with contributions by

Brian D. Andresen
William F. Balistreri
Edward W. Bermes, Jr.
Robert V. Blanke
Robert A. Blouin
Larry D. Bowers
Gregory J. Buffone
Carl A. Burtis
Wendell T. Caraway
Sati C. Chattoraj
Robert H. Christenson
Walter J. Decker
Franklin R. Elevitch
Emanuel Epstein
Virgil F. Fairbanks
Montserrat deM. Fencl
Donald Fraser
Esther F. Freier
Robert S. Galen
Gregor H. Grant
Michael F. Greene
Mary Alice Heard
A. Ralph Henderson
G. Phillip Hicks
Bethany W. Hurtuk
Robert A. Jacob
C. Darrell Jennings
Glenville Jones
John F. Kachmar
George G. Klee
Sang Whay Kooh
Robert F. Labbe
Joel M. Lamon
H. Peter Lehmann
Nancy M. Logan
Ernest Maclin
Donald B. McCormick
Donald W. Moss
Thomas P. Moyer
Dorice M. Czajka-Narins
Kwokie J. Ng
John F. O'Brien
Theodore Peters, Jr.
Charles E. Pippenger
Edward R. Powsner
Elizabeth L. Pruden
Ingeborg C. Radde
Marilyn R. Rettaliata
Alan D. Rinker
Robert C. Rock
Leslie M. Shaw
Ole Siggaard-Andersen
Lawrence M. Silverman
Helge Erik Solberg
Evan A. Stein
Nancy L. Strang
Thomas O. Tiffany
Dan Tulchinsky
M. David Ullman
W. Gordon Walker
Nelson B. Watts
James O. Westgard
Donald S. Young

Fundamentals of Clinical Chemistry

THIRD EDITION

EDITOR

Norbert W. Tietz, Ph.D.

Professor, Department of Pathology,
Director of Clinical Chemistry,
College of Medicine,
University of Kentucky,
Lexington, Kentucky

EDITORIAL COMMITTEE

Joan Aldrich, M.S., M.T.(ASCP)
N.V. Bhagavan, Ph.D.
Diane Coiner, M.S., CLS
Rex B. Conn, M.D.
Richard S. Kowalczyk, Ph.D.
H. Peter Lehmann, Ph.D.
Elizabeth L. Pruden, Ph.D., M.T.(ASCP)
Ronald J. Whitley, Ph.D.

1987
W.B. Saunders Company
PHILADELPHIA □ LONDON □ TORONTO □ MEXICO CITY
RIO DE JANEIRO □ SYDNEY □ TOKYO □ HONG KONG

W. B. Saunders Company: West Washington Square
Philadelphia, PA 19105

Library of Congress Cataloging-in-Publication Data

Fundamentals of clinical chemistry.

Includes bibliographies and index.
1. Chemistry, Clinical. I. Tietz, Norbert W.,
1926- . [DNLM: 1. Chemistry, Clinical.
QY 90 F981]
RB40.F84 1987 616.07′56 86-1937
ISBN 0-7216-8862-4

Editor: Dean Manke
Designer: Terri Siegel
Production Manager: Frank Polizzano
Manuscript Editor: Linda Mills
Illustrator: Philip Ashley
Illustration Coordinator: Walt Verbitski

Fundamentals of Clinical Chemistry ISBN 0-7216-8862-4

Last digit is the print number: 9 8 7 6 5 4 3 2 1

In Thankfulness Dedicated to
MY WIFE GERTRUD
and to my children
MARGARET ANN, KURT RICHARD
ANNETTE MARIE *and* MICHAEL GERHARD

CONTRIBUTORS AND REVIEWERS

JOAN ALDRICH, M.S., M.T.(ASCP)
Assistant Professor, Department of Medical Laboratory Sciences, School of Allied Health Sciences, University of Texas, Dallas, TX
Editorial Committee

BRIAN D. ANDRESEN, Ph.D
Director, Mass Spectrometry Facilities, Biomedical and Environmental Science Division, University of California, Lawrence Livermore National Laboratory, Livermore, CA
Mass Spectrometry

WILLIAM F. BALISTRERI, M.D.
Professor, Departments of Pediatrics and Medicine, College of Medicine, University of Cincinnati; Director, Division of Pediatric Gastroenterology and Nutrition, Children's Hospital Medical Center, Cincinnati, OH
Liver Function

EDWARD W. BERMES, Jr., Ph.D.
Professor, Departments of Pathology and Biochemistry, Associate Chairman, Department of Pathology, Stritch School of Medicine, Loyola University; Director, Clinical Chemistry, Loyola University Medical Center, Maywood, IL
General Laboratory Techniques and Procedures; Specimen Collection and Processing; Sources of Biological Variation

N. V. BHAGAVAN, Ph.D.
Professor and Chairman, Department of Biochemistry and Biophysics, John A. Burns School of Medicine, University of Hawaii; Consultant Biochemist, Kaiser Medical Center, Honolulu, HI
Editorial Committee

ROBERT V. BLANKE, Ph.D.
Professor, Departments of Pathology, Pharmacology, and Toxicology; Director, Hospital Toxicology Laboratory, Medical College of Virginia, Virginia Commonwealth University, Richmond, VA
Therapeutic Drug Monitoring; Analysis of Toxic Substances

ROBERT A. BLOUIN, Pharm.D.
Associate Professor, College of Pharmacy, University of Kentucky; Director, Clinical Pharmacokinetics Program, University of Kentucky Medical Center, Lexington, KY
Therapeutic Drug Monitoring

LARRY D. BOWERS, Ph.D.
Associate Professor, Department of Laboratory Medicine and Pathology, University of Minnesota; Director, Drug Analysis Section, University of Minnesota Hospitals, Minneapolis, MN
Chromatography

GREGORY J. BUFFONE, Ph.D.
Associate Professor, Department of Pathology, Baylor College of Medicine; Director, Clinical Chemistry, Department of Pathology, Texas Children's Hospital, Houston, TX
Principles of Immunochemical Techniques

CARL A. BURTIS, Ph.D.
Chief, Clinical Chemistry, Health Division, Chemical Technology Division, Oak Ridge National Laboratory, Oak Ridge, TN
Coordinator: Analytical Procedures and Instrumentation; Chromatography

WENDELL T. CARAWAY, Ph.D.
Clinical Biochemist (Retired), McLaren General Hospital and St. Joseph Hospital, Flint, MI
Photometry; Carbohydrates

SATI C. CHATTORAJ, Ph.D.
Associate Professor, Department of Biochemistry; Director, Research Laboratories, Department of Obstetrics and Gynecology, Boston University Medical Center; Director, Gynecologic Endocrine Lab, Boston University School of Medicine, Boston, MA
Chromatography; Endocrinology

ROBERT H. CHRISTENSON, Ph.D.
Assistant Professor, Department of Pathology, Duke University Medical Center; Chief, Clinical Chemistry, Durham Veterans Administration Medical Center, Durham, NC
Amino Acids and Proteins

DIANE COINER, M.S., CLS
Associate Professor, Department of Medical Technology, School of Allied Health Sciences, University of Texas Medical Branch, Galveston, TX
Editorial Committee

REX B. CONN, M.D.
Professor, Department of Pathology and Laboratory Medicine, Emory University School of Medicine; Clinical Professor of Medical Technology, Georgia State University; Associate Chief, Pathology and Laboratory Medicine, and Director, Clinical Pathology Laboratories, Emory University Hospital, Atlanta, GA
Editorial Committee

WALTER J. DECKER, Ph.D.
Adjunct Associate Professor, Departments of Pharmacology, Toxicology, and Pediatrics, University of Texas Medical Branch, Galveston, TX; Toxicology Consultant Services, Inc., El Paso, TX
Analysis of Toxic Substances

FRANKLIN R. ELEVITCH, M.D.
Associate Clinical Professor, Department of Laboratory Medicine, University of California, San Francisco Medical Center; Director, Clinical Laboratories, El Camino Hospital, Department of Pathology, Mountain View, CA
Microprocessors and Computers in the Clinical Laboratory

EMANUEL EPSTEIN, Ph.D
Associate Clinical Professor, Department of Pathology, College of Medicine, Wayne State University, Detroit, MI: Medical Staff, William Beaumont Hospital, Royal Oak, MI
Electrophoresis

VIRGIL F. FAIRBANKS, M.D.
Professor, Department of Laboratory Medicine, Mayo Medical School; Consultant, Department of Laboratory Medicine, Mayo Clinic, Rochester, MN
Biochemical Aspects of Hematology

MONTSERRAT deM. FENCL, Ph.D.
Associate Director, Laboratory of Reproductive Endocrinology, Brigham and Women's Hospital; Assistant Professor, Department of Obstetrics and Gynecology, Harvard Medical School, Boston, MA
Biochemical Aspects of Pregnancy

DONALD FRASER, M.D., Ph.D.
Professor, Departments of Paediatrics and Physiology, University of Toronto; Senior Physician, Department of Endocrinology, Hospital for Sick Children, Toronto, Ontario, Canada
Calcium and Phosphate Metabolism

ESTHER F. FREIER, M.S., M.T.(ASCP)
Professor, Department of Laboratory Medicine and Pathology, University of Minnesota Medical School; Co-Director, Clinical Chemistry Laboratories, University of Minnesota Hospitals, Minneapolis, MN
Osmometry

ROBERT S. GALEN, M.D., M.P.H.
Chairman, Department of Biochemistry, The Cleveland Clinic Foundation, Cleveland, OH
Analytical Goals and Clinical Relevance of Laboratory Procedures

GREGOR H. GRANT, B.M., F.R.C.Path.
Consultant Chemical Pathologist (Retired), Royal Shrewsbury Hospital, Shrewsbury, Shropshire, England
Amino Acids and Proteins

MICHAEL F. GREENE, M.D.
Instructor, Department of Obstetrics and Gynecology, Harvard Medical School, Cambridge, MA; Obstetrician Gynecologist, Brigham and Women's Hospital, Boston, MA
Biochemical Aspects of Pregnancy

MARY ALICE HEARD, M.A.
Data Coordinator, Department of Pathology, University of Kentucky Medical Center, Lexington, KY
Editorial Assistant

A. RALPH HENDERSON, M.B., Ch.B., Ph.D.
Professor, Department of Biochemistry, University of Western Ontario; Chief, Department of Clinical Biochemistry, University Hospital, London, Ontario, Canada
Enzymes; Gastric, Pancreatic, and Intestinal Function

G. PHILLIP HICKS, Ph.D.
Adjunct Associate Professor, Department of Medicine, University of Wisconsin Medical School; President, Laboratory Consulting, Inc., Madison, WI
Microprocessors and Computers in the Clinical Laboratory

BETHANY W. HURTUK, Ph.D., M.T.(ASCP), CLS
Clinical Assistant Professor, Department of Chemistry, Cleveland State University; Medical Technologist-Biochemistry, Cleveland Clinic Foundation, Cleveland, OH
Mass Spectrometry

ROBERT A. JACOB, Ph.D.
Project Leader and Research Chemist, Bioanalytical Research Unit, USDA-ARS Western Human Nutrition Research Center, Presidio of San Francisco, CA
Trace Elements

C. DARRELL JENNINGS, M.D.
Assistant Professor, Department of Pathology and Co-Director, Clinical Immunology Laboratory, University of Kentucky Medical Center, College of Medicine; Assistant Chief, Laboratory Service, Veterans Administration Hospital, Lexington, KY
Nitrogen Metabolites and Renal Function

GLENVILLE JONES, Ph.D.
Associate Professor, Departments of Biochemistry and Medicine, Queen's University, Kingston, Ontario. Formerly: Associate Professor, Department of Biochemistry, University of Toronto and Research Institute, Hospital for Sick Children, Toronto, Ontario, Canada
Calcium and Phosphate Metabolism

JOHN F. KACHMAR, Ph.D.
Senior Attending Scientist, Presbyterian-St. Luke's Hospital; Emeritus Associate Professor of Biochemistry, Rush University, Chicago, IL
Enzymes

GEORGE G. KLEE, M.D., Ph.D.
Associate Professor, Department of Laboratory Medicine, Mayo Medical School; Director, Thyroid Function and Radioimmunoassay Laboratory, Mayo Clinic, Rochester, MN
Quality Assurance; Biochemical Aspects of Hematology

SANG WHAY KOOH, M.D., Ph.D., F.R.C.P.(C)
Associate Professor, Departments of Paediatrics and Physiology, University of Toronto; Senior Physician, Hospital for Sick Children, Toronto, Ontario, Canada
Calcium and Phosphate Metabolism

RICHARD S. KOWALCZYK, Ph.D., F.A.C.B.
Clinical Biochemist, Northern Michigan Hospitals, Petrosky, MI
Editorial Committee

ROBERT F. LABBE, Ph.D.
Professor, Department of Laboratory Medicine, School of Medicine; Head, Clinical Chemistry Division, University of Washington, Seattle, WA
Porphyrins and Disorders of Porphyrin Metabolism

JOEL M. LAMON, M.D.
Clinical Associate Member, Division of Hematology-Oncology; Adjunct Assistant Member, Department of Clinical Research, Scripps Clinic and Research Foundation, La Jolla, CA
Porphyrins and Disorders of Porphyrin Metabolism

H. PETER LEHMANN, Ph.D.
Professor, Department of Pathology, Louisiana State University Medical Center, New Orleans, LA
Editorial Committee

NANCY M. LOGAN, B.A.
Research Associate, Division of Clinical Chemistry, University of Kentucky Medical Center, Lexington, KY
Appendix

ERNEST MACLIN, P.E.
Vice President, Research and Development, Electro-Nucleonics, Inc., Fairfield, NJ
Automation in the Clinical Laboratory

DONALD B. McCORMICK, Ph.D.
Fuller E. Callaway Professor and Chairman, Department of Biochemistry, Emory University School of Medicine, Atlanta, GA
Vitamins

DONALD W. MOSS, Ph.D., D.Sc.
Professor of Clinical Enzymology, Department of Chemical Pathology, Royal Postgraduate Medical School, University of London; Honorary Biochemist, Hammersmith Hospital, London, England
Enzymes

THOMAS P. MOYER, Ph.D.
Associate Professor, Department of Laboratory Medicine, Mayo Medical School; Director, Toxicology Laboratory, Mayo Clinic, Rochester, MN
Therapeutic Drug Monitoring

DORICE M. CZAJKA-NARINS, Ph.D.
Professor, Department of Human and Family Resources, Northern Illinois University, DeKalb, IL; Guest Lecturer, Chicago Medical School, North Chicago, IL
Nutrition

KWOKIE J. NG, Ph.D.
Senior Scientist, Department of Analytical Biochemistry and Mass Spectrometry, Schering Plough Corporation, Bloomfield, NJ
Mass Spectrometry

JOHN F. O'BRIEN, Ph.D.
Associate Professor, Department of Laboratory Medicine, Mayo Medical School; Director, Enzyme Laboratory, Section of Clinical Chemistry, Mayo Clinic, Rochester, MN
Lysosomal Enzymes

THEODORE PETERS, Jr., Ph.D.
Adjunct Associate Professor, Department of Biochemistry, College of Physicians and Surgeons, Columbia University; Research Biochemist, The Mary Imogene Bassett Hospital, Cooperstown, NY
Analytical Goals and Clinical Relevance of Laboratory Procedures; Evaluation of Methods

CHARLES E. PIPPENGER, Ph.D.
Head, Section of Applied Clinical Pharmacology, Department of Biochemistry, Cleveland Clinic Foundation, Cleveland, OH
Therapeutic Drug Monitoring

EDWARD R. POWSNER, M.D.
Director, Division of Nuclear Medicine, Department of Pathology, St. John Hospital, Detroit, MI
Basic Principles of Radioactivity and its Measurement

ELIZABETH L. PRUDEN, Ph.D., M.T.(ASCP)
Associate Professor, Department of Pharmacology and Therapeutics, School of Medicine in Shreveport; Co-Director, Clinical Toxicology Laboratory, Louisiana State University, Shreveport, LA
Editorial Committee; Electrolytes, Blood Gases, and Acid-Base Balance

INGEBORG C. RADDE, M.D., Ph.D., F.R.C.P.(C)
Associate Professor, Departments of Paediatrics and Pharmacology, University of Toronto; Associate Professor and Senior Staff Physician, Research Institute, Hospital for Sick Children, Toronto, Ontario, Canada
Calcium and Phosphate Metabolism

MARILYN R. RETTALIATA, B.S., M.T.(ASCP)
Consultant, Bangor, PA
Invited Reviewer

ALAN D. RINKER, M.S.
Supervisor, Radioimmunoassay and Immunochemistry Laboratory, Division of Clinical Chemistry, University of Kentucky Medical Center, Lexington, KY
Gastric, Pancreatic, and Intestinal Function; Invited Reviewer

ROBERT C. ROCK, M.D.
Associate Professor, Department of Pathology and Laboratory Medicine, Johns Hopkins University School of Medicine; Director, Department of Laboratory Medicine, Johns Hopkins Hospital, Baltimore, MD
Nitrogen Metabolites and Renal Function

LESLIE M. SHAW, Ph.D.
Associate Professor, Department of Pathology; Director, Toxicology Laboratory, Division of Laboratory Medicine, William Pepper Laboratory, Hospital of the University of Pennsylvania, Philadelphia, PA
Liver Function

OLE SIGGAARD-ANDERSEN, M.D., Ph.D.
Professor, Department of Clinical Chemistry, University of Copenhagen; Director, Clinical Chemistry, Herlev Hospital, Herlev, Denmark
Electrochemistry; Electrolytes, Blood Gases, and Acid-Base Balance

LAWRENCE M. SILVERMAN, Ph.D.
Associate Professor, Department of Pathology and Biochemistry, University of North Carolina; Director, Special Chemistry Laboratory; Associate Director, Clinical Chemistry Laboratories, Department of Hospital Laboratories, North Carolina Memorial Hospital, Chapel Hill, NC
Amino Acids and Proteins

HELGE ERIK SOLBERG, M.D., Ph.D.
Consultant, Department of Clinical Chemistry, Rikshospitalet, University of Oslo, Oslo, Norway
Establishment and Use of Reference Values

EVAN A. STEIN, M.D., Ph.D.
Professor, Department of Pathology and Laboratory Medicine; Associate Professor, Internal Medicine, University of Cincinnati Medical Center; Director, Division of Chemical Pathology, University Hospital, Cincinnati, OH
Lipids, Lipoproteins, and Apolipoproteins

NANCY L. STRANG, M.S.
Data Coordinator, Department of Pathology, University of Kentucky Medical Center, Lexington, KY
Editorial Assistant

NORBERT W. TIETZ, Ph.D.
Professor, Department of Pathology, College of Medicine; Director, Clinical Chemistry, University of Kentucky Medical Center, Lexington, KY
Editor; Electrolytes, Blood Gases, and Acid-Base Balance; Gastric, Pancreatic, and Intestinal Function; Appendix

THOMAS O. TIFFANY, Ph.D.
Director, Research and Development, Instrumentation Laboratory, Inc., Spokane, WA
Fluorometry, Nephelometry, and Turbidimetry

DAN TULCHINSKY, M.D.
Associate Professor, Department of Obstetrics and Gynecology, Harvard Medical School; Director, Reproductive Endocrine Laboratory, Brigham and Women's Hospital, Boston, MA
Biochemical Aspects of Pregnancy

M. DAVID ULLMAN, Ph.D.
Chief, Special Research Laboratory, Veterans Administration Hospital, Bedford, MA; Associate Research Professor, Department of Psychiatry, Boston University School of Medicine, Boston, MA
Chromatography

W. GORDON WALKER, M.D.
Professor, Department of Medicine, Johns Hopkins University School of Medicine; Director, Renal Division, Department of Medicine; Director, Adult Clinical Research Center, and Physician, Johns Hopkins Hospital; Director, Nephrology and Physician, Good Samaritan Hospital, Baltimore, MD
Nitrogen Metabolites and Renal Function

NELSON B. WATTS, M.D.
Assistant Professor of Medicine, Emory School of Medicine and Emory University Clinic, Section of Internal Medicine, Department of Endocrinology, Atlanta, GA
Carbohydrates; Endocrinology

JAMES O. WESTGARD, Ph.D.
Professor, Department of Pathology and Laboratory Medicine, University of Wisconsin; Associate Director, Clinical Laboratories–Quality Control, University of Wisconsin Hospitals and Clinics, Madison, WI
Coordinator: Acquisition, Management, and Application of Laboratory Data; Evaluation of Methods; Quality Assurance

RONALD J. WHITLEY, Ph.D.
Associate Professor, Department of Pathology, College of Medicine; Assistant Director, Clinical Chemistry, University of Kentucky, Lexington, KY
Editorial Committee

DONALD S. YOUNG, M.B., Ph.D.
Professor, Department of Pathology and Laboratory Medicine; Head, Division of Laboratory Medicine, University of Pennsylvania; Director, William Pepper Laboratory, Hospital of the University of Pennsylvania, Philadelphia, PA
General Laboratory Techniques and Procedures; Automation in the Clinical Laboratory; Specimen Collection and Processing; Sources of Biological Variation

FOREWORD TO THE FIRST EDITION

By Michael Somogyi, Ph.D.

The clinical chemistry laboratory a few decades ago was a modest handmaiden supplying a few chemical tests that appeared serviceable in diagnostic procedures. These tests were developed in Otto Folin's laboratory at Harvard University, in D. D. Van Slyke's laboratory in the hospital of the Rockefeller Institute, and in a number of biochemistry departments, which were themselves new in some medical schools. These methods, which filled small booklets, were in most instances like cookbook recipes, and were frequently in the hands of semiskilled persons with no great competence in any branch of chemistry. A great change was inaugurated when a few progressive hospitals appointed to their laboratories highly trained professional chemists, men who were qualified to carry on investigative work in addition to supplying reliable data for diagnostic use. These chemists faced great challenges in refining existing methods and in the development of a rapidly increasing number of new tests and techniques. Today, in place of small "cookbooks," the chemist is confronted with a burgeoning amount of literature on clinical chemistry, the "fundamentals" of which are presented in this volume by the cooperative and coordinated effort of a group of competent and erudite clinical chemists.

With the well organized, rich content of this volume before me, and with the picture of instrumentation and equipment of an up-to-date clinical laboratory in mind, I cannot help but hark back to the humble beginnings in this field. In 1926, when I moved from academic work to the newly constructed Jewish Hospital of St. Louis, this was the first hospital in this city (priding itself on two medical schools!) to place a chemist in charge of its clinical laboratory. The space originally allotted to chemistry consisted of a single room of moderate size, with $1200 available for equipment. As my technical assistant I trained a young, small-town school teacher with no previous experience in chemistry. True, the routine work comprised at that time but a dozen or so analytical procedures, involving exclusively simple manual operations. Thus we found time for some investigative and educational activities, which year after year attracted the interest of the interns, residents, and younger members of the medical staff. They learned to use and appreciate the services of the laboratory so that the erstwhile modest handmaiden has developed into "laboratory medicine." My participation in this process, my collaboration with progressive, spirited, mostly youthful, clinicians was the most rewarding experience in my professional life. I am confident that these progressive young clinicians will find the study of the *Fundamentals of Clinical Chemistry* as useful as will workers in the laboratory. By so doing they will narrow the gap between medical science and medical practice.

Michael Somogyi

(1883–1971)

PREFACE

Since the publication of the second edition of *Fundamentals of Clinical Chemistry*, many new tests and some remarkable new technologies have been introduced. In addition, the practice of clinical chemistry has changed drastically as new versions of reagent packaging and novel procedures have been developed, along with innovative instrumentation. These developments have led to an improvement in quality and an increase in the amount of information made available by manufacturers.

The growth of the field of clinical chemistry and the change in its practice have required significant adjustments in the presentation of material included in *Fundamentals of Clinical Chemistry*. A new chapter has been included for the discussion of immunochemical principles on which many of the newer procedures are based, and an instrumentation chapter discusses the various steps of automation and uses existing instrumentation to illustrate these processes. Furthermore, chapters on therapeutic drug monitoring, biochemistry of pregnancy, and clinical relevance of laboratory procedures have been added to address these newly developed areas. To avoid duplication with kit inserts and instrument manuals, many of the laboratory tests are discussed in principle only, without listing procedural details. However, those tests not available as kits but considered important for the practice of clinical chemistry are included in detail; other methods are included that describe important analytical principles that the student must learn.

Selection of material for a textbook in a field as diversified as clinical chemistry is always difficult because of differences in teaching approach and varying interests of the readership. The selection of methods for inclusion was made by a group of medical technology educators with many years of experience in teaching clinical chemistry. Readers interested in more detailed methodology will find references to the original literature in the text. In addition, the *Quick Index to Methods* will also direct the student to procedures that are listed in detail in the *Textbook of Clinical Chemistry*, N. W. Tietz, Editor, W. B. Saunders, 1986. Since the organization of both textbooks is very similar, it is easy for the more interested or more advanced student to find in the *Textbook* additional information on methodology or on other aspects of clinical chemistry.

As with the second edition of *Fundamentals of Clinical Chemistry*, the third edition is written predominantly for undergraduate students in medical technology and clinical chemistry. *Fundamentals* attempts to provide a useful balance between basic biochemistry, the application of laboratory tests in clinical diagnosis, and a discussion of methods performed in the laboratory. It is hoped that this coverage of traditional as well as new areas of clinical chemistry will fill not only contemporary but also future needs of the student.

The third edition of *Fundamentals of Clinical Chemistry* uses a second color for major headlines and to highlight definitional equations, designate important aspects of chemical equations, and draw attention to important aspects of some illustrations. It is hoped that this functional use of color not only will add to the attractiveness of the book but will increase its teaching effectiveness.

We would like to express to all readers of the second edition of *Fundamentals of Clinical Chemistry* our sincere thanks for the extensive use of the textbook and its enormous success. We hope that the third edition will be even more effective in meeting the needs of instructors and students. Your suggestions and thoughts are most welcome.

ACKNOWLEDGMENTS

I am very grateful to all contributors for the preparation of their manuscripts and for cooperating so well with the editorial review process. My special thanks go to the members of the Editorial Committee for their endurance and for the numerous contributions they made to the editing process. Nancy L. Strang, M.S., and Mary Alice Heard, M.A., assisted in editing manuscripts, and Tammy I. Rayburn and Susan C. Blandford assisted with typing manuscripts; their contributions and dedication are gratefully acknowledged.

The cooperation of the staff of W. B. Saunders Company—especially of Dean J. Manke, Medical Editor; Frank Polizzano, Production Manager; Linda Mills, Copy Editor; and Terri Siegel, Designer—is also gratefully acknowledged.

NORBERT W. TIETZ
Editor

CONTENTS

We will be
not what we are —
but what we aim to be.

I

LABORATORY PRINCIPLES AND INSTRUMENTATION

GENERAL LABORATORY TECHNIQUES AND PROCEDURES

1

EDWARD W. BERMES, Jr., Ph.D.
DONALD S. YOUNG, M.B., Ph.D.

The function of the clinical chemistry laboratory is to perform qualitative and quantitative analyses on body fluids such as blood, urine, and spinal fluid as well as on feces, tissue, calculi, and other materials. If the results are to be useful to the physician in the diagnosis and treatment of disease, the tests must be performed as accurately as possible. To achieve accurate and precise results requires the use of sound analytical methods and good instrumentation. However, underlying all standards of analytical quality are *basic principles and procedures*. Essential factors include the purity of reagent solutes and solvents, quality of containers, reliability and quality of measuring devices and methods, appropriate choice of separative methods and devices, and observance of established safety procedures. There should be an understanding of the chemical reaction involved in each test and the effect of physical variables on the procedures.

CHEMICALS AND RELATED SUBSTANCES

Laboratory chemicals are available in a variety of grades (see page 7). The solutes and solvents used in analytical work are *reagent grade chemicals*, among which water is a solvent of primary importance. The *classification* of reagent grade chemicals is quite unofficial, and there is no agreement among manufacturers concerning the designation of the various degrees of purity. The IUPAC (International Union of Pure and Applied Chemistry) has established criteria for "primary standards." The National Bureau of Standards (NBS) has a number of Standard Reference Materials (SRMs) available for the clinical chemistry laboratory, and the College of American Pathologists (CAP) also supplies some Certified Reference Materials (CRMs). In addition, the National Committee for Clinical Laboratory Standards (NCCLS) has published a standard for reagent grade water and has "Guidelines for the Development of Certified Reference Materials for use in Clinical Chemistry" in the "proposed standard" stage.

Reagent Grade Water

The preparation of most reagents and solutions used in the clinical laboratory requires "pure" water.

Although in many laboratories the term *deionized water* may have replaced *distilled water*, both terms should be discouraged, since they describe methods of preparation and do not in themselves reflect the quality of the final product. NCCLS specifications for reagent grade water (Table 1-1) are the current standards for laboratory use. The term *reagent grade water*, followed by designation of the Type (I through III) better defines the specifications of the water and is independent of the method of preparation.

In general, no single process of purification can produce water that meets the rigid specifications for reagent grade water Type I set forth by NCCLS. The following processes or any combination of these may be used in the preparation of reagent grade water as long as the final product meets the specifications. The combination selected is usually dictated in part by the quality of the source water and in part by the intended use of the water.

Table 1-1. NCCLS SPECIFICATIONS FOR REAGENT GRADE WATER[8]

	Type I	Type II	Type III
Microbiological content, * colony forming units per mL, cfu/mL (maximum)	10	10^3	N.A.
pH	N.A.	N.A.	5.0–8.0
Resistivity, † Mohm centimeter (MΩ-cm), 25 °C	10 (in-line)	2.0	0.1
Silicate, mg SiO_2/L (maximum)	0.05	0.1	1.0
Particulate matter ‡	Water passed through 0.2-μm filter	N.A.	N.A.
Organics §	Water passed through activated carbon	N.A.	N.A.

* *Microbiological content.* The microbiological content of viable organisms, as determined by total colony count after incubation at 36 ± 1 °C for 14 h, followed by 48 h at 25 ± 1 °C, and reported as colony forming units per mL (cfu/mL).

† *Specific resistance or resistivity.* The electrical resistance in ohms measured between opposite faces of a 1.00 cm cube of an aqueous solution at a specified temperature. For these specifications, the resistivity will be corrected for 25 °C and reported in megohm centimeters (MΩ-cm). The higher the amount of ionizable materials, the lower the resistivity and the higher the conductivity.

‡ *Particulate matter.* When water has been passed through an intact membrane filter with a mean pore size of 0.2 μm, it is considered to be free of particulate matter.

§ *Organic material.* When water has been passed through a bed of activated carbon, it is considered to contain minimum organic material.

N.A. Not applicable.

Preparation of Reagent Grade Water

Filtration. Three types of filtration equipment may be used to purify water.

Prefilters are usually composed of glass or cotton microfibers that remove 98% or more of the particulate matter and thus protect the whole water flow and containment system. These filters can usually be cleaned and reused many times.

A second type of filter is a bed of *activated carbon*. The bed is effective in removing organic materials and chlorine. A combination of prefiltration and carbon bed filtration is extensively used to increase the life expectancy of an ion-exchange resin used later in the purification process.

The third type of filtration employs a *submicron filter* that removes all particles or microorganisms larger than the membrane's pore size (usually 0.2 μm) without releasing contaminants that could alter the quality of the water. A submicron filter is usually used as the last stage of a system (i.e., near the point of delivery).

Distillation. The oldest method of water purification is distillation. During vigorous boiling a small number of water droplets entrapped in vapor may carry impurities such as sodium, potassium, manganese, carbonates, and sulfates to the distillate. As a result, water treated by distillation alone will not meet the specific conductivity requirement of Type I water. Distillation also involves a large expenditure of energy and time to clean and maintain the still.

Deionization. Deionized water is water from which the ions have been removed by the process of ion exchange. It is most conveniently prepared by using commercial deionizing resins. Deionization is accomplished by passing source water over insoluble resin polymers that exchange H^+ and OH^- ions for the ionized impurities present in the water. These insoluble resin polymers are prepared with acid or amine functional groups, which make them cation-exchangers or anion-exchangers; "mixed-bed resins" are made by placing a mixture of cation- and anion-exchange resins into the same container.

A typical cation-exchange resin reacts as follows:

$$(RSO_3)H + Na^+ \rightarrow (RSO_3)Na + H^+$$

whereas an anion-exchange resin with a quaternary ammonium type of structure acts as follows:

$$(RNR_3')OH + Cl^- \rightarrow (RNR_3')Cl + OH^-$$

A mixed-bed deionizer generally is capable of producing water with specific resistance in excess of 1 MΩ-cm. When connected in series, mixed-bed deionizers usually produce water with a specific resistance that exceeds 10 MΩ-cm.

Reverse osmosis is a process by which water is forced through a membrane that acts as a molecular filter to remove 95–99% of organics, bacteria, and other particulate matter; 90–97% of all ionized and dissolved minerals; but less of the gaseous impurities. Although the process is not adequate to produce reagent grade water for the laboratory, it may be used as a preliminary purification method.

Quality, Use, and Storage of Reagent Grade Water

Type III water may be used for washing glassware; however, final rinsing should be done with water of a grade suitable for the intended use of the glassware. This Type III water may also be used for certain qualitative procedures, such as urinalysis.

Type II water is used for general laboratory testing not requiring Type I water. Storage time should be kept to a minimum, and storage as well as delivery systems should be constructed so as to ensure a minimum of chemical or bacterial contamination.

Type I water should be used in test methods requiring minimum interference and maximum precision and accuracy. Such procedures might include trace metal determination, enzyme measurements, electrolyte measurements, and the preparation of all standards. This water should be used immediately after production because it loses its high resistivity after it is drawn off and stored.

Specific Systems for the Preparation of Reagent Grade Water

Any system may be used for preparing reagent grade water as long as the desired specifications are attained. Considerations in the selection of a system to produce Type I reagent water include quality of feed water, amount of water needed, availability of space near the area of use, and electrical energy cost in the geographic area. In general, if the feed water contains more than 1000 ppm of total dissolved solids (TDS), reverse osmosis should be considered as a pretreatment.

Type I reagent water with neutral pH and <170 ppm TDS can be produced from municipal feed water by first using a primary dual filtration cartridge to remove particulate matter and organic materials and then treating it with two tanks containing a mixed-bed ion-exchange resin. The second tank "polishes" the water to give water with a resistance of up to 15 Mohm-cm, which is then filtered through a 0.2-μm membrane filter. Water is delivered for use from a spigot that is within a few feet of the final filter.

Testing for Water Purity

The specific methods used for testing water purity are not covered here. At a minimum, water should be tested for microbiological content, resistivity, and soluble silica. The reader is referred to one of several references for specifics.[1,2,8]

Reagent Grade or Analytical Reagent Grade (AR) Reagents

Chemicals that meet specifications of the American Chemical Society (ACS) are described as reagent grade or analytical reagent grade. These are available in two forms: (1) lot-analyzed reagents, in which each individual lot is analyzed and the actual amount of impurity is reported (e.g., lead—0.0005%) and (2) maximum impurities reagents, for which maximum impurities are listed (e.g., lead—maximum 0.002%). Both types of these reagent grade chemicals are of very high purity and are recommended for quantitative or qualitative analyses.

Ultra Pure Reagents

Gas chromatography, high performance liquid chromatography, fluorometry, and trace element analyses frequently call for reagents whose purity exceeds the specifications of the two types described previously. Manufacturers now offer selected chemicals that have been especially purified to meet specific needs. There is no uniform designation for these chemicals and organic solvents. Terms such as *spectrograde, nanograde,* and *HPLC pure* have been used. Data of interest to the user (e.g., absorption at a specific ultraviolet wavelength) are supplied with the reagents.

Other Designations of Chemical Purity

There are several other designations of chemical purity. In general, however, chemicals with these designations are not of sufficient purity for use as reagents. *Chemically pure* (CP) is a designation that fails to reveal the tolerance limits of impurities, and the practice followed by different manufacturers in the use of this designation is not uniform. The term *highest purity* is used by manufacturers for organic chemicals that they have purified to as great a degree as they find practical. *USP* and *NF grade* indicate chemicals produced to meet specifications set down in the United States Pharmacopoeia (USP) or the National Formulary (NF). These designations specify impurity tolerances that will not be injurious to health. Chemicals labeled *purified, practical, technical* or *commercial grade* should not be used in clinical chemistry analysis without further purification.

Purity of Organic Reagents

The purity of commercially obtained organic reagents for clinical chemistry purposes is generally inferior to that of inorganic reagents. The majority of impurities in these compounds will have been introduced in the synthesis either in the starting materials or as by-products of the synthesis. Such impurities are presumably more difficult or more expensive to remove than are impurities in inorganic substances. In addition, some organic compounds oxidize or decompose on standing; the quantity of impurities resulting from this process will depend on how long the reagent has been opened or stored.

The presence of impurities in an organic reagent may be a source of difficulty when it is used. If the contaminant does not react with the substance being determined, no interference will occur as long as there are no interfering side reactions and there is enough of the original reactant remaining. On the other hand, if a reagent is impure, the final color developed in a colorimetric reaction may be considerably suboptimal because of a high blank caused by the impurity. The presence of isomers in a particular lot of an organic reagent may also be a cause of analytical problems, since in some reactions only certain isomers may produce the desired colored product or be suitable as substrate.

Organic solvents are available in several grades of purity. The common grades are listed in Table 1-2. The required degree of a solvent's purity depends on its intended use. The properties of some of the organic solvents commonly used in the clinical laboratory are listed in the *Appendix*, Table 29-4. Organic solvents should generally be considered potentially hazardous to health, and thus inhalation of fumes should be avoided. Unless a solvent is known to be nonflammable, it should be considered a potential fire hazard. The maximum quantities of solvents that can be stored in particular

Table 1-2. COMMON GRADES OF ORGANIC SOLVENTS

Practical Grade	Contains some impurities but usually adequate for most organic preparations
USP Grade and NF Grade	Meets standards established by the US Pharmacopeia (USP) or National Formulary (NF); may contain impurities which have not been tested for
Chemically Pure (CP)	Almost as pure as reagent grade chemicals
Spectroscopic Grade	Spectrally pure in the visible, ultraviolet, and near and mid-infrared ranges
Chromatographic Grade	Minimum purity of greater than 99% as determined by gas chromatography; no single impurity exceeding 0.2%
Reagent Grade	Certified to contain impurities below levels established by the Committee on Analytical Reagents of the American Chemical Society

(Modified from Shugar, G. J., Shugar, R. A., Bauman, L., et al.: Chemical Technicians' Ready Reference Handbook. 2nd ed. New York, McGraw-Hill, 1981.)

types and sizes of containers are described by OSHA regulations (see *Safety*, below). Pure solvents are best stored sealed in glass containers under an inert gas or under positive pressure. With some solvents, sealing the stopper with paraffin prolongs their storage life.

Standards

Primary standards are highly purified chemicals that can be directly weighed or measured to produce a solution whose concentration is exactly known. These chemicals may be weighed directly for the preparation of solutions of selected concentration or for the standardization of solutions of unknown strength. Companies supply primary standards with certificates of analysis for each lot. These chemicals must be stable substances of definite composition that can be dried, preferably at 104–110 °C, without a change in composition. They must not be hygroscopic (i.e., they must not absorb water during weighing).

Secondary standards are solutions whose concentration cannot be exactly known by weighing the solute and dissolving a known amount into a volume of solution. The concentration of secondary standards is usually determined by analysis of an aliquot of the solution by an acceptable reference method, using a primary standard to calibrate the method.

Standard Reference Materials (SRMs) for clinical laboratories are available from the National Bureau of Standards (Table 1-3). Not all standard reference materials have the properties and the degree of purity specified for a primary standard, but the NBS has described each one with regard to certain chemical or physical properties and issues for each a certificate that gives the results of the characterization. These standards may then be used to characterize other materials.

Desiccants and Use of Desiccators

The choice of the drying agent required for the absorption of moisture depends on the composition of the material to be dried and on the desiccant's convenience of use, efficiency, and sometimes cost. For example, certain drying agents are deliquescent, and when liquefaction of the drying agent occurs, a decline in drying efficiency results. Most published information concerning the comparative efficiency of drying agents is based upon experiments that measured the amount of moisture absorbed from air flowing through a bed of desiccant. Desiccants that produce dust should also be avoided; granular calcium chloride, for instance, frequently carries a large amount of dusty "fines" when fresh. Drying agents that incorporate cobalt chloride or some other moisture-sensitive salt to indicate exhaustion, although more expensive than agents without an indicator, are much preferred. Silica gel and anhydrous calcium sulfate (Drierite) are examples of agents that contain indicators. Some drying agents can easily be regenerated by heat, an important consideration when comparing costs. The properties of various desiccants are described in Table 1-4.

Desiccators should be opened slowly. Desiccators often contain air at less than atmospheric pressure as a result of the cooling of hot samples. If such a desiccator is not opened slowly, the inrush of air may create drafts sufficient to dislodge materials from open vessels or to stir up dust particles from the drying agent. Vacuum desiccators should be provided with a curved inlet tube to deflect incoming air against the lid, and the stopcocks on these containers should be opened very carefully when the internal pressure is restored to that of the atmosphere.

GENERAL LABORATORY SUPPLIES

Most of the laboratory ware used in clinical chemistry today is made of either glass or plastic. The advantages as well as the limitations imposed by the properties of these materials should be appreciated in light of their possible effects upon analyses performed in such containers.

Table 1-3. SOME STANDARD REFERENCE MATERIALS (SRMs) FOR CLINICAL CHEMISTRY

Analyte	Material	SRM Number
Albumin (bovine)	Powder	926
Angiotensin I (human)	Powder	998
Bilirubin	Powder	916
Blood gas, 10% CO_2 + N_2	Gas in cylinder	1700
Blood gas, 5% CO_2; 12% O_2 + N_2	Gas in cylinder	1701
Blood gas, 5% CO_2; 20% O_2 + N_2	Gas in cylinder	1702
Blood gas, 10% CO_2; 7% O_2 + N_2	Gas in cylinder	1703
Cadmium*	Human serum, lyophilized	909
Calcium*	Human serum, lyophilized	909
Calcium carbonate	Powder	915
Chloride*	Human serum, lyophilized	909
Cholesterol*	Human serum, lyophilized	909
Cholesterol	Powder	911A
Chromium*	Human serum, lyophilized	909
Copper*	Human serum, lyophilized	909
Cortisol	Powder	921
Creatinine	Powder	914
Ethosuximide	Human serum, lyophilized	900
Glucose*	Human serum, lyophilized	909
Glucose	Powder	917
Iron*	Human serum, lyophilized	909
Iron	Metal	937
Lead, 20 ng/mL*	Human serum, lyophilized	909
Lead	Porcine blood	955
Lead nitrate	Powder	928
Lithium*	Human serum, lyophilized	909
Lithium carbonate	Powder	924
Magnesium*	Human serum, lyophilized	909
Magnesium gluconate	Powder	929
Mannitol	Powder	920
p-Nitrophenol	Powder	938
Phenobarbital	Human serum, lyophilized	900
Phenytoin	Human serum, lyophilized	900
Potassium*	Human serum, lyophilized	909
Potassium chloride	Powder	918
Potassium dichromate	Crystalline	935
Primidone	Human serum, lyophilized	900
Quinine sulfate	Powder	936
Sodium chloride	Powder	919
Trace elements	Bovine serum	8419
Tripalmitin	Crystalline	1595
Urea	Powder	912
Uric acid*	Human serum, lyophilized	909
Uric acid	Powder	913
Vanadium*	Human serum, lyophilized	909
VMA (4-hydroxy-3-methoxymandelic acid)	Powder	925
Glass filters		930D
Spectrophotometer wavelength standard	Holmium oxide solution	2034
Thermometers, clinical laboratory		933
		934

*All analytes are contained in same lyophilized serum.

(Modified from Seward, R. W., and Mavrodineanu, R.: Standard Reference Materials: Summary of the Clinical Laboratory Standards Issued by the National Bureau of Standards. Washington, D.C., National Bureau of Standards, 1981.)

Glassware

Glass is a complex silicate whose properties depend on the type of silicate anion in the structure as well as its cation content. The addition of metal cations such as Fe(III) or Ni(II) to the basic structure of the glass can alter its color. The thermal properties of glass can be significantly improved by the addition of boron oxide (B_2O_3). This type of glassware, known as *borosilicate glass,* is extensively used in clinical laboratories.

Table 1-4. PROPERTIES AND ACTIVITIES OF DESICCANTS

Drying Agent	Activity*	Capacity	Deliquescence	Easy Regeneration	pH
Phosphorus pentoxide	0.02	Very low	Yes	No	Acidic
Barium oxide	0.6–0.8	Moderate	No	No	Alkaline
Alumina	0.8–1.2	Low	No	≥175 °C	Neutral
Magnesium perchlorate (anhydrous)	1.6–2.4	High	Yes	No	Neutral
Calcium sulfate (Drierite)	4–6	Moderate	No	≥275 °C	Neutral
Silica gel	2–10	Low	No	≥120 °C	Neutral
Potassium hydroxide (stick)	10–17	Moderate	Yes	No	Alkaline
Calcium chloride (anhydrous)	330–380	High	Yes	No	Neutral

* Micrograms of residual water per liter of air at 30 °C.

Glass with *high thermal resistance* is usually a borosilicate glass with low alkali content. It is free from zinc-group elements and heavy metals (arsenic and antimony) and resists heat, corrosion, and thermal shock. Because this glass has a relatively low coefficient of expansion, its dimensions change very little with temperature changes. Therefore, this type of glassware should be used whenever heating or sterilization by heat is to be performed. Some borosilicate glassware, if properly supported and not under internal pressures, can be heated to about 600 °C for a relatively short period of time. If the glass is cooled too quickly, however, it will acquire strains that may affect its future serviceability. The highest safe operating temperature for this glass is called its *strain point*, or the temperature at which deformation may result due to heat stress. *Pyrex* (Corning) and *Kimax* (Kimble) glass are the most common thermal-resistant borosilicate glassware found in the laboratory. Most clinical laboratory apparatus such as beakers, flasks, and pipets are made of this type of glass. Because this glassware contains so few elements, it contaminates liquids minimally, even when they are hot.

Several years ago a special alumina-silicate glass was developed that is at least six times stronger than borosilicate glass. This *Corex* (Corning) laboratory glassware has been strengthened chemically. Corex laboratory glassware is harder than conventional borosilicates and is better able to resist scratching and clouding due to alkali. This glass is also used for high-temperature thermometers (250 °C), graduated cylinders, and centrifuge tubes.

Vycor (Corning) laboratory glassware is recommended for use in applications involving high temperatures, drastic heat shock, and extreme chemical treatment with acids and dilute alkalis. This transparent glassware is resistant to attack by all acids except hydrofluoric, and even in the upper temperature range it is more resistant to alkali than is borosilicate glass. Vycor ware is used primarily in ashing and ignition techniques. It can be heated to 900 °C and can withstand the shock of temperature drop from 900 to 0 °C.

The high silica content (over 96%) of *high-silica glass* makes it comparable to fused quartz in its thermal endurance, chemical stability, and electrical characteristics. The glass is made by removing almost all elements except silica. It has good optical and temperature characteristics; it is used for high-precision analytical work and can also be used for optical reflectors and mirrors.

Boron-free glassware has high resistance to alkali and was developed particularly for use with strongly alkaline solutions and for digestions involving strong alkali. Its thermal resistance is much less than that of borosilicate glass, and it must therefore be heated and cooled very carefully. This glass is often referred to as "soft glass."

Low-actinic glassware contains materials that impart an amber or red color to the glass, thus reducing light exposure of its contents. It was developed to provide a highly protective laboratory glassware for handling materials sensitive to light in the 300–500 nm range (e.g., bilirubin, carotene, and vitamin A).

Flint glass is a soda-lime glass composed of a mixture of oxides of silicon, calcium, and sodium. This type of glass is lowest in cost of all glasses and is readily fabricated in a wide variety of shapes. Such glass has poor resistance to high temperatures and sudden changes of temperature, and its resistance to attack by chemicals is only fair. Since this glass is relatively easy to melt and shape, it is used for bottles and for some disposable laboratory glassware such as pipets. If not rinsed after manufacture, soda-lime pipets may release alkali into the pipetted liquid and will cause considerable errors. Organic contaminants found in disposable pipets made of soft or borosilicate glass can also interfere with analytical procedures using UV absorption or fluorescence techniques.

Colored and opal glasses, used in filters and light bulbs, are made by adding small amounts of metallic oxides to the glass.

Coated glass has a thin, metallic oxide permanently fire-bonded to its surface; it can conduct electricity, and it has electronic applications as a heat shield to protect against infrared light and as an electrostatic shield to carry off charges.

Optical glass is made of soda lime, lead, and borosilicate; because of its high optical purity, it is used in making prisms, lenses, and optical mirrors.

Glass ceramics (Pyroceram) have high thermal resistance, chemical stability, and corrosion resistance, and they are useful for making hot plates, table tops, and heat exchangers.

Radiation-absorbing glass is made of soda lime and lead and is useful in preventing transmission of high-energy radiation (e.g., γ- and X-rays).

Plasticware

The introduction of plasticware has greatly enhanced laboratory analysis. Plastic beakers, bottles, flasks, graduated cylinders, funnels, centrifuge tubes, tubing, and pipets have unique qualities that make them ideal for use when high corrosion resistance and high-impact and tensile strength are required. Table 1-5 lists the resistance of plasticware to various chemicals.

The *polyolefins* (polyethylene and polypropylene) are a unique group of resins that are chemically relatively inert. Although concentrated sulfuric acid slowly attacks polyethylene at room temperature, the polyolefins as a group are unaffected by acids, alkalis, salt solutions, and most aqueous solutions. Aromatic, aliphatic, and chlorinated hydrocarbons cause moderate swelling in these plastics at room temperature, whereas organic acids and halogens slowly penetrate them. Strong oxidizing agents attack this group of resins only at elevated temperatures. Polyethylene is less expensive than polypropylene and is used in most disposable plasticware. Polypropylene has a distinct advantage in that it withstands higher temperatures and can be sterilized; however, it absorbs pigments and tends to become discolored.

Polycarbonate is twice as strong as polypropylene and may be used at temperatures ranging from −100 to +160 °C; however, its chemical resistance is not as wide as that of the polyolefins. This resin is unsuitable for use with strong acids, bases, or oxidizing agents, and it is dissolved by chlorinated aliphatic and aromatic hydrocarbons. Polycarbonate resin is insoluble in aliphatic hydrocarbons, some alcohols, and dilute aqueous acids and salts. Since labware molded from this resin is glass-clear and shatterproof, polycarbonate is used extensively in centrifuge tubes and graduated cylinders.

Fluorocarbon resins (fluorinated ethylene propylene, Teflon) have unique qualities that make them almost chemically inert and therefore ideal when high corrosion resistance at extreme temperatures is essential. Because Teflon resists temperatures ranging from −270 to +255 °C, bottles and beakers made of this material are suitable for use in cryogenic experiments. Labware made of fluorocarbon resins is translucent white and is inert to such corrosive reagents as boiling aqua regia, nitric and sulfuric acids, boiling hydrocarbons, ketones, esters, and alcohols. Because of its unique antiadhesive properties and its nonwettable surface, Teflon is used for nonlubricated stopcocks, stirring bars, bottle cap liners, and tubing. It is also quite easy to clean and is fast drying, but it can easily be scratched and warped.

Use and care of plastic laboratory ware. Whenever possible, plastics should be used in place of glass, because plasticware has the advantage of being unbreakable and it does not release ions into a solution as does glass. However, polyethylene is permeable to water vapor, and thus evaporation may occur. This phenomenon does not occur if

Table 1-5. CHEMICAL RESISTANCE OF PLASTICS AT 20 °C

Classes of Substances	Acids, Dilute or Weak	Acids,* Strong and Concentrated	Alcohols, Aliphatic	Bases	Hydrocarbons, Aliphatic	Hydrocarbons, Aromatic	Hydrocarbons, Halogenated	Ketones	Oxidizing Agents, Strong
Low-density polyethylene	E	E	E	E	F	F	N	G	F
High-density polyethylene	E	E	E	E	G	G	F	G	F
Polypropylene/polyallomer	E	E	E	E	G	F	F	G	F
Polymethylpentene ("TPX")	E	E	E	E	F	F	N	G	F
Teflon	E	E	E	E	E	E	E	E	E
Polycarbonate	E	N	G	N	F	N	N	N	N
Polysulfone	E	G	G	E	G	N	N	N	G
Polyvinyl chloride	E	E	E	E	E	N	N	N	G
Polystyrene	E	F	E	E	N	N	N	N	N
Nylon	F	N	G	F	E	E	G	E	N

E = Excellent; G = Good; F = Fair; N = Not recommended.
*Except for oxidizing acids; for oxidizing acids, see "Oxidizing Agents, Strong."
(From Nalgene Labware, Nalge Co., Rochester, NY.)

containers are filled with liquid. Therefore, small volumes of reagent should never be stored in oversized plastic bottles for long periods of time. Furthermore, polyethylene may bind or adsorb proteins, dyes, stains, iodine, and picric acid. Colorless reagents may bind to plasticware without being detected, only to manifest themselves in subsequent chemical analysis as a cause of erratic results.

Linear polyethylene, polypropylene, Teflon, polymethylpentene, and polycarbonate plastics can be cleaned in ordinary glassware washing machines. Ultrasonic cleaners may also be used, provided the plasticware does not rest directly upon the transducer diaphragm. The use of abrasive cleaners and strong oxidizing agents should be avoided.

Polypropylene, Teflon, and polymethylpentene may be repeatedly autoclaved under normal conditions. Polycarbonate shows some loss of mechanical strength when autoclaved; thus, this procedure should be limited to 20 min at 121 °C. Plasticware should be well cleaned and rinsed with deionized water before it is sterilized, since certain chemicals that have no appreciable effect on plastics at room temperature may cause deterioration at autoclave temperatures. Polystyrene, polyvinyl chloride, styrene acrylonitrile, and conventional polyethylene are not autoclavable, but they may be gas-sterilized with ethylene oxide or chemically sterilized by rinsing with benzalkonium chloride. With the exception of Teflon, none of the listed resins should be hot-air sterilized because of the potential for accelerated oxidative degradation.

Some of the transparent plastics such as polycarbonate and polystyrene may absorb minute quantities of water vapor and appear cloudy after autoclaving. This clouding effect is transient and disappears as the plastic dries. Clearing of the plastic may be accelerated by drying the plasticware in an oven at 110 °C.

Cleaning of glass- and plasticware. Laboratory glassware should be rinsed immediately after use and placed in a weak solution of a detergent until it is washed. Hazardous materials should be safely discarded, and their containers should be washed immediately and not be left with other vessels to be picked up by personnel who wash glassware. In many institutions, the glassware is washed in an automatic glassware washer, rinsed in a rinse cycle, and then placed in an automatic dryer. The manufacturers of automatic washing equipment recommend detergents compatible with the equipment and effective with the local water supply. Some residual detergents may be detected by measuring the pH of water added to the glassware or by using a dilute solution of an acid-base indicator that will detect an alkaline residue.

For general washing without an automatic washer, most laboratories prefer *detergents* that are not highly alkaline, are nonionic, and are metal free. Special care must be taken to ensure adequate rinsing. Volumetric glassware and glass apparatus must be absolutely clean to ensure that measured volumes are accurate and that chemical reactions are therefore not adversely affected. One gross method that may be used to test for cleanliness is to fill the vessel with distilled water and then empty it and examine the walls to see whether they are covered by a continuous, thin film of water. Imperfect wetting or the presence of discrete droplets of water indicates that the vessel is not sufficiently clean. Obviously this criterion cannot be used on nonwettable plastics. The number of pieces of glassware checked depends on the work load and number of pieces cleaned, but at least one piece of randomly selected glassware should be checked each day.

Of the various cleaning agents in common usage, the National Bureau of Standards prefers fuming sulfuric acid and a chromic–sulfuric acid mixture (see below). However, in general the method used to clean glassware should be as mild as possible and appropriate to the type of contamination present.

Ultrasonic cleaners may be used to supplement the action of detergent. In these baths, where very-high-frequency vibrations cause the soil to be broken free from the walls of the vessel, neutral detergents are usually preferred.

A *cleaning solution* that consists of sodium dichromate in sulfuric acid is a common cleaning agent. It is prepared by stirring about 20 g of powdered, technical grade sodium dichromate with just enough water to make a thick paste. Slowly and carefully add 300 mL of technical grade concentrated sulfuric acid, with stirring. This preparation is best carried out in a sink. Clear supernatant solution should be decanted for use. Because of its strong oxidizing power, this material removes organic material as well as inorganic material; it should not be used on any plasticware except Teflon. Scrupulous rinsing with tap water followed by reagent grade water must follow the use of this strong chemical cleaner. Personnel handling this solution must wear safety glasses, rubber gloves, and a rubber or plastic apron because of the corrosive properties of the cleaning solution. The solution may be used until the reddish color of dichromate has been replaced by the green color of the chromic ion.

Nitric acid may also be used as a cleaning agent, with the same safety precautions. In addition, strong nitric acid solutions should be handled only in a chemical hood. Although nitric acid effectively removes trace metals, relatively dilute solutions of hydrochloric acid ($\leq$1 mol/L) are equally effective and are more easily used, since they do not require the use of a hood. Personnel should take care not to reintroduce trace metals when rinsing.

Air drying or oven drying at temperatures below 100 °C, with the laboratory ware placed bottom-up, is the preferred drying method. Occasionally it is desirable to rinse the glassware with a water-miscible organic solvent and then to expose it to a stream of air or nitrogen. To prevent contamination, the solvent must be of high quality and the gas must be pure. Storage should be such that the laboratory ware is protected from dust.

Synthetic and Rubber-Base Tubing

Tygon is a modified polyvinyl plastic substance that is clear and resistant to chemical attack and is used extensively for the manufacture of tubing. It is inert to most chemicals and comes in a variety of sizes, which are specified by inside diameter (ID), by outside diameter (OD), and by wall thickness. Tygon tubing is flexible. It can be used with a variety of positive-action (peristaltic) fluid-pumping systems. Because it is available in continuous lengths, Tygon requires few joints or couplings. It is readily cut with a sharp blade and can be coupled to other tubings by using a small amount of solvent (cyclohexanone) as a bond. Careful heating of the ends of the tube with a hot metal probe can also assist in softening the plastic for coupling of tubing. Tygon will sometimes be difficult to slip over metal or glass in making a connection. A soapy solution or a synthetic wetting agent will lubricate the tubing, allowing it to slip on more easily. In contact with solutions containing zinc or copper, Tygon can become discolored at high temperatures. Teflon tubing can be used in place of Tygon in circumstances requiring its special properties.

Amber *latex* rubber tubing is translucent and is recommended for glass connections in which a highly elastic tubing with a long life is required. Neoprene is a synthetic rubber that is soft and pliable with a smooth finish. Neoprene is not affected by oil, alkalis, hot water, and many corrosive substances. However, it is not recommended for use with chlorinated or aromatic hydrocarbons.

Volumetric Equipment and Its Calibration

Clinical chemistry procedures require exact volumetric measurements to ensure accurate results. For precise work, only Class A glassware should be used. Class A glassware is certified to conform to the specifications outlined in NBS circular C-602.

Pipets

Two types of pipets commonly used in clinical chemistry are transfer pipets and measuring pipets (Figure 1-1). *Transfer pipets* include volumetric and Ostwald-Folin pipets and are designed to deliver a fixed volume of liquid. A *volumetric transfer pipet* is calibrated to deliver accurately a fixed volume of a dilute aqueous solution such as a standard, a filtrate, or a nonviscous sample. These pipets are allowed to drain by gravity. For complete drainage the flow rate should be adjusted to NBS flow rates and the tip

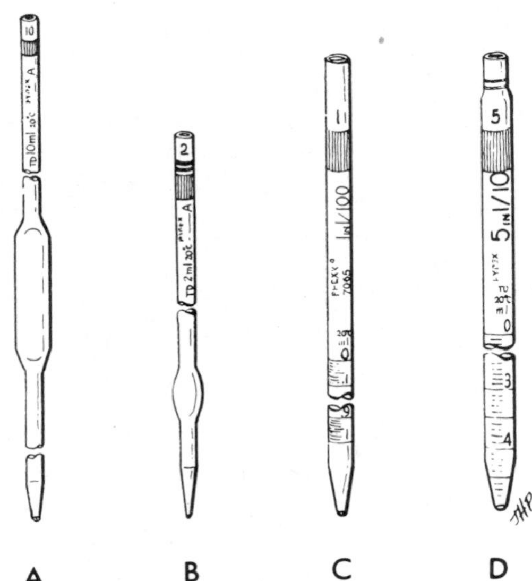

Figure 1-1. Pipets. *A,* Volumetric (transfer). *B,* Ostwald-Folin (transfer). *C,* Mohr (measuring). *D,* Serologic (graduated to the tip).

must be placed against the side of the accepting vessel but should not touch the liquid in the vessel. Table 1-6 shows examples of tolerances of various types of pipets. Since reliability of the calibration diminishes with a decrease in size, special micropipets have been developed.

Ostwald-Folin pipets (Figure 1-1,*B*) are similar to volumetric pipets but have their bulb closer to the delivery tip, thereby reducing the surface area in contact with the liquid. These pipets are used for the accurate measurement of viscous fluids such as blood or serum. In contrast to a volumetric pipet, an Ostwald-Folin pipet has an etched ring near the mouthpiece, indicating that it is a blow-out pipet. The liquid is blown out of the pipet only after the blood or serum has drained freely to the last drop in the delivery tip. Controlled, slow drainage is required with all viscous solutions so that no residual film is left on the walls of the pipet. The top of the meniscus must be read when the pipet is filled with opaque fluids such as blood.

The second type of pipet is the *graduated or measuring pipet* (Figure 1-1,*C* and *D*). Two kinds are available: the Mohr pipet is calibrated between two marks on the stem, and the other, the serologic pipet, has graduation marks down to the tip. The serologic pipet (Figure 1-1,*D*) has an etched ring (or pair of rings) near the mouth end of the pipet, signifying that it is a blow-out pipet. Mohr pipets deliver between their calibration marks. Serologic pipets have a larger orifice than the Mohr pipets and thus drain faster. Measuring pipets are principally used for the transfer of reagents and are not generally considered accurate enough to pipet samples and standards.

Borosilicate glass pipets require frequent inspection to be certain that pipets with broken tips and those that have become badly etched are discarded. Pipets that are made from special chemically tempered glass are now available. Although they cost

Table 1-6. TOLERANCES OF VARIOUS TYPES OF PIPETS

Type	Size, mL	Class	Limit of Error, mL
Transfer	5	A	0.0100
Ostwald-Folin	5	A	0.0100
Serologic	5		0.0400
Mohr	5		0.0400

more, they are less likely to break than their borosilicate counterparts. In addition, they are less susceptible to scratching and chipping because the glass is harder. Plastic pipets are generally not considered to be sufficiently accurate for quantitative chemical analysis.

There are certain common techniques that apply to all pipets described above. First, mouth pipetting should never be allowed. Instead, pipetting bulbs should be used. Pipets must be held in a vertical position when the liquid level is adjusted to the calibration line and during delivery. The lowest part of the meniscus, when it is sighted at eye level, should be level with the calibration line on the pipet. The flow of the liquid should be unrestricted when volumetric pipets are used, and the tips must be touched to the inclined surface of the receiving container until two seconds after the liquid has ceased to flow.

With measuring pipets, the flow of liquid may have to be slowed during delivery. *Serologic pipets* are calibrated to the tip, and the etched glass ring on top of the pipet signifies that it is to be blown out. However, the final drop of liquid is blown out only after the fluid in the pipet has drained completely.

Verification of pipet calibration. The following procedure is used to verify calibration of a to-deliver pipet. Water, weighing vial, and pipet must all be at room temperature at the time of the procedure.

1. Determine and record the temperature (t) of a large volume of Type I water to the nearest whole degree.
2. Weigh a clean, dry, empty, stoppered weighing vial. Handle the vial without touching it with bare fingers. Record its weight to the nearest 0.1 mg (W_v).
3. Transfer a sample of water to the weighed vial with the pipet to be tested.
4. Restopper and reweigh the vial. Record the weight to the nearest 0.1 mg (W_f).
5. Refer to the table entitled "True capacity of glass vessels from the weight of the contained water or mercury when weighed in air" in *The Handbook of Chemistry and Physics*[9] to obtain the correction factor (F_t) for the water temperature.
6. Calculate:

$$(W_f - W_v) \times (F_t) = \text{actual capacity, mL at temperature of manufacturer's calibration}$$

Example for a 10 mL pipet:
W_f = 31.9961 g
W_v = 22.0391 g
t = 24 °C
F_t = 1.003 771 (from table)[9]

$$(31.9961 - 22.0391) \times (1.003\ 771) = 9.9945 \text{ mL, actual capacity}$$

The deviation, or error, from stated capacity is

$$\frac{(10 - 9.9945)}{10} \times 100 = 0.055\%$$

In routine analysis, an error of 0.1% or less may be ignored when using the pipet. With an error in excess of 0.1%, depending upon the application, the stated capacity of a given pipet may require correction to the actual capacity found.

Micropipets. In micro work, the volume of fluid that coats the inner wall of a pipet after dispensing can cause significant error. For this reason, most micropipets are calibrated "to contain" (TC) the stated volume rather than to deliver it. Proper use requires rinsing the pipet with the final solution up to the mark but not beyond it after the contents are delivered into the diluent. Volumes are expressed in microliters (μL). Micropipets are generally available in small sizes, ranging from 1 to 500 μL.

Semiautomatic pipets and dispensers. Semiautomatic versions of pipets and dispensers are available in sizes from 1 μL to 20 mL. Figure 1-2,*A* illustrates a positive-displacement pipet. The capillary attachments into which the sample is drawn may be siliconized glass, glass, or plastic. Models that allow for digital adjustment of volume are also available.

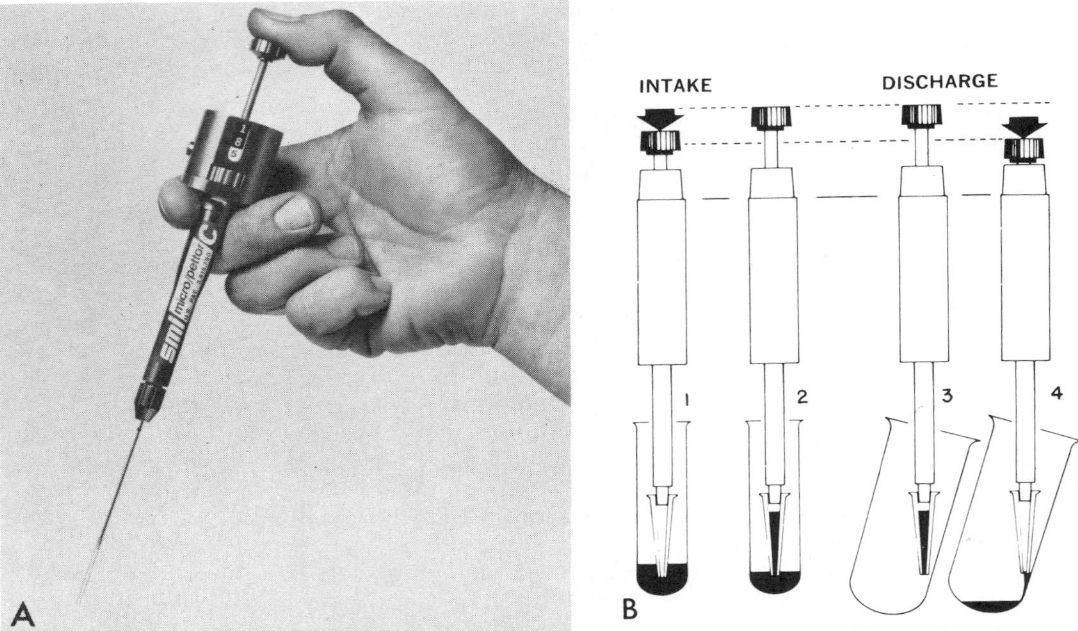

Figure 1-2. *A,* Multirange micropipetting device. (Courtesy of American Dade, Miami, FL 33152) *B,* A semiautomatic micropipetting device that is designated as a TD pipet.

Figure 1-2,*B* illustrates another type of single or multirange micropipetting device. This device draws up and dispenses its predefined volume when the plunger is moved through a complete cycle. The fluid containment tip is made of a disposable plastic material that tends to retain less inner surface film than does glass. Such tips avoid the risk of cross contamination between samples, eliminate the need for washing between samples, and improve the precision of measurements. A "macro" version of this device can be used to dispense or transfer larger volumes. The useable range is 1–5 mL, and the device may be adjusted for any volume within those limits.

Figure 1-3,*A* shows an example of a diluter that is a semiautomatic sampling, diluting, and dispensing apparatus that measures and dispenses preset volumes of solutions by means of two motor-driven syringes, one for metering the sample and one for metering the diluent. The volumes of sample aspirated and diluent dispensed are adjustable. The reproducibility of the metered pumps is claimed to be ±0.5% of delivered volume. A more automatic and versatile piece of equipment is shown in Figure 1-3,*B*. This is really a complete liquid handling system that can aspirate up to five different fluids and deliver them with diluent. The equipment is controlled by a microprocessor and is easily programmed. Accuracy is claimed to be better than 0.5% of dispensed volume and reproducibility better than 0.1% when at least 10% of syringe volume is dispensed.

Calibration of micropipets, dispensers and dilutors. Accurate calibration of micropipets, dispensers, and dilutors is as important as the calibration of volumetric macropipets. The classical procedure for checking the calibration of micropipets employs gravimetric measurement of mercury measured with the micropipet.

Gravimetric calibration. Determine the weight of mercury contained or delivered by the micropipet and divide by the density of mercury at its observed temperature to find the volume of the pipet. The density of mercury at 25 ± 5 °C is 13.5340 ± 0.0123 (± 0.091%); therefore, for temperatures between 20 and 30 °C it is sufficiently accurate to use the density at 25 °C. The weights needed and the calculations performed are analogous to the procedure for verification of pipet calibration (page 17).

Spectrophotometric calibration of micropipets. As an alternative to the gravimetric measurement of mercury, a colored compound of known absorptivity may be

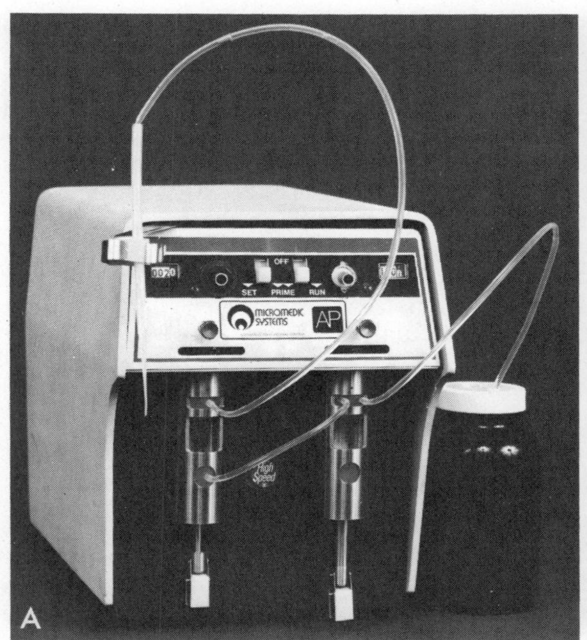

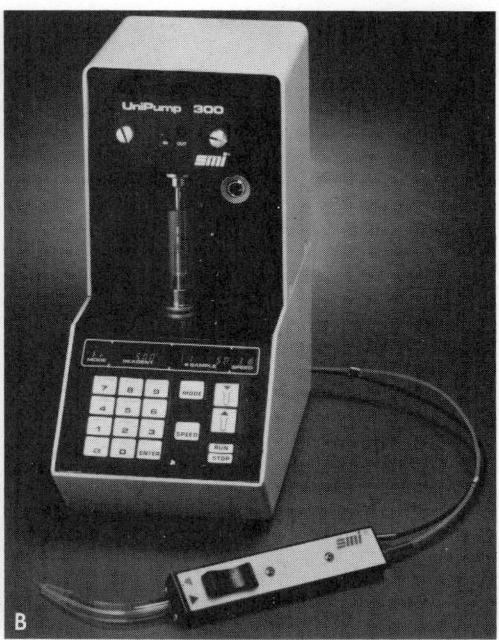

Figure 1-3. *A,* Semiautomatic sampling, diluting, and dispensing apparatus. (Courtesy of Micromedic Systems, Inc.) *B,* An automatic and versatile sampling, diluting, and dispensing apparatus. (Courtesy of American Dade)

used to verify the calibration of micropipets. The following is a procedure using 4-nitrophenol in a spectrophotometric verification of the calibration of a 10-μL pipet.

1. All glassware used should meet Class A specifications.
2. Reagents required:
   a. NaOH, 0.01 mol/L
   b. 4-Nitrophenol, 105 mg/dL. Dissolve 105 mg of high purity 4-nitrophenol (NBS SRM 938) in Type I deionized water in a 100-mL volumetric flask. Fill to the mark.
3. Preparation of dilutions:

Reference dilutions: Fill three 250-mL volumetric flasks to the mark with NaOH, 0.01 mol/L. To each add 1.0 mL of the 4-nitrophenol solution, using a different volumetric pipet for each addition.

Test dilutions: Add to each of five test tubes 2.5 mL NaOH, 0.01 mol/L, using either a calibrated pipet or calibrated dispenser. Add 10 μL of 4-nitrophenol solution to each tube, using the technique that is appropriate for the type of micropipet tested.

4. Read the absorbance of each of the reference and test dilutions, in 10-mm cuvets, at 401 nm on a narrow-bandpass spectrophotometer, using NaOH, 0.1 mol/L, as a blank. For NBS 4-nitrophenol with a "certified specific absorbance" of 131.48 $L \times g^{-1} \times cm^{-1}$ in NaOH, 0.01 mol/L, A_{401} should be 0.550 for an exactly prepared reference dilution. Average the readings of the three reference dilutions; the mean value (A_1) should be close to 0.550. Average the five test dilution readings to obtain a mean value (A_2).
5. Calculation:

$$A_2/A_1 \times \mathrm{D} \times \mathrm{V} = \text{Volume, in } \mu\text{L, delivered by pipet}$$

where D is the dilution of test solution (1/251 in this example), and V is the final volume in μL of the test dilution (2510 μL in this example).

For this example, then, if A_1 were 0.550 and A_2 were 0.561, $0.561/0.550 \times 1/251 \times 2510\ \mu L = 10.20\ \mu L$ delivered by pipet with a nominal capacity of 10 μL. This is a 2% error. The analyst must make a decision as to the acceptability of this error. As supplied by the manufacturer, micropipets of this capacity are capable of delivering between 0.5

and 1% of nominal capacity, depending on the type of pipet. Automatic dilutors can be calibrated by using similar techniques.

Calibration of micropipets and dilutors by using radioisotopes. Another technique to verify the calibration of micropipets or dilutors and to estimate the precision of their performance is available to laboratories routinely using gamma-emitting radioisotopes. Because of its availability and low radiation hazard, the isotope of choice is ^{125}I, and any water-soluble, labeled material of high specific activity is acceptable.

To achieve a valid estimate of precision, (1) the total volumes of the replicates must be identical in order to avoid variations in counting geometry, (2) the count rate (counts per minute) must be at least 50 times the background count, and (3) the total counts must be at least 40 000 to result in a coefficient of variation (CV) of 0.5%. The minimal counting time to achieve this total count may be calculated from

$$T = N/R$$

or

$$T = 40\,000/R$$

where N is the net total count, R is the rate, and T is the time.

The accuracy of the micropipet or dilutor may be ascertained by comparing it with a "reference" or "master" pipet whose calibration has been verified with a gravimetric technique.

Burets

Standard burets vary in size from 1 to 100 mL. Burets having a capacity of 10 mL or less are classified as microburets.

The outflow of liquid from the buret is usually controlled by an all-glass or all-Teflon stopcock. The latter type does not require any lubricant and is especially useful when the titrant is an alkali. An all-glass stopcock should be lightly greased with petrolatum-like lubricants.

A buret calibration is verified by accurate delivery of carbon dioxide–free water to a tared vessel, which is then stoppered and weighed. The delivered volume is then corrected for temperature (see *Verification of pipet calibration*, page 17, for method and calculations). As is the case with graduated pipets, the accuracy of the overall calibration and mark-to-mark calibration decreases with a decrease in buret size.

Microburets. Some microburets are constructed like a syringe. The plunger and the ID of the syringe can be machined to an exact diameter. The displacement caused by advancing the plunger is indicated on a micrometer dial and is proportional to the volume delivered. A typical model is shown in Figure 1-4. The tip of the pipet must be placed beneath the surface of the solution to be titrated to permit even dispensing of reagent and to avoid discrete drop formation. The tip must have a fine point to prevent diffusion of liquid back into the tip.

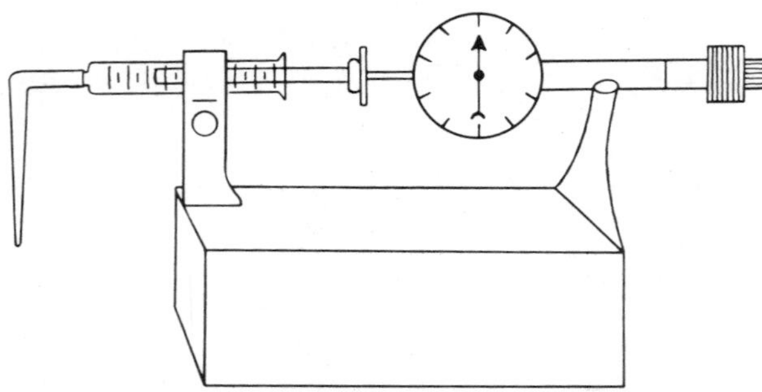

Figure 1-4. Syringe-type microburet.

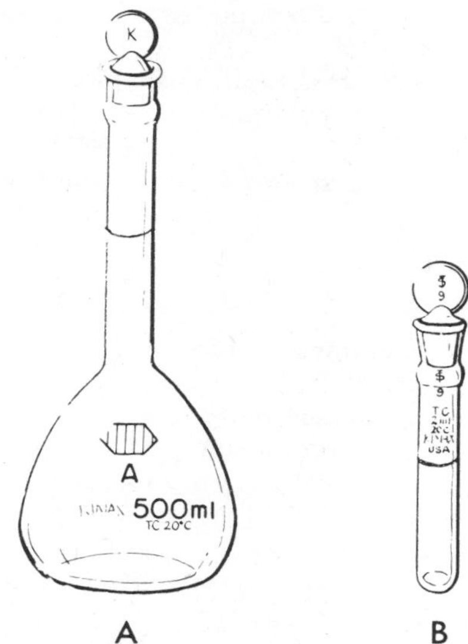

Figure 1-5. Volumetric flasks. *A,* Macro. *B,* Micro.

Volumetric Flasks

Volumetric flasks (Figure 1-5), commonly found in sizes varying from 1 to 4000 mL, are primarily used in preparing solutions of known concentration. They are available in various grades. The most accurate are certified to meet standards set forth by the National Bureau of Standards.

An important factor in the use of volumetric apparatus is the accurate sighting of the meniscus. A small card that is half black and half white is most useful. The card is placed 1 cm behind the apparatus with the white half uppermost and the top of the black area about 1 mm below the meniscus. The meniscus then appears as a clearly defined, thin black line. This device is also useful in reading the meniscus of a buret.

Volumetric equipment should be used with solutions equilibrated to room temperature. Solutions diluted in volumetric flasks should be repeatedly mixed during dilution so that the contents are homogeneous before the solution is made up to final volume. Errors due to expansion or contraction of liquids upon mixing are thereby minimized.

Calibration. For small volumetric flasks, a flask is weighed and then filled with carbon dioxide–free Type I deionized water to just above the graduation mark. The meniscus mark is set at the graduation line by removing excess water and the flask is reweighed. The final weight is corrected for the temperature to obtain the volume of the flask (page 17). Large-volume volumetric flasks may be calibrated by the spectrophotometric technique previously described (page 18).

Centrifuges

A centrifuge is a device to accelerate gravitational separation of substances differing significantly in their masses. In the clinical laboratory centrifugation is used:

1. To separate particles from a solution in which they are suspended. Examples of this application are (a) separating cells from blood to provide cell-free plasma or serum for analysis; (b) concentrating cellular elements and other components of biological fluids for microscopic examination or chemical analysis; (c) eliminating chemically precipitated protein from an analytic specimen; (d) separating protein-bound or antibody-bound ligand from free ligand in immunochemical and other assays; and

2. To separate two liquid phases of different densities (e.g., an aqueous phase from

an organic solvent) or to separate lipid components from plasma or serum, or from each other.[5]

Despite years of experience with centrifuges, there is no specific recommendation for the RCF or the time used for centrifugation of blood specimens. NCCLS standard H18-P proposes an RCF of 1000–1200 $\times g$ for 10 $\pm$ 5 min, but clearly considerable deviations from the recommendation allow adequate separation. No standards have been established for centrifugation of other specimens, such as serum to which a protein precipitant has been added.

Types of Centrifuges

Centrifuges may be classified generally into three types: horizontal-head or swinging-bucket, fixed-angle or angle-head, and ultracentrifuges.

Horizontal-head or *swinging-bucket centrifuges* allow the tubes placed in the cups of the rotor to assume a horizontal plane when the rotor is in motion and a vertical position when it is at rest. During centrifugation, particles travel in a constant manner along the tube while the tube is at a right angle to the shaft of the centrifuge. Thus, the sediment is distributed uniformly against the bottom of the tube. The surface of the sediment is flat (parallel to the shaft of the centrifuge) and remains so, with a column of liquid on top of it when the rotor stops.

In an *angle-head rotor*, tubes are held at a fixed angle from 25 to 40 degrees to the vertical axis of rotation. As in the horizontal-head centrifuge, particles are driven outward horizontally but strike the side of the tube so that the sediment packs against the side and bottom of the tube and the surface of the sediment is parallel to the shaft of the centrifuge. As the rotor slows down and then stops, gravity may cause the sediment to slide down the tube; generally a poorly packed pellet is formed.

The shape of a *fixed-angle rotor* allows more rapid sedimentation of small particles than is usually possible with a horizontal-head rotor. Fixed-angle rotors can be run at a higher speed than swinging-bucket rotors, which offer considerable resistance to rotation and generate heat as a result of air friction. The sedimentation of large particles is efficient at low speed, so the horizontal-head rotor is adequate for the separation of erythrocytes from plasma or of a protein precipitate from a supernatant.

Ultracentrifuges are high-speed centrifuges that use mainly fixed-angle rotors. The most common application of an ultracentrifuge in the clinical laboratory is the separation of lipoproteins. Because the separation may require hours or days and may generate considerable heat, ultracentrifugation requires a refrigerated chamber. The tabletop model Airfuge Ultracentrifuge (Beckman Instruments, Inc.) is a miniature air turbine with a small rotor capable of achieving a centrifugal force of 165 000 times gravity ($\times$ g). It is used in the clinical laboratory to clear serum of chylomicrons so that accurate analyses may be performed on the infranatant.

Components of a Centrifuge

All centrifuges contain a rotor or centrifuge head, drive shaft, and motor. The rotor is enclosed in a chamber that is provided with a cover and latch. Most centrifuges also have a power switch, timer, speed control, tachometer, and brake. Some are equipped with a refrigerator to reduce the temperature within the chamber. Centrifuges may also have audible or visible alarms to indicate malfunctions such as an imbalance of the rotor. Some models of centrifuges are available with a plastic cover over the rotor chamber or over each swinging bucket to reduce aerosol formation when a tube breaks during centrifugation.

The typical design of a horizontal-head rotor uses pairs of buckets that swing freely. The buckets accept a variety of racks or adapters into which the tubes to be centrifuged are inserted. Most tube holders contain cushioning pads to lessen the possibility of tube breakage during centrifugation.

The drive motor of a large centrifuge is usually a high-torque, heavy-duty, direct current (DC) electric motor. Alternating current (AC) motors are usually used with small centrifuges. Power is transmitted to the rotors through a commutator and brushes. Generally the rotor shaft is driven directly or through a gyro system, but occasionally a pulley system is used. In most centrifuges sealed bearings are used to reduce the need for lubrication and to minimize vibration. The speed of the centrifuge rotor is governed by a potentiometer, which changes the voltage supplied to the motor. However, differences in the load cause the rotor to turn at different speeds, even if the potentiometer setting is the same. The tachometer indicates the speed of the rotor (revolutions per minute, rpm), not the gravitational force applied. The brake operates by reversing the polarity of the current to the motor, thus causing rapid deceleration. The timer on a centrifuge allows a rotor to reach a preprogrammed speed under optimized conditions and then to decelerate after the set time without braking.

Centrifugation generates heat as a result of friction. This may affect the stability of the material centrifuged and may also cause evaporation. The temperature in the chamber of many centrifuge models may increase by as much as 5 °C after a single run to separate serum or plasma from cells. The change in temperature depends on the initial ambient temperature in the centrifuge chamber, on the rotor speed, and on the duration of centrifugation. A large chamber around the rotor dissipates heat. When the material to be centrifuged is temperature labile, a refrigerated centrifuge should be used.

Principles of Centrifugation

The speed of centrifugation, measured in rpm, does not describe the force required to separate two phases in a centrifuge. The correct term is *relative centrifugal force* (RCF), also called relative centrifugal field. Units are expressed as number of times greater than gravity (e.g., $5000 \times g$).

Maximum relative centrifugal force is calculated as follows:

$$\mathrm{RCF} = 1.118 \times 10^{-5} \times \mathrm{r} \times \mathrm{n}^2$$

where 1.118×10^{-5} is an empirical factor; r is the horizontal distance (i.e., the radius in centimeters) from the center of rotation to the bottom of the tube in the rotor cavity or bucket during centrifugation; and n is the speed of rotation of the rotor in rpm.

The RCF achieved with a centrifuge at a given rpm may also be determined from a nomogram (see *Appendix*, Figure 29-3). RCF is derived from the distance from the rotor center to the bottom of the tube, whether the tube is horizontal to, or at an angle to, the rotor center. The minimum RCF, calculated in the same way but from the center of rotation to the *surface* of the liquid, may be several hundred *g* less than the maximum value. Note also that the RCF applied to a tube in a fixed-angle head may be much less than that applied to the same tube in a horizontal-head rotor, because the tube is unable to swing outward, which reduces the effective radius.

Duplication of the conditions of centrifugation is often desirable. The following is a useful formula for calculating speed required of a rotor whose radius differs from that with which a prescribed RCF was originally defined.

$$\text{rpm (alternate rotor)} = 1000 \times \sqrt{\frac{\text{RCF (original rotor)}}{11.18 \times \text{r (cm, alternate rotor)}}}$$

The time required to sediment particles depends on the rotor speed, the radius of the rotor, and the effective pathlength traveled by the sedimented particles (i.e., the depth of the liquid in the tube). Length of time for centrifugation can also be calculated so that centrifugation with an alternate rotor of a different size gives an RCF equivalent to the original.

$$\text{Time (alternate rotor)} = \frac{\text{Time} \times \text{RCF (with original rotor)}}{\text{RCF (with alternate rotor)}}$$

Note, however, that the time for each rotor to reach operating speed (which may vary from instrument to instrument) and the deceleration time, during which sedimentation is still occurring (but less efficiently), cannot be accounted for.

Operation of the Centrifuge

For proper operation of a centrifuge, only those tubes recommended by its manufacturer should be used. Polypropylene tubes are generally capable of withstanding RCFs of up to 5000 × *g*. The tubes should have a tapered bottom and should be of a size to fit securely into the rack to be centrifuged. The top of the tube should not protrude so far above the bucket that the swing into a horizontal position is impeded by the rotor.

For smooth operation of the centrifuge, the rotor must be properly balanced. The weight of racks, tubes, and their contents on opposite sides of a rotor should not differ by more than 1% or by an acceptable limit established by the manufacturer. The smaller the difference, the smoother the centrifugation. The placement of tubes should be symmetrical and opposite weights be equal. The total weight of each rack should not exceed the limits stated by the centrifuge manufacturer for a rated speed.

Imbalance of the rotor causes vibration. Even when an imbalance is not great enough to trigger the imbalance detector, tubes break more frequently, and wear on the centrifuge is increased. A slight vibration can normally be expected when the rotor is accelerating or decelerating through slow speeds. Vibration may be amplified at a critical range of speed during the acceleration or deceleration. Continuous operation of a centrifuge in this critical range should be avoided. Although the rotor must pass through this range as it decelerates, centrifugal force is still enough to prevent disturbance of a well-packed sediment.

Tubes of collected blood should be centrifuged prior to being unstoppered to reduce the probability that an aerosol will be produced during centrifugation. If a specimen needs to be recentrifuged after removal of the stopper, the tube should be restoppered or covered with Parafilm. After centrifugation the lid of the centrifuge should not be opened until the rotor has come to a complete stop.

Operating Practice

The cleanliness of a centrifuge is important in minimizing the possible spread of infectious agents. When sample tubes break, the racks and chamber of the centrifuge must be carefully cleaned. Any spillage should be considered a possible hazard. Gray dust arising from the sandblasting of the chamber by fragments of glass indicates tube breakage and possible contamination, necessitating cleaning of the chamber. Broken glass embedded in the cushions of tube holders may be a continuing cause of breakage if cushions are not inspected and replaced in the cleanup procedure.

The speed of a centrifuge should be checked at least once every three months with the use of an external tachometer of known accuracy. The measured speed should not differ by more than 5% from the rated speed under specified conditions. All the speeds at which the centrifuge is commonly operated should be checked. The centrifuge timer should be checked weekly against a reference timer such as a stop watch and should not be more than 10% in error. The temperature of a refrigerated centrifuge should be measured monthly under reproducible conditions and should be within 2 °C of the expected temperature. Commutators and brushes should be checked at least every three months, and brushes should be replaced when they show considerable wear.

Criteria of satisfactory performance should be established for every centrifuge, just as for other laboratory instruments. Maintenance and repair records should be carefully kept.

Balances and Weighing

Mass is an invariant property of matter. Weight is a function of mass under the influence of gravity, a relationship expressed in the equation

$$\text{Weight} = \text{Mass} \times \text{gravity}$$

Two substances of equal mass and subject to the same gravitational force have equal weights. A balance compares the mass of an unknown against a known mass. In practice, the terms *mass* and *weight* are used synonymously.

The classic form of a balance is a beam poised on an agate knife-edged fulcrum, a pan hanging from each end of the beam, and a rigid pointer hanging from the beam at the poise point. When the object to be weighed is placed on one pan and weights of equal mass are on the other pan, the pointer will come to rest at an equilibrium point (i.e., will indicate "balance.")

More than one balance is required for a clinical laboratory, since there is a need to weigh, for example, 3 kg of a timed urine specimen as well as microgram amounts of a drug for a standard solution. Coarse balances of large capacity (up to 5 kg) have a sensitivity of 0.1 g. The typical analytical balance has a capacity of 200 g with a sensitivity of ~10 μg. Microbalances may have a maximum capacity of as little as 5 g and a sensitivity of 0.1 μg.

All balances require a vibration-free location. The more sensitive a balance, the more protection it needs, not only from vibration but also from air currents that can disturb the equilibrium between the weighed object and the weights. Loose crystals of chemicals or liquids with corrosive vapors should not be permitted to remain on or around the immediate area of the pans. Good weighing technique at sensitivities under 1 g calls for handling weights and weighed objects with forceps to avoid deposition of moisture, oils, or salts from an analyst's skin.

Principles of weighing. There are three general principles of weighing. In weighing by *direct comparison*, weights are added to one side of the beam to counterbalance the weight of the object on the other side. In weighing by *substitution*, weights are removed from the side of a balance to which the object to be weighed has been added, to restore equilibrium. Electronic balances use an *electromagnetic force* to balance the weight of the object.

Types of balances. A balance may have one or two pans. *Double-pan* balances have a single beam with arms of equal length. Standard weights are added manually to the right-side pan to counterbalance the weight of the object on the other; in some models a dial or vernier with chain is used to make fine adjustments to the mass associated with the right-side pan. In *single-pan* balances, the arms are of unequal length. The object to be weighed is placed on the pan attached to the shorter arm. A restoring force is applied mechanically to the other arm to return the beam to its null position. Double- and triple-beam balances are forms of the unequal-arm balance that are generally designed to weigh bulk reagents and have relatively low accuracy. Single-pan analytical balances that use the weighing by substitution principle are also available.

The *electronic single-pan* balance is a common balance in the clinical laboratory. When a load on the pan causes the beam to tilt downward, a null detector senses that the beam has deviated from the equilibrium point. An electromagnetic force is applied to return the balance beam to its null position. The electromagnetic force takes the place of weights in a two-pan balance. The restoring force is proportional to the weight on the pan and is applied through a solenoid or torque motor. The current required to produce the force is displayed digitally in a form equivalent to the mass on the balance pan. The accuracy of an electronic balance depends on the linearity of both the digital voltmeter and the torque motor. Most electronic balances have a built-in provision for taring so that the weight of the container can be subtracted easily from the total weight measured.

Analytical weights. Analytical weights must be used to counterbalance the weight of objects weighed on two-pan balances and to verify the performance of both single and two-pan balances. The National Bureau of Standards defines several classes of analytical weights. *Class M weights* are of primary standard quality and are used only to calibrate other weights. *Class S weights* are used for calibrating balances. In the clinical laboratory, balances should be calibrated at least monthly as well as prior to very accurate analytical work. *Class S-1 weights* have greater tolerance limits than class S weights and are used for routine analytical work. Tolerances of different weights have been defined by the NBS. For class S weights from 5 to 1 g, the tolerance is ±0.054 mg; from 500 to 100 mg, ±0.025 mg; and from 50 to 1 mg, ±0.014 mg.

Concept of Solute and Solvent

In clinical chemistry most measurements are concerned with the concentration of substances in solutions such as blood, serum, urine, spinal fluid, or other body fluids. The substance that is dissolved in a solution is a *solute*. Thus, the compounds that are typically measured are solutes. In clinical chemistry these are frequently referred to as *analytes*. The substance in which the solute is dissolved is a *solvent*. A *solution* is correctly defined as a physically homogeneous mixture of two or more substances. In clinical chemistry we are concerned primarily with the measurement of solids in liquids, when there is a relatively large amount of solvent in comparison to the amount of solute.

When a solution holds as much of a dissolved solute as it can at a specific temperature, it is said to be *saturated*. Saturation is influenced by the temperature of the solution, atmospheric pressure, and the nature of the solute and solvent. Solvents generally dissolve more solute at higher temperatures, whereas the solubility generally decreases with decrease in temperature. An *unsaturated* solution is one that contains less solute than the solvent is capable of holding. In a *supersaturated* solution (which contains more solute than it can hold when the solution is saturated), the addition of undissolved substance, jarring, or stirring will cause precipitation of the excess solute and will produce a saturated solution.

When two liquids dissolve in each other in any proportion, they are completely *miscible*. If the liquids do not dissolve at all in each other, they are completely *immiscible*. When each of two liquids is partially soluble in the other, they are said to be *partially miscible*.

Solutions

The concentration of a solute in a solution may be expressed in different ways. Traditionally, in clinical chemistry concentration has most often been expressed as mass of solute per volume of solution (i.e., mass concentration). However, the *Système International* (SI, see page 33) recommends the use whenever possible of moles of solute per volume of solution (i.e., substance concentration) and the use of liter as the reference volume (Table 1-7).

Gases dissolve in liquids to form solutions. *Henry's law*, which is applicable to gases at low pressures, states that the solubility of a gas in a liquid is directly proportional to the pressure of the gas above the liquid. The relationship between pressure and solubility varies with the nature of the gas. When several gases are dissolved at the same time in a single solvent, the solubility of each gas is proportional to its partial pressure in the mixture. The solubility of most gases in liquids decreases with an increase in temperature, and indeed boiling a liquid will frequently drive out all dissolved gases. Traditionally, the unit used to describe the concentration of gases in liquids has been per cent by volume (v/v). Using the SI, gas concentrations are expressed in moles per cubic meter (mol/m^3).

Dilute, concentrated, and standard solutions. Many measurements in clinical chemistry require the preparation of dilute working solutions (standards or reagents) from concentrated stock solutions.

Making dilutions

Most dilutions are made by transferring a precise volume of a concentrated solution into an appropriate flask, adding water or other diluent to the required volume, and mixing appropriately to ensure homogeneity. With the use of proper techniques, errors in the preparation of dilutions should always be $<0.1\%$. Errors may be of two types—random and determinate. *Random errors* vary nonreproducibly from one measurement to another and may thus be minimized by careful technique. A *determinate error* affects every one of a set of measurements, remaining fixed for all measurements. Such errors may be avoided by using correctly calibrated equipment and careful laboratory techniques.

If a solution to be pipetted does not have the same density, viscosity, or surface tension as water, the volume of solution delivered may not be the same as it would be

Table 1-7. CONCENTRATION QUANTITIES AND UNITS

Name of Quantity	Definition	Unit
Substance concentration [of a given solute component]	Amount of substance of a solute divided by volume of solution	mol/m^3; mol/L
Molality [of a given solute component]	Amount of substance of a solute divided by mass of solvent	mol/kg
Mole fraction (or substance fraction) [of a given component]	Amount of substance of a component divided by amount of substance of mixture (i.e., all components of the system)	mol/mol
Mole ratio (or substance ratio) [of a given solute component]	Amount of substance of a solute divided by amount of substance of the solvent	mol/mol
Mass concentration [of a given component]	Mass of a component (e.g., solute) divided by volume of system (e.g., solution)	kg/m^3; kg/L
Mass fraction [of a given component]	Mass of a component divided by mass of system (mixture)	kg/kg
Volume fraction [of a given component]	Volume of a component divided by volume of system (mixture)	m^3/m^3; L/L
Number concentration	Number of specified particles or elementary entities divided by volume of system (mixture)	m^{-3}; L^{-1}
Number fraction	Number of specified particles or elementary entities divided by total number of particles or entities in the system (mixture)	1 (a ratio)
Substance content [of a given component]	Amount of substance of a component divided by mass of system (mixture)	mol/kg

(From The SI for the Health Professions, World Health Organization, 1977.)

for water, with which pipets are usually calibrated. For dilute solutions, this error is usually so small that it can be ignored. To avoid errors that arise when two liquids of very different viscosity are mixed, the technique of diluting to volume is used. Instead of adding 90 mL of water to 10 mL of concentrated solution, the 10 mL of concentrated solution should be pipetted into a 100-mL volumetric flask and water added to bring the volume to 100 mL.

When a solution is diluted, its volume is increased and its concentration decreased, but the total amount of solute remains unchanged. The following formula can be used to determine the volume of a concentrated solution that is required to make a known volume of a less concentrated solution.

$$\text{Concentration}_1 \times \text{volume}_1 = \text{Concentration}_2 \times \text{volume}_2$$

Physical and Chemical Units

Generally, laboratory data are reported in the United States in terms of mass concentration (i.e., the mass of solute per unit volume of solution). Although the deciliter (dL) has been the traditional reference volume, SI recommends the liter (L). An exception to the general expression of analyte concentrations in terms of volume of solution is the measurement of osmolality, in which concentrations are expressed in terms of mass of solvent (kg water).

When the solution and solvent are both liquids, as in alcohol solutions, the concentration of such a solution is frequently expressed in terms of volume per volume (v/v). By adding 70 mL alcohol to a volumetric flask and diluting it to 100 mL with water, a solution whose concentration is 700 mL/L would be achieved. The expression, 700 mL/L, is preferred to the alternatives of 70 volumes per cent or 70% (v/v).

A *molar* solution contains 1 mole of solute in one liter of solution (1 mol/L); for example, since the molecular weight of H_2SO_4 is 98.08, a 1-molar solution of H_2SO_4

contains 98.08 g H_2SO_4 per liter of solution. The symbol M, to denote molarity, is no longer preferred and has been replaced by mol/L. A *molal* solution contains 1 mole of solute in 1 kg of solvent.

A *normal* solution contains 1 gram-equivalent weight of solute in one liter solution; i.e., 1 mol HCl, 0.5 mol H_2SO_4, and 0.33 mol H_3PO_4, each in one liter of solution, are one normal solutions. The normality scale is limited in that a given solution may have more than one normality, depending on the type of reaction (acid-base or oxidation-reduction) for which the solution is used. Normality is no longer recommended to express concentrations. Nevertheless, the term is included here because it remains in common usage and is related to the concept of equivalents, still favored for serum electrolyte concentrations in the United States and some other countries.

The following equations define the expressions of concentrations:

$$\text{Mole} = \frac{\text{Weight (g)}}{\text{Gram molecular weight}}$$

$$\text{Molarity of a solution} = \frac{\text{Number of moles of solute}}{\text{Number of liters of solution}}$$

$$\text{Molality of a solution} = \frac{\text{Number of moles of solute}}{\text{Number of kilograms of solvent}}$$

$$\text{Normality of a solution} = \frac{\text{Number of gram equivalents of solute}}{\text{Number of liters of solution}}$$

$$\text{Normality (in oxidation-reduction reaction)} = \text{Molarity} \times \text{difference in oxidation state}$$

$$\text{Gram-equivalent weight (as oxidant or reductant)} = \frac{\text{Gram molecular weight}}{\text{Difference in oxidation state}}$$

A milliequivalent of a substance is its equivalent weight expressed in milligrams. The equivalent weight of H_2SO_4 is 49.04 g. Then, one milliequivalent of H_2SO_4 is 49.04 mg H_2SO_4. Since substances may react on the basis of their valence, one mole of calcium (atomic weight = 40), which is bivalent, has twice the combining power of one mole of sodium (atomic weight = 23). Forty mg Ca^{2+} is therefore as reactive as two times 23 mg Na^+.

The unit of measurement commonly used to express the concentration of electrolytes in plasma is the milliequivalent (mEq), which is one-thousandth of an equivalent.

$$\text{Milliequivalents (mEq)} = \frac{\text{Weight (g)}}{\text{Milliequivalent mass (g)}}$$

Milligrams per 100 mL (dL) can be converted to mEq per liter by the following formula:

$$\text{mEq/L} = \frac{\text{mg/dL} \times 10 \times \text{valence}}{\text{mg atomic mass}}$$

Example: If the serum sodium concentration is 322 mg/dL, then the serum contains 3220 mg/L. The equivalent mass of sodium is 23, and the valence is 1; therefore,

$$\text{mEq/L} = \frac{322 \times 10 \times 1}{23} = 140$$

The recommended units for sodium concentration in plasma are millimoles per liter (mmol/L). In the example above, the concentration of sodium is

$$\text{mmol/L} = \frac{\text{mg/L}}{\text{mg molecular mass}} = \frac{322 \times 10}{23} = 140$$

In chemical terms the *titer* of a solution is the mass of a substance equivalent to a unit volume of the solution.

$$\text{Titer} = \frac{\text{Grams of substance}}{\text{Liters of solution}} = \frac{\text{mg substance}}{\text{mL solution}}$$

However, in clinical laboratory practice a titer is more commonly thought of as the lowest dilution at which a particular reaction takes place. Titer is customarily expressed as ratio (e.g., 1:10 or 1 to 10).

Procedures for Concentrating Solutions

Evaporation is often used to concentrate solutes in a solution. A technique frequently used for the evaporation of volatile solvents is to direct a stream of air through a fine nozzle onto the surface of the liquid in a tube to flush out vapors in equilibrium with the solution. This procedure can be used on many tubes simultaneously by using a manifold, and it can be speeded up by placing the tubes in a heated water bath. However, this procedure should be carried out in a fume hood to avoid a health hazard from evaporating solvents. If compressed air is used, it should be cleansed by passing it through a filter and over a desiccating agent.

The rapid removal of large volumes of solvent is best effected through the use of a combination of reduced pressure and heat. The solution to be evaporated is placed in a round-bottomed flask of a capacity no larger than 100 mL to avoid the danger of implosion. Vacuum is applied to the flask through a safety trap, and the flask is rotated in a heated water bath. Swirling creates a large surface area, which speeds evaporation and reduces bumping. The use of a suitable trap to collect the vapors drawn off is essential. To concentrate a large volume of solution, a rotary evaporator is used. Multiple-unit evaporators are available to concentrate several specimens at once.

Freeze-drying may be used for the preparation of control materials, unstable reagents, or individual specimens for analysis. More than 99% of the water in a substance is removed in the typical freeze-drying process. The composition of the freeze-dried material is generally unchanged except for the loss of water or some volatile organic solvents. Freeze-drying entails first freezing a material at −40 °C or less and then subjecting it to a high vacuum. The very low pressure causes the ice to sublime while the nonsublimable material remains behind in a dried state.

Extraction Procedures

Extraction is the process used to separate compounds on the basis of their different solubilities in different solvents. The technique may be used either to purify or to concentrate a material. Application of the technique and the principle on which it is based are discussed in Chapter 2, Section Six, and in Chapter 18.

The choice of a solvent for extraction depends on the solute to be recovered. Generally, it is desirable to use a solvent that is easily evaporated and does not form emulsions. An emulsion consists of a colloidal suspension of an organic solvent in an aqueous solution or vice versa. Emulsification occurs most often when the solute acts as a detergent or when viscous solutes are present. Emulsification is less likely to occur when the two solvents are swirled together rather than shaken vigorously. An emulsion may be broken by salting out—adding an inorganic salt soluble in water—which causes the solubility of organic solvents in water to be reduced. Other techniques that may be effective include the addition of a silicone defoamer, a detergent, an acid, or several drops of ethanol or diethyl ether; the last method is especially useful when the organic layer is chloroform. An emulsion may also be resolved by allowing it to stand, by centrifugation, or by cooling the mixture.

Separatory funnels are used in simple extraction procedures. A known volume of solution is brought into contact with a known volume of solvent, which is immiscible in the solution, and the funnel is shaken until equilibrium is attained. The liquid layers can be separated by draining the lower layer through a stopcock at the bottom of the separatory funnel. This method of separation is rapid and simple.

Phase separating paper (Whatman No. 1 PS) is a silicone-impregnated paper that removes aqueous solutions from water-miscible solutions. Because the paper is hydrophobic the organic layer passes through it but the aqueous layer is retained. Solvents heavier than water flow through the apex of the filter cone, whereas solvents lighter than water pass through the wall of the cone.

Filtration

Two types of filtration are used to separate particles from liquids: surface and depth filtration. *Surface filtration* (screen filtration), common in the clinical laboratory, is performed with filter papers, membranes, or sieves. The process is one of exclusion of the particles having a larger diameter than the pores of the separation medium. In *depth filtration*, a thick layer of separation medium allows particles to be retained in the body of the filter as well as on the surface. Depth filters are usually made of cotton, fiberglass, asbestos, or other materials such as diatomaceous earth. Depth filters usually consist of a matrix of fibers arranged in a randomized manner to eliminate channels through the filters but still allow the passage of fluid.

Filtration may be carried out by gravity or with the aid of vacuum. The latter may be used to accelerate the filtration rate by combining the influences of gravity, suction, and capillary attraction. The factors affecting gravity filtration include the nature of the filter medium, its surface area, and the size of its pores. The number and size of particles in the solution and its pH also may affect the filtration rate.

Filter paper is used in many of the filtrations carried out in the clinical laboratory. Different types of filter paper include low-ash or ashless paper as well as paper graded according to thickness. Vacuum filtration requires hardened grade paper with high, wet tensile strength. Fiberglass papers have the greatest strength and combine the advantages of retention of fine particles with fast filtration. Table 1-8 describes the grade, flow rate, and retention characteristics of commonly used filter papers.

Much of the filtration in clinical laboratories is now carried out with *membranes* of controlled pore size. These filters are made from homogeneous polymeric materials such as polyvinylidene fluoride, cellulose esters, cellulose acetate, polytetrafluoroethylene, and polyvinyl chloride, depending on the intended application of the filter. The most widely used filter is composed of cellulose acetate and cellulose nitrate. These filters contain no loose fibers or particulate matter and are manufactured in a variety of sizes up to 30 cm in diameter and with pore sizes that vary from 10 to 0.025 μm, the latter size being suitable for sterilization of a liquid. Eighty per cent of the surface area of a typical membrane filter is occupied by pores, so that high flow rates occur even with pores of the smallest diameter. The basic structure of membrane filters is hydrophobic, although the surface of the membrane can be modified chemically to make it hydrophilic. Membrane filters may be used under vacuum, with positive pressure, or with gravity.

Table 1-8. GRADE, FLOW RATE, AND RETENTION CHARACTERISTICS OF FILTER PAPERS

Grade	Thickness, mm	Ash, %	Wet Strength*	Initial Filtration Speed†	Particle Size Retained, μm
40	0.20	0.010	L	75	8
41	0.21	0.010	L	12	20-30
42	0.20	0.010	L	240	2.5
43	0.21	0.010	L	40	16
44	0.17	0.010	L	175	3
50	0.12	0.025	H	250	2.7
52	0.17	0.025	H	55	7
54	0.18	0.025	H	10	20-25
540	0.16	0.008	H	55	8
541	0.16	0.008	H	12	20-25
542	0.15	0.008	H	250	2.7

*L = Low; H = High.

† Time taken for 100 mL of water to pass through 15-cm quadrant-folded circle, according to a modified ASTM method D981–56.

(Information from Whatman Laboratory Products, 9 Bridewell Place, Clifton, NJ.)

Ultrafiltration is a technique for removing dissolved particles by an extremely fine filter. It is used to concentrate macromolecules such as proteins, because smaller dissolved molecules pass through the filter.

Many specimens of small volumes of biological fluids may be filtered and concentrated at the same time by using porous *cellulose acetate filter* cones and a centrifuge. The filter cones fit inside centrifuge tubes and centrifugal force is used to drive water and molecules of molecular weight $<50\,000$ through the anisotropic membrane filter. This approach is widely used to concentrate proteins in urine or cerebrospinal fluid for electrophoresis.

Buffer Solutions and Their Action

Buffers are defined as substances that resist changes in the pH of a system. All weak acids or bases form buffer systems in the presence of their salts. The action of buffers and their role in maintaining the pH of a solution can best be explained by the Henderson-Hasselbalch equation, which is derived as follows:

The ionization of a weak acid, HA, and of a salt of that acid, BA, can be represented as

$$HA \leftrightarrows H^+ + A^-$$

$$BA \longrightarrow B^+ + A^-$$

The dissociation constant for a weak acid may be calculated from the following equation:

$$\frac{[H^+][A^-]}{[HA]} = K_a$$

Thus,

$$[H^+] = K_a \times \frac{[HA]}{[A^-]}$$

or

$$\log [H^+] = \log K_a + \log \frac{[HA]}{[A^-]}$$

Now, multiplying throughout by -1,

$$-\log [H^+] = -\log K_a - \log \frac{[HA]}{[A^-]}$$

Since, by definition, $pH = -\log [H^+]$ and $pKa = -\log K_a$, we may write:

$$pH = pK_a + \log \frac{[A^-]}{[HA]}$$

Since A^- is derived principally from the salt that is completely dissociated, the equation may, for practical purposes, be written:

$$pH = pK_a + \log \frac{[\text{salt}]}{[\text{undissociated acid}]}$$

or simply:

$$pH = pK_a + \log \frac{[salt]}{[acid]}$$

where [salt] = $[A^-]$ = concentration of salt, and [acid] = [HA] = concentration of weak acid.

Consequently, the pH of the system is determined by the p*K* of the acid and the ratio of $[A^-]$ to [HA]. The buffer has its greatest buffer capacity at its p*K*, i.e., that pH at which the $[A^-]$ = [HA]. Entered into the preceding equation, this gives

$$pH = pK_a + \log 1$$
$$pH = pK_a + 0$$

The capacity of the buffer decreases as the ratio deviates from 1. In general, buffers should not be used at a pH outside the range of $pK_a \pm 1$. If the ratio of [salt] to [acid] is beyond 50/1 or 1/50, the system is considered to have lost its buffering capacity. This point is ~1.7 pH units to either side of the p*K* of the acid, since

$$pH = pK_a + \log 50/1 \qquad \text{or} \qquad pH = pK_a + \log 1/50$$
$$pH = pK_a + 1.7 \qquad\qquad pH = pK_a - 1.7$$

For acetate buffer, the relationship between pH and the ratio of CH_3COONa to CH_3COOH, ($[A^-]/[HA]$), is shown in Table 1-9.

The chemical mechanisms by which buffers exert their effect may be seen by considering the reactions involved when base is added to a buffer solution containing acetate ions, CH_3COO^-, and acetic acid molecules, CH_3COOH. On addition of NaOH:

$$\begin{matrix} CH_3COO^-H^+ \\ CH_3COO^-\ Na^+ \end{matrix} + Na^+ + OH^- \longrightarrow \begin{matrix} CH_3COO^-\ Na^+ \\ CH_3COO^-\ Na^+ \end{matrix} + HOH$$

Base (OH^-) is removed by combining with the hydrogen ion dissociated from acetic acid, thus minimizing pH changes. The addition of alkali decreases $[CH_3COOH]$ in the buffer and increases $[CH_3COONa]$. The pH of the solution increases in proportion to the change in the logarithm of the ratio of the concentration of salt to acid in the buffer solution.

On addition of HCl:

$$\begin{matrix} CH_3COO^-\ H^+ \\ CH_3COO^-\ Na^+ \end{matrix} + H^+ + Cl^- \longrightarrow \begin{matrix} CH_3COO^-\ H^+ \\ CH_3\ COO^-H^+ \end{matrix} + Na^+ + Cl^-$$

H^+ is removed by combining with acetate to form poorly dissociated acetic acid. In this case, the addition of HCl acts to decrease $[CH_3COONa]$ and increase $[CH_3COOH]$ in the

Table 1-9. CONCENTRATION, SALT/ACID RATIO, AND pH OF ACETATE BUFFER

CH_3COONa, mol/L	CH_3COOH, mol/L	Ratio Salt/Acid	pH
0.00	0.20	0.00	2.7
0.01	0.20	0.05	3.4
0.05	0.20	0.25	4.1
0.10	0.20	0.50	4.4
0.20	0.20	1.00	4.7
0.40	0.20	2.00	5.0
1.00	0.20	5.00	5.4
2.00	0.20	10.00	5.7

buffer. The pH of the solution falls in proportion to the change in the logarithm of the ratio of salt to acid in the solution; thus only a small change in pH occurs.

Tables for the preparation of buffers commonly used in clinical chemistry laboratories are given in Tables 29-11 to 29-14 in the *Appendix*.

Units of Measurement

A meaningful measurement requires both a number and a unit. The unit identifies the dimension—mass, volume, or concentration—of a measured property. The number indicates how many units are contained in the property. Traditionally, measurements in the clinical laboratory have been made in metric units. In the early development of the metric system, units were referenced to length, mass, and time. The first system was based on the centimeter, gram, and second (CGS), and then later, the meter, kilogram, and second (MKS) formed the basis. The current system is called the *Système International d'Unités* (SI) and was accepted internationally in 1960. The *units of the system* are called SI units.

International System of Units[7]

There are three classes of SI units—base, derived, and supplemental units.

A *base unit* was decreed by the Conférence Générale des Poids et Mesures (CGPM) for each of seven fundamental but dimensionally independent physical quantities. These units, and the provisionally accepted katal, are listed in Table 1-10. A *derived unit* is derived mathematically by multiplication or division of one or more base units. The SI is coherent since derived units relate unequivocally to base units and contain no numerical factor other than one. A *supplemental unit* is a unit that conforms to the SI but that has not been classified as either base or derived. At present, only the radian (for plane angles) and the steradian (for solid angles) are classified this way. The kilogram is the only SI base unit with a prefix. Multiples and submultiples of all other base units are denoted by an appropriate prefix to the base unit.

Because derived units consist of base units linked by multiplication or division, some are complex and have been given special names and symbols to simplify their use. Some examples of derived units appear in Table 1-11. For measurements of properties that are dimensionless, such as refractive index or relative density, the appropriate SI unit is 1 (one) because the derived unit is a ratio between two identical base units.

Some units outside the SI continue to be important and useful in particular applications; an example is the liter as the reference volume in clinical chemistry. Liter is the name of the submultiple (cubic decimeter) of the SI unit of volume, the cubic meter. The CGPM recommends that such exceptional units as the liter should not be combined with SI units and should be replaced with SI units whenever possible.

The minute, hour, and day have had such longstanding use in everyday life that it is unlikely that new SI units derived from the second could supplant them. Some other non-SI units are still accepted because they are very important in a few specialized fields. Examples of such units are illustrated in Table 1-12.

Table 1-10. SI BASE UNITS

Quantity	Name	Symbol
Length	meter	m
Mass	kilogram	kg
Time	second	s
Electric current	ampere	A
Thermodynamic temperature	kelvin	K
Amount of substance	mole	mol
Luminous intensity	candela	cd
Catalytic amount	katal	kat

Table 1-11. EXAMPLES OF SI-DERIVED UNITS IMPORTANT IN CLINICAL MEDICINE, EXPRESSED IN TERMS OF BASE UNITS

Quantity	Name	SI Symbol	Expression in Terms of Other SI Units	Expression in Terms of SI Base Units
Volume	cubic meter	m^3		m^3
Mass density	kilogram per cubic meter	kg/m^3		kg/m^3
Concentration of amount of substance	mole per cubic meter	mol/m^3		mol/m^3
Frequency	hertz	Hz		s^{-1}
Force	newton	N		$m \cdot kg \cdot s^{-2}$
Pressure	pascal	Pa	N/m^2	$m^{-1} \cdot kg \cdot s^{-2}$
Energy, work, quantity of heat	joule	J	$N \cdot m$	$m^2 \cdot kg \cdot s^{-2}$
Power	watt	W	J/s	$m^2 \cdot kg \cdot s^{-3}$
Electric potential, potential difference, electromotive force	volt	V	$W \cdot A^{-1}$	$m^2 \cdot kg \cdot s^{-1} \cdot A^{-1}$

Definition of Base Units

Length. The unit of length is the *meter*. It is defined as "the length equal to 1 650 763.73 wavelengths in vacuum of the radiation corresponding to the transition between the levels $2p_{10}$ and $5d_5$ of the krypton-86 atom."

Mass. The *kilogram* is the unit of mass and is equal to the mass of the international prototype of the kilogram. Mass refers to the quantity of matter and is used in place of the ambiguous term "weight," which reflects the force of gravity.

Time. The *second*, the unit of time, is "the duration of 9 192 631 770 periods of the radiation corresponding to the transition between two hyperfine levels of the ground state of the cesium-133 atom."

Electric current. The *ampere* is that constant current which, if maintained in two straight parallel conductors of infinite length and of negligible circular cross section, and placed one meter apart in vacuum, would produce between these conductors a force equal to 2×10^{-7} newtons per meter of length.

Temperature. The triple point of water is the fundamental fixed point for temperature measurements. The temperature of 273.16 K (kelvin) was assigned to this point. The *kelvin* is defined as the fraction 1/273.16 of the triple point of water. The unit "degree Celsius" (°C) is exactly equivalent to the unit kelvin.

Luminous intensity. The *candela* is the luminous intensity, in the perpendicular direction, of a surface of 1/600 000 square meter of a blackbody at the temperature of freezing platinum under a pressure of 101 325 newtons per square meter.

Amount of substance. The *mole* is the amount of substance that contains as many elementary entities as there are atoms in 0.012 kilogram of ^{12}C. When the mole is used, the elementary entities must be specified and may be atoms, molecules, ions, electrons, other particles, or specified groups of such particles.

Table 1-12. NON-SI UNITS RETAINED FOR USE WITH THE SI

Quantity	Unit	Symbol	Value in SI Units
Time	minute	min	1 min = 60 s
	hour	h	1 h = 3 600 s
	day	d	1 d = 86 400 s
Plane angle	degree	°	1° = π/180 rad
	minute	′	1′ = π/10 800 rad
	second	″	1″ = π/648 000 rad
Volume	liter	L	1 L = 1 dm^3 or 10^{-3} m^3
Mass	tonne	t	1 t = 1000 kg

Table 1-13. PREFIXES AND THEIR SYMBOLS USED TO DESIGNATE DECIMAL MULTIPLES AND SUBMULTIPLES

Prefix	Symbol	Factor	Numerical Value
tera	T	10^{12}	1 000 000 000 000
giga	G	10^{9}	1 000 000 000
mega	M	10^{6}	1 000 000
kilo	k	10^{3}	1000
hecto*	h	10^{2}	100
deka*	da	10^{1}	10
deci*	d	10^{-1}	0.1
centi*	c	10^{-2}	0.01
milli	m	10^{-3}	0.001
micro	μ	10^{-6}	0.000 001
nano	n	10^{-9}	0.000 000 001
pico	p	10^{-12}	0.000 000 000 001
femto	f	10^{-15}	0.000 000 000 000 001
atto	a	10^{-18}	0.000 000 000 000 000 001

Prefixes marked with * do not conform to the systematic pattern with exponents that are simple multiples of three and should be avoided in scientific practice.

Decimal Multiples and Submultiples of SI Units

In the practical application of SI units, certain values are too large or too small to be expressed conveniently in the base or derived units. Numeric values can be brought to convenient size when the unit is appropriately modified by prefixes. Generally, the prefixes are such that the value of the unit changes a thousandfold. However, certain common previously accepted multiples or submultiples, such as deci- and hecto-, are still accepted within the SI framework. The SI prefixes are listed in Table 1-13 together with their symbols.

Applications of SI in Laboratory Medicine

Use of the liter as a reference volume causes little confusion in interpretation of test results. However, introduction of the concept of amount of substance expressed as mole (mol) and its decimal fractions, such as mol, μmol, or nmol, could lead to confusion because of a general lack of familiarity with these units in comparison with more traditional units. However, in those countries in which the SI has been introduced into medical practice, amounts of constituents of body fluids have generally been reported as substance concentrations rather than mass concentrations (e.g., 2.5 mmol/L instead of 10.0 mg/dL for calcium and 3.9 mmol/L instead of 70 mg/dL for glucose). A comparison of results of some of the commonly measured serum constituents, at concentrations found in healthy individuals, is shown in Table 1-14.

The proposed base unit katal (symbol: kat), mol/s, is the catalytic amount of any

Table 1-14. TYPICAL VALUES FOR ANALYTES AND REPORTING INCREMENTS

	Conventional Units	Recommended Units	Rounded Recommended Unit	Smallest Recommended Reporting Increment
Albumin	3.8 g/dL	550.6 μmol/L	550 μmol/L	10 μmol/L
Bilirubin	0.2 mg/dL	3.42 μmol/L	3 μmol/L	2 μmol/L
Calcium	9.8 mg/dL	2.45 mmol/L	2.45 mmol/L	0.02 mmol/L
Cholesterol	200 mg/dL	5.17 mmol/L	5.2 mmol/L	0.05 mmol/L
Creatinine	0.8 mg/dL	90.48 μmol/L	90 μmol/L	10 μmol/L
Glucose	90 mg/dL	5.00 mmol/L	5.0 mmol/L	0.1 mmol/L
Phosphorus	3.0 mg/dL	0.97 mmol/L	1.0 mmol/L	0.05 mmol/L
Thyroxine	7.0 μg/dL	90.09 nmol/L	90 nmol/L	10 nmol/L
Triglycerides	100 mg/dL	1.14 mmol/L	1.15 mmol/L	0.05 mmol/L
Urea nitrogen*	10 mg/dL	3.57 mmol/L	3.5 mmol/L	0.05 mmol/L
Uric acid	5.0 mg/dL	297 μmol/L	300 μmol/L	10 μmol/L

*Urea nitrogen is reported as urea (mmol/L) when SI units are used.

catalyst, including *enzymes*, that catalyzes a reaction rate of one mole per second in an assay system. There is a constant relationship between the International Unit (1 μmol/min) and the katal (1 mol/s); to convert a value in International Units to nmol/s, it is multiplied by 16.67. Note, however, that the same dependency on reaction conditions applies to SI units as to International Units, so that data reported in the same units but obtained under different conditions may not be comparable. Although the katal has been approved by the IUB-IUPAC Joint Commission on Biochemical Nomenclature of the International Union of Pure and Applied Chemistry, it has yet to receive official sanction by CGPM.

For some *proteins* whose molecular weight is uncertain, debate continues as to the appropriate unit for reporting test results. Nevertheless, as substance concentration gives a better indication of the relative amount of a protein, substance concentration units are still recommended, even if the molecular weight of the protein is not precisely known. The molecular weight used for the substance concentration should be included as part of the test description. The concentration of *hormones* should be reported as substance concentration; units of activity, such as international units related to a consensus standard, should be avoided. Again, when uncertainty in the correct molecular weight of the hormone exists, an approximate value may be used.

For *drugs*, it is essential that units be clearly stated to avoid dangerous misinterpretations of test values that might occur when drug concentrations are expressed in different units. The use of substance concentration allows the concentration of a drug to be understood in comparison with the concentration of proteins, to which drugs are largely bound, and of other compounds with which drugs compete for protein-binding sites.

The osmole is not an SI unit and therefore theoretically is not suitable for reporting the results of osmotically active particles. The appropriate alternatives would be to record the depression of freezing point, whose unit is kelvin or degree Celsius, or the change in vapor pressure in pascals. But because the units would be method dependent, some of the benefits of using the osmole would be lost. The current practice is to report osmolality in osmoles per kilogram; however, the recommended units, mol/kg, will eventually be phased in.

The SI unit for reporting *pressure*, whether it is the partial pressure of blood gases or blood pressure, is the pascal. Because the numerical values expressed in pascals are too large to be used for most clinically important pressure measurements, the kilopascal is a more convenient unit. Thus, kilopascal replaces mm Hg or torr. The SI unit for gas concentrations in blood is mmol/L.

Reporting Increments

Although rounding of numerical values is necessary, this should be deferred to the last step to minimize errors. When conversions are made, no greater precision should be implied than was present in the original measurement. Thus, the data expressed in the second column of Table 1-14 imply a greater degree of precision than is warranted. The usual increment for reporting albumin in traditional units is 0.1 g/dL. The comparable increment for albumin in SI units is 10 μmol/L. Although the current reporting increment for calcium is typically 0.1 mg/dL, the increment with the SI should be 0.02 mmol/L, even though it implies greater precision in measurement.

Standardized Reporting of Test Results

To describe test results properly it is important that all necessary information be included in the test description. The International Federation of Clinical Chemistry (IFCC) and the International Union of Pure and Applied Chemistry (IUPAC) recommend that the following items be included:

1. The name of the system or its abbreviation
2. A dash (two hyphens)
3. The name of the component (never abbreviated) with an initial capital letter
4. A comma
5. The quantity name or its abbreviation

6. An equal sign
7. The numerical value and the unit or its abbreviation.

The example that follows illustrates a correct description of a test result in complete and abbreviated forms.

Plasma from fasting patient—Glucose, substance concentration = 5.5 millimoles per liter
(fPt)P—Glucose, substc. = 5.5 mmol/L

The *system* represents the material that is examined. It should be described to the extent necessary to make interpretation of results unambiguous. The system may be the patient, a body fluid from the patient, or a component of a body fluid from a patient. Although spelling out the system is preferable, certain abbreviations have been recommended by IFCC and IUPAC (Table 1-15).

The *component* that is measured should be described in complete terms. The component corresponds to what was previously described as the test name.

The *quantity* is the property that is measured. The different kinds of quantities and their accepted abbreviations are listed in Table 1-16. Each quantity is associated with a specific unit. The relation between the concentration quantities and their units is illustrated in Table 1-7.

Rules for Conversion from One Set of Units to Another

To convert from conventional units to SI units, the following scheme may be used:

1. Mass concentration (mass units/dL) to mass concentration (mass units/L):

$$\text{Numerical value in mass units/dL} \times 10 = \text{Mass units/L}$$
$$\text{e.g., Serum albumin, } 4.0 \text{ g/dL} \times 10 = 40 \text{ g/L}$$

2. Mass concentration to substance concentration:

$$\frac{\text{Numerical value in mass concentration/dL}}{\text{Molecular mass}} \times 10 = \text{Substance units/L}$$

$$\text{e.g., Serum albumin, } 4.0 \text{ g/dL} = \frac{4.0 \times 10}{69\ 000} \text{ mol/L} = 580 \text{ mol/L}$$

3. Mass amount to amount of substance:

$$\frac{\text{Numerical value in mass units}}{\text{Molecular mass}} = \text{Amount of substance}$$

$$\text{e.g., Urine albumin, } 150 \text{ mg/d} = \frac{150}{69\ 000} \text{ mmol/d} = 2.2\ \mu\text{mol/d}$$

$$\text{Urine urate, } 600 \text{ mg/d} = \frac{600}{168} = 3.6 \text{ mmol/d}$$

Table 1-15. ABBREVIATIONS FOR SYSTEMS AND PREFIXES RECOMMENDED BY IFCC AND IUPAC

a	arterial	Lkc	leukocyte
c	capillary	Lkcs	leukocytes
v	venous	P	plasma
d	24 hours (diem)	Peritf	peritoneal fluid
f	fasting	Plf	pleural fluid
Amf	amniotic fluid	Pt	patient
B	blood	S	serum
Df	duodenal fluid	Semf	seminal fluid
Erc	erythrocyte	Sf	spinal fluid
Ercs	erythrocytes	Synf	synovial fluid
F	feces	T	tissue
Gf	gastric fluid	U	urine

Table 1-16. ABBREVIATIONS FOR NAMES OF QUANTITIES

ams	amount of substance
diff	difference
equil	equilibration
massc*	mass concentration
massfr†	mass fraction
molal	molality
numc	number concentration
numfr	number fraction
rel	relative
substc	substance concentration
substfr	substance fraction
vol	volume
volfr	volume fraction

* Previously masc.
† Previously masfr.

In some cases a factor of 1000 may be required so that values in SI units fall between 1 and 1000.

The factors for converting from traditional to SI units for many of the constituents of body fluids commonly measured in the clinical laboratory are listed in Tables 29-18 and 29-19, *Reference Ranges*, included in the *Appendix*.

Safety

The federal Occupational Safety and Health Act (OSHA) of 1970 was the beginning of the formal regulatory oversight of employee safety. Consideration for the health and safety of the employees is now accepted as an obligation of all employers and laboratory directors. Clinical laboratories are generally considered safe places to work because necessary safety rules and procedures have been incorporated into the daily activities of the workers.

Every clinical laboratory must have a formal safety program. A specific individual should be appointed as "Safety Officer" with the duties of administering the safety program and keeping it current. One part of such a program consists of educating all clinical laboratory employees. During orientation, each employee should be given a copy of the general laboratory safety manual. Each employee should know the location of available fire-fighting equipment, how to use it, and where the evacuation routes are. The continuing education program of the laboratory should include periodic programs on safety.

The second part of a safety program should ensure that the laboratory environment meets accepted safety standards. This effort includes, but should not be limited to, types and location of fire extinguishers, hoods that are in good working order, proper labeling of chemicals and grounding of electrical equipment, and proper disposal of biological specimens and hazardous solvents.

Identification of Hazards

Clinical laboratories deal with each of the nine classes of hazardous materials in the United Nations classifications: (1) explosives, (2) compressed gases, (3) flammable liquids, (4) flammable solids, (5) oxidizer materials, (6) toxic materials, (7) radioactive materials, (8) corrosive materials, and (9) miscellaneous hazardous materials not elsewhere classified.

Warning labels aid in the identification of chemical hazards during shipment. The US Department of Transportation (DOT) placards or labels are diamond-shaped with a digit imprinted on the bottom corner that identifies the UN hazard class (1 to 9). The hazard is identified more specifically in printed words placed along the horizontal axis of the diamond. Color-coding and a pictorial art description of the hazard supplement the identification of the hazardous material on the label; the artwork appears in the top half of the diamond (Figure 1-6,*A*). When a hazardous material reaches its destination and is removed from the shipping container, its identification as a hazard is lost. The laboratory must then label each individual container. Usually the information needed to classify the contents of the container appropriately is contained on the shipping label and should be noted. Important first-aid information is also usually provided on this label.

Even though OSHA prescribes the use of labels or other appropriate warnings, at present no single uniform labeling system for hazardous chemicals exists for clinical laboratories. The National Fire Protection Association (NFPA) developed the 704-M Identification System, which classifies hazardous material from 0 to 4 (most hazardous) according to flammability and reactivity (instability). This system also uses diamond-shaped labels, which are color coded and divided into quadrants. Each of three quadrants has a characteristic color and represents a type of hazard. A number in the quadrant indicates the degree of the hazard. The fourth (lower) quadrant contains information of special interest to firemen; for example, "W" indicates a water-ignitable material (Figure 1-6,*B*). Some chemicals require labels utilizing two or three of the quadrants to convey the necessary information.

Descriptive labels such as "corrosive," "flammable," "poison," and "explosive" may also be used. Obviously, some containers may require more than one label, since several types of hazard are involved. Safety precautions and labeling criteria for radioisotopes are discussed in Chapter 2, Section 7.

Safe Handling of Chemicals

The storage and use of chemicals may be attended by a variety of dangers, such as burns, explosions, fires, and toxic fumes. A knowledge of the properties of the chemicals in use and of proper handling procedures will greatly reduce any danger. Bottles of chemicals and solutions should be handled carefully, and chemical containers made of glass should be transported in rubber or plastic protective holders that will protect them from breakage and, in the event of breakage, contain the spill. A manual on the handling of chemicals should be available in each laboratory.

Spattering from acids, caustic materials, and strong oxidizing agents probably represents the greatest hazard to clothing, skin, and eyes and is a potential source of chemical burns. Bottles must never be grasped by the neck but should be held firmly around the body. Acids must be diluted by slow addition of the acid to the water with mixing; *water should never be added to concentrated acid*. When working with acid or alkaline solutions, laboratory personnel should wear safety glasses. Thought should be

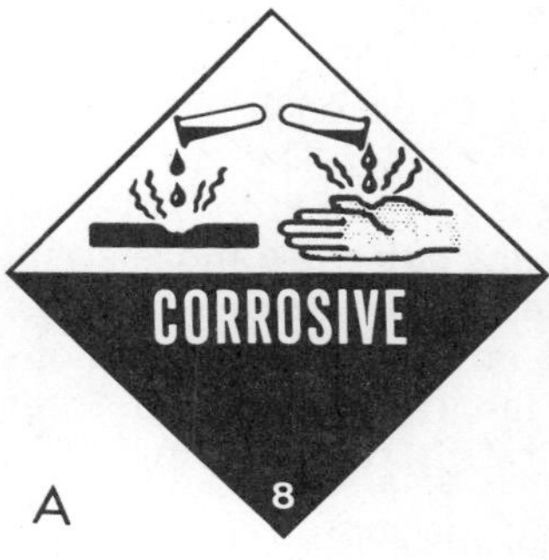

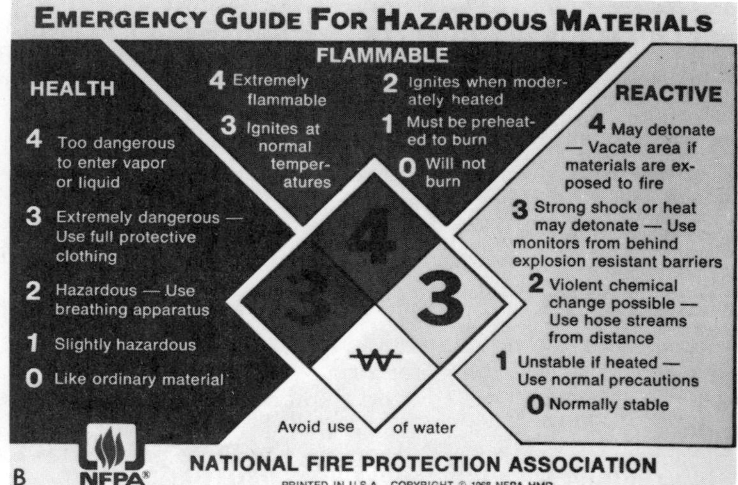

Figure 1-6. *A*, Department of Transportation label for corrosives. *B*, Labeling identification system of National Fire Protection Association.

given to the possibility of breakage and if possible mixing of acid should be done in a sink, which would provide water for cooling as well as for confinement of the acid in the event the container were to break. Strong acids, caustic materials, and strong oxidizing agents should be dispensed with automatic dispensing devices. Under no circumstances is mouth pipetting permitted.

All bottles containing reagents must be properly labeled, preferably before the reagent is added. The label should bear the name and concentration of the reagent, the initials of the person who made up the reagent, and the date on which the reagent was prepared. When appropriate, the expiration date should also be included. The labels should be color coded or an additional label added to designate specific storage instructions, such as the need for refrigeration or special storage related to a potential hazard.

All reagents found in unlabeled bottles should be discarded using rigid precautions. The disposal method used will depend on the type of chemical, the amount to be discarded, and the local laws. Some chemicals may be burned in a chemical incinerator; others may have to be buried in a chemical landfill. In the clinical chemistry laboratory small amounts of relatively dilute solutions can be disposed of down the drain after prior neutralization. The reagent should be flushed down the drain with large amounts of water. Acids must be diluted sufficiently in order not to harm the plumbing. Large amounts of water must be used to flush the drain completely of all reagents, particularly those that present a danger, either actual or potential. Although federal standards for all types of hazardous waste disposal are not available, many municipalities have their own regulations. At this time, most clinical laboratories can dispose of aqueous solutions of chemicals down the drain after proper treatment. Local authorities should be contacted for specific information.

Most of the general precautions just mentioned also hold for automated chemistry analyzers. In some instances, all waste materials cannot be collected in a single container. With certain pieces of equipment, strong acids or other hazardous materials are pumped directly into the drain. This flow should always be accompanied by a steady flow of a large amount of water from the faucet. Safety glasses should be used by instrument operators when acids are pumped under pressure.

Perchloric acid is potentially explosive when in contact with organic materials and requires careful handling. Perchloric acid should not be used on wooden bench tops, and bottles of this acid should be stored on a glass tray. If large amounts of this acid are used, special perchloric acid hoods, with special wash-down facilities, should be installed.

Special care is needed in dealing with mercury. Even small drops of mercury on bench tops and floors may poison the atmosphere in a poorly ventilated room. The element's ability to amalgamate with a number of metals is well known. Therefore, after an accidental mercury spill, the surrounding area should be cleaned carefully until there are no droplets remaining. All containers of mercury should be kept well stoppered.

Hazards from volatiles. The use of organic solvents in a clinical laboratory represents a potential fire hazard as well as a hazard to health from the inhalation of toxic vapors or from skin contact. These solvents should be used in a fume hood and stored in an OSHA-approved metal storage cabinet that is properly vented. The maximum volume of flammable solvents that can be stored in a building is regulated by rules set down by OSHA (Table 1-17); however, some local fire department rules are more stringent. The maximum working volume of any flammable solvent allowed outside of storage cabinets is five gallons per room. CAP requirements state that no more than 60 gallons of flammable liquids may be stored per 5000 square feet of laboratory space. Larger amounts than shown in Table 1-17 must be kept in special refrigerated rooms or in outside storage buildings.

Vaporization is the major problem in the ignition and spread of fires. Flammable and combustible liquids and solids are those whose vapors form a flammable mixture with air. They are characterized by their flash point, defined as the lowest temperature at which a solvent will give off flammable vapors in the close vicinity of its surface. The mixture will ignite at its flash point when it is exposed to a source of ignition. At temperatures below the flash point, the vapor given off is too lean for ignition. The flash points of solvents commonly found in clinical laboratories are listed in Table 1-18.

Table 1-17. MAXIMUM ALLOWABLE SIZE OF CONTAINERS FOR STORAGE OF SOLVENTS INSIDE BUILDINGS (OSHA)

Container Type	Maximum Allowable Size of Containers			
	Class 1A	***Class 1B***	***Class 1C***	***Classes II and III***
Glass or approved plastic	1 pt	1 qt	1 gal	1 gal
Safety cans*	1 gal	2 gal	5 gal	5 gal
Metal drums (DOT)	60 gal	60 gal	60 gal	60 gal
Definition of Classes:				
IA	Flash point below 22.8 °C and boiling point below 37.8 °C			
IB	Flash point below 22.8 °C and boiling point above 37.8 °C			
IC	Flash point between 22.8 °C and 37.8 °C			
II	Flash point between 37.8 °C and 60 °C			
III	Flash point above 60 °C			

*Safety cans are metal or plastic containers having a spring-closing spout cover designed so that it will relieve internal pressure when subjected to the heat of fire and will prevent leakage if tipped over. They are also equipped with a flame-arrester screen. Portable safety cans range in size from 1 pint to 5 gallons and are made in numerous styles with faucets, pouring spouts, or dispensing hoses. Only safety cans recognized by FM (Factory Mutual Engineering Corporation of the Factory Mutual System) or UL (Underwriters Laboratories, Inc.) should be used, because these have been tested and approved by these agencies. OSHA = Occupational Safety and Health Act.

In general, disposal of flammable solvents in sewers is not allowed. Exceptions are small amounts of those materials that are miscible with water, but even these should be followed by large amounts of cold water. Other solvents should be collected in safety cans. Separate cans should be used for ether and for chlorinated solvents; all other solvents may be combined in a third can. The cans should be stored in a safety cabinet, in keeping with storage quantity rules, until pickup by a waste-disposal firm.

Hazards from compressed gases. Regulations of the US Department of Transportation cover the labeling of cylinders of compressed gases that are transported by interstate carriers. The diamond-shaped labels described previously are used on all large cylinders and on any boxes containing small cylinders. The OSHA regulations regarding compressed gas in cylinders are based on publications of the Compressed Gas Association, Inc.[3]

Some general rules for handling large cylinders of compressed gas are:

1. Always transport cylinders on a hand truck to which the cylinder is secured.
2. Leave the valve cap on the cylinder until the cylinder is ready for use, at which time the cylinder should have been secured by a support around the upper third of its body. Shut off the valve of the tank that is to be replaced, disconnect the hose or regulator,

Table 1-18. FLASH POINTS OF SOLVENTS COMMONLY FOUND IN CLINICAL LABORATORIES

	°F	°C
Acetone	−4	−20
Butanol	82	28
Cyclohexane	−4	−20
Ethyl alcohol	55	13
Ethyl ether	−49	−45
Ethyl acetate	24	−5
Heptane	25	−4
Hexane	−7	−22
Isopropanol	53	12
Petroleum ethers	−70	−56
Propanol	77	26
Toluene	40	14
Xylene	81	27

and replace the cap before the cylinder is completely empty to avoid the possibility of the development of a negative pressure. Place an "empty" sign or label on the cylinder.

3. Chain or secure cylinders at all times, even when empty.
4. Always check cylinders for the composition of contents before they are connected.
5. Never force threads; if the regulator does not thread readily, something is wrong.

Empty cylinders are usually picked up by gas suppliers for return to the manufacturer for refilling.

Small cylinders. The precautions cited for large refillable gas cylinders apply also to small cylinders that are not refillable, such as propane cylinders of the type frequently secured to a flame photometer and cylinders of calibrating gases for blood gas equipment. Cylinders in floor-standing base supports require the additional security of attachment to a wall or fixed piece of furniture. Local fire department regulations govern disposal of these cylinders when they are exhausted.

Electrical Hazards

Wherever there are electrical wires or connections, there is a potential shock or fire hazard. Worn wires on all electrical equipment should be replaced immediately; all equipment should be grounded by three-pronged plugs. OSHA regulations stipulate that laboratories should meet the requirements for grounding electrical equipment suggested by the National Electrical Code published by the National Fire Protection Association. Some local codes are more stringent than OSHA requirements.

The use of extension cords is prohibited by the 1980 NCCLS standard on power requirements for clinical laboratories.[6] This standard is more stringent than any other existing regulation. In some instances, an extension cord may have to be used temporarily. In such cases, the cord should be <12 feet in length, have at least 16-AWG (American Wire Gauge) wire, be approved by the Underwriters Laboratory (UL), and have only one outlet at the end. If several outlets are needed in an area, a power strip with its own fuse or circuit breaker may be installed at least three inches above benchtop level. Several manufacturers now sell devices to check for high resistance in neutral or ground wiring or for excess voltage in the neutral wiring.

Electrical equipment and connections should not be handled with wet hands, nor should electrical equipment be used after liquid has been spilled on it. The equipment must be turned off immediately and dried thoroughly; a fan or hair dryer will speed up the drying process. In case of a wet or malfunctioning electrical instrument that is used by several people, the plug should be pulled and a note cautioning coworkers against its use should be left on the instrument.

Electrical apparatus that are operated in an area where there are flammable vapors must be explosion-proof; induction-driven motors are well suited for these areas.

Fire Safety

The ideal solution to the problem of fires and indeed to all laboratory accidents is prevention. Since all fires cannot be prevented, provisions must be made for those that do occur. NFPA (National Fire Protection Association) and OSHA publish standards covering subjects from emergency exits to safety and firefighting equipment. NFPA also publishes the National Fire Codes. Many state and local agencies have adopted these codes (some of which are more stringent than OSHA requirements), which makes them legally enforceable.

Every laboratory should have the necessary equipment to put out or to confine a fire in the laboratory, as well as to put out a fire on the clothing of an individual. Easy access to safety showers is essential. A safety shower should have a pull chain with a large ring attached to it so that the shower may be easily activated, even with eyes closed. Fire blankets for smothering clothing fires should be available in an easily accessible, wall-mounted case. The location of this equipment as well as the locations of fire alarms and maps of evacuation routes is usually dictated by the local fire marshal.

Various types of *fire extinguishers* are available. The type to use depends upon the type of fire. Because it is impractical to have several types of fire extinguishers present in every area, dry chemical fire extinguishers are the best all-purpose extinguishers for laboratory areas. An extinguisher should be provided near every laboratory door and, in a large laboratory, also at the end of the room opposite the door. Everyone in the laboratory should be instructed in the use of these extinguishers and of any other firefighting equipment. All fire extinguishers should be tested by qualified personnel at intervals specified by the manufacturer. The three classes of fires and the type of fire extinguisher to be used for each are listed in Table 1-19. Every fire extinguisher is labeled as to the type of fire it should be used to extinguish.

Many clinical laboratories have a computer that is housed in a temperature- and humidity-controlled room. The most popular automatic fire control system in use for these rooms is the Halon 1301 (bromotrifluoromethane). Although this is the least toxic of the halons, NFPA regulations require a warning sign at the entrance to the room as well as the availability of self-contained breathing equipment.

Biological Hazards

Among the most frequent biohazards are: (1) accidental puncture with hypodermic needles, (2) spraying, spilling, or splattering infectious materials on bench tops or floors, (3) centrifuge accidents, and (4) cuts or scratches from contaminated glassware.

Viral hepatitis represents a special health hazard for clinical laboratory personnel and other health care individuals. Although in the past hepatitis B virus presented the greatest risk, non-A, non-B hepatitis is now the most common form of post-transfusion hepatitis and is seen frequently among renal dialysis patients. The risk of infection from specimens of patients with acquired immune deficiency syndrome (AIDS) is under investigation. At present, these specimens should be handled with the same precautions that are used for specimens from patients with hepatitis.

Techniques and Procedures to Minimize Laboratory Infections

Sound fundamental laboratory techniques, well supervised and conscientiously carried out, can do much to achieve environmental control and reduce the hazards of infection. A list of rules that are widely applicable in clinical laboratories follows:

1. Never perform direct mouth pipetting in the laboratory.
2. Do not blow out pipets that contain infectious material.
3. Do not mix infectious materials by bubbling air through the liquid.

Table 1-19. CLASSIFICATION OF FIRES AND FIRE EXTINGUISHER REQUIREMENTS

Type of Hazard	Class of Fire	Recommended Extinguisher Agents
Ordinary combustibles: Wood, cloth, paper	A	Water, dry chemical foam, loaded steam
Flammable liquids and gases: Solvents and greases; Natural or manufactured gases	B	Dry chemical, carbon dioxide, loaded steam, Halon 1211 or 1301 foam
Electrical equipment: Any energized electrical equipment. If electricity is turned off at source this reverts to a Class A or B	C	Dry chemical, carbon dioxide, Halon 1211 or 1301
Combinations of: Ordinary combustibles and flammable liquids and gases	A & B	Dry chemical, loaded steam, foam
Combinations of: Ordinary combustibles and electrical equipment	A & C	Dry chemical
Combinations of: Flammable liquids and gases and electrical equipment	B & C	Dry chemical, carbon dioxide, Halon 1211 or 1301
Combinations of: Ordinary combustibles, flammable liquids and gases, and electrical equipment	A, B & C	Triplex dry chemical

4. Use only needle-locking hypodermic syringes. Avoid using syringes whenever possible. Dispose of needles in special containers designed for that purpose.

5. Expel excess fluid and bubbles from a syringe into a cotton pledget moistened with disinfectant.

6. Sterilize reusable glassware immediately after use by placing it into a pan containing a phenolic compound such as Staphene (Vestal Laboratories, St. Louis, MO 63110).

7. Before centrifugation, inspect tubes for cracks. Inspect the inside of the trunnion cup for rough walls caused by erosion or adhering matter. Be sure that rubber cushions are free from all bits of glass. A germicidal solution added between the tube and the trunnion cup not only disinfects the surfaces of both but also provides an excellent cushion against shocks that otherwise might break the tube.

8. Avoid decanting from centrifuge tubes; if you must do so, wipe off the outer rim afterward with a disinfectant. Avoid filling a tube to the point where the rim becomes wet with potentially contaminated specimen.

9. Never leave a discarded tube or infected material unattended or unlabeled.

10. Periodically clean out deep freeze and dry-ice chests to remove broken ampules or tubes of biological specimens stored there. Use rubber gloves and respiratory protection during this cleaning.

11. Label all samples known or strongly suspected to have come from a patient with hepatitis or AIDS. The label should be placed on all tubes containing aliquots (e.g., all serum aliquots removed from the clot and used in various sections of the laboratory).

12. Handle with disposable gloves serum specimens that carry a risk of infectious hepatitis or AIDS.

13. As part of the disposal process, sterilize all biological specimens (except urine) by autoclaving or incineration. Urine may be poured into a drain.

14. Do not allow smoking or consumption of food or beverages in the laboratory. Food and beverages are never to be stored in any laboratory refrigerator (or freezer) unless it has been specifically and exclusively designated for such purpose.

15. Encourage frequent hand washing; this practice is mandatory before an employee leaves the laboratory.

16. To reduce the possibility of self-inoculation, develop the habit of keeping your hands away from your mouth, nose, eyes, and any other mucous membranes.

Safety Equipment for Personnel

A large number of safety items are available for the laboratory. Most of these are shown and described in the Fisher Manual.[4] A few important safety items will be mentioned briefly here. Eyewashers or facewashers should be available in every chemistry laboratory. A hand-held eye and face safety spray is a convenient safety device that should be consistently placed in a position next to each sink. Safety showers strategically located in the laboratory should be tested on a regular schedule.

Safety goggles, glasses, and visors are available in many sizes and shapes. Personnel wearing contact lenses should be aware of the danger of irritants getting under a lens, thus making it difficult to irrigate the eye properly after an accident. Shatterproof safety shields should be used in front of systems posing a potential danger because of implosion (vacuum collapse) or pressure explosion. Heat-resistant (nonasbestos) gloves should be available for handling hot glassware and dry ice, and hot beakers should be handled with tongs. Acids should be dispensed with polyethylene pumps. Spill kits for acids, caustic materials, or flammable solvents should be located at convenient sites in the laboratory.

A chemical fume hood is a necessity for every clinical chemistry laboratory. The fume hood is the only safe place to open any container from which harmful vapors may escape, to prepare reagents that produce fumes, or to use flammable solvents.

Safety Inspections

Safety is a concern of regulatory and accrediting agencies. OSHA inspectors have the authority to enter a clinical laboratory unannounced and, upon presentation of credentials, inspect it. The inspection may be regular or prompted by a complaint. The Commission on Inspection and Accreditation of CAP inspects clinical laboratories and uses various safety check lists (available to the laboratory prior to inspection) when evaluating a laboratory for accreditation. These inspections are carried out on a regular basis. The Joint Commission on Accreditation of Hospitals (JCAH) does not publish specific safety guidelines for the clinical laboratory; the subject is covered very generally and briefly in its manual. Although JCAH will accept CAP accreditation of a laboratory, it may still conduct a safety inspection of the laboratory when it inspects the hospital. Inspections may also be made on a regular basis by state or local health departments or by local fire departments to determine conformity to their requirements. At the present time, a laboratory that meets federal or state OSHA requirements is likely to satisfy the standards of any other inspecting agency.

References

1. American Society for Testing and Materials: Water, Annual Book of ASTM Standards, Part 31, Philadelphia, 1981.
2. Commission on Laboratory Inspection and Accreditation: Reagent Water, Skokie, IL, College of American Pathologists, 1978.
3. Compressed Gas Association, Inc.: Handbook of Compressed Gases. 2nd ed. New York, Reinhold Publishing Corporation, 1981.
4. Fisher Scientific Company: Fisher Safety Manual, Pittsburgh, Fisher Scientific Co., 1982.
5. National Committee for Clinical Laboratory Standards: Centrifugation of Blood, Proposed Standard H18-P. Villanova, PA, NCCLS, 1981.
6. National Committee for Clinical Laboratory Standards: Power Requirements for Clinical Laboratory Instruments and for Laboratory Power Sources, Approved Standard I5-A, Villanova, PA, NCCLS, 1980.
7. National Committee for Clinical Laboratory Standards: Quantities and Units: SI; Committee Report C11-CR. Villanova, PA, NCCLS, 1983.
8. National Committee for Clinical Laboratory Standards: Preparation and Testing of Reagent Water in the Clinical Laboratory. 2nd ed. Proposed guideline C3-P2. Villanova, PA, NCCLS, 1985.
9. Weast, R. C. (Ed.): The Handbook of Chemistry and Physics, 58th ed. Cleveland. The Chemical Rubber Co., 1980–81, p. D-144.

Additional Reading

College of American Pathologists: Standards, Reference Materials, and Methods. A Practical Guide for the Medical Laboratory, Skokie, IL, College of American Pathologists, 1983.

Committee on Hazardous Substances in the Laboratory: Prudent Practices for Handling Hazardous Chemicals in Laboratories: National Research Council. Washington, D.C., National Academy Press, 1981.

Gordon, A. J., and Ford, R. A.: The Chemist's Companion, New York, Wiley-Interscience, 1972.

National Bureau of Standards: Standard Reference Materials: Summary of the Clinical Laboratory Standards, NBS Special Publication 260-71. Washington, D.C., U.S. Department of Commerce, 1981.

Shugar, G. J., Shugar, R. A., Bauman, L., et al.: Chemical Technicians' Ready Reference Handbook. 2nd ed. New York, McGraw-Hill, 1981.

World Health Organization: The SI for the Health Professions. Geneva, Switzerland, 1977.

ANALYTICAL PROCEDURES AND INSTRUMENTATION

2

Coordinated by CARL A. BURTIS, Ph.D.

SECTION ONE

Photometry

WENDELL T. CARAWAY, Ph.D.

Many determinations made in the clinical laboratory are based upon measurements of radiant energy emitted, transmitted, absorbed, or reflected under controlled conditions. The principles involved in such measurements will be considered in this section. Details of operation and maintenance for a particular instrument are typically supplied by the manufacturer and should be consulted for further information.

The term "photometric measurement" was defined originally as making a measurement of light intensity independently of wavelength. Most instruments used at present, however, have some means of isolating a narrow wavelength range of the spectrum for measurements. Those that use filters for this purpose are referred to as *filter photometers*, whereas those that use prisms or gratings are called *spectrophotometers*. Both types will be considered.

Nature of Light

Electromagnetic radiation includes radiant energy from short-wavelength gamma rays to long-wavelength radio waves. The term "light" will be used to describe radiant energy with wavelengths visible to, and bordering on those visible to, the human eye. The wavelength of light is defined as the distance between two peaks as the light travels in a wavelike manner. This distance is expressed in nanometers (nm) for wavelengths commonly used in photometry. Other (obsolete) units that may be encountered are Ångstroms (Å) and millimicrons (mμ):

$$1 \text{ nm} = 1 \text{ m}\mu = 10 \text{ Å} = 10^{-9}\text{m}$$

The human eye responds to radiant energy with wavelengths between ~380 and 750 nm, but modern instrumentation permits measurements at both shorter-wavelength (ultraviolet, UV) and longer-wavelength (infrared, IR) portions of the spectrum.

Sunlight, or light emitted from a tungsten filament, is a mixture, or spectrum, of radiant energy of different wavelengths that the eye recognizes as "white." Table 2-1 shows approximate relationships between wavelengths and color characteristics for the UV, visible, and short IR portions of the spectrum. Thus, a solution will appear green when viewed against white light if it transmits light maximally between 500 and 580 nm, but the solution absorbs light at other wavelengths. Similarly, a solid object appears green if it reflects light in this region (500–580 nm) but absorbs light at other portions of the spectrum. In general, if we compare the intensity of light transmitted by a colored solution with that of a blank or reference solution over the entire spectrum, we obtain a typical spectral-transmittance curve that is characteristic for that solution. Such curves are shown in Figure 2-1 for solutions of nickel sulfate (*A*) and potassium permanganate (*B*). Inspection of the curves should lead us to predict that the color of solution *A* is

Table 2-1. UV, VISIBLE, AND SHORT IR SPECTRUM CHARACTERISTICS

Wavelength (nm)	Region Name	Color Observed*
<380	Ultraviolet†	Not visible
380–440	Visible	Violet
440–500	Visible	Blue
500–580	Visible	Green
580–600	Visible	Yellow
600–620	Visible	Orange
620–750	Visible	Red
750–2000	Short IR	Not visible

*Owing to the subjective nature of color, the wavelength ranges shown are only approximations.

† The ultraviolet portion of the spectrum is sometimes further divided into "near" UV (220–380 nm) and "far" UV (<220 nm). This arbitrary distinction has a practical basis since standard silica used to make cuvets transmits light effectively at wavelengths ≥220 nm.

green inasmuch as light is transmitted maximally near the green portion of the spectrum. Curve *B*, on the other hand, illustrates the spectrum of a solution that transmits light maximally in the blue, violet, and red portions of the spectrum. The eye recognizes this mixture of colors as purple.

The chief advantage of filter photometry or spectrophotometry is that we can isolate and use discrete portions of the spectrum for purposes of measurement. Thus, in the case of the permanganate solution, we may use relatively pure green light, which will be absorbed selectively, rather than white (composite) light. This use of pure light results in improved specificity, sensitivity, and linearity of spectrophotometric measurements.

Beer's Law

Consider an incident light beam with intensity I_0 passing through a square cuvet containing a solution of a compound that absorbs light of a certain wavelength (lambda, λ) (see Figure 2-2). The intensity of the transmitted light beam I_S will be less than I_0, and we define the transmittance (*T*) of light as I_S/I_0. Some of the incident light, however, may be reflected by the surface of the cell or absorbed by the cuvet wall or by the solvent. In order to focus attention on the compound of interest, it is necessary to eliminate these factors. This is done by using a reference cell identical to the sample cell, except that the compound of interest is omitted from the solvent. The transmittance through this

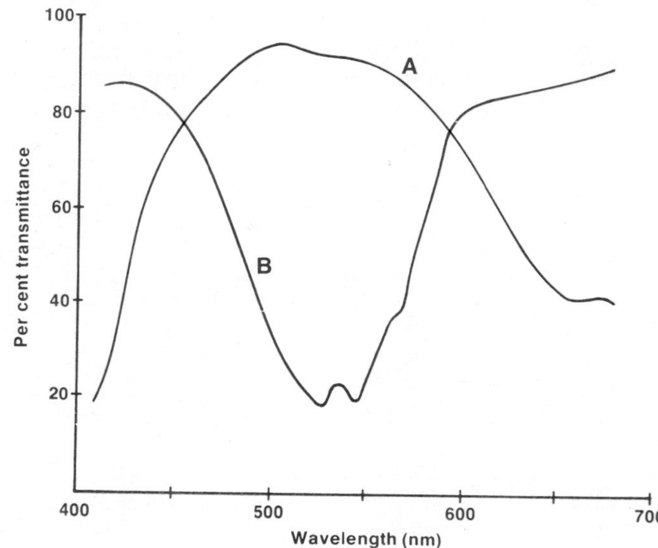

Figure 2-1. Spectral-transmittance curves of nickel sulfate (*A*) and potassium permanganate (*B*). Arbitrary concentrations, read versus water as a blank (Beckman DB-G spectrophotometer).

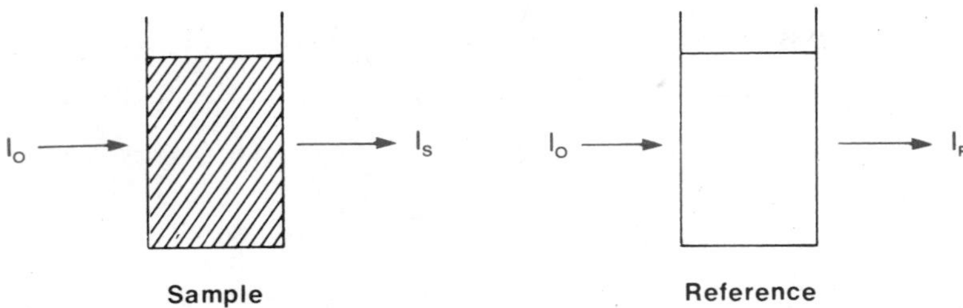

Figure 2-2. Transmittance of light through sample and reference cells. Transmittance of sample versus reference = I_S/I_R.

reference cell is I_R/I_O; the transmittance for the compound in solution is then defined as I_S/I_R. In practice, this reference cell is inserted and the instrument adjusted to an arbitrary scale reading of 100 (corresponding to 100% transmittance), following which the per cent transmittance reading is made on the sample. As we increase the concentration of the compound in solution, we find that transmittance varies inversely and logarithmically with concentration. Consequently, it is convenient to define a new term, absorbance (A), that will be directly proportional to concentration. (The term optical density, OD, sometimes used for absorbance, is now obsolete.) Hence

$$A = -\log I_S/I_R = -\log T = \log \frac{1}{T} = \log \frac{100\%}{\text{per cent } T}$$
$$= \log 100 - \log \text{per cent } T = 2 - \log \text{per cent } T$$

To clarify these relationships more fully, assume that we have a solution of a compound at a concentration of 1 g/L that transmits only half of the incident light ($T = 0.5$) compared with a reference setting of 100. Therefore,

$$T = 50\% \text{ and } A = 2 - \log 50 = 0.301$$

If we increase the concentration to 2 g/L, the light transmitted becomes $0.5 \times 0.5 = (0.5)^2 = 0.25 = 25\%\ T$. At 3 g/L, transmittance $= (0.5)^3 = 0.125 = 12.5\%\ T$, and so on. It can be seen that as the concentration increases linearly, the per cent T decreases *geometrically*. On the other hand, absorbance is *linearly* related (directly proportional) to the concentration (Figure 2-3, *A* and *C*).

Data can also be plotted on semilogarithmic paper, as shown in Figure 2-3, *B*. When per cent T is plotted on the log scale, a straight line with negative slope is obtained. Some instruments have scales that show per cent T and absorbance. Most instruments now provide direct digital readouts of absorbance.

Experience has shown that with most photometers the response of the detector to a signal of transmitted light is such that any uncertainty in T is constant over the entire

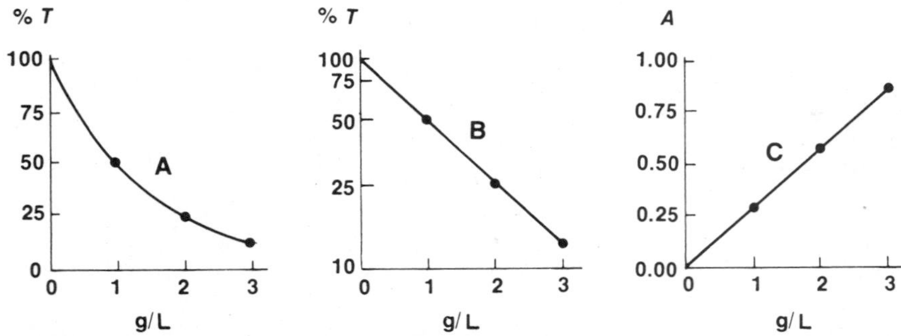

Figure 2-3. Transmittance and absorbance as a function of concentration. *A,* Per cent *T,* linear scale. *B,* Per cent *T,* logarithmic scale. *C,* Absorbance, linear scale.

T scale. The uncertainty derives from electrical and mechanical imperfections in the instrument and from individual variations in the use of the instrument.

A fixed distance on the linear scale (e.g., 1% T) represents a greater change in absorbance for low values of per cent T than for high values of per cent T. For this reason, the *absolute* concentration error or uncertainty is greater when readings are taken at high absorbance. However, the *relative* concentration error is greater for readings at either low *or* high absorbances. An uncertainty of 1% in readings on the linear T scale results in changes in absorbance (and in apparent concentration) as follows:

Per Cent T	Absorbance	Absolute Error	Per Cent Relative Error
10	1.000		
11	0.959	0.041	4.1
45	0.347		
46	0.337	0.010	2.9
90	0.046		
91	0.041	0.005	10.9
95	0.022		
96	0.018	0.004	18.2

The relative error is actually minimal at 36.8% T, corresponding to an absorbance of 0.434. Consequently, methods should be designed such that readings fall near the center of the scale, preferably between 20 and 80% T (A between about 0.7 and 0.1). Judgment must be exercised as to the importance of absolute and relative errors in a particular determination.

When we hold the concentration constant (say, at 1 g/L) and double the inside diameter of the cell, the effect on absorbance is the same as doubling the concentration, since we have introduced twice as many absorbing molecules in the light path. From this example it follows that absorbance is also directly proportional to the light path through the cell. This relationship is often referred to as *Bouguer's law* or *Lambert's law.*

The overall equation relating these variables may be expressed as

$$A = abc \tag{1}$$

where A is the absorbance, a is a proportionality constant defined as absorptivity, b is the light path in cm, and c is the concentration of the absorbing compound. This equation is called *Beer's law* and forms the basis of quantitative analysis by absorption photometry. Absorbance values have no units; hence, the units for a are the reciprocal of those for b and c.

When $b = 1$ cm and c is expressed in moles per liter, the symbol ϵ (epsilon) is substituted for the constant a. The value for ϵ is a constant for a given compound at a given wavelength under prescribed conditions of solvent, temperature, pH, and so forth, and is called the *molar absorptivity*. The nomenclature of spectrophotometry is summarized in Table 2-2. Values for ϵ are used to characterize compounds, to establish their purity, and to compare sensitivities of measurements obtained on derivatives. Bilirubin, for example, when dissolved in chloroform at 25 °C, should have a molar absorptivity

Table 2-2. SPECTROPHOTOMETRY NOMENCLATURE

Name	Symbol	Definition
Absorbance	A	$2 - \log T$ or $\log I_0/I$
Absorptivity	a	A/bc (c in g/L)
Molar absorptivity	ϵ	A/bc (c in mol/L)
Path length	b	Internal cell or sample length in cm
Transmittance	T	I/I_0
Wavelength unit	nm	10^{-9}m
Absorption maximum	λ_{max}	Wavelength at which maximum absorption occurs

of 60 700 ± 1600 at 453 nm. The molecular weight of bilirubin is 584. Hence, when a solution containing 5 mg/L (0.005 g/L) is read in a 1-cm cell, it should have an absorbance of

$$A = (60\ 700)(1)(0.005/584) = 0.520$$

Conversely, a solution of this concentration showing an absorbance of 0.490 could be assumed to have a purity of 0.490/0.520 or 94%.

The molar absorptivity of the complex between ferrous iron and *s*-tripyridyltriazine is 22 600, whereas that with 1,10-phenanthroline is 11 000. These values illustrate that, for a given concentration of iron, the former reagent produces a complex that has an absorbance about twice that of the complex with the latter reagent. Hence, it imparts greater sensitivity to the measurement of iron.

In toxicologic work, it is customary to list constants based on concentrations in g/dL rather than in mol/L. This practice is also necessary when the molecular weight of a substance is unknown. For $b = 1$ cm and $c = 1$ g/dL (1%), A can be written as $A_{1cm}^{1\%}$. This constant is called the *absorption coefficient.* An older symbol, $E_{1cm}^{1\%}$, now obsolete, was called the *extinction coefficient*.

The direct proportionality between absorbance and concentration must be established experimentally for a given instrument under specified conditions. Frequently there is a linear relationship up to a certain concentration or absorbance. We then say that the solution obeys Beer's law up to this point. Within this limitation, a calibration constant (K) may be derived from formula (1) and used to calculate the concentration of an unknown solution by comparison with a standard.

$$A = abc$$

Therefore,

$$c = A/ab = A \times 1/ab \tag{2}$$

The absorptivity (a) and light path (b) remain constant in a given method of analysis; hence, $1/ab$ can be replaced by a constant (K) and

$$c = A \times \mathrm{K} \tag{3}$$

where $\mathrm{K} = c/A$. The value of the constant K is obtained by measuring the absorbance (A) of a standard of known concentration (c).

Certain precautions must be observed in using such calibration constants. Under no circumstances should the constant be used when either the standard or unknown readings exceed the linear portion of the calibration curve (i.e., if the curve no longer obeys Beer's law). A standard should be included in each series of determinations to permit direct comparison of unknown to standard or to calculate the calibration constant, since variations in reagents, working conditions, cell diameters, deterioration or change in instruments, and so on, may result in day-to-day changes in the absorbance value for the standard. A nonlinear calibration curve may be used if a sufficient number of standards of varying concentrations are included to cover the entire range encountered for readings of unknowns.

In some cases, pure standards may not be readily available and constants may be provided that were obtained on pure standards by others and reported in the literature. In general, the use of published constants should be discouraged unless the method is followed in detail and readings are made on a spectrophotometer capable of providing light of high spectral purity at a verified wavelength. Use of broad-band light usually leads to some decrease in absorbance. The absorbance of NADH at 340 nm, for example, is frequently used as a reference for the determination of enzyme activity, based on an assigned molar absorptivity of 6.22×10^3. This value is acceptable only under carefully controlled conditions and should not be used unless these conditions are met.

Components of Spectrophotometers

The major components of a *single-beam spectrophotometer* are shown schematically in Figure 2-4. Light is passed through a monochromator to provide selection of the desired region of the spectrum to be used for measurements. Slits are used to isolate a narrow beam of the light and improve its chromatic purity. The light next passes through an absorption cell (cuvet), where a portion of the radiant energy is absorbed, depending on the nature and concentration of the solution. Any light not absorbed is transmitted to a detector (photocell or phototube), which converts light energy to electrical energy that can be registered on a meter or digital readout.

In operation, an opaque block is substituted for the cuvet so that no light reaches the photocell, and the meter is adjusted to read 0% *T*. Next, a cuvet containing a reagent blank is inserted and the meter is adjusted to read 100% *T* (i.e., zero absorbance). The composition of the reagent blank should be identical to that of standard or unknown solutions except for the substance to be measured. Standard solutions containing various known concentrations of the substance are inserted and readings are recorded. Finally, a reading is made on the unknown solution, and its concentration is determined by comparison with the readings obtained on the standards.

For instruments that provide readouts in absorbance only, it is usually not possible to make an adjustment to 0% *T* (infinite absorbance) when the light path is blocked. These instruments usually have a method of compensating for this signal so that the detector registers zero at zero light input; hence, the absorbance is simply set to zero with the reagent blank in the cuvet prior to taking readings on standards and unknowns.

Double-beam spectrophotometers are also widely used. In one such type of double-beam system, shown schematically in Figure 2-5, all components are duplicated except the light source. This system is referred to as a double-beam-in-*space* spectrophotometer. Another approach is to use a light-beam chopper (a rotating wheel with alternate silvered sections and cut out sections) inserted after the exit slit (Figure 2-6). A system of mirrors passes the portions of the light reflected off the chopper alternately through the sample and a reference cuvet onto a common detector. This system is referred to as a double-beam-in-*time* spectrophotometer. The chopped-beam approach, using one detector, compensates for light source variation as well as for sensitivity changes in the detector.

The Light Source

The light source for measurements in the visible portion of the spectrum is usually a *tungsten lamp*. This light source is acceptable for making measurements of moderately dilute solutions in which the change in color intensity varies significantly with small changes in concentration. A potential source of error with some photometers is the heat generated by the light source; this heat may cause problems in measurement either by changing the geometry of the optical system or by changing the sensitivity of the photocell.

At the temperature of the tungsten filament, some of the metal vaporizes and condenses on the cooler glass surface of the bulb, thus reducing the intensity of radiant energy. Such coatings may also change the spectrum sufficiently to alter instrument response. Similar changes may be brought about by deposits of chemical fumes or oily particles on the outside of the bulb. These coatings are often uneven and are particularly likely to cause trouble in the double-beam-in-space spectrophotometers. Periodic inspection and replacement of lamps are necessary to avoid these difficulties.

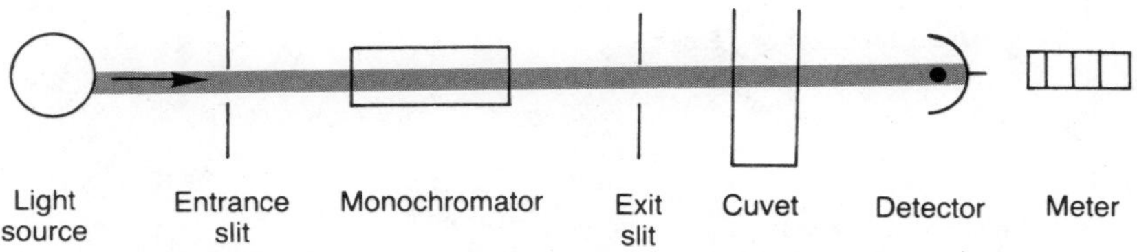

Figure 2-4. Major components of a single-beam spectrophotometer.

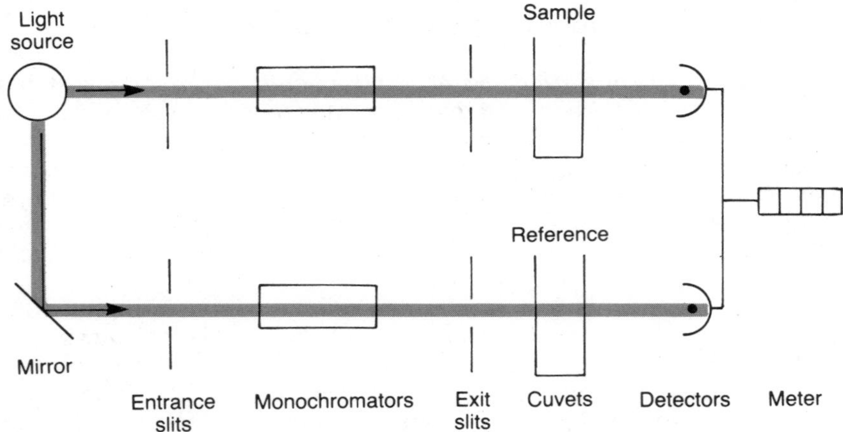

Figure 2-5. Double-beam-in-space spectrophotometer.

A tungsten light source does not supply sufficient radiant energy for measurements below 320 nm. *Hydrogen* and *deuterium lamps* provide sources of continuous spectra in the ultraviolet region with some sharp emission lines, as do high-pressure mercury and xenon arc lamps. A deuterium lamp is more stable and has a longer life than a hydrogen lamp. Electric current is the energy source for the lamp. Because this current is subject to voltage fluctuations, variation may occur in the light intensity from the lamp, resulting in unstable meter readings. Therefore, suitable voltage-regulation devices should be installed for optimal stability.

Spectral Isolation

A system for isolating radiant energy of a desired wavelength and excluding that of other wavelengths is called a *monochromator*. Spectral isolation can be accomplished by various means, including the use of filters, prisms, and diffraction gratings. Combinations of lenses and slits may be inserted before or after the monochromatic device to render light rays parallel or to isolate narrow portions of the light beam. Variable slits may be used to adjust the total radiant energy that reaches the photocell.

The simplest type of filter is a thin layer of colored glass. Certain metal complexes or salts, dissolved or suspended in glass, absorb specific wavelengths, thus leaving the desired wavelength of light to be transmitted through the filter.

Strictly speaking, a glass filter is not a true monochromator since it transmits light over a relatively wide range of wavelengths. The spectral purity of a filter or other monochromator is usually described in terms of its *spectral bandwidth*. This is defined

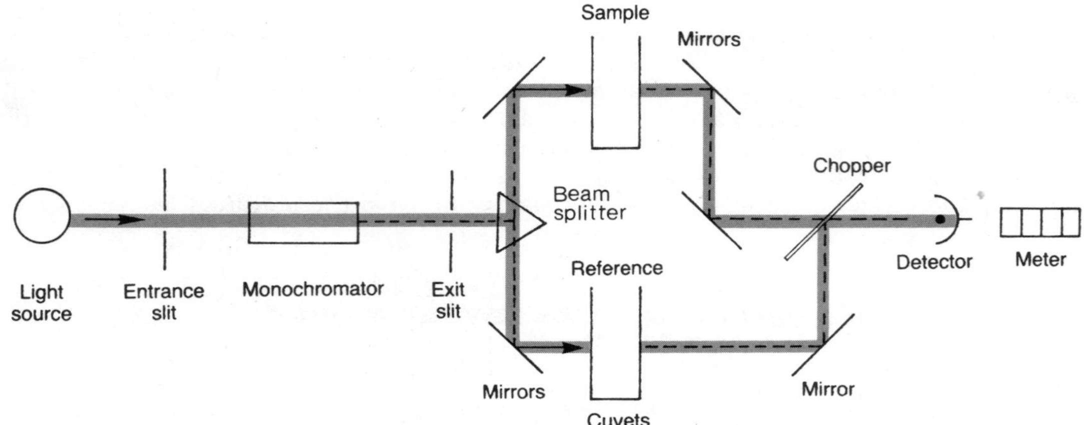

Figure 2-6. Double-beam-in-time spectrophotometer.

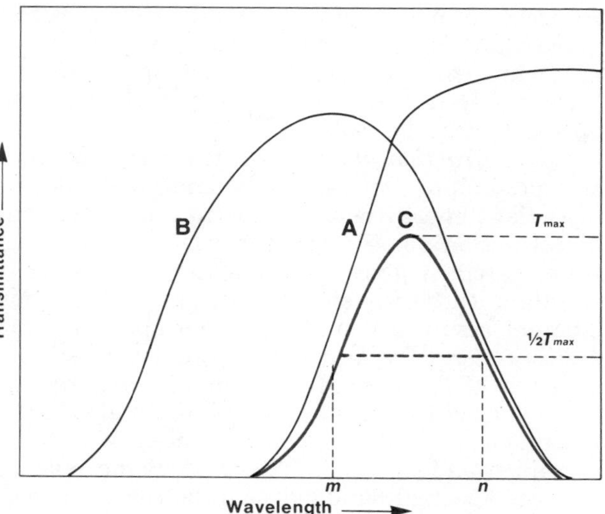

Figure 2-7. Spectral characteristics of a sharp-cutoff filter (*A*) and a wide-bandpass filter (*B*). The narrow-bandpass filter (*C*) is obtained by combining filters *A* and *B*. The spectral bandwidth of filter *C* (distance $n - m$) is defined as the width in nanometers of the spectral-transmittance curve at a point equal to one-half of maximum transmittance.

as the width, in nm, of the spectral transmittance curve at a point equal to one-half the peak transmittance (Figure 2-7). Commonly used glass filters have spectral bandwidths of ~50 nm and are entirely adequate for many purposes. These are referred to as wide-bandpass filters.

Other glass filters include the narrow-bandpass and sharp-cutoff types (Figure 2-7). As implied, the latter filter typically shows a sharp rise in transmittance over a narrow portion of the spectrum and is used to eliminate light below a given wavelength. Narrow-bandpass filters may be constructed by combining two or more sharp-cutoff filters or regular filters but require a high-intensity light source.

Narrow-bandpass filters are constructed from a dielectric material of controlled thickness sandwiched between two thin-silvered pieces of glass. The thickness of the layer determines the wavelength of energy transmitted. Energies of wavelengths that are multiples of this thickness stay in phase as they reflect back and forth between the silvered surfaces and finally emerge, whereas other wavelengths will cancel owing to phase differences. Light striking the semitransparent film at a given angle θ will be partly reflected and partly passed (Figure 2-8, point *a*). The same process occurs at *a′*, *b*, *b′*, and so on. For reinforcement to occur at point *b*, the distance traveled in the medium by the beam reflected at *a′* must be some multiple of its wavelength. Filters so constructed have narrow spectral bandwidths, usually from 5 to 10 nm, and are referred to as *interference filters*. Since these interference filters also transmit harmonics, or multiples, of the desired wavelength, accessory glass filters are required to eliminate the undesired wavelengths. Thus, an interference filter designed for 620 nm will also transmit some radiation at 310 and 1240 nm unless accessory cutoff filters are provided to absorb this undesired stray light.

Prisms and diffraction gratings are widely used as monochromators because they produce complete spectra instead of a single wavelength. A *prism* separates white light

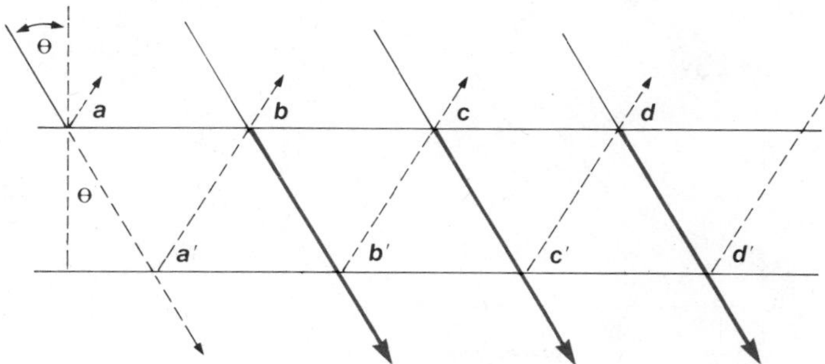

Figure 2-8. Passage of light through an interference filter. The red lines indicate reinforcement of entering light rays.

into a continuous spectrum by refraction; i.e., shorter wavelengths are bent, or refracted, more than longer wavelengths as they pass through the prism. This results in a nonlinear spectrum with the longer wavelengths close together and shorter wavelengths more widely spaced. With the use of suitable accessories, a narrow-bandwidth portion of the spectrum may be isolated.

A *diffraction grating* is a thin layer of aluminum-copper alloy on the surface of a flat glass plate that has many small parallel grooves ruled into the metal coating. Rays of radiant energy bend (refract) around the sharp edges of the grooves, and the extent of refraction varies with the wavelength. Thus, when each groove on the grating is illuminated, it gives rise to a tiny spectrum. Wave fronts formed from these spectra reinforce those wavelengths in phase with each other and cancel those not in phase. The net result is a uniform linear spectrum. Diffraction gratings may be either of the *transmission* or *reflection* type; the latter is more common. A grating may also be ruled at a specified angle, so that a maximum fraction of the radiant energy is directed into wavelengths diffracted at a selected angle. This type of grating is called an *echelette* and is said to have been given a *blaze* at a particular angle or to have been blazed at a certain wavelength (e.g., 250 nm). Some instruments contain diffraction gratings that produce spectral bandwidths of 20 nm or more; higher-priced instruments may have a resolution of 0.5 nm or less. Good gratings are generally better than prisms for spectral isolation.

The type of monochromator chosen will depend on the objectives of the analyst. Narrow-bandpass filters are highly desirable in flame photometry in order to isolate emission energy at a given wavelength. Narrow-spectral bandwidths are required if one is interested in resolving and identifying sharp absorption peaks that are closely adjacent.

Some increase in absorbance as well as improved linearity with concentration is usually observed with instruments that operate at narrow bandwidths of light. This is especially true for substances that exhibit a sharp peak of absorption. Spectral absorbance curves for a solution of coproporphyrin I (Figure 2-9) demonstrate the marked decrease in maximum absorbance as the spectral bandwidth is increased from 1 to 20 nm.

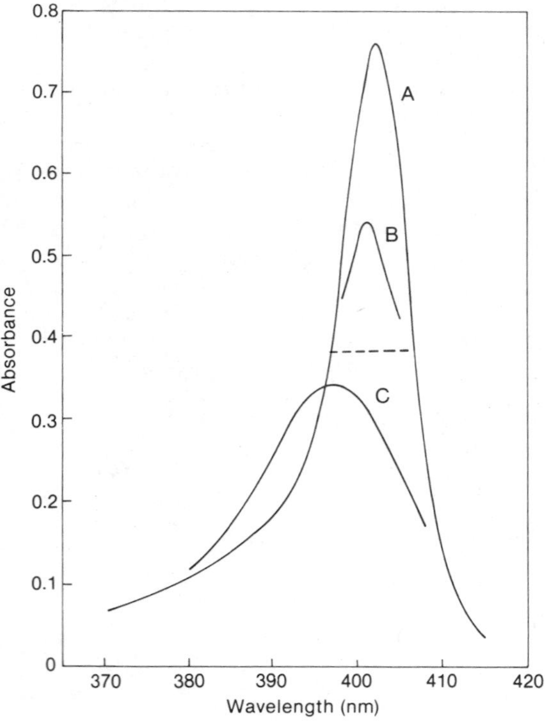

Figure 2-9. Effect of spectral bandwidth (SBW) on the absorption spectrum of coproporphyrin I. Nominal concentration, 1 μg/mL in HCl, 0.1 mol/L. Curve *A*, SBW = 1 nm, Beckman DB-G spectrophotometer; curve *B*, SBW = 10 nm, and curve *C*, 20 nm, Beckman DB spectrophotometer. The red horizontal line shows a natural bandwidth of 10 nm for coproporphyrin I when scanned at a spectral bandwidth of 1 nm. The shift of A_{max} to lower wavelengths, as SBW is increased, is related to skewness of the absorption spectrum to the left.

The *natural bandwidth* of an absorbing substance is defined as the bandwidth of the spectral absorbance curve at a point equal to one-half of the maximum absorbance. Curve A in Figure 2-9, scanned at a *spectral* bandwidth of 1 nm, shows a *natural* bandwidth of 10 nm. As a general rule, for peak absorbance readings to be within 99.5% of true values, the spectral bandwidth should not exceed 10% of the natural bandwidth. It follows that an instrument capable of measuring 1-nm bandwidths is necessary for an accurate measurement of coproporphyrin I.

Coproporphyrin is an unusual example of a compound showing a very narrow natural bandwidth. Most common clinical chemistry laboratory procedures produce a color for which the natural bandwidth ranges from 40 to over 200 nm.[2] The natural bandwidth of NADH is 58 nm (λ_{max} = 339 nm). Ideally, for accurate measurements of this compound, we should use a spectral bandwidth of 6 nm or less. Actual studies have shown that use of a 10-nm spectral bandwidth at 340 nm produces absorbance values approximately 98% of those obtained with a 1-nm spectral bandwidth.[5]

The wavelength selected is usually at the peak of maximum absorbance in order to achieve maximum sensitivity; however, it may be desirable to choose another wavelength to minimize interfering substances. For example, turbidity readings on a spectrophotometer are greater in the blue region than in the red region of the spectrum, but the latter region is chosen for turbidity measurements to avoid absorption of light by bilirubin (460 nm) or hemoglobin (417 and 575 nm). The color developed in the alkaline picrate procedure for creatinine produces a relatively flat peak in the visible region at ~480 nm, but the reagent blank itself absorbs light strongly below 500 nm. A compromise is made by selecting a wavelength at 520 nm to minimize the contribution of the blank. Blank readings should, of course, be kept to a minimum, since a small difference between two large numbers is subject to greater uncertainty; hence, minimizing absorbance of the blank improves precision and accuracy. The linear working range of a method can be expanded by not measuring at the peak absorbance, but measurements on the steep slope of an absorption curve should be avoided, since a slight error in wavelength adjustment would introduce a significant error in absorbance readings.

Multiple Wavelength Readings

Background interference can often be eliminated or minimized either by the inclusion of blanks or by reading absorbance at two or three wavelengths. In one approach, absorbance is measured at two wavelengths, one corresponding to peak absorbance and another at a point near the base of the peak to serve as a baseline. The difference in absorbance at the two wavelengths is related to concentration. In effect, this approach (called bichromatic analysis) provides a blank reference point for each individual specimen. Another method to correct for background interference is to measure absorbance at the peak wavelength and at two other wavelengths equidistant from the peak. Values for the latter are averaged to obtain a baseline under the peak, which is then subtracted from the peak reading. The value thus obtained is known as a "corrected" absorbance and can be related to the concentration, provided that the background absorbance is linear with the wavelength over the region in which readings are made. This technique of making corrections for interfering substances is called the *Allen correction* and is illustrated in Figure 2-10, where the corrected absorbance at 300 nm is obtained from the Allen equation:

$$A_{corr} = A_{300} - \frac{(A_{280} + A_{320})}{2}$$

Similar corrections are applied in procedures for spectrophotometric determinations of salicylates, porphyrins, steroids, and other compounds. The correction must be applied similarly to standards, since the Allen corrected absorbance is typically less than the true absorbance, even in the absence of interfering background.

Before the Allen correction is used, knowledge of the interference and of the shape of the absorption curve for the substance of interest is required. The linearity of the baseline shift should be verified by measuring the absorption spectra of commonly

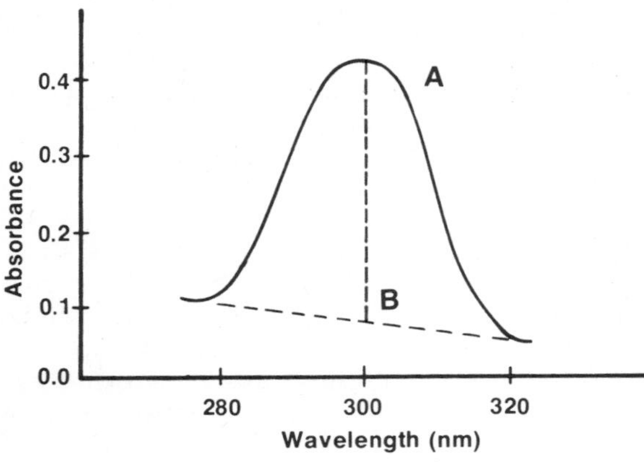

Figure 2-10. Hypothetical example of an Allen correction. *A* is the absorption curve of the test mixture. Absorbance readings are taken at the absorption peak (300 nm) and at equidistant points from the peak (280 and 320 nm). *B* is background absorbance, which is linear over the measured range.

encountered interferences. Care should be exercised in the use of the Allen correction, because if it is not properly used, it may introduce larger errors than would be observed without correction. Such a situation may occur if the background reading is not linear over the region measured.

Cuvets

Cuvets, also known as absorption cells, may be round, square, or rectangular and may be constructed from glass, silica (quartz), or plastic. Depending on the material used, the amount of light transmitted by the cell may vary significantly. For routine colorimetric manual procedures, the round, glass test tube is most common. These are inexpensive and satisfactory, provided they are matched; i.e., to be used interchangeably, the cuvets must be of uniform inside diameters so that the absorbance of a solution will be within specified tolerances when measured in different cuvets. This criterion is readily checked by preparing a stable solution, such as cyanmethemoglobin, to read approximately 50% *T* at 540 nm against a reference blank. This solution is then added to each dry cuvet to be tested and readings are taken. Cuvets with readings that fall within selected tolerances are reserved for use. Since test tube–type cuvets may not be perfectly round, they should be rotated in the well to observe any changes in readings with position; those showing changes should be either discarded or etched to indicate the position of use.

Square or rectangular cuvets have plane-parallel optical surfaces and a constant light path. The most popular have a 1.0-cm light path, held to close tolerances. Compared with round cuvets, they have few of such optical aberrations as lens effect and refractive errors. On the other hand, these cuvets are more expensive and are less convenient to handle. Ordinary borosilicate glass cuvets are suitable for measurements in the visible portion of the spectrum. For readings below 340 nm, however, silica absorption cells are usually required. Some plastic cells have good clarity in both the visible and ultraviolet range but can present problems relating to tolerances, cleaning, etching by solvents, and temperature deformations. Many of the plastic cuvets are designed for disposable, single-use applications.

Cuvets must be clean and optically clear. Etching or deposits on the surface will obviously affect absorbance values. Round cuvets, used in the visible range, are cleaned by copious rinsings with tap water and distilled water. Alkaline solutions should not be left standing in cuvets for prolonged periods since alkali slowly dissolves glass and produces etching. Both round and square cuvets may be cleaned in mild detergent or soaked in a mixture of concentrated HCl-water-ethanol (1:3:4). Cuvets should never be soaked in dichromate cleaning solution since the solution tends to adsorb onto and discolor the glass.

Cuvets used for measurements in the ultraviolet region should be handled with special care. Invisible scratches, fingerprints, or residual traces of previously measured substances may be present and may absorb significantly. A good practice is to fill all such cuvets with distilled water and measure the absorbance for each against a reference blank over the wavelengths to be used. Any cuvet showing an absorbance greater than zero should be cleaned to eliminate falsely elevated readings or should be discarded.

Detectors

The two most commonly used devices for measuring light intensity in the UV and visible regions of the spectrum are barrier layer cells and photomultiplier tubes. The barrier layer cells (also known as photovoltaic cells) are rugged and are used in inexpensive instruments; photomultipliers are almost always used in higher quality, more expensive spectrophotometers.

Barrier layer cells. The barrier layer cell consists of a thin, semitransparent layer of silver on a layer of the semiconductor selenium. The silver and selenium metals are then mounted on an iron backing or support (Figure 2-11). When light passing through the transparent metal layer falls upon the selenium surface, electrons are released in proportion to the intensity of the light and are collected on the silver layer to produce a negative charge. The iron base plate becomes the positive pole. When this cell is connected to a galvanometer, a current will flow that will vary with the intensity of the incident light.

Photomultiplier tubes. A photomultiplier is an electron tube that is capable of significantly amplifying a current. The photomultiplier tube has as its cathode a light-sensitive metal that absorbs light and emits electrons in proportion to the radiant energy that strikes the surface of the light-sensitive material. The electrons produced by this first stage go to a secondary stage (dynode), where each electron produces between four and six additional electrons. Each of these electrons from the second stage goes on to another stage, again producing four to six electrons. Each electron cascades through the photomultiplier stages; thus, the final current produced by such a tube may be one million times as much as the initial current (Figure 2-11). As many as 10 to 15 stages or dynodes are present in common photomultipliers.

Photomultiplier tubes have extremely rapid response times, are very sensitive, and do not show as much fatigue as other detectors. Because the tubes have excellent sensitivity and rapid response, they must be carefully shielded from all stray light and

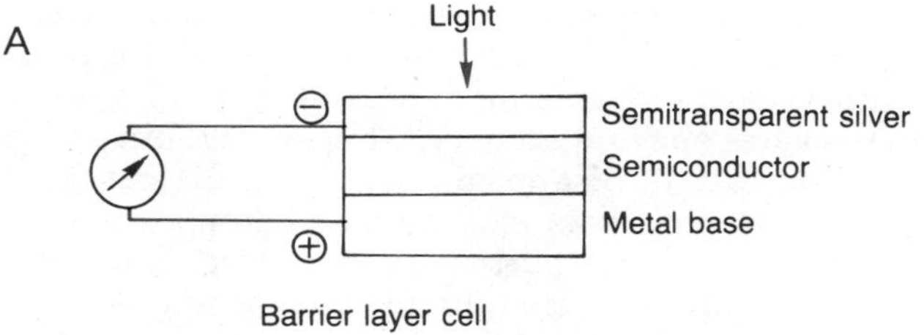

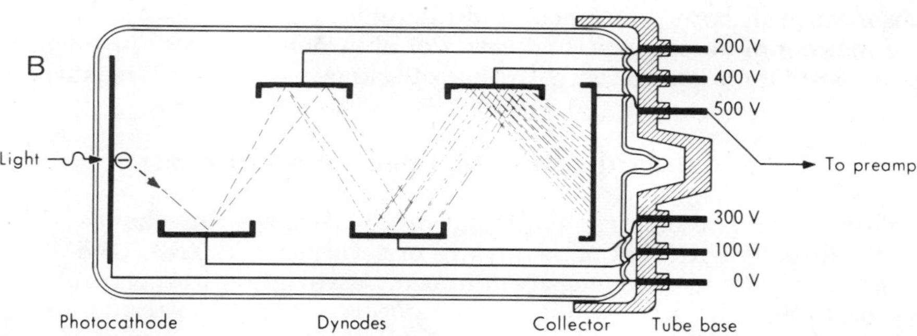

Figure 2-11. *A,* Schematic diagram of a barrier layer cell. See text. *B,* Schematic diagram of a photomultiplier tube. (From Powsner, E. R., and Raeside, D. E.: Diagnostic Nuclear Medicine. New York, Grune & Stratton, 1971.)

daylight. A photomultiplier with the voltage applied should never be exposed to room light because it may burn out. Because of the fast response time of the photomultiplier, this detector is readily used with interrupted light beams such as those produced by choppers, and it therefore has significant advantages when used in double-beam-in-time spectrophotometers.

When voltage is applied to photomultipliers and all light has been blocked from them, some current usually will be produced. This current is called *dark current*. It is desirable to have the dark current of photomultipliers at their lowest level, since this current is also amplified and appears as background noise.

Readout Devices

Electrical energy from a detector is displayed on some type of meter or readout system. In *direct-reading systems*, the output of the detector is used to drive a sensitive meter directly with no further amplification. Other instruments utilize an amplifier to increase the output of the detector. In the *null point system*, the output of the detector is balanced against the output of a reference circuit. The meter may be replaced by a servomotor, which is activated by an imbalance of current and which stops when the two circuits are balanced. Direct digital readouts are thus obtained.

Also popular are *digital readout devices* that provide a numerical display of absorbance or converted values of concentrations. These operate on the principle of selective illumination of portions of a bank of light-emitting tubes or diodes, controlled by the voltage signal generated. Typical examples include visible light-emitting diodes (LEDs).

Microprocessors are now widely incorporated into many instruments. Signal output from a standard or calibrator can be stored, signals from blanks can be subtracted from both standards and unknowns, and the concentration of unknown can be readily calculated analogous to the application of Beer's law. Data from multiple standards may be used to store a complete calibration curve, display or print out the curve for inspection, and calculate results of unknowns based on the curve or some mathematical transformation of it.

Recorders

Spectrophotometers may be equipped with recorders in addition to or instead of the digital display of values. Recorders are synchronized to provide line traces of transmittance or absorbance, as a function of either time or wavelength. Change of absorbance with time is widely used to measure enzyme activity. Thus, in the kinetic determination of a dehydrogenase enzyme, the rate of disappearance of NADH can be monitored at 339 nm. Multiple readings at short intervals may also be taken electronically and fed into a microprocessor to identify the linear portion of the curve, calculate the slope as ΔA/min, and multiply by an appropriate constant to provide readout and printout of the final result.

When a continuous tracing of absorbance versus wavelength is recorded, the resultant figure is called an *absorption spectrum*. This type of procedure is especially useful for the identification of drugs that absorb in the ultraviolet range. Several criteria are used, including the determination of those wavelengths showing maximum and minimum absorbance in both dilute acid and alkaline solutions, absorptivity at the wavelength of maximum absorbance, and ratios of absorbance at two wavelengths. Finally, the entire spectrum is compared with that of a known sample of the suspected drug.

Standardization of Spectrophotometers

In most spectrophotometric analytical procedures, the absorbance of an unknown is compared directly with that of a standard or series of standards. Under these circumstances, minor errors in wavelength calibration, a variation in spectral bandwidth, the presence of stray light, and so on, do not usually contribute serious errors. Use of a series of standards covering a wide range of concentrations also provides a measure of linearity (i.e., agreement with Beer's law for a given procedure and instrument). When calculations are based on published or previously determined values for molar absorptivities or absorption coefficients, however, the spectrophotometer must be checked more rigorously. Standardization of spectrophotometers on a periodic basis also improves reliability of routine comparative analyses.[1]

Wavelength Calibration

For many analytical purposes, the wavelength chosen may be satisfactory if it is close to the λ_{max} of the measured chromogen and if the wavelength is reproducible. Use of filters can be satisfactory because unknowns are compared with standards at a fixed wavelength and spectral bandwidth. With prisms and diffraction gratings, however, a continuous choice of wavelengths is available, and it becomes necessary to verify their accuracy and reproducibility. Knowledge of the exact wavelength becomes critical when using published molar absorptivities, for identification of substances in toxicological studies, and in the use of differential absorption techniques. Enzyme assays employing the NAD-NADH reaction, for example, are based on a molar absorptivity constant for NADH of 6.22×10^3 at 339 nm. (Recently 6.23×10^3 has been suggested.) It is therefore necessary that the wavelength setting be accurate and reproducible and that the instrument show spectrophotometric accuracy if the constant is to be used in calculating results.

For instruments of narrow spectral bandwidths, a *holmium oxide glass* may be scanned over the range of 280–650 nm. This material shows very sharp absorbance peaks at well-defined wavelengths, and the operator may compare the wavelength scale readings that produce maximum absorbance with established values. Should these not coincide, a calibration curve can be constructed to relate scale readings to true wavelengths. A typical spectral-transmittance curve for holmium oxide glass is shown in Figure 2-12. Selected absorption peaks for this filter, suitable for calibration purposes, occur at 279.3, 287.6, 333.8, 360.8, 418.5, 536.4, and 637.5 nm. Solutions of holmium oxide in dilute perchloric acid have also been recommended and may be used with any spectrophotometer.[1]

With broad-bandpass instruments a didymium filter may be used to verify wavelength settings. This filter should show a sharp minimum per cent transmittance at 530 nm (depending on the instrument) against an air blank (Figure 2-13). Since didymium has several absorption peaks, the setting should be verified grossly by visual examination of transmitted light. This light should appear green at 530 nm.

Spectral Bandwidth

The spectral bandwidth (SBW) of a spectrophotometer is customarily specified by the manufacturer. This value is generally accepted without verification because of the lack of a convenient method for directly verifying it. The SBW can be measured by use

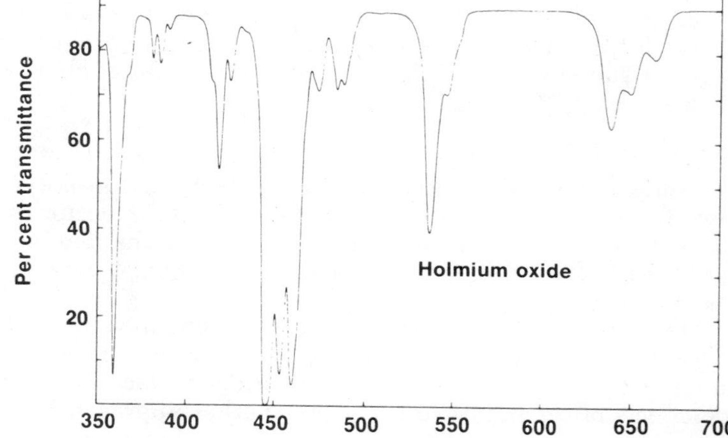

Figure 2-12. Spectral-transmittance curve of holmium oxide filter. (Courtesy of Beckman Instruments, Inc.)

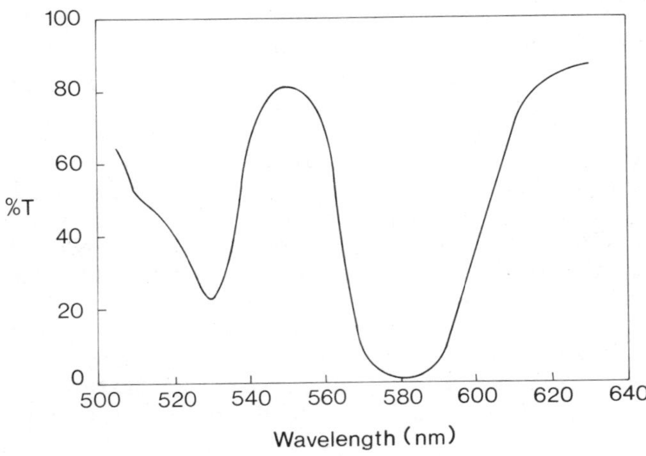

Figure 2-13. Spectral-transmittance curve of a didymium filter. (Perkin-Elmer Model 35 spectrophotometer, 8-nm nominal spectral bandwidth.)

of a mercury vapor lamp, which shows a number of sharp, well-defined emission lines between 250 and 580 nm. The apparent width of an emission band at half-peak height is taken to be the spectral bandwidth of the instrument (see Figure 2-7). The SBW may also be calculated from the manufacturer's specifications.[4] Interference filters with spectral bandwidths of 1–2 nm are available and may be used to check those instruments with a nominal SBW of 8 nm or more.

Stray Light

Stray light, in general terms, is radiation of wavelengths outside the narrow band nominally transmitted by the monochromator. A perfect monochromator would transmit light only within its bandpass. In practice, scattering and diffraction inside the monochromator introduce light of other wavelengths into the exit beam. This light is further modified by other components of the spectrophotometer and by the sample itself. Stray light is usually defined as a ratio or per cent of the stray light to the total detected light.

Other sources of unwanted light include light leaks and fluorescence of the sample. Light leaks should be excluded by covering the cell compartments. Light arising from fluorescence can increase the signal to the detector and cause an apparent decrease in absorbance. These sources of light are not included in the usual definition of stray light.

Cutoff filters are satisfactory for the detection of stray light. These may be of glass similar to the composition of stray-light filters discussed above, which produce a sharp cut in the spectrum with almost complete absorption on one side and high transmittance on the other. Liquid cutoff filters are satisfactory and convenient in the UV range, where stray light is usually more of a problem. An aqueous solution of sodium nitrite, 50 g/L, should show essentially zero transmittance when read against water over the range of 300–385 nm. Acetone, read against water, should show zero transmittance over the range of 250–320 nm. With either solution, readings above ~2% *T* indicate the presence of stray light.

Linearity

Linearity of calibration curves has been discussed earlier with reference to Beer's law. In order for a spectrophotometer to give accurate absorbance measurements throughout its absorbance range, its response to changes in light intensity must be linear. This means that a linear relationship should exist between the light absorbed and the instrument readout.

Liquid solutions of various concentrations have been used to check instrument linearity. However, several sources of error may be encountered in this type of standard; among them are errors caused by dilution, by lack of stability, by shifts in pH, and by the effects of temperature changes. Another approach is to use a solid glass filter that, at a given wavelength, has an absorbance that is a small fraction of the total linear range of the instrument.[3] The didymium filter used for Figure 2-13 has an absorbance of ~0.09 at 550 nm and is satisfactory for establishing linearity. Nickel sulfate solution ($NiSO_4 \cdot 6H_2O$, 12 g/dL, in sulfuric acid, 0.05 mol/L) may also be used at 550 nm. When the filter is used, zero absorbance is set with the cell compartment empty. For the nickel sulfate solution, water is used as a reference. For filter or solution, the procedure is as follows:

1. Set the wavelength at 550 nm and set absorbance to zero.
2. Read the absorbance of the filter or solution and record the value.
3. Remove the filter or solution and set absorbance to 0.25.
4. Read the absorbance of the filter or solution and record the value.
5. Repeat by setting the absorbance to 0.50, 0.75, 1.00, and 1.25, and read the absorbance of the filter or solution as above.
6. Calculate the ΔA for each increase in settings.

Linearity is satisfactory as long as the ΔA increments remain constant and within specified tolerance limits. The above procedure may be used on spectrophotometers in which the blank reference settings can be adjusted in steps up to a total of at least 1 A.

Photometric Accuracy

Solutions of potassium dichromate are recommended for overall checks on photometric accuracy.[1] These also provide an indication of wavelength calibration, linearity, cuvet light path, and freedom from stray light in the UV region. Analytical reagent grade potassium dichromate ($K_2Cr_2O_7$) is dried at 110 °C for 1 h. Exactly 0.0500 g is dissolved in, and diluted to 1 L with sulfuric acid, 0.005 mol/L. The absorbance is read at 350 nm in 10-mm cells with the temperature controlled in the range of 15–25 °C; sulfuric acid, 0.005 mol/L, is used as the reference. The absorbance should read 0.536 ± 0.005. Since the natural bandwidth of the dichromate solution at 350 nm is ~63 nm, the value given applies strictly to spectrophotometers with a spectral bandwidth that is ≤ 6 nm. The solution should also show an absorbance peak at 350 nm.

Glass filters made from a moderately neutral glass may also be used to check photometric accuracy. These are available from the National Bureau of Standards (Schott NG-4 SRM 930) and are calibrated on the NBS high-accuracy spectrophotometer at levels of approximately 10, 20, and 30% transmittance. A certificate is supplied that describes the filters and gives the calibration values at four wavelengths, ranging from 440 to 635 nm.

Flame Photometry

Flame emission photometry is most commonly used for the quantitative measurement of sodium and potassium in body fluids. Lithium, although not normally present in serum, may also be measured in connection with the therapeutic use of lithium salts in the treatment of some psychiatric disorders.

Atoms of many metallic elements, when excited by a hot flame, will emit energy at wavelengths characteristic for the element. A specific amount or quantum of thermal energy is absorbed by an orbital electron. The electrons, unstable in this high-energy (excited) state, release their excess energy as photons of a particular wavelength as they change from the excited to their previous or ground state. The light emitted may consist of more than one energy level and therefore of different wavelengths. This line spectrum is characteristic for each element. Sodium, for example, emits energy primarily at 589 nm along with other, much less intense emissions. The wavelength to be used for the measurement of an element depends upon the selection of a line of sufficient intensity to provide adequate sensitivity as well as freedom from interference by other lines at or near the selected wavelength.

Some metals are comparatively easy to excite in the flame of an ordinary laboratory burner. Lithium produces a red, sodium a yellow, potassium a violet, rubidium a red, and magnesium a blue color in the flame. These colors are characteristic of the metal atoms that are present as cations in solution. Under constant and controlled conditions, the light intensity of the characteristic wavelength produced by each of the atoms is directly proportional to the number of atoms that are emitting energy, which in turn is directly proportional to the concentration of the substance of interest in the sample. Of the more easily excited alkali metals, such as sodium, only 1–5% of those atoms present in solution become excited in a flame. Even with this small percentage of excited atoms, the method has adequate sensitivity for the measurement of alkali metals for most bioanalytical applications. Thus, flame photometry lends itself well to direct concentration measurements of some metals.

Other cations, such as calcium, are less excited in a propane-air flame. In these cases, the amount of light given off may not provide adequate sensitivity for analysis by flame emission methods. The sensitivity can be improved slightly by using high-temperature flames. Most other metal ions are not as easily excited in a flame, and flame emission methods are not as suitable for their measurement.

Components of a Flame Photometer

Figure 2-14 shows a schematic diagram of the basic parts of a flame photometer. The sample is sprayed as fine droplets into the flame through the atomizer. The monochromator, entrance and exit slits, and detectors are similar to those discussed previously for spectrophotometers. In effect, the light source for the spectrophotometer has been replaced with an atomizer-flame combination, and emission of light is measured rather than absorption.

Various combinations of gases and oxidants are used in flame photometry. The choice of flame depends on the temperature desired; for sodium and potassium determinations, a propane–compressed air flame is entirely adequate.

The *atomizer* and the *flame* are critical components in a flame photometer. The atomizer provides a means of drawing the sample through the aspirator and converting it into a fine mist, which then enters the flame. The most important variable in the flame itself is the temperature. Frequent standardization of flame photometers is essential because thermal changes affect the number of excited atoms. In addition, temperature changes affect the output of photocell detectors; for this reason, a period of warm-up, which includes aspiration of water and standards, is required to establish thermal equilibrium for the flame and the atomizer chamber.

Ideally, monochromators in flame photometers should be of higher quality than those found in absorption spectrophotometers. When nonionic materials are burned, light of varying wavelengths is given off. This light is known as *continuous emission* and will be added to the *line emission* of the element being measured. For this reason, the narrowest bandpass that is achievable should be used to eliminate as much of the extraneous, continuous emission as possible but still permit a maximum of the line emission to pass through to the detector. The detectors used in flame photometers operate by the same principle and in the same way as those described for spectrophotometers.

Direct and Internal Standard Flame Photometry

In some of the instruments of earlier designs, diluted solutions of sodium or potassium were aspirated directly into the flame to provide a series of meter readings. This approach, referred to as the *direct reading method*, presents certain problems; these include fluctuations in air and gas pressures that result in imprecise and inaccurate

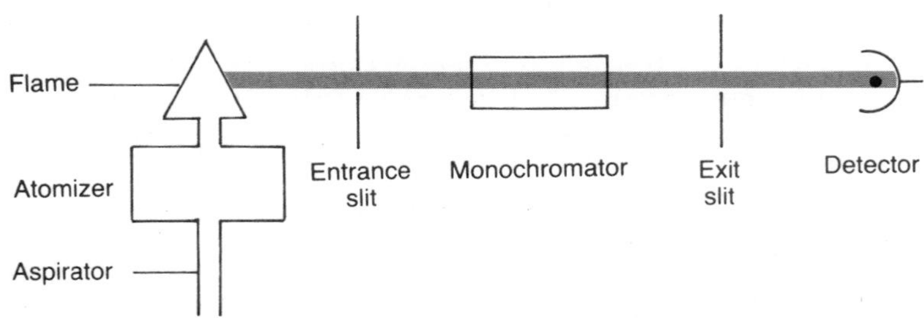

Figure 2-14. Essential components of a flame photometer.

data and enhancement of the potassium signal by the sodium present, the so-called mutual excitation.

In the *internal standard* method, lithium or cesium is added to all standards, blanks, and unknowns in equal concentrations. Lithium has a high-emission intensity, is normally absent from biological fluids, and emits at a wavelength sufficiently removed from sodium and potassium to permit spectral isolation. The flame photometer makes a comparison of the emission of the desired element (sodium or potassium) with the emission of the reference lithium element. When the ratios of emissions are measured in this way, small variations in atomization rates, flame stability, and solution viscosity are compensated for. Lithium does not function as a "standard" under these conditions but as a reference signal for analytical variation. Known concentrations of sodium and potassium in the lithium diluent must be used to establish calibration curves or to verify linearity of response.

Serum lithium concentrations in patients receiving lithium salts are maintained at ~1 mmol/L. However, this amount is insignificant compared with the high amount added as internal standard. The large concentration of lithium in the diluent also acts as a *radiation buffer* to minimize the effects of mutual excitation, i.e., the transfer of energy from one excited atom to another (as noted above from excited sodium to potassium atoms).

A wetting agent is frequently recommended for inclusion in standards and sample dilutions. This agent minimizes changes in atomizer flow rates due to differences in the viscosity of the samples. Viscosity effects are further reduced by diluting samples 100- to 200-fold.

Atomic Absorption Spectrophotometry

Atomic absorption spectrophotometry in some respects is the inverse of flame emission photometry. In all emission methods, the sample is excited in order to measure the radiant energy given off as the element returns to its lower energy level. Extraneous radiation must be filtered out from the energy of interest if interference by these signals is to be avoided.

In atomic absorption spectrophotometry the element is not appreciably excited in the flame but is merely dissociated from its chemical bonds and placed in an unexcited or ground state. This means that the atom is at a low energy level in which it is capable of absorbing radiation at a very narrow bandwidth corresponding to its own line spectrum. A hollow cathode lamp, made of the material to be analyzed, is used to produce a wavelength of light specific for the kind of metal in the cathode. Thus, if the cathode were made of sodium, sodium light at predominantly 589 nm would be emitted by the lamp. When the light from the hollow cathode lamp enters the flame, some of it is absorbed by the ground-state sodium atoms in the flame, resulting in a net decrease in the intensity of the beam arriving at the detector. This process is referred to as atomic absorption.

The process is analogous to absorption spectrophotometry. A specific hollow cathode lamp serves as the light source, and the sample heated in the flame replaces the sample in the cuvet. The pathlength of the flame is analogous to the light path through the cuvet. As noted previously, only a small fraction of the sample in the flame contributes emission energy, and only a fraction of this will be transmitted to the detector. Hence, most of the atoms are in the ground state and are able to absorb light emitted by the cathode lamp. In general, atomic absorption methods are approximately 100 times more sensitive than flame emission methods. In addition, owing to the unique specificity of the wavelength from the hollow cathode lamp, these methods are highly specific for the element being measured.

Components of an Atomic Absorption Spectrophotometer

Figure 2-15 shows the basic components of an atomic absorption spectrophotometer. The hollow cathode lamp is the light source. A nebulizer sprays the sample into the flame; the monochromator, slits, and detectors have the functions described for spec-

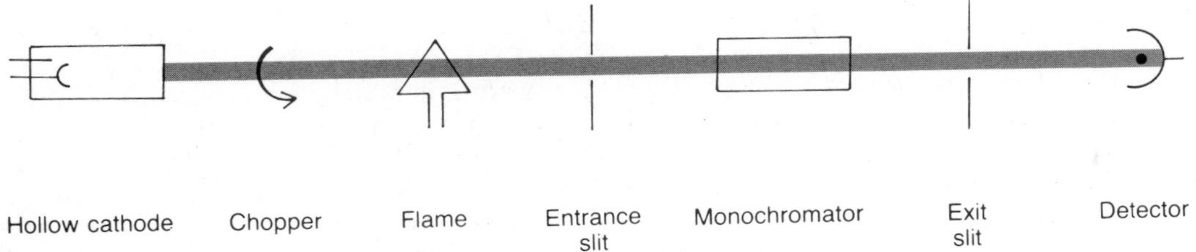

Figure 2-15. Essential components of an atomic absorption spectrophotometer.

trophotometry. The cathode is made of the metal to be analyzed and is different for each metal analysis. In some cases, when an alloy is used to make the cathode, a multielement lamp is produced.

The hollow cathode lamp usually contains argon or neon gas at a pressure of a few millimeters of mercury. An argon-filled lamp produces a blue-to-purple glow during operation, and the neon produces a reddish-orange glow inside the hollow cathode lamp. Quartz or special glass that allows transmission of the proper wavelength is used as a window. A current is applied between the two electrodes inside the hollow cathode lamp, and metal is sputtered from the cathode into the gases inside the glass envelope. When the metal atoms collide with the neon or argon gases, they lose energy and emit their characteristic radiation. Calcium has a sharp, intense, analytical emission line at 422.7 nm, which is most frequently used for calcium analysis. In an interference-free system, only calcium atoms will absorb the calcium light from the hollow cathode as it passes through the flame.

Two kinds of burners have been used in most clinical applications. One is a *total consumption burner*, illustrated in Figure 2-16, in which the gases (hydrogen and air) mix with the sample within the flame. One disadvantage of the total consumption burner is that relatively large droplets produced in the flame scatter light and thus cause signal noise. An advantage of this type of burner is that the flame is more concentrated and it can be made hotter than that of other burners, causing molecular dissociation that may be desirable for some chemical systems.

Figure 2-16 also shows a *premix burner* (laminar flow burner) in which the gases are mixed and the sample is atomized before it is burned. An advantage of this system is that the larger droplets go to waste while only the fine mist enters the flame; thus, a less noisy signal is produced. In addition, the pathlength through the flame of the burner is longer than that of the total consumption burner. This longer pathlength

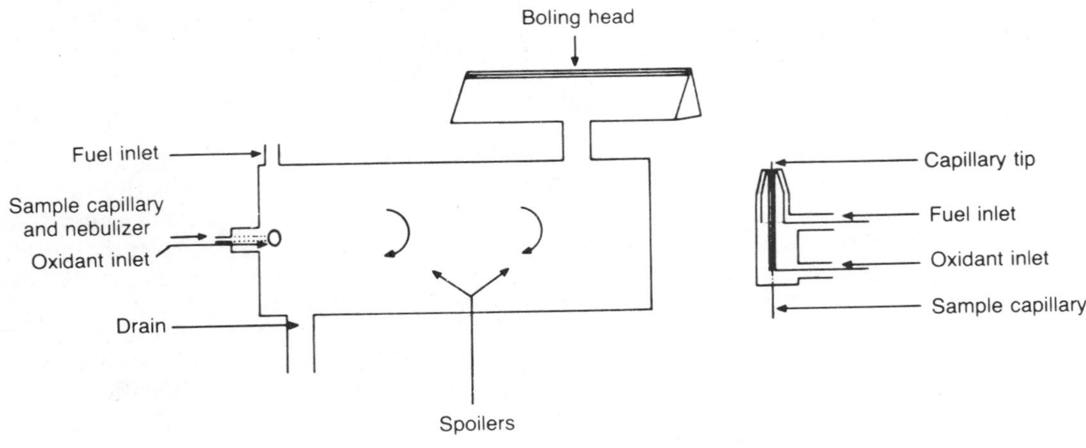

Figure 2-16. Schematic diagrams of a total consumption burner and a laminar flow burner.

produces a greater absorption and increases the sensitivity of the measurement. A disadvantage of the premix burner is that the flame is usually not as hot as that of the total consumption burner, and thus it cannot adequately dissociate certain metal complexes (e.g., calcium phosphate complexes) in the flame. *Nitrous oxide premix burners* produce higher temperatures and will dissociate some calcium complexes; however, at these higher temperatures, calcium becomes excited to a significant extent and emits in the flame, thus introducing another problem.

In *flameless atomic absorption* techniques (carbon rod or "graphite furnace"), the sample is placed in a depression on a carbon rod in an enclosed chamber. Strips of tantalum metal may also be used. In successive steps, the temperature of the rod is raised to dry, char, and finally atomize the sample into the chamber. The atomized element then absorbs energy from the corresponding hollow cathode lamp. This approach is more sensitive than the conventional flame methods and permits determination of trace metals in small samples of blood or tissue.

Interferences in Atomic Absorption Spectrophotometry

There are three general types of interferences in atomic absorption spectrophotometry: chemical, ionization, and matrix effects.

Chemical interference refers to the situation when the flame cannot dissociate the sample into free atoms so that absorption can occur. An example is the phosphate interference in the determination of calcium, caused by the formation of calcium phosphate complexes. These complexes do not dissociate in the flames unless a special high-temperature burner is used. The phosphate interference is overcome by adding a cation that will compete with calcium for the phosphate. Usually, in the determination of calcium, lanthanum or strontium is added to the dilute sample to replace and release calcium from its phosphate complex. The freeing of calcium occurs because lanthanum and strontium form more stable complexes with phosphate than does calcium. The free calcium is then capable of absorbing the calcium light from the hollow cathode lamp.

Ionization interference results when atoms become excited (instead of only dissociated) by the flame and then emit energy of the same wavelength that is being measured. This effect can be overcome by adding an excess of a more easily ionized substance that will absorb most of the flame energy so that the substance of interest will not become excited. Ionization interference can also be decreased by reducing the flame temperature.

A third type of interference is *matrix interference*. One example of a matrix effect is caused by organic solvents, which cause enhancement of light absorption. An atom may absorb between two and five times more energy when dissolved in an organic solvent than when dissolved in an aqueous solvent. A second kind of matrix effect is the light absorption caused by the formation of solids from sample droplets as the solvent is evaporated in the flame. This will usually occur only with solutions of concentrations >0.1 mol/L. Refractory oxides of metals formed in the flame can also lead to matrix interferences.

Atomic absorption spectrophotometry is sensitive, accurate, precise, and highly specific. One of the reasons for these advantages is that the method does not require excitation of the element and thus is less affected than flame photometry by temperature variations in the flame and by the transfer of energy from one atom to another. The high specificity results from the fact that the light used has an extremely narrow bandwidth (0.01 nm) and is selectively absorbed by the atoms being measured. The most significant disadvantage is the problem of interferences. This is being solved, however, by extraction techniques and the introduction of competing cations to release the element to be measured from complexing or chelating anions.

References

1. Burgess, C., and Knowles, A., Eds.: Standards in Absorption Spectrometry. New York, Chapman and Hall, 1981.
2. James, G. P., and DJang, M. H.: Evaluation of clinical laboratory instruments. Part III. Spectral bandwidth and wavelength accuracy. Am J. Med. Technol., *47*:477–483, 1981.

3. Lucas, D. H., and Blank, R. E.: Spectrophotometric standards in the clinical laboratory. Am. Lab., *9*:77–83, 1977.
4. Passey, R. B., Gillum, R. L., and Fuller, J. B.: Measurement of spectral bandwidth, as exemplified with the Beckman "Enzyme Analyzer System TR Spectrophotometer." Clin. Chem., *21*:1582–1584, 1975.
5. Surles, T., and Erickson, J. O.: Absorbance measurements at various spectral bandwidths. Clin. Chem., *20*:1243–1244, 1974.

Additional Reading

Bauer, H. H., Christian, G. D., and O'Reilly, J. E., Eds.: Instrumental Analysis. Boston, Allyn and Bacon, 1978.
Ewing, G. W.: Instrumental Methods of Chemical Analysis. 3rd. ed. New York, McGraw-Hill, 1969.
Knowles, A., and Burgess, C., Eds.: Practical Absorption Spectrometry. New York, Chapman and Hall, 1984.

SECTION TWO

Fluorometry, Nephelometry, and Turbidimetry

THOMAS O. TIFFANY, Ph.D.
CARL A. BURTIS, Ph.D.

Fluorescence and Light Scattering Measurements

The interaction of radiant energy with molecules or particles in solution can result in fluorescence or light scattering. *Fluorescence* occurs when a molecule (i.e., a fluorophor) absorbs light at one wavelength and re-emits light at a longer wavelength. *Light scattering* occurs when radiant energy passing through a solution encounters a molecule in an elastic collision, resulting in the light being scattered in all directions. Unlike fluorescence emission, the scattered light is of the same frequency as the incident light.

Approaches to Fluorescence Measurements

The Nature of Fluorescence

Luminescence is the emission of light or radiant energy when an electron returns from an excited or higher energy level to a lower energy level. Types of luminescence phenomena include fluorescence, phosphorescence, and chemiluminescence. Although these phenomena differ in how an electron is activated to the excited state, they result in similar emissions of radiant energy. Fluorescence luminescence and phosphorescence luminescence occur when radiant energy is absorbed by a molecule. Absorption of light energy by a molecule takes place in discrete units, or quanta. The energy of a quantum of light is expressed by the formula $E = h\nu$ or $E = hc/\lambda$, where E is the energy, h is Planck's constant, c the velocity of light, ν the frequency, and λ the wavelength. Figure 2-17 shows diagrammatically the relationship between absorption, fluorescence, and phosphorescence. Absorption of a quantum of light energy by a molecule causes the transition of an electron from the singlet ground state to one of a number of possible vibrational levels of its first excited singlet state. Once the molecule is in an excited state it can return to its original energy state by (1) radiationless vibrational equilibration, (2) the fluorescence process, (3) quenching of the excited singlet state, (4) radiationless crossover to a triplet state, (5) quenching of the first triplet state, or (6) the phosphorescence process.

Fluorescence is the most probable process to occur under the typical reaction conditions used by the clinical chemist, and certain behavior is characteristic of fluorescence. Vibrational equilibration before fluorescence results in some loss of the exci-

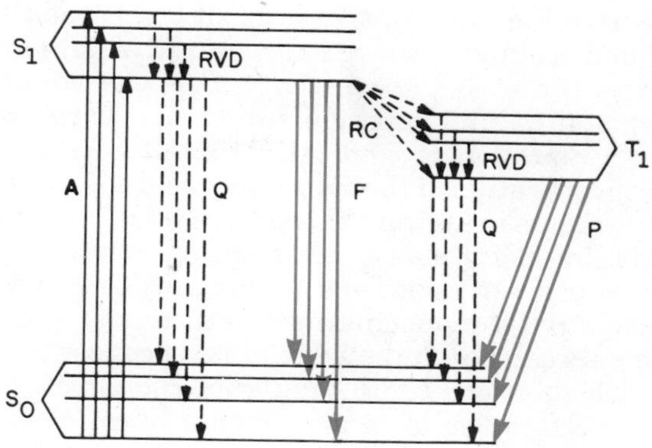

Figure 2-17. Luminescence energy level diagram of typical organic molecule. S_0 is the ground level singlet state; S_1 is the first excited singlet state; A is the absorption process; T_1 is the first excited triplet state; and RVD is the radiationless vibrational deactivation. Q is quenching of the excited singlet or triplet state. F is the fluorescence process from the first excited singlet state. P is the phosphorescence process from the first excited triplet state. RC is the radiationless crossover from the first excited singlet state to the first excited triplet state.

tation energy (Figure 2-17). As a result, the emitted fluorescent light has lower energy and a longer wavelength than the excitation light. The difference between the maximum wavelength of the excitation light and the maximum wavelength of the emitted fluorescent light is a constant referred to as the "Stokes shift." This constant is a measure of the energy lost during the lifetime of the excited state (radiationless vibrational deactivation) before return to the ground singlet level (fluorescence emission). Emission of fluorescent light is characterized by a rapid (10^{-8} s) decay time.

Phosphorescence shows a larger shift in emitted-light wavelength than does fluorescence, and the decay time of emission of phosphorescence is longer (10^{-4} to 100 s) than that of fluorescence.

Chemiluminescence and ***bioluminescence*** differ from the other luminescence phenomena in that the excitation event is caused by a chemical reaction. The physical event of the light emission in chemiluminescence and bioluminescence is similar to fluorescence; it occurs from an excited singlet state and the light is emitted when the electron returns to the ground state. *Chemiluminescence* involves the oxidation of an organic compound, such as luminol, by an oxidant (e.g., hydrogen peroxide, hypochlorite, or oxygen), and the emission of light occurs from the excited product formed in the chemical reaction. *Bioluminescence* is the name given to a special form of chemiluminescence found in biological systems in which a catalytic protein, e.g., luciferase, increases the efficiency of the luminescent reaction. Chemiluminescence and particularly bioluminescence assays have a sensitivity greater than most direct fluorescence assays; the sensitivity is comparable to the sensitivity of radioimmunoassays.

The Relationship of Concentration to Intensity of Fluorescence Emission

The relationship of concentration to intensity of fluorescence emission may be derived from Beer-Lambert's law:

$$I/I_O = 10^{-abc} \tag{1}$$

where I is the intensity of light transmitted, I_O the intensity of the incident light, a the absorptivity, b the pathlength, and c the concentration of the absorbing species. The amount of light absorbed is obtained by rearrangement of equation (1):

$$I_O - I = I_O(1 - 10^{-abc}) \tag{2}$$

The fluorescence intensity F is proportional to the quanta of light absorbed and to the efficiency of the emitted light as expressed by equation (3):

$$F = \phi I_O(1 - 10^{-abc}) \tag{3}$$

leading, through mathematical rearrangement, to:

$$F = \phi I_O(2.303\, abc) \tag{4}$$

where F is the relative intensity, ϕ is the fluorescence efficiency (i.e., the ratio between quanta of light emitted and quanta of light absorbed), I_0 is the initial excitation intensity, a is the molar absorptivity, b is the volume element defined by the geometry of the excitation and emission slits, and c is the concentration in mol/L.

Equation (4) indicates that fluorescence intensity is directly proportional to the concentration of the fluorophor and the excitation intensity. However, fluorescence measurements are often expressed in relative intensity units. The word *relative* is used because the intensity measured is not an absolute quantity. It is a small part of the total fluorescence emission, and its magnitude is defined by the instrument slit widths, detector sensitivity, monochromator efficiency, and excitation intensity. Fluorescence measurements can be expressed in units of concentration, provided the system has been calibrated with solutions containing known quantities of the fluorescing reference material.

Examples of important analytes in clinical chemistry that can be determined by the use of fluorescence measurements are given in Table 2-3. As indicated by this wide range of assays, fluorometry is a versatile analytical technique since it combines the simplicity of photometry with high specificity and sensitivity. However, factors limiting fluorescence measurements include concentration effects on fluorescence (inner filter effect), background effects due to Rayleigh and Raman scattering, solvent effects (e.g., interfering nonspecific fluorescence and quenching), and sample matrix effects (e.g., light scattering, interfering fluorescence, adsorption, and photodecomposition).

The inner filter effect. The inner filter effect is caused by a loss of excitation intensity across the cuvet pathlength as the excitation light is absorbed by the fluorophor. At higher concentrations of the fluorophor, the absorbance of the excitation beam increases, and the loss of the excitation light as it travels through the cuvet increases correspondingly. Therefore, the linear relationship between concentration and fluorescence emission (equation 4) holds only as long as solutions are used that absorb $<2\%$ of the exciting light.

Light scattering. Light scattering occurs when radiant energy collides with molecules in solution, thereby scattering light in all directions. A portion of this scattered light may enter the photodetector and therefore contribute to the background "noise"; this interference leads to loss in sensitivity. For fluorophors with small Stokes shifts, the excitation and emission spectra overlap and are particularly susceptible to loss of sensitivity due to background light scatter.

Both elastic and nonelastic collisions can occur, leading to what is termed Rayleigh and Raman scattering. Rayleigh scattering can be controlled by the use of well-defined emission and excitation interference filters or by appropriate monochromator settings and by the use of polarizers. Raman light scattering can be controlled by setting the excitation and emission wavelengths far enough apart to avoid the Raman scatter. It can also be controlled by narrowing the slit width on the excitation monochromator. However, both options tend to decrease sensitivity.

Table 2-3. EXAMPLES OF ANALYTES MEASURED BY FLUORESCENCE

Analytes	*Enzymes*
Glucose	Oxidoreductases
Bilirubin	Glucosidases
Protoporphyrin	Hydrolases
Magnesium	Proteases
Calcium	*Therapeutic drugs*
Catecholamines	Phenytoin
Bile acids	Phenobarbital
Estrogens (urinary)	Amikacin
*Coagulation**	Tobramycin
Heparin	Gentamicin
Antithrombin III	Theophylline
Plasminogen	Primidone

*Using specific fluorescent substrates for proteolytic enzymes.

Solvent and cuvet material effects. Certain quartz glass and plastics contain UV-absorbing materials that will fluoresce. Some solvents, such as ethanol, can also cause appreciable fluorescence. It is therefore important when developing a fluorescence assay to check the background fluorescence of all components of the reaction mixture. Fluorescence-grade solvents and cuvets with low fluorescence emission are commercially available. Quenching by the solvent can also be a problem and should be investigated when setting up a new analytical method. *Quenching* is related to the interaction of the fluorophor with the solvent or with a solute. Such interaction results in a loss of fluorescence due to energy transfer or other mechanisms, but there is no effect on the absorbance spectrum of the fluorophor.

Sample matrix effects. A serum or urine sample contains many compounds that can fluoresce. Thus, the sample matrix is a potential source of unwanted background *fluorescence* and must be examined when new methods are developed. Examples of serious contributors to unwanted fluorescence are proteins and bilirubin. *Light scattering* by proteins and other macromolecules in the sample matrix can also cause unwanted background. Lipemic samples, for example, are noted for their intense light scattering.

Instrumentation for Measurement of Fluorescence

Basic Instrumentation

The instruments used to measure fluorescence are fluorometers or spectrofluorometers. The fluorometer uses interference filters or glass filters to produce monochromatic light for sample excitation and for isolation of fluorescence emission, whereas the spectrofluorometer uses a grating or prism monochromator for this purpose. The basic components required for fluorescence spectrometry are (1) the excitation source, (2) the excitation monochromator, (3) the sample cell, (4) the emission monochromator, and (5) the detector.

The excitation source. The absorption spectra of most fluorescent compounds of interest are in the range of 300–550 nm. An intense lamp capable of emitting radiant energy over a large spectral region is desirable. The light source that best fits these criteria is the xenon arc lamp, but a quartz halogen cycle lamp of high intensity may also be used for excitation in special applications. A mercury lamp provides high-intensity line spectra at several wavelengths, but the most useful lamp is one that provides a continuum over the spectral range of interest.

The xenon lamp is popular because it provides a continuum of relatively high-intensity radiant energy over the spectral range of 250–800 nm. A limitation of xenon lamps for analytical use is arc wandering or flicker. The arc between the internal electrodes of the xenon lamp tends to move slightly, causing some random variation in the intensity of the light reaching the cuvet. The use of current-stabilized power supplies has improved the performance of fluorescence instrumentation using xenon lamps.

Excitation and emission monochromator. A monochromator is defined as a device for isolating a narrow portion of the spectrum. Monochromators used in fluorescence instrumentation are interference filters, colored glass filters, gratings, and prisms (see also Section One, *Photometry*). Most modern analytical instruments using *interference filters* use the all-dielectric multicavity filter or a hybrid Fabry-Perot coupled-dielectric filter (i.e., a filter with metal reflective layers). *Colored glass filters* selectively absorb certain wavelengths of light. These filters have been used for both excitation and emission wavelength selection, but they are more susceptible to transmitting stray light and they can exhibit unwanted fluorescence. *Grating monochromators* are devices that isolate narrow regions of the spectrum (see *Photometry*). The spectral resolution of the light at the slit is a function of the slit width and the resolution of the grating. Spectrofluorometers generally use larger slit widths than absorbance spectrophotometers in order to obtain higher excitation intensities. The advantage of the grating monochromater is its ability to select excitation and emission wavelengths. Such selectivity is required when working with new fluorophors with absorbance and emission maxima for which specially fabricated interference filters may not exist.

The ***sample cell compartment*** defines the geometry the instrument uses for fluorescence measurements. Fluorescent light is emitted in all directions from a molecule, and there are several excitation/emission geometries that can be used to measure fluorescence (Figure 2-18). Most commercial spectrofluorometers and fluorometers use the *right-angle-detector approach* because it minimizes the background signal that limits analytical sensitivity. The *end-on approach* allows the adaptation of a fluorescence detector to existing 180° absorption instruments. The sensitivity of this approach is limited by the quality of the excitation/emission interference filter pair, the excitation/emission spectral band overlap, and the inner filter effect. The *front-surface approach* provides the greatest linearity over a broad range of concentrations because it minimizes the inner filter effect and it has been widely applied to heterogeneous solid-phase fluorescence immunoassay systems (see item 5 in Table 2-4). The material of the sample cell is generally fused silica or quartz. Plastic cuvets can be used for certain applications, but some plastics will fluoresce, causing unwanted background signal and loss of sensitivity.

Photodetectors. Since the human eye is a sensitive detector with a wide range of spectral recognition, qualitative fluorescent thin-layer methods in the clinical laboratory make use of an ultraviolet lamp source and visual observation. For quantitative assays, the most commonly used detector in fluorometers and spectrofluorometers is the photomultiplier tube (PM), which has been described in the photometry chapter. The important features of the PM tube for fluorescence measurements are a wide choice of spectral responses, rapid (nanosecond) photon response time, and high sensitivity.

Manual Fluorometers and Spectrofluorometers

There are many commercially available fluorometers and fluorescence spectrophotometers offering a variety of features. These features can include ratio referencing, microprocessor-controlled excitation and emission monochromators, pulsed xenon light sources, photon counting, rhodamine cells for corrected spectra, polarizers, flow cells, front-surface viewing adapters, multiple cell holders, and microprocessor-based data-reduction systems. Regardless of the complexity of these instruments at first appearance, they all contain the basic instrumental components discussed in the previous section.

A spectrofluorometer consists of a light source, a sample compartment, excitation and emission filters or monochromators, a detector, and a simple digital or meter display. The operation of this device requires that the fluorometer display first be set to zero by the use of a sample blank. A calibrating solution is then placed into the fluorometer and the range is adjusted. Once the zero and standard settings have been made, unknown

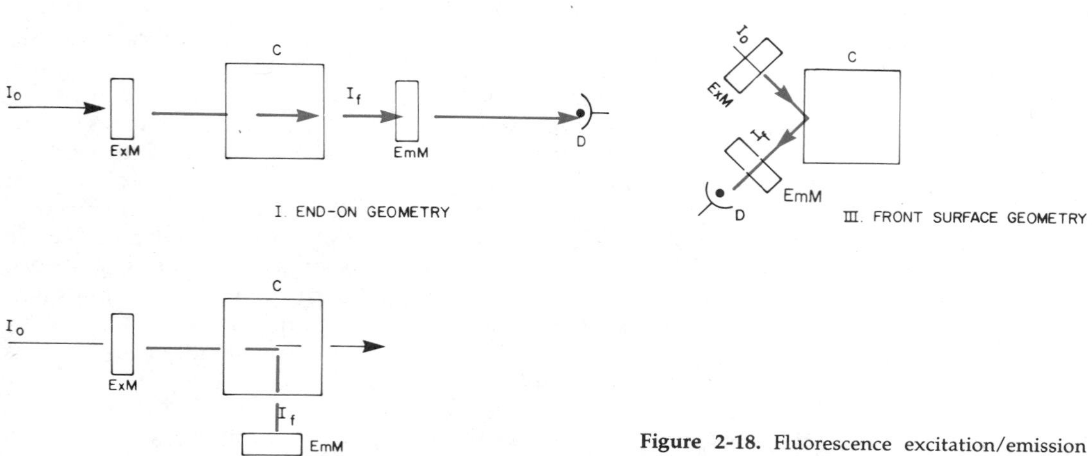

Figure 2-18. Fluorescence excitation/emission geometries: I_0 is the initial excitation energy, ExM is the excitation monochromator, C is the sample cuvet, I_f is the fluorescence intensity, EmM is the emission monochromator, and D is the detector.

Table 2-4. SPECIAL APPLICATIONS FLUOROMETERS

Instrument	Application	Manufacturer
1. Protopath Fluorometer	Coagulation: specific factor assays	Dade Division, American Hospital Supply Corporation
2. ESA 4000 Hemato-Fluorometer	Bilirubin; protoporphyrin	Environmental Sciences Associates
3. TDx-Fluorescence Polarization Analyzer	Therapeutic drugs	Abbott Laboratories, Diagnostics Division
4. Ortho Spectrum III	Cellular immunofluorescence testing	Ortho Diagnostics
5. Fluorometer-FIAX	Antinuclear antibodies, immunoglobulins (IgG, IgA, IgM), and antibodies to rubella G, cytomegalovirus, herpes simplex virus, rheumatoid factor, *Toxoplasma gondii*	International Diagnostic Technology, Inc.
6. Photon Counting Fluorometer	Thyroxine, triiodothyronine	BioRad Laboratories

solutions can be read. This simple single-beam fluorometer can measure concentrations over three or four orders of magnitude.

A typical ratio-referencing spectrofluorometer is a right-angle instrument that uses two monochromators, two photomultiplier detectors, and a xenon lamp source (Figure 2-19). The light from the excitation monochromator is split and a small portion (10%) is directed to the reference photomultiplier for ratio-referencing purposes. The remaining excitation light is focused into the sample cuvet. Emission optics are positioned at a right angle to the excitation optics. An emission monochromator is used to select or scan the desired portion of the emission spectra, which is directed to the sample detector tube for measurement of the emission intensity. The output signals from the reference and the sample photomultipliers are amplified, and a ratio of the sample to the reference signal is provided by a digital display or a chart recorder. A ratio-referencing spectrofluorometer provides two advantages over single-beam spectrofluorometers. First, it eliminates short-term and long-term xenon lamp energy fluctuations (i.e., arc flicker and

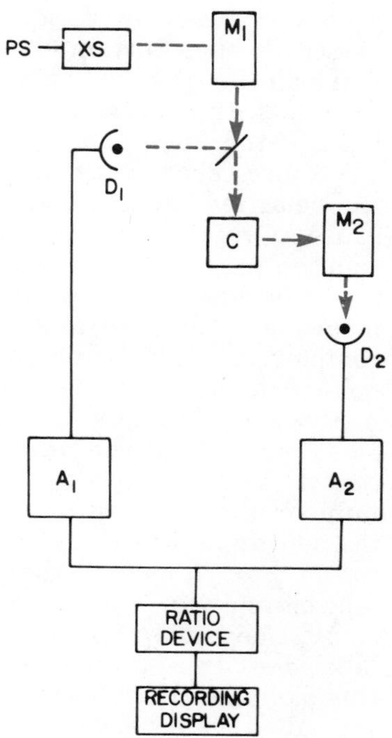

Figure 2-19. Block diagram of a typical spectrofluorometer: XS is the xenon source; PS is the power supply; M_1 is the excitation monochromator; C is the sample cell; M_2 is the emission monochromator. D_1 and D_2 are detectors; D_1 monitors the variation in excitation intensity and D_2 measures fluorescence emission intensity. A_1 and A_2 are excitation signal and emission signal amplifiers, respectively.

lamp decay) and thus minimizes the need for frequent standardization of the instrument during analysis. Second, it provides corrected excitation spectra by compensating for wavelength-dependent energy fluctuations.

Some spectrofluorometers have the capability of synchronous scanning by both the excitation and emission monochromators. Synchronous scanning is performed at a constant wavelength increment of 5–50 nm or more, but intervals in the range of 15–30 nm yield the most distinctive spectra. This technique has been valuable in the area of forensic chemistry for qualitative and quantitative analyses of mixtures of fluorophors, such as drugs and their metabolites in solution.

Automated Fluorometers and Spectrofluorometers

There are several automated analyzers with fluorometric and absorbance measurement capabilities for use in the clinical chemistry laboratory. These systems include a centrifugal analyzer that uses a xenon light source and a monochromator to isolate the excitation spectrum. The excitation light is directed into the spinning reaction cuvets at a 90° angle by use of a quartz fiberoptic bundle. Interference filters are used to isolate the emission spectrum. A photomultiplier tube is used to measure the emission intensity. A fluorescence compound in the reference cuvet of the rotor is used to provide a stable reference similar to that in manual spectrofluorometers.

Fluorometers for Special Applications

Several automated analyzers have been developed for special fluorescence applications in clinical chemistry. These include a fluorescence polarization spectrometer and an enzyme-linked fluoroimmunoassay system for measurement of therapeutic drugs; a hematofluorometer for quantitation of protoporphyrin, bilirubin, and reserve bilirubin-binding capacity; a front-surface solid-phase instrument for quantitation of immunoglobulins and antinuclear antibodies; and a flow-cell fluorometer with photon counting for quantitation of thyroxine, free thyroxine, and triiodothyronine. Each of these automated fluorescence instruments makes use of a special principle or process in fluorescence measurements to accomplish a specific type of analysis. The special instruments and their manufacturers are listed in Table 2-4.

Fluorescence polarization and fluorescence polarization spectrometer. Light waves produced by standard excitation sources are oriented randomly. Light waves passed through certain crystalline substances (polarizers) become oriented in a single plane and are said to be plane polarized. When a fluorophor is excited by polarized light, the size of the molecule will determine if the fluorophor emits polarized or depolarized light. If the fluorophor is large, the rotational relaxation time is long compared with the fluorescence decay time, and the emitted light will be plane polarized, parallel to the polarized excitation light. A small molecular fluorophor causes depolarization because its rotational relaxation is faster than the fluorescence decay time, and thus its rotation may change between excitation and emission; light will therefore be emitted in many different directions (i.e., light will be depolarized). However, if such a small molecule is attached to a macromolecule, as in a fluorophor-labeled antigen-antibody complex, the small fluorophor will behave similarly to a large fluorophor and emit polarized light. This phenomenon can be used analytically, for example, in immunoassay methods for the quantitation of a number of therapeutic drugs, drugs of abuse, and other small molecules. The central principle of these methods is competition between the analyte in the sample and a known amount of fluorophor-labeled analyte for a limited number of binding sites on an antibody to the analyte. The amount of analyte present in the sample will therefore have a direct effect on the change in fluorescence polarization. For example, if the amount of analyte in the sample is small, more fluorophor-labeled analyte- (antigen-) antibody complex forms and the degree of depolarization is small. Conversely, if the analyte concentration in the sample is high, less fluorophor-labeled analyte- (antigen-) antibody complex forms, and depolarization is significant. In this assay, bound fluorophor does not need to be separated from free; such techniques are called homogeneous immunoassays (see also Chapter 3).

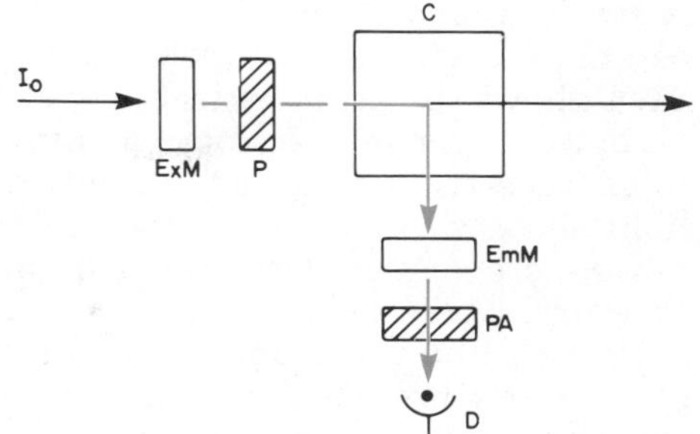

Figure 2-20. Schematic of a fluorescence polarization analyzer. P is the excitation polarizer to provide polarized excitation light. PA is the emission polarizer, which is rotated to provide the measurement of parallel and perpendicular polarized fluorescence-emission intensity. ExM is the excitation monochromator. EmM is the emission monochromator, D is the detector, and C is reaction cell or cuvet.

The schematics of one of the instrument types used to measure fluorescence polarization is shown in Figure 2-20. In the normal operating mode, the sample is excited with polarized light provided by the light source and the excitation polarizer. The emission polarizer is first positioned to measure the intensity of the emitted fluorescent light in the vertical plane (I_v) and is then rotated by 90° to measure the emitted fluorescent light intensity in the horizontal plane (I_h). Fluorescence polarization (P) is then calculated by use of equation (5).

$$P = (I_v - I_h)/(I_v + I_h) \tag{5}$$

The Abbott TDx represents a similar approach, but instrument configuration differs somewhat from that in Figure 2-20. In the TDx, the polarizer in the excitation beam is electronically positioned to produce first vertically polarized, then horizontally polarized light. The polarizer in the emission path is instead stationary; only vertically polarized light reaches the photodetector and is measured.

The ***hematofluorometer*** uses a front-surface cell for quantitation of bilirubin and protoporphyrin in erythrocytes by fluorescence measurement (see also Chapter 25). With this instrument the inner-filter effect is minimized by using front-surface fluorescence (Figure 2-18) and short-pathlength cuvets.

Flow cytometry is used for the differentiation of lymphocytes from other peripheral leukocyte types. Flow cytometers have been built with an argon laser that provides excitation light at 488 nm and with optics to measure right-angle fluorescence, right-angle light scatter, and forward light scatter. This new instrument is capable of analyzing a flowing stream of blood cells for size and, by prior admixture of fluorescence-labeled monoclonal antibodies, for subclasses of specific cells such as T-lymphocytes.

Light Scattering Measurements

Light scattering occurs when radiant energy collides with molecules in solution. The relationship between scattering intensity and the variables that influence it was shown by Rayleigh to be

$$\frac{i_S}{I_O} = \frac{4\pi^2(dn/dc)^2\, Mc \sin^2\theta}{N_a\lambda^4 r^2} \tag{6}$$

where i_S is the intensity of scattered light from small particles excited by polarized light, I_O is the incident intensity, dn/dc is the change in refractive index of the solvent with respect to change in solute concentration, N_a is Avogadro's number, M is the molecular weight (g/mol), c is the concentration (g/mL) of the particles, λ is the wavelength of the incident light, θ is the angle of observation, and r is the distance from the point of light scattering to the detector.

The expression for light scattering given in equation (6) holds for dilute solutions of small particles whose largest dimension is less than one-tenth of the wavelength of the incident light. Thus, the upper limit on size of particles exhibiting Rayleigh scattering is about 40 nm when visible light at 400 nm is used. Many of the plasma proteins such as immunoglobulins, β-lipoproteins, and albumin have sizes below this limit. Figure 2-21,*A* shows the symmetrical light scattering caused by these molecules. As the particle becomes larger in size from ~40 nm to 400 nm, the angular dependence of the scattered light loses the symmetry around the 90° axis as seen in Figure 2-21*B*, and scattered light shows an increase in forward scattering. Some plasma proteins of the IgM class, chylomicrons, and aggregating immunoglobulin/antigen complexes fall into this latter category. The scattering from particles in this size range is known as *Rayleigh-Debye scattering,* and the equation for this type of scattering becomes more complex. Particles such as erythrocytes and bacteria are even larger, 7000–40 000 nm. These particles show a complex, narrow angular dependence of light scattering in the forward direction, and this type of scattering from very large particles is called *Mie scattering*.

The Rayleigh and Rayleigh-Debye expressions are important in the optimization of analytical instrumentation for measuring light scattering and provide useful information about the scattering of light by small and intermediate-sized particles. This information includes the wavelength dependence, the angular dependence, the relationship between intensity of scattering and the detector, and the effect of polarization on the scattered light at a right angle to the incident light. Furthermore the Rayleigh expression, equation (6), shows a direct relationship between the concentration of particles and scattered light at a fixed angle of observation provided that the particles are uniform in molecular weight and are of the proper size. However, the relationship between light scattering intensity and concentration becomes more complex when applied to aggregating systems such as antigen-antibody reactions, because these reactions involve a change in concentration, a change in molecular weight, and a change in size. Although more difficult to describe by theory, light scattering measurements have proved to be useful for the quantitation of serum proteins as discussed in Chapters 3 and 11.

Nephelometric and Turbidimetric Measurements of Scattered Light

Turbidity causes the attenuation (decrease) of the intensity of the incident beam of light as it passes through a solution of particles. The measurement of this decrease in intensity of the incident light beam that is caused by scattering, reflectance, and ab-

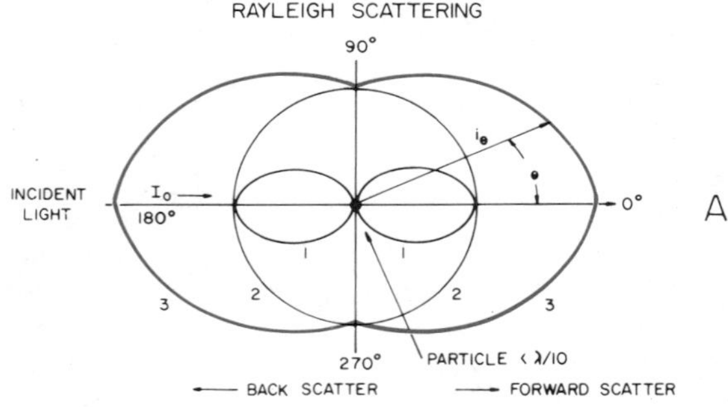

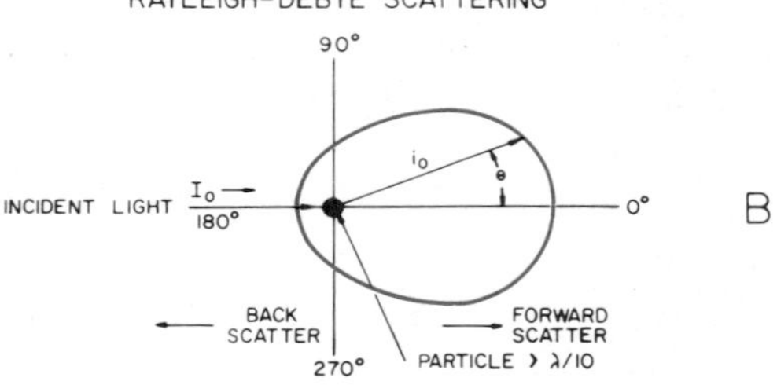

Figure 2-21. The angular dependence of light scattering intensity with nonpolarized and polarized incident light for small particles (*A*) and the angular dependence of light scattering with nonpolarized light for larger particles (*B*).

sorption of the light is called *turbidimetry*. Analogous to absorption spectroscopy, turbidity can be defined as:

$$I = I_0 e^{-bt} \quad \text{or} \quad t = (1/b)\ln I_0/I \tag{7}$$

where t is the turbidity and b is the pathlength of the incident light through the solution of light scattering particles. Turbidity is measured at 180° from the incident beam, that is, in the same manner as absorbance measurements are made in a spectrophotometer. Turbidity can be measured on most spectrophotometers and automated clinical chemistry analyzers.

Nephelometry is defined as the detection of light energy scattered or reflected toward a detector that is not in the direct path of the transmitted light. Common nephelometers measure scattered light at right angles to the incident light. The ideal nephelometric instrument would be free of stray light; neither light scatter nor any other signal would be seen by the detector when no particles are present in solution in front of the detector. However, due to stray light–generating components in the optics path as well as in the sample cuvet or sample itself, a truly dark field is difficult to obtain when making nephelometric measurements.

The choice between turbidimetry and nephelometry depends upon the application and the available instrumentation. Until recently, the statement was often made that for relatively clear solutions, where the transmission of light in the forward direction is > 95%, small changes in absorption due to turbidity were difficult to measure with precision and nephelometry was the method of choice. However, with the advent of stable high-resolution photometric systems, turbidimetric measurements have become competitive in sensitivity with nephelometric methods for immunological quantitation of serum proteins. Nephelometry, however, still offers some advantage in sensitivity when low-level antigen-antibody reactions are measured.

Limitations of Light Scattering Measurements

Antigen excess. Antigen-antibody reactions are complex and appear to result in a mixture of aggregate sizes. As the turbidity increases during the addition of antigen to antibody, the signal increases to a maximum value and then decreases. The point at which the decrease begins marks the beginning of the phase of antigen excess; the phenomenon is explained in Chapter 3, *Principles of Immunochemical Techniques*. Light scattering methods for quantitation of antigen-antibody reactions, to be fully acceptable, must provide a method for detecting antigen excess.

Matrix effects. Particles and solvents as well as all serum macromolecules scatter light. Lipoproteins and chylomicrons in lipemic serum provide the highest background turbidity or intensity. With appropriate dilutions, the relative intensity of light scattering from a lipemic sample is less than that of the antiserum blank. However, as the concentration of the antigen in serum decreases and correspondingly less dilute samples are used, the background interference from lipemic samples becomes greater. Large particles, such as suspended dust, can also cause significant background interference, which can be controlled by filtering all buffers and diluted antisera before analysis.

Instrumentation for the Measurement of Light Scattering

Light scattering can be measured with standard clinical fluorometers or photometers. Although intensity of light scattering can be thought of as analogous to fluorescence intensity, the angular dependence of light scattering intensity has resulted in the design of special nephelometers. These devices place the photomultiplier detector at an appropriate angle to the excitation light beam. The design principle of a nephelometer is similar to the design principle applied in fluorescence measurements. The major operational difference between the fluorometer and the nephelometer is that the excitation and detection wavelengths will be set to the same value. The principal concerns of light scatter instrumentation are excitation intensity, wavelength, distance of the detector from the sample cuvet, size of sample slits, and minimization of external stray light. A

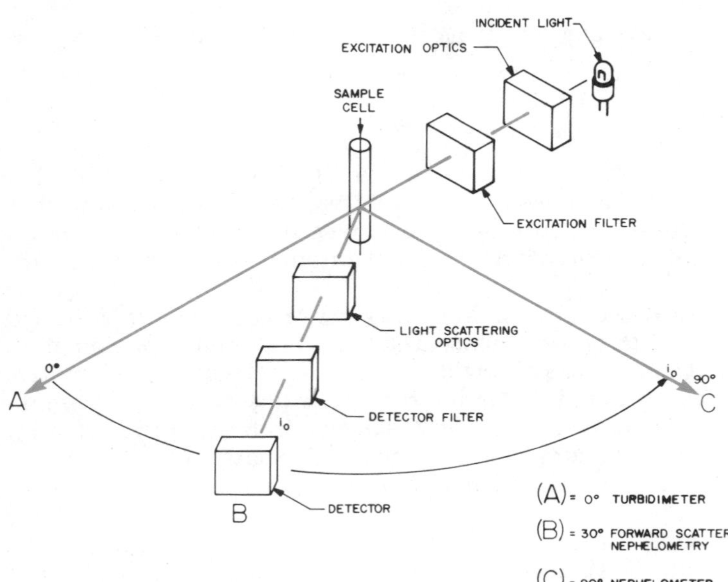

Figure 2-22. Schematic diagram of light scattering instrumentation showing: *A,* the optics position for a turbidimeter; *B,* the optics position for a forward-scattering nephelometer; and *C,* the optics position for a right-angle nephelometer.

schematic diagram of the basic components of an instrument for measuring light scattering is shown in Figure 2-22. It consists of a light source, collimating optics, a sample cell, and the collection optics, which include light scattering optics, detector filter, and a detector. The schematic also shows the different angles from the incident light beam where the detector, filter, and optics can be placed to measure light scattering. Figure 2-22,*A* is the straight-through arrangement for turbidimetry, whereas Figure 2-22,*B* and 2-22,*C* are two possible arrangements frequently found in nephelometers. Some high-precision research instruments for measuring light scattering provide detectors that are movable about the sample cell to take advantage of the angular dependence of light scatter intensity. However, most clinical instruments capable of measuring light scattering have a fixed optical arrangement.

The optical components used for the nephelometer or turbidimeter are similar to those in fluorometers or photometers. The light sources commonly used are quartz halogen lamps, xenon lamps, and lasers. The laser beam is used specifically in some nephelometers because of its high intensity; in addition, the coherent nature of laser light makes it ideally suited for nephelometric applications.

Turbidimetric measurements are easily carried out on photometers or spectrophotometers and require little optimization. The principal concern of turbidimetric measurements is signal-to-noise ratio. Photometric systems with electro-optical noise in the range of ±0.0002 absorbance unit or less are useful for turbidity measurements. Ratio-referencing fluorometers and specifically designed nephelometers are well suited for nephelometric measurements.

Several automated instruments capable of light scattering intensity or turbidimetry measurements are commercially available for the clinical chemistry laboratory. Light scattering measurements are best applied to immunoassays of immunoglobulins, specific proteins, and haptens such as therapeutic drugs.

Additional Reading

Jolley, M. E., Stroupe, S. D., Schwenzer, K. S., et al.: Fluorescence polarization immunoassay. III. An automated system for therapeutic drug determination. Clin. Chem., *27*:1575–1579, 1981.

Schulman, S. .G.: Fluorescence and Phosphorescence Spectroscopy: Physiochemical Principles and Practice. Oxford, Pergamon Press, 1977.

Soini, E., and Hemmila, I.: Fluoroimmunoassay: Present status and key problems. Clin. Chem., *25*:353–361, 1979.

Tiffany, T. O., Huey, E. E., Manning, G. B., et al.: The application of fluorescence measurements to centrifugal analyzers. *In*: Centrifugal Analyzers in Clinical Chemistry. C. Price and K. Spencer, Eds. Eastbourne, East Sussex, Praeger Publishers, 1980.
Wampler, J. E.: Measurements and physical characteristics of luminescence. In: Bioluminescence in Action. P. J. Herring, Ed. New York, Academic Press, 1978.
Wehry, E. L.: Modern Fluorescence Spectroscopy, Vol. 1. New York, Plenum Press, 1976.
Wehry, E. L.: Modern Fluorescence Spectroscopy, Vol. 2. New York, Plenum Press, 1976.
Whitehead, T. P., Kricka, L. J., Carter, T. J. N., et al.: Analytical luminescence: Its potential in the clinical laboratory. Clin. Chem., *25*:1531–1546, 1979.

SECTION THREE

Electrophoresis

EMANUEL EPSTEIN, Ph.D.

Electrophoresis refers to the migration of charged solutes or particles in a liquid medium under the influence of an electrical field. Although the expression *electrophoresis* applies to migration of all species of particles, the term *iontophoresis* is specifically limited to the migration of small ions. The first electrophoresis method used in the study of proteins was the free solution or moving boundary method devised by Tiselius in 1937. This technique is still used in research for measurement of electrophoretic mobility and for study of protein-protein interaction but is not used in clinical laboratories for routine work. A complex apparatus is needed, the technique is difficult, and samples of ~0.5 mL are required.

The term *zone electrophoresis* refers to the migration of charged macromolecules in a porous supporting medium such as cellulose paper, cellulose acetate sheets, or agarose gel film. Zone electrophoresis generates an *electrophoretogram*, a display of protein zones, each one sharply separated from neighboring zones on the electrophoretic support material. Solutes analyzed by this technique in clinical chemistry are mainly macromolecular in size and colloidal in nature; they include proteins in serum, urine, cerebrospinal fluid, and other biological fluids, as well as proteins in erythrocytes and tissues.

Theory of Electrophoresis

Chemical species carrying an electric charge by virtue of ionization will move either to the cathode or to the anode in an electrophoresis system, depending on the kind of charge on the molecules. In a solution more acid than the isoelectric point (pI) of the solute, an *ampholyte* (a solute molecule that can be either positively or negatively charged, also called a zwitterion) takes on a positive charge by binding protons and migrates toward the cathode (i.e., the negatively charged electrode). Conversely, the ampholyte that has lost protons is in the anionic form and migrates toward the anode (positive electrode). The amphoteric nature of proteins can be illustrated as

$$H_2N-R-C\begin{matrix}{=}O\\ {-}OH\end{matrix} \xrightarrow{OH^-} H_2N-R-C\begin{matrix}{=}O\\ {-}O^-\end{matrix} \quad \text{(Anodic form)}$$

$$H_2N-R-C\begin{matrix}{=}O\\ {-}OH\end{matrix} \xrightarrow{H^+} H_3\overset{+}{N}-R-C\begin{matrix}{=}O\\ {-}OH\end{matrix} \quad \text{(Cathodic form)}$$

(At isoelectric point)

where the protein R is shown with terminal amino ($-NH_2$) and carboxyl ($-COOH$) groups. Since proteins contain many ionizable groups, this equation should be seen as a simplification.

The rate of migration is dependent on such factors as (1) net electrical charge of the molecule; (2) size and shape of the molecule; (3) strength of the electrical field; (4) properties of the supporting medium; and (5) temperature of operation. The equation expressing electrophoretic mobility is a general formula that derives from two formulas, one describing the forward motion toward the electrode of charge opposite to that on the molecule, and the other describing the retarding force due to frictional resistance:

$$\mu = \frac{Q}{kr\eta}$$

where

μ = electrophoretic mobility, cm/(V)(s)
Q = the net charge on the molecule
k = a constant
r = the ionic radius of the migrating particle
η = the viscosity of the solution in which migration is occurring

Clearly, electrophoretic mobility is directly proportional to the net charge as the protein migrates under the influence of an applied electric current, and inversely proportional to the size of the protein ion and the viscosity of the medium (electrophoretic buffer).

A schematic illustration of a typical electrophoretic apparatus is shown in Figure 2-23. An electrophoretic support such as agarose or cellulose acetate is placed between electrode chambers. The chambers are filled with a buffer of the desired pH; the electrophoretic support is also saturated with the same buffer. Electrodes in each chamber extend into the buffer solutions. A source of constant current or voltage is connected to the electrodes, and the circuit is made complete by the contact of the electrophoretic support with each buffer chamber. A protein sample placed on the electrophoretic support will migrate when current is applied.

Buffer ions have a twofold purpose in electrophoresis; they carry the applied current, and—by fixing the pH at which electrophoresis is carried out—they determine the electrical charge on the solute. The ionic strength of the buffer determines the thickness of the ionic cloud surrounding the charged molecule and thus also the rate of migration and the sharpness of the zones. Increasing the ionic strength by increasing the concentration of buffer ions causes an increase in the ionic cloud around the charged molecule, and the molecule becomes more hindered in its movement. Buffers of high ionic strength yield sharper separations of bands, but the benefits of sharper resolution are diminished by the Joule (heat) effect that leads to denaturation of heat-labile proteins. Many buffer systems have been used in electrophoretic procedures, but those most widely used are barbital buffers (ionic strength = 0.025–0.075) and tris buffers (ionic strength =

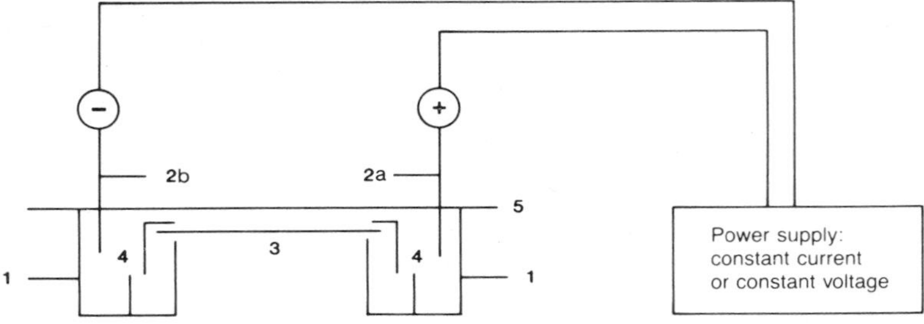

Figure 2-23. A schematic of a typical electrophoresis apparatus. Two electrode chambers with baffle plates (1) contain the buffer. In each chamber is an electrode (2), either platinum or carbon, the sign of which is fixed by the mode of connection to the power supply (2a, anode; 2b, cathode). The electrophoresis support on which separation takes place (3) is in contact with buffer by means of wicks (4) or it dips directly into the buffer solutions. The whole apparatus is covered (5) to minimize evaporation and protect the operator. Direct-current power supply may be either constant current or constant voltage.

0.03−0.12). Because the base in both of these buffers is monovalent, any association between buffer and protein is minimal.

A General Method for Electrophoresis

Principle

First, the hydrated support material (agarose gel or previously wetted cellulose acetate) is placed into the electrophoresis chamber. Care must be taken that excess buffer is removed from the support surface and that bubbles are not present. The support is put into contact with buffer previously placed in the electrode chambers. Sample is applied to the support, and by using either constant voltage or constant current, electrophoresis is conducted for a determined length of time. The support is then removed from the electrophoresis cell and rapidly dried or placed in a fixative to prevent diffusion of sample components. Fixation is followed by treatment with a staining reagent to locate and reveal the individual protein zones. After excess dye is washed out, the support either is dried immediately or is first placed in a clearing agent.

The result of electrophoresis of serum proteins is an electrophoretogram in which five major zones of proteins or protein groups can be seen. A typical electrophoretogram of a normal serum is illustrated in Figure 11-6.

Protein Stains for Electrophoresis Procedures

Stains used to locate and reveal the separated protein fractions of the sample differ in accordance with the type of application and the personal choice of the analyst. Stains most commonly utilized for specific applications are described in connection with particular procedures, given in Chapter 11. Table 2-5 lists common stains and wavelengths used for quantitative examination of electrophoretograms.

The amount of dye bound by each protein zone is related to the amount of protein present in the zone but is affected by many other factors, such as the type of protein present and the degree of denaturation caused by the fixing agent. Recently, silver nitrate has become popular because of its great sensitivity for staining of proteins and polypeptides (see Chapter 11).

Quantitation of Protein Zones

Customarily, results of electrophoretic separation are reported in terms of the percentage of each fraction present or in terms of absolute concentration when the concentration of total protein is known. Quantitation of the amount of dye in individual

Table 2-5. STAINS FOR VISUALIZATION OF SERUM PROTEIN ZONES AND WAVELENGTHS FOR QUANTITATION BY DENSITOMETRY OR ELUTION SPECTROPHOTOMETRY

Application	Stain	Nominal Wavelength, nm	
		Densitometry	*Elution*
Staining of serum proteins in general	Amido Black (Naphthol Blue Black)	640	600
	Bromophenol Blue	600	595
	Coomassie Brilliant Blue G-250 (Brilliant Blue G)	595	610
	Coomassie Brilliant Blue R-250 (Brilliant Blue R)	560	580
	Nigrosin	540	560
	Ponceau S	520	565 (alkaline) 520 (neutral)
Staining of isoenzymes	Nitrotetrazolium Blue (as the formazan)	570	—
Staining of lipoproteins	Fat Red 7B (Sudan Red 7B)	540	—
	Oil Red O	520	—
	Sudan Black B	600	—

zones can be conveniently accomplished either by *direct densitometry* or by *elution of dye* from individual zones and by subsequent spectrophotometric measurement of eluted dye. In densitometry, the fixed and cleared electrophoretic medium is moved past the optical system of an instrument, called a densitometer; the location and intensity of zones are indicated as successive peaks on a recorder chart. In some instruments, the area under a peak is automatically integrated. The elution method involves cutting the support into separate zones and eluting the adsorbed dye in each with suitable solvents, such as basic buffers or alcoholic solutions.

Densitometry has become the method of choice in most clinical laboratories. The following features should be sought when choosing a densitometer: (1) ability to scan electrophoresis supports with lengths of 25–150 mm; (2) automatic gain control to prevent the response for the most intense peak or zone from going off scale; (3) automatic zeroing capability for background correction (this feature allows the instrument to choose the lowest background point in the electrophoretogram for a baseline so that minor peaks are not lost or cut off); (4) a variable-wavelength control, either as a continuously variable monochromator or as variably selected interference filters, to allow operation in the range of 400–700 nm; (5) variable slits from 0.1 × 2.0 to 0.2 × 10.0 mm (this feature allows the instrument to discriminate between individual narrow bands); and (6) an integrating device.

The simplest *integrator* provides zigzag integration, requiring the operator to count the number of sawteeth and to compute manually the area of the zones integrated. Microprocessor-fitted models, however, rapidly and automatically compute areas of peaks and usually provide printouts. When automatic computation is desired, one should consider a densitometer capable of integrating up to 30 peaks rather than one limited to the five zones associated with conventional serum protein electrophoresis. Such a choice allows the analyst to apply the instrument to those high-resolution electrophoretic techniques that fractionate serum proteins more completely. Other desirable features are an ultraviolet light source and the ability to measure fluorescence. Additional features, not essential but helpful, are automatic indexing (a feature that automatically advances an electrophoretic strip containing multiple samples from one sample to the next); built-in diagnostics for instrument troubleshooting; and a choice of several scanning speeds.

The linearity of response of a densitometer may be tested by standards available from various instrument manufacturers. Standards have been designed with several separate zones of increasing density or with gradually increasing density. The first type causes the recorder pen to return to baseline between zones. By evaluating the behavior of peak tracing and baseline return, the operator can check the optical, mechanical, and electrical functions of the densitometer. Densities of zones in the standard are compared with their expected values. With modern cellulose acetate and agarose gel techniques, the use of very small sample size, as well as excellent transparency when strips are scanned, has substantially eliminated the shortcomings of densitometry observed with opaque cellulose paper strips. Nevertheless, problems persist because of differences in affinity by individual proteins for stain and differences in size among protein zones.[10,11]

Power Supply

Commercially available power supplies allow operation at either constant current or constant voltage. The flow of current through a medium that offers electrical resistance is associated with the production of heat:

$$\text{Heat} = (E)\,(I)\,(t)$$

where E is electromotive force in volts (V); I is current in amperes (A); and t is time in seconds (s). Heat evolved during electrophoresis increases the conductance of the system (i.e., resistance decreases) as a result of the increase in thermal agitation of all dissolved ions. When a constant-voltage power source is used, the resultant rise in current causes a progressive increase in the migration rates of proteins (Ohm's law, $E = IR$). The rate of evaporation of water from the support medium is also increased, and this loss causes an increase in ion concentration and thus a further decrease in resistance (R). These effects on the migration rate are best minimized by the use of a constant-current power supply. In this case, a decrease in R due to heat also decreases V, and the current remains constant (see Ohm's law). As a result, the heat effect is decreased and the migration rate is kept relatively constant.

Types of Electrophoresis

Paper Electrophoresis (PE)

Paper electrophoresis, once widely used in clinical laboratories for the separation of serum proteins, has largely been replaced by cellulose acetate or agarose gel electrophoresis. Significant disadvantages in the use of paper electrophoresis include the long separation time (14–16 h), excessive background (tailing), and decreased resolution. The advantages of paper as a support medium are high tensile strength, low cost, and ease of handling.

Agarose Gel Electrophoresis (AGE)

Agar gel electrophoresis has been successfully applied to the analysis of serum proteins, hemoglobin variants, lactate dehydrogenase isoenzymes, lipoprotein fractions, and other substances. This medium parallels cellulose acetate in versatility, convenience, and applicability to routine clinical demands. However, agar contains at least two fractions: agaropectin and agarose. Agaropectin has acid sulfate and carboxylic acid groups and accounts for the considerable endosmosis and background staining that are observed on electrophoretograms made on unfractionated agar. *Agarose*, which is essentially free of ionizable groups,[4,7] exhibits little endosmosis and has replaced agar in routine clinical laboratory procedures.[8] Occasionally lots of commercially available agarose, however, contain some residual charged groups.

Agarose gel has a low affinity for protein and has very little effect on migration rate. Because it is clear after drying, it permits excellent densitometric examination. Usually 0.5–1.0 g of agarose/dL of buffer provides a gel with the desired strength and with good migration properties. Unmodified serum or serum dissolved in warmed agarose is applied directly into a precut or precast well. Although the latter technique is less convenient, it has the advantage that the agarose-sample solution solidifies to become part of the agarose support. In contrast, the direct application of serum to a precut slot causes an uneven surface at the sample application point and an artifactual peak in densitometry. Another alternative uses a thin plastic template that has small slots corresponding to sample application points. The template is placed upon the agarose surface and 5-μL samples are introduced into each slot. The serum is allowed to diffuse into the agarose for 5 min; the excess is removed by blotting, and the template is then removed from the agarose surface. Sample size for AGE is relatively small (0.6–3.0 μL), and electrophoresis time is relatively short (30–90 min).

Cellulose Acetate Electrophoresis (CAE)

When the hydroxyl groups of cellulose are reacted with acetic anhydride, cellulose acetate is formed. This acetylated cellulose forms the raw material for cellulose acetate membranes. The membranes commercially available contain about 80% air space in the form of pockets within the interlocking cellulose acetate fibers. The membranes as purchased are dry, opaque, brittle films that crack easily if not handled gently. When the film is placed in buffer, the air spaces fill with liquid and the film becomes quite pliable. Characteristics of specific membranes vary with the extent of acetylation, the prewashing procedure employed by the manufacturer, and the additives used, as well as with the pore size and thickness of the membrane. Cellulose acetate that has been especially prepared to reduce electroendosmosis is also commercially available.

Serum samples (0.3–2.0 μL) are generally applied with a twin-wire applicator to cellulose acetate strips that have been presoaked in buffer. Cellulose acetate membranes may be made transparent (cleared) for densitometry by treatment with a solvent mixture containing one solvent that dissolves cellulose acetate and a second solvent that promotes miscibility between the buffer and the first solvent. A typical clearing solution consists of 95 parts methanol and 5 parts glacial acetic acid. The cellulose acetate fibers partially dissolved by the action of the solvent coalesce, thus eliminating the air spaces responsible for the original opacity.

Advantages of cellulose acetate electrophoresis are the speed of separation (20 min to 1 h) and the stability of the cleared membranes when stored for long periods. Applications of CAE are discussed in Chapter 11.

Polyacrylamide Gel Electrophoresis (PAGE)

Protein electrophoresis on paper, cellulose acetate, or some agarose gels yields only five to seven zones owing to diffusion during the electrophoretic procedure. Polyacrylamide, or starch gel, electrophoresis commonly yields 20 or more fractions (Figure 2-24) and is therefore widely used to study individual serum proteins, particularly genetic variants and isoenzymes.

The polyacrylamide gel electrophoresis (PAGE) technique employs layers of gel that differ in composition and pore size. Because of the *discontinuities* of the electrophoretic matrix and the *discoid* shape of the separated zones of protein, this method is sometimes referred to as *disc* electrophoresis. This technique, which was introduced in 1964,[2,9] minimizes band broadening due to diffusion.

The individual layers are prepared in situ in glass tubes by polymerizing a gel monomer and a cross-linking agent with the aid of an appropriate catalyst. The first gel to be poured into the tubular-shaped electrophoresis cell is the small-pore separation gel. After gelation takes place (usually 30 min), a large-pore gel (the spacer gel) is cast on top of the separation gel. Then a large-pore monomer solution containing ~3 μL of serum is polymerized above the spacer gel, so that the finished product is composed of three different gel layers. When electrophoresis begins, all protein ions migrate unimpeded through the large-pore gels and stack up along the end of the separation gel in a very thin zone. This process serves to concentrate protein components at the border (or starting) zone, so that the preconcentration of specimens with low-protein content (e.g., CSF) may not be necessary. Separation of the individual protein ions then occurs in the bottom separation gel, not only on the basis of their charges but also on the basis of their molecular sizes. The average pore size in a typical 7.5% PAGE separation gel is ~5 nm (50 Å). Proteins whose molecular radius or length (or both) exceeds critical limits will be more or less impeded in their migration. Since both net charge and molecular size affect migration rate, serum proteins are fractionated into many zones.

Acrylamide gel is thermostable, transparent, strong, and relatively chemically inert; it can be made in a wide range of pore sizes to optimize particular separations. Furthermore, these gels are uncharged, thus eliminating electroendosmosis. In recent years, simplifications of the original technique have been introduced. In one such procedure, the use of spacer and sample gels has been eliminated, a continuous buffer system replaces the discontinuous buffer system, and undiluted serum is delivered directly onto the top surface of the separation gel.[5,6] Polyacrylamide gel is also used in isoelectric focusing.

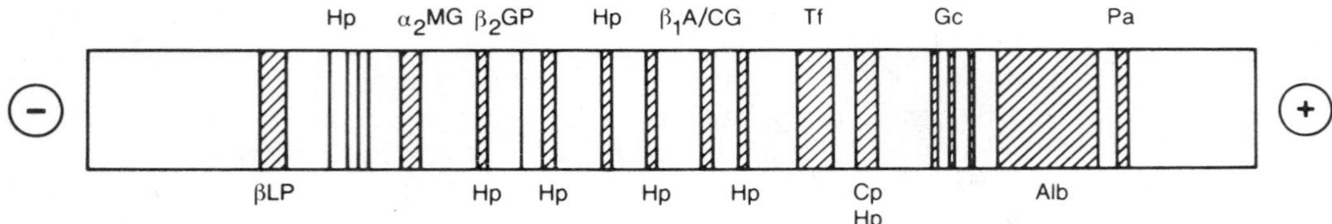

Figure 2-24. A simplified schematic of a protein pattern from the serum of a subject with haptoglobin Type 2-1, separated by PAGE. Immunological techniques demonstrate that some zones contain more than the one protein indicated. Abbreviations: Pa, prealbumin; Alb, albumin; Gc, Gc-globulin; Cp, ceruloplasmin; Tf, transferrin; β_1A/CG, β_1-A/C-globulin; Hp, haptoglobin(s); β_2GP, β_2-glycoprotein; α_2MG, α_2-macroglobulin; βLP, β-lipoprotein.

Starch Gel Electrophoresis

Starch gel electrophoresis, like PAGE, separates macromolecular ions on the basis of both surface charge and molecular size. Partially hydrolyzed starch is used, since native starch does not gel. Starch gel may be used in a horizontal process or with migration taking place in the vertical direction. Proper preparation of gels is relatively difficult and requires considerable skill. The starch concentration is 10–16 g/dL, and the pH of the buffer varies according to the specific application.

Isoelectric Focusing (IEF)

Isoelectric focusing is a further advance in the electrophoretic separation of proteins into discrete bands. The medium provides a stable pH gradient by distributing carrier ampholytes along its axis. The compounds to be separated (e.g., proteins) migrate to the zone in the medium where the pH is equal to the isoelectric point (pI) of each compound. In that zone the charge on the protein becomes zero and migration ceases. Figure 2-25 illustrates the process and shows the electrophoretic conditions before and after current is applied. With IEF, separated protein zones are very sharp because the pI of a protein is confined to a narrow pH range and because diffusion of the protein is counteracted by acquisition of charge as it moves away from its pI position. Electrical forces then drive it back to its pI position (Figure 2-26). Proteins that differ in pI value by only 0.02 pH unit have been separated by IEF.

The pH gradient is created by the use of amphoteric polyaminocarboxylic acids (carrier ampholytes), a group of compounds with molecular weights from 300 to 1000. The carrier ampholytes are used in mixtures containing 50–100 different compounds; under electrophoretic conditions, the many different carrier ampholytes generate a pH gradient as each of them reaches its individual isoelectric point. The anode is surrounded by a dilute acid solution and the cathode by a dilute alkaline solution. Since carrier ampholytes are generally used in relatively high concentrations, a high voltage (up to 2000 *V*) is necessary. As a result, the electrophoretic matrix must be cooled. Depending on analytical conditions, the power required may be in the vicinity of 250 *W* and use of a constant-power power supply is advisable. If IEF is carried out with constant voltage, frequent adjustments of the voltage are necessary because of a drop in current as electrophoresis proceeds. The drop in current is due to lower conductivity of carrier ampholytes at their isoelectric points and to the creation of zones of pure water as electrophoresis progresses. With a constant-power power supply, manual resetting of voltage is avoided or minimized. Constant-current power supplies are not customarily used for IEF.[1,3]

Several support media have been used for IEF, including polyacrylamide, agarose, and cellulose acetate. **Polyacrylamide gel-isoelectric focusing (PAGE-IEF)** is widely used in analytical work. The polyacrylamide gel must be optically clear and have a large

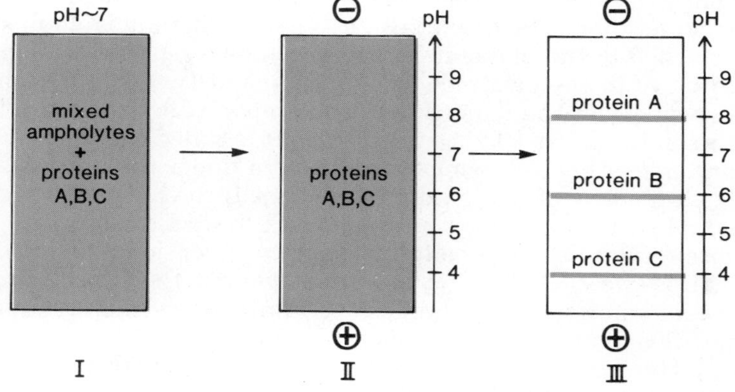

Figure 2-25. Schematic of an IEF procedure. *I,* Homogeneous mixture of carrier ampholytes, pH range 3–10, to which proteins A, B, and C with pI 8, 6, and 4, respectively, were added. *II,* Current is applied and each carrier ampholyte rapidly migrates to the pH zones where its net charge is zero. *III,* Proteins A, B, and C, sharing a high molecular weight, migrate more slowly to their respective pI zones, where migration ceases. The high buffering capacity of the carrier ampholyte creates stable pH zones.

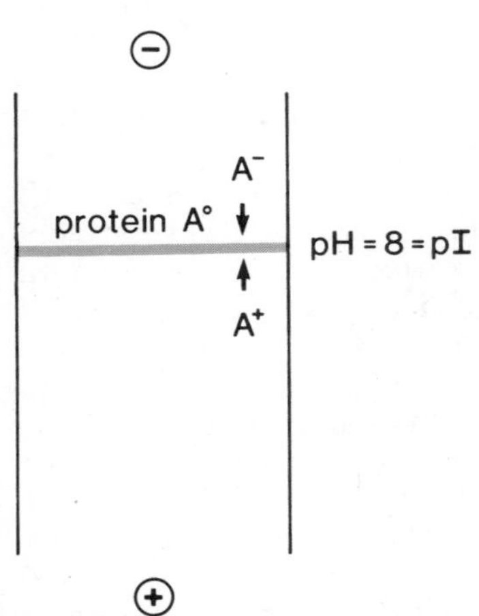

Figure 2-26. After attaining the pH at which protein A has a net charge of zero (A^*), diffusion toward the cathode bestows a negative charge on A (A^-), and migration in the electric field forces A^- back to A^0. Diffusion toward the anode causes A to take on the opposite charge (A^+), and migration is toward the cathode and to the point at which A^* exists. IEF processes of this kind cause sharp zones to form; i.e., the protein is focused.

enough pore size so that, ideally, protein migration is unaffected by molecular sieving. In practice, impediment of migration of some proteins, characteristically of IgM, is unavoidable. Agarose and cellulose acetate products that are free of electroendosmotic effects have been adapted for AGE-IEF and CAE-IEF; these adaptations have the advantage of simple operating conditions, and their media have a large enough pore size to make the exclusion of proteins on the basis of molecular weight and size unlikely.

Limitations and Errors in Routine Electrophoresis

Electroendosmosis and "wick flow" are the major factors that perturb electrophoretic procedures. Satisfactory electrophoretograms are also dependent on proper preparation and storage of buffer and stain solutions and on correct application of sample.

Wick flow. As electrophoresis proceeds, heat is generated and causes evaporation of solvent from the electrophoretic support. The drying effect causes buffer to rise into the electrophoresis support from both buffer compartments. This flow of buffer from both directions affects protein migration. Wick flow can be minimized by cooling the electrophoresis chamber and by keeping the chamber closed to preserve a high humidity within it.

Electroendosmosis. Some electrophoretic support media in contact with water take on a negative charge because of adsorption of hydroxyl ions. Since the ions are fixed to the surface of the electrophoretic support, they are rendered immobile relative to the other ions in solution. Positive ions in solution cluster about the fixed negative-charge sites, forming an ionic cloud of mostly positive ions. The number of mobile negative ions increases with the distance away from the fixed, negative-charge sites until, eventually, positive and negative ions are present in equal concentration (Figure 2-27). The potential that exists between the fixed ions and the associated cloud of ions is termed the electrokinetic or zeta potential (ζ). When a current is applied to such a system, charges attached to the immobile support remain fixed, but the cloud of ions in solution is free to move to the electrode of opposite polarity. Since the ions in solution are hydrated, movement of the ionic cloud is accompanied by solvent. This migration of solvent and its solutes relative to the fixed support is called *endosmosis*. Macromolecules in solution, and migrating in the opposite direction under the influence of the electrical field, must buck the flow of hydrated positive ions. If the molecules are insufficiently charged, they stand still or may even be swept back toward the opposite pole. In media where endosmosis is strong (e.g., cellulose paper, conventional cellulose acetate, and conventional agarose gel), γ-globulins are swept behind the line of application. In electrophoretic media where surface charges are minimal, as in starch, polyacrylamide gel, or highly purified agarose, endosmosis is minimal.

Buffers. Once a buffer has been made up with water to the desired pH and ionic strength, it should be stored refrigerated to inhibit the growth of microorganisms. Fur-

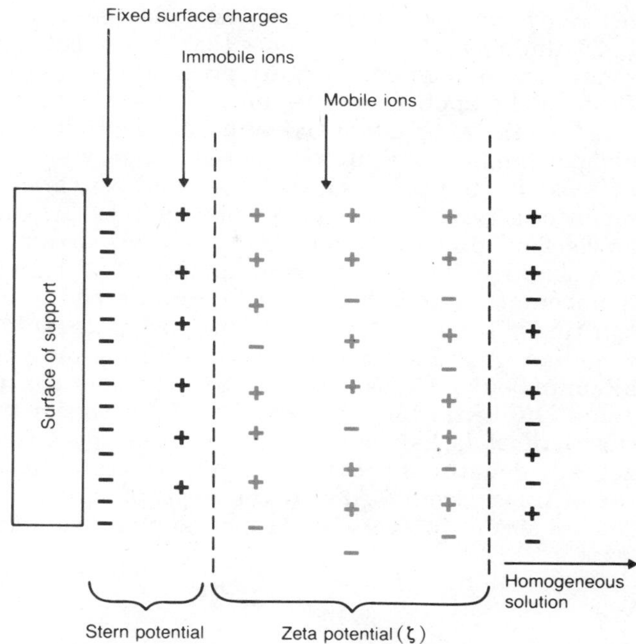

Figure 2-27. Distribution of positive (+) and negative (−) ions around the surface of an electrophoretic support. Fixed on the surface of the solid is a layer of − ions. (These may be + ions under suitable conditions.) A second layer of + ions is attracted to the surface. These two layers compose the Stern potential. The large, diffuse layer containing mostly + ions is the electrokinetic or zeta (ζ) potential; extending further from the surface of the solid is homogeneous solution. The Stern potential plus the zeta potential equal the electrochemical or epsilon (ϵ) potential.

thermore, cold buffer in an electrophoretic run improves resolution and lessens evaporation from the electrophoresis support. Buffer used in an apparatus with small-volume buffer compartments should be discarded after each run because pH change occurs as a result of the electrolysis of water during electrophoresis. When compartment volumes are >700 mL, switching polarity of the electrophoretic cell after each run allows use of the same buffer for up to four runs. Alternatively, buffer from the two compartments can be poured into a common container, stored at 4 °C between runs, and used up to four times.

Stain solution. A typical stain solution for proteins may be used several times. A good rule of thumb is that 100 mL of stain solution may be used for a total of 387 cm^2 (60 in^2) of cellulose acetate or agarose film. Deterioration of the stain should be suspected if leaching of the stained zones occurs in the 5% acetic acid wash step or (with cellulose acetate only) in the clearing step. Stain solutions should be stored tightly closed to avoid evaporation.

Sample application. In order to obtain satisfactory electrophoretograms, the amount of serum applied must be optimal. The concentration of albumin in serum is ~10 times greater than that of α_1-globulins. Therefore, the amount of sample applied should be small enough to avoid overloading with albumin but large enough to allow quantitation of α_1-globulin. When, as in isoenzyme analysis, a procedure calls for multiple applications of sample, albumin overload is not a concern.

Problem Solving in Protein Electrophoresis

Problems may arise from sample application or from the character of the support materials. Application problems are especially prominent in CAE and AGE.

Discontinuities in sample application may be due to a dirty or deformed applicator. Because applicators are easily bent, caution is necessary in cleaning them. Twin-wire applicators are best cleaned merely by dipping and agitating them in water, and then removing water and residue by gently pressing the applicators against absorbent paper. Manual wiping is inadvisable.

Distortion of protein zones may be due to bent applicators or overloading of the support with sample. Care should be taken not to overload the twin-wire applicator with serum or to overfill the sample well in AGE. Excessive drying of a cellulose acetate film before or during electrophoresis may also cause distorted zones. Improper tension of the CAE film may often cause zones that look "bent over." Irregularities other than broken zones in sample application are probably due to excessively wet cellulose acetate films. Parts of the applied samples may look washed out.

Certain artifacts cause unusual bands that may be easily recognized. Since free hemoglobin migrates in the β-globulin zone, hemolysis is a frequent cause of an apparent increase in that zone. An unusual band between the α_2- and β_1-globulins may be the result of a hemoglobin-haptoglobin complex. A band occurring at the starting point of an electrophoretogram may be fibrinogen; the sample should be verified as serum before this band is considered to be an abnormal protein. Splitting of α_1-, α_2-, and β-globulins is not unusual and should not be considered an error in electrophoresis. Occasionally splitting of the albumin zone is observed as a result of a rare, benign, genetically related condition. A grossly widened albumin zone may be the result of a medication that is albumin-bound. Occasionally an irregular but sharp protein zone is seen at the starting point. Unlike fibrinogen or other proteins that may be seen at about the same location, this artifact lacks the somewhat diffuse appearance that proteins usually show; it is actually denatured protein from a deteriorated serum or is evidence of damage done to the cellulose acetate film by the applicator. It is good laboratory practice to include a control serum in every electrophoretic run as an aid to evaluating the quality of the run.

References

1. Catsimpoolas, N.: Isoelectric focusing and isotachophoresis of proteins. 13th Eastern Analytical Symposium Articles. Separation Sci., *8*:71–121, 1973.
2. Davis, B. J.: Disc electrophoresis—II. Method and application to human serum proteins. Ann. N.Y. Acad. Sci., *121*:404–427, 1964.
3. Drysdale, J. W.: Isoelectric focusing in polyacrylamide gel. *In*: Methods of Protein Separation, Vol. 1. N. Catsimpoolas, Ed. New York, Plenum, 1975.
4. Duckworth, M., and Yaphe, W.: Preparation of agarose by fractionation from the spectrum of polysaccharides in agar. Anal. Biochem., *44*:636–641, 1971.
5. Epstein, E., Wolf, P. L., Horwitz, J. P., et al.: An indigogenic reaction for alkaline phosphatase in disk electrophoresis. Am. J. Clin. Pathol., *48*:530–534, 1967.
6. Epstein, E., and Zak, B.: Separation of haptoglobins by disc electrophoresis. Ann. Clin. Lab. Sci., *2*:191–197, 1972.
7. Hegenauer, J. C., and Nace, G. W.: An improved method for preparing agarose. Biochim. Biophys. Acta, *111*:334–338, 1965.
8. Jeppsson, J.-O., Laurell, C.-B., and Franzen, B.: Agarose gel electrophoresis. Clin. Chem., *25*:629–638, 1979.
9. Ornstein, L.: Disc electrophoresis—I. Background and theory. Ann. N.Y. Acad. Sci., *121*:321–349, 1964.
10. Zak, B., Baginski, E. S., and Epstein, E.: Associated problems of protein electrophoresis, staining and densitometry. Ann. Clin. Lab. Sci., *8*:385–395, 1978.
11. Zak, B., Epstein, E., and Watkins, R.: Postulated flaw in densitometry. Microchem. J., *23*:226–236, 1978.

Recommended Reading

Bier, M., Ed.: Electrophoresis: Theory, Methods and Applications. New York, Academic Press, Vol. 1, 1959, Vol. 2, 1967.

Block, R. J., Durrum, E. L., and Zweig, G.: Paper Chromatography and Paper Electrophoresis. 2nd ed. New York, Academic Press, 1958.

Deyl, Z., Ed.: Electrophoresis: A Survey of Techniques and Applications. Amsterdam, Elsevier, 1979.

SECTION FOUR

Electrochemistry

OLE SIGGAARD-ANDERSEN, M.D., Ph.D.

Several types of analytical methods are based on electrochemical phenomena. Those which have been applied to clinical chemical procedures are potentiometry, polarography, amperometry, coulometry, and conductometry.

Potentiometry

Potentiometry is the measurement of the electric potential difference between two electrodes in an electrochemical cell. An *electrochemical* (galvanic) *cell* always consists of two electrodes (electron or metallic conductors) connected by an electrolyte solution (ion conductor). An *electrode* or *half-cell* consists of a single metallic conductor in contact with an electrolyte solution. The ion conductors can be composed of one or more phases in direct contact with each other, or they can be separated by membranes that are permeable only to specific cations or anions (Figure 2-28). One of the electrolyte solutions is the unknown or test solution, which may be replaced by an appropriate standard solution for calibration or quality control purposes. A salt solution or bridge may be interposed in the cell to reduce any liquid-liquid junction potential that is present. By convention, the cell is shown so that the left electrode (M_L) is the *reference electrode*, whereas the right electrode (M_R) is the *indicator (measuring) electrode.*[1] *The electromotive force* (EMF, E) is defined as the maximal difference in potential between the two electrodes (right minus left) obtained when the current drawn from the cell is zero.

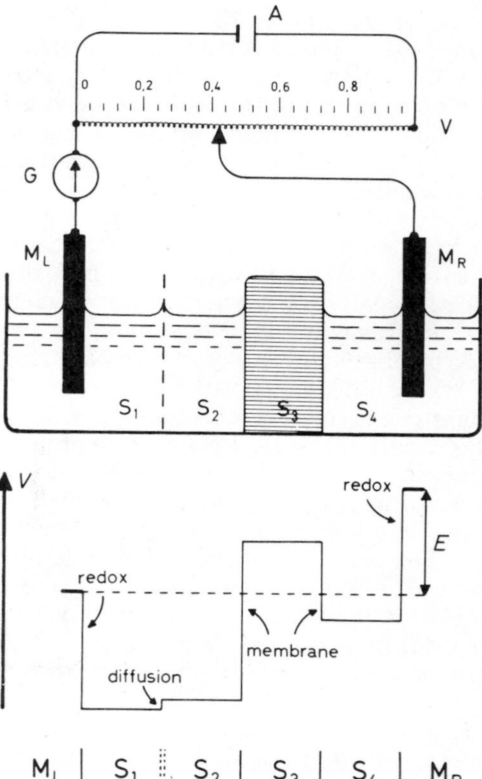

Figure 2-28. Schematic diagram of an electrochemical cell. M_L and M_R are two metallic conductors. S_1, S_2, S_3, and S_4 symbolize a variable number of *ion-conductive* phases. S_1 may be saturated KCl solution; S_2, a given test solution; S_3, an ion-selective membrane (which need not necessarily be a thin membrane); and S_4, a given reference solution. The liquid-liquid junction between S_1 and S_2 may be an open contact between the two solutions, or a porous membrane, or a fiber junction. The two electrodes (M_L and M_R) are connected externally via a potentiometer (e.g., a null-point potentiometer) consisting of a voltage source (A), a voltage divider (V), and a sensitive galvanometer (G). The voltage divider is adjusted so that the galvanometer shows zero current. The EMF of the cell is then read directly on the calibrated scale.

The lower half of the figure illustrates the potential jumps at the phase boundaries caused by (1) redox potentials, (2) membrane potentials, and (3) diffusion potentials (liquid-liquid junction potentials). At the bottom, the components of the cell are presented as symbols; vertical lines indicate phase boundaries, and a double dotted line indicates the liquid-liquid junction.

Measurement of the cell potential is carried out with a *potentiometer*, of which the common pH meter is a special type.

Within any one given conductive phase, the potential is constant as long as the current flow is zero. Between two phases, however, a potential difference arises. The overall potential of an electrochemical cell is the sum of all the potential jumps existing between different phases of the cell (Figure 2-28). The potential of a single electrode with respect to the surrounding electrolyte and the absolute magnitude of the individual potential jumps between the phases are actually unknown and cannot be measured. We can measure only the *potential differences* between two electrodes (half-cells). The potential jumps can be classified as (1) redox potentials, (2) membrane potentials, and (3) diffusion potentials. Generally, it is possible to devise a cell in such a manner that all the potential jumps except one are constant. This potential can then be related to the *activity* of some specific ion of interest (H^+, Na^+, and so forth).

Measurement of Ion Activity Versus Ion Concentration

Most methods, such as flame photometry, atomic absorption spectrometry, and titration, provide the concentration of total ions. Direct potentiometric measurement on an undiluted sample provides a means of determining the concentration of *free unbound ions* or, more precisely, the *activity of the ions*. This is the relevant quantity for most purposes because chemical equilibria and biological phenomena are dependent on the activity of the ions rather than on the concentration of total ions. But for some purposes (e.g., for balance studies, when the input and output of the total amount of substance are of interest), the concentration of *total ions* is more relevant.

The *activity* (a) equals the numerical value of the *molality* of free ion (m; (mol/kg))* times the *activity coefficient* (γ). Both the activity and the activity coefficient are dimensionless. The activity coefficient depends on the *ionic strength* (I) of the solution ($I = \frac{1}{2} \Sigma m \cdot z^2$, where z is the charge number of the ions) and generally decreases with increasing ionic strength.

The decrease in the activity coefficient with increasing ionic strength is more pronounced the higher the charge number of the ion, and hence activity coefficients are generally much lower for divalent than for monovalent ions. For most biological fluids it is difficult to calculate the activity coefficients accurately because the contribution of the protein ions to the ionic strength is highly uncertain. Examples of estimated activity coefficients for normal human blood plasma are Na^+, 0.75; K^+, 0.74; and Ca^{2+}, 0.32.

Results of potentiometric measurements on undiluted whole blood, plasma, or serum are usually reported not as activity but as the activity multiplied by an appropriate factor, so that the normal reference interval becomes the same as the conventional normal reference interval for substance concentration. In actual practice the factor is incorporated in the calibration by using standards with known activity and correcting electronically for the activity coefficient.

In this way major discrepancies between the results obtained by flame photometry and by direct potentiometry for Na^+ and K^+ will occur only when the mass concentration of water deviates significantly from normal. In cases of severe hyperlipemia, when the mass concentration of water may be as low as 0.80 kg/L (normally about 0.93 kg/L), the result for Na^+ by flame photometry may be 119 mmol/L, whereas direct potentiometry gives 140 mmol/L. The flame photometric value may be misinterpreted as indicating Na^+ deficiency or water intoxication or both (also called pseudo-hyponatremia). The potentiometric result shows that the Na^+ activity is normal.

The *precision* of potentiometric measurements is limited by the standard deviation for the measured potentials (~50 μV). This uncertainty corresponds to a coefficient of variation for the activity of a monovalent ion of ~0.2%. For divalent ions, the coefficient of variation will be about twice as high. The *accuracy* of direct potentiometric measurement is limited by the uncertain liquid junction potential, which may vary about

*Notice that the symbol m may be employed for *molality* as well as for *mass*. In order to distinguish these, the symbol for mass may be printed as an italic boldface $\boldsymbol{m}$.

±0.5 m*V* with the composition of the serum. The variation corresponds to a bias in the measured concentration of about ±2% for monovalent ions.

If potentiometric measurements are employed for *the determination of the concentration of total ions* (free + bound), the sample must be diluted by a suitable diluent that liberates the complex-bound ion from its binding agent (e.g., by a pH adjustment). The dilution should at the same time serve to establish a constant ionic strength so that a constant activity coefficient is obtained independent of variations in ionic strength of the original sample. This method is called *indirect potentiometry* as opposed to *direct potentiometry*, which involves direct measurement in the undiluted sample. A different approach consists of *titration* utilizing a potentiometric end point detection. In this case, the electrode is used only to sense the sudden change in activity as the end point is reached. Titration is generally considered among the most accurate and precise analytical methods available.

REDOX ELECTRODES

Redox potentials are due to chemical equilibria involving electron transfer reactions:

$$\text{Oxidized form (Ox)} + ze^- \rightleftharpoons \text{Reduced form (Red)} \tag{1}$$

e.g.,

$$Fe^{3+} + e^- \rightleftharpoons Fe^{2+}$$

or

$$2\,H^+ + 2\,e^- \rightleftharpoons H_2$$

where z symbolizes the number of electrons involved in the reaction. Any substance that binds (accepts) electrons is an oxidant (Ox), and a substance that gives off electrons is a reductant (Red). The two forms, Ox and Red, represent a redox couple (conjugate redox pair). Redox processes take place only between two redox couples; the electrons are transferred from a reductant (Red_1) to an oxidant (Ox_2). In this process, Red_1 is oxidized to its conjugate Ox_1, while Ox_2 is reduced to Red_2:

$$Red_1 + Ox_2 \rightleftharpoons Ox_1 + Red_2 \tag{2}$$

e.g.,

$$2\,Fe^{2+} + 2\,H^+ \rightleftharpoons 2\,Fe^{3+} + H_2$$

The *electrode potential* (reduction potential) for a redox couple is defined as its potential measured with respect to the standard hydrogen electrode, whose potential is defined as zero (see later). This potential, by convention, is the **EMF** of a cell, where the standard hydrogen electrode is the reference electrode and the given half-cell is the indicator electrode. The reduction potential for a redox couple is given by the *Nernst equation*:

$$E = E^\circ - \frac{N}{z} \cdot \log \frac{a\text{Red}}{a\text{Ox}} = E^\circ - \frac{59.2\text{ m}V}{z} \cdot \log \frac{a\text{Red}}{a\text{Ox}} \tag{3}$$

where

E = electrode potential of the half-cell

E° = standard electrode potential when $a\text{Red}/a\text{Ox} = 1$

z = number of electrons involved in the reduction reaction

R = gas constant ($8.3143\ J \cdot K^{-1} \cdot mol^{-1}$)

T = absolute temperature (unit: K, kelvin)

F = Faraday constant (96487 $C \cdot mol^{-1}$), (C = coulomb = joule $\cdot$ $volt^{-1}$)

$N = R \cdot T \cdot \ln 10/F$ (the Nernst factor if $z = 1$)*, $N = 0.0592$ V if $T = 298.15$ K (25 °C)

ln 10 = natural logarithm of 10 = 2.303

a = activity

aRed/aOx = product of mass action for the reduction reaction

The redox electrodes presently in use can be divided into two major groups or classes, namely, (1) inert metal electrodes immersed in solutions containing redox couples, and (2) metal electrodes in which the metal functions as a member of the redox couple.

Inert Metal Electrodes

Platinum and gold electrodes are examples of inert metal electrodes used to record the redox potential of a redox couple dissolved in an electrolyte solution. Not all dissolved redox couples, however, are able to equilibrate with an inert metal such as platinum or gold. In many cases the presence of a catalyst is required to establish a reproducible potential. The catalyst can be a small amount of another redox couple, a so-called mediator, that readily equilibrates with both the metal electrode and the more sluggish redox couple of interest.

The ***hydrogen electrode*** is a special redox electrode for pH measurement. It consists of a platinum or gold electrode that is coated with highly porous platinum (platinized) to catalyze the electrode reaction

$$H^+ + e^- \rightleftharpoons \tfrac{1}{2} H_2$$

The electrode potential is given by

$$E = E^\circ - N \cdot \log \frac{(aH_2)^{1/2}}{aH^+}$$

where

E° = 0 at all temperatures (by convention)

aH^+ = activity of hydrogen ions

$-\log aH^+$ = negative log of the H^+ activity (paH^+ or pH)

When the partial pressure of H_2 in the solution (and hence aH_2) is maintained constant by bubbling H_2 through the solution, the potential is a linear function of log aH^+ (= −pH). In the *standard hydrogen electrode* the electrolyte consists of an aqueous solution of HCl with aHCl = 1.000 (or cHCl ≃ 1.2 mol/L) in equilibrium with a gas phase with aH_2 = 1.000 (or pH_2 = 101.3 kPa = 1 atm).

Metal Electrodes Participating in Redox Reactions

The ***silver electrode*** consists of a silver wire immersed in a solution containing silver ions. The electrode process consists of the reduction of silver ions to metallic silver:

$$Ag^+ + e^- \rightleftharpoons Ag$$

* N should not be mistaken for the Avogadro constant.

The expression for the electrode potential reduces to:

$$E = E^{\circ} + N \cdot \log a\mathrm{Ag}^{+}$$

because the activity of pure silver is unity. According to this equation, the electrode measures the silver ion activity in the solution. An application for this electrode is the determination of chloride by titration with $AgNO_3$. In this method the silver electrode is coupled with a mercurous sulfate reference electrode ($\frac{1}{2}\ Hg_2^{2+} + e^- \rightleftharpoons Hg$) for the end point determination. When all Cl^- ions are precipitated as AgCl, the excess of Ag^+ causes a sudden change in the electrode potential, which indicates the end point of the reaction.

In an analogous fashion, other metal electrodes (e.g., Zn, Cu) measure the activity of their respective ions. When the metal is chemically unstable in pure form, an electrode consisting of an amalgam of the metal (e.g., calcium amalgam) can sometimes be used. Unfortunately, such electrodes are of little value for direct measurements in biological fluids because proteins and lipids tend to coat the amalgam surface, and various redox couples present in biological fluids affect the electrode potential.

The ***silver/silver chloride electrode*** consists of a silver wire, electrolytically coated with AgCl, which dips into a solution containing chloride ions. The electrode process is

$$\mathrm{AgCl\ (solid)} + e^- \rightleftharpoons \mathrm{Ag\ (metal)} + \mathrm{Cl}^-$$

Since aAgCl and aAg are both unity (because both components are present as pure substance), the expression for the electrode potential reduces to

$$E = E^{\circ} - N \cdot \log a\mathrm{Cl}^{-}$$

which shows that the electrode measures chloride activity. This electrode has been utilized for direct measurement of the chloride activity in serum. By placing it directly on the skin surface, the electrode can also be used for the measurement of the chloride activity of sweat in connection with the diagnosis of cystic fibrosis.

When the chloride activity is kept constant (e.g., with cKCl fixed at 0.1 mol/L), the electrode potential is constant, and therefore the electrode is frequently employed as a reference electrode or "inner electrode" in membrane electrodes (Figure 2-29).

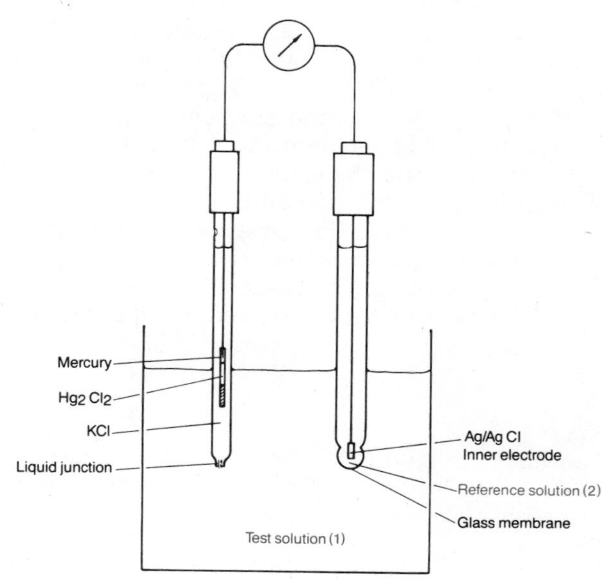

Figure 2-29. Electrochemical cell for measuring pH. The reference calomel electrode is on the left. KCl diffuses slowly out, forming a liquid-liquid junction with the test solution. The indicator pH electrode with glass membrane is on the right. The Ag/AgCl inner electrode makes contact with a sealed-in reference solution. The potential difference between the test solution (1) and reference solution (2) is read on a voltmeter calibrated in pH units.

The ***calomel electrode*** consists of mercury covered by a layer of calomel (Hg_2Cl_2), which is in contact with an electrolyte solution containing chloride. The electrode process is

$$\frac{1}{2} Hg_2Cl_2 \text{ (solid)} + e^- \rightleftharpoons Hg \text{ (metal)} + Cl^-$$

Since aHg and $a Hg_2Cl_2$ are both unity (present as pure substance), the electrode potential reduces to

$$E = E^\circ - N \cdot \log a\mathrm{Cl}^-$$

This equation shows that the calomel electrode also functions as a chloride electrode, and its potential varies with the chloride activity. The chloride activity is generally maintained constant, either as saturated KCl or at a concentration of 3.5 mol/L. A calomel electrode is frequently employed as a reference electrode together with a glass electrode for pH measurement.

ION-SELECTIVE MEMBRANE ELECTRODES[3,4]

Membrane potentials are due to permeability of certain types of membranes to selected anions or cations. Biological membranes (e.g., the glomerular membrane in the kidney) are often impermeable to the high-molecular protein ions; this fact gives rise to the Donnan potential, which in turn leads to an uneven distribution of the diffusible ions on both sides of the membrane. For analytical applications, membranes are required that possess a selective permeability for a single ion species.

The ion-selective membrane, separating the solution on the left side (1) from the solution on the right side (2), can be illustrated as

1 | Membrane | 2

The membrane potential is conventionally defined as the potential on the right side minus the potential on the left. The potential difference between the two solutions is compared by using a suitable reference electrode. The pH electrode in Figure 2-29 is an example of the use of a glass membrane. The potential of the reference solution (2) is generally detected by means of an Ag/AgCl electrode, whereas the potential of the test solution (1) is detected by means of a calomel electrode via a liquid-liquid junction between the saturated KCl and the test solution.

If the activity of a diffusible cation is higher in solution 1 than in solution 2, a positive membrane potential develops. In most cases the mechanism is as follows: on the left side, where the cation activity is high, cations are bound to the membrane surface by specific binding groups, and the membrane is thereby charged positively. This charging of the membrane causes a dissociation of cations from the other side of the membrane into the solution on the right side, which is thereby charged positively with respect to the solution on the left side. Thus, it *appears* as if the membrane were permeable to the cation. This process proceeds until an equilibrium is established, which occurs when the electric potential difference across the membrane matches the difference in activity of the diffusible ions on the two sides of the membrane.

The membrane potential, V, is given by the Nernst equation:

$$V = \frac{\mathrm{R} \cdot T}{z \cdot F} \cdot \ln \frac{a(2)}{a(1)} \quad (4)$$

where z is the ion charge number (positive for cations, negative for anions), and a is the activity of the diffusible ion (cation or anion). If 2 is a reference solution with constant activity of the diffusible ion [constant $a(2)$], the equation reduces to

$$V = V' + \frac{N}{z} \cdot \log a(1) \quad (5)$$

where V' is constant. In other words, the *membrane potential is directly proportional to the logarithm of the activity of the diffusible ion in the test solution*.

The *sensitivity* of the electrode, $\Delta V/\Delta \log a$, is given by the (theoretical) Nernst slope, N/z. However, the actual sensitivity often deviates slightly, being $s \cdot N/z$, where s is the *relative sensitivity*. For most pH-glass electrodes s is 0.98–1.00. For other ion-selective electrodes s may be as low as 0.90. Lower values indicate poor electrode performance due to malfunction or poor selectivity. It should be noted that the theoretical sensitivity for divalent ions is only half the value for monovalent ions.

The *selectivity* of the electrode for a single ion species is seldom absolute. If the membrane is "permeable" to several different ions, a diffusion potential may arise, and the theory for the membrane potential becomes more complicated.

The ***design of membrane electrodes*** is as multifaceted as the human imagination. Electrodes may be bulb-shaped, inverted bulb-shaped, capillary type, needle type, and flow-through. Many attempts are being made to miniaturize electrodes for measurements in the blood stream (catheter-tip electrodes) or measurements in single tissue cells (glass capillary microelectrodes).

Available ion-selective membrane electrodes can be classified arbitrarily as (1) glass electrodes, (2) solid-state electrodes, and (3) liquid ion-exchange electrodes. By means of additional membranes these electrodes may be modified into (4) gas electrodes or (5) enzyme electrodes.

Glass Electrodes

Glass electrodes are made from specially formulated glass consisting of a melt of SiO_2 with added oxides of various metals. Membranes with varying composition of the glass have been prepared with directed selectivity for H^+, Na^+, K^+, Li^+, Rb^+, Cs^+, Ag^+, Tl^+, and NH_4^+.[3] The membranes generally have a thickness of 10–100 μm.

Hydrogen-ion selective glass electrodes (pH electrodes) can be manufactured from the classic Corning 015 glass, consisting of SiO_2, Na_2O, and CaO in the molar ratio of 72.2:21.4:6.4, but many other compositions have also been found suitable. With the older glass, significant error in measurement (erroneously low pH values) occurred in the presence of Na^+ at pH values above 8.0. The sodium error, which is due to a loss in specificity in the electrode, increases greatly with temperature. This error is greatly reduced with newer types of glass that allow pH measurements in NaOH solutions up to about pH 13 with a pH error of <0.02.

Na^+-selective glass electrodes can be prepared from glass consisting of SiO_2, Na_2O, and Al_2O_3 in the ratio of 71:11:18. Lithium-aluminum silicates have also been found suitable. These electrodes are insensitive to H^+ in the pH range 6–10.

Solid-State Electrodes

Solid-state membranes can be either homogeneous membranes consisting of a "single" crystal, or heterogeneous membranes consisting of an active substance embedded in an inert matrix.[10]

The ***homogeneous-membrane electrodes*** include those for F^- (lanthanum fluoride crystal), Cl^- (AgCl crystal), Br^- (AgBr crystal), I^- (AgI crystal), S^{2-} (Ag_2S), Cu^{2+} (cupric selenide crystal), and others. The silver salt solid-state membrane electrodes are less susceptible to interference from redox systems than are the equivalent silver/silver salt redox electrodes. The AgCl membrane electrode is used for measurement of the activity of chloride in sweat by direct measurement on the skin surface (Orion electrode; Orion Co., Boston, MA). The fluoride electrode has been utilized for measurements of the fluoride concentration in saliva.

Ion-Exchange Electrodes

Liquid ion-exchange membranes consist of an *inert solvent* in which an *ion-selective carrier* substance is dissolved. Both solvent and carrier should be insoluble in water. The membrane solution can be separated from the test solution by means of a collodion membrane, or a porous matrix can be soaked by the membrane solution. The inert solvent and the ion-selective carrier are often embedded in a matrix of polyvinylchloride (PVC). Many different membranes have been prepared with selectivity for specific cations or anions (e.g., K^+, NH_4^+, Ca^{2+}) and recently also for H^+, Cl^-, and HCO_3^-.

Potassium-selective membranes can be made by dissolving the antibiotic valinomycin in a suitable solvent. Valinomycin is a neutral carrier that binds K^+ in the center of a ring of oxygen atoms (Figure 2-30). This membrane is highly selective for potassium.

The ammonium ion–selective membrane is based on a mixture of the antibiotics nonactin and monactin, which are also neutral carriers.

A calcium-selective membrane can be made by dissolving the calcium salt Ca^{2+}-bis(di-*p*-octylphenyl phosphate) in PVC.[9,10] Two di-*p*-octylphenyl phosphate ions bind a Ca^{2+} ion and act as the Ca^{2+} carrier. Di-octylphenyl phosphonate is added as a plasticizer to the PVC. *A neutral carrier* has also been developed.[12]

Protein interference. A protein coating on the ion-selective membrane may cause a lack of reproducibility when measurements are performed alternately in protein-free and protein-containing solutions. This source of bias may be avoided by covering the ion-selective membrane by a dialysis membrane, which also prolongs the lifetime of the electrode. A disadvantage of using a cellophane membrane is a significant prolongation of the response time of the electrode from <1 s to >20 s for 98% response.

GAS ELECTRODES

Gas electrodes are specially designed for the measurement of specific gases in gas mixtures or in solutions. Examples are the CO_2 electrode and the NH_3 electrode, which are both based on potentiometric measurement using a pH glass electrode. The pO_2 electrode is based on amperometric measurement using a polarized platinum cathode (see below).

Usually the gas electrode is separated from the test solution by means of a thin, gas-permeable membrane (e.g., polyethylene, polypropylene, Teflon, silicone rubber).

The ***pCO_2 electrode*** represents a special application of a pH glass electrode (Figure 2-31). The sample in this case is in contact with a membrane that is permeable to gas but not to solutions. This membrane (e.g., silicone rubber) is separated from the actual glass electrode by a thin film of bicarbonate solution (5 mmol/L). The CO_2 gas diffuses from the sample (or test gas) through the membrane and rapidly enters equilibrium with the bicarbonate solution altering its pH.

$$CO_2 + H_2O \rightleftharpoons H_2CO_3 \rightleftharpoons H^+ + HCO_3^-$$

The pH of the bicarbonate solution is a simple function of the pCO_2. The pCO_2 electrode

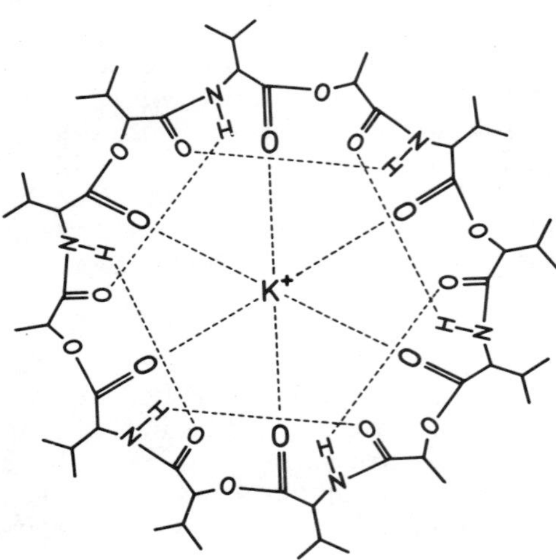

Figure 2-30. Valinomycin has a heterocyclic structure consisting of alternating peptide and ester linkages. The molecule is cylindrical and stabilized by six hydrogen bridges (from =NH to C=O). The outer surface of the cylinder is strongly hydrophobic and the interior is strongly hydrophilic. The size of the central cavity nearly equals the diameter of an unhydrated potassium ion.

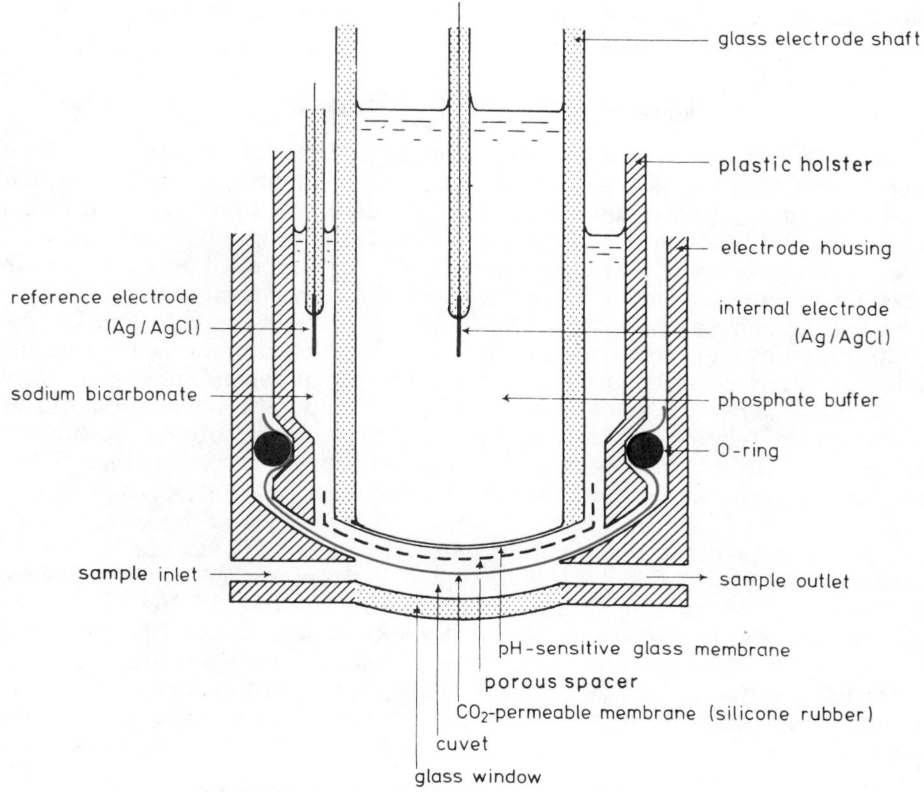

Figure 2-31. Schematic illustration of a pCO_2 electrode.[13]

has been used extensively for measurements in arterial blood samples. More recently the application has been extended to *transcutaneous measurements*.[7,11]

The ***NH_3* gas electrode** is similar to the pCO_2 electrode except that the bicarbonate solution is replaced by an ammonium chloride solution. Again the measured pH of the ammonium chloride solution behind the gas-permeable membrane varies linearly with the logarithm of the partial pressure of NH_3.

ENZYME ELECTRODES[5]

The various electrodes may be covered by a layer of immobilized enzymes that catalyze a chemical reaction that can be monitored by the electrode. The enzymes may be immobilized on the electrode surface by simple adsorption, by entrapment behind a dialysis membrane, or by covalent linkage. This principle may be applied to redox electrodes, ion-selective membrane electrodes, and also to amperometric electrodes (see below).

The response of enzyme electrodes is complex because it is dependent on the rate of diffusion of substrates into the enzyme layer, the rate of the enzyme-catalyzed reaction, and the rate of back diffusion of the reaction products. It should be noted that enzyme electrodes are much less durable than H^+, K^+, or Ca^{2+} electrodes, since immobilized enzymes gradually lose their activity owing to surface contamination and aging.

Urea electrodes. Urea from the sample diffuses to the membrane consisting of polyacrylamide matrix containing immobilized urease fixed on a Dacron net. The urea is then hydrolyzed to NH_4^+, which is measured by a NH_4^+-selective membrane electrode, or the rise in pNH_3 is sensed by a NH_3 gas electrode, or the rise in pH is sensed with a pH glass electrode.

A ***glucose electrode*** based on the use of glucose oxidase, in combination with a pH electrode for detection of gluconic acid formed in the enzyme-catalyzed reaction, was the first enzyme electrode proposed. However, amperometric glucose electrodes (see later) have proved to be superior.

Liquid-Liquid Junction Potentials

The variable and generally unknown liquid-liquid junction potential constitutes the most important source of bias in potentiometric measurements of ion activities. Diffusion potentials arise at a liquid-liquid junction (1⫶2) where two solutions of different ionic composition are in direct contact. If cations diffuse faster from left to right than anions, a positive liquid junction potential develops at the interface. Generally, the aim is to reduce the liquid junction potential to a value as low and as reproducible as possible. This is achieved by using a concentrated KCl solution as the bridge solution. The potential is then dominated by the large excess of K^+ and Cl^- diffusing at almost the same rate into the test solution, and the potential becomes largely independent of the composition of the test solution. The difference between the junction potential with test solution (e.g., plasma) and the calibration solution (e.g., phosphate buffer) is called the *residual liquid junction potential*. In order to reduce the residual liquid junction potential, the composition of the calibration solution should match that of the test solution as closely as possible.

Erythrocytes present at the junction, with saturated KCl as the bridge solution, increase the liquid junction potential ($\simeq$+0.6 mV when comparing whole blood and plasma). They do so because precipitated proteins at the liquid junction act as an ion exchanger influencing the diffusion rate of the ions, slowing down Cl^- more than K^+. The effect is reduced to about one third when a flowing junction is employed. It is also possible to use a different bridge solution. Thus, with sodium formate, 4 mol/kg, as bridge solution, the effect of erythrocytes on the junction potential is negligible, and measurements on whole blood give the same result as measurements on the corresponding plasma.

Polarography

Polarography is based on the dual measurement of the current flowing through an electrochemical cell and the electrical potential between the two electrodes when the potential is gradually increased at a constant rate by means of an external voltage source.[6] The function between current and potential is called a *polarogram* (Figure 2-32).

A polarographic cell generally consists of a polarizable indicator electrode and a nonpolarizable reference electrode. The indicator electrode can be a dropping mercury

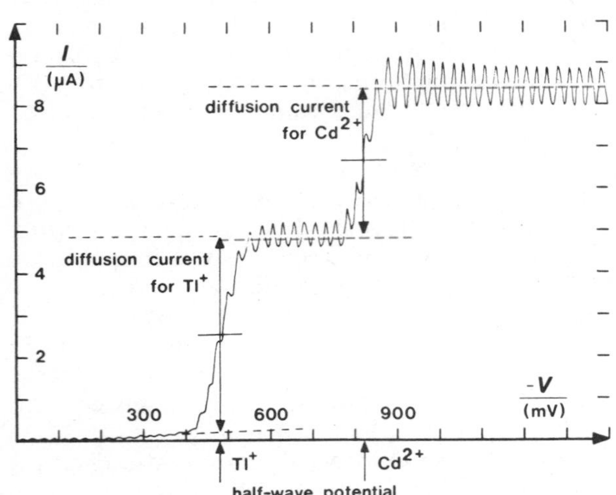

Figure 2-32. Polarogram of a solution containing TlCl (1.0 mmol/L) and $CdCl_2$ (0.5 mmol/L) in an ammonia/ammonium chloride buffer (1 mol/L). The abscissa shows the potential of the indicator electrode. The ordinate shows the current through the cell. The current pulsates owing to the dropping mercury cathode.

electrode, i.e., a glass capillary tube (50 μm ID) filled with mercury that slowly drips into the test solution. In this way the electrode surface is constantly renewed. The reference electrode can be a large pool of mercury in the bottom of the vial.

The cell may be symbolized as follows, where the indicator electrode is the cathode:

Hg | Test solution | Hg
(**Anode**) (**Cathode**)

When an increasingly negative potential is applied to the cathode at a constant rate, the current at first is almost zero, because the cathode is electrolytically polarized (i.e., the cell becomes asymmetrical and therefore generates a counterelectromotive force that balances the applied potential). If the test solution is oxygen-free water, the electrode processes are

Cathode $H^+ + e^- \rightleftharpoons \frac{1}{2} H_2$

Anode $2\,Hg \rightleftharpoons Hg_2^{2+} + 2e^-$

Hydrogen is formed at the cathode, and mercury is oxidized to Hg_2^{2+} at the anode. However, owing to the large surface of the anode, the changes in concentration at the anode are small and insignificant in relation to the concentration changes at the cathode. The anode is therefore considered to be nonpolarized. The maximal counterelectromotive force that can be obtained is reached when pH_2 at the cathode has reached atmospheric pressure.

If the test solution contains substances that are reduced more easily than H^+, the cathode will be depolarized by these substances and a current will flow through the cell. For example, if the test solution contains oxygen, oxygen will be reduced at the cathode according to the overall reaction

$$O_2 + 2\,H_2O + 4e^- \rightarrow 4OH^-$$

When the decomposition potential for this reaction (about -0.3 to -0.4 V) is reached, the current abruptly increases until it reaches a plateau, called the diffusion current. The height of the plateau is dependent on the rate at which O_2 can diffuse from the surrounding solution to the surface of the cathode, where pO_2 is zero. The diffusion current is therefore directly proportional to the pO_2 of the test solution. Any substance that is reduced at the cathode is characterized by two values: (1) the *half-wave potential* (i.e., the potential at which the current is equal to one-half the plateau current) characterizes the kind of substance reduced, and (2) the *diffusion current* is proportional to the concentration of the substance in the test solution. Polarography may therefore be used for identification as well as for quantitation. In medicine, polarographic analysis is used for the determination of some metals (e.g., copper, lead, and zinc) in addition to the determination of oxygen (see *Amperometry*).

Anodic stripping voltammetry and *potentiometric stripping analysis*[8] are variants of polarography. Both methods employ an initial plating step whereby trace metal ions are reduced and concentrated on the surface of a cathode consisting of a mercury-coated graphite rod. The plating time is usually from 1 to 30 min, depending on the concentration of the trace metal. Extremely dilute samples may be analyzed because of this initial plating step. During the second phase, the metals concentrated on the graphite rod are successively oxidized, and this process is followed by measuring the stripping current as a function of the stripping potential (anodic stripping voltammetry) or by measuring the stripping potential as a function of time (potentiometric stripping analysis).

Amperometry

This technique is based on measurement of the current flowing through an electrochemical cell when a constant electric potential is applied to the electrodes. Some examples of this analytical principle are described in the following.

The ***p*O$_2$ electrode** (Clark electrode,[2] Figure 2-33) is actually a complete electrochemical cell consisting of a small platinum cathode (area about 300 μm^2) and a Ag/AgCl anode in phosphate buffer with added KCl. The platinum cathode, covered by a thin film of electrolyte, is separated from the test solution by a gas-permeable membrane (e.g., polypropylene). The cathode potential is adjusted to -0.65 *V*. In the absence of oxygen in the test solution, the current is almost zero because the cathode is polarized. In the presence of oxygen, a current is observed that is due to a diffusion of O_2 from the test solution through the membrane to the cathode, where it is reduced. The current is directly proportional to the pO_2 in the test solution. The sensitivity of commercial pO_2 electrodes is on the order of $\Delta I/\Delta pO_2 = 10^{-14}$ A/Pa, depending on the size of the cathode area and the thickness of the gas-permeable membrane.

The purpose of the membrane is twofold: first, to prevent proteins and other (dissolved) oxidants from gaining access to the cathode surface, and second, to limit the diffusion zone to the membrane and hence prevent variations in the diffusion coefficient of O_2 in the test solution (or gas) from influencing the result. Nevertheless, pO_2 electrodes generally read higher values in gases than in liquids, and it is necessary to take this gas/liquid ratio into account when calibrating the electrodes.

The pO_2 electrode has found widespread application for measurements of pO_2 in arterial or capillary blood. Estimation of the arterial pO_2 by *transcutaneous measurement* is also possible[7] (see Chapter 19, Section Two). In addition, the pO_2 electrode has been applied to the measurement of the concentration of *total oxygen* in the blood after liberation of hemoglobin-bound oxygen with ferricyanide or carbon monoxide (by forming methemoglobin and carboxyhemoglobin, respectively).

Glucose determination by means of the pO_2 electrode is based on the glucose oxidase–catalyzed reaction:

$$\text{Glucose} + O_2 \rightarrow \text{D-Glucono-}\delta\text{-lactone} + H_2O_2$$

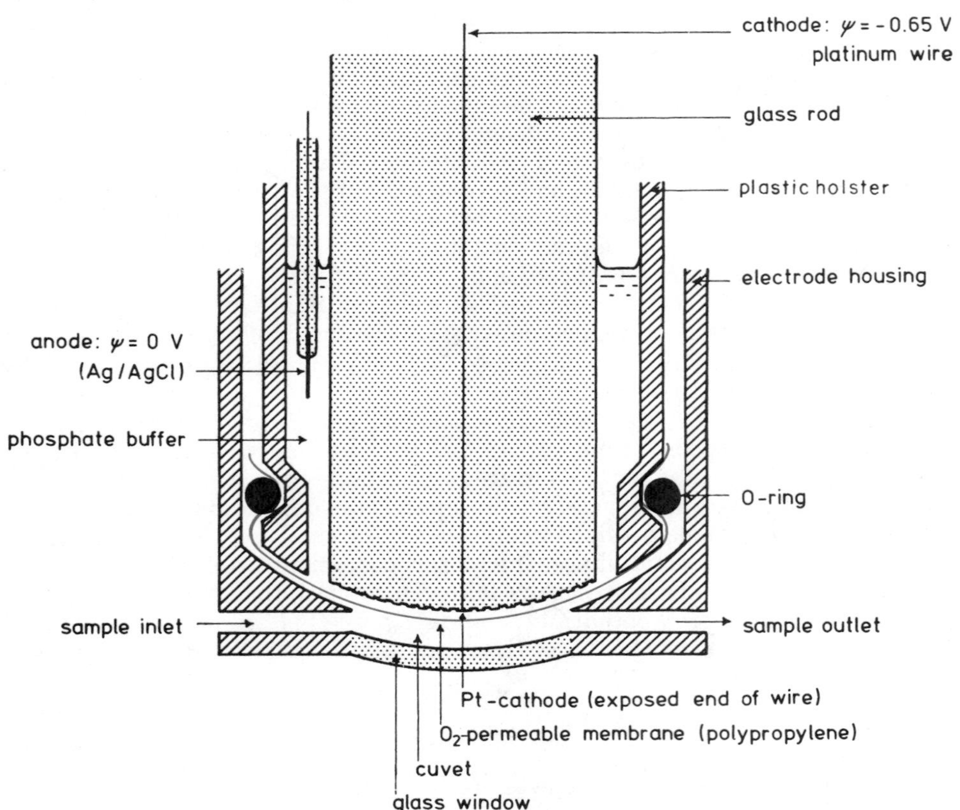

Figure 2-33. Schematic illustration of a pO_2 electrode.[13]

The rate of fall in pO_2 under standardized conditions is a measure of the glucose concentration. The glucose oxidase may be in the test solution or it may be trapped in a gel on the outside of the O_2-permeable membrane in front of the cathode so that a diffusion equilibrium involving both glucose and oxygen is established. Under standardized conditions, the response of the electrode is then directly related to the glucose concentration.

The ***peroxidase electrode*** consists of a *polarized platinum anode* and a nonpolarized Ag/AgCl cathode. The anode potential is +0.6 *V*. In the presence of peroxide, a current flows owing to oxidation of peroxide at the anode:

$$H_2O_2 \rightarrow 2\,H^+ + 2\,e^- + O_2$$

The peroxidase electrode covered by an enzyme membrane or enzyme layer has been proposed by Clark[2] for the determination of any substance for which a suitable oxygen oxidoreductase is available. Important examples are glucose, lactate, uric acid, and ethanol, in which case the membrane contains glucose oxidase, lactate oxidase, uricase, and alcohol oxidase, respectively. The substrate diffuses to the enzyme layer, where it is dehydrogenated, and H_2O_2 is formed. The H_2O_2 diffuses to the anode and causes a current proportional to the rate of formation of H_2O_2 (i.e., proportional to the diffusion of substrate to the membrane).

Amperometric end point determination may be applied in the titration of Cl^- with Ag^+. The sample is acidified by means of nitric acid (or nitric plus acetic acid), and the chloride ions are titrated with Ag^+. The silver ions may be added as a solution of $AgNO_3$ (volumetric titration), or Ag^+ may be generated from a silver electrode by means of an electric current (coulometric titration; see later). During titration, the Ag^+ concentration remains low owing to the reaction $Ag^+ + Cl^- \rightarrow AgCl$, which causes the precipitation of AgCl. At the end point, Ag^+ appears in excess, and the increase in Ag^+ activity may be detected either potentiometrically (with an Ag electrode and a mercury/mercurous sulfate reference electrode) or amperometrically. In the latter case, two silver electrodes are employed and a negative potential of 0.15–0.25 *V* is applied to the cathode. During titration, the cathode is polarized and the current is very low. At the end of the titration, the excess of Ag^+ depolarizes the cathode by the reaction $Ag^+ + e^- \rightarrow Ag$, and a current flows that is proportional to the excess of Ag^+. When the current has reached a preset value, the titration can be automatically stopped. This indicator principle is applied in the Cotlove chloride titrator, which in addition uses coulometric generation of Ag^+.

Coulometry

Coulometry is the technique used to measure the *amount* of electricity passing between two electrodes in an electrochemical cell. The amount of electricity is directly proportional to the amount of substance produced or consumed by the redox process at the electrodes. This is called Faraday's first law and may be expressed as

$$z \cdot n \cdot F = Q \tag{6}$$

where z is the number of electrons involved in the reduction (or oxidation) reaction, n is the amount of substance reduced or oxidized (mol), F is the Faraday constant (96 487 $C \cdot mol^{-1}$), and Q is the amount of electricity (coulomb = ampere • second) passing through the cell.

An example of an application is the coulometric titration of chloride (Cotlove titrator), in which silver ions are generated by electrolysis from a silver wire used as an anode. The anode potential must be high enough to exceed the decomposition potential and force the reaction $Ag \rightarrow Ag^+ + e^-$. At the cathode, H^+ is reduced to H_2 and the amount of silver ions generated is measured coulometrically. When the current is kept constant, the measurement is reduced to a measurement of time according to

$$Q = I \cdot t \tag{7}$$

where Q is the amount of electricity (coulomb), I is the electric current (ampere), and t is the time (second). The current may also be gradually decreased as the titration approaches the end point. In this case the amount of electricity is calculated as the integral: $Q = \int_0^t I \cdot dt$, a calculation performed electronically by the coulometer. The end point of the titration may be detected either amperometrically or potentiometrically.

Acid-base titrations can also be performed coulometrically by using a platinum generator electrode in the test solution separated from the other electrode by a sintered glass filter. If the generator electrode is the cathode, H^+ is removed ($H^+ + e^- \rightarrow 1/2\, H_2$), which is equivalent to the addition of base. If the generator electrode is the anode, H^+ is added ($H_2O \rightarrow 1/2\, O_2 + 2\, H^+ + 2\, e^-$). In either case, two sets of electrodes are necessary: the generator electrodes and the indicator electrodes. The latter may be used for direct potentiometric determination of the end point or in connection with amperometric end-point determination. Coulometric titrations are among the most accurate analytical determinations available.

Conductometry

Conductometry is the measurement of the current flow between two nonpolarized electrodes, between which a known electrical potential is established. In order to avoid polarization of the electrodes, an alternating potential is applied with a frequency between 100 and 3000 Hz. The resulting current (also alternating) is directly proportional to the conductivity* of the solution.

The conductivity of aqueous solutions is dependent on the concentration of electrolytes and is closely related to the ionic strength. In the purest available water, the conductivity is $\kappa = 4.9 \cdot 10^{-6}\ S \cdot m^{-1}$ at 18 °C. In ordinary distilled or deionized water the value is $\kappa < 2 \cdot 10^{-4}\ S \cdot m^{-1}$. A higher conductivity indicates the presence of electrolytes, and conductivity measurements therefore serve to monitor the performance of deionizers and to provide a warning that the ion-exchange resin should be regenerated.

Some chemical reactions are associated with a change in conductivity of the reaction medium. For example, urea can be measured rapidly and accurately by the change or the initial rate of change in conductivity that accompanies the urease-catalyzed formation of ammonium ions and bicarbonate ions. However, the high conductivity of biological fluids (owing to the presence of electrolytes) limits the use of this technique to analyses of low concentrations in which the change in conductivity is too small against a high background. Conductivity measurement can also be used for end-point detection in many kinds of titrations (acid-base, precipitation, compleximetric).

References

1. Bates, R. G.: Determination of pH. Theory and Practice. 3rd ed. New York, John Wiley & Sons, Inc., 1973.
2. Clark, L. C., Jr.: A family of polarographic enzyme electrodes and the measurement of alcohol. Biotechnol. Bioeng., *3*:377, 1972.
3. Freiser, H., Ed.: Ion-Selective Electrodes in Analytical Chemistry, Vol. 1. New York, Plenum Press, 1978.
4. Freiser, H., Ed.: Ion-Selective Electrodes in Analytical Chemistry, Vol. 2. New York, Plenum Press, 1980.
5. Guilbault, G. G.: Analytical uses of immobilized enzymes. Biotechnol. Bioeng., *3*:361, 1972.
6. Heyrovsky, J., and Kuta, J.: Principles of Polarography. Prague, Publishing House of the Czechoslovak Academy of Science, 1965.
7. Huch, R., and Huch, A., Eds.: Continuous Transcutaneous Blood Gas Monitoring. New York, Marcel Dekker, 1983.
8. Jagner, D., and Graneli, A.: Potentiometric stripping analysis. Anal. Chim. Acta, *83*:19–26, 1976.
9. Ross, J. W.: Calcium-selective electrode with liquid ion exchange. Science, *156*:1378, 1967.
10. Růžička, J., Hansen, E. H., and Tjell, C.: Selectrode™, the universal ion selective electrode. VI. The

* *Conductance* (symbol: G) is defined as the current (A) divided by the potential difference (V), when no EMF is present, and the unit therefore is ohm^{-1} = siemens ($\Omega^{-1} = S$). *Conductivity* (symbol: κ) is defined as the current density ($A \times m^{-2}$) divided by the electric field strength ($V \cdot m^{-1}$), when no EMF is present, and the unit therefore is: $\Omega^{-1} \cdot m^{-1} = S \cdot m^{-1}$.

calcium(II) selectrode employing a new ion exchanger in a non-porous membrane and a solid state reference system. Anal. Chim. Acta, *67*:155, 1972.
11. Severinghaus, J. W., Stafford, M., and Bradley, A. F.: tcP_{CO_2} electrode design, calibration and temperature gradient problems. Acta Anaesth. Scand., *68*(Suppl.):118–122, 1978.
12. Simon, W., Ammann, D., Oehme, M., et al.: Calcium-selective electrodes. Ann. N.Y. Acad. Sci., *307*:52, 1978.
13. Siggaard-Andersen, O.: The Acid-Base Status of the Blood. 4th ed. Baltimore, Williams & Wilkins, 1974.

SECTION FIVE

Osmometry

ESTHER FREIER, M.S., M.T. (ASCP)

Osmometry is a technique for measuring the concentration of solute particles that is directly related to the osmotic pressure of the solution. Osmotic pressure governs the movement of water (or solvent) across membranes separating two solutions. Membranes vary in pore size and thus in their ability to select molecules of different sizes and shapes. Those membranes that are permeable only to water are described as *strictly semipermeable*. The term *partially semipermeable* is applied to membranes that permit passage of small molecules or ions in addition to water.

Osmosis

Osmosis refers to the movement of water across membranes in response to differences in osmotic pressure between the two sides of the membrane. Water generally migrates from the side of the membrane with the more dilute solute to that containing the more concentrated solute.

Consider an aqueous solution of sucrose placed in a sac made of a strictly semipermeable membrane, and an open, vertical, glass-tube manometer attached to this sac. If this device is placed into a beaker of distilled water, water will move across the membrane into the sucrose solution in an attempt to equalize the chemical potential of the solvent on both sides of the membrane. The sucrose solution will rise some distance up the manometer tube. At equilibrium, the gravitational pressure of the column of solution in the manometer equals the osmotic pressure and thus will prevent further net movement of water. The osmotic pressure of this solution is defined as equal to the pressure that would have to be exerted on the sucrose side of the membrane to prevent the flow of water across the membrane.

If the sucrose solution within the membrane sac were replaced with a sodium chloride solution of the same molarity, the solution in the manometer would reach equilibrium at a point almost twice as high as that previously observed. Sodium chloride dissociates into two ions per molecule. The total number of individual solute particles present in the solution per given mass of solvent, regardless of their molecular nature (i.e., nonelectrolyte, ion, or colloid), determines the total osmotic pressure of the solution. In blood plasma, for example, osmotic pressure is the sum of effects of nonelectrolytes such as glucose and urea, of electrolytes present as free ions or in undissociated form, and of proteins.

Colligative Properties

The *colligative properties* of solutions—increase in osmotic pressure, increase in boiling point, lowering of vapor pressure, and decrease in freezing point—are directly related to the total number of solute particles per mass of solvent. The term *osmolality* expresses concentrations in terms of *mass* of solvent (1 osmolal solution contains 1 osmol/kg H_2O). In contrast, the term *osmolarity* expresses concentrations per volume of solution

(1 osmolar solution contains 1 osmol/L solution). Osmolality (osmol/kg H_2O) is a thermodynamically more precise expression because solution concentrations expressed on a weight basis are temperature independent, whereas those based on volume will vary with temperature in a manner dependent on the thermal expansion of the solution. Since the osmolality of physiological fluids is relatively low, the milliosmol (mOsm) is used as a more convenient unit.

In the ideal case (i.e., where there is no association between molecules of any species), 1 mole of a nondissociated solute dissolved in 1 kg of water results in a 1-molal solution that contains 6.023×10^{23} particles (Avogadro's number) per kg of water. This solution boils at a temperature 0.52 °C higher and freezes at a temperature 1.858 °C lower than pure water. The vapor pressure of this solution is 0.3 mm Hg lower than the vapor pressure of pure water, which is 23.8 mm Hg at 25 °C. The osmotic pressure of the same solution is increased from zero to 17 000 mm Hg or to 22.4 atmospheres.

A solution of an electrolyte dissociates into two (in the case of NaCl) or three (in the case of $CaCl_2$) particles, and therefore the colligative effects of such solutions are multiplied by the number (n) of dissociated ions formed per molecule. Because of incomplete electrolyte dissociation, as well as association between molecules, many solutions do not behave as expected in the ideal case, and a 1-molal solution may give an osmotic pressure lower than that theoretically expected. The osmotic activity coefficient is a factor used to correct for the deviation from the "ideal" behavior of the system:

$$\text{Osmolality} = \text{osmol/kg } H_2O = \phi(\text{n})(C)$$

where

ϕ = osmotic coefficient (experimentally derived from a solution of known concentration)
n = number of particles into which each molecule in the solution potentially dissociates
C = concentration in mol/kg H_2O

The total osmolality of a solution is equal to the sum of the osmolalities of all solute species present.

Because of their high molecular weights, serum proteins contribute less than 0.3% of the total osmolality of serum. The electrolytes Na^+, Cl^-, and HCO_3^-, which are present in relatively high concentrations, make the greatest contribution to serum osmolality, whereas nonelectrolytes such as glucose and urea, which are present normally at low concentrations, contribute only to a small extent. For example, the theoretical molality of osmolutes (osmotically active particles) in blood serum is about 307 mmol/kg water, divided among the various osmolutes approximately as follows: Na^+, 148; K^+, 4; Ca^{2+}, 2; Mg^{2+}, 1; Cl^-, 108; HCO_3^-, 27; HPO_4^{2-}, 2; SO_4^{2-}, 1; organic anions, 5; glucose, 5; and urea, 4 (units: mmol/kg H_2O). The serum proteins contribute only 1 mmol/kg H_2O. The total measured osmolality of this mixture is, however, only 290 mOsm/kg H_2O, because the osmotic coefficients are <1.

The Freezing Point Osmometer

Theoretically, any of the four colligative properties could be used as a basis for the measurement of osmolality. The freezing point depression is most commonly used because its measurement is simple and convenient. The instrument used is a freezing point depression osmometer, usually simply called an osmometer or, less commonly, a cryoscope. The components of the cryoscope (see Figure 2-34) are:

1. A thermostatically controlled cooling bath set at −7 °C.
2. A rapid-stir mechanism to initiate ("seed") freezing of the sample.
3. A thermistor probe connected to a Wheatstone bridge circuit to measure the temperature of the sample. (The thermistor is a metal stem whose resistance varies rapidly and predictably with temperature.)
4. A galvanometer indicating current flow.

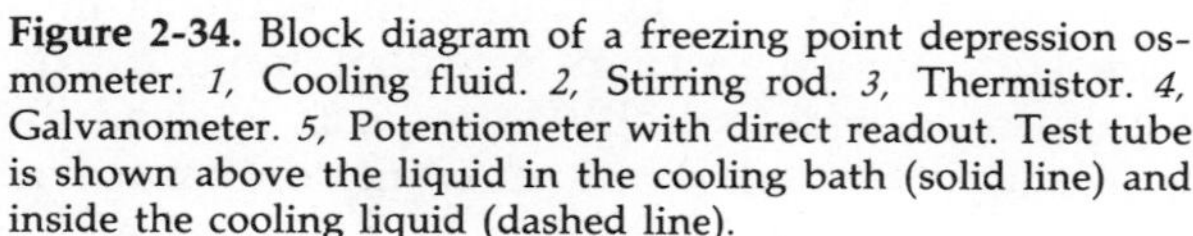

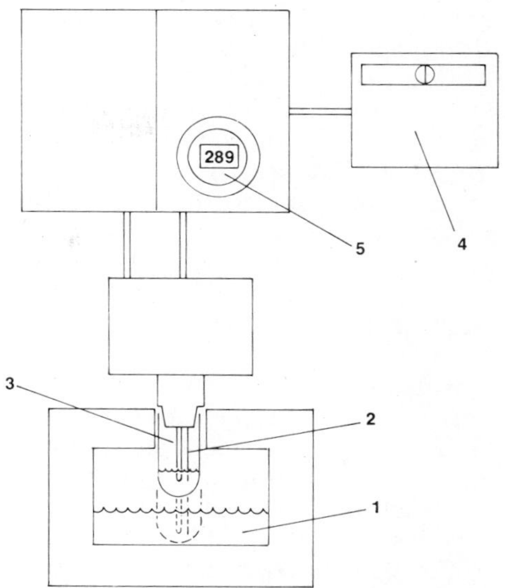

Figure 2-34. Block diagram of a freezing point depression osmometer. *1,* Cooling fluid. *2,* Stirring rod. *3,* Thermistor. *4,* Galvanometer. *5,* Potentiometer with direct readout. Test tube is shown above the liquid in the cooling bath (solid line) and inside the cooling liquid (dashed line).

5. A measuring potentiometer (variable resistor) used to null (balance) the current in the Wheatstone circuit. Automatic instruments exist in which components 4 and 5 are replaced by a light-emitting diode (LED) display, which both indicates the time course of the freezing curve and presents the final result in a numerical display.

Operation of a Freezing Point Osmometer[1]

The tube with sample (in which the thermistor probe and stirring wire have been centered) is lowered into the bath and with gentle stirring is supercooled to a temperature (−7 °C) several degrees below its freezing point. When movement of the galvanometer or LED display indicates that sufficient supercooling has occurred, the wire stirrer is changed from a gentle stir to a momentary (1 s) vigorous rate with increased amplitude, which initiates freezing of the supercooled solution. This freezing is only to the slush stage; about 2–3% of the solvent solidifies. The galvanometer reverses direction, as the released heat of fusion warms the solution, and then remains stationary, indicating the equilibrium temperature at which both freezing and thawing of the solution occur. During this equilibrium period of 2–3 min, the balancing potentiometer is adjusted, thus changing the variable resistance of the Wheatstone bridge to bring the galvanometer to the null position. At the end of the plateau caused by the temperature equilibrium, the galvanometer again indicates decreasing temperature as the sample freezes solid. The balancing potentiometer readings can be related to the degree of freezing point depression in °C, but commonly they are calibrated with suitable standards to read directly in mOsm/kg.

Available freezing point osmometers are designed for use with samples varying from 2.0 mL to 20 μL. After analysis, the thawed samples may be used for other tests. Day-to-day precision of ±2 mOsm/kg H_2O can be obtained in a service laboratory. The precision of the automated and manual instruments is essentially the same.

Reference ranges for osmolality in body fluids, as well as physiological and pathological variation of osmolality in body fluids, are discussed in Chapters 19 and 20.

The Vapor Pressure Osmometer

Another type of osmometer intended to measure total osmolality has been introduced for use in the clinical laboratory by Wescor Inc. (Logan, UT). Though the instrument is called a vapor pressure osmometer, the osmolality measurement is not related directly to the change in vapor pressure (in mm Hg) but to the decrease in the *dew point temperature* of the pure solvent (water), caused when the vapor pressure of the solvent is decreased by the solutes. In this instrument, the osmolality measurement is made by introducing a 7-μL sample into a vaporization-condensation chamber on a 5-mm disk of filter paper. The chamber is sealed and time is allowed for thermal and vapor equilibration at room temperature. The thermocouple is cooled to a temperature below the dew point so that water condenses from the air enclosed in the chamber and forms a thin, liquid film on the thermocouple junction. Cooling is then stopped, and the chamber temperature increases to the dew point temperature. Temperature remains constant since water is in equilibrium between the liquid and vapor phases. This temperature is measured by means of a thermocouple. The dew point temperature depression is proportional to the decreased water vapor pressure, which in turn is proportional to the solute concentration. The instrument is calibrated with suitable standards to read directly in milliosmols. Early models were sensitive to changes in ambient temperature, but a temperature-compensating circuit has reduced such influence to <0.1 mOsm/°C, provided the instrument is not subjected to rapid changes in ambient temperature. The instrument is compact and does not need a refrigerated bath or test tubes. When the chamber and thermocouple hygrometer are kept scrupulously clean, the instrument is capable of a within-run precision of ± 3 mOsm/kg water at serum levels of 290.

In clinical laboratories, the vapor pressure osmolality technique has been reported to be less precise than the freezing point depression method, both in the interlaboratory proficiency surveys conducted by the College of American Pathologists (CAP) and in an individual intralaboratory comparison.[2,3] More than 80% of the laboratories in the CAP surveys use freezing point depression osmometers. For serum samples, the coefficients of variation obtained for the vapor pressure osmometer are about twice those obtained for the cryoscopic osmometer. An important difference between the vapor pressure technique and the cryoscopic osmometer is the failure of the former to include in the measurement of the total osmolality the volatile solutes present in the serum. Substances such as ethanol, methanol, ethylene glycol, and even dissolved CO_2 are volatile and thus escape from the solution. As a result, the dew point of the remaining solution is lowered to a lesser degree than would have been the case without loss of solute. With either technique, serum specimens should be kept stoppered, because loss of CO_2 from bicarbonate may occur.

The Colloid Osmotic Pressure Osmometer

Colloid osmotic pressure (COP) is a measure of the contribution of the serum proteins to the serum osmolality. In the COP osmometer, also called a membrane osmometer (Wescor, Inc., Logan, UT, and Instrumentation Laboratories, Lexington, MA), the measuring cell consists of two fluid chambers separated by a semipermeable membrane. The sample that contains protein is placed on one side of the membrane, and a colloid-free physiological saline solution is placed on the reference side. This saline solution is in contact with a pressure transducer. Fluid moves by osmosis from the saline side across the membrane, causing a negative pressure to develop on the saline (reference) side equal to the COP of the sample. The membrane commonly used for serum is impermeable to particles with molecular weights $\geq 30\,000$.

The reference range for serum is 21 ± 2 mm Hg in supine individuals and 25 ± 2 mm Hg in ambulatory individuals. The measurement of colloid osmotic (also called oncotic) pressure is useful in guiding postoperative intravenous fluid therapy.

References

1. Johnson, R. B., Jr., and Hoch, H.: Osmolality of serum and urine. *In*: Standard Methods of Clinical Chemistry, Vol. 5, S. Meites, Ed. New York, Academic Press, 1965.
2. Juel, R.: Serum osmolality, A CAP survey analysis. Am. J. Clin. Pathol., *68* (Suppl.):165–169, 1977.
3. Mercier, D. E., Feld, R. D., and Witte, D. L.: Comparison of dewpoint and freezing point osmometry. Am. J. Med. Tech., *44*:1066–1069, 1978.

SECTION SIX

Chromatography

CARL A. BURTIS, Ph.D. • LARRY D. BOWERS, Ph.D.
SATI C. CHATTORAJ, Ph.D.
M. DAVID ULLMAN, Ph.D.

Chromatography is a versatile and powerful analytical tool for separation, detection, and quantitation of a wide variety of analytes. Among these analytes are gases and inorganic ions; amino acids, sugars, lipids, vitamins, drugs, and steroids; macromolecular species such as proteins, polysaccharides, nucleic acids, and soluble viruses; and particulate materials such as subcellular components and bacteria.

Chromatographic separation of a mixture of analytes is based on the differences in mass transport rates that are displayed by the individual substances when they are dissolved in a liquid or gas (mobile phase) that is flowing through a selected medium (stationary phase). Separation is contingent upon different degrees of physical interaction of the individual substances with the stationary phase of the system. A substance more attracted to the stationary phase than the mobile phase progresses more slowly through the system than does a substance more attracted to the mobile phase. Characteristics of mobile and stationary phases, characteristics of the analytes in the mobile phase, characteristics of the porous medium, and flow dynamics therefore determine the rates at which particular substances introduced into a chromatographic system will pass through it and will be separated from one another during passage. Appropriate choice of sample preparation, conditions for separation, and mode of detection of separated compounds will result in qualitative and quantitative chromatographic techniques that are especially useful in the clinical laboratory.

Separation Mechanisms

The physicochemical processes that control the interactions of solutes with mobile and stationary phases are surface adsorption, liquid partition, ion exchange, steric exclusion, and molecular affinity.

Surface Adsorption Chromatography

Surface adsorption chromatography, also known as liquid-solid chromatography (LSC), is the oldest form of the chromatographic modes. The Russian botanist Twsett used this technique in 1906 to separate plant pigments into a number of colored bands by using a column filled with calcium carbonate, and he gave it the name *chromatography*.

The electrostatic, hydrogen-bonding, and dispersive interactions between a molecule and a solid support or adsorbent are the basis for adsorption chromatography. In gas chromatography the adsorption mode is restricted to the analysis of low-molecular-weight species (e.g., methyl, ethyl, and isopropyl alcohols) and of compounds that are normally gases at room temperature (e.g., CO_2, N_2, and O_2), by using stationary phases such as "molecular sieve," alumina, and Porapak polymers. In liquid chromatography (LC), three types of adsorbents are generally used—nonpolar, acidic polar, and basic polar. The nonpolar adsorbents include charcoal and polystyrene-divinylbenzene. The main acidic polar adsorbent is silica gel. Alumina is the main basic adsorbent; it preferentially retains acidic substances.

Competition between the mobile phase and the solutes for the adsorption sites on the support is characteristic of LSC. As a result, a scale of solvent strength, or eluotropic series, can be developed based on the strength of solvent adsorption. For a polar adsorbent such as silica gel, one would expect either methanol or water to be strongly adsorbed and therefore to serve as a strong solvent. Pentane would be poorly adsorbed and would thus act as a weak solvent. In contrast, for a nonpolar adsorbent such as charcoal, water would be a weak solvent and hexane would be a strong solvent.

Adjustment of the elution properties of the analytes is accomplished by judicious choice, usually with the assistance of solvent strength tables, of a mixture of a weak and a strong solvent. The main drawback to adsorption chromatography is the difficulty in preparing a support with a homogeneous distribution of adsorption sites. Technical problems associated with the reproducibility of retention have reduced the popularity of this mode of chromatography. Nevertheless, it has been used, for example, to separate steroids or vitamins prior to radioimmunoassay.

Partition Chromatography

Separation by this mode is based on differences in the solubility of individual solute molecules in two immiscible liquids that are in contact with one another. Therefore, this technique is also called liquid-liquid chromatography (LLC). The degree to which the individual solutes distribute between the two solvents is expressed by the partition (distribution) coefficient (K_d). This coefficient is defined as the ratio of the solubility of the solute in the extracting solvent (stationary phase) to its solubility in the original solvent (mobile phase). In general, polar solutes extract better into polar solvents, and nonpolar solutes move preferentially into nonpolar solvents.

For *normal-phase partition chromatography,* a polar solvent is used as the stationary phase and a nonpolar solvent mixture is used as the mobile phase. A partition system in which the stationary phase is less polar than the mobile phase is called *reversed-phase partition chromatography.* With the advent of chromatographic supports that have a nonpolar stationary phase chemically bonded to an inert support, as much as 80% of all LC separations are carried out by reversed-phase partition. However, the mechanisms of a separation are often not clear cut; both adsorptive and partitioning forces may be involved.

Ion-Exchange Chromatography

In ion-exchange chromatography, solutes in a sample mixture are separated by virtue of their differences in sign and magnitude of ionic charge. This technique is especially applicable to compounds such as inorganic ions, amino acids, proteins, and nucleotides.

A cation-exchange resin has many negatively charged functional groups (Figure 2-35) that are covalently bound to it. Associated with these groups are loosely bound cations, such as Na^+, that are available for exchange with cations of the solute. The functional moieties can be either strongly acidic groups, such as sulfonate ions, or weakly acidic groups, such as carboxylate ions or carboxymethyl, phosphate, sulfomethyl, sulfoethyl, or sulfopropyl groups.

Anion-exchange resins are characterized by the presence of strongly basic quaternary amines such as triethylaminoethyl groups or weakly basic groups such as aminoethyl, diethylaminoethyl, or guanidoethyl groups that can bear a positive charge. Loosely bound anions such as OH^- exchange with analyte anions.

Ionic compounds can be selectively eluted from ion-exchange resins by varying the pH and ionic strength of the mobile phase. The role of ionic strength can be viewed as a competition between the salt ion and the analyte ion for the exchange sites. For example, a positively charged amino acid can compete better with one sodium atom for a cation binding site than it can with ten (see Figure 2-35). However, some ion exchange selectivity exists for inorganic ions even when the net charge on them is the same. For example, the binding of the monovalent cations to a cation exchange resin increases in the order:

$$Li^+ < H^+ < Na^+ < NH_4^+ < K^+ < Rb^+ < Cs^+ < Ag^+$$

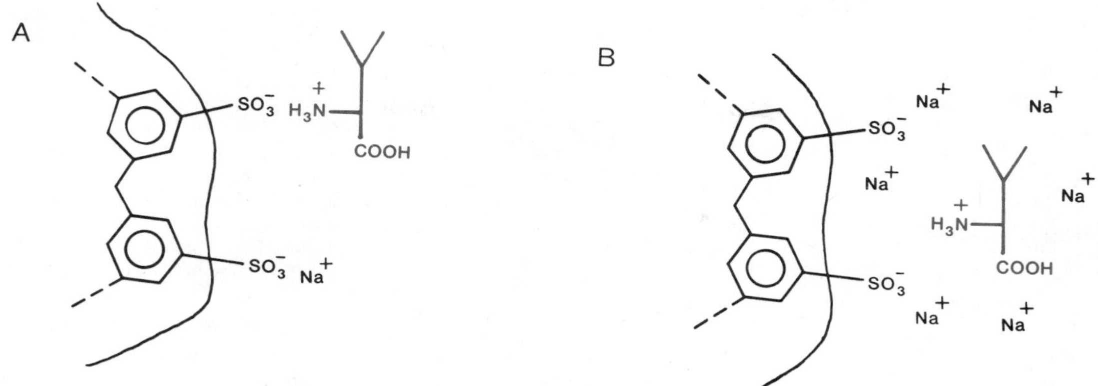

Figure 2-35. Retention (*A*) and elution (*B*) of an amino acid with a cation-exchange resin as a function of sodium ion concentration. A higher ionic strength results in decreased retention. Variation of pH can also be used to adjust retention by changing the charge on the analyte molecule.

For the anion exchange resins of the strong type, the selectivity would be:

$$F^- < OH^- < CH_3COO^- < HCOO^- < H_2PO_4^- < HCO_3^- < Cl^- < NO_3^- < HSO_4^- < I^-$$

This is mainly due to the size of the solvated ion and its ability to penetrate the resin. Thus, a separation with insufficient resolution using hydrogen ions in the eluent may benefit from the use of a lithium eluent, because the lithium displaces the analytes less vigorously and may allow a difference in their binding to appear.

Ion-exchange resins have been used for a spectrum of applications in the clinical laboratory—from mixed-bed resins for removal of ions from water (deionization) to chromatographic separation of hemoglobin variants, isoenzymes of creatine kinase and lactate dehydrogenase, and amino acids. Ion exchangers may also be used to remove an interfering ion. For example, urinary porphobilinogen can be retained on an anion exchange resin while other compounds that might interfere with the color reaction are eluted. Porphobilinogen can then be eluted with acid and reacted with Ehrlich's reagent to obtain a selective quantitative test.

Steric Exclusion Chromatography

Steric exclusion chromatography (formerly known as gel-filtration, gel-permeation, size exclusion, molecular exclusion, or molecular sieve chromatography) separates solutes on the basis of the differences in their molecular size; however, molecular shape and hydration are also factors in the process. The types of stationary phase used in this mode of chromatography are gel-forming hydrophilic beads, such as cross-linked dextran (Sephadex), polyacrylamide (Bio-Gel), and agarose (Sepharose). The beads are porous and contain openings and microchannels. The size of these channels and pores can be designed to pass or retard molecules of different sizes by varying the degree of cross-linking in the course of manufacture. Large solute molecules that cannot enter the beads flow along the spaces between them, thus eluting first (Figure 2-36). The solvent and all solute molecules with diameters less than that of the pores will enter the gel beads, be entrapped within the channels, and therefore will elute later.

Steric exclusion chromatography can be used for the separation of either hydrophilic or hydrophobic materials, depending on the type of gel used. Hydrophilic gels are particularly suitable for the separation of analytes in aqueous solutions (e.g., enzymes, antibodies, proteins, hemoglobins, polysaccharides, nucleic acids). Hydrophobic gels, such as methylated Sephadex polystyrene beads (Styragel) and cross-linked polymethacrylate, have been developed for the separation of solutes that are soluble only in organic solvents. These hydrophobic gels have been useful in the separation of triglycerides and fatty acids.

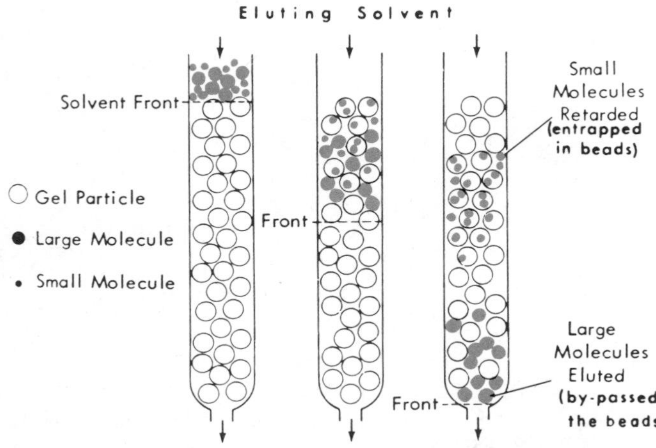

Figure 2-36. Representation of gel-filtration chromatography by column technique. (Modified from Bennett, T. P.: Graphic Biochemistry, Vol. 1: Chemistry of Biological Molecules. New York, The Macmillan Company, 1968.)

This chromatographic mode is useful for obtaining data on molecular weights and is also a convenient technique for separating molecules that differ significantly in molecular size (such as inorganic ions) and biopolymers (such as proteins). An important application is the desalting of protein-containing solutions, which is fast by this technique and is therefore preferred over dialysis, especially when labile substances are studied.

Affinity Chromatography

The term *affinity chromatography* has been used to cover a myriad of separation mechanisms involving some aspect of an interaction that occurs between biochemical species. These interactions can range from highly specific enzyme-substrate, hormone-receptor, or antigen-antibody complexes, to less specific charge-transfer complexes.

The stationary phase in affinity chromatography is prepared by immobilizing a molecule (called a ligand because it participates in binding) on a support, either directly or through a spacer (see Figure 2-37). Because of the special stereochemical aspects of the interaction, the length and charge of the spacer may be critical to the ability of the stationary phase to bind the analyte specifically and to release it. The support must be as inert as possible to prevent nonspecific adsorption. Adjustments of pH and ionic strength are required to achieve a strong binding of the analyte to the ligand. If the interaction between analyte and ligand is specific, the analyte may be displaced in a single step by the addition of a substrate or an inhibitor or, alternatively, by a change in pH or ionic strength or by the addition of a hydrogen bond–breaking agent such as guanine hydrochloride, urea, or sulfite. The latter compounds are also known as chaotropic agents.

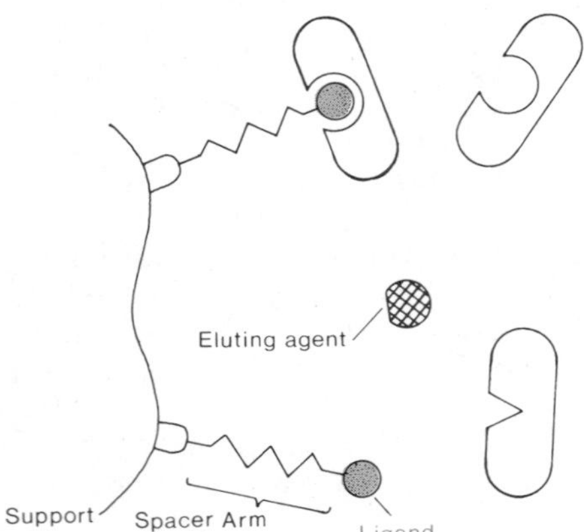

Figure 2-37. Principle of affinity chromatography separation. The analyte (e.g., enzyme, antibody, antigen, tissue receptor) binds to the support-bound ligand. It can then be eluted with a general eluent such as a chaotropic agent, with a pH change, or with a biospecific eluent such as an inhibitor or substrate.

The power of affinity chromatography lies in its selectivity. In the clinical laboratory, the use of affinity chromatography is limited at present to preparative separations of proteins and antibodies. Lectin columns have also been used to separate cells with different surface carbohydrate moieties. More recently, heparin has been used as a ligand to specifically retain low-density and very-low-density lipoproteins. The use of phenyl boronate affinity supports to separate glycosylated hemoglobin from other hemoglobins may be the first widespread analytical use of affinity chromatography.

Types of Chromatography

Chromatographic separations can also be classified according to the configuration of the chromatographic bed and the physical state of the stationary and mobile phases.

Paper Chromatography

The adsorbents used in this technique are special grades of filter paper selected for homogeneity, wet strength, thickness, level of impurities, and mobile phase migration rate. This technique is still used when inexpensive and rapid separations are needed. However, it is not widely used in the clinical laboratory and will not be discussed further.

Thin-Layer Chromatography (TLC)

Thin-layer and paper chromatography are closely related techniques using similar procedures for application, development, detection, and analysis. In TLC, a thin layer of a sorbent such as silica gel (usually only 0.2 mm in thickness) is spread uniformly on a glass plate or a plastic sheet with applicators designed for this purpose. Prepared plates coated with a variety of sorbents (e.g., silica gel, microcellulose, alumina, and Sephadex) are commercially available. The sample to be analyzed is added as a concise spot or streak near an edge of the plate, which is then placed in a closed glass container with the lower edge of the plate in contact with the mobile phase (Figure 2-38). As is the case with other types of chromatography, solvents and sorbents can be chosen from one of the four types of separatory processes. After the solvent front has reached a desired height, the plate is removed from the tank and dried. The separated sample components are located and identified by appropriate procedures such as ultraviolet illumination, spraying with specific color reagents, or autoradiography. If it is necessary to recover the separated substances for quantitation or for further experiments, the appropriate sorbent region is scraped from the plate and extracted to remove the solute. To aid in the identification of the compounds that give rise to the chromatographic spots, standards are usually chromatographed with the unknown sample. Compounds in the spots are usually identified by their R_f values (R_f = distance of spot migration ÷ distance of mobile phase migration). If the migrations of the unknown compound and standard are

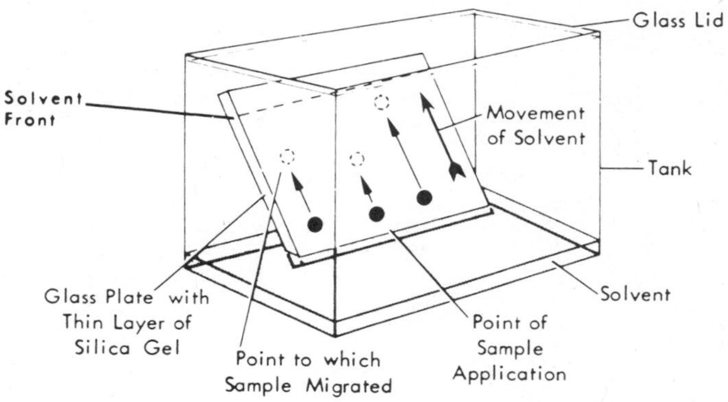

Figure 2-38. Illustration of thin-layer chromatography (TLC). The solvent is drawn up the thin layer of adsorbent by capillary action. Ascending development is commonly used in TLC. (Modified from Bennett, T. P.: Graphic Biochemistry, Vol. 1: Chemistry of Biological Molecules. New York, The Macmillan Company, 1968.)

not identical, the compounds are not the same. The reverse is not true, however, since more than one compound can have the same R_f value in a particular chromatographic system. Absolute quantitation by TLC is generally difficult, because the size and density of the spots depend on a number of factors, including the application technique. Nevertheless, densitometry has been successfully used to quantitate drugs, lipids from serum and amniotic fluid, and a number of other compounds of clinical interest.

Gas Chromatography (GC)

Principle. Gas chromatography is a process by which a mixture of compounds in volatilized form is separated into its constituent components by moving a mobile (gas) phase over a stationary phase. Under controlled conditions, the individual components of the sample, in accordance with their vapor pressure, will be present partially in the stationary and partially in the mobile phases. A compound with a high vapor pressure (i.e., which vaporizes readily) will be eluted more rapidly than another compound with a lower vapor pressure. If one compound has a selective interaction with the stationary phase and another has no or a lesser degree of interaction, the difference in the rate of elution and the efficiency of separation is increased.

Gas chromatography is divided into two major categories: (1) gas-solid chromatography (GSC), in which the sorbent is a solid of large surface area, and (2) gas-liquid chromatography (GLC), in which a nonvolatile liquid, the stationary phase, is coated on an inert solid support. GLC is most commonly used in medical laboratories. The mobile phase in both cases is an inert gas such as nitrogen, helium, or argon. The gas carries the solute molecules through the gas chromatographic column, hence the name *carrier gas*. The effluent from the column carries the separated sample constituents to the detector in the order of their elution. The response signal of the detector, after amplification in an electrometer, is fed into a recorder for display. The graphic representation is usually a straight baseline upon which are superimposed a number of peaks, one for each of the components of the sample. Retention time, the elapsed time between injection of sample and the appearance of a given peak maximum at the detector, is a distinctive property of a particular component in a particular separation system and is therefore a means of qualitative identification of individual components of the mixture. Peak size (area or height) is proportional to the amount of component detected and is a means for quantitation of the separated components.

Instrumentation

A gas chromatograph consists of six basic parts: (1) a carrier gas supply with flow control; (2) an injector; (3) a chromatographic column located in an oven with temperature control; (4) a detector; (5) an electrometer (amplifier); and (6) a data recorder (Figure 2-39).

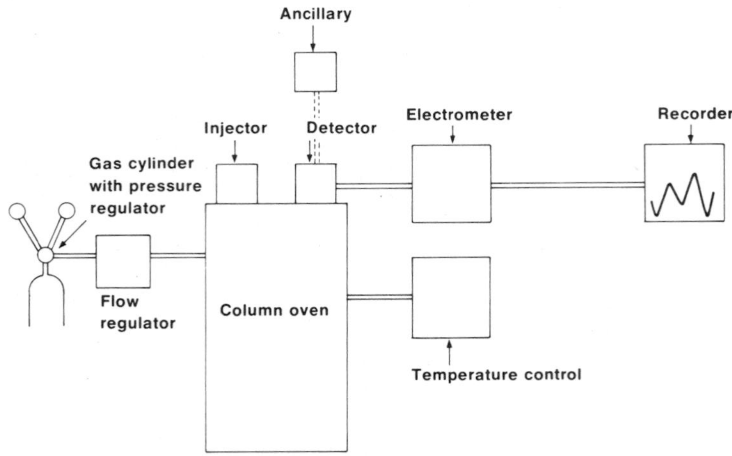

Figure 2-39. Gas chromatographic system.

Ancillary attachments include devices for collecting the eluate and instruments such as mass spectrometers and spectrophotometers for identification of the eluted components of the sample mixture. In recent years, microprocessor-controlled electronics have made automation possible for the entire chromatographic process from injection and separation through peak measurement, calculation, and final report, and they have thereby improved the overall accuracy, precision, and utility of gas chromatographic analysis. The computer system offers, in addition to data handling, the capability of recognizing and quantitating unresolved peaks and correcting for baseline drift and nonzero baseline. Automated gas chromatographs are now available from many commercial sources.

Carrier gas supply with flow control. The type of carrier gas is determined by the type of detector utilized in the system. The most frequently used carrier gas is nitrogen, which can be used with flame ionization, electron capture, or thermal conductivity detectors. Helium may be used with flame ionization and thermal conductivity detectors, and nitrogen admixed with argon-methane is used with the electron capture detector.

Precise control of the flow rate of carrier gas through the column is important, since the retention time of individual sample components is in part determined by the gas flow rate. Thus, any variations in flow rate will lead to a change in retention time and uncertainty as to the identity of the eluting compounds.

Injector. The sample to be introduced may be a solid, liquid, or gas. Gases are usually introduced by gas-tight syringes or by gas-sampling valves. Liquids are injected into the carrier gas stream with a microliter syringe through a self-sealing septum in the apparatus. Solids may be dissolved in a suitable solvent and injected in liquid form. Injection of liquid sample is the most common introduction mode for clinical use of GC; devices for the direct injection of solids have also become commercially available.

The sample introduction system is kept at a temperature that will cause quick vaporization of the sample components so that the sample mixture moves onto the column in the form of a narrow band. Excessive dead volume in the injection port results in diffusion of the sample and therefore in increased peak width and tailing. Efficiency of separation is also affected by the sample volume, which should be kept as small as possible (2–10 μL).

Formation of derivatives. The high boiling point of some compounds requires that they be injected in the form of a more volatile derivative so that the desired instantaneous vaporization can be accomplished. Additional reasons for forming derivatives may include (1) decrease of polarity, which minimizes "tailing" and adsorption of the solute compound on the column; (2) stabilization of thermally reactive structures of the compound; and (3) alteration of retention behavior, which improves separation of compounds in specific mixtures.

Column oven and its temperature control. The column oven of the gas chromatograph is designed to maintain the constancy and uniformity of the column temperature, which are necessary for reproducibility of retention times and for maintaining a constant bleed rate of the column (evaporation of components of the liquid stationary phase itself). Any "hot spots" along the column may cause localized deterioration of the column packing and undesirable adsorption of compounds. Any temperature drop either in the column itself or in the connection between column and detector may lead to partial condensation and, consequently, to reduced efficiency of separation and to decreased sensitivity.

Isothermal (constant temperature) operation limits gas chromatographic analysis to mixtures of a few components, all of whose boiling points are within a relatively narrow range. In an isothermal run early peaks, representing low-boiling substances, may emerge as sharp overlapping peaks, whereas higher-boiling components emerge as flat peaks that are difficult to measure. The use of temperature programming (controlled gradual or stepwise change of column temperature) allows the analysis of complex mixtures containing analytes with a wide range of boiling points. At the start, a lower temperature is used and substances with low boiling peaks are well separated. With the rising temperature, each higher boiling component is eluted as a sharp peak, and total analysis time is shortened.

Chromatographic column. Separation of the sample mixture takes place in the gas chromatographic column. *Packed analytical columns* vary from 0.1 to 0.4 cm in inner diameter (ID), and 1 to 4 meters in length. Longer columns provide more efficiency and better resolution, whereas an increase in column diameter increases its capacity for sample. On the other hand, very long columns require very high carrier-gas pressures at the inlet; these pressures pose problems in injection technique and in avoiding gas leaks. Very wide columns are less efficient in separating compounds.

The materials used for the construction of columns are usually glass or stainless steel. The choice of material depends upon the substance analyzed. Cholesterol, and particularly samples containing halogens, may react with hot metal surfaces; thus, the use of a glass column in such cases is essential. *Stainless steel* columns are inexpensive and durable and have a nonadsorptive surface with excellent heat-transfer properties. *Glass columns*, on the other hand, allow visual inspection of the support during the packing process and detection of discontinuity of packing material or deposition of nonvolatile carbonaceous residues in the top of the column. However, the fragility of the column and the need to deactivate the active sites inside the tubing (silanization) are disadvantages of glass columns.

Capillary columns are long tubes of small diameter. They range from 0.02 to 0.05 cm (ID) and from 10 to 150 meters in length. The inside wall of the tubing is coated with a thin film of liquid phase. Capillary columns have very high efficiencies but low sample capacity. They are most suitable for the separation of complex mixtures and the resolution of closely related isomers such as saturated hydrocarbons and aromatic compounds.

Column support material. Columns are packed with particles of a solid support material onto which a liquid stationary phase has been coated or chemically bonded. The support material should be inert and have a high crushing strength, large surface area, and uniform size. Its sole purpose is to hold the thin layer of the stationary phase. If the support is not inert, interactions of the vapor solute with the support will occur, and asymmetrical peaks (tailing) or partial loss of the injected material due to irreversible adsorption or decomposition will result. Silica is the most widely used support material.

The efficiency of the column is in part determined by the size and distribution of the supporting particles. By decreasing the particle size of a support of uniform distribution, the multiple-path effect (eddy diffusion) on the transport of the solute is minimized, and thus the separatory power of the column is enhanced. Solid supports in the range of 80–100 mesh and 100–120 mesh are generally used for routine analysis. Supports are generally sized by screening through standard screens. Mesh numbers refer to the number of openings per linear inch. Particles that will pass through 60-mesh but not 80-mesh are referred to as 60/80-mesh.

Stationary phase (liquid phase). The proper choice of a liquid phase is of great importance for the separation of the sample compounds. The nature of the stationary phase to be used is primarily determined by the physical and chemical properties of the compounds to be analyzed. The amount of stationary liquid phase used in relation to the amount of inert support (loading) may vary from <1 to 50% by weight. In general, diffusion phenomena are reduced in columns with relatively low amounts of a liquid phase. The volatility of the sample needs to be considered when choosing the amount of liquid phase. As a rule, compounds of low volatility, such as steroids, are best analyzed on relatively low-loaded columns (3% or less), whereas very volatile materials such as light hydrocarbons, because of their poor solubility, require high concentrations (20–30%) of liquid phase. The retention time is proportional to the amount of liquid phase present; the higher the liquid load, the longer the retention time for a compound.

Although a wide variety of compounds have been used as liquid phases, a few carefully selected compounds probably suffice to fill the needs of analytical procedures carried out in a clinical laboratory. The ideal liquid phase, under experimental conditions, should be nonvolatile, thermally stable, and chemically inert toward the solutes under study. Methyl silicone polymers, substituted silicone polymers, and silicone polyesters fulfill these criteria fairly well. These compounds are either coated or chemically bonded onto the surface of the column support material. Although more expensive, the bonded materials are preferred because of their durability and stability.

Stationary phases can be separated into nonselective and selective phases. The former types are suggested for the separation of compounds with differing molecular weights and differing boiling points and for the separation of samples of a relatively nonpolar nature. The selective type phases are more frequently employed for the separation of isomers and compounds of varying polarities. In general, the selective phases are less thermostable than the nonselective phases.

Column conditioning. Freshly prepared gas chromatographic columns and, especially, their stationary liquid phases may contain impurities such as low-molecular-weight compounds (shorter polymers) of the liquid phase. Such impurities are removed by heating the column, usually to 30–40 °C above the intended operating temperature for the column. This procedure is called column conditioning. In order to avoid contamination of the detector with impurities, detector lines should be disconnected during the process. The conditioning period varies with different liquid phases but is ordinarily 12–24 h. During this time the carrier gas is permitted to sweep through the column to carry off the impurities. Satisfactory conditioning of the column has been accomplished if a gas chromatogram shows a steady baseline without any peaks or spikes.

Detectors. The column effluent, containing the individual separated compounds, enters the detector through the detector lines. As the compounds pass through the detector, an electrical signal is generated proportional to the amount of substance present in the carrier gas stream. Many types of detectors have been invented. Among these, the detectors for thermal conductivity, electron capture, flame ionization, and alkali flame ionization (recently introduced) are most noteworthy.

Thermal conductivity (TC) detector. As the name implies, the operational principle of this detector is based on the changes in thermal conductivity (ability to carry heat) of the carrier gas caused by the admixture of sample components. Changes in conductivity are proportional to the amount of substance eluted from the column. The detector consists of a metal block with two separate channels through which the column effluent and the pure gas, respectively, flow (Figure 2-40). Each channel has a wire of tungsten or tungsten-rhenium alloy; these wires are connected to opposite arms of a Wheatstone bridge circuit that is fed from a constant current supply. Both filaments are heated electrically to raise their temperature above that of the detector block.

Since most substances analyzed have low thermal conductivities, sensitivity of the detector is increased by using a gas with high thermal conductivity. Helium is most frequently used. As the column effluent containing the mixture of carrier gas and separated sample components passes through the sample side of the detector, the temperature and therefore the electrical resistance of the sample filament rise, while resistance of the reference filament remains unchanged. Difference in resistance causes a voltage change that is amplified and transmitted to the recorder for graphic display.

The TC detectors are nondestructive, and thus sample components may be collected for further studies. Other advantages of TC detectors are their ruggedness and their ability to respond to a wide variety of samples, including respiratory gases such as O_2, N_2, CO_2, CO, and others that cannot be detected with the flame ionization detector. Limitations of TC detectors are their relatively low sensitivity and their tendency to give baseline drifts.

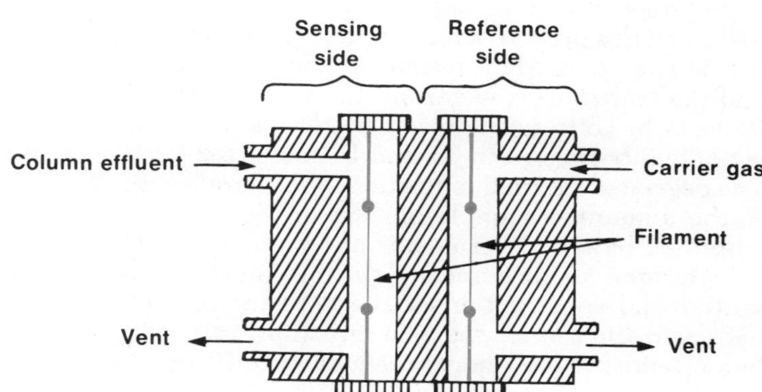

Figure 2-40. Schematic diagram of a thermal conductivity detector.

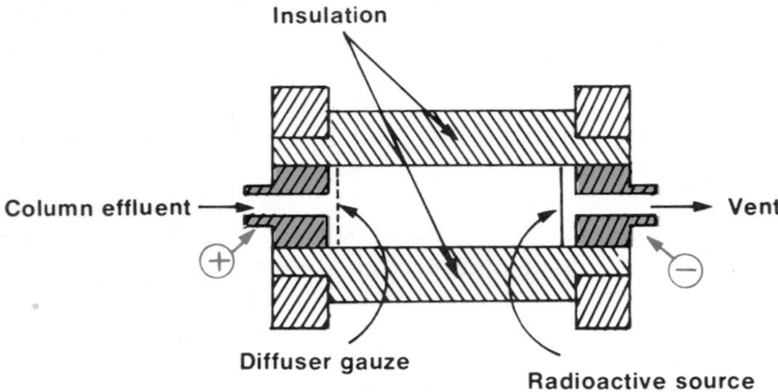

Figure 2-41. Schematic diagram of an electron capture detector.

Electron capture (EC) detector. These detectors are the most sensitive but they are also the most selective, since they detect only compounds with affinity for electrons (i.e., those compounds that capture electrons). Thus, the use of EC detectors is essentially limited to halogenated compounds. If the compound itself does not contain halogens, appropriate derivatives of the sample compounds can be formed.

The EC detectors contain two electrodes separated by an insulator (Figure 2-41). A radioactive source such as tritium (^3H) or ^{63}Ni is attached to the cathode. The column effluent containing carrier gas and separated sample components enters the detector through a hole in the cathode. As the carrier gas (e.g., nitrogen admixed with argon-methane) goes through the detector, it is ionized by the radioactive source. A fixed potential, or a short pulse of potential, just sufficient to collect all ions and electrons is applied to the electrode, yielding a very small but constant current (standing current) across the detector. If compounds having an electron-capturing ability enter the detector, some electrons are captured, and as a result the standing current is diminished. The changes in current can be amplified in the electrometer, and polarity can be reversed so that a positive peak is depicted on the recorder chart.

EC detectors have the great advantage of sensitivity and selectivity, but unfortunately they are also extremely sensitive to impurities in the sample and to changes in the operating conditions. Thus, their use should be considered only when the flame ionization detector proves inadequate.

Flame ionization (FI) detector. This is the most widely used detector because of its simple construction, reliable performance, wide range of linear response, high sensitivity, and ease of operation. FI detectors are more sensitive than TC detectors, are not subject to corrosion, and can be operated at high temperatures. FI detectors can essentially sense all organic substances and are therefore highly suitable for analytical work in a clinical laboratory. It should be noted, however, that not all compounds give the same response in an FI detector. In addition, some compounds, such as inorganic gases and water, cannot be detected with an FI detector. Also, the sample is destroyed in the FI detector, and therefore sample collection is possible only if the column effluent is split prior to entering the detector.

In most FI detectors (Figure 2-42), the effluent from the column is mixed with hydrogen gas prior to its admission to the detector. The mixture then enters a jet, where it is burned to produce a small flame. As a sample burns in the flame, ionization occurs and the transmitted electrons are collected either by a collector loop placed around the flame or by collector plates on both sides of the flame. A potential difference (e.g., 300 V) is applied between the jet and the collector loop, each of which serves as an electrode. The degree of thermal ionization, and therefore the current flow, is directly proportional to the amount of sample burned in the flame. The current thus produced is linearly amplified by the electrometer and transmitted to a strip chart recorder for display.

Optimal performance of the FI detector is dependent not only on the applied potential and construction of the detector but also on the ratio between flow rates of hydrogen and carrier gas. The ratio (flow rate of H_2 to flow rate of carrier gas) determines flame temperature and, consequently, efficiency of ionization.

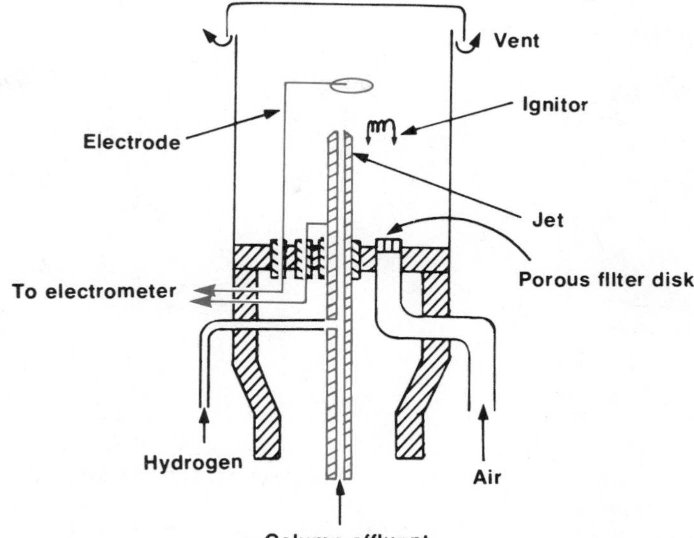

Figure 2-42. Schematic diagram of a flame ionization detector.

Alkali flame ionization detector (nitrogen-phosphorus detector—thermionic specific detector). The alkali flame ionization detectors (AFID) are very similar in operation and construction to the FID, except that the burner is fitted with a cap of compressed alkali metal salt, such as rubidium sulfate. In the presence of alkali atoms in the flame, nitrogen-containing compounds give a 15 times greater, and phosphorus-containing compounds a 300 times greater, response than the corresponding FID response, hence the name nitrogen-phosphorus detector. The principle of the response of the AFID has not yet been fully explained. It is generally considered that the presence of halogens, phosphorus, nitrogen, sulfur, arsenic, tin, silicon, or lead increases the volatility of the alkali metal salt and thus also the concentration of the alkali metal in the flame. This increased concentration results in an increase in ionization current and consequent enhanced response for the compounds containing the above elements.

High Performance Liquid Chromatography (HPLC)

Liquid chromatography (LC) is a process by which a mixture of compounds is separated into its constituent compounds by flowing a liquid (i.e., mobile phase) over a stationary phase. HPLC is a type of LC that separates components of a mixture more efficiently.

Instrumentation

Usually, there are six components in the HPLC instrument: (1) a solvent reservoir, (2) a pump, (3) an injector, (4) a chromatographic column, (5) a detector, and (6) a data recorder (Figure 2-43). Since the availability of microprocessors, these devices are now generally included as a seventh component of HPLC instruments.

Solvent reservoir. The solvent reservoir is usually a brown glass bottle or flask that contains the mobile phase.

Pump. The pump drives the mobile phase from the reservoir and through the injector, the column, and one or more detectors. In the past, several different types of pumps have been used. The type most widely used today is the reciprocating pump, in which an asymmetric cam drives pistons into and out of two pumping chambers (Figure 2-44). In the fill stroke, the piston is withdrawn from the chamber and a volume of liquid proportional to the diameter and stroke of the piston is drawn from the solvent reservoir, through the open inlet check valve, into the pumping chamber. In the pumping stroke, the inlet check valve is closed, the outlet valve is opened, the piston is driven into the chamber, and liquid is pumped into the column.

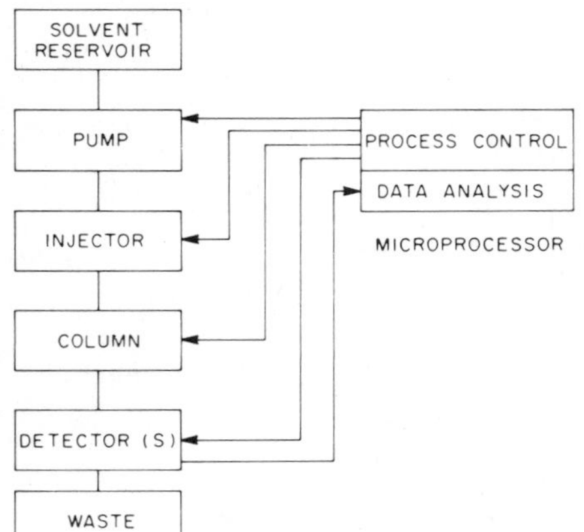

Figure 2-43. Functional schematic of a high performance liquid chromatograph.

Although the reciprocating pump has definite advantages over other types of pumps, a major disadvantage is that, owing to its reciprocating action, the mobile phase pulsates and causes variations in the flow rate. This phenomenon is known as "pump noise" and is manifested as fluctuations in the output of the detectors. Thus, most reciprocating pumps use either a pulse damper or a multihead reciprocating pump. In the multihead type of pump, two or more reciprocating pistons (Figure 2-44) operate out of phase to minimize the flow pulsations. Reciprocating pumps operate at up to 6000 psi and generate flow rates between 1 and 10 mL/min.

The pump in a liquid chromatograph operates in two different modes: isocratic or gradient. In the *isocratic mode* the mobile phase composition remains constant throughout the chromatographic run. An isocratic mobile phase can be a single solvent, such as methanol, or it can be premixed from several different solvents (e.g., methanol-water) and delivered from a single solvent reservoir.

In the *gradient mode*, the mobile phase composition is changed, either in a stepwise or in a continuous fashion, throughout the run. In one technique, two pumps are used in parallel. A wide variety of gradient profiles can then be generated by programming the output of each one. Alternatively, the mobile phase can be proportioned on the inlet side of a single pump. For example, three different solvent reservoirs can be connected via proportioning valves to the inlet check valve of a single pump. The composition of the mobile phase can then be varied by programming the time during which solvent is delivered through each of the proportioning valves.

Injector. An aliquot of sample is introduced into a liquid chromatograph via some type of sample injector. The most widely used type is the loop injector. As shown in Figure 2-45, in the fill position, an aliquot of sample is introduced with a syringe into an external loop of the injector at atmospheric pressure. In the inject mode, the valve is activated, thereby rotating the sample loop into the flowing stream of mobile phase, which results in the aliquot of sample being flushed into the chromatographic column.

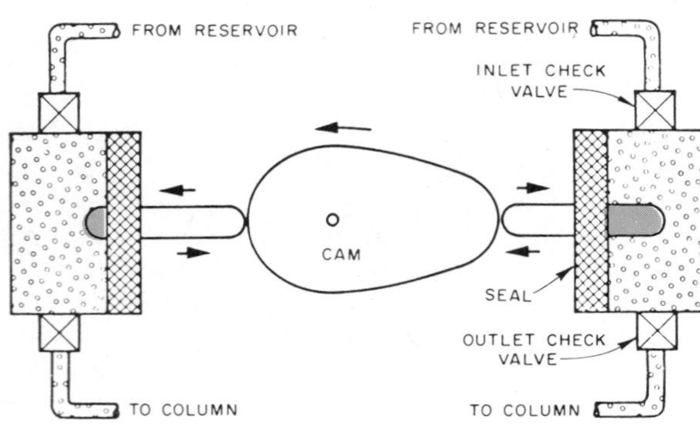

Figure 2-44. Cross-sectional view of a dual-piston reciprocating pump. (From Analysis Instrumentation, Vol. 12. Reprinted by permission. Copyright © Instrument Society of America 1974.)

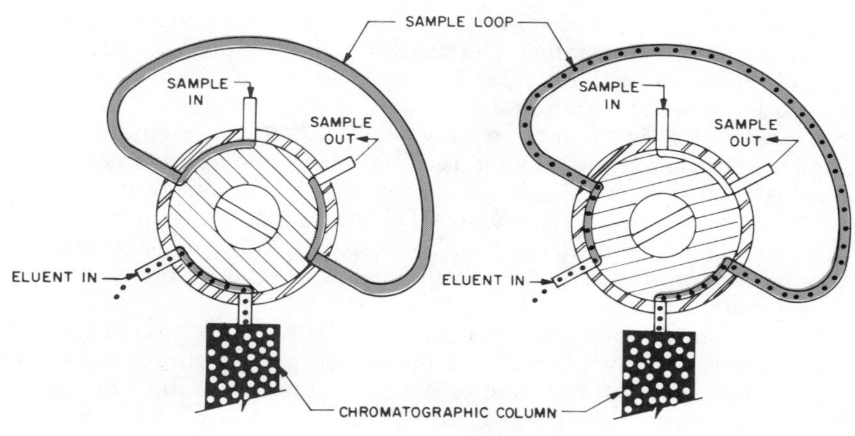

Figure 2-45. Cross-sectional view of a sample loop injector that is commonly used in a liquid chromatograph.

Although they are more expensive, loop injectors are precise and can be used at higher pressures. In addition, they can be automated, and consequently multisample loaders that incorporate a loop injector are commercially available and have been integrated into automated liquid chromatographs.

Chromatographic column. The chromatographic column consists of a stainless steel tube that contains the packing (stationary phase) and the ancillary fittings used to connect it to the chromatographic system.

Column packings. Numerous materials are used as column packing, including silica, alumina, charcoal, and organic polymers. Silica is the most widely used material and is produced in three principal types of particles: macroparticulate, microparticulate, and pellicular (Figure 2-46). Macroparticulate supports have a mean particle diameter larger than 40 μm and are completely porous. Microparticulate supports have a mean particle diameter between 3 and 10 μm, are completely porous, and have either a spherical or irregular shape. Pellicular packing has a thin pellicle or porous layer of stationary phase coated on 20–60 μm solid glass beads. Microparticulate packings have the highest efficiencies as well as the greatest loading capacities and are the most widely used.

In *liquid-solid HPLC*, in which the separation depends on the differential adsorption and desorption of solutes on the surface of the stationary phase, the particles of silica are used directly. In *partition or liquid-liquid HPLC*, the solutes differentially distribute themselves between two immiscible liquids; one of these, attached to particles of silica, serves as the stationary phase. In the past, the attachment process consisted of physically coating the particles with the desired liquid. However, columns packed with coated column packings have short lifetimes and consequently are not widely used in modern-day HPLC.

A very significant advancement made in column technology has been the development of the "bonded-phase" column packings. In these packings the molecules of the liquid used as the stationary phase are chemically bonded to the surfaces of the silica particles through a silica ester or, preferably, a silicone polymeric linkage. The resultant

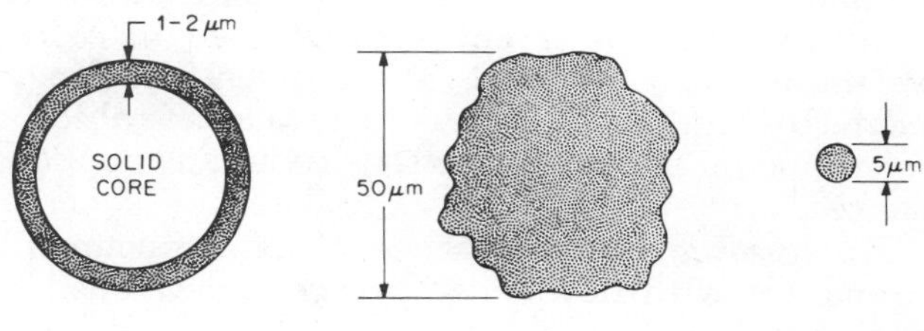

Figure 2-46. Various types of chromatographic packings used in liquid chromatography.

packings are mechanically and chemically stable, have long lifetimes, and provide excellent chromatographic performance. The capability of chemically bonding specific functional groups has resulted in the availability of a variety of versatile column packings that can be used for both partition and ion-exchange applications.

Reversed-phase HPLC is the most widely practiced form of liquid chromatography, and it requires a nonpolar stationary phase. The most popular reversed-phase packing is the octadecyl type, in which a C_{18} hydrocarbon is bonded to silica particles. A column with octadecyl silane packing is called an ODS column. Reversed-phase column characteristics can be altered by attaching different alkyl chains (e.g., octyl) to the silica. Mobile phases such as acetonitrile, aliphatic alcohols, tetrahydrofuran, water, or mixtures of such solvents, are used.

"Ion-pair" HPLC is a version of reversed-phase HPLC and is particularly useful for biological separations. Generally, reversed-phase columns are most effective in separating nonionic compounds; but most biological compounds have ionic functional groups. In ion-pair HPLC, these ionic groups are neutralized by adding a second ion (i.e., counter ion) to the mobile phase. This counter ion combines with the sample ions and forms a neutral ion pair. The neutral ion pair will then undergo the normal partitioning process. In this manner, strongly ionic compounds can be separated by reversed-phase HPLC. When the sample is a strong acid, a basic compound (such as tetrabutyl ammonium phosphate) is added to the mobile phase as the counter ion. For strong bases, sodium lauryl sulfonate or another sulfonic acid is used as the counter ion.

The various column packings are contained in a stainless steel tube, which ranges in length from 5 to 150 cm and has an internal diameter from 1 to 8 mm. These tubes usually have a straight configuration, although both "U" and coiled shapes have been used. The tubes are packed by either dry- or wet-packing techniques.

In some cases, a *precolumn* is used to protect the column packing from stripping of the stationary phase or from extraneous materials, such as protein, found in the sample. Most chromatographic packings will irreversibly adsorb protein with a concurrent loss in both resolution and the life of the column.

Detectors. As the name implies, the function of the detector is to detect compounds as they elute from the chromatographic column. Ultraviolet and visible photometers are the most popular HPLC detectors, although fluorometers and electrochemical detectors are becoming more widely used for clinical applications.

Ultraviolet and visible photometers. Ultraviolet (UV) and visible photometers measure the absorption of radiant energy by the compounds as they elute from the chromatographic column. The magnitude of the absorption and the wavelengths at which it occurs are functions of the molecular structure and concentration of the compounds. Most organic compounds will absorb in the ultraviolet and a few in the visible region of the electromagnetic spectrum.

Photometers operate in either fixed- or variable-wavelength modes. The fixed-wavelength UV photometer is the most widely used of all LC detectors, and most utilize the intense 254-nm resonance line produced by a mercury arc lamp (Figure 2-47). The radiation from such a lamp is directed through the reference and sample compartments of the flow cell, and the resultant transmitted radiation is detected and measured by a photodetector. The output from the photodetector is then displayed on a strip-chart recorder or a printer plotter. This type of detector is very sensitive but does require that the compounds of interest absorb energy at the photometer's fixed wavelength. To provide a fixed-wavelength detector with more flexibility, a phosphor can be placed between the lamp and the flow cell. The resultant emitted fluorescent light can then be used as a light source with a different fixed wavelength. With a variable-wavelength detector, a wavelength can be selected from a given wavelength range (e.g., 190–650 nm). Thus, the detector can be "tuned" to operate at the absorbance maximum for a compound or set of compounds and thus the sensitivity and applicability of the detector can be greatly enhanced.

Fluorescence detectors. Since many compounds have fluorescent properties, small fluorometers with flow cells have been developed and are increasingly being used as

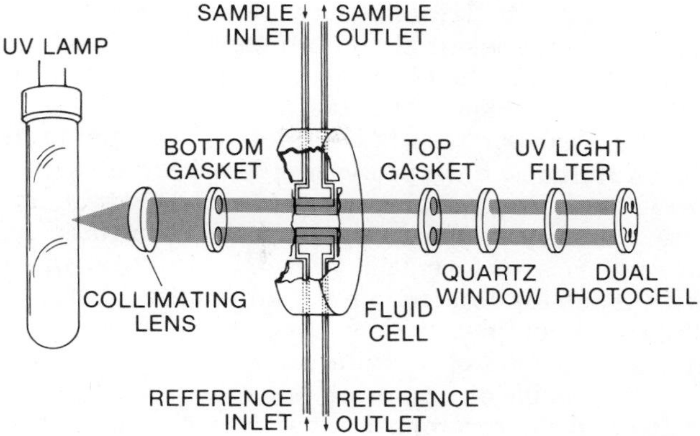

Figure 2-47. Optical schematic of a simple photometer and flow cell (fluid cell), which constitute the most widely used detector in liquid chromatography. (Adapted from Johnson, E. L., and Stevenson, R.: Basic Liquid Chromatography. Palo Alto, CA, Varian Associates, 1978.)

detectors in liquid chromatography. Most are relatively simple in design and construction and are very selective and sensitive for compounds that fluoresce within the range of their operating wavelengths. In addition, either a pre- or postcolumn reactor can be added for the purpose of chemically tagging a compound with a fluorescent label. Amino acids or other primary amines are often labeled with either a dansyl or fluorescamine tag before HPLC separation and fluorometric monitoring of the column eluent.

Electrochemical detectors. In this type of detector, the sample is either oxidized or reduced at an electrode surface under a constant potential. For such a detector to be useful, the compounds in question must be electroactive at the available potentials or be made electroactive by adding electrochemically active tags (e.g., bromine).

Data recorder. A liquid chromatograph produces data in a response-versus-time (y vs x) format. The time required for a compound to pass through a liquid chromatograph is called the retention time, and under a given set of chromatographic conditions this time can be used to identify eluting compounds. The detector response is used for quantitative purposes, since the magnitude of the response is proportional to the quantity of the compound passing through it.

A strip-chart recorder, one of the earliest data-recording devices used in chromatographs, is still widely used for this purpose. A strip-chart recorder displays chromatographic data as a sequential series of peaks, called a chromatogram (see Figures 2-48 and 27-6). To utilize quantitatively the data contained in a chromatogram, either the area or height of each individual peak is measured.

Peak areas can be calculated manually by the triangle formula area (A) = $HW_{1/2}$, where H is the height of the peak and $W_{1/2}$ is its width at half height. The area can also

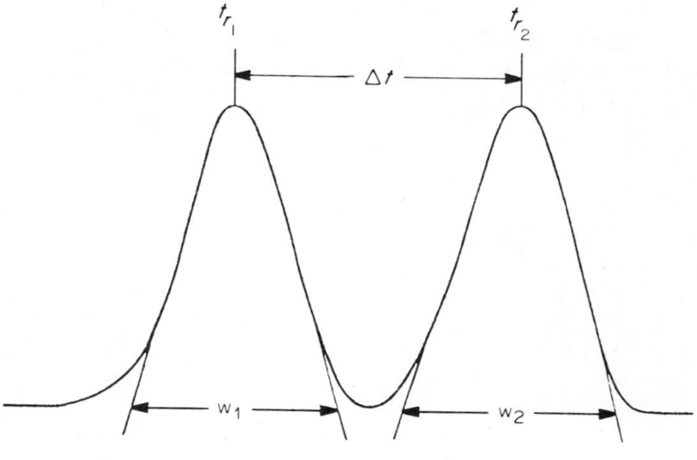

Figure 2-48. A two-component separation showing the chromatographic values that are measured and used to compute resolution.

be calculated by triangulation with the formula A = ½ BH, where B is the peak width measurement taken at the intersection of the baseline with two lines drawn tangent to the side lines of the peak. Although these manual techniques are simple and inexpensive, they are laborious, and coefficients of variation for results are 2–4%, considerably greater than for results obtained by automated data processing.

Automatic data processing is achieved by connecting the output of the detector in series with the digital integrator and a strip-chart recorder. As solutes pass through the detector, they produce a signal that is monitored by the integrator and digitally converted to numerical information. Advanced digital integrators provide the chromatographer with a means to compensate automatically for many chromatographic anomalies, including incompletely separated peaks, tailing peaks, and drifting baselines. The analytical precision of a digital integrator is ±0.5%.

To provide computation and interpretation capabilities with on-line data acquisition, mini- and microcomputers have been combined with HPLC systems. These interfaced systems are very useful, since they provide the chromatographer with the practical and analytical advantages that result from a fully automatic data system.

Applied Theory

The primary goal of any chromatographic process is to separate a mixture into its individual components, called solutes. For an effective separation, the individual solutes must emerge from the column at different times and their band profiles must be narrow enough to minimize overlap (Figure 2-48). Resolution (R_s) is a measure of the degree of separation of two solute peaks. By definition, R_s is the distance between the peak centers divided by the average of the peak widths. This is expressed mathematically as:

$$R_s = \frac{t_{r_2} - t_{r_1}}{\frac{w_1 + w_2}{2}} = \frac{2\Delta t}{w_1 + w_2} \quad (1)$$

where t_{r_1} and t_{r_2} are retention times for solutes 1 and 2, respectively, and w_1 and w_2 are peak widths in units of time measured at the base for solutes 1 and 2, respectively.

Inadequate separations usually occur when the calculated value for R_s is <0.8, whereas baseline separation is obtained when R_s is >1.25 (Figure 2-49). In actual practice, it is seldom necessary to calculate an R_s value, since by simple inspection of the shapes of peaks in a chromatogram, one can determine if the resolution is adequate. Unsatisfactory R_s can be increased by improving the column capacity factor, column selectivity, or column efficiency.

Column Capacity Factor

The resolution provided by a chromatographic column is a function of the ratio of the number of solute molecules on the column packing to the number of those in the mobile phase. This ratio is called the capacity factor (k') and can be computed from the

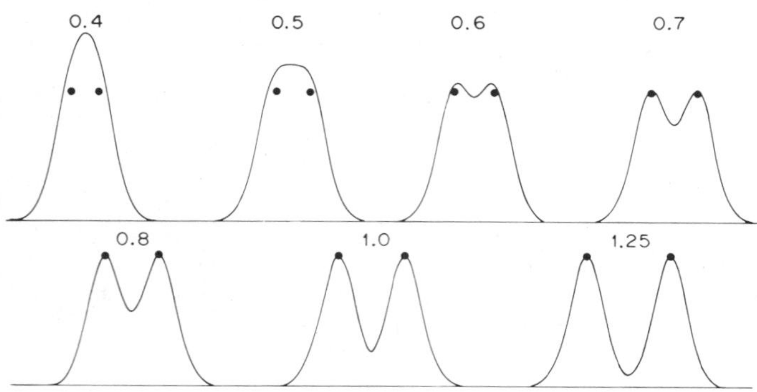

Figure 2-49. Separation of chromatographic peaks as a function of resolution (R_S). (From Snyder, L. R.: J. Chromatogr. Sci., *10:*202, 1972. Reproduced with permission.)

following relationship:

$$k' = \frac{t_r - t_m}{t_m} \quad (2)$$

where t_r is retention time of retained solute and t_m is retention time of unretained solute (time to appearance of solvent or injection peak). The k' is small when few solute molecules are associated with the column packing. Solutes with a $k' < 2$ are only slightly retained by the column packing and emerge from the column soon after the unretained materials. Resolution is poor at these conditions because the unretained components may trail through the solute band of interest.

When k' is large, the number of solute molecules that are associated with the column packing is large and the solute band migrates slowly. Large values of k' improve separation but at the expense of longer elution times and increased broadening of the solute bands. Optimal values of k' lie between 2 and 6, although in actual practice k' values between 1 and 10 can be used.

Column Selectivity

Column selectivity, α, is a measure of the relative separation between solute peak centers. It is a function of the column packing, mobile phase, and solute chemistry. It can be calculated from:

$$\alpha = \frac{t_{r_2} - t_m}{t_{r_1} - t_m} = \frac{k'_2}{k'_1} \quad (3)$$

where t_{r_1} and t_{r_2} are retention times of solutes 1 and 2, respectively; t_m is retention time of unretained solute (often called solvent or injection peak); and k'_1 and k'_2 are capacity factors of solutes 1 and 2, respectively. Since the k' of a solute equals its ratio of molecules on the column packing to those in the mobile phase, α is improved by changing that ratio for the solutes of interest (Figure 2-50). This change can be induced by altering (1) the composition of the mobile phase; (2) the stationary phase (e.g., the polarity of a reversed-phase column can be changed by using C_8 instead of C_{18} column packing); (3) the sample chemistry, through derivatization; and (4) the separation temperature.

When calculating α, the larger capacity factor is placed in the numerator; thus, α is always ≥ 1.0. When two solutes have the same k', α will be equal to 1.0 and no separation will occur. In general, solutes are relatively easy to separate if α is ≥ 1.1.

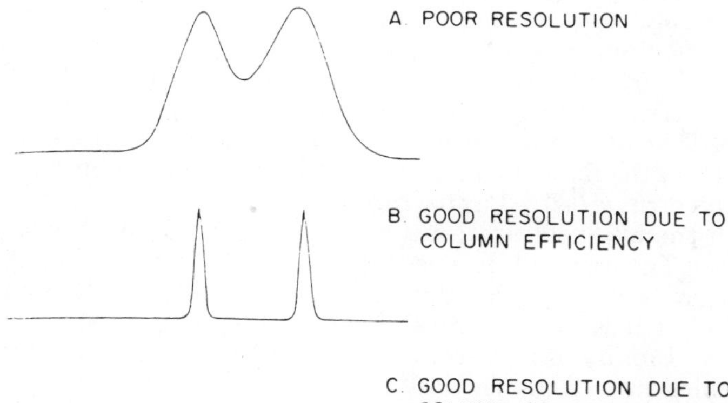

Figure 2-50. Demonstration of the effect of selectivity and efficiency on chromatographic resolution. (From Johnson, E. L., and Stevenson, R.: Basic Liquid Chromatography. Palo Alto, CA, Varian Associates, 1978.)

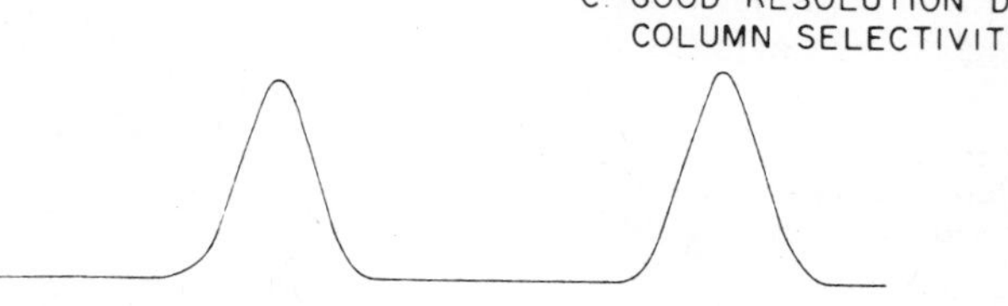

Column Efficiency

A very efficient column is characterized by the appearance of narrow peaks in the chromatogram (Figure 2-50). The efficiency of a chromatographic column is indicative of its effectiveness in minimizing the spreading of solute bands. Column efficiency is expressed as the number of theoretical plates, N.* Each plate corresponds to one equilibrium separation in a sequence of multiple extractions. N can be calculated from:

$$N = \left(\frac{t_r}{\sigma}\right) \quad \text{or} \quad N = 5.5\left(\frac{t_r}{w_{1/2}}\right)^2 \tag{4}$$

where t_r is the retention time, σ is one quarter of the peak width at the baseline, and $w_{1/2}$ is the peak width at half peak height. The measurement of $w_{1/2}$ is usually more accurate because it is less arbitrary than baseline measurements. Efficient columns frequently have N values of 5000 or greater.

Efficiency is also expressed as the number of theoretical plates per unit length (N/L) or the reciprocal (L/N) of this relationship, which is called the height equivalent to one theoretical plate (HETP). Thus, smaller HETPs signify more efficient columns. It becomes obvious that N (and therefore column efficiency) can be increased by increasing the length of the column. Other factors that affect efficiency are characteristics of the column packing particles, mobile phase viscosity, column temperature, and noncolumn contributions. In liquid chromatography, the most efficient column packings are composed of small (3–10 μm), regularly shaped particles.

It should be noted that the high efficiency of any chromatographic system will be lost if noncolumn contributions to solute band spreading are not properly controlled. Pre- and postcolumn tubing, connectors, column end fittings, and the detector cell must be of minimum diameter, length, and volume.

Improvement of resolution. To improve the resolution for separation, it is important to understand the relationship of resolution to the factors that control it; namely, capacity, selectivity, and efficiency.

With the appropriate substitution and mathematical manipulation, equations 2, 3, and 4 can be combined to disclose this relationship as:

$$R_s = \left(\frac{k'}{k' + 1}\right)\left(\frac{\alpha - 1}{\alpha}\right)\left(\frac{\sqrt{N}}{4}\right) \tag{5}$$

where k' is the capacity factor for the peak of interest, α is the selectivity factor for the peaks to be resolved, and N are the theoretical plates. Thus, resolution can be improved through an increase in k', α, or N. The most cost- and time-effective approach to improvement of resolution is to adjust first the capacity factor (k'), then the selectivity factor (α), and then the efficiency factor (N).

The capacity factors for the solutes of interest are first adjusted to between 2 and 6. This adjustment is easy to make, since it entails a simple modification of mobile phase strength. R_s will be appreciably improved when the initial value of k' is small, since the term $k'/(k' + 1)$ is asymptotic with 1.0. However, an increase in k' requires an increase in the analysis time, a sometimes unacceptable trade-off.

The α value is evaluated after k' is acceptable. If α is <1.1, it will have to be increased. The most dramatic improvement in R_s can be obtained with an increase in α, but it is also the most difficult and time-consuming factor to vary.

Finally, if necessary, the chromatographic efficiency should be adjusted. Three variables have a major influence on chromatographic efficiency: column length, column packing characteristics, and mobile phase flow rate. The column length required to give

---

*One theoretical plate is defined as a region on a column with perfect equilibrium of the solute between mobile and stationary phases.

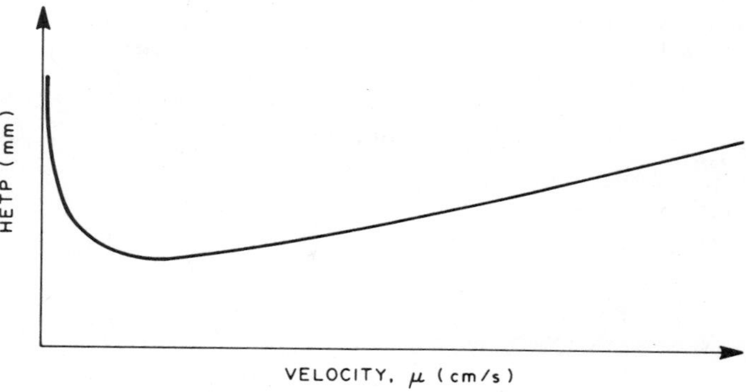

Figure 2-51. Typical van Deemter plot of mobile phase velocity versus efficiency (HETP).

the desired resolution can be approximated from the following relationship:

$$\text{Column lengths required} = \left(\frac{\text{Desired resolution}}{\text{Observed resolution}}\right)^2$$

For example, if the desired resolution is 1.2 and the observed resolution is 0.6, then the column length must be increased by at least a factor of 4. However, an increase in column length increases the analysis time. Increasing column length is also expensive, since additional columns are needed. The efficiency of the chromatographic system can also be improved through a decrease in the mean particle diameter of the column packing.

Once the appropriate column length and the packing material have been selected, the efficiency of a column can be improved by decreasing the flow rate of the mobile phase. The relationship between efficiency and flow rate is shown with the van Deemter plot (Figure 2-51).

Summary

Chromatography is a powerful analytical technique that is widely used to separate or purify two or more compounds that are indistinguishable by other analytical methods. The objective of the chromatographer, then, is to select those conditions that achieve the separation in the most reproducible and controlled manner available. If one has specimens to run on a repeated basis, which is the usual case in the clinical laboratory, one can afford to devote significant time to the development of chromatographic methods. This requires knowledge of the physical properties of a compound (e.g., volatility, solubility, ionicity) as well as the chemical properties that make its detection possible (e.g., UV absorption, fluorescence, electrochemical activity, electronic capture capability). With this information and the various modes of chromatography that are available, one can then optimize a separation scheme either in preparation for another analytical technique such as radioimmunoassay or for direct quantitation with, for example, a flame ionization detector or a UV-visible detector.

Although chromatography is used to separate and quantitate a wide variety of compounds including proteins, peptides, amino acids, carbohydrates, lipids, hormones, and nucleic acids, its greatest application is in the measurement of therapeutic drugs and drugs of abuse (see Chapter 26).

Recommended Reading

Colin, H., Krstulovic, A. M., Excoffier, J. L., and Guiochor, G.: A Guide to the HPLC Literature, Vol. 1: 1966–1979. New York, John Wiley & Sons, Inc., 1984.

Hancock, W. S., and Sparrow, J. T.: HPLC Analysis of Biological Compounds: A Laboratory Guide. Chromatographic Science Series, Vol. 26. New York, Marcel Dekker, Inc., 1984.

Hearn, M. T. W., Regnier, F. E., and Wehr, C. T.: High Performance Liquid Chromatography of Proteins and Peptides. New York, Academic Press, 1983.

Johnson, E. L., and Stevenson, R.: Basic Liquid Chromatography. Walnut Creek, CA, Varian Associates, 1978.
Kabra, P. M., and Marton, L. J.: Critical Liquid Chromatography, Vol. 1: Analysis of Exogenous Compounds; Vol. 2: Analysis of Endogenous Compounds. Boca Raton, FL, CRC Press, 1984.
Moyer, T. P., and Boeckx, R. R., Eds.: Applied Therapeutic Drug Monitoring, Vol. 1: Fundamentals. Washington, D.C., AACC Press, 1982.
Snyder, L. R., and Kirkland, J. J.: Introduction to Modern Liquid Chromatography, 2nd ed. New York, Wiley-Interscience, 1979.

SECTION SEVEN

Basic Principles of Radioactivity and its Measurement

EDWARD R. POWSNER, M.D.

Atomic Structure, Radiation, and Radioactivity

The Atom

Atomic nucleus. The atomic model is a small, central nucleus of positive charge surrounded by "orbital" electrons, each carrying a single negative charge. In this planetary model, the mass of the atom is concentrated in the dense, central nucleus. Although the composition of the nucleus and the nature of its internal "glue" or binding energy are still not well understood, we assume that the nucleus is composed of *protons* and *neutrons*, collectively referred to as *nucleons*. The proton carries a single positive charge; the neutron carries no charge.

The masses of atoms and other particles are expressed in *atomic mass units* (amu). By definition, 1 amu is 1/12 of the mass of one atom of carbon-12, or about 1.6604×10^{-27} kg. The mass of a nucleon is slightly more than 1 amu; in contrast, an electron has a mass of about 1/1900 amu.

Atomic species. The *atomic number* (Z) is the number of protons in the nucleus; the total number of nucleons (protons plus neutrons) is the *mass number* (A). A *nuclide* is an atomic species with a given atomic number and a given mass number. The generally accepted way of representing a nuclide is to place its mass number as a superscript to the left of its chemical symbol and its atomic number as a left subscript (e.g., ${}^{32}_{15}P$, ${}^{14}_{6}C$). Since both atomic number and chemical symbol identify the chemical species, the subscript is often omitted (e.g., ${}^{32}P$, ${}^{14}C$).

Isotopes are nuclides with the same atomic number but different mass numbers. These represent various nuclear species of the same element. Most elements occurring in nature are mixtures of isotopes. For example, natural carbon is predominantly ${}^{12}C$ with about 1% ${}^{13}C$ and a trace of ${}^{14}C$. The naturally occurring isotopes of the lighter elements such as carbon and iron are stable, but among the heavy elements, such as lead and bismuth, some of the naturally occurring isotopes are unstable. These unstable nuclides undergo spontaneous decay to stable nuclides by a process referred to as radioactive decay. Decay is typically accompanied by the emission of energy in the form of radiation.

In the electrically neutral, un-ionized atom, the number of orbital electrons equals the number of protons. Because chemical properties depend upon the number and arrangement of the orbital electrons and only slightly on the atomic mass, the chemical properties of all isotopes of an element are virtually identical. This identity of chemical properties is the basis for isotopic tracer methods, since the living system does not differentiate between isotopes of the same element.

Nuclear stability. In the neutron-proton model of the nucleus, protons contribute mass and positive charge, whereas neutrons contribute only mass; the combination of protons and neutrons provides stability. However, only a few possible proton-neutron combinations are stable. Binding energy and nuclear stability depend upon the number of neutrons relative to the number of protons (n/p ratio) in the nucleus. An unstable nucleus can adjust its n/p ratio by the emission of particles of matter. This adjustment, known as radioactive decay, is discussed below.

A solitary neutron does not form a stable nuclide; in contrast, a solitary proton is the nucleus of ordinary hydrogen ($^{1}_{1}H$). The one-proton combinations with neutrons start with stable deuterium ($^{2}_{1}H$), one proton with one neutron. The next member of this series is unstable, radioactive tritium ($^{3}_{1}H$), one proton with two neutrons. Combinations containing two protons are called helium. There is no stable nuclide of two protons alone; at least one neutron is required. The two-proton combinations with one and two neutrons are $^{3}_{2}He$ and $^{4}_{2}He$, respectively; both are stable, but helium isotopes with more than two neutrons are unstable.

As atomic numbers increase, the difference between the number of neutrons and protons increases. At the end of the list, bismuth has a stable isotope with 126 neutrons for its 83 protons. No stable nuclides are found with more than 83 protons.

Radiation

Radiations from the atom and its nucleus may be classified as particulate or electromagnetic. *Particulate radiation* consists of small bits of high-velocity matter, either the so-called elementary particles, such as the electron, positron, proton, neutron, and neutrino (discussed later), or the more complex particles, such as the helium nucleus and larger nuclear fragments. *Electromagnetic radiation* has been assigned special names on the basis of its energy, wavelength, or frequency (Table 2-6). Thus, electromagnetic radiations with very low energies, low frequencies, and long wavelengths are called radiowaves, whereas those with higher energies and frequencies and shorter wavelengths are called light. Those with the highest energy are called X-rays and gamma (γ-) rays. Gamma- and X-rays are distinguished as follows: γ-ray implies origin in the de-excitation of an atomic nucleus, and X-ray implies origin from the acceleration of orbital or other electrons. Nevertheless, both are electromagnetic radiation, and, once emitted, a 35-keV γ-ray is physically indistinguishable from a 35-keV X-ray.

A major tenet of modern physics is that all radiation has properties resembling those of both particles and waves. For electromagnetic radiation, *photon* is the name for the particle or quantum of radiation. The photon has no rest mass; indeed, it does not exist except in motion. This property is more easily understood in terms of waves because waves imply motion. Wavelength is inversely proportional to the energy of the photon.

$$E = 1.24/\lambda$$

where energy (E) is expressed in million electron volts (MeV) and wavelength (λ) in picometers (pm or 10^{-12} m). For example, the 0.364-MeV gamma radiation emitted after iodine-131 decay has a wavelength of 1.24/0.364 or 3.4 pm. For comparison, a photon of

Table 2-6. A CLASSIFICATION OF ELECTROMAGNETIC RADIATION

Name	Frequency (Hz)
Radiowaves	10^{4} –10^{11}
Infrared light	10^{12}–10^{14}
Visible light	10^{14}–10^{15}
Ultraviolet light	10^{15}–10^{17}
Gamma and X-rays	10^{17}–10^{20}

green light with a wavelength of 530 nm (530 000 pm) has an energy of 1.24/530 000 or 2.3×10^{-6} MeV (2.3 eV).

Radioactive Decay

Energetics. The radioactive decay of an atomic nucleus is evidence of its instability. By radioactive decay the nucleus alters its composition or configuration to increase its stability. Decay is manifested by spontaneous loss of mass and emission of energetic radiations.

Alpha decay. To achieve stable configurations, heavy elements, particularly those with atomic numbers above 70, may shed some of their nuclear mass by emitting a two-proton, two-neutron fragment identifiable after emission as a helium nucleus. Because nuclear radiations were observed before their identity was known, this fragment was called an *alpha* (α-) *particle* and its emission was termed α-decay. These names are still used. For example, radium-226 decays by α-emission to produce radon-222 as follows:

$$^{226}_{88}Ra \rightarrow ^{222}_{86}Rn + ^{4}_{2}He$$

The α-particle after it acquires two orbital electrons becomes a helium atom. Most α-emitters are naturally occurring radioisotopes of the heavy elements and have little clinical application.

Beta decay. For some heavy nuclides and for almost all those with atomic numbers below 60, stability is achieved by a rearrangement of the nucleus in which the total number of nucleons and the mass number are unchanged. In terms of the neutron-proton model of the nucleus, this rearrangement is the conversion of a neutron to a proton or vice versa. During such conversions, the nucleus emits either a negative electron or its positive equivalent, a *positron*. The emission of the negative electron, named the *beta* (β-) *particle*, is β-decay.

The emission of a β-particle leaves the nucleus with one additional positive charge. From our viewpoint, a neutron is converted to a proton, and the nucleus assumes the next higher atomic number. Beta-emission is characteristic of a nucleus that has more neutrons than required by its protons for stability. Examples of nuclides that decay by negative β-emission are carbon-14, hydrogen-3 (tritium), iodine-131, and iron-59. Emission of a positive β-particle (positron) has the opposite effect. A proton becomes a neutron, and the atomic number decreases by one. This process is characteristic of nuclei that have an excess of protons over neutrons. Positron emitters include carbon-11, fluorine-18, and iron-52. Some radionuclides used in clinical pathology are listed in Table 2-7.

Table 2-7. RADIATION PROPERTIES OF COMMONLY USED RADIONUCLIDES*

Nuclide	Half-Life	Type of Decay	Maximum Energy of Radiation, MeV†	
			Beta	***Gamma***
^{3}H	12.3 y	$\beta-$	0.018	
^{14}C	5730 y	$\beta-$	0.156	
^{32}P	14.3 d	$\beta-$	1.71	
^{35}S	87.1 d	$\beta-$	0.167	
^{51}Cr	27.8 d	EC‡		0.32
^{59}Fe	45 d	$\beta-, \gamma$	0.273, 0.475	1.095, 1.292
^{57}Co	270 d	EC		0.122, 0.136
^{58}Co	71.3 d	EC, $\beta+, \gamma$	0.47	0.81, 1.67§
^{99m}Tc	6 h	IT‡		0.141
^{125}I	60 d	EC		0.036
^{131}I	8.1 d	$\beta-, \gamma$	0.61	0.364

*Data taken in part from Dillman, L. T.: Radionuclide decay schemes and nuclear parameters for use in radiation-dose estimation, J. Nucl. Med., Supplement Number 2, March, 1969.

†Only the principal energies are given here.

‡EC refers to electron capture. IT refers to isomeric transition.

§1.1 MeV annihilation radiation is also produced.

Electron capture. An alternative decay process to the emission of positive β-particles is the capture of an electron. In this process, an orbital electron is absorbed by the nucleus. The end effect on nuclear structure is the same; a proton appears to have changed into a neutron and the atomic number decreases by one. As an example, Figure 2-52 shows the decay scheme for iodine-125, which decays exclusively by electron capture. A simple explanation of electron capture is that electrons in the low-energy s orbital have some probability of overlapping the nucleus and of permitting the nucleus to capture the electron. Although the n/p ratio is stable following electron capture, the atom is left with a vacancy in its s orbital. This is an unstable or excited state for the atom. *Atomic de-excitation* is accomplished by rearrangement of the orbital electrons, usually by a jump of an electron from a higher-energy orbital to the inner-orbital vacancy. As the electron falls to the inner orbital, energy is released in the form of an X-ray, or the energy may be absorbed by, and cause ejection of, a more weakly bound orbital electron (*Auger electron*).

Gamma emission and internal conversion. The original observations of naturally occurring radioactive nuclides disclosed a third radiation, the *gamma* (γ-) *ray*. This ray is distinguished by its ability to penetrate materials that block both α- and β-radiation. Gamma-radiation is high-energy electromagnetic radiation emitted during nuclear de-excitation. After either α- or β-decay or after electron capture, the nucleus may be left in an excited state. In these decays, the daughter nuclide goes from the excited to the ground state by shedding the excess energy either through the emission of a γ-photon or through the ejection of an orbital electron.

The energy of the emitted γ-ray is equal to the energy released in de-excitation. When de-excitation occurs in a single step, a single photon is produced. When it occurs in multiple steps, each produces a photon of corresponding energy. For every radionuclide, the de-excitation steps, and therefore the energies of the γ-rays, are unique; indeed, the spectrum of γ-ray energies may be used to identify unknown nuclides.

Rate of Radioactive Decay

Random nature of decay. The rate of decay is a characteristic of an individual radionuclide; the rate is a constant that is not affected by temperature, pressure, concentration, or any other chemical or physical condition. The number of disintegrations

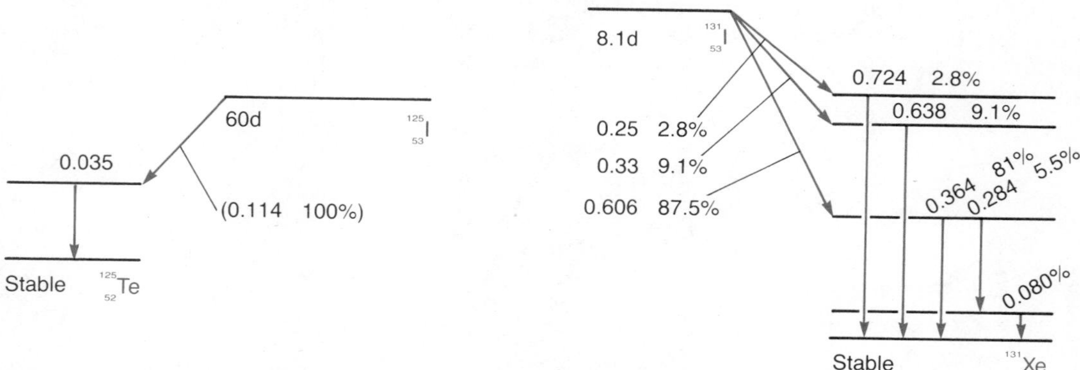

Figure 2-52. Atomic mass-energy diagram for the decay of two iodine isotopes. In these diagrams, the symbol for the nuclide is given just below the lowest or only horizontal line of each stack. The time to the left of the symbol for a radionuclide is its half-life; nonradioactive nuclides are simply labeled "stable." The height of a horizontal line is proportional to its energy with respect to the ground state of the stable nuclide. The relative position of a line from right to left indicates relative atomic number. ^{125}I (atomic number 53) decays to ^{125}Te (atomic number 52). ^{131}I (53), on the other hand, decays by β-emission to the higher atomic number ^{131}Xe (54), which is therefore placed to the right.

Arrows sloping down to the right indicate negative β-emission, vertical arrows indicate γ-emission, and arrows sloping down to the left show electron capture. The number assigned to each of the arrows is the energy in million electron volts released during the transition; the second number, where given, is the frequency in percent. For example, the 0.364 MeV γ-ray is emitted following 81% of ^{131}I decays. (Based on data of Way, K., et al.: Nuclear Data Sheets. New York, Academic Press, 1966; from Powsner, E. R., and Raeside, D. E.: Diagnostic Nuclear Medicine. New York, Grune & Stratton, 1971.)

in a short time interval (the rate of disintegration) is proportional to the number of atoms present at the beginning of the time interval. The proportionality constant, referred to as the decay constant, is a measure of the intrinsic probability of decay. The strength or rate of decay of a radioactive source is called its *activity*, A, and is simply the rate at which the radioactive atoms change to a lower energy atom.

It is convenient to describe the rate of decay in terms of *half-life*, $t_{½}$, the time required for sample activity to decrease to half its initial value. The relationship between activity, in terms of the half-life, and the elapsed time, t, is

$$A = A_0 \, e^{-(0.693/t_{½})t} \tag{1}$$

where A_0 is the initial activity at zero time, e is the base of natural logarithms, and 0.693 is the natural logarithm of two. This relationship is conveniently pictured by plotting the logarithm of the per cent activity remaining against time (Figure 2-53). Each half-life reduces the activity by half; the effect is cumulative. In general,

$$A_n = A_o \times 2^{-t/t_{½}} \tag{2}$$

or

$$A_n = A_o \times 2^{-n} \tag{3}$$

where n is the number of half-lives. Equation (3) is useful in planning experiments and in the disposal of radioactive waste. For disposal a rule of thumb is that seven half-lives reduce the activity to $<1\%$ of its original value ($2^{-7} = 1/128 = 0.78\%$) and ten half-lives reduce it to $<0.1\%$.

Units. The conventional unit of radioactivity is the *curie* (Ci), defined as 3.7×10^{10} disintegrations per second (dps) or 2.22×10^{12} disintegrations per minute (dpm). This number was originally selected because it approximates the activity of 1 g of radium-226 in equilibrium with its daughters. The new SI unit of activity is the *becquerel* (Bq) which is equal to 1 dps. One curie equals 3.7×10^{10} Bq; 1 μCi equals 37 kBq.

Specific activity. The term "specific activity" has several meanings; it may refer to any one of the following: radioactivity per unit mass of an element, radioactivity per mass of labeled compound, or radioactivity per unit volume of a solution. The reference used for the denominator must be specified. In terms of radioactivity per unit mass, the maximum specific activity attainable for each radionuclide is that for the pure radionuclide.

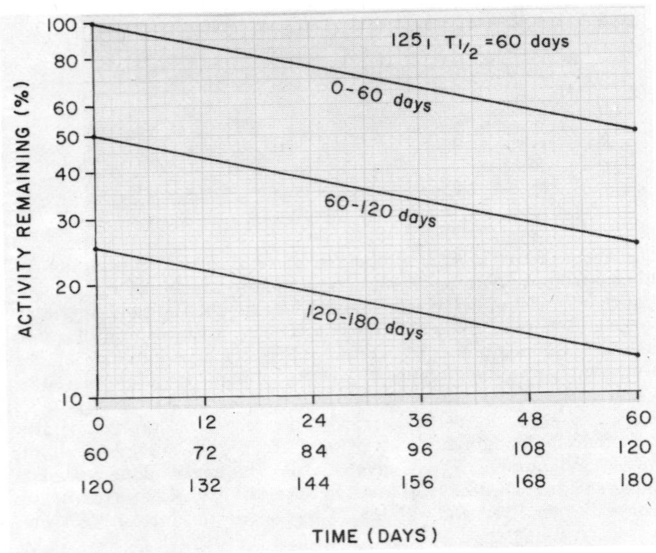

Figure 2-53. Radioactivity versus time for ^{125}I ($t_{½}$ = 60 d). The logarithm of activity remaining is plotted against elapsed time over a period of three half-lives.

Interaction of Radiation with Matter

Radioactive emissions possess energy either in the form of kinetic energy of motion, as in the case of α- and β-particles, or in the form of electromagnetic radiation, as in the case of γ-rays. In passing through matter these radiations transfer energy to the atoms and molecules encountered, chiefly through excitation and ionization. Alpha- and β-particles are charged and transfer energy by the interaction of their electric fields with other molecules. Gamma rays, which are uncharged, interact by other processes that will be noted later. The ability of radiation to produce excitation and ionization is one of its most important properties. This property is the basis for the detection of radioactivity and is responsible for the biologic effects of radiation.

Excitation and Ionization

Excitation. In this process, the energy of the incident radiation is transferred to matter by raising the electrons of the irradiated material to higher energy levels.

Ionization. If the energy absorbed from the radiation completely removes an electron from its atom or molecule, the process is called ionization. The resulting positive ion and negative electron are referred to as an *ion pair*. The ejected electron itself is called a *secondary electron*.

Particulate Radiation

The most important interactions are those of the charged particles, α and β. Both α- and β-particles cause ionization and excitation as a result of their velocity and charge. In mechanical terms, α- and β-particles are rapidly moving, charged bodies capable of forcing electrons from the atoms they pass. At each encounter the particles lose energy, and after many encounters they finally come to rest.

Electromagnetic Radiation

The nature of the initial interaction of photons, including γ-rays, is energy dependent. At the low end of the energy spectrum, particularly if the energy is insufficient to cause ionization, photons induce electronic and molecular motions, which register as *heat*. This is the predominant effect of the radiowaves used in microwave or radar heaters. Photons that are higher on the energy spectrum can eject electrons from matter by several mechanisms, of which the most important are the photoelectric and Compton interactions.

Photoelectric effect. The ejection of an atomic electron (Figure 2-54) is the most likely interaction of a moderately energetic photon. The photoelectric effect is possible only if the energy of the incoming photon at least equals the binding energy of the electron. Photon energy in excess of the binding energy imparts kinetic energy to the ejected *photoelectron*; the photon itself is absorbed.

Compton effect. The Compton effect (Figure 2-54) predominates at energies greatly in excess of the binding energy of atomic electrons. In a Compton interaction, the incoming photon interacts with a free or weakly bound valence electron as though the photon and electron were each billiard balls in collision. The electron is propelled in one direction while the photon is scattered in another direction, simultaneously losing

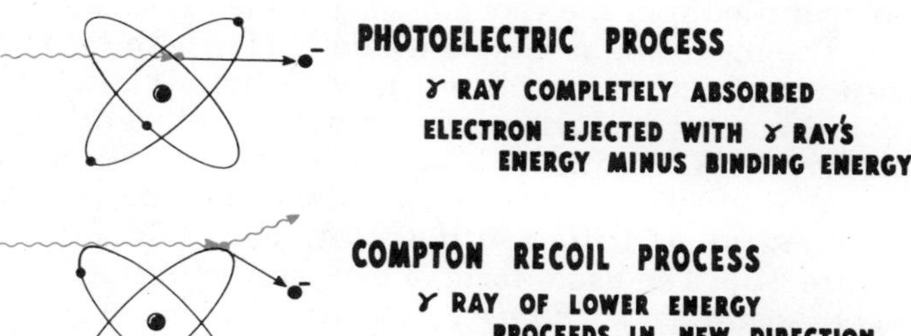

Figure 2-54. Interactions of γ-rays with matter (see text). US Atomic Energy Commission, ID2165A.

some of its energy to the electron. The electron deposits its energy in the surrounding matter by ionization and excitation of other atoms. The scattered photon has lower energy than the original photon.

Units of Radiation Exposure and Radiation Dose

Exposure. X-ray or γ-ray exposure is defined in terms of the ionization produced by the radiation. The conventional unit is the *roentgen* (R); the SI unit is the *coulomb per kilogram* (C/kg). The roentgen is the amount of radiation required to produce ions carrying one electrostatic unit of electric charge per cubic centimeter of dry air at standard temperature and pressure. In SI units, 1 R produces ionization equal to 0.258 mC/kg.

Radiation dose. Dose is described by the amount of energy absorbed. The conventional unit is the *rad* (radiation absorbed dose), which is defined as 100 ergs absorbed per gram of matter. Exposure of dry air to 1 R results in a dose of 0.87 rad; during the exposure of soft tissue, 1 R equals 0.93–0.97 rad, depending on photon energy. In part, the rad was chosen because the number of rads is close to the number of roentgens to which soft tissue is exposed. The SI unit of radiation dose is the *gray* (Gy), defined as one joule absorbed per kilogram; 1 Gy equals 100 rad. The relationship between radiation dose and biological effect is presented in a later section.

Detection and Measurement of Radioactivity

The darkening of photographic emulsion, ionization of gas, or fluorescent scintillation is the basis for most modern detection techniques. The underlying physical process in all of these methods is the excitation or ionization caused by radiation. In radiology, exposure and observation of a photographic film are the principal procedures. Measurement of ionization in gas or, more recently, in solids is widely used for environmental monitoring. In the clinical chemistry laboratory almost all measurements make use of scintillation detectors.

Gas-Filled Detectors

Detectors filled with an inert gas capture and measure the ions produced by radiation within the detector. The positive and negative ions are moved through the gas in opposite directions by a low-intensity electric field; the resulting current is amplified and measured. If the applied field is moderately high (e.g., several hundred volts over a few centimeters), the electrons move with enough velocity to ionize additional gas molecules. This process is referred to as *gas amplification*, which is advantageous because the current it generates is easier to measure than the small current of the primary ions alone.

The *Geiger counter* is the type of gas-filled counter commonly seen in the clinical chemistry laboratory, where it is used as a portable radiation monitor.

Scintillation Detectors

The principal types of scintillation detectors found in the clinical chemistry laboratory are the sodium iodide crystal scintillation detector and the organic liquid scintillation detector. The crystal detector is the more widely used because of the ease of its operation and the economy of sample preparation.

The operating principle is similar for both types of scintillation detectors. The absorption of radiation within the scintillator causes excitation and ionization. This absorbed energy in turn produces a flash of light, or scintillation. The distinguishing property of a scintillator is its capacity to convert a significant fraction of the absorbed energy into visible or ultraviolet light that can be detected and quantitated.

Crystal scintillation detector. This solid scintillation detector is diagrammed in Figure 2-55. The usual form of a crystal scintillation detector encountered in the chemistry laboratory is the *well detector*. This has a hole drilled in the end or side of the cylindrical crystal to accept a test tube (Figure 2-55). Because it is hygroscopic, the crystal is hermetically sealed in an aluminum can with a transparent quartz window

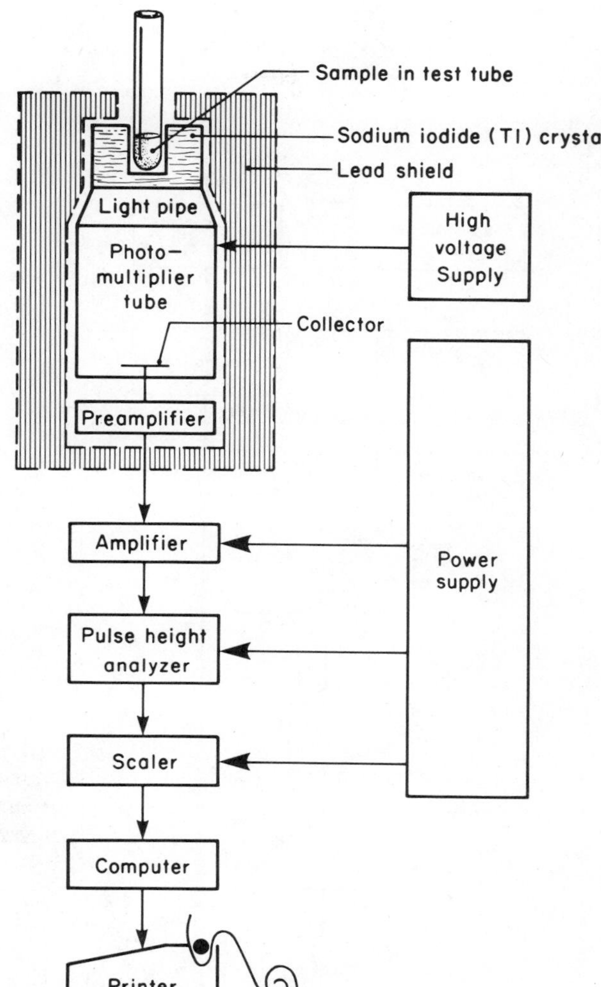

Figure 2-55. Crystal scintillation well detector. Thallium (Tl) activates the NaCl crystal.

at one end through which the blue-violet (420 nm) scintillations can be detected. The photons of iodine-125 in the sample easily penetrate the specimen test tube and aluminum can to enter the crystal, where they are likely to be absorbed in the thick, high-density sodium iodide. Of course, only the energy that is actually absorbed in the crystal can be detected.

The well detector is the instrument of choice for such γ- and X-ray–emitting nuclides as chromium-51, cobalt-57, iron-59, iodine-125, and iodine-131; for this reason it is often referred to as a gamma-counter. This detector is not suitable for β-radiation, which usually cannot penetrate the sample container and aluminum lining of the well. For a typical well detector, the counting efficiency for iodine-125, expressed as the percentage of decays that produces counts, approximates 70%.

Liquid scintillation detector. As shown in Figure 2-56, this detector measures radioactivity by recording scintillations occurring within a transparent vial that contains the unknown sample and the liquid scintillator. Because the radionuclide is intimately mixed with, or actually dissolved in, the liquid scintillator, the technique is ideal for the pure β-emitters of low energy and short range such as tritium (maximum β-energy is 19 keV and maximum range is <10 μm in water) and carbon-14 (156 keV maximum energy and <300 μm range). Typical efficiencies for liquid scintillation counting in the absence of significant quenching are 60% for tritium and 90% for carbon-14.

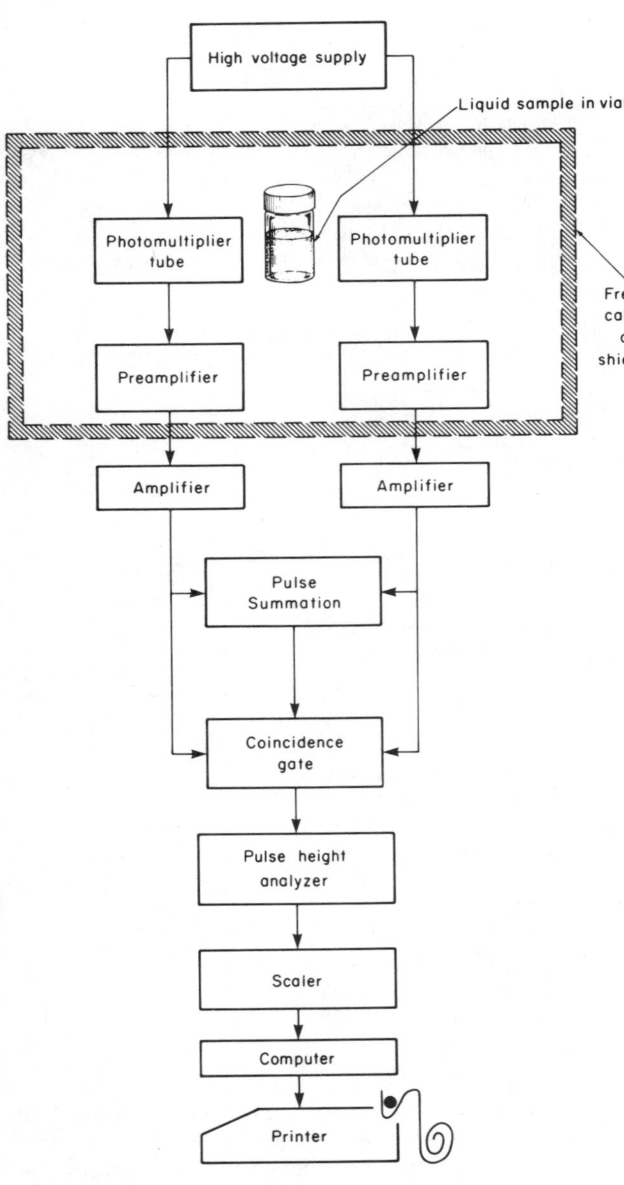

Figure 2-56. Liquid scintillation detector.

The liquid scintillator itself, the so-called *scintillation cocktail*, contains at least two components. The first is the *primary solvent*, usually an aromatic hydrocarbon such as toluene, xylene, or pseudocumene (1,2,4-trimethyl benzene). The second necessary component is a *primary scintillator*, which is dissolved in the primary solvent, absorbs energy from it, and converts the energy into light. A typical scintillator is PPO (2,5-diphenyloxazole), which emits ultraviolet light at 380 nm. In addition to the primary solvent and scintillator, the cocktail may contain:

1. A *secondary solvent* to improve the solubility of aqueous samples or a surfactant to stabilize or emulsify the sample;

2. A *secondary scintillator*, sometimes referred to as a wavelength shifter, to absorb the ultraviolet photons of the primary scintillator and re-emit the energy at a longer wavelength to facilitate measurement;

3. One or more *adjuvants*, such as suspension agents, solubilizers for biologic tissues, and antifreezes to prevent freezing and separation of water at low temperatures.

Electronics of Scintillation Counting[4]

After each burst of scintillation, the light is collected and converted to an electric pulse; the pulses are amplified, sorted by size, and counted. The components that perform these functions in the crystal and liquid scintillation detectors are shown in Figures 2-55 and 2-56, respectively. The more important components are discussed here.

Photomultiplier. This tube converts scintillations to electrical pulses (see Chapter 2, Section One). For well detectors, light is collected through the transparent window in the crystal cover and transmitted to the photomultiplier through a short *light pipe*. For liquid scintillators, the sample vial is placed in a lightproof opening between two photomultipliers. The photons of light enter the *photomultiplier* through its quartz window. Within the photomultiplier the photons strike the *photosensitive layer* (cathode) and eject outer-shell electrons from it. These electrons are accelerated toward the first of several intermediate electrodes, called *dynodes*, each of which is maintained at a progressively higher positive potential relative to the photosensitive cathode. As each electron strikes the dynode, it ejects several additional electrons from the dynode surface, thus multiplying the number of electrons traveling down the tube. These likewise are accelerated by the increasing potential differences to successive dynodes, where the multiplication is repeated.

Preamplifier. Despite the multiplication of electrons that occurs in the photomultiplier tube, the output of the tube still requires amplification. A preamplifier, typically placed close to the photomultiplier, is used to boost the current before it is analyzed.

Pulse-height analyzer. Pulses are sized electronically by an analyzer; its circuit passes only pulses within an acceptable *window* or *channel*. The analyzer for a well detector is usually supplied with one or more windows, often factory set, each corresponding to the narrow, γ-energy spectrum of a specific nuclide. Generally, the window is centered on the photopeak; when multiple photopeaks are used, the window may be set wider to include more than one peak, thus resulting in increased counting efficiency. An adjustable window may also be supplied. In liquid scintillation counters, the windows must be wider to accommodate the wider spectrum of β-energies, and the channel width is usually adjusted by the user because the intensity of the scintillation is affected by the composition of sample as well as by the β-energy.

Efficiency of Scintillation Counting

General factors affecting efficiency. The overall efficiency of a scintillation detector can be defined as the ratio of count rate to decay rate. The factors that affect the count rate and, hence, the efficiency are (1) the fraction of all decays that yield useful radiation (i.e., radiation capable of affecting the scintillator), (2) the fraction of potentially useful radiations that are directed into the scintillator, (3) the fraction of photons entering the scintillator that deposit energy therein, (4) the scintillation efficiency (i.e., the ratio of energy emitted as light to energy deposited as radiation), and (5) the detector threshold (i.e., the number of light photons, usually 10–20, required to trigger a count in the electronic circuits monitoring the scintillator).

Of the five factors affecting the count rate, scintillation efficiency is the one most likely to require the attention of the laboratory chemist. For solid crystal scintillation detectors, a decrease in efficiency may be an indication that the crystal has cracked or otherwise failed and that it should be replaced. For liquid scintillation detectors, changes in efficiency are more likely to be related to the sample or its preparation than to the detector. Monitoring the efficiency of a scintillation detector is an important part of the quality control program.

The term "quenching" is often used to refer to all the factors that reduce scintillation efficiency. Quenching lowers efficiency by absorbing part of the energy of a photon emitted in the scintillator. Quenching can cause batch-to-batch or sample-to-sample variations in efficiency. It is important to monitor and, when necessary, to correct for quenching in every assay. In some instruments, this is done automatically; alternatively, internal standards may be used.

Counting Statistics

Poisson distribution. Rutherford observed that the number of decays of a radioactive sample in successive, equal, time intervals follows a Poisson distribution. For a typical assay in which thousands of decays are counted, the Poisson distribution closely resembles the bell-shaped normal distribution, with one provision: the standard deviation of the Poisson distribution is always equal to the square root of its mean, $s = \sqrt{\bar{x}}$. The standard deviation is of importance in estimating the uncertainty in the apparent value of any statistical datum. For practical purposes, if the count measured for a specimen is 10 000 counts, the standard deviation is 100 counts. Notice that what has been written here is "counts," not "count rate"; the calculation of standard deviation is based on counts. As in Gaussian distribution, the confidence limit at ±1 SD is 68% and at ±2 SD is 95.5%. Thus, in the above example, at the 95.5% level there is a 2% error. If only 100 counts are collected, the SD is 10, but the error is 20% at the same confidence limit.

Effect of background count. The count referred to above is the actual count as read from the scintillation counter. This is sometimes referred to as the gross count. In practice, samples are always counted in the presence of background radiation. This background activity originates from a variety of sources including the ever-present, naturally occurring radionuclides in the environment and the radioactive materials in the laboratory, as well as electronic "noise." Unless the background is negligible, a clinical assay will include a correction for the background count. The corrected or net count is obtained by subtraction; thus, if sample and background are counted equal times,

$$C_s = C_{s+b} - C_b$$

where counts, C, are specified by the subscripts: s for sample net, b for background, and s + b for sample with background (gross counts).

The background count must be considered when the uncertainty in the count is calculated. This uncertainty is usually referred to as counting error, and its calculation follows the general rules for propagation of errors. For the simple case of a sum or difference, the rule states that the square of the error in the result equals the sum of the squares of the individual errors. Following this rule,

$$s_s^2 = s_{s+b}^2 + s_b^2$$

where the standard errors (sample standard deviations), s, are specified by the subscripts as before.

For example, if the mean value of the background count over 10 minutes is 2500 counts and the mean value of the gross count of the sample for 10 minutes in the presence of background is 10 000, the standard error of the net count of the sample is given by the following:

$$s^2 = 2500 + 10\,000 = 12\,500$$

or,

$$s = 111.8$$

Thus the standard error of the net count rate is higher than that for the gross count rate even though the net count rate (10 000 − 2500 = 7500 counts) is lower.

Radiation Safety

All individuals working in or frequenting any portion of a restricted area where radioactivity is present are required to have instructions regarding the health problems associated with radiation exposure.[1]

Absorbed energy. All of the effects of radiation on tissue begin with the absorption of radiation energy and the resultant disruption of chemical bonds. If a large amount of energy is absorbed, cells are killed and the tissue is destroyed almost immediately. If less energy is absorbed, the damage to cells and tissue may not be clinically apparent for hours or even years. At all lower levels there may be no discernible damage either because the damage to cells is too slight to detect or because biological repair has been complete.

Health Risks from Radiation

Types of risk. For the small amounts of radionuclide used and the low exposures received in the clinical chemistry laboratory, risk is limited to the possible effects of low-level, long-term exposure. *Acute radiation sickness* with immediate organ damage may follow acute exposure of >25 rem (0.25 Sv)* and usually follows doses >100 rem (1.0 Sv). Doses as large as 25 rem are not encountered in clinical laboratories.

It is illustrative to compare estimates of risk for persons exposed at the maximum rate permitted by regulations with the risks incurred by persons not exposed to radiation. The maximum permissible dose equivalent for the whole body of a worker exposed in restricted areas is 1.25 rem for each calendar quarter. Over an adult career of 30 years, the dose will be 150 rem. This exposure has been estimated to entail a risk of about 3.6 cancer deaths per thousand exposed workers. For comparison, the risk of developing cancer spontaneously is about 20 times greater. From another perspective, the risk of cancer associated with 30 years of exposure at the maximum permissible rate is lower than the risk of a fatal automobile accident but higher than the risk of a fatal electric shock.[3,5]

Regulatory Requirements

The requirements of state and federal agencies must be followed when radioactive materials are used. States differ substantially in their regulations; some have no regulatory agency and depend entirely on federal regulations, whereas others require the user to report to several different agencies. The federal regulations are covered in Title 10 of the Code of Federal Regulations (CFR),[1] which discusses radiation protection and licensing requirements. The Nuclear Regulatory Commission (NRC) controls the possession, use, and transfer of licensed radionuclides; the Department of Transportation regulates the shipping of radioactive materials as described in CFR Title 49;[2] and the Food and Drug Administration regulates the manufacture of pharmaceuticals and diagnostic kits, including those containing radionuclides. Users should obtain a current copy of the Code of Federal Regulations and consult a radiation health physicist regarding their own unique requirements for occupational protection from radiation.

In general, the user is required to maintain a level of occupational exposure as low as reasonably achievable (ALARA). Most large institutions employ radiation health officers or radiation health physicists to provide guidance in radiation protection and ensure that the user conforms to current regulations. There is no absolute lower threshold to occupational radiation exposure below which danger of possible injury can be positively excluded. However, existing guidelines for handling radioactive materials will result in risks as low as, and in most cases a risk much lower than, those accepted by workers in other well-regulated industries. This risk is low enough to be acceptable to most persons working with ionizing radiations. *Dosage limits*, as specified in the CFR, represent doses of radioactivity that are not expected to cause appreciable bodily injury to the individual at any time during a lifetime. Doses to workers in clinical laboratories are normally a small fraction of those specified in the CFR.

---

*The rem (roentgen equivalent in humans) is the conventional unit for the so-called dose equivalent or biologically "weighted" absorbed dose. Thus, 1 rem of any radiation has the same biological effect as 1 rad of standard X-rays (200–250 keV) (see page 130). The sievert (Sv) is the corresponding SI unit and is defined as 1 joule per kilogram; 1 Sv = 100 rem.

Safe Handling of Radionuclides

Radiation monitoring. For individuals working with radioactive materials, it is a good safety precaution to wear some form of *personal radiation monitor*. The most popular devices are the film badge and the thermoluminescent dosimeter, both of which measure absorbed dosage in millirems.

For the laboratory, an *area monitor* should be in continuous operation. The usual area monitor is a Geiger counter with an audible signal and a loud alarm to warn of high count rates. Since Geiger counters will not detect low-level radiation, all laboratory work areas should be checked regularly for surface contamination, preferably by a *wipe test*. This consists of wiping each surface to be examined with a separate, moistened absorbent material (the swipe) and counting each swipe in a gamma-counter for γ-radiation or in a liquid scintillation counter for β-radiation.

Safety precautions. General safety precautions must be practiced in any clinical laboratory where radioisotopes are used. Some of these are as follows:

1. Accurately record the receipt, transfer, and disposal of radioactive materials.
2. Label prominently all substances containing or contaminated with radioactivity.
3. Protect sources and post adequate warning signs in areas in which sources are stored. Do not leave sources of radioactivity unattended; lock unattended storage facilities.
4. Wear a protective laboratory coat at all times in the radiation area.
5. Wear disposable gloves, and thoroughly wash hands after handling radioactive substances and before leaving the work area.
6. Do not eat, store, or prepare food; smoke; or apply cosmetics in a work area.
7. Use and store radioactive materials only in designated, posted, and protected work areas.
8. Plan experiments carefully to minimize the time spent working with high-energy β- or γ-emitting materials.
9. Work behind lead or other shielding and use tongs or remote devices to handle high-energy β- or γ-emitting sources that emit >100 μCi. Radiation intensity decreases as the square of the distance from a "point" source.
10. Handle millicurie quantities of radioiodine only in a fume hood approved and tested for this purpose.
11. Handle unsealed radioactive materials over removable absorbent covers to simplify decontamination should spillage occur.
12. Pipet radioactive materials only with a pipet bulb or syringe. Do not pipet by mouth.
13. Contain all spills of radioactive materials. When spills are detected, immediately notify the proper safety officers; begin decontamination promptly.
14. Store radioactive waste, liquid or solid, in labeled containers until disposed of by the proper agencies. Do not allow radioactive waste to enter normal routes of trash or sewage disposal.

References

1. Code of Federal Regulations, Title 10 Energy. Office of the Federal Register, General Services Administration, Washington, D.C., January 1, 1983.
2. Code of Federal Regulations, Title 49 Transportation, Subchapter C: Hazardous Materials Regulations, Parts 171–179. Washington, D.C., Office of the Federal Register, General Services Administration, October 1, 1982.
3. International Commission on Radiological Protection: Problems Involved in Developing an Index of Harm. Publication No. 27 (Annals of the ICRP Vol. 1, No. 4). New York, Pergamon, 1977.
4. Krugers, J.: Instrumentation in Applied Nuclear Chemistry. New York, Plenum, 1973.
5. Whipple, G. H.: Unpublished lecture notes, 1978.

Additional Reading

Bushong, S. C.: Radiologic Science for Technologists. 2nd ed. St. Louis, C. V. Mosby Company, 1980.
Hendee, W. R.: Radioactive Isotopes in Biological Research. New York, John Wiley & Sons, 1973.

Howard, P. L., and Trainer, T. D.: Radionuclides in Clinical Chemistry. Boston, Little, Brown & Company, 1980.

Lederer, C. M., Hollander, J. M., and Perlman, I.: Table of Isotopes. 6th ed. New York, John Wiley & Sons, 1967.

Noz, M. E., and Maguire, G. Q.: Radiation Protection in the Radiologic and Health Sciences. Philadelphia, Lea & Febiger, 1979.

Peng, C.-T., Horrocks, D. L., and Alpen, E. L.: Liquid Scintillation Counting, Recent Applications and Development, Vol. 1, Physical Aspects; Vol. 2, Sample Preparation and Applications. New York, Academic Press, 1980.

Pizzarello, D. J., and Witcofski, R. L.: Medical Radiation Biology. Philadelphia, Lea & Febiger, 1982.

Powsner, E. R., and Raeside, D. E.: Diagnostic Nuclear Medicine. New York, Grune & Stratton, 1971.

SECTION EIGHT

Mass Spectrometry

BRIAN D. ANDRESEN, Ph.D.
BETHANY W. HURTUK, Ph.D., M.T. (ASCP)
KWOKIE J. NG, Ph.D.

Mass spectrometry (MS) is a technique that can provide information concerning the elemental composition and structure of organic compounds. The technique determines the arrangement of functional groups of a molecule, and the molecular weight may be measured to ten-thousandths of an atomic mass unit. Submicrogram quantities of an unknown substance in either a solid, liquid, or gaseous state may be completely characterized and quantitated with extreme specificity and sensitivity. Hence, mass spectrometry has gained increased use in clinical chemistry laboratories for the identification of compounds present in body fluids in health and disease and for the identification of ingested drugs and their metabolites.

In order to perform an analysis by mass spectrometry, molecules are ionized and then manipulated (filtered) through an electrostatic or magnetic field, or both, and positively identified according to their mass (m) to charge (z) ratio (m/z). Furthermore, sufficient excess energy can be imparted to the molecular ion to generate many fragment ions that can also be separated, measured, and recorded, from the smallest fragment to the intact molecular ion, to produce a mass spectrum (Figure 2-57). A mass spectrum is usually displayed with the most abundant fragment (i.e., the most stable and most often recorded ion) normalized to an arbitrary 100% relative intensity. All other fragment ion and molecular ion intensities are present relative to this major fragment ion, which is termed the "base peak."

The mass spectrum of a compound is specific for that particular compound. Reference libraries of mass spectral data are available for use in the identification of an unknown substance. If the compound has not been analyzed previously by mass spectrometry, a search through the library collection of mass spectral data may reveal many compounds that produce similar fragment ions of unique m/z values. Such information identifies specific structural features and aids in supporting a reasonable proposal for the complete identification of the unknown compound.

The Instrument

The essential components of a mass spectrometer are the vacuum system, ion source, mass filter, and detector, as illustrated in Figure 2-58.

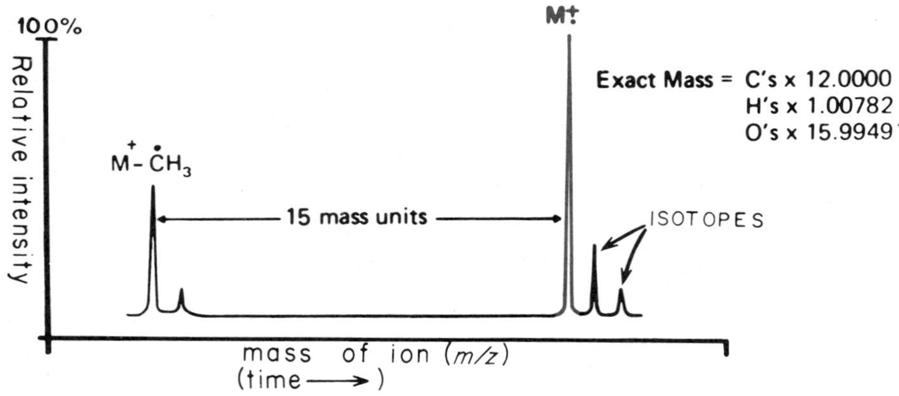

Figure 2-57. Analog mass spectral data produced by a mass spectrometer. The molecular ion, M^+, is equal to the molecular weight of a substance and is the precursor for all the ions seen in the spectrum. Stable isotopes of individual elements can also be observed. The x-axis of the plot shows mass, whereas the y-axis reveals abundance of ion formation.

Vacuum System

Mass spectrometers operate under conditions of very high vacuums, in the range of 10^{-5} to 10^{-7} torr (1 torr = 1.0 mm Hg). At this pressure there are ~10^8 molecules/cm^3 of background air molecules. If more background air molecules were present in the instrument, ions generated in the ion source would collide with the background molecules and scatter, resulting in decreased sensitivity. To avoid contamination and background noise caused by the condensation of previous samples on the interior of the instrument, parts of the mass spectrometer are continuously heated to 200–320 °C.

Ion Source

In mass spectrometry the sample must be pure and dry to avoid simultaneous detection of water or other solvents. The sample to be analyzed is either fractionally distilled, sublimed from a probe inserted into the ion source, or eluted from a gas chromatographic column. The purified components are directed into the ion-source chamber, where they are ionized before analysis. The most common ionization techniques are electron impact, chemical ionization, and field desorption.

Electron impact is the most widely used ionization technique; it uses an electron gun precisely positioned in the ionization chamber to bombard the molecules from the sample (Figure 2-59). As the molecules enter the ion source they are struck by an energetic electron beam, which causes ionization and fragmentation. The molecular ion (parent compound minus an electron) as well as fragments are produced. These fragments

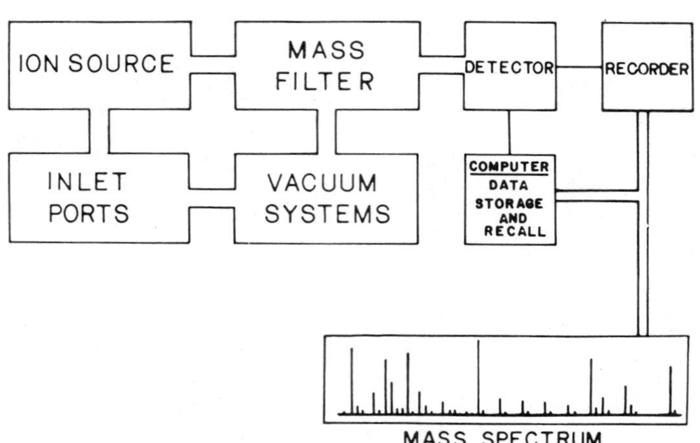

Figure 2-58. Block diagram of the essential components of a mass spectrometer.

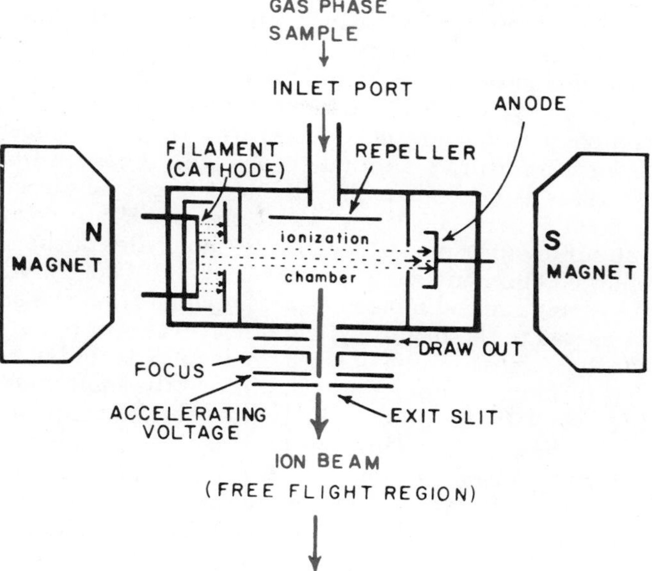

Figure 2-59. Electron-impact ion source. The small magnets are used to collimate a dense electron beam, which is drawn from a heated filament placed at a negative potential. The electron beam is positioned in front of a repeller, which is at a slight positive potential compared with the ion source. The repeller sends any positively charged fragment ions toward the opening at the front of the ion source. The accelerating plates strongly attract the positively charged fragment ions.

possess either positive (+), negative (−), or neutral (·) charges. Although abundant, neutral fragments are not detected by mass spectrometers. For most compounds negative ions are generated approximately 1/100th as often as positive ions. Therefore, most commercial mass spectrometers are designed to detect and record positively charged molecular and fragment ions. If an opening in the ion source is near a negatively charged draw-out plate, positive molecular ions and positively charged fragment ions will rush from the ion source in an attempt to neutralize their charge. If the draw-out plate is designed with an escape (opening), a great many of the positively charged fragment ions can travel out of the ion source in free flight. Focusing plates near the exit of the ion source allow intense total ion beams to enter the analyzer section of the mass spectrometer.

Chemical ionization uses a reagent gas that ionizes the sample more gently than electron impact. The reagent gas is continuously leaked into the ion source, raising the source pressure to a substantially higher value (0.05–0.1 torr) than exists in the rest of the mass spectrometer. Initially, the reagent gas is ionized by an electron gun in a manner similar to the electron impact technique. Because of the higher source pressure, the ionized reagent-gas molecules react chemically in the gas phase with other nonionized reagent molecules to form a highly activated intermediate, which is used to transfer a charge to the sample molecules. The sequence of gas-phase chemical events leading to the transfer of a proton from methane (reagent gas) to the sample molecule is shown below:

$$CH_4 \text{ (bombarded with electron beam)} + e^- \rightarrow CH_4^{+\bullet} + 2e^- \quad \text{(Ionization)}$$

$$CH_4^{+\bullet} + CH_4 \rightarrow CH_5^+ + \dot{C}H_3 \quad \text{(Activation)}$$

$$CH_5^+ + \text{Analyte} \rightarrow \underset{(M+1)^+}{\text{Analyte-H}^+} + CH_4 \quad \text{(Charge transfer)}$$

Chemical ionization provides a unique method for the determination of the molecular weights of compounds that normally do not form stable molecular ions under electron impact conditions. Substances analyzed under chemical ionization techniques produce "quasi" molecular ions (i.e., $M^+ + 1$) and only a few fragment ions. This technique is especially useful for determining the molecular weight of an unknown substance.

Chemical ionization techniques have been used clinically to screen complex biological samples rapidly in order to identify the molecular weights of unknown drugs. A probe distillation of a crude serum or urine sample extract that has been volatilized directly into the chemical ionization source permits the identification of all compounds by their $M^+ + 1$ ions. Relative concentrations of individual drugs can also be determined rapidly by comparisons of peak heights of individual molecular ions.

Field desorption is used to produce molecular ions of large, polar, and nonvolatile compounds that are not ionized by either electron impact or chemical ionization. Field desorption utilizes a unique probe with a heated filament upon which the sample to be analyzed is deposited. The filament is placed directly in front of a negatively charged (−8000 to −10 000 V) draw-out plate. When a positively charged voltage is applied to the sample probe, electrons in the molecules of the sample are drawn into the positively charged filament surface (+8000 to +10 000 V), which is simultaneously heated. The positively ionized molecules are repelled from the surface of the probe and are strongly attracted to the negative field of the draw-out plate, which is constructed with an orifice to allow most of the ionized molecules to enter the mass filter section of the mass spectrometer. Under these conditions the molecular ion leaves the ion source at a very high velocity. Because of the limited resident time of the molecular ion in the ion source, only a few fragment ions are formed before the molecular ion strikes the detector.

Figure 2-60 contrasts the fragment ions generated for tetrahydrocortisone, by using the three different ionization techniques. Field desorption clearly produces the simplest mass spectrum and most abundant molecular ion.

Mass Filters

A total ion (TI) beam containing the molecular and fragment ions of the compounds under investigation continuously emerges in free flight from the ion source. To obtain a complete mass spectrum, the ion beam must be separated into its individual ion components according to their masses (m) and charges (z). Two different methods, one using a magnetic sector analyzer and the other a quadrupole analyzer, are commonly found to separate the various fragment and molecular ions.

Magnetic sector instruments. As charged particles travel through a magnetic field, they are deflected at right angles to the lines of magnetic flux. If the total ion beam travels in free flight through a strong magnetic field, the fragment ions are deflected differently, depending upon their mass (m) and charge (z). For ions with the same charge,

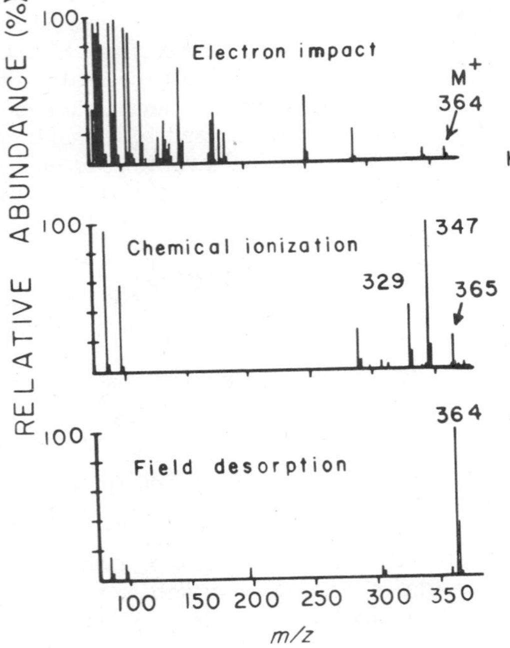

Figure 2-60. Mass spectral comparisons of tetrahydrocortisone under electron impact, chemical ionization, and field desorption. The molecular ion is clearly evident when field desorption is used.

the lightest fragment ions are deflected to the greatest degree and the heavier fragment ions are deflected to a lesser amount. All fragment ions with identical charge and mass are deflected to the same degree and are recorded as a unique signal. The intensity of this signal is dependent upon the number of fragment ions recorded, which in turn is dependent upon the stability of the fragment ion. The more stable an ion is in the gas phase, the more often it is recorded.

Figure 2-61 shows the ion paths generated for acetone. The molecular ion (M^+) is deflected the least in a fixed magnetic field, whereas the lightest fragment ion shown (CH_3^+) is deflected the most. In this example, only the fragment ion at m/z 43 ($CH_3—C{\equiv}O^+$) is of the proper mass and charge to travel through the flight path (with a radius of curvature R) and strike the detector. To allow observation of all fragment and molecular ions (i.e., a complete mass spectrum of the compound of interest), the magnetic sector instrument may (1) utilize a fixed magnetic field, (2) sweep the magnetic field to "fan" individual fragments past the detector, or (3) uniformly increase or decrease the accelerating voltage. Most magnetic sector instruments use the second method to scan the magnetic field and obtain the full spectrum as each unique mass ion passes by the entrance to the detector. Magnetic sector instruments can usually detect high-mass fragment ions over 2000 amu and possess great resolving power. High-resolution instruments are capable of measuring exact masses and with computer support can calculate the molecular formula of each ion recorded, a capability that is often very important for the characterization of a completely unknown compound.

Quadrupole analyzer. The quadrupole instrument, as its name implies, uses four metallic rods, which are held in place by insulators. During operation, an oppositely charged potential is placed on the rods, and DC voltage is increased while the polarity of each rod is constantly alternated at a high radio frequency (RF, Figure 2-62).

Ions leaving the ion source travel into the center of the oscillating electrostatic field produced by the quadrupoles. Mass filtering is accomplished as ions are attracted to the oscillating, charged surface of the rods. The separation of ions is determined by the optimal path that a certain mass ion will take as it travels through the length of the quadrupole.

The paths that the ions follow through the mass filter are very contorted. Each ion with a unique m/z value will follow a specific path as determined by the voltage and

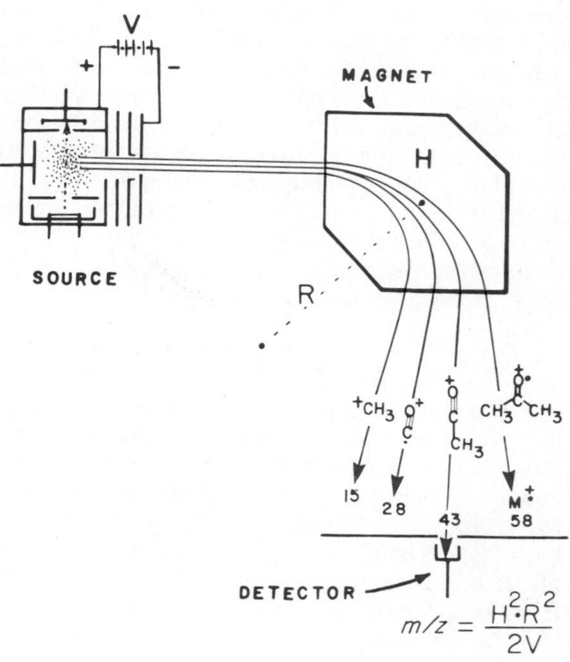

Figure 2-61. Ion path through a strong external magnetic field. Ion trajectories and radius of curvature, R, are related to the accelerating voltage, V, magnetic field strength, H, and m/z ratio. Only when specific V, H, and m/z conditions are met will an ion follow the unique path of radius R and strike the detector.

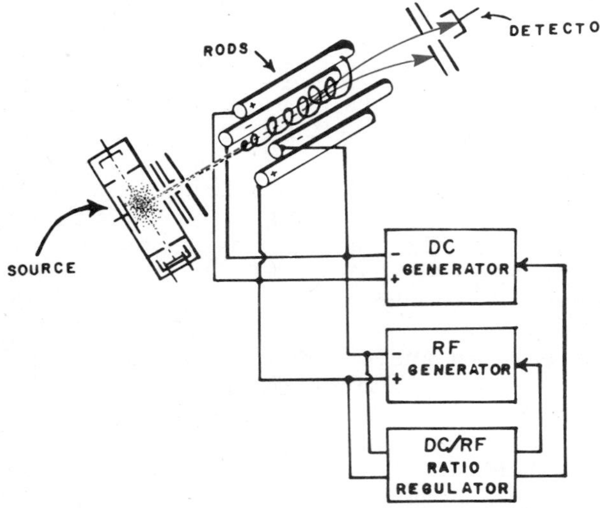

Figure 2-62. Quadrupole mass spectrometer with associated electronic controls for scanning the DC voltage and radio frequency (RF) of the mass filter rods. Ions in a quadrupole mass filter follow a contorted path, and only those that have a specific m/z ratio that matches the applied voltage and radio frequency will be detected.

frequency applied on the rods. Only when specific conditions are satisfied will an ion emerge at the end of the quadrupole and strike the detector. A complete mass spectrum is obtained by uniformly scanning the quadrupoles through the whole range of voltages and frequencies.

Quadrupole analyzers can produce unit mass resolution of fragment ions to ~800–1000 amu, which is adequate for most analytical purposes. The small size of these analyzers, the lack of large, magnetic power supplies, and their low maintenance requirements make them suitable as routine instruments. Although they do not have great resolving power, quadrupole instruments are capable of scanning rapidly and therefore have been used extensively for combined gas chromatographic/mass spectrometric and computer analyses.

Detector

One of the main attributes of mass spectrometry, its great sensitivity, is achieved because fragment ions are recorded with an electron multiplier. The device resembles a photomultiplier tube (see Figure 2-11,*B*) except that ions, rather than photons, set off a large cascade of electron discharges from the surfaces of metal dynodes specifically arranged with high potential differences between them. The electrons boil off the surface of the electron multiplier and amplify the incoming ion signal. The cascade of electrons generated in this manner increases the detected signal 10^6 to 10^7 times. With an output signal of great intensity from the electron multiplier, mass spectral data can be recorded directly on an analog recording device, or the signal itself can be passed through an analog-to-digital converter for computer acquisition and storage. High-gain electron multipliers permit analyses of compounds present in complex mixtures in amounts ranging from picograms to femtograms.

Gas Chromatography and Mass Spectrometry

The resolving power of gas chromatography (GC), when coupled with the sensitivity and specificity of mass spectrometry (MS), forms a unique tool for the analysis of complex biological samples. Because packed gas chromatographic columns require helium flow rates of 30–40 mL/min, much of the carrier gas must be removed before an eluent from the column can be introduced into the ion source. This is achieved with lipophilic membranes, fritted glass, or jet separators that take advantage of the fact that helium (because of its inertness, small molecular size, and high kinetic energy) can be stripped away from larger organic compounds (Figure 2-63).

As an alternative to packed columns, capillary columns with flow rates of only

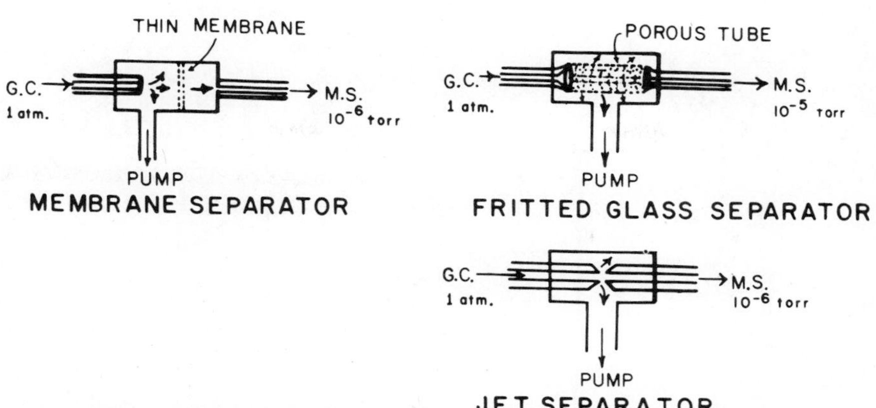

Figure 2-63. Techniques to remove helium (carrier gas) and enrich the sample as it elutes from the end of the gas chromatographic column. The membrane, fritted glass, and jet separators all remove the helium with an external vacuum pumping system that draws the helium preferentially away from the sample.

1–4 mL/min are often inserted directly into the ion source. This flow rate is small enough to be handled easily by the vacuum system of the mass spectrometer. All interface lines between the gas chromatograph and mass spectrometer are heated to avoid sample condensation.

Sample Preparation

Both gas chromatography and mass spectrometry are generally applicable only to volatile compounds. Consequently, many compounds must be derivatized to make them sufficiently volatile for GC, MS, or combined GC-MS analysis. Derivatization increases the sensitivity of the assay for the individual components and generates unique mass-spectral data that aid in characterizing substances in complex biological mixtures. For most biological samples, a derivatization procedure is followed that produces silylated or methylated esters or ethers.

Conclusion

Mass spectrometry is one of the most comprehensive analytical techniques now available. The mass spectrometer is no longer a specialized instrument used only by a few research laboratories. User-oriented mass spectrometers are available that enable the practicing clinical chemist to utilize this technique for numerous applications such as identification and quantitation of drugs as well as abnormal metabolites associated with inborn errors of metabolism.

Recommended Reading

Gudzinowicz, B. J., and Gudzinowicz, M. J.: Analysis of Drugs and Metabolites by Gas Chromatography-Mass Spectrometry, Vols. 1–5. New York, Marcel Dekker, 1977.

McLafferty, F. W.: Interpretation of Mass Spectra. Reading, MA, W. A. Benjamin, 1973.

Roboz, J.: Introduction to Mass Spectrometry, Instrumentation and Techniques. New York, John Wiley and Sons, 1968.

Waller, G. A., Ed.: Biochemical Applications of Mass Spectrometry. New York, Wiley Interscience, 1972.

Watson, T.: Introduction to Mass Spectrometry. New York, Raven Press, 1976.

3 PRINCIPLES OF IMMUNOCHEMICAL TECHNIQUES

GREGORY J. BUFFONE, Ph.D.

Any discussion of immunochemistry—the branch of chemistry concerned with detection and quantitation of chemical substances by measurement of antigen-antibody interactions—requires knowledge of the basic "reactants." Most immunochemical analyses performed in the chemistry laboratory use antibody as a reagent to detect the chemical substance (antigen) of interest. The exquisite specificity and the high affinity of antibodies for specific antigens, coupled with the ability of antibodies to crosslink antigens, allow their identification and quantitation by a variety of methods. Principles of the methods most commonly used in the clinical laboratory are discussed in this chapter. This introduction is intended to acquaint the reader with the structure and function of antibodies (immunoglobulins) in relation to their use as reagents in immunoanalyses. Immunoglobulins in health and disease are discussed in Chapter 11.

Antibodies are capable of binding specifically to a wide variety of natural and synthetic antigens, including proteins, carbohydrates, nucleic acids, lipids, and other molecules. Immunoglobulins consist of five general classes, designated IgG, IgA, IgM, IgD, and IgE. Immunoglobulin G is the most prevalent immunochemical reagent in use. Immunoglobulins A, M, D, and E, although essential in the host immune response, do not play an important role in immunochemical analysis and will not be discussed here. A schematic diagram of the IgG molecule is shown in Figure 3-1. IgG is a glycoprotein with a MW of ~158 000 and is classified as a 7S globulin by sedimentation velocity analysis. The protein is composed of two duplex chains; each set is made up of a heavy (γ) and a light (κ or λ) chain joined by disulfide bonds. Interchain disulfide bonds hold the duplex chains together. For more information on the structure and classification of immunoglobulins, see Chapter 11.

The variable amino acid sequence at the amino terminal end of each chain determines the antigenic specificity of a particular antibody. Each unique amino acid sequence is a product of a single plasma cell line or clone. The normal host response to an immunogen (any chemical substance capable of inducing an immune response) results in the stimulation of lymphocytes that are able to divide and form plasma cells capable of secreting antibody. Each plasma cell line produces antibodies with a single specificity. A complex antigen is capable of eliciting a multiplicity of antibodies with different specificities that are derived from different cell lines. Antibodies derived in this manner are termed *polyclonal* and exhibit diverse specificities in their reactivity with the immunogen. Each unique region of the molecular antigen that will bind complementary antibody is termed an *epitope*, the more contemporary term for an antigenic determinant. Both terms define a specific physical site on the molecule that reacts with the antibody. The circulating immunoglobulins are collectively referred to as *humoral antibodies*.

Antigens and immunogens. It is extremely difficult to provide a rigid definition of these terms, and almost any definition becomes circular. Thus, an *immunogen** (or an immunogenic material) is either a protein or a substance coupled to a carrier, usually a protein, that when introduced into a foreign host is capable of inducing the formation

*The term *immunogen* (instead of the term *antigen*) is now used when referring to materials capable of eliciting antibody formation when injected into a host. The term *antigen* is used for any material capable of reacting with an antibody, without necessarily being capable of inducing antibody formation.

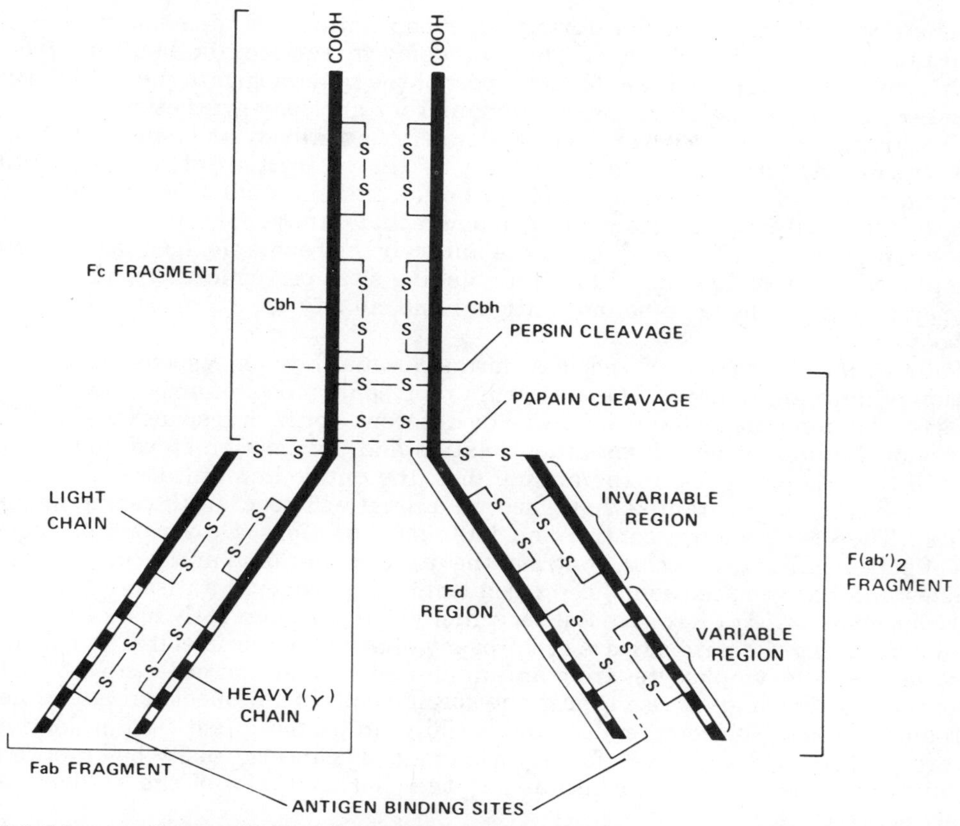

Figure 3-1. Schematic diagram of IgG antibody molecule, showing carbohydrate (Cbh), disulfide bonds (—S—S—), and major fragments produced by proteolytic enzyme treatment [$F(ab')_2$, Fc, Fab, Fd].

of an antibody in the host. The definition is neither precise nor complete. The route of introduction of the immunogen is usually, but not always, intradermal, and the antibody produced may be either circulating (humoral) or tissue-bound (cellular).

The types and sources of immunogens vary greatly. If we include those molecules (haptens) that are capable of inducing specific antibody formation when coupled to a carrier protein, even molecules such as L-tyrosine, penicillin, D-glucose, and arsanilic acid are immunogenic. A *hapten* is a chemically defined determinant that, when conjugated to an immunogenic carrier, stimulates the synthesis of antibody specific for the hapten. It is capable of binding antibody but cannot by itself stimulate an immune response. Natural immunogens are usually more complex than haptens and have several epitopes. Such immunogens are capable of reacting with several antibodies, allowing crosslinking and the formation of large immune complexes. In addition to proteins and polypeptides, lipids, nucleic acids, and many other materials can function as antigens. Some general properties requisite for immunogenicity are as follows: (1) areas of structural stability and chemical complexity within the molecule, (2) randomness of structure (i.e., lack of repeating units), (3) a minimal MW of 8000–10 000, (4) ability to be metabolized (a necessary but not sufficient criterion for some classes of antigens), (5) accessibility of a particular immunogenic configuration to the antibody-forming mechanism, and (6) foreignness of structure to the host. The overall shape and charge of the molecule (in most cases) have very little influence on the ability to induce antibody formation.

Injection of an immunogen into a host animal induces a series of biologic changes leading to markedly increased production of certain serum globulins with specific antibody activity—the immunoglobulins—and to changes in the capability of certain cellular elements to give an immune response. After the first exposure to an immunogen, a latent period (induction) occurs, during which no antibody is present in serum; this period may last from 5 to 10 days. The long delay in antibody build-up in vivo is apparently due to the time required for transport of the immunogen to the site of antibody production, for transmission of information, and for antibody synthesis.

The strength or energy of interaction between the antibody and antigen is described by two terms. *Affinity* refers to the energy of the interaction of a single antibody-combining site and its corresponding epitope on the antigen. *Avidity* refers to the overall strength of binding of antibody and antigen and includes the sum of the binding affinities of all the individual combining sites on the antibody. For example, IgG has two affinity-binding sites, whereas IgM has 10 affinity-binding sites per molecule. Thus, affinity is a property of the substance bound (antigen) and avidity is a property of the binder (antibody).

Polyclonal antiserum is raised in a normal animal host in response to the administration of pure immunogen. The antiserum represents a pool of antibodies with specificities for the immunogen but is produced by different clones in response to the different epitopes of the immunogen. Even after isolation and purification steps, the antiserum may still be heterogeneous. If the animal dies, the entire immunization process must be repeated, with no guarantee that a second animal will react in the same manner as the first. Therefore, quality control and lot-to-lot variations are serious problems for users of polyclonal antiserum. In contrast, the use of monoclonal antiserum avoids these problems and has replaced some of the conventional polyclonal antisera.

Monoclonal antisera have absolute specificity and are completely homogenous, since all cells producing the antibody are the progeny of a single hybrid cell. The hybridomas are formed from B lymphocytes from an immunized animal and are fused with plasma myeloma cells, which gives rise to just one specific cell line. Monoclonal antibodies can be produced from hybridoma cells in potentially unlimited quantities, as long as the cells are properly fed and cared for. Hybridoma cells can also easily be cryopreserved and cultured as needed. Thus, continuity in the characteristics of the antibody is reasonably assured.

Monoclonal antibodies have the following limitations for use in diagnostic assays.

1. Monoclonal antibodies react with only a single epitope on a multivalent antigen and therefore will not crosslink to form precipitates of macromolecular antigen. This characteristic makes them undesirable for precipitation or hemagglutination assays.

2. The avidity of monoclonal antibodies cannot be manipulated because the antibodies react with only a single antigen epitope. In contrast, avidity of polyclonal antisera may sometimes be increased by affinity purification to concentrate antibodies with high avidity for a specific antigen epitope.

3. Some subclasses of immunoglobulins do not fix complement; in these cases monoclonal antisera cannot be used in complement fixation assays.

In spite of these limitations, monoclonal antibodies have been adapted to many immunological binding assays, including immunoradiometric assay (IRMA), enzyme-linked immunosorbent assay (ELISA), and immunofluorometry. Examples are the IRMA or ELISA procedures in which a solid phase antibody (i.e., bound to a solid phase) specific for a unique epitope and another enzyme- or radiolabeled antibody specific for a different epitope can be reacted with antigen in a single step. This approach eliminates the two-step sequential addition of antigen and labeled antibody to the solid phase and avoids one incubation step and one washing step, which would both be necessary when polyclonal antibodies to both sites are used.

Antigen-Antibody Binding—Theory of Interaction

Several forces act cooperatively to produce antigen-antibody binding. The major contributing forces are van der Waals–London dipole-dipole interaction, ionic coulombic bonding,[5,11] hydrogen bonding, and hydrophobic interaction. These are all weak forces; affinity is due to a close geometric fit and many interaction points.

Reaction Mechanism for Antigen and Antibody Binding

The binding of antigen to antibody is not static; it is instead an equilibrium reaction that proceeds in three phases. The initial reaction of a multivalent antigen (Ag_n) and a bivalent antibody (Ab), called phase 1, occurs very rapidly in comparison with the subsequent growth of the complexes (phase 2) and is depicted by the following equation:

$$Ag_n + Ab \underset{k_{-1}}{\overset{k_1}{\rightleftharpoons}} Ag_nAb \underset{k_{-2}}{\overset{k_2}{\rightleftharpoons}} Ag_aAb_b \quad (1)$$

where k_1 is much greater than k_2, n is the number of epitopes per molecule, and a and b are the number of antigen and antibody molecules per complex. Depending on the relative concentrations of Ag_n and Ab, the complexes may crosslink in the third phase to form larger complexes, which then precipitate out of the solution. The speed of these reactions can be influenced by such factors as electrolyte concentration, pH, and temperature, as well as by antigen and antibody types and the binding affinity of the antibody.

The Precipitin Reaction

If the number of antibody combining sites, [Ab], is significantly greater than the antigen binding sites [Ag], then antigen binding sites are quickly saturated by antibody before crosslinking between the antigens can occur, and the formation of small antigen-antibody complexes of the composition AgAb results (Figure 3-2,*A*). For the case in which antibody is in moderate excess, the probability of crosslinking of Ag by Ab is more likely, and hence large complex formation is favored (Figure 3-2,*B*). When [Ag] is in great excess, large complexes are again less probable (Figure 3-2,*C*). The curve shown in Figure 3-3 is a schematic diagram of the classical precipitin curve illustrating the results observed when antigens and antibodies are mixed in various concentration ratios. Although the concentration of total antibody is constant, the concentration of free antibody, $[Ab]_f$ (i.e., not bound to antigen), and free antigen, $[Ag]_f$, varies throughout the range for any given Ag/Ab ratio. A low Ag/Ab ratio exists in section *A* of Figure 3-3 (the zone of antibody excess). Under these conditions $[Ab]_f$ exists in solution but $[Ag]_f$ does not. As total antigen increases, the size of the immune complexes increases up to equivalence (Figure 3-3,*B*) where little or no $[Ab]_f$ or $[Ag]_f$ exists. This is the zone of

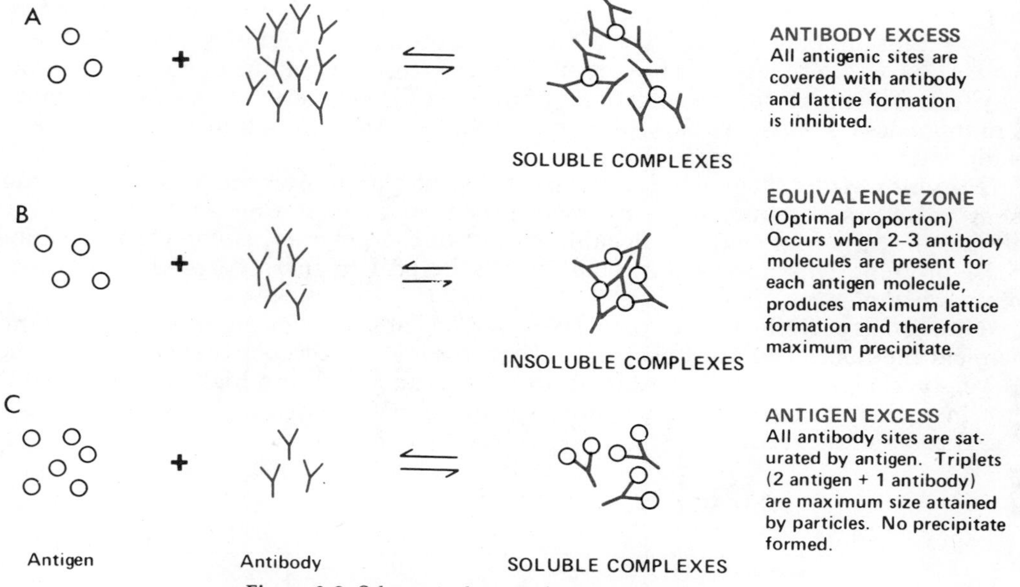

Figure 3-2. Schematic diagram for precipitin reaction.

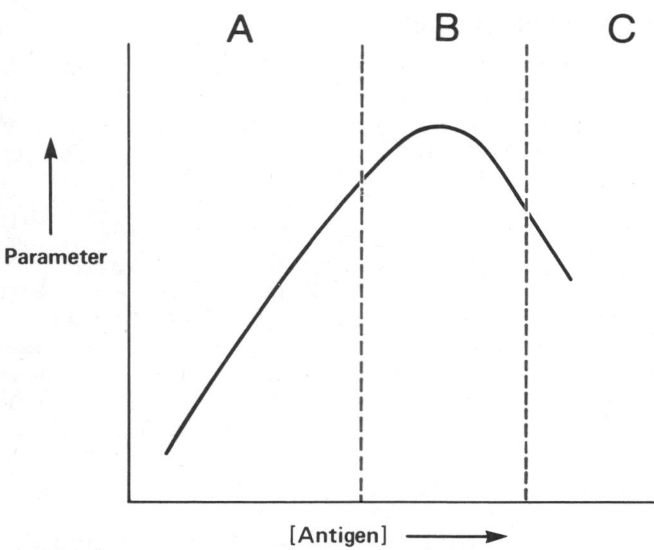

Figure 3-3. Schematic diagram of precipitin curve, illustrating zones of antibody excess (*A*), equivalence (*B*), and antigen excess (*C*). The parameter measured may be quantity of protein precipitated, light scattered, or another measurable value. Antibody concentration is held constant in this example. The same relationship is true for an agglutination reaction.

maximum size for the immune complex. This *equivalence zone* represents the optimal combining ratio for crosslinking in the particular system under examination. As Ag/Ab increases (Figure 3-3,*C*), the size of the immune complex will decrease and $[Ag]_f$ will increase (the zone of antigen excess). Immunoprecipitation in the "equivalence zone" is the basis for detection and quantitation of analytes by many immunoassays.

The Effect of Ionic Species, Ionic Strength, and Soluble Linear Polymers on Antigen-Antibody Binding

Ionic species and ionic strength affect the binding of antigen and antibody.[4,10] When a cationic hapten is used, anionic salts have little or no influence on binding, in contrast to cationic salts that produce an inhibition of the antibody-hapten binding. The order of inhibition by various cations is $Cs^+ > Rb^+ > NH_4^+ > K^+ > Na^+ > Li^+$. This order corresponds to the decreasing ionic radius and the increasing radius of hydration. Presumably, the lesser degree of hydration permits greater interaction of the salt with an anionic group located in the antibody combining site. Similar results have been found for anionic haptens and anionic salts. The order of inhibition of binding is $CNS^- > NO_3^- > I^- > Br^- > Cl^- > F^-$, again in the order of decreasing ionic radius and increasing radius of hydration. Therefore, the selection of buffers and polymers (see below) for inclusion in immunoassays must be carefully considered in order to obtain the most desirable conditions for specific assays.

Polymer effect.[8,9] In general, the solubility of a protein in the presence of different linear polymers is found to be inversely proportional to the molecular weight of the polymer; i.e., the higher the molecular weight of the polymer, the lower the solubility of the protein. This decrease in solubility is helpful to increase precipitation of the immune complex.

Numerous polymer species have been tested (Table 3-1) for applications in immunochemical methods. The most desirable characteristics of the polymer are high molecular weight, a high degree of linearity (minimum branching), and high aqueous solubility. Most investigators have found polyethylene glycol 6000 (PEG) in concentrations of 30–50 g/L to be most useful.

Table 3-1. LINEAR POLYMERS USED TO ENHANCE ANTIGEN-ANTIBODY REACTIONS

Polyethylene glycol (Carbowax)	Polypropylene glycol
Dextran	Polyvinyl alcohol
Modified cellulose	Polyvinyl pyrrolidone

Qualitative Methods for Detecting Proteins

Passive Gel Diffusion

Many qualitative and quantitative immunochemical methods are carried out in a semisolid medium such as agar or agarose. The primary advantages of a gelatinous medium are that it stabilizes the diffusion process with regard to mixing caused by vibration or convection and that it allows visualization of precipitin bands for qualitative and quantitative evaluation of the reaction. Antigen-antibody ratio, salt concentration, and polymer enhancement have the same influence on the antigen-antibody reaction in gels as they have on reactions in solution.

If the matrix does not interact with the molecular species under investigation, passive diffusion of reactants in a semisolid matrix can be described by Fick's equation:

$$\frac{dQ}{dt} = -DA\frac{dC}{dx} \tag{2}$$

where dQ is the amount of diffusing substance that at time t passes through the area A, dC/dx is the concentration gradient, and D is the diffusion coefficient. The amount of diffusing species transferred from the origin to a distant point (over the migration distance) is dependent on the length of time diffusion is allowed to occur; dQ/dt is clearly a function of dC/dx.

The initial concentration of antigen and antibody is critical. Each molecule in the system will achieve a unique concentration gradient with time. When the leading fronts of antigen and antibody diffusion overlap, the reaction will begin but formation of a precipitin line will not occur until moderate antibody excess is achieved. A precipitin band may form and be dissolved many times by incoming antigen before equilibrium is established.

Two basic approaches to passive diffusion are in common use today. *Single immunodiffusion*, in which a concentration gradient is established for only a single reactant, depends on the diffusion of an antigen into agar impregnated with antibody. A quantitative technique based on this principle is radial immunodiffusion (RID), which will be discussed later in this section.

The second approach is *double diffusion*, in which a concentration gradient is established for both reactants (antigen and antibody). Double immunodiffusion (in two dimensions; Ouchterlony technique) allows comparison of two or more test materials and provides a direct method for determining if the antigens in the test specimens are identical, cross-reactive, or nonidentical. The simplest method uses a standard Petri dish filled with agar. Holes are cut in the agar plate, and the bottom of each well is sealed to prevent diffusion under the agar. Antibody is placed in one well, and the two antigens to be tested are placed in two other wells. When the same antigen is in both wells, the lines of precipitation fuse and are continuous—the reaction of identity (Figure 3-4,*A*). When the precipitin bands cross each other, the reaction is one of nonidentity (Figure 3-4,*B*). If the two antigens are related but not identical, a reaction of partial identity is observed (Figure 3-4,*C*). Here, the cardinal point is that the precipitate does not act as a barrier to block unrelated, diffusing species. As shown in Figure 3-4,*D*, when two related antigens, Ag and Ag_1, are in separate wells and the respective antibodies, Ab and Ab_1, are in the third well, an AgAb precipitate forms on one side and blocks further diffusion of Ab from the antibody well; on the other side, the Ag_1Ab precipitate does not stop Ab from migrating further and from forming an AgAb spur.

Double diffusion can easily be done on a microscale by using small glass slides. Only a small sample (1 μL) is required, and the reaction time is markedly decreased from as long as a week to 24 h or less in some cases. Sharp lines are produced because the agar layer is thin and the area from which the reactants diffuse is small. A negative reaction does not necessarily imply absence of antibody or antigen but may result from the use of amounts of material too small for the sensitivity of the method or from a nonprecipitating antibody.

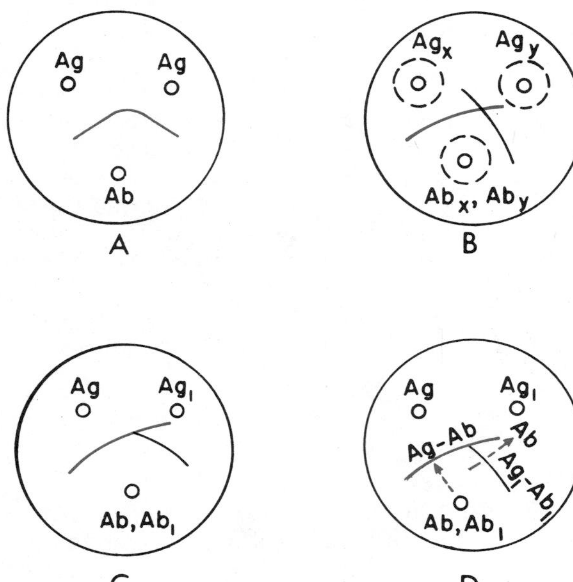

Figure 3-4. Double immunodiffusion in two dimensions by Ouchterlony technique. In order to identify an Ag, the well containing the unknown must be adjacent to an antigen of known quantity. Ag = antigen: Ab = antibody. *A,* Reaction of identity. *B,* Reaction of nonidentity. *C,* Reaction of partial identity. *D,* Scheme for spur formation. Ab specificity can be identified in the same manner by placing Ag in a single well and Ab of known and unknown specificity in adjacent wells.

Immunoelectrophoresis

If several antigens of interest exist in a solution (e.g., proteins in spinal fluid or serum), the various species can be separated and identified by immunoelectrophoresis (IEP). The procedure is carried out on an agarose gel medium poured onto a thin plastic sheet or onto glass. The sample to be analyzed is placed in a well in the gel, and an electrical field is applied across the gel surface. During electrophoresis the proteins in the serum are separated according to their electrophoretic mobilities (Figure 3-5). After electrophoresis, an antiserum to the protein of interest is placed in a trough parallel and adjacent to the electrophoresed sample. Simultaneous diffusion of the antigen from the separated sample and the antibody from the trough results in the formation of precipitin arcs whose shape and position are characteristic of the individual separated proteins in the specimen. The pattern is interpreted after formation of the precipitate. By comparison with a known control separated on the same plate, tentative indentification of individual proteins can be made. In some cases, such characteristics of the precipitin arc as bowing, brushing at the end of the arc, or thickening may suggest to the experienced observer the presence of special types of proteins. A permanent record can be made either by photographing the electrophoresis film or by staining and retaining the film. For an illustration of this technique see Figure 11-8 on page 326.

Crossed immunoelectrophoresis (CRIE), also known as two-dimensional immunoelectrophoresis, is a variation of IEP wherein electrophoresis is also used in the second dimension to drive the antigen into a gel containing antibodies specific for the antigens of interest (Figure 3-6).[1,6] This technique is more sensitive and produces higher resolution than is possible with immunoelectrophoresis.

In ***counterimmunoelectrophoresis (CIE)***[2] two parallel lines of wells are punched in the agar. Antigen solution is placed in a well adjacent to a well filled with antibody solution (Figure 3-7). If the solutions were allowed to diffuse passively, a precipitin line would form between the opposing wells where antigen reacted with specific antibody. In CIE, this process is made to occur more rapidly by applying a voltage across the gel so that the antigen and antibody move toward each other at a faster rate. The antigen migrates toward the anode, and the antibody moves in the opposite direction as a result of electroendosmosis; a precipitin line is formed where they meet. This method has found application in the detection of bacterial antigens in blood, urine, and CSF.

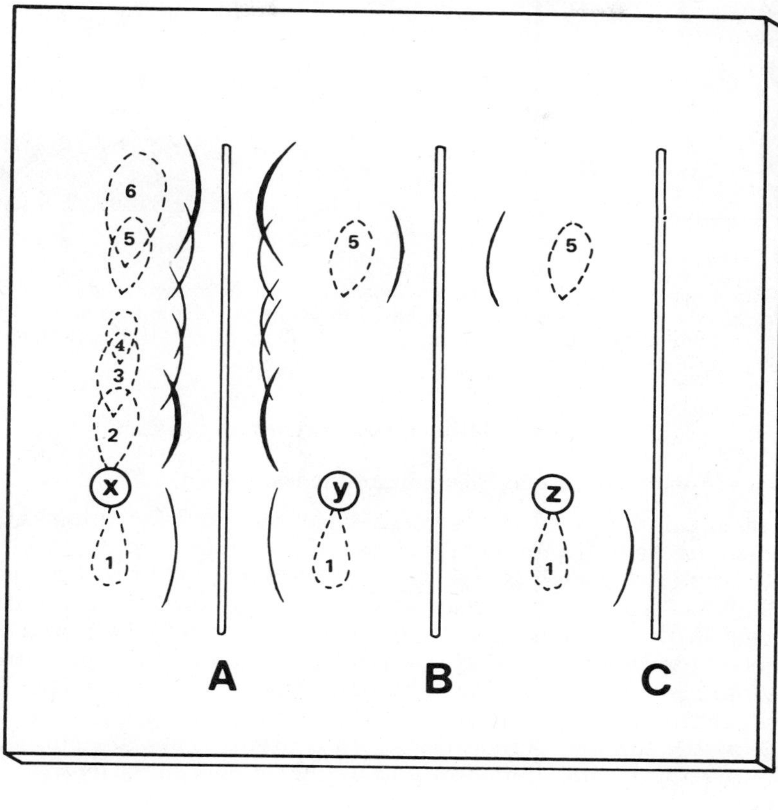

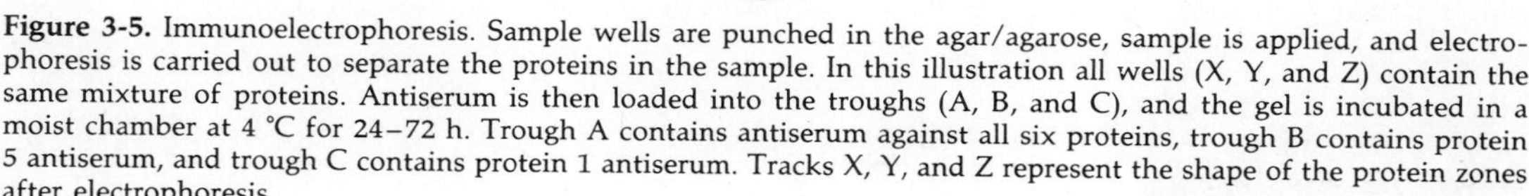

Figure 3-5. Immunoelectrophoresis. Sample wells are punched in the agar/agarose, sample is applied, and electrophoresis is carried out to separate the proteins in the sample. In this illustration all wells (X, Y, and Z) contain the same mixture of proteins. Antiserum is then loaded into the troughs (A, B, and C), and the gel is incubated in a moist chamber at 4 °C for 24–72 h. Trough A contains antiserum against all six proteins, trough B contains protein 5 antiserum, and trough C contains protein 1 antiserum. Tracks X, Y, and Z represent the shape of the protein zones after electrophoresis.

In ***immunofixation (IF)*** as in IEP and crossed immunoelectrophoresis, a first-dimension electrophoresis is performed in agarose gel to separate the proteins in the mixture. Subsequently, various antisera spread directly on different areas of the gel will precipitate the proteins of interest. The immune precipitate is trapped within the gel matrix, and all other nonprecipitated proteins can be removed by washing the gel. The gel may then be stained for observation and identification of the proteins. For a more detailed description of this procedure, see Chapter 11, page 325. Crossed immunoelectrophoresis is more sensitive than IF in terms of detection limit and also shows better resolution. Proteins of closely related or identical electrophoretic mobility can be better separated into individual bands by crossed immunoelectrophoresis, whereas in IF they may appear as a single band. In practice, IF is technically more efficient than either IEP or crossed immunoelectrophoresis, and it produces patterns that are more easily interpreted.

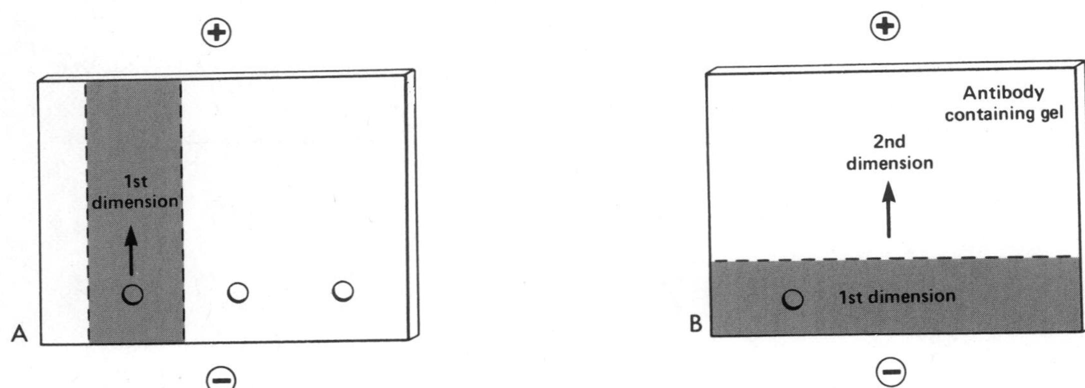

Figure 3-6. Two-dimensional (CRIE) electrophoresis. Diagram *A* shows the configuration for the first dimension of crossed immunoelectrophoresis. The segment of the gel denoted by the red color in diagram *A* is cut out and placed on a second plate, and antibody-containing gel is poured next to this segment (*B*). Electrophoresis is then carried out at 90° relative to the first-dimension run.

Quantitative Methods for Measuring Proteins

Radial Immunodiffusion (RID) and Electroimmunoassay (EIA)

The two most commonly encountered gel-based methods for quantitative immunochemical studies are radial immunodiffusion and electroimmunoassay ("rocket" technique). *Radial immunodiffusion* is a passive diffusion method in which a concentration gradient is established for a single reactant, usually the antigen. The antibody is uniformly dispersed in the gel matrix. Antigen is allowed to diffuse from a well into the gel until antibody excess exists and immune precipitation occurs; the antigen-antibody interaction is manifest as a well-defined ring of precipitin around the well. The diameter of the ring will continue to increase until equilibrium is reached. Standards are run at the same time as the sample. Standard antigen concentrations are plotted against the squares of the precipitin ring diameter, and unknown concentrations are interpolated from this curve.

In *electroimmunoassay*, as in RID, a single concentration gradient is established for the antigen, but in this case an applied voltage is used to drive the antigen from the application well into a homogeneous suspension of antibody in the gel (Figure 3-8).[7] This

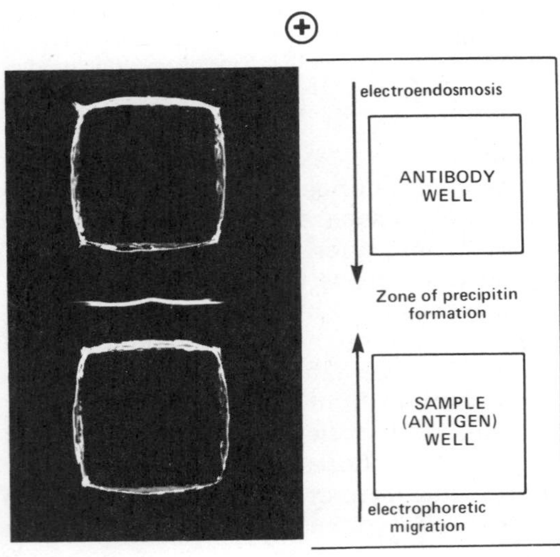

Figure 3-7. Counterimmunoelectrophoresis showing positive reaction between anti-*H. influenzae B* (upper well) and a CSF sample containing *H. influenzae B* (lower well).

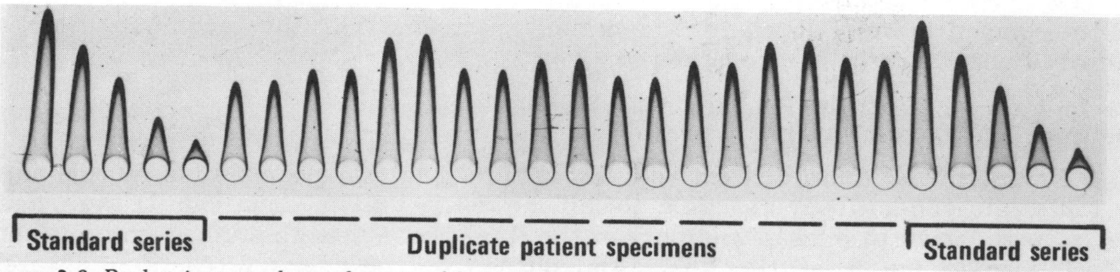

Figure 3-8. Rocket immunoelectrophoresis of human serum albumin. Patient samples were applied in duplicate. Standards were placed at opposite ends of the plate.

produces a rocket-shaped precipitin line whose size is proportional to the antigen concentration. Quantitation is effected by using standards on the same plate along with the unknowns and then estimating the concentrations of unknowns from the heights of the "rockets" obtained. The calibration curve is linear only over a narrow concentration range, so samples may have to be diluted or concentrated as needed.

In general, gel-based methods for quantitation of antigens require excellent technical skill, as well as tedious preparation and execution, to produce good results. In many clinical laboratories today, these methods are restricted to qualitative studies or are used as reference methods. Quantitative data are more commonly obtained by turbidimetric and nephelometric methods, radioimmunoassays, enzyme immunoassays, and fluorometric immunoassays.

Turbidimetric and Nephelometric Assays

Turbidimetry and nephelometry are convenient techniques for measuring the rate of formation of immune complexes in solution. Instrumental principles for these methods are described in Chapter 2, Section Two. Turbidimetric and nephelometric immunochemical methods using both rate and equilibrium protocols have been described for proteins, antigens, and haptens. In rate assays, measurements are usually made within the first few minutes of the reaction, since the largest change in intensity of scattered light (I_S) with respect to time (dI_S/dt) is obtained during this time interval. For so-called pseudoequilibrium assays, it is necessary to wait 30–60 min so that the dI_S/dt is small relative to the time required to make the necessary measurements. Such assays are referred to as "pseudoequilibrium" because true equilibrium is not reached within the time allowed for these assays.

Nephelometric methods in general are more sensitive than turbidimetric assays and have an average sensitivity of ~1–10 mg/L. Better sensitivities are obtained in fluids such as CSF and urine than in serum because their lower lipid and protein concentrations result in a better signal-to-noise ratio.

Nephelometric and turbidimetric assays also have been applied to the measurement of haptens (e.g., drugs) by using *inhibition techniques*. The reagent is prepared by attaching the drug of interest to a carrier molecule such as bovine serum albumin. The hapten-bound albumin then competes with free hapten (drug introduced in the sample) for the hapten-antibody. In the presence of free hapten, immune complex formation is decreased, since more antibody sites are saturated; thus, light scattering is decreased. The decrease of light scattering is related to the concentration of free hapten.

Particle Counting Immunoassay (PACIA)

PACIA was developed to increase the detection sensitivity of the antigen-antibody reaction and can be used for quantitation of either antibodies or antigens. For quantitating antigens in patient specimens, the appropriate antibody is bound to latex particles that are then mixed with the patient specimen. Antigen in the specimen binds to the antibodies on the latex particles and causes them to agglutinate, forming particles with a wide range of sizes. After the reaction is complete, the latex aggregates are counted in a darkfield illuminated cell counter with a size threshold set at ~1.2 μm. Only the unbound latex particles are counted, and their concentration is inversely proportional to the concentration of antigen in the patient specimen. With this technique, detection limits of about 1 ng/L can be achieved.

Techniques for Detecting Primary Association Between Antigen and Antibody

The methods discussed so far rely on examining immune complex formation as an index of antigen-antibody reaction. Recall, however, that in equation (1) the overall reaction is shown in phases, and only in the final phase are crosslinked immune complexes formed. For measuring the initial phase (i.e., the primary reaction), different techniques are required (Table 3-2). The following section will discuss only those methods appropriate for clinical chemistry laboratory use.

Competitive immunoassays share a common principle: labeled and unlabeled antigen compete for a limited number of protein binding sites (Figure 3-9). This technique is commonly used to measure the concentrations of hormones or drugs (antigen). The label on the antigen may vary, as does the manner of measuring the label. Also, the binding protein may be an antibody, serum or tissue receptor protein, serum binding protein, or a cell membrane receptor. Radioimmunoassay, using a radionuclide as label, is the oldest of these methods and will be discussed first.

Radioimmunoassay (RIA) and immunoradiometric assay (IRMA) methods are capable of measuring the primary reaction between hapten or antigen and a single antibody. In RIA, the hapten or antigen is labeled with a radioactive isotope, whereas in IRMA, the antibody is labeled. In RIA or IRMA, we are concerned only with the primary combination of antigen and antibody.

$$\mathrm{Ab} + \mathrm{Ag} \underset{k_{-1}}{\overset{k_1}{\rightleftharpoons}} \mathrm{AbAg} \tag{3}$$

$$K = \frac{(\mathrm{AbAg})}{(\mathrm{Ab})(\mathrm{Ag})} \tag{4}$$

where k_1 is the rate constant for the forward reaction, k_{-1} is the rate constant for the reverse reaction, and K is the equilibrium constant for the overall reaction. As would be predicted from the law of mass action, the relative concentrations of Ab, Ag, and AbAg will be dependent on the magnitude of k_1 and k_{-1}.

The two standard procedures for RIA are the competitive and the sequential assays. In a *competitive* RIA, all reactants are mixed together simultaneously. Labeled antigen (Ag*) and unlabeled antigen (Ag) compete for binding to the antibody. In such a system, it is very desirable for the avidity of the antibody for both the labeled and unlabeled antigen to be the same. Under these conditions, the probability of the antibody binding the labeled antigen is inversely proportional to the concentration of unlabeled antigen; hence, bound counts (Ag*Ab) are inversely proportional to unlabeled antigen concentration (see Figure 3-9). For some antisera, a competitive RIA is not as sensitive as sequential RIA, especially in cases where the antiserum has different avidities for the labeled and unlabeled antigen.

In the *sequential* approach, unlabeled antigen is first mixed with excess antibody

Table 3-2. METHODS FOR MEASURING THE PRIMARY REACTION OF ANTIGEN AND ANTIBODY

Competitive Immunoassays	Noncompetitive Immunoassays
RIA	IRMA
EMIT	ELISA
FPIA	

Ag	+	Ag*	+	Ab	⇌	AgAb	+	Ag*Ab	+	Ag*	% Ag* Bound	% Ag* Free
0		80		40		0		40		40	40/80=50%	40/80=50%
40		80		40		13		27		53	27/80=34%	53/80=66%
80		80		40		20		20		60	20/80=25%	60/80=75%
120		80		40		24		16		64	16/80=20%	64/80=80%
160		80		40		27		13		67	13/80=16%	67/80=84%

Figure 3-9. Principles of competitive protein binding. Ag = unlabeled antigen, Ag* = labeled antigen, Ab = antibody or binding protein. The numbers refer to relative concentrations of each reaction component. In an assay, Ag* must be greater than Ab, and concentrations of Ag* and Ab are kept constant so that unlabeled Ag is the only variable. Under these conditions, as unlabeled Ag increases, bound Ag* (Ag*Ab) decreases due to increased displacement of unlabeled Ag from Ab binding sites; conversely, free Ag* increases.

and binding is allowed to achieve equilibrium. Next, labeled antigen is added and allowed to equilibrate, and then after separation, the bound and free counts are determined.

Step 1 $$\text{Ag} + \text{Ab} \underset{k_{-1}}{\overset{k_1}{\rightleftharpoons}} \text{AgAb} + \text{Ab}$$

Step 2 $$\text{AgAb} + \text{Ab} + \text{Ag}^* \rightleftharpoons \text{AgAb} + \text{Ag}^*\text{Ab} + \text{Ag}^*$$

With this approach, a higher fraction of the unlabeled antigen can be bound by the antibody than in a competitive assay, especially at low antigen concentrations. Sequential assays can provide a two- to fourfold increase in sensitivity compared to a competitive assay, provided k_1 is significantly greater than k_{-1}. The increase in sensitivity results from an increase in AgAb binding (and thus in a decrease in Ag* binding), which is favored by the sequential addition of Ag and Ag*. If k_1 is slightly greater than or equal to k_{-1}, sensitivity decreases, because dissociation of AgAb becomes more likely and competition between Ag* and Ag results.

Radiolabeling of antigen with an isotope can affect the affinity of an antigen for an antibody and the avidity of the antibody. Therefore, to assure that the antibody reacts equally with each form, labeled and unlabeled antigens should always be evaluated when a competitive assay is used.

In order to quantitate the amount of unlabeled antigen (Ag), the next step in the RIA technique is to separate Ag* (free) from Ag*Ab (bound). Separation relies upon some physical method of separation (Table 3-3). Three general types of separation are commonly used. In the first type, retention of the free low-molecular-weight antigen is achieved by *adsorption or steric exclusion.* Adsorbents used for this purpose include activated charcoal, dextran-coated charcoal, ion exchange resin, magnesium silicate (Florisil), fuller's earth, and talc. The disadvantage of this method is that the time of contact between the adsorbent and the incubation mixture is critical, especially for the more active adsorbents such as activated charcoal. In some cases timing has to be controlled to within seconds in order to get reproducible results.

The second type of separation involves *precipitation of the bound antigen* from the solution by using a protein precipitant such as ammonium sulfate, ethanol, dioxane, or polyethylene glycol (polymer precipitation). The bound antigen can also be precipitated immunologically with a second antibody. For example, if the primary antibody is derived

Table 3-3. PHASE SEPARATION METHODS FOR IMMUNOASSAYS

Adsorption	Double antibody precipitation
Ion exchange	Solid phase antibodies
Gel filtration	Electrophoresis
Solvent or salt precipitation	Protein A
Polymer precipitation	Biotin-avidin

from rabbits, the second antibody can be an antiserum against rabbit γ-globulin derived from goats or sheep. This method has the advantage that it can be used for practically any assay. It has the disadvantage, however, that it usually requires longer assay times and additional steps.

The third type of method utilizes *solid phase antibodies*. In this method, the binding of antigen occurs on a solid surface to which the antibody is attached. The solid surface may be the inside surface of a plastic tube, a polymerized antibody, or antibody conjugated by a covalent bond to insoluble inert materials, such as cellulose, Sephadex, glass beads, or iron particles. After incubation, the solid phase antibody along with bound antigen can be sedimented to the bottom of the tube by centrifugation or, in the case of iron particles, separated by use of magnets. The supernatant containing free antigen can then easily be siphoned off or decanted.

The use of solid phase separation makes automation easier. In addition, when the antibody is bound to a solid phase, the protein exists in a microenvironment that is different from that of a protein that is free in solution. The water surrounding the protein is more highly ordered near the surface of the solid phase, which is more favorable for van der Waals–London dipole-dipole interaction and coulombic bonding. This situation favors the formation of antigen-antibody complexes and hence can provide better sensitivity for analytical applications.

After separation, either the free or bound Ag* is counted in a scintillation counter (see Chapter 2, Section Seven). Standard solutions are run in parallel with patient samples for use in the preparation of a standard curve. Typical standard curves of raw data are shown in Figure 3-10,*A* and *B*. Numerous mathematical transformations such as log and logit functions are available to obtain a linearized relationship; two such relationships are shown in Figure 3-10,*C* and *D*.

Although RIA can be used for the measurement of protein antigens, its primary application has been for quantitating haptens or peptides, particularly hormones or drugs. In most cases, these molecules have only one epitope and therefore are more suitable for RIA. (See applications in Chapters 18, 26, and 28.)

IRMA assays, using labeled antibodies, have the advantage of not requiring purified labeled antigen as a reagent. This obviates potential problems that may be caused by iodination of labile antigens. Antibodies are more stable proteins and can be labeled without damaging the protein's function. Typically, a "sandwich" or two-site IRMA method is used. In such a system, antibody is first attached to a solid phase by passive adsorption or by covalent binding of the antibody to the solid phase. Antigen from the sample is then allowed to react with the solid phase antibody, other protein is washed away, and a radionuclide-labeled antibody is added that reacts with the bound antigen through a second and distinct epitope. After a second washing, the bound counts are determined and are directly proportional to the concentration of antigen. IRMA is a common approach for protein antigen measurement owing to its simplicity and ease of application in the routine laboratory.

Indirect IRMAs have also been described. In such a method, the labeled antibody is reacted with the antigen of interest, which is bound to a solid phase (e.g., cell mem-

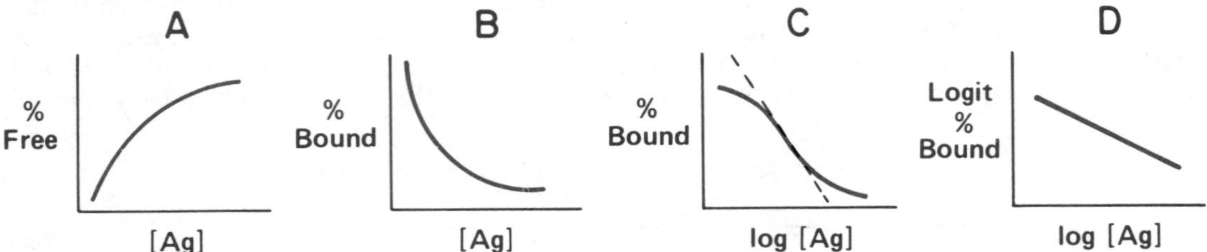

Figure 3-10. Schematic diagrams of dose-response curves using data from a competitive protein binding assay. Logit %B (bound) = $\text{Log}\left[\frac{\%B}{100 - \%B}\right]$.

brane). After removal of the bound antibody, the excess labeled antibody, which remains in solution, can then be measured and is inversely related to the antigen concentration.

Nonisotopic immunoassays. The use of nonisotopic immunoassays in the last few years has become increasingly common in both research and clinical laboratories. Although these immunoassays do not employ radioactive labels, they may possess sensitivities similar to those of classic RIA. The immunochemical mechanisms and principles are the same as for RIA; only the label and method of measurement are different. EMIT (Enzyme-Multiplied Immunoassay Technique) is one example of a nonisotopic immunoassay currently used primarily for drug assays. In this procedure antibody against the drug to be measured is added to the patient's sample. Enzyme-labeled drug is then added that reacts with remaining antibody. The binding of antibody to the drug-enzyme molecule affects enzyme activity by physically blocking access of the substrate to the active site or by changing the conformation of the enzyme molecule. In the final step an appropriate substrate is added to measure the enzyme activity. The decrease in enzyme activity is proportional to the drug concentration in the patient's sample. Some EMIT systems also work in the reverse order; i.e., the enzyme activity is enhanced by binding of the antigen-enzyme molecule to the antibody. The EMIT system is considered a *homogeneous* system, since quantitation can be accomplished without separating the bound and free fractions of the antibody. One of the major goals in the design of all nonisotopic immunoassays has been to provide a homogeneous system in which no separation of bound and free antibody or antigen is necessary and, therefore, to make these assays technically easier and faster.

Enzyme-linked immunosorbent assay (ELISA) is a heterogeneous enzyme immunoassay technique. The technique is similar to the IRMA assays except that the antibody has an enzyme label instead of a radionuclide. In the most common approach, antibody is nonspecifically adsorbed to a solid phase, such as a microtiter tray or a plastic bead. Sample or standard containing the antigen to be quantitated is added to, and allowed to bind to, the solid phase antibody. After washing, enzyme-labeled antibody (different from the solid phase antibody) is added and forms a "sandwich complex" of solid phase Ab-Ag-Ab′-enzyme. Excess enzyme-labeled antibody is then washed away and enzyme substrate is added; the enzyme hydrolyzes (or oxidizes) the substrate, causing color formation proportional to the amount of antigen in the sample. Antibodies in a sample can also be quantitated by an ELISA system in which antigen instead of antibody is bound to a solid phase and the second reagent is an enzyme-labeled antibody specific for the sample antibody. The procedure is otherwise identical to that described above for the measurement of antigens.

The choice of an *enzyme label* in any nonisotopic immunoassay depends primarily on the application. Alkaline phosphatase, β-glucuronidase, glucose-6-phosphate dehydrogenase, and horseradish peroxidase are commonly used. Beta-glucuronidase and alkaline phosphatase can be used with either a colorimetric or fluorometric substrate.

Fluorescence polarization immunoassay (FPIA) is discussed in Chapter 2, Section Two.

Other Immunochemical Techniques

Immunocytochemistry

The use of labeled antibody reagents as specific probes for protein and peptide antigens allows the investigator to examine single cells for their ability to synthesize specific markers of cell lines. Enzyme labels provide several advantages over fluorescent labels. They permit the use of fixed tissues embedded in paraffin, which provides excellent preservation of cell morphology and eliminates the problems of autofluorescence from tissue. In addition, immunoperoxidase stains are permanent, and only a standard light microscope is needed to identify labeled features. The peroxidase-antiperoxidase and direct immunoperoxidase methods are the most versatile and generally the most sensitive.[3]

Agglutination Assays

Agglutination assays have been used for many years for the qualitative and quantitative measurements of antigens and antibodies. In an agglutination method, the visible clumping of particles such as cells, bacteria, and latex particles is used as an indicator of the primary reaction of antigen and antibody. Agglutination methods require stable and uniform particles, pure antigen, and specific antibody. IgM antibodies are more likely to produce complete agglutination than are IgG antibodies because of the size and valence of the IgM molecule. As with all immunochemical reactions in which aggregation is the measured endpoint, the ratio of antigen to antibody is critical. Extremes in antigen or antibody concentration will result in inhibition of aggregation.

Hemagglutination refers to agglutination reactions in which the antigen is located on an erythrocyte. Erythrocytes are not only good passive carriers of antigen but are also easily coated with foreign proteins and are easily obtained and stored. For an application of this technique, see Chapter 28, page 909.

Direct testing of erythrocytes for blood group, Rh, and other antigenic types is widely used in blood banks; specific antisera, such as anti-A, anti-C, and anti-Kell, are used to detect such antigens on the erythrocyte surface. In *indirect* or *passive hemagglutination*, the erythrocytes are used as a particulate carrier of foreign antigen (and, in some tests, of antibody); this technique has wide applications. Other materials available in the form of fine particles, such as bentonite and latex, also have been used as antigen carriers, but they are more difficult to coat, standardize, and store. A variation of this technique, known as *hemagglutination inhibition*, determines the ability of antigens, haptens, or other substances to inhibit hemagglutination of sensitized cells (cells coated with antibody). An application of this technique is a pregnancy test described in Chapter 28.

The classic agglutination test can be conducted either on a slide or in a test tube. Microprocedures that require only small quantities of costly reagents are being used with increasing frequency. These methods are easier to set up, are more sensitive, and produce settling patterns that can be read easily. The choice of animal for the source of erythrocytes is important; ideally one should use the species that produces the least degree of cross reaction. In practice, sheep erythrocytes are used most frequently. Erythrocytes are relatively unstable after depletion of endogenous substrates and accumulation of metabolic products. However, if the cells are collected directly into Alsever's solution (acid-citrate-dextrose solution, pH 6.1, sometimes containing a broad-spectrum antibiotic) and stored at 4 °C, the erythrocytes are usable for up to 30 days. Cells treated with tannic acid or formalin (formalinized cells) can be used for months and may be stored for even longer periods in the frozen and lyophilized state. However, settling patterns may not be as satisfactory as with cells remaining in an aqueous medium and stored in the cold.

In general, the agglutination methods are more sensitive than precipitation techniques but are not as quantitative as other immunochemical methods discussed thus far. Isotopic and nonisotopic immunoassays, especially enzyme immunoassays, are often as convenient as agglutination reactions and are therefore slowly replacing agglutination methods.

References

1. Clarke, H. G. M., and Freeman, T.: A quantitative immunoelectrophoresis method (Laurell electrophoresis). *In*: Protides of the Biological Fluids, 14th colloquium. H. Peeters, Ed. Amsterdam, Elsevier, 1967, pp. 503-509.
2. Daniels, J. C.: Practical applications of gel diffusion tests. *In*: Immunoassays in the Clinical Laboratory, R. M. Kakamura, W. M. Dito, and E. S. Tucker III, Eds. New York, Alan R. Liss, 1979, pp. 52–54.
3. Giegel, J. L., Brotherton, M. M., Cronin, P., et al.: Radial partition immunoassay. Clin. Chem., *28*:1894–1898, 1982.
4. Grossberg, A. L., Chen, C. C., Rendina, L., et al.: Specific cation effects with antibody to a hapten with a positive charge. J. Immunol., *88*:600–603, 1962.
5. Kabat, E. A.: Structural Concepts in Immunology and Immunochemistry. 2nd ed. New York, Holt, Rinehart and Winston, 1976.
6. Laurell, C.-B.: Antigen-antibody crossed electrophoresis. Anal. Biochem., *10*:358–361, 1965.

7. Laurell, C.-B.: Electroimmunoassay. Scand. J. Clin. Lab. Invest., *29* (Suppl. 124): 21–37, 1972.
8. Laurent, T. C., and Ogston, A. G.: The interaction between polysaccharides and other macromolecules. 4: The osmotic pressure of mixtures of serum albumin and hyaluronic acid. Biochem. J., *89*:249–253, 1963.
9. Ogston, A. G., and Phelps, C. F.: The partition of solutes between buffer solutions containing hyaluronic acid. Biochem. J., *78*:827–833, 1961.
10. Pressman, D., Nisonoff, A., and Radzimski, G.: Specific anion effects with antibenzoate antibody. J. Immunol., *86*:35–41, 1961.
11. Steward, M. W.: Immunochemistry. New York, John Wiley and Sons, 1977, p. 233.

4 AUTOMATION IN THE CLINICAL LABORATORY

ERNEST MACLIN, P. E.
DONALD S. YOUNG, M. B., Ph.D.

Automation, as it is traditionally understood in the clinical laboratory, is the mechanization of an analytical process. When initially introduced, automation mimicked manual test procedures and was applied to those tests requested most often. Precision and sometimes accuracy were improved. Low-cost microprocessors have now been incorporated into most analyzers and are used not only to direct the analytical process but also to manipulate quality control and statistical data, to generate instrument diagnostics, and to collate reports.

Despite the applicability of the same analytical principles to the assay of both serum and urine, most automated systems are designed primarily for serum analysis. Only a few automated systems are provided with operating modes specifically defined for the analysis of urine, since ranges of analyte concentration in urine are very wide and different degrees of dilution of sample are required to stay within the linear range of the chemical system.

In this chapter we discuss the principles that are applied in individual steps of automated analysis, and we illustrate applications of these principles to some representative instruments.*

Instrumental Concepts

It is difficult to classify automated systems. Although individual manufacturers typically use one major approach to automation, many systems incorporate several different concepts into a single instrumental unit. The following definitions are commonly used to describe features of automated analysis in the clinical chemistry laboratory.

Definitions

Batch analysis: A number of specimens are processed in the same analytical session or "run."

Sequential analysis: Specimens in the batch enter the analytical process one after another, and each result or set of results emerges in the same order as the specimens that are entered.

Continuous-flow analysis: Each specimen in the batch passes sequentially through the same continuous stream and is subjected to the same analytical reactions and at the same rate as every other specimen.

Discrete analysis: Each specimen in the batch has its own physical space, and chemical reagents are added to it separate from every other specimen.

Single-channel analysis (also known as single-test analysis): Each specimen is subjected to a single process so that results for a single analyte are produced.

Multiple-channel analysis (also known as multitest analysis): Each specimen is subjected to multiple analytical processes so that a set of test results is obtained for it.

*To avoid repetition, the names and locations of manufacturers of instruments cited in the text are given in full in Appendix 4-II, at the end of the chapter.

Parallel analysis: All specimens are subjected to a series of analytical processes at the same time in a parallel fashion.

Discretionary multiple-channel analysis: Specimens in sequence can be analyzed by any one or more than one of the available processes (methods, channels) as a result of a specific command to the processing system for that specimen.

Random access analysis: Any specimen, by a command to the processing system, can be analyzed by any available process, in or out of sequence with other specimens, and without regard to their initial order.

Centrifugal analyzers incorporate several process concepts. All are single-test batch analyzers; analysis is sequential, discrete, and parallel since specimens are individually added into successive and separate reaction vessels, and the chemical reaction and detection of products for all samples proceed at the same time. Results for an entire batch of specimens are available in 2–13 min. When different batches are performed sequentially in batch mode, 200–450 results may be produced per hour, depending on the system and the length of each type of analysis. Given that appropriate reagents for a particular test have been mounted, keyboard entry (COBAS-BIO, MULTISTAT, or ENCORE) or a card reader (GemENI) designates the operational parameters of a specific analysis.

Several benchtop analyzers use a batch, sequential, discrete approach, with specimens first presented in a platter or tray on a *batch* basis; each is then entered into the analyzer sequentially, with analyses performed on a *discretionary* basis. The widely used discrete analyzers, ASTRA 4 and 8, implement *discrete* pipetting of specimens and reagents with a discretionary selection of tests from an array of several possible tests (multiple channels) and with *sequential* specimen presentation from a batch of specimens. Keyboard commands are used to select the appropriate tests for each specimen.

The *AutoAnalyzer* is a single-channel, continuous-flow, batch analyzer providing one result per analyte for each specimen at a rate of 40–60 specimens per hour. The latest multichannel version of this system is the SMAC. This is an analyzer capable of performing 23 simultaneous analyses on one specimen in a batch of as many as 144 specimens presented to the analyzer. This instrument is nondiscretionary; that is, all 23 tests are done on every specimen.

Random access analyzers such as the *aca*, Ektachem, and DACOS perform analyses on a batch of specimens, and each specimen is analyzed sequentially but only for tests specified. The broad array of tests available gives the instruments some of the same capabilities as the multichannel analyzers, but the tests in the random access analyzers are not preordained, since they are selectable through the use of different vials, liquid reagents (as in the DACOS), reagent packs (*aca*), or thin-film slides (Ektachem). This approach permits the measurement of a variable number and variety of analytes in each specimen. As with discretionary multichannel systems, profiles can be defined for a specimen at the time of specimen entry into the analyzer by means of a keyboard (as used in most systems) or by operator selection of appropriate reagent packs (as required for the *aca*).

The *specimen throughput rate* in most multichannel and random access systems is independent of the reaction times of the chemistry procedures; instead, the rate of analysis depends on a fixed sequence of events involving mechanical motions and optical measurement cycles. This is in contrast to batch analyzers, in which throughput rate is affected by the assays performed.

Steps in Automated Analysis

The component steps in automated systems include:

1. Specimen identification
2. Specimen preparation
3. Specimen transport and delivery
4. Removal of protein and other interferents
5. Sample measurement, transport, and delivery
6. Reagent handling

7. Reagent measurement and delivery
8. Chemical reaction phase
9. Measurement phase
10. Signal processing and data handling.

These steps are usually performed sequentially, but in some instruments they may be combined and occur in parallel.

Specimen Identification

Typically, the identifying link between patient and specimen is made at the patient's bedside, and the link between the specimen and test result is made in the laboratory. Linkage in the laboratory is done by a variety of techniques ranging from manual transcription of all information to the generation of labels and printing of reports by the instrument itself. See Chapter 9 for a description of computer-assisted labeling and acquisition as well as load list (worklist) generation. Once the specimen is labeled with its unique accession number, it enters the analytical process. A load list usually associates accession numbers in sequence with the physical positions in the loading zone of the analyzer. This loading zone may be a revolving tray or turntable, a mechanical belt, or a set of racks by which specimens are delivered to the sample aspiration station of the analyzer. In those analyzers that do not automatically link specimen identity and the position where the sample is aspirated, the sequence of results produced must be linked manually with the sequence of specimens entered. Some analyzers print out or transmit to a host computer each result or set of results from a specimen either by its position in the loading zone or by its accession number programmed to that position. When results are printed out only by position number, they must subsequently be hand-transcribed and correlated with the work list.

Clearly, many opportunities arise for mismatching specimens and results.[6] The risks begin at the bedside and are compounded at every processing step that a specimen undergoes between collection from the patient and the analysis by the instrument. Automated analyzers reduce these errors by (1) reading the identification label in the specimen loading zone (as in the SMAC, Figure 4-1,*A*); (2) generating coded labels at the time of test request entry (as in the Paramax, Figure 4-1,*B*); (3) providing the specimen label with the test requisition (as in the Parallel); or (4) reading codes by a separate, off-line light pen in conjunction with separate coded labels that are affixed to the specimen container (as in the ACCLAIM). In all instruments with the capability of scanning coded labels, results are matched with a patient's identification number.

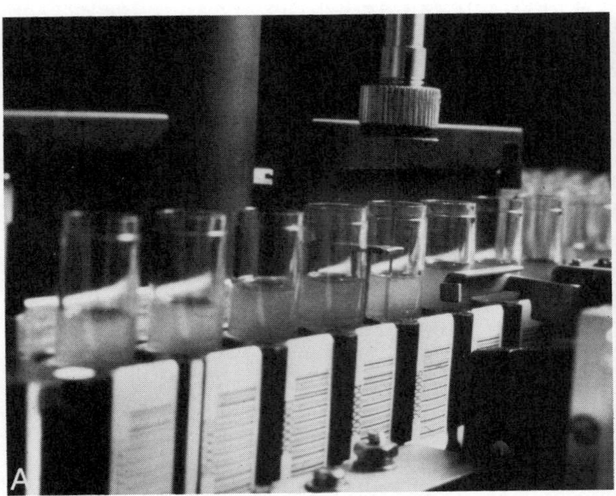

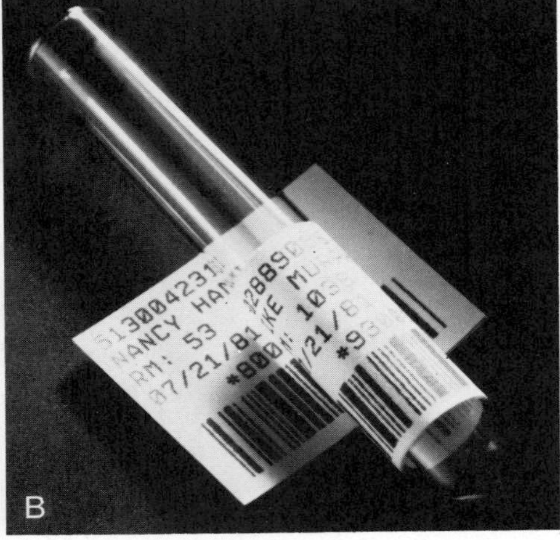

Figure 4-1. Examples of codes used on labels for specimen identification. Labels, as shown, have human-readable as well as machine-readable coding. *A,* Codabar intermixed on SMAC II. (Courtesy of Technicon Instruments Corp.) *B,* Codell on Paramax. (Courtesy of American Dade, Division of American Hospital Supply Corporation.)

Several codes have been used on labels in automated systems. Examples of bar codes that are used in chemistry analyzers are illustrated in Figure 4-1. All labels must be readable by humans as well as machines so that specimens can be located rapidly for emergency tests, or, from storage, for retesting.

Specimen Preparation

The clotting of blood in the specimen collection tubes and the centrifuging of these tubes, with subsequent transfer of serum to secondary tubes, cause delays and expense in the preparation of a specimen for analysis. To date, no system for automatic preparation of original specimens has been incorporated into any analyzer. However, three ongoing developments are noteworthy.

1. *Use of whole blood for analysis*. Automated or semiautomated ion-selective electrodes, which measure ion activity in whole blood, can be incorporated into automated systems to provide test results within minutes of a specimen's being drawn. This approach is already available for the measurement of sodium, potassium, and calcium, and extended application of it to other analytes holds considerable promise for more rapid testing. Another approach involves manual or automated application of whole blood to dry reagent films and visual or instrumental observation of a quantitative change; this approach is exemplified by systems available from Boehringer Mannheim Diagnostics (BMD) and Ames Division, Miles Laboratories, Inc.[14] When whole blood can be added directly to an assay system, specimen preparation time is essentially eliminated.

2. *Automation of specimen preparation*. Instrumentation Laboratory, Inc., has developed a membrane filtration device in a disposable cartridge (Statsep) that produces nonhemolyzed plasma from a specimen of whole blood in 60 s. No clinically significant difference in the concentration of 18 commonly measured analytes has been observed between plasma obtained by this device and by centrifugation. This approach has the potential to reduce the delay between specimen receipt in the clinical laboratory and the production of test results, but it has not yet been integrated into any automated system.

3. *Use of robotics*. Robots can be used to mechanize the specimen preparation steps in a manner analogous to those automated systems that implement manual analytical procedures. A modular robotics system that has "hands" and "fingers," positions tubes and syringes relative to various specimen racks and centrifuges, and can pipet, pour, and mix solutions has been offered by Zymark Corporation.

Specimen Handling, Presentation, and Transport

In most situations, the specimen presented to an automated analyzer is serum. A few analyzers (e.g., SMAC and Parallel) have been designed to sample a supernatant serum from the original collection tube. In these cases, collecting tubes frequently contain a separator material that forms a barrier between supernatant and cells (see Chapter 10). Most analyzers, however, sample from a cup or tube filled with serum that was transferred from the original specimen tube. The design of the cups should minimize dead volume—the excess serum that must be present in a cup to permit aspiration of the full volume required for testing. Specimen containers should be made of glass or polyvinyl; if made of polystyrene, it must be demonstrated that they do not add material that is a positive interferent and that they do not remove analyte from the specimen. Specimen cups should also be disposable to lower cost and minimize cross contamination.

Evaporation of specimen from cups in the loading zone may cause analytical errors as great as 50% over 4 h.[3] The extent of fluid loss depends on environmental factors (temperature, relative humidity, and air flow) and the shape of the cup. A fluid surface area that is small in relation to the depth of the specimen reduces the rate of evaporation. Cups for standards and controls should be freshly filled for every run to avoid errors due to evaporation. Many manufacturers of automated analyzers provide covers for individual cups or for part or all of the loading zone to reduce evaporation.

Temperature-labile analytes may undergo degradation if held at ambient temperature. Constituents that are photolabile, such as bilirubin, are destroyed by ultraviolet light. The adverse effects of ambient temperature could be avoided by refrigeration of both specimens and standards in the loading zone, but no automated analyzer yet includes such refrigeration as an integral unit. Photodegradation is reduced by the use of semiopaque cups and smoke- or orange-colored plastic covers placed over the specimen cups. Such covers are often an integral part of the analyzer, as in the SMAC and ASTRA 8.

The *loading zone* of an analyzer is the area in which specimens are held before they are analyzed. The holding area may be in the form of a circular tray (Hitachi 705 and ASTRA), a rack or series of racks in the form of a cassette (ACCLAIM, SMAC), or a serpentine chain of containers (DEMAND) into which individual tubes are inserted. When specimens are not identified automatically, they must be presented to the sampling device in the correct sequence. The sampling mechanism determines the exact volume of sample removed from the specimen.

For most analyzers, specimens for a second run may be prepared on a separate tray while one run is in progress. This permits machine operation and human actions to proceed in parallel for optimal efficiency. In some analyzers, specimens may be added continuously by the operator as they become available. A desirable feature of any automated analyzer is the ability to insert new specimens ahead of specimens already in place in the loading zone to allow the timely analysis of specimens with a high medical priority.

Removal of Proteins and Other Interferents

Removal of proteins and other interferents is often necessary to assure specificity of an analytical method. Several approaches have been used.

Dialysis is a process by which constituents of low molecular weight are separated across a semipermeable membrane from compounds with a high molecular weight (e.g., proteins). Dialysis is used in some continuous-flow analyzers to deproteinize serum; deproteinization prevents precipitation of proteins by denaturing reagents in chemical reactions, removes optical interference caused by macromolecules, and extends the life of reusable immobilized enzyme reagents.

The quantity of solute that passes through a dialysis membrane is determined by the concentration gradient across the membrane, the duration of contact of the two solutions, the area of contact, and the temperature, as well as by the thickness and porosity of the membrane. As solutes cross the membrane, their concentration in the donor stream decreases as does their rate of transfer, so that even if the membrane area is doubled, the quantity of solute transferred is not proportionally increased. Other factors that influence the rate of transfer are the size and shape of the molecules, their electrical charge, and the composition of the fluids on either side of the membrane. When specimens differ in character, as is the case with diluted serum or aqueous standards as opposed to serum specimens with relatively high protein content, the rate of dialysis of solutes may also differ.[1]

Column chromatography is used in the DuPont *aca* to remove compounds that would adversely affect test reactions. Gel filters, ion exchangers, or resins are used to eliminate organic salts, proteins, and interfering compounds and to isolate isoenzymes.

Other separation systems are used in "dry chemistry" analyzers. In the Kodak Ektachem, the top spreading layer of the test slide separates compounds of low molecular weight from interferents of large molecular weight. In the BMD and Ames dry chemistry devices, a film membrane is used to exclude blood cells that interfere with the analytical procedures.

Sample Transport and Delivery

The method of specimen transport and delivery into the analyzer is the major difference between continuous-flow and discrete systems. In continuous-flow systems, the sample is aspirated through a probe into a continuous reagent stream, whereas in discrete analyzers the sample is aspirated into a probe and then delivered, often with reagent, through the same orifice into a reaction vessel or other container. *Note*: When an exact aliquot of specimen is taken, the term for this aliquot is "sample."

Continuous-Flow Analyzers

Technicon Instruments Corporation pioneered the use of peristaltic pumps and plastic tubing to move sample and reagents. Peristaltic pumps are also used in many analyzers with ion-selective electrodes. Peristaltic pumps trap a "slug" of fluid between two rollers that occlude the tubing. As the rollers travel over the tubing, the trapped fluid is pushed forward and, as the leading roller lifts from the tubing, is added to the fluid beyond the pump. To assure proportionality between calibrators, controls, and specimens, the pump must act uniformly on the sample tube and the roller speed must remain constant.

For uniformity of sample aspiration, the length of time the sample probe remains in each specimen cup (dwell time) must be the same, and the wash time, or time between specimen aspirations, must also remain constant. An air bubble is aspirated into the sample line as the sample probe lifts out of the specimen. In all continuous-flow systems, the air bubble between specimens and additional bubbles introduced through a separate tubing scour the inside of the tubing, separating specimens and reducing carryover from one specimen to the next. In the SMAC series of continuous-flow analyzers, the sample probe dips three times into a water reservoir between serum specimens, aspirating a bubble of air between each dip. This introduction of air bubbles is very effective at reducing carryover between specimens.

Sample dilution. The volume of all solutions pumped in a continuous-flow analyzer is determined by the inside diameter of the tubing in the pump. Inside diameters of the available pump tubings range from 0.005 to 0.110 inch (0.13–2.79 mm) to give flow rates from 0.015 to 3.90 mL/min. In AutoAnalyzers, a *manifold* is assembled from tubings of the desired diameters to assure proportional mixing of sample and reagents. The manifold is stretched and secured between plastic holders (end-blocks) and fittings. Excessive stretching of the tubings, which occurs as they age, is most marked in large-diameter tubings; for a time the proper tension can be restored by adjusting the end-blocks of the manifold. When this adjustment no longer assures proportionality, the tubings must be replaced, preferably all at one time. Although standard solutions frequently lack protein, and though the viscosity of consecutive fluids in a tube (e.g., standards and serum specimens) may be different, flow rates are basically constant since the peristaltic pumps used are positive displacement systems.

Discrete Processing Systems

Positive liquid-displacement pipets are used for sample handling in most discrete automated systems, including centrifugal analyzers. In most batch and random access systems and in some multitest units, specimens, standards, and controls are delivered by a single pipet to the next stage in the analytical process.

A positive-displacement pipet may be designed for one of two operational modes—either to dispense aspirated sample only into the reaction receptacle or to flush out the sample together with the diluent. These devices require careful control of the speed for aspiration and delivery and for the transition between the two steps, because abrupt changes in velocity lead to inaccuracy and imprecision. Nevertheless, the cycle time may be as short as 3 s. The cycle time of the sample pipet largely governs the throughput of the analyzer. When a liquid (e.g., diluent or reagent) is used as a displacement medium, these pipets are capable of highly reproducible measurement of serum. However, when air is the displacement medium, viscous fluids may be measured with less accuracy. This inaccuracy becomes apparent when sera containing large amounts of lipids or proteins are pipetted and the analytical results are compared against standards that are water based.

Pipets may be categorized as fixed, variable, or selectable volume. Selectable-volume pipets allow the selection of a limited number of predetermined volumes. The size of the sample has an important influence on the reagent volume that is used in a system since the ratio of sample to reagent typically affects the analytical range (linearity limit) for a particular analyte. Generally, pipets with selectable volumes are used in systems that allow many different applications, whereas fixed-volume pipets are usually used in instruments dedicated to performing only a limited variety of tests.

Inaccuracy and imprecision for delivery of samples from pipets should not exceed 1%. Similar performance is desired for reagent delivery. The accuracy of the pipetting must be verified periodically, and such checks are a key factor in assessing the quality of instrument performance. The accuracy and reproducibility of pipets in automated instruments are checked in the same way as they are for pipets in manual systems (see Chapter 1).

Carryover

Carryover between specimens should be kept as small as possible. Most manufacturers of discrete systems reduce the carryover by setting an adequate flush-to-specimen ratio or by incorporating wash stations for the sample probe. The ratio of flush to specimen may be as much as 4:1 to limit carryover to <1%,[4] although recent advances in materials and control of dispenser velocity have permitted lower ratios. The outside of the sample probe is wiped in some instruments to prevent transfer of a portion of the previous specimen into the next specimen cup. A recent innovation in the Technicon RA-1000 uses minute amounts of a fluorocarbon fluid to renew a coating on the probe tip. This coating prevents the sample from adhering to the underlying material.

In discrete systems with disposable reaction vessels and measuring cuvets, any carryover is entirely caused by the pipetting system. In instruments with reusable cuvets or flow cells, carryover can occur at every point through which samples pass sequentially. Disposable sample-probe tips can eliminate between-specimen carryover.

Reagent Handling

Most automated systems use liquid reagents stored in plastic or glass containers. The volumes of reagents stored depend on the number of tests to be performed without operator intervention to refill. Where possible, manufacturers use single reagents for test procedures, although two or more reagents are required for some tests. Prepackaging of liquid reagents provides convenience and independence from variations in water quality. Some of the newer analyzers, such as *aca* (Figure 4-2) and Paramax, use reagents in dry pill form, and others, such as Ektachem and Seralyzer, use reagent-impregnated slides or sticks.

For many analyzers in which specimens are not processed continuously, reagents are stored in laboratory refrigerators and introduced into the instruments as required. In the larger systems, as in the SMAC II and Parallel, sections of the reagent storage

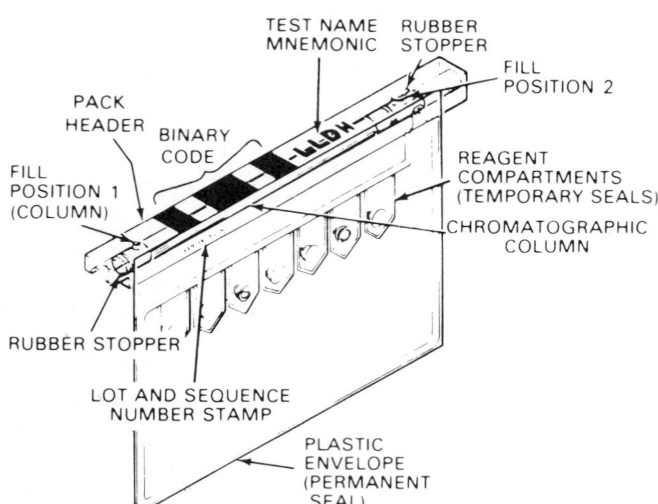

Figure 4-2. Unit analytical test pack used for *aca* discrete clinical analyzer. Reagents in compartments are in the form of liquids and tablets. (Courtesy of DuPont Company.)

compartments are maintained at 4–8 °C. Reagents prepared manually or by automated reconstitution from dry pill or lyophilized powder typically are stable after reconstitution for 2–8 d when stored at 4–8 °C. Some reagents that contain enzymes are supplied as single component liquids to eliminate the need for reconstitution; more often two separate reagents, each with one year's stability, are provided. When the reagents are mixed, they are usually stable for 10 d. This development provides an economical choice in a laboratory where test volume is low, since the long-term stability of reagents yields significant cost savings.

Liquid Reagent Systems

The ASTRA, Parallel, and SMAC continuous-flow analyzers are examples of systems using liquid reagents. Large volumes of reagents, adequate for operation of the analyzer for hours or days, are stored in the systems. The containers are reservoirs that provide reagent continuously, as in the SMAC, or sample by sample, as in the ASTRA and Parallel. Reagents with limited stability may be prepared during the set-up phase and stored during a run in refrigerated space adjoining the system. The DEMAND conserves storage space by placing concentrated reagents in small containers and diluting them from a central reservoir at the time of dispensing.

Dry Reagent Systems

Reagents for the *aca* and Paramax are initially in a dry tablet form. For the *aca*, tablets are in the compartments of a test pack. After sample and diluent have been injected into the pack, mechanical breaker-mixers crush the compartments holding the reagent tablets, thus allowing the reagents to mix with the diluted serum to start the reaction. The Paramax tablet is held in a plastic dispenser and then dispensed into a one-test reaction vessel; fluid is added, and an ultrasonic horn aids in mixing and dissolving the tablet. Sample is then added. Both analyzers illustrate the "unit dose" concept. Unit dose reagents use little storage space and operator time and avoid test-to-test reagent carryover. Dry reagents, although frequently more costly than liquid reagents, have the advantage of long stability and are less likely to be wasted.

Nonliquid Reagent Systems

Unit dose application is carried to a further level of sophistication in systems where no, or very little, liquid is used in the analytic process, as in the Ektachem, Seralyzer, and Stratus. For "color reactions," the Ektachem uses a multilayered, 16-mm square slide (Figure 4-3) in which reagents dispersed in emulsions are activated by diffusion of the sample fluid into the layers. From three to seven layers containing reagents are used for each of the different tests. For electrometric assays the Ektachem also uses slides that incorporate miniature ion-selective electrodes. On these slides, a reference solution and patient specimen provide fluid that turns the electrodes into electrochemical half-cells (see Chapter 19, Section One). Slides of this type that measure sodium, potassium, carbon dioxide, and chloride are in current use.

The Seralyzer system has dry reagents impregnated in a spongelike matrix at the end of a plastic strip. Serum, which is diluted for some tests, activates the reaction when it is deposited on the fiber matrix. The fiber sticks contain little residual water, so they can be packaged in glass or plastic jars without special precautions until opened. The immunoassays of the Stratus use dry reagents in a fiber filter paper matrix in the form of "tabs."

Systems Using Recyclable Reagents

The protein immobilization system used in the Stratus has an analogy in the immobilization of reagent enzymes in continuous-flow (SMAC) and discrete (YSI) systems and of reagent antibodies in continuous-flow immunoassays (ARIA II). In these systems, the analyte in each successive sample is presented to the fixed enzyme or antibody reagent. In the SMAC glucose assay, hexokinase and glucose-6-phosphate dehydrogenase are bound to the inner wall of a coiled tube. Several thousand tests can be performed with one coil during a four- to five-week period. Coils for enzymatic measurement of

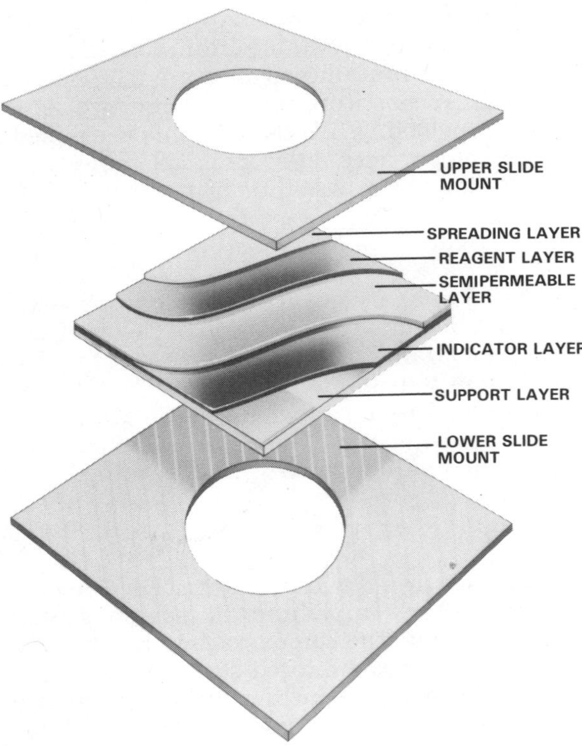

Figure 4-3. Schematic, exploded view of the multilayer dry-film reagent used in Kodak Ektachem analyzers. (Courtesy of Eastman Kodak Company.)

uric acid are also available. In the YSI Glucose Analyzer, the enzyme for the measurement of glucose is immobilized on a membrane of a sensing electrode. The reaction products are then measured by the sensing device. In the ARIA II, antibody is covalently bound to a chamber through which antigen (analyte) and competitive antigen pass. The separation of free from bound antigen is accomplished by a buffer in the first phase; then the immobilized antibody is cleaned of bound antigen by another buffer before the next sample enters the chamber.

Reagent Delivery

Liquid reagents are taken up and delivered to mixing and reaction chambers either by pumping through tubes or by positive-displacement syringes. In continuous-flow systems such as SMAC, a peristaltic pump and manifold both deliver fluids and determine the proportion of reagents to sample. In the ASTRA, a stepper-motor is used to drive a pump that functions in the same way as the peristaltic pump of the AutoAnalyzer by alternately compressing and releasing a single reagent tube. In most high-throughput automated analyzers, reagents and diluent are drawn up from bulk containers through tubes. Syringe devices are common to many automated systems for both reagent and sample delivery. They are usually positive-displacement devices with selectable volumes.

Chemical Reaction Phase

The chemical reaction phase and the measurement of the reaction are incorporated into the design of every automated analyzer, but the manner in which each aspect is handled varies among instruments and depends on other factors. Five of these factors are (1) the container in which the reaction occurs, (2) the mixing and transport of reactants, (3) thermal conditioning of fluids, (4) timing of reactions, and (5) measurement of the reaction.

In **continuous-flow systems**, the reactants are contained in tubing where *mixing* is achieved by tumbling air-segmented slugs of fluids as they pass through helical coils. *Thermal conditioning* occurs as the stream passes through tubing enclosed in baths set at particular temperatures, and *timing of the reaction period* is determined by the distance the stream travels through the tubing and coils between the reagent-sample mixing and the measurement station. *Measurement* of the reaction occurs in a stationary flow cell. Serum blanking, if required, is done simultaneously in a parallel channel.

Discrete analyzers may use as *reaction containers* either (1) individual (disposable or reusable) reaction vessels that are transported through the system after the sample and reagent have been dispensed (*aca*, DEMAND, Parallel) or (2) a stationary reaction chamber, as in the ASTRA. In current models of centrifugal analyzers, the reaction containers become, in effect, a multicuvet insert to the rotor. The *mixing of reactants* is accomplished by a variety of means—forceful dispensing (ABA 100, Abbott VP), magnetic stirring (ASTRA or ACCLAIM), vigorous lateral displacement (RA-1000), or vibrating rod (DEMAND, Hitachi 705) or probe (DACOS). The *aca* utilizes breaker-mixers that knead the fluid or solid chemical in the internal reagent compartments of its flexible test pack and mix it with the diluted serum. The nature of such systems as Ektachem, Seralyzer, and Stratus obviates the need for mixing since the serum completely interacts with the dry chemicals as it flows through the matrix of the reaction unit. *Thermal conditioning* is achieved by temperature-controlled air or liquid baths or by heating blocks placed around vessels or chambers. In the *aca* and DEMAND, *reaction time* is dependent on the rate of transport of the reaction mixture through the system to the measurement station, on timed events of reagent addition (or activation) relative to measurement, or on both. For the ASTRA, reaction time is either a function of length of stay in the chamber for the sample-reagent mixture prior to measurement or of the time interval over which continuous measurement is made. In transported-vessel systems, measurement may be made on fluid in the original vessel that has been transported into the measurement station or in a stationary flow cell at the measurement station to which the fluid has been transported. If the first approach is used, the optical quality of the reaction vessel is an important consideration; if the reaction vessel is reusable, it must be cleaned well before it is recycled. Obviously, cleansing of a flow cell or reaction chamber is important to minimize carryover.

Mixing and Incubation in Centrifugal Analyzers

In centrifugal analyzers, the functions of mixing, thermal equilibration, transport, and measurement are integrated into a rotor assembly. The basic principle of operation is illustrated in Figure 4-4. Within the rotor, a set of cavities radially aligned to a corresponding cuvet creates a set of discrete reaction units. Each specimen is processed in a separate reaction unit. A light source and a monochromator are arranged on one side of the rotor so that a stationary light passes through the cuvets to a photodetector on the other side while the cuvets spin through the light beam (Figure 4-5). One cuvet with a water or reagent blank permits setting to 100% transmission, and the solid portion of the rotor between every cuvet allows setting of 0% transmission. Absorbance measurements in each cuvet are acquired (e.g., every 100 milliseconds) and averaged to enhance precision.

Serum and reagents are placed in the inner cavities of the rotor disk. In most systems, fluids are pipetted into compartments of the rotor by a separate instrument module, and the rotor is subsequently mounted into the instrument. As the disk is rotated rapidly about its axis in the analyzer, the reagents and the sample flow over the walls of their cavities and come together in the outer chamber of the rotor. Partial mixing occurs with this transfer of fluids; when the rotor is then abruptly braked and accelerated again, the fluids are mixed completely by agitation.

Methods of thermal equilibration differ in each system. To ensure rapid transfer of heat into the fluids, some rotors are manufactured of aluminum or graphite covered with Teflon. The usual heat sources are air baths or electric heaters mounted directly on the rotor.

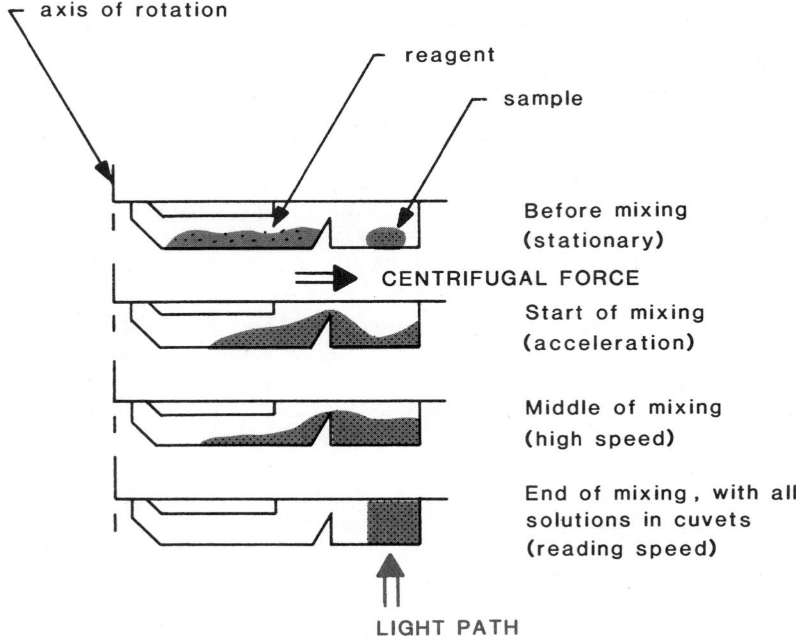

Figure 4-4. Sequence of transfer of solutions and mixing in a disposable rotor of a centrifugal fast analyzer. (Courtesy of Electro-Nucleonics, Inc.)

Reaction Vessels and Cuvets in Discrete Systems

In some discrete systems, reaction vessels are reused; in others they are discarded after use. In some systems, the reaction vessels serve as cuvets, whereas in others, solutions are aspirated from the reaction vessels into separate cuvets for measurement. The use of inexpensive disposable cuvets followed the development of improved plastics (notably acrylic and polyvinylchloride). The cuvets must be transparent in the spectral range of interest. The use of disposable cuvets has simplified automation by eliminating carryover and the maintenance of flow cells. Examples of systems using reusable or disposable reaction vessels and cuvets are discussed below.

In the ACCLAIM, reagents are dispensed into polyvinyl reaction cups in a circular table sitting in a heated water bath. Undiluted serum specimens are added to the reaction cups from 45 s to 11 min prior to aspiration into a series of four photometer flow cells. In each cup, a magnet is used to mix the reagents. Several reagents can be added to

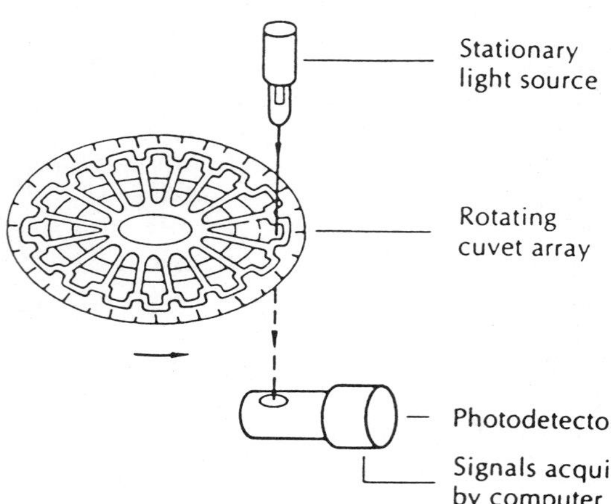

Figure 4-5. Schematic view of operation of centrifugal analyzer. The cuvet array rotates through a stationary light beam that monitors absorbance of each cuvet at every revolution. One cuvet with a water blank permits setting to 100% transmission while the solid portion of the rotor between every cuvet allows setting of 0% transmission. Absorbance measurements in each cuvet are acquired every 100 milliseconds and averaged to enhance precision. (Courtesy of Electro-Nucleonics, Inc.)

each reaction cup so that substrate-initiated reactions can be performed. After the reaction solution has been aspirated into a cuvet, a wash device descends into the reaction vessel to flush it with a high-pressure water jet oriented to scour the sides of the cup. The wash fluid is aspirated and the cup is air-dried, allowing its reuse. Reaction vessels are also reused in the DACOS, but here they are also used as measuring cuvets. Reaction cups or cuvets are arranged in a large, circular rotor that sits in an air bath at 37 °C. The probe used to dispense reagents into the cuvet is oscillated to mix the reactants.

In the SBA 300 batch analyzer the reaction fluids are aspirated into the flow cell of a spectrophotometer. Carryover is minimized by passing a portion of the reaction mixture, equal to 5–10 times the internal volume of the flow cell, through the cell prior to the actual photometric measurement.

Figure 4-6 illustrates the Technicon RA-1000 in which an array of cuvets is used. The unit is slued rapidly to mix reagents in the cuvets. Each cuvet is designed so that the fluorocarbon fluid dispensed through the sample probe to prevent carryover (see page 166) does not interfere with the optical path. At the end of each group of reactions, the rotor is manually removed and replaced. Other instruments, such as the Abbott VP, also use a disposable circular array of reaction cuvets. In these, high-velocity injection of a reagent into the cuvets is used to mix solutions, and optical measurements are made through the water bath in which the cuvet holder sits. The water-cuvet interface reduces reflections that might increase imprecision.

The unit test pack of the *aca*, made of a flexible ionomer (Surlyn), acts as a reagent container, mixing vessel, and measuring cuvet. The sample is flushed into the pack by diluent dispensed at the initial filling station. The pack is then attached to a continuous chain to transport it through the instrument. A breaker-mixer breaks the sealed reagent compartments and kneads the pack to mix reagents with sample and diluent. The pack is transported to a measurement station, where a cuvet with a precise 1-cm pathlength is formed by hydraulic pistons that squeeze the flexible pack between quartz windows. A wetting solution applied to the surface of the windows eliminates distortion of the optical interface at the pack walls. After the photometric measurement is made, the entire pack is discarded automatically.

The dry reagent systems such as the Ektachem and the Seralyzer eliminate dispensing and mixing of liquid reagents. Nevertheless, these instruments still require a mechanism to maintain a stable temperature and to provide accurate positioning of the reaction unit for optical measurements.

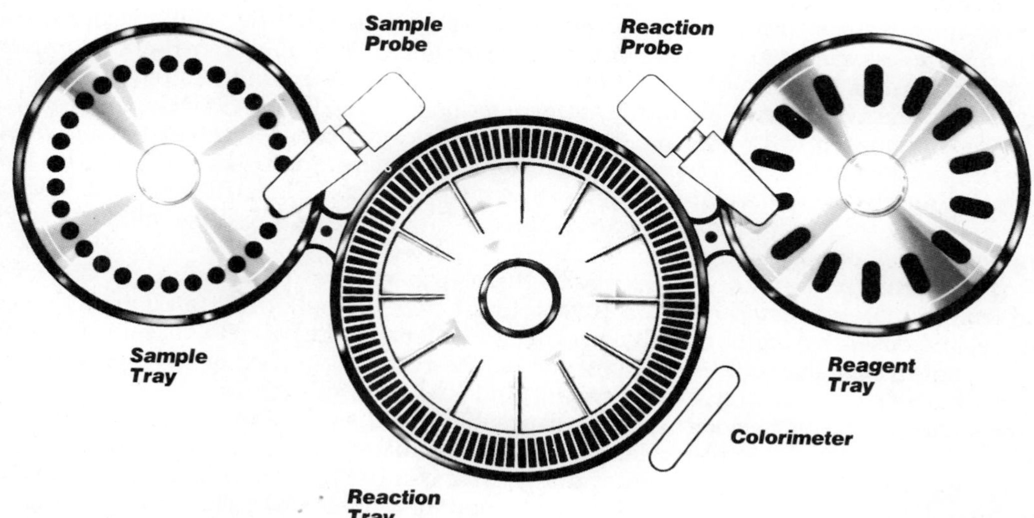

Figure 4-6. Schematic plan of the RA-1000, a random access instrument that employs a 100-cuvet disposable reaction tray. (Courtesy of Technicon Instruments Corp.)

Measurement Approaches

Automated analyzers have traditionally relied on principles of absorbance and transmittance photometry for measurements of analytes. However, in recent years several alternative photometric approaches have been adapted to automated analysis: reflectance photometry (Ektachem), fluorometry (Optimate, Advance), nephelometry (ICS), and fluorescence polarization (TDx). Nonoptical analysis by means of ion-selective electrodes and other electrochemical techniques is becoming increasingly popular. Principles of these measurement techniques are discussed elsewhere (Chapter 2, Sections Two and Four). Here, the special features and applications of the various approaches to automated analysis are reviewed.

Systems Using Absorbance and Transmittance Photometry

The measurement of absorbance requires three basic components—a radiant energy source, a means of spectral isolation, and a detector (see also Chapter 2, Sections One and Two).

The ***radiant energy sources*** used in automated systems include tungsten, quartz halogen, deuterium, mercury, and xenon lamps, as well as lasers. The spectrum produced by a quartz halogen lamp includes wavelengths from approximately 300 to 700 nm. In the COBAS-BIO, a pulsed xenon lamp is used to provide a high-intensity, long-lived source. Some nephelometric instrument systems use lasers to provide monochromatic light at discrete wavelengths (see Appendix 4-I,*B*). The ICS uses a tungsten lamp.

Spectral isolation is most commonly achieved with interference filters in automated systems. In several of the multitest analyzers in which a single cuvet is used for all photometric measurements, the required filters are mounted in a wheel and the appropriate filter is moved into place under command of the microprocessor.

The most popular ***detector*** used in automated systems is the photomultiplier tube (PMT), although photodiodes have adequate sensitivity for measuring most reactions. PMTs, however, will probably continue to be required to give sufficient sensitivity for fluorescent and chemiluminescent reaction measurements and for measurements that require a fast detector response time.

Several approaches have been used for the *electro-optical integration and packaging* in different analyzers. In centrifugal analyzers, a double-beam spectrophotometric system is, in effect, created by the constant monitoring of a reference solution (usually water) set at 100% transmittance against the zero transmittance obtained as the solid portions of the rotor block the light path during rotation. Absorbance readings are made at very short intervals, such as every 0.2 ms. A logarithmic amplifier or a microprocessor software routine converts transmittance to absorbance. The rapid, continuous sampling of optical signals in centrifugal and other analyzers allows optimal timing of measurements, especially when following kinetic reactions.[12] The DACOS (Discrete Analyzer with Continuous Optical Scanning) uses a similar concept. However, instead of the cuvets' being rapidly rotated with respect to a stationary light beam, in the DACOS, the light beam is rapidly rotated relative to slowly rotating cuvets. Eight filters, each with its own detector, rotate around the quartz halogen light source, creating, in effect, eight double-beam photometers.

The Paramax uses liquid-filled light guides (fiberoptics) to convey light energy to and from remotely located reaction cuvets. Light from a single tungsten quartz halogen source is focused by eight radially arranged condensing lenses onto a spinning filter wheel containing several filters (Figure 4-7). The resulting pulses of light of seven different wavelengths are conveyed by light guides to each of eight photometric read positions. During the 5-s period that each disposable cuvet is at the read station, it receives 150 pulses of light of each wavelength. The transmitted energy is directed through a second spinning filter wheel to eight radially arranged PMTs. For each measured analyte the microprocessor chooses the appropriate pair of wavelengths that have previously been defined for the bichromatic reading and then processes the signals.

Systems that use time-sharing of components of the optical system tend to be less costly and show good performance and reliability. In general, an electro-optical package

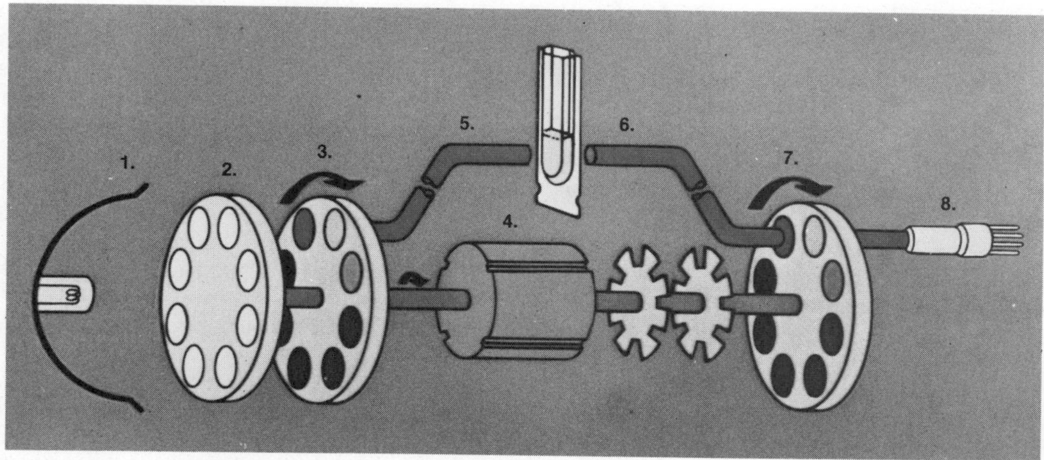

Figure 4-7. Optical arrangement for Paramax. (1) Source with reflector, (2) fixed focusing lens, (3) filter wheel, (4) double-shafted motor, (5) fiber optic bundle, (6) cuvet, (7) filter wheel, (8) photomultiplier tube. (Courtesy of American Dade, Division of American Hospital Supply Corporation.)

should offer an absorbance range of up to 2.5 A in order to permit the extended linearity desirable for some reactions. Detection sensitivity should be 0.001 absorbance unit at 1 A, and the noise level should not exceed 0.0005 A at 1 A, in order to ensure precision adequate for measuring kinetic reactions. The spectral range typically required is 340–650 nm. For enzyme assays, a narrow bandpass is required, but a bandpass that is $\leq$20 nm is satisfactory for most other assays.

Proper alignment of cuvets with the light paths is as important in automated as in manual photometry; stray energy and internal reflections must be kept to acceptable levels. If the light path is not perpendicular to the cuvet, inaccuracy and imprecision may occur, particularly in kinetic analyses, since these are not referenced to a standard.

Other Forms of Photometry

Several forms of photometry other than absorbance photometry are now widely used in automated systems. Among these techniques are reflectance photometry, fluorometry, fluorescence polarization, and nephelometry.

In ***reflectance photometry***, diffuse reflected light is measured. The intensity of the reflected light from the reagent carrier is compared with the intensity of light reflected from a reference surface. Intensity of reflected light is not linear with the concentration of the analyte; two algorithms are commonly utilized to make the relation of reflectance to concentration linear: the Kubelka-Munk equation or the Clapper-Williams transformation.[14] The electro-optical components used in reflectance photometry are essentially the same as those required for absorbance photometry. The Ektachem (see page 180) and Seralyzer employ reflectance photometry with their dry, unit dose reagent systems.

Fluorescence is the emission of electromagnetic radiation by a species that has absorbed exciting radiation from an outside source. Intensity of emitted (fluorescent) light is directly proportional to concentration of the excited species. (see Chapter 2, Section Two). Fluorometry is widely used for automated immunoassay. Sensitivity is about 1000 times greater than for comparable absorbance spectrophotometry, but background interference due to the native fluorescence of some sera may be a major problem. Examples of different optical arrangements are found in the Optimate, Stratus, and the VP series of analyzers.

Right-angle fluorescence measurement is a feature of the Optimate. The system can be used for both absorbance spectrophotometry and fluorometry. The primary path, used for photometric measurements, consists of a quartz halogen lamp, interference filters mounted in a stepper motor–driven filter wheel, the flow cell, and a photomultiplier. A secondary light path, used in fluorometric measurements, starts at right angles to the flow cell, passes through another interference filter mounted in a four-position wheel, and goes to a second photomultiplier.

Front-surface reflectance fluorometry is used in the Stratus, where light passed through a 360-nm filter is reflected from a dichroic mirror onto the front of the reagent tab on which the reaction of the antigen-conjugated enzyme occurs. The light emitted from the front surface of the tab travels back through the dichroic mirror and to a photomultiplier through an interference filter that selects only 450-nm light. Fluorescence intensity is monitored.

Forward-emitted fluorescence is measured in the VP series II-F analyzer with a tungsten-halogen lamp, a photomultiplier tube, and four filters (excitation, reference, emission, reference) mounted on a rapidly spinning filter wheel.[8] As with the Optimate, the approach combines elements for both absorbance and fluorescence detection in a single instrument.

In *fluorescence polarization*, the radiant energy is in the form of polarized light. Measurement is then made of the change in the angle of rotation of polarized light emitted by fluorescent molecules (see Chapter 2, Section Two). Fluorescence polarization is applied in the TDx for homogeneous assays of compounds of small molecular weights.

Turbidimetry and nephelometry are particularly applicable to methods measuring the precipitate formation in antigen-antibody reactions. Details of turbidimetry and nephelometry have been discussed in Chapter 2, Section Two.

The ICS nephelometric system employs a tungsten-iodide source lamp, and a series of focusing lenses and a photomultiplier tube are located at a 70° forward angle to detect scattering of light in the 400–500 nm range. The rate of change of scattered light intensity is monitored; the nonlinear signal is converted mathematically in the microprocessor to achieve a linear relation of rate to concentration.

In the PDQ Laser Nephelometer, the highly collimated beam from a 0.5-m*W* helium-neon laser is directed to the antigen-antibody complexes in the cuvet. A photomultiplier tube measures the scattered-light intensity passing through apertures placed at an angle of 31° above the incident-light beam.

Systems Using Electrochemical Measurements

A variety of electrochemical methods have been incorporated into automated systems. The most widely used electrochemical approach involves ***ion-selective electrodes***. These electrodes have replaced flame photometry for the determination of electrolytes such as sodium and potassium in many analyzers and have lately found direct application in the measurement of other serum constituents such as glucose and urea. The operating principle of ion-selective electrodes is given in some detail in Chapter 2, Section Four. The relationship between ion activity and the concentration of ions in the specimens must be established with standard solutions, and frequent recalibration must be done to compensate for alterations of electrode response.

Electrodes in automated systems may be either of *direct-dip* or *flowthrough design*. In the direct-dip design, as used in the DYNA-LYTE and the IONETICS Dual Analyzer, miniature electrodes are clustered around a reference electrode, and when dipped into the specimen the whole assembly is agitated to ensure the contact of each electrode with sample. In the more widely used flowthrough systems (ORION 1020, NOVA), peristaltic pumps are used to move the sample into chambers containing fixed sample and reference electrodes. Whatever the design, the electrodes must remain in contact with the specimen for a sufficiently long time (typically, 20–30 s) to reach steady-state conditions. The response time of the electrodes, therefore, determines throughput of an automated system, and care must be taken to monitor response time and to restore it, when it degrades, by appropriate maintenance procedures.

Ion-selective electrodes coupled with immobilized enzymes represent an extended application. In such systems, the enzymatic reaction with the analyte involves a product that can be measured electrochemically. For example, in the YSI Glucose Analyzer glucose oxidase is immobilized between two membrane layers and catalyzes the oxidation of glucose in the fluid, which is in contact with the membrane, to produce hydrogen peroxide. The hydrogen peroxide in turn is oxidized at a platinum anode, which causes an electric current directly proportional to the hydrogen peroxide concentration and therefore proportional to glucose concentration.

Coulometry is used in the ASTRA to measure chloride concentration. The specimen is introduced into an acid solution in the reaction cup, which contains a silver ion–generating electrode and detector electrodes. Solid-state electronic circuits monitor the amount of silver ion required to titrate the chloride in solution, and the amount of silver consumed is directly proportional to the concentration of chloride.

Conductimetry is another electrochemical approach used in the ASTRA systems for the measurement of urea nitrogen. The specimen is introduced into the reaction cup along with a buffered urease solution. As urea is converted to ammonium and carbonate ions, the rate of change of conductivity across gold electrodes embedded in the cup is proportional to the concentration of urea nitrogen.

Signal Processing, Data Handling, and Microprocessors

The inclusion of digital computers in automated analyzers has had a major impact on the acquisition and processing of analytical data. Analog signals from detectors are converted rapidly (10^{-3} to 10^{-5} s) to digital form by analog-to-digital converters. The computer then processes the digital data, by means of algorithms, into immediately useful and meaningful output. Data processing can be as sophisticated as computer hardware and software allow. Real-time acquisition and processing of data by microprocessors have allowed the automation of such procedures as nonisotopic immunoassays and reflectance spectrometry because algorithms can readily transform complex, nonlinear standard responses into linear calibration curves. The nature and character of data output can also be as varied as analytical purposes require. Some functions performed by microprocessors incorporated into automated analyzers are listed in Table 4-1.

Examples of Automated Clinical Chemistry Systems

The field of automation is changing rapidly. New instruments are introduced continuously, replacing older systems. Any attempt, therefore, to describe all currently used automated systems is unrealistic. However, to provide a perspective on different approaches to the implementation of automation, six representative systems are described. These systems exemplify the various processes required for automation that have been

Table 4-1. SOME SIGNAL AND DATA PROCESSING FUNCTIONS PERFORMED BY COMPUTERS OF AUTOMATED ANALYZERS

DATA ACQUISITION AND CALCULATION
Acquisition of response signal and signal averaging
Subtraction of blank response
Correction of response of unknown for interferences (e.g., Allen-type corrections)
First-order linear regression for determining slope ($\Delta A/\Delta t$) of rate reactions; ($\Delta A/\Delta C$) of absorbance/concentration relation; ($\Delta R/\Delta C$) of any response measurement to concentration
Statistics (mean, SD, CV) on patient or control values
Mathematical transformation of nonlinear relations to linear counterparts
MONITORING
Test for fit of data to linearity criteria for calibration curves or reaction rates
Test of patient result against reference range criteria
Test of control result against criteria of a quality control standard of performance
DISPLAY
Accumulation of sets of patient results
Collation of results for patient-oriented printout
Display of warning messages to alert operator to instrument malfunction, need for maintenance, or unusual clinical situation

previously discussed. These examples include instruments that employ continuous-flow, parallel, batch, discrete, random access, and unit-dose concepts in their design. Note, however, that these concepts are not exclusive, and some manufacturers utilize several of these characteristics in one system.

From 23 to 61 assays can be performed by the six systems described. Most measure the same analytes, and their methods are quite similar. The key differences are the throughput, response time, and operating economics of the systems and the way in which the test assays are automated, factors that are briefly discussed in a later section on instrument selection.

ASTRA

The ASTRA systems are discrete, multitest, selective, sequential analyzers. The ASTRA 4 is a benchtop unit with slots for four analytical modules, whereas the ASTRA 8 has eight available slots. The ASTRA 8 consists of a sampler, sample assay compartment, video display unit (VDU), printer, programming keyboard, and microprocessor with a disk operating system (Figure 4-8). Thirty-eight specimens and two calibrators are placed in disposable cups on the 40-position specimen changer for presentation to a dual sample probe. One of three possible operating modes is selected at the keyboard: (1) a basic mode in which all specimens are assayed on all channels; (2) a program mode in which the operator selects any combination of assays on an individual specimen or group of specimens; and (3) a stat mode that allows interruption for analysis of a specimen with high priority. The specimen identification system is specimen oriented rather than patient oriented.

A plastic cover placed over the specimen tray reduces evaporation and protects specimens and calibrators from foreign particles. The tray accommodates specimen cups

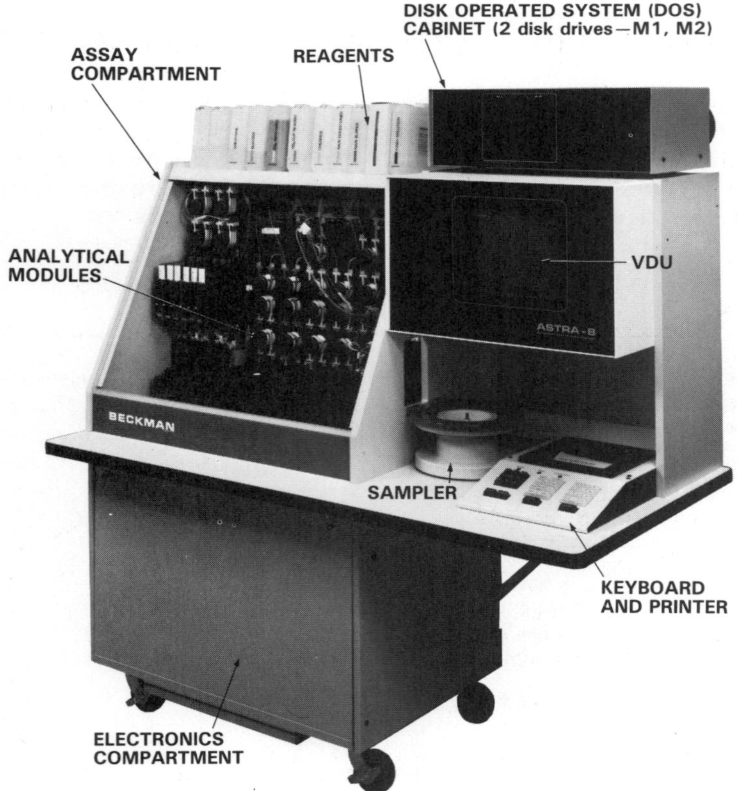

Figure 4-8. The ASTRA 8 is a discrete, discretionary multitest analyzer in which replaceable analytical modules with similar components are used. Detection methods of modules in the assay compartment differ for different analytes. (Courtesy of Beckman Instruments, Inc.)

with capacities of 0.25 to 4.0 mL. A test of conductivity between the two elements of the dual sample probe detects missing cups or insufficient specimen.

Two probes, moving as a single unit, drop through a hole in the specimen tray cover to acquire specimens. Motor-driven syringes sip the required volume of specimen for the tests programmed plus a small excess, as well as two air slugs that are used to minimize cross contamination. All of the sample required for all tests is aspirated at the same time. The time required for sample acquisition and dispensing is reduced by the use of the dual probes. The dual probes move into the analytical section of the instrument and separate to dispense the required specimen volume into the individual reaction chambers without flush. The probes return to a wash cup, and pumps purge the inside and the outside of the sample probes with deionized water containing cleaning solution. The reaction chambers are fixed, cuplike elements that are repetitively filled and drained by peristaltic pumps for successive assays.

Mixing in the reaction chamber is accomplished by stirring with a Teflon-coated magnetic bar. For some assays a heater under the cup provides thermal control of 37 ± 0.1 °C.

The modular design of analytical (and other) units of the ASTRA systems allows the rapid replacement of a malfunctioning element with a spare. The chemistry measurement modules, with the exception of the integrated sensor devices and reaction chambers, are essentially similar. Each has three to four peristaltic pumps: one or two to fill the reaction chamber with reagents, one to sip excess reagent, and one to drain the chamber after the measurement is completed.

Six of the 23 assays available are performed in modules using electrochemical principles; the remaining assays are spectrophotometric.

The video display provides communication between the operator and the microprocessor and is used in conjunction with the system keyboard to display instructions as they are programmed. It is at this keyboard that test selection and numerical patient data are entered. Patient results are displayed at the video display and are printed on a tape.

COBAS-BIO

The COBAS-BIO is a discrete, batch, centrifugal, parallel analyzer. The instrument is a tabletop system that provides a throughput of 200–400 results per hour. It utilizes 30 disposable reaction cuvets assembled in a rotor, a grating monochromator, and a microprocessor to control the total system (Figure 4-9).

Specimens are loaded in sequence, and a printout identifies their positions in the circular specimen tray. Translucent polyethylene cups have a snap top that prevents evaporation of specimen; the sampling probe penetrates the top. A regulated volume of deionized water flushes the sample from the probe into the rotor.

Tests are selected at the keyboard from a list of up to 40 tests previously programmed by the user. When a specific test is selected from a menu, parameters for that test are displayed on the front panel; they include linearity limits, operating temperature, incubation time, pipettor settings, and one of seven report formats.

The loader unit is an integral part of the system. Although this arrangement has some advantages in regard to convenience and space saving, it has the disadvantage that only one tray can be loaded in advance of the analytical run.

The 30-position disposable cuvet ring is manually mounted on the rotor in the analysis station. Twenty-eight of these positions are available for unknowns, standards, or controls, whereas the other two cuvets are reserved for water and reagent blanks. Inside the analysis station, an air bath heats and cools the metal surface on which the cuvet ring is mounted. The temperature of the reaction mixture is maintained to ±0.1 °C. The photodetector is on a hinged arm at the center of rotation in the analysis station.

Reagents and standards are drawn up from a storage tray by two positive-displacement Hamilton syringes. To minimize carryover during fluid aspiration, an air bubble

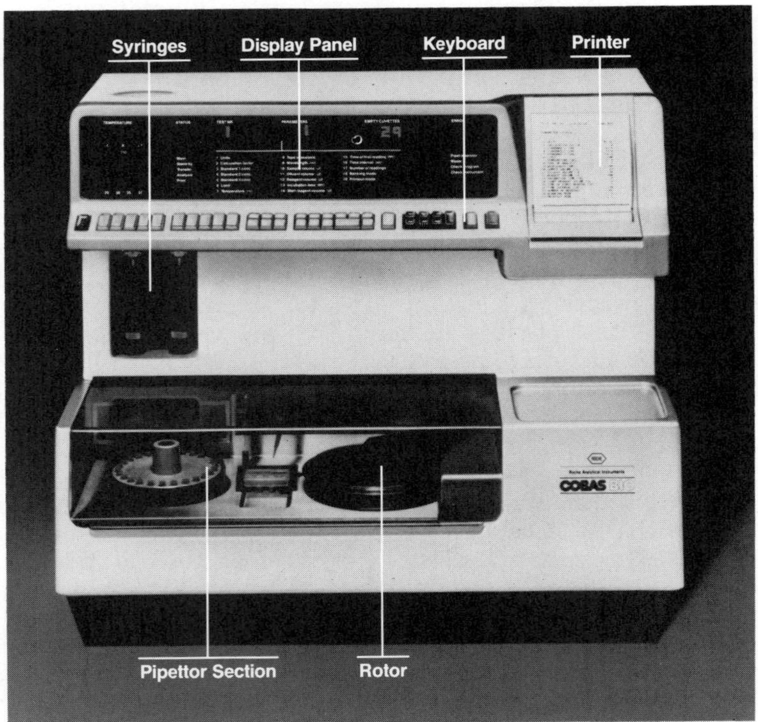

Figure 4-9. COBAS-BIO—a centrifugal, batch, parallel, fast analyzer. (Reprinted with permission of Roche Diagnostic Systems.)

is maintained between the sample and a column of deionized water. Sample washout occurs with the delivery of deionized water, and external cleaning is accomplished by rinsing the sample probe at a wash station. Prior to the transfer of fluid from the specimen ring and reagent trays, the cuvet ring is spun briefly, and satisfactory optical quality of the cuvets is verified.

Wavelength selection is provided by a holographic grating monochromator with a spectral range of 285–750 nm. A pulsed, xenon flash tube provides a high-intensity, long-lived light source; pulsing is phased to those times when readings are required. A semiconductor photodiode serves as detector.

Measurement in the COBAS does not depend on a fixed optical pathlength; the optical path is through the long axis of the fluid contained in the cuvet (Figure 4-5) so that the length is defined by the contained volume. The plane of the meniscus is made flat by the centrifugal field and is parallel to the axis of rotation. In the application of Beer's law that is invoked, the absorbance is independent of the effective length of the light path. Thus, absorbance does not change with a change in reagent volume.[2]

The microprocessor monitors many system features and alerts the operator to errors or problems. Coded computer messages are used for trouble-shooting in conjunction with the service manual and a test board.

Results are printed on a 48-column, thermal strip-printer. The computer has a 64K memory for function control and programming of the system. An RS-232C port allows communication to a central computer.

DEMAND

The DEMAND system is a random access analyzer that utilizes disposable reaction cuvets and concentrated liquid reagents. The system, housed in a floor-model console (Figure 4-10), performs up to 23 different assays at a throughput of 400 results per hour (up to 600 per hour with the ion-selective electrode module).

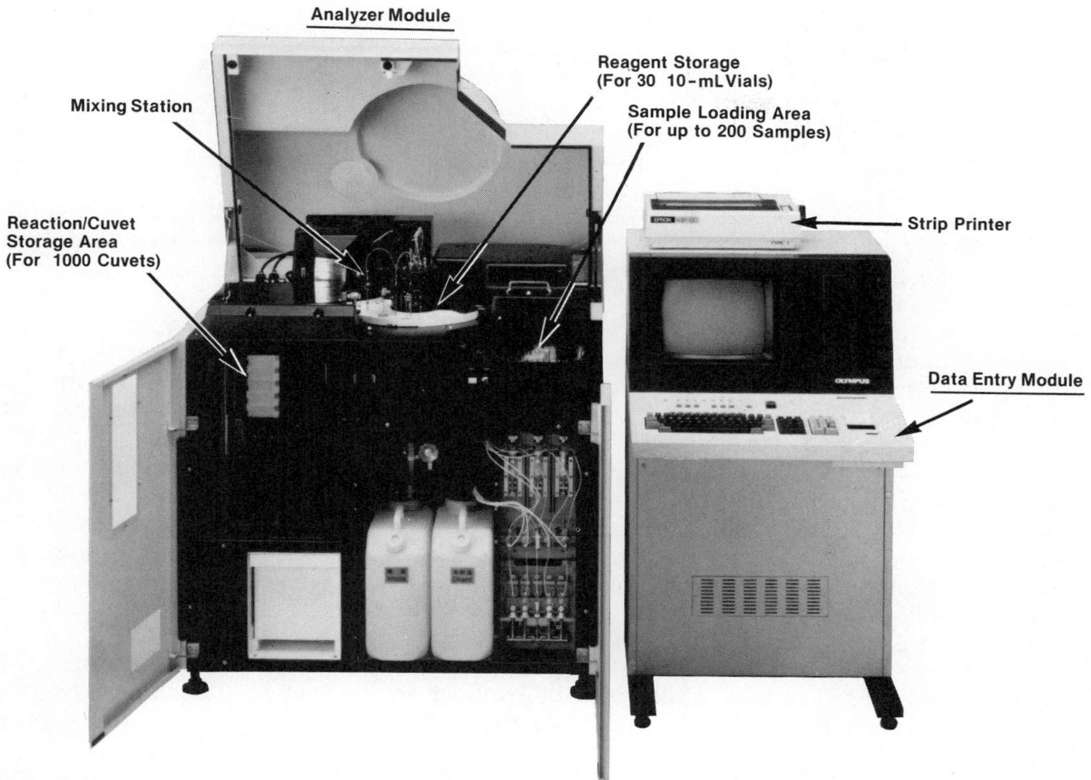

Figure 4-10. DEMAND—a random access analyzer, with major system components shown. (Courtesy of Cooper Biomedical, Inc.)

More than 100 specimens may be presented for sampling by placing them in cups in numbered positions on a serpentine chain. The chain is enclosed in a compartment kept at high relative humidity to reduce evaporation.

The system is "specimen oriented" in that every test requested is performed on each specimen before the next specimen is sampled and tested. Profiles with 2–23 components can be preprogrammed. Emergency tests can be performed by placing the stat specimen in a special position and then instructing the system, by means of the keyboard, to sample and test it next.

Disposable, 10-mm rectangular reaction cuvets are loaded into the lower section of the console. Cuvets are fed by an electrically driven push-bar onto a rotating reaction ring that holds 72 cuvets and indexes every 9 s. At the end of one rotation, used cuvets drop through a chute into a disposal bag. A water bath at 37 ± 0.1 °C maintains the temperature of an aluminum ring in which the reaction cuvets sit.

Sample volumes range from 5 to 25 μL and are metered by a digitally controlled positive-displacement syringe. A separate wash position located on the path from sample cup to cuvet ring supplies deionized water for internal and external washing of the probe. Delivery volumes for concentrated reagents are also set at the keyboard for each of two dispensers. In the dispensing step, the concentrate is diluted 3–5 times with prewarmed, deionized water. Dilution of the reagent with preheated diluent provides reaction aliquots at operating temperatures. Oscillating metal rods provide mixing after each reagent and sample delivery.

Fifteen fixed photometer stations located around the rotating cuvet ring acquire absorbance data on each cuvet every 36 s. The output of a single lamp, located above the plane of the turntable, is directed down through one of ten selectable filters in a wheel. The monochromatic beams are sequentially reflected through a cuvet to each of the 15 silicon photodiode detectors by a digitally controlled, rotating, reflecting prism.

When it passes each of the 15 photometer stations, each cuvet receives light through each of the ten filters every 9 s. Bichromatic absorbance readings are selected from all of those acquired to correct for several types of interference, such as hemolysis, lipemia, and icterus.

EKTACHEM 700

The Ektachem 700 is a random access system that integrates multilayer, film-slide techniques for end point, electrochemical, and enzyme rate measurements within one housing. One slide is used per assay and then discarded. The system has a maximum throughput of 540 results per hour.

Ten specimens are placed in a tray quadrant, and four of these quadrants are assembled to make a complete unit. Specimens are identified at the CRT screen by their position. Each quadrant contains disposable pipet tips mounted concentrically with the specimens. An arm in the metering tower that contains the fluid dispensers for specimens picks up each pipet tip as required (Figure 4-11).

A positive-displacement pump in the metering tower aspirates as many aliquots of specimen as are required for the number of tests requested. Assays require 10 or 11 μL of serum. The aspirator dispenser maintains an air space between the piston and the fluid so that the piston never touches the specimen. Diffusion of the specimen into the film layers accomplishes mixing and initiates the reactions. Each specimen cup is covered with a flexible plastic cap to prevent evaporation. The cap is cross-cut so that the tip of the sample pipet penetrates easily.

The slide supply compartments provide space for 60 cartridges, each containing 50 slides, so that a total of 3000 slides may be in place in the system. The cartridges are bar-coded and are automatically identified by an optical reader in the slide supply compartments.

As individual tests are called up for each specimen, a slide is drawn from the cartridge in the housing and placed in a distributor arm that moves the slide to a metering station below the specimen tray; a metered drop of sample is then injected onto the slide. The potentiometric slides receive, in addition, a 10-μL drop of reference fluid for

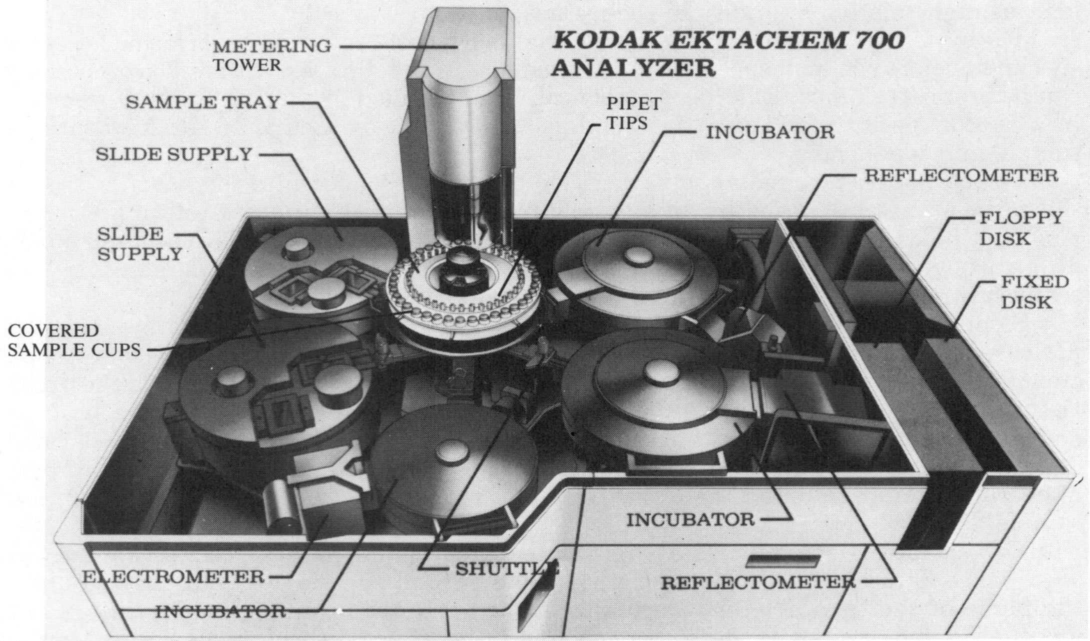

Figure 4-11. Analyzer module for the Ektachem 700. (Courtesy of Eastman Kodak Company.)

a reference electrode; the reference fluid is deposited by a separate dispenser in the metering tower.

After metering, the slide is transferred into the incubator by the distributor arm. The time in each incubator is set to be consistent with the analytical requirements. Thermal control for all slides is obtained through contact of the slide with the rotating thermal mass of the incubator; the enzyme and endpoint measurement disks are maintained at 37 ± 0.1 °C, and slides that have built-in ion-selective electrodes are maintained at 25 °C and monitored by an electrometer.

Endpoint and kinetic enzyme measurements are monitored by separate reflectometers, each with a permanent reflectance standard for a reference. Slides are removed from the colorimetric incubator and placed into the reflectometer so that readings occur at 5 min and every 12 s thereafter. For some endpoint reactions, such as albumin, readings are taken at 2.5 min. A two-point rate determination for amylase uses readings at 2.5 and 5 min.

The operator uses a touch-sensitive CRT to activate the instrument controls and to order tests. Up to 14 test panels can be selected by the operator. The computer monitors instrument functions and environmental conditions and provides self-diagnostic capability in the event that repair is necessary. Diagnostic messages are displayed on the CRT screen to guide the operator through the steps necessary to identify the problem.

PARALLEL

The Parallel analytical system is a floor-model, 30-channel, selective and discrete analyzer (Figure 4-12). It has multiple analytical stations and generates results at a maximum rate of 7200 per hour. As with many discretionary systems, actual throughput depends on the tests performed.

The system consists of five major components: (1) the chemistry console, (2) the electronics console, (3) the database console, (4) the video terminals, and (5) a printer. Each of the individual channels on the chemistry console contains its own sample and

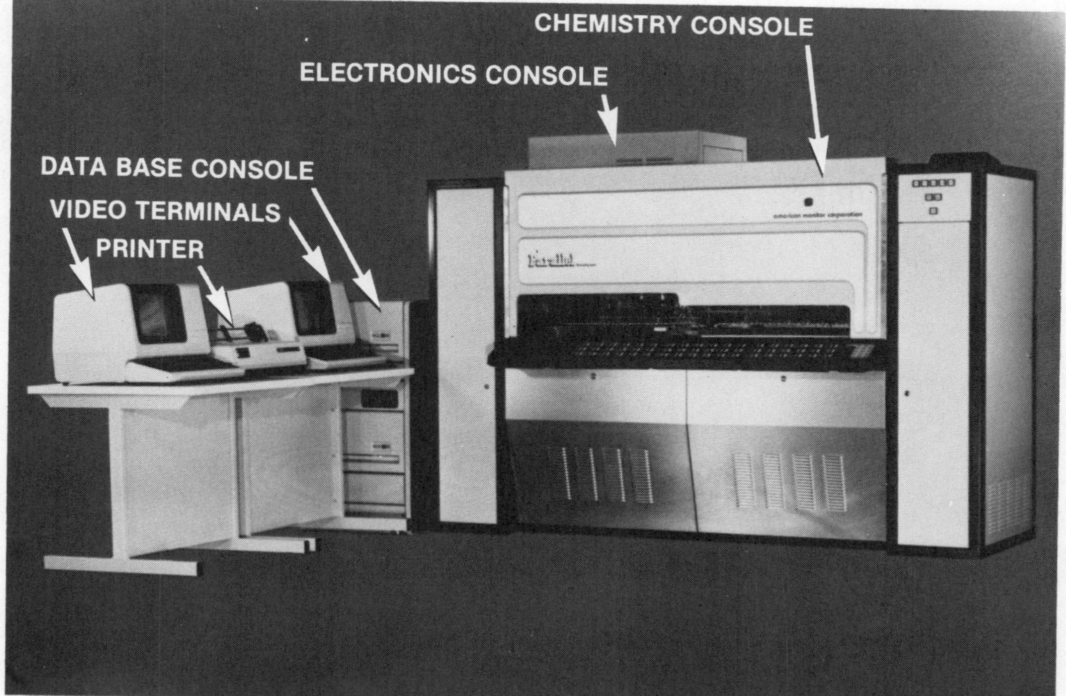

Figure 4-12. Main modules of the Parallel, a multichannel, discretionary, highly computerized system. (Courtesy of American Monitor Corp.)

reagent dispensers, mixers, measurement stations, and wash position. The system has substantial computer capacity to accumulate, process, and display data. User-friendly menu screens contribute to ease of operation.

A key feature of the analyzer is its test selectivity. Any one or a combination of the 30 available tests can be requested as a test panel. A desired test panel can be entered into the system by one of four options: (1) readout of the test schedule and patient's accession number from a bar-code label affixed to the specimen; (2) download of accession numbers and corresponding test schedules from an external laboratory computer; (3) direct entry of the test schedule from the chemistry console's STAT pad; or (4) keyboard entry of the test schedule at the video display unit.

The specimen chain accommodates 150 specimens in a continuous loop. For each assay, an aliquot of the specimen is transferred to one of 30 washable, polypropylene reaction vessels, each with a 2.0-mL capacity. Each group of 30 reaction vessels is carried by one of 68 carrier bars on the system so that a total of 2040 reaction vessels are available. A schematic of the main elements of the analyzer's chemistry console is shown in Figure 4-13.

Sampling for all or any of the 30 channels occurs according to a fixed cycle time of 15 s. Sampling is by positive displacement; a deionized water column is present in the probe between the syringe piston and the aspirated sample. Dispensing is accomplished by passing deionized water through a bypass valve and flushing the specimen from the sample aspiration tip. This sample-dispense design reduces syringe motion and avoids serum contact with the syringe piston. Sample volumes $\leq 100\ \mu L$ can be dispensed.

A fiberglass platen contains fluid-dispensing heads, motorized stirrers, transfer probes, photometers, and wash probes, all of which move up and down above the plane of the reaction vessels once every index cycle. Up to six reagents can be dispensed per channel; the volume of reagents dispensed ranges from 50 to 1300 μL. Bulk liquid reagents are stored in 32 containers at ambient temperature and in 28 containers at 10–14 °C.

A motorized stirrer for each channel is mounted on the platen between the reagent dispenser and the photometer. Up to three mixers can be placed in one channel, although for most procedures only one is required. Twenty-nine transfer probes aspirate reaction mixture to the 10-mm flow cells of 29 photometers. The location of a transfer probe is determined by the reaction requirements of a channel; e.g., one channel is devoted to monitoring specimen absorbance at 500, 546, and 650 nm for subsequent correction of icterus, hemolysis, or turbidity, and another probe leads to a flame photometer for the determination of sodium and potassium. Each flow cell has a measurement volume of 76 μL and is flushed continuously as fresh reagents are aspirated. If no test has been

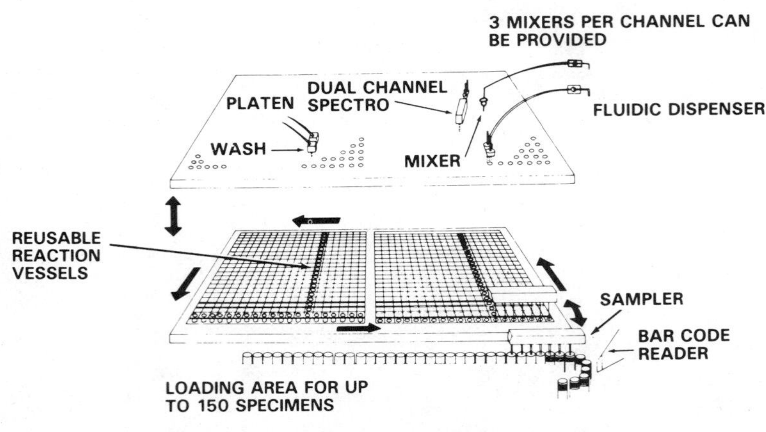

Figure 4-13. Schematic of the main elements of the Parallel chemistry console. (Courtesy of American Monitor Corp.)

requested on a particular channel, its reaction vial is filled with deionized water which, when aspirated into the flow cell, ensures that it is never empty.

Each optical measuring station is a dual-channel filter photometer with a water-cooled tungsten-halogen lamp and an interference filter. The interference filters cover the range of 340-750 nm; they have bandpasses of 7–19 nm. Within each photometer a beam-splitter creates two separate light paths. One beam illuminates a photodetector that regulates output of the light source through a negative feedback network. The other beam passes through an interference filter and then through the flow cell to a second detector for measurement of the absorbance of the reaction mixture.

The spectrophotometric process begins after transfer of the reaction mixture is complete. Readings are processed through a log amplifier and A/D converter and sent to the database console, where various mathematical routines correct for carryover and spectral interference and convert absorbance data to concentration values. Analytical modes for standardizing include linear regression, third-degree polynomial fits, and multiple kinetic standardization relative to a specific reference material. Log-logit transformations for nonisotopic immunoassays are also performed.

The last step in the assay sequence is the wash station, where reaction vessels are cleansed.

The chemistry console is directed by the electronics console, a 64K minicomputer that controls not only the mechanical functions but also the transfer of spectrophotometric data to the database console. The database console consists of a 256K minicomputer with two ten-megabyte hard-disk drives; it executes all mainline programs and provides order entry, result reporting, and data storage and retrieval.

SMAC II

The Technicon SMAC II system (Sequential Multiple Analysis plus Computer) is the most recent model of the continuous-flow systems applying principles devised by Skeggs in the 1950s. Samples, sequentially presented, are split automatically into multiple parallel analytical pathways for the simultaneous assay of 23 serum components.

The system consists of two major subsystems. The analytical processor (AP) instructs and monitors analysis; the results processor (RP) provides data handling of results and release of an edited report, complete with appropriate demographic information. The operating units of the system are shown in Figure 4-14.

The operator initiates analysis through the keyboard of the AP control terminal (Figure 4-14,*A*) by use of command codes and dedicated keys. The operator also defines, through keyboard entries, the cycles for calibration and the frequency of control specimens and intervals for readjustment of reagent baselines. During the run, the operator can display analytical curves of 16 channels at one time on the CRT or can call up management reports of system parameters.

The sampler (Figure 4-14,*B*) has the capacity for 18 carrier blocks, each with space for eight specimen collection tubes or polystyrene specimen cups. Each position in a block may be assigned to a specimen by the operator at the keyboard, or the position may be identified by a bar-coded label affixed to a specimen collection tube or to a specimen position in a block. A fixed module in the center of the sampler holds wash solution, control solutions, calibrators (in a refrigerated portion), and a position for placing a priority (STAT) specimen. Sampling of the STAT specimen is directed from the keyboard and occurs without affecting the sequence of specimens already in the loading zone. Calibrators and controls are sampled in the pattern previously defined by the operator.

Samples are taken into the analytical console (Figure 4-14, *C*) by a probe that dips into and aspirates aliquots from specimens. Approximately 600 μL of specimen is required for the sampling process. The probe is connected by tubing to the dilution cartridge on the analytical console. The major portion of aspirated sample is diluted with distilled water and segmented with air bubbles to minimize sample-to-sample interaction. Diluted sample then travels up a riser in the center of the console past a series of analytical cartridges.

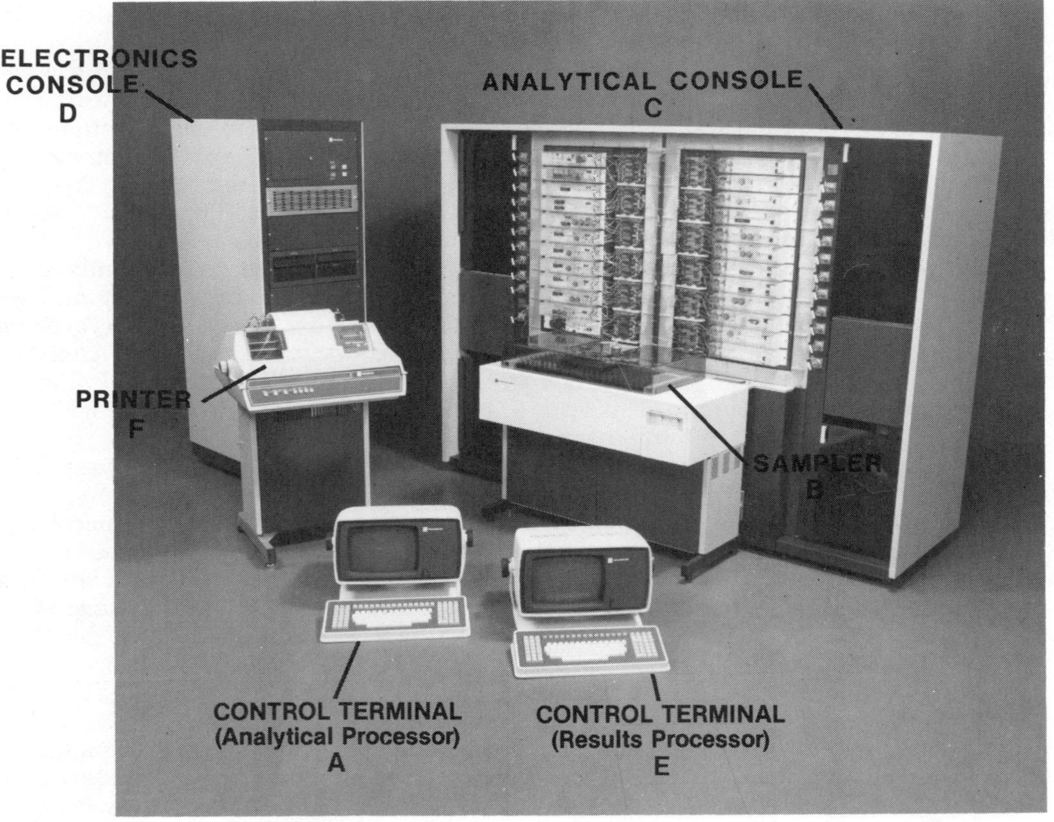

Figure 4-14. Main modules of the SMAC II, a sequential, multichannel, nondiscretionary, continuous-flow analyzer. (Courtesy of Technicon Instruments Corp.)

Each cartridge is a miniature, single-channel analyzer made of the components required for one analysis. A typical cartridge includes a peristaltic pump, pump tubes, mixing coils, injection blocks, and flow cells. When required, an on-line dialyzer for protein separation or an on-line heating bath for color development or both are present. Each cartridge draws from the central riser an aliquot of diluted sample. Reagents are drawn through tubes from an array of reagent containers mounted on both sides of the analytical console.

Analytes are measured by spectrophotometry except for sodium and potassium, which are measured by in-line ion-selective electrodes. Blanking, when necessary, is achieved by a parallel channel on another analytical cartridge. Methods for glucose and uric acid use specific enzymes immobilized in nylon coils and use spectrophotometry for quantitation. Kinetic assays are made by placing multiple flow cells on an analytical cartridge and making multipoint measurements.

Optical fibers transmit light from one of two light sources (UV and visible) to each flow cell and from each flow cell to a single photomultiplier tube detector. A computer synchronizes selected wavelengths provided by filters on a rotating wheel with the appropriate flow cell. The transmittance signal of the photomultiplier is converted to absorbance by a logarithmic amplifier, and the analog signal is digitized through an analog-to-digital converter. Atypical curve shapes due to insufficient sample, poor mixing, or irregularities of flow are detected by the AP, which also alerts the operator to problems and presents a message defining probable cause and suggested remedial action.

Both the AP and the RP interact with the electronics console (Figure 4-14, *D*). Data released from the AP system flow, together with identification number or specimen number, to the RP, where the identification number is the key to linking analytical data to demographic information. The RP terminal and its subsystem manage the review and editing of patient and control data received from the AP. Quality control programs apply Westgard multirule algorithms[15] for testing control data (see Chapter 8).

Selection of Instruments

Automated systems need to be judiciously selected to fit into that combination of automated, semiautomated, and manual modes of analysis that is optimal for the kind and volume of work performed in any one laboratory. Currently, there are no standardized approaches to the selection and evaluation of instrumentation, but useful checklists, workshop manuals, and charts have been published.[5,9,11] A two-phase effort is suggested: the first is selection; the second is evaluation. In both phases the steps are the same, but their sequences differ.

In the *selection* phase, the priority is: first, systematic analysis of the role of an automated system in the workflow and identification of candidate systems; second, detailed analysis of cost-effectiveness of rival candidate systems; third, assessment of acceptability to operators ("user friendliness," serviceability, and similar factors); fourth, characterization of the analytical performance of instruments being considered. At this stage, manufacturers' claims may be assumed to be valid because of today's environment of competitive pressures and government regulations. In the *evaluation* phase, the priority is exactly reversed.

A preliminary screening method for candidate systems has been proposed.[7] In this approach a matrix is set up with one axis listing features to be considered and the other listing candidate systems. The list of features may include any operational, technical, or financial elements of interest. Table 4-2 is an example of such a matrix.[7] In the table are shown values assigned on an arbitrary scale of 0 to 2 that indicate relative importance of an evaluation category according to individual judgment or to a consensus. This approach enables a semiquantitative estimate of a candidate system's agreement with targeted goals. This screening mechanism forces the systematic definition of areas of concern in the selection process and provides a first estimate of the strongest contenders among competing systems.

When the range of choice among systems has been narrowed by the screening process, detailed analyses of costs for acquiring and running the candidate systems are

Table 4-2. PRELIMINARY "WANT" LIST FOR THE EVALUATION OF AUTOMATED ANALYZERS

Instrument Designation	Evaluation Categories*													Total Points
	1	*2*	*3*	*4*	*5*	*6*	*7*	*8*	*9*	*10*	*11*	*12*	*13*	
A	1.0	0.0	0.0	1.0	1.0	1.0	0.5	0.0	0.0	1.0	0.5	0.0	2.0	8.0
B	1.5	0.0	0.0	1.5	1.5	1.0	0.5	0.5	2.0	2.0	2.0	0.0	1.5	14.0
C	0.5	0.0	0.0	0.0	2.0	0.5	1.0	2.0	0.5	2.0	2.0	1.5	2.0	14.0
D	1.0	2.0	2.0	1.5	2.0	0.5	2.0	1.5	0.0	0.0	0.5	0.0	2.0	15.0
E	1.0	2.0	0.0	1.5	2.0	1.5	2.0	0.0	2.0	2.0	0.5	1.0	1.5	17.0
F	2.0	1.5	0.0	1.5	2.0	1.5	1.5	1.0	1.5	2.0	0.5	0.0	2.0	17.0
G	2.0	1.5	0.0	1.5	2.0	2.0	1.5	1.5	0.0	0.0	0.5	0.0	2.0	14.5
H	2.0	2.0	0.0	1.5	1.0	2.0	2.0	0.5	1.5	0.0	0.5	0.0	2.0	15.0
I	1.0	2.0	2.0	0.5	1.0	1.0	2.0	1.5	0.0	0.0	0.5	0.0	2.0	13.5
J	2.0	1.0	0.0	1.5	1.0	0.5	1.0	0.0	0.5	2.0	0.5	2.0	2.0	14.0
K	2.0	0.0	0.0	1.0	1.5	1.0	1.5	0.0	0.0	2.0	0.5	0.0	2.0	11.5

*Categories: 1. sample consumed per test; 2. throughput (testing rate); 3. stat capability; 4. methodology; 5. accuracy and precision; 6. reagent cost/test; 7. manpower requirements; 8. personnel training requirements; 9. computer compatibility; 10. capability to flag instrumental anomalies; 11. automatic deproteinization; 12. sample identification system; 13. selectable tests for profiling.

(Reproduced with permission from Gudzinowicz, B. J., et al.: Chem. Biomed. Environ. Instrument., *11*:107–126, 1981, by courtesy of Marcel Dekker, Inc.)

Table 4-3. ECONOMIC ASPECTS OF INSTRUMENT EVALUATION

Yearly projections for a given test over a specific amortization period for the instrument*:
- Workload
- Manpower needs
- Manpower costs
- Preventive maintenance
- Down-time costs
- Reagent and disposable item costs
- Service costs
- Instrument cost
- Cost adjustments for wage-scale or cost-of-living changes, for effect of inflation on cost of reagents and other consumables

Total costs for a given test over the amortization period of the instrument, which is the sum of:
- Direct costs (operating)
- Indirect costs (overhead)

*For multicomponent analyzers, analysis of costs of a given test will differ from that made for a single-element analyzer, since some costs (e.g., for maintenance and service) must be prorated either over the spectrum of all tests available on the multicomponent analyzer or relative to variable usage of certain tests within the spectrum.

Summarized from Gudzinowicz, B. J., et al.: Chem. Biomed. Environ. Instrument., *9*:315–333, 1979.

required. Several references[5,7,11] are useful in systematizing the process of cost analysis for capital equipment. Table 4-3 lists the economic considerations involved in a selection process.[7] *Acquisition costs* must be considered in terms of possible options of capital investment (purchase), lease, straight rental, or reagent rental. *Operating costs* are another aspect of the analysis. The chief determinants are the labor demand of a system in both operation and maintenance; supply cost, which is a function of both work volumes and the nature of assays on the system; and service requirements, which may be supplied in-house or by a service contract. Systems with an attractively low cost per test may have unattractive total operating costs if using them is labor-intensive in setup and maintenance or if service arrangements are expensive.

The success of many automated systems depends heavily on the interface between the machine and the human operator. Those systems deemed "user friendly" obtain greatest acceptance by laboratory workers; acceptance, in turn, leads to more rapid and less disruptive introduction into the laboratory workflow. It is highly desirable for technologists who will eventually use the systems under consideration to operate them for some period of time in other laboratories, in a demonstration session, or in a preliminary on-site trial.

Once a system has been delivered, *performance evaluation* must ensure that the sensitivity, specificity, accuracy, and precision claimed by the manufacturer are achieved. The NCCLS user-evaluator protocol EP5-T provides a standardized approach to assess precision and carryover for a wide range of clinical values and within a reasonable time period.[10] The use of NCCLS evaluation protocols will not, however, identify specificity problems or the effects of different lots of calibrating materials, nor will it detect variations in operator skills, preanalytical error,[13] or environmental conditions. A major concern is the generation of a falsely negative result that fails to identify patients with an existing disease. Gudzinowicz and coworkers[7] deal with this concern by testing the ability of the overall system (instrument and reagents) to identify the maximum separation between false-negative and false-positive test results.

Instrument systems in use should be demonstrably rugged enough to withstand continual use. Sufficient training and adequate manuals should be provided for operators so that they can detect ordinary malfunctions and make minor on-site repairs. Microprocessor technology has reduced the dependency on manufacturer's service engineers either by incorporating diagnostic routines on software and making available 24-h telephone consultation, or by speedy dispatch of replacement or substitution boards. Overall, the best recommendation for equipment reliability is a well-founded reputation for satisfying past and current users.

Appendix 4-I. SELECTED CHARACTERISTICS OF REPRESENTATIVE INSTRUMENTS

A. PHOTOMETRIC INSTRUMENTS

Name and Categories	Manufacturer	Specimen Loading Zone Capacity	Number of Channels	Approximate Throughput (Results/h)	Cabinet Type	Type of Reagents Used	Measurement Method(s)	Comments
ABAVP Batch, sequential, discrete	Abbott Laboratories	32	N/A	300–465	Table top	Liquid, bulk	Photometric	Disposable cuvet array
aca Discrete, selective random access	DuPont Company	34	N/A	100	Floor console	Tablet and liquid	Photometric and ion-selective electrodes	Disposable, unit dose, cuvet pack; integral columns for separation for some assays
ACCLAIM Discrete, selective random access	Electro-Nucleonics, Inc.	150	N/A	360	Floor console	Liquid	Photometric	Reusable reaction vessels, flow-through cuvets
ASTRA 8 Multichannel, discretionary, discrete	Beckman Instruments, Inc.	40	8	580–720	Floor console	Liquid, bulk	Photometric, ion-selective electrode, coulometric, and conductometric	Reusable reaction vessels
AUTOANALYZER I Single channel, sequential batch, continuous-flow	Technicon Instruments Corp.	40	N/A	60	Table top	Liquid, bulk	Photometric	The basic, initial, continuous-flow, single-channel instrument
COBAS-BIO Discrete, batch, centrifugal, parallel, fast analyzer	Roche Diagnostic Systems	28	N/A	200–400	Table top	Liquid	Photometric, fluorescence polarization	Disposable cuvets
DACOS Discrete, selective random access	Coulter Electronics, Inc.	40+	N/A	600	Floor console	Liquid	Photometric	Semireusable cuvets
DEMAND Discrete, selective random access	Cooper Biomedical, Inc.	100	N/A	400	Floor console	Liquid	Photometric	Disposable cuvet packets
EKTACHEM 400 Discrete, selective random access	Eastman Kodak Company	40	N/A	300	Floor console	Film, unit dose	Reflectometric	Disposable slides
ENCORE Batch, centrifugal, fast analyzer	Baker Instruments Corp.	96	N/A	400	Floor console	Liquid, bulk	Photometric, fluorometric	Reusable rotor; six separate analytes may be run simultaneously on one rotor

Appendix 4-I. SELECTED CHARACTERISTICS OF REPRESENTATIVE INSTRUMENTS (*Continued*)

A. PHOTOMETRIC INSTRUMENTS

Name and Categories	Manufacturer	Specimen Loading Zone Capacity	Number of Channels	Approximate Throughput (Results/h)	Cabinet Type	Type of Reagents Used	Measurement Method(s)	Comments
GemENI Batch, centrifugal, fast analyzer	Electro-Nucleonics, Inc.	20	N/A	200–400	Table top	Liquid, bulk	Photometric	Disposable rotor
HITACHI 705 Discrete, selective random access	Boehringer Mannheim	40	N/A	180 (300 with ISE)	Floor console	Liquid	Photometric; ion-selective electrodes	Reusable cuvets
HITACHI 737 Discrete, selective random access	Boehringer Mannheim	60	N/A	1200	Floor console	Liquid	Photometric	Reusable cuvets
MULTISTAT III PLUS Batch, contrifugal, fast analyzer	Instrumentation Laboratory, Inc.	20	N/A	400	Floor console	Liquid, bulk	Photometric, fluorometric, and nephelometric	Disposable rotor
IMPACT 400 E Batch, sequential, discrete	Gilford Instruments	60	N/A	218–480	Table top	Liquid, bulk	Photometric	
PARALLEL Multichannel, discretionary, discrete	American Monitor Corp.	150	30	7200	Floor console	Liquid, bulk	Photometric and ion-selective electrode	
PARAMAX Discrete, selective random access	American Dade	—	N/A	720	Floor console	Tablet, unit dose	Filter photometer	Disposable cuvets in reels of 2100
RA-1000 Discrete, selective random access	Technicon instruments Corp.	30	N/A	240	Table top	Liquid	Photometric	Disposable cuvet ring
SMA 12/60 Multichannel, sequential, nondiscretionary, continuous-flow	Technicon Instruments Corp.	40	12	720	Floor console	Liquid, bulk	Photometric	
SMAC Multichannel, sequential, nondiscretionary, continuous-flow	Technicon Instruments Corp.	144	23	3400	Floor console	Liquid, bulk	Photometric	

SPIRIT Discrete, selective random access	Baker Instruments Corp.	40	N/A	320	Table top	Liquid	Photometric and ion-selective electrodes	Disposable reaction vessel; reusable flow-through cuvet; maximum of 12 reagents can be used at one time
System 508 Multichannel, discretionary, discrete	Instrumentation Laboratory, Inc.	100	8	800	Floor console	Liquid, bulk	Photometric, ion-selective electrodes, and mano-metric	Reusable reaction vessels

B. INSTRUMENTS WITH NONPHOTOMETRIC MEASUREMENT PRINCIPLES

Name and Categories	Manufacturer	Measurement Techniques, Energy Source, and Details	Approximate Throughput (Results/h)	Comments
Auto ICS Sequential, batch	Beckman Instruments, Inc.	Rate nephelometry; quartz iodide lamp, 450–620 nm; PMT at 70° to incident light	25–55	Reusable cuvet; automated 40-position sample
IL System 504 Discrete 4-channel (selectable for 3 channels)	Instrumentation Laboratory, Inc.	Ion-selective electrodes for Na, K; mercuric thiocyanate for Cl (photometric); direct pressure for tCO_2 (manometric)	400	Automated 100-position, linear sampler
Model 420 Sequential, batch, discrete	Baker Instruments Corp.	Rate and endpoint nephelometry; helium-neon laser at 632.8 nm; dual diode mounted at 37.8° to incident light	25–30	Flow-through cuvet; automated 25-specimen sampler
Model 1020 Semiautomated discrete, dual channel	Orion Research, Inc.	Ion-selective electrodes for Na, K	60	Calibration is automatic; specimen: whole blood, plasma, serum
NOVA 4 + 4 Discrete, 4-channel, nondiscretionary	NOVA Biomedical	Ion-selective electrodes for Na, K, Cl, and CO_2	85	Automated 40-position sampler
PDQ Laser Nephelometer Sequential, batch, discrete	Hyland Diagnostics	Endpoint nephelometry; using a helium-neon laser at 632.8 nm, with PMT mounted at optimum forward detection angle of 31°	20–30	Automated pipettor-dilutor available for up to 150 specimens
YSI Model 23A Blood Glucose Analyzer Discrete, single channel	Yellow Springs Instrument Co., Inc.	Immobilized enzyme (glucose oxidase) on membrane converts glucose to H_2O_2, monitored amperometrically	60	Specimen: whole blood, plasma

Appendix 4-II. NAMES AND ADDRESSES OF MANUFACTURERS OR DISTRIBUTORS

Abbott Laboratories 1921 Hurd Drive Irving, TX 75061	ABA 100™, TDx™, VP™
American Can—Parafilm Div. American Lane Greenwich, CT 06830	Parafilm®
American Dade Division American Hospital Supply Corp. P.O. Box 520672 Miami, FL 33152	Paramax®, Stratus®
American Monitor Corp. P.O. Box 68505 Indianapolis, IN 46268	Parallel®
Ames Company Division of Miles Laboratories Elkhart, IN 46514	Optimate®, Seralyzer®
Baker Instruments Corp. 100 Cascade Drive Allentown, PA 18103	Encore™, Spirit™, Model 420™
Beckman Instruments, Inc. Healthcare Products 200 South Kramer Blvd. Brea, CA 92621	ASTRA™, ICS™
Becton Dickinson Rutherford, NJ 07070	ARIA II™
Boehringer Mannheim Diagnostics 9115 Hague Road Indianapolis, IN 46250	Hitachi 705, Hitachi 737
Cooper Biomedical, Inc. One Technology Court Malvern, PA 19355	DEMAND™
Coulter Electronics, Inc. 590 West 20th Street Hialeah, FL 33010	DACOS™
DuPont Instruments Quillen Bldg./Concord Plaza Wilmington, DE 19898	*aca*™ (Teflon®, Surlyn®)
Eastman Kodak Co. 343 State Street Rochester, NY 14650	Ektachem
Electro-Nucleonics, Inc. Separation and Analytical Systems Div. 368 Passaic Avenue Fairfield, NJ 07006	ACCLAIM™, GemENI®
Gilford Instruments 132 Artino Street Oberlin, OH 44074	SBA 300, IMPACT® 400E
Hyland Diagnostics Division of Cooper Biomedical, Inc. One Technology Court Malvern, PA 19355	PDQ™

Appendix 4-II. NAMES AND ADDRESSES OF MANUFACTURERS OR DISTRIBUTORS (*Continued*)

Instrumentation Laboratory, Inc. 113 Hartwell Avenue Lexington, MA 02173	MULTISTAT III PLUS®, STATSEP™, SYSTEM 504™, SYSTEM 508™
Ionetics, Inc. 3020 Enterprise Street Costa Mesa, CA 92626	IONETICS Dual Analyzer
Nova Biomedical 20 Ossipee Road Newton, MA 02164	NOVA 4 + 4
Orion Research, Inc. 840 Memorial Drive Cambridge, MA 02139	ORION Model 1020©
Roche Diagnostic Systems 340 Kingsland Nutley, NJ 07110	COBAS-BIO®
Sequoia-Turner Corp. 755 Ravendale Avenue Mountain View, CA 94043	DYNA-LYTE 100, 200
Syva Corporation 3181 Porter Drive Palo Alto, CA 94304	Advance™
Technicon Instruments Corp. 511 Benedict Avenue Tarrytown, NY 10591	AutoAnalyzer®, RA-1000®, SMA®, SMAC®
Yellow Springs Instrument Co. P.O. Box 279 Yellow Springs, OH 45387	Model 23A Blood Glucose Analyzer
Zymark Corporation Zymark Center Hopkinton, MA 01748	Zymate

References

1. Babson, A. L., and Kleinman, N. M.: A source of error in an AutoAnalyzer determination of serum iron. Clin. Chem., *13*:163–166, 1967.
2. Billmeyer, F. W.: Principles of Color Technology. New York, Interscience Publishers, 1966, pp. 61–63.
3. Burtis, C. A.: Factors influencing evaporation from sample cups, and assessment of their effect on analytical error. Clin. Chem., *21*:1907–1917, 1975.
4. Burtis, C. A., Johnson, W. F., Mailen, J. C., et al.: Automated sample-reagent loader for use with the GeMSAEC fast analyzer. Clin. Chem., *18*:433–439, 1972.
5. Gochman, N., and Bowie, L. J.: Selection factors for automated analytical instrumentation. Workshop No. 201, American Association for Clinical Chemistry National Meeting. Clin. Chem., *28*:1675, 1982.
6. Grannis, G. F., Grümer, H.-D., Lott, J. A., et al.: Proficiency evaluation of clinical chemistry laboratories. Clin. Chem. *18*:222–236, 1972.
7. Gudzinowicz, B. J., Driscoll, J. L., Martin, H. C., et al.: Economic and technical aspects of instrument evaluation. Chem. Biomed. Environ. Instrument., *4*:1-25, 107–166, 315–333, 1980.
8. Khalil, O. S., Routh, W. S., Lingenfelter, K., et al.: Automated in-line ratio-correcting filter fluorometer. Clin. Chem., *27*:1586–1591, 1981.
9. Krieg, A. F., and Baer, D.: Sharpen your instrument-shopping savvy. Med. Lab. Obs., April 1974, pp. 70–82.
10. National Committee for Clinical Laboratory Standards: User evaluation of precision performance of clinical chemistry devices. NCCLS Standard EP5-T. Villanova, Pa., National Committee for Clinical Laboratory Standards, 1984.
11. Oszustowicz, R. J.: Following the equipment acquisition process. J. Hosp. Fin. Manage. Assoc., April 1982, pp. 12–33.

12. Renoe, B. W., Savory, J., and Wills, M.: Measurement principles for centrifugal analyzer methods. J. Clin. Lab. Auto., *1*:47–54, 1981.
13. Statland, B. E., and Winkel, P.: Physiologic variation of the concentration values of selected analytes as determined in healthy young adults. Proceedings of the 1976 Aspen Conference on Analytical Goals in Clinical Chemistry. Chicago, College of American Pathologists, 1976, pp. 94–101.
14. Walter, B.: Dry reagent chemistries in clinical analysis. Anal. Chem., *55*:498A–514A, 1983.
15. Westgard, J. O., Barry, P. L., Hunt, M. R., et al.: A multi-rule Shewhart chart for quality control in clinical chemistry. Clin. Chem., *27*:493–501, 1981.

II

ACQUISITION, MANAGEMENT, AND APPLICATION OF LABORATORY DATA

ESTABLISHMENT AND USE OF REFERENCE VALUES

With an Introduction to Statistical Techniques

5

HELGE ERIK SOLBERG, M.D.

Introduction to Statistical Terms and Techniques

Statistical Concepts

When we measure a quantity (for example, the serum triglyceride concentration [S-Triglyceride] on several individuals), we collect *data*. Statisticians have developed techniques to analyze data. The following is a very brief review of basic statistical concepts and methods to refresh the reader's memory. More details can be found in textbooks of statistics.[7,8] The rationale, calculations, and use of specific statistics are discussed later in this chapter and in Chapters 7 and 8.

The type of quantity studied (in our example, S-Triglyceride) is called a *variable* and each result is a *value*, a data point, or an observation. The common mathematical symbol for a variable is the letter x and the individual values are symbolized x_1, x_2, ..., x_i, ..., x_n. The symbol x_i denotes the *i*th data value; the letter n is the symbol for the number of values or the sample *size*.

It is usually not feasible to study all possible individuals of a certain category of the general *population* (for example, S-Triglyceride of all healthy males aged 20–40 years living in the USA). We therefore have to select for study a smaller group, a *sample*. The individuals in the sample should preferably be selected at *random* from those in the complete population. Otherwise the sample data may not be representative of the population. In this use, "random" means that each individual has an equal chance of being selected.

Note: There is a distinction between the census bureau's meaning of "population" (a group of people within some geographical or demographical limits) and its connotation in statistics (an infinite or finite set of values that conform to a stated description).

Statistical methods allow us to make two types of inferences from sample data.

1. We can *estimate* properties of the population. Our sample data may, for example, allow us to estimate the mean S-Triglyceride in the complete population. The sample mean is more or less close to the true population mean, depending on the size of the sample and how representative the sample is. It is also possible to estimate a *confidence interval* for the property studied (for example, the mean). The confidence interval is bounded by limits within which the true value is located with a specified confidence. The confidence is expressed as a number in the interval 0–1 (or 0–100%), indicating its degree on the scale between "never" and "always."

2. Another inference is to *test hypotheses* regarding properties of the population. We may, for example, measure the S-Triglyceride concentration in a sample of females and in a sample of males from the same population. If the observed mean concentrations are different for the two sexes, we can test whether this difference is real or only a result of chance (random variation).

Statistical tests provide quantitative approaches to making such decisions. The first step in the process is to state a *null hypothesis* ("no difference"); for example, "there is no difference between the S-Triglyceride means for females and males in the population."

This hypothesis has to be rejected if the statistical test shows that the probability that it is true is less than a stated *significance level*; for example, <0.05 (5%). The *probability* is a number in the interval 0–1 (or 0–100%) that indicates the degree on the scale between "impossible" and "certain." If the computed probability is less than the stated significance level, we have to reject the null hypothesis and accept the *alternative hypothesis* that the true means are different. We then say that the difference is *statistically significant* at the 0.05 level; i.e., there is <5% probability (or less than a 5% chance) that the true means of S-Triglyceride of the two sexes in the population are equal (see example later).

Estimation

We may display our data graphically. A convenient graph is the *histogram*, which gives a picture of the distribution of values (Figure 5-1,*a*). The height of the bars located on the horizontal axis represents the number (or frequency) of observations within small intervals of the measurement scale.

We can also describe the distribution by computed values. The arithmetic ***mean*** is a measure of the location of the distribution on the measurement scale. This sample mean is designated by the symbol $\bar{x}$ (pronounced "x-bar") and is calculated by the formula:

$$\bar{x} = \frac{\Sigma x_i}{n}$$

The meaning of the symbol Σ (sigma) is "the sum of." To calculate the mean we add all the individual values (x_i) and divide the sum ($\Sigma x_i = x_1 + x_2 + ... + x_n$) by the number of values (n). The calculation may be simplified if the individual x-values are classified into k groups. If m_j designates the midpoint value of the *j*th group, which contains f_j of the original values (j = 1, 2, ..., k), then $n = f_1 + f_2 + ... + f_k = \Sigma f_j$, and $\Sigma x_i = f_1m_1 + f_2m_2 + ... + f_km_k = \Sigma f_jm_j$.

The ***standard deviation*** is a measure of the dispersion of the values (i.e., how much they vary around the mean). The sample standard deviation is designated by the symbol s (or more precisely, s_x) and is defined by the formula:

$$s = \sqrt{\frac{\Sigma(x_i - \bar{x})^2}{n - 1}}$$

Or in words, compute the difference between each value and the mean ($x_i - \bar{x}$), square the differences, and add all the squared differences [$\Sigma(x_i - \bar{x})^2$]. Then divide the sum

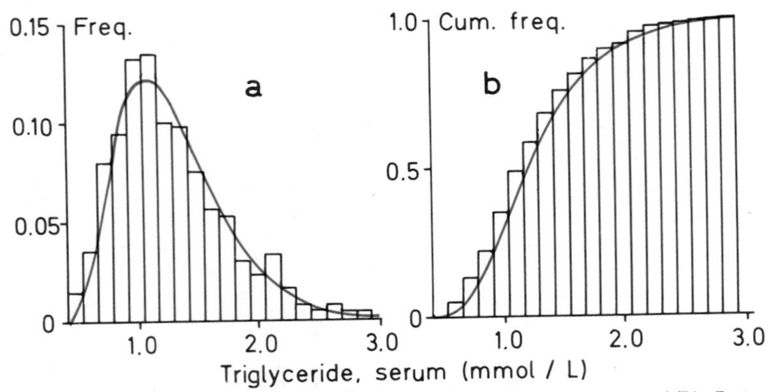

Figure 5-1. Observed and hypothetical distributions of 500 S-Triglyceride values (mmol/L). *Part a:* The vertical bars of the histogram show the frequency, i.e., the number of observations in each interval divided by the total number of observations. The curve is the estimated probability distribution of the population, assuming random sampling and a log-Gaussian distribution. *Part b* shows the cumulated frequencies (bars) and the estimated cumulative probability distribution (curve). Freq. denotes frequency; cum. freq. refers to cumulative frequency.

by one less than the number of values (n − 1) and take the square root. The calculation is made easier by the equivalent formula:

$$s = \sqrt{\frac{\Sigma x_i^2 - \frac{(\Sigma x_i)^2}{n}}{n - 1}}$$

Then we only need to count the values (n) and to compute the sum of the values (Σx_i, see above) and the sum of the squared values $\Sigma x_i^2 = x_1^2 + x_2^2 + \ldots + x_n^2$ or, for grouped data (see above), $\Sigma x_i^2 = f_1 m_1^2 + f_2 m_2^2 + \ldots + f_k m_k^2 = \Sigma f_j m_j^2$. The mean and the standard deviation both have the same unit as the data values (e.g., mmol/L).

It is often convenient to relate the standard deviation to the mean. The ***coefficient of variation*** (CV) is the standard deviation expressed as a percentage of the mean:

$$CV = \frac{s}{\bar{x}}(100\%)$$

Other statistical measures may be necessary to describe the distribution completely; for example, to indicate how asymmetrical it is (its skewness).[7,8] Such measures are not presented, however, in this introduction.

The mean and standard deviation, as defined above, are computed from sample data. They are estimates of the corresponding properties of the population. Sample properties, called *statistics*, are given lower-case, Latin letters or symbols (e.g., $\bar{x}$ and s). The corresponding properties of the population, called *parameters*, are symbolized by lower-case, Greek letters, such as μ (mu) and σ (sigma), for the population mean and standard deviation, respectively.

If we select several random samples from the same population, measure the same quantity in all cases, and compute, for example, the mean ($\bar{x}$) for each sample, we obtain several estimates with some variation around the unknown population mean (μ). A statistical measure of this sampling variation of means is provided by the ***standard error of the mean***, that is, the standard deviation of the distribution of mean values. An estimate ($s_{\bar{x}}$) of the population standard error of the mean ($\sigma_{\bar{x}}$) is calculated from a single sample by the formula:

$$s_{\bar{x}} = \frac{s}{\sqrt{n}}$$

That is, divide the sample standard deviation by the square root of the sample size.

This measure is useful because it allows us to estimate the ***confidence interval*** of the mean. The approximate 0.95 confidence interval of the mean is bounded by a lower limit ($\bar{x} - 2s_{\bar{x}}$) and an upper limit ($\bar{x} + 2s_{\bar{x}}$). The population mean is located within these two limits with a confidence of ~95%. A more accurate estimate is obtained by replacing the factor 2 by the quantity $t_{0.05}$ found in a Student's t table at $\nu = n - 1$ degrees of freedom (Table 29-16 in the *Appendix*).

Distributions

As described above, we may display a sample distribution as a histogram. The frequency of observations constituting a bar of the histogram—i.e., the number of values in the bar divided by the total number of values (the sample size)—is an estimate of the probability of finding a random population value in the interval of the measurement scale covered by the bar. The sample distribution is therefore an estimate of the population ***probability distribution***. In Figure 5-1,*a* the estimated population probability distribution is superimposed on the histogram as a continuous curve. The area under this curve is a graphical representation of probability, as will be discussed below.

We can also display the sample data as a cumulative distribution. In this case, we consecutively add the frequencies of values in each bar of the histogram, starting with the leftmost bar, and make another graph where the heights of the bars show the cumulative sum (Figure 5-1,*b*). The continuous curve in this figure represents the estimate of the population's *cumulative probability distribution*.

There are many types of distributions. Some are symmetric; i.e., the areas of the histogram on each side of a vertical line drawn at the mean value are equal in size and shape (mirror images of each other). This is often the case with data obtained in quality control. Distributions of biological data are, however, often asymmetric (skewed).

The population probability distribution behind symmetric sample distributions may be close to the *Gaussian distribution* (or "normal" distribution) shown in Figure 5-2. Table 29-15 in the *Appendix* gives upper tail areas under the standard Gaussian curve for positive values of $z_i = (x_i - \bar{x})/s$. Several important statistical methods are based on the assumption that population data have this distribution. As indicated in the figure, there are important relations between the mean, the standard deviation, and probabilities when data have a Gaussian distribution. For example, the area under the curve between the vertical lines drawn at two standard deviations below and above the mean is always 95.5% of the total area. The probability that a random population value is located within two standard deviations from the mean is therefore 0.955. However, this simple relationship does not hold for other types of distributions.

Testing of Hypotheses

The concepts of null hypothesis and significance testing were introduced above. Let us apply them to testing the significance of the difference between two sample mean values. Other statistical tests are presented in Chapter 7.

Table 5-1 shows the data obtained by measuring S-Triglyceride in two samples, one consisting of females and the other of males, drawn randomly from the same general population. Calculated sample statistics are also shown. As an exercise, you may try to verify the calculations using the formulas given above. Remember that $\Sigma x_i = \Sigma f_j m_j$ and $\Sigma x_i^2 = \Sigma f_j m_j^2$ when data are grouped into classes with midpoint values m_j, as is the case with the data in the table.

The question arises: Is the difference between the two means significant? We begin to resolve the question by stating the null hypothesis that the two sexes in the population have equal means of S-Triglyceride and then test whether the observed difference between the two sample means can be ascribed to chance (sampling variation) alone. A first approximation may be obtained by comparing the confidence intervals for the two means: 0.916–1.056 for males and 1.031–1.187 for females (Table 5-1). The difference would have been declared significant if the confidence limits did not overlap. In our example, however, we have a small overlap, which causes some doubt about the conclusion.

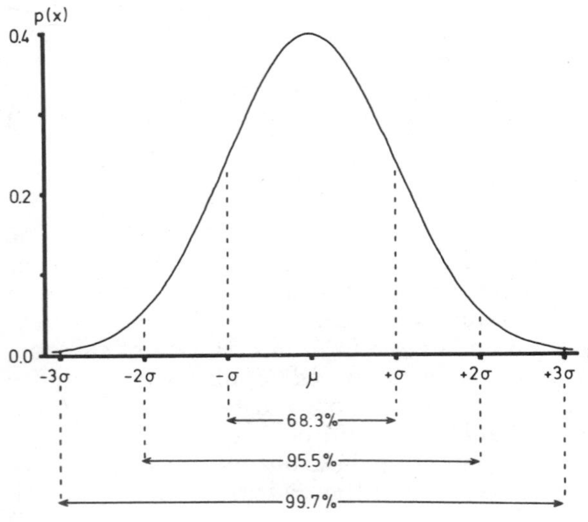

Figure 5-2. The Gaussian distribution. The percentage of the area under the curve that corresponds to ±1, ±2, and ±3 standard deviations (±σ, ±2σ, ±3σ) from the mean (μ) are indicated below the figure.

Table 5-1. AN EXAMPLE OF SIMPLE STATISTICAL CALCULATIONS AND STUDENT'S *t*-TEST*

Data		
S-Triglyceride Class Midpoints (m_j), mmol/L	***Number of Observations (f_j)***	
	Females	*Males*
0.5	1	1
0.6	3	2
0.7	6	4
0.8	8	6
0.9	10	4
1.0	8	8
1.1	7	8
1.2	5	7
1.3	3	5
1.4	2	6
1.5	1	3
1.6	1	2
1.7	0	1
1.8	1	0
1.9	0	1
Sample Statistics		
Sample sizes	$n_1 = 56$	$n_2 = 58$
Means	$\bar{x}_1 = 0.986$	$\bar{x}_2 = 1.109$
Standard deviations	$s_1 = 0.262$	$s_2 = 0.298$
Standard error of the means	$s_{\bar{x}_1} = 0.035$	$s_{\bar{x}_2} = 0.039$
Confidence interval of the means	0.916–1.056	1.031–1.187
***t*-Test**		
Difference of means		$d = 0.123$
Standard error of difference		$s_d = 0.052$
t-Test criterion		$t = 2.34$
Number of degrees of freedom		$df = 112$
Probability		$0.01 < P < 0.025$

*The table shows calculations based on unpaired observations of S-Triglyceride (mmol/L) for females and males from the same general population.

A more formal and reliable test is provided by the ***Student's t-test for unpaired observations***. This test is based on the *t*-value, which is calculated as the ratio between the absolute difference between the two means (d) and the standard error for that difference (s_d):

$$t = \frac{|d|}{s_d}$$

where $d = \bar{x}_1 - \bar{x}_2$ and

$$s_d = \sqrt{\left(\frac{n_1 + n_2}{n_1 n_2}\right)\left(\frac{(n_1 - 1)s_1^2 + (n_2 - 1)s_2^2}{n_1 + n_2 - 2}\right)}$$

In addition, we need the number of degrees of freedom (df)*:

$$df = n_1 + n_2 - 2$$

To complete the test, we decide to reject the null hypothesis if the probability that the difference (d) is due to chance alone is less than 0.05 (5% significance level). We

*Degrees of freedom is generally notated as df, but in statistical tables the same expression is often referred to as ν.

then compare our calculated test criterion $t = 2.34$ (Table 5-1), with critical values given in a Student's t table at $\nu = 112$. Table 29-16 in the *Appendix* gives probability points of the Student's t distribution with ν degrees of freedom. For our example we enter the table at $\nu = 120$ (as there is no line for $\nu = 112$). Our calculated criterion falls between the values in the columns for 0.01 and 0.025 probability. Therefore, the probability that the calculated difference is due only to chance is between 0.01 and 0.025, which obviously is less than the stated significance level. We therefore reject the null hypothesis and declare the difference between the two means significant at the 0.05 (5%) level.

Comments

1. The calculation of s_d is based on the assumption that the variances s_1^2 and s_2^2 (squared sample standard deviations) are not significantly different. A test for the difference between variances, the F-test, is presented in Chapter 7. If they are significantly different, consult textbooks of statistics for a variant of the t-test designed for that situation.[7,8]

2. The t-test also becomes unreliable if the data distributions deviate too greatly from the Gaussian distribution. In such cases one should use nonparametric tests (for example, the Wilcoxon's rank sum test, which can be found in textbooks of statistics).[7]

3. The concept of *degrees of freedom* (df) also needs clarification. It may be defined as the number of data values that carry new information when a statistical calculation is performed. When the mean is computed, all values carry information (i.e., $df = n$). But if we already know the mean, we can compute the remaining value if we know $n - 1$ data values. Since the mean is part of the definition of the standard deviation, the number of degrees of freedom for the standard deviation is one less than the sample size (i.e., $df = n - 1$). Likewise, the standard error of the difference between two means (s_d) is based on two different standard deviations. The number of degrees of freedom for the s_d is accordingly:

$$df = (n_1 - 1) + (n_2 - 1) = n_1 + n_2 - 2.$$

Establishment and Use of Reference Values

The Concept of Reference Values

Medicine is an art and a science in the service of fellow human beings. To increase the health of their patients, physicians collect empirical data, interpret these data using scientific knowledge and professional experience, make decisions concerning diagnoses, recommend preventive measures, and execute therapeutic actions.

Data collected during the medical interview, clinical examination, and supplementary investigations must be interpreted by *comparison* with reference data. If the condition of the patient resembles what is considered typical of a particular disease, the physician may base his diagnosis on this observation (positive diagnosis). This diagnosis is made more likely if observed symptoms and signs do not fit the patterns characterizing a set of alternative diseases (diagnosis by exclusion).

The interpretation of medical laboratory data is just a special case of decision-making by comparison. For this decision process, we need *reference values* for all tests performed in the laboratory, not only from healthy individuals, but also from patients with relevant diseases.[2,4,10] Such persons are called *reference individuals*, i.e., individuals selected according to stated criteria with the purpose of producing reference values.[4] Ideally, observed values should be related to several collections of reference values: values from healthy persons, from the undifferentiated hospital population, from persons with typical diseases, and from ambulatory individuals, as well as previous values from the same individual.

Certain conditions are mandatory before comparison of a patient's laboratory result with reference values will be possible and valid:

1. All groups of reference individuals should be clearly defined.

2. The patient examined should resemble the reference individuals (in all groups selected for comparison) in all respects other than those under investigation.

3. The conditions under which the specimens were obtained and processed for analysis should be known.

4. All laboratory results should be produced by adequately standardized methods under sufficient analytical quality control (see Chapter 7).

The term *normal values* has frequently been used in the past. It is roughly equivalent to "reference values obtained from healthy individuals." Confusion arises, however, because the word *normal* has several very different connotations. Three important and very different meanings of the word *normal* are given below.

1. Values are often referred to as "normal" if their observed distribution seems to follow closely the theoretical "normal distribution" of statistics, that is, the Gaussian probability distribution (Figure 5-2).

2. Another meaning of *normal* is met in the following statement: it is "normal" to find that the concentration of S-Triglyceride is between 0.6 and 2.4 mmol/L, whereas it is considered "abnormal" to have a serum triglyceride concentration outside these limits. Alternative terms for *normal* in this sense are *common*, *frequent*, *habitual*, *usual*, and *typical*.

3. The term *normal* is also often used to indicate that values show the absence of certain diseases or the lack of risks for the development of these diseases. In this sense a "normal value" is considered a sign of health. Better terms are *healthy* or *nonpathological values*.

The International Federation of Clinical Chemistry (IFCC) recommends the concept of *reference values* and related terms to avoid the problems related to the term *normal values*.[4] This concept was purposely introduced as a very vague term to force us to define in each case what is actually meant. Reference values in clinical chemistry are the results of quantitative analysis of a certain analyte obtained from an individual or a group of individuals who are selected according to clearly delineated criteria, which must be stated so that others can also use these values.

It is convenient, and frequently sufficient, to give a short description in connection with the term *reference values*, such as *health-associated reference values*. Other examples of such qualifying words are: *diabetic*, *hospitalized diabetic*, and *ambulant diabetic*. These short descriptions prevent the common misunderstanding that reference values are necessarily associated with health.

We may further distinguish between subject-based and population-based reference values. *Subject-based reference values* are previous values from the same individual, obtained when the individual was in a defined state of health. *Population-based reference values* are those obtained from a group of well-defined reference individuals and are usually the type of values referred to when the term *reference values* is used without qualifying words.

Each analyte may have one or more collections of *univariate reference values*. The term *multivariate reference values* denotes that results of two or more analytes obtained from the same set of reference individuals are treated in combination.

Selection of Reference Individuals

A set of *selection criteria* determines who should be included in or excluded from the group of reference individuals; these standards describe the source population and specify criteria for health or for the diseases of interest.[2,4] The required characteristics of the reference values determine which criteria should be used in the selection process. Table 5-2 shows a list of important exclusion criteria to use when the production of health-associated reference values is the aim.

Ideally, the group of reference individuals should be a *random* sample of all the individuals in the parent population who fulfill the selection criteria. A strictly random sampling scheme is in most cases impossible for practical reasons. It would imply the examination of, and application of selection criteria to, the entire population (thousands or millions of persons) and the random selection (for example, by raffling) of a subset of individuals from among those accepted.

Table 5-2. EXAMPLES OF EXCLUSION CRITERIA FOR HEALTH-ASSOCIATED REFERENCE VALUES*

Risk factors	***Disease***
Obesity	***Intake of pharmacologically active agents***
Hypertension	Drug treatment for disease or suffering
Risks from occupation or environment	Oral contraceptives
Genetically determined risks	Drug abuse, alcohol, tobacco
Specific physiological states	
Pregnancy	
Stress	
Excessive exercise	

*The table lists only some major classes of criteria. It should be supplemented with other relevant criteria based on known sources of biological variation.

We therefore have to use the best reference sample we can possibly get after all practical considerations are taken into account. Data should then be used and interpreted with due caution, remembering the possible bias introduced by the nonrandomness of the sample selection process.

Often separate reference values for each sex, different age groups, and other criteria are needed. For this purpose we need to specify the *partition criteria*, i.e., rules for the division (Table 5-3).[4] The number of partition criteria should usually be kept as small as possible so that sufficient sample sizes are obtained to derive reliable estimates.

Age and *sex* are the most frequently used criteria for subgrouping. Age may be categorized by equal intervals (for example, by decades), or the intervals may be narrower if greater variation is observed in some periods of life. It is often convenient to use such qualitative classifications as postnatal, infancy, childhood, prepubertal, pubertal, adult, premenopausal, menopausal, and geriatric. Height and weight can be used as criteria for categorizing the reference values of children.

Specimen Collection

Preanalytical standardization of the preparation of individuals before sample collection, of the procedure of sample collection itself, and of the handling of the specimen prior to analysis may eliminate or minimize biological "noise" that might otherwise conceal important biological "signals" of disease, risk, or effect of treatment.[2,4] (See also Chapter 10.)

The magnitudes of preanalytical sources of variation are clearly not equal for different analytes. It could therefore be argued that we should only consider those factors that cause unwanted variation for the biological quantity for which we want to produce reference values. Body posture during specimen collection, for instance, is highly relevant for the establishment of reference values for nondiffusible analytes such as albumin in serum but irrelevant for serum sodium.

On the other hand, several constituents are usually analyzed in the same clinical specimens. It would therefore be impractical to devise special systems for every analyte. For that reason, standardized general procedures for blood specimen collection by venipuncture and skin puncture (capillary blood) have been recommended.[2,6]

A special problem is caused by drugs taken by individuals prior to specimen collection. We may distinguish between indispensable and dispensable medication. The

Table 5-3. EXAMPLES OF PARTITION CRITERIA TO BE USED FOR POSSIBLE SUBGROUPING OF THE REFERENCE GROUP

Age (not necessarily categorized by equal intervals)	***Physiological factors***
Sex	Stage in menstrual cycle
Genetic factors	Stage in pregnancy
Race (ethnic origin)	Physical condition
Blood groups (ABO)	***Other factors***
Histocompatibility antigens (HLA)	Socioeconomic
	Environmental
	Chronobiological

latter type of drugs should always be avoided for at least two days prior to obtaining a biological specimen. The use of indispensable drugs such as contraceptive pills or essential medication may be a criterion for exclusion or partition.

Analytical Procedures and Quality Control

Essential components of the required definition of a set of reference values are specifications concerning the following:

1. Analysis method, including information on equipment, reagents, calibration standards, type of raw data, and calculation method.
2. Quality control (see Chapter 8).
3. Reliability criteria (see Chapter 7).

The specifications should be so thoroughly described that another investigator can reproduce the study and the user of reference values can evaluate their comparability with values obtained by the methods used for producing the patient's values in the routine laboratory.

Statistical Analysis of Reference Values[2,4,10]

Partitioning of reference values. The sample of reference individuals and the corresponding reference values may be partitioned according to sex, age, and other characteristics. The aim of the partitioning is to reduce, if possible and necessary, variation among subjects so that there wil be less biological "noise." Minimizing the intraclass variation gives narrower and more sensitive reference intervals.

The distribution of reference values in the individual classes may show different locations (mean values may vary) or different intraclass variations (standard deviations may vary). Both contribute to the larger total variation of reference values that is present before partitioning. In general, reference values should be partitioned when statistical tests show that the differences between class means or class standard deviations, or both, are significant.[7,8]

In the following sections we assume that we have a homogeneous reference distribution, either the complete sample distribution (if partitioning has been shown to be unnecessary) or the distribution of a subclass after partitioning.

Inspection of the distribution. It is always advisable to display the reference distribution graphically in a histogram (Figure 5-1,*a*) and to inspect it. We should look for the following characteristics of the distribution:

1. Highly deviating values (*outliers*) may represent erroneous values.
2. *Bimodal* or *polymodal distributions* have two or more peaks and may indicate that the distribution is nonhomogeneous. If there is more than one peak, the criteria used to select reference individuals should be re-evaluated, or partitioning of the values according to age, sex, or other relevant factors should be tried.
3. The shape of the distribution should be observed. It may be asymmetrical (*skewness*) or more or less peaked than the Gaussian distribution (non-Gaussian *kurtosis*).[7,8,10] Asymmetry and non-Gaussian peakedness may also be present together.
4. The visual inspection may also provide *initial estimates* of the location of reference limits that are useful as checks on the validity of computations.

Identification and handling of outliers.[4] Erroneous values sometimes can be traced to gross deviations from the prescribed procedure for the establishment of reference values. They may either deviate significantly from the proper reference values (*outliers*) or they may be hidden in the reference distribution.

Visual inspection of a histogram is a reliable method for identification of possible outliers. It is important to keep in mind, however, that values far out in the long tail of a skewed distribution may easily be misinterpreted as outliers. If the distribution has a long right tail, a histogram of the logarithms of the values may aid in the identification of outliers.

Some outliers may also be identified by statistical tests, but no single method is capable of detecting outliers in every situation that may occur. The two main problems encountered are:

1. Many tests for outliers assume that the type of the true distribution is known prior to using the test. Some specifically require that the distribution be Gaussian, which

often is not the case with biological data. The following range test is relatively insensitive to departures from Gaussian behavior.[4] Identify the extreme value as an outlier if the difference between the two highest (or two lowest) values in the distribution exceeds one-third of the range of all values.

2. Several tests for outliers assume that only one outlier is present in the data. The range test, for example, usually fails in the presence of several outliers.

Deviating values that are identified as possible outliers should not be discarded automatically. Values should be included or excluded on a rational basis (for example, by checking the analytical records and correcting any errors).

Determination of reference limits.[2,4,10] In clinical practice one usually compares an observed patient's value with the corresponding *reference interval* (sometimes called *reference range*). The interval is bounded by a pair of *reference limits*.

Note: Reference limits should not be confused with *clinical decision limits*, the limits used for clinical judgments concerning diagnosis, prognosis, and therapy. The basis for the estimation of clinical decision limits may be several collections of reference values. The actual calculation, based on Bayes' law or other statistical theory, is, however, not discussed here.

Three kinds of reference intervals have been suggested: tolerance interval, prediction interval, and interpercentile interval. The IFCC has recommended the use of the most common type, the simple *interpercentile interval*, which is bounded by an upper and a lower percentile of the reference distribution.[4] A ***percentile*** denotes a value that divides the reference distribution such that a specified percentage of its values have magnitudes less than or equal to the limiting value. For example, if 2.4 mmol/L is the 97.5 percentile of S-Triglyceride, 97.5% of the concentration values are equal to or below this value.

It is a common convention to define the reference interval as the central 95% interval bounded by the 2.5 and 97.5 percentiles (Figure 5-3); that is, 2.5% of the values are cut off in each tail of the reference distribution. Another size or an asymmetric location of the reference interval may be more appropriate in particular cases.

The precision of a sample percentile as an estimate of a population value is dependent on the size of the sample; it is less precise when there are few values. We may determine the *confidence interval* of the percentile, that is, the limits within which the true percentile is located with a specified degree of confidence (Figure 5-3). The 0.90 confidence interval of the 97.5 percentile (upper reference limit) for S-Triglyceride may, for example, be 2.3–2.5 mmol/L. We would expect to find the true percentile in this interval with a confidence of 90% if we measured the S-Triglyceride of all individuals in the total reference population. The theoretical minimum sample size required for the estimation of the 2.5 and 97.5 percentiles is 40 values, but at least 120 reference values are required to obtain reliable estimates.

The interpercentile interval is easily determined by either parametric or nonparametric statistical techniques. We are using a *parametric method* when we use estimates of the mean, standard deviation, and other distribution parameters for the calculation of a reference interval. Such methods are usually based either on the Gaussian distribution or on a mathematical transformation of data to fit a Gaussian distribution. The

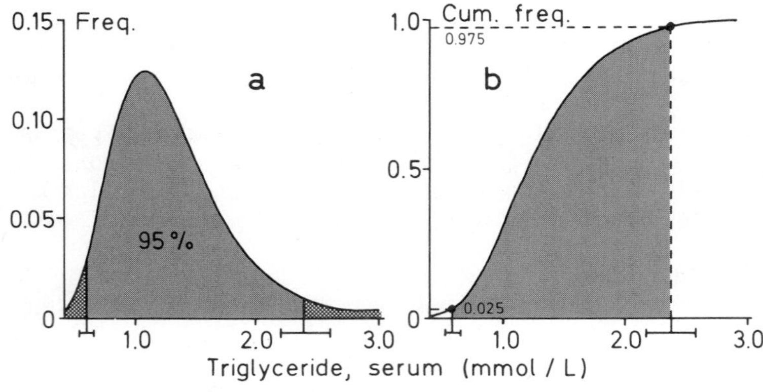

Figure 5-3. Central 95% reference interval. The 2.5 and 97.5 percentiles of the 500 S-Triglyceride values shown in Figure 5-1 and their 0.90 confidence intervals (horizontal lines below the x-axis) were determined by the parametric method. The curves are the estimated probability distributions.

nonparametric method makes no assumption concerning the type of distribution and does not use estimates of distribution parameters. The percentiles are determined simply by cutting off the required percentage of values in each tail of the sample reference distribution.

These two methods may produce similar estimates of the percentiles. In most cases, the simple nonparametric method is preferred.

Nonparametric method. Several methods are available, but those based on rank numbers are simple and reliable and allow the estimation of the confidence interval of the percentiles. Computer programs for the method are available.[9] Table 5-4 shows an example of the nonparametric determination of percentiles using the S-Triglyceride values shown in Figure 5-1. The procedure is as follows:

1. Sort the n reference values in ascending order of magnitude and rank the values. The minimum value has rank number 1, the next value has rank number 2, and so on. Consecutive rank numbers should be given to two or more values that are equal. When large amounts of data are analyzed, considerable labor can be saved by identifying and sorting only values in the tails of the distribution.
2. Compute the rank number of the 2.5 and 97.5 percentiles as 0.025 (n + 1) and 0.975 (n + 1), respectively.
3. Determine the percentiles by finding the original reference values that correspond to the computed rank numbers. If the rank numbers are not integers, interpolate between the two limiting values.
4. Finally, determine the confidence interval of each percentile by entering Table 5-5. Look up the two rank numbers for each percentile and find the corresponding original reference values.

Parametric method. The parametric method is much more complicated and often requires a computer when large samples are to be processed. Computer programs are, however, available.[9]

Table 5-4. NONPARAMETRIC DETERMINATION OF REFERENCE INTERVAL*

Sorted and ranked serum triglyceride values in the left tail of the distribution:

Values:	0.41	0.43	0.45	0.46	0.47	0.49	0.51	0.55	0.55	0.55
Ranks:	1	2	3	4	5	6	7	8	9	10
Values:	0.56	0.58	0.58	0.61	0.62	0.62	0.64	0.64	0.65	0.65
Ranks:	11	12	13	14	15	16	17	18	19	20

Sorted and ranked serum triglyceride values in the right tail of the distribution:

Values:	2.21	2.22	2.26	2.27	2.27	2.28	2.30	2.31	2.34	2.35
Ranks:	481	482	483	484	485	486	487	488	489	490
Values:	2.48	2.50	2.55	2.62	2.63	2.65	2.72	2.78	2.90	2.91
Ranks:	491	492	493	494	495	496	497	498	499	500

Calculation of rank numbers of the percentiles:

Lower: 0.025 · (500 + 1) = 12.5
Upper: 0.975 · (500 + 1) = 488.5

Finding the original values corresponding to these rank numbers:

Lower reference limit (2.5 percentile): 0.58
Upper reference limit (97.5 percentile): 2.32 (by interpolation)

Rank numbers (see Table 5-5) and values of the 0.90 confidence limit of the lower reference limit:

Rank numbers: 7 and 19
Confidence limits: 0.51 and 0.65

Rank numbers (see Table 5-5) and values of the 0.90 confidence limit of the upper reference limit:

Rank numbers: 500 + 1 − 19 = 482 and 500 + 1 − 7 = 494
Confidence limits: 2.22 and 2.62

Summary:

Lower reference limit: 0.58 (0.51–0.65) mmol/L
Upper reference limit: 2.32 (2.22–2.62) mmol/L

*The table shows a worked-out example using the 500 serum triglyceride concentrations displayed in Figure 5-1. See the text for a description of the nonparametric method. The unit of all concentrations in the table is mmol/L.

Table 5-5. NONPARAMETRIC CONFIDENCE INTERVALS OF REFERENCE LIMITS

Sample Size	Rank Numbers Lower	Rank Numbers Upper	Sample Size	Rank Numbers Lower	Rank Numbers Upper
119–132	1	7	556–574	8	22
133–160	1	8	575–598	9	22
161–187	1	9	599–624	9	23
188–189	2	9	625–631	10	23
190–218	2	10	632–665	10	24
219–248	2	11	666–674	10	25
249–249	2	12	675–698	11	25
250–279	3	12	699–724	11	26
280–307	3	13	725–732	12	26
308–309	4	13	733–765	12	27
310–340	4	14	766–773	12	28
341–363	4	15	774–799	13	28
364–372	5	15	800–822	13	29
373–403	5	16	823–833	14	29
404–417	5	17	834–867	14	30
418–435	6	17	868–871	14	31
436–468	6	18	872–901	15	31
469–470	6	19	902–919	15	32
471–500	7	19	920–935	16	32
501–522	7	20	936–967	16	33
523–533	8	20	968–970	17	33
534–565	8	21	971–1000	17	34

The table shows the rank numbers of the 0.90 confidence interval of the 2.5 percentile for samples with 119–1000 values. To obtain the corresponding rank numbers of the 97.5 percentile, subtract the rank numbers in the table from (n + 1), where n is the sample size.

The table is taken from Part 5 of the series of recommendations on reference values prepared by the IFCC[4] (with permission from the IFCC).

The parametric method for estimating percentiles assumes that the true distribution is Gaussian. Ignoring this fact often results in seriously biased estimates of reference limits. Therefore, a critical phase in the parametric method is to *test the goodness-of-fit* of the reference distribution to a Gaussian distribution. A simple test is obtained by plotting the cumulative distribution on Gaussian probability paper, which has a nonlinear vertical axis based on the Gaussian distribution. The plot will be close to a straight line if the distribution is Gaussian. However, it is very difficult to evaluate visually deviations from the straight line because of the nonlinearity of the vertical distances in the graph. Many statistical computer programs have fine goodness-of-fit tests, for example, tests based on coefficients of skewness and kurtosis, the Kolmogorov-Smirnov test, or the Anderson-Darling test.[4,9]

If the reference distribution is not significantly different from the Gaussian distribution, the *2.5 and 97.5 percentiles* can be estimated by the values approximately two standard deviations (2s) on each side of the mean, or more accurately:

$$2.5 \text{ percentile} = \bar{x} - 1.96s$$
$$97.5 \text{ percentile} = \bar{x} + 1.96s$$

The *0.90 confidence interval* of each percentile is estimated by the following two limits:

$$\text{Lower confidence limit} = \text{Percentile limit} - 2.81s/\sqrt{n}$$
$$\text{Upper confidence limit} = \text{Percentile limit} + 2.81s/\sqrt{n}$$

If the reference distribution is non-Gaussian, mathematical *transformation of data* may give a distribution close to the Gaussian distribution. It is frequently observed that logarithmically transformed values of a distribution with a long right tail (positively skewed) fit the Gaussian distribution rather closely. In other cases, square roots of the values result in a better approximation to the Gaussian distribution. This is the basis for the common use of the logarithmic and square root transformations when reference limits are estimated as described below. If these two functions fail to transform data to fit a Gaussian distribution, more general transformations can be used. Such functions are described elsewhere.[2,4,9,10] To apply the parametric procedure, follow these steps:

1. Transform data by the logarithmic function $y = \log(x)$ and test the fit to the Gaussian distribution using the methods described above. (You can use either natural logarithms, $\log_e(x)$, or common Briggsian logarithms, $\log_{10}(x)$.) If the logarithmic transformation is unsuccessful, try the square root function $y = \sqrt{x}$. If both of these functions fail, use more general functions (complicated) or use the above nonparametric method (simple).

2. Compute the mean ($\bar{y}$) and the standard deviation (s_y) of the transformed data. Then estimate the percentiles and their confidence intervals in the transformed scale by the formulas given above, using $\bar{y}$ for $\bar{x}$ and s_y for s.

3. The final step is reconversion of the percentiles and their confidence intervals to the original data scale by using the inverse functions: antilogarithms or squares, respectively.

Example: The mean and standard deviations of the S-Triglyceride values of Figure 5-1 following logarithmic transformation (natural logarithms) are $\bar{y} = 0.172$ and $s_y = 0.357$. The 2.5 percentile is:

$$0.172 - 1.96 \cdot 0.357 = -0.528$$
$$\text{2.5 percentile} = e^{-0.528} = 0.59$$

The lower reference limit of serum triglyceride is thus 0.59 mmol/L. The 0.90 confidence interval of this percentile is:

$$-0.528 - 2.81 \cdot 0.357/\sqrt{500} = -0.573$$
$$\text{Lower confidence limit} = e^{-0.573} = 0.56$$
$$-0.528 + 2.81 \cdot 0.357/\sqrt{500} = -0.483$$
$$\text{Upper confidence limit} = e^{-0.483} = 0.62$$

that is, 0.56–0.62 mmol/L. The 97.5 percentile (and its 0.90 confidence interval) is, by the same method, found to be 2.39(2.29–2.50) mmol/L. As an exercise, verify the latter results.

Comparison with Table 5-4 shows that the nonparametric and parametric methods result in very similar estimates of reference limits (percentiles). The parametric confidence intervals are, however, somewhat narrower than the nonparametric ones.

Presentation of an Observed Value in Relation to Reference Values[2,4]

An *observed value* (patient's result) is interpreted by comparison with reference values. This comparison often looks like hypothesis testing, but it is seldom a statistical test in the strict sense. It is advisable to consider reference values as the yardstick for a less formal assessment than hypothesis testing.

The clinician should always be supplied with as much information about the reference values as needed for the interpretation, including information about the analysis methods and their imprecision.

A convenient presentation of an observed value in relation to reference values may be a great help for the busy clinician. The *reference intervals* may be preprinted on report forms, or the computer system may select the appropriate age- and sex-specific reference intervals from a file and print it next to the test result or in graphical form. Figure 5-4 shows a combined numerical and graphical presentation used by the author's laboratory computer system.

An observed value may be classified as "low," "usual," or "high" depending on its location in relation to the reference interval. Reports printed by a computer conveniently *flag unusual results* with symbols, such as L (for low) and H (for high).

SPECIMEN DATE AND TIME: 20/09/84 09:7 RESULT REFERENCE INTERVAL

S- Sodium	145.0 mmol/L	135.0-148.0	R
S- Potassium	4.5 mmol/L	3.7-4.6	R
S- Chloride	105.0 mmol/L	100.0-109.0	R
S- Calcium	2.3 mmol/L	2.2-2.6	R
S- Urea	7.4 mmol/L	2.8-7.7	R
S- Creatinine	105.0 μmol/L	71.0-124.0	R
S- Protein Total	64.0 g/L	63.0-80.0	R
S- AST	138.0 U/L	13.0-38.0	R
S- ALT	69.0 U/L	10.0-41.0	R
S- LD	1038.0 U/L	227.0-452.0	R

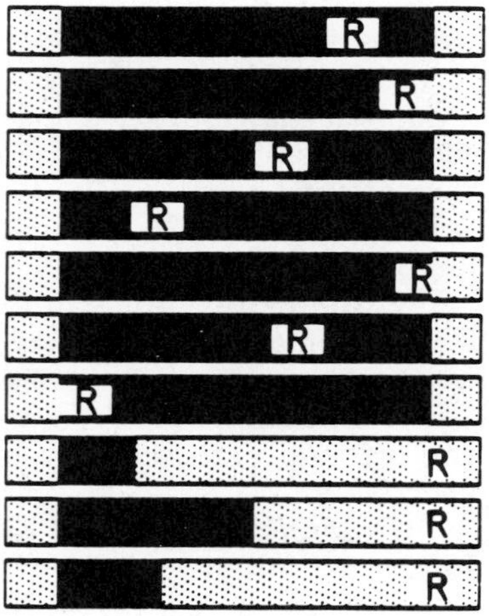

Figure 5-4. Combined numerical and graphical presentation of the observed results of a patient in relation to reference intervals. The figure shows an example from the CLDS computer program (Clinical Laboratory Display System, Department of Clinical Chemistry, Rikshospitalet, Oslo, Norway). The original display on a screen terminal is in color to enhance the interpretation: results are shown in red and reference intervals are displayed in green. The left numeric column contains results. The right numeric column lists the age- and sex-specific reference interval for each value. The graphical display shows both the reference intervals (linear scale, horizontal bars on the screen) and the location of the result (the letter R).

Another popular method for presenting an observed value is to express it with a mathematical *distance measure*, for example the *standard deviation (SD) unit*, which is the difference between the observed value and the mean of the reference values divided by their standard deviation.[2]

Multivariate Reference Regions

The previous sections have concerned univariate population-based reference values and quantities derived from them. Such values do not, however, fit the common clinical situation in which observed values of several different laboratory tests are available for interpretation and decision-making. For example, the mean number of individual laboratory tests requested on one sample received in the author's laboratory is 5.4 (range from one to more than 30 tests). There are two models for interpretation by comparison in this situation.

1. We can compare each observed value with the corresponding reference values or interval; that is, we perform *multiple, univariate comparisons*. Five per cent of the reference values are, by definition, located outside the 95% reference interval for each test. When we use the reference intervals for several tests simultaneously, the total fraction of reference values located outside the reference intervals is $100\,(1 - 0.95^{m})$ per cent (m represents the number of laboratory tests performed on the specimen). We would, for example, expect to find $100\,(1 - 0.95^{10})$, or 40%, "false positives" when interpreting 10 laboratory tests in this way.

2. A better method is to consider the set of observed values as a single multivariate observation and to interpret it in relation to a *multivariate reference region.*[1,2]

First, a few words on the multivariate concept. A univariate observation, such as a single laboratory result, may be represented graphically as a point on a line, the axis or scale of values. The results obtained by two different laboratory tests performed on the same specimen, a bivariate observation, may be displayed as a point in a plane defined by two perpendicular axes. With three results we have a trivariate observation and a point in a space defined by three perpendicular axes, and so on. We lose the possibility for direct graphical representation of multivariate observation with more

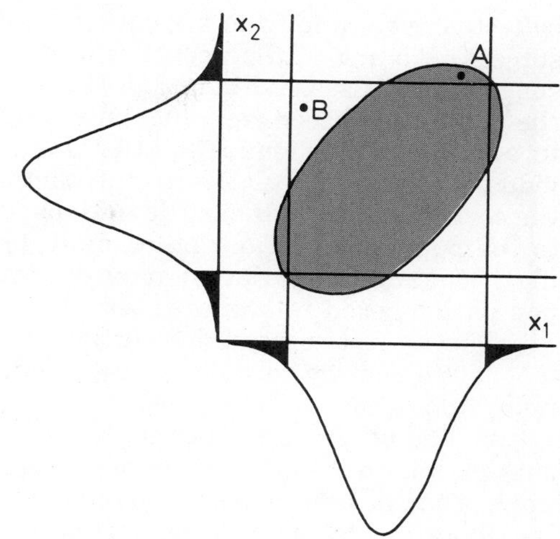

Figure 5-5. Bivariate reference region (ellipse) compared with the region defined by the two univariate reference intervals (box). The point A is outside the univariate reference interval of x_2 but inside the bivariate reference region. The point B is within both univariate reference intervals but outside the bivariate reference region.

than three dimensions. Still, we can consider the multivariate observation as a point in a multidimensional hyperspace with as many mutually perpendicular axes as there are results of different tests (*hyper* here designates "more than three dimensions"). Multivariate observations are also called *patterns* or *profiles*. A distribution of many multivariate observations is thus represented by a cluster of points in the space or hyperspace. Several multivariate statistical methods are straightforward extensions of univariate methods to the multivariate situation. A number of good and not-too-difficult textbooks on multivariate statistical methods exist.[5]

The multivariate reference region is based on the joint distribution of the reference values for two or more laboratory tests. This region may be an ellipse in the plane (for two tests), as shown in Figure 5-5, or an ellipsoid body in space (for three or more tests). The multivariate reference region is a straightforward extension of the univariate 95% interval to the multivariate situation; it may be set to enclose 95% of the central multivariate reference data points.[1,2] In that case we would expect to find only 5% "false positives."

The establishment of multivariate reference regions usually requires the assistance of a computer. The interpretation of a multivariate observation is the task of comparing a calculated index with a critical value established on the basis of a corresponding set of reference values.[1,2]

Subject-Based Reference Values

Figure 5-6 depicts the inherent problem associated with population-based reference values. It shows two hypothetical reference distributions, one based on single specimens obtained from a group of reference individuals and the other based on several specimens

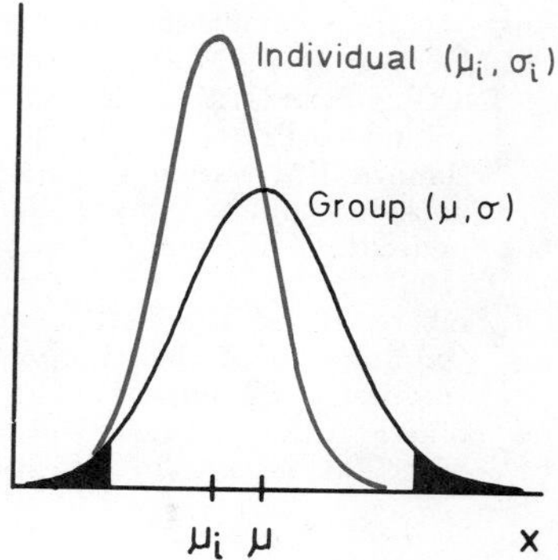

Figure 5-6. The relationship between population-based and subject-based reference distributions and reference intervals. The example is hypothetical, and the two distributions are, for simplicity, Gaussian. (With permission after Harris, E. K.: Effects of intra- and interindividual variation on the appropriate use of normal ranges. Clin. Chem., *20:*1535–1542, 1974.)

collected over time from a single individual. If an observed value is located outside the subject's 2.5 and 97.5 percentiles, the cause may be a change in the biochemical status, suggesting a disease. Figure 5-6 shows that such an observed value may still be within the population-based reference interval. The sensitivity of the latter interval to changes in a subject's biochemical status is accordingly dependent on the location of the individual's mean (μ_i) relative to the common mean (μ) and to the relative magnitudes of the corresponding standard deviations (σ_i and σ, respectively). An individual mean close to the population mean and a small individual standard deviation relative to the population standard deviation may conceal the individual's changes entirely within the population-based reference interval.

There are two possible solutions to this problem:

1. We can try to reduce the variation in the reference values by *partitioning* into more homogeneous subclasses.

2. The other possibility is to use the subject's previous values, obtained when the subject was in a known state of health, as the reference for any future value.[2,3,10] The application of *subject-based reference values* becomes more feasible as "health-screening" by laboratory tests and computer storage of results become available to large segments of the general population.

Transferability of Reference Values

The determination of reliable reference values for each test in the laboratory's repertoire is a major task, often far beyond the capabilities of an individual laboratory. It would therefore be convenient if reference values generated in another laboratory could be used. This is possible if certain conditions are fulfilled:

1. The populations should be adequately described and matched.

2. Subsets of data from both laboratories should be compared to check for bias arising from analytical factors.

3. Analytical performance in both laboratories should agree.

4. Preparation of individuals before specimen collection and the specimen collection itself should follow a standardized scheme in both laboratories.

References

1. Boyd, J. C., and Lacher, D. A.: The multivariate reference range: An alternative interpretation of multiple profiles. Clin. Chem., *28*:259–265, 1982.
2. Gräsbeck, R., and Alström, T., Eds.: Reference Values in Laboratory Medicine. Chichester, England, John Wiley & Sons, Ltd., 1981.
3. Harris, E. K., Cooil, B. K., Shakarji, G., et al.: On the use of statistical models of within-person variation in long-term studies of healthy individuals. Clin. Chem., *26*:383–391, 1980.
4. IFCC, Expert Panel on Theory of Reference Values: The theory of reference values. *Part 1*. The concept of reference values. Clin. Chem., *25*:1506–1508, 1979. *Part 2*. Selection of individuals for the production of reference values. J. Clin. Chem. Clin. Biochem., *22*:203–208, 1984. *Part 3*: Preparation of individuals and collection of specimens for the production of reference values. In preparation. *Part 4*: Control of analytical variation in the production, transfer and application of reference values. In preparation. *Part 5*: Statistical treatment of collected reference values. Determination of reference limits. J. Clin. Chem. Clin. Biochem., *21*:749–760, 1983. *Part 6*: Presentation of observed values related to reference values. J. Clin. Chem. Clin. Biochem., *20*:841–845, 1982.
5. Morrison, D. F.: Multivariate Statistical Methods. New York, McGraw-Hill Book Co., 1976.
6. NCCLS: Standard Procedures for the Collection of Diagnostic Blood Specimens by Venipuncture (H3-A). Standard Procedures for the Collection of Diagnostic Blood Specimens by Skin Puncture (H4-A). Villanova, PA, National Committee for Clinical Laboratory Standards, 1980 and 1982, respectively.
7. Sachs, L.: Applied Statistics. A Handbook of Techniques. New York, Springer-Verlag, 1982.
8. Snedecor, G. W., and Cochran, W. G.: Statistical Methods. 6th ed. Ames, Iowa State University Press, 1967.
9. Solberg, H. E.: REFVAL. Technical Report, February 1983. Department of Clinical Chemistry, Rikshospitalet, N-0027 Oslo 1, Norway. (The technical report, including computer programs, is available on request to the author.)
10. Solberg, H. E.: Establishment and use of reference values. *In*: Textbook of Clinical Chemistry. N. W. Tietz, Ed. Philadelphia, PA, W. B. Saunders Co., 1986.

ANALYTICAL GOALS AND CLINICAL RELEVANCE OF LABORATORY PROCEDURES

6

ROBERT S. GALEN, M.D., M.P.H.
THEODORE PETERS, JR., Ph.D.

Analytical goals are those standards that the laboratory adopts for its performance of various analyses. However, for an analytically acceptable result to be useful to the physician, information must also be relevant to the clinical problem; i.e., procedures must be examined not only for their analytical but also for their diagnostic attributes. In other chapters of Part II, preanalytical factors such as patient preparation and specimen handling; analytical factors such as accuracy, precision, specificity, and sensitivity of assays; and postanalytical factors such as reporting and record keeping are described. In this chapter, we discuss the manner in which analytical goals are determined and the clinical value of laboratory procedures is examined.

Laboratory tests are requested by the physician to obtain specific and usually quite limited information about the patient. The questions that are asked reflect the testing rationale: (1) Is disease present but not apparent? (Implies that a screening test may be useful.) (2) What is the nature of an apparent disease? (Implies definition of the pathophysiology of the suspected disease.) (3) Can a tentative diagnosis be confirmed? (Implies that additional evidence is needed to increase confidence in an original impression.) (4) How severe is a disease process? (Implies evaluation relative to prognosis.) (5) Has a change occurred since the last observation? (Implies monitoring of a continuing disease process or its treatment.) Obviously, the same laboratory test may be requested in different circumstances to serve different clinical needs.

Analytical attributes of a test are critically important in determining whether the test will be useful in answering one or all of the questions; analytical precision will directly affect our ability to distinguish a normal from an abnormal result or to determine whether a day-to-day change in test results is significant. The diagnostic attributes of a test, diagnostic sensitivity and diagnostic specificity (which will be discussed later in this chapter), are important in assessing whether a test is useful (e.g., relevant in screening, differential diagnosis, or confirmation of a tentative diagnosis).

Analytical Goals

Analytical goals should be appropriate for their intended medical use and hence will vary in different clinical situations as well as among different analytes.

Although one intuitively assumes that a laboratory test is usually obtained for the purpose of diagnosing an illness, surveys show that most laboratory tests are in fact obtained to monitor treatment. There is a basic distinction in performance requirements for these two purposes. For monitoring therapy or following the course of an illness, the concentration of one or more substances is compared with previous results on the same patient. In this case, day-to-day observations must be reliable, and precision (reproducibility) is the prime concern. For diagnosis, on the other hand, the concentration of one or more substances is compared with the established reference range or discrimination value to determine if the value for the subject is indicative of a suspected illness. Thus, in a diagnostic application, both precision and accuracy are important.

When laboratory analysis is used to screen for disease, an abnormal value may lead to immediate therapy or it may be the signal for further, more extensive testing. Therefore, results must be both accurate and precise in order to minimize the expense of the additional tests and the adverse effect on the subject of a false-positive finding. At the same time, the chance that the presence of a disease would be missed (false-negative result) must be minimized.

In emergencies, merely knowing whether a substance is present may justify initiation of appropriate treatment. In some situations, the speed of a report may be of such importance that a somewhat lower performance in terms of accuracy and precision is acceptable. For example, various "quick" tests for blood glucose, ketones, amylase, and urea, despite their lower precision than conventional methods, have proved useful for certain applications. However, the accuracy and precision of these tests must still be known and the analytical runs must be under control.

Some guidelines for establishing analytical goals have been formulated.[4]

1. Analytical goals can only be defined in terms of the needs for patient care; any other basis is irrelevant.

2. Goals for precision and accuracy of quantitative methods, where the analyte is well defined, should be at least as stringent as the current performance by well-managed laboratories, or the "state of the art."

3. For group screening, in which an individual is to be selected from a population, a goal for an analytic coefficient of variation (CV_a) is defined as:

$$CV_a \leq \frac{1}{2}\sqrt{(CV_{Intraindividual})^2 + (CV_{Interindividual})^2}$$

4. For testing to evaluate diagnosis or monitor treatment of an individual, a goal for an analytic coefficient of variation is defined as:

$$CV_a \leq \frac{1}{2}\, CV_{Intraindividual}$$

Goals for Analytical Accuracy

For many analytes there is no agreement on the "true value" among laboratories or on reference methods to obtain such values. Studies of patients often extend over long periods of time, and patients may shift geographical locations; therefore results must be comparable from lab to lab. Thus, a growing need exists for standardization of both analytical methods and reference ranges among laboratories.

There are few published analytical goals for accuracy of clinical laboratory tests. Gilbert, in 1975, proposed that accuracy goals be expressed as percentages; he cited his experience with the evaluation of the College of American Pathologists' (CAP) survey results (Table 6-1).[10] Gilbert's approach reflects confidence in the mean of values from many participating laboratories as an approach to the establishment of the "true" value for an analyte. This confidence has been strengthened by agreement between such interlaboratory means and values established by definitive methods.

Goals for Analytical Precision

Two basic approaches have been used to set analytical goals in terms of medical needs—questionnaires and consensus. The more widely used approach is to poll a large number of practicing physicians concerning how large a change in the concentration of an analyte would be considered medically significant (i.e., real and requiring action). Precision criteria derived in this fashion are listed in Table 6-2.

Precision, expressed as coefficient of variation, generally varies with the concentration of the analyte. Hence, goals should be specified at several levels, particularly at the medically important "decision levels." Precision criteria based on biological variations are listed in Table 6-2. These values were calculated as one-half of the measured day-to-day or intraindividual precision. The most stringent of the precision goals set by clinical surveys (column 3, Table 6-2) were for calcium, chloride, osmolality, and sodium (CVs ≤1.4%). These substances are under tight homeostatic control; note that they show day-to-day variability within the same individual of ≤1.1% (column 4). Day-to-day individual variability is also small for albumin, alkaline phosphatase, cholesterol, creat-

Table 6-1. ACCURACY GOALS FOR CLINICAL LABORATORY TESTS AND PERFORMANCE OF CLINICAL LABORATORIES

Substance (in Serum)	Definitive Value*	Accuracy Goal, %[c]	Clinical Laboratory Values	
			Bias, %	CV, %
Calcium[a]	10.0 mg/dL	2.0	−1.1	4.2
Chloride[a]	105 mmol/L	1.0	−0.8	2.4
Iron[a]	80 μg/dL	5.0	−3.5	12.7
Lithium[a]	1.00 mmol/L		+0.2	13.6
Magnesium[a]	2.00 mg/dL	5.0	−0.7	12.9
Potassium[a]	4.0 mmol/L	2.0	+1.0	2.4
Sodium[a]	145 mmol/L	0.7	−0.5	1.5
Cholesterol[b]	173 mg/dL	4.0	+1.1	5.8
Glucose[b]	104 mg/dL	5.0	−1.2	7.0
Urea nitrogen[b]	17 mg/dL	5.0	+2.8	7.8
Urate[b]	4.8 mg/dL	3.8	−1.8	4.7
Creatinine[b]	0.89 mg/dL	6.7	+17.8	11.2
Cortisol[b]	15.8 μg/dL		+29.1	14.3

*Definitive values on two serum pools were obtained by isotope dilution–mass spectrometry at the US National Bureau of Standards or at the Swedish Karolinska Institute[b] and compared with the mean values found by clinical laboratories on the same pools.

[a] Mean of 1100–4200 US laboratories in CAP survey.[11]

[b] Mean of 18–64 Swedish clinical laboratories.[3]

[c] From Gilbert.[10]

inine, glucose (fasting), phosphorus, potassium, total protein, thyroxine, and uric acid (column 4). A weakness of this approach is that these values are set from observations on healthy individuals; day-to-day variability may be appreciably higher in pathological states.

Analytical Performance Compared with Analytical Goals

Comparison of the accuracy of averaged results of surveys with values obtained by definitive methods is shown in Table 6-1. The bias in the averaged results of surveys for many frequently assayed substances is small—less than the CV of the assay. This also suggests that the mean value obtained by a large number of participating laboratories for a common pool is a reasonable estimate of the true value for these substances. This conclusion cannot be applied as yet to other analytes, particularly enzymes, for which reaction conditions vary among laboratories, nor can it be extended as yet to specimens containing interfering substances. Furthermore, the encouragingly low figures for bias shown by survey results do not necessarily apply to the performance of individual laboratories; the spread of individual results can be quite large. Laboratories should determine their own bias for each test from their survey results.

Results of several large quality assurance programs provide a basis for assessing precision of participating laboratories in the USA (Table 6-2). When compared with requirements of columns 3 and 4, results for albumin, alkaline phosphatase, amylase, calcium, chloride, cholesterol, creatine kinase, creatinine, glucose, pO_2, protein, sodium, and thyroxine fall short of one or more of the criteria. However, the trend toward improved precision over the last decade is encouraging.[18] CVs for determination of glucose, calcium, cholesterol, electrolytes, urea, creatinine, uric acid, and protein have fallen from about 8% to 3%. Even CVs for enzymes have improved markedly, to the 4–10% range. The improvement is the result of both the increasing use of automated equipment and the increased quality and growth of quality control programs.

Clinical Relevance of Laboratory Procedures

To be relevant in the patient care process, a laboratory test must provide information that cannot be obtained by other means at comparable cost and risk to the patient. In addition, a useful abnormal value for a laboratory test must have a high association with the disorder in question and should be infrequently encountered in other conditions.

Table 6-2. ANALYTICAL PERFORMANCE AND MEDICAL NEEDS FOR FREQUENTLY ANALYZED SERUM CONSTITUENTS

		Estimate of Precision Required for Medical Needs (CV, %)		"State-of-the-Art" Analytical Precision (CV, %)	
Analyte	**Mean Value Found for Analyte**	***Based on Clinical Surveys***	***Based on Biological Variability***	**Average Laboratory**	***Laboratory in 0-10th Percentile***
1	*2*	*3*	*4*	*5*	*6*
Acid phosphatase	1.2 U/L			9.3	
Alanine amino-transferase	32 U/L	10.3		7.9	
Albumin	3.5 g/dL	3.7	1.4	3.4	5.7
Alkaline phosphatase	280 U/L	4.3	1.8	3.4	
Amylase	117 U/L	6.0		7.6	
Aspartate amino-transferase	35 U/L			7.5	
Bilirubin, direct	0.3 mg/dL			22.7	
	1.0 mg/dL	10.5		9.8	15.4
Bilirubin, total	1.0 mg/dL			8.3	17.9
Calcium, total	11.0 mg/dL	0.8	0.9	2.0	3.5
Carbon dioxide, total	20 mmol/L	9.0		4.9	
Chloride	90 mmol/L	1.3	1.1	1.5	2.8
Cholesterol	250 mg/dL	2.8	2.4	3.2	6.5
Cortisol	20 μg/dL		13.0	10.1*	16.1*
Creatine kinase	100 U/L	5.1	13.0	9.0	
Creatinine	1.5 mg/dL	15.4	2.2	5.6	10.0
Digoxin	2.5 ng/mL			7.0	10.4
Fibrinogen	200 mg/dL			6.0	10.0
Glucose	50 mg/dL	3.2		3.9	6.0
	120 mg/dL	2.7	2.2	2.8	4.9
γ-Glutamyl transferase	30 U/L			5.0	
Iron	80 μg/dL		13.0	4.0	8.6
Iron binding capacity, total	250 μg/dL			6.2*	10.4*
Lactate	1.5 mmol/L			11.6	
Lactate dehydrogenase	155 U/L	7.9		3.5	
Lipase	10.1 U/L			27.5	
Lithium	1.5 mmol/L			3.8	6.9
Magnesium	1.5 mg/dL			9.0	14.8
Osmolality, total	280 mOsmol/kg	1.3		1.8*	3.6*
pCO_2	32 mm Hg	4.4		5.0	
pO_2	75 mm Hg	4.8		7.2	
pH	7.30	2.8		2.7	
Phosphorus	4.5 mg/dL	4.0	2.9	3.0	5.3
Potassium	3.0 mmol/L	3.7	2.2	2.4	4.0
Protein, total	6.0 g/dL	2.9	1.5	2.2	3.7
Salicylate	15.0 mg/dL			3.5	5.4
Sodium	130 mmol/L	1.4	0.4	1.1	1.8
Thyroxine	12 μg/dL	3.7	3.8	7.2	10.1
Triglycerides	130 mg/dL	10.5	13.0	5.2	9.2
Urea nitrogen	27 mg/dL	15.0	6.2	3.5	6.1
Uric acid	6.0 mg/dL	3.1	3.7	3.0	5.3

*Manual methods.

Estimates of medical needs were obtained either from questionnaires received from clinicians (column 3)[1,5] or from the biological variability of the concentration of an analyte as evaluated at a conference on analytical goals held in Aspen, CO, in 1976 (column 4).[4,10]

State-of-the-art figures (columns 5 and 6) were obtained from cumulative results of large-scale survey and quality control programs.[13,14,15,18]

Many laboratory tests do not meet these conditions. A variety of statistical methods can be used to assess relevancy in these terms. The most widely used approach is the predictive value model. It should be emphasized that although this type of analysis is valuable, even essential, in evaluating the general clinical usefulness of a laboratory test, the techniques are of limited applicability to a specific patient because they only give the probability that a patient has the disease under consideration.

The Predictive Value of Laboratory Tests[9,19,20]

The diagnostic performance of a laboratory test can be defined by its diagnostic sensitivity, diagnostic specificity, predictive value, and efficiency.

Diagnostic sensitivity is a measure of the frequency of a positive test when a particular disease is present; it is the percentage of *true-positive* results. ***Diagnostic specificity*** is a measure of the frequency of a negative test in the absence of a particular disease; it is the percentage of *true-negative* results. A perfect laboratory test would have 100% sensitivity and 100% specificity, and an abnormal result would unfailingly identify those with the disease in question.

When a test is applied to a population, not all of whose members have the disorder being considered, some patients having the disorder will have negative test results (false negatives) and some not afflicted with the disorder will have positive test results (false positives). The probabilities of these outcomes can be calculated for any defined population and are expressed as predictive values. The ***predictive value*** of a positive result is the percentage of all positive results that are true positives, that is, the frequency of the disease in all patients with positive test results. The predictive value of a negative result is the percentage of all negative results that are true negatives, that is, the frequency of patients not having the disease in all patients with negative test results. The ***efficiency*** of a test indicates the percentage of patients who are correctly classified by the test result as having or not having the disorder. The relationships of these terms are shown in detail in Table 6-3.

In most diagnostic applications, we are interested primarily in the predictive value of the positive result. This value identifies the percentage of patients who are presumptive positives for a particular disease and who, after detailed diagnostic work-ups,

Table 6-3. PREDICTIVE VALUE OF TEST APPLIED TO HEALTHY AND DISEASED POPULATIONS[8]

	Number with Positive Test Result	Number with Negative Test Result	Totals
Number with disease	TP	FN	TP + FN
Number without disease	FP	TN	FP + TN
Totals	TP + FP	FN + TN	TP + FP + TN + FN

Definitions:

TP = True positives: number of diseased patients correctly classified by the test.
FP = False positives: number of nondiseased patients misclassified by the test.
FN = False negatives: number of diseased patients misclassified by the test.
TN = True negatives: number of nondiseased patients correctly classified by the test.

Sensitivity = Positivity in disease, expressed as per cent = $\frac{TP}{TP + FN} \times 100.$

Specificity = Negativity in health, or absence of a particular disease, expressed as per cent = $\frac{TN}{FP + TN} \times 100.$

Predictive value of positive test =

Per cent of patients with positive test results that are diseased = $\frac{TP}{TP + FP} \times 100.$

Predictive value of negative test =

Per cent of patients with negative test results that are nondiseased = $\frac{TN}{TN + FN} \times 100.$

Efficiency of test =

Per cent of patients correctly classified as diseased and nondiseased = $\frac{TP + TN}{TP + FP + FN + TN} \times 100.$

will be found to have the disease. In the discussion that follows, predictive value will always be used to refer to the predictive value of the positive test result.

The Effect of Disease Frequency

The *incidence* of a disease is the number of cases of the disease that arise during a specified period of time (e.g., number of patients per 100 000 population who develop the disease in a given year). The *prevalence* is the frequency of the disease at a designated point in time. The particular application determines whether prevalence or incidence rates are more appropriate for calculating the predictive value of a test.

A marked change in predictive value occurs when there is a change in the prevalence of the disease in the population under study. The relationship of prevalence and predictive values is shown by a formula derived from Bayes's theorem.[2]

$$\text{Predictive value} = \frac{(\text{Prevalence})\,(\text{sensitivity})}{(\text{Prevalence})\,(\text{sensitivity}) + (1 - \text{prevalence})\,(1 - \text{sensitivity})}$$

Table 6-4 demonstrates the change in predictive value that occurs for a hypothetical test with changing prevalence of disease. A particular test will have a higher predictive value when it is applied to a population with a higher prevalence of the disease being studied, for example, those populations selected because of symptoms or physical examination. But a good diagnostic test frequently fails as a screening test when applied to the general population, where the prevalence of a disease is very low. If a laboratory test is useful, its predictive value will be much greater than the disease prevalence.

It is essential that laboratorians and clinicians be able to use the predictive value model for evaluating the diagnostic accuracy of laboratory tests as well as other procedures. This will facilitate rational test selection and anticipation of outcomes.

Example: A hypothetical serum enzyme determination used in the diagnosis of myocardial infarction (MI) has 95% sensitivity for patients with MIs and 95% specificity for non-MI patients. Let us assume that the prevalence of MI in the outpatient clinic patients with "chest pain" was 5%, whereas it was 50% in patients admitted to the coronary care unit (CCU) in order to rule out MI. When used in the clinic, the enzyme test would have a predictive value of 50%; i.e., 50% of the patients with positive results would have MIs and the remaining 50% would not and would therefore be false positives. This same enzyme test when used in the CCU would have a predictive value of 95%; i.e., 95% of the patients with positive results would have MIs, and only 5% of positive results would be false positives.

The Effect of Cut-Off Point

Although an abnormal laboratory test result can be defined as one falling outside the limits of a predefined reference range, this range may not be the most effective one for making medical decisions since it may produce an unacceptably high frequency of false-positive or false-negative results. By using a different medical decision level or "cut-off point," it is possible to eliminate or decrease false-positive tests (at the cost of more false-negative tests) or to eliminate or decrease false-negative results (at the cost

Table 6-4. PREDICTIVE VALUE AS A FUNCTION OF DISEASE PREVALENCE[8] FOR A LABORATORY TEST WITH 95% SENSITIVITY AND 95% SPECIFICITY

Prevalence of Disease, %	Predictive Value, %
1	16.1
2	27.9
5	50.0
10	67.9
15	77.0
20	82.6
25	86.4
50	95.0

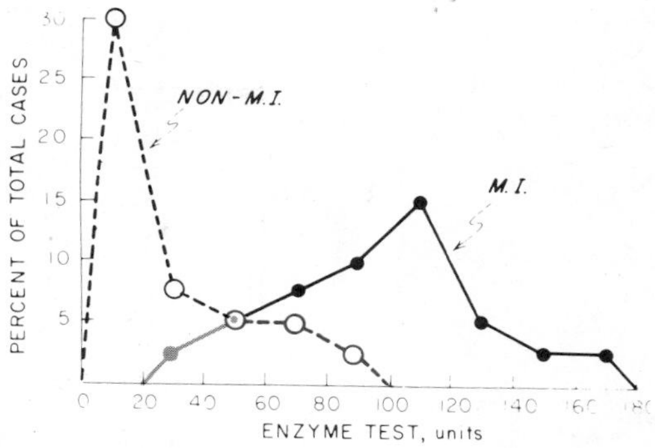

Figure 6-1. Selection of a cut-off point based on the sensitivity and specificity of a laboratory test.[8] Prevalence of myocardial infarction = 50%. Mean enzyme activity, MI = 97 U/L. Mean enzyme activity, non-MI = 27 U/L. Upper limit of normal = 20 U/L, sensitivity = 100%, specificity = 60%. Upper reference limit = 100 U/L, sensitivity = 50%, specificity = 100%. (From Galen, R. S.: Orthop. Clin. North Am., *10:*293, 1979.)

of more false-positive results). It is impossible to select a cut-off point that would afford complete discrimination between the two groups. At any cut-off point, one must sacrifice sensitivity for specificity or vice versa. Herein lies the major flaw of diagnoses from laboratory tests, in that tests are not sensitive and specific at the same time. Use of multiple laboratory tests improves the predictive value, but the trade-off between sensitivity and specificity always remains.

Figure 6-1 illustrates how sensitivity and specificity are altered by the selection of a cut-off point for a particular test. If we grouped the patients admitted to the CCU with the possible diagnosis of myocardial infarction into those with MI and those without MI, based on history and electrocardiographic (ECG) findings, and then expressed the enzyme results for each group as a frequency distribution, we would find the classical overlapping distributions seen in Figure 6-1. We could choose any point in the overlapping region as the cut-off point, but each choice would affect both the sensitivity and specificity of the enzyme result. Several cut-off points are listed in Table 6-5 with the sensitivity, specificity, and predictive value associated with each. For any laboratory test, the predictive value of the test in diagnosing a particular disease can be altered by changing the cut-off point.

There are several caveats in using the predictive value model. The population on which sensitivity is calculated should unequivocally have the disease and yet should not have the disease in a far-advanced stage. If this is not the case, many false-negative results will be obtained when the test is applied to detect the disease in its early stages.

Application of the Model—Single Test

In 1977, Foti and coworkers[7] reported on a radioimmunoassay (RIA) procedure for prostatic acid phosphatase (PAP). Foti's findings are summarized in Table 6-6 (stages I through IV describe the extent of spread of the tumor). Foti's study, and an editorial by Gittes[12] that followed, suggested that acid phosphatase be used not only as a tumor marker but also as a cancer screening test.

Table 6-5. CUT-OFF POINTS AND ASSOCIATED PREDICTIVE VALUES IN ENZYME TESTING[8]

Cut-off Point, U/L	Sensitivity, %	Specificity, %	Predictive Value of a Positive Test, %	Predictive Value of a Negative Test, %	Efficiency, %
20	100.0	60.0	71.4	100.0	80.0
40	95.0	75.0	79.2	93.8	85.0
50	90.0	80.0	81.8	88.9	85.0
60	85.0	85.0	85.0	85.0	85.0
80	70.0	95.0	93.3	76.0	82.5
100	50.0	100.0	100.0	66.7	75.0
Chance Test	50.0	50.0	50.0	50.0	50.0
Perfect test	100.0	100.0	100.0	100.0	100.0

Disease prevalence = 50%.
Cut-off points and associated predictive values and efficiencies for data illustrated in Figure 6-1.

Table 6-6. SENSITIVITY AND SPECIFICITY OF RIA TEST FOR PROSTATIC ACID PHOSPHATASE*

Group	Number of Patients	Sensitivity, %
Patients with prostate cancer	113	70
Stage I	24	33
Stage II	33	79
Stage III	31	71
Stage IV	25	92
	Number of Patients	**Specificity, %**
Patients without prostate cancer	217	94
Normal controls	50	100
Benign prostatic hyperplasia	36	94
Total prostatectomy	28	96
Other cancers	83	89
Gastrointestinal disorders	20	95

* Data summarized from Table 2, Foti et al.[7]

Foti and coworkers determined the sensitivity of the radioimmunoassay of PAP (based on current clinical staging criteria) to be 33, 79, 71, and 92% in stages I to IV, respectively. The specificity of radioimmunoassay of PAP has been measured in specific subgroups but is not available for the at-risk population as a whole. However, on the basis of specificity of 100% in normal subjects, 94% in patients with benign prostatic hypertrophy, and 89% in patients with other carcinomas, a reasonable estimate of specificity would be 95%. The third factor, prevalence, is a function of age and the method of diagnosis. An overall estimate of prevalence of 25% for men over age 60 seems more than reasonable.

Finally, the cases must be staged. Clinical data, probably biased toward advanced stages, combined with autopsy data, were used to estimate that 15, 25, 35, and 25% of the patients with carcinomas would fall into stages I to IV, respectively.[6] Data showing how this test can be evaluated by using the predictive value model are presented in Table 6-7. On the basis of these estimates, the predictive value of the radioimmunoassay for PAP can be calculated to be 83% with an overall sensitivity of 73%.

Table 6-7. TESTING FOR CARCINOMA OF THE PROSTATE: DISTRIBUTION OF RESULTS IN A HYPOTHETICAL UNSCREENED POPULATION, n = 10 000; PREVALENCE = 25%; SPECIFICITY = 95%[6]

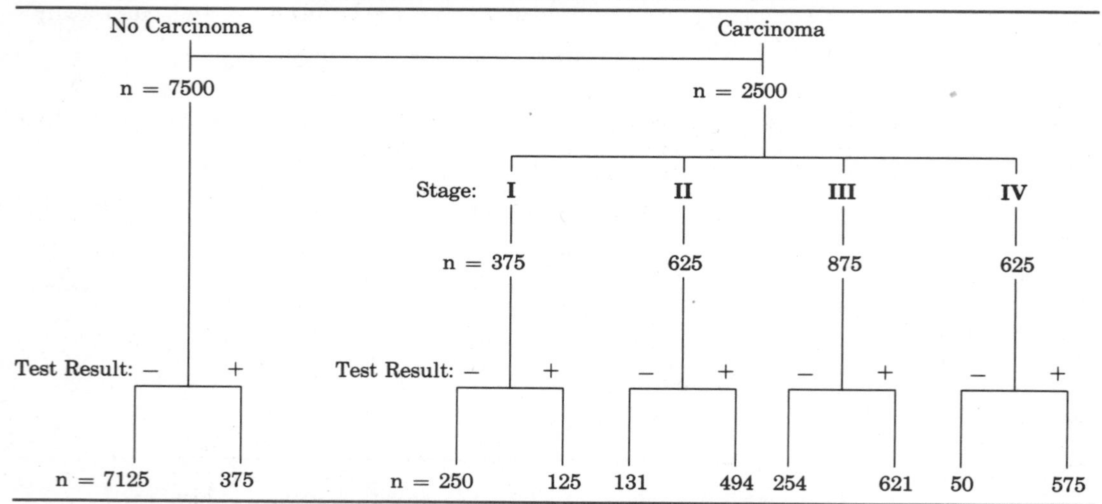

These calculations are representative of an unscreened population. If the population had been previously screened for prostatic carcinoma, the test would be detecting the incident rather than the prevalent cases. An incidence rate of 5% per year (5–10 times the maximal reported clinical incidence rate) is more than adequate to explain the increase in prevalence with age. This rate yields a predictive value of a positive test of 43%. Therefore, follow-up studies (perhaps needle biopsy of the prostate) on 57% of the patients would be done unnecessarily. With this frequency of false positives, it would be neither medically prudent nor cost-effective to use this test for screening purposes. With a lower incidence, the test would perform even worse. With a 1% incidence, the predictive value would be only 13%.

These considerations demonstrate the importance of evaluating laboratory tests in terms of sensitivity, specificity, and predictive values in the population in which they will be used.

Predictive Value of Laboratory Diagnosis—Combination Testing

The result of a single test is rarely used as the final arbiter of a medical decision. Therefore, let us explore the benefit of using more than one test to diagnose a clinical problem.

Two independent tests in a screening or diagnostic situation can be used in different ways. In the *series approach*, test A is applied first and all those specimens with a positive result are retested with test B, or vice versa. In the *parallel approach*, tests A and B are used together and all those specimens with positive results for either or both tests are considered to be positives.

With parallel testing, the combined sensitivity is greater than the individual sensitivities of the contributing tests, but the specificity is low, whereas series testing results in lower sensitivity but higher specificity. If two tests are going to be used in series, the optimal sequence can be determined by using the predictive value model.

Application of the Model—Multiple Tests

The above relationship can be demonstrated with actual data from a study of haptoglobin, total LDH, and LDH isoenzymes in the diagnosis of hemolytic disorders.[16] One hundred hospitalized patients were evaluated based on the physician's request for a serum haptoglobin. The laboratory then performed LDH and LDH isoenzyme determinations. Without knowledge of the haptoglobin, LDH, or LDH isoenzyme data, a hematologist classified these 100 cases into two major groups: 24 patients with hemolytic disease and 76 patients with other (nonhemolytic) diseases. Table 6-8 summarizes the sensitivity, specificity, predictive value, and efficiency of these procedures used as single tests and in combination with each other. Haptoglobin was determined nephelometrically and considered positive for hemolytic disease at a concentration of ≤ 25 mg/dL (3 μmol/L). LDH activity was considered abnormal at ≥ 250 U/L. LDH isoenzymes were separated on cellulose acetate and scanned fluorometrically. LDH isoenzymes were considered positive when LDH-1 activity exceeded LDH-2 activity (a "flipped" LDH pattern).

Table 6-8. HEMOLYTIC DISEASE STUDY[16]

Test	Sensitivity, %	Specificity, %	Predictive Value, %	Efficiency, %
Haptoglobin	83	96	87	93
Lactate dehydrogenase	83	61	40	66
Lactate dehydrogenase isoenzymes	58	93	74	85
Haptoglobin *and* LDH isoenzymes*	50	100	100	88
Haptoglobin *or* LDH isoenzymes†	92	89	73	90

* Series interpretation.
† Parallel interpretation.

Table 6-9. HEMOLYTIC DISEASE STUDY[16]

	Haptoglobin Results		
	≤25 mg/dL	**>25 mg/dL**	**Total**
Hemolytic disease	20	4	24
Other	3	73	76
Totals	23	77	100

Sensitivity = 83%; specificity = 96%; predictive value = 87%; efficiency = 93%.

The highest efficiency (overall correct classification) is achieved by simply using haptoglobin alone. The predictive value for haptoglobin is presented in Table 6-9. It is interesting to explore the four cases of hemolytic disease that were "false negatives," that is, patients without depressed haptoglobin levels. These turned out to be two cases of hypersplenism and two cases of pernicious anemia. On the other hand, there were three "false positives," that is, nonhemolytic patients with depressed haptoglobin levels. These turned out to be cases of ovarian carcinoma, cirrhosis, and sarcoidosis.

The contrast between series and parallel testing is demonstrated by the combined use of haptoglobin and LDH isoenzymes. Although the efficiency is essentially the same, a series approach requiring both tests to be positive has a predictive value of 100%. The sensitivity, however, is only 50%, and half of the cases would go undetected. A parallel approach requiring either test to be positive actually has the highest sensitivity of the tests listed (see Table 6-8).

The effect of increasing specificity with the series approach can be demonstrated with actual data from a study of enzymes and isoenzymes in the diagnosis of acute myocardial infarction.[10] By using the final clinical diagnosis as the ultimate criterion, the sensitivity and specificity of single and multiple enzyme determinations were calculated. Positive results were defined for creatine kinase and aspartate aminotransferase on day one, and for lactate dehydrogenase on day two, following chest pain; the laboratory's upper limit was used for each test. The results are summarized in Table 6-10.

It can be seen that interpreting results in a series fashion always decreases sensitivity and increases specificity relative to single tests or smaller test combinations. The overall effect of the series approach is that more cases will be missed, but there will be a higher predictive value of the positive result (i.e., fewer false-positive results).

Biochemical profiles or batteries of laboratory tests interpreted in a parallel fashion have extremely high sensitivity but low specificity. Because of the high sensitivity, these profiles, when negative, have a very high negative predictive value. For that reason, they are used by clinicians to exclude or rule out a variety of diagnoses with a high degree of certainty.

Later on in the diagnostic work-up, series testing is performed. This approach is highly predictive of the presence of a particular disease because of its high specificity.

Table 6-10. SENSITIVITY AND SPECIFICITY OF SINGLE AND COMBINATION* ENZYME TESTS PERFORMED ON PATIENTS IN THE CCU[8]

Test	**Sensitivity, %**	**Specificity, %**
CK	95.7	57.1
AST	91.4	73.5
LDH	87.2	90.5
AST, then LDH	82.2	92.9
CK, then LDH	82.2	92.9
CK, then AST	87.2	83.7
CK, AST, then LDH	77.7	95.2

* All combination results interpreted in series fashion as listed.

As the clinician follows his own algorithm for testing a patient, the probability of a correct diagnosis continues to increase as he receives more information.

Evaluation of Test Results by the Laboratory

In addition to adhering to quality control procedures (see Chapter 8), the laboratory staff has the responsibility to evaluate test results with regard to their clinical relevance. In particular, the need for timeliness in reporting a result and the consistency of a result with the patient's condition should be checked.

Critical values. Results should be screened at the earliest opportunity for values that are so far from the norm as to indicate a potentially dangerous condition requiring immediate attention by the clinician. Such critical results should be reported immediately even if they were not requested on a priority basis. Table 6-11 lists some typical "critical values."

Cumulative reports. If serial analyses are performed on specimens taken at different times from the same patient, it is helpful to the clinician if these results are presented as a table on a single sheet. Such cumulative reports, often generated by a computer, enable the physician to see at a glance the patient's progress (see also Chapter 9) and allow the laboratory staff to make additional checks on the validity of values. Random errors such as albumin values greater than total protein values or accidental

Table 6-11. EXAMPLES OF CRITICAL VALUES AND ALLOWED DIFFERENCES FROM PRIOR VALUES*

	Critical Values		Allowed Differences from Prior Values		
Analyte	*Lower*	*Upper*	*Low Range*	*Division Point*	*High Range*
Glucose (serum), mg/dL	40	400	100	300	50% of prior value
Glucose (CSF), mg/dL	40	—	—	—	—
Urea-N, mg/dL	3	50	10	25	20% of prior value[a]
Creatinine, mg/dL	0.3	7.5	0.5	2	25% of prior value
Calcium, total, mg/dL	6.0	13.0	1.5	—	1.5
Calcium, ionized, mg/dL	3.0	6.5	—	—	—
Magnesium, mg/dL	1.0	—	—	—	—
Phosphate-P, mg/dL	1.0	8	1.0	—	1.0
Uric acid, mg/dL	1.0	12	2.0	12	25% of prior value
Bilirubin (adult), mg/dL	—	—	2.0	5	50% of prior value
Bilirubin (neonatal), mg/dL	—	12	2.5	10	25% of prior value
Albumin, g/dL	1.5	6	1.0	—	1.5
Protein, g/dL	4	9	1.5	—	1.5
Protein (CSF), mg/dL	—	100	—	—	—
Sodium, mmol/L	120	160	10	—	10
Potassium, mmol/L	2.5	6.5	1.2	—	1.2
Lithium, mmol/L	—	2.0	—	—	—
Chloride, mmol/L	85	115	10	—	10
CO_2 content, mmol/L	10	40	6	—	6
Alkaline phosphatase, U/L	—	—	50[b]	125	50% of prior value
Aspartate aminotransferase, U/L	—	—	100[b]	500	40% of prior value
Alanine aminotransferase, U/L	—	—	100[b]	500	40% of prior value
Creatine kinase, U/L	—	—	200[b]	600	40% of prior value
Lactate dehydrogenase, U/L	—	—	200[b]	600	40% of prior value
Amylase, U/L	—	—	100[b]	500	25% of prior value
pH	7.20	7.65	—	—	—
pO_2, mm Hg	40	—	—	—	—
pCO_2, mm Hg	—	50	—	—	—

*From references 21, 22, and the authors' experience.

[a] Not applicable if patient is undergoing renal dialysis.

[b] Method-dependent values.

Critical values are those which should be reported immediately to the clinician responsible for the patient as well as verified in the laboratory. Values showing changes greater than the *allowed differences from prior values* should be verified and then flagged to indicate their possible importance to the clinician.

transposition of urea and creatinine results can readily be noticed by an experienced laboratorian (internal consistency). That the pattern of results is reasonable, considering the suspected diagnosis (biological consistency), can also be judged—low levels for both total protein and cholesterol in malnutrition, or elevated sodium and chloride in dehydration, for example.

Review of calculated data such as urea nitrogen/creatinine ratio, anion gap, and base excess also helps detect inconsistencies (see also Chapters 8, 9, and 20).

Checks with prior values. An additional check compares the most recent value for an analyte with the previous value obtained for the same analyte. Such comparisons are termed *prior value tests* or *delta checks*. (See also Chapter 8.) Examples of the allowable differences between two consecutive test results are given in Table 6-11. A difference greater than the figures given implies either that the patient has experienced an acute change in condition or that there is a laboratory error in one of the results. Typically, 5–25% of results failing a delta check are attributable to laboratory errors.[21]

References

1. Barrett, A. E., Cameron, S. J., Fraser, C. G., et al.: A clinical view of analytical goals in clinical biochemistry. J. Clin. Pathol., *32*:893–896, 1979.
2. Bayes, T.: An essay toward solving a problem in the doctrine of change. Philos. Trans. R. Soc. Lond., *53*:370–418, 1763.
3. Björkhem, I., Bergman, A., Falk, O., et al.: Accuracy of some routine methods used in clinical chemistry as judged by isotope dilution–mass spectrometry. Clin. Chem., *27*:733–735, 1981.
4. Elevitch, F. R., Ed.: Proceedings of the 1976 Conference on Analytical Goals in Clinical Chemistry, College of American Pathologists. Skokie, IL, College of American Pathologists, 1977.
5. Elion-Gerritzen, W. E.: Analytic precision in clinical chemistry and medical decisions. Am. J. Clin. Pathol., *73*:183–197, 1980.
6. Fink, D. J., and Galen, R. S.: Immunologic detection of prostatic acid phosphatase: Critique II. Hum. Pathol., *9*:621–623, 1978.
7. Foti, A. G., Cooper, J. F., Herschmann, H., et al.: Detection of prostatic cancer by solid-phase radioimmunoassay of serum prostatic acid phosphatase. N. Engl. J. Med., *297*:1357–1361, 1977.
8. Galen, R. S.: Selection of appropriate laboratory tests. *In*: Clinician and Chemist. D. S. Young, J. Hicks, H. Nipper, et al., Eds. Washington, D.C., American Association for Clinical Chemistry, 1979, pp. 69–105.
9. Galen, R. S. and Gambino, S. R.: The Predictive Value and Efficiency of Medical Diagnoses. New York, John Wiley & Sons, 1975.
10. Gilbert, R. K.: Progress and analytic goals in clinical chemistry. Am. J. Clin. Pathol., *63*:960–973, 1975.
11. Gilbert, R. K.: Accuracy of clinical laboratories studied by comparison with definitive methods. Am. J. Clin. Pathol., *70*:450–470, 1978.
12. Gittes, R.: Acid phosphatase reappraised. N. Engl. J. Med., *297*:1398–1399, 1977.
13. Hyland SCAN Service, Hyland Diagnostics, Deerfield, IL 60015, January–December, 1982.
14. Itano, M.: CAP blood gas survey—first year's experience. Am. J. Clin. Pathol., *74*:535–541, 1980.
15. Lohff, M. R., DiSilvio, T. V., Ross, J. W., et al.: Analytic clinical laboratory precision. State of the art for selected enzymes. Am. J. Clin. Pathol., *78*(Suppl.):634–643, 1982.
16. Marchand, A., Galen, R. S., and Van Lente, F.: The value of serum haptoglobin in hemolytic disease. JAMA, *243*:1909–1911, 1980.
17. Ransohoff, D. F., and Feinstein, A. R.: Problems of spectrum and bias in evaluating the efficiency of diagnostic tests. N. Engl. J. Med., *299*:926–930, 1978.
18. Ross, J. W., and Fraser, M. D.: Clinical laboratory precision. The state of the art and medical usefulness based internal quality control. Am. J. Clin. Pathol, *78*(Suppl.):578–586, 1982.
19. Vecchio, T. J.: Predictive value of a single diagnostic test in unselected populations. N. Engl. J. Med., *274*:1171–1173, 1966.
20. Weinstein, M. C., and Fineberg, H. V.: Clinical Decision Making. Philadelphia, W. B. Saunders Company, 1980.
21. Wheeler, L. A., and Sheiner, L. B.: A clinical evaluation of various delta check methods. Clin. Chem., *27*:5–9, 1981.
22. Whitehurst, P., DiSilvio, T. V., and Boyadjian, G.: Evaluation of discrepancies in patients' results—an aspect of computer-assisted quality control. Clin. Chem., *21*:87–92, 1975.

EVALUATION OF METHODS

7

THEODORE PETERS, JR., Ph.D.
JAMES O. WESTGARD, Ph.D.

Method selection and evaluation are important steps in establishing high quality laboratory services. Before a new or improved method is introduced into a laboratory, it must be selected with care and its performance evaluated under laboratory conditions. The purpose of this chapter is to outline some of the techniques that are utilized for selecting and evaluating analytical methods, whether these are manual or automated, and whether the reagents are prepared in the laboratory or obtained from a manufacturer.

The selection and evaluation of a new method require considerable judgment. The analyst must be able to understand the claims of manufacturers and the studies reported in the literature. He must also be able to perform evaluation studies to make sure that the desired performance is achievable under the conditions prevailing in his laboratory.

Definitions and Goals

The first step in selecting a method is to confirm the need for the assay and to define the goals that are to be accomplished. Pertinent properties relate to medical usefulness (see Chapter 6) and include the choice of analyte to be measured, the diagnostic sensitivity and specificity of the procedure, and the precision and accuracy required for effective performance. Other related characteristics are the type of specimen to be analyzed, the hours the test should be available, and required turnaround time for results. The importance of a careful definition of requirements should be apparent. It is this list of requirements that guides the whole process of selecting and evaluating analytical methods.

Characteristics of Analytical Methods

Practicality characteristics are factors (other than analytical performance) that determine whether the method can be implemented in the laboratory. They include equipment, workload, specimen handling, run size, personnel skill, cost per test, methods of standardization and quality control, space needs (including reagent storage), and safety precautions and procedures that are required for a specific method. Most practicality requirements can be defined by the laboratory personnel and can often be assessed from the technical literature describing the analytical method.

Reliability characteristics relate to the performance of the method. They include such factors as precision, accuracy, analytical sensitivity, analytical specificity, recovery, interference, blank readings, linear range, sample interaction, and reagent stability. The assessment of reliability characteristics requires experimental studies to estimate the method's performance, followed by judgments on whether the observed performance is satisfactory for the medical use of the test results. The definition of "medical usefulness requirements" may require conferences between laboratory and clinical personnel.

Analytical range is the "range of concentration or other quantity in the specimen over which the method is applicable without modification."[2] It is tested by a "linearity experiment" in which a series of solutions, usually standards representing a wide concentration range, are analyzed by the method. The analytical range should be wide enough to include most (e.g., 95%) of the expected clinical specimens without dilution.

Ideally the standard curve (plot of response versus analyte concentration) should be linear and pass through the origin. When linearity is assumed in the calibration process, the linear range of the standard curve is the critical characteristic. If linear response is not obtained, calibration procedures must employ a higher number of standard solutions to define the response curve adequately, and unknowns should be bracketed by the standard solutions.

Analytical sensitivity is a measure of "the ability of an analytical method to detect small quantities of the measured component."[2] Sensitivity is related to precision. When the concern is performance at concentrations above the detection limit, it is best to determine precision at the particular concentrations of interest. When the concern is performance at very low concentrations (for example, in measuring trace materials), then it is useful to determine the detection limit.

Detection limit is defined as "the smallest single result which, with stated probability (commonly 95%), can be distinguished from a suitable blank. The limit may be a concentration or an amount and defines the point at which the analysis becomes just feasible."[2] The detection limit is related to precision and also depends on the amplitude of blank readings.

Blank readings are responses observed by the measurement procedure due to reagent and sample constituents, exclusive of the desired analyte. It may be useful to quantitate the blank readings directly by making measurements of the reagent solutions without sample present (*reagent blanks*) and of sample dilutions without at least one reagent that initiates the reaction (*sample blanks*). Low blank readings are of course desirable and provide the best possibilities for precision and accuracy. Despite high blank readings, the performance of a method may still be satisfactory if the blank measurements are stable and precise, therefore permitting small differences between measurements to be meaningful.

Analytical specificity is "the ability of an analytical method to determine solely the component(s) it purports to measure."[2] Specificity is related to accuracy. Lack of crossreactivity with closely related substances is one aspect of specificity; other common examples are the potential problems due to frequently encountered substances such as bilirubin, hemoglobin, and lipids, which may cause errors by virtue of their color, turbidity, or other characteristics.

Interference relates to specificity and is defined as "the effect of a component, which does not by itself produce a reading, on the accuracy of measurement of another component."[2] Interferences may be very subtle, such as the unsuspected presence of a drug that exhausts a reagent.

Recovery is "the ability of an analytical method to correctly measure pure analyte when added to the samples routinely analyzed."[2] Recovery is related to accuracy and it tests whether the analytical method can measure the desired substance in the presence of all of the other materials appearing in the matrix of the real specimen. Addition of the analyte in vitro is obviously an artificial approach; it cannot be assured that the physical state of the analyte, its solubility, and its binding to serum proteins are the same as would be attained in vivo. Nor can the possibility of a change in chemical form in vivo (such as conjugation or modification in the liver) be taken into account. But in the absence of analytical reference methods and reference materials, recovery experiments may be one of the few ways to assess accuracy.

Precision refers to the "agreement between replicate measurements"[2] and is estimated by a "replication experiment" in which the same sample material is analyzed several times (usually a minimum of 10–20 measurements). *Within-run precision* is the variability found when the same material is analyzed repeatedly in one analytical run (or, alternatively, when duplicate analyses are made within a run for a series of clinical specimens and the standard deviation of the duplicates is calculated). It usually provides an optimistic estimate of the expected performance since there has been minimal opportunity for conditions to change during the time between the replicate analyses. *Within-day precision* is estimated when the same material is analyzed repeatedly in several different runs on the same working day. This variability is usually somewhat higher than observed for within-run replicates. *Day-to-day precision* is the variability found when the same material is analyzed repeatedly on different days. This last estimate is the most realistic assessment of the performance that will be observed by the clinicians in their use of the test.

The systematic error concept of accuracy is a measure of the "agreement between the best estimate of a quantity and its true value."[2] It is usually estimated by "comparison of methods" experiment in which clinical specimens are analyzed by the method under evaluation (*test method*) and by another method whose accuracy has already been established (*reference method*). The disagreement between methods can be described as *systematic error*—"an error that is always in one direction."[6] Systematic errors can be further divided into two types—constant and proportional. *Constant systematic error* is "an error that has always the same direction and magnitude even as the concentration of analyte changes."[6] *Proportional systematic error* is "an error which is always in one direction and whose magnitude is a percentage of the concentration of analyte being measured."[6]

In addition to these types of errors, there is *random bias*, which is a systematic error that varies from sample to sample due to interferences present in some but not all samples. Its magnitude may depend on the patient population being tested.

The nature of these different types of errors can be seen by plotting the results of a test method versus a reference method for a group of specimens. As shown in Figure 7-1, random error appears as scattered data points about the line of best fit. Constant error causes the line to shift in one direction and is most easily seen at the point where the line intersects the ordinate (y-intercept). Proportional error causes the angle (or slope) of the line to change. Careful inspection of this plot can provide information about the random, constant, and proportional components of analytical error.

Another useful graph is a plot of the differences between the results of the test and reference methods on the y-axis versus the values obtained by the reference method on the x-axis. This "scatterplot" permits closer inspection of the differences for individual specimens.

The total error concept of accuracy considers all types of errors, both random and systematic, as causing an observed measurement to deviate from its true value. The total error concept and its relationship to the random and systematic components are illustrated in Figure 7-2. The distribution of measurements ($\pm 2s$) around a central value ($\bar{x}$) represents random error. The amount of shift of that central value from the true value (μ) represents sytematic error ($\bar{x} - \mu$). The total error ($2s + |\bar{x} - \mu|$) shows how large the errors can be when the random and systematic components occur in the same direction. This "worst-case" estimate of the errors may be larger than either the random or systematic components alone, and therefore it has been argued to be a "safer" estimate for judging the quality of analytical methods.[5]

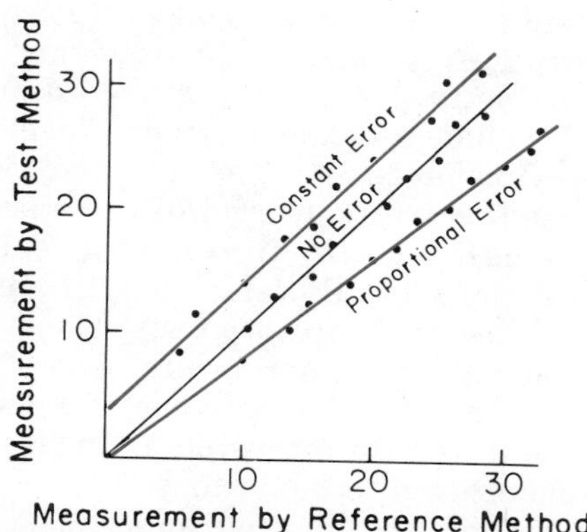

Figure 7-1. Appearance of different types of analytical errors when observed values (test method measurements) are plotted on the y-axis, versus some estimate of the true values (measurements by a reference method) plotted on the x-axis. Note that random error here could include both the imprecision of the measurements and the random bias due to interferences. See text for more explanation. (From Westgard, J. O., de Vos, D. J., Hunt, M. R., et al.: Method Evaluation. Houston, American Society for Medical Technology, 1978.)

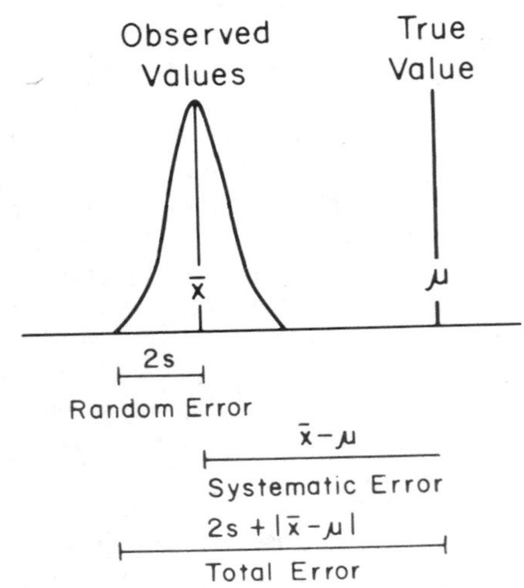

Figure 7-2. Illustration of concepts of accuracy: random error, systematic error, and total error. (From Westgard, J. O., de Vos, D. J., Hunt, M. R., et al.: Method Evaluation. Houston, American Society for Medical Technology, 1978.)

Setting Quality Goals for Analytical Performance

Performance goals should be expressed as specifications for the amount of analytical error that can be permitted without invalidating the medical usefulness of test results. Separate specifications can be defined as goals for precision and accuracy, or a single specification can be defined as a goal for total error.

A ***precision goal*** can be stated as an allowable standard deviation (s_A). For a urea method, for example, the allowable standard deviation could be set as 1.0 mg/dL at a concentration of 15 mg/dL and as 5.0 mg/dL at a concentration of 100 mg/dL.

An ***accuracy goal*** can be stated as an allowable systematic error or allowable bias (B_A). For a urea method, the allowable bias could be defined as 1.0 mg/dL at a concentration of 15 mg/dL and 4.0 mg/dL at 100 mg/dL.

A ***total error goal*** (TE_A) can be stated as a 95% limit of the allowable analytical error (i.e., a limit of two standard deviations) which is to include the effects of both random error (RE) and systematic error (SE), where TE = RE + SE, and RE = 2 × the day-to-day standard deviation. Only one patient test result out of twenty would be permitted to have an error greater than the amount specified. For the urea example, the total analytical error could be specified as 3.0 mg/dL (2 × 1.0 + 1.0) at 15 mg/dL and 14 mg/dL (2 × 5 + 4) at 100 mg/dL.

A ***medical decision level*** (X_C) specifies a concentration at which test results are critically interpreted for purposes of diagnosis, monitoring, or therapeutic decisions. Several decision levels can be defined when there are different medical applications. Quality specifications should include a medical decision level to indicate the concentration at which the desired performance is to be achieved. Method evaluation studies should be designed to estimate the errors at each of these critical concentrations.

Recommendations for quality goals are found in Tables 6-1 and 6-2. It should be recognized that recommendations for allowable analytical errors are judgments based on experience with the medical use of laboratory tests. There are no error specifications that are universally applicable. What is appropriate depends on the medical purposes of the health care facility, the population being served, the particular application of the test, and the way in which the physician interprets the test result. Therefore, performance goals may differ from laboratory to laboratory.

It is difficult to make judgments on allowable analytical errors, but these judgments are made by laboratory personnel who are knowledgeable about the subject matter. The alternative is to try to make judgments on what statistical results are acceptable. Historically, statistics such as *t*-value, F-value, and the correlation coefficient (r) have been used to judge the acceptability of analytical performance. Such judgments are also difficult and usually not very objective.[7] Therefore, it is preferable for laboratory personnel to define allowable analytical errors and to judge analytical performance relative to medical needs.

Judging Performance Based on Quality Goals

When a specification for the allowable analytical error is given, the acceptability of the analytical performance of a method can be judged by the following steps:

1. The observed errors are compared with the specified allowable errors.
2. When an observed error is less than the allowable error, performance is acceptable and the method may be implemented for routine service.
3. When an observed error is greater than the allowable error, performance is not acceptable. The errors of the method must be reduced by modification of the procedure, or else the method must be rejected.

Quantitative criteria can be developed for each experiment to compare the observed errors with the specified allowable errors. The observed errors are estimates based on the use of certain statistics. For example, the standard deviation is used to estimate precision or random error; a criterion for judging precision is to compare the observed standard deviation with the specified allowable standard deviation. When a total error goal has been specified, the observed standard deviation can be multiplied by 2 to provide a 95% limit for random error, which can then be compared with the allowable total error. Similar criteria can be developed for interference, recovery, and comparison-of-methods experiments[5]; these criteria are shown in Table 7-1.

Selecting an Analytical Method

After goals for a method are set, the next step is to survey the technical literature and decide which procedures best fulfill the requirements. This literature includes both professional journals and manufacturers' reports and brochures. A preliminary judgment can often be made from a review of this information, and it is also possible to assign weights or points to the different requirements based on their relative importance, thus providing a more quantitative selection process.

Some guidelines for reviewing prospective methods have been established by the American Association for Clinical Chemistry.[1] These are oriented to kit methods, but provide recommendations that are useful in general. Attention is directed to the following

Table 7-1. CRITERIA FOR JUDGING THE ACCEPTABILITY OF ERRORS AS ESTIMATED FROM DIFFERENT EVALUATION EXPERIMENTS

Type of Error	Experiment	Criteria
Random error	Replication	$s_{obs} < s_A$, or $2s_{obs} < TE_A$
Proportional error	Recovery	$\left\lvert\frac{(\overline{R} - 100)}{100}\right\rvert X_C < B_A$
Constant error	Interference	$\lvert Bias\rvert < B_A$
Systematic error	Comparison	$\lvert(a + bX_C) - X_C\rvert < B_A$
Total error	Replication and comparison	$2s_{obs} + \lvert(a + bX_C) - X_C\rvert < TE_A$

s_{obs} is the standard deviation determined in the replication experiment.
$\overline{R}$ is the average recovery (in per cent) determined in the recovery experiment.
s_A, B_A, and TE_A are the allowable standard deviation, allowable bias, and allowable total error, respectively.
X_C is the decision level concentration at which medical interpretation is critical.
Bias is the average difference determined in the interference experiment.
a and b are the y-intercept and slope, respectively, determined by regression analysis of the comparison of methods data.

features, which should be considered in relation to the laboratory situation:

1. Principle of the assay, with original reference.
2. Composition of reagents and standards, quantity provided (equivalent number of tests), and storage requirements (space, temperature, light, humidity) applicable both before and after opening the original container.
3. Stability of reagents and standards (shelf life).
4. Possible hazards and appropriate safety precautions.
5. Specimen requirements, i.e., conditions for collection, specimen volume, anticoagulants, preservatives, and storage.
6. Anticipated analytical performance, i.e., accuracy, precision, analytical range, linear range, and specificity with biological samples.
7. Reference range, with information on how it was derived; typical values obtained in health and disease; whether there will be a need to determine a reference range for one's own institution.
8. Complete and detailed protocol for performing a test.
9. Instrumental requirements and limitations.
10. Availability of technical support, supplies, and service.

Evaluating an Analytical Method

After the definition of needs and quality goals as well as the careful selection of an analytical method, evaluation studies are performed to test critical reliability characteristics. In general, a preliminary period should be allowed for familiarization with the analytical method, equipment, and procedural steps. This may involve trial runs to determine within-run precision and analytical range. Once the analyst is satisfied that the method can be performed without any procedural difficulties, studies on recovery and interference can be started. Day-to-day precision should be determined by using at least a four-week period for the experiment. Accuracy should be tested by a comparison-of-methods experiment, using a reference method with established and satisfactory accuracy and precision. The acceptability of the new method is determined from the experimental results by estimating the observed analytical errors, then comparing these error estimates with the defined allowable limits for the errors. When the observed errors are less than the defined allowable errors, the method can be implemented for routine service.

Step-by-Step Outline of a Method Evaluation Study

1. Plan in advance to assign several technologists, not just the most capable individual, to carry out the testing procedure. If the test is not frequently performed, it will save time later to stockpile frozen clinical specimens that have been analyzed by the existing procedure; specimens with possible interference from bilirubin, hemolysis, or turbidity are of special interest. If the sample volume required per test is high, prepare pools with high and low concentrations from clinical specimens for use when testing precision. Any frozen specimens must be thawed and mixed completely before use.
2. Obtain the reagents, kit, instrument, and all necessary materials and supplies. Study the instructions, then set up the method and perform some initial runs on control materials and random clinical specimens by carefully following the directions of the literature reference or of the manufacturer. Repeat these steps for the next several days to become familiar with the method.
3. Determine the within-run precision by analyzing 10–20 aliquots of a control material or a single clinical specimen. Alternatively, perform duplicate analyses on a variety of clinical specimens. Select control materials that have concentrations corresponding to the different decision-level concentrations of interest, or select groups of clinical specimens near the different decision levels of interest. Calculate the mean and standard deviation for each data set.

When data are assembled, results that are questionable should be excluded only under the following conditions: (a) those related to documented errors, such as results obtained when an automated instrument is indicating "Error" status; (b) those that can arbitrarily be excluded as undocumented errors, such as decimal point or transcription errors; and (c) statistical outliers, defined as values lying more than three standard deviations from the mean. After excluding the outliers, the mean and standard deviation are recalculated for each data set.

To judge acceptability, compare the observed standard deviation (s_{obs}) with the allowable standard deviation (s_A) or compare $2 \times s_{obs}$ with TE_A. If $s_{obs} < s_A$ (or $2s_{obs} < TE_A$), within-run precision is acceptable; proceed to further testing as described below. If not, attempt to locate the sources of imprecision, or abandon the analytical method in favor of another. Steps at which imprecision frequently arises are the dispensing of specimens or reagents, mixing operations, and reading of instrument outputs such as optical absorbance.

4. Determine the analytical range by analyzing a set of standards or a series of dilutions of either control materials or an elevated specimen pool. In some assays, the choice of diluent may be important; for example, a constant protein matrix or a serum that is known to be free of the analyte of interest may be necessary. Plot the observed value versus the concentrations of the samples or versus the relative dilutions for the series of samples (Figure 7-3). Estimate the analytical range by visual inspection of the graph. When applicable, also include measurements of the reagent blanks and sample blanks (if any) versus water, and note the magnitude of the blank values compared with the instrument reading for a low concentration of analyte.

5. Determine recovery by adding known amounts of the analyte in concentrated form to actual clinical specimens to prepare a new test sample. Add the same volume of solvent without analyte to provide a new baseline sample. Analyze the test and baseline samples, calculate the difference to determine the amount recovered, then calculate recovery by dividing the amount recovered by the original amount added, multiplying by 100 to express the result in per cent. Repeat this procedure for several specimens, then average the per cent recoveries to obtain the final estimate. The difference between this average recovery and the ideal 100% recovery provides an estimate of proportional systematic error, which can be multiplied by a decision-level concentration of interest to determine the magnitude of the error in concentration units. Compare this estimate of systematic error with the allowable bias (B_A) or with the allowable total error (TE_A) to judge acceptability (see Table 7-1).

6. Test the interference of common substances, such as bilirubin, hemoglobin, lipids, anticoagulants, and common drugs, and of specific substances known to cause problems

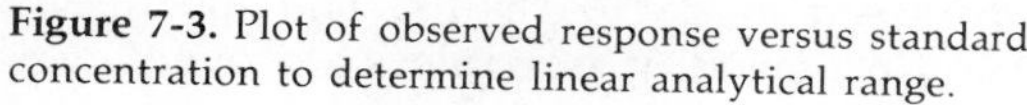
Figure 7-3. Plot of observed response versus standard concentration to determine linear analytical range.

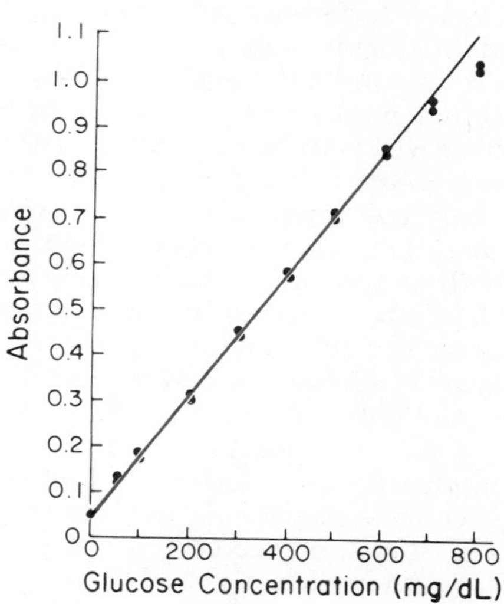

in the determination of the particular analyte being measured. The concentrations of added drugs should be within the therapeutic or toxic range. The effects of anticoagulants are best tested on paired specimens from a blood sample. Interference can be tested by an experiment analogous to the recovery experiment, except that the material added is the suspected interfering substance rather than the analyte to be determined. The results can be expressed in terms of the difference in concentration between the sample containing the interferent and the baseline sample. Interference can also be tested by comparing results with an analytical method known to be free of interferences, taking the difference between the values determined by the two methods as the estimate of interference. This difference, or bias, can then be compared with the allowable bias (B_A) or the allowable total error (TE_A) to judge its acceptability (see Table 7-1).

7. Determine other specific characteristics that may be critical for the particular test method, such as sample interaction due to inadequate washout of pipetting devices or cuvets on certain automated analytical instruments. The experimental design will depend on the type of instrument system but generally will involve introduction of a high-concentration specimen followed by two or more aliquots of a low-concentration specimen—for example, the series H_1, L_1, L_2. The difference between the results of L_1 and L_2 is taken as the error due to interaction. This is a systematic error, at least in the sense that it will be the same when the same concentrations of materials are involved. Acceptability of performance can be judged by comparing the observed error with the allowable bias (B_A).

8. Determine day-to-day precision by analyzing serum pools or control materials each working day over a one-month period. Select materials at two or three decision levels, including one near the upper limit of the analytical range. Calculate the mean and standard deviation for each data set. To judge acceptability, compare the observed standard deviation (s_{obs}) with the allowable standard deviation (s_A) or compare $2 \times s_{obs}$ with the allowable total error, TE_A.

9. Compare the results by the test method with those obtained by a reference analytical method. The differences between methods can be interpreted as analytical errors, but there are several factors that are critical for the performance of the experiment and the interpretation of the results:

(a) *Reference analytical method.* To assign the cause of a systematic error to the test method assumes that the comparison method is free from analytical errors. There are some high quality analytical methods that have been established as reliable for use in validating new analytical methods,[4] but they are not readily available. Most often the comparison method chosen is one that is in routine use in the laboratory. In such a case, there exists the possibility that any differences observed may actually be caused by limitations of the comparison method. When these differences are large, additional data from interference and recovery experiments may be necessary to determine which method is at fault.

(b) *Clinical specimens.* Specimens from 40–200 patients should be analyzed. The actual number of specimens is less important than the concentration range that is covered by the specimens and the variety of diseases and medical problems that are represented. The concentration of the specimens should be distributed over the entire analytical range of interest. Specimens should be fresh, if possible, or preserved under conditions known not to alter the results. It is desirable to obtain duplicate results by both analytical methods, preferably from two different analytical runs. The specimens should be analyzed by the two methods within a 4-h period, if possible, to prevent any changes due to instability of the specimens themselves. Specimens should then be analyzed in small groups over a period of several days to provide data that are representative of the performance of both analytical methods.

(c) *Data analysis.* The results should first be displayed graphically with the test method results plotted on the y-axis and the comparative results on the x-axis (Figure 7-4). Then statistical calculations are performed, usually least squares (regression) analysis, *t*-test, and the correlation coefficient.

It is generally best to perform the regression calculations to determine slope, y-intercept, and standard deviation of the residuals or points around the regression line.

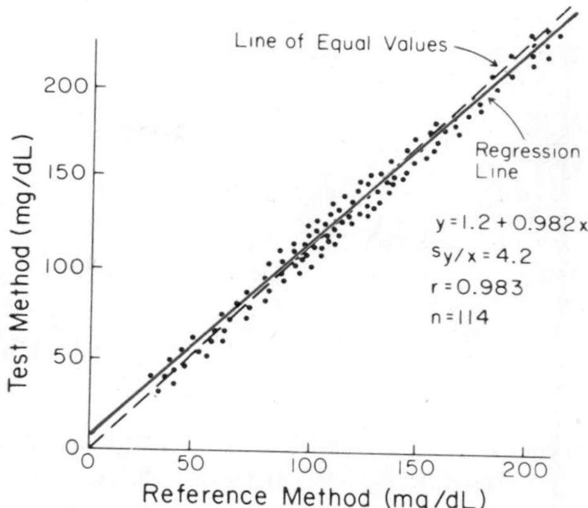

Figure 7-4. Comparison of serum glucose values obtained with a test method versus those obtained with a reference method. From statistical calculations on the 114 pairs of results, the slope is 0.982, the y-intercept is 1.2 mg/dL, the standard error about the regression line is 4.2 mg/dL, and the correlation coefficient is 0.983.

The slope (b) provides an estimate of proportional analytical error, and when it is near 1.00, the y-intercept (a) provides an estimate of constant analytical error. The standard deviation ($s_{y/x}$) provides an estimate of the random error between the methods. The regression equation ($Y = a + bX$) is used to estimate the systematic error at a decision-level concentration as follows. The Y-value (Y_C) corresponding to the critical X-value (X_C) is calculated ($Y_C = a + bX_C$) and the systematic error is estimated as the difference between X_C and Y_C ($SE = X_C - Y_C$). SE can be compared with the specification for allowable bias (B_A) or allowable total error (TE_A) to judge acceptability (see Table 7-1). A total error criterion can be employed by adding the observed sytematic error to two times the observed standard deviation ($TE = SE + 2s_{obs}$) and then comparing this estimate with the allowable total error (TE_A; see Table 7-1).

10. Determine the stability and uniformity of reagents and standards by comparing results of different batches (lots) of materials, and by using batches (lots) that have been stored for varying periods of time. The variation of results should be much less than the defined error specifications.

11. Establish or verify each reference range or interval. If one accurate method is selected to replace another, there should be no change from the previously established reference intervals. In other instances, it will be necessary to analyze a group of specimens from selected reference populations, following the guidelines in Chapter 5.

12. Prepare a report that summarizes the experimental results and data analysis. Carefully review and interpret the data and determine whether or not the method is acceptable. If no method for measuring the analyte has been previously available in your laboratory, a conference with clinicians may be useful to help assess the clinical utility of the method.

13. Document the procedure to be followed in the routine operation of the analytical method. Prepare in-service training materials and instruct personnel in the performance of the test. Implement quality control procedures and establish control limits and corresponding actions.

14. Prepare documentation for the clinical staff, with instructions on specimen collection, hours of test availability, expected turnaround time, reference ranges, charge for the test, and a short description of its clinical utility.

15. Inform the clinical staff of the availability of the test. Encourage staff feedback during the introductory phase of the new test. Monitor performance carefully during the first several months of use.

Statistical Concepts and Calculations

Mean, standard deviation, coefficient of variation. The nature and properties of a Gaussian distribution were discussed in Chapter 5. The same concepts apply here; however, the focus is on the distribution of errors from the analytical method, not the distribution of values from a healthy or patient population. The equations for mean ($\overline{x}$), standard deviation (s), and coefficient of variation (CV) are:

$$\overline{x} = \frac{\Sigma x_i}{n}$$

$$s = \sqrt{\frac{\Sigma x_i^2 - \frac{(\Sigma x_i)^2}{n}}{n - 1}}$$

$$CV = \frac{s(100\%)}{\overline{x}}$$

where x_i is an individual measurement and n is the number of measurements.

Standard deviation of duplicates. When samples are analyzed in duplicate, the differences between duplicates can be used to estimate the standard deviation:

$$s_{dup} = \sqrt{\frac{\Sigma d^2}{2n}}$$

where d is the difference between the duplicate aliquots of a patient specimen and n is the number of different patient samples analyzed. If the duplicates are performed in one run, s_{dup} estimates within-run precision or random error; if the duplicates are analyzed in different runs, between-run variation will be included in the estimate.

Paired t-test. The use of the "unpaired" *t*-test was described in Chapter 5 to test the difference in means from two different populations. For comparison-of-methods studies, a different form of the *t*-test, called a "paired" *t*-test, is used to test the average difference of the pairs of values obtained by the test and reference methods. Note that a *t*-test is applied to means or average differences and therefore provides a test for accuracy or the presence of systematic errors of a method. The null hypothesis being tested is that there is no systematic difference or bias between the two analytical methods.

Three statistics are actually calculated. The average difference or "bias" is calculated by the following equation:

$$\text{bias} = \overline{y} - \overline{x}$$

where $\overline{y}$ is the mean of the test method values and $\overline{x}$ is the mean of the reference method values. Next, the standard deviation of the differences (s_d) is calculated:

$$s_d = \sqrt{\frac{\Sigma[(y_i - x_i) - \text{bias}]^2}{n - 1}}$$

where y_i and x_i are the values for individual patient samples as measured by the reference and test methods, respectively; bias is from the previous equation, and n is the number of patient samples tested. The *t*-value is obtained as follows:

$$t = \frac{|\text{bias}|\sqrt{n}}{s_d}$$

To interpret the *t*-test, the calculated *t*-value is compared with a "critical" *t*-value obtained from a statistics table such as shown in Table 29-16. The critical values are given for different probabilities (p) and degrees of freedom (df*). A probability of 0.05 is usually selected, which allows a 5% chance that the calculated *t*-value will exceed the critical *t*-value owing to chance alone. Degrees of freedom here are equal to n − 1.

* Also frequently referred to as ν, especially in statistical tables.

If the calculated *t*-value is greater than the critical *t*-value, the difference (bias) is interpreted as being statistically significant, meaning that a real difference has been observed between the two methods. If the calculated *t*-value is less than the critical *t*-value, then the bias is interpreted as not being statistically significant; the data do not support a conclusion that a difference exists between the two methods.

For example, if a calculated *t*-value of 1.80 was obtained based on a comparison study that included 41 different specimens, the critical *t*-value can be found in Table 29-16 by looking in the row for 40 degrees of freedom (df = n − 1) and the column for p = 0.025 (use this column because the table shows the probability in the upper tail only; there would also be a probability of 0.025 in the lower tail, for a total of 0.05 or p = 0.05). The critical *t*-value is 2.02. Thus, the calculated value is less than the critical value (1.80 < 2.02), meaning that the original null hypothesis stands. The experimental data do not support a conclusion that there is any bias between the two analytical methods.

Interpretation of the *t*-test is more difficult when the question is to determine whether a method's accuracy or systematic error is acceptable. For example, although the *t*-value can be seen to increase directly as the bias increases, it also increases as the number of specimens increases and as s_d decreases. Thus, a very small bias may be statistically significant when a large comparison study (large n) has been performed between two methods, both having very good precision (small s_d). On the other hand, a large bias may not be statistically significant when a small number of specimens have been compared between two imprecise methods.

To judge acceptability, it is better to interpret the bias as an estimate of systematic error and compare its size to the defined allowable bias (B_A) or to the defined allowable total error (TE_A). When bias is smaller than either B_A or TE_A, the observed systematic error is less than the defined allowable error and performance can be judged acceptable. When bias is larger than either B_A or TE_A, the observed systematic error is larger than the allowable error and performance is not acceptable.

F-test. To compare the precision of two analytical methods, the standard deviations of the methods can be compared by use of the F-test. The F-value is calculated from the following equation:

$$F = \frac{(s_1)^2}{(s_2)^2}$$

where s_1 is the larger standard deviation and s_2 the smaller standard deviation. The null hypothesis being tested is that there is no difference between the two standard deviations.

As an example, consider a test method having a standard deviation of 2 mg/dL (n = 31) and a reference method having a standard deviation of 1 mg/dL (n = 21). The calculated F-value would be 4.00 ($F = 2^2/1^2$).

To interpret the F-test, the calculated F-value is compared with a critical F-value. When the calculated value is greater than the critical value, a statistically significant difference has been observed, meaning that the standard deviation of one method is larger than that of the other. When the calculated value is less than the critical value, no statistically significant difference has been observed. Therefore, the original null hypothesis stands: based on the study data used, there is no evidence for concluding that there is any difference between the two standard deviations.

The critical F-value can be found in Table 29-17. For the values given in this table, there is only a 5% chance (p = 0.05) that the critical F-value will be exceeded due to chance alone. To look up a critical value, the degrees of freedom (df) of the numerator and denominator are needed, which are each equal to n − 1. For example, there were 31 samples analyzed by the test method and 21 samples analyzed by the reference method. To find the critical F-value, first calculate the degrees of freedom for the numerator and denominator (df of numerator = 31 − 1 = 30; df of denominator = 21 − 1 = 20). The critical F-value is found in Table 29-17 at the intersection of the column (numerator) having 30 df and the row (denominator) having 20 df (i.e., a value of 2.04). Since the observed F-value of 4.00 is greater than the critical F-value, there is a statistically significant difference (at $p = 0.05$) between the standard deviations of the two methods. It is therefore concluded that the reference method is more precise than the test method.

Whether the precision of the test method is acceptable is a different question and is not answered by the F-test. To judge acceptability, the observed standard deviation needs to be compared with the defined allowable standard deviation or with the defined allowable total error. If $s_{obs} < s_A$ or if $2s_{obs} < TE_A$, precision is acceptable. If $s_{obs} > s_A$ or $2s_{obs} > TE_A$, the observed precision is not acceptable.

Linear regression or least squares analysis. To describe the plot of test values versus reference values from the comparison-of-methods experiment, the best straight line through the data can be determined from linear regression or least squares analysis. A straight line is given by an equation of the form:

$$y = a + bx$$

where a is the y-intercept and b is the slope of the regression line describing how estimates of the test method values (y) are related to reference method values (x).

The *slope* (b) is calculated from the following equation:

$$b = \frac{n\Sigma xy - \Sigma x \Sigma y}{n\Sigma x^2 - (\Sigma x)^2}$$

The *y-intercept* (a) is calculated from the equation:

$$a = \bar{y} - b\bar{x}$$

where $\bar{y}$ and $\bar{x}$ are the mean values from the test and reference method samples, and b is the slope calculated from the previous equation.

The *standard deviation about the regression line* ($s_{y/x}$) can be calculated from the equation:

$$s_{y/x} = \sqrt{\frac{\Sigma(y_i - Y_i)^2}{n - 2}}$$

where y_i is the value observed for a sample by the test method and Y_i is the value calculated from the regression equation, given the value by the reference method (x_i). The denominator gives the number of degrees of freedom, which is equal to $n - 2$ because the slope and intercept have already been calculated, thereby imposing two restrictions on the data set (leaving $n - 2$ independent comparisons or degrees of freedom).

Ideally, the slope would be 1.00; the y-intercept, 0.0; and $s_{y/x}$, 0.0. Deviations from these ideal values are caused by different kinds of analytical errors.[7] Proportional errors cause changes in the slope, constant errors cause changes in the y-intercept, and random errors cause $s_{y/x}$ to increase.

The systematic error at any concentration can be determined by substituting X_C, a given concentration (e.g., a medical decision level), for x_i in the regression equation, then calculating the corresponding Y_C. Systematic error is estimated by the difference between these two ($SE = Y_C - X_C$).

For example the comparison data shown in Figure 7-4 give a regression line of $y = 1.2 + 0.982x$. The y-intercept is therefore 1.2 mg/dL, meaning that a constant error of 1.2 mg/dL has been observed. The slope is 0.982, meaning that a proportional error of 1.8% has been observed ($1.000 - 0.982 = 0.018$ or 1.8%). The standard deviation about the regression line is 4.2 mg/dL, meaning that it is expected that 95% of the data points will fall within 8.4 mg/dL (±2s) of the line.

To judge the acceptability of accuracy or systematic error, it is best to determine the magnitude of the error at critical medical decision levels. If the allowable error were 5.0 mg/dL at a decision level of 100 mg/dL (X_C), the systematic error should be calculated for a concentration of 100 mg/dL. The test method value corresponding to 100 mg/dL by the reference method can be calculated from the regression line [$Y_C = 1.2 + 0.982(100) = 99.4$]. The difference between Y_C and X_C provides an estimate of the systematic error at the critical concentration ($|99.4 - 100| = 0.6$). If the observed error is less than the defined allowable error, then the accuracy of the method is acceptable. If it is greater than the defined allowable error, then the accuracy of the method is not acceptable. In the case above, the accuracy would be judged acceptable because the observed systematic error is 0.6 mg/dL, which is less than the defined allowable error of 5.0 mg/dL.

In applying linear regression to data from a comparison-of-methods experiment, it is important to graph the data for visual inspection. Check to see that the response between the two methods is linear, that a wide analytical range has been studied, and that there are no widely discrepant values (outliers). Estimates of slope and intercept will not be reliable if there is nonlinearity, a narrow range of concentrations, or outliers. When nonlinearity is present, restrict the comparison data to the linear range. If the range of data is narrow (also indicated by a correlation coefficient <0.95), an alternative statistical technique such as Deming's regression may be necessary.[3] If outliers are present, remeasure those samples if possible. Otherwise, determine whether the outliers affect the decision on acceptability by analyzing the data with and without the outlying values; if the decision changes, the presence of the outliers is critical and it is best to extend the study to obtain more reliable data.

Correlation coefficient. Although it is not a necessary calculation, the correlation coefficient (r) is very commonly used in comparison-of-methods studies. It can be calculated from the equation:

$$r = \frac{n\Sigma xy - \Sigma x \Sigma y}{\sqrt{[n\Sigma x^2 - (\Sigma x)^2]\,[n\Sigma y^2 - (\Sigma y)^2]}}$$

Ideally, the value for r should be 1.00. Values less than 1.00 are due to random error within or between the methods. Systematic errors have no effect on r, and therefore r can be near its ideal value even when a method is inaccurate.[7] The acceptability of the accuracy of a method cannot be judged on the value of the correlation coefficient. However, the correlation coefficient may be helpful in assessing the reliability of the linear regression estimates of slope and y-intercept. When $r > 0.95$, the estimates of slope and y-intercept are not very much affected by variability in the comparative method.

References

1. American Association for Clinical Chemistry Committee on Standards: AACC policy regarding reagent sets and kits. Clin. Chem., *12*:43, 1966.
2. Büttner, J., Borth, R., Boutwell, J. H., et al. (IFCC Committee on Standards): Provisional recommendation on quality control in clinical chemistry. Part I. General principles and terminology. Clin. Chem., *22*:532–540, 1976.
3. Cornbleet, P. J. and Gochman, N: Incorrect least-squares regression coefficients in method-comparison analysis. Clin. Chem., *25*:432–438, 1979.
4. Tietz, N. W. : A model for a comprehensive measurement system in clinical chemistry. Clin. Chem., *25*:594, 1979.
5. Westgard, J. O., Carey, R. N., and Wold, S.: Criteria for judging precision and accuracy in method development and evaluation. Clin. Chem., *20*:825–833, 1974.
6. Westgard, J. O., de Vos, D. J., Hunt, M. R., et al.: Method Evaluation. Houston, American Society for Medical Technology, 1978.
7. Westgard, J. O., and Hunt, M. R.: Use and interpretation of common statistical tests in method comparison studies. Clin. Chem., *19*:49–57, 1973.

8 QUALITY ASSURANCE

JAMES O. WESTGARD, Ph.D.
GEORGE G. KLEE, M.D., Ph.D

All laboratory personnel must be conscious of how the quality of their work affects the medical diagnosis and treatment of patients. Along with a general concern for accuracy, there must also be a thorough understanding of the essential quality characteristics of the analytical services they provide. Laboratory analysts, supervisors, and directors must have concern for the scope of testing that is offered, for the delivery of analytical services in a useful and convenient manner, and for ensuring that the analytical quality is appropriate for the medical application and interpretation of the test results.

It is important for laboratories to define their service objectives and to establish goals for quality. Without quality goals, there is no objective way to determine whether acceptable quality is being achieved, to plan effective strategies for improving quality, or to design procedures that assure that a specified level of quality will be attained.

Elements of a Quality Assurance Program

The attainment of quality goals in a service laboratory requires a comprehensive quality assurance program. "Quality assurance" is used here to represent those practices that are generally recommended for assuring that desired quality goals are achieved. Quality assurance is a broad spectrum of plans, policies, and procedures that together provide an administrative structure for a laboratory's efforts to achieve quality goals. The term *quality control (QC)* is often used to represent those techniques and procedures that monitor performance characteristics. Generally these are quantitative techniques that monitor particular sources of errors, estimate the magnitude of the errors, and alert laboratory personnel when there are indications that quality has deteriorated.

A quality assurance program involves virtually everything and everybody in the clinical laboratory. An error in any step during the acquisition, processing, and analysis of a specimen and the reporting of a laboratory test result can invalidate the quality of the analysis and cause the laboratory to fall short of its quality goals. There are several essential elements of a quality assurance program.

Commitment. Dedication to quality service must be central. Quality must be a major consideration in all management decisions because any single decision may compromise other plans and practices for attaining quality goals.

Facilities and resources. Laboratories must have adequate space, equipment, materials, supplies, staff and supervisory personnel, and budgetary resources. These facilities and resources are the basis upon which quality services can be developed and maintained.

Technical competence. High-quality personnel are essential for high-quality services. The educational background and experience of all personnel are important, as is the capability for providing in-service training to develop and maintain skills and knowledge.

Quality assurance procedures. Three groups of procedures are discussed in this chapter:

1. The control of preanalytical variables such as test request, patient preparation and identification, specimen acquisition and transport, specimen processing and distribution, preparation of worklists and logs, and maintenance of records.

2. The control of analytical variables, which include analytical methodology, standardization and calibration procedures, documentation of analytical protocols and procedures, and the monitoring of critical equipment and materials.

3. The monitoring of analytical quality by the use of statistical methods and control charts.

Problem-solving mechanism. A particularly critical element in a quality assurance program is a mechanism for solving problems and implementing change. Such a mechanism provides the link between the identification of a problem and the implementation of a solution to that problem. For problems with individual methods or instrument systems, the delegation of responsibility for each method or system may provide the corrective mechanism because specialized trouble-shooting skills can be developed and improved and preventive maintenance programs can be instituted. For more generally occurring problems, the in-service training program can be an important part of the problem-solving mechanism but will often require additional input from a quality control technologist or supervisor to initiate the corrective training.

Problem-solving teams can also be established. These are small groups of people who meet to discuss problems and suggest solutions. "Quality-improvement teams" can be organized to focus on selected problems. Personnel are generally appointed to provide specific skills and background related to the problem to be solved. "Quality circles" can be implemented to provide a more general forum in which to identify problems. Participation is usually voluntary and limited to peers. By more widespread involvement of personnel, quality circles heighten their interest and commitment to quality.

Control of Preanalytical Variables

Monitoring preanalytical variables requires the coordinated efforts of many individuals and hospital departments, each of which must recognize the importance of these efforts in maintaining a high quality of service.

Test utilization. Current concerns with the cost of medical care have led to government regulations that place new emphasis on efficient utilization of resources. Careful monitoring of the appropriateness of test requests is likely to become increasingly important. The laboratory should play a role in identifying situations in which test utilization can be improved.

Patient identification. Correct identification of patients and specimens is a major concern for laboratories. The highest frequency of errors occurs with the use of handwritten labels and request forms. The use of plastic-embossed patient identification cards to imprint the patient's name on test request forms and on blood collection labels is recommended to eliminate transcription and identification errors. However, it is still the responsibility of the person who obtains a specimen to verify that the patient's name on the label correctly identifies the donor of the specimen.

Turnaround time. Delayed and lost test requisitions, specimens, and reports can cause major problems in diagnosis and treatment. An essential procedure for identifying the causes of delays is the recording of the actual times of specimen collection, receipt in the laboratory, and reporting of test results.

Patient preparation. Laboratory tests are affected by many factors, such as recent intake of food, alcohol, or drugs, as well as by smoking, exercise, stress, sleep, posture during specimen collection, and other variables (see Chapter 10 on *Specimen Collection*). Therefore, proper patient preparation is essential for the test results to be meaningful. The laboratory must define the instructions and procedures for patient preparation and specimen acquisition, which should be included in hospital procedure manuals and should be transmitted to patients by both oral and written instructions.

Specimen collection. The techniques used to acquire a specimen may affect many laboratory tests. Therefore, the best mechanism to monitor and control this aspect of laboratory processing is to have a specially trained team assigned to specimen collection. The identity of the person collecting a specimen should be maintained so that any collection problems can be reviewed and recurrence of the problem prevented. Individuals who process specimens should be trained to look for and document collection problems.

Specimen transport. The stability of specimens during transport from the patient to the laboratory may be critical for some tests when performed locally and for most tests when sent to regional centers and commercial laboratories. In controlling specimen transport, the laboratory that performs the analysis must retain the authority to reject unsatisfactory specimens.

Specimen separation and aliquoting. The main variables are the centrifuges, containers, and personnel.

Centrifuge performance. The calibration and maintenance schedules of centrifuges are discussed in Chapter 1. Special attention must be paid to monitoring the speed, time, and temperature.

Containers. Collection tubes, glass beads, or other materials added to specimens during processing, as well as pipets and aliquot tubes are possible sources of contamination. Some of the materials used in making plastic containers interfere with drug analyses, whereas others adsorb some analytes, making trace analysis difficult. Trace metal analyses are particularly vulnerable to contamination problems. Any changes in supplies and processing procedures must be carefully evaluated (see also Chapter 10).

Personnel should be thoroughly trained and supervised with regard to all techniques and protocols used in the laboratory. A written procedure manual is an essential part of this requirement. Also, training and equipment should be available to limit the exposure of personnel to hazards in the laboratory. A continual program of in-service training ensures knowledge of and aids compliance with these aspects of quality assurance.

Laboratory logs. When specimens arrive in the laboratory, the information (patient's name, identification number, tests requested) on the request forms or collection logs should be matched against the information on the labels of the specimen containers. A master log containing the patient's name, identification number, and the tests requested should be maintained. When aliquots are distributed, additional logs or worksheets should include the test results and quality control information. Whenever there are manual transfers of test results, a second person should check for transcription errors. Some computer-assisted laboratory report systems help to control errors by requiring special release codes or by comparing current results with previous results for a patient.

Control of Analytical Variables

There are certain variables that should be monitored on a laboratory-wide basis, since they will affect many of the methods in the laboratory. These include water quality, calibration of analytical balances and volumetric glassware, stability of electrical power, and the temperature of heating baths, refrigerators, freezers, and centrifuges. Specific methods and criteria for these variables are discussed in other chapters. In addition, there are certain variables that relate directly to individual analytical methods, and these require that procedures be developed to deal specifically with the characteristics of those methods.

Choice of analytical methodology. The selection and evaluation of analytical methods are discussed in Chapter 7. It is important to recognize that the initial evaluation of a method often takes place in a setting somewhat more ideal than the production setting. It is therefore desirable to have a start-up period in the service laboratory before test results are reported, in order to discover any additional problems and to have time to train a sufficient number of analysts to support the routine service operation.

Standards and calibration. The reliability of the analytical values obtained from a particular procedure often depends on the quality of the standards and the calibration procedure employed. Some knowledge of the overall structure of an accuracy-based measurement system is necessary to appreciate fully the problems in standardization and calibration.[12] The highest-quality methods, *definitive methods*, are to be used to validate reference methods and primary reference materials. The highest-quality reference materials, namely *primary reference materials*, should be used in the development and validation of reference methods, the calibration of definitive and reference methods, and the production of secondary reference materials. *Reference methods* should be used to validate field (routine) methods. *Secondary reference materials* should be used to provide working standards for field methods and to assign values to control materials. *Control materials* are used only to monitor field methods.

To help implement this conceptual structure in clinical chemistry, the following definitions of reference materials have been proposed:[9]

Reference material (RM): "A material or substance, one or more properties of which are sufficiently well established to be used for the calibration of an apparatus or for the verification of a measurement method."[9]

Certified reference material (CRM): This is a primary standard described as a "reference material accompanied by, or traceable to, a certificate stating the property value(s) concerned, issued by an organization, public or private, which is generally accepted as technically competent" (for example, standard reference materials, SRMs, from the US National Bureau of Standards).[9]

Calibration and test material (CTM): This is a secondary material described as "a reference material or solution with which the test sample is compared in order to determine the concentration of analytes or other quantities."[9]

Control material: A "specimen, or solution, which is analyzed solely for quality control purposes and is not used for calibration purposes."[9]

To illustrate the use of these materials: a field (routine) method is traceable to a reference method when it has been evaluated against that method or when it has been calibrated with a secondary calibration and test material (CTM) whose value was established by analysis with a reference method. The reference method itself should have been validated against a definitive method or calibrated with a primary certified reference material (CRM) whose value was established by a definitive method.

Documentation of analytical protocols. Step-by-step descriptions of analytical procedures are critical if methods are to provide the same results when used by different analysts over a long period of time. To maintain such consistency requires written protocols in the form of method or procedure manuals. The essential information to be contained in such documents is outlined in Table 8-1. Method manuals should be revised and updated whenever changes occur.

Verification of reference intervals. The details for establishing reference values and intervals are presented in Chapter 5. Such studies are generally performed for methods during their development, and the results are available in the technical literature. Local verification of reference ranges is strongly encouraged.

Table 8-1. OUTLINE FOR A PROCEDURE MANUAL

A. *Procedure name:* List principal name of procedure first, alternative names next. List commonly used abbreviations.
B. *Clinical significance:* Give a brief explanation of how the test is used in clinical medicine. Include reference intervals for specific diseases and recommended diagnostic and therapeutic action limits.
C. *Principle of method:* Give a brief statement of the principles on which the method is based.
D. *Specimen:* List type of specimens that can be used, recommended volume, as well as minimum volume. Indicate conditions that render the specimen unacceptable, such as hemolysis or lipemia. List patient preparation procedures. Provide instruction for specimen handling prior to testing.
E. *Reagents and equipment:* Provide a list of reagents in order of their use, including standards. Indicate the names and addresses of suppliers and detailed instructions for preparation, including checks to be performed prior to use. List equipment used and special precautions required.
F. *Procedure:* This should be a step-by-step description, including calibration and quality control procedures. There should be enough detail so that the assay could be performed by a person unfamiliar with the test. The necessary calculations should be included.
G. *Reference values:* List the reference ranges for healthy subjects. Indicate factors such as age, sex, or race that affect reference values. Include the nature of the population studied, the number of subjects, and the date or reference for the original work.
H. *Comments:* Include any special analytical variables affecting the test, such as pH or temperature. Include effects of commonly used drugs, any dangers or personal hazards in the procedure, and any special safety precautions and procedures.
I. *References:* Give the primary literature references which describe the method, or the references upon which the method is based.

Monitoring technical competence. The personal characteristics and techniques of individual analysts may greatly affect certain analytical methods, particularly manual methods. Proper training of laboratory personnel to achieve uniformity in technique is important, as is scheduling sufficient routine service to maintain proper techniques. When personnel are trained on new analytical methods, it is helpful to have written objectives that outline the critical tasks and knowledge that must be mastered. Before analyses for clinical use are performed, the technical competence of the personnel should be checked and practice runs should be performed. In-service and continuing education programs help maintain and improve competence.

Inventory control of materials. Stable operation of laboratory methods is vitally dependent on the materials utilized with the methods. In general, when materials are stable and changes in product lots (lot numbers) cause problems, as large a stock as feasible should be maintained. Adequate inventory should be maintained to allow time for shipment and testing of additional supplies if the current shipment should not be satisfactory.

Input control of materials. Along with inventory management, the quality of materials should be monitored when they are received. Reagents and materials should be labeled with the name, lot number (particular manufacturing batch), concentration, date received, date prepared, date opened, date of expiration, and recommended storage conditions. Interruptions in service can be avoided by testing the adequacy of materials prior to their introduction for routine use. New lots of blood-drawing tubes, processing tubes, reagents, and other supplies should receive parallel or comparison testing prior to their routine use.

Monitoring method changes. An essential tool for solving analytical problems is a record or log of all changes and problems occurring with a method. Logs should include the date and time the problem was discovered, a description of the problem that occurred, and the name of the analyst, and they should identify the lot numbers of reagents, materials, and standards used. The actions that resolved the problem are also recorded. These logs may be used to identify in-service training needs or to monitor the shelf life of reagents.

All instrument maintenance should be recorded, whether it is performed by laboratory personnel or by service personnel from outside the laboratory. Periodic review of these logs may be used as a basis for the revision of preventive maintenance procedures and schedules.

Control of Analytical Quality Using Stable Control Materials

The performance of analytical methods can be monitored by analyzing specimens whose concentrations are known and then comparing the observed values with the known values. The known values are usually represented by a range of acceptable values, or upper and lower limits for a control specimen (control limits). When the observed values fall within the control limits, this should assure the analyst that the analytical method is working properly. When the observed values fall outside the control limits, the analyst should be alerted to the possibility of problems in the analytical determination.

Control Materials

The known specimens that are analyzed for quality control purposes are called "control materials." They need to be stable materials, available in aliquots or vials that can be analyzed periodically over a long period of time. There should be little vial-to-vial variation so that differences between repeated measurements can be attributed to the analytical method alone. A control material should preferably have the same matrix as the test specimens of interest; for example, a protein matrix may be best when serum is the test material to be analyzed by the analytical method. The concentration of analyte in different control materials should be in the normal and abnormal ranges, corresponding to the concentrations that are critical in the medical interpretation of the test results (medical decision levels).

Control materials can be prepared in the laboratory from unused sera,[2] but for safety, stability, and economy, most laboratories choose to purchase control sera or "control products." Commercial products are generally supplied as lyophilized materials that are reconstituted by adding water or a specific diluent solution. Materials are also available that have matrices representing urine, spinal fluid, and whole blood.

In the selection of commercial control materials, there are several considerations besides the matrix of the material. Stability is critical because it is often desirable to purchase a year's supply of one manufacturing batch or lot. Different batches (or lot numbers) of the same material will have different concentrations, which require new estimates of the mean and standard deviation for each of the analytes of interest. The size of the aliquots or vials must be adequate for the analytical methods to be monitored.

Control products may be purchased as assayed or unassayed materials. Assayed materials are accompanied by a list of values (mean and standard deviation) for the concentrations that are expected for that material. Values may be specified for several of the common analytical methods and preferably for a reference method for each analyte. Assayed materials are more expensive because of the work required to establish the values. Even when assayed materials are selected, it is advisable to determine the mean and standard deviation in the user's own laboratory.

General Principles of Control Charts

The most common way of comparing the observed values for control materials with their known values is to use *control charts*. These are graphical displays in which the observed values are plotted versus the time when the observations were made. The acceptable range of values is indicated on the chart by lines for upper and lower control limits. When the plotted points fall within the control limits, the method is judged to be performing properly. When points fall outside the control limits, problems may be developing.

The control limits are calculated from the mean and standard deviation obtained from repeated measurements on the known specimens by the particular analytical method that is to be controlled. The mean ($\bar{x}$) and standard deviation (s) are calculated from the following equations:

$$\bar{x} = \frac{\Sigma x_i}{n}$$

$$s = \sqrt{\frac{\Sigma x_i^2 - \frac{(\Sigma x_i)^2}{n}}{n - 1}}$$

where x_i is an individual control observation and n is the number of observations in the time period being monitored. The initial estimate should be based on measurements over a time period of at least four weeks (or 20 measurements) when the method is working properly. The initial estimate may not be entirely reliable, owing to the low number of data points and possible outliers in the data. The estimates can be revised when more data have been accumulated by recording n and the summation of x_i and x_i^2, then using the cumulative totals in the equations above to give cumulative means and standard deviations.

It is assumed that the error distribution is Gaussian. The control limits are selected to include most of the control values, usually 95 to 99.7%, which correspond to $\bar{x} \pm 2s$ or $\bar{x} \pm 3s$. Since it should be a relatively rare occurrence to observe a value in the tails of the distribution (only 1 out of 20 values for 2s limits, 3 out of 1000 for 3s limits), such an observation is suspect and suggests that something may have happened to the analytical method. A shift in the mean could have occurred (an accuracy problem), which would result in a higher probability for exceeding one of the limits; or an increase in the standard deviation could have occurred (a precision problem), widening the distribution and increasing the probabilities for exceeding both limits.

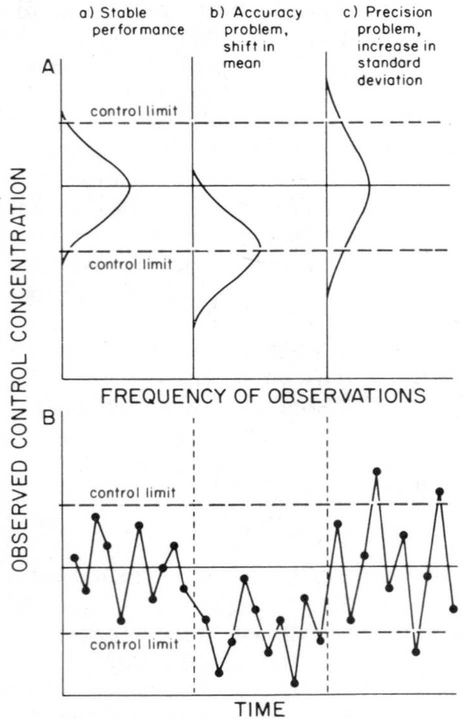

Figure 8-1. Conceptual basis of control charts. *A,* Frequency distributions of control observations for different error conditions. *B,* Display of control values versus time on a control chart.

Figure 8-1,*A* illustrates how the distributions of control values appear for three different situations: (a) stable performance in which only an occasional observation exceeds the control limits; (b) occurrence of a systematic error, which shifts the mean of the distribution and causes a much higher expectation or probability of observing control values outside one of the control limits; and (c) occurrence of an increase in random error, which widens the distribution and causes a much higher probability of observing a control value outside either of the control limits.

Control charts are used to compare the observed control values with the control limits and to provide a visual display that can be quickly inspected and reviewed. The concentration or observed value is plotted on the y-axis versus time of observation on the x-axis. It is common practice to plot one month's data on a chart, usually one or two points a day, but the time axis should be appropriately chosen for the method being monitored. An example of a control chart is shown in Figure 8-1,*B*, where the control values represent the three situations discussed above. When the analytical method is operating properly, the control values fall predominantly within the control limits. When there is an accuracy problem, the control values are shifted to one side of the mean, and there may be several values in a row falling outside one of the limits. When there is a precision problem, the control values fluctuate widely, and there may be values exceeding both the upper and lower control limits.

Interpretation of control data to judge whether a run is "in control" or "out of control" can be guided by certain decision criteria or control rules. These rules can be symbolized as A_L, where A is the abbreviation for a statistic or the number of control observations exceeding control limits, and L refers to the control limits. For example, 1_{3s} refers to a control rule in which one observation exceeding $\overline{x} \pm 3s$ control limits is the criterion for rejecting an analytical run.

Different control procedures have different capabilities, depending on the control rules used and the number of control observations per run (N). There should be some consideration of the number of "false alarms" versus "true alarms" when selecting control procedures. Information is available in the form of probabilities for rejection.[16] *Probability* refers to the likelihood that an event will occur; in this case, the event is the rejection of an analytical run. Ideally, there should be no rejections of analytical runs when the analytical method is performing properly (no false alarms or false rejections). On the other hand, when analytical errors occur, the control procedure should always provide a rejection signal (all true alarms or true rejections). This corresponds to having a probability for false rejections (false alarms) of 0.00 and a probability of error detection (true alarms) of 1.00.

The critical performance characteristics for most commonly used control procedures are described in the literature.[14] The probabilities for rejecting runs having different sizes of errors have been summarized by "power function" graphs, which are plots of probability for rejection versus the size of analytical errors. The ability to detect errors depends on each control rule selected, the number of control observations used, and the size of the error occurring.

Levey-Jennings Control Chart

Control charts were introduced into the clinical chemistry laboratory in the 1950s by Levey and Jennings.[8] They demonstrated how the industrial control procedures developed by Shewhart[11] could be used with the mean and range of duplicate measurements. In 1952 Henry and Segalove[7] described an alternate chart in which single control values were plotted directly on a graph. This "single value" chart was adopted by most laboratories because data calculations were not required prior to plotting the control results. Today this single value chart is commonly known as a "Levey-Jennings" chart.

To use a Levey-Jennings control chart, follow these steps:

1. Analyze samples of the control material by the analytical method to be controlled on at least 20 different days. Calculate the mean and standard deviation for those results.

2. Construct a control chart on graph paper. Label the y-axis "control value" and set the range of concentrations to include the $\bar{x} \pm 4s$. Draw horizontal lines for the mean and the upper and lower control limits. Set the control limits as $\bar{x} \pm 3s$ when the number of control observations, N, is 2 or greater. When N is 1, control limits may be set as the $\bar{x} \pm 2s$. Label the x-axis in terms of time, using day, run number, control observation number, or whatever is appropriate for recording the relative time of control observations.

3. Introduce control samples into each analytical run, record the values, and plot each value on the control chart.

4. When the control values fall within the control limits, interpret the run as being "in control" and report the patient's results. When a single control value exceeds the control limits, do not report the patient's results. Inspect the method to determine the cause for the errors. Solve the problem, then repeat the entire run of patient and control samples. Determine the control status for the new run in the same manner.

An example of a Levey-Jennings chart is shown in Figure 8-2, where control limits have been set as the $\bar{x} \pm 3s$. This control procedure will have a false rejection rate of only 1% or so, but its error detection will also be low when N is only 2–4. There is only

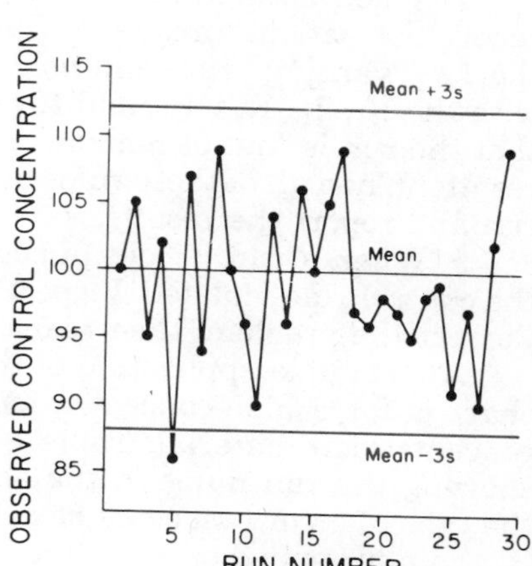

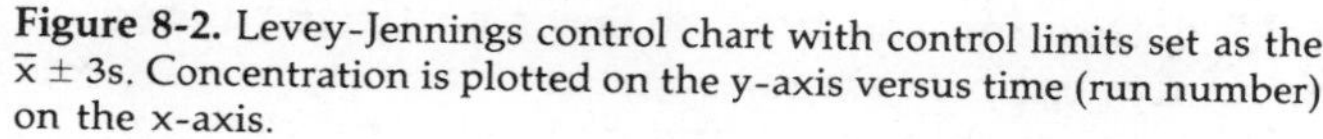
Figure 8-2. Levey-Jennings control chart with control limits set as the $\bar{x} \pm 3s$. Concentration is plotted on the y-axis versus time (run number) on the x-axis.

a 20–40% detection of shifts equivalent to 2s or increases in random error equivalent to a doubling of s. If instead 2s control limits were used with N = 1, the error detection would be a little higher but the false rejections would also be higher, about 5%. With 2s limits and N > 1, the false rejections will increase, for example, 9% with N = 2 and 18% with N = 4. Because of these limitations, alternative control procedures using different rules have been developed.

Westgard Multi-Rule Control Chart

The multi-rule procedure developed by Westgard and coworkers[13] uses a series of control rules for interpreting control data. The probability for false rejection is kept low by selecting only those rules whose individual probabilities for false rejection are very low (≤ 0.01). The probability for error detection is improved by selecting those rules that are particularly sensitive to random and systematic errors. The procedure requires a chart with control limits drawn at $\bar{x} \pm 3s$, $\bar{x} \pm 2s$, and $\bar{x} \pm 1s$, and it can be adapted to existing Levey-Jennings charts by adding the missing limits.

The following control rules are used:

1_{2s}, *one control observation exceeding the* $\bar{x} \pm 2s$, a "warning" rule to initiate further testing of the control data by other control rules;

1_{3s}, *one control observation exceeding the* $\bar{x} \pm 3s$, a rejection rule that is primarily sensitive to random error;

2_{2s}, *two consecutive control observations exceeding the same* $\bar{x} + 2s$ *or* $\bar{x} - 2s$ *limit*, a rejection rule sensitive to systematic error;

R_{4s}, *one observation exceeding the* $\bar{x} + 2s$ *and another exceeding the* $\bar{x} - 2s$, a rejection rule sensitive to random error;

4_{1s}, *four consecutive control observations exceeding the same* $\bar{x} + 1s$ *or* $\bar{x} - 1s$ *limit*, a rejection rule sensitive to systematic error;

$10_{\bar{x}}$, *ten consecutive control observations falling on one side of* $\bar{x}$ (above or below, with no other requirement on the size of the deviation), a rejection rule sensitive to systematic error.

The use of the multi-rule procedure is similar to the use of a Levey-Jennings chart, but the data interpretation is more structured. To employ the multi-rule procedure, follow these steps:

1. Analyze samples of the control material by the analytical method to be controlled on at least 20 different days. Two different materials having appropriate concentrations are recommended, but a single material can be used. Calculate the mean and standard deviation from the results for each control material.

2. Construct a control chart for each control material used. Label the y-axis "control value" and set the range of concentrations to include $\bar{x} \pm 4s$. Draw horizontal lines for $\bar{x}$, $\bar{x} \pm 1s$, $\bar{x} \pm 2s$, and $\bar{x} \pm 3s$. Label the x-axis for time, day, or run number.

3. Introduce one sample of each of the two control materials into each analytical run. Record the control values and plot them on their respective control charts.

4. When both control values fall within 2s limits, accept the analytical run and report the patient's results. When one of the control values exceeds a 2s limit, apply the 1_{2s} "warning" rule and hold the patient's results while inspecting the control data with the 1_{3s}, 2_{2s}, R_{4s}, 4_{1s}, and $10_{\bar{x}}$ rules. When any one of these additional rules indicates that the run is "out of control," reject the analytical run and do not report the patient's results. When all of these rules indicate that the run is "in control," accept the analytical run and report the results.

5. When a run is "out of control," determine the type of error occurring based on the control rule violated. Inspect the analytical process for sources of that type of error. Correct the problem, then reanalyze the entire run including controls and patients.

An example application of the multi-rule procedure is shown in Figure 8-3; the top chart is for a high-concentration control material and the bottom chart is for a low-concentration material. Table 8-2 summarizes the interpretation of the charted data, showing the run number, the accept or reject decision, the control rules violated, and the type of error suspected based on the rule violations. It is important to note that the R_{4s} rule is applied only within a run, so that between-run systematic changes or errors

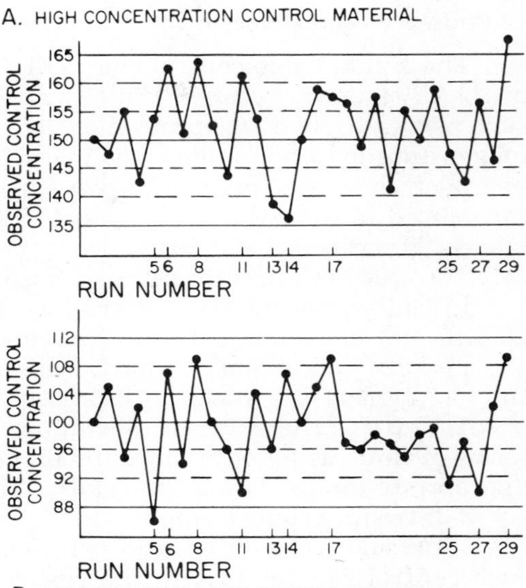

Figure 8-3. Westgard multi-rule control chart with control limits drawn at the $\bar{x} \pm 1s$, $\bar{x} \pm 2s$, and $\bar{x} \pm 3s$. Concentration is plotted on the y-axis versus time (run number) on the x-axis. *A,* Chart for high-concentration control material. *B,* Chart for low-concentration control material. (From Westgard, J. O., et al., Clin. Chem., *27*:493–501, 1981.)

are not wrongly interpreted as random errors. However, the rule may be applied "across" materials, meaning that one of the observations can be on the low-concentration material and the other on the high-concentration material, as long as they are within the same run. On the other hand, the 2_{2s}, 4_{1s}, and $10_{\bar{x}}$ rules can be applied across runs and materials, which effectively increases N and improves the error-detection capabilities of the procedure.

Comparison of the error detection of the Westgard multi-rule procedure with the Levey-Jennings chart having 3s control limits shows improved error detection by the multi-rule procedure. The R_{4s} rule improves the detection of random error, and the 2_{2s}, 4_{1s}, and $10_{\bar{x}}$ rules improve the detection of systematic errors. Elimination of the $10_{\bar{x}}$ rule does not cause much loss in error detection but reduces considerably the amount of control data that must be inspected; this simplification may make the multi-rule procedure easier to use. The 4_{1s} rule could possibly be eliminated to further simplify the procedure, but this would cause an appreciable loss in error detection. Another alternative is to interpret the $10_{\bar{x}}$ and 4_{1s} rules as "prevention" rules that initiate inspection and preventive maintenance of the analytical method prior to performing additional runs.

Table 8-2. INTERPRETATION OF EXAMPLES OF CONTROL DATA USING THE WESTGARD MULTI-RULE PROCEDURE

Run Number	Decision on Run		Control Rule Violated					Type of Error	
	Accept	*Reject*	1_{3s}	2_{2s}	R_{4s}	4_{1s}	$10_{\bar{x}}$	*RE*	*SE*
5		x	x					x	
6	x								
8		x		x					x
11		x			x			x	
13	x								
14		x		x					x
17		x				x			x
25	x								
27		x					x		x
29		x	x	x					x

Cumulative Sum Control Chart

The single value control charts discussed in the previous sections provide a display of the differences between the observed values and the expected mean. Control rules such as 2_{2s}, 4_{1s}, and $10_{\bar{x}}$ provide one way of determining when successive differences no longer appear to be random (too many in a row on one side of a control limit). Another approach is to use the cumulative sum control procedure, or cusum chart, which was introduced in industry in the late 1950s[10] and which has found limited use in clinical laboratories since the 1960s.

The steps required to set up a cusum chart are as follows.

1. Analyze the control material by the analytical method to be controlled on at least 20 different days, and calculate the mean and standard deviation of those results.

2. Construct a control chart. Label the y-axis "cusum." Draw a horizontal line at the midpoint of the y-axis to represent a cusum of zero. Set the range of values to be about 10 times the standard deviation, both above and below zero. Label the x-axis in terms of time, using day, run number, control observation number, or whatever is most appropriate for recording the relative time of the control observations.

3. Introduce control samples into each analytical run and record the values obtained.

4. Calculate the differences between the observed values and the expected mean for each control sample. Obtain the cusum by adding each of these differences to the cumulative sum of the previous differences. Plot the cusum on the control chart and inspect the plot.

5. Interpret the charted data by examining the angle of the cusum line. A steep angle suggests that a systematic error is present and that the analytical method is "out of control."

Example calculations are shown in Table 8-3, and the cusum values are plotted versus observation number in Figure 8-4. When control values scatter on both sides of the mean, giving both positive and negative differences, the cusum will alternate in sign and the plotted values will wander back and forth across the zero line on the control chart. When the control values fall predominately on one side of the mean so that most of the differences have the same sign, the cusum value increases in magnitude and the plotted values will move away from the zero line of the control chart.

Because of the difficulties in interpreting the angle of the cusum line, an alternative form of cusum, called decision limit cusum,[15] may be found in clinical laboratories, particularly when microcomputer quality control programs are being used. In this form,

Table 8-3. EXAMPLE OF CUSUM CALCULATIONS (FOR CONTROL MATERIAL WITH $\bar{x}$ = 100, s = 5.0)

Control Observation Number	Control Value	Difference from $\bar{x}$, d_1	Cumulative Sum of Differences, CS_1
1	110	+10	+10
2	100	0	+10
3	108	+8	+18
4	105	+5	+23
5	105	+5	+28
6	101	+1	+29
7	96	−4	+25
8	105	+5	+30
9	101	+1	+31
10	101	+1	+32
11	111	+11	+43
12	102	+2	+45
13	110	+10	+55
14	107	+7	+62
15	107	+7	+69
16	107	+7	+76

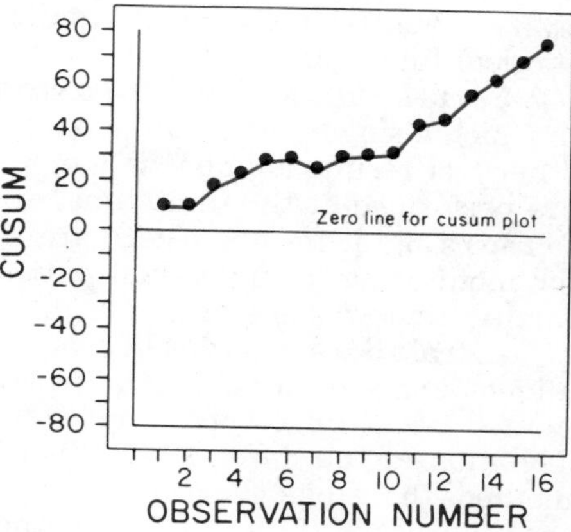

Figure 8-4. Cumulative sum control chart for example data is in Table 8-3. The cumulative sum of the differences from the mean is plotted on the y-axis versus time (control observation number) on the x-axis.

the cusum value is compared directly with a numerical limit, or control limit, making the determination of control status easier.

Cusum procedures are very sensitive to systematic errors because they effectively increase N by including all past data. However, they do not provide very good detection of random error; thus, a cusum procedure should not be used alone but should be included with a Levey-Jennings procedure on a combined chart[15] or employed as a separate chart along with a Levey-Jennings chart.

Shewhart Mean and Range Control Charts

When several values are obtained on the same control material, the mean (average) and range (difference between high and low values) can be determined and used as control statistics. The mean is plotted on an "x-bar" chart and the range is plotted on an "R-chart." Detailed directions on the use of mean and range charts have been prepared by Hainline.[6] These charts have not been widely used in clinical laboratories because of the low numbers of controls analyzed in each run; however, they should be useful for assessing the performance of analytical methods over longer time periods when N is accumulated over time.

Related procedures such as moving averages or moving standard deviations[4] have been implemented in microcomputer quality control programs (under the name of "trend analysis" procedures) and are available for use in clinical laboratories.

Control of Analytical Quality Using Data From Patients

Procedures Using Individual Laboratory Results

The patient's test results are the final product of most laboratory procedures, and the monitoring of these results is the most direct form of quality control. Unfortunately, procedures for monitoring these results are not very sensitive and have low probabilities for error detection. The most effective procedure is the clinical correlation of test results with other information related to the patient, especially surgical findings, response to therapy, and autopsy data. Less sensitive but easier to implement are comparisons with previous test values and correlation with related test results. The easiest procedure is the comparison of test results with physiological or probabilistic limits.

Clinical correlation. In modern high-volume laboratories, it is impractical to correlate all test results with the clinical status of patients. In general, the clinicians ordering the tests are in a better position to examine the appropriateness of test results. They should be encouraged to report discrepancies to the laboratory, and a mechanism should be available to investigate these problems. However, it should be recognized that clinicians often order a test because they are uncertain of the exact clinical status of their patient. In such cases, it may be circular reasoning to correlate clinical diagnoses with laboratory test results. Clinical correlation studies are best performed retrospectively, when laboratory test results can be directly related to other evidence, such as surgical findings.

Correlation with other laboratory tests. There are a limited number of situations in which a single test result is plausible, but the combination of that result with certain other test results is impossible or very unlikely. If the tests involved in the correlation are performed at the same time, errors can be identified and problems corrected prior to reporting of results. There are several relationships that provide some possibilities for monitoring an individual patient's test results, at least for purposes of initiating further review; some examples are anion gap, acid-base balance, and thyroxine-TSH.

Intralaboratory duplicates. Specimens can be divided into two aliquots and each aliquot analyzed for all tests ordered. The differences between the duplicates can be plotted on a range type of control chart having limits calculated from the standard deviation of the differences. When duplicates are obtained from the same analytical method, the range chart monitors only random error and is not adequate for assuring the accuracy of the analytical method. When duplicates are obtained from two different analytical methods, the range chart monitors both random and systematic errors but cannot separate the two types of errors.

Delta checks with previous tests. Certain errors, particularly errors in specimen identification, can be detected by comparing laboratory test results with values obtained on previous specimens from the same patient. The expected variability of test results depends on both the analyte and the time interval between determinations. See Table 6-11 for recommendations for specific tests. In identifying specimen mix-ups, it is helpful to examine several tests, including hematological tests, since differences in many tests will make a mix-up more apparent.

Limit checks. A patient's test results should be checked to be sure they are within the physiological ranges compatible with life. See Table 6-11 for recommendations for specific tests. Limit checks are helpful for detecting clerical errors such as transposed digits or misplaced decimal points. Somewhat narrower limit checks can be used as warning limits for detecting and verifying possible but infrequently occurring test results. Limit checks have a relatively low chance for detecting errors; thus they cannot replace conventional quality control procedures and should be used as supplemental procedures.

Procedures Based on Results from Many Patients

"Mean of normals." Statistics based on distributions of test results from large numbers of patients are useful for detecting systematic errors (shifts and drifts), but are of no value for detecting random errors (increased variability). They are useful adjuncts to conventional control procedures but should not be substituted for them. Values from patients include numerous sources of variation—demographical, biological, pathological, and preanalytical—in addition to any analytical variation caused by the analytical method. Individual patient test values are too variable to have any utility for quality control; however, the mean of the test values from a group of patients is more stable and may be useful for control purposes.

The "mean of normals" approach calls for establishing truncation limits, usually near the limits for reference intervals, for "trimming" patient data. Values outside the truncation limit are eliminated from the calculations to reduce the effect of outliers and subpopulations. The mean of the truncated results is calculated and compared with the control limits. The details of the procedure and guidelines for optimizing its performance have been presented by Cembrowski.[3]

Although control statistics based on a patient's results can be readily available in computerized laboratories, their utility is still controversial. These procedures detect only systematic error and their sensitivity depends on the variation of a patient's test values relative to analytical variation and on the number of values included in the calculation.

In summary, quality control procedures based on data from patients can provide information that is useful in monitoring the quality of laboratory analyses, but these procedures should not generally be substituted for conventional procedures unless stable control materials are unavailable. Procedures for data from patients can detect quality control problems that may not be detected by conventional control procedures and therefore should be used to supplement conventional control procedures when appropriate.

Role of External Quality Assurance (QA) Procedures

All control procedures described so far have focused on monitoring a single laboratory; they compose what is called *internal quality assurance.* In contrast, *external quality assurance* procedures compare the performance of a laboratory with the performance of a group of different laboratories. The two are complementary activities; internal QA is necessary for the daily monitoring of the precision and accuracy of the analytical method, and external QA is important for maintaining the long-term accuracy of the analytical method.

Limitations of internal QA. The problems detected are only the changes in performance between the present operation and the "stable" operation that was characterized during the baseline period when the analytical method was thought to be working properly. Initial method evaluation studies are essential to assure that systematic errors have been eliminated prior to the baseline period when control limits are established. Ongoing comparison-of-methods studies are desirable to assure that systematic errors do not slowly increase and go undetected by internal QA procedures. Such studies are provided by external QA procedures.

Features of external QA programs. Several programs are available to laboratories, some sponsored by professional societies and others by manufacturers of control materials. These programs have the participating laboratories analyze the same lot of control material, usually as part of the daily internal QA activities. The results are tabulated monthly and sent to the sponsoring group for data analysis. Summary reports are prepared by the program sponsor and are distributed to all participating laboratories, usually a few weeks after the data were submitted. Thus, the reports are useful for monthly reviews and periodic problem-solving activities.

The reports often include extensive data analysis, statistical summaries, and plots. The results from individual laboratories are compared with the "true" or correct values, which are usually estimated from the overall mean of all the laboratories in the program, or from the mean of values of a selected group of "reference" laboratories, or sometimes from the results of reference or definitive analytical methods. Different programs present the data in different ways. For example, the statistical significance of the difference between an individual laboratory's observed mean and the group mean can be tested by use of a t-test. When the difference is significant, the laboratory is alerted that its results are biased as compared with the results of the other laboratories. Another approach is to divide the difference by the survey standard deviation and to express the difference in terms of the number of standard deviations:

$$\text{SDI} = \frac{\text{Lab mean} - \text{group mean}}{\text{Survey standard deviation}}$$

where SDI is the abbreviation for *standard deviation interval* or *index*, and the survey standard deviation may be for the whole group, a selected subgroup, or even the individual laboratory. The calculation of SDIs reduces all the test results to the same values, making it possible to interpret the data without reference to the exact mean and standard deviation for each analytical method. For example, an SDI value of +2.0 has the same meaning for any test, indicating that the value is two standard deviations above its established mean. Differences >2 indicate that a laboratory result is not in good agreement with the results from the other laboratories in the program.

Information about the nature of the systematic error can be obtained when two different control materials have been analyzed by each laboratory. The laboratory's observed mean for material A can be plotted on the y-axis versus its observed mean for material B on the x-axis. Such graphs are called Youden plots.[5] Ideally, the point for a laboratory should fall at the center of the plot. Points falling away from the center but on the 45-degree line suggest a proportional analytical error. Points falling away from the center but not on a 45-degree line suggest either an error that is constant for both materials or an error that occurs with just one material.

Identifying the Sources of Analytical Errors

The control procedures discussed in the previous sections provide a way of alerting the analyst to analytical problems that may cause the quality of analytical performance to fall short of the quality goals set for the laboratory. However, these control procedures do not identify the sources of the analytical errors or solve the control problems. Additional procedures are often needed to identify the sources.

Physical inspection. When the analyst is alerted to a control problem, the first step is careful inspection of the analytical method, equipment, reagents, and specimens. An inspection may seem to be a very qualitative and sensory technique, but it can be very powerful when used with checklists developed for specific analytical methods. The inspection should include a review of logs and records that document changes occurring with the instrument and reagents. Extra instrument function checks can often be performed to verify proper system performance and to separate chemical from instrumental sources of errors.

Relationship of type and source of error. The type of error itself provides a clue about the source of the error. For example, systematic errors are often related to calibration problems, such as impure calibration materials, improper preparation of standard solutions, erroneous set point and assigned values, unstable or contaminated standard solutions, inadequate calibration techniques, nonlinear or unstable calibration functions, unstable reagent blanks, and inadequate sample blanks. Random errors are more likely due to lack of reproducibility in: the pipetting of samples and reagents, dissolving of reagent tablets, mixing of samples and reagents, and the lack of stability of temperature baths, timing regulation, and photometric or other sensors.

Knowing to which errors a control procedure responds offers information about the type of error occurring. Different control rules have different sensitivities to random and systematic errors, as is illustrated by the multi-rule control procedure. It is often possible to make a good judgment based on the control procedure or control rule that provides the alert as to the type of error. Control procedures that employ samples from a patient rather than stable control materials can help identify preanalytic sources of errors such as specimen handling and processing. External control procedures may provide information about systematic errors. The information from all of these procedures is complementary and when used together provides a more complete assessment of the types of errors occurring and their possible sources.

Method evaluation experiments. In cases in which the type of error cannot be easily identified and quantitated from available control information, it may be necessary to utilize the experiments employed in method evaluation studies. Recovery, interference, linearity, and comparison-of-methods experiments can be used to assess systematic errors. Random error can be assessed from replication experiments, particularly those designed to estimate the components of variance.[1] For example, when duplicates are analyzed within a run for several different runs, the data can be used to determine the components of within-run and between-run variation. This procedure isolates the errors occurring in different time periods, which in turn may help to identify the sources of random error.

References

1. Bauer, S., and Kennedy, J. W.: Applied statistics in the clinical laboratory: II. Variability over time. J. Clin. Lab. Autom., *2*:35–40, 1982.
2. Bowers, G. N., Burnett, R. W., and McComb, R. B.: Preparation and use of human serum control materials for monitoring precision in clinical chemistry. *In:* Selected Methods in Clinical Chemistry, Vol. 8. Washington, D.C., Am Assoc. Clin. Chem., 1977, pp. 21–27.
3. Cembrowski, G. S., Chandler, E. P., and Westgard, J. O.: Assessment of "average of normals" quality control procedures and guidelines for implementation. Am. J. Clin. Pathol., *81*:492–499, 1984.

4. Cembrowski, G. S., Westgard, J. O., Eggert, A. A., et al.: Trend detection in control data: Optimization and interpretation of Trigg's technique for trend analysis. Clin. Chem., *21*:1396–1405, 1975.
5. Elion-Gerritzen, W. E.: Quality control in clinical chemistry—the two sample plot and improvement in laboratory performance. Am. J. Clin. Pathol., *67*:91–96, 1977.
6. Hainline, A.: Quality assurance: Theoretical and practical aspects. Selected Methods in Clinical Chemistry, *9*:17–31, 1982.
7. Henry, R. J., and Segalove, M.: The running of standards in clinical chemistry and the use of the control chart. J. Clin. Pathol., *5*:305–311, 1952.
8. Levey, S., and Jennings, E. R.: The use of control charts in the clinical laboratory. Am. J. Clin. Pathol., *20*:1059–1066, 1950.
9. National Committee for Clinical Laboratory Standards: Nomenclature and definitions for use in the national reference system in clinical chemistry. NCCLS Document PSC-13. Villanova, PA, 1979.
10. Page, E. S.: Continuous inspection schemes. Biometrika, *41*:100–114, 1954.
11. Shewhart, W. A.: Economic Control of Quality of the Manufactured Product. New York, Van Nostrand, 1931.
12. Tietz, N. W.: A model for a comprehensive measurement system in clinical chemistry. Clin. Chem., *25*:594, 1979.
13. Westgard, J. O., Barry, P. L., Hunt, M. R., et al.: A multi-rule Shewhart chart for quality control in clinical chemistry. Clin. Chem., *27*:493–501, 1981.
14. Westgard, J. O., and Groth, T.: Power functions for statistical control rules. Clin. Chem., *25*:863–869, 1979.
15. Westgard, J. O., Groth, T., Aronsson, T., et al.: Combined Shewhart-cusum control chart for improved quality control in clinical chemistry. Clin. Chem., *23*,1881–1887, 1977.
16. Westgard, J. O., Groth, T., Aronsson, T., et al.: Performance characteristics of rules for internal quality control: Probabilities for false rejection and error detection. Clin. Chem., *23*:1857–1867, 1977.

MICROPROCESSORS AND COMPUTERS IN THE CLINICAL LABORATORY

FRANKLIN R. ELEVITCH, M.D.
G. PHILLIP HICKS, Ph.D.

To appreciate fully the computer's impact on and benefits to the clinical laboratory operation, one must thoroughly understand the laboratory's problem of managing specimens and data. The increase in the number of laboratory tests performed, in addition to the trend to centralize laboratory services, has led to significant management problems in handling and identifying specimens submitted for analysis. Furthermore, the critical nature of the data involved requires an accurate flow of laboratory information between other areas of the hospital, including nursing services, the business office, the medical record room, and physicians' offices. Thus, it is necessary to consider the entire patient-care process and the effect the computer has on all associated areas, as well as the place it occupies in the entire hospital communications network (Figure 9-1).

This chapter provides only an introduction to basic computer information and concepts and an overview of the application of the computer in the laboratory.

The Computer System

There are two components of a computer system, *hardware* and *software*. These two components are analogous to the division of the analytical process into instrumentation and methodology. When a large-scale integration (LSI) chip is used as a component of equipment such as a home appliance or a laboratory instrument, it is referred to as a *microprocessor*. A system designed as a general-purpose computer that uses the same chip is called a *microcomputer*. The *minicomputer* is manufactured from the same technological base as the microcomputer but is generally designed to utilize several terminals and programs simultaneously (multi-user environment), whereas the microcomputer is generally intended for one user at a time (single-user environment). Some general-purpose microcomputers now rival the capacity of the commonly known minicomputers, and the distinction between them is becoming less clear.

Computer Hardware

The computer hardware can be divided into two categories—the *central processing unit* (CPU), which is the computer proper, and peripheral devices (peripherals), which are external to the CPU.

Peripheral Devices

The various peripheral devices provide the means for transferring information into and out of the CPU and for storing data and programs for use in the CPU. Peripherals combined with the CPU make up what is known as a *configuration*. There are two types of peripheral devices: (1) those that allow the transfer of information between the computer and the laboratory technologist (user) or laboratory instrumentation and (2) those that store and retrieve information.

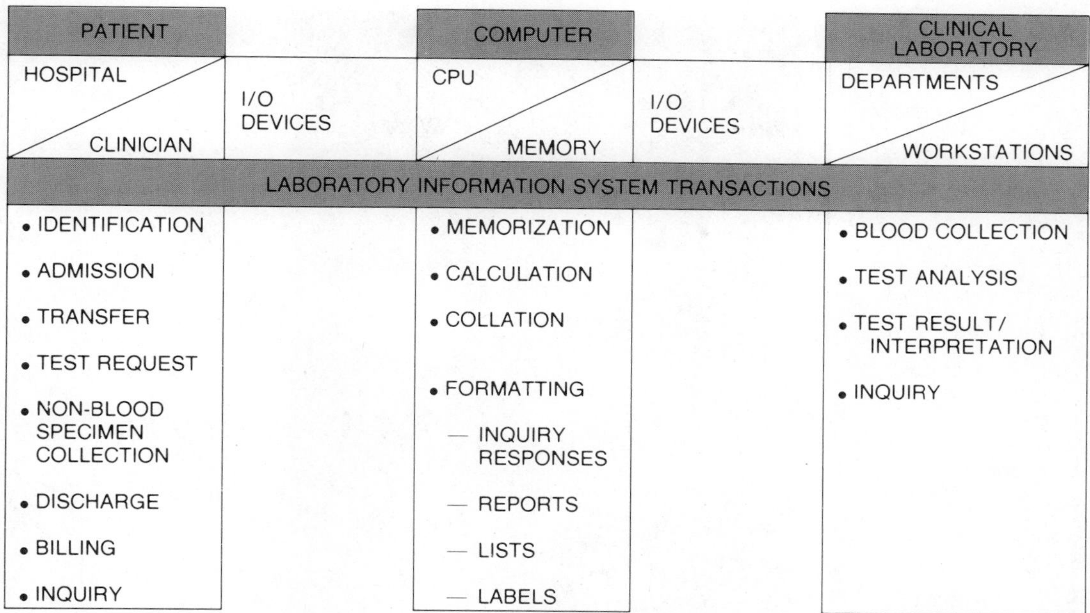

Figure 9-1. Overview of a computerized laboratory information system.

Input and Output Devices

Cathode ray tube (CRT). Devices that communicate with the user or instrumentation are referred to as Input/Output (I/O) devices. One of the most common peripheral I/O devices is the cathode ray tube (CRT), or monitor, with a keyboard. The CRT terminal is used to transfer information between the user and the computer. It can be used in a conversational mode at a high speed of output from the computer to the screen (30–9600 characters/s). The CRT allows for observation on the screen of the keyboard entries as well as for program-controlled editing of the data entered. However, as the screen fills up, the information is either erased or disappears off the top of the screen. Consequently, there is no permanent record (audit trail) of the transaction between the user and the computer. For this reason laboratories use the CRT for interactive functions such as inquiry and use printer terminals for the output of data such as laboratory reports.

CRTs may be considered "intelligent" or "dumb." An *intelligent terminal* has one or more internal microprocessors with programs to perform specific tasks at the work station. The program may be permanently contained in the terminal or it may be loaded from the main computer. The latter technique is called *down loading.* The intelligent terminal performs functions that would otherwise be required of the main computer. Input to the intelligent terminal can continue even when the main computer is not operating or when it is not available to receive the data. In this case, the "intelligent" terminal stores the information until the main computer is operational and ready to receive the information. A *dumb terminal*, in contrast, requires that the main computer be operating to accept each character or each line and to control the input process.

Printer terminals. The second major type of I/O terminal is a printer with a keyboard similar to a typewriter and that allows entry of information or queries. The computer in turn responds by typing information back on either the same or another terminal. Output from any printing terminal is referred to as a *hard copy* and has the advantage of providing a permanent record (called *audit trail*) of every transaction, thus allowing retrospective examination of entries.

Card reader. Another type of device, which is suitable for input of information only, is the card reader, which can "read" either pencil marks on or holes in a card

Figure 9-2. Handheld microcomputer with optical scanner.

(thus the name *mark-sense reader*) by measuring the light reflected from or transmitted through the card.

Optical scanners. These devices use optical character recognition, OCR, to read and recognize either a bar code or a printed character directly. The optical scanner shown in Figure 9-2 is a hand-held microcomputer-based device that is widely used in supermarkets and industry inventory systems. Optical bar-code scanners that utilize laser beams are used in blood banks, for example, to read identification labels.

Digital matrix devices. The incorporation of the microprocessor into peripheral devices has provided many options for data input that are beyond the scope of this chapter, but a few are worth mentioning here. *Digital pads* allow the direct input of a "drawing" or "graphic" into the computer through a pen. *Light pens* and *touch-sensitive screens* can be used to select options on a CRT by pointing to the location on the screen or to enter graphic data by "drawing" on the CRT screen.

Printers. There are a variety of flexible and inexpensive *dot matrix* and *character* printers available as output devices with speeds of 30–240 characters/s. Despite their low speed, these *low-speed printers* can be an effective backup to higher-speed printers.

The *high-speed printer* is usually capable of printing at speeds from 300–1100 lines/min. This device is often referred to as a *line printer* because it prints one whole line at a time. There is a complete character set for each column space on the line printer, and when the printer has received a complete line from the CPU, the entire line is printed in one process. Because of their high printing speed, line printers are suitable for the output of bulky reports such as cumulative reports for all patients in the hospital and long administrative reports such as patient directories and master work logs for the entire laboratory.

Instrument interfaces. The instrument interface is a peripheral device that allows the direct electronic connection of a laboratory instrument to the CPU. There are two major types of interfaces, *analog* and *digital* (Figure 9-3). An *analog signal* varies continuously with time. A recorder tracing of a chromatographic separation provides a good example of a graphic demonstration of an analog signal. Since a digital computer can only process discrete numbers, it is necessary to convert the continuously changing

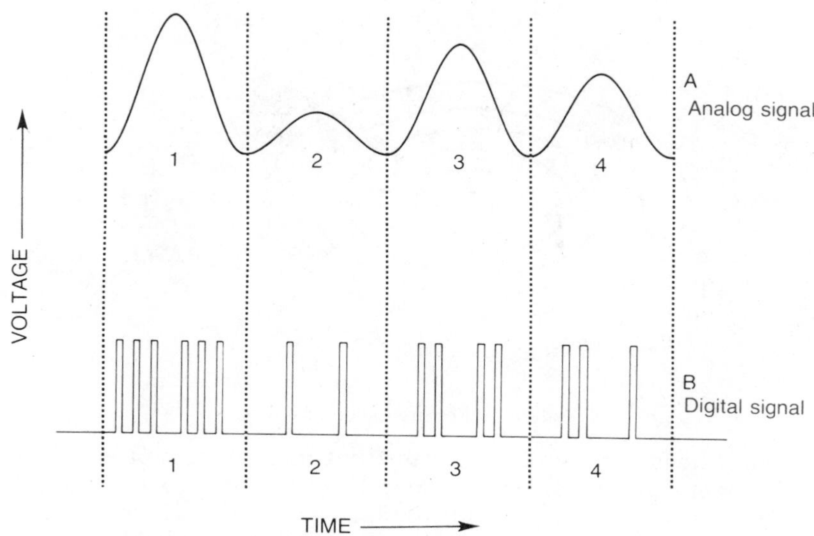

Figure 9-3. Representation of analog and digital signals.

voltage generated by the detector into a series of discrete numbers. This conversion is made by an *analog-to-digital converter* (ADC), which converts the value of the analog signal height into a discrete number proportional to the voltage of the original signal. A series of numbers taken at sufficiently frequent intervals (e.g., each second) permits the computer to construct a corresponding curve and to find the highest point on this curve.

A *digital signal* typically consists of a series of equal-height voltage "pulses." The magnitude of the signal is indicated by the presence or absence of pulses in the pulse pattern, or sometimes by the interval between pulses, but not by the height of the voltage pulses. Figure 9-3 illustrates four sets of digital pulses that correspond to the four analog peaks. Since the digital pulses are already discrete, a digital interface needs only to "decode" the pulse pattern to obtain the discrete number representing the signal value.

Storage Devices

Perhaps the most common storage device external to the CPU is the ***disk.*** This is a platter on which magnetized spots or *bits** can be deposited or erased. The heads within the disk drive read and write information on the disks.

Figure 9-4 illustrates magnetic bits of data on the tracks of each disk. Each bit produces an electronic pulse when it passes the read head. A bit of data is determined to be "0" or "1" by the presence or absence of an electronic pulse. A number of bits together produce a pattern of pulses called a *word.* For example, a 16-bit word consists of a pattern of 16 pulses that give the numeric value of the word in a manner similar to that described for digital signals in Figure 9-3. Each word has a specific location on the disk assigned to it; this location is called the *address* of the word. One-half of a word, generally 8 bits, is called a *byte.* A byte or one-half word is generally used to store the equivalent of a single character such as "A" or "1" and has become one of the most common ways of expressing memory and file sizes. Thus, 64 000 bytes of memory would be 32 000 words or 64 000 characters.

The disk platter is divided into *tracks,* and the tracks are divided into *sectors* and *blocks,* which may be designated to store specific files of information or programs. Some disk drives have for each track a read head, called a "fixed head," but most disk drives

---

*All information handled by the computer is contained in a numerical code whose fundamental unit is called a *bit.* Since the values of a bit are limited to zero and one, all calculations and data are contained as coded numbers in base 2 or "binary." A computer recognizes bits in a grouping that contains a constant number of bits; this grouping is called a word. Word sizes for typical laboratory computers range from 8 to 32 bits.

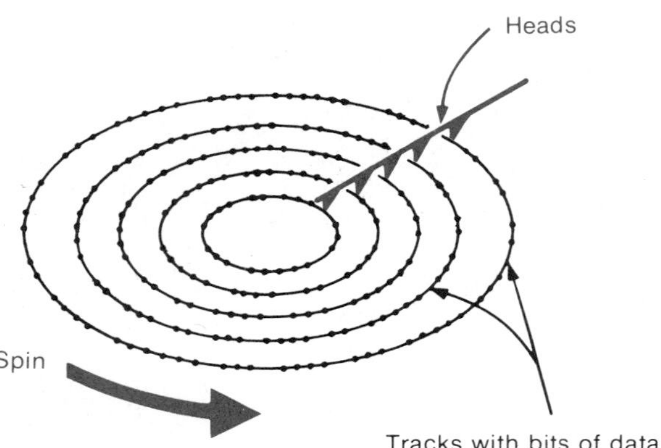

Figure 9-4. Representation of a disk.

have a single read head for each disk platter that must move to a track and then read the block of interest. Since this type of head has to move to the appropriate location and then read the sector to find the information, its average access time is typically 50–80 ms. The moving-head disk, however, usually has a high capacity—in the range of 2.5–50 million words per platter.

Moving-head disks are further categorized as *hard* and *floppy*. The *hard disk* is made of a rigid aluminum platter that has a magnetized coated surface. The *floppy disk* is similar to magnetic tape except that it is flexible and is contained in a rigid jacket for protection. Although both floppy and hard disks may be used on all types of computers—from the microcomputer to the largest mainframes—the floppy disk is used more commonly with the microcomputer. Floppy disks are relatively inexpensive, but hard disks generally have greater storage capacity.

The storage device with the fastest access time is the *mass memory*. This form of memory is accessible at the same speed as the main memory of the computer and may be constructed by LSI chip technology. Mass memories are provided in modules of 250 000 to 500 000 bytes and are usually "plug-to-plug compatible" with the disk controllers provided by major manufacturers. Large systems use up to four million bytes of this type of storage. Typical access times are 3 μs, a thousand times faster than fixed-head disks, which have access times of 4 ms.

Magnetic tape is the storage medium with the slowest access time. Like a disk, it retains information in the form of magnetic bits. Small tapes used on laboratory computers typically contain only 100 000 to 200 000 words, and the computer can scan such a tape within 20–30 s. Although this is extremely slow compared with the speed of a disk, it may be quite acceptable for storing information that is to be retrieved infrequently. Large amounts of information may be stored on "industry-compatible tape." These tapes may be 2400 feet in length and may have 20 times or more the capacity of small tapes, but the access time may be as long as 20 min. Use of industry-compatible tape has the great advantage that information generated by the laboratory computer can be transferred to any other computer equipped with an industry-compatible tape drive.

A new development is the *video disk* that uses a laser beam in a manner analogous to the stylus of a phonograph. It has the potential for storage of billions of bytes of data. Video disks currently allow for information retrieval (read) only; however, research is under way to develop a video disk that will have capabilities for both reading and writing.

Central Processing Unit (CPU)

The CPU is the "electronic logic" of the computer system. In the case of the microcomputer, the CPU might be a single chip on the main electronic board. Each CPU has its own set of instructions that it executes in the appropriate order as specified by the software. The number of different instructions in an *instruction set* ranges from six to several hundred. The CPU has its own storage device, the main memory, which may contain from 4000 (4 K) to one million (1 mega) or more bytes or instructions, depending upon the specific requirements.

The fast access time of the main memory and the high-speed circuitry of the CPU in general are very seldom the limiting factors in the speed of the overall system. Peripheral devices, such as disks, terminals, and tape units are many orders of magnitude slower than the CPU and therefore determine the overall speed of the system. The central processor is capable of executing many thousands of instructions in the same time that it takes a terminal to print or display one character.

Computer Software

Software Programming Languages

Computer software programs provide detailed instructions to the electronic hardware and thus enable the overall system to function. Programs can be constructed in several "levels" of languages. Those languages relating to the detailed machine structure (e.g., binary words, bytes, bits) are called *machine* or *assembly languages* and are considered to be "low level." Languages relating to commands that provide operations directly for the user (e.g., C = A + B, PRINT) are referred to as "high level."

Machine instruction is the most fundamental means of communicating to the computer what function to perform. An instruction is expressed as one or more discrete numbers that are decoded by the electronic logic of the computer. For example, the number coded 2000 may instruct the computer to ADD a number (xx) to another number (xy). The address (location) of the number to be added (xy) might also be included within the basic instruction. Thus, the number coded 2100 may instruct the computer to ADD the number (xy) in location 100 to the other number (xx). A series of such discrete instructions that are executed in sequence constitutes a *program.*

Techniques have been developed to allow programmers to substitute English-type *mnemonic codes* for instructions, such as ADD, LOAD, MULT, or STORE, in place of their numeric equivalents. The computer program that converts the mnemonic codes to equivalent numeric instructions that are decoded by the computer is called an *assembler.* Programming in numeric instructions is called *machine language programming,* whereas programming in mnemonic codes is called *assembly language programming.*

High-level programming allows the programmer to construct programs more easily and at a "higher level" through a series of high-level mnemonics referred to as *high-level language.* Examples of some well-known high-level languages are FORTRAN (FORmula TRANslation), which is especially useful for mathematics; COBOL (COmmon Business Oriented Language), useful for business applications; BASIC (a simple beginning language); PASCAL (structured language); and MUMPS (designed for medical applications).

High-level and machine-level programming both have advantages and disadvantages. A well-designed system usually will mix the higher-level functions with specific machine instructions to optimize the advantages of both types of programming. Since memory is becoming less expensive, more programming is being done in high-level languages. However, certain applications such as interfacing to instrumentation or communications require the use of machine language to attain sufficient speed and control.

For functional purposes, a program is classified either as an *operating system* or as an *application program.* The group of programs called the operating system is analogous to the central nervous system. Written for specific central processing units, these programs "manage" all of the jobs and the operation of the peripheral devices of the computer. Application programs, which are often written in higher-level languages, are designed to interrelate the computer and its user.

Database

The organization of data in the computer ultimately determines the function and flexibility of the system software. New approaches for structuring data have recently been developed and promise to increase greatly the utility of computer information systems. A detailed discussion of this new approach, called database, is beyond the scope of this chapter; therefore, only key features of database systems are mentioned. In contrast to database systems, programs using structured file systems typically are "data dependent." This means that each program is written to utilize the files exactly as they are designed by referring to the specific location and structure of data within each file. With this approach, it is necessary to rewrite every program each time the data files are altered by adding new types of data or by changing its structure. Thus, data-dependent programs have contributed to the inflexibility of laboratory systems.

Database designs promise to provide dramatic improvements in flexibility. Briefly stated, a database system is designed to file or store data without a programming structure related to the specific process it initially intended to serve; i.e., all data are accessible without limitations set by previous assumptions regarding the use of data. Instead, the data are organized in logical groupings that provide the computer programs essentially equal access to all data. This allows new programs to use data in ways not anticipated by the designer of the system, and makes the programs *data independent.*

Computerized Laboratory Information System

In a computerized laboratory information system (LIS), the flow of information passes through the central processing unit and then into memory, where it is available for inquiry and reports (see Figure 9-1). All information is entered into the computer through terminals or machine-readable forms. In this interactive mode the computer is used to manage the data transactions through appropriate software-controlled prompts and edits. Requisitions, logs, and workbooks are used as a backup on those occasions when the computer is inoperative for prolonged periods of time.

In an LIS, entry of information related to patient's identification, admission, transfer, discharge, test requests, and specimen collection is a function performed by nonlaboratory personnel. In institutions that have a hospital information system (HIS), this information is automatically transmitted to the LIS through an electronic interface and communications protocol. Usually test results are returned through LIS to HIS for inquiry and reporting on the nursing stations, whereas billing information is transferred to the hospital financial management system (FMS).

Planning a Laboratory Information System

The LIS integrates all aspects of the laboratory, medical staff, and hospital that are concerned with laboratory testing.[3] Since the installation of an LIS is a major commitment of effort and money, a feasibility study should first be conducted to determine the technical and economic practicality of replacing the existing manual system with a computer system.[2] The feasibility study may show that improvements in the manual system are sufficient to correct existing problems and bring about enhanced service, in which case the plans for an LIS would be abandoned. On the other hand, if an LIS appears preferable, the next step is writing specifications for the LIS. These specifications must be detailed and specific, addressing the needs of the laboratory as well as the rest of the hospital and medical staff. If an HIS already exists (or if one is anticipated imminently), a computer interface specification is included in the planning process. The laboratory may then choose to develop its own LIS or purchase one from the LIS vendor that meets the required specifications. In either case, a lifelong relationship (at least for the life of the LIS) is established with those who provide the computer, its peripheral devices, and its programs. The mutual obligations of this relationship are described as clearly as possible in a legal contract signed by the laboratory and the LIS vendor.

Data Transactions in the Laboratory

For purposes of this discussion, the flow of clinical laboratory information has been divided into seven steps. Each step is characterized by a *data transaction* in which there is a unidirectional transfer of data from a *data provider* to a *data receiver.* Although laboratory personnel are involved at each step, the patient, medical staff, and five other hospital services participate in the overall process (Table 9-1).

Table 9-1. TRANSACTION OF CLINICAL LABORATORY INFORMATION AND DATA

Transaction	Data Provider	Data Receiver
Patient identification admission transfer	Patient Nurse	Admission clerk ER clerk Nurse Lab clerk
Test request	Clinician	Lab phlebotomist Nurse
Specimen collection	Lab phlebotomist Nurse	Lab technologist Nurse Clinician
Test analysis	Lab instrument	Lab technologist
Test result/interpretation	Lab technologist Lab director	Clinician Nurse
Patient discharge	Clinician	Admission clerk Business clerk Lab clerk Medical records Nurse
Patient/guarantor billing	Lab clerk	Business clerk

The Laboratory Database

A major advantage of the LIS in a medical facility is that it establishes standardized laboratory nomenclature and standardized information processing procedures. The basic unit of information in an LIS file is called a *data element*. All the aspects of clinical laboratory testing are described by the appropriate combinations of fewer than 50 data elements grouped to define a patient, a specimen, or a test (Table 9-2). Actually, all of the transactions shown in Table 9-1 translate into data-element processing in LIS: adding, deleting, editing. Data elements are organized in LIS into two functional types of files: definitional and descriptive. *Definitional files* or dictionaries are intended for long-term repetitious use. They include data elements that constitute the basic services of the

Table 9-2. DATA ELEMENTS OF A LABORATORY INFORMATION SYSTEM

Patient	Specimen	Test
1. Number	1. Number	1. Number
2. Name	2. Status	2. Name(s)
3. Birth date	a. Ordered	3. Mnemonic
4. Sex	b. Collected	4. Source
5. Admission date	c. Received	5. Turnaround time
6. Admission status	d. Reported	6. Container/volume
7. Location	3. Date	7. Special requirements
8. Room/bed	a. Year	8. Department
9. Physician(s)	b. Month	9. Work list
10. Diagnosis(es)	c. Day	10. Method
11. Discharge date	4. Time	11. Units of measurement
12. Financial status	a. 12-hour clock	12. Numeric?
13. Guarantor(s)	b. 24-hour clock	13. Required?
14. Billing number	5. Phlebotomist identification	14. Delta check
	6. Ordering physician	15. Controls/standards
		16. Reference intervals
		17. Panic value(s)
		18. Charge code
		19. Charge
		20. Work units

laboratory and whose definitions can be changed only by management. These files include (1) test list; (2) worklist formats; (3) standards and controls; (4) coded comments; (5) laboratory departments; (6) hospital, emergency room, and clinic locations; (7) physicians' names, addresses, and phone numbers; (8) diagnostic classifications; and (9) automated instrument interface specifications. *Descriptive files* retain the historical events in LIS, such as the patient's identification; admission, discharge, and transfer information; test requests; and specimens. Relevant definitional data elements (Table 9-2) are included in descriptive files that make up each patient's record.

For every transaction in LIS, the appropriate data elements are incorporated into a descriptive file. Table 9-3, the LIS database, illustrates the logical accumulation of data elements on patients (P), specimens (S), and tests (T) from patient identification through billing. With each new transaction, data elements are added. Because of the cumulative process, it is unnecessary to store repeatedly data elements that are already

Table 9-3. LABORATORY INFORMATION SYSTEM DATABASE

Transaction		Data Element
Patient identification admission transfer	P	1. Patient number
		2. Patient name
		3. Birth date
		4. Sex
		5. Admission date
		6. Admission status
		7. Location
		8. Room/bed
		9. Clinician(s)
		10. Diagnosis(es)
Test request	S	1. Specimen number
		2. Specimen status
		3. Date
		4. Time
		5. Phlebotomist ID
	T	1. Test number
		2. Test name(s)
		3. Mnemonic
		4. Source
		5. Turnaround time
Specimen collection		6. Container volume
		7. Special requirements
Specimen analysis		8. Department
		9. Work list
		10. Method
		11. Units of measure
		12. Numeric?
		13. Required?
		14. Delta check
		15. Controls/standards
Test result/interpretation		16. Reference intervals
		17. Panic value(s)
Patient discharge	P	11. Discharge date
Patient/guarantor billing	T	18. Charge code
		19. Charge
		20. Work units
	P	12. Financial status
		13. Guarantor(s)
		14. Billing number
		15. Addresses

filed with each new test request. All transactions relate specific test numbers to specific specimen numbers for a specific patient number.

Patient identification. All hospital services for each patient are related to a unique patient number, which should be permanent for the lifetime of the patient. Sometimes the prefix "I" or "O" may be used for quick recognition of inpatient or outpatient status. At other times a numerical suffix can identify each of multiple admissions, or an additional separate admission number may be assigned from a nonrepeating serial sequence of numbers. In an emergency room admission of a patient in a life-threatening circumstance, time may be insufficient to obtain a formal patient number. In such cases, a separate manual system consisting of unique, nonrepeating numbered labels and matching wristbands that can be worn by the patient is essential. The temporary number is transcribed into LIS so that testing can proceed without delay. When a permanent patient number is assigned by LIS, the information under the temporary number is merged into the file under the permanent number.

Age is usually derived from the patient's birth date and is used in LIS for assigning age-related reference intervals to the respective test results. The birth date may also help to discriminate between patients with identical names. Sex-related reference intervals may be stratified by gender.

Admission, transfer, and discharge. A patient's *demographic data* elements often differ from one hospital stay to another. These data elements include (1) admission date, (2) admission status, (3) location, (4) room and bed, (5) each clinician, (6) each diagnosis, (7) discharge date, (8) financial status, (9) each guarantor, and (10) billing number. *Admission statuses* include outpatient, inpatient, preadmission, and ambulatory surgery categories, among others. In LIS, these classifications initiate report routing, define the length of time that data are retained in LIS, and organize pricing schedules. *Location,* such as room and bed, is used for inpatient specimen collection routes and report delivery. Outpatient reports are routed to the emergency room, clinics, or clinicians' offices.

Responsibility for documenting the patient's *diagnosis* (or diagnoses) is shared between the hospital staff and the medical staff. While physicians make diagnoses, hospitals maintain the medical records. Physicians' diagnoses follow a variety of classification systems, which hospitals have the responsibility to translate into a single uniform system of coding. Hospitals currently use a coding system called ICD9-CM (International Classification of Diseases, Clinical Manifestations). These codes are collated on a regular basis into reports of the hospital's activities.

In LIS, patient identification as well as admission and transfer data elements should be entered before tests are ordered and processed. If LIS communicates with HIS, these data elements are transmitted automatically to the LIS.

The billing process is initiated by acquisition of the patient's financial status and guarantor as specified on admission. The guarantor is the specific person, insurance company, or government agency to which the statement of charges will be sent.

Test request. Nursing service is the communications hub for patient care. All orders from clinicians, including those for laboratory testing, pass through nursing service, which has the responsibility to assure that all specimens are collected and that all results return to the patient's chart. Nonblood specimens (e.g., urine, sputum) are collected by the nursing service. The specimen together with a requisition (transmittal slip) is then sent to the laboratory. For routine and timed specimens collected by the laboratory, either collection lists and labels are printed in the laboratory through HIS or manual requisitions are sent to the laboratory for transcription into LIS. When tests are ordered through the HIS, a specimen number is assigned by the computer to all of the tests ordered at that time. This number is the unique identifier that both HIS and LIS use for inquiry and reporting. If this specimen number is also used as an accession number in the laboratory, the numerical series in each department will not be continuous. This makes it somewhat difficult at a later time to look for retained specimens, but with specimen lists and machine-readable labels, the specimens are identifiable.

Specimen collection. The receipt of a collected specimen in the laboratory is recorded in LIS by changing the specimen status from "ordered" or "collected" to "received." This change in specimen status automatically completes the order in the

HIS record and can be used to assign an accession number in LIS. Both quality control specimens and standards should have an accession numbering sequence analogous to but distinguishable from patient specimens. Identification of the phlebotomist and nursing service should also be stored in LIS.

Specimen analysis. For specimen processing, LIS groups tests with the appropriate standards and controls by work station or reference laboratory. These groupings are called work lists and may be reviewed on either a CRT terminal or printed report. The format for work lists is specifically designed for each work area. A work list may be a loading list for an automated instrument or it may provide a format for recording data for tests that are done manually.

Test results are produced by a variety of laboratory instruments. With built-in microprocessor capabilities, many instruments display preformatted results together with limit checks and results of control samples. Once matched with the proper patient identifiers, these values are transmitted directly to LIS and HIS. LIS stores all control results and uses these results for several *quality control routines,* including the multi-rule control procedures (see Chapter 8). A cumulative summary of observed control values can be displayed in a standard graphical format, together with a display of the expected mean and standard deviation and a list of observed means and their standard deviations for the last several months.

Test results. Laboratory technologists review and verify all results entered manually or through on-line instrument interfaces before the results are available for inquiry or reporting. The identification of the laboratory technologists who performed and verified the results is retained. Free text and coded comment entries may be included with results. The LIS test data elements (Table 9-2) provide on-line editing capabilities for numeric and nonnumeric results, required results, and results outside of reference, panic, or delta check limits. (See also Chapter 8, *Quality Assurance.*)

Test interpretation. Computer-generated *interpretive reports* assist clinicians in evaluating patient problems and arriving at medical decisions with accuracy and efficiency. The first step in report design is to format data in such a way that abnormal results and pathophysiological relationships are highlighted. This can be accomplished by flagging abnormal and critical values with distinctive symbols. For analytes whose trends may be meaningful on a daily basis (e.g., calcium), a horizontal format with analytes listed in rows and dates in columns is appropriate. For analytes in which date and time are more meaningful (e.g., blood gases or isoenzymes), a vertical format is preferable. For interpretive reporting, an *explanatory comment* can be added to emphasize an abnormal finding or to caution against false-positive results. An elaboration of this is the categorical listing of disease entities that might be associated with abnormalities.

Computer-assisted interpretive reports have taken several different approaches.[1] The report may list possible causes of a single abnormal result, differential possibilities offered by multiple abnormal results, and suggestions for further laboratory testing. This approach has been useful for analytes such as calcium, thyroid hormones, and serum proteins, in which the pathophysiological aspects are defined.

Patient or guarantor billing. The hospital business office is responsible for processing laboratory charges. The laboratory, the business office, and hospital administration collaborate to develop a price list, which includes test name, charge code, and charge. The charge code may be associated with a single dollar value and it may identify one or more tests in LIS and the hospital financial management system (FMS). For example, blood and urine sodium will have the same charge code because the amount of work involved in these analyses is comparable. On the other hand, the charge code for electrophoresis of protein in serum and cerebrospinal fluid may be different because of the extra concentrating step required for spinal fluid.

If a laboratory test, once requested, cannot be performed, proper credit can be given to the patient or the charge can be deleted automatically if the billing report has not yet been issued. Certain codes, such as "QNS" (quantity not sufficient) or "ACC" (laboratory accident), are "crediting statements," which cause the computer to generate automatically a credit to the patient.

Future Use of Computers

Many hospitals and companies are in the process of developing total information systems in which computers will control the information flow throughout the entire hospital. Such information systems can be interfaced with the laboratory computer, so that laboratory data can be transmitted automatically to the central communications system. Such development is certainly technically feasible, and routine implementation can be expected in the next decade. It is likely that large central computers will be connected to various configurations of smaller laboratory computers. The performance of functions unique to the laboratory and the integration of laboratory instrumentation with small computers make the presence of the latter in the laboratory highly desirable, if not essential.

References

1. Elevitch, F. R.: Computer-assisted clinical laboratory interpretive reports. Lab. Med., *13*:45–47, 1982.
2. Elevitch, F. R., and Aller, R.: Planning for a Computer in Your Laboratory. Chicago, American Society of Clinical Pathologists, 1983, p. 208.
3. Grams, R., Ed.: MedSy Report: Review of Laboratory Systems. Gainesville, FL, University of Florida, 1981.

10 SPECIMEN COLLECTION AND PROCESSING; SOURCES OF BIOLOGICAL VARIATION

DONALD S. YOUNG, M.B., Ph.D.
EDWARD W. BERMES, JR., Ph.D.

Physicians use laboratory tests to diagnose disease, to evaluate its severity, to monitor its progress or response to treatment, and to screen for disease in seemingly healthy individuals. Many factors besides disease affect the composition of body fluids; these factors may be either preanalytical or analytical. The variability of test results due to biological factors is often greater than the variability due to analytical factors. The control of biological variability begins with proper preparation of an individual prior to specimen collection. When factors are not controllable, as is the case with genetic and other long-term influences, it is important to recognize and consider the effects of these factors when interpreting laboratory data. The influence of biological variables on test values has been discussed in depth by Ladenson[8] and Young.[25]

Specimen Collection

Collection of Blood

Blood for analysis may be obtained from veins, arteries, or capillaries. Venous blood is usually the specimen of choice and *venipuncture* is the method for obtaining this specimen. In young children *skin puncture* is frequently used to obtain what is predominantly capillary blood. *Arterial puncture* is used mainly for blood gas analyses. An evacuated blood tube or a needle and syringe are usually used to obtain blood specimens, although an indwelling intravenous line with a tap or an indwelling butterfly-winged needle may occasionally be used.

Venipuncture

Before performing a venipuncture, the phlebotomist should verify the patient's identity and that the patient is fasting, should fasting be necessary to ensure medically useful results. The patient should be comfortably seated, or supine if sitting is not feasible, and should have been in this position for twenty minutes before the specimen is drawn. Standardization of the patient's position minimizes differences in concentrations of blood constituents caused by postural variations in blood volume (hemoconcentration). The median cubital vein in the antecubital fossa, or crook of the elbow, is the preferred site for collecting venous blood in adults, since the vein is both large and close to the surface of the skin. An arm containing a cannula or arteriovenous fistula should not be used without consent of the patient's physician. If fluid is being infused intravenously into a limb, the fluid should be shut off for 3 min before a specimen is obtained and a suitable note made in the patient's chart and on the result report form. Specimens obtained from the opposite arm or below the infusion site in the same arm may be satisfactory for most tests except for those analytes that are contained in the infused solutions (e.g., glucose or electrolytes).[15]

After the skin is cleaned with 70% isopropanol and allowed to dry, a tourniquet is applied 4 to 6 inches (10–15 cm) above the intended puncture site to obstruct the return of venous blood to the heart and to distend the veins. Tourniquets are best made from precut soft-rubber strips about 1 inch wide and 15 inches long that are commercially available, or from Velcro-type bands designed for this purpose. Alternatively, a blood-

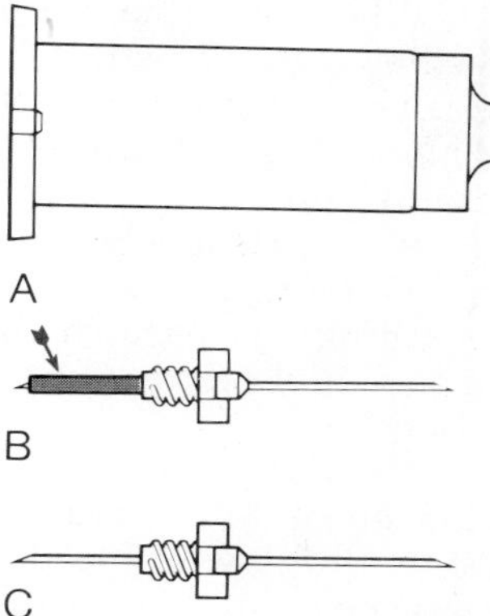

Figure 10-1. *A*, Blood collection tube holder (Becton Dickinson, Rutherford, NJ 07070). *B*, A needle for use with multiple tubes. *C*, A needle for use with a single tube. Arrow indicates rubber sleeve (red).

pressure cuff may be used, but it should be inflated only enough to impede venous return to the heart.

Typical systems for collecting blood in evacuated tubes are shown in Figure 10-1. There are two main classes of evacuated blood tubes: those containing a serum-separating material and those without. Various additives, including anticoagulants and preservatives, may be used with these tubes. The stopper color denotes the type of additive, as shown in Table 10-1. The additives must fall within certain tolerances as listed in NCCLS-approved standard, ASH-1.[13]

Serum separator tubes contain an inert, thixotropic, polymer gel material with a specific gravity of ~1.04. Silica or glass particles that accelerate clotting may be associated with it. Aspiration of blood into the tube and subsequent centrifugation displace the gel, which settles like a disk between cells and serum when the tube is centrifuged. Relative centrifugal force (RCF) must be at least 1100 × *g* for gel release and barrier formation. The release of intracellular components into the serum is prevented by the barrier for several hours or, in some cases, for a few days. Several studies indicate that the separation material does not contaminate the serum, but others demonstrate that free gel particles may persist and can affect dialysis membranes or ion-selective electrodes in some analyzers. As evacuated tubes age, the vacuum fails and their effective draw diminishes. The silicone coat that lines the tube and stopper also tends to break down with age. Therefore, the stock of these tubes should be rotated and careful attention paid to the expiration date.

Before performing a venipuncture the phlebotomist selects the appropriate tubes and needle for the tests to be done. The most commonly used needle sizes are gauges 19–22 (1.06–0.71 mm) in outside diameter. All needles must be sterile, sharp, and without barbs. Blood drawn for trace element measurements requires specially cleaned equipment (see Chapter 17).

The time at which a specimen is obtained is important for those blood constituents that undergo marked diurnal variation (e.g., corticosteroids and iron) and for those used to monitor drug therapy (e.g., digoxin or prothrombin time) because the length of time since drug administration affects the drug concentration. Furthermore, special precautions in the collecting and handling of specimens are required in tests for alcohol and in tests of medicolegal importance for which a chain of custody for the specimen must be established.

Table 10-1. CODING OF STOPPER COLOR TO INDICATE ADDITIVE IN EVACUATED BLOOD TUBE[13]

Color	Use	Additive
Gray	Plasma or whole blood with glycolysis inhibition	Oxalate (Na or K), fluoride (Na), or iodoacetate (Na)
Yellow	Sterile interior of tube	None
Green	Plasma or whole blood	Heparin (Na, Li, or NH_4)
Red	Serum	None
Blue	Plasma or whole blood	Citrate (Na)
Lavender	Plasma or whole blood	EDTA (Na_2 or K_2)

When blood collection is complete, the tourniquet should be released before the needle is withdrawn. When the needle has been removed, the patient should hold a dry gauze pad over the puncture site, with the arm raised to lessen the likelihood of leakage of blood. The pad can subsequently be held in place by a bandage, which can be removed after 15 min.

A comprehensive description of the proper technique for venipuncture is contained in NCCLS-approved standard ASH-3.[14] The standard contains additional information regarding the performance of difficult venipunctures, collection procedures for blood culture and other sterile techniques, and for specimen collections from patients in isolation.

Venous occlusion. It is rarely necessary to leave a tourniquet in place for longer than one minute, but even within this short time the composition of blood changes. When the flow of blood in the large veins is obstructed by a tourniquet, the filtration pressure across the capillary walls is increased. The increased filtration pressure causes fluid and compounds with molecular weights <5000 to pass through the capillary wall. The situation is analogous to that caused by the change of posture from lying to standing (i.e., a relative hemoconcentration). Although the changes that occur in one minute are slight, marked changes have been observed after 3 min (see Table 10-2). Note that these changes often exceed the analytical variability of the methods used to measure the constituents of the blood.

The composition of blood drawn first will be most representative of the composition of circulating blood. The first-drawn specimen should therefore be used for those tests pertinent to critical medical decisions (e.g., calcium). Blood drawn later, which was contained in the small veins and capillaries when the tourniquet was applied, shows a greater effect of venous stasis.[10] In general, the concentrations of proteins and protein-bound constituents are most influenced by stasis, and indeed, one study has demonstrated that 3-min stasis may be enough to increase the concentration of protein or protein-bound constituents by 15%. A uniform procedure for "order of draw" for tests should therefore be established.

Stresses associated with blood collection can have effects in patients at any age. Plasma concentrations of cortisol and growth hormone may increase. Particularly in young children who are frightened, struggling, and held in physical restraint, collection may cause adrenal stimulation leading to an increased plasma glucose concentration or create increases in the serum activities of enzymes that originate in skeletal muscle. Vigorous suction on a syringe during collection or forceful transfer from the syringe to the receiving vessel may cause hemolysis of blood. Hemolysis is generally less when blood is drawn through a small-bore needle because turbulence of the blood is less than when a larger-bore needle is used.

Skin Puncture

If only a small volume of blood is required for a blood test, venipuncture is unnecessary. However, a skin puncture is time consuming and there is a greater risk of infection than from a venipuncture, since the fingertip is more difficult to sterilize than the antecubital fossa. To improve circulation of the blood, the site of puncture may be

Table 10-2. CHANGES* IN COMPOSITION OF SERUM WHEN VENOUS OCCLUSION IS PROLONGED FROM 1 MIN TO 3 MIN[18]

Increase	%	Decrease	%
Total protein	4.9	Potassium	6.2
Iron	6.7		
Total lipids	4.7		
Cholesterol	5.1		
Aspartate aminotransferase	9.3		
Bilirubin	8.4		

* In order to evaluate the magnitude of the effect on results, relate the per cent increase or decrease shown in the table to the analytical variation (±CV) routinely found for analytes.

Mean values obtained from 11 healthy individuals.

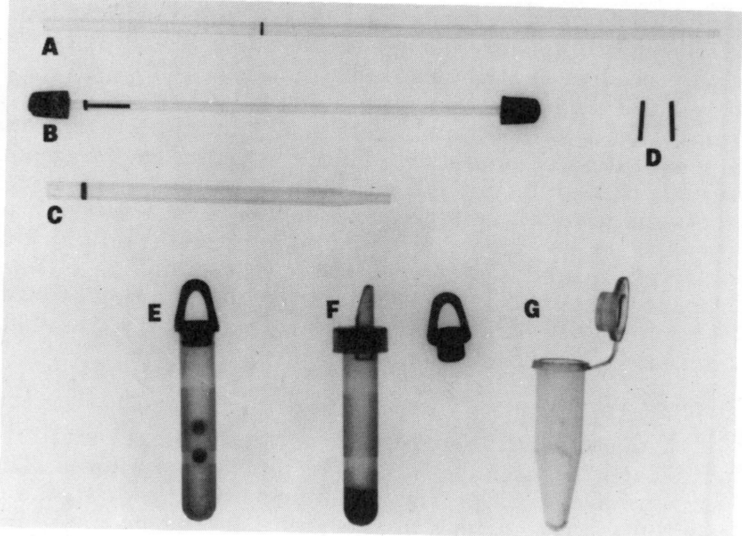

Figure 10-2. Representative devices for collection of capillary specimens. *A,* Natelson microcollection tube. *B,* Heparinized capillary tube containing metal mixing bar, used for specimens for blood gas analysis. *C,* Caraway tube. *D,* Metal mixing bar for mixing blood in capillary tubes. *E,* Plastic microtube containing anticoagulant and two inert mixing beads. *F,* Plastic microtube with serum separator gel and collecting device to improve blood flow into tube. *G,* Plastic microcentrifuge tube.

warmed by application of a warm, wet washcloth for three minutes prior to pricking. The skin is cleaned as for a venipuncture, and when dry it is quickly punctured by a sharp stab with a lancet. The depth of the incision should be <2.5 mm to avoid contact with bone.[4] To minimize the possibility of infection, a different site should be selected for each finger puncture. The finger should be held in such a way that gravity assists the collection of blood on the finger tip. Massage of the finger to stimulate blood flow should be avoided, because it causes the outflow of debris or of tissue fluid, which does not have the same composition as plasma. The first drop of blood is wiped off and subsequent drops are transferred to the appropriate collection tube held at a slight angle to the skin.

Blood may be collected into capillary blood tubes by capillary attraction. A variety of collection tubes is commercially available (Figure 10-2). The Natelson tubes (280-μL capacity, 1.5–1.6 mm ID) and Caraway tubes (350-μL, 2.5–2.7 mm ID) are widely used. Such containers can be purchased with different anticoagulants, such as sodium and ammonium heparin, and some can be obtained in brown glass for collection of light-sensitive analytes such as bilirubin.

In an *infant* less than one year old, the lateral or medial plantar surface of the foot should be used for skin puncture; suitable areas are illustrated in Figure 10-3. In older children the plantar surface of the big toe may also be used, although blood collection on ambulatory patients from anywhere on the foot should be avoided. Blumenfeld and coworkers[4] have described in detail the suitable locations for drawing blood from the feet of infants. The proper procedure for collecting blood from infants is also contained in an NCCLS standard.[12]

For the collection of blood specimens on filter paper for neonatal screening, the skin is cleaned and punctured as described previously. Then the filter paper is gently touched

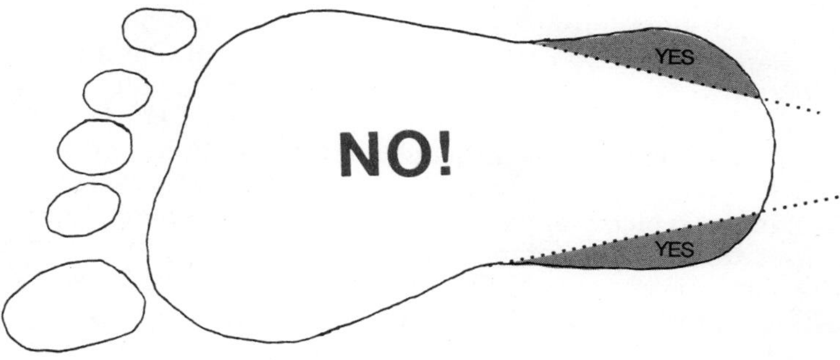

Figure 10-3. Acceptable sites for skin puncture to collect blood from an infant's foot. (Modified from Blumenfeld, T. A., Turi, G. K., and Blanc, W. A.: Lancet, *1*:230–233, 1979.)

against a large drop of blood, which is allowed to soak into the paper to fill each marked circle and completely penetrate the paper. Avoid milking or squeezing the foot, since this procedure contributes tissue fluids.

Blood from skin puncture is more like arterial than venous blood; also, it is contaminated to some extent by interstitial and intracellular fluids. Although the potassium concentration of interstitial fluid is as much as 0.3 mmol/L less than that of venous serum, potassium in capillary blood may be higher than in venous blood because of contamination with tissue debris. Total protein concentration in plasma is more than three times the concentration found in interstitial fluid. The low capillary blood protein observed in analyses may therefore reflect mixing of interstitial fluid with capillary blood. Such dilution also reduces the concentration of protein-bound constituents such as calcium and bilirubin. Therefore, variations in concentration of an analyte will depend on relative amounts of contamination.

Arterial Puncture

Arterial punctures require considerable skill and are usually only performed by physicians or specially trained technicians or nurses. The preferred sites of arterial puncture are, in order, the radial artery at the wrist, the brachial artery in the elbow, and the femoral artery in the groin. In the neonate, an indwelling catheter in the umbilical artery is best to obtain arterial specimens (e.g., for blood gas analysis).

In patients in whom it is impossible to perform an arterial puncture, *arterialized capillary blood* may be obtained from a skin puncture after warming the puncture site with a hot, moist towel to improve the circulation. Such a specimen yields acceptable values for pH and pCO_2, but not always for pO_2. Capillary blood specimens are particularly inappropriate when the blood circulation is poor and thus should be avoided when a patient has reduced cardiac output, hypotension, or vasoconstriction. The puncture itself should be performed as described previously. A free flow of blood is essential, and heparinized capillary tubes containing a small metal bar should be used to collect the blood. The tubes should be sealed quickly and the contents mixed well by using a magnet to move the metal bar up and down in the tube so that a uniform specimen is available for analysis (see Figure 10-2).

Anticoagulants and Preservatives for Blood

If whole blood or plasma is desired for testing, an anticoagulant must be added to the specimen during the collection procedure. Whole blood is rarely required for clinical chemistry tests; only for blood gas and lactate determinations is it the preferred specimen. Serum from coagulated blood is the specimen of choice for many assay systems, but plasma obtained with an appropriate anticoagulant may be an equally valid specimen, and, in certain circumstances, preferable to serum. Because harvest of serum must wait 15–30 min for completion of coagulation before centrifugation, the use of plasma expedites analysis in medical emergencies. Furthermore, plasma yield from a given volume of whole blood is always greater than the yield of serum. A disadvantage is the formation of fibrin clots or fragments during plasma storage and the subsequent risk of clogging the sample probes of automated analytical instruments.

Heparin is the most widely used anticoagulant for clinical chemical tests and it causes the least interference. This anticoagulant probably acts to prevent the transformation of prothrombin into thrombin and thus the formation of fibrin from fibrinogen. Although ammonium heparin has some effect on erythrocyte volume, other heparin salts appear to have none. Either lithium or ammonium heparin is required when plasma is obtained for the analysis of sodium and potassium. Although lithium is often used as an internal standard for flame photometry, the amount of lithium added as lithium heparinate has an insignificant effect on sodium and potassium measurements.

Ethylenediaminetetraacetic acid (EDTA), a chelating agent, in concentrations of 1–2 mg/mL blood, is particularly useful for hematological examinations because it preserves the cellular components of blood. It is used as the disodium or dipotassium salt; the latter is more soluble. EDTA prevents coagulation by binding calcium, which is essential for the clotting process.

Sodium fluoride is usually considered as a preservative for blood glucose, together with an anticoagulant such as potassium oxalate; however, fluoride also acts as a weak anticoagulant. It exerts its preservative action by inhibiting the enzyme systems involved in glycolysis. Fluoride is also a potent inhibitor of urease and of many serum enzymes. Fluoride is the best available preservative for glucose. Most specimens will be preserved at 25 °C for 24 h or at 4 °C for 48 h, whereas without an antiglycolytic agent, the blood glucose concentration decreases as much as ~10 mg/dL (0.56 mmol/L) per hour at 25 °C. The rate of decrease is even faster in newborns because of the increased metabolic activity of their erythrocytes and in leukemic patients because of the high metabolic activity of their leukocytes.

Sodium citrate solution has little application in clinical chemistry. It is widely used for coagulation studies since its effect is easily reversible by the addition of Ca(II) and it appears to preserve labile procoagulants.

Sodium, potassium, ammonium, and lithium *oxalates* inhibit blood coagulation by forming insoluble complexes with calcium ions. Although potassium oxalate ($K_2C_2O_4 \cdot H_2O$) is the most widely used oxalate, it causes a reduction in hematocrit because of fluid loss and the exchange of electrolytes across the erythrocyte membranes. In addition, oxalate inhibits several enzymes, including acid and alkaline phosphatases, amylase, and lactate dehydrogenase.

Sodium iodoacetate at a concentration of 2 mg/mL blood is an effective antiglycolytic agent and a substitute for sodium fluoride. Since it has no effect on urease, it can be used when glucose and urea are performed on a single specimen. It inhibits creatine kinase but appears to have no significant effects on other clinical chemistry tests.

Hemolysis

Serum shows visual evidence of hemolysis when the hemoglobin concentration exceeds 20 mg/dL. Slight hemolysis has little effect on most values. Severe hemolysis causes a slight dilutional effect on those constituents present at a lower concentration in the erythrocytes than in plasma. However, a marked effect may be observed on those constituents that are present at a higher concentration in erythrocytes than in plasma. Thus, the concentrations or activities of aldolase, total acid phosphatase, lactate dehydrogenase, isocitrate dehydrogenase, potassium, and magnesium are particularly affected. Furthermore, the inorganic phosphate in serum increases rapidly as the organic esters released from the cells are hydrolyzed. Aspartate aminotransferase activity increases by 2% with each 10 mg hemoglobin/dL. In colorimetric procedures for cholesterol without extraction, 10 mg hemoglobin/dL will raise the apparent cholesterol concentration by 5 mg/dL. Ten milligrams hemoglobin/dL will increase serum lactate dehydrogenase by ~10% and serum potassium by ~ 0.6%. Serum protein electrophoresis may show an additional band due to hemoglobin.

Although the amount of free hemoglobin could be measured and a calculation made to correct test values affected by hemoglobin,[5] this practice is undesirable because factors other than hemoglobin could contribute to the altered test values and it would be impossible to assess their effect.

Urine Collection

The type of urine specimen to be collected is dictated by the tests to be performed. Generally, urine specimens for quantitative analyses must be collected over an interval of time, such as 1, 4, or 24 h. Untimed or random specimens are suitable for only a few chemical tests. A *clean, early morning, fasting specimen* is generally the most concentrated specimen and thus is preferred for microscopic examinations and for the detection of abnormal amounts of constituents such as proteins, or for quantitation of unusual compounds such as β-chorionic gonadotropin. The *clean, timed specimen* is one obtained at specific times of the day or during certain phases of micturition. The *double-voided specimen* is the urine excreted during a timed period following a complete emptying of the bladder; it is used, for example, to assess glucose excretion during a glucose tolerance test. Its collection must be timed in relation to the ingestion of glucose.

Catheter specimens, used for microbiological examination in critically ill patients or in those with urinary tract obstruction, should not normally be obtained just for examination of chemical constituents. The *suprapubic tap specimen* is a useful alternative, because the tap is unlikely to cause infection. Details of the collection of urine specimens are contained in an NCCLS standard currently in preparation.

The collection period for *timed urine specimens* must be long enough to minimize the influence of short-term biological variations. When specimens are to be collected over a specified period of time, the patient's close adherence to instructions is important. Before beginning a timed collection a patient should be given written instructions with regard to diet or medication, if appropriate, to avoid the interference of ingested compounds with analytical procedures. Appropriate information regarding the collection, including warnings with respect to handling of the specimen, should appear on the bottle label.

Urine cannot be collected at the same time for two or more tests for which different preservatives are required. Aliquots for an analysis such as a microscopic examination may not be removed while a 24-h collection is in progress. The removal of aliquots is not permissible even when the volume removed is measured and corrected for, since the excretion rate of most compounds varies throughout the day, and the effect of removal of some specimen on the total concentration cannot be corrected.

Urine preservatives. The most common preservatives and the tests for which the preservatives are required are listed in Table 10-3. Preservatives have different roles but are usually added to reduce bacterial action or chemical decomposition, or to solubilize constituents that might otherwise precipitate out of solution. Another use of preservatives is to decrease atmospheric oxidation of unstable compounds. Some specimens should not have *any* preservatives added because of the possibility of interference with analytical methods.

One of the most satisfactory forms of preservation of urine specimens is *refrigeration* during collection. *Acidification* to below pH 3 is widely used to preserve 24-h specimens and is particularly useful for specimens for calcium, steroid, and vanillylmandelic acid determinations. However, precipitation of urates will occur, thereby rendering a specimen unsuitable for measurement of uric acid. *Sulfamic acid* and *boric acid* have been used, but they also cause precipitation of urates. Although *thymol* and *chloroform* were widely used in the past to preserve specimens for routine qualitative and microscopic urinalysis, it is now recognized that specimens for these tests should be analyzed immediately and that the addition of preservatives is both largely ineffective and a source of interference with several methods. *Toluene* is the only organic solvent that is still used as a preservative. When present in a large enough amount, it acts as a barrier between the air and the surface of the specimen. Toluene, however, will not prevent the growth of anaerobic microorganisms, and because of its flammable nature it is a safety hazard. *Sodium bicarbonate* is used to preserve porphyrins and urobilinogen.

When a timed collection is complete, the specimen should be delivered without delay to the clinical laboratory, where the volume should be measured. This may be done by using graduated cylinders or, when preweighed or uniform containers are used, by weighing the container and urine. The mass in grams (g) may be reported as if it were the volume in milliliters (mL). There is rarely a need to measure the specific gravity of a weighed specimen since errors in analysis usually exceed the error arising from failure to correct the volume of urine for its mass.

Before a specimen is transferred into small containers for each of the ordered tests, it must be thoroughly mixed to assure homogeneity. The small container into which an aliquot is transferred should not be a plastic bottle if toluene or another organic compound has been used as preservative; metal-free containers must be used for trace metal analyses.

Collection of Feces

Small aliquots of feces are frequently analyzed to detect the presence of *occult (hidden) blood,* which is recognized as one of the most effective clues to the presence of a bleeding ulcer or malignant disease in the gastrointestinal tract (see Chapter 24). Tests

Table 10-3. COMMONLY USED URINE PRESERVATIVES

Analyte	**None**	**Freeze**	**Glacial Acetic Acid, 33% (20 mL/24 h)**	**Boric Acid (5 g/24 h)**	**Toluene (cover surface of specimen)**	**Concentrated Hydrochloric Acid, 6 mol/L (10 mL/24 h)**	**Sodium Bicarbonate (5 g/24 h)**	**Petroleum Ether (100 mL/24 h)**
Aldosterone		X	X	X				
Amino acids					X	X		
δ-Aminolevulinic acid			X			X		
Calcium						X		
Catecholamines			X			X		
β-hCG	X			X				
Citrate				X	X			
Copper	X					X		
Cortisol (free)	X		X	X				
Cystine		X			X	X		
Estrogens			X	X				
Homogentisic acid	X							
Homovanillic acid			X	X		X		
5-Hydroxyindoleacetic acid				X		X		
Hydroxyproline					X	X		
17-Ketosteroids			X	X				
Magnesium	X					X		
Mercury	X					X		
Metanephrines			X			X		
Nitrogen	X					X		
Osmolality	X	X						
Oxalate	X					X		
Porphyrins							X	
Pregnanetriol			X	X				
Uric acid	X				X			
Urobilinogen	X	X						X
Vanillylmandelic acid				X		X		
Zinc	X					X		

for occult blood should be done on aliquots of excreted stool rather than on material obtained on the glove of a physician doing a rectal examination, since this procedure may cause enough bleeding to produce a positive result, or, conversely, the specimen may not be representative of the whole, so that bleeding may not be recognized.

Collection of Spinal Fluid

Spinal fluid is normally obtained from puncture of the lumbar region, although occasionally fluid obtained during surgery from the cervical region or from a cistern or ventricle of the brain may be submitted for analysis. Spinal fluid is examined when there is suspicion of a cerebrovascular accident, meningitis, demyelinating disease, or meningeal involvement in malignant disease. Lumbar punctures are always performed by a physician.

Up to 20 mL of spinal fluid can be safely removed from an adult, although this amount is not usually required. Antiglycolytic agents are generally not added to the tube for glucose measurement, but since rapid processing of specimens is a clinical requirement for tests on spinal fluid, little metabolism of glucose occurs even in the presence of many bacteria. To allow proper interpretation of a spinal fluid glucose value, a blood specimen should be obtained simultaneously.

Collection of Other Fluids and Tissues for Analysis

Synovial fluid aspiration. The technique of obtaining synovial fluid for examination is called *arthrocentesis.* Synovial fluid is withdrawn from joints to aid characterization of the type of arthritis and to differentiate noninflammatory effusions from inflammatory fluids. Normally, only a very small amount of fluid is present in any joint, but this volume is usually very much increased in the presence of inflammatory conditions. Arthrocentesis is performed by a physician. Sterile, plain tubes should be used for culture and for glucose and protein measurements; an EDTA tube is needed for a total leukocyte, differential, and erythrocyte count. Microscopic slides are prepared for staining with Gram's or other stains and for visual inspection.

Amniocentesis. Amniocentesis is performed by a physician for prenatal diagnosis of congenital disorders, to assess fetal maturity, or to detect Rh isoimmunization or intrauterine infection.

If the lecithin/sphingomyelin (L/S) ratio is to be measured, the container is immediately placed in ice. If spectrophotometric analysis of bilirubin is to be performed, the specimen should be transferred to a brown tube or the tube should be wrapped in foil to prevent photodegradation.

Pleural, pericardial, and ascitic fluid collection. The pleural, pericardial, and peritoneal cavities normally contain a small amount of serous fluid that lubricates the opposing parietal and visceral membrane surfaces. Inflammation or infections affecting the cavities cause fluid to accumulate. The fluid may be removed to determine if it is an effusion or an exudate, a distinction made possible by protein or enzyme analysis. The collection procedure is called *paracentesis.* When specifically applied to the pleural cavity, the procedure is a *thoracentesis;* when applied to the pericardial cavity, a *pericardiocentesis.* Pericardiocentesis has now been largely supplanted by echocardiography. Paracenteses are performed only by physicians.

Collection of saliva. Although measurements of concentrations of certain analytes in saliva have been advocated, there is presently no clinical justification for such assays, with the possible exception of blood group substances to determine secretor status and Le genotype. The measurement of a drug in saliva has been suggested to estimate the free, pharmacologically active concentration of the drug in serum. There is, however, a considerable difference in pH between saliva and serum, and ratios of bound drug to free drug would not be the same.

Preparation of solid tissue. The solid tissue most often analyzed in the clinical chemistry laboratory is malignant tissue from the breast for determination of estrogen and progesterone receptors. During surgery, at least 0.5–1 g tissue is removed and trimmed of fat and nontumor material. The tissue should be frozen, preferably in liquid nitrogen or in a mixture of dry ice and alcohol, within 20 min of collection. A histological section of the specimen should always be examined to confirm that it is indeed malignant tissue. The same procedure may be used to obtain and prepare solid tissues for toxicological analysis. When trace element determinations are to be made, all materials used in the collection or handling of the tissue should be made of plastic or materials known to be free of contaminating trace elements (see also Chapter 17).

Handling of Specimens for Testing

Maintenance of Specimen Identification

Valid test results require a representative and properly collected, preserved, and identified specimen. Every specimen container must be adequately labeled even if the specimen must be placed in ice or if the container is small, such as a capillary blood tube. Direct labeling of a capillary blood tube by folding the label like a flag around the tube is preferred; a less satisfactory alternative is to label a larger container into which the capillary tube can be placed. The minimum information on a label should include the patient's name, location, and identifying number, as well as the date and time of collection. A system of color-coded tags should also be used to identify tubes that contain specimens known to come from patients with infectious diseases.

Preservation of Specimens in Transit

The specimen must be properly treated both during its transport to the laboratory and from the time the serum has been separated until it is analyzed. For some tests, specimens must be kept at ~4 °C from the time the blood is drawn until the specimens are analyzed or until the serum or plasma is separated from the cells. Examples are specimens for ammonia (Chapter 23) and blood gas determinations such as pCO_2, pO_2, and blood pH (see Chapter 19, Section Two). Transfer of these specimens to the laboratory must be done by placing the specimen container in ice water. Specimens for acid phosphatase, lactate and pyruvate, and certain hormone tests (e.g., gastrin, renin activity, and parathyroid hormone) should be treated the same way.

For all test constituents that are thermally labile, serum and plasma should be separated from cells in a refrigerated centrifuge. Specimens for bilirubin or carotene must be protected from both daylight and fluorescent light to avoid photodegradation. Some special handling requirements are listed in Table 10-4. If blood tubes are well filled they may be sent to a clinical laboratory via pneumatic tube, without producing significant hemolysis.

Table 10-4. SELECTED BLOOD CONSTITUENTS REQUIRING SPECIAL COLLECTION AND STORAGE CONDITIONS[23]*†

Constituent	Anticoagulant	Handling Requirements
S-Acetone, acetoacetate†‡		Freeze; stopper
S-Acid phosphatase		Add citrate (10 mg/mL); freeze
P-ACTH (corticotropin)	Heparin, 0.2 mg/mL	Freeze within 15 min of collection
S-Alcohol		Add NaF (10 mg/mL) to serum
S-Aldolase		Freeze
P,S-Aldosterone		Add boric acid (25 mg/mL) or freeze
P-Amino acids	Heparin, 0.2 mg/mL	Freeze
S-Androstenedione		Draw sample in A.M. (0000–1200 h)
S-Ascorbic acid		Freeze
S-Barbiturates		Do not use heparin
S-Bile acids		Freeze
S-C-Peptide		Draw from fasting patient; freeze
S-Calcitonin		Freeze

Continued on following page

Table 10-4. SELECTED BLOOD CONSTITUENTS REQUIRING SPECIAL COLLECTION AND STORAGE CONDITIONS[23]*†(*Continued*)

Constituent	Anticoagulant	Handling Requirements
P-Carcinoembryonic antigen (CEA)	EDTA, 2 mg/mL	
B-Cholinesterase	Heparin, 0.2 mg/mL	
S-Citric acid		Freeze
S-Complement		Freeze
P-Cortisol	Heparin, 0.2 mg/mL	Separate immediately
S-Creatine		Freeze
S-Creatine kinase isoenzymes		Freeze
S-Creatinine		Freeze
S-Cryoglobulins		Keep above 20 °C
P-11-Deoxycortisol	Heparin, 0.2 mg/mL	Separate immediately
P,S-Digitoxin		Draw 6–12 h postadministration
S-Digoxin		Draw 8 h postadministration
P-Estradiol	Heparin, 0.2 mg/mL	Freeze
P-Fatty acids (free and esterified)	Heparin, 0.2 mg/mL	Freeze
P-Fibrinogen	Citrate, ~4 mg/mL	Do not use heparin
S,B-Fluoride		Do not collect in glass container
P-Folate (tetrahydrofolate)		Freeze
B-Galactose	Heparin, 0.2 mg/mL	NaF (10 mg/mL) added with heparin
S-Gastrin†		Collect from fasting patient; freeze
B-Glucose-6-phosphate dehydrogenase†	EDTA, 2 mg/mL	
P-HDL-cholesterol	EDTA, 2 mg/mL	Collected after 12–14 h fast; freeze
B-Hemoglobins (for quantitation and electrophoresis)	EDTA, 2 mg/mL	
S-Histidine		Freeze
S-17-Hydroxyprogesterone		Draw between 0900 and 1100 hours
S-Insulin		Collect from fasting patient; freeze
S-Isocitrate dehydrogenase		Freeze
B-Lactate		Immediately dilute with an equal volume of 5% (50 g/L) perchloric acid; shake and mix
B-Lead	Heparin, 0.2 mg/mL	Collect in lead-free tube containing heparin
S-Lipoprotein phenotyping		Freeze
S-Lysozyme		Freeze
S-Magnesium		Separate immediately
S-Parathyroid hormone (PTH)		Freeze
S-Pepsinogen		Freeze
S-Placental lactogen		Freeze
S-Prolactin		Freeze
S-Prostaglandin $F_{2\alpha}$		Freeze
P-Pyridoxal phosphate (vitamin B_6)	EDTA, 2 mg/mL	Protect from light
B-Pyruvate		As B-lactate
P-Renin	EDTA, 2 mg/mL	Chill during collection, centrifugation
S-Vitamin A		Protect from light
S-Zinc		Use only acid-washed glass; avoid hemolysis

P = plasma, S = serum, B = whole blood, EDTA = ethylenediaminetetraacetate.

*All specimens listed, except those to be assayed for cryoglobulins, should be transported at temperatures below 15 °C, except as noted.

†A sealable Styrofoam container with "freezer packs" is usually sufficient to keep specimen frozen for 12 h. Solid CO_2 is necessary for longer periods.

‡There is some question as to whether this constituent remains stable even with this procedure.

Separation and Storage of Specimens

Plasma or serum should be separated from cells as soon as possible, and certainly within 2 h. If this is impossible, the specimen should be held at room temperature rather than at 4 °C to decrease hemolysis. If the specimens cannot be analyzed at once, the separated serum should generally be stored in capped tubes at 4 °C until analysis, both to maintain stability of the specimen and to reduce evaporation. If a specimen for a particular test is sufficiently unstable at 4 °C, the serum specimen should be held at −20 °C in a freezer capable of maintaining this temperature. Note, however, that neither 4 °C nor −20 °C is an optimal storage temperature for all tests; some lactate dehydrogenase isoenzymes, for instance, are more stable at room temperature than at 4 °C.

Specimen tubes should be centrifuged with their stoppers in place to reduce evaporation and prevent aerosolization of infectious particles. Specimen tubes containing volatiles such as ethanol *must* be stoppered while they are spun. Centrifuging specimens with the stopper in place maintains anaerobic conditions that are important in the measurement of carbon dioxide, acid phosphatase, and ionized calcium. Removal of the stopper before centrifugation allows loss of carbon dioxide and an increase in blood pH.

Transport of Specimens to Referral Laboratories

It should be assumed that transport from a referring laboratory to a referral laboratory may take as long as 72 h. A referring laboratory should retain enough specimen under optimal conditions for retesting, should an unanticipated problem arise during shipment. The tube used for holding a specimen (*primary container*) should be so constructed that the contents will not escape if the container is exposed to extremes of heat, cold, or sunlight.[24] Reduced pressure of 0.50 atmosphere may be encountered during air transportation, together with vibration, and specimens should be protected by a suitable container from these adverse effects. Variability of temperature is the biggest cause of instability of test constituents.[6]

The shipping or *secondary container* used to hold one or more specimen tubes or bottles must be constructed to prevent the tubes from knocking against each other. A padded shipping envelope provides adequate protection for shipping single specimens. When specimens are shipped as drops of blood on filter paper (e.g., for neonatal screening), the paper should be enclosed in a plastic bag, which then can be mailed in a regular envelope.

For transportation of *frozen* or *refrigerated* specimens, a Styrofoam container should be used. Dry ice is the most convenient refrigerant material for maintaining specimens frozen, and temperatures as low as −70 °C can be achieved. The amount of dry ice required in a container depends on the size of the container and the efficiency of its insulation, as well as the length of time for which the specimens must be kept frozen. A piece of dry ice about 3 inches × 4 inches × 1 inch in a container with 1-inch Styrofoam walls and a volume of 125 cubic inches will maintain a specimen frozen for 48 h.

Various laws and regulations apply to the shipment of biological specimens. Although these rules theoretically apply only to etiologic agents (known infectious agents), all specimens should be transported as if the same regulations applied. Airlines have rigid regulations covering the transport of specimens and consider dry ice a hazardous material, so the transport of most clinical laboratory specimens requires accommodation to their requirements.

Physiological Factors Affecting the Composition of Body Fluids

Standardization of specimen collection practices minimizes the variables that cause changes in test values within one day or from one day to another and thereby reduces difficulty in interpretation of values. In hospital practice, however, standardization is not always possible; thus, one must understand the effects of both controllable and uncontrollable variables on the composition of body fluids.

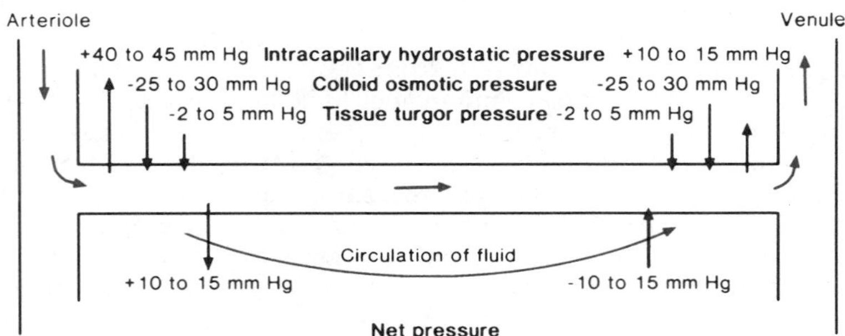

Figure 10-4. Pressure relationships across blood capillary affecting the transfer of fluids and solutes. (Reproduced with permission from Pitts, R. F.: Physiology of the Kidney and Body Fluids, 3rd ed. Chicago, Year Book Medical Pub., Inc., 1974.)

Controllable Biological Variables

Posture. The blood volume of an adult in an upright position is typically 600–700 mL less than when the person is lying down. The reduced plasma volume is associated with a comparable increase in the plasma protein concentration. The concentrations of all proteins, including enzymes and hormones, and of protein-bound compounds such as drugs, calcium, and bilirubin, are also affected.

The transfer of fluid and solute across a capillary wall depends on the interaction of hydrostatic and osmotic pressures of the capillary and interstitial fluids. Fluid moves into the interstitial space at the arteriolar end of the capillary and returns to the capillary at the venular end. A greater volume of fluid leaves the capillary at the arteriolar end than is returned to it at the venous end. The excess drains into the lymphatic system. When an individual lies down, more fluid returns to the capillaries because the capillary pressure is reduced. The pressure relationships involved in fluid transfer across capillaries are illustrated in Figure 10-4. Changes in the concentration of some major serum constituents brought about by change in posture are listed in Table 10-5.

Hospitalization and immobilization. With prolonged bed rest, fluid retention occurs and serum concentrations of protein and albumin and protein-bound constituents may be decreased. Calcium is mobilized from bones and the increased ionized fraction

Table 10-5. CHANGE IN CONCENTRATION OF SERUM CONSTITUENTS WITH CHANGE FROM LYING TO STANDING[7]*

Constituent	Average Increase, %
Alanine aminotransferase	7
Albumin	9
Alkaline phosphatase	7
Amylase	6
Aspartate aminotransferase	5
Calcium	3
Cholesterol	7
IgA	7
IgG	7
IgM	5
Thyroxine	11
Triglycerides	6

* In order to evaluate the magnitude of the effect on results, relate the per cent increase shown in the table to the analytical variation (±CV) routinely found for analytes.

compensates for the reduced protein-bound calcium, so serum total calcium is less affected. The serum activity of enzymes arising from skeletal muscle is usually slightly less in individuals confined to bed than in those undertaking normal physical activity. Prolonged bed rest is also associated with increased urinary nitrogen and calcium excretion. When an individual becomes active after a period of bed rest, more than three weeks are required before calcium excretion reverts to normal.

Exercise. The influence of exercise on the composition of body fluids is related to the duration and intensity of the activity. Only the effect of moderate exercise will be described here, as patients are unlikely to have exercised strenuously. With moderate exercise, the provoked stress-response causes an increase in blood glucose, which stimulates insulin secretion. The arteriovenous difference in glucose concentration is increased by the greater tissue demand for glucose. Lactate concentration may be increased twofold by the increased metabolic activity of skeletal muscle. Arterial pH and pCO_2 are reduced by exercise. Competition between uric acid and lactate and products of increased tissue catabolism for renal excretion causes the serum urate concentration to increase. Exercise causes a reduction of cellular ATP, which increases cellular permeability, causing slight increases in the serum activities of enzymes originating from skeletal muscle, such as aspartate aminotransferase, lactate dehydrogenase, creatine kinase, and aldolase.[19] As little as five minutes of walking will increase the activity of these enzymes in plasma.

Physical training. Athletes generally have a higher serum activity of enzymes of skeletal muscular origin than do nonathletes. However, the release of these enzymes in response to exercise is less in athletes than in other individuals. Serum concentrations of urea, urate, creatinine, and thyroxine are higher in athletes than in comparable untrained individuals.[25]

The total serum lipid concentration is reduced by physical conditioning. Total serum cholesterol may be lowered by as much as 25%, whereas HDL-cholesterol increases. Thus, the decrease in total cholesterol is mostly due to reduction in LDL-cholesterol. The serum triglyceride concentration may be reduced by up to 20 mg/dL (0.23 mmol/L), but the free fatty acid concentration is higher in fit individuals than in those who do not exercise.

Circadian variation. Many constituents of body fluids exhibit cyclical variations throughout the day. Factors contributing to such variations include posture, activity, food ingestion, stress, daylight or darkness, and sleep or wakefulness. Some of these cyclical variations may be quite large. The concentration of serum iron, for example, may change by as much as 50% from 0800 to 1400 h, and that of cortisol by a similar amount from its peak around 0600–0800 h to its trough around 1600 h.[20] The typical total variation of several commonly measured serum constituents over 6 h is illustrated in Table 10-6. The total variation is contrasted with analytical error.

Maximum renin activity normally occurs early in the morning during sleep; the minimum occurs late in the afternoon. Glomerular filtration rate varies inversely with the secretion of renin; GFR is least at the time of maximum renin secretion and ~20% greater in the afternoon, when renin activity is at a minimum.

The excretion of 17-ketosteroids and 17-hydroxycorticosteroids is low at night and reaches a maximum about midafternoon. A 20 to 40% increase of plasma testosterone occurs during the night. Prolactin concentration is also greatest during sleep. The serum TSH is at a maximum between 0200 and 0400 h and at a minimum between 1800 and 2200 h. The variation is of the order of 50%. There are also variations in the serum thyroxine concentration, but these appear to be related to the changes in concentration of binding protein brought about by changes in posture.

Growth hormone secretion is greatest shortly after sleep commences. Conversely, basal plasma insulin is higher in the morning than later in the day, and its response to glucose is also greatest in the morning and least about midnight. Thus, when a glucose tolerance test is given in the afternoon, higher glucose concentrations occur than when the test is given early in the day. The higher plasma glucose occurs in spite of a great insulin response which, however, is delayed and less effective.

Table 10-6. TOTAL AND ANALYTICAL VARIATION FOR SERUM TESTS ON SPECIMENS OBTAINED AT 0800 AND 1400 h[22]*

Constituent	Mean	Total Variation, %†	Analytical Variation, %
Sodium, mmol/L	141	1.9	1.8
Potassium, mmol/L	4.4	7.1	2.8
Calcium, mg/dL	10.8	3.2	2.7
Chloride, mmol/L	102	3.8	3.4
Phosphate, mg/dL	3.8	10.7	2.4
Urea N, mg/dL	14	22.5	2.5
Creatinine, mg/dL	1.0	14.5	6.3
Uric acid, mg/dL	5.6	11.5	2.6
Iron, μg/dL	116	36.6	3.4
Cholesterol, mg/dL	193	14.8	5.7
Albumin, g/dL	4.5	5.5	3.9
Total protein, g/dL	7.3	4.8	1.7
Total lipids, g/L	5.3	25.0	3.6
Aspartate aminotransferase, U/L	9	25	6
Alanine aminotransferase, U/L	6	56	17
Acid phosphatase, U/L	3	15	8
Alkaline phosphatase, U/L	63	20	3
Lactate dehydrogenase, U/L	195	16	12

* In order to evaluate the magnitude of the effect on results, relate the per cent increase or decrease shown in the table to the analytical variation (±CV) found for analytes.

† 11 male subjects, aged 21–27 years, studied at 0800, 1400 h.

Influence of Food, Drugs, and Stimulants

Recent food ingestion. The concentration of certain plasma constituents is affected by the ingestion of a meal. The biggest increases in serum concentrations or activities occur for glucose, iron, total lipids, and alkaline phosphatase. The increase in alkaline phosphatase (mainly intestinal isoenzyme) is greater when a fatty meal is ingested and is influenced by the blood group of the individual and the substrate used for the enzyme reaction. Lipemia may affect some analytical methods, but ultracentrifugation or the use of serum blanks can reduce the adverse analytical effects of lipemia.

The effects of a meal may be long lasting. Thus, ingestion of a protein-rich meal in the evening may cause increases in the serum urea nitrogen, phosphorus, and urate concentrations that are still apparent 12 h later. Nevertheless, these changes may be less than the typical intraindividual variability. Large protein meals at lunch or in the evening also increase the serum cholesterol and growth hormone concentrations for at least 1 h after a meal. The effect of carbohydrate meals on blood composition is less than that of protein meals. Glucagon and insulin secretions are stimulated by a protein meal, and insulin secretion is stimulated by carbohydrate meals.

Ingestion of specific foods. *Caffeine,* which is contained in many beverages including coffee, tea, and colas, has considerable effect on the concentration of blood constituents. Caffeine stimulates the adrenal medulla, causing an increased excretion of catecholamines and their metabolites, a slight increase in the plasma glucose concentration, and impairment of glucose tolerance.[2] The adrenal cortex is also affected; plasma cortisol is increased, accompanied by excretion of free cortisol and 11-hydroxycorticoids. The effect of caffeine may be so marked that the normal diurnal variation of plasma cortisol may be eliminated. Caffeine has a marked effect on lipid metabolism. Ingestion of two cups of coffee may increase the plasma free fatty acid concentration by as much as 30%. Caffeine is also a potent stimulant of gastric juice, hydrochloric acid, and pepsin secretion. Coffee has a diuretic effect and also increases the excretion of erythrocytes and renal tubular cells in the urine.

Drug administration. It is rare for a patient to be hospitalized without receiving some medication. Drugs may have both in vivo and in vitro effects on laboratory tests.

Comprehensive listings of the effects of drugs on laboratory tests have been published by Young and coworkers[27] and Young.[26] Only a few representative effects are discussed here.

Many drugs when administered intramuscularly cause sufficient muscle irritation to increase the release into the serum of enzymes such as creatine kinase, aldolase, and the skeletal muscle isoenzyme of lactate dehydrogenase. The increased activities may persist for several days after a single injection, and consistently high values may be observed during a course of treatment. Penicillin derivatives given intramuscularly are particularly likely to increase the activity of these enzymes. Opiates such as morphine or meperidine can cause spasm of the sphincter of Oddi. The spasm transmits pressure up the bile and pancreatic ducts, causing release of liver and pancreatic enzymes into the serum. Aspartate aminotransferase activity may be so increased that the level may suggest a myocardial infarction.

Oral contraceptives affect many different constituents measured in the clinical laboratory. Tests are affected by both the progestin and estrogen components. The overall effect depends on the proportion of the two components.

Diuretic drugs often cause a mild reduction of the plasma potassium concentration; hyponatremia may be observed. Hypercalcemia may occur with hemoconcentration. Thiazides cause hyperglycemia and reduce glucose tolerance, especially in diabetics. Thiazides may also cause prerenal azotemia with hyperuricemia due to decreased renal blood flow and GFR as a result of reduced blood volume.

Anticonvulsants have widespread effects on clinical laboratory tests, largely as a result of hepatic microsomal enzyme induction. The tests affected include many liver function tests as well as tests for adrenal function and bone metabolism. Erythrocyte metabolism may also be affected.

Effect of Underlying Medical Conditions on Laboratory Values

Some general clinical conditions have an effect per se on the composition of body fluids. These conditions may exist in addition to the primary disorder that prompted a patient's admission to the hospital.

Fever. Fever provokes many hormone responses.[16] Hyperglycemia occurs early and stimulates the secretion of insulin, which improves glucose tolerance; however, insulin secretion does not necessarily reduce the blood glucose concentration, since increased secretion of growth hormone and glucagon also occurs. Fever results in reduced serum thyroxine, as does acute illness when fever is not present. In response to increased corticotropin secretion, the plasma cortisol concentration is increased and its normal diurnal variation may be abolished. The urinary excretion of free cortisol, 17-hydroxycorticosteroids, and 17-ketosteroids is increased.

Glycogenolysis and a negative nitrogen balance occur with the onset of fever because of the decreased food intake and wastage of skeletal muscle that typically accompany fever. Although there is usually an increase in the blood volume with fever, the serum concentrations of creatinine and uric acid are usually increased. Aldosterone secretion is increased, which results in retention of sodium and chloride. The secretion of antidiuretic hormone also contributes to the retention of water by the kidneys.

Fever accelerates lipid metabolism. The serum concentrations of cholesterol, nonesterified fatty acids, and the other lipids may decrease initially, but within a few days the free fatty acid concentration may increase markedly. Fever is often associated with a respiratory alkalosis caused by hyperventilation. This pH increase causes a reduction of the plasma phosphate concentration, accompanied by an increased excretion of phosphate and other electrolytes.[1]

Shock and trauma. Regardless of the cause of shock or trauma, certain characteristic biochemical changes ensue. The typical stress response is an increased secretion of cortisol, aldosterone, renin, and catecholamines, as well as of growth hormone, glucagon, and insulin. Plasma lipid and glucose concentrations are also increased.

Immediately after an injury there is loss of fluid to extravascular tissues with the result of a decreased plasma volume. If the decrease is enough to impair circulation, glomerular filtration is reduced. Diminished renal function will lead to the accumulation of urea and other end products of protein metabolism in the circulation. In burned patients, serum total protein concentration falls by as much as 0.8 g/dL because of both loss to extravascular spaces and the catabolism of protein. Serum α_1-, α_2- and γ-globulin concentrations increase but not enough to compensate for the reduced albumin concentration. The plasma fibrinogen concentration responds dramatically to trauma and may double in two to eight days after surgery.

With tissue destruction, there is increased urinary excretion of major components of skeletal muscle. The muscle damage associated with trauma can markedly increase the activity of serum enzymes originating in skeletal muscle. Increased tissue catabolism requires increased oxygen consumption and also leads to the production of acid metabolites. Thus, blood lactate may increase two- to threefold. With tissue anoxia and impairment of renal and respiratory function, a metabolic acidosis develops.

Long-Term Biological Influences on Body Fluids

Published reference ranges are usually broad; however, the intraindividual variability is significantly less than the group or population variability. Indeed, the values obtained from one individual may be very different from those from another individual, yet both sets of values could fall within a predetermined reference range. Not only do the set points for different individuals appear to be different, but the extent by which values change from one occasion to another, even when specimens are obtained under standardized conditions, appears to differ among individuals. Both of these factors are probably determined genetically. The difference in set points underscores the fallibility of using a population reference range to associate small changes in laboratory data with a deviation from health for a given individual.

The influences of gender and age are discussed with the individual laboratory tests in various chapters of this book. The most important influences that determine the overall effect of age are the degree of sexual maturity and the amount of skeletal muscle mass of the individual. Generally, individuals can be considered in four groups—the newborn, the older child to puberty, the sexually mature adult, and the elderly adult.

Influences of Age, Gender, and Race

Newborn. The composition of the newborn's blood is affected by the maturity of the infant at birth. In the mature infant most of the hemoglobin is the adult form, hemoglobin A, whereas in the immature infant much of the hemoglobin may be the fetal form, hemoglobin F. Arterial blood oxygen saturation is very low initially. Newborns develop a metabolic acidosis that results from the accumulation of organic acids, especially lactic acid. The acid-base status, however, reverts to normal within 24 h.

In infants, even in the absence of disease, the concentration of bilirubin rises following birth and peaks about the third to fifth day of life. This physiological jaundice of the newborn rarely produces serum bilirubin values >5 mg/dL (85 μmol/L). Distinguishing this naturally occurring phenomenon from other conditions that produce neonatal hyperbilirubinemia may be difficult, and the chronological course of the hyperbilirubinemia is important.

The blood glucose concentration is low in newborns because of their small glycogen reserves and adrenal immaturity. Plasma calcium is high initially but falls by as much as 1.4 mg/dL (0.35 mmol/L) during the first day of life. The plasma urea nitrogen concentration decreases after birth as the infant synthesizes new protein, and the concentration does not begin to rise until tissue catabolism becomes prominent. The plasma amino acid concentration is low as a result of synthesis of tissue protein, although urinary excretion of amino acids may be quite high because of immaturity of the tubular reabsorptive mechanisms.

The serum thyroxine concentration of the healthy newborn, like that in the pregnant woman, is considerably higher than in the nonpregnant adult. After its birth, an infant secretes thyroid-stimulating hormone (TSH), which causes a further increase in the serum thyroxine concentration. This physiological hyperthyroidism gradually declines over the first year of life.

Childhood to puberty. Plasma protein concentrations increase after infancy, and adult concentration values are attained by the age of 10 y. The serum activity of most enzymes decreases during childhood to adult values by puberty or earlier, although the activity of alanine aminotransferase may continue to rise, at least in men, until middle age. Serum alkaline phosphatase activity is high in infancy but decreases during childhood before rising again with growth. The activity of the enzyme is better correlated with skeletal growth and sexual maturity than with chronological age; it is greatest at the time of maximal osteoblastic activity occurring with bone growth. The activity decreases rapidly after puberty, especially in girls.

The serum creatinine concentration increases steadily from infancy to puberty parallel to the development of skeletal muscle; until puberty there is little difference in the concentration between sexes. The serum uric acid concentration decreases from its high at birth until age 7–10 y, at which time it begins to increase, especially in boys, until about age 16 y.

The adult. The concentrations of most test constituents remain quite constant between puberty and menopause in women and between puberty and middle age in men. During the midlife years, serum total protein and albumin concentrations decrease slightly. There may be a slight decrease in the serum calcium concentration in both sexes. In men, the serum phosphorus decreases after age 20 y; in women, the phosphorus also decreases until menopause, when a marked increase takes place. The serum alkaline phosphatase begins to rise in women at the menopause so that in elderly women activity of this enzyme may actually be higher than it is in men.

Serum uric acid concentrations peak in men in their twenties and in women during middle age. Urea concentration increases in both sexes in middle age. The serum total cholesterol and triglyceride concentrations increase in both men and women at a rate of ~2 mg/dL (0.02 mmol/L) per year to a maximum between ages 50 and 60 y. The concentration of glucose in plasma 1 h after a loading dose rises ~8 mg/dL (0.44 mmol/L) per decade.

The elderly adult. Significant increases in the plasma concentrations of many constituents occur in women after menopause.[21] Renal concentrating ability is reduced in the elderly adult, so that creatinine clearance may decline by as much as 50% by the ninth decade. This decreased clearance is more related to decreased urinary creatinine associated with reduced lean body mass with age than to renal problems. The tubular maximum capacity for glucose is reduced. The blood urea concentration rises with age, as does the urinary excretion of protein.

Hormone concentrations are also affected by aging. However, changes in concentration are much less important than an endocrine organ's response to stimuli. Triiodothyronine concentration decreases by up to 40% in people over 40 years of age. Plasma parathyroid hormone concentration decreases with age. Cortisol secretion is reduced, although the serum concentration may not be affected. 17-Ketosteroid excretion in the elderly adult is about half of that of the younger adult. The secretion and metabolic clearance of aldosterone are decreased with a reduction of ~50% in the plasma concentration. The aldosterone response to sodium restriction is diminished. Basal insulin concentration is unaffected by aging, but the insulin secretion in response to glucose is reduced. In the male, the secretion rate and concentration of testosterone are reduced after age 50 y. In elderly women, the concentration of pituitary gonadotropins is increased in the blood and urine.

Influence of gender. Until puberty there are few differences in laboratory data between boys and girls. After puberty, the serum activities of alkaline phosphatase, the aminotransferases, creatine kinase, and aldolase are greater in men than in women. The higher activity of enzymes originating from skeletal muscle in men is related to their greater muscle mass.

Serum iron is low during a woman's fertile years, and her plasma ferritin may be only one-third the concentration in males. The reduced iron concentration in women is attributable to menstrual blood loss. Cholesterol concentration is typically higher in

Table 10-7. INFLUENCE OF GENDER ON COMPOSITION OF SERUM AT DIFFERENT AGES[9]*†

Constituent	Male-Female Difference†				
	< 29 y	30–39 y	40–49 y	50–59 y	60–69 y
Albumin, g/dL	0.1	0.1	0	0	−0.1
Alkaline phosphatase, U/L	14	12	−8	2	−1
Aspartate aminotransferase, U/L	5	8	8	1	−1
Bilirubin, mg/dL	0.1	0.1	0.1	0.1	0.1
Calcium, mg/dL	0.1	0.1	0.1	−0.1	−0.2
Cholesterol, mg/dL	−14	2	6	−16	−34
Creatinine, mg/dL	0.2	0.2	0.2	0.2	0.1
Glucose, mg/dL	5	3	6	0	6
Phosphorus, mg/dL	0.1	0.1	0	−0.1	−0.2
Total protein, g/dL	−0.1	−0.1	−0.1	−0.1	−0.2
Urea N, mg/dL	3	3	3	2	0
Uric acid, mg/dL	1.5	1.7	1.7	1.0	0.5

*In order to evaluate the magnitude of the effect on results, relate the per cent increase or decrease shown in the table to the analytical variation (±CV) routinely found for analytes.

†Male values are higher than female values except where indicated by a minus sign.

men than women, whereas the α-lipoprotein concentration is less. The plasma amino acid concentration, as well as the concentrations of creatinine, urea, and uric acid, is higher in males than in females. The combined influences of age and gender on the concentrations of serum constituents are illustrated in Table 10-7.

Influence of race. Differentiation of the effects of race from those of socioeconomic conditions is often difficult. The activities of creatine kinase and lactate dehydrogenase in serum are usually much higher in black men and women than in whites. This is presumed to be related to the amount of skeletal muscle, which tends to be greater in blacks than whites. Due to their greater skeletal development, black children generally have a higher serum alkaline phosphatase at puberty than do white children.

Carbohydrate and lipid metabolisms differ among races.[3] Glucose tolerance is less in blacks, Polynesians, American Indians, and Eskimos than in comparable age- and sex-matched whites. After age 40 y, the serum cholesterol and triglyceride concentrations are consistently higher in both white men and women than in blacks. These may be dietary rather than racial factors, since the concentration of plasma lipids has been shown to be different in the same racial group in different parts of the world.

Long-Term Cyclical Changes

Seasonal influences. Seasonal influences on the composition of body fluids are small in comparison to those related to changes in posture or the misuse of a tourniquet. Dietary changes as different foods become available or altered physical activity as more or different forms of exercise become feasible contribute to the seasonal influence. Evaluations of seasonal variation are difficult, since they depend on the definition of a season and on the magnitude of temperature change from one season to another.

Influence of menstrual cycle. The plasma concentrations of many female sex hormones as well as other hormones are affected by the menstrual cycle (see also Chapter 18). Other plasma constituents that have a cyclic change dependent on the hormone variation include cholesterol, total protein, albumin, and fibrinogen. Plasma iron, magnesium, sodium, chloride, and phosphate also vary during the cycle. Enzyme activity appears to be little affected.

Body Habitus

The serum concentrations of cholesterol, triglycerides, and very-low-density lipoproteins are positively correlated with obesity.[17] The serum urate concentration is also correlated with body weight, especially in individuals weighing over 80 kg.[11] Serum lactate dehydrogenase activity and glucose concentration increase in both sexes with increasing body weight.[17] In men, serum aspartate aminotransferase, creatinine, and total protein increase with increasing body weight, as does blood hemoglobin concentration. In women, serum calcium increases with increasing body weight. In both sexes, serum phosphate decreases with increased body mass.

Cortisol production is increased in obese individuals, and urinary excretion of 17-hydroxycorticosteroids and 17-ketosteroids is increased as well. Plasma insulin concentration is also increased in the obese, but glucose tolerance is impaired (see Chapter 13). The serum triiodothyronine correlates significantly with body weight and increases further with overeating. In obese men, the serum testosterone concentration is reduced.

References

1. Beisel, W. R., Goldman, R. F., and Joy, R. J. T.: Metabolic balance studies during induced hyperthermia in man. J. Appl. Physiol., *24*:1–10, 1968.
2. Bellet, S., Roman, L., DeCastro, O., et al.: Effect of coffee ingestion on catecholamine release. Metabolism, *18*:288–291, 1969.
3. Benedek, T. G., and Sunder, J. H.: Comparison of serum lipid and uric acid content in white and Negro men. Am. J. Med. Sci., *260*:331–340, 1970.
4. Blumenfeld, T. A., Turi, G. K., and Blanc, W. A.: Recommended site and depth of newborn heel skin punctures based on anatomic measurements and histopathology. Lancet, *1*:230–233, 1979.
5. Caraway, W. T.: Chemical and diagnostic specificity of laboratory tests. Am. J. Clin. Pathol., *37*:445–464, 1962.
6. Felding, P., Hyltoft Petersen, P., and Horder, M.: The stability of plasma and serum constituents during simulated transport. Scand. J. Clin. Lab. Invest. *41*:35–40, 1981.
7. Felding, P., Tryding, N., Hyltoft Petersen, P., et al.: Effects of posture on concentrations of blood constituents in healthy adults: practical application of blood specimen collection procedures recommended by the Scandinavian Committee on Reference Values. Scand. J. Clin. Lab. Invest., *40*:615–621, 1980.
8. Ladenson, J. H.: Nonanalytical sources of variation in clinical chemistry results. *In:* Gradwohl's Clinical Laboratory Methods and Diagnosis. A. C. Sonnenwirth and L. Jarett, Eds. 8th ed. St. Louis, Mosby, 1980.
9. Leonard, P. J.: The effect of age and sex on biochemical parameters in blood of healthy human subjects. *In:* Reference Values in Human Chemistry. G. Siest, Ed. Basel, Karger, 1973.
10. McNair, P., Nielsen, S. L., Christiansen, C., et al.: Gross errors made by routine blood sampling from two sites using a tourniquet applied at different positions. Clin. Chim. Acta, *98*:113–118, 1979.
11. Munan, L., Kelly, A., PetitClerc, C., et al.: Association with body weight of selected chemical constituents in blood. Clin. Chem., *24*:772–777, 1978.
12. National Committee for Clinical Laboratory Standards: Approved standard procedure for the collection of diagnostic blood specimens by skin puncture. NCCLS standard H4-A. Villanova, PA, National Committee for Clinical Laboratory Standards, 1982.
13. National Committee for Clinical Laboratory Standards: Standard for evacuated tubes for blood specimen collection. 2nd ed. NCCLS Standard ASH-1. Villanova, PA, National Committee for Clinical Laboratory Standards, 1980.
14. National Committee for Clinical Laboratory Standards: Standard procedure for the collection of diagnostic blood specimens by venipuncture. NCCLS standard ASH-3. Villanova, PA, National Committee for Clinical Laboratory Standards, 1977.
15. Ong, Y. Y., Boykin, S. F., and Barnett, R. N.: You can draw blood from the "IV arm" below the intravenous needle if you put a tourniquet in between. Am. J. Clin. Pathol., *72*:101–102, 1979.
16. Rayfield, E. J., Curnow, R. T., George, D. T., et al.: Impaired carbohydrate metabolism during a mild viral illness. N. Engl. J. Med., *289*:618–621, 1973.
17. Siest, G., Henny, J., and Schiele, F., Eds.: Interpretation des Examens de Laboratoire. Basel, Karger, 1981.
18. Statland, B. E., Bokelund, H., and Winkel, P.: Factors contributing to intraindividual variation of serum constituents: 4. Effects of posture and tourniquet application on variation of serum constituents in healthy subjects. Clin. Chem., *20*:1513–1519, 1974.
19. Thomson, W. H. S., Sweetin, J. C., and Hamilton, I. J. D.: ATP and muscle enzyme efflux after physical exertion. Clin. Chim. Acta, *59*:241–245, 1975.
20. Weitzman, E. D.: Circadian rhythms and episodic hormone secretion. Ann. Rev. Med., *27*:225–243, 1976.
21. Wilding, P., Rollason, J. G., and Robinson, D.: Pattern of change for various biochemical constituents detected in well-population screening. Clin. Chim. Acta, *41*:375–387, 1972.
22. Winkel, P., Statland, B. E., and Bokelund, H.: The effects of time of venipuncture on variation of serum constituents. Am. J. Clin. Pathol., *64*:433–447, 1975.
23. Winsten, S., and Gordesky, S. E.: Transportation of specimens. *In:* Selected Methods of Clinical Chemistry. W. R. Faulkner, Ed. Washington, D.C., AACC, 1982, Vol. 9, pp. 11–15.
24. Wollner, G. C.: Handling, storing and transporting diagnostic specimens. Lab. Med., *11*:87–91, 1980.
25. Young, D. S.: Biological variability. *In:* Chemical Diagnosis of Disease. S. S. Brown, F. L. Mitchell, and D. S. Young, Eds. Amsterdam, Elsevier, 1979, pp. 1–13.

26. Young, D. S.: The effects of frequently prescribed drugs on common laboratory procedures. *In*: Practice of Medicine. J. A. Spittell, Jr., Ed. Philadelphia, J. B. Lippincott Co., 1984. Vol. 1, Chapter 26, pp. 1–21.
27. Young, D. S., Pestaner, L. C., and Gibberman, V.: Effects of drugs on clinical laboratory tests. Clin. Chem., *21*:1D–432D, 1975.

Additional Reading

Brown, S. S., Mitchell, F. L., and Young, D. S., Eds.: Chemical Diagnosis of Disease. Amsterdam, Elsevier, 1979.

Calam, R. R.: Reviewing the importance of specimen collection. J. Am. Med. Technol., *39*:297–302, 1977.

Meites, S., Ed.: Pediatric Clinical Chemistry. Washington, D.C., AACC. 2nd edition, 1981.

Slockbower, J. M., and Blumenfeld, T. A., Eds.: Collection and Handling of Laboratory Specimens. Philadelphia, J. B. Lippincott Co., 1983.

Statland, B. E., and Winkel, P.: Effects of preanalytical factors on the intraindividual variation of analytes in the blood of healthy subjects: consideration of preparation of the subject and time of venipuncture. CRC Crit. Rev. Clin. Lab. Sci., *8*:105–144, 1977.

Tietz, N. W., Ed.: Clinical Guide to Laboratory Tests. Philadelphia, W. B. Saunders Co., 1983.

Wilding, P., Zilva, J. F., and Wilde, C. E.: Transport of specimens for clinical chemistry analysis. Ann. Clin. Biochem., *14*:301–306, 1977.

III

ANALYTES, METHODS, PATHOPHYSIOLOGY, AND INTERPRETATION

AMINO ACIDS AND PROTEINS

11

GREGOR H. GRANT, B. M., F.R.C. Path.
LAWRENCE M. SILVERMAN, Ph.D.
ROBERT H. CHRISTENSON, Ph.D.

BASIC CHEMISTRY OF AMINO ACIDS AND PROTEINS

Amino acids are organic compounds containing both an amino group and a carboxyl group. Those occurring in proteins are called α-amino acids and have the empirical formula $RCH(NH_2)COOH$. Beta-amino acids and γ-amino acids also occur in nature but not as components of proteins. As shown below, the core of an α-amino acid is the carbon atom next to the carboxylic acid group, the α-carbon. With the exception of glycine, all α-amino acids are asymmetric; i.e., four different groups are bonded to the α-carbon—a hydrogen atom, a carboxyl group, an amino group, and a distinctive R group. These four groups, which are mirror images of one another and rotate polarized light in opposite directions, can be bonded in two different ways. They are called D and L optical isomers; in natural proteins of higher organisms, only the L-isomer is found. In the illustration below, *A* is the conventional way to depict the L-configuration, but *B* is the more convenient form for writing and printing.

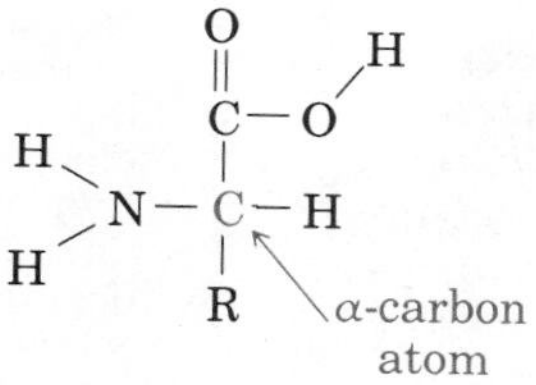

```
          O                        H  H
          ||    H                   \ /
          C—O /                      N      O
  H       |                          |     //
    N— C—H                      R— C— C
  H       |  ↖                       |     \
          R   α-carbon               H      O—H
                atom
          A                          B
```

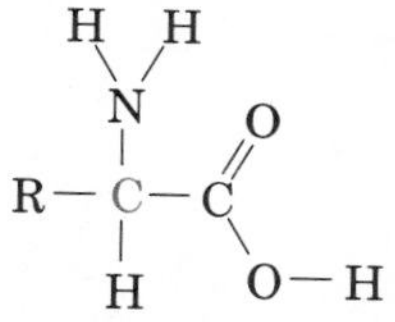

Only 20 L-α-amino acids listed in Table 11-1 are used to build the enormous number of biologically active peptides and proteins that exist in nature. However, there are many other amino acids of biological significance, such as thyroxine and triiodothyronine (the thyroid hormones) and β-alanine, a constituent of the vitamin pantothenic acid. What makes peptides and proteins so versatile in structure and function are the characteristic acid-base properties of individual amino acids and the variety of possible R group interactions. The acid-base properties depend on the amino and carboxyl groups attached to the α-carbon and on the basic or acidic functional groups that may be present in an R group. In aqueous media, very little amino acid exists in an un-ionized form. The acidic carboxyl group is a proton donor and the basic amino group is a proton acceptor. In the pH range of 7.35–7.45, the carboxyl group is dissociated and the amino group is protonated to give the following structure:

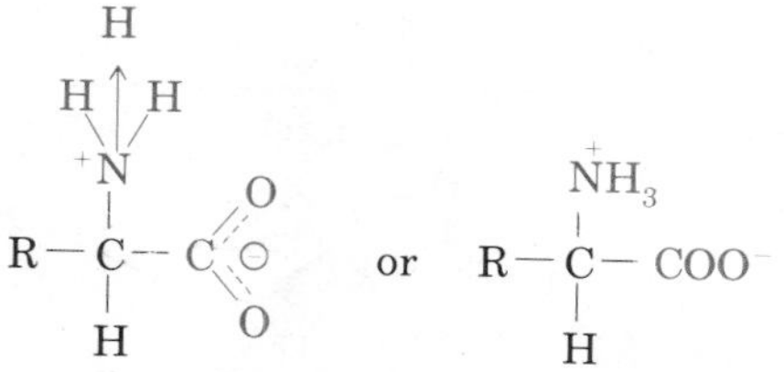

```
        H
      H ↑ H
        \|/
        +N        O                 +
         |       ⁄⁄                NH3
    R— C-- C ⊖        or     R— C— COO⁻
         |       \\                |
         H        O                H
```

Table 11-1. AMINO ACIDS

Name and Abbreviation	MW	Structure at pH 6–7	Comments
1. Hydrophobic Amino Acids; Nonpolar R Groups			
Alanine Ala	89	$CH_3-CH(\overset{+}{N}H_3)-C(=O)O^-$	Substrate for ALT (alanine transaminase); least hydrophobic of the group
Leucine Leu	131	$(H_3C)_2CH-CH_2-CH(\overset{+}{N}H_3)-C(=O)O^-$	Branched-chain R group; essential; ketogenic; metabolism is faulty in maple syrup urine disease
Isoleucine Ile	131	$CH_3-CH_2-CH(CH_3)-CH(\overset{+}{N}H_3)-C(=O)O^-$	Essential; partly ketogenic; see Leu
Valine Val	117	$(H_3C)_2CH-CH(\overset{+}{N}H_3)-C(=O)O^-$	Essential; partly ketogenic; see Leu
Proline Pro	115	ring: $H_2C-CH_2(C\,H_2)-CH(COO^-)-\overset{+}{N}H_2-CH_2$	Important constituent of connective tissue proteins (e.g., collagen and elastin); some hydroxylated to Hyp during collagen synthesis; destabilizes α-helical and β-structures; contains an α-imino group
Methionine Met	149	$CH_3-S-CH_2-CH_2-CH(\overset{+}{N}H_3)-C(=O)O^-$	Essential; important in transfer of methyl groups; provides sulfur for other sulfur-containing compounds
Phenylalanine Phe	165	$C_6H_5-CH_2-CH(\overset{+}{N}H_3)-C(=O)O^-$	Essential; excess in phenylketonuria
Tryptophan Trp	204	(indol-3-yl, N–H)$-CH_2-CH(\overset{+}{N}H_3)-C(=O)O^-$	Essential; metabolites found in carcinoid disease; contains indole ring system; precursor of serotonin and melatonin
2. Hydrophilic Amino Acids; Uncharged Polar R Groups			
Glycine Gly	75	$H-CH(\overset{+}{N}H_3)-C(=O)O^-$	Simplest amino acid; optically inactive; placed in this group because its R group (single H) is unable to affect

Table 11-1. AMINO ACIDS (*Continued*)

Name and Abbreviation	MW	Structure at pH 6-7	Comments
			polarity of the rest of the molecule; used in biosynthesis of purines and porphyrins; used in vitro as a buffer
Serine Ser	105	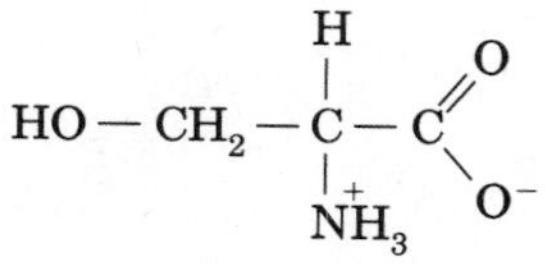 $HO-CH_2-CH(NH_3^+)-COO^-$	Constituent in active center of many enzymes; hydroxyl group can be phosphorylated
Threonine Thr	119	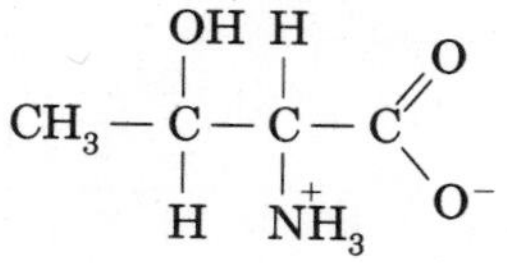 $CH_3-CH(OH)-CH(NH_3^+)-COO^-$	Essential
Cysteine Cys	121	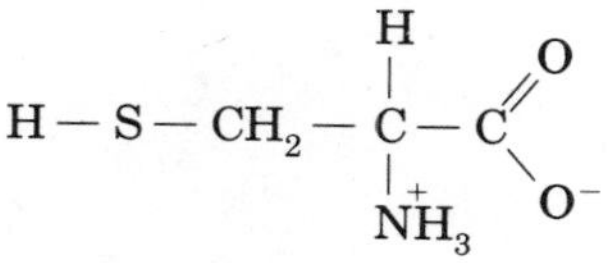 $H-S-CH_2-CH(NH_3^+)-COO^-$	Sulfhydryl group functional in the activity of many enzymes; is responsible for disulfide bridges in peptides and proteins; cystine is dicysteine, Cys−S−S−Cys; homocysteine has one carbon more than cysteine and forms homocystine (dihomocysteine)
Tyrosine Tyr	181	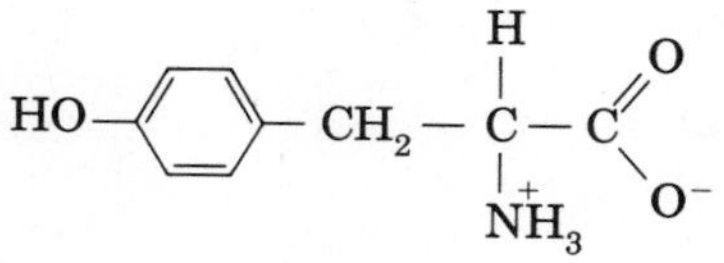 $HO-C_6H_4-CH_2-CH(NH_3^+)-COO^-$	Usually nonessential; intermediate in synthesis of catecholamines, thyroxine, and melanin; functional phenolic group; reacts with Folin's reagent in quantitative protein assay
Glutamine Gln	146	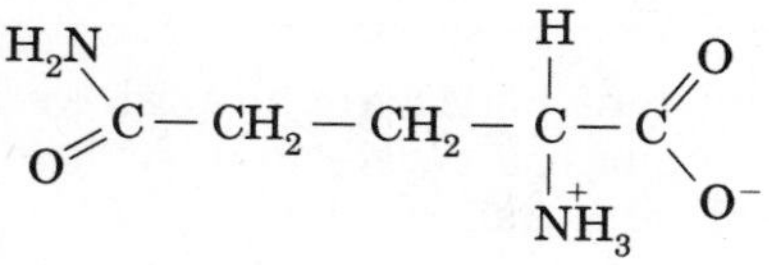 $H_2N-C(=O)-CH_2-CH_2-CH(NH_3^+)-COO^-$	Storage form of ammonia in tissue; supplies the amido nitrogen used in purine and pyrimidine biosynthesis
Asparagine Asn	132	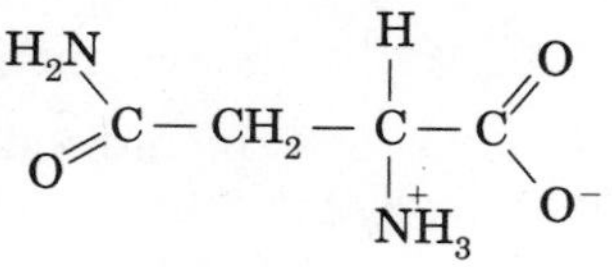 $H_2N-C(=O)-CH_2-CH(NH_3^+)-COO^-$	Storage form of ammonia in tissues.
Hydroxy-proline Hyp	131	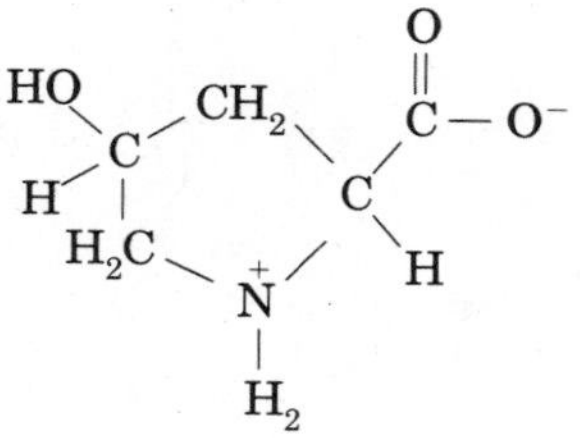 Ring: $HO(H)C-CH_2-CH(COO^-)-\overset{+}{N}H_2-H_2C$ (closing to $HO(H)C$)	Constituent of collagen—the only human protein to contain appreciable amounts; urinary output is used as an indicator of bone matrix metabolism; contains an α-imino group

Table continued on following page

Table 11-1. AMINO ACIDS (*Continued*)

Name and Abbreviation	MW	Structure at pH 6-7	Comments
		3. Dicarboxylic Amino Acids; Acidic R Groups	
Aspartic acid Asp	133	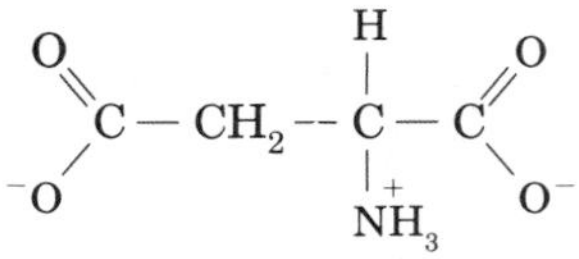 $^{-}O(O{=})C-CH_2-CH(\overset{+}{N}H_3)-C({=}O)O^{-}$	Co-substrate with Glu for AST; used in pyrimidine biosynthesis
Glutamic acid Glu	147	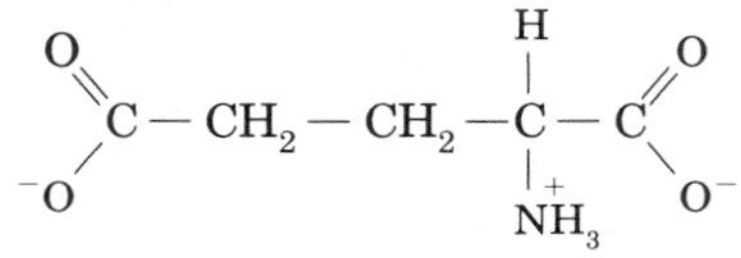 $^{-}O(O{=})C-CH_2-CH_2-CH(\overset{+}{N}H_3)-C({=}O)O^{-}$	Co-substrate with Ala for ALT and with Asp for AST
		4. Basic Amino Acids; Basic R Groups	
Lysine Lys	147	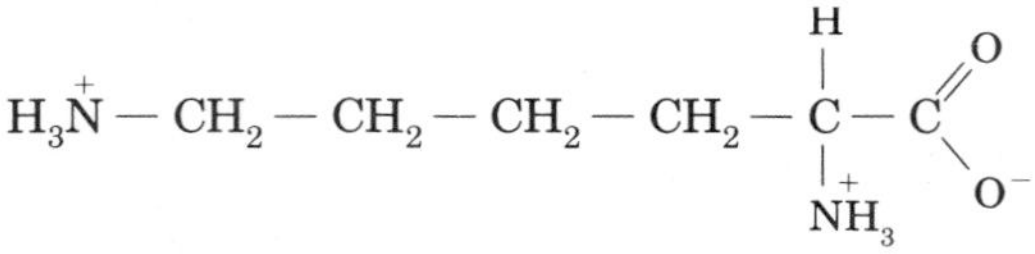 $H_3\overset{+}{N}-CH_2-CH_2-CH_2-CH_2-CH(\overset{+}{N}H_3)-C({=}O)O^{-}$	Essential; terminal NH_2 called ϵ-amino
Arginine Arg	174	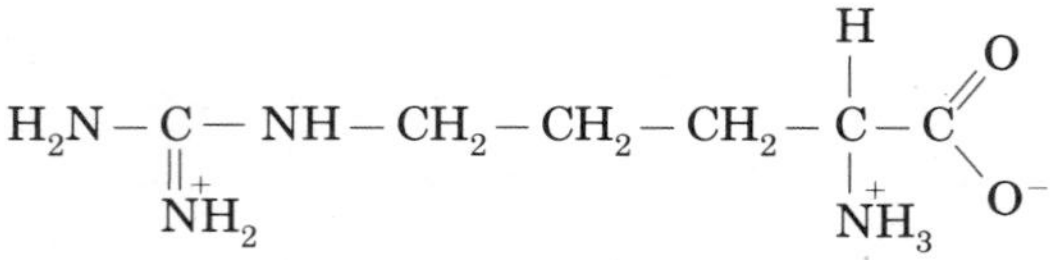 $H_2N-C({=}\overset{+}{N}H_2)-NH-CH_2-CH_2-CH_2-CH(\overset{+}{N}H_3)-C({=}O)O^{-}$	Involved in urea synthesis; the basic group is a guanidinium group
Histidine His	155	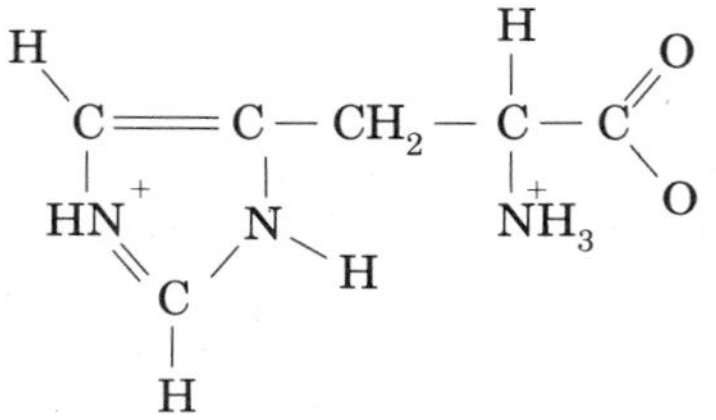 Imidazolium ring (H–C═C–, $H\overset{+}{N}$, N–H, C–H) $-CH_2-CH(\overset{+}{N}H_3)-C({=}O)O$	The imidazole group of histidine is the most important buffer group in the physiological pH range

This kind of ionized molecule with coexistent negative and positive charges is called a dipolar ion or ampholyte (in older terminology, a zwitterion). At low pH, an amino acid is in its cationic form; i.e., both the amino and carboxyl groups are protonated as $-\overset{+}{N}H_3$ and —COOH. As the pH rises, the carboxyl group loses its proton, and the ampholyte form appears at about pH 6. With further increases in pH, the amino $-\overset{+}{N}H_3$ is also deprotonated and the anionic form of the molecule is formed. This process can be shown for a monoamino and monocarboxylic amino acid as follows:

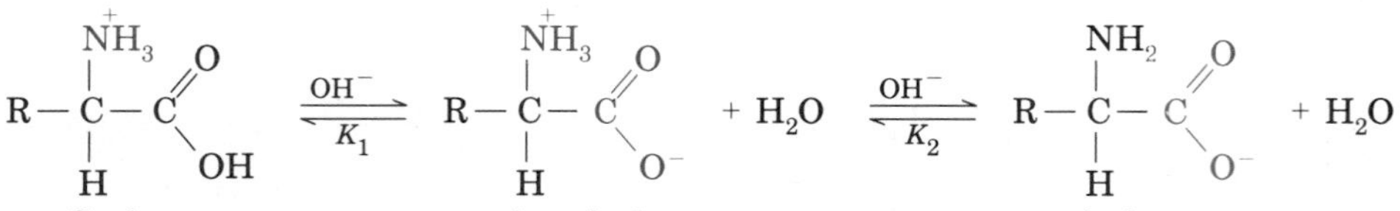

$$R-CH(\overset{+}{N}H_3)-C({=}O)OH \underset{K_1}{\overset{OH^-}{\rightleftharpoons}} R-CH(\overset{+}{N}H_3)-C({=}O)O^- + H_2O \underset{K_2}{\overset{OH^-}{\rightleftharpoons}} R-CH(NH_2)-C({=}O)O^- + H_2O$$

Cation at pH < pI   **Ampholyte at pH = pI**   **Anion at pH > pI**

The dissociation constants, K_1 and K_2, are usually expressed logarithmically as pK_1 and pK_2, where $pK = -\log K$, in a manner analogous to the notation for pH. A pK is the pH at which equal quantities of the protonated (associated) and unprotonated (dissociated) forms are present. The isoelectric point, pI, is the pH at which all molecules exist in the ampholyte forms; i.e., with net charge = 0. At pH > pI, the amino acid is negatively charged, and at pH < pI, it is positively charged. The isoelectric point of a neutral amino acid can be calculated from the pKs of its amino and carboxyl groups; i.e., $pI = \frac{1}{2}(pK_1 + pK_2)$. The concept of an ampholyte and its dissociation characteristics is also applicable to proteins.

The R groups of individual amino acids are responsible for their special properties. Note that Table 11-1 is organized in terms of the differences in R groups. Some R groups are nonpolar and therefore hydrophobic; others, although uncharged, are polar and hydrophilic. Still others can become charged, either negatively (the acidic amino acids) or positively (the basic amino acids). R groups may be linear as in alanine or cyclic as in proline, small as in glycine or bulky as in tryptophan. Electron density may be low as in aliphatic chains or high as in aromatic rings. This variation in R group structure makes possible several kinds of interaction between R groups. Such interactions are critical in determining the structures and functions of proteins into which the amino acids are incorporated.

R groups with ionizable substituents have their own pK values, so that at physiological pH some of the amino acids exist primarily either as anions or cations. At pH 7, the second carboxyl group of the acidic amino acids glutamine and aspartic acid is fully ionized, and therefore these amino acids are negatively charged. The basic amino acids lysine and arginine are present almost entirely as positive ions, but <10% of histidine is in the cationic form. Glycine, with pK_1 2.34 and pK_2 9.60, has at pH 6–7 a net charge near zero and exemplifies acid-base behavior of those amino acids whose R groups have no ionizable substituents.

The differing solubilities and acid-base properties of amino acids provide the basis for their separation by partition chromatography, ion-exchange chromatography, or electrophoresis. Differences in the chemical nature of R groups may permit their identification or measurement by color reactions.

Polymerization of amino acids. Amino acids can be linked to one another by a *peptide bond*, indicated in color below, between the α-amino group of one and the α-carboxyl group of another.

$$R_1-\underset{H}{\overset{NH_2}{C}}-C\begin{matrix}\diagup\!\!\!\!\diagup O\\ \diagdown O-H\end{matrix} + \begin{matrix}H\diagdown\\ H\diagup\end{matrix}N-\underset{R_2}{\overset{\overset{O}{\|}C-O-H}{C}}-H \longrightarrow R_1-\underset{H}{\overset{NH_2}{C}}-\overset{O}{\overset{\|}{C}}-\underset{H}{N}-\underset{R_2}{\overset{\overset{O}{\|}C-O-H}{C}}-H + H_2O$$

First amino acid **Second amino acid** **Dipeptide**

The peptide bond may be formed between the $-NH_2$ of the first and the $-COOH$ of the second or between the $-COOH$ of the first and the $-NH_2$ of the second. For example, glycine and alanine can react to form two different dipeptides:

glycyl-alanine, $NH_2CH_2-CO \cdot NH-CH(CH_3)COOH$

or

alanyl-glycine, $CH_3CHNH_2-CO \cdot NH-CH_2COOH$

In glycyl-alanine, alanine is called the *C*-terminal residue of the peptide because its *carboxyl* group is free; glycine is the *N*-terminal residue because its *amino* group is free. In alanyl-glycine, the designations are reversed. The *C*- and *N*-terminal designations for dipeptides apply also to polypeptides and proteins that have amino acids linked by peptide bonds to form chains. Very short chains are often designated as tri-, tetra-, or pentapeptides and so on. Chains up to five residues are called oligopeptides. Longer chains (6–30 residues) are referred to as *polypeptides*. When the number of amino acids linked together exceeds 40 (MW ~5000), the chain takes on the physical properties associated with proteins. The terms *proteose* and *peptone* refer to protein breakdown products; they are large polypeptides that differ from true proteins in that they are not coagulated by heat. A wide variety of peptides are formed when proteins are hydrolyzed in the digestive tract or in the laboratory.

From the structures of the amino acids in Table 11-1, it is clear that, in an appropriate three-dimensional orientation, possible interactions of one R group with another include disulfide bond formation, hydrogen bond formation, hydrophobic interaction, interaction between charged R groups, and steric effects. The disulfide bond is the only covalent bond, other than the peptide bond, that contributes to polypeptide structure, RSH + HSR → R—S—S—R. Other types of interactions are weaker, but are often so numerous that their collective strength is large. *Hydrogen bonds* result from the sharing of a hydrogen atom between two electronegative atoms (e.g., N or O) which have unbonded electrons. In proteins, groups having a hydrogen atom that can be shared include $>$N—H (peptide nitrogen, imidazole, and indole); —OH (serine, threonine, tyrosine, and hydroxyproline); $—NH_2$ and $—\overset{+}{N}H_3$ (arginine, lysine, and α-amino); and —CONH (carbamino). Groups that can accept the sharing of a hydrogen atom include $—COO^-$ (aspartate, glutamate, and α-carboxylate); —S—S— (disulfide); and $>$C=O (in peptides and ester linkages). Hydrophobic interactions result from the tendency of nonpolar groups (e.g., methyl or phenyl) to associate with one another in aqueous or other polar solutions. In proteins, this association serves to bend and fold a molecule in such a way as to bring nonpolar R groups inside to the less polar interior and polar R groups outside to the more polar aqueous environment.

Proteins may be classified as fibrous or globular. The basic structural organizations of fibrous proteins are the α-helix and the β-sheet. The molecules of fibrous proteins tend to be longer than they are wide, and they tend to be insoluble in physiologic fluids. Globular proteins usually include mixtures of helical and sheet conformation; they tend to be compact and roughly spherical and are soluble in physiological fluids. These distinctions and details of protein structure are discussed at length in texts of biochemistry.

Nearly all proteins of interest in clinical chemistry are soluble globular proteins such as hemoglobin (see Figures 24-3 and 24-4), enzymes, and plasma proteins (with the exception of fibrinogen). The complex bending and folding of their polypeptide chains are a result of the numerous interactions of their R groups. Globular proteins are compact with little or no space for water in the interior of the molecule where most of the hydrophobic R groups are located. The polar R groups are located on the surface of the protein, where they exert a substantial influence on protein solubility, acid-base behavior, and electrophoretic mobility.

Most globular proteins retain their biological activity only within very limited ranges of temperature and pH, at which their native structural arrangements are the most stable. Even short periods of exposure to temperatures of 65–70 °C or to extremes of pH will cause "denaturation" with loss of solubility and biological activity; enzymes, for example, lose their catalytic powers after denaturation occurs (see Chapter 12). The amino acid sequence remains unchanged, but the polypeptide chain becomes unfolded, or the quaternary structure is lost.

After mild denaturation, many protein molecules are able to revert spontaneously to their most stable or native globular forms and regain their biological functions when physiologic conditions are restored.

When describing the structure of a protein, particularly a globular protein, it is convenient to consider its complexity in four stages.

1. *Primary structure* refers to the identity and specific order of amino acid residues in the polypeptide chain. This sequence, which depends exclusively on covalent (peptide) bonds, is predetermined by the DNA coding; the remarkable fact is that the three-dimensional structure and any special biological properties of the protein follow automatically from this amino acid sequence, which folds itself into the most stable structure possible under physiological conditions (e.g., of temperature, pH).

2. *Secondary structure* may be defined as a regular recurring arrangement in space of the primary structure extending along one dimension. The secondary structure of a globular protein usually has stretches of α-helix, β-pleated sheet, and random coils, all dependent on numerous hydrogen bonds and occasional disulfide covalent bonds.

3. *Tertiary structure* involves the intramolecular folding of the polypeptide chain into a compact three-dimensional structure with a specific shape. This structure is maintained by electrovalent linkages, hydrogen bonds, disulfide bridges, van der Waals forces, and hydrophobic interactions. Hydrophobic interactions are considered to be a major force in maintaining the unique tertiary structure of proteins.

4. *Quaternary structure* refers to the association of several polypeptide chains or subunits into a larger "oligomeric" aggregate unit. This structure depends on the close fit of the polypeptide subunits through interactions at their contact surfaces and with any prosthetic groups.

Proteins containing nonamino acid components (*prosthetic groups*) are often referred to as *conjugated proteins* and are classified according to the nature of their prosthetic groups as metalloproteins, lipoproteins, glycoproteins or mucoproteins, phosphoproteins, and so on. Glycoproteins and mucoproteins both have covalently linked carbohydrate prosthetic groups, but the amount of carbohydrate varies—in glycoproteins it may be ~5–15%, in mucoproteins ~15–75%. Conjugated proteins freed of their prosthetic groups are called *apoproteins*.

Biological functions of proteins are numerous, varied, and important. Enzymes catalyze biochemical transformations essential to metabolism; protein, polypeptide, and oligopeptide hormones regulate metabolism; antibodies and components of the complement system protect against infection. Plasma proteins maintain the oncotic pressure of plasma; they transport hormones, vitamins, metals, and drugs, often serving as reservoirs for their release and utilization; apolipoproteins solubilize lipids; hemoglobin carries oxygen; protein coagulation factors affect hemostasis. In addition, there are structural proteins such as the keratins, collagen, and elastin; contractile proteins such as myosin; storage proteins such as ferritin; chromosomal proteins such as the basic histones; and many others.

Properties that proteins have in common, such as the ability to react with biuret reagent, the UV absorption characteristics of peptide bonds, or the ease of their precipitation or dye binding, provide tools for determining *total* protein. Properties of *individual* proteins utilized for their separation, identification, and distinctive assay include:

1. Molecular size. A majority of proteins are macromolecules, compounds of large size and consequently of high molecular weight. Hence they can be separated from one another or from smaller molecules by dialysis or ultrafiltration across membranes of specified pore size, or by molecular exclusion chromatography (gel filtration). Because of their differing molecular weights, they can be separated by density-gradient ultracentrifugation.

2. Differential solubility. Loss of solubility is essentially a loss of affinity of the solute for the solvent. The solubility of a protein is affected by the pH, ionic strength, temperature, and dielectric constant of the solvent in which it is placed. Changing the pH of the solvent affects the surface charges of the polar R groups exposed to the solvent; at its isoelectric point (net charge zero), a protein in a polar solvent is in its least soluble state because it has the least affinity for the solvent. Manipulation of ionic strength of the solvent affects the degree of hydration of the protein; in high concentrations, ionic salts compete with protein for water and thus decrease the amount of water available for hydration of the protein molecules. This behavior is the basis of "salting-in" and "salting-out" procedures, which historically were the earliest methods for separating and characterizing different proteins and groups of proteins. Raising the temperature affects solubility by decreasing the intramolecular bonding of a protein; loss of secondary and tertiary structure exposes the less polar interior of the protein molecule to the polar medium. Water-miscible, neutral organic solvents have dielectric constants that are less than that of water; they suppress the ionization of the surface R groups on the protein and thereby decrease its solubility. Manipulation of solvent character and composition in such ways is often useful in analyses of biological, protein-containing fluids.

3. Electrical charge. The ability of pH to confer, enhance, or change the surface charges on a protein can create, in a mixture of different proteins, a variety of species of different charge/mass ratios, which migrate at different rates in an electrical field; electrophoresis and isoelectric focusing are based on this behavior. Ion-exchange chromatography is based on electrostatic interactions of charged proteins with oppositely charged solid media.

4. Adsorption on finely divided inert material. These materials offer large surface areas for interaction with protein. Adsorption of protein on charcoal, a nonpolar substance, occurs presumably by hydrophobic interaction and can be used to separate proteins from a fluid matrix. Adsorption on polar substances such as silica, alumina, or hydroxyapatite depends on ionic interactions or hydrogen bonding; these substances can be used with buffer *elution* in chromatographic separations.

5. Specific binding to antibodies, co-enzymes, or hormone receptors. The unique property of a protein to recognize and bind to a complementary compound with high specificity has been valuable in immunoassays and other competitive protein binding assays. Proteins can also be separated by affinity chromatography, where a ligand attached to a solid medium provides high selectivity for a protein that binds to it. For the determination of individual proteins in serum, use of specific antisera as reagents obviates the need for preliminary isolation or purification of most proteins from the mixture present in serum.

Amino Acid Metabolism

In health, the primary source of amino acids for endogenous protein synthesis is dietary protein. Eight to ten of the twenty common amino acids are considered "essential" dietary components because they cannot be synthesized by the body rapidly enough to maintain health or growth, or both. Proteolytic enzymes in the GI tract, acting on proteins, release amino acids, which are then absorbed from the jejunum into the blood and subsequently become part of the body pool of amino acids. The liver and other tissues draw on the pool for synthesis of plasma, intracellular, and structural proteins. The liver and kidneys are active in interconverting amino acids by transamination and in degrading them by deamination. Deamination produces ammonium ions that are consumed in the synthesis of urea; urea is then excreted by the kidneys.

Amino acids in body fluids.[71] Plasma amino acid concentrations vary during the day by about 30%; they are highest in midafternoon and lowest in early morning, a finding that must be remembered when investigating heterozygotes of the inherited disorders of amino acid metabolism. Levels are high during the first days of life, especially in premature neonates, but tend to be low in babies with birth weights low for their gestational age. Maternal values are low in the first half of pregnancy.

Amino acids in the blood are filtered through the glomerular membrane but are normally reabsorbed in the renal tubules by saturable transport mechanisms. The reabsorption mechanism is still obscure, although it is known to be an active transport system dependent upon membrane-bound carriers.

Three types of *aminoaciduria* may be distinguished: (1) *overflow aminoaciduria*, in which the plasma level of one or more amino acids exceeds the renal threshold, or tubular capacity for reabsorption (this may also occur transiently in normal subjects); (2) *renal aminoaciduria*, in which plasma levels are normal but there is a congenital or acquired defect of the renal transport system; and (3) *no-threshold aminoaciduria*, which is similar to the overflow type except that plasma levels are essentially normal because all of the amino acid is excreted. Note that no-threshold aminoacidurias such as homocystinuria are not due to congenital or acquired kidney defects but are solely due to inefficient reabsorption by normal renal tubular mechanisms.

In a chromatogram of *normal adult urine*, glycine is usually the dominant fraction, with alanine, serine, glutamine, and, in heavy meat eaters, histidine and 1-methyl histidine present in smaller quantities; in some normal urines, taurine is prominent; in others, β-aminoisobutyric acid is seen. The renal threshold for many substances is lowered during pregnancy, and amino acids such as histidine, phenylalanine, lysine, and tyrosine are commonly seen in urine. Premature babies, especially during the first week, have a physiological generalized renal aminoaciduria; even at full term, aminoaciduria is marked when compared with values for normal adults.

CSF concentrations of most amino acids are less than in plasma. *Cells* have amino acid concentrations about ten times those of plasma.

THE AMINOACIDURIAS

Aminoacidurias[2,5,28,78] may be primary or secondary. Primary disease is due to an inherited enzyme defect, also called an inborn error of metabolism. The defect is located either in the pathway by which a specific amino acid is metabolized or in the specific renal tubular transport system by which the amino acid is reabsorbed. Descriptions of the many acidurias are given elsewhere.[78] Secondary aminoaciduria is due to a disease of an organ such as the liver, which is an active site of amino acid metabolism, or to generalized renal tubular dysfunction, or to protein-energy malnutrition.

A ***primary aminoaciduria*** results from a mutation in the DNA base sequence that codes for the amino acid sequence of a particular enzyme. This mutation produces a defective enzyme that causes a block in the metabolic pathway of its substrate; consequently, the substrate accumulates or is diverted into alternative paths.

It should be possible to diagnose such an inherited disease at three levels: (1) the DNA abnormality, (2) the enzyme defect, and (3) the metabolic abnormalities due to the defect. At present, diagnosis commonly depends on recognizing suggestive clinical signs and symptoms and then characterizing the metabolic abnormalities. Demonstration of an enzyme defect is restricted to prenatal diagnosis or to confirmation of a postnatal diagnosis by testing cultured cells. Techniques for demonstrating DNA abnormalities, capable of application early in pregnancy, are now being developed.

Although an enormous number of congenital errors exist, many of them are harmless and often go unrecognized; nevertheless, more than 1350 human diseases, including the primary aminoacidurias, are now known to be caused by alteration of one single gene out of the 50 000 or more in each human cell. Most of the serious inherited diseases are rare, but collectively they pose an increasingly important diagnostic problem because some are now treatable, if diagnosed sufficiently early.

The symptoms and prognosis of the primary aminoacidurias that cause disease may vary from being almost benign, as in alkaptonuria, to being lethal, as in maple syrup urine disease. Their causes are usually a block in a major catabolic pathway, and their expression often indicates the diversion of substrate to an alternative, metabolically ineffective minor path. As a result, substrates and precursor substrates behind the block accumulate in the blood. When renal transport mechanisms for these substances are saturated, these products spill into the urine. Accumulation of keto acids in maple syrup urine disease and of phenylalanine in phenylketonuria (PKU) are examples of precursors and substrates with toxic effects. Past the block, products and intermediates are decreased. If an alternative path is used, products and intermediates of the alternative path are increased in blood and urine.

Occasionally, an enzyme block occurs in an anabolic pathway, as in the synthesis of melanin (where the consequence is albinism). In some cases, substrates do not accumulate in the blood because there is no reabsorption in the kidneys (no-threshold aminoaciduria) and urine levels are increased.

The *renal* types of primary aminoacidurias (e.g., cystinuria) are not characterized by high levels of amino acids in blood because the defective protein coded by the altered gene is an element of the tubular reabsorption mechanism for that amino acid.

A few of the primary aminoacidurias are listed in Table 11-2; all are inherited as autosomal recessive disorders. The overflow types (Table 11-2,*A*) probably involve a change in only one amino acid residue of the enzyme concerned. For an autosomally recessive disease to occur in an offspring, the gene defect must be present in the chromosomes of both parents (i.e., the offspring's genome must be homozygous for the defect). The homozygote therefore fails to make the critical protein with a correct amino acid

Table 11-2,*A*. PRINCIPAL PRIMARY OVERFLOW AMINOACIDURIAS (AUTOSOMAL RECESSIVE DISORDERS*)

Disorder	Incidence	Abnormal Enzyme(s) (or other defects)	Excesses in Blood	Excesses in Urine	Clinical Features	Treatment
Hyperphenylalaninemia						
Classic PKU (Type I)	1:10 000	Phenylalanine hydroxylase (absent)	Phenylalanine	Phenylalanine and its metabolites (phenylpyruvate, phenyllactate, *o*-hydroxyphenylacetate)	Mental retardation; seizures, eczema	Dietary restriction of Phe
Variant PKU (Type II)	1:14 000	Phenylalanine hydroxylase (deficient)	Phenylalanine	Variable	Mild retardation	Dietary restriction of Phe
Tyrosinemia						
Type I (Tyrosinosis)	1:100 000	Fumarylacetoacetate hydrolase? (absent)	Tyrosine; methionine	Tyrosine and its metabolites (PHPPA, PHPLA, PHPAA); DOPA; generalized aminoaciduria (e.g., Fanconi's syndrome)	Hepatic cirrhosis; renal damage	Dietary restriction of Phe, Tyr, Met (treatment does not cure liver disease)
Alkaptonuria	1:250 000	Homogentisic acid oxidase (absent)	Homogentisic acid (slight)	Homogentisic acid	Degenerative arthritis, cartilage pigmentation	None available
Homocystinuria	1:200 000	Cystathionine β-synthase (absent or deficient)	Homocystine; methionine	Homocystine; methionine and its sulfoxide	Ocular, skeletal, vascular effects	Pyridoxine; low methionine diet supplemented with cystine
Branched-chain ketoaciduria (Maple syrup urine disease)	1:250 000	Branched-chain ketoacid decarboxylase (deficient)	During acute attacks: leucine, isoleucine, alloisoleucine, valine, and corresponding keto acids	During acute attacks: as in blood; also odor	Overwhelming acidosis; vomiting and CNS symptoms; sometimes mental retardation; respiratory failure; may be fatal	Dietary restriction of Leu, Ile, Val

Table 11-2,*B*. PRINCIPAL PRIMARY RENAL AMINOACIDURIAS (AUTOSOMAL RECESSIVE DISORDERS*)

Disorder	Incidence	Excess Amino Acids in Urine	Clinical Symptoms†	Treatment
Cystinuria, classic	1:13 000	Lysine; ornithine; arginine; cystine	Cystine renal calculi	High fluid intake; keep urine alkaline; D-penicillamine
Hartnup disease	1:18 000	All neutral amino acids	May be symptomless; usually pellagra-like dermatitis; neurological and psychiatric symptoms	Adequate protein diet; nicotinamide

*In some probable, rather than proved.

†Affected subjects usually have a transport defect in the jejunal mucosa, as well as in the proximal renal tubules.

sequence, and normal enzyme activity is either absent or significantly reduced. The defective protein is usually detectable immunologically; occasionally it is completely absent, as in half the cases of homocystinuria. In the heterozygote, enzyme activity is decreased (e.g., half normal) and the block in the metabolic path is only partial.

Secondary aminoacidurias affect many amino acids simultaneously and may be of the "overflow" or "renal" type. Examples of disorders that result in secondary *overflow aminoaciduria* are acute hepatic failure and acetaminophen poisoning. Generalized secondary *renal aminoaciduria* has a variety of causes, acquired or inherited, all of which cause proximal renal tubular dysfunction. If rickets is also present, the condition is called Fanconi's syndrome;[68] in childhood it may be secondary to cystinosis, an autosomal recessive disorder causing widespread intracellular deposits of cystine but without cystinuria. Generalized renal aminoacidurias may be caused by poisons, especially heavy metals; by wasting from starvation or disease; by acute tubular necrosis; or by congenital diseases such as galactosemia and Wilson's disease. In some patients the tubular defect involves amino acids only. In others, all the functions carried out by the proximal tubule are affected, as in Fanconi's syndrome, where glucosuria and hyperphosphaturia are also common.

The diagnosis of the inherited aminoacidurias. In the study and treatment of primary aminoacidurias, there are three separate problems: the diagnosis of the sick infant or child, routine neonatal screening, and the possibility of prenatal diagnosis.

The sick infant or child.[10] Inherited aminoacidurias are uncommon and their symptoms are often nonspecific, but they must be diagnosed rapidly before irreparable damage is done. These diseases should be suspected when an infant or young child suffers from persistent vomiting, failure to thrive, or neurological or liver abnormalities, especially if a baby appears well initially but deteriorates after a few days as the protein intake is increased.

The determination of *total* amino acids in plasma or urine is seldom useful clinically. Instead, separation, detection, and estimation of *individual* amino acids are required. Thin-layer chromatography (TLC) is ideal for this purpose since it is relatively simple and inexpensive, and a large number of specimens can be processed simultaneously. This technique, however, is only semiquantitative and is difficult to automate. Simple chemical or microbiological tests for individual amino acids and metabolites are often used to supplement TLC.

An attempt should be made to reach, as soon as possible, a provisional diagnosis that can be confirmed later if necessary. Failure to do so carries the risk that the child may die or develop irreversible problems before any information is obtained from the laboratory. Ideally the laboratory should carry out a battery of tests to exclude inherited metabolic diseases in general, and not just the aminoacidurias.[27]

Neonatal screening is commonly carried out in order to make the earliest possible diagnosis in serious treatable aminoacidurias and especially phenylketonuria. The most widely used procedure is the Guthrie test,[25] a semiquantitative microbiological assay that will be discussed later.

Prenatal diagnosis of an inherited aminoaciduria is particularly desirable in the case of a mother who has had a previous child with a severe inherited defect. Preliminary study of parents, affected relatives, or the previous affected child, preferably by skin fibroblast cultures, is essential for defining the defect to be excluded. The usual chemical tests for metabolites are not applicable to prenatal diagnosis. Although amniotic fluid is easily obtained, it may not reflect a metabolic defect in the fetus. Prenatal diagnosis, therefore, must be based on identification either of primary enzyme deficiency or of abnormal DNA in fetal cells. The most common approach is to collect amniotic fluid by amniocentesis at 14–16 weeks of gestation, culture the fetal cells, and then examine the cells for enzyme activity. Unfortunately, 3–6 weeks may be required to grow enough cells for the enzyme assays. For the earliest diagnosis, sufficient fetal cells would have to be obtained by biopsy of trophoblastic villi to make culture unnecessary and yet supply enough fetal cells for gene analysis. These techniques are now being developed. Direct gene analysis involves cutting up DNA with restriction endonucleases, separating the fragments by gel electrophoresis, and identifying the genes with radioactive gene probes.[20]

Amino Acid Analysis

Three groups of tests for amino acid analysis may be considered.

1. The initial diagnostic screening tests referred to above—thin-layer chromatography, urine color tests, and Guthrie microbiological tests.

2. Quantitative tests for monitoring treatment or confirming the initial diagnosis. For these estimations, ion-exchange or high performance liquid chromatography is preferred; these procedures are complex and expensive, but can be automated. Some individual amino acids may also be determined by quantitative chemical methods.

3. Exact identification of an unknown amino acid or metabolite. Since chromatographic separation gives only one identifying characteristic, the retention time, other properties such as the infrared spectrum may need to be examined; the most discriminating automated system is a combination of gas-liquid chromatography and mass spectrometry, available in selected medical centers.

Screening Tests

Specimen. The child should be on a normal diet, and blood and urine specimens for chromatography should be collected simultaneously. Plasma should be harvested promptly and hemolysis avoided, because the concentration of amino acids is much greater in erythrocytes than in plasma. Plasma, urine, and CSF should be stored frozen if analysis is delayed.

Thin-Layer Chromatography (TLC)

There are three stages in any chromatographic analysis of amino acids: preparation of the sample, chromatographic separation, and identification of the separated amino acids. Chromatography is a versatile technique, and details can be varied to suit specific problems (see Chapter 2, Section Six). For the analysis of amino acids in body fluids and tissues, pretreatment of the sample is often necessary in order to remove proteins, lipids, inorganic salts, or other substances that will interfere with chromatographic resolution. Most biological fluids are first deproteinized with sulfosalicylic acid or some other appropriate protein-precipitating agent. Following centrifugation to remove precipitated proteins, the supernatant is then desalted by means of solvent extractions, ion-exchange resins, or electrodialysis. Relatively dilute specimens such as CSF or urine may not require extensive desalting pretreatment.

The amount of amino acids visible in a chromatogram is influenced not only by the disease process but also by the volume of fluid applied to the chromatogram. The sample volume is therefore standardized by reference to its total nitrogen content or the amount of creatinine in a specified volume of the specimen.

Of the possible choices of *stationary phases*, paper and thin layers of cellulose continue to be favored by many investigators. Procedures using cellulose have the advantage of giving superior chromatographic resolution and reducing the time required for solvent development. Procedures using paper, however, are very useful when blood or urine samples are collected on filter paper disks.

A large number of *solvent systems*[77] have been proposed for the separation of amino acid mixtures. One-dimensional chromatography is favored by some laboratories because of its simplicity—multiple standards and samples can be easily run on a single plate. Two-dimensional TLC is similar to one-dimensional methods except that, after the first migration, the chromatogram is rotated 90° and then transferred to another solvent system for a second migration. As a result, amino acids are separated in two dimensions. When selective staining reagents are used in conjunction with two-dimensional solvent systems, identification of more than 75 compounds of biochemical interest is possible.

Many *staining reagents* can be used to reveal amino acids separated by TLC. The most widely used is ninhydrin. A number of colored products are formed, but the major one is thought to result from deamination and condensation of the amino acid with ninhydrin:

$$\text{Ninhydrin} + R-\underset{H}{\overset{NH_2}{C}}-COOH \longrightarrow \text{Colored Product}$$

Ninhydrin **Amino acid** **Colored Product**

Note that only the α-nitrogen from the original amino acid molecule is ultimately part of the measured product.

Most amino acids react with ninhydrin at ambient temperatures to form a blue color that becomes purple upon heating. However, proline and hydroxyproline yield yellow compounds that are less satisfactory for visual observation; consequently, additional stains such as isatin (indole-2,3-dione) are often used. Isatin converts proline and hydroxyproline to blue compounds that are detectable against a yellow background. The addition of various divalent metal ions to ninhydrin or isatin staining solutions is frequently used to help stabilize the color complexes that are formed. Addition of an organic base, such as collidine, to ninhydrin solutions produces a more discriminating polychromatic stain; the reader is referred to other publications for descriptions of stains.[69,76]

One-dimensional chromatography. The principle of one-dimensional chromatography is described in Chapter 2, Section Six. One-dimensional thin-layer chromatography of plasma, serum, or urine can detect most aminoacidopathies. Relatively inexpensive commercial kits (e.g., Helena Laboratories, Beaumont, TX 77704, catalog no. 544) are available and well suited for use in smaller laboratories. In these methods, specimens, standards, and controls are applied to precoated cellulose thin-layer plates. Development is carried out in an appropriate solvent-vapor system,[5] followed by drying and staining of the chromatogram. The standard is a mixture of amino acids of known concentration. Examples of one-dimensional chromatograms are shown in Figure 11-1. By running several samples of the same specimen on a single plate and then staining each with different specific stains, it is often possible to identify an individual amino acid without resorting to the two-dimensional technique.

Two-dimensional thin-layer chromatography.[69,76] The samples, usually fresh urine or serum, are purified, standardized, and analyzed on precoated cellulose thin-layer plates. One sample is applied at one corner of each plate, and after separation of the amino acids in one direction the plate is then turned at a right angle and the amino

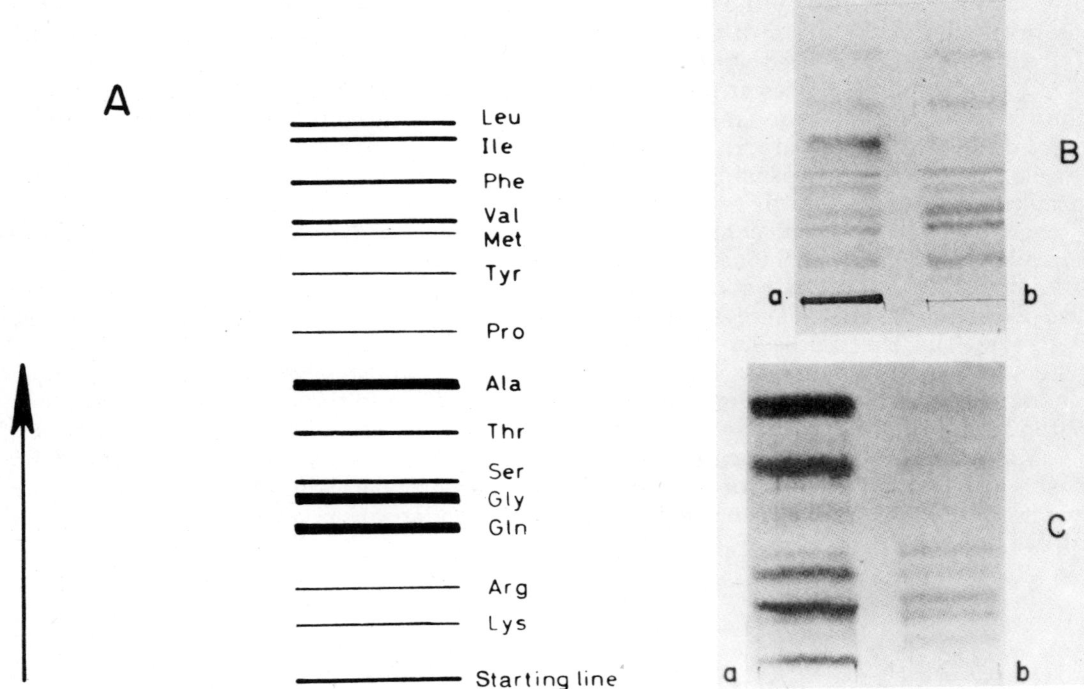

Figure 11-1. One-dimensional chromatography of amino acids in a solvent system consisting of *n*-butanol: acetone:acetic acid:water (35:35:10:20 by volume). *A,* Positions of several amino acids after their separation from a standard mixture. *B,* Separation of plasma amino acids in transient neonatal tyrosinemia. Note tyrosine band **a**; **b** = amino acid standard. *C,* Separation of plasma amino acids in untreated maple syrup urine disease; **a** = sample; the most prominent bands are those of isoleucine, leucine, and valine; **b** = amino acid standard. (From Bremer, H. J., Duran, M., Kamerling, J. P., et al.: Disturbances of Amino Acid Metabolism: Clinical Chemistry and Diagnosis. Baltimore, Urban and Schwarzenberg, Inc., 1981.)

acids are further separated by a second solvent mixture, as in Figure 11-2. The first solvent mixture contains ammonia to increase the mobility of basic amino acids; the second contains formic acid to increase the mobility of dicarboxylic amino acids. After development, the plate is dried and stained with ninhydrin-collidine. The colors and R_f values of the spots are compared with those obtained with standard mixtures separated on the same type of plate at the same time; each amino acid can be identified, and its relative depth of staining assessed, by tracing the patient's chromatogram onto a clear vinyl sheet that is then superimposed onto the standard's chromatogram.[76] Note that R_f values may vary slightly between runs. A complete map is presented in Figure 11-2. A *normal* urine chromatogram is characterized by the presence of five prominent amino acids: alanine, glutamine, glycine, histidine, and serine. In pediatric urines moderate amounts of β-aminoisobutyric acid, ethanolamine, glutamic acid, lysine, taurine, and threonine also occur. Traces of asparagine, cystine, or methylhistidine are only infrequently found.

Colorimetric Screening Tests for Urine

A variety of qualitative tests (described later in detail) can be employed for screening, spot checking, or supplemental information. These tests are summarized in Table 11-3.

Guthrie Microbiological Test for Neonatal Screening

Bacterial spores, usually *B. subtilis*, are added to an agar medium containing a competitive growth inhibitor specific for the amino acid to be determined and usually of similar molecular structure (Table 11-4). Blood or urine from the patient is spotted onto a piece of soft filter paper; a standardized portion of the paper with the specimen is punched out and laid on the agar surface. The agar plate is then incubated and later

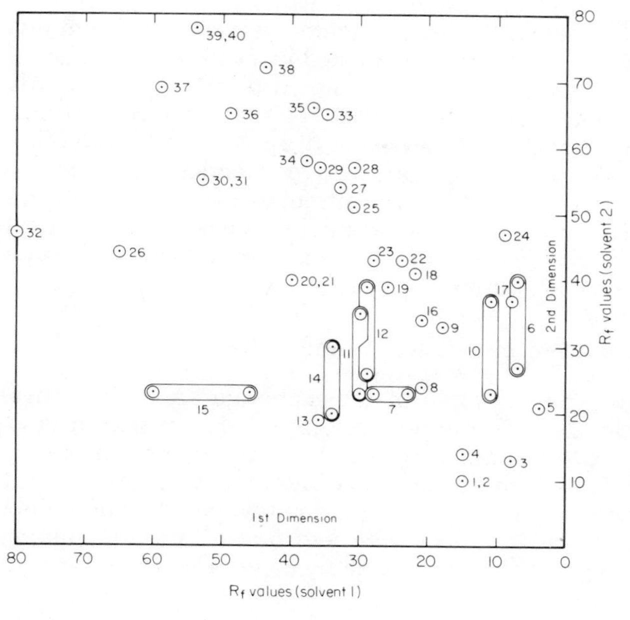

Figure 11-2. Composite map of R_f values (×100) of amino acids obtained with two-dimensional chromatography on cellulose. First dimension solvent—pyridine:acetone:concentrated ammonium hydroxide:water (40:30:5:20 by volume). Second-dimension solvent—2-propanol:concentrated formic acid:water (66:15:15 by volume). 0 = origin.

Amino Acid	Color with Ninhydrin-Collidine Stain
1. Cystine	Gray
2. Cysteine	Gray
3. Cystathionine	Violet
4. Cysteic acid	Violet
5. Phosphoethanolamine	Violet
6. Argininosuccinic acid	Violet
7. Ornithine	Violet
8. Asparagine	Beige
9. Homocystine	Violet
10. Arginine	Violet
11. Lysine	Violet
12. 1-Methylhistidine	Gray
13. Histidine	Gray
14. Carnosine	Tan
15. Taurine	Violet
16. Glutamine	Violet
17. Aspartic acid	Aquamarine
18. Citrulline	Violet
19. Methionine sulfoxide	Violet
20. Methionine sulfone	Violet
21. Serine	Violet
22. Glycine	Rose
23. Hydroxyproline	Orange
24. Glutamic acid	Violet
25. Sarcosine	Gray-pink
26. Threonine	Violet
27. β-Alanine	Aquamarine
28. Alanine	Violet
29. Proline	Yellow
30. Tyrosine	Gray-brown
31. Tryptophan	Yellow-gray
32. Ethanolamine	Violet
33. γ-Aminobutyric acid	Violet
34. α-Aminobutyric acid	Violet
35. β-Aminoisobutyric acid	Violet
36. Methionine	Violet
37. Phenylalanine	Gray
38. Valine	Violet
39. Leucine	Violet
40. Isoleucine	Violet

Table 11-3. COLOR TESTS ON URINE

Test and Color	Metabolites	Disorder
Ferric chloride* (p. 308)		
Dark blue-green (persistent)	Phenylpyruvate	Phenylketonuria
Green (transient)	*p*-Hydroxyphenylpyruvate	Tyrosinemia
Blue (transient)	Homogentisic acid	Alkaptonuria
Gray-green	Imidazolepyruvate	Histidinemia
Gray-blue	Branched-chain keto acids	Maple syrup urine disease
Blue-green	5-Hydroxyindoleacetic acid	Carcinoid
Purple	Salicylates	(Interferent)
Purple-brown	Phenothiazines	(Interferent)
Cyanide/nitroprusside (p. 311)		
Cherry-red	Cystine	Cystinuria; generalized aminoacidurias
	Homocystine	Homocystinuria
	Cysteine-homocysteine disulfide	Homocystinuria
	Penicillamine-cysteine disulfide	(Treatment)
2,4-Dinitrophenylhydrazine (p. 312)		
Yellow-white (ppt)	Branched-chain keto acids; phenylpyruvate; *p*-hydroxyphenylpyruvate	Maple syrup urine disease; phenylketonuria; tyrosinosis
Nitrosonaphthol (p. 310)		
Orange-red	Tyrosine and its metabolites (e.g., *p*-hydroxyphenylpyruvate, -lactate, -acetate)	Tyrosinemia

*Phenistix (Ames Co., Division of Miles Laboratories, Elkhart, IN 46514) is a commercial version of the ferric chloride test (p. 308).

observed for bacterial growth. In the presence of elevated concentrations of the amino acid of interest, the effect of the growth inhibitor is diminished or overcome, and zones of bacterial growth will be observed. The test system is designed to show growth only when the concentration of the amino acid of interest exceeds its upper reference limit. To avoid false-negative results, the specimen must be free of drugs such as antibiotics. Every positive result should be confirmed with a quantitative chemical test for the specific amino acid or by chromatography. Note that positive results on premature infants may be due to immaturity of the liver. Specimens should not be obtained until 48 h after birth, since 2 days of protein intake are required to accumulate an abnormal concentration of an amino acid (e.g., phenylalanine).

Quantitative Separation Procedures for Amino Acids

Ion-exchange chromatography. The principle of ion-exchange and other chromatographic separations of amino acids has been extensively applied in research laboratories (see Chapter 2, Section Six). After column separation of the compounds, ninhydrin or some other indicator is combined with the column effluent so that amino acids can be detected by using an on-line spectrophotometer, fluorometer, or other detection device. The amino acids are identified by relating the *retention times* of the components in the specimen to those of standards. Quantitation is made by comparison of specimen peak areas with those from sets of standards.

Ion-exchange chromatography, however, is not suitable for the average clinical laboratory. The technique requires a large, expensive instrument dedicated solely to assays that are infrequently needed. Furthermore, completion of an analysis of one specimen requires at least 3 h.

High performance liquid chromatography (HPLC). HPLC[37] has emerged as an attractive means of amino acid analysis because of its great sensitivity, high resolution, and relatively short analysis time. The major advantage of HPLC over gas-liquid chromatography is the avoidance of amino acid decomposition at high temperatures. Detection of amino acids separated by HPLC is ordinarily achieved by ultraviolet or visible spectrophotometry, fluorometry, or electrochemical detection. Initial sample purification and prederivatization may be required, since a number of interfering non-amino acid substances also absorb light in the UV range.

Gas-liquid chromatography (GLC). Advantages of GLC include small sample size, sensitivity, and speed, but a major limitation is the relatively low volatility of amino acids at temperatures conventionally used in this technique; this necessitates multiple derivatization to increase volatility and to enhance chromatographic and detection characteristics and often involves formation of amino acid alkyl esters with subsequent acetylation reactions.[75]

Selected Disorders of Amino Acid Metabolism

Hyperphenylalaninemias

The hyperphenylalaninemias are a group of disorders resulting from impaired conversion of phenylalanine (Phe) to tyrosine. Normally, the major metabolic pathway for phenylalanine involves an enzyme complex known as phenylalanine hydroxylase (PH), present in liver and kidneys.

Hyperphenylalaninemia is nearly always due to deficiency of PH, although some 1–3% of cases are due to a deficiency of tetrahydrobiopterin. A PH defect causes Phe to accumulate in blood and CSF; at a plasma level of 12–15 mg/dL, the minor transamination pathway is stimulated, and increased production of phenylpyruvate and its metabolites leads to increased renal excretion of these substances. The term *phenylketonuria* (PKU) indicates increased urinary concentration of phenyl compounds that react as ketones do in the classical ferric chloride tests; actually, phenylpyruvic acid is the only true ketone involved.

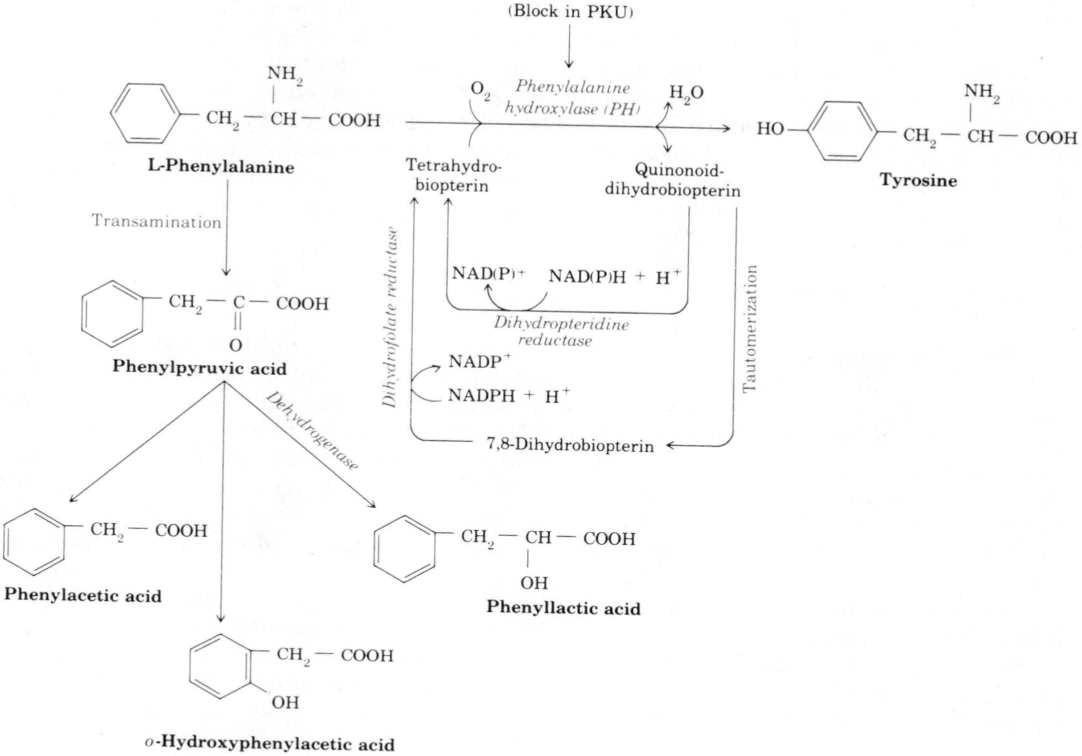

(Block in PKU)

L-Phenylalanine → (Phenylalanine hydroxylase (PH); O_2 → H_2O) → Tyrosine

Tetrahydro-biopterin → Quinonoid-dihydrobiopterin

$NAD(P)^+$ ← $NAD(P)H + H^+$ (Dihydropteridine reductase)

Tautomerization → 7,8-Dihydrobiopterin

$NADP^+$ ← $NADPH + H^+$ (Dihydrofolate reductase)

Transamination → Phenylpyruvic acid

Phenylpyruvic acid → Phenylacetic acid; o-Hydroxyphenylacetic acid; Dehydrogenase → Phenyllactic acid

In about half the cases of *primary phenylalanine hydroxylase deficiency*, PH is almost totally absent. Less severe forms of PKU, often called PKU variants, result from partial deficiencies of PH. Heterozygote carriers of the defective gene compose about 2% of the Caucasian population in the United States. They usually have normal blood levels of Phe, but their reduced capacity to metabolize Phe can be demonstrated with a load test. Delay in hepatic synthesis of PH may, in normal neonates, cause a *transient neonatal hyperphenylalaninemia*, but the initial high blood level gradually falls to normal. In infants with the inborn error, the Phe level continues to rise and may reach 10 times or more the normal level for unaffected adults.

PKU, unless treated, causes severe mental retardation. Affected children appear normal at birth and the earliest symptoms are usually nonspecific—delayed development, feeding difficulties, and vomiting sometimes sufficiently severe to suggest pyloric stenosis. Brain injury from the high concentration of Phe in CSF begins within the second or third week of life and becomes maximal in 8–9 months. Other symptoms are a musty odor from the phenylpyruvate in urine and sweat, and hypopigmentation, a consequence of phenylalanine's activity as a competitive inhibitor of tyrosinase (the enzyme that initiates melanogenesis).

Early diagnosis is essential if treatment is to be started in time to avoid mental retardation. Hence, routine neonatal screening by the Guthrie test is widely practiced. Treatment consists of restricting dietary Phe to keep plasma Phe concentration at 3–8 mg/dL. Even if the diagnosis is made as late as 4–6 months, dietary therapy decreases the rate of further mental deterioration. The patient is monitored best by quantitative blood analysis, but urine tests for phenylketones (see below) are helpful as early warning of breakdown of dietary control. Monitoring of blood Phe and dietary control should continue indefinitely, especially for females, since a high serum Phe in pregnancy has disastrous effects on the fetus (cardiac defects, anencephaly, retardation). To avoid *maternal hyperphenylalaninemia*, which is becoming more common as successfully treated children grow up, dietary Phe (including aspartame, the synthetic peptide sweetener) must be restricted before conception and throughout pregnancy.

Hyperphenylalaninemia due to *tetrahydrobiopterin deficiency* arises from a defect in either dihydropteridine reductase or an enzyme involved in biopterin synthesis. Tetrahydrobiopterin is also required as coenzyme in the hydroxylation of tryptophan and tyrosine. The products of these reactions are precursors of neurotransmitters, and therefore deficiencies result in severe, progressive neurological disease that is not prevented by dietary control of Phe. Replacement therapy with precursors of the neurotransmitters has some clinical benefit.

Methods for Determination of Serum Phenylalanine

The Guthrie qualitative microbiological screening test is used to identify neonates with high serum Phe (page 304). For quantitative estimation of phenylalanine, there is a choice of fluorometric, chromatographic, and spectrophotometric methods. A capillary blood specimen sufficient to provide 50 μL of serum is required for fluorometric procedures. If analysis is delayed, plasma or serum should be separated and stored frozen. In one fluorometric method[19] Phe reacts with ninhydrin at pH 5.88 in the presence of the dipeptide L-leucyl-L-alanine to form a product whose fluorescence is proportional to the concentration of Phe. Alkaline copper tartrate is present to stabilize the fluorescent product.[76]

Alternative methods are automated ion-exchange column chromatography,[47] gas liquid chromatography, and spectrophotometry. A contemporary spectrophotometric method uses the enzyme phenylalanine-ammonia lyase to convert Phe to transcinnamic acid.[74] Older methods that use L-amino acid oxidase[41] are subject to interferences by tyrosine, tryptophan, and histidine.

Ferric Chloride Test for Phenylpyruvic Acid (PPA) in Urine

This test is both qualitative and nonspecific and is useful in monitoring dietary therapy in PKU. It is not an alternative to blood Phe estimations for diagnosis, since PPA usually cannot be detected in infant urines until the serum phenylalanine concentration has reached levels between 12 and 15 mg/dL; this does not usually occur until 10–14 d after birth. Affected infants will excrete up to 2 g PPA/g creatinine, and urinary concentrations will vary from 50 to 100 mg/dL.

Reagents

1. $FeCl_3$ solution. Dissolve 10 g $FeCl_3 \cdot 6\ H_2O$ in water to give 100 mL solution.
2. Phosphate precipitating agent. Dissolve 2.2 g $MgCl_2 \cdot 6\ H_2O$, 1.4 g NH_4Cl, and 2.0 mL concentrated NH_4OH in water to give 100 mL of solution.

Procedure

Add, with mixing, 1.0 mL of phosphate precipitant to 4.0 mL of fresh urine. Filter the mixture and acidify the filtrate with 2–3 drops of concentrated HCl. Add 2–3 drops of $FeCl_3$ solution; observe the filtrate after each drop. A dark green to blue-green color that persists for 2–4 min indicates a positive test. Very rapidly fading greens suggest homogentisic acid (HGA) or *p*-hydroxyphenylpyruvic acid (PHPPA). Imidazolepyruvic acid gives a positive test in histidinemia, identical with that for PPA, but it is encountered only rarely. Bilirubin may give a false-positive reaction. Other colors than those given in Table 11-3 are read as negative. The sensitivity for PPA is about 10 mg/dL. By using serial dilutions of the urine filtrate, a semiquantitative measure of the PPA concentration may be made. A commercial product Phenistix (Ames Co., Division of Miles Laboratories, Elkhart, IN 46515), based on the ferric chloride reaction, may also be used to test for PPA. It changes the color of the reagent-impregnated paper from a light gray to a gray-green or green-blue. Light, medium, and high readings correspond to PPA levels of 15, 40, and 100 mg/dL, respectively.

Test for Heterozygote or Carrier State

The suspected carrier is given an oral dose of 100 mg of phenylalanine per kg body weight. Blood specimens are then obtained at hourly intervals for 4 h and assayed for phenylalanine. In a noncarrier, phenylalanine concentrations will rise from a fasting value of about 1.4 mg/dL to a value of 9 mg/dL at 1 and 2 h, and then drop to 5 mg/dL at 4 h. In a heterozygote, the phenylalanine concentration will rise from normal to a value of about 19 mg/dL at the first hour and fall much more slowly than in the normal person. A phenylalanine tolerance index is calculated by summing the 1-, 3-, and 4-h values. The values of the index for normal individuals and carriers are sufficiently disparate to distinguish clearly between the two.

Tyrosinemia and Related Disorders

There are several forms of tyrosinemia, each of which is accompanied by tyrosinuria and phenolic aciduria. Tyrosine is a precursor of thyroxine, melanin, and the catecholamines. An outline of its metabolism is shown in Figure 11-3. The major path is by way of tyrosine aminotransferase to *p*-hydroxyphenylpyruvic acid (PHPPA) and homogentisic acid (HGA). HGA oxidase then opens the ring.

The pigment melanin is produced by the melanocytes in the skin. Many of the clinical syndromes associated with hypomelanosis, known collectively as *albinism*, result from inherited defects in melanin synthesis, in which tyrosinase activity is absent or deficient. In contrast, melanomas (melanocyte tumors) of the skin or retina, especially when complicated by liver metastases, may cause overproduction of melanin and its colorless precursors and result in *melanuria*. The melanogens in urine darken on exposure to air and sunlight.

Inherited tyrosinemia is of two types. *Type I* (tyrosinosis), probably due to defective FAA hydrolase, causes liver damage and generalized renal tubular failure (Fanconi's syndrome, page 301). Most cases of Type I have been found in an isolated French-Canadian population in Quebec. The rare *Type II*, due to defective tyrosine aminotransferase, affects the eyes and the skin of the palms and soles. In both types there are high levels of tyrosine in blood and urine. Patients with Type I also have high blood levels of methionine and α-fetoprotein and a urinary excretion of PHPPA up to 1.6 g/d, about 25 times normal.

Transient neonatal tyrosinemia occurs in premature infants and in full-term infants of low birth weight; it is caused by immaturity of the liver and its decreased ability to synthesize enzymes. As the liver matures, the accumulated tyrosine is metabolized, and serum levels fall to adult levels in 4–8 weeks.

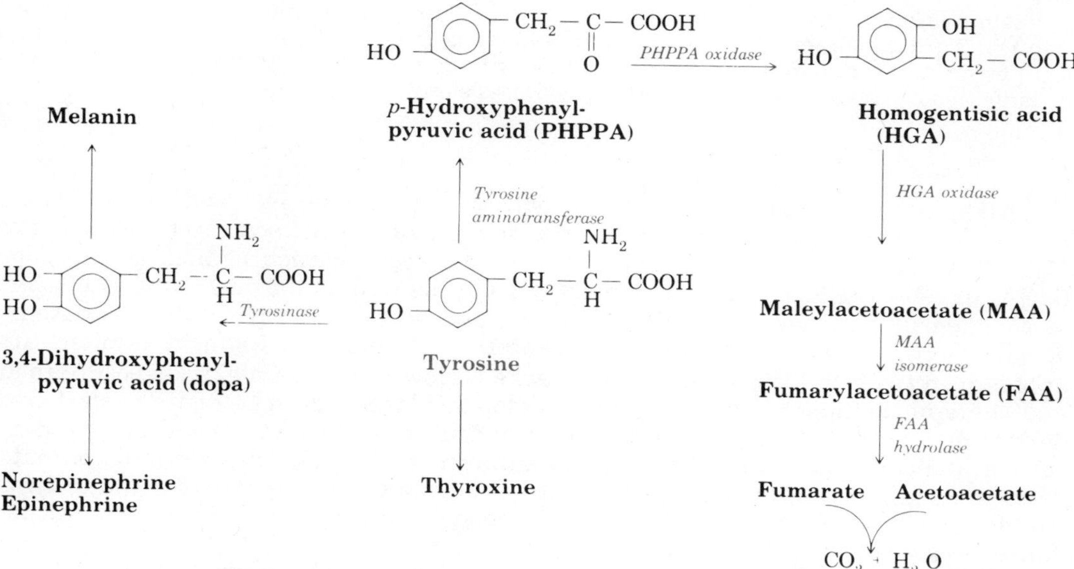

Figure 11-3. Metabolism of tyrosine with important intermediary and end products.

Methods for the Determination of Tyrosine in Serum

The most convenient method is the fluorometric procedure of Udenfriend and Cooper,[86] but it is unsuitable for urine, where interfering compounds are present in significant concentrations. Ion-exchange column chromatography is regarded as a reference method.

Enzymatic methods employ tyrosinase from mushrooms, L-amino acid oxidase from snake venom,[41] or ammonia lyase from yeast.[74] Tyrosinase catalyzes the oxidation of tyrosine to dopaquinone, and consumption of oxygen can be measured amperometrically.

Detection of PHPPA in Urine

Type I tyrosinemia can be diagnosed most reliably through isolation and identification of PHPPA from urine by *chromatography*. More conveniently, the Millon reaction or the nitrosonaphthol test can be used to detect a wide variety of substituted phenolic compounds in urine. These tests, however, are not specific for PHPPA; positive reactions are also given by tyrosine and tyrosine metabolites, such as *p*-hydroxyphenyllactate (PHPLA) and *p*-hydroxyphenylacetate (PHPA), that accompany the increased urinary output of PHPPA.

Modified Millon reaction. Mix 2.5 mL of urine and 2.0 mL of a solution of 15 g $HgSO_4$ in H_2SO_4, 2.5 mol/L; allow to stand for 1 h and then centrifuge. Transfer the supernatant to a 50-mL beaker, add 10 mL of H_2SO_4, 1 mol/L, and gently boil the mixture for 10–15 min. Add an additional 2 mL of sulfuric acid, 1 mol/L, and cool for 30 min. Then add 1 mL of fresh $NaNO_2$, 2 g/dL. The formation of an orange color indicates a positive test.

Nitrosonaphthol test. To 1 mL nitric acid (one volume of concentrated acid plus five volumes H_2O), add 1 drop of a freshly prepared aqueous solution of sodium nitrite, 2.5 g/dL. Mix well. Then add 10 drops nitrosonaphthol solution, 0.1 g/dL, in 95% ethanol, and 3 drops of urine. Mix well. Formation of an orange-red color within 3–5 min is reported as a positive result.

Tests for Melanogens in Urine

Thormählen test. Add 5–6 drops of fresh sodium nitroferricyanide solution (5.0 g/dL in water) to 5 mL of urine in a test tube, then add 0.5 mL of NaOH, 10 mol/L, and mix vigorously. Rapidly cool the tube under cold tap water and acidify with acetic acid solution (glacial acetic acid, 33 mL/dL). Normal urine gives an olive or brownish-green color; if melanogens are present, the color varies from greenish-blue to bluish-black, depending on the amount present. Although high levels of creatinine give a brown color, the Thormählen test is probably the most specific and sensitive test available for melanin or melanogens.

Ammoniacal silver nitrate test. Add 0.5 mL of urine to 5.0 mL of an aqueous $AgNO_3$ solution, 3 g/dL, followed by dilute ammonia solution (2 mL concentrated NH_4OH/dL) until the AgCl precipitate is almost dissolved. The solution will darken as a result of the formation of both melanin and colloidal silver. The reaction progresses slowly.

Alkaptonuria

Alkaptonuria results from deficient activity of homogentisic acid oxidase in the major catabolic pathway that converts tyrosine to fumarate and acetoacetate (see Figure 11-3). Reduced activity of this oxidase causes the accumulation of homogentisic acid (HGA) in cells and body fluids. Polymers of HGA, as well as the compound itself, bind to collagen in cartilage, intervertebral disks, and other connective tissue, eventually causing degenerative arthritis and pigmentation of cartilage. Pigment is often first apparent in the ears (ochronosis). Urine darkens slowly on standing, on exposure to air and sunlight, or upon addition of alkali; color formation must be distinguished from darkening due to melanuria, phenols, gentisic acid (a salicylate metabolite), or indoxyl sulfate (indican, a metabolite of tryptophan, formed in stagnant intestinal contents). Alkaptonuria is not usually diagnosed until middle age, when ochronosis and arthritis lead to suspicion, but it can be diagnosed in neonates if the dark stain in an unwashed diaper is noticed and investigated.

There is no satisfactory treatment for alkaptonuria. If the diagnosis is made early, dietary restriction of tyrosine or its precursor phenylalanine may be beneficial.

Reduction Tests for Homogentisic Acid

Ammoniacal silver nitrate test. If HGA is present, a brown-black to black precipitate of elemental silver is formed immediately, often even before the addition of ammonia. Melanogens react slowly and only if there is an excess of ammonia.

Benedict's qualitative glucose reagent, if heated with urine containing HGA, develops a dark supernatant and a yellow precipitate of cuprous oxide. Melanogens do not reduce the reagent unless present in very large quantity; even then the supernatant is not dark.

Homocystinuria

Cystathionine β-synthase deficiency is the usual cause of homocystinuria. This deficiency is caused by a block in the conversion of homocysteine to cystathionine with the result that homocysteine and methionine accumulate in body fluids while their normal metabolites, cysteine and cystine, decrease. Methionine increases up to 30 mg/dL in plasma, and homocystine is detectable. The urine contains homocystine, methionine, and other sulfur-containing amino acids.

There are no symptoms in the newborn, but ocular, skeletal, and vascular symptoms gradually develop. The most serious is thrombosis, which may be lethal at an early age or may cause mental retardation. Early diagnosis is possible by neonatal screening for increased plasma methionine levels. Treatment by dietary restriction of methionine, supplements of cystine, or large doses of the cofactor pyridoxine can be started early. Note that increased plasma methionine may also be transient or due to other causes such as tyrosinosis.

Deficiencies of N^5-methyltetrahydrofolate or its transferase cause a block in the normal conversion of homocysteine back to methionine and cause rare forms of homocystinuria, distinguished by the absence of an increased level of plasma methionine.

Cyanide-nitroprusside test for cystine and homocystine is based on the reaction of sodium nitroprusside with sulfhydryl compounds to produce a red-purple color.

To 5 mL urine add 5 drops of concentrated NH_4OH followed by 2 mL of an aqueous solution of NaCN, 5 g/dL. *Caution*: NaCN is poisonous. Allow the mixture to stand for 10 min. Then add, dropwise, a freshly prepared aqueous solution of sodium nitroprusside, 5 g/dL. Normal urines show a persistent yellow or pale brown color. Formation of a red-purple color is a positive result. The use of NH_4OH instead of NaOH prevents interference from creatinine. Acetoacetic acid and acetone give negative results with this particular version of the nitroprusside test.

Silver nitroprusside test for homocystine differentiates between cystine and homocystine. By substituting silver nitrate for sodium cyanide as the reducing agent, homocystine can be reduced to its thiol form (homocysteine) while cystine remains in the nonreactive oxidized form. Homocysteine then reacts with sodium nitroprusside to yield a pink-purple color.[76]

Cystinuria

Classic cystinuria is the most common inborn error of *amino acid transport*. It is characterized by massive urinary excretion of the dibasic amino acids: cystine, lysine, arginine, and ornithine. Normally these amino acids are filtered by the glomeruli and reabsorbed in the proximal renal tubules. In cystinuria reabsorption fails because a carrier system that transports all four amino acids is defective.

Because cystine is the least soluble of the naturally occurring amino acids, its overexcretion often leads to the formation of cystine calculi in the renal pelves, ureters, and bladder; obstruction, infection, and renal insufficiency occasionally result. Some 1–2% of all urinary tract stones are composed of cystine. Treatment involves reducing the concentration of cystine in urine by drinking large amounts of water, increasing cystine solubility by maintaining an alkaline urine, and, if necessary, reducing cystine excretion by using D-penicillamine. A test for cystine is given above.

Table 11-4. GUTHRIE TEST SYSTEMS

L-Amino Acid	Inhibitor	Disease	Reference Limits, mg/dL
L-Phenylalanine	β-2-Thienylalanine	Phenylketonuria	4
L-Leucine	4-Azaleucine	Maple syrup urine disease	4
L-Methionine	Methionine sulfoximine	Homocystinuria	2
L-Tyrosine	D-Tyrosine	Tyrosinemia	8
L-Lysine	S-(β-Aminoethyl) cysteine	Hyperlysinemia	4

Maple Syrup Urine Disease (MSUD)

The branched-chain amino acids, leucine, isoleucine, and valine, are normally converted by transamination to their corresponding α-keto acids, which, by oxidative decarboxylation, are then converted to acyl-CoA derivatives. An inherited defect of the decarboxylase step results in the accumulation of the branched-chain amino acids and their corresponding α-keto acids in blood, urine, and CSF. An affected infant appears normal at birth; then, as in other organic acidurias,[18,81] acute ketoacidosis develops and causes vomiting, lethargy, seizures, coma, and, in survivors, mental retardation. Amino acid analyses of blood and urine show high levels of leucine, isoleucine, and valine, and the presence of allo-isoleucine. The urine contains increased keto acids, which are responsible for its "maple syrup" odor, positive reaction in the dinitrophenylhydrazine (DNPH) test, and gray-blue color in the ferric chloride test. Neonatal screening is possible by a Guthrie test (Table 11-4) and prenatal diagnosis by an assay of decarboxylase in cultured amniotic cells. Treatment is by dietary restriction of leucine, isoleucine, and valine, monitored by daily DNPH urine tests and monthly amino acid assays.

2,4-Dinitrophenylhydrazine test for keto acids. Mix equal quantities of filtered urine and DNPH solution, 0.4 g/dL in HCl, 1 mol/L. A yellow to yellow-white precipitate of hydrazone forming within 5 min indicates excess keto acids. Interference by acetone is eliminated by first boiling the urine briefly.

PROTEINS OF BODY FLUIDS

The human body contains countless different proteins, some soluble in fluids within or outside the cells, some insoluble and structural. The only ones easily available for investigation are the soluble proteins of the extracellular fluids. In plasma, more than 300 proteins have been identified, but they differ greatly in concentration and only the most abundant will be considered here. The plasma proteins move between the blood and other extracellular fluids by active transport as well as by passive diffusion. When cells are damaged, soluble proteins normally in cells or on their surfaces may also be present in extracellular fluid. Most of the proteins of urine and spinal fluid are derived from plasma.

Table 11-5 lists the principal plasma proteins.[67] These proteins are described below,[89] following the discussion of protein analysis. Other chapters describe enzymes, lipoproteins, and hormones, as well as hemoglobin, fibrinogen, and other coagulation factors. The column headed "Acute Phase Behavior" in Table 11-5 refers to the "acute phase reaction"[42,50] (the response to acute inflammation, for example, after surgery, myocardial infarction, and infections, or in the presence of tumors). In this plasma protein reaction, more of the "acute phase reactants" (APRs) and less of the "negative acute phase reactants" are synthesized by the liver, the site of synthesis of most plasma proteins. Levels of the individual APRs increase at different rates and times—first, C-reactive protein and α_1-antichymotrypsin; then, within the first 12 h, α_1-acid glycoprotein; then α_1-antitrypsin, haptoglobin, C4, and fibrinogen; and finally C3 and ceruloplasmin. All reach their maxima within 2–5 d (Figure 11-4). These changes in APRs do not aid in diagnosing the cause of the inflammation, but measurements of those APRs

Table 11-5. PROPERTIES OF SELECTED PLASMA PROTEINS

Protein	Adult Reference Ranges, mg/dL	$t_{1/2}$	pI	MW	Carbohydrate Content, %	Acute Phase Behavior	Comments
Prealbumin (PA)	20–40	12 h	4.7	54 400	0	Neg APR*	Indicator of nutrition
Albumin	3500–5000	15–19 d	4–5.8	66 000	0	Neg APR	Reduced in many diseases: a general transport protein
Electrophoresis, α_1-region							
α_1-Antitrypsin (AAT)	78–200	4 d	4.8	55 000	12	APR	Congenital deficiency may cause emphysema or cirrhosis
α_1-Acid glycoprotein (AAG, orosomucoid)	50–150	5 d	2.7–4	40 000	45	APR	Function obscure
α_1-Lipoprotein (Apolipoprotein A)	170–325			200 000			Transports lipids
α_1-Fetoprotein (AFP)	0.003			69 000			Principal fetal protein
Electrophoresis, α_2-region							
Haptoglobin (HAP, Hp)	30–215	2 d	4.1†	85 000 >1 000 000	12	APR	Binds Hb; reduced by hemolysis
α_2-Macroglobulin (AMG)	125–410	5 d	5.4	800 000	8		Increased in nephrotic syndrome
Ceruloplasmin (CER)	20–50	4.5 d	4.4	160 000	7	APR	Decreased in Wilson's disease; contains Cu
Electrophoresis, β_1-region							
Transferrin (TRF, siderophilin)	200–350	7 d	5.7	77 000	6	Neg APR	Transports Fe, increased in hypochromic anemia
Hemopexin (Hx, Hpx)	50–115			57 000			Binds heme
β-Lipoprotein (Apolipoprotein B)	60–155			~3 000 000			Transports lipids
C4	10–40			206 000	7	APR	Complement factor
Electrophoresis, β_2-region							
Fibrinogen	200–400	2.5 d	5.5	340 000	3	APR	Precursor of fibrin clot
C3	70–150			180 000	2	APR	Complement factor
β_2-Microglobulin (BMG)	0.1–0.2			11 800			Used to test renal tubular function
Electrophoresis, γ-region							
IgG	525–1650	24 d	6–7.3	160 000	3		Antibodies increase in immune reactions; monoclonal increases in B cell tumors
IgA	40–390	6 d		170 000	8		
IgM	25–310	5 d		900 000	12		
C-reactive protein (CRP)	<0.8		6.2	~120 000	0	APR ++	Probably a nonadaptive defense protein

* APR: Acute phase reactant.

† For HAP 1-1 phenotype.

++ Reactant with largest relative increase.

Note: The adult reference ranges reported are established by nephelometry and may vary with differences in technique. Other reference ranges in this chapter may differ according to ages tested and methods used.

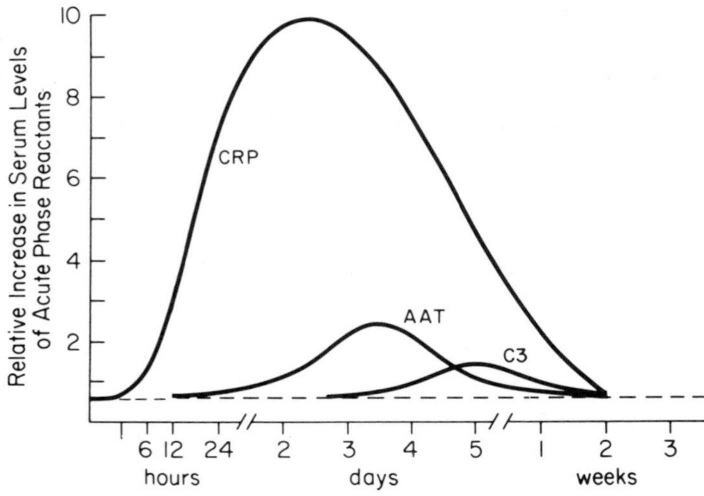

Figure 11-4. Relative increases of acute phase reactants: C-reactive protein (CRP), α_1-antitrypsin (AAT), and complement factor 3 (C3). Concentrations are expressed as multiples of the upper limit of the reference range. The dotted line represents the upper reference limit.

with the largest and earliest rises (e.g., C-reactive protein) are useful in monitoring the progress of the inflammation or its response to treatment.

Rates of hepatic synthesis of many plasma proteins are also affected by the patient's endocrine status, which may alter the protein levels characteristic of a specific disease.[89]

The total concentration of all proteins in plasma may vary because of changes in the volume of plasma water or changes in the amounts of individual proteins. Decrease in the volume of plasma water (*hemoconcentration*) causes the concentrations of all the proteins to be increased to the same degree. This *hyperproteinemia* is seen in dehydration due either to inadequate water intake or to excessive loss of water, as in severe vomiting, Addison's disease, or diabetic acidosis. Increase in plasma water volume (*hemodilution*), on the other hand, causes a decrease in the concentrations of all proteins. This *hypoproteinemia* occurs in water intoxication or salt retention syndromes, during massive intravenous infusions, and physiologically when a person is recumbent. A recumbent position decreases the total protein concentration by 0.3–0.5 g/dL.

Of the individual serum proteins, albumin forms such a large proportion of the total protein that low albumin levels often cause hypoproteinemia. Mild hyperproteinemia may be caused by increased concentrations of proteins normally present in relatively low concentrations (e.g., increases in APRs and immunoglobulins during infections). Marked hyperproteinemia can be caused by high levels of monoclonal immunoglobulins.

Analysis of Proteins of Body Fluids

Methods for the analysis of body fluid proteins fall into four groups:

1. Quantitative measurements of total protein and albumin.

2. Separation of proteins by electrophoresis, which acts as a screening test for many diseases and allows semiquantitative estimations of the more abundant proteins (Chapter 2, Section 3).

3. Specific quantitative assays of particular proteins by immunochemical methods using specific antisera and measurement of the antigen-antibody (Ag-Ab) complexes by nephelometry, turbidimetry, radial immunodiffusion, or electroimmunoassay; or, for proteins in very low concentrations, by radioimmunoassay (RIA) or enzyme immunoassay (EIA) (Chapter 2, Section Two, and Chapter 3).

4. Detection and identification of abnormal proteins by radial immunodiffusion (RID), immunoelectrophoresis (IEP), and immunofixation electrophoresis (IFE) (see Chapter 3).

Methods for the Determination of Total Protein

Serum Total Protein

When the total serum protein is measured, it is arbitrarily assumed that: (1) all protein molecules are pure polypeptide chains, containing on the average 16% by weight of nitrogen; and (2) each of the several hundred individual proteins present in serum reacts chemically in the same way as every other protein. Clearly, the first assumption is not true at all, and the second cannot always be true. Nevertheless, these simplifying assumptions make measurement of total protein a practical, though empirical, procedure.

Historically, total protein was first determined by the ***Kjeldahl method***, in which acid digestion was used to convert nitrogen in the protein to ammonium ion.[66] The concentration of ammonia nitrogen was then evaluated by titration or nesslerization, a correction was made for nitrogen contributed by nonprotein compounds, and the ammonia nitrogen value was multiplied by 100%/16% (i.e., the factor 6.25) to express protein nitrogen as total protein. The method is well defined and reproducible but impractical for routine use. It remains, however, a means of defining reference standards (see page 317).

The ***biuret method*** depends on the presence of peptide bonds in all proteins. When a solution of protein is treated with Cu(II) ions in a moderately alkaline medium, a colored chelate is formed between the Cu(II) ion and the peptide bond.[40] An analogous reaction occurs between cupric ion and the organic compound biuret,

```
 H       H       H
 |       |       |
HN — C — N — C — NH
     ‖       ‖
     O       O
```

hence the name. The reaction occurs with any compound containing at least two —HN—CO—, —HN—CH_2—, —HN—CS—, or similar groups joined together directly or through a carbon or nitrogen atom. One copper ion is probably linked to six nearby peptide linkages by coordinate bonds. Amino acids and dipeptides do not react, but tri-, oligo-, and polypeptides do react to give pink to reddish-violet products. The intensity of the color produced is proportional to the number of peptide bonds that are reacting, and therefore to the number of protein molecules present in the reaction system. Thus, the biuret reaction is suitable for the quantitative determination of total protein by spectrophotometry.

Although small peptides react, their concentration in serum is so low that they contribute little to the biuret color. Ammonium ions interfere but not at concentrations that can occur in serum. Most biuret methods can detect between 1 and 15 mg of protein in the aliquot being measured, an amount present in 15–200 μL of a serum containing protein at 7 g/dL. Numerous versions of the biuret method have been reported. All methods are simple, can be automated, and are sufficiently precise for clinical use.[58] The biuret reagent contains sodium potassium tartrate to complex cupric ions and maintain their solubility in alkaline solution; potassium iodide is included as an antioxidant, and the absorbance of the colored product is measured spectrophotometrically at 540 nm. Serum samples should not show hemolysis or gross lipemia.

Refractometry. If a solute is added to water, the refractive index of the solution increases in proportion to the concentration of solute. Since the solutes present in greatest concentration in serum are proteins, the refractive index of serum provides a rapid approximate measure of serum total protein concentration within the range of 3.5–10 g/dL. Clinical refractometers are calibrated to read protein in g/dL as well as the specific gravity of urine. They compensate for temperature changes in the range of 15–37 °C, but significant errors may be caused by alterations in albumin/globulin ratio, azotemia, hyperglycemia, hyperbilirubinemia, and especially lipemia.

Reference ranges. The total protein concentration of serum obtained from healthy ambulatory adults is 6.3–8.3 g/dL and from an adult at rest, 6.0–7.8 g/dL.

Methods for the Determination of Total Protein in Other Fluids

The methods discussed here are dependent on four general properties of proteins, namely:

1. Reaction of tyrosine and tryptophan residues with the Folin-Ciocalteu (phenol) reagent.
2. Characteristic absorption of ultraviolet (UV) light at 200–225 and 270–290 nm.
3. Precipitation (for measurement by turbidimetry or nephelometry).
4. Ability to bind colored dyes.

Although any of these methods could be and have been used for serum protein determinations, none is more convenient or practical than the biuret method. The biuret method, however, is not sufficiently sensitive to determine the smaller concentrations of protein in urine or spinal fluid.

Tyrosine and tryptophan residues occur in most proteins, but each protein contains a unique proportion of them. Albumin, for instance, has only 0.2% tryptophan by weight, whereas the tryptophan content of individual globulins varies between 2 and 3%. These amino acids, either free or in an unfolded polypeptide chain, reduce phosphotungstic-phosphomolybdic acid (Folin-Ciocalteu) reagent to give a blue color. This property is better suited for assays of a pure protein such as fibrinogen than for mixtures of proteins with different proportions of tyrosine and tryptophan. However, it has been applied to the assay of total protein in the *Lowry method*,[30] which combines the biuret and Folin-Ciocalteu reactions. This method can detect proteins in concentrations as low as 10–60 μg/mL and is widely used in research to measure tissue and enzyme proteins in purified preparations. The method is not fully satisfactory for the determination of total protein in urine or spinal fluid because reactions with nonprotein compounds, particularly in spinal fluids, give a positive error of 3–9 mg/dL.

Absorption of UV light at 280 nm is chiefly the result of the high electron density of aromatic rings of tyrosine and tryptophan in solution at pH 8. Because these amino acids are unevenly distributed among individual proteins, and because free tyrosine and tryptophan, uric acid, and bilirubin—all of which absorb significantly at or near 280 nm—are also present in body fluids, this method suffers in respect to specificity. At 200–225 nm, however, where peptide bonds are chiefly responsible for UV absorption (70% at A_{205}), specific absorption by proteins is 10–30 times greater than at 280 nm.

Precipitation of protein for turbidimetric or nephelometric assays can be achieved with sulfosalicylic acid alone, with sulfosalicylic acid in combination with sodium sulfate or trichloroacetic acid, or with trichloroacetic acid alone. For the light scattering to be reproducible, the precipitate must consist of fine, uniform particles that are homogeneously distributed. In addition, the principal proteins must be precipitated equally efficiently, and the standard used must behave like normal and abnormal mixtures of plasma proteins. Turbidimetric methods are often used for urine and CSF.

Dye-binding methods are based on the ability of proteins to bind dyes such as Amido Black 10B and Coomassie Brilliant Blue. This property is used for staining protein bands after electrophoresis, assaying serum albumin, and determining total protein by spectrophotometric methods. The unequal affinities and dye-binding capacities of individual proteins[64] and the lack of a satisfactory standard are limitations. The dye-binding method of greatest interest, particularly for assay of total protein in urine, employs Coomassie Brilliant Blue G-250, which binds to protonated amine groups of amino acid residues in the polypeptide chain. The absorbance is measured at 595 nm (the absorbance maximum for the bound dye) and at 465 nm, where the free dye shows maximum absorption and the bound dye shows a decrease in absorbance (Figure 11-5).

Standardization of Total Protein Methods

The use of bovine or human albumin to standardize *biuret methods* is well established. Albumin can be prepared consistently in high purity, it contains only amino acids, its nitrogen content is a constant fraction of its molecular mass, and the number

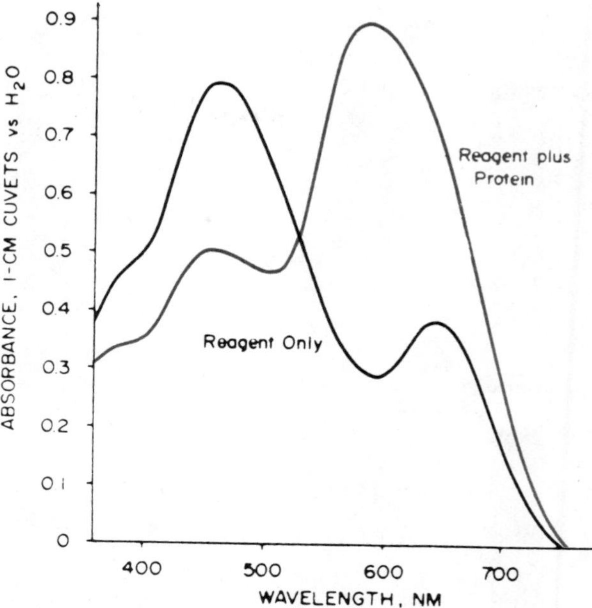

Figure 11-5. Absorption spectrum of Coomassie Brilliant Blue with and without added protein. (From Lott, J. A., Stephan, V. A., and Pritchard, K. A.: Evaluation of the Coomassie Brilliant Blue G-250 method for urinary protein. Clin. Chem., *29*:1946–1950, 1983.)

of peptide bonds per molecule is known. Since the peptide bond is the biuret-reacting unit in all proteins, and the number of peptide bonds determines the absorbance of the colored product, albumin is a reasonable peptide-bond standard for all the proteins in the mixture. Albumin should be standardized by unequivocal means such as Kjeldahl nitrogen content (Armour Pharmaceutical Co., Chicago, IL 60690) or weighed in and verified by reference laboratories (College of American Pathologists, Skokie, IL 60076). A standardized bovine serum albumin solution, Standard Reference Material (SRM) #927, is available from the National Bureau of Standards, Washington, D.C. It may also be used for assigning protein concentrations to secondary standards, such as pooled sera or bovine albumin solutions. Standardization with bovine albumin is less costly than with human albumin; its use by a majority of laboratories is an advantage in quality control programs and proficiency surveys.

For the standardization of precipitation and dye-binding methods other than those using trichloroacetic acid precipitation, pure albumin standards are unsatisfactory because albumin behaves differently from other proteins both with sulfosalicylic acid and with dyes. The standard must contain a mixture of proteins similar to that of the sample; such a standard can be obtained by diluting a normal serum (or serum pool) and determining its total protein with a correctly standardized and well-controlled Kjeldahl or biuret method.

Serum Protein Electrophoresis (SPE)

Separation of proteins on cellulose acetate or agarose gel is almost always performed on serum in order to avoid the complication of a fibrinogen band in the $\beta\gamma$-region. SPE is often performed by using systems that integrate apparatus, materials, and reagents from a single supplier (Beckman Instruments, Inc., Brea, CA 92621; Helena Laboratories, Beaumont, TX 77704). The standard buffer is barbital with an ionic strength of 0.05 and a pH of 8.6. The usual sample is 3–5 μL, applied with a mechanical device to obtain an even stripe of sample across the width of the support medium or over a track of the medium. Figure 11-6 illustrates some SPE separations.

Normal SPE usually produces only five bands (albumin, α_1, α_2, β, and γ), whose principal components are listed in Table 11-5. A sixth (β_2, C3 complement component) is seen if the serum is fresh and the buffer contains Ca(II) ions. Many of the individual protein concentrations are too low to manifest as distinct stained bands, or they are

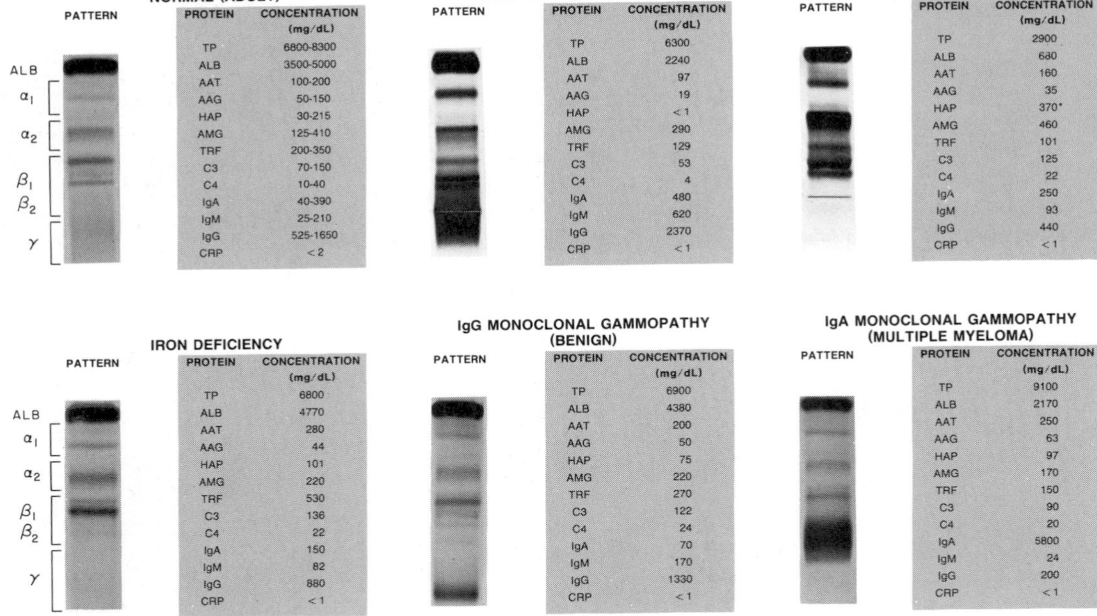

Figure 11-6. Electrophoretic patterns typical of normal and of some pathological conditions (agarose gel). Information in the screened areas gives examples of changes in protein composition in each of the listed diseases. *Hp 2-2 phenotype.

overshadowed by proteins of higher concentrations that migrate near them (e.g., ceruloplasmin is masked by haptoglobin and α_2-macroglobulin). Some proteins stain poorly because they contain high proportions of lipid (lipoproteins) or carbohydrate (α_1-acid glycoprotein). A variety of stains may be used (Table 2-5). Amido Black and Ponceau S have been popular in the past; Coomassie Brilliant Blue is more sensitive. Fat stains are needed for visualizing the lipoproteins (see Chapter 14). The pattern produced by a patient's serum is inspected for the following changes[34] by comparing it with that of a normal control serum run in parallel.

Sharp, narrow bands occurring from the α- to γ-regions suggest monoclonal immunoglobulins (paraproteins).

Multiple bands, absent bands, or different mobility of normal bands can be due to genetic variants such as haptoglobin in the α_1-region and transferrin in the β-region.

Other causes of altered mobility. Increased mobility of albumin occurs when it is bound to penicillin or salicylates, or to greater than normal amounts of bilirubin or fatty acids. Decreased mobility of α_1-antitrypsin occurs when it binds thiol groups or Bence Jones protein.

Appearance of a band not normally seen. The concentration of a normal protein may increase to such a level as to become visible as a line. For instance, a faint, sharp band may appear between albumin and the α_1-region as a result of a hundredfold increase of α-fetoprotein from certain tumors. Similarly, a large increase of C-reactive protein (CRP) in a severe acute phase reaction may generate a faint band in the γ-region, or an increase of lysozyme in monocytic leukemia may produce a band in the post-γ-region.

Changes in the relative staining intensity of normal bands. Decreased albumin- and γ-bands in conjunction with an increased α_2-band suggest selective proteinuria such as is observed in the *nephrotic syndrome*. Note, however, that the albumin concentration must fall by at least one third of its normal level before its decrease will be evident on the electrophoretic strip.

An increase in both the *α_1-band* (α_1-antitrypsin and α_1-acid glycoprotein) and *α_2-band* (haptoglobin) suggests an acute phase reaction. An increase in α_1-components alone may be seen in chronic hepatitis and in acute phase reactions accompanied by hemolysis, as well as during estrogen therapy or pregnancy. In diseases characterized by vasculitis (such as rheumatoid arthritis) or in immune-complex diseases, there may be a predominant increase in the α_2-band. An increase in the *β_1-band* suggests iron deficiency anemia (transferrin increase) or high levels of estrogen. Fusion or bridging of β- and γ-bands suggests an increase in IgA such as occurs in cirrhosis, respiratory or skin infections, and rheumatoid arthritis. An increase in the γ-band suggests a polyclonal γ-globulin increase associated with an immune reaction, chronic inflammatory disease, liver disease, or disseminated neoplasms. Oligoclonal bands are occasionally seen in chronic aggressive hepatitis and chronic viral infections. A decrease or absence of the γ-band suggests immune deficiency.

Any such abnormalities of the electrophoretic pattern can be investigated further by immunochemical determination of appropriate individual proteins. Also, the electrophoretic strips can be cleared and their bands evaluated by densitometry. However, only albumin has a band sufficiently separated from other proteins for it to be reliably estimated by densitometry.

Immunochemical Methods for the Determination of Individual Proteins

Any single protein in serum can be determined, without prior separation and at any concentration down to ~1 ng/mL, if a sufficiently specific antiserum is available. The methodological principles are described in Chapter 3. Nephelometry and turbidimetry, because of their speed and ease, are most widely used[21] to measure either the amount of Ag-Ab complex formed (end-point methods) or the rate of complex formation (kinetic methods). Several factors must be considered before deciding which quantitative method is most appropriate:

Sensitivity. Nephelometry, turbidimetry, and radial immunodiffusion (RID) are all sufficiently sensitive to measure protein concentrations down to ~20 μg/mL. Turbidimetry is slightly less sensitive (20–30 μg/mL) than nephelometry (~10 μg/mL). The sensitivity of RID (10–20 μg/mL) can be enhanced by using "low level" plates (Behring Corp., La Jolla, CA 92037). RIA or EIA is required for determining ng/mL levels (e.g., for serum IgE).

Precision. With nephelometry, within-run CVs of <5% are usual; the end point methods, however, tend to be less precise than the kinetic methods because sample blanks are required. RID and EIA systems have higher run-to-run CVs, usually in the range of 5–15%, but the concentrations measured are much lower.

Turnaround time. Nephelometry is fast; kinetic methods give results within minutes[7,79] whereas end point methods take up to 1 h. RID usually requires 24–48 h of incubation, an unacceptable period for urgent tests; RIA requires several hours.

Kinetic nephelometric methods, although somewhat less sensitive than end point methods, are more rapid, and, because sample blanks are unnecessary, more precise.[76] In such methods, specific antiserum against the protein to be determined is added to a reaction cuvet containing a serum sample in a buffered solution; this allows the formation of antigen-antibody complexes. The measurement is made by passing a light beam through the cuvet containing the reaction mixture. The beam is scattered by the suspended antigen-antibody complexes; the intensity of scattered light, measured at an angle of 70° to the incident beam, is proportional to the number of suspended particles in solution. As antigen and antibody react, the rate of formation of particles progressively increases until it reaches a maximum level. The *rate of change* of the light scatter is monitored and the maximum rate during the reaction selected; it is directly proportional to the concentration of antigen in the reaction mixture. When standards of known concentrations are used, maximum rate versus concentration can be plotted to form a standard curve. The concentration of antigen in an unknown sample is then calculated from the standard curve.

A kinetic method can be performed on the Beckman Immunochemistry System (ICS). A calibration curve consisting of 8–15 points is prepared by the reagent manufacturer for each lot of calibrator serum and specific antiserum for each analyte. These values are coded onto the antibody program card, which provides the microprocessor with all the measurements needed to perform a specific protein assay and calculate results for patients' specimens. The derived constants define the shape of the calibration curve and are used by the microprocessor to convert peak rate values for unknown samples to actual concentration values. Since the microprocessor recognizes the calibration curve by reading the antibody card, it is necessary to measure and verify only *one* point on the curve. Once the instrument verifies the known target value (calibration point), all unknown samples may be assayed and their results interpolated from the calibration curve for that particular antiserum.

Standardization of Immunochemical Methods

Reference materials containing accurately determined quantities of purified individual proteins seldom exist. For many proteins of interest, there are also no well-defined purified protein preparations available. Depending on their sources, antisera may vary greatly in specificity and sensitivity. Their inconsistent character and the lack of well-defined reference materials make it imperative that each laboratory establish its own reference ranges for its own reagent systems and methods.

The World Health Organization (WHO) has available reference preparations for IgG, IgA, IgM, IgD, IgE, α-fetoprotein, carcinoembryonic antigen (CEA), albumin, C3, ceruloplasmin, and transferrin. The values assigned are, however, in International Units (IU); factors have not been defined for conversion of IU to mass units. Another preparation, the United States National Reference Preparation (USNRP) for Specific Human Proteins, is available from the Centers for Disease Control, Atlanta, GA; it contains 13 individual proteins, and the values in IU are assigned against WHO standards. A reference preparation whose assigned values are in mass units has been developed by the College of American Pathologists (CAP). The 12 individual proteins in this preparation were assayed by a number of experts using specific protein assays and by manufacturers of diagnostic products for such assays. The methods used include nephelometry, RID, electroimmunodiffusion, and turbidimetry. An arithmetic mean value for each protein was calculated from the results of all the methods and defined as the assigned value for that protein.

Individual Plasma Proteins

The immunoglobulins and complement components, as distinct groups of closely related proteins, are considered first; the other proteins are then addressed in the same order as shown in Table 11-5.

IMMUNOGLOBULINS (HUMORAL ANTIBODIES)

Immunoglobulins are unique in their heterogeneity, in their sites of synthesis, and in the fact that their synthesis is an adaptive response to antigenic stimulation. They have two functions: recognition of antigens and initiation of effector mechanisms to destroy antigens. Their basic characteristics and the definitions of some terms commonly used in immunology are described in Chapter 3.

Whereas most plasma proteins are synthesized in the liver, immunoglobulins are synthesized in and secreted by plasma cells, which are the end products of a line of cell types that start with the B lymphocyte stem cell in bone marrow.[76,89] More mature (early and late) B lymphocytes, found mainly in lymph nodes and in blood, develop numerous receptor immunoglobulins on their surface membranes. Upon encountering antigen, these B lymphocytes proliferate and develop into plasma cells, each of which secretes into the blood a highly specific antibody capable of binding antigen. Each immunoglobulin (Ig) molecule consists of two or more basic units built of two identical heavy (H) chains and two identical light (L) chains. The variable regions at the *N*-terminal ends of the four chains provide the antigen-binding sites (see Figure 3-1), and their amino acid sequences determine the antigenic specificity characteristic of the Ig molecules (antibody). The remainder, the "constant" part, is the same for every immunoglobulin molecule of a given class—IgM, IgG, IgA, IgD, or IgE. Their respective heavy chains are called μ, γ, α, δ, and ϵ. They carry the effector sites that interact with cells (e.g., mast cells) and with complement. The light chains, which are produced independently and then assembled with the heavy chain, are of two types, κ and λ. They occur in all classes of immunoglobulins in the proportion $\kappa:\lambda = 2:1$; the two halves of a given molecule always have the same type.

Properties of Immunoglobulin Classes

IgM is the most primitive and least specialized immunoglobulin and the only one that a newborn synthesizes. In adult serum, IgM accounts for 5–10% of the total serum Ig. Most of the IgM is a pentamer of five IgM monomers, each very similar to the IgG molecule, except that IgM has attached to it a small glycopeptide, the J chain. The large MW of IgM (900 000) prevents its passage into extravascular spaces. B lymphocytes initially have IgM surface receptors and in the first or primary response to an antigen they secrete IgM; then as the cells change into plasma cells, a second exposure to the same antigen causes a larger secondary response, now of IgG secretion. IgM continues, however, to be synthesized against antigens confined to the blood, such as erythrocyte surface antigens and tropical parasites. IgM is not transported across the placenta and is therefore not involved in hemolytic disease of the newborn. It is an efficient complement activator.

IgG (MW 160 000) is the most abundant immunoglobulin of extravascular fluids (65%) and of serum (70–75%). IgG antibodies are produced in response to the antigens of most bacteria and viruses and to small soluble protein antigens such as toxins in tissue spaces. The "clones" or subsets of plasma cells generated in response to these different antigens each secrete IgG molecules differing slightly in the structure of their variable regions and hence in their electrophoretic mobility. Because of this "polyclonal" response, IgG appears as a diffuse band from the γ- to the α_2-regions in SPE.

IgG has four subclasses: IgG_1, IgG_2, IgG_3 and IgG_4. Structurally they differ in their hinge regions,[61] which join the Fc fragment to the two Fab fragments (see Figure 3-1). In IgG_3, the hinge is extended by up to 15 half-cystines, which allows binding of C1q. IgG_1 and IgG_3 both bind firmly to the Fc receptors of phagocytic cells, activate killer monocytes (K cells), and cross the placenta by an active transport process dependent on Fc binding. Maternal IgG protects the newborn for the first three months (Figure 11-7).

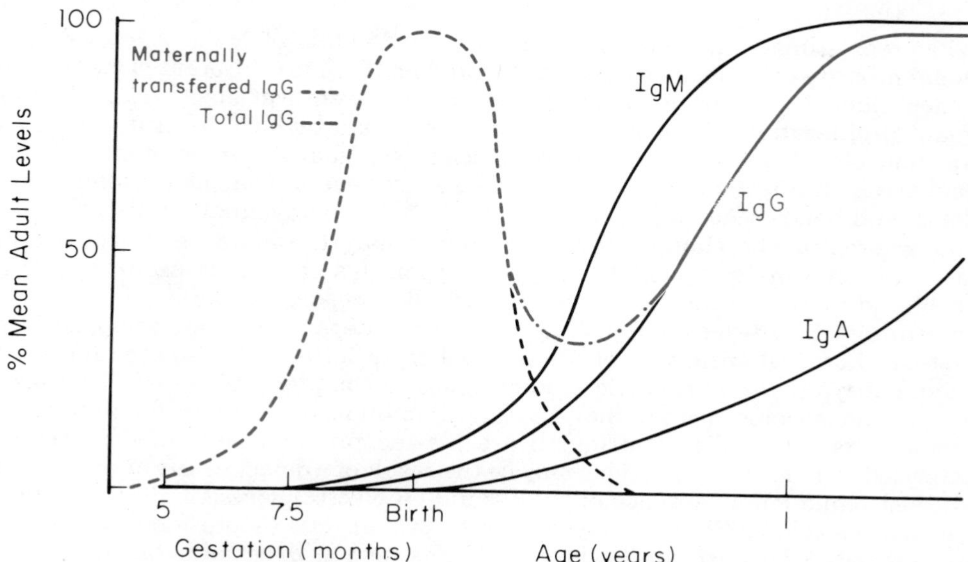

Figure 11-7. Serum immunoglobulin levels as per cent of adult levels before birth and for the first year of life.

IgA (MW 160 000) composes ~10–15% of serum Ig. In its monomeric form, its structure is similar to that of IgG, but 10–15% of IgA in serum is polymeric, particularly IgA_2, which is more resistant to destruction by some pathogenic bacteria than IgA_1. On electrophoresis, IgA migrates in the $\beta\gamma$-region, ahead of most of the IgG. IgA can activate complement by the alternative pathway but its exact role in serum is not clear.

Another and probably more important form of IgA is called *secretory IgA*. It is found in tears, sweat, saliva, milk, and colostrum, and in gastrointestinal and bronchial secretions. Secretory IgA has a MW of 380 000 and consists of two molecules of IgA, a secretory component whose MW is 70 000, and a J chain of MW 15 600. It is synthesized mainly by plasma cells in the mucous membranes of the gut and bronchi and in the ductules of the lactating breast. The secretory component makes secretory IgA more resistant to enzymes and more able to protect the mucosa from bacteria and viruses. Its presence in colostrum and milk probably protects the newborn from intestinal infections.

IgD accounts for <1% of serum Igs. It is monomeric, contains about 12% carbohydrate, and has a MW of 184 000. Its structure is similar to that of IgG. Like IgM, IgD is a surface receptor for antigen in B lymphocytes, but its primary function is unknown.

IgE is so rapidly and firmly bound to mast cells that only trace amounts of it are normally present in serum. IgE has a MW of 188 000 and a structure similar to IgG. Many IgE molecules with different variable regions (i.e., different antibodies) may be attached by their Fc chains to each mast cell. When antigen (allergen) combines with one of the attached IgE molecules, the mast cell is stimulated to release histamine and other vasoactive amines responsible for the vascular permeability and smooth muscle contraction characteristic of allergic reactions such as hay fever, asthma, urticaria, and eczema.

"Normal" Immunoglobulin Levels

Reference ranges for IgG and IgM in various human adult populations differ around the world because levels depend on the extent of antigenic stimulation coming from the environment. IgA levels, however, are unaffected by environmental factors. Normal newborn infants do not have significant rates of Ig synthesis. The IgG level in a neonate represents IgG transferred across the placenta from the mother. Immunoglobulin synthesis is stimulated by environmental antigens so that serum IgM reaches adult levels at about 9 months (see Figure 11-7), IgG in 3 years, IgA, IgE, and IgD in about 14 years. IgA in saliva, however, reaches adult levels in about 6 weeks.

Clinical Significance of Immunoglobulins

When we estimate the concentration of a single class of Ig, we are evaluating a polyclonal mixture of antibodies that are the products of many different clones of plasma cells, each clone producing Ig molecules with a different variable region. Benign or malignant proliferation of one such clone will produce a high concentration of a single type (a monoclonal antibody) that may appear as a sharp, narrow band on protein electrophoresis. If a few clones proliferate, there are several sharp bands (for example, the oligoclonal bands seen on electrophoresis of CSF in demyelinating diseases such as multiple sclerosis). The changes found in disease may therefore be a decrease or an increase in the normal polyclonal mixture of serum Igs or an increase in one or more monoclonal idiotypes (in multiple myeloma usually one, rarely two).

Immunoglobulin deficiency. Immune defense depends on four complex, interactive systems: humoral antibodies of the B lymphocyte series; cell-mediated immunity of the T lymphocytes; the phagocytic system; and the complement system. The last two systems are nonspecific in that they have no immunologic memory for antigen. Only the components of the first and fourth are plasma proteins. Immunodeficient states characterized by recurrent infections may be the result of a defect in any of these systems.

Marked reduction or absence of the γ-band on electrophoresis indicates deficiency of Ig antibodies. The deficiency may be due to protein loss (nephrotic syndrome), to a primary inherited defect in synthesis, or to toxins such as drugs or substances retained in renal failure, or it may be secondary to lymphoid malignancy such as multiple myeloma. Diagnosis is important, since replacement therapy is available. Some primary deficiencies involve only one or two immunoglobulin classes; if the total Ig level is not greatly affected, the deficiency (e.g., IgM) may not be suspected from the electrophoretic pattern.

Immunodeficiency is a risk for infants. Figure 11-7 shows how levels of maternal IgG, transferred across the placenta, rise in the fetus during the last three months of pregnancy. Contact of the neonate with environmental antigens causes B lymphocytes to begin to multiply, IgM levels to start to rise, and plasma cells producing IgG and IgA to increase in number. These developments are paralleled by a decrease of maternal IgG, so that in the infant's blood, IgG falls to a minimum at about three months of age. Premature babies, because they start with less than the full-term amount of maternal IgG, and babies in whom initiation of IgG synthesis is transiently delayed, are at risk. IgG determinations are invaluable in these cases since levels may fall dangerously low if the baby is not treated. Rising serum IgM and normal salivary IgA concentrations at six weeks of age suggest a good prognosis.

Polyclonal hyperimmunoglobulinemia. Polyclonal increase in serum Igs is the normal response to infections. IgG tends to predominate in autoimmune responses; IgA in skin, gut, respiratory, and renal infections; and IgM in primary viral infections and blood stream infections such as malaria. *Chronic bacterial infections* cause an increase in serum levels of all Ig classes. In such cases, estimations of the individual classes seldom provide more information than protein electrophoresis. They are of value, however, in the differential diagnosis of liver disease and of intrauterine infections. In *primary biliary cirrhosis*, the IgM level is markedly increased; in chronic active hepatitis, IgG and sometimes IgM are increased; and in *portal cirrhosis*, IgA and sometimes IgG are increased. In *intrauterine infections*, production of IgM by the fetus increases and at birth the IgM level in cord blood is increased. Estimations of IgE are used in the management of asthma and other allergic conditions, especially in children.

Monoclonal Immunoglobulins (Paraproteins)

A single clone of plasma cells produces immunoglobulin molecules with identical structures. If the clone is permitted to multiply, the concentration of its particular protein in the patient's serum becomes so great that on electrophoresis it produces a narrow, sharply discrete spike. These monoclonal immunoglobulins, which are also called paraproteins, may be polymers, monomers, or fragments of Ig molecules; if fragments, they are usually light chains (Bence Jones proteins) or, rarely, heavy chains or half molecules.

Multiple myeloma is a malignant neoplasm of a single clone of plasma cells of the bone marrow. The plasma cells most often proliferate diffusely throughout the marrow but occasionally they form a solitary tumor called a *plasmacytoma*. Osteolytic bone lesions are produced and the other bone marrow cells are reduced so that thrombocytopenia, anemia, and leukopenia develop. Development of normal clones of plasma cells is inhibited; consequently, synthesis of the other Ig classes is reduced and a syndrome of recurrent infections occurs. The incidence of multiple myeloma is low in individuals under age 60 but rises rapidly with age. The patient may present with local symptoms of bone lesion but more often presents with nonspecific symptoms such as weight loss, obscure anemia, hemorrhages, repeated infections, or renal failure. A highly suggestive laboratory finding is a normal serum alkaline phosphatase level in a patient with destructive bone lesions. Cardinal diagnostic features of the disease are the findings of neoplastic plasma cells in bone marrow aspirate, radiologic demonstration of osteolytic lesions, and identification of a paraprotein in serum or in concentrated urine. All patients who could conceivably have the disease should be screened for paraproteins, since $<1\%$ with the disease fail to have detectable paraproteins. Table 11-6 lists the paraproteins that may be associated with multiple myeloma and shows some characteristic findings for them.

Should a paraprotein be identified in blood, or urine, or both, its heavy and light chains should be typed and the concentrations of polyclonal IgG, IgA, and IgM determined. These studies confirm whether the spike on the electrophoretic pattern is indeed a paraprotein, they help to decide the probable prognosis, and they show whether the

Table 11-6. MONOCLONAL IMMUNOGLOBULINS IN MULTIPLE MYELOMA

Plasma, Paraprotein	Incidence,* %	Age of Occurrence,* Mean	Average Doubling Time for Concentration, Months	Incidence of Bence Jones Proteinuria, %	Comments
IgG	50	65	10	60	Patients more susceptible to immunodeficiency
IgA	25	65	6	70	Tend to have hypercalcemia and amyloidosis
Bence Jones protein only	20	56	3.5	100	Often renal failure; bone lesions; amyloidosis. Poor prognosis.
IgD	2	57	—	100	90% λ type. Often have extraosseous lesions, amyloidosis, renal failure; 50% have enlarged lymph nodes, liver, spleen. Poor prognosis.
IgM	1	—	—	100	May or may not have viscosity syndrome
IgE	0.1	—	—	Most	—
Biclonal	1	—	—	—	—
None detected	<1	—	—	0	Usually reduction of normal immunoglobulins

*Approximate values

polyclonal Igs are so low as to make the patient vulnerable to infections. If the paraprotein proves to be IgM, the diagnosis is probably Waldenström's macroglobulinemia and not multiple myeloma. Prognosis is based on the class of the paraprotein found, its concentration at the time of diagnosis, and the rate at which its concentration increases.

Lymphoid tumors, presenting as lymphomas or chronic lymphocytic leukemias, arise from less mature stages in B lymphocyte development; about 1 in 5 produce paraproteins, usually of the IgM class. *Waldenström's macroglobulinemia* arises from the most mature B lymphocytes and invariably produces IgM; this very high-molecular-weight protein produces an increase in viscosity of the blood and hence the symptoms. Bence Jones proteinuria occurs in 80% of these cases, but the condition is much less malignant than multiple myeloma. Symptoms are treatable by exchange transfusion. Rarely, unusual forms of IgG, IgA, or light chains polymerize and cause a similar syndrome of high blood viscosity. *Heavy-chain diseases* are rare conditions in which the paraprotein consists only of a heavy chain, usually incomplete.

Two other disorders are sometimes characterized by the presence of paraproteins: *cryoglobulinemia* and *amyloid disease*. Cryoglobulin is a serum protein that precipitates at temperatures lower than body temperature. Most cryoglobulins are polyclonal immunoglobulin complexes, but nearly half are monoclonal, usually IgM. For cryoglobulin examinations, a temperature of 37 °C must be maintained for blood collection and for serum separation in order to keep the cryoglobulin from precipitating out of the serum.

Methods for Analysis of Immunoglobulins[91]

Only immunochemical methods are sensitive enough to quantitate individual Igs at normal levels. Although RID or electroimmunoassay gel techniques may be used, nephelometry or turbidimetry is preferred now that very specific antisera with high titer and affinity are available. The latter methods require few manipulations and are more rapid and precise. They are applicable to serum and CSF IgG, IgA, and IgM assays, but for IgD and IgE the greater sensitivity of radioimmunoassay or enzyme immunoassay is required. Remarkably good precision can be obtained for Ig assays, but because reference materials containing precisely determined quantities of purified proteins are not currently available, it is impossible to say how accurate the methods are.

Assays of Ig classes involve determining the concentration of a mixture of protein molecules having similar constant regions but different variable regions (idiotypes). Reagent antisera and reference Ig standards used in most immunochemical assays have been generated against normal human sera containing a mixture of Ig subclasses. Despite the uncertainties inherent in these assay systems, determination of individual polyclonal Igs is quite reliable. By contrast, the determination of monoclonal proteins is not. A monoclonal immunoglobulin will have only some of the determinants with which the antibodies in the antiserum usually react and thus the standard dilution curve will not apply. The few antibodies in the antiserum that *do* react are soon precipitated and the paraprotein is overestimated. Furthermore, if a new batch of antiserum is introduced, the relationship between paraprotein concentration and the standard curve is likely to change. For these reasons, the absolute concentration of a paraprotein must be estimated by electrophoresis and densitometry, but subsequent *changes* of concentration of the *same* paraprotein can be determined immunochemically, provided that the *same* batch of antiserum is used for all determinations.

Reference ranges are related to age, geographic, and methodological differences and vary widely in different circumstances (see *Appendix*, Table 29-18).

Investigation of Paraproteins

The first step consists of agarose electrophoresis of serum and of concentrated urine. Serum should be collected and separated at 37 °C to avoid loss of cryoglobulins. An estimate of the paraprotein should then be made from the spike identified on the electrophoretic plate by densitometric scan and by determination of total protein in the sample. Bence Jones protein, when it is present only in urine, is the paraprotein most often missed.

The second step is to identify the class of paraprotein with specific antisera against heavy and light chains by immunoelectrophoresis[85] (IEP) or by immunofixation electrophoresis (IFE).[65] (Commercial kits are available for each.) Reaction of the protein causing the spike with only one heavy-chain and one light-chain antiserum confirms that it is a paraprotein and that the spike is not caused by denaturation, fibrinogen, hemoglobin, or some other artifact. If a paraprotein does not react with any of these heavy-chain antisera, it should be tested with anti-delta and anti-epsilon before assuming that it is a Bence Jones protein. The subsequent changes in paraprotein concentration are monitored by immunochemical assay.

Immunofixation electrophoresis (IFE). Blood is collected in the fasting state without anticoagulants and centrifuged immediately for 15 min at 900 $\times$ *g*. The serum ($\geq$100 μL) is best used fresh but can be stored at 2–8 °C up to 5 d or at −20 °C for longer periods. Spinal fluid and urine may require concentrating. Aliquots of the same specimen are placed in six wells on an agarose gel and their major protein groups are separated by electrophoresis (Figure 11-8). One track is then treated with a chemical fixative solution to fix all the proteins and create an electrophoresis reference pattern. The other five tracks are treated with antisera against heavy chains γ(IgG), α(IgA), and μ(IgM), and κ and λ light chains. They react with individual immunoglobulins and immunofix them. All unreacted proteins in these five tracks are then washed out of the gel, and all six tracks are stained. When the locations of the stained, immunofixed bands are compared with bands of the same location in the reference pattern, specific proteins can be identified.

In monoclonal gammopathies, the IFE patterns yield a distinct, sharply defined, precipitin band with one heavy-chain and one light-chain antiserum. These bands match the location of the paraprotein spike in the reference pattern (see Figure 11-8). In polyclonal gammopathies, there are diffuse precipitin bands with *both* light-chain antisera and one or more heavy-chain antisera.

IFE patterns should always be confirmed by quantitating the immunoglobulins (IgG,

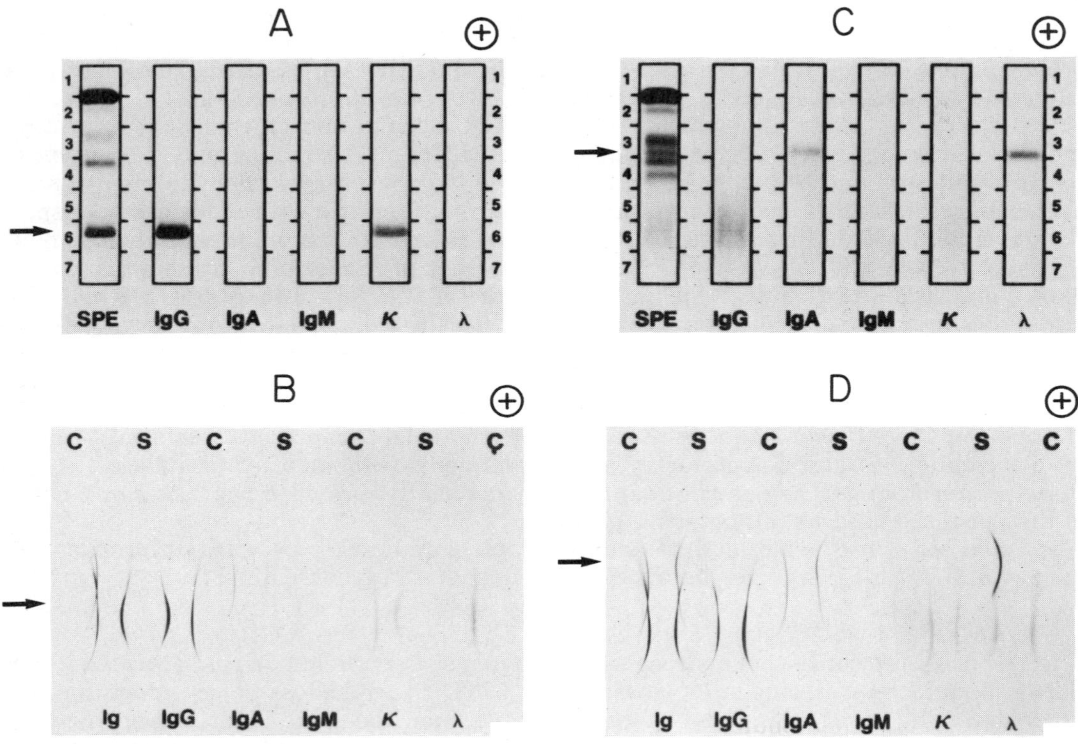

Figure 11-8. Comparison of IFE and IEP for two patients with monoclonal gammopathies.

A, Patient specimen with an IgG (kappa, κ) monoclonal protein as identified by IFE. Arrow indicates position of monoclonal protein. After electrophoresis, each track except SPE is reacted with its respective antiserum, then all tracks are stained to reveal the respective protein bands. SPE = chemically fixed serum protein electrophoresis; IgG, IgA, IgM, κ, and λ indicate antiserum used on each track.

B, Same specimen as in *A,* with proteins identified by IEP. Arrow indicates position of monoclonal protein. Normal control (C) and patient serum (S) are alternated. After electrophoresis, antiserum is added to each trough as indicated by the labels Ig (polyvalent Ig antiserum), IgG, IgA, IgM, κ, and λ. The antisera react with separated proteins in the specimens to form precipitates in the shape of arcs. The IgG and kappa arcs are shorter and thicker than those in the normal control, showing the presence of the IgG (kappa) monoclonal protein. The concentrations of IgA, IgM, and λ-light chains are also reduced.

C, Patient specimen with an IgA (lambda, λ) monoclonal protein identified by IFE procedure as described in *A.*

D, Same specimen as in *C* with proteins identified by IEP as described in *B.* The abnormal IgA and λ-arcs for the patient specimen indicate an elevated concentration of a monoclonal IgA (lambda) protein. All separations were performed using the Beckman Paragon system.

IgA, IgM) in the specimen. Elevations of specific immunoglobulins should correspond to more intensely stained bands on the IFE pattern, but if a monoclonal protein is present, the result will be an approximation only. These quantitations, if performed first, will suggest suitable serum dilutions for IFE.

Reagents. Agarose gels for IFE may be prepared by the laboratory or purchased (e.g., Beckman Instruments, Inc., Brea, CA 92621; or Worthington Diagnostics, Freehold, NJ 07728). These gels are often supplied as part of a kit that includes antisera, stains, and buffer. The antisera may also be obtained as individual reagents from Atlantic Antibodies, Scarborough, MA 04074 or from Kallestad Laboratories, Inc., Austin, TX 78701. Barbital, 0.05 mol/L, pH 8.6, is commonly used for the electrophoresis buffer, Coomassie Brilliant Blue as the protein stain, and acetic acid, 5 mL/dL, in aqueous ethanol, 60 mL/dL, as the protein fixative.[76] Further details are provided in kit inserts.

COMPLEMENT PROTEINS

The complement proteins are a group of at least 20 immunologically distinct components in blood and tissue fluids.[1,13,49] They are able to interact sequentially with antigen-antibody complexes, with each other, and with cell membranes in a complex but adaptable way to destroy viruses, bacteria, and, pathologically, even the host's own cells. Complement proteins are synthesized by the liver and are normally present in the blood as functionally inactive molecules. A sequence of interactions can be activated by antigen-antibody complexes through the "classical" pathway from C1 to C5, or even by nonantibody-initiating factors through the "alternative" or nonclassical pathway from C3 to C5.[76] Either path leads to the final membrane attack sequence (C5 to C9) and lysis. The whole process is controlled by the very short life of many of the intermediates and by specific inhibitor proteins.

Activation of the classical pathway is initiated by IgG or IgM bound to antigen, or by complexed C-reactive protein. The alternative pathway provides an immediate nonspecific defense against invading organisms, since antibody is not essential to activate it even though IgA and IgG can do so. Subsequent pathway steps include proteolytic cleavage of C2, C4, C3, and C5; in each case a major and a minor fragment result. The major fragment has an enzymic site for attacking the next component and a binding site for attachment to a cell membrane or to an immune complex, so that subsequent steps are localized. The minor peptide fragments act on various cells, which take part in the very complex process of inflammation.

Since pure preparations and specific antisera are now available, complement components can be determined either immunologically or by a functional assay (hemolytic complement assay). Here we shall consider C1, C1 esterase inhibitor (C1 inhibitor, C1INA), C3, C3 proactivator (Factor B), and C4, since estimations of them and their breakdown products have proved most useful clinically.

C1 consists of three distinct protein molecules—C1q, C1r, and C1s—combined in a molar ratio of 1:2:2. These components are held together by a Ca(II)-dependent interaction and dissociate only under pathological conditions. C1q has binding sites for the Fc regions of IgG and IgM. When bound with an antigen-antibody complex, the C1 complex starts the complement cascade.

C1 esterase inhibitor (C1INH) is an α_2-globulin that controls the first stage of the classical pathway and also inhibits plasmin, thrombin, and kallikrein. Its absence causes inherited angioedema.

C3 and ***C4*** are β-glycoproteins with similar properties. In fresh serum, the concentration of C3 is sufficient to produce a β_2-band on electrophoresis; C4 is usually not visible. Both molecules are polymers; C3 contains two subunits and C4 three. Both polymers dissociate upon storage, altering their electrophoretic mobilities and causing the β_2-band to disappear; immunochemical recognition, however, is retained. Both C3 and C4 are acute phase reactants, and both take part in the classical pathway, but only C3 participates in the alternative pathway.

C3 proactivator (Factor B), a β_2-globulin, is a component of the alternative pathway.

Clinical Significance

Abnormal serum levels may be due to either inherited or acquired diseases.[22] ***Inherited deficiencies*** of most complement proteins have been described.[23] Deficiency of C1 esterase inhibitor causes *inherited angioedema*, an autosomal dominant disease characterized by subcutaneous, bronchial, and gastrointestinal edema. Laboratory findings include low levels of C4 and C1 esterase inhibitor, but normal levels of C3. In about 15% of cases, levels of C1 inhibitor are normal, but the protein is nonfunctional. In such cases, a functional assay for the inhibitor is essential to diagnosis. Diagnosis is important since mortality is high if the disorder is not treated.

Inherited deficiencies of C1, C4, and, most commonly, *C2* also occur. These deficiencies are associated with increased incidence of immune complex diseases. *Deficiency of C3* results in recurrent infections and impaired leukocyte mobilization. *Deficiency of C3b inhibitor* has a similar outcome, since activation of the alternative pathway is not controlled and C3 and Factor B deficiencies ensue. *Deficiencies in components C5 to C9* are associated with recurrent and persistent Neisserial infections.

Secondary changes in serum levels of complement components may occur in either direction; increases occur in acute phase reactions, decreases in immune complex diseases. Hence, levels of C3 and C4 reflect a balance between complement consumption due to immune complex formation and increased synthesis due to the acute phase response.

Immune complex diseases are disorders in which antigen-antibody complexes are either formed and deposited in tissues or formed in the circulation and then attached to tissue cells. Measurement of complement components can be useful in the diagnosis of these diseases, because the Ag-Ab complexes activate the complement system. Hence, changes in the concentration of the complement components reflect the activity of the deposition process.

Methods for the Determination of Selected Complement Proteins

Specimen. Serum is preferred; however, plasma from EDTA-anticoagulated blood is often suitable. The most important requirements are prompt separation of plasma or serum from cells and prompt analysis or storage at −70 °C, in order to preserve the very labile components.

Immunochemical assays. Nephelometry or RID can be used for assays of C3, C4, and Factor B, although there may be overestimation of C3 and C4 by RID when breakdown products are present. Breakdown products can be readily detected and estimated by immunofixation.[90] C1q and C1 esterase inhibitor are also estimated immunochemically. *Reference ranges*[84] for complement factors are age and method dependent.

Functional assays depend on observation of the final stage of the final pathway of complement activity as indicated by the lysis of erythrocytes coated with antibody. Individual components are tested by adding all the other components in excess. Test results are often expressed as the amount of factor required to cause 50% lysis (CH50).

Functional assays are used for evaluating patients whose symptoms imply deficiency of a complement factor but whose serum levels of complement factors are apparently normal by immunochemical assays. In such cases, these assays are essential. They are time-consuming, expensive, and best performed in a specialized laboratory.

ALBUMIN

Albumin, from week 20 of gestation and throughout life, is the most abundant protein in human plasma, representing 55–65% of the total protein.[57] It is synthesized in the liver at a rate that is dependent on protein intake but subject to feedback regulation by the plasma albumin level. The molecule contains no carbohydrate and is not stored to any extent in parenchymal cells. Little albumin is filtered through the kidney glomeruli, and most of that is reabsorbed by the proximal tubule cells and degraded by their lysosomal enzymes into fragments that are returned to the circulation. Plasma albumin is catabolized in various tissues, where cells remove it by pinocytosis. Its constituent amino acids are released by intracellular proteolysis and returned to the body pool.

Albumin has a MW of ~66 000; this relatively small size makes it a useful indicator of the integrity of glomerular and other membranes. Because its pI lies between 4 and 5.8, albumin is an anion at pH 7.4 with >200 negative charges per molecule. Its chief biological functions are to transport and store a wide variety of ligands, to maintain the plasma oncotic pressure, and to serve as a source of endogenous amino acids. The vast capacity of albumin for ligand binding is due to the large number of charges on each molecule as well as the very large number of molecules available. For example, albumin binds (and solubilizes) nonpolar compounds such as plasma bilirubin and long-chain fatty acids. It also binds a proportion of many hormones, among them thyroxine, triiodothyronine, cortisol, and aldosterone, thus acting as a reservoir in which these potent compounds are stored in an inactive form, but from which they are readily mobilized. Some 45% of serum calcium is bound to albumin, and many drugs are also strongly bound.

Clinical significance. Albumin levels, although important for management and follow-up, have very little value in diagnosis. *Hyperalbuminemia* is of little diagnostic significance except in dehydration. *Hypoalbuminemia*, however, is very common in many illnesses and results in most instances from one or more of the following factors: (1) impaired synthesis, either primary as in liver disease or secondary to diminished protein intake (hence its use to assess nutritional status); (2) increased catabolism as a result of tissue damage and inflammation; (3) reduced absorption of amino acids caused by malabsorption syndromes or malnutrition; (4) protein loss: in urine, due to nephrotic syndrome, chronic glomerulonephritis, diabetes, or systemic lupus erythematosus; in feces, due to protein-losing enteropathy arising from inflammatory or neoplastic disease; or from the skin through burns; and (5) altered distribution that may sequester large amounts of albumin in an extravascular compartment, as for instance in ascites where high pressure in the portal circulation drives albumin into the peritoneal fluid.

The most severe hypoalbuminemia is caused by protein loss by way of urine or feces; when plasma albumin levels are below 2.5 g/dL, the low plasma oncotic pressure allows water to move out of the blood capillaries into the tissues (edema). Changes in plasma albumin concentration affect the relative amounts of the bound and free concentrations of the ligands it carries; since free ligands are those that interact with tissue receptor sites and that can be excreted, plasma albumin levels have important influences on the metabolism of endogenous substances such as calcium, bilirubin, and fatty acids, and on the effects of drugs and hormones.

There are more than 20 *genetic variants* of albumin that are not associated with disease. Most commonly they cause a widening of the albumin band on electrophoresis or two bands (*bisalbuminemia*). There is also a transient form of the latter, sometimes caused by drugs.[82] Surprisingly, congenital absence of albumin, *analbuminemia*, is asymptomatic except for occasional slight edema.

Methods for the Determination of Albumin

Determination of albumin[58] in serum or plasma is usually based on its binding with the anionic dyes bromcresol green (BCG) or bromcresol purple (BCP) in a manual or automated procedure. The requirements of dye-binding methods include specific binding of the dye to albumin in the presence of other plasma or serum proteins; high binding affinity between the dye and albumin so that small changes in ionic strength and pH or the presence of competing ligands will not break the dye-protein complex; a substantial shift in the absorption wavelength of the dye in the bound form so that it will be spectrally distinct from the free form present in excess; and an absorption maximum for the bound form at a wavelength distinct from those where bilirubin and hemoglobin can interfere.

The BCG method almost meets these requirements. Bilirubin and hemoglobin do not interfere, and absorbance is linear for albumin concentrations from 1.0–6.0 g/dL. The dye, however, binds slowly to other serum proteins; hence, the absorbance has to be read at a short and standardized interval after mixing, and the method is unreliable for very low concentrations of albumin.

Albumin at pH 4.2 is sufficiently cationic to bind the anionic dye, and the absorption of the BCG-albumin complex is measured spectrophotometrically at 628 nm.[76] Acceptable within-run CV is $<3\%$; between-run, $<4\%$. If there is marked lipemia, the method must be modified to correct for its interference. Venostasis must be avoided during venipuncture because hemoconcentration increases the concentration of plasma proteins; also, the posture of the patient should be recorded since albumin concentration is ~0.3 g/dL higher when the patient is ambulatory.

An alternative dye method employing BCP[45] in place of BCG is not yet as widely used but is claimed to bind more specifically to albumin. Either method may be manual or automated.[29,59] Dipstick tests for *urinary* albumin are probably performed more often than any other test and are discussed later in the section on *Proteins in Urine*.

Electrophoresis separates albumin as a discrete band that, after staining (page 317), can be estimated by densitometry or cut out, eluted, and measured spectrophotometrically.

Immunochemical methods such as RID and nephelometry are more specific and sensitive. Highly purified albumin is available for standards and for preparing good, consistent antisera. These methods, however, are expensive. They, and electrophoresis after concentration, are suitable for quantitating albumin in solutions with low albumin content, such as urine or CSF.

α_1-ANTITRYPSIN

Alpha$_1$-antitrypsin (AAT) is an acute phase reactant with antiprotease activity,[43] whose inherited deficiency is associated with lung and liver disease. On electrophoresis AAT accounts for ~90% of the α_1-band because the other two components of this band stain only faintly (α_1-acid glycoprotein because of its high carbohydrate content and α_1-lipoprotein because of its high lipid content). AAT is responsible for 90% of serum antitrypsin activity, although it is relatively inactive toward trypsin. Its main function is to neutralize lysosomal elastase released upon phagocytosis of particles by polymorphonuclear leukocytes. AAT, a relatively small molecule, can pass from capillaries into tissue fluid, bind protease, and pass back into the intravascular fluid.

Genetic variants[35] are numerous. Some 33 allotypes have been described based on differences in electrophoretic mobility. The protein produced by 95% of the US population, whose genotype is Pi^{MM} (homozygotic for M protease inhibitor), is called M protein. Two other proteins, Z and S, are found in genotypes Pi^{ZZ}, Pi^{SS}, Pi^{SZ}, Pi^{MZ}, and Pi^{MS}; taken all together, these genotypes have a prevalence of <1% in the US population. Ability to inhibit proteases is directly related to the circulating level of M protein. If normal mean activity in the MM phenotype is taken as 100%, the level of activity in ZZ is 15%, in SS 60%, in MZ 57.5%, and in MS 80%. A rare null genotype (Pi^-) produces no AAT at all.[80]

Clinical significance. Low levels of AAT are found in neonatal respiratory distress syndrome, in severe protein-losing disorders, and in congenital deficiency. Increased levels are more common since AAT is an acute phase reactant. AAT deficiency of the ZZ, SS, and possibly MS phenotypes is associated with pulmonary emphysema of early onset (20–30 years). Inhaled particles and bacteria are continually removed from the lungs by polymorphonuclear leukocytes. This phagocytosis gives rise to the release of elastase. When the M protein of AAT is deficient, the uninhibited enzyme attacks the elastin of the alveolar wall, whose springy recoil after distention by inhaled air normally drives out exhaled air. The loss of elasticity of the lung tissue results in emphysema with impaired ventilation and vulnerability to serious respiratory infections. The liver is also affected, most seriously in the ZZ phenotype, where the Z protein accumulates in hepatic cells. Whether SZ, MS, and SS phenotypes are also at risk is not yet clear.

Methods of analysis. The functional assay, called trypsin inhibitory capacity (TIC),[16] has been virtually replaced by immunochemical (chiefly nephelometric) assays for the M protein. In the absence of an acute phase reaction to inflammation, a level of <50 mg/dL is presumptive evidence for a variant phenotype; testing should be pursued with typing by isoelectric focusing[33] or starch gel electrophoresis. In the presence of clinical infection and demonstrable increases in other acute phase reactants, a level >50 mg/dL does not exclude deficiency.

α_1-ACID GLYCOPROTEIN

Alpha$_1$-acid glycoprotein (AAG, orosomucoid)[70] is unique among the plasma proteins because of its low pI (2.7–3.5) and large carbohydrate content. It is very soluble in water and other polar solvents and has a high-negative-charge density at pH 7.4. Its electrophoretic mobility may vary slightly but this polymorphism has no known clinical significance. AAG is an acute phase reactant whose exact role is unknown. A certain degree of homology exists among the amino acid sequences of AAG, the immunoglobulins, and the haptoglobin α-chain, suggesting a common ancestry or role in the immune system.

Early work indicated that the liver was the only site of synthesis of AAG, but more recent studies have shown that some tumors are also able to synthesize it. Clinically it is used for monitoring the acute phase reaction.

Methods of analysis are immunochemical, such as RID or nephelometry; the antisera react with the peptide part of the molecule common to all its polymorphic forms.

α_1-FETOPROTEIN

Alpha$_1$-fetoprotein (AFP), the principal fetal protein, is described in Chapter 28, where the value of its determination in amniotic fluid and in maternal serum for the antenatal diagnosis of neural tube defects is discussed.

Later in life gross elevations of AFP serum levels are found in ~80% of patients with *hepatocellular carcinoma*, in 50% of those with *germ cell tumors*, and in all children with *hepatoblastoma*. Sequential assays are particularly useful for prognosis and for monitoring treatment. A lesser elevation of AFP occurs in some cases of carcinoma of the pancreas, gut, or lung, and in some nonmalignant diseases such as cirrhosis of the liver, tyrosinemia, and ataxia telangiectasia.

Methods of analysis.[72] For amniotic fluid, RID or electroimmunoassay (rocket technique) is adequate, but radioimmunoassay on a suitable dilution of fluid is widely used. It is essential to avoid contamination of the sample with fetal blood, which has an AFP concentration up to 200 times that of amniotic fluid. For serum, radioimmunoassay or enzyme-labeled immunoassay is required. Counterimmunoelectrophoresis may be used to screen patients with hepatic disease for increased AFP due to hepatocellular carcinoma. This application is useful in those parts of the world where hepatitis B, an etiological factor in the occurrence of hepatocellular carcinoma, is prevalent.

Reference ranges for amniotic fluid and maternal serum vary with weeks of gestation.[84]

HAPTOGLOBIN[14,62]

Haptoglobin (HAP, Hp) binds free hemoglobin in serum and is an acute phase reactant. It is an α_2-glycoprotein consisting of two pairs of nonidentical chains, α and β, linked by disulfide bonds. The subunit structure can therefore be designated as $\alpha_2\beta_2$. Three main phenotypes can be identified by electrophoresis, but they are of no clinical importance. A small percentage of Black and Oriental adults have no detectable HAP (anhaptoglobinemia), and this may be an unrecognized factor in the diagnosis of diseases that deplete HAP plasma levels.

The primary function of HAP appears to be the irreversible binding of free oxyhemoglobin in plasma, which occurs between HAP and the α-chain of the globin portions of hemoglobins A, F, S, or C. The complex is removed from the plasma within minutes by the reticuloendothelial system, where its components are metabolized to free amino acids and iron within a few hours. HAP thus prevents loss of hemoglobin to urine and conserves iron. In an acute episode of intravascular hemolysis, the total circulating HAP pool is capable of binding about 3 g of Hb. A single severe episode of hemolysis may deplete HAP entirely. After such depletion plasma levels return to normal in about one week. HAP does not bind methemoglobin, heme, or unusual forms of Hb in which the α-chain is missing, but a β_1-globulin, *hemopexin,* binds free heme. In conditions where hemolysis and an acute phase reaction are concurrent, it may be preferable to follow the changes in hemopexin concentration rather than those of HAP.

Clinical significance. Isolated estimations of plasma HAP have limited value because of the wide normal range. Serial determinations, however, are used to detect and monitor acute phase reactions and hemolytic states. *Increases in plasma HAP* concentrations occur in acute phase reactions and in conditions such as burns and nephrotic syndrome, when large amounts of fluid and lower-molecular-weight plasma proteins have been lost. *Low serum HAP* levels are most frequently associated with conditions of increased intravascular hemolysis or Hb turnover such as occur in hemolytic anemias, transfusion reactions, and malaria. Chronic intravascular hemolysis causes a small but continuing drop in HAP levels.

Methods of analysis. Immunochemical methods are simplest; all three phenotypes react with antisera raised against HAP 1-1. Nephelometry[87] is preferable to RID because the phenotype with the greatest molecular weight diffuses so slowly that RID takes several days for completion. Haptoglobin phenotypes may be separated by starch gel or polyacrylamide gel electrophoresis (PAGE).

CERULOPLASMIN

Ceruloplasmin (CER),[26,60] a late acute phase reactant, is the principal copper-containing protein of plasma. It is an α_2-glycoprotein with a single polypeptide chain containing 6–7 copper atoms per molecule. Because of its copper content, the pure protein has a blue color, hence the name. With a pI of 4.4, CER migrates on electrophoresis in the fast α_2-region, but it makes no visible contribution to the α_2-band except when the concentration in plasma is greatly increased. Its chief function appears to be as an enzyme. CER was originally called "copper oxidase"; its oxidase behavior (reduction of O_2 to water) has been used for its identification and assay. CER rapidly catalyzes oxidation of Fe(II) to Fe(III), and this "ferroxidase" activity is essential to transform Fe(II) at the cell surface, before iron can be bound to transferrin.

Ceruloplasmin is not, as originally thought, a transport protein for copper in plasma. Once synthesized, it neither gains nor loses copper unless metabolized. Plasma copper consists of a nondialyzable fraction (95%) attached to CER and of a dialyzable fraction (5%) loosely bound to albumin and histidine. Copper is believed to be transported in the dialyzable form from the gut to the liver; it is incorporated into the CER apoprotein, which is released into the blood stream. Animal experiments have shown that copper cannot combine with apoprotein in the circulation. Increased absorption of copper leads to increased synthesis of CER and increased excretion of nonreabsorbable copper-protein complexes in the bile. Synthesis of CER thereby provides a first-line reaction to potential copper toxicity.

A possible role for CER is as an antioxidant. Spontaneous oxidation of many organic substances in contact with oxygen is a constant danger to life but is prevented by the presence of antioxidants in plasma and tissues; this is perhaps the role of CER in acute phase reactions.

Clinical significance. The most important clinical application of CER assay is in the diagnosis of *Wilson's disease.*[15] Typically the plasma CER is reduced, whereas dialyzable copper concentration is increased. Unless treated with copper chelators such as D-penicillamine, the disease is always progressive and fatal. Prompt diagnosis is important since the treatment takes 3–6 months to have the desired effect. When treated, most patients lead a normal life. A majority of cases present with liver dysfunction; any patient between the ages of 5 and 50 years with unexplained liver disease, enlarged liver and spleen, a history of attacks of jaundice, or signs of unexplained brain damage should be screened for Wilson's disease.[52] Neurological symptoms are more common in adults, whereas liver dysfunction is more common in children.

The disease is rare (incidence 1:50 000–1:100 000) and is due to an autosomal recessive defect in the regulation of copper metabolism. The two fundamental disturbances are a gross decrease in the rate of incorporation of copper into apoprotein and a marked reduction in the biliary excretion of copper. Although the exact defect is unknown, it is probably lack of an enzyme or a carrier protein required for the incorporation of copper into proteins. Copper is deposited in the kidneys, in the liver (where it causes cirrhosis), and in the brain (where it damages the basal ganglia). The combination of liver and neuron damage gives the disease its alternative name, hepatolenticular degeneration. Copper also deposits in the cornea, giving rise to the characteristic Kayser-Fleischer rings. Their presence, together with plasma CER <10 mg/dL, high plasma levels of dialyzable copper, and increased urinary copper, is diagnostic in all cases with neurological involvement but is diagnostic in only 80% of the cases with liver disease. The most reliable test for Wilson's disease is injection of radioactive copper, followed by timed observations of plasma radioactivity. In normal patients, the radioactivity first increases, then falls, then rises again as CER is synthesized. The secondary rise will not occur in Wilson's disease because of the defect in incorporating copper into CER.

Low plasma levels of CER are also found in malnutrition, malabsorption, nephrosis, and severe liver disease, particularly primary biliary cirrhosis. In the last instance, the radiocopper test will probably be required to rule out Wilson's disease.

Methods of analysis.[79] CER is most readily measured immunochemically by nephelometry. RID and spectrophotometric measurement of oxidase activity are alternatives.

Reference ranges. CER levels vary with age. Exercise, pregnancy, and estrogen administration increase CER concentration (see *Appendix*, Table 29-18).

TRANSFERRIN

Transferrin (TRF, siderophilin)[6,63] is the principal plasma protein for transport of iron. Its concentration correlates with the total iron-binding capacity (TIBC) of serum (see Chapter 24). TRF is a glycoprotein with a single polypeptide chain. It reversibly binds numerous polycations—iron, copper, zinc, cobalt, and calcium—although only iron and copper binding appear to have physiological significance. One molecule of TRF binds two ferric ions and an associated anion that is, in vivo, usually bicarbonate. There are two binding sites for iron, probably with different affinities; they are strong at physiological pH but weaken with decreasing pH.

TRF is synthesized in the liver and to a small extent in the reticuloendothelial system and in endocrine glands such as testes and ovaries. Plasma levels are apparently regulated by availability of iron, since in iron deficiency states plasma levels rise and, upon successful treatment with iron, return to normal levels. Iron metabolism is discussed in Chapter 24.

The TRF apoprotein binds iron released from catabolism of hemoglobin or absorbed from the intestine. The TRF-Fe(III) complex in plasma then carries iron to storage sites (the liver and reticuloendothelial system) where it is incorporated into ferritin and hemosiderin, and to cells that synthesize the iron-containing compounds such as hemoglobin, myoglobin, and cytochromes.

Measurement of plasma transferrin levels is useful for the differential diagnosis of anemia and for monitoring its treatment. In common iron deficiency or hypochromic anemia, the transferrin level is increased due to increases in synthesis, but the protein is less saturated with iron because plasma iron levels are low. On the other hand, if the anemia is due to a failure to incorporate iron into erythrocytes, the transferrin level is normal or low but the protein is highly saturated with iron. In iron overload, transferrin concentration is normal but saturation (normally 30–38%) is >55%, and may be as high as 90%.

Transferrin is a negative acute phase reactant, and *low levels* will occur, along with low levels of albumin, prealbumin, and β-lipoprotein, in inflammation or malignancy. Other causes of decreased synthesis and low plasma levels are chronic liver disease, malnutrition,[11] and protein loss, as in the nephrotic syndrome or protein-losing enteropathies. In congenital atransferrinemia, a very low level of transferrin is accompanied by iron overload and a severe hypochromic anemia resistant to iron therapy. *High levels* of transferrin occur in pregnancy and during estrogen administration.

Methods of analysis. Immunochemical methods—nephelometry or RID—are preferred. Abnormally low levels may first have been suspected from a routine electrophoretic separation. Because of the convenience of simultaneous measurement of serum iron and TIBC and the desirability of knowing per cent saturation of TRF, TRF is sometimes estimated indirectly from the TIBC value by the equation:

$$\text{TRF, mg/dL} = 0.70 \times \text{TIBC, } \mu\text{g/dL}, \quad \text{or} \quad \text{TIBC} = 1.43 \times \text{TRF}$$

This procedure, however, overestimates the TRF concentration by 16–20% because it assumes that all plasma iron is bound to TRF. In fact, when TRF is more than half saturated, some of the iron is bound to other plasma proteins such as albumin.

Adult reference ranges for nephelometry are 220–400 mg/dL.[84]

C-REACTIVE PROTEIN[51,55]

A substance present in the sera of acutely ill patients and able to bind the C-polysaccharide on the cell wall of *Streptococcus pneumoniae* was first described in 1930. It was shown to be a protein and was given the name C-reactive protein (CRP). It has five identical polypeptide subunits and is synthesized by the liver. It migrates on electrophoresis anywhere from the slow γ- to the mid-β-region, depending on factors such as the concentration of calcium ions in the buffer.

CRP binds not only the polysaccharides present in many bacteria, fungi, and protozoal parasites but also phosphorylcholine, phosphatidylcholines such as lecithin, and polyanions such as nucleic acids. Once complexed, CRP becomes an activator of the classical complement pathway. Like antibodies, CRP can initiate opsonization, phagocytosis, and lysis of invading cells. Perhaps its main role is to recognize potentially toxic autogenous substances released from damaged tissues, to bind them, and then to detoxify them or clear them from the blood. CRP, which was the first acute phase reactant to be discovered, is also the one exhibiting the most dramatic increases in concentration following myocardial infarction, trauma, infection, surgery, or neoplastic proliferation. The increase occurs within 24–48 h and the level may be 2000 times normal. For example, cord blood normally has low CRP concentrations (1–35 μg/dL), but in intrauterine infection levels may be as high as 26 000 μg/dL. Determination of this nonspecific increase of CRP is clinically useful for screening for organic disease; assessing the activity of an inflammatory disease such as rheumatoid arthritis; detecting intercurrent infections in systemic lupus erythematosus, in leukemia, or after surgery (secondary rise in plasma level); detecting rejection in renal allograft recipients; and managing neonatal septicemia and meningitis when the collection of specimens for bacteriological investigations may be difficult.

Methods of analysis.[9,79] Since CRP is normally present in blood at <800 μg/dL, immunochemical methods are used, such as RID, nephelometry, or the more sensitive RIA or homogeneous enzyme immunoassay (Syva Corp., Palo Alto, CA 94303).

OTHER PLASMA PROTEINS OF CLINICAL INTEREST

Amyloid and its precursors.[76] Amyloid is a pathological extracellular deposit sometimes seen in biopsy sections. Deposits may exert pressure on vital structures and eventually cause death. They are characterized by staining with iodine and certain dyes, in particular Congo Red, which give them a green fluorescence in polarized light. Under the electron microscope deposits have a fibrillar appearance. Such "amyloidosis" is found in various organs in a variety of apparently unrelated diseases—a situation recently explained by the discovery that amyloid has a variable protein composition but always a β-pleated sheet structure, which gives it its characteristic insolubility and staining properties.[24] Its components all appear to have soluble plasma protein precursors; a minor one always present is the "P-component," a close relative of CRP. Work on identifying the major components, their precursors, and the diseases associated with them has progressed far enough to draw up a partial classification of amyloid.[76] For example, the amyloid found in some cases of multiple myeloma has Ig L-chain precursors; that associated with chronic inflammation has *serum amyloid A protein*, an acute phase protein, as its main precursor; and for some other clinical types, prealbumin is the plasma protein precursor.

Prealbumin (PA) and retinol-binding protein (RBP) are both synthesized by the liver. During electrophoresis, both migrate ahead of albumin. The $t_{1/2}$ for each is extremely short, probably <12 h; consequently, measurement of these serum levels may provide a more timely and sensitive assessment of protein malnutrition or liver dysfunction than measurement of transferrin or albumin levels.

Both PA and RBP are transport proteins. PA binds thyroxine and triiodothyronine. RBP, in the form of a 1:1 complex with PA, transports vitamin A (retinol). PA is a negative acute phase reactant; owing to decreased synthesis, the serum level falls in inflammation, malignancy, cirrhosis of the liver, and protein wasting diseases of the gut or kidney. Zinc is required for RBP synthesis, and zinc deficiency states are characterized by low serum levels of both RBP and total vitamin A.

Determination of PA and RBP can be made by nephelometry[79] or RID. The adult reference range for PA measured by RID is 10–40 mg/dL and for RBP, 3.5–9.0 mg/dL. The level in children is about half the adult level; sharp increases occur at puberty.[84]

Alpha$_1$-antichymotrypsin (α_1AC), inter-α-trypsin inhibitor (IαTI), and vitamin D-binding protein (DBP; Gc globulin) form two very faint bands between the α_1- and α_2-regions on electrophoresis. The first two are protease inhibitors. The α_1AC is a rapidly responsive acute phase protein and occurs normally in relatively high concentrations in bronchial secretions. IαTI probably serves as a parent of protease inhibitors of smaller size. DBP has a MW of 59 000 and functions in vitamin D transport; it is measured by RIA. Both protease inhibitors can be assayed by RID.

Alpha$_2$-macroglobulin (AMG) is so large a molecule that it tends to remain intravascular. It is capable of binding a wide variety of substances, including many proteases whose active sites remain unmasked.[4,43] Its role is unknown, but it is probably vital in inflammation.[32] At present, measuring the plasma level of AMG is of little clinical value. In states such as the nephrotic syndrome, in which selective loss (sieving) of proteins occurs, AMG is increased because its large size aids in its retention. Nephelometry and RID are acceptable methods for the determination of AMG.

Beta$_2$-microglobulin (BMG) is a low-molecular-weight (11 800) protein found on the cell surfaces of all nucleated cells and shed into the blood, particularly by lymphocytes and tumor cells. The protein is the light- or β-chain of the human leukocyte antigen (HLA) on cell surfaces and consists of a single polypeptide chain with one intrachain disulfide bridge. The small size of the molecule allows BMG to pass through the glomerular membrane, but normally $<1\%$ of the filtered BMG is excreted in the urine; the remainder is reabsorbed and catabolized in the proximal tubules of the kidney.

High plasma levels occur in renal failure, inflammation, and neoplasms, especially those associated with B lymphocytes. The principal clinical value of BMG assay is to test renal tubular function,[92] particularly in kidney transplant patients in whom rejection of the allograft will manifest as diminished tubular function. Serial assays of β_2-microglobulin are also useful to monitor B-cell tumors. Radioimmunoassay is the usual method of analysis for BMG.

Lysozyme (muramidase) occurs in lysosomes intracellularly and in most extracellular fluids, especially in exocrine secretions. It is bactericidal and is produced by granulocytes and monocytes but not by lymphocytes. The plasma level rises in tuberculosis, sarcoidosis, and other granulomatous diseases. The enzyme is filtered by the glomerulus, but almost all of it is reabsorbed and catabolized by the tubules; hence, its measurement in urine may be employed to test renal tubular function. Lysozyme is produced in such large quantities in monocytic leukemia that it may be visible as a post-γ-band in an electrophoretic separation of serum or urine concentrate. Its concentration is determined by nephelometry.[84]

Proteins in Urine[12,39]

The glomeruli of the kidney behave as ultrafilters for plasma proteins. The degree to which individual proteins are normally filtered through the membrane is dependent on both their molecular size and their plasma concentration. In general, transport of protein molecules through the glomerular membrane progressively diminishes as protein size increases. Normally, high-molecular-weight proteins such as IgM (MW 900 000) do not appear in glomerular filtrate except in trace amounts. Relatively small yet significant amounts of albumin (MW 66 000) are passed into the filtrate as a result of its high plasma concentration and relatively low molecular weight. Proteins with MW 15 000–

40 000 filter more readily but in lesser quantities because of their low plasma concentrations. The proportions of individual proteins *excreted* in the urine depend on the extent of their reabsorption by the renal tubules. The lower-molecular-weight proteins, including some albumin, are actively reabsorbed from the filtrate and catabolized by the proximal tubules. Consequently, the total protein excreted is only 20–150 mg/d; about half is albumin and the remainder almost entirely Tamm-Horsfall protein (uromucoid), a constituent of urinary casts and probably secreted by the distal tubules. Increased permeability of the glomeruli is therefore first indicated by increased amounts of albumin in the urine, and increasing permeability is indicated by the appearance in urine of proteins with increasingly greater molecular weights. Diminished or diminishing tubular reabsorption is indicated by the appearance of increasing concentrations of small-molecular-weight proteins in urine (see also Chapter 20).

Clinical Significance

There are four ways in which *proteinuria* (increased amounts of protein in urine) can occur: increased glomerular permeability (*glomerular proteinuria*), in which the urinary protein is mainly albumin; defective tubular reabsorption (*tubular proteinuria*), in which urinary proteins are mainly normal, low-molecular-weight plasma proteins; overflow from plasma to urine of an abnormal low-molecular-weight protein (*overload proteinuria*); and production of protein by the urinary tract (*postrenal proteinuria*).

Glomerular proteinuria is by far the most common and serious type of proteinuria. Patients are routinely screened for this disorder by a simple dipstick test for albumin. If the dipstick test is negative, clinically significant glomerular proteinuria is excluded; if the test is positive, further investigation such as a quantitative evaluation of protein excretion is indicated. Since most of the excreted protein is albumin, glomerular proteinuria is often labeled "albuminuria." Increase in glomerular permeability occurs in numerous conditions characterized by diffuse injury to the kidney (see also Chapter 20). In diabetes, vascular permeability increases and albuminuria appears when metabolic regulation is poor.[53] Immune complex disease often affects the glomerulus; in nephritis associated with systemic lupus erythematosus and in most cases of glomerulonephritis, granular deposits containing immunoglobulins and complement components can be demonstrated on the glomerular membrane. In such progressive diffuse renal diseases, the ability to restrict filtration of the smallest of the larger proteins is lost first; thus, albumin appears first in the urine. As long as the molecular weight determines which proteins are filtered, the proteinuria is called *selective*. Progressively severe glomerular lesions produce less selective proteinuria until proteins of all sizes pass the glomerulus (*non-selective proteinuria*). In the final stages of disease, as glomeruli are destroyed and large numbers become functionless, proteinuria decreases and renal failure is established.

Protein excretion rate is ordinarily determined from a 24-h urine collection because random specimens vary considerably in protein concentration. When protein excretion is >1 g/d, the proteinuria is likely to be glomerular. When excretion is >2 g/d, heavy loss of plasma albumin may cause generalized edema (e.g., in the nephrotic syndrome).

Functional or benign proteinuria is a mild form of glomerular proteinuria that is probably due to changes of blood flow through the glomeruli. It is associated with exercise, pyrexia, exposure to cold, congestive heart failure, hypertension, or arteriosclerosis. Protein excretion rates are <1 g/d. *Postural or orthostatic proteinuria*, associated with the upright position, is also a functional proteinuria but excretion may exceed 1 g/d. It complicates assessment of proteinuria in otherwise symptomless patients. If transient, it is probably benign; but if chronic or not entirely related to posture, protein excretion rates should be checked at six-month intervals. Persistence of proteinuria suggests underlying renal disease. In normal pregnancy, protein excretion may increase harmlessly to 200–300 mg/d. This slight increase is in contrast to the proteinuria of pre-eclamptic toxemia, a glomerular proteinuria of up to 3 g/d, and to the proteinurias of latent renal disease or of urinary infection.

Tubular proteinuria results from defective reabsorption of the smaller plasma protein molecules that always pass through the glomerular filters, such as β_2-microglobulin (MW 11 800), lysozyme (MW 14 500), retinol-binding protein (MW 21 000),

α_1-microglobulin (MW 27 000), and α_1-acid glycoprotein (MW 40 000). Some albumin is also excreted, but not enough to give a positive dipstick test.

The sizes of the protein molecules found in the urine indicate the type of proteinuria—small molecules in tubular, mainly albumin in early glomerular, larger ones in later glomerular, and all sizes in mixed tubular and glomerular types. Hence a technique such as SDS-PAGE that separates the proteins purely according to molecular size reveals most clearly the nature of a proteinuria. (The proteins are given a uniform negative charge by sodium dodecyl sulfate and then separated by electrophoresis on 10% polyacrylamide gel. See Chapter 2, Section Three.)

In practice, tubular proteinuria is usually diagnosed by determining the urinary concentration of a small protein molecule, usually β_2-microglobulin or lysozyme. If, as is common, there is also glomerular disease, the dipstick test will be positive.

Acute tubular proteinuria may occur in severe disturbances of metabolism that are associated with burns, acute pancreatitis, or administration of nephrotoxic drugs; it may later clear up completely. ***Chronic tubular proteinuria*** is usually irreversible. Its origin may be hereditary as in Fanconi's syndrome, or a kidney disease such as chronic pyelonephritis, or a systemic disease such as cirrhosis or sarcoidosis, or acquired owing to drugs such as phenacetin. In such instances, slight tubular proteinuria may be the only sign of progressive renal damage. Tests for tubular proteinuria are therefore now used to monitor renal allograft rejection,[92] aminoglycoside and cadmium toxicity, and chronic pyelonephritis.

Overload proteinuria includes hemoglobinuria, myoglobinuria, and Bence Jones proteinuria, due to high plasma concentrations of immunoglobulin light-chain paraproteins (e.g., in multiple myeloma). Detection of light chains depends on electrophoretic and immunochemical testing. Hemoglobin and myoglobin can be detected by tests discussed in Chapter 24.

Post-renal proteinuria refers to protein arising from the urinary tract below the kidneys and is usually due to inflammation or malignancy. It can be diagnosed by microscopic examination of the urinary sediment for inflammatory (pus) cells and malignant cells. The presence of erythrocytes or pus cells in casts in such a centrifuged urinary sediment is valuable proof that their origin is from the kidney and is *not* extrarenal.

Laboratory Studies of Proteinuria

Qualitative detection of excess protein in urine is largely based on use of dipsticks. The reactive portion of the stick is coated with a buffered indicator that changes color in the presence of protein. The principle is often called the "protein error of indicators." A typical example is Albustix (Ames, Elkhart, IN 46514), in which bromphenol blue, buffered to pH 3 with citrate, is present predominantly in the protonated, yellow form. When protein is added, the affinity of the anionic form of the indicator dye for protein causes a shift of the equilibrium between anionic and protonated forms of the indicator toward formation of the blue anionic species. The intensity of blue color produced is proportional to the concentration of protein in the specimen. The detection limit of the Albustix test is 7 mg/dL.

Like all dye-binding techniques, the dipstick methods are more sensitive to albumin than to other plasma proteins. They are therefore excellent screening tests for glomerular proteinuria but unsatisfactory for detection of tubular proteinuria or overload proteinuria of the Bence Jones type. They are only roughly quantitative, and their use should be limited to screening and to estimates required before concentrating a specimen for electrophoresis or diluting it for quantitative assay. A first morning urine specimen is best, since it tends to be concentrated and unaffected by postural factors.

Quantitative assay for total protein (or for individual proteins) is usually performed on timed collections. Periods of 4, 8, and 12 h may be appropriate for following a renal transplant patient or a patient whose acute renal losses of albumin are being compensated with closely regulated replacement therapy. In most cases, however, a 24-h collection is chosen, both for quantitative total or specific protein assay and for electrophoretic separation. An alternative approach is to measure the protein/creatinine ratios of random specimens.[3,73] Creatinine concentration is relatively constant in any one subject, and the ratio correlates well with the 24-h total excretion of protein.

Quantitative assay of specific proteins is necessary for estimates of glomerular selectivity and for evaluation of tubular proteinuria. Nephelometry, radial immunodiffusion, and rocket immunoelectrophoresis are sufficiently sensitive for most of the analytes. For quantitating β_2-microglobulin, whose excretion normally is ~0.1 mg/d, radioimmunoassay is required. Because this protein is susceptible to degradation at low pH, specimens should be stored at a pH >6.0 during the collection period. In *assessment of the nephrotic syndrome*, glomerular selectivity is estimated usually by measuring two individual proteins, one that in health is just retained by the glomerulus (e.g., albumin) and another that is well retained (e.g., IgG). A random urine sample and a corresponding serum are collected and analyzed for the two proteins. The clearances of each protein are calculated by using the classical formula in which U and P represent the concentrations of the compound in urine and plasma, respectively, and V is the volume of excreted urine per minute. The ratio of the two clearances is then calculated.

$$\text{Ratio} = \left(\frac{\text{UV}}{\text{P}}\text{ for IgG}\right) \Big/ \left(\frac{\text{UV}}{\text{P}}\text{ for albumin}\right)$$

Note that V cancels out, an advantage with children. A ratio higher than 0.16 implies some loss of selectivity in that larger molecules are being lost. A ratio >0.30 indicates low selectivity. To evaluate selectivity by using a random specimen, one would calculate the protein-creatinine indices for the two proteins in the urine specimen and substitute them for urine concentration terms in the clearance equation.

For evaluation of *tubular proteinuria*, one may calculate the excretion rate of the protein per day or measure in a random specimen the concentrations of both creatinine and the specific protein (e.g., lysozyme) and determine their ratio:

$$\text{Lysozyme-creatinine Index} = \frac{[\text{Lysozyme}]}{[\text{Creatinine}]}$$

Methods for the Determination of Total Protein in Urine

Recent comparisons of methods have been reported in the literature.[17,46] The biuret method applied to acid-precipitated protein[85] or to a concentrate obtained by membrane filtration has the advantage of equal sensitivity to each of the individual proteins in the mixture. Turbidimetric and dye-binding methods are faster and simpler. Of the dye-binding methods, Ponceau S and Coomassie Brilliant Blue (CBB)[44] are the most popular; the CBB method is probably the better of the two. It is sensitive and has a precision of 6–7% (CV) but, like turbidimetric methods, has limited linearity so that urines must be diluted to varying degrees for assay. Like turbidimetric and other dye-binding methods, it has unequal sensitivities for individual proteins and underestimates low-molecular-weight proteins in tubular proteinuria and immunoglobulin light chains in overload proteinuria. No method is wholly satisfactory. A method using a TCA-Ponceau dye reagent has been published by Pesce and Strande.[56]

In one method evaluated and reported by the AACC as a Selected Method in Clinical Chemistry,[58,76] proteins in urine are precipitated as fine particles with trichloroacetic acid and quantitated by turbidimetry at 620 nm. Standardization is made against serum that has been analyzed by a properly standardized and well-controlled biuret method and then diluted appropriately. A similar method, in which TCA is replaced by benzethonium chloride, has been automated on the *aca* (DuPont Instruments, Wilmington, DE 19898).

Electrophoretic separation of urinary proteins. The procedure for electrophoresis of urine on agarose gel or cellulose acetate is identical to that for serum, with the exception that urine must be concentrated prior to application. Concentration is most conveniently achieved with the Minicon B15 concentrators (Amicon Corporation, Danvers, MA 01923). Up to 5 mL of a sample is placed in a chamber of the concentrator.

The inner surface of the chamber is a membrane of selective permeability that is backed by absorbent pads. Water from the specimen crosses the membrane and is absorbed by the pads; proteins left behind on the specimen side of the membrane are concentrated to a degree dependent on the time allowed for the process. Up to 100-fold concentration is possible. This particular concentrator retains in the specimen proteins with MWs $>15\,000$, which includes Bence Jones proteins and albumin.

Simple electrophoresis of urine concentrated to a protein concentration of ~3 g/dL may suggest the presence of immunoglobulin light chains (Bence Jones protein) or other low-molecular-weight proteins typical of tubular proteinuria. Comparison of a urine separation with a corresponding serum separation may also indicate the degree of selectivity in glomerular proteinuria. Separations of urine concentrates on SDS-PAGE are particularly informative.

Proteins in Cerebrospinal Fluid[39,88]

Cerebrospinal fluid (CSF) is secreted by the choroid plexuses, around the cerebral vessels, and along the walls of the ventricles of the brain. It fills the ventricles and cisternae, bathes the spinal cord, and is reabsorbed into the blood through the arachnoid villi. More than 80% of CSF protein content originates from plasma by ultrafiltration through the walls of capillaries in the meninges and choroid plexuses; the remainder originates from intrathecal synthesis. The lowest concentration of total protein and the smallest proportion of the larger protein molecules are in the ventricular fluid; as the CSF passes down to the lumbar spine (from which site specimens are usually collected), the protein concentration increases. In CSF from neonates, levels of total protein are considerably higher than in older children and healthy adults.

Because CSF is mainly an ultrafiltrate of plasma, low-molecular-weight plasma proteins such as prealbumin, albumin, and transferrin normally predominate. No protein with a molecular weight greater than that of IgG is present in sufficient concentration to be visible on electrophoresis. The electrophoretic pattern of normal CSF after concentrating the fluid has two striking features—a prominent prealbumin band and two transferrin bands. The second form of transferrin bands, the τ band, is made or transformed intrathecally.

Clinical Significance of CSF Protein Analysis

Examination of CSF total protein and specific proteins is used chiefly to detect increased permeability of the blood/brain barrier* to plasma proteins or to detect increased intrathecal secretion of immunoglobulins (Igs).

The permeability of the blood/brain barrier to plasma proteins is increased by high intracranial pressure due to brain tumor or intracerebral hemorrhage, or by inflammation, especially bacterial meningitis. Lumbar CSF protein is increased when the CSF circulation is mechanically obstructed above the puncture site (as by a spinal cord tumor) and plasma proteins equilibrate across the walls of meningeal capillaries into the stagnant CSF. The effect of any of these conditions is that the proportions of specific proteins in CSF increasingly resemble those characteristic of serum.[8] The degree of permeability of the blood/brain barrier can be evaluated by immunochemical measurements of albumin in CSF and in serum specimens obtained at the same time. Albumin is a particularly suitable indicator protein because it is neither synthesized nor metabolized intrathecally. In CSF that is free of contaminating blood, albumin must necessarily have come from plasma through the blood/brain barrier.

---

*The blood/brain barrier is a concept rather than an anatomical structure. The barrier is defined by the many complex factors that govern the distribution of compounds other than water, CO_2, and O_2 between the blood and the extracellular fluid of the brain and its accessory structures. In the present discussion, the term *blood/brain barrier* is used, admittedly inaccurately, as a synonym for the capillary endothelium of vessels of the central nervous system.

Based on the results of the immunochemical albumin assays, a ratio named the CSF/Serum Albumin Index is arbitrarily calculated as follows:

$$\text{CSF/Serum Albumin Index} = \frac{\text{Albumin}_{\text{CSF}},\ \text{mg/dL}}{\text{Albumin}_{\text{SER}},\ \text{g/dL}}$$

An index value <9 is considered normal. Values of 14–100 suggest moderate to severe impairment of the barrier; values >100 indicate its complete breakdown.

Demonstration of increased *intrathecal synthesis of immunoglobulins*, particularly IgG, has great importance in the diagnosis of demyelinating diseases of the central nervous system (CNS), especially multiple sclerosis.[88] In multiple sclerosis, patchy deterioration of myelin sheaths of axons in the CNS profoundly affects conduction of nerve impulses. The sites of the lesions are unpredictable, and symptoms vary widely. B lymphocytes infiltrate the lesions and synthesize IgG and occasionally other Igs, which then appear in the CSF.

Increased intrathecal synthesis of IgG causes an increase in the CSF/serum immunoglobulin concentration ratio and the appearance of oligoclonal Ig bands on electrophoretic separations of CSF.[83] The investigation of these changes has now become routine when multiple sclerosis is suspected, but their presence is not diagnostic. They may also be found in some other chronic inflammatory diseases of the CNS, which must be ruled out clinically. Finding a CSF total protein >100 mg/dL, a CSF leukocyte count >50/μL, or a positive test for neurosyphilis tends to exclude multiple sclerosis. The CSF albumin concentration in 70% of cases of multiple sclerosis is within the reference range.

A complicating factor in testing for intrathecal IgG is that an increase in CSF IgG concentration, or in the CSF/serum IgG concentration ratio, may occur because of increased permeability of the blood/brain barrier, or increased local production of IgG, or both. In order to identify intrathecal production specifically, one of the following tests can be performed, using the CSF albumin concentration to correct for increased permeability.

1. Concentrations in CSF of IgG and albumin are measured, and a ratio is calculated:

$$\text{Ratio} = \frac{\text{IgG}_{\text{CSF}},\ \text{mg/dL}}{\text{Albumin}_{\text{CSF}},\ \text{mg/dL}}$$

A ratio >0.27, indicating increased synthesis, is found in about 70% of cases of multiple sclerosis.

2. Concentrations in CSF *and* serum of IgG and albumin are measured, and the "CSF immunoglobulin index" is calculated from the CSF/serum IgG concentration ratio divided by the CSF/serum albumin concentration ratio so that

$$\text{Index} = \frac{\text{IgG}_{\text{CSF}},\ \text{mg/dL} \times \text{Albumin}_{\text{SER}},\ \text{g/dL}}{\text{Albumin}_{\text{CSF}},\ \text{mg/dL} \times \text{IgG}_{\text{SER}},\ \text{g/dL}}$$

The *reference range* for the index is 0.3–0.77. Values >0.77 are considered evidence of increased IgG synthesis; in over 90% of cases of multiple sclerosis, the index exceeds 0.77. This estimator is the one most often used.

Methods for the Determination of Total Protein in CSF

A CSF specimen is very precious since only a small volume is generally obtained (5–10 mL or less) and it must often be shared with the microbiology and cytology departments. Repeated spinal punctures should be avoided because the procedure is invasive, may be painful, and can produce CSF that merely reflects trauma created by the first puncture rather than a change in the patient's condition. Assay techniques should be chosen to give maximum information on the least volume of sample and any sample remaining should be refrigerated or frozen in case additional testing is required. Blood in a CSF specimen, present because of bleeding within the cord, ventricles, or cisternae, or as the result of a "bloody tap," invalidates the protein values, since the protein concentration of whole blood is ~1000 times that of normal CSF. Total protein measurement, even on clear supernatant of a bloody specimen, inevitably reflects contamination with plasma proteins.

The small volumes of CSF available limit the choice of methods; turbidimetric methods and versions of the Coomassie Brilliant Blue (CBB) dye-binding method[31,36] remain the most popular. The most serious disadvantage of turbidimetric methods is the requirement for 0.2–0.5 mL of sample. CBB methods require only 25 μL, but they underestimate globulins. However, since albumin is the predominant protein of CSF, this underestimation is often not serious. A *turbidimetric method* in which proteins in CSF are precipitated by a combination of sulfosalicylic acid and sodium sulfate has been shown to be equally efficient in precipitating albumin and globulins.[54] It seems probable, however, that immunochemical methods that use small sample volumes and determine individual proteins rapidly, specifically, precisely, and quantitatively may eventually replace determinations of total protein in CSF.

Methods for Specific Proteins in CSF

Rate nephelometry, electroimmunodiffusion, and radial immunodiffusion are most often used for measurements of albumin and IgG. When diffusion methods are employed, CSF and serum concentrations of a protein can be directly compared on the same plate; electroimmunodiffusion even allows all four estimations needed for calculating the CSF/serum immunoglobulin index to be performed on the same plate, since albumin and IgG migrate in opposite directions. Apparent absence of IgG may be due to its degradation by proteinases in the specimen. Radioimmunoassay (RIA) will be required for determination of specific proteins present in very low concentrations (e.g., IgM).

Reference ranges.[84] The range for albumin levels in lumbar spinal fluid by RID is 17.7–25.1 mg/dL. In normal CSF, IgA, IgD, and IgM, measured by RIA, are each less than ~0.2 mg/dL. Reference ranges for IgG, also measured by RIA, are age related and their means rise from 3.5 mg/dL in the 15–20-year age group to 5.8 in adults 60 years or older.

Electrophoretic Separation of CSF Proteins

The electrophoretic procedure used for serum can be employed for CSF if the CSF has been sufficiently concentrated, but the procedure requires too large a specimen and always involves some loss of proteins. More sensitive modifications require 100-fold concentration but only 3 μL of concentrate, and they use high-voltage electrophoresis on thin agarose gels with concurrent cooling. Commercial procedures are available (e.g., Beckman Instruments, Brea, CA 92621; Helena Laboratories, Beaumont, TX 77704). Coomassie Brilliant Blue is suitable for staining the bands, but *silver staining* is 20–50 times more sensitive, allowing unconcentrated CSF to be used.[48] The stain binds to specific amino acid groups in proteins, is independent of the net protein charge, and detects proteins in concentrations as low as 100 μg/dL; however, the procedure is more difficult to perform. (Details of the technique are included with the reagent kits.) Comparisons of silver and CBB staining are shown in Figure 11-9. If immunofixation electrophoresis is followed by silver staining, IgG oligoclonal bands can be identified with certainty.

Polyacrylamide gel electrophoresis is another technique applicable to unconcentrated CSF and suitable for routine work. A sample of 100–200 μL is applied directly to the gel, and concentration occurs in the medium. After electrophoresis and staining, oligoclonal bands can be clearly seen. If the sample is pretreated with sodium dodecyl sulfate to invoke separation by molecular size, then the SDS-PAGE procedure can be used for evaluating the permeability of the blood/brain barrier in a manner analogous to its use for investigating proteinuria.

CSF and serum for electrophoresis should be collected from the patient at the same time and run in parallel after dilution of the serum to a protein concentration similar to that of CSF. Although the main purpose is to detect CSF oligoclonal IgG bands, a number of other abnormalities may be seen and matched to the serum pattern. Some

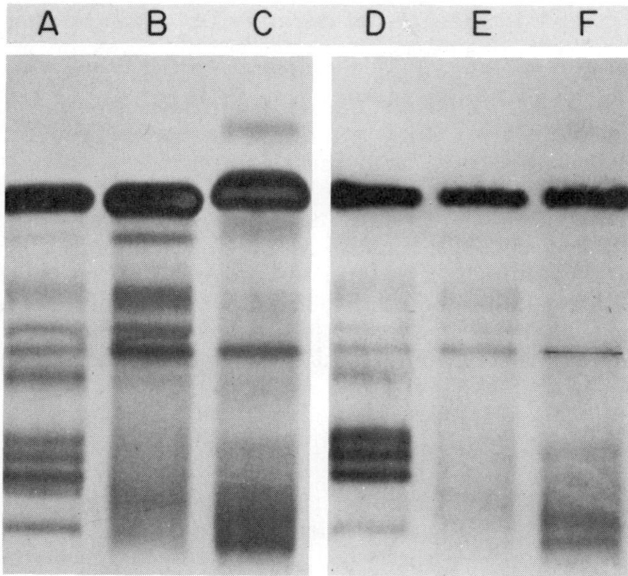

Figure 11-9. Comparison of staining methods for cerebrospinal fluid proteins. A–C, Coomassie Brilliant Blue stain. D–F, Silver stain. Samples were run for 40 min at 100 *V*, on Beckman SPE-11 agarose gel in a Paragon electrophoresis system. *Lanes A, D*: Control Mixture—serum from mulitple myeloma patients, diluted 100-fold, 2μL applied. *Lanes B, E*: Serum from a multiple sclerosis patient, diluted 1:100, 2 μL applied. *Lane C*: CSF from a multiple sclerosis patient, concentrated 80-fold in an Amicon B-15 concentrator, 1 μL applied. *Lane F*: Same CSF, unconcentrated, 1 μL applied. (Photograph courtesy of Dennis B. Lubahn, Ph.D., Department of Hospital Laboratories, North Carolina Memorial Hospital, Chapel Hill, NC.)

are secondary to changes in the serum pattern due, for example, to an acute phase reaction or to the presence of Bence Jones proteins in plasma. Others are the result of intrathecal desialization of glycoproteins.

Proteins in Other Body Fluids

Investigation of proteins in other body fluids is seldom of clinical value. *Amniotic fluid* is analyzed for α-fetoprotein in the detection of neural tube defects. *Saliva* is tested for secretory IgA in infantile hypogammaglobulinemia and for β_2-microglobulin in Sjögren's syndrome. Assay of *feces* for α_1-antitrypsin is sometimes used in the diagnosis of exudative enteropathy, although a better test is intravenous injection of ^{52}Cr-labeled albumin, followed by measurement of the radioactive label in feces.

Pathological accumulations of *fluid in the peritoneal and pleural cavities* or elsewhere vary greatly in protein content; they may be ultrafiltrates with low protein concentrations and scant amounts of high-molecular-weight proteins, or they may be serous fluids with high protein concentrations and significant amounts of large proteins. These fluids are divided arbitrarily according to their protein concentration into *transudates* with total protein below ~3 g/dL and *exudates* with total protein concentrations above ~3 g/dL. Transudates ordinarily reflect changes in permeability of filtering membranes or in hydrostatic pressure across membranes, whereas exudates usually result from infection or malignancy; the latter may contain large numbers of leukocytes or malignant cells. Both transudates and exudates usually contain enough protein to allow the biuret method to be used for the determination of total protein.

References

1. Alper, C. A.: Complement. *In*: Structure and Function of Plasma Proteins. A. C. Allison, Ed., VI. New York, Plenum Publishing Corp., 1974.
2. Ampola, M. G.: Metabolic Diseases in Pediatric Practice. Boston, Little, Brown and Company, 1982.
3. Barratt, M.: Proteinuria. Br. Med. J., *287*:1489–1490, 1983.
4. Barrett, A. J., Starkey, D. M., and Munn, E. A.: The unique nature of the interaction of α_2-macroglobulin with proteases. *In*: Bayer Symposium V Protease Inhibitors. H. Fritz, H. Tschesche, and L. J. Greene, Eds. Berlin, Springer-Verlag, 1974, p. 72.
5. Bremer, H. J., Duran, M., and Kamerling, J. P., et al.: Disturbances of Amino Acid Metabolism: Clinical Chemistry and Diagnosis. Baltimore, Urban and Schwarzenberg, Inc., 1981.

6. Buffone, G. J., Lewis, S. A., Iosefsohn, M., et al.: Chemical and immunochemical measurement of total iron-binding capacity. Clin. Chem., *24*:1788–1791, 1978.
7. Buffone, G. J., Savory, J., Cross, R. E., et al.: Evaluation of kinetic light scattering as an approach to the measurement of specific proteins with the centrifugal analyzer. I. Methodology. Clin. Chem., *21*:1731–1734, 1975.
8. Christenson, R. H., Behlmer, P., Howard, J. F., et al.: Interpretation of cerebrospinal fluid protein assays in various neurological diseases. Clin. Chem., *29*:1028–1030, 1983.
9. Claus, D. R., Osmand, A. P., and Gewurz, H.: Radioimmunoassay of human C-reactive protein and levels in normal sera. J. Lab. Clin. Med., *87*:120, 1976.
10. Clayton, B. E., and Round, J. M.: Chemical Pathology and the Sick Child. Oxford, Blackwell, 1984.
11. Collins, J. A.: Clinical judgment versus the laboratory. N. Engl. J. Med., *306*:987, 1982.
12. Cooper, E. H., and Morgan, D. B.: Proteinuria. Am. Assoc. Clin. Chem., *1*(10):1–11, 1984.
13. Cooper, N. R.: The complement system. *In*: Basic and Clinical Immunology. 4th ed. D. P. Stites, J. D. Stobo, H. H. Fudenberg, et al., Eds. Los Altos, CA, Lange Medical Publishers, 1982, pp. 124–135.
14. Daniels, J. C.: Carrier protein abnormalities. *In*: Serum Protein Abnormalities: Diagnostic and Clinical Aspects. S. E. Ritzmann, and J. C. Daniels, Eds., Boston, Little, Brown and Company, 1975, pp. 213–242.
15. Danks, D. M.: Hereditary disorders of copper metabolism in Wilson's disease and Menkes' disease. *In*: The Metabolic Basis of Inherited Disease. 5th ed. J. B. Stanbury, J. B. Wyngaarden, D. S. Fredrickson, et al., Eds. New York, McGraw-Hill, 1983, pp. 1251–1268.
16. Dietz, A. A., Rubinstein, H. M., and Hodges, L.: Measurement of α_1-antitrypsin in serum by immunodiffusion and by enzymic assay. Clin. Chem., *20*:396–399, 1974.
17. Dilena, B. A., Penberthy, L. A., and Fraser, C. G.: Six methods for determining urinary protein compared. Clin. Chem., *29*:1533–1557, 1983.
18. Duran, M., and Wadman, S. K.: Organic acidurias. *In*: Recent Advances in Clinical Biochemistry, Number 2. K. G. M. Alberti and C. P. Price, Eds. Edinburgh, Churchill Livingstone, 1981, pp. 103–127.
19. Faulkner, W. R.: Phenylalanine. *In*: Standard Methods of Clinical Chemistry, Vol. 5. S. Meites, Ed. New York, Academic Press, 1965, pp. 199–209.
20. Fincham, J. R. S.: Genetics. Bristol, John Wright, 1983.
21. Finley, P. R.: Nephelometry: Principles and clinical laboratory applications. Lab. Management, *20*:34–45, 1982.
22. Gaither, T. A., and Frank, M. M.: Complement, *In*: Clinical Diagnosis and Management by Laboratory Methods. 17th ed. J. B. Henry, Ed. Philadelphia, W. B. Saunders Co., 1984, pp. 879–892.
23. Glass, D. N., Fearon, D. T., and Austen, K. F.: Inherited abnormalities of the complement system. *In*: The Metabolic Basis of Inherited Diseases. 5th ed. J. B. Stanbury, J. B. Wyngaarden, D. S. Fredrickson, et al., Eds. New York, McGraw-Hill, 1983, pp. 1934–1955.
24. Glenner, G. G.: Amyloid deposits and amyloidosis. N. Engl. J. Med., *302*:1283–1292, 1333–1343, 1980.
25. Guthrie, R., and Susi, A.: A simple phenylalanine method for detecting phenylketonuria in large populations of newborn infants. Pediatrics, *32*:338–343, 1963.
26. Gutteridge, J. M. C., and Stocks, J.: Ceruloplasmin: Physiological and pathological perspectives. CRC Crit. Rev. Clin. Lab. Sci., *14*:257–329, 1981.
27. Haan, E. A., and Danks, D. M.: Clinical investigation of suspected metabolic disease. *In*: Laboratory Investigation of Fetal Disease. A. J. Basson, Ed. Bristol, John Wright, 1981, pp. 410–428.
28. Harrison, T. R.: Principles of Internal Medicine. 8th ed. New York, McGraw-Hill, 1977.
29. Haythorn, P., and Sheehan, M.: Improved centrifugal analyzer assay of albumin. Clin. Chem., *25*:194, 1979. Letter.
30. Henry, R. J., Cannon, D. C., and Winkelman, J. W., Eds.: Clinical Chemistry: Principles and Techniques. 2nd ed. New York, Harper and Row, 1974, pp. 424–428.
31. Hische, E. A. H., van der Helm, H. J., van Meegen, M. T., et al.: Protein estimations in cerebrospinal fluid with Coomassie Brilliant Blue. Clin. Chem., *28*:1236–1237, 1982.
32. James, K.: Alpha$_2$-macroglobulin and its possible importance in the immune system. *In*: Trends in Biochemical Sciences, 1980, pp. 43–47.
33. Jeppsson, J. O., and Franzen, B.: Typing of genetic variants of α_1-antitrypsin by electrofocusing. Clin. Chem., *28*:219–225, 1982.
34. Jeppsson, J. O., Laurell, C. B., and Franzen, B.: Agarose gel electrophoresis. Clin. Chem., *25*:629–638, 1979.
35. Johnson, A. M.: Genetically determined variations in plasma proteins. *In*: Protein Abnormalities: Proteins in Body Fluids, Amino Acids, and Tumor Markers: Diagnostic and Clinical Aspects, Vol. 1. S.E. Ritzmann and L. M. Killingsworth, Eds. New York, Alan R. Liss, Inc., 1983, pp. 53–100.
36. Johnson, A., and Lott, J. A.: Standardization of the Coomassie Brilliant Blue method for cerebrospinal fluid proteins. Clin. Chem. *24*:1931–1933, 1978.
37. Jones, B. N., and Gilligan, J. P.: Amino acid analysis by *o*-phthaldialdehyde precolumn derivatization and reversed-phase HPLC. Am. Lab., *15*(12):46–51, 1983.
38. Kachmar, J. F., and Grant, G. H.: Proteins and amino acids. *In*:Fundamentals of Clinical Chemistry, 2nd ed. N. W. Tietz, Ed. Philadelphia, W.B. Saunders Co., 1976.
39. Killingsworth, L. M.: Clinical applications of protein determinations in biological fluids other than blood. Clin. Chem., *28*:1093–1103, 1982.
40. Kingsley, G. R.: The direct biuret method for the determination of serum proteins as applied to photoelectric and visual colorimetry. J. Lab. Clin. Med., *27*:840–845. 1942.
41. La Du, B. N., and Michael, P. J.: An enzymatic spectrophotometric method for the determination of phenylalanine in blood. J. Lab. Clin. Med., *55*:491–496, 1960.

42. Laurell, C. B.: Acute phase proteins—a group of protective proteins. *In*: Recent Advances in Clinical Biochemistry, No. 3. K. G. Alberti and C. P. Price, Eds. Edinburgh, Churchill Livingstone, 1985.
43. Laurell, C. B., and Jeppsson, J. O.: Protease inhibitors in plasma. *In*: The Plasma Proteins, Vol. 1. 2nd ed. F. Putnam, Ed. New York, Academic Press, 1975, p. 229.
44. Lott, J. A., Stephan, V. A., and Pritchard, K. A.: Evaluation of the Coomassie Brilliant Blue G-250 method for urinary protein. Clin. Chem., *29*:1946–1950, 1983.
45. Louderback, A., Mealy, E. H., and Taylor, N. A.: A new dye-binding technique using bromcresol purple for determination of albumin in serum. Clin. Chem., *14*:793–794, 1968. Abstract.
46. McEldeny, L. A., Tarbit, I. F., and Cassells-Smith, A. J.: Six methods for urinary protein compared. Clin. Chem., *28*:256–260. 1982.
47. Mechanic, G., Efron, M. L., and Shih, V. E.: A rapid quantitative estimation of tyrosine and phenylalanine by ion exchange chromatography. Anal. Biochem., *16*:420–426, 1966.
48. Mehta, P. D., Mehta, S. P., and Patrick, B. A.: Silver staining of unconcentrated cerebrospinal fluid in agarose gel (Panagel) electrophoresis. Clin. Chem., *30*:735–736, 1984.
49. Müller-Eberhard, H. J.: Complement. Annu. Rev. Biochem., *45*:697–724, 1975.
50. Nagashima, M., and Schreiber, G.: Changes in specific proteins during acute inflammation. Am. Assoc. Clin. Chem., *1*(9):1–8, 1984.
51. Osmand, A. P., Friedenson, B., Gewurz, H., et al.: Characterization of C-reactive protein and the complement subcomponent C1t as homologous proteins displaying cyclic pentameric symmetry (pentraxins). Proc. Natl. Acad. Sci. USA, *74*:739–743, 1977.
52. Parkes, D.: Wilson's disease. Br. Med. J., *288*:1180–1181, 1984.
53. Parving, H. H., Noer, I., Decrert, T., et al.: The effect of metabolic regulation on microvascular permeability to small and large molecules in short-term diabetes. Diabetologia, *12*:161–166, 1976.
54. Pennock, C. A., Passant, L. P., and Bolton, F. G.: Estimation of cerebrospinal fluid protein. J. Clin. Pathol., *21*:518–520, 1968.
55. Pepys, M. B.: C-reactive protein. Fifty years on. Lancet, *1*:653–657, 1981.
56. Pesce, M. A., and Strande, C. S.: A new micromethod for determination of protein in cerebrospinal fluid and urine. Clin. Chem., *19*:1265–1267, 1973.
57. Peters, T., Jr.: Serum albumin. *In*: The Plasma Proteins, Vol. 1. 2nd ed. F. Putnam, Ed. New York, Academic Press, 1975, p. 133.
58. Peters, T., Jr., Biamonte, G. T., and Doumas, B. T.: Protein (total protein) in serum, urine, and cerebrospinal fluid; albumin in serum. *In*: Selected Methods of Clinical Chemistry, Vol. 9. W. R. Faulkner, and S. Meites, Eds. Washington, D.C., American Association for Clinical Chemistry, 1982.
59. Pinnel, A. E., and Northam, B. E.: New automated dye-binding method for serum albumin determination with bromcresol purple. Clin. Chem., *24*:80–86, 1978.
60. Poulik, M. D., and Weiss, M. L.: Ceruloplasmin. *In*: The Plasma Proteins, Vol. II. 2nd ed. F. Putnam, Ed. New York, Academic Press, 1975, pp. 52–109.
61. Pumphrey, R. S. H.: Structure and function of immunoglobulin. *In*: Immunochemistry in Clinical Laboratory Medicine. A. M. Ward and J. T. Whicher, Eds. Lancaster, England, M.T.P. Press, 1979, pp. 85–98.
62. Putnam, F. W.: Haptoglobin, *In*: The Plasma Proteins, Vol. 2. 2nd ed. F. W. Putnam, Ed. New York, Academic Press, 1975, pp. 2–51.
63. Putnam, F. W.: Transferrin. *In*: The Plasma Proteins, Vol. 1. 2nd ed. F. W. Putnam, Ed. New York, Academic Press, 1975, pp. 265–316.
64. Ramakers, J. M.: Coomassie Blue: An alternative procedure for proteins. Clin. Chem., *30*:1433–1434, 1984.
65. Ritchie, R. F., and Smith, R.: Immunofixation III. Application to the study of monoclonal proteins, Clin. Chem., *22*:1982–1985, 1976.
66. Ritzmann, S. E., and Daniels, J. C.: Introduction *In*: Serum Protein Abnormalities: Diagnostic and Clinical Aspects. S. Ritzmann and J. Daniels, Eds. Boston, Little, Brown and Company, 1975.
67. Ritzmann, S. E., and Finney, M. A.: Proteins—synopsis of characteristics and properties. *In*: Proteins in Body Fluids, Amino Acids and Tumor Markers. Diagnostic and Clinical Aspects. New York, Alan R. Liss, Inc., 1983, pp. 415–455.
68. Roth, K. S., Foreman, J. W., and Segal, S.: The Fanconi syndrome and mechanisms of tubular transport dysfunction. Kidney Int., *20*:705–716, 1981.
69. Saifer, A.: Rapid screening methods for the detection of inherited and acquired aminoacidopathies. Adv. Clin. Chem., *14*:145–218, 1971.
70. Schmid, K.: Alpha$_1$-acid glycoprotein. *In*: The Plasma Proteins, Vol. 1. 2nd ed. F. W. Putnam, Ed. New York, Academic Press, 1975, pp. 184–228.
71. Scriver, C. R., and Rosenberg, L. E.: Amino Acid Metabolism and Its Disorders. Philadelphia, W. B. Saunders Co., 1973.
72. Sell, S., and Becker, F. F.: Alpha-fetoprotein, I. Natl. Cancer Inst. Monogr., *60*:19–26, 1978.
73. Shaw, A. B., Risdon, P., and Lewis-Jackson, J. D.: Protein creatinine index and Albustix in assessment of proteinuria. Br. Med. J., *287*:929–932, 1983.
74. Shen, R. S., and Abell, C. W.: Phenylketonuria: A new method for the simultaneous determination of plasma phenylalanine and tyrosine. Science, *197*:665–667, 1977.
75. Siezen, R. J., and Mague, T. H.: Gas-liquid chromatography of the *N*-heptafluorobutyryl isobutyl esters of fifty biologically interesting amino acids. J. Chromatogr., *130*:151–160, 1977.
76. Silverman, L. M., Christenson, R. H., and Grant, G. H.: Amino acids and proteins. *In*: Textbook of Clinical Chemistry. N. W. Tietz, Ed. Philadelphia, W. B. Saunders Co., 1986.

77. Smith, I.: Chromatographic and Electrophoretic Techniques, Vol. 1. 3rd ed. London, Heinemann Medical Publishers, 1969.
78. Stanbury, J. B., Wyngaarden, J. B., Fredrickson, D. S., et al.: The Metabolic Basis of Inherited Disease. 5th ed. New York, McGraw-Hill, 1983.
79. Sternberg, J. C.: A rate nephelometer for measuring specific proteins by immunoprecipitation reactions. Clin. Chem., *23*:1456–1464, 1977.
80. Talamo, R. C., Langley, C. E., and Reed, C. E.: Alpha$_1$-antitrypsin deficiency: A variant with no detectable α_1-antitrypsin. Science, *181*:70, 1973.
81. Tanaka, K.: Disorders of organic acid metabolism. *In*: Biology of Brain Dysfunction, Vol. 3. G. E. Gaull, Ed. New York, Plenum Press, 1975, pp. 145–214.
82. Tarnoky, A. L.: Genetic and drug-induced variation in serum albumin. *In*: Advances in Clinical Chemistry, *21*:101–146, 1980.
83. Thompson, E. J., Kaufmann, P., Shortman, R. C., et al.: Oligoclonal immunoglobulins and plasma cells in spinal fluid of patients with multiple sclerosis. Br. Med. J., *1*:16–17, 1979.
84. Tietz, N. W., Ed.: Clinical Guide to Laboratory Tests. Philadelphia, W. B. Saunders Co., 1983.
85. Tietz, N. W., Ed.: Fundamentals of Clinical Chemistry. 2nd ed. Philadelphia, W. B. Saunders Co., 1976.
86. Udenfriend, S., and Cooper, J. R.: Assay of L-phenylalanine as phenylethylamine after enzymatic decarboxylation; application to isotope studies. J. Biol. Chem., *203*:953–960, 1953.
87. Van Lente, F., Marchand, A., and Galen, R. S.: Evaluation of a nephelometric assay for haptoglobin and its clinical usefulness. Clin. Chem., *25*:2007–2010, 1979.
88. Walsh, M. J., and Tourtellotte, W. W.: The cerebrospinal fluid in multiple sclerosis. *In*: Multiple Sclerosis, J. F. Hallpike, C. W. M. Adams, and W. W. Tourtellotte, Eds. London, Chapman and Hall, 1983, pp. 275–358.
89. Whicher, J. T.: Abnormalities of plasma proteins. *In*: Biochemistry in Clinical Practice. D. L. Williams and V. Marks, Eds. London, William Heinemann, 1983, pp. 221–250.
90. Whicher, J. T., Higginson, J., Riches, P. G., et al.: Clinical applications of immunofixation: detection and quantitation of complement activation. J. Clin. Pathol., *33*:781–785, 1980.
91. Whicher, J. T., Warren, C., and Chambers, R. E.: Immunochemical assays for immunoglobulins. Ann. Clin. Biochem., *21*:78–91, 1984.
92. Woo, J., Floyd, M., and Cannon, D. C.: Albumin and β_2-microglobulin radioimmunoassays applied to monitoring of renal-allograft function and in differentiating glomerular and tubular diseases. Clin. Chem., *27*:709–713, 1981.

Additional Reading

Dickerson, R. E., and Geis, I.: Proteins, Structure, Function and Evolution. 2nd ed. Menlo Park, CA Benjamin/Cummings Publishing Company, 1983.

Ritzmann, S. E.: Protein Abnormalities. Volume 1: Physiology of Immunoglobulins: Diagnostic and Clinical Aspects. New York, Alan R. Liss, Inc., 1982.

Ritzmann, S. E.: Protein Abnormalities. Volume 2: Pathology of Immunoglobulins: Diagnostic and Clinical Aspects. New York, Alan R. Liss, Inc., 1982.

Ritzmann, S. E., and Killingsworth, L. M.: Protein Abnormalities. Volume 3: Fluids, Amino Acids, and Tumor Markers: Diagnostic and Clinical Aspects. New York, Alan R. Liss, Inc., 1983.

Stanbury, J. B., Wyngaarden, J. B., Fredrickson, D. S., et al.: The Metabolic Basis of Inherited Disease. 5th ed. New York, McGraw-Hill Book Company, 1983.

12 ENZYMES

DONALD W. MOSS, Ph.D., D.Sc.
A. RALPH HENDERSON, M.B. Ch.B., Ph.D.
JOHN F. KACHMAR, Ph.D.

Clinical enzymology is the application of the science of enzymes to the diagnosis and treatment of disease. Enzymes are proteins with catalytic properties due to their powers of specific activation of their substrates. This definition indicates the characteristic properties of enzymes, which in turn govern the principles of methods of enzyme analysis. This chapter surveys these principles, outlines the main diagnostic interpretations of enzyme assays, and describes in principle the most common and clinically useful methods for the determination of enzyme activities.

Enzyme Nomenclature

The first enzymes to be discovered were given empirical names, such as pepsin, but by the turn of the century it had become customary to identify individual enzymes by using the name of the substrate or group on which the enzyme acts and then adding the suffix *-ase*. Thus the enzyme hydrolyzing urea was ure*ase*; the one acting on starch, amyl*ase* (from *amylum*, starch); and the one acting on phosphate esters, phosphat*ase*. Later, for purposes of clarity, the type of reaction involved was also identified, as in carbonic anhydrase, D-amino acid oxidase, and succinic dehydrogenase. As the science of enzymology grew, a definitive and standardized system of identifying enzymes became available as a result of the work of the Enzyme Commission of the International Union of Biochemistry (IUB).[80]

Two names are provided for each enzyme: (1) a systematic name, which clearly describes the nature of the reaction catalyzed and with which is associated a unique numerical code designation, and (2) a trivial or practical name, which may be identical with the systematic name but is often a simplification of it, suitable for everyday use. The unique numerical designation for each enzyme consists of four numbers, separated by periods (for example, 2.2.8.11). The number is prefixed by the letters *EC*, denoting "Enzyme Commission." The first three numbers indicate the class, subclass, and sub-subclass to which the enzyme is assigned. The last number is the specific serial number given to each enzyme within its sub-subclass.

It is a common and convenient practice in clinical chemistry to use capital letter abbreviations for the names of certain enzymes, such as ALT (formerly GPT) for alanine aminotransferase (EC 2.6.1.2); other examples are AST (GOT) for aspartate aminotransferase, LD (or LDH) for lactate dehydrogenase, and CK for creatine kinase.

Enzymes as Proteins

All enzyme molecules possess the three levels of structure characteristic of proteins in general. The *primary structure* is the linear sequence of amino acids linked through their α-carboxyl and α-amino groups by peptide bonds, and it is unique for each type of enzyme molecule. Each polypeptide chain is coiled up into three-dimensional secondary and tertiary levels of structure. *Secondary structure* refers to the conformation of limited segments of the polypeptide chain, namely, α-helices, β-pleated sheet, random coils, and β-turns. The complex, three-dimensional structure assumed by the whole chain is referred to as its *tertiary structure*. The amino acid sequence, the primary structure, determines the conformation of the higher levels of molecular structure that are assumed spontaneously by the growing chain. Therefore, the secondary and tertiary structures are as characteristic of a particular protein molecule as is its primary structure.

A further level of structure exists in protein molecules that contain two or more polypeptide chains (subunits). This *quaternary* level consists of the association of small groups of subunits (also called protomers), each consisting of one polypeptide chain in its characteristic conformation, to make up an oligomeric protein molecule; e.g., creatine kinase is a dimer, and lactate dehydrogenase a tetramer. The catalytic activity of enzymes is often found to be a property of the oligomeric molecule, so that activity is lost under conditions in which the subunits separate from each other. The protomers that compose a particular type of enzyme molecule are frequently identical (as in the MM isoenzyme of CK, or the H_4 lactate dehydrogenase isoenzyme). However, association of unlike subunits can occur and gives rise to the heteropolymeric isoenzymes described later.

The biological activity of a protein molecule—its catalytic activity in the case of an enzyme molecule—depends generally on the integrity of its structure. Therefore, any disruption of the structure is accompanied by a loss of activity, a process that is known as *denaturation*. If the process of denaturation has not gone too far, it may be reversed, and activity may be recovered when the denaturing agent is removed. Renaturation occurs because of the tendency of the enzyme molecule to resume its usual conformation. However, prolonged or severe denaturing conditions result in an irreversible loss of activity. Denaturing conditions include elevated temperatures, which weaken the stabilizing bonds by increasing the vibration of the constituent atoms. The rate of inactivation increases rapidly as temperature rises, becoming almost instantaneous in most cases above ~60 °C. Low temperatures are therefore needed to preserve enzyme activity, especially in aqueous solutions such as serum. Extremes of pH also cause unfolding of enzyme molecular structures and, except in a few cases, should be avoided when enzyme samples are preserved. Urea and related compounds disrupt hydrogen bonds and hydrophobic interactions, so that exposure of enzymes to strong solutions of these reagents results in inactivation.

Specificity and the Active Center

Interaction between the enzyme and its substrate involves the combination of one molecule of enzyme with one substrate molecule (or two, in the case of bisubstrate reactions). The reaction involves the attachment of the substrate molecule to a specialized region of the enzyme molecule, its *active center*. The various chemical groups that are important in substrate-binding are brought together at the active center, where the processes of activation and transformation of the substrate take place. The composition and spatial arrangement of the active center are the basis of what is one of the outstanding characteristics of enzymes, their specificity.

Each enzyme catalyzes only one reaction or, at most, a limited range of chemical reactions. The degree of specificity varies from one enzyme to another. Some enzymes show *absolute specificity*; i.e., they catalyze a unique reaction and no others. Pyruvate kinase, for example, mediates the transfer of a phosphate group between phosphoenol pyruvate and ADP and can function in no other reaction. A somewhat lesser degree of substrate specificity is found for hexokinase. This enzyme transfers a phosphate group from ATP to D-glucose, but it will also phosphorylate D-fructose, D-mannose, and 2-deoxy-D-glucose at almost equivalent rates. The phosphatases are examples of enzymes with *group specificity*. These enzymes split phosphate from any of a large variety of organic phosphate esters, although at somewhat different rates. Substances as varied as glucose-6-phosphate, phenyl phosphate, and β-glycerophosphate can serve as substrates. The esterases have even lower specificity and hydrolyze esters to alcohols and carboxylic acids.

Stereoisomeric specificity is characteristic of many enzymes. The enzymes involved in glycolysis act only on the D-stereoisomers of glucose and its derivatives and never on the L-forms. The transaminases convert oxo-acids only to the L-isomers of the amino acids. Human α-amylase hydrolyzes only the linear segments of starches in which the

D-glucose residues are linked by α-1,4-linkages and is inactive toward the branch points (α-1,6-linkages) of glycogen and amylopectin.

Enzymes as Antigens

Like other proteins, enzymes usually elicit the production of antibodies when they are injected into animals of a species other than that from which they originate. The availability of enzyme-specific antisera opens up a wide range of methods of enzyme analysis, some of which, such as radioimmunoassays, do not depend on the catalytic activity of the enzyme molecules that are being determined and thus can measure inactive enzyme protein. Immunochemical methods are particularly important in the analysis of isoenzyme mixtures.

Enzymes as Catalysts

A catalyst increases the rate of a particular chemical reaction without itself being consumed or permanently altered; at the end of a catalyzed reaction the catalyst appears unchanged in form and quantity, whereas the main reaction materials have undergone transformation into new products. Enzymes are protein catalysts of biological origin, and virtually all the chemical reactions that take place in living matter are catalyzed by specific enzymes.

A small number of enzyme molecules are able to convert an enormously greater number of substrate molecules to products within a short period of time. Therefore, the appearance of increased amounts of enzymes in the blood stream can be detected with great sensitivity, even though the amount of enzyme protein released from damaged cells is negligible compared with the background level of nonenzymic proteins in blood. Since enzymes are specific in their actions, a particular enzyme can be recognized by its characteristic effect on a given chemical reaction, in spite of the presence of a vast excess of other proteins.

Like other catalysts, an enzyme changes only the rate at which equilibrium is established between reactants and products; it does not alter the equilibrium constant of the reaction. In a reaction in which only one set of products is chemically possible, the catalyst cannot effect any change in the nature of the products. Furthermore, when different possible pathways exist by which a substrate can be converted to different products, an enzyme will direct the reaction along only one pathway.

The Enzyme-Substrate Complex

Enzymes act through the formation of an enzyme-substrate complex, *ES*, in which a molecule of substrate is bound to the specialized substrate-binding region (the *active center*) of the enzyme molecule. The binding process transforms the substrate molecule to an activated state; the energy required for this transformation is provided by the free energy of binding of S and E. Therefore, the energy barrier to the reaction is lowered and breakdown to products is speeded up. The *ES* complex breaks down to give the reaction products (P) and free enzyme (E):

$$E + S \rightleftharpoons ES \rightarrow P + E \tag{1}$$

Factors Governing the Rate of Enzyme-Catalyzed Reactions

Enzyme concentration. In the enzymatic reaction represented in equation (1), it is assumed that the equilibrium reaction between E and S is very rapid compared with the breakdown of *ES* into free enzyme and products. The overall rate of the reaction under otherwise constant conditions is therefore considered to be proportional to the concentration of the *ES* complex. Provided that an excess of free substrate molecules is maintained, addition of more enzyme molecules to the reaction system will increase the concentration of the *ES* complex and thus the overall rate of reaction. This accounts for the observation that the rate of reaction is generally proportional to the concentration of enzyme present in the system and is the basis for the quantitative determination of enzymes by measurement of reaction rates. Reaction conditions are selected to ensure that the observed reaction rate is proportional to enzyme concentration over as wide a range as possible.

Substrate concentration. In addition to explaining the dependence of reaction rate on enzyme concentration under conditions in which excess substrate is present, the formation of an enzyme-substrate complex also accounts for the typical hyperbolic relationship between reaction velocity and substrate concentration (Figure 12-1). Such curves are usually known as Michaelis-Menten curves, after the workers who first emphasized their importance.

If the enzyme concentration is fixed and the substrate concentration is varied, the rate of reaction is almost directly proportional to substrate concentration at low values of the latter; i.e., the reaction is essentially first order with respect to substrate concentration. At low concentrations of substrate, only a fraction of the enzyme is associated with substrate, and the rate observed reflects the low concentration of the enzyme-substrate complex. At high substrate concentrations, on the other hand, variation in substrate concentration has no effect on rate, and the reaction is zero order with respect to substrate concentration. Under these conditions, all the enzyme is bound to substrate and a much higher rate of reaction is obtained. Moreover, since all the enzyme is now present in the form of the complex, no further increase in complex concentration and no further increment in reaction rate are possible. The maximum possible velocity for the reaction has been reached.

A typical Michaelis-Menten curve can be described by the equation:

$$v = \frac{V_{max}[S]}{K_m + [S]} \tag{2}$$

V_{max} is the velocity that the observed value of v approaches at high values of $[S]$. It increases with increasing enzyme concentration. K_m, the Michaelis constant, is the substrate concentration at which $v = V_{max}/2$, and it is a constant for a given enzyme acting under given conditions. If an equilibrium is set up between E and S, as is usually the case, K_m is the equilibrium constant of this reaction. However, the symbol K_s (substrate constant) is now usually used if this meaning is intended, and K_m is reserved for the experimentally determined value of $[S]$ at which $v = V_{max}/2$.

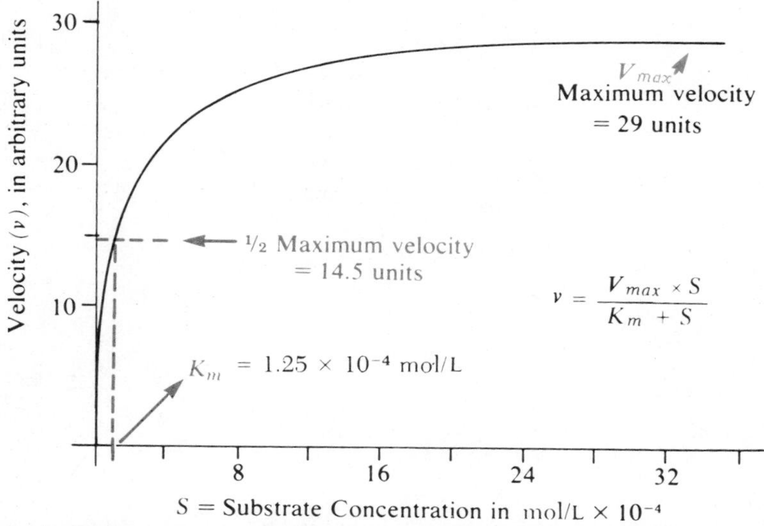

Figure 12-1. Michaelis-Menten curve relating velocity (rate) of an enzyme-catalyzed reaction to substrate concentration. The value of K_m is the substrate concentration at which one-half of the maximum velocity is obtained.

Although it is quite simple to set up an experiment to determine the variation of v with $[S]$, the exact value of V_{max} cannot be evaluated easily from the hyperbolic curve. Furthermore, many enzymes deviate from ideal behavior at high substrate concentrations and indeed may be inhibited by excess substrate, so that the calculated value of V_{max} cannot be achieved in practice. The Michaelis-Menten equation (2) is therefore usually transformed into one of several reciprocal forms (for example, equation 3) when enzyme constants are experimentally measured.

$$\frac{1}{v} = \left(\frac{K_m}{V_{max}} \times \frac{1}{[S]}\right) + \frac{1}{V_{max}} \tag{3}$$

Equation (3), when plotted, gives a straight line with intercepts at $1/V_{max}$ on the ordinate and $-1/K_m$ on the abscissa. The graph is usually referred to as the *Lineweaver-Burk plot*. The data of Figure 12-1 are shown in Lineweaver-Burk form in Figure 12-2.

It is necessary to explore the relationship between reaction velocity and substrate concentration over a wide range, when setting up methods of enzyme assay, to determine K_m, and to detect any inhibition at high substrate concentrations. Zero-order kinetics will be maintained if the substrate is present in large excess; i.e., at concentrations at least 10 and preferably 100 times that of the value of K_m. When $[S] = 10 \times K_m$, v is ~91% of the theoretical V_{max}.

Two-substrate reactions. Although the effect of changes in the concentration of only a single substrate on the rate of reaction has been discussed so far, most enzymatic reactions are of the type:

$$\underset{S_1}{\text{Substrate 1}} + \underset{S_2}{\text{Substrate 2}} \underset{}{\overset{E}{\rightleftharpoons}} \underset{P_1}{\text{Product 1}} + \underset{P_2}{\text{Product 2}} \tag{4}$$

Among the bisubstrate reactions important in clinical enzymology are the reactions catalyzed by dehydrogenases in which the second substrate is a specific coenzyme,

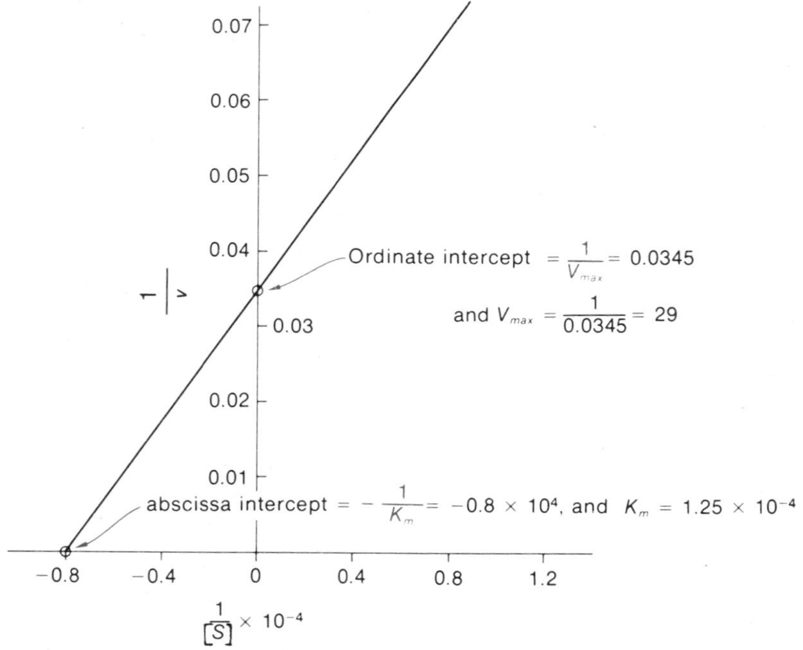

Figure 12-2. Lineweaver-Burk transformation of the curve in Figure 12-1, with $1/v$ plotted on the ordinate (Y-axis), and $1/[S]$ on the abscissa (X-axis). The indicated intercepts permit calculation of V_{max} and K_m. The units of v and $[S]$ are those given in Figure 12-1.

NAD(H) or NADP(H), or by aminotransferases. The concentrations of both substrates affect the rates of two-substrate reactions. Values of K_m and V_{max} for each substrate can be derived from experiments in which the concentration of the first substrate is held at saturating levels while the concentration of the second substrate is varied, and vice versa. There is of course no reason why the K_m values for the two substrates should be the same, or even similar.

The selection of reaction conditions for the measurement of enzymatic activity involving two substrates can be approached empirically by varying the concentration of the first substrate while keeping the concentration of the second substrate constant, until maximum activity is reached. The process is then repeated with the concentration of the first substrate held at the value thus determined, while the concentration of the second substrate is varied. In recent years the traditional empirical approach to optimization has been replaced by newer techniques of simplex co-optimization[52] or response surface methods.[85]

Consecutive enzymatic reactions. All reactions catalyzed by enzymes are in theory reversible, the enzymes catalyzing the forward and backward reactions equally. However, in practice, the reaction is usually found to be more rapid in one direction than the other, so that an equilibrium is reached in which the product of either the forward or the backward reaction predominates, sometimes so markedly that the reaction is virtually irreversible.

If the product of the reaction in one direction is removed as it is formed, e.g., because it is the substrate of a second enzyme present in the reaction mixture, the equilibrium of the first enzymatic process will be displaced so that the reaction may go to completion in that direction. Reaction sequences in which the product of one enzyme-catalyzed reaction becomes the substrate of the next enzyme and so on, often through many stages, are characteristic of biological processes. In the laboratory also, several enzymatic reactions may be linked together in order to provide a means of measuring the activity of the first enzyme or the concentration of the initial substrate in the chain. For example, the activity of creatine kinase is usually measured by a series of linked reactions, and the concentration of glucose can be determined by consecutive reactions catalyzed by hexokinase and glucose-6-phosphate dehydrogenase.

When a secondary enzyme-catalyzed reaction, known as an *indicator reaction*, is used to determine the activity of a different enzyme, the primary reaction (i.e., that catalyzed by the enzyme to be determined) must be the rate-limiting step. Conditions are chosen to ensure that the rate of reaction catalyzed by the indicator enzyme is directly proportional to the rate of product formation in the first reaction.

Effect of pH on Reaction Rate

The rate of enzyme-catalyzed reactions typically shows a marked dependence on pH (Figure 12-3). Many of the enzymes in blood plasma show maximum activity in vitro somewhere in the pH range from 7 to 8. However, activity can be observed at pH values as low as 1.5 (pepsin) and as high as 10.5 (alkaline phosphatase). The optimal pH for a given forward reaction may be different from the optimal pH found for the corresponding reverse reaction. The form of the pH-dependence curve is a result of a number of separate effects. The enzyme may catalyze the reaction only if the substrate is in either the undissociated or dissociated form, and the extent of dissociation is dependent on the pH of the reaction system and the pK of the dissociating acid or base group of the substrate. The activity of the enzyme itself is also affected by the extent of dissociation of certain key amino acid side-chains in the protein molecule, both at the active center and elsewhere in the molecule. Both pH and ionic environment will have an effect on the three-dimensional conformation of the protein and therefore on enzyme activity to such an extent that enzymes may be irreversibly denatured at extreme pH values.

The pronounced effects of pH on enzyme reactions emphasize the need always to control this variable by means of adequate buffer solutions. Enzyme assays should be carried out at the pH of optimal activity, because the pH-activity curve has its minimum slope near this pH, and a small variation in pH will cause a minimal change in enzyme activity. The buffer system must be capable of counteracting the effect of adding the specimen (e.g., serum itself is a powerful buffer) to the assay system, as well as the

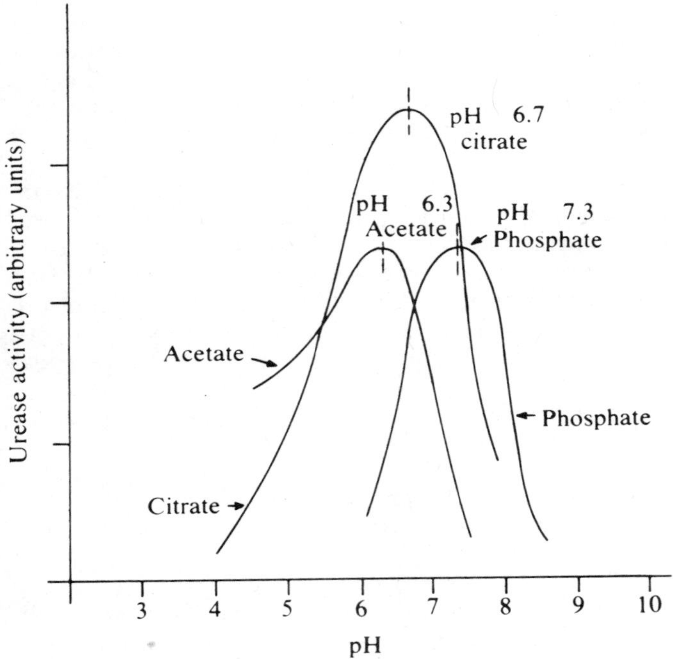

Figure 12-3. pH activity curves for urease, showing effect of buffer species on pH optimum. (Adapted from Howell, S. F., and Sumner, J. B.: J. Biol. Chem., *104*:619–626, 1934.)

effects of acids or bases formed during the reaction (e.g., formation of fatty acids by the action of lipase). Since buffers have their maximum buffering capacity close to their pK_a values, whenever possible a buffer system should be chosen with a pK_a value within 1 pH unit of the desired pH of the assay. Interaction between buffer ions and other components of the assay system (e.g., activating metal ions) may eliminate certain buffers from consideration.

Effect of Temperature on Reaction Rate

The rate of any chemical reaction increases as the temperature at which the reaction is taking place increases. Enzymatic reactions are no exception. For most enzymatic reactions, values of Q_{10} (the relative reaction rates at two temperatures differing by 10 °C) vary from 1.7 to 2.5. However, an increase in the rate of the catalyzed reaction is not the only effect of increasing temperature on an enzymatic reaction. At least in theory, the initial rate of reaction measured instantaneously goes on increasing with rising temperature. In practice, however, a finite time is needed to allow the components of the reaction mixture, including the enzyme solution, to reach temperature equilibrium and to permit the formation of a measurable amount of product. During this period the enzyme is undergoing thermal inactivation and denaturation, a process that has a very large temperature coefficient for most enzymes and thus becomes virtually instantaneous at temperatures of 60–70 °C.

The counteracting effects of the increased rate of the catalyzed reaction and more rapid enzyme inactivation as the temperature is raised account for the existence of an apparent *optimum temperature* for enzyme activity (Figure 12-4). The apparent optimum temperature depends on the time taken to make the activity measurement. With older fixed-time assay methods that require lengthy periods of incubation of enzyme with substrate, enzyme inactivation takes effect at lower temperatures and the phenomenon is more easily seen.

The choice of temperature for the assay of enzymes of clinical importance has been the subject of extensive debate.[24,64] Current opinion favors either 30 °C or 37 °C, each temperature having advantages in particular circumstances. Whatever temperature is chosen, precise temperature control to within ±0.1 °C during the enzymatic reaction is essential.

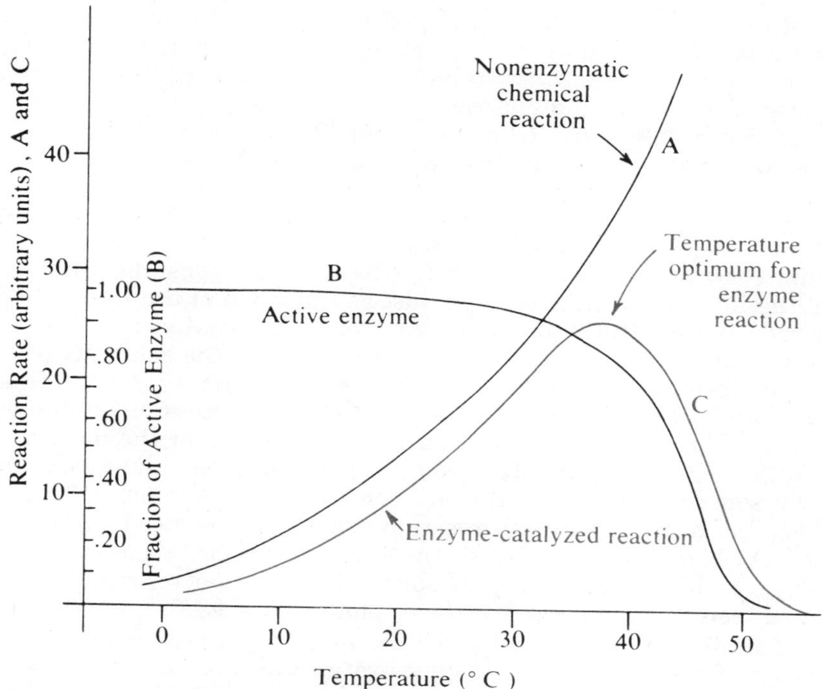

Figure 12-4. Schematic diagram showing effect of temperature on rate of non-enzyme-catalyzed and enzyme-catalyzed reactions.

Inhibition and Activation of Enzyme Activity

The rates of enzymatic reactions are often found to be affected by changes in the concentrations of substances other than the enzyme or substrate. These modifiers may be activators (i.e., they increase the rate of reaction) or they may be inhibitors (i.e., their presence reduces the reaction rate). Activators and inhibitors are usually small molecules (compared with the enzyme itself) or even ions. They vary in specificity from modifiers that exert similar effects on a wide range of different enzymatic reactions to substances that affect only a single reaction. Some phenomena of enzyme activation or inhibition are due to interaction between the modifier and a nonenzymatic component of the reaction system, such as the substrate. In most cases, however, the modifier combines with the enzyme itself in a manner analogous to the combination of enzyme and substrate.

Inhibition of Enzyme Activity

Inhibitors may be reversible or irreversible. *Reversible inhibition* implies that the activity of the enzyme is fully restored when the inhibitor is removed from the system by some physical separation process such as dialysis or gel filtration. An *irreversible inhibitor*, on the other hand, combines covalently with the enzyme so that physical methods are ineffective in separating the two.

Characteristics of irreversible inhibition. An irreversible inhibitor is not in equilibrium with the enzyme. Its effect is progressive with time and becomes complete if the amount of inhibitor present exceeds the total amount of enzyme. The rate of the reaction between enzyme and inhibitor is expressed as the fraction of the enzyme activity that is inhibited in a fixed time by a given concentration of inhibitor. When the inhibitor is added to the enzyme in the presence of its substrate, the reaction between the enzyme and inhibitor may be delayed because some of the enzyme molecules will be combined with the substrate and therefore protected from reacting with the inhibitor. However, as the substrate molecules are broken down, the active centers become available for combination with the inhibitor. Thus, inhibition will eventually become complete even though an excess of substrate may initially have been present. Furthermore, addition of more substrate is ineffective in reversing the inhibition, in contrast to its effect on reversible competitive inhibition discussed later.

Kinetics of reversible inhibition. Reversible inhibition is characterized by the existence of an equilibrium between enzyme, E, and inhibitor, I:

$$E + I \rightleftharpoons EI \tag{5}$$

The equilibrium constant of the reaction, K_i (the inhibitor constant), is a measure of the affinity of the inhibitor for the enzyme, just as the value of the substrate constant, K_s, generally reflects the affinity of the enzyme for its substrate.

A *competitive* inhibitor is usually a structural analog of the substrate and binds to the enzyme at the substrate-binding site, but because it is not identical with the substrate, breakdown into products does not take place. When the process of inhibition is fully competitive the enzyme can combine with either the substrate or the inhibitor, but not with both simultaneously. At low substrate concentrations, the binding of substrate is reduced because some enzyme molecules are combined with inhibitor. Thus, the concentration of ES and hence the overall reaction velocity are reduced, and K_m is apparently increased. At high $[S]$, however, all the enzyme molecules can combine to form ES, so that V_{max} is unaffected by the inhibitor. These characteristics of competitive inhibition can be seen in the Lineweaver-Burk plot (Figure 12-5).

Competitive inhibition is responsible for the inhibition of some enzymes by excess substrate, because of competition between substrate molecules for a single binding site. In two-substrate reactions, high concentrations of the second substrate may compete with the binding of the first substrate. Competitive inhibition can also contribute to the reduction of the rate of an enzymatic reaction with time. For example, in a freely reversible reaction, a rate reduction can occur because increasing concentrations of reaction products tend to drive the reaction backward. A product of a reaction may itself be an inhibitor of the forward reaction so that, even if the reaction is not readily reversible, it proceeds against an increasing concentration of inhibitor. Product inhibition is thus one cause of nonlinearity of reaction progress curves.

A *noncompetitive* inhibitor is usually unlike the substrate in structure. It is assumed to bind at a site on the enzyme molecule that is different from the substrate-binding site; thus, there is no competition between inhibitor and substrate, and a ternary enzyme-inhibitor-substrate complex can form. Attachment of the inhibitor to the enzyme does

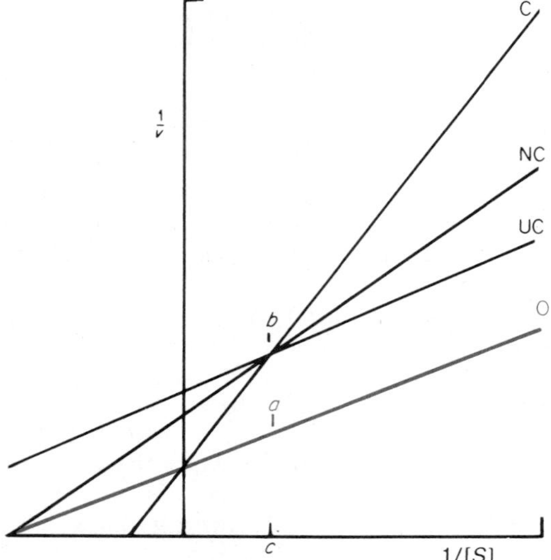

Figure 12-5. Effects of different types of inhibitors on the double-reciprocal plot of $1/v$ against $1/[S]$. Each of the inhibitors has been assumed to reduce the activity of the enzyme by the same amount, represented by the change in $1/v$ from a to b at a substrate concentration of c. Line O is the plot for enzyme without inhibitor, C with a competitive inhibitor, NC with a noncompetitive inhibitor, and UC with an uncompetitive inhibitor. (From Moss: D. W.: Measurement of enzymes. *In*: Enzymes in Cardiology: Diagnosis and Research. D. J. Hearse and J. de Leiris, Eds. Copyright 1979, John Wiley & Sons. Reprinted by permission of John Wiley & Sons, Ltd.)

not alter the affinity of the enzyme for its substrate (i.e., K_m is unaltered) but the enzyme-inhibitor-substrate complex does not break down to give products. Since the substrate does not compete with the inhibitor for binding sites on the enzyme molecule, increasing the substrate concentration does not overcome the effect of a noncompetitive inhibitor. Thus, V_{max} is reduced in the presence of such an inhibitor, whereas K_m is not altered, as the Lineweaver-Burk plot shows (Figure 12-5).

In a rather unusual type of reversible inhibition, *uncompetitive* inhibition, parallel lines are obtained when plots of $1/v$ against $1/[S]$ with and without the inhibitor are compared (Figure 12-5); i.e., both K_m and V_{max} are decreased. Uncompetitive inhibition is due to combination of the inhibitor with the *ES* complex. It is more common in two-substrate reactions, in which a ternary *ESI* complex can form after the first substrate has combined with the enzyme.

Enzyme Activation

Activators are considered to increase the rates of enzyme-catalyzed reactions by promoting formation of the most active state either of the enzyme itself or of other reactants such as the substrate. Many enzymes contain as an integral part of their structures metal ions that appear to play a direct part in catalysis, in addition to any possible structural role that they may fulfill. When the activator ion is an essential part of the functional enzyme molecule, it is usually incorporated quite firmly into the enzyme molecule. Therefore, it is not usually necessary to add the activator to reaction mixtures, and excess of the ion may even have an inhibitory effect. However, in many cases the activating ion is attached only weakly or transiently to the enzyme (or its substrate) during catalysis. Common activating cations are Mg(II), Fe(II), Ca(II), Zn(II), and K^+. More rarely, anions may act as activators. Amylase functions at its maximal rate only if Cl^- or other monovalent anions such as Br^- or NO_3^- are present. Some enzymes require the presence of two activating ions. The velocity of the reaction depends on the concentration of a reversible activator in a fashion similar to its dependence on substrate concentration. It is important that activator-dependent enzyme reactions be performed in the presence of the optimal concentration of activator. Serum specimens may be deficient in an activator such as a metal ion, so the addition of such an ion increases the reaction rate.

Coenzymes and Prosthetic Groups

Coenzymes are usually more complex molecules than activators, although they are smaller molecules than the enzyme proteins themselves. Some compounds that are classified as coenzymes (e.g., the dinucleotides NAD^+ and $NADP^+$) are specific substrates in two-substrate reactions, and their effect on the rate of reaction follows the Michaelis-Menten pattern of dependence on substrate concentration. The structures of these two coenzymes are identical except for the presence of an additional phosphate group in $NADP^+$; nevertheless, individual dehydrogenases, for which these coenzymes are substrates, are predominantly or even absolutely specific for one or the other form.

Coenzymes such as NAD^+ and $NADP^+$ are bound only momentarily to the enzyme during the course of reaction, as is the case for substrates in general. Therefore, no reaction takes place unless the appropriate coenzyme is present (e.g., if added to the reaction mixture). In contrast to these entirely soluble coenzymes, a number of coenzymes are more or less permanently bound to the enzyme molecules, where they form part of the active center and undergo cycles of chemical change during the reaction. The active holoenzyme results from the combination of the inactive apoenzyme with the *prosthetic group*, as such a bound coenzyme is called. An example of a prosthetic group is pyridoxal phosphate, a component of aspartate and alanine aminotransferases. The pyridoxal phosphate prosthetic group undergoes a cycle of conversion of the pyridoxal moiety to pyridoxamine and back again during the transfer of an amino group from an amino acid to an oxoacid. Prosthetic groups often do not have to be added to an assay mixture to elicit full catalytic activity of the enzyme, unless previous treatment has caused the prosthetic group to be lost from some enzyme molecules. However, both normal and pathological serum specimens contain appreciable amounts of apoaminotransferases, which can be converted to the active holoenzymes by a suitable period of incubation with pyridoxal phosphate. This treatment increases the aspartate aminotransferase activity of serum specimens by an average of about 50%, although the effect on alanine aminotransferase levels is considerably less. There is considerable variation from sample to sample in the degree of activation and also, in the case of aspartate aminotransferase, among sera from patients with different diseases. In order to ensure that assay methods measure only differences in the total enzyme content of specimens, the most recent descriptions of methods for aminotransferase assay specify conditions for activation of any apoenzyme that may be present. This principle applies also to known cofactors in other enzyme assays.

Progress of Enzymatic Reactions and the Measurement of Reaction Rates

Since the rate of an enzyme-catalyzed reaction is directly proportional to the amount of active enzyme present in the system, determination of the rate of reaction under defined and controlled conditions provides a very sensitive and specific method for the measurement of enzymes in specimens such as serum. The progress of conversion of the substrate into products in the presence of an enzyme can be followed by measuring the decreasing concentration of the substrate or the increasing concentration of a product. Both principles are found among assays of clinically important enzymes.

All determinations of reaction rate involve the measurement of the amount of change produced in a defined time interval, i.e., all measurements of reaction rate are kinetic measurements. This is just as true when the amount of change produced by the enzyme is measured after stopping the reaction at the end of a fixed-time interval as it is when the progress of the reaction is monitored continuously. The term *kinetic method* has often been limited erroneously to the latter procedure. These two analytical approaches are referred to respectively as *fixed-time* and *continuous-monitoring* methods.[74] They have different advantages and limitations. To appreciate these it is necessary to consider the way in which the rate of an enzymatic reaction varies with time.

At the moment when the enzyme and substrate are mixed, the rate of the reaction is zero or virtually so. Typically, the rate then rises rapidly to a maximum value, which remains constant for a period of time (Figure 12-6). During the period of constant reaction rate, which may be short or long in different circumstances, the rate depends only on enzyme concentration and is completely independent of substrate concentration. The reaction is said to follow zero-order kinetics; i.e., its rate is proportional to the zero power of the substrate concentration. Ultimately, however, as more and more substrate is consumed, the reaction rate declines and enters a phase of first-order dependence on substrate concentration. Other factors that contribute to the decline in reaction include accumulation of products that may be inhibitory, the growing importance of the reverse reaction, and even enzyme denaturation.

Although the rates of reaction produced by different amounts of an enzyme can be compared under first-order conditions, it is easier to standardize such comparisons when the enzyme concentration is the only variable that influences the reaction rate. Therefore, enzyme assays are almost always made under conditions that are initially saturating with respect to substrate concentration. The rate of reaction during the zero-order phase can be determined by measuring the product formed during a fixed period of incubation, but this assumes that the rate has, in fact, remained constant during this period. If it has not, the apparent rate of reaction will no longer be proportional to enzyme concentration, as illustrated in Figure 12-7. Measurement of reaction rates over any portion of curve A will give results that will be identical to the true "initial rate." However, curve B deviates from linearity over its entire course, and rates fall off with time. At no time will it give a measure of the "initial rate." From curve C, correct results cannot be obtained if the rate is measured during the lag phase (I) or during during phase III.

Careful selection of reaction conditions, such as the concentrations of substrates and cofactors, can often improve the reaction progress curves, eliminating lag phases and

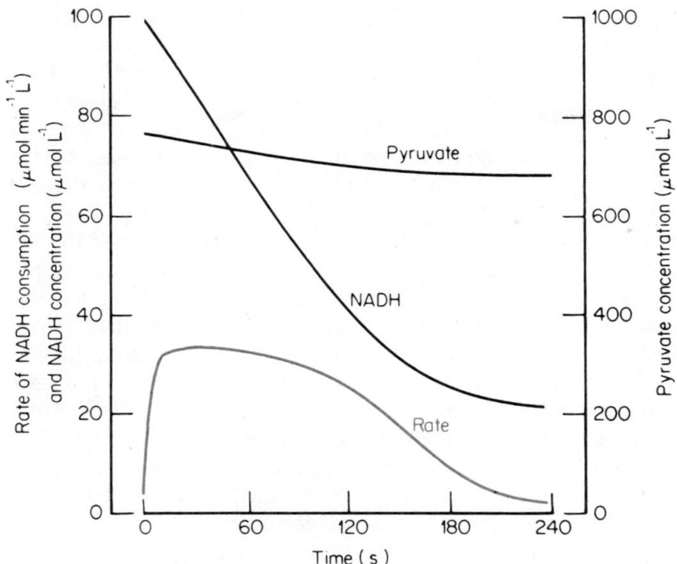

Figure 12-6. Changes in substrate concentrations and rate of reaction during an assay of lactate dehydrogenase activity at 37 °C in phosphate buffer, with pyruvate and NADH as substrates. The reaction is followed by observing the fall in absorbance at 340 nm as NADH is oxidized to NAD^+. The rate of reaction rises rapidly to a maximum value, from which it declines only slightly until about half the NADH has been used up. During this phase of the reaction, the rate is essentially zero order with respect to substrate concentration. At the point at which the rate falls below about 90% of its maximum value, NADH concentration is approximately $10 \times K_m$. The K_m for NADH is of the order of 5×10^{-6} mol $\times$ L^{-1}, whereas for pyruvate it is 9×10^{-5} mol $\times$ L^{-1}. Thus an initial pyruvate concentration approximately 10 times that of NADH is used. (Concentrations are per liter of reaction mixture.) (From Moss, D. W.: Measurement of enzymes. *In*: Enzymes in Cardiology: Diagnosis and Research. D. J. Hearse and J. de Leiris, Eds. New York, John Wiley & Sons, 1979.)

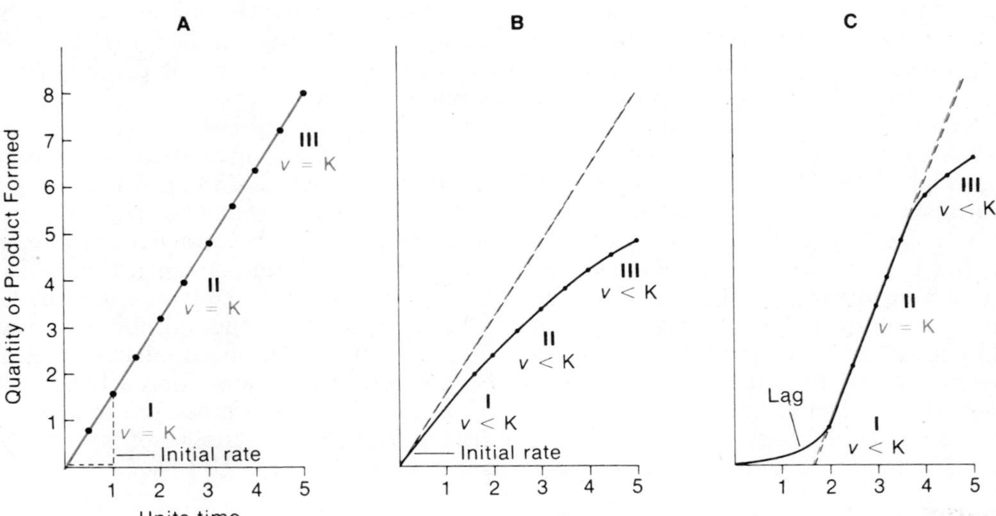

Figure 12-7. Enzyme reaction rate as a function of time. In *A*, the rate is constant during the entire time, and rates calculated as I, II, and III will be identical to the initial rate. In *B*, the rate falls off continuously; rates calculated at I, II, and III will be different and less than the true initial rate. In *C*, a measurement at II will be representative of the maximal rate, but at I (lag period) and III (substrate depletion) it will be less than at II.

prolonging the period of linearity so that fixed-time methods of analysis become feasible. Improvements in photometry leading to more sensitive measurement of product formation have also allowed the duration of incubation to be shortened compared with older assays. These improvements have resulted in a corresponding increase in the range over which enzyme activity can be measured. Nevertheless, there is an upper limit of activity in all fixed-time methods, above which progress curves will no longer be linear and the amount of change measured over the fixed-time interval no longer represents the true zero-order rate.

The existence of an upper limit of proportionality with enzyme activity implies that samples with activities above this limit must receive special attention. The upper limit of activity acceptable in the unmodified method must be chosen so that samples with activities below it can be presumed with a high degree of certainty to give linear progress curves; on the other hand, if the limit is set too low, many samples will be reanalyzed unnecessarily. Samples that are above the limit should ideally be assayed by shortening the incubation period until a constant reaction rate is obtained. However, this is difficult or impossible in some automated methods, in which the duration of incubation is fixed by the configuration of the apparatus. It then becomes necessary to dilute the specimen, but dilution may not always result in a proportionate change in activity.

An upper limit of activity that can be accepted without modification of the assay procedure exists even in continuous-monitoring methods, but this limit is usually much higher than that applicable in corresponding fixed-time methods. Fewer samples therefore require special treatment. *Continuous-monitoring methods* (or methods in which the progress of reaction is monitored at frequent intervals) therefore *possess a decisive advantage in enzyme assay and should be used whenever possible*.

Units for Measuring Enzyme Activity

Since enzymes are almost always measured by their catalytic activities, the results of such determinations are expressed in terms of the number of activity units present in a convenient volume or mass of specimen. The unit of activity is the measure of the rate at which the reaction proceeds (e.g., the quantity of substrate consumed or product formed in a chosen unit of time). The quantity of substrate or product may be given in any convenient unit—milligrams, micromoles, change in absorbance, change in viscosity, or microliters of gas formed; time may be expressed in seconds, minutes, or hours. The activity concentration is expressed in terms of some convenient volume of sample (e.g., the liter). Since the rate of the reaction depends on variables such as pH, type of buffer, temperature, nature of substrate, ionic strength, and concentration of activators, these parameters must be specified in the definition of the unit. Any methodological modification, however slight, that involves a change in the values chosen for these variables may alter the significance of the units associated with that method.

In the course of many decades a multiplicity of units for expressing enzyme activity have been introduced, often specific to a particular method and identified by the names of the originators. To bring order into this situation, the Commission on Enzymes of the IUB proposed that the unit of enzyme activity be defined as that quantity of enzyme that will catalyze the reaction of one micromole (μmol) of substrate per minute, and that this unit be termed the *International Unit* (U). Catalytic concentration is to be expressed in terms of U/L, or mU/L, whichever gives the more convenient numerical value. Although the basis of expressing enzyme activity is thus made uniform, it cannot be emphasized too strongly that methods with different reaction conditions will give different numerical values for the activity of a given sample, even when all values are expressed in International Units. A newer unit, the *katal*, expresses enzyme activity in mol/s, in conformity with the Système International (SI) scheme of units. Thus, $1\ \text{U} = 10^{-6}\ \text{mol}/60\ \text{s} = 16.7 \times 10^{9}\ \text{mol/s}$, i.e., $1\ \text{nkat/L} = 0.06\ \text{U/L}$.

Measurement of Chemical Changes in Enzymatic Reactions

The amount of substrate transformed into products during an enzyme-catalyzed reaction can be measured with any appropriate analytical method. However, the method most commonly used is photometric analysis. The reaction may be accompanied by a change in the absorbance characteristics of some component of the assay system, either in the visible or ultraviolet spectrum; this change can be observed while it is proceeding. "Self-indicating" reactions of this type are particularly valuable since they make possible continuous monitoring in a recording spectrophotometer. Important examples of self-indicating reactions are the determination of dehydrogenase activity, by following the change in absorbance at 339 (340)* nm of the coenzymes NAD(H) or NADP(H) during oxidation or reduction, and the measurement of alkaline phosphatase activity by the liberation of the yellow 4-nitrophenoxide ion from the substrate 4-nitrophenyl phosphate in alkaline solution. The advantages of continuous monitoring are so marked that coupled reactions are frequently used to provide an observable change in absorbance accompanying a primary reaction in which such a change is not present, as in aminotransferase assays, for example. The photometric requirements of enzymatic analysis are not different from those of photometric quantitative analysis in general. However, the photometer used must be provided with means to maintain the contents of the cuvet at a constant temperature during the reaction, and provision of a chart recorder or computer simplifies monitoring of the progress of reaction.

Optimization, Standardization, and Quality Control of Enzyme Assays

In all analytical methods, the quantity of the analyte being measured should be the only variable that causes an observable difference in results. In the case of enzyme activity measurements, it is obvious that all the factors that affect the reaction rate, other than the concentration of active enzyme, must be rigidly controlled. However, some slight variation in such factors as pH or temperature between samples or between analytical runs is inevitable, and this contributes to the imprecision of the analysis. The effect of these slight variations is least under optimal conditions (e.g., of pH or substrate concentration), where the dependence of reaction velocity on these variables is least marked. Furthermore, since the reaction velocity is at or near its maximum under optimal conditions, a larger analytical signal is obtained, which can be measured more easily and precisely than the smaller signal obtained under suboptimal conditions. Much effort has therefore been devoted to optimizing conditions for measuring the activities of enzymes of clinical importance. However, optimal conditions are not necessarily those that provide the maximum attainable rate of reaction but are those that are practicable in the routine laboratory and that combine a high rate of reaction with the least marked dependence on variables other than enzyme concentration.

Standardization. The large number of methods used for the measurement of the activities of enzymes in clinical laboratories and the consequent multitude of different units of activity and reference ranges form a serious barrier to the interpretation of data obtained in different laboratories, both in the records of individual patients and in reports in the scientific literature.

Two approaches to standardization of enzyme assays are possible. The first is to provide enzyme preparations of stated catalytic activity that can be used as calibration standards (calibrators). However, this approach requires the availability of enzyme preparations of defined and reliable properties, particularly with regard to stability. The second approach is to standardize the conditions of assay, thus reducing the number of methods in use. Recommended methods range from those intended for use in small groups of neighboring laboratories to rigorously specified reference methods intended to provide a criterion of analytical performance against which routine methods can be judged. Between these extremes are the methods recommended by national or international enzyme committees that combine analytical reliability with applicability to current standards of instrumentation and skill in well-equipped laboratories. Increased use of recommended methods has led to a marked improvement in interlaboratory agreement as shown in external quality assessment schemes.

Quality control. The application of quality control procedures to ensure that satisfactory analytical performance of enzyme assays is maintained on a day-to-day basis

---

*Most recent data give 339 nm as the absorbance maximum of NAD(P)H, instead of the often-quoted 340 nm.[66,114]

is complicated by the tendency of enzyme preparations to undergo denaturation with loss of activity. This may make it difficult to distinguish between poor analytical performance and denaturation as possible causes of a low result obtained for a control sample introduced into a batch of analyses. Assured stability within a defined usable time-span is therefore the prime requirement for enzyme control materials. It is the function of a *control* to reveal significant variations in reaction conditions; therefore, it must be at least as sensitive to these variations as the samples analyzed. For this reason, the chosen control enzyme must not be very different in its properties from the analogous human enzyme.

Lyophilized preparations containing various enzymes are available from commercial sources and these have a useful function in quality control. Serum pools that are prepared in the laboratory (with care to exclude specimens contaminated with hepatitis or other viruses), then assayed for enzyme activity and stored frozen in small portions for daily use, are also valuable. However, the use of both types of control materials is attended with uncertainties for the reasons mentioned above. Furthermore, it is often difficult to reproduce the values for activity stated by the manufacturers for freeze-dried sera, even when the same methods are apparently in use, so that recalibration in the laboratory is necessary. Another useful procedure in the control of enzyme analysis is to carry forward one or two samples from one batch to the next, so that the successive results can be compared. Some loss of activity usually takes place between the first and second analyses, but for many enzymes this is $<10\%$ with suitable interim storage.

The systematic application of quality control programs is as essential in enzyme analysis as it is in other forms of clinical analysis. When well-designed assay methods are in use, the coefficient of variation for reproducibility of results of enzyme assays on a day-to-day basis is usually less than $\pm10\%$ for activities within the normal range and less than $\pm5\%$ for increased activities.

Enzymes as Analytical Reagents

The use of enzymes as analytical reagents offers the advantage of great specificity for the substance being determined. This high specificity typically reduces the need for preliminary separation or purification stages, so that the analysis can be carried out directly on complex mixtures such as serum. Enzymes with absolute specificity for the substance to be estimated are clearly preferable for analytical use; such enzymes include uricase (urate oxidase), urease, and glucose oxidase. However, this ideal cannot always be achieved in practice, and knowledge of the substrate specificities of reagent enzymes is therefore essential to allow possible interferences with the assay to be anticipated and avoided. Coupled reactions are often used to construct an enzymatic analytical system for determining a particular compound, and the specificity of the coupled reactions may modify the specificity of the overall process. An example of this is the determination of glucose by the hexokinase reaction. Hexokinase will convert sugars other than glucose to their 6-phosphate esters. However, the indicator reaction used to monitor this change is catalyzed by glucose-6-phosphate dehydrogenase, an enzyme that is highly specific for its substrate, so that the overall process is highly specific for glucose.

The principle that is most widely used to determine the amount of a substance enzymatically is to allow the reaction to go to completion, so that all the substrate has been converted into a measurable product. These methods are called "end point" or, more correctly, *equilibrium methods*, since the reaction ceases when the equilibrium is reached. Reactions in which the equilibrium point corresponds to virtually complete conversion of the substrate are obviously preferable for this type of analysis. However, unfavorable equilibria can often be displaced in the desired direction by additional enzymatic or nonenzymatic reactions that convert or "trap" a product of the first reaction; e.g., in measuring lactate with lactate dehydrogenase, the pyruvate formed can be trapped by the addition of hydrazine, with which it forms a hydrazone. A further important consideration is that the time required to transform a fixed quantity of substrate into products is inversely proportional to the amount of enzyme present. Equilibrium methods may therefore require the use of appreciable amounts of enzyme for each sample to avoid inconveniently long incubation periods. As the substrate concentration falls to low levels toward the end of the reaction, the K_m of the enzyme becomes important in determining the reaction rate. Enzymes with high affinities for the substrates (low K_m values) are therefore most suitable for equilibrium analysis. Equilibrium methods are largely insensitive to minor changes in reaction conditions. It is not necessary to have exactly the same amount of enzyme in each reaction mixture or to maintain the pH or temperature absolutely constant, provided that the variations are not so great that the reaction is not completed within the fixed time allowed.

The amount of reagent enzyme required for each analysis can be reduced and the time shortened by the use of methods in which the amount of change produced in a fixed time interval is measured. As already described, the rate of an enzyme-catalyzed reaction follows first-order kinetics with respect to substrate concentration when the substrate concentration is low (typically when $[S] < 0.2 \times K_m$). It is a property of first-order reactions that the change in concentration of the substrate over a fixed time interval is directly proportional to its initial concentration.

Methods in which some property that is related to substrate concentration (e.g., light absorbance) is measured at two fixed times during the course of the reaction are known as "two-point" kinetic methods. They are theoretically the most accurate for the enzymatic determination of substrates. However, these methods are technically more demanding than equilibrium methods. Since the reaction rate is measured in the two-point methods, all factors that affect reaction rate such as pH, temperature, and amount of enzyme must be kept constant from one assay to the next, as must the timing of the two measurements. These conditions can now readily be achieved in automatic analyzers. To ensure first-order reaction conditions the substrate concentration must be low, of the order of $< 0.2 \times K_m$. Enzymes with high K_m values are therefore preferred for kinetic analysis to give a wider usable range of substrate concentration. A standard solution of the analyte (substrate) must be used for calibration.

Isoenzymes and Other Multiple Forms of Enzymes

Improved techniques for analyzing mixtures of proteins, developed over the past twenty years, have shown that a particular type of catalytic activity within a single individual is frequently due to the existence of several distinct forms of an enzyme rather than to only one type of molecule.[76] These enzyme variants may occur within a single individual, a single organ, or even within a single type of cell. The forms can be distinguished on the basis of differences in various physical properties, such as electrophoretic mobility or resistance to chemical or thermal inactivation. Often there are significant quantitative differences in catalytic properties between them. However, all forms of a particular enzyme retain the ability to catalyze its characteristic reaction. The multiple molecular forms of an enzyme are often described as *isoenzymes* (or isozymes), although it is now recommended by IUB that this term should be restricted to those forms that originate at the level of the genes that encode the structures of the enzyme proteins in question.

The existence of multiple forms of enzymes in human tissues has important implications in the study of human disease. The presence of isoenzymes with distinctive properties in different organs helps in understanding organ-specific patterns of metabolism, whereas genetically determined variations in enzyme structure between individuals account for such characteristics as differences in sensitivity to drugs, as well as those differences in metabolism that manifest themselves as hereditary metabolic diseases. For the diagnostic enzymologist, the existence of multiple forms of enzymes, whether due to genetic or nongenetic causes, provides opportunities to increase the diagnostic specificity and sensitivity of enzyme assays carried out on specimens of body fluids.

Genetic Origins of Enzyme Variants

True isoenzymes may be due to the existence of more than one gene locus coding for the structure of the enzyme protein. A substantial proportion of human enzymes (perhaps more than one third) seems to be determined by more than one structural gene locus. The structural genes at the separate loci have undergone different modifications during the course of evolution, so that the enzyme proteins coded by them no longer have identical structures, although they are recognizably similar.

Multiple gene loci have become disseminated throughout a whole species during the course of evolution, so that these genes and their resultant isoenzymes typically are present in all individuals of that species. However, a large number of human enzymes exist in multiple molecular forms that differ in type from one individual to another. Family studies show that the various forms of a given enzyme are inherited according to Mendelian laws. In these cases the different enzyme forms originate from modified genes, or alleles. Allelic genes are alternative forms of the gene that occur at a particular locus and that give rise to gene products with the same function. The isoenzymes that result from the existence of allelic genes are termed *allelozymes*. The proportion of human gene loci subject to allelic variation is considerable, and the probability that individuals will differ to some degree in their allelozyme patterns is correspondingly high. The likelihood of different allelozyme patterns is further increased by the possibility that individuals may be heterozygous for particular alleles; i.e., they have inherited different alleles at a particular locus.

In many cases the products of allelic genes are functionally adequate and their inheritance causes no adverse effects. In other cases the individual who inherits particular allelozymes may experience illness when exposed to certain drugs or foodstuffs, because of functional abnormalities of the allelozymes. In extreme cases, the allele may produce no active isoenzyme at all, so that metabolic disease of varying degrees of severity results.

Another category of multiple molecular forms can arise in the case of enzymes that are oligomeric (i.e., that consist of molecules made up of subunits). The association of different types of subunits in various combinations gives rise to a group of active enzyme molecules. When the different subunits are derived from distinct structural genes, either multiple loci or multiple alleles, the hybrid molecules so formed are included within the definition of isoenzymes and are called *hybrid isoenzymes*. The ability to form hybrid isoenzymes is evidence of considerable structural similarities between the different subunits. Hybrid isoenzymes can be formed in vitro, but they are also formed in vivo in cells in which the different types of constituent subunits are present in the same subcellular compartment.

The number of different hybrid isoenzymes that can be formed from two nonidentical protomers depends on the number of subunits in the complete enzyme molecule. For a dimeric enzyme, one mixed dimer (hybrid isoenzyme) can be formed (in addition to two homopolymeric isoenzymes). If the enzyme is a tetramer, three heteropolymeric isoenzymes may be formed (Figure 12-8).

Nongenetic Causes of Multiple Forms of Enzymes

Many different types of post-translational modification of enzyme molecules can give rise to multiple forms. Several of these processes have been shown to cause the heterogeneity of various enzymes, either in living matter or as a result of changes taking place during extraction or storage. Modification of residues in the polypeptide chains of enzyme molecules (e.g., by removal of amide groups) can take place in living cells to give multiple forms. Modification can also take place as a result of procedural manipulations; for example, several erythrocyte enzymes contain sulfhydryl groups that are susceptible to oxidation in hemolysates. Thus, variant enzyme molecules with altered molecular charge may be generated.

Modifications affecting nonprotein components of enzyme molecules may also lead to molecular heterogeneity. Many enzymes are glycoproteins, and variations in carbohydrate side-chains are a common cause of nonhomogeneity of preparations of these enzymes. Some carbohydrate moieties, notably *N*-acetyl neuraminic acid (sialic acid), are strongly ionized and consequently have a profound effect on such properties of enzyme molecules as electrophoretic mobility.

Aggregation of enzyme molecules with each other or with nonenzymatic proteins

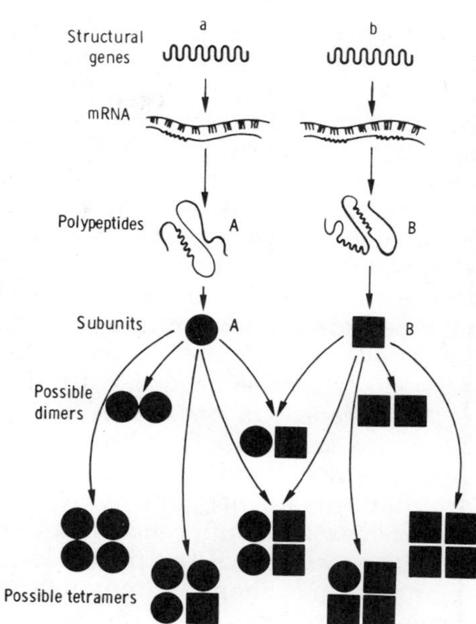

Figure 12-8. Diagram showing the origin of isoenzymes, assuming the existence of two distinct gene loci. When the active enzymes contain more than one subunit, hybrid isoenzymes consisting of mixtures of different subunits may be formed. One such isoenzyme can be formed in the case of a dimeric enzyme (e.g., creatine kinase) and three if the enzyme is a tetramer (e.g., lactate dehydrogenase). In both cases two homopolymeric isoenzymes can also exist. (From Moss, D. W.: Isoenzyme Analysis. London, The Chemical Society, 1979.)

may give rise to multiple forms that can be separated by techniques that depend on differences in molecular size. A specific form of interaction between enzymatic and nonenzymatic proteins is the cause of unusual enzyme components seen when some samples of human plasma are fractionated by electrophoresis or chromatography. These components are due to combination of apparently normal enzyme or isoenzyme molecules with plasma immunoglobulins. The enzyme-protein complexes thus formed may themselves be heterogeneous. Since the identification of "macroamylase," the first such enzyme-immunoglobulin complex to be identified, similar complexes involving lactate dehydrogenase, alkaline phosphatase, and creatine kinase have been observed.

Distribution of Isoenzymes and Other Multiple Forms of Enzymes

The existence of multiple gene loci and the isoenzymes derived from them has presumably conferred an evolutionary advantage on the species and has thus become part of its normal metabolic pattern. Some of these adaptations are related to the division of function between and within different types of specialized cells and tissues. It is to be expected, therefore, that the distribution of isoenzymes will not be uniform throughout the body, and wide variations in the activity of different isoenzymes do indeed occur between organs, between the cells that compose a particular organ, and even between the structures that constitute a single cell.

Certain gene loci may be expressed almost exclusively in a single tissue or only at a particular stage in development. A particularly striking example of the localized expression of multiple gene loci is provided by distinct isoenzymes that occur exclusively in specific subcellular organelles. Differences between mitochondrial isoenzymes and their functionally analogous counterparts in the cytoplasm have been demonstrated in several cases. Tissue-specific differences in the distributions of some multiple isoenzymes and other multiple forms of enzymes can form the basis for organ-specific diagnosis through isoenzyme measurements.

Differences in Properties Between Multiple Forms of Enzymes

The structural differences between the multiple forms of an enzyme give rise to greater or lesser differences in physicochemical properties such as electrophoretic mobility, resistance to inactivation, or solubility, or in catalytic characteristics such as ratio of reaction with substrate analogs or response to inhibitors. Methods of isoenzyme analysis therefore depend on a wide range of catalytic and structural properties of enzyme molecules.[75]

Isoenzymes that derive from the existence of multiple gene loci usually differ quantitatively in catalytic properties such as molecular activity, Michaelis constants, sensitivity to various inhibitors, and relative rates of activity with substrate analogs. In contrast, multiple enzyme forms that arise by such post-translational modifications as aggregation usually have similar catalytic properties. Multilocus isoenzymes also usually differ in antigenic specificity, although immunological cross-reaction is also not uncommon among multilocus isoenzymes when polyclonal antisera are used. Multiple enzyme forms due to postsynthetic modification frequently share antigenic determinants. Isoenzymes derived from allelic genes ("allelozymes") are often antigenically similar, even to the extent that substances that cross-react with antisera to the common isoenzyme may be detectable when mutation has abolished enzyme activity altogether. However, the greater specificity of monoclonal antibodies seems capable of revealing differences between otherwise antigenically similar isoenzyme molecules.

Differences in resistance to denaturation (e.g., by heat or concentrated urea solutions) are commonly found between true isoenzymes, whether these are the products of multiple loci or multiple alleles. Other multiple forms of enzymes often do not differ, or differ only slightly in this respect. The most commonly exploited difference between isoenzymes is the difference in net molecular charge that results from the altered amino acid compositions of the molecules; this forms the basis of separation by zone electrophoresis, ion-exchange chromatography, or isoelectric focusing. Separation methods that depend on differences in molecular size, such as gel filtration, are important in the detection of multiple forms that involve aggregation or association of enzyme molecules with other proteins.

Zone electrophoresis is the most generally useful technique of isoenzyme analysis. When suitable, sensitive methods are used to detect and to measure the separated enzyme zones, electrophoresis requires only small volumes of the specimen. Certain supporting media, notably starch and polyacrylamide gels, have pore sizes that approach the molecular dimensions of protein molecules. This introduces an element of segregation according to size into the resolution of protein mixtures on these media. Isoenzyme zones are usually located after electrophoresis by reactions that make use of the enzyme's activity to form a colored precipitate (or otherwise detectable product) at the site of the isoenzyme zone.

When the purpose of isoenzyme separation is to distinguish between possible alternative sources of increased enzyme activity, as for instance between heart and liver as sources of lactate dehydrogenase in serum, inspection of the relative intensity of the colored isoenzyme zones may provide sufficient information. However, in order to exploit fully the potential diagnostic sensitivity of isoenzyme separation, quantitative determination of the relative proportions of the different fractions is necessary. When isoenzymes are well separated by zone electrophoresis, densitometric scanning of stained electrophoretic strips gives a good estimate of the activities of the individual isoenzymes, provided the total activity applied to the strip and the staining conditions generate an intensity of the color proportional to the enzymatic activity. The need to detect and measure low activities of some isoenzymes calls for the use of sensitive techniques, such as measurement of the fluorescence emission. Elution of electrophoretically separated isoenzyme zones, followed by measurement of the enzymatic activities of the eluates, provides an alternative to densitometric methods of isoenzyme quantitation. However, sensitive methods are needed to assay the activity eluted from the strip. Recoveries from some media are low or variable.

Ion-exchange chromatography also makes use of differences in net molecular charge at a given pH to separate isoenzymes. In general, the technique is not as highly resolving of closely similar proteins as is zone electrophoresis, but relatively large amounts of proteins can be separated with good recoveries of enzymatic activity, so that the method is of great value in enzyme purification. Measurement of the enzymatic activity of each of the multiple fractions eluted by the gradient method from an ion-exchange column is laborious. When the isoenzymes to be separated are quite distinct in their elution characteristics, stepwise elution or batch processing may be used. The simplicity of stepwise ion-exchange methods makes them suitable for routine analysis of serum specimens. Other forms of chromatography that have been applied to fractionation of isoenzyme mixtures include high performance liquid chromatography and affinity chromatography.

Selective inactivation under controlled conditions is a technique of isoenzyme characterization based on differences in stability that result from small changes in the structure of protein molecules. Elevated temperatures or concentrated solutions of urea or other reagents are frequently chosen to denature the enzyme. Rates of enzyme inactivation by these agents are critically dependent on the conditions of the experiment, which must therefore be strictly controlled if reliable comparisons between samples are to be made. Selective inactivation is a useful semiquantitative or quantitative method of determining the composition of isoenzyme mixtures in serum. However, the results are less easy to interpret when more than two isoenzymes are present, especially when hybrid isoenzymes form part of the mixtures, since these forms possess stability characteristics intermediate between those of their parent homopolymeric isoenzymes.

Differences in catalytic properties, such as Michaelis constants, relative rates of reaction with substrate analogs (when the specificity of the enzyme allows for variation in the structure of the substrate), pH optima, and response to inhibitors, typically exist between isoenzymes that are the products of multiple gene loci. These differences can be made the basis for identification and measurement of particular isoenzymes. Under the most favorable circumstances, this approach to isoenzyme measurement has the advantage that only slight changes in the usual method of measuring enzyme activity may be needed.

Immunochemical methods of isoenzyme analysis are particularly applicable to isoenzymes derived from multiple gene loci, since these are usually antigenically distinct. Some of these methods make use of catalytic activity of the isoenzymes. For example, residual activity may be measured after reaction with antiserum. Alternatively, enzymatic activity may be used to locate the enzyme-antigen precipitate in immunodiffusion or "rocket" electroimmunoassays if the reaction with antibody does not inhibit activity. Radioimmunoassays have also been applied to isoenzyme measurement; these methods do not depend on the catalytic activity of the isoenzyme being determined.

Principles of Diagnostic Enzymology

All of the hundreds of different enzymes present in the human body are synthesized intracellularly, and most of them carry out their functions within the cells in which they are formed. However, certain enzymes are secreted, either in an active or an inactive form and, after activation, function within the extracellular fluids. The most obvious examples of these enzymes are the proteases and other hydrolases secreted into the gastrointestinal tract, some of which (e.g., pancreatic amylase) may find their way into the blood. Other examples of secreted enzymes include those of the blood-clotting mechanism and complement system and those concerned with fibrinolysis, which become functionally active in the plasma; these are referred to as plasma-specific enzymes (Table 12-1).

Table 12-1. CLASSIFICATION OF ENZYMES IN BLOOD

Classification	Examples
Plasma-specific enzymes	Serine protease procoagulants: thrombin, Factor XII (Hageman factor), Factor X (Stuart-Prower factor), and others Fibrinolytic enzymes or precursors: plasminogen, plasminogen proactivator
Secreted enzymes	Lipase, α-amylase (from salivary glands and pancreas), trypsinogen, cholinesterase, prostatic acid phosphatase
Cellular enzymes	Lactate dehydrogenase, aminotransferases, alkaline phosphatases

With a few exceptions, the clinical chemist is principally concerned with changes in the activity in serum or plasma of enzymes that are predominantly intracellular and that are normally present in the serum in low activities only. By measuring changes in the activities of these enzymes in disease, it is possible to infer the location and nature of pathological changes in the tissues of the body. Therefore, it is necessary to understand the factors that affect the rate of release of enzymes from their cells of origin and the rate at which they are cleared from the circulation, so that changes in activity in disease can be interpreted correctly.

Factors Affecting Enzyme Levels in Plasma or Serum

The measured level of activity of an enzyme in blood is the result of the balance between the rate at which it enters the circulation from its cells of origin and the rate at which it is inactivated or removed.

Entry of enzymes into the blood. By far the most important factors that affect enzyme activities in serum or plasma are those that influence the rate at which enzymes enter the circulation from the cells. These factors can be divided into two main categories: those that affect the rates at which enzymes leak from cells, and those that reflect altered rates of enzyme production, due either to increased synthesis of a particular enzyme by individual cell types or to proliferation of a particular type of enzyme-producing cell.

Leakage of enzymes from cells. Enzymes are retained within their cells of origin by the plasma membrane surrounding the cell. The plasma membrane is a metabolically active part of the cell, and its integrity depends on the cell's energy production. Any process that impairs energy production, either by depriving the cell of oxidizable substrates or restricting access of oxygen needed for energy production, will promote deterioration of the cell membrane. The membrane will become leaky and, if cellular injury becomes irreversible, the cell will die. Small molecules are the first to leak from damaged or dying cells, followed by larger molecules such as enzymes; ultimately the whole contents of necrotic cells are discharged.

Direct attack on the cell membranes by such agents as *viruses* or *organic chemicals* is an obvious cause of enzyme release and one that is particularly important in the case of the liver. A reduction in the supply of oxygenated blood perfusing *any* tissue will promote enzyme release. The most obvious clinical condition in which such a reduction occurs is myocardial infarction. The cells of the affected region rapidly begin to deteriorate and die, releasing their enzyme contents. Transfer of the enzymes to the extracellular fluid and lymphatics and then to the systemic circulation accounts for the rapid rise in serum enzyme activity that is characteristic of this condition. The liver is very sensitive to *hypoxia*, which can result from diminished cardiac output (heart failure). Increased activities of hepatocellular enzymes in the blood accompany a wide variety of conditions such as *congestive heart failure*, *shock*, and *hypoxemia*.

Skeletal muscle can also contribute enzymes to blood. Again, the cause may be poor perfusion, hypothermia, or direct trauma to the muscles (crush injuries). Infection, inflammation (polymyositis), degenerative changes (dystrophies), drugs, and alcohol (alcoholic myopathy) will cause enzyme leakage from myocytes. Enzyme release from muscles and other tissues also occurs as a result of anesthesia. Table 12-2 summarizes causes of cell damage or death.

Because of the very high concentrations of enzymes within the cells—thousands or even tens of thousands times greater than the concentrations in extracellular fluid—and because of the sensitivity with which small amounts of enzyme can be detected by their catalytic activity, an increase of enzyme activity in the extracellular fluid or plasma is an extremely sensitive indicator of even minor cellular damage.

Enzyme changes reflecting altered enzyme production. The small amounts of intracellular enzymes normally present in the plasma can be assumed to result from turnover of cells or leakage of enzymes from healthy cells. This contribution of enzymes to the circulating blood may *decrease*, either as the result of a genetic deficiency of enzyme production or when enzyme production is depressed as a result of disease. However, cases in which enzyme production is *increased* are of more general interest in diagnostic enzymology.

Table 12-2. CAUSES OF CELL DAMAGE OR DEATH

Category	Examples
Hypoxia	Loss of blood supply due to narrowing (atheromatous plaques) or blocking (thrombosis) of artery or vein; inadequate oxygenation due to cardiorespiratory failure; loss of oxygen-carrying capacity (anemia)
Chemicals and drugs	Environmental pollutants (lead, mercury); drugs (use and abuse); alcohol; tobacco
Physical agents	Trauma, extremes of heat and cold; radiation; electrical energy
Microbiological agents	Bacteria, viruses, fungi, protozoa, and helminths
Immune mechanisms	Immune disorders can cause tissue damage by a number of mechanisms: 1. Anaphylaxis (causing release of vasoactive amines) 2. Cytotoxicity (causing the target cell to be lysed) 3. Immune complex disease (leading to release of lysosomal enzymes) 4. Cell-mediated hypersensitivity (leading to cytotoxicity)
Genetic defects	Disorders with polygenic inheritance (diabetes mellitus, gout) Mendelian disorders (X-linked disorders, autosomal dominant and recessive disorders); disorders with variable modes of transmission
Nutritional disorders	Protein-calorie malnutrition, vitamin deficiencies, mineral deficiencies

After the classification of Robbins, S. L., Angell, M., and Kumar, V., *In*: Basic Pathology, 3rd ed. Philadelphia, W. B. Saunders Co., 1981.

An increase in the number and activity of the alkaline phosphatase–producing osteoblasts of bone is responsible for the increased level of alkaline phosphatase in the serum of normally growing children. Increased osteoblastic activity also accounts for the increased levels of this enzyme in serum in various types of bone disease. Toward the end of normal pregnancy, the placenta constitutes a new source of alkaline phosphatase and contributes its characteristic isoenzyme to the maternal circulation.

Enzyme induction may increase enzyme production; an example of such induction is the increased activity of γ-glutamyl transferase in serum, which may result from the administration of drugs such as barbiturates or phenytoin and from intake of ethanol. Biliary obstruction causes increased synthesis of alkaline phosphatase in the liver.

Proliferation of acid phosphatase–producing cells accounts for the increased levels of prostatic-type acid phosphatase that are found in about three-quarters of all cases of metastatic cancer of the prostate. As the malignant cells spread beyond the capsule of the prostate gland, they are more favorably placed to contribute their characteristic prostatic acid phosphatase isoenzymes to the circulation. However, in this case, the amount of enzyme produced by each cell is often less than that produced by a normal prostatic cell and indeed may decline to zero as the cancer cell becomes less and less like the prostatic cell from which it has originated.

Clearance of enzymes.[27,53] Little is known about the way in which enzymes are cleared from the circulation. Few enzyme molecules are small enough to pass through the healthy glomeruli of the kidney, and therefore urinary excretion is not a major route for elimination of enzymes from the circulation. An exception to this is amylase; increased levels of this enzyme in the blood (e.g., following acute pancreatitis) are accompanied by increased excretion in the urine. Present evidence suggests that enzyme inactivation begins in the plasma and that inactivated enzymes are rapidly removed, probably by the reticuloendothelial system. The half-life ($t_{1/2}$) of an enzyme in plasma may be a few hours or several days, but in most cases an average $t_{1/2}$ of 24–48 h can be expected.

The existence of *circulating inhibitors* or *activators* of enzymes has little effect on activities measured in the laboratory. When these inhibitors are of a reversible nature, the relatively high dilutions of serum used in most modern enzyme assays are sufficient to eliminate any possible effects from this cause.

Selection of Enzyme Tests

The selection of which enzyme to measure in serum for diagnostic or prognostic purposes depends on a number of factors. An important one is the distribution of enzymes among the various tissues (shown, for example, for aspartate and alanine transaminase and creatine kinase in Figure 12-9). The main enzymes of established clinical value

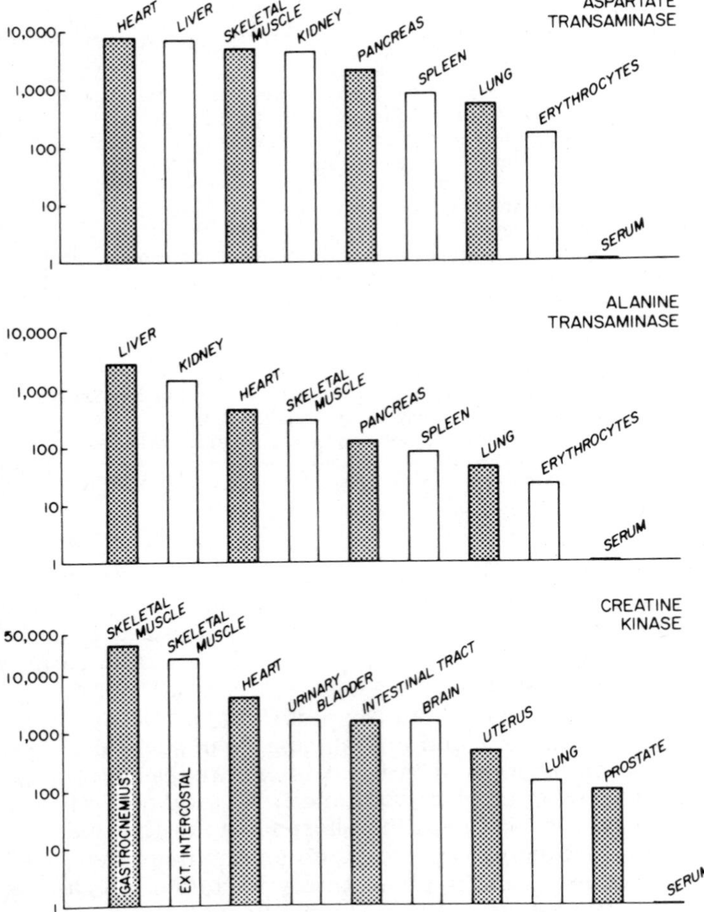

Figure 12-9. The concentration gradients between some human tissues and serum for aspartate transaminase, alanine transaminase, and creatine kinase. The concentration gradient axis is logarithmic.

together with their tissues of origin and their clinical applications are listed in Table 12-3.

Not all intracellular enzymes are equally valuable as indicators of cellular damage. For example, isocitrate dehydrogenase activity is high in heart muscle but, after a myocardial infarction, it is rapidly inactivated on entering the vascular compartment. Ornithine carbamyltransferase is an enzyme of the urea cycle and it is therefore almost totally liver-specific, with a liver:blood ratio of about 10^5:1; thus, even minor degrees of hepatocyte damage should create a readily detectable elevation in blood levels. Yet this enzyme has not found particular favor in clinical enzymology, possibly because of the inconvenience of the assay and a consequent lack of wide clinical experience with its results.

The mass of the damaged or malfunctioning organ, together with the enzyme's cell:blood gradient, obviously has a profound influence on the resulting elevation of enzyme activity in blood. Thus, the gradient of activity of prostatic acid phosphatase between prostate and blood is about 10^3:1, and the mass of that organ is 20 g. By contrast, the cell:blood gradient of alanine aminotransferase in the liver cell is 10^4:1, and the mass of the liver can exceed 1000 g. Obviously, fewer cells have to be damaged in the liver than in the prostate for the abnormality to be detected by an enzyme elevation in the blood. If, on the other hand, there is total organ involvement, then clearly the vast number of affected liver cells will elevate blood levels of any liver enzyme quite markedly. It has been estimated that if only one liver cell in every 750 were damaged, there would be a detectable elevation in the blood level of alanine aminotransferase.

Pathological damage to a tissue can embrace a wide spectrum of effects. Thus, a mild, reversible, viral inflammation of the liver—such as a mild attack of viral

Table 12-3. DISTRIBUTION OF DIAGNOSTICALLY IMPORTANT ENZYMES

Enzyme	Principal Sources	Principal Clinical Applications
Acid phosphatase	Prostate, erythrocytes	Carcinoma of prostate
Alanine aminotransferase	Liver, skeletal muscle, heart	Hepatic parenchymal disease
Aldolase	Skeletal muscle, heart	Muscle diseases
Alkaline phosphatase	Liver, bone, intestinal mucosa, placenta, kidney	Bone diseases, hepatobiliary diseases
Amylase	Salivary glands, pancreas, ovaries	Pancreatic diseases
Aspartate aminotransferase	Liver, skeletal muscle, heart, kidney, erythrocytes	Myocardial infarction, hepatic parenchymal disease, muscle disease
Cholinesterase	Liver	Organophosphorus insecticide poisoning, suxamethonium sensitivity, hepatic parenchymal diseases
Creatine kinase	Skeletal muscle, brain, heart, smooth muscle	Myocardial infarction, muscle diseases
Glutamate dehydrogenase	Liver	Hepatic parenchymal diseases
γ-Glutamyltransferase	Liver, kidney	Hepatobiliary disease, alcoholism
Lactate dehydrogenase	Heart, liver, skeletal muscle, erythrocytes, platelets, lymph nodes	Myocardial infarction, hemolysis, hepatic parenchymal diseases
5′-Nucleotidase	Hepatobiliary tract	Hepatobiliary disease
Sorbitol dehydrogenase	Liver	Parenchymal hepatic diseases
Trypsin(ogen)	Pancreas	Pancreatic diseases

hepatitis—is likely to increase only the permeability of the cell membrane and allow cytoplasmic enzymes to leak out into the blood, whereas a severe attack causing cell necrosis will also disrupt the mitochondrial membrane and both cytoplasmic and mitochondrial enzymes will be detected in blood. Knowledge of the intracellular location of enzymes can therefore assist in determination of the nature and severity of a pathological process if suitable enzymes are assayed in the blood. However, this possibility remains in an exploratory stage.

Finally, in the selection of a suitable enzyme in blood to assay for diagnostic purposes, the rate at which its activity will disappear from blood is of some significance. An enzyme with a very short $t_{1/2}$ in blood has little or no diagnostic value, and the most commonly assayed enzymes are those with $t_{1/2}$ in the range of 6 h or greater.

It must be emphasized that diagnostic decisions are made by integrating the history of the patient's illness, the results of clinical examination, and the analytical findings. Therefore, the results of enzyme analyses should always be assessed alongside the results of other analyses (such as bilirubin in acute hepatitis) and the clinical findings.

The Aminotransferases (Transaminases)

ASPARTATE AMINOTRANSFERASE; ASPARTATE TRANSAMINASE

(EC 2.6.1.1; L-Aspartate:2-Oxoglutarate Aminotransferase; AST or AspAT; formerly Glutamate Oxaloacetate Transaminase, GOT)

ALANINE AMINOTRANSFERASE; ALANINE TRANSAMINASE

(EC 2.6.1.2; L-Alanine:2-Oxoglutarate Aminotransferase; ALT or AlaAT; formerly Glutamate Pyruvate Transaminase, GPT)

The aminotransferases are a group of enzymes that catalyze the interconversions of the amino acids and α-oxoacids by transfer of amino groups. Distinct isoenzymes of AST are present in the cytoplasm and the mitochondria of cells. For example, in conditions associated with a mild degree of liver tissue injury, the predominant isoenzyme form in serum is that from the cytoplasm, although some mitochondrial isoenzyme is also present. Severe tissue damage results in the release of much mitochondrial isoenzyme as well.

The α-oxoglutarate/L-glutamate couple serves as one amino group acceptor and donor pair in all amino-transfer reactions; the specificity of the individual enzymes derives from the particular amino acid that serves as the other donor of an amino group. Thus, aspartate aminotransferase catalyzes the reaction shown in equation (6).

$$\begin{array}{ccccccc} COO^- & & COO^- & & COO^- & & COO^- \\ | & & | & & | & & | \\ H-C-NH_2 & + & C=O & \underset{}{\overset{AST,\ P\text{-}5'\text{-}P}{\rightleftharpoons}} & C=O & + & H-C-NH_2 \\ | & & | & & | & & | \\ CH_2 & & CH_2 & & CH_2 & & CH_2 \\ | & & | & & | & & | \\ COO^- & & CH_2 & & COO^- & & CH_2 \\ & & | & & & & | \\ & & COO^- & & & & COO^- \\ \textbf{L-Aspartate} & & \boldsymbol{\alpha}\textbf{-Oxoglutarate} & & \textbf{Oxaloacetate} & & \textbf{L-Glutamate} \end{array} \tag{6}$$

Alanine aminotransferase catalyzes the analogous reaction presented in equation (7). The reactions are reversible, but the equilibria of the AST and ALT reactions favor formation of aspartate and alanine, respectively.

$$\begin{array}{ccccccc} COO^- & & COO^- & & COO^- & & COO^- \\ | & & | & & | & & | \\ H-C-NH_2 & + & C=O & \underset{}{\overset{ALT,\ P\text{-}5'\text{-}P}{\rightleftharpoons}} & C=O & + & H-C-NH_2 \\ | & & | & & | & & | \\ CH_3 & & CH_2 & & CH_3 & & CH_2 \\ & & | & & & & | \\ & & CH_2 & & & & CH_2 \\ & & | & & & & | \\ & & COO^- & & & & COO^- \\ \textbf{L-Alanine} & & \boldsymbol{\alpha}\textbf{-Oxoglutarate} & & \textbf{Pyruvate} & & \textbf{L-Glutamate} \end{array} \tag{7}$$

Pyridoxal-5′-phosphate (phosphopyridoxal; P-5′-P or PLP) functions as a prosthetic group in the amino transfer reactions. The pyridoxal-5′-phosphate bound to the apoenzyme accepts the amino group from the first substrate, aspartate or alanine, to form enzyme-bound pyridoxamine-5′-phosphate and the first reaction product, oxaloacetate or pyruvate, respectively. The coenzyme, in amino form, then transfers its amino group to the second substrate, α-oxoglutarate, to form the second product, glutamate. Pyridoxal-5′-phosphate is thus regenerated.

Both holoaminotransferases and the coenzyme-deficient apoenzymes may be present in serum. Therefore, addition of pyridoxal phosphate under conditions that allow recombination with the enzymes usually produces a marked increase in aminotransferase activity. This increase ranges from zero to three- or fourfold and averages ~50% for AST in normal serum and ~20% for ALT. It has not been customary to add pyridoxal phosphate when assaying aminotransferase activity for routine diagnostic purposes, but the reference method drawn up by the Expert Panel on Enzymes of the International Federation of Clinical Chemistry (IFCC) specifies this addition, in accordance with the principle that all factors affecting the rate of reaction must be optimized and controlled. Addition of pyridoxal phosphate is also recommended in some national methods intended for routine use.

Clinical Significance

Transaminases are widely distributed in human tissues. Both AST and ALT are normally present in human plasma, bile, cerebrospinal fluid, and saliva, but none is found in urine unless a kidney lesion is present.

In *viral hepatitis* and other forms of liver disease associated with hepatic necrosis, levels of serum AST and ALT are elevated even before the clinical signs and symptoms of disease, such as jaundice, appear. Activities for both enzymes may reach values as high as 100 times the upper reference limit, although 20- to 50-fold elevations are most frequently encountered. Peak values of transaminase activity are seen between the seventh and twelfth days; values return to normal levels by the third to fifth week if recovery is uneventful. In toxic or viral hepatitis, ALT is characteristically as high as or higher than AST, and the ALT/AST (De Ritis) ratio, which normally is <1, approaches or becomes greater than unity.

Moderately increased levels of AST and ALT activity may also be observed in *extrahepatic cholestasis*. Levels observed in *cirrhosis* vary with the status of the cirrhotic process; they range from upper normal to some four to five times normal, with the level of AST activity higher than that of ALT activity. Five- to tenfold elevations of the two enzymes may occur in patients with primary or metastatic *carcinoma of the liver*, with AST usually being higher than ALT, but levels are often normal in the early stages of malignant infiltration of the liver. Slight or moderate elevations of both AST and ALT activities may be observed after intake of alcohol, during delirium tremens, and after administration of drugs such as opiates, salicylates, or ampicillin.

Although serum levels of both AST and ALT become elevated whenever disease processes affect liver cell integrity, ALT is the more liver-specific enzyme.[22] Serum elevations of alanine transaminase activity are rarely observed in conditions other than parenchymal liver disease. Moreover, elevations of ALT activity persist longer than do those of AST activity.

After *myocardial infarction* an increased level of AST activity appears in serum. These changes are discussed in a later section. ALT levels are within normal limits, or are only marginally increased, in uncomplicated myocardial infarction, since the concentration of ALT in heart muscle is only a fraction of that of AST. However, ALT is increased in liver damage secondary to heart failure.

AST and occasionally ALT activity levels are increased in *progressive muscular dystrophy* and *dermatomyositis*, reaching levels up to eight times normal; they are usually normal in muscle diseases of neurogenic origin. *Pulmonary emboli* can raise AST levels to two to three times normal, and slight to moderate elevations (two to five times normal) are seen in *acute pancreatitis, crushed muscle injuries*, *gangrene*, and *hemolytic disease*.

Methods for the Measurement of Transaminase Activity[42,86]

The assay system for measuring transaminase activity contains two amino acids and two oxoacids (equations 6, 7, 8). There is no convenient method for assaying either of the amino acids in the reaction system. Therefore, formation or consumption of the oxoacids is measured.

Continuous Monitoring Method for the Measurement of Aminotransferase Activity

It is not possible to monitor transaminase reactions directly, but the advantages of continuous-monitoring assays can be obtained by coupling the transaminase reactions to specific dehydrogenase reactions. The oxoacids formed in the transaminase reaction are measured indirectly by enzymatic reduction to the corresponding hydroxyacids, and the accompanying change in NADH concentration is monitored spectrophotometrically. Thus, oxaloacetate, formed in the AST reaction, is reduced to malate in the presence of malate dehydrogenase (MDH), as shown in equation (8);

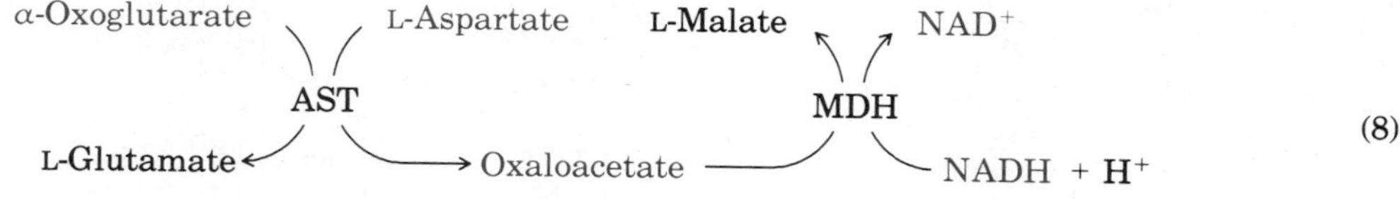

Aminotransferase reaction
(Formation of oxaloacetate)
Assay reaction

Dehydrogenase reaction
(Quantitation of oxaloacetate)
Indicator reaction

pyruvate formed in the ALT reaction is reduced to lactate by lactate dehydrogenase (LDH). The substrate, NADH, and the auxiliary enzymes, MDH and LDH, must be present in sufficient quantity so that the reaction rate is limited only by the amounts of AST and ALT, respectively. As the reactions proceed, NADH is oxidized to NAD^+. The disappearance of NADH is followed by the decrease in absorbance at 339 (340) nm, which is measured for several minutes either continuously or at frequent intervals. The change in absorbance per minute (ΔA/min) is related directly to micromoles of NADH oxidized and, in turn, to micromoles of substrate transformed per minute (International Units).

Measurements should be made by using a spectrophotometer with established spectrophotometric and wavelength accuracy and with good resolution at 339 (340) nm. The reaction mixture in the cuvet must be maintained at the selected temperature by means of a thermostatically controlled cuvet compartment. A preliminary incubation period is necessary to ensure that NADH-dependent reduction of endogenous oxoacids in the sample is completed before 2-oxoglutarate is added to start the transaminase reaction. Most modern methods include pyridoxal-5′-phosphate to reactivate any apotransaminase in the sample. This process also takes place during the preincubation period. After a brief lag phase, seen with most coupled reactions, ΔA is monitored, preferably with a recorder; alternatively, five to eight readings are taken at 1-min intervals to establish the linear portion of the curve. Because the reaction mixture has a rather high initial absorbance, it is customary to use a solution such as potassium dichromate as a reference blank or to use the zero offset (zero suppression) function of the spectrophotometer.

Determination of Aspartate Transaminase (AST) Activity[86]

Specimens. Serum is the specimen of choice. Hemolysis should be avoided, since AST and ALT activities in erythrocytes are some fifteen and seven times higher, respectively, than those in normal sera. Specimens are best stored frozen if they are to be kept more than 3–4 d. Minimal loss of activity occurs at 0–4 °C over 1–3 d.

Procedure. Prepackaged reagents are now invariably used in clinical laboratories. These are based on national or international recommendations and should incorporate the following features. Tris is used as a buffer in place of the earlier choice of phosphate because phosphate appears to increase the rate of NADH decomposition and to inhibit association of P-5′-P with the transaminase apoenzyme. The coenzyme may be added to ensure that all transaminase is fully active, although this recommendation is not yet universally followed. Addition of lactate dehydrogenase to the coupled enzyme system accelerates the endogenous side reactions and thus shortens the preincubation period. A pH of 7.8 is used because the pH optimum for the coupled enzyme system is between 7.7 and 7.9 and the stability of NADH is greater at this pH than at pH 7.4. Stock preparations of both MDH and LDH are diluted with glycerol rather than with $(NH_4)_2SO_4$; this procedure avoids introducing ammonium ions, thereby eliminating a possible side reaction catalyzed by L-glutamate dehydrogenase (GLDH, EC 1.4.1.3), in which NADH is consumed. Increased concentrations of GLDH may be seen in parenchymal liver disease. The reaction temperature is either 30 or 37 °C.

The concentrations of the various components in the complete reaction mixture are as follows: Tris buffer, pH 7.8, 90 mmol/L; L-aspartate, 175 mmol/L; 2-oxoglutarate, 15 mmol/L; NADH, 0.15 mmol/L; MDH, 600 U/L (25 °C); LDH, 1200 U/L (25 °C); and P-5′-P, 0.12 mmol/L. The ratio of serum volume to total reaction volume (serum dilution) is 1/15. The substrate concentrations are in agreement with those recommended by the Enzyme Study Group of the American Association for Clinical Chemistry (AACC).[86] The substrate concentrations chosen by the Expert Panel on Enzymes are slightly different: aspartate, 240 mmol/L; 2-oxoglutarate, 12 mmol/L. However, in both methods the substrate concentrations support a reaction rate equivalent to ~96% of the theoretical V_{max} value.

Reference ranges[105] at 30 °C are 25–75 U/L for newborns, 15–20 U/L for infants, and 8–20 U/L for adults. Serum values for adults with addition of P-5′-P to the reaction mixture are 10–30 U/L. The average rate of the AST reaction in human serum at 37 °C is 1.59 times the rate at 30 °C.

The distribution of values in the normal population is skewed to the right; the skewness may be real or may reflect the inclusion of sera with nonvisible hemolysis and sera from persons with possible occult tissue damage. Individuals in the fasting state may show slightly lower values. Values for men are slightly higher than those for women.

Determination of Alanine Transaminase (ALT) Activity[42]

The procedure used is identical to that for measuring AST activity, except that L-alanine replaces aspartate as the amino group donor and LDH replaces MDH as the indicator enzyme. The concentration of alanine needed to give an optimal rate of ALT activity is much greater than that of aspartate in measuring AST activity. The added LDH both speeds up the side reaction and serves as the indicator enzyme.

The composition of a typical, complete reaction mixture is as follows: Tris buffer, 90 mmol/L, pH 7.3; L-alanine, 500 mmol/L; oxoglutarate, 15 mmol/L; NADH, 0.15 mmol/L; LDH, 2400 U/L (25 °C); and P-5′-P, 0.12 mmol/L. The serum dilution is 1/15. The substrate concentrations are in agreement with recommendations of the Enzyme Study Group of the AACC and the Expert Panel on Enzymes of the IFCC and support a reaction rate of ~92% of the theoretical V_{max} value.[42]

Reference ranges at 30 °C are 5–28 U/L for newborns and infants and 8–20 U/L for adults.[105] Serum values for adults with addition of P-5′-P to the reaction mixture are up to 30 U/L. Values in men are slightly higher than in women.

Colorimetric Methods

The oxoacids can be assayed colorimetrically by coupling with 2,4-dinitrophenylhydrazine. However, in each aminotransferase reaction, two oxoacids are present, one is a substrate and the other a product, and both can react to give dinitrophenylhydrazones. As the reaction proceeds, one oxoacid is increasing in concentration while the other is decreasing. Furthermore, oxoacids that are normal components of serum, such as pyruvate, also produce phenylhydrazones and contribute to the high blanks. Despite these limitations, the colorimetric approach is still feasible, because the phenylhydrazones of the products of both the AST and ALT reactions (oxaloacetate and pyruvate, respectively) are considerably more chromogenic than is the phenylhydrazone of α-oxoglutarate. Colorimetric methods based on dinitrophenylhydrazone formation are relatively simple and have limited but acceptable accuracy. Assay methods based on this approach must now be considered obsolete.

CREATINE KINASE[54]

(EC 2.7.3.2; Adenosine Triphosphate: Creatine *N*-Phosphotransferase; CK)

Creatine kinase, also incorrectly referred to as creatine phosphokinase (CPK), catalyzes the reversible phosphorylation of creatine by adenosine triphosphate (ATP), as shown in equation (9).

$$\underset{\text{Creatine (Cr)}}{{}^{-}O-\overset{O}{\overset{\|}{C}}-CH_2-N(CH_3)-\underset{\underset{NH}{\|}}{C}-NH_2} + \underset{\text{Adenosine triphosphate (ATP)}}{AMP-O-\underset{O^-}{\overset{O}{\overset{\|}{P}}}-O-\underset{O^-}{\overset{O}{\overset{\|}{P}}}-O^-} \underset{\text{pH 6.7}}{\overset{\text{pH 9.0}}{\underset{CK,\ Mg^{2+}}{\rightleftharpoons}}} \tag{9}$$

$$\underset{\text{Phosphocreatine (PCr, Creatine phosphate)}}{{}^{-}O-\overset{O}{\overset{\|}{C}}-CH_2-N(CH_3)-\underset{\underset{NH}{\|}}{C}-\overset{H}{\overset{|}{N}}-\underset{O^-}{\overset{O}{\overset{\|}{P}}}-O^-} + \underset{\text{Adenosine diphosphate (ADP)}}{AMP-O-\underset{O^-}{\overset{O}{\overset{\|}{P}}}-O^-} + H^+$$

Phosphocreatine, the major phosphorylated compound in muscle, is present in about an eightfold excess over ATP. When muscle contracts, ATP is consumed (to form ADP) and creatine kinase catalyzes the rephosphorylation of ADP to form ATP, using phosphocreatine. The optimal pH values for the forward (Cr + ATP → ADP + PCr) and reverse (PCr + ADP → ATP + Cr) reactions are 9.0 and 6.7, respectively. The equilibrium position for the reaction is dependent on pH. At neutral pH, PCr has a much higher phosphorylating potential than does ATP; this higher potential favors the reverse reaction. The reverse reaction proceeds two to six times faster than the forward reaction, depending on the reaction conditions.

As is true for all kinases, Mg(II) is an obligate activating ion, functioning with ADP and ATP. The optimal concentration range for Mg(II) is quite narrow, and excess Mg(II) is inhibitory. Many metal ions, such as Mn(II), Ca(II), Zn(II), and Cu(II), inhibit enzyme activity, as do iodoacetate and other sulfhydryl-binding reagents. Activity is also inhibited by excess ADP, urate, and cystine. The enzyme in serum is relatively unstable; activity is lost as a result of sulfhydryl group oxidation at the active site of the enzyme. Activity can be partially restored by incubating the enzyme preparation with sulfhydryl compounds such as *N*-acetylcysteine, thioglycerol, dithioerythritol, dithiothreitol (Cleland's reagent), thioglycolic acid, mercaptoethanol, or cysteine. The extent of activity recovered by this procedure is inversely proportional to the time delay in adding these reagents after collection of the blood specimen and depends on the type of sulfhydryl reagent. The current agents of choice are *N*-acetylcysteine (NAC) or thioglycerol at a final concentration of 20 mmol/L. The former is soluble, stable, odorless, and inexpensive; *N*-acetylcysteine can be lyophilized, whereas thioglycerol cannot.

CK activity is greatest in striated muscle, brain, and heart tissues, which contain some 2500, 550, and 470 U/g protein, respectively. Other tissues, such as the kidney and the diaphragm, contain significantly less activity, and the liver and erythrocytes are essentially devoid of activity.

Creatine kinase is a dimer composed of two subunits, each with a MW of ~40 000. The subunits (B, or brain, and M, or muscle) are the products of two distinct structural genes. Three different pairs of subunits can exist: BB (or CK-1), MB (or CK-2), and MM (or CK-3). The Commission on Biochemical Nomenclature has recommended that isoenzymes should be numbered on the basis of their electrophoretic mobility, with the most anodal form receiving the lowest number. Accordingly, the CK isoenzymes are numbered CK-1, CK-2, and CK-3. CK-1 predominates in brain, prostate, gut, lung, bladder, uterus, placenta, and thyroid, whereas CK-3 predominates in skeletal and cardiac muscle. CK-2 is present, to varying degrees, in heart muscle (25–46% of CK activity) and also to a minor degree in skeletal muscle (<5%). There is presently wide divergence in published data on tissue composition of CK.

All three of the isoenzymes are found in the cell cytosol or are associated with myofibrillar structures. However, there exists a fourth form that differs from the others both immunologically and in electrophoretic mobility. This isoenzyme, *CK-Mt*, is located between the inner and outer membranes of mitochondria, and in the heart it constitutes up to 15% of the total CK activity. CK activity may also be found in two macromolecular forms, *macro CK Types 1 and 2*.[55] Type 1 is CK-1 associated with IgG or CK-3 with IgA, and Type 2 is oligomeric CK-Mt. The M subunit undergoes post-translational modification in serum and thus there exist at least two other M subunits, each of which is capable of hybridizing with other M or B subunits to form active isoenzyme species with electrophoretic mobilities slightly different from the original, unmodified subunit.

Serum CK activity is subject to a number of physiological variations. For example, the activity in serum appears to be a function of the muscle mass of the individual. This is presumably the basis for the finding that females have lower serum activity than males and that slightly built individuals often have lower serum CK activity than more muscular members of the same sex. Ethnic origin is also important; thus, a healthy black North American female may have higher serum CK activity than a white male.

Clinical Significance[30,54,55]

Diseases of skeletal muscle. In all types of muscular dystrophy, and especially in the Duchenne type (in which levels up to 50 times the upper limit of normal may be encountered), serum CK activity is greatly elevated at some time during the course of the disease. In *progressive muscular dystrophy* (particularly the Duchenne sex-linked muscular dystrophy), enzyme activity in serum is highest in infancy and childhood (7–10 years of age) and may be elevated long before the disease is clinically apparent. Serum CK activity characteristically falls as the patient gets older and as the mass of functioning muscle diminishes with the progression of the disease. About 50–80% of the asymptomatic female carriers of Duchenne dystrophy show three- to sixfold elevations of CK activity, but values may be normal following a period of physical inactivity. Quite high values of CK are also seen in a wide variety of *muscle diseases* including malignant hyperthermia; serum enzyme activity is normal in *neurogenic muscle diseases* such as myasthenia gravis, multiple sclerosis, poliomyelitis, and Parkinsonism.

Diseases of the heart. After a *myocardial infarction*, CK activity in serum is invariably elevated. The use of total CK and CK-2 in the diagnosis of myocardial infarction is the most important single application of CK measurements in clinical chemistry and is discussed more fully later (page 383). Obviously, *cardiac trauma* following heart surgery (including transplantation) will also cause an elevation of total serum CK and CK-2 sufficient to mask elevations due to intraoperative myocardial infarction.

Other cardiac conditions —angina pectoris, cardiogenic shock, electrical countershock, tachycardia, myocarditis, and congestive cardiac failure—usually do not have elevated total serum CK or CK-2. In the few positive cases, particularly those with CK-2 increases, it seems reasonable to suspect cardiac damage severe enough to cause changes in the cell membranes of the myocytes and to allow intracellular CK to diffuse into the extracellular space. The mere presence of CK-2 activity in serum does not necessarily indicate myocardial damage. Many assay methods are sensitive enough to detect serum CK-2 in a healthy individual, and CK-2 can arise from tissues other than the myocardium. In fact, CK-2 activity (<6% of total CK activity) can be detected in serum in inflammatory and degenerative muscle diseases, traumatic lesions including shock, intoxications, delirium tremens, hypothyroidism, and acute psychosis, and in women immediately after obstetric delivery. Thus, the diagnosis of myocardial damage must be based on both clinical findings and a consideration of the extent of CK-2 elevation. Myocardial damage is usually excluded if an increased CK-2 activity is <5 or 6% of the total CK activity.

Diseases of the central nervous system. Serum CK activity may increase after cerebral ischemia, acute cerebrovascular disease (including subarachnoid hemorrhage), and head injury. In *Reye's syndrome* (a childhood disorder characterized by acute brain swelling with fatty infiltration and nonicteric dysfunction of the liver) there may be up to a 70-fold increase in serum CK activity, indicating the severity of the encephalopathy.

Diseases of the thyroid. Serum CK activity demonstrates an inverse relationship with thyroid activity. About 60% of *hypothyroid* subjects show an average elevation of CK activity of five times the upper reference limit. The major isoenzyme present is CK-3. Hypothyroidism predisposes a patient to ischemic heart disease, in which CK-2 may account for up to 13% of total CK activity. By contrast, in *hyperthyroid* subjects, the serum CK activity tends to be at the low end of the reference range.

Clinical Conditions Associated with the Presence of CK-1 in Serum

During *normal childbirth*, there can be a sixfold elevation in maternal total serum CK activity, with up to 10% CK-1 from the uterus and possibly the placenta. Surgical intervention during labor will further increase the serum activity of CK-1. CK-1, usually at low levels, has also been reported in a wide variety of disorders such as *multiorgan involvement*, in many *critical care patients*, following *aortocoronary bypass* operations, and in *hypothermia*. *Gastrointestinal infarction* or *adenocarcinoma*, as well as *lung tumors*, may cause elevated serum CK-1 activities. *Tumors* of the prostate, bladder, kidney, breast, and ovary have been found to be associated to varying degrees with elevations of serum CK-1; these findings suggest that serum CK-1 may be of use as a tumor marker.

Methods for the Determination of Creatine Kinase Activity[30]

Numerous coupled enzyme methods have been developed for the assay of CK activity by using either the forward (Cr → PCr) or the reverse (PCr → Cr) reaction. The reverse reaction is preferred, despite the greater cost of the starting chemicals, phosphocreatine and ADP, because it proceeds faster than the forward reaction. ATP produced in the reverse reaction is measured by coupling the hexokinase (HK) and the glucose-6-phosphate dehydrogenase (GPD) reactions to the CK reaction [equation (10)].

$$\begin{gathered}\text{Phosphocreatine} + \text{ADP} \xrightarrow[\text{pH 6.8}]{CK} \text{Creatine} + \text{ATP} \\ \text{ATP} + \text{glucose} \xrightarrow{HK} \text{Glucose-6-phosphate} + \text{ADP} \\ \text{Glucose-6-phosphate} + \text{NADP}^+ \xrightarrow{GPD} \text{6-Phosphogluconate} + \text{NADPH} + \text{H}^+\end{gathered} \tag{10}$$

The HK catalyzes phosphorylation of glucose by ATP to form glucose-6-phosphate (G-6-P) and regenerates ADP for the CK reaction. The G-6-P is then oxidized with $NADP^+$ to form 6-phosphogluconic acid and NADPH. The rate of NADPH formation is a measure of the CK activity, provided that the concentrations of all other components in the three-enzyme system are present in suitable excess so that the CK activity is the only limiting factor. This coupled enzyme system is completely "downhill" (i.e., all reactions proceed in a favorable direction). The pH optimum for the entire system is 6.8–6.9.

There are a number of sources of interference in this assay system. The *adenylate kinase (AK) effect* is caused by an enzyme found in fairly high concentration in nearly all tissues, including erythrocytes. AK catalyzes the reaction,

$$2\ \text{ADP} \xrightarrow{AK} \text{ATP} + \text{AMP} \tag{11}$$

which in the presence of ADP (in the CK assay mixture) will produce ATP; ATP is also the product of the CK reaction, as shown in equation (10). Serum AK activity results in an apparent increase in CK activity because more ATP is produced than CK activity alone produces. AK activity can be inhibited by adding AMP or diadenosine pentaphosphate (Ap_5A) to the CK assay. AMP (competitive inhibitor of AK) causes product inhibition of reaction (11). Ap_5A competitively inhibits the AK of muscle and erythrocytes but has less effect on the liver and kidney enzymes. The combination in an assay mixture of AMP, 5 mmol/L, and Ap_5A, 10 μmol/L, is extremely effective in inhibiting AK activity. Alternatively, AK activity can be measured directly by omitting phosphocreatine from the CK assay to obtain a "blank" activity, which is then subtracted from the value for total CK activity obtained in a complete assay system.

CK assays demonstrate a *lag phase*; i.e., a slow increase in reaction rate after the addition of substrate or serum to the assay mixture. Typical lag phases in an optimized assay system are 90 s at 30 °C and 60 s at 37 °C. This phenomenon has important practical implications, since inappropriate timing of the measurement period may cause incorrect and low estimations of the reaction rate.

The presence of *endogenous inhibitors* in serum has been noted earlier. Ca(II) is an important competitive inhibitor for Mg(II) which is required by CK for full catalytic activity. The Ca(II) effect can be nullified by the addition of EDTA, 2 mmol/L, provided the Mg(II) concentration is increased to 10 mmol/L to compensate for some binding of Mg(II) by the EDTA. Because of the presence of endogenous inhibitors in serum, the dilution of serum in the CK assay mixtures has a significant effect on the apparent CK activity. Decreasing the serum volume fraction may result in a significant increase in the observed CK activity.

Specimen. Serum is the preferred specimen, although heparinized plasma may be used. CK activity in serum is unstable and is rapidly lost during storage. Full activity may persist at ambient temperatures for 4 h, at 4 °C for 8–12 h, and for 2–3 d when frozen; the degree of stability varies with the individual specimen and also with the activating efficiency of the assay mixture. The serum specimen should be chilled to 4 °C as soon as practicable after collection. It is probably not necessary to add any sulfhydryl agent for storage because the optimized assay formulation will reactivate CK in serum up to 99% even after it has been stored for one week at 4 °C. A slight degree of hemolysis can be tolerated because erythrocytes contain no CK activity. However, moderately or severely hemolyzed specimens are unsatisfactory because enzymes and intermediates (adenylate kinase, ATP, glucose-6-phosphate) liberated from the erythrocytes may affect the lag phase and the side reactions occurring in the assay system.

Procedure

Principle. The reverse reaction [see equation (10)] is used almost universally. In the Oliver-Rosalki assay, which was subsequently further optimized by Szasz, the hexokinase and glucose-6-phosphate dehydrogenase reactions are coupled to the production of ATP. *N*-acetylcysteine (NAC) is included to activate CK, EDTA to bind Ca(II) and to increase the stability of the reaction mixtures, and Ap_5A together with AMP to inhibit adenylate kinase (the CK-NAC-EDTA method). The reaction is initiated by adding phosphocreatine and the absorbance is recorded continuously, after allowing for a 2-min lag phase. The assay is linear up to at least 3000 U/L. If dilution of sera with very high CK activities is necessary, use NaCl, 150 mmol/L. A twofold dilution will produce an apparent increase of CK activity of ~10%; the increase may be much greater with less optimized methods.

A narrow-bandpass spectrophotometer equipped with a device to maintain the temperature of the cuvet compartment at 30 (or 37) °C ± 0.1 °C is required; wavelength is set at 339 (340) nm.

Reagents. In view of the complexity of the CK-NAC-EDTA mixture and the different stability of its components, prepackaged reagents are widely used. The most satisfactory of the formulations currently available is that recommended by the Scandinavian Society for Clinical Chemistry and Clinical Physiology.[16,30] Other formulations exist;[20] for example, the DuPont *aca* assay was used by 40% of all participants in one of the 1984 College of American Pathologists' Special Enzyme Surveys.

The concentrations of the reagents in the complete reaction mixture in the Scandinavian formulation are: imidazole acetate buffer, 100 mmol/L; phosphocreatine, 30 mmol/L; ADP, 2 mmol/L; D-glucose, 20 mmol/L; $NADP^+$, 2 mmol/L; hexokinase, 3500 U/L (at 25 °C); D-glucose-6-phosphate dehydrogenase, 2000 U/L (at 25 °C); magnesium acetate, 10 mmol/L; AMP, 5 mmol/L; Ap_5A, 10 μmol/L; *N*-acetylcysteine, 20 mmol/L; EDTA, 2 mmol/L. The volume fraction of serum is 0.043 and the pH is 6.5 at 37 °C.

Reference ranges.[105] The CK levels in the sera of normal newborns are elevated during the first 24-h postpartum to about three times adult values, and a slight elevation remains throughout the first year of life. In the male, serum CK activity remains constant after the first year; it rises with increase in muscle mass in puberty, then decreases slightly until age 20. In the female, activity is stable from the first year until the first menses, at which time a decrease occurs. CK activity decreases in pregnancy and rises after the menopause. In both sexes CK activities are higher in summer months. Blacks have slightly higher CK activity values than whites.

Exercise and *muscle trauma* can elevate serum CK values. Serum CK-2 activity is also significantly elevated by jogging. Many *drugs* (including alcohol) and exposure to *carbon monoxide* can cause elevations of serum total CK activity because of their effect on skeletal muscle. Bed rest, even overnight, can lower serum creatine kinase levels by 20% or more.

Examples of adult reference ranges (U/L) for serum CK, using the CK-NAC-EDTA method described above, are for males, at 25 °C, 10–65; at 30 °C, 15–105; and at 37 °C, 38–174. For females they are, at 25 °C, 7–55; at 30 °C, 10–80; and at 37 °C, 26–140.

Caution is necessary when published data regarding reference ranges or physiological variation are interpreted, because there are wide variations in activators, inhibition, and serum volume fractions. In addition, the physiological state of the subjects in many studies is ill defined.

Methods for the Separation and Quantitation of Creatine Kinase Isoenzymes[30,54,55]

The three techniques most commonly used for the determination of CK isoenzymes are electrophoresis, ion-exchange chromatography, and immunological methods.

Electrophoresis on agar, agarose, or cellulose acetate can separate the CK isoenzymes satisfactorily. The isoenzyme bands are then visualized by incubating the support with a concentrated CK assay mixture that uses the reverse reaction. The NADPH formed in this reaction may be detected by observing the bluish-white fluorescence after excitation by long-wave (360 nm) ultraviolet light or quantitated by fluorescent densitometry, which is capable of detecting activities of 2–10 U/L. Typical examples of results are shown in Figure 12-10, *A*. Alternatively, the NADPH can reduce a tetrazolium salt (NBT, nitroblue tetrazolium; MTT, 3-(4,5-dimethylthioazolyl-2)-2,5-diphenyl-2H-tetrazolium bromide; or INT, 3-(4-iodophenyl)-2-2(4-nitrophenyl)-5-phenyl-2H-tetrazolium chloride) to form a colored formazan. The discriminating power of electrophoresis allows the detection of many abnormal bands, some of which are shown in Figure 12-10, *B*. In patients on maintenance hemodialysis for end-stage renal disease, albumin becomes endogenously fluorescent and may simulate the CK-1 band in the fluorescent assay. This possibility can be readily excluded by examining the electrophoretic strip, under fluorescent light, before and after development. Albumin fluorescence does not interfere when strips are stained. The macro-CK forms have been described earlier (page 374). The postsynthetic bands result from post-translational modifications to the M subunit occurring in blood; these sub-bands will appear between CK-3 and CK-2 and between CK-2 and CK-1 (Figure 12-10) but are not readily detectable by routine methods.

Ion-exchange techniques can be used to separate the CK isoenzymes either by batch adsorption or by column chromatography. DEAE-Sephadex A50, usually in the form of a minicolumn (0.5 × 6.0 cm) holding about 60 mg of gel, has been most favored.

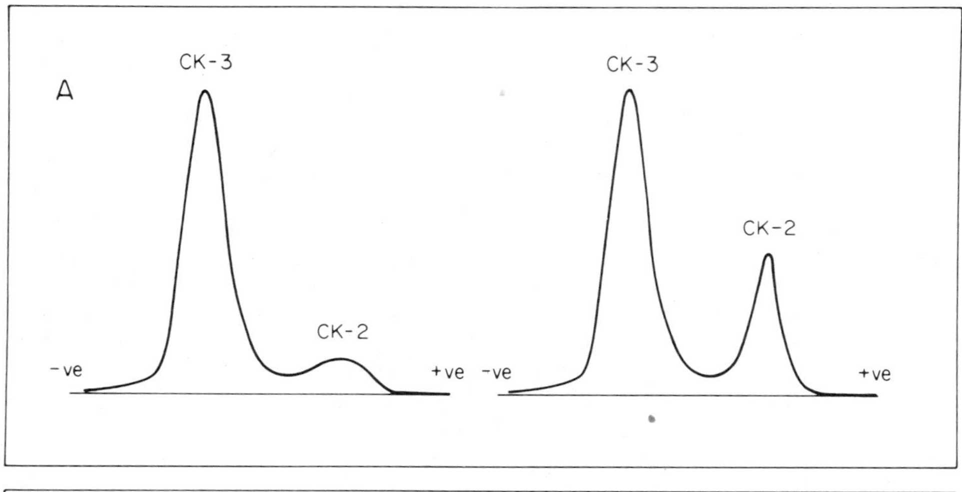

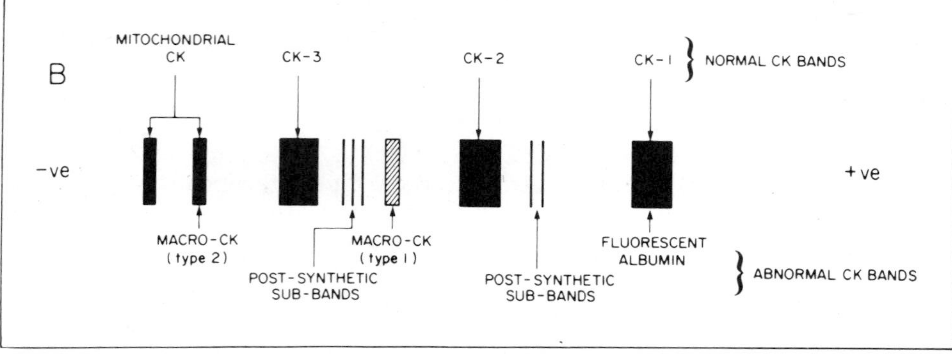

Figure 12-10. *A*, The separation of serum CK isoenzymes from a healthy adult (left) and from a patient (right) who had a myocardial infarction 24 h previously. *B*, A diagrammatic representation of the naturally occurring CK isoenzymes (some of which are seen only in disease) and some of the reported anomalous isoenzymes seen on agarose gels.

The CK isoenzymes are adsorbed onto the gel, the gel is then washed, and the CK-3 is eluted with Tris buffer, 50 mmol/L, pH 8.0, containing NaCl, 100 mmol/L. CK-2 is next eluted with the same Tris buffer containing NaCl, 200 mmol/L. Finally, CK-1 is eluted with Tris buffer, 50 mmol/L, pH 7.0, containing NaCl, 500 mmol/L. The lower limit of detection is 1–5 U/L. The elution process considerably dilutes the CK isoenzymes. It is therefore important to reduce this dilutional effect by using small volumes of eluting buffer, or by concentrating each fraction after collection, or by using the eluate as part of the solvent used for the reconstitution of the CK reagents. There are several published procedures and many commercially available kits based on this technique. The method is relatively simple and rapid to perform, but precautions must be taken to ensure that there is no carryover between the different isoenzymes eluted from the gel. Carryover is the major limitation of this technique; this risk is greatest for samples of high CK-3 activities.

Immunological methods for measuring the CK isoenzymes require specific antisera against the M and B subunits. These are obtained by preparing monoclonal or polyclonal antibodies. Rabbit antisera can be used to precipitate specific isoenzymes out of solution (immunoprecipitation) or they can competitively bind specific isoenzymes (immunoassay). Goat antisera appear to be more effective inhibitors of subunit activity (immunoinhibition). In contrast with immunoprecipitation and immunoinhibition techniques, which measure the isoenzymes by determination of residual enzyme activity, immunoassays such as RIA or immunoenzymometric assays (using poly- or monoclonal antibodies) measure the enzyme mass, whether or not the enzyme molecule is catalytically active. Current practice uses the "sandwich" technique in which the B subunits of CK-2 (and CK-1, if present) are bound by anti-CK-B sera rendered immobile on a matrix (such as the test-tube wall or an insoluble bead), and then any M subunits associated with these bound B subunits (MB or CK-2) are estimated with the use of an anti-CK-M antibody. This antibody may either be labeled with a radioactive tracer (such as ^{125}I) or be conjugated to an enzyme (such as bovine alkaline phosphatase); the antibody-bound CK-2 is thus measured by comparing the amount of ^{125}I or alkaline phosphatase activity against a calibration curve. These "sandwich" techniques have a higher analytical sensitivity than methods that measure CK activity and are therefore preferred.

A wide spread of ***reference ranges*** have been obtained by these various immunological assays: for serum CK-3, the values for adult males are 102–1688 μg/L and for adult females, 36–487 μg/L; for CK-2, 2–49 μg/L; and for CK-1, <6.2 μg/L. The detection limit for these assays varies between 0.5 and 10 μg/L and the precision is between 5 and 11%.

LACTATE DEHYDROGENASE

(EC 1.1.1.27; L-Lactate:NAD$^+$ Oxidoreductase; LDH; LD)

Lactate dehydrogenase is a hydrogen transfer enzyme that catalyzes the oxidation of L-lactate to pyruvate with the mediation of NAD$^+$ as hydrogen acceptor [equation (12)]. The reaction is reversible and the equilibrium strongly favors the reverse reaction, namely, the reduction of pyruvate to lactate (P→L).[65]

$$\underset{\textbf{L-Lactate}}{\mathrm{H{-}C(CH_3)(COO^-){-}O{-}H}} + \mathrm{NAD^+} \underset{\text{pH 7.4–7.8}}{\overset{\textit{LDH,}\ \text{pH 8.8–9.8}}{\rightleftharpoons}} \underset{\textbf{Pyruvate}}{\mathrm{CH_3{-}C(=O){-}COO^-}} + \mathrm{NADH} + \mathrm{H^+} \quad (12)$$

The pH optimum for the lactate to pyruvate (L→P) reaction is 8.8–9.8, and an optimal assay mixture at 30 °C contains NAD$^+$, 5 mmol/L, and L-lactate, 50 mmol/L; for the P→L assay, at 30 °C, the pH optimum is 7.4–7.8, and optimal concentrations for reactants are NADH, 150 μmol/L, and pyruvate, 1.5 mmol/L. The optimal pH varies among the different isoenzymes and depends on the temperature as well as on substrate and buffer concentrations. The specificity of the enzyme extends from L-lactate to a variety of related α-hydroxy acids and α-oxo acids; for example, catalytic oxidation of α-hydroxybutyrate to α-oxobutyrate is referred to as α-hydroxybutyrate dehydrogenase (HBDH) activity. LDH does not act on D-lactate, and only NAD^+ will serve as coenzyme. The enzyme has a MW of 134 000 and is composed of four peptide chains (subunits) of two types: M and H, which are determined by loci on human chromosomes 11 and 12, respectively. The subunit compositions of the five isoenzymes, in order of decreasing anodal mobility in an alkaline medium, are LDH-1 (HHHH; H_4); LDH-2 (HHHM; H_3M); LDH-3 (HHMM; H_2M_2); LDH-4 (HMMM; HM_3); LDH-5 (MMMM; M_4). A different sixth isoenzyme, LDH-X composed of four X subunits, is present in the postpubertal human testicle.

Lactate dehydrogenases are inhibited by reagents such as mercuric ions and *p*-chloromercuribenzoate that react with thiol groups; the inhibition is reversed by the addition of thiol reagents such as cysteine or glutathione. Borate and oxalate inhibit by competing with lactate for its binding site on the enzyme; similarly, oxamate competes with pyruvate for its binding site. Both pyruvate and lactate in excess inhibit enzyme activity, although the effect of pyruvate is greater. Inhibition by either substrate is greater for the H form than for the M form, and substrate inhibition decreases with increase in pH. EDTA inhibits the enzyme, perhaps by binding Zn(II); however, the postulated activator role for zinc ions is not fully established.

Distribution and Clinical Significance[110]

LDH content in various tissues. LDH activity is present in all cells of the body and is invariably found only in the cytoplasm of the cell. Enzyme levels in various tissues (in U/g) are very high, compared with those in serum: liver, 145; heart, 124; kidney, 106; skeletal muscle, 147; and erythrocytes (U/g hemoglobin), 36. Thus, tissue levels are about 500-fold higher than those normally found in serum, and leakage of the enzyme from even a small mass of damaged tissue can increase the observed serum level of LDH to a significant extent.

In addition to their higher enzyme concentration, many of these tissues show different isoenzyme composition. In cardiac muscle, kidney, and erythrocytes the electrophoretically faster-moving isoenzymes LDH-1 and LDH-2 predominate, whereas in liver and skeletal muscle the more cathodal LDH-4 and LDH-5 isoenzymes predominate. Isoenzymes of intermediate mobility are found in many tissues, such as endocrine glands, spleen, lung, lymph nodes, platelets, and nongravid uterine muscle.

Serum Levels of Lactate Dehydrogenase in Disease States

Myocardial infarction is usually associated with three- to fourfold elevations of total LDH, but values may be as high as ten times the upper reference limit (see page 384). Values may be moderately elevated in *myocarditis* and in cardiac failure with *hepatic congestion* but are normal in *angina* and in *pericarditis*. Enzyme levels may be moderately elevated in *severe shock* and *anoxia*.

Almost any cause of *hemolysis*, if sufficiently severe, can produce an LDH isoenzyme pattern similar to that seen in myocardial infarction. An increasingly important cause of hemolysis is the presence of a prosthetic heart valve. In *megaloblastic anemias* due to a deficiency of folate or vitamin B_{12}, breakdown of the erythrocyte precursor cells in the bone marrow (ineffective erythropoiesis) causes release of large quantities of LDH-1 and LDH-2 isoenzymes. Elevations of total serum LDH activity up to 50 times the upper reference limit may be observed, but these elevations rapidly return to normal after appropriate treatment.

Elevations of LDH activity are also observed in *liver disease*, but these elevations are not as great as the increases seen in aminotransferase activity. Elevations are especially high (ten times normal) in toxic hepatitis with jaundice; slightly lower values are observed in viral hepatitis and in infectious mononucleosis, which are often associated with elevations of LDH-3 (Figure 12-11). Total LDH activity is normal, or at most twice

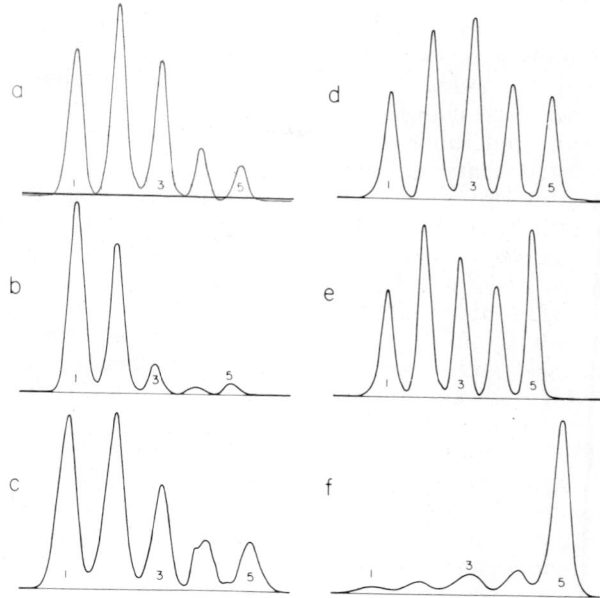

Figure 12-11. Serum LDH isoenzyme patterns obtained with a thin-layer agarose gel electrophoresis system. The isoenzymes were detected by incubating the gel with a liquid overlay of NAD^+ and L-lactate. The generated NADH was detected by fluorescent scanning. Pattern *a*, normal serum. Pattern *b*, acute myocardial infarction (common pattern showing "flipped" LDH-1). Pattern *c*, acute myocardial infarction (showing elevated LDH-1 that has not "flipped"). Pattern *d*, involvement of platelets or lymphatic tissue. This pattern was obtained from a patient with infectious mononucleosis. Pattern *e*, congestive cardiac failure showing elevated LDH-5 as a result of hepatic anoxia. Pattern *f*, acute circulatory shock showing very severe hepatic anoxia.

the upper limit of normal, in cirrhosis and in obstuctive *jaundice*. Serum LDH-5 is often markedly elevated in patients with either *primary liver disease* or *liver anoxia* secondary to decreased oxygen perfusion (Figure 12-11).

Patients with *malignant disease* show increased LDH activity in serum; the LDH pattern usually shows a nonspecific increase in LDH-4 and LDH-5. One unusual exception to this observation is the markedly elevated LDH-1 seen in *germ cell tumors* such as teratomas, seminoma of the testis, and dysgerminoma of the ovary.

Moderately increased LDH activity is found in the sera of all patients affected by *progressive muscular dystrophy*, especially in the early and middle stages of the disease. The increase is confined to LDH-5, the isoenzyme form characteristic of striated muscle. In the later stages of the disease, after a large mass of tissue containing the LDH-5 has been lost, the total LDH levels in serum may even drop to normal levels, with LDH-1 and LDH-2 predominating. Only occasionally is an elevated serum LDH level found in persons with other forms of neuromuscular disorders.

Alpha-Hydroxybutyrate Dehydrogenase[38]

As an alternative to LDH isoenzyme determinations, Rosalki and Wilkinson advocated measurement of *α-hydroxybutyrate dehydrogenase* (HBDH) activity and calculation of the LDH/HBDH ratio. LDH-1 and LDH-2 (the two isoenzyme forms with greatest proportions of monomer H) are relatively more active with α-oxobutyrate as substrate than with pyruvate. In serum from *healthy individuals*, the LDH/HBDH ratio varies from 1.2 to 1.6; in *parenchymal liver disease* with greater LDH-5 activity, the ratio is increased and ranges from 1.6 to 2.5. In *myocardial infarction*, with elevated LDH-1 and LDH-2 activity, the ratio is decreased to between 0.8 and 1.2. The exact ratios will depend on the conditions chosen for the assay of the two activities.

Methods for Lactate Dehydrogenase Activity Measurements[65,92,109]

A multiplicity of procedures have been introduced over the last 25 years; the procedures use the forward (L→P) and reverse (P→L) reactions in almost equal numbers. The L→P assay has the following advantages: substrate inhibition by lactate is less than that produced by pyruvate, NAD^+ preparations used in the L→P reaction appear to contain fewer endogenous LDH inhibitors than NADH preparations used in the P→L reaction, and the reaction linearity of the L→P assay is more prolonged than that of the P→L assay. However, the advantages of the P→L assay include a less expensive assay formulation because of the much lower concentration of reactants; greater change in absorbance with time, thus allowing more precise measurements; and greater stability of the working reagents once they are prepared in assay solutions. Both assays enjoy significant popularity, and reagent kits are widely available. The usual concentrations of the reactants are given on page 380.

Specimens. Serum or heparinized plasma specimens are satisfactory; plasma containing other anticoagulants, especially oxalate, should not be used. Serum or plasma should be separated from the clot as soon as possible after the specimen has been obtained. Hemolyzed serum or plasma must not be used, since erythrocytes contain 150 times more LDH activity (particularly LDH-1 and LDH-2) than serum.

The different isoenzymes vary in their sensitivity to cold; LDH-4 and LDH-5 are especially labile. Loss of activity may be prevented by addition of NAD^+ or glutathione. Both H and M monomers bind a molecule of NAD^+, but the binding of NAD^+ to the M form is weaker and exposes the sulfhydryl groups to oxidation. In serum, the sulfhydryl groups in albumin and other proteins retard inactivation of the M-rich isoenzymes (LDH-4 and LDH-5); therefore, serum specimens should be stored at room temperature, at which no loss of activity will occur for 2–3 d. If serum specimens must be stored for longer periods, they should be kept at 4 °C with NAD^+ (10 mg/mL) or glutathione (3.1 mg/mL) added to retard the inactivation of LDH-4 and LDH-5. Note that there is considerable disagreement over the essential storage conditions for the LDH isoenzymes; therefore, each laboratory should determine suitable storage procedures.

Reference ranges.[105] Values for LDH activity in *serum* vary considerably, depending on the direction of the enzyme reaction and the experimental variables. For the P→L reaction at 30 °C and at pH 7.4, a range of 95–200 U/L represents the experience of most workers. With the Scandinavian assay,[84] reference ranges from 200 to 380 U/L (at 37 °C) are obtained. For the L→P reaction at pH 8.8–9.0 (at 30 °C), the range of 35–88 U/L represents the consensus of values proposed by various investigators. The reference values for *spinal fluid* LDH are 7–30 U/L for the P→L reaction at 30 °C. In *urine*, a reference range of 42–90 U/L has been reported for the L→P reaction at 25 °C.

Methods for Measuring LDH Isoenzymes[38]

Electrophoresis. Electrophoretic separation on agarose gels or cellulose acetate membranes is most commonly used to demonstrate LDH isoenzymes. Many procedures and specialized pieces of equipment are available commercially for this technique. The serum specimen is inserted into a well in the gel surface or is touched onto the surface of the membrane and electrophoresis is performed. After the isoenzymes have been separated by electrophoresis, a reaction mixture is layered over the separation medium. The mixture (typically L-lactate, 500 mmol/L, and NAD^+, 13 mmol/L, often dissolved in a suitable pH 8.0 buffer) may be applied as a liquid or in a gel. The overlay and medium are incubated at 37 °C. The NADH generated over the LDH zones is detected either by its fluorescence when excited by long-wave ultraviolet light (360 nm) or by its reduction of tetrazolium salt (NBT: nitroblue tetrazolium; MTT: 3-(4,5-dimethylthiazolyl-2)-2,5-diphenyl-2H-tetrazolium bromide; INT: 3-(4-iodophenyl)-2-(4-nitrophenyl)-5-phenyl-2H-tetrazolium chloride) to form a colored formazan. Figure 12-11 shows examples of the separations obtained on agarose gel. Ion-exchange chromatographic separation of the LDH isoenzymes has also been used with considerable success.[69]

Reference ranges.[105] With an agarose gel technique and fluorometric densitometry, the following reference ranges for a healthy population (expressed as per cent of total LDH) have been determined: LDH-1, 14–26; LDH-2, 29–39; LDH-3, 20–26; LDH-4, 8–16; LDH-5, 6–16. The LDH-1/LDH-2 ratio is 0.45–0.74. The clinical sensitivity and specificity of this ratio test, as determined in a coronary cardiac care unit, are 100 and 90%, respectively. A cellulose acetate method using NBT staining produced the following reference ranges (as per cent of total LDH): LDH-1, 27–35; LDH-2, 34–44; LDH-3, 16–22; LDH-4, 4–8; LDH-5, 3–7.

Immunoprecipitation. An important recent advance in LDH-1 measurement is an immunoprecipitation technique[108] using goat antiserum against purified rhesus monkey LDH-5. This antiserum binds all human LDH isoenzymes containing one or more

M subunits (i.e., LDH-2 through LDH-5). When added to a patient's serum, the antiserum combines with all LDH isoenzymes except LDH-1. A second antibody, conjugated to polyvinylidene fluoride particles, is added and reacts with the immune complexes of M-containing isoenzymes. The particles are then separated by centrifugation and the supernatant, containing only the LDH-1 fraction, is assayed for LDH activity. LDH-1 results by this method have a high correlation with those obtained by electrophoresis, and the diagnostic sensitivity for myocardial infarction has been reported as 94%. This assay is more convenient and less demanding than electrophoretic and chromatographic assays.

Enzyme Tests in the Diagnosis of Myocardial Infarction[1]

The key roles played by CK, AST, and LDH in the investigation of suspected myocardial infarction have already been mentioned. In the investigation of this disorder, it is customary to measure activities of two or even all three of these enzymes as well as the isoenzymes of CK and LDH. This section considers the changes in activities of these enzymes collectively and discusses the relative advantages and disadvantages of each type of measurement.

The first symptom of a myocardial infarction is usually chest pain (due to ischemia of the cardiac muscle), often severe and frequently described as "crushing" or "tightness." Characteristically, the patient is sweating markedly, and nausea as well as vomiting occur frequently. Blood pressure or heart rhythm may be abnormal, and the patient may be in circulatory shock. The patient may die if the abnormality of heart rhythm or the shock-like state remains untreated; this is the reason for admission to special coronary (cardiac) care units (CCU), where intensive and continuing electocardiographic monitoring and therapy for arrhythmias are readily available. Chest pain, per se, may be due to causes other than myocardial infarction. Among them are aortic dissection, mitral valve prolapse, or pericarditis; pulmonary embolism, pneumothorax, or pneumonia; esophageal spasm or rupture or gastric ulcer; degenerative arthritis of cervical or thoracic vertebrae; and herpes zoster. Chest pain may also have an emotional origin (depression, anxiety, malingering).

About a quarter of all myocardial infarctions may be clinically "silent," associated either with atypical symptoms or with no symptoms at all. Many patients with silent infarcts are diabetic, and their lack of pain is attributed to the autonomic neuropathy that accompanies longstanding diabetes. Often it is difficult, after general surgery, to diagnose a myocardial infarction because of the effect of surgery, the anesthetic, and the patient's antecedent illness. In the elderly, a myocardial infarction often presents with the onset of sudden breathlessness, acute confusion, fainting, or even a stroke.

With many patients, the examining physician is fairly certain after obtaining the patient's history and completing a physical examination that a myocardial infarction has occurred. In these circumstances the existence of either typical electrocardiographic (ECG) changes or typical serum enzyme changes may be sufficient for establishing the diagnosis. The ECG can show changes that are quite characteristic of myocardial infarction (pathologic changes in ST segment and Q- and T-waves), provided the ECG had previously been normal. These specific patterns can sometimes be obscured by previous heart block, cardiac muscle hypertrophy, or infarction, or they may not appear at all (Q waves do not develop in subendocardial infarctions). These facts account for the diagnostic sensitivity of the ECG ranging from only 63 to 82%, although its diagnostic specificity is nearly 100%. If the ECG pattern is equivocal, then the physician must depend on the serum enzyme patterns.

Total Creatine Kinase and CK-2[63,98]

After a myocardial infarction there is an initial period during which all serum enzyme activities remain within their reference ranges. The duration of this phase is about 6 h but may be less, particularly when the infarct is large. Thereafter, activities rapidly rise to peaks whose magnitudes are proportional to the extent of myocardial necrosis (Figure 12-12).

The rise and subsequent fall of both CK-3 and CK-2 is nearly always indicative of

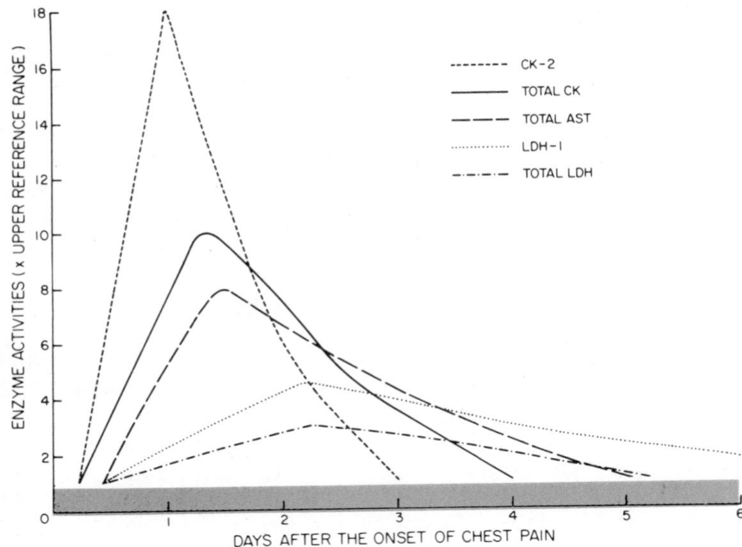

Figure 12-12. Typical pattern of changes in serum enzyme activities following an uncomplicated myocardial infarction. These patterns and the magnitude of the peaks can vary quite markedly among patients, depending on the location and size of the infarct and the enzyme assay methods in use. The area of color indicates the reference range of all enzymes.

a myocardial infarction. The first enzyme to rise is CK-3. The increase is roughly equivalent to the total CK activity, which exceeds the upper limit of the reference range in 6–8 h after the onset of chest pain, reaches a peak between 18 and 30 h, and returns to within the reference range by the third or fourth day. CK-2 rises about the same time but it reaches a peak up to 12 h earlier than CK-3. Because of its shorter half-life in blood, CK-2 returns to preinfarction levels by or before the third day. Serum CK-2 values may be reported as a percentage of total CK activity; preinfarction values are usually <6%, but following an infarction values can increase up to 30% depending on the methods used for analysis, the extent of myocardial damage, and the location of the infarct. Diagnostic sensitivity of total CK elevation is 93–98%, and of CK-2, nearly 100%. Diagnostic specificity for total CK elevation is 75–85%; for CK-2 >6% of total CK, nearly 100%.

Aspartate Aminotransferase

Aspartate aminotransferase (AST) activity rises 6–8 h after the onset of chest pain, peaks at ~18–24 h, and returns to preinfarction levels by the fourth or fifth day. AST has poorer diagnostic performance because it is less cardiac specific. Recall that an AST rise may reflect diseases of liver, lung, and skeletal muscle as well as heart.

Total LDH and LDH-1[60,63]

LDH-1, often paralleled by total LDH activity, rises 8–12 h after the onset of chest pain although it may rise concomitantly with CK-3. The isoenzyme peaks at 24–48 h and returns to preinfarction levels after 7–12 days. LDH-1 is therefore an extremely useful marker of myocardial infarction because of its prolonged presence in serum. The diagnostic sensitivity of LDH-1 estimations in myocardial infarction approaches 90%, and diagnostic specificity is 90–99%. The test sensitivity may be further increased by using the LDH-1/LDH-2 ratio. More than 80% of all myocardial infarctions show, at some time, a "flipped" ratio (i.e., LDH-1 > LDH-2), although greater sensitivity may be attained by comparing the patient's LDH-1/LDH-2 ratio with the laboratory's reference range since the frequency of a flipped ratio in myocardial infarction tends to alter with different methods. A common cause of false-positive results with LDH-1 or LDH-1/LDH-2 determinations is *hemolysis*, which produces exactly the same effect on LDH-1 and LDH-1/LDH-2 values. Hemolysis can usually be ruled out, but hemolysis due to a prosthetic heart valve should always be considered whenever appropriate.

Test Sequence

Appropriate timing of blood sampling for the cardiac enzymes and isoenzymes is essential (Figure 12-12), because the demonstration of both a rise and a subsequent fall in the activities of these enzymes increases the predictive value of these tests for myocardial infarction. Clearly, the most effective test sequence for confirming or excluding a diagnosis of myocardial infarction are serial measurements of CK and LDH isoenzymes. A variety of sampling sequences have been suggested: every 12 h for the first 48 h of hospital admission; three samples within the first 36 h of admission (at admission, 6–12 h later, and 24–36 h later); sampling every 6–12 h (as an aid in the detection of even small subendocardial infarcts); every 8 h for the first 48 h. In practice, use depends on the local availability of service, since isoenzyme tests are rarely available at night—except under special circumstances—and specimens obtained for these tests will be held until the next working day; by contrast, total CK, LDH, and AST estimations are often available around the clock.

A patient with a suspected myocardial infarction is admitted to the CCU and treated as such until infarction can be ruled out. The greater expense of a CCU bed compared with the cost of a bed on a general medical floor mandates that a diagnosis of myocardial infarction be confirmed or excluded as soon as possible. To rule out the diagnosis requires tests with high diagnostic sensitivity. Undoubtedly, CK-2 and, later in time after admission, LDH-1 and the LDH-1/LDH-2 ratio determinations are most useful in this respect. Alternatively, to rule in (confirm) the diagnosis requires tests with high diagnostic specificity; again measurements of CK-2, LDH-1, and the LDH-1/LDH-2 ratio are all valuable. However, the use of these tests to rule-in or rule-out an MI presupposes that the physician is knowledgeable about the need to establish prior decision thresholds for these purposes; in this regard the laboratory can provide valuable advice.

ALKALINE PHOSPHATASE[67]

(EC 3.1.3.1; Orthophosphoric-Monoester Phosphohydrolase (alkaline optimum); ALP)

The isoenzymes of alkaline phosphatase exhibit optimal activity in vitro at a pH of about 10, although the optimal pH varies with the nature and concentration of the substrate acted upon, the type of buffer or phosphate acceptor present, and, to some extent, the nature of the isoenzymes. Alkaline phosphatases act on a large variety of naturally occurring and synthetic substrates, but the natural substrates on which they act in the body are not known. Some divalent ions such as Mg(II), Co(II), and Mn(II) are activators of the enzyme, and Zn(II) is a constituent metal ion. The correct ratio of Mg(II)/Zn(II) ions is necessary to obtain optimal activity. Phosphate, borate, oxalate, and cyanide ions are inhibitors of all forms of the enzyme. The individual multiple forms are inhibited to different degrees by L-phenylalanine, urea, excess Zn(II), or AsO_4^{3-}. Variations in Mg(II) and substrate concentrations change the pH optimum. The type of buffer present may affect the rate of enzyme activity. Buffers can be classified as inert (carbonate; barbital), inhibiting (glycine; propylamine), or activating (2-methyl-2-aminopropanol-1, AMP; tris(hydroxymethyl)aminomethane, Tris; and diethanolamine, DEA).

Alkaline phosphatase is present in practically all tissues of the body especially at or in the cell membranes, and it occurs at particularly high levels in intestinal epithelium, kidney tubules, bone (osteoblasts), liver, and placenta. Although the precise metabolic function of the enzyme is not yet understood, the enzyme is closely associated with the calcification process in bone. Alkaline phosphatase displays considerable inter- and intratissue heterogeneity, but there are rarely more than two or three forms in any one serum specimen. The forms present in sera from patients with various diseases have the characteristics of the specific forms present in liver, bone, intestinal, placental, or, very rarely, renal tissue. The forms present in the sera of normal adults originate in the liver or the biliary tract, and up to half the total activity comes from the skeleton. The relative contributions of bone and liver isoenzymes to the total activity are markedly age-dependent. There is also a significant difference between the sexes in serum alkaline phosphatase activity at some ages, although this is less important for interpretation of values than the dependence on age. A small amount of intestinal alkaline phosphatase may also be present, particularly in the sera of individuals of blood groups B or O who are secretors of blood-group substances. The enzyme found in urine is probably derived from renal tissue.

Clinical Significance[67]

Serum alkaline phosphatase measurements are of particular interest in the investigation of *hepatobiliary disease* and *bone disease* associated with increased osteoblastic activity.

The response of the liver to any form of *biliary obstruction* is to synthesize more alkaline phosphatase; i.e., the effect is one of enzyme induction. The main sites of new enzyme syntheses are the hepatocytes adjacent to the biliary canaliculi. Some of the newly formed enzyme enters the circulation to raise the enzyme level in serum. The elevation tends to be more marked (greater than threefold) in *extrahepatic obstruction* (e.g., by stone or by cancer of the head of the pancreas) than in intraheptic obstruction and is greater the more complete the obstruction. Serum enzyme activities may reach 10–12 times the upper limit of normal and may return to normal on surgical removal of the obstruction. *Intrahepatic obstruction* of the bile flow (e.g., by invading cancer tissue or by drugs such as chlorpromazine that affect the biliary tree) also raises serum alkaline phosphatase, but usually to a lesser extent than extrahepatic obstruction. Liver diseases that principally affect parenchymal cells, such as *infectious hepatitis*, typically also show only moderately (less than threefold) elevated or even normal serum alkaline phosphatase levels. These distinctions between extra- and intrahepatic obstructive jaundice and parenchymal jaundice are nevertheless subject to exceptions in individual instances.

Among the *bone diseases*, the highest levels of serum ALP activity are encountered in Paget's disease (osteitis deformans) as a result of the action of the osteoblastic cells as they try to rebuild bone that is resorbed by the uncontrolled activity of osteoclasts. Values from 10 to 25 times the upper limit of normal are not unusual. Levels are generally normal in osteoporosis. Moderate rises are observed in osteomalacia. In rickets, levels two to four times the normal may be observed, and these fall slowly to normal upon treatment with vitamin D. Slight to moderate elevations are seen in *Fanconi's syndrome*. *Primary* and *secondary hyperparathyroidism* may be associated with slight to moderate elevations of alkaline phosphatase activity in serum, reflecting the presence and extent of skeletal involvement. Very high enzyme levels are present in patients with *osteogenic bone cancer*. Transient elevations may be found during healing of *bone fractures*. Physiological *bone growth* elevates alkaline phosphatase in serum to 1.5–2.5 times that present in normal adult serum.

An increase of up to two to three times normal may be observed in women in the the third trimester of *pregnancy*; however, levels may not exceed the upper limit of the reference range in some cases. The additional enzyme is of placental origin. Although upward or downward trends in placental ALP levels may presage complications of pregnancy, such as *hypertension* or *pre-eclampsia*, single estimations are of little diagnostic value.

Forms of the enzyme that are essentially identical with the normal placental isoenzyme may appear in the sera of some patients with malignant diseases. These carcinoplacental or "Regan" isoenzymes appear to result from the derepression of the placental phosphatase gene. In some cases, modification of the gene or its product may cause the appearance of isoenzymes that differ in some respects from the normal placental isoenzyme while still retaining its general characteristics. Tumors may also produce alkaline phosphatases that appear to be modified forms of nonplacental isoenzymes.

Methods for the Determination of Alkaline Phosphatase Activity[107]

The most widely used substrate for assay of alkaline phosphatase is 4-nitrophenyl phosphate (usually abbreviated to 4-NPP or PNPP from the older name, para-nitrophenyl phosphate). This ester is colorless, but 4-nitrophenol (4-NP), the product of enzyme action, is colored at the pH of the reaction; thus the enzyme reaction can be followed continuously by observing the rate of formation of yellow color [equation (13)]. This reaction forms the basis of current recommended and standard methods of alkaline phosphatase assay.

Other self-indicating substrates include phenolphthalein monophosphate, thymolphthalein phosphate, and α-naphthyl phosphate. However, methods based on these substrates are used relatively little. Fixed-time methods for alkaline phosphatase assay are still carried out in considerable numbers because of their suitability for automatic analyzers. 4-NPP is increasingly used as the substrate in these assays also.

4-Nitrophenyl phosphate (colorless) + HOH $\xrightarrow[\text{pH 10.3}]{ALP,\ Mg^{2+}}$ HPO_4^{2-} + 4-Nitrophenoxide (colorless benzenoid form) $\xrightarrow{\text{Rearranges at alkaline pH}}$ 4-Nitrophenoxide (yellow, quinonoid form)

4-Nitrophenyl phosphate (colorless)

4-Nitrophenoxide (colorless benzenoid form)

4-Nitrophenoxide (yellow, quinonoid form)

In all older alkaline phosphatase methods the liberated phosphate group is transferred to water; i.e., the reaction is hydrolytic. The rate of phosphatase action is much enhanced, however, if certain amino alcohols are used as buffers. Among these apparent activators are compounds such as 2-amino-2-methyl-1-propanol (AMP), diethanolamine (DEA), and tris(hydroxy)methylaminomethane (Tris). These materials act as buffers by binding protons at the nitrogen atom and as phosphate acceptors by reacting with phosphate through the hydroxyl group. Enzyme activity in the presence of optimal concentrations of these buffers can be two to six times greater than in the presence of a nonactivating buffer such as carbonate. The ALP methods outlined here use such phosphate-accepting buffers.

Determination of Alkaline Phosphatase Activity by Continuous-Monitoring Procedures

4-Nitrophenyl phosphate is now almost invariably chosen as the substrate in reagent kits. However, these formulations, and the national and international recommendations on which they are based, differ in the choice of buffer. The Expert Panel on Enzymes of IFCC and some other authorities have expressed a preference for AMP.[25,107]

In a typical method the reaction mixture has the composition: 4-NPP, 14 mmol/L; AMP, 750 mmol/L; Mg(II), 0.1 mmol/L; pH 10.30 at 30 °C. Production of 4-NP is monitored at 405 nm in a spectrophotometer with a thermostat in the cuvet compartment. However, the method can readily be adapted to a two-point procedure in which the reaction is allowed to proceed for a predetermined time (e.g., 15 min) and is then stopped by addition of 10 mL NaOH, 0.05 mol/L, to a 1-mL reaction mixture. The added alkali raises the pH to 11.5–12.0, which inactivates the enzyme, converts 4-NP to the colored quinonoid form, and dilutes the yellow color to a measurable intensity.

Specimens. Serum or heparinized plasma, free of hemolysis, should be used. Complexing anticoagulants such as citrate, oxalate, and EDTA must be avoided. Freshly collected serum samples should be kept at room temperature (20–26 °C) and assayed as soon as possible, but preferably not later than 4 h after collection. Sera kept at room temperature usually show a slight but real increase in activity, which varies from 1% over a 6-h period to 3–6% over a 1- to 4-d period. Even in sera stored at refrigerator temperature, activity increases slowly (2%/d). Freezing causes a loss of activity, which is slowly recovered after the sera are thawed. A similar enhancement of activity, but of greater magnitude, occurs with reconstituted lyophilized control sera or reference materials. Frozen specimens should be thawed and kept at room temperature for 18–24 h before measurement to achieve full enzyme reactivation. In reconstituted materials stored at 37 °C, the increase in activity of some materials can be as high as 50–100%

over a 24-h period, and the increases with storage at 4 °C and 20 °C are about 10 and 30%, respectively. Enhancement of activity continues for several days but at a decreasing rate. The cause of this phenomenon is not known.

Reference ranges (30 °C; AMP buffer) are <350 U/L at 1–12 years of age. In females older than 15 years, values are 25–100 U/L, whereas values for males are <500 U/L at 12–15 years of age and 25–100 U/L for those over 16 years of age. When enzyme activity at 30 °C is taken as 1.0, relative enzyme activities have been reported as 0.76 at 25 °C and 1.47 at 37 °C. The Q_{10} value is 1.7.

IFCC and AACC Method

The method developed by the Expert Panel on Enzymes of the IFCC and the Alkaline Phosphatase Study Group of AACC uses 4-NPP as substrate and AMP as phosphate-acceptor buffer. The concentrations of Mg(II) and Zn(II), excesses of which are inhibitory, are controlled at optimal levels by addition of Mg(II) and Zn(II) and of the chelating agent, *N*-hydroxyethylethylenediaminetriacetic acid (HEDTA). Although Zn(II) is present in a total concentration of 1 mmol/L, most is bound to HEDTA, leaving only a small, experimentally determined, optimal concentration of free ions. A similar situation exists for Mg(II). Thus, HEDTA acts as a metal ion buffer, maintaining optimal concentrations of the two ions. The temperature range of measurement is 30 ± 0.1 °C, and the acceptable wavelength range is 405 ± 2 nm.

Optimal conditions for measurement confirmed by the techniques of response surface methods are as follows: 2-amino-2-methyl-1-propanol (AMP), 0.35 mol/L; 4-nitrophenyl phosphate, 16.0 mmol/L; magnesium acetate, 2.0 mmol/L; zinc sulfate, 1.0 mmol/L; *N*-hydroxyethylethylenediaminetriacetic acid (HEDTA), 2.0 mmol/L; and volume fraction of sample, 0.0196 (1:51).

Reference ranges. By this method, the following reference ranges in U/L have been established for adults in the fasting state: males, 35–94; females, 25–80. Values for females older than 60 years of age increase to 40–98.

Multiple Forms of Alkaline Phosphatase[65]

Alkaline phosphatases prepared from different human tissues are not identical in all properties, although they are similar in many respects. The alkaline phosphatase activity in tissues such as liver, bone, and kidney seems to be due almost entirely to the presence of a form of the enzyme that is characteristic of that tissue. Criteria that have been used to differentiate the isoenzymes of alkaline phosphatase include differences in relative rates of reaction with various substrates, differences in response to selected inhibitors, variations in stability to denaturation by heat or urea, and differences in electrophoretic mobility and immunochemical characteristics. The tissue-specific characteristics of the alkaline phosphatase isoenzymes are retained when the enzymes are released into the circulation and can be used to identify the tissue responsible for the elevation of the serum level of the enzyme. However, the differences in properties between alkaline phosphatase isoenzymes are in some cases quite small (this is particularly true for bone and liver phosphatases), so that a combination of techniques may be needed to ensure reliable discrimination.

Electrophoretic separation. When serum specimens are separated by electrophoresis at alkaline pH, liver phosphatase moves most rapidly toward the anode (Figure 12-13). Bone phosphatase, which typically gives a more diffuse zone than the liver isoenzyme, has a slightly lower anodal mobility, although the two zones usually overlap to some extent. Intestinal phosphatase also migrates diffusely but more slowly than the bone enzyme, and kidney phosphatase, which occurs very rarely in serum, migrates even more slowly. The placental isoenzymes have mobilities of the same order as those of liver and bone, depending on the phenotype.

Additional minor phosphatase zones are also present in tissue extracts and occasionally in serum. One such zone, named the "fast liver" fraction, migrates more anodally than the main liver zone on media such as cellulose acetate and has been observed more frequently in hepatobiliary disease. Complexes between alkaline phosphatase and

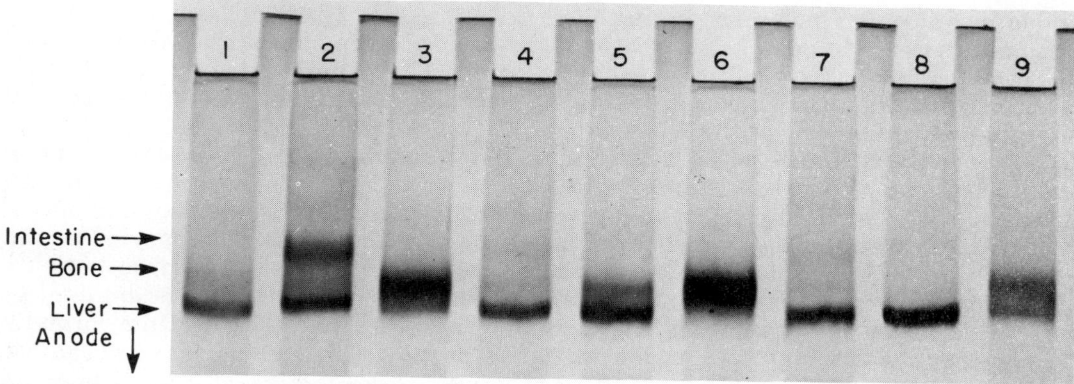

Figure 12-13. Alkaline phosphatase zones in human serum separated by vertical polyacrylamide gel electrophoresis at pH 9.5. The compact zone of liver phosphatase is prominent in (from the left) samples 1, 2, 4, 7, and 8. The more diffuse, less anodal zone of bone phosphatase can be seen in samples 3 and 6. Bone and liver phosphatase zones are both present in samples 5 and 9. Sample 2 displays a prominent zone of intestinal phosphatase, migrating more slowly than either the liver or bone forms. The anodal direction is downward.

immunoglobulins occur occasionally in serum, giving rise to abnormally migrating zones; however, in the present state of knowledge, these do not provide specific diagnostic information.

Although the difference in electrophoretic mobility between bone and liver phosphatase zones is small, it is sufficient to distinguish between these two tissues as the possible source of increased phosphatase levels (hyperphosphatasemia). However, it is advisable to run reference sera (markers) of known isoenzyme composition, e.g., specimens from cases of proven bone or liver disease, along with the unknown specimens on the same support medium at the same time, e.g., on gel slabs rather than in individual tubes. It is also useful to compare the electrophoretic patterns obtained on specimens before and after they have been subjected to heat denaturation, inhibition, or neuraminidase treatments (see below).

Recently, two methods have been reported to greatly improve the electrophoretic separation between bone and liver alkaline phosphatases and allow quantitative estimations to be made by densitometric scanning. Both methods exploit differences in the carbohydrate portions of the two forms of alkaline phosphatase. In one method, electrophoretic separation is carried out in the presence of wheat-germ lectin, which retards bone alkaline phosphatase to a greater extent than the liver enzyme.[89] An alternative method is to treat the serum *briefly* with neuraminidase.[77] Brief treatment with neuraminidase removes sialic acid residues of bone phosphatase more readily than those of liver phosphatase, so that the mobility of bone phosphatase is reduced more than that of liver phosphatase. Treatment with 1 U of neuraminidase per 10 μL of serum for 10 min at 37 °C gives patterns in which the bone and liver alkaline phosphatase zones are sufficiently well separated to allow densitometric scan of the pattern. Measurements of liver phosphatase made in this way agree well with results of measurements by a quantitative, selective heat-inactivation technique and show good reproducibility.

Starch, agar, and polyacrylamide gels, as well as cellulose acetate, have been used for the separation of alkaline phosphatase isoenzymes.[75] The choice of a particular system depends on the resources of individual laboratories, but it is important that the chosen method be used regularly and consistently to ensure repeatable and reliable results. Careful attention to cooling (e.g., by carrying out the procedure in a cold room) is needed to minimize enzyme inactivation. After electrophoresis, alkaline phosphatase zones are made visible by incubating the gel in a solution of buffered substrate to which a chromogenic system is added. A suitable staining mixture consists of 1-naphthyl phosphate (50 mg) and Fast Blue BB salt (70 mg) in either Tris-borate-$MgCl_2$ buffer (70 mL) at pH 9.5 or carbonate-bicarbonate buffer at pH 10.0.

When electrophoresis suggests the presence of intestinal phosphatase, confirmatory evidence can be obtained by repeating the separation after *prolonged* incubation of the sample with neuraminidase. This treatment, by completely removing negatively charged sialic acid residues, reduces the anodal mobility of all phosphatase isoenzymes except intestinal phosphatase, which is resistant to neuraminidase since terminal sialic acid residues are not present in the molecule.

Heat-inactivation.[78,79] The most remarkable property of placental alkaline phosphatase is its pronounced resistance to heat. Incubation of the enzyme at a temperature as high as 65 °C for 30 min therefore provides a convenient test for the presence of this isoenzyme. Pronounced heat stability is also shown by the "Regan" isoenzyme, a placental-like form of alkaline phosphatase that occurs in 5–15% of specimens from patients with cancers of various types. Thus, heating the sample at 65 °C can also be used to detect this abnormal isoenzyme. The similarity between these two enzyme forms is evidenced by the fact that the Regan isoenzyme will react with antibodies raised against the placental isoenzyme.

Other isoenzymes of alkaline phosphatase can also be differentiated on the basis of their stability at temperatures lower than 65 °C. At 56 °C, for example, liver ALP is more stable than the bone isoenzyme. Thus, if $<20\%$ of the serum alkaline phosphatase activity remains after 10 min at 56 °C, the enzyme present is probably largely of bone type; residual activities of between 25 and 55% support electrophoretic evidence that liver phosphatase is the predominant isoenzyme. For this test to be meaningful, very careful control of the experimental conditions is essential because the temperature coefficients for inactivation of enzymes by heat are high. At the end of the heating period, specimens must be immediately immersed in ice water.

Procedure. The following procedure[79] for heat inactivation of alkaline phosphatase in serum has been found to give reproducible results, provided the precautions mentioned above are followed carefully:

1. Place ~0.5 mL of serum in a small, thin-walled glass tube (Dreyer tube) and seal with plastic film (Parafilm).
2. Place the tube into a thermostatically controlled water bath that is already stabilized at 56 °C, ensuring that the surface of the serum in the tube is below the level of the water, and simultaneously start the stopwatch. The water bath should have good temperature stability, be of large volume, and be well stirred to minimize temperature fluctuations on introducing the specimens.
3. After *exactly* 10 min, remove the serum tube and rapidly place it in ice water.
4. Determine the alkaline phosphatase activity of the heated specimen; express it as a percentage of the activity of an unheated portion of the same serum that has been kept in ice water but not frozen.

Residual activities of 20% or less suggest that the predominant isoenzyme is of the bone type. Values between 25 and 55% are associated with sera in which the predominant isoenzyme(s) is of liver or intestinal origin, or both. If the presence of placental or Regan isoenzymes is suspected, the incubation should be done at 65 °C for 30 min. Any remaining activity is due to one of these two isoenzymes.

Chemical inhibition. *Urea* at high concentrations inhibits the enzyme. Inhibition is irreversible and varies with the tissue of origin of the ALP. This urea inhibition provides an alternative to heat inactivation in isoenzyme analysis. The bone enzyme is most readily inactivated, liver phosphatase has intermediate resistance, and placental phosphatase is most resistant to urea.

Specific chemical inhibitors have also been applied to the characterization of alkaline phosphatase isoenzymes in serum. L-*Phenylalanine* markedly inhibits intestinal, placental, and Regan isoenzymes when present at a concentration of 5 mmol/L, but it has less effect on the isoenzymes of bone or liver. Levamisole preferentially inhibits liver and bone alkaline phosphatases and is effective at much lower concentrations.

Immunochemical techniques. Monospecific antisera to placental and intestinal alkaline phosphatase have been prepared, and immunochemical methods provide the best measurements of these isoenzymes. However, polyclonal antisera to either liver or bone phosphatases cross-react completely.

GAMMA-GLUTAMYLTRANSFERASE[33,88]

(EC 2.3.2.2; 5-Glutamyl-Peptide:Amino-Acid 5-Glutamyltransferase; γ-Glutamyltransferase; GGT)

Peptidases are enzymes that catalyze the hydrolytic cleavage of peptides to form amino acids or smaller-sized peptides, or both. They constitute a broad group of enzymes of varied specificity, and some individual enzymes catalyze the transfer of amino acids from one peptide to an amino acid or another peptide; i.e., they act as amino acid transferases. Included among the latter is γ-glutamyltransferase (GGT), which transfers the γ-glutamyl group from peptides and other compounds that contain it to an acceptor. The glutamyl acceptor can be the substrate itself, some amino acid or peptide, or even water, in which case a simple hydrolysis takes place. The enzyme acts only on peptides or peptide-like compounds containing a terminal glutamate residue joined to the remainder of the compound through the terminal (5- or γ-) carboxyl [see equation (14)]. This is the reaction used in measuring enzyme activity.

$$O{=}C(NH{-}C_6H_4{-}NO_2){-}{}_{\gamma}CH_2{-}{}_{\beta}CH_2{-}{}_{\alpha}CHNH_2{-}COOH + H_2N{-}CH_2{-}C{=}O{-}NH{-}CH_2{-}COOH \xrightarrow{GGT,\ pH\ 8.2} O{=}C(NH{-}CH_2{-}C{=}O{-}NH{-}CH_2{-}COOH){-}CH_2{-}CH_2{-}CHNH_2{-}COOH + NH_2{-}C_6H_4{-}NO_2 \qquad (14)$$

γ-Glutamyl-*p*-nitroanilide Substrate (donor) + **Glycylglycine** Acceptor ⇌ **γ-Glutamylglycylglycine** Transfer product + ***p*-Nitroaniline** Donor residue

GGT is present in serum and in all cells except those in muscle. It is predominantly located in the cell membrane and may act to transport amino acids and peptides into the cell in the form of γ-glutamyl peptides. It may also be involved in some aspects of glutathione metabolism.

Clinical Significance

GGT present in serum appears to originate primarily from the hepatobiliary system, and GGT activity is elevated in all forms of liver disease. It is highest in cases of *intra-* or *posthepatic biliary obstruction*, reaching levels 5–30 times normal. GGT is more sensitive than alkaline phosphatase in detecting obstructive jaundice, cholangitis, and cholecystitis, and its rise occurs earlier and persists longer. Only moderate elevations (2–5 times normal) are seen in *infectious hepatitis*, and in this condition GGT determinations are less useful diagnostically than are measurements of the transaminases. High levels of GGT are also seen in patients with either primary or secondary (metastatic) liver cancer; again changes occur earlier and are more pronounced than those with the other liver enzymes. Small increases (2–5 times normal) of GGT activity are observed in patients with *fatty livers*, and similar but transient increases are seen in cases of drug intoxication. In acute and chronic *pancreatitis* and in some *pancreatic malignancies* (especially if associated with hepatobiliary obstruction), enzyme activity may be 5–15 times the upper limit of normal. In summary, GGT is the most sensitive enzymatic indicator of hepatobiliary disease available at present; normal values are rarely found in the presence of liver disease. However, GGT is of little value in attempting to discriminate between different kinds of liver disease. *Normal levels* of the enzyme are seen in cases of *skeletal disease*, in *children* older than one year of age, and in healthy *pregnant women*—conditions in which alkaline phosphatase is elevated. Thus, measurement of GGT levels in serum can be used to ascertain whether observed elevations of ALP are due to skeletal disease or reflect the presence of hepatobiliary disease.

Increased levels of GGT are seen in the sera of heavy drinkers and of patients with alcoholic cirrhosis. In patients receiving anticonvulsant *drugs* such as phenytoin and phenobarbital, increased levels of the enzyme in serum may reflect induction of new enzyme activity and the toxic effects of alcohol and other drugs on microsomal structures in liver cells. Hepatic complications occurring in cystic fibrosis also lead to elevations of GGT.

High levels of GGT are present in the *prostate*, which may account for the fact that the activity of GGT in sera of males is ~50% higher than that in females. Prostatic malignancy may at times be the source of an elevated GGT activity in serum. The *irradiation of tumors* in cancer patients may be accompanied by a rise in GGT activity. However, in malignant disease in general, an increased serum GGT activity must arouse suspicion that the disease is metastatic to the liver.

Specimens. Serum free from hemolysis is the preferred specimen, but EDTA-plasma (up to 1 mg/mL blood) can be used. Heparin produces turbidity in the reaction mixture; citrate, oxalate, and fluoride depress activity by 10–15%.

Methods for the Determination of Gamma-Glutamyltransferase[17,45]

GGT assays were based on the use of L-γ-glutamyl-*p*-nitroanilide as the donor substrate and glycylglycine as the glutamyl acceptor [equation (14)]. This combination is still in widespread use. However, L-γ-glutamyl-*p*-nitroanilide has limited solubility in the reaction mixture, and it is therefore difficult to obtain saturating substrate concentrations. More concentrated solutions, used to initiate the reaction, can be prepared in dilute HCl. The *p*-nitroaniline produced in the reaction is determined by its yellow color, which is monitored at 405 nm.

An alternative procedure uses L-γ-glutamyl-3-carboxy-4-nitroanilide as substrate. This compound is readily soluble in water, and activities in serum are higher with the carboxyl derivative than with the noncarboxylated substrate; this is partly due to the higher substrate concentrations attainable with the former. Reagent kits using either of these substrates are available.

An IFCC reference method for GGT is at present available as a consultative document.[45] It proposes the use of L-γ-glutamyl-3-carboxy-4-nitroanilide as the donor substrate and retains glycylglycine as the acceptor and also as buffer. The conditions in the reaction mixture are: temperature, 30 °C; pH, 7.90; L-γ-glutamyl-3-carboxy-4-nitroanilide, 6 mmol/L; glycylglycine, 150 mmol/L; and volume fraction of serum, 0.091. The wavelength of measurement is 410 nm, at which the product, 5-amino-2-nitrobenzoate, has an absorption coefficient of 7.908 $L \times mmol^{-1} \times cm^{-1}$.

Typical upper ***reference range*** limits using the *p*-nitroanilide substrate at 30 °C are <40 U/L for males and <25 U/L for females; at 37 °C they are <50 U/L for males and <30 U/L for females.

Digestive Enzymes of Pancreatic Origin

Assays of serum amylase, lipase, and trypsin are almost exclusively applied to the investigation of pancreatic disease.[32,70] Although pancreatic function is treated in Chapter 23, a brief outline is included here to orient the reader to clinical applications for individual digestive enzymes.

Pancreatitis is classified as acute or chronic by clinical or pathological criteria. Recovery from an attack of acute pancreatitis leaves the patient asymptomatic. Chronic pancreatitis leaves the patient with persistent pain or evidence of endocrine or exocrine insufficiency of the pancreas. *Relapsing* is a term characterizing recurrent attacks of either acute or chronic pancreatitis. The milder form of acute pancreatitis is edematous; mortality in the edematous form, though significant, is much less than in the hemorrhagic form, where it has been reported to be from 20 to 50%. Either ascites or pleural effusion (usually left-sided), or both, are frequent complications of acute pancreatitis. Persistent ascites may cause elevation of digestive enzymes in serum as a result of absorption from the ascitic fluid into the circulation. Some 50% of all cases of acute pancreatitis go on to form a pseudocyst. The pseudocyst (i.e., not contained by a true capsule) is a collection of enzyme-rich pancreatic fluid encapsulated by inflammatory tissue and occurs as a consequence of autodigestion and liquefaction of pancreatic tissue after hemorrhage and necrosis. Pseudocyst formation is suspected if, one week after onset of an attack, clinical improvement has not occurred. In 50–85% of patients with pseudocyst, serum levels of digestive enzymes are increased.

In *chronic pancreatitis*, marginal elevations of serum digestive enzymes may be observed during attacks when the disease is in its early stages; eventually severe destruction of acinar tissue reduces the amount of enzymes released into the circulation, and normal or even low levels of serum enzymes can be expected as the disease progresses. *Trauma to the pancreas*—as from automobile accidents, physical assaults, surgical procedures, and diagnostic radiologic procedures such as endoscopic retrograde cholangiopancreatography (ERCP)—may cause transient or persistent elevation of serum enzymes, depending on the severity of the injury. *Carcinoma* of the pancreas with obstruction of the pancreatic duct makes elevation of serum enzymes likely; in the absence of obstruction, the carcinoma is "silent" in terms of altered serum enzyme levels or may even be associated with decreased serum enzyme levels.

AMYLASE

(EC 3.2.1.1; 1,4-α-D-Glucan Glucanohydrolase)

Amylases are a group of hydrolases that split complex carbohydrates constituted of α-D-glucose units linked through carbon atoms 1 and 4 located on adjacent glucose residues. Both straight-chain (linear) polyglucans such as amylose and branched polyglucans such as amylopectin and glycogen are hydrolyzed, although at different rates. In the case of amylose, the enzyme splits the chains at alternate α-1,4-hemiacetal (—C—O—C—) links, forming maltose and some residual glucose; maltose, glucose, and a residue of limit dextrins are formed if branched-chain polyglucans are used as substrate. The α-1,6-linkages at the branch points are not attacked by the enzyme.

Two types of amylases are recognized. *Beta-amylase* (e.g., plant and bacterial exoamylase) acts only at the terminal reducing end of a polyglucan chain; it splits off two glucose units (maltose) at a time. Human amylases are α-amylases; they are also called endoamylases because they can attack the α-1,4-linkage in a random manner anywhere along the polyglucan chain. Large polysaccharide molecules are thus rapidly broken down into small units (e.g., dextrins, maltose, and some glucose units). Since both maltose and glucose are reducing sugars, the course of the hydrolytic reaction is paralleled by an increase in soluble reducing materials.

Amylase in human serum has a moderately sharp pH optimum at 6.9–7.0. The enzyme is customarily assayed at 37 °C, although it is active at 50 °C, and some automated procedures employ this higher temperature. The temperature coefficient is 6%/°C increase in temperature ($Q_{10} = 1.6$). Alpha-amylases are calcium metalloenzymes, and calcium is absolutely required for functional integrity; saturation with Ca(II) is obtained at ~1.0 mmol/L. Full activity is displayed only in the presence of anions such as chloride, bromide, nitrate, cholate, or HPO_4^{2-}. The first two anions are the most effective activators; for example, optimal activity is obtained at chloride ion concentrations of 10 mmol/L. The amylases normally occurring in human plasma are small molecules with molecular weights varying from 40 000 to 50 000. The enzyme is thus small enough to pass through the glomeruli of the kidney, and amylase is the only plasma enzyme normally found in urine. Serum and urine amylases migrate electrophoretically with the β- and γ-globulins.

Sources of amylase. Amylase is present in a number of organs and tissues. The greatest concentration is present in the pancreas, where the enzyme is synthesized by the acinar cells and then secreted into the intestinal tract by way of the pancreatic duct system. The salivary glands also secrete a potent amylase to initiate hydrolysis of starches while the food is still in the mouth and esophagus. The action of the salivary enzyme is terminated by acid in the stomach. At the alkaline conditions in the duodenum, pancreatic and intestinal amylase act on dietary polyglucans to produce maltose. Intestinal maltase then further hydrolyzes maltose to glucose, which is absorbed into the blood stream. Most of the pancreatic amylase is destroyed by trypsin activity in the lower portions of the intestinal tract, although some amylase activity is present in feces.

Amylase activity is found in extracts from semen, testes, ovaries, fallopian tubes, striated muscle, lung, and adipose tissue. The enzyme is also found in colostrum, tears, and milk. There is little or no amylase activity in the liver. Tumors of lung and ovary may contain considerable amylase activity, and ascitic and pleural fluids may contain amylase as a result of the presence of a tumor or pancreatitis. The enzyme present in normal serum is predominantly of pancreatic (P-type) and salivary gland (S-type) origin, and the enzyme found in urine is derived from the plasma. Serum amylase activity is not markedly diminished after pancreatectomy. Thus, the remaining activity is derived from the salivary glands and perhaps from some as yet unidentified source (see also later section, *Amylase Isoenzymes*).

Macroamylases[26] are sometimes present in sera; these rare forms are probably complexes between ordinary amylase (usually S-type) and IgA, IgG, or other normal or abnormal high-molecular-weight plasma proteins. Macroamylases cannot be filtered through the glomeruli of the kidney because of their large size (MW >200 000) and are thus retained in the plasma, where their presence may increase amylase activity to some six- to eightfold over that observed in health. In macroamylasemia, amylase activity in the urine is lower than normal, since less amylase is cleared by the kidneys. No clinical symptoms are associated with this disorder; however, many of these cases have been detected after investigation of abdominal pain.

Clinical significance.[91] Assays of amylase activity in serum and urine are used in the diagnosis of diseases of the pancreas and in the investigation of pancreatic function. In *acute pancreatitis*, there is a transient rise in serum amylase activity within 2–12 h of the onset with a peak at 12–48 h; return to normal levels occurs by the third or fourth day. There is usually a four- to sixfold elevation in amylase activity above the reference limit (although a significant number of subjects show a lesser elevation and sometimes none). The magnitude of the elevation of serum enzyme activity is not related to the severity of pancreatic involvement; however, the greater the rise, the greater the probability of acute pancreatitis. In acute pancreatitis associated with hyperlipemia, serum amylase activity may be normal. A significant amount of the serum amylase is excreted in the urine and, therefore, elevation of serum activity is reflected in the rise of urinary amylase activity. Urine amylase, as compared with serum amylase, appears to be more frequently elevated, it reaches higher levels, and its elevations persist for longer periods.

Amylase assays are also of value in detecting the development of complications, such as *pseudocyst*, *ascites*, and *pleural effusion* following acute pancreatitis. Amylase activity in pleural fluid, after an attack, can be 100 times the serum reference range. In an *acute relapse of chronic pancreatitis*, the serum amylase may be minimally elevated; more often the activity remains within the reference range.

The diagnosis of acute pancreatitis is sometimes difficult to make, since it must be differentiated from other acute intra-abdominal disorders (see *Lipase*, page 399) and since an increase in serum amylase activity may not necessarily be due to pancreatitis. Table 12-4 lists some of the many causes of hyperamylasemia and hyperamylasuria. The organ source can sometimes be identified by determining whether the major isoenzyme present is of the P-type or S-type. In some important groups of *nonpancreatic disorders* (e.g., some neoplasms), reasons for hyperamylasemia are not known.

Salivary gland lesions produce a significant S-type hyperamylasemia. Mumps (infective parotitis) and maxillofacial surgery can cause a twofold elevation, whereas salivary gland irradiation can produce a transient 9- to 18-fold elevation of serum amylase activity. *Biliary tract disease* such as cholecystitis and a variety of *intra-abdominal events* can cause up to fourfold elevations of the serum amylase activity, due to either primary or secondary pancreatic involvement. *Postoperative hyperamylasemia* occurs in about 20% of all patients subjected to a wide variety of surgical interventions, including extra-

Table 12-4. CAUSES OF HYPERAMYLASEMIA AND HYPERAMYLASURIA*

Pancreatic disease (P-type↑)
Pancreatitis
Acute
Chronic
Complications
Pseudocyst
Ascites and pleural effusion
Abscess
Pancreatic trauma, including investigative maneuvers
Pancreatic carcinoma
Disorders of nonpancreatic origin (mechanism unknown)
Renal insufficiency (mixed↑)
Neoplastic hyperamylasemia—usually bronchogenic or ovarian (usually S-type↑)
Salivary gland lesions, e.g., mumps, calculus disease (S-type↑)
Macroamylasemia (predominantly S-type)
Disorders of complex origin (mechanism unknown or uncertain)
Biliary tract disease
Intra-abdominal disease (other than pancreatic diseases; see above)
Perforated peptic ulcer (P-type↑)
Intestinal obstruction (P-type↑)
Mesenteric infarction (P-type↑)
Peritonitis (mixed↑; depends on cause)
Acute appendicitis
Ruptured ectopic pregnancy (S-type↑)
Aortic aneurysm with dissection
Cerebral trauma (type depends on other organ damage)
Burns and traumatic shock
Postoperative hyperamylasemia (usually S-type↑)
Diabetic ketoacidosis (mixed↑)
Renal transplantation (S-type↑)
Acute alcoholism (mixed↑)
Drugs
Medicinal opiates (P-type↑)
Heroin addiction →? heroin lung (S-type↑)

*After Salt and Schenkers' classification. Medicine, *55*:269–289, 1976.

Predominant isoenzyme type is shown in parentheses: P-type (pancreatic); S-type (salivary); mixed (either or both isoenzymes may be present).

abdominal procedures. The elevations can be about fourfold, and the principal isoenzyme is usually of the S-type, less often of the P-type.

Amylase clearance. Comparison of renal clearance of amylase with clearance of creatinine, the amylase creatinine clearance ratio (ACCR), has been found useful in diagnosis. The ACCR is readily calculated from amylase activity and creatinine concentration determined from the same urine and from a single serum specimen obtained at the time of urine collection. The calculation simplifies to:

$$\text{ACCR (\%)} = \frac{\text{Urine amylase (U/L)} \times \text{serum creatinine (mg/L)}}{\text{Serum amylase (U/L)} \times \text{urine creatinine (mg/L)}} \times 100 \qquad (15)$$

Because volume and time are identical, they cancel out. Therefore, timed urine collection is unnecessary and random or short (2–4 h) collections are adequate. The reference range for ACCR is ~2–5%, but it is affected by the type of amylase assay used. Thus, it is imperative to establish a reference range for the assay method currently in use.

In *acute pancreatitis* tubular reabsorption of amylase and other proteins is reduced (probably due to competition from other low-molecular-weight proteins) and ACCR is increased; values >8% are not uncommon. Caution must be exercised in interpreting increased ACCR values, because elevations have been observed also in burns, ketoacidosis, renal insufficiency, myeloma, light-chain proteinuria, and march hemoglobinuria, and following extracorporeal circulation, large intravenous doses of corticosteroids, duodenal perforations, and extraperitoneal surgical procedures. In *macroamylasemia* ACCR is usually $<2\%$.

Methods for the Determination of Amylase Activity[4,62]

Amylase is quite stable in serum; loss of activity is negligible at room temperature in the course of a week or at 4–8 °C over a two-month period. In urine, an acid pH may make the enzyme less stable; therefore, pH should be adjusted to ~7.0 before storage. With the exception of heparin, all common anticoagulants inhibit amylase activity because they chelate Ca(II). As a consequence, amylase assays should be performed only on serum or heparinized plasma.

Starch-Substrate Methods

Until a decade ago, three starch-based methods (each with several variations) predominated throughout the world; these were the *saccharogenic*, the *amyloclastic*, and the *chromolytic* methods.

Saccharogenic assays. In saccharogenic assays the course of the enzyme reaction is followed by measuring the quantity of reducing materials (sugars, dextrins) formed. Any of the common procedures for measuring reducing substances, such as the Folin-Wu or Somogyi-Nelson methods, may be used. The chief reducing sugar present is maltose, which, mass for mass, has ~40% of the reducing capacity of glucose. These assays are rarely used today.

Amyloclastic assays. Amylase activity can also be evaluated by following the decrease in starch substrate concentration rather than by measuring the product formed. Methods based on this approach are called *amyloclastic methods*. In the Wohlgemuth method, serial dilutions of the enzyme preparation are added to a fixed quantity of starch, and that dilution is found that is just able to hydrolyze all the starch present in a fixed time period. The *amylometric* procedures of Van Loon and others measure the amount of starch hydrolyzed in a fixed period of time by using the intensity of the blue starch-iodine color as the means for quantitating unhydrolyzed starch. Linear starch chains in helical form react with molecular iodine to form a deep blue starch-iodine complex. The endoamylolytic hydrolysis (dextrinization) of starch to dextrins and oligosaccharides is thus paralleled by a gradual loss of ability to bind iodine; the hue of the glucan-iodine complex slowly changes to light blue, then to violet, and finally to red. No color is formed when the chain size is six or fewer glucose units. These assays have been largely discontinued.

Chromolytic assays. The *chromolytic* assays use dye-labeled amylase substrate materials. Most of these substrates are synthesized by linking amylose or amylopectin to a variety of dyes. These dye substrates are usually water insoluble; after amylase attacks the α-1,4-bonds, water-soluble dye-containing fragments are produced that can be measured colorimetrically after they are separated from the insoluble, unreacted substrate.

Defined-Substrate Methods for Amylase Activity Measurements[50]

The recent use of both well-defined substrates and of auxiliary and indicator enzymes in amylase assays has improved the reaction stoichiometry and has led to more controlled, and therefore more consistent, hydrolysis conditions. Amylolytic hydrolysis of small oligosaccharides gives better-defined products than hydrolysis of starch substrates. Both maltopentaose and maltotetraose are excellent α-amylase substrates; they have good stability, consistent hydrolysis products, and unambiguous reaction stoichiometry. Other useful substrates are small oligosaccharides (of 4–7 glucose units) that have the 4-nitrophenyl group covalently bound to their reducing ends. Examples of these different substrate formulations will be described. The use of a coupled enzyme assay (i.e., an assay using auxiliary and indicator enzymes) was first proposed in practical form by Tietz and his colleagues, who used α-glucosidase (EC 3.2.1.20; maltase) to hydrolyze to glucose the di- and trisaccharides that are produced by the action of amylase; the technique is illustrated for a starch substrate [equation (16)]. The rate of glucose production was determined with glucose oxidase (EC 1.1.3.4) and an oxygen electrode to measure the decrease in pO_2.

$$\begin{gathered}\text{Starch} \xrightarrow{\alpha\text{-Amylase}} \text{Maltose} + \text{maltotriose} + \text{dextrins} \\ \text{Maltose} + \text{maltotriose} \xrightarrow{\alpha\text{-Glucosidase}} \text{Glucose} \\ \text{Glucose} + \tfrac{1}{2}O_2 \xrightarrow{\text{Glucose oxidase}} H_2O_2 + \delta\text{-gluconolactone}\end{gathered} \tag{16}$$

For each mole of glucose oxidized, 0.5 mole of O_2 is consumed. Thus, the rate of oxygen consumption is directly related to the amylase activity in the sample.

An alternative method uses the same principle but requires the addition of peroxidase (EC 1.11.1.7) in an indicator reaction using the diammonium salt of 2,2′-azine-di-(3-ethylbenzthiazoline)-6-sulfonic acid (ABTS); ABTS has an absorbance maximum of 410 nm when oxidized.

$$H_2O_2 + \text{ABTS (reduced: colorless)} \xrightarrow{\textit{Peroxidase}} H_2O + \text{ABTS (oxidized: colored)} \quad (17)$$

The increase in absorbance at 410 nm is proportional to the amylase activity of the specimen.

The DuPont *aca* Amylase Assay

The basic reactions of this assay are:

$$\begin{aligned} &\text{Maltopentaose} \xrightarrow{\alpha\text{-Amylase}} \text{Maltotriose} + \text{maltose} \\ &\text{Maltotriose} + \text{maltose} \xrightarrow{\alpha\text{-Glucosidase}} 5\ \text{Glucose} \\ &\text{Glucose} + \text{ATP} \xrightarrow{\textit{Hexokinase}} \text{G-6-P} + \text{ADP} \\ &\text{G-6-P} + \text{NAD}^+ \xrightarrow{\text{G-6-P Dehydrogenase}} \text{6-P-Gluconolactone} + \text{NADH} + \text{H}^+ \end{aligned} \quad (18)$$

Endogenous glucose in the sample is removed before the reaction by a gel-filtration column in the test pack. Hydrolysis of maltopentaose produces five molecules of glucose (and hence NADH) for every single saccharide bond hydrolyzed by the amylase. This assay has a rather substantial blank reaction due to contamination by trace amounts of shorter-chain oligosaccharides, and the blank reaction displays nonlinear kinetics, as do patient samples. However, measurements are reproducible because of the accurate and consistent timing of the preincubation and measurement intervals on the DuPont analyzer.

The Beckman DS Amylase Assay

This method uses maltotetraose as substrate and a more complex auxiliary and indicator enzyme system for measuring the hydrolysis products:

$$\begin{aligned} &\text{Maltotetraose} \xrightarrow{\alpha\text{-Amylase}} 2\ \text{Maltose} \\ &\text{Maltose} + \text{P}_i \xrightarrow{\text{Maltose phosphorylase}} \text{Glucose} + \text{glucose-1-P} \\ &\text{G-1-P} \xrightarrow{\beta\text{-Phosphoglucomutase}} \text{G-6-P} \\ &\text{G-6-P} + \text{NAD}^+ \xrightarrow{\text{G-6-P Dehydrogenase}} \text{6-P-Gluconolactone} + \text{NADH} + \text{H}^+ \end{aligned} \quad (19)$$

For each bond hydrolyzed by amylase, two molecules of NADH are produced. Endogenous glucose does not interfere with this measurement of amylase activity. However, contaminating lactate dehydrogenase (from both the specimen and the assay reagents) reacts with any pyruvate present, causing negative interference by removing NADH. The degree of this interference depends on the lactate dehydrogenase activity, the pyruvate concentration, and the amylase activity of the individual specimen.

Amylase Assays Using 4-Nitrophenyl–Glucoside Substrates[36]

Substrates for this assay are prepared by bonding 4-nitrophenol (4-NP) to the reducing end of a defined oligosaccharide. If the oligosaccharide is maltoheptaose (G7), the substrate is then 4-NP–G7. Amylase splits this substrate to produce oligosaccharides (G5, G4, G3) and 4-NP–G2 (9%), 4-NP–G3 (31%), and 4-NP–G4 (60%). Alpha-glucosidase added to the system hydrolyzes some of the 4-NP–G4 to free 4-NP and free oligosaccharides; 4-NP–G3 and 4-NP–G2 will be hydrolyzed to free 4-NP and glucose.

$$\begin{aligned} &\text{4-NP-(glucose)}_7 \xrightarrow{\alpha\text{-Amylase}} \text{4-NP-(glucose)}_{4,3,2} + (\text{glucose})_{5,4,3} \\ &\text{4-NP-(glucose)}_{4,3,2} \xrightarrow{\alpha\text{-Glucosidase}} \text{4-NP-(glucose)}_4 + \text{x-glucose} + \text{4-NP} \end{aligned} \quad (20)$$

The result of the sequential hydrolysis by amylase and α-glucosidase is that >30% of the product is free 4-NP, which is detected by its absorbance at 405 nm. Alpha-glucosidase does not react with the original substrate or with any oligosaccharide containing more than four glucose molecules in the chain; G4 is hydrolyzed only very slowly.

The Boehringer Mannheim Diagnostics assay utilizes this principle. The mixture is preincubated at the assay temperature, 50 or 100 μL of specimen is added, and after a further incubation of 2 or 3 min, ΔA/min is determined at 405 nm. The ΔA/min is proportional to the amylase activity of the specimen. Since only one out of the three substrate molecules actually cleaved produces free 4-nitrophenol, a stoichiometric factor of three is used to express amylase activity in International Units.

Other procedures utilize 4-NP–G5 and 4-NP–G6 as substrates. In the presence of α-glucosidase, these substrates are also hydrolyzed to free 4-NP– and 4-NP–glucosides with short chain length. An example is the method by Behring Diagnostics (La Jolla, CA 92037). In other respects, the method is similar to that described above.

Reference ranges for the DuPont *aca* amylase assay at 37 °C are 28–85 U/L for serum and 4–37 U/2 h for urine. Values for the Beckman DS amylase assay at 37 °C are 25–125 U/L for serum and 1–17 U/h for urine. The upper reference value for serum for the Boehringer Mannheim Diagnostics method at 30 °C is 130 U/L and 200 U/L at 37 °C; the upper value for urine at 30 °C is 650 U/L.

Amylase Isoenzymes

Analytical methods for human amylase isoenzymes (isoamylases) include electrophoresis, ion-exchange chromatography, isoelectric focusing, and selective inhibition of the S-type isoamylase by a wheat germ inhibitor. Up to three pancreatic and three salivary isoenzymes can be demonstrated after separation on agarose or cellulose acetate media.[59,97] A minicolumn ion-exchange technique has been described that separates P-type from S-type isoamylases.[100] Isoelectric focusing can be carried out on thin-layer polyacrylamide gels.[61] S-type amylase is preferentially inhibited by a protein isolated from wheat.[43,81] In a method based on this principle, the amylase activity of the serum specimen is determined in the presence and absence of the inhibitor. The extent of inhibition is proportional to the S-type isoenzyme content of the specimen. A monoclonal antibody that specifically inhibits the S-type isoenzyme has recently been reported.[71]

Reference ranges. After birth, serum S-type amylase activity increases steadily with age and reaches normal adult levels of 82 U/L (range, 35–155 U/L) or 48% of total activity at five years of age.[97] Serum P-type amylase activity is not demonstrable in most children younger than one year, but activity rises slowly thereafter to reach adult levels (88 U/L, with a range of 45–180 U/L, or 52% of total activity) at 10–15 years of age. Štěpán and Škrha,[100] using a minicolumn ion-exchange technique, found similar ratios (48% and 52%) for the serum S- and P-type activities in sera from adults. Levitt's group[61] obtained values of 68% and 32% for the S- and P-type isoamylases in serum with their isoelectric focusing method. O'Donnell and colleagues,[81] using the wheat inhibitor technique, obtained the following values: males, 38% S-type and 62% P-type; females, 34% S-type and 66% P-type. Huang and Tietz,[43] with a modified technique, found values for P-type amylase of 27–70% of total amylase activity.

LIPASE[11]

(EC 3.1.1.3; Triacylglycerol Acylhydrolase)

Lipases are defined as enzymes that hydrolyze glycerol esters of long-chain fatty acids. Only the ester bonds at carbons 1 and 3 (α- and α_1-positions) are attacked, and the products of the reaction are two moles of fatty acids and one mole of 2-acylglycerol (β-monoglyceride) per mole of substrate. The latter is resistant to hydrolysis, due to steric hindrance, but it can spontaneously isomerize to the α-form (3-acylglycerol). This isomerization permits the third fatty acid to be split off, but at a much slower rate. A scheme for the steps in the complete hydrolysis of a fat to glycerol and three fatty acids is given in equation (21).

$$\begin{array}{ccccccccc} CH_2OFA' & & CH_2OH & & CH_2OH & & CH_2OH & & CH_2OH \\ | & & | & & | & & | & & | \\ CHOFA'' & \xrightarrow[\text{HOH, OH}^-]{\textit{Lipase}} & CHOFA'' & \xrightarrow[\text{HOH, OH}^-]{\textit{Lipase}} & CHOFA'' & \xrightarrow{\textit{Isomerization}} & CHOH & \xrightarrow[\text{HOH, OH}^-]{\textit{Lipase}} & CHOH \\ | & & | & & | & & | & & | \\ CH_2OFA''' & & CH_2OFA''' & & CH_2OH & & CH_2OFA''' & & CH_2OH \end{array} \quad (21)$$

Triacylglycerol (Triglyceride)	**α,β-Diacylglycerol + FA′OH (Fatty acid I)**	**β-Monoacylglycerol + FA‴OH (Fatty acid III)**	**α-Monoacylglycerol**	**Glycerol + FA″OH (Fatty acid II)**

Lipase acts only at the interface between water and the substrate, i.e., when the substrate is present in an emulsified form. The rate of lipase action depends on the surface area of the dispersed substrate. Bile acids appear to be necessary to ensure that the surface of the dispersed substrate remains free of other proteins, including some lipolytic enzymes (e.g., lipoprotein lipase). Lipase seems to gain access to the substrate surface in the following manner: colipase (secreted by the pancreas) attaches to a micelle of bile salt, thus forming a colipase-bile salt complex that attaches to the substrate surface. This complex binds lipase with high affinity, thus allowing lipase action to proceed. Human lipase can also be fully activated by colipases from other species (e.g., by porcine colipase).

Most of the lipase found in serum is produced in the pancreas, but some lipase is also secreted by the tongue and by gastric, pulmonary, and intestinal mucosa. Lipase activity can also be demonstrated in leukocytes, in adipose tissue cells, and in milk.

Clinical significance. Lipase measurements on serum, plasma, ascitic, and pleural fluids are used exclusively to investigate pancreatic disorders. After an attack of *acute pancreatitis*, the serum lipase activity rises within 2–12 h to two to four times (sometimes as much as 10 times) the upper reference limit with the two-point assay and 2 to 50 times with the multipoint assay. Values may return to normal within 48–72 h but may stay elevated for considerably longer periods (10–14 d) than amylase values. Lipase elevations usually parallel those of amylase, but increases in lipase activity may occur sooner or persist longer than increases in amylase activity and lipase may rise to a greater extent. For this reason it has been suggested that both enzymes be assayed. The increase in serum lipase activity is not necessarily proportional to the severity of the attack.

Acute pancreatitis may produce ascitic fluid or pleural fluid (usually on the left side), or both, which may contain lipase activity. About 50% of patients with acute pancreatitis develop a *pseudocyst*; its presence is usually suspected when there is no clinical improvement within a week of the acute attack. About 50% of patients with pseudocyst formation show persistent serum amylase elevations, and lipase assays may also be useful under these circumstances.

The diagnosis of acute pancreatitis is sometimes difficult since this disorder must be differentiated from other acute intra-abdominal disorders with similar clinical findings, such as perforated gastric or duodenal ulcer, intestinal obstruction, and mesenteric vascular obstruction. Elevation of serum lipase activity is probably a more specific diagnostic test in these cases than serum amylase activity, because many of these conditions are less likely to cause increases in lipase activity than in amylase activity.

Serum lipase assays may also be of value in the diagnosis of *chronic pancreatitis*, but in the later stages of the disease there is severe destruction of the acinar tissue, resulting in a reduction of the amount of enzyme that can enter the circulation. Marginal or no increase of serum lipase activity is therefore not unusual in this disease. *Obstruction* of the pancreatic duct by a calculus or by carcinoma of the pancreas may cause an increase in serum lipase activity, depending on the location of the obstruction and the amount of remaining functioning tissue. In *acute* and *chronic renal disease*, serum lipase activity may be slightly increased.

In *mumps* (acute parotitis) without pancreatic involvement, serum lipase activity is usually not elevated, although serum amylase activity is. Investigation of the biliary tract by *retrograde pancreatography* (ERCP) or treatment with *opiates* (which cause the sphincter of Oddi to contract) may cause elevation of serum lipase activity.

Methods for Measuring Lipase Activity[111]

In 1932 Cherry and Crandall first recognized the clinical value of blood lipase measurement in following pancreatic injury. The method bearing their names used a buffered, stabilized 50% emulsion of olive oil as substrate; lipase activity was defined by the amount of free fatty acids released in a 24-h incubation at 37 °C and estimated by titration to neutrality with dilute alkali. Since 1932 many lipase methods have been described, including titrimetric, turbidimetric, spectrophotometric, fluorometric, and immunologic techniques. Over the years, methods using long-chain substrates have displayed a correlation of results with the clinical state that is superior to that seen with other methods.

Titrimetric methods. Olive oil is a mixture of long-chain triglycerides that has been widely used as a substrate. Two-point methods using this substrate suffer from the required length of incubation period of 24, 6, 4, 3, or 1 h, depending on the degree of optimization of reaction conditions. Phenolphthalein, the original acid-base indicator of the titration step, has been replaced by thymolphthalein, whose end point at pH 10.5 confers superior stoichiometry. Dependence on a visual end point was avoided by Tietz and Fiereck by the use of potentiometric titration.[104] The character of the olive oil emulsion has been considerably improved in respect to stability, uniformity, and droplet size. Triolein, a pure long-chain triacylglycerol, has been shown to be an alternative substrate. Although triolein has the advantage of being a single molecular species, the low cost and availability of olive oil, specially purified to remove endogenous fatty acids and other impurities, will no doubt contribute to its continued use for routine analyses.

The relatively long incubation times of the titrimetric methods have always posed a risk of enzyme denaturation due to long exposure of enzyme to the relatively high reaction temperature. The much greater sensitivity of multipoint methods using a pH-Stat makes shorter incubation periods possible. The method of Tietz and Repique[106] is an example of such a continuous-monitoring technique. During the reaction period, the pH is kept constant at 8.8 at 30 °C (or 8.6 at 25 °C) by the addition of dilute alkali. The amount of alkali used is recorded as a function of time and serves as a measure of fatty acid produced during the reaction. Good precision is obtained with a serum sample of 200–500 μL and a reaction time of 3–8 min. When olive oil substrate is used at 30 °C, the reported reference range is <160 U/L; with triolein substrate, slightly higher activities are obtained and the upper limit of the reference range increases to 200 U/L.

Until recently, *colipase*[10] has not been added to reaction mixtures. Older methods therefore determine activities of both lipase and lipoprotein lipase (also known as clearing factor), which hydrolyze protein-bound triacylglycerols, particularly in postheparin specimens. Recently, however, Hockeborn and Rick have described a continuous-monitoring (pH-Stat) method with addition of colipase and bile salts to triolein substrate.[40] But a subsequent publication advised against the use of colipase at this time. Tietz and Astles, in a modification of their original pH-Stat method, use sodium glycocholate, 30 mmol/L, and colipase at 3 mg/L, and are able to stabilize the enzyme and eliminate postheparin lipase activity.[103]

Multipoint titrimetric assays using a pH-Stat are reliable but unfortunately demanding of special skills and equipment and thus not well suited to routine laboratory use. They are mainly reference methods for evaluating other lipase assays or for research purposes.

Turbidimetric methods. Emulsions of fat in water have a milky appearance because the micelles absorb and scatter incident light. Clearing occurs as lipase hydrolyzes the triacylglycerols in the micelles and micellar disruption takes place. The rate of micellar disintegration as measured by decreases in turbidity or light scattering, however, is not necessarily equal to the rate of triglyceride hydrolysis as measured by the appearance of free fatty acids. A turbidimetric method must therefore be calibrated against a titrimetric method or with a serum calibrator that has a lipase activity assigned to it by a titrimetric technique. Stable, reproducible substrates with appropriate initial turbidity are difficult to prepare and store; high readings for initial absorbance result in a high degree of analytical imprecision. Turbidities have occasionally been observed to increase rather than to decrease during the reaction period. Increase may be due to an interaction of paraproteins and the micelles of emulsified lipids; such increases in turbidity have frequently been observed with specimens containing rheumatoid factor. Despite these theoretical and practical objections, turbidimetric assays for lipase activities are popular, especially because of recent improvements in techniques as described below. Recent improvements of the turbidimetric assay incorporate deoxycholate and porcine colipase[41] (see below).

The method of Vogel and Zieve uses an olive oil emulsion and measures change in turbidity at 400 nm over a 20-min reaction period. An optimal deoxycholate concentration and pH are defined, but the method is relatively insensitive. Shihabi and Bishop have modified the procedure by increasing the reaction temperature and by reading absorbance at a lower wavelength. The DuPont *aca* lipase assay is adapted from both methods but uses triolein as substrate.

Determination of Lipase Activity in Serum

At present it is difficult to suggest one particular assay for the determination of lipase activity in the service laboratory. Accordingly, the principles of titrimetric and turbidimetric assays are briefly outlined.

Specimen. Lipase activity in *serum* is stable at room temperature for one week, for three weeks at 4–8 °C, and for several months if frozen. Bacterial contamination of the specimen may result in an increase in lipase activity.

Titrimetric assay.[47,104] An aliquot of serum is incubated in a stabilized olive oil emulsion at pH 7.8 for 3 h at 37 °C. The liberated fatty acids are then titrated to pH 10.5 with NaOH, 50 mmol/L, either potentiometrically or to a light blue color with thymolphthalein as indicator. Considerable enzyme inactivation may be expected over a 3-h incubation period at 37 °C. The enzyme reaction is not linear with time over the 3-h period, but lack of linearity is significantly less than in methods using 6- or 12-h incubation periods. This assay does not include deoxycholate or colipase. Therefore, it is likely to be less specific for pancreatic lipase than assay formulations that incorporate these reagents into the reaction mixture. The *reference range* is 28–280 U/L or up to 1.0 (conventional) unit; the CV of the assay is 12% at the upper reference limit but approaches 4% at higher lipase activities.

Turbidimetric assay.[41,113] An aliquot of serum is added to a preincubated stabilized triolein emulsion containing Na deoxycholate (17 mmol/L), $CaCl_2$ (0.1 mmol/L), porcine colipase (3 mg/mL) with Tris buffer, 25 mmol/L, pH 9.2, and triolein, 0.29 mmol/L. Assay temperature can be 25 or 30 °C. Absorbance at 340 nm is read after 4 min and again after an additional 5 min. The ΔA/min is taken as a measure of lipase activity. This method requires the use of a standard lipase preparation as calibrator. Addition of deoxycholate and colipase increases the specificity and sensitivity of this assay. However, the assay still has the disadvantage of giving negative results (increase in absorbance) with some sera, predominantly from patients with rheumatoid factor and disorders of active immune processes. These difficulties can generally be overcome by monitoring the reaction with a chart recorder, by using a bile salt concentration to 35 mmol/L,[103] and by selecting the linear segment of the reaction rate curve (between 4 and 19 min, depending on the sample). This method is available as a kit from Boehringer Mannheim Diagnostics, Indianapolis, IN 46250. The *reference range* is 30–190 U/L.

TRYPSIN

(EC 3.4.21.4; no systematic name)

Trypsin is a serine proteinase that hydrolyzes the peptide bonds formed by the carboxyl groups of lysine or arginine with other amino acids, although esters and amides

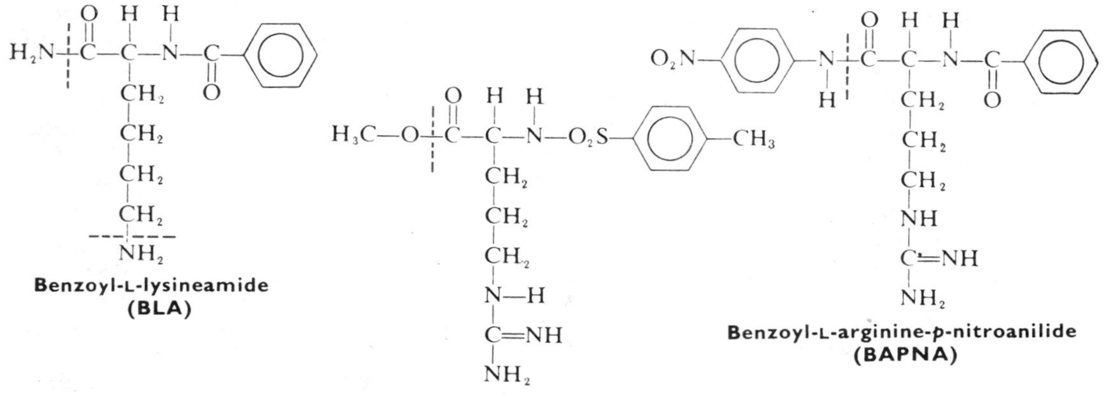

Benzoyl-L-lysineamide (BLA)

p-Toluenesulfonyl-L-arginine methyl ester (TAME)

Benzoyl-L-arginine-p-nitroanilide (BAPNA)

Figure 12-14. Formulas for three synthetic substrates for trypsin, illustrating the bond specificity of the enzyme, which attacks bonds involving the —COOH of lysine and arginine. The dashed line indicates the bond that is hydrolyzed. Hydrolysis of TAME is an example of esterase activity; hydrolyses of BLA and of BAPNA are examples of amidase activity.

involving these amino acids are actually split more rapidly than peptide bonds. This specificity is illustrated in Figure 12-14, which presents the structures of three synthetic substrates, benzoyl-L-lysineamide (BLA), *p*-toluenesulfonyl-L-arginine methyl ester (TAME), and benzoyl-L-arginine-*p*-nitroanilide (BAPNA); all are hydrolyzed rapidly by trypsin.

The acinar cells of the human pancreas synthesize two different forms of trypsin (I and II) as inactive proenzymes (or zymogens), trypsinogens I and II. These are stored in zymogen granules and are secreted into the duodenum under the stimulus of either the vagus nerve or the intestinal hormone cholecystokinin-pancreozymin. The two trypsinogens represent ~19% of the total protein in pancreatic juice; trypsinogen I is present at about twice the concentration of trypsinogen II. In the intestinal tract, the trypsinogens are converted to the active enzyme trypsin by the intestinal enzyme enteropeptidase (EC 3.4.21.9), or by preformed trypsin (autocatalysis).

Trypsin I is also described as cationic and trypsin II as anionic because of their differing electrophoretic mobilities; the cationic form predominates and is the better documented enzyme. The pH optimum for trypsin I for natural substrates is in the range of 8.0–9.0, but for synthetic substrates such as TAME or BAPNA it is 7.8; the pH optimum for trypsin II is 8.0–10.0. Trypsin II differs from trypsin I in that it undergoes autolysis rapidly at neutral or alkaline pH values and it is not stabilized against autolysis by Ca(II). Because the two trypsins show little immunological cross-reactivity, specific immunoassay of each of them is possible.

Trypsin activity is stimulated by calcium and magnesium ions and, to a lesser extent, by cobalt and manganese ions and by aliphatic alcohols. Cyanide, sulfide, citrate, fluoride, and heavy metals inhibit activity, as do those organic phosphorus compounds that combine with serine at the active site.

Nondialyzable trypsin inhibitors are present in pancreatic juice, serum, and urine. Methods for determining the plasma inhibitors, α_1-antitrypsin and α_2-macroglobulin are discussed in Chapter 11. These inhibitors protect plasma and other proteins against hydrolysis by trypsin and other proteases if for some reason any appreciable quantity of the enzyme enters the vascular system. The absence of α_1-antitrypsin is associated with an increased tendency toward panlobular emphysema in early life, illustrating the effects of uninhibited proteases on organ function.

Trypsin in duodenal content and feces.[70] Determinations of trypsin in gastrointestinal secretions are used in the evaluation of pancreatic function and in the diagnosis of *chronic pancreatitis* and *cystic fibrosis*, as discussed in Chapter 23. Trypsin assays can be performed either on aspirated specimens of duodenal content or on fresh feces. Determinations of trypsin in feces are of limited value, since much of the pancreatic trypsin may be destroyed by proteases during passage through the tract. In addition, the remaining activity cannot easily be differentiated from protease activity associated with intestinal bacteria.

Cystic fibrosis in children is accompanied by deficient secretion of trypsin by the pancreas. The disease is diagnosed on clinical grounds and by the measurement of sweat electrolytes. Either at the time of presentation or at the stage when pancreatic insufficiency develops, there is little or no tryptic activity in fecal specimens or in duodenal content.

Immunoreactive trypsin in serum.[9,28,32,58,99] Radioimmunoassay detects trypsinogen I, trypsin I, and the trypsin I:α_1-antitrypsin complex; it does not detect the trypsin I:α_2-macroglobulin complex.[12] Free trypsin I is not usually found in serum. The diagnostic use of measurements of immunoreactive trypsin is discussed in Chapter 23. Trypsin II exists, in healthy persons, as trypsinogen II, and in individuals with disease it is also found complexed to α_1-antitrypsin and α_2-macroglobulin. Assays of trypsin II have been reported, but there is little clinical experience with this enzyme assay.

In *cystic fibrosis*, plasma immunoreactive trypsin levels have been reported to be high in neonates; as the disease progresses, the level falls. Dried blood specimens have been suggested for use in screening tests. In both Type I and Type II *diabetes*, significantly lower than normal plasma immunoreactive trypsin levels have been reported with a graded reduction that is related to the degree of insulin failure.

Methods for the Determination of Trypsin in Duodenal Fluid and Stool

Semiquantitative Determination of Trypsin by the X-Ray Film Test[19]

The test specimen may be either duodenal fluid or a fresh stool specimen. Serial, twofold dilutions of the specimen in barbital buffer, 100 mmol/L (pH 8.0), are prepared. For a duodenal fluid specimen, the dilution may range from 1/2 to 1/256, and for a stool specimen, from 1/5 to 1/320. A drop of each dilution is spotted onto a small piece of unexposed X-ray film, along with a drop of barbital buffer to act as a control. The X-ray film is incubated for 30 min in a Petri dish at 37 °C, cooled in a refrigerator to harden the gelatin, and washed with a slight stream of water to remove any loose or hydrolyzed gelatin. The film is then examined for clear (digested) areas, which indicate hydrolysis of the gelatin by the enzyme; the control spot should show no evidence of digestion. The greatest dilution of the specimen giving a cleared area is a measure of trypsin activity present.

Reference values. The stool of a normal infant under one year of age will show tryptic activity through a dilution of 1/80 or higher. With older children, activity may be evident only through a dilution of 1/20 or 1/40. Infants with cystic fibrosis disease will rarely give positive tests beyond the 1/10 dilution. In the case of duodenal contents, normal infants give positive tests through a dilution of 1/32 or 1/64; in children with cystic fibrosis the test is usually positive in dilutions of 1/4 or less.

Determination of Trypsin in Stool or Duodenal Fluid Using TAME as Substrate[37]

The trend is now toward the use of synthetic peptide substrates such as BAEE (benzoyl-L-arginine ethyl ester), BLA, BAPNA, and TAME, which are specific for trypsin and are not attacked by other proteases found in duodenal contents and stool specimens. In addition, these synthetic substrates have only one bond that can be split so that expression of activity in international units is possible. Procedures using these well-defined substrates are simple and convenient, and enzyme concentration is expressed in mass units. With the TAME substrate, *reference ranges* for normal children and adults have values of 40–760 μg trypsin/g stool; adults with (alcoholic) pancreatic insufficiency have values of 0–33 μg/g. Children with cystic fibrosis have concentrations of <20 μg/g stool.

CHYMOTRYPSIN

(EC 3.4.21.1; no systematic name)

Chymotrypsin, like trypsin, is a serine proteinase. It hydrolyzes peptide bonds that involve carboxyl groups of tryptophan, leucine, tyrosine, or phenylalanine, with preference for the aromatic residues. Specificity of chymotrypsin therefore contrasts with that of pepsin, which splits bonds involving amino groups of the aromatic amino acids. Chymotrypsin also demonstrates hydrolytic activity for other types of bonds, including esters (especially *N*-substituted tyrosine esters) and amide peptides. *N*-acetyl-L-tyrosine ethyl ester (ATEE) is an example of a synthetic substrate that is not significantly hydrolyzed by trypsin:

$$H_3C-C(=O)-NH-CH(COOCH_2CH_3)-CH_2-C_6H_4-OH \xrightarrow{Chymotrypsin} H_3C-C(=O)-NH-CH(COOH)-CH_2-C_6H_4-OH + C_2H_5OH \quad (22)$$

N-acetyl-L-tyrosine ethyl ester **N-acetyl-L-tyrosine** **Ethanol**

The acinar cells of the human pancreas synthesize two different chymotrypsins (I and II, the latter being the major species) in the form of the inactive proenzymes (or zymogens) chymotrypsinogens I and II. These zymogens are stored in granules and are secreted into the pancreatic duct in the same manner as trypsinogen. In the intestinal tract, the chymotrypsinogens are converted to chymotrypsin by the action of trypsin. Chymotrypsin is more resistant than trypsin to degradation in the intestine; it is therefore the enzyme of choice for assay in feces.

Chymotrypsin I is also described as anionic and chymotrypsin II as cationic because of their differing electrophoretic mobilities; the cationic form predominates. The molecular weight of both forms is ~25 000. There is close immunological identity between the chymotrypsins and chymotrypsinogens; immunoassay for chymotrypsin II reacts totally with chymotrypsinogen II and strongly with both chymotrypsin I and chymotrypsinogen I. Chymotrypsin, like trypsin, is bound in blood plasma by α_1-antitrypsin and α_2-macroglobulin. The complex of chymotrypsin with α_1-antitrypsin (but not the chymotrypsin I or II:α_2-macroglobulin complex) is detected by the immunossay.

Clinical significance.[2,56] The predominant form of chymotrypsin found in blood is chymotrypsin II. Although RIAs for immunoreactive chymotrypsin II were described in the 1970s, little clinical experience with these serum assays is available. The major application of these assays is in the investigation of pancreatic disease. In the newborn, chymotrypsin II in cord blood is 2.2–7.5 times higher in infants with *cystic fibrosis* than in healthy infants. In the adult, *acute pancreatitis* can elevate immunoreactive chymotrypsin II to 35 times the levels found in health. In *pancreatic carcinoma*, levels may be normal or slightly elevated. Following total *pancreatectomy*, no chymotrypsin II is detectable in serum. In *gastric carcinoma*, levels may be slightly elevated but are more usually within the reference range. In *hepatobiliary disease*, elevations up to two times the upper reference range are found. In *renal failure*, as with trypsin and amylase, there is elevation of serum chymotrypsin II up to eight times the reference range.

In *duodenal aspirates* (or pancreatic juice) and in *feces*, chymotrypsin levels are variably reduced in pancreatic dysfunction (see also Chapter 23).

Methods for the Determination of Chymotrypsin in Serum

Immunoassay.[11,29,46] Methods for the determination of chymotrypsin that use synthetic substrates or peptides are being replaced in the clinical laboratory by immunoassays. Immunoassays detect chymotrypsinogen II, chymotrypsin II, and, to a lesser extent, chymotrypsin II bound to α_1-antitrypsin. There is a significant cross reaction with chymotrypsinogen I (55–85%), chymotrypsin I (25–50%), and chymotrypsin I:α_1-antitrypsin (20–32%) in these assays; however, the complex between either form of chymotrypsin and macroglobulin is not detected. This type of assay is not yet commercially available.

Methods for the Determination of Chymotrypsin in Duodenal or Pancreatic Aspirates and Feces[32]

Synthetic substrates are commonly used for these applications. A sensitive kinetic assay using Succ-Ala-Ala-Pro-Phe-*p*-nitroanilide has been described. Duodenal or pancreatic aspirates are obtained following stimulation with test meals or by administration of secretin or secretin and cholecystokinin-pancreozymin (see Chapter 23). Stool specimens are either random or 24-h collections. When chymotrypsin is measured in stool specimens, prior treatment of the specimen with detergent to release particle-bound chymotrypsin is essential.[49]

An indirect estimate of chymotryptic activity in the gut can be made by giving the compound *N*-benzoyl-L-tyrosyl-*p*-aminobenzoic acid (Bz-Tyr-PABA) orally, usually with a Lundh meal. Bz-Tyr-PABA is specifically hydrolyzed by chymotrypsin in the gut, and free *p*-aminobenzoic acid (PABA) is released. PABA is absorbed, conjugated in the liver, and excreted in the urine, where it can be readily measured. Details of these assays can be obtained from current reviews[2,56] (see also Chapter 23). The obvious applications of chymotrypsin immunoassay to duodenal or pancreatic aspirate and to feces have not yet been reported. Whether immunoassay will prove superior to current approaches remains to be seen.

Reference ranges. In serum from cord blood, total chymotrypsin is <24 μg/L. In adults, total serum chymotrypsin ranges from 15 to 78 μg/L (mean of 37.5 μg/L), although an earlier report suggested ~10 μg/L. Because of the many and varied techniques used for assay of duodenal aspirates, pancreatic juice, and feces, reference ranges should be obtained from current reviews or should be established in the laboratory performing the test.

CHOLINESTERASES[13]

(EC 3.1.1.7; Acetylcholine Acetylhydrolase)
(EC 3.1.1.8; Acylcholine Acylhydrolase, SChE)

There are two related enzymes that have the ability to hydrolyze acetylcholine. One is *acetylcholinesterase*, which is called "true" cholinesterase, or cholinesterase I. True cholinesterase is found in erythrocytes, lung, spleen, nerve endings, and the gray matter of the brain. It is responsible for the prompt hydrolysis of acetylcholine, which is released at the nerve endings and mediates transmission of the neural impulse across the synapse. The degradation of acetylcholine is necessary to the depolarization of the nerve so that it can be repolarized in the next conduction event.

The other cholinesterase is *acylcholine acylhydrolase* (SChE); it is also called pseudocholinesterase, benzoyl cholinesterase, or cholinesterase II. This enzyme is found in liver, pancreas, heart, the white matter of the brain, and serum, but its biological role is unknown. This serum enzyme is the one whose assay is clinically useful.

The type of reaction catalyzed by both cholinesterases is:

$$\underset{\textbf{Acetylcholine bromide}}{(CH_3)_3\overset{+}{N}-CH_2CH_2-O-\overset{O}{\overset{\|}{C}}-CH_3\ Br^-} + H-O-H \xrightarrow[\textit{Cholinesterase}]{} \underset{\textbf{Choline bromide}}{(CH_3)_3\overset{+}{N}-CH_2CH_2-O-H\ Br^-} + \underset{\textbf{Acetate ion}}{CH_3\overset{O}{\overset{\|}{C}}-O^-} + \underset{\textbf{Hydrogen ion}}{H^+} \quad (23)$$

The two enzymes differ in specificity toward some substrates and behave similarly toward others. The serum enzyme acts on benzoylcholine but cannot hydrolyze acetyl-

β-methylcholine; the erythrocyte enzyme acts on the latter but not on the former. Both enzymes are inhibited by the alkaloids prostigmine and physostigmine, which are typical competitive inhibitors, and both enzymes are irreversibly inhibited by some organic phosphorus compounds, such as diisopropylfluorophosphate. The phosphoryl group binds very tightly to the enzyme site at which binding of the acyl group normally occurs, thus preventing attachment of the acetylcholine. Both enzymes are also inhibited by a large variety of other compounds, among which are morphine, quinine, tertiary amines, phenothiazines, pyrophosphate, bile salts, citrate, fluoride, and borate.

Atypical genetic variants of the enzyme, characterized by diminished activity against acetylcholine and other substrates, are found in the sera of a small fraction of apparently healthy persons. The gene (E_1) controlling the synthesis of SChE can exist in at least four allelic forms, designated as E_1^u, E_1^a, E_1^f, and E_1^s. Several other allelic forms may also exist, and another gene locus is recognized (E_2). The normal, most common phenotype is designated as $E_1^uE_1^u$, or UU. The gene E_1^a is referred to as the atypical gene; the sera of persons homozygous for this gene ($E_1^aE_1^a$ or AA) are only weakly active toward most substrates for cholinesterase and possess increased resistance to inhibition of enzyme activity by dibucaine. The E_1^f gene also gives rise to a weakly active enzyme but with increased resistance to fluoride inhibition. The E_1^s gene (*s* for *silent*) is associated with absence of enzyme or the presence of enzyme protein with minimal or no catalytic activity. The mutations that give rise to the atypical and fluoride-resistant SChE variants involve a change in the structure of the active center that reduces both the affinity for substrates (i.e., K_m is increased) and for competitive inhibitors such as dibucaine or fluoride. This gives rise to the characteristic dibucaine- or fluoride-resistant properties of the genetic variants that are exploited in their characterization. The homozygous forms, AA or FF, are found in only 0.3–0.5% of the white population; their incidence among blacks is even lower. Inheritance of increased serum cholinesterase activity has also been reported in a few families. This is apparently due to increased production of the usual allelozyme.

Clinical significance.[13] Cholinesterase levels in serum are useful as an indicator of possible *insecticide poisoning*, for the detection of patients with *atypical forms* of the enzyme, or as a test of *liver function*. Among the organic phosphorus compounds that inhibit cholinesterase activity are many insecticides, such as Parathion, Sarin, and tetraethyl pyrophosphate. Workers in agriculture and in organic chemical industries are subject to *poisoning* by inhalation of these materials or by contact with them. Obviously, if enough material is absorbed to inactivate all the acetylcholinesterase of nervous tissue, death will result. Both cholinesterases are inhibited, but the activity of the serum enzyme falls more rapidly than does that of the erythrocyte enzyme. A 40% drop in serum enzyme activity occurs before the first symptoms are felt, and a drop of 80% is required before neuromuscular effects become apparent. Near-zero levels of enzyme activity require emergency treatment of the patient with such enzyme reactivators as pyridine-2-aldoxime.

Succinyldicholine (suxamethonium), a drug used in surgery as a muscle relaxant, is also hydrolyzed by cholinesterase, and its pharmacological effect persists only long enough to meet the needs of the surgical procedure. In patients with low levels of enzyme activity or in those with a weakly active variant, destruction of the drug will not occur rapidly enough, and the patient may enter a period of *prolonged apnea* requiring mechanical ventilation until the drug is eliminated by other routes. Preoperative screening has been advocated to identify patients in whom suxamethonium administration may lead to complications. The degree of sensitivity varies with the phenotype of the patient. The phenotypes most susceptible to apnea after succinyldicholine administration are AA, AS, FF, FS, SS, AF, and, to some extent, UA.[21] Measurements of total serum cholinesterase activity as well as determination of the "dibucaine number" and "fluoride number" are needed to characterize cholinesterase variants fully. The latter values indicate the percentage inhibition of enzyme activity toward specified substrates in the presence of standard concentrations of dibucaine or fluoride. Measurement of serum cholinesterase activity can also serve as a sensitive measure of the *synthetic capacity of the liver* if the patient's normal (baseline) level is known. In the absence of genetic causes or known inhibitors, any decrease in activity in serum reflects impaired synthesis of the enzyme by the liver. A 30–50% decrease in level is observed in acute hepatitis and in chronic hepatitis of long duration. Decreases of 50–70% occur in advanced cirrhosis and carcinoma with metastases to the liver. Essentially normal levels are seen in chronic hepatitis, mild cirrhosis, and obstructive jaundice.

Methods for the Determination of Cholinesterase Activity[3,96]

The hydrolysis of acetylcholine, the substrate used in older methods, results in the formation of one mole of hydrogen ion for each mole of substrate reacted, and analyses were based on measurement of the resulting pH change.

Contemporary methods employ acylthiocholine esters as substrates. These substrates are hydrolyzed at approximately the same rate as choline esters, and the thiocholine formed can be measured by reaction with chromogenic disulfide agents such as DTNB [5-5′-dithiobis(2-nitrobenzoic acid); Ellman's reagent] or 4,4′-dithiodipyridine.

In a typical procedure, activity of SChE is determined from the rate of hydrolysis of butyryl- or propionylthiocholine in the presence of DTNB. The reaction of the thiocholine product with colorless DTNB [equation (24)] forms a colored product that is measured spectrophotometrically at 410 nm. Dibucaine or fluoride inhibition can be estimated by performing concurrent assays in which dibucaine or fluoride is present in the substrate mixture. Per cent inhibition is evaluated by comparison of activity in the inhibited system with that in the uninhibited system.

Specimen. Serum is the specimen of choice. Enzyme activity in serum is stable for several weeks whether the specimen is stored at room temperature or under refrigeration. Moderate hemolysis does not interfere if separated serum has been centrifuged to remove erythrocyte ghosts. Specimens from patients who have displayed apnea after succinylcholine treatment should not be obtained until after paralysis has passed; metabolites of the drug appear to interfere with the assay.

$$(CH_3)_3N^+\text{—}(CH_2)_2\text{—S—C(=O)—}(CH_2)_2\text{—}CH_3 + HOH \xrightarrow{SChE} (CH_3)_3N^+\text{—}(CH_2)_2\text{—SH} + {}^-O\text{—C(=O)—}(CH_2)_2\text{—}CH_3 + H^+$$

Butyrylthiocholine ion → Thiocholine ion + Butyrate + Hydrogen ion

$$(CH_3)_3N^+\text{—}(CH_2)_2\text{—SH} + \text{DTNB} \rightleftharpoons (CH_3)_3N^+\text{—}(CH_2)_2\text{—S—S—}C_6H_3(COOH)(NO_2) + HOOC(O_2N)C_6H_3\text{—SH} \quad (24)$$

Thiocholine ion + DTNB (colorless) ⇌ Mixed disulfide (colored) + 5-Mercapto-2-nitrobenzoic acid; 5-MNBA (colorless)

ACID PHOSPHATASE

(EC 3.1.3.2; Orthophosphoric-Monoester Phosphohydrolase (acid optimum); ACP)

Under the name of acid phosphatase are included all phosphatases with optimal activity below a pH of 7.0. Thus, the name refers to a group of similar or related enzymes rather than to one particular enzyme species.[7,112] The acid phosphatase of greatest clinical importance, namely, that derived from the prostate, has a pH optimum in the range of pH 5–6.

Acid phosphatase is present in lysosomes, which are organelles present in all cells except erythrocytes. Extralysosomal acid phosphatases are also present in many cells. The greatest concentrations of acid phosphatase activity occur in liver, spleen, milk, erythrocytes, platelets, bone marrow, and the prostate gland. The last is the richest source, and it contributes about one-third to one-half of the enzyme present in sera from healthy males. The source of the remainder of the acid phosphatase in sera from healthy males and females is not known, but there is evidence that it derives from the osteoclasts of bone.

The optimal pH for the individual acid phosphatases varies, depending on the tissues from which they originate and the substrate on which the enzyme acts; the more acidic the substrate, the lower the pH at which maximum activity is obtained. The enzymes can hydrolyze a variety of phosphate esters, and indeed every substrate utilized in measuring alkaline phosphatase activity in serum has also been used to determine acid phosphatase activity.

The acid phosphatases are unstable, especially at temperatures above 37 °C and at pH levels above 7.0. Some of the enzyme forms in serum (especially the prostatic enzyme) are particularly labile, and over 50% of the ACP activity may be lost in 1 h at room temperature. Acidification of the serum specimen with citrate to a pH below 6.5 aids in stabilizing the enzymes.

Because of the clinical importance of increased serum acid phosphatase levels in the diagnosis and monitoring of prostatic cancer, it is desirable to be able to differentiate specifically between the prostatic and nonprostatic forms. The prostatic enzyme is strongly inhibited by dextrorotatory tartrate ions, whereas the erythrocyte isoenzyme is not. Erythrocyte acid phosphatase is inhibited by formaldehyde and by cupric ions, to which prostatic acid phosphatase is resistant. Thus these inhibitors, particularly tartrate, allow a distinction to be made between prostatic and erythrocyte acid phosphatases. However, interference by the erythrocyte isoenzyme in the measurement of prostatic acid phosphatase in serum is not a significant problem in blood specimens collected with precautions against hemolysis and from which serum has been separated without delay.

An alternative approach to increase the specificity of the assay is the use of substrates that are hydrolyzed rapidly by the prostatic enzyme but at a significantly slower rate by the other forms of the enzyme. For example, α-naphthyl phosphate and thymolphthalein monophosphate are relatively less sensitive to the action of nonprostatic acid phosphatases than are such substrates as phenyl phosphate and 4-nitrophenyl phosphate.

Clinical significance.[35] Determination of acid phosphatase activity in serum is almost always directed toward the prostatic enzyme with the intent of detecting or monitoring *carcinoma of the prostate*. This important disease is currently the second most frequent cause of cancer deaths (17 000/year) in American males. Carcinoma of the prostate is graded by clinical examination, supplemented by a variety of radiological examinations and histological examination of biopsy tissue.[39]

Elevations of the enzymatic activity of prostatic acid phosphatase (and thus, generally, of total acid phosphatase activity also) are found in the sera of ~75% of males with prostatic cancer with metastases. Total activities may reach 40–50 times the upper limit of normal, and nearly all of the activity is inhibited by tartrate. However, when the carcinoma remains localized in the prostate gland, normal or only slightly increased levels of enzyme activity are found. After surgery or estrogen therapy, the levels slowly approach normal; there is a subsequent rise if the treatment is unsuccessful. In patients with benign (nodular) hypertrophy of the prostate, the enzyme level in serum is within normal limits, even after prostatic manipulation, except in rare instances.

In recent years several immunological methods based on the use of specific antisera to prostatic acid phosphatase have been introduced. The potentially greater sensitivity and specificity of these methods, especially RIA, have prompted a re-examination of the value of acid phosphatase measurements in the determinations of the various stages of prostatic cancer.[35,87,95] A slight advantage in diagnostic sensitivity is gained when analytically sensitive and specific immunologic methods are compared with enzyme activity measurements. The advantage is, however, marginal, particularly in the early stages of prostatic carcinoma.[39]

Slight or moderate elevations in total ACP activity often occur in *Paget's disease*, in *hyperparathyroidism* with skeletal involvement, and in the presence of *malignant invasion* of the bones by cancers such as female breast cancer. The enzyme in these cases typically is not inhibited by tartrate and is thought to come from osteoclasts. In some cases of metastatic breast cancer, however, elevations of the tartrate-inhibited ACP have been observed. The osteoclasts are also the probable source of the increased tartrate-resistant acid phosphatase activity of growing children. Increased concentrations of nonprostatic acid phosphatase have also been observed in patients with *Gaucher's* and *Niemann-Pick diseases*, in *myelocytic leukemia*, and in some other *hematological disorders*. Acid phosphatase is present in very high concentration in *semen*, a fact utilized in forensic medicine investigations of rape and similar offenses.[31,57]

Methods for the Determination of Acid Phosphatase Activity in Serum

Methods for acid phosphatase activity measurements are essentially adaptations of those developed for alkaline phosphatase. Phenyl phosphate and 4-nitrophenyl phosphate, for example, have frequently been used in an acid pH medium for acid phosphatase estimation. However, when these methods are used to measure acid phosphatase they must generally be performed as fixed-time procedures. This is because the spectral differences between the phosphate ester and its parent phenol, which are exploited in continuous-monitoring procedures for the assay of alkaline phosphatase, are distinctive only at alkaline pH. Therefore, the acid phosphatase reaction must first be stopped by making the reaction mixture alkaline before the amount of product formed can be measured.

Specimen. Serum is separated immediately from erythrocytes and stabilized by the addition of disodium citrate monohydrate at a level of 10 mg/mL of serum. Pellets (10 mg) are available from several commercial sources. Alternatively, 50 μL of acetate buffer (5 mol/L) per mL of serum may be added to lower the pH to 5.4, at which the enzyme is stable at room temperature for several hours and for up to a week if the serum is refrigerated. Although the substrate is relatively insensitive to erythrocyte acid phosphatase, hemolyzed serum specimens are contaminated with considerable amounts of this isoenzyme and should be rejected. Chylous sera should be avoided because of possible interference with measurement due to turbidity. In the sera of healthy adult men, the *reference range* is from 0.5 to 1.9 U/L (mean $\pm$2 SD).

Continuous-monitoring methods for assay of acid phosphatase activity can be based on the principle introduced by Hillmann, in which α-naphthol released from its phosphate ester forms a colored product with the stabilized diazonium salt of 2-amino-5-chlorotoluene-1,5-naphthalene disulfonate (Fast Red TR). Fast Red TR does not react significantly with bilirubin, and coupling with α-naphthol takes place at the pH of enzyme action. The method is not without practical difficulties; the rate of coupling is dependent on pH, and pronounced lag-phases may occur before zero-order kinetics are established. In addition, the colored product is not very stable. Nevertheless, the method has been successfully used in manual and automated procedures. A recent development is the introduction of alcohols, such as 1,5-pentanediol, which accelerate the reaction and increase sensitivity by acting as a phosphate acceptor in transfer reactions.

As already mentioned, nonprostatic acid phosphatases hydrolyze certain substrates less readily than does the prostatic enzyme. Since the object of diagnostic assays is almost always to determine the prostatic phosphatase, the substrates preferred by that isoenzyme are now usually chosen for the assay system. The two most popular of these are thymolphthalein monophosphate and α-naphthyl phosphate; the latter is favored for continuous-monitoring procedures. Although neither substrate is completely specific, tartrate inhibition to enhance specificity is seldom considered advantageous for methods in which they are used. With the less specific 4-nitrophenyl phosphate substrate, the assay system can be set up to determine total and tartrate-*un*inhibitable phosphatase activity and consequently to determine tartrate-inhibitable prostatic enzyme activity by difference.

A widely used assay method derives its specificity for prostatic acid phosphatase from the use of *thymolphthalein monophosphate* as substrate.[23,90] This substrate is hydrolyzed by prostatic acid phosphatase at pH 5.4 and 37 °C. The reaction is stopped after 30 min by addition of $NaOH\text{-}Na_2CO_3$ solution. At this alkaline pH, the color of the liberated thymolphthalein is measured at 595 nm.

The conditions for the complete incubation mixture are reaction temperature, 37 °C; pH 5.4; acetate buffer, 0.15 mol/L; thymolphthalein monophosphate (TMP), 1.0 mmol/L; Brij-35, 1.5 g/L; and volume fraction (sample/total), 0.083 (1:12).

Immunological methods.[95] RIA, counterimmunoelectrophoresis, and immunoprecipitation methods derive their specificity from the high specificity of the antiserum. An assay using specific antiserum may be used in combination with a colorimetric assay of activity; if the antiserum is used to "capture" or precipitate prostatic acid phosphatase activity, determination of total and residual activity again provides measurement, by difference, of the prostatic isoenzyme activity. RIA determines enzymatically inactive as well as active molecules, provided that they are immunologically recognizable. The antigenic identity of acid phosphatase molecules is lost at a slower rate than catalytic activity. Nevertheless, since loss of antigenic identity takes place at appreciable rates at room temperature in nonacidified sera, the precautions of acidification and refrigeration should be applied to specimens for the determination of acid phosphatase by RIA as well as to those for enzymatic methods.

Results of RIA or other immunologic methods such as counterimmunoelectrophoresis for the determination of prostatic acid phosphatase are usually expressed in mass concentrations. *Reference ranges* are method dependent; representative values are <3 μg/L for RIA or <20 μg/L for counterimmunoelectrophoresis.

Enzymes Less Frequently Applied to Diagnosis

The enzymes already described are those most often assayed for diagnostic purposes. The diagnostic potential of many more enzymes has been explored, but assays for some enjoy only regional popularity and assays for others that were once very popular are now only occasionally used. The properties and applications of four of these enzymes are summarized below.

ISOCITRATE DEHYDROGENASE[110]

(EC 1.1.1.42; Isocitrate:NADP$^+$ Oxidoreductase, Decarboxylating; ICD)

Isocitrate dehydrogenase catalyzes the oxidative decarboxylation of isocitrate to α-oxoglutarate [equations (25) and (26)]. The enzyme is substrate specific; only D_s-isocitrate is acted on by the enzyme (both *cis*-aconitate and L-isocitrate are inert), and only NADP$^+$ can serve as the hydrogen transfer coenzyme in the reaction. Mn(II) is an activator. Genetically distinct cytoplasmic (soluble) and mitochondrial isoenzymes exist. ICD activity can be assayed by monitoring the increase in NADPH concentration (A_{340}) during the course of the reaction.[34]

$$\underset{\textbf{D}_s\textbf{-Isocitrate}}{\begin{array}{c} COO^- \\ | \\ H-C-OH \\ | \\ {}^-OOC-C-H \\ | \\ CH_2 \\ | \\ COO^- \end{array}} + NADP^+ \xrightleftharpoons{\textit{Isocitrate dehydrogenase}} NADPH + H^+ + \underset{\textbf{Oxalosuccinate}}{\begin{array}{c} COO^- \\ | \\ C=O \\ | \\ {}^-OOC-C-H \\ | \\ CH_2 \\ | \\ COO^- \end{array}} \quad (25)$$

Hydrogen Transfer Reaction

$$\begin{array}{c} COO^- \\ | \\ C{=}O \\ | \\ {}^-OOC{-}C{-}H \\ | \\ CH_2 \\ | \\ COO^- \end{array} + H^+ \xrightleftharpoons{\text{Isocitrate dehydrogenase, } Mn^{2+}} \begin{array}{c} COO^- \\ | \\ C{=}O \\ | \\ H{-}C{-}H \\ | \\ CH_2 \\ | \\ COO^- \end{array} + CO_2 \quad (26)$$

Oxalosuccinate **α-Oxoglutarate** **Carbon dioxide**

Decarboxylation Reaction

The ICD of diagnostic interest must be distinguished from the NAD$^+$-dependent isocitrate dehydrogenase (EC 1.1.1.41), which participates in the Krebs tricarboxylic acid cycle and is a distinct enzyme with quite different properties. The NADP$^+$-dependent enzyme is found in high concentrations not only in the liver but also in heart, skeletal muscle, kidney, adrenal tissue, platelets, and erythrocytes.

Elevations of ICD activity are very sensitive indicators of parenchymal liver disease and can be seen even in the early incubation phase of the disease. The highest values are found in viral hepatitis (10 to 40 times the upper limit of normal). Changes in serum ICD in other hepatobiliary diseases are generally similar to those of AST or ALT. However, normal ICD values are encountered in a variety of diseases involving the heart, lungs, kidney, skeletal muscle, and other tissues. The greater organ specificity of ICD compared with that of the transaminases accounts for the clinical interest in ICD. However, this advantage is not sufficiently marked to offset the greater difficulties of ICD assay, and therefore interest in ICD remains limited.

GLUTAMATE DEHYDROGENASE[93]

(EC 1.4.1.3; L-Glutamate:NAD(P)$^+$ Oxidoreductase, Deaminating; GLDH or GMD)

Glutamate dehydrogenase is a mitochondrial enzyme found mainly in the liver, heart muscle, and the kidney, but small amounts occur in other tissues including brain, skeletal muscle, and leukocytes. It is a zinc-containing enzyme. The smallest active molecule is of MW ~350 000, consisting of six polypeptide chains, but larger polymers are also found. The enzyme catalyzes the removal of hydrogen from L-glutamate to form the corresponding imino acid, which undergoes spontaneous hydrolysis to 2-oxoglutarate:

$$\begin{array}{l} CH_2{-}COO^- \\ | \\ CH_2 \\ | \\ CH{-}NH_2 \\ | \\ COO^- \end{array} + NAD^+ \rightleftharpoons \begin{array}{l} CH_2{-}COO^- \\ | \\ CH_2 \\ | \\ C{=}NH \\ | \\ COO^- \end{array} + NADH + H^+$$

$$\updownarrow +H_2O \quad (27)$$

$$\begin{array}{l} CH_2{-}COO^- \\ | \\ CH_2 \\ | \\ CO \\ | \\ COO^- \end{array} + NH_3$$

L-Glutamate **2-Oxoglutarate**

Although NAD$^+$ is the preferred coenzyme, NADP$^+$ can also act as the hydrogen acceptor.[94]

In normal serum, glutamate dehydrogenase is present in only trace amounts, but increased activities are observed in liver disease in which hepatocellular damage is present. However, the degree of elevation is relatively much less than that of the transaminases in uncomplicated viral hepatitis. Four- or fivefold elevations are seen in chronic hepatitis (similar to those for transaminases); lesser increases occur in cirrhosis. Large increases in serum glutamate dehydrogenase occur in halothane toxicity and in response to some other hepatotoxic agents. Compared with other hepatocellular enzymes such as the transaminases, increases in the activity of glutamate dehydrogenase in serum are disproportionately large and may reach 10 to 20 times normal.

Glutamate dehydrogenase can thus offer some potential for differential diagnosis in the investigation of liver disease, particularly when interpreted in conjunction with other enzyme tests (e.g., by calculating the transaminase:glutamate dehydrogenase ratio). The difficulties of making use of the additional diagnostic information potentially offered by glutamate dehydrogenase are similar to those discussed under isocitrate dehydrogenase: absolute levels of activity in serum are low; precision of assay, and particularly the precision of ratios, is correspondingly difficult to achieve and to maintain. The use and interpretation of this enzyme test remain mainly in specialist laboratories of diagnostic enzymology.

Continuous-monitoring methods for the determination of glutamate dehydrogenase can be based upon either the forward or reverse reactions[94] and measure the formation or consumption of NADH at 339 nm.

ALDOLASE

(EC 4.1.2.13; D-Fructose-1,6-bisphosphate: D-Glyceraldehyde-3-phosphate-lyase; ALD)

Aldolase catalyzes the splitting of D-fructose-1,6-diphosphate (FDP) to D-glyceraldehyde-3-phosphate (GLAP) and dihydroxyacetone-phosphate (DAP), an important reaction in the glycolytic breakdown of glucose to lactate. The reaction equilibrium favors the formation of fructose diphosphate.

Aldolase is a tetrameric enzyme with subunits determined by three separate gene loci. Only two of these loci, those producing A and B subunits, appear to be active simultaneously in most tissues, so that the most common isoenzyme pattern consists of varying proportions of the components of a five-membered set of isoenzymes; i.e., A_4, A_3B, A_2B_2, AB_33, B_44. The locus that determines the structure of the C subunit is active in brain tissue, as is the A locus, so that this tissue contains aldolases A_4 and C_4 together with the three corresponding heteropolymers. Aldolase A is the predominant isoenzyme in skeletal muscle, in which it cleaves fructose-1,6-diphosphate to triosephosphate, a key reaction in glycolysis. Aldolase B, the main isoenzyme in liver, is better adapted to utilization of fructose and to gluconeogenesis.

Serum aldolase determinations have been of greatest clinical interest in primary diseases of skeletal muscle.[102] In general, however, measurement of aldolase activity in serum does not provide information that is not available more readily by measurement of other, more easily assayed enzymes, especially CK.

All aldolase assay methods are based on the forward reaction; both colorimetric and continuous-monitoring procedures have been developed.[83,84]

5′-NUCLEOTIDASE

(EC 3.1.3.5; 5′-Ribonucleotide Phosphohydrolase; 5′NT, NTP)

5′-Nucleotidase is a phosphatase that acts only on nucleoside 5′-phosphates, such as adenosine 5′-phosphate (adenosine 5′-monophosphate, AMP; adenylic acid), to form adenosine and release inorganic phosphate.[8] The enzyme is widely distributed throughout the tissues of the body and appears to be localized in the plasma membrane of the cells in which it occurs. Its pH optimum is between 6.6 and 7.0.

The activity of 5′-nucleotidase in serum is increased two to six times in those hepatobiliary diseases in which there is intra- or extrahepatic interference with the secretion of the bile.

In skeletal disease, increases in the activity of 5′-nucleotidase in serum are rarely observed, and, when they do occur, they are marginal. Thus, assays of 5′-nucleotidase are valuable in differentiating between elevations in alkaline phosphatase caused by hepatobiliary diseases and those caused by diseases involving the skeletal system. In fact, any rise in NTP activity that is more than marginal is virtually specific for hepatobiliary disease. 5′-Nucleotidase activity is normal in late pregnancy and during childhood, conditions in which alkaline phosphatase levels are generally elevated.

The most often used substrate of 5′-nucleotidase, adenylate (adenosine 5′-monophosphate, AMP) is also hydrolyzed to an appreciable degree by other nonspecific (alkaline) phosphatases, even at a pH as low as 7.5. Methods for the estimation of 5′-nucleotidase in serum must therefore incorporate some means for correcting for the hydrolysis of the substrate by the nonspecific phosphatases. 5′-Nucleotidase differs from alkaline phosphatase in being inhibited by nickelous ions, and this property can be used to distinguish the two enzymes.[14] In another approach, a large excess of some phosphate ester, such as β-glycerophosphate or phenyl phosphate, is added to the reaction mixture.[8,82] These are substrates for alkaline phosphatase but not for 5′-nucleotidase, and by forming substrate complexes with the former enzyme, they reduce the proportion of the total alkaline phosphatase activity that is directed to the hydrolysis of adenosine 5′-monophosphate. Since inorganic phosphate is a product of both enzymatic reactions, adenosine (the second product of the 5′-nucleotidase reaction) is determined colorimetrically or spectrophotometrically by measuring the amount of ammonia formed by the action of added adenosine deaminase.

TERMINAL DEOXYNUCLEOTIDYL TRANSFERASE*

(EC 2.7.7.31; DNA Nucleotidylexotransferase, TdT)

Terminal deoxynucleotidyl transferase (TdT) is a unique DNA polymerase in that DNA is not required as a template for nucleoside triphosphate addition to the 3′ terminal end of a primer DNA chain. The biological distribution of TdT in normal individuals is generally restricted to thymocytes and to precursor cells of lymphocytes in bone marrow. An ontological association between the appearance of TdT and immunocompetent T cells has been described by Sujimoto and Bollum.[101] Thus, TdT seems to be present early in lymphocyte differentiation and may persist somewhat longer in those precursors destined to mature as T cells. The physiological function of this enzyme is not known.

Clinical significance. The presence of TdT in T-cell precursors suggests that it should be found in T-cell–derived lymphoblastic leukemias and lymphomas. In fact, the most significant elevations of TdT are found in *acute lymphoblastic leukemia*. Virtually all patients with "T" and "null" cell acute lymphoblastic leukemia (ALL) and *lymphomas*, as well as many patients with pre-B cell lymphoblastic leukemia, will have TdT in their tumor cells. Some adult non-Hodgkin's lymphoma lymphocytes show lesser, yet significant elevations. Approximately 50% of patients with *acute undifferentiated leukemia* and 30% with *chronic myelogenous leukemia* in blast crisis will have elevated TdT activity in their leukemic blasts. In such cases, the acute phase of chronic granulocytic leukemia is a lymphoblastic leukemia. A small number of other unusual acute leukemias may also demonstrate TdT activity.

Because of the association of TdT with immature T-cell malignancies, assay of TdT activity is a useful adjunct for determining the type of leukemia, for selecting appropriate treatment, and for monitoring the effectiveness of the treatment.

TdT may not be a dependable predictor of relapse in children with common ALL. This is because TdT-positive cells may constitute up to 10% of the nucleated cells from normal pediatric marrows and because *marrow regeneration*, *idiopathic thrombocytopenic purpura*, and *neuroblastoma* may be causes for TdT-positive marrows.[44]

Specimen. Bone marrow aspirates and peripheral blood are the usual specimens used for TdT assays, although pleural fluid, spinal fluid, lymph nodes, and testicular tissue may also be used. For semiquantitative methods, a slide smear prepared from the specimen is necessary, whereas, quantitative methods require separation and concentration as well as lysing of lymphocytes or tissue cells to release TdT. Preparation of specimens may vary; therefore, appropriate references should be consulted.[6,15,18,48,72]

Principles of TdT methods. TdT catalyzes the addition of deoxynucleoside triphosphates to the 3′-OH end of a polydeoxynucleotide primer. TdT activity is determined by incubating the enzyme with ^{3}H-deoxynucleoside triphosphate, a divalent cation such as Mg(II), and an appropriate primer DNA polymer in a suitable buffer system. The extent of polymerization is determined by isolating the newly formed polymer and measuring its radioactivity. The amount of radioactivity incorporated in the polymer during a defined reaction time is used to calculate TdT enzyme activity. The general reaction catalyzed by TdT is:

$$\text{n(Deoxynucleoside triphosphate)} + \text{(oligodeoxynucleotide)}_{m} \xrightarrow{TdT} \text{(Oligodeoxynucleotide)}_{m+n} + \text{n(pyrophosphate)}$$

*Contributed by John F. O'Brien, Ph.D.

where m is the number of nucleotides in the primer DNA chain and n is the number of nucleotides added to the DNA chain.

An immunofluorescent method for the *semiquantitative* determination of terminal deoxynucleotidyl transferase has also been described[15] that requires fewer cells for detection of TdT and no preliminary preparation of lymphocytes. However, this method is not as sensitive as the radiometric method, and problems with preparation of slide smears do occur. For cases in which a positive qualitative result is sufficient, this shorter, more convenient method may be the assay of choice. Immunological methods that yield *quantitative* results are the solid-phase enzyme immunoassay described by Kaneda and co-workers[48] and, more recently, an ELISA method that is sensitive enough for measurement of TdT in whole blood.[51] Also, an enzyme immunoassay for TdT activity based on the "sandwich" principle is now commercially available (Abbott Laboratories, North Chicago, IL 60064).

References

1. Alpert, J. S., and Braunwald, E.: Acute myocardial infarction: pathological, pathophysiological, and clinical manifestations. *In*: Heart Disease: A Textbook of Cardiovascular Medicine. 2nd ed. E. Braunwald, Ed. Philadelphia, W. B. Saunders Co., 1984, pp. 1262–1300.
2. Arvanitakis, C., and Cooke, A. R.: Diagnostic tests of exocrine pancreatic function and disease. Gastroenterology, *74*:932–948, 1978.
3. Augustinsson, K. B.: Determination of activity of cholinesterases. *In*: Methods of Biochemical Analysis, Vol. 19. D. Glick, Ed. New York, John Wiley & Sons, 1971, pp. 217–273.
4. Bayse, D. D., Duncan, P. H. and Mather, A.: Amylase in serum and urine. *In*: Clinical and Analytical Concepts in Enzymology. H. A. Homburger, Ed. Skokie, IL, College of American Pathologists, 1983, pp. 173–194.
5. Belfield, A., and Goldberg, D. M.: Inhibition of the nucleotidase effect of alkaline phosphatase by β-glycerophosphate. Nature (London), *219*:73–75, 1968.
6. Beutler, E., and Kuhl, W.: An assay for terminal deoxynucleotidyl transferase in leukocytes and bone marrow. Am. J. Clin. Pathol., *70*:733–737, 1978.
7. Bodansky, O.: Acid phosphatase. Adv. Clin. Chem., *15*:44–136, 1972.
8. Bodansky, O., and Schwartz, M. K.: 5′-Nucleotidase. Adv. Clin. Chem., *11*:277–328, 1968.
9. Borgström, A., and Ohlsson, K.: Radioimmunological determination and characterization of cathodal trypsin-like immunoreactivity in normal human plasma. Scand. J. Clin. Lab. Invest., *36*:809–814, 1976.
10. Borgström, B., and Brockman, H. L., Eds.: Lipases. Amsterdam, Elsevier, 1984.
11. Borgström, A., Sveger, T., Lindberg, T., et al.: Immunoreactive trypsin, chymotrypsin, and pancreatic secretory trypsin inhibitor in cord blood from infants with cystic fibrosis. Acta Pediatr. Scand., *70*:619–621, 1981.
12. Broderick, J. W., Geokas, M. C., Largman, C., et al.: Molecular forms of immunoreactive pancreatic cationic trypsin in pancreatitis patient sera. Am. J. Physiol., *237*:E474–E480, 1979.
13. Brown, S. S., Kalow, W., Pilz, W., et al.: The plasma cholinesterases: A new perspective. Adv. Clin. Chem., *22*:1–123, 1981.
14. Campbell, D. M.: Determination of 5′-nucleotidase in blood serum. Biochem. J., *84*:34P, 1962.
15. Cibull, M. L., Coleman, M. S., Nelson, O., et al.: Evaluation of methods of detecting terminal deoxynucleotidyl transferase in human hematologic malignancies. Am. J. Clin. Pathol., *77*:420–423, 1982.
16. Committee on Enzymes of the Scandinavian Society for Clinical Chemistry and Clinical Physiology: Recommended method for the determination of creatine kinase in blood modified by the inclusion of EDTA. Scand. J. Clin. Lab. Invest., *39*:1–5, 1979.
17. Committee on Enzymes of the Scandinavian Society for Clinical Chemistry and Clinical Physiology: Recommended method for the determination of γ-glutamyl transferase in blood. Scand. J. Clin. Lab. Invest., *36*:119–125, 1976.
18. Deibel, M. R., Jr., and Coleman, M. S.: Biochemical properties of purified human terminal deoxynucleotidyl transferase. J. Biol. Chem. *255*:4206–4212, 1980.
19. Demetriou, J. A., Drewes, P. A., and Gin, J. B.: Enzymes. *In*: Clinical Chemistry, Principles and Technics. 2nd ed. R. J. Henry, D. C. Cannon, and J. W. Winkelman, Eds. Hagerstown, MD, Harper and Row, 1974, pp. 815–1101.
20. Desjarlais, F., Morin, L. G., and Daigneault, R.: In search of optimum conditions for the measurement of creatine kinase activity: A critical review of nineteen formulations. Clin. Biochem., *13*:116–121, 1980.
21. Dietz, A. A., Rubinstein, H. H., and Lubrano, T.: Colorimetric determination of serum cholinesterase and its genetic variants by the propionylthiocholine-dithiobis(nitrobenzoic acid) procedure. Clin. Chem., *19*:1309–1313, 1973.
22. Ellis, G., Goldberg, D. M., Spooner, R. J., et al: Serum enzyme tests in diseases of the liver and biliary tree. Am. J. Clin. Pathol., *70*:248–258, 1978.
23. Ewen, L. M., and Spitzer, R. W.: Improved determination of prostatic acid phosphatase (sodium thymolphthalein monophosphate substrate). Clin. Chem., *22*:627–632, 1976.

24. Expert Panel on Enzymes, Committee on Standards (IFCC): Approved recommendations of IFCC methods for the measurement of catalytic concentrations of enzymes. Part I. General Considerations. Clin. Chim. Acta, *98*:163F–174F, 1979.
25. Expert Panel on Enzymes, IFCC: IFCC methods for the measurement of catalytic concentration of enzymes. Part 5. IFCC method for alkaline phosphatase. Clin. Chim. Acta, *135*:339F–367F, 1983; and J. Clin. Chem. Clin. Biochem., *21*:731–748, 1983.
26. Fridhandler, L., and Berk, J. E.: Macroamylasemia. Adv. Clin. Chem., *20*:267–286, 1978.
27. Friedel, R., Diederichs, F., and Lindena, J.: Release and extracellular turnover of intracellular enzymes. *In*: Advances in Clinical Enzymology. E. Schmidt, F. W. Schmidt, I. Trautschold, et al., Eds. Basel, Karger, 1979, pp. 70–105.
28. Geokas, M. C., Largman, C., Brodrick, J. W., et al.: Determination of human pancreatic cationic trypsinogen in serum by radioimmunoassay. Am. J. Physiol., *236*:E77–E83, 1979.
29. Geokas, M. C., Largman, C., Brodrick, J. W., et al.: Immunoreactive forms of human pancreatic chymotrypsin in normal plasma. J. Biol. Chem., *254*:2775–2781, 1979.
30. Gerhardt, W., and Wulff, K.: Creatine kinase. *In*: Methods of Enzymatic Analysis. 3rd ed. H. U. Bergmeyer, J. Bergmeyer, and M. Grassl, Eds. Weinheim, Verlag-Chemie, 1983, Vol. 3, pp. 508–539.
31. Gohara, W. F.: Rate of decrease of glutamyltransferase and acid phosphatase activities in the human vagina after coitus. Clin. Chem., *26*:254–257, 1980.
32. Goldberg, D. M.: Enzymes and isoenzymes in the evaluation of diseases of the pancreas. *In*: Clinical and Analytical Concepts in Enzymology. H. A. Homburger, Ed. Skokie, IL, College of American Pathologists, 1983, pp. 31–55.
33. Goldberg, D. M.: Structural, functional and clinical aspects of gamma-glutamyltransferase. CRC Crit. Rev. Clin. Lab. Sci., *12*:1–58, 1980.
34. Goldberg, D. M., and Ellis, G.: Isocitrate dehydrogenase. *In*: Methods of Enzymatic Analysis. 3rd ed. H. U. Bergmeyer, J. Bergmeyer, and M. Grassl, Eds. Weinheim, Verlag-Chemie, 1983, Vol. 3, pp. 183–190.
35. Griffiths, J. C.: The laboratory diagnosis of prostatic adenocarcinoma. CRC Crit. Rev. Clin. Lab. Sci., *19*:187–204, 1983.
36. Hägele, E-O., Schaich, E., Rauscher, E., et al.: Mechanism of action of human pancreatic and salivary α-amylase on α-4-nitrophenyl maltoheptaoside substrate. Clin. Chem., *28*:2201–2205, 1982.
37. Haverback, B. J., Dyce, B. J., Gutentag, P. J., et al.: Measurement of trypsin and chymotrypsin in stool. Gastroenterology, *44*:588–597, 1963.
38. Henderson, A. R.: Lactate dehydrogenase isoenzymes. *In*: Methods of Enzymatic Analysis. 3rd ed. H. U. Bergmeyer, J. Bergmeyer, and M. Grassl, Eds. Weinheim, Verlag-Chemie, 1983, Vol. 3, pp. 138–155.
39. Henderson, A. R., and Nealon, D. A.: Enzyme measurements by mass: An interim review of the clinical efficacy of some mass measurements of prostatic acid phosphatase and the isoenzymes of creatine kinase. Clin. Chim. Acta, *115*:9–32, 1981.
40. Hockeborn, M., and Rick, W.: Determination of lipase by continuous titration. J. Clin. Chem. Clin. Biochem., *20*:773–785, 1982.
41. Hoffmann, G. E., and Weiss, L.: Specific serum pancreatic lipase determination, with use of purified colipase. Clin. Chem., *26*:1732–1733, 1980.
42. Hørder, M., and Rej, R.: Alanine transaminase. *In*: Methods of Enzymatic Analysis. 3rd ed. H. U. Bergmeyer, J. Bergmeyer, and M. Grassl, Eds. Weinheim, Verlag-Chemie, 1983, Vol. 3, pp. 444–456.
43. Huang, W. Y., and Tietz, N. W.: Determinations of amylase isoenzymes in serum by use of a selective inhibitor. Clin. Chem., *28*:1525–1527, 1982.
44. Hutton, J. J., Coleman, M. S., Moffit, S., et al: Prognostic significance of terminal transferase activity in childhood acute lymphoblastic leukemia: A prospective analysis of 164 patients. Blood, *60*:1267–1276, 1982.
45. IFCC Expert Panel on Enzymes: IFCC methods for the measurement of the catalytic concentration of enzymes. Part 4. IFCC method for γ-glutamyl transferase. J. Clin. Chem. Clin. Biochem., *21*:633–646, 1983.
46. Iwaki, K., Ogawa, M., Tanaka, S., et al.: Radioimmunoassay for human pancreatic chymotrypsin and measurement of serum immunoreactive chymotrypsin contents in various diseases. Res. Commun. Chem. Pathol. Pharmacol., *40*:489–496, 1983.
47. Kachmar, J. F., and Moss, D. W.: Enzymes. *In*: Fundamentals of Clinical Chemistry. 2nd ed. N. W. Tietz, Ed. Philadelphia, W. B. Saunders Co., 1976, pp. 565–698.
48. Kaneda, T. Kuroda, S., Hirota, Y., et al.: Highly sensitive solid-phase enzyme immunoassay for terminal deoxynucleotidyl transferase. Anal. Biochem., *126*:327–334, 1982.
49. Kaspar, P., Moeller, G., Wahlefeld, A. W., et al.: A new photometric method for the determination of chymotrypsin in stool. Fresenius Z. Anal. Chem., *311*:391–392, 1982.
50. Kaufman, R. A., and Tietz, N. W.: Recent advances in measurement of amylase activity—a comparative study. Clin. Chem., *26*:846–853, 1980.
51. King, W. J., Fairbanks, T. R., Manderino, G. L., et al.: Measurement of terminal deoxynucleotidyl transferase (TdT) antigen by solid-phase enzyme immunoassay. Published with comments in Transactions of the 1984 Children's Cancer Study Group Proceedings.
52. Krause, R. D., and Lott, J. A.: Use of the simplex method to optimize analytical conditions in clinical chemistry. Clin. Chem., *20*:775–782, 1974.
53. Kristensen, S. R., and Hørder, M.: Release and turnover of intracellular enzymes. *In*: Clinical and Analytical Concepts in Enzymology. H. A. Homburger, Ed. Skokie, IL, College of American Pathologists, 1983, pp. 1–13.

54. Lang, H., Ed.: Creatine Kinase Isoenzymes. Berlin, Springer-Verlag, 1981.
55. Lang, H., and Wurzburg, U.: Creatine kinase, an enzyme of many forms. Clin. Chem., *28*:1439–1447, 1982.
56. Lankisch, P. G.: Exocrine pancreatic function tests. Gut, *23*:777–798, 1982.
57. Lantz, R. K., and Eisenberg, R. B.: Preservation of acid phosphatase activity in medico-legal specimens. Clin. Chem., *24*:486–488, 1978.
58. Largman, C., Brodrick, J. W., Geokas, M. C., et al.: Demonstration of human pancreatic anionic trypsinogen in normal serum by radioimmunoassay. Biochim. Biophys. Acta, *543*:450–451, 1978.
59. Legaz, M. E., and Kenney, M. A.: Electrophoretic amylase fractionation as an aid in diagnosis of pancreatic disease. Clin. Chem., *22*:57–62, 1976.
60. Leung, F. Y., and Henderson, A. R.: Thin-layer agarose electrophoresis of lactate dehydrogenase isoenzymes in serum: A note on the method of reporting and on the lactate dehydrogenase isoenzyme-1/isoenzyme-2 ratio in acute myocardial infarction. Clin. Chem., *25*:209–211, 1979.
61. Levitt, M. D., Ellis, C., and Engel, R. R.: Isoelectric focusing studies of human serum and tissue isoamylases. J. Lab. Clin. Med., *90*:141–152, 1977.
62. Lorentz, K.: α-Amylase assay: Current state and future development. J. Clin. Chem. Clin. Biochem., *17*:499–504, 1979.
63. Lott, J. A., and Stang, J. M.: Serum enzymes and isoenzymes in the diagnosis and differential diagnosis of myocardial ischemia and necrosis. Clin. Chem., *26*:1241–1250, 1980.
64. Mangum, B. W., and Thornton, D. D., Eds.: Standard Reference Materials: The Gallium Melting Point Standard. Washington, D.C., NBS Special Publication 481, 1977.
65. McComb, R. B., Bond, L. W., Burnett, R. W., et al.: Determination of the molar absorptivity of NADH. Clin. Chem., *22*:141–150, 1976.
66. McComb, R. B.: The measurement of lactate dehydrogenase. *In*: Clinical and Analytical Concepts in Enzymology. H. A. Homburger, Ed. Skokie, IL, College of American Pathologists, 1983, pp. 157–171.
67. McComb, R. B., Bowers, G. N., Jr., and Posen, S.: Alkaline Phosphatase. New York, Plenum Press, 1979.
68. McKenzie, D., and Henderson, A. R.: Electrophoresis of lactate dehydrogenase isoenzymes. *In*: Selected Methods of Clinical Chemistry. G. R. Cooper, Ed. Washington, D.C., American Association for Clinical Chemistry, 1983, Vol. 10, pp. 58–67.
69. Mercer, D. W.: Improved column method for separating lactate dehydrogenase isoenzymes 1 and 2. Clin. Chem., *24*:480–482, 1978.
70. Meyer, J. H.: Pancreatic physiology. *In*: Gastrointestinal Disease: Pathophysiology, Diagnosis, Management. 3rd ed. M. H. Sleisenger and J. S. Fordtran, Eds. Philadelphia, W. B. Saunders Co., 1983, pp. 1426–1436.
71. Mifflin, T. E., Benjamin, D. C., and Bruns, D. E.: Rapid quantitative, specific measurement of pancreatic amylase in human serum with use of a monoclonal antibody. Clin. Chem., *31*:1283–1288, 1985.
72. Modak, M. J., et al.: A micro method for determination of terminal deoxynucleotidyl transferase (TdT) in the diagnostic evaluation of acute leukemias. J. Cancer. Res. Clin. Oncol. *98*:91–104, 1980.
73. Moss, D. W.: Alkaline phosphatase isoenzymes. Clin. Chem., *28*:2007–2016, 1982.
74. Moss, D. W.: Automatic enzyme analyzers. Adv. Clin. Chem., *19*:1–56, 1977.
75. Moss, D. W.: Isoenzyme Analysis. London, The Chemical Society, 1979.
76. Moss, D. W.: Isoenzymes. London and New York, Chapman and Hall 1982.
77. Moss, D. W., and Edwards, R. K.: Improved electrophoretic resolution of bone and liver alkaline phosphatases resulting from partial digestion with neuraminidase. Clin. Chim. Acta, *143*:177–182, 1984.
78. Moss, D. W., Shakespeare, M. J., and Thomas, D. M.: Observations on the heat-stability of alkaline phosphatase isoenzymes in serum. Clin. Chim. Acta, *40*:35–41, 1972.
79. Moss, D. W., and Whitby, L. G.: A simplified heat-inactivation method for investigating alkaline phosphatase isoenzymes in serum. Clin. Chim. Acta, *61*:63–71, 1975.
80. Nomenclature Committee, IUB: Enzyme Nomenclature, 1978. Recommendations of the Nomenclature Committee of IUB on the Nomenclature and Classification of Enzymes. New York, Academic Press, 1979.
81. O'Donnell, M. D., Fitzgerald, O., and McGeeney, K. F.: Differential serum amylase determination by use of an inhibitor, and design of a routine procedure. Clin. Chem., *23*:560–566, 1977.
82. Persijn, J. P., van der Slik, W., Kramer, K., et al: A new method for the determination of serum nucleotidase. Z. Klin. Chem. Klin. Biochem., *6*:441–446, 1968.
83. Pinto, P. V. C., Kaplan, A., and Van Dreal, P. A.: Aldolase. II. Spectrophotometric determination using an ultraviolet procedure. Clin. Chem. *15*:349–360, 1969.
84. Pinto, P. V. C., Van Dreal, P. A., and Kaplan, A.: Aldolase. I. Colorimetric determination. Clin. Chem., *15*:339–348, 1969.
85. Rautela, G. S., Snee, R. D., and Miller, W. K.: Response-surface co-optimization of reaction conditions in clinical chemical methods. Clin. Chem., *25*:1954–1964, 1979.
86. Rej, R., and Hørder, M.: Aspartate aminotransferase. *In*: Methods of Enzymatic Analysis. 3rd ed. H. U. Bergmeyer, J. Bergmeyer, and M. Grassl, Eds. Weinheim, Verlag-Chemie, 1983, Vol. 3, pp. 416–433.
87. Romas, N. A., Rose, N. R., and Tannenbaum, M.: Acid phosphatase: New developments. Hum. Pathol., *10*:501–512, 1979.
88. Rosalki, S. B.: Gamma-glutamyl transpeptidase. Adv. Clin. Chem., *17*:53–107, 1975.
89. Rosalki, S. B., and Foo, Y. A.: Two new methods for separating and quantifying bone and liver alkaline phosphatase isoenzymes in plasma. Clin. Chem., *30*:1182–1186, 1984.
90. Roy, A. V., Brower, M. E., and Hayden, J. E.: Sodium thymolphthalein monophosphate: A new acid

phosphatase substrate with greater specificity for the prostatic enzyme in serum. Clin. Chem., *17*:1093–1102, 1971.
91. Salt, W. B., and Schenker, S.: Amylase—its clinical significance: A review of the literature. Medicine, *55*:269–289, 1976.
92. Scandinavian Society for Clinical Chemistry and Clinical Physiology: Recommended Methods for the Determination of Four Enzymes in Blood. Scand. J. Clin. Lab. Invest., *33*:291–306, 1974.
93. Schmidt, E., and Schmidt, F. W.: Enzyme diagnosis in diseases of the liver and biliary system. *In*: Advances in Clinical Enzymology. F. W. Schmidt, E. Schmidt, I. Trautschold, et al., Eds. Basel, Karger, 1979, pp. 239–293.
94. Schmidt, E., and Schmidt, F. W.: Glutamate dehydrogenase. *In*: Methods of Enzymatic Analysis. 3rd ed. H. U. Bergmeyer, J. Bergmeyer, and M. Grassl, Eds. Weinheim, Verlag Chemie, 1983, Vol. 3, pp. 216–227.
95. Shaw, L. M.: The specificity of the immunochemical determination of prostatic acid phosphatase. *In*: Clinical and Analytical Concepts in Enzymology. H. A. Homburger, Ed. Skokie, IL, College of American Pathologists, 1983, pp. 57–68.
96. Silk, E., King, J., and Whittaker, M.: Assay of cholinesterase in clinical chemistry, Ann. Clin. Biochem., *16*:57–75, 1979.
97. Skude, G.: Human amylase isoenzymes. Scand. J. Gastroent., *12*(Suppl. 44): 1–36, 1977.
98. Speicher, C. E., and Smith, J. W.: Cardiovascular Subproblems. *In*: Choosing Effective Laboratory Tests. Philadelphia, W. B. Saunders Co., 1983, pp. 155–182.
99. Stagg, B. H., and Wood., T. P.: Radioimmunoassay of trypsin: A new aid in the assessment of pancreatic function. Ann. Clin. Biochem., *16*:147–151, 1979.
100. Štěpán, J., and Škrha, J., Measurement of amylase isoenzymes in human sera and urine using a DEAE-cellulose minicolumn method. Clin. Chim. Acta, *91*:263–271, 1979.
101. Sujimoto, M., and Bollum, F. J.: Terminal deoxynucleotidyl transferase (TdT) in chick-embryo lymphoid tissues. J. Immunol., *122*:392–397, 1979.
102. Thompson, W. H. S.: The clinical chemistry of muscular dystrophies. Adv. Clin. Chem., *7*:138–197, 1964.
103. Tietz, N. W., Astles, J. R., and Shuey, D.: Personal communication, 1985.
104. Tietz, N. W., and Fiereck, E. A.: Measurement of lipase activity in serum. *In*: Standard Methods of Clinical Chemistry. G. R. Cooper, Ed. New York, Academic Press, 1972, Vol. 7, pp. 19–31.
105. Tietz, N. W., and Finley, P. R., Eds.: Clinical Guide to Laboratory Tests, Philadelphia, W. B. Saunders Co., 1983.
106. Tietz, N. W., and Repique, E. V.: Proposed standard method for measuring lipase activity in serum by a continuous sampling technique. Clin. Chem., *19*:1268–1275, 1973.
107. Tietz, N. W., and Rinker, A. D.: The measurement of alkaline phosphatase (ALP). *In*: Analytical Concepts in Enzymology. H. A. Homberger, Ed. Skokie, IL, College of American Pathologists, 1983, pp. 195–203.
108. Usategui-Gomez, M., Wicks, R. W., and Warshaw, M.: Immunochemical determination of the heart isoenzymes of lactate dehydrogenase (LDH_1) in human serum. Clin. Chem., *25*:729–734, 1979.
109. Vassault, A., Wahlefeld, A. W., and Deneke, U.: Lactate dehydrogenase. *In*: Methods of Enzymatic Analysis, Vol. 3. 3rd ed. H. U. Bergmeyer, J. Bergmeyer, and M. Grassl, Eds. Weinheim, Verlag-Chemie, 1983, pp. 118–138.
110. Wilkinson, J. H.: The Principles and Practice of Diagnostic Enzymology. London, Edward Arnold, 1976.
111. Williamson, T.: The estimation of pancreatic lipase—a brief review. Med. Lab. Sci., *33*:265–279, 1976.
112. Yam, L. T.: Clinical significance of the human acid phosphatases: A review. Am. J. Med., *56*:604–616, 1974.
113. Ziegenhorn, J., Neumann, U., Knitsch, K. W., et al.: Determination of serum lipase. Clin. Chem., *25*:1067, 1979.
114. Ziegenhorn, J., Senn, M., and Bücher, T.: Molar absorptivities of β-NADH and β-NADPH. Clin. Chem., *22*:151–160, 1976.

The Lysosomal Enzymes

by JOHN F. O'BRIEN, Ph.D.

Inborn errors of metabolism is a general term that is applied to numerous genetic disorders. The lysosomal storage diseases are a distinct subset of the inborn errors of metabolism. These diseases result from accumulation in lysosomes of metabolites that would normally be degraded by one of the many hydrolytic enzymes that reside in these subcellular organelles. The lysosomal enzymes are unique inasmuch as they have acid pH optima. The specific enzyme deficiencies of the lysosomal storage diseases have been elucidated by identifying products stored in tissues and metabolites found in urine. The nomenclature of enzymes, substrates, storage products, and metabolites is not always systematic. The general names for the storage products are *sphingolipids* and *glycosaminoglycans*; the latter have been previously known as *mucopolysaccharides*.

In most cases, demonstration of enzyme deficiency by an assay using an appropriate nonphysiological substrate is sufficient for diagnostic purposes. Some lysosomal enzyme deficiencies can be detected by assay of a serum specimen; others require a specimen consisting of cells that contain lysosomes, such as leukocytes, fibroblasts, or cells from amniotic fluid. Erythrocytes are unsuitable for assay purposes. In affected cells, the deficiency of a lysosomal enzyme is constant; it is not inducible or episodic. Multiple allelic forms exist for many of the lysosomal enzymes and these may have differing kinetic or other functional properties that may be associated with variable expression of the clinical disease. One must expect that for each of these proteins there may be many variants and that the severity of disease may vary accordingly. The rare nature of the lysosomal storage diseases has made it impractical for most clinical laboratories to acquire the technical expertise for performance of all the lysosomal enzymes.

Among all the lysosomal storage diseases, *Tay-Sachs disease (TSD)* has been the most thoroughly studied. The biochemical defect responsible for TSD is a nearly total deficiency of the enzyme *N*-acetyl-β-hexosaminidase A (β-NAGS-A; Hex A).[14] The absence of enzyme activity results in the neuronal storage of G_{M2} ganglioside (Figure 12-15) and the attendant clinical symptoms. Because of the storage of G_{M2} gangliosides, this disease, along with at least three other variants, is classified as a G_{M2} gangliosidosis. The isoenzymes of β-NAGS that are present in highest concentration in human tissues and blood are commonly designated A and B. Deficiency of both A and B isoenzymes results in the *Sandhoff variant* of G_{M2} gangliosidosis, which is more rare than TSD. The difference in structure of β-NAGS isoenzymes allows the separation of the two isoenzymes by electrophoresis, ion-exchange chromatography, and heat stability. It is the last difference that is most often used to estimate the activity of each of the isoenzymes in a mixture.

In the autosomal recessive inherited disorders, clinical manifestation is apparent only in homozygotes. However, unaffected heterozygotes can often be detected by enzyme assay. It is more difficult to identify heterozygotes of X-chromosome–linked disorders because of the random inactivation of one or the other of the two X-chromosomes in each cell nucleus. The only method by which such heterozygotes might be detected is

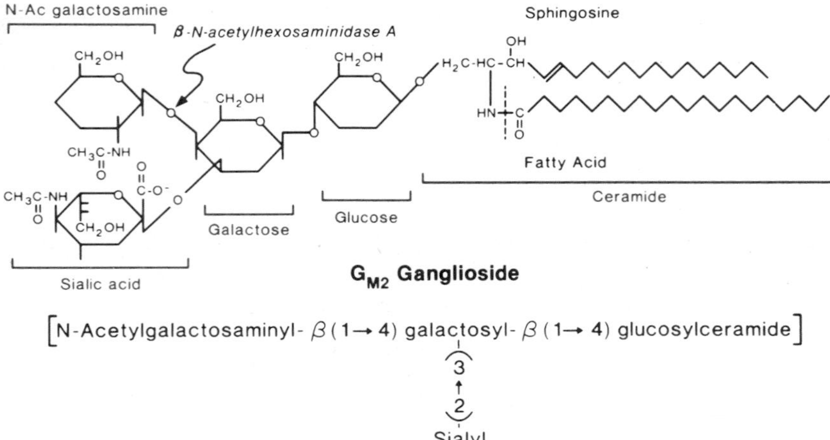

Figure 12-15. The structure of G_{M2} ganglioside that accumulates in neurons of the central nervous system in Tay-Sachs disease. The terminal monosaccharides are *N*-acetylgalactosamine and sialic acid. Both of these monosaccharides are linked to galactose, which is, in turn, linked to the ceramide moiety through a glucose unit. Ceramide is a combination of sphingosine and an amide-linked fatty acid.

by examining clones of cells. Carrier identification of any of the lysosomal storage diseases may be advanced in the future by use of specific gene probes.

Clinical significance. Most of these disorders are expressed early in life and may include delayed or regressing psychomotor performance, abnormal bone development, and ophthalmological abnormalities. Table 12-5 gives a summary of these disorders. The lysosomal enzyme deficiencies are reviewed in greater detail in reference 17.

Table 12-5. LYSOSOMAL STORAGE DISORDERS

Storage Disease	Deficient Enzyme	Inheritance
I. SPHINGOLIPIDOSES		
Niemann-Pick disease (*Sphingomyelin lipidosis*)	Sphingomyelinase[6]	Autosomal recessive
Gaucher's disease (*Glucosyl ceramide lipidosis*)	β-Glucocerebrosidase[16]	Autosomal recessive
Krabbe's disease (*Galactosylceramide lipidosis, globoid cell leukodystrophy*)	Galactosylceramide-β-galactosidase[18]	Autosomal recessive
Metachromatic leukodystrophy (*Sulfatide lipidosis*)	Arylsulfatase A[1,2]	Autosomal recessive
Fabry's disease (*Angiokeratoma corporis diffusum universale*)	α-Galactosidase A[4]	X-linked
G_{M2} Gangliosidoses:		
Tay-Sachs disease (*Infantile onset*)	β-*N*-Acetylglucosaminidase A	Autosomal recessive
Bernheimer-Seitelberger disease (*Juvenile onset*)	β-*N*-Acetylglucosaminidase A	Autosomal recessive
Adult—Chronic gangliosidosis (*Juvenile–adult onset*)	β-*N*-Acetylglucosaminidase A	Autosomal recessive
Sandhoff's disease (*A B variant*)	β-*N*-Acetylglucosaminidase A *and* B[12]	Autosomal recessive
Generalized gangliosidosis (*G_{M1} Gangliosidosis*)	β-Galactosidase[9,11]	Autosomal recessive
Ganglioside sialidase deficiency	Ganglioside sialidase (tentative)	Autosomal recessive
***II. MUCOPOLYSACCHARIDOSES (MPS)**[5]		
Hurler's syndrome (*MPS I*)	α-L-Iduronidase[10]	Autosomal recessive
Scheie's syndrome (*MPS Ia*)	α-L-Iduronidase[10]	Autosomal recessive
Hunter's syndrome (*MPS II*)	Iduronate sulfatase[10]	X-linked
Sanfilippo's syndrome (*MPS IIIa*)	Heparan *N*-sulfatase[8]	Autosomal recessive
Sanfilippo's syndrome (*MPS IIIb*)	α-*N*-Acetylglucosaminidase (α-hexosaminidase)[19]	Autosomal recessive
Sanfilippo's syndrome (*MPS IIIc*)	α-Glucosaminide-*N*-acetyltransferase	Autosomal recessive
Sanfilippo's syndrome (*MPS IIId*)	*N*-Acetylglucosamine-6-sulfate sulfatase	Autosomal recessive
Morquio's disease (*MPS IVa*)	Galactosamine-6-sulfate sulfatase	Autosomal recessive
Morquio's disease (*MPS IVb*)	β-Galactosidase[9]	Autosomal recessive
Maroteaux-Lamy disease (*MPS VI*)	Arylsulfatase-B[1]	Autosomal recessive
Glucuronidase deficiency (*MPS VII*)	β-Glucuronidase[7]	Autosomal recessive
III. DISORDERS OF UNDEFINED GLYCOPROTEIN AND/OR LIPID STORAGE PRODUCTS		
Fucosidosis	α-Fucosidase[3]	Autosomal recessive
Mannosidosis	α-Mannosidase[7]	Autosomal recessive
IV. DISORDERS PRESUMED TO INVOLVE MULTIPLE STORAGE PRODUCTS		
Mucolipidosis I (*ML I; sialidosis*)	Neuraminidase[13]	Autosomal recessive
Mucolipidosis II (*ML II; I cell disease*)	Biochemical defect in post-transcriptional modification that leads to absence of mannose-6-phosphate on a number of acid hydrolases, which is required for proper sequestration into lysosomes	Autosomal recessive
Mucolipidosis III (*ML III; pseudo-Hurler polydystrophy*)	See ML II	Autosomal recessive
Multiple sulfatase deficiency (*MSD*)	Deficiency of all lysosomal sulfatases to greater or lesser degrees (arylsulfatase A < arylsulfatase B)[1]	Autosomal recessive

*All the mucopolysaccharidoses (except for Morquio's disease) and mucolipidoses II and III show delayed breakdown of mucopolysaccharides in cultured skin fibroblasts. This is a helpful screening adjunct, which can be followed by specific enzyme assays if available.

Mucopolysaccharides are now called glycosaminoglycans. However, the term *mucopolysaccharidosis* (MPS) is retained because of its frequent use in the clinical description of these diseases.

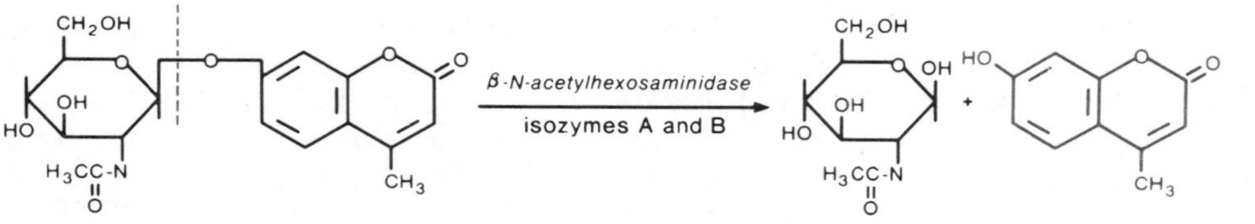

4-methylumbelliferyl-2-acetamido-2-deoxy-β-D-glucopyranoside

2-acetamido-2-deoxy-β-D-glucopyranose

4-methylumbelliferone (Fluorescent at 446 nm upon excitation at 365 nm)

Figure 12-16. The most frequently used reaction for the determination of β-*N*-acetylhexosaminidases involves hydrolysis of the 4-methylumbelliferyl derivative of *N*-acetylglucosamine (shown) or *N*-acetylgalactosamine. The free 4-methylumbelliferone product is quantitated by spectrophotofluorometry.

Methods

Lysosomal enzyme quantitation most often uses artificial glycosides or sulfate derivatives of 4-methylumbelliferone or *p*-nitrophenol as substrates. The former substrate yields a fluorescent product (Figure 12-16), and the latter yields a colored *p*-nitrophenol product. In addition, the demonstration of elevated urinary glycosaminoglycans and thereafter the delayed turnover of ^{35}S-labeled glycosaminoglycans in cultured skin fibroblasts can be used.

It is beyond the scope of this text to include all of the many enzyme assay methods. These are referenced in Table 12-5. Leukocytes can be used for most lysosomal enzyme assays. They are often requested by reference laboratories, and shipment of a preparation made from freshly drawn blood is superior to shipment of the whole blood specimen. For this reason, the method for purification of leukocytes is presented here.

Preparation of Leukocytes (Percy and Brady,[15] Modified)

Principle. Whole blood is allowed to sediment through a dense medium (e.g., dextran). Recovery of the supernatant and centrifugation at low centrifugal force result in a pellet made up largely of leukocytes. Contaminating erythrocytes are eliminated by selective shock treatment, since they are less resistant than leukocytes to lysis in hypotonic solutions.

Specimen. Ten milliliters of blood collected in acid-citrate-dextrose usually provide an adequate supply of leukocytes. Heparinized blood may also be used but gives considerably lower yield unless isolation is completed within a few hours after the specimen is drawn.

Reagents

1. Dextran solution, 5 g/dL. Into 16 × 125-mm tubes, weigh 0.5-g aliquots of dextran, grade HH, average MW 254 000 (Pharmachem Corp., Bethlehem, PA 18017). Immediately before use, dissolve one aliquot in 10.0 mL of NaCl, 0.7 g/dL.
2. NaCl, 0.7 g/dL. Place 0.7 g NaCl into a 100-mL volumetric flask and dilute to mark with water. Stable for two months at 4 °C.
3. NaCl, 0.9 g/dL. Place 0.9 g NaCl into a 100-mL volumetric flask and dilute to mark with water. Stable for two months at 4 °C.
4. NaCl, 1.8 g/dL. Place 1.8 g NaCl into a 100-mL volumetric flask and dilute to mark with water. Stable for two months at 4 °C.
5. ISOTON II (Coulter Electronics, Inc., Hialeah, FL 33010).

Procedure

1. Place 10 mL of anticoagulated blood into a 15-mL plastic centrifuge tube.
2. Add 2 mL of a freshly prepared solution of dextran, 5 g/dL, in NaCl, 0.7 g/dL. Mix gently by inversion.
3. Allow to stand for 45 min for sedimentation of cells.
4. Transfer supernatant with a plastic disposable pipet into another plastic centrifuge tube.

5. Centrifuge at 500 × g for 10 min in the cold.
6. Draw off supernatant and discard.
7. Resuspend the button of leukocytes in a 1.0 mL of *cold* NaCl, 0.9 g/dL.
8. Shock treat the cells as follows: Add 3.0 mL of ice-cold distilled H_2O and mix gently for 45 s. Immediately add 3.0 mL of *cold* NaCl, 1.8 g/dL, and mix.
9. Centrifuge at 500 × g for 10 min.
10. Draw off supernatant and discard.
11. Follow steps 7 through 10 for second shock treatment.

At this point, it is possible either to count the leukocytes or to determine the protein concentration in the preparation. These determinations are eventually used to determine the specific activity of the lysosomal enzyme. Circumstances in each laboratory may dictate which of these alternatives to use. If a Coulter Counter is readily available and the leukocyte preparation is freshly prepared, counting the cells provides narrower reference limits, probably because small cells and membrane contaminants are excluded from the count. When frozen pellets of leukocytes are used for the assay, counting is precluded, and the protein concentration must be determined. If the sample is to be sent out, the cells can be frozen as a pellet and the protein can be determined in the reference laboratory.

References

1. Baum, H., Dodgson, K. S., and Spencer, B.: The assay of arylsulphatases A and B in human urine. Clin. Chim. Acta, *4*:453–455, 1959.
2. Beratis, N. G., Aron, A. M., and Hirschhorn, K.: Metachromatic leukodystrophy: Detection in serum. J. Pediatr., *83*:824–827, 1973.
3. Beratis, N. G., Turner, B. M., Labadie, G., et al.: α-L-Fucosidase in cultured skin fibroblasts from normal subjects and fucosidosis patients. Pediatr. Res., *11*:862–866, 1977.
4. Desnick, R. J., Allen, K. Y., Desnick, S. J., et al.: Fabry's disease: Enzymatic diagnosis of hemizygotes and heterozygotes. J. Lab. Clin. Med., *81*:157–171, 1973.
5. Fratantoni, J. C., Hall, C. W., and Neufeld, E. F.: The defect in Hurler's and Hunter's syndromes: Faulty degradation of mucopolysaccharide. Proc. Natl. Acad. Sci. USA, *60*:699–706, 1968.
6. Gal, A. E., Brady, R. O., Hibbert, S. R., et al.: A practical chromogenic procedure for the detection of homozygotes and heterozygous carriers of Niemann-Pick disease. New Engl. J. Med., *293*:632–636, 1975.
7. Gehler, J., Cantz, M., Tolksdorf, E., et al.: Mucopolysaccharidosis VII: β-Glucuronidase deficiency. Humangenetik, *23*:149–158, 1974.
8. Hall, C. W., Liebaers, I., Di Natale, P., et al.: Enzymic diagnosis of the genetic mucopolysaccharide storage disorders. *In*: Methods in Enzymology, V. Ginsburg, Ed. New York, Academic Press, 1978, Vol. 50.
9. Ho, M. W., and O'Brien, J. S.: Differential effect of chloride ions on β-galactosidase isoenzymes: A method for separate assay. Clin. Chim. Acta, *32*:443–450, 1971.
10. Hopwood, J. J., Muller, V., Smithson, A., et al.: A fluorometric assay using 4-methylumbelliferyl α-L-iduronide for the estimation of α-L-iduronidase activity and the detection of Hurler and Scheie syndromes. Clin. Chim. Acta, *92*:257–265, 1979.
11. Kint, J. A.: Fabry's disease: Alpha-galactosidase deficiency. Science, *167*:1268–1269, 1970.
12. O'Brien, J. S., Okada, S., Chen, A., et al.: Tay-Sachs disease: Detection of heterozygotes and homozygotes by serum hexosaminidase assay. New Engl. J. Med., *283*:15–20, 1970.
13. O'Brien, J. S., and Warner, T. G.: Sialidosis: Delineation of subtypes by neuraminidase assay. Clin. Genet., *17*:35–38, 1980.
14. Okada, S., and O'Brien, J. S.: Tay-Sachs disease: Generalized absence of a beta-D-*N*-acetylhexosaminidase component. Science, *165*:698–700, 1969.
15. Percy, A. K., and Brady, R. O.: Metachromatic leukodystrophy: Diagnosis with samples of venous blood. Science, *161*:594–595, 1968.
16. Raghaven, S. S., Topol, J., and Kolodny, E. H.: Leukocyte β-glucosidase in homozygotes and heterozygotes for Gaucher's disease. Am. J. Hum. Genet., *32*:158–173, 1980.
17. Stanbury, J. B., Wyngaarden, J. B., Fredrickson, D. S., et al., Eds.: Part 5: Disorders of lysosomal enzymes. *In*: The Metabolic Basis of Inherited Disease. 5th ed. New York, McGraw-Hill, 1983.
18. Suzuki, K.: Enzymatic diagnosis of sphingolipidoses. *In*: Methods in Enzymology. V. Ginsburg, Ed. New York, Academic Press, 1978, Vol. 50.
19. von Figura, K., Logering, M., Mersmann, G., et al.: Sanfilippo B disease: Serum assays for detection of homozygous and heterozygous individuals in three families. J. Pediatr., *83*:607–611, 1973.

13 CARBOHYDRATES

WENDELL T. CARAWAY, Ph.D.
NELSON B. WATTS, M.D.

Glucose is the major energy source for the human body. The glucose level in blood is kept within a narrow range through a variety of influences. Although there is some variation in blood glucose as circumstances change (feeding, prolonged fasting), levels above or below the normal range usually indicate disease. Glucose determinations on serum or blood are ordered for the diagnosis and follow-up of abnormalities of carbohydrate metabolism (diabetes mellitus, hypoglycemia), and on cerebrospinal fluid glucose determinations are performed in conditions such as suspected meningitis.

Chemistry of Carbohydrates

Carbohydrates are the aldehyde and ketone derivatives of polyhydric alcohols. The simplest carbohydrate is glycol aldehyde, the derivative of ethylene glycol. The aldehyde and ketone derivatives of glycerol are, respectively, glyceraldehyde (glycerose) and dihydroxyacetone (Figure 13-1).

Monosaccharides

Sugars containing three, four, five, and six or more carbon atoms are known, respectively, as trioses, tetroses, pentoses, and hexoses (and so on) and are classified as monosaccharides. Aldehyde derivatives are called *aldoses* and ketone derivatives are called *ketoses*. The most common hexoses are shown in Figure 13-2. The carbon atoms in the chain are numbered 1 to 6 as shown by the numbers at the left of the formula for D-glucose. The designation D- or L- refers to the position of the hydroxyl group on the carbon atom next to the last (bottom) CH_2OH group. By convention, D-sugars are written with the hydroxyl group on the right and L-sugars are written with the hydroxyl group on the left. Most of the sugars occurring in the body are of the D-configuration. A number of different structures exist, depending on the relative positions of hydroxyl groups on the carbon atoms.

The formula for glucose can be written in either aldehyde or enol form. A shift to the enol anion is favored in an alkaline solution.

$$\underset{\text{Aldehyde}}{\begin{array}{l} H-C=O \\ \quad\;\, | \\ H-C-OH \\ \quad\;\, | \end{array}} \;\leftrightharpoons\; \underset{\text{Enol}}{\begin{array}{l} H-C-OH \\ \quad\;\, \| \\ \quad\;\, C-OH \\ \quad\;\, | \end{array}} \;\xrightarrow{OH^-}\; \underset{\text{Enol anion}}{\begin{array}{l} H-C-O^- \\ \quad\;\, \| \\ \quad\;\, C-OH \\ \quad\;\, | \end{array}} \;+ H_2O$$

The presence of a double bond and a negative charge in the enol anion form makes glucose an active reducing substance and provides one basis for its analytical determination. Thus, glucose in hot alkaline solution readily reduces metallic ions such as cupric to cuprous ions, and the color change can be used as a presumptive indication for the presence of glucose. Sugars capable of reducing cupric ions in alkaline solution are commonly known as reducing sugars.

Aldehyde and alcohol groups can react to form hemiacetals. In the case of glucose, the aldehyde group reacts with the hydroxyl group on carbon number 5 (C-5), as shown in Figure 13-3. With this ring structure the hydroxyl group on the first carbon can be written to the right or to the left. By convention, the form with the hydroxyl group on

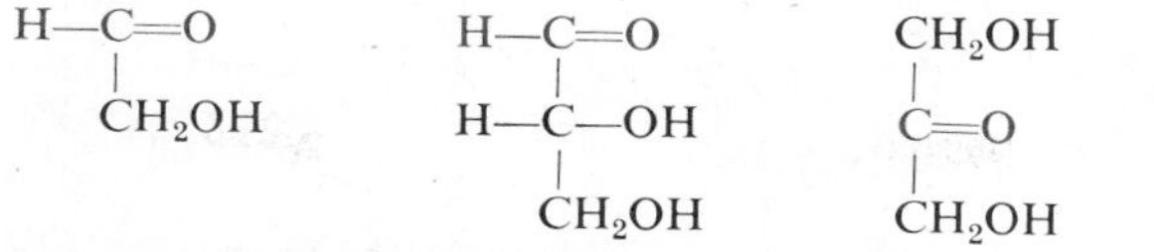

Glycol aldehyde	Glyceraldehyde (glycerose)	Dihydroxyacetone
H—C=O	H—C=O	CH_2OH
CH_2OH	H—C—OH	C=O
	CH_2OH	CH_2OH

Figure 13-1. Examples of 2- and 3-carbon carbohydrates.

the right is called α-D-glucose and the form with the hydroxyl group on the left is called β-D-glucose. The common anhydrous crystalline glucose is in the α-D-form. As a result of mutarotation, either form in an aqueous solution gives rise to an equilibrium mixture that has about 36% of the glucose in the α-form and 64% in the β-form; only a trace amount remains in the free aldehyde form. Since the enzyme glucose oxidase reacts only with β-D-glucose, standard solutions used in glucose oxidase methods for glucose determinations should stand at least 2 h to obtain equilibrium comparable with that in the test samples to be analyzed.

Disaccharides

Chemical interactions between two monosaccharides, with loss of a molecule of water, lead to the formation of disaccharides. The chemical bond between the saccharides always involves the aldehyde or ketone group of one monosaccharide joined to an alcohol group (e.g., in maltose) or to an aldehyde or ketone group (e.g., in sucrose) of the other monosaccharide. The linkage of oxygen on C-1 of sugars (C-2 in the case of fructose) with an alcohol is called a glycosidic linkage.

The most common disaccharides are maltose (glucose + glucose), lactose (glucose + galactose), and sucrose (glucose + fructose). If the linkage between two monosaccharides is between the aldehyde or ketone group of one molecule and a hydroxyl group of another molecule (as in maltose and lactose), there remains one potentially free ketone or aldehyde group on the second monosaccharide. Consequently, this sugar will be a reducing sugar. The reducing power is ~40% of that of the two single monosaccharides added together, since one of the reducing groups is not available. If the linkage between two monosaccharides involves the aldehyde or ketone groups of both molecules (as in sucrose), a nonreducing sugar results, since no free aldehyde or ketone group remains.

Polysaccharides

The linkage of many monosaccharide units results in the formation of polysaccharides. Each molecule of starch or glycogen, the chief carbohydrate reserve of plants and animals, respectively, typically contains 25–2500 glucose units. The suffix *-an* attached to a name of monosaccharide indicates the main type of sugar present in the polysaccharide. Starch and glycogen, for example, are *glucosans*, since they are composed of a series of individual glucose molecules.

Nearly all starches are composed of a mixture of two kinds of glucosans called amyloses and amylopectins. The proportions of these glucosans in starch vary from ~20%

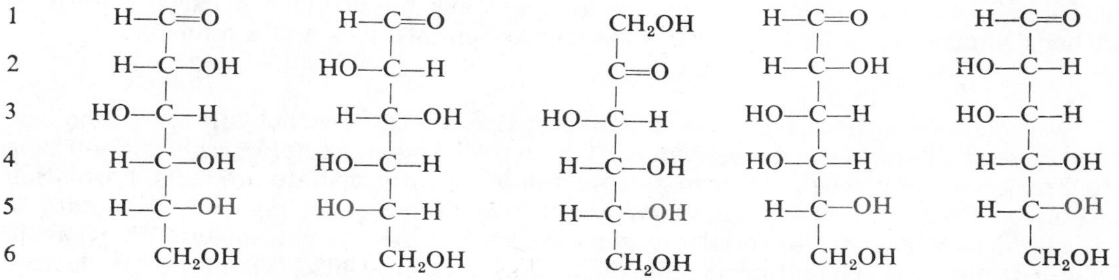

	D-Glucose	L-Glucose	D-Fructose	D-Galactose	D-Mannose
1	H—C=O	H—C=O	CH_2OH	H—C=O	H—C=O
2	H—C—OH	HO—C—H	C=O	H—C—OH	HO—C—H
3	HO—C—H	H—C—OH	HO—C—H	HO—C—H	HO—C—H
4	H—C—OH	HO—C—H	H—C—OH	HO—C—H	H—C—OH
5	H—C—OH	HO—C—H	H—C—OH	H—C—OH	H—C—OH
6	CH_2OH	CH_2OH	CH_2OH	CH_2OH	CH_2OH

Figure 13-2. Typical 6-carbon sugars.

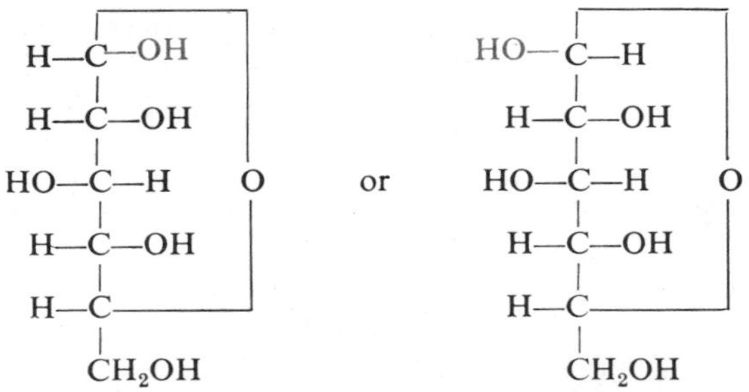

```
    H—C—OH                  HO—C—H
    H—C—OH                   H—C—OH
   HO—C—H      O      or    HO—C—H      O
    H—C—OH                   H—C—OH
    H—C—                     H—C—
      CH2OH                    CH2OH
```

α-D-**Glucose** β-D-**Glucose**

Figure 13-3. Structure of D-glucose (hemiacetal form).

amylose and ~80% amylopectin in wheat, potato, and ordinary corn starch to nearly 100% amylopectin in the starch of waxy corn. Although both amylose and amylopectin are made up of glucose molecules, there is one significant difference in their structure. Amylose consists of one long unbranched chain of 25–300 units of glucose, whose units are linked together by α-1,4-linkages with only the terminal aldehyde group free. In amylopectin the majority of the units are also connected with α-1,4-links but, in addition, there are α-1,6-glycosidic bonds that form side chains. Glycogens have structures similar to amylopectins except that branching is more extensive. The average length of a branch in a glycogen molecule is usually 12 or 18 D-glucose units, compared with about 25 units in amylopectin.

Metabolism of Carbohydrates

Starch and glycogen ingested as food are partially digested by the action of salivary amylase to form intermediate-sized glucosans, called dextrins, and maltose. Amylase activity is inhibited by the acid pH of the stomach. In the small intestine, where pH is increased by alkaline pancreatic juice, the amylase of the pancreas completes digestion of starch and glycogen to remnants called "limit dextrins" and to maltose. The latter, along with any ingested lactose and sucrose, is split by the disaccharidases (maltase, lactase, and sucrase) from the intestinal mucosa to form the monosaccharides glucose, galactose, and fructose. Absorption of these monosaccharides is fairly complete and appears to occur by an active carrier-mediated transfer process.

After absorption into the portal vein, hexoses are transported to the liver. Depending on the needs of the body, carbohydrates are converted to liver glycogen and stored; or metabolized completely to carbon dioxide and water to provide immediate energy; or converted to keto acids, amino acids, and proteins; or converted to fat (triglyceride) and stored in adipose tissue. Fructose and galactose are phosphorylated and eventually enter the same metabolic pathway as glucose.

Of clinical interest are metabolic disease states based on a deficiency of one or more enzymes in the intermediary metabolism of carbohydrates. A number of these are listed in Table 13-1. The complete picture of intermediary metabolism of carbohydrates is rather complex and interwoven with the metabolism of lipids and amino acids.

Regulation of Blood Glucose Concentration

Various terms are used to describe general processes in carbohydrate metabolism. *Glycogenesis* refers to the conversion of glucose to glycogen, and *glycogenolysis* refers to the breakdown of glycogen to form glucose and other intermediate products. Formation of glucose from noncarbohydrate sources, such as amino acids, glycerol, or lactate, is called *gluconeogenesis*. Conversion of glucose or other hexoses into lactate or pyruvate is called *glycolysis*. Oxidation of glucose to CO_2 and H_2O also occurs via the hexose monophosphate shunt pathway. The net result of all factors affecting these various processes determines the concentration of glucose in the blood.

Table 13-1. ENZYME DEFICIENCIES AND METABOLIC DISEASE STATES

Enzyme Deficiency	Disease State
Glucose-6-phosphatase	Type I GSD* (von Gierke's disease)
Muscle phosphorylase	Type V GSD* (McArdle's disease)
Hepatic phosphorylase	Type VI GSD* (Hers' disease)
Galactose-1-Puridyl transferase	Galactosemia
Galactokinase	Galactosemia
Hepatic fructokinase	Essential fructosuria
Hepatic fructose-1-P aldolase	Hereditary fructose intolerance
Pyruvate kinase	Hemolytic anemia
Glucose-6-P dehydrogenase	Hemolytic disease

*GSD = Glycogen storage disease.

During a brief fast, a precipitous drop in the level of blood glucose is avoided by drawing upon the glycogen stores of the liver, which contains the specific enzyme glucose-6-phosphatase, necessary for the conversion of glucose-6-phosphate (derived from either gluconeogenesis or glycogenolysis) to glucose. In contrast, glycogen stored in skeletal muscle cannot directly contribute glucose to the blood since muscle lacks this enzyme. During long-term fasting, gluconeogenesis plays a significant role in maintaining plasma glucose levels. On the other hand, as blood glucose levels increase, glycogenolysis is replaced by glycogenesis, whereby blood glucose is converted into glycogen in liver and muscle; thus, excessive blood glucose is avoided.

Hormones that Affect Blood Glucose Concentration

The concentration of glucose in blood is normally maintained in a narrow range despite changes in feeding and fasting. This is accomplished by the coordinated effects of insulin on one hand, which tends to lower blood glucose, and several "counterregulatory" hormones (glucagon, cortisol, epinephrine, and growth hormone) on the other hand, which tend to elevate glucose levels (Figure 13-4). ***Insulin***, produced by the B (beta)* cells of the pancreatic islets of Langerhans, enhances the entry of glucose into

*It has become customary to denote alpha, beta, and delta cells as A, B, and D cells, respectively, but the Greek letter designations are still widely used.

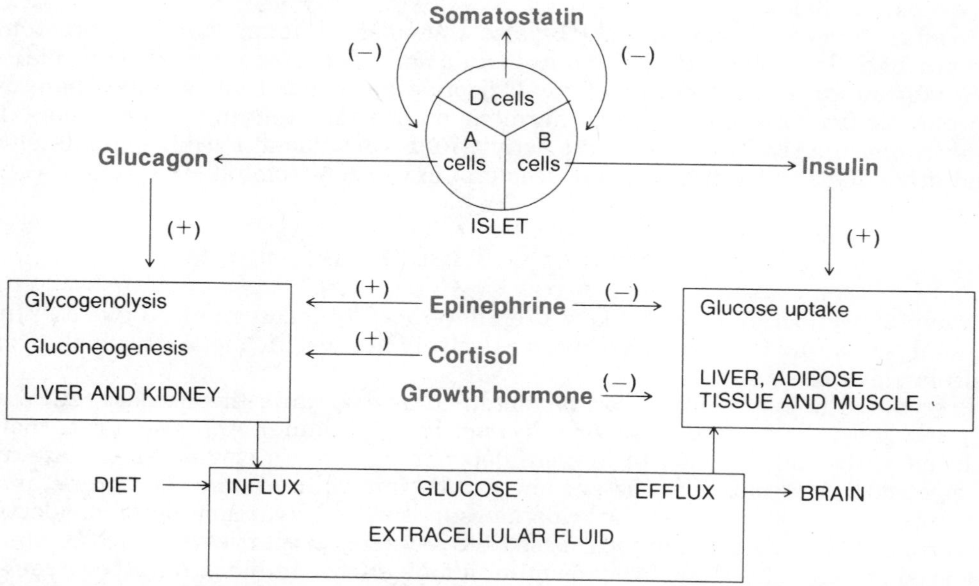

Figure 13-4. Hormonal influences on glucose homeostasis (see text). (+) Stimulation. (−) Inhibition.

liver, muscle, and adipose tissue and promotes the storage of energy substrate in the form of glycogen, fat, and protein.[39]

The absolute concentration of insulin required for a specific decrease of blood glucose varies considerably, depending on circumstances. Overweight individuals with normal carbohydrate metabolism require much higher levels of insulin than do normal-weight individuals with the same carbohydrate response, both in the basal state and after eating. The problem does not seem to be with the insulin molecule itself; there appears to be a resistance to the effect of insulin in the overweight individual that is possibly the result of fewer binding sites.[29]

Somatostatin is a polypeptide found in several areas of the body but is concentrated in the hypothalamus and in the D (delta) cells of the pancreatic islets. It inhibits secretion of glucagon and insulin, thus modulating the reciprocal relationship of these two hormones.

"Counterregulatory" hormones serve as substrate providers, enhancing breakdown of glycogen to glucose (glycogenolysis) and stimulating synthesis of glucose from glycerol and propionic acid derived from fat and from protein (gluconeogenesis). Changes in blood glucose, as well as other stimuli, will trigger the appropriate hormone response.

Growth hormone and adrenocorticotropic hormone (ACTH) are polypeptides secreted by the anterior pituitary. Both have an antagonistic action to insulin and tend to raise blood glucose concentration.

Cortisol, a hormone secreted by the adrenal cortex, stimulates gluconeogenesis. Therefore, individuals with Cushing's syndrome frequently develop hyperglycemia. Conversely, individuals with Addison's disease (destruction or atrophy of the adrenal cortex) have moderate hypoglycemia.

Epinephrine, secreted by the adrenal medulla, stimulates glycogenolysis. Physical or emotional stress causes increased production of epinephrine and an immediate increase in production of glucose for energy. Tumors of adrenal medullary tissue, known as pheochromocytomas, secrete excess epinephrine or norepinephrine and produce moderate hyperglycemia for as long as glycogen stores are available in the liver.

Glucagon, a polypeptide secreted by the A (alpha) cells of the pancreatic islets, increases blood glucose by stimulating hepatic glycogenolysis and gluconeogenesis. Glucagon has no effect on muscle glycogen.

Thyroxine, secreted by the thyroid gland, stimulates glycogenolysis, often leading to depletion of glycogen stores in the liver. The rate of glucose absorption from the intestine may be increased. These factors often contribute to a mild glucose intolerance in thyrotoxic individuals.

Human placental lactogen (hPL), also known as a human chorionic somatomammotropin (hCS), is a polypeptide with anti-insulin activity that is secreted by the placenta.

Somatomedins are a family of small peptide hormones that mediate the growth-promoting actions of human growth hormone on the skeleton and other tissues. They also have nonsuppressible insulin-like activity (NSILA). Somatomedin A, somatomedin C, and other IGFs (insulin-like growth factors) have been isolated.

Determination of Glucose in Body Fluids

Many analytical procedures have been developed to measure blood glucose. In the past, analyses were often performed with relatively nonspecific methods, which resulted in falsely elevated values.

Procedures in common use can be classified as *enzymatic* (hexokinase, glucose oxidase, and glucose dehydrogenase) or *colorimetric* (*o*-toluidine). An oxygen rate method, which measures consumption of oxygen, depends on the use of glucose oxidase and a pO_2 electrode. Alternatively, the enzyme can be immobilized on a membrane and the rate of production of hydrogen peroxide measured with a hydrogen peroxide electrode, as described below. Older methods, based on reducing properties of glucose, are now considered obsolete. Methods for determination of glucose in blood have been reviewed by Cooper.[4] Ten glucose methods have been evaluated and compared with a generally accepted hexokinase reference method.[32]

Collection and Handling of Specimens

When *blood* is drawn and permitted to clot and to stand uncentrifuged at room temperature, the average decrease in serum glucose is ~7% in 1 h (5–10 mg/dL).[45] This decrease is the result of glycolysis. Plasma removed from cells after moderate centrifugation contains leukocytes that also metabolize glucose. In separated, unhemolyzed *serum*, the glucose concentration is generally stable up to 8 h at 25 °C, if kept free of bacterial contamination, and up to 72 h at 4 °C. Without preservative, plasma or serum must be separated from the cells or clot within 1 h after blood is drawn if glucose values within 10 mg/dL of the original value are to be obtained consistently.

Glycolysis can be prevented and glucose in whole blood stabilized for up to 24 h at room temperature by adding sodium fluoride to the specimen. Fluoride ions prevent glycolysis by inhibiting enolase. Fluoride ions also inhibit coagulation by binding calcium; however, clotting may occur after several hours. It is therefore advisable to use a combined fluoride-oxalate mixture, such as 2 mg of potassium oxalate (or $EDTA \cdot Na_2$) and 2.5 mg of sodium fluoride per mL of blood, to prevent late clotting. Fluoride ions in high concentration inhibit the activity of urease and certain other enzymes; consequently, specimens are unsuitable for determination of urea in some procedures that require urease and for direct assay of some serum enzymes.

Cerebrospinal fluid frequently contains bacteria or other cellular constituents and should be analyzed for glucose without delay. Glucose may be preserved in 24-h collections of *urine* by adding 5 mL of glacial acetic acid to the container before starting the collection. The final pH of the urine is usually between 4 and 5, and bacterial activity is inhibited at this level of acidity. The use of 5 g of sodium benzoate per 24-h specimen is also effective in preserving the urine.

Hexokinase Methods

Principle. Hexokinase is an enzyme that catalyzes phosphorylation of glucose by adenosine triphosphate (ATP) to form glucose-6-phosphate and adenosine diphosphate (ADP). To follow the reaction, a second enzyme, glucose-6-phosphate dehydrogenase (G-6-PD), is used to catalyze oxidation of glucose-6-phosphate by nicotinamide adenine dinucleotide phosphate ($NADP^+$) to form NADPH in proportion to the glucose originally present. The increase in absorbance at 339 (340) nm due to NADPH provides a measure of glucose content. When G-6-PD derived from yeast is used as the enzyme, $NADP^+$ is the cofactor. NAD^+ is used instead of $NADP^+$ if the source of the dehydrogenase is bacterial (*Leuconostoc mesenteroides*).

$$\text{Glucose} + \text{ATP} \xrightarrow{\textit{Hexokinase}} \text{Glucose-6-phosphate} + \text{ADP}$$

$$\text{Glucose-6-phosphate} + \text{NAD(P)}^+ \xrightarrow{\textit{G-6-PD}} \text{6-Phosphogluconate} + \text{NAD(P)H} + \text{H}^+$$

A generally accepted reference method based on this principle has been developed and validated.[26] Serum or plasma is deproteinized by adding solutions of barium hydroxide and zinc sulfate. The clear supernatant is mixed with a reagent containing ATP, NAD^+, hexokinase, and glucose-6-phosphate dehydrogenase. The mixture is incubated at 25 °C until the reaction is complete and the absorbance related to the production of NADH is measured at 340 nm. Standards and blanks are carried through the entire procedure, including the deproteinization step.

Although capable of high accuracy and precision, the reference method is too exacting and time-consuming to serve as a routine method in the clinical laboratory. An *alternative approach* is to apply the reaction directly to serum or plasma and use a specimen blank to correct for interfering substances that absorb at 340 nm. Since most methods in use rely on commercially prepared reagents supplied in lyophilized form, only a general discussion of the direct procedure will be presented.

Either serum or plasma may be used. Hemolyzed specimens containing >0.5 g hemoglobin/dL are unsatisfactory owing to the release from erythrocytes of phosphate esters and enzymes that interfere.

Nineteen substances were tested for interference with the reaction. Only fructose affected the blanked direct hexokinase method in serum. However, only small amounts of fructose are usually present in normal fasting serum.

Absorbances of samples or standards are measured against corresponding blanks after the reaction has gone to completion. Although glucose concentrations may be calculated directly, based on the molar absorptivity of NAD(P)H, inclusion of a set of standards is recommended to detect possible deterioration of enzymes, ATP, $NADP^+$, or NAD^+, all of which are unstable. Reagents may also contain substances that react with coenzymes. Presence of these substances can be evaluated by measuring the increase in absorbance observed in a reagent blank. The highest standard provides a check on linearity of response and the adequacy of the enzyme reagent.

Hexokinase procedures are available in which indicator reactions produce colored products so that absorbance may be measured in the visible range. An oxidation-reduction system containing phenazine methosulfate (PMS) and a substituted tetrazolium compound, 2-(*p*-iodophenyl)-3-*p*-nitrophenyl-5-phenyltetrazolium chloride (INT), is reacted with NADPH formed in the reaction. The reduced INT is colored with maximum absorbance at 520 nm. The PMS-INT color developer must be refrigerated when not in use and must be protected from exposure to light to retard autoreduction. Co-immobilized hexokinase and G-6-PD on the inner surface of small-bore plastic tubing are used in certain continuous-flow automated systems with considerable savings in reagent cost.

Glucose Oxidase Methods

Principle. The enzyme glucose oxidase catalyzes the oxidation of glucose to gluconolactone and hydrogen peroxide:

$$\text{Glucose} + O_2 \xrightarrow{\textit{Glucose oxidase}} \text{D-Glucono-}\delta\text{-lactone} + H_2O_2$$

Addition of the enzyme peroxidase and a chromogenic oxygen acceptor, such as *o*-dianisidine, results in formation of color that can be measured:

$$\underset{\text{(colorless)}}{o\text{-Dianisidine}} + H_2O_2 \xrightarrow{\textit{Peroxidase}} \underset{\text{(colored)}}{\text{Oxidized } o\text{-dianisidine}} + H_2O$$

Glucose oxidase is highly specific for β-D-glucose. As noted, glucose in solution exists as ~36% α- and ~64% β-form. Complete reaction of glucose, therefore, requires mutarotation of the α- to β-form. Some commercial preparations of glucose oxidase contain an enzyme, mutarotase, that accelerates this conversion. Otherwise, extended incubation time allows for spontaneous conversion.

The second step, involving peroxidase, is much less specific than the glucose oxidase reaction. Various substances such as uric acid, ascorbic acid, bilirubin, and glutathione inhibit the reaction, presumably by competing with the chromogen for hydrogen peroxide, which results in lower values. Some glucose oxidase preparations contain catalase as a contaminant; catalase activity decomposes hydrogen peroxide and decreases the final color obtained. Standards and unknowns should be analyzed simultaneously under conditions such that the rate of oxidation is proportional to glucose concentration.

In some methods the final mixture is acidified slightly to stop the reaction, and the yellow color is measured at 400 nm. In a stronger acid solution the color becomes pink, with maximum absorbance at 540 nm; both sensitivity and stability are improved. Other approaches to measurement of hydrogen peroxide produced include peroxide-mediated oxidative coupling of 3-methyl-2-benzothiazolinone hydrazone (MBTH) with *N,N*-dimethylaniline (DMA), catalyzed by peroxidase,[9] or oxidative coupling of *p*-aminophenazone (PAP) to phenol.[43] Both procedures have been automated. MBTH/DMA and PAP procedures are not affected by high concentrations of creatinine, uric acid, or hemoglobin and are performed directly on serum.

Glucose oxidase methods are suitable for measuring glucose in *cerebrospinal fluid* but are not directly applicable to urine specimens, owing to the high concentration of substances that interfere with the peroxidase reaction and produce falsely low results. Uric acid is the major interfering substance in urine.

Some instruments employ a *polarographic oxygen electrode*, which measures the rate of oxygen consumption after addition of the sample to a solution containing glucose oxidase.[17] Since this measurement involves only the first reaction shown above, interferences encountered in the peroxidase step are eliminated. To prevent formation of O_2

from H_2O_2 by catalase present in some preparations of glucose oxidase, H_2O_2 is removed by inclusion of two additional reactions:

$$H_2O_2 + C_2H_5OH \xrightarrow{Catalase} CH_3CHO + 2\,H_2O$$

Ethanol Acetaldehyde

$$H_2O_2 + 2\,H^+ + 2\,I^- \xrightarrow{\text{Molybdate}} I_2 + 2\,H_2O$$

The latter reaction is effective even when catalase activity has diminished upon storage of reagents. The procedure can be applied directly to urine, serum, plasma, or cerebrospinal fluid.

In the YSI Model 23A (Yellow Springs Instrument Co., Yellow Springs, OH 45387), glucose oxidase is immobilized in a thin layer of resinous material sandwiched between two membranes. When a buffered sample is introduced, glucose diffuses through the first polycarbonate membrane and reacts with the enzyme to produce H_2O_2. The latter then diffuses through the second smaller-pore cellulose acetate membrane and contacts a platinum anode, where oxidation takes place. The current thus created is directly proportional to the glucose concentration in the diluted sample.

$$H_2O_2 \rightarrow 2\,H^+ + O_2 + 2\,e^-$$

The circuit is completed by a silver cathode, at which oxygen is reduced to water.

$$4\,e^- + 4\,H^+ + O_2 \rightarrow 2\,H_2O$$

Any H_2O_2 diffusing back into the sample chamber is destroyed by catalase to prevent interference with the analysis. Determinations may be performed on 25 μL of plasma, serum, or whole blood. Good precision and correlation with an oxygen consumption rate analyzer have been reported.[3]

The Kodak Ektachem system (Eastman Kodak Co., Rochester, NY 14650) makes use of *dry multilayer films* for chemical analyses.[5] Glucose is measured by a glucose oxidase procedure. A 10-μL sample of serum is placed on a porous film on top of the layer containing the reagents. Glucose diffuses through the film and reacts with the reagents to produce a colored end product or dye. The intensity of this dye is measured through a lower transparent film by reflectance spectrophotometry. Advantages of this system include small sample size, elimination of liquid reagents, and improved storage requirements.

Reference ranges. The normal range for fasting adult *plasma* or *serum* glucose, determined by specific methods, is 70–105 mg/dL. For whole blood the range is 60–95 mg/dL. There is no sex difference and the ranges given are applicable to children after the first few weeks of life. In the premature newborn, plasma levels may range from 25 to 80 mg/dL; in the full-term infant, expected values are 30–90 mg/dL. Fasting plasma glucose values increase ~2 mg/dL per decade in the adult years. Postprandial values increase at the rate of 4 mg/dL per decade and those following a glucose challenge, as much as 8–13 mg/dL per decade.[30]

Cerebrospinal fluid glucose values are usually ~60% of the plasma values (fasting range 40–75 mg/dL) and should be compared with concurrent plasma values for adequate clinical interpretation. The amount of glucose excreted in the *urine*, determined by highly specific enzymatic methods, is <500 mg/d. Random specimens show an upper limit of normal of ~30 mg/dL. Less specific methods for measuring glucose in urine are discussed in the section on *Urinary Sugars*.

Glucose Dehydrogenase Methods

Principle. The enzyme glucose dehydrogenase (GDH; β-D-glucose:NAD oxidoreductase, EC 1.1.1.47) catalyzes oxidation of glucose to gluconolactone:

$$\text{Glucose} + NAD^+ \xrightarrow{\textit{Glucose dehydrogenase}} \text{D-Glucono-}\delta\text{-lactone} + NADH + H^+$$

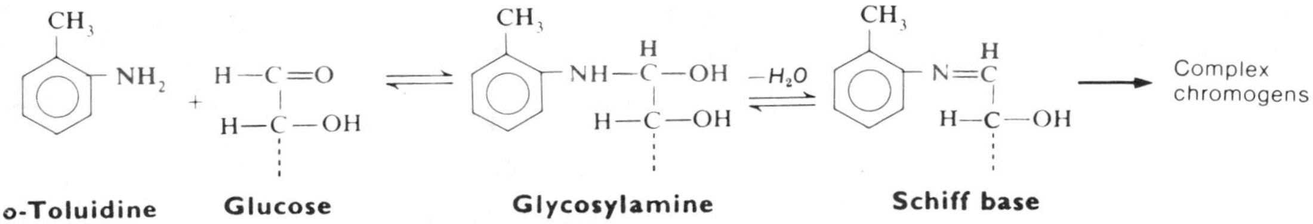

Figure 13-5. Reaction of glucose with *o*-toluidine.

Mutarotase is added to shorten the time necessary to reach the end point. The amount of NADH generated is proportional to the glucose concentration.

Glucose dehydrogenase, for this assay, is isolated from *Bacillus cereus*. The reaction appears to be highly specific for glucose, shows no interference from common anticoagulants and substances normally found in serum, and provides results in good agreement with hexokinase procedures. Methods have been adapted to continuous-flow analyzers, including the use of immobilized enzymes, and to centrifugal analyzers.[22] At this writing, the glucose dehydrogenase procedure is not widely used, although some commercial reagent sets are available.

Relative rates of reaction of other sugars with GDH compared with that of glucose are: glucose, 100%; D-glucosamine, 31%; D-xylose, 15%; D-mannose, 8%; fructose, none; and galactose, none. Concentrations of other reacting sugars in serum should be negligible; however, the procedure should not be used for blood glucose determinations during oral xylose absorption tests.

o-Toluidine Methods

Principle. Various aromatic amines react with glucose in hot acetic acid solution to produce colored derivatives. Among those used are aniline, benzidine, 2-aminobiphenyl, and *o*-toluidine. Since the others are (or are suspected) carcinogens, only *o*-toluidine finds wide use. The latter condenses initially with the aldehyde group of glucose to form an equilibrium mixture of a glycosylamine and the corresponding Schiff base, as illustrated in Figure 13-5. Further rearrangements and reactions take place after the original condensation to produce a mixture of green chromogens with an absorption maximum at 630 nm.[46]

Sugars other than glucose produce variable amounts of color in the reaction, which appear to depend somewhat on reaction conditions and should be determined by the individual laboratory. Of the sugars that occur naturally in serum, only galactose produces an absorbance comparable with that of glucose. Pentoses react with *o*-toluidine to produce an orange color with maximum absorbance near 480 nm.

Moderate hemolysis of the specimen does not interfere significantly with the *o*-toluidine method for measuring glucose. Bilirubin contributes some increase in absorbance because it is partially converted to the green pigment biliverdin. This limitation of the method must be considered when measuring glucose in serum of infants with elevated bilirubin levels. EDTA in concentrations >1 mg/mL and sodium fluoride at levels >5 mg/mL in the specimen will cause some increase in color. Thymol preservative should be avoided since it inhibits color formation.

Reference ranges for the *o*-toluidine method are essentially the same as those for the glucose oxidase and hexokinase procedures, since all three methods give similar results. In patients with uremia, somewhat higher values are obtained with the *o*-toluidine method.

Home Monitoring of Blood Glucose

Persons with diabetes, especially those with Type I insulin-dependent diabetes, require some type of monitoring to improve control of blood glucose concentrations. Testing urine for glucose is a simple but unreliable guide since the renal threshold must be exceeded before glucose appears in the urine. Renal thresholds vary considerably and tend to be higher in diabetic individuals, especially in elderly persons who have decreased glomerular filtration rates associated with arteriosclerosis and glomerulosclerosis. Insulin pumps, which now replace daily injections for many patients, should not be used without frequent monitoring of blood glucose. It is impractical for patients themselves to perform blood or plasma glucose determinations by the methods given above, but simple test strips are now available that permit rapid and reasonably accurate measurements on a drop of whole blood. Colors developed on the strips may be evaluated visually or measured quantitatively on meters designed especially for such strips.[41]

Home monitoring of blood glucose is useful for all patients with diabetes, but is especially recommended for the following:[40] diabetic women during pregnancy, users of insulin infusion pumps, patients with renal disease, and patients with unstable, insulin-dependent diabetes. Two methods are described briefly; both depend on the glucose oxidase-peroxidase chromogenic reaction. In both methods, a small surface area of the test strips is impregnated with combined reagents in dry form.

With Dextrostix (Ames Co., Elkhart, IN 46514) a drop of blood is placed on the reagent pad, let stand exactly 60 s, then washed off for 2 s with water from a wash bottle. The strip is read on a reflectance colorimeter (several models are available), which displays results on a digital display screen with a range of 0–399 mg/dL. The meter must be calibrated and controlled daily prior to use. To obtain valid results, the test must be performed exactly as directed. Both the 60-s incubation and 2-s wash times are crucial. The strips may not be held under tap water, since this tends to result in over-washing, removal of chromogen, and thus in falsely low results. Readings must be made immediately, since the color is unstable; hence, the strip may not be kept for later validation.[35] Visual reading of Dextrostix with a color chart is not accurate enough for most clinical circumstances.[41]

The bG Chemstrips (Bio-Dynamics, Indianapolis, IN 46250) have two adjacent reagent zones. Each zone is covered simultaneously with one drop of blood. After 60 s, the blood is wiped off with a cotton ball. After a further 60 s, colors are compared with a color chart. The two zones appear as different colors and provide a range of 20–800 mg/dL of blood glucose. The final color is usually stable for several days, depending on temperature and humidity; thus, results can be checked later by a trained observer. The Accucheck bG meter (Bio-Dynamics) can be used for a digital reading with bG Chemstrips.

With careful attention to instructions, results between 80 and 240 mg/dL obtained by self-monitoring compare reasonably well with those obtained by conventional laboratory procedures.[41] Values <80 mg/dL tend to be lower with strip tests, whereas values >240 mg/dL can be very erratic. Results with the strip tests obtained on whole blood are 5–10% less than results obtained on plasma or serum. In a report of May 31, 1984, Dr. Donald Widmann, Director of Standards for the Joint Commission on Accreditation of Hospitals (JCAH), released minimum requirements for quality control procedures for glucose monitoring using these techniques.

DIABETES MELLITUS

Diabetes mellitus, as currently understood, is almost certainly more than a single disorder. Of particular importance, yet poorly understood, is the propensity for persons with diabetes mellitus to develop specific complications (retinopathy with blindness, kidney failure with uremia, nerve damage, and circulatory problems that predispose to tissue damage and may lead to amputations, heart disease, and stroke). Diabetes mellitus is a common disorder, but the actual prevalence of this disease depends on the criteria used for its diagnosis. According to criteria that will be discussed later in this section, there are perhaps as many as 10 million persons with diabetes mellitus in the United States; half of them are not aware that they have the disease.

Hyperglycemia is the hallmark of diabetes mellitus. The common underlying defect is a *deficiency of insulin secretion or action, which leads to the development of hyperglycemia*. This may be an absolute deficiency of insulin production, typical of Type I diabetes, or a relative deficiency of insulin, commonly seen in Type II diabetes. In both types, insulin production by the B cells of the pancreas is impaired. In most persons with Type II diabetes, there is also resistance to the peripheral action of insulin. This insulin resistance may be due to changes in insulin receptor number or affinity, or in postreceptor events.

Classification of Hyperglycemic Disorders

In an effort to standardize terminology applied to diabetes mellitus, the National Diabetes Data Group has proposed a classification for hyperglycemic disorders that has been widely adopted.[25]

Insulin-dependent diabetes mellitus (IDDM, Type I). Persons with this type of diabetes mellitus are usually acutely ill at the time of diagnosis with severe hyperglycemia and other metabolic derangements. These individuals produce very little insulin and thus have an *absolute deficiency of insulin*. Without insulin treatment, these metabolic changes will cause death in a short time. With insulin treatment, satisfactory control of blood glucose can be obtained, yet many of these individuals will develop significant microvascular changes with time. Diagnosis of this type of diabetes is fairly straightforward when the patient presents in the acute stages. Type I, insulin-dependent diabetes mellitus, makes up about 10% of all patients with diabetes mellitus.

IDDM appears to be due to autoimmune destruction of B cells. Several studies have shown that circulating antibodies to pancreatic islet cells are present in patients for months or years prior to the development of overt diabetes. Even when glucose and insulin responses to an oral glucose tolerance test are normal, these "prediabetics" have impaired early insulin release after intravenous glucose. In a few centers, clinical trials are underway using cyclosporine, an immunosuppressive drug, in patients with new-onset IDDM, with hopes of preventing or minimizing B-cell destruction. Preliminary data are encouraging.

Non-insulin-dependent diabetes mellitus (NIDDM, Type II). NIDDM is the most common form of diabetes mellitus (80–90% of patients). Most individuals with NIDDM have plasma insulin concentrations that are higher than normal but not sufficient to maintain a normal blood glucose concentration in the face of insulin resistance. Thus, most persons with this form of diabetes mellitus have a *relative deficiency of insulin activity*. Obesity is commonly associated with this form of diabetes, and a reduction in body weight usually brings about an improvement in hyperglycemia and, often, normal glucose concentrations. This is not always the case, however, and some persons with NIDDM may require insulin or an oral hypoglycemic agent to control symptomatic hyperglycemia. Patients with NIDDM are subdivided into (1) those patients who are not obese and (2) those who are obese; these subgroups are further divided into (a) those who require insulin for control of hyperglycemia and (b) those who do not require insulin. Early diagnosis of this type of patient is difficult, since the disease is usually slow to develop and progress, and precise separation of all persons with NIDDM from those who are normal may not be possible.

Impaired glucose tolerance (IGT). This category identifies individuals who are not clearly normal at the time of testing but are not sufficiently abnormal to be certain that they have diabetes mellitus. They have a greater than usual risk for the development of diabetes mellitus at a later age, and there is a slow rate of progression to overt diabetes (1–5% a year). In this group, there is also a greater than normal risk for the development of cardiovascular disease. Microvascular disease is quite rare in this group, even in patients in whom definite hyperglycemia appears with time. This category is useful for identifying persons whose glucose tolerance is not clearly normal, but it avoids the inappropriate label of diabetes mellitus for persons in this group.

Gestational diabetes mellitus (GDM). Gestational diabetes is diagnosed when an abnormal glucose concentration is discovered for the first time during pregnancy; this is not the same as a known diabetic who becomes pregnant. Early diagnosis of gestational diabetes is important so that the patient can be monitored and treatment can be instituted when necessary. There is a high probability (30–60%) that a woman with gestational diabetes will develop diabetes mellitus later in life, but carbohydrate tolerance often reverts to normal after delivery and may remain normal.

Previous abnormality of glucose tolerance (Prev. AGT). This category includes persons known to have had an abnormality of glucose tolerance at one time but who have been shown to have normal results on repeat testing. For example, this would include someone who had Type II diabetes mellitus, was obese, was not treated with insulin, and after weight reduction had a normal glucose tolerance test.

Potential abnormality of glucose tolerance (Pot. AGT). This includes persons at greater than the average theoretical risk for the later development of diabetes mellitus, for example, the identical twin of a patient with Type II diabetes or the child of parents who each had Type II diabetes.

Glucose intolerance associated with certain conditions and syndromes. This category includes the uncommon patient in whom hyperglycemia is due to a specific underlying disorder (such as chronic pancreatitis with endocrine and exocrine insufficiency, Cushing's disease, acromegaly, or glucagon-producing islet-cell tumor of the pancreas), to other diseases, or to the use of certain drugs that are known to cause hyperglycemia.

The previously used terms *prediabetes, latent diabetes, chemical diabetes, subclinical diabetes, borderline diabetes*, and *asymptomatic diabetes* do not appear in this classification, but these conditions are covered in the categories of Pot. AGT, Prev. AGT, and IGT. Juvenile-onset, or ketosis-prone, diabetes is now classified as Type I diabetes, and maturity-onset, or ketosis-resistant, diabetes as Type II diabetes. Although insulin-dependent diabetes is most commonly seen in young persons and non-insulin-dependent diabetes usually develops in middle age, these disorders are not age specific.

Diagnostic Strategies

The demonstration of significant hyperglycemia is the key to the diagnosis of diabetes mellitus. For Type I diabetes, the diagnosis is usually simple, since hyperglycemia appears abruptly, is severe, and is accompanied by serious metabolic derangements (e.g., metabolic ketoacidosis). It is in Type II diabetes mellitus (NIDDM) that early diagnosis becomes troublesome.

The risk for the later development of microvascular disease makes it important to identify patients with Type II diabetes. Although it seems that bringing glucose levels to normal or near normal will help prevent or minimize microvascular disease, the evidence is inconclusive.

Defining normal. Some difficulty exists in setting limits of hyperglycemia that differentiate normal individuals from those with diabetes mellitus, either in the fasting state or after a glucose challenge is administered orally. Most populations show a unimodal distribution for plasma glucose results. Early population studies, which were used to set standards for normal and abnormal responses to an oral glucose load, were done on young healthy individuals—a group not particularly likely to develop Type II diabetes mellitus. For this reason, some well-accepted standards for interpreting an oral glucose tolerance test are unreasonably low when dealing with a population that is at high risk for the development of Type II diabetes mellitus (i.e., middle-aged and overweight individuals).

Effect of age. Postprandial and postchallenge glucose values tend to increase with age.[30] For 2-h values, this increase may be as high as 10 mg/dL per decade after age 40. Thus, values >200 mg/dL may be found in elderly individuals who do not have diabetes mellitus.

Postprandial plasma glucose. Demonstration of unequivocal hyperglycemia (plasma glucose >200 mg/dL) 2 h or more after a mixed meal is considered diagnostic for diabetes mellitus. A person with a clear elevation of plasma glucose needs no further diagnostic testing. A person with a value <140 mg/dL (though most people usually have <120 mg/dL) is unlikely to have diabetes mellitus. Thus, the postprandial plasma glucose results either will be normal and exclude the diagnosis of diabetes mellitus or will quickly establish the diagnosis where it is clear cut, leaving only a small number of individuals with borderline values who need to be studied further.

Post-challenge plasma glucose. Administering a standard oral glucose load* (75 g is recommended, though 50- and 100-g loads have been used)[25] and determining the plasma glucose concentration 2 h later allow for more objective conditions of testing than the postprandial method and are therefore a more precise comparison of one individual's response with that of a large reference group. This test has the advantage of being more precise than that for postprandial glucose, where the meal content, the time

---

*A commercially prepared and flavored carbohydrate load equivalent to this glucose dose is also acceptable.[25]

used to consume a meal, and absorption of the meal cannot be completely controlled. However, the procedure requires preparation: i.e., unrestricted activity and a diet containing ≥150 g of carbohydrates daily for three days before the test, fasting for 10–16 h prior to the test, testing in the morning, and sampling 2 h after ingestion of the glucose load. For the 2-h postchallenge glucose, sustained values >200 mg/dL are considered diagnostic of diabetes mellitus.

Fasting plasma glucose. An elevation of fasting plasma glucose values is highly specific for diabetes mellitus. Values >140 mg/dL on two or more occasions would be diagnostic. Fasting hyperglycemia, however, occurs relatively late in Type II diabetes mellitus, so that waiting for fasting hyperglycemia to appear delays the time of diagnosis.

Oral glucose tolerance test (GTT). Serial measurement of plasma glucose before and after glucose is given orally should provide a standard method to evaluate individuals and establish values for normal and disease states. There are a number of factors that may affect glucose tolerance and that should be controlled or eliminated before such a test is performed; other conditions altering the glucose response need to be controlled during the test (Table 13-2). As a result, abnormal values may occur in the absence of diabetes mellitus, and repeat testing in the same individual does not always give reproducible results, even under the same test conditions. When an oral glucose tolerance test is ordered, the following conditions should be met: omit medications known to affect glucose tolerance; perform the test in the morning after 3 d of unrestricted diet and activity; and perform the test after a 10–16 h fast. Plasma glucose should be measured fasting, then every 30 min for 2 h after an oral glucose load. For adults, the recommended load is 75 g and for children, 1.75 g/kg, up to 75 g maximum. There is still some question as to what constitutes the ideal glucose load. In adults, 50 g is suboptimal, whereas 100 g frequently causes nausea and vomiting; 75 g is the consensus but may not be a maximal stimulus.[25]

The National Diabetes Data Group has proposed a standard set of criteria for use in the diagnosis of diabetes mellitus.[25] According to these criteria, no GTT is necessary if fasting plasma glucose is >140 mg/dL or the 2-h postprandial value is >200 mg/dL. Diabetes is confirmed if the 2-h plasma glucose *and* at least one other glucose value in a GTT are >200 mg/dL. According to these criteria, diabetes mellitus in its early stages may not be identified but persons with abnormal results can be diagnosed with relative certainty. These criteria should be used for interpretation of glucose tolerance test results.

An oral glucose tolerance test is not always necessary for the diagnosis of diabetes mellitus, particularly in a person with classic symptoms of diabetes mellitus (thirst, frequent urination, unexplained weight loss). The oral glucose tolerance test is indicated in the following situations:

1. Diagnosis of gestational diabetes mellitus.
2. Further evaluation of an individual with a borderline elevation of fasting or postprandial plasma glucose.

Table 13-2. FACTORS THAT AFFECT GLUCOSE TOLERANCE

Factors During the Test That Affect Glucose Tolerance	
Posture	Cigarettes
Nausea	Time of day
Anxiety	Activity
Coffee	Amount of glucose ingested
Factors Before the Test That Affect Glucose Tolerance	
Carbohydrate intake	Propranolol
Time of previous food intake	Corticosteroids
GI surgery and malabsorption	Age
Thiazides	Inactivity
Estrogens	Weight
Diphenylhydantoin	Stress (surgery, infection)

From Watts, N. B. (used with permission).[44]

Table 13-3. DIAGNOSIS OF GESTATIONAL DIABETES (GDM)

Oral Glucose Tolerance Test (OGTT) for GDM
Omit medications known to affect glucose tolerance. Perform in A.M. after 3 days of unrestricted diet and activity and after 10–16 h fast. Measure fasting plasma glucose. Give 100 g of glucose or equivalent orally. Measure plasma glucose hourly for 3 h. At least 2 values must exceed the following values: Fasting: 105 mg/dL 1 h: 190 mg/dL 2 h: 165 mg/dL 3 h: 145 mg/dL If results are normal in a clinically suspicious situation, repeat OGTT during the third trimester.

3. Risk counseling in an individual with a previously abnormal glucose tolerance test under suboptimal conditions (repeat results will often be normal).

4. Risk counseling in persons at high risk for the development of diabetes mellitus, but with normal fasting and postprandial plasma glucose values (a normal result, however, does not mean that diabetes mellitus will not develop in the future).

5. Evaluation of a patient with unexplained nephropathy, neuropathy, or retinopathy (abnormal results in this setting do not necessarily indicate a cause-and-effect relationship, and further studies will be needed to exclude conclusively other diseases).

6. Population studies for epidemiologic data.

The role of the glucose tolerance test in the diagnosis of hypoglycemia is discussed later in this chapter.

Evaluation for gestational diabetes mellitus (GDM). Clues from the history of a patient that would raise the suspicion of gestational diabetes include a strong family history of diabetes mellitus, a history of unexplained stillbirth or neonatal death, a history of an infant with a congenital anomaly, the delivery of an infant weighing 4000 g or more, or a poor reproductive history. Suspicious clinical findings include obesity, recurrent monilial infections, hydramnios, or glycosuria. Pregnant women with symptoms of hyperglycemia (thirst, frequent urination, unexplained weight loss) should have serum glucose measured without delay to look for clear-cut and serious hyperglycemia; pregnant women with some other reason to suspect gestational diabetes can be evaluated in a more systematic way.

The O'Sullivan *screen for gestational diabetes* is a simple way to evaluate the significance of a suspicious history or clinical presentation.[31] The woman is given a glucose load after fasting 10–16 h, and plasma glucose is measured afterward. A plasma glucose >150 mg/dL 1 h after a 50-g glucose load, or a plasma glucose >140 mg/dL 2 h after a 100-g glucose load indicates a high probability of gestational diabetes, and these women should be tested with an oral glucose tolerance test. If results of the initial screening are normal, the screening test should be repeated in the third trimester, since carbohydrate tolerance deteriorates in the second half of pregnancy.

When an oral glucose tolerance test is performed for diagnosis of gestational diabetes, the conditions and criteria for diagnosis are different from those when the test is done for the diagnosis of idiopathic diabetes mellitus (Table 13-3).[31] A 3-h glucose tolerance test using a 100-g glucose load is used for the diagnosis of gestational diabetes.

GLYCOHEMOGLOBINS

In adults, circulating hemoglobin (Hb) typically consists of Hb A (97% of the total), Hb A_2 (2.5%) and Hb F (0.5%). Chromatographic analysis of Hb A has shown that it contains a number of minor hemoglobins, identified as Hb A_{1a}, Hb A_{1b}, and Hb A_{1c}. Collectively, these are referred to as Hb A_1, "fast hemoglobins," glycosylated hemoglobins, glycated hemoglobins, or simply glycohemoglobins. Hemoglobin A consists of four polypeptide chains, two α-chains and two β-chains. Hemoglobin A_{1c} is formed by the condensation between glucose and the *N*-terminal valine amino acid of each β-chain to

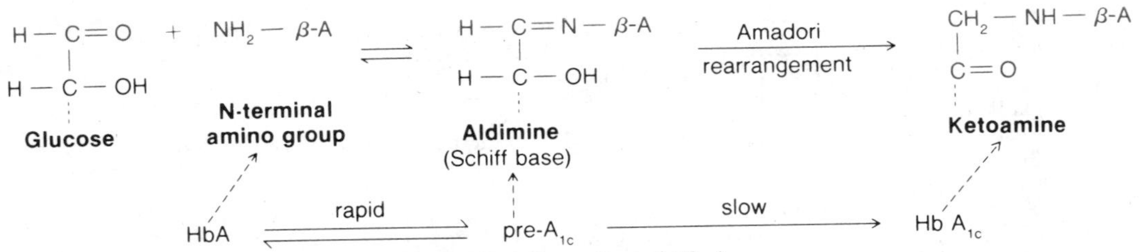

H—C=O + NH_2—β-A ⇌ H—C=N—β-A —Amadori rearrangement→ CH_2—NH—β-A
H—C—OH | H—C—OH | C=O
Glucose | N-terminal amino group | Aldimine (Schiff base) | Ketoamine
HbA ⇌(rapid) pre-A_{1c} —slow→ Hb A_{1c}

Figure 13-6. Formation of Hb A_{1c}.

form an unstable Schiff base (aldimine), which then undergoes an Amadori rearrangement to form a stable ketoamine (Figure 13-6). Hb A_{1c} is the major fraction of the "fast hemoglobins."

The formation of the glycohemoglobins is nonenzymatic and occurs over the life span of the erythrocyte (average 120 days). The amount of total Hb A_1 or Hb A_{1c} depends on the time-averaged glucose concentration over the two or three months prior to the measurement. This value provides a single objective means of reflecting control of diabetes, unrelated to short-term fluctuations in plasma glucose levels. Thus, measurement of glycohemoglobins supplements other more traditional methods of assessing control. It becomes a means for judging control of diabetes in a new patient or when urine glucose records are inadequate, when blood glucose levels vary markedly within one day or from day to day, and in patients during pregnancy, when close control is especially important. Measurement of Hb A_{1c} also provides additional confirmation of the clinical impression of control.

The interpretation of Hb A_{1c} values is based on the assumption of a normal erythrocyte life span. Patients with hemolytic disease, or other conditions with shortened erythrocyte survival, have a significant reduction in Hb A_{1c}.

Most methods for measurement of glycohemoglobins yield values for total Hb A_1, rather than Hb A_{1c} alone. This is of little clinical significance, since the two values have a high degree of correlation.[14] Of more concern is the inclusion of the labile intermediate (pre-A_{1c}) in most assays, including those by chromatography, HPLC, and electrophoresis.[24] The labile fraction changes rapidly with acute changes in blood glucose concentration; hence, methods for Hb A_1 or Hb A_{1c}, which include pre-A_{1c}, may show changes not truly reflecting longer time-averaged glucose concentrations. The labile fraction amounts to 5–8% of total Hb A_1 in normal individuals and ranges from 8 to 30% in patients with diabetes, depending on the degree of control of blood glucose levels. Figure 13-6 suggests that pre-A_{1c} should revert to glucose and Hb A in the absence of excess glucose. This reversion to Hb A is the basis for some procedures in which washed erythrocytes are incubated in saline solution to eliminate the labile fraction. Thus, if the analytical method includes both fractions in the measurement, the labile pre-A_{1c} should first be removed to avoid factitious results.

Nonenzymatic attachment of sugars to amino groups of other proteins also occurs. A stable glycosylated form of serum albumin has been found to be elevated in humans with diabetes.[6] Since the circulating half-life of human serum albumin is ~19 days, this suggests that monitoring of glycosylated albumin would provide a more recent time-averaging of blood glucose than would determination of Hb A_1.

Methods for the Determination of Glycohemoglobins

Currently available methods for the determination of glycohemoglobins include ion-exchange chromatography, high performance liquid chromatography, colorimetry, radioimmunoassay, electrophoresis, isoelectric focusing, and affinity chromatography.

Ion-exchange chromatography. Short columns are used that contain weakly acidic cation-exchange or carboxymethyl cellulose resin. Glycosylated hemoglobins at a selected ionic strength and pH of the eluent buffer are less positively charged than Hb A and elute from the column first. A second buffer may be used to elute Hb A, and the absorbance of the two eluates is used to calculate per cent of total glycosylated hemoglobin. Alternatively, only the Hb A_1 is eluted, and a separate dilution of the original hemolysate is made, against which the Hb A_1 is compared. A number of commercial modifications of the procedure are available.

In all ion-exchange column methods it is important to control the temperature of reagents and columns to obtain accurate and reproducible results. In addition, rigid control of pH and ionic strength must be maintained. Results with three commonly used, commercially available column-chromatographic methods were found to correlate satisfactorily with a high performance liquid chromatographic method.[11]

The labile fractions elute with the stable forms and will produce elevated results unless pretreatment of the erythrocytes is included to remove the labile form (see below). Hemoglobin F (fetal hemoglobin) typically elutes with Hb A_1 and produces falsely elevated results. On the other hand, Hb S, Hb C, and their glycosylated derivatives, if present, do not elute. Hence, misleading low values for Hb A_1 are obtained in the presence of Hb S and Hb C.

High performance liquid chromatography (HPLC). HPLC has been introduced as a method for separating and quantitating Hb A_{1c} as well as other hemoglobin fractions and has since been recommended as a reference method.[28]

Colorimetry. A colorimetric method is based on the observation that Hb A_{1c}, when subjected to mild acid hydrolysis, releases 5-hydroxymethylfurfural (5-HMF). Oxalic acid is added to an aliquot of the hemolysate, the mixture is heated at 100 °C for 5 h and cooled, and proteins are precipitated with trichloroacetic acid. The supernatant is reacted with thiobarbituric acid to produce a colored solution whose absorbance is measured at 443 nm. Because the reaction is specific for ketoamine-linked glucose, it is unaffected by the presence of Hb F, hemoglobin variants, and labile Hb pre-A_{1c}. However, the test is difficult to standardize because the yield of 5-HMF from Hb A_{1c} is only about 30%. An evaluation of the colorimetric assay indicates that, under controlled conditions, the assay provides reliable results.[33]

Radioimmunoassay. Sheep antiserum against Hb A_{1c} has been prepared and used as the basis for a radioimmunoassay.[15] Partial cross-reactivity with Hb A_{1a} was noted. At present, the antiserum is not commercially available, and the method has not been fully evaluated.

Electrophoresis. Agar gel electrophoresis on whole blood hemolysates at pH 6.3 has been shown to provide good resolution of Hb A and Hb A_1. In addition, Hb S and Hb C and their glycosylated components are resolved and do not interfere. Hb F migrates to the same region as Hb A_1 and would cause a falsely elevated Hb A_1 value. This method includes the labile aldimine as part of the total Hb A_1 fraction.[24]

Isoelectric focusing. Hemolysates are prepared from washed erythrocytes and diluted with a phosphate-glycerol reagent. Ampholines in the pH range of 6–8 are used to establish the gradient in 1-mm-thick acrylamide gel slabs. On completion of isoelectric focusing, the gels are fixed in trichloroacetic acid and then scanned on a high-resolution, integrating microdensitometer. Hemoglobin A_{1c} is adequately resolved from hemoglobins A_{1a}, A_{1b}, S, and F.[38]

Affinity chromatography. Affinity gel columns are used to separate glycosylated hemoglobin from the nonglycosylated fraction. The gel contains immobilized *m*-aminophenylboronic acid on crosslinked, beaded agarose. Boronic acid reacts with the *cis*-diol groups of glucose bound to hemoglobin to form a reversible 5-membered ring complex, thus selectively holding the glycosylated hemoglobin on the column. Nonglycosylated hemoglobin is eluted. The complex is next dissociated by sorbitol, which permits elution of glycosylated hemoglobin. Absorbances of the two fractions, measured at 415 nm, are used to calculate the per cent of glycosylated hemoglobin.

Advantages of affinity chromatography are (1) no interference from nonglycosylated hemoglobins, (2) negligible interference from the labile intermediate form of Hb A_{1c}, and (3) minimal dependence on variations in ambient temperature in contrast to ion-exchange chromatography. The method has been evaluated in some detail.[20] Values in the reference (normal) range tend to agree well with those of other procedures. At higher values the results are generally more increased, presumably owing to the inclusion of glucose attached to free amino groups in hemoglobin in addition to the *N*-terminal amino groups.

Removal of labile glycohemoglobin from erythrocytes. The concentration of the labile form of Hb A_{1c} fluctuates rapidly in response to changes in plasma glucose levels. To reflect accurately long-term blood glucose regulation, especially in poorly controlled diabetics, the labile form should be removed prior to assay. This may be accomplished by incubating erythrocytes in saline or buffer solutions at pH 5–6. Either procedure results in a suspension of erythrocytes suitable for use in all procedures discussed above.

Incubation with saline: Collect blood with EDTA anticoagulant, centrifuge, and remove the plasma and buffy coat. To 0.5 mL of packed cells add 5 mL of saline (9 g NaCl/L), mix, and centrifuge. Aspirate and discard the supernatant. Add 5 mL of saline, mix, and incubate in a 37 °C water bath for 5 h. The mixture may be incubated overnight at room temperature (~14 h at 21–25 °C), if this is more convenient. Centrifuge and discard the supernatant. Add 0.5 mL of saline to the cells, mix, and assay the mixture for glycohemoglobin as if it were whole blood.

Incubation at pH 5: Collect blood with EDTA anticoagulant, centrifuge, and remove the plasma and buffy coat. To 0.5 mL of packed cells add 5 mL of citrate buffer (0.1 mol/L, pH 5.0), mix, and incubate in a 37 °C water bath for 15 min. Centrifuge and discard the supernatant. Add 0.5 mL of saline to the cells, mix, and assay the mixture for glycohemoglobin as if it were whole blood.

To prepare the citrate buffer, dissolve 8.62 g of citric acid monohydrate and 17.35 g of trisodium citrate dihydrate in water and dilute to 1 L.

Reference Ranges

Values for glycohemoglobins are usually expressed as a percentage of total blood hemoglobin. Reference ranges vary depending on the method, the subfractions measured (Hb A_1 or Hb A_{1c}), and whether the labile fraction is included in the assay. A consensus, based on several studies of normal subjects, suggests the following reference ranges.

	Mean, %	Range, %
Hb A_1 (A_{1a+b+c})	6.5	5.0–8.0
Hb A_{1c} only	4.5	3.0–6.0

Reference ranges show some increase with age, in agreement with similar observations of fasting blood glucose levels. In poorly controlled patients with diabetes mellitus, values may extend to twice the upper limit of normal or more, but rarely exceed 20%.

KETONE BODIES

The metabolism of fatty acids results in the formation of a small amount of acetoacetate, which is subsequently metabolized in peripheral tissues. When there is carbohydrate deprivation (e.g., starvation) or decreased utilization of carbohydrates (e.g., diabetes mellitus), there is increased production of acetoacetate, which may exceed the capacity of peripheral tissues to metabolize this compound. Thus, acetoacetate accumulates in blood, where a small part is converted to acetone by spontaneous decarboxylation and the greater part is converted to β-hydroxybutyrate in accordance with the following reactions:

$CH_3-CO-CH_2-COO^-$
Acetoacetate

β-Hydroxybutyrate dehydrogenase (in liver): NADH + H^+ → NAD^+

Spontaneous: H^+ → CO_2

$CH_3-CH(OH)-CH_2-COO^-$
β-Hydroxybutyrate

$CH_3-CO-CH_3$
Acetone

The relative proportions in which the three ketone bodies are present in blood vary; average figures are 78% β-hydroxybutyrate, 20% acetoacetate, and 2% acetone. None of the most commonly used methods for the detection and determination of ketone bodies in serum or urine react with all three ketone bodies. Tests employing nitroprusside are 15–20 times more sensitive for acetoacetate than for acetone and give no reaction at all with β-hydroxybutyrate. Thus, the tests to be described essentially detect or measure acetoacetate only. Tests for β-hydroxybutyrate are indirect; they require brief boiling of the urine to remove acetone and acetoacetate by evaporation (acetoacetate first breaks down spontaneously to acetone), followed by gentle oxidation of β-hydroxybutyrate to acetoacetate and acetone with peroxide, ferric ions, or dichromate. Acetoacetate thus formed may be detected with one of the procedures employing nitroprusside. Determination of β-hydroxybutyrate in urine is not considered to be a routine procedure.

Clinical significance. Excessive formation of ketone bodies results in increased blood concentrations (ketonemia) and increased excretion in the urine (ketonuria). This process is observed in conditions associated with a decreased intake of carbohydrates, such as starvation, digestive disturbances, dietary imbalance, and frequent vomiting. The most frequent cause of excessive ketone formation is decreased utilization of carbohydrates, as in diabetes mellitus.

Semiquantitative determination of ketone bodies in plasma or urine is a helpful guide in the treatment of diabetic ketoacidosis. Excess production of ketone bodies in the diabetic leads to ketoacidosis with a decrease in plasma pH and bicarbonate. Treatment of such patients must be directed to the metabolic abnormalities and to the associated acid-base and electrolyte problems.

Determination of Ketone Bodies in Serum and Urine

Although a number of quantitative and semiquantitative methods for the estimation of ketone bodies have been devised, the semiquantitative Acetest and Ketostix reagents (Ames Co., Division of Miles Laboratories, Elkhart, IN 46514) offer information sufficient for clinical purposes.[8]

Specimens. Serum or plasma should be free of visible hemolysis, since discoloration of the tablet or reagent strip may occur if an excessive amount of hemoglobin is present. Urine specimens should be fresh. All specimens should be refrigerated if not tested within 30 min.

Detection of Ketone Bodies by Acetest

Principle. Acetest tablets contain a mixture of glycine, sodium nitroprusside, disodium phosphate, and lactose. Acetoacetate or acetone in the presence of glycine will form a lavender-purple complex with nitroprusside. Disodium phosphate provides an optimal pH for the reaction, and lactose enhances the color.

Procedure. Acetest was designed mainly for detection of ketone bodies in urine. Directions of the manufacturer should be closely followed. If serum is used, the tablet should be crushed and a drop of serum should be added to the powder. Failure to do so will result in falsely low results.

Urine specimens containing very high quantities of phenylketones may give false positive results, as will specimens preserved with 8-hydroxyquinoline. L-Dopa metabolites may produce an atypical reaction, which could be misinterpreted as a positive result.

A positive reaction (appearance of purple-lavender color) indicates the presence of 5–10 mg or more of ketone bodies per dL. A color chart provided with the package may be used to estimate the actual concentration of the ketone bodies. If desirable, dilutions of serum with saline can be prepared to measure levels of ketone bodies >80 mg/dL. A "large" reaction in an undiluted sample corresponds to ~80 mg/dL; hence, a "large" reaction in a twofold dilution corresponds to ~160 mg/dL. Since the reaction is affected by proteins, any dilution with saline introduces a certain amount of error.

Detection of Ketone Bodies by Ketostix

Ketostix is a modification of the nitroprusside test, in which a reagent strip is used instead of a tablet. Ketostix gives a positive reaction within 15 s with 5–10 mg acetoacetate per dL. Approximate acetoacetate values are assigned to the color blocks. Ketostix does not react with acetone.

HYPOGLYCEMIA

Hypoglycemia as a laboratory diagnosis has no specific defined limits. It is not uncommon for an isolated plasma glucose concentration to be as low as 50 mg/dL several hours after the ingestion of an oral glucose load; even in the fasting state, there may be an occasional extremely low blood glucose value without concurrent symptoms or evidence of underlying disease. An accurate interpretation of a single low blood glucose value is impossible without knowledge of the clinical setting in which it occurs.

There are no symptoms that are specific for hypoglycemia. A rapid fall in plasma glucose will usually trigger the release of epinephrine, and it is epinephrine that accounts for the signs and symptoms which are most commonly attributed to "hypoglycemia" (i.e., weakness, shakiness, sweating, nausea, rapid pulse, lightheadedness, hunger, and epigastric discomfort). Identical signs and symptoms often occur with other conditions, such as hyperthyroidism, pheochromocytoma, and anxiety. The symptoms may occasionally be triggered by a rapid drop in blood glucose concentration even though the blood glucose itself may not drop below the reference range. These symptoms can be termed "adrenergic" symptoms and may occur 2–3 h after eating.

Often, a gradual fall in plasma glucose is not accompanied by epinephrine release or associated adrenergic symptoms. A very low level of plasma glucose (<20–30 mg/dL) causes impairment of central nervous system function, such as confusion, lethargy, seizures, or loss of consciousness; these symptoms are known collectively as *neuroglycopenia*. It is extremely important to differentiate adrenergic signs and symptoms from those of neuroglycopenia and to realize that adrenergic symptoms are often caused by problems other than low blood glucose.

When a low blood glucose level is detected after fasting, there is usually serious underlying organic disease. Such an individual should be studied intensively; the work-up is best done in a hospital because serious permanent neurological damage may occur from prolonged low blood glucose levels.

Hypoglycemia in Infancy and Childhood

It is generally accepted that values for plasma glucose <30 mg/dL in a term infant and <20 mg/dL in a premature infant are abnormal; some authorities feel that values <50 mg/dL are abnormal and medical attention is required. Oral glucose tolerance testing is not generally indicated in childhood. If spontaneous hypoglycemia can be demonstrated, appropriate studies to determine the diagnosis should be undertaken.

Hypoglycemia in Adults

When planning a hypoglycemia evaluation, it is helpful to consider whether hypoglycemia occurs spontaneously in the fasting state or seems to be triggered by the ingestion of food, because appropriate diagnostic strategies for these two situations are considerably different.

Fasting hypoglycemia. Fasting hypoglycemia in adults is rare. A precise point for separation between the low range of normal and abnormally low is not possible; values as low as 30 mg/dL may be seen in healthy premenopausal women after a 72-h fast. However, fasting plasma glucose concentrations ≤50–60 mg/dL are uncommon and should be investigated.

Patients with hypoglycemia in the fasting state will generally have serious organic disease. The patient should be allowed a liberal intake of non-glucose-containing fluids, and specimens should be drawn frequently for plasma glucose during such a fast (at least every 4 h) to anticipate a dangerously low value before it occurs. If a fast has progressed to 48 h with no hypoglycemia, the patient should be exercised and a final specimen drawn before the fast is terminated. It is uncommon to find significant hypoglycemia in a patient who tolerates a 48-h fast and exercise without a fall in blood glucose.

Table 13-4. CAUSES OF SPONTANEOUS (FASTING) HYPOGLYCEMIA IN ADULTS

Medications (insulin, oral hypoglycemic agents, alcohol)
Severe hepatic dysfunction
Deficiency of adrenal glucocorticoids or other counterregulatory hormones
Extrahepatic neoplasms
Insulin-producing pancreatic tumors
Sepsis

Conditions that should be considered in patients with fasting hypoglycemia are shown in Table 13-4. A low plasma glucose (<30 mg/dL) with an inappropriately high plasma insulin (>15 μU/mL) is strong evidence for the presence of an insulin-producing tumor.[7] Measurement of C-peptide or proinsulin may establish the diagnosis in the rare insulinoma patient in whom the tumor produces more proinsulin than insulin.[1,13]

Appropriate historical information should be taken to exclude the possibility of drug-induced hypoglycemia. The C-peptide assay can help in establishing surreptitious injection of insulin (C-peptide should be suppressed).[13] Sepsis as a cause of hypoglycemia should be readily apparent, and laboratory tests can confirm the suspicion of hepatic dysfunction or glucocorticoid deficiency.

Nonpancreatic neoplasms that cause hypoglycemia represent a heterogeneous group.[18] Often, these neoplasms are extremely large mesenchymal tumors that may overutilize glucose and also may inhibit glucose mobilization. Occasionally, small tumors have been reported that cause hypoglycemia, apparently through the production of substances known as insulin-like growth factors or nonsuppressible insulin-like activity (NSILA).[10] Locating the tumor is the most direct way to arrive at this diagnosis.

The oral glucose tolerance test is not an appropriate method to evaluate a patient suspected of having fasting hypoglycemia.

Postprandial hypoglycemia. In most cases, the question of hypoglycemia will be raised when a patient complains of adrenergic symptoms occurring ~2 h after eating and seems to obtain relief by food intake. There are specific situations in which these symptoms are likely to be related to low blood glucose; for example, in early diabetes mellitus, when insulin release is delayed and exaggerated, or after gastrointestinal surgery with rapid gastric emptying. However, many individuals with adrenergic complaints in the postprandial setting exhibit these symptoms when their blood glucose is clearly normal. A 5- or 6-h glucose tolerance test has been the standard procedure to establish the presence of postprandial hypoglycemia, but it is not an ideal test because symptoms and blood glucose levels often do not correlate.[16]

The best diagnostic strategy is to obtain a blood specimen at the time the patient is experiencing the symptoms. Finding a normal blood glucose at a time when symptoms are present is strong evidence that the symptoms are not causally related to hypoglycemia. If it is not possible to obtain a blood glucose at the time symptoms occur spontaneously, performance of a 5-h meal tolerance test[2,19] has the advantage of simulating normal circumstances. A blood glucose concentration <50 mg/dL with associated adrenergic symptoms would suggest a cause-and-effect relationship.

Other Tolerance Tests

Intravenous glucose tolerance test. Poor absorption of oral glucose may result in a "flat" tolerance curve. Some patients are unable to tolerate a large carbohydrate load orally (vomiting) or may have altered gastric physiology (e.g., gastric resection). In these cases an intravenous glucose tolerance test may be performed. Plasma insulin assays may also be requested in conjunction with the test.

The dose of glucose is 0.5 g/kg of body weight, given as a 25 g/dL solution administered intravenously within 2–4 min. Blood is collected every 10 min for 1 h. If insulin assays are performed, a specimen is also obtained 5 min after the start of the injection.

Tolbutamide test. Tolbutamide, or 1-butyl-3-(*p*-tolylsulfonyl)urea (Orinase), stimulates the normal pancreas to produce insulin. After intravenous injection of this drug, the normal response is a decrease in blood glucose to ~50% of the fasting level by 30 min, followed by a return to normal. The test is sometimes useful in evaluating hypoglycemic states caused by insulinomas, when a marked decrease in blood glucose to values in the range of 20–30 mg/dL and hypoglycemia persist up to 3 h. If the insulin response to tolbutamide is measured, diagnostic data can usually be obtained by 60 min, thus reducing the risk of severe hypoglycemia. Patients must be watched carefully for hypoglycemic reactions, in which case the test is terminated by intravenous administration of glucose.

Epinephrine tolerance test. This test is used to evaluate one form of glycogen storage disease (Type I, von Gierke's disease), a condition in which there is a deficiency or absence of the enzyme glucose-6-phosphatase in the liver. This enzyme is the catalyst for the final step in the formation of blood glucose from hepatic glycogen, which is normally stimulated by epinephrine. Individuals with von Gierke's disease have low blood glucose, increased liver glycogen, but decreased *availability* of liver glycogen as shown by less than normal or no increase in blood glucose following intravenous administration of epinephrine.

Lactose tolerance test. A lactose tolerance test can be used to evaluate deficiency of small bowel mucosal lactase,[27] which may be associated with intolerance to lactose manifested by diarrhea and other symptoms following ingestion of milk. This test and its interpretation are discussed in Chapter 23.

LACTATE AND PYRUVATE

Clinical significance. Lactic acid, present in *blood* entirely as lactate ion (pK 3.86), is an intermediary product of carbohydrate metabolism and is derived mainly from muscle cells and erythrocytes. It is normally metabolized by the liver. The blood lactate concentration is affected by its rate of production as well as its rate of metabolism. During exercise, lactate levels may increase significantly from an average normal concentration of ~0.9 mmol/L to ~12 mmol/L. However, pyruvate levels increase under these conditions as well, and the normal ratio of lactate to pyruvate remains approximately 6:1 or 7:1.

Severe oxygen deprivation of tissues results in a blockage of aerobic oxidation of pyruvic acid in the tricarboxylic acid cycle and subsequent glycolytic reduction of pyruvate to lactate. This leads to a severe acidosis, called lactic acidosis, which is associated with an increase in the lactate:pyruvate ratio in blood and lactate levels to as high as 25 mmol/L (225 mg/dL) or more. Such extreme findings are associated with marked hyperpnea, weakness, fatigue, stupor, and finally coma. Clinical examples are irreversible stage of shock, diabetic coma without ketosis, and a variety of illnesses in the terminal stage (see also Chapter 19, Section Three). Hypoxia seen in shock, cardiac decompensation, hematological disorders, and pulmonary insufficiency are associated with less extreme increases in blood lactate.

The liver can normally metabolize more lactate than is produced. In the case of decreased perfusion of the liver, however, removal of lactate by the liver may be significantly reduced.

The amount of lactate in cerebrospinal fluid normally parallels blood levels. In cases of biochemical alterations in the central nervous system, however, CSF lactate values change independently of blood values. Increased CSF levels are detected in cerebrovascular accidents, intracranial hemorrhage, bacterial meningitis, epilepsy, and other CNS disorders.

Determination of Lactate in Whole Blood[21,23]

Principle. Lactate in the presence of NAD^+ and lactate dehydrogenase (LDH) is oxidized to pyruvate; the NADH formed is measured spectrophotometrically at 340 nm and serves as a measure of the lactate concentration.

$$\text{L-Lactate} + NAD^+ \xrightleftharpoons{LDH,\ \text{pH } 9.0\text{–}9.6} \text{Pyruvate} + NADH + H^+$$

The equilibrium of the reaction normally lies far to the left. However, by using a pH of 9.0–9.6 and an excess of NAD^+ and by trapping the reaction product pyruvate with hydrazine, the equilibrium can be shifted to the right. Use of a Tris buffer results in faster completion of a side reaction between NAD^+ and hydrazine. Pyruvate can also be removed by reacting it with L-glutamate in the presence of alanine aminotransferase (ALT).

Metaphosphoric acid is preferred by some as a protein-precipitating agent, since perchloric acid does not precipitate mucoproteins and it interferes with the enzymatic method for pyruvate (if this analyte is determined on the same filtrate as lactate). Also, the enzymatic LDH reaction proceeds much more slowly in the presence of perchloric acid.

Because of its high specificity and simplicity, the enzymatic method is the method of choice for measuring lactate, although other methods may also be used (e.g., gas chromatography and colorimetry).

Specimens. Collection of a satisfactory specimen for lactate analysis requires special procedures to prevent changes in lactate both while and after the specimen is drawn. The patient should be fasting and at complete rest. A venous specimen is best drawn without a tourniquet or immediately after the tourniquet has been applied. If there is any delay in obtaining the specimen, the tourniquet should be removed after the puncture has been performed, and blood should be allowed to circulate for at least 2 min before the specimen is withdrawn.

Venous or arterial blood may be collected conveniently in heparinized syringes and immediately delivered into a premeasured amount of chilled metaphosphoric acid. The clear supernatant, after centrifugation, is stable at 4–8 °C for up to 8 d. If blood is not preserved, lactate will increase rapidly as a result of glycolysis. Increases may be as great as 20% within 3 min and 70% within 30 min at 25 °C. Specimens collected as described are also suitable for the determination of pyruvate.

If plasma is required as a specimen, blood is best collected in a container with 10 mg sodium fluoride and 2 mg potassium oxalate per mL of blood, followed by *immediate* chilling of the specimen in ice water and separation of the cells *within 15 min*.

Reference ranges. Fasting *venous* blood has a lactate concentration of 0.5–1.3 mmol/L (5–12 mg/dL). *Arterial* blood contains 0.36–0.75 mmol/L (3–7 mg/dL) if the patient is in a state of complete rest. Patients under usual hospital conditions show a higher range, up to 0.9–1.7 mmol/L (8–15 mg/dL) for venous blood and up to 1.25 mmol/L for arterial blood. Sudden, severe exercise increases lactate levels dramatically. Even movement of leg muscles of patients at bed rest may result in significant increases. *Plasma* values are ~7% higher than those seen in whole blood, although differences are dependent on the procedure employed. *CSF* values are normally close to those found in blood but may change independently in CNS disorders. Normal 24-h *urine* output of lactate is 5.5–22 mmol/d.

Determination of Pyruvate in Whole Blood[23,42]

Principle. The reaction involved in the determination of pyruvate is essentially the reverse of the reaction used in the lactate procedure.

$$\text{Pyruvate} + \text{NADH} + H^+ \xrightleftharpoons{\textit{LDH}, \text{ pH 7.5}} \text{Lactate} + \text{NAD}^+$$

At about pH 7.5, the equilibrium constant strongly favors the reaction to the right. The method is very specific, and α-oxoglutarate, oxaloacetate, acetoacetate, and β-hydroxybutyrate do not interfere, as is the case with colorimetric methods.

Specimen. Pyruvate in blood is extremely unstable and the same precautions detailed for lactate should be observed. A protein-free filtrate prepared with metaphosphoric acid (MPA) is suitable for both lactate and pyruvate determinations. Collection of blood and the preparation of a protein-free filtrate are discussed under lactate.

Reference ranges. Fasting *venous blood*, drawn with the patient at rest, has a pyruvate concentration of 0.03–0.10 mmol/L (0.3–0.9 mg/dL). *Arterial* blood contains 0.02–0.08 mmol/L (0.2–0.7 mg/dL). Values for *cerebrospinal fluid* are 0.06–0.19 mmol/L (0.5–1.7 mg/dL).[36] Normal 24-h *urine* output of pyruvate is ~1 mmol/d or less.

URINARY SUGARS

Occurrence of Sugars in Urine

Urine is examined routinely to detect the presence or determine the amount of *glucose*; the test is done as a screening procedure as a guide to insulin therapy. Other

Table 13-5. REDUCING SUBSTANCES IN URINE

Fructose	Ketone bodies
Lactose	Sulfanilamide
Galactose	Oxalic acid
Maltose	Hippuric acid
Arabinose	Homogentisic acid
Xylose	Glucuronic acid
Ribose	Formaldehyde
Uric acid	Isoniazid
Ascorbic acid	Salicylates
Creatinine	Cinchophen
Cysteine	Salicyluric acid

sugars may also appear in the urine in certain conditions. Those of clinical interest are all reducing sugars; that is, they readily reduce cupric ions in hot alkaline solution. Except for galactose, glucose is the only sugar found in urine that is of pathological significance.

Galactose appears in the urine of infants with galactosemia, a condition characterized by an inability to metabolize galactose. Such infants fail to thrive on milk, since half of the milk sugar (lactose) is galactose. Failure to discover and control this condition can result in liver disease, mental retardation, and cataracts. *Lactose* is sometimes found in urine of women during lactation and occasionally toward the end of pregnancy. The laboratory may be required to differentiate this sugar from glucose. *Fructose* may appear in the urine after eating fruits, honey, and syrups, but has no clinical significance. Fructosuria also occurs in a rare and harmless congenital defect that should not be confused with diabetes. *Pentoses* may occur in urine after eating such fruits as cherries, plums, or prunes, or as a harmless congenital anomaly and, as with fructose, must be distinguished from glucose. *Maltose* has been reported to occur along with glucose in the urine of some patients with diabetes.

Many *reducing substances* other than sugars may also occur in urine. A partial list of the more important reducing substances is shown in Table 13-5. Ascorbic acid, especially, may be ingested in large amounts or be present in antibiotic preparations administered intravenously. In either case, excess concentrations usually appear in the urine and contribute significantly to the total reducing substances present.

Qualitative Methods for Total Reducing Substances

Principle. Benedict's qualitative reagent contains cupric ions complexed with citrate in alkaline solution. Glucose or other reducing substances reduce cupric ions to cuprous ions with resultant formation of yellow cuprous hydroxide or red cuprous oxide. A convenient adaptation of this procedure is marketed in tablet form (Clinitest; Ames Co., Elkhart, IN 46514). The tablets contain anhydrous cupric sulfate, sodium hydroxide, citric acid, and sodium bicarbonate. Five drops (0.25 mL) of urine are mixed with 10 drops of water in a test tube. One tablet is added and the mixture is allowed to stand undisturbed for 15 s; it is then remixed and observed for color. A chart provided by the manufacturer is used to interpret the result. Heat is generated by contact of sodium hydroxide with water and speeds up the reaction. The initial reaction between citric acid and sodium bicarbonate causes the release of carbon dioxide, which blankets the mixture and reduces contact with oxygen from the air to prevent reoxidation of cuprous ions.

Quantitative Methods for the Determination of Glucose in Urine

Applications of various procedures for the quantitative determination of glucose in urine have been discussed under *Determination of Glucose in Body Fluids.* Hexokinase or glucose dehydrogenase procedures are recommended for greatest accuracy and specificity. Glucose oxidase procedures that depend only on consumption of oxygen or production of hydrogen peroxide are also reliable. Glucose oxidase procedures that include the hydrogen peroxide–peroxidase reaction are not acceptable. The *o*-toluidine procedure is a satisfactory method for most routine applications.

Separation and Identification of Sugars

Techniques for separating and identifying sugars have included fermentation, optical rotation, osazone formation with phenylhydrazine, specific chemical tests, and paper or thin-layer chromatography. The availability of glucose oxidase test strips has greatly simplified the differentiation of glucose from many other reducing substances. Galactose, pentoses, and other reducing sugars do not react with the glucose oxidase strips. Urine from infants and children should be tested routinely by both glucose oxidase and copper reduction tests to identify individuals with congenital anomalies who might otherwise be diagnosed as diabetics and to detect substances that interfere with the glucose method (see below). Non-glucose reducing substances should be further identified by chromatographic procedures.

Qualitative Tests for Individual Sugars

Glucose. A convenient paper test strip is commercially available (Clinistix; Ames Co., Elkhart, IN 46514). The filter paper is impregnated with glucose oxidase, peroxidase, and the dye *o*-tolidine and provides a simple color test according to principles discussed earlier. The test end is moistened with urine and examined after 10 s. A blue color develops if glucose is present. Sensitivity of the strip has been adjusted to take into account the presence of enzyme inhibitors normally occurring in urine. Thus, a positive test will be obtained with lower concentrations of glucose in water as compared with urine. For the same reason, misleading high results may be obtained with very dilute specimens.

False-negative results with enzyme paper tests may be caused by ascorbic acid. Contamination of urine with hydrogen peroxide or a strong oxidizing agent such as hypochlorite (bleach) produces false-positive results. For routine examinations, a negative result using the strip test is usually interpreted to mean that the urine specimen is negative for glucose.

Another strip test (Ketodiastix; Ames Co., Elkhart, IN 46514) is designed for the semiquantitative estimation of both glucose and ketone bodies. The glucose portion of the strip utilizes the glucose oxidase–peroxidase method. The peroxide produced oxidizes iodide to iodine, yielding varying intensities of a brown color that correspond to the concentration of glucose in the urine. Compared with Clinistix, the Ketodiastix glucose test is much less subject to inhibition by ascorbic acid. Other glucose test strips are also available (e.g., Chemstrip; Bio-Dynamics/bmc, Indianapolis, IN 46250).

Identification of Urinary Sugars by Paper Chromatography

Principle. Sugars can be separated by ascending or descending chromatography on paper and located after color development with dinitrosalicylic acid.[37] The variable rates of migration depend upon the solubility of the sugars in the particular solvent system. Presumptive identification is made by comparison of the R_f value of the unknown with those of standards.

$$R_f = \frac{\text{Distance traveled by solute spot}}{\text{Distance traveled by solvent front}}$$

The R_f values vary slightly from run to run, and for this reason known reference samples should be included each time.

The unknown sugar in the urine is presumed to be the same as a known reference standard when both migrate the same distance under the same test conditions. For confirmation, the urine specimen may be mixed with an equal volume of known standard and rechromatographed. Only one sugar spot will appear on the paper if the two sugars are identical.

Dinitrosalicylic acid is a highly specific reagent for reducing sugars. Other reagents to detect reducing substances in urine have also been described.[12] Phloroglucinol reagent, for example, produces a faint brown color with galactose, orange-brown with fructose, and greenish-blue with pentoses. No color is produced with lactose, maltose, or glucose.

Identification of urinary sugars can also be made by using thin-layer chromatographic techniques as described by Young and Jackson.[47] When frequent chromatographic separations are necessary, these methods are preferred over paper chromatography because of the shorter time required.

MUCOPOLYSACCHARIDES

The mucopolysaccharide storage diseases are clinically progressive, hereditary disorders marked by accumulation of mucopolysaccharides (MPS, glycosaminoglycans) in various tissues. This is due to a deficiency of one of a group of enzymes that degrade three classes of mucopolysaccharides: dermatan sulfate, heparan sulfate, and keratan sulfate. The general MPS phenotype includes coarse facies, corneal clouding, hepatosplenomegaly, joint stiffness, hernias, dysostosis multiplex, MPS excretion in urine, and metachromatic staining of leukocytes in peripheral blood and bone marrow.

The mucopolysaccharides are complex molecules that contain repeating disaccharide units. *Dermatan sulfate* is composed of alternating sulfated *N*-acetylgalactosamine and uronic acid residues. *Heparan sulfate* is composed of alternating glucosamine and uronic acid residues. *Keratan sulfate* is composed of alternating *N*-acetylglucosamine and galactose residues. *Chondroitin sulfates* are composed of alternating sulfated *N*-acetylgalactosamine and galactose residues.

The sulfated mucopolysaccharides are covalently linked to proteins to form macromolecules, which may be parts of larger aggregates in tissues. The mucopolysaccharidoses are inherited by a pattern of simple Mendelian genetics and, with the exception of Hunter's syndrome (which is an X-linked recessive disease), are autosomal recessive diseases. The major MPS excreted by normal individuals is chondroitin-6-sulfate.

Laboratory diagnostic aids for the detection and evaluation of MPS storage diseases have been reviewed by Pennock.[34] A simple screening test for urinary MPS should be performed first. If the result is positive, further tests could include the quantitative estimation of MPS by measurement of hexuronic acid content and identification of the MPS excretion pattern by electrophoresis.

Screening Test for Mucopolysaccharides in Urine

Principle. Under controlled conditions of pH and electrolyte concentration, cetylpyridinium chloride (CPC) reacts with MPS to form an insoluble precipitate. In the presence of citrate buffer, pH 4.8, this precipitate is sufficiently stabilized and dispersed so that absorbance of the turbid solution may be compared with a similarly treated standard of chondroitin sulfate. The procedure is described elsewhere.[42]

Specimen. Early-morning specimens should not be used for screening tests, and bacterially infected specimens are unsuitable for analysis. Single specimens passed between 0900 and 1800 h or 24-h specimens should be collected without preservative and stored frozen until analyzed.

References

1. Alsever, R. N., Roberts, J. P., Gerber, J. G., et al.: Insulinoma with low circulating insulin levels: The diagnostic value of proinsulin measurements. Ann. Intern. Med., *82*:347–350, 1975.
2. Charles, M. A., Hofeldt, F., Shakelford, A., et al.: Comparison of oral glucose tolerance tests and mixed meals in patients with apparent idiopathic postabsorptive hypoglycemia. Absence of hypoglycemia after meals. Diabetes, *30*:465–470, 1981.
3. Chua, K. S., and Tan, I. K.: Plasma glucose measurement with the Yellow Springs Glucose Analyzer. Clin. Chem., *24*:150–152, 1978.
4. Cooper, G. R.: Methods for determining the amount of glucose in blood. Crit. Rev. Clin. Lab. Sci., *4*:101–145, 1973.
5. Curme, H. G., Columbus, R. L., Dappen, G. M., et al.: Multilayer film elements for clinical analysis: General concepts. Clin. Chem., *24*:1335–1342, 1978.
6. Dolhofer, R., and Wieland, O. H.: Increased glycosylation of serum albumin in diabetes mellitus. Diabetes, *29*:417–422, 1980.
7. Fajans, S. S., and Floyd, J. C., Jr.: Fasting hypoglycemia in adults. N. Engl. J. Med., *294*:766–772, 1976.
8. Fraser, J., Fetter, M. C., Mast, R. L., et al.: Studies with a simplified nitroprusside test for ketone bodies in urine, serum, plasma, and milk. Clin. Chim. Acta, *11*:372–378, 1965.
9. Gochman, N., and Schmitz, J. M.: Application of a new peroxide indicator reaction to the specific automated determination of glucose with glucose oxidase. Clin. Chem., *18*:943–950, 1972.
10. Gordon, P., Hendricks, C. M., Kahn, C. R., et al.: Hypoglycemia associated with non-islet-cell tumor and insulin-like growth factors. N. Engl. J. Med., *305*:1452–1455, 1981.
11. Hammons, G. T., Junger, K., McDonald, J. M., et al.: Evaluation of three minicolumn procedures for measuring hemoglobin A_1. Clin. Chem., *28*:1775–1778, 1982.
12. Horrocks, R. H., and Manning, G. B.: Partition chromatography on paper. Identification of reducing substances in urine. Lancet, *1*:1042–1045, 1949.
13. Horwitz, D. L., Kuzuya, H., and Rubenstein, A. H.: Circulating serum C-peptide. A brief review of diagnostic implications. N. Engl. J. Med., *295*:207–209, 1976.
14. James, T. M., Davis, J. E., McDonald, J. M., et al.: Comparison of hemoglobin A_{1c} and hemoglobin A_1 in diabetic patients. Clin. Biochem., *14*:25–27, 1981.

15. Javid, J., Pettis, P. K., Koenig, R. J., et al.: Immunologic characterization and quantification of hemoglobin A_{1c}. Br. J. Haematol., *38*:329–337, 1978.
16. Johnson, D. D., Dorr, K. E., Swenson, S. M., et al.: Reactive hypoglycemia. JAMA, *243*:1151–1155, 1980.
17. Kadish, A. H., Little, R. A., and Sternberg, J. C.: A new and rapid method for the determination of glucose by measurement of rate of oxygen consumption. Clin. Chem., *14*:116–131, 1968.
18. Kahn, C. R.: The riddle of tumor hypoglycemia revisited. Clin. Endocrinol. Metab., *9*:335–360, 1980.
19. Kansal, P. C., Buss, R. W., Pino, J. A., et al.: Glucose tolerance test (GTT) vs. meal test (MT) in reactive hypoglycemia. Diabetes, *26*(Suppl. 2):403, 1977.
20. Klenk, D. C., Hermanson, G. T., Krohn, R. I., et al.: Determination of glycosylated hemoglobin by affinity chromatography: Comparison with colorimetric and ion-exchange methods, and effects of common interferences. Clin. Chem., *28*:2088–2094, 1982.
21. Livesley, B., and Atkinson, L.: Accurate quantitative estimation of lactate in whole blood. Clin. Chem., *20*:1478, 1974.
22. Lutz, R. A., and Flückiger, J.: Kinetic determination of glucose with the GEMSAEC (ENI) centrifugal analyzer by the glucose dehydrogenase reaction, and comparison with two commonly used procedures. Clin. Chem., *21*:1372–1377, 1975.
23. Marbach, E. P., and Wells, M. H.: Rapid enzymatic measurement of blood lactate and pyruvate. Clin. Chem., *13*:314–325, 1967.
24. Nathan, D. M.: Labile glycosylated hemoglobin contributes to hemoglobin A_1 as measured by liquid chromatography or electrophoresis. Clin. Chem., *27*:1261–1263, 1981.
25. National Diabetes Data Group: Classification and diagnosis of diabetes mellitus and other categories of glucose intolerance. Diabetes, *28*:1039–1057, 1979.
26. Neese, J. W., Duncan, P., Bayse, D., et al.: Development and evaluation of a hexokinase/glucose-6-phosphate dehydrogenase procedure for use as a national reference method. HEW Publication No. (CDC) 77-8330. Atlanta, Centers for Disease Control, 1976.
27. Newcomer, A. D., McGill, D. B., Thomas, P. J., et al.: Prospective comparison of indirect methods for detecting lactase deficiency. N. Engl. J. Med., *293*:1232–1236, 1975.
28. Niederau, C. M., and Reinauer, H.: Analyseverfahren für glykosidierte Hämoglobine. Ein Methodenvergleich. J. Clin. Chem. Clin. Biochem., *19*:1097–1101, 1981.
29. Olefsky, J. M.: Insulin resistance and insulin action. Diabetes, *30*:148–162, 1981.
30. O'Sullivan, J. B.: Age gradient in blood glucose levels. Diabetes, *23*:713–715, 1974.
31. O'Sullivan, J. B., Mahan, C. M., Charles, D., et al.: Screening criteria for high-risk gestational diabetic patients. Am. J. Obstet. Gynecol., *116*:895–900, 1973.
32. Passey, R. B., Gillum, R. L., Fuller, J. B., et al.: Evaluation and comparison of 10 glucose methods and the reference method recommended in the proposed product class standard (1974). Clin. Chem., *23*:131–139, 1977.
33. Pecoraro, R. E., Graf, R. J., Halter, J. B., et al.: Comparison of a colorimetric assay for glycosylated hemoglobin with ion-exchange chromatography. Diabetes, *28*:1120–1125, 1979.
34. Pennock, C. A.: A review and selection of simple laboratory methods used for the study of glycosaminoglycan excretion and the diagnosis of the mucopolysaccharidoses. J. Clin. Pathol., *29*:111–123, 1976.
35. Plasse, N. J.: Monitoring blood glucose at home. A comparison of three products. Am. J. Nursing, *81*:2028–2029, 1981.
36. Pryce, J. D., Gant, P. W., and Saul, K. J.: Normal concentrations of lactate, glucose, and protein in cerebrospinal fluid, and the diagnostic implications of abnormal concentrations. Clin. Chem., *16*:562–565, 1970.
37. Sophian, L. H., and Connolly, V. J.: Chromatographic identification of reducing sugars in urine. Am. J. Clin. Pathol., *22*:41–45, 1952.
38. Spicer, K. M., Allen, R. C., and Buse, M. G.: A simplified assay of hemoglobin A_{1c} in diabetic patients by use of isoelectric focusing and quantitative microdensitometry. Diabetes, *27*:384–388, 1978.
39. Steiner, D. F.: Insulin today. Diabetes, *26*:322–340, 1977.
40. Stevens, A. D.: Monitoring blood glucose at home. Who should do it. Am. J. Nursing, *81*:2026–2027, 1981.
41. Symposium on blood glucose self-monitoring. Diabetes Care, *4*:392–426, 1981.
42. Tietz, N. W., Ed.: Textbook of Clinical Chemistry. Philadelphia, W. B. Saunders Company, 1986.
43. Trinder, P.: Determination of glucose in blood using glucose oxidase with an alternative oxygen acceptor. Ann. Clin. Biochem., *6*:24–27, 1969.
44. Watts, N. B.: Oral glucose tolerance test: Nondiagnostic. Lab '79, *2*:28–33, 1979.
45. Weissman, M., and Klein, B.: Evaluation of glucose determinations in untreated serum samples. Clin. Chem., *4*:420–422, 1958.
46. Yee, H. Y., and Goodwin, J. F.: Evaluation of some factors influencing the *o*-toluidine reaction with glucose. Anal. Chem., *45*:2162–2165, 1973.
47. Young, D. S., and Jackson, A. J.: Thin-layer chromatography of urinary carbohydrates. A comparative evaluation of procedures. Clin. Chem., *16*:954–959, 1970.

14 LIPIDS, LIPOPROTEINS, AND APOLIPOPROTEINS

EVAN A. STEIN, M.D., Ph.D.

Of the numerous different lipids known to exist in humans, only a limited number are of clinical and analytical importance. Lipids are ubiquitous in the body tissues and play an important role in virtually all aspects of biological life; they serve as hormones or hormone precursors, aid in digestion, provide energy storage and metabolic fuels, act as functional and structural components in biomembranes, and form insulation to allow nerve conduction or to prevent heat loss. In the last decade, however, lipids in clinical chemistry have become associated with lipoprotein metabolism and a form of arterial disease known as atherosclerosis (see page 460).

Chemistry and Physiology of Lipids

The term *lipid* applies to a class of compounds that are soluble in organic solvents and nearly insoluble in water. Chemically, lipids include compounds that yield fatty acids on hydrolysis and complex alcohols that can combine with fatty acids to form esters. Some lipids are more complex, containing groups that tend to increase the solubility of lipids in polar solvents. Lipids can be broadly subdivided into five groups based on their chemical structure (Table 14-1).

CHOLESTEROL

Cholesterol is found exclusively in animals and humans, where it is also the main sterol. Virtually all human cells and body fluids contain some cholesterol. Like other sterols, cholesterol is a solid alcohol of high molecular weight and possesses the tetracylic perhydrocyclopentanophenanthrene skeleton. The molecule contains 27 carbon atoms, numbered as shown in Figure 14-1.

Absorption. Cholesterol in the intestine comes from diet, bile, intestinal secretions, and cells. Animal products, especially meat, egg yolk, seafood, and whole-fat dairy products, provide the bulk of dietary cholesterol. The average American diet is estimated to contain ~400–700 mg of cholesterol per day. A similar amount of cholesterol is present in the gut from biliary secretion and the turnover of mucosal cells. Practically all cholesterol in the intestine is present in the unesterified (free) form, since esterified cholesterol is rapidly hydrolyzed in the intestine by cholesterol esterases in pancreatic and small intestinal secretions.

In order to be absorbed, cholesterol is solubilized by formation of mixed micelles containing unesterified cholesterol, fatty acids, monoglycerides (monoacylglycerols), phospholipids (lysolecithin), and conjugated bile acids. These micelles also facilitate cholesterol transport across the luminal cell surfaces. In the absence of bile acids, digestion and absorption of both cholesterol and triglycerides are severely impaired. On the average, 30–60% of dietary and intestinal cholesterol is absorbed daily, to a maximum of ~1 g/d when the oral intake reaches 3 g/d.

Helpful contributions by Ronald J. Whitley, Ph.D., University of Kentucky Medical Center, are gratefully acknowledged.

Table 14-1. CLASSIFICATION OF CLINICALLY IMPORTANT LIPIDS

Sterol derivatives	***Glycerol esters***
Cholesterol and cholesteryl esters	Triglycerides (triacylglycerols)
Steroid hormones	Phosphoglycerides
Bile acids	***Sphingosine derivatives***
Vitamin D	Sphingomyelin
Fatty acids	Glycosphingolipids
Short chain (2–4 carbon atoms)	***Terpenes (isoprene polymers)***
Medium chain (6–10 carbon atoms)	Vitamin A
Long chain (12–26 carbon atoms)	Vitamin E
Prostaglandins	Vitamin K

Maximum absorption of cholesterol occurs in the middle and terminal ileum, where micelles contain significant amounts of fatty acids and monoglycerides. As absorption of fat and cholesterol occurs, the micelles break up, thus reducing further cholesterol absorption.

In addition to animal cholesterol, ~200–300 mg of plant sterols, such as β-sitosterol, are ingested daily, but they are poorly absorbed. When plant sterols are administered in amounts of 5–15 g/d, absorption of cholesterol is significantly inhibited. Therefore, plant sterols have been used therapeutically to reduce cholesterol absorption in patients with increased plasma cholesterol levels.

After its absorption into the mucosal cell, cholesterol is reassembled along with triglycerides, phospholipids, and specific apolipoproteins into a large micelle called a chylomicron. The apolipoprotein component is vital to the formation of chylomicrons, and, in subjects with a rare deficiency of apolipoprotein B, chylomicron formation and consequently fat and cholesterol absorption are severely impaired. Chylomicrons enter the lymphatics, which empty into the thoracic duct, and eventually enter the systemic venous circulation.

Synthesis. Almost 90% of the body's cholesterol is synthesized by the liver and gut from simpler molecules, particularly acetate. Hepatic synthesis of cholesterol appears to be inhibited by newly absorbed dietary cholesterol, which reaches the liver in the chylomicron remnant. It is uncertain whether cholesterol produced by endogenous synthesis has a similar feedback effect on the liver.

Esterification. Once synthesized, cholesterol is released into the circulation for transport with specific apolipoproteins in complexes known as lipoproteins. Minimal cholesterol esterification occurs within the liver prior to its release, and cholesterol is mainly esterified within the vascular compartment, where cholesteryl esters account for about 70% of the total cholesterol. Esterification is important, since it serves to enhance the lipid-carrying capacity of the lipoproteins. The reaction is catalyzed by the enzymes lecithin-cholesterol acyltransferase (LCAT) in the plasma and acyl-cholesterol acyltransferase (ACAT) intracellularly. In plasma, LCAT activity is responsible for nearly all the esterified cholesterol. LCAT is synthesized in the liver, is released into the circulation, and appears to be activated by apolipoprotein AI and perhaps apolipoprotein CI.

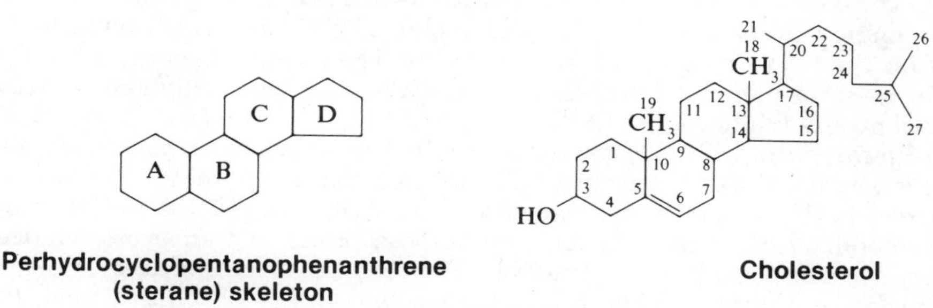

Perhydrocyclopentanophenanthrene (sterane) skeleton

Cholesterol

Figure 14-1. Structure of cholesterol.

Catabolism. After cholesterol enters the cell, the esters are hydrolyzed by the action of specific lysosomal esterases. The lack or malfunction of these lysosomal enzymes results in intracellular accumulation of cholesteryl esters and produces a clinical disorder known as *cholesteryl ester storage disease*.

Cholesterol that reaches the liver is either secreted unchanged into bile or metabolized to form bile acids (~30 %/d). The bile acid synthesis rate averages 200–400 mg/d. Two bile acids, cholic and chenodeoxycholic, constitute the primary bile acids (see Chapter 22). They are conjugated with either glycine or taurine and enter the bile canaliculi. Conjugated bile acids play an active part in cholesterol and fat absorption in the small intestine, as has been previously discussed. The cholesterol, excreted directly into the biliary system, is solubilized by mixed micelles of bile acids and phospholipids. If the amount of cholesterol exceeds the capacity of the solubilizing agents, a supersaturated state can occur (lithogenic bile) that can result in the formation of gallstones that contain precipitated cholesterol. In Western societies, ~80% of all gallstones are cholesterol stones.

FATTY ACIDS

Fatty acids (FA) are simple molecular forms of lipids depicted by the chemical formula RCOOH, where R stands for an alkyl chain. The most common classification is based on chain length: 2–4 carbon atoms (short chain), 6–10 carbon atoms (medium chain), and 12–26 carbon atoms (long chain). The fatty acids that are important in human nutrition and metabolism are of the long-chain class and contain even numbers of carbon atoms.

Fatty acids can be further classified according to their degree of saturation. Saturated fatty acids contain an alkyl chain without any double bonds between carbon atoms, monounsaturated fatty acids contain one double bond, and polyunsaturated fatty acids have more than one double bond. In mammals, all naturally occurring unsaturated fatty acids are of the *cis*- variety. *Trans*- fatty acids are a result of catalytic hydrogenation, a process used to "harden" fats in the manufacture of certain foods such as margarine. Most fats in the human body are derived from our diet, which on the average contains 40% fat, 90% of which is triglyceride. In addition, humans can synthesize most fatty acids. Linoleic acid ($C_{18}:2^{9,12}$), however, cannot be synthesized by mammals; it is found only in plants. This fatty acid is vital for maintenance of health, as well as for growth and development, and is termed an *essential fatty acid*. Linoleic acid plays an important role in prostaglandin synthesis and perhaps in myelinization of the central nervous system.

The carboxyl group of a fatty acid has a pK_a of ~4.8; thus, free fatty acid molecules in both plasma and intracellular fluid exist in an ionized form. However, most fatty acids exist as esters with cholesterol or glycerol or are bound to albumin and prealbumin.

Catabolism. Mitochondria oxidize long-chain fatty acids for energy production by a series of reactions known as β-oxidation. These reactions operate in a repetitive manner to shorten the chains by two carbon atoms at a time. The acetyl-CoA that is produced does not normally accumulate in the cell but is enzymatically condensed with oxaloacetate, a substance derived largely from carbohydrate metabolism (Figure 14-2). The product of the condensation reaction is citrate, which is a major component of the Krebs cycle (the citric acid or tricarboxylic acid cycle). The Krebs cycle serves as a common pathway for the final oxidation of carbon chains. The chemical energy from β-oxidation is thus released for metabolic processes or stored in the form of high-energy compounds, such as adenosine triphosphate (ATP).

Ketone formation. During prolonged starvation or when carbohydrate metabolism is severely impaired, as in uncontrolled diabetes mellitus, the amount of acetyl-CoA formed exceeds the supply of oxaloacetate. The abundance of acetyl-CoA results from excessive mobilization of fatty acids from adipose tissue and an excessive degradation of the fatty acids by β-oxidation in the liver. The resulting excess of acetyl-CoA is diverted to an alternative pathway in the mitochondria that forms acetoacetic acid, β-hydroxybutyric acid, and acetone; these three compounds are known collectively as *ketone bodies* (see also Chapter 13). The presence of ketone bodies is frequently found in severe cases

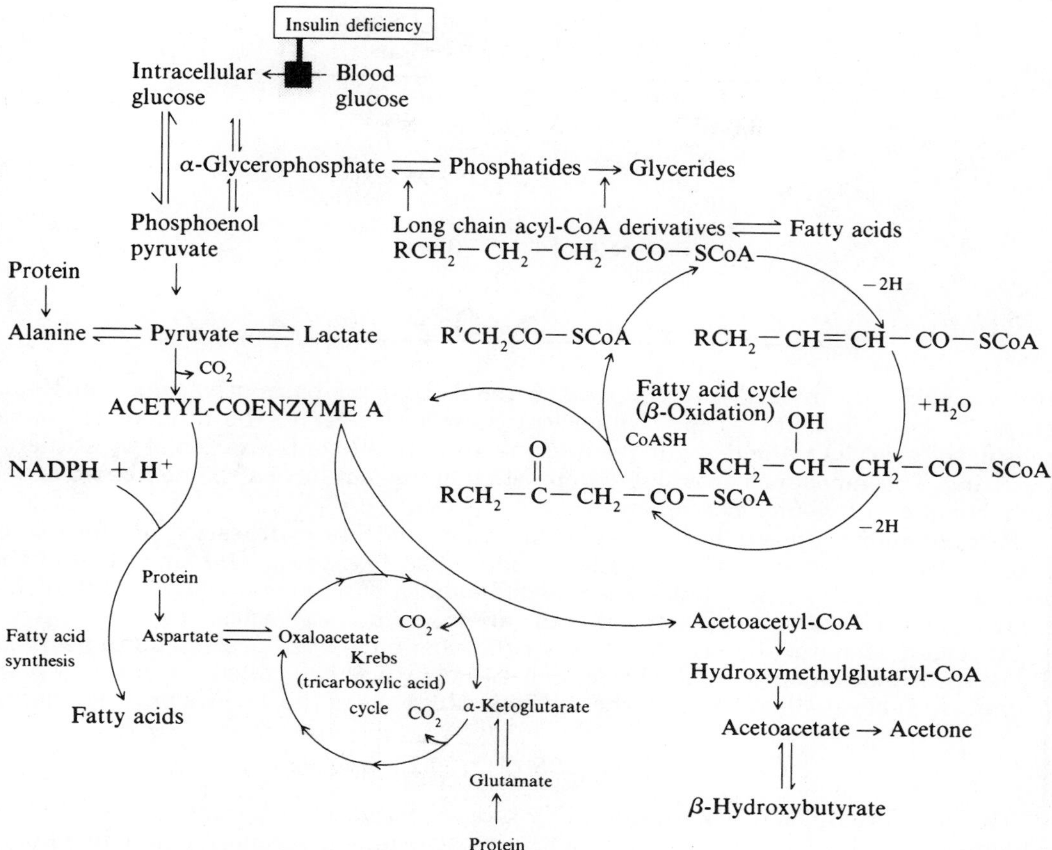

Figure 14-2. Metabolic relations among intermediates of carbohydrate, fat, and protein metabolism. Note that acetyl-CoA is produced from both carbohydrate and fat. The glucogenic amino acids, derived from protein metabolism, enter glycolytic paths as α-keto acids. Ketogenic amino acids enter as acetyl-CoA.

of uncontrolled diabetes mellitus, when ketosis develops from excessive production of acetyl-CoA as the body attempts to obtain necessary energy from stored fat in the absence of an adequate supply of carbohydrate metabolites.

The entire process of ketosis can be reversed by restoring an adequate level of carbohydrate metabolism. In starvation, restoration consists of adequate carbohydrate ingestion; in diabetes mellitus, ketosis can be reversed by insulin administration, which permits circulating blood glucose to be taken up by the cells. With production of oxaloacetate, normal metabolism is restored, and the release of fatty acids from adipose tissues slows and is finally reversed. These metabolic reactions are outlined in Figure 14-2, which shows the interrelationship between carbohydrate, fatty acid, and protein metabolism.

Prostaglandins

Prostaglandins and related compounds, such as thromboxanes and leukotrienes, are derivatives of fatty acids. Although their full physiological role is not completely known, these compounds are extremely potent, producing diverse physiological actions at concentrations as little as 1 μg/L.

The prostaglandins (PG) are a series of C_{20} unsaturated fatty acids containing a cyclopentane ring; the parent fatty acid is prostanoic acid.

COO^-
9 8 7 5 3
10 12
11 13 15 17 19 CH_3

Table 14-2. NATURALLY OCCURRING PROSTAGLANDINS (PG)

Primary PG	Other PG
PGE_1	PGA_1
$PGF_{1\alpha}$	PGA_2
PGE_2	19α-OH-PGA_1
$PGF_{2\alpha}$	19α-OH-PGA_2
PGI_2	PGB_1
Thromboxane A_2	PGB_2
Thromboxane B_2	19α-OH-PGB_2
	PGE_3
	$PGF_{3\alpha}$

Prostaglandins are divided into classes A through I on the basis of various functional groups in the cyclopentane ring. Numerical subscripts refer to the number of double bonds in the side chains, and α- and β-subscripts refer to the configuration of substituents in the ring. Thromboxanes are related to prostaglandins but have a six-membered oxane ring instead of a five-membered ring.

Prostaglandins appear hormone-like in action and are synthesized at the site of action and in almost all tissues. Linoleic acid ($C_{18}:2^{9,12}$) is the precursor of two of the three 20-carbon fatty acids that form prostaglandins. The other precursor is linolenic acid ($C_{18}:3^{9,12,15}$). Interestingly, both of these are *essential* fatty acids.

Although 16 naturally occurring prostaglandins have been described (Table 14-2), only seven are commonly found throughout the body and are called the *primary prostaglandins*. Table 14-3 lists some of the reported functions of the various prostaglandins.

GLYCEROL ESTERS (ACYLGLYCEROLS)

The complex lipids are virtually all fatty acid derivatives; in most cases they are covalently linked to an alcohol. One of the most common alcohols found in human metabolism is glycerol, a three-carbon molecule containing three hydroxyl groups.

$$\begin{array}{c} H_2-C_{\alpha}-OH \\ | \\ H-C_{\beta}-OH \\ | \\ H_2-C_{\alpha'}-OH \end{array}$$

The two terminal carbon atoms in the molecule are chemically equivalent and are designated α and α' (α prime), whereas the center carbon is labeled β. An alternate labeling system uses the numerals 1 for the α-carbon, 2 for the β-carbon, and 3 for the α'-carbon. Esterification of fatty acids with glycerol produces acylglycerols (glycerides), which are classified as mono-, di-, or triglycerides according to their number of fatty acids (Figure 14-3). ***Triglycerides*** are the most prevalent dietary glycerol esters encountered. They also constitute 95% of the fat stored in tissue and are the predominant form of glycerol ester found in plasma. Their fatty acid residues vary considerably but usually include combinations of the long-chain fatty acids. Triglycerides from plants tend to have large amounts of $C_{18}:2$ or linoleic residues. They are termed *polyunsaturated*

Table 14-3. PROSTAGLANDIN-MEDIATED EFFECTS

Site of Action	Physiological Response
Arterial smooth muscle	Alterations of blood pressure
Uterine muscle	Induces labor, therapeutic abortion
Lower gastrointestinal tract	Increases motility
Bronchial smooth muscle	Bronchospasm
Platelets	Increases coagulability
Stomach	Enhances gastric acid secretion
Capillaries	Increased permeability with flushing
Adipose tissue	Inhibits triglyceride lipolysis

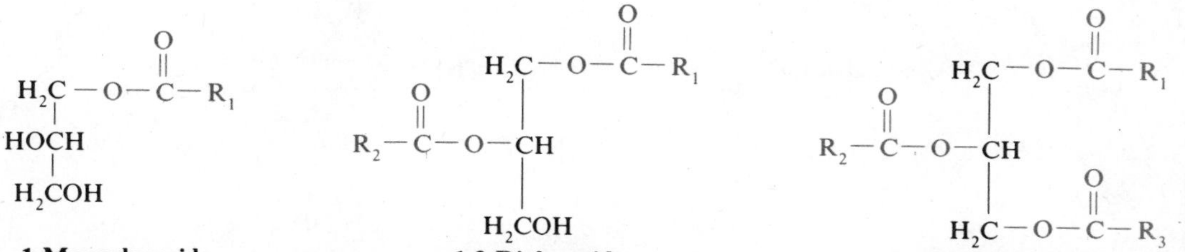

$H_2C{-}O{-}C(=O){-}R_1$ / $HOCH$ / H_2COH

1-Monoglyceride

$H_2C{-}O{-}C(=O){-}R_1$ / $R_2{-}C(=O){-}O{-}CH$ / H_2COH

1,2-Diglyceride

$H_2C{-}O{-}C(=O){-}R_1$ / $R_2{-}C(=O){-}O{-}CH$ / $H_2C{-}O{-}C(=O){-}R_3$

Triglyceride

Figure 14-3. Structure and classification of glycerol esters (acylglycerols). R_1, R_2, and R_3 are fatty acid residues of varying carbon atom lengths.

fats and are liquid even at 4 °C. Triglycerides from animals tend to have C_{12}:0 through C_{18}:0 fatty acid residues (saturated fats) and are solid even at room temperature. Triglycerides such as eicosapentaenoic acid (EPA) from fish, especially cold-water, fatty fish, have multiple unsaturated bonds and are now becoming recognized as beneficial in lowering plasma triglycerides.

Triglycerides undergo digestion in the duodenum and proximal ileum. Through the action of lipases and bile acids, triglycerides are hydrolyzed into glycerol, fatty acids, and monoglycerides. After absorption, triglycerides are resynthesized in the epithelial cells and form chylomicrons, as explained earlier.

Phosphoglycerides are another major class of glycerol esters. They have fatty acids esterified at the α- and β-carbon but contain phosphoric acid at the third carbon atom (Figure 14-4). Usually, an additional alcohol-derived group such as choline, serine, inositol, or ethanolamine is substituted on the phosphoric residue. If choline is the attached group, the compound is referred to as phosphatidylcholine; if ethanolamine is substituted, the compound is phosphatidylethanolamine, and so on. A more common term for phosphatidylcholine is *lecithin*, whereas phosphatidylethanolamine, -serine, and -inositol are referred to as *cephalins*. Each of these terms includes a group of compounds with different fatty acids (R_1 and R_2 in Figure 14-4).

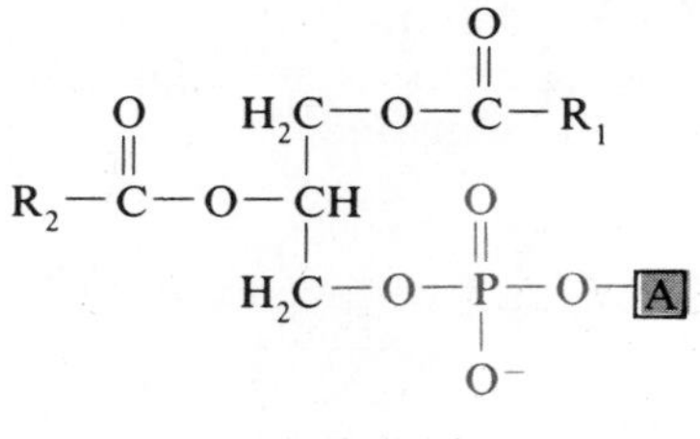

$H_2C{-}O{-}C(=O){-}R_1$ / $R_2{-}C(=O){-}O{-}CH$ / $H_2C{-}O{-}P(=O)(O^-){-}O{-}$[A]

When [A] =	Phosphoglyceride
$-H$	= Phosphatidic acid
$-CH_2CH_2-\overset{+}{N}H_3$	= Phosphatidyl ethanolamine*
$-CH_2CH_2-\overset{+}{N}(CH_3)_3$	= Phosphatidyl choline (lecithin)
$-CH_2CH(\overset{+}{N}H_3)-COO^-$	= Phosphatidyl serine*
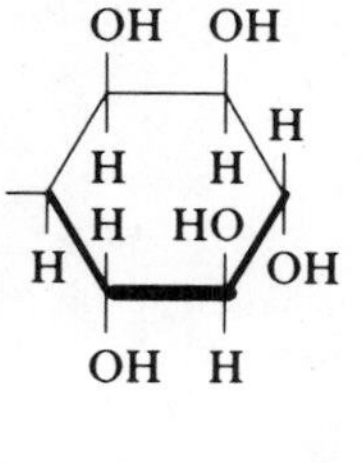 inositol ring (OH, OH, H, H, H, H, HO, H, OH, OH, H)	= Phosphatidyl inositol*

*Commonly known as cephalins.

Figure 14-4. Structures of phosphoglycerides and common alcohol groups associated with them. R_1 and R_2 are fatty acid residues of varying carbon atom lengths.

```
      OH H
      |  |
      HC—C=C—(CH2)12CH3
      |     |
      |     H
H2N— CH
      |
      |
      CH2OH
```

Sphingosine

```
                OH H
                |  |
                HC—C=C—(CH2)12CH3
     O          |     |
     ||         |     H
R—C—NH— CH
                |
                |
                CH2OH
```

Ceramide

```
                OH H
                |  |
                HC—C=C—(CH2)12CH3
     O          |     |
     ||         |     H
R—C—NH— CH
                |        O
                |        ||
                H2C—O—P—O—CH2—CH2—N+(CH3)3
                         |
                         O-
```

Sphingomyelin

Figure 14-5. Structures of some sphingolipids.

Sphingolipids

The fourth class of lipids found in humans is derived from the amino alcohol sphingosine. Sphingosine binds to one fatty acid containing 18 or more carbon atoms via the amino group, forming a ceramide. Additional substitutions at the terminal hydroxyl group form sphingomyelin, glycosyl ceramides, and other complex sphingolipids (Figure 14-5). These complex sphingolipids form the major lipids of cell membranes and the central nervous system. Membrane glycosphingolipids play a major role in cell recognition and blood typing, whereas gangliosides, which contain more complex substituted glycosyl groups, are particularly prevalent in the gray matter of the brain. A number of inherited lipid storage diseases known as sphingolipidoses have been characterized and their basic biochemical defects identified.[42]

Terpenes

These are polymers of isoprene, a five-carbon unit, and include vitamins A, E, and K (see Chapter 16).

APOLIPOPROTEINS

Lipids are synthesized in the intestine or liver but must be transported to many distant tissues and organs in order to complete their varied metabolic functions. Given the hydrophobic nature of the neutral fats (triglycerides and cholesteryl esters), lipid transport and delivery via plasma would not be possible without some form of hydrophilic adaptation. The lipids are consequently transported by means of a series of rather complex micellar structures that consist of an outer monolayer of protein (an apolipoprotein) and polar lipids (phospholipid and unesterified cholesterol), and an inner core of neutral lipids (triglycerides and cholesteryl esters) (Figure 14-6). The core neutral

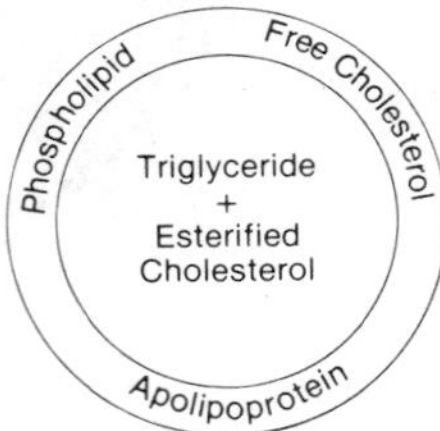

Figure 14-6. Schematic diagram of a lipoprotein. (Modified and published with permission from M. L. Kashyap, M.D.)

Table 14-4. MAJOR AND MINOR APOLIPOPROTEINS

Apolipoprotein	Function	MW	Site of Synthesis
AI	LCAT activation	28 300	Liver, intestine
AII	LCAT inhibition; lipid transport	17 000	Liver, intestine
AIV	Chylomicron-triglyceride transport		Intestine
B-100	Lipid transport and clearance	8000–275 000	Liver
B-48	? Chylomicron transport		Intestine
CI	? LCAT activation	6 331	Liver
CII	Lipoprotein lipase activation	8 837	Liver
CIII	? Lipoprotein lipase inhibition	8 764	Liver
D	? LCAT activation; lipid transfer	22 100	?
EII	?	38 000	Liver
EIII	?	38 000	Liver
EIV	IDL clearance	39 500	Liver

lipids are largely inactive components, whereas the apolipoproteins, which interact with a series of enzymes or tissue receptors, are mainly responsible for further metabolism and catabolism of the particle. The initial classification divided lipoproteins into five main density classes: chylomicron, very-low-density lipoprotein (VLDL), intermediate-density lipoprotein (IDL), low-density lipoprotein (LDL), and high-density lipoprotein (HDL). This nomenclature is currently in common use, but recent insights into the importance of apolipoproteins may alter the classification system. At least five major apolipoproteins and a number of subgroups have been described in association with these lipoproteins (Table 14-4). Apolipoproteins differ in their primary, secondary, and tertiary structures, in their physicochemical behavior, and in their function and distribution in the various lipoprotein classes mentioned above. *Reference ranges* for apolipoproteins are given in Table 14-5.

Apolipoprotein A (Apo A)

The A apolipoproteins[22,39] form the major proteins found in HDL, but they are also found in chylomicrons. Approximately 50% of HDL mass is protein, and Apo AI and AII constitute nearly 90% of it. Although the exact site of *synthesis* is still unclear, experimental evidence suggests that Apo AI and AII originate in the intestine or liver or both. After they enter the circulation as a component of chylomicrons, Apo AI and AII transfer to and accumulate in HDL.

Apolipoprotein AI plays a *role* in the activation of LCAT and removal of free cholesterol from extrahepatic tissues. Although the function of Apo AII is not entirely clear, it appears to play a structural role in HDL and may inhibit LCAT. Apo AII may also play a role in HDL catabolism, perhaps by stimulating hepatic lipase.

Since it appears that apolipoproteins exchange between lipoprotein classes, it has been difficult to study apolipoprotein A *catabolism*. The tissue site of degradation is probably the liver or kidney or both. A number of disease states have been associated with deficiencies of one or more apolipoproteins. The clinical classification of disease states will be discussed later.

Table 14-5. REFERENCE RANGES FOR PLASMA APOLIPOPROTEINS

Apolipoprotein	Mean, mg/dL	Mean ± 2 SD
AI	121	73–169
AII	37	19–55
B	98	58–138
CI	7	3–11
CII	3.7	0.5–6.9
CIII	13	3–23
E	4	2–6

Modified from Norum, R. A., Lakier, J. B., Goldstein, S., et al. N. Engl. J. Med., *306*:1513–1519, 1982.

Apolipoprotein B (Apo B)

Apo B is the major protein moiety of all lipoproteins other than HDL. Two forms of Apo B are found in humans. The most abundant form is known as large B or B-100 and constitutes the Apo B found in lipoproteins synthesized in the liver. The other form of Apo B, small B or B-48, is believed to be synthesized in the intestinal wall. In normal subjects, very little if any B-48 is found in fasting plasma; however, in subjects with defects in clearance of chylomicron remnants (e.g., with chronic renal failure), levels of B-48 are increased.[34]

Synthesis of Apo B is thought to occur on membrane-bound polyribosomes. Incorporation of Apo B into chylomicrons and VLDL is essential to the formation and release of these lipoproteins into the plasma. Although Apo B is a component of the VLDL molecule, crucial parts of the Apo B are masked, and the apolipoprotein plays a passive role. With delipidation of VLDL and gradual conversion to the relatively triglyceride-poor LDL molecule, Apo B assumes the role of a recognition protein on the surface of the lipoprotein, allowing the binding of LDL to specific high-affinity receptors on the cell membrane.

The *catabolism* of Apo B occurs mainly via the receptor-mediated pathway. Once internalized in the cell, Apo B is degraded to its constituent amino acids by lysosomal enzymes. Some Apo B is also catabolized by non-receptor-mediated processes that include a scavenger pathway. These latter processes may be important in the development of atherosclerosis. Some Apo B is also catabolized as part of VLDL and IDL. Both excess and absence of Apo B in the circulation have been associated with various diseases.

Apolipoprotein C (Apo C)

This group of low-molecular-weight apolipoproteins[22,39] is soluble in an aqueous solution and binds avidly to phospholipids to form protein-lipid complexes. Although the site of *synthesis* is not known with certainty, current evidence suggests that the liver is the primary organ. In fasting subjects, most of the C apolipoproteins are found in the VLDL and HDL fractions. The *function* of Apo CII has been conclusively demonstrated to be that of a cofactor of extrahepatic lipoprotein lipase, the enzyme responsible for hydrolysis of triglyceride-rich lipoproteins such as chylomicrons and VLDL. Apo CII also inhibits binding of VLDL and chylomicrons to hepatic receptors, thereby targeting them for extrahepatic clearance. In addition, the presence of a lipoprotein lipase activated by Apo CI has been reported, and CI has been implicated in the activation of LCAT. The third Apo C, CIII, is also thought to play a role in triglyceride hydrolysis, possibly as an inhibitor of lipoprotein lipase. Little is known about the *catabolism* of the C apolipoproteins. There is virtually no Apo C found in IDL or LDL. There is, however, well-documented transfer and redistribution of the C apolipoproteins between VLDL, HDL, and chylomicrons. Apo C from HDL is transferred to newly secreted chylomicrons and VLDL, whereas chylomicron and VLDL hydrolyses result in movement of Apo Cs, especially CII, from these lipoproteins to HDL.

Apolipoprotein D (Apo D)

Virtually nothing is known of Apo D synthesis or catabolism. It is suggested that Apo D may function as a transfer protein, assisting in the movement of cholesteryl esters and triglycerides between lipoprotein species, especially from VLDL to HDL and vice versa.

Apolipoprotein E (Apo E)

Apo E[22,39] is a glycoprotein with at least four polymorphic forms that differ from one another by one charge only. Apo E is most likely synthesized in the liver and initially enters the plasma as part of nascent HDL. Under the influence of LCAT, HDL accumulates cholesterol and Apo E is rapidly transferred to VLDL and chylomicrons, where it remains while VLDL is catabolized to IDL. Apo E plays a significant *role* in the recognition and catabolism of chylomicron remnant and IDL via specific receptors in hepatic cells. In subjects who produce only EII, there is a strong association with impaired clearance of chylomicron remnants and IDL, producing a disorder known as dys-β-lipoproteinemia or Type III hyperlipoproteinemia.

Lipoprotein Little A [Lp(a)] Antigen

In addition to the above apolipoproteins, another apolipoprotein known as Lp(a) antigen occurs in a variant form of LDL termed *sinking pre-β-lipoprotein*. Its function is not known, but it has been associated with increased atherosclerosis risk.

LIPOPROTEINS

The rapidly changing knowledge of the apolipoproteins and their functional interrelationships among the various lipoproteins does not permit a definitive classification of the lipoproteins at present. However, the varied physical and chemical properties of these lipoproteins do allow for an operational definition based on their sizes, floating densities in an ultracentrifugal field, or electrophoretic mobilities. For purposes of clarity and consistency, the classification based on ultracentrifugal flotation density will be used (Table 14-6).

The *physiological function* and *degradation* of lipoproteins are not only restricted to the transport of water-insoluble lipids; lipoproteins also play an important role in cholesterol and triglyceride metabolism. Unlike many other serum components that are principally transferred from one point to another as metabolites, lipoproteins undergo a series of complex metabolic processes in which changes and exchanges occur continuously in and between the various lipoproteins. These processes are referred to as the *lipoprotein cascade* (see below).

Chylomicrons

Chylomicrons, responsible for transporting dietary fat, are synthesized by and released from the intestinal epithelial cells. The lipid content is derived predominantly

Table 14-6. CLASSIFICATION, PROPERTIES, AND COMPOSITION OF HUMAN SERUM LIPOPROTEINS

Measurement	Chylomicron	VLDL	IDL	LDL	HDL
Hydrated density, g/mL	0.93	0.97	1.003	1.034	1.121
Solvent density for isolation, g/mL	<1.006	<1.006	1.006–1.019	1.019–1.063	1.063–1.21
Molecular weight	$(0.4–30) \times 10^9$	$(5–10) \times 10^6$	$(3.9–4.8) \times 10^6$	2.75×10^6	$(1.75–3.6) \times 10^5$
Diameter, nm	>70.0	25.0–70.0	22.0–24.0	19.6–22.7	4–10
Electrophoretic mobility (paper, agarose)	Origin	Pre-β	Broad β (between β and pre-β)	β	α
Composition, % by weight					
Cholesterol, unesterified	2	5–8	8	13	6*
Cholesterol, esterified	5	11–14	22	49	13*
Phospholipid	7	20–23	25	27	28*
Triglyceride	84	44–60	30	11	3*
Protein	2	4–11	15	23	50*
Apoproteins, % total apolipoprotein					
AI	7.4	Trace	—	—	67
AII	4.2	Trace	—	—	22
B-100	Trace	36.9	50–70	98	Trace
B-48	22.5	Trace	Trace	—	—
CI, CII, CIII	66	49.9	5–10	Trace	5–11
EII, EIII, EIV	—	13.0	10–20	Trace	1–2
D	—	—	—	—	Trace
Synthesis	Intestine	Liver, intestine	Intravascular	Intravascular	Intestine, liver

*From Jackson, R. L., Morrisett, J. D., and Gotto, A. M. *In*: Hyperlipidemia: Diagnosis and Therapy. B. M. Rifkind, and R. I. Levy., Eds., New York, Grune & Stratton, 1977.

from the alimentary tract and is termed *exogenous*. The major lipid fraction is triglyceride, which constitutes over 80% of the total particle by weight. Chylomicrons contain a variety of apolipoproteins, but only 1–2% of the total particle is made up of apolipoproteins (Table 14-6). The most notable apolipoproteins found are B-48, AI, AII, AIV, and the C apolipoproteins. Chylomicrons traverse the lymphatics into the thoracic duct, eventually entering the main systemic circulation. While traveling through the lymphatic system, chylomicrons acquire significant quantities of the C apolipoproteins, presumably as a result of transfer from HDL that has filtered into the lymph from plasma. After entry into the systemic circulation, chylomicrons interact with other triglyceride-rich lipoproteins and are eventually acted on by a number of lipoprotein lipases. These enzymes hydrolyze the triglyceride component to monoglycerides, glycerol, and free fatty acids, which can then be taken up at the cellular level for energy metabolism or for resynthesis of triglyceride for storage. The Apo AI, AII, and C are transferred to HDL together with a small amount of lipid. The remainder of the chylomicron, known as chylomicron *remnant*, now falls into the density region between VLDL and IDL; its main apolipoproteins are B-48 and E. The remnant particle is 30–80 nm in diameter and retains most of the cholesteryl esters found in the chylomicron. Chylomicron remnants are rapidly removed from the circulation by hepatic lysosomes and are quickly internalized in the liver by Apo E receptor–mediated endocytosis and degraded, thus delivering dietary cholesterol to the liver.

Very-Low-Density Lipoproteins

The other triglyceride-rich lipoprotein, very-low-density lipoprotein (VLDL), is synthesized in and released from the liver. VLDL transports hepatic-synthesized triglyceride and cholesterol, which are therefore called *endogenous lipids*. A number of important apolipoproteins are contained in VLDL, the most notable being Apo B-100 and Apo CII. A schematic diagram of VLDL catabolism is shown in Figure 14-7.

Intermediate-Density Lipoproteins

After VLDL hydrolysis by lipoprotein lipase, a short-lived intermediate-density lipoprotein (IDL), partly depleted of TG, is formed. IDL contains about equal amounts

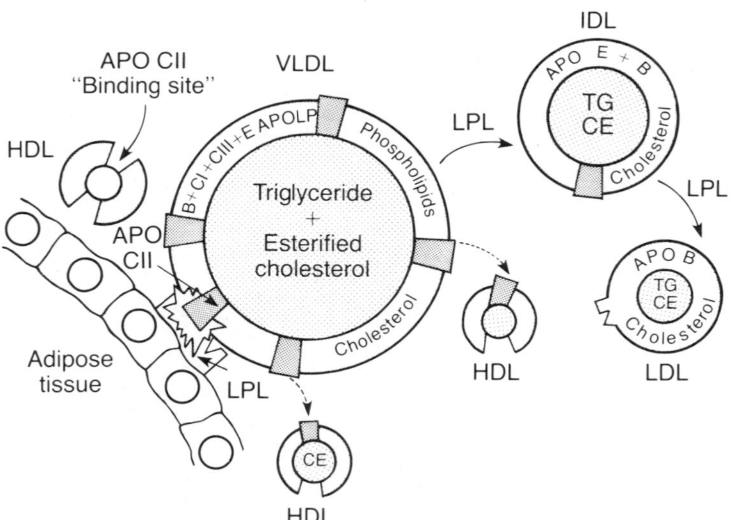

Figure 14-7. Catabolism of VLDL to LDL. Activation of lipoprotein lipase (LPL) by Apo CII leads to hydrolysis of VLDL triglycerides by LPL and formation of a short-lived lipoprotein called *intermediate-density lipoprotein* (IDL). IDL is further catabolized by LPL to form the cholesterol-rich LDL particle. During VLDL catabolism, Apo CII is transferred to HDL; all the remaining apolipoproteins except Apo B are lost as LDL is formed. The Apo C on HDL is later transferred back to chylomicrons and VLDL to complete the cycle. (Modified and published with permission from M. L. Kashyap, M.D.)

of cholesterol and triglycerides (see Table 14-6), and its major apolipoproteins are Apo B and E. It is Apo E that determines the continued catabolic process (cascade) of IDL to LDL for hepatic uptake and degradation. The role of Apo E here may be viewed as similar to the role it plays in respect to chylomicron remnants.

Low-Density Lipoproteins

Low-density lipoproteins (LDL) are normally formed by hepatic delipidation of IDL by hepatic lipoprotein lipase. As delipidation occurs, Apo B becomes unmasked, thus allowing LDL to bind to specific membrane receptors. LDL catabolism takes place in the liver and the peripheral tissues. In the proposed sequence of events, LDL interacts with high-affinity receptor sites; the bound LDL is internalized and subjected to lysosomal degradation that ultimately hydrolyzes the Apo B to amino acids, and the esterified cholesterol is hydrolyzed to free cholesterol, which enters the cytoplasm.

The release of free cholesterol is responsible for three regulatory responses that assist in cholesterol homeostasis: (a) suppression of the rate-limiting enzyme, HMG-CoA reductase, and thus of new cholesterol synthesis; (b) activation of ACAT activity to esterify excess cholesterol for intracellular storage as cholesteryl ester droplets; and (c) modulation of the number of LDL receptors on the plasma membrane to prevent the over-accumulation of intracellular cholesterol through the receptor pathway.

In addition to catabolism via the high-affinity Apo B LDL–receptor pathway, plasma LDL can be degraded by nonspecific pathways not subject to feedback regulation. These modes achieve significant rates of removal only at elevated plasma levels of LDL.

Patients with the genetic disorder *familial hypercholesterolemia* (FH) have a defect in the gene that codes for the LDL receptor. The receptor is absent or deficient, or it is present and binds normally but is unable to internalize the LDL particle. In these persons, elevated plasma levels of LDL-cholesterol (LDL-C) cause increased uptake of LDL by macrophages and smooth muscle cells, cholesterol deposits in arterial walls, and, consequently, premature atherosclerosis.

High-Density Lipoproteins

Although the metabolism of chylomicrons, remnants, VLDL, IDL, and LDL is reasonably well understood, our knowledge about high-density lipoproteins (HDL) is relatively new and still growing. HDL consists of a number of heterogeneous particles that vary with respect to size and content of lipid and apolipoprotein (see Table 14-6). The major apolipoproteins found in HDL are AI and AII, and they constitute about 90% of total HDL protein.

Both the liver and the intestine are involved in the *production* of HDL, although the exact roles and relative importance of each are not fully understood. Current evidence indicates that the mature spherical form of HDL is not secreted directly into the plasma but is derived from a nascent precursor form. Transformation of this nascent form of HDL into the spherical lipid-transporting moiety depends to a great extent on the action of LCAT (Figure 14-8). HDL components may also arise from the catabolism of VLDL and chylomicrons, since transfer of both apolipoproteins and lipids from VLDL and chylomicrons to HDL has been demonstrated. Little is known about the sites of HDL catabolism; however, the liver and kidney are probably involved.

Although our knowledge is still limited, HDL appears to be important in cholesterol efflux from tissues, thereby reducing the amount of cholesterol stored there. HDL is also involved in returning cholesterol from the periphery to the liver for removal as bile acids, a process known as reverse cholesterol transport. In addition, HDL plays a major role as a scavenger of lipid and apolipoprotein during the normal catabolism of chylomicrons and VLDL. HDL is also known to play an important part as a plasma reservoir for Apo CII.

Figure 14-8 summarizes the complex interrelationships between the various lipoproteins and provides a starting point from which a coherent and systematic *classification of lipoprotein disorders* can be derived.

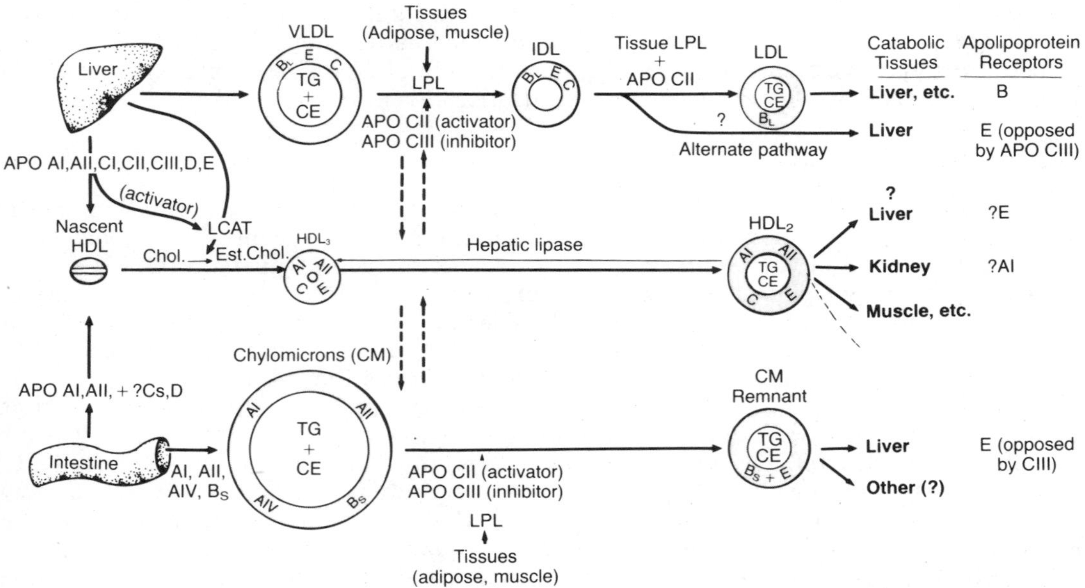

Figure 14-8. Overview of lipoprotein interrelationships.

Upper path: Endogenous lipoprotein metabolism of VLDL. Hepatic-synthesized triglyceride and cholesterol are transported by VLDL. Apo CII catalyzes and Apo CIII inhibits lipoprotein triglyceride hydrolysis by LPL. As VLDL is catabolized to IDL and then to LDL, the C-apoproteins are transferred to HDL (indicated by the broken arrow). IDL may also be catabolized via the Apo E receptor pathway in the liver. LDL is largely catabolized via the Apo B receptor. Degradation of LDL results in the release of esterified cholesterol, which is later hydrolyzed to free cholesterol.

Middle path: HDL formation and lipid transfer. Apo AI and AII are synthesized in the liver and intestine, enter the plasma (probably as part of VLDL or chylomicrons), and then come together to form nascent HDL in plasma, where Apo AI activates LCAT. Free cholesterol is then esterified and enters the core of the HDL particle. Thus, removal of cholesterol from cell surfaces and other lipoproteins is facilitated. These esters are later transferred from HDL to VLDL for transport to the liver for catabolism. Little is known about the sites of HDL catabolism. As indicated by the broken arrows, the Apo C on HDL is transferred to VLDL and chylomicrons.

Lower path: Exogenous lipoprotein metabolism of chylomicron. Dietary triglyceride constitutes over 80% of these particles. As chylomicrons are catabolized, Apo AI, II, and C are transferred to HDL (indicated by the broken arrow). The remaining chylomicron remnant is removed by the Apo E receptor. Degradation of these remnants results in delivery of dietary cholesterol to the liver.

TG = triglycerides; CE = cholesteryl ester; B_L = large form of Apo B; B_S = small form of Apo B.

(From Vascular Medicine, *2*:16–21, 1984. Reproduced with permission of copyright owner: Westminster Publications, Inc., Roslyn, New York 11576. All rights reserved.)

Clinical Interrelationships of Lipids, Lipoproteins, and Apolipoproteins

Epidemiology of Lipoprotein Disorders

An increase in excess circulating levels of any lipoprotein can be caused by either excess production or decreased catabolism; to date, most abnormalities have been related to the latter. Lipoprotein metabolism is of particular interest to clinicians concerned with the diagnosis and treatment of atherosclerosis, a vascular disease characterized by progressive thickening of the inner layer of the arterial wall. This thickening is caused by localized deposition of lipid and cellular material, resulting in reduced blood flow and weakening of affected arteries. Atherosclerotic coronary artery disease (CAD) accounts for more than 50% of all deaths in the US.[27] Of the many risk factors so far implicated in causing CAD, three are accepted as being of major importance: dyslipoproteinemia, hypertension, and cigarette smoking. There are numerous other risk factors, including diabetes, inactivity, obesity, and family history. The importance of serum lipoprotein disturbances as an etiological factor in atherosclerosis is supported by evidence that was amassed from epidemiological and population studies.[2,8,29,47] Many studies, using different approaches, have suggested that elevated HDL levels are "antiatherogenic," whereas reduced levels are associated with increased risk for coronary artery disease.[4,16,35]

Altered metabolism of chylomicrons and VLDL does not appear directly related to atherosclerosis. It is not known how chylomicron remnants, IDL, and LDL relate to the development of the disease or how HDL prevents it. It is known, however, that the disease involves more than just circulating lipoproteins and that cellular enzyme systems (such as LPL), platelets, arterial smooth muscle and endothelium, and macrophages play significant roles. In addition to these cellular and humoral factors, a host of genetic, environmental, and local factors have been implicated.

Apolipoproteins

In recent years evidence has accumulated to suggest that increased levels of Apo B may be a better marker of atherosclerosis in patients than are decreased HDL-cholesterol (HDL-C) and increased LDL-cholesterol (LDL-C) levels. In addition, plasma Apo AI may be a very good predictor of CAD in subjects requiring angiography.[31] Apo A levels have also been shown to be better correlated with peripheral vascular disease (PVD) than the lipid values presently measured.[13]

The diagnosis of Type III disease (dys-β-lipoproteinemia), which is due to improper catabolism of IDL and remnant particles, currently requires difficult, tedious, and expensive ultracentrifugal isolation and measurement of IDL. Although Type III hyperlipidemia is relatively rare, it must still be ruled out in all patients who display combined hyperlipidemia (increased cholesterol and triglyceride), a lipid phenotype found in 2–3% of the population. It appears that measurement of Apo E levels and Apo E isomorphic forms may be more specific and perhaps less difficult.[54]

Classification of Clinical Lipoprotein Disorders

The contemporary classification of dyslipoproteinemias shown in Tables 14-7 and 14-8 is an extension of the initial classification, known as the *Fredrickson* or *NHLBI classification*.[14] This classification was adopted by the World Health Organization (WHO) in 1970 and describes hyperlipoproteinemias as plasma phenotypes, i.e., particular patterns of lipid and lipoprotein levels with strong associations to clinical disease. Fredrickson's types continue to be important tools, but their major limitation is that they "lump" etiologically heterogeneous disease states together. For example, hyperchylomicronemia (Fredrickson Type I) can be due to either LPL deficiency or to Apo CII deficiency. This classification also does not take into consideration HDL, individual apolipoproteins, diet, drugs or alcohol, or general health.

A more logical pathophysiological classification based on the lipoprotein metabolic "cascade" is beginning to emerge. For instance, hyper-β-lipoproteinemia is characterized by elevations of circulating LDL that result from a number of defects in the high-affinity receptor-mediated LDL pathway. Subclassification of this Fredrickson Type II disorder would then include both a binding defect and an internalization defect.

Lipoprotein disorders may also be classified as primary or secondary. Primary disorders are genetic (familial) or nongenetic (sporadic), whereas secondary disorders are caused by diet; by use of alcohol or drugs; or by diseases of metabolic, hormonal, infectious, or malignant causes. Secondary causes are much more frequent and should be ruled out before the presence of a much less common primary disorder is sought.

HYPERLIPOPROTEINEMIAS

Hyperlipoproteinemia disorders are listed and the findings are summarized in Table 14-7.

Increased Chylomicrons (Hyperchylomicronemia; Fredrickson Type I Hyperlipoproteinemia)

In most cases, this extremely uncommon disorder is diagnosed before the patient reaches 10 years of age. The most frequent presentation is recurrent abdominal pain, often accompanied by splenomegaly or hepatomegaly or both, and eruptive xanthomas. Pancreatitis is the most serious complication and may be the cause of death.

Table 14-7. CLASSIFICATION OF HYPERLIPOPROTEINEMIAS

Lipoprotein Abnormality (Increase)	Plasma Appearance*	Total Cholesterol	Triglyceride	LDL-Cholesterol	HDL-Cholesterol	Apolipoprotein	Lipoprotein Electrophoresis†	Lipoprotein Phenotype	Clinical Association
Chylomicrons	Creamy layer, infranate clear or slightly turbid	Normal to moderately elevated	Markedly elevated	Normal	Normal to decreased	↑B-48 ↑A-IV ↓↑CII	Intense band at origin	Type I‡	Acute abdomen, pancreatitis
Low-density lipoprotein (LDL)	Clear, possible increase in yellow-orange tint	Usually elevated, occasionally within normal range	Normal	Elevated	Normal to decreased	↑B-100	Increased band in β-region	Type IIA‡	Markedly increased risk of CAD
Low-density lipoprotein (LDL); very-low-density lipoprotein (VLDL)	Clear to slightly turbid	Elevated, occasionally marginally	Elevated	Elevated	Normal to decreased	↑B-100	Increased β- and pre-β-band	Type IIB‡	Increased risk of CAD
Intermediate-density lipoprotein (IDL)	Turbid to opaque with thin creamy layer occasionally present	Elevated	Elevated	Normal to decreased	Normal to decreased	↑EII ↓EIII ↓EIV	Broad β-band	Type III‡	Increased risk of CAD
Very-low-density lipoprotein (VLDL)	Turbid to opaque	Normal to slightly elevated	Moderately to markedly elevated	Normal	Normal to decreased	↓↑CII ↑B-100	Increased pre-β-band	Type IV‡	Increased risk of CAD
Very-low-density lipoprotein (VLDL); chylomicrons	Creamy layer, infranate turbid to opaque	Slightly to moderately elevated	Markedly elevated	Normal	Normal to decreased	↓↑CII ↑B-48 ↑B-100	Intense band at origin plus increased pre-β-band	Type V‡	Pancreatitis, increased risk of CAD
High-density lipoprotein (HDL)	Clear	Normal to moderately elevated	Normal	Normal	Elevated	↑AI ↑AII	Increased α-band	Hyper-α-lipoproteinemia	Decreased risk of CAD

* After 16 h at 4 °C.
† Agarose.
‡ Fredrickson type.

Table 14-8. CLASSIFICATION OF HYPOLIPOPROTEINEMIAS

Lipoprotein Abnormality (Decrease)	Plasma Appearance*	Total Cholesterol	Triglyceride	LDL-Cholesterol	HDL-Cholesterol	Apolipoprotein	Lipoprotein Electrophoresis	Lipoprotein Phenotype	Clinical Association
Low-density lipoprotein (LDL)	Clear	Markedly decreased	Decreased	Absent	Normal	↓↓B-100	Absent β-band and decreased pre-β-band	A-β-lipoproteinemia	Malabsorption; mental retardation; growth failure
Low-density lipoprotein (LDL)	Clear	Mildly to markedly decreased	Normal	Decreased	Normal	↓B-100	Decreased β-band	Hypo-β-lipoproteinemia	Decreased risk of CAD
High-density lipoprotein (HDL)	Clear	Normal to decreased	Normal	Normal	Absent	↓↓↓AI ↓↓AII ?↓CIII	Absent α-band	A-α-lipoproteinemia	Increased risk of hypersplenism, CAD
High-density lipoprotein (HDL)	Clear	Normal, decreased, or increased	Normal to increased	Normal to increased	Decreased	↓AI ↓AII	Decreased α-band	Hypo-α-lipoproteinemia	Increased risk of CAD

* After 16 h at 4 °C

Adapted from Stein, E. A., and Glueck, C. J.: Hyperlipoproteinemia: Implications, diagnosis and therapy. *In*: Cardiac Diagnosis and Treatment, 3rd ed. N. O. Fowler, Ed. New York, Harper and Row, 1980, pp. 677–695.

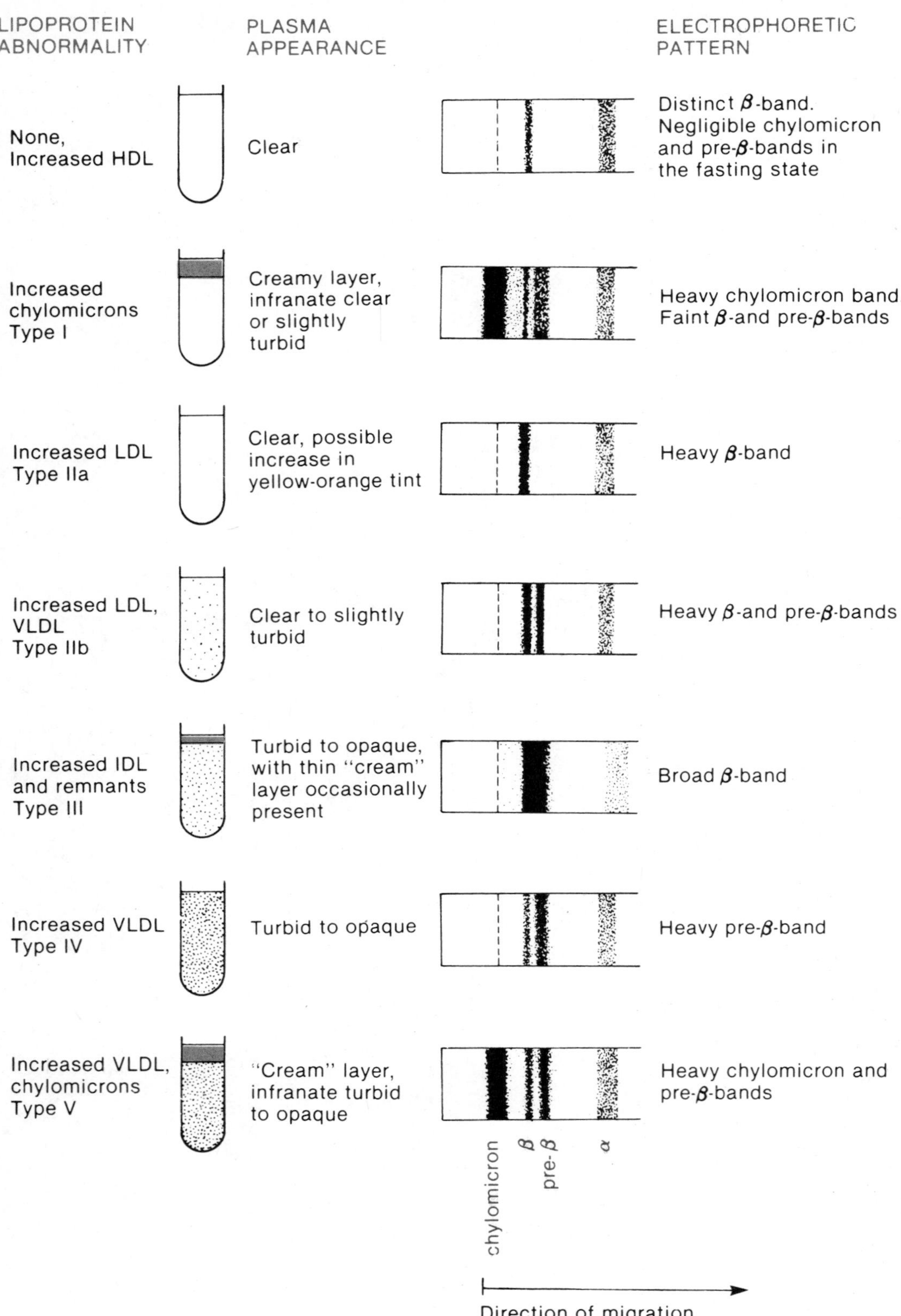

Figure 14-9. Hyperlipoproteinemias. Plasma appearance after 16 h at 4 °C. Electrophoretic migration is from left to right. Dotted line = origin. Relative separations are those typically found on agarose or paper. On polyacrylamide gel, the pre-β-lipoprotein band migrates slower than the β-lipoprotein band. (Modified from Ellefson, R. D., and Caraway, W. T.: Lipids and lipoproteins. *In*: Fundamentals of Clinical Chemistry, 2nd ed. N. W. Tietz, Ed., Philadelphia, W. B. Saunders Co., 1976.)

When diagnosed in a child, the disease is usually primary or familial. However, Type I phenotypes have been observed, often transiently, in older patients with a number of underlying disorders.

Biochemical findings. Hyperchylomicronemia is associated with some of the highest triglyceride levels reported and is easily diagnosed on the basis of the creamy layer of chylomicrons present in a fasting sample (Figure 14-9). A deficiency in the activity of the enzyme lipoprotein lipase (LPL) is thought to be the major cause of the impaired chylomicron clearance. LPL is responsible for initial triglyceride hydrolysis, and its inactivity prevents the removal of ingested fat from the circulation. Lack of LPL activity may be due either to its absence or to the lack of Apo CII, its activator.

Increased Low-Density Lipoprotein (Familial Hypercholesterolemia; Fredrickson Type II Hyperlipoproteinemia)

A number of disease entities are associated with or may cause Type II hyperlipoproteinemia (Table 14-9). The familial disorder, also known as familial hypercholesterolemia (FH), is inherited as an autosomal dominant gene.

Type II hyperlipoproteinemia is a metabolic problem of major clinical interest. This disorder is characterized by a strikingly high incidence of premature atherosclerosis, especially among young persons with FH, and tends to cause disease and death at a young age. In subjects homozygous for FH, very marked elevations of cholesterol (usually 800–1200 mg/dL) and major elevations of LDL are observed even with severe reductions in dietary fat and cholesterol. The disorder is usually manifest in infancy, with tuberous and plantar xanthomas, and is characterized by the early appearance of tendon xanthomas and of aortic stenosis. Often in childhood, and without exception by early adulthood, clinical symptoms and electrocardiographic signs of CAD develop. Death due to severe atherosclerosis in the untreated homozygote usually occurs before age 30.

Biochemical findings. The hallmark of Type II is an elevated level of LDL, which is usually measured indirectly by quantitating or calculating the LDL-C content. This Fredrickson classification is subdivided into Type IIa, with normal VLDL and triglyceride concentrations, and Type IIb, with elevation of VLDL and triglycerides. A greater prevalence of Type IIb is noted with increasing age and obesity. For a patient with values of cholesterol and triglyceride both above 300 mg/dL, it is often necessary to resort to ultracentrifugal and Apo E studies to differentiate between Types IIb and III. In familial hypercholesterolemia, patients have a deficiency in the number of functional LDL receptors.

With lack of uptake of LDL via the receptor-dependent pathway, LDL levels increase significantly. The scavenger pathway, which usually accounts for only 15% removal of the LDL pool daily, continues to function. Consequently, there is a large accumulation of cholesterol in tendon macrophages (xanthomas), splenic macrophages, hepatic Kupffer cells, bone marrow histiocytes, and similar scavenger cells in many organs.

Increased Intermediate-Density Lipoproteins (Dys-β-lipoproteinemia; Fredrickson Type III Hyperlipoproteinemia)

This disease is clinically manifest in about 1 in 10 000 persons, but it is rarely observed in children and is rarely acquired secondary to other causes. Most of the patients detected are asymptomatic. When physical signs of the disease are present, the most

Table 14-9. CAUSES OF SECONDARY INCREASES IN LDL

Diet high in cholesterol and total and saturated fat	Diabetes
Hypothyroidism	Chronic renal failure
Nephrotic syndrome	Drugs:
Dys-γ-globulinemia, multiple myeloma	Estrogens
Hepatic obstruction, hepatic disease	Androgens (anabolic steroids)
Porphyria	β-Blockers
Pregnancy	Carbamazepine
Anorexia nervosa	Progestins

characteristic is a skin lesion, xanthoma striata palmaris, which ranges from a yellowish discoloration to elevation and even obliteration of the palmar creases. Cerebral and peripheral vascular disease are as common as CAD.

Biochemical findings. Type III hyperlipoproteinemia is characterized by elevations of both cholesterol and triglyceride levels as well as an abnormal lipoprotein, which floats with the VLDL fraction on preparative ultracentrifugation. This floating β-lipoprotein or β-migrating VLDL often merges with the normal pre-β-band on electrophoresis, giving rise to a broad β-band. For a definitive diagnosis of Type III, an ultracentrifugal study with measurement of the lipid composition in the separated fractions is necessary. Type III lipoproteinemia should be suspected when the ratio of VLDL-cholesterol (VLDL-C) to total triglyceride (TG) exceeds 0.25, and it should be considered almost definite if the ratio exceeds 0.35. Confirmation of Type III should be done by studying the Apo E isoforms.

In dys-β-lipoproteinemia, hepatic receptors, which usually have high affinity and specificity for Apo E, do not bind or remove chylomicron remnants and IDL in a normal manner, due to a lack of the isoform known as Apo EIV.

Increased Very-Low-Density Lipoprotein (Fredrickson Type IV Hyperlipoproteinemia)

A high proportion of adults with hyperlipoproteinemia have increased levels of VLDL. Type IV is generally inherited as a recessive disease and is less penetrant in children. A significant number of persons with hypertriglyceridemia have the disease secondarily to obesity. It is rare to observe xanthoma, corneal arcus, or xanthelasma in young patients with primary Type IV; however, the last two signs may be present at a later age. Glucose intolerance, hyperuricemia, hypertension, premature CAD, and peripheral vascular disease are associated with both primary and secondary Type IV.

VLDL elevation often occurs secondary to a wide variety of diseases, drugs, and dietary habits (Table 14-10). Because of the numerous acquired hypertriglyceridemias, an approach to the management of the hypertriglyceridemic patient involves a careful distinction between primary and secondary Type IV.

Biochemical findings. Type IV hyperlipoproteinemia is characterized by an increase in triglyceride and VLDL levels; plasma cholesterol levels range from normal to moderately increased, LDL-C levels are normal or low, and HDL-C levels are often lower than in normotriglyceridemic populations. Chylomicrons are not present in fasting

Table 14-10. CAUSES OF SECONDARY CHANGES IN CHYLOMICRONS, VLDL, HDL, AND LDL

Disorder	Effect
Alcoholism	↑ Chylomicrons, VLDL
Regular alcohol intake	↑ Chylomicrons, VLDL, HDL
Excess calories	↑ Chylomicrons, VLDL
Obesity	↑ Chylomicrons, VLDL; ↓ HDL
Diabetes mellitus	↑ Chylomicrons, VLDL; ↓ HDL
Hypothyroidism	↑ Chylomicrons, VLDL; ↓ HDL
Hyperthyroidism	↓ HDL, LDL
Chronic renal disease	↑ Chylomicrons, VLDL
Pancreatitis	↑ Chylomicrons, VLDL
Intravenous hyperalimentation	↓ LDL, HDL
Malnutrition	↓ LDL
End-stage liver disease	↓ LDL, HDL
Chronic anemia	↓ LDL
Drugs	
Probucol	↓ HDL
Androgens	↑ VLDL; ↓ HDL
Estrogen-progestins	↑ VLDL
Beta-blockers	↑ VLDL; ↓ HDL
Thiazides	↑ VLDL; ↓ HDL
Corticosteroids	↑ VLDL
Pregnancy	↑ VLDL

plasma. Increased VLDL production by the liver or reduced VLDL removal are important factors in the development of hypertriglyceridemia.

Increased Chylomicrons and Very-Low-Density Lipoproteins (Fredrickson Type V Hyperlipoproteinemia)

Hypertriglyceridemia is the most common lipid abnormality encountered. The majority of hypertriglyceridemic patients have Type IV hyperlipoproteinemia; Type V is encountered less often. Expression of Type V in childhood or adolescence is uncommon. In addition, the disorder is commonly secondary to other disease states (see Table 14-10).

Biochemical findings. Type V is characterized by elevation of triglyceride levels due to increase in both VLDL and chylomicrons. The increased chylomicrons form a creamy surface layer when a plasma specimen is allowed to stand for 16 h at 4 °C. The turbid to lactescent layer below the surface layer is due to the VLDL increase. Excessive endogenous triglyceride synthesis or a deficiency of Apo CII is thought to be the major cause of impaired clearance of these triglyceride-rich lipoproteins.

Increased High-Density Lipoproteins

Familial hyper-α-lipoproteinemia appears to be transmitted as an autosomal dominant trait.[6] The condition is fully expressed in children and may also be diagnosed in neonates. Secondary hyper-α-lipoproteinemia associated with use of estrogenic preparations or alcohol has been described. No clinical features have been associated with hyper-α-lipoproteinemia.

Biochemical findings. Hyper-α-lipoproteinemia is characterized by distinct elevations of HDL-C, a slight elevation of total cholesterol, and normal triglyceride concentrations. At present it is merely a biochemical description associated with a less than average cardiovascular disease risk. No hypothesis for the cause of the increased HDL levels has been made.

HYPOLIPOPROTEINEMIA

Disorders and findings are summarized in Table 14-8.

Reduced Low-Density Lipoproteins (Hypo-β-Lipoproteinemia)

Familial hypo-β-lipoproteinemia is inherited as an autosomal dominant trait. Reduced plasma LDL-C is associated with significant increases in life expectancy in both males and females and a relative diminution of morbid and lethal myocardial infarctions. A number of other disorders are associated with secondary hypo-β-lipoproteinemia (see Table 14-10).

The syndrome is arbitrarily defined as an LDL-C level below the fifth percentile for a given sex- and age-matched population. The condition should not be confused with a-β-lipoproteinemia, which is a rare and severe disorder characterized by a total lack of Apo B–containing lipoproteins and by fat malabsorption, failure to thrive in infancy, and mental retardation.

Absent Low-Density Lipoproteins (A-β-lipoproteinemia)

A-β-lipoproteinemia is a rare autosomal recessive disorder characterized by an absence of the lipoprotein species that contain Apo B. Thus, there is total absence of LDL and, consequently, very low levels of plasma total cholesterol. The underlying biochemical defect probably involves an abnormality in the synthesis or secretion of the lipoproteins containing Apo B. The heterozygotes have no known clinical or metabolic abnormality but may have low LDL levels (hypo-β-lipoproteinemia).

Clinically, the features of a-β-lipoproteinemia are variable. Malabsorption of dietary fat is present in infancy with poor weight gain and failure to thrive. Steatorrhea is present even in individuals who consume a normal fat diet. Growth retardation is a common feature. Jejunal biopsy shows villi with extensive vacuolated mucosal cells that are packed with lipid droplets. The malabsorption of fat is associated with severe malabsorption of fat-soluble vitamins. There is progressive degeneration of the central nervous system.

Reduced High-Density Lipoproteins (Hypo-α-Lipoproteinemia)

Studies have indicated a significant increase in the risk of atherosclerotic persons with lower HDL-C levels compared with those with higher HDL-C concentrations.[37] The differences in HDL-C levels between various risk groups are small (5–10 mg/dL), and extreme caution must be exercised when HDL deficiency states are diagnosed.

A number of underlying disease states or environmental influences are associated with reduced levels of HDL (see Table 14-10); one of the most common associations is with hypertriglyceridemia. Familial hypo-α-lipoproteinemia appears to follow Mendelian dominant transmission.[17] Identification of individuals with familial hypo-α-lipoproteinemia is important in order to treat other associated risk factors. Currently a number of drugs are being investigated for therapeutic use to increase HDL concentrations.

Absent High-Density Lipoproteins (Tangier Disease)

The best-known disorder of HDL metabolism is Tangier disease. It is a rare autosomal recessive disorder characterized by accumulation of cholesteryl esters in the reticuloendothelial system and in other tissues. The disorder is associated with the virtual absence of HDL in plasma. The diagnosis is suspected when extremely low HDL-C levels are detected by conventional precipitation techniques, and it is confirmed by the absence of an α-lipoprotein band on electrophoresis. The homozygote is found to have undetectable amounts of HDL-C and extremely low levels of Apo AI and AII. Plasma total and LDL-C concentrations are usually low, and mild hypertriglyceridemia is often observed. The genetic defect most likely involves altered synthesis or catabolism of AI or AII.

The clinical findings of Tangier disease reflect the sites of accumulation of the cholesteryl esters and most notably include enlarged, lobulated orange-yellow tonsils and adenoids. Mucosa of the pharynx and rectum may also show orange discoloration. Recurrent peripheral neuropathy, muscle weakness and atrophy, and depressed tendon reflexes are also found. Splenomegaly and mild thrombocytopenia are common. The course of the disease is extremely variable; it has been diagnosed in patients who ranged in age from 3 to 69 years. The disorder is relatively benign since the lipid accumulation in the reticuloendothelial tissue does not usually affect any organ function. A recent review of subjects with Tangier disease has suggested that atherosclerosis may occur at a younger age than usual, mainly between the ages of 35 and 50 years. Unlike findings in LDL abnormalities, the clinical expression of Tangier disease is predominantly cerebrovascular in nature.

Investigation of Plasma Lipoprotein Disorders

Factors affecting diagnostic tests. The three most common errors leading to incorrect diagnosis are improper sampling, unreliable laboratory analysis, and failure to repeat sampling when indicated. The usual routine screening tests require a fast of at least 12 h (overnight), during which time only water should be consumed.

Analyzing a baseline blood sample while the patient is on a habitual diet is particularly important. Weight loss acutely lowers triglyceride concentrations and may transiently elevate cholesterol and low-density lipoprotein levels. Any acute febrile illness, trauma, or recent surgical procedure may affect plasma lipids, often by elevating triglyceride levels and depressing total cholesterol and LDL-C concentrations. Chronic and debilitating illness may severely decrease LDL and HDL concentrations. In addition, the patient may be taking prescription drugs that either lower lipid levels or substantially affect lipid metabolism. Often, young women do not consider estrogens or estrogen-progestin oral contraceptives as drugs and will not volunteer the information unless specifically asked about these medications. Similarly, young males participating in contact sports may illegally take anabolic steroids and hesitate to reveal this information even on direct questioning. Ideally, a specimen for the determination of plasma lipids should be obtained when the patient is not taking any drugs; if this is not possible, the use of drugs should be noted and taken into account when interpreting the laboratory data. If an abnormal lipid or lipoprotein concentration is detected, it is extremely important that the biochemical finding be confirmed on at least two subsequent occasions, preferably two to four weeks apart.

Diagnostic Tests for Lipoprotein Disorders

Plasma Appearance

The plasma appearance test is simple, convenient, inexpensive, and often diagnostically useful. If the plasma is clear, the triglyceride level is most likely normal or near normal (<200 mg/dL). When the triglyceride level increases to ~300 mg/dL, the plasma is usually hazy to turbid in appearance and is not translucent enough to allow clear reading of newsprint through the tube. When the plasma triglyceride level is >600 mg/dL, the plasma is usually opaque and milky (lipemic, lactescent). If chylomicrons are present, a thick, homogeneous, creamy layer may be observed floating at the plasma surface after it stands for several hours at 4 °C. As summarized in Figure 14-9, a uniformly opaque plasma sample usually denotes a Type IV pattern, whereas an opaque plasma sample with a creamy layer on top usually indicates a Type V pattern. Plasma with a thick chylomicron creamy layer and generally clear infranate is consistent with the Type I pattern. In nonfasting subjects, a chylomicron layer above a clear infranate may also be found, but this does not constitute an abnormal finding unless the blood specimen was drawn after more than 12 h of fasting. In patients with hypercholesterolemia due only to elevated LDL concentrations, the plasma is clear but may have an orange-yellow tint, since carotenoids are carried in LDL.

Total Cholesterol and Triglyceride Concentrations

After visual observation of the specimen, the next most useful and reliable tests are determinations of total cholesterol and triglyceride concentrations. A single cholesterol and triglyceride value more than 20% below the suggested age- and sex-adjusted upper reference limits virtually eliminates a diagnosis of hyperlipoproteinemia; it does not, however, address the equally important lipoprotein deficiency states (see Table 14-8). It is important that no dietary or drug intervention be instituted while evaluation is taking place.

Determination of High-Density Lipoprotein Cholesterol (HDL-C)

Currently, it is extremely difficult and impractical to quantitate HDL directly. Most methods depend on the measurement of the plasma content of HDL-C after selective precipitation of VLDL and LDL with various polyanions (e.g., heparin and manganese chloride). Although the precipitation stage is acceptably accurate and precise, the measurement of the supernatant cholesterol is subject to significant variation because of the low HDL-C concentration.

Some authors believe ratios of various lipoproteins (e.g., LDL-C:HDL-C) can be correlated with coronary artery disease risk, but these ratios contribute little to the understanding of the underlying disease and may prove misleading. For example, incorrect diagnostic or "risk" assessment may occur if the total cholesterol:HDL-C ratio is used without also measuring triglycerides and estimating LDL-C.

Determination of Low-Density Lipoprotein Cholesterol (LDL-C)

Traditionally, LDL-C measurements have required the use of a preparative ultracentrifuge. However, LDL-C can be fairly accurately estimated using the Friedwald formula:[15]

$$\text{LDL-C} = \text{Total cholesterol} - \text{HDL-C} - \text{triglyceride}/5$$

This formula is based on the assumption that VLDL-C is present in a concentration equal to one-fifth of the triglyceride concentration. This assumption is usually valid for triglyceride concentrations of <400 mg/dL; at higher triglyceride concentrations, inconsistencies in the VLDL triglyceride/cholesterol ratio occur, and the formula *must not* be used.

Lipoprotein Electrophoresis (LPE)

The value of LPE as part of the routine lipid or lipoprotein profile has been in question for a number of years, and its use in primary assessment of lipid or lipoprotein disorders is now discouraged. Instead, quantitative assays for triglycerides and for total and HDL-C, calculations of VLDL-C and LDL-C, inspection of plasma, and follow-up of abnormal findings with ultracentrifugation to establish the phenotype are now recommended. In this context, LPE remains valuable as a supplementary technique for characterization of uncommon disorders such as a-β-lipoproteinemia, Tangier disease, and Type III hyperlipoproteinemia (broad or floating β-disease). LPE also continues to be important for assessing post-heparin lipolytic activity.

There are two instances in which LPE serves useful purposes by first intent. One is when a patient is suspected of being nonfasting; in this case a small chylomicron band at the origin will alert the physician to retest the patient after proper preparation. The second instance is for the detection of Lp-X, an abnormal lipoprotein that is a marker in obstructive jaundice (see page 471).

Ultracentrifugation

In many research centers that specialize in lipoprotein abnormalities, preparative ultracentrifugation is used to separate the various lipoprotein classes. The most common method for expressing the concentration of the isolated lipoprotein fractions is by their cholesterol or apolipoprotein content. Clinically, when cholesterol and triglycerides are both elevated above 300–350 mg/dL, the presence of IDL (Type III) is suspected and should be confirmed by ultracentrifugation.

Determination of HDL Subfractions

In some clinical cases, distinction of subclasses of HDL may be useful. Separation and analysis of the HDL subclasses have, for the most part, been carried out by ultracentrifugal techniques. However, polyanion-precipitating reagents have also been used. HDL_2, which has a lower protein to lipid ratio than HDL_3, is selectively precipitated, and the cholesterol content of the HDL_3 fraction remaining in solution is measured. HDL_2-C is calculated by subtracting HDL_3-C from total HDL-C.

Methods for the Determination of Lipids and Lipoproteins

Separation and Isolation of Lipoproteins

Isolation of lipoproteins. Lipoproteins may be separated in at least five different ways: (1) by ultracentrifugal techniques; (2) by electrophoresis; (3) by precipitation as insoluble complexes between lipoproteins, polyanions, and divalent cations; (4) by gel or membrane filtration; or (5) by precipitation with antibodies to their apolipoproteins.

Separation of plasma lipoproteins by ultracentrifugation[18] is currently the reference method, and preparative ultracentrifugation is the most common technique used. It allows lipoprotein separation from other plasma proteins and from each other on the basis of their different individual densities. Ultracentrifugation, however, requires expensive instrumentation and significant technical skills and therefore is used mainly in specialized laboratories. The availability of a relatively inexpensive tabletop air-driven ultracentrifuge (Beckman Airfuge, Beckman Instruments, Inc., Palo Alto, CA 94304) permits the ultracentrifugal separation of the lipoproteins in 175 μL plasma in 2.5 h.[6]

Lipoprotein electrophoresis. The major classes of lipoproteins separate electrophoretically on the basis of their charge, size, shape, and interaction with the supporting medium. Agarose gel and paper electrophoresis produce similar separations of the lipoproteins, and the agarose gel offers increased resolution and, occasionally, increased separation within classes. Cellulose acetate may be inadequate to detect chylomicrons that co-migrate with VLDL, and this technique is not recommended. Most agarose gel electrophoresis methods are based on the procedure developed by Noble,[36] which needs about 90 min of electrophoresis in a barbital buffer followed by fixing, drying, and staining with Fat Red 7B or Sudan Black B.

Lipoproteins separated by LPE are identified according to the serum protein band that has similar mobility. On agarose and paper, HDL is conventionally associated with the α_1-globulin region, LDL with the α_2-globulin region, and VLDL with the β-globulin region; chylomicrons remain at the origin (see Figure 14-9). However, the β-, pre-β, α_2-, and α_1-bands that are observed on the electrophoretic strip cannot always be assigned with certainty to classes of lipoproteins defined by their density characteristics. Nevertheless, LPE of native serum is helpful in detecting an abnormal serum lipoprotein called lipoprotein-X (Lp-X), which on most support media migrates with the β-lipoproteins but in agar gels moves cathodically behind the origin. Lp-X is found in sera of patients with obstructive jaundice.[40] LPE of native serum is also helpful in evaluating post-heparin lipolytic activity and in distinguishing Lp(a) antigen, a minor β-lipoprotein that is also called "sinking pre-β-lipoprotein." LPE is most useful when applied to fractions of serum isolated by differential-density preparative ultracentrifugation. However, even when indicated in lipid studies, lipoprotein electrophoresis is recommended only as a qualitative supplement to quantitative lipid measurements. Problems of incomplete resolution and differences in staining sensitivity of lipoproteins prevent accurate quantitation of their concentrations on the basis of electrophoresis alone.

Procedure for the Separation of Lipoproteins by Electrophoresis

Specimen. After a 12-h fast, collect blood by using an evacuated blood collection tube containing dry Na_2EDTA (1 mg/mL). Store plasma at 4 °C until analysis. Whenever possible, lipoprotein electrophoresis should be performed on the same day the specimen is collected. Mix the sample well before performing the assay.

Principle. Lipoproteins are separated electrophoretically on the basis of their differing surface electrical charges. The rate of migration is also affected by the type of supporting medium (e.g., paper, agarose gel, polyacrylamide gel, or starch). The last two support media also provide size separation; the first two do not. The HDL (α-lipoprotein) migrates most rapidly toward the anode. With starch and polyacrylamide gel, the LDL (β-band) is the next fastest band, followed by VLDL (pre-β-lipoprotein). With paper and agarose, the β- and pre-β-migrations are reversed. On all media except cellulose acetate, chylomicrons do not migrate well and generally remain at the origin.

Interpretations. Visual interpretation of lipoprotein electrophoresis separations after staining is preferred (see Tables 14-7 and 14-8) and should be made in conjunction with quantitative lipid and lipoprotein measurements. The control sample should demonstrate distinct α- and β-bands and a faint pre-β-band. Densitometric quantitation is not recommended.

Methods for the Determination of HDL-Cholesterol (HDL-C)

Numerous studies[52] have compared the different methods for quantitating HDL-C. Isolation procedures involving preparative ultracentrifugation are considered reference methods. Electrophoretic techniques lack precision and accuracy in the range of 20–40 mg/dL, where greatest clinical interest lies.[44] Ion-exchange and gel-permeation chromatography are too complex and time-consuming for routine analysis. Therefore, precipitation techniques are recommended.

The major lipoproteins are precipitated by divalent cations and sulfated polysaccharides or by sodium phosphotungstate. For clinical purposes, a reagent precipitating system must form an insoluble complex with all the plasma lipoproteins except HDL so that HDL remaining in the supernatant after centrifugation can be quantitated by its cholesterol content. One of the frequently used and well-studied methods incorporates heparin-Mn(II) as the polyanion-divalent cation pair.[49] Other precipitating reagents include dextran sulfate (MW 50 000 or 500 000) or sodium phosphotungstate with Mg(II),[11,30] heparin with Ca(II),[52] concanavalin A, and polyethylene glycol 6000.[48] Reagent systems containing sodium phosphotungstate-Mg(II) and dextran sulfate-Mg(II) have been reported as more stable and more compatible with enzymatic assays for cholesterol.[30] Although these reagents produce a better sediment of VLDL and LDL, especially in lipemic samples, they do precipitate small but significant amounts of HDL;[52]

consequently, lower HDL-C values are obtained. The presence of Mn(II) and its interaction with phosphate buffer in an enzymatic cholesterol assay may cause an overestimation of the cholesterol content of HDL separated by heparin-Mn(II) precipitation.[43] The interference is totally eliminated by reconstituting the cholesterol reagent with EDTA solution[43] to chelate the Mn(II) or by using a reagent that does not contain phosphate. When very lipemic samples are precipitated with heparin-Mn(II), the insoluble complex has a low density and may not sediment completely. Ultrafiltration or high-speed centrifugation[51] may then be used to achieve separation of the precipitate from the HDL-containing supernatant. The heparin-Mn(II) procedure is effective even if the heparin concentration is varied over a wide range.[21]

Specimen. After a 12-h fast, collect blood by using an evacuated blood collection tube containing dry Na_2EDTA (1 mg/mL). Store plasma at 4 °C prior to analysis. HDL is stable for 4 d at 4–6 °C; slight but significant changes occur by 7 d. Specimens frozen at −20 °C show statistically but not clinically significant decreases in HDL measured at 7–14 d. Whenever possible, HDL separation should be performed on the day of specimen collection.

Principle.[52] HDL is isolated from the other major classes of plasma lipoproteins by the formation of insoluble complexes of lipoproteins, polyanions, and divalent cations. In the presence of Mn(II) and heparin, chylomicrons, VLDL, and LDL are selectively precipitated, leaving only HDL in solution. The precipitated lipoproteins are sedimented by centrifugation and the clear, HDL-containing supernatant is recovered for cholesterol analysis.

Reagents

1. Sodium heparin, 40×10^6 USP units/L.
2. Manganese chloride, 1.06 mol/L. Dissolve 20.98 g of $MnCl_2 \cdot 4\ H_2O$ in distilled water and dilute to 100 mL (stable for three months at 4 °C).
3. Working precipitation reagent. Add 0.6 mL of the sodium heparin solution to 10 mL $MnCl_2 \cdot 4\ H_2O$, 1.06 mol/L (stable for one month at 4 °C).

Procedure

1. Pipet 1.0 mL of specimens and controls into disposable 10 × 75-mm glass tubes.
2. Add 0.1 mL of the working precipitation reagent to each tube and mix thoroughly on a vortex mixer.
3. Allow the tubes to stand for 10 min at room temperature.
4. Centrifuge at 1500 × *g* for 30 min at 4 °C.
5. Transfer the clear supernatant carefully to a clean tube for use in cholesterol analysis. (See section on *Cholesterol Methods.*)

Calculations. The result of the quantitative analysis for cholesterol in the supernatant is multiplied by 1.1 to account for dilution with the precipitating reagent.

Comments

1. Samples anticoagulated with liquid EDTA will be somewhat diluted; thus, results of all lipids, including HDL-C, may be low.
2. Lipoproteins precipitated from lipemic samples may not sediment. The supernatant remains or becomes turbid, contains non-HDL lipoproteins, and is not suitable for further analysis unless subjected to one of the following procedures:
   a. Ultrafiltration of the turbid supernatant through a 0.22-nm filter in a 25-mm filter holder. Aspirate the turbid supernatant into a syringe which is then used to express the fluid through the filter. Collect the *clear* filtrate in a small, clean disposable glass tube.[51]
   b. Dilution of lipemic samples with an equal volume of NaCl, 0.15 mol/L, prior to precipitation. Dilution decreases the density of the solution and facilitates sedimentation of the precipitate. However, HDL-C, which is usually already low in lipemic samples, is also diluted and analytical error is increased.
   c. Ultracentrifugation for 18 h at 10 000 × *g*, to remove chylomicrons and VLDL, prior to performing the precipitation.
3. Supernatants should be mixed well before analysis for cholesterol.

4. EDTA plasma is the preferred specimen and is required for the reagent formulation specified above. Serum may be used in the same procedure but only if the Mn(II) concentration in the precipitation mixture is reduced from 92 to 46 mmol/L (by preparing Reagent 2 at half concentration, i.e., 0.53 mol/L). The higher concentration of Mn(II) used with plasma counteracts the EDTA it contains, but if the higher concentration of Mn(II) is used with serum, small amounts of HDL are precipitated by the excess Mn(II), and HDL-C is subsequently underestimated.

Reference ranges.[28] HDL-C levels vary considerably with age and sex. The overall mean is 45 mg/dL for males and 55 mg/dL for females. Systematic differences between different methods of separating HDL are sufficiently great to recommend caution in evaluating results. HDL-C can also be expressed as per cent of total cholesterol. For average risk of coronary heart disease, reference ranges are 15–25% for males and 18–27% for females.

Standing Plasma Test (Plasma Appearance)

Specimen. Collect blood in evacuated tubes containing dry Na_2EDTA (1 mg/mL). Although serum can be used, EDTA plasma is convenient since it can be used for other lipoprotein analyses when the standing plasma test has been completed. The test should be carried out on *fresh* specimens and never on previously frozen plasma.

Principle. Chylomicrons in plasma have a density of <1.006 and insufficient protein-protein interaction to keep them in solution. After plasma is left standing, chylomicrons float to the uppermost surface and form a distinct opaque layer. The VLDL molecule is large enough to cause dispersion of light (turbidity) but will remain uniformly distributed in plasma on standing.

Procedure. A 500-μL aliquot of fresh, well-mixed plasma is placed in a 6 × 50-mm disposable-glass culture tube, covered with Parafilm, and allowed to remain overnight (16–18 h) at 4 °C. The tube is then carefully examined in a strong light against a black or dark background.

Report. Chylomicrons, lactescence, and turbidity are noted if present (see also Figure 14-9). Chylomicrons appear as a thick, creamy layer at the top of the sample; the infranate may be clear or turbid. The sample should also be assessed for color; increased concentrations of LDL may be associated with an orange color. Hemolysis and jaundice should also be noted if present.

Comments

1. Chylomicrons should not be confused with the white material that floats to the top of hypertriglyceridemic plasma that has been stored for several days in the refrigerator.

2. Either a creamy layer or uniform turbidity of a sample indicates a need for dilution or other special handling for analysis of triglycerides and for isolation of HDL.

Methods for the Determination of Cholesterol

Cholesterol may be quantitated either by enzymatic or by chemical methods. The methods are either *direct*, utilizing serum or plasma directly for the assay, or *indirect*, treating the sample with solvent extraction or other isolation procedures prior to the assay of cholesterol. The direct methods are simple, convenient, and readily adapted to automated analysis but have a higher risk of error than indirect methods.

In the *chemical methods*, cholesterol reacts as a typical alcohol with strong, concentrated acids; the products are colored substances, chiefly cholestapolyenes and cholestapolyene carbonium ions (Lieberman-Burchard reaction). In virtually all of these procedures, acetic acid and acetic anhydride are used as solvents and dehydrating agents and sulfuric acid is used as a dehydrating and oxidizing reagent. The reaction is enhanced by addition of various metal ions (e.g., iron salts).

Direct *enzymatic procedures* have virtually replaced chemical methods in the clinical laboratory. The initial reaction steps, common to all enzymatic procedures, are the hydrolysis of cholesteryl esters at C-3 to form free cholesterol, and the subsequent oxidation step to produce H_2O_2.

$$\text{Cholesteryl esters} \xrightarrow[\text{esterase}]{\text{Cholesterol}} \text{Cholesterol} + \text{fatty acids}$$

$$\text{Cholesterol} + O_2 \xrightarrow[\text{oxidase}]{\text{Cholesterol}} \text{Cholest-4-ene-3-one} + H_2O_2$$

Some enzymatic methods measure the amount of O_2 consumed amperometrically by using an oxygen-sensing electrode. Other methods involve quantitation of the H_2O_2 by formation of a colored oxidation product or a reduced pyridine nucleotide, for example:

$$2\,H_2O_2 + \text{4-aminoantipyrine} + \text{phenol} \xrightarrow{\text{Peroxidase}} \text{Quinoneimine dye} + 4\,H_2O$$

or,

$$H_2O_2 + \text{ethanol} \xrightarrow{\text{Catalase}} \text{Acetaldehyde} + 2\,H_2O$$

$$\text{Acetaldehyde} + \text{NAD(P)}^+ \xrightarrow[\text{dehydrogenase}]{\text{Aldehyde}} \text{Acetate} + H^+ + \text{NAD(P)H}$$

The quinoneimine product absorbs light of 510 nm, and NAD(P)H absorbs at 340 nm.

Interferences. In direct methods, interference observed with turbid, lipemic, icteric, or hemolyzed specimens is an important consideration. Bilirubin causes positive interference with the Lieberman-Burchard and iron-salt methods by forming biliverdin, which absorbs in the spectral region of the cholesterol reaction product. It causes negative interference in the enzymatic methods that produce a colored chromophore because bilirubin reacts with hydrogen peroxide, thus reducing the amount of peroxide available for formation of the colored complex. Bilirubin can also interfere directly due to its absorbance around 500 nm. Effects of bilirubin interference are minimized in enzymatic assays in which oxygen consumption is measured electrochemically.[10]

Blanking of cholesterol assays is a complex and unresolved issue because of the varied nature of the interferents and their effects. Problems caused by interferents can be reduced by using methods that partially purify the specimen by means of solvent extraction of cholesterol.[1,32] In the Abell-Kendall method,[1] cholesterol and cholesteryl esters are first extracted from serum; the esterified cholesterol is then enzymatically hydrolyzed or chemically saponified, and then the Lieberman-Burchard reagent is added for color development.

Determination of Cholesterol by an Enzymatic Method

Specimen. After a 12-h fast, collect blood by using an evacuated tube containing dry Na_2EDTA (1 mg/mL). Separated plasma can be stored at 4 °C for up to 4 d prior to analysis. The choice between plasma and serum is controversial. EDTA-plasma has the advantages that specimens can be separated immediately and that lipoproteins have enhanced stability during storage at 4 °C. Specimens should preferably be analyzed on the day of collection. Plasma is stable for three months at −20 °C and for many years at −70 °C. Thawed specimens should be mixed well and brought to room temperature prior to analysis.

Principle. All cholesteryl esters present in plasma are quantitatively hydrolyzed into free cholesterol and fatty acids by cholesterol esterase. In the presence of oxygen, free cholesterol is then oxidized by cholesterol oxidase to cholest-4-ene-3-one and hydrogen peroxide (H_2O_2). The hydrogen peroxide reacts with phenol and 4-aminophenazone (4-aminoantipyrine) in the presence of peroxidase (POD) to form an *o*-quinoneimine dye (Trinder reaction). The intensity of the color formed is proportional to the cholesterol concentration and can be measured photometrically between 480 and 520 nm. The assay may be run in end point or kinetic modes and is adaptable to automated continuous-flow and centrifugal analyzers. Reagents are available in kit form (Boehringer Mannheim Diagnostics, Indianapolis, IN 46250).

Calibration. The method should be standardized with a plasma or serum secondary calibrator previously assigned a target value by an independent reference technique. The optimal target value for a total cholesterol calibrator is between 250 and 300 mg/dL. If standardization uses a primary standard containing only unesterified cholesterol, patient values are underestimated, probably due to incomplete hydrolysis of esters.[3] Recently an esterase has become available that is claimed to achieve total and rapid hydrolysis and that appears to eliminate the difference between the primary standard and the secondary calibrator. A second problem with primary standards is slow reaction time due to inhibition of cholesterol oxidase by the detergents used to solubilize the free cholesterol; this latter point may be especially important in the kinetic procedure.

Comments

1. The method shows linearity up to 500 mg/dL. Specimens with values above 500 mg/dL should be reanalyzed after dilution with an equal volume of NaCl, 0.15 mol/L.
2. The method also demonstrates good linearity at low cholesterol concentrations, an important feature for the measurement of HDL-C. For measurement of HDL-C, a secondary calibrator with a target value between 45 and 60 mg/dL should be used.
3. For total cholesterol measurements, a normal (170–190 mg/dL) and an elevated (280–320 mg/dL) control should be included in each run. For HDL-C measurements, at least four controls are required (e.g., serum pools). Two controls should have *total* cholesterol concentrations of 25–35 and 60–70 mg/dL, respectively; these are used only in the run for quantitating cholesterol. The other two should have HDL-C concentrations of 25–35 and 60–70 mg/dL; these should be carried through the HDL separation step *and* the cholesterol quantitation.
4. If a secondary serum calibrator is used, primary cholesterol standards with values of 100, 200, and 400 mg/dL should be included in the cholesterol assay to check linearity. For the HDL-C determination, primary standards of 25, 50, and 100 mg/dL should be used.
5. Commercial lyophilized or stabilized serum-based preparations are not suitable for the calibration of enzymatic cholesterol methods. The turbidity of these products is not cleared by the surfactant present in the reagent, as is turbidity of patient specimens. Therefore, an attempted blank correction would be erroneous. If possible, a frozen (−70 °C) plasma or serum calibrator should be used.
6. Total cholesterol and HDL-C measurements should be carried out by the same procedure, particularly if LDL-C is to be calculated by the Friedewald formula.[15]
7. The end point assay is slightly more affected by interfering substances, such as bilirubin, hemoglobin, and ascorbic acid, than is the kinetic procedure.

Reference ranges.[28] Significant changes in total, LDL, and HDL cholesterol occur with age, and there are also significant differences between the sexes. It is therefore vital that laboratories provide reference values for specific ages and sexes if clinically significant lipid abnormalities are not to be overlooked or overestimated. The practice of citing a single reference value for cholesterol with the upper limit usually between 250 and 300 mg/dL, regardless of age and sex, is strongly discouraged. On the basis of results obtained both from epidemiological studies of populations in which atherosclerosis is uncommon and from recent clinical trials that have unequivocally demonstrated a reduced risk of atherosclerotic complications with reduced cholesterol levels, it appears prudent to aim at keeping cholesterol levels close to the 50th percentile.

Methods for the Quantitation of Triglyceride

Methods for the quantitation of plasma triglycerides include chemical and enzymatic methods.[5,12,50] The chemical methods require solvent extraction of the plasma to solubilize triglycerides and to denature and remove protein. The extract is treated with an adsorbent material to remove phospholipids and interfering substances; isopropanol extracts are treated with a zeolite mixture or with alumina, and chloroform extracts are treated with silicic acid. Alternatively, an extraction system that employs nonane and isopropanol may be used to separate triglycerides into the upper, nonane layer. Contaminating phospholipids, free glycerol, and glucose are left in the lower isopropanol layer, which is discarded. Once isolated and purified, triglycerides are quantitated by either chemical or enzymatic reactions directed against their glycerol component.

In the ***chemical methods***, glycerol is released from triglycerides in the purified extract by saponification with potassium hydroxide or by transesterification with alkoxides. The glycerol is then oxidized to formaldehyde by sodium periodate. The formaldehyde is either reacted with a chromotropic–sulfuric acid mixture[7] to form a product that absorbs at 570 nm or is treated with acetylacetone and ammonium acetate to produce a fluorescent condensation product (Hantzsch condensation reaction), which may be quantitated by either fluorometry or colorimetry.[25]

$$\underset{\textbf{Formaldehyde}}{H-\overset{\overset{\displaystyle O}{\|}}{C}-H} + NH_4^+ + 2\ \underset{\substack{\textbf{Acetylacetone}\\ \textbf{(2,4-pentanedione)}}}{CH_3-\overset{\overset{\displaystyle O}{\|}}{C}-CH_2-\overset{\overset{\displaystyle O}{\|}}{C}-CH_3} \longrightarrow$$

$$\textbf{3,5-Diacetyl-1,4-dihydrolutidine} + 3\ H_2O + H^+$$

The multiple steps and complexities of the chemical methods complicate their adaptation to automation.

In the ***enzymatic methods***, hydrolysis of triglycerides to form glycerol may be achieved by saponification or by lipase.

$$\text{Triglyceride} + 3\ H_2O \xrightarrow{\textit{Lipase}} \text{Glycerol} + 3 \text{ free fatty acids}$$

Glycerol produced by hydrolysis may be assayed by a variety of coupled-enzyme approaches. In each of these methods, the first step is conversion of glycerol to glycerol-3-phosphate.

$$\text{Glycerol} + \text{ATP} \xrightarrow{\textit{Glycerol kinase}} \text{Glycerol-3-phosphate} + \text{ADP}$$

The final measured product of subsequent coupled reactions is proportional to the amount of glycerol, which in turn is proportional to the amount of triglycerides.

One method uses the ADP of the first step to generate pyruvate, which is consumed by an indicator reaction:

$$\text{ADP} + \text{phosphoenolpyruvate} \xrightarrow{\textit{Pyruvate kinase}} \text{ATP} + \text{pyruvate}$$

$$\text{Pyruvate} + \text{NADH} + H^+ \xrightarrow{\textit{Lactate dehydrogenase}} \text{Lactate} + \text{NAD}^+$$

The disappearance of NADH is followed spectrophotometrically at 340 nm. Other methods utilize the glycerol-3-phosphate. The reaction

$$\text{Glycerol-3-phosphate} + \text{NAD}^+ \xrightarrow{\textit{Glycerol phosphate dehydrogenase}} \text{Dihydroxyacetone phosphate} + \text{NADH} + H^+$$

allows the rate of appearance of NADH to be monitored at 340 nm. An alternative to glycerol phosphate dehydrogenase is glycerol phosphate oxidase, which produces hydrogen peroxide:

$$\text{Glycerol-3-phosphate} + O_2 \xrightarrow{\textit{Glycerol phosphate oxidase}} \text{Dihydroxyacetone phosphate} + H_2O_2$$

The H_2O_2 is then reacted by means of peroxidase to give a red quinoneimine dye, which can be measured spectrophotometrically at 505 nm.

$$H_2O_2 + \text{4-chlorophenol} + \text{4-aminophenazone} \xrightarrow{Peroxidase} HCl + 2\,H_2O + \text{4-(}p\text{-benzoquinoneimino)-phenazone}$$

Potassium ferrocyanide → Potassium ferricyanide

This last approach confers a greater degree of specificity by avoiding oxidative/reductive side reactions of NAD^+. Improved linearity and sensitivity of the peroxidase method has been described when 2-hydroxy-3,5-dichlorobenzenesulfonate is substituted for 4-chlorophenol.[33] The red chromophore produced in this modification is measured at 510 nm.

The presence of *free glycerol* in patient specimens is a source of error in both chemical and enzymatic methods. Increased concentrations of free glycerol may occur in plasma with greatly increased triglyceride concentrations, in plasma from patients taking glycerol-containing medications, or in specimens that are not fresh. Except for the last, errors due to free glycerol can be eliminated in both colorimetric and enzymatic assays by measuring a specimen blank in which the initial hydrolysis step is omitted. For the majority of fresh refrigerated samples with triglycerides below 300 mg/dL, the contribution to the triglyceride measurement caused by free glycerol and other interfering substances is relatively low and constant and is of minimal clinical importance. When false elevations of triglycerides are suspected, a sample blank should be determined.

Specimen. A specimen should not be obtained for triglyceride determination unless the subject has been fasting for 10–14 h. Collect blood in an evacuated tube containing dry Na_2EDTA (1 mg/mL) and store the plasma at 4 °C until analysis. Specimens can be stored at 4 °C for seven days, frozen at −20 °C for three months, or frozen at −70 °C for years. Lipemic specimens may require warming to 37 °C and vigorous mixing to attain homogeneity prior to analysis, especially if they have been frozen.

Calibration. Aqueous primary standards of glycerol can be used; however, an aqueous primary standard of glycerol is not identical to a patient specimen or a plasma calibrator containing triglyceride. Triolein, the triglyceride that has been used as a primary standard in some chemical methods, cannot be used in this method because it is not water soluble; when triolein is dissolved in an organic solvent and used as standard, the solvent will interfere in the enzymatic assay. A plasma- or serum-based secondary standard should therefore be prepared from a frozen pool of plasma with a triglyceride value of about 200 mg/dL. The pool must first be repeatedly assayed by the reference method to assign its secondary standard value. Although they cannot be used for calibrators, primary standards of glycerol at 100, 200, 400, and 600 mg/dL should be included in each run to monitor linearity.

Comments

1. Blood collection tubes whose stoppers have been lubricated with glycerol should not be used for collection of specimens for triglyceride assay. Such tubes, usually for collection of 2-mL specimens, are sometimes found in use for pediatric and micro specimen collection.

2. Controls at both normal and abnormal triglyceride levels (90–110 and 190–200 mg/dL, respectively) should be included in each assay run.

3. For specimens with triglyceride levels below 200 mg/dL, preformed glycerol content is almost always low, and blanking of the assay is not necessary. Larger amounts of preformed glycerol occur when triglycerides exceed 300 mg/dL but are probably not clinically significant; elevated glycerol levels are found in subjects who have undergone prolonged starvation or who have acute or severe illnesses that are characterized by excessive release of fatty acids from tissue storage depots. However, in samples taken post mortem, glycerol blanking should always be done, since triglycerides break down rapidly after death. Frozen or lyophilized controls may contain measurable amounts of free glycerol and should have a blank run with each assay.

4. When enzymatic assays are performed on automated instruments, stability of reagents must be carefully monitored. Some reagents have extremely short stability after reconstitution (e.g., one hour).

5. Various kits for the determination of triglycerides are commercially available.

Reference ranges. Many laboratories set an upper reference limit for triglycerides somewhere between 130 and 200 mg/dL, using the 95th percentile value for a given population distribution. Reference ranges based on an "apparently healthy population" are generally higher than the recommended (desirable) levels for adults: 40–160 mg/dL (males) and 35–135 mg/dL (females).

Simultaneous Determination of Cholesterol and Triglyceride, Using Continuous-Flow Analysis (AutoAnalyzer II)[32]

Principle. This semiautomated method is used by the Lipid Research Clinics and differs from the Abell-Kendall reference procedure only by omitting the step that hydrolyzes esterified cholesterol. After the plasma is manually extracted with isopropanol in the presence of a zeolite mixture, the specimen is aspirated into the cholesterol assay manifold (containing the Lieberman-Burchard reagent) and incubated at 60 °C; absorbance of the color developed is measured at 630 nm. Another aliquot of the extract enters a triglyceride assay stream of 80% isopropanol, to which is added a stream of KOH. After incubation at 50 °C for triglyceride saponification, periodate reagent enters the stream to oxidize the glycerol to formaldehyde, followed by acetylacetone in ammonium acetate to condense with the formaldehyde. Condensation takes place in a second heating bath (50 °C) to form the fluorescent product 3,5-diacetyl-1,4-dihydrolutidine, which is measured in a fluorometer.

Measurement of Apolipoproteins

Immunochemical measurements of apolipoproteins are rapidly gaining in popularity as specific antisera and various purified apolipoproteins become more widely available. Nonimmunological procedures such as molecular-sieve chromatograpy, isoelectric focusing, and polyacrylamide gel electrophoresis are laborious and imprecise but are still used as a reference to validate immunological assays. In general, immunochemical assays are more sensitive, specific, reproducible, and potentially more adaptable to automation than are nonimmunological procedures. There are, however, significant problems in producing antisera and in standardizing apolipoprotein assays. Polymorphism is a characteristic of apoproteins AI, B, and E, and an antiserum produced to one of them may not react with all the isoforms of the specific apoprotein. Masking of antigenic sites on the apoprotein while it is residing in the lipoprotein is a source of analytical error; e.g., 90–95% of the Apo AI in HDL is not antigenically available.[41] The tendency of apoproteins to self-associate, especially when purified and in dilute solutions, also poses a problem in certain immunochemical approaches.

A number of established immunochemical techniques are used: radial immunodiffusion (RID),[9] electroimmunoassay (EIA) in agar or agarose gel,[9] radioimmunoassay (RIA),[23] immunonephelometry (INA),[53] enzyme-linked immunoassay (ELISA),[38,39,77] and fluorescence immunoassay (FIA).[26]

RID for apolipoprotein analysis is slow (2 or 3 d are required for the diffusion-precipitation process) and calls for pretreatment of sample. Assays are critically dependent on antibody characteristics, antibody-antigen ratio, and the properties of the gel. The sensitivity of RID procedures is generally lower than that of EIA and INA and considerably less than that of FIA, ELISA, and RIA.

EIA, or "rocket" assay, is a more rapid and more sensitive immunoprecipitation technique than RID. In EIA, rocket formation depends on the low intrinsic charge of the agarose support gel, the type of buffer, the concentration of polyethylene glycol (PEG), and the duration of electrophoresis.

RIA[24,38,46] is the most traditional approach to apolipoprotein measurement but has fallen out of favor in many clinical laboratories.

In ***INA***,[38,46] the antigen-antibody interaction results in turbidity that is quantitated by measuring changes in light-scattering intensity. The procedure can be performed in either end point or kinetic mode. Moreover, the procedure is adaptable to automation, requires a small sample volume, and has a high sample throughput.

ELISA systems[19,20,45] in general require minimal upkeep once they have been developed. Only small amounts of antiserum are required, assay time is short, and assay precision is acceptable. Other than INA, only ELISA methods have potential for automation and high throughput.

FIA has the same advantages as ELISA in terms of analytical specificity and reagent stability, but it has greater sensitivity than ELISA.

Commercial assays for *Apo AI* and *B* have recently become available in the US. These kits use the principles of RID (Tago Laboratories, Burlingame, CA 94010) and INA (Hyland Diagnostics, Deerfield, IL 60015). An RIA procedure for Apo AI is also available (Ventrex, Portland, ME 04103).

References

1. Abell, L. L., Levey, B. B., Brodie, B. B. et al.: A simplified method for the estimation of total cholesterol in serum and demonstration of its specificity. J. Biol. Chem., *195*:357–366, 1952.
2. Arteriosclerosis. A Report by the National Heart and Lung Institute Task Force on Arteriosclerosis. Department of Health, Education, and Welfare publication (NIH) 72-137. Washington D.C., National Institutes of Health, 1971, Vol. 1.
3. Bacharik, P. S., and Wood, P. D. S.: Laboratory considerations in the diagnosis and management of hyperlipoproteinemia. *In*: Hyperlipidemia: Diagnosis and Therapy. B. M. Rifkind and R. I. Levy, Eds. New York, Grune and Stratton, 1977.
4. Barr, D. P., Russ, E. M., and Eder, H. A.: Protein-lipid relationships in human plasma. II. In atherosclerosis and related conditions. Am. J. Med., *11*:480–493, 1951.
5. Bartlett, G. R.: Phosphorus assay in column chromatography. J. Biol. Chem., *234*:466, 1959.
6. Bronzert, T. J., and Brewer, H. B.: New micromethod for measuring cholesterol in plasma lipoprotein fractions. Clin. Chem., *23*:2089–2098, 1977.
7. Carlson, L. A., and Wadström, L. B.: Determination of glycerides in blood serum. Clin. Chim. Acta, *4*:197, 1959.
8. Coronary Drug Project Research Group: The Coronary Drug Project: Clofibrate and niacin in coronary heart disease. JAMA, *231*:360–381, 1975.
9. Curry, M. D., Gustafson, A., Alaupovic, P., et al: Electroimmunoassay and radial immunodiffusion assay evaluated for quantification of human apolipoprotein B. Clin. Chem., *24*:280–286, 1978.
10. Dietschy, J. M., Weeks, L. E., and Delente, J. J.: Enzymatic measurement of free and esterified cholesterol levels in plasma and other biological preparations using the oxygen electrode in a modified glucose analyzer. Clin. Chim. Acta, *73*:407, 1976.
11. Finley, P. R., Schifman, R. B., Williams, R. J., et al.: Cholesterol in high-density lipoprotein: Use of Mg^{2+}/dextran sulfate in its enzymic measurement. Clin. Chem., *24*:931, 1978.
12. Folch, J., Lees, M., and Sloane-Stanley, G. H.: A simple method for the isolation and purification of total lipids from animal tissues. J. Biol. Chem., *266*:497, 1957.
13. Franceschini, G., Bondioli, A., Mangero, M., et al.: Increased lipoprotein B in very low density lipoproteins of patients with peripheral vascular disease. Arteriosclerosis, *2*:74–80, 1982.
14. Fredrickson, D. S., and Levy, R. I.: Familial hyperlipoproteinemias. *In*: The Metabolic Basis of Inherited Disease 3rd ed. J. B. Stanbury, J. B. Wyngaarden, and D. S. Fredrickson, Eds. New York, McGraw-Hill, 1972, pp. 545–614.
15. Friedewald, W. T., Levy, R. I., and Fredrickson, D. S.: Estimation of the concentration of low-density lipoprotein cholesterol in plasma without use of the preparative ultracentrifuge. Clin. Chem., *18*:499, 1972.
16. Glueck, C. J., Fallat, R. W., Millet, F., et al.: Familial hyperalphalipoproteinemia. Metabolism, *24*:1243, 1975.
17. Glueck, C. J., Mellies, M. J., Tsang, R. C., et al: Low and high density lipoprotein cholesterol interrelationships in neonates with low density lipoprotein cholesterol less than or equal to the 10th percentile and in neonates with high density lipoprotein cholesterol greater than or equal to the 90th percentile. Pediatr. Res., *11*:957, 1977.
18. Havel, R. J., Eder, H. A., and Bragdon, J. H.: Distribution and chemical composition of ultracentrifugally separated lipoproteins in human serum. J. Clin. Invest., *34*:1345–1354, 1955.
19. Holmquist, L: Quantitation of human serum apolipoprotein B by enzyme immunoassay. Clin. Chim. Acta, *121*:327–336, 1982.

20. Holmquist, L.: Quantitation of human very low density apolipoproteins Cl, C2, C3 and E by enzyme immunoassay. J. Immunol. Methods, *34*:234–251, 1980.
21. Ishikawa, T. T., Brazier, J. B., Steiner, P. M., et al.: A study of the heparin-manganese chloride method for determination of plasma alpha-lipoprotein cholesterol concentration. Lipids, *11*:628, 1976.
22. Jackson, R. L.: Lipid transport and atherosclerosis: Role of apolipoproteins in vascular smooth muscle. *In*: Topics in Physiology; Metabolic, Ionic, and Contractile Mechanisms. M. F. Krass III and C. D. Barnes, Eds., New York, Academic Press, 1982, pp. 163–196.
23. Karlin, J. B., Juhn, D. J., Goldberg, R., et al.: Recent progress in the development of radioimmunoassays for human serum lipoproteins. Am. Clin. Lab. Sci., *8*:142–154, 1978.
24. Kashyap, M. L., Srivastava, L. S., Chen, C. Y., et al.: Radioimmunoassay of human apoliprotein CII: A study of normal and hypertriglyceridemic subjects. J. Clin. Invest., *60*:171–180, 1977.
25. Kessler, G., and Lederer, H.: Fluorometric measurement of triglycerides. *In*: Automation in Analytical Chemistry. L. T. Skeggs, Jr., et al., Eds. Technicon Symposia 1965. New York, Mediad, 1966, p. 341.
26. Lee, J. K. and Carrico, R. T.: Substrate labeled fluorescence immunoassay for the determination of serum protein. *In*: Proceedings of the Workshop on Apolipoprotein Quantification. K. Lippel, Ed. U.S. Department of Health and Human Services, National Institutes of Health, Publication No. 83-1266, 1983, pp. 332–342.
27. Levy, R. I.: Declining mortality in coronary heart disease. Arteriosclerosis, *1*:312–325, 1981.
28. Lipid Research Clinics Program Epidemiology Committee: Plasma lipid distributions in selected North American population. The Lipid Research Clinics Program Prevalence Study. Circulation, *60*:427–439, 1979.
29. Lipid Research Clinics Program: The Lipid Research Clinics Coronary Primary Prevention Trial Result. II. The relationship of reduction in incidence of coronary heart disease to cholesterol lowering. JAMA, *251*:365–374, 1984.
30. Lopes-Virella, M. F., Stone, P., Ellis, S., et al.: Cholesterol determinations in high-density lipoproteins separated by three different methods. Clin. Chem., *23*:882, 1977.
31. Maciejko, J. J., Holmes, D. R., Kottke, B. A., et al.: Apolipoprotein A-I as a marker of angiographically assessed coronary artery disease. N. Engl. J. Med., *309*:385–389, 1983.
32. Manual of Laboratory Operations, Lipid Research Clinics Program. Lipid and lipoprotein analysis. U.S. Department of Health, Education and Welfare, National Institutes of Health, Publication No. 75-628, 1974.
33. McGowan, M. W., Artiss, J. D., Strandbergh, D. R., et al.: A peroxidase-coupled method for the colorimetric determination of serum triglycerides. Clin. Chem., *29*:538–542, 1983.
34. Nestel, P. J., Fridge, N. H., and Tan, M. H.: Increased lipoprotein-remnant formation in chronic renal failure. N. Engl. J. Med., *307*:329–333, 1982.
35. Nikkila, E.: Studies on the lipid-protein relationships in normal and pathological sera and the effect of heparin on serum lipoproteins. Scand. J. Clin. Lab. Invest., *5*:1–101, 1953.
36. Noble, R. D.: Electrophoretic separation of plasma lipoproteins in agarose gel. J. Lipid Res., *9*:693, 1968.
37. Rhoads, G. C., Gulbrandsen, D. L., and Kagan, A.: Serum lipoproteins and coronary heart disease in a population of Hawaii Japanese men. N. Engl. J. Med., *294*:293, 1976.
38. Rosseneu, M., Vercaemst, R., Steinberg, K. K., et al.: Some consideration of methodology and standardization of apolipoprotein B immunoassays. Clin. Chem., *29*:427–433, 1983.
39. Schaefer, E. J., Eisenberg, S., and Levy, R. I.: Lipoprotein apoprotein metabolism. J. Lipid Res., *19*:667–687, 1978.
40. Seidel, D., Alaupovic, P., Furman, R. H., et al.: A lipoprotein characterizing obstructive jaundice. II. Isolation and partial characterization of the protein moieties of low density lipoproteins. J. Clin. Invest., *49*:2396, 1970.
41. Shonfeld, F., and Pfleger, B.: The structure of human high density lipoprotein and levels of apolipoprotein AI in plasma as determined by radioimmunoassay. J. Clin. Invest., *54*:235, 1974.
42. Stanbury, J. B., Wyngaarden, J. B., Fredrickson, D. S., et al.: Metabolic Basis of Inherited Disease, 5th ed. New York, McGraw-Hill, 1983.
43. Steele, B. W., Kochler, D. F., Azar, M. M., et al.: Enzymatic determinations of cholesterol in high-density lipoprotein fractions prepared by a precipitation technique. Clin. Chem., *22*:98, 1976.
44. Stein, E. A., McNeely, S., and Steiner, P.: Electrophoretic separation of high-density lipoprotein cholesterol evaluated and compared with the modified lipid research clinic procedure. Clin. Chem., *24*:1934, 1979.
45. Stein, E. A., and Pesce, A.: Enzyme linked immunoassays for apolipoproteins: Advantages, problems and prototype assay. *In*: Proceedings of the Workshop on Apolipoprotein Quantification. K. Lippel, Ed. U.S. Department of Health and Human Services, National Institutes of Health, Publication No. 83-1266, 1981, pp. 319–331.
46. Steinberg, K. K., Cooper, G. R., Graiser, S. R., et al.: Some considerations of methodology and standardization of apolipoprotein A-I immunoassays. Clin. Chem., *29*:415–426, 1983.
47. Turpeinen, O.: Effect of cholesterol lowering diet on mortality from coronary heart disease and other causes. Circulation, *59*:1–7, 1979.
48. Viikari, J.: Precipitation of plasma lipoproteins by PEG-6000 and its evaluation with electrophoresis and ultracentrifugation. Scand. J. Clin. Lab. Invest., *36*:265, 1976.
49. Warnick, G. R., and Albers, J. J.: A comprehensive evaluation of the heparin-manganese precipitation procedure for estimating high density lipoprotein cholesterol. J. Lipid Res., *19*:65, 1978.
50. Warnick, G. R., and Albers, J. J.: Evaluation of quality control materials for high-density lipoprotein cholesterol separation. Clin. Chem., *26*:998, 1980.

51. Warnick, G. R., and Albers, J. J.: Heparin-Mn^{2+} quantitation of high-density lipoprotein cholesterol: An ultra-filtration procedure for lipemic samples. Clin. Chem., *24*:900, 1978.
52. Warnick, G. R., Cheung, M. C., and Albers, J. J.: Comparison of current methods for high-density lipoprotein quantitation. Clin. Chem., *25*:596, 1979.
53. Weinstock, N., Bartholome, M., and Seidel, D.: Determination of apolipoprotein A-I by kinetic nephelometry. Biochim. Biophys. Acta, *663*:279–288, 1981.
54. Zannis, V. I., and Breslow, J. L.: Apolipoprotein E. Mol. Cell. Biochem., *42*:3–20, 1982.

15 NUTRITION

DORICE M. CZAJKA-NARINS, Ph.D.

With the progress of science and technology, the cellular role of nutrients has become more clearly delineated and the importance of nutrition, both as a vital component of health and in minimizing the effects of disease, has become more apparent. As our understanding of nutrition has expanded, so has the need for methods to assess early nutritional depletion and to monitor recovery.

Approximately forty dietary components and elements are recognized as essential for humans; additional minerals may be required in such small amounts that their essentiality has not yet been recognized. In general, *macronutrients* (proteins, carbohydrates, and lipids) constitute structural and functional components of the cells and are sources of energy. *Micronutrients* (vitamins and minerals) are components of enzymes or cofactors needed for metabolic reactions. *Conditionally essential nutrients* are nonessential in the healthy person but are required by certain sick individuals who have lost the ability to synthesize such compounds at an adequate rate.

Recommended dietary allowances (RDA) have been established as goals for good nutrition. As originally defined, the recommended dietary allowances are "goals at which to aim in providing the nutritional needs of groups of people."* Except for energy,† the RDA are set at levels sufficiently above the average physiological requirements to cover the variability that arises from differences in genetic makeup and to ensure that the needs of nearly all are met. The energy allowance is set at a level believed to be consistent with the good health of average persons.

RDA for energy are categorized by age and sex. Energy needs per kilogram of body weight are highest for the young infant; after age ten they are generally higher for males than females. For young adults, the allowances are set for individuals doing light work; for example, for males 19–22 years of age, the RDA is 2900 ± 400 kcal. The range emphasizes the variation in energy intake appropriate for any group of people. Allowances for older groups are set to compensate for a 2% decrease in basal metabolic rate (BMR) per decade and for a reduction in activity. (The basal metabolic rate is the sum of the energy needed for all involuntary activities.)

Proteins, carbohydrates, and lipids are dietary sources of energy. Proteins and carbohydrates yield 4 kcal/g (17 kJ/g), and triglycerides yield 9 kcal/g (38 kJ/g). Foods typically contain all three macronutrients in varying proportions. Carbohydrates are the primary source of energy for the body. Although proteins and amino acids are needed as the precursors of structural proteins, enzymes, antibodies, some hormones, and other metabolically active compounds, they can also be used for energy. Lipids, the source of essential fatty acids, are also a concentrated source of energy. Lipids function as components of all cell membranes and as precursors of prostaglandins and steroid hormones.

RDA have been established for those vitamins and minerals for which sufficient information is available. For another group of vitamins and minerals, there is less information available, and for these only "estimated safe and adequate daily dietary intakes" have been developed.

---

* Food and Nutrition Board, National Academy of Sciences: The Recommended Dietary Allowances, 9th ed. Washington, D.C., National Research Council, National Academy of Sciences, 1980.

† In the nutritional sciences the word *energy* refers to food energy: the unit is the calorie or, in the SI, the joule (J).

Factors Altering Nutritional Requirements

Several physiological and pathological conditions affect nutritional requirements. Anabolic processes such as growth increase energy and nutrient needs. Conditions such as those associated with reduced gastrointestinal absorption, trauma and burns, or increased renal excretion result in the need to increase dietary intake. Drug-nutrient interactions and nutrient-nutrient interactions can also result in deficiency.

Physiological Factors

Growth. Physiological conditions characterized by the expansion of tissue mass (e.g., growth and pregnancy) result in increased requirements. Requirements are highest in the newborn. As the growth rate slows, the requirements decrease and do not rise again until the adolescent growth spurt.

Lactation. Lactation is nutritionally demanding; the energy required is proportional to the volume of milk produced. The vitamin content of milk appears to be more sensitive to maternal intake than is the mineral content.

Gender. Energy requirements per kilogram body weight are higher in males than in females due to different proportions of body fat and the effects of hormones. Iron requirements are substantially higher in females from menarche to menopause. On the other hand, several vitamins in the B group are required in higher amounts by the male than by the female because of higher caloric intake by males.

Age. Lean body mass and basal metabolic rate, as well as caloric need, decline with age; however, requirements for many essential nutrients do not. Information on the requirements of elderly individuals is generally inadequate; most available information relates to young adults.

Nutrient-Nutrient Interactions

Nutrients in food can interact with each other or with non-nutrient components. The most thoroughly studied of these interactions is the enhancement of iron absorption by ascorbic acid. Meat, poultry, or fish proteins exert a similar effect, but the specific compound responsible has not been identified. Another example of a nutrient-nutrient interaction is the decrease in absorption of several trace minerals caused by the phytic acid or fiber present in cereals that bind these minerals.

A similar interaction is that between calcium and protein. Increased fecal excretion of calcium occurs when intake of some proteins increases. Since the protein intake of most Americans is higher than the RDA, the requirements for calcium in the US are also higher than the standards established for other countries.

Drug-Nutrient Interaction

Drugs affect nutritional status in a variety of ways. Some drugs alter food intake and others interact to reduce absorption or to alter distribution, transport, utilization, storage, or excretion (Table 15-1). Antineoplastic drugs may cause nausea or gastrointestinal damage and thus reduce both intake and absorption. Insulin, steroids, some psychotropic drugs, and certain antihistamines increase food requirements. Tetracyclines and other broad-spectrum antibiotics inhibit vitamin K synthesis by microorganisms in the gastrointestinal tract. In addition, malabsorption can be caused by laxatives, hypocholesterolemic drugs, antibiotics, and anticonvulsants. These examples of drug influences demonstrate the necessity of including information on the use of over-the-counter and prescription medication as part of the diet history.

Adaptation to Altered Nutrient Intake

In the face of decreased intake or increased need for nutrients, physiological adjustments take place, particularly if the change occurs slowly and is not too severe. For example, the consumption of pharmacological doses of vitamin C by pregnant women can lead to rebound scurvy in their formula-fed infants, since formulas were developed for infants of mothers with an average intake of vitamin C. Once the tissues of the fetus

Table 15-1. MAJOR DRUG EFFECTS ON NUTRIENTS

Therapeutic Class	Major Drugs	Effect on Nutritional Status
Alcohols	Ethanol	Vitamin B_1 deficiency; impaired vitamin B_6 activation; folic acid deficiency; magnesium excretion
Antacids*	Those containing Al or Ca	Reduced PO_4 absorption
Anticonvulsants and sedatives	Diphenylhydantoin Phenobarbital Glutethimide	Accelerated vitamin D metabolism; accelerated vitamin K metabolism; folic acid deficiency; vitamin B_6 deficiency
Anti-inflammatory drugs	Aspirin	Iron loss caused by GI bleeding
Antitubercular drugs	Isoniazid	Vitamin B_6 deficiency; niacin deficiency
Corticosteroids	Cortisone Prednisone	Accelerated vitamin D metabolism; accelerated vitamin C excretion; increased vitamin B_6 requirement; increased zinc excretion
Diuretics*	Chlorothiazide	Increased K and Mg excretion
	Spironolactone	Reduced K excretion
Hypocholesterolemic agents	Cholestyramine	Malabsorption of fat, vitamins A, K, D, and B_{12}
Hypotensive drugs	Hydralazine	Vitamin B_6 depletion
Laxatives*	Mineral oil	Malabsorption of vitamins A, D, and K
	Phenolphthalein	Malabsorption of vitamin D and calcium
Nonabsorbed antibiotics	Neomycin Kanamycin	Lower absorption of vitamins A, D, K, B_{12}, carotene, protein
Oral contraceptive agents	Mestranol Ethinyl estradiol Conjugated estrogens	Folic acid deficiency; reduced calcium excretion

* Categories of drugs which may cause significant mineral depletion in the elderly.

are saturated with vitamin C, the excess is metabolized and excreted. The greater the excess, the more efficient the excretion. When intake of the newborn drops suddenly, the high rate of excretion continues and depletes the body of vitamin C. An example of protective adaptation is the gastrointestinal absorption of some minerals, such as iron, that becomes more efficient in individuals consuming diets low in that mineral.

Effects of Disease on Nutrient Requirements

RDA are inadequate for patients with serious acute diseases or chronic conditions. Some data on patients with cancer suggest an increased metabolic rate; other data suggest no change. Other examples of diseases that alter nutrient requirements are measles, infections, and burns. Measles affects the wall of the gastrointestinal tract and causes leakage of albumin and an associated drop in serum albumin concentration. Infection causes a decrease in the serum concentrations of zinc and iron by redistribution; deficiency is relative rather than absolute. The most dramatic increase in requirements occurs in patients with extensive burns (see *Nutritional Support in Special Clinical Situations,* below).

Development of a Nutritional Problem

Deficiencies can develop within a few weeks or over several years, depending on the nutrient in question. A deficiency develops by degree in a series of steps; there may or may not be laboratory tests available to detect each step of deficiency. For example, although iron and vitamin A are both stored in the body, our ability to monitor the development of deficiency in each case is quite different. For iron, measurement of serum ferritin provides an estimate of reserves, whereas for vitamin A there is no reliable measure of the amount stored.

The rapidity with which a deficiency develops depends on whether or not the nutrient is stored in the body. *Stores* are reserves of nutrients that can be mobilized in times of need without diminishing the ability of the body to do work. For those nutrients that are stored, the initial response to insufficient intake is utilization of body stores. For those nutrients that are not stored, deficiency is accompanied by some impairment of health or function.

As the deficient intake for any stored nutrient continues, either in severity or duration, changes in serum concentration or altered urinary excretion of the nutrient or its metabolites may occur. For example, at an intermediate stage of iron deficiency, serum iron begins to decrease and erythrocyte protoporphyrin begins to increase. In vitamin A deficiency, a moderate decrease in the serum concentration reflects depletion of stores. These changes suggest that the risk of deficiency is now much greater than at the earlier stages, and functional changes may now also appear. If the depletion is sufficiently severe or continues for a longer time, first general and then more specific physical signs of deficiency will appear.

Malnutrition (i.e., poor nutrition) can result from either deficient or excessive intake. For many essential nutrients, such as most of the water-soluble vitamins, any excess is merely excreted; for others, such as vitamins A and D, intake of large amounts may result in toxicity. Excessive intake is generally due to the use of high-potency vitamin-mineral supplements.

Nutritional Assessment

A nutritional assessment is made by observing physical signs, making anthropometric measurements, collecting dietary information, taking medication and medical histories, measuring metabolic and immunological factors, and evaluating these data in relation to reference standards.

Physical Examination

Physical signs are the last evidence of deficiency to develop and the first to disappear when correction of the deficiency has begun. Physical signs are rarely specifically associated with the deficiency of a single nutrient and may vary with age. For example, in children who have scurvy, one symptom is painful, swollen joints due to hemorrhages under the periosteum; in the elderly, ecchymoses appear. Bleeding gums, the classic sign of ascorbic acid deficiency, do not occur in young infants nor in adults who have lost their teeth.

Anthropometry

Anthropometric measurements can provide information on general nutritional status. Among the large number of anthropometric measurements, four are most useful: height, weight, skinfold measurements, and arm circumference.

Height for age reflects total previous history as well as genetic potential. Changes in body *weight*, on the other hand, generally reflect caloric intake. A comparison of the adult patient's weight with those on standard tables as a measure of appropriate body weight may be misleading; instead, a comparison with the usual weight of the patient should be used to assess weight loss (or gain). The rapidity of weight loss is as important as the actual amount, but a 10% loss of body weight is always clinically significant.

Skinfold measurements, used to estimate body fat, are based on the assumption that approximately one-half of the total body fat is deposited as subcutaneous tissue loosely attached to underlying muscles. To make the measurements, the skin and underlying fat are lifted up with the fingers into a fold and the thickness of the fold is measured with a caliper. An objective measurement of thinness is the determination of mid-upper *arm circumference*, which correlates with weight.

Energy Needs

Calorie balance is achieved when intake equals expenditure. If caloric intake exceeds expenditure, the patient is in positive balance. Energy derived from excess intake is converted to fat for storage. If caloric intake is less than expenditure, the patient is in negative balance. Stored fat and body protein are then mobilized and metabolized to provide the energy needed by the body. Energy needs are the sum of the energy required for basal needs, physical activity, and heat production. An estimate of basal energy expenditure (BEE) in kcal/d can be made using the Harris-Benedict equations.[6,15]

For men,

$$BEE = 66 + (13.7 \times \text{wt in kg}) + (5 \times \text{ht in cm}) - (6.8 \times \text{age in years})$$

For women,

$$BEE = 655 + (9.6 \times \text{wt in kg}) + (1.7 \times \text{ht in cm}) - (4.7 \times \text{age in years})$$

The estimate is increased by 10–30% for patients with multiple fractures, 20–50% for those with sepsis, and 90–110% for those with burns. For most hospitalized patients, activity is limited and the above equations are sufficiently accurate.

The amount of energy stored as fat is also important. Body weight does not accurately reflect fat stores. Ultrasound, computerized tomography, neutron activation, total body electrical conductivity, and underwater weighing have been used to provide accurate estimates of muscle and fat reserves or both, but at present these are primarily research techniques. Ultrasound has the greatest potential for widespread use. Skinfold measurements, although not as accurate, are currently the most frequently used clinical method.

Nutritional History

A nutritional history involves the evaluation of food intake as well as social, cultural, economic, and medical data. These data can be used to determine the patient's risk of deficiency and to provide information that can serve as the basis of a care plan.

Two methods of evaluating *intake* are diaries and recalls. Diaries of food intake are maintained for varying intervals. The patient is asked to record everything eaten or drunk. Serving sizes may be estimated, or the patient may be asked to use measuring cups and spoons to measure the servings. Recalls aim at eliciting information on actual intake, as remembered at the time of the interview. Food composition tables are used to estimate the intake for nutrients of interest, and then the total intake is compared with the RDA.

A method for evaluating the *adequacy* of intake is to categorize the patient's intake by food groups and to compare servings within groups with the number of servings recommended. Grouping of specific foods is determined by nutrient content. The basic groups are (1) milk and milk products, (2) meat or foods containing protein, (3) fruits and vegetables, (4) grains and pulses, and (5) others (i.e., foods that may be added to meet total energy needs).

Frequency of food intake provides another bit of information concerning relative risk and a long-term pattern of intake. The nutritional history should also include questions about prescription and nonprescription drugs currently in use, recent weight loss or gain, the presence of chronic illness, dietary habits, and allergic reactions to food.

Biochemical Data

Biochemical markers of food intake are needed to confirm dietary information. These markers are available for some nutrients; for example, urinary excretion of vitamin C indicates adequacy of intake. For many other nutrients markers are not available, and there is no single master test to determine nutritional status. Multitest systems should be used, but as yet there are few data to define these systems. Biochemical procedures are used to measure (1) nutrient concentrations in blood, (2) the urinary excretion rate of nutrients, (3) abnormal metabolic products in blood or urine, (4) changes in blood components or enzyme activities, and (5) changes after a test load or the administration of stable isotopes. The clinical chemist is of paramount importance in this aspect of nutritional assessment and should work with nutritionists to develop the chemical markers needed to confirm nutritional status.

Protein-calorie malnutrition (PCM). Of the two main components of protein reserves, skeletal muscle (approximately 60% of the body cell mass) is estimated from anthropometric measurements and visceral protein (20–30% of the body cell mass) is estimated from biochemical measurements. The adequacy of protein intake is determined by estimating nitrogen balance (nitrogen intake minus nitrogen losses via urine, feces, and skin).[1] Nitrogen intake is calculated by dividing daily protein intake in grams by 6.25, a factor based on the assumption that all protein contains 16% nitrogen. The major route of nitrogen loss is urine (70–90%). The major catabolite of protein in urine is urea, and the amount of urea nitrogen is closely correlated with total urinary nitrogen and protein intake. Ammonia and creatinine excretion are fairly constant in adults eating either high- or low-protein diets. There are changes, however, during fasting.

For precise determinations of nitrogen balance, fecal and urine nitrogen losses are measured directly. In most hospitals, a factor of 4 g/d is used as an estimate of fecal and other losses.[15] If no stool was passed, the factor is taken as 3 g/d.

$$\text{Nitrogen balance} = \underset{(\textit{Intake})}{\frac{\text{Protein intake, g/d}}{6.25}} - \underset{(\textit{Excretion})}{[\text{urinary urea N, g/d} + 3 \text{ (or 4)}]}$$

A positive nitrogen balance is indicative of net protein synthesis; a negative balance indicates net protein catabolism. In adults, negative nitrogen balance is associated with starvation, fever, and illness. Negative nitrogen balance can be more serious in children than in adults because growth can be compromised.

There are sources of error in estimating nitrogen balance. Exudation from burns, oozing from wounds, or copious sputum can result in significant additional losses. Severely malnourished patients receiving diet supplements can be in strong positive nitrogen balance before improvement in other measurements becomes evident. Nitrogen balance is therefore useful for monitoring nutritional therapy. The goal of therapy for depleted patients is a positive nitrogen balance of 4–6 g/d.

Urinary urea nitrogen can also be used as an indicator of the level of catabolic stress. As the severity of stress increases and catabolism becomes pronounced, the urinary excretion of urea nitrogen increases. Bistrian[5] has proposed a "Stress Index":

$$\text{Stress Index (SI)} = \text{24-h urinary urea nitrogen} - (\tfrac{1}{2} \text{ nitrogen intake} + 3)$$

In patients with an SI <1 there is no significant stress. An SI between 1 and 5 denotes moderate stress, and >5, severe stress.

Visceral protein mass. Visceral proteins are depleted as the protein status of the patient worsens. Protein undernutrition (kwashiorkor in children and kwashiorkor-like syndrome in adults) is characterized by depletion of the visceral protein mass. Marasmus, a severe deficiency in energy stores, is characterized by depletion of the skeletal muscle mass with relative preservation of the visceral component.

Concentrations of the various serum transport proteins are used to indicate adequacy of the visceral component. The concentration of a serum protein is the result of the rates of synthesis, utilization, and excretion of the protein, as well as of the hydration of the patient. Administration of whole blood or certain fractions of blood also affects the concentration of serum proteins.

Although serum albumin is widely measured to assess protein nutrition, it is a poor indicator of marginal protein depletion since it is depressed only in severe undernutrition. Clinical nutritionists generally agree that in the absence of liver or kidney disease or infection, <2.1 g/dL serum albumin is indicative of severe kwashiorkor-like depletion.[11] Patients who have low serum albumin at admission have frequently been shown to stay in the hospital longer than those with normal values.[2] Serum albumin is lower in the elderly than in young adults by a slight but significant amount.[10]

Plasma transferrin has been proposed as a more sensitive indicator of protein status than albumin because of its smaller pool and shorter half-life (8.8 d). However, in the first week after an injury, measurement of transferrin concentration is not useful because transferrin is an acute-phase reactant. Plasma transferrin concentration falls during catabolic stress and rises in patients with iron-deficiency anemia, during pregnancy, and after blood loss. On the basis of limited testing, a weighted index of serum albumin and serum transferrin appears to be useful for selecting candidates for nutritional support.

Prealbumin and retinol-binding protein have also been proposed as indicators of protein status. However, prealbumin concentration drops whenever there is a sudden increase in protein synthesis. Changes in retinol-binding protein (RBP) occur with minor stress, and RBP concentration in blood is affected by kidney function, thus limiting its use in nutritional assessment.

The status of other nutrients also affects serum proteins. Concentrations of albumin, transferrin, and prealbumin are depressed in patients with zinc deficiency, but values rise after a short period of zinc supplementation.[3]

Somatic protein mass. The most widely used biochemical marker for the estimation of body muscle mass is 24-hour urinary creatinine excretion. Undernourished patients who have lost muscle mass excrete less creatinine than do well-nourished patients. Renal function must be normal to obtain valid results.

Urinary excretion of hydroxyproline and 3-methylhistidine. Hydroxyproline derived from collagen catabolism and 3-methylhistidine (3MH) from myofibrillar catabolism have been suggested as indicators of protein status.[7,23] Conditions other than protein nutrition affect both of these metabolites, so values must be interpreted carefully.[4,20,24] Serial measurements of 3MH are used to monitor patients receiving total parenteral nutrition.[17]

Host defenses. Infection is a leading cause of the morbidity and mortality of nutritionally debilitated individuals. Mucocutaneous integrity is impaired by deficiencies of several vitamins and minerals, some of which are commonly consumed in less than adequate amounts by certain segments of the population.

A significant impairment of cell-mediated immunity (CMI) occurs in patients with protein-calorie malnutrition (PCM). PCM is in actuality a multiple deficiency syndrome; therefore, the specific cause of impaired CMI might be one or another of the deficiencies that occur concurrently. Iron deficiency can account for most of the changes, although zinc deficiency also produces some of the aberrations in CMI. For nutritional assessment, the most common measures of CMI are the total number of lymphocytes and skin testing. In patients with kwashiorkor-like syndrome, impairment of delayed hypersensitivity can be expected when the serum albumin falls below 3.5 g/dL. In patients with marasmus, immune function is typically intact until their weight falls below 85% of the reference standard. With nutritional repletion, the defects are reversed.

Serum immunoglobulin levels are variable in undernourished children. Antibody responses to some antigens, such as typhoid and influenza, are depressed but are normal or increased for other antigens, such as tetanus toxoid, and for attenuated measles virus.[22] Nutritional supplementation and clinical improvement are associated with rapid recovery of immune responsiveness except in the small-for-gestation-age newborn, in whom the defect persists for several months and perhaps years.[8]

Lipids

Essential fatty acid deficiency can be established by determining the concentration of 5,8,11-eicosatrienoic acid in serum or plasma lipids. A deficiency of essential fatty acids has been seen only in patients who are not consuming food by mouth. Serum lipid profiles are frequently performed to measure cholesterol, triglycerides, and apolipoproteins A and B and to identify types of hyperlipidemia, some of which are treated by dietary management.

Vitamins

Biochemical tests of vitamin status may reflect recent dietary intake or they may show the metabolic changes that occur in a deficiency state. For instance, plasma vitamin A concentration simply reflects dietary intake or indicates that the nutrient is in transit from one organ to another. On the other hand, measurement of erythrocyte transketolase activity reflects adequacy or deficiency of thiamin. Usually the determination of vitamin status is performed only during in-depth assessment of the individual and is based on clues provided by the patient's history or by clinical and dietary information. For example, the presence of certain diseases or the chronic use of certain drugs might suggest a potential for deficiencies. Roe[21] recommends that the nutritional profile of alcoholics include assessment of niacin, ascorbic acid, pyridoxine, and vitamin B_{12}.

Minerals

Assessment of the nutritional status of trace minerals is still in the early stages of development. Of all the essential minerals, only four are usually measured in surveys—calcium, phosphorus, magnesium, and iron. Of these, calcium and iron intakes are frequently significantly less than the RDA.

Biopsy Samples

Until iron stores could be estimated from serum ferritin values, staining a bone marrow sample to determine the amount of iron present was the only way to estimate iron stores. This remains the definitive method, but it is not used in nutritional assessment or in the diagnosis of simple iron deficiency. Megaloblastic anemia due to deficiency of vitamin B_{12} or folate can be confirmed by bone marrow biopsy, which shows the presence of erythroid hyperplasia and megaloblastic changes in the erythroid precursors.

Hair. Many constituents of hair have been studied. Microscopic examination is tedious and time consuming, taking 1–2 h per subject. Many trace minerals can be measured in small samples of hair, but analysis is complex and still in the developmental stage. Hair analysis appears to have promise for assessment of zinc status.[14]

Nutritional Support in Patient Management

Enteral feeding. Patients who cannot eat or drink amounts adequate for their needs should receive enteral feeding by nasogastric or nasoduodenal tubes. This method of management is used for low-birth-weight infants and for patients with severe illnesses that reduce appetite or make the patient too weak to eat.

Patients receiving enteral feedings should be monitored for bloating, nausea, fluid and electrolyte balance, glucosuria, and hyperglycemia, as well as for hepatic and hematological status. Many formulas used for tube feeding do not contain sufficient water to allow for the renal excretion of urea and electrolytes; thus, there is the possibility that patients can become dehydrated, hyperosmolar, and hyperglycemic. Blood glucose should be monitored several times a week. In patients with liver dysfunction, possible complications are hyperammonemia, hepatic encephalopathy, and edema. Formulas should be started at a low concentration by slow-drip infusion and should gradually become more concentrated.

Metabolic alterations in patients receiving enteral nutrition are generally less severe than in patients receiving total parenteral nutrition (TPN). Major advantages are the avoidance of possible sepsis from the catheter required for TPN and having the gastrointestinal (GI) tract as a buffer.

Parenteral nutrition. Patients with GI disease can be put into positive nitrogen balance by infusion of hypertonic nutrient solution through a catheter inserted into the subclavian vein and threaded to the superior vena cava. Hospitals where enteral and parenteral feeding are extensively used now have a metabolic support team of clinicians, scientists, dietitians, pharmacists, and nurses to provide the expertise needed to manage the patients successfully on either regimen.

Total parenteral nutrition is often indicated as primary therapy for patients with such conditions as acute renal failure, inflammatory bowel disease, and radiation enterocolitis. TPN is supportive nutritional therapy for comatose patients; for patients with neoplasms or anorexia nervosa; for infants with failure to thrive; and for burn patients. TPN is contraindicated when the patient is in good nutritional status and will be able to eat in a few days or when he or she is in the terminal stages of illness. The use of TPN is also questionable in patients with systemic sepsis or those with fluid overload.

Glucose is the primary source of energy in the nutrient solution. Other components are *amino acids*, available either as protein hydrolysates or as pure amino acids. Protein hydrolysates may cause mild hypersensitivity reactions, whereas pure amino acids occasionally cause metabolic acidosis; hyperammonemia can occur in patients receiving either. Both the absolute amounts of essential amino acids and the total amount of amino acids are important for protein metabolism. Approximately half the total nitrogen should be in the form of essential amino acids to meet the increased needs of protein synthesis.

Several specialized amino acid solutions have been developed for use in patients with clinical conditions in which amino acid metabolism is changed or mechanisms of nitrogen excretion are impaired. *Fat* emulsions are now used for supplying energy and for eliminating deficiencies of essential fatty acids.

Vitamins are either added to the solution or given intramuscularly. There are several preparations available that must be combined or given on alternate days to meet the requirements. Vitamin K is usually omitted from preparations of multivitamins because of its incompatibility with the solutions.

Potassium and *phosphate* are required in higher than normal amounts in TPN solutions to satisfy the body's needs for protein synthesis. *Calcium* gluconate and *magnesium* are added to prevent the neuromuscular irritability, tremor, and seizures seen in chronically malnourished patients. Recently, other minerals such as *zinc, chromium,* and *copper* have been added because of reported cases of deficiency. Preterm infants receiving TPN present special problems. If they are maintained on TPN for long periods, they develop pathological fractures and rickets. Infants given large amounts of calcium develop fewer fractures.

Insulin is frequently added to the TPN solution or given subcutaneously to maintain serum glucose concentrations within or slightly above normal limits. Heparin is added in low concentrations to prevent occlusion of the subclavian catheters used for TPN.

Osmotic diuresis, sepsis, and metabolic changes are the main complications of TPN (Table 15-2). Hyperosmolar, hyperglycemic, nonketotic dehydration can result if hyperglycemia is not detected soon enough and the infusion rate is not adjusted. In more severe cases, hypotonic saline, 5% dextrose in water, and insulin should be given to the patient instead of the TPN solution. Middle-aged or elderly patients who are obese, physically inactive, and possibly mildly diabetic are more likely to develop this problem.

Hypophosphatemia is a serious problem that results when a semistarved patient deriving energy from fat metabolism (which is less dependent on phosphate) is suddenly given glucose as the predominant fuel. Serum glucose concentration rises and causes a

Table 15-2. POSSIBLE METABOLIC COMPLICATIONS OF PARENTERAL NUTRITION

Complication	Symptom	Cause
Hyperglycemic, hyperosmolar nonketotic coma	Hyperglycemia; dehydration; increase in serum osmolality and serum sodium; somnolence; seizures	Carbohydrate intolerance; too rapid initiation of PN; infection; latent diabetes mellitus; pancreatic disease
Hypoglycemia	Hypothermia; somnolence; lethargy; peripheral vasoconstriction	Rapid cessation of infusion
Hyperchloremic metabolic acidosis	Decrease in blood pH and serum bicarbonate	Excessive renal or GI losses of base; concentration of cationic amino acids greater than anionic amino acids
	Increase in serum chloride and sodium	Excessive chloride in solution
Hyperammonemia	Elevated blood ammonia; somnolence; coma; lethargy; seizures	Hepatic dysfunction; deficiency in urea cycle amino acids
Hypophosphatemia	Paresthesia; mental confusion; hyperventilation; lethargy; decreased RBC function	Inadequate inorganic phosphate in solution; concentrated glucose infus may precipitate syndrome, redistribution into cells or bone
Hypokalemia	Muscular weakness; cardiac arrhythmias; altered digitalis sensitivity	Excessive gastrointestinal or urinary losses of potassium
Hyponatremia	Lethargy; confusion	Excess GI or urinary losses of sodium; water intoxication

Table 15-3. VALUES USUALLY MONITORED DURING PARENTERAL NUTRITION*

Frequency	Values
Every six hours	Urinary glucose
	Vital signs (initially)
Daily	Weight
	Input and output (nitrogen balance)
	Vital signs after initial period
Daily until stable, then twice weekly	Serum electrolytes and acid-base status
	Serum glucose
	Blood urea nitrogen
	Serum calcium, phosphorus, magnesium, transferrin, and albumin
Twice a week	Hemoglobin or hematocrit
	WBC
	Platelet count
Weekly	Plasma zinc and copper; prothrombin time
	Blood ammonia
	3-Methylhistidine
	Height in infants and children
	Head circumference in infants
	Liver function tests
Monthly	Serum folate and B_{12}
Four hours after lipid infusion	Serum triglycerides and cholesterol
As needed	Serum iron, reticulocyte count

*If patient is a child, values are measured less frequently.[12]

release of insulin, which not only promotes uptake of glucose and amino acids by tissues but also favors storage of organic phosphorus in the liver. The clinical changes are complex and can result in cerebral hypoxia, acute cardiopulmonary decompensation, and greater susceptibility to infection.

Most complications of parenteral nutrition can be avoided or minimized by monitoring the patient at regular intervals. Table 15-3 lists the values that should be monitored and the suggested frequency of monitoring. The frequency for infants and small children must also be evaluated in terms of the total amount of blood required for the tests.

Nutritional Support in Special Clinical Situations

Most of what we know about nutrition relates to the nutrition of the basically healthy individual. We have some understanding of the effects of specific diseases on nutritional status, but more information is needed to provide the best possible nutritional care.

Sick patients frequently have anorexia or are physically unable to eat so that their nutritional status is altered not only by the disease but also by starvation. The patient's body adapts through a series of steps in order to best utilize reserves and to meet the needs of the body during relative or total energy restriction. *Fat,* stored as triglycerides, is by far the largest reservoir of energy in the body (~130 000 kcal). In obese individuals, the value is higher.

The major body reservoir of *protein* is muscle tissue (~40 000 kcal). All body proteins are either functional or structural and as such are not reserves in the same sense as fat. During the initial response to protein deprivation, somatic protein is lost. As depletion continues and more crucial functions are compromised, progressively more visceral protein is lost; the loss leads eventually to death. The magnitude of protein loss cannot be measured with precision.

Carbohydrate storage is limited. If fasting continues for more than 18 h, liver glycogen is depleted. The endocrine metabolic response to starvation mainly involves changes in the insulin/glucagon ratios that enable the patient to provide adequate substrates for brain metabolism and to minimize the loss of body protein.

The normal response to starvation is a continuum that can be divided into three stages: postabsorptive (6–12 h after eating), short-term starvation (up to 3–7 d), and prolonged starvation (2 weeks or longer). During an overnight fast, 75% of the glucose released is derived from glycogen; the rest is from gluconeogenesis. Much of this glucose is used by the brain, an organ with a high metabolic rate. In the adult the liver represents a larger proportion of the body weight and the brain a smaller proportion than in the child, and thus in adults the liver is able to provide sufficient glucose to meet the needs of the brain during starvation. Children, who have relatively larger brains and smaller livers, place more demand on the glycogen stores during a short fast. As fasting continues, the body metabolism shifts to the utilization of lipids, and only 13% of the energy requirements come from the catabolism of tissue protein. The relatively small muscle mass in children limits the amount of amino acids that can be mobilized, and ketosis may develop sooner than in adults.

Obesity. Many weight-reduction diets result in weight loss; unfortunately, most patients gain the weight back. Modification of eating behavior, modification of diet composition, and an understanding of factors that are important in the causes of obesity may help the patient maintain a lower body weight once reduction has been achieved.

A balanced diet is the safest, most effective way to lose weight; however, weight loss is gradual. Any diet that advocates the consumption of no more than 900 kcal/d does result in rapid weight loss, but it leads to dangerous losses of potassium and to utilization of protein from muscle. All fad diets produce an initial rapid weight loss from the diuresis that occurs when intake is suddenly reduced. Fortunately, most individuals cannot tolerate these diets for more than a few days, and thus they avoid potentially dangerous or fatal consequences.

Any diet that restricts the caloric intake makes it difficult for the patient to consume the requirements for critical nutrients such as iron, zinc, and some of the water-soluble vitamins. Starvation should not be used as a treatment unless the patient is carefully monitored, especially for potassium losses. Other treatments of obesity can also result in complications. Jejunoileal bypass surgery may result in several metabolic derangements such as renal oxalate stones and renal failure and sometimes fatal hepatic disease. Gastroplasty has not yet been shown to be as damaging metabolically. Drug therapies are effective on a short-term basis, but effects may not extend beyond the period of ingestion.

Anorexia nervosa is a psychological disorder of eating behavior. It results in a severe weight loss, which frequently endangers the life of the patient. Nutritional treatment is the same as for other starving patients.

Physical stress. In patients who have undergone severe physical stress, such as injury or infection, there is extensive loss of tissue protein. The requirements of a patient under stress differ quantitatively rather than qualitatively from those of the healthy individual. There is a range of response, depending on the extent and type of trauma. Uncomplicated surgical procedures increase energy requirements the least. Multiple fractures, severe infection (sepsis), and third-degree burns produce progressively greater increases in energy requirements. Patients with extensive thermal injury may require up to 8000 kcal/d.

Skeletal muscle can be depleted of 25–30% of its protein after seven days of stressful illness. Protein synthesis is more sensitive to severe stress than is protein breakdown. During moderate to severe stress, urinary excretion of zinc, copper, magnesium, potassium, and calcium increases. Blood levels of vitamin A, ascorbic acid, zinc, and iron are decreased, and sodium and water are retained.

Diseases That Produce Nutritional Problems

Diseases of the liver and biliary systems. In patients with *cholestasis*, the quantity of bile salts in the intestine decreases and the absorption of fat diminishes. Dietary fats stimulate gallbladder contraction and the relaxation of the sphincter of Oddi; the reduction of dietary fat will result in less stimulation. After an episode of cholestasis, the diet initially should be limited to small amounts of carbohydrate food. Later, cereals, skim milk, broiled fish, and lean, broiled meats can be added.

In patients with *liver disease of non-nutritional origin*, dietary management is limited to improving the patient's nutritional status and enabling the liver to function as efficiently and easily as possible. During the acute stage of hepatitis, patients may require 10% glucose intravenously. When appetite returns, simple foods, given as frequent small feedings, are tolerated best.

Patients with *cirrhosis* may develop hidden water retention, frank edema, or ascites associated with nutritional, endocrine, or secondary metabolic disorders. Initial treatment of bed rest and a low-salt diet may be sufficient to cause diuresis. In patients with persistent ascites, a low-sodium diet is used in combination with diuretics.

For patients with *hepatic failure*, the amount of protein needed to sustain liver function while minimizing protein intolerance is not well defined. When hepatic cells are damaged, increased amounts of ammonia enter the systemic blood stream and may precipitate hepatic coma (see also Chapter 22). Ammonia is readily absorbed from the GI tract, where it is formed by the action of intestinal bacteria on protein foods or on blood (in the case of GI bleeding).

Diseases of the GI tract. Nutrition is recognized as an integral part of the therapy of patients with *cystic fibrosis* or *Crohn's disease*. Children with cystic fibrosis who are adequately nourished are better able to resist infection and to gain weight. Most of these patients have pancreatic insufficiency and malabsorption; thus, pancreatic enzymes are given with each feeding. The use of medium-chain triglycerides in the diet helps decrease the steatorrhea and increase weight gain. Starches are poorly tolerated; therefore carbohydrates should be consumed in the form of simple sugars. Protein is usually consumed in amounts well in excess of the RDA, and vitamin supplements are given to satisfy requirements.

During exacerbations, patients with Crohn's disease are usually in negative nitrogen balance, which can be reversed by parenteral nutrition or the use of an elemental diet. At other times, the diet should be low in residue and high in calories, animal protein, vitamins, and minerals.

Food intolerances. Gluten in the diet of individuals with celiac disease damages the intestinal mucosa, which leads to malabsorption. If gluten is removed from the diet of these patients, improved absorptive function returns as the mucosa of the small intestine recovers. The anemia seen in these patients is caused by an unknown interference with folate absorption. Oral supplements of folate may not alleviate the problem, and intramuscular injections may be necessary.

Disaccharide intolerance, of which lactose intolerance is the most common, occurs with high frequency among blacks, Arabs, Japanese, and other groups (see also Chapter 23). In these individuals, ingested lactose is not absorbed but is acted upon by intestinal bacteria, which produces bloating, cramps, and diarrhea. Affected individuals can frequently tolerate small amounts of milk or milk products as sources of calcium. The analysis of breath for hydrogen production after an oral load of lactose appears to be more sensitive for the detection of deficiency than is absence of a rise in serum glucose.[19] The definitive test is peroral intestinal mucosal biopsy and demonstration of decrease or absence of the disaccharidase involved, but this is rarely necessary.

A gastroenteropathy is occasionally seen in younger infants fed cow's-milk products and in older infants fed whole cow's milk. In young infants the symptoms are diarrhea, steatorrhea, growth retardation, anemia, hypoproteinemia and edema, respiratory or skin allergies, and peripheral eosinophilia. Older infants are most often asymptomatic, and anemia is detected only during routine testing. GI bleeding is usually not apparent but can be detected by testing the stool for occult blood to distinguish this condition from strictly dietary causes of anemia. Tolerance for milk is regained with maturity.

Renal disease. Patients with chronic renal failure retain nitrogenous waste products and phosphate, have a low serum calcium, and may develop edema. Restriction of protein intake results in a trend toward normalization of serum urea values. Complicating the nutritional status of the patient with renal failure is the loss during hemodialysis of nutrients (free amino acids, glucose, vitamins, and minerals). Approximately 40% of the amino acids lost during hemodialysis are essential; therefore, the patients need to consume high-quality proteins such as meat. Patients treated with peritoneal dialysis require more protein because of the greater losses of amino acids during this procedure. Circulating levels of folic acid, pyridoxine, and ascorbic acid are often decreased; therefore, a supplement of these compounds is advised.

Most adult patients with renal disease maintain sodium and water balance with an intake of 2–8 g of salt and 1.5–3 L of water per day. These patients should avoid diets high in potassium and magnesium and should take calcium supplements. Their absorption of calcium is impaired, and their diets usually include few calcium-rich dairy products because of the need to restrict protein. In renal failure the kidney's inability to transform 25-hydroxycholecalciferol to 1,25-dihydroxycholecalciferol causes vitamin D deficiency and contributes to the deranged calcium metabolism. A patient's response to vitamin D therapy can be determined by the observation of improved bone structure and the return of serum calcium concentrations toward normal.

Cancer. The debilitation of patients with cancer is due to the effects of treatment as well as to the disease itself. Weight loss may occur in patients with a variety of malignancies. One-third of the patients with cancer of the esophagus, stomach, or pancreas will have lost more than 10% of their weight before the institution of treatment. Because of increased energy and protein requirements, surgery, radiation, and chemotherapy tend to accelerate weight loss.

Tumors of the GI tract may cause problems by obstruction. Absorption is altered even before obstruction is complete. Resection of the GI tract for removal of the tumor, together with the early and late effects of therapy, may cause destruction of some cells and atrophy of the villi.

Anorexia in patients with cancer is due to anxiety, depression, and metabolic changes. Certain patients develop protein aversion and alteration of taste preferences, and will avoid meat and protein products. Consequently, both protein and energy malnutrition are seen frequently in patients with cancer.

Diabetes mellitus. For the insulin-dependent diabetic patient, the goal of diet therapy is the balance of food intake with exercise and insulin administration. The most common method for dietary management uses exchange lists, which group foods according to carbohydrate, protein, and fat values. The patient with diabetes should limit consumption of foods high in simple sugars and increase the amount of complex carbohydrates.

Non-insulin-dependent diabetes occurs mainly in adults, particularly in those who are obese. For many, diet management or weight reduction results in a correction of the diabetes. Frequent urine testing for glucose is a necessary part of the management of all patients with diabetes. Blood glucose is also monitored, but less frequently. Diabetes in some elderly patients may actually be due to chromium deficiency. Although chromium appears to have a role in glucose metabolism, its role in diabetes is not yet defined.

The fiber content of the diet has been suggested to be a factor in the treatment of diabetes, although its importance has not been unequivocally demonstrated. Patients with non-insulin-dependent diabetes will consume additional fiber if they increase their intake of complex carbohydrates.

Hypertension. Although sodium restriction is recommended for patients with hypertension, the main mode of treatment is medication. If oral diuretics are prescribed, the side effects are hypokalemia, hyperuricemia, and hyperglycemia. Hypokalemia is particularly common and can be treated with diet or a potassium supplement.

Hyperlipidemia. Dietary treatment of patients with hyperlipoproteinemia varies with the phenotype. Type I patients are treated with a diet low in fat; Type II with restricted cholesterol intake and an increase in the ratio of polyunsaturated to saturated fats; and Type III with weight monitoring, limited cholesterol intake, controlled carbohydrate intake, and a diet consisting of fats whose ratio of polyunsaturated to saturated fatty acids is increased. Type IV patients are treated by weight monitoring, controlled carbohydrate intake, moderate cholesterol restriction, and an increase in the ratio of polyunsaturated to saturated fats. Type V patients should restrict their intake of total fat and, to some extent, of carbohydrate (see Chapter 14, *Lipids, Lipoproteins, and Apolipoproteins*).

Inborn errors of metabolism. For patients with some *inborn errors of carbohydrate metabolism*, the dietary treatment is avoidance or decreased intake of the offending carbohydrate. Treatment of patients with *inborn errors of amino acid metabolism* involves bypass of the block by supplying the offending amino acid in minimal amounts and giving all other amino acids as needed. Dietary management of the individual with phenylketonuria is currently the best defined. Lofenalac is a commercially available casein hydrolysate from which 95% of the phenylalanine has been removed. The use of this product minimizes increase in the blood levels of phenylalanine and prevents the resulting derangements of brain metabolism, impaired mental development, aggressive behavior, hyperactivity, neuromuscular instability, seizures, and eczema.

Maple syrup urine disease is more complex to treat because three essential amino acids (leucine, isoleucine, and valine) must be monitored. Mental retardation is severe if the diet is not restricted. Dietary management of patients with other less common metabolic disorders is also difficult.

Physical handicaps. Nutrition for the handicapped is particularly important during the period of growth and development. Children with motor dysfunction severe enough to prevent ambulation require only 75% of the calories of active children. Those children with less severe motor dysfunction have caloric needs that are unaffected by their handicap.

The handicapped adult compromised by limb injury, progressive neurologic disease, or dysphagia has such symptoms as fatigue, loss of food from the mouth, and a greater length of time required for eating. Other conditions, such as arthritis, may make food preparation and consumption painful. When the total effect of a clinical condition on a patient is determined, economics, physical environment, and other factors must be taken into account in the evaluation of the patient and the implementation of dietary change.

Conclusion

Nutritional assessment of individual patients is a relatively new concept in health care. The role of the clinical chemist in such assessment is largely that of an expert on the biochemical aspect. A broad general understanding is also needed so that the clinical chemist can interact effectively in team care of nutritionally compromised patients and can, furthermore, work with physicians and nutritionists to develop better methods of determining body composition, the functional status of many vitamins and minerals, and biochemical markers that will confirm dietary intake data. More complete discussions of nutrition are available in *Human Nutrition: Clinical and Biochemical Aspects,*[9] *Food, Nutrition and Diet Therapy,*[18] and *Modern Nutrition in Health and Disease.*[12] Specific texts and review articles are also listed under *Additional Reading.*

References

1. Allison, J. B., and Bird, J. W. C.: Nitrogen excretion from the body. *In*: Mammalian Protein Metabolism, Vol. 1. H. N. Munroe and J. B. Allison, Eds., New York, Academic Press, 1964.
2. Anderson, C. F., and Wochos, D. N.: The utility of serum albumin values in the nutritional assessment of hospitalized patients. Mayo Clin. Proc., *57*:181–184, 1982.
3. Bates, J., and McClain, C. J.: The effect of severe zinc deficiency on serum levels of albumin, transferrin and prealbumin in man. Am. J. Clin. Nutr., *34*:1655–1660, 1981.
4. Bilmazes, C., Uauy, R., Haverberg, L. N., et al.: Muscle protein breakdown rates in humans based on N′-methylhistidine (3-methylhistidine) content of mixed proteins in skeletal muscle and urinary output of N′-methylhistidine. Metabolism, *27*:525–530, 1978.
5. Bistrian, B. R.: A simple technique to estimate severity of stress. Surg. Gynecol. Obstet., *148*:675–678, 1979.
6. Blackburn, G. L., Bistrian, B. R., Maini, B. S., et al.: Nutritional and metabolic assessment of the hospitalized patient. J. Parent. Ent. Nutr., *1*:11–22, 1977.
7. Cabacungan, N. B., Miles, C. W., Abernathy, R. P., et al.: Hydroxyproline excretion and nutritional status of children. Am. J. Clin. Nutr., *26*:173–176, 1973.
8. Chandra, R. K.: Fetal malnutrition and postnatal immunocompetence. Am. J. Dis. Child., *129*:450–454, 1975.
9. Garry, P. J., Ed.: Human Nutrition: Clinical and Biochemical Aspects. Washington, D.C., American Association for Clinical Chemistry, 1982.

10. Gersovitz, M., Munro, H. N., Udall, J., et al.: Albumin synthesis in young and elderly subjects using a new stable isotope methodology: Response to level of protein intake. Metabolism, *29*:1075–1086, 1980.
11. Golden, M. H.N.: Transport proteins as indices of protein status. Am. J. Clin. Nutr., *35*:1159–1165, 1982.
12. Goodhart, R. S., and Shils, M. E.: Modern Nutrition in Health and Disease, 6th ed. Philadelphia, Lea & Febiger, 1980.
13. Grant, J. P.: Handbook of Total Parenteral Nutrition. Philadelphia, W. B. Saunders Co., 1980.
14. Hambidge, K. M., Hambidge, C., Jacobs, M., et al.: Low levels of zinc in hair, anorexia, poor growth and hypogeusia in children. Pediat. Res., *6*:868–874, 1972.
15. Harris, J. A., and Benedict, F. A.: A biometric study of basal metabolism in man. Washington, D. C., Carnegie Institute, Publication No. 120, 1919.
16. Kerner, J. A., Jr., Ed.: Manual of Pediatric Parenteral Nutrition. New York, John Wiley & Sons, 1983.
17. Kim, C. W., Okada, A., Itakura, T., et al.: Urinary excretion of 3-methylhistidine in patients receiving parenteral nutrition. J. Parent. Ent. Nutr., *3*:255–257, 1979.
18. Krause, M., and Mahan, L. K.: Food, Nutrition and Diet Therapy. 7th ed. Philadelphia, W. B. Saunders Co., 1984.
19. Levitt, M. D., and Donaldson, R. M.: Use of respiratory hydrogen (H_2) excretion to detect carbohydrate malabsorption. J. Lab. Clin. Med., *75*:937–945, 1970.
20. Long, C. L., Birkhahn, R. H., Geiger, J. W., et al.: Urinary excretion of 3-methylhistidine: An assessment of muscle protein catabolism in adult normal subjects and during malnutrition, sepsis and skeletal trauma. Metabolism, *30*:765–776, 1981.
21. Roe, D. A.: Nutritional assessment of alcoholics. *In*: Alcohol and the Diet. Westport, CT, AVI Publishing Company, 1979.
22. Stiehm, E. R.: Humoral immunity in malnutrition. Fed. Proc., *39*:3093–3097, 1980.
23. Young, V. R.: The role of skeletal and cardiac muscle in the regulation of protein metabolism. *In*: Mammalian Protein Metabolism, Vol. 4. H. N. Munro, Ed. New York, Academic Press, 1970.
24. Young, V. R., and Munro, H. N.: N'-methylhistidine (3-methylhistidine) and muscle protein turnover: An overview. Fed. Proc., *37*:2291–2300, 1978.

Additional Reading

Chernoff, R.: Enteral feedings. Am. J. Hosp. Pharm., *37*:65–74, 1980.
Deitel, M., Ed.: Nutrition in Clinical Surgery. Baltimore, Williams & Wilkins, 1980.
Newell, G. R., and Ellison, N. M.: Cancer and Nutrition: Etiology and Treatment. New York, Raven Press, 1981.
Schneider, H. A., Ed.: Nutritional Support of Medical Practice, 2nd ed. Philadelphia, J. B. Lippincott Company, 1983.

VITAMINS

16

DONALD B. McCORMICK, Ph.D.

Among factors that affect the health and development of humans, none is more essential than vitamins. Although deficiency of a single vitamin is relatively uncommon, it can occur as a result of an inborn error of metabolism, unusual restriction in dietary intake, or failure to absorb a single vitamin. More frequent are complex deficiencies that arise as a result of food fads; as complications of certain diseases, especially those affecting food absorption; as a result of massive losses of blood or from hemodialysis; and with the use of certain drugs. There are instances, more common in the Western world, where excessive use of some vitamins is encountered. These and other influences that can cause vitamin imbalance, together with advances in knowledge of what constitutes imbalance, have increased the demand for laboratory tests for the evaluation of the vitamin status of individuals.

Definition of vitamins. Vitamins are organic compounds required in trace amounts (μg to mg quantities per day) for health, growth, and reproduction. Vitamins are natural materials that can be isolated from animal or vegetable sources or that can be chemically synthesized. Only small amounts of vitamins are required for the functional, often catalytic (coenzymatic) roles they serve, in contrast to the relatively large amounts of such macronutrients as proteins, lipids, and carbohydrates, which constitute the bulk of the diet and which serve primarily as sources for energy and the reconstitution of body mass. Several naturally occurring compounds serve the defined vitaminic roles: after ingestion they are either metabolically interconverted to a common functional form (e.g., the vitamin B_6 group of pyridoxine, pyridoxal, and pyridoxamine to the coenzyme pyridoxal-5′-phosphate), or they directly substitute for each other, such as phylloquinones (K_1) and menaquinones (K_2). Such vitamins are reasonably subclassed under the parent common or chemical name for the respective vitamin group.

Classification of vitamins. Historically, vitamin groups such as A, B, and D bear an Arabic subscript number following the letter. The number may designate structural and functional similarity, such as A_1 (retinol) and A_2 (3-dehydroretinol), or may indicate the approximate order in which they are identified as the members of the so-called B-complex, e.g., B_1 (thiamin) and B_2 (riboflavin). Common chemical names, which are receiving greater usage, give a better indication of the types of compounds involved. These often reflect the presence of some specific atom (*thi*amin), a prime functional group (pyridox*amine*), or an even larger portion of the molecular structure (phyllo*quinone*). Parts of some names reflect functional properties (chole*calciferol*).

Another classification pertains to relative solubility of vitamins. The "fat-soluble" group (i.e., A, D, E, and K) are more soluble in organic solvents, whereas the B-complex group and vitamin C are "water-soluble." This classification is useful not only for purposes of noting gross physical properties but as a reminder that the fat-soluble vitamins are absorbed, transported, and stored for longer times and in a manner similar to fats. Most water-soluble vitamins are not retained for a long time in the body and are lost through urinary excretion.

Vitamins Essential for Humans

Table 16-1 provides a current list of those 13 known vitamins and vitameric groups essential to humans. For ten of these, well-defined requirements have led to recommended daily dietary allowances (RDA) in the US; for three (K, biotin, and pantothenic acid), only "estimated safe and adequate daily dietary intakes" have been published.[29]

Table 16-1. VITAMINS REQUIRED BY HUMANS*

Common Name	Trivial Chemical Name	General Roles	Deficiency Symptoms or Disease	Direct and Indirect Assays
Fat-Soluble				
Vitamin A_1	Retinol	Vision, growth, reproduction	Nyctalopia, xerophthalmia, keratomalacia	Colorimetric, fluorometric, dark adaptation, RIA, HPLC
A_2	3-Dehydroretinol			
Vitamin D_2	Ergocalciferol	Modulation of Ca(II) metabolism, calcification of bone and teeth	Rickets (young), osteomalacia (adult)	CPB, HPLC
D_3	Cholecalciferol			
Vitamin E	Tocopherols, α, β, γ, δ	Antioxidant for unsaturated lipids	Lipid peroxidation including red cell fragility, hemolytic anemia (premature, newborn)	Colorimetric, HPLC, erythrocyte hemolysis
Vitamin K_1	Phylloquinones	Blood clotting, osteocalcins	Increased clotting time, hemorrhagic disease (infant)	Colorimetric, HPLC, prothrombin time, RIA for abnormal prothrombin
K_2	Menaquinones			
Water-Soluble				
Vitamin B_1	Thiamin	Carbohydrate metabolism, nervous function	Beriberi, Wernicke-Korsakoff syndrome	Fluorometric, microbiological, transketolase
Vitamin B_2	Riboflavin	Oxidation-reduction reactions	Angular stomatitis, dermatitis, photophobia	Fluorometric, microbiological, glutathione reductase
Vitamin B_6	Pyridoxine, pyridoxal, pyridoxamine	Amino acid, phospholipid, and glycogen metabolism	Epileptiform convulsions, dermatitis, hypochromic anemia	Microbiological, tyrosine decarboxylase
Niacin	Nicotinic acid	Oxidation-reduction reactions	Pellagra	Microbiological, fluorometric
Niacinamide	Nicotinamide			
Folacin	Pteroylglutamic acid	Nucleic acid and amino acid biosynthesis	Megaloblastic anemia	CPB, microbiological
Vitamin B_{12}	Cyanocobalamin	Amino acid and branched-chain keto acid metabolism	Pernicious and megaloblastic anemia, neuropathy	CPB, microbiological
Biotin	—	Carboxylation reactions	Dermatitis	Microbiological, colorimetric, carboxylases
Pantothenic acid	—	General metabolism	Burning feet syndrome	Microbiological
Vitamin C	Ascorbic acid	Connective tissue formation	Scurvy	Colorimetric

*Only one vitamer from a vitamin group (e.g., A_1 for vitamin A or pyridoxine for vitamin B_6) is sufficient for that vitamin.

R = CH_2OH for retinols, CHO for retinals, CO_2H for retinoic acids:

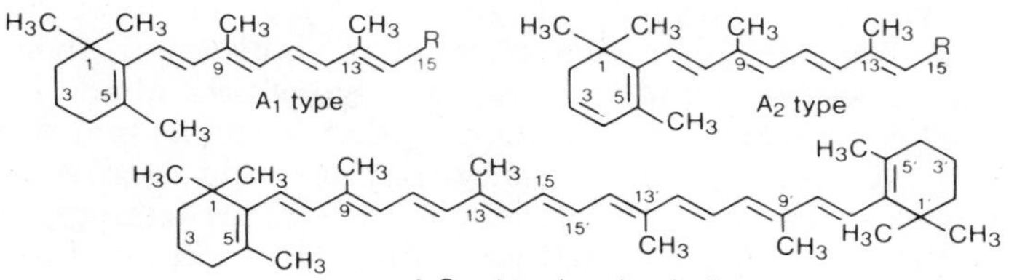

A_1 type

A_2 type

β-Carotene (provitamin A)

Figure 16-1. Vitaminic forms of A_1, A_2, and β-carotene.

VITAMIN A

Chemistry and sources. The two natural forms of vitamin A, *retinol* (A_1) and *3-dehydroretinol* (A_2), are C_{15}-isoprenoid alcohols, as shown in Figure 16-1. These compounds are yellowish oils or low-melting solids that are practically insoluble in water but soluble in organic solvents and mineral oil. Vitamin A is sensitive to oxygen and ultraviolet light. Vitamin A_1 predominates, especially as long-chain fatty acid esters, in the liver of mammals and salt-water fish (e.g., cod), whereas the biologically less active vitamin A_2 is found in fresh-water fish oils. Higher animals are unable to synthesize the β-ionone type ring structure but can derive the aldehyde retinal from provitamin carotenes and cryptoxanthin of plants. Retinal is then reversibly reduced to retinol.

The structure for the most common and effective provitamin A, *β-carotene*, is also given in Figure 16-1. The other carotenes yield less vitamin A_1 activity. Such carotenoid compounds, which constitute the yellow-to-orange pigments of most vegetables and fruits, vary considerably in their availability to the human; they often constitute the main dietary source of what ultimately becomes vitamin A.[9]

Requirements and allowances. Recommended daily allowances (RDA) for adult males are 1000 and for females 800 retinol equivalents, where 1 retinol equivalent is 1 μg retinol, or 6 μg β-carotene, or 12 μg of other provitamin A carotenoids. In terms of international units (IU), 1 retinol equivalent is 3.33 IU of retinol or 10 IU of β-carotene. The RDA for infants, based on the content and volume of human milk, is 420 retinol equivalents until the age of six months and 400 retinol equivalents from then until the first year. Higher levels are recommended to satisfy growth needs of children and adolescents. An additional 200–400 retinol equivalents are recommended for pregnant and lactating women.

Absorption, transport, and metabolism. The emulsification to micelles of forms of vitamin A and provitamin A by bile salts enhances their uptake by mucosal cells of the small intestine. Absorbed free retinol is re-esterified with long-chain fatty acids within the cell. Retinyl esters in association with chylomicrons then pass via the lymphatic system to the liver and are stored in a lipoglycoprotein complex. Retinol associated with plasma retinol-binding protein is released from hepatocytes to form, with circulating prealbumin, a molecular aggregate of sufficient size to avoid loss through glomerular filtration.

Functions. The best-understood physiological function of vitamin A is its participation as retinal in vision. Retinol and retinal are involved in a reversible cycle with opsin and rhodopsin, resulting in optic nerve transmission. Other broader functions of vitamin A occur in reproduction and growth. Systemic effects of an optimal level of vitamin A are the stabilization of cellular and intracellular membranes, maintenance of the integrity of epithelial tissue, and synthesis of glycoproteins.

Vitamin A deficiency. Clinically, degenerative changes in the eyes and skin are commonly observed in vitamin A deficiency.[49] Poor adaptation to darkness, or *night blindness* (nyctalopia), is an early symptom that is followed by degenerative changes in the retina. Associated skin changes include dryness, roughness, papular eruptions, and follicular hyperkeratosis.

Vitamin A toxicity. Toxic effects occur with hypervitaminosis A from ingestion of excess vitamin or from inappropriate therapy.[5,6,49] One of the most important factors in the development of toxicity is the form of vitamin A administered.[17] Symptoms can appear more rapidly after administration of aqueous emulsions rather than oily solutions. Hypervitaminosis A occurs after the liver storage of retinol and its esters exceeds 10 000 IU/g tissue, a level ten times the estimated RDA for adult males, or if circulating retinyl esters exceed 140 μg/dL. Acute toxicity from a single massive dose presents as abdominal pain, nausea, vomiting, severe headaches, dizziness, sluggishness, and irritability, followed within a few days by desquamation of the skin and recovery. Chronic toxicity from moderately high doses taken for protracted periods is characterized by bone and joint pain, hair loss, dryness and fissures of the lips, anorexia, benign intracranial hypertension, weight loss, and hepatomegaly. Carotenemia results from chronic excessive intake of carotene-rich foods, principally carrots. This condition, in which yellowing of skin is observed, is benign, since the excess carotene is deposited rather than converted to vitamin A.

Methods. In the *Neeld-Pearson method,*[32] trifluoroacetic acid is reacted with the conjugated double-bond system of the organic solvent–extracted compounds to produce a blue color (A_{max} = 620 nm). A correction for the color contributed by carotenes is necessary. The preparation of standards and all analytical operations must be performed in low-actinic glassware or under very-low-intensity light.

Blood is collected from the patient in the fasting state. The specimen should be free from hemolysis and protected from light. Serum is separated promptly and frozen at −20 °C if not analyzed immediately; vitamin A and β-carotene are stable for at least two weeks if frozen and protected from light.

The Neeld-Pearson procedure is not affected by moisture and is more sensitive than the classic Carr-Price method using antimony trichloride in chloroform. Nevertheless, this method is still relatively imprecise, tedious, and nonspecific. Several colorimetric/spectrophotometric and fluorometric methods applied after solvent extractions with and without additional separation techniques have been used to quantitate carotenoids and retinoids in numerous biological specimens.[22,25] The HPLC method by Bieri and colleagues[2] is an example of newer methods that are gaining rapid acceptance. This method is rapid and specific for retinol as well as for individual carotenes. It also allows for the simultaneous determination of vitamins A and E.

The serum or plasma concentration of vitamin A in a fasting subject is presently the only practical biochemical indicator of vitamin A status. The *reference ranges* for children and adults are 30–80 and 30–65 μg/dL, respectively. Values for men are generally about 20% higher than those for women. Values above 100 μg/dL indicate toxicity. The reference range for serum *carotenoids* (mainly β-carotene and xanthophyll) is 60–200 μg/dL in petroleum ether extracts and 50–300 μg/dL in isooctane extracts. Elevated levels are found in hypothyroidism, in which conversion to vitamin A is decreased, and in patients with hyperlipemia associated with diabetes mellitus.

Plasma ***retinol-binding protein***, usually measured by RIA, has been used to assess the nutritional status of vitamin A; values normally range between 40 and 50 μg/mL.[15] See also the discussion in Chapters 11 and 15.

VITAMIN D

Chemistry and sources. The two forms of vitamin D, *ergocalciferol* (D_2) and *cholecalciferol* (D_3), are steroid-derived compounds. D_2 does not occur naturally but is produced by ultraviolet irradiation of ergosterol, which occurs in molds, yeast, and higher plants. Vitamin D_3 in our bodies derives from irradiation of provitamin 7-dehydrocholesterol in skin. As shown by structures in Figure 16-2, the only chemical difference between the two vitamin and provitamin forms is in the side chain. There are at least ten compounds known to yield vitamin D–active compounds upon irradiation. Vitamin D forms white, odorless crystals that are soluble in fats and organic solvents. It is generally stable to heat, alkalis, acids, and oxidation.

Requirements and daily allowances. Ten μg (400 IU) is recommended as the daily allowance from infancy through 18 years of age. As the rate of skeletal growth and calcium needs decrease, the daily allowances are reduced to 5 μg after age 22. An additional 5 μg/d is recommended for pregnant and lactating women. In such diseases as uremic osteodystrophy and with autosomal recessive vitamin D dependency, requirements are increased to the pharmacological dose range of 30–50 μg (1200–2000 IU).

Absorption, transport, and metabolism. The extent to which ultraviolet irradiation of 7-dehydrocholesterol in skin provides vitamin D_3 varies with seasonal changes

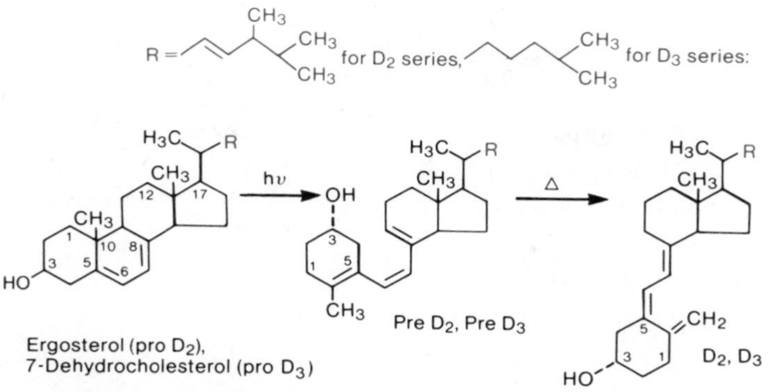

Figure 16-2. Provitamin, previtamin, and vitamin D_2 and D_3.

in sunlight, amount of clothes worn, degree of skin pigmentation, and other factors. The availability and consumption of such vitamin D_3-containing foods as fatty fish, eggs, liver, and butter also varies. Hence, widespread fortification of food has been adopted to ensure an adequate supply. Because dairy foods in the United States supply about 75% of dietary calcium in a calcium/phosphorus ratio near optimal for bone growth, milk and infant formulas are desirable vehicles for vitamin D fortification.[1] Though a poor source of vitamin D itself, milk now marketed contains 10 μg (400 IU) of added vitamin D per quart.

Vitamin D is efficiently absorbed from the gastrointestinal tract unless there is fat malabsorption due to lack of bile salts, pancreatic insufficiency, or a defect of intestinal mucosal function. When absorbed, the vitamin is bound directly to chylomicrons. Much of the vitamin initially supplied to the liver is hydroxylated at the terminal side chain position to yield 25-hydroxy-D. This 25-hydroxycholecalciferol, which represents the major metabolite of D_3 in plasma, circulates bound to a vitamin D-binding α-globulin (MW ~52 000). In kidneys, a second hydroxylation occurs at the 1 α-position. This change is regulated by vitamin D status and by calcium/phosphate levels, which are controlled by parathormone (PTH) and calcitonin.

The 1α,25-dihydroxycholecalciferol formed in the kidney complexes with vitamin D-binding protein for transfer via blood to target tissues (e.g., small intestine and bone). In the intestinal mucosal cell, the hormonally active dihydroxy-D mediates synthesis of a specific calcium-binding protein.

An outline of the transport and metabolism of vitamin D is given in Figure 16-3. The regulatory interplay of parathormone and calcitonin, the clinical significance of insufficient and excess vitamin D,[5,33] and suitable analytical methods for its determination[5,22,25] are found in Chapter 21.

VITAMIN E

Chemistry and sources. The form of vitamin E that is biologically most active, and upon which units are based, is *d-α-tocopherol*. There are eight related natural compounds in this vitaminic group, which is biosynthesized in plants and is especially abundant in vegetable oils.[16] Structures are shown in Figure 16-4. They are soluble in fat solvents and insoluble in water, stable to acid and heat in the absence of oxygen, but labile to oxygen in alkaline solutions and to ultraviolet light.

Requirements and allowances. The requirement for vitamin E is related to the polyunsaturated fatty acid content of cellular structures and, therefore, depends upon the nature and

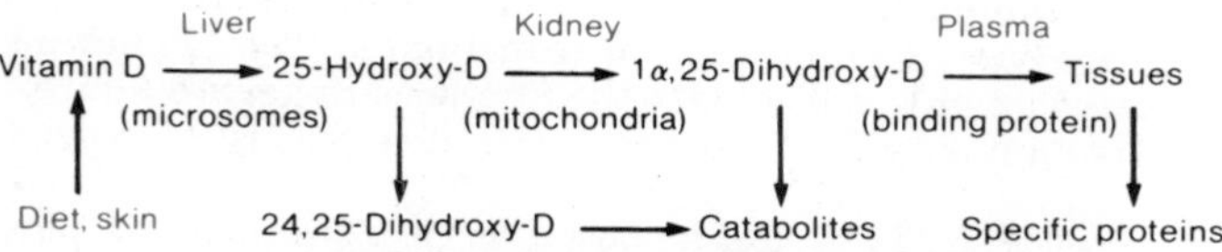

Figure 16-3. Transport and metabolism of vitamin D.

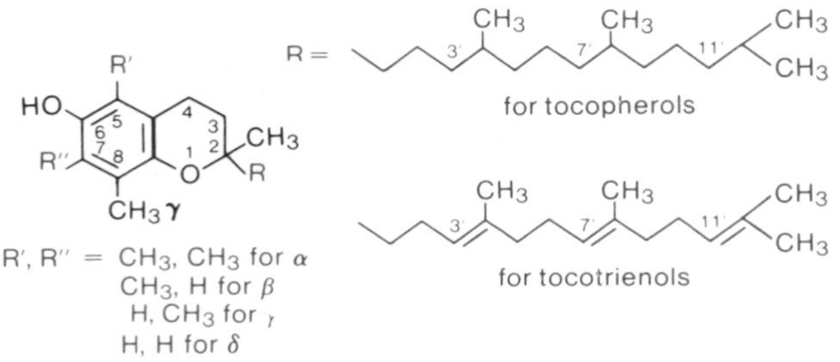

R = for tocopherols

for tocotrienols

R′, R″ = CH_3, CH_3 for α
CH_3, H for β
H, CH_3 for γ
H, H for δ

Figure 16-4. Vitaminic forms of E.

quantity of dietary fats that affect such composition. Hence, the minimum adult requirement for vitamin E is not certain but is probably not more than 3–4 mg (4.5–6 IU) of *d*-α-tocopherol per day.[29] When vitamin E activity of tocopherols and tocotrienols in usual diets is calculated, an estimated range of 7–13 mg of α-tocopherol equivalents (10–20 IU) can be expected. This intake will maintain plasma concentrations of total tocopherols within the reference range of 0.5–1.2 mg/dL, which also ensures an adequate concentration in all tissues.

Recommended daily dietary intakes, in α-tocopherol equivalents, are 10 mg for adult males and 8 mg for adult females and boys 11–14 years old. During pregnancy and lactation, intake should be increased by 2–3 mg. Infants should receive 3 mg/d during the first half year, and children 4–8 mg/d.

Absorption, transport, and metabolism. Vitamin E is absorbed from the small intestine in the presence of bile. Most tocopherol enters the blood stream via lymph, where the tocopherol is associated with chylomicrons and very-low-density lipoproteins. The vitamin is stored in most tissues, with the largest amount stored in adipose tissue. Rapid exchange of tocopherol occurs between the erythrocyte membranes and plasma lipoproteins. When physiological amounts are administered, only a small fraction of the dose appears in urine.

Functions. The best-defined role for vitamin E is as an antioxidant for unsaturated fatty acyl moieties of lipids within membranes. Oxidative damage to polyunsaturated fatty acyl parts of membrane phospholipids can occur as a result of hydrogen peroxide production by flavoprotein oxidases. Concurrent free-radical damage can ensue.

Deficiency. Premature and low-birth-weight infants are particularly susceptible to the development of vitamin E deficiency, since placental transfer is poor and they have limited adipose tissue, where much of the vitamin is normally stored.[50] Signs of deficiency include irritability, edema, and hemolytic anemia. The anemia reflects the shortened life span of erythrocytes with fragile membranes; it does not respond to iron therapy, which may aggravate the condition. Deficiency symptoms are rare in children and adults except in cases of severe malabsorption, as for example in cystic fibrosis.

Excess. Toxicity from chronic high intake of vitamin E has been noted in animals, in which competition for absorption may increase requirements for other fat-soluble vitamins, notably D and K. This problem appears to be the case for humans who already have limited vitamin K because of anticoagulant therapy following coronary infarcts. In general, however, relatively high doses of vitamin E (e.g., 300 mg/d) appear to be tolerated.

Methods. These involve solvent extraction of vitamin E forms after saponification of samples, most commonly blood plasma.[22,42] Molecular distillation and two-dimensional reversed-phase paper, thin-layer, or column chromatography have been used for quantitation. More recently *gas chromatography* and *high performance liquid chromatography* (HPLC) permit rapid separation of the different tocopherols and tocotrienols. The method by Bieri and coworkers,[2] which was mentioned in the section on *Vitamin A,* is specific for α- and λ-tocopherol and is recommended.

Chemical methods are based on an oxidation-reduction reaction. Following elution by specific techniques, fractions are commonly subjected to the Emmerie-Engel procedure, in which tocopherol is oxidized to tocopheryl quinone by $FeCl_3$, and the Fe(II) produced is complexed with α,α′-dipyridyl to give a red color.

Evaluation of vitamin E status should be based on measurements of plasma tocopherol and

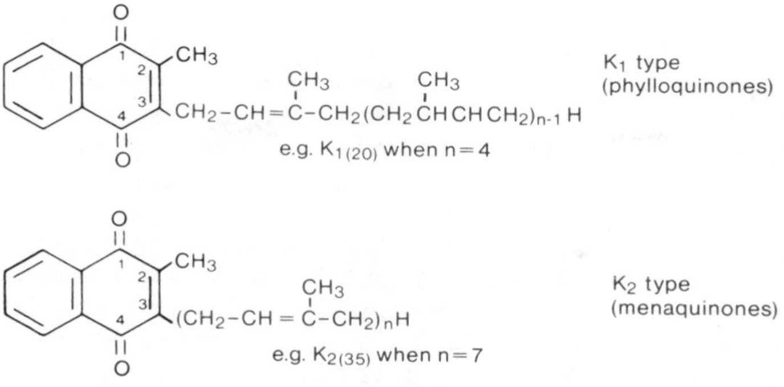

K₁ type (phylloquinones): $CH_2-CH=C(CH_3)-CH_2(CH_2CH(CH_3)CHCH_2)_{n-1}H$ side chain at position 3, CH_3 at position 2; e.g. $K_{1(20)}$ when $n=4$

K₂ type (menaquinones): $(CH_2-CH=C(CH_3)-CH_2)_nH$ side chain at position 3, CH_3 at position 2; e.g. $K_{2(35)}$ when $n=7$

Figure 16-5. Vitaminic forms of K.

on the degree of *erythrocyte hemolysis* after treatment of cells with H_2O_2 (preferred), dialuric acid, or isotonic saline–phosphate buffer.[42]

VITAMIN K

Chemistry and sources. Compounds in the vitamin K series, shown in Figure 16-5, are the *phylloquinones* (K_1 type) synthesized in plants and the *menaquinones* (K_2 type) of bacterial origin. Several synthetic analogs and derivatives have been used; most relate to or derive from *menadione* (K_3), which lacks a side-chain substituent at position 3. The K vitamins are insoluble in water but dissolve in organic fat solvents. They are destroyed by alkaline solutions and reducing agents and are also sensitive to ultraviolet light.

Estimated adequate intake. No specific RDA are presently given for vitamin K, since intestinal bacteria in normal individuals synthesize the menaquinones, which are partially absorbed. However, because the sufficiency of this source of K is uncertain over long periods, an estimated adequate range of dietary intake has been suggested.[29] The intake for adults, 70–140 $\mu g/d$, is easily supplied by an average mixed diet.[34] The intake for infants, 12 $\mu g/d$, is within the range supplied by mature breast milk.

Absorption, transport, and metabolism. The absorption of natural vitamin K from the small intestine is facilitated by bile. Vitamins K_1 and K_2 are bound to chylomicrons for transport from mucosal cells to the liver. Menadione (K_3) is more rapidly and completely absorbed. Vitamin K in the blood stream is associated with circulating β-lipoproteins for transport. Whereas only traces of urinary metabolites of vitamins K_1 and K_2 appear in urine, a considerable portion of vitamin K_3 (menadione hydroquinone) is conjugated to β-glucuronide and sulfate esters, which are excreted.

Function. Initially, vitamin K was recognized as a dietary antihemorrhagic factor. It has been found to be necessary for liver syntheses of plasma clotting factors II (prothrombin), VII (proconvertin), IX (plasma thromboplastin component), and X (Stuart factor). It is now known that reduced K (hydroquinone) participates in the oxygen-dependent incorporation of CO_2 into specific L-glutamyl residues of prothrombin and other plasma proteins.[45] The formation of bone Gla protein (γ-carboxylglutamyl proteins) or osteocalcin, possible regulators of mineralization, also depends on vitamin K.

Bis-4-hydroxycoumarin (dicumarol), the anticlotting compound from spoiled sweet clover, and synthetic 4-hydroxycoumarins such as warfarin act as anticoagulants by interfering with the conversion of provitamin K to the functional hydroquinone.[14] Hence, 4-hydroxycoumarins are able to suppress the formation of prothrombin and other vitamin K–dependent Ca(II)-binding proteins.

Deficiency. *Hemorrhagic disease* of the newborn can develop readily, since the menaquinone-synthesizing intestinal flora have not become established within the first week postpartum, and early breast milk is low in vitamin K. Prothrombin levels during this period are only about 25% of the adult level. Severe diarrhea and antibiotics used to suppress diarrhea readily exacerbate the situation, so that prothrombin levels can drop below 5% of the adult level and bleeding can occur. Development of vitamin K deficiency in the adult may require both reduction of dietary intake and antibiotic inhibition of intestinal microflora or 4-hydroxycoumarin-type anticoagulant therapy. Deficiency is thus relatively uncommon and is found in cases with chronic malabsorption of fats, including fat-soluble vitamins, or during long-term antibiotic or anticoagulant treatments.[49] Defective blood coagulation and demonstration of abnormal noncarboxylated prothrombin are at present the only well-established signs of vitamin K deficiency.

Excess. The use of high doses of naturally occurring vitamin K (K_1 and K_2) appears to have no untoward effect; however, menadione (K_3) treatment can lead to the formation of erythrocyte cytoplasmic inclusions known as Heinz bodies and to hemolytic anemia. With severe hemolysis, increased bilirubin formation and an undeveloped capacity for its conjugation may produce kernicterus in the newborn.

Methods. Direct methods for the quantitation of the different forms of vitamin K following extraction and chromatographic separations[22] involve spectrophotometry.[25] *Colorimetric determinations* based on reactivity of the quinoid nucleus of K vitamins include reactions with acidic phenylhydrazine, 2,6-dichloroindophenol, active methylene compounds, alkali, and piperidine. Also, spectrophotometric determination of chemically reduced vitamin K has been used.

Conventional assessment of vitamin K status relies on the clotting ability of plasma as reflected by the *prothrombin time*. A deficiency of vitamin K can be distinguished from the hypoprothrombinemia of liver disease by measurement with RIA of the non-γ-carboxylated prothrombin precursor (i.e., "abnormal prothrombin") that accumulates in plasma when the vitamin is deficient. Alternatively, if a vitamin K deficiency exists, parenteral administration of the vitamin will result in prompt correction of an abnormal prothrombin or clotting time. Routine determination of prothrombin time before all surgical procedures and deliveries has been suggested.[49]

THIAMIN

Chemistry and sources. *Thiamin* (thiamine, vitamin B_1) is a pyrimidyl-substituted thiazole (Figure 16-6). The principal if not sole coenzyme form is the pyrophosphate ester (TPP) also shown in Figure 16-6. Monophosphate and triphosphate esters occur naturally. Small amounts of thiamin and its phosphates are present in most plant and animal tissues, but more abundant sources are unrefined cereal grains, liver, heart, kidney, and lean cuts of pork. Thiamin is somewhat heat labile, particularly in alkaline solutions.

Requirements and allowances. The requirements for thiamin vary considerably, since they directly correlate with the amount of metabolizable carbohydrate intake. There is generally an increased requirement during situations in which metabolism is heightened, as in increased muscular activity, pregnancy, and lactation, or in protracted fever and hyperthyroidism.[13] With considerations of average caloric intakes and activities in different age groups, the RDA range from 1.0 mg/d for older women to 1.5 for young men. An additional allowance of 0.3 to 0.5 mg/d is suggested for pregnant and lactating women and growing children.

Absorption, transport, and metabolism. Thiamin is readily absorbed in the small intestine by an active transport process when intake is <5 mg/d; at higher levels of intake, passive diffusion increasingly contributes to absorption.[15,49] Phosphorylation takes place in the jejunal mucosa to yield thiamin pyrophosphate. Thiamin is carried by the portal blood to the liver. The free vitamin occurs in the plasma, but the coenzyme, TPP, predominates in the cellular components. About half of the body stores are found in skeletal muscles, with much of the remainder in heart, liver, kidneys, and nervous

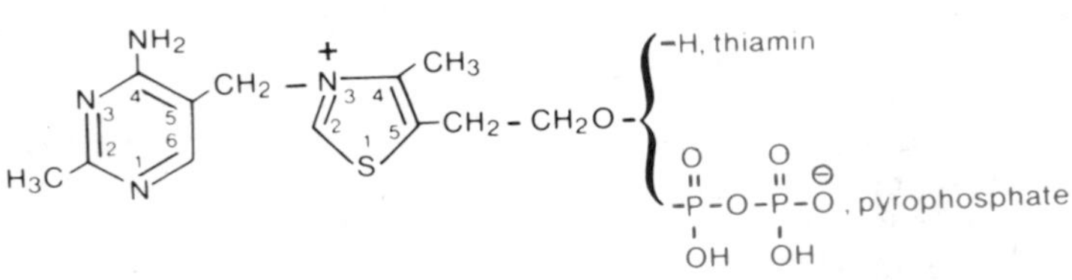

Figure 16-6. Thiamin and the pyrophosphate coenzyme.

tissue, including brain. Thiamin, as well as several of its catabolites, is excreted into the urine by the renal tubules.[7,15]

As a thiamin deficiency develops, there is a rather rapid loss of the vitamin from all tissues except the brain. The decrease of TPP in the erythrocyte roughly parallels the decrease of this coenzyme in other tissues. During this time, the thiamin in urine falls to near zero; the urinary metabolites remain high for some time before decreasing.

Functions. TPP functions as the Mg(II)-coordinated coenzyme for so-called active aldehyde transfers in the oxidative decarboxylation of α-keto acids and the formation of α-ketols (ketoses) catalyzed by transketolase.[48]

Deficiency. *Beriberi* is the disease resulting from thiamin deficiency. The causes for deficiency[15,29] include inadequate intake owing to diets largely dependent on milled, nonenriched grains such as rice and wheat, or to the ingestion of raw fish containing microbial thiaminases,[10] which hydrolytically destroy the vitamin in the gastrointestinal tract. Chronic alcoholism is a common contributor to deficiency because of both a low intake of thiamin (and other B vitamins) and impaired absorption and storage. There are several thiamin-responsive inborn errors of metabolism;[40,43] these include a megaloblastic anemia of unknown mechanism, lactic acidosis due to low or defective pyruvate decarboxylase, branched-chain ketoaciduria with poor activity of the keto acid dehydrogenase system, and subacute necrotizing encephalomyelopathy, where there is a lack of thiamin triphosphate in neural tissues. Therapeutic doses of 5–20 mg of thiamin daily have proved beneficial in some cases.[5] Other at-risk patients are those undergoing long-term renal dialysis or intravenous feedings and even those with chronic febrile infections.[29]

Clinical signs primarily involve the nervous and cardiovascular systems.[49] In adults, symptoms most frequently observed are mental confusion, anorexia, muscular weakness, ataxia, peripheral paralysis, ophthalmoplegia, edema (wet beriberi), muscle wasting (dry beriberi), tachycardia, and an enlarged heart. In infants, symptoms appear suddenly and severely and often involve cardiac failure and cyanosis.

Methods. The most useful and reliable method presently used for assessing thiamin status is a measurement of whole blood or *erythrocyte transketolase*. This enzyme catalyzes two reactions in the pentose phosphate pathway with TPP as coenzyme.

Xylulose-5-phosphate + D-ribose-5-phosphate →
Sedoheptulose-7-phosphate + glyceraldehyde-3-phosphate

Xylulose-5-phosphate + erythrose-4-phosphate →
Fructose-6-phosphate + glyceraldehyde-3-phosphate

Transketolase activity in blood is usually measured by determining the rate of disappearance of D-ribose-5-phosphate by using orcinol and ferric chloride in concentrated hydrochloric acid. Transketolase activity in blood can also be measured by determining the amount of fructose-6-phosphate formed in the second reaction above, utilizing the anthrone reagent.

The enzyme assay is performed by incubating a sample of heparinized blood (previously hemolyzed in a phosphate buffer containing sodium chloride and magnesium sulfate) with ribose-5-phosphate and removing aliquots of the incubation mixture at three successive time intervals. Trichloroacetic acid is added to the aliquots to stop the reaction. After centrifugation, the supernatant is analyzed for the remaining ribose-5-phosphate by the orcinol reaction. The product of the transketolase reaction, sedoheptulose-7-phosphate, also reacts with the orcinol reagent to yield a colored product with an absorption spectrum (λ_{max} = 670 nm) that overlaps with the spectrum of the colored product of ribose-5-phosphate with the orcinol reagent (λ_{max} = 580 nm). By measuring the absorbances of samples at both 580 nm and 670 nm and solving simultaneous equations, one can measure transketolase activity in terms of both rate of substrate (ribose-5-phosphate) disappearance and rate of product (sedoheptulose-7-phosphate) formation.[47]

The xylulose-5-phosphate in the reaction is not included in the assay mixture because erythrocytes contain nonlimiting amounts of two extremely active enzymes that can convert ribose-5-phosphate to xylulose-5-phosphate. Ribose-5-phosphate added to an incubation mixture containing these enzymes is rapidly converted to an equilibrium mixture of ribose-5-phosphate, ribulose-5-phosphate, and xylulose-5-phosphate in a 1:9:1 ratio. There is a rapid decrease in ribose-5-phosphate during the first five minutes of incubation due to the formation of this equilibration mixture of pentoses. The decrease in ribose-5-phosphate concentration after the five-minute equilibration period is linear with time and is presumed to be due only to transketolase activity. Some previously proposed assays of transketolase activity measure ribose-5-phosphate disappearance before the equilibrium of the pentoses takes place. Such methods give falsely high values for transketolase activity.

The sensitivity of the procedure to measure thiamin deficiency can be improved by performing the assay before and after addition of thiamin pyrophosphate (TPP). The presence of TPP results in optimal enzyme activity. This increase in activity is known as the *TPP effect.* Results are interpreted as follows: 0–14%, normal; 15–24%, marginal deficiency; over 25%, deficiency.

The TPP effect is generally eliminated within 2–4 h after the parenteral administration of 50 mg of thiamin to the patient. Transketolase activity responds slowly to oral therapy. Occasionally, however, patients with very low transketolase activity may be encountered in whom the TPP effect is eliminated by the parenteral therapy, but whose total enzyme activity may not return to normal values for 1–14 days. These are generally very malnourished individuals, often with liver disorders, who may have an apotransketolase deficit in addition to a coenzyme depletion.

RIBOFLAVIN

Chemistry and sources. *Riboflavin* (vitamin B_2) is a yellow fluorescent compound that is widely distributed throughout the plant and animal kingdoms. Numerous naturally occurring flavins are now known to have the vitamin or closely related derivatives as an integral part of their structures.[9] Principal among these are the coenzymes flavin mononucleotide (FMN) and flavin adenine dinucleotide (FAD). The components of the naturally predominant FAD are indicated in Figure 16-7. Such coenzyme forms are rich sources of the vitamin in liver, kidney, and heart. Many vegetables are also good sources, but cereals are rather low in flavin content. Raw milk is a good source of the vitamin, but considerable loss can occur from exposure to light during pasteurization and bottling or as a result of irradiation to increase the vitamin D content. Flavins are stable to heat but are decomposed by light.

Requirements and allowances. Assessment of riboflavin status has been made on the basis of the relationship of dietary intake to overt signs of hyporiboflavinosis, urinary excretion of the vitamin, riboflavin content of erythrocytes, and, more recently, erythrocyte glutathione reductase activity.[29] Calculations have been based on protein allowances, energy intakes, and metabolic body size, but these do not differ significantly, since they are interdependent. Suggested daily allowances range from 1.2 to 1.7 mg/d, with an additional intake of 0.3–0.5 mg/d for pregnant or lactating women.

Absorption, transport, and metabolism. The process by which riboflavin and lesser amounts of natural derivatives are released by digestion of complexes with food proteins and then absorbed, transported, and metabolically altered has been reviewed in a fairly comprehensive manner.[9] A salient feature is release of coenzyme forms of the vitamin (mainly FAD and less FMN) as a consequence of gastric acidification. After hydrolysis of coenzymes, the vitamin is primarily absorbed in the proximal small intestine by a saturable, Na^+-dependent transport system that is rapid and proportional to intake before leveling off at doses near 25 mg riboflavin per day. Bile salts appear

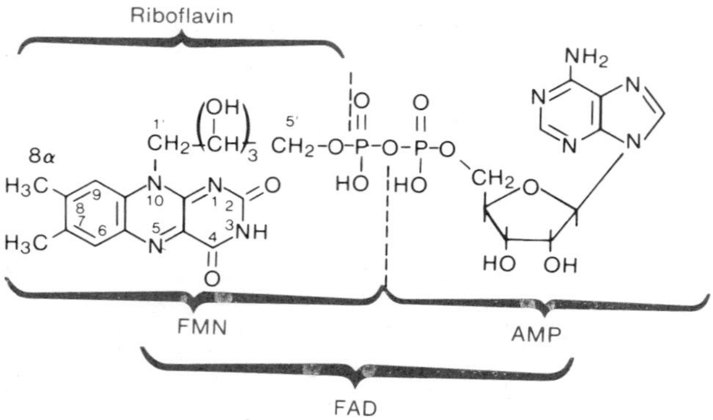

Figure 16-7. Riboflavin and FMN as components of FAD.

to facilitate the uptake. Much of the riboflavin circulating in blood plasma is loosely associated with albumin; lesser amounts complex with other proteins. One subfraction of IgG avidly binds a small portion of the total flavin in blood.[27] Conversion of riboflavin to the coenzymes FMN and FAD occurs within the cell cytoplasm of most tissues, but particularly in the small intestine, liver, heart, and kidneys.[9,15] Thyroxine and triiodothyronine stimulate FMN and FAD synthesis in mammalian systems.[37,38] FAD is the predominant flavocoenzyme present in tissues, where it is mainly complexed with numerous flavoprotein dehydrogenases and oxidases. Since there is little storage of riboflavin as such, the urinary excretion reflects dietary intake.

Functions. In bound coenzymatic form, riboflavin participates in oxidation-reduction reactions in numerous metabolic pathways and in energy production via the respiratory chain.[18] Flavins serve as redox carriers upon differential binding to proteins, participate in both one- and two-electron transfers, and, in reduced (1,5-dihydro) form, react rapidly with oxygen.

Deficiency. Although riboflavin is widely distributed in foodstuffs, many people live for long periods on low intakes, and consequently minor signs of deficiency are common in many parts of the world.[50] Deficiency is encountered almost invariably in combination with a deficit of other water-soluble vitamins. The deficiency syndrome is characterized by sore throat, hyperemia and edema of the pharyngeal and oral mucous membranes, cheilosis, angular stomatitis, glossitis (magenta tongue), seborrheic dermatitis, and normochromic, normocytic anemia associated with pure red cell aplasia of the bone marrow.[49] Severe riboflavin deficiency can affect also the conversion of vitamin B_6 to its coenzyme[11] and can even curtail conversion of tryptophan to niacin.[48]

Methods. There are numerous biochemical methods aimed at the separation and quantitation of the diverse natural flavins.[21,24] Among the more sensitive are those utilizing specific binding, such as riboflavin with egg white riboflavin-binding protein, FMN with apoflavodoxin, and FAD with apoproteins for D-amino acid oxidase or glucose oxidase. Nutritional status is commonly assessed by measuring urinary excretion of the vitamin in fasting, random, or 24-h specimens or by load return tests, measurement of erythrocyte riboflavin concentration, or determination of the erythrocyte glutathione reductase activity coefficient.[5,31]

Urinary riboflavin can be measured by *fluorometric* as well as by microbiological procedures. Under conditions of adequate intake, the amount excreted is >120 μg/d or 80 μg/g creatinine. The rate of excretion expressed as μg/g creatinine is greater for children than for adults. Conditions causing negative nitrogen balance and the administration of antibiotics and certain psychotropic drugs (phenothiazine) increase urinary riboflavin as a consequence of tissue displacement and depletion. A load return test augments the reliability in a given case.

Erythrocyte riboflavin can also be determined by either fluorometric or microbiological means. Since changes observed are rather small, there is some problem with sensitivity and interpretation of results. Values <15 μg/dL of cells reflect low or deficient status.

Currently, the most commonly used method for assessing riboflavin status is the determination of FAD-dependent glutathione reductase activity in freshly lysed erythrocytes as described by Sauberlich and colleagues.[41] It has, however, one drawback:[31] the test cannot be used in persons with glucose-6-phosphate dehydrogenase deficiency because of an increased avidity of the glutathione reductase for FAD in this disease.

Reference ranges. The results of erythrocyte glutathione reductase assays are expressed in terms of "activity coefficients" (AC), which represent the degree of stimulation of apoenzyme resulting from addition in vitro of FAD. An AC of 1.0 indicates no stimulation; i.e., that only holoenzyme is present because FAD (and riboflavin) levels in the erythrocytes are adequate. Suggested guidelines for interpretation of such coefficients are <1.2, acceptable; 1.2–1.4, low; >1.4, deficient.

VITAMIN B_6

Chemistry and sources. The vitamin B_6 group comprises three natural forms: *pyridoxine* (pyridoxol), *pyridoxamine*, and *pyridoxal*, shown in Figure 16-8. Both pyridoxamine-5′-phosphate and pyridoxal-5′-phosphate (PLP, P-5′-P) interconvert as coenzyme forms during aminotransferase- (transaminase-) catalyzed reactions. Vitamin B_6 is widely distributed in animal and plant tissues, where the phosphorylated forms, and particularly PLP, predominate. Meats, poultry, and fish are good sources, as are yeast,

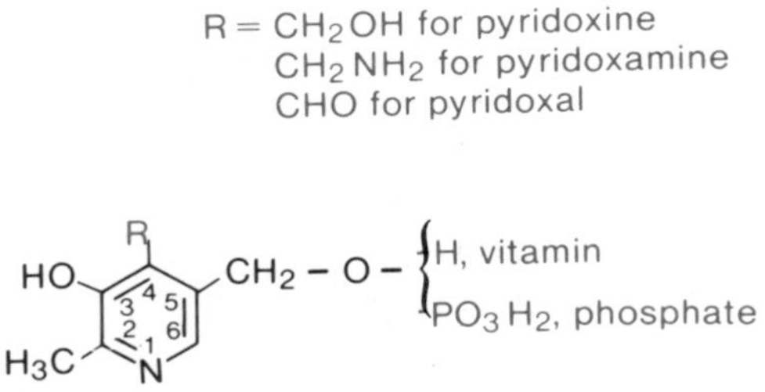

R = CH_2OH for pyridoxine
CH_2NH_2 for pyridoxamine
CHO for pyridoxal

HO, R, $CH_2 - O -$ {H, vitamin; PO_3H_2, phosphate}; H_3C, N

Figure 16-8. Free and phosphorylated forms of vitamin B_6.

certain seeds, and bran; somewhat more limited sources are milk, eggs, and green leafy vegetables.[13,29,49,50] The common commercial form of the vitamin is pyridoxine hydrochloride, which is a water-soluble, white, crystalline solid. Solutions of the B_6 vitamins are decomposed by light, especially in the UV region, at neutral to alkaline pH. There is significant loss of PLP during thermal processing of foods.[12]

Requirements and allowances. Requirements for vitamin B_6 are complicated by differences in protein intake, by the probable provision of a fraction of the needed quantity through bacterial synthesis in the intestinal tract, and by the use of alcohol or oral contraceptives.[13,29,49] Estimates of requirements with some margin of safety are 2.2 mg/d for adult males and 2.0 mg/d for adult females, with an additional 0.6 mg/d for pregnant women and during lactation.

Absorption, transport, and metabolism. The three B_6 vitamers are readily absorbed by the mucosal cells. PLP binds for catalytic function with numerous specific apoenzymes throughout the cell. The erythrocyte, in addition, traps PLP as a conjugate Schiff base with hemoglobin.[26] Glycogen phosphorylase contains most of the PLP in skeletal muscle.[3]

The release of free vitamin, mainly pyridoxal, when physiological nonsaturating levels of vitamin are absorbed, occurs when the phosphates are hydrolyzed by nonspecific alkaline phosphatase located in the plasma membrane of cells. Some PLP is also released into the circulation by the liver.[46] Although PLP is the principal tissue form of vitamin B_6 and pyridoxal constitutes much of the circulating vitamin, the main catabolite excreted in urine is 4-pyridoxic acid.

Functions. As coenzyme PLP, vitamin B_6 functions in the metabolism of proteins, carbohydrates, lipids, and amino acids.[48] A few examples of PLP coenzyme–dependent reactions include formation of amines such as epinephrine and norepinephrine, formation of aminolevulinate in heme synthesis, and phosphorolysis of the α-1,4-linkages of glycogen.

As with other water-soluble vitamins that function as coenzymes, the rate at which symptoms become worse when there is deficiency of the vitamin depends both on the relative affinity of the coenzyme for a given apoenzyme and on the essentiality of the catalyzed reaction.

Deficiency. A deficiency of vitamin B_6 alone is uncommon; it usually occurs in association with deficits of other vitamins of the B-complex. Chemotherapy or fortuitous ingestion of antagonists has led to hypovitaminosis B_6.[15,49] Antagonists include the tuberculostatic drug isoniazid[46] and several naturally occurring substituted hydrazines and hydroxylamines,[23] among them D-cycloserine.[8] Penicillamine (β-dimethyl cysteine), used in treatment of patients with Wilson's disease to decrease the levels of copper in the liver, inactivates PLP. There are several genetic conditions in which abnormalities in the function of vitamin B_6 occur, but they usually respond to 5–50 mg/d of administered vitamin B_6.[40]

Biochemical changes occur early and become more marked as deficiency of B_6 progresses.[13,15,49] Plasma levels of PLP and urinary output of B_6 and 4-pyridoxic acid decrease within a week of removal of the vitamin from the diet. There is increased xanthurenic acid in urine. Transaminase activity in serum and erythrocytes also decreases. Electroencephalographic abnormalities appear within three weeks. Epileptiform convulsions are a common finding in young vitamin B_6–deficient subjects. In addition, skin changes include a dermatitis; hematological manifestations may include a decrease in circulating lymphocytes and possibly a normocytic, microcytic, or sideroblastic anemia.

Methods. Direct assessment of all or separate vitaminic forms of B_6 in urine and blood has been done in the past by using *microbiological assays*. Levels of 20 μg vitamin B_6/g creatinine in urine are considered indicative of marginal or inadequate dietary intake of the vitamin. Fluorometric assays of *urinary 4-pyridoxic acid* and blood PLP after conversion of the latter to the cyanide complex or condensation with a fluorophore, such as methyl anthranilate, followed by reduction, have also found application. 4-Pyridoxic acid is now best determined by HPLC. During deficiency, the level of 4-pyridoxic acid in urine will drop well below the normal of at least 0.8 mg/d. The *PLP concentration in plasma*, usually measured by using radioactive tyrosine and the apodecarboxylase, has been judged the most reliable indicator of B_6 status. The reported *reference ranges* are 5–23 ng/mL and 13–16 ng/mL (53–65 pmol/mL) of plasma.[36]

Erythrocyte levels of aspartate and alanine aminotransferases provide a good reflection of vitamin B_6 status. Enzymatic assays run after and before in vitro addition of PLP yield an activity coefficient ratio. Ratios of <1.5 for aspartate aminotransferase and <1.2 for alanine aminotransferase are considered normal but may depend somewhat on the assay method used.[5] This ratio is invalid if pyridoxal phosphate is already a component of the reaction mixture.

Measurement of urinary tryptophan metabolites, particularly *xanthurenic acid*, following an oral load of 2–5 g of L-tryptophan, is one of the most common indices used in studies of vitamin B_6 nutriture, because changes can be recognized early and measured relatively easily. Amounts of xanthurenate well above the normal of ~25 mg/d are seen in vitamin B_6 deficiency.

NIACIN AND NIACINAMIDE

Chemistry and sources. Though the term *niacin* is chemically synonymous with nicotinic (pyridine-3-carboxylic) acid, it is now used as the generic name for the specific compound as well as for derivatives exhibiting qualitatively the biological activity of niacinamide (nicotinamide, nicotinic acid amide). Thus "niacin activity" and "niacin deficiency" carry this broader meaning in nutritional literature. Structures of both vitamers and the two coenzyme forms containing the nicotinamide moiety are given in Figure 16-9. Nicotinamide adenine dinucleotide (NAD; diphosphopyridine nucleotide, DPN) and nicotinamide adenine dinucleotide phosphate (NADP; triphosphopyridine nucleotide, TPN) represent most of the niacin activity found in food sources, which include yeast, lean meats, liver, and poultry. Milk, canned salmon, and several leafy green vegetables have lesser amounts. Protein provides a considerable portion of niacin equivalent because of the tryptophan content. As much as two-thirds of niacin required by adults can be derived from tryptophan metabolism via nicotinic acid ribonucleotide to NAD and NADP. Free forms of the vitamin are white, stable solids that are quite soluble in water. The oxidized coenzymes are labile to alkali, whereas the reduced (dihydro) coenzymes are labile to acid. Reduction of the oxidized coenzymes commonly occurs by addition of a hydride ion to the *para*(4) position of the nicotinamide ring, with

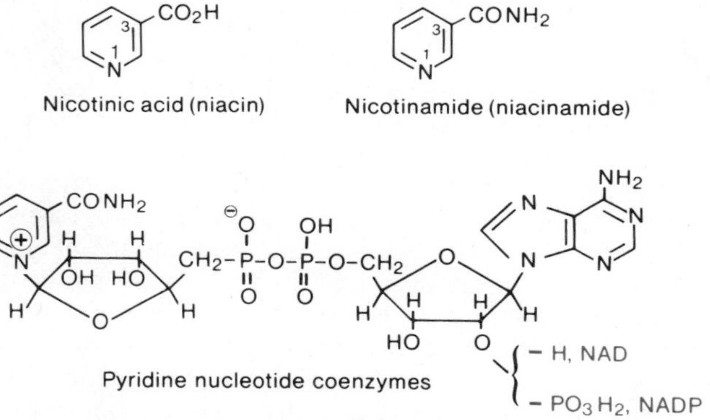

Nicotinic acid (niacin)

Nicotinamide (niacinamide)

Pyridine nucleotide coenzymes

– H, NAD

– PO_3H_2, NADP

Figure 16-9. Niacin, niacinamide, and coenzymes.

simultaneous formation of a solvated proton. NADH and NADPH (but not NAD and NADP) absorb light in the near ultraviolet region (339 nm).

Requirements and allowances. Estimations of niacin requirements take into account the contribution of tryptophan derived from protein. Average diets in the US supply 0.5–1 g of tryptophan plus 8–17 mg of niacin, for a total of 16–34 mg of niacin equivalents (NE).[29] The recommended allowance range is 9–19 NE/d for children and adults, with an increase of 2 NE/d during pregnancy and 5 NE/d for lactation. RDA are set at 6–8 NE/d for infants.

Absorption, transport, and metabolism. The coenzymes are hydrolyzed in the intestinal tract, and both the acid and amide forms of the vitamin are readily absorbed. Nicotinic acid and nicotinamide are both present in blood and move between blood and cerebrospinal fluid.[44] Both compounds are converted to the coenzyme forms in blood cells, kidneys, brain, and liver. In the tissues most of the vitamin is present as nicotinamide in NAD and NADP, although liver may contain a significant fraction of the free vitamin. There is little storage of niacin as such.

Although nicotinamide can be converted to nicotinic acid by a rather widespread microsomal deamidase, there is no direct reamidation of nicotinic acid. Humans excrete 1-methylnicotinamide and 1-methyl-3-carboxamido-6-pyridone, which are primary urinary metabolites.[13,15,29,48]

Functions. There are hundreds of enzymes that require the nicotinamide moiety within either NAD or NADP. Most of these oxidoreductases function as dehydrogenases and catalyze such diverse reactions as the conversion of alcohols (often sugars and polyols) to aldehydes or ketones, hemiacetals to lactones, aldehydes to acids, and certain amino acids to keto acids. Generally, most NAD-dependent enzymes are involved in catabolic reactions, whereas NADP systems are more common to biosynthetic reactions.

Deficiency. *Pellagra* is a classic deficiency disease most often found among people who subsist chiefly on corn.[39] Though the pathogenesis has been attributed to a deficiency of niacin (and tryptophan), other associated complicating factors may be an imbalance of amino acid intake, particularly the ingestion of high levels of leucine,[15,49] and the presence of mycotoxins elaborated by mold infestations.[49] Pellagra is also an occasional secondary manifestation of two disorders that profoundly affect tryptophan metabolism:[49] *carcinoid syndrome*, in which up to 60% of tryptophan is catabolized by what is ordinarily a minor pathway, and *Hartnup disease*, an autosomal recessive disorder in which several amino acids, including tryptophan, are poorly absorbed.

The typical presentation of pellagra is that of a chronic wasting disease associated with dermatitis, dementia, and diarrhea. The characteristic erythematous dermatitis occurs on skin areas exposed to sunlight. Mental changes include fatigue, insomnia, and apathy, which precede an encephalopathy and, eventually, frank organic psychoses. The diarrhea reflects a widespread inflammation of the intestinal mucous surfaces.

Excess. Although relatively large daily intakes of niacin (40–200 mg) may be required in treatment of Hartnup disease and carcinoid syndrome,[49] the use of pharmacologic doses is of doubtful value for other dysfunctions and may even prove harmful.[6,15,29] Massive (gram) doses of nicotinic acid (but not the amide) produce vasodilation or "flushing," with an accompanying sensation of burning or stinging of the face and hands. Pruritus, nausea, vomiting, and diarrhea have been commonly reported but often abate with continued therapy. Varying degrees of hyperpigmentation occur in rare cases. Additional effects are abnormal glucose tolerance, hyperuricemia, peptic ulcer, hepatomegaly, jaundice, and increased serum transaminases.

Methods. The measurement of the excretion of *N*(1)-methylnicotinamide and *N*(1)-methyl-3-carboxamide-6-pyridone (also named *N*(1)-methyl-2-pyridone-5-carboxamide) has received continued use in the biochemical assessment of niacin nutriture. Normally, adults excrete 20–30% of their niacin in the form of methylnicotinamide and 40–60% as the pyridone.[42] An excretion ratio of pyridone to methylnicotinamide of 1.3–4.0 is acceptable, but latent niacin deficiency is indicated by a value below 1.0. As depletion occurs, the pyridine is absent for weeks before clinical signs are noted, and the methylnicotinamide excretion falls to a minimum at about the time clinical signs are evident. The pyridone/methylnicotinamide ratio is best determined with *HPLC* *methods*. There are several efficient means for separating and quantitating both vitaminic and coenzymic forms.[21,24] Some of these rely on the absorbance of reduced pyridine nucleotides at 340 nm in coupled-enzyme systems, and others utilize fluorescent properties, especially of addition products such as methyl ethyl ketone.

FOLIC ACID AND VITAMIN B_{12}

Discussion of these vitamins, which interrelate in one-carbon metabolism, is covered in Chapter 24, *Biochemical Aspects of Hematology*.

BIOTIN

Chemistry and sources. *Biotin* (vitamin H) is shown in Figure 16-10. The vitamin in most organisms occurs mainly bound to protein. Good sources of biotin include liver, kidney, pancreas, eggs, yeast, and milk.

Estimated adequate intakes. Intestinal microfloras make a significant contribution to the body pool of available biotin, making determination of the dietary requirement difficult.[29] An intake of 100–200 μg/d for young infants is believed to be more than adequate.

Absorption, transport, and metabolism. Digestion of dietary proteins containing bound biotin yields considerable biocytin (ϵ-*N*-biotinyl lysine). Biocytin and biotin are readily absorbed. Biotin is cleared from the circulating blood and is taken up by such tissues as liver, muscle, and kidney. Careful balance studies in humans, in whom perhaps only 1 mg is the total body content, show that the urinary excretion of biotin often exceeds dietary intake and that in all cases fecal excretion is as much as three to six times greater than dietary intake because of microfloral biosynthesis.[13,29]

Functions. Four biotin-dependent enzymes have been found in human tissues.[19] They are carboxylases for acetyl-CoA, propionyl-CoA, β-methylcrotonyl-CoA, and pyruvate.

Deficiency. Dietary deficiency of biotin is infrequently seen and has been produced in adults only after ingestion of large amounts of raw egg white that contains the protein avidin, which complexes biotin.[29,42] Symptoms include anorexia, nausea, vomiting, glossitis, pallor, depression, and a dry, scaly dermatitis. A seborrheic dermatitis in infants under six months of age can also be caused by inadequate biotin, but the condition responds promptly to biotin therapy. Significantly lowered urinary excretion or circulating blood levels have also been found in pregnant women, alcoholics, and patients with achlorhydria, as well as among the elderly.[4] Finally, there are rather rare genetic enzyme defects such as those in holoenzyme synthetase and in propionyl-CoA carboxylase.[19]

Methods. Biotin can be quantitated by using *microbiological assays*.[42] Bound biotin is first liberated by proteolytic digestion of the sample using, for example, papain with whole blood. Then aliquots are added to a biotin-deficient medium inoculated with a test organism such as *Lactobacillus plantarum*. Standard curves are derived from growth in controls containing known amounts of biotin. Other methods include isotopic dilution assays[23] and even a colorimetric determination with acidic *p*-dimethylaminocinnamaldehyde, which forms a red Schiff base.[20]

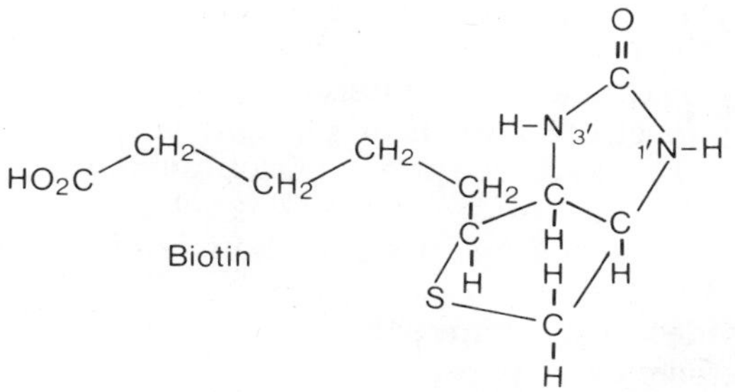

Figure 16-10. Biotin.

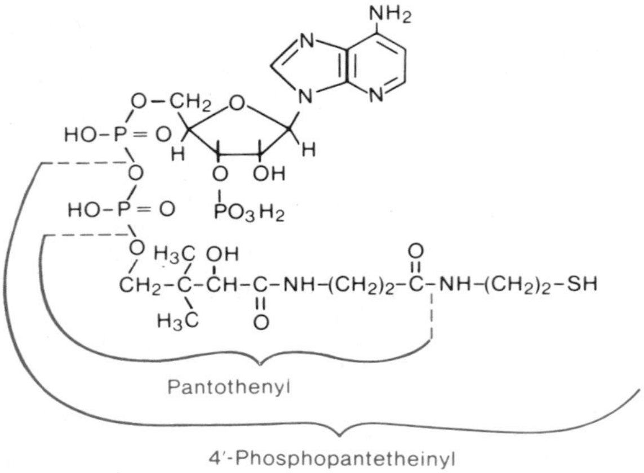

Figure 16-11. Pantothenate and 4′-phosphopantetheine as components of CoA.

PANTOTHENIC ACID

Chemistry and sources. *Pantothenic acid*, ubiquitous in nature, is synthesized by most microorganisms and plants from pantoic acid. The vitamin is an integral part of a covalently attached prosthetic group of acyl carrier proteins and is within the structure of coenzyme A, as shown in Figure 16-11. The vitamin is widely distributed in foods and is particularly abundant in animal sources, legumes, and whole grain cereals.[13,29] Excellent food sources include egg yolk, kidney, liver, and yeast. Fair sources include broccoli, lean beef, skimmed milk, sweet potatoes, and molasses. Over half the pantothenate in wheat may be lost during the manufacture of flour, and up to a third is lost during the cooking of meat. The most common commercial synthetic form is the calcium salt.

Estimated adequate intake. An intake of pantothenic acid of 4–7 mg/d seems sufficient for adults. Urinary excretion, which generally correlates with dietary intake, is 2–7 mg/d in adults. A slightly higher intake may be warranted for pregnant or lactating women. Adequate intakes for younger age groups are based on proportional energy needs and range from a suggested 2 mg/d for young infants to 4–5 mg/d for adolescents.

Absorption, transport, and metabolism. Coenzyme A, the form in which much of the pantothenic acid is ingested, is hydrolyzed by intestinal pyrophosphatase and phosphatase to pantetheine, which together with pantothenate is absorbed into the portal circulation. Within cells resynthesis of coenzyme compounds occurs. About 80% of the vitamin in animal tissues is in CoA form. Cleavage enzymes operate during turnover and release of the vitamin, which is excreted in the urine. Only a small fraction of pantothenate is secreted into milk, and even less into colostrum.

Functions. The myriad acyl thiol esters of CoA, of which pantothenic acid is a constituent, are central to the metabolism of numerous compounds (especially lipids) and the ultimate catabolic disposition of carbohydrates and ketogenic amino acids.[13,48] For example, acetyl-CoA, which derives from the metabolism of carbohydrates, fats, and amino acids, can acetylate compounds such as choline and hexosamines to produce essential biochemicals; it can also condense with other metabolites such as oxaloacetate or with itself to supply citrate and cholesterol, respectively. Another essential role of pantothenic acid is its participation in the 4′-phosphopantetheine moiety of acyl carrier protein (ACP), where it facilitates chain elongation during fatty acid biosynthesis. Although the reactive thiol function of CoA and ACP is not an integral part of pantothenate, the steric and chemical properties conferred by the vitamin structure are important for enzymic recognition.

Deficiency. The widespread occurrence of pantothenic acid in foods makes an uncomplicated dietary deficiency of pantothenate unlikely in the human.[15,29] Symptoms have been produced in a few volunteers who have received an antagonist, and more recently in persons fed semisynthetic diets virtually free of pantothenate. Subjects became irascible and developed postural hypotension and rapid heart rate on exertion, epigastric distress with anorexia and constipation, numbness and tingling of the hands and feet, hyperactive deep-tendon reflexes, and weakness of finger extensor muscles. The eosinopenic response to ACTH was impaired. More severe deficiency in animals leads to failure of the adrenal cortex.

Methods. Pantothenic acid is usually measured by microbiological procedures. A radioimmunoassay recently developed is also being used.[51] Gas chromatographic methods are employed for pharmaceutical preparations. Enzymatic assays can be used for determining the quantity of CoA and ACP.[20,23]

Since urinary output of pantothenate is directly proportional to dietary intake, present assessment relies mainly on this index.[5,42] Urinary excretion of <1 mg/d is considered abnormally low. Suspicion of inadequate intake is further supported if whole blood values are <100 μg/dL.

ASCORBIC ACID

Chemistry and sources. L-*Ascorbic acid* (*vitamin C*) is a white, crystalline solid that is readily soluble in water. Ascorbic acid is a relatively strong reductant with an E'_0 (pH 7) of +0.58 volt. It is reversibly oxidized to dehydroascorbic acid (ascorbone), also shown in Figure 16-12. The dehydro form is more labile than the reduced form.[29] Some mammals, including the human, lack the enzyme that catalyzes the formation of 2-keto-L-gulonolactone, which spontaneously tautomerizes to L-ascorbic acid. The best sources of the vitamin are citrus fruits, berries, melons, tomatoes, green peppers, raw cabbage, and leafy green vegetables.[13] Losses during processing, especially with heat and aerobic conditions, can be considerable.

Requirements and allowances. The amount of vitamin C sufficient to alleviate and cure the clinical signs of scurvy is only 10 mg/d.[29] This amount, however, does not provide for any reserves. The RDA for adults is 60 mg. An additional 20 mg/d is recommended for the pregnant woman, and an additional 40 mg/d for lactating women. A daily recommendation of 35 mg has been made for breast-fed infants, and an RDA of 100 mg is suggested for premature and newborn babies during the first week to protect against possible tyrosinemia. The RDA for children is 45 mg.

Absorption, transport, and metabolism. Absorption of vitamin C occurs readily, mostly from the stomach, where some of the ascorbic acid is converted to the dehydro form. At physiological pH, the uncharged dehydroascorbic acid passes across cell membranes faster than the anionic L-ascorbate. Passive diffusion of vitamin C may largely account for entry into some cells, such as leukocytes and erythrocytes, but an active transport mechanism may also operate, especially for platelets, adrenals, and retina. Vitamin C is found in most tissues, but glandular tissues such as the pituitary, adrenal cortex, corpus luteum, and thymus have the highest amounts,[13] and the retina has 20–30 times the plasma concentration. The $t_{1/2}$ for vitamin C in humans is only ~16 days. Ascorbate, dehydroascorbic acid, and lesser amounts of a number of catabolites are present in urine.

Functions. The most clearly established and critical functional role for ascorbic acid is as a cofactor for protocollagen hydroxylase, the enzyme responsible for hydroxylation of prolyl and lysyl residues within nascent peptides in connective tissue proteins.

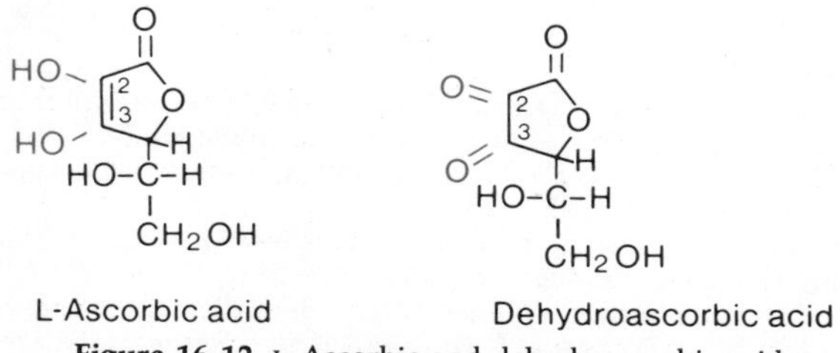

L-Ascorbic acid   Dehydroascorbic acid

Figure 16-12. L-Ascorbic and dehydroascorbic acids.

Among these are collagen and related proteins that comprise intercellular material of cartilage, dentin, and bone.[13,48] Vitamin C may also be involved in tyrosine metabolism, microsomal drug metabolism, synthesis of epinephrine and anti-inflammatory steroids by the adrenals, folic acid metabolism, and leukocyte functions.[9] Absorption of Fe(II) is enhanced by simultaneous ingestion of the vitamin.

Deficiency. Protracted deficiency of vitamin C leads to the classic disease called *scurvy*. Inability to form adequate intercellular substances in connective tissue is reflected in swollen, tender, and often bleeding or bruised loci at joints and in other areas where structurally weakened tissue cannot withstand stress.[50] Infantile scurvy, also known as Barlow's disease, exhibits a bayonet-rib syndrome. The gums are livid and swollen. Cutaneous bleeding often begins on the lower thighs as perifollicular hemorrhages, which may then spread to the buttocks, abdomen, legs, and arms. Petechial hemorrhages due to the rupture of capillaries often appear. Thereafter, large spontaneous bruises (ecchymoses) may arise almost anywhere on the body. Ocular hemorrhages, drying of salivary and lacrimal glands, parotid swelling, femoral neuropathy, edema of the lower extremities, and psychological disturbances have also been described. Some scorbutic patients may develop anemia, may display radiological changes characteristic of osteoporosis, or may die suddenly from heart failure.

Excess. The use of "megadoses" of vitamin C, as suggested by some for the prevention or amelioration of the common cold, does not appear warranted on the basis of present information. Large doses of ascorbic acid have generally been considered nontoxic, except for gastrointestinal symptoms, which are experienced by some subjects; however, more serious adverse effects have been observed and suspected as potential hazards.[6,13,15,29] These include reductive destruction of concomitantly ingested vitamin B_{12}, too much iron absorption, increased production of oxalate favoring deposition of calcium oxalate stones in kidney and bladder (particularly in congenital, familial hyperoxalurics), and uricosuria. Both benefits and hazards from high levels of vitamin C seem minimal for most people, who simply excrete the excess.

Methods. Assessment of vitamin C status is generally done by measuring serum (plasma) and leukocyte levels of the vitamin. Urinary excretion and erythrocyte concentrations are neither specific nor useful indices of vitamin C status; however, urinary levels of ascorbic acid, especially following a load test, can be helpful in the clinical diagnosis of scurvy.

Ascorbic acid in plasma is oxidized by Cu(II) to form dehydroascorbic acid, which reacts with acidic 2,4-dinitrophenylhydrazine to form a red *bis*-hydrazone, which is measured at A_{520}. Heparinized blood is collected and centrifuged to obtain plasma, which should be analyzed immediately, and not later than 3 h if the specimen is refrigerated.

Plasma (or serum) levels of ascorbate show a linear increase with dietary intake of the vitamin up to a level of 1.2–1.4 mg/dL, beyond which urinary excretion rapidly increases.[5,13,42] Plasma values >0.3 mg/dL are considered acceptable, those from 0.2 to 0.29 are "at risk," and those <0.2 indicate deficiency. In persons with adequate intake of vitamin C, plasma concentrations of total vitamin (ascorbic acid plus dehydroascorbic acid) are between 0.6 and 2.0 mg/dL. The lower limit value may be seen in some cases with subclinical vitamin C deficiency and in older individuals. *Leukocyte* ascorbate levels are somewhat more difficult to measure but are more representative of tissue stores. Values ranging from 0 to 7 mg/dL of leukocytes are suggestive of deficiency, whereas values >15 indicate satisfactory vitamin C status. Values may also be expressed as 20–50 $\mu g/10^8$ WBC. The *urinary* excretion of ascorbic acid by the normal adult ranges from 8 to 27 mg/d; this declines to undetectable levels after depletion.[42]

References

1. American Medical Association, Council on Foods and Nutrition: Improvement of the nutritive quality of foods. JAMA, *225*:1116, 1973.
2. Bieri, J. G., Tolliver, T. J., and Catignani, G. L.: Simultaneous determination of α-tocopherol and retinol in plasma or red cells by high pressure liquid chromatography. Am. J. Clin. Nutr., *32*:2143, 1979.
3. Black, A. L., Guirard, B. M., and Snell, E. E.: The behavior of muscle phosphorylase as a reservoir for vitamin B_6 in the rat. J. Nutr., *108*:670, 1978.
4. Bonjour, J. P.: Biotin in man's nutrition and therapy—a review. Internat. J. Vit. Nutr. Res., *47*:107, 1977.
5. Briggs, M., Ed.: Vitamins in Human Biology and Medicine. Boca Raton, FL, CRC Press, Inc., 1981.
6. Campbell, T. C., Allison, R. G., and Carr, C. J.: Feasibility of Identifying Adverse Health Effects of Vitamins and Essential Minerals in Man. Bethesda, MD, Life Sciences Research Office, Federation of American Societies for Experimental Biology, 1980.

7. Catabolites of thiamine from the rat. Nutr. Rev., *29*:120, 1971.
8. Cohen, A. C.: Pyridoxine in the prevention and treatment of convulsions and neurotoxicity due to cycloserine. Ann. N.Y. Acad. Sci., *166*:346, 1969.
9. Darby, W. J., Ed.: Annual Review of Nutrition, Vol. 1. Palo Alto, CA, Annual Reviews, Inc., 1981.
10. Evans, W. C.: Thiaminases and their effects on animals. *In*: Vitamins and Hormones, Vol. 33. P. L. Munson, J. Glover, E. Diczfalusy, et al., Eds. New York, Academic Press, 1975.
11. Genetic determination of coenzyme synthesis in red cells. Nutr. Rev., *39*:331, 1981.
12. Gregory, J. F., and Kirk, J. R.: Vitamin B_6 in foods: assessment of stability and bioavailability. *In*: Human Vitamin B_6 Requirements. Washington, D.C., National Academy of Sciences, 1978.
13. Harper, H. A.: The water-soluble vitamins. *In*: Review of Physiological Chemistry. H. A. Harper, V. A. Rodwell, and P. A. Mayes, Eds. Los Altos, CA, Lange Medical Publications, 1979.
14. Hauschka, P. V., Lian, J. B., and Gallop, P. M.: Vitamin K and mineralization. Trends Biochem. Sci., *3*:75, 1978.
15. Hegsted, D. M., Chichester, C. O., Darby, W. J., et al., Eds.: Present Knowledge in Nutrition, 4th ed. Washington, D.C., The Nutrition Foundation, Inc., 1976.
16. Janiszowska, W., and Pennock, J. F.: The biochemistry of vitamin E in plants. *In*: Vitamins and Hormones, Vol. 34. P. L. Munson, J. Glover, E. Diczfalusy, et al., Eds. New York, Academic Press, 1976.
17. Körner, W. F., and Völlm, J.: New aspects of the tolerance of retinol in humans. Internat. J. Vit. Nutr. Res., *45*:363, 1975.
18. Massey, V., and Williams, C. H., Jr., Eds.: Flavins and Flavoproteins. New York, Elsevier Biomedical, 1982.
19. McCormick, D. B., and Olson, A. E.: Biotin. *In*: Present Knowledge in Nutrition, 5th ed. R. E. Olson, et al., Eds. Washington, D.C., The Nutrition Foundation, Inc., 1984.
20. McCormick, D. B., and Wright, L. D., Eds.: Vitamins and coenzymes. Methods in Enzymology, Vol. 18, part A. New York, Academic Press, 1970.
21. McCormick, D. B., and Wright, L. D., Eds.: Vitamins and coenzymes. Methods in Enzymology, Vol. 18, part B. New York, Academic Press, 1971.
22. McCormick, D. B., and Wright, L. D., Eds.: Vitamins and coenzymes. Methods in Enzymology, Vol. 18, part C. New York, Academic Press, 1971.
23. McCormick, D. B., and Wright, L. D., Eds.: Vitamins and coenzymes. Methods in Enzymology, Vol. 62, part D. New York, Academic Press, 1979.
24. McCormick, D. B., and Wright, L. D., Eds.: Vitamins and coenzymes. Methods in Enzymology, Vol. 66, part E. New York, Academic Press, 1980.
25. McCormick, D. B., and Wright, L. D., Eds.: Vitamins and coenzymes. Methods in Enzymology, Vol. 67, part F. New York, Academic Press, 1980.
26. Mehansho,H., and Henderson, L. M.: Transport and accumulation of pyridoxine and pyridoxal by erythrocytes. J. Biol. Chem., *255*:11, 901, 1980.
27. Merrill, A. H., Jr., Froehlich, J. A., and McCormick, D. B.: Isolation and identification of alternative riboflavin-binding proteins from human plasma. Biochem. Med., *25*:198, 1981.
28. National Dairy Council: Recent developments in vitamin D. Dairy Council Digest, *47*:13, 1976.
29. National Research Council, Committee on Dietary Allowances: Recommended Dietary Allowances, 9th revised ed. Washington, D.C., National Academy of Sciences, 1980.
30. Neal, R. A.: Vitamin deficiencies: Thiamin. *In*: Proc. Workshop on Problems of Assessment and Alleviation of Malnutrition in the US. R. G. Hansen and H. N. Munro, Eds. Nashville, GSMHA and National Institutes of Health, 1970.
31. Nichoalds, G. E.: Riboflavin. Clin. Laboratory Med., *1*:685, 1981.
32. Nino, H. V., and Shaw, W.: Vitamins. *In*: Fundamentals of Clinical Chemistry, 2nd ed. N. W. Tietz, Ed. Philadelphia, W. B. Saunders Co., 1976.
33. Norman, A. W., and Henry, H.: 1,25-Dihydroxy D_3. A hormonally active form of Vitamin D_3. Recent Prog. Horm. Res., *30*:431, 1974.
34. Olson, R. E.: Vitamin K. *In*: Modern Nutrition in Health and Disease. R. S. Goodhart and M. E. Shils, Eds. Philadelphia, Lea & Febiger, 1973.
35. Pahuja, D. N., and DeLuca, H. F.: Stimulation of intestinal calcium transport and bone calcium mobilization by prolactin in vitamin D–deficient rats. Science, *214*:1038, 1981.
36. Reynolds, R. D.: Nationwide assay of vitamin B_6 in human plasma by different methods. Fed. Proc., *42*:665 (Abst. No. 2185), 1983.
37. Rivlin, R. S.: Hormones, drugs, and riboflavin. Nutr. Rev., *37*:241, 1979.
38. Rivlin, R. S., Ed.: Riboflavin. New York, Plenum Press, 1975.
39. Roe, D. A.: A Plague of Corn. Ithaca, NY, Cornell University Press, 1973.
40. Rosenberg, L. E.: Vitamin-responsive inherited diseases affecting the nervous system. *In*: Brain Dysfunction in Metabolic Disorders; Research Publication for the Association for Research in Nervous and Mental Disease, Vol. 53. F. Plum, Ed. New York, Raven Press, 1974.
41. Sauberlich, H. E., Judd, J. H., Jr., Nichoalds, G. E., et al.: Application of the erythrocyte glutathione reductase assay in evaluating riboflavin nutritional status in a high school student population. Am. J. Clin. Nutr. *25*:756, 1972.
42. Sauberlich, H. E., Skala, J. H., and Dowdy, R. P.: Laboratory Tests for the Assessment of Nutritional Status. Boca Raton, FL, CRC Press, Inc., 1974.
43. Scriver, C. R.: Vitamin-responsive inborn errors of metabolism. Metabolism, *22*:1319, 1973.
44. Spector, R.: Niacin and niacinamide transport in the central nervous system. In vivo studies. J.Neurochem., *33*:895, 1979.

45. Suttie, J. W., Ed.: Vitamin K Metabolism and Vitamin K–Dependent Proteins. Baltimore, University Park Press, 1979.
46. Tryfiates, G. P., Ed.: Vitamin B_6 Metabolism and Role in Growth. Westport, CT, Food and Nutrition Press, Inc., 1980.
47. Warnock, L. G.: A new approach to erythrocyte transketolase measurement. J. Nutr., *100*:1057, 1970.
48. White, A., Handler, P., Smith, E. L., et al.: Principles of Biochemistry. New York, McGraw-Hill Book Company, 1978.
49. Wilson, J. D.: Disorders of vitamins—deficiency, excess and errors of metabolism. *In*: Harrison's Principles of Internal Medicine, 10th ed. R. G. Petersdorf, R. D. Adams, E. Braunwald, et al., Eds. New York, McGraw-Hill Book Company, 1982.
50. Winick, M.: Nutrition in Health and Disease. New York, John Wiley & Sons, Inc., 1980.
51. Wyse, B. W., Wittmer, C., and Hansen, R. G.: Radioimmunoassay for pantothenic acid in blood and other tissues. Clin. Chem., *25*:108, 1979.

TRACE ELEMENTS

17

ROBERT A. JACOB, Ph.D.

Major elements in biological tissues are those that are present in mg/g amounts, whereas *trace element concentrations* are those present in μg/g or smaller amounts. Correspondingly, human intake requirements for major elements are expressed in grams or fractions of a gram per day, whereas intakes of trace elements are in mg or μg/d. Improved sensitivities of analytical methods have allowed the study of *ultratrace elements*, those present in ng/g amounts or less.

Essential Trace Elements

The capability of reliably quantitating levels of trace elements in biological specimens has led to discoveries of the vital roles that trace elements play in human and animal metabolism. An *essential element* is one that is uniquely required for growth or for the maintenance of life or health. A deficiency of the element consistently produces a functional impairment that is only alleviated by physiological supplementation of that element. A biochemical basis for the element's essential functions must be demonstrated. A simple model of the relationship between tissue concentration of an essential nutrient and a dependent biological function is shown in Figure 17-1. The plateau of the curve reflects the operation of a homeostatic regulation system that ensures optimal health over a range of nutrient levels. At low nutrient levels, dependent biological functions are impaired, thus defining a deficiency state. Very high nutrient levels may result in another impairment, i.e., a toxicity state. Each nutrient has its own characteristic curve, and since pathological effects due to deficiency are often different from those due to toxicity, such curves may be asymmetric.

Over the past several decades, as analytical detection limits have been lowered, the list of trace elements claimed to be essential for humans or animals has grown steadily. A complete list of essential trace elements is somewhat controversial because of fragmentary data on elements currently under study. Also, comprehensive criteria that define the essentiality of trace elements are not universally accepted. The elements currently accepted as essential in animals or humans are shown in a periodic table

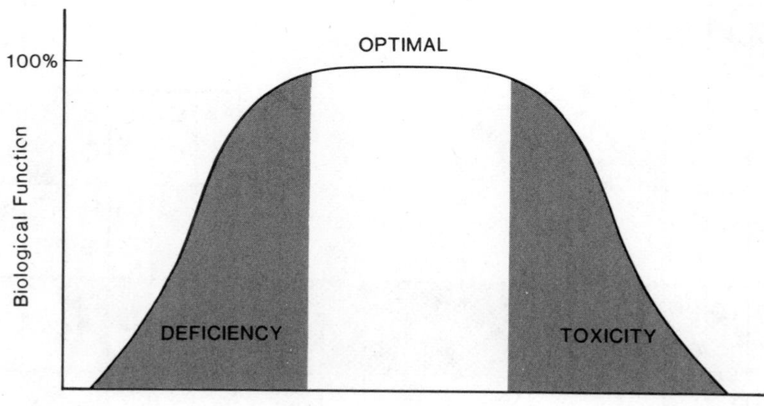

Figure 17-1. Model of the relationship between tissue concentration or intake of an essential nutrient and dependent biological function.

format in Figure 17-2. All macro and trace (shaded) elements are considered essential for humans, with the exception of nickel, vanadium, arsenic, and silicon, for which essentiality has been demonstrated only in animals. Trace elements that are consistently present in human tissues but are considered nonessential and can be toxic are aluminum, cadmium, lead, and mercury.

Currently, perhaps because of available analytical methods, most of the trace elements characterized here are metals. The emphasis of this chapter is therefore on trace metals. Iodine metabolism is discussed in Chapter 18 and fluorine is discussed at the end of this chapter.

Characteristics of Trace Element Functions

Although many details of trace element functions are not yet understood, some general characteristics are well known. These are *amplification* of trace element action, *specificity, homeostasis*, and *interactions.*[16]

Amplification of Trace Element Action

Very small amounts of trace elements are necessary for optimal performance of an organism. Lack of even a small amount of a trace element (e.g., iron) can result in disease (anemia) seemingly disproportionate to the amount of element missing. The basis for *amplification* of trace element action is that trace elements are constituents of, or interact with, enzymes and hormones that regulate the metabolism of much larger amounts of biochemical substrates. If the substrates are also regulatory, the effect is even further amplified.

Specificity of Trace Elements

Essential trace elements are *specific* for their in vivo functions; they cannot be effectively replaced by chemically similar elements. The essential trace metals interact with atoms such as nitrogen, sulfur, and oxygen; the types of interaction are dependent on configurational preferences and bond types. Certain trace metals are stable in more than one valence state (e.g., Fe, Cu, Mo), allowing biological redox function, whereas others are stable in only a single state [e.g., Zn(II), Ni(II)]. Metals with partially filled *d* electron orbitals (e.g., Fe, Cu, Co) tend to coordinate with a larger number of atoms than metals with filled *d* orbitals (Zn, Se), and they exhibit qualitatively different preferences for the atoms. Specificity of trace element function is also promoted by specific carrier and storage proteins such as transferrin and ferritin for iron, albumin and α_2-macroglobulin for zinc, ceruloplasmin for copper, transmanganin for manganese,

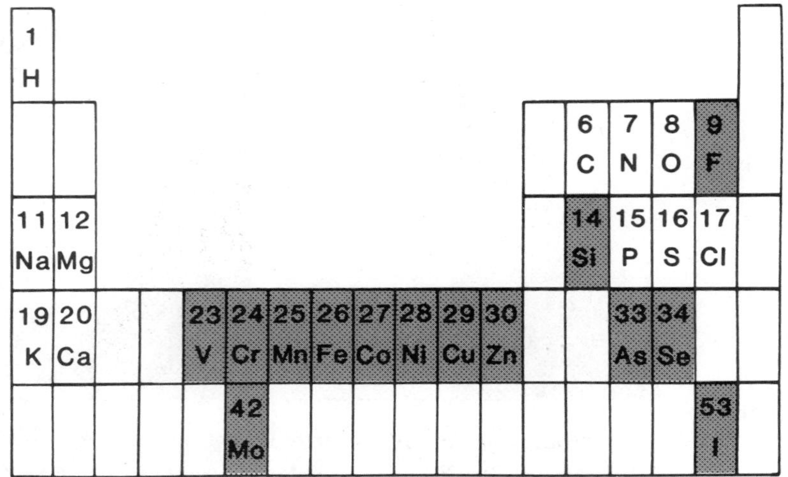

Figure 17-2. Chemical elements accepted as essential for human or animal nutrition (trace elements shaded).

and nickeloplasmin for nickel. These carrier proteins recognize and bind specific metals and transport them to, or store them at, specific sites within the organism.

Homeostasis of Trace Elements

Mechanisms that ensure optimal body distribution of an element over a range of intake levels constitute a system of *homeostatic regulation* for that element and include absorption, storage, and excretion. Although many details of trace element absorption are still unknown, the rate of absorption of a trace element generally decreases with its increasing concentration in the intestinal lumen. Active transport mechanisms involving specific binders, feedback inhibition, or both, have been postulated. Carrier and storage proteins have some capability to buffer against excess free metal levels, but they generally are not as important for regulation as are absorption control mechanisms.

The principal excretory route for elimination of trace metals is through the feces. Relatively small amounts of trace metals are excreted in the urine, but halides (iodide and fluoride) as well as selenium and chromium are effectively eliminated in the urine. The loss of trace elements through hair, skin cell desquamation, and sweat is generally of a minor degree.

Interactions of Trace Elements

An overabundance of one trace element can interfere with the metabolic utilization of another element present in normal or marginal concentrations.[13,17] Alternatively, the effect of a toxic trace element may be ameliorated by another "protective" trace element. The addition of large amounts of zinc to a diet interferes with the intestinal copper absorption to cause copper deficiency in spite of an otherwise adequate copper intake. Hence, zinc is said to *antagonize* copper absorption. Copper deficiency, in turn, is known to produce iron deficiency and anemia. Interactions that involve toxic elements include the increased retention of cadmium and lead that is associated with iron deficiency.

LABORATORY ASSESSMENT OF TRACE METAL STATUS

No completely satisfactory laboratory methods for the clinical assessment of trace metal status have yet been established. Currently the only definitive test of human trace metal deficiency is the clinical response to therapeutic supplementation with the trace metal in question. Measurements of metalloenzyme activities have been proposed as useful assessment tests because plasma trace metal concentrations are often affected by factors unrelated to the status of metal in the entire body. Although hair is relatively easy to analyze for trace metal content, its chief problem as a specimen for assessing metal nutriture is its susceptibility to environmental contamination. In the clinical setting it is most practical to assess trace metal status from analysis of a battery of tests performed on a single blood specimen.

Sample Collection and Testing

Accurate determination of trace metals in biological specimens presents special analytical difficulties and requires precautions. *Sampling* procedures must be carefully considered since heterogeneity of trace metal distributions in specimens is the rule rather than the exception.

Hair samples should be cut with stainless steel scissors to avoid contamination. The best source for hair is the occipitonuchal region, since this area is low in the back of the head, is less visible, and the hair is probably less contaminated. The proximal (scalp-end) segments of the hairs should be taken for analysis. Washing procedures have been developed to remove oils and exogenous surface metal contamination and leave all or most of the endogenous metal.[9]

The chief analytical problem encountered is external contamination. Many trace metals are prevalent in the laboratory environment in ng and even μg amounts; hence, an appreciable portion of the result of a trace metal analysis may be caused by contamination. Precautions must be taken through all sampling and analytical procedures to minimize contamination.

Common sources of trace metal contamination in analytical work include rubber, wood, paper products, metal surfaces, skin, dandruff, and hair. Equipment made from plastic, borosilicate glass, or stainless steel is best suited for trace metal analysis and, of these, the fluorocarbon, polypropylene, and polyethylene plastics are generally best. Stainless steel knives or scissors are suited for cutting specimens for determination of iron, zinc, and copper. Water should meet ACS specifications for Type I water. Glassware should be cleaned by soaking it overnight in dilute nitric acid or in diluted, commercial metal-scavenging solutions marketed for radioactive cleanup. The latter solutions are effective and are considerably safer and easier to handle than large volumes of dilute acids. Only evacuated blood collection tubes specified for trace metal analysis should be used for blood collection; other tubes may contaminate the specimen owing to the leaching of metals from the rubber stoppers. Plastic disposable syringes with stainless steel needles are also suitable. Other materials used in specimen collection or processing, such as clotting aids or gel serum separators, should be avoided or checked before use for absence of the trace metals of interest. The level of trace metals should be checked in all anticoagulants used, particularly heparin, which is derived from animal tissue sources. Common laboratory practices that can contribute to contamination include the use of wooden applicator sticks to ring blood clots off tube walls, the wiping of pipet tips with tissues, and the mixing or storing of specimens in rubber-stoppered tubes. Skin contact with any surface that will be exposed to specimens or solutions must be avoided.

Analytical Methods

Analytical methods for the determination of trace metals in biological specimens must be sensitive, specific, and accurate. Since these method requirements are not of the same importance for all metals and for all specimens, the analytical technique of choice is dependent on the specimen type and the metal to be determined.

The most popular of currently employed techniques for trace metal analysis include atomic absorption spectrophotometry (AAS), emission spectroscopy (ES), neutron activation analysis (NAA), colorimetry, mass spectrometry (MS), and, to a lesser extent, X-ray fluorescence spectrometry (XRF) and electrochemical techniques such as anodic stripping voltammetry (ASV). The chief drawbacks to ES, NAA, MS, and XRF techniques are the complexity and expense of equipment and materials and the high level of operator skill required. For this reason, these techniques are generally not practical for most clinical diagnostic laboratories. Electrochemical and ASV techniques are highly sensitive but have been applied to only a few metals in biological specimens (primarily to zinc, copper, cadmium, and lead). The matrix-sensitivity of these techniques complicates determinations in biological specimens, and this reduces sample throughput considerably. These constraints leave AAS and colorimetry as alternative methods for trace metal determinations. However, because nearly all *d*-type transition elements are active in forming colored complexes, colorimetric methods are often not completely specific for a particular metal.

Atomic Absorption Spectrophotometry

Atomic absorption spectrophotometry (AAS) has supplanted colorimetric techniques as the method of choice for trace metal determination. The AAS technique detects only one metal at a time, and some specimen preparation work is often necessary. Metal determination, however, is quick and accurate once the specimen is in the proper form. Besides offering good sensitivity and specificity, AAS instruments are readily available and relatively simple to operate. Well-tested AAS procedures for the determination of most trace metals in biological specimens are available from either instrument manuals or the general scientific literature. The principles of AAS are detailed in Chapter 2, Section One and other sources.[4,24,25]

Methods in which diluted serum or plasma is aspirated directly into the AAS flame are effective for the determination of magnesium, zinc, and copper. Discrete nebulization techniques minimize sample aspiration waste and allow metal determinations on microliter volumes of serum.[14,30] Electrothermal or "flameless" AAS micromethods, using as little as 10 μL of serum, have also been reported.[7,29] Serum dilutions of tenfold or more decrease the problems of clogging nebulizers and burners, but they result in poor sensitivity. Therefore, sample dilutions of fivefold or less are recommended. Differences in aspiration rates due to viscosity differences in aqueous standard solutions versus diluted serum specimens can be overcome by the use of glycerol diluents for the standards.[3]

In order to obtain reproducible AAS results it is most important to maintain consistently optimal aspiration and flame conditions. This is especially true when solutions with high solid concentrations (rather than protein-free or digested specimens) are aspirated. Dilute hydrochloric acid should be periodically aspirated to help keep the nebulizer system clean. The nebulizer system should be taken apart periodically and cleaned. The burner head should be soaked overnight in a nonacidic cleaner and cleaned thoroughly before every run. The inside of the burner orifice should be scraped with a thin metal file to dislodge aggregated solids and maintain smooth surfaces on the orifice. An extra burner head and nebulizer system should be kept on hand for installation when unexplained variability in results occurs.

The sensitivity of the AAS technique for determining various metals depends on the particular metal of interest and on the technique used. The flame AAS mode is simpler and less tedious to perform than the flameless mode, although it is less sensitive. Generally, if metal concentrations of a specimen are below 50 ng/g, flameless AAS techniques are necessary. *Atomic fluorescence spectrometry*, a technique somewhat analogous to AAS, can be performed on many commercial AAS instruments with certain optional equipment. The technique is considerably more sensitive than AAS for some metals, such as zinc and especially cadmium.

Quality Control of Trace Metal Determinations

In general, control or reference materials that contain certified levels of trace elements are not widely available. A convenient and reliable approach for control of biological fluid assays (blood, urine, saliva) is the use of frozen aliquots of a specimen pool. Aliquots should be stored frozen in tightly capped polypropylene tubes with a minimum of head space. Frozen pool aliquots can be used to assess the day-to-day precision of methods. Recovery studies of metal from specimens containing known quantities of added analyte are useful initially for assessing accuracy and linearity of the method.

Because standard hair reference material is not available, several laboratories have prepared homogeneous hair "pools" that can be used as controls for hair analysis and for assessing precision and accuracy through standard recovery tests. A hair pool can be prepared by cutting several grams of hair as finely as possible with stainless steel scissors, washing successively with acetone-ether-detergent, drying, and storing in a desiccator.

For quality control of trace metal determinations in tissue and serum specimens, some NBS Standard Reference Materials are available (US Department of Commerce, National Bureau of Standards, Office of Standard Reference Materials, Washington, D.C. 20234).

MAGNESIUM

Magnesium (Mg) is not a true trace element; along with potassium, magnesium is a major intracellular cation.

Biochemistry and metabolism. As an alkaline earth metal, magnesium has chemical properties distinctly different from those of the transition metals. Unlike the transition metals, magnesium interacts with other chemical species with much stronger electrostatic bonding and a relative preference for oxygen over nitrogen atoms. Magnesium is an essential factor in many important enzymatic reactions, either as an integral part of a metalloenzyme or as an activator. Magnesium is mostly in bone (~55%, which is available for mobilization during deficiency) and muscle (~27%).

The Recommended Dietary Allowance (RDA) for magnesium is listed in Table 17-1. The magnesium content of food varies widely. Seafood, meats, nuts, and grains contain appreciable amounts, whereas oil, fats, and sugars contain very little.

Magnesium is absorbed in the upper intestine. The major proportion of ingested magnesium (60–70%) is not absorbed but is excreted in the feces. Unlike calcium absorption, magnesium absorption is not vitamin D dependent, but it can be affected by the amount of calcium, phosphate, protein, lactose, or alcohol ingested. Intestinal absorption, together with an effective renal regulation of urinary excretion, allows homeostasis of magnesium over a wide range of dietary intakes. The major part of magnesium in plasma (~65–70%) exists as free ions or in the form of small-molecular-weight complexes; the remainder is protein bound, mostly to albumin.

Clinical significance. The best-defined manifestation of magnesium deficiency is impairment of neuromuscular function (e.g., hyperirritability, tetany, convulsions, and electrocardiographic changes). In deficiency states, levels of serum magnesium have been found to be reduced much sooner than erythrocyte magnesium levels. Symptoms of magnesium *deficiency* usually do not occur until serum magnesium levels decrease below

Table 17-1. RECOMMENDED DIETARY INTAKES OF ESSENTIAL TRACE ELEMENTS*

A. Recommended Dietary Allowances (RDA) †

	Age, Years	Magnesium, mg	Iron, mg	Zinc, mg	Iodine, μg
Infants	0.0–0.5	50	10	3	40
	0.5–1.0	70	15	5	50
Children	1–3	150	15	10	70
	4–6	200	10	10	90
	7–10	250	10	10	120
Males	11–14	350	18	15	150
	15–18	400	18	15	150
	19–22	350	10	15	150
	23–50	350	10	15	150
	51+	350	10	15	150
Females	11–14	300	18	15	150
	15–18	300	18	15	150
	19–22	300	18	15	150
	23–50	300	18	15	150
	51+	300	10	15	150
Pregnant		+150	‡	+5	+25
Lactating		+150	‡	+10	+50

B. Estimated Adequate and Safe Daily Dietary Intakes §

	Age, Years	Copper, mg	Manganese, mg	Fluoride, mg	Chromium, mg	Selenium, mg	Molybdenum, mg
Infants	0–0.5	0.5–0.7	0.1–0.5	0.1–0.5	0.01–0.04	0.01–0.04	0.03–0.06
	0.5–1	0.7–1.0	0.7–1.0	0.2–1.0	0.02–0.06	0.02–0.06	0.04–0.08
Children and adolescents	1–3	1.0–1.5	1.0–1.5	0.5–1.5	0.02–0.08	0.02–0.08	0.05–0.10
	4–6	1.5–2.0	1.5–2.0	1.0–2.5	0.03–0.12	0.03–0.12	0.06–0.15
	7–10	2.0–2.5	2.0–3.0	1.5–2.5	0.05–0.20	0.05–0.20	0.10–0.30
	11+	2.0–3.0	2.5–5.0	1.5–2.5	0.05–0.20	0.05–0.20	0.15–0.50
Adults		2.0–3.0	2.5–5.0	1.5–4.0	0.05–0.20	0.05–0.20	0.15–0.50

*From Food and Nutrition Board, National Research Council; Recommended Dietary Allowances, 9th ed. Washington, D.C., National Academy of Sciences, 1980.

†The RDA are the daily levels of intake of essential nutrients considered to be adequate to meet the known nutritional needs of practically all healthy persons.

‡Supplemental iron (30–60 mg) is recommended during pregnancy and for 2–3 months after parturition.

§Estimated ranges of adequate but safe intakes of nutrients for which RDA cannot be established on the basis of present knowledge. Since toxic levels for some of the elements are only several times more than usual intakes, the upper levels of estimated intakes should not be greatly and habitually exceeded.

0.5 mmol/L. Human magnesium deficiency, as indicated by reduced serum magnesium levels (hypomagnesemia), has been shown to occur with either normal or reduced serum calcium levels. A hypomagnesemic-normocalcemic tetany has been described and is effectively treated with magnesium supplementation alone. During this type of tetany, serum magnesium levels of 0.15–0.5 mmol/L, accompanied by normal serum calcium concentrations and pH, have been reported. Tetany due to hypocalcemia *and* hypomagnesemia cannot be treated optimally with calcium alone. Decreased serum potassium concentrations (hypokalemia) have also been found to accompany magnesium depletion, and the occurrence of otherwise unexplained hypokalemia and hypocalcemia should suggest magnesium deficiency.

Conditions that have been associated with *hypomagnesemia* include chronic alcoholism, childhood malnutrition, lactation, malabsorption, acute pancreatitis, hypoparathyroidism, aldosteronism, digitalis intoxication, and prolonged intravenous feeding. Magnesium depletion occurs in conditions that disrupt the normal renal conservation of magnesium, such as in patients with renal tubular reabsorption defects and those taking chlorothiazides, ammonium chloride, or mercurial diuretics for congestive heart failure.

Increased serum magnesium concentrations have been observed in dehydration, severe diabetic acidosis, and Addison's disease. Conditions that interfere with glomerular filtration (e.g., renal failure) result in retention of magnesium and hence elevation of serum levels.

Determination of Magnesium in Body Fluids

Specimens. Serum magnesium determinations should be performed on specimens drawn without venous stasis. Since the magnesium concentration in erythrocytes is substantially greater than that in serum, the specimen should be separated from the erythrocytes as soon as possible, and hemolyzed specimens should not be used for analysis. Magnesium levels in serum are stable for several days if the specimen is stored in the refrigerator, after it has been separated from the erythrocytes.

Methods. Atomic absorption spectrophotometry (AAS) is the preferred technique for the determination of magnesium in biological specimens. Some of the analytical difficulties are similar to those discussed under calcium analysis (see Chapter 21). La(III) or Sr(III) is added to the sample diluent to bind with phosphates that otherwise would form refractory calcium and magnesium compounds in the AAS flame and result in falsely low values. Some methods involve release of magnesium from proteins by treatment with hydrochloric or trichloroacetic acid followed by centrifugation.

Direct dilution techniques are fast and have been found to be precise and accurate, provided proper care is taken to maintain burner and nebulizer performance. Because relatively large sample dilutions are made for magnesium determinations, the problem of nebulizer or burner clogging is minor. Serum specimens are diluted 50-fold with an acidic lanthanum chloride or oxide diluent and aspirated directly into the AAS flame without further treatment. Serum magnesium levels are calculated from a linear calibration curve of standards made up with the same lanthanum diluent. Sodium and potassium salts are added to the standard solutions and the baseline-zeroing solution to compensate for spectral interferences contributed by the same salts in the diluted serum samples.

The above method is primarily that of Bowers and Pybus[2] for determining serum calcium, but the method is equally applicable to serum magnesium. Hence, both calcium and magnesium can be determined on the same serum dilution.

Alternative methods. Many *fluorometric* (8-hydroxyquinoline) and *colorimetric* methods have been used for determining serum magnesium.[26] For laboratories without an AAS instrument, the Calmagite colorimetric method has been adapted for both manual[1] and automated[5] use. The colored Calmagite-Mg complex is formed immediately upon mixing of Calmagite reagent with plasma and is stable for over 30 min. This method is simple and rapid. Between-batch precision (CV) is 3%, and method results correlate well with results from an atomic absorption procedure.

Reference ranges. The reference range for magnesium in *serum* with the atomic absorption method is 0.6–1.1 mmol/L. Serum levels in newborns are lower than in adults (0.5–0.9 mmol/L).[27] Infants older than five months, children, and adolescents have levels essentially the same as adults. Levels do not change appreciably throughout the day, but in females the highest serum values are obtained at the time of menses.

Erythrocytes contain about 1.7–2.7 mmol/L. The concentration range in *cerebrospinal fluid* is 1.0–1.4 mmol/L. The *urinary excretion* is 3.0–5.0 mmol/d; values for males are slightly higher than those for females.

IRON

Iron (Fe) is the most important of the essential trace metals. An appreciable number of human diseases are related to iron deficiency or disorders of iron metabolism. These topics as well as methods for the determination of iron are covered in Chapter 24, *Biochemical Aspects of Hematology*. The RDA for iron is listed in Table 17-1.

ZINC

Zinc (Zn) is second only to iron in importance as an essential trace metal and is present in abundance in the body. Since zinc deficiency was first described in the early 1960s, research has elucidated many specific metabolic interactions of zinc. The discovery of a variety of zinc-related clinical disorders has directly demonstrated the importance of zinc in human nutrition.

Biochemistry and metabolism. The metabolic functions of zinc are largely due to its presence in zinc metalloenzymes. Important zinc metalloenzymes in humans include carbonic anhydrase, alkaline phosphatase, RNA and DNA polymerases, thymidine kinase, carboxypeptidases, and alcohol dehydrogenase. The zinc atom is an integral, firmly bound part of the metalloprotein molecule and is often directly involved in the active catalytic site; it contributes also to the structural stability of metalloenzymes.

Zinc absorption occurs mostly in the duodenum and proximal jejunum. The absorption process is active, energy dependent, and apparently mediated by specific zinc transport (binding) ligands. Tissues and fluids especially rich in zinc are prostate, semen, liver, kidney, retina, bone, and muscle.[8] The zinc concentration of erythrocytes is about ten times that of plasma because of their rich content of carbonic anhydrase and other zinc metalloenzymes.

High-protein foods such as meat, fish, and dairy products are good sources of available zinc. The *bioavailability* of zinc from vegetables and cereal grains is reduced because phytates (inositol phosphates) and other dietary fiber components inhibit zinc absorption. Western diets generally supply adults with 10–15 mg Zn/d.

Clinical significance. Human *zinc deficiency* is due to low-zinc diet in which a high fiber content has decreased the availability of zinc for intestinal absorption. Primary clinical features include retardation of growth and skeletal maturation, testicular atrophy, hepatosplenomegaly, and reduced taste acuity. Other manifestations of human zinc deficiency include susceptibility to infection, impaired healing of wounds, sores and ulcers, scaly dermatitis, and diarrhea. In most reported clinical cases, the physical sequelae of zinc deficiency are rapidly cleared by administration of dietary zinc sulfate supplements, not to exceed two or three times the 15 mg/d RDA.

Many diseases and medical treatments may lead to *conditioned zinc deficiency*. Among these are hepatic cirrhosis, gastrointestinal disorders, renal disease, neoplastic and inflammatory disease, and severe burns.[21] *Iatrogenic causes* of conditional zinc deficiency include administration of anabolic and metal-chelating drugs such as corticosteroids and penicillamine, respectively, and synthetic diet therapies. Synthetic oral diets and total parenteral alimentation fluids are sometimes deficient in trace metals and have been shown to produce zinc and other trace metal deficiencies in patients receiving long-term treatment. Adequate zinc nutrition is especially important for the postsurgical patient because zinc is needed for optimal wound healing and immune function.

Pregnant women are at higher risk of acquired zinc deficiency due to the uptake of zinc by the fetus and associated tissues. The use of oral contraceptives produces a decrease in plasma zinc and an increase in erythrocytic zinc.

The most clearly defined genetic disorder of zinc metabolism is *acrodermatitis enteropathica*. Manifestations of the disease are those of zinc deficiency. The patients exhibit lowered plasma zinc levels, and symptoms are completely cleared with zinc sulfate supplementation.

Laboratory Assessment of Zinc Nutriture

Laboratory tests for assessing zinc nutriture can be classified into two groups: those involving the determination of zinc in a body tissue or fluid and those testing a *zinc-dependent function*. Useful tests in the former group are determinations of zinc in plasma or serum, hair, urine, erythrocytes, and saliva, whereas functional tests include measurement of zinc metalloenzyme activities and assessment of taste acuity.[23] All of the indices decrease with zinc deficiency except ribonuclease activity, which increases.

Although the useful tests mentioned above have been shown to reveal zinc deficiency in animals and humans, no one test has proved to be a definitive indicator of zinc status. Plasma zinc determination is the best single choice, whereas serum alkaline phosphatase or erythrocyte zinc measurements may also be useful. However, test results must be interpreted with caution, since they may be confounded by clinical conditions unrelated to the subject's zinc status per se.

Circulating zinc levels in plasma and serum have often been shown to indicate human zinc deficiency, but they do not accurately reflect whole body zinc status in all cases. Circulating zinc levels closely correlate with the major carrier protein, albumin. Thus, lowered plasma zinc levels observed in hypoalbuminemic conditions, such as hepatic cirrhosis and malnutrition, may reflect depressed plasma protein binding of zinc. Circulating zinc levels are also decreased by the action of bacterial endotoxin and by various forms of steroid regimens (e.g., oral contraceptives).

Of the many zinc-dependent enzymes, alkaline phosphatase, carbonic anhydrase, lactate dehydrogenase, and ribonuclease are the most consistently useful indices of zinc deficiency. Depression of serum alkaline phosphatase activity has been observed in a variety of human zinc-deficient conditions. However, a decrease in alkaline phosphatase levels is nonspecific and is affected by many metabolic conditions unrelated to zinc status.

Methods for the Determination of Zinc in Plasma or Serum

Specimens. Precautions must be taken throughout all procedures to avoid zinc contamination during specimen collection and from reagents. Collect at least 3 mL of blood by venipuncture; avoid hemolysis since erythrocytes contain at least ten times more zinc than plasma. The specimens may be stored in the refrigerator for several days but should be frozen at −20 °C for long-term storage.

Methods. The determination of plasma or serum zinc concentration by *atomic absorption spectrophotometry* (AAS) is the simplest and analytically most reliable test for the routine assessment of zinc nutriture. A fivefold diluted specimen is aspirated and the zinc level is determined by comparing the signal from the diluted specimen with the signal from aqueous standards prepared in a matrix of glycerol (5 mL/dL) to simulate the viscosity of diluted plasma. The method, which requires a 500-μL specimen, has been published as a proposed selected method.[22] Routine quality control of plasma zinc measurements is accomplished by including aliquots of a frozen plasma pool in each analytical batch.

Colorimetric methods are also available. The method using 4-(2-pyridylazo)-resorcinol is simple and rapid and gives results comparable with those obtained by AAS.[11] The day-to-day coefficient of variation is 2%. A specimen of 200 μL is required. Metals are released from serum proteins with guanidine hydrochloride, and cyanide is added to complex all metals except zinc, which is selectively demasked by adding chloral hydrate.

Other methods include anodic stripping voltammetry, emission spectroscopy, and neutron activation analysis.

The *reference range* for zinc in plasma is 70–150 μg/dL. Plasma zinc levels exhibit both circadian and postprandial fluctuations. The levels decrease after eating and from morning to evening. Fasting morning values below 70 μg/dL represent zinc-deficient status.

COPPER

Biochemistry and metabolism. Copper (Cu) is an integral component of many metalloenzymes, including cytochrome oxidase, superoxide dismutase, uricase, dopamine-β-hydroxylase, lysyl oxidase, ceruloplasmin, and tyrosinase. The important protein metallothionein binds copper as well as other heavy metals. A variety of pathological conditions have been attributed to loss of cuproenzyme activity; they include failure of pigmentation, connective tissue cross-linking defects, ataxia, and abnormal catecholamine conversions. Copper deficiency impairs iron absorption, and anemia accompanies severe copper deficiency. *Ceruloplasmin*, the major plasma copper protein, is a ferroxidase that oxidizes ferrous iron to the ferric state prior to its binding by plasma transferrin (see also Chapters 11 and 24).

Copper absorption in humans occurs primarily in the stomach and duodenum. The amount of orally ingested copper that is absorbed is variable. Copper absorption may be impaired in patients with diffuse small bowel disease. Absorbed copper binds to albumin and is transported to the liver, where it is stored. Most copper is released as ceruloplasmin, which contains 90–95% of circulating plasma copper. Since nearly all plasma copper is present as ceruloplasmin, levels of plasma ceruloplasmin and copper almost always change proportionately.

Copper is excreted primarily in the feces, mostly from unabsorbed dietary copper and from biliary and gastrointestinal secretions. A small amount of copper, 0.3–3% of dietary intake, is excreted in the urine.

Most populations consume an adequate amount of copper, and reported cases of frank dietary copper deficiency in humans are rare. The copper content of foods is variable. Liver, crustaceans, and shellfish contain large amounts of copper. Cow's milk has a low copper content; therefore, infants solely on a milk diet may develop copper deficiency and anemia. Western diets generally supply 2–4 mg/d. Estimated "adequate and safe" intakes of copper are shown in Table 17-1.

Clinical significance. Copper deficiency in infants has been observed in prematurity, malnutrition, malabsorption, chronic diarrhea, hyperalimentation, and prolonged feeding with low-copper, total-milk diets. Symptoms of copper deficiency include (1) neutropenia and hypochromic anemia in the early stages, both of which are responsive to oral copper but not iron; (2) osteoporosis and various bone and joint abnormalities; (3) decreased pigmentation of the skin and general pallor, attributed to loss of tyrosinase required for melanin synthesis; and (4) in the later stages, possible neurological abnormalities. Copper deficiency symptoms in infants are generally reversed by copper supplementation. The levels of plasma copper and ceruloplasmin directly reflect the degree of copper deficiency and repletion. The monitoring of body copper status or copper intake or both in the premature infant is warranted.

Although the need for trace metals in parenteral fluids is obvious, recommendations on the optimal amounts have not been established. A minimum parenteral intake of copper of 0.3 mg/d for adults is recommended.[15] Long-term hyperalimentation with infusates deficient in trace metals has been shown to produce copper deficiency in both infants and adults. Observed symptoms are those usually seen in copper deficiency. Human copper deficiency has also been associated with zinc therapy and treatment with copper-chelating agents such as penicillamine.

Menkes' syndrome. A genetically determined defect in copper transport and storage underlies the rare Menkes' *steely-hair syndrome*. Clinical manifestations include kinky or steely hair, depigmentation of the skin and hair, hypothermia, seizures, cerebral degeneration, and vascular defects. The formation of steely hair is attributed to loss of copper-catalyzed disulfide bond formation. The affected infants have been shown to be copper deficient and to have low serum, hepatic, and cerebral copper levels; low ceruloplasmin; and little or no cytochrome oxidase activity in nerve tissue. Symptoms appear usually by three months of age, and death occurs by five years.

Wilson's disease (hepatolenticular degeneration) is a genetically determined copper accumulation disease that usually becomes manifest between 6 and 20 years of age; it has a prevalence of about 1:200 000 (see also Chapter 22). Copper accumulates in the liver, brain, kidney, and cornea, and urinary excretion of copper is increased. The basic defect in Wilson's disease is unknown. Plasma levels of ceruloplasmin are decreased; however, some patients exhibit normal levels. There appears to be a reduction in the rate of incorporation of copper into ceruloplasmin and in biliary excretion of copper. Although free and albumin-bound copper is increased, total serum copper is generally decreased because of low ceruloplasmin levels. Thus, simultaneous determinations of serum ceruloplasmin, serum copper, and urinary copper are useful in the diagnosis of this disease. Disease symptoms include neurological disorders, cirrhosis of the liver, and Kayser-Fleischer rings (green-brown deposits) in the cornea. The disease can be controlled, and in some cases halted, by early and persistent treatment with copper chelators.

Copper toxicity, characterized by nausea, vomiting, epigastric burning, and diarrhea, can occur subsequent to ingestion of copper-contaminated solutions, the use of copper-containing intrauterine devices, the use of copper salts in animal feeds, and exposure to copper-containing fungicides. Systemic toxic effects include hemolysis, hepatic necrosis, gastrointestinal bleeding, oliguria, azotemia, hemoglobinuria, hematuria, proteinuria, hypotension, tachycardia, convulsions, coma, and death.

Abnormal Copper Metabolism Associated with Other Conditions

Estrogens increase serum copper levels, probably by increasing hepatic ceruloplasmin synthesis. Serum copper is normally higher in women than in men, and this difference increases further during pregnancy and in women taking estrogenic oral contraceptives. Abnormal copper levels seen in diseases of the pituitary-adrenal axis include elevated plasma copper and decreased urinary copper in Addison's disease and hypopituitarism. The reverse is found in Cushing's syndrome and adrenal cortical carcinomas. During infections or inflammatory stress, serum copper levels rise owing to the acute-phase reaction.

Since the liver is the important organ for copper storage and regulation of homeostasis, abnormal copper metabolism is common with liver disease. Elevated serum copper levels are found in cases of hepatitis and biliary tract disease; hypocupremia has been observed in patients with hemolytic jaundice, hemochromatosis, and some types of hepatic cirrhosis.

Laboratory Assessment of Copper Nutriture

Measurements of plasma or serum copper levels provide the best routine test for the clinical assessment of copper nutriture. As in the case of zinc, factors not related to copper nutriture can affect plasma copper levels, including the amount of binding protein (apoceruloplasmin) and the effect of hormones, both endogenous (as in pregnancy) and exogenous. Plasma copper levels exhibit diurnal variation, with the highest levels in the morning.

Measurements of hair copper, urinary copper, and cuproenzyme activities are also useful for indexing copper deficiency.[23] However, urinary copper levels in infants with Menkes' syndrome may be normal. Only the most proximal hair segments should be analyzed as a test for copper nutriture. The best cuproenzyme tests for assessing copper status are erythrocyte superoxide dismutase and leukocyte cytochrome oxidase.

Methods for the Determination of Copper

Specimens. Plasma or serum may be used; hemolysis is not a great concern, since levels of copper in plasma and erythrocytes are nearly equal.

Method. *Atomic absorption spectrophotometry (AAS)* is the method of choice for determining serum copper. Specimens diluted with an equal volume of deionized water are aspirated directly into the AAS flame, and the copper concentrations are calculated against copper standards in a glycerol matrix (10 mL/dL) that approximates the viscosity of the diluted specimens. The procedure is similar to that for plasma zinc determination by AAS; however, because the sensitivity of AAS for copper is less than for zinc, a twofold rather than a fivefold dilution of serum is recommended. Aspiration of the more concentrated sample solution requires more frequent cleaning of the burner orifice or use of a high-solids (Boling) burner to avoid clogging and erratic absorbance readings.

Despite claims that *urinary* copper can be determined by direct aspiration of acidified (or even diluted) specimens into the AAS flame, the sensitivity of flame AAS is not sufficient to differentiate between normal and subnormal urinary copper levels. Colorimetric methods for serum copper are also available, such as those using bis-cyclohexanone oxaldihydrazone (Cuprizone)[20] or bathocuproine[12,31] as chromogens.

Reference ranges for serum are higher in pregnancy (120–300 μg/dL), in children 6–12 years (80–190 μg/dL), and in adult females (80–155 μg/dL) than in adult males (70–140 μg/dL) and in infants (20–70 μg/dL).

Serum ceruloplasmin can be determined by a colorimetric copper oxidase reaction, by immunonephelometry, or by radial immunodiffusion (see also Chapter 11). The *reference range* for adults is 15–60 mg/dL.

MANGANESE

Biochemistry and metabolism. The biochemical basis for manganese (Mn) essentiality is its role as a cofactor in certain enzymatic reactions. Manganese metalloproteins include avimanganin, manganin, and concanavalin A as well as the enzymes pyruvate carboxylase and superoxide dismutase. The estimated adequate and safe intake of manganese is 2.5–5.0 mg/d for adults (see Table 17-1). *Chronic manganese poisoning* has occurred in workers with chronic exposure. The symptoms are schizophrenia-like psychiatric effects and neurological disorders clinically similar to those of Parkinson's disease.

Unlike iron, zinc, and copper, which are considered essential because clinical evidence of deficiency exists, manganese is accepted as essential for humans on the basis of its proved role in manganese-dependent enzymes rather than on direct evidence of human deficiency. Serum manganese levels are increased after industrial exposure, acute hepatitis, and myocardial infarction. Increased erythrocyte levels have been found in patients with rheumatoid arthritis.

Laboratory assessment of manganese status. Laboratory tests that reliably assess body manganese status have not been established. Widely varying levels of manganese in blood have been reported. Adult *reference ranges* for manganese are 0.4–2.4 μg/dL for whole blood, 0.04–1.4 μg/dL for serum, 0.5–4.6 μg/dL for erythrocytes, 1–50 μg/L for urine, and 0.2–2.1 μg/g for hair.[8,27] The large variation in reported manganese levels can be partly attributed to specimen contamination during collection or processing, the use of older analytical methods that were relatively insensitive and nonspecific, the use of different analytical methods, or any combination of these factors.

The most popular analytical techniques for determining manganese in biological specimens are neutron activation analysis (NAA), emission spectroscopy (ES), and atomic absorption spectrophotometry (AAS). Spectrophotometric and electrochemical methods have also been used. For the clinical laboratory, the AAS technique is the method of choice. Because of the low levels of manganese in blood, hair, and urine, the more sensitive flameless AAS mode should generally be used.[25,32]

MOLYBDENUM

The essentiality of molybdenum (Mo) for animals and humans is based on its incorporation into three metalloenzymes: xanthine oxidase, aldehyde oxidase, and sulfite oxidase.

As with other ultratrace metals, the wide variation in reported molybdenum levels for tissues and body fluids may be due largely to contamination and the use of varied analytical techniques, rather than to true biologic variation. Reported *reference ranges* for healthy adults are 1–15 ng/mL for whole blood, 18 ng/mL (mean) for erythrocytes, 0.1–6 ng/mL for plasma or serum, 10–16 μg/L for urine excretion, and 0.06–0.20 μg/g for hair.[8]

No well-defined cases of human molybdenum deficiency have been reported. Molybdenum is relatively nontoxic to humans. High activities of xanthine oxidase associated with elevated molybdenum levels may lead to high serum uric acid levels and a predisposition to gout. Though accepted as an essential element, molybdenum does not appear to be directly related to any clinically observed conditions. Methods for the determination of molybdenum in biological specimens use emission spectroscopy, neutron activation analysis, and atomic absorption spectrophotometry techniques.

CHROMIUM

Biochemistry. Chromium (Cr) is accepted as essential only because insulin resistance is a consequence of chromium deficiency, and insulin appears to be ineffective as a glucose regulator without chromium. The agent that potentiates the action of insulin has been isolated from brewer's yeast and kidney powder and has been termed *glucose tolerance factor* (GTF). GTF is an organic low-molecular-weight complex containing trivalent chromium. The biosynthesis and structure of GTF have not been elucidated.

Metabolism and clinical significance. The site and mechanism of intestinal chromium absorption in humans have not been determined. A small portion of circulating blood chromium is apparently bound to a protein in the β-globulin fraction, whereas the major portion is presumed to be bound to GTF. The regulation of chromium homeostasis is still obscure. Most orally absorbed chromium is excreted by the kidneys in an unknown form.

The total content of chromium in foods has been studied but is of limited value, since chromium absorption and its biological activity are highly dependent on the chemical form. Estimated adequate and safe dietary intakes of chromium are 50–200 μg/d for adults (see Table 17-1).

Trivalent chromium has a low order of *toxicity*. An apparently wide margin of safety exists between toxicity and supplementation to cure deficiency. Chromium(VI) is considerably more toxic than Cr(III). Acute exposure may produce allergic reaction, conjunctivitis, nasal septum ulceration, dermatitis, edema, and ulcer. Chronic exposure may produce gastrointestinal symptoms, hepatitis, and lung cancer.

The clinical significance of chromium rests primarily on its relation to glucose metabolism. In the US, simple addition of 150 μg of Cr(III) to the daily diet improved glucose tolerance of maturity-onset diabetics and middle-aged and elderly subjects. The potentiation of insulin action due to dietary supplementation with GTF (as a component of brewer's yeast) has been demonstrated in several human studies.[18] Deficiencies of chromium have been documented in two cases of total parenteral feeding.[6,10]

Methods for the Determination of Chromium Status

No laboratory tests that reliably define body chromium status have been established. The quantitation of chromium in human tissues and fluids has proved to be among the most difficult of trace metal determinations. *Reported levels* of chromium in human specimens vary widely: 0.08–45 μg/L in serum, 19–66 ng/mL in erythrocytes, 0.7–28 ng/mL in whole blood, 0.1–3.6 μg/g in hair, and <1 μg/d in urine.[8,27] The commonly used instrumental techniques for determining chromium in biological specimens are neutron activation as well as emission and atomic absorption spectrophotometry (AAS). The flameless AAS technique is recommended as the most practical for clinical laboratories because of the very low levels of chromium in biological specimens. A National Bureau of Standards (NBS) Reference Material (SRM #1659), brewer's yeast, contains a volatile organochromium component and has a certified value for chromium content.

This material is recommended as a reference standard for assessing accuracy of methods for chromium determination.

COBALT

Cobalt (Co) is essential for humans only as an integral part of vitamin B_{12} (cobalamin); no other function for cobalt in the human body is known. Details of vitamin B_{12} biochemistry and function in humans are discussed in Chapter 24. Microflora of the human intestine cannot utilize cobalt to synthesize physiologically active cobalamin. The human vitamin B_{12} requirement must be supplied from the diet. Free (nonvitamin B_{12}) cobalt does not interact with the body vitamin B_{12} pool. Because cobalt itself is nonessential, mechanisms for homeostatic control of body cobalt levels apparently have not evolved.

Relatively large doses of cobalt can be ingested without appreciable ill effects. However, *toxic effects* due to excess cobalt intake have been noted in renal patients receiving cobalt as an erythropoietic agent and in heavy drinkers of beer containing cobalt as a foam stabilizer. In both cases, cardiomyopathy resulting in death was observed.

Determination of ultratrace levels of cobalt in human body fluids and tissues has proved to be fraught with analytical difficulty. Emission spectroscopy, neutron activation analysis (NAA), and flameless AAS provide the required sensitivity and have been the most commonly used analytical techniques.

SELENIUM

Biochemistry and physiology. The human essentiality for selenium (Se) is based on its incorporation into the enzyme *glutathione peroxidase* (GSH-Px). Selenium has no other confirmed function in humans. Glutathione peroxidase catalyzes the breakdown of hydrogen peroxides and lipid hydroperoxides in body tissues and fluids, thereby protecting against oxidative damage to body tissues. High activity of GSH-Px in phagocytic cells, such as leukocytes and macrophages, helps to protect the phagocytic cell itself from peroxides it generates during oxidative destruction of foreign matter. Reduced GSH-Px activity in platelets has been linked to bleeding disorders; reduced activity in plasma has been linked to edema due to peroxidation of capillary membranes. Selenium, vitamin E, glutathione, and sulfur-containing amino acids all provide some protection against hepatic damage due to oxidative stress, such as occurs in carbon tetrachloride poisoning.

Metabolism and clinical significance. Little is known about the site and mechanism of gastrointestinal absorption of selenium. Urinary excretion appears to correlate well with dietary intake and plays a major role in regulation of selenium homeostasis. In blood, selenium is preferentially taken up by erythrocytes, whereas plasma selenium is bound less avidly by protein sulfhydryl groups.

Selenium content in foods is directly related to the protein content and the selenium content in the soil from which the food was produced. The estimated adequate and safe dietary intakes of selenium are listed in Table 17-1. Effective selenium nutriture is also dependent on the bioavailability of the dietary selenium.

Selenium *poisoning* is well known in animals, but few cases of human toxicity have been reported. There are no well-defined cases of human *disease* due to selenium deficiency except in China, where it has been associated with Keshan's disease, a specific type of cardiomyopathy. A number of clinical disorders have been shown to involve glutathione peroxidase (GSH-Px) deficiency, among them hemolytic anemia, a platelet disorder, a chronic granulomatous disease, and carcinoma.[19]

Determinations of urinary and blood selenium are useful measures of human selenium status. GSH-Px activity in erythrocytes has been shown to correlate with blood selenium up to 100 ng/mL and is a useful functional test of selenium status. The selenium *concentrations* reported in specimens from healthy adults are 57–340 ng/mL in whole blood, 71–340 ng/mL in erythrocytes, 78–320 ng/mL in serum or plasma, 5–100 μg/L in urine, and 0.6–2.6 μg/g in hair.[8,27] *Low* blood selenium concentrations have been observed in patients with gastrointestinal cancer, during pregnancy, and in protein-calorie malnutrition, whereas *high* values have been reported in patients with reticuloendothelial neoplasia. Recommended methods for determining selenium in biological specimens are flameless AAS[28] and spectrofluorometry.

FLUORIDE

Interest in human fluoride (F) status first resulted from the discovery in 1931 of chronic, endemic fluoride toxicity in several countries. Earlier, it was observed that children with mottled enamel (a sign of fluoride excess) had very little dental decay. Since then, numerous clinical and epidemiological studies have demonstrated the unique property of fluoride in preventing tooth decay. The exact form of fluoride treatment that provides maximal anticariogenic effect and the mechanism for the effect are not definitely known. Supplemental fluoride treatment is generally most effective during the period of tooth eruption and immediately thereafter.

Inorganic fluoride is readily absorbed in the small intestine and distributed almost entirely to bone and teeth. Fluoride replaces hydroxyl groups in tooth apatite. Renal excretion is most important for regulation of body fluoride levels.

The *reported concentrations* of fluoride in specimens from healthy adults are 10–370 ng/mL in serum or plasma, 450 ng/mL in erythrocytes, and 0.2–1.9 mg/L in urine.[8,27] Concentrations of fluoride in these specimens vary widely, depending on whether or not the water in an area is fluoridated. The estimated safe and adequate dietary intakes of fluoride are listed in Table 17-1. Fluoridated drinking water contributes appreciably to total fluoride intake. Seafoods and tea contain *high levels* of fluoride.

Some evidence suggests that *low levels* of fluoride intake are associated with osteoporosis and that sodium fluoride and calcium supplementation may stimulate bone formation. *Excessive fluoride* intake produces mottled or pitted enamel in growing teeth. Chronic fluorosis may produce osteosclerosis, calcification of ligaments and tendons, and crippling deformities such as kyphosis, stiffness of the spine, and bony exostoses. Fluoride assay *methods* are discussed in Chapter 27, *Analysis of Toxic Substances*.

References

1. Abernethy, M. H., and Fowler, R. T.: Micellar improvement of the calmagite compleximetric measurement of magnesium in plasma. Clin. Chem., *28*:520, 1982.
2. Bowers, G. N., and Pybus, J.: Total calcium in serum by atomic absorption spectrophotometry. *In*: Standard Methods of Clinical Chemistry, Vol. 7., S. Meites, Ed. New York, Academic Press, 1972, p. 143.
3. Butrimovitz, G. P., and Purdy, W. C.: The determination of zinc in blood plasma by atomic absorption spectrometry. Anal. Chim. Acta, *94*:63, 1977.
4. Dean, J. A., and Rains, T. C.: Flame Emission and Atomic Absorption Spectrometry, Vols. 1–3. New York, Marcel Dekker, Inc., 1971.
5. Fowler, R. T., Abernethy, M. H., Walmsley, T. A., et al.: Measurement of magnesium by continuous-flow colorimetry. Clin. Chem., *28*:523, 1982.
6. Freund, H., Atamain, S., and Fisher, J. E.: Chromium deficiency during total parenteral nutrition. JAMA, *241*:496, 1979.
7. Issaqui, H. J.: Effects of matrix on the determination of volatile metals in biological samples by flameless atomic absorption spectrometry with the graphite tube atomizer. Anal. Chem., *51*:657, 1979.
8. Iyengar, I. Y., Kollmer, W. E., and Bowert, H. J. M.: The Element Composition of Human Tissues and Body Fluids. Weinheim, Verlag Chemie, 1978.
9. Jacob, R. A.: Hair as a biopsy material. *In*: Systemic Aspects of Biocompatibility. D. F. Williams, Ed. Boca Raton, FL, CRC Press, Inc., 1981.
10. Jeejeebhoy, K. N., Chu, R. C., Marliss, E. B., et al.: Chromium deficiency, glucose intolerance, and neuropathy reversed by chromium supplementation in a patient receiving long-term total parenteral nutrition. Am. J. Clin. Nutr., *30*:531, 1977.
11. Johnson, D. J., Djuh, Y. Y., Bruton, J., et al.: Improved colorimetric determination of serum zinc. Clin. Chem., *23*:1321, 1977.
12. Kossman, K. T.: Copper in serum measured with the Cobas-Bio centrifugal analyzer. Clin. Chem., *29*:578, 1983.
13. Levander, O. A., and Cheng, L., Eds.: Micronutrient interactions: Vitamins, minerals, and hazardous elements. Ann. N.Y. Acad. Sci., *355*:1, 1980.
14. Makino, T., and Takahara, K.: Direct determination of plasma copper and zinc in infants by atomic absorption with discrete nebulization. Clin. Chem., *27*:1445, 1981.

15. Mason, K. E.: Copper metabolism and requirements of man. J. Nutr., *109*:1979, 1979.
16. Mertz, W.: The essential trace elements. Science, *213*:1332, 1981.
17. Mills, C. F.: Interactions between elements in tissues. Studies in animal models. Fed. Proc., *40*:2138, 1981.
18. Prasad, A. S.: Chromium. *In*: Trace Elements and Iron in Human Metabolism. New York, Plenum Medical Book Company, 1978, pp. 3–15.
19. Prasad, A. S.: Selenium. Ibid. pp. 215–250.
20. Rice, E. W.: Principles and Methods of Clinical Chemistry. Springfield, IL, Charles C Thomas, 1960, pp. 157–159.
21. Sandstead, H. H., Vo-Khactu, K. P., and Solomons, N.: Conditioned zinc deficiencies. *In*: Trace Elements in Human Health and Disease, Vol. 1. A. S. Prasad, Ed., New York, Academic Press, 1976, pp. 33–49.
22. Smith, J. C., Jr., Butrimovitz, G. P., and Purdy, W. C.: Direct measurement of zinc in plasma by atomic absorption spectroscopy. Clin. Chem., *25*:1487, 1979.
23. Solomons, N. W.: On the assessment of zinc and copper nutriture in man. Am. J. Clin. Nutr., *32*:856, 1979.
24. Sunderman, F. W., Jr.: Atomic absorption spectrometry of trace metals in clinical pathology. Hum. Pathol., *4*:549, 1973.
25. Sunderman, F. W., Jr.: Electrothermal atomic absorption spectrometry of trace metals in biological fluids. Ann. Clin. Lab. Sci., *5*:421, 1975.
26. Thiers, R. E.: Magnesium (fluorometric) and magnesium (titan yellow). *In*: Standard Methods of Clinical Chemistry, Vol. 5. S. Meites, Ed. New York, Academic Press, 1965, pp. 131–142.
27. Tietz, N. W., Ed.: Clinical Guide to Laboratory Tests. Philadelphia, W. B. Saunders Company, 1983.
28. Tulley, R. T., and Lehmann, H. P.: Flameless atomic absorption spectrophotometry of selenium in whole blood. Clin. Chem., *28*:1448, 1982.
29. Vieir, N. E., and Hansen, J. W.: Zinc determined in 10-μL serum or urine samples by flameless atomic absorption spectrometry. Clin. Chem., *27*:73, 1981.
30. Weinstock, N., and Uhlemann, M.: Automated determination of copper in undiluted serum by atomic absorption spectroscopy. Clin. Chem., *27*:1438, 1981.
31. Williams, H. L., Johnson, D. J., and Haut, M. J.: Simultaneous spectrophotometry of Fe^{2+} and Cu^{2+} in serum denatured with guanidine hydrochloride. Clin. Chem., *23*:237, 1977.
32. Wilson, D. L.: Separation and concentration techniques for atomic absorption spectrometry: A guide to the literature. Atomic Absorption Newsletter, *18*:13, 1971.

Additional Reading

Hook, G. R., Hosseini, J. M., and Elin, R. J.: Analytical approaches for biomedical elemental analysis. J. Am. Coll. Nutr., *4*:599, 1985.

Prasad, A. S., Ed.: Clinical, Biochemical, and Nutritional Aspects of Trace Elements, Current Topics in Nutrition and Disease, Vol. 6. New York, Alan R. Liss, Inc., 1982.

Versieck, J.: Trace element analysis—a plea for accuracy. Trace Elements in Medicine, *1*:2, 1984.

Versieck, J., and Cornelis, R.: Normal levels of trace elements in human blood plasma or serum. Anal. Chim. Acta, *116*:217, 1980.

ENDOCRINOLOGY

18

SATI C. CHATTORAJ, Ph.D.
NELSON B. WATTS, M.D.

Endocrine glands are tissues that release their secretory products into extracellular and interstitial fluids. The secretory products are called hormones, and endocrinology is the study of hormones and their actions. Hormones may be transported in the bloodstream from their place of synthesis to a distant location where they exert their action, but some hormones also act in close proximity to their site of release. Figure 18-1 shows the location of endocrine glands in the human body, and Table 18-1 summarizes the types of hormone actions.

Most hormones can be grouped into three general classes: (1) steroids, (2) polypeptides or proteins, and (3) hormones derived from amino acids. Steroid hormones are generally hydrophobic. Many steroid hormones circulate in plasma bound to high-affinity plasma proteins, leaving only a small percentage of the hormone free for biologic activity. The half-lives of these hormones vary from 60 to 100 min. In contrast, polypeptide and protein hormones are water soluble, circulate unbound in plasma, and may undergo rapid fluctuations. Their half-lives vary from 5 to 60 min. Half-lives of amino acid–derived hormones differ depending on whether or not they are protein bound. Thyroxine, for example, circulates bound to three binding proteins and has a half-life of almost one week, whereas epinephrine, which is not protein bound, has a half-life of <1 min.

Hormones possess a high degree of structural specificity. A slight alteration in molecular composition may bring significant changes in physiological activity. For example, estradiol and estriol differ by only an α-hydroxyl group, but estradiol is the most potent estrogen and estriol is almost inert. Classifications of hormones, their sources, and a brief description of their actions are given in Table 18-2.

Actions of Hormones

Actions of hormones may be broadly divided according to three general aspects:

Regulatory function. One of the major functions of the endocrine system is to maintain constancy of chemical composition of extracellular and intracellular fluids. Homeostasis is maintained through action of hormones that closely regulate the metabolism of salt, water, carbohydrate, fat, and protein. Other important regulatory functions of hormones are the responses to demands of starvation, infection, trauma, psychological stress, and the process of sexual reproduction.

Morphogenesis. Some hormones play an important part in controlling the growth and development of an organism. The development of male and female sex characteristics under the influence of testosterone and estradiol is perhaps the best example.

Integrative action. This aspect of hormonal function is the most complex and the least understood. In general, each hormone has a specific function. However, hormones produced by several endocrine glands may be important for regulating a single function. For example, normal carbohydrate metabolism involves the concerted actions of glucagon and insulin (from the pancreas), growth hormone (pituitary), glucocorticoids and epinephrine (adrenals), thyroxine (thyroid), and estrogens (gonads). These interrelations are not limited to the endocrine glands but extend to the nervous system as well. For example, mineralocorticoids (deoxycorticosterone, aldosterone) have a profound influence on the maintenance of salt and water balance that is correlated to simultaneous adjustment of the rate of blood flow, blood pressure, and vasoconstriction by the autonomic nervous system.

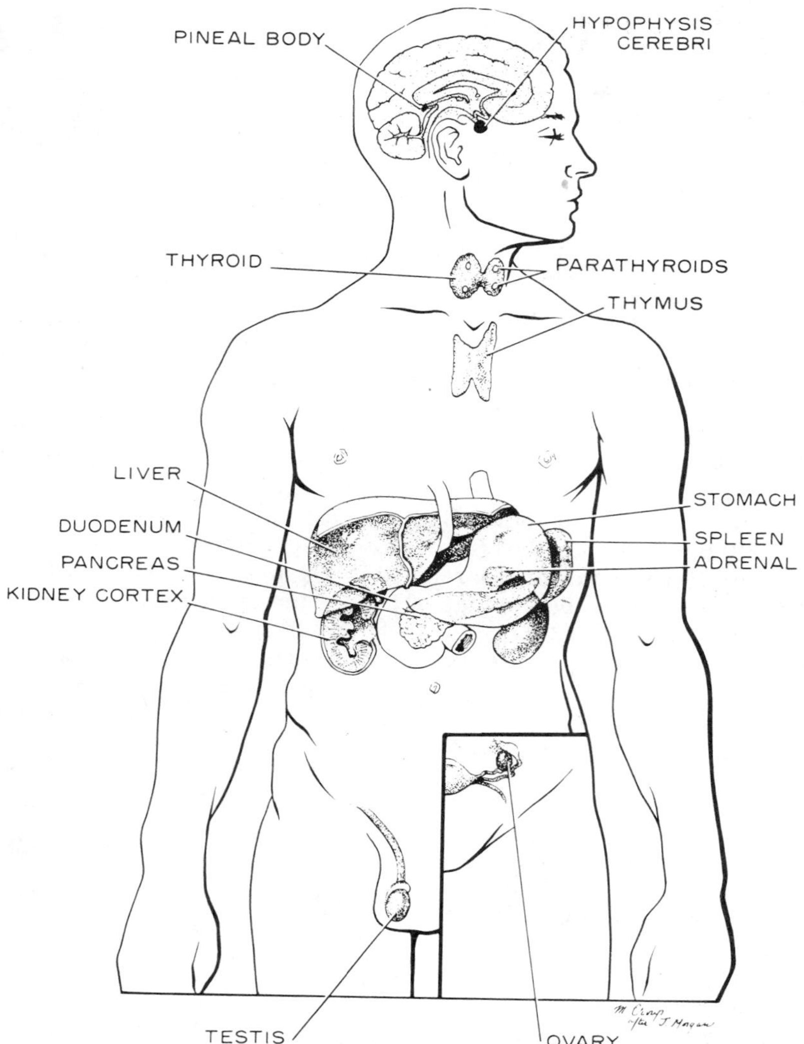

Figure 18-1. Location of the endocrine glands in humans. (From Turner, C. D.: General Endocrinology. 4th ed. Philadelphia, W.B. Saunders Co., 1966.)

Control of Hormone Secretion

The following covers only salient features of the several mechanisms involved in maintaining the balance between production of hormones and the needs of the organism for hormones.[123]

The *anterior pituitary* occupies a central role in the control of hormone secretion (*see Anterior Pituitary Hormones*, page 546) through the production of tropic hormones that stimulate and maintain certain other endocrine glands. In the absence of these tropic hormones, the target glands are unable to maintain a normal rate of secretion. The main target organs of the pituitary tropic hormones are the thyroid gland, adrenal cortex, and gonads.

Feedback regulation is important in the control of hormone secretion. A feedback system is one in which the product of a system affects an earlier part of the system and thus alters the rate of production of the product itself. If an increase in the product causes an increase in the activity of the system, the relationship is described as *positive feedback*. If an increase in the product decreases the activity of the system, this is *negative feedback*, the principal mechanism for the control of hormone secretion. Indirect negative

Table 18-1. TYPES OF HORMONE ACTION

Endocrine:	Hormone synthesized in one location and released into plasma; binds to specific receptors in cells at a distant site to elicit characteristic response. *Example*: The action of TSH from the anterior pituitary on the thyroid.
Neuroendocrine:	Hormone synthesized in nerve ending and released into extracellular space; interacts with receptors of cells at distant site. *Example*: The action on the heart of norepinephrine synthesized in splanchnic nerve ending.
Neurocrine:	Hormone synthesized in neurons and released into extracellular space; binds to receptor in nearby cell and affects its function. *Example*: The action on cardiac muscle cells of norepinephrine synthesized in nerve endings in heart.
Neurotransmission:	Hormone synthesized in neurons and released from nerve endings; crosses synapse and binds to specific receptor in another neuron, affecting its action. *Example*: Release of acetylcholine from preganglionic nerve fibers in sympathetic ganglia and binding to receptor in postganglionic neuron with liberation of norepinephrine.
Paracrine:	Hormone synthesized in endocrine cells and released into extracellular space; binds to specific receptor of nearby cell and affects its function. *Example*: Release of somatostatin from islet D cells and its subsequent action on nearby A and B cells in the same pancreatic islet.
Exocrine:	Hormone synthesized in endocrine cells and released into lumen of gut; binds to cells lining the gut at varying distances from endocrine cell, thereby affecting their function. *Example*: The release of gastrin by mucosal cells and its action on the gastric acid secretion by the stomach.

Modified from Williams, R. H., Ed.: Textbook of Endocrinology. 6th ed. Philadelphia, W.B. Saunders Co., 1981.

feedback regulation via tropic hormones is exemplified in the control of adrenocortical, gonadal, and thyroidal hormone secretions.

As opposed to negative feedback, positive feedback does not operate in isolation but constitutes an integral part of a control system. Positive feedback regulation is primarily observed in the hormonal control of the menstrual cycle (see *The Normal Menstrual Cycle*, page 581).

Control by the central nervous system (CNS). Hormones secreted by the neurohypophysis or posterior pituitary (e.g., oxytocin, vasopressin) and the adrenal medulla (e.g., epinephrine) are controlled primarily by the nervous system. The importance of the relationship between the nervous system and endocrine system is apparent through various feedback control mechanisms that involve the hypothalamus. Since the hypothalamus receives signals from higher centers in the brain, one feature of the regulation of pituitary hormones is the ability of the brain to override or to fine-tune other control mechanisms. This type of control by the higher centers of the brain is called an *open-loop control system*; the hormonal response to stress is an example of this system.

Hormone-Receptor Interactions

The biological response to a hormone is initiated by the binding of the hormone to target cell receptors. The receptor provides the target cell with a mechanism for recognizing and concentrating the hormone, and the hormone-receptor complex activates the target cell to begin the chain of events that constitutes the biological effect(s) of that hormone. The hormone-receptor complex has several characteristics:

1. It is highly specific; i.e., the target tissue accumulates only the active hormone from among the myriad of other substances to which the cells are exposed.
2. It is an equilibrium system in which hormone and receptor react reversibly to form the hormone-receptor complex (HR), which is similar to an enzyme-substrate complex; the reaction is assumed to reach equilibrium.
3. It is saturable due to the presence of a finite number of receptors in cells. The degree of biological response of a target tissue is directly proportional to the number of hormone-receptor complexes; a maximal response is obtained when all sites are filled.
4. It has high affinity, which allows hormone-receptor complexes to form in the presence of very low circulating levels of hormone.

Table 18-2. CLASSIFICATION OF HORMONES, THEIR SOURCES, AND THEIR ACTIONS

Endocrine Gland and Hormone	Nature of Hormone	Site of Action	Principal Actions
Hypothalamus			
Thyrotropin-releasing hormone (TRH)	Peptide (3 aa)*	Anterior pituitary	Release of TSH and PRL
Gonadotropin-releasing hormone (GnRH) or LH-releasing hormone (LHRH)	Peptide (10 aa)	Anterior pituitary	Release of LH and FSH
Corticotropin-releasing hormone (CRH)	Polypeptide (41 aa)	Anterior pituitary	Release of ACTH and β-LPH
Growth hormone–releasing hormone (GHRH)	Polypeptide (40 aa)	Anterior pituitary	Release of GH
Somatostatin† (SS) or growth hormone–inhibiting hormone (GHIH)	Peptide (14 aa)	Anterior pituitary	Suppression of GH and TSH; inhibition of gastrin, VIP, GIP, secretin, motilin, and insulin
Prolactin-releasing factor (PRF)	Peptide?	Anterior pituitary	Release of PRL
Prolactin-inhibiting factor (PIF)	Dopamine?	Anterior pituitary	Suppression of PRL
Anterior pituitary			
Thyrotropin or thyroid-stimulating hormone (TSH)	Glycoprotein‡ (α, 89 aa; β, 112 aa)	Thyroid	Stimulation of thyroid hormone formation and secretion
Follicle-stimulating hormone (FSH)	Glycoprotein‡ (α, 89 aa; β, 115 aa)	Ovary	Growth of follicles and, with LH, secretion of estrogens and ovulation
		Testis	Development of seminiferous tubules, spermatogenesis
Luteinizing hormone (LH)	Glycoprotein‡ (α, 89 aa; β, 115 aa)	Ovary	Ovulation, formation of corpora lutea, secretion of progesterone
		Testis	Stimulation of interstitial tissue; secretion of androgens
Prolactin (PRL)	Protein (198 aa)	Mammary gland	Proliferation of mammary gland; initiation of milk secretion; antagonist of insulin action
Growth hormone (GH) or somatotropin	Protein (191 aa)	Body as a whole	Growth of bone and muscle
β-Lipotropin (β-LPH)	Polypeptide (91 aa)	Unknown	Precursor of β-MSH and the endorphins
Corticotropin or adrenocorticotropin (ACTH)	Polypeptide (39 aa)	Adrenal cortex	Stimulation of adrenocortical steroid formation and secretion

β-Endorphin (β-END)[†¶]	Polypeptide (31 aa)	Brain	Endogenous opiate; raising of pain threshold and influence on extrapyramidal motor activity
α-Melanocyte-stimulating hormone (α-MSH)	Peptide (13 aa)	Skin	Dispersion of pigment granules, darkening of skin
Leu-enkephalin (LEK)[†¶] and metenkephalin (MEK)[†¶]	Peptide (5 aa)	Brain	Same as β-endorphin
Posterior pituitary			
Vasopressin or antidiuretic hormone (ADH)	Peptide (9 aa)	Arterioles Renal tubules	Elevation of blood pressure Water reabsorption
Oxytocin	Peptide (9 aa)	Smooth muscles (uterus, mammary gland)	Contraction, action in parturition and in sperm transport, ejection of milk
Pineal			
Serotonin or 5-hydroxytryptamine (5-HT)	Indoleamine	Cardiovascular, respiratory, and gastrointestinal systems, brain	Neurotransmitter; stimulation or inhibition of various smooth muscles and nerves; possible role in mental illness
Melatonin	Indoleamine	Hypothalamus	Suppression of gonadotropin and GH secretion; induction of sleep
Thyroid			
Thyroxine (T_4) and triiodothyronine (T_3)	Iodoamino acids	General body tissue	Stimulation of oxygen consumption and metabolic rate of tissue
Calcitonin or thyrocalcitonin	Polypeptide (32 aa)	Skeleton	Inhibition of calcium resorption; lowering of plasma calcium and phosphate
Parathyroid			
Parathyroid hormone (PTH) or parathormone	Polypeptide (84 aa)	Skeleton, kidney, gastrointestinal tract	Regulation of calcium and phosphorus metabolism
Adrenal cortex			
Cortisol	Steroid	General body tissue	Metabolism of carbohydrates, proteins, and fats; inflammation, resistance to infection; hypersensitivity
Aldosterone	Steroid	Kidney	Salt and water balance

Table continued on following page

Table 18-2. CLASSIFICATION OF HORMONES, THEIR SOURCES, AND THEIR ACTIONS (*Continued*)

Endocrine Gland and Hormone	Nature of Hormone	Site of Action	Principal Actions
Adrenal medulla			
Norepinephrine and epinephrine	Aromatic amines	Sympathetic receptors	Stimulation of sympathetic nervous system
Epinephrine		Liver and muscle	Glycogenolysis
		Adipose tissue	Lipolysis
Ovary			
Estrogens	Phenolic steroids	Female accessory sex organs	Development of secondary sex characteristics
Progesterone	Steroid	Female accessory reproductive structure	Preparation of the uterus for ovum implantation, maintenance of pregnancy
Relaxin	Polypeptide	Uterus	Inhibition of myometrial contraction
Inhibin	Polypeptide	Hypothalamus	Suspected role in the control of FSH secretion
Testis			
Testosterone	Steroid	Male accessory sex organs	Development of secondary sex characteristics, maturation, and normal function
Inhibin	See above	See above	See above
Placenta			
Estrogens	See above	See above	See above
Progesterone	See above	See above	See above
Relaxin	See above	See above	See above
Human chorionic gonadotropin (hCG) or choriogonadotropin	Glycoprotein‡ (α, 92 aa; β, 144 aa)	Same as LH	Same as LH; prolongation of corpus luteal function; suspected role in steroidogenesis during fetal life
Human chorionic somatomammotropin (hCS) or human placental lactogen (hPL)	Protein (191 aa)	Same as PRL	Same as PRL
Pancreas			
Insulin	Polypeptide§	Most cells	Regulation of carbohydrate metabolism; lipogenesis
Glucagon	Polypeptide (29 aa)	Liver	Glycogenolysis

Pancreatic polypeptide (PP)	Polypeptide (36 aa)	Gastrointestinal tract	Increased gut motility and gastric emptying; inhibition of gallbladder contraction
Gastrointestinal tract			
Gastrin[¶]	Peptide (17 aa)	Stomach	Secretion of gastric acid, gastric mucosal growth
Secretin	Polypeptide (27 aa)	Pancreas	Secretion of pancreatic bicarbonate and digestive enzymes
Cholecystokinin-pancreozymin (CCK-PZ)[¶]	Polypeptide (33 aa)	Gallbladder and pancreas	Stimulation of gallbladder contraction and secretion of pancreatic enzymes
Motilin	Polypeptide (22 aa)	Gastrointestinal tract	Stimulation of gastrointestinal motility
Vasoactive intestinal peptide (VIP)[¶]	Polypeptide (28 aa)	Gastrointestinal tract	Neurotransmitter; relaxation of smooth muscles of gut and of circulation; increase of release of hormones and secretion of water and electrolytes from pancreas and gut
Gastric inhibitory polypeptide (GIP)	Polypeptide (42 aa)	Gastrointestinal tract	Inhibition of gastric secretion and motility; increase of insulin secretion
Bombesin[¶]	Peptide (14 aa)	Gastrointestinal tract	Stimulation of release of various hormones and pancreatic enzymes, smooth muscle contractions and hypothermia, changes in cardiovascular and renal function
Neurotensin[¶]	Peptide (13 aa)	Gastrointestinal tract and hypothalamus (gut and brain)	Uncertain
Substance P (SP)[¶]	Peptide (11 aa)	Gastrointestinal tract and brain	Sensory neurotransmitter, analgesic; increase in contraction of gastrointestinal smooth muscle; potent vasoactive hormone; promotion of salivation, increased release of histamine

Table continued on following page

Table 18-2. CLASSIFICATION OF HORMONES, THEIR SOURCES, AND THEIR ACTIONS (*Continued*)

Endocrine Gland and Hormone	Nature of Hormone	Site of Action	Principal Actions
Kidney			
1,25-$(OH)_2$ Vitamin D	Sterol	Intestine	Facilitation of calcium and phosphorus absorption
		Bone	Increase in bone resorption in conjunction with PTH
		Kidney	Increase in reabsorption of filtered calcium
Erythropoietin	Glycoprotein	Bone marrow	Stimulation of erythrocyte formation

*aa = Amino acid residues.

†Also produced by gastrointestinal tract.

‡Glycoprotein hormone composed of two dissimilar peptides. The α-chain is similar in structure or identical; the β-chain differs for each hormone and confers specificity.

§Two chains linked by disulfide bonds: A, 21 aa; B, 30 aa.

¶Also produced in the brain.

For most hormones the *equilibrium dissociation constant*, K_d, falls approximately in the range of circulating hormone concentrations.

Receptor Regulation

The number of receptors may increase or decrease in response to various stimuli, and receptor properties may change with varying physiologic conditions. The exposure of responsive cells to high concentrations of hormone will decrease the number of surface membrane receptors. In addition, the binding of a hormone molecule to its receptor may influence the affinity of neighboring binding sites. Such a mechanism modulates hormone action by enhancing cell sensitivity at low hormone concentrations and reducing sensitivity at high hormone concentrations.

Receptors may also be modulated by other hormones. Furthermore, one hormone may have strong affinity for its own receptor and some affinity for the receptor of another hormone. Such a "spillover" generally occurs between hormones with structural similarities. At physiological levels of hormones this spillover is of no consequence. However, a hormone produced in excess may show not only an excessive biological effect of its own but also the response of a structurally similar hormone.

Dysfunction of receptors may be the cause of a variety of diseases.[90] For example, in pseudohypoparathyroidism, parathyroid hormone fails to elicit the normal responses in its target organs.

Mechanism of Action of Hormones

Currently, two general models of hormone action prevail, one for water-soluble hormones—peptides, proteins, and catecholamines—and one for lipid-soluble hormones such as steroids, 1,25-$(OH)_2$-vitamin D, and iodothyronines.

Mechanism of protein, peptide, and catecholamine hormone action. Protein hormones and catecholamines are too polar to diffuse passively through lipoprotein cell membranes; protein hormones are also too large to pass through membrane pores. These hormones initiate their response by binding to receptors located on or in the cell membrane. For many hormones, combination with their respective receptors activates adenylate cyclase, a membrane protein that catalyzes the formation of cyclic AMP (cAMP) from ATP at the inner membrane surface. Cyclic AMP diffuses through the cell and activates a group of closely related enzymes known as cAMP-dependent protein kinases. These kinases phosphorylate intracellular proteins, especially other enzymes, and thereby regulate (activate or inactivate) the action of these intracellular enzymes or proteins. The functional relationships of these components are shown in Figure 18-2. Several hormones (e.g., insulin, prolactin, angiotensin) that bind to membrane receptors do not activate adenylate cyclase but appear to work through other chemical messengers.

Mechanism of steroid, 1,25-$(OH)_2$-vitamin D, and thyroid hormone actions. These hormones act by regulating gene expression to induce synthesis of specific proteins or enzymes in target cells. These hormones are lipid soluble; they passively diffuse across cell membranes and then bind to specific receptor proteins in the cytoplasm. The hormone-receptor complex is translocated into the nucleus (Figure 18-3).

Thyroid hormones (T_4, T_3) do not fit into the conventional model for hormone action but instead enter cells and bind directly to receptors in nuclear chromatin. Nuclear T_3 is derived both from circulating T_3 and from intracellular deiodination of T_4. The T_3-nuclear receptor complex acts on DNA to increase the synthesis of messenger RNA (mRNA) and ribosomal RNA, resulting in synthesis of proteins. Some of these induced proteins are enzymes that modify cell function.

PROTEIN HORMONES

Protein and peptide hormones vary in their chemical structures, from a simple tripeptide like thyrotropin-releasing hormone to complex glycoprotein molecules such as FSH, LH, and TSH. Since these hormones have no distinctive chemical groups toward

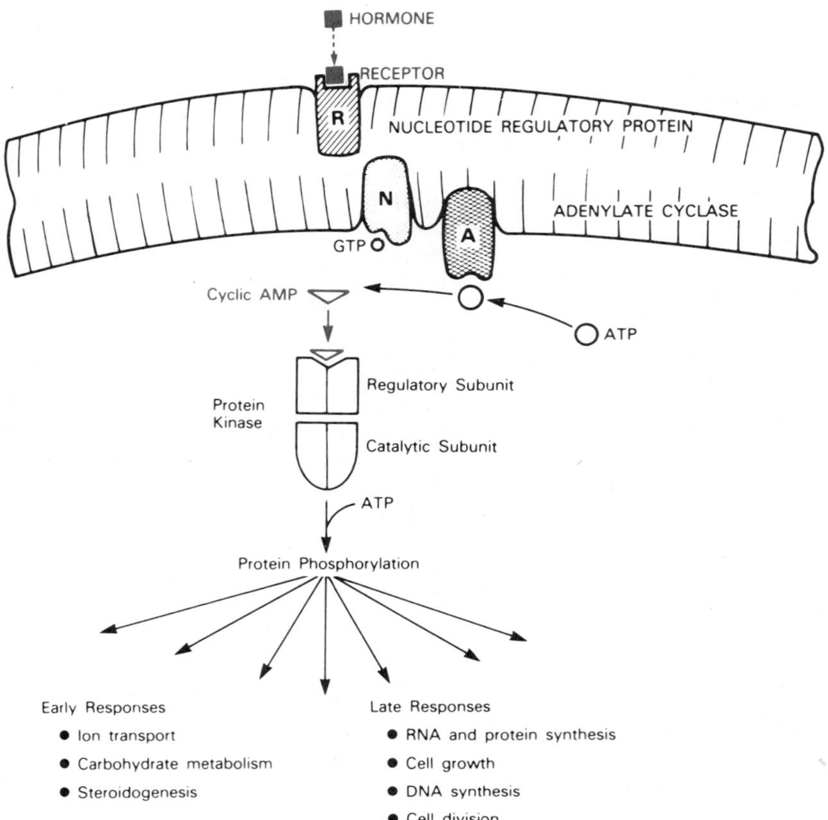

Figure 18-2. Cyclic AMP as a "second messenger" mediating the effects of extracellular chemical messengers such as hormones and neurotransmitters. The ligand (hormone) binds to the receptor protein (R), and this reaction activates adenylate cyclase (A) through the nucleotide regulatory protein (N), which binds GTP. Activation of adenylate cyclase converts ATP to cAMP. The cyclic AMP activates protein kinase, which in turn phosphorylates enzymes and other proteins to produce a variety of effects. (Reproduced with permission from Catt, K. J., et al.: Recent Prog. Horm. Res., *36*:557, 1980.)

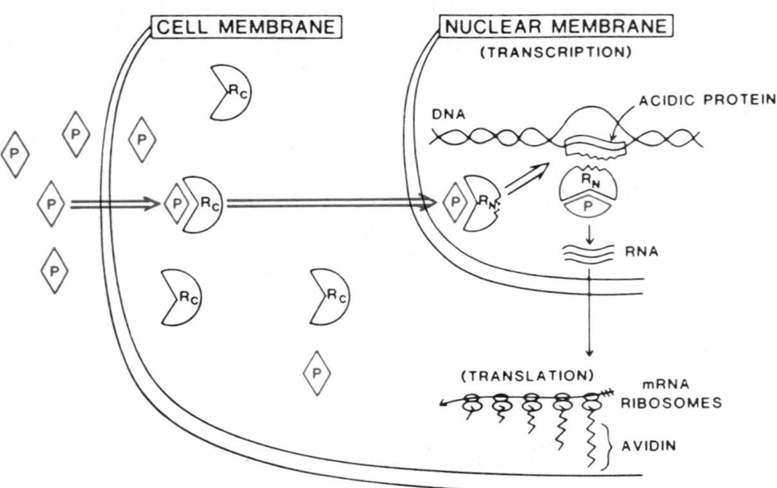

Figure 18-3. General model of steroid hormone action. Binding of hormone, in this case progesterone (P), to the cytoplasmic receptor (R_c) is followed by translocation (arrow) and induction of gene transcription by nuclear receptor (R_N) to produce messenger RNA (mRNA). mRNA is translated on cytoplasmic ribosomes to synthesize new protein, in this case avidin, a protein whose synthesis is induced by progesterone in the chick oviduct. (Reproduced with permission from O'Malley, B. W., et al.: Endocrine Rev., *3*:141, © by The Endocrine Society, 1982.)

which a specific color reaction may be directed as a means of assay, measurements of protein hormones depend on other techniques.

Bioassay Techniques

Bioassay[56] is based on the observation of a physiological response specific for the hormone being measured in a suitably prepared animal. Magnitudes of test and standard responses are compared to estimate the amount of hormone present in the unknown. This technique is fraught with problems: variability of response in the experimental animals; the presence of substances, other than the hormone, that alter the response; and chemical change in the hormone before or after injection. At present, bioassays are primarily employed as reference methods during the process of purifying hormones.

Radioreceptor Assays

Radioreceptor assays[56] depend on an in vitro interaction of hormone with its biological receptor. Unlabeled hormone displaces trace amounts of radioactively labeled hormone from receptor sites. An advantage of these assays is that they measure biologically active hormone; by contrast, radioimmunoassays may measure inactive hormone derivatives that have the same antigenic determinants. Radioreceptor assays have some important disadvantages: the binding sites provided in the assay may not be the true biological receptor (e.g., talcum in insulin assay); interference from binding by other substances may lead to spurious results; or enzymes (peptidases and proteases) in the sample may destroy labeled tracer or alter the membrane that contains the receptors and thus decrease specific binding.

Immunoassay Techniques

Of all the immunological assay techniques for the quantitation of protein hormones, radioimmunoassay (RIA) is used most often. The sensitivity of RIA has made possible the measurement of most protein and polypeptide hormones without prior extraction. Since the use of commercial kits from many different manufacturers[29] has become almost universal in clinical laboratories, only a general discussion on the assay of protein hormones follows.

Although RIA procedures for different hormones vary in details of sample preparation and separation of bound from free hormone, the basic steps are similar. Standards or samples containing unlabeled hormone are added to predetermined amounts of radiolabeled hormone and antibody. The unlabeled and radiolabeled hormones compete for binding sites on the antibodies during incubation. The antibody-bound and free hormones are then separated so that the radioactivity of either fraction can be measured. A calibration curve is constructed by plotting the radioactivity in either fraction against the standard concentrations.

The selection of specimen for assay depends on the sensitivity of the assay, the ease of sample collection, and the physiological characteristics of the hormone to be measured. If fluctuations of the hormone in blood are extreme or rapid, measurements of timed urine collections may be more useful than measurements on a single blood specimen.

Concentration or extraction procedures may be necessary if levels of the hormones to be determined are below the sensitivity of the assay, or if the concentration of degradative enzymes in plasma samples is high enough to cause significant in vitro inactivation, or if there is a need to separate the hormones from interfering substances.

INSULIN

Human insulin (MW 6000) is a protein hormone produced by the B (β)* cells of the pancreatic islets of Langerhans; it consists of 51 amino acids in two polypeptide chains (A and B) joined by two disulfide bridges. The amino acid sequence of human insulin differs from that of insulin of other species. Insulin was the first protein hormone to be

---

*It has become customary to change the designation of α-, β-, δ-cells to A, B, and D cells, respectively, but the Greek letter designations are still widely used.

sequenced, the first substance to be measured by radioimmunoassay, and the first compound produced by recombinant DNA technology for practical use. Insulin is clinically important in disorders of carbohydrate metabolism (e.g., diabetes mellitus and hypoglycemia), which are discussed further in Chapter 13, *Carbohydrates*.

Synthesis, Release, and Degradation

Preproinsulin, a protein of ~100 amino acids, is formed in the rough endoplasmic reticulum of the pancreatic B cells. Preproinsulin is not detectable under normal conditions because it is rapidly converted to proinsulin. ***Proinsulin*** (MW 9000) has little biological activity and is stored in secretory granules in the Golgi complex, where proteolytic cleavage to insulin and C-peptide occurs. At the cell membrane, insulin and C-peptide are released into the portal circulation in equimolar amounts. Small amounts of minimally active proinsulin and intermediate cleavage forms are also released.[95]

Insulin release is stimulated by glucose, amino acids, other pancreatic and gastrointestinal hormones (e.g., glucagon, gastrin, secretin, pancreozymin, gastrointestinal polypeptide), and some medications (e.g., sulfonylureas and β-adrenergic stimulators such as isoproterenol). Insulin release is inhibited by hypoglycemia, by somatostatin (which is produced in the pancreatic D cells), and by a variety of drugs (α-adrenergic stimulators, β-adrenergic blocking agents, diazoxide, phenytoin, phenothiazines, and nicotinic acid).[86] Fifty to 70% of secreted insulin is extracted by the liver on the first pass through the portal circulation and is degraded there. The kidney is also a site of insulin degradation. The half-life of insulin in the circulation is between 5 and 10 min.[105]

C-peptide. Proinsulin is cleaved to form a 31–amino acid connecting (C) peptide and insulin. C-peptide (MW 3600) is devoid of biological activity. Although insulin and C-peptide are secreted into the portal circulation in equimolar amounts, the circulating concentration of C-peptide is greater than that of insulin due to its longer half-life of about 20 min. C-peptide is removed from circulation by the kidney and degraded. A fraction of unchanged C-peptide is excreted in the urine.[26]

C-peptide measurements in blood or urine are particularly useful in assessing B cell function in the presence of exogenous insulin or circulating antibodies to insulin. However, the primary diagnostic use of C-peptide measurement has been in patients with fasting hypoglycemia and suspected of having an insulin-producing B cell tumor, in which concentrations of insulin, C-peptide, and proinsulin are all likely to be high. In some of these subjects, particularly if hyperinsulinism is intermittent, C-peptide concentration may be diagnostically elevated when the insulin concentration is not. The 24-h urine C-peptide content correlates well with fasting serum C-peptide concentration or with the sum of C-peptide concentrations after a glucose load. When surreptitious insulin injection causes hypoglycemia, insulin levels will be high but C-peptide levels will be low, since C-peptide is not found in commercial insulin preparations and exogenous insulin suppresses B cell function.

Insulin receptors and insulin action. Insulin acts on its target tissue cells (primarily liver, muscle, and adipose tissue) by combining with specific *insulin receptors* on the plasma membrane. The insulin-receptor combination initiates changes within the cell as discussed in Chapter 13. The end result is anabolism—facilitation of the entry of glucose into cells for storage as glycogen and the inhibition of protein breakdown and lipolysis.

Clinical significance. Patients with insulin-producing islet-cell tumors usually present with fasting hypoglycemia and elevated plasma insulin levels. Decreased levels of plasma insulin occur in Type I diabetes mellitus or after pancreatectomy.

Elevated concentrations of insulin associated with fasting or postprandial hyperglycemia suggest insulin resistance (see Chapter 13). Such resistance is commonly seen in obesity, Type II diabetes mellitus, uremia, and hypercortisolism and may be caused by a decrease in insulin receptors or by post-receptor defects.

Antibodies to insulin. Antibodies to insulin develop in almost all patients who are treated with exogenous bovine or porcine insulin. These antibodies are usually of the IgG class and are rarely of sufficient quantity or affinity to cause clinical problems other than interference with the radioimmunoassay for insulin. Occasionally, however, binding of insulin to circulating high-affinity antibodies will interfere with the action of exogenously administered insulin; fluctuations of plasma glucose may then occur as free insulin levels rise and fall due to erratic release from the antibody-bound form. Beef insulin is more immunogenic than pork insulin.

The method for detection of such antibodies is similar to RIA for insulin, except that the patient's serum is used as the source of antiserum. Different dilutions of the patient's serum are incubated with labeled porcine or bovine insulin. The dilution of the patient's serum at which the binding of the labeled material is equivalent to that of a negative control serum is considered the end point of the titration. The result is expressed as the antibody titer.

Nonsuppressible insulin-like activity (NSILA). Insulin accounts for only ~7% of the hypoglycemic effect of compounds in peripheral blood as determined by bioassay. The remainder of the effect is due to "insulin-like" activity of certain other compounds. Since insulin activity is variable and suppressible—for instance, it can be blocked by antibodies to insulin—and activity of these other compounds is rather constant and nonsuppressible, the term *NSILA* has come into use. Such NSILA compounds are heterogeneous and include the somatomedins, responsible for 4–8% of total insulin-like activity, and a high-molecular-weight protein called NSILAP, which accounts for 85–89% of total insulin-like activity.[87] The somatomedins are also known as NSILAs, where the *s* stands for soluble fraction. The significance of these compounds in normal carbohydrate metabolism is not known. They are occasionally produced in excess by extrapancreatic neoplasms and cause fasting hypoglycemia. NSILA may also be a tumor marker in some patients who do not have hypoglycemia.

Methods for the Determination of Insulin, Proinsulin, and C-Peptide

Radioimmunoassay of Insulin[124]

Because of limitations of biological methods,[56] RIA has become the technique of choice for measurement of insulin in biological fluids.

Principle. Various RIA methods and commercial kits for insulin measurement are now available. The principle and general procedures of RIA have been described previously (see Chapter 3).

The antibody is produced against porcine insulin in guinea pigs. (Antisera raised against porcine insulin react identically with porcine and human insulin.) Samples and dilutions of an insulin standard are incubated with antiserum and ^{125}I-insulin. Antibody-bound insulin is then separated and the radioactivity of this fraction is determined. The amount of insulin in the sample is calculated by comparing the percent of bound radioactivity in the sample with the standard curve.

Specimen collection and storage. The specimen should be collected from a fasting patient except when otherwise indicated. Measurements of insulin may be carried out on serum or plasma. However, the use of heparin as an anticoagulant should be avoided because high concentrations of heparin interfere with the antigen-antibody reaction. Plasma is best collected in the presence of 1.5 mg EDTA/mL of whole blood. Hemolysis can cause falsely low insulin values. Insulin is stable in whole blood at room temperature for several hours. However, plasma and serum should be separated promptly and stored at −20 °C; repeated thawing and freezing should be avoided. *Urine* may be assayed for insulin content, but albumin must be added to the collection vessel to minimize denaturation of insulin in the specimen.

Reference range. After an overnight fast, plasma insulin levels in healthy, normal, subjects range from 2 to 25 μU/mL.

Comments. Antisera raised against insulin show some degree of cross-reactivity with proinsulin. Under most circumstances, proinsulin is present in blood in relatively small amounts and does not appreciably affect the absolute value of immunoreactive insulin. In certain situations (e.g., some islet-cell tumors) proinsulin may be the major component of circulating insulin. In such instances, direct assay of plasma may falsely estimate the true insulin concentration. The magnitude of the error will depend on the concentration of proinsulin and on the relative reactivity of the antiserum with insulin and proinsulin.

Radioimmunoassay of Proinsulin[104]

Because antisera specific for proinsulin are not available, an anti-insulin antibody must be used. However, separation of insulin from proinsulin prior to immunoassay by chromatography or incubation with insulin-specific protease provides an assay that is specific for proinsulin. The *reference range* for proinsulin in fasting plasma is 0.05–0.5 ng/mL.

Radioimmunoassay of C-Peptide[104]

C-peptide was initially difficult to measure by RIA since it does not react with anti-insulin antibodies, it is a poor immunogen, it is species specific, and it does not contain tyrosine needed for iodination. These difficulties have been overcome, and reliable commercial procedures are now available for C-peptide assay.[104] The *reference range* for serum from fasting individuals is approximately 1.0–4.0 ng/mL. However, reference ranges vary with the methods used.

ANTERIOR PITUITARY HORMONES

The pituitary gland (hypophysis) lies in a bony cavity, the sella turcica, located at the base of the skull. The gland is ~1 cm in diameter and weighs ~500 mg. Anatomically, the gland is divided into two lobes, anterior (adenohypophysis) and posterior (neurohypophysis). Blood reaches the pituitary from two sources: arterial blood is supplied by a branch of the internal carotid, and venous blood carrying neurosecretory material from the hypothalamus reaches the pituitary through a venous portal system.

Until recently, human anterior pituitary cells were categorized on the basis of their histologic staining reactions, as *acidophils* (about 40% of the cells), *basophils* (about 10% of the cells), and *chromophobes* (the remaining 50% of the cells). The anterior pituitary contains five types of cells: somatotropes, which secrete growth hormone (GH); mammotropes, which secrete prolactin (PRL); thyrotropes, which secrete TSH; gonadotropes, which secrete both luteinizing hormone (LH) and follicle-stimulating hormone (FSH); and corticotropes, which secrete both adrenocorticotropic hormone (ACTH) and β-lipotropin (β-LPH).

The pituitary gland secretes at least eight hormones, all of which are proteins or peptides. Of the eight hormones, six—GH, PRL, TSH, ACTH, FSH, and LH—are produced in the anterior pituitary. The anterior pituitary also secretes β-lipotropin and a number of smaller peptides of undetermined significance. GH and PRL act primarily on diffuse target tissues; the other tropic hormones act primarily on target endocrine glands. The other two hormones, vasopressin (also known as antidiuretic hormone, ADH) and oxytocin, are produced in the hypothalamic nuclei and are carried through the neurohypophyseal nerve axons to the posterior pituitary, where they are stored and released.

Regulation of Anterior Pituitary Function

The activity of the anterior pituitary is under the control of the hypothalamus, which synthesizes small peptides (neuropeptides) known as *releasing* or *inhibiting hormones* or *factors*. The existence of at least seven different and specific hypothalamic substances has been established. They are corticotropin-releasing hormone (CRH), thyrotropin-releasing hormone (TRH), growth hormone–releasing hormone (GRH), somatostatin (also called growth hormone–inhibiting hormone, GHIH), gonadotropin-releasing hormone (GnRH), prolactin-releasing factor (PRF), and prolactin-inhibiting factor (PIF, thought to be dopamine). GnRH is sometimes referred to as LHRH (luteinizing hormone–releasing hormone). The releasing hormones are not species specific. The therapeutic use of releasing hormones may be indicated in conditions where insufficient secretion of anterior pituitary hormones is due to hypothalamic dysfunction.

The functional relationship between the pituitary and its target glands is based on feedback regulation, primarily negative feedback from the target gland to the pituitary and hypothalamus. The effect of negative feedback control is opposite to that of the initial stimulus. The importance of such feedback control lies in the maintenance of an optimal concentration of hormones in the blood under a variety of physiological circumstances.

GROWTH HORMONE AND SOMATOMEDINS

Growth hormone (GH, somatotropin, STH), the most abundant hormone in the anterior pituitary, is a single-chain polypeptide with two disulfide bridges. It bears a marked structural resemblance to prolactin and to the placental hormone chorionic somatomammotropin (CS or placental lactogen), with which it shows overlapping biological activities.

Plasma levels of GH in normal adults remain stable and relatively low with four or five sharp spikes occurring 3–4 h after meals and a marked rise an hour after the onset of sleep.

Regulation of Secretion

The release of GH appears to be under the control of two hypothalamic factors. One is GH-releasing hormone (GRH) and the other is somatostatin or GHIH, which inhibits GH release. Somatostatin is also found in the D cells of the pancreatic islets of Langerhans and in other sites in the digestive tract, where it has inhibiting effects on gastrointestinal hormones and inhibits insulin and glucagon release. The hypothalamic influence on GH release seems to be predominantly stimulatory. The release of the two hypothalamic factors is, in turn, controlled by the higher centers of the brain. Thus, various stimuli such as exercise, physical and emotional stress, hypoglycemia, increased amino acid levels, and hormones such as testosterone, estrogens, and thyroxine stimulate GH secretion. The presence of high levels of glucocorticoids suppresses GH secretion.

Physiological Action

The overall action of GH is to promote growth of cartilage, bone, and many soft tissues by stimulating protein synthesis. Increased growth is accompanied by changes in electrolyte metabolism, including positive nitrogen and phosphorus balance, increase in intestinal absorption of calcium, and decrease in urinary excretion of sodium and potassium. GH also stimulates uptake of nonesterified fatty acids (NEFA) by muscle and accelerates fat mobilization from adipose tissues to the liver. In addition, GH stimulates hepatic glycogenolysis and antagonizes the effect of insulin on glucose uptake by peripheral cells (see Chapter 13), so that blood glucose concentrations increase. GH and insulin are similar in inducing growth, since they both have protein anabolic effects and stimulate the transport of amino acids into peripheral cells. However, their respective effects on the blood glucose level oppose each other. Most GH effects are delayed rather than immediate.

Somatomedins

GH stimulates secretion by the liver of peptides called *somatomedins*. Their effects include enhanced incorporation of sulfate into cartilage and stimulation of synthesis of DNA and RNA for collagen formation. These peptides (MW ranging from 6000 to 8500) include somatomedin A, somatomedin C, and insulin-like growth factors I and II [IGF I and II, which were formerly known as nonsuppressible insulin-like activity (NSILA)]. They show insulin-like activity in some tissues—they inhibit lipolysis and increase glucose oxidation in adipose tissue and stimulate glucose and amino acid transport into diaphragm and heart muscle. Unlike most other peptide hormones, somatomedins circulate in blood complexed with a plasma-binding protein.

Clinical Significance

Clinically important states of growth hormone excess or deficiency are rare. Growth hormone levels are quite variable under normal circumstances, so that measurement of GH under random conditions is generally not of value. GH measurements are best done as part of dynamic tests that involve pharmacologic or physiologic stimuli for release or suppression of GH.[27]

Somatomedin levels are fairly constant; a single measurement of somatomedin C can be taken as an accurate reflection of somatomedin production.[87] Somatomedin concentrations are increased in conditions of growth hormone excess and low in states of GH deficiency. However, somatomedin levels may also be elevated or depressed in conditions in which GH levels are normal.

Growth hormone excess. Excess growth hormone production is usually seen in association with eosinophilic or chromophobic adenomas of the pituitary. Prolonged excess of GH results in overgrowth of the skeleton and soft tissues. This most commonly occurs in adults and is known as *acromegaly*. In advanced cases, there is coarsening of facial features and enlargement of the hands and feet. The most important requirement for diagnosis is demonstration of inappropriate GH secretion. When GH excess occurs before long-bone growth is complete, there is a striking acceleration of linear growth, and the condition is termed *pituitary gigantism*.

Essentially all patients with acromegaly can be shown to have an abnormal GH response to an oral glucose load. The response in acromegaly may be either no change in the basal level or paradoxical increase of GH; normal individuals, on the other hand, will suppress GH levels to <5 ng/mL and usually to <1 ng/mL 60–120 min after a 100-g oral glucose load. Somatomedin C levels are usually elevated in active acromegaly, but the degree of clinical correlation is variable.

Growth hormone deficiency states and growth retardation. Children who lack sufficient GH production or action will not grow normally. GH deficiency may be congenital or acquired, idiopathic or due to anatomic damage to the pituitary or hypothalamus, or isolated or associated with deficiencies of other pituitary hormones. Although GH deficiency is not a common cause of growth retardation, children with growth retardation or dwarfism that has no clear explanation should at least be screened for GH deficiency.

GH deficiency in adults is probably the most common demonstrable abnormality in patients who have large pituitary adenomas or who have received pituitary irradiation, but this condition is rarely of clinical significance.

When GH production is evaluated, the use of provocative tests with arginine, L-dopa, or other agents is useful. Although a normal GH response to a provocative test points strongly against GH deficiency, no single test can be considered diagnostic. In many centers, testing for GH deficiency is done by obtaining frequent (every 20–30 min) blood specimens for GH; lack of normal secretory episodes after meals and during sleep is taken as evidence of GH deficiency.

Somatomedin levels are low in patients with growth hormone deficiency and growth failure. However, patients with growth failure due to other endocrine diseases or to nonendocrine organic diseases also have low levels of somatomedin; thus a low level of somatomedin is not a specific test for GH deficiency.

Methods for the Determination of Growth Hormone[82] and Somatomedins[20]

Radioimmunoassay is the method of choice for the analysis of GH, and reliable kits are now available. Somatomedin assays on plasma or serum by in vitro bioassay, using cartilage radioreceptor assay or radioimmunoassay, have been reported.[20,87] Although these assays have provided some overall information about the concentrations of somatomedins in health and disease, they have limitations. For example, radioligands and antisera used in radioimmunoassays do not interact equally with all forms of somatomedins, resulting in poor precision and specificity. Bioassays lack precision and are affected by the presence in serum of inhibitors of biological activity. However, immunoassay and bioassay techniques may give complementary results when levels of growth factors in the presence of suspected inhibitors are evaluated.

PROLACTIN

Prolactin (PRL) is similar in structure to growth hormone. PRL is secreted by the pituitary mammotropic cells. A biologically inactive high-molecular-weight "big prolactin" may be produced by pituitary adenomas. PRL secretion is pulsatile and diurnal; values are slightly higher in the afternoon than in the morning, and there is a surge after the onset of sleep.

Regulation of Secretion

PRL secretion is controlled by an inhibitory influence from the hypothalamus that is thought to be dopamine. The physiologic stimulus for PRL release is suckling; maternal plasma levels increase within minutes after initiation of breast feeding, and following delivery they remain elevated longer if the newborn is breast fed. Many other stimuli induce PRL release; stress is one factor. The most important influence on PRL regulation appears to be exerted by PRL itself by way of the short feedback loop (feedback regulation between the pituitary and the hypothalamus). Short feedback loop inhibition of GnRH secretion by PRL has also been suggested as being responsible for the inhibition of gonadotropin secretion that occurs in women who are nursing and in patients with PRL-secreting adenomas of the pituitary.

Physiological Action

Initiation and maintenance of lactation are the primary functions of PRL. However, breast tissue must first be primed by estrogens, progestins, corticosteroids, and insulin for prolactin to act. PRL induces ductal growth, development of the lobular alveolar system, and synthesis of specific milk proteins.

PRL, like other pituitary hormones, has a receptor on the cell membrane of its target organs—the breast, adrenal, ovary, testis, kidney, and liver. However, the exact mechanism of PRL action at the molecular level has not been delineated.

Clinical Significance

Now that assays for PRL are widely available, patients with PRL elevations are identified more frequently. However, there is still a lack of agreement as to what constitutes a significant PRL elevation, what (if any) stimulus for PRL release or suppression gives helpful diagnostic information, and what tests help explain the cause of PRL elevations. PRL deficiency does not appear to have any clinical importance.

Hyperprolactinemia. PRL levels may be elevated in patients who have variable alterations of fertility such as anovulation with or without menstrual irregularity, amenorrhea and galactorrhea, or galactorrhea alone. PRL excess in males may present as oligospermia or impotence or both. As many as 30% of patients with clinically silent pituitary adenomas will have elevated PRL levels. Microadenomas of the pituitary that produce PRL may present as galactorrhea-amenorrhea or simply as infertility. A marked increase (>5 times the upper reference limits) in the basal level of PRL is relatively strong evidence for a pituitary tumor. Intermediate elevations of PRL may be seen with pituitary macroadenomas, microadenomas, or with no demonstrable pathological endocrine disorder. Thyrotropin-releasing hormone (TRH) will stimulate PRL in normal individuals. Failure of TRH to stimulate PRL is seen with pituitary adenomas, but this finding is not specific.

If only a slight elevation of PRL is present in an initial sample, it is advisable to repeat the measurement on at least two other occasions. The specimen should be obtained with minimal trauma, no breast stimulation, and in the morning; the patient should not take medication that stimulates PRL release.

Methods for the Determination of Prolactin[99]

The isolation and purification of human PRL have made possible the development of homologous radioimmunoassays,[99] and a number of commercial kits are available. Like other protein hormones, PRL has also been measured by radioreceptor assays.[31]

Radioiodinated PRL is allowed to bind to specific membrane receptors prepared from rabbit or mouse mammary glands or from rabbit liver. Biological activity is determined by the ability of PRL in the sample to inhibit the binding of radioiodinated PRL to these receptors. This method does not discriminate between PRL and growth hormone or chorionic somatomammotropin.

Reference ranges vary with the kits employed. Typical normal levels (in ng/mL) for PRL in premenopausal females are <20; for postmenopausal females, <12; and for males, <15.

CORTICOTROPIN (ACTH) AND RELATED PEPTIDES

Corticotropic cells of the anterior pituitary synthesize a large precursor protein called *pro-opiomelanocortin* or "31 K precursor" that is cleaved to form ACTH (amino acids 1–39) and β-lipotropin (amino acids 1–91, β-LPH). Regulation of secretion, physiological action, and clinical significance of ACTH are described in the section on *Adrenocortical Steroids*, page 561.

β-LPH contains the amino acid sequences of the endogenous opioid peptides *α-endorphin, γ-endorphin, β-endorphin,* and *enkephalin*, although there is evidence to indicate that enkephalin arises independently, presumably from another prohormone. Their physiological actions include sedation, an increase in the threshold of pain, changes in extrapyramidal motor activity (e.g., postural control), and possibly changes in behavior. Endorphins and enkephalins may also form a major peptidergic system in the brain that is involved in modifying endocrine responses to stress.

Determination of ACTH by RIA[80]

RIA is the method of choice for the measurement of plasma ACTH.[56] Commercial reagent kits are now available.

Specimen collection and storage. ACTH is adsorbed onto glass surfaces, is very unstable in whole blood, and is destroyed by proteolytic enzymes that are activated during freezing and thawing of the sample. Physiological factors that influence plasma ACTH levels, such as prior administration of corticosteroids, time of the day at which the sample is taken (diurnal variation), and stress resulting from a poorly performed venipuncture, must be taken into account.

Blood samples for ACTH should be drawn by an experienced phlebotomist using a lightly heparinized plastic syringe at a specified time. If repeated serial samples are required, these should be obtained via an indwelling venous needle or catheter. High concentrations of heparin may interfere with antigen-antibody binding, and EDTA concentrations >1 mmol/L may interfere with RIA. Blood should be collected into a plastic tube chilled in ice water, then centrifuged for 15 min at 1000–1500 × *g* at 4 °C. The supernatant should then be recentrifuged at 6000 × *g* at 4 °C for another 10 min. The second centrifugation is very important since it removes additional formed elements that, if frozen and then thawed for assay, apparently rupture and release proteolytic enzymes that damage ACTH. Plasma is stored at −20 °C.

Reference ranges are 25–100 pg/mL at 0800 h, and <25 pg/mL at 1800 h.

PITUITARY GLYCOPROTEIN HORMONES

The glycoprotein hormones of the pituitary (LH, FSH, TSH) and placenta (hCG; see Chapter 28) are composed of two nonidentical, noncovalently bound subunits, α and β, each with a carbohydrate substituent group attached. The carbohydrate moiety, accounting for 15–31% of the molecular weight, includes fucose, mannose, galactose, glucosamine, galactosamine, and sialic acid. The α-subunits of these hormones are interchangeable and are devoid of biological activity. The β-subunits confer hormonal and immunological specificity. Isolated β-subunits have little intrinsic biologic activity, but full activity is attained when α- and β-subunits are recombined. The availability of specific antisera for β-subunits has led to the development of radioimmunoassay techniques for measurements of these hormones (see *Human Chorionic Gonadotropin,* Chapter 28).

GONADOTROPINS

The gonadotropic cells of the anterior pituitary secrete two gonadotropins, follicle-stimulating hormone (FSH) and luteinizing hormone (LH). Since these two hormones control the functional activity of gonads, they are grouped together under the generic term *gonadotropins.*

Physiological action. In females, FSH stimulates growth of ovarian follicles and, in the presence of LH, promotes secretion of estrogens by maturing follicles. LH in females causes release of ova from ovarian follicles that have previously been ripened by FSH. After ovulation, transformation of the ruptured follicle into the corpus luteum and secretion of progesterone by the corpus luteum are primarily under the influence of LH. In males, FSH stimulates spermatogenesis, and LH is responsible for production of testosterone by the Leydig cells of the testes. Regulation of LH and FSH secretion and their clinical significance in reproductive endocrinology are discussed under *Reproductive Endocrinology,* page 580.

Methods for the Determination of LH and FSH[75]

Reliable commercial radioimmunoassay kits for both blood and urine samples are available. Plasma, serum, or urine specimens can be utilized for LH and FSH measurements. However, because of episodic, circadian, and cyclic variations in secretion of gonadotropins, a meaningful clinical evaluation of these hormones often requires determinations in pooled blood samples, in multiple serial blood samples, or in a timed sample of urine.

Reference Ranges

	FSH, mIU/mL	LH, mIU/mL
Female,		
Follicular phase:	2–10	0–14
Midcycle peak:	9–18	20–70
Luteal phase:	0–9	0–16
Postmenopause:	20–100	20–70
Male:	2–10	0–9

Mean FSH and LH levels in prepubertal boys and girls are approximately half those of adults.

THYROTROPIN

Thyrotropin (thyroid-stimulating hormone, TSH) is a glycoprotein secreted by the thyrotropic cells of the anterior pituitary. The α subunit of TSH is similar to the α-chain of LH, FSH, and hCG. TSH stimulates the growth of the thyroid follicular cells and practically every step in thyroid hormone synthesis, including the uptake of iodine, organic binding, coupling of tyrosines, and release of thyroid hormones. Regulation of secretion of TSH, its clinical significance, its relation to thyroid disease, methods for its determination, and reference ranges, are discussed under *Thyroid Function,* page 584.

POSTERIOR PITUITARY HORMONES (NEUROHYPOPHYSEAL HORMONES)

The posterior pituitary is part of a neurosecretory system consisting of the supraoptic and paraventricular nuclei in the hypothalamus, the tracts of nerve fibers from the cell bodies in the hypothalamic nuclei, and the posterior lobe of the pituitary. *Vasopressin* (antidiuretic hormone, ADH) is synthesized mainly in the supraoptic nuclei, whereas *oxytocin* is produced in the paraventricular nuclei. After synthesis, both oxytocin and ADH become bound to specific proteins called *neurophysins*. There are separate and specific neurophysins for both ADH and oxytocin. These molecular complexes are incorporated into granules that migrate down the nerve axons to the posterior pituitary, where they are stored. The nerve endings are intimately associated with posterior pituitary capillaries, which carry the hormone into the blood stream.

Regulation of Secretion

ADH. The principal stimulus for ADH secretion is an increase in plasma osmolality. Osmoreceptors located in the hypothalamus and probably in the posterior pituitary respond to changes in plasma osmolality. Other important stimuli for ADH release include a reduction in plasma volume or arterial pressure (or both), pain, stress, sleep, exercise, and drugs such as nicotine, morphine, and barbiturates. Hyposmolality, volume expansion, hypertension, exposure to cold, alcohol, and some drugs (e.g., phenytoin, glucocorticoids) inhibit the release of ADH.

Oxytocin. The primary stimulus for oxytocin release is suckling. Stretch receptors in the uterus and possibly in the vaginal mucosa may also stimulate release of oxytocin. Estrogens enhance the response of oxytocin to these stimuli. Emotional stress inhibits lactation, possibly by inhibiting oxytocin release.

Physiological Action

ADH. The principal physiological function of ADH is stimulation of water reabsorption from the renal tubular fluid. The blood level of ADH directly regulates water balance, which indirectly regulates the concentration of osmotically active solutes in extracellular fluid. ADH increases the volume of water reabsorbed by the renal tubules; consequently, urine becomes concentrated and its volume decreases. A further action of ADH is expressed by its alternative name, vasopressin. In sufficient quantity, ADH induces generalized vasoconstriction that leads to a rise in arterial blood pressure. ADH is believed to play an important role in maintaining arterial blood pressure during acute blood loss.

Oxytocin is present in both males and females but its physiological effects are known only in females. Oxytocin stimulates contraction of the estrogen-primed uterus. It is a useful therapeutic agent in inducing labor, but the physiological mechanism whereby oxytocin induces uterine contractions remains obscure. Oxytocin also stimulates lactation by the mammary gland. Progestins are believed to inhibit the action of oxytocin.

Clinical Significance

Polyuric states. Deficient production or action of ADH results in polyuria due to the failure of the renal tubule to reabsorb solute-free water. When urine output is >2.5 L/d, further investigation is usually indicated; with complete deficiency of ADH, urine output may approach 1 L/h. If thirst is normal, increased ingestion of fluid (polydipsia) will follow; if access to water is not restricted, serum osmolality and electrolytes will usually remain normal.

Polyuric states can be divided into three general categories: (1) deficient ADH production (central diabetes insipidus), (2) decreased response of the kidney to the action of ADH (nephrogenic diabetes insipidus), and (3) excessive water intake (psychogenic polydipsia).

Central diabetes insipidus may be idiopathic or due to destructive or infiltrative lesions involving the hypothalamus or pituitary. The deficiency of ADH may be partial or complete. *Nephrogenic diabetes insipidus* may be due to a variety of conditions affecting the kidney, such as amyloidosis, multiple myeloma, and chronic electrolyte abnormalities (hypokalemia, hypercalcemia), as well as side effects from certain medications, including lithium, demeclocycline, and methoxyflurane. Nephrogenic diabetes insipidus may also be idiopathic or inherited, partial or complete, transient or permanent. This form of diabetes insipidus is due to an inability of the renal tubule to respond normally to ADH.

Psychogenic polydipsia may be difficult to distinguish from diabetes insipidus. Patients with psychogenic polydipsia generally have significant psychiatric illness and may deny that their intake of liquids is excessive. In many cases, it may be difficult to determine whether the primary problem is due to excessive fluid intake or to uncontrolled water loss as a result of deficient ADH action.

Diagnostic studies. Since patients with diabetes insipidus may be extremely sensitive to water restriction, diagnostic studies are best performed in the hospital under close supervision. To document polyuria prior to hospitalization, the patient should collect urine for two separate days on an ad lib fluid intake. Creatinine in these collections should be measured to provide an index of the adequacy of the collection. Substances such as nicotine, alcohol, and caffeine that influence the action of ADH must be avoided.

The traditional approach to the diagnosis of polyuric disorders is the water deprivation test.[73] During a period of water deprivation, patients with central or nephrogenic diabetes insipidus lose free water; the result is weight loss, hypernatremia, and hyperosmolality of plasma. Urine osmolality remains low, often less than that of plasma. If maximum urine osmolality after water deprivation is <400 mOsm/kg, some form of diabetes insipidus is likely. At this point, an injection of vasopressin may be given and urine osmolality measured 60 min later. An increase in urine osmolality of >10% after vasopressin administration indicates central diabetes insipidus, whereas failure to increase urine osmolality after ADH injections indicates nephrogenic diabetes insipidus. Patients with psychogenic polydipsia will usually produce a normal urine osmolality after water deprivation, but occasionally patients with psychogenic polydipsia fail to produce a concentrated urine and show a variable response to ADH injection.

Measurement of serum ADH concentration will help to separate the difficult cases.[127] Patients with central diabetes insipidus show low or inappropriately "normal" values for ADH in the presence of hyperosmolality induced by water deprivation. Patients with nephrogenic diabetes insipidus will have high plasma ADH levels but will continue to produce hypotonic urine. Patients with psychogenic polydipsia will have low or normal levels of plasma ADH, and their plasma will not be hyperosmolal. Since the best discrimination occurs when the patient is hyperosmolal, blood for ADH can be obtained after hypertonic saline is given to produce a hyperosmolal state.

Syndrome of inappropriate ADH secretion (SIADH). Excess ADH causes retention of free water by the kidney, which leads to volume expansion and dilutional hyponatremia. SIADH may be caused by a variety of acute and chronic disorders of the central nervous system or by pulmonary disorders, malignancies, and other illnesses. Other conditions that cause dilutional hyponatremia and that must be differentiated from SIADH include congestive heart failure, renal insufficiency, chronic liver disease, glucocorticoid deficiency, and hypothyroidism as well as excessive administration of hypotonic fluids, diuretics, and drugs that stimulate or potentiate ADH. Hyponatremia may also occur from depletion of total body sodium content as a result of mineralocorticoid deficiency, renal sodium wasting, or excessive gastrointestinal losses of sodium.

The clinical manifestations of hyponatremia are nonspecific. These include weakness and central nervous system changes such as lethargy, confusion, coma, and convulsions. There are no signs or symptoms that are specific for SIADH. History, physical examination, and routine laboratory tests often indicate whether hyponatremia is dilutional or depletional and may indicate the presence of one of the causes of hyponatremia.

Measurement of sodium and osmolality in plasma and urine will usually permit this differentiation. The typical patient with SIADH will have hyposmolal plasma (<275 mOsm/kg), urine osmolality that slightly exceeds plasma (300–400 mOsm/kg), and a urine sodium concentration that is neither high nor low (40–80 mEq/L). Patients with hyponatremia from excess water intake will have hypotonic plasma, urine sodium concentration usually >80 mEq/L, and urine osmolality less than plasma. Patients with hyponatremia from total body sodium depletion have hypotonic plasma, a low urine sodium concentration (usually <20 mEq/L), and urine osmolality greater than serum (often >600 mOsm/kg).

If the cause for hyponatremia remains unclear after determination of sodium and osmolality on plasma and urine, a water loading test may be done. Patients with SIADH show a reduced ability to excrete a water load. Measurement of plasma ADH is not usually needed in these patients, but plasma ADH would be expected to be high relative to the patient's hyposmolar state.

Methods for the Determination of ADH and Oxytocin[47]

Numerous RIA methods for measuring ADH or oxytocin have been described. However, their routine clinical applications have been hampered because of their complexity and their lack of sensitivity and specificity. Measurements of serum and urine osmolality are sufficient in most clinical situations for evaluating patients with ADH problems. The clinical significance of oxytocin measurements is not established.

STEROID HORMONES

Structure and Nomenclature

Steroids are compounds containing the cyclopentanoperhydrophenanthrene ring system (Figure 18-4). Three six-membered rings (A, B, C) constitute the phenanthrene nucleus to which is attached a five-membered ring (D), cyclopentane. The prefix "perhydro" refers to the fact that all the necessary hydrogen atoms are present to make it fully saturated. Examples of this class of compounds include sterols (e.g., cholesterol), bile acids (e.g., cholanic acid), sex hormones (e.g., estrogens, androgens), adrenocortical hormones (corticosteroids), cardiac glycosides (e.g., digitoxigenin, the aglycone of digitoxin), sapogenins (e.g., tigogenin), and some alkaloids (e.g., solasodine). Steroid hormones contain up to 21 carbon atoms (C_{21} steroids), numbered as shown in Figure 18-4. Each carbon atom of a ring bears two hydrogen atoms, except when it is common to two rings or the side chain at C-17, in which case it bears only one hydrogen atom. The carbon atoms composing the rings and the hydrogen atoms attached to them are not usually written into the structure unless they are required to draw special attention to configuration. Furthermore, in all naturally occurring steroid hormones, the solid line projecting from the carbon atom at position 10 or 13 designates the presence of an angular methyl group, unless otherwise indicated.

Steroids are three-dimensional, and thus the constituent atoms lie in different planes, giving rise to isomers. If a hydrogen atom points in the same direction as the angular methyl groups, the compound is said to be the *cis* or *normal* form. If, however, they are on opposite sides, the compound is said to be the *trans* or *allo* form. Although the rings A and B may be either cis or trans, the rings B/C and C/D have trans configuration in all naturally occurring steroid hormones. Substituents on the same side as these two methyl groups are said to possess a β-configuration, which is indicated by a solid line (—) joining them to the appropriate carbon atoms in the nucleus. Substituents on the opposite side are attached by a broken line (- - -) to denote an α-configuration.

The innumerable steroids containing the cyclopentanoperhydrophenanthrene nucleus differ from one another by double bonds between certain pairs of carbon atoms, by substituents for the hydrogen atoms, or by a specific type of side chain attached to

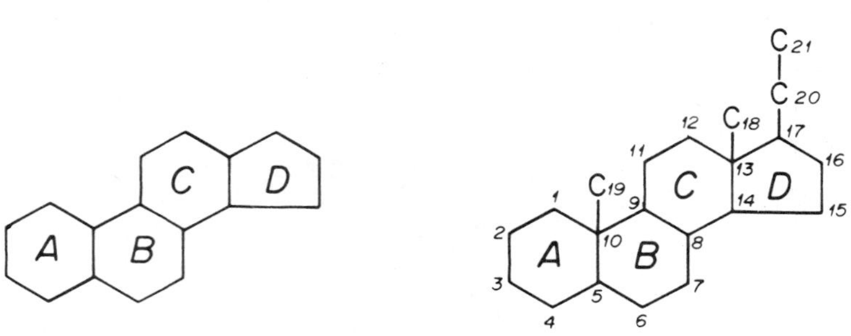

A B C D

Cyclopentanoperhydrophenanthrene

C_{21} C_{20} 12 C_{18} 17 11 16 13 C_{19} C D 1 9 14 15 2 10 8 A B 3 5 7 4 6

Figure 18-4. Numbering system of steroids.

Table 18-3. COMMON SUFFIXES AND PREFIXES FOR STEROIDS

Suffix or Prefix	Definition
Suffix	
-al	Aldehyde group
-ane	Saturated hydrocarbon
-ene	Unsaturated hydrocarbon
-ol	Hydroxyl group
-one	Ketone group
Prefix	
hydroxy- (oxy-)	Hydroxyl group
keto- (oxo-)	Ketone
deoxy- (desoxy-)	Replacement of hydroxyl group by hydrogen
dehydro-	Loss of two hydrogen atoms from adjacent carbon atoms
dihydro-	Addition of two hydrogen atoms
cis-	Spatial arrangement of two substituents on the same side of the molecule
trans-	Spatial arrangement of two substituents on opposite sides of the molecule
α-	Substituent which is *trans* to the methyl group at C-10
β-	Substituent which is *cis* to the methyl group at C-10
epi-	Isomeric in configuration at any carbon atom except at the junction of two rings
Δ^n-	Position of unsaturated bond

C-17. On the basis of such structural characteristics, steroidal compounds are classified as derivatives of *estrane* (for estrogens), *androstane* (for androgens), and *pregnane* (for corticosteroids and progestins). In addition, a variety of suffixes and prefixes are used to describe a chemical compound (Table 18-3).

General Aspects of Biosynthesis and Metabolism

Cholesterol is the precursor of steroid compounds and may be synthesized from acetic acid by steroidogenic cells or taken up from circulating lipoproteins (see Chapter 14). In the adrenals, ovaries, placenta, and possibly the testes, lipoprotein cholesterol is the major source of steroidogenic substrate. With the exception of the placenta, lipoprotein cholesterol uptake is coordinated with intracellular cholesterol synthesis and mobilization of cholesteryl esters to ensure a continuous supply of free cholesterol for steroid synthesis. When the rate of cholesterol uptake exceeds the rate of steroidogenesis, intracellular cholesterol synthesis is suppressed and cholesterol in excess of cellular needs is esterified and stored for future use.

The adrenals, ovaries, and testes all utilize the same precursors, acetate and cholesterol, but the nature and the quantity of steroid hormones produced by each gland depend upon the degree of activity of certain enzyme systems. For example, the enzymes 11β-hydroxylase and 21-hydroxylase, present only in the adrenals, synthesize the unique hormones of the adrenal glands (cortisol and aldosterone). Similarly, the enzymatic distinction between the ovaries and the testes lies in the fact that, in addition to the enzymes found in the testes, the ovaries contain an active aromatizing enzyme system that converts male sex hormones to female sex hormones (e.g., testosterone to estradiol).

The enzymes of steroid hormone synthesis may be broadly classified into the following functional groups:

1. Hydroxylases. These enzymes catalyze the substitution of a hydroxyl group (—OH) for hydrogen (—H). For example, 21-hydroxylase introduces a hydroxyl group at C-21. The required cofactors are NADPH and molecular oxygen; the reaction is irreversible.

$$\mathrm{H{-}{}^{21}C(H)_2{-}{}^{20}C{=}O} \xrightarrow[\text{21-Hydroxylase}]{\text{NADPH, } O_2} \mathrm{HO{-}{}^{21}C(H)_2{-}{}^{20}C{=}O}$$

2. *Desmolases.* These enzymes are required for cleaving the side chain. There are two desmolases: 20,22-desmolase, which participates in the conversion of C_{27} cholesterol to a C_{21} compound, pregnenolone; and 17,20-desmolase, which transforms C_{21} steroid hormones to C_{19} steroid hormones. The required cofactors are thought to be NADPH and molecular oxygen.

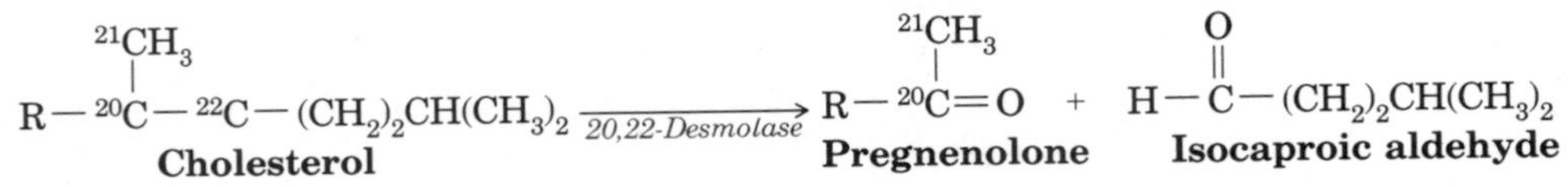

$$R-{}^{20}C(^{21}CH_3)-{}^{22}C-(CH_2)_2CH(CH_3)_2 \xrightarrow{\text{20,22-Desmolase}} R-{}^{20}C(^{21}CH_3)=O + H-C(=O)-(CH_2)_2CH(CH_3)_2$$

Cholesterol → **Pregnenolone** + **Isocaproic aldehyde**

3. *Dehydrogenases.* This group of enzymes catalyzes the transfer of hydrogen (oxidation and reduction). The reaction is generally reversible. The cofactor is either the oxidized or reduced form of NAD^+ or $NADP^+$, depending on the direction of the reaction. An example is 17β-hydroxysteroid dehydrogenase.

NAD⁺ or NADP⁺ → *17 β-Dehydrogenase* ← *NADH or NADPH*

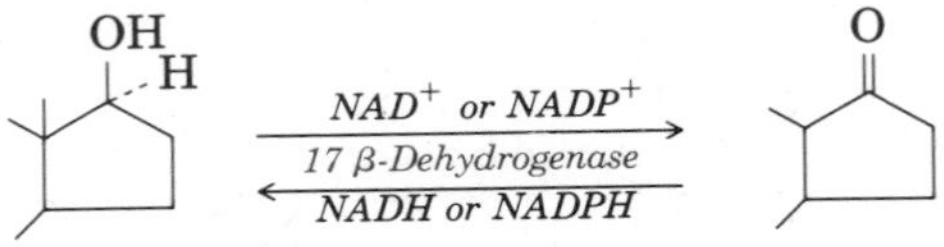

4. *Isomerases.* These enzymes catalyze the migration of a double bond. The most important enzyme of this group is Δ^5-ketosteroid isomerase (Δ^{4-5}-isomerase). The concerted action of Δ^5-β-hydroxysteroid dehydrogenase and Δ^5-ketosteroid isomerase on pregnenolone produces progesterone.

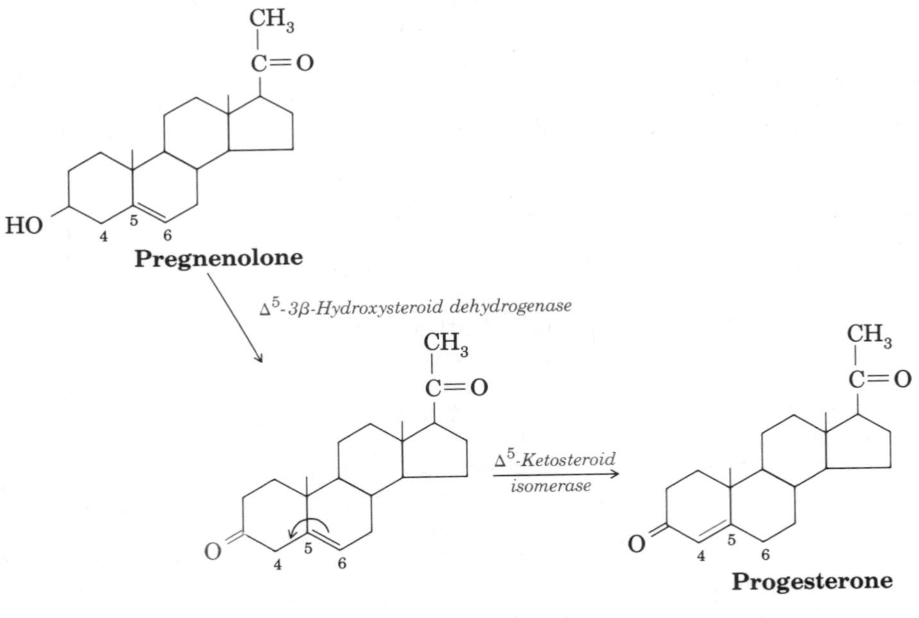

Pregnenolone

Δ^5-3β-Hydroxysteroid dehydrogenase

Δ^5-Ketosteroid isomerase

Progesterone

The liver is the major site of steroid catabolism; however, the kidney and the gastrointestinal tract may also carry out some of the metabolic transformations. Important biochemical steps for neutralizing the biological activity of hormones and for their rapid elimination from the systemic circulation are (1) the introduction of a new hydroxyl group (e.g., estradiol to estriol); (2) dehydrogenation (e.g., testosterone to androstenedione); (3) the reduction of a double bond (e.g., cortisol to dihydrocortisol); or (4) conjugation of one or more essential hydroxyl groups with a chemical moiety such as glucuronic acid (e.g., testosterone to testosterone glucuronide). The conjugation of these hormones and their metabolites with sulfuric or glucuronic (glucosiduronic) acid is by far the most efficient single metabolic process for their excretion in the urine.

Methods for the Determination of Steroid Hormones

Specimen. The *urinary* excretion of a hormone, its metabolites, or both, represents a proportion of the amount secreted during the period of urine collection. However, such factors as incompleteness of collection, altered renal function, and contribution by more than one gland to the total excretion of hormone warrant special precaution in the interpretation of urinary values. Analyses of total urinary metabolites have further shortcomings, since they measure only the fraction of a hormone that is metabolized; the quantity and nature of these metabolites may, in turn, be dependent not only on pathologic conditions but also on the intake of drugs and diet. Other arguments against urinary assay of metabolites include the delay and inconvenience encountered in the collection of 24-h urine specimens. In contrast, *plasma* samples are more convenient for the RIA of specific plasma steroids. Moreover, plasma assays are more appropriate than urine measurements in clinical investigations that dictate rapid and repeated analyses, such as dynamic tests, control of ovulation induction, and monitoring of fetoplacental function. However, plasma estimations may be limited in value because there are rapid fluctuations in hormone levels, and thus a plasma sample is representative only of the concentration at the time of sampling. Rapid separation of erythrocytes from the specimen is important since erythrocytes at room temperature can alter plasma concentrations of active steroid hormones by degradation.

It has been suggested that the measurement of *salivary* steroids provides information similar to that derived from urinary free steroids.[94] Patients find little difficulty in salivating directly into disposable tubes, thus providing an adequate volume in ~10 min. Assays of samples collected at 1- to 2-h intervals provide an accurate assessment of baseline endocrine activity. Since smaller aliquots (500 μL) can be collected at 15- or even 10-min intervals, salivary samples could well be more useful than either plasma or urine samples in short-term dynamic tests. Although results appear promising, the clinical usefulness of salivary measurements of steroids is not yet fully established.

Hydrolysis, Extraction, and Separation

Chemical assay procedures for steroids have many similarities in requirements for hydrolysis, extraction, purification and separation, and final quantitation. Techniques such as HPLC and RIA can often be applied directly to unmodified samples.

Hydrolysis. Steroid hormones and their metabolites are present in blood and urine as water-soluble conjugates of glucuronic (glucosiduronic) acid and sulfuric acid. Two general types of procedures for hydrolysis are available; namely, acid hydrolysis and enzymatic hydrolysis. In *acid hydrolysis*, an aliquot of urine is boiled in the presence of a mineral acid. For *enzymatic hydrolysis*, a portion of a urine or plasma sample is adjusted with buffer to the optimal pH for the enzyme used; after addition of an adequate amount of β-glucuronidase and sulfatase, the test sample is incubated for 18 to 72 h.

Acid hydrolysis is preferred because of simplicity, speed, and completeness of reaction. Enzymatic hydrolysis, on the other hand, requires special attention to optimal concentration and type of enzyme, pH, temperature, and duration of incubation. In addition, the variable presence of enzyme inhibitors casts some doubt on the degree of completeness of enzyme hydrolysis. However, enzymatic hydrolysis is obligatory for those steroids that are labile in strong acid solution; it is also used to avoid interfering substances produced by acid hydrolysis.

Extraction. Following hydrolysis, deconjugated steroids are sparingly soluble in aqueous solution. Thus, when the hydrolyzed sample is shaken with an immiscible organic solvent in which steroids are highly soluble, the vast majority of free and deconjugated steroids are extracted into the organic solvent. Repeating the extraction process with a fresh volume of organic solvent increases the recovery of steroids. Note that the addition of an organic solvent to plasma denatures the binding proteins. Thus, the extracted steroid represents both protein-bound and unbound fractions.

Selection of the extraction solvent is based on the polarity of the steroid hormone to be measured. The nonpolar tetracyclic ring system is common to all steroids, but polarity increases as the number of oxygen groups (i.e., ketone and hydroxyl groups) and double bonds increases. Steroids with three or more oxygens (e.g., corticosteroids and their metabolites) are quite polar; for their extraction, polar organic solvents are most suitable.

In addition, the improved recovery of aqueous compounds such as urinary corticosteroids and estriol from the polar phase can be achieved by adding ammonium sulfate or sodium chloride prior to extraction.

Purification and separation. Although the proper choice of solvent improves the selectivity of extraction, a large number of closely related steroids, chromogenic substances, and other nonspecific materials are also extracted. The removal of such contaminants, especially those that will interfere in the final estimation, is very important.

The *solvent partition* method is the simplest and most suitable method for a clinical laboratory. Steroids and other urinary impurities contained in the organic solvent are treated with a weakly basic solution that partitions strongly acidic components into the basic aqueous layer. For example, phenolic steroids (estrogens) can be readily extracted from an organic phase with an aqueous solution of sodium hydroxide. Most often, the organic extract is washed to neutrality with water to ensure complete removal of alkali which, if allowed to remain, might interfere with the subsequent analytical process.

The degree of purification and separation needed will depend on the method used for final quantitation. For example, if the final mode of estimation is a very specific color reaction or an RIA that uses fairly specific antiserum, the quantitation may be carried out without further purification of organic extracts. However, specific measurement of an individual steroid or group of steroids may necessitate further purification and separation of the extract. Various techniques available for this purpose are based either on some selective chemical reaction (e.g., digitonide formation to separate 3β-hydroxy- from 3α-hydroxysteroids) or on physical techniques such as countercurrent distribution, paper and column-partition chromatography, column and thin-layer adsorption chromatography, gel filtration, gas chromatography (GC), or high performance liquid chromatography (HPLC).

Quantitation

Methods for quantitative estimation of steroids may be divided into five categories: colorimetric, fluorometric, chromatographic (GC and HPLC), radioisotope (double-isotope derivative formation), and saturation analysis (radioimmunoassay). Radioimmunoassay is by far the most sensitive and widely applied assay method at present.

Colorimetric estimation. In these methods a specific functional group of a steroid is reacted with a reagent to form a colored product. For example, steroids containing a keto ($>$C=O) group in position 17 (17-ketosteroids) react with *m*-dinitrobenzene in alcoholic alkali to produce a reddish-purple compound (the Zimmermann reaction). The intensity of the color is proportional to the concentration of the steroid. The principal source of error in this type of procedure is the interference from nonspecific chromogens derived from other steroidal and nonsteroidal components of biological extracts.

Fluorometric estimation. Many steroids produce characteristic fluorescence in a medium such as sulfuric acid or phosphoric acid. Although this technique increases sensitivity and specificity, it has many drawbacks, including nonspecific fluorescence and quenching effects caused by solvent residues and inadequately purified reagents. The need for controlled reaction conditions and the instability of fluorescence have further limited the usefulness of this technique.

Gas chromatographic estimation. This technique offers speed, sensitivity, accuracy, precision, and specificity. It has been found to be quite suitable for routine measurement of urinary pregnanediol, pregnanetriol, 17-ketosteroids, and estriol.[30,125]

High performance liquid chromatography (HPLC).[49] HPLC is the fastest-growing technique in clinical chemistry. Its advantages include high resolution that is superior even to GLC; increased speed, since analysis times of <1 h are common; increased sensitivity, since UV absorption detectors can detect nanogram quantities and the detection sensitivity can be increased to picogram quantities by the use of fluorescence and electrochemical detectors; reusable columns; and easy sample recovery, since most detectors are nondestructive. In contrast to GLC, nonderivatized compounds and compounds that are thermally labile or nonvolatile can also be analyzed.

Both adsorption and reversed-phase partition chromatography using isocratic or gradient elution modes have also been employed. Steroids are usually co-assayed with an internal standard so that their elution times, relative to the internal standard, can be used for identification of individual steroids and that corrections can be made for losses due to extraction inefficiency. Quantitation is based on measuring the peak heights or areas of the eluted steroids. Most published methods for steroid analyses by HPLC are still investigative analytical procedures and are less popular than other routine clinical techniques such as radioimmunoassay.

Double-isotope derivative assay. In the double-isotope derivative assay, a steroid labeled with an isotope of high specific activity and negligible mass (e.g., ^{3}H or ^{14}C) is added to the specimen to be analyzed. Following extraction and purification of the sample, a suitable reagent labeled with a second isotope (e.g., ^{35}S) is added to yield chemically stable derivatives of the analyte. After further purification, measurement of the isotope ratio of the derivative provides the amount of steroid originally present in the sample. Two isotopes that emit different energy spectra are chosen so that the radioactivity of both isotopes can be measured simultaneously in a liquid scintillation counter. The radioactive steroid added prior to processing of the samples serves as an indicator for calculating losses during the extraction and purification, whereas the formation of a derivative with a radiolabeled reagent aids in the specific detection and the quantitation. Double-isotope derivative assays are considered reference methods because of their high sensitivity, specificity, and accuracy.[25] Their complexity and expense generally preclude their routine use.

Determination of Steroid Hormones by Radioimmunoassay

The measurement of steroid hormones by radioimmunoassay is the most significant advance in steroid methodology of the past decade. Because of its speed, sensitivity, specificity, simplicity, and application to simultaneous assays of a large number of samples, the technique has been applied to the determinations of virtually every steroid of physiological importance.

Steroid hormones are not naturally immunogenic; antibodies are produced by immunization of animals with steroids covalently linked to carrier proteins. The most commonly used carrier protein is bovine serum albumin (BSA). Antiserum tends to be specific for that part of the steroid molecule farthest removed from the site of conjugation.

Recently, the use of monoclonal antibodies has received considerable attention.[43] The hybridoma technique makes it feasible to select an antibody with desired specificity and sensitivity from a large number of antibodies produced by numerous clones. The availability of such standardized antibodies holds a great potential for achieving significant improvement in overall assay specificity.

Clinical laboratories generally prefer to use steroids labeled with γ-emitters because they usually have short half-lives and high specific activity, resulting in short counting times and great assay sensitivity. The most widely used γ-emitting isotope is ^{125}I, which has an adequate half-life (60 d). It is also less hazardous (has lower energy of emitted radiation) than other γ-nuclides. Steroids labeled with ^{125}I can be prepared at high specific activities of ~1000 Ci/mmol. However, with the exception of phenolic estrogens, direct radioiodination of steroid molecules is not possible. Steroids must first be conjugated to moieties such as tyramine, histamine, or tyrosine methyl esters that can be easily radioiodinated. There are, however, drawbacks to using iodinated steroid ligand, notably the fact that incorporation of an iodine atom whose size is approximately the same as that of the A ring may adversely affect binding between antibody and labeled steroid.

In recent years commercial kits for RIA of almost all steroid hormones have become available.

General Procedures for Steroid RIA

Reagents

Assay buffer. Binding of steroids to antisera is influenced by the pH and ionic strength of the incubation medium. Sodium phosphate buffer, ranging in molarity from

0.01 to 0.1 and with pH near neutrality (7.4), is found to be optimal for most steroid assays. Protein incorporated into the buffer prevents adsorption of steroids onto the surfaces of glassware, minimizes nonspecific binding, and stabilizes the overall assay system.

Radioactive tracer. Steroids labeled with ^{125}I are usually stable for 6–8 weeks when stored in ethanol at 2–4 °C. ^{125}I-labeled or tritium-labeled steroids sold commercially as components of kits should be stored in the manner suggested by the supplier. The stability of labeled steroid varies with the nature of the steroid derivative used for iodination.

Standard solutions. An assay standard set is usually provided by the kit manufacturer. For direct (nonextraction) assay, all dilutions of working standard should be made with serum or plasma (free from the steroid being measured) so that the standard reflects the biological matrix of the sample.

Reconstitution of antiserum. The procedure for reconstitution is usually prescribed by the supplier. It is convenient to make whatever dilution is appropriate for the assay and then to freeze aliquots suitable for daily use. Repeated freezing and thawing must be avoided because of the risk of denaturing the antibody protein.

The optimal dilution of an antiserum depends on the affinity and concentration of its antibody. The optimal titer is usually defined as the dilution that binds 50% of a standard amount of radiolabeled ligand under specified technical conditions. In practice, analysis can be performed with good precision and sensitivity when per cent binding ($\%B_0$) of radiolabeled ligand is 30–50%. Note, however, that when $\%B_0$ is low, sensitivity is comparatively high but precision is poor. The reverse is true when $\%B_0$ is high.

Separation agents. A variety of separation techniques are used in steroid assays to separate steroid-antibody complexes from unbound steroids. The most common techniques are double antibody, solid-phase antibody, and dextran-coated charcoal methods. For more detail on these techniques, see Chapter 3.

Specimen Collection and Storage

Most steroids are very stable, and no special precaution is necessary during collection of biological samples. Steroid levels can be determined either in plasma (heparin or EDTA) or serum. However, erythrocytes, which metabolize steroid hormones, should be separated rapidly. Repeated freezing and thawing of samples may induce error.

For urinary measurements, a 24-h specimen is collected in the presence of a preservative such as boric acid (1 g/L). Bacteria can degrade steroids and hydrolyze steroid conjugates. Plasma, serum, or urine without preservative may be stored overnight at 2–8 °C or for longer periods (up to 2 years) at −20 °C in airtight containers. Nonpolar steroids such as progesterone and androstenedione are very unstable when stored at −20 °C in an aqueous solution unless protein is added.

Sample Preparation

The mode and nature of specimen preparation depend primarily on the affinity constant and cross-reactivity of the antiserum. The higher the affinity constant the greater the sensitivity. Cross-reactivity dictates whether or not samples need to be purified before assay. In those assays in which extraction remains necessary, an aliquot of the organic extract is evaporated to dryness at 40 °C under a gentle stream of nitrogen to preclude the oxidation and decomposition of steroids. The dried residue is then dissolved in assay buffer.

Procedure

Incubation. Aliquots of unknown samples, controls, and standards are incubated with antiserum and radioactive ligand. An additional tube containing only tracer and assay buffer (substituted for antibody) gives information on the nonspecific binding (NSB).

For some steroid immunoassays, incubation at 4 °C is preferred because the binding affinity and stability of the assay system are enhanced. In many commercially available kits for steroid RIA, the incubation period has been shortened to 30–60 min at room temperature or 37 °C. Note, however, that in such nonequilibrium conditions small inequalities in incubation times reduce precision of the assay.

Separation of free from antibody-bound steroid. After incubation, free steroid is separated from the bound form and the radioactivity of the centrifuged sediment or the supernatant is then measured. For direct RIA of steroids using ^{125}I-labeled ligand, techniques that use a mixture of polyethylene glycol and second antibody as precipitating reagent or that use an antibody-coated tube (solid-phase) are widely employed.

Calculations

Patient results are determined from a standard curve; therefore, all conditions of the assay must be identical for both the unknowns and the standards. A standard curve is formed by plotting the radioactivity of the antibody-bound fractions against the standard concentrations used in the assay. Bound radioactivity is most commonly expressed as %B, where %B = (cpm of standard or unknown/cpm of zero standard) × 100. When %B versus standard concentrations are plotted on logarithmic graph paper, a sigmoidal curve is obtained. A patient result is then interpolated from this curve using its measured %B. Linear curves can be derived by converting %B to its logit:

$$\text{logit } \%B = \ln[\%B/(100 - \%B)]$$

and then plotting the logit versus standard concentrations on logarithmic graph paper. Alternatively, %B can be plotted on logit-log graph paper whose ordinate is already in logit divisions. The logit transform requires subtraction of nonspecific counts (NSB) from all values before the logit relation is calculated. Software is now available for logit-log data reduction by the programmable microprocessors of automated counters.

The logit-log transform has the following advantages: (1) within-assay variations between duplicates are readily obtained from it; (2) between-assay variations may be assessed by comparing standard curves of successive runs; (3) accuracy may be assessed by measuring parallelism between sample dilutions and the standard curves; and (4) the characteristic slope and intercept of each standard curve can be calculated. The one drawback of the transform is the increase of error that occurs from extrapolation at either end of the curve.

Competitive Protein Binding (CPB) and Radioreceptor Assays (RRA)

In addition to RIA procedures, CPB and RRA have been used for the measurement of steroid hormones. The important difference between these assays is the type of binding protein. For steroid CPB assays the binding agent is a naturally occurring protein found in blood. Until recently, CPB assays were the most sensitive methods for steroid determination. Plasma proteins widely used as binding proteins include sex hormone–binding globulin (SHBG) and corticosteroid-binding globulin (CBG, transcortin). Plasma binding proteins have advantages in that they are readily available and large quantities can be obtained at relatively low cost. However, the major disadvantage of this assay technique is its lack of specificity; as many as eight different corticosteroids can be measured using CBG. Consequently, CPB assays have largely been replaced by specific RIA methods.

Target tissues are characterized by the presence of high-affinity hormone receptors that interact only with biologically active steroid hormones. In addition, tissue receptors have high affinities, resulting in highly sensitive assays. However, the unstable nature and the complexity of preparation and purification of these receptors have limited their routine use.

ADRENOCORTICAL STEROIDS

Corticosteroids

The corticosteroids are physiologically and quantitatively the most important group of adrenal steroids. The major corticosteroids, cortisol and aldosterone, are secreted by the adrenals at the rate of approximately 25 mg/d and 200 μg/d, respectively.

Functionally, corticosteroids may be subdivided into glucocorticoids and mineralocorticoids. The influence of *glucocorticoids* tends to elevate blood glucose through promotion of gluconeogenesis and decreased carbohydrate utilization. In addition, glucocorticoids inhibit amino acid uptake and protein synthesis in peripheral tissues (muscle, skin, bone) and affect fat metabolism. Large quantities of circulating glucocorticoids also have anti-inflammatory, immunosuppressive, and antiallergic properties. Of the naturally occurring corticosteroids, only cortisol and corticosterone possess glucocorticoid activity. Of the two, cortisol is the more active.

The term *mineralocorticoids* refers to those steroids that regulate salt metabolism (sodium conservation and loss of potassium) and water metabolism (control of extracellular fluid volume). The most potent naturally occurring mineralocorticoid is aldosterone.

The structural formulas of some of the most significant biologically active corticosteroids are included in Figure 18-5. These compounds all possess a Δ^4-3-keto group (unsaturation between carbon atoms 4 and 5 and a keto group at carbon atom 3); a side chain ($\overset{\text{OH}}{\underset{|}{\text{CH}_2}}-\overset{\text{O}}{\underset{\|}{\text{C}}}-$) substituted at C-17 in the β-position (above the plane of the paper); and, with the exception of 11-deoxycortisol and deoxycorticosterone, an oxygen function (keto- or β-hydroxyl) at C-11. Cortisone and cortisol also have a 17-α-hydroxyl group (below the plane of the paper). Structural alterations, especially the reduction of the Δ^4-3-keto group, eliminate biologic activity.

Other Adrenal Steroid Hormones

Besides corticosteroids, the adrenals also secrete androgens, progesterone, and estrogens, all of which are produced by the gonads as well. Androgens include androstenedione, testosterone, dehydroepiandrosterone (DHEA), and 11β-hydroxyandrostenedione. From the quantitative standpoint, androstenedione and DHEA are the most important adrenal androgens. Estrogens (e.g., estrone) and progesterone of adrenal origin are quantitatively insignificant.

Biosynthesis of Adrenal Corticosteroids

The rate-limiting step in the biosynthesis of steroids is the conversion of cholesterol to pregnenolone (see *General Aspects of Biosynthesis and Metabolism*, page 555). Important and characteristic biochemical events involved in the formation of adrenal steroids include the introduction of hydroxyl groups at C-21, C-17, and C-11. The biosynthesis of aldosterone has not yet been completely elucidated. Under normal circumstances, cortisol, aldosterone, and DHEAS are quantitatively the most significant secretory products.

The adrenal cortex consists of three layers of cells (zones), each having different secretory products: the zona glomerulosa, the zona fasciculata, and the zona reticularis. The *zona glomerulosa* produces aldosterone. The principal secretory products of the *zona fasciculata* in humans are cortisol and, to a lesser extent, corticosterone and 11-deoxycorticosterone. The *zona reticularis* is the principal site of secretion of androgens such as androstenedione and dehydroepiandrosterone sulfate and, to a very limited extent, estrogens. When stimulated by ACTH, the fascicular and reticular zones are able to transform cholesterol to glucocorticoids and androgens. ACTH stimulates the zona glomerulosa to synthesize steroids from cholesterol to progesterone and corticosterone. Subsequent transformation of corticosterone to aldosterone is controlled by the renin-angiotensin system (see *Regulation of Aldosterone Secretion*). In the absence of ACTH, the fascicular and reticular zones atrophy; however, the zona glomerulosa, which is under the control of the renin-angiotensin system, is relatively preserved.

Metabolism of Adrenal Corticosteroids

The major site of corticosteroid catabolism is the liver. Reduction of the double bond between carbon atoms 4 and 5 and of the ketone group at C-3 produces tetrahydro compounds, which are the major excretory metabolites. The metabolism of cortisol is shown in Figure 18-6. Corticosterone, aldosterone, and other C_{21} steroids follow the same catabolic sequences as cortisol. Ninety per cent of cortisol metabolites are conjugated at C-3 and excreted as conjugates of glucuronic acid. Under normal circumstances about 1% of the total amount of cortisol secreted appears unchanged in urine (urinary free cortisol).

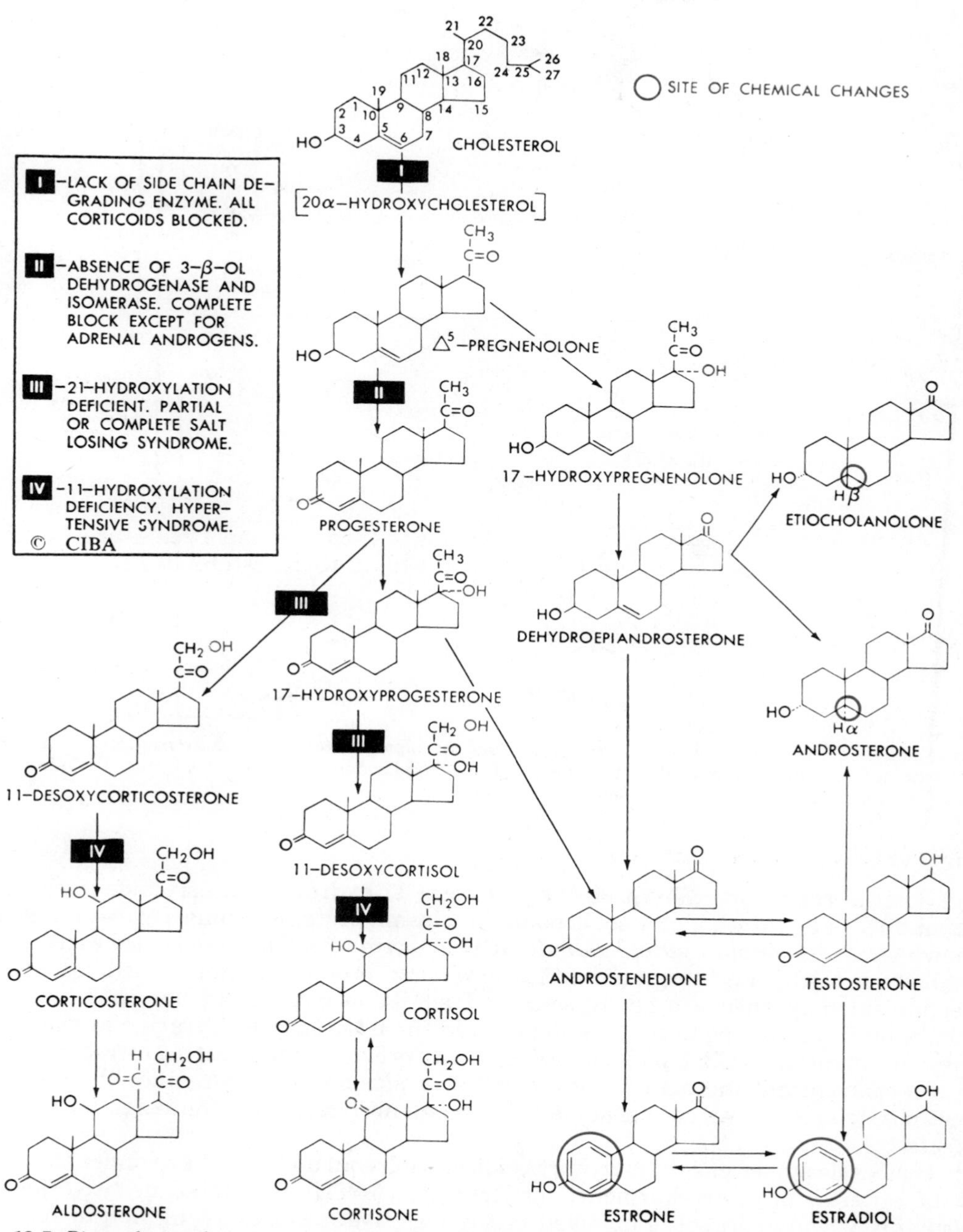

Figure 18-5. Biosynthesis of corticosteroids. I, II, III, and IV indicate sites of major blocks causing adrenogenital syndromes. (Modified from *The CIBA Collection of Medical Illustrations* by Frank H. Netter, M.D. © Copyright 1959 CIBA Pharmaceutical Company, Division of CIBA-GEIGY Corporation.)

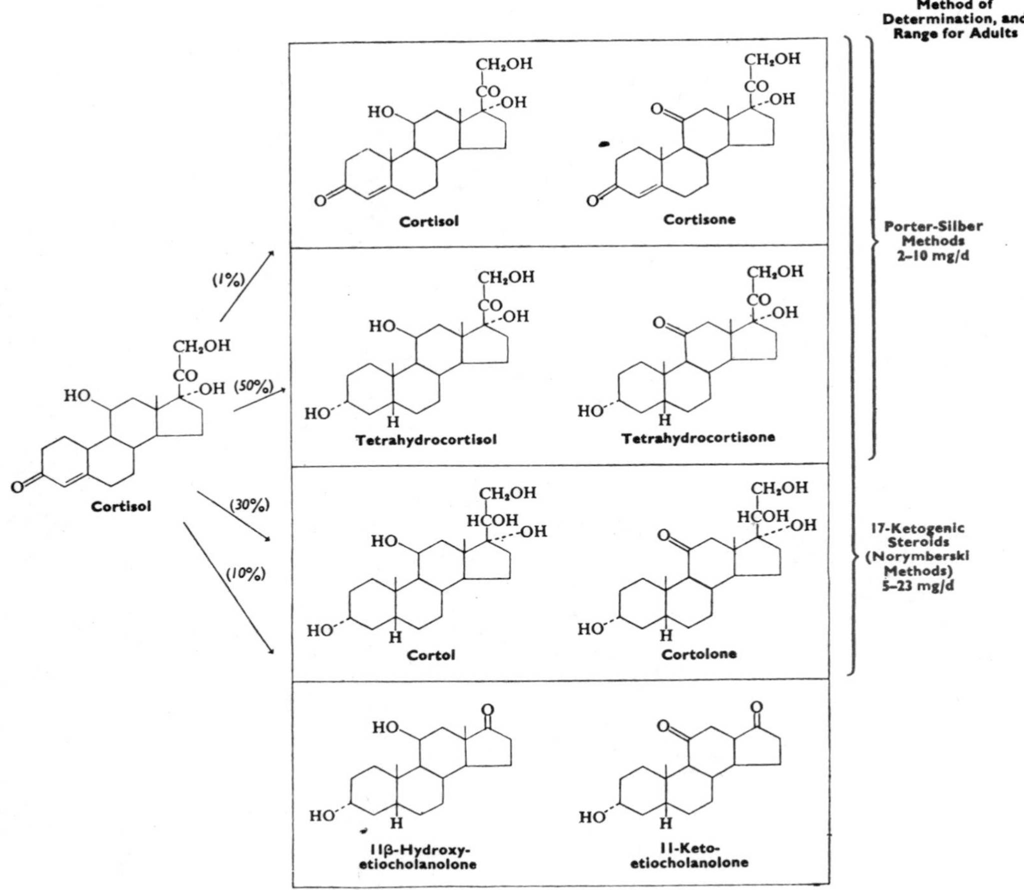

Figure 18-6. Major urinary metabolites of cortisol showing the approximate extent of conversion and the metabolites as determined by different methods. (After James, V. H. T., and Landon, J.: *In:* Recent Advances in Endocrinology. 8th ed. Boston, Little, Brown and Co., 1968.)

Regulation of Cortisol Secretion

Cortisol represents almost 80% of the total 17-hydroxycorticosteroids in the blood. About 90% of circulating cortisol is bound to plasma proteins, primarily to an α-globulin known as *corticosteroid-binding globulin* (CBG), or *transcortin;* there is also weak binding to albumin. Although the precise function of such protein binding is still obscure, this mechanism may ensure a ready source of available circulating hormone as well as protection from inactivation by conjugation in the liver or from filtration by the kidney. The concentration of CBG in plasma rises during pregnancy and during estrogen therapy, with a concomitant increase of total cortisol in plasma. However, since the amount of free, physiologically active hormone remains at normal levels, there are no untoward effects.

Synthesis and release of glucocorticoids and adrenal androgens are under the control of the anterior pituitary hormone, corticotropin (ACTH), through a negative feedback mechanism. Only cortisol is involved in the negative feedback regulation of ACTH secretion. As the level of free (non-protein-bound) cortisol in blood rises, the release of ACTH is inhibited. Conversely, if cortisol levels are subnormal, negative feedback is diminished, ACTH levels rise, and the adrenal cortex is stimulated until normal blood levels of cortisol are restored. The release of ACTH itself is under the control of hypothalamic corticotropin-releasing hormone (CRH).

There is diurnal variation in the secretion of cortisol and ACTH. In the early morning hours, blood levels of ACTH and cortisol start to rise, reaching a maximum between

0600 and 0800 h, after which there is a gradual decline to the lowest level between early evening and midnight. The level at 2000 h is normally two-thirds the level at 0800 h. The regularity of this rhythm is a function of sleep-wake habits and can be altered by consistent revision of the time and duration of the sleep-wake schedule. The normal regulatory system can be superseded by stimulatory factors such as trauma and stress; regardless of the time of day, and independent of the plasma level of cortisol, the response of normal individuals to stressful situations is a sharp rise in ACTH secretion and a consequent increase in cortisol secretion. In certain adrenal diseases (e.g., Cushing's syndrome), the diurnal rhythm of cortisol secretion may be absent.

Regulation of Aldosterone Secretion

The control mechanism for the secretion of aldosterone involves the renin-angiotensin system. Renin is a proteolytic enzyme synthesized and stored in juxtaglomerular cells located along the afferent arterioles of the renal glomeruli. Primary stimuli for renin release are a decrease in perfusion pressure to the juxtaglomerular apparatus and negative sodium balance. Following stimulation, renin is released into the blood, where it hydrolyzes its substrate, angiotensinogen (a globulin that is synthesized in the liver), to produce angiotensin I. Angiotensin I is then rapidly converted to angiotensin II by a circulating angiotensin-converting enzyme that is abundant in the lung. Angiotensin II, a potent vasoconstrictor, stimulates the cells of the zona glomerulosa to produce aldosterone.

Clinical Significance of Glucocorticoid Measurements

Cortisol measurements are the key to evaluating suspected abnormalities of glucocorticoid production. However, a single cortisol measurement under random conditions is rarely of value, since cortisol varies under normal circumstances through the course of a day and with stress.

Glucocorticoid deficiency (Addison's disease). Deficient glucocorticoid production may be due either to destructive processes that involve both adrenal glands (such as autoimmunity, infection, or neoplasm) or to deficient pituitary production of ACTH secondary to disorders of the pituitary or hypothalamus. Pharmacologic doses of glucocorticoids may also cause sufficient suppression of ACTH such that cessation of treatment will be followed by a period of hypoadrenocortical function.

When adrenal insufficiency is present in early or mild forms, symptoms or signs of adrenal hypofunction may not be evident unless the patient is under stress. In this case, failure to increase cortisol and aldosterone secretion appropriately may lead to hypotension, hyponatremia, and hyperkalemia. Chronic cortisol deficiency has less definitive symptoms, such as easy fatigue and weakness. With primary adrenal insufficiency, deficient mineralocorticoid production may cause postural hypotension and electrolyte imbalances; excess ACTH release, unchecked by negative feedback, may cause darkening of the skin and mucous membranes through a melanocyte-stimulating hormonal effect.

In healthy individuals the adrenal gland will increase cortisol production rapidly in response to ACTH stimulation. Administration of cosyntropin, a synthetic analog of ACTH, should be followed by a ≥ 7 μg/dL rise in plasma cortisol to a peak value of ≥ 20 μg/dL.[103] A normal response establishes that the adrenal cortex is capable of releasing cortisol in a normal fashion in response to ACTH and excludes a diagnosis of primary adrenal insufficiency. Primary and secondary adrenal insufficiency are differentiated by further testing. Failure to respond to a single dose of cosyntropin strongly suggests adrenal insufficiency, either primary or secondary to disease of the pituitary or hypothalamus. Measurement of ACTH in plasma will usually allow the definitive diagnosis of primary adrenal insufficiency in the unstressed patient (elevated ACTH level, low cortisol) or secondary adrenal insufficiency (inappropriately low ACTH level in relation to a low cortisol level).[10]

If secondary adrenal insufficiency is suspected and the diagnosis is still in doubt after initial studies,[54] additional tests may be done. Metyrapone is an inhibitor of the 11β-hydroxylase enzyme that converts 11-deoxycortisol to cortisol. In a normal person, the fall in plasma cortisol after metyrapone stimulates pituitary ACTH, and cortisol precursors such as 11-deoxycortisol accumulate. In secondary adrenal insufficiency, the pituitary is incapable of responding to the stimulus.

Glucocorticoid excess (Cushing's syndrome). Excessive cortisol production results in the clinical picture known as Cushing's syndrome; truncal obesity, hypertension, hypokalemic metabolic alkalosis, carbohydrate intolerance, disturbance of reproductive function, and neuropsychiatric symptoms are seen. Cortisol excess may be due to excess ACTH production by a pituitary adenoma or a benign or malignant adrenal cortical neoplasm, or it may be due to the production of ACTH by a nonendocrine neoplasm. A simple and sensitive test to screen for Cushing's syndrome is needed since the symptoms are also commonly noted in patients with normal adrenal function; measurements of cortisol under conditions of suppression and stimulation are needed to provide a differential diagnosis.

Testing for Cushing's syndrome. Urinary free cortisol is high in most patients with Cushing's syndrome.[28] This test, which requires a 24-h urine collection, appears to be the most specific and sensitive one for initial screening. A normal value is strong evidence against Cushing's syndrome. When a high value is found, repeat testing may be indicated since transient increases can occur as a result of stress. Urinary free cortisol values are not increased in obesity. Values are spuriously decreased, however, in patients with renal disease.

Blood samples may be more easily obtained than timed urine samples. Because plasma cortisol levels vary with time of day, standardized collection times and standardized conditions are necessary. Dexamethasone is a potent glucocorticoid that, in a dose of 1 mg at bedtime, suppresses the nocturnal rise in ACTH level and thus suppresses the normal morning rise in cortisol. In normal individuals the cortisol level at 0800 h will be suppressed to <5 μg/dL, whereas values >10 μg/dL will be found in patients with Cushing's syndrome. Borderline values (0800-h levels between 5–10 μg/dL) and false-positive responses make the overnight dexamethasone suppression test somewhat less specific and less sensitive than urinary free cortisol measurement. An equivocal result on a screening test can be pursued with a two-day *low-dose dexamethasone test.* The dose is sufficient to suppress cortisol production into the normal range in persons whose elevations are due to stress but should not suppress hypercortisolism in patients with Cushing's syndrome.

Differential diagnosis of Cushing's syndrome. Most commonly, Cushing's syndrome is due to overproduction of ACTH by a pituitary adenoma. However, the possibilities of a benign adrenal adenoma, adrenocortical carcinoma, or ectopic ACTH syndrome (ACTH production by a nonendocrine malignant tumor) must also be considered. Laboratory results that supplement methods of anatomic localization, such as CT scans of adrenals and pituitary, are listed in Table 18-4. Differentiation of pituitary Cushing's syndrome from other causes, such as adrenal adenoma, is possible because with the high-dose dexamethasone suppression test, suppression of urine and plasma cortisol occurs only in pituitary Cushing's syndrome. Low or undetectable ACTH levels are virtually diagnostic of Cushing's syndrome of adrenal origin. Plasma ACTH values >200 pg/mL are highly predictive of ectopic Cushing's syndrome. Values between 75 and 200 pg/mL are consistent with, but not highly predictive of, pituitary Cushing's disease. Since none of these tests is totally specific for any of the causes of Cushing's

Table 18-4. DIFFERENTIAL DIAGNOSIS OF CUSHING'S SYNDROME

	Plasma ACTH	Suppression with High-Dose Dexamethasone	Response to Cosyntropin or Metyrapone
Pituitary Cushing's syndrome (bilateral adrenal hyperplasia)	Normal or slight elevation	Yes	Yes
Adrenal adenoma	Low	No	Variable
Adrenal carcinoma	Low	No	No
Ectopic ACTH production	Very high	No	Variable

syndrome, a number of chemical and radiologic studies may be needed to make a differential diagnosis with confidence.

Clinical Significance of Measurements of Renin and Aldosterone

The interaction of renin, angiotensin, and aldosterone is important in the regulation of extracellular fluid volume, blood pressure, and the balance of sodium and potassium. A change in one of these variables leads to changes in the others. A decrease in effective plasma volume or mean arterial pressure leads to the release of renin from juxtaglomerular cells of the kidneys; more angiotensin is formed, which increases production of aldosterone by the adrenal glands. The result is retention of water and sodium, an increase in extracellular volume, and a decrease in serum potassium. This state of *secondary hyperaldosteronism* is commonly present in congestive heart failure, nephrotic syndrome, cirrhosis of the liver, other hypoproteinemic states, or any condition of chronic depletion of plasma volume. Secondary hyperaldosteronism can be inferred when one of these clinical conditions is observed in patients with volume depletion, edema, and hypokalemic alkalosis. Measurements of renin and aldosterone are seldom needed in these cases.

Primary hyperaldosteronism. Overproduction of aldosterone may be due to an adenoma of one adrenal gland (APA, or Conn's syndrome), to adrenal carcinoma, or to idiopathic hyperplasia of aldosterone-producing cells in both adrenal glands (idiopathic adrenal hyperplasia; IAH). Primary aldosteronism is characterized by hypertension and hypokalemia.[71] However, since $<1\%$ of hypertensive patients have primary aldosteronism, indications that should lead to further evaluation include hypokalemia without an apparent cause, development of profound or refractory hypokalemia upon treatment with diuretics, or lack of response to combination antihypertensive drug treatment. In primary hyperaldosteronism, high aldosterone levels and low renin levels are expected. However, many factors influence basal secretion of renin and aldosterone, and these factors must be controlled prior to measurement. For example, diuretic medications and upright posture spuriously raise renin, and hypokalemia lowers aldosterone. Once these factors are standardized, measurement of plasma renin activity (PRA) is a good initial screening test for primary hyperaldosteronism. Plasma renin levels vary with sodium balance; thus, it is often helpful to compare an individual's ambulatory PRA with his level of sodium excretion.

Another approach to screening patients suspected of having primary hyperaldosteronism is the furosemide stimulation test. Furosemide is a potent diuretic that will stimulate plasma renin production in the normal individual. A low or blunted PRA response therefore is highly indicative of primary aldosteronism. A high value, on the other hand, might suggest renovascular disease or pheochromocytoma (secondary hyperaldosteronism).

The demonstration of an elevated level of aldosterone in blood or urine in a patient with a low plasma renin is presumptive evidence for primary hyperaldosteronism. Since hypokalemia has a suppressive effect on aldosterone secretion, potassium replacement should be considered before aldosterone measurements are undertaken. To establish autonomous aldosterone overproduction, one may also attempt to suppress aldosterone production with a high-salt intake (i.e., saline infusion) or with a mineralocorticoid such as fludrocortisone. Failure to suppress aldosterone under these conditions would confirm primary aldosteronism. Differential diagnosis of primary hyperaldosteronism is important, since most patients with APA respond well to surgical removal of the tumor, whereas patients with IAH do not. X-ray localization can help, but the tumors may be too small to be detected in this way. Aldosterone secretion and plasma renin suppression are usually greater with APA, and patients with APA will usually show no change or a fall in plasma aldosterone with sodium depletion and upright posture; patients with IAH will show a rise in plasma aldosterone after sodium restriction and upright posture.

States of low renin or low aldosterone. Deficient production of aldosterone causes hyponatremia and hyperkalemia, often with hypochloremic acidosis. Mild or moderate volume depletion may also occur. Hyporeninemic hypoaldosteronism is most commonly seen in patients with diabetes mellitus and mild or moderate renal insufficiency. Finding low plasma renin and aldosterone levels under circumstances that should cause an increase (furosemide stimulation, upright posture) would be diagnostic.

Plasma renin in renovascular hypertension. Plasma renin activity that is elevated after furosemide stimulation or in relation to urinary sodium excretion is suggestive of renal artery stenosis as the cause of hypertension.

Congenital Adrenal Hyperplasia (CAH; Adrenogenital Syndrome)

CAH is characterized by an inherited absence or deficiency of any of the biosynthetic enzymes that lead to cortisol production.[71] As noted in Figure 18-5, defective or deficient enzymes may block corticoid synthesis at four sites. Low levels of plasma free cortisol then provide a stimulus for ACTH secretion, which results in hyperplasia of the adrenal cortex and accumulation of intermediate compounds behind the block. A partial block may cause either marked or subtle clinical manifestations; a complete block is incompatible with life.

In female neonates, the diagnosis is commonly suggested by ambiguous genitalia. In males, the abnormality may not be suspected until there are signs of precocious puberty or accelerated growth. Because mineralocorticoid production may be affected by diversion of intermediates, hypertension and salt-wasting may also be present.

The most frequently encountered enzyme defect causing congenital adrenal hyperplasia is 21-hydroxylase deficiency. Other enzyme defects include 11β-hydroxylase and 3β-hydroxysteroid dehydrogenase deficiencies (Figure 18-5). Measurement of specific precursor steroids in blood—17-hydroxyprogesterone in 21-hydroxylase deficiency, 11-deoxycortisol in 11β-hydroxylase deficiency, and 17-hydroxypregnenolone in 3β-hydroxysteroid dehydrogenase deficiency—have largely replaced less specific determinations of urinary steroids.[57,67] When congenital adrenal hyperplasia manifests in an adolescent or an adult, basal levels of these precursors may not be clearly elevated but will rise to abnormal levels after ACTH stimulation.

Methods for the Determination of Cortisol in Blood and Urine[56]

There are three general methods for the estimation of blood cortisol in a routine clinical laboratory. They are based on the Porter-Silber color reaction,[91] measurements of the sulfuric acid–induced fluorescence,[64] or ligand assays (e.g., RIA, CPB assay, and enzyme immunoassay).[9] The Porter-Silber methods are now obsolete, and fluorometric and CPB methods have been largely replaced by RIA.

Fluorometric assay requires a relatively small volume of plasma and is free of interference from compounds such as prednisone, prednisolone, dexamethasone, ketones, hexoses, and drugs such as propoxyphene. However, fluorometry overestimates plasma cortisol by ~2.5 μg/dL and does not measure 11-deoxycortisol, a steroid intermediate that is normally stimulated in a metyrapone challenge test.

CPB has been widely used because it is fast, simple to perform, and sufficiently sensitive for measurement of plasma cortisol. Synthetic analogs of steroids (spironolactone, dexamethasone) do not interfere. The binder employed is CBG from human plasma; separation of bound and free tracer is achieved with Florisil.

Determination of Cortisol by RIA[43]

Principle. Most cortisol assays performed in clinical laboratories are direct assays and require no initial extraction. In these direct assays, cortisol is quantitatively displaced from endogenous binding proteins by protein-binding agents such as salicylate, by low pH, or by heat treatment.[43] Total cortisol (protein-bound and free) is measured. Procedures that involve organic solvent extraction or precipitation of protein prior to assay also measure total cortisol.

^{125}I-labeled cortisol is used as tracer. Popular commercial cortisol kits use cortisol antibody, which is immobilized onto the lower inner wall of polyethylene tubes. During incubation, cortisol is bound to antibody; the contents of the tubes are then decanted, and the radioactivity remaining in the tube (^{125}I-cortisol bound to antibody) is counted. This relatively quick and simple procedure avoids a centrifugation step.

Other commercial kits employ double-antibody separation or solid-phase antibody suspension. In double-antibody systems, a second antibody is used to precipitate the primary antibody, and, after centrifugation and aspiration of the supernatant, the antibody-bound tracer is counted. Solid-phase suspension systems employ antibodies bound to small particles that remain suspended in the reaction medium. After antigen-antibody reactions have occurred, the suspended particles are separated by centrifugation, the supernatant is aspirated, and particles in the precipitate are counted to measure bound radioactivity.

Most commercial kits for plasma cortisol can also be used for measurement of *urinary free cortisol* after extraction of the sample with an appropriate organic solvent, such as methylene chloride. Extraction is required because of the large number of water-soluble cortisol metabolites and conjugates in urine that cross-react with the cortisol antibody in the assay. An aliquot of the extract is evaporated to dryness and redissolved in a buffered solution. This solution is then used in the assay for serum cortisol. Extraction efficiency can be monitored by including a recovery tube containing a known standard in each batch.

Comments. In general, direct assays for cortisol are precise, and results show high correlation with those obtained by reference methods. Accuracy is improved by dissolving standards in analyte-free plasma or serum to mimic the biological matrix of the samples.

Because of low cross-reactivity of the antiserum with corticosterone and 11-deoxycortisol, these compounds do not normally interfere. However, when extremely high plasma concentrations of 11-deoxycortisol are found (such as in 11β-hydroxylase deficiency or after administration of metyrapone), preliminary extraction with carbon tetrachloride to remove this steroid is important.

Reference ranges. Plasma cortisol reference values are 5–23 μg/dL (mean, 12 μg) between 0800 and 1000 h and 2–15 μg/dL (mean, 8 μg) at 1600 h. Urinary excretion is 10–100 μg/d. In adults, there is no significant dependence of values on age or sex.

Methods for the Determination of Aldosterone in Plasma and Urine[122]

Until recently, estimations of aldosterone in plasma and urine have been beyond the scope of most routine clinical laboratories. Simple and reliable RIA methods, including commercial kits, are now available.[47] RIA methods differ primarily in the types of aldosterone-protein conjugates of antigens used and in the specificity of the antisera raised against these antigens. New radioimmunoassays use highly specific antisera that avoid organic solvent extraction of aldosterone and tedious purification procedures.[43]

Determination of Aldosterone by RIA[47]

Most direct RIA methods for plasma aldosterone employ highly specific antisera. After incubation of aldosterone standards or plasma aldosterone with ^{125}I-labeled aldosterone and antisera, the separation of antibody-bound from free fractions is achieved by adding dextran-coated charcoal suspension, followed by centrifugation. The supernatant is aspirated and the free fraction (charcoal sediment) counted in a gamma counter. Concentrations of aldosterone in samples are calculated from the standard curve by interpolation.

The *reference range* for plasma aldosterone for individuals on ad libitum sodium intake varies from 48 to 148 pg/mL when the patient is recumbent and from 120 to 300 pg/mL after the patient has been standing for 4 h. Urinary excretion is 6–8 μg/d.

Determination of 11-Deoxycortisol in Plasma[47]

Measurement of plasma 11-deoxycortisol (compound S) is required for the metyrapone test. Because metyrapone inhibits 11β-hydroxylase, an enzyme that catalyzes the conversion of 11-deoxycortisol to cortisol, a large increase in 11-deoxycortisol is expected in patients with adequate pituitary-adrenal reserve. Consequently, immunoassay methods for 11-deoxycortisol need not be highly specific or sensitive. Even though a method for direct analysis of plasma 11-deoxycortisol has been reported,[43] very few reliable commercial kits are yet available. Several RIA methods involving extraction are available.[47] The method of Mahajan and colleagues[60] appears suitable for clinical use.

Reference ranges. In normal subjects receiving a single dose of metyrapone, values for 11-deoxycortisol are 8.0–22.3 μg/dL. In patients with hypopituitarism or Addison's disease, response values are <1.0 μg/dL.

Determination of 17-Hydroxyprogesterone in Plasma[56]

The RIA method described by Sommerville and his colleagues[126] appears to be simple, rapid, specific, and suited for use in routine clinical laboratories. Recently, commercial kits have also become available for the determination of this hormone. CPB methods are now obsolete.

Reference Ranges

	ng/mL
Males, 2–40 years:	0.3–3.2
Females, days 1–10 of cycle:	0.2–0.7
day 18 to end of cycle:	0.4–2.2
pregnant 12th week to term:	2.2–12.9

Determination of Plasma Dehydroepiandrosterone Sulfate (DHEAS)[13]

The clinical value of plasma assays of DHEA or its sulfated conjugate (DHEAS) is in the diagnosis of adrenal hyperplasia and the differential diagnosis of hirsutism. Since DHEAS predominates in the circulation and is a more specific product of the adrenals than DHEA, determination of this steroid is widely used in clinical practice. Methods based on RIA[13] are used in most clinical laboratories and a variety of commercial kits are now available.

Reference Ranges

	μg/mL
Males:	2.0–3.3
Females,	
Premenopausal:	0.8–3.4
Term pregnancy:	0.2–1.2
Postmenopausal:	0.1–0.6
Newborns, both sexes:	1.7–3.6

Methods for the Determination of Corticosteroids in Urine[110]

Colorimetric methods based on the Porter-Silber reaction for steroids containing the dihydroxyacetone side chain are still used in a few laboratories. This color reaction, however, does not include all the C_{21} metabolites. By contrast, the assay based on the oxidation of cortisol and its metabolites to 17-ketosteroids (i.e., the method for 17-ketogenic steroids) measures all the major C_{21} metabolites (see Figure 18-6).

17-Ketogenic Steroids

The determination of 17-ketogenic steroids (17-KGS) in urine includes all C_{21} 17-hydroxysteroids that also have hydroxyl (—OH) or keto ($>C{=}O$) groups at the C-20 position. When treated with oxidizing agents such as sodium metaperiodate ($NaIO_4$), these ketogenic steroids are subject to oxidative cleavage of the C-17—C-20 bond, giving rise to 17-ketosteroids (17-KS). Preliminary reduction with sodium borohydride ($NaBH_4$) will convert any 20-ketosteroid to the corresponding 20-hydroxysteroid, thus facilitating subsequent periodate oxidation.

$^{21}CH_3(CH_2OH)$ — $^{20}C{=}O$ — OH (C-17) $\xrightarrow{NaBH_4}$ $CH_3(CH_2OH)$ — HCOH — OH $\xrightarrow{NaIO_4}$ O= (C-17) $\xrightarrow[\text{benzene}]{m\text{-Dinitro-}}$ Purple complexes

$A_{max} = 520$ nm

17α-Hydroxy-20-ketosteroids | **17α,20(21)-Hydroxysteroids** | **17-Ketosteroids**

Newly formed 17-ketosteroids are extracted into dichloromethane and measured by the Zimmermann color reaction using dehydroepiandrosterone (DHEA) as the reference standard. 17-KS originally present in the urine are converted to C_{19} 17β-hydroxysteroids by the preliminary borohydride reduction. These C_{19} 17β-hydroxysteroids are not reoxidized by periodate and will not react with *m*-dinitrobenzene.

$$\text{17-Ketosteroids} \xrightarrow{NaBH_4} \text{17-Hydroxysteroids} + NaIO_4 + m\text{-dinitrobenzene} \longrightarrow \text{No reaction}$$

17-Ketosteroids **17-Hydroxysteroids**

In addition to splitting the C-17—C-20 bonds, periodate also oxidizes glucuronides at C-3 to the free steroids or to their alkali-sensitive formates, thus eliminating the need for harsh acid hydrolysis.

Methods for the Determination of Renin and Angiotensins[32]

Several commercial RIA kits for the measurement of renin are available. However, none of the kits measure renin concentration; instead, they measure plasma renin activity (PRA) by assaying angiotensin I. Most radioimmunoassays for angiotensins do not differ significantly in their essential characteristics. Like steroid hormones, angiotensins are poor immunogens because of their small molecular size; thus, they must first be conjugated to proteins for the production of antisera.

Determination of Plasma Renin Activity (PRA)

PRA is defined as the rate of production of angiotensin I from angiotensinogen by renin in a patient's plasma. PRA is expressed in nanograms angiotensin I produced per milliliter of plasma per hour and is determined by assaying angiotensin I before and after incubation of plasma at 37 °C.

Specimen collection and storage. Angiotensins are very labile in plasma and are generated in vitro in untreated plasma. Draw blood into EDTA anticoagulant, centrifuge quickly at room temperature, then freeze the plasma and store at −20 °C or lower until analysis.

Reference ranges. Reference ranges vary depending on sodium intake, posture, hydration state, and other factors. For patients on a low-sodium (10 mmol/d) diet, a reference range of 5–24 ng angiotensin I/mL/h is frequently observed. High values may be seen in patients receiving diuretics, which may cause volume and sodium depletion, or in those who receive estrogen-containing medications, which stimulate hepatic synthesis of angiotensinogen. Most antihypertensive drugs alter plasma renin activity; therefore, such drugs should be discontinued 1–2 weeks before the test is performed.

Androgens

Androgens are a group of C_{19} steroids. They are required for differentiation of the male genital tract and development and maintenance of male secondary sex characteristics; they also influence muscle bulk, bone mass, sex drive, and sexual performance in males. In normal females, androgens serve as precursors for estrogens but have no other well-defined function.

In males, androgen excess is rarely seen; androgen deficiency results in infertility and diminished libido and potency. In females, androgen deficiency has no known clinical significance; mild or moderate androgen excess disturbs the normal menstrual cycle and leads to excessive facial and body hair. Severe androgen excess in females causes virilization. Biologically active androgens include testosterone, dihydrotestosterone (DHT), androstanediol, and androstenediol; plasma concentrations of the latter two compounds are too low to be significant in normal physiology.

Testosterone is the most important androgen in blood. In males, testosterone is secreted primarily by the Leydig cells of the testes; in females, 50% of circulating testosterone is derived from peripheral conversion of androstenedione, 25% from the ovaries, and 25% from the adrenal glands (except at midcycle, when the ovarian contribution increases by 10–15%). The structural characteristics of this steroid are an

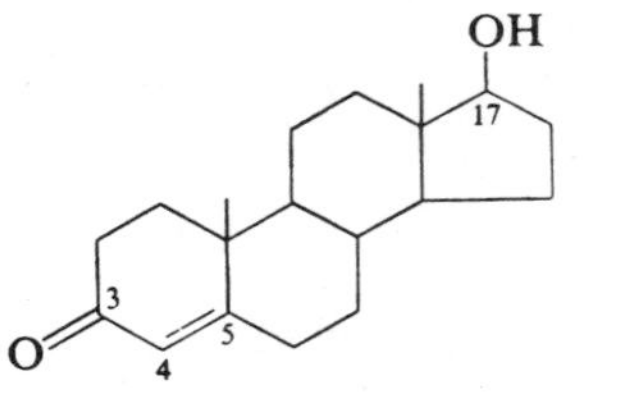

Testosterone
(17β-hydroxyandrost-4-en-3-one)

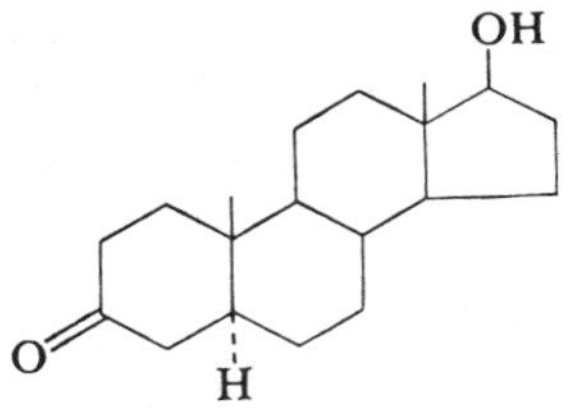

Dihydrotestosterone
(17β-hydroxy-5α-androstan-3-one)

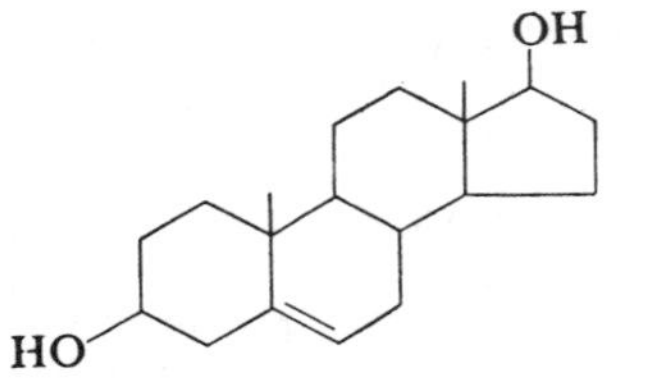

Androstenediol
(3β,17β-dihydroxyandrost-5-ene)

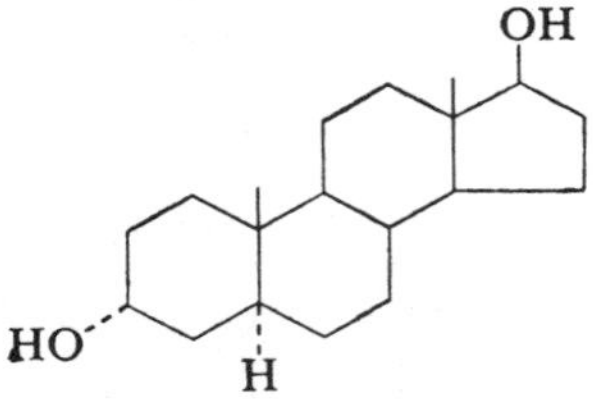

Androstanediol
(3α,17β-dihydroxy-5α-androstane)

Figure 18-7. Chemical structure of androgens.

unsaturated bond between C-4 and C-5, a ketone group at C-3 (Δ^4-3-keto), and a hydroxyl group in the β-position at C-17 (Figure 18-7). Testosterone is converted to dihydrotestosterone (DHT) by the enzyme 5α-reductase, found in cells in the prostate, skin (particularly genital skin), and seminal vesicles. Data from bioassays indicate that DHT is a much more potent androgen than testosterone. The affinities of DHT and testosterone for their common receptor are about the same. Their respective effects on a tissue will therefore depend on the rate of conversion of testosterone to DHT in that tissue.

Androstanediol is produced from testosterone in peripheral tissues and passes into the blood. Androstanediol may be secreted by the adrenals or ovaries or may be derived from peripheral conversion of plasma dehydroepiandrosterone (DHEA). Androstenedione and DHEA are weakly androgenic; they are much more important as prehormones of testosterone.

The mean production rates of testosterone and DHT in normal males are 700 and 300 μg/d, respectively, whereas the corresponding production rates in females are 300 and 60 μg/d. The production rate of androstenedione in healthy men is about 2.5 and in healthy women about 3.5 mg/d.

Control of the testicular hormone secretion. Testicular function is regulated by means of the hypothalamic-pituitary-testicular axis (Figure 18-8) in a manner similar to the regulation of other endocrine systems. Hypothalamic gonadotropin-releasing hormone (GnRH) mediates pituitary release of LH (also known as ICSH, interstitial cell–stimulating hormone) and FSH. The primary effect of LH is to stimulate the synthesis and secretion of testosterone by Leydig cells; this effect is enhanced by FSH, which increases the number of LH receptors on the cells. Increasing levels of free testosterone, in turn, inhibit LH secretion. The negative feedback of testosterone may possibly be carried out through its conversion to estradiol within the hypothalamus. In contrast to the effects on LH, testosterone at physiologic levels does not suppress FSH. This phenomenon led to the discovery of *inhibin,* a polypeptide that specifically inhibits FSH. Inhibin is produced by the Sertoli cells of the seminiferous tubules (see Figure 18-8). When the seminiferous tubules fail, FSH secretion increases, presumably the result of lack of inhibin. Inhibin reduces FSH secretion both by direct action on the pituitary and through inhibition of GnRH release.

Biosynthesis. There are two pathways for the formation of androgens from pregnenolone (see Figure 18-5). The Δ^5-pathway includes the transformation of pregnenolone to 17-hydroxypregnenolone to DHEA and then to androstenedione and testosterone. The

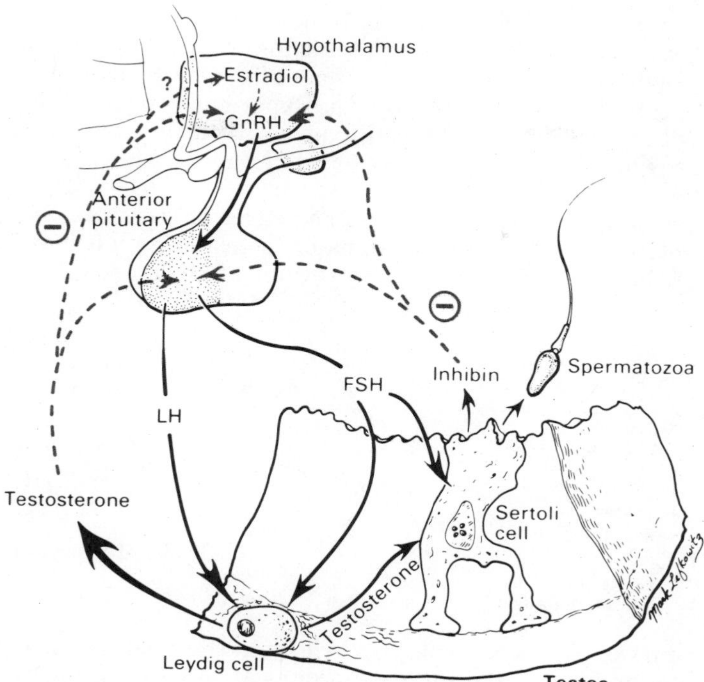

Figure 18-8. Control of testicular function. See text for explanation.

Δ^4-pathway leads from pregnenolone to progesterone and then to androstenedione. Although both pathways function in the testes and adrenals, the Δ^5-route is probably more active. The transformation of androstenedione into testosterone is a reversible reaction. In the ovaries, the equilibrium lies far toward the formation of androstenedione, which is secreted. In the testes, production and secretion of testosterone are favored.

Metabolism. After entering the bloodstream, testosterone is bound to plasma proteins, particularly one with high affinity, called sex hormone–binding globulin (SHBG) or testosterone-estradiol–binding globulin (TeBG) to indicate its capacity for binding estradiol. SHBG is distinct from CBG, the major transport protein for cortisol. In women, about 80% of circulating testosterone is bound to SHBG, 19% to albumin, and 1% remains unbound. In men, 3% of total testosterone is free; 78% is bound to SHBG, and 19% is bound loosely to albumin. Only the unbound fraction is biologically active and available for metabolism. Increases or decreases in the metabolic clearance rate (MCR) of testosterone are primarily due to changes in the concentration of SHBG.

The main metabolites of androstenedione and testosterone are produced by reduction of the double bond of the ketone group at C-3 and reduction of the ketone group at C-17. DHEAS is catabolized similarly following its irreversible conversion to androstenedione. Quantitatively, the 3α-hydroxysteroids predominate in urine, and they constitute a group of steroids known as 17-ketosteroids (Figure 18-5). Conjugation of these steroids may occur with sulfuric acid and glucuronic acid.

Determination of Plasma Testosterone by RIA[47]

Earlier methods for measuring testosterone by fluorometry, double-isotope derivative, electron-capture gas-liquid chromatography, and competitive protein-binding assays have been largely replaced by radioimmunoassay using highly specific antisera.[56] Various RIA methods differ mainly in the mode of sample preparation, the quality of the antiserum used, and the nature of the steroid-protein conjugate used to generate the antiserum.[47] Several commercial assay kits for the direct assay of testosterone have recently become available.

Most testosterone antisera currently available lack the specificity required for direct assay. According to one report,[92] overestimations of testosterone by direct methods are caused by the presence of one or more water-soluble cross-reacting substances in the specimen. Clinical laboratories using direct assay kits must be aware that an occasional sample may yield completely spurious results. Experience suggests that the validity of results obtained with kit methods is best confirmed by tests for parallelism of the response curves for standards and several randomly selected plasma samples obtained from both males and females.

Almost all testosterone antisera show some degree of cross-reactivity with DHT. In most clinical situations, however, estimations of testosterone without prior separation of DHT are acceptable, since plasma concentrations of DHT are only 10–20% of those for testosterone.

Reference Ranges

	ng/dL
Males,	
Prepubertal:	10–20
Adult:	300–1000
Females,	
Prepubertal:	10–20
Adult:	20–75 (higher at midcycle peak)
Pregnant:	60–300
Postmenopausal:	8–35

In males, serum testosterone levels rise briefly in the twelfth week of gestation and again shortly after birth. A sharp increase occurs at puberty. Subsequently, the level remains relatively constant except for slight diurnal variation. The highest level occurs in the morning and gradually decreases to 25% of the peak by evening. A slow and progressive decline begins after the fifth decade. In females, there is an increase in plasma testosterone at puberty and a cyclic pattern of elevation for 1–2 days at midcycle.

Determination of Free (Non-Protein-Bound) Testosterone in Plasma

Only a small fraction of unconjugated testosterone exists free (i.e., non-protein-bound) in plasma. None of the conventional assay methods, including RIA, are sufficiently sensitive to quantitate the free steroid directly in a protein-free ultrafiltrate of plasma. Instead, one can estimate free steroid in plasma by adding a known amount of radiolabeled compound to the plasma and allowing labeled compound and unlabeled compound to reach equilibrium in their competition for the same binding sites on the proteins. Bound and free radioactive labels are then separated, and the ratio of free label to total label is determined and is taken as a measure of the ratio of free to total endogenous testosterone. Most problems with this procedure have involved the separation of bound and free label. Several separation techniques have been used, including equilibrium dialysis, membrane ultrafiltration, and steady-state gel filtration. Equilibrium dialysis has been used most often in the past, but more recent work[115] has pointed out some of the errors arising from the dilution of sample required by this method. A recent method of ultrafiltration[38,40] appears to overcome the problems listed above and to obviate errors due to dilution.

The mean *reference value* (±SD) for free testosterone as per cent of total in healthy, non-pregnant women is 1.09 ± 0.25% and for healthy men is 2.47 ± 0.61%. These values compare favorably with those reported for membrane ultrafiltration, gel filtration, and equilibrium dialysis methods.

Determination of 17-Ketosteroids in Urine[101]

Clinical significance. *Decreased values* of 17-ketosteroids are generally found in males with primary hypogonadism (Klinefelter's syndrome, castration), in secondary hypogonadism (panhypopituitarism), and in women with primary hypoadrenalism (Addison's disease). *Increased values* are obtained in patients with testicular tumors (interstitial cell tumor, chorioepithelioma), adrenal hyperplasia, and adrenal carcinoma, and in some women with hirsutism.

Principle. 17-ketosteroids are excreted as water-soluble conjugates of glucuronic acid and sulfuric acid. Cleavage of these conjugates with acid is followed by extraction, washing with alkali, and finally the color reaction. Although estrone is a 17-ketosteroid, it is removed by alkali treatment prior to the colorimetric reaction of the "neutral" 17-ketosteroids because of its phenolic nature. The color formation is based on the reaction of 17-ketosteroids with *m*-dinitrobenzene in alcoholic potassium hydroxide (Zimmermann reaction) to produce a reddish-purple color with maximum absorption at 520 nm.

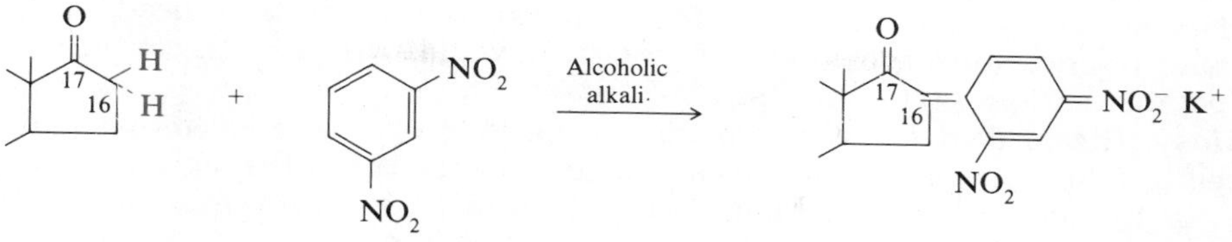

17-Ketosteroid **m-Dinitrobenzene** **Purple compounds**

When the ketone group is situated at other positions (e.g., Δ^4-3-keto in testosterone, progesterone, cortisol) the color development is less intense and the absorption maxima differ. Quantitation is carried out by comparison with a reference standard such as dehydroepiandrosterone.

Reference Ranges

	ng/d
Up to a year:	<1
1–4 years:	<2
5–8 years:	<3
8–12 years:	3–10
13–16 years:	5–12
Young adult male:	9–22
Adult male:	8–20
Adult female:	6–15

Excretion is the same for both sexes throughout childhood. After about age 60, the rate of excretion declines progressively in both sexes. Note that the measurement of DHEAS in blood serves as a more convenient marker for adrenal androgen production than does urinary 17-ketosteroid excretion.[55] Thus, many clinicians now prefer plasma DHEAS to urinary 17-ketosteroid measurements.

PROGESTERONE

During the menstrual cycle, progesterone, in conjunction with estrogens, regulates the functions of the sex organs. This hormone is especially important in preparing the uterus for implantation of a blastocyst and in maintaining pregnancy. In nonpregnant women, progesterone is secreted mainly by the corpus luteum, a yellow glandular mass in the ovary formed by an ovarian follicle following the discharge of its ovum. During pregnancy the placenta becomes the major source. Minor sources are the adrenal cortex in both sexes and the testes in males.

Progesterone (pregn-4-ene-3,20-dione) contains a keto group (at C-3) and a double bond between C-4 and C-5 (Figure 18-9); both structural characteristics are considered

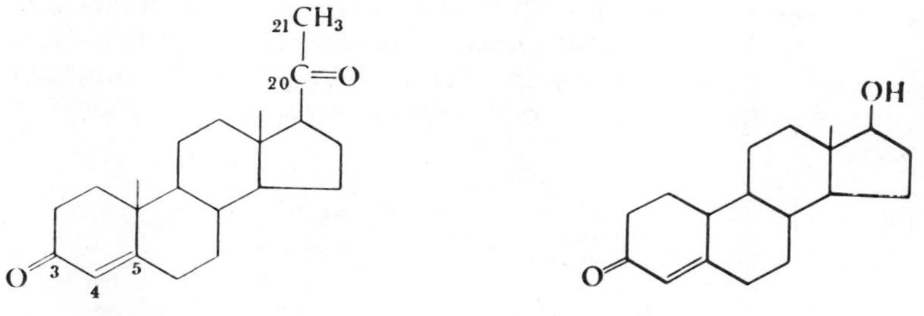

Progesterone (Pregn-4-ene-3,20-dione)

19-Nortestosterone 17β-Hydroxy-19-norandrost-4-en-3-one)

Figure 18-9. Structural formulas of progesterone and 19-nortestosterone.

to be essential for progestational activity. The 2-carbon side chain ($CH_3-\overset{O}{\overset{\|}{C}}-$) on C-17 does not seem to be important for its physiological action. Indeed, the synthetic compound 19-nortestosterone and its derivatives, widely used as oral contraceptives, are more potent than progesterone itself.

Biosynthesis and metabolism. Biosynthesis of progesterone in ovarian tissue is believed to follow the same path, from acetate to cholesterol through pregnenolone, as it does in the adrenal cortex (Figure 18-5). In luteal tissue, however, low-density-lipoprotein cholesterol is thought to serve as the preferred precursor. Initiation and control of luteal secretion of progesterone are regulated by LH. The interrelationship of progesterone and gonadotropins will be discussed under *Female Reproductive Endocrinology,* page 581.

Progesterone does not have a specific plasma-binding protein but, like cortisol, is bound to CBG. Reported values for plasma free progesterone vary from 2 to 10% of total concentration, and the percentage of unbound progesterone remains constant throughout the normal menstrual cycle. The production rate at the luteal peak may reach 30 mg/d, whereas the production rate of progesterone by the placenta during the third trimester of pregnancy is ~300 mg/d. The important metabolic events leading to the inactivation of progesterone are reduction and conjugation. Reduced metabolites (mainly pregnanediol) are eventually conjugated with glucuronic acid and excreted as water-soluble glucuronides.

Urinary pregnanediol (5β-pregnane-3α,20-α-diol) is an index of endogenous production of progesterone.

Clinical significance. During the follicular phase of the menstrual cycle, plasma levels are <1 ng/mL. After ovulation, production of progesterone by the corpus luteum increases rapidly, reaching a maximal concentration of 10–20 ng/mL in 4–7 d. Progesterone concentration remains elevated for 4–6 d and then falls abruptly to baseline levels ~24 h before the onset of menstruation. Since the rise and fall of progesterone parallels the activity of the corpus luteum, measurements of plasma progesterone are used clinically to confirm ovulation and normal function of the corpus luteum in nonpregnant women. If ovulation does not occur, the corpus luteum is not formed, and no cyclical rise of plasma progesterone is observed.

During pregnancy, progesterone is produced by the placenta. Maternal plasma progesterone rises steadily, reaching values as high as 400 ng/mL at term.

Progesterone concentrations vary considerably among individuals and in the same individual from day to day or even hour to hour. Consequently, serial rather than single measurements of plasma progesterone are preferred. Abraham and co-workers[3] found that the functional status of the corpus luteum is normal in females with normal menstrual cycles if the sum of three plasma progesterone values, obtained 4–11 d prior to menstruation, is ≥15 ng/mL. In females with abnormal cycles, the sum was <15 ng/mL.

Methods for the Determination of Progesterone in Plasma[47]

Direct measurement of progesterone in plasma is considered to be the most reliable way to assess its rate of production. Techniques based on gas-liquid chromatography and competitive protein binding (CPB) have been described,[9] but these assays have now been largely replaced by RIA. With the availability of specific antisera, direct measurement of progesterone in serum or plasma is now feasible.[16,93] Several commercial kits are available.

Reference Ranges

	ng/mL
Male:	0.12–0.3
Female,	
Follicular phase:	<1
Luteal phase:	5–20
Pregnancy	
1st trimester:	20–50
2nd trimester:	50–100
3rd trimester:	100–400

ESTROGENS

Estrogens are responsible for the development and maintenance of the female sex organs and secondary female sex characteristics. They also participate in regulation of the menstrual cycle and in the maintenance of pregnancy. In normal women, most estrogens are secreted by the ovarian follicles, the corpus luteum, and, during pregnancy, the placenta. Adrenals and testes are also believed to secrete estrogens, but only in minute quantities. Estradiol is by far the most potent naturally occurring estrogen. The names and structural formulas of the principal estrogens are shown in Figure 18-10.

Structurally, estrogens consist of 18 carbon atoms and possess the following characteristic features: (1) aromatic ring A; (2) the presence of a ketone (estrone) or a hydroxyl group (estradiol) at C-17 and frequently at C-16 (estriol); (3) the phenolic hydroxyl group at C-3, which gives the compounds acidic properties; and (4) the absence of the methyl group at C-10. The phenolic ring (A) and oxygen function at C-17 are essential for biological activity. Substituents at other positions in the molecule diminish feminizing potency.

Biosynthesis. The normal human ovaries produce all three classes of sex steroids: estrogens, progestins, and androgens. Estradiol and progesterone, however, are their primary secretory products. Unlike the testes, the ovaries possess a highly active aromatase system that rapidly converts androgens to estrogens.

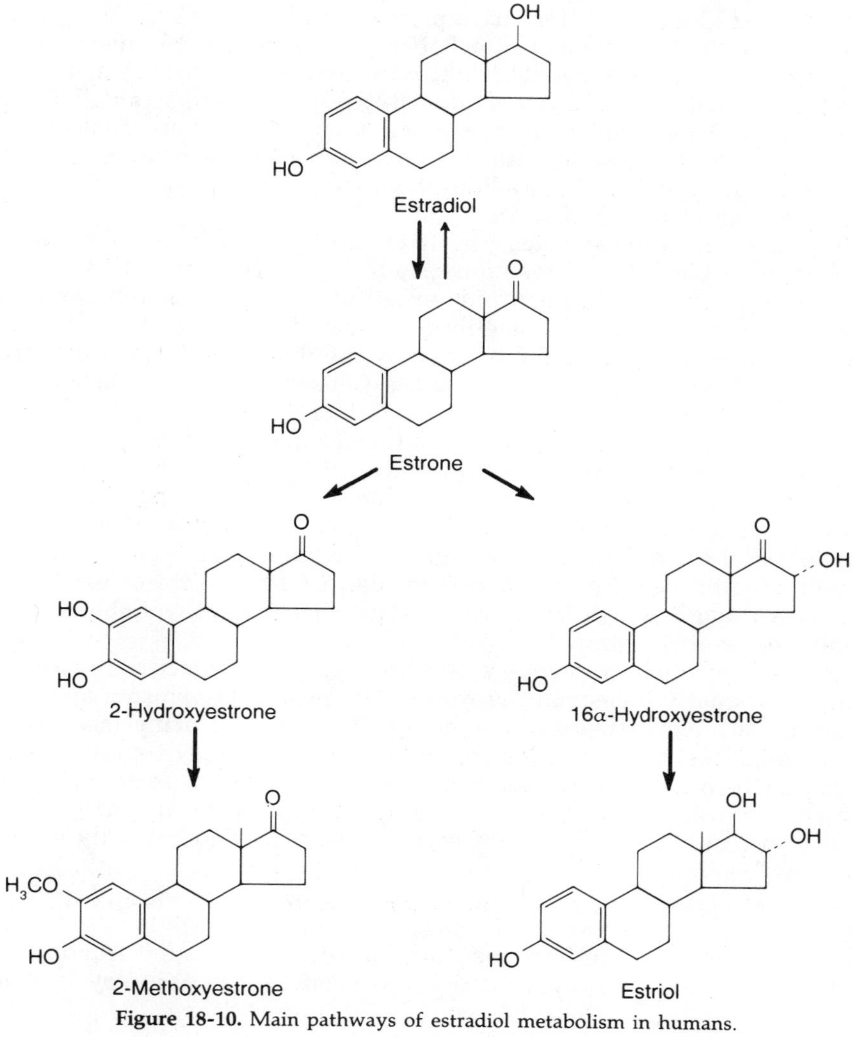

Figure 18-10. Main pathways of estradiol metabolism in humans.

Estrogens are also produced by peripheral aromatization of androgens, primarily androstenedione. In normal men and women, ~1% of secreted androstenedione is converted to estrone. Although the ovaries of menopausal women do not secrete estrogens, these women have significant levels of estrone originating from peripheral conversion of androstenedione secreted by the adrenal. Since a major site of this conversion is adipose tissue, estrone will increase in obese postmenopausal women, sometimes to the point of producing uterine bleeding.

Circulating estradiol is largely unconjugated; it is strongly bound to a specific globulin, sex hormone–binding globulin (SHBG), and loosely bound to albumin. Only 2–3% of total estradiol circulates in the biologically active unbound form. In contrast, estrone and estrone sulfate circulate almost exclusively bound to albumin. In the normal nonpregnant female, estradiol is produced at the rate of 100–300 μg/d. The production of estrone ranges from 100 to 200 μg/d.

Biosynthesis of estriol during pregnancy. During pregnancy, the major source of estrogens is the placenta, whereas in nonpregnant women the ovaries are the main site of synthesis. In comparison with microgram quantities excreted by nonpregnant women, the amount of estrogen excreted during pregnancy increases to milligram amounts. The major estrogen secreted by the ovaries is estradiol, whereas the major product secreted by the placenta is estriol. For further discussions and the clinical significance of estriol, see Chapter 28, *Biochemical Aspects of Pregnancy.*

Metabolism. Estrogen catabolism is very complex. In the normal course of estrogen metabolism, estradiol is believed to form a reversible redox system with estrone. Estrone is then metabolized along two alternative pathways (Figure 18-10): the 2-hydroxylation pathway leading to the formation of catechol estrogens and the 16α-hydroxylation pathway leading predominantly to estriol.[41] Recent evidence indicates that in nonpregnant women the 2-hydroxylation pathway is quantitatively more important.[17] The dominant pathway of estradiol metabolism is dependent upon the pathophysiological state. In hyperthyroidism and anorexia nervosa the 2-hydroxylation pathway is increased at the expense of 16-hydroxylation of estradiol. Hypothyroidism, liver damage, and obesity reverse the direction of metabolism.[63]

As is the case for other steroids, the liver is the primary site for inactivation of estrogens. The main biochemical reactions are hydroxylation, oxidation, reduction, and methylation. Conjugation with glucuronic or sulfuric acid is the final step in the metabolic process that allows them to be eliminated rapidly through the kidney.

Clinical significance. Estradiol is not only the most potent natural estrogen, it is also derived almost exclusively from the ovaries. Consequently, its measurement is often considered sufficient to evaluate ovarian function. The secretory patterns of estrogen during the menstrual cycle, as well as their clinical significance in menstrual disorders, are described under *Female Reproductive Endocrinology* (page 581). The clinical value of estrone determinations is limited to the diagnosis of postmenopausal bleeding and to menstrual dysfunction due to extraglandular estrone production. Normally, blood estrone levels parallel estradiol levels throughout the menstrual cycle but at one-third to one-half their magnitude.[2] Except in pregnancy, measurements of estriol have little clinical value. Although assays for catechol estrogens are now available,[17] their clinical value remains to be established.

Both plasma estradiol and urinary estrogen excretion reflect the secretion pattern of estrogens in normally menstruating women. In most cases, plasma and urine values correlate quite well. Nevertheless, one should recall that in urine most estrogens are present as metabolites, and therefore changes in excretion may represent altered liver and renal metabolism as well as changes in estrogen production. Certain clinical entities, such as polycystic ovary disease, exhibit significant extra-ovarian estrogen production. In these cases, measurement of urinary estrogen may be diagnostically more reliable.

Methods for the Determination of Estrogens in Plasma

Most plasma estradiol methods measure both free and bound fractions of the unconjugated steroid. Because plasma estradiol concentrations are very low throughout the menstrual cycle, earlier methods have been superseded by radioimmunoassays, which combine advantages of sensitivity, reliability, and practicality.[1]

Determination of Plasma Estradiol by RIA[78]

Highly specific antisera permit direct quantitation of estradiol in plasma or serum.[48] Commercial kits employing ^{125}I-labeled tracer for direct assay of estradiol are available.

Antisera commonly used for estradiol do not cross-react with any other steroids or estrogens except the 6-oxygenated estradiols. Since the 6-oxygenated compounds are absent or very low in concentration relative to that of estradiol, direct quantitation without extraction is possible. Problems of accuracy in direct assays for plasma or serum estradiol are similar to those described earlier for cortisol and testosterone. Experience suggests that extraction as well as purification using a Sephadex LH-20 column is necessary for accurate measurement of estradiol in men or in prepubertal children.

Reference Ranges

	pg/mL
Female,	
Prepubertal:	4–12
Early follicular phase:	30–100
Late follicular phase:	100–400
Luteal phase:	50–150
Postmenopausal:	5–18
Male,	
Prepubertal:	2–8
Adult:	10–60

Methods for the Determination of Estrogens in Urine

Urine is the principal route of estrogen excretion; the urinary determination of three estrogens—estrone, estradiol, and estriol, either individually or together—renders adequate information about the endogenous production of estrogenic hormones. The Kober reaction is the most widely accepted principle used in the quantitative colorimetric determination of estrogens. The specific functional groups that are responsible for the development of the pink color in the Kober reaction include phenolic or phenolic ether groups at C-3 of ring A and an intact ring D oxygenated at C-17. Although the Kober reaction is specific for estrogens, urinary contaminants produce a yellow-brown color, which decreases sensitivity. For this reason, extensive purification of the urinary extract prior to color development, as well as a background color correction, becomes necessary for reliable measurement of estrogens.

The main steps in this method[11] are acid hydrolysis, extraction, separation of neutral from phenolic fractions, conversion of the three estrogens into their respective 3-methyl ethers, purification and separation by alumina chromatography, and final quantitation by the Kober color reaction using the Allen correction.

To avoid the contributions of nonspecific chromogens, Ittrich[46] suggested that the color complex be extracted into chloroform containing 2% *p*-nitrophenol and then measured colorimetrically or fluorometrically. This procedure has been incorporated into various methods for the estimation of estrogens in urine and blood.[56]

The quantitative determination of urinary total estrogens may also be achieved by RIA.[78] The specific steps for a procedure depend upon the characteristics of antisera. If the antiserum is directed only toward nonconjugated estrogens, then an aliquot of 24-h urine requires hydrolysis (acid or enzyme) prior to RIA (see *Determination of Steroid Hormones by Radioimmunoassay,* page 559). However, if an antibody reacts with conjugated estrogens as well, then direct assay of estrogens (following pH adjustment of urine) is possible.

Determination of Urinary Total Estrogens in Nonpregnant Subjects[12]

Estrogen glucuronides are acid hydrolyzed; free estrogens are then extracted into diethyl ether, washed with bicarbonate buffer, and partitioned between a sodium hydroxide solution and the organic extract. Following re-extraction of phenolic steroids with diethyl ether, the Kober color is developed using the Ittrich modification and is measured fluorometrically.

The excretion of estrogens in children is generally <1 μg/d. Men excrete a constant amount of estrogen derived from the adrenals and probably also from the testes. The average daily value is ~11 μg, with a range from 5 to 18 μg.

Brown[11] has extensively studied the excretion of estrone, estradiol, and estriol during the normal menstrual cycle. The following are excretion values of total estrogens computed from his data:

	μg/d	
Onset of menstruation:	4–25	(mean, 13)
Ovulation peak:	28–99	(mean, 56)
Luteal peak:	22–105	(mean, 43)
Menopausal women:	1.4–19.6	(mean, 6.4)

Measurements of Estrogen and Progesterone Receptors in Breast Cancer[50]

An *estrogen receptor* is a specific cytoplasmic protein with high affinity and great specificity for estrogen. The estrogen receptor protein is found in target tissue cells such as the uterus, pituitary, hypothalamus, and breast. Since estrogen stimulates biochemical processes in target cells that normally contain estrogen receptor protein, a reduction in blood estrogen levels would be expected to reduce the biochemical activity of these cells. This has been the rationale for using endocrine therapy in women with breast carcinoma. Approximately one-third of women with metastatic breast carcinoma obtain an objective remission following various types of endocrine therapy directed at lowering their estrogen levels. Such therapy includes oophorectomy, hypophysectomy, and adrenalectomy (ablative therapy) as well as administration of anti-estrogens and androgens (additive therapy).

Cytoplasmic estrogen receptors are now routinely measured in samples of breast tissue after surgical removal of the tumor.[65] Generally there is a highly significant association between the presence of estrogen receptors and the response to endocrine therapy. Approximately three-fourths of patients with estrogen receptor–positive tumors respond to endocrine therapy; 95% of the patients with estrogen receptor–negative tumors fail to respond. The greater the estrogen receptor content of the tumor, the higher the response rate to endocrine therapy.

Progesterone receptor assay[66] is a useful adjunct to assay of estrogen receptors. Since progesterone receptor synthesis appears to be dependent upon estrogen action, measurement of progesterone receptor activity provides confirmation that all the steps of estrogen action are intact. Indeed, patients with metastatic breast cancer and with both estrogen receptor– and progesterone receptor–positive tumors have a response rate to endocrine therapy of 75%, whereas those with estrogen receptor–positive and progesterone receptor–negative tumors have only a 30% response rate.

Reproductive Endocrinology

Puberty

During childhood, circulating levels of sex steroids and gonadotropins are low and are similar for both sexes. The transition from sexual immaturity appears to begin with a diminished sensitivity of the pituitary or hypothalamus, or both, to the negative feedback effect of the sex steroids. Values for LH, FSH, and gonadal steroids rise gradually over several years before stabilizing in the adult range, when full sexual maturity is reached. In girls, precocious puberty is defined as the onset of pubertal development before the age 8½, and puberty is considered delayed if there has been no development by the age of 13½ or if menarche has not occurred by age 16½. In boys, puberty is considered precocious before the age of 10 and delayed when the onset is later than age 16.

True precocious puberty is more common in girls. The physical and hormonal changes are generally identical with those of normal puberty. Sex hormone–producing tumors of the ovaries or adrenal glands or gonadotropin-secreting tumors may occur, but these are rare. When puberty is delayed, measurement of gonadotropins may be useful; elevated levels would indicate gonadal failure. Screening for chromosome abnormalities is particularly important in a girl with delayed puberty. Pituitary function testing and X-rays may give helpful data if gonadotropin levels are not elevated. Often, delayed puberty is not due to specific organic disease but is simply the unusual end of the normal spectrum of maturation.

Female Reproductive Endocrinology

The normal menstrual cycle. During a normal menstrual cycle, there is a closely coordinated interplay of feedback effects between the hypothalamus, the anterior pituitary gland, and the ovaries. There are also cyclic hormone changes that lead to functional and structural changes in the ovaries (follicle maturation, ovulation, and corpus luteum development), uterus (preparation of the endometrium for possible implantation of the fertilized ovum), cervix (to permit transport of sperm), and vagina.

The menstrual cycle is usually counted as beginning with day 1 on the first day of menstrual bleeding. The cycle is divided into three phases. The *follicular phase,* which is about the first half of the cycle (14 days of a 28-day cycle) actually begins a few days before day 1. During the early follicular phase, the level of estrogen produced by the developing follicles is low. This causes a diminished negative feedback effect on the pituitary-hypothalamic axis, with consequent greater secretion of LH and FSH. Progesterone secretion is also low during this phase. Toward the end of the follicular phase, estradiol level rises; the higher estrogen level triggers positive feedback centers in the pituitary and hypothalamus, resulting in an increase in GnRH. This begins the *ovulatory phase,* a period of about 48 h at midcycle, when there is a rise in estradiol, LH, and FSH; this is followed by ovulation. The last half of the cycle, the *luteal phase,* is characterized by increasing production of progesterone and estrogen with consequent gradual lowering of LH and FSH levels. A schematic representation of the hormone changes of a normal menstrual cycle is shown in Figure 18-11.

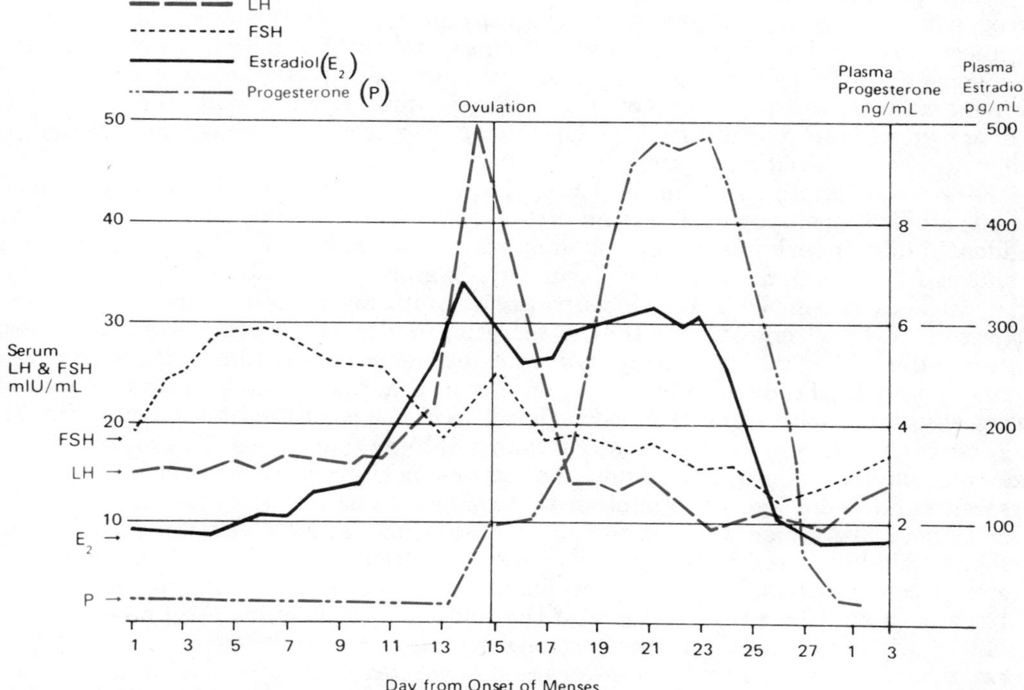

Figure 18-11. Composite of hormone changes during normal menstrual cycle. (From Watts, N. B., and Keffer, J. H.: Practical Endocrine Diagnosis. 3rd ed. Philadelphia, Lea & Febiger, 1982.)

There are changes in androgen production during the menstrual cycle; a peak at midcycle possibly serves to increase libido at the time. Estradiol is not the only estrogen produced; estrone, mainly from peripheral sources, also rises through the cycle. Estrogen and progesterone have visible effects on vaginal cytological characteristics and cervical mucus, and progesterone elevates body temperature.

Infertility and irregular menses. In normal ovulatory menstrual cycles, bleeding usually occurs 14 or 15 days after the midcycle LH surge. Normal women display a fair amount of variation in cycle length due to variations in the length of the follicular phase. In infertile women, the luteal phase may be shortened; in this case, progesterone secretion is low, the corpus luteum regresses early, and menses occur only 5–10 d after the midcycle peak. In some abnormalities, the cycle length itself may be normal but anovulatory; a record of basal body temperature or a profile of the pattern of gonadotropin release may elucidate the problem.

A diagnosis of *primary amenorrhea* should be considered if menses have not begun by age 16½. Sex chromosome abnormalities, such as in Turner's syndrome, constitute the most common cause of primary amenorrhea. Typical hormone findings in gonadal failure are high levels of LH and FSH, low levels of estrogens, and low effects of estrogens on target tissues. *Secondary amenorrhea* is diagnosed when menses have stopped after months or years of cyclic bleeding. Depending on the clinical findings, evaluation might focus on the exclusion of serious underlying disease such as hypopituitarism or the evaluation of reproductive hormone function. If fertility is important, a more detailed evaluation is necessary.

The physical effects of gonadal steroids, or the lack of them, are easily read from evaluations of vaginal cells, cervical mucus, basal body temperature, and endometrial biopsy. The laboratory measurement of specific estrogens or total estrogens, though readily made, is relatively insensitive and rarely adds much to these other clinical maneuvers.[71] However, progesterone measurement is helpful in the diagnosis of the short luteal phase syndrome. Menstrual disturbance due to androgen excess will be considered later; prolactin excess as a cause of infertility and amenorrhea has been discussed previously (see *Prolactin,* page 549). Anatomical abnormalities that cause infertility may occur with or without endocrine abnormalities.

Menopause. With advancing age, the ovaries fail to produce adequate amounts of estrogen and progesterone, ovulation ceases, and cycle length shortens. Cycles then become irregular, due in part to irregular maturation of the follicles, which results in anovulatory bleeding. FSH begins to rise, then fluctuates irregularly; both FSH and LH then increase as the failure of ovarian estrogen production eliminates the negative feedback effect on the pituitary. Ovarian failure may occur at any age, but failure prior to age 40 is considered premature.

Hirsutism. Excess of facial and body hair (hirsutism), sometimes a side effect of various medications, may be due to an increase in androgen production or may not have an identifiable underlying cause. It is important to establish that hirsutism is not due to androgen production from a neoplasm.[62] Such androgen excess is usually severe and often causes virilism. DHEAS concentrations in plasma of >700 μg/dL are strongly suggestive of an adrenal tumor that is secreting androgens. Total serum testosterone concentrations of >200 ng/dL suggest a testosterone-producing tumor; these tumors are usually found in the ovaries but may arise in the adrenal glands. Signs of androgen excess associated with signs of glucocorticoid excess suggest Cushing's syndrome.

There are a variety of nontumorous causes of androgen excess. The source of excess androgen may be the adrenal glands, the ovaries, peripheral conversion of steroid precursors to androgens by skin and other tissues, or a combination of these. Selective use of hormone measurements is appropriate in evaluating patients with hirsutism to detect hormone imbalance and to follow the response to treatment; however, no matter how precise or controlled the measurement, determination of basal androgen levels in blood or urine does not identify the source of the androgen production. An elevated level of DHEAS[52] with a normal total testosterone value suggests an adrenal origin for androgen excess, whereas elevated total testosterone but normal DHEAS levels indicate ovarian androgen excess. Elevation of both DHEAS and total testosterone levels indicates a mixed source of the androgens. Usually, it is not important to determine the specific defect to manage the patient.

Amounts of androstanediol and its glucuronide conjugate are markedly elevated in almost all women with hirsutism, regardless of the specific androgen excess involved. Androstanediol is derived primarily from the metabolism of dihydrotestosterone in target tissues; its measurement seems promising as a means to identify women with androgen excess and thus to select them for further study.

Male Reproductive Endocrinology

Infertility is rarely the sole presentation of primary or secondary hypogonadism in men; other clinical features (particularly impotence) generally point to that diagnosis. Azoospermia or oligospermia should always be considered as a possibility in men who have no obvious clinical manifestations of androgen deficiency. Hyperprolactinemia may cause oligospermia along with secondary hypogonadism. Unless evidence of androgen deficiency is present, endocrine profiles in patients with oligospermia are not generally helpful.

Impotence is the persistent inability to develop or maintain a penile erection sufficient for intercourse and ejaculation. If there is no obvious explanation for troublesome impotence, LH and FSH measurements are indicated; elevated values establish a diagnosis of primary hypogonadism. Total and even free testosterone levels may be in the normal range yet may still be subnormal for a specific individual, as indicated by elevated LH or FSH values. Hyperprolactinemia is an infrequent cause of impotence but should be considered in unusual situations.

The pituitary-gonadal axis is affected by a number of systemic diseases. Chronic renal failure causes a decrease in testosterone and sperm count, but LH and FSH levels are not elevated. With chronic cirrhosis, testosterone concentration tends to be low and estrogen levels elevated, but the plasma LH and FSH values are in the normal range. Even in the absence of liver disease, alcohol may have a direct effect on the testes, impairing spermatogenesis. With severe malnutrition, plasma testosterone declines but LH levels are variably affected. Cortisol excess and deficiency states, thyroid hormone excess and deficiency states, or any chronic debilitating illness may affect reproductive hormone levels and sexual function.

Gynecomastia. In adult males, the growth of glandular breast tissue should be viewed with suspicion. Gynecomastia may be unilateral or bilateral; unilateral gynecomastia has no specific diagnostic significance. Gynecomastia may develop due to excessive use of alcohol, marijuana, or certain medications, or it may be associated with hyperthyroidism or hypothyroidism. Chromosomal abnormalities such as Klinefelter's syndrome are often accompanied by gynecomastia, but hypogonadism is usually obvious in these patients. Germinal-cell or nonendocrine tumors that produce chorionic gonadotropin; estrogen-producing tumors of the adrenals, testes, or liver; and prolactin-producing pituitary adenomas all may cause gynecomastia. These are uncommon disorders, and the cause of gynecomastia may remain unknown in many cases. When history and physical examination do not point to a specific disorder, measurements of hCG, plasma estradiol (or total estrogens), and prolactin are appropriate. The presence of glandular breast tissue in pubertal males is quite common and almost always resolves spontaneously.

Hypogonadism. Generally, the cause of male hypogonadism can be classified by using hormone measurements. A simplified classification includes (1) abnormalities in hypothalamic or pituitary function (secondary hypogonadism); (2) primary gonadal abnormalities; and (3) chemical or cellular defects in androgen synthesis, androgen action, or both.

Measurement of serum testosterone, LH, and FSH provides the necessary initial data in individual cases. In primary hypogonadism, LH and FSH levels are elevated but those of testosterone are low. When hypogonadism is secondary to pituitary or hypothalamic hypofunction, levels of LH, FSH, and testosterone are all low. Less commonly, germ cell failure will be found, with elevated amounts of FSH but normal amounts of LH and testosterone. In syndromes of partial androgen resistance,[39] LH concentration is increased, FSH is normal, and testosterone is normal or increased. Chromosome analysis is useful for providing specific differential information.

THYROID FUNCTION

The thyroid gland is composed of two lobes connected by a thin band of tissue, the isthmus, which gives the gland the appearance of a butterfly. Each lobe measures ~2.0–2.5 cm in both thickness and width and 4.0 cm in length.

The secretory units of the gland are the follicles, which consist of an outer layer of epithelial cells that enclose an amorphous material called *colloid.* Colloid is mainly composed of an iodinated glycoprotein, thyroglobulin, and small quantities of iodinated thyroalbumin. Important reactions of thyroid hormone synthesis, such as iodination and the initial phase of hormone secretion (colloid resorption), are believed to take place at or near the epithelial cell surface.

The thyroid also contains another type of cell known as parafollicular or C cell. These cells produce the polypeptide hormone *calcitonin* (see Chapter 21) and are confined within the follicular basement lamina or exist in clusters in the interfollicular spaces.

THYROID HORMONES

The thyroid gland secretes two hormones, thyroxine (3,5,3′,5′-L-tetraiodothyronine) and triiodothyronine (3,5,3′-L-triiodothyronine), that are commonly known as T_4 and T_3. In addition, the thyroid secretes minute amounts of biologically inactive 3,3′,5′-L-triiodothyronine (reverse T_3 or rT_3). At least 80% of T_3 and essentially all rT_3 result from the deiodination of T_4 by the liver and other peripheral tissues rather than from direct secretion by the thyroid. T_4 was once considered to be the principal biologically active hormone, but it has been shown that T_3 is 4–5 times more potent than T_4; thus, T_3 is biologically the most important thyroid hormone.

Although thyroid hormones have many actions,[106] the primary one is to increase energy expenditure in many tissues (as indicated by oxygen consumption). Thyroid hormones are indispensable for growth, development, and sexual maturation. Other actions include stimulation of heart contractions, maintenance of body weight, stimulation of protein synthesis and carbohydrate metabolism, increase in the synthesis and degradation of cholesterol and triglycerides, increase in vitamin requirements, and enhancement of sensitivity of the β-receptors to catecholamines.

Biosynthesis, Secretion, and Metabolism

The important element involved in the biosynthesis of thyroid hormones is iodine, which is normally ingested in the form of iodides. Its transport to the follicles is the first and rate-limiting step in the synthetic process. Iodide is then oxidized in the thyroid gland, probably to the free radical of iodine, I*, and bound to tyrosine molecules, which are attached to a thyroidal protein called *thyroglobulin.* The enzyme responsible for oxidation and binding of iodide (organification) is thyroid peroxidase. Monoiodotyrosine (MIT) and diiodotyrosine (DIT) are thus formed. Two DIT molecules then condense, forming T_4, still in peptide linkage to thyroglobulin. Similarly, T_3 and a small amount of rT_3 result from condensation of MIT and DIT. The condensation reaction is aerobic, requires energy, and is also catalyzed by thyroid peroxidase (Figure 18-12).

Synthesis of T_3, T_4, DIT, and MIT in the thyroglobulin molecules occurs mainly at the follicular cell/colloid interface but also within the colloid. Thyroglobulin is present in highest concentrations within the colloid, where it is stored. Follicular cells engulf colloid globules by endocytosis. These globules then merge with lysosomes in the follicular cell, where lysosomal proteases break the peptide bonds between iodinated residues and thyroglobulin; T_4 and T_3 thus formed diffuse into the systemic circulation, whereas DIT and MIT remain in the follicular cells, where they are deiodinated and the freed iodide is re-utilized for hormonal synthesis.

Each step in the synthesis of thyroid hormones is regulated by pituitary thyrotropin, TSH. In addition, TSH induces an increase in the size and number of thyroidal follicular

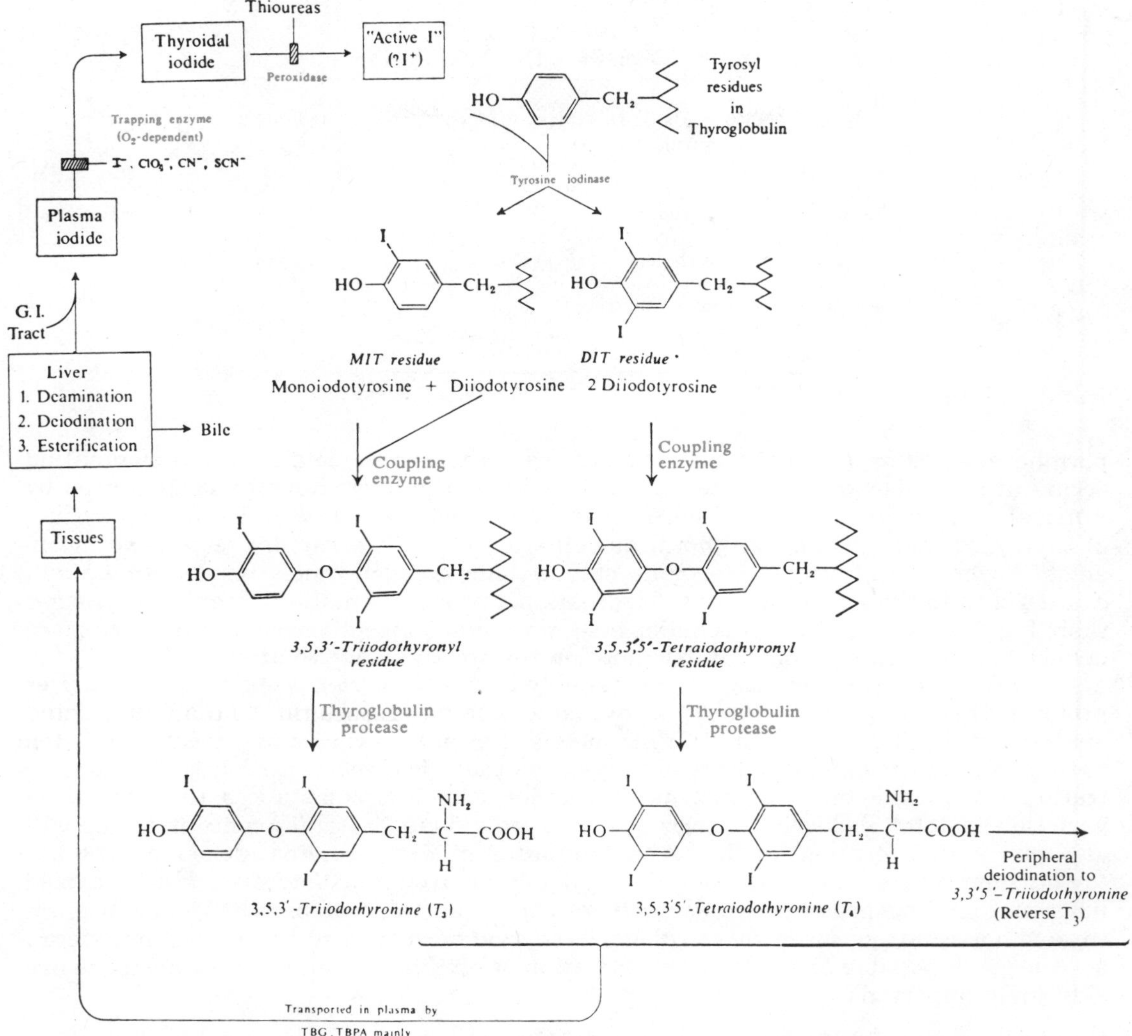

Figure 18-12. The metabolism of iodine, emphasizing formation and secretion of the thyroid hormones. Hatched areas indicate block in the pathway. Iodide transport is inhibited by monovalent anions such as thiocyanate (SCN^-), perchlorate (ClO_4^-), and pertechnetate (TcO_4^-). The oxidation and organic binding of iodide to thyroglobulin is blocked by thiourylenes, sulfonamides, and high concentrations of iodide. (Modified from Berger, S., and Quinn, J. L., *In:* Tietz, N. W.: Fundamentals of Clinical Chemistry, 2nd ed. Philadelphia, W.B. Saunders, Co., 1976.)

cells. Prolonged TSH stimulation leads to increased vascularity and eventual hypertrophic enlargement (goiter) of the gland.

In the liver, T_3 and T_4 are conjugated to form sulfates and glucuronides. These conjugates enter the bile and pass into the intestine. The thyroid hormone conjugates are hydrolyzed, some are reabsorbed through the enterohepatic circulation, and some are excreted in the stool. A thorough discussion of thyroid metabolism, secretion, and synthesis is available.[45]

Normal thyroid physiology. There is a tightly coordinated feedback relationship between the thyroid gland, the hypothalamus, and the pituitary gland. Thyrotropin-releasing hormone (TRH), a tripeptide, is produced in the hypothalamus and acts on the pituitary to cause the synthesis and release of thyroid-stimulating hormone (TSH). A rise in thyroid hormone level inhibits the pituitary response to TRH (negative feedback). A fall in thyroid hormone causes an increase in TRH and TSH secretion.

T_4 is the primary secretory product of the normal thyroid gland. T_4 undergoes

Table 18-5. ALTERATIONS IN PLASMA CONCENTRATION OF THYROID HORMONE–BINDING PROTEINS

Decreased levels in:
Treatment with anabolic steroids, androgens, diphenylhydantoin
Major illness or surgical stress
Nephrotic syndrome
Active acromegaly
Genetic (inherited) deficiency
Increased levels in:
Treatment with estrogens, perphenazine
Pregnant or newborn state
Acute intermittent porphyria
Infectious hepatitis
Genetic (inherited) increase in synthesis

peripheral deiodination of the outer ring at the 5′ position to yield T_3. This deiodination occurs at a number of tissue sites, but primarily in the liver. Reverse T_3, produced by removal of one iodine from the inner ring, is metabolically inactive and is an end-product of thyroxine metabolism. Peripheral deiodination of T_4 is a rapidly responsive mechanism for control of thyroid hormone balance. Acute or chronic stress or illness will cause a shift in the direction of this deiodination, favoring formation of reverse T_3 rather than T_3, whereas the T_4 level remains essentially unchanged. Various medications (see below) will also shift peripheral deiodination toward the inactive product rT_3.

T_4 and T_3 in the circulation are reversibly bound to carrier proteins. These carrier proteins (thyroxine-binding globulin, thyroxine-binding prealbumin, and albumin) bind ~99.97% of T_4 and ~99.7% of T_3. This means that only a very small fraction of each hormone is free for biological activity. Because there is large variation in the concentration of thyroxine-binding proteins even under normal circumstances, there is a wide variation in total T_4 levels among euthyroid individuals. Total T_3 concentrations will also vary with alterations in the concentrations of binding proteins, although not quite to the same degree as T_4 levels will vary. Common circumstances in which thyroid hormone–binding proteins are altered are shown in Table 18-5. When one relies on thyroid hormone measurements to diagnose thyroid hormone excess or deficiency states, knowledge of possible changes in concentration of binding proteins or free hormone are extremely important.

Diagnosis of Thyroid Hormone Excess and Deficiency States

Laboratory tests most commonly used to evaluate patients for thyroid function disorders are listed in Table 18-6. Clinical signs and symptoms of thyroid hormone excess or deficiency are nonspecific. When hypothyroidism or hyperthyroidism is suspected, confirmation with laboratory tests is generally necessary. However, normal thyroid hormone levels do not exclude thyroid disease, and diffuse or nodular thyroid enlargement may be seen in euthyroid patients.

Patients can be screened by using one of three methods: (1) a total thyroxine measurement, (2) calculation of a free thyroxine index (FT_4I) from total T_4 and T_3 uptake ratio measurements, or (3) one of the newer assays for free thyroxine (FT_4) measurement.[69] T_4 measurement is a good reflection of thyroid hormone production. However, changes in serum thyroid hormone–binding proteins also affect total T_4 measurements. Even the most reliable thyroxine index or free T_4 measurement may not completely correct for extreme variations in binding proteins. Also, because acute changes in T_4 conversion to T_3 and rT_3 may occur with nonthyroidal illnesses, relationships between these hormones may not be totally predictable. Therefore, for individual patient testing, total T_4, T_3 uptake, and calculated FT_4I provide more reliable information than a total T_4 value alone.

Hypothyroidism. Primary hypothyroidism is a common disease, affecting 1–2% of the population. Primary hypothyroidism results when the thyroid gland is damaged,

Table 18-6. ABBREVIATIONS AND NOMENCLATURE FOR THYROID TESTS

Hormone concentration	
T_4(RIA)	Total serum thyroxine (radioimmunoassay)
T_3(RIA)	Total serum triiodothyronine (radioimmunoassay)
FT_4D	Free thyroxine (dialysis method)*
FT_3D	Free triiodothyronine (dialysis method)
rT_3	Reverse 3,3′,5′-triiodothyronine
TSH	Thyroid-stimulating hormone, thyrotropin
Indirect measure of binding sites	
T_3U	T_3 resin uptake†
Indirect estimates of free hormone concentrations	
FT_4 index (FT_4I)	Free thyroxine index [T_4(RIA) × T_3U]
FT_3 index (FT_3I)	Free triiodothyronine index [T_3(RIA) × T_3U]
Serum binding proteins	
TBG	Thyroxine-binding globulin‡
TBPA	Thyroxine-binding prealbumin

*Any method other than dialysis for measurement of "free" thyroxine must be critically evaluated; results are not always comparable with dialysis techniques.

†The recommended test is the T_3 uptake ratio, where the T_3U of the patient is divided by the T_3U from a pool of normal standard reference serum included in the same assay run. This ratio is highly preferred and should be used in calculating the free thyroxine index. The normal range for T_3U ratio, although centered on 1.0, is not necessarily the same for each method.

‡Thyroxine-binding globulin, the protein that binds thyroxine in serum, is distinct from thyroglobulin, the intracellular carrier protein that is involved in thyroid synthesis. Thyroglobulin concentrations in serum may be elevated in well-differentiated thyroid carcinoma and other conditions.

such as in chronic lymphocytic thyroiditis; or in inherited conditions where thyroid hormone synthesis is inefficient, such as in enzyme abnormalities associated with dyshormonogenesis; or after ablation or removal of thyroid tissue with radioactive iodine or surgery. The fall in T_4 and T_3 leads to an increase in TSH, which confirms a diagnosis of primary hypothyroidism. In early or mild hypothyroidism, symptoms are subtle and the thyroid hormone levels may still be above the lower limit of the reference range, but TSH levels will be elevated. In mild hypothyroidism, TSH measurement is invaluable in establishing the diagnosis.

Congenital hypothyroidism may be due to absence of the thyroid gland (athyreosis) or to defects of thyroid hormone synthesis. Screening programs for congenital hypothyroidism[116] have been established in most states. This disorder occurs once in every 3500 to 4000 live births, and early treatment is critical in preventing mental retardation. For a primary screening test, T_4 is measured on blood eluted from a punched-out filter-paper spot collected from the heel of a newborn infant. When results are outside the lower 2–3 SD of the normal distribution, a TSH determination from a second punched-out spot from the same filter paper is indicated. In many instances, the diagnosis of primary hypothyroidism can be made in the neonate; in some cases, further testing is needed to confirm the diagnosis.

Secondary hypothyroidism is due to pituitary or hypothalamic disease; serum thyroid hormone concentrations are low but the TSH value is either within the normal range (and inappropriately low for the low T_4) or only very slightly elevated. When T_4 and TSH levels are both low, a TRH stimulation test is helpful to distinguish hypothalamic from pituitary disease. In patients with destructive lesions of the pituitary that cause TSH deficiency, no response of TSH to TRH is expected. In patients with hypothalamic abnormalities affecting TRH release, the peak TSH response to TRH may be normal but is generally delayed until 45 or 60 min after TRH administration.

Hyperthyroidism. This disorder can be caused by a number of conditions, including diffuse toxic goiter, toxic multinodular goiter, solitary toxic nodule, and damage to the thyroid gland from thyroiditis. Hyperthyroidism is generally easier to diagnose on clinical grounds than is hypothyroidism. If a patient has diffuse thyroid enlargement, as well as rapid pulse, tremor, nervousness, weight loss, increased sweating, easy fatigue, heat intolerance, and exophthalmos, the diagnosis of hyperthyroidism may require only a confirmatory determination of T_4. If the clinical picture is not clearly that of hyperthyroidism, or if the T_4 level is not clearly elevated, measurement of serum T_3 is helpful. In hyperthyroidism, T_3 is often elevated earlier and to a greater degree than T_4. The use of a total T_3 measurement with a T_3 uptake ratio (FT_3I) may be helpful in adjusting the total T_3 (RIA) for abnormalities in binding proteins. In a patient with transient lowering of T_3 due to medication or nonthyroidal illness, the diagnosis of hyperthyroidism using total T_4, total T_3, and T_3 uptake may not be possible.

Hyperthyroidism may be seen with normal total T_4 and FT_4 but with an elevated T_3; this syndrome of *T_3 toxicosis* is most commonly seen following previous radioiodine treatment or surgery for hyperthyroidism, or after treatment with thyroid hormone preparations that contain both T_4 and T_3. *T_4 toxicosis,* high total T_4 and normal total T_3, is not a common situation. In many patients with high T_4 but low T_3 levels and clinical hyperthyroidism, the low T_3 level is probably the result of the transient influence of drugs or stress. These patients will usually have high T_3 levels if tested again in a basal state.

Occasionally, increases in T_4 and T_3 may occur due to release of thyroid hormones from damaged thyroid parenchyma associated with subacute thyroiditis or chronic lymphocytic thyroiditis ("hyperthyroiditis"). The increase in T_4 and T_3 may be associated with clinical findings suggestive of hyperthyroidism. Low radioactive iodine uptake (per cent of orally administered radioactive iodine taken up by the gland at 6 or 24 h) accompanies the hyperthyroidism of thyroiditis. In most cases, this is a self-limiting condition that will resolve without residual thyroid function abnormality.

The most conclusive test for hyperthyroidism is the TRH stimulation test. High levels of thyroid hormone at the pituitary will block release of TSH from the pituitary. In a patient with equivocal but suggestive symptoms of hyperthyroidism and T_4 or T_3 levels slightly above the normal range, failure of TRH to stimulate a rise in TSH establishes the diagnosis of hyperthyroidism. This test, however, is not needed in most hyperthyroid patients.

Effects of Drugs and Nonthyroidal Illness on Thyroid Tests

Peripheral deiodination of T_4 is the primary source of T_3, the more potent of the two thyroid hormones in blood. This enzymatic conversion is inhibited by a number of medications, by acute and chronic stress, and in a variety of acute and chronic nonthyroidal illnesses.[22,117] As conversion of T_4 to T_3 is reduced, conversion of T_4 to rT_3 increases, resulting in an increase in serum rT_3. Common test abnormalities accompanying nonthyroidal illnesses in euthyroid patients include an increase in rT_3, lowering of total T_3(RIA) and FT_3I, and lowering of TBG and other binding protein levels. Low T_4 levels are often also seen in sick euthyroid patients. Less frequently, the T_4 concentration is increased.[35] Free T_4 may be high with acute illness or low with severe chronic illness. TSH and TRH responses are rarely affected. Thus, in nonthyroidal illness, abnormal thyroid tests are not necessarily suggestive of thyroid disease. Hyperthyroid patients with acute or chronic nonthyroidal illness may have T_3 levels that are lowered into the normal range, and these patients sometimes show transient lowering of T_4 levels as well. TSH and TRH responses are usually valid despite nonthyroidal illness; these tests should be used if hyper- or hypothyroidism is suspected.

Almost all laboratory tests for thyroid function are now commercially available in kit forms. A comparison of methods and kits for assessing thyroid function has been published.[69]

Determination of Total T_4 by Radioimmunoassay[21] [T_4(RIA)]*

Current RIA methods do not require extraction of T_4 and T_3 from serum but do require dissociation of the hormones from serum binding proteins prior to assay. Barbital buffer is used to dissociate iodothyronines from thyroxine-binding prealbumin (TBPA). Other blocking agents, such as 8-anilino-1-naphthalene-sulfonic acid (ANS), salicylate,

*The assistance of Joseph D. Musto, M.Sc., Ph.D., Newton-Wellesley Medical Laboratory, Inc., Natick, MA, 01760, in the preparation of this section is gratefully acknowledged.

thimerosal, or phenytoin, inhibit the binding of iodothyronines to TBG. These blocking agents effectively displace T_4 and T_3 from TBG without affecting the binding of these thyroid hormones to their specific antibodies in the radioimmunoassays.

A typical method[21] for T_4 consists of incubating serum specimens and standards with antiserum and ^{125}I-T_4 in ANS-barbital buffer. After incubation, the bound and free labeled T_4 fractions are separated, and the radioactivity of the bound label in the precipitate is determined. T_4 values of specimens are calculated by comparing their bound radioactivity with that of standards of known concentrations. Intra- and interassay reproducibilities, expressed as CV, are 4.3 and 7.1%, respectively. This type of method is sufficiently sensitive to allow the use of samples as small as 2.5 μL, such as those obtained from infants by heel puncture and collected in capillary tubes or on filter paper. Dried blood specimens on filter paper are stable and easily transported and are ordinarily obtained for purposes of screening neonates for congenital hypothyroidism. When this collection technique is used, a ⅛-inch dot is punched out from the blood-saturated paper and is extracted into buffer prior to T_4 assay.

Many commercial kits with good analytical reliability are available. They differ primarily in methods of separation of antibody-bound and unbound fractions. Some utilize a double antibody technique with polyethylene glycol (PEG); some use a solid-phase technique with T_4-specific antibody bound to beads, magnetizable particles, or tubes. Use of solid-phase systems often simplifies the procedure by eliminating the centrifugation step for separation of bound and free label.

Nonisotopic immunoassays are also commercially available. As in RIA, the principle is based on competition of labeled and unlabeled analyte for an antibody and subsequent measurement of a property of the labeled analyte that is bound to antibody. The label may be an enzyme or a fluorophore. In an *enzyme-linked immunosorbent assay* (ELISA), offered by ICL Scientific (Fountain Valley, CA 92708), a serum sample is incubated with peroxidase-labeled T_4 in a test tube coated with T_4 antibody. The endogenous T_4 competes with the peroxidase-labeled T_4 for the binding sites on the antibody. After incubation, the contents of the tube are aspirated, and the inside of the tube is washed to remove any unbound labeled T_4. Enzyme activity due to enzyme-labeled T_4 bound to antibody on the tube is determined by photometric measurement of color intensity following incubation with a suitable substrate. The color intensity is inversely proportional to the T_4 concentration in serum. The *enzyme-multiplied immunoassay* technique (EMIT) for T_4 developed by Syva (Palo Alto, CA 94303) is a homogeneous immunoassay system; i.e., separation of unbound from antibody-bound T_4 is not required. In this system, malate dehydrogenase covalently linked to T_4 is used as the enzyme label. T_4 bound to this enzyme blocks the enzyme's active site. (This is in contrast to other EMIT assays using different enzyme labels.) This inhibition is reversed when the T_4-enzyme complex binds to a T_4 antibody, allowing access of a suitable substrate to the catalytic site. Endogenous T_4 competes with the enzyme-labeled T_4 for antibody-binding sites. Consequently, as the concentration of endogenous T_4 increases, less of the enzyme-labeled T_4 becomes bound to the antibody, and enzyme activity decreases. In the case of malate dehydrogenase, the reaction involves reduction of NAD^+ and can be observed photometrically. Since enzyme activity in this case is a property only of antibody-bound enzyme, separation of bound from free enzyme-labeled T_4 is not required.

Fluorescent labels can also be used in T_4 immunoassays. In the *homogeneous fluorescence immunoassay* developed by Syva, fluor-labeled T_4 is added to the specimen and competes with endogenous T_4 for binding to a T_4-specific antibody, which is labeled with a fluor-acceptor compound such as rhodamine. When the fluor-labeled T_4 is bound to the antibody, its fluorescence intensity is reduced. The change in fluorescence of the reaction mixture is directly proportional to the amount of unlabeled T_4 in the sample. *Heterogeneous fluorescence immunoassay* (Bio-Rad Laboratories, Richmond, CA 94804) uses a solid-phase antibody technique. Fluor-labeled T_4 competes with T_4 in the sample for binding to antibody, which is covalently linked to polyacrylamide beads. After incubation, the beads are sedimented and the supernatant discarded. The beads are then resuspended, and the fluorescence intensity of the suspension is measured in a fluorometer.

In some cases, radioimmunoassays and nonisotopic immunoassays have been partially or fully automated.[19] A technique based on fluorescence polarization immunoassay developed for the Abbott TDx has been introduced.

Specimen collection and storage. The preferred sample is serum; plasma with EDTA or heparin as anticoagulant may also be used. Note, however, that plasma tends to form fibrin clots after freezing and thawing. T_4 in serum is quite stable; storage of serum at room temperature for up to 7 d results in no appreciable loss of T_4. However, storage at 2–8 °C or freezing is recommended. Frozen specimens are stable for at least 30 d. Repeated freezing and thawing of the sample should be avoided. Grossly lipemic specimens should not be used; fatty acids compete with T_4 for binding sites on TBG. If patients have received diagnostic or therapeutic radionuclides within two weeks prior to the T_4 determination, the radioactivity of serum samples should be checked in a gamma counter to determine whether the radioactivity in the sample could significantly affect RIA results.

Reference Ranges[112]

	μg/dL
Cord Blood:	8.0–13.0
Newborn:	11.5–24.0
(Lower in low-birth weight infants)	
Newborn screen (filter paper):	6.2–22.0
Neonate:	9.0–18.0
Infant:	7.0–15.0
1–5 years:	7.3–15.0
5–10 years:	6.4–13.3
10–60 years:	5.0–12.0
> 60 years, male:	5.0–10.0
female:	5.5–10.5

Determination of T_3 by Radioimmunoassay [T_3(RIA)][21]

Principle. T_3 measurement is similar to that described for serum T_4(RIA). In a typical assay the reagents are added to tubes in the following order: (1) barbital buffer, (2) unknown samples or varying concentrations of standards, (3) ANS solution, (4) T_3-antiserum diluted so as to bind about 30 to 50% of $^{125}I-T_3$ at zero concentration of standard, (5) $^{125}I-T_3$ in barbital buffer. The reagents are mixed and incubated at room temperature. A second antibody is used to precipitate antibody-bound T_3. Bound and free labeled T_3 are separated, and the radioactivity of the bound label in the precipitate is counted.

As with T_4(RIA), numerous commercial RIA kits for measuring serum T_3 are available. Most of these procedures can be performed within 4 h with incubation periods ≤2 h. A variety of separation techniques for the bound and unbound fractions are available and are similar to those described for commercial T_4 methods. The *specimen collection and storage* are the same as described for T_4.

Reference ranges. A typical adult reference range for serum T_3 is 110–190 ng/dL.

Methods for the Determination of Thyroid Hormone Binding

Total serum T_4 and T_3 concentrations are dependent on their production rates and on the level of circulating thyroxine-binding globulin. Thus, tests measuring thyroid hormone binding are performed to determine whether a change in the total concentration of the hormone is due to a change in the level of binding protein or to a change in hormone production. Furthermore, such tests provide evidence of whether an abnormality in the total concentration of serum T_3 or T_4 is associated with alterations in the free (non-protein bound) fraction of these hormones.

Several approaches have been taken to assess thyroid hormone binding to proteins, including direct estimation of free thyroxine concentration (FT_4), measurement of the concentration of unoccupied binding sites on TBG (T_3U), and the direct determination of TBG concentration by RIA.

Determination of Free Thyroxine (FT_4) in Serum[44,108]

The interaction between serum FT_4 and the unoccupied or available binding sites on TBG conforms to the law of mass action and can be described by the following equation:

$$[FT_4] \times [TBG^0] = K \times [T_4 \cdot TBG]$$

where $[FT_4]$ = concentration of free T_4 in serum
$[TBG^0]$ = concentration of unoccupied binding sites on serum TBG
K = an association constant
$[T_4 \cdot TBG]$ = concentration of T_4-occupied binding sites on serum TBG

The relationship among these entities has important diagnostic implications. A primary increase in either $[FT_4]$ or $[TBG^0]$ drives this reaction to the right, increasing serum $[T_4 \cdot TBG]$. Since 99.97% of the total thyroxine is bound to TBG, the value of $[T_4 \cdot TBG]$ is considered equivalent to that of total T_4. Hyperthyroidism produces a primary increase in FT_4, whereas high estrogen levels and idiopathic or genetic conditions may produce a primary increase in TBG. In both cases there is an increase in $[T_4 \cdot TBG]$, but in the first case, the patient is ill and requires treatment, whereas in the latter case, the patient is euthyroid. Likewise, a low serum $[T_4 \cdot TBG]$ may be due to a primary decrease in $[FT_4]$ or to a primary decrease in [TBG]. It is therefore clinically important to differentiate between changes in $[T_4 \cdot TBG]$ that are due to primary changes in $[FT_4]$, such as hyper- or hypothyroidism, and those due to primary changes in $[TBG^0]$.

In normal serum the amount of FT_4 averages 0.03% of the total. The percentage of free T_3 is approximately ten times that of T_4 (i.e., 0.3%) and reflects the lesser affinity of TBG for T_3.

The concentrations of FT_4 and FT_3 in serum are exceedingly low. Measurement techniques such as equilibrium dialysis, ultrafiltration, and gel chromatography are too cumbersome and time-consuming for routine clinical use and, except for equilibrium dialysis, have found little application. Recently a number of commercial kits have become available for the measurement of FT_4 and FT_3. These assays are designed so that the original equilibrium between free and protein-bound T_4 in the patient sample remains undisturbed. The results may, however, differ from FT_4 determined by the standard equilibrium dialysis technique because of the influence of extremely high or low TBG concentrations on the commercial methods.

Specimen collection and storage should be carried out as described for T_4(RIA). The typical *reference range* for FT_4 in adults is 0.7–1.9 ng/dL and for FT_3, 230–660 pg/dL.

Determination of T_3 Uptake (T_3U) by Radioassay[107]

Determinations of total T_3 and T_4 concentrations in serum are clinically more meaningful if the functional levels (capacity to bind hormone) of thyroid hormone–binding protein in blood are known. The T_3 uptake (T_3U) assay is an indirect measure of the number of available free binding sites on TBG.

Note that the T_3U test is not an estimation of endogenous T_3; rather, radioactive T_3 is used as a reagent in the assay. $^{125}I\text{-}T_3$ is added to serum followed by the addition of a resin that binds T_3. Some of the $^{125}I\text{-}T_3$ binds to the unoccupied binding sites on TBG, and the remainder binds to the resin. The percentage of added radioactivity taken up by the resin (i.e., the T_3 resin uptake) is inversely proportional to the number of available free binding sites on TBG in the test serum.

A "normalized" per cent T_3 uptake for each unknown sample may be determined by comparison with a euthyroid reference serum of known per cent uptake.

$$\%T_3U = \frac{cpm_x}{cpm_r} \times \%T_3U_r$$

where cpm_x = cpm for each patient tube
cpm_r = cpm for the reference tube
$\%T_3U_r$ = % uptake of a normal (euthyroid) reference serum

Table 18-7. RELATIONSHIPS OF FINDINGS IN THYROID FUNCTION TESTS IN VARIOUS CLINICAL ENTITIES

	Thyroid Function Tests					
Clinical Entity	T_4	T_3	FT_4	T_3U	FT_4I	TBG
Normal	N	N	N	N	N	N
Hyperthyroidism	↑	↑	↑	↑	↑	N,↓
Hypothyroidism	↓	↓	↓	↓	↓	N,↑
Increased TBG	↑	↑	N	↓	N	↑
Decreased TBG	↓	↓	N	↑	N	↓
Displacement of hormone by drugs	↓	↓	N	↑	N	N

Procedures for *collection and storage of serum specimens* are similar to those described for T_4(RIA). Repeated freezing and thawing, mixing by vortexing, or vigorous agitation of serum samples should be avoided since such treatment causes denaturation of proteins. The use of hemolyzed or lipemic samples is not recommended.

Many commercial T_3U kits are currently available. These kits utilize a wide range of secondary binders, such as dextran-coated charcoal, talcum, Sephadex, or high-affinity antiserum. A convenient procedure that yields accurate, sensitive, and precise results is based on the solid-phase RIA technique. Kits utilizing charcoal and antibody-coated tubes, however, are particularly sensitive to variations in incubation time. Charcoal kits also exhibit marked temperature-dependent variations in the uptake values. In most kits there is a pronounced change in the T_3U value at TBG concentrations <10 mg/L and very little change in the value at TBG concentrations >40 mg/L, which indicates that the T_3U test has limited accuracy at very high TBG concentrations. Typical intra- and interassay CVs are 4 and 5%, respectively.

Comments. Interrelationships between serum TBG, T_3U, and other functional tests are illustrated in Table 18-7. The concentration of TBG is frequently normal in patients with thyroid dysfunction but may be slightly low in hyperthyroidism and slightly high in hypothyroidism. However, in hyperthyroidism more binding sites, and in hypothyroidism fewer binding sites on TBG are occupied by T_4. As a result, the uptake of $^{125}I\text{-}T_3$ by resin is high in hyperthyroidism and low in hypothyroidism. In contrast, free thyroid hormone levels are normal in patients with a primary increase in serum TBG, but the total T_4 is increased since the total number of binding sites is increased. The increased number of TBG binding sites will bind more $^{125}I\text{-}T_3$, and thus less $^{125}I\text{-}T_3$ will remain free to bind with resin; therefore, T_3U is low. The opposite changes occur in conditions of decreased TBG. Certain drugs, such as phenytoin (Dilantin), dicumarol, heparin, and salicylates, compete with thyroid hormones for the TBG binding sites. This situation is reflected by normal FT_4, low total T_4, and high T_3U values (Table 18-7).

Reference ranges. Since reference ranges vary considerably, each laboratory must establish its own. A typical reference range is 25–35%; variations stem from the fact that the tracer uptake depends on assay conditions such as time, temperature, pH, quantity of serum added, and nature and quantity of the secondary binder.

Free Thyroxine Index (FT_4I)[23] and FT_3 Index (FT_3I)

Total T_4 levels in serum are not wholly indicative of thyroid function since it is the level of *free* hormone that determines clinical status. An *indirect* assessment of free hormone concentration can be obtained by computing a free thyroxine index (FT_4I), which usually correlates with FT_4 concentrations; it is calculated by multiplying the value of total T_4 by the value of T_3U.

The calculation of FT_4I is based on the equilibrium relationship that exists between bound T_4 ($TBG \cdot T_4$) and free T_4 (FT_4). If K is an equilibrium constant, then

$$K = \frac{[TBG \cdot T_4]}{[FT_4] \times [TBG^0]}$$

Since 99.97% of the total thyroxine is bound to TBG, the term $[TBG \cdot T_4]$ can be essentially replaced by the total thyroxine concentration. Furthermore, since $[TBG^0]$, the concentration of unoccupied TBG binding sites, is inversely proportional to the $\%T_3U$, then

$$K \times [FT_4] = [T_4] \times \frac{1}{[TBG^0]} = [T_4] \times \%T_3U$$

If FT_4I is defined as $K \times [FT_4]$, then

$$FT_4I = [T_4] \times \%T_3U$$

T_4 is expressed in μg/dL and T_3U is expressed as percentage of a euthyroid control. Thus, like T_4, FT_4I has units of μg/dL. Furthermore, the *reference range* for FT_4I will vary with reference ranges established for T_4 and T_3U. The upper and lower limits of the FT_4I range can be calculated by multiplying the upper and lower limits of the T_3U and T_4 ranges.

In recent years, a number of other designations for FT_4I have appeared in the literature and in commercial kits. Effective thyroxine ratio (ETR), T-7, and T-12 are only a few of the examples. These tests differ either in the mode of calculating results or in the mode of performing the tests. The underlying intent of all these tests is to evaluate the FT_4 indirectly. The Committee on Nomenclature of the American Thyroid Association[102] has recommended that all tests designed to estimate FT_4 indirectly be designated as FT_4I.

In acute medical illness, FT_4I may be a better index of thyroid status than FT_4, whereas in chronic nonthyroidal illness FT_4 seems to be better than FT_4I in resolving problem cases. In the final analysis, neither FT_4I nor FT_4 can be counted on to indicate thyroid status in all circumstances; the choice of test depends on the clinical situation. With either test, supplementary data (e.g., T_3, TSH, TRH stimulation test, or even rT_3 measurement) may be required to evaluate a particular case.

The FT_3I is calculated, as is the FT_4I, as $T_3 \times T_3U$, and is used in the same way as an estimate for FT_3. FT_3I may be helpful in some patients to exclude a diagnosis of T_3 toxicosis. In general, however, the FT_3I offers no advantage over the FT_4I and is utilized less frequently in clinical practice.

Determination of Thyroxine-Binding Globulin (TBG)[36]

Tests described above are quite adequate in most clinical situations for the assessment of thyroid hormone–binding to TBG. However, direct measurement of TBG is occasionally required to confirm the results of the free hormone assay in serum. For example, in patients with familial deficiency or absence of TBG, low T_4 values and high T_3U ratios are conveniently clarified by direct measurement of TBG. Similarly, during pregnancy, elevated T_4 levels accompanied by a less than normal T_3U value are confirmed by the observation of an elevated TBG.

TBG can be measured indirectly in terms of its T_4-binding capacity (T_3U) or directly by RIA.[15] Direct measurement of TBG by RIA[36] is more popular because it is technically less complex. The antiserum is generated against purified TBG, which is also used for the preparation of the standard and the radioiodinated ligand. Bound and free fractions are separated by a double-antibody technique using goat-antirabbit antibody. An example of this assay is the two-site immunoradiometric assay (IRMA) or "sandwich" assay now commercially available (Corning Medical, Medfield, MA 02052). The adult *reference range* for this procedure is 1.5–3.4 mg/dL. For *specimen collection and storage*, see requirements listed for T_4(RIA).

T_4/TBG and T_3/TBG Ratios

These ratios are indices derived to correct for the hormonal variation that is a function of TBG concentration or binding capacity. They give essentially the same information as the FT_4I and FT_3I. There is debate whether FT_4I and FT_3I or T_4/TBG and T_3/TBG ratios correlate better with clinical status;[120a] the calculation of choice seems to depend on the particular clinical situation. The reference range for T_4/TBG ratios is

0.35–0.42 when reference values for T_4 and TBG are 4.5–12.5 μg/dL and 13–30 μg/mL, respectively.

Determination of Thyrotropin (TSH) by RIA[1]

Both conventional RIA and immunoradiometric assay (IRMA) are used in kit methods for the measurement of TSH.[72] Principles of conventional RIA have been described in Chapter 3. In a typical IRMA procedure, the patient sample is first incubated with a ^{125}I-labeled antibody to TSH to allow the binding of the patient's TSH to this antibody. Then a second unlabeled antibody, which is specific for a different antigenic site on the patient's TSH, is added. This second antibody is usually immobilized on a solid phase (beads). During a second incubation, a "sandwich" complex is formed, consisting of labeled antibody–TSH–solid phase antibody. After centrifugation and decanting, the radioactivity in the solid-phase pellet is counted in a gamma counter. Bound radioactivity is directly proportional to TSH concentration in the sample, and patient TSH values are determined from a standard curve. Sensitivities for most IRMA TSH procedures are ≤0.5 μIU/mL. Typical CVs for TSH determinations within the reference range are ~5% intra-assay and ~8% inter-assay.

The assay for TSH with sensitivities ≤0.1 μIU/mL has become available for general use. The improved sensitivity may permit use of the TSH assay to differentiate hyperthyroid from euthyroid subjects without the need to perform a TRH test.

Specimen collection and storage. The preferred specimen is serum free of hemolysis and lipemia. Serum is stable for 7 d at 2–8 °C and for at least one month frozen.

Reference range. The lower limit is presently not well defined but is ~1 μIU/mL; it should be established in each laboratory for the method in use. The upper reference limit is ~5.0 μIU/mL.

Determination of Thyroglobulin (Tg) by RIA[114]

Two types of commercial kits are currently available. One is a competitive double-antibody RIA technique and the other is a solid phase, two-site immunoradiometric assay (IRMA). Before Tg is measured, the presence of anti-Tg in the serum should be determined. If such antibodies are detected, further analysis of Tg in the serum cannot be performed, since anti-Tg autoantibodies will interfere with the immunoassay.

Reference ranges. Tg is present in 75% of normal sera at a concentration of ~5 ng/mL. Tg has also been detected in healthy pregnant women and in cord blood. Tg is elevated in patients with thyroid carcinoma, thyrotoxicosis, and subacute thyroiditis. Measurement of Tg is most useful in detecting recurrence of thyroid carcinoma following surgical resection or radioactive iodide ablation.

Human Thyroid-Stimulating Immunoglobulins (hTSI)[120]

The indication that autoantibodies might play a role in the pathogenesis of Graves' disease came with the discovery of *long-acting thyroid stimulator* (LATS) in the serum of some patients with this disease. The generic term for these IgG autoantibodies is *thyroid-stimulating immunoglobulins* (TSI). TSI displace TSH from thyroid gland receptor sites and block TSH from binding to its receptor on human thyroid cell membranes. These immunoglobulins appear to be antibodies to the TSH receptor that interact with the receptor to activate the adenylate cyclase system and stimulate the thyroid in a manner similar to TSH, but they do not respond to the negative feedback system. TSI demonstrate substantial heterogeneity. Some interfere with the binding of TSH to its receptor, some are agonists of the adenylate cyclase system, and some stimulate the thyroid of other species (e.g., mouse).

Two types of assays are available for detection of TSI. One radioreceptor method[68] is based on the fact that TSI interfere in a dose-dependent manner with the binding of radioiodinated TSH to a crude preparation of human thyroid membranes. The method is rapid, simple, and suitable for analyzing large numbers of specimens. The other assay is based on the ability of TSI to stimulate adenylate cyclase activity in human thyroid preparations.

TSI are undetectable in normal human serum but are present in 83–93% of patients with hyperthyroidism due to Graves' disease. High titers of TSI in Graves' hyperthyroidism carry an unfavorable prognosis for spontaneous remission. The presence of high levels of TSI in pregnant women is predictive of neonatal hyperthyroidism.

Determination of Antithyroglobulin and Antimicrosomal Antibodies[7]

Principle. In the tanned erythrocyte hemagglutination method for antithyroglobulin antibodies, an aliquot of a patient's serum is mixed with erythrocytes that have been treated with tannic acid and then coated with thyroglobulin. When antibodies in the patient's serum combine with antigen (i.e., thyroglobulin) bound to erythrocytes, agglutination of the erythrocytes occurs. The hemagglutination reactions are much more sensitive than most other antigen-antibody reactions. Serial dilutions of the patient's serum are employed to establish the thyroglobulin antibody titer. A similar procedure is followed for the measurement of antimicrosomal antibodies, in which the erythrocytes used have been coated with microsomal antigen derived from human hyperplastic thyroid glands.

One or both of these antibodies are detected in 5–10% of normal patients. However, in patients with primary myxedema and Graves' disease, and strikingly in Hashimoto's thyroiditis, reactivity occurs much more frequently and to a higher titer. An agglutination test that is positive at a dilution of 1:2500 is equivalent to histologic proof for Hashimoto's disease. However, in patients who have multinodular goiter, thyroid adenomas, and thyroid carcinomas, antibody titers are not found with any greater frequency than in normal control subjects.

Commercial kits for both antithyroglobulin and antimicrosomal antibodies are available. The preferred *specimen* for testing is a serum specimen that has been deactivated at 56 °C for 30 min. Heat inactivates complement and TBG, which otherwise would interfere with the assay. Serum specimens should be kept frozen if the test is not performed the day blood is drawn.

CATECHOLAMINES AND METABOLITES

Catecholamines are benzene rings bearing two hydroxyl groups (catechol) and an alkylamine chain. The most important endogenously produced compounds of this group are epinephrine (adrenalin), norepinephrine (noradrenalin), and dopamine (3,4-dihydroxyphenylethylamine). The structural formulas and the numbering system of these compounds are shown in Figure 18-13. Dopamine and norepinephrine are primary amines, whereas epinephrine is a secondary amine. Epinephrine and norepinephrine exhibit the chemical properties of phenols, alcohols, and amines. Epinephrine is rapidly

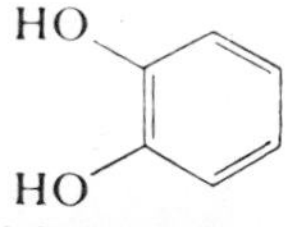

HO, HO

Catechol (Dihydroxybenzene)

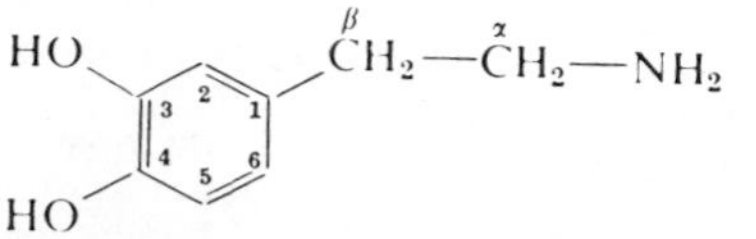

HO, HO, $\overset{\beta}{C}H_2-\overset{\alpha}{C}H_2-NH_2$, 1, 2, 3, 4, 5, 6

Dopamine [β(3,4-dihydroxyphenyl)ethyl amine]

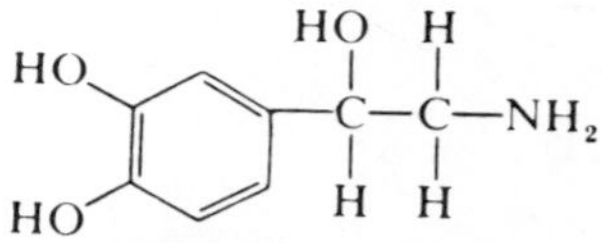

HO, HO, HO H, C—C—NH_2, H H

Norepinephrine (Noradrenaline)

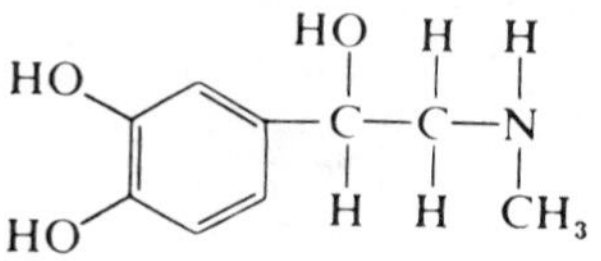

HO, HO, HO H H, C—C—N, H H CH_3

Epinephrine (Adrenaline)

Figure 18-13. Structure and numbering system of catecholamines.

oxidized in neutral and alkaline solutions; norepinephrine is more resistant to oxidation. This difference in properties of the two compounds has been utilized for their individual estimations in biological fluids.

The main sites of production of catecholamines are the brain, chromaffin cells of the adrenal medulla, and sympathetic neurons. Epinephrine is quantitatively the most important substance produced by the adrenal medulla. Norepinephrine is the major substance liberated by the postganglionic sympathetic nerves.

Each of the catecholamine hormones has characteristic physiological functions and pharmacologic actions.[123] Dopamine and norepinephrine are important neurotransmitters in the central nervous system (the brain and the spinal cord). Both have a marked influence on the vascular system, whereas epinephrine influences metabolic processes, especially carbohydrate metabolism. There are certain tissues in which epinephrine and norepinephrine produce divergent effects. The existence of two types of adrenergic receptors, α and β, with different sensitivities for the various catecholamines, accounts for the varying responses. The α-receptors interact with both epinephrine and norepinephrine; the β-receptors respond to epinephrine but are relatively insensitive to norepinephrine.

Biosynthesis, storage, and release. The L-amino acid tyrosine is the precursor of the catecholamines (Figure 18-14). The first and rate-limiting step in the biosynthetic path of catecholamines is hydroxylation of tyrosine in the mitochondria by tyrosine hydroxylase to dihydroxyphenylalanine (dopa). Dopa is converted to dopamine in the cytoplasm through the action of aromatic L-amino acid decarboxylase. Dopamine is then transported by an energy-dependent mechanism into storage granules (vesicles) that are present both in the sympathetic nerve endings and in the adrenal medulla. These granules contain dopamine-β-oxidase, an enzyme that is responsible for final conversion of dopamine to norepinephrine, which is then stored in the granules. In the adrenal medulla, norepinephrine released from storage granules becomes the substrate for another cytoplasmic enzyme, phenylethanolamine-*N*-methyl transferase, which catalyzes the formation of epinephrine. Biosynthesis and release of catecholamines are regulated both by intracellular compartmentalization and by feedback inhibition on rate-limiting steps

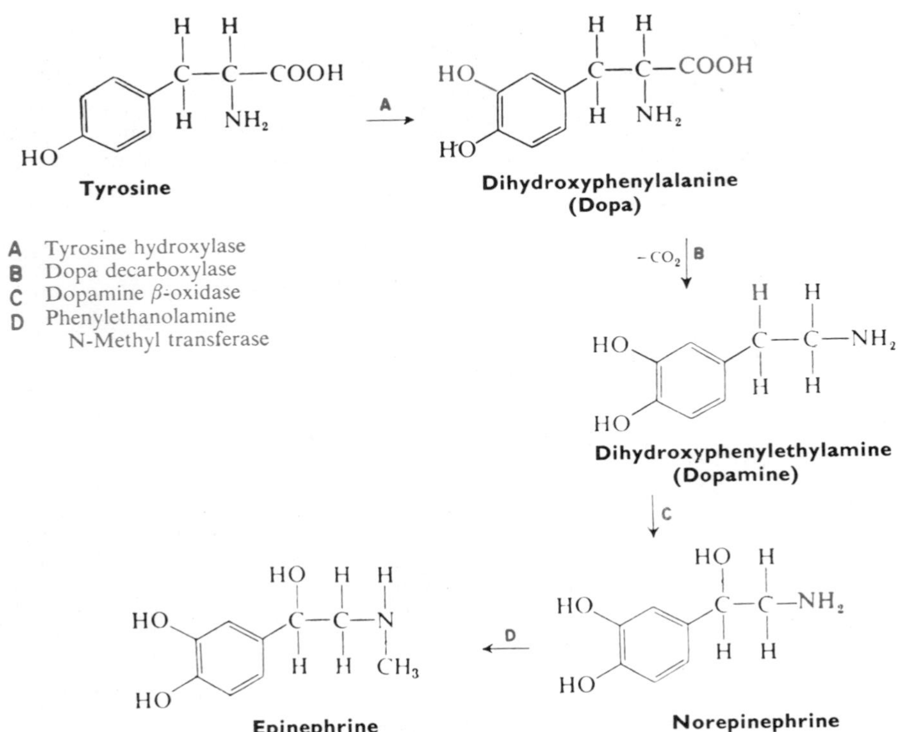

Figure 18-14. Biosynthesis of catecholamines. Red components of structure indicate location of structural change.

by the products. Nerve stimulation is necessary for the release of catecholamines from their vesicles by the process of exocytosis, which is aided by contraction of the microfibrils of the cells.

Metabolism. Upon release from storage vesicles, catecholamines act on effector sites. Catecholamines are then rapidly inactivated via reuptake by storage granules, conversion to metabolites, or excretion. Catecholamines are transported free in blood and have a very short half-life of ~2 min. Two important enzymatic events for inactivation of catecholamines involve catechol-O-methylation and oxidative deamination. Catechol-O-methyl transferase (COMT) is present in most tissues, especially in liver, kidney, and erythrocytes. This cytoplasmic enzyme methylates the C-3 hydroxyl group of norepinephrine and epinephrine, resulting in the formation of normetanephrine and metanephrine, respectively (Figure 18-15). A certain portion of these methylated products is excreted either in the free form or as conjugates of sulfuric or glucuronic acid. However, most of these 3-methoxy derivatives undergo deamination by monoamine oxidase to 3-methoxy-4-hydroxymandelic aldehyde (MHM aldehyde). This latter intermediate is

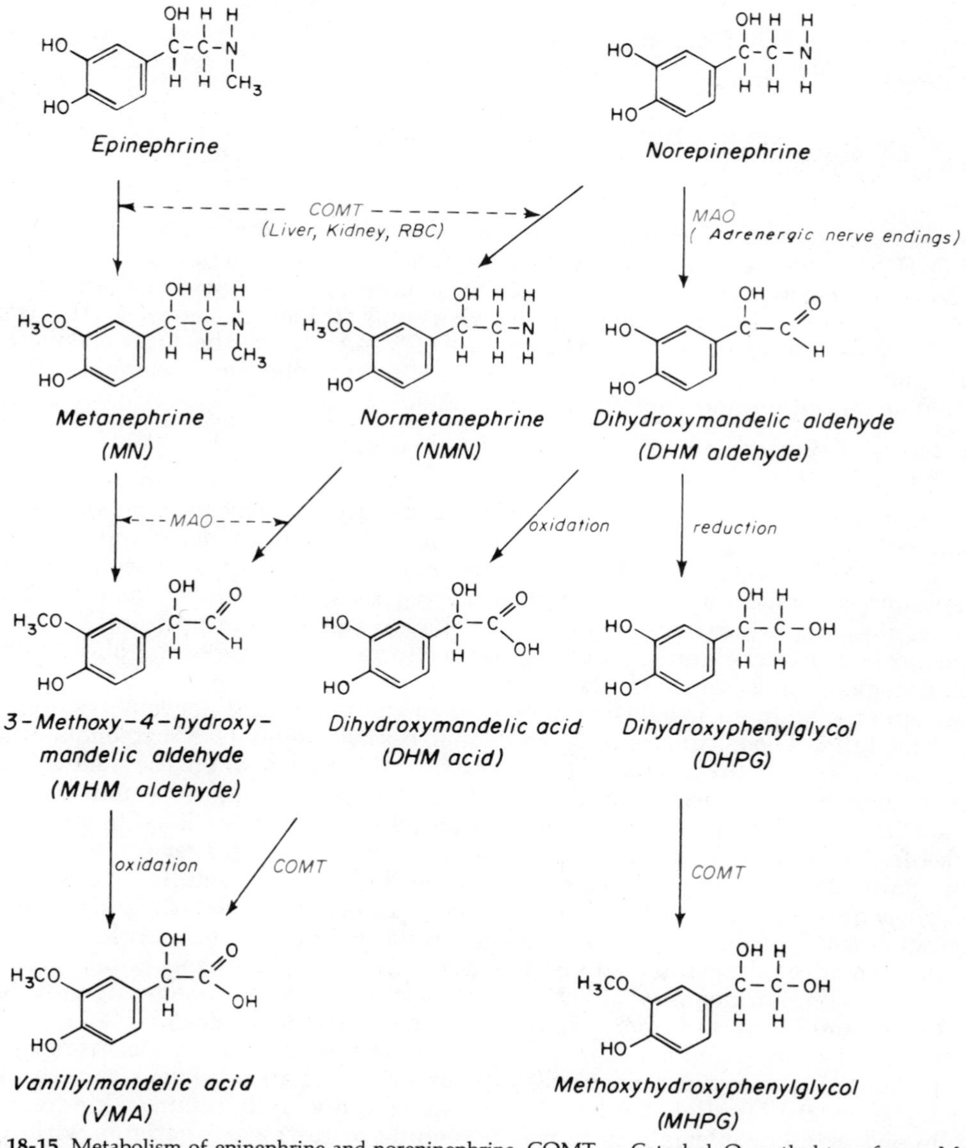

Epinephrine

Norepinephrine

COMT (Liver, Kidney, RBC)

MAO (Adrenergic nerve endings)

Metanephrine (MN)

Normetanephrine (NMN)

Dihydroxymandelic aldehyde (DHM aldehyde)

MAO

oxidation

reduction

3-Methoxy-4-hydroxymandelic aldehyde (MHM aldehyde)

Dihydroxymandelic acid (DHM acid)

Dihydroxyphenylglycol (DHPG)

oxidation

COMT

COMT

Vanillylmandelic acid (VMA)

Methoxyhydroxyphenylglycol (MHPG)

Figure 18-15. Metabolism of epinephrine and norepinephrine. COMT = Catechol-O-methyl transferase; MAO = Monoamine oxidase.

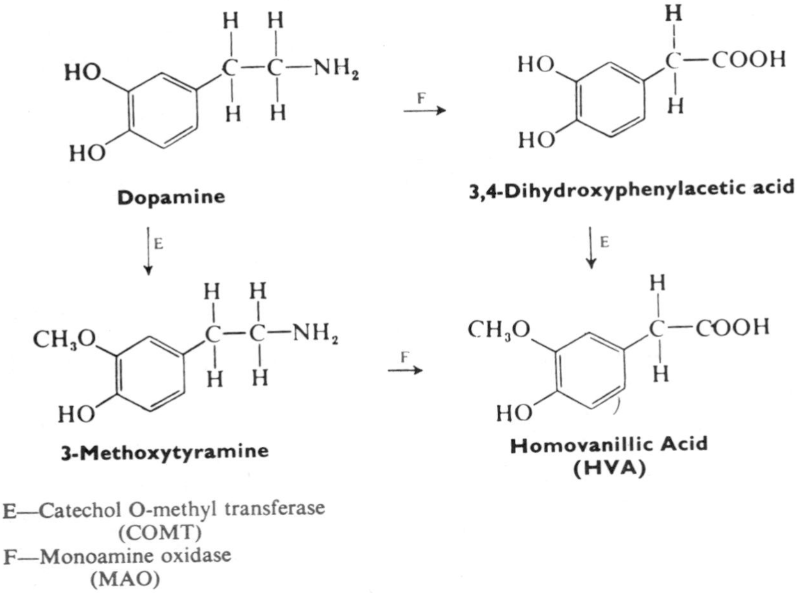

Dopamine → (F) 3,4-Dihydroxyphenylacetic acid

Dopamine → (E) 3-Methoxytyramine

3,4-Dihydroxyphenylacetic acid → (E) Homovanillic Acid (HVA)

3-Methoxytyramine → (F) Homovanillic Acid (HVA)

E—Catechol O-methyl transferase (COMT)
F—Monoamine oxidase (MAO)

Figure 18-16. Metabolism of dopamine.

either oxidized to vanillylmandelic acid (VMA) or reduced to methoxyhydroxyphenylglycol (MHPG). These end products are subsequently excreted. In adrenergic nerve endings these enzyme reactions occur in a different order, but the principal end products of both norepinephrine and epinephrine are VMA and, to a lesser degree, MHPG (Figure 18-15). For this reason urinary measurements of VMA reflect the total production of norepinephrine and epinephrine in the body. The final metabolite of dopamine following oxidative deamination and O-methylation is homovanillic acid, HVA (Figure 18-16).

Clinical Significance of Epinephrine and Norepinephrine and Their Metabolites

Catecholamines play an important role in health and disease. Increase in catecholamines is associated with stress, a fall in blood pressure or blood volume, thyroid hormone deficiency, congestive heart failure, and arrhythmias; decreased amounts of catecholamines are seen in idiopathic postural hypotension. However, laboratory measurements of epinephrine, norepinephrine, or their metabolites are primarily useful for the diagnosis of catecholamine-secreting neurochromaffin tumors (i.e., pheochromocytomas, paragangliomas, or neuroblastomas).

Benign or malignant neoplasms may arise in the autonomic nervous system or the adrenal medulla. These tumors may produce excessive amounts of catecholamines or catecholamine metabolites. Such tumors are rare. However, when the classic clinical picture of catecholamine excess occurs—sustained or paroxysmal hypertension, weight loss, "spells" of sweating, headache, palpitations, and anxiety—it is quite striking. Although life-threatening, this condition is usually curable with surgery.

In adults, 90% of these rare tumors occur in the adrenal medulla, where they are called pheochromocytomas. The 10% that occur in extra-adrenal sites are known as paragangliomas. The clinical picture of pheochromocytoma may be seen in the absence of tumor with adrenal medullary hyperplasia.[96] Nine out of ten pheochromocytomas in adults are benign; hormonally "silent" tumors are more likely to be malignant. Malignant tumors and bilateral tumors occur more frequently in children.[33]

Although hypertension is a common symptom of patients with pheochromocytoma, this catecholamine-secreting tumor accounts for $<0.5\%$ of all hypertensive cases. Pheochromocytomas are familial in about 10% of cases and usually occur in kindreds with multiple endocrine neoplasia syndromes. Screening or testing all patients with hypertension for pheochromocytoma is not cost effective. However, patients with clinical features that suggest adrenergic excess, particularly those with hypertension, should be tested. Patients in kindreds known to have multiple endocrine neoplasia syndrome should also be tested, even when they are asymptomatic. Despite selectivity in testing, false-positive test results are frequent and are much more common than true-positive tests.

A number of aspects of normal physiology make it difficult for one test of catecholamine production or action to be satisfactory for screening or for definitively diagnosing pheochromocytoma. Norepinephrine levels in plasma usually reflect only a small fraction of the concentration of norepinephrine at the synaptic nerve ending. The half-life of catecholamines in plasma is brief. Rapid and striking increases in plasma catecholamine concentrations may be induced by stress, upright posture, exercise, hypoglycemia, hypovolemia, cold, hypoxia, hypercapnia, or the mental states of anxiety or anger. Also, there is a diurnal variation of catecholamines; the zenith is in the morning and the nadir is at night. In addition, medications may change catecholamine concentrations. Alpha- and beta-adrenergic blockers, vasodilators, and theophylline increase norepinephrine levels; adrenergic neuron blockers, clonidine, α-methyl dopa, bromocriptine, and phenothiazines decrease the levels of norepinephrine. Other factors that further complicate the selection of these tests include drug interference with specific assay procedures and the conditions of sample collection. Furthermore, patients with pheochromocytoma may release catecholamines intermittently, or the uptake and degradation of catecholamines by the tumor may be so great that normal or near-normal concentrations of catecholamines will be seen in blood and urine despite striking increases in urinary metanephrines and VMA.

Measurement of urinary metanephrine is recommended as the initial screening test for pheochromocytoma.[37] Measurement of urinary metanephrines produces almost no false negatives (nearly 100% sensitivity). False-positive tests are often seen in time of severe stress, but results are likely to be normal when the test is repeated. A 24-h urine collection is desirable and provides an "integrated" picture of catecholamine production. A 12-h overnight sample can be satisfactory, provided that urinary metanephrine is expressed per gram of creatinine; a normal metanephrine/creatinine ratio is $<2.2\ \mu g$ metanephrine/g creatinine.[37,109]

Measurement of VMA or catecholamines or both in urine is best applied to follow-up testing in patients with an elevated urinary metanephrine/creatinine ratio or in subjects with normal urinary metanephrine levels but highly suspect for pheochromocytoma.[37] Urinary VMA and catecholamine determinations are more specific (fewer false positives) but are less sensitive (more false negatives) and are more subject to drug interferences than are urinary metanephrine measurements.

Some clinicians advocate measurement of plasma catecholamines as an alternative initial test to measuring urinary metanephrines. Cases of pheochromocytoma with normal levels of catecholamines, metanephrine, and VMA in urine but with definite elevations of norepinephrine or epinephrine in plasma have been reported.[8] Conditions of sampling, however, are most important; blood should be drawn in the morning after 30 min of rest in the recumbent position and ideally from an indwelling needle placed 20–30 min beforehand. Venipuncture may cause stress and increase plasma levels of catecholamines. Plasma must be promptly separated from blood. If measurements of basal levels of catecholamines and metabolites fail to establish or exclude the diagnosis of pheochromocytoma, pharmacologic tests to suppress or stimulate catecholamines may be considered.[18]

The clinical significance of determining norepinephrine and its metabolites, specifically 3-methoxy-4-hydroxy phenylglycol (MHPG) in depressed patients, stems from the hypothesis that a norepinephrine deficiency at critical CNS synapses underlies the cyclic emergence of depression in many patients with affective disorders.[59] Although urinary norepinephrine, normetanephrine, metanephrine, and perhaps VMA originate in catecholamine pools external to the CNS, there is evidence that a significant fraction of the MHPG excreted in the urine has its origin in the metabolism of brain norepinephrine. Several investigators have demonstrated that patients with low urinary excretion of MHPG have a favorable therapeutic response to antidepressants known to inhibit selectively the reuptake of norepinephrine from the synaptic cleft into the presynaptic neuron.

Methods for the Determination of Catecholamines and Metabolites

Among the numerous methods for measurement of catecholamines and their metabolites, fluorometric, spectrophotometric, radioenzymatic, and HPLC methods with electrochemical (EC) detection are widely used in clinical laboratories.

Determination of Plasma Catecholamines

The concentrations of epinephrine and norepinephrine in plasma are normally very low, necessitating the use of highly sensitive assay techniques. Fluorometry has been the classic method, but HPLC methods are the latest and possibly most promising[42] because of their high specificity and sensitivity. Extremely low plasma levels of epinephrine and dopamine require that HPLC systems operate close to their limit of detectability, resulting in occasional unreliable estimations. Until further improvements in the methods are achieved, routine application of HPLC is largely limited to the analyses of urinary catecholamines. Recently a radioenzymatic assay has found increasing application in the clinical field.

Fluorometric Determination[61]

Principle. Fluorometric methods for the determination of epinephrine and norepinephrine in plasma depend upon the conversion of the amines into characteristic derivatives that emit more intense fluorescence than does the parent amine. In the ethylenediamine (EDA) method,[61] epinephrine and norepinephrine are first oxidized in dilute acetic acid to the corresponding adrenochromes, which are then condensed with ethylenediamine.

Norepinephrine reacts with 2 mol of EDA and loses its side chain, whereas epinephrine reacts with 1 mol of EDA and retains its side chain as part of the final product. The fluorescent products of the two catecholamines have different emission spectra, which allow simultaneous measurement of each of the two hormones in the presence of the other without prior separation. The reaction is performed in red glassware, since the condensation product of norepinephrine is unstable to light.

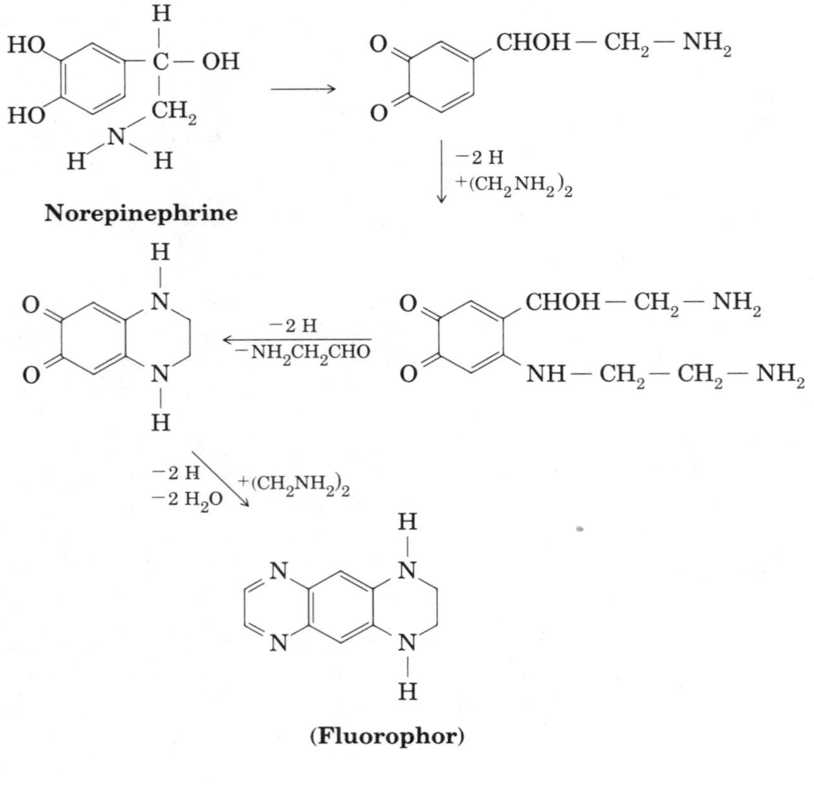

HO, HO, H, C—OH, CH_2, N, H, H

Norepinephrine

O, O, $CHOH—CH_2—NH_2$

$-2\ H$
$+(CH_2NH_2)_2$

O, O, $CHOH—CH_2—NH_2$, $NH—CH_2—CH_2—NH_2$

$-2\ H$
$-NH_2CH_2CHO$

O, O, H, N, N, H

$-2\ H$
$-2\ H_2O$
$+(CH_2NH_2)_2$

N, N, H, N, N, H

(Fluorophor)

Specimen collection and storage. Venous blood is collected in a tube containing EDTA and sodium thiosulfite sufficient to give a final concentration of 10 mg EDTA and 20 mg thiosulfite per milliliter of blood. Heparin is an acceptable alternative to EDTA. Specimens should be immediately chilled in ice water and plasma separated within 30 min of collection. The plasma must be frozen at once in a plastic vial at −70 °C. Collection and processing to storage must be quick, because even small delays cause significant losses in catecholamine content. Prolonged storage of frozen plasma samples and repeated thawing and freezing also cause losses.

The patient should be properly prepared for testing. Certain common drugs (ampicillin, methyldopa, promethazine, protamine, sulfonamides, vitamin B complex) and beverages (cocoa, coffee, tea) interfere with the fluorometric assay. Care should be taken with venipuncture since physical or emotional stress can cause temporary elevation of plasma catecholamines. Patients should be supine for 30 min preceding sampling, because catecholamines increase significantly in the upright posture. When multiple samples are necessary, they should be taken after a period of rest following placement of an indwelling venous catheter. Failure to standardize rigorously such factors as posture or failure to avoid stress and interfering substances may invalidate the most careful biochemical analysis and confound interpretation of the data.

Reference ranges. The ranges for plasma epinephrine are 140–300 and for norepinephrine, 360–800 pg/mL.

Radioenzymatic Assay[85]

The enzyme catechol-O-methyl transferase (COMT) is used as a reagent to transfer a tritiated methyl group from S-adenosylmethionine (^{3}H-SAM) to epinephrine, norepinephrine, and dopamine in the sample (plasma or cerebrospinal fluid). The radiolabeled derivatives are then separated by thin-layer chromatography. Dopamine is directly quantitated from the radioactivity of the ^{3}H-3-methoxytyramine spot. The epinephrine and norepinephrine derivative spots are converted with periodate to ^{3}H-vanillin; the radioactivity of the vanillin is measured and calculated to give the concentration of original compounds. The method, without chromatography, can be applied to determine total plasma catecholamines. A commercial kit is available (CAT-A-KIT, Upjohn Co, Kalamazoo, MI 49001).

Reference ranges for normotensive, supine adults, in pg/mL, are 111–603 for norepinephrine, 0–62 for epinephrine, and 0–83 for dopamine.

Determination of Catecholamines and Their Metabolites in Urine

Current methods of choice for the determination of urinary catecholamines are fluorometry and HPLC. For the analysis of metabolites such as metanephrines and VMA, colorimetric procedures are widely used. However, numerous HPLC methods are also available[5,34,74,79,98] and are replacing the cumbersome and less specific fluorometric and colorimetric techniques for the analysis of catecholamines and their metabolites.

Specimen collection and storage. All antihypertensive medications should be withheld from the patient for at least 2 d prior to and during specimen collection. If the patient cannot be completely removed from a drug regimen, analysis may still be performed as long as appropriate judgment is applied when results are interpreted.

A complete 24-h urine is collected in a container with 10 mL HCl, 6 mol/L, as a preservative. The specimen should be refrigerated during collection. Upon receipt of the specimen, adjust pH to between 2 and 5 by adding HCl, 6 mol/L. The total urine volume is measured and recorded; a 100-mL aliquot is stored in the refrigerator until the test is performed or may be frozen and stored indefinitely.

Determination of Metanephrine and Normetanephrine

The *colorimetric method* described by Pisano[88] has found wide application in clinical laboratories. After acid hydrolysis, metanephrines are adsorbed on Amberlite CG-50 and eluted with NH_4OH, 3 mol/L. The eluted compounds are then converted to vanillin by periodate oxidation (see *Determination of VMA in Urine*, page 603). The resulting compound is measured spectrophotometrically at 360 nm. This method does not distinguish between metanephrine and normetanephrine but serves as a rapid, easy, and reliable method for screening urine for pheochromocytoma. The reference value for total metanephrines is reported to be 0.6 ± 0.3 mg/d (SD).

The *gas-chromatographic method*[14] allows estimation of metanephrines and VMA in the same urine sample. Acidified urine (pH 1.0) is extracted with ethyl acetate; the organic extract is saved for VMA estimation. The aqueous phase is subjected to hydrolysis by hydrochloric acid, followed by oxidation of the metanephrines with periodate to vanillin at pH 10–11. Vanillin is extracted with methylene dichloride containing the internal standard, *p*-hydroxybenzoic ethyl ester, and is then quantitated as a trimethylsilyl (TMS) derivative by gas chromatography using a 3% OV-1 column. The ethyl acetate phase containing VMA is extracted with acetate buffer, pH 6.2, and VMA is oxidized to vanillin by sodium periodate at pH 11–12. After extraction into methylene dichloride, the vanillin product is estimated by gas chromatography. Values for VMA are similar to those obtained by the colorimetric technique, but the quantities of metanephrines measured by GLC are substantially lower than those obtained by chemical methods. Thus, Pisano's colorimetric procedure appears to overestimate metanephrines because of the presence of interfering contaminants.

Several highly selective methods based on *HPLC* with electrochemical detection have been reported.[5,79,98] Most of these procedures measure not only metanephrines but also homovanillic acid and 3-methoxytyramine. In principle, these methods consist of isolating the urinary metabolites of interest by passing the hydrolyzed urine through small ion-exchange columns. Samples are then concentrated by solvent extractions and injected onto the liquid chromatographic column.

The ***reference ranges*** are 105–354 μg/d for normetanephrine, 74–297 μg/d for metanephrine, and 66–222 μg/d for homovanillic acid.

3-Methoxy-4-Hydroxy Phenylglycol (MHPG)

MHPG is an important metabolite of norepinephrine and may have particular relevance to the disposition of norepinephrine in the brain. In humans, MHPG is found in the CSF, urine, and plasma. MHPG may be assayed by different methods, including HPLC[97] and GLC.[24]

Reference ranges. The mean 24-h MHPG urinary excretion for normal males is 1600 ± 380 μg (±SD) and for normal females is 1320 ± 360 μg (±SD).

Homovanillic Acid (HVA)

HVA has been identified as the principal urinary metabolite of dopa and dopamine. The lack of suitable methods for measuring dopamine in blood and urine has aroused interest in the estimation of urinary HVA, which is excreted in free form in relatively large amounts. Both gas chromatography[121] and HPLC[34] methods allow simultaneous determination of both HVA and VMA, the major terminal metabolites of dopamine and norepinephrine, respectively. Since dopamine or norepinephrine may be elevated in patients with neuroblastoma, it is diagnostically important to determine both HVA and VMA.

Reference ranges are 1–40 μg HVA/mg creatinine and 1.5–7.1 μg VMA/mg creatinine. Urinary excretion is expressed in terms of μg/mg creatinine to allow random specimens of urine to be used when complete collections are difficult to obtain. In addition, use of this ratio minimizes variations due to sex and age.

Determination of Urinary Free Catecholamines (Fractionated) by HPLC-EC[76]

Until recently, measurements of urinary total catecholamines[119] and VMA[89] were common routine clinical procedures. With the advent of HPLC with electrochemical detection, rapid and reliable determinations of individual catecholamines are possible. The following is a general description of a method for urinary free catecholamines based on high performance liquid chromatography with electrochemical detection.

Principle. Following protein precipitation with perchloric acid, an aliquot of a 24-h urine collection (preserved in acid) is first applied to a weak-acid cation exchange resin. Unconjugated catecholamines are selectively adsorbed at pH 6.5 and then eluted with dilute boric acid, pH 4.0. An intermediate water wash removes interfering urine impurities. Subsequent resolution of the individual catecholamines in the boric acid eluate is achieved by reversed-phase, paired-ion high performance liquid chromatography (HPLC) under optimized isocratic conditions. Alkyl-bonded silica is used as the nonpolar stationary phase, and an organic/aqueous buffer mixture, pH 2.8, is used as the polar mobile phase. To enhance the affinity of the polar catecholamines for the hydrophobic stationary phase, an ion of opposite charge (octyl sodium sulfonate) is also included in the mobile phase. This "counter-ion" is capable of forming uncharged ion-pair conjugates with catecholamine cations prior to partitioning into the lipophilic stationary phase. This separation mechanism may also be described by postulating that the counter-ion itself partitions into the stationary phase with its ionic groups oriented at the surfaces. The reversed-phase column then has the physical characteristics of a conventional ion-exchange resin.

A thin-layer glassy carbon or carbon-paste working electrode, in conjunction with an Ag/AgCl reference electrode and a stainless steel auxiliary electrode, is utilized as the amperometric detection system. Each catecholamine passing through the detector cell undergoes a rapid two-electron oxidation at a fixed potential to form an *o*-quinone:

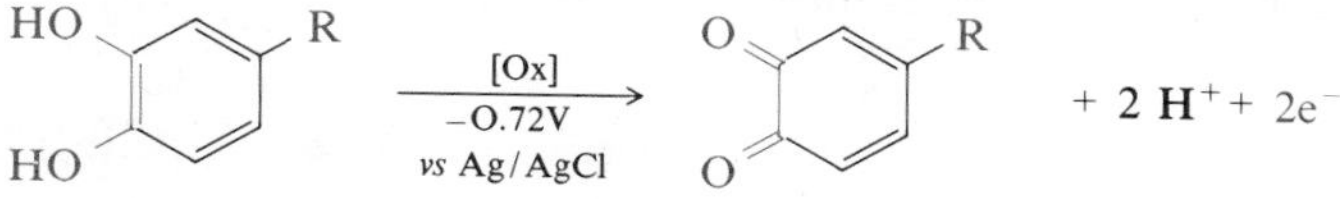

$$\text{(HO)}_2\text{C}_6\text{H}_3\text{–R} \xrightarrow[\text{-0.72V } vs \text{ Ag/AgCl}]{[\text{Ox}]} \text{(O=)}_2\text{C}_6\text{H}_3\text{–R} + 2\,H^+ + 2e^-$$

R = — $CH_2CH_2NH_2$, dopamine
= — $CH(OH)CH_2NH_2$, norepinephrine
= — $CH(OH)CH_2NH(CH_3)$, epinephrine

The current resulting from this reaction is converted to a voltage signal and monitored as a function of retention time. At a constant temperature and flow rate, this oxidation current is directly proportional to the concentration of the analyte.

Catecholamine standards are used to calibrate the system on the basis of peak heights and retention times. To calculate sample concentrations, peak height ratios relative to an internal standard, dihydroxybenzylamine, are compared with those of standards.

Reference Range

	Urinary Excretion Rate, µg/d		
Age, years	***Norepinephrine***	***Epinephrine***	***Dopamine***
0–1	0–10	0.0–2.5	0–85
0–2	0–17	0.0–3.5	10–140
1–4	4–29	0.0–6.0	40–260
4–7	8–45	0.2–10	65–400
7–10	13–65	0.5–14	65–400
10–15	15–80	0.5–20	65–400
>15	14–80	0.5–20	65–400

Determination of VMA (Vanillylmandelic Acid; 3-Methoxy-4-Hydroxymandelic Acid) in Urine[89,111]

Principle. VMA and other phenolic acids are extracted from acidified urine with ethyl acetate and then re-extracted into aqueous potassium carbonate solution. The addition of sodium metaperiodate to the carbonate extract oxidizes VMA to vanillin.

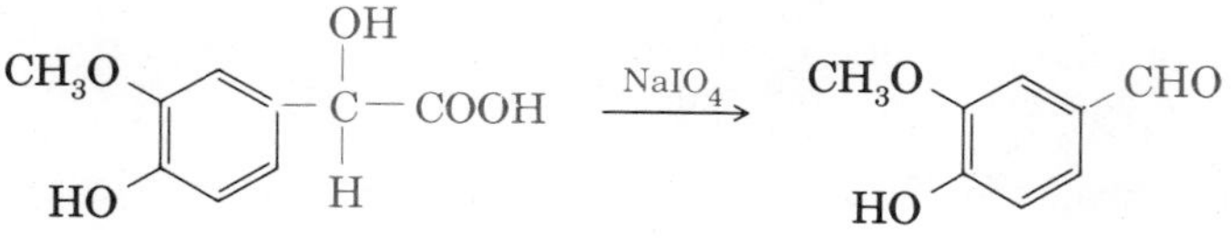

$$\underset{\textbf{VMA}}{\text{CH}_3\text{O(HO)C}_6\text{H}_3\text{–CH(OH)–COOH}} \xrightarrow{\text{NaIO}_4} \underset{\textbf{Vanillin}}{\text{CH}_3\text{O(HO)C}_6\text{H}_3\text{–CHO}}$$

Vanillin is then separated from contaminating phenolic acids by selective extraction into toluene and back-extraction into carbonate. The vanillin concentration is determined spectrophotometrically at a wavelength of 360 nm.

Specimen collection and storage. The intake of chocolate, coffee, bananas, foods containing vanilla, citrus fruits, and drugs such as aspirin and antihypertensive agents (e.g., Aldomet) may cause falsely high values for urinary VMA with some methods. No dietary restrictions during urine collection are necessary if the VMA method described below is used.

The pH of the urine should be kept at ~2 during the collection by placing 10 mL of HCl, 6 mol/L, into a suitable container (dark-brown bottle). After measurement of the total volume, aliquots may be stored at 4 °C and are stable for several weeks.

Reagents

All reagents should be of reagent grade purity.

1. Hydrochloric acid, 6 mol/L. Slowly add 500 mL concentrated HCl to a 1-L volumetric flask containing ~300 mL water and dilute to the mark with water. Water to be used for the preparation of all reagents should be distilled twice in an all-glass apparatus.
2. Sodium chloride.
3. Ethyl acetate.
4. Potassium carbonate, 1 mol/L. Dissolve 138 g potassium carbonate in 1 L of distilled water. Store at room temperature and discard after one month.
5. Sodium metaperiodate, 2 g/dL, in distilled water. Prepare fresh daily.
6. Sodium metabisulfite, 10 g/dL, in distilled water. Prepare fresh daily.
7. Acetic acid, 5 mol/L. Dilute 286 mL glacial acetic acid with distilled water to 1 L.
8. Phosphate buffer, 1 mol/L, pH 7.5. *Solution A*: dissolve 268.07 g disodium phosphate ($Na_2HPO_4 \cdot 7H_2O$) in distilled water and dilute to 1 L. Store in a refrigerator. *Solution B*: dissolve 27.22 g potassium dihydrogen phosphate (KH_2PO_4) in 200 mL of distilled water. Mix 168.2 mL of solution A with 31.8 mL of solution B. Check with pH meter and add acid or base to obtain a pH of 7.5.
9. Hydrochloric acid, 0.01 mol/L. Dilute 0.83 mL of concentrated HCl to 1 L with distilled water.
10. Standard solutions. *VMA stock solution,* 1 mg/mL: Accurately weigh 100 mg of VMA and dissolve in 100 mL of HCl, 0.01 mol/L, in a volumetric flask. The solution is stable for approximately three months when under refrigeration. *VMA working solution,* 10 μg/mL: Dilute 1 mL of the stock solution to 100 mL with HCl, 0.01 mol/L. Prepare fresh before use.

Procedure

1. Pipet 0.2% of the 24-h urine volume into 50-mL glass-stoppered (or Teflon-lined screw-capped) centrifuge tubes marked "tests," "internal standards," and "unoxidized blanks" in duplicate. To the internal standard tubes also add 1 mL of the working standard.
2. Dilute the contents of all tubes to 5.5 mL with distilled water, and further acidify with 0.5 mL of HCl, 6 mol/L.
3. Add a saturating amount of sodium chloride (~3 g), mix, and extract with 30 mL of ethyl acetate by shaking on a mechanical shaker for 30 min. Centrifuge for 5 min.
4. Transfer 25 mL of the organic extract (upper layer) to a second glass-stoppered centrifuge tube containing 1.5 mL of potassium carbonate, 1 mol/L. Shake mechanically for 3 min and centrifuge for 5 min. Aspirate and discard the upper organic phase.
5. Pipet 1 mL of the carbonate phase (lower layer) to a third glass-stoppered centrifuge tube.
6. To the test and standard tubes add 0.1 mL of sodium metaperiodate, 2 g/dL, mix, and stopper loosely; place all tubes, including tubes marked "unoxidized blank" (metaperiodate solution is omitted at this stage), into a water bath at 50 °C for 30 min.
7. At the end of the incubation period, cool the tubes to room temperature.
8. To the "unoxidized blank" tubes add 0.1 mL of sodium metaperiodate and mix.
9. Without delay add 0.1 mL of metabisulfite solution to all tubes to reduce residual periodate.
10. Neutralize all solutions by adding 0.3 mL of acetic acid, 5 mol/L. Allow to stand for 10 min.
11. Add 0.6 mL of phosphate buffer, 1 mol/L, pH 7.5. (The pH can be checked at this point by adding one drop of aqueous cresol red, 0.04 g/dL. The solution should be yellow, indicating a pH of <8.8.)
12. Shake mechanically for 3 min with 20 mL of toluene to extract vanillin, the oxidized product of VMA.

13. Centrifuge for 5 min, and transfer 15 mL of the toluene extract into a fourth glass-stoppered centrifuge tube containing 4.0 mL of potassium carbonate, 1 mol/L. Shake mechanically for 3 min and centrifuge for 5 min.

14. Transfer the carbonate layer containing vanillin into a microcuvet, and determine the absorbance at 360 nm against a water blank.

Calculations

$$\text{VMA, mg/d} = \frac{A_t - A_b}{A_{st} - A_t} \times \frac{10}{1000} \times \frac{100}{0.2} = \frac{A_t - A_b}{A_{st} - A_t} \times 5$$

where A_b = absorbance of "unoxidized" urine blank
A_t = absorbance of test
A_{st} = absorbance of internal standard (standard + test)
10 = concentration of standard
1000 = converts standard from μg to mg
0.2 = per cent of total urine volume used

Comments. An unoxidized blank is prepared for every sample to correct for the presence of vanillin in urine. The internal standard compensates for procedural losses, for decomposition of vanillin, and for the relative inhibition of its formation by unknown urinary factors. At room temperature the oxidation of VMA by periodate proceeds smoothly in pure solutions, whereas an elevated temperature (50 °C) is required for the oxidation of VMA in urinary extracts. In some urine specimens, the oxidation may be strongly inhibited even at 50 °C.[119]

The oxidation of VMA to vanillin is also sensitive to hydrogen ion concentration. In neutral and acidic solutions, oxidation forms a yellow pigment; strongly alkaline solutions, on the other hand, delay the formation of vanillin and cause its decomposition. Optimal conditions are obtained in the presence of sodium or potassium carbonate solution/dL urine to give an approximate pH of 11.[89] The maximum absorption of vanillin occurs at 348 nm. However, at this wavelength there is considerable absorbance of the oxidation product (*p*-hydroxybenzaldehyde) of *p*-hydroxymandelic acid, a normal constituent of urine. For this reason, the measurement is carried out at 360 nm, where the absorbance of vanillin is 80% of its peak value and interference is minimal. The absorbance of vanillin drops sharply between 350 and 380 nm, so the wavelength setting must remain exactly at 360 nm.[89]

Reference Ranges

	mg/d
Newborn and neonate:	<1.0
Infant:	<2.0
Child:	1–5
Adolescent:	1–5
Thereafter:	2–7 (or 1.5–7.0 μg/mg creatinine)

Serotonin and 5-Hydroxyindoleacetic Acid

Biosynthesis and Metabolism

Serotonin (5-hydroxytryptamine, 5-HT), a powerful smooth-muscle stimulant and vasoconstrictor, is a derivative of the amino acid tryptophan. It is transported in the blood by the platelets. The formation and breakdown of serotonin is depicted in Figure 18-17. Approximately 1–3% of dietary tryptophan is normally metabolized by this pathway. The enzymatic decarboxylation (step B in figure) is very active in carcinoid tumors. Serotonin may rapidly undergo oxidative deamination in a tumor or in the blood after release from a tumor. The oxidative deamination of serotonin by the enzyme monoamine oxidase (MAO) leads to the formation of 5-hydroxyindoleacetic acid (5-HIAA), which is quantitatively the most significant metabolite. The majority of the 5-HIAA is excreted in the free form, although a small amount may be conjugated as the O-sulfate ester before excretion.

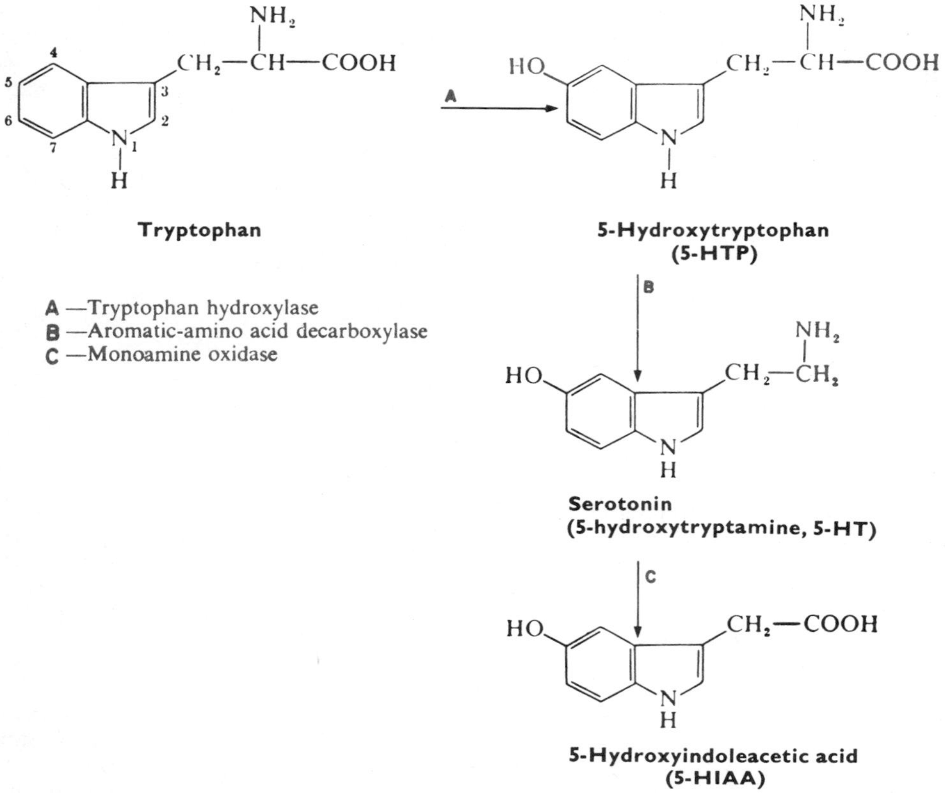

Figure 18-17. Biosynthesis and metabolism of serotonin. Red components of structure indicate location of structural change.

Clinical Significance

Depression. The most important physiological function of serotonin is that of a transmitter in neurons (serotoninergic neurons) within the brain. In humans, serotonin has been implicated in a variety of behavioral patterns including sleep, perception of pain, social behavior, and mental depression.[6]

Carcinoid tumors. Serotonin is produced in excess by carcinoid tumors,[83] which develop from enterochromaffin cells. These cells are widely distributed throughout the gastrointestinal tract, biliary tract and gallbladder, pancreatic ducts, and bronchial tree and are also found in the thymus, thyroid, ovary, uterus, and salivary glands. Carcinoid tumors usually arise in the small intestine, appendix, or rectum, and the patients usually present with bleeding, obstruction, or metastases, without humoral manifestations. The syndrome associated with the humoral manifestations of these tumors is quite striking but is seen only rarely. The tumors may be malignant and often metastasize early.[70]

The production and metabolism of serotonin differ relative to the tissue of origin of the tumor. Tumors from mid-gut cells usually contain and release large quantities of 5-HT; this large amount may not be fully reflected in the amount of the metabolite 5-HIAA in urine, because little is metabolized. Tumors from fore-gut cells produce large amounts of 5-HT, which is oxidized within the tumor to 5-HIAA. With these tumors, urinary excretion of 5-HIAA is often much higher than would be expected from the clinical presentation. Tumors from hind-gut cells only rarely produce excess 5-HT or 5-HIAA.

The classic clinical presentation of carcinoid syndrome includes pronounced flushing, bronchial constriction, diarrhea, and cardiac valvular lesions, often associated with right-sided heart failure.[70] Not all of these clinical manifestations of carcinoid syndrome can be explained by serotonin excess, but carcinoid tumors often produce excesses of other substances as well, among them histamine, catecholamines, prostaglandins, vasoactive peptides, kallikrein, ACTH, growth hormone, and insulin. They may also be associated with multiple endocrine neoplasia.

Patients with functioning carcinoid tumors usually have striking increases in urinary excretion of 5-HIAA (> 25 mg/d when normal is < 9 mg/d). If a borderline elevation of 5-HIAA is found (6–25 mg/d), repeat collections should be made and care taken to avoid the patient's use of food and medications that might elevate 5-HIAA (e.g., pineapples, avocados, bananas, walnuts, chocolate, guaifenesin, and reserpine). Nontropical sprue may cause a slight increase in urinary 5-HIAA.[53] 5-HIAA levels are lowered by phenothiazines, in renal insufficiency, and after small bowel resection.

When 5-HIAA in urine is normal, assay for blood levels of 5-hydroxytryptophan (5-HTP) or 5-HT is sometimes needed to document the diagnosis of carcinoid syndrome.

Methods for Determination of Serotonin and Its Metabolite[58]

Serotonin and its metabolite in body fluids and tissues have been measured by spectrophotometry, fluorometry, and gas chromatography. A specific and sensitive radioenzymatic method for the determination of serotonin has been reported, and HPLC with electrochemical detection has also been applied to the analysis of serotonin and 5-HIAA in human body fluids, including cerebrospinal fluid.[84]

Determination of 5-Hydroxyindoleacetic Acid in Urine

Estimation of the parent hormone, 5-HT, in blood and urine has been severely limited because of its very low concentration and because of methodological complications. As a result, the urinary determination of 5-HIAA continues to be the most practical means for the diagnosis of carcinoid tumors. This metabolite of serotonin is excreted in very large amounts, often exceeding 350 mg/d, and a positive result is obtained with simple qualitative (screening) tests. However, for early diagnosis, when the tumor is small and has not metastasized, and in some carcinoid tumors in which the excretion values barely exceed 8 mg, the more sensitive and specific quantitative test is required.

Specimen collection and storage. A random specimen is usually suitable for the screening test. For quantitative analysis, a 24-h urine specimen is collected without preservatives, and the specimen should be refrigerated during collection. When the specimen is received in the laboratory, the pH of the urine should be adjusted between 2 and 3 by addition of HCl, 6 mol/L. Acidified urine can be kept at 4 °C for two weeks or at −20 °C for longer periods of time. False-negative results may occur in patients taking phenothiazine drugs. Ingestion of bananas, avocados, red plums, eggplant, tomatoes, or cough medication containing glycerol guaiacolate may produce false-positive results. Therefore, these drugs and diets should be restricted three to four days prior to and during the collection.

Principle of screening test.[100] This test is based on the development of a purple color, specific for 5-hydroxyindoles, upon the addition of 1-nitroso-2-naphthol and nitrous acid. Other interfering chromogens are extracted into ethylene dichloride.

Reagents

1. 1-Nitroso-2-naphthol, 0.1 g/dL in 95% ethanol.
2. Sulfuric acid, 1 mol/L.
3. Sodium nitrite, 2.5 g/dL in water. Prepare fresh at frequent intervals. Refrigerate.
4. Nitrous acid reagent. Prepare fresh before use by adding 0.2 mL sodium nitrite, 2.5 g/dL, to 5 mL of sulfuric acid, 1 mol/L.
5. Ethylene dichloride, redistilled.

Procedure

1. Pipet into a test tube 0.2 mL of urine, 0.8 mL of distilled water, and 0.5 mL of 1-nitroso-2-naphthol. Mix. Prepare another tube in the same way with normal urine to serve as a negative control.
2. Add 0.5 mL of freshly prepared nitrous acid reagent to both tubes and mix again. Let tubes stand at room temperature for 10 min.

3. Add 5 mL of ethylene dichloride and shake. If turbidity results, centrifuge. A positive test shows a purple color in the top aqueous layer. The negative control with normal urine produces a slight yellow color.

Comments. A purple color (positive test) will be seen at levels of 5-HIAA excretion as low as 40 mg/d. At higher levels, the color is more intense, and it is almost black at levels > 300 mg/d. A positive result should be verified with a quantitative method. Observance of dietary and drug restrictions as outlined are important if false negative or positive results are to be avoided. The substance *p*-hydroxyacetanilide derived from acetanilide or related drugs reacts in the same manner as 5-HIAA and adds to the color. Color formation may be inhibited in specimens from patients who excrete large amounts of keto acids.

Principle of quantitative test.[113] 5-Hydroxyindoleacetic acid (5-HIAA), other phenolic acids, and drug metabolites are extracted into diethyl ether from acidified urine; a saturating amount of sodium chloride is added to promote quantitative transfer into the ether phase. The 5-HIAA is then back-extracted into a phosphate buffer (0.1 mol/L, pH 7) and reacted with nitroso-naphthol and nitrous acid (HONO) at 37 °C to form a violet color.

Phosphate buffer at pH 7 is chosen for efficient extraction, since 5-HIAA becomes progressively more unstable at higher pH values. Urinary phenols lacking an acid group are not ionized at pH 7 and remain behind in the ether layer, thereby providing a relatively clean extract. An intense blue chromophore is rapidly formed upon subsequent addition of 2-mercaptoethanol:

$$\text{5-HIAA (HO, } CH_2-COOH\text{, N–H)} + \text{Nitrosonaphthol (N=O, OH)} \xrightarrow[37\ ^\circ C]{HONO} \text{Violet chromophore } (\lambda_{max} = 535\text{ nm}) \xrightarrow[37\ ^\circ C]{HSCH_2CH_2OH} \text{Blue chromophore } (\lambda_{max} = 645\text{ nm})$$

5-HIAA **Nitrosonaphthol** **Mercaptoethanol**

Extraneous colors caused by reactive phenols and indoleacetic acid are removed by treatment with the mercaptoethanol followed by extraction into ethyl acetate. The absorbance maximum of the blue solution occurs at 645 nm, but measurements are generally made at 590 nm, where Beer's law is obeyed throughout a suitable range of 5-HIAA concentrations.

Comments. Acetic acid should not be used as a urine preservative since it decreases the extraction efficiency of 5-HIAA into the phosphate buffer. Preservatives are not required for 24-h specimens, provided the urine is kept refrigerated during the collection. The reaction of 5-HIAA with nitroso-naphthol to form a violet chromophore is claimed to be very specific. Serotonin and 7-hydroxyindoles do not react.[113]

Ectopic Hormones and Paraneoplastic Syndromes

Malignant neoplasms may produce a variety of ectopic products[77] that may exhibit striking clinical effects. These products are termed *ectopic* because they are not normally derived from the tissue affected by the tumor. The ability of neoplasms to produce hormones ectopically is due in part to the frequent occurrence of tumor cells derived from neural ectoderm. These are small cells that have *a*mine *p*recursor *u*ptake and *d*ecarboxylase activity (APUD cells).[4,118] Although not all hormone-producing malignancies are derived from APUD cells, probably all cancers can produce and secrete one or more proteins or peptides. Not all of the products are biologically active, and none are specific for a particular cell type. As tumor markers, however, these substances attract great interest.

Hormones are among the markers produced by neoplasms (Table 18-8). In many instances, these markers lack biological activity, but their measurement may be useful in the diagnosis of certain malignancies or in following the response to treatment.[51] Some neoplasms may exhibit these markers inconsistently, not at all, or variably with the natural progression of the tumor.

Most frequently, these ectopically produced substances are immunochemically similar to the naturally occurring hormones and are therefore detectable with standard radioimmunoassay techniques. However, immunochemical differences between the ec-

Table 18-8. ECTOPIC HORMONES PRODUCED BY MALIGNANT TUMORS

ACTH, proACTH	Corticotropin-releasing hormone (CRH)
Lipotropin	Calcitonin
Chorionic gonadotropin (intact and β-subunit)	Pituitary glycoprotein hormones (α-subunit)
Somatomedins	Antidiuretic hormone (vasopressin)
Parathyroid hormone	Prostaglandins
Prolactin, "big" prolactin	Growth hormone
Gastrin	Secretin
Glucagon	Somatostatin
Chorionic somatomammotropin	Vasoactive intestinal polypeptide
Erythropoietin	

topic and natural hormones are usually insufficient for most assays to distinguish normal from ectopic production.

Common Endocrine Syndromes Associated with Malignancies[81]

The most commonly encountered instances of ectopic hormone production are ectopic ACTH associated with oat-cell carcinomas of the lung, bronchial and other carcinoid tumors, and with medullary carcinomas of the thyroid; ectopic ADH (SIADH) associated with oat-cell and other malignancies of the lung, carcinomas of the colon, carcinoid tumors, and lymphomas; and humoral hypercalcemia of malignancy due to renal adenocarcinomas, squamous cell and other carcinomas of the lung, and carcinomas of the ovary. In most cases of humoral hypercalcemia, the exact mediator of hypercalcemia is not yet known; ectopic production of parathyroid hormone has only rarely been demonstrated. With ectopic production of ACTH, the typical picture of cortisol excess (Cushing's syndrome) is often overshadowed by wasting associated with the tumor.

References

1. Abraham, G. E., Ed.: Radioassay Systems in Clinical Endocrinology. New York, Marcel Dekker, Inc., 1981.
2. Abraham, G. E., and Chakmajian, Z. H.: Serum steroid levels during the menstrual cycle in bilaterally adrenalectomized women. J. Clin. Endocrinol. Metab., *37*:581–587, 1973.
3. Abraham, G. E., Maroulis, G. B., and Marshall, J. R.: Evaluation of ovulation and corpus luteum function using measurement of plasma progesterone. Obstet. Gynecol., *44*:522–525, 1974.
4. Baylin, S. B., and Mendelsohn, G.: Ectopic (inappropriate) hormone production by tumors: Mechanisms involved and the biological and clinical implications. Endocr. Rev., *1*:45–77, 1980.
5. Bertani-Dziedzic, L. M., Krstulovic, A. M., Dziedzic, S. W., et al.: Analysis of urinary metanephrines by reversed-phase high performance liquid chromatography and electrochemical detection. Clin. Chim. Acta, *110*:1–8, 1981.
6. Boullin, D. J., Ed.: Serotonin in Mental Abnormalities. New York, John Wiley & Sons, 1978.
7. Boyden, S. V.: Adsorption of proteins on erythrocytes treated with tannic acid and subsequent hemagglutination by antiprotein sera. J. Exp. Med., *93*:107–120, 1951.
8. Bravo, E. L., Tarazi, R. C., Gifford, R. W., et al.: Circulating and urinary catecholamines in pheochromocytoma. N. Engl. J. Med., *301*:682–686, 1979.
9. Breuer, H., Hamel, D., and Krüskemper, H., Eds.: Methods of Hormone Analysis. New York, John Wiley & Sons, 1976.
10. Broughton, A.: Application of adrenocorticotropin assays in a routine clinical laboratory. Am. J. Clin. Pathol., *64*:618–624, 1975.
11. Brown, J. B.: A chemical method for the determination of oestriol, oestrone and oestradiol in human urine. Biochem. J., *60*:185–193, 1955.
12. Brown, J. B., MacNaughton, C., Smith, M. A., et al.: Further observations on the Kober colour and Ittrich fluorescence reactions in the measurement of oestriol, oestrone, and oestradiol. J. Endocrinol., *40*:175–188, 1968.
13. Buster, J. E., and Abraham, G. E.: Radioimmunoassay of plasma dehydroepiandrosterone sulfate. Anal. Letters, *5*:543–551, 1972.
14. Calseyde, J. F. van de, Scholtis, R. J., Schmidt, N. A., et al.: Gas chromatography in the estimation of urinary metanephrines and VMA. Clin. Chim. Acta, *32*:361–366, 1971.
15. Cavalieri, R. R., and Ingbar, S. H.: Methods for measuring the thyroxine-binding proteins and free thyroid hormone concentration in serum. Methods Enzymol., *36*:126–132, 1975.

16. Chattoraj, S. C., Rankin, J. L., Turner, A. K., et al.: Urinary progesterone as an index of ovulation and corpus luteal function. J. Clin. Endocrinol. Metab., *43*:1402–1405, 1976.
17. Chattoraj, S. C., Fanous, A., Cecchini, D., et al.: A radioimmunoassay method for urinary catechol estrogens. Steroids, *31*:375–391, 1978.
18. Chattoraj, S. C., and Watts, N. B.: Endocrinology. *In*: Textbook of Clinical Chemistry. N. W. Tietz, Ed. Philadelphia, W.B. Saunders Co., 1986.
19. Chen, I. -W.: Commercially available fully automated systems for radioligand assays. Part I. Overview. Ligand Rev., *2*(2):46–50, 1980. Part II. Performance characteristics. Ligand Rev., *2*(3):46–48, 1980.
20. Chochinov, R. H., and Daughaday, W. H.: Somatomedin A, Somatomedin C., and NSILA-s. *In*: Methods of Hormone Radioimmunoassay. B. M. Jaffe and H. R. Behrman, Eds. New York, Academic Press, 1979, pp. 960–977.
21. Chopra, I. J.: Radioimmunoassay of iodothyronines. *In*: Handbook of Radioimmunoassay. G. E. Abraham, Ed. New York, Marcel Dekker, Inc., 1977, pp. 679–703.
22. Chopra, I. J., Hershman, J. H., Pardridge, W. M., et al.: Thyroid function in nonthyroidal illness. Ann. Intern. Med., *98*:946–957, 1983.
23. Clark, F., and Horn, D. B.: Assessment of thyroid function by the combined use of the serum protein-bound iodine and resin uptake of ^{131}I-triiodothyronine. J. Clin. Endocrinol. Metab., *25*:39–45, 1965.
24. Dekirmenjian, H., and Maas, J. W.: An improved procedure of 3-methoxy-4-hydroxyphenyl-ethylene glycol determination by gas-liquid chromatography. Anal. Biochem., *35*:113–122, 1970.
25. Dorfman, R. I., Ed.: Methods in Hormone Research, Vol. I. 2nd ed. New York, Academic Press, 1968.
26. Duckworth, W. C., and Kitabchi, A. E.: Insulin metabolism and degradation. Endocrine Reviews, *2*:210–233, 1981.
27. Eddy, R. L., Gilliland, P. F., Ibarra, J. D., et al.: Human growth hormone release: Comparison of provocative test procedures. Am. J. Med., *56*:179–185, 1974.
28. Eddy, R. L., Jones, A. L., and Gilliland, P. F.: Cushing's syndrome: A prospective study of diagnostic methods. Am. J. Med., *55*:621–630, 1973.
29. Edwards, L., Ed.: Product guide for radioassay and nonisotopic assays. Clin. Chem., *29*:889–986, 1983.
30. Eik-Nes, K. B., and Horning, E. C.: Gas Phase Chromatography of Steroids. New York, Springer-Verlag, 1968.
31. Frantz, W. L., MacIndoe, J. H., and Turkington, R. W.: Prolactin receptors: Characteristics of the particulate fraction binding activity. J. Endocrinol., *60*:485–497, 1974.
32. Freedlender, A. E., and Goodfriend, T. L.: Renin and the angiotensins. *In*: Methods of Hormone Radioimmunoassay. B. M. Jaffe and H. R. Behrman, Eds. New York, Academic Press, 1979, pp. 889–907.
33. Freier, D. T., Tank, E. S., and Harrison, T. S.: Pediatric and adult pheochromocytoma. Arch. Surg., *107*:252–255, 1973.
34. Fujita, K., Maruta, K., Ito, S., et al.: Urinary 4-hydroxy-3-methoxymandelic (vanillylmandelic) acid, 4-hydroxy-3-methoxyphenylacetic (homovanillic) acid, and 5-hydroxy-3-indoleacetic acid determination by liquid chromatography with electrochemical detection. Clin. Chem., *29*:876–878, 1983.
35. Gavin, L. A., Rosenthal, M., and Cavalieri, R. R.: The diagnostic dilemma of isolated hyperthyroxinemia in acute illness. JAMA, *242*:251–253, 1979.
36. Gershengorn, M. C., Larsen, P. R., and Robbins, J.: Radioimmunoassay for serum thyroxine-binding globulin: Results in normal subjects and in patients with hepatocellular carcinoma. J. Clin. Endocrinol. Metab., *42*:907–911, 1976.
37. Gitlow, S. E., Mendlowitz, M., and Bertrani, I. M.: The biochemical techniques for detecting and establishing the presence of a pheochromocytoma. Am. J. Cardiol., *26*:270–279, 1970.
38. Green, P. J., and Yucis, M. J.: Free testosterone determination by ultrafiltration and comparison with dialysis. Clin. Chem., *28*:1237–1238, 1982.
39. Griffin, J. E., and Wilson, J. D.: The syndromes of androgen resistance. N. Engl. J. Med., *302*:198–209, 1980.
40. Hammond, G. L., Nisker, J. A., Jones, J. A., et al.: Estimation of the percentage of free steroid in undiluted serum by centrifugal ultrafiltration dialysis. J. Biol. Chem., *255*:5023–5026, 1980.
41. Hellman, L., Bradlow, H. L., and Zumoff, B.: Recent advances in human steroid metabolism. Adv. Clin. Chem., *13*:1–35, 1970.
42. Holly, J. M. P., and Makin, H. L. J.: The estimation of catecholamines in human plasma. Anal. Biochem., *128*:257–274, 1983.
43. Hunter, W. M., and Corrie, J. E. T., Eds.: Immunoassays for Clinical Chemistry. 2nd ed. Edinburgh, Churchill Livingstone, 1983.
44. Ingbar, S. H., Braverman, L. E., Dawber, N. A., et al.: A new method for measuring the free thyroid hormone in human serum and an analysis of the factors that influence its concentration. J. Clin. Invest., *44*:1679–1689, 1965.
45. Ingbar, S. H., and Woeber, K. A.: The thyroid gland. *In*: Textbook of Endocrinology. R. A. Williams, Ed. Philadelphia, W. B. Saunders Co., 1981, pp. 117–247.
46. Ittrich, G.: Eine neue Methode zur chemischen Bestimmung der oestrogenen Hormone in Harn. Hoppe-Seylers Z. Physiol. Chem., *312*:1–14, 1958.
47. Jaffe, B. M., and Behrman, H. R., Eds.: Methods of Hormone Radioimmunoassay. New York, Academic Press, 1979.
48. Jurjens, H., Pratt, J. J., and Woldring, M. G.: Radioimmunoassay of plasma estradiol without extraction and chromatography. J. Clin. Endocrinol. Metab., *40*:19–25, 1975.
49. Kautsky, M. P., Ed.: Steroid Analysis by HPLC. New York, Marcel Dekker, Inc., 1981.

50. Kelsey, J. L.: A review of the epidemiology of human breast cancer. Epidemiol. Rev., *1*:74–109, 1979.
51. Klee, G. G., and Go, V. L. W.: Serum tumor markers. Mayo Clin. Proc., *57*:129–132, 1982.
52. Korth-Schultz, S., Levine, L. S., and New, M. E.: DHEA-S levels. A rapid test for abnormal adrenal androgen secretion. J. Clin. Endocrinol. Metab., *42*:1005–1013, 1976.
53. Kowlessar, O. D., Williams, R. C., Law, D. H., et al.: Urinary excretion of 5-hydroxyindoleacetic acid in diarrheal states, with special reference to nontropical sprue. N. Engl. J. Med., *259*:340–341, 1958.
54. Lindholm, J., Kehlet, H., Blickert-Toft, M., et al.: Reliability of the 30-minute ACTH test in assessing hypothalamic-pituitary-adrenal function. J. Clin. Endocrinol. Metab., *47*:272–279, 1978.
55. Lobo, R. A., Paul, W. L., and Goebelsmann, U.: Dehydroepiandrosterone sulfate as an indicator of adrenal androgen function. Obstet. Gynecol., *57*:69–73, 1981.
56. Loraine, J. A., and Bell, E. T., Eds.: Hormone Assays and Their Clinical Application. 4th ed. Edinburgh, Churchill Livingstone, 1976.
57. Loriaux, D. L., Ruder, H. J., and Lipsett, M. D.: Plasma steroids in congenital adrenal hyperplasia. J. Clin. Endocrinol. Metab., *39*:627–630, 1974.
58. Lovenberg, W., and Engelman, K.: Assay of serotonin, related metabolites and enzymes. *In*: Methods of Biochemical Analysis. (Suppl. volume on biogenic amines). D. Glick, Ed. New York, Interscience Publishers, 1971, pp. 1–34.
59. Maas, J. W.: Biogenic amines and depression. Arch. Gen. Psychiatry, *32*:1357–1361, 1975.
60. Mahajan, D. K., Wahlen, J. D., Tyler, F. H., et al.: Plasma 11-deoxycortisol radioimmunoassay for metyrapone tests. Steroids, *20*:609–620, 1972.
61. Manger, W. M., Steinsland, O. S., Nahas, G. G., et al.: Comparison of improved fluorometric methods used to quantitate plasma catecholamines. Clin. Chem., *15*:1101–1123, 1969.
62. Maroulis, G. B.: Evaluation of hirsutism and hyperandrogenemia. Fertil. Steril, *36*:273–305, 1981.
63. Martucci, C., and Fishman, J.: Direction of estradiol metabolism as a control of its hormonal action—uterotrophic activity of estradiol metabolites. Endocrinology, *101*:1709–1715, 1977.
64. Mattingly, D.: A simple fluorimetric method for the estimation of free 11-hydroxycorticoids in human plasma. J. Clin. Pathol., *15*:374–379, 1962.
65. McGuire, W. L., Carbone, P. P., and Vollmer, E. P., Eds.: Estrogen Receptors in Human Breast Cancer. New York, Raven Press, 1975.
66. McGuire, W. L., Raynaud, J.-P., and Baulieu, E. E., Eds.: Progesterone Receptors in Normal and Neoplastic Tissues. New York, Raven Press, 1977.
67. McKenna, T. J., Jennings, A. S., Liddle, G. W., et al.: Pregnenolone, 17–OH pregnenolone, and testosterone of patients with congenital adrenal hyperplasia. J. Clin. Endocrinol. Metab., *42*:918–925, 1976.
68. Mehdi, S. Q., and Nussey, S. S.: A radioligand receptor assay for the long-acting thyroid stimulator. Inhibition by the long-acting thyroid stimulator of the binding of radioiodinated thyroid stimulating hormone to human thyroid membranes. Biochem. J., *145*:105–111, 1975.
69. Melmed, S., Geola, F. L., Reed, A. W., et al.: A comparison of methods for assessing thyroid function in nonthyroidal illness. J. Clin. Endocrinol. Metab., *54*:300–306, 1982.
70. Melmon, K. L.: The endocrinologic function of selected autacoids: Catecholamines, acetylcholine, serotonin, and histamine. *In*: Textbook of Endocrinology. 6th ed. R. H. Williams, Ed. Philadelphia, W.B. Saunders Co., 1981, pp. 515–588.
71. Mikhail, G.: Hormone assays and the gynecologist. Fertil. Steril., *27*:229–237, 1976.
72. Miles, L. E. M.: Immunoradiometric assay (IRMA) and two-site IRMA systems (assay of soluble antigens using labeled antibodies). *In*: Handbook of Radioimmunoassay. G. E. Abraham, Ed. New York, Marcel Dekker, 1977, pp. 131–177.
73. Miller, M., Dalakos, T., Moses, A. M., et al.: Recognition of partial defects in antidiuretic hormone secretion. Ann. Intern. Med., *73*:721–729, 1970.
74. Moleman, P., and Borstrok, J. J. M.: Determination of urinary vanillylmandelic acid by liquid chromatography with electrochemical detection. Clin. Chem., *29*:878–881, 1983.
75. Moudgal, N. R., Muralidhar, K., and Madhwa Raj, H. G.: Pituitary gonadotropins. *In*: Methods of Hormone Radioimmunoassay. B. M. Jaffe and H. R. Behrman, Eds. New York, Academic Press, 1979, pp. 173–198.
76. Moyer, T. P., Jiang, N.-S., Tyce, G. M., et al.: Analysis for urinary catecholamines by liquid chromatography with amperometric detection: Methodology and clinical interpretation of results. Clin. Chem., *25*:256–263, 1979.
77. Odell, W. D., and Wolfsen, R. A.: Hormones from tumors: Are they ubiquitous? Am. J. Med., *68*:317–318, 1980.
78. Orczyk, G. P., Caldwell, B. V., and Behrman, H. R.: Plasma estradiol, estrone, estriol and urinary estriol glucuronide. *In*: Methods of Hormone Assay. B. M. Jaffe and H. R. Behrman, Eds. New York, Academic Press, 1979, pp. 675–700.
79. Orsulak, P. J., Kizuka, P., Grab, E., et al.: Determination of urinary normetanephrine and metanephrine by radial compression liquid chromatography and electrochemical detection. Clin. Chem., *29*:305–309, 1983.
80. Orth, D. N.: Adrenocorticotropic hormone. *In*: Methods of Hormone Radioimmunoassay. B. M. Jaffe and H. R. Behrman, Eds. New York, Academic Press, 1979, pp. 245–284.
81. Orth, D. N.: Ecotopic hormone production. *In*: Endocrinology and Metabolism. P. Felig, J. D. Baxter, A. E. Broadus, et al., Eds. New York, McGraw-Hill, Inc., 1981, pp. 1191–1217.
82. Peake, G. T., Morris, J., and Buckman, M. T.: Growth hormone. *In*: Methods of Hormone Radioimmunoassay. B. M. Jaffe and H. R. Behrman, Eds. New York, Academic Press, 1979, pp. 223–244.

83. Pernow, B., and Waldenström, J.: Determination of 5-hydroxytryptamine, 5-hydroxyindoleacetic acid and histamine in thirty-three cases of carcinoid tumor (argentaffinoma). Am. J. Med., *23*:16–25, 1957.
84. Petrucelli, B., Bakris, G., Miller, T., et al.: A liquid chromatographic assay for 5-hydroxytryptophan, serotonin and 5-hydroxyindoleacetic acid in human body fluids. Acta Pharmacol. Toxicol., *51*:421–427, 1982.
85. Peuler, J. D., and Johnson, G. A.: Simultaneous single isotope radioenzymatic assay of plasma norepinephrine, epinephrine and dopamine. Life Sci., *21*:625–636, 1977.
86. Pfeifer, M. A., Halter, J. B., and Porte, D.: Insulin secretion in diabetes mellitus. Am. J. Med., *70*:579–588, 1981.
87. Phillips, L. S., Vassilopoulou-Sellin, R.: Somatomedins (in two parts). N. Engl. J. Med., *302*:371–380, 438–446, 1980.
88. Pisano, J. J.: A simple analysis for normetanephrine and metanephrine in urine. Clin. Chim. Acta, *5*:406–414, 1960.
89. Pisano, J. J., Crout, R. J., and Abraham, D.: Determination of 3-methoxy-4-hydroxymandelic acid in urine. Clin. Chim. Acta, *7*:285–291, 1962.
90. Pollet, R. J., and Levey, G. S.: Principles of membrane receptor physiology and their application to clinical medicine. Ann. Intern. Med., *92*:663–680, 1980.
91. Porter, C. C., and Silber, R. H.: A quantitative color reaction for cortisone and related 17,21-dihydroxy-20-ketosteroids. J. Biol. Chem., *185*:201–207, 1950.
92. Ratcliffe, W. A.: Direct (non-extraction) serum assays for steroids. *In*: Immunoassays for Clinical Chemistry. W. M. Hunter and J. E. T. Corrie, Eds. Edinburgh, Churchill Livingstone, 1983, pp. 401–409.
93. Ratcliffe, W. A., Corrie, J. E. T., Dalziel, A. H., et al.: Direct ^{125}I-radioligand assays for serum progesterone compared with assays involving extraction of serum. Clin. Chem., *28*:1314–1318, 1982.
94. Riad-Fahmy, D., Read, G. F., Walker, R. F., et al.: Steroids in saliva for assessing endocrine function. Endocr. Rev., *3*:367–395, 1982.
95. Rubenstein, A. H., Kuzuya, H., and Horwitz, D. L.: Clinical significance of circulating C-peptide in diabetes mellitus and hypoglycemic disorders. Arch. Intern. Med., *137*:625–632, 1977.
96. Rudy, F. R., Bates, R. D., Cimorelli, A. J., et al.: Adrenal medullary hyperplasia: A clinicopathologic study of four cases. Hum. Pathol., *11*:650–657, 1980.
97. Shipe, J. R., Savory, J., and Wills, M. R.: Improved liquid-chromatographic determination of 3-methoxy-4-hydroxyphenylethyleneglycol in urine with electrochemical detection. Clin. Chem., *30*:140–143, 1984.
98. Shoup, R. E., and Kissinger, P. T.: Determination of urinary normetanephrine, metanephrine and 3-methoxytyramine by liquid chromatography with amperometric detection. Clin. Chem., *23*:1268–1274, 1977.
99. Sinha, Y. N., Selby, F. W., Lewis, U. J., et al.: A homologous radioimmunoassay for human prolactin. J. Clin. Endocrinol. Metab., *36*:509–516, 1973.
100. Sjoerdsma, A., Weissbach, H., and Udenfriend, S.: Simple test for diagnosis of metastatic carcinoid (argentaffinoma). JAMA, *159*:397, 1955.
101. Sobel, C. S., Golub, O. J., Henry, R. J., et al.: Study of the Norymberski methods for determination of 17-ketogenic steroids (17-hydroxycorticosteroids) in urine. J. Clin. Endocrinol. Metab., *18*:208–221, 1958.
102. Solomon, D. H., Benotti, J., DeGroot, L. J., et al.: Revised nomenclature for tests of thyroid hormones in serum. J. Clin. Endocrinol. Metab., *42*:595–598, 1976.
103. Speckart, P. F., Nicoloff, J. T., and Bethune, J. E.: Screening for adrenocortical insufficiency with cosyntropin (synthetic ACTH). Arch. Intern. Med., *128*:761–763, 1971.
104. Starr, J. I., Horwitz, D. L., Rubenstein, A. H., et al.: Insulin, proinsulin and C-peptide. *In*: Methods of Hormone Radioimmunoassay. 2nd ed. B. M. Jaffe and H. R. Behrman, Eds. Academic Press, Inc., 1979, pp. 613–642.
105. Steiner, D. E.: Insulin today. Diabetes, *26*:322–340, 1977.
106. Sterling, K.: Thyroid hormone action at the cell level (in two parts). N. Engl. J. Med., *300*:117–123, 173–177, 1979.
107. Sterling, K., and Tabachnik, M.: Resin uptake of ^{131}I-triiodothyronine as a test of thyroid function. J. Clin. Endocrinol. Metab., *21*:456–464, 1961.
108. Sterling, K., and Brenner, M. A.: Free thyroxine in human serum: Simplified measurement with the aid of magnesium precipitation. J. Clin. Invest., *45*:153–163, 1966.
109. Sullivan, J. M., and Solomon, H. S.: The diagnosis of pheochromocytoma: Overnight excretion of catecholamine metabolites. JAMA, *231*:618–619, 1975.
110. Sunderman, F. W., Jr.: Measurement of urinary corticosteroids; modified Porter-Silber method. *In*: Lipids and Steroid Hormones in Clinical Medicine. F. W. Sunderman and F. W. Sunderman, Jr., Eds. Philadelphia, J.B. Lippincott Co., 1960, pp. 162–164.
111. Thomasson, C. G., Blijenberg, B. G., Eilers, G. A., et al.: A comparative study of five different methods for the determination of 3-methoxy-4-hydroxymandelic acid in urine. J. Clin. Chem. Clin. Biochem., *21*:417–427, 1983.
112. Tietz, N. W., Ed.: Clinical Guide to Laboratory Tests. Philadelphia, W. B. Saunders Co., 1983.
113. Udenfriend, S., Titus, E., and Weissbach, H.: The identification of 5-hydroxy-3-indoleacetic acid in normal urine and a method for its assay. J. Biol. Chem., *216*:499–505, 1955.
114. Van Herle, A. J., Uller, R. P., Matthews, N. I., et al.: Radioimmunoassay for measurement of thyroglobulin in human serum. J. Clin. Invest., *52*:1320–1327, 1973.
115. Vermeulen, A., Stoica, T., and Verdonck, L.: The apparent free testosterone concentration, an index of androgenicity. J. Clin. Endocrinol. Metab., *33*:759–767, 1971.

116. Walfish, P. G.: The best way to screen for neonatal hypothyroidism. Diag. Med., *7*(2):67–75, 1984.
117. Wartofsky, L., and Burman, K. D.: Alterations in thyroid function in patients with systemic illness: The "euthyroid sick syndrome." Endocr. Rev., *3*:164–217, 1982.
118. Weichert, R. F.: The neural ectodermal origin of the peptide-secreting endocrine glands. Am. J. Med., *49*:232–241, 1970.
119. Weil-Maherbe, H.: The estimation of total (free and conjugated) catecholamines and some catecholamine metabolites in human urine. *In*: Methods of Biochemical Analysis. D. Glick, Ed. New York, Interscience Publishers, 1968. Vol. 16, pp. 293–326.
120. Werner, S. C.: Immune System, III. Role in thyroid disease. *In*: The Thyroid. 4th ed. S. C. Werner and S. H. Ingbar, Eds. Hagerstown, MD, Harper & Row, 1978, pp. 615–623.
120a. Wilke, T. J.: Free thyroid hormone index, thyroid hormone/thyroxine-binding globulin ratio, triiodothyronine uptake and thyroxine-binding globulin compared for diagnostic value regarding thyroid function. Clin. Chem., *29*:74–79, 1983.
121. Williams, C. M., and Greer, M.: Estimation by gas chromatography of urinary homovanillic acid and vanilmandelic acid in neuroblastoma. Methods Med. Res., *12*:106–114, 1970.
122. Williams, G. H., and Underwood, R. H.: Mineralocorticoids: Aldosterone, deoxycorticosterone, 18-hydroxydeoxycorticosterone and 18-hydroxycorticosterone. *In*: Methods of Hormone Radioimmunoassay. B. M. Jaffe and H. R. Behrman, Eds. New York, Academic Press, 1979, pp. 743–766.
123. Wilson, J. D., and Foster, D. W., Eds.: Williams Textbook of Endocrinology. 7th Ed. Philadelphia, W.B. Saunders Co., 1985.
124. Wilson, M. A., and Miles, L. E.: Radioimmunoassay of insulin. *In*: Handbook of Radioimmunoassay. G. E. Abraham, Ed. New York, Marcel Dekker, Inc., 1977, p. 275.
125. Wotiz, H. H., and Chattoraj, S. C.: The role of gas-liquid chromatography in steroid hormone analysis. J. Chromatogr. Sci., *11*:167–174, 1973.
126. Youssefnejadian, E., Florensa, E., Collins, W. P., et al.: Radioimmunoassay of 17-hydroxyprogesterone. Steroids, *20*:773–788, 1972.
127. Zerbe, R. L., and Robertson, G. L.: A comparison of plasma vasopressin measurements with a standard indirect test in the differential diagnosis of polyuria. N. Engl. J. Med., *305*:1539–1546, 1981.

19 ELECTROLYTES, BLOOD GASES, AND ACID-BASE BALANCE

SECTION ONE

Electrolytes

NORBERT W. TIETZ, Ph.D.
ELIZABETH L. PRUDEN, Ph.D.
OLE SIGGAARD-ANDERSEN, M.D., Ph.D.

Electrolytes are anions or cations, depending on whether they move in an electrical field toward the anode or toward the cathode, that is, whether they have a negative or positive charge. They are essential components of all living matter and include the major electrolytes Na^+, K^+, Ca^{2+}, Mg^{2+}, Cl^-, HCO_3^-, $HPO_4^{-,2-}$, SO_4^{2-}, and lactate, as well as the trace elements. Although amino acids and proteins in solution also carry an electrical charge, in clinical chemistry they are usually considered separately from electrolytes. The major electrolytes occur primarily as free ions; the trace metals occur primarily in combination with proteins.

The dietary requirements for electrolytes vary widely; most need to be consumed only in small amounts or at rare intervals and are retained when in short supply. Some, like calcium and phosphorus, are continuously excreted and must be ingested regularly in order to prevent deficiency.

The role of electrolytes in the human body is manifold. There are almost no metabolic processes that are not dependent on or affected by electrolytes. Among functions of the electrolytes are maintenance of osmotic pressure and water distribution in the various body fluid compartments, maintenance of the proper pH, regulation of the proper function of the heart and other muscles, involvement in oxidation-reduction (electron transfer) reactions, and participation in catalysis as cofactors for enzymes. Thus, abnormal levels of electrolytes may be either the cause or the consequence of a variety of disorders, and determination of electrolytes is one of the most important functions of the clinical laboratory.

In this section, we discuss sodium, potassium, chloride, and bicarbonate in body fluids as they are commonly grouped in the familiar test order for an "electrolyte profile." Other substances that are also electrolytes, but that have special functions in particular contexts, are discussed elsewhere in the book.

SODIUM

Sodium ions are the major cations of extracellular fluid, representing ~90% of the 154 mmol of inorganic cations per liter of plasma water. Thus, sodium ions are responsible for almost half the osmolality of the plasma and play a central role in maintaining the normal distribution of water and the osmotic pressure in the extracellular fluid compartment. The average daily diet contains 130–260 mmol (8–15 g) of sodium chloride, which is nearly completely absorbed from the gastrointestinal tract. Since the body requirement is only 1–2 mmol/d and the excess is excreted in the urine, the kidneys are the ultimate regulators of the amount of sodium and potassium in the body. Sodium is initially filtered by the glomeruli, but 60–70% of the filtered load is reabsorbed in the proximal tubules along with bicarbonate and water. Another 25–30% is reabsorbed in the loop of Henle along with chloride and more water. In the distal tubules, interaction of the adrenocortical hormone aldosterone with the coupled systems of Na^+–K^+ and Na^+–H^+ exchange directly modulates the reabsorption of sodium from the remaining 5–15% of the filtered load. It is the management of this fraction of the filtered sodium that sets the renal threshold* for sodium at 110–130 mmol/L and determines the amount of sodium excreted in the urine (see also Section Three of this chapter).

Hyponatremia (decreased plasma sodium concentration) is a predictable consequence of decreased intake of sodium, particularly when low intake is complicated by unusual losses. *Depletional* hyponatremia may be the result of excessive sweating, prolonged vomiting, persistent diarrhea, or salt-losing enteropathies (i.e., conditions in which sodium loss is greater than water loss). Renal loss due to diminished tubular reabsorption may be caused by inappropriate choice, dose, or use of diuretics; by primary or secondary deficiency of aldosterone and other mineralocorticoids; or by severe polyuria. Hyponatremia is common in metabolic acidosis (e.g., diabetic ketoacidosis) in which cations are lost by co-excretion with large amounts of organic anions, and in renal tubular acidosis in which impaired Na^+–H^+ exchange may be the cause. Any form of alkalosis or any condition associated with alkalinized urine is also a factor contributing to increased renal loss of sodium. The *dilutional* type of hyponatremia is secondary to excessive retention of water; total body sodium may be normal or increased. It occurs with edema, ascites in chronic cardiac failure, uncontrolled diabetes, hepatic cirrhosis, nephrotic syndrome, and malnutrition. In SIADH (syndrome of inappropriate antidiuretic hormone secretion, Chapter 18) hyponatremia is seen because of the expansion of extracellular fluid in conjunction with diminished Na^+ or increased water reabsorption in the distal tubules. It is important to recognize a factitious hyponatremia due to hyperlipidemia or hyperproteinemia (see discussion on page 620.)

Hypernatremia (increased plasma sodium concentration) is often attributable to excessive loss of sodium-poor body fluids, i.e., to water loss disproportionate to sodium loss, as may occur in profuse sweating, prolonged hyperpnea, vomiting or diarrhea, or polyuria. Other causes include decreased production of antidiuretic hormone, decreased tubular sensitivity to the hormone, and osmotic diuresis. Other cases of hypernatremia are due to accumulation of sodium as in inappropriate forms of parenteral therapy with saline solutions or high salt intake without corresponding intake of water; hyperaldosteronism; hyperadrenocorticism (Cushing's syndrome), in which increased production of mineralocorticoids causes increased tubular reabsorption of sodium; certain types of brain injury; and response to insulin treatment of uncontrolled diabetes, in which falling plasma concentration of glucose causes decreases in plasma osmolality and contraction of the extracellular fluid volume. Hypernatremia is often associated with hypercalcemia and hypokalemia.

Hypernatriuria (increased urinary sodium excretion) is a frequent corollary to initial stages of development of hyponatremia; it is observed in hypoaldosteronism, adrenal failure, salt-losing nephritis, diuretic therapy, and SIADH. Physiologic causes are increased dietary intake and postmenstrual diuresis. Conversely, *hyponatriuria* is associated with low dietary intake of sodium and premenstrual retention of sodium and water. It occurs pathologically in adrenocortical hyperfunction, hyperaldosteronism, conditions with decreased glomerular filtration rates, acute oliguria, and prerenal azotemia. In these instances, hyponatriuria is a corollary to sodium and water retention (i.e., to expansion of extracellular fluid volume).

Specimens. Serum, heparinized plasma, sweat, urine, feces, or gastrointestinal fluids may be assayed. Timed collections of urine, feces, or gastrointestinal fluids are needed to allow for comparison of values with reference ranges or for the determination of rates of electrolyte loss. Serum, plasma, and urine may be stored at 2–4 °C or frozen for delayed analysis. Erythrocytes contain only one tenth the sodium present in plasma;

---

*The renal threshold of a substance is the plasma concentration at which the rate of glomerular filtration of the substance begins to exceed the rate at which the tubules can reabsorb it (see also Chapter 20).

thus, hemolysis of blood does not cause a significant decrease in observed sodium values for serum or plasma unless hemolysis is severe, in which case there is a dilutional effect. If plasma is used for the assay, the lithium or ammonium salt of heparin should be used as anticoagulant. Urine should be collected without addition of a preservative.

Fecal analysis is justified only when stools are liquid, since only then are fecal losses of electrolytes significant. Liquid stool specimens should be clarified by filtration through gauze or filter paper or by centrifugation.

Determination of Sodium in Body Fluids

Sodium is currently most often determined by flame emission spectrophotometry (FES) or electrochemically with an ion-selective electrode (ISE). Excellent accuracy and coefficients of variation of $<1.5\%$ are readily achieved.

Reference ranges.[13] The range of *serum* sodium is 136–146 mmol/L* from infancy throughout life. *Urinary* sodium excretion varies with dietary intake; for persons on an average diet, a range of 40–220 mmol/d is typical. There is a large diurnal variation in sodium excretion. The sodium concentration of *cerebrospinal fluid* is 138–150 mmol/L.

POTASSIUM

Potassium is the major intracellular cation. In tissue cells, its average concentration is 150 mmol/L. In erythrocytes the concentration is 105 mmol/L, ~23 times the concentration in plasma. High intracellular concentrations are maintained because K^+ diffuses only slowly outward through the cell membrane whereas the Na^+, K^+–ATPase pump continually transports K^+ into the cells. Skeletal muscle activity, because it draws its energy from anaerobic glycolysis, causes K^+ efflux from muscle cells into plasma.

The body requirement for potassium is satisfied by dietary intake of 50–150 mmol/d. Potassium absorbed from the gastrointestinal tract is rapidly distributed; a small amount is taken up by cells but most is excreted by the kidneys. Potassium filtered through the glomeruli is almost completely reabsorbed in the proximal tubules and is then secreted in the distal tubules. The amount of potassium excreted in urine varies relative to intake. The renal secretory system reponds almost immediately to K^+ loading with an increase in K^+ output, and urine may have a potassium concentration as high as 100 mmol/L.[3] But tubular secretory response to conserve K^+ in instances of total body depletion is very slow in the initial stages of depletion; the normal kidneys continue to excrete potassium. Unlike the prompt response of the tubules to conserve Na^+ in deficit states, 1–2 weeks may be required for the tubules to reduce K^+ excretion to 5–10 mmol/d. Intracellular stores of K^+ maintain the K^+ concentration in the extracellular compartment at a normal level until K^+ depletion is severe.

Factors that regulate distal tubular secretion of potassium are intake of Na^+ and K^+, flow rate in the distal tubules, plasma level of mineralocorticoids, and acid-base balance. Because the conservation mechanism is slow to respond, potassium depletion is an early consequence of restricted K^+ intake or unusual losses of K^+ by extrarenal routes. Diminished glomerular filtration rate is typical of renal failure, and the consequent decrease in distal tubular flow rate is an important factor in the retention of potassium. Aldosterone enhances K^+ secretion in portions of the distal tubules and collecting ducts; the effect is interrelated with the control of Na^+ reabsorption by aldosterone and with Na^+–K^+ exchange. Renal tubular acidosis, as well as metabolic and respiratory acidosis and alkalosis, affects renal regulation of potassium excretion. General aspects of these topics are discussed in Section Three and in Chapter 20.

Disturbance of potassium homeostasis has serious consequences. For example, decrease of extracellular K^+ is characterized by muscle weakness, irritability and paralysis; fast heart rate and specific conduction effects apparent by electrocardiographic examination; and eventual cardiac arrest. Because extracellular potassium concentrations are maintained at the expense of intracellular supply, plasma potassium and neuromuscular function can be normal even with substantial total body deficit of K^+; plasma

---

*In many laboratories, mEq/L remains the unit for expressing electrolyte concentrations; mmol/L is the SI unit.

potassium <3.0 mmol/L is associated with marked neuromuscular symptoms and is evidence of a critical degree of intracellular depletion. Abnormally high extracellular potassium produces symptoms of mental confusion; weakness, numbness, and tingling of the extremities; weakness of the respiratory muscles; flaccid paralysis of the extremities; and slowed heart rate, typical conduction defects on the electrocardiogram, and eventually peripheral vascular collapse and cardiac arrest. The symptoms are apparent at potassium levels >7.5 mmol/L; levels >10.0 mmol/L are in most cases fatal.

Causes of *hypokalemia* (decreased plasma potassium concentration) can be grouped into decreased intake, redistribution of extracellular K^+ into intracellular fluid, and increased loss of potassium-rich body fluids. Situations of decreased intake include chronic starvation and postoperative therapy with potassium-poor fluids. Redistribution is illustrated by the fall in plasma potassium when insulin therapy of diabetic hyperglycemia is instituted; cellular uptake of glucose is accompanied by uptake of potassium and water. Hypokalemia is also a feature of alkalosis, in which K^+ moves from the extracellular fluid into the cell as H^+ moves into the extracellular fluid; thus, all other things being equal, alkalosis of itself causes hypokalemia. On the other hand, intracellular potassium depletion may be a cause of alkalosis, as will be explained in Section Three of this chapter. Gastrointestinal loss of potassium occurs with fluids lost because of vomiting, diarrhea, or intestinal fistulas; renal losses may be the result of renal tubular acidosis, primary or secondary aldosteronism, or Cushing's or Bartter's syndromes. Aldosterone decreases renal tubular potassium reabsorption, in contrast to the increase in sodium reabsorption caused by aldosterone.

Hyperkalemia (increased plasma potassium concentration) may be precipitated by intravenous infusion of K^+ at a rate in excess of 20 mmol/h or by treatment using K^+ solutions with concentrations >40 mmol/L or doses of >80 mmol/d. Overtreatment is unlikely to produce hyperkalemia as long as renal function is normal because excess K^+ is readily excreted in the urine. Transfer of intracellular K^+ into extracellular fluid may occur in dehydration and shock with tissue hypoxia, diabetic ketoacidosis, massive intravascular or extracorporeal hemolysis, severe burns, and with violent muscular activity such as that in status epilepticus. Decreased excretion of potassium in acute renal failure or endstage renal failure with oliguria or anuria and acidosis is a common cause of hyperkalemia. In renal failure due to shock and in renal tubular acidosis, interference with Na^+–H^+ exchange in the tubules leads to retention of potassium. The hyperkalemia in acidosis is the result of K^+ moving from intracellular fluids into the plasma as H^+ moves into the cells from extracellular fluid. Hyperkalemia occurs along with sodium depletion in adrenocortical insufficiency because, in the absence of adequate amounts of aldosterone and other mineralocorticoids, diminished Na^+ reabsorption and Na^+–K^+ exchange and decreased K^+ secretion lead to retention of K^+.

Hyperkaliuria, increased urinary excretion of potassium, occurs at the onset of starvation, in primary and secondary aldosteronism, in primary renal diseases, in renal tubular syndromes, and during the recovery phases of acute tubular necrosis, metabolic acidosis, and metabolic alkalosis. Although *hypokaliuria* is eventually a mark of body depletion of potassium, its occurrence is far less worthy of note than the hypokalemia and symptoms of depletion that appear much earlier in a disease process.

Specimens for serum or plasma assay of potassium must be collected in such a way as to minimize hemolysis; release of potassium from as few as 0.5% of the erythrocytes can increase the serum level by 0.5 mmol/L. Extracorporeal circulation may cause hemolysis, and the amount of potassium added to the plasma by the procedure is important clinical information.

Potassium levels in plasma have been shown to be 0.1–0.7 mmol/L lower than in serum. The difference is due to release of potassium from platelets ruptured in the coagulation process. These considerations now make plasma the specimen of choice and emphasize the necessity of noting on reports whether serum or plasma was assayed and whether the specimen was visibly hemolyzed.

Plasma potassium levels may be increased by 10–20% as a result of muscle activity if the patient opens and closes his fist repeatedly prior to venipuncture. Increases can also occur as a result of K^+ leakage from erythrocytes when plasma or serum is not promptly separated from cells after collection, or if the whole blood specimen is chilled prior to separation. The increase in serum is of the order of 0.2 mmol/L in 1.5 h at 25 °C; at 4 °C, the increase is 1.5 times greater.[5] Extreme thrombocytosis or leukocytosis can also cause marked increases in measured potassium levels even up to an apparent doubling of serum K^+ in thrombocytosis.

Determination of Potassium in Body Fluids

The most popular analytical methods are currently flame emission spectrophotometry and potentiometry with an ion-selective electrode. With either approach, coefficients of variation of <2% are readily achieved.

The ***reference range***[13] for *serum* of adults is 3.5–5.0 mmol/L; for *plasma* the range is 3.5–4.5 mmol/L. Serum values for newborns are 3.7–5.9 mmol/L. *Cerebrospinal fluid* levels are ~70% of those determined simultaneously in serum. *Urinary* excretion of potassium varies with dietary intake, but a typical range observed for persons on an average diet is 25–125 mmol/d. With severe diarrhea, *fecal* loss may be as much as 60 mmol/d.

Methods for the Determination of Sodium and Potassium

Flame Emission Spectrophotometry

Principle. Sample is diluted with a solution containing an internal standard (lithium or cesium) and aspirated into a propane-air flame. Na^+, K^+, Li^+, and Cs^+, when excited, emit spectra with sharp, bright lines at 589, 768, 671, and 852 nm, respectively. As the thermally excited ions return to the ground state, light is emitted and is directed through separate interference filters to corresponding photodetectors. The internal standard signal is taken as a reference against which the analytical signal (sodium, potassium, or lithium) is compared. The system is standardized relative to low and high concentrations of each analyte, and the relation of signal to concentration is computed by an associated microprocessor. Upon aspiration of controls and patient samples into the standardized instrument, results are presented in mmol/L on a digital readout device (see also Chapter 2, Section One).

Reagents. Commercially prepared standards of NaCl and KCl as well as diluents are available, often as concentrates, from instrument manufacturers. Standard reference materials (SRMs, see Table 1-3) for Na and K are available from the National Bureau of Standards (Washington, DC 20334) and can be used when needed. Dilution of standard or diluent concentrates for use should be made in Type II water. Nonionic, surface-active agents are added to diluted solutions of standards and unknowns to facilitate uniform aspiration and atomization of the fluids into the flame.

Quality control. Reliable assay of samples depends on meticulous maintenance of instrumentation and on close adherence to operating instructions, as specified by the manufacturer. Careful cleaning of the atomizer-burner, sample receptacle, and glass chimney, as well as the proper adjustment of flame size and aspiration rate of the atomizer, are of primary importance for all analytical systems. A warm-up period, after ignition of the flame and before standardization, is required. Initial evaporation of water in the diluent decreases the temperature of the burner body and the air in the burner chamber, and thermal equilibrium must be established with the flame alight and diluent being aspirated so that flame and optics are stabilized and reproducible readings can be obtained on standards and unknowns. Good practice also requires dilution of unknowns in the same batch or lot of diluent as is used for dilution of standards.

Safety. Propane, commercially supplied in small, easily replaceable tanks, is currently an almost universal fuel for flame photometers in the US. This circumstance has been largely dictated by OSHA (Occupational Safety and Health Administration) regulations for safety in the work place and by policies adopted by instrument manufacturers. Small tanks are often usable to exhaustion of their contents without any particular effect on the flame quality. With large tanks, however, flame quality generally begins to deteriorate before the tank is empty. The liquified propane gas (LPG) in the larger tanks consistently contains appreciable quantities of other hydrocarbons such as butane. As the tank empties, the relative proportions of the other hydrocarbons in the fuel increase, and flame characteristics deteriorate.

Most contemporary flame photometers are fitted with safety valves that open to allow fuel and oxidant to flow only when the igniter is energized and close promptly if ignition does not occur at once. Some instruments emit audible warnings when the fuel valve is open but the flame is not lit. Every flame photometer should be lighted according to prescribed, safe procedures. Leakage of propane from tanks, instrument fittings, or valves is always possible but may be detected with the aid of soap solution (e.g., SNOOP). Because propane is heavier than air, it readily collects at low spots in the laboratory, thus posing an explosion hazard. Work areas should therefore be well ventilated.

Methods Using Ion-Selective Electrodes (ISE)[11]

Analyzers fitted with ISEs usually have Na^+ electrodes with glass membranes and K^+ liquid ion-exchange membranes that incorporate valinomycin (see Chapter 2, Section Four). The principle of potentiometry can be simply stated as the determination of change in electromotive force (*E*, potential) in the potential-measuring circuit between a measurement electrode (the ISE) and a reference electrode, as the selected ion interacts with the membrane of the ISE. The measuring system is calibrated with solutions containing Na^+ and K^+, and the unknown concentration is interpolated from their measured potentials.

There are two types of ISE methods. In the *indirect methods*, sample is introduced into the measurement chamber and mixed with a rather large volume of diluent of high ionic strength, as in the ASTRA systems (Beckman Instruments, Brea, CA 92621). In the *direct methods*, sample is presented to the electrodes without dilution; the NOVA systems (NOVA Biomedical, Newton, MA 02164) exemplify this approach. In some direct method applications, either serum, plasma, or whole blood may be used. Single-use, thin-film ion-selective electrodes for Na^+, K^+, Cl^-, and CO_2 determinations are unique applications of a direct method (EKTACHEM, Eastman Kodak Co., Rochester, NY 14650; see also Chapter 2, Section Four). Figure 19-1 illustrates a potassium slide; slides for the other analytes differ chiefly in terms of the ion-selectivity of the membrane upon

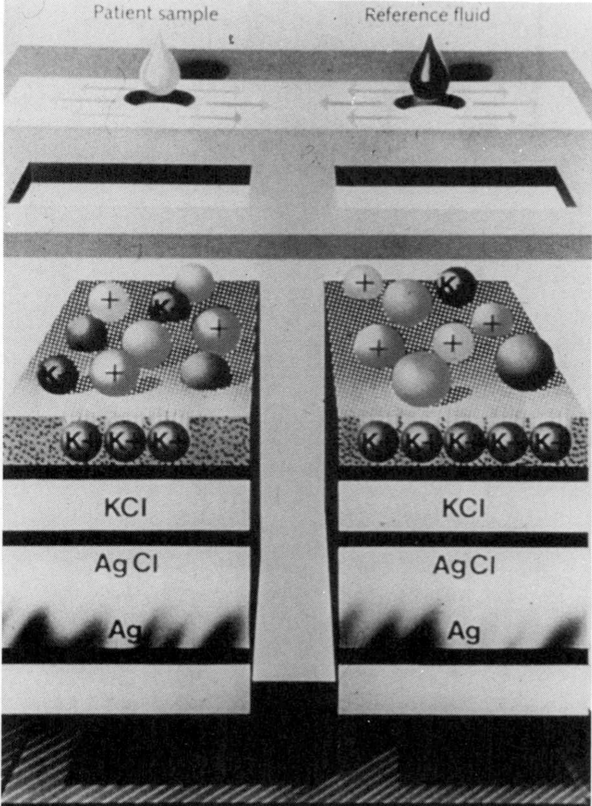

Figure 19-1. A thin-film ion-selective electrode slide (Kodak EKTACHEM) for the determination of potassium. (Courtesy of Eastman Kodak Co., Rochester, NY 14650.)

which sample and reference solution are dispensed. The analyte ions selected at the membrane diffuse to the silver/silver chloride electrode and generate a potential. The difference between potentials of the reference and sample electrodes is used as a measure of ion concentration in the sample.

Errors observed in the use of ISEs fall into two categories. One includes the obvious errors due to lack of selectivity, protein coating of the ion-sensitive membrane, or contamination of the membrane by ions that compete or react with the selected ion and thus alter electrode response to the selected ion. The second category, the "electrolyte exclusion effect," applies chiefly to indirect methods and is mainly due to extreme concentrations of lipid and protein in the sample. (This effect is also referred to as the "exclusion error.")

Electrolyte exclusion effect. The volume fraction of total solids in plasma is ordinarily ~8%. The bulk of the total solids is protein and lipid; 92% of the plasma is water. Free electrolyte ions are almost entirely associated with plasma water; i.e., they are not evenly distributed throughout the whole volume of plasma. For this reason, sodium concentration determined as 145 mmol/L in whole plasma containing 92% water represents a concentration of 145 × (100/92) or 158 mmol/L in plasma water. This negative "error" in plasma electrolyte analysis has been recognized for many years and, in certain applications, corrected. But in practical experience, electrolyte concentrations have been defined in terms of whole plasma rather than plasma water. Except in cases of extreme hypo- or hyperproteinemia[7,10] or extreme hyperlipidemia,[2] electrolyte concentrations of whole plasma reliably parallel concentrations in plasma water, where the electrolytes play their special physiological roles. The magnitude of error is <5% when the triglyceride concentration is <2500 mg/dL.

The electrolyte exclusion effect arises most commonly when plasmas with very high total solids content are analyzed by certain analytical approaches. Those approaches that are subject to the effect are flame emission and atomic absorption spectrophotometry, amperometry/coulometry, and potentiometry by the indirect ISE methods. The methods have in common dilution of the plasma sample in a large volume of ionic solution prior to the measuring step, followed by definition of concentration as a simple function of the volume of whole plasma diluted times the dilution factor. Analytes usually involved are Na^+, K^+, Li^+, Ca^{2+}, Mg^{2+}, and Cl^-. The "effect" consists of concentration values for whole plasma that do not parallel the concentrations in plasma water and that in fact underestimate them to a degree dependent upon the relative volume fractions of solids and water in the plasma. In certain clinical situations, notably diabetic ketoacidosis with severe hyperlipidemia and multiple myeloma with severe hyperproteinemia, the effect may be so large that laboratory results lead attending physicians to believe that electrolyte concentrations are normal or low when the concentration in the water phase may actually be dangerously high or, in less extreme cases, to believe them low when concentrations in plasma water are actually normal. The clinical significance of the electrolyte exclusion effect is therefore the risk it raises that patients may be treated with inappropriate amounts or formulations of IV fluids.

Blood gas and pH measurement, osmolality by freezing point depression, and potentiometry of electrolytes by direct ISE methods are analytical approaches that are free of the exclusion effect. When sodium values obtained by direct and indirect ISE methods are compared, those from the direct method are higher. The explanation lies in the elimination of dilution in the direct methods. *Indirect ISE methods*, because they dilute plasma in a diluent of fixed high ionic strength, control the activity coefficient of individual ions and set it to a virtual constant; that for sodium approaches a value of 1. Under these circumstances, measurement of activity, a (where $a = \gamma \times$ concentration, and γ is the activity coefficient), is tantamount to measurement of concentration. In the *direct ISE methods,* sample is not diluted. The concentration is still determined relative to activity, but the activity is directly proportional to the concentration of ion in the water phase. The activity coefficient, γ, is about 0.7 in normal plasma; in any one plasma, the value depends slightly on ionic strength, I, which is in turn a function of all the ions in the plasma water. The volume fraction of water in whole plasma may vary widely as a result of variation of quantities of lipid or protein, but measured activity of the ions in the water phase, and consequently the estimate of their concentration, becomes independent of the relative proportions of water and total solids.

CHLORIDE

Chloride is the major extracellular anion and represents the largest fraction of the plasma total inorganic anion concentration of ~154 mmol/L. Chloride is therefore significantly involved in the maintenance of water distribution, osmotic pressure, and anion-cation balance in the extracellular fluid compartment. The concentration of chloride in intracellular fluid of erythrocytes is 45–54 mmol/L and in intracellular fluid of tissue cells, ~1 mmol/L. In both gastric and small and large intestinal secretions, chloride is the most abundant anion.

Chloride ions in food are almost completely absorbed from the intestinal tract. They are filtered from plasma at the glomeruli and passively reabsorbed, along with Na^+, in the proximal tubules. In the thick ascending limb of the loop of Henle, Cl^- is actively reabsorbed by the so-called chloride pump, whose action promotes reabsorption of Na^+ as well. Loop diuretics such as furosemide and ethacrynic acid inhibit the chloride pump. Surplus chloride is excreted in the urine; excessive losses in sweat, as can occur in hot weather, are normally minimized by the secretion and action of aldosterone, which causes the sweat glands to reabsorb more Na^+ and Cl^-.

Hypochloremia (decreased plasma chloride concentration) is observed in salt-losing nephritis as associated with chronic pyelonephritis. The loss is probably due to defective tubular reabsorption despite a body deficit of chloride. In Addison's disease, chloride levels are usually maintained close to normal except in Addisonian crisis, when chloride as well as sodium levels may drop significantly. Hypochloremia may also be seen in those types of metabolic acidoses that are caused by increased production or diminished excretion of organic acids (e.g., diabetic ketoacidosis or renal failure). In such cases, the fraction of total anion concentration represented by chloride is diminished to compensate for the increased fraction of β-hydroxybutyrate, acetoacetate, lactate, and phosphate. Prolonged vomiting results in significant loss of chloride, and ultimately in hypochloremia and depletion of total body chloride. In metabolic alkalosis, plasma chloride tends to fall as bicarbonate increases.

Hyperchloremia (increased plasma chloride concentration) occurs with dehydration, renal tubular acidosis, acute renal failure, metabolic acidosis associated with prolonged diarrhea and loss of $NaHCO_3$, and salicylate intoxication. A slight rise in chloride is seen in respiratory alkalosis. Hyperchloremic acidosis may be a sign of severe pathological disorders of the renal tubules. Extremely high dietary intake of salt and overtreatment with saline solutions are also causes of hyperchloremia. In patients with primary hyperparathyroidism, mean serum chloride concentrations are ~3 mmol/L higher than normal. This is probably due to the effect of parathyroid hormone on distal tubular function.[16]

Urinary excretion of chloride normally approximates dietary intake; a physiologic increase occurs with postmenstrual diuresis and a decrease with premenstrual salt and water retention, in parallel with an increase and decrease of urinary sodium. Massive diuresis of any cause is accompanied by increased chloride excretion. Urinary excretion of chloride decreases when losses by other routes are increased.

Methods for Determination of Chloride in Body Fluids

Specimens. Chloride is measured in serum or plasma, urine, and sweat. Cells should be separated from plasma soon after blood is exposed to air since loss of gaseous CO_2 alters distribution of Cl^- between cells and plasma. With anaerobic separation, chloride in venous samples is found to be 3–4 mmol/L less than in arterial samples. Measurement of chloride lost in gastric aspirates or intestinal drainages is an adjunct to parenteral replacement therapy. Fecal chloride determination is clinically useful for the diagnosis of congenital hypochloremic alkalosis with hyperchlororrhea (increased excretion of chloride in stool). In this condition, concentration of Cl^- in feces may reach 180 mmol/L, but almost no Cl^- is found in urine. A filter paper test has been described for rapid semiquantitative estimate of fecal chloride.[15]

Mercurimetric Titration (Schales and Schales,[12] Modified)

Principle. A tungstic acid protein-free filtrate of specimen is titrated with mercuric nitrate solution in the presence of diphenylcarbazone added as indicator. Free mercuric ions combine with chloride ion to form soluble but essentially un-ionized mercuric chloride:

$$2Cl^- + Hg(NO_3)_2 \rightarrow HgCl_2 + 2NO_3^-$$

Excess Hg^{2+} reacts with diphenylcarbazone to form a blue-violet color complex. The first appearance of this color is considered the end point of the titration.

Spectrophotometric Methods

Mercurimetric methods are based on the reaction of chloride with mercuric thiocyanate.

$$Hg(SCN)_2 + 2\ Cl^- \rightarrow HgCl_2 + 2\ SCN^-$$

$$3\ (SCN)^- + Fe^{3+} \rightarrow Fe(SCN)_3$$

Chloride ions react with undissociated mercuric thiocyanate to form undissociated mercuric chloride and free thiocyanate ions. The thiocyanate ions react with ferric ions to form highly colored, reddish ferric thiocyanate with an absorption peak at 480 nm. Perchloric acid increases the intensity of the red color. High levels of globulins in the serum produce turbidity that interferes in methods in which serum is added directly to the reagent.[8]

Use of mercurimetric methods presents the problem of disposal of reagent waste containing a significant amount of toxic mercury. Substituting TPTZ, mercuric-2,4,6-tri(2-pyridyl)-5-triazine for $Hg(SCN)_2$ reduces waste mercury by about half.[4] TPTZ reacts with Cl^- to produce an intense blue color that is stable for 5–30 min and can be measured at 560–600 nm.

The ferric perchlorate method[9] uses no mercurial reagent; chloride ions react with the reagent in dilute perchloric acid to form a color complex with absorption maxima at 344 and 562 nm. The absorbance/concentration relationship is linear and the reaction is more specific for Cl^- than are most others.

The analytical range of spectrophotometric automated methods is, more often than not, limited to concentrations of chloride that can be expected in serum, plasma, or spinal fluid. The methods are therefore often not applicable to chloride analysis of other body fluids in which chloride concentrations may be much more or much less than 80–120 mmol/L.

Coulometric-Amperometric Titration[1]

In coulometric-amperometric determinations of chloride, Ag^+ generated from a silver electrode at a constant rate reacts with chloride ion in the sample to form insoluble AgCl:

$$Ag^+ + Cl^- \rightarrow AgCl$$

As soon as the stoichiometric point is reached, continued generation of Ag^+ triggers shutdown of the system. A timing device records the elapsed time between start and stop of silver ion generation. (A more detailed description of this principle can be found in Chapter 2, Section Four.) The time interval (t) is proportional to the amount of chloride ion present in the sample; thus:

$$\text{Chloride, mmol/L} = \frac{t_{\text{unknown}} - t_{\text{blank}}}{t_{\text{standard}} - t_{\text{blank}}} \times C_s$$

BICARBONATE (TOTAL CARBON DIOXIDE)

Total carbon dioxide ($ctCO_2$) in plasma consists of CO_2 in physical solution, the CO_2 bound loosely to amine groups of proteins (carbamino compounds), the CO_2 existing as HCO_3^- or CO_3^{2-} ions and other undissociated bicarbonates and carbonates, as well as carbonic acid ($HHCO_3$). Bicarbonate ions make up most of the total carbon dioxide of plasma.[6] Therefore, measurement of the total carbon dioxide concentration (or CO_2, as we can conveniently call it henceforth) is useful chiefly to evaluate bicarbonate concentration. Thus, in the following discussions, *(total) CO_2* and *bicarbonate* are used interchangeably.

Clinical significance. Alterations of bicarbonate and CO_2 dissolved in plasma are characteristic of acid-base imbalance. The nature of the imbalance cannot, however, be inferred from the CO_2 value by itself; its value has significance in the context of other electrolytes and pH that have been determined simultaneously. When acid-base imbalance is suspected, evaluation of blood gases and pH is required to provide a definitive picture of the overall pattern of imbalances. *Increases* in $ctCO_2$ occur in metabolic alkalosis due to severe vomiting, hypokalemic states, or excessive intake of alkali, as well as in respiratory acidosis, especially when compensated. *Decreases* in $ctCO_2$ are seen in renal failure, renal tubular acidosis with coexisting hyperchloremia, diarrhea, states of poor tissue perfusion, and respiratory alkalosis, especially when compensated. The clinical significance of changes in total carbon dioxide (bicarbonate) is discussed in Section Three.

Specimens. Either serum or heparinized plasma from venous or capillary blood is the usual specimen. For a specimen drawn in an evacuated tube, the concentration of plasma total CO_2 is most accurately determined when the specimen in the unopened tube is centrifuged as promptly as possible after collection and then analyzed as promptly as possible after the tube is opened and the plasma is separated. CO_2 will escape from a specimen exposed to ambient air; the value may decrease by 6 mmol/L in the course of an hour.[14] In practical terms, most measurements of plasma CO_2 are made on specimens that have lost some or most of the dissolved, gaseous CO_2, simply as a result of exposure to air during separation of plasma and delay prior to sampling. In this case, bicarbonate may be a better description of the analyte than is $ctCO_2$. Analysis of an arterial sample collected anaerobically for blood gas analysis preserves dissolved CO_2 content of the sample until the time of sampling; in these circumstances, the analyte is really total CO_2.

Methods for the Determination of Serum or Plasma Total CO_2

The first step in automated methods is the acidification of the sample to convert the various forms of CO_2 in plasma to gaseous CO_2. The step is ordinarily accomplished by dilution with an acid buffer.

In *continuous-flow analysis* (SMAC, Technicon Instruments Corp., Tarrytown, NY 10591) the gaseous CO_2 is allowed to diffuse across a silicone membrane into a recipient solution buffered at pH 9.2 and containing the indicator phenolphthalein. As CO_2 is absorbed by the alkaline medium to form HCO_3^-, the decrease in pH of the recipient stream intensifies the red color. The increase in color over the baseline is determined spectrophotometrically and is proportional to the amount of CO_2 released from the sample. Baseline drift is the most frequent evidence of trouble with the method. The fault can usually be traced either to the color reagent, whose pH must be maintained within 0.2 unit, or to exhaustion of the CO_2 trap that is fitted to the air lines that supply the air segments for the reagent stream.

In the *ASTRA method* (Beckman Instruments, Brea, CA 92621), the released gaseous CO_2, is determined by a pCO_2 electrode (of the type illustrated in Figure 2-31, page 95) that is set in the reaction chamber of the CO_2 module. The *rate* of change of pH of the buffer inside the membrane of the measuring electrode is taken as the measure of total CO_2 present in the sample.

An *enzymatic method* for CO_2 is illustrated by the method used in the *aca* (DuPont, Wilmington, DE 19898). The specimen is first alkalinized to convert all forms of CO_2 to HCO_3^-. The enzymatic reactions are as follows (PEPC is phosphoenolpyruvate carboxylase and MDH is malate dehydrogenase):

$$HCO_3^- + \text{phosphoenolpyruvate} \xrightarrow{PEPC} \text{Oxaloacetate} + P_i$$

$$\text{Oxaloacetate} + NADH + H^+ \xrightarrow{MDH} \text{Malate} + NAD^+$$

Decrease in absorbance of NADH at 340 nm is proportional to the total carbon dioxide content.

Reference ranges generally are instrument dependent. Those representative of anaerobic venous plasma samples are 22–29 mmol/L. Values for newborns and infants are lower.

Quality Control of Concurrent Electrolyte Measurements

An estimator called the anion gap can be calculated from concentrations of Na^+, Cl^-, and total CO_2 concentration determined on the same plasma or serum. The value of the anion gap can be used clinically to assess unmeasured anions and in the laboratory as a quality control estimate. Both aspects of anion gap are discussed in Section Three under *Estimates Derived from Electrolyte Measurements.*

References

1. Cotlove, E.: Determination of chloride in biological materials. *In*: Methods of Biochemical Analysis, Vol. 12. D. Glick, Ed. New York, Interscience Publishers Inc., 1964.
2. Creer, M. H., and Ladenson, J.: Analytical errors due to lipemia. Lab. Med., *14*:351–355, 1983.
3. Fernandez, P., and Cox, M.: Basic concepts of renal physiology. Int. Anesthesiol. Clin., *22*:1–34, 1984.
4. Fried, Von R., Hoeflmayr, J., and Velösy, G.: Eine neue, hochempfindliche Methode zur Bestimmung des Chlorids in Körperflüssigkeiten ohne Eiweissfällung. Z. Klin. Chem. Klin. Biochem., *10*:280, 1972.
5. Goodman, J. T., Vincent, J., and Rosen, I.: Serum potassium changes in blood clots. Am. J. Clin. Pathol., *24*:111–113, 1954.
6. Hall, W. D.: Serum total CO_2 content. *In*: Clinical Methods: The History, Physical, and Laboratory Examinations. H. K. Walter, W. D. Hall, and J. W. Hurst, Eds. Boston, Butterworth Publisher, 1980.
7. Ladenson, J. H., Apple, F. S., Aguanno, J. J., et al.: Sodium measurements in multiple myeloma: Two techniques compared. Clin. Chem., *28*:2383–2386, 1982.
8. Landowne, R. A.: Immunoglobulin interference with phosphorus and chloride determinations with the Coulter Chemistry. Clin. Chem., *25*:1189–1190, 1979.
9. Law, W. T., and Ertingshausen, G.: A novel reagent and method for determination of chloride in serum with a centrifugal analyzer. Clin. Chem., *26*:1874–1877, 1980.
10. Levy, G. B.: Determination of sodium with ion-selective electrodes. Clin. Chem., *27*:1435–1438, 1981.
11. Maas, A. H. R., Kofstad, J., Siggaard-Andersen, O., et al., Eds.: Ionized Calcium, Sodium, and Potassium by Ion Selective Electrodes. Vol. 5. Proceedings of the first meeting of the European working group on ion selective electrodes. IFCC Workshop, Oslo, 1983. Copenhagen, Private Press, 1984.
12. Schales, O., and Schales, S. S.: A simple and accurate method for the determination of chloride in biological fluids. J. Biol. Chem., *140*:879–884, 1941.
13. Tietz, N. W., Ed.: Clinical Guide to Laboratory Tests. Philadelphia, W. B. Saunders Co., 1983.
14. Tietz, N. W., Ed.: Fundamentals of Clinical Chemistry, 2nd ed. Philadelphia, W. B. Saunders Co., 1976.
15. Tomaszewski, L., and Konarska, L.: A simple, rapid, and accurate screening test for hyperchlororrhoea. J. Clin. Chem. Clin. Biochem., *15*:385–388, 1977.
16. Wills, M. R.: Value of plasma chloride concentration and acid-base status in the differential diagnosis of hyperparathyroidism from other causes of hypercalcaemia. J. Clin. Pathol., *24*:219–227, 1971.

SECTION TWO

Blood Gases and pH

ELIZABETH L. PRUDEN, Ph.D.
OLE SIGGAARD-ANDERSEN, M.D., Ph.D.
NORBERT W. TIETZ, Ph.D.

Clinical management of respiratory and metabolic disorders and of patients undergoing surgical correction of cardiopulmonary disorders has been greatly improved by the development of instrumentation for rapid, accurate measurements of oxygen and carbon dioxide in blood and expired air. The determination of blood gases and pH also plays an important part in the detection of acid-base imbalance and in following the effect of therapy. Table 19-1 addresses the nomenclature used in the following discussion and alternative terminology that may be encountered in practice.[6]

Table 19-1. CONVERSION FACTORS, PREFIXES, SYMBOLS, AND DESCRIPTORS USED IN DISCUSSIONS OF GASES MEASURED IN BLOOD*

Conversion Factors

1 mm Hg = 0.133 kPa; 1 kPa = 7.5 mm Hg

kPa: 1 kilopascal = 1000 pascal. The pascal is the SI-derived unit of pressure; 1 pascal = 1 Newton/m^2 (see also Chapter 1).

General Prefixes

p: partial pressure or tension
Usage: pO_2, pCO_2, pH_2O; alternative: P_{O_2}

s: saturation fraction
Usage: sO_2; alternative: S_{O_2}

c: substance concentration (trivial, "content")
Usage: ctO_2 for concentration of total oxygen; $ctCO_2$ for concentration of carbon dioxide; $cHCO_3^-$ for concentration of bicarbonate. Alternative: C. Note, however, risk of confusion when used as C_{O_2} for oxygen concentration.

d: dissolved gas, used with substance concentration (c); e.g., $cdCO_2$

t: total, used with substance concentration (c); e.g., $ctCO_2$

Specimen origin is indicated by lower-case letters. Whole blood and plasma are distinguished by capitals.

a: arterial; v: venous; c: capillary

B: blood; P: plasma
Usage: pO_2(aB) for partial pressure of oxygen in arterial blood

Other Descriptors

BTPS: *B*ody *T*emperature (37 °C or 310.16 K) and ambient *P*ressure, fully *S*aturated (pH_2O = 47 mm Hg)

STPD: *S*tandard *T*emperature (0 °C or 273.16 K) and standard *P*ressure (760 mm Hg) of *D*ry gas

SVP: *S*aturated *V*apor *P*ressure of water = pH_2O at 37 °C = 47 mm Hg

*This list is not complete and not all the terms and quantities shown are in the forms recommended by the IFCC.[8] The list is presented to facilitate interpretation of terms used in the text and to illustrate some alternative forms.

Respiration

Respiration is responsible for supplying oxygen needed by tissue cells for their metabolic processes and for removing the carbon dioxide produced by these processes. *External respiration* is the name given to the exchange of O_2 and CO_2 in the lungs between alveolar air and blood (in contrast to *internal respiration*, occurring at the tissue level). The exchange takes place across the membranes of the pulmonary capillaries and sac-like alveoli in the lungs, as air moves in and out in response to the contraction of the diaphragm and the elastic rebound of the lungs and chest wall. Loss of elasticity of pulmonary parenchyma, restriction of thoracic expansion, and damage to alveolar membranes are basic causes of pulmonary distress.

The amount of gas exchanged is a function of the rate and depth of respiration. In a resting state, the respiration rate is normally 12–15/min. CO_2 is exchanged at 0.2 L/min and O_2 at 0.25 L/min. Physical activity increases ventilation. Voluntary efforts can increase the rate of ventilation 20–30 times over the resting level and increase exchange of CO_2 to a maximum of 8.0 L/min and exchange of O_2 to a maximum of 4.0 L/min, but only briefly.

Regulation of respiration is mediated by neurochemical mechanisms. Involuntary increases in rate and depth of respiration are directed by the medullary respiratory center in the brain stem, which in turn is stimulated by central chemoreceptors located on the anterior surface of the medulla oblongata, and by peripheral chemoreceptors that include the carotid and aortic bodies. Peripheral chemoreceptors are stimulated by a fall in pH of arterial blood as a result of accumulation of CO_2. The central chemoreceptors are stimulated by a decrease in the pH of cerebrospinal fluid. Decrease of arterial pO_2 stimulates peripheral but not central chemoreceptors. Increased temperatures also accelerate the respiratory rate. The importance of the regulatory system will be explored in Section Three. The measurement of blood gases and pH becomes especially important when the clinical management of various disorders of cardiopulmonary function and metabolism overrides the physiological regulatory system. In assisted ventilation, for instance, mechanical means are used to establish respiratory rate, and selected and varying mixtures of respiratory gases are provided intermittently through an endotracheal tube.

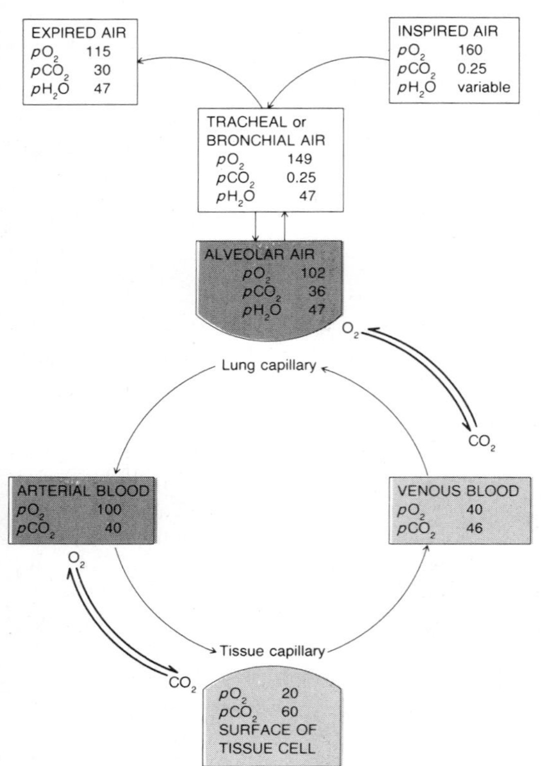

Figure 19-2. Partial pressures of oxygen and carbon dioxide in air, blood, and tissue. Values shown are approximations in mm Hg. Heavy arrows show directions of gradients.

Exchange of Gases in the Lungs and Peripheral Tissues

The movement of oxygen and carbon dioxide across alveolar and peripheral cell membranes is governed by gradients in the partial pressure of each gas. Dry air inspired at a pressure of one atmosphere (760 mm Hg) consists of 20.95% O_2 (~159 mm Hg), 0.03% CO_2 (~0.25 mm Hg), 78.1% nitrogen, and ~0.1% other inert gases. As inspired air passes over the moist mucous membranes of the upper respiratory tract, it is warmed to 37 °C, becomes saturated with water vapor (which also exerts a partial pressure), and mixes with air in the respiratory tree and in the alveoli; mixing decreases pO_2 and increases pCO_2. Venous blood on the opposite side of the alveolar membrane contains oxygen at an average partial pressure of ~40 mm Hg and carbon dioxide at ~46 mm Hg; the gradient for oxygen is toward the blood and for carbon dioxide away from the blood (Figure 19-2). Carbon dioxide removal is so efficient that the pCO_2 of expired air is >100 times the pCO_2 in inspired air.

At the arterial end of the capillaries of peripheral tissues, partial pressures of oxygen at ~100 mm Hg are substantially higher than the average pO_2 at the surface of the tissue cell (20 mm Hg), and the partial pressure of carbon dioxide (35–46 mm Hg) is substantially lower than that in the cell (50–70 mm Hg). In the tissue capillary, the gradient for oxygen is therefore toward the cells, and for carbon dioxide it is toward the blood. The arteriovenous difference is ~60 mm Hg for O_2 and $\leq$6 mm Hg for CO_2; the arteriovenous difference for oxygen is an indicator of the efficiency of oxygen extraction in the passage of blood through the capillary.

Internal respiration of tissue cells consumes oxygen and produces carbon dioxide and organic acid metabolites. The acid products of metabolism are buffered or removed by various mechanisms so that close control (homeostasis) of hydrogen ion concentration (and therefore pH) is maintained in the extracellular fluid of the body (see Section Three).

CO_2 and O_2 in Blood

Carbon dioxide is transported in the blood in several forms; the three most abundant and important are bicarbonate (HCO_3^-) in the plasma and erythrocytes, carbamino-hemoglobin ($HbCO_2$) in erythrocytes, and dissolved CO_2 (dCO_2) in fluids of erythrocytes and plasma. Oxygen is transported to the tissues in two forms: one, reversibly bound to hemoglobin in the erythrocytes (HbO_2); the other, dissolved O_2 (dO_2), physically dissolved in fluids of the erythrocytes and plasma.

The various forms of CO_2 deserve special notice, since it is on an understanding of these forms that comprehension of acid-base regulation is based. In *plasma*, a small portion of dCO_2 reacts with plasma water to form carbonic acid and bicarbonate:

$$dCO_2 + H_2O \rightleftharpoons \underset{\text{Carbonic acid}}{HHCO_3} \rightleftharpoons \underset{\text{Bicarbonate}}{HCO_3^-} + H^+$$

Another small amount reacts with amino groups of plasma proteins to form carbamino compounds. At a blood pH of 7.4 the equilibrium of the carbonic acid/bicarbonate dissociation favors the bicarbonate ion. As a result, the concentrations of undissociated carbonic acid and carbonate ion at blood pH 7.4 are minuscule. In *erythrocytes*, the same reactions occur between dCO_2 and intracellular water and between dCO_2 and amino groups of hemoglobin and other proteins. The pCO_2 of *whole blood* therefore reflects the equilibrium and the concentrations of forms of CO_2 that are present in both erythrocytes and plasma. A number of derived quantities, defined by means of calculation and experimental observations, are in use to describe the relationships of forms of CO_2 and pH in ways useful in acid-base evaluation. These quantities include buffer base, standard bicarbonate, actual bicarbonate, base deficit or base excess, and others; they are explained in Section Three. Contemporary blood gas instrumentation is commonly fitted with software for their calculation. But to study acid-base relationships in whole blood and plasma, as described by the Henderson-Hasselbalch equation, we are concerned only with the basic quantities of total concentration of CO_2 ($ctCO_2$), bicarbonate concentration ($cHCO_3^-$), concentration of dissolved CO_2 ($cdCO_2$), and pH.

Physical Principles

Behavior of Gases

Determination of partial pressures of gases in expired air or blood depends on certain familiar physical principles (Table 19-2). In blood gas analysis, however, many of the applications and assumptions are quite different from those made in basic sciences. For example, gas laws are defined in terms of closed systems but in the context of blood gases are applied to open systems (i.e., spaces whose components can move into other spaces). For this reason, the gas in blood is conceptualized as if the gas in the liquid phase were in equilibrium with an imaginary gas phase in contact with the liquid phase. Furthermore, assumptions are made that the partial pressure of the gas is the same in

Table 19-2. PHYSICAL PRINCIPLES APPLIED IN BLOOD GAS MEASUREMENTS

Boyle's law The volume of an ideal gas at a constant temperature varies inversely with the pressure exerted to contain it.	$V \propto 1/P$
Charles's (Gay-Lussac's) law The volume of an ideal gas at a constant pressure varies directly with its absolute temperature.	$V \propto T$
Avogadro's hypothesis Equal volumes of different ideal gases at the same temperature and pressure contain the same number of molecules.	$n_i/V_i = n_j/V_j$
Dalton's law The total pressure exerted by a mixture of ideal gases is the sum of the partial pressures of each of the gases in the mixture.	$P = \Sigma p_i$
Henry's law The amount of a sparingly soluble gas dissolved in a liquid is proportional to the partial pressure of the gas over the liquid.	$c = a \times p$

the liquid phases of erythrocytes and plasma, and that the partial pressure of the gas in whole blood is equal to its partial pressure in plasma. The term *equilibrium* means an even balance of forces that results in a stable condition. "Equilibration" of blood with a gas mixture of known composition is a procedure of exposing the blood to the mixture such that each gas in it is partitioned to an equilibrium state between liquid and gaseous phases; that is, the partial pressure of each gas in liquid is brought to equilibrium with its partial pressure in the gas phase. In the laboratory, equilibration is a routine procedure known as tonometry (see page 638).

Atmospheric (barometric) pressure, p(Atm), is assumed to be the prevailing pressure in the room, in the patient's tracheobronchial tree and alveoli, and in the measuring chamber of an instrument. Partial pressures of each of the gases present in these spaces are assumed to add up to the value of p(Atm). However, no such assumption can be or is made for partial pressures described for gases in circulating blood.

By scientific convention, *STPD* defines standard conditions (see Table 19-1) for dry gas. In respiratory physiology an additional and different definition, *BTPS*, is used. BTPS specifies that measurements of partial pressure in biological specimens are always made at a temperature of 37 °C in the presence of a partial pressure of water vapor at 47 mm Hg and at local atmospheric pressure (Table 19-1). The use of this convention has the following practical effects:

1. It relates laboratory data for blood gases strictly to the geographic location of the patient, so that reference ranges are altitude dependent.
2. It assumes that human body temperature is exactly 37 °C and that the measuring device also holds the blood sample at exactly 37 °C. This assumption requires thermal stability of instrumentation and, in instances in which a patient's temperature is not 37 °C, consideration of possible correction of measured values.
3. It recognizes that the partial pressures of measured gases in blood coexist with a partial pressure of water (SVP, saturated vapor pressure), which is a constant at a given temperature and must therefore be identical for both calibration of the instrument and measurements made on the blood sample.

Boyle's and *Charles's laws* and *Avogadro's hypothesis* are combined in what is called the general gas equation:

$$P = (nRt)/V$$

where P = pressure in units of mm Hg, pascals (Pa) or kilopascals (kPa)*
V = volume, in L, in which an ideal gas is contained
t = temperature in K (0 °C = 273.16 K)
n = number of moles (or amount of substance) of gas
R = universal gas constant: 62.36 mm Hg $\times$ L^{-1} $\times$ K^{-1} $\times$ mol^{-1}, or 8.31 kPa $\times$ L^{-1} $\times$ K^{-1} $\times$ mol^{-1}

The general gas equation is the justification for accepting partial pressures of gases in blood as estimators of their concentrations, since a partial pressure (p) as well as P, pressure, at a constant temperature is directly proportional to n/V, a concentration term in units of mol/L. Pressure, P or p, may mean either total pressure, as in the expression p(Atm) for the mixture of gases in room air, or partial pressure in blood, as in pO_2(aB).

Dalton's law may be written for room air as:

$$p(\text{Atm}) = pO_2 + pCO_2 + pN_2 + pH_2O + pX$$

where pH_2O is the partial pressure of water vapor and pX is that of any other gas or gases in the sample. For venous blood in a measuring device, the equation can be written

$$p(\text{vB}) = pO_2 + pCO_2 + pX'$$

* Use of Pa (or kPa), the SI unit, has the advantage that one atmosphere (760 mm Hg) is almost equal to 100 kPa (1 atm = 101.325 kPa). Partial pressures in kPa for gases in a mixture at 1 atm are therefore easily converted to quite close estimates of percentages of gases in the mixture.

by assuming that pX' is a constant (i.e., the total of partial pressures of all the gases present in the blood except oxygen and carbon dioxide). Dalton's law of partial pressures is important for the standardization and control of measuring devices.

Consider the example of a calibrator gas that contains 15% O_2 (L/L or mol/mol) and 5% CO_2, the remainder being N_2. This mixture, after saturation with water vapor at 37 °C (to mimic a patient's blood or alveolar air), is introduced into the instrument's measuring chamber (held at 37 °C to mimic a patient's body temperature) for the purpose of standardizing the instrument. If the local barometric pressure p(Atm) on this occasion is 747 mm Hg, then the humidified calibrator gas is present in the chamber at ambient barometric pressure, such that

$$p(\text{Atm}) = 747 \text{ mm Hg} = pO_2 + pCO_2 + pN_2 + pH_2O$$

To set the instrument to the pO_2 and pCO_2 of the calibrator gas, we must first account for pH_2O at 37 °C, which we know is 47 mm Hg.

$$p(\text{Atm}) - pH_2O = pO_2 + pCO_2 + pN_2$$
$$747 - 47 = 700 \text{ mm Hg}$$

Then p(Atm) corrected for pH_2O represents the sum of partial pressures for the dry gases. The exact partial pressures of oxygen and carbon dioxide for calibration of the instrument are:

$$\mathbf{pO_2} = 700 \times 0.15 = 105 \text{ mm Hg}$$
$$\mathbf{pCO_2} = 700 \times 0.05 = 35 \text{ mm Hg}$$

Henry's law (Table 19-2) predicts the amount of dissolved gas in a liquid in contact with a gaseous phase. For gas in blood at 37 °C, the solubility coefficient for oxygen, αO_2(B), is 0.00140 mol/L per mm Hg. For plasma the coefficient is 0.00126 mmol/L per mm Hg. When pO_2(aB) is ~98 mm Hg, cO_2(aB) is 0.00140 × 98 or 0.137 mmol/L—a very small proportion of total oxygen content (ctO_2). Increasing the O_2 fraction of inspired air to 100%, or increasing the pressure of inspired air, as can be done in hyperbaric chambers, forces more O_2 into solution. In therapy with pure O_2, when pO_2 may rise to 640 mm Hg, cdO_2(aB) could be as high as 0.9 mmol/L. In hyperbaric treatment, a pO_2(aB) of 2500 mm Hg (~3.2 atmospheres) is equivalent to cdO_2(aB) of 3.5 mmol/L. These therapies, and calculation of cdO_2, become important when hemoglobin-mediated delivery of oxygen is impaired.

The concentration of dissolved CO_2 can be calculated in the same way. The αCO_2(P) at 37 °C is 0.0306 mmol/L per mm Hg. At a pCO_2 of 40 mm Hg therefore, $cdCO_2$ is 40 × 0.0306 = 1.224 mmol/L. Modern blood gas instrumentation calculates $cdCO_2$ from measured pCO_2 as a preliminary to calculation of derived values such as bicarbonate concentration and base excess or deficit.

The Interrelationship Between $ctCO_2$, $cHCO_3^-$, $cdCO_2$, and pH

The Henderson-Hasselbalch Equation

Carbon dioxide and water react to form carbonic acid, which in turn dissociates to hydrogen and bicarbonate ions:

$$CO_2 + H_2O \underset{}{\overset{K_{\text{hydration}}}{\rightleftharpoons}} H_2CO_3 \overset{K_{\text{dissociation}}}{\rightleftharpoons} H^+ + HCO_3^-$$

According to the law of mass action,

$$K_{\text{hydration}} = \frac{aH_2CO_3}{aCO_2 \times aH_2O} = 2.29 \times 10^{-3} \qquad (pK = 2.64 \text{ at } 37\ ^\circ C)$$

and

$$K_{\text{dissociation}} = \frac{aH^+ \times aHCO_3^-}{aH_2CO_3} = 2.04 \times 10^{-4} \qquad (pK = 3.69 \text{ at } 37\text{ °C})$$

These can be combined to give

$$K_{\text{combined}} = \frac{aH_2CO_3}{aCO_2 \times aH_2O} \times \frac{aH^+ \times aHCO_3^-}{aH_2CO_3} = \frac{aH^+ \times aHCO_3^-}{aCO_2 \times aH_2O} = 4.68 \times 10^{-7} \qquad (pK = 6.33)$$

In the classical formulation Henderson (1908) used concentrations (c) rather than activities (a) for bicarbonate, CO_2, and H^+; the concentration of water was assumed to be constant and was therefore incorporated into the constant K'.

$$K' = \frac{cH^+ \times cHCO_3^-}{cdCO_2}$$

The symbol $cdCO_2$ stands for the concentration of dissolved CO_2 *including* the small amount of undissociated H_2CO_3. It can be expressed as $cdCO_2 = \alpha \times pCO_2$, where α is the solubility coefficient for CO_2. The symbol $cHCO_3^-$ represents the concentration of total CO_2 ($ctCO_2$) minus the concentration of dissolved CO_2 ($cdCO_2$), which includes H_2CO_3.

$$cHCO_3^- = ctCO_2 - (a \times pCO_2)$$

The "bicarbonate" concentration by definition includes undissociated $NaHCO_3$, carbonate (CO_3^{2-}), and carbamate (carbamino-CO_2), which are present only in small amounts in plasma.

If the Henderson equation is rearranged and $cdCO_2$ is replaced by $a \times pCO_2$, the following equation results:

$$cH^+ = K' \times \frac{a \times pCO_2}{cHCO_3^-}$$

Hasselbalch (1916) showed that a logarithmic transformation of the equation was a more useful form and employed the symbols pH ($= -\log cH^+$) and $pK'(= -\log K')$. Currently, the pH is defined as the negative log of the activity of H^+ (aH^+), which is the entity actually measured with pH meters. The resulting Henderson-Hasselbalch equation becomes:

$$pH = pK' + \log \frac{cHCO_3^-}{cdCO_2} = pK' + \log \frac{ctCO_2 - (a \times pCO_2)}{a \times pCO_2}$$

K' is the first, apparent, overall (combined) dissociation constant for carbonic acid—apparent, because concentrations are employed rather than activities and because it includes both the hydration and dissociation constants; overall, because both the concentration of dissolved CO_2 and the true concentration of H_2CO_3 are used. K' depends not only on the temperature but also on the ionic strength of the solution. For an aqueous sodium bicarbonate solution at 37 °C the following approximate relationship exists between K' and ionic strength (I) measured in mol/kg H_2O:

$$pK' = 6.33 - 0.5\sqrt{(I)}$$

For blood plasma at 37 °C the normal mean value is $pK'(P) = 6.103$ with a normal biological standard deviation of about ± 0.0015, mainly due to normal variations in ionic strength. In pathological cases with markedly deviant ionic strength, the standard deviation for pK' may be significantly greater. The solubility coefficient for CO_2 gas, α, for normal plasma at 37 °C is 0.0306 mmol $\times$ L^{-1} $\times$ mm Hg^{-1} with a biological SD of about ± 0.0003 $mmol/L^{-1} \times mm\ Hg^{-1}$. In lipemic plasma the value of α may be 0.033 or even higher.

Inserting pK' and α for normal plasma at 37 °C, the Henderson-Hasselbalch equation takes the following form:

$$pH = 6.103 + \log \frac{cHCO_3^-}{0.0306 \times pCO_2}$$

or

$$pH = 6.103 + \log \frac{ctCO_2 - 0.0306 \times pCO_2}{0.0306 \times pCO_2}$$

where pCO_2 is measured in mm Hg, and $cHCO_3$ and $ctCO_2$ are measured in mmol/L.

Taking the antilogarithm and combining constants, the equation becomes:

$$aH^+ = 24.1 \times \frac{pCO_2}{cHCO_3^-} \times 10^{-9}$$

or, taking the activity coefficient of hydrogen ions in plasma to be 1.00 (the true value is probably ~0.80), the equation can be written:

$$cH^+ = 24.1 \frac{pCO_2}{cHCO_3^-}$$

where cH^+ is now given in nmol/L, pCO_2 in mm Hg, and $cHCO_3^-$, in mmol/L. If normal values are substituted in the equation,

$$cH^+ = 24.1 \times \frac{40}{25.4} \text{ nmol/L} = 38.0 \text{ nmol/L}$$

Application of the Henderson-Hasselbalch Equation in Blood Gas Measurements

By measuring any two of the four parameters, pCO_2 or $cdCO_2$, pH, $ctCO_2$, $cHCO_3^-$, and using the Henderson-Hasselbalch equation with appropriate values for pK' and α, the other values may be calculated. Microprocessors in state-of-the-art blood gas instrumentation use measured pH and measured pCO_2 in the equation

$$pH = pK' + \log \frac{cHCO_3^-}{(\alpha \times pCO_2)}$$

to calculate $cHCO_3^-$. pK' is taken as 6.103 and α as 0.0306 mmol/L per mm Hg. Should pH be 7.42 and pCO_2 37.5 mm Hg, the equation is thus evaluated

$$7.42 = 6.103 - \log cHCO_3^- - \log(0.0306 \times 37.5)$$

Upon rearrangement to solve for log $cHCO_3^-$ and taking the antilog,

$$\mathbf{cHCO_3^-} = 23.97 \text{ mmol/L}$$

Dissolved CO_2 concentration is the product of α and pCO_2,

$$\mathbf{cdCO_2} = 0.0306 \times 37.5 = 1.1 \text{ mmol/L}$$

and total CO_2 concentration

$$\mathbf{ctCO_2} = cdCO_2 + cHCO_3^- = 25.1 \text{ mmol/L}$$

Oxygen in Blood

Most of the oxygen in blood is transported as HbO_2 in erythrocytes. Uptake of O_2 by the blood in the lungs is governed primarily by the pO_2 of alveolar air and by the ability of O_2 to diffuse freely across the alveolar membrane into the blood, and—to a lesser extent—by the affinity of deoxyhemoglobin (HHb) in the erythrocytes for O_2. At the pO_2 normally present in aveolar air, and with a normal membrane and normally functional hemoglobin protein, >95% of the hemoglobin binds oxygen. At a pO_2 above 100–120 mm Hg, 100% of the hemoglobin binds oxygen; i.e., hemoglobin is *saturated* with O_2. When all hemoglobin capable of binding O_2 is saturated, further increase in the pO_2 of alveolar air simply increases cdO_2 of arterial blood. Delivery of O_2 by the blood to the tissues is governed by the large gradient between pO_2 of the blood and that of the tissue cells, by the working of the isohydric and chloride shift, and by the dissociation behavior of oxyhemoglobin (HbO_2) in the erythrocytes at the lower pO_2 of the blood-tissue cell interface.

Oxygen Saturation

Traditionally, sO_2, the oxygen saturation of blood, has been defined as

$$sO_2, \% = \frac{\text{Oxygen content}}{\text{Oxygen capacity}} \times 100\%$$

in terms of determination of O_2 by manometric methods. But now that spectrophotometric methods are used to determine oxyhemoglobin and reduced hemoglobin (along with carboxyhemoglobin and methemoglobin) specifically, sO_2 is calculated according to the equation

$$sO_2, \% = \frac{cHbO_2, \text{g/dL}}{cHb, \text{g/dL}} \times 100\%$$

where $cHbO_2$ is the concentration of oxyhemoglobin and cHb is the concentration of total functional hemoglobin (sum of oxy- and deoxyhemoglobin).* sO_2 is usually expressed in per cent in the US but may also be expressed as a decimal fraction of 1.00. Mathematical approximations calculated by microprocessors of blood gas instruments should be interpreted with reservations since the algorithmic approach assumes normal oxygen affinity of the hemoglobin.

Decrease of oxygen saturation of arterial blood below the level critical for adequate oxygenation of tissue is a grave clinical situation. Such decrease indicates either a low pO_2 or seriously impaired ability of hemoglobin to serve as an oxygen carrier. The amount of oxygen that the blood can carry is determined by three major factors: the amount of functional hemoglobin available within erythrocytes; the pO_2 that determines how much oxygen will dissolve in the blood; and the affinity of the available hemoglobin for oxygen. Decreases in the amount of hemoglobin can arise from a decreased number of erythrocytes that contain a normal concentration of hemoglobin (normochromic anemia) or from a decreased concentration of hemoglobin in the erythrocytes (hypochromic anemia). Decreased amounts of functional hemoglobin can occur as a result of poisonings that convert large quantities of hemoglobin into nonfunctional carboxyhemoglobin, met-

---

* sO_2 is sometimes calculated as

$$sO_2, \% = \frac{cHbO_2}{ctHb} \times 100\%$$

where ctHb includes carboxy-, met-, and sulfhemoglobin. With this equation, the value of sO_2 can never reach 100% if any of the nonfunctional hemoglobin derivatives is present. For example, should carboxyhemoglobin amount to 10% of ctHb, sO_2 could be no more than 90% in fully saturated blood. This finding would be misleading, indicating increased shunting of blood in the lungs. Therefore, this calculation for sO_2 should be abandoned.

hemoglobin, sulfhemoglobin, or cyanmethemoglobin. Impaired diffusion of oxygen across the alveolar membrane or inadequate circulation due to cardiac insufficiency results in decrease of pO_2(aB) and may affect oxygen delivery to tissues. Compensatory factors that govern affinity of hemoglobin for oxygen may intervene to keep delivery at an adequate level. Clinically, it is important to distinguish between *hypoxia* (decreased pO_2 and decreased sO_2 because of decreased availability of O_2) and *cyanosis* (decreased sO_2) because of abnormally high concentrations of reduced hemoglobin or chemically altered hemoglobin incapable of carrying O_2.

Hemoglobin-Oxygen Dissociation

The degree of association or dissociation between oxygen and hemoglobin is determined by pO_2 and by the affinity of hemoglobin for oxygen. When the saturation, sO_2, is determined over a range of pO_2 and plotted against pO_2, a sigmoidal curve called the oxygen dissociation curve is obtained. The *shape* of this curve is due to the increasing efficiency with which deoxyhemoglobin molecules bind more oxygen once the first oxygen molecule has been bound. The *location* of the curve, relative to the pO_2 required to achieve a particular degree of saturation, is a function of the affinity of the hemoglobin for oxygen. Changes in affinity cause the curve to shift to the right or to the left.

The affinity of hemoglobin for oxygen depends on five factors: temperature, pH, pCO_2, concentration of 2,3-diphosphoglycerate (DPG), and the type of hemoglobin protein. Variants of hemoglobin are discussed in Chapter 24; the effect of changes in DPG concentration in shifting the dissociation curve is shown in Figure 24-5. Figure 19-3, top left, illustrates the effect of plasma pH on the sO_2/pO_2 relationship. If similar graphs were made for variation of pCO_2 and for variation of temperature, the shifting locations of the curves could be described as in the chart in the legend of Figure 19-3. The shifts develop because of deviations of values from reference points—temperature from 37 °C; pCO_2 from 40 mm Hg; pH(P) from 7.40; and cDPG/cHb from some value defined as "normal." The sigmoidal curves are difficult to use when quantitative estimates of the amount of shift are desired. The Hill transformation (on the right side of Figure 19-3) converts the curvilinear dissociation function into a linear function. The slope of the linear function, called the Hill slope, or n_{Hill}, is essentially constant at 2.7 for sO_2 of 30–70%. Linear transforms allow determination of coefficients of change (Figure 19-3) that can be used for correcting measured pO_2 or calculated $pO_2(0.5)$ for temperature, pCO_2, pH(P), and cDPG effects.

$pO_2(0.5)$ or P_{50}

$pO_2(0.5)$ is defined as the value of pO_2 for a given blood at which hemoglobin of the blood is half-saturated with O_2. The $pO_2(0.5)$ represents what the pO_2 *would* be if its sO_2 *were* equal to 0.5. The measured value of $pO_2(0.5)$ differs from the standard value of $pO_2(0.5)$ by some amount determined by the effect that nonstandard pH, pCO_2, temperature, or cDPG has on hemoglobin affinity.

Determination of $pO_2(0.5)$

The simplest method consists of measuring pO_2 potentiometrically and sO_2 spectrophotometrically on an anaerobically collected venous whole blood sample. The following calculation is made:

$$\log pO_2(0.5) = \log pO_2 - \frac{\text{logit } sO_2}{2.7}$$

where logit $sO_2 = \log [sO_2/(1 - sO_2)]$ and 2.7 is the Hill slope, which is assumed to be independent of the other variables. The antilog of the value obtained is for $pO_2(0.5)$ at 37 °C. Corrected values, obtained with coefficients of change shown in Figure 19-3, may be calculated for various clinical purposes.

Reference ranges. For adults, the 95% limits for $pO_2(0.5)$, measured at 37 °C and corrected to pH(P) of 7.4, are 25–29 mm Hg. For newborn infants, the limits are 18–24 mm Hg.

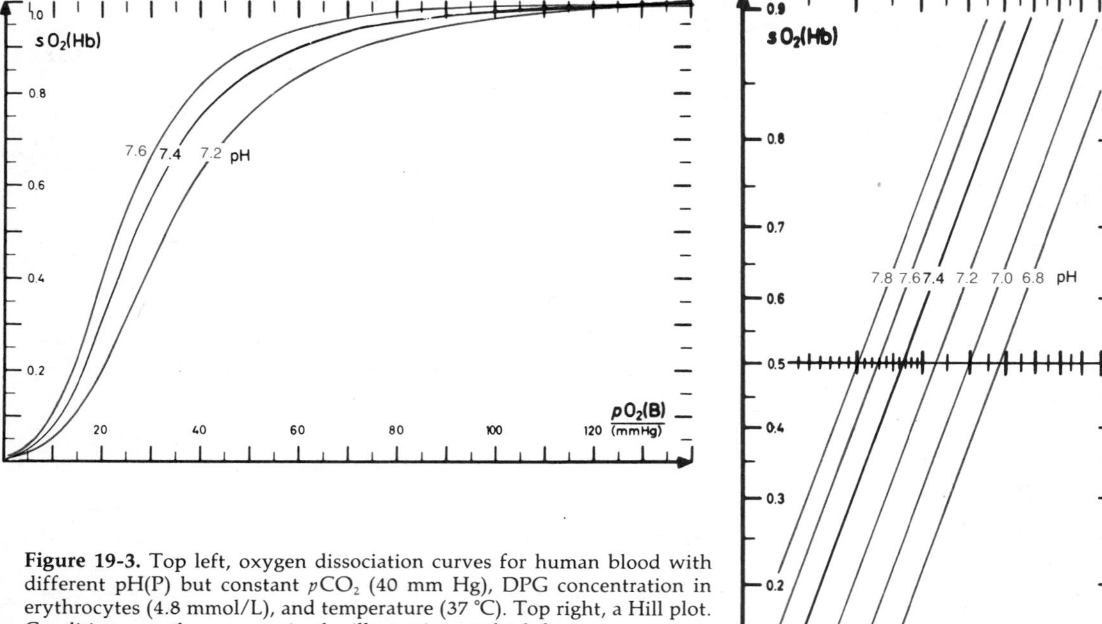

Figure 19-3. Top left, oxygen dissociation curves for human blood with different pH(P) but constant pCO_2 (40 mm Hg), DPG concentration in erythrocytes (4.8 mmol/L), and temperature (37 °C). Top right, a Hill plot. Conditions are the same as in the illustration at the left.

Deviation from Standard Conditions	Shift in Dissociation Curve	Affinity of Hemoglobin for Oxygen	Coefficient of Change*
pH(P) >7.4	←	↑	$\frac{\Delta \log pO_2}{\Delta pH(P)} = -0.38$
pH(P) <7.4	→	↓	
Temperature			
>37°C	→	↓	$\frac{\Delta \log pO_2}{\Delta T} = +0.024\ K^{-1}$
<37°C	←	↑	
pCO_2 >40 mm Hg	→	↓	$\frac{\Delta \log pO_2}{\Delta \log pCO_2} = +0.10$
pCO_2 <40 mm Hg	←	↑	
DPG/Hb ratio			
>normal	→	↓	$\frac{\Delta \log pO_2}{\Delta(cDPG/cHb)} = +0.60$
<normal	←	↑	

*The coefficients given in this chart are the basis for the correction of measured pO_2. The effect of pH(P) to shift the dissociation curve is called the Bohr effect; the coefficient above for $\Delta \log pO_2/\Delta pH(P)$ applies to conditions when the pCO_2 is 40 mm Hg and changes in pH(P) are due to changes in concentrations of noncarbonic acids and bases. If, however, the changes in pH(P) are caused by changes in pCO_2, then the absolute value of the coefficient is greater (i.e., $\Delta \log pO_2/\Delta pH(P) = -0.53$). The coefficients for the Bohr effect are specified for pO_2 of whole blood but utilize the pH of plasma, pH(P). The coefficient for DPG effect utilizes the common form of expression of DPG concentration, ratio of DPG concentration to hemoglobin concentration.

Clinical significance. *Increased values* for $pO_2(0.5)$ indicate displacement of the oxygen dissociation curve to the right (i.e., a decreased affinity of the hemoglobin for oxygen). The chief causes are hyperthermia, acidemia, hypercapnia, high concentrations of DPG, or the presence of an abnormal hemoglobin with decreased oxygen affinity. DPG concentrations tend to be increased in chronic alkalemia and in anemia. An example of hemoglobins with decreased O_2 affinity is hemoglobin Seattle. Physiological effects of decreased affinity of hemoglobin are small.

Low values for $pO_2(0.5)$ signify displacement of the oxygen dissociation curve to the left (i.e., increased affinity of hemoglobin). The main causes are hypothermia, acute alkalemia, hypocapnia, low concentration of DPG, or an abnormal hemoglobin. Decreases of DPG are commonly observed in acidemic states which have persisted for more than a few hours; the initial increase in $pO_2(0.5)$ caused by the acidemia is gradually compensated by decrease in DPG so that $pO_2(0.5)$ then falls to lower than normal values. The physiological consequence of increased affinity of hemoglobin for oxygen is less efficient dissociation of HbO_2 at the peripheral tissues.

Determination of pCO_2, pO_2, and pH

Specimens

Arterial or venous blood is the usual specimen to be analyzed in a clinical laboratory, although in some facilities expired gases may also be submitted for analysis. Blood specimens may be obtained from chambers of the heart and great vessels of the chest during heart catheterization as well as from the vessels of the extremities. The volume of specimen collected for analysis, especially from premature neonates being monitored by frequent, repetitive testing, needs to be as small as is possible without sacrifice of specimen quality. Quality control of blood analysis for gases and pH is highly dependent on control of preanalytical error (i.e., proper collection and handling of specimen).

Arterial puncture carries a risk, and under no circumstances should it be undertaken by anyone who has not been properly trained to perform it. The patient should be spared, insofar as possible, the pain and anxiety that may cause hyperventilation. Timing of collection should be appropriate to treatment regimens. Arterial puncture is always done with a syringe and needle. No tourniquet is used and no pull applied to the plunger of the syringe; the arterial blood pressure pushes blood into the syringe.

Venous blood is best collected with a syringe, but specimens drawn to complete fill of an evacuated tube containing a dry heparin salt are also acceptable. When an arm vein is used, the specimen should be obtained within the first few seconds after application of the tourniquet. The patient should not be allowed to flex his fingers or clench his fist. Prolonged application of the tourniquet or muscular activity decreases venous pO_2 and allows accumulation of acid metabolites. Indwelling cannulas with heparin locks for short- and long-term therapies are now fairly common; the cannula serves often as a port for specimen collection. The lock must be thoroughly flushed with blood before a specimen for blood gases and pH (or any other assay) is drawn.[3] Failure to flush the lock properly is usually indicated by bizarre test results.

Arterial or venous specimens are best collected anaerobically with liquid heparin anticoagulant in all-glass, sterile syringes with capacities of 1–5 mL.* A syringe with a Luer-Lok hub is best for collecting arterial specimens. Glass is preferred to plastic to avoid loss of gases through the syringe wall, and in arterial collections to avoid friction on the plunger as arterial blood pressure fills the syringe. The size of the syringe, the concentration of heparin, and the maximum volume of blood drawn into the syringe are important. For syringes of 1–5 mL capacity, dead space is 0.1–0.2 mL and holds more than enough heparin for anticoagulation of 5 mL of blood. Sterile heparin solution is drawn into the syringe in such a way as to wet the interior of the barrel over the maximum capacity of the syringe; the heparin is then ejected so that the dead space is filled with heparin. Increasing the dead-space volume of the syringe by addition of a mixing bead, or drawing blood to less than full capacity, increases the proportion of heparin solution to blood and the error due to dilution of the specimen. Furthermore, certain lots of heparin solution are acidic and an increase in the ratio of heparin to blood decreases the value of measured pCO_2 and leads to error in values calculated from pCO_2.[5] Mixing specimen with heparin in the syringe is efficiently accomplished by rolling the syringe between the palms.

The *anaerobic technique* for blood collection has as its guiding principle either no or minimal exposure of specimen to atmospheric air. The pCO_2 of air is much less than

---

*Packaged, disposable sets for collection of blood gases are commercially available, as are cold-packs for subsequent transport of specimens.

that of blood, so that exposure of blood to air allows CO_2 content and pCO_2 to decrease and pH (a function of pCO_2) to rise. The pO_2 of air is higher than that of blood, and venous and arterial blood with normal oxygen content, when exposed to air, gains O_2. But blood with pO_2 >158 mm Hg, as might be obtained from patients undergoing oxygen therapy, will lose O_2 upon exposure to air. In routine arterial or venous puncture, depending on whether the needle is full of heparin or empty, a single small bubble of air may appear in the blood in the syringe. The error, however, will be minimal if the bubble is ejected immediately upon removing the needle from the puncture site and before sealing the syringe tip. Sealing the sample by sticking the needle into a cork is not recommended; snug-fitting caps for the syringe hub are commercially available.

Arterialized capillary blood is sometimes an acceptable alternative to arterial blood when specimen volumes must be minimized, repeated arterial puncture must be avoided, or an arterial cannula is not available. Arterialized capillary blood is not acceptable, however, when systolic blood pressure is <95 mm Hg, in cases of vasoconstriction, in patients on oxygen therapy, in newborns in the first few hours or days of life, or in newborns with respiratory distress syndrome. In these situations, there is a particular risk of admixture with blood from the venules that causes erroneously low pO_2 values. Freely flowing cutaneous blood originates in the arterioles and corresponds to arterial blood in composition. The selected skin puncture site must be warmed for 10 min to achieve vasodilation and good blood flow through the local capillaries. For collection from a child's or adult's finger or an infant's heel, warming may be accomplished by immersing the arm or foot in water warmed to 45 °C. The limb should be rosy and warm to the touch before the skin is punctured. The first drop of blood to appear should be wiped away, and subsequent free-forming drops should be taken up in a capillary collection tube containing dried heparin. Appropriate tubes hold 70–300 μL of blood. They are allowed to fill, and then a flea (a slender wire ~5 mm long) is immediately inserted into the tube, a cap is fitted to the filled end, and a magnet is used to move the flea to and fro in the blood for mixing it with heparin coating the tube walls.

Transport and analysis of specimens should be prompt even when there is no urgency in returning test results. Specimens should never be stored, although results on properly preserved specimens may remain valid up to 1 h. The pH of freshly drawn blood decreases on standing at a rate of 0.04–0.08 unit/h at 37 °C and 0.008 unit/h at 4 °C. The decrease in pH is accompanied by a corresponding decrease in glucose and an equivalent increase in lactate. The primary cause of these changes is thought to be glycolysis by leukocytes. In leukemia, the pH drop during the first 30 min after specimen collection may be as high as 0.6 pH unit. pCO_2 increases in freshly drawn blood by ~5 mm Hg/h at 37 °C but only about one-tenth as fast at 2–4 °C. If blood with normal pO_2 is kept anaerobic, glycolysis causes pO_2 to decrease at a rate of 2–6 mm Hg/h at room temperature and twice as rapidly at 37 °C. Adverse effects of glycolysis on pH, $ctCO_2$, pO_2, and pCO_2 are best avoided by immersing the specimen in a mixture of ice and water immediately upon collection and analyzing it within 30 min. Under these conditions, changes in blood pH or blood gases are negligible.

Instrumentation

A schematic diagram characteristic of contemporary instrumentation is shown in Figure 19-4. Electrochemical principles and structural features of electrodes are discussed in Chapter 2, Section Four. The operation of a blood gas instrument begins with the operator's presenting a blood specimen at the sample probe and initiating its uptake by a keyboard command. The pump device then draws the sample through the probe and into the measuring chamber (C), bringing the sample into contact with the electrodes. The pump then pauses to allow the sample to reside in the chamber for temperature equilibration prior to measurement. Upon completion of measurement, the pump pushes the sample to waste, while digital output is made available on a display, printed tape, or a printed multi-copy requisition/report ticket. The instrument is designed so that a manually or electronically actuated valve (V) admits calibrator gases, standard buffers, or sample to a small chamber (C) holding 60–150 μL and maintained by a fluid or metal bath (B) at a constant temperature of 37 ± 0.1 °C. Measuring and reference electrodes

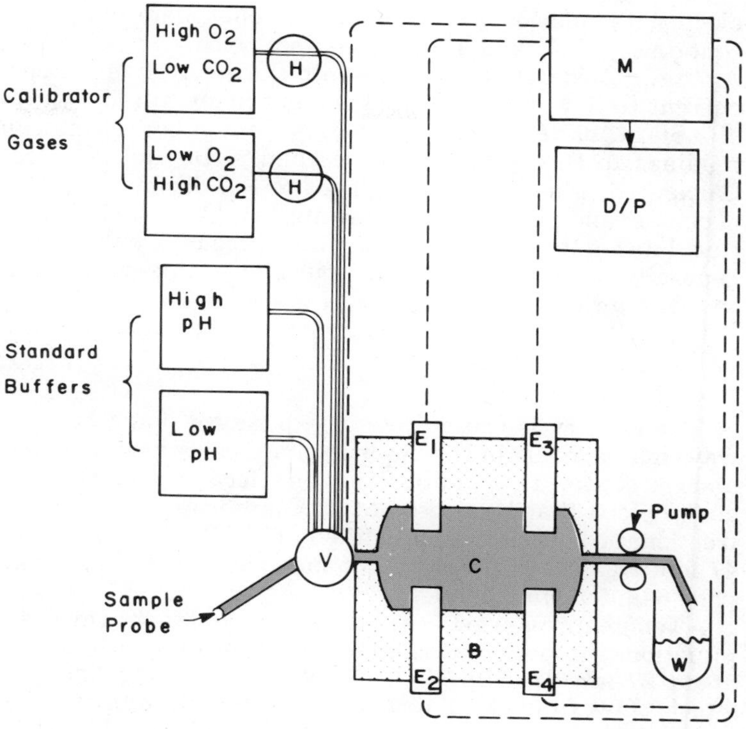

Figure 19-4. Diagram of blood gas instrumentation. H, humidification device; V, valve; C, chamber; B, constant temperature bath at 37 °C; W, waste; M, microprocessor; D/P, display/printer. E (electrodes) where E_1 is pO_2; E_2, pCO_2; E_3, pH; and E_4, reference for pH.

(E) protrude into the chamber. In the pH calibration phase of the instrument, high pH standard buffer and low pH standard buffer are alternately admitted into the chamber, and electronic responses of the upper and lower limits of a linear pH curve are established. In the gas calibration phase, gas mixtures with high and low fractional concentrations of O_2 and CO_2 are alternately admitted into the chamber and electronic responses of the upper and lower limits of linear pO_2 and pCO_2 curves are set. In the measurement phase, an anaerobically collected blood sample is admitted.

Most instruments are designed to be self-calibrating. Under the command of the microprocessor (M), calibrator gases and buffers are cycled at short intervals through the chamber. Electronic responses are continually monitored and reset to the constants initially entered for high and low pCO_2 and pO_2 and high and low pH of the calibrator materials.

Electrodes. Electrodes are typically constructed with threaded collars or bayonet locks for a leak-proof fit when they are installed in the chamber. They are positioned to bring their tips into contact with the contained sample. The tip of the pH-measuring electrode is made of H^+-sensitive glass. The pH reference electrode is a calomel or silver/silver chloride electrode filled with saturated KCl that contacts the sample through a port, thus forming a salt bridge.

The construction of pO_2 and pCO_2 electrodes is shown in Figures 2-31 and 2-33. The gas electrodes are fitted at the tip with gas-permeable membranes held in place with O-rings. Inside these membranes are electrolyte solutions in contact with the measuring and reference elements of the electrode. The membrane of the pCO_2 electrode usually consists of Teflon or silicone rubber ~25 μm thick. The electrolyte solution is a thin film containing $NaHCO_3$ at 0.005 mol/L and NaCl at 0.1 mol/L and saturated with AgCl. A spacer of nylon net or cellophane lies between the solution and the H^+-sensitive glass of the measuring element proper. The electrolyte also contacts the silver/silver chloride reference electrode. As CO_2 diffuses from the sample into the electrolyte, the slight rise in $[H^+]$ from its hydration reaction is measured as ΔpH by an especially sensitive potentiometer and electronically transformed to $\Delta \log pCO_2$.

The membrane of a pO_2 electrode is usually of polypropylene ~20 μm thick. The electrolyte solution is a thin film of phosphate buffer saturated with AgCl and also containing KCl, and it is in contact with the polarized platinum cathode and the Ag/AgCl anode. As O_2 diffuses into the electrolyte, it reacts with the cathode to cause current to flow; the generated current is measured.

Significant figures of results. Contemporary instruments commonly display pH readings to three decimal places and pCO_2 and pO_2 readings to one decimal place. Calculated values may be displayed to 1–3 decimal places. pH should be reported to two decimals, and pCO_2 and pO_2 as integers. The number of significant digits reported for calculated data takes precision into account. Values with two significant digits are generally adequate for clinical purposes; imprecision and inaccuracy often lie in the second digit.

Tonometry

Tonometry is the process of exposing a liquid to an ambient gas phase in such a way that each gas in the gaseous phase partitions to an equilibrium between the liquid and gas phases. This equilibration, in effect, imparts the pCO_2 and pO_2 of an equilibrating gas to blood that is exposed to it. Equilibration by tonometry employs gases of known fractional composition, humidified at 37 °C to give a saturated water vapor pressure of 47 mm Hg. The pCO_2 or pO_2 of such gases is calculated according to Dalton's law (see page 628, for an example).

Tonometry is used to treat blood samples for various purposes. For $pO_2(0.5)$ determination, the patient's sample is equilibrated with a gas mixture with known pO_2 of about 27 mm Hg and pCO_2 of about 40 mm Hg, prior to measurements of sO_2 from which $pO_2(0.5)$ is calculated. For determination of standard bicarbonate, the patient's sample is equilibrated with gas of pCO_2 40 mm Hg and pO_2 >100 mm Hg (i.e., pO_2 sufficient to guarantee sO_2 >0.90) prior to measurement of pH and pCO_2. Other applications include preparation of a blood sample for quality control of blood gas measurements, for determining the liquid-gas difference for a pO_2 electrode, and for verifying accuracy and linearity of pO_2 and pCO_2 electrodes.

The principles that should be observed in tonometric equilibration of blood are rapid equilibration, preferably within 3 min, to minimize glycolysis in the blood sample; temperature control at 37 ± 0.1 °C; humidity and temperature control to prevent evaporation of water from the blood or condensation of water into the blood; and a mode of exposure of gas to blood that causes no more hemolysis than 50 μmol Hb/L of blood plasma. Tonometry may be achieved by simple, homemade assemblies or by commercially available equipment.

Removal of tonometered blood from a treatment container should be made with a syringe that has been thoroughly flushed with the equilibrating gas. The orifice through which the syringe is filled should be as large as possible and the blood should be drawn up with care to avoid creating a vacuum ahead of the plunger. For most purposes, the sample should be analyzed immediately; otherwise, it should be tightly sealed for delayed analysis.

Calibration

pH measurement. The pH measurement system is calibrated against primary standard buffers. The buffers are phosphate solutions with a temperature coefficient, $\Delta pH/\Delta T$, of $-0.0015 \times K^{-1}$ (or $°C^{-1}$) and should meet NBS specifications (National Bureau of Standards, Washington, DC). Such buffers are prepared from Standard Reference Materials, SRM #186Ic (potassium dihydrogen phosphate) and SRM #186IIc (disodium monohydrogen phosphate), according to directions furnished with the materials, to give two buffer solutions: one at ionic strength of 0.1 mol/kg and pH 7.386 at 37 °C; the other at ionic strength of 0.1 mol/kg and pH 6.841 at 37 °C. A Sørensen buffer of ionic strength of ~0.17 mol/kg and pH 7.383 at 37 °C is sometimes used.

Calibration buffers meeting NBS specifications are commercially available, usually in containers of appropriate size and shape for mounting as a reservoir on the instrument. At pH 6.8 or 7.4 at 37 °C, the tolerance (SD) allowed should be less than or equal to

±0.003 in order to achieve SDs of ±0.005 to ±0.01 in measuring blood pH. Another independent buffer of as good or better quality should always be kept on hand and sampled as part of routine quality control in order to assure that buffers mounted on the instrument have not degraded because of heat or contamination with dust and bacteria. Unopened containers should be stored at room temperature.

***p*CO_2 and *p*O_2 measurement.** Calibration of the gas measurement system is made against gases of known O_2 and CO_2 composition admitted into the sample chamber. Compressed gases, with certificate of analysis provided by the manufacturer, are employed as primary standards. Pure O_2, CO_2, and N_2 may be obtained in individual tanks and mixed as desired with a precision gas mixer in the blood gas laboratory; gas mixtures of various fractional compositions may also be purchased. The "low gas" mixture for calibration usually has a fractional composition of 5% CO_2, 0% O_2, and 95% N_2; the "high gas" mixture has a fractional composition of 10% CO_2, 20% O_2, and 70% N_2. These compositions correspond roughly to a calibration range of 38–76 mm Hg for pCO_2 and 0–152 mm Hg for pO_2. When specimens with expected pO_2 of >200 mm Hg must be analyzed, 99.8% pure (medical) oxygen should be available for establishing calibration over the interval from 150 to ~700 mm Hg.

The mode of calibration is determined by the design of the instrument. Some instruments contain a barometer or a transducer responsive to ambient atmospheric pressure so that barometric pressure is always known to the microprocessor. With such instruments, only a keyboard entry of the fractional composition of O_2 and CO_2 in low- and high-calibrator gas mixtures needs to be made; the microprocessor then calculates the values of pO_2 and pCO_2, according to Dalton's law, for gases saturated with water vapor at 37 °C. For other instruments, the operator must read a free-standing barometer, manually calculate the set points for the calibrator gases, and then enter them into the instrument.

Liquid-gas difference of a p*O_2 *electrode. When calibration gases are used to standardize an instrumental system for the measurement of gases in blood samples, a particular property of the pO_2 electrode needs to be considered. This property is called the liquid-gas or blood-gas difference. pO_2 and pCO_2 electrodes are alike in that gas diffusing from liquid or gaseous sample passes through a gas-permeable membrane into the enclosed electrolyte solution in contact with measuring and reference elements. In both, the rate of diffusion of a gas through their membranes is slower from a liquid phase than from a gaseous phase. But pCO_2 and pO_2 electrodes use different principles of measurement (see Chapter 2, Section Four). CO_2 entering bicarbonate-electrolyte solution enters into a reversible equilibrium in which the change of H^+ activity (pH) in the electrolyte becomes a measure of pCO_2. The CO_2 is not consumed; therefore, the rate of its diffusion through the membrane is not limiting on the measurement. The pCO_2 electrode senses pCO_2 to the same extent whether the sample is liquid and diffusion is relatively slow or the sample is gaseous and diffusion relatively fast.

By contrast, O_2 entering the electrolyte solution enters an irreversible reaction at a polarized cathode so that current generated by the reaction is proportional to the amount of O_2 consumed at the cathode. The amount of O_2 available for consumption is dependent on the rate of diffusion of O_2 through the membrane; operation of a pO_2 electrode is therefore diffusion-limited. The electrode responds to a greater degree to O_2 diffusing from a gaseous phase than from a liquid phase.

The liquid-gas difference for a pO_2 electrode becomes significant when the electrode calibrated on gas is used to measure pO_2 of blood. The difference is usually expressed as a ratio of pO_2 (gas sample) to pO_2 (liquid sample). For contemporary electrodes, the ratio is commonly 1.02–1.06. For routine clinical work, a ratio of 1.04 is frequently assumed rather than determined. But for highest accuracy of blood pO_2 analysis, the ratio should be determined and applied as a correction factor to measured pO_2.

The factor is determined by measuring the pO_2 of a gas with a known content of O_2, and the pO_2 of a normal blood sample equilibrated against the same gas, and by calculating

$$\text{Factor} = \frac{pO_2\text{, equilibrating gas}}{pO_2\text{, equilibrated blood}}$$

The factor is then applied to correct measured blood pO_2, thus:

$$\text{Blood } pO_2\text{, corrected} = \text{Blood } pO_2\text{, measured} \times \text{factor}$$

Quality Control

Maintenance of instrumentation. The sophistication of contemporary equipment and availability of high-quality calibrator materials have made reliable and accurate determination of blood pH and gases primarily a matter of meticulous maintenance of the equipment and meticulous control of specimens. In many instances, monitor programs of the microprocessor display warning and diagnostic routines that alert the operator and then assist in the troubleshooting process. Nevertheless, regular maintenance of all aspects of the instrumentation, with close adherence to the manufacturer's recommended procedures, is essential to satisfactory operation. The frequency with which maintenance should be scheduled is in direct proportion to the number of blood gas analyses performed in the laboratory; in very active laboratories, certain maintenance procedures may have to be performed more than once in a 24-h period, but in every laboratory some maintenance is required every 24 h. The manufacturer's suggested schedule should be considered a guideline; often, experience will indicate the need to speed up the schedule.

The cleanliness of the sample chamber and path is especially important. Automatic flushing to cleanse the sample chamber and path after each blood sample measurement is a feature of many instruments. Despite proper flushing, however, clogging of chamber or path may occur. The frequency of clogging is related to the number of heparinized capillary blood samples that are analyzed. Fibrin threads and small clots may be present in the specimen or may form while the sample resides in the warm chamber. If allowed to remain, they will affect subsequent measurements or calibrations by interfering with contact of blood, buffers, or gases with electrode membranes or with KCl from the pH reference electrode, or by causing segmentation in the stream as the next blood sample is admitted. Cleaning should be performed by procedures recommended in the instrument's manual; the procedure may be as simple as passing a horsehair through the sample path or as drastic as dismantling the chamber and its accessory structures.

Control Materials

Quality control materials can be classed as tonometered whole blood for pCO_2 and pO_2 control, independent standard buffers for pH control, and commercial blood-base or blood-base-like fluids and aqueous fluids for pH, pCO_2 and pO_2 control. Pure and mixed gases are also needed for some control procedures.

Tonometered whole blood. A fresh blood specimen with a normal hemoglobin is preferred. When equilibrated against gas mixtures of primary standard quality, a tonometered sample measured on a properly calibrated and properly operating instrument should give measured values for pCO_2 and pO_2 that are within 1–4% of those calculated for the equilibrating gas. Quality control of pO_2 and pCO_2 by tonometered blood is considered by many workers the method of choice, since the control material most nearly approximates patients' samples in its interaction with gas electrodes. Tonometered blood is acknowledged to have greater sensitivity for detecting deterioration of the gas-permeable membranes of electrodes than do aqueous fluid controls. The disadvantages are the time required for tonometry, particularly for two or three levels of control; difficulties with availability of fresh, normal blood samples; the necessity for repeated calculation (and risk of miscalculation) of pCO_2 and pO_2 of equilibrating gases; the need to keep special gas mixtures for equilibration; and the inapplicability to direct control of pH measurement.

Standard buffers. The buffers used for control of pH measurement should be of a quality equal to or greater than that of the calibrator buffers. Buffer salts such as TES (2-[tris(hydroxymethyl)methylamino]-1-ethanesulfonic acid) or HEPES (N-2-hydroxyethylpiperazine-*N′*-2-ethanesulfonic acid) may be used to detect the failure of temperature control of the sample chambers.[4] Tris(hydroxymethyl)aminomethane and its hydrochloride are available as SRMs #922 and #923 from NBS. The temperature coefficient of a Tris buffer is $-0.026 \times K^{-1}$ (or $°C^{-1}$), almost 20 times that of phosphate buffers and almost 200 times that of whole blood. Desirable precision of pH measurements on control material is ±0.003 pH unit.

Blood-base (or blood-base-like) control material is a relatively recent development. One type of material (Quantra Plus, American Dade, Miami, FL 33152) consists of aldehyde-tanned human erythrocytes suspended in buffered medium and sealed in vials with a gas mixture of known O_2 and CO_2 content. Another (Prime, Fisher Scientific, Pittsburgh, PA 15219) contains human hemoglobin in a similar medium and package. A nonblood fluorocarbon material, with oxygen-carrying properties similar to those of blood, is also available (abc, Instrumentation Laboratories, Lexington, MA 02173). These products are made at three levels of pH, pCO_2, and pO_2. Prior to sampling, Quantra Plus must be incubated in the vials at 37 °C; all of them must be shaken vigorously for a prescribed length of time in order to equilibrate the gas and fluid phases. Unopened, these types of control materials have the advantage of long shelf life, and the buffered medium allows for control of pH along with that of pCO_2 and pO_2. Within-day CVs have been reported at 1–2% for both gases and pH.

Aqueous fluid control materials consist of buffered media sealed in vials with gas mixtures; the fluid is equilibrated with the gas by vigorous shaking by hand for a prescribed length of time immediately before the vial is opened and the contents sampled.

Gases: pure and mixed. Keeping a supply of gases on hand is a necessary part of quality control procedures such as tonometry. The quality of pure or mixed gases for these purposes should be as good as that of gases used for calibration. The kinds of pure gases or mixtures kept depend on their applications. Some laboratories invest in a precision gas mixer, but most purchase selected mixtures with certified composition and in small tanks.

Data obtained from assays of the various control materials may be handled in the same way as data from other clinical chemistry determinations; i.e., by calculation of mean, SD, and CV, and of control and confidence limits for construction of Levy-Jennings plots. Intangible but important features of quality control are the sixth sense of practiced operators for detecting subtle manifestations of deterioration of instrument performance and the suspicion of trouble expressed by physicians closely supervising care of individual patients monitored with frequent blood gas analysis.

Linearity of Electrodes

The linearity of new electrodes should be verified upon their installation. For the pCO_2 electrode, linearity can be tested after calibration with 5% and 10% CO_2, by determination of pCO_2 on a gas containing 7% CO_2. A similar test of an intermediate point of the calibration range for a pO_2 electrode calibrated with 0 and 20% O_2 might use 10% O_2, or for a pO_2 electrode calibrated in the extended range, 50% O_2. The gases used in the initial verification can subsequently be used for periodic checks in the quality control program. In some laboratories, checking linearity with tonometered whole blood samples is preferred to using gases.

Barometer

A barometer, whether built into the instrument or free-standing, should be periodically checked for accuracy. The most reliable reference is local atmospheric pressure as recorded by the nearest official meteorological station.

Temperature Control

Because exact temperature at 37 °C is essential to accurate measurement of blood gases and pH, state-of-the-art instrumentation has thermal sensors embedded in the heat sink around the measuring chamber and audible or visible alarms. Deviation of temperature outside of the preset tolerance (usually ±0.1 °C) is readily detected and must be corrected promptly.

Sources of Error

If analytical error is under control, the most probable source of error is in the collection and handling of the specimen. Among general causes of analytical error are calibration of the instrument with incorrect set points for pH buffers or calibrator gases, failure of temperature control of the measuring chamber, a dirty sample chamber or path, and inadequate mixing of the blood prior to sampling.

A failure of temperature control may occur when delivery tubing is pinched or kinked in instruments whose heat sinks are maintained by a circulating fluid system; temperature equilibration of the sample may become slow or chamber temperature may be slightly low. Dirt or clogs in sample spaces may also cause slow equilibration of the temperature of the sample but are more likely to interfere with the contact of sample with electrodes. Segmenting of the sample stream upon admission may place voids opposite the electrodes. Homogeneous distribution of blood cells through the sample is necessary because junction potentials of the electrodes are sensitive to the presence of erythrocytes.

A source of error for the pO_2 electrode lies in using it for measurements above ~500 mm Hg when it has not been calibrated over the high range. pO_2 measurements are particularly sensitive to temperature error; in order to keep systematic error to 1–2%, temperature control at 37 °C must be within ±0.1 °C. Failure to account for the liquid-gas difference of a pO_2 electrode can introduce an error of 2–6%. The performance of the electrode may be affected by gases other than O_2 present in the sample. The anesthetic gases halothane and nitrous oxide can both be reduced at the polarized cathode in competition with O_2; the effect is, however, small and ordinarily ignored.

Temperature Correction of Measured pH and pCO_2

In the Henderson-Hasselbalch equation, pK' and α are used as constants for a temperature of 37 °C. But the body temperature of a patient may, because of fever, be elevated to 40–41 °C, or the temperature of a patient made hypothermic for cardiopulmonary bypass surgery may decrease to as low as 26 °C. Most blood gas instruments, upon keyboard entry of the patient's actual temperature, can calculate and present temperature-corrected pH and pCO_2, as well as values derived from the temperature-corrected primary data. Algorithms used in some contemporary instruments have been listed in a recent publication.[2] For the constants used in the equations, $\Delta pH/\Delta T = 0.0147 \times K^{-1}$ (or $°C^{-1}$) and $\Delta \log pCO_2/\Delta T = 0.021 \times K^{-1}$ (or $°C^{-1}$) appear to be reasonable estimates over the range of 25–42 °C.[2] Their validity for calculating corrected values is not controversial.

There is a general consensus that routine correction of pH and pCO_2 values in hyperthermic states is unnecessary. The issue of correction for hypothermic states remains unresolved. Anesthesiologists, by choosing ventilation conditions and gas mixtures for the hypothermic patient, can force the pH and pCO_2 of the patient at 26–30 °C to the same values that are considered ideal at 37 °C. Clinical disagreement is focused on whether pH and pCO_2 values that are ideal at 37 °C are ideal at 26–30 °C. Prudent policy for the laboratory might be to generate and report temperature-corrected results only upon specific request of the physician. At no time, however, should temperature-corrected results be released unaccompanied by the original data acquired by the measurement at 37 °C.

Temperature Correction of Measured pO_2

There is no controversy with respect to desirability of correcting the pO_2 measured at 37 °C to the patient's body temperature. Correction should be made if the patient's temperature differs by more than 1 °C from 37 °C. The equations used are complex,[1,2] because at pO_2 < 100 mm Hg ($sO_2 \leq 0.95$) the hemoglobin-oxygen dissociation curve is shifted to the left by decrease in temperature as well as by the concomitant rise in pH (see Figure 19-3). At pO_2 > 100 mm Hg, the hemoglobin is fully saturated ($sO_2 = 1.00$) whatever the temperature.

Transcutaneous Monitoring of Blood Gases

Obtaining arterial, venous, or capillary blood is an invasive procedure. Repetitive sampling in intensive acute care management carries risks of infection and vascular complications. In premature infants particularly, repeated sampling imposes an undesirable blood loss and can alter blood gas values sufficiently to make them unreliable for clinical decisions. Transcutaneous (or percutaneous) monitoring of pCO_2 and pO_2 is noninvasive and can be continuously maintained for several hours at a time. Although its use does not totally replace measurements on blood, it allows real-time bedside management with fewer blood specimens and consequently with less risk and trauma. It has been widely and successfully utilized, primarily in neonatal care.[9]

Oxygen tension at the surface of the skin is very low, 2–3 mm Hg; carbon dioxide tension is even lower. But if the skin is warmed to 42–44 °C, permeability of the skin increases, and pO_2 and pCO_2 at the surface approach those of arterial blood. Electrodes and equipment are now available for percutaneous monitoring of both pCO_2 and pO_2 (Radiometer America, Westlake, OH 44145; Kontron Medical Instruments, Everett, MA 02149). Designs incorporate heating elements that warm the skin to 44 °C and that surround an electrode body whose gas-permeable membrane is held directly against the skin by self-adhesive portions of the assembly. Thermocouples in the assembly monitor the temperature of the heater and of the skin surface. Gases pass from the warmed surface through the membrane into an electrolyte solution in contact with the measuring and reference elements of the potentiometric circuit. The membrane for the O_2 electrode consists of two layers, one 12-μm-thick Teflon, the other 15-μm-thick polypropylene; the membrane for the CO_2 electrode is a single Teflon layer. Although the electrodes differ considerably in appearance from electrodes used on blood gas instrumentation, they operate on exactly the same electrochemical principle. The electrodes are usually applied to the chest or abdomen. Calibration is made against dry calibrator gases prior to applying the electrodes to the skin; because the current generation of electrodes may display a drift of up to 5%/h, recalibration at 3-h intervals has been recommended.[7] The microprocessor-based display monitor used with the electrodes can be fitted with a strip chart recorder for continuous recording of results.

Transcutaneous pCO_2 ($tc\text{-}pCO_2$) values can be up to 80% higher than corresponding capillary or arterial blood values;[7] a difference of 40–50% can be largely explained by the difference in the temperatures of measurement (44 vs. 37 °C). In infants, $tc\text{-}pO_2$ does not differ by more than 5% from arterial pO_2 or by more than 8% from capillary pO_2. In adults, $tc\text{-}pO_2$ is more dependent on cutaneous O_2 consumption and blood flow than it is in infants. A recent study,[10] however, shows ~5% greater $tc\text{-}pO_2$ than capillary pO_2. Transcutaneous gas monitoring becomes unreliable when blood pressure falls below −2 SD of normal and also when tolazoline (a pulmonary vasodilator) or anesthetics such as halothane or nitrous oxide are administered.

Clinical Significance of pO_2 Levels in Blood

Increased arterial oxygen tension, pO_2(aB), occurs upon breathing oxygen-enriched air; with administration of 100% O_2, the pO_2 may be as high as 640 mm Hg. Exercise causes an increase over the resting level.

Decreased arterial oxygen, *hypoxemia*, may occur at high altitudes, with exposure to carbon monoxide, or with improper anesthesia. General causes of hypoxemia can be classified as:

1. Decreased capacity for pulmonary diffusion, as for example due to respiratory distress syndrome in adults or newborns.

2. Decreased surface area of alveolocapillary membranes as the result of resection or compression of the lung.

3. Inequalities of ventilation/perfusion in the cardiopulmonary system because of bronchitis, asthma, emphysema, or obstruction of the airway by a neoplasm, foreign body, or secretions (e.g., croup).

4. Generalized alveolar hypoventilation of *peripheral origin* (e.g., as a result of suffocation or inadequate expansion of the chest or neuromuscular abnormalities that affect respiration, as for example phrenic nerve paralysis) or of *central origin* because of depression of the respiratory center by drugs such as barbiturates or morphine.

Clinical Significance of pCO_2 Levels in Blood

Increased blood pCO_2, *hypercapnia*, causes respiratory acidosis. pCO_2 rises with decreased alveolar ventilation due to diseases of the lungs or bronchial tree, to obstructions of the airway, or to breathing CO_2-enriched air. Depression of the respiratory center by drugs or impairment of the neuromuscular respiratory apparatus by decreasing the rate and depth of respiration may lead to retention of CO_2. In metabolic alkalosis, compensatory hypoventilation is responsible for CO_2 retention and increased pCO_2.

Decreased blood pCO_2, *hypocapnia*, is the cause of respiratory alkalosis; hypocapnia occurs upon increased alveolar ventilation as a result of too rapid rates of mechanically assisted respiration or of stimulation of the respiratory center to increase the rate and depth of respiration. The effect of altered CO_2 content of the blood on acid-base status is discussed in Section Three.

Clinical Significance of pH Levels in Blood

The pH range of blood compatible with life is 6.80–7.80. The normal arteriovenous difference is 0.01–0.03 but may be greater in cases of congestive heart failure or shock. Increase in blood or plasma pH above 7.45 is called *alkalemia*. Decrease in blood or plasma pH below 7.35 is called *acidemia*. Metabolic and respiratory alkalosis and acidosis are acid-base disorders in which blood pH may or may not be affected, depending on the degree of compensation.

Alterations of blood pH are discussed in conjunction with acid-base disorders in Section Three.

Reference Ranges

Reference ranges for ***arterial blood*** pO_2, sO_2, pCO_2, and pH are described in Chapter 29. Arterial blood pO_2 is low at birth but rises to an adult level of 83–108 mm Hg within 2 d. Values decrease with age.

Saturation fraction, sO_2(aB), may be as low as 0.40 at birth but thereafter is 0.95–0.98. $pO_2(0.5)$ corrected to pH 7.40 is 18–24 mm Hg for newborns and 25–29 mm Hg for adults.

Arterial blood $\boldsymbol{pCO_2}$ ranges, at sea level, are somewhat lower for infants than for adults; the range for adult males is 35–48 mm Hg, for adult females, 32–45 mm Hg. Values decrease with altitude above sea level at a rate of 3 mm Hg/km (5 mm Hg/mile). A physiologic change occurs with change in posture; pCO_2(aB) is 2–4 mm Hg higher for a sitting or standing subject than for one in the supine position. During pregnancy, pCO_2 falls gradually to a mean of ~28 mm Hg just prior to term.

Arterial blood pH, in the first few hours of life, may vary normally over a range of 7.09–7.50 but thereafter is 7.35–7.45.

References

1. Adams, A. P., and Hahn, C. E. W.: Principles and Practice of Blood Gas Analysis. London, Franklin Scientific Projects, Ltd., 1979.
2. Ashwood, E. R., Kost, G., and Kenny, M.: Temperature correction of blood gas and pH measurements. Clin. Chem., *29*:1877–1885, 1983.
3. Eichhorn, J. H.: Inaccuracy in blood gas/pH measurements caused by the blood sample. J. Med. Technol., *2*:23–26, 1985.
4. Elser, R. C., Stiler, J., and Garver, C.: A flexible and versatile program for blood-gas quality control. Am. J. Clin. Pathol., *78*:471–478, 1982.
5. Hutchison, A. S., Ralston, S. H., Dryburgh, F. J., et al.: Too much heparin: Possible source of error in blood gas analysis. Br. Med. J., *287*:1131–1132, 1983.
6. Joint Committee on Pulmonary Nomenclature, American College of Chest Physicians and American Thoracic Society: Pulmonary terms and symbols. A report of the AACP-ATS. Chest, *67*:583–593, 1967.
7. Kost, G. J., Chow, J. L., and Kenny, M. A.: Transcutaneous carbon dioxide for short-term monitoring of neonates. Clin. Chem., *29*:1534–1536, 1983.
8. Siggaard-Andersen, O. S., Durst, R. A., and Maas, A. H. J.: Physico-chemical quantities and units in clinical chemistry. International Federation of Clinical Chemistry and International Union of Pure and Applied Chemistry. Pure Appl. Chem., *56*:567–594, 1984.
9. Townsend-Collymore, L. A., and Buffone, G. T.: Transcutaneous pO_2 and pCO_2 monitoring in the newborn. Laboratory Management, April, 1985, pp. 41–46.
10. Wimberly, P. D., Pederson, K. G., et al.: Transcutaneous and capillary p_{CO_2} and p_{O_2} measurements in healthy adults. Clin. Chem., *29*:1471–1473, 1983.

Additional Reading

Maas, A. H. J., Kofstad, J., Siggaard-Andersen, O., et al., Eds.: Physiology and Methodology of Blood Gases and pH, Vol. 4. Proceedings of the 9th meeting of the IFCC expert panel on pH and blood gases, IFCC Workshop, Oslo, 1983. Copenhagen, Private Press, 1984.

National Committee for Clinical Laboratory Standards (NCCLS). Devices Measuring P_{O_2} and P_{CO_2} in Blood Samples (C21-P); Definitions of Quantities and Conventions Related to Blood pH and Gas Analysis (C12-T); Service of Clinical Laboratory Instruments (I6-A); Establishing Performance Claims for Clinical Chemical Methods, Introduction and Performance Check Experiment (EP2-T); Establishing Performance Claims for Clinical Chemical Methods, Replication Experiment (EP3-T); Establishing Performance Claims for Clinical Chemical Methods, Comparison of Methods Experiment (EP4-T); User Evaluation of Precision Performance of Clinical Chemistry Devices (EP5-T); National Committee for Clinical Laboratory Standards. Villanova, PA 19805.

SECTION THREE

Acid-base Balance and Acid-base Disorders

NORBERT W. TIETZ, Ph.D.
OLE SIGGAARD-ANDERSEN, M.D., Ph.D.
ELIZABETH L. PRUDEN, Ph.D.

Metabolic processes in the body result in the production of relatively large amounts of carbonic, sulfuric, phosphoric, and other acids. A person weighing 70 kg disposes daily of ~20 moles of carbon dioxide through the lungs and 70–100 millimoles of nonvolatile acids, such as sulfuric and phosphoric acids, through the kidneys. Lactic acid, β-hydroxybutyric acid, and acetoacetic acid are intermediary products that are normally metabolized to carbon dioxide and water before excretion. However, in some abnormal conditions such as diabetes mellitus, they may accumulate to a significant extent. The products of metabolism are transported to the excretory organs (lungs and kidneys) via the extracellular fluid and the blood with only minimal change in the plasma pH. This feat is accomplished by the combined functions of the *buffer systems* of the blood and mechanisms of the *respiratory* and *renal systems.*

Acid-base balance. In physiology, a component is *in balance* if the rates of input and output of this component are equal for a given time interval. A *positive balance* indicates that over a period of time there is a net gain of the component in the body, whereas a *negative balance* indicates that there is a net loss. Evaluation of the net balance must take into account all input (intake plus production) and all output (excretion plus metabolic conversion) over a given time interval. *Acid-base balance* involves CO_2 balance as well as the balance of noncarbonic acid and base and affects the *acid-base status* of the body fluids (i.e., the pH, the pCO_2, the concentration of titratable acid or base, and other acid-base variables).

Whole blood, plasma, serum, and urine are the main specimens used to evaluate the acid-base and electrolyte status of the body. It must be kept in mind, however, that the composition of the various body fluid compartments is different and that the movement of ions and other solutes between these compartments is not necessarily free. Results obtained on plasma or serum reflect predominantly the composition of extracellular fluid.

The pH of plasma is a function of two independent variables: the pCO_2, which is regulated by the respiratory mechanism, and the concentration of titratable base (base excess or deficit), which is regulated by the renal mechanism. Plasma bicarbonate concentration is often taken as a measure of the base excess or deficit in plasma and extracellular fluid.

The H^+ balance is closely associated with the balance of other electrolytes because H^+ cannot be introduced without a concomitant anion (e.g., Cl^-) or in exchange for a cation (e.g., Na^+). Table 19-3 gives the concentrations of the cation and anion charges for normal plasma. Note that there must be an exact electrical neutrality (i.e., the sum

Table 19-3. CONCENTRATIONS OF CATION AND ANION CHARGES IN PLASMA (EXPRESSED IN mmol/L)

Cation charges		Anion charges	
Na^+	142	Cl^-	103
K^+	4	HCO_3^-	27
Ca^{2+}	5	HPO_4^{2-}	2
Mg^{2+}	2	SO_4^{2-}	1
Others (trace elements)	1	Organic acids⁻	5
	154	Protein⁻	16*
			154

*The ion charge, mmol/L, for protein in serum of a healthy individual can be calculated by multiplying the total protein, g/dL, by the factor 2.41 (Van Slyke, D. D., Hastings, A. B., Hiller, A., et al.: Studies of gas and electrolyte equilibrium in blood. XIV. The amounts of alkali bound by serum albumin and globulin. J. Biol. Chem., *79*:769–780, 1928). Van Leeuwen (Acta Med. Scand. Suppl., *422*, 1–212, 1964), however, has found the following relationship: net negative charge (mEq/100 g protein) = 6.9 (pH − 4.96).

of negative and positive charges is zero). Any increase or decrease in the concentration of one anion is accompanied either by a compensating change of other anions or by a change of one or more cations, or both, so that total electrical neutrality is invariably maintained.

An acid is a substance that can donate protons (H^+) and a base is a substance that can accept protons (H^+).

Examples:

$$\textit{Acid} \rightleftharpoons H^+ + \textit{conjugate base}$$

$$HCl \rightleftharpoons H^+ + Cl^-$$

$$NH_4^+ \rightleftharpoons H^+ + NH_3$$

$$\text{Glycine} \rightleftharpoons H^+ + \text{glycinate}^-$$

$$\text{Glycinium}^+ \rightleftharpoons H^+ + \text{glycine (electrically neutral)}$$

Thus, HCl, NH_4^+, and amine ions (glycinium$^+$) are acids, whereas NH_3, glycinate$^-$, and free (electrically neutral) amines are bases. Some hydrogen-containing anions, such as HPO_4^{2-}, amino acids, and proteins, can act as acid or base:

$$\underset{\text{(acid)}}{H_2PO_4^-} \rightleftharpoons H^+ + \underset{\text{(conjugate base)}}{HPO_4^{2-}}$$

$$\underset{\text{(acid)}}{HPO_4^{2-}} \rightleftharpoons H^+ + \underset{\text{(conjugate base)}}{HPO_4^{3-}}$$

Concentrations of electrolytes in body fluids are commonly expressed in units of millimoles per liter (mmol/L).* In the case of polyvalent ions it is important to distinguish between the (substance) concentration of the ion (cI) and the (substance) concentration of the ion charge ($c\mathrm{I}_{1/z}$). Thus, when the concentration (c) of the total calcium ions in normal plasma is 2.5 mmol/L, the concentration of the total calcium ion charge is 5.0 mmol/L.

Electrolyte Composition of Body Fluids

The body has two main fluid compartments, the *intracellular* and the *extracellular*, the latter being further divided into blood plasma (vascular compartment) and interstitial

*It is still the custom in the United States to express electrolyte concentrations in terms of mEq/L (in the case of calcium, also in mg/dL). The use of these units is now discouraged.

Table 19-4. MASS (*m*) AND VOLUME (V) OF VARIOUS BODY COMPARTMENTS IN A 70-kg ADULT

	m, kg	V, L	*m*H_2O, kg	*m*Bone, kg	*m*Fat, kg
Interstitial fluid, cerebrospinal fluid, and bone	19	17	14	4.5	
Plasma	3.4	3.2	3.0		
Erythrocyte fluid	2.3	2.1	1.5		
Intracellular fluid	45.5	44	24.5		12
Total	70	66	43		

fluid. These compartments differ in composition in regard to water, electrolytes, and other components. The masses and volumes of the different body compartments are summarized in Table 19-4 (see also Figure 19-5).

Plasma has a volume of 1300–1800 mL/m² of body surface and constitutes ~5% of the body volume. Depending upon the protein and lipid content, the mass concentration of water in normal plasma is about 0.933 kg/L. The electrolyte composition of plasma is summarized in Table 19-3 and Figure 19-5.

Interstitial fluid is essentially an ultrafiltrate of blood plasma. Its volume makes up ~15% of the total body volume; however, when all extracellular spaces except plasma are included, the volume accounts for about 26% (17/66) of the total body volume (see Table 19-4). Plasma is separated from the interstitial fluid by the capillary endothelium,

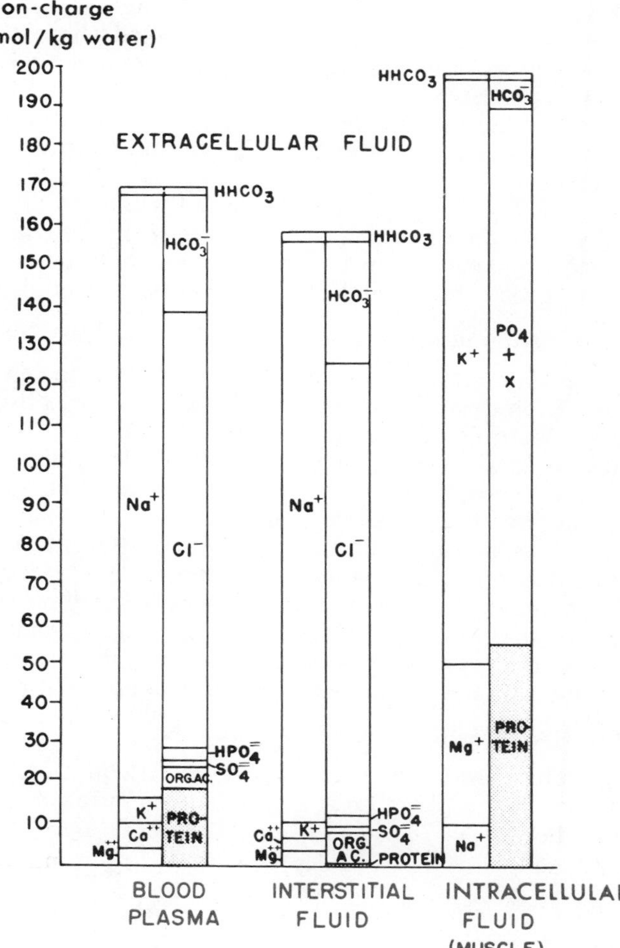

Figure 19-5. Electrolyte composition of blood plasma, interstitial fluid, and intracellular fluid. Note that the solutes are expressed in molality (i.e., amount of substance of solute divided by the mass of solvent). This form of expression is most meaningful when concentrations of solute in two aqueous solutions of widely differing water content are compared, since it is only in solvent that solutes are free to move and be osmotically active. When values are expressed in molality, the mass of solvent is kept constant at 1 kg. (After Gamble, J.: Chemical Anatomy, Physiology, and Pathology of Extracellular Fluid. Cambridge, Harvard University Press, 1950.)

which acts as a semipermeable membrane and allows passage of water and diffusible solutes, but not of compounds of large molecular mass such as proteins. However, this "impermeability" is not absolute.

The exact composition of ***intracellular fluid*** varies, depending on cell specialization, and is extremely hard to measure. Data on cell composition, therefore, are considered only approximations. The volume of intracellular fluids contributes ~66% to the total body volume.

Gibbs-Donnan equilibrium. Two solutions separated by a semipermeable membrane will establish an equilibrium in such a way that all ions are equally distributed in both compartments, provided that the solutions contain only solutes that can freely move through the membrane. At the state of equilibrium the total ion concentration and therefore the total concentration of osmotically active particles (osmolutes) are the same on both sides of the membrane.

If the solution on one side of a membrane contains ions that cannot move freely through the membrane (e.g., proteins), distribution of the diffusible ions at the steady state is unequal. However, *the product of the concentrations of the diffusible ions in one compartment is equal to the product of the concentrations of the diffusible ions in the other compartment (Gibbs-Donnan Law).* Furthermore, the law of electrical neutrality is obeyed for both compartments; i.e., the number of positive and negative charges in each compartment are in balance. The forces governed by the Gibbs-Donnan Law are responsible, for example, for the unequal distribution of chloride ions in serum versus spinal fluid.

Since the concentration of total particles in the compartment containing nondiffusible ions is greater than that of the other compartment, water tends to diffuse toward the more concentrated solution. This increase in pressure is due to the presence of the nondiffusible ions (colloids) and is called the *colloidal osmotic pressure.* The colloidal osmotic (oncotic) pressure is especially important in the body because it aids in the proper distribution of water throughout the various fluid compartments. Decrease in the colloidal osmotic pressure of plasma (e.g., owing to loss of protein in renal disease) results in a loss of water from the vascular compartment to the interstitial fluid, causing edema.

Body cells, which contain nondiffusible protein anions, can withstand only a limited difference in osmotic pressure across the cell membrane. Their cell membranes can correct concentration differences by excluding some small ions by active processes. If these processes cease, the cells gradually swell and eventually burst.

Distribution of Ions by Active and Passive Transport

Examination of Figure 19-5 reveals that the electrolyte compositions of blood plasma and interstitial fluid (both of which are extracellular fluids) are similar, but their compositions differ markedly from that of intracellular fluid. The major extracellular ions are Na^+, Cl^-, and HCO_3^-; in intracellular fluids the predominant ions are K^+, Mg^{2+}, organic phosphates, and proteins. The mechanism for this unequal distribution has so far not been adequately explained.

The Na^+ distribution is greatly different from the distribution predicted by the membrane potential and is due to an active transport of Na^+ from inside to outside the cell against an electrochemical potential gradient, a process that requires energy supplied by the metabolic processes in the cell (e.g., glycolysis). An active sodium pump deriving its energy from ATP is now thought to be present in most cell membranes, frequently coupled with a transport of K^+ in the opposite direction.

The Henderson-Hasselbalch Equation

The Henderson-Hasselbalch (H-H) equation aids in the understanding and explanation of the pH of body fluids. The equation derived on pages 629–630 can be written as follows:

$$\text{pH} = \text{p}K' + \log \frac{c\text{HCO}_3^-}{c\text{dCO}_2}$$

where pK' in plasma at 37 °C is 6.1 and $cdCO_2$ is the concentration of dissolved gas, which by definition includes also the low concentration of undissociated carbonic acid. The $cdCO_2$ is equal to $\alpha \times pCO_2$, where α is the solubility coefficient for CO_2. Dissolved CO_2 itself does not change the pH. However, dissolved CO_2 produces H^+ through this equilibrium:

$$dCO_2 + H_2O \rightleftharpoons H_2CO_3 \rightleftharpoons H^+ + HCO_3^-$$

The average normal ratio of the concentrations of bicarbonate to dissolved carbon dioxide in plasma is 25 (mmol/L)/1.25 (mmol/L) = 20/1, and the $\log_{10}$ of 20 is 1.3. The H-H equation applied to normal plasma can then be written:

$$pH = 6.1 + 1.3 = 7.4$$

It follows that any change in the concentration of either bicarbonate or dissolved CO_2 and, therefore, in the ratio of $cHCO_3^-/cdCO_2$ must be accompanied by a change in pH. Such changes in the ratio can occur through a change either in the numerator ($cHCO_3^-$) or in the denominator ($cdCO_2$).

Clinical conditions characterized as *metabolic* disturbances of acid-base balance are classified as primary disturbances in HCO_3^- concentration. Those characterized as *respiratory* disturbances are classified as primary disturbances in $cdCO_2$. Various compensatory mechanisms attempt to reestablish the normal ratio of $cHCO_3^-/cdCO_2$ and hence normal pH. The compensatory mechanisms may result in changes in the bicarbonate or dissolved CO_2 concentration. The application of the Henderson-Hasselbalch equation can be illustrated by the lever-fulcrum (teeter-totter) diagram (Figure 19-6).

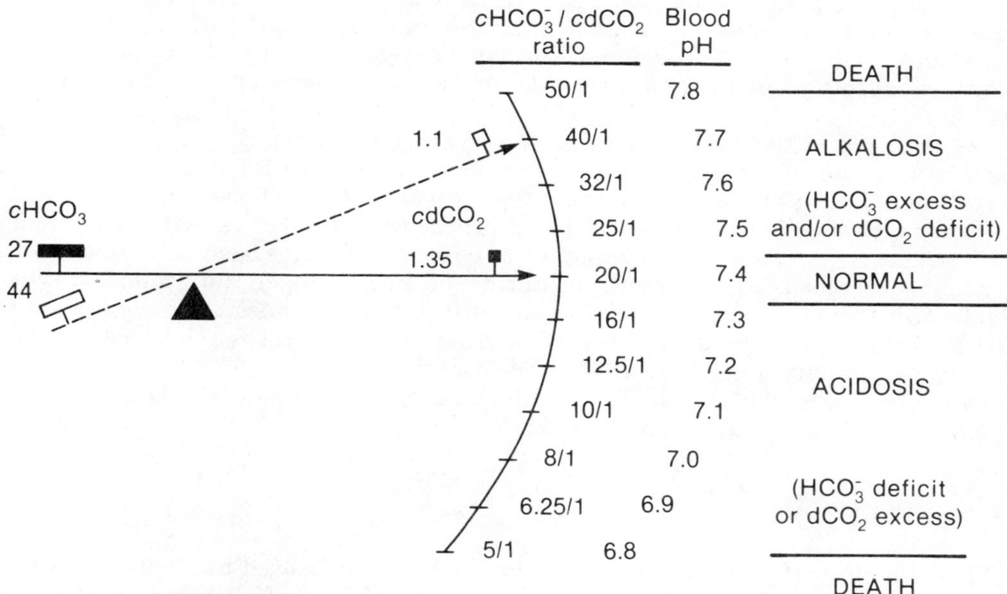

Figure 19-6. Scheme demonstrating the relation between pH and ratio of bicarbonate concentration to the concentration of dissolved CO_2. If the ratio in blood is 20/1 ($cHCO_3^-$ = 27 mmol/$cdCO_2$ = 1.35 mmol/L), the resultant pH will be 7.4 as demonstrated by the solid beam. The dotted line shows a case of uncompensated alkalosis (bicarbonate excess) with a bicarbonate concentration of 44 mmol/L and a dCO_2 concentration of 1.1 mmol/L. The ratio therefore is 40/1 and the resultant pH is 7.7. In a case of uncompensated acidosis, the pointer of the balance indicates a pH between 6.8 and 7.35, depending on the $cHCO_3^-/cdCO_2$ ratio. pH values below 6.8 or above 7.8 are incompatible with life. (After Weisberg, H. F.: Surg. Clin. N. Amer., *39*:93, 1959; Snively, W. D., and Wessner, M.: J. Ind. State Med. Assn., *47*:957, 1954.)

Acid-Base Parameters—Definitions and Abbreviations

Bicarbonate. Bicarbonate is the second largest fraction of the anions of the plasma (~25 mmol/L). Usually plasma "bicarbonate" includes true bicarbonate, carbonate, and CO_2 bound in plasma carbamino compounds. True bicarbonate is the largest component of the three, since at the pH of blood, the plasma carbonate concentration is only ~25 μmol/L and that of plasma carbamino compounds is 0.2 mmol/L. (In erythrocytes, it is 1.5 mmol/L.)

Dissolved CO_2 (dCO_2). The dissolved CO_2 fraction is defined to include both undissociated carbonic acid and physically dissolved, free CO_2. At the pH of the blood, the amount of dissolved CO_2 is 700–1000 times greater than the amount of carbonic acid, and therefore $cdCO_2$ is the term used to express their combined concentration. $cdCO_2$ is calculated from the solubility coefficient of CO_2 in blood at 37 °C ($\alpha = 0.0306 \times$ mmol $\times$ L^{-1} $\times$ mm Hg^{-1}). At a pCO_2 of 40 mm Hg, $cdCO_2$ is 1.224 mmol/L.

pCO_2, the partial pressure of CO_2. In the US, pCO_2 is generally expressed in mm Hg, but the IFCC recommends the use of kPa (kilopascal): 1 kPa = 7.5 mm Hg.

Total CO_2. The concentration of total CO_2 (formerly called CO_2 content) of blood consists of the sum of bicarbonate and dissolved CO_2 concentrations, as each of them is defined above. In equation form:

$$ctCO_2, \text{mmol/L} = cHCO_3^- + cdCO_2$$

Standard bicarbonate of blood. This term describes plasma $cHCO_3^-$ in blood that has been equilibrated with a gas mixture having pCO_2 = 40 mm Hg and pO_2 > 100 mm Hg at 37 °C. It is a reference value with which bicarbonate calculated from the actual pCO_2 of a blood sample can be compared.

Base excess. The base excess concentration is defined as the concentration of titratable base when the blood or plasma is titrated with a strong acid or base to a plasma pH of 7.40 at a pCO_2 of 40 mm Hg at 37 °C. Positive values (base excess) indicate a relative deficit of noncarbonic acid and suggest the presence of a metabolic alkalosis; negative values (base deficit) indicate a relative excess of noncarbonic acids and suggest the presence of a metabolic acidosis. The base excess value can be obtained from a nomogram or from algorithms programmed in the microprocessor of a blood gas instrument.

pH. The pH is the negative log of the hydrogen ion activity (pH = $-\log aH^+$). Thus, the average pH of blood (7.40) corresponds to a hydrogen ion concentration of 4×10^{-8} mol/L = 40 nmol/L, assuming that the activity coefficient of H^+ is 1.

Potentiometric determinations of blood pH measure the H^+ activity (aH^+) and not the H^+ concentration, and cH^+ is assumed to equal aH^+. It has been recommended that the acidity of blood be expressed in terms of its hydrogen ion concentration in nmol/L. This form of expression has the advantage that one does not have to deal with logarithms when performing acid base calculations (see below). The relationship between hydrogen ion activity and pH is illustrated in Figure 19-7.

To approximate cH^+ in blood, the following equation has been derived from the Henderson-Hasselbalch equation (page 629):

$$cH^+ = 24 \times \frac{pCO_2}{cHCO_3^-}$$

This is a clinically useful approximation of cH^+ and is used to identify the nature of acid-base disorders, since it relates pCO_2 and $cHCO_3^-$ to cH^+. However, calculated estimates are not a substitute for the actual determination of the blood acid-base values.

Buffer Systems and Their Role in Regulating the pH of Body Fluids

The action of buffers in the regulation of body pH can be explained by using the bicarbonate buffer system as an example. (For definition of buffers, see Chapter 1.) If

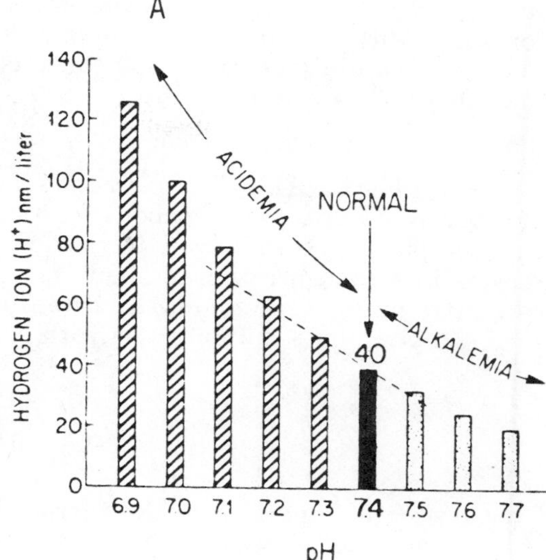

Figure 19-7. Relationship of pH to hydrogen ion concentration. *A,* Broken line is drawn to emphasize the (approximately) linear relationship between hydrogen ion concentration and pH over the pH range of 7.2–7.5. (Used by permission from Narins, R. G., and Emmett, M.: Simple and mixed acid-base disorders; a practical approach. Medicine, *59*:161–187, 1980.) *B,* Scale for the interconversion of pH and hydrogen ion concentration.

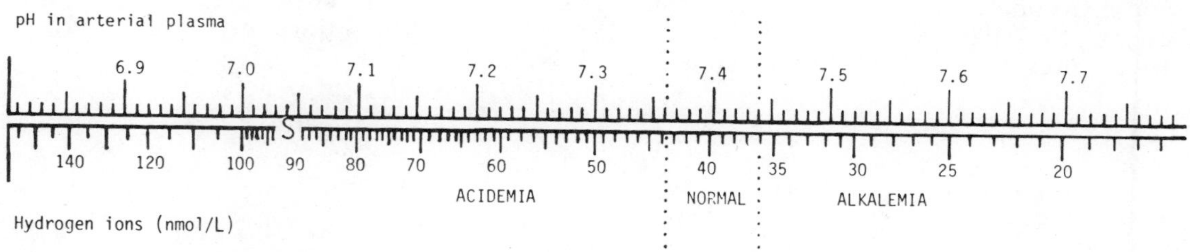

we add a strong acid to a solution containing HCO_3^-, the H^+ will react with HCO_3^- to form $HHCO_3$ and subsequently CO_2. The hydrogen ions are thereby bound, and the increase in the H^+ concentration will be minimized:

$$HCO_3^- + H^+ \rightarrow H_2CO_3 \rightarrow CO_2 + H_2O$$

On the other hand, if we add a base to the same buffer solution, the base will react with H_2CO_3 to form bicarbonate. The pH change, therefore, will be small:

$$HPO_4^{2-} + H_2CO_3 \rightleftharpoons H_2PO_4^- + HCO_3^- \qquad (4)$$

(Weak base)

$$OH^- + H_2CO_3 \rightleftharpoons H_2O + HCO_3^- \qquad (5)$$

(Strong base)

The buffer systems of greatest interest in regulation of the pH of body fluids are those of plasma and erythrocytes.

The Bicarbonate/Carbonic Acid Buffer System

The most important buffer of plasma is the bicarbonate/carbonic acid pair, whose effectiveness is based on its high concentration and on the fact that CO_2 can readily be disposed of or retained by the lungs. In addition, the renal tubules can increase or decrease the reclamation of bicarbonate from the glomerular filtrate (see page 656). In such an *open system,* where pCO_2 is constant, the buffer value* is normally ~57 mmol/L, compared with a value of ~2.7 mmol/L for a closed system in which $ctCO_2$ is constant. These values illustrate the importance of the bicarbonate buffer in an open system as it exists in the living organism.

The Phosphate Buffer System

At a plasma pH of 7.4, the ratio $cHPO_4^{2-}/cH_2PO_4^-$ is 80/20 ($pK' = 6.8$). Inorganic phosphate accounts for only about 5% of the total nonbicarbonate buffer value of plasma. Organic phosphate, however, in the form of 2,3-diphosphoglycerate (present in erythrocytes in a concentration of about 4.5 mmol/L) accounts for about 16% of the nonbicarbonate buffer value of erythrocyte fluid.

The phosphate buffer reacts with acids and with bases as follows:

$$HPO_4^{2-} + H^+ \rightarrow H_2PO_4^-$$

$$H_2PO_4^- + OH^- \rightarrow HPO_4^{2-} + H_2O$$

This system is important in the excretion of acids in the urine, as will be explained in the section on *The Renal Compensatory Mechanism.*

The Plasma Protein Buffer System

The proteins, especially albumin, account for the greatest portion (95%) of the nonbicarbonate buffer value of the plasma. The most important buffer groups of proteins in the physiological pH range are the imidazole groups of histidines (pK ~7.3), of which 16 are present for each albumin molecule:

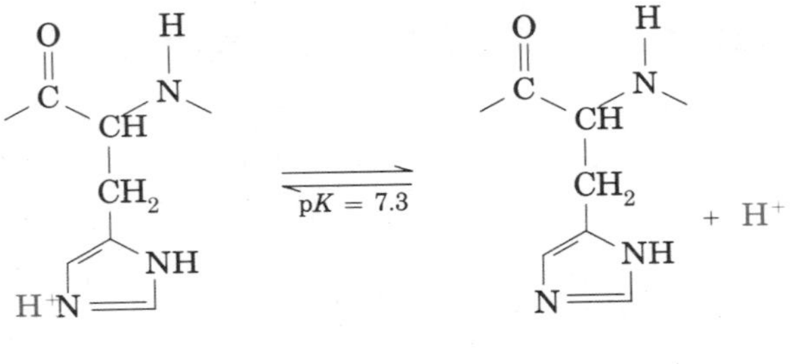

O H
C N
CH
CH₂ ⇌ (pK = 7.3) ... + H^+

(Structure: histidine residue with protonated imidazole ring (H⁺N, NH) in equilibrium, $pK = 7.3$, with deprotonated imidazole ring (N, NH) $+ H^+$)

The Hemoglobin Buffer System

Hemoglobin accounts for the major part of the buffering capacity of erythrocytes (the remainder is due mainly to 2,3-diphosphoglycerate). The imidazole groups of hemoglobin are quantitatively the most important buffer groups.

The buffer capacity of hemoglobin varies with pH and with oxygenation because the pK values of oxygen-linked acid-base groups decrease upon oxygenation of hemoglobin. This pK shift also causes a liberation of H^+ upon oxygenation of hemoglobin, a phenomenon called the *Haldane effect.*

The Isohydric and Chloride Shift

Because of the continuous production of carbon dioxide within the tissue cells, there is a downhill concentration gradient for carbon dioxide from cells to the plasma and the erythrocytes. This gradient causes a shift of physically dissolved carbon dioxide from the tissue cells into the plasma and erythrocytes (Figure 19-8). A small portion of the carbon dioxide entering the plasma stays as dissolved carbon dioxide, and another small portion reacts with water to form carbonic acid. The increased amount of H^+ is buffered by the plasma buffers, including the proteins (Figure 19-8, reaction 1). Another small

*The buffer value is defined as the amount of base required to cause a change in pH of one unit.

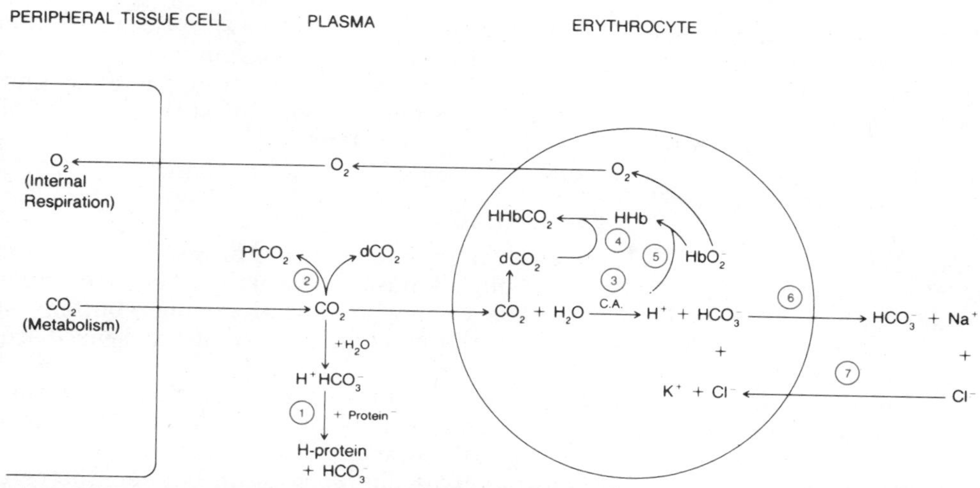

Figure 19-8. Scheme demonstrating the isohydric and chloride shift. The encircled numbers refer to the reactions described in the text. For details, see text.

portion combines with the amino groups of proteins and forms carbamino compounds ($PrCO_2$; Figure 19-8, reaction 2). Most of the carbon dioxide enters the erythrocytes and reacts with water to form carbonic acid. This reaction is catalyzed by the enzyme carbonic anhydrase (C.A.) and therefore proceeds at a relatively high speed (Figure 19-8, reaction 3). Some CO_2 remains as dissolved CO_2, and some combines with deoxyhemoglobin (HHb) to form carbamino hemoglobin ($HHbCO_2$; Figure 19-8, reaction 4).

The carbonic acid formed in reaction 3 would contribute to the H^+ concentration, but the pH change is fully or partially compensated by the release of oxygen from oxyhemoglobin (HbO_2^-) as the oxygen-linked acid-base groups accept the H^+ (Figure 19-8, reaction 5). This facilitates the release of O_2 from oxyhemoglobin. For each mole of O_2 given off, the hemoglobin binds about 0.5 mol of H^+. Furthermore, deoxyhemoglobin binds more CO_2 in the form of carbamino-CO_2 than does oxyhemoglobin, and thus a small but significant fraction of CO_2 is transported in this form. The oxygen released from HbO_2^- moves from the erythrocytes through the plasma into the peripheral tissue cells. These transformations (Figure 19-8, reactions 1–5) are called the *isohydric shift* (the shift in concentrations of components such that the concentration of H^+ remains essentially unchanged).

The remainder of the hydrogen ions formed in reaction 3 are buffered by the non-bicarbonate buffers of the erythrocyte fluid, whereas the concentration of HCO_3^- increases to the same extent that the concentration of Hb anions falls. The concentration of HCO_3^- increases relatively more in the erythrocytes than in the plasma, the pH of plasma falls relatively more than the pH of the erythrocytes, and, most importantly, the protein anion concentration in the erythrocytes falls owing to the decrease in protonation of proteins and hemoglobin. The membrane potential of the erythrocytes therefore becomes less negative and the distribution of all diffusible ions must change in accordance with the new membrane potential. HCO_3^- moves out of the erythrocytes and Cl^- moves into the erythrocytes to provide electrochemical balance. This shift of chloride ions is referred to as the *chloride shift* (Figure 19-8, reactions 6 and 7). There is a net increase of osmolutes inside the erythrocyte; therefore, a small amount of H_2O will pass from the plasma into the erythrocytes. As a result of these ion and water fluxes, the concentration of chloride in venous plasma is about 1 mmol/L lower than it is in arterial plasma, and the mean volume of an erythrocyte is about 1 femtoliter (fL) greater in venous blood than it is in arterial blood. This increase in erythrocyte volume increases the packed cell volume of arterial blood by ~0.4%, but the hematocrit of venous blood increases by 2–3% owing to loss of plasma water.

The venous blood is transported to the pulmonary capillaries, where the comparatively low pCO_2 in the alveoli causes a shift of carbon dioxide from the erythrocytes and the plasma into the alveoli. On the other hand, the high pO_2 in the alveoli causes a shift of oxygen into the plasma and the erythrocytes. This exchange causes a reversal of reactions 1–7 in Figure 19-8. The removal of carbon dioxide from the blood and the oxygenation of the blood are the reactions that convert venous blood into arterial blood.

The Respiratory Compensatory Mechanism in the Regulation of Acid-Base Balance

The respiratory mechanism contributes to the maintenance or restoration of normal body pH through retention of CO_2 in metabolic alkalosis or through increased elimination of CO_2 in metabolic acidosis. These processes also support the bicarbonate buffer system, as was explained earlier. The regulation of the respiratory system has been discussed in Section Two of this chapter.

The respiratory system responds immediately to a change in acid-base status, but about 3–6 h are required for the central and peripheral chemoreceptors to be fully stimulated. For example, in the early stages of metabolic acidosis plasma pH decreases, but because ions equilibrate rather slowly across the blood brain barrier, the cerebrospinal fluid pH remains nearly normal. The peripheral chemoreceptors are stimulated by the decrease in plasma pH, hyperventilation occurs, and plasma pCO_2 equilibrates rapidly across the blood brain barrier. Therefore, initially the pH of the extracellular fluid of the brain tends to rise and the central chemoreceptors are inhibited. But as plasma bicarbonate continues to fall, bicarbonate concentration in extracellular fluid of the brain also falls, and over the subsequent 3–6 h the extracellular fluid pH decreases to below normal. It is at this point that stimulation of respiration becomes maximal. The reverse is true when a patient is treated with HCO_3^-.

The Renal Compensatory Mechanism

The average pH of the glomerular filtrate, like that of plasma, is ~7.4; the average urinary pH of fasting individuals is ~6.0. This drop in pH is brought about by the kidneys, which excrete the nonvolatile acids produced by metabolic processes. The ability to excrete variable amounts of acid or base makes the kidney the final defense mechanism against any changes in body pH. The various acids produced during metabolic processes are buffered in the extracellular fluid at the expense of HCO_3^- (see *Bicarbonate/Carbonic Acid Buffer System,* page 651). Renal excretion of acid and conservation of HCO_3^- occur through several mechanisms: (1) reclamation of bicarbonate, (2) the Na^+–H^+ exchange, and (3) production of ammonia and excretion of NH_4^+.

Excretion of Acids

Strong acids such as sulfuric, hydrochloric, and phosphoric acids are fully ionized at the pH of urine and are excreted only after the H^+ derived from these acids reacts with a buffer base (e.g., HPO_4^{2-}, Figure 19-9,1; or NH_3, Figure 19-9,2), as will be explained below. Excretion of the anions of these acids is accompanied by the simultaneous removal of an equal number of cations, such as Na^+, K^+, or NH_4^+, to provide electrochemical balance. Some acids, such as acetoacetic acid ($pK = 3.58$) and β-hydroxybutyric acid ($pK = 4.7$), are present in blood almost entirely in ionized form but are partially nondissociated at the acid pH frequently prevailing in urine. The nondissociated form of the acid can be excreted as such (Figure 19-9,3).

Na^+–H^+ Exchange

In the renal tubules, (predominantly the proximal but also the distal tubules), hydrogen ions from the tubular cells can be exchanged for sodium ions from the tubular fluid—an energy-dependent process involving an ATPase (Figure 19-9,4). This process is enhanced in acidosis and decreased in alkalosis. The proximal tubules, however, cannot maintain a H^+ gradient of more than ~1 pH unit and the distal tubules cannot maintain one of more than ~3 pH units. Thus, maximum urine acidity is reached at pH ~4.4. Hydrogen ions of the glomerular filtrate or those that may have been transported into

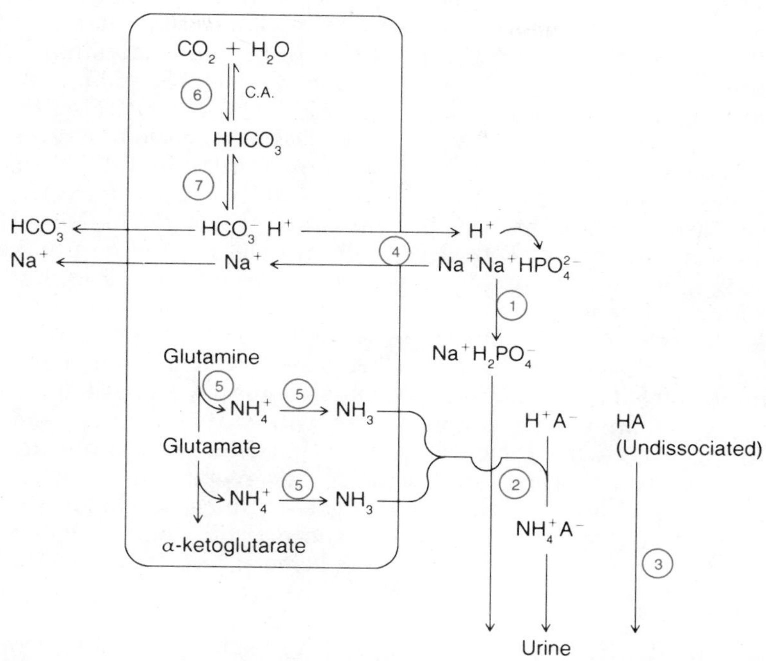

Figure 19-9. Hydrogen ion excretion, sodium hydrogen ion exchange, and ammonia production in the renal tubules. 1, Conversion of HPO_4^{2-} to $H_2PO_4^-$; 2, reaction of hydrogen ions with NH_3; 3, excretion of undissociated acids; 4, Na^+–H^+ exchange; 5, NH_3 production; 6 and 7, synthesis of carbonic acid from CO_2.

the tubular fluid through the Na^+–H^+ exchange may react with NH_3 or HPO_4^{2-}, as shown in Figure 19-9,1 and 2. In some forms of renal tubular acidosis, this exchange process is defective and may lead to a decrease in blood pH and an increase in urinary pH, as will be explained later.

Potassium ions compete with hydrogen ions in the Na^+–H^+ exchange. If the intracellular K^+ level of renal tubular cells is high, more K^+ and less H^+ are exchanged for Na^+; therefore, the urine becomes less acid and the acidity of body fluids increases. If there is K^+ depletion, more H^+ ions are exchanged for Na^+, and the urine becomes more acid and the body fluids more alkaline. Since the body's compensatory mechanism against metabolic alkalosis is relatively ineffective, K^+ depletion frequently results in a metabolic alkalosis (see *Metabolic Alkalosis,* page 661). Remember that serum K^+concentrations are not always an accurate indicator of the intracellular K^+ concentrations; in potassium depletion and in acidosis there is a tendency for K^+ to move from the cells into the serum. This movement will, at least temporarily, maintain normal serum K^+ concentrations despite intracellular K^+ depletion.

Renal Production of Ammonia and Excretion of Ammonium Ions

The renal tubular cells are able to generate ammonia from glutamine and other amino acids derived from muscle and liver cells according to the following reaction:[5]

$$\text{Glutamine} \xrightarrow[H_2O \;\to\; NH_4^+]{\textit{Glutaminase}} \text{Glutamate} \xrightarrow[NAD^+ \;\to\; NADH\; NH_4^+]{\textit{Glutamate dehydrogenase}} \alpha\text{-Ketoglutarate}$$

$$NH_4^+ \rightleftharpoons NH_3 + H^+$$

The ammonium ion produced dissociates into ammonia and hydrogen ions to a degree dependent on the pH (Figure 19-9,5). Ammonia, being a gas, diffuses readily across the cell membrane into the tubular lumen, where it combines with hydrogen ions to form ammonium ions (Figure 19-9,2). The NH_4^+ formed in the tubular lumen cannot easily cross cell membranes and thus is trapped in the glomerular filtrate and excreted with anions such as phosphate, chloride, or sulfate. In normal individuals, NH_4^+ production in the tubular lumen accounts for the excretion of ~60% (30–60 mmol) of the hydrogen ions associated with nonvolatile acids. The amount of H^+ excreted bound to NH_3 can be measured as NH_4^+. The H^+ required for NH_4^+ formation may be present in the glomerular filtrate or may be generated within the tubular cell through the synthesis of carbonic acid from CO_2, catalyzed by carbonic anhydrase (Figure 19-9,6 and 7). These hydrogen ions are secreted into the tubular lumen through the Na^+–H^+ exchange. In systemic acidosis, glutamine is released from muscle, glutamine uptake by the liver is decreased, and glutamine release is increased. More glutamine is therefore present in the circulation, and renal production of ammonia increases. In fact, under these circumstances, NH_4^+ excretion accounts for by far the greatest net excretion of H^+ by the kidneys. However, the maximum rate of glutamine release and therefore of NH_3 production is not achieved until acidosis has persisted for 3 d. In patients with chronic renal insufficiency, the kidneys are unable to generate sufficient NH_3 and this defect contributes significantly to the acidosis in such patients. Potassium depletion increases ammonia genesis; K^+ overload decreases it. This reciprocal relationship between the availability of K^+ and NH_3 production is of great clinical importance in those acid-base disturbances that are associated with mineralocorticoid deficiency or excess and in the pathogenesis of some forms of renal tubular acidosis (see Chapter 20).

Excretion of H^+ as $H_2PO_4^-$

H^+ secreted into the tubular lumen by the Na^+–H^+ exchange mechanism may react with HPO_4^{2-} to form $H_2PO_4^-$ (see Figure 19-9,1). This process depends on the amount of phosphate filtered by the glomeruli and on the pH of urine. Under normal physiological conditions, ~30 mmol of H^+ is excreted per day as $H_2PO_4^-$, and this amount accounts for ~90% of the titratable acidity of urine. In case of high-protein intake, phosphate production and subsequent filtration may be increased; after low-protein intake they may be decreased. Acidemia increases phosphate excretion and thus provides additional buffer for reaction with H^+. A decrease in the glomerular filtration rate, as is observed in renal disease, may result in a decrease of $H_2PO_4^-$ excretion.

The secretion of H^+ and its subsequent reaction with HPO_4^{2-} allow the removal of one H^+ without any significant decrease in urinary pH. As a result, the pH gradient is not greatly affected, more H^+ can be secreted into the tubules, and more Na^+ is reabsorbed and conserved. The HCO_3^- formed in the tubular cells as the result of H^+ secretion (Figure 19-9,7) reclaims the HCO_3^- consumed in buffering the body burden of acid.

Reclamation of Filtered Bicarbonate

The unmodified glomerular filtrate has the same concentration of HCO_3^- as does plasma; however, with increasing acidification of the proximal tubular urine, the urine pCO_2 increases and the HCO_3^- concentration decreases. It is believed that these changes are triggered by the excretion of H^+ by the Na^+–H^+ exchange mechanism, which results in a decrease in urinary pH. The H^+ thus excreted reacts with HCO_3^- to form H_2CO_3 and subsequently CO_2 and H_2O (catalyzed by carbonic anhydrase, C.A., in the brush border of the proximal tubular cells; Figure 19-10,1).

The increase in urinary pCO_2 causes carbon dioxide to diffuse across the tubular wall into the tubular cell, where it reacts with H_2O in the presence of carbonic anhydrase to form H_2CO_3 and subsequently H^+ and HCO_3^- (see Figure 19-10,2). Thus, reclamation of bicarbonate is in fact diffusion of CO_2 into tubular cells and its subsequent conversion to HCO_3^-. (The term *bicarbonate reabsorption,* which is often used for this process, is inappropriate.) This process helps to reclaim the filtered HCO_3^- and therefore to maintain or restore a normal pH in the general circulation. Normally nearly 90% of the HCO_3^- (or about 4500 mmol/d) is reclaimed in the proximal tubules, and the extent of HCO_3^-

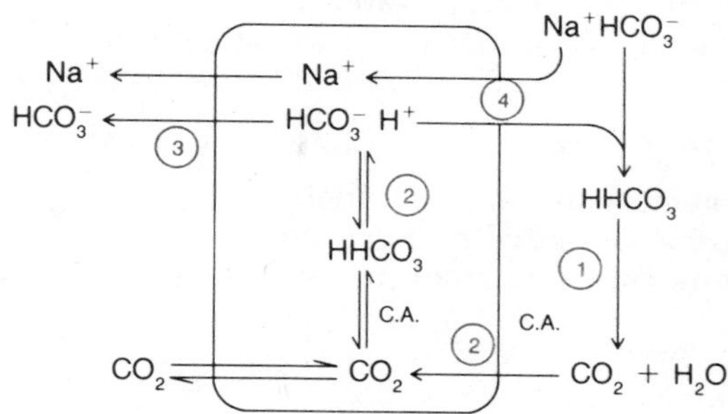

Figure 19-10. 1, Formation of CO_2 from bicarbonate in the tubular fluid; 2, formation of H^+ and HCO_3^- from CO_2 in the tubular cell; 3, new generation of HCO_3^-; and 4, NA^+–H^+ exchange.

reclamation parallels the Na^+ reabsorption. Thus, for each H^+ secreted into the tubular fluid, one Na^+ enters the tubular cell, one HCO_3^- is formed, and both return to the general circulation. In addition, about 50 mmol HCO_3^- per day is newly formed, mainly in the distal tubules.

The reclamation of HCO_3^- in the proximal tubules is a high-capacity, low-gradient process. Probably because of the fast dehydration of carbonic acid, the pH of tubular fluid at the ends of the proximal tubules is only about pH 6.7. When plasma HCO_3^- concentration increases above 26 mmol/L, the capacity of the proximal and distal tubules to reclaim HCO_3^- is exceeded and HCO_3^- is excreted in the urine. The process of bicarbonate reclamation is enhanced in acidosis (and is decreased in alkalosis), most likely as a result of increased Na^+–H^+ exchange.

Conditions Associated with Abnormal Acid-Base Status and Abnormal Anion-Cation Composition of the Blood

Many pathological conditions are accompanied by or caused by disturbances of the acid-base balance and electrolyte composition of the blood. These changes are usually reflected in the acid-base and anion-cation patterns of extracellular fluid. Because blood is the only sample of extracellular fluid that can be readily obtained, acid-base status is evaluated from measurements of blood pH and pCO_2 and determination of $cHCO_3^-$ or $ctCO_2$. It is important to note, however, that results obtained on blood or plasma may not always be a true indication of the acid-base status of *intra*cellular fluid.

Abnormalities of acid-base balance of the blood are usually accompanied by characteristic changes in electrolyte concentrations in the plasma, especially in the case of the metabolic acid-base disorders. Acid-base disturbances are classified into one of the following four types: metabolic acidosis, metabolic alkalosis, respiratory acidosis, or respiratory alkalosis.

Metabolic Acidosis (Primary Bicarbonate Deficit)

Metabolic acidosis is usually caused by one or more of the following processes:

1. Production of organic acids that exceeds the rate of elimination (e.g., production of acetoacetic acid and β-hydroxybutyric acid in diabetic acidosis and of lactic acid in lactic acidosis). Acidosis may be accompanied by loss of cations that are excreted with the anions.
2. Reduced excretion of acids (e.g., in renal failure, tubular acidosis).
3. Excessive loss of bicarbonate (base) because of excessive loss of duodenal fluid (as in diarrhea). Plasma $cHCO_3^-$ falls and is replaced by anions such as chloride, phosphate, sulfate, or organic anions.

When any of these conditions exists, the ratio of $cHCO_3^-/cdCO_2$ is decreased because of the primary decrease in bicarbonate. Compensatory mechanisms then restore the ratio and the pH to normal. If this restoration is complete (or nearly complete so that the pH remains between 7.35 and 7.45), the condition is called *fully compensated metabolic acidosis.* If, in spite of the compensatory mechanisms, the pH stays below 7.35, the condition is called *partially compensated metabolic acidosis.*

Classification and Selected Examples of Metabolic Acidosis

The various forms of metabolic acidosis can be classified as *normal anion gap acidosis*, which is characterized by hyperchloremia, or *increased anion gap acidosis*, in which the level of chloride is normal (normochloremic metabolic acidosis) or at times low.

Normal Anion Gap Acidosis

Normal anion gap acidosis can be divided into normokalemic and hypokalemic acidosis.

Normokalemic acidosis is seen in early renal failure; in hydronephrosis; after treatment with NH_4Cl, arginine-HCl, or similar therapeutic agents; and in hypoaldosteronism or after administration of aldosterone antagonists such as spironolactone, triamterene, or amiloride. (After amiloride, a hyperkalemia may be present.) ***Hypokalemic acidosis*** includes renal tubular acidosis caused either by proximal or distal tubular dysfunction or by buffer deficiency; acidosis caused by gastrointestinal loss of Na^+, K^+, and HCO_3^-; acidosis due to diuretics that are carbonic anhydrase inhibitors, such as acetazolamide and mafenide; and acidosis caused by ureteral diversions such as ureterosigmoidostomy, or obstructed ileal bladder or ureter.

Diarrhea may cause acidosis as a result of loss of Na^+, K^+, and HCO_3^-; the resulting hyperchloremia is due to the replacement of lost bicarbonate with Cl^- to maintain electrochemical balance.

Renal tubular acidosis is predominantly characterized by loss of bicarbonate due to decreased tubular secretion of H^+ (see Chapter 20). Lost HCO_3^- is replaced by chloride, resulting in hyperchloremic acidosis.

Carbonic anhydrase inhibitors cause wasting of Na^+, K^+, and HCO_3^-. Posthypocapnic acidosis may develop, but it rarely lasts more than 6–12 h and rarely has severe consequences.

Hyperkalemic acidosis inhibits renal ammonia formation and therefore elimination of H^+. If associated with increased extracellular fluid (ECF) volume, HCO_3^- reclamation in the tubules may also be depressed. Failure of the kidneys to synthesize renin, failure of the adrenal cortex to secrete aldosterone, and renal tubular resistance to aldosterone are the most common causes of this type of acidosis.

Increased Anion Gap Acidosis

Renal failure may result in decreased glomerular filtration rate (GFR) and loss of functional renal tubular mass with consequent decrease in ammonia formation and Na^+–H^+ exchange. All result in decreased acid excretion (see Chapter 20). Acidosis usually develops if GFR falls below 20 mL/min and when urea nitrogen increases to >40 mg/dL and creatinine to >4 mg/dL.

Ketoacidosis. The pathogenesis of this disorder has been discussed in Chapter 13. Recall that the accumulation of ketone bodies causes a decrease in HCO_3^-, a normal or low serum chloride, and a high anion gap.

Salicylate intoxication generally occurs with blood salicylate concentrations above 30 mg/dL. Initially the drug stimulates the respiratory center to increase the rate and depth of respiration, resulting in a low pCO_2, low HCO_3^-, and respiratory alkalosis. Salicylates, however, also alter peripheral metabolism, causing production of various organic acids without dominance of any specific acid. The processes eventually lead to a metabolic acidosis with a high anion gap. In adults, mixed respiratory alkalosis and metabolic acidosis are more common, whereas in children metabolic acidosis predominates.

Lactic acidosis. Lactic acid ($pK_1 = 3.86$) is an intermediate of carbohydrate metabolism derived mainly from muscle cells and erythrocytes and is normally metabolized by the liver. It is present in blood entirely as lactate ion (see also Chapter 13). Any increase in the concentration of lactate and the associated H^+ to >2 mmol/L after resting leads to a condition called *lactic acidosis.* During exercise, lactate levels may increase significantly; however, pyruvate levels increase under these conditions as well, and the ratio of lactate/pyruvate remains ~6 or 7/1 (see below).

Type A lactic acidosis is caused by severe tissue hypoxia, secondary to severe anemia, shock, cardiac decompensation, or pulmonary insufficiency. Severe oxygen deprivation of tissues is associated with a significant increase in the lactate/pyruvate ratio in blood, marked hyperpnea, weakness, fatigue, stupor, and finally coma. Conditions at these stages are frequently irreversible.

Type B lactic acidosis is caused by drugs such as biguanides and streptozotocin; by acquired and hereditary defects in enzymes involved in gluconeogenesis; by disorders such as severe acidosis, uremia, liver failure, tumors, and seizures; and during anesthesia.

Toxins and other chemical agents can also be the cause of metabolic acidosis, as seen in the following examples. *Methanol,* although not toxic itself, is metabolized to formaldehyde and formic acid. Accumulation of this acid leads to metabolic acidosis with a high anion gap. *Ethylene glycol,* if ingested, is metabolized to glycolic and oxalic acids as well as other acidic metabolites, any of which lead to an acidosis with a high anion gap. Accumulation of the metabolites may lead to lactic acid production and further contribute to the acidosis. Precipitation of calcium oxalate and hippurate crystals in the urinary tract may lead to acute renal failure. *Paraldehyde* toxicity may develop after chronic paraldehyde ingestion. The pathogenesis is ill defined, but the acidosis may actually be a ketosis with β-hydroxybutyric acid as the main acidic product.

Compensatory Mechanisms in Metabolic Acidosis

Buffer systems. Bicarbonate is the principal buffer involved in minimizing changes in pH. Bicarbonate is consumed in its reaction with hydrogen ions, and the decrease in its concentration reduces the ratio of $cHCO_3^-/cdCO_2$ to <20/1.

Respiratory compensatory mechanisms. The decrease in pH in metabolic acidosis stimulates the respiratory mechanism and produces hyperventilation, which results in the elimination of carbonic acid as CO_2, a decrease in pCO_2 (hypocapnia), and consequently a decrease in $cdCO_2$. Thus, the ratio of $cHCO_3^-/cdCO_2$ approaches its normal value of 20/1 (e.g., $cHCO_3^-/cdCO_2$ before compensation = 15/1.2 or 12.5/1 → pH 7.2; after partial compensation = 14.5/0.86 or 16/1 → pH 7.3).

Renal compensatory mechanism. The kidneys partially restore the normal pH by increased excretion of acid and conservation of base (increased rate of Na^+–H^+ exchange, increased ammonia formation, and increased reclamation of bicarbonate). As a result, the primary deficit of $cHCO_3^-$ will be ameliorated (e.g., to 22/1.1 or 20/1 → pH 7.40). This is a fully compensated metabolic acidosis. It is so named because the pH has returned to normal; however, acidosis still exists because a primary $cHCO_3^-$ deficit persists.

Laboratory Findings in Metabolic Acidosis

Plasma bicarbonate, total carbon dioxide content, and blood pCO_2 are decreased in metabolic acidosis (Table 19-5). In the uncompensated state, blood pH is decreased to a degree depending on the ratio of $cHCO_3^-/cdCO_2$. Note that in a mixed acid-base disorder such as a respiratory acidosis superimposed on a pre-existing metabolic acidosis, the pCO_2 will be higher than expected for a given degree of metabolic acidosis.

Electrolytes are also altered in various ways, depending on the cause of metabolic acidosis. In *diabetic ketoacidosis* (Figure 19-11,*B*) the increased organic acids (ketone bodies) are reflected by a decrease in plasma bicarbonate and sometimes in chloride. Plasma Na^+ and K^+ are also decreased because of the associated polyuria and coexcretion of these cations with acetoacetate and β-hydroxybutyrate. Furthermore, there is a dilutional effect as a result of the osmotically induced increase in the vascular volume. When glucose concentration decreases as a result of insulin treatment, water leaves the vascular compartment and Na^+ concentration increases. Serum potassium levels, however, may be normal or even high, despite severe total body depletion of K^+; the serum level represents a balance struck between the amount of K^+ lost in the urine,

Table 19-5. CLASSIFICATION AND CHARACTERISTICS OF SIMPLE ACID-BASE DISORDERS

	Primary Change	Compensatory Response	Expected Compensation
METABOLIC			
Acidosis	↓↓↓$cHCO_3^-$	↓↓pCO_2	pCO_2 = 1.5 ($cHCO_3^-$) + 8 ± 2 pCO_2 falls by 1–1.3 mm Hg for each mmol/L fall in $cHCO_3^-$ Last 2 digits of pH = pCO_2 (e.g., if pCO_2 = 28, pH = 7.28) $cHCO_3^-$ + 15 = last 2 digits of pH ($cHCO_3^-$ = 15, pH = 7.30)
Alkalosis	↑↑↑$cHCO_3^-$	↑↑pCO_2	pCO_2 increases 6 mm Hg for each 10 mmol/L rise in $cHCO_3^-$ $cHCO_3^-$ + 15 = last 2 digits of pH ($cHCO_3^-$ = 35, pH = 7.50)
RESPIRATORY			
Acidosis			
Acute	↑↑↑pCO_2	↑$cHCO_3^-$	$cHCO_3^-$ increases by 1 mmol/L for each 10 mm Hg rise in pCO_2
Chronic	↑↑↑pCO_2	↑↑$cHCO_3^-$	$cHCO_3^-$ increases by 3.5 mmol/L for each 10 mm Hg rise in pCO_2
Alkalosis			
Acute	↓↓↓pCO_2	↓$cHCO_3^-$	$cHCO_3^-$ falls by 2 mmol/L for each 10 mm Hg fall in pCO_2
Chronic	↓↓↓pCO_2	↓↓$cHCO_3^-$	$cHCO_3^-$ falls by 5 mmol/L for each 10 mm Hg fall in pCO_2

Modified from Narins, R. G., and Gardner, L. B.: Simple acid-base disturbances. Med. Clin. North Am., *65*:321–346, 1981.

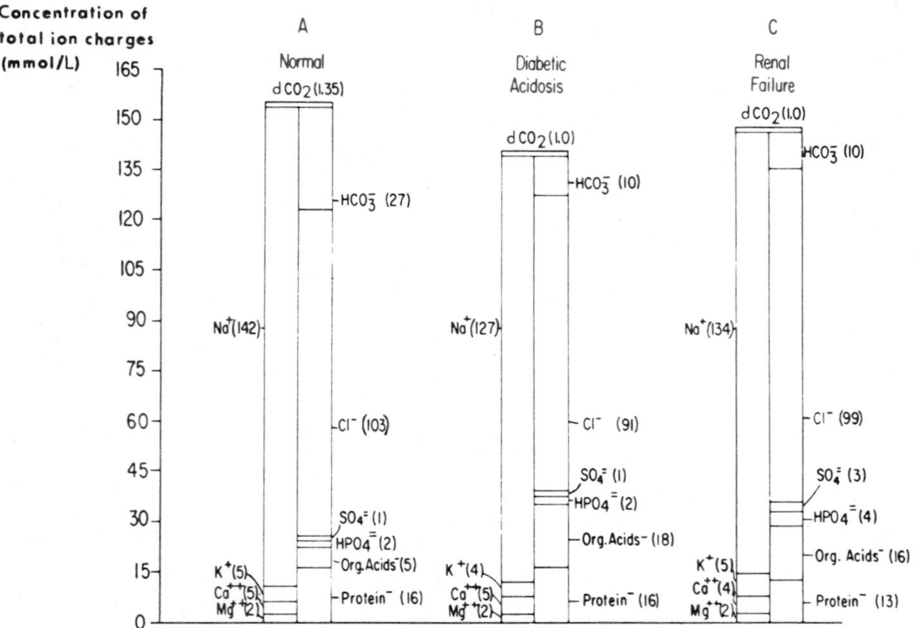

Figure 19-11. *A,* Gamblegram illustrating normal electrolyte composition of plasma. *B,* Example of anion-cation pattern as may be found in diabetic acidosis. Na^+ and Cl^- are decreased because of polyuria and osmotically induced hypervolemia due to increased plasma glucose. The organic acid fraction is increased because of excessive formation of ketone bodies. The ratio of HCO_3^-/dCO_2 in this example is 10/1; thus, the plasma pH must be 7.10. *C,* Example of anion-cation pattern as may be seen in renal failure. Organic acids as well as phosphates and sulfates are retained owing to decreased renal function. (From Tietz, N. W.: The Chicago Medical School Quarterly, *22*:156, 1962.)

the amount of K^+ shifted from cells into extracellular fluid, and the degree of dehydration. In *renal failure* (Figure 19-11,*C*) organic acids, phosphate, and sulfate are increased because of retention. Chronic metabolic acidosis also enhances mobilization of calcium from bone, and the decrease in plasma pH increases dissociation of plasma protein–bound calcium. Thus, more Ca^{2+} is filtered through the glomeruli and less is reabsorbed in the tubules. Other types of electrolyte changes are described below in the discussions of specific acidoses and in the section on renal tubular acidosis in Chapter 20. As long as the renal compensating mechanisms are functioning, urinary acidity and levels of urinary ammonia are increased.

Metabolic Alkalosis (Primary Bicarbonate Excess)

Metabolic alkalosis is most frequently caused by one of the following:

1. Administration of excess alkali, such as $NaHCO_3$, citrate (in transfused blood), or some antacids.
2. Excessive loss of hydrochloric acid from the stomach after prolonged vomiting, in pyloric or high intestinal obstruction, and after gastric suction.
3. Potassium depletion in Cushing's syndrome, hyperaldosteronism, after intake of licorice and carbenoxolone, or because of low K^+ intake. In these conditions, renal Na^+ reabsorption, H^+ secretion, and NH_3 production are enhanced secondary to K^+ depletion. This enhanced secretion of H^+ secondarily increases HCO_3^- reclamation. Potassium plays a less critical role in non-mineralocorticoid–induced alkalosis since hypokalemia suppresses aldosterone secretion and thus removes its stimulatory effect on H^+ secretion.
4. Renal bicarbonate retention, as seen in hypovolemia-induced aldosterone stimulation, which causes an increased reabsorption of Na^+ together with HCO_3^- or Cl^-.
5. Prolonged administration of certain diuretics. Diuretics causing alkalosis block sodium absorption by the various portions of the renal tubules. They include organomercurials; carbonic anhydrase inhibitors such as acetazolamide (Diamox) and dichlorphenamide; and both furosemide (Lasix) and ethacrynic acid. The resulting increase in the Na^+ concentration in the tubular fluid reaching the distal convoluted tubule stimulates aldosterone secretion and therefore secretion and loss of K^+.
6. After administration of laxatives, carbenoxolone, and potassium-poor intravenous fluids.

In these conditions, the ratio of $cHCO_3^-/cdCO_2$ becomes $>20/1$ because of a primary increase in bicarbonate. The compensatory mechanisms of the body then act to restore the normal plasma pH. With progression of the disturbance, the compensatory mechanisms are not effective enough, and the pH will increase. If the increase in pH is great enough, increased neuromuscular activity may be seen as ionized calcium becomes bound to protein or other anions; above pH 7.55, tetany may develop.

Compensatory Mechanisms in Metabolic Alkalosis

Buffer systems. Excess base reacts with the $HHCO_3$ of the $HHCO_3/HCO_3^-$ buffer system to form an increased amount of HCO_3^-, thereby minimizing pH change.

Respiratory mechanism. The increase in pH depresses the respiratory center, causing a retention of carbon dioxide (hypercapnia). Thus, the ratio of $cHCO_3^-/cdCO_2$ approaches its normal value although the actual levels of both $cHCO_3^-$ and $cdCO_2$ are increased. The respiratory response to metabolic alkalosis is erratic, and increases in pCO_2 are variable.

Renal mechanism. The kidneys respond to the state of alkalosis by decreased Na^+–H^+ exchange, decreased formation of ammonia, and decreased reabsorption of bicarbonate.

Laboratory Findings in Metabolic Alkalosis

Blood plasma values for $cHCO_3^-$, $cdCO_2$, and pCO_2 and therefore the plasma total CO_2 concentration are increased, and the ratio of $cHCO_3^-/cdCO_2$ is high. In uncomplicated metabolic alkalosis, the pCO_2 increases by ~6 mm Hg for each 10 mmol/L rise in $cHCO_3^-$. A higher than expected pCO_2 may indicate superimposed respiratory acidosis.

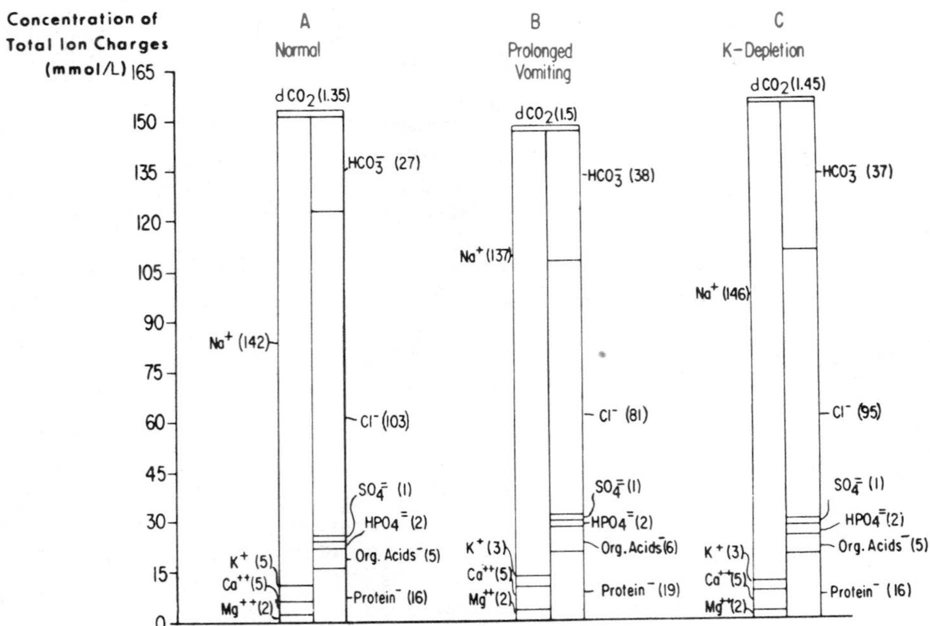

Figure 19-12. *A*, Electrolyte composition of normal plasma. *B*, Example of plasma electrolyte composition after prolonged vomiting, showing the decrease in K^+ and Cl^- and the increase in protein$^-$, organic acids$^-$, HCO_3^-, and dCO_2. The ratio of the last two is 38/1.5 = 25.3. The pH is therefore 7.50. *C*, Example of typical plasma electrolyte composition in a patient with intracellular K^+ depletion. There is a decrease in K^+ and Cl^- and an increase in HCO_3^- and dCO_2, resulting in a ratio of 37/1.45 = 25.4 (pH = 7.50). (From Tietz, N. W.: The Chicago Medical Quarterly, *22*:156, 1962.)

Other electrolyte ions vary, depending on the condition. In cases of prolonged vomiting (Figure 19-12, *B*), Cl^- and possibly K^+ levels are low because of the loss of these ions through the vomitus. Protein values may be increased owing to dehydration. If food intake is inadequate, formation of ketone bodies may increase the organic acid fraction. In cases of excessive administration of $NaHCO_3$, Na^+ levels are increased. In K^+ depletion (Figure 19-12,*C*) decreased concentrations of Cl^- are common. Serum K^+ concentrations are generally but not necessarily low, since loss of about 100–200 mmol is required before serum K^+ concentrations decrease below normal.

Urinary pH values are usually increased because of the decreased excretion of acid and increased excretion of bicarbonate. Urinary ammonia values are decreased because of decreased formation of ammonia in the tubules. In K^+ depletion, H^+ is preferentially exchanged for Na^+ and the pH of the urine may be low in spite of a metabolic alkalosis; this is called paradoxical aciduria.

Respiratory Acidosis (Primary dCO_2 Excess)

Any condition that decreases elimination of carbon dioxide through the lungs results in an increase of pCO_2 (hypercapnia) and respiratory acidosis. Increase in pCO_2 results in an increase of $cdCO_2$, which in turn causes a decrease in the $cHCO_3^-/cdCO_2$ ratio. Causes of decreased CO_2 elimination can be classified as those caused by factors that directly depress the respiratory center (such as centrally acting drugs, CNS trauma, and infections) or those that affect the respiratory apparatus or cause mechanical obstruction of the airways. Cardiac disease may also cause respiratory acidosis, although generally it causes a slight alkalosis because the hypoxemia stimulates hyperventilation. Rebreathing or breathing of air high in CO_2 content may also cause high pCO_2.

Compensatory Mechanisms in Respiratory Acidosis

Buffer systems. Carbonic acid present in blood in excess is buffered to a great extent by the hemoglobin and protein buffer systems. This process results in a slight rise in $cHCO_3^-$.

Respiratory compensatory mechanisms. The increase in pCO_2 stimulates the respiratory center and results in increased respiratory rate and depth of respiration, provided that the primary defect is not in the respiratory center. The respiratory response is proportional to the degree of acidosis and is therefore fairly predictable.

Renal compensatory mechanisms. The kidneys respond to respiratory acidosis in the same way that they do to metabolic acidosis; namely, with increased Na^+–H^+ exchange, increased ammonia formation, and increased reclamation of bicarbonate. In a chronic respiratory acidosis in steady state, the normal renal compensation returns the plasma pH about halfway toward normal. A so-called fully compensated chronic respiratory acidosis, in which the pCO_2 is high but the pH is within the reference range, is actually not the result of a true compensation but is generally the result of a chronic respiratory acidosis with a superimposed metabolic alkalosis arising from some other cause. Renal compensation is not effective before 6–12 h and is not optimal before 2–3 d.

Laboratory Findings in Respiratory Acidosis

Before compensation, pCO_2 and therefore $cdCO_2$ are increased. After compensation, $cHCO_3^-$ and therefore $ctCO_2$ are increased as well (Figure 19-13, *B*). Due to an increase in $cdCO_2$, the ratio of $cHCO_3^-/cdCO_2$ is decreased, resulting in a decreased pH. In the acute phase, $cHCO_3^-$ will increase by about 1 mmol/L for each 10 mm Hg rise in pCO_2. If respiratory acidosis persists, the change will be 3.5 mmol/L, mainly as a result of renal compensation.

Uncomplicated acute respiratory acidosis will generally not increase $cHCO_3^-$ to >32 mmol/L or decrease the pH to <7.20. Greater decreases are generally due to a mixed acidosis. The plasma chloride decreases as plasma bicarbonate increases. Hyperkalemia may occur but is not as predictable as in some forms of metabolic acidosis. This increase in K^+ is mainly due to the movement of K^+ from the cells into the plasma in exchange for the movement of H^+ into the cells. Urinary acidity and ammonia content are increased.

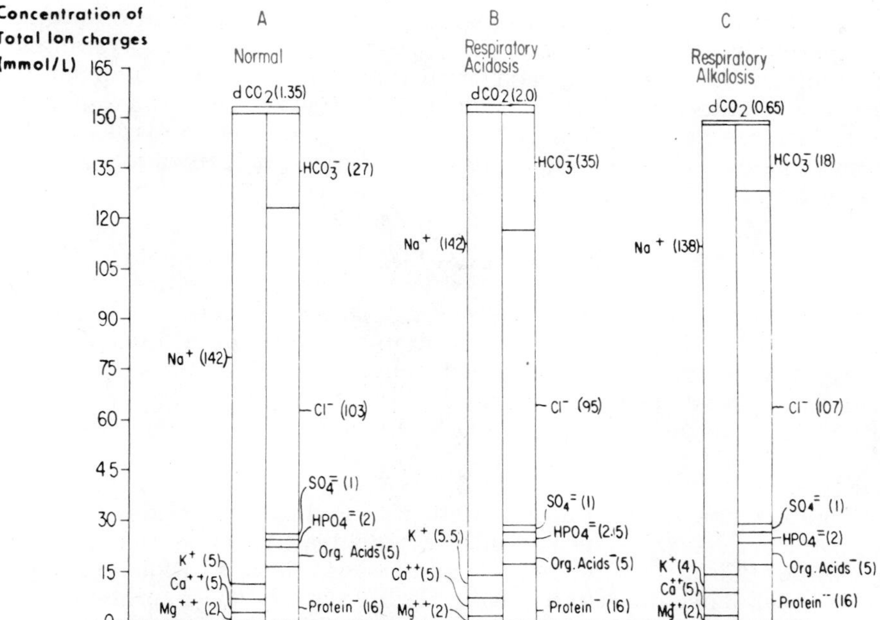

Figure 19-13. *A,* Electrolyte composition of normal plasma. *B,* Example of possible electrolyte pattern in a patient with respiratory acidosis. Note the increase in HCO_3^- and dCO_2. The increase in the latter fraction is more pronounced than that of HCO_3^-. The ratio therefore is decreased, and the chloride fraction shows a decrease. *C,* Theoretical electrolyte pattern of patient in respiratory alkalosis. There is a decrease in the HCO_3^- and especially the dCO_2 fraction. Therefore, the ratio of HCO_3^-/dCO_2 and the pH are increased. The Na^+ is at the lower limit of normal. (From Tietz, N. W.: The Chicago Medical School Quarterly, *22*:156, 1962.)

Respiratory Alkalosis (Primary dCO_2 Deficit)

A decrease in pCO_2 (hypocapnia) is caused by an increased rate or depth of respiration, or both. Excessive elimination of carbon dioxide reduces the pCO_2, which causes an increase in the $cHCO_3/cdCO_2$ ratio and increases pH. In respiratory alkalosis the compensation is very efficient and may return the pH almost to the original value.

Causes of respiratory alkalosis, analogous to causes of respiratory acidosis, can be classified as those with a direct stimulatory effect on the respiratory center (hysteria, fever, CNS infections, hypoxia, and drugs such as salicylates and catecholamines) and those due to effects on the pulmonary mechanism (such as pneumonia, asthma, pulmonary emboli, and congestive heart failure).

Compensatory Mechanisms in Respiratory Alkalosis

The compensatory mechanisms respond to respiratory alkalosis in two stages. In the first stage, erythrocyte and tissue buffers provide H^+ ions that consume a small amount of HCO_3^-. The second stage becomes operational in prolonged respiratory alkalosis and depends on the renal compensation as outlined for metabolic alkalosis.

Laboratory Findings in Respiratory Alkalosis

In chronic respiratory alkalosis, the $cdCO_2$, pCO_2, $cHCO_3^-$, and thus the total CO_2 concentration are decreased (Figure 19-13,*C*). The ratio of $cHCO_3^-/cCO_2$ is increased and causes an increase in pH which, however, rarely exceeds 7.60. Greater increases are usually due to a mixed alkalosis. During the *acute* phase, only the $cdCO_2$ and pCO_2 may show significant changes; $cHCO_3^-$ falls by 2.0 mmol/L for each 10 mm Hg decrease in pCO_2. In *chronic* respiratory alkalosis renal compensation further decreases HCO_3^-, the $cHCO_3^-$ falls 5 mmol/L for every 10 mm Hg decrease in pCO_2, and the pH may return to near normal.

Individuals living at high altitudes chronically hyperventilate owing to hypoxia and have pCO_2 values lower than those seen at sea level. In one study, the average pCO_2 was 31 mm Hg, but the blood pH was within reference range as a result of the compensatory mechanism.

Movement of K^+ into cells in exchange for H^+ that moves into plasma results in mild K^+ wasting by the kidneys and mild hypokalemia. Chloride is retained to replace the decreasing HCO_3^-. There may be a small increase in the anion gap due to enhanced glycolysis and lactate formation. Lactate levels may increase 2–4 mmol/L, probably owing to a decrease in hepatic blood flow. In prolonged severe alkalosis there may be an increase of ketone bodies due to decreased carbohydrate utilization. Phosphate levels may be significantly decreased.

Estimates Derived from Electrolyte Measurements

Electrolyte analysis is usually defined as concurrent determinations of Na^+, K^+, Cl^-, and total CO_2. The set of results is often clinically more informative than each result by itself. Supplemental data derived from the set of results, such as the calculated anion gap and calculated osmolality, can have additional clinical value. In calculations of this kind, the concentration of total CO_2 is taken as the concentration of HCO_3^-.

Anion Gap

Anion gap is a calculation used to detect altered concentrations of ions other than Na^+, K^+, Cl^-, and HCO_3^-. The law of electroneutrality demands that the number of positive charges (Na^+, K^+, Ca^{2+}, Mg^{2+}, and traces of other cations) in plasma exactly equal the number of negative charges (Cl^-, HCO_3^-, SO_4^{2-}, $HPO_4^{-,2-}$, protein anions, organic anions, and traces of other anions). Therefore, the term *anion gap* refers to the differences in charge concentration between unmeasured anions (cUA^-) and unmeasured cations (cUC^+) as shown also in the nonshaded area in Table 19-6.

$$\text{Anion gap, mmol ion charge/L} = cNa^+ - (cCl^- + cHCO_3^-) \quad (6)$$

Table 19-6. ILLUSTRATION OF GAP BETWEEN UNMEASURED ANIONS (UA^- = 24 mmol ION CHARGE/L) AND UNMEASURED CATIONS (UC^- = 12 mmol ION CHARGE/L)

Na^+	142		Cl^-	103	
K^+	4	12	HCO_3^-	27	
Ca^{2+}	5		HPO_4^{2-}	2	24
Mg^{2+}	2		SO_4^{2-}	1	
Others	1		Org. anions	5	
			Protein	16	
	154			154	

Red indicates measured cation and anions.

or alternatively,

$$\text{Anion gap, mmol ion charge/L} = (c\text{Na}^+ + c\text{K}^+) - (c\text{Cl}^- + c\text{HCO}_3^-) \quad (7)$$

Both definitions are acceptable. Note, however, that the reference ranges for the two equations are different.

Clinical significance.[2] The value of the calculated anion gap may be affected in five situations.[4]

1. Anion gap values may be *increased* due to retention of one or more of the unmeasured anions, which causes a decrease in $cHCO_3^-$ or cCl^- or both. Such increases include increased β-hydroxybutyrate and acetoacetate in diabetic, alcoholic, or starvation ketoacidosis; increased lactate in lactic acidosis; increased sulfate and phosphate in renal failure; the presence of salicylate or acidic metabolites such as formate, oxalate, hippurate, or glycolate accumulating in poisonings; increased negative charge of protein anions as a result of increased blood pH in respiratory or metabolic alkalosis; or increased number of protein anions because of dehydration (hemoconcentration).

2. Anion gap values may also be slightly *increased* due to an *apparent* increase in unmeasured anions that is actually a real decrease in unmeasured cations (hypocalcemic, hypomagnesemic, or hypokalemic states).

3. Anion gap values may be *decreased* due to a real decrease in the number of negatively charged protein anions (hypoalbuminemia), or to a real increase of special proteins that carry relatively few negative charges (hypergammaglobulinemia), or to an increase in plasma water (hemodilution).

4. Anion gap values may also be *decreased* due to an increase in unmeasured cations (hypercalcemic, hypermagnesemic, or hyperkalemic states, or in lithium toxicity) or in the presence of abnormal proteins (e.g., myeloma proteins), carrying a net positive charge.

5. Anion gap values can be affected by analytical error in estimation of Na^+, Cl^-, or HCO_3^-. Analytical error may also develop when a particular sample contains an interferent that reacts with the method available for the assay. Thus, the calculation may be used to detect some gross analytical errors.

The finding of an abnormal anion gap is nonspecific. In the absence of laboratory errors, however, a high value, taken together with patient history, physical examination, and other laboratory data, is very useful for the detection of organic acidoses in which real increases occur in anions that are difficult or impossible to measure, especially on a "stat" basis.[2,3] Low and high values are considered less useful in detection of altered levels of unmeasured cations or proteins.[3]

Use of Anion Gap in Quality Control of Electrolyte Analysis

On the assumption that unmeasured ion concentrations lie within their reference ranges, an anion gap calculated with equation (6) that exceeds 12–20 mmol/L indicates a need for review of calculations and procedures for Na^+, Cl^-, and HCO_3^-. If results are confirmed, reasons for a high anion gap may be sought; for example, a high glucose or a high urea nitrogen would suggest accumulation of organic anions above normal concentrations (ketoacidosis and renal acidosis, respectively).

Reference ranges. For anion gap calculated without the inclusion of K^+ [equation (6)], the reference range is 8–16 (mean, 12) mmol/L. For anion gap calculated with

Table 19-7. METABOLIC ACIDOSES WITH HIGH ANION GAP

Etiology	Acids Retained	Anion Gap, mmol/L	Osmolal Gap, mOsm/kg	Other Laboratory Findings
Renal failure	Sulfuric, phosphoric, organic	15–20 (usually)	<10	Increased plasma urea nitrogen, creatinine
Ketoacidosis				
Diabetic Alcoholic Starvation	Acetoacetate, β-hydroxybutyrate	>25 (often)	<10	Increased urine and plasma ketone bodies and hyperuricemia
Poisoning				
Aspirin	Salicylate	>25 (often)	<10	Respiratory alkalosis
Methanol	Formate		>15	
Ethylene glycol	Hippurate, glycolate, oxalate		>15	Urine oxalate crystals
Lactic acidosis	Lactate	>25 (often)	<10	Normal arterial pO_2 (for Type B lactic acidosis)

Modified from Narins, R. G., and Gardner, L. B.: Simple acid-base disturbances. Med. Clin. North Am., *65*:321–346, 1981.

inclusion of K^+ [equation (7)], the range is 10–20 (mean, 16) mmol/L. Low values are characteristic of hypoproteinemia or hyperproteinemia of the cationic type. The highest values are associated with forms of acidosis summarized in Table 19-7. When using electrolyte analysis and anion gap calculated without K^+ to screen for acid-base imbalance, most physicians react to an anion gap of <5 or >19 mmol/L, since there is a 2–4 mmol/L random variation at the upper and lower limits of the reference range.[2]

Osmolal Gap

The osmolal gap is a mathematical construct similar to the anion gap, but it applies to osmolute (osmotically active solute) concentrations rather than to concentrations of ion charges. The value of the gap is:

$$\text{Osmolal gap, mOsm/kg} = \text{Measured osmolality, mOsm/kg} - \text{calculated osmolality, mOsm/kg}$$

Calculation of the osmolal gap requires measurement of plasma or serum osmolality by freezing point depression osmometry and measurement of the plasma or serum concentrations of Na^+, glucose, and urea (nitrogen). Osmolality measured by freezing point depression, as opposed to vapor pressure measurements, reflects the total concentration of osmolutes in the specimen, including volatiles such as alcohols or acetone. Since total osmolute concentration is essentially determined by the five major osmolutes—Na^+, Cl^-, HCO_3^-, urea, and glucose—and since each Na^+ ion can be assumed to be accompanied by an anion, only sodium, glucose, and urea need to be measured in order to calculate the osmolality of total solutes in plasma. The best-documented equation for calculating plasma osmolality is:

$$\text{Calculated osmolality, mOsm/kg} = 1.86\,(Na^+, \text{mmol/L}) + \text{glucose, mmol/L} + \text{urea, mmol/L} + 9$$

or

$$= 1.86\,(Na^+, \text{mmol/L}) + \frac{\text{glucose, mg/dL}}{18} + \frac{\text{urea N, mg/dL}}{2.8} + 9$$

or

$$= 1.86\,(Na^+, \text{mmol/L}) + 0.056\,(\text{glucose, mg/dL}) + 0.36\,(\text{urea N, mg/dL}) + 9$$

The equation is empirical. It was developed by regression analysis to solve for the factor 1.86 (slope for the relation of Na^+ concentration in mmol/L to the calculated osmolality) and the composite intercept represented by the last three terms in the equation. The equation was validated by comparison of calculated osmolality with measured osmolality; the mean difference was approximately zero, and the standard deviation of the differences ~6 mOsm/kg. The factors applied to glucose and urea N convert concentration expressed in mg/dL to mmol/L:

For Glucose,

$$0.056 = \frac{1}{18} = \frac{1\ \text{mg/dL} \times 10\ \text{dL/L}}{180\ \text{mg/mmol glucose}}$$

For urea N,

$$0.36 = \frac{1}{2.8} = \frac{1\ \text{mg/dL} \times 10\ \text{dL/L}}{28\ \text{mg N/mmol urea}}$$

A number of other equations can be used; the above equation, as well as any of the others, may be applied on electrolyte analyzers that display the calculated osmolality together with the results of a test panel that includes sodium, urea nitrogen, and glucose. The test of suitability of an equation used to calculate osmolality is whether calculated osmolality of almost all specimens containing a normal amount of plasma water compares closely with measured osmolality on the same specimens.

The ability of contemporary analyzers to calculate and display a result for calculated osmolality is a mixed blessing. On the one hand, the calculated osmolality may lead some physicians and laboratorians to believe that whatever equation has been used for the algorithm in instrument software is the only or the best equation for calculating osmolality, or to believe that calculated osmolality is a uniformly valid substitute for measured osmolality. On the other hand, it provides informed laboratorians and physicians with the opportunity to calculate the osmolal gap more conveniently.

Reference ranges and clinical utility of osmolal gap. The mean for the plasma osmolal gap calculated as described above is approximately zero, with a standard deviation of ~6 mOsm/kg. The value of the gap is increased in hyperosmolal states due to increase in osmolutes other than the five major ones. Values up to 10 (i.e., very moderate increases) are associated with ketoacidosis, renal acidosis, and lactic acidosis. Values frequently fall within the reference range and thus are not very helpful in these clinical situations. However, the osmolal gap is most effectively used to detect and follow poisonings with nonelectrolytes such as acetone, ethanol, methanol, isopropanol, ethylene glycol, diethyl ether, paraldehyde, or trichloroethane. High concentrations of both parent compounds and metabolites cause a hyperosmolality and a large increase in the osmolal gap that are often correlated with the severity of the intoxication. Such increase in osmolality is not detected by calculating the osmolality on the basis of the Na^+, glucose, and urea concentrations. The osmolal gap therefore is valuable in detecting and following these particular hyperosmolal states and avoids difficult or specific assays for individual, and perhaps unknown, analytes. Osmolal gaps expected in some clinical conditions are shown in Table 19-7.

References

1. Dorwart, W. V., and Chalmers, L.: Comparison of methods for calculating serum osmolality from chemical concentrations, and the prognostic value of such calculations. Clin. Chem., *21*:190–194, 1975.
2. Emmett, M., and Narins, R. G.: Clinical use of the anion gap. Medicine, *56*:38–54, 1977.
3. Gabow, P. A., Kaehny, W. D., Fennessey, W. D., et al.: Diagnostic importance of an increased serum anion gap. N. Engl. J. Med., *303*:854–858, 1980.
4. Gambino, R.: The anion gap in '82. Lab Report for Physicians, *4*:28–30, 1982.
5. McGilvery, R. W.: Biochemistry. Philadelphia, W. B. Saunders Co., 1983, pp. 778–782.

Additional Reading

Bockelman, H. W., Cembrowski, G. S., Kurtycz, D. F. I., et al.: Quality control of electrolyte analyzers: Evaluation of the anion gap average. Am. J. Clin. Pathol., *81*:219–223, 1984.

Cembrowski, G. S., Westgard, J. O., and Kurtycz, D. F. I.: Use of anion gap for the quality control of electrolyte analyzers. Am. J. Clin. Pathol., *79*:688–696, 1983.

Narins, R. G., and Emmett, M.: Simple and mixed acid-base disorders: A practical approach. Medicine, *59*:161–187, 1980.

Narins, R. G., and Gardner, L. B.: Simple acid-base disturbances. Med. Clin. North Am., *65*:321–346, 1981.

NITROGEN METABOLITES AND RENAL FUNCTION

20

ROBERT C. ROCK, M.D.
W. GORDON WALKER, M.D.
C. DARRELL JENNINGS, M.D.

The kidneys are a paired organ system located in the retroperitoneal space; the right kidney is situated slightly lower than the left (Figure 20-1, *A*). Each kidney weighs about 150 g and is bean shaped with a central hilus on the medial side, where vessels, lymphatics, and the renal pelvis join (Figure 20-1, *B*). The kidney parenchyma is divided into lobes that consist of an outer cortex, which covers a pyramidal medulla. This cortex is composed mostly of glomeruli and the proximal and distal convoluted tubules. The medulla consists mostly of loops of Henle, vasa recta (elongated vessels accompanying the loops of Henle), and collecting ducts. The collecting ducts drain through the rounded apex of the pyramid into branches of the renal pelvis called calyces. From here urine collects in the pelvis and flows through the ureteropelvic junction into the ureter. The ureters carry urine from each kidney into the bladder, where it is stored until voided through the urethra. The renal blood supply represents ~25% of the cardiac output. The renal artery divides into posterior and anterior portions with subsequent divisions that ultimately terminate in the afferent arterioles, each supplying a single glomerulus.

The short course of the renal artery and rapid branching leading into the afferent arterioles are very important for renal function. This vascular arrangement allows transmittance of adequate hydrostatic pressure to the glomerular capillary bed for glomerular ultrafiltration. The oncotic pressure difference between the capillary lumen and Bowman's space opposes filtration. Since most proteins are not normally filtered, the oncotic pressure in the glomerular capillary progressively rises until filtration ceases. This limits the amount of filtrate that can be obtained from a given volume of plasma. Thus, maintenance of glomerular filtration rate is dependent upon an adequate renal

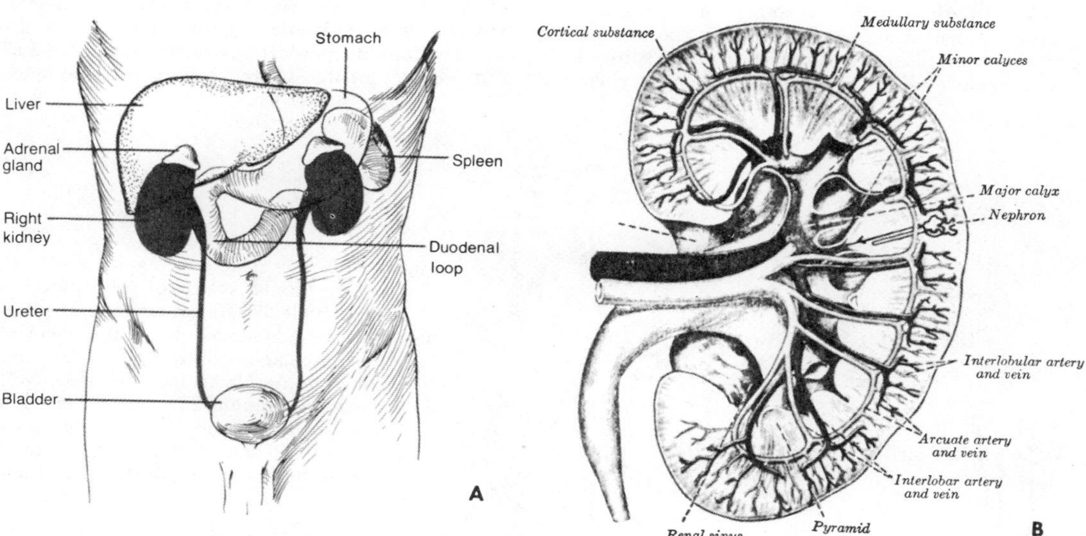

Figure 20-1. *A,* Schematic drawing of right and left kidneys with respect to other abdominal viscera. *B,* Diagram of vertical section through the kidney. Nephron and blood vessels are greatly enlarged. (From Goss, C. M., Ed.: Gray's Anatomy. 28th ed. Philadelphia, Lea & Febiger, 1966.)

blood flow rate as well as adequate perfusion pressure. The normal renal blood supply is well designed to meet these needs.

Renal tubular function is also dependent upon the anatomy of the vessels providing the blood supply. Efferent arterioles form capillary plexuses as well as elongated vessels accompanying the loops of Henle (vasa recta), which provide the rich, intimate perfusion necessary for the highly active tubular cells to reabsorb various urine constituents. Further, the flow through the vasa recta is vital for the steady-state operation of the countercurrent exchange mechanism that allows formation of the medullary osmotic gradient that is so important in renal processing of water.

The nephron. The nephron is the functional renal unit. It consists structurally of (1) the glomerulus, a spherical epithelial space invaginated by a capillary tuft that connects the afferent and efferent arterioles, and (2) a tubule of epithelial cells, continuous with the glomerular epithelial space and ultimately leading to collecting ducts that empty into the renal pelvis. Figure 20-2 schematically depicts the functional regions of the nephron. Note that the efferent arteriole is located in close association with the exterior surface of the nephron.

Formation of urine. The first step in urine formation is filtration at the glomerulus of plasma water driven by a net filtration pressure of about 15 mm Hg. The filtrate is called an ultrafiltrate because its composition is essentially the same as that of plasma, with the notable absence of molecules of MW above ~15 000. Each kidney contains about one million nephrons, and ~170–200 L of ultrafiltrate pass through the glomeruli in 24 h. In the passage of ultrafiltrate through the tubules, reabsorption of solutes and water in various regions of the tubules reduces the total volume, which ranges between 0.4 and 2 L of urine per day.

In the tubules, the solute composition of the ultrafiltrate is altered by the processes of reabsorption and secretion, so that the urine excreted may have a very different composition from that of the original filtered fluid. Different regions of the tubule have been shown to specialize in certain functions. In the *proximal tubule*, 60–80% of the ultrafiltrate is reabsorbed in an obligatory fashion along with sodium, chloride, bicarbonate, calcium, phosphate, and other ions. The threshold substance,* glucose, is also

*Certain substances, almost completely reabsorbed by the tubule when their concentrations in plasma are relatively low, appear in the urine when their plasma levels are above a certain set point or threshold level. High-threshold substances, such as glucose and amino acids (whose conservations are important to the body economy), are almost completely reabsorbed by means of specific transport systems in the tubular cells. The appearance of a high-threshold substance in the urine is evidence that the filtered load of the substance is exceeding the maximal reabsorption rate of its transport system. This may occur owing either to elevated plasma concentration or to a defect in the transport system. The renal threshold of a filterable compound is therefore that plasma concentration of the substance at which the tubular reabsorptive rate is not sufficiently fast to remove all of the substance from the plasma ultrafiltrate, so that the substance spills into the urine.

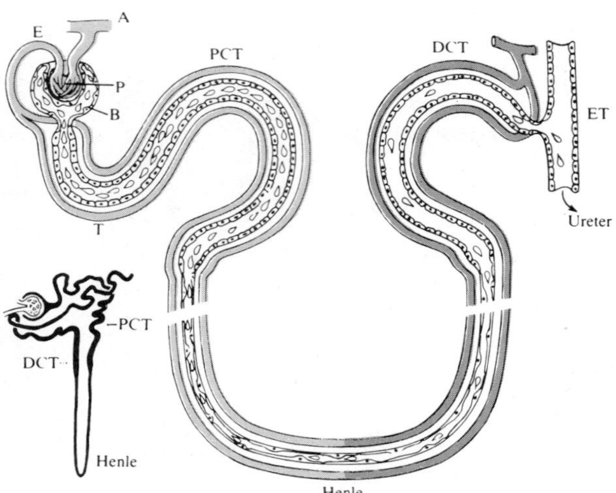

Figure 20-2. Schematic drawing of the glomerulus and the tubular system of the nephron. A, afferent arteriole; E, efferent arteriole; P, plexus of capillaries (glomerular tuft); B, Bowman's capsule; T, tubular blood supply; PCT, proximal convoluted tubule; Henle, loop of Henle; DCT, distal convoluted tubule; ET, excretory tubule or duct. The blood capillaries shown along the tubular system (T) gradually change to venous capillaries as they pass down the tubular system.

reabsorbed in the proximal tubule. Certain nonbiological compounds, such as phenolsulfonphthalein (PSP), *p*-aminohippurate (PAH), and iodopyracetate (Diodrast), are secreted by the proximal tubule and have been used for the evaluation of renal tubular secretory capacity. When blood levels of creatinine increase above normal, creatinine is secreted in this region of the nephron. In the *loop of Henle*, chloride and more sodium without water are reabsorbed, generating a dilute urine. Water reabsorption in the more distal tubules and collecting ducts is then regulated by the antidiuretic hormone (ADH). In the *distal tubule*, secretion is the prominent activity. It is also this region that secretes H^+ and reabsorbs Na^+ and reclaims HCO_3^- in aid of acid-base regulation (see *Excretion of Acids*, Chapter 19, Section Three).

Homeostasis. The obvious role of glomerular filtration and tubular secretory functions is to eliminate waste products of metabolism. Another equally important role is to maintain the constant optimal chemical composition of the blood and thereby to maintain the constant optimal chemical composition of the interstitial and intracellular fluids throughout the body (homeostasis). The kidney must therefore be viewed not only as an excretory organ but also as a regulatory organ.

FUNCTIONS OF THE KIDNEY

The functions of the kidney may be characterized as excretory, regulatory, and endocrine. The ***excretory function*** serves to rid the body of most of the undesirable end products of metabolism, as well as any excess of inorganic substances ingested in the diet. Waste products include the nonprotein nitrogenous compounds urea, creatinine, and uric acid; a number of other organic acids, including amino acids, are excreted in small quantities. Diet contains a variable and usually excessive supply of sodium, potassium, chloride, calcium, phosphate, magnesium, sulfate, and bicarbonate. The dramatic efficiency of the excretory function is illustrated by sodium; body sodium content is maintained essentially constant, regardless of whether daily sodium intake is 1 or 150 mmol or more. In order to achieve the excretion of the metabolic wastes and the ingested surpluses without disrupting homeostasis, the kidney must exercise both its excretory and regulatory functions.

The ***regulatory function*** of the kidney plays a major role in homeostasis. The mechanisms of differential reabsorption and secretion, located in the tubule of the nephron, are the effectors of regulation. Further discussion of the mechanism appears under *Urine Production* in a later section of this chapter and in Chapter 19, Section Three.

The ***endocrine functions*** of the kidney may be regarded either as primary, because the kidney is an endocrine organ producing hormones, or as secondary, because the kidney is a site where hormones produced elsewhere are activated. In its *primary endocrine function*, the kidney produces renin, prostaglandins, and erythropoietin. Erythropoietin acts on the bone marrow to stimulate proliferation and maturation of erythrocyte precursors. Anemia, therefore, is a prominent feature of renal insufficiency. In the anephric state, the hematocrit can be expected to fall as low as 10–12% before stabilizing.[38] Renin and prostaglandins affect the vascular system. The renin-angiotensin system is an important regulator of blood pressure; it is responsive to plasma sodium and potassium concentrations and regulates the aldosterone effect on K^+–Na^+ exchange in the distal tubule (see also Chapter 18).

In the *secondary* but no less important aspect of its endocrine function, the kidney is a site of degradation of insulin, glucagon, and aldosterone. Furthermore, the kidney is the location of the important 1-hydroxylation of 25-$(OH)_2D_3$ that produces 1,25-$(OH)_2$-D_3, the most active known form of vitamin D. Progressive, chronic renal failure is regularly associated with a combination of osteomalacia (inadequate mineralization of bone), due to altered vitamin D metabolism, and osteitis fibrosa cystica, caused by secondary elevation of PTH levels.

Table 20-1. BIOCHEMICAL MONITORING OF HEMODIALYSIS

Disorder	Tests* (Plasma or Serum)
Fluid and electrolyte imbalance	Sodium, potassium, chloride, bicarbonate, osmolality
Nitrogen imbalance	Total protein, albumin, urea, creatinine, uric acid
Osteodystrophy	Calcium, phosphorus, alkaline phosphatase, magnesium, parathyroid hormone
Hepatitis	Hepatitis B antigen, bilirubin, aspartate aminotransferase, alanine aminotransferase

*None of these tests provide direct data bearing on balance but must be interpreted in relation to changes in body weight, fluid intake, and food intake as well as changes attributable directly to dialysis.

Renal Dialysis

The development of techniques for hemodialysis (treatment with the "artificial kidney") and peritoneal dialysis* has changed end-stage renal disease from an illness that once terminated fatally within a few months to a chronic illness with an average survival time of 10–15 years. Hemodialysis is generally more efficient than peritoneal dialysis in removal of waste compounds in uremic patients, although peritoneal dialysis is of value in situations in which arteriovenous fistulas ("shunts") fail.

Complications of hemodialysis include mechanical difficulties such as air embolism, hemorrhage, or obstruction of the shunt, as well as toxicity caused by the dialysate itself. The syndrome of "dialysis dementia" in chronic renal dialysis patients has, for example, been attributed to high levels of aluminum in dialysate.[23] Fluid and electrolyte shifts during dialysis produce symptoms at the time and in the immediate postdialysis period. Rapid removal of urea, sodium, and other osmotically active substances causes "dialysis disequilibrium," with symptoms ranging from lethargy and headache to stupor and seizures. Rapid lowering of potassium during dialysis may lead to myocardial toxicity. Vitamins such as folate and pyridoxine are removed by dialysis and must be replaced.

Viral hepatitis (B and non-A, non-B) is a particular problem for dialysis services. Renal osteodystrophy, with secondary hyperparathyroidism and vitamin D deficiency resulting in osteomalacia, persists despite hemodialysis. Control of hyperparathyroidism occasionally requires subtotal parathyroidectomy.

Laboratory tests routinely used to monitor patients on hemodialysis[14] are listed in Table 20-1. Effectiveness of dialysis is usually monitored by measurement of creatinine, serum potassium, and urea nitrogen. Measurement of body weight, blood pressure, and serum sodium assesses regulation of body hydration and sodium balance; measurement of bicarbonate and arterial pH indicates acid-base status and buffering capacity; and serum phosphate, calcium, and periodically PTH (and possibly hand X-rays) assess the effects upon bone metabolism. The most reliable measurements for following residual renal function are urine volume and a reliably measured glomerular filtration rate (GFR). However, in the 24-h period immediately after a dialysis session the GFR has limited usefulness, since GFR shows a transient decrease during this period. GFR is evaluated from clearance of creatinine or ^{125}I-iothalamate.[30] Several therapeutic drugs, especially the nephrotoxic aminoglycosides, should be monitored by measuring serum levels in dialysis patients, since the effects of poor renal function and dialysis are not always predictable. Table 20-1 makes clear the full range of concern that the laboratory data must address in order to support the total care of the end-stage renal disease patient undergoing dialysis.

**Hemodialysis*, the general term, refers to a procedure in which blood solutes are allowed to diffuse down a concentration gradient, across a semipermeable membrane, and into a recipient fluid, the dialysate. The artificial kidney is an application of extracorporeal dialysis, since an extracorporeal circulation is set up via an implanted arteriovenous cannula to lead the patient's blood to the membrane cartridge immersed in the dialyzer bath. Peritoneal dialysis is intracorporeal, so to speak; the patient's abdominal cavity is perfused with (or filled with and then emptied of) the recipient fluid, and dialysis occurs across the peritoneal membrane. Clearance or dialysance achievable by peritoneal dialysis is $<15–20\%$ of that achieved by hemodialysis. It is especially slow at removing K^+. Hence, when hyperkalemia is the prime indication for dialysis, hemodialysis is the preferred and more efficient means of treatment unless contraindicated.

Transplantation

Transplantation of kidneys into immunologically suitable recipients corrects many of the abnormalities of chronic renal failure that hemodialysis can only partially correct. But immunosuppression, which is required to forestall rejection of the transplanted kidney, causes new problems for the recipient. The most immediate danger is infection by opportunistic organisms such as fungi, protozoa, certain bacteria, and certain viruses. Long-range problems relate to the development of malignancies in immunosuppressed recipients of renal allografts.[25]

Biochemical monitoring of renal transplant patients is designed to detect deterioration of function in the transplanted kidney, especially as it relates to delayed rejection of the graft. Serial monitoring of plasma creatinine is most useful in the early recognition of decline of glomerular filtration. Detection of tubular dysfunction, by following urinary excretion of markers such as β_2-microglobulin (low molecular weight) or *N*-acetylglucosaminidase (high molecular weight),[27] may allow earlier detection of graft rejection. Beta$_2$-microglobulin may also be helpful in the oliguric patient with a recent transplant. Here it may help differentiate acute tubular necrosis from early acute rejection. The patient with acute renal failure will have higher urine values of this protein than will the patient with acute humoral rejection (see also Chapter 11).

TESTS OF RENAL FUNCTION

The kidney's excretory, regulatory, and endocrine roles display complex interactions. Not surprisingly, the composition of urine, as well as plasma, reflects not only functional disorders of the nephrons but also various systemic disorders. The practical evaluation of the status of the kidney in renal disease is more or less limited to examining the nephron's functions of glomerular filtration, secretory capacity for particular endogenous and exogenous compounds, and reabsorptive capacity for water (i.e., concentrating ability).

Quantitation of overall renal function is based on a simplifying assumption, the *intact nephron hypothesis*. This hypothesis holds that each individual and intact nephron is normally functional and that decline in renal function is due to loss of functional nephrons. Thus, in nearly all types of diffuse renal disease, impaired function of the kidney is attributed to a diminished number of functioning nephrons rather than to the diminished function of individual nephrons. Since glomerular filtration is the initiating phase of all nephron functions, measurement of glomerular filtration rate provides the most useful general index for the physician in his assessment of the severity and progress of renal damage.

Specific defects in particular functions of the nephron can also be identified and evaluated. For instance, measurement of *titratable acidity* of urine (*net acid secretion*) can identify defective function of the distal tubule with respect to hydrogen ion secretion; assessment of the *maximal concentrating capacity* of the kidney gives an estimate of ADH-controlled reabsorption of solute-free water in the distal portion of the tubule. "Pinpoint defects," such as aminoacidurias due to genetically determined deficiency of specific tubular transport systems or toxic and pharmacologic lesions of such transport systems, may be evaluated by *determining quantities of endogenous marker compounds* appearing in urine (see Chapter 11). *Secretory capacity (tubular secretory mass)* may be estimated by the *p*-aminohippurate (PAH) load test. Most of the tests that measure localized functions of the nephron are not commonly performed; the reader is referred to the literature.[5,27,34,42]

Of possible tests for evaluating renal function, however, the most practical on a routine basis are clearances of various compounds (creatinine, in particular) to estimate glomerular filtration rate, assessment of glomerular permeability by establishing the types of proteins appearing in urine, measurement of plasma creatinine, and measurement of the concentrating ability of the tubules.

Renal Clearance and Glomerular Filtration Rate

Most clinical laboratory information used to assess kidney function is derived from or related to measurement of "clearance" of some substance by the kidney. Clearance is defined as the (hypothetical) quantity of blood or plasma completely cleared of a substance per unit of time, and it is expressed mathematically as:

$$C_s = (U_s \times V)/P_s$$

where U_s is the urinary concentration of the substance, s; P_s is its plasma concentration; V is urine flow rate in mL/min; and C_s is the clearance in units of mL of plasma cleared of the substance per min. C_s is proportional to the rate of solute mass excretion ($U_s \times V$).

Para-aminohippurate (PAH) is a substance almost completely cleared from the blood by the kidney tubules, and thus its clearance may be used to measure tubular excretory capacity. The PAH clearance test is also a useful means of determining total renal plasma flow if tubular function is known to be normal. Renal *plasma* flow determined by PAH clearance is ~574 mL/min per 1.73 m^2 of body surface area.

Clearance of substances that are filtered exclusively or predominantly by the glomerulus but neither reabsorbed nor secreted by other regions of the nephron can be used to measure the glomerular filtration rate. Inulin (a polyfructan) is such a substance, and thus the GFR may be determined by the inulin clearance. Furthermore, when a substance is freely filterable through the glomerulus, it is possible to assess quantity reabsorbed from or secreted into the tubule by multiplying its plasma concentration, P_s, times GFR. If the product (quantity filtered per min) is greater than $U_s \times V$ (excretion rate), then the substance must be secreted by the tubule. The magnitude of the difference between filtration and excretion rates is a quantitative measure of tubular reabsorption or secretion of the substance.

The primary or reference substance for measuring GFR is inulin, and PAH is the reference substance for measurement of renal plasma flow. Both substances have been extensively used in clinical and experimental studies. The necessary plasma and urine assays for inulin and PAH are too difficult and too time consuming to be practical in routine clinical laboratories; thus, creatinine clearance is almost universally used for the clinical assessment of GFR. Plasma and urine creatinine levels are easily and conveniently measured by either manual or automated methods whose precision and reliability are well documented. These methods will be discussed below.

Glomerular permeability. The glomerular capillary wall is a special membrane uniquely designed to support rapid ultrafiltration and to retain nearly all proteins within the plasma. This wall (Figure 20-3) is composed of three structural elements: (1) a layer of endothelial cells, (2) the glomerular basement membrane that separates the endo-

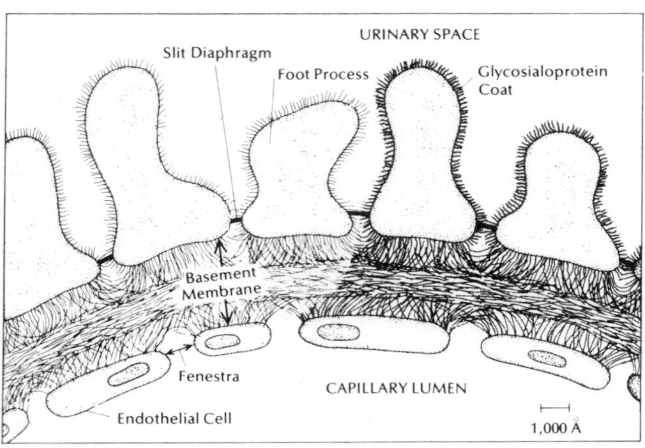

Figure 20-3. Glomerular capillary wall. In glomerular filtration, filtered fluid is believed to traverse the capillary wall via an extracellular route (i.e., through endothelial fenestrae, basement membrane, and slit diaphragms). Circulating polyanions (e.g., albumin) are thought to be retarded by the rich distribution in inner barriers of negatively charged glycosialoproteins (shaded area in schematic diagram). (Used with permission from Brenner, B. M., and Beeuwks, R., III: The kidney in health and disease. III. The renal circulations. Hosp. Prac., *13* [July]:35–46, 1978. Artist: Nancy Lou Gaban Mahrís.)

thelial cells from the next layer, and (3) the outer epithelial cell layer that is attached to the basement membrane by discrete cytoplasmic extensions (pedicels or foot processes). Its structure allows for ultrafiltration at a rate that removes ~20% of the plasma water (with small-molecular-weight solutes) during transit of plasma through the glomerular capillaries but that restricts passage of protein molecules with MW above ~15 000.

Molecular shape and charge also affect glomerular filtration. Clearance of globular proteins is less than that of similar-sized but linear molecules, which presumably have the ability to turn end-on during passage through the filtration barrier.[31] Also important is the dependence on molecular charge. Molecules that are highly negatively charged exhibit much lower glomerular permeance than do neutral molecules of the same size. This is strong direct evidence that the filtering membrane has a negative charge that tends to repel negatively charged molecules and thus decrease the effective radius of "pores" in the membrane as molecules with like charges approach it. A major form of injury to the glomerulus in human renal disease may be the loss of glomerular polyanionic charge, perhaps because of decrease or alteration of sialoprotein content of the basement membrane.[3]

The glomerular filtration apparatus may become grossly distorted during the course of numerous diffuse renal diseases, and the primary derangement is loss of the ability to retain selectively the large protein molecules. The result is loss of large quantities of low- and high-molecular-weight proteins in the urine. Three types of measurements provide diagnostic and prognostic information when proteinuria is a feature of renal disease; they are (1) measurement of excretion rate of total urinary protein, (2) electrophoresis of urinary proteins, and (3) measurement of selective clearance of proteins of different molecular sizes (see also Chapter 11).

Measurement of the excretion rate of total urinary protein is the standard means to quantitate the severity of proteinuria. The magnitude of protein loss generally directs the physician toward selection of an appropriate diagnostic category. Some diseases rarely have protein excretion rates of >1 g/d; others may have rates of >3 g/d. The latter category includes patients who have persistent or intermittent clinical manifestations of the nephrotic syndrome. A characteristic means of evaluating the effectiveness of therapy (e.g., when corticosteroids are used to treat some forms of the nephrotic syndrome) is to follow the protein excretion rate sequentially. The excretion rate is ordinarily determined on 24-h specimens since it tends to fluctuate during the day with changes of activity and posture.

Qualitative evaluation of urine proteins can be valuable in the differential diagnosis of diseases producing proteinuria. The relatively benign cause of nephrotic syndrome, minimal change disease (lipoid nephrosis), produces little alteration in glomerular structure other than loss of glomerular basement membrane charge characteristics and hence produces a selective proteinuria consisting almost entirely of albumin. More severe glomerulonephritis causes significant glomerular injury with resulting loss of larger-molecular-weight globulins into the urine and hence an increased globulin/albumin or IgG/albumin ratio that can readily be measured (see also Chapter 11). Tubular diseases result in decreased reabsorption of those small proteins that are normally filtered freely. Thus, one can detect increased urinary levels of lysozyme, β_2-microglobulin, and immunoglobulin light chains in such conditions as acute tubular necrosis (shock kidney), heavy metal poisoning, chronic transport rejection, and nephrotoxicity after intake of certain drugs.

Nonprotein Nitrogen-Containing Compounds

Catabolism of proteins and nucleic acids results in formation of so-called NPN compounds (*n*on*p*rotein *n*itrogenous compounds). The principal nonprotein nitrogenous substances, their biochemical origins, and situations in which their assay has significant clinical utility are summarized in Table 20-2. Several of these metabolic products are sequentially derived from the catabolism of either exogenous (dietary) or endogenous (tissue) proteins.

Table 20-2. URINARY NITROGENOUS METABOLITES*

Metabolite	Biochemical Origin	Clinical Utility of Measurement
Amino acids	Proteins, endogenous and exogenous	Liver disease; inborn errors of metabolism; tubular disorders
Ammonia†	Amino acids	Liver disease; renal disease, congenital and acquired; inborn errors of metabolism
Urea	Ammonia	Liver disease; renal disease
Creatinine	Creatine	Renal function
Uric acid	Purine nucleosides	"Marker" for cell turnover; disorders of purine synthesis

*These compounds account for >90% of urinary nonprotein nitrogen.
†Present in urine as NH_4^+.

UREA

Biochemistry and physiology. Urea is the major nitrogen-containing metabolic product of protein catabolism in humans, composing over 75% of the nonprotein nitrogen eventually excreted. The biosynthesis of urea from amino nitrogen–derived ammonia is carried out exclusively by hepatic enzymes of the urea cycle.

Over 90% of urea is excreted through the kidney; losses through the gastrointestinal tract and skin account for most of the remaining minor fraction. Urea is filtered freely by the glomeruli. In the normal kidney, 40–70% of the highly diffusible urea moves passively out of the renal tubule and into the interstitium, ultimately to re-enter plasma. The back-diffusion of urea is dependent on urine flow rate, and more enters the interstitium in slow-flow states. Consequently, urea clearance underestimates GFR.

Clinical utility of plasma urea measurement. A wide variety of renal diseases can cause an increase in plasma urea concentration. Unfortunately, the usefulness of urea as an independent indicator of renal function is limited by the variability of its blood levels as a result of nonrenal factors. Mild dehydration, high-protein diet, increased protein catabolism, muscle wasting as in starvation, reabsorption of blood proteins after a GI hemorrhage, treatment with cortisol or its synthetic analogs, or decreased perfusion of the kidneys may cause azotemia (increased blood urea), which is called *prerenal azotemia.* Impaired perfusion may be due to decreased cardiac output or shock due to blood loss or other causes. The key to identifying the azotemia as prerenal is documentation of an increase of plasma urea without concomitant increase of plasma creatinine. *Postrenal azotemia* has its cause in conditions that obstruct urine outflow through the ureters, bladder, or urethra; examples of these conditions are nephrolithiasis, prostatism, and tumors of the genitourinary tract. With obstruction, plasma urea and creatinine both rise, but there is a disproportionately greater rise of urea than of creatinine, because the obstruction of urine flow causes a back-pressure on the tubule and back-diffusion of urea into the blood from the tubule.

Thus, the principal clinical utility of plasma urea determination lies in its measurement in conjunction with the measurement of plasma creatinine and in discrimination between prerenal and postrenal azotemia. Clinicians frequently calculate a convenient relationship, the *urea nitrogen/creatinine ratio.* For a normal individual on a normal diet, the ratio ranges between 12 and 20; values for most individuals lie between 12 and 16. Significantly lower ratios usually denote acute tubular necrosis, low protein intake, starvation, or severe liver disease (decreased urea synthesis). High ratios with *normal* creatinine levels may be seen with catabolic states of tissue breakdown; prerenal azotemia; high protein intake, especially in uremic patients; and after gastrointestinal hemorrhage. High ratios associated with *elevated* creatinine concentrations may denote either postrenal obstruction or prerenal azotemia superimposed on renal disease. The ratio may also be affected by the degree of specificity of the method used for urea and creatinine determination. Thus, the urea nitrogen/creatinine ratio can show great variability and is only a rough guide to the nature of the underlying abnormality.

Methods for the Determination of Urea[7,21]

Indirect methods are based on preliminary hydrolysis of urea with urease followed by some process that quantitates the ammonium ion.[13]

$$\underset{\textbf{Urea}}{(H_2N)_2C{=}O} + H_2O \xrightarrow{Urease} 2\,NH_3 + CO_2 \xrightarrow{H_2O} 2\,NH_4^+ + CO_3^{2-}$$

Spectrophotometric approaches to ammonium quantitation include the *Berthelot reaction* and the *enzymatic assay* with glutamate dehydrogenase.

The method using the **Berthelot reaction** is based on the following principle:[7,32] Urea is hydrolyzed with urease, and the ammonium ion formed is reacted with phenol and hypochlorite in alkaline medium to form indophenol.

$$NH_4^+ + 5\,NaOCl + 2\,\underset{\textbf{Phenol}}{C_6H_5OH} \xrightarrow{OH^-} \underset{\textbf{Indophenol}}{O{=}C_6H_4{=}N{-}C_6H_4{-}O^-} + 5\,NaCl + 5\,H_2O$$

Nitroprusside is used to catalyze the reaction. Absorbance of dissociated indophenol, a blue chromophore, is measured at 560 nm.

For the enzymatic assay using glutamate dehydrogenase (GluDH), the principle is as follows:

$$\underset{\textbf{2-Oxo-glutarate}}{HOOC{-}(CH_2)_2{-}\overset{O}{\overset{\|}{C}}{-}COO^-} + NH_4^+ + NAD(P)H + H^- \underset{}{\overset{GluDH}{\rightleftarrows}}$$

$$NAD(P)^+ + H_2O + \underset{\textbf{Glutamate}}{HOOC{-}(CH_2)_2{-}\overset{\overset{+}{N}H_3}{\underset{H}{C}}{-}COO^-}$$

For serum or plasma assays, the reaction system for both approaches is usually formulated with urease so that the addition of sample containing urea starts the reaction. Decrease in absorbance, resulting from the glutamate dehydrogenase reaction, is monitored at 340 nm in either an endpoint or kinetic mode.[33] *Electrochemical* approaches to quantitation of ammonium ions involve conductimetry and potentiometry. The conductimetric method has been automated on the ASTRA instruments (Beckman Instruments, Brea, CA 92621), where a sample and a urease-containing reagent are introduced into a conductivity cell. The rate of change of the conductivity of the mixture, as nonionized urea is transformed to an ionic species, is measured. Potentiometry using an ammonium ion-selective electrode fitted with a membrane on which urease has been immobilized,[15] although promising, has not yet come into widespread use.

Indirect methods may be *standardized* with either urea or ammonium salt solutions, but urea is much to be preferred. Any of these methods are at risk of ammonia contamination of reagents or glassware. Some caution must be observed in selection of an indirect method for assay of urine specimens in which significant amounts of preformed ammonia may be present. Either the preformed ammonia must be removed with Permutite prior to assay or the method must prescribe a step that compensates for it before urease is added. Use of the enzymatic assay would require blanking by running sample in the reaction system both before and after urease addition. The automated conductimetric method noted above measures the *rate* of change in conductivity due to ammonia production, and thus preformed ammonia does not interfere.

Direct methods for urea are usually variations on the Fearon reaction, the condensation of diacetyl with urea to form the chromogen diazine. Since diacetyl is unstable, it is usually generated in the reaction system from diacetyl monoxime and acid.

$$\underset{\text{Diacetyl monoxime}}{CH_3-C(=O)-C(=NOH)-CH_3} + H_2O \xrightarrow{H^+} \underset{\text{Diacetyl}}{CH_3-C(=O)-C(=O)-CH_3} + \underset{\text{Hydroxylamine}}{HONH_2}$$

The reaction of diacetyl and urea gives diazine

$$\underset{\text{Urea}}{(H_2N)_2C{=}O} + CH_3-C(=O)-C(=O)-CH_3 \xrightarrow[\text{Heat}]{H^+} \underset{\text{Diazine}}{CH_3-C{=}N-C(=O)-N{=}C-CH_3\ (\text{ring})} + 2\,H_2O$$

which absorbs strongly at 540 nm. Thiosemicarbazide and Fe(III) are added to the system to enhance and stabilize the color. The method may be used manually and has been automated on various systems, among them the Technicon continuous-flow systems.[24] Urea must be used as standard for a direct method. The Fearon reaction is applicable to both plasma and urine assays.

Comments. Contemporary methods for serum or plasma, whether direct or indirect, rarely call for deproteinization of sample. The need has been obviated by selection of small specimen volumes and by optimization of reagent systems.

Although BUN (*b*lood *u*rea *n*itrogen) continues to be the terminology for ordering the plasma or serum urea nitrogen test, analysis of *blood* for urea has essentially been discontinued.

The long-established habit of reporting and expressing results of a urea assay in units of urea nitrogen appears to be strongly entrenched in the US, although the SI system recommends use of urea, expressed in mmol/L. Thus, it behooves the student of clinical chemistry to have in mind the conversion factors for urea to urea nitrogen. Since 60 g (one gram molecular weight) of urea contain 28 g (two gram atomic weights) of nitrogen, the factor is 0.467 for converting urea mass units to those of urea nitrogen, and 2.14 for converting urea nitrogen mass units to those of urea.

Specimen. Either serum or urine is suitable for the assay. Plasma is also suitable, provided that the anticoagulant used contains neither ammonium nor fluoride salts. Fluorides in high concentration inhibit urease. Urea in serum is stable for up to 24 h at room temperature, for at least several days at 4–6 °C, and for at least 2–3 months when frozen.

Reference ranges. The reference range for *plasma* of healthy ambulatory adults is 7–18 mg/dL as urea nitrogen (15–39 mg/dL or 2.5–6.4 mmol/L expressed as urea). Levels in the neonatal period are 3–6 mg/dL lower than in adults; in adults over 60 years of age, levels are 1–4 mg/dL higher than in younger adults.[40] Plasma concentrations also tend to be slightly higher in males than in females. High-protein diet causes significant increases in plasma urea concentrations and urinary excretion. On an average protein diet, *urinary excretion* expressed as urea nitrogen is 17–20 g/d.

CREATININE AND CREATINE

Biochemistry and physiology. Creatine is synthesized in the kidneys, liver, and pancreas and is then transported in blood to other organs such as muscle and brain, where it is phosphorylated to phosphocreatine, a high-energy compound. Interconversion of phosphocreatine and creatine is a particular feature of metabolic processes of muscle contraction; some of the free creatine in muscle spontaneously converts to creatinine, its anhydride. Between 1 and 2% of muscle creatine is converted to creatinine daily. Because the amount of endogenous creatinine produced is proportional to muscle mass, the production varies with age and sex: nonobese adult males excrete ~1.5 g/d, females, 1.2 g/d. Most of the interindividual variations of creatinine excretion in healthy subjects are attributable to age, sex, and lean body mass; intraindividual variation (<15% from day to day) is partly due to dietary fluctuations.

$$\text{Creatine} + ATP \xrightarrow{\text{Creatine kinase}} \text{Phosphocreatine} + ADP$$

Creatine: $^{-}O-C(=O)-CH_2-N(CH_3)-C(=NH)-NH_2$

Phosphocreatine: $^{-}O-C(=O)-CH_2-N(CH_3)-C(=NH)-NH-P(=O)(=O)-O^{-}$

Creatine → (Spontaneous, nonenzymatic cyclization) → H_2O + Creatinine

Phosphocreatine → (Spontaneous) → Creatinine + P_i

Creatinine: cyclic $H_2C-N(CH_3)-C(=NH)-NH-C(=O)$

Clinical Utility of Creatinine Clearance and Determination of Plasma Creatinine

Constancy of endogenous creatinine production and its release into the body fluids at a constant rate, and constancy of plasma levels of creatinine over the 24 h of a day, make creatinine a useful endogenous substance whose clearance may be measured as an indicator of glomerular filtration rate. However, a small quantity of creatinine is reabsorbed by the tubules and a small quantity of creatinine appearing in the urine (7–10%) is due to tubular secretion.

Two factors influence measurement of creatinine clearance and thus its correct interpretation. First, the most common methods for measuring creatinine employ the nonspecific alkaline picrate reaction, and thus noncreatinine chromogens in plasma increase the apparent plasma concentration by as much as 30% if serum values are <1.0 mg/dL and by ~10% if values are >1.0 mg/dL. The per cent increase is progressively less with higher creatinine concentrations. (Urine contains considerably less noncreatinine chromogens.) This overestimate of plasma creatinine concentration results in an underestimate of creatinine clearance and partially offsets the apparent high clearance of creatinine that is due to tubular secretion. As a result, the endogenous creatinine clearance, by coincidence, agrees closely with the inulin clearance over a substantial range of clearances.[26] If highly specific methods are used for assay of plasma creatinine, the GFR estimated by creatinine clearance may not be as well correlated with the GFR estimated by inulin clearance. Second, GFR measured by creatinine clearance and GFR measured by inulin clearance in the same patient progressively diverge as renal failure progresses and plasma creatinine rises. The greater apparent GFR found by creatinine clearance may be due to an increase in tubular secretory activity for creatinine when plasma levels increase much above normal and to the relatively smaller contribution of noncreatinine chromogens in a nonspecific assay of plasma creatinine.

The creatinine clearance as a measure of GFR has several unique advantages over the clearance of inulin or other exogenous substances. Not only are plasma and urine creatinine easily measured, but extensive data are available for reference ranges of both plasma creatinine levels and creatinine clearances in individuals of all age groups. The creatinine clearance—flawed as it is—remains the most practical and popular method for estimating GFR in most hospitals.

The linear decrease of creatinine clearance over time, as renal function fails, has been thoroughly documented for several forms of chronic glomerulonephritis (including chronic membranous glomerulonephritis), diabetic nephropathy, chronic interstitial nephritis, and at least some cases of chronic pyelonephritis. Moreover, graphic representation of creatinine clearance as a function of time has some prognostic value, since it may allow the physician to estimate when the severity of renal failure may make dialysis mandatory. The reciprocal of plasma creatinine concentration has also been used as a guide in following and documenting the rate of loss of renal function in progressive, diffuse renal disease. As the creatinine clearance of a patient increases (or decreases), the reciprocal of his plasma creatinine concentration increases (or decreases) linearly.

Estimation of Creatinine Clearance from Plasma Creatinine Level

Clinical pharmacologists, pharmacokineticists, and clinicians, when supervising the administration of potentially toxic drugs cleared by the kidneys, commonly use creatinine clearance as an adjunct to dose management. Among the drugs of concern are the nephrotoxic aminoglycosides and also digoxin, guanethidine, sulfonamides, methotrexate, and 5-fluorouracil. Because plasma creatinine determinations are quicker and easier than the creatinine clearance procedure, a one-time plasma creatinine level is sometimes used, together with the patient's body weight, age, and sex, to arrive at an estimate of clearance.[8] The estimate may be based on a calculation[2,9,20] or derived from a nomogram (Figure 20-4).[35]

Estimations of clearance made by these techniques must be interpreted with great caution! The nomogram is based on the assumption that an increase in body weight represents mainly increase in lean body mass. Therefore, obesity of the patient is a source of potentially large error in the clearance estimate even if correction for weight is made, and thus the nomogram should not be used for obese individuals. The random error of the plasma creatinine determination is another source of error in estimating clearance.

Determination of Creatinine

The methods most widely used today are based on the *Jaffe reaction*. This reaction occurs between creatinine and the picrate ion formed in alkaline medium; a red-orange adduct develops, but in spite of extensive study its structure remains uncertain. Concentration of alkali in the reagent system is critical. The observed rate constant of adduct formation is a linear function of the hydroxyl ion concentration over a broad range of picric acid concentrations.[6] Not only does hydroxyl ion concentration determine the rate of reaction, it also specifies to a large extent the behavior of the spectral absorbance curve of the adduct over the wavelengths 485–520 nm.[28] Modern methods usually employ a sodium hydroxide concentration below 1.0 mol/L (in most instances below 0.5 mol/L), and picric acid in excess of stoichiometric amounts. In most methods, wavelengths of 505–520 nm are specified for spectrophotometry, and a narrow-bandpass spectrophotometer is required to provide appropriate analytical sensitivity and specificity.

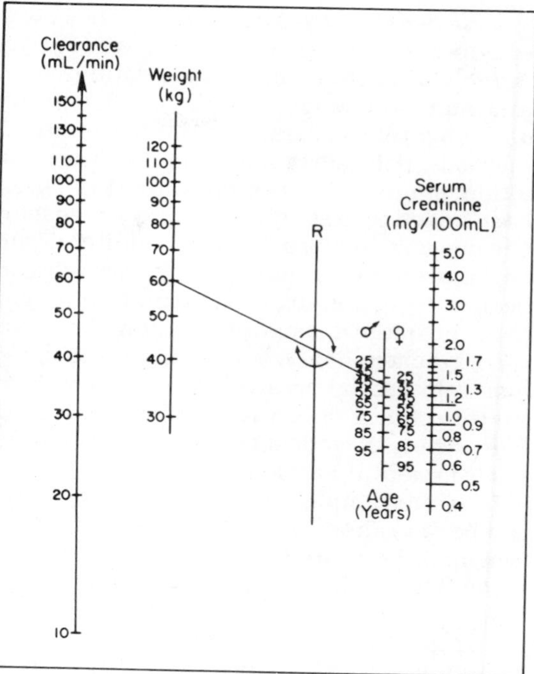

Figure 20-4. The Siersbaek-Nielsen nomogram for estimation of endogenous creatinine clearance. A point is first defined on the *R* line by the patient's body weight, age, and sex. The point is then lined up with the observed value for plasma creatinine (right-hand scale), and the estimate of clearance is read from the left-hand scale. (Drawn with permission from Siersbaek-Nielsen, K., Molholm-Hansen, J., Kampmann, J., et al.: Rapid evaluation of creatinine clearance. Lancet, *1*:1133–1134, 1971.)

The Jaffe reaction has long been known to be nonspecific when applied to the measurement of creatinine in plasma. Among the noncreatinine Jaffe-reacting chromogens are protein, glucose, ascorbic acid, guanidine, acetone, cephalosporins, and α-ketoacids such as acetoacetate and pyruvate.[37] Depending on the method, these "quasi-creatinine" compounds cause an overestimate of plasma "true" creatinine by 0.2–0.4 mg/dL. The numerous modifications of Jaffe-reaction methods have had as their chief purpose improvement of specificity for creatinine in plasma; urine has a relatively small content of the same interferents. Protein effect has been eliminated in automated continuous-flow systems by dialysis, in manual methods by a protein precipitation step (as in the method discussed below), or by adding sodium dodecyl sulfate to the reagent in order to form nonreactive protein complexes. Glucose and ascorbic acid have been complexed with borate in a buffered system.[17]

Another approach to eliminate interferences in the Jaffe reaction has been acid-blanking, i.e., addition of acid to the reaction mixture after color development and subtraction of this value from the absorbance value obtained without acid addition. Treating a protein-free filtrate with Lloyd's reagent (aluminum silicate)[16] or with ion-exchange resin to adsorb and remove noncreatinine chromogens, although generally successful in improving specificity, is considered an excessively time-consuming procedure and tends to increase method imprecision.[28]

Manual methods have traditionally been endpoint methods, with 10–15 min allowed for color development at room temperature. *Kinetic assays* (i.e., those measuring the rate of the pseudo first-order reaction that forms adduct), were developed in a quest both for specificity and for fast, automatable analyses. In these assays, temperature control is important in order to ensure reproducibility of rate measurements.[4,10] Timing of measurements becomes critical to specificity since two noncreatinine chromogens can be distinguished by rate methods. One kind reacts rapidly only in the first 20 s after reagent addition; the other does not react until 80 s or more after reagent addition. Therefore, the "window" between 20 and 80 s after addition and mixing of reagent is a period in which the observed rate of adduct formation can be attributed mainly to creatinine. A variety of automated microprocessor-controlled instruments implement the kinetic Jaffe reaction by measuring the rate 25–60 s after reagent addition and at a controlled temperature.

Enzymatic or partially enzymatic assays for creatinine have had a long history. Two enzymes are of particular interest: (1) creatininase (EC 3.5.4.21), also called creatinine *imino*hydrolase or deiminase, catalyzes creatinine degradation to *N*-methylhydantoin and ammonium ion; and (2) creatinine hydrolase (EC 3.5.2.10), also called creatinine *amido*hydrolase, catalyzes the hydrolysis of creatinine to creatine. Partially enzymatic methods, both early manual and recent automated ones, have used these two enzymes, either singly or in combination, to achieve specificity of analysis. The enzymes have been added to destroy creatinine, so measurement of the difference of total Jaffe-reacting chromogens (before enzyme addition) and of residual, noncreatinine Jaffe-reacting chromogens (after enzyme addition) represents the true creatinine. Creatininase has also been used in conjunction with an ammonium ion–selective electrode or with indicator dyes for quantitating ammonium. An example of the latter approach is the method on the Ektachem analyzer (Eastman Kodak Co., Rochester, NY 14650), in which ammonium ions released by creatininase react with bromphenol blue and the color is determined by reflectance spectrophotometry. Methods using creatininase are subject to risks of high levels of endogenous ammonia and contamination of the reaction mixture with environmental ammonia.

In one completely enzymatic assay, creatinine is hydrolyzed to form creatine, which can be measured by using creatine kinase. The indicator reaction in this coupled-enzyme system is monitored spectrophotometrically at 340 nm in either an end point or a kinetic mode. The method is highly specific and has been adapted for both manual and automated use.[19]

High performance liquid chromatography (HPLC) has also been described for creatinine assay.[36] Although the HPLC assay is impractical for routine analysis of creatinine, its specificity recommends it as a reference method and for validating other methods.

Importance of Specificity

Currently popular methods for assaying plasma creatinine have varying degrees of specificity. Laboratory workers who serve active renal disease centers must recognize how the specificity of the method used in their facility affects the clinical application of results. Reference ranges for creatinine clearance, the reciprocal of plasma creatinine concentration, and plasma urea nitrogen/creatinine ratio—because each of these is a calculated value—will vary depending on how "true" is the creatinine value used in the calculation (i.e., how specific is the method used to obtain the result). The more a method overestimates true creatinine, the greater the underestimation of these three values.

As methods of more uniform and greater specificity are developed and utilized, reference intervals for plasma creatinine levels probably will become less method dependent and will be 0.1–0.3 mg/dL (10–25 μmol/L) lower than many of those currently quoted.

Determination of Creatinine by Using the Jaffe Reaction[13]

Specimen. Either plasma or serum is suitable. Creatinine in the sample is stable for at least 7 d in the refrigerator and indefinitely when frozen. Significant hemolysis of a blood specimen may cause spurious elevations of creatinine values for serum or plasma since rupture of erythrocytes increases the concentration of noncreatinine chromogens.

Creatinine in urine is stable for 2–3 d at room temperature and for at least 5 d refrigerated. If urine must be collected with a preservative for other analytes, only thymol or toluene may be used.

Principle. Creatinine in a protein-free supernatant of plasma or serum is reacted with alkaline picrate to form a color complex whose intensity is measured at 510 nm.[39]

Comments. The random error of manual methods for creatinine may be significantly greater than that of automated methods. Recommended practice is to determine reproducibility for whatever method is used, particularly for plasma or serum, and to inform clinicians who use plasma creatinine values for supervision of their renal patient of the magnitude of random error. Standard deviation for day-to-day reproducibility should be of the order of 0.10–0.15 mg/dL so that the physician can rely on a change of 0.2–0.3 mg/dL as signifying real change in the patient's plasma creatinine level.

Reference ranges[40] for *serum* or *plasma* creatinine, measured by methods similar to the one described above, are 0.9–1.5 mg/dL (80–133 μmol/L) in men and 0.7–1.3 mg/dL (62–115 μmol/L) in women. In specific methods, values for males are 0.72 – 1.22 mg/dL (64–108 μmol/L) and for females, 0.53–1.00 mg/dL (47–90 μmol/L). Values in children are lower.

Urinary creatinine excretion is 14–26 mg/kg/d (124–230 μmol/kg/d) in adult men and 11–20 mg/kg/d (97–177 μmol/kg/d) in adult women. There is some intraindividual variability in renal excretion of creatinine that can be assigned to the amount of dietary (exogenous) creatinine ingested in meats. This variability may account for the 15–20% between-day variation in calculated creatinine clearances for a given individual. Creatinine excretion decreases with age.

Endogenous Creatinine Clearance

Principle. Renal clearance of creatinine is calculated from plasma creatinine concentration and mass of creatinine excreted in the urine over a fixed period of time (see also page 674).

Procedure

1. Hydrate the patient with at least 600 mL of water. Note that special orders may have been given by the attending physician of a patient with documented renal disease; these special orders may override recommendations made here with respect to water intake or withholding of medication.
2. Withhold tea, coffee, and drugs on the day of the test.
3. Have the patient void and discard the specimen. Note time, and from then on collect all urine passed for 4, 12, or 24 h. Keep patient well hydrated during collection period in order to assure a urine flow rate of ≥2 mL/min.
4. Collect a blood specimen. Because plasma creatinine concentration is relatively constant, the collection can be made at any time during the urine collection period. For the sake of consistency, however, blood collection at midpoint of the urine collection period is recommended.
5. In the laboratory, measure the volume of the total urine collected and record both volume and minutes (h × 60) of the period in which it was collected. Perform the assays of plasma (serum) and urine creatinine.
6. Calculate clearance:

$$\text{Plasma cleared, mL/min, per standard surface area} = \frac{U_{Cr} \times V}{P_{Cr}} \times \frac{1.73}{A}$$

where U_{Cr} = concentration of creatinine in urine, in mL/min
P_{Cr} = concentration of creatinine in plasma in the same units as for urine
V = volume of urine flow in mL/min
A = body surface area in square meters

The factor 1.73/A normalizes clearance for average body surface, correcting as it does for variation in creatinine excretion relative to lean body (muscle) mass. Nomograms for determining body surface area from height and weight are found in the *Appendix* (Figures 29-1 and 29-2).

Sources of Error

1. Errors in recording the timing of collection period or loss of a portion of the urine during collection are the most common sources of error.
2. Vigorous exercise during the urine collection period may alter clearance.

3. Proper hydration of the patient to assure urine flow rate of ≥2 mL/min improves the accuracy of the measurement of filtration rate and tends to eliminate retention of urine in the bladder as a source of negative error.

Reference Ranges[40]

With the creatinine method described in this chapter, clearance for males is 105 ± 20 mL/min; for females it is 95 ± 20 mL/min. When a more specific method for creatinine is used, clearance is 117 ± 20 mL/min for males and 108 ± 20 mL/min for females. Clearance rates decrease with age at a rate of ~6.5 mL/min/decade.

Intraindividual variability in creatinine clearance is approximately ±15%; sequential determinations of creatinine clearance and averaging of values are required to reduce this variation appreciably.

Creatine in Serum and Urine

The determinations of creatine levels, once used in diagnosis of muscle dystrophic and atrophic diseases, have been superseded for that purpose by assays for activity of the muscle enzymes, particularly creatine kinase.

URIC ACID

Biochemistry and physiology. In humans, uric acid is the major product of the catabolism of the purine nucleosides, adenosine and guanosine. Purines from catabolism of dietary nucleic acid are converted to uric acid directly. However, the bulk of purines ultimately excreted as uric acid in the urine arises from degradation of endogenous nucleic acids. The catabolism of purines is illustrated in outline in Figure 20-5.

Reutilization of the major purine bases, adenine, hypoxanthine, and guanine, is achieved through "salvage" pathways in which phosphoribosylation of the free bases causes resynthesis of the respective nucleotide monophosphates.

Lower primates and mammals other than humans carry purine catabolism one step further with the formation of allantoin from uric acid, a step mediated by uricase. Approximately 75% of uric acid excreted is lost in the urine; most of the remainder is

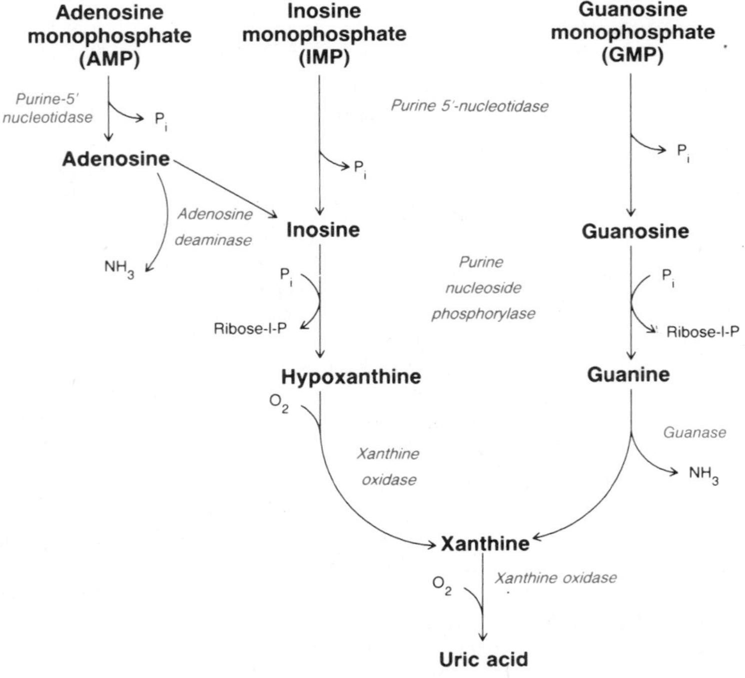

Figure 20-5. Catabolism of purines.

secreted into the gastrointestinal tract, where it is degraded to allantoin and other compounds by bacterial enzymes. Renal handling of uric acid involves glomerular filtration, reabsorption in the proximal convoluted tubule, subsequent secretion of uric acid into the distal portion of the tubule, and further reabsorption in the distal tubule. The net urinary excretion of uric acid is 6–12% of the amount filtered.

The physicochemical properties of uric acid are important in considering uric acid concentrations in the circulation, in tissues, and in the kidney. The first pK_a of uric acid is 5.75; above this pH, uric acid exists chiefly as urate ion, which is more soluble than uric acid. At urine pH below 5.75, uric acid is the predominant form.

Clinical significance. *Hyperuricemia* is most commonly defined by serum or plasma uric acid concentrations >7.0 mg/dL in men or >6.0 mg/dL in women (if specific methods are used to measure uric acid). The major causes of hyperuricemia are summarized in Table 20-3. Asymptomatic hyperuricemia is frequently detected through biochemical screening; long-term follow-up of asymptomatic hyperuricemic subjects is undertaken because many are at risk for renal disease that may develop as a result of hyperuricemia and hyperuricosuria.

Gout occurs when monosodium urate precipitates from supersaturated body fluids; the deposits of urate are responsible for the clinical signs and symptoms. Gouty arthritis may be associated with urate crystals in joint fluid as well as with deposits of crystals (tophi) in tissues surrounding the joint. The deposits may occur in other soft tissues as well, and wherever they occur they elicit an intense inflammatory response consisting of infiltration by polymorphonuclear leukocytes and macrophages. Renal disease associated with hyperuricemia may take one or more of several forms: (1) gouty nephropathy with urate deposition in renal parenchyma, (2) acute intratubular deposition of urate crystals, and (3) urate nephrolithiasis.

Gout may be classified as primary or secondary. *Primary gout* is associated with essential hyperuricemia; i.e., that due to metabolic overproduction of purines or to underexcretion of uric acid. *Secondary gout* is a result of hyperuricemia attributable to several identifiable causes. Renal retention of uric acid may occur in acute or chronic renal disease of any type or as a consequence of administration of drugs; diuretics, in particular, are implicated. Organic acidemia may interfere with tubular secretion of urate. Increased nucleic acid turnover and consequent increase in catabolism of purines may be seen in rapid proliferation of tumor cells as well as in massive destruction of tumor cells upon therapy with certain chemotherapeutic agents.

Hyperuricemia is also attributable to primary defects of enzymes in the pathways of purine metabolism. The *Lesch-Nyhan* syndrome is characterized by complete deficiency of hypoxanthine-guanine phosphoribosyl transferase (HGPRT), the major enzyme of the purine salvage pathways. Glucose-6-phosphatase deficiency also leads to hyperuricemia due to both overproduction and underexcretion of uric acid.

Quantitation of urinary uric acid excretion is an aid in selecting the appropriate

Table 20-3. CAUSES OF HYPERURICEMIA

"Essential" hyperuricemia
Overproduction (associated with hyperuricaciduria)
Underexcretion (associated with normal or decreased renal excretion of uric acid)
Renal retention
Renal failure
Drug therapy: diuretics, salicylates, pyrazinamide, ethambutol
Poisons: lead, alcohol
Organic aciduria: acetoacetate, lactate
Endocrinopathies: hypothyroidism, hyperparathyroidism
Increased turnover of nucleic acids
Myeloproliferative syndromes
Chemotherapy of malignant tumors, especially leukemias and lymphomas
Specific enzyme defects
Deficiency of hypoxanthine-guanine phosphoribosyl transferase
Complete (Lesch-Nyhan syndrome)
Partial
Abnormal phosphoribosyl pyrophosphate synthetase

ment for asymptomatic hyperuricemia. Hyperuricemic patients excreting <600 mg uric acid daily are candidates for treatment with uricosuric drugs, which enhance renal excretion of uric acid by blocking the carriers in the tubular cells that mediate reabsorption. Patients excreting >600 mg/d are candidates for treatment with allopurinol, a drug that suppresses degradation of hypoxanthine to uric acid.

About one in five patients with clinical gout will also have urinary tract uric acid stones. In about 50% of patients, uric acid stones are relatable to either hyperuricosuria or excretion of a persistently acid urine, or both. Undissociated uric acid (pK_a 5.75) is relatively insoluble, whereas urate at pH 7.0 is more than ten times as soluble. Thus, in patients with urinary pH persistently <6.0, relatively small amounts of uric acid in urine may produce supersaturation. Measurement of both urine pH and uric acid excretion are important in the investigation of uric acid urolithiasis. In any patient with urolithiasis, identification of crystals present in urine may provide a significant clue as to the nature of stones present. Pure uric acid stones represent 5–10% of all urinary tract stones.

Hypouricemia, often defined as serum urate concentrations below 2.0 mg/dL, is much less common than hyperuricemia. It may be secondary to any one of a number of underlying conditions. Severe hepatocellular disease with reduced purine synthesis or xanthine oxidase activity is one possibility. Another is defective renal tubular reabsorption of uric acid, either congenital, as in generalized Fanconi's syndrome, or acquired.

Methods for the Determination of Uric Acid

Methods in current use for assay of uric acid fall into two groups: phosphotungstic acid (PTA) methods and uricase methods.

Phosphotungstic acid methods rely on the development of a blue color ("tungsten blue") as PTA is reduced by urate in alkaline medium; the color is read by spectrophotometry at wavelengths of 650–700 nm. The presence of protein during the color development causes turbidity and quenching of the absorbance. In methods for plasma uric acid, protein removal is therefore an obligatory step. Removal has been accomplished in automated continuous-flow methods by dialysis and in manual methods by preparation of a $ZnSO_4/Ba(OH)_2$ protein-free filtrate or supernatant. These approaches avoid pH values of <6.0 and consequent negative error due to co-precipitation of urate as poorly soluble uric acid. Another approach uses PTA in excess as the precipitant in the protein removal step; the color reaction is then initiated by adding base to the supernatant that contains the PTA-urate complex. However, even in the absence of protein, the color system may develop turbidity.

PTA methods are subject to many interferences, including *endogenous* compounds such as glucose and ascorbic acid in plasma or urine and glutathione, ergothionine, and cysteine spilled into plasma from hemolyzed erythrocytes; and *exogenous* compounds such as acetaminophen, acetylsalicylic acid, gentisic acid (a salicylate metabolite), and the substituted purines such as caffeine, theobromine, and theophylline. All of these compounds also reduce phosphotungstic acid and thus introduce a positive error. Efforts to modify PTA methods have had little success in improving their specificity.

Uricase methods are inherently more specific because they have, either as a single step or as the initial step, urate oxidation catalyzed by the enzyme uricase. Preliminary precipitation of protein is not required. The reaction common to all enzymatic methods is

$$\text{Urate} + O_2 \xrightarrow{\text{Uricase}} \text{Allantoin} + H_2O_2 + CO_2$$

Urate **Allantoin**

Some uricase methods are based on this reaction after the plasma or urine is diluted with a borate or Tris buffer. At an alkaline pH, gaseous CO_2 is converted to bicarbonate, which prevents bubbles from interfering with spectrophotometry. The reaction can be observed in either the kinetic or the end point mode. The decrease of absorbance as urate is converted may be monitored spectrophotometrically at 282–292 nm.

Coupled uricase methods as a rule utilize the hydrogen peroxide produced by an initial uricase reaction as substrate for a chemical indicator reaction involving peroxidase or catalase.[41] An example of an automated uricase/catalase system with a chemical indicator reaction is the following:

$$H_2O_2 + \underset{\textbf{Methanol}}{CH_3OH} \xrightarrow{Catalase} \underset{\textbf{Formaldehyde}}{H_2C{=}O} + 2\,H_2O$$

$$H_2C{=}O + 3\,\underset{\textbf{Acetylacetone}}{H_3C{-}\overset{O}{\overset{\|}{C}}{-}CH_2{-}\overset{O}{\overset{\|}{C}}{-}CH_3} + NH_3 \longrightarrow 3\,H_2O + \textbf{3,5-Diacetyl-1,4-dihydrolutidine}$$

The chromophore absorbs at 410 nm.

Another method uses an enzymatic indicator system to follow H_2O_2 production.

$$H_2O_2 + \underset{\textbf{Ethanol}}{CH_3CH_2OH} \xrightarrow{Catalase} \underset{\textbf{Acetaldehyde}}{CH_3CHO} + 2\,H_2O$$

$$CH_3CHO + NAD^+ \xrightarrow[H_2O]{\text{Aldehyde dehydrogenase}} \underset{\textbf{Acetate}}{CH_3COO^-} + NADH + 2\,H^+$$

Increase in absorbance is followed spectrophotometrically at 340 nm.

High performance liquid chromatographic methods on reversed-phase columns are sufficiently specific to serve as reference methods.[29]

Reference ranges.[40] Reference intervals are method dependent; studies using PTA methods show higher values than those using enzymatic assays. For PTA methods, ranges are 4.5–8.2 mg/dL for males and 3.0–6.5 mg/dL for females. For uricase methods the corresponding values are 3.5–7.2 and 2.6–6.0 mg/dL.

Renal Excretion of Amino Acids

Amino acids in plasma are filtered by the glomeruli and appear in the glomerular filtrate in the same proportions as they do in plasma. A great portion of amino acids is reabsorbed by the renal tubular cells. In some congenital disorders there is a defect in the reabsorption of amino acids that results in aminoaciduria. An example of such a condition is cystinuria, in which there is a failure to reabsorb dibasic amino acids (cystine, lysine, arginine, and ornithine). In Fanconi's syndrome there is a failure to reabsorb a wide variety of amino acids. These and other forms of aminoaciduria are discussed in more detail in Chapter 11.

In conditions associated with decreased glomerular filtration, amino acids are retained, together with other nonprotein nitrogenous compounds, and contribute to the acidity and the organic anion fraction of plasma.

Renal Tubular Acidosis (RTA)

RTA is a result of decreased reabsorption of bicarbonate by the proximal renal tubules, or the result of insufficient acidification of the distal tubular fluid due to different causes. According to the specific defect, renal tubular acidosis is classified as proximal RTA, distal RTA, a combination of both, or RTA due to selective aldosterone deficiency. RTA is usually associated with hyperchloremia, a normal anion gap, and urinary

HCO_3^- or H^+ excretion inappropriate for the plasma pH. Hyperchloremia is caused by enhanced Cl^- reabsorption stimulated by contraction of the extracellular volume. A summary of the most important laboratory findings in RTA is given in Table 20-4. The interested reader is referred to more detailed discussions of the topic in references 1 and 11.

Proximal renal tubular acidosis (RTA Type II). Primary or idiopathic proximal RTA (PRTA) can be either sporadic or genetically transmitted. It occurs chiefly in male infants and is commonly associated with growth retardation. Proximal RTA may also occur secondary to inherited systemic diseases, such as cystinosis, tyrosinosis, hereditary fructose intolerance, Wilson's disease, and Lowe's syndrome. Drugs or toxic agents known to produce PRTA are carbonic anhydrase inhibitors, heavy metals, outdated tetracyclines, acetazolamide, and streptozotocin. PRTA is usually not an isolated defect but is accompanied by other tubular dysfunctions that result in hyperaminoaciduria, glycosuria, phosphaturia, and uricosuria (Fanconi's syndrome). The renal loss of ions and compounds shown in Table 20-4 may lead to chronic volume depletion with secondary hyperaldosteronism, malnutrition, failure to thrive, and osteomalacia.

Distal renal tubular acidosis (DRTA). Classic DRTA (DRTA Type I) occurs most often in infants (sometimes transiently) or young children, but it may also be seen in adults. The condition may occur: (1) as an autosomal dominant condition with higher incidence in females; (2) as a sporadic, nonfamilial disease; (3) as the result of administration of drugs such as amphotericin B, gentamicin, phenacetin, and lithium or after inhaling toluene (e.g., glue sniffing); (4) as a consequence of a variety of autoimmune disorders; and (5) in low renin/low aldosterone states, resulting in low H^+ and low K^+ excretion, as in diabetes (RTA Type IV).

The condition presents frequently with growth retardation; rickets and osteomalacia resulting from increased excretion of phosphate and hypophosphatemia; hypercalciuria often associated with nephrocalcinosis, renal calculi, renal colic, and pyelonephritis; and muscle weakness often caused by hypokalemia. The observed characteristic bone changes in children are termed *renal rickets.*

Combined proximal and distal RTA (RTA Type III). Certain features of both proximal and distal RTA have been reported in the same individual. The clinical and laboratory findings depend on the specific combination of defects that are present.

Selective aldosterone deficiency (RTA Type IV). In this form of RTA, hyperchloremia is present in about 75% of cases and hyperkalemia, although mild, is a usual manifestation. The hyperchloremic, hyperkalemic acidosis is primarily caused by a decreased glomerular filtration rate; it does not appear to be the result of impaired proximal bicarbonate reabsorption. The symptoms of this disorder are similar to those seen in *aldosterone resistance.*

Water Homeostasis

The kidney is the principal organ for regulating both the total water content and the total solute content of the body. It maintains a precise balance between water and solute, and it controls plasma osmolality within an extremely close tolerance (283 ± 3 mOsm/kg). In humans, the kidney can vary the rate of excretion of water and solute independently to achieve urine osmolalities as low as 50 mOsm/kg and as high as 1200 mOsm/kg or greater.

Urine volume. In health, water homeostasis is determined by several interrelated processes: (1) water intake and water formed through oxidation of foodstuffs; (2) extrarenal water loss through the feces, insensible water loss, and sweating; (3) a solute load to be excreted that is derived from ingested minerals and nitrogenous substances; and (4) the ability of the kidney to produce a concentrated or dilute urine. Other factors such as vomiting and diarrhea become important in various disease states. Loss of ability to produce a concentrated urine is a hallmark of virtually all types of chronic renal diseases.

To maintain water homeostasis, the kidney must produce urine in a volume that precisely balances water intake and production with water loss through extrarenal

Table 20-4. CHARACTERIZATION OF RENAL ACIDIFICATION DEFECTS

	Minimal Urine pH	Renal NH_4^+ Excretion	Reabsorption of Filtered HCO_3^-	Plasma Cl^-	Plasma K^+	Other Frequent Laboratory Findings
End-stage renal disease	<4.8	Markedly reduced	Reduced	Normal or low	Normal or increased	Azotemia, hyperuricemia
Proximal renal tubular acidosis	>5.5, but <5.5 during acidosis	Low at urine pH >6.0; normal at urine pH <5.5	Severely impaired; near normal at low plasma HCO_3^-; fractional HCO_3^- excretion $>15\%$	Increased	Low or normal	Hypocalcemia, hypophosphatemia, hypouricemia; volume depletion, increased aldosterone; aminoaciduria, glycosuria; increased phosphate and uric acid clearance; lysozyme and immunoglobulin light chains in urine; Na and K wasting
Distal renal tubular acidosis	>5.5	Reduced	Slightly reduced; fractional HCO_3^- excretion 3–5%	Increased	Low	Hypophosphatemia, hypokalemia; hyperchloremic acidosis; increased PTH if hypovolemia present; increased excretion of Ca, Na, K, uric acid; decreased titratable acidity; no increase in urine pCO_2 during alkaline diuresis
Hyperkalemic DRTA	>5.5	Reduced	Reduced	Increased	Increased	K excretion not lowered in response to Na_2SO_4
Selective aldosterone deficiency (SAD; RTA IV)	>5.5	May be reduced due to hyperkalemia		Increased in 75%	Increased (modestly)	Urine pCO_2 normal during alkaline diuresis

routes. Minimum urine volume may be as low as 400 mL/d, whereas maximum urine volume may be as high as 20–25 L/d. The kidney is more proficient in excreting excess water than in conserving water. Fluid intake that is inadequate to provide the volume necessary to excrete the solute load will result in a net loss of body water and will produce prerenal azotemia. This condition is frequently seen in elderly patients with compromised renal concentrating ability and can be treated simply by increasing fluid intake.

Solute load. The solute load that must be excreted by the kidney is determined by the mineral and nitrogenous content of the diet or, in the fasting state, by the amounts of these materials produced by breakdown of body tissues. The usual American diet produces a solute load of 100–1200 mOsm/d, although the load will be higher in those who consume large meat meals and who use excessive amounts of salt. Carbohydrates and triglycerides produce no metabolic end-products that require renal excretion; however, metabolism of proteins and nucleic acids results in formation of urea, creatinine, uric acid, and various inorganic ions that contribute to the solute load. Thus, the solute load can be reduced by providing a high-carbohydrate, high-fat, low-protein, low-salt diet, which is usually prescribed for patients in acute or chronic renal failure.

In certain disease states the oxidation of carbohydrates and triglycerides results in the production of large quantities of organic acids (e.g., lactate, acetoacetate, β-hydroxybutyrate), and the excretion of these compounds by the kidneys requires both water and Na^+. The highest rate of solute excretion encountered in disease is in diabetes mellitus, when solute excretion rates may approach 5000 mOsm/d. This rate of solute excretion can lead very rapidly to profound water deficits, particularly in elderly diabetics.

Control of Osmolality

All cell membranes are freely permeable to water, and osmolality is the same in all fluid compartments of the body (intracellular, extracellular, intravascular). Measurement of plasma osmolality therefore provides an indication of the total body fluid osmolality. An increase in plasma osmolality initiates two responses: (1) thirst, which stimulates an increase in water intake, and (2) secretion of antidiuretic hormone (ADH), which stimulates renal tubular reabsorption of water. The threshold for thirst is higher than for ADH secretion, and body osmolality is controlled largely by the effects of ADH on renal water conservation and excretion.

Antidiuretic hormone is synthesized in the hypothalamus and then transported complexed with a carrier protein in the neural tracts connecting the hypothalamus with the posterior portion of the pituitary (the neurohypophysis). Release of ADH from its storage site in the posterior pituitary is controlled by osmoreceptors located in the hypothalamus. Plasma hyperosmolality stimulates release of ADH, while contraction of extracellular fluid volume can also stimulate secretion of ADH and can apparently override the stimulus from hyperosmolality; however, this override does not occur until there has been a 10% loss in fluid volume. ADH acts on the cortical collecting tubules and the collecting ducts of the nephrons to make them permeable to water. The degree of permeability is determined by the rate of ADH secretion. In the absence of ADH the tubules are impermeable to water. The mechanism of action of ADH consists of activation of adenylate cyclase with the production of cAMP and increased phosphorylation of key proteins that increase the permeability to water.

The Countercurrent Multiplier System

The countercurrent multiplier system is the mechanism through which a large osmolal gradient is established in the interstitial tissue between the corticomedullary junction and the tips of the renal papillae (Figure 20-6). The medullary collecting ducts course through this hyperosmolal interstitium. In the presence of ADH, which renders the collecting ducts permeable to water, fluid is reabsorbed until the osmolality within the collecting ducts is equal to that of the interstitium, thus producing a hyperosmolal urine. In the absence of ADH, the collecting ducts are impermeable to water and the dilute tubular fluid traverses the medulla unchanged, resulting in a hyposmolal urine.

The urine concentrating/diluting system involves the loops of Henle, the distal

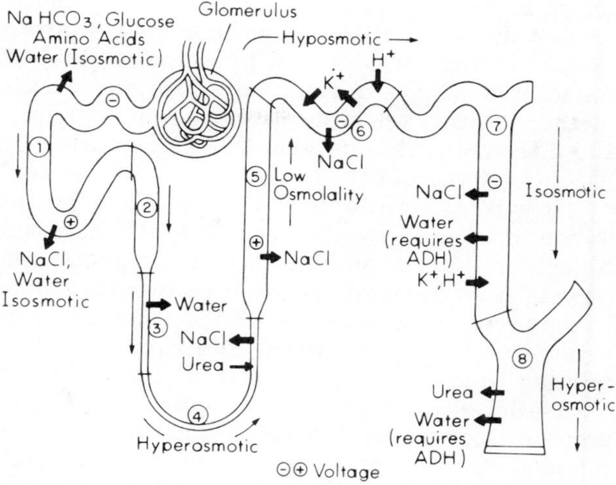

Figure 20-6. Schematic representation of the principal processes of transport in the nephron. In the convoluted portion of the proximal tubule (1) salt and water are reabsorbed at high rates in isotonic proportions. Bulk reabsorption of most of the filtrate (65–70%) and virtually complete reabsorption of glucose, amino acids, and bicarbonate take place in this segment. In the pars recta (2) organic acids are secreted and continuous reabsorption of sodium chloride takes place. The loop of Henle comprises three segments: The thin descending (3) and ascending (4) limbs and the thick ascending limb (5). The fluid becomes hyperosmotic, because of water abstraction, as it flows toward the bend of the loop, and it becomes hyposmotic, because of sodium chloride reabsorption, as it flows toward the distal convoluted tubule (6). Active sodium reabsorption occurs in the distal convoluted tubule and in the cortical collecting tubule (7). This latter segment is water-impermeable in the absence of ADH, and the reabsorption of sodium in this segment is increased by aldosterone. The collecting duct (8) allows equilibration of water with the hyperosmotic interstitium when ADH is present. (Adapted from Burg, M. B.: The kidney in health and disease. V. The nephron in transport of sodium, amino acids, and glucose. Hosp. Pract., *13* (Oct.): 100, 1978. Artist: Alan D. Iselin.)

convoluted tubules, the cortical collecting tubules, the collecting ducts, and the associated capillary vasa recta. It is important to recognize that in humans there is no physiological mechanism for active transport of water. Solutes, primarily sodium and chloride, can be actively "pumped" from the tubular lumen into the interstitium, and water will follow passively if the wall of the tubule is permeable to water.

Failure of ADH production results in marked polyuria (diabetes insipidus), which stimulates thirst and greatly increases water intake. Lack of response of the kidney to ADH has a similar effect (nephrogenic diabetes insipidus). A number of other conditions are also characterized by diminished ability to produce a concentrated urine (Table 20-5).

Urine Production

Consideration of the volumes and osmolalities of fluid during the sequential steps of urine production will illustrate the function of the concentrating/diluting mechanism of the kidney. Renal blood flow results in a renal plasma flow of 600 mL/min. One fifth of the renal plasma flow, or 120 mL/min, is filtered through the glomeruli as isosmotic fluid of very low protein content. About 60–80% of the glomerular filtrate is reabsorbed isosmotically in the proximal convoluted tubules. In the loops of Henle, reabsorption of

Table 20-5. DISEASES ASSOCIATED WITH DISTURBANCES IN THE RENAL CONCENTRATING MECHANISM

Marked polyuria and hypotonic urine after water deprivation
Diabetes insipidus
Hereditary nephrogenic diabetes insipidus
Chronic lithium toxicity
Sickle cell nephropathy
Hypokalemia (rarely)
Moderate polyuria and inability to produce hypertonic urine
Hypercalcemia
Hypokalemia
Chronic pyelonephritis
End-stage renal disease
Amyloidosis
Interstitial nephritis

solute and water does not occur isosmotically; about 25% of the sodium chloride and 15% of the water filtered by the glomeruli are reabsorbed. The higher solute reabsorption, of course, results in a hyposmolar fluid (100–150 mOsm/kg) entering the distal convoluted tubules. Additional solute is reabsorbed in the distal convoluted tubules and in the cortical collecting tubules and further lowers the osmolality of the fluid (50–100 mOsm/kg). It is only on those structures beyond the distal convoluted tubules that ADH exerts its effect.

It is important to note that the major portion of water reabsorption by the tubules is not influenced by the rate of ADH secretion. In the absence of ADH the cortical collecting tubules and the collecting ducts remain impermeable to water, and fluid flowing from the distal convoluted tubules is excreted as a large volume of dilute urine. In the presence of ADH the cortical collecting tubules and the medullary collecting ducts become permeable to water, and rapid osmolal equilibration takes place between the tubular fluid and the interstitial fluid. Water is reabsorbed in the cortex until the osmolality of the tubular fluid approximates that of plasma (~300 mOsm/kg), and this process accounts for the major part of water conservation during antidiuresis. It is in the medullary collecting ducts that the final concentration of tubular fluid takes place, increasing the osmolality of the tubular fluid from that of the cortical interstitium to the osmolality of the interstitium of the papillae.

Quantitation of Water Excretion and Conservation

The quantitative description of water conservation or excretion by the kidneys can be derived from clearance concepts previously presented. The kidneys are required to excrete a given amount of solute each day as determined by the daily intake of minerals and nitrogenous substances. The solute excretion rate can be calculated from the solute concentration in the urine (U_{Osm}) expressed in mOsm/kg and the urine flow rate (V) expressed in mL/min. (The calculations assume that 1 kg = 1 L of urine.)

$$\text{Solute excretion rate, mOsm/min} = U_{Osm} \times V \times 1000$$

Water excretion with varying solute loads can be expressed as the osmolal clearance (C_{Osm}), which is the hypothetical volume of plasma completely cleared of solute per unit time expressed in mL/min.

$$C_{Osm}, \text{mL/min} = \frac{U_{Osm} \times V}{P_{Osm}}$$

At a low rate of solute excretion, the ability of the kidney to reabsorb solute-free water is limited by its capacity to increase the osmolality of urine compared with plasma, the U:P osmolal ratio. This ratio varies between 3.5 and 4.0 in normal individuals. In advanced renal failure, the kidney is unable to produce a significantly hypertonic urine, and water conservation is seriously impaired.

Assessment of Renal Concentrating Ability

The capacity of the kidney to conserve water can be assessed simply by demonstrating that the solute concentration of the urine approaches the maximal range that can be achieved in health. The usual test consists of withholding fluids overnight, obtaining the first voided specimen in the morning, and measuring osmolality. After fluid deprivation of 18 h or more, a urine osmolality >850 mOsm/kg is considered to reflect a normal renal concentrating mechanism.

In patients with either hypothalamic or pituitary disorders causing complete ADH deficiency or in patients who lack the normal renal response to ADH (nephrogenic diabetes insipidus), urine osmolality rarely exceeds 300 mOsm/kg. However, partial defects in ADH secretion or renal response to ADH will result in urine osmolalities between 300 and 800 mOsm/kg. The assessment of partial deficiencies of ADH secretion or deficient renal response to ADH requires measurement of urinary osmolality or

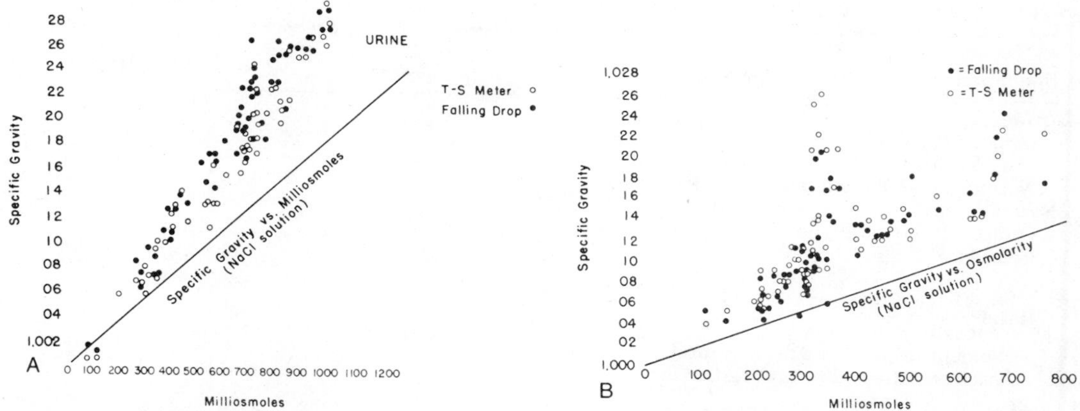

Figure 20-7. *A,* Comparison of the urinary specific gravity and urinary osmolality in a series of urines obtained from healthy medical students. The straight line represents comparative readings on various concentrations of sodium chloride solutions. *B,* Comparison of the urinary specific gravity and urinary osmolality in a series of unselected urines obtained from patients on the renal service. (From Holmes, J. H.: Workshop on Urinalysis and Renal Function Studies. Chicago, American Society of Clinical Pathologists, © 1962. Used by permission.)

measurement of plasma ADH levels or both during specific manipulations of plasma osmolality through dehydration of the patient or intravenous osmotic loading

A useful guide to the adequacy of the renal concentrating mechanism is the measurement of urine specific gravity, as long as the limitations of this procedure are recognized. This simple test measures the density of urine relative to the density of water, and in most individuals density bears a constant relationship to osmolality. This relationship is illustrated in Figure 20-7. In some conditions there is a remarkable divergence between osmolality and specific gravity. Urine collected after intravenous administration of iodine-containing radiopaque compounds for radiological studies may give extraordinarily high values, since heavy molecules affect density more than osmolality. Glucose and protein may also contribute substantial increments to the density of urine, and semiquantitative determination of these substances is necessary for valid interpretation or correction of urine specific gravity measurements. Diabetic patients with uncontrolled hyperglycemia and glucosuria may have a high urine specific gravity, even when the normal renal concentrating function is seriously impaired.

Determination of Specific Gravity with the Urinometer

Specimen. Specific gravity is most often determined as a part of a routine urinalysis on a random urine specimen. Less frequently, however, specific gravity is measured on timed specimens after water restriction, in which case more exact information is derived.

Reference range. Specific gravity values observed in healthy individuals vary greatly with fluid intake and the state of hydration. Thus, the normal values for a 24-h specimen are usually considered to be from 1.015 to 1.025. Values after fluid restriction are >1.025.

Principle. The urinometer is a hydrometer designed for the measurement of urinary specific gravity. When placed in the specimen contained in a cylinder, it sinks to the level characteristic of the specific gravity of the specimen. The value may then be read directly from the calibrations on the stem.

Sources of error. Readings may be in error if:

1. The urinometer is not allowed to float freely. It must not adhere to the sides of the tube, and no bubbles should cling to the stem.

2. There is failure to compensate for temperature or for gross proteinuria and glycosuria. Subtract 0.003 from the reading for each 1 g of protein/dL of urine; subtract 0.004 for each 1 g of glucose/dL of urine.

The urine specific gravity and osmolality of normal individuals may vary widely depending on the state of hydration (see Table 20-6).

Table 20-6. VALUES FOR OSMOLALITY OBSERVED UNDER VARIOUS CONDITIONS

Specimen	mOsm/kg H_2O
Urine	
Osmotic limits of renal dilution and concentration	50–1200
Normal random specimen (average fluid intake)	300–900
Normal range during maximum urine concentration (fluid restriction for 12 h)	850–1200
Serum	
Osmotic limits observed	230–490
Reference range	278–298 (mean: 283)
Ratio $\frac{\text{Urine osmolality}}{\text{Serum osmolality}}$	
Random urine (average fluid intake)	1.0–3.0
Specimen after 12-h fluid restriction	3.0–4.7
Ratio $\frac{\text{Serum Na}}{\text{Serum osmolality}}$	0.43–0.50

Determination of Specific Gravity by Refractometry

Specimen. See under *Determination of Specific Gravity with the Urinometer.*

Principle. The refractive index and the specific gravity of a urine specimen are both related functions of the quantity and type of dissolved substance in the specimen. Each substance contributes differently to the refractive index and also to the specific gravity; however, because various urine specimens are likely to contain dissolved substances of similar types and proportions, the refractive index and the specific gravity may be correlated. Increased amounts of abnormal substances such as glucose and protein may partially invalidate the correlation and give specific values that are misleading.

Osmolality of Serum and Urine

Measurements of the osmotic concentration of urine are considered more valid than specific gravity measurements in assessing the concentrating ability of the kidney, since the regulation of water excretion is, in part, determined by the osmolality of the fluid compartments of the body. Consequently, measurement of the urine osmolality, especially as part of a concentration test, is preferred. Holmes[18] performed a comparative study of specific gravity versus osmolality values in normal individuals and in unselected patients with renal disease. Although in normal individuals there appears to be a reasonably close relationship between specific gravity and osmolality, the relationship is maintained to a much lesser degree in patients with renal disease (see Figure 20-7). This lack of correlation can be explained at least partially by the fact that the presence of heavy molecules, such as protein, glucose, or iodine-containing compounds, affects the specific gravity of urine substantially more than its osmolality.

In *chronic progressive renal failure*, the concentrating ability of the tubules is diminished, which may be readily and reliably shown by means of urine osmolality measurements. In acute tubular necrosis, the urine osmolality, if there is urine output at all, approaches the osmolality of the glomerular filtrate.

In *polyuria due to diabetes insipidus*, the urine osmolality is extremely low, and this may serve as a means to differentiate this condition from polyuria due to neurogenic origin (see also *Ratio of Urine Osmolality to Serum Osmolality*). Also, if the fluid intake of patients with diabetes insipidus is restricted, only slight changes in osmolality are noted, whereas in the latter type of disease, urine will be concentrated to above 850 mOsm/kg (see Table 20-6).

Ratio of Urine Osmolality to Serum Osmolality

In normal individuals on an average fluid intake, this ratio is most frequently between 1.0 and 3.0. After fluid restriction, as in the case of the concentration tests, the ratio is 3.0 or above and may rise in extremes to 4.7. In patients with *renal tubular deficiency*, the ratio will be below that observed in normal individuals. In polyuria of *diabetes insipidus*, the ratio will be between 0.2 and 0.7 even after fluid restriction. This allows for differentiation of this type of polyuria from that seen in other conditions such as diabetes mellitus. In polyuria of neurogenic origin, the ratio may be normal without fluid restriction and will increase after fluid restriction.

Determination of Osmolality in Serum and Urine

Specimen. Blood should be collected by venipuncture with a minimum of stasis, and the serum should be separated by centrifugation soon after collection. In order to lessen the possible presence of particulate matter, a second centrifugation is recommended. If the serum is not to be analyzed soon after centrifuging, it should be refrigerated or frozen. Heparinized plasma is also satisfactory, but oxalated plasma is not.

Urine should be collected in clean, dry containers without preservatives and centrifuged at sufficiently high speed to remove all gross particulate material. If the analysis cannot be carried out soon after collection, the specimen should be refrigerated. Before analysis, refrigerated specimens should be warmed to aid the complete solution of any precipitated substances.

Several different osmometers are commercially available, and each one must be operated in a manner appropriate to the particular instrument, as described in its corresponding manual (see also Chapter 2, Section Five).

Comments and sources of error. It has been reported that a reproducibility of ±1 mOsm/kg H_2O with an equal accuracy can be obtained; however, under routine conditions an accuracy and precision of ±2 mOsm/kg H_2O is quite acceptable. Faulty standardization, faulty use of the osmometer, and the presence of particulate matter in the specimen are the most common sources of inaccurate results.

Renal Disease and the Role of the Laboratory

There are few renal diseases for which we know the entirety of etiology, pathophysiology, structural change, and clinical presentation, and many renal diseases are defined in terms of their clinical presentation and structural change alone. Consequently, many names for renal diseases actually represent clinical syndromes and their structural basis as revealed by renal biopsy. An additional difficulty in classifying renal diseases is the tendency of distinct primary diseases to merge with time into a single clinical entity, the end-stage kidney, regardless of their initial etiology or anatomic region of injury. This is primarily due to the intimate relationship between the anatomy and function of renal structures and their blood supply, as discussed below. This delicately balanced system is vulnerable to disease processes at many points. A diminution of the kidney's share of the normal cardiac output, about one-fifth of the total, may impair all the functions of the nephron and produce the clinical syndrome of prerenal azotemia. If severe, it may lead to ischemic injury of renal tubules and the syndrome of acute renal failure. Sustained renal ischemia triggers renin release from the kidney with resulting hypertension. Hypertension, in fact, is a complication of many renal diseases of varied etiologies. Additionally, severe hypertension itself can damage renal arterioles, further compromising nephron function. The complex interdependence of various nephron functions and their interrelated blood supply makes loss of entire individual nephrons, together with hypertrophy of those remaining, the common denominator of most chronic renal diseases. Whether the initial insult is immunologic injury of the glomerulus, infection of the interstitium, or toxic injury to tubular cells, the ultimate result is loss of the complete nephron. The loss of >90% of nephrons, resulting in the end-stage kidney, produces a remarkably consistent clinical and laboratory picture. Once chronic renal disease has reached this point, it may be impossible to determine the site and nature of the initial injury.

End-stage renal disease and the pathophysiology of the uremic syndrome. The uremic syndrome, the terminal clinical manifestation of kidney failure, is the constellation of symptoms, physical signs, and abnormal diagnostic studies that results from the failure of the kidneys to maintain adequate excretory, regulatory, and endocrine

Table 20-7. STAGES OF CHRONIC PROGRESSIVE RENAL DISEASE

Stage	Renal Function Remaining, %	Serum Creatinine, mg/dL	Serum Urea N, mg/dL
1. Decreased renal reserve	50–75	1.0–2.5	15–30
2. Renal insufficiency	25–50	2.5–6.0	25–60
3. Renal failure	10–25	5.5–11.0	55–110
4. Uremic syndrome	0–10	>8.0	>80

functions. Classic signs of uremia include progressive weakness and easy fatigability, loss of appetite followed by nausea and vomiting, muscle wasting, tremors, abnormal mental function, frequent but shallow respirations, and metabolic acidosis. The syndrome progresses to produce stupor, coma, and ultimately death unless support is provided by hemodialysis or successful renal transplantation. It is helpful to characterize progressive renal disease as occurring in four stages, defined by the percentage of renal function remaining and by the plasma concentrations of creatinine and urea nitrogen (Table 20-7), although there are no well-defined clinical syndromes that necessarily characterize these four stages.

The most characteristic *laboratory findings* are increased concentrations of nitrogenous compounds in plasma; e.g., urea nitrogen and creatinine (azotemia), as a result of reduced glomerular filtration rate and decreased tubular function. Retention of these compounds and of metabolic acids is followed by progressive hyperphosphatemia, hypocalcemia, and potentially dangerous hyperkalemia. Although most patients eventually become acidemic, respiratory compensation by elimination of CO_2 is extremely important. In addition, there is reduced endocrine function manifested by inadequate synthesis of erythropoietin and vitamin D with resulting anemia and osteomalacia. Disordered regulation of blood pressure generally leads to hypertension. Biochemical characteristics of the uremic syndrome are summarized in Table 20-8.

In addition to the consequences of reduced excretory, regulatory, and endocrine function of the kidney, there are several systemic manifestations of the uremic syndrome—among them pericarditis, pleuritis, disordered platelet and granulocyte function, and encephalopathy—that have been difficult to explain. Many retained metabolites have been implicated in the systemic toxicity of the uremic syndrome. The

Table 20-8. BIOCHEMICAL CHARACTERISTICS OF THE UREMIC SYNDROME

Retained nitrogenous metabolites
- Urea
- Cyanate
- Creatinine
- Guanidine compounds
- "Middle molecules"
- Uric acid

Fluid, acid-base, and electrolyte disturbances
- Fixed urine osmolality
- Metabolic acidosis (decreased blood pH, bicarbonate)
- Hypo- or hypernatremia
- Hypo- or hyperkalemia
- Hyperchloremia
- Hypocalcemia
- Hyperphosphatemia
- Hypermagnesemia

Carbohydrate intolerance
- Insulin resistance
  - Plasma insulin normal or increased
  - Delayed response to carbohydrate loading
  - Hyperglucagonemia

Abnormal lipid metabolism
- Hypertriglyceridemia
- Decreased HDL-cholesterol
- Hyperlipoproteinemia

Altered endocrine function[22]
- Secondary hyperparathyroidism
- Osteomalacia (secondary to abnormal vitamin D metabolism)
- Hyperreninemia and hyperaldosteronism
- Hyporeninemia
- Hypoaldosteronism
- Decreased erythropoietin production
- Altered thyroxine metabolism
- Gonadal dysfunction (increased prolactin and luteinizing hormone, decreased testosterone)

major nitrogenous metabolites are listed in Table 20-2. The syndrome itself is probably a result of the cumulative effect of many retained compounds acting as toxins and having an effect on metabolism in general.

Acute renal failure. This clinical syndrome occurs when there has been a rapid and severe reduction of GFR to <10 mL/min. Most of these patients are oliguric with urine outputs of <400 mL/d after fluid challenge. During this oliguric phase one of the most significant alterations is the inability to excrete sodium and water. This retention leads to expansion of extracellular fluid with resulting peripheral edema, hypertension, and congestive heart failure. In many patients, water is retained in excess of salt, producing hyponatremia. If the hyponatremia is severe, central nervous system (CNS) symptoms may result, beginning with somnolence and leading to seizures and coma.

As in chronic renal failure, the inability to excrete potassium may result in life-threatening hyperkalemia manifested by cardiac arrhythmias. This situation may be aggravated by metabolic acidosis due to the failure to excrete endogenous acid. Retention of phosphate leads to hyperphosphatemia, hypocalcemia, and later to secondary hyperparathyroidism. Hypermagnesemia may contribute to CNS depression and coma, as well as to suppression of cardiac pacemakers. As in chronic renal failure, there is an accumulation of the metabolic products creatinine and urea nitrogen. When acute renal failure is due to underperfusion of the kidney without intrinsic renal disease (prerenal azotemia), urea nitrogen will accumulate in excess of creatinine (high urea/creatinine ratio). The rise in serum creatinine and, less often, the increase in urea nitrogen are used to monitor the degree of renal impairment.

Patients with acute renal failure may develop all the symptoms of the uremic syndrome. In general, however, the anemia and metabolic bone disease seen in chronic renal failure are not as prominent in acute renal failure. Patients with acute renal failure are particularly at risk for gastrointestinal hemorrhage and systemic infection, which are the major causes of death.

Laboratory monitoring is crucial for assessing the severity of renal impairment and for the management of the fluid and electrolyte disorders that accompany acute renal failure.[12] Patients with reversible acute renal failure generally remain oliguric for about two weeks and then enter the diuretic phase in which there is gradually increasing urine output as the GFR begins to rise. Small changes in GFR may give rise to significant changes in urine volume and urine output may rise while urea nitrogen and creatinine are still being retained. Recovery of tubular function may lag behind the recovery of GFR so that regulation of sodium, potassium, and water reabsorption may be inadequate, with resulting hypovolemia and hypokalemia. Serial monitoring of creatinine, GFR, serum electrolytes, and urine volume may show more rapidly changing values than in chronic renal failure; such changes may require prompt attention.

A complete discussion of the causes of acute renal failure is beyond the scope of this chapter. In the majority of cases the initial event is either ischemic or toxic injury to the kidney. Inadequate perfusion of the kidney results in reduced GFR and may mimic acute renal failure. Judicious use of the chemistry laboratory is essential in recognizing and monitoring acute renal failure and in distinguishing it from inadequate perfusion of an intrinsically normal kidney. With normal glomerular and tubular function, the kidney responds to the underperfusion with avid reabsorption of sodium and water to produce a highly concentrated urine containing little sodium. Following ischemic or toxic damage to renal tubules, the ability to concentrate urine as well as maximally reabsorb sodium is lost. Thus, there is frequently an isosmotic urine with higher urine sodium values.

In addition to the urine chemistry tests discussed above, examination of the urinary sediment can also be of great value. Moderate numbers of hyaline and finely granular casts are typical of prerenal azotemia. In contrast, acute tubular injury produces free renal epithelial cells as well as epithelial cell casts and fatty casts. The presence of erythrocyte casts strongly suggests acute glomerular injury, whereas leukocyte casts are most often seen in pyelonephritis.

Although the preceding tests are very useful in distinguishing underperfusion of the kidney from acute tubular injury, they may be misleading in certain patients. Prior treatment with diuretics may make interpretation of urine chemistry values impossible. Acute oliguric renal failure due to urinary tract obstruction does not have a characteristic picture and may require the use of other techniques such as radiographic studies for proper diagnosis.

Glomerular diseases. Diseases that cause direct injury to the renal glomeruli while preserving tubular function may initially resemble prerenal azotemia. Eventually chronic progressive glomerular disease will affect the blood supply of tubules and result in the loss of complete nephrons to produce chronic renal disease and the uremic syndrome. The glomerular diseases give rise to particular syndromes in which laboratory evaluation is a key procedure for diagnosis and follow-up. Examples of each syndrome will be discussed to illustrate the use of the laboratory in their evaluation.

The ***acute nephritic syndrome*** (acute glomerulonephritis) is characterized by the rapid onset of hematuria, proteinuria, reduced GFR, and sodium and water retention with resulting hypertension and sometimes localized peripheral edema. Congestive heart failure and oliguria may also be seen. Renal biopsy shows enlarged, inflamed glomeruli with narrowed capillary lumina.

In a number of patients with the acute nephritic syndrome, the pathologic process is related to recent group A β-hemolytic streptococcal infection. Glomerular injury with damage of the glomerular basement membrane leading to reduction of GFR is thought to be due to activation of the inflammatory response by immune complexes. Abnormal *laboratory results* are usually present early in the course of acute nephritis. Hematuria, gross or microscopic, and proteinuria, usually <3 g/d, are almost always present. Red cell casts are highly suggestive of glomerulonephritis. They are commonly present in urine but will be observed only if the specimen is fresh and acidic, centrifugation is light, and sediment (after decantation) is resuspended gently. Large numbers of hyaline and granular casts are also common; waxy casts suggest a chronic process and should raise a question of acute exacerbation of a pre-existing disease. The proteinuria is progressively more nonselective, with larger-molecular-weight species having greatly increased clearance (see also Chapter 11).

In patients suspected of acute poststreptococcal glomerulonephritis, evidence of recent infection may be found in increased titers of antibodies to streptococcal extracellular products; i.e., antistreptolysin O (ASO), antihyaluronidase (AHase), and antideoxyribonuclease-B (ADNase-B). Most patients will have moderate reductions in total hemolytic complement activity (CH_{50}) and in the C3 component of the complement cascade. Persistent and severe depression of C3 levels should suggest membranoproliferative glomerulonephritis, systemic lupus erythematosus (SLE), endocarditis, or other forms of sepsis.

Other causes of acute nephritis are reactions to drugs, acute infection of the kidney, systemic diseases with immune complexes such as SLE, bacterial endocarditis, and possibly immunologic injury related to antecedent viral infections.

Rapidly progressive glomerulonephritis (RPGN) refers to a heterogeneous group of disorders characterized by a fulminant clinical course, leading to renal failure in weeks to a few months. This group of diseases is characterized by a consistent histologic picture of glomerular crescent formation. The glomerular epithelial crescent is a proliferation of blood-derived macrophages and parietal epithelial cells lining Bowman's capsule. Characteristically, the urine contains many RBCs and RBC casts as well as other casts and protein.

Chronic glomerulonephritis refers to the clinical syndrome that may result from a number of glomerular diseases that have a prolonged downhill course with progressive loss of nephron mass. Many of these diseases may remain entirely asymptomatic except for mild hematuria, proteinuria, and slightly reduced renal function. In some cases, the first indication of disease is the gradual onset of the uremic syndrome. Many of the glomerular diseases discussed under other headings may lead to chronic glomerulonephritis, or chronic glomerulonephritis may occur as an idiopathic process. Differentiation from nonglomerular diseases is not always possible, but heavy proteinuria or red cell casts suggest a glomerular process.

The ***nephrotic syndrome*** consists of massive proteinuria of ≥ 3.5 g/d, hypoalbuminemia of <3 g/dL, generalized edema, and hyperlipidemia. This syndrome results from glomerular injury that produces excessive permeability to plasma proteins and massive proteinuria. GFR is usually depressed. The loss of albumin in urine overwhelms hepatic synthetic capacity and hypoalbuminemia results. Loss of intravascular oncotic pressure results in edema, which is enhanced by sodium and water retention. This retention is secondary to aldosterone secretion stimulated by intravascular hypovolemia. The genesis of the hyperlipidemia is not clear.

Leakage of lipoproteins into the urine results in lipiduria with oval fat bodies (degenerated tubular cells that have reabsorbed lipoproteins).

The nephrotic syndrome can be seen with several systemic diseases that involve the kidney, such as diabetes mellitus, amyloidosis, and SLE, or it may be due to primary glomerular disease. In children between 2 and 6 years of age, *minimal change disease* (a synonym for lipoid nephrosis) accounts for over 90% of cases of primary nephrotic glomerular disease. Systemic disease is a more important cause of nephrotic syndrome in adults.

In addition to massive proteinuria, hypoalbuminemia, and hyperlipidemia, there are other important *laboratory changes* in the nephrotic syndrome as well. Serum levels of α_2- and β-globulins are usually increased, whereas α_1-globulins are normal to slightly decreased. IgG levels may be dramatically decreased although other immunoglobulins are usually normal. Factor B of the alternative pathway of complement activation may also be significantly decreased and, along with depressed IgG, may give rise to the significant risk of infection in these patients. Other complement components are not significantly decreased. There is an increase in the levels of fibrinogen and several other coagulation factors along with a decrease in antithrombin III. These changes, together with increased platelet aggregation, hypovolemia, and edema, contribute to the occurrence of thromboembolic problems in the nephrotic patient. There is significant alteration of several important binding proteins including depression of serum levels of transferrin, transcortin, and thyroxine-binding globulin. Thus, tests such as total iron-binding capacity, total serum iron, total T_4, and T_3 resin uptake will be affected. Total cholesterol and triglycerides will be elevated secondary to changes in lipoproteins. Finally, patients may have reduced serum concentrations of vitamin D, copper, and zinc due to loss of binding proteins.

Tubular diseases. Two important types of tubular disease, ischemic and toxic tubular necrosis, were discussed previously under acute renal failure because this is their usual clinical presentation. Another group of diseases in which primary tubular dysfunction affects acid-base balance is discussed separately under renal tubular acidosis. Here we limit discussion to diseases affecting predominantly the tubules and interstitium and, in particular, those that present clinically as interstitial nephritis rather than acute renal failure. This group of diseases is characterized in the acute stage by leukocytic infiltration of the interstitium and tubules. The causes of the initial inflammatory response can be quite varied and include bacterial, fungal, or viral infection; hypersensitivity reactions to drugs such as methicillin; toxic effects of analgesic drugs; radiation toxicity; and immunologic rejection of a transplanted kidney.

Hallmarks of glomerular disease such as severe proteinuria, nephrotic syndrome, and red cell casts are usually missing in the tubulointerstitial nephritides. Clinical presentation is marked by reduced GFR associated with tubular dysfunction and loss of concentrating ability, reduced acid excretion, and inadequate regulation of sodium balance.

The most clear-cut instance of acute tubulointerstitial disease is acute pyelonephritis, which is easily recognized by the accompanying clinical manifestations, including acute febrile episodes associated with more or less severe flank pain and tenderness. Acute pyelonephritis is usually associated with urinary symptoms of frequency and urgency. Characteristically, the urine contains numerous leukocytes and frequent leukocyte casts. The diagnosis is nearly always readily made by culture.

In contrast to the characteristic clinical features of acute or recurrent pyelonephritis, noninfectious tubulointerstitial disease is virtually never symptomatic until the symptoms of uremia appear. In most instances, the findings on urinalysis are quite benign, though on occasion modest proteinuria with moderate numbers of leukocytes in the urine is seen. The great majority of these cases can be attributed to drug toxicity. Careful examination may reveal impaired acidification and other urinary abnormalities such as the presence of glucose and amino acids in the urine; more commonly, reduced concentrating ability and slight proteinuria are the only manifestations.

Urinary tract obstruction. Urinary tract obstruction may cause disease, either indirectly by predisposing to urinary tract infection or directly by raising tubular pressure that destroys entire nephrons and leads to chronic renal failure. Obstruction may be at any level in the urinary tract from the renal tubule to the distal urethra. Obstruction is frequently complicated by urinary tract infection. In the progression of obstructive disease of one or both kidneys, the earliest impairment is loss of concentrating ability followed by reduced acid excretion, and ultimately by reduction of GFR and renal blood flow. The most useful *laboratory tests* are routine urinalysis, urine culture, CBC, serum urea nitrogen, and creatinine. Diagnosis is generally based on radiological evaluation.

Two very important systemic diseases, diabetes mellitus and hypertension, account for a sizable portion of the chronic renal disease in the western world. Both produce renal disease primarily by means of a vascular mechanism that causes progressive renal ischemia and consequently gradual loss of nephrons. Both diseases and their laboratory findings are discussed more fully elsewhere.

Use of the Laboratory for the Patient with Renal Disease

The diagnostic process in nephrology is heavily dependent upon the clinical laboratory because a majority of the diseases of the kidney produce few or no clinical manifestations until 50–75% of the kidney is destroyed. Laboratory tests often provide initial recognition of renal impairment and indicate the direction for further productive laboratory investigations.

The initial step of diagnosis is to determine biochemical or physiological parameters that characterize, in a general manner, a relatively small number of subgroups of the myriad diseases possible in the kidney. The next step is to determine specific parameters that focus on particular diseases within an indicated subgroup. Table 20-9 shows laboratory findings in conjunction with clinical syndromes. The initial laboratory data base needed to identify one of the clinical syndromes consists of a complete urinalysis, complete blood count, and the following serum chemistries: urea nitrogen, creatinine, sodium, potassium, chloride, CO_2, phosphorus, calcium, and uric acid. Supplemental data that aid in establishing the identification are 24-h protein excretion, urine protein electrophoresis, urine culture, and serial measurements of serum creatinine, of GFR, and of 24-h urine volume.

Presumptive diagnosis is therefore made by history and physical examination in conjunction with the data base provided by the laboratory. The diagnosis is refined by selective testing, which is also used to follow the progress of the disease. Although biochemical studies are important in the initial detection process, their use is overshadowed by the history and physical findings. But in the subsequent differentiation and in monitoring the disease process, the role of biochemical studies dominates that of clinical observations.

Renal Calculi

Most kidney stones are composed of one or more of the following substances: calcium oxalate, calcium phosphate, uric acid, cystine, or a mixture of these with magnesium ammonium phosphate (struvite). These substances crystallize within an organic matrix, the nature of which is not well understood. All of the substances listed above are poorly soluble, and, for some, solubility is influenced to a major degree by the urinary pH. The mechanisms responsible for the multiple recurrences of renal stones in only certain individuals are not completely understood but probably involve a multitude of factors,

Table 20-9. LABORATORY FINDINGS IN TEN CLINICAL SYNDROMES (AFTER COE,[9] MODIFIED)

Clinical Syndrome	Characteristic Laboratory Findings	Other Possible Laboratory Findings
Acute nephritis	Hematuria or RBC casts and documentation of one or more of following: decreased GFR, oliguria, edema, or hypertension	S: ↑ Creatinine, urea N, K^+ S: ↓ Na^+, CO_2, CH_{50}, C3 U: Proteinuria usually <3 g/d
Nephrotic syndrome	Proteinuria (>3.5 g/d)/1.75 m^2	S: ↑ Triglyceride, cholesterol, α_2- and β-globulin S: ↓ Albumin, IgG U: Oval fat bodies, free fat, ↓ Na^+ excretion
Asymptomatic urinary abnormalities	Hematuria or proteinuria without the presence of one of the other nine syndromes	
Acute renal failure	Oliguria, anuria, or reduced GFR and documentation of one of the following: normal GFR <3 months previously, GFR falling in days to weeks, creatinine rising daily, or the absence of anemia	S: ↑ Urea N, creatinine, K^+, PO_4^{2-}, Mg^{2+} S: ↓ Na^+, Ca^{2+}, CO_2, pH U: Epithelial cells and epithelial casts
Chronic renal failure	Reduced GFR stable for >3 months, uremic symptoms for >3 months, renal osteodystrophy, or demonstration of small kidneys	See Table 20-1
Urinary tract infection	Bacteriuria $>10^5$ colonies/mL or documentation of other infectious agent	B: Leukocytosis U: Pyuria, leukocyte casts, positive nitrite (some organisms); positive urine leukocyte esterase (dipstick)
Urinary tract obstruction	Obstruction visualized directly or seen on X-ray, or documentation of urinary retention or residual volume	S: ↑ Urea N, creatinine U: ↓ GFR, osmolality following fluid restriction, acid excretion Findings of urinary tract infection
Renal tubule defect	Documentation of actual tubular defects such as depressed secretion or reabsorption, impaired concentrating ability, renal tubular acidosis	S: ↑ Urea N, creatinine, K^+ U: ↓ GFR, osmolality after fluid restriction Proteinuria, glucosuria, phosphaturia, aminoaciduria, uricosuria, natriuria, renal tubular acidosis
Hypertension	Increased blood pressure of $>145/95$, documented on three separate occasions	Renal vascular disease S: ↑ Renin, aldosterone, Na^+, renal vein renin differential S: ↓ K^+ U: ↑ K^+ For hyperaldosteronism, see Chapter 18 For pheochromocytoma, see Chapter 18
Nephrolithiasis	Stone passed, removed, or seen by X-ray	Depends on underlying metabolic defect: S: ↑ Uric acid in gout, ↑ PTH and calcium in primary hyperparathyroidism

S: serum; U: urine; B: blood; ↑ increase; ↓ decrease.

including (1) urine flow (fluid intake), (2) excretion of excess quantities of the relatively insoluble substances listed above, and perhaps (3) the absence of a substance in the urine that, under normal circumstances, inhibits precipitation of some of these nearly insoluble agents. Certain types of stones are recognized as associated with specific disorders, and these are listed in Table 20-10.

Calcium oxalate stones are perhaps the most common stones that are encountered, and they occur in a worldwide stone belt that is mainly confined to the tropical and subtropical regions of the globe. They may be associated with either a persistently concentrated urine or a consistently increased excretion of urinary calcium or oxalate. Most commonly this type of stone is not composed of pure calcium oxalate, but calcium oxalate predominates, with small quantities of calcium phosphate and uric acid also present. In many instances it is not possible to find any abnormality in the stone-forming

Table 20-10. COMMON VARIETIES OF RENAL STONES

Type of Stone	Urine pH	Cause(s)

Calcium oxalate	Variable	Concentrated urine
		Hypercalciuria
		  Vitamin D intoxication
		  Hyperparathyroidism
		  Sarcoidosis
		Milk-alkali syndrome
		Osteoporosis
		Renal tubular acidosis
		Idiopathic hypercalciuria
		Hyperoxaluria

Calcium phosphate	Alkaline	Renal tubular acidosis Alkali ingestion Infection with urea-splitters (produces mixed stones)
Magnesium ammonium phosphate	Alkaline	Infection

Uric acid	Acid	Hyperuricaciduria
		  Hyperuricemia
		  Gout
		  High purine diet
		Urinary hyperacidity

Cystine stone	Acid	Cystinuria

individual beyond a persistently small urine volume; however, it is important to obtain data on calcium and oxalate output in the urine and to ascertain that the mechanism of urinary acidification is normal.

Hypercalciuria, which can be documented in perhaps half of the patients with calcium oxalate stones, can be categorized by appropriate studies into absorptive, resorptive, or renal hypercalciuria (Table 20-11). As the name implies, *absorptive hypercalciuria* results from increased gastrointestinal absorption of calcium; it is documented by demonstrating that oral calcium administration leads to an increase in urinary calcium excretion. Levels of serum calcium and phosphorus are normal, and dietary reduction of calcium reduces the rate of urinary calcium excretion. *Resorptive hypercalciuria* represents excretion of calcium that has been resorbed from bone. Most commonly, this resorption is caused by hyperparathyroidism or neoplastic bony metastases. Most of these patients have elevated serum calcium and their conditions are thus easily distinguished from the other forms of hypercalciuria. Patients with either primary or secondary hyperparathyroidism exhibit elevated levels of urinary cyclic AMP and elevations in circulating parathormone. *Renal hypercalciuria* represents persistent loss of calcium via the urine as a result of a renal leak, as in renal tubular acidosis. Details of other clinical entities associated with stones are found in Table 20-12.

Laboratory Analysis of Kidney Stones

Both qualitative and quantitative analysis of the chemical constituents of kidney stones may be useful in establishing their origin and in planning rational therapy. Analysis is not needed for every urinary calculus when other laboratory and clinical findings are diagnostic. In some patients, however, two or more possible etiologies exist,

Table 20-11. DIFFERENTIATION OF TYPES OF HYPERCALCIURIA

Type	Serum [Ca]	Serum [P]	Bone Density	Serum PTH	Urinary cAMP
Absorptive	Normal	Normal	Normal	↓	↓
Resorptive	↑ or Normal	Normal or ↓	↓	↑	↑
Renal	Normal	Normal	↓	↑	↑

Table 20-12. FREQUENCY, CHARACTERISTICS, AND ETIOLOGIES OF KIDNEY STONES

Chemical Type	Approximate Frequency	Characteristics	Etiologies
Calcium	65% Oxalate 30% Phosphate 10% Mixed 25% (oxalate-phosphate)	Hard, small to medium-sized, often multiple X-ray: radiopaque; may have diffuse renal calcium deposits (nephrocalcinosis)	Idiopathic lithiasis (30%) Normocalcemic hypercalciuria (30%) Hyperuricosuria (20%) Primary hyperparathyroidism (7%) Other causes, e.g., hyperoxaluria (5%), renal tubular acidosis (3%), medullary sponge kidney, sarcoidosis, immobilization
Magnesium ammonium phosphate ($Mg\ NH_4\ PO_4$)	25%	May be large ("staghorn") X-ray: radiopaque	Urinary tract infection (with hydrolysis of urea by organisms such as *Proteus*)
Uric Acid	5%	Yellow, friable, may be large ("staghorn") X-ray: radiolucent	Idiopathic hyperuricosuria Primary or secondary gout (e.g., malignancies)
Cystine	2%	X-ray: radiolucent May be large ("staghorn"); brownish color ("maple sugar")	Primary cystinuria

and analysis of the calculus may help distinguish between different diagnoses. For example, patients with hypercalciuria and calcium oxalate stones may develop renal infections with deposition of magnesium ammonium phosphate upon the calcium oxalate stones, to form "mixed stones." Chemical analysis of these stones is needed to confirm this sequence of events.

Specialized physical techniques such as infrared spectroscopy and X-ray diffraction are gradually replacing less specific, qualitative chemical methods for stone analysis.

References

1. Batlle, D., and Kurtzman, N. A.: Acid-base physiology and pathophysiology. *In*: Contemporary Nephrology, Vol. 2. S. Klahr and S. G. Massry, Eds. New York, Plenum Publishing Corp., 1983.
2. Bjornsson, T. D.: Use of serum creatinine concentrations to determine renal function. Clin. Pharmacokinet., *4*:200–222, 1979.
3. Blau, E. B., and Haas, J. E.: Glomerular sialic acid and proteinuria in human renal disease. Lab. Invest., *28*:477–481, 1973.
4. Bowers, L. D., and Wong, E. T.: Kinetic serum creatinine assays. II. A critical evaluation and review. Clin. Chem., *26*:555–561, 1980.
5. Brenner, B. M., and Rector, F. C., Jr.: The Kidney. 2nd ed. Philadelphia, W. B. Saunders Co., 1981.
6. Butler, A. R.: The Jaffe reaction. Identification of the coloured species. Clin. Chim. Acta, *59*:227–232, 1976.
7. Chaney, A. L., and Marbach, E. P.: Modified reagents for determination of urea and ammonia. Clin. Chem., *8*:130–132, 1962.
8. Cockcroft, D. W., and Gault, M. H.: Prediction of creatinine clearance from serum creatinine. Nephron, *16*:31–41, 1976.
9. Coe, F. L.: Clinical and laboratory assessment of the patient with renal disease. *In*: The Kidney, Vol. 1. 2nd ed. B. M. Brenner and F. C. Rector, Jr., Eds. Philadelphia, W. B. Saunders Co., 1981.
10. Cook, J. G. H.: Factors influencing the assay of creatinine. Ann. Clin. Biochem., *12*:219–232, 1975.
11. DeFronzo, R. A., and Thier, S. O.: Inherited disorders of renal tubular function. *In*: The Kidney, Vol. 2. B. M. Brenner and F. C. Rector, Jr., Eds. Philadelphia, W. B. Saunders Co., 1981.
12. Del Greco, F., and Krumlovsky, F. A.: Role of the laboratory in management of acute and chronic renal failure. Ann. Clin. Lab. Sci., *11*:283–291, 1981.
13. DiGiorgio, J.: Nonprotein nitrogenous constituents. *In*: Clinical Chemistry: Principles and Technics. 2nd ed. R. J. Henry, D. C. Cannon, and J. W. Winkelman, Eds. New York, Harper & Row, 1974.
14. Gardner, M. D., and Henderson, A. R.: Test-request patterns for clinical chemistry in a British and a Canadian dialysis unit. Clin. Chem., *29*:1546–1548, 1983.

15. Georges, J.: Determination of ammonia and urea in urine and of urea in blood by use of an ammonia-selective electrode. Clin. Chem., *25*:1888–1890, 1979.
16. Haeckel, R.: Assay of creatinine in serum, with use of fuller's earth to remove interferents. Clin. Chem., *27*:179–183, 1981.
17. Heinegård, D., and Tiderström, G.: Determination of serum creatinine by a direct colorimetric method. Clin. Chim. Acta, *43*:305–310, 1973.
18. Holmes, J. H.: Measurement of osmolality in serum, urine, and other biological fluids by the freezing point determination. *In*: Workshop on Urinalysis and Renal Function Studies. Commission on Continuing Education. American Society of Clinical Pathologists, 1962.
19. Jaynes, P. K., Feld, R. D., and Johnson, G. F.: An enzymic, reaction-rate assay for serum creatinine with a centrifugal analyzer. Clin. Chem., *28*:114–117, 1982.
20. Kampmann, J., Siersbaek-Nielsen, K., Kristensen, M., et al.: Rapid evaluation of creatinine clearance. Acta. Med. Scand., *196*:517–520, 1974.
21. Kaplan, A.: Urea nitrogen and urinary ammonia. *In*: Standard Methods of Clinical Chemistry, Vol. 5. S. Meites, Ed. New York, Academic Press, 1965, pp. 245–256.
22. Knochel, J. P., and Seldin, D. W.: The pathophysiology of uremia. *In*: The Kidney, 2nd ed. B. M. Brenner and F. C. Rector, Jr., Eds. Philadelphia, W. B. Saunders Co., 1981.
23. Lazarus, J. M.: Complications in hemodialysis: An overview. Kidney Internat., *18*:783–796, 1980.
24. Marsh, W. H., Fingerhut, B., and Miller, H.: Automated and manual direct methods for the determination of blood urea. Clin. Chem., *11*:624–627, 1965.
25. Merrill, J. P.: Dialysis versus transplantation in the treatment of end-stage renal disease. Ann. Rev. Med., *29*:343–358, 1978.
26. Miller, B. F., and Winkler, A. W.: The renal excretion of endogenous creatinine in man. Comparison with exogenous creatinine and inulin. J. Clin. Invest., *17*:31–40, 1938.
27. Morgan, D. B.: Assessment of renal tubular function and damage and their clinical significance. Ann. Clin. Biochem., *19*:307–313, 1982.
28. Narayanan, S., and Appleton, H. D.: Creatinine: A review. Clin. Chem., *26*:1119–1126, 1980.
29. Nissen, P.: Simultaneous determination of allopurinol, oxipurinol, and uric acid in human plasma by high performance liquid chromatography. J. Chromatogr., *228*:382–386, 1982.
30. Ott, N. T., and Wilson, D. M.: A simple technique for estimating glomerular filtration rate with subcutaneous injection of [^{125}I] iothalamate. Mayo Clinic Proc., *50*:664–668, 1975.
31. Rennke, H. G., and Venkatachalam, M. A.: Glomerular permeability of macromolecules: Effect of molecular configuration on the fractional clearance of uncharged dextran and neutral horseradish peroxidase in the rat. J. Clin. Invest., *63*:713–717, 1979.
32. Rock, R. C., Walker, W. G., and Jennings, C. D.: Nitrogen metabolites and renal function. *In*: Textbook of Clinical Chemistry, N. W. Tietz, Ed. Philadelphia, W. B. Saunders Co., 1986.
33. Sampson, E. J., Baird, M. A., Burtis, C. A., et al.: A coupled-enzyme equilibrium method for measuring urea in serum: Optimization and evaluation of the AACC Study Group on urea candidate reference method. Clin. Chem., *26*:816–826, 1980.
34. Scriver, C. R., Clow, C. L., and Lamm, P.: On the screening, diagnosis and investigation of hereditary aminoacidopathies. Clin. Biochem., *6*:142–188, 1973.
35. Siersbaek-Nielsen, K., Molholm-Hansen, J., Kampmann, J., et al.: Rapid evaluation of creatinine clearance. Lancet, *1*(Part 2):1133–1134, 1971.
36. Smith, C. H., Landt, M., Steelman, M., et al.: The Kodak Ektachem 400 Analyzer evaluated for automated enzymic determination of plasma creatinine. Clin. Chem., *29*:1422–1425, 1983.
37. Soldin, S. J., Henderson, L., and Hill, J. G.: The effect of bilirubin and ketones on reaction rate methods for the measurement of creatinine. Clin. Biochem., *11*:82–86, 1978.
38. Stenzel, K. H., Cheigh, J. S., Sullivan, J. F., et al.: Clinical effects of bilateral nephrectomy. Am. J. Med., *58*:69–75, 1975.
39. Taussky, H. H.: Creatinine and creatine in urine and serum. *In*: Standard Methods of Clinical Chemistry, Vol. 3. D. Seligson, Ed. New York, Academic Press, 1961, pp. 99–113.
40. Tietz, N. W., Ed.: Clinical Guide to Laboratory Tests. Philadelphia, W. B. Saunders Co., 1983.
41. Watts, R. W. E.: Determination of uric acid in blood and urine. Ann. Clin. Biochem., *11*:103–111, 1974.
42. Wesson, L. G., Jr.: Physiology of the Human Kidney. New York, Grune & Stratton, 1969.

Additional Reading

Brenner, B. M., and Rector, F. C., Jr., Eds.: The Kidney. 2nd ed. Philadelphia, W. B. Saunders Co., 1981.
Ganong, W. F.: Review of Medical Physiology. 10th ed. Los Altos, CA, Lange Med. Pub., 1981.
Klahr, S., Ed.: The Kidney and Body Fluids in Health and Disease. New York, Plenum Medical Book Co., 1983.

CALCIUM AND PHOSPHATE METABOLISM

21

DONALD FRASER, M.D., Ph.D.
GLENVILLE JONES, Ph.D.
SANG WHAY KOOH, M.D., Ph.D., F.R.C.P.(C)
INGEBORG C. RADDE, M.D., Ph.D., F.R.C.P.(C)

Calcium and inorganic phosphate in the mineral phase of the hard tissue—bone, dentin, and enamel—co-exist in relatively fixed proportions. The calcium and inorganic phosphate of bone exchange with those in the tissue fluids. The concentrations of calcium and inorganic phosphate in plasma are dependent in large part on the balance between rates of bone mineral deposition and bone resorption; at the same time, the plasma concentration of each element has an important bearing on the bone mineralization process itself. The three main regulators of calcium and inorganic phosphate homeostasis—parathyroid hormone, vitamin D hormone, and calcitonin—each exert independent effects on both ions.

CALCIUM

Distribution. Calcium is the fifth most common element in the body.[9] Practically all of the ~25 mol (1 kg) of calcium in the human adult is extracellular, and 99% of it exists as crystalline hydroxyapatite in bones and teeth. The extracellular fluid contains ~25 mmol (1 g) of calcium; blood contains 7.5 mmol (300 mg), virtually all of which is in the plasma. Approximately 125 mmol (5 g) of calcium is in the exchangeable calcium pool, located mainly on bone-forming surfaces and in soft tissues, where the calcium is readily accessible for physiological functions.

Calcium exists in plasma in three physicochemical states: protein-bound (~45%); complexed with small diffusible ligands such as citrate, lactate, phosphate, and bicarbonate (~10%); and ionized (~45%). Eighty per cent of the protein-bound fraction is bound to albumin and 20% to globulins. Each fraction is in dynamic equilibrium with the other forms. The ionized fraction (Ca_I) is the physiologically active form; its magnitude is affected by pH and plasma protein concentration. Acidosis favors dissociation, which raises the portion existing as Ca_I; alkalosis has the opposite effect.

In contrast to the large quantity of extracellular calcium, only 1 mmol (40 mg) is present in the cytosol. This intracellular calcium serves a vital role in controlling the function of all cells.

Physiology. Intracellularly, calcium is the prime inorganic messenger for regulation of cell functions. It determines activity of several enzymes, notably adenylate cyclase and phosphodiesterase, through reversible combination with calmodulin,[13] a small protein found in all nucleated cells; it has four calcium-binding sites for each molecule. In striated muscle, calcium activates contraction of the myosin fibril through combination with troponin, a calcium-binding protein similar, if not identical, to calmodulin. Calcium also functions at the plasma membrane. It serves to regulate membrane permeability, to cause neurotransmitter release, and to diminish neuromuscular excitability. Secretions of several endocrine glands are controlled by the extracellular Ca_I concentration at the cell surface. Examples include the parathyroid glands, thyroid C cells, and pancreatic B (beta) cells. Ca_I plays a vital role in blood coagulation and is involved in cell-to-cell adhesion and possibly communication. As a principal constituent of hydroxyapatite, calcium serves to confer rigidity to bones and teeth.

Calcium is constantly lost from the body through excretion in feces, urine, and, to a small extent, sweat. The dietary calcium requirement of humans depends on age, growth, pregnancy, and lactation. Dairy products contribute 70% of the calcium in the North American diet. The US recommended dietary allowance (RDA) for calcium is 800 mg/d for adults and children above one year of age. An additional 400 mg/d calcium is recommended during adolescence, pregnancy, and lactation. Net absorption of ingested calcium in the healthy adult is about 25–30%.

PHOSPHATE

Distribution. The body of a human adult contains ~20 mol (620 g) of phosphorus, entirely in the form of phosphate.[9] Phosphate is distributed fairly equally between extracellular and intracellular compartments. Intracellularly, phosphate occurs mainly as a component of macromolecules such as phospholipids and phosphoproteins. This fraction is termed *organic phosphate.* A small but extremely important fraction exists as inorganic phosphate and participates in high-energy transfer reactions.

About 85% of extracellular phosphate occurs in inorganic form as hydroxyapatite. In plasma or serum, most phosphate exists in the inorganic form (P_i); this fraction is present as the mono- and dihydrogen forms, the relative proportions varying with the pH. At pH 7.40, the HPO_4^{2-}:$H_2PO_4^-$ ratio is close to 80:20 (see Chapter 19, Section Three). Approximately 15% of plasma P_i is bound to protein; the remainder exists in complexed and free forms and is ultrafiltrable. Total plasma and total urinary P_i concentrations are not altered by pH changes.

Physiology. Phosphate in the form of phospholipid is a structural component of cell membranes. It participates in many important cell functions, including energy metabolism (high-energy nucleotides such as ATP) and gene replication (nucleic acids). Intracellularly, inorganic phosphate participates in many reactions concerned with the generation of metabolic energy. P_i is a substrate for oxidative phosphorylation. Glycogen breakdown in liver and kidney requires P_i to produce free glucose and in muscle to provide glycolytic intermediates. In erythrocytes, production of 2,3-diphosphoglycerate from glyceraldehyde-3-phosphate consumes P_i to generate the reducing equivalents required to maintain integrity of the cell membrane.

Extracellular inorganic phosphate is a major component of hydroxyapatite. The concentration of P_i in plasma and extracellular fluid is an important determinant of bone mineral turnover. Indeed, chronic hypophosphatemia is associated with rickets and osteomalacia.

Phosphorus is present in most foods. The RDA for phosphorus (in mg/d) is the same as that for calcium.

Hormones that Regulate Calcium and Inorganic Phosphate Transport[3,17,56]

The three primary hormonal regulators of calcium and inorganic phosphate are parathyroid hormone (PTH), calcitonin (CT), and 1,25-dihydroxyvitamin D (1,25-$(OH)_2D$; 1α,25-dihydroxyvitamin D).

PARATHYROID HORMONE (PTH)

PTH is synthesized by the parathyroid glands. In humans there are usually four glands situated close to or on the posterior surface of the thyroid gland. However, additional parathyroid glands may be located elsewhere in the neck or within the thymus in the superior mediastinum. There are two types of cells in the gland: the chief cells, responsible for the production, storage and secretion of PTH, and the oxyphil cells, whose function is unknown.

Parathyroid hormone is a single-chain polypeptide of 84 amino acids and a MW of 9500.[14,33,62] The precursor of PTH, pre-proparathyroid hormone, is a 115-amino acid polypeptide synthesized on the ribosomes within the chief cells. It is rapidly converted by proteolysis to pro-PTH (90 amino acids) and then to intact PTH. Intact PTH 1–84 is the form stored in the glands and the principal form secreted into the blood stream. It is metabolized in peripheral tissues, principally the liver and kidney, to an amino-terminal (*N*-terminal) fragment of at least 34 amino acids, MW ~2000, and to several carboxy-terminal (*C*-terminal), and midmolecule fragments (e.g., amino acids 53–84, 44–68, 35–64, and sequences that overlap these).

The biological activity of PTH resides in the first 29 and maximum activity resides in the first 34 amino acids of the *N*-terminal portion of the molecule. Thus, intact PTH 1–84 and PTH 1–34 both possess full biological activity. The *C*-terminal and midmolecule fragments are biologically inert. Intact PTH and the *N*-terminal fragment are removed from the plasma by attachment to hormone-specific receptors on the plasma membranes of target cells, by renal excretion, and by degradation in the liver, kidney, and possibly bone. The inactive midmolecule and *C*-terminal fragments are removed exclusively by renal excretion. The plasma $t_{1/2}$ of PTH 1–84 is 15–30 min, and the $t_{1/2}$ of the mid- and *C*-terminal fragments are 2–4 h.

Because of these differences in metabolism and excretion, the plasma concentrations and relative distributions of PTH-derived peptides cover a wide range and vary according to the physiological or pathological status. The concentration of biologically active hormones is very low in normal plasma, whereas that of the inert midregion and *C*-terminal fragments is 5–10 times greater.

Function. The primary role of PTH is regulation of the extracellular fluid (ECF) calcium concentration. Synthesis and secretion of PTH are under precise feedback regulation by the concentration of ionized calcium (Ca_I) in the circulation. PTH synthesis and secretion are stimulated within a few minutes by low concentration of Ca_I; high Ca_I concentration suppresses secretion. Although PTH also affects the concentration of plasma P_i (see below), the plasma P_i concentration does not control PTH secretion directly.

PTH activity is mediated by cAMP and acts on target organs in several ways:

1. It stimulates osteoclasts and probably osteocytes to mobilize calcium from bone. This action requires the synergistic action of 1,25-$(OH)_2$D. Mobilization of calcium, if persistent, leads to demineralization of bone and to release of phosphate and hydroxyproline into the plasma.
2. It increases renal tubular reabsorption of calcium.
3. It inhibits proximal renal tubular reabsorption of inorganic phosphate.
4. It inhibits proximal tubular reabsorption of HCO_3^-, Na^+, K^+, H_2O, and amino acids.
5. It stimulates renal hydroxylation of 25-(OH)D to 1,25-$(OH)_2$D by increasing the synthesis of renal 1-hydroxylase.
6. It indirectly induces an increase in intestinal calcium absorption by stimulating 1,25-$(OH)_2$D synthesis.
7. It promotes bone formation in certain circumstances after PTH binds to membrane receptors of target cells.

The net effects of PTH are increases in plasma Ca_T *and* Ca_I *and a decrease in* P_i. Renal phosphate and cAMP excretion rise. The urinary excretion of calcium is usually increased because the effect of PTH to increase tubular calcium reabsorption is offset by its effect on plasma calcium concentration and the consequent increase in the filtered load of calcium. PTH causes varying degrees of generalized aminoaciduria and metabolic acidosis.

CALCITONIN

Calcitonin (CT) is produced mainly by C cells distributed thoughout the thyroid gland.[4] CT consists of a 32–amino acid polypeptide chain, MW 3500, that is derived from a large precursor polypeptide chain. Interspecies differences in amino acid sequence cause differences in potency and immunospecificity (e.g., patients treated with salmon or porcine CT may develop antibodies to the agent).

Secretion of CT is regulated by the level of Ca_I in the circulation. Healthy individuals normally have low plasma concentrations of CT. Hypercalcemia stimulates CT release, whereas hypocalcemia suppresses secretion.

Function. Calcitonin binds to specific membrane receptors of target cells and stimulates cAMP production. The effect of small doses of CT is to decrease resorption of calcium and inorganic phosphate from bone by inhibiting the action of osteoclasts and probably osteocytes, leading to lower plasma Ca_T, Ca_I, and P_i concentrations.

Pharmacological doses of CT (1) decrease renal tubular reabsorption of calcium, phosphate, sodium, potassium, and magnesium; (2) decrease gastrin secretion, and (3) increase secretion of sodium, potassium, chloride, and water by the small intestine.

CT has its greatest physiological effect in the fetus and in young children, probably because bone calcium turnover is very rapid at this stage of development. Plasma CT levels are higher in neonates and in pregnant and lactating women than in normal adults. CT may have a function in promoting deposition of calcium in the skeleton of the human fetus and may be a causative factor in neonatal hypocalcemia.

VITAMIN D AND ITS METABOLITES

There are two families of vitamin D (see Chapter 16 for structures, sources, and dietary allowances).[17] Vitamin D_3 (cholecalciferol) is the natural form produced in the skin from the action of sunlight. Vitamin D_2 (ergocalciferol) is manufactured commercially from precursors of plant origin. In nature, 7-dehydrocholesterol is present in skin and is converted by certain ultraviolet wavelengths (290–310 nm) of sunlight to provitamin D_3 and thence by the heat of the body to vitamin D_3. Skin exposure to sunlight is a very efficient means of acquiring vitamin D.[1] Neither vitamin D_2 nor vitamin D_3 is active as such.[17] Before the D vitamins can exert physiological action, each must undergo two hydroxylations, both of which are catalyzed by specific hydroxylases (OHase) (Figure 21-1). The first metabolic step takes place in the liver and converts vitamin D to 25-hydroxyvitamin D [25-(OH)D]. The latter is transported by vitamin D–binding protein (DBP) to the kidney, where it undergoes further hydroxylation to either 1,25-dihydroxyvitamin D [1,25-$(OH)_2$D] or 24,25-dihydroxyvitamin D [24,25-$(OH)_2$D]. 1,25-$(OH)_2$D disappears quickly from the plasma ($t_{1/2}$ = 0.5 h), and its usual concentration in the plasma is low. 24,25-$(OH)_2$D disappears more slowly ($t_{1/2}$ = 10 h) and its concentration is considerably greater.

Function. 1,25-$(OH)_2$D is the most potent of the known metabolites. It has a rapid and powerful stimulatory action on membrane transport of calcium. When calcium mobilization is required, 1-OHase activity is stimulated and synthesis of 1,25-$(OH)_2$D takes place. Other factors that stimulate 1-OHase activity are estrogens, progestins,

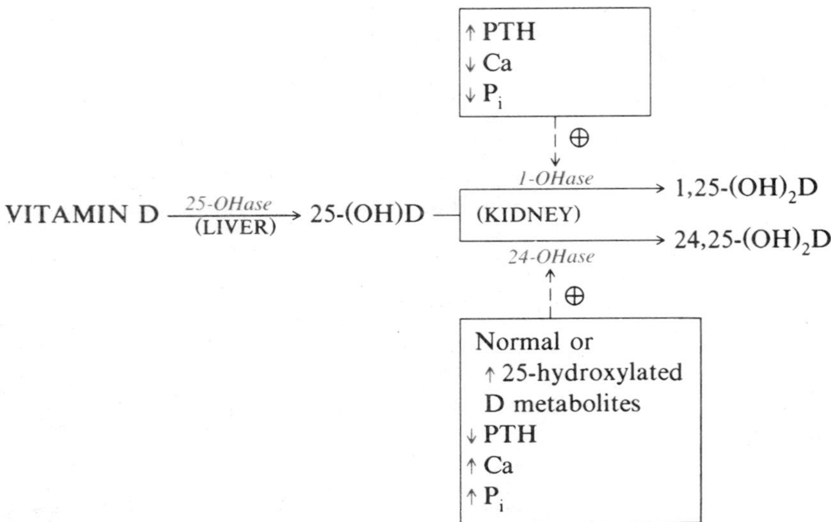

Figure 21-1. Simplified diagram to illustrate the pathways of vitamin D metabolism and the main factors that stimulate 1,25-$(OH)_2$D and 24,25-$(OH)_2$D synthesis.

testosterone, growth hormone, and prolactin. When calcium mobilization is unnecessary or would be detrimental to the individual, 1-hydroxylation ceases and 25-(OH)D is converted to inactive 24,25-$(OH)_2$D.

1,25-$(OH)_2$D has the following biological actions:

1. It stimulates calcium absorption by the small intestine.
2. It stimulates intestinal absorption of phosphate (independent of its action on intestinal calcium absorption). However, it probably does not have a significant effect on renal tubular phosphate reabsorption.
3. It stimulates calcium resorption from bone. In this action, PTH acts synergistically.
4. It increases calcium reabsorption by the distal renal tubules.
5. It stimulates differentiation of a variety of cells. It is suggested that 1,25-$(OH)_2$D stimulates transformation of progenitor cells into osteoclasts.

1,25-$(OH)_2$D-receptors have been identified not only in intestinal mucosa, bone, and renal tubules but also in a large variety of organs. However, a physiological action of 1,25-$(OH)_2$D has not been proved at sites other than intestine, bone, and kidney. 1,25-$(OH)_2$D does not stimulate mineralization of bone directly.

The net effect of 1,25-$(OH)_2$D is an increase in plasma Ca_T and Ca_I concentrations. Plasma P_i concentration is usually increased also. Urinary calcium excretion increases because, as with PTH, the effect of 1,25-$(OH)_2$D on plasma Ca and hence on the filtered load of calcium overrides its effect of increasing tubular reabsorption of calcium.

Regulation of Ca and P_i Metabolism

The concentrations of Ca and P_i in plasma are determined by the balance between absorption and excretion in the kidneys, the intestine, and to a small extent the skin, and by exchanges between plasma, extracellular fluid (ECF), and tissues including bone. These dynamic processes are determined by many factors, principally PTH, CT, and vitamin D. The actions of these three hormones (Figure 21-2) result in a remarkably precise control of the plasma Ca_I concentration in the healthy individual. Prostaglandin E_2,

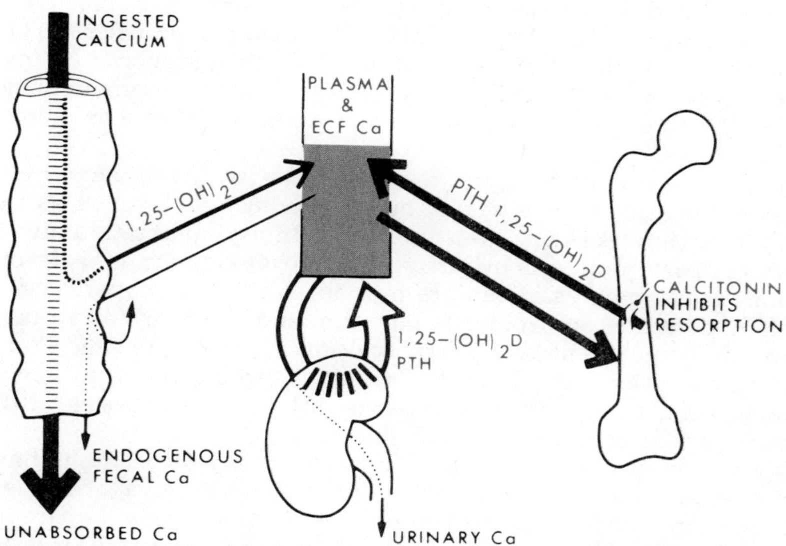

Figure 21-2. Diagram showing the role of 1,25-$(OH)_2$D, PTH, and CT in maintaining calcium homeostasis. The relative magnitudes of the various homeostatic processes are indicated by the thickness of the arrows. In the normal adult, rates of calcium deposition and resorption are equal. The processes stimulated by 1,25-$(OH)_2$D, PTH, and CT are indicated. (From Fraser, D., and Kooh, S. W., *In*: Textbook of Paediatrics, Vol. 2, 3rd ed. J. O. Forfar and G. C. Arneil, Eds. Edinburgh, Churchill Livingstone, 1984, pp. 1157–1172.)

Table 21-1. PLASMA CHEMISTRY FINDINGS IN CONDITIONS ASSOCIATED WITH HYPERCALCEMIA

	P_i	ALP	PTH	25-(OH)D	1,25-(OH)$_2$D
Primary hyperparathyroidism	↓ or N	↑ or N	↑	N	↑
Malignancy					
Tumor producing PTH-like factor	↓ or N	↑ or N	↓ or N		
Skeletal metastasis of tumor	↑ or N	↑ or N	↓ or N		
Multiple myeloma, lymphoma, leukemia	↑ or N	↑ or N	↓ or N		
Prostaglandin-producing tumor	↓ or N	↑ or N	↓ or N		
Local osteolytic metastasis	N	N	↓ or N		
Vitamin D toxicity	↑ or N	↓ or N	↓ or N	↑	N
Drug-induced hypercalcemia					
Thiazides, Ca-containing antacids, parenteral Ca	↓ or N	N	↓ or N		
Sarcoidosis	↓ or N	↑ or N	↓ or N	N	↑
Thyrotoxicosis	↑ or N	↑ or N	↓ or N		
Chronic renal failure	↑	↑ or N	↑	N	N or ↓
Adrenal insufficiency	N	N	↓ or N		
Immobilization	N	N	↓ or N		
Familial hypocalciuric hypercalcemia	↓ or N	N	↑ or N	N	N or ↑
Infantile hypercalcemia syndrome	↑ or N	↓ or N	↓	N	↓

Abbreviations: N, normal range; ↓, significantly decreased or undetectable; ↑, significantly increased; ALP, alkaline phosphatase.

osteoclast activating factor (OAF), and other local bone-resorbing factors, as well as adrenal corticosteroids, estrogen, and growth hormone influence calcium homeostasis secondarily. In the case of plasma P_i regulation, PTH exerts a particularly important effect; by inhibiting phosphate reabsorption by the kidney tubules, PTH lowers the plasma P_i concentration.

Diseases Involving Calcium and Phosphate Homeostasis

Diseases Associated with Hypercalcemia

Diseases associated with hypercalcemia and their important biochemical features are listed in Table 21-1. Primary hyperparathyroidism and hypercalcemia of malignancy are discussed below because they account for the great majority of hypercalcemic patients. Primary hyperparathyroidism is caused by persistent semiautonomous hypersecretion of PTH,[2,62] which causes hypercalcemia. Adenoma of one or more parathyroid glands is the major cause.

Primary hyperparathyroidism. Most patients with *primary hyperparathyroidism* are either asymptomatic or have vague, nonspecific symptoms such as weakness and fatigue. Nowadays, the disease is often uncovered in asymptomatic individuals during routine laboratory screening tests by virtue of an elevated plasma Ca_T level. In patients with symptoms, most manifestations are attributable to the hypercalcemia resulting from the PTH hypersecretion. Nocturia, polyuria, and thirst are early signs; anorexia, nausea, and vomiting may be present if hypercalcemia is protracted and severe. Increased urinary calcium excretion can sometimes result in urolithiasis and nephrocalcinosis. Renal colic occurs in about one-third of patients with primary hyperparathyroidism, but chronic renal failure is an uncommon complication.

Patients with longstanding hyperparathyroidism may occasionally have bone pain and may show generalized demineralization of the bones by X-ray, but the diagnosis of hyperparathyroidism is usually made before bone lesions become apparent.

Laboratory findings. Hypercalcemia is the hallmark of hyperparathyroidism. Those patients with so-called normocalcemic hyperparathyroidism (normal plasma Ca_T) usually have increased Ca_I concentrations[46] or their hypercalcemia may occur intermittently as a result of episodic PTH hypersecretion or after consumption of a calcium-rich meal. True normocalcemia can occur in primary hyperparathyroidism only if the patient has co-existing vitamin D deficiency, some other condition that blunts the bone action of the excessive PTH secretion, or depleted calcium stores. The best means of discriminating hyperparathyroid patients from euparathyroid individuals is by measuring Ca_I.[7,44]

Hypophosphatemia is present in about 50% of patients with hyperparathyroidism. However, plasma P_i may be normal, particularly in patients with diminished renal function, and it may be elevated when severe renal failure has occurred. Hyperchloremic acidosis occurs in about one-third of patients due to the inhibitory action of PTH on renal tubular reclamation of bicarbonate. In contrast, patients with hypercalcemia from other causes tend to develop metabolic alkalosis and depressed serum chloride along with low plasma P_i concentration. The plasma urate concentration may be increased. The plasma alkaline phosphatase (ALP) activity is usually normal. However, if bone lesions have developed, ALP tends to be increased roughly in proportion to the degree of skeletal involvement.

Urinary excretion of calcium is usually increased due to an increased filtered load of calcium. The magnitude of *hypercalciuria* is less in hyperparathyroidism than in hypercalcemia from other causes because PTH enhances tubular reabsorption of calcium. If urinary calcium excretion is low in the presence of hypercalcemia, the diagnosis of *familial hypocalciuric hypercalcemia (FHH)*[50] should be considered. Tubular reabsorption of phosphate (TRP) is usually decreased in primary hyperparathyroidism, and about 80% of patients excrete increased amounts of cAMP in the urine. However, these features are not sufficiently consistent to be helpful in establishing the diagnosis of primary hyperparathyroidism. The plasma 1,25-$(OH)_2D$ concentration tends to be elevated but levels may overlap with normal values.

The plasma PTH concentration is raised in primary hyperparathyroidism, and this determination is extremely important in establishing the diagnosis and in differentiating primary hyperparathyroidism from hypercalcemia of malignancy and other hypercalcemic conditions (see also page 722). The discrimination is improved when the PTH value is correlated with the corresponding serum calcium value (Figure 21-3).

Hypercalcemia of malignancy. Hypercalcemia occurs in 10–20% of patients with malignant disease.[3,55] Malignancies associated with hypercalcemia are classified into three categories:

1. Solid tumors secreting a humoral factor with PTH-like activity (10% of patients). These tumors secrete a humoral factor that activates PTH receptors but does not cross-react with RIAs for PTH. Most patients in this group have increased urinary excretion of nephrogenous cAMP, suppressed plasma PTH level, low plasma 1,25-$(OH)_2D$ concentration, and decreased TRP.

2. Solid, nonsecreting tumors with skeletal metastases (70% of patients). In this condition, the most common category of hypercalcemia of malignancy, the metastases result in hypercalcemia by causing bone resorption.

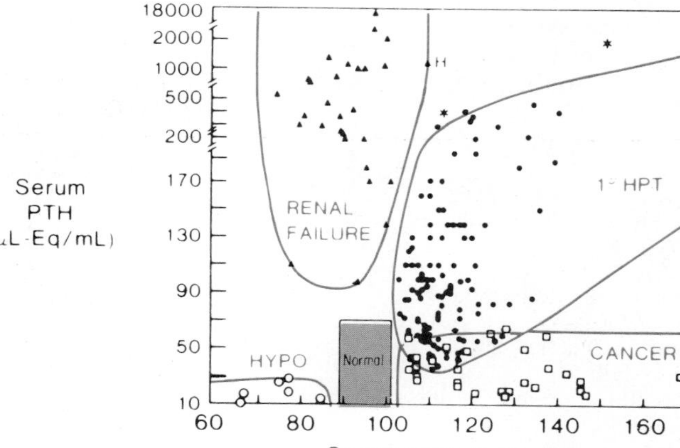

Figure 21-3. Graph showing correlations of serum PTH and serum Ca concentrations in patients with primary hyperparathyroidism (1° HPT) (●), cancer (□), hypoparathyroidism (hypo) (○), chronic renal failure (▲), parathyroid carcinoma (*), and hypernephroma (H). (From Kao, P. C., Jiang, N., Klee, G. G., et al.: Clin. Chem., *28*:69–74, 1982.)

3. Hematologic malignancy (20% of patients). Multiple myeloma, lymphoma, and leukemia are the most common neoplasms in this group. Resorption of bone mineral by invasive tumor tissue is the suggested mechanism for hypercalcemia.

In addition to the effects of the tumor per se, the clinical symptoms are those of hypercalcemia—mainly gastrointestinal, renal, and neurological. Hypercalcemia of malignancy runs a more rapidly progressive course than primary hyperparathyroidism; the levels of plasma Ca_T and Ca_I are higher. Patients tend to show symptoms of hypercalcemia at lower levels of plasma Ca but are less likely to have renal calculi.

Laboratory findings. The plasma PTH concentration is the laboratory feature that best distinguishes hypercalcemia of malignancy from primary hyperparathyroidism (see Table 21-1). Typically, plasma PTH concentrations are elevated in primary hyperparathyroidism and low in hypercalcemia of malignancy and other forms of hypercalcemia. Certain other features suggest hypercalcemia of malignancy. In the early stages of multiple myeloma, the plasma Ca_T may be elevated without any increase in the plasma Ca_I concentration because of increased binding with an abnormal plasma protein. In other forms of malignancy, plasma Ca_T and Ca_I concentrations tend to be more elevated and more variable than in primary hyperparathyroidism. The plasma P_i concentration in hypercalcemia of malignancy may be elevated, normal, or decreased, whereas it is likely to be decreased in primary hyperparathyroidism. ALP is often elevated in hypercalcemia of malignancy, but is normal in primary hyperparathyroidism unless, uncommonly, bone lesions have occurred. The ratio of plasma Cl (mmol/L) to plasma P_i (mmol/L) is >90 in over half the patients with hypercalcemia of malignancy, whereas the ratio is likely to be <90 in primary hyperparathyroidism. The plasma HCO_3^- content is elevated or normal in hypercalcemia of malignancy; it is normal or decreased in primary hyperparathyroidism. The 1,25-$(OH)_2$D level tends to be low in hypercalcemia of malignancy; this is in contrast to elevated levels observed in primary hyperparathyroidism. Measurements of TRP and urinary cAMP excretion are usually not helpful in the differential diagnosis of hypercalcemia of malignancy, although the diagnosis of primary hyperparathyroidism is unlikely with a TRP value $>90\%$.

Diseases Associated with Hypocalcemia

Diseases associated with hypocalcemia and their important biochemical features are listed in Table 21-2. Some of the conditions are discussed in greater detail below.

Hypoparathyroidism and pseudohypoparathyroidism. Hypoparathyroidism is due to deficient secretion of PTH, and pseudohypoparathyroidism results from impaired response of the target organs to the hormone.[16,27] In both conditions, hypocalcemia and hyperphosphatemia are present.

The pathogenesis of *hypoparathyroidism* includes congenital aplasia of the parathyroid glands, postnatal atrophy, inadvertent removal or damage during thyroidectomy, and, in a significant proportion of patients, autoimmune destruction of the glands. Often,

Table 21-2. PLASMA CHEMISTRY FINDINGS IN CONDITIONS ASSOCIATED WITH HYPOCALCEMIA

	P_i	ALP	PTH	25-(OH)D	1,25-$(OH)_2$D
Hypoparathyroidism	↑	↓ or N	↓	N	↓
Pseudohypoparathyroidism					
Type 1	↑	N	↑ or N	N	↓
Type 2	↑	N	↑ or N	N	
Neonatal hypocalcemia	↑	N	Variable	N	
Hypomagnesemia	N or ↑	N	↓	N	
Vitamin D deficiency and malabsorption syndromes	↓	↑	↑	↓	N or ↓
Renal failure	↑	↑ or N	↑	N	N or ↓
Hereditary vitamin D–dependency rickets					
Type I	↓	↑	↑	N	↓
Type II	↓	↑	↑	N	↑

Abbreviations: N, normal range; ↓, significantly decreased or undetectable; ↑, significantly increased.

patients with isolated pathological disorders of the parathyroid glands have symptoms of hypocalcemia as the only clinical manifestation, but in hypoparathyroidism resulting from autoimmune disease, additional endocrine deficiences (e.g., Addison's disease, diabetes mellitus, ovarian failure) and other manifestations of autoimmunity (alopecia, candidiasis, pernicious anemia) are likely to develop. The plasma PTH concentration is low or undetectable in all these forms of hypoparathyroidism.

In *pseudohypoparathyroidism*, of which there are three genetically distinct forms, hypocalcemia and hyperphosphatemia are caused by failure of the target tissues to respond to endogenous PTH. The plasma PTH concentration is often elevated, but values within the normal range are sometimes found. Half the patients with pseudohypoparathyroidism have one or more associated abnormalities (obesity, short stature, short metacarpals, mental defect) that help to distinguish pseudohypoparathyroidism from hypoparathyroidism. In some patients, differentiation between the two diseases requires the PTH response test. In this procedure,[16,27] the rates of urinary cAMP excretion are measured sequentially before and immediately after a standardized intravenous injection of PTH. Little or no increase in cAMP excretion occurs in pseudohypoparathyroidism, whereas in the various forms of hypoparathyroidism, as in normal individuals, cAMP increases more than 50-fold during the half-hour after injection.

Vitamin D deficiency. Hypocalcemia may develop as a result of vitamin D deficiency. In North America, dietary deficiency of vitamin D is uncommon because of the widespread use of vitamin D–enriched milk. However, full-blown vitamin D deficiency rickets is occasionally observed in breast-fed infants, because the vitamin D content of mother's milk is insufficient to meet the requirement.

Three stages of vitamin D deficiency can be distinguished.[28] *Stage I* deficiency involves impaired intestinal calcium absorption and diminished calcium mobilization from bone, with resultant hypocalcemia. Plasma P_i is not usually affected at this stage, and skeletal lesions on radiographs are minimal. In *Stage II* deficiency, hypocalcemia stimulates PTH secretion, which returns the plasma Ca_T and Ca_I to normal or nearly normal, but the secondary hyperparathyroidism results in hyperphosphaturia, hypophosphatemia, and aminoaciduria. The inadequate concentrations of plasma calcium and phosphate contribute to classical rachitic and osteomalacic lesions. In *Stage III* deficiency, vitamin D stores are severely depleted; the consequent widespread rachitic lesions and absence of vitamin D make the patient less sensitive to the actions of PTH. Hypocalcemia is present in this advanced stage of rickets, and PTH-stimulated hypophosphatemia and aminoaciduria persist.

The plasma 25-(OH)D level, which is the most practical measure of the state of vitamin D nutrition, is low or undetectable in all stages of vitamin D deficiency. In response to the stimulus of secondary hyperparathyroidism, plasma 1,25-$(OH)_2D$ concentrations are often within the normal range or even increased during the early stages of vitamin D deficiency; the level becomes low or undetectable as vitamin D deficiency advances.

Hereditary vitamin D–dependency rickets Type I. This condition is caused by a recessively inherited defect in renal 25-(OH)D-1-hydroxylase.[70] The plasma 25-(OH)D concentration is normal but the production of the 1,25-$(OH)_2D$ is blocked and the concentration of the active vitamin D hormone is extremely low. Complete healing of the biochemical and skeletal lesions occurs with administration of very small amounts of synthetic 1,25-$(OH)_2D_3$ (calcitriol). Clinically, the disease resembles Stage III vitamin D deficiency.

Hereditary vitamin D–dependency rickets Type II. This is a rare autosomal recessive form of severe rickets[21] in which target organs are deficient in cytosolic receptors for 1,25-$(OH)_2D$. The plasma and urine findings are the same as with Type I vitamin D dependency except that the plasma 1,25-$(OH)_2D$ concentration is raised, often to extremely high levels.

Intestinal malabsorption. Rickets and osteomalacia sometimes complicate chronic hepatobiliary and gastrointestinal diseases because bile salts and an intact intestinal mucosa are essential for efficient intestinal absorption of vitamin D. The severe bone disease that occurs in longstanding liver disease appears to be caused more by malabsorption of vitamin D than by defective hepatic conversion of vitamin D to 25-(OH)D.[42]

Neonatal hypocalcemia. During the first days of life, about 1% of full-term infants, and a higher percentage of premature infants and infants with perinatal complications, have transient symptomatic hypocalcemia. However, neonatal hypocalcemia may also be the first sign of more serious permanent forms of parathyroid gland hypofunction.

Diseases Associated with Hyperphosphatemia

Conditions associated with hyperphosphatemia are listed in Table 21-3.

Renal osteodystrophy. The term renal osteodystrophy (ROD) refers to the skeletal lesions that occur in chronic renal failure. Rickets, osteomalacia, hyperparathyroid bone changes (osteitis fibrosa), osteosclerosis, and osteoporosis are found in varying proportions.[14] Hyperphosphatemia plays a pivotal role in the pathogenesis of ROD.

In chronic renal failure the ability of the kidneys to excrete P_i is greatly impaired

Table 21-3. CONDITIONS ASSOCIATED WITH HYPO- AND HYPERPHOSPHATEMIA

Hyperphosphatemia	Hypophosphatemia
Intake	
Poisoning by phosphate-containing enema or laxative	Deficiency of dietary phosphate Phosphate malabsorption in dialyzed patients, alcoholics Overuse of phosphate-binding agents Total parenteral alimentation (TPN)
Redistribution	
Respiratory acidosis Lactic acidosis Diphosphonate therapy	Glucose infusion Treatment of diabetes mellitus Respiratory alkalosis β-Adrenergic agents Increased skeletal uptake in healing phase of rickets Osteoblastic metastases of cancer Nutritional recovery syndrome Diuretic phase of severe burns
Renal Causation	
Acute and chronic reduction in glomerular filtration rate Hypoparathyroid state Primary hypoparathyroidism Pseudohypoparathyroidism Suppression of PTH secretion from any hypercalcemic condition	Specific phosphate transport defect: X-linked dominant hypophosphatemia Autosomal dominant hypophosphatemia Multiple renal tubular transport defects Idiopathic Fanconi's syndrome Cystinosis Hereditary fructose intolerance Hereditary tyrosinemia Galactosemia Wilson's disease Oculocerebrorenal (Lowe's) syndrome Phosphaturia due to primary or secondary hyperparathyroidism Primary hyperparathyroidism Secondary hyperparathyroidism due to hereditary vitamin D dependency, Types I and II Hypocalcemia from any cause, provided parathyroid glands are intact
Miscellaneous Causes	
Vitamin D intoxication Acromegaly Tumoral calcinosis Cortical hyperostosis	Tumor-induced hypophosphatemia Post-transplantation hypophosphatemia Hypercalciuric nephrolithiasis

and hyperphosphatemia occurs. Although the level of plasma P_i does not regulate PTH secretion directly, hyperphosphatemia causes movement of calcium into bone, decreases resorption of mineral from bone, and increases formation of undissociated calcium-phosphate complexes in plasma. The resultant decrease of plasma Ca_I indirectly stimulates PTH hypersecretion. Impaired synthesis of 1,25-$(OH)_2$D by the damaged kidneys diminishes intestinal calcium absorption to further aggravate the bone lesions.

Diseases Associated with Hypophosphatemia

Conditions associated with hypophosphatemia are listed in Table 21-3.

Familial hypophosphatemic syndromes (Vitamin D–resistant rickets). Chronic hypophosphatemia from any cause can eventually result in rickets and osteomalacia. This mechanism is considered the main factor in the pathogenesis of two well-recognized genetic forms of chronic rickets not caused by deficiency of vitamin D.

X-linked hypophosphatemia[70] results from a selective inborn defect in renal phosphate reabsorption located in the proximal renal tubule cells. The main biochemical abnormality is hypophosphatemia. Plasma Ca_T and Ca_I are normal, and increased plasma PTH and generalized aminoaciduria, hallmarks of hereditary vitamin D dependency Type I, are absent, suggesting that hypophosphatemia is a primary abnormality. Plasma ALP is frequently but not always increased.

Defective renal tubular reabsorption of P_i is indicated by abnormally high urinary P_i clearance, diminished tubular reabsorption of P_i (low TRP), and a low ratio of transport maximum for P_i to glomerular filtration rate (TmP_i/GFR), even at low plasma P_i concentrations. Plasma 25-(OH)D concentrations are normal, and 1,25-$(OH)_2$D concentrations are also normal or low.

Autosomal dominant hypophosphatemia[70] is a less common condition in which the pathology of the defect in renal phosphate conservation is less well understood. On the whole the rickets is less severe than in x-linked hypophosphatemia.

Bone Diseases with Normocalcemia and Normophosphatemia

Osteoporosis. Osteoporosis is a general term used to describe a group of diseases of diverse causes that are characterized by reduction of bone mass below that required for normal mechanical function even though the ratio of mineral to organic matrix in bone is normal. Osteoporosis can be secondary to well-defined conditions such as Cushing's disease, glucocorticoid administration, and chronic immobilization. More often it is primary and not associated with any other definable disease. Idiopathic osteoporosis is the most common skeletal abnormality in postmenopausal women, and it is frequently seen in aged individuals.

Laboratory findings. Characteristically, the usual laboratory test results are normal. Hypercalcemia may occur transiently when the patient is immobilized. Increased ALP may occur if a patient has fractured a bone. Decreased levels of plasma 1,25-$(OH)_2$D have been reported in some osteoporotic individuals.

Paget's disease of bone. Paget's disease is among the most common of the non-metabolic bone diseases. It is characterized by regions of excessive bone resorption and repair, increased turnover of bone minerals and matrix, and formation of abnormal bone. Evidence for a viral origin has recently been presented.[51] The condition is often discovered incidentally during radiographic examination or by the finding of increased serum ALP during biochemical screening.

Laboratory findings. The plasma Ca and P_i concentrations are usually normal. However, hypercalcemia develops occasionally and, when present, is related to immobilization or fracture. A higher than usual incidence of primary hyperparathyroidism and hypercalcemia has been noted. Serum ALP and urinary hydroxyproline are usually markedly increased; the elevations tend to correlate with the extent and activity of the disease. The urinary calcium excretion rate varies; when bone resorption predominates, it tends to be increased, and this may account for the high incidence of urolithiasis in the disease.

Diseases Associated with Hypercalciuria

The diagnosis of ***idiopathic hypercalciuria*** is important because approximately 70% of kidney stones are caused by this condition. In ***absorptive hypercalciuria,*** the primary abnormality is considered to be overabsorption of calcium from the gut. This causes the rate of urinary calcium excretion to increase after a meal because of an increased filtered load of calcium and the absence of PTH action to increase tubular calcium reabsorption. Fasting urinary calcium levels are normal. In ***renal hypercalciuria,*** the primary abnormality is thought to be a selective defect in the ability of the renal tubules to reabsorb calcium. The urinary loss of calcium elicits secondary hyperparathyroidism, which in turn stimulates synthesis of 1,25-$(OH)_2D$ to maintain normocalcemia.

The initial investigation of ***kidney stone disease*** should include the following determinations: plasma Ca_T, Ca_I, P_i, creatinine, uric acid, electrolytes, blood pH, pCO_2, and HCO_3^-. The 24-h urinary excretion rates of calcium, creatinine, oxalate, and uric acid should be measured. A fresh urine sample should be examined for specific gravity and pH. Microscopy and a bacterial culture should be performed to rule out urinary tract infection. The plasma PTH concentration and the urinary excretion of nephrogenous cAMP should be measured if hyperparathyroidism is suspected. Assessment of the urinary amino acid pattern and a qualitative test for cystine should be carried out routinely on a random sample to exclude the diagnosis of cystinuria.

ANALYTICAL METHODS

Analysis of Calcium in Biological Fluids

Atomic absorption spectrophotometry (AAS) and spectrophotometry using calcium-dye complexes are most often used for measuring Ca_T. Ca_I measurement has become feasible on a routine basis through development of calcium ion–selective electrodes (ISE) and sensitive potentiometers.

Determination of Ionized Calcium (Ca_I)

Several analytical systems are available for the determination of ionized calcium. The potentiometric determination of Ca_I, strictly speaking, does not measure Ca ion concentration but rather the activity of the ionic species in the sample. The term *concentration* will be used subsequently in this qualified sense.

Principle. A sensitive potentiometer measures the voltage differential (E, EMF) developed between a reference electrode and a calcium ion–selective electrode exposed to a liquid containing the calcium ion. The potential established by the unknown solution is compared with the potential developed by a standard solution of known ionic strength and ion composition and containing calcium ions in exactly known concentration (see also Chapter 2, Section Four).

Instruments

Several commercial instruments are available for measuring Ca_I. Each consists of a sampler pump, a calcium ion–selective electrode, a reference electrode, a potentiometer, and electronic components to compute and display results. These instruments differ in sample volume requirements, length of time required for measurement, whether the electrode dips into the sample or the sample flows through the electrode, and the type and frequency of standardization required. Particular considerations govern the choice of an instrument. A vertical flow path has the advantage of easier elimination of inadvertently introduced air bubbles and of less frequent clogging with small fibrin clots. Automatic self-standardization simplifies the operation during long sample runs and allows the instrument to be used at any time without delay. Ability to accept small sample volumes is of particular importance for pediatric applications.

Calcium Ion–Selective Electrodes

Calcium-selective electrodes are designed around organophosphate or other organic ion exchanger compounds that have been modified for selective interaction with calcium. The reaction process is analogous to the interaction of hydrogen ions with the glass electrode.

Selectivity of an electrode is defined in terms of relative electrode response to calcium ions compared with the response to competing cations such as Mg^{2+}. The extent of interference of a competing ion is given by its selectivity coefficient, K.[66,67] The smaller the selectivity coefficient of the competing ion, the greater the selective response of the electrode to calcium ion.

Temperature has an effect both on electrode response and on the calcium-binding equilibrium with plasma proteins and small-molecule ligands. Instruments for measuring Ca_I are therefore designed to control and maintain temperature constant at 37 °C.

Chemical interference with the ISE may be caused by contamination of the specimen with surfactants having anionic properties or by added or intrinsic chemicals in the sample.[15] Ethanol, when present at high blood levels, may interfere directly with the Ca_I measurement by affecting the ion-exchanger liquid membrane. EGTA, EDTA, and other strong calcium chelators do not interfere directly with the operation of newer electrodes. However, they chelate calcium in the plasma sample and thus lower the Ca_I value. They should not be used as anticoagulants in specimens intended for Ca_I determination.

Heparin is the only suitable anticoagulant for whole blood or plasma samples intended for Ca_I determination. However, heparin causes a 0.01 mmol/L decrease in the Ca_I value for each unit of heparin added per mL blood. For this reason it must be stressed that the amount of added heparin should be kept constant and at the lowest concentration that prevents clotting (about 2 U/mL blood). Intravenous injection of 10 000 units of heparin has been shown, in adults, to cause a decrease in Ca_I of about 0.03 mmol/L.[6]

Several anions, such as phosphate, citrate, lactate, sulfate, and oxalate, can form complexes or other nonionized compounds and thus decrease the concentration of Ca_I in the specimen.

Specimen

Venous, arterial, and capillary blood specimens are suitable for Ca_I determination. Whole blood, plasma, or serum can be used. However, because pH has a great effect on the concentration of Ca_I in plasma, it is necessary that the pH of the specimen at the time of measurement be the same as that of the patient's blood at the time of sampling. This is most readily achieved by collecting and processing the specimen anaerobically. When the specimen is not collected anaerobically, its pH should be adjusted to the pH of the patient's blood at the time of sampling—a complicated and time-consuming procedure. The practice of adjusting the pH of specimens to a standard pH 7.4 is not recommended because values of Ca_I obtained in this manner do not necessarily reflect the in vivo situation.

The preferred sample for measurement of Ca_I is heparinized whole blood, collected anaerobically. Whole blood has the following advantages over plasma: (1) minimal handling, since the specimen can be introduced directly into the electrode; (2) smaller sample volume requirement; (3) greater ease of maintaining anaerobic conditions; and (4) rapid acquisition of test results (i.e., within a few minutes of drawing the blood). The whole blood specimen can be stored anaerobically at 4 °C for 48 h.

A drawback of using whole blood is the effect of erythrocytes on the junction potential of the ISE, an effect that cannot be compensated for by using standards. Ca_I determination in whole blood with a normal hematocrit gives a value 1.6% higher than in plasma or serum of the same specimen.

Concentrations of Ca_T and protein in plasma and serum are affected by posture of the patient and by venous stasis at the time of blood sampling;[65] Ca_I concentrations are less affected. The level of Ca_I in individuals confined to bed is 0.10–0.15 mmol/L lower than in ambulatory individuals.[67] Exercise or muscular activity increases plasma lactate, Ca_T, and protein, and decreases blood pH,[65] but the net effect on Ca_I is unpredictable.

The recommended collection procedure is to keep the patient recumbent and at rest for 30 min before the blood is drawn and to obtain the specimen without the use of a tourniquet.

Reagents

Prepared standards are available commercially or may be prepared in-house. In some instruments, the 1.0 and 2.0 mmol/L solutions are used for continuous self-standardization. Standards are prepared with NaCl, 150 mmol/L, to "normalize" ionic strength. For standardization of some systems, KCl, 4 mmol/L, is also incorporated into the standard solutions; in that circumstance, the NaCl concentration is reduced to 146 mmol/L. As long as the sum of the patient's sodium and potassium concentrations lies between 120 and 160 mmol/L, differences in the ionic strengths of the unknowns and standards will not cause clinically significant errors. If the sum of the Na and K concentrations in the unknown is <120 or >160 mmol/L, accurate determination of Ca_I by ISE requires a comparison of the unknown with standards containing NaCl concentrations that mimic the ionic strength of the unknown sample. Such action may be necessary when Ca_I is measured in dialysis fluids or in urine.[74]

Reference Ranges

Because of the relative novelty of the technique and differences in instrumentation and procedure, reference ranges reported in the literature are somewhat varied, particularly at the upper limit. In 42 healthy term infants, the range for whole blood Ca_I was 1.18–1.47 mmol/L (4.72–5.88 mg/dL).[43] In children, the reference range for whole blood Ca_I in the authors' laboratory is 1.10–1.50 mmol/L (4.40–6.00 mg/dL); another report[26] gives the range for serum Ca_I as 1.10–1.27 mmol/L (4.40–5.08 mg/dL) in 24 children under 16 years of age. For a series of reference ranges for serum or whole blood in adults,[8,26,43,71,76] the lowest reported limit is 1.09 mmol/L (4.36 mg/dL), and the highest limit is 1.33 mmol/L (5.32 mg/dL). Use of an internationally standardized reference serum should reduce interlaboratory differences.

Determination of Total Calcium (Ca_T)

The definitive method for Ca_T is isotope dilution (ID) followed by mass spectrometry (MS);[52] however, it is not used for routine analysis. The ID-MS method also serves as reference for atomic absorption spectrophotometry (AAS), the secondary reference method.[10] Atomic absorption spectrophotometry is well within the scope of many laboratories; however, a commonly used routine method employs a simple spectrophotometric measurement based on formation of a color complex between calcium and *o*-cresolphthalein complexone (CPC).

Determination of Ca_T in Plasma, Serum, or Urine by Atomic Absorption Spectrophotometry[63]

Principle. When calcium is introduced into a flame of appropriate temperature, free calcium atoms are formed, a majority of which are in the ground state. The atoms at ground state absorb incident light emanating from a calcium hollow-cathode lamp at the resonance wavelength of 422.7 nm. The degree of absorption is concentration dependent (see Chapter 2, Section One). Manufacturers of atomic absorption spectrophotometers provide specific directions for use of their instruments and preparation of required reagents.

Specimen. Serum, heparinized plasma, and urine are suitable specimens. Serum and urine, but not plasma, may be stored at 4 °C for weeks or months. Pyrex vials with air-tight Teflon-lined screw caps are recommended for specimen storage. Plastic containers should be avoided because calcium may be lost by adsorption to the walls; this loss is particularly serious for specimens of small volume.[54] Use of acid-etched glassware may lead either to contamination with calcium because the etched areas cannot be cleaned thoroughly or to calcium loss because of adsorption of calcium onto the damaged surface. The risk of calcium loss is high when heparinized plasma is stored, since any calcium that coprecipitates with the fibrin that forms during storage will be removed by centrifugation.

Timed (24-h) urine specimens should be collected in acid-washed bottles containing 5 mL of HCl, 6 mol/L. If the urine specimen is collected without acid, the pH should be adjusted to 3–4 with no more than 10 mL of HCl, 6 mol/L, at the time of receipt in the laboratory and before an aliquot is taken for assay or storage. As for serum, aliquots of urine should be stored in Pyrex vials.

Comments

1. Good *analytical technique* includes the following precautions. Aspirate deionized water frequently, preferably between successive samples. Keep the burner clean and replace it when corrosion appears at the edges of the slot or when the signal-to-noise ratio decreases. Bracket unknowns between standards and repeat the aspiration of blank and standards often, readjusting the instrument as often as necessary. Use a standardized control solution with each analytical run.

2. *Pre-analytical errors* may arise from unreliable standards, contamination of specimen or reagents with calcium, or loss of calcium from stored specimens or standard solutions by adsorption to the container. Standards should be "certified" for AAS or checked against a primary standard such as SRM #915 obtained from the National Bureau of Standards, Washington, D.C. 20234. Calcium contamination of glassware can be eliminated or minimized by acid-wash (HCl, 0.1–0.3 mol/L), followed by copious rinses with deionized water. Corks and rubber stoppers, including those used with some commercially available evacuated tubes, and reused plastic bottles or pipet tips are sources of calcium contamination and should be avoided. Calcium losses from stock and working standards can be minimized by storage in tightly-sealed borosilicate glass containers.

3. *Analytical interferences. Anionic interference* can arise from phosphate, sulfate, and citrate, which are co-constituents with calcium in biological materials. Phosphate is the most serious interferent since it forms a complex with calcium that is not readily dissociated in an air-acetylene flame and thus reduces absorption of light. La^{3+} and Sr^{2+} incorporated into the sample diluent dissociate calcium from anion complexes and reduce the depressant effect of anions on light absorption by atomic calcium.[77] *Cationic interference* of Na^+, K^+, and Mg^{2+} is less troublesome than anionic interference and can be obviated by including these cations in standards at ionic stengths similar to those of plasma. *Protein interference* may occur due to calcium-protein complex formation or to the mechanical effect of viscosity on aspiration and atomization of the sample.[24] Addition of TCA to the sample diluent acidifies the sample, breaks the calcium-protein complex, and precipitates the protein.

4. Calcium assays of *feces, tissue,* or *food* are achieved by taking weighed aliquots of homogenized specimens, ashing them, and dissolving the ash in dilute nitric acid. The acid digest is then analyzed for calcium or magnesium by AAS and for phosphate, Na, and K by other methods.

Reference Ranges

Mean *plasma* or *serum* Ca_T concentration for adults is 2.35 mmol/L (9.4 mg/dL) with 95% confidence ranges from 2.20 to 2.55 mmol/L (8.8–10.2 mg/dL). Five per cent lower means have been reported for persons over 50 y.[40] The 95% confidence range for Ca_T in children, ages 4–20 y, is 2.30–2.75 mmol/L (9.2–11.0 mg/dL).[11,12,31]

Urinary Ca_T is usually expressed as a 24-h excretion rate; the excretion depends on many variables including dietary intake of calcium.[41] Ninety per cent of healthy men excrete <7.5 mmol (300 mg) per day, and 90% of healthy women excrete <6.25 mmol (250 mg) per day.[5,36,41,57,69] Children excrete <0.15 mmol (6 mg) per kg body weight per day.[41]

In a voided urine sample, the ratio of the concentration of calcium to creatinine (Ca:Cr) provides a convenient and fairly reliable index of the rate of calcium excretion when muscle mass is constant (i.e., not during growth). The ratio of urinary Ca:Cr for adults is normally <0.40 when both analytes are measured in mmol/L, or <0.14 when both are measured in mg/dL. A value >0.57 (using mmol/L) or >0.20 (using mg/dL) in a fasting specimen signifies hypercalciuria.[59] The lower limit of normal calcium excretion in urine is about 1.8 mmol (75 mg) per day.

Fecal Ca_T. Assuming an average intake of 17.5 mmol (700 mg) calcium per 24 h and net absorption of 20–40%, the fecal calcium excretion rate in healthy adults ranges from 10.5 to 14 mmol/d (420–560 mg/d).

Spectrophotometric Determination of Ca_T in Plasma or Serum with *o*-Cresolphthalein Complexone[45]

Principle. The metal-complexing dye *o*-cresolphthalein complexone (CPC) forms a chromophore with calcium in alkaline solution; the color produced is measured spectrophotometrically at 578 nm. The chromogen reagent contains 8-hydroxyquinoline to mask interfering cations (mainly Mg^{2+}), urea to diminish the turbidity of lipemic specimens and to enhance metal-dye complex formation, and ethanol to inhibit color development in the blank. After the developed color is measured, addition of an excess of the calcium chelator EDTA allows correction for light absorption by hemolysis. This treatment dissociates the calcium-chromogen complex but not the hemoglobin-chromogen complex; the absorbance due to this hemoglobin-chromogen complex is subtracted from the total absorbance to give the corrected Ca_T absorbance.

Reagents

1. Buffer. Diethanolamine, 2 mol/L, pH 11.7; urea, 5 mol/L. Dissolve 210 g of diethanolamine and 300 g of urea in about 500 mL of water. Adjust pH to 11.7 with acetic acid, and dilute to 1 L. Stable for two months.
2. Chromogen. *o*-Cresolphthalein complexone, 100 μmol/L; 8-hydroxyquinoline, 8 mmol/L; urea, 5 mol/L; acetic acid, 43 mmol/L; and ethanol, 25% (v/v). Dissolve 64 mg of CPC, 1.16 g of 8-hydroxyquinoline, and 2.5 mL acetic acid in 250 mL ethanol in a 1-L volumetric flask. Add 300 g urea and dilute to mark with deionized water. Stable for three months.
3. Color reagent. Mix equal volumes of buffer and chromogen reagent. Use within one day.
4. Calcium standard, 2.5 mmol/L. Weigh 125.1 mg $CaCO_3$, reagent grade, and dissolve in 500 mL of trichloroacetic acid, 0.1 mol/L.
5. Ethylenediaminetetraacetic acid (EDTA), 150 mmol/L. Dissolve 558 mg $Na_2EDTA \cdot 2\,H_2O$ in 10 mL of deionized water.

Procedure

1. Add 1.0 mL of color reagent to test tubes labeled blank, unknown, and standard.
2. To the blank tube add 20 μL of water, to the standard tube 20 μL of standard, and to the unknown tube 20 μL of serum or heparinized plasma.
3. Mix and let stand for 5–60 min.
4. Read at 578 nm and record absorbance of standard (A_{std}) and unknowns (A_{tot}) against the blank set to zero absorbance.
5. To the blank and unknowns, add 10 μL of EDTA. Mix.
6. Read and record the absorbance of the unknown (A_{hem}) with the blank set to zero absorbance at 578 nm as in step 4.

Calculations

$$A_{unknown} = A_{tot} - A_{hem}$$

$$Ca_T,\ mmol/L = \frac{A_{unknown}}{A_{std}} \times 2.5$$

Comments

1. The method has been adapted to many automated analyzers. Details of particular applications may be obtained from the manufacturers.
2. The intensity of color developed by the Ca-CPC complex increases proportionately with an increase in pH and is also dependent on the buffer species employed. Diethanolamine, 1 mol/L

of final mixture, enhances the color intensity; the concentration of this buffer is adequate to maintain the pH at 11.7 without increasing the viscosity or absorbance excessively. 2-Amino-2-methyl-1-propanol (AMP) is used as an alternative buffer in some adaptations.[53]

3. 8-Hydroxyquinoline in the color reagent almost completely eliminates interference from Mg^{2+} at concentrations encountered clinically, although it decreases the color intensity by 25–40%.

4. Bilirubin in high concentrations introduces a significant negative error in Ca_T results.[30] Acetaminophen and the antihypertensive agent hydralazine cause positive interference. A positive bias from hemolysis has been reported for some automated methods (5% at a hemoglobin value of 2.5 g/L),[61] but addition of EDTA or EGTA compensates for this error.[32]

5. *Reference ranges* are identical to those obtained by AAS.[68]

Ektachem Slide Method for Ca_T

Principle. A sample (10 μL) is deposited on multilayered slides (Eastman Kodak Company, Rochester, NY 14650). The calcium in the sample reacts with an indicator dye (Arsenazo III) at pH 7.8 to form a colored complex, the amount of which is proportional to the calcium present in the sample. Intensity of color is measured by reflectance spectrophotometry at 680 nm.

Methods for the Determination of Inorganic Phosphate*

Since the method of Fiske and SubbaRow was introduced in 1925,[25] almost all methods for assay of inorganic phosphate (P_i) have relied on the formation of a complex of phosphate ion with a molybdate compound. In the more commonly used methods, serum or plasma proteins are precipitated and inorganic phosphate is liberated from ligands by trichloroacetic acid. The protein-free filtrate is mixed with ammonium molybdate in acid solution to form ammonium phosphomolybdate.

$$7\ H_3PO_4 + 12\ (NH_4)_6Mo_7O_{24} \cdot 4\ H_2O \rightarrow 7\ (NH_4)_3[PO_4(MoO_3)_{12}] + 51\ NH_4^+ + 51\ OH^- + 33\ H_2O$$

A reducing agent is then added to convert the hexavalent molybdenum-phosphate complex to the pentavalent blue-colored complex. The intensity of the blue color is measured spectrophotometrically at a wavelength near 700 nm.

Many reducing agents are suitable for producing the colored pentavalent molybdenum-phosphate complex. The HCl salt of *N-phenyl-p-phenylenediamine (semidine)* provides good sensitivity for phosphate and good color stability.[20,29]

Other reducing agents have disadvantages worthy of note. *Aminonaphtholsulfonic acid* (ANSA) is relatively unstable; it tends to precipitate, and the color intensity of the reduced molybdate increases over several hours so that strict timing of standard and sample readings is required. Although *stannous chloride* gives greater color intensity than ANSA, the color is unstable, its formation is very sensitive to small differences in acid concentrations, and it deviates from Beer's law at concentrations observed in clinical measurement.

Some methods do not employ reduction of the phosphomolybdate complex. A *molybdivanadophosphate complex* with a stable yellow color is formed by addition of excess molybdate to an acid solution of vanadate and orthophosphate.[72] The chief disadvantage of this method is acid hydrolysis of labile phosphate esters and hence overestimation of P_i in biological specimens. Complex formation of certain *basic dyes* with *phosphomolybdate* will cause a marked shift in the absorbance maximum of the uncomplexed dye. The dye method based on the triphenylmethane dye Malachite Green is considerably more sensitive than the Fiske-SubbaRow procedure.[37]

Specimens. Serum or heparinized plasma is suitable. Citrate, oxalate, and EDTA should not be used as anticoagulants because they interfere with the formation of the phosphomolybdate complex. Serum or plasma should be separated from cells within

* Standard methods applied to plasma and serum measure only the inorganic phosphate fraction. However, by tradition, the result is expressed as phosphorus; hence, the term *phosphorus* is still frequently employed, although its use is unjustified from the physiological standpoint. When results are expressed in mmol/L rather than in mg/dL, the need to distinguish between "phosphate" and "phosphorus" is obviated.

1 h after collection of the specimen. Hemolysis must be avoided because phosphate may be split off from labile esters in the erythrocytes. Plasma or serum may be stored at 4 °C for several days or in the frozen state for several months.

Ideally, specimens should be obtained from a recumbent fasting patient without the use of a tourniquet. Venous stasis, because it causes hemoconcentration, affects plasma P_i in a manner analogous to its effect on Ca_T, with the added effect of hypoxia, which increases the breakdown of phosphate esters. Food ingestion causes a transient decrease in plasma P_i; the decrease occurs because the increase in blood pH after meals (alkaline tide) enhances in vivo complex formation of calcium with phosphate and its subsequent deposition in the bone mineral phase, and because glycogen deposition in muscle and liver draws on the P_i pool for formation of glucose-phosphate intermediates.

Urine for P_i determination should be collected in acid-washed, detergent-free bottles. If urinary calcium is to be determined on the same specimen, the bottle should contain 5 mL of HCl, 6 mol/L.

Comments

1. The order of addition of reagents—semidine before molybdate—must be observed; the excess semidine-HCl immediately and preferentially reduces the complex formed of phosphate with endogenous ascorbic acid before other side reactions with the complex can occur. Reversing the order can lead to 8% overestimation of P_i.

2. Interference by lipemia is not found with the semidine method and others that use protein separation. However, "creamy" serum causes a positive error and hemoglobin causes a negative error with the DuPont *aca* adaptation of the semidine method. Hemoglobin causes a positive error with the manual and dialysis methods.[29] Neither ascorbic acid (up to 25 mg/dL) nor bilirubin interferes in any of the methods. Mannitol interferes by decreasing the rate of formation of the phosphomolybdate complex.[18] Although citrate and oxalate do not interfere at concentrations usually found in the circulation and in urine,[75] they prevent full color development; neither should be used as a plasma anticoagulant.

Reference Ranges. The reference ranges for fasting plasma or serum P_i vary greatly with the age of the patient, and this observation must be taken into consideration in the diagnosis of hypophosphatemia and hyperphosphatemia. During the first six months of life, P_i values are 1.6–2.5 mmol/L (5.0–7.8 mg/dL).[58] Thereafter, values decrease significantly. At 5 years of age the P_i concentration range is 1.2–2.0 mmol/L (3.7–6.2 mg/dL) and at 10 years of age it is 1.1–1.8 mmol/L (3.4–5.6 mg/dL).[12] No sex difference is observed in childhood, but during puberty the P_i concentration decreases abruptly to the range observed in young adults: 0.8–1.45 mmol/L (2.5–4.5 mg/dL) in males and 0.9–1.5 mmol/L (2.8–4.7 mg/dL) in females.[40] Levels diminish slightly in adults of both sexes until middle age; thereafter they increase in elderly females and decrease slightly in elderly males.[40]

Measurement of PTH in Plasma

Radioimmunoassays are currently the only widely available procedures for measuring PTH. Causes of increased and decreased plasma PTH concentrations are listed in Table 21-4. The measurement of PTH by RIA is complicated by several factors. Plasma contains not only intact PTH 1–84 but also *N*-terminal, *C*-terminal, and midmolecule fragments, the relative concentrations of which vary from condition to condition and from patient to patient. Furthermore, radioimmunoassays differ in their ability to detect the various PTH fragments. Presently, no one radioimmunoassay is considered ideal in all disorders and there is lack of agreement about the best method to use in various clinical situations.

In employing a PTH radioimmunoassay system, it is essential that the characteristics of the antibody for recognition of the various fragments be accurately established and that the reference range of the particular assay be known. The choice of the particular radioimmunoassay system depends partly on the tentative diagnosis of the patient. Assays directed toward the *C*-terminal and midmolecule fragments presently

Table 21-4. CAUSES OF INCREASED AND DECREASED PLASMA PTH CONCENTRATIONS

Conditions in which plasma PTH concentration is elevated:
Primary hyperparathyroidism
Hyperplasia
Adenoma
Carcinoma
Familial hypocalciuric hypercalcemia, familial neonatal primary hyperparathyroidism
Medullary thyroid carcinoma
MEN: Types I, IIa, and IIb
Secondary hyperparathyroidism (in response to hypocalcemia of various causes)
Vitamin D deficiency
Hereditary vitamin D dependency rickets: Types I and II
Chronic renal failure
Pseudohypoparathyroidism
Dietary calcium deficiency
Conditions in which plasma PTH concentration is diminished or undetectable:
Hypoparathyroidism
Isolated
Autoimmune (MEDAC syndrome)
DiGeorge syndrome (sometimes temporary)
Injury of parathyroid glands (hemochromatosis, thyroid or parathyroid surgery)
Secondary to severe hypomagnesemia
Neonatal
Transient neonatal hypocalcemia
Maternal hyperparathyroidism
Secondary to hypercalcemia of various causes:
Hypercalcemia of malignancy
Infantile hypercalcemia syndrome
Hypervitaminosis D
Sarcoidosis
Sudden immobilization
Hyperthyroidism

have the greatest general use.[47,49] Midmolecule procedures tend to have a shorter assay time (<8 h) than do *C*-terminal procedures (~24 h).

Principle. Specific antiserum is incubated with individual plasma (serum) specimens or standards. ^{125}I-PTH is then added, and the reaction mixture is incubated to allow ^{125}I-PTH to compete with unlabeled PTH for binding sites on the antiserum. After incubation, a second antibody directed specifically against the first antibody is added to precipitate the antibody-bound PTH. The mixture is centrifuged, the supernatant discarded, and the bound ^{125}I-PTH of the precipitate is measured in a gamma-counter. The concentration of PTH in samples is determined by interpolation using a standard curve.

Reference ranges. Reference ranges common to all methods cannot be presented because of differences in sequence specificity of the antisera employed, in the proportions of the various PTH fragments in the plasma of individuals under investigation, and in the modes of expressing results. The reference range for a typical midmolecule radioimmunoassay is 29–85 pmol/L (87–225 ng/mL). PTH values of infants, children, and pregnant women tend to be somewhat higher than those of other healthy subjects.

Measurement of Calcitonin in Plasma

The clinical application of the plasma calcitonin assay is in the diagnosis of *medullary carcinoma of the thyroid* (MCT). This condition is a malignancy of the calcitonin-secreting cells (C cells).[4] Despite excessive secretion of CT, the plasma Ca_T and Ca_I concentrations are usually normal. If hypocalcemia does occur, it is related to diarrhea, a commonly associated feature. The plasma P_i concentration is usually normal.

Several techniques have been described that follow the established principles for radioimmunoassay of peptide hormones.[34,35] Suitable dilutions of human calcitonin (hCT) standard, serum, or plasma are preincubated with antibody for 3–7 d at 4 °C. ^{125}I-hCT is added and the mixture incubated for an additional 3–6 d at 4 °C. Antibody-bound and free fractions are separated by precipitation or adsorption and the radioactivities of the bound and free fractions are determined. The concentration of CT in the patient's sample is determined by interpolation from a standard curve.

Reference ranges. Basal (i.e., unstimulated) CT values vary depending on sequence specificity of the antiserum and heterogeneity of the secreted hormone. Thus, results of assays must be interpreted in the light of reference data of the same laboratory. Basal levels in adults are generally <100 pg/mL. The plasma CT level is higher in newborn infants than in adults and decreases gradually in the elderly. The plasma CT level is also somewhat higher during pregnancy and lactation. Plasma CT does not appear to be strongly influenced by meals. Values of CT in serum are somewhat higher than in plasma. Hemolysis causes artifactually high values.

Comments. Despite the high sensitivity of recently described radioimmunoassays, the basal plasma CT level is less than the detection limit in an appreciable proportion of normal subjects tested. Provocative secretion tests usually cause a significant increase of plasma CT, and the hormone becomes readily detectable in nearly all normal individuals. The most commonly used secretagogues for provocative tests are pentagastrin, glucagon, and calcium salts, each of which is administered intravenously.[34,60] Measurements of plasma CT levels are of particular value in the diagnosis of medullary thyroid carcinoma (MTC), the identification of asymptomatic affected relatives of patients with MTC and multiple endocrine neoplasia (MEN, Types IIa and IIb), and the monitoring of surgically treated MTC patients for possible recurrence of the tumor.

Methods for the Detection of Vitamin D and Its Metabolites

Vitamin D and its metabolites can be separated from the bulk of plasma components by solvent extraction techniques. All vitamin D compounds possess a *cis*-triene bond that gives rise to an unusually intense UV-absorption with λ_{max} = 265 nm, λ_{min} = 228 nm, and ϵ_{molar} = 18 300. These features are utilized in designing methods for separation and quantitation.

A number of natural binding proteins and receptors for vitamin D and its metabolites can be utilized in radioligand assays. Vitamin D–binding protein (DBP) has strong binding affinities for 25-(OH)D and 24,25-$(OH)_2D$ and is a basic component of a popular form of 25-(OH)D assay. The same protein can be used, though less easily, for measuring vitamin D. The cytosolic binding protein of chick intestine, the probable intestinal receptor for 1,25-$(OH)_2D$ in that species, facilitates quantitation of extremely small amounts of 1,25-$(OH)_2D$. Other tissues from mammalian sources (e.g., bovine thymus and rat osteosarcoma) are also useful sources of the 1,25-$(OH)_2D$ receptor. Recently, specific antibodies have been generated for measuring individual vitamin D metabolites; however, these antibodies are not widely available.

In an individual receiving both dietary vitamin D_2 supplements and exposure to sunlight, the plasma will contain vitamin D_2 and D_3 metabolites in proportion to the amounts of the parent vitamins acquired. Thus, methods designed to measure metabolites of vitamins D_2 and D_3 must have equal sensitivity for both types. Most procedures involve three basic steps: (1) extraction, (2) purification, and (3) radioligand or UV assay. Internal standards must be used to assess losses of vitamin D compounds during the purification process. In most cases, the corresponding radioactive vitamin D metabolite is used as an internal standard.[73]

Determination of Total 25-(OH)D* in Plasma by Competitive Binding Assay[38]

Principle. 25-(OH)D along with other plasma lipids is extracted from serum or heparinized plasma, partially purified by Sephadex LH-20 chromatography or by HPLC on microparticulate silica, and assayed by virtue of its strong binding affinity for rat vitamin D–binding protein (DBP). The ability of nonradioactive 25-(OH)D from the plasma extract to compete with $[^3H]$25-$(OH)D_3$ for DBP binding is compared with the competition by known amounts of pure 25-$(OH)D_3$ standard.

---

*Total 25-(OH)D = 25-$(OH)D_2$ plus 25-$(OH)D_3$; total 1,25-$(OH)_2D$ = 1,25-$(OH)D_2$ plus 1,25-$(OH_2)D_3$; and total vitamin D = vitamin D_2 plus vitamin D_3 forms.

Table 21-5. REFERENCE RANGES FOR VITAMIN D AND ITS METABOLITES IN SERUM

	Mean (± SD)		Range	
Vitamin D[38]	5.7 ± 2.9 nmol/L	(2.18 ± 1.1 ng/mL)	2.1– 12.2 nmol/L	(0.80– 4.68 ng/mL)
25-(OH)D[38]				
Summer	55.3 ± 17.8 nmol/L	(22.1 ± 7.1 ng/mL)	20.0– 82.5 nmol/L	(8.0–33.0 ng/mL)
Winter	40.0 ± 9.8 nmol/L	(16.0 ± 3.9 ng/mL)	22.8– 59.8 nmol/L	(9.1–23.9 ng/mL)
1,25-$(OH)_2D$*	88.9 ± 30.5 pmol/L	(36.9 ± 12.6 pg/mL)	40.9–141.8 pmol/L	(17.0–58.9 pg/mL)
24,25-$(OH)_2D_3$ [16]	5.8 ± 2.6 nmol/L	(2.4 ± 1.0 ng/mL)	1.9– 10.3 nmol/L	(0.8– 4.2 ng/mL)

*Jones, G., unpublished.

Determination of Total 25-(OH)D in Plasma by Ultraviolet Detection[23,38]

Principle. 25-(OH)D is extracted along with plasma lipids from serum or heparinized plasma. 25-$(OH)D_2$ and 25-$(OH)D_3$ are separated from each other and from the other UV-absorbing substances by two HPLC steps, and the two vitamin D compounds are measured independently by using their strong UV-absorption characteristic at 265 or 254 nm.

Comments. The UV-detection method separates 25-$(OH)D_2$ and 25-$(OH)D_3$, and it is faster and more accurate than the competitive protein-binding assay, is more readily automated, and permits analysis of 25-(OH)D over a concentration range of 12–2500 nmol/L (5–1000 ng/mL). To encompass such a range with the competitive protein-binding assay requires initial dilutions of the plasma extract. On the other hand, the UV-detection method has a lower sensitivity than the competitive protein-binding assay and larger sample volumes must be used when plasma contains <10 ng 25-(OH)D/mL.

Determination of Total 1,25-$(OH)_2D$* in Plasma by Competitive Protein-Binding Assay[22,48,64]

Principle. 1,25-$(OH)_2D$ is selectively extracted from plasma by use of reverse-phase cartridges (Extrelut, Merck Darmstadt, West Germany; Bondelut-C_{18},[39] Analytichem International, Harbor City, CA 90710; or Sep Pak-C_{18}, Waters Associates, Inc., Milford, MA 01757) at pH 10, partially purified on silica cartridges (Sep Pak-SIL, Waters; Bondelut-SIL, Analytichem) or HPLC, and assayed by virtue of its strong binding affinity for cytosolic vitamin D–binding receptors in homogenates of chick intestinal mucosa,[22] bovine thymus[64] or rat osteosarcoma.[48] The extracted 1,25-$(OH)_2D$ is compared with a standard curve obtained with pure 1,25-$(OH)_2D_3$ with respect to its ability to compete with [^{3}H]1,25-$(OH)_2D_3$ for tissue cytosol-binding sites.

Determination of 24,25-$(OH)_2D_3$ in Plasma by Competitive Protein-Binding Assay[19]

Principle. 24,25-$(OH)_2D_3$ is extracted from plasma along with total lipids, including 25-(OH)D; the metabolite is then separated from 25-(OH)D by chromatography on HPLC and assayed by virtue of its strong binding affinity for DBP. The affinity of the 24,25-$(OH)_2D_3$ for DBP is equal to that of 25-$(OH)D_3$; thus, a standard reagent of 24,25-$(OH)_2D_3$ is not needed. Per cent bound 24,25-$(OH)_2D_3$ is compared with a standard curve of pure 25-$(OH)D_3$ in respect to its ability to compete with [^{3}H]25-$(OH)D_3$ for DBP binding.

Determination of Vitamin D* in Plasma by Ultraviolet Detection[38]

Principle. Vitamin D is extracted from plasma along with total lipids, separated from the bulk of the lipids and the vitamin D metabolites by HPLC on microparticulate silica, then further purified by a second HPLC step on octadecasilane-bonded silica prior to quantitation by ultraviolet spectrophotometry. This method is suitable for measurement of vitamin D in plasma of individuals receiving therapeutic doses of vitamin D_2 or D_3.

Reference ranges for vitamin D and metabolites in serum are shown in Table 21-5.

*Total 25-(OH)D = 25-$(OH)D_2$ plus 25-$(OH)D_3$; total 1,25-$(OH)_2D$ = 1,25-$(OH)D_2$ plus 1,25-$(OH_2)D_3$; and total vitamin D = vitamin D_2 plus vitamin D_3 forms.

References

1. Adams, J. S., Clemens, T. L., Parrish, J. A., et al.: Vitamin D synthesis and metabolism after ultraviolet irradiation of normal and vitamin D-deficient subjects. N. Engl. J. Med., *306*:722–725, 1982.
2. Arnaud, C. D.: The parathyroid glands. *In*: Cecil Textbook of Medicine. 16th ed. J. B. Wyngaarden and L. H. Smith, Jr., Eds. Philadelphia, W. B. Saunders Co., 1982, pp. 1286–1302.
3. Aurbach, G. D., Marx, S. J., and Spiegel, A. M.: Parathyroid hormone, calcitonin and the calciferols. *In*: Textbook of Endocrinology. 6th ed. R. H. Williams, Ed. Philadelphia, W. B. Saunders Co., 1981, pp. 922–1031.
4. Austin, L. A., and Heath, H., III: Calcitonin. Physiology and pathophysiology. N. Engl. J. Med., *304*:269–278, 1981.
5. Bijvoet, O. L. M.: Kidney function in calcium and phosphate metabolism. *In*: Metabolic Bone Disease, Vol. 1. L. V. Avioli and S. M. Krane, Eds., New York, Academic Press, 1977, pp. 49–140.
6. Biswas, C. K., Ramos, J. M., and Kerr, D. N. S.: Heparin effect on ionised calcium concentration. Clin. Chim. Acta, *116*:343–347, 1981.
7. Boyd, J. C., Lewis, J. W., Slatopolsky, E., et al.: Parathyrin measured concurrently with free or total calcium in the differential diagnosis of hypercalcemia. Clin. Chem., *27*:574–579, 1981.
8. Brauman, J., Delvigne, C., and Brauman, H.: Measure of blood ionized calcium versus total calcium in normal man in renal insufficiency and in hypercalcemia of various origins. Scand. J. Clin. Lab. Invest., *43* (Suppl. 165):75–78, 1983.
9. Brinkhurst, F. R., and Potts, J. T., Jr.: Calcium and phosphate distribution, turnover, and metabolic actions. *In*: Endocrinology, Vol. 2. L. J. DeGroot et al., Eds. New York, Grune & Stratton, 1979, pp. 551–585.
10. Cali, J. P., Bowers, G. N., Jr., and Young, D. S.: A reference method for the determination of total calcium in serum. Clin. Chem., *19*:1208–1213, 1973.
11. Cheng, M. H., Lipsey, A. I., Blanco, V., et al.: Microchemical analysis for 13 constituents of plasma from healthy children. Clin. Chem., *25*:692–698, 1979.
12. Cherian, A. G., and Hill, J. G.: Percentile estimates of reference values for fourteen chemical constituents in sera of children and adolescents. Am. J. Clin. Pathol., *69*:24–31, 1978.
13. Cheung, W. Y.: Calmodulin. Sci. Am., *246*:62–70, 1982.
14. Coburn, J. W., and Slatopolsky, E.: Vitamin D, parathyroid hormone, and renal osteodystrophy. *In*: The Kidney, Vol. 2. B. M. Brenner and F. C. Rector, Jr., Eds. Philadelphia, W. B. Saunders Co., 1981, pp. 2213–2305.
15. Craggs, A., Moody, G. J., Thomas, J. D. R., et al.: Effect of anionic surfactants on calcium ion-selective electrodes. Analyst, *105*:426–431, 1980.
16. Daneman, D., Kooh, S. W., and Fraser, D.: Hypoparathyroidism and pseudohypoparathyroidism in childhood. Clin. Endocrinol. Metab., *11*:211–231, 1982.
17. Deluca, H. F., and Schnoes, H. K.: Vitamin D: Recent advances. Ann. Rev. Biochem., *52*:411–439, 1983.
18. Donhowe, J. M., Freier, E. F., Wong, E. T., et al.: Factitious hypophosphatemia related to mannitol therapy. Clin. Chem., *27*:1765–1769, 1981.
19. Dreyer, B. E., and Goodman, D. B. P.: A simple direct spectrophotometric assay for 24,25-dihydroxyvitamin D_3. Anal. Biochem., *114*:37–41, 1981.
20. Dryer, R. L., and Routh, J. I.: Determination of serum inorganic phosphorus. Stand. Methods Clin. Chem., *4*:191–195, 1963.
21. Eil, C., Libermann, U. A., Rosen, J. F., et al.: A cellular defect in hereditary vitamin D-dependent rickets Type II: Defective nuclear uptake of 1,25-dihydroxyvitamin D in cultured skin fibroblasts. N. Engl. J. Med., *304*:1588–1591, 1981.
22. Eisman, J. A., Hamstra, A. J., Kream, B. E., et al.: A sensitive, precise and convenient method for determination of 1,25-dihydroxyvitamin D in human plasma. Arch. Biochem. Biophys., *176*:235–243, 1976.
23. Eisman, J. A., Shepard, R. M., and DeLuca, H. F.: Determination of 25-hydroxyvitamin D_2 and 25-hydroxyvitamin D_3 in human plasma using HPLC. Anal. Biochem., *80*:298–305, 1977.
24. Fassel, V. A., and Becker, D. A.: Chemical or solute vaporization interference in flame atomic emission and absorption spectrometry. Anal. Chem., *41*:1522–1526, 1969.
25. Fiske, C. H., and SubbaRow, Y.: The colorimetric determination of phosphorus. J. Biol. Chem., *66*:375–400, 1925.
26. Fogh-Andersen, N., McNair, P., Møller-Petersen, J., et al.: Lowered serum ionized calcium in insulin treated diabetic subjects. Scand. J. Clin. Lab. Invest., *43* (Suppl. 165):93–97, 1983.
27. Fraser, D., and Kooh, S. W.: Disorders of the endocrine glands: Disturbances of parathyroid hormone and calcitonin. *In*: Textbook of Paediatrics, Vol. 2. 3rd ed. J. O. Forfar and G. C. Arneil, Eds. Edinburgh, Churchill Livingstone, 1984, pp. 1157–1172.
28. Fraser, D., Kooh, S. W., and Scriver, C. R.: Hyperparathyroidism as the cause of hyperaminoaciduria and phosphaturia in human vitamin D deficiency. Pediatr. Res., *1*:425–435, 1967.
29. Garber, C. C., and Miller, R. C.: Revisions of the 1963 semidine HCl standard method for inorganic phosphorus. Clin. Chem., *29*:184–188, 1983.
30. Gitelman, H. J.: An improved automated procedure for the determination of calcium in biological specimens. Anal. Biochem., *18*:521–530, 1967.
31. Gomez, P., Coca, C., Vargas, C., et al.: Normal reference-intervals for 20 biochemical variables in healthy infants, children, and adolescents. Clin. Chem., *30*:407–412, 1984.

32. Haas, R. G., and Mushel, S.: Modified DuPont *aca* calcium method for hemolyzed specimens. Am. J. Clin. Pathol., *77*:216–219, 1982.
33. Haefner, J. F., and Potts, J. T., Jr.: Biosynthesis of parathyroid hormone. N. Engl. J. Med., *299*:580–585, 635–644, 1978.
34. Heath, H., III: Tests of parathyroid function: Utility and limitations. Endocrinol. and Metab., *2*:1–10, 1984.
35. Hillyard, C. J., Cooke, T. J. C., Coombes, R. C., et al.: Normal plasma calcitonin: Circadian variation and response to stimuli. Clin. Endocrinol. (Oxf.), *6*:291–298, 1977.
36. Hodgkinson, A., and Pyrah, L. N.: The urinary excretion of calcium and inorganic phosphate in 344 patients with calcium stone of renal origin. Br. J. Surg., *46*:10–18, 1958.
37. Itaya, K., and Ui, M.: A new micromethod for the colorimetric determination of inorganic phosphate. Clin. Chim. Acta, *14*:361–366, 1966.
38. Jones, G.: Assay of vitamin D_2 and D_3 in human plasma by high performance liquid chromatography. Clin. Chem., *24*:287–298, 1978.
39. Kao, P. C., and Heser, D. W.: Simultaneous determination of 25-hydroxy- and 1,25-dihydroxyvitamin D from a single sample by dual-cartridge extraction. Clin. Chem., *30*:56–61, 1984.
40. Keating, F. R., Jones, J. D., Elveback, L. R., et al.: The relation of age and sex to distribution of values in healthy adults of serum calcium, inorganic phosphorus, magnesium, alkaline phosphatase, total proteins, albumin, and blood urea. J. Lab. Clin. Med., *73*:825–834, 1969.
41. Knapp, E. L.: Factors influencing the urinary excretion of calcium. I. In normal persons. J. Clin. Invest., *28*:182–196, 1947.
42. Kooh, S. W., Jones G., Reilly, B. J., et al.: Pathogenesis of rickets in chronic hepatobiliary disease in children. J. Pediatr., *94*:870–874, 1979.
43. Larsson, L., Finnstrom, O., Nilsson, B., et al.: Evaluation of Radiometer ICA 1 as a routine instrument for serum ionized calcium and its application for whole blood capillary samples from newborn infants. Scand. J. Clin. Lab. Invest., *43*(Suppl. 165):21–26, 1983.
44. Larsson, L., and Öhman, S.: Serum ionized calcium and corrected total calcium in borderline hyperparathyroidism. Clin. Chem., *24*:1962–1965, 1978.
45. Lorentz, K.: Improved determination of serum calcium with *o*-cresolphthalein complexone. Clin. Chim. Acta, *126*:327–334, 1982.
46. Low, J. C., Schaaf, M., Earl, J. M., et al.: Ionic calcium determination in primary hyperparathyroidism. JAMA, *223*:152–155, 1973.
47. Mallette, L. E., Tuma, S. N., Berger, R. E., et al.: Radioimmunoassay for the middle region of human parathyroid hormone using a homologous antiserum with a carboxyl-terminal fragment of bovine parathyroid hormone as radioligand. J. Clin. Endocrinol. Metab., *54*:1017–1024, 1982.
48. Manolagas, S. C., and Deftos, L. J: Cytoreceptor assay for 1,25-dihydroxyvitamin D_3: A novel radiometric method based on binding of the hormone to intracellular receptors in vitro. Lancet, *2*:401–402, 1980.
49. Martin, K. J., Hruska, K., Freitag, J., et al.: Clinical utility of radioimmunoassays for parathyroid hormone. Min. Electr. Metab., *3*:283–290, 1980.
50. Marx, S. J., Spiegel, A. M., Brown, E. M., et al.: Divalent calcium metabolism: Familial hypocalciuric hypercalcemia versus typical primary hyperparathyroidism. Am. J. Med., *65*:235–242, 1978.
51. Mills, B. G., Singer, F. R., Weiner, L. P., et al.: Immunohistological demonstration of respiratory syncytial virus antigens in Paget's disease of bone. Proc. Natl. Acad. Sci. USA, *78*:1209–1213, 1981.
52. Moore, L. J., and Machlan, L. A.: High accuracy determination of calcium in blood serum by isotope dilution mass spectrometry. Anal. Chem., *44*:2291–2296, 1972.
53. Moorehead, W. R., and Biggs, H. G.: 2-Amino-2-methyl-1-propanol as the alkalinizing agent in an improved continuous-flow cresolphthalein complexone procedure for calcium in serum. Clin. Chem., *20*:1458–1460, 1974.
54. Muhlert, M., Julita, M., and Quamme, G.: Disappearance of calcium and other electrolytes from microvolume samples. Am. J. Physiol., *242*:F202–F206, 1982.
55. Mundy, G. R., and Martin, J.: The hypercalcemia of malignancy: Pathogenesis and management. Metabolism, *31*:1247–1277, 1982.
56. Neer, R. M.: Calcium and inorganic phosphate homeostasis. *In*: Endocrinology, Vol. 2. L. Groot et al., Eds. New York, Grune and Stratton, Publishers, 1979, pp. 669–692.
57. Nordin, B. E. C., Peacock, M., and Wilkinson, R.: Hypercalciuria and calcium stone disease. Clin. Endocrinol. Metabol., *1*:169–183, 1972.
58. Owen, G. M., Garry, P., and Fomon, S. J.: Concentrations of calcium and inorganic phosphorus in serum of normal infants receiving various feedings. Pediatrics, *31*:495–498, 1963.
59. Pak, C. Y. C., Britton, I., Peterson, R., et al.: Ambulatory evaluation of nephrolithiasis. Am. J. Med., *69*:19–30, 1980.
60. Parthemore, J. G., and Deftos, L. H.: Calcitonin secretion in normal human subjects. J. Clin. Endocrinol. Metab., *47*:184–188, 1978.
61. Porter, W. H., Carroll, J. R., and Roberts, R. E.: Hemoglobin interference with the DuPont Automatic Clinical Analyzer procedure for calcium. Clin. Chem., *23*:2145–2147, 1977.
62. Potts, J. T., Jr.: Disorders of parathyroid glands. *In*: Harrison's Principles of Internal Medicine. 10th ed. R. G. Peterdorf, R. D. Adams, E. Braunwald, et al., Eds. New York, McGraw-Hill Book Co., 1983, pp. 1929–1943.
63. Pybus, J., Feldman, F.J., Bowers, G. N., Jr.: Measurement of total calcium in serum by atomic absorption spectrophotometry with use of a strontium internal reference. Clin. Chem., *16*:998–1007, 1970.

64. Reinhardt, T. A., Horst, R. L., Orf, J., et al.: A microassay for 1,25-dihydroxyvitamin D which does not require HPLC. J. Lab. Clin. Med., *58*:91–98, 1984.
65. Renoe, B. W., McDonald, J. M., and Ladenson, J. H.: The effects of stasis with and without exercise on free calcium, various cations, and related parameters. Clin. Chim. Acta, *103*:91–100, 1980.
66. Robertson, W. G.: Measurement of ionized calcium in body fluids—a review. Ann. Clin. Biochem., *13*:540–548, 1976.
67. Robertson, W. G., and Marshall, R. W.: Ionized calcium in body fluids. CRC Crit. Rev. Clin. Lab. Sci., *15*:83–125, 1981.
68. Robertson, W. G., and Marshall, R. W.: Calcium measurements in serum and plasma—total and ionized. CRC Crit. Rev. Clin. Lab. Sci., *13*:271–304, 1979.
69. Robertson, W. G., and Morgan, D. B.: The distribution of urinary calcium excretion in normal persons and stone formers. Clin. Chim. Acta, *37*:503–508, 1972.
70. Scriver, C. R., Fraser, D., and Kooh, S. W.: Hereditary rickets. *In*: Clinical Endocrinology 2, Calcium Disorders. D. A. Heath and S. J. Marx, Eds. London, Butterworths' Intl. Med. Rev. 1982, pp. 1–46.
71. Smith, S. C. H., Buckley, B. M., Wedge, G., et al.: An evaluation of the ICA 1 ionized calcium analyzer in a clinical chemistry laboratory. Scand. J. Clin. Lab. Invest., *43*(Suppl. 165):33–37, 1983.
72. Taussky, H. H., and Shorr, E.: A microcolorimetric method for the determination of inorganic phosphorus. J. Biol. Chem., *202*:675–685, 1953.
73. Tietz, N. W., Ed.: Textbook of Clinical Chemistry, Philadelphia, W. B. Saunders Company, 1986.
74. Thode, J., Wandrup, J., Aas, F., et al.: Evaluation of a new semiautomatic electrode system for simultaneous measurement of ionized calcium and pH. Scand. J. Clin. Lab. Invest., *42*:407–415, 1982.
75. Vreman, H. J., and Jöbsis, F. F.: Interference by mannitol and other compounds with phosphate determinations. Anal. Biochem., *17*:108–118, 1966.
76. Wandrup, J., and Kancir, C.: The concentration of free ions in whole blood. Scand. J. Clin. Lab. Invest., *43* (Suppl. 165):47–48, 1983.
77. Zettner, A., and Seligson, D.: Application of atomic absorption spectrophotometry in the determination of calcium in serum. Clin. Chem., *10*:869–890, 1964.

LIVER FUNCTION

22

WILLIAM F. BALISTRERI, M.D.
LESLIE M. SHAW, Ph.D.

The human liver performs multiple functions essential for life. Hepatocytes have a unique regenerative capability as well as a marked capacity to respond to increased metabolic demands. The liver receives, processes, and stores materials absorbed from the digestive tract such as amino acids, carbohydrates, fatty acids, cholesterol, and vitamins, and is capable of releasing metabolites of these compounds on demand. The liver synthesizes multiple plasma proteins including albumin, α- and β-globulins, clotting factors, transport proteins, and contributes to the body's immune system. The liver is the main organ of detoxification, is the site of metabolic conversion of endogenous and exogenous compounds, and synthesizes bile acids from cholesterol. These bile acids are secreted into the intestine, generating bile flow and facilitating dietary fat emulsification and absorption. Because the liver is a major site of catabolism of hormones, it participates in regulation of plasma hormone levels. The liver also responds to hormonal and neural stimuli to regulate the blood glucose concentration.

Assessment of Liver Function

Tests of liver function are used for: (1) *screening* for abnormalities in liver function, (2) *documenting* an abnormality, (3) identifying the *type* and *site* of injury, as well as (4) *prognostication* and follow-up of patients with hepatic disease. Because a single laboratory test will not satisfy all of the postulated purposes, tests must be used in combination. It is, therefore, important to indicate the value of each test in regard to (1) diagnostic *sensitivity* in screening for hepatic dysfunction, (2) *specificity* for liver disease, and (3) *selectivity* in differentiating these disorders. Radiographic, endoscopic, ultrasonic, and scintigraphic techniques provide sophisticated second-line diagnostic modalities; however, laboratory assessment of hepatic dysfunction is a valuable initial discriminator and complements findings from the patient's history and physical examination prior to selection of more definitive or invasive tests in the patient with suspected or proven hepatic disease.

Liver Anatomy

Gross Anatomy

The liver, the largest organ in the human body (weighing ~1.2–1.5 kg in the human adult), is located in the right upper quadrant of the abdomen, beneath and attached by ligaments to the diaphragm. The liver has a dual blood supply: (1) the portal vein, which carries blood from the alimentary tract and is rich in nutrients absorbed from the gastrointestinal tract, and (2) the hepatic artery, which carries oxygenated blood to the liver. Venous drainage is via the right and left hepatic veins, which enter into the inferior vena cava near the right atrium.

The *portal triad* consists of bile ducts that accompany the hepatic artery and portal vein (Figure 22-1). Bile flows from the hepatocyte into the bile canaliculi and ductules, to the larger intrahepatic bile ducts, and finally to the left and right hepatic bile ducts, which emerge from the liver to form the common hepatic duct. The hepatic duct is joined by the cystic duct from the gallbladder to form the common bile duct. The common bile duct and pancreatic duct enter the duodenum at the ampulla of Vater.

Microscopic anatomy. The liver is composed of polyhedral prisms of tissue (lobules) demarcated by connective tissue septa and vascular and biliary vessels (Figure 22-1).

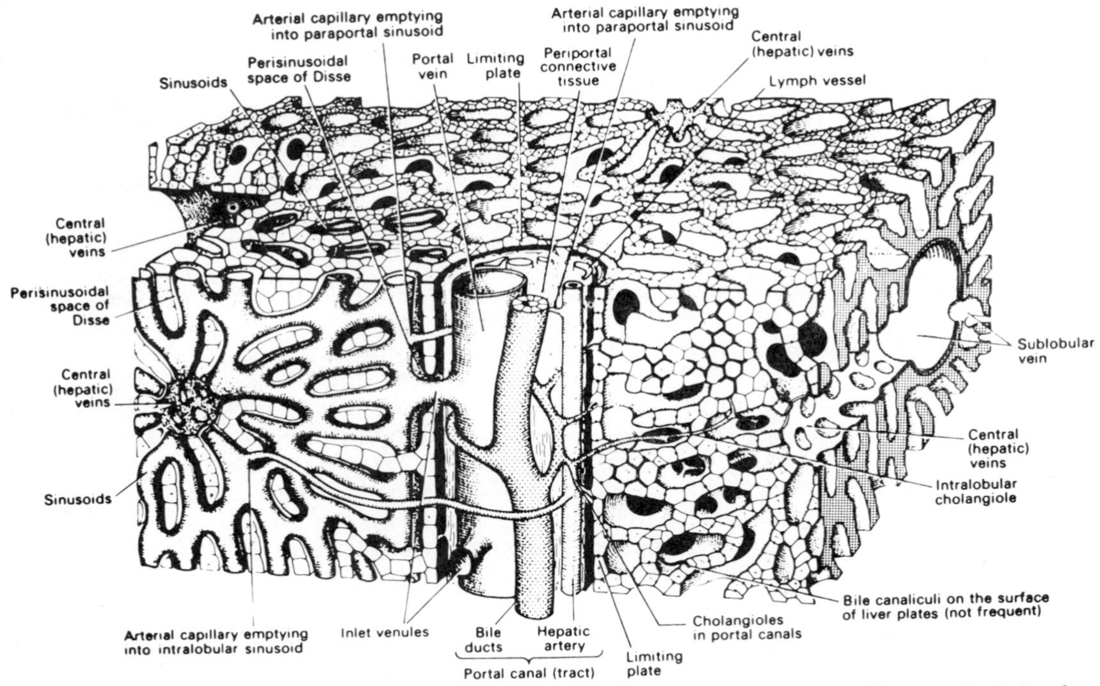

Figure 22-1. The classic liver lobule. (From Sherlock, S.: Diseases of the Liver and Biliary System. 6th ed. London, Blackwell Scientific Publications, 1981.)

The central vein (terminal hepatic venule) is in the center of the lobule, and single cell plates of parenchymal epithelial cells radiate from the central vein.

The *sinusoids* are blood-carrying vascular channels on either side of the liver cell plates; they form a rich, intralobular vascular network that converges toward the central vein. The sinusoidal lining cells are either endothelial cells or Kupffer cells; the latter constitute the largest group of fixed macrophages in the body. The tissue space between the endothelial cells and hepatocytes (i.e., the space of Disse) contains interstitial fluid, across which the transfer of nutrients and waste products from the blood occurs.

Hepatocytes, which constitute ~60% of the liver mass, are polygonal cells ~30 μm in diameter. The surface of the cell is specialized: (1) the sinusoidal surface, which faces the incoming blood (sinusoid and space of Disse); (2) the intercellular surface, contiguous with the sinusoidal surface; and (3) the canalicular surface.

Ultrastructure. Hepatic parenchymal cells have a well-developed organellar substructure (Figure 22-2) that includes:

1. *Mitochondria,* which participate in energy generation (oxidative phosphorylation and fatty acid oxidation).
2. *Lysosomes,* which have specific degradative functions (hydrolytic enzymes).
3. The *endoplasmic reticulum* (ER), which is the site of many functions, including drug metabolism and bile acid synthesis. The *smooth* ER assumes the form of tubules and vesicles and is the site of bilirubin conjugation, drug detoxification, and cholesterol synthesis. The *rough* ER forms lamellar profiles lined with ribosomes, which are the sites of synthesis of specific proteins, such as albumin and coagulation factors.
4. The multifunctional *Golgi complex,* which produces lipoproteins (VLDL) and is involved in glycoprotein synthesis and albumin secretion (a "packaging" process).
5. *Microtubules and microfilaments,* which maintain cell shape and provide contractile force.

Clinical Manifestations of Liver Disease

Jaundice. The classic clinical manifestation of hepatic disease is yellow discoloration of the plasma, skin, sclerae, and mucous membranes that is caused by bilirubin

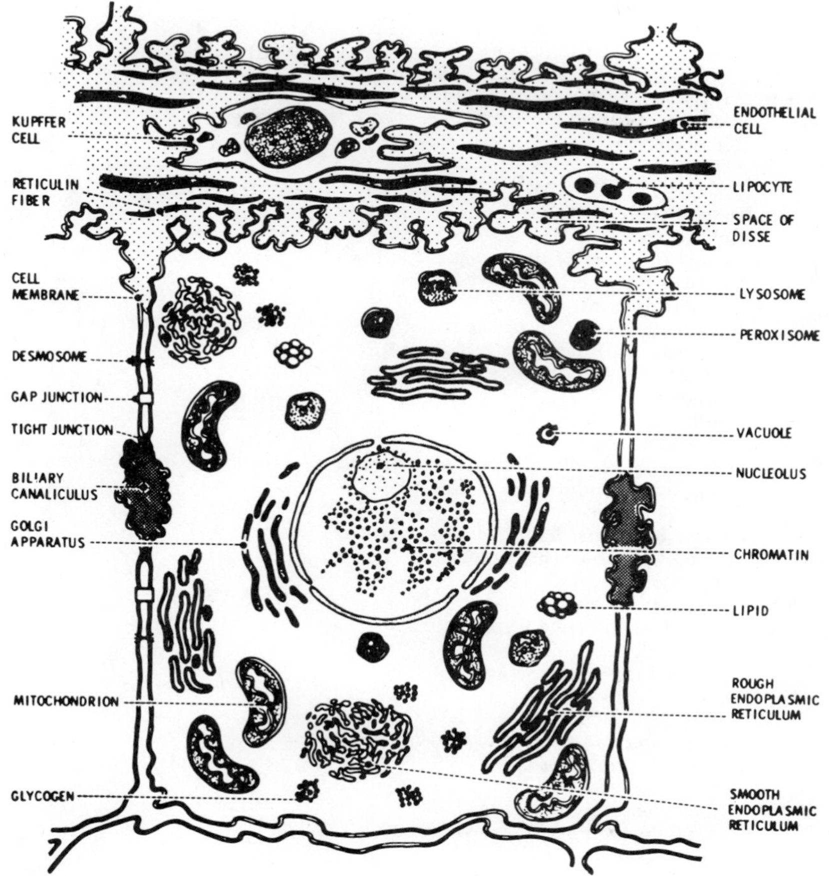

Figure 22-2. Hepatic ultrastructure. (From Sherlock, S.: Diseases of the Liver and Biliary System. 6th ed. London, Blackwell Scientific Publications, 1981.)

accumulation. Jaundice, or icterus, may be the first, and often the only manifestation of liver disease. The yellow discoloration is clinically apparent when the serum total bilirubin concentration reaches 2–3 mg/dL. There may be other signs of disturbances in bilirubin metabolism, such as the passage of dark urine or acholic (white) stools. The clinical finding of jaundice is not specific and may indicate a variety of diseases. Initial evaluation of the jaundiced patient should focus upon whether the hyperbilirubinemia arises from the predominance in serum of the unconjugated form of bilirubin or whether there is a significant increase of the conjugated form (Table 22-1).

Portal hypertension. Blood entering the liver via the portal vein represents the influx from the abdominal portion of the digestive tract, pancreas, and spleen. Under normal conditions, the portal venous system differs from other venous blood systems in that the pressure is slightly higher in order to overcome the resistance of the liver sinusoidal system. In the face of sinusoidal infiltration, scarring, or hepatic vein (outlet) obstruction, the normal portal pressure of ~5–10 mm Hg will be increased to varying degrees, causing portal hypertension, operationally defined as >20 mm Hg. When the resistance to the flow of portal venous blood is increased, there is dilatation of the venous blood system behind the obstruction and enlargement of the spleen. With time, there is an increase in collateral flow around the liver and the development of variceal vessels in the esophagus and at other locations that are potential sites of bleeding.

In addition to hepatosplenomegaly and the presence of collateral vessels, ascites and the hepatorenal syndrome are major complications of cirrhosis and portal hypertension. *Ascites* represents the accumulation of free fluid within the peritoneal cavity and is a consequence of sinusoidal hypertension, excess lymph formation, and hypoalbuminemia.

Table 22-1. CLASSIFICATION OF JAUNDICE (BASED ON TYPE AND SITE)

- **Unconjugated hyperbilirubinemia**
  - *Increased* ***production*** *of unconjugated bilirubin from heme*
    - Hemolysis
      - Hereditary
      - Acquired
    - Ineffective erythropoiesis
    - Rapid turnover of increased erythrocyte mass (neonate)
  - *Decreased* ***delivery*** *of unconjugated bilirubin (in plasma) to hepatocyte*
    - Right-sided congestive heart failure
    - Portacaval shunt
  - *Decreased* ***uptake*** *of unconjugated bilirubin across hepatocyte membrane*
    - Competitive inhibition
      - Drugs
      - Others?
      - Gilbert's syndrome?
      - Sepsis, fasting?
  - *Decreased* ***storage*** *of unconjugated bilirubin in cytosol ($\downarrow$ Y and Z proteins)*
    - Competitive inhibition
    - Fever
  - *Decreased* ***biotransformation*** *(conjugation)*
    - Neonatal jaundice (physiologic)
    - Inhibition (drugs)
    - Hereditary (Crigler-Najjar)
      - Type I (Complete enzyme deficiency)
      - Type II (Partial deficiency)
    - Hepatocellular dysfunction
    - Gilbert's syndrome?
- **Conjugated hyperbilirubinemia** (cholestasis)
  - *Decreased* ***secretion*** *of conjugated bilirubin into canaliculi*
    - Hepatocellular disease
      - Hepatitis
      - Cholestasis (intrahepatic)
    - Dubin-Johnson and Rotor syndromes
    - Drugs (estradiol)
  - *Decreased* ***drainage***
    - Extrahepatic obstruction
      - Stones
      - Carcinoma
      - Stricture
      - Atresia
      - Primary sclerosing cholangitis
    - "Intrahepatic" obstruction
      - Drugs
      - Granulomas
      - Primary biliary cirrhosis
      - Tumors

Hepatic failure and encephalopathy. Neuropsychiatric disturbances (altered mentation and neuromuscular dysfunction, as well as altered consciousness and coma) may result from the metabolic abnormalities that occur in the presence of chronic liver disease.

The surgical creation of a portal-systemic shunt may precipitate encephalopathy, which can be attributed to the shunting of blood from the intestine to the systemic circulation, bypassing the liver. Therefore, substances such as ammonia or mercaptans, which are normally metabolized by the liver, accumulate and exert a cerebrotoxic effect.

Multiple metabolic abnormalities are seen in the patient with liver failure and hepatic coma; some are primary, and others represent secondary compensatory or decompensatory mechanisms. The major findings are increased ammonia concentrations in blood and the accumulation of ammonia, glutamine, and α-ketoglutarate in brain and CSF. There is an increase in short- and medium-chain fatty acids and amino acids in the plasma; mercaptans may account for the characteristic odor known as "fetor hepaticus" in these patients.

Altered drug metabolism. The liver plays a major role in enzymatic transformation and disposition of drugs.[4] During the course of diseases causing hepatic dysfunction, many drugs will undergo limited or aberrant transformation, will accumulate, and will exert unwanted effects upon the host.

Endocrine abnormalities. Hormone imbalances are another manifestation of hepatic disease. In addition to serving as a target tissue for the action of some hormones, the liver plays a key role in the uptake and metabolism of others. Among the common manifestations of chronic liver disease in men is feminization (i.e., gynecomastia and testicular atrophy) due to increased estrogen effects.

Nutritional and metabolic abnormalities. Severe metabolic and nutritional derangements may occur in the cirrhotic patient, such as alterations in glucose metabolism due to insulin resistance and hypokalemia secondary to excess losses of potassium as a result of vomiting or diarrhea. Secondary hyperaldosteronism may occur due to hyponatremia. Hypoalbuminemia is frequently present due to decreased synthesis. Inefficient delivery of bile acids to the duodenum impairs absorption of fats and fat-soluble vitamins and leads to chronic deficiencies of vitamins A, D, E, and K and to malnutrition and steatorrhea.

Immunoglobulin abnormalities. A number of observations implicate the liver in immunological responses; for example, the liver efficiently sequesters dimeric IgA and secretes it into the bile. In liver disease, however, there are increased serum concentrations of IgA due to defects in the uptake process.

Disordered hemostasis in liver disease. Chronic liver disease produces alterations in hemostasis and a generalized hemorrhagic tendency; there may be decreased synthesis of specific clotting factors and production of qualitatively abnormal proteins. There may also be alterations in platelet number and function due either to hepatic disease or to associated hypersplenism. In addition, endogenous anticoagulant substances may be present, fibrinolytic systems may be aberrant, and there may be intravascular coagulation.

BIOCHEMICAL ASSESSMENT OF LIVER FUNCTION

Hepatic Excretory Function

Organic anions of both endogenous and exogenous origin are removed from the sinusoidal blood and excreted into the bile. The most frequently used means for assessment of this excretory function are measurements of serum concentrations of endogenously produced compounds such as bilirubin and bile acids and the rate of clearance of an administered exogenous dye such as indocyanine green.

Bilirubin Metabolism

Bilirubin IXα, the orange-yellow bile pigment, is produced from protoporphyrin IX (Figure 22-3). For each mole of heme catabolized by this pathway, one mole each of carbon monoxide, bilirubin, and ferric ion are produced. In humans, daily bilirubin production from all sources averages 250–300 mg, of which 85% is derived from the heme moiety of the hemoglobin released from senescent erythrocytes that are destroyed in the reticuloendothelial system. The remaining 15% of bilirubin is produced from erythrocyte precursors destroyed in the bone marrow and from the catabolism of other heme-containing proteins such as myoglobin, cytochromes, and peroxidases.

After it is produced in peripheral tissues, bilirubin is transported to the liver in association with albumin. Bilirubin is then rapidly taken up by hepatocytes across the sinusoidal membrane by a carrier-mediated, active-transport process (Figure 22-4). Once inside the liver cells, bilirubin is tightly, although reversibly, bound to ligandin and Z protein, which accounts for most of the bilirubin binding.

Inside the hepatocytes, bilirubin is rapidly conjugated with glucuronic acid to produce bilirubin mono- and diglucuronide, which are then excreted into bile (see Figure 22-4). The excretion of conjugated bilirubin into bile against a significant concentration

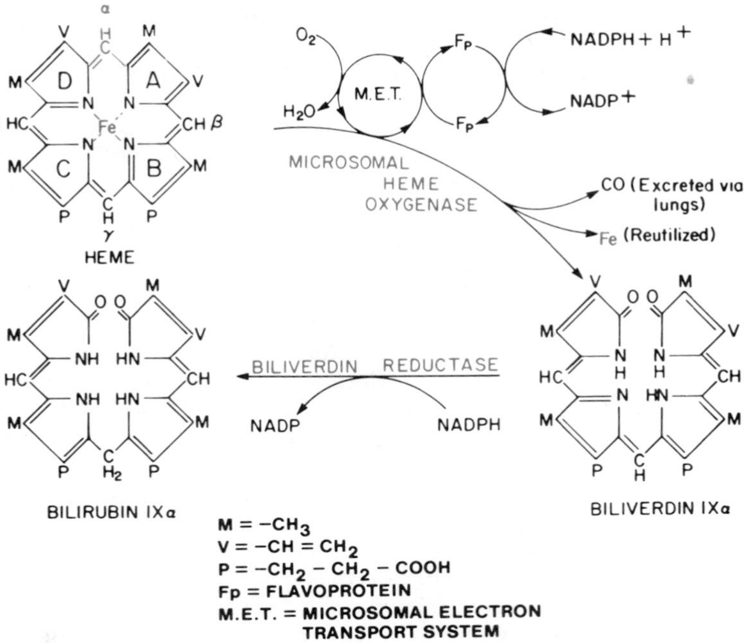

M = $-CH_3$
V = $-CH=CH_2$
P = $-CH_2-CH_2-COOH$
Fp = FLAVOPROTEIN
M.E.T. = MICROSOMAL ELECTRON TRANSPORT SYSTEM

Figure 22-3. Catabolism of heme to bilirubin IXα. (From Berlin, N. I., and Berk, P. D.: Blood, *57*:983–999, 1981.)

gradient is an energy-dependent, active-transport process. In human adults, virtually all bilirubin excreted into bile is in the form of glycosidic conjugates, glucuronides (diglucuronide ~90%; monoglucuronide ~10%), glucosides, and xylosides.

Once secreted into the upper small intestine, bilirubin glucuronides are not reabsorbed to any great extent. In the intestinal tract, bilirubin glucuronides are hydrolyzed to the unconjugated pigment at the alkaline pH of the upper portion of the small intestine by β-glucuronidase from liver, intestinal epithelial cells, and intestinal bacteria. The unconjugated bilirubin is then reduced by the anaerobic intestinal microbial flora to

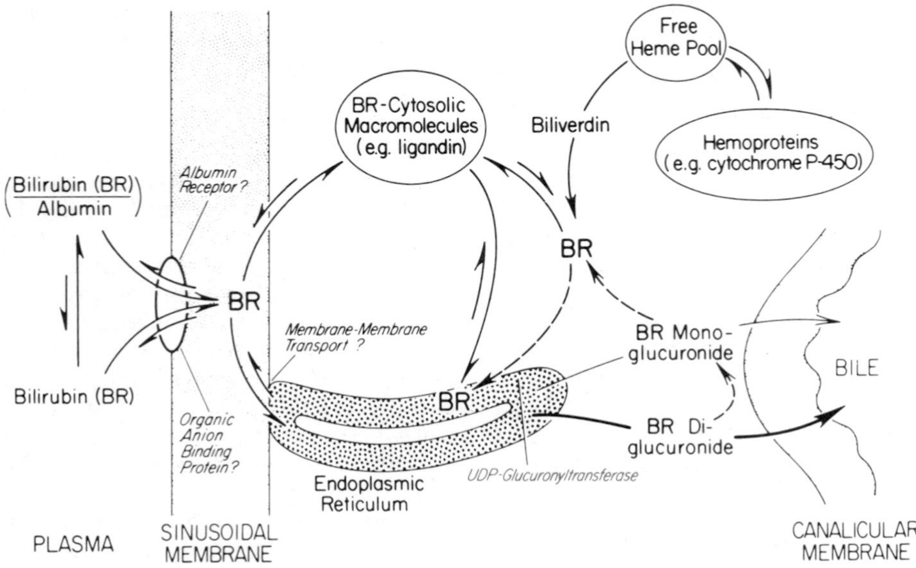

Figure 22-4. Bilirubin uptake, metabolism, and transport in the hepatocyte. (From Gollan, J. L., and Schmid, R.: Progress in Liver Diseases, Vol. VII. Chap. 15, 1982. By permission.)

form a group of three colorless tetrapyrroles collectively called *urobilinogens* (*stercobilinogen, mesobilinogen,* and *urobilinogen*). Up to 20% of the urobilinogens produced daily are reabsorbed from the intestine and enter the enterohepatic circulation. The majority of the reabsorbed urobilinogens are taken up by the liver and are re-excreted in the bile; 2–5% enter the general circulation and appear in the urine. In the lower intestinal tract the three urobilinogens spontaneously oxidize to produce the bile pigments *stercobilin, mesobilin,* and *urobilin,* which are orange-brown and are the major pigments of stool.

Disturbances of bilirubin metabolism. A number of diseases affect one or more of the steps involved in the production, uptake, storage, metabolism, and excretion of bilirubin. Depending on the disorder, unconjugated or conjugated bilirubin or both are major contributors to the hyperbilirubinemia resulting from these disturbances.

Unconjugated hyperbilirubinemia. The most common form of unconjugated hyperbilirubinemia is that seen in newborns and referred to as *physiological jaundice.* All newborns have serum unconjugated bilirubin concentrations greater than values in the healthy adult population; unconjugated bilirubin values rise to 40–50 mg/L (in a small percentage of cases, to as high as 100 mg/L) by 48 h and decrease to normal values by 7–10 d. This temporary increase is caused by increased bilirubin production as a result of hemolysis of erythrocytes and by incomplete maturation of several steps involved in bilirubin metabolism and excretion.

The increased production of bilirubin that accompanies the premature *breakdown of erythrocytes* (hemolytic anemia) and *ineffective erythropoiesis* results in hyperbilirubinemia in the absence of any liver abnormality. Total bilirubin concentrations are usually not >40 mg/L in these conditions; unconjugated bilirubin is the predominant form.

A common feature of the rare genetic *Crigler-Najjar Type I and Type II syndromes* is an absent or lower than normal activity of hepatic bilirubin uridine diphosphate-(UDP-)glucuronyl transferase. A more common inherited disorder, *Gilbert's syndrome,* is a heterogeneous condition that is usually ascribed to decreased UDP-glucuronyl transferase activity but may also be caused by defects in membrane transport. In the case of Crigler-Najjar Type I syndrome, bilirubin UDP-glucuronyl transferase activity is totally absent; reduced but measurable bilirubin UDP-glucuronyl transferase activity is present in liver tissue of patients with Crigler-Najjar Type II syndrome. Since bilirubin cannot be conjugated at the normal rate in patients with these disorders, the rate of excretion of bilirubin is significantly reduced and the serum concentration of unconjugated bilirubin rises. Serum total bilirubin concentrations of 200–500 mg/L are commonly seen in Crigler-Najjar Type I syndrome. Patients with this disease usually die in early life due to the development of kernicterus (bilirubin staining of the basal ganglia of the brain). Total bilirubin concentrations of <200 mg/L and of <30 mg/L are the usual ranges in patients with Crigler-Najjar Type II and Gilbert's syndromes, respectively. These disorders have a benign prognosis.

Conjugated hyperbilirubinemia. In hepatobiliary diseases bilirubin uptake, storage, and excretion are impaired to varying degrees. Both conjugated and unconjugated bilirubin are retained in these disorders, and a wide range of abnormal serum concentrations of each form of bilirubin may be observed. When any portion of the biliary tree becomes partially or totally blocked, biliary passage of all other components of bile is retarded and these substances are retained. As a result, plasma concentrations of conjugated bilirubin, cholesterol, γ-glutamyltransferase, alkaline phosphatase, and bile acids increase to abnormal values. In addition, obstruction of the biliary tree also leads to an increased synthesis (induction) of enzymes such as γ-glutamyltransferase and alkaline phosphatase that increases blood levels further. Drug- and steroid hormone-induced cholestasis and, occasionally, alcoholic hepatitis and acute viral hepatitis are examples of diseases that may cause *intrahepatic cholestasis.* Bile canaliculi and bile ductules are injured by the disease process, but larger biliary channels are not.

Mechanical obstruction of the biliary tree produces the increased concentrations of serum conjugated bilirubin observed in such disorders as carcinoma of the head of the pancreas, common bile duct, or ampulla of Vater; choledocholithiasis; and fibrosis of the head of the pancreas.

The *Dubin-Johnson syndrome* is a rare, benign genetic disorder of bilirubin excretion. In patients with this disorder, hepatic uptake and storage are normal but bilirubin excretion into bile is markedly impaired. In these patients serum total bilirubin concentrations are usually in the range of 20–50 mg/L and the predominant form is conjugated bilirubin.

Delta-hyperbilirubinemia. Delta-bilirubin is a form of bilirubin that is covalently bound to albumin. This form is not detectable in the serum of normal volunteers, neonates with physiologic jaundice, and patients with Gilbert's disease or hemolytic disorders. This albumin-bound bilirubin fraction appears only in the serum of adult, pediatric, and neonatal patients when hepatic excretion of conjugated bilirubin is impaired.[27] As recovery begins, δ-bilirubin becomes a larger percentage of total bilirubin owing to its smaller clearance compared with the clearances of the other bilirubin subfractions. This prolonged half-life and the fact that δ-bilirubin is not excreted in urine explains why urinary bilirubin is normal in certain patients with liver disease and hyperbilirubinemia (see also page 741).

Disturbances of Urobilinogen Metabolism

The formation of urobilinogen is decreased in all conditions in which liver secretion or the biliary drainage of bilirubin is impaired. As biliary obstruction progresses, delivery of bilirubin to the gut becomes limiting and excretion of urine as well as stool urobilinogen decreases to very low concentrations. The clay-colored or chalky white stool in obstructive jaundice reflects this impairment.

Chemical Properties of Bilirubins

An important chemical property of the bilirubin molecule is its insolubility in water and its ready solubility in a variety of nonpolar solvents. Recent X-ray crystallographic studies of crystalline bilirubin confirmed two important structural features of bilirubin: (1) a "Z-Z" (trans) conformation for the double bonds between carbons 4 and 5 as well as 15 and 16, and (2) an involuted hydrogen-bonded structure in which the propionic acid carboxylic acid groups are hydrogen bonded to the pyrrole ring nitrogens (Figure 22-5). When exposed to light, bilirubin in the "Z-Z" configuration is converted to the "E-E" (cis) conformation as well as to other combinations. Bilirubin in the "E-E" conformation and other E-containing isomers do not permit the degree of internal hydrogen bonding that occurs in the "Z-Z" conformation and are therefore more water soluble

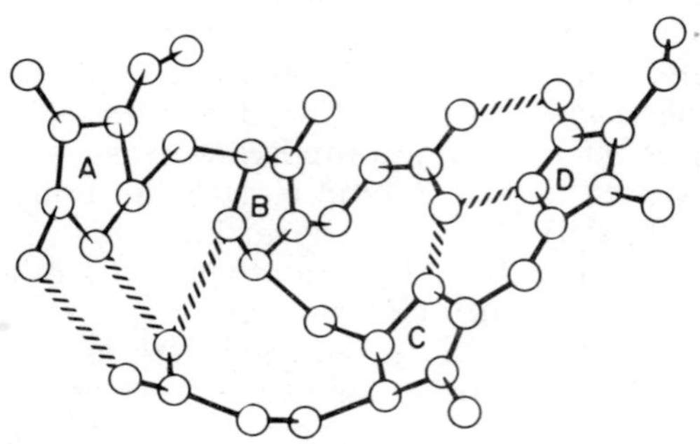

Figure 22-5. Bilirubin IXα structure. *Top,* The unfolded or linear tetrapyrrole structure showing the "Z" configuration at the C-4 to C-5 and C-15 to C-16 double bonds. *Bottom,* The folded conformation showing extensive internal hydrogen bonding. (Reprinted with permission from Elsevier Science Publishing Co., Inc., from Schmid, R.: Gastroenterology, *74*:1307–1312, 1978. Copyright 1978 by the American Gastroenterological Association.)

(and therefore more readily excreted) than is bilirubin in the "Z-Z" conformation.[23] This is the rationale for exposing to light neonates who have clinically significant jaundice in order to reduce plasma unconjugated bilirubin concentrations.

The occurrence of a hydrogen-bonded structure in aqueous solution explains some of the unique chemical properties of bilirubin IXα. For example, the addition of hydrogen bond–breaking chemicals such as caffeine, methanol, ethanol, or urea, 6 mol/L, is required for unconjugated bilirubin to react with the diazo reagent in the van den Bergh reaction. These reagents probably act by breaking internal hydrogen bonds of the bilirubin molecule to allow it to react with diazotized sulfanilic acid. On the other hand, bilirubin IXα diglucuronide is quite soluble in water and readily reacts with the diazo reagent. Presence of the bulky glucuronic acid moiety blocks the formation of internal hydrogen bonds.

Methods for the Determination of Bilirubin

The HPLC method of Lauff and colleagues,[12] which uses unconjugated, monoconjugated, and diconjugated bilirubin standards, is a rigorous reference method against which to compare routine laboratory bilirubin assays for accuracy. HPLC methods, however, are presently not technically suitable for measurements of bilirubin and its metabolites in the routine laboratory.

The most widely used methods for bilirubin measurement are those based on the diazo reaction. In this reaction diazotized sulfanilic acid (the diazo reagent) reacts with bilirubin to produce two azodipyrroles (Figure 22-6), which are reddish purple at neutral pH and blue at low or high pH values.[21] It has been established that alcohol or some other previously mentioned agents accelerate the diazotization reaction of unconjugated bilirubin. The fraction of bilirubin that reacts with the diazo reagent in the absence of alcohol is therefore also referred to as the "direct" bilirubin fraction. The term "indirect" bilirubin is used for the difference between total bilirubin, found after the addition of alcohol to the reaction mixture, and the so-called direct bilirubin fraction. It has long been thought that direct-reacting bilirubin is primarily conjugated bilirubin, that total bilirubin consists of both the conjugated and unconjugated species, and that indirect bilirubin is the unconjugated species. In light of recent work, we now know that there are at least four bilirubin species in serum. Direct reacting bilirubin consists of mono- and diconjugated bilirubin and the δ-fraction, which is bilirubin tightly bound to albumin. Unconjugated bilirubin, which is water insoluble and is associated with albumin by noncovalent bonding, only reacts with the diazo reagent after the addition of an "accelerator" such as alcohol or caffeine.

The diazo method, originally described by Jendrassik and Grof, gives results for serum bilirubin that are reliable, as judged by the HPLC method of Lauff and colleagues.[12] In this procedure an aqueous solution of caffeine and sodium benzoate serves as the accelerator.

Determination of Conjugated and Total Bilirubin in Serum (Jendrassik and Grof Method)

Specimen

Either serum or plasma may be used as the specimen; a morning fasting specimen is preferred in order to avoid lipemia. Hemolysis should be avoided, since hemoglobin produces falsely low values with diazo methods. Because both conjugated and unconjugated bilirubin are photo-oxidized when exposed to white or UV light, specimens should be protected from direct exposure to either artificial light or sunlight. The sensitivity to light is temperature dependent; for optimal stability, store specimens in the dark and at low temperatures. When specimens are stored in the refrigerator, stability is maintained for 3 d. Specimens are stable for three months when stored frozen at −70 °C in the dark.

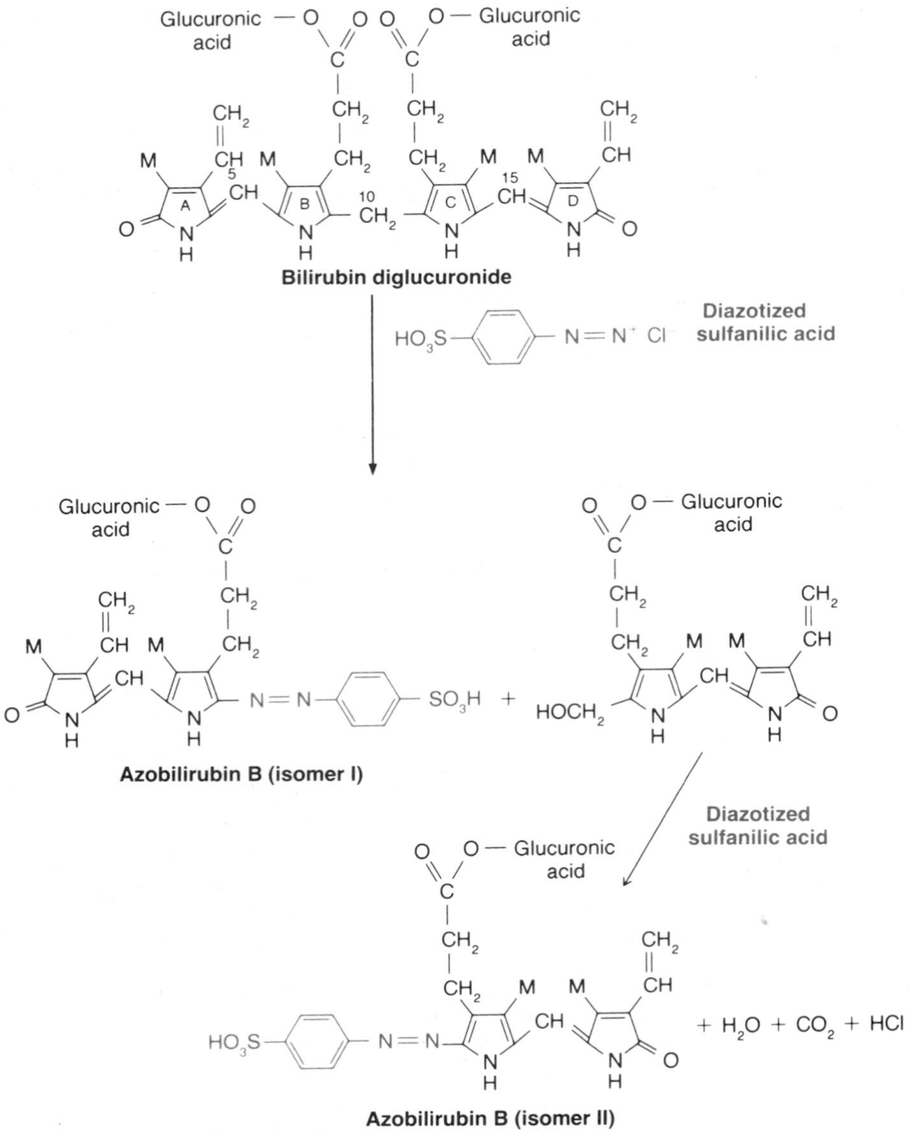

Bilirubin diglucuronide

Diazotized sulfanilic acid

Azobilirubin B (isomer I)

Diazotized sulfanilic acid

$+ H_2O + CO_2 + HCl$

Azobilirubin B (isomer II)

Figure 22-6. The reaction of bilirubin glucuronide with diazotized sulfanilic acid to produce isomers I and II of azobilirubin B. Unconjugated bilirubin reacts in the same way to produce isomers I and II of azobilirubin A.

Principle[17]

For the measurement of *conjugated bilirubin,* the serum or plasma is acidified with dilute HCl and then mixed with diazotized sulfanilic acid to produce azobilirubin. Only the conjugated forms of bilirubin will react with the diazo reagent in the absence of the accelerator caffeine-benzoate. The reaction is stopped by the addition of an ascorbic acid solution; then an alkaline tartrate solution is added to the reaction mixture, followed by the addition of an aliquot of the caffeine reagent. The tartrate reagent provides an alkaline pH to produce the blue and more intense color of azobilirubin, which has an absorbance peak at 600 nm in the presence of caffeine.

The measurement of *total bilirubin* in serum or plasma is achieved by adding caffeine reagent (accelerator) to the specimen followed by the addition of diazotized sulfanilic acid. During the incubation period both conjugated and unconjugated bilirubin react with the diazo reagent to produce azobilirubin. Ten minutes after the addition of diazotized sulfanilic acid, solutions of ascorbic acid, alkaline tartrate, and dilute hydrochloric acid are added to the reaction mixture. The absorbance of the resulting blue azobilirubin solution is measured at 600 nm.

Reagents

1. Caffeine-benzoate reagent. Dissolve 56 g anhydrous sodium acetate, 56 g sodium benzoate, 1 g Na_2EDTA, and 37 g caffeine in 700 mL of reagent grade water. Dilute to 1 L. This reagent is stable for at least six months at room temperature.
2. Hydrochloric acid, 0.05 mol/L.
3. Sulfanilic acid, 5 g/L. Add 5 g of sulfanilic acid to 700 mL of reagent grade water. Add 15 mL of concentrated hydrochloric acid. Dilute to 1 L. Store at room temperature.
4. Sodium nitrite, 5 g/L. Dissolve 0.5 g of sodium nitrite in 70 mL of reagent grade water and dilute to 100 mL. Store at 4 °C. Prepare fresh every two weeks.
5. Diazotized sulfanilic acid. Prepare by mixing 20 mL of the stock sulfanilic acid solution (reagent 3) with 0.5 mL of a stock solution of sodium nitrite (reagent 4). Prepare fresh daily and store in a refrigerator.
6. Ascorbic acid solution, 40 g/L. Dissolve 200 mg ascorbic acid in 5 mL water. Prepare fresh daily and store in a refrigerator.
7. Alkaline tartrate solution. Dissolve 75 g sodium hydroxide and 320 g of sodium potassium tartrate in 700 mL of reagent grade water. After the solution has cooled, dilute to 1 L. This reagent is stable for at least six months at room temperature.

Procedure

Conjugated Bilirubin

1. Add 1.0 mL of HCl, 0.05 mol/L, to each of two glass test tubes, one labeled C (conjugated) and the other B (blank). Add 2.0 mL of caffeine-benzoate reagent to tube B only.
2. Add 0.2 mL of sample to each tube.
3. Add 0.5 mL of sulfanilic acid solution to tube B only and mix.
4. Add 0.5 mL of freshly prepared, diazotized sulfanilic acid reagent to tube C and mix.
5. Exactly 10 min later, add 0.1 mL ascorbic acid solution to tubes B and C, followed immediately by the addition of 1.5 mL of alkaline tartrate solution to tubes B and C, and mix.
6. Add 2.0 mL of caffeine-benzoate reagent to tube C and mix.
7. Read the absorbance of tube C at 600 nm against tube B set at zero absorbance.

Total Bilirubin

1. Add 2.0 mL caffeine-benzoate reagent to a glass test tube labeled T (Total).
2. Add 0.2 mL of sample to the tube.
3. Add 0.5 mL of freshly prepared diazotized sulfanilic acid reagent to the tube and mix.
4. Exactly 10 min after the addition of diazotized sulfanilic acid in step 3, add, in succession, 0.1 mL of ascorbic acid solution, 1.5 mL of alkaline tartrate, and 1.0 mL of HCl, 0.05 mol/L, to the tube and mix.
5. Read the absorbance of the solution in tube T at 600 nm against that of tube B (from step 5 in the conjugated bilirubin method) set at zero absorbance.

Calibration

Only unconjugated bilirubin of acceptable purity should be used for establishing calibration curves. Such preparations of bilirubin should have a molar absorptivity of 60 700 ± 1600 at 453 nm in chloroform at 25 °C. An acceptable diluent for the unconjugated bilirubin is bovine serum albumin (BSA) Cohn Fraction V, 40 g/L. The bilirubin standard to be used for the preparation of the calibration curve should be prepared as follows.

1. Weigh out 20.0 mg of bilirubin of acceptable purity, transfer to a 100-mL volumetric flask, and dissolve by adding 1.0 mL of dimethyl sulfoxide and 2.0 mL of Na_2CO_3, 0.1 mol/L. Weigh bilirubin on a plastic weighing dish, not weighing paper. When transferring the bilirubin to the 100-mL volumetric flask, use the dimethyl sulfoxide for quantitative transfer of any bilirubin that sticks to the weighing dish. Dilute the solution to 100 mL with the bovine serum albumin diluent solution (40 g/L) previously adjusted to pH 7.4. Since bilirubin is light sensitive, the flask should immediately be wrapped with aluminum foil to protect the solution from light as much as possible. Stability of the standard is dependent on the storage temperature. Deterioration is about 1.5% per month at −20 °C and about 1% in six months at −70 °C.

2. Standard solutions for the preparation of the calibration curve are prepared as follows:

Total bilirubin, mg/L	40 g/L BSA, mL	200 mg/L Standard, mL
0	4.0	0
20	9.0	1.0
50	3.0	1.0
100	2.0	2.0
150	1.0	3.0
200	0	4.0

Note: Minimize exposure to light.

3. Determine the absorbance values of azobilirubin produced by each of the standard solutions using the procedure for the analysis of serum bilirubin. The procedure is linear to 200 mg/L.

Comments

1. It is important to determine the molar absorptivity of the bilirubin standard solution after color development; it should fall within the range of 75 080 ± 1520.

2. Check the acceptability of the bilirubin stock standard solution by centrifuging 5–10 mL of the standard in a table-top centrifuge at 1500 × *g* for 10 min and inspecting the tube. If there is any orange sediment on the wall or the bottom of the tube, the solution is unacceptable for use.

Reference Ranges

Reference values for serum total bilirubin of newborns, infants, and adults are summarized below.

Age	Premature		Full-Term Newborn	
	mg/L	*μmol/L*	*mg/L*	*μmol/L*
Up to 24 h:	10–80	17–137	20–60	34–103
Up to 48 h:	60–120	103–205	60–100	103–171
3–5 days:	100–140	171–239	40–80	68–137
Infants >1 month and adults,				
Conjugated:	0–2.0	0–3.4		
Total:	2.0–10.0	3.4–17		

Determination of Bilirubin Using Direct-Reading Bilirubinometers

Specimen. Draw a capillary specimen from the heel or fingertip; avoid hemolysis; separate serum (plasma) and protect it from light.

Principle. The absorbance of bilirubin in serum at 454 nm is proportional to its concentration. The serum of newborn infants does not contain carotene and other pigments that increase the absorbance at 454 nm. However, these pigments may be present in serum from older children and adults, and thus, use of the direct spectrophotometric method should be restricted to newborns. The contribution of hemoglobin to the absorbance at 454 nm is corrected by subtracting the absorbance at 540 nm.

The principle of the direct spectrophotometric method for measuring total bilirubin in serum has been applied in the design of direct-reading bilirubinometers. These instruments are differential spectrophotometers, either with two light paths passing at right angles to each other through the same cuvet or with a single light path that is split into two beams after passing through the cuvet. The resulting light beams are passed through two individual narrow-bandpass filters to separate photodetectors. One filter has a wavelength of 454 (or 461) nm, which is the absorption peak wavelength for bilirubin. The other filter wavelength is 540 (or 561) nm, chosen as the point at which oxyhemoglobin has the same absorbance as at 454 (or 461) nm.

When bilirubin is present in the sample, the absorbance at 454 (or 461) nm increases and unbalances the photodetection system. The bilirubin concentration in the sample is read directly on a meter, or the null point needle is returned to zero and the concentration is read on a scale calibrated in mg/dL or mg/L. If oxyhemoglobin is present in the sample no interference will occur, because the absorbances of oxyhemoglobin at 540 (or 561) nm and at 454 (or 461) nm are equal and will therefore cancel.

The instrument can be standardized with bilirubin solutions of known concentrations or with secondary standards, such as a methyl orange solution at pH 7.4, or a multilayered colored glass standard (see manufacturer's instructions).

Determination of Bilirubin Using Reflectance Spectrophotometry

Method for neonates. A new thin-film technique (Kodak Ektachem) that utilizes reflectance spectrophotometry for the determination of total bilirubin in the serum of neonates has been developed.[28] The thin film has three main components. The top layer, or spreading layer, contains caffeine and sodium benzoate to dissociate bilirubin from albumin; the second layer, the screen layer, contains gelatin, which traps and holds serum proteins in place; the third layer, the reaction layer, contains a cationic polymer, called a mordant, that binds the bilirubin. The interaction of both unconjugated and conjugated bilirubin with the mordant produces spectral shifts to lower wavelengths and increases molar absorptivity values. The reflectance spectra of the mordant-diconjugated bilirubin and mordant-monoconjugated bilirubin complexes are similar, but they differ from that of the unconjugated bilirubin-mordant complex at 460 nm. However, the reflectance densities of all three bilirubin species are comparable at 400 nm, which is the measuring wavelength for this method. Substances such as hemoglobin, chylomicrons, lipoproteins, and pigmented proteins that interfere with other direct spectrophotometric methods are trapped and retained in the screen layer. This reflectance technique measures only bilirubin or other species that absorb light at 400 nm and that are located in the reaction layer. The δ-bilirubin fraction is also retained in the screen layer and thus will not be detected by this pediatric bilirubin technique. This is not a significant problem for bilirubin determinations in sera of neonates, since the δ-bilirubin fraction from neonates up to 14 d of age is $<5\%$ of the total bilirubin. In contrast, adult specimens can contain up to 90% of the total bilirubin as the δ-fraction; therefore, a modification of the technique is required for the analysis of such specimens (see below).

The Kodak system requires a total of about 35 μL of serum, 10 μL of which is dispensed onto the spreading layer from a specimen cup. For neonatal sera the method agrees very well with the modified Jendrassik-Grof diazo method.[7,11]

Method for total and individual fractions of bilirubin. A bilirubin procedure that uses thin-film technology and reflectance spectrophotometry for the determination of total bilirubin and the individual bilirubin fractions—unconjugated bilirubin, conjugated bilirubin (mono- plus diconjugates), and δ-bilirubin—is also available.[26] This procedure takes advantage of the difference in reflectance spectra of unconjugated bilirubin and that of monoconjugated and diconjugated bilirubin. Reflectance densities are measured at two wavelengths, 400 and 460 nm, using the Kodak film system described above. The concentrations of unconjugated and conjugated bilirubin are then derived from these measurements, taking advantage of the difference in reflectance spectra between unconjugated and conjugated bilirubin (see above). A separate determination of total bilirubin is made with a different thin-film slide. In this thin-film slide, all of the bilirubin fractions react in the top layer, or the spreading layer, with a diazonium salt in the presence of an accelerator (dyphylline) and a surfactant (Triton X-100) to produce the corresponding azo pigments. The reflectance densities of the azo derivatives of all bilirubin fractions are equivalent at 540 nm, the measuring wavelength for this method. The concentration of the δ-bilirubin is derived by subtraction of the unconjugated and conjugated bilirubin values from the total bilirubin result.

Methods for the Determination of Urine Bilirubin

The qualitative detection of bilirubin in urine is a useful test in the differential diagnosis of jaundice. Because conjugated, but not unconjugated, bilirubin is excreted in urine, the urine test can demonstrate whether a patient with hyperbilirubinemia has increased concentrations of conjugated bilirubin. The most commonly used test method for detecting bilirubin in urine, which involves the use of a "dipstick" impregnated with a diazo reagent, can detect concentrations ≥5 mg/L.

Specimens. A fresh urine specimen is required since bilirubin is very unstable when exposed to light and room temperature. If the test must be delayed, protect the specimen from light and store in a refrigerator at 2–8 °C for not more than 24 h.

Procedure. Dip the reagent strip into the urine specimen for 1 s. Allow 60 s for the bilirubin to react with 2,6-dichlorobenzene-diazonium-tetrafluoroborate (Chemstrip, Bio-Dynamics/bmc, Indianapolis, IN 46250) at an acidic pH. A pink to red-violet color is produced, the intensity of which is proportional to the bilirubin concentration. The reaction mechanism for urinary conjugated bilirubin is the same as that described in Figure 22-6 with diazotized sulfanilic acid except for the different diazo reagent. Another commonly used dipstick is the Ictotest procedure (Ames Co., Division of Miles Laboratories, Elkhart, IN 46514). In this test *p*-nitrobenzenediazonium *p*-toluenesulfonate is the diazo reagent.

Sources of error. The Chemstrip test for bilirubin in urine is a highly specific test; it has a low incidence of false-positive results. Medications that color the urine red or give a red color in an acid medium, such as phenazopyridine, can produce a false-positive reading. Also, large quantities of ascorbic acid or of nitrite lower the sensitivity of the test.

Urobilinogen in Urine and Feces

Any pathological process that leads to increased concentrations of urobilinogen in the gastrointestinal tract results in an increase in the quantity of urobilinogen excreted in urine. As noted in the discussion of disturbances of urobilinogen metabolism, when biliary obstruction becomes more complete, urinary excretion of urobilinogen decreases due to the very limited delivery of bilirubin to the gut and the extremely low rate of urobilinogen production. The clay-colored stool in patients with obstructive jaundice results from the very low quantities of the bile pigment metabolites of urobilinogen (urobilin) produced in this condition. Urobilinogen measurement is of little help in the evaluation of liver disease, since it adds no information to that obtained from the commonly measured tests of liver disease.

BILE ACIDS

The regulation of bile acid metabolism is a major function of the liver, and detectable derangements in bile acid metabolism are a reflection of liver dysfunction. Through the process of bile acid synthesis, conjugation, and secretion, the liver serves to maintain cholesterol balance and to provide surface-active detergent molecules that facilitate both hepatic excretion of cholesterol and intestinal absorption of lipids. The proper functioning of this system involves an efficient ileal conservation mechanism through which bile acid molecules are rescued from fecal loss. Therefore, alterations in hepatic bile acid synthesis, intracellular metabolism, excretion, intestinal absorption, or plasma extraction are reflected in derangements in bile acid metabolism.

Bile Acid Chemistry and Metabolism

Cholesterol is continually synthesized by all tissues. Some is utilized unchanged as a structural component of membranes; some is excreted unchanged by the liver into bile. A small amount serves as precursor for steroid hormone synthesis; a large amount is converted in the liver to the highly polar bile acids, which are subsequently secreted in bile. Transformation of cholesterol to bile acids and the ability of bile acids to solubilize additional cholesterol in bile are the major mechanisms of cholesterol elimination from the body. Prior to secretion into the bile canaliculi, both of the primary bile acids (cholic acid and chenodeoxycholic acid) are conjugated at the carboxylic acid carbon with an amino acid (either glycine or taurine, Figure 22-7) that increases polarity, and hence water solubility, and leads to formation of four primary bile acids (cholyltaurine, cholylglycine, chenodeoxycholyltaurine, and chenodeoxycholylglycine). In health, the glycine conjugates predominate in a ratio of approximately 3:1 to 4:1, whereas unconjugated (free) bile acids are not present in bile.

During passage through the small intestine and colon, bile acids are subject

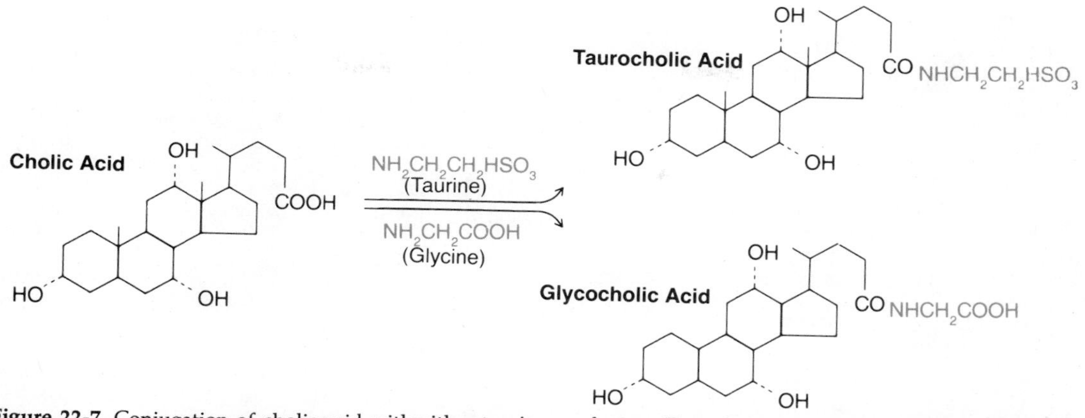

Figure 22-7. Conjugation of cholic acid with either taurine or glycine. (From Balistreri, W. F., and Soloway, R. D.: Clinical Guide to Bile Acid Physiology and Alterations in Disease States. Reproduced with the permission of Abbott Laboratories, Diagnostics Division, North Chicago, IL. © 1979 by Abbott Laboratories.)

to alterations by enzymes produced by the indigenous bacterial flora, giving rise to the *secondary bile acids,* deoxycholic acid and lithocholic acid (Figure 22-8). Secondary bile acids are also conjugated in the liver with glycine or taurine and join the primary bile acids as components of bile.

The Role of Bile Acids in Hepatic Bile Formation

Transport of bile acids into the bile canaliculi generates osmotic water flow and is a major factor regulating bile formation and secretion. Transport of these organic anions also influences secretion of the remainder of the major components of bile, such as bilirubin, cholesterol, and phospholipids. There is no secretion of the latter two compounds in the absence of bile acid secretion.

The ability of bile acids to solubilize biliary lipids is due to their amphophilic nature; i.e., they possess polar and nonpolar regions in the same molecule and are able to align at water-lipid interfaces, reduce surface tension, and therefore act as detergents. In an aqueous solution, bile acids aggregate to form small polymolecular aggregates called micelles, which are capable of incorporating cholesterol and phospholipids. Formation of mixed micelles of bile salts and phospholipids enhance the aqueous solubility of cholesterol, a weakly polar compound, and permit cholesterol excretion in bile, an aqueous polar medium.

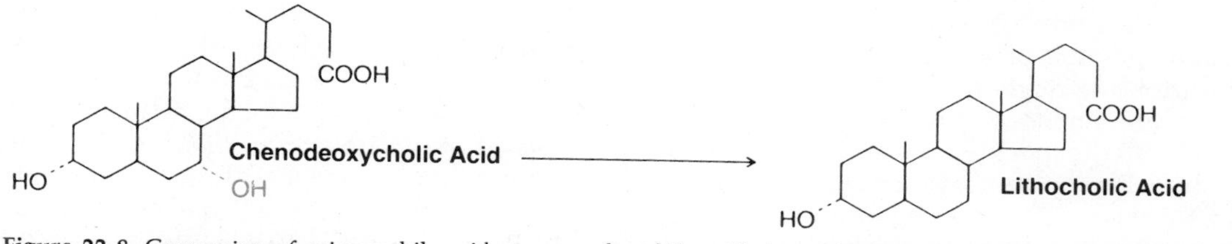

Figure 22-8. Conversion of primary bile acids to secondary bile acids by endogenous microflora. (From Balistreri, W. F., and Soloway, R. D.: Clinical Guide to Bile Acid Physiology and Alterations in Disease States. Reproduced with the permission of Abbott Laboratories, Diagnostics Division, North Chicago, IL. © 1979 by Abbott Laboratories.)

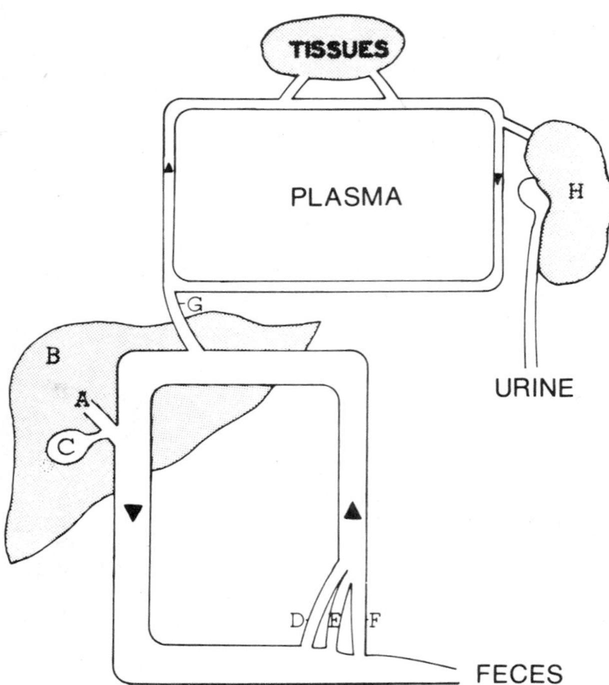

Figure 22-9. Fasting bile salt kinetics in a normal adult with estimated degree of compartmentation in the various organs involved in the enterohepatic circulation. Hepatic synthesis (**A**) is followed by delivery from the liver (**B**) to the gallbladder (**C**); intestinal reabsorption occurs in several sites (**D** = Jejunal *passive* transport; **E** = Ileal *active* transport; **F** = colonic *passive* transport); **G** = Portal-systemic shunting; **H** = Kidney.

Kinetic data: Pool size = 2–4 g; jejunal concentration = 2–3 mmol/L; synthesis rate = 0.3 g/d; fecal loss = 0.3 g/d; peripheral plasma level = 3–14 μmol/L. With ingestion of a *meal,* jejunal concentration will increase to 5–10 mmol/L owing to gallbladder contraction. An increase in portal venous levels will follow owing to ileal reabsorption; there will be "spillover" into peripheral plasma of this increased load. (Modified from Balistreri, W. F., and Soloway, R. D.: Clinical Guide to Bile Acid Physiology and Alterations in Disease States. Reproduced in part with the permission of Abbott Laboratories, Diagnostics Division, North Chicago, IL. © 1979 by Abbott Laboratories.)

The Enterohepatic Circulation of Bile Acids

The body conserves the bile acid pool through the enterohepatic circulation (EHC). The components of the EHC are the liver, biliary tract, intestine, and portal venous system. The restriction of the bulk of the bile acid pool to this compartment is made possible by efficient intestinal reabsorption and hepatic extraction of these molecules.

During fasting, bile acids pass down the biliary tree to enter the gallbladder (Figure 22-9). Here they undergo a tenfold increase in concentration due to reabsorption of water and electrolytes. After an overnight fast, 95% of the bile acid pool may be sequestered in the gallbladder. resulting in low levels of bile acids in the intestine, portal vein, and liver and a very low concentration of bile acids in the plasma.

In response to a meal, hormones such as cholecystokinin cause relaxation of the sphincter of Oddi and contraction of the gallbladder. This allows a concentrated solution of mixed micelles of bile salts and the phospholipid lecithin to enter the intestine, carrying ~1 g of cholesterol daily. In the intestinal lumen, endogenous (biliary) phospholipid and cholesterol leave the micelles and are replaced by dietary cholesterol and the products of triglyceride digestion, predominantly free fatty acids and monoglycerides. Micelles serve to facilitate fat absorption in the jejunum by accelerating the hydrolytic action of pancreatic lipase on triglycerides and by solubilizing the hydrolytic products. In order to carry out these functions, a minimal concentration of bile acids is necessary; below this critical concentration micelles are no longer formed, and fat digestion and absorption are impaired.

Bile acids are subsequently reabsorbed in the intestine by active transport in the distal portion of the ileum. Selective reabsorption of less polar, freely diffusible, nonionized bile acids takes place by passive transport in the jejunum and colon.

Bile acids are carried in blood tightly bound to protein. The concentration of bile acids in the portal vein is high; however, because of the efficiency (~80%) of hepatic extraction, systemic blood levels remain low. The portal vein concentration of bile acids is increased after a meal, systemic plasma levels rise owing to spillover, and the plasma pool of bile acids is increased.

Bile acid concentrations in serum increase as a result of decreased liver uptake or altered metabolism and therefore measurement of serum bile acid levels is a sensitive

Table 22-2. DISTURBANCES IN BILE ACID METABOLISM

Defective bile acid *synthesis*	**Bile acid *malabsorption***
Congenital impairment of hepatic synthesis	Primary bile acid malabsorption
Specific defects in bile acid synthesis as seen in:	(absent or inefficient ileal active transport)
Cerebrotendinous xanthomatosis	Intractable diarrhea (infancy)
Intrahepatic cholestasis	Irritable bowel (adults)
Qualitative	Secondary bile acid malabsorption
Quantitative	Ileal disease or resection
Acquired defects in bile acid synthesis, as observed in liver diseases such as hepatitis and cirrhosis	Crohn's disease
Abnormalities of bile acid *delivery* to the bowel	Ileal resection
Celiac sprue	Ileal bypass
Extrahepatic bile duct obstruction due to:	Radiation enteritis
Congenital biliary atresia	Postinfectious enteritis
Stricture	Exogenous bile acid administration (e.g., gallstone dissolution)
Stone	Cystic fibrosis
Carcinoma	Tertiary bile acid malabsorption
***Interruption* of the enterohepatic circulation of bile acids**	Postcholecystectomy
External bile fistula	Renal failure
Ileojejunal exclusion for exogenous obesity or hypercholesterolemia	Drugs
Cystic fibrosis	**Defective *uptake* or altered intracellular metabolism**
Contaminated small bowel syndrome (with bile acid precipitation, increased jejunal absorption, and "short circuiting")	Parenchymal disease (acute hepatitis, cirrhosis) associated with regurgitation from cells or portosystemic shunting
Entrapment of bile acids in intestinal lumen by:	Cholestasis
Cholestyramine	
Trivalent cations	
Fiber	

indicator of liver dysfunction. In many patients with liver disease, increase of bile acids in 2-h postprandial blood is the only detectable abnormality.

Abnormalities of Bile Acid Metabolism

There are several potential sites for primary or secondary disturbances of the described physiologic sequence of bile acid metabolism (Table 22-2).

Defective bile acid synthesis. The most frequently encountered abnormalities are *acquired* defects that have been noted in hepatitis and cirrhosis; serum bile acid levels are also elevated. In acute hepatitis, alterations of bile acid synthesis and conjugation occur due to changes in hepatocyte function, loss of liver cell mass, and presence of portosystemic shunting. The severity of cirrhosis correlates closely with loss of cholic acid synthetic capacity. The decreased ratio of the trihydroxy bile acid (cholic acid) to the dihydroxy bile acid (chenodeoxycholic acid) in serum has been used as a diagnostic test of liver cell dysfunction.

Abnormalities of bile acid delivery to the bowel. Delayed emptying or mechanical obstruction of the gallbladder or common bile duct can prevent bile acids from reaching the duodenal lumen. Extrahepatic bile duct obstruction results in bile acid retention and regurgitation from the liver cell into plasma (thereby increasing serum levels), as well as decreased delivery to the intestine.

Interruption of the enterohepatic circulation of bile acids. Approximately 95% of the bile acids that are secreted during a single enterohepatic cycle are recirculating bile acids. Therefore, a significant interruption of this cycle leads to a decrease in hepatic bile acid secretion. Increased bile acid synthesis caused by negative feedback can only partially compensate for the losses. The active transport system for bile acid absorption, located in the terminal ileum, has a major input to the enterohepatic circulation. Therefore, resection, inflammation, or bypass of the ileum will be associated with specific clinical symptoms, disturbances of bile acid metabolism, and low levels of bile acids in serum, especially after a meal.

Disturbances of bile acid metabolism in cholestasis. In cholestasis of any cause, biliary excretion of bile acids is decreased, hepatic accumulation occurs, and there is a shift of the bile acid pool to nonintestinal compartments such as plasma and peripheral tissues. Elevated peripheral serum bile acid levels are present.[1]

Methods for the Measurement of Bile Acids in Serum

GLC. This technique requires extensive preparation of each specimen; i.e., removal of the glycine or taurine moiety by means of alkaline or enzymatic hydrolysis and preparation of a volatile derivative. The method is time consuming and tedious, and its use in routine clinical practice is not recommended.

HPLC. Several HPLC methods have been developed for analysis of duodenal bile acids; these techniques permit quantitation with minimal derivatization. The major problem in the assay of serum is limited sensitivity of currently available detectors.

Enzymatic method. An enzyme produced by *Pseudomonas testosteroni*, 3α-hydroxysteroid dehydrogenase, catalyzes the oxidation of the 3α-hydroxyl group of all bile acids, producing a 3-keto group in the presence of NAD^+. The assay is usually carried out as a single-point determination by measuring the change in absorbance at 340 nm as NADH is formed. Sensitivity is increased by measuring NADH fluorometrically or by coupling the reaction to a fluorescent dye or by using luciferase to produce chemiluminescence. If individual bile acid quantitation is needed, the various species can be separated by thin-layer chromatography into classes and quantitated by the enzymatic method. The major disadvantages at present are the cost of the commercial assay kits and the limitation on the number of specimens that can be processed in a single day.

Immunoassays. Antibodies have been prepared that have high specificity for the steroid moiety of bile acids (and to some extent the conjugated moiety), permitting individual bile acids to be measured by radioimmunoassay or enzyme-linked immunoassay. Several standardized kits for measuring individual bile acids in serum are available and the assays are easy to perform.

Clinical Utility of Serum Bile Acid Measurements

Increased serum bile acid concentrations in the *fasting* state suggest impaired hepatic uptake or secretion, or portosystemic shunting. Thus, such measurements may be used as a sensitive endogenous clearance test. The *2-h postprandial* serum bile acid determination may be a more sensitive test of liver function than a test performed in the fasting state. Stimulation of gallbladder contraction, followed by bile acid determination in serum, may serve as a "stress test" and as a clearance test to provide further diagnostic information. Furthermore, since cholate conjugates are absorbed exclusively in the ileum, the postprandial increase in the serum bile acid level may be an index of ileal absorption. Ileal disease or dysfunction would be manifest by the lack of a rise in the postprandial serum bile acid levels.

Hepatic Synthetic Function

The liver has an extensive synthetic function and plays a major role in the regulation of carbohydrate, lipid, and protein metabolism. Normal blood glucose concentrations are maintained during short fasts by the breakdown of hepatic glycogen and during prolonged fasts by hepatic gluconeogenesis. Protein, triglyceride, fatty acid, cholesterol, and bile acid synthesis also occur in the liver.

Protein Synthesis

The liver is the primary site for the synthesis of many plasma proteins. Severe or long-standing hepatic disease will lead to *decreased synthesis* and decreased plasma levels of many proteins normally synthesized in the liver: albumin, α_1-antitrypsin, fibrinogen, ceruloplasmin, haptoglobin, transferrin, the coagulation proteins, and others. The pattern of plasma protein alterations depends on the type, severity, and duration of liver injury or disease. For example, in acute hepatic dysfunction, there is usually very little change in the plasma protein profile or the total plasma protein concentration. In chronic liver disease γ-globulin levels increase, whereas other proteins, such as albumin, decrease. For a detailed discussion of the individual proteins, refer to Chapter 11. However, in the context of liver disease, some additional information on selected proteins is of interest:

The liver responds to inflammation with *increased synthesis* of the acute-phase reactants (C-reactive protein, α_1-antichymotrypsin, α_1-acid glycoprotein, α_1-antitrypsin, haptoglobin, complement C3 and C4, fibrinogen, and ceruloplasmin).

Prealbumin has an extremely short half-life (~1.9 d), and the concentration of prealbumin is therefore a very sensitive index of alterations in hepatic synthetic or catabolic function.

Serum levels of ***immunoglobulins*** indirectly reflect impairment of hepatic uptake and that of the reticuloendothelial system of the hepatic sinusoid in its function to filter incoming intestinal tract antigens from the portal venous blood. The antigens stimulate the extrahepatic reticuloendothelial system and elicit the immune response (i.e., increased antibody production). In acute viral hepatitis there is often an increase in serum γ-globulin concentrations. Persistence of hypergammaglobulinemia suggests the presence of chronic active liver disease.

Ceruloplasmin levels increase in liver diseases such as chronic active liver disease, biliary cirrhosis, and hemochromatosis. Of specific interest are the decreased levels that have been noted in Wilson's disease (see Chapter 11).

Serum levels of ***β_2-microglobulin*** can be increased in hepatobiliary disease complicated by renal disease[19] and in chronic active hepatitis and alcohol-induced liver cirrhosis without renal disease.

Ligandin is one of a class of multifunctional proteins known as glutathione transferase; its name is derived from its ability to bind a wide range of ligands (small molecules). Some of the roles postulated for ligandin are (1) detoxification by binding certain toxic chemicals; (2) storage or transport of organic ions; and (3) enzymatic degradation of some compounds.

Most ***coagulation proteins*** are synthesized in the liver. These proteins interact in a cascade fashion to produce a fibrin clot. Two fibrinolytic proteins, *plasminogen* and *α_2-antiplasmin,* and the anticoagulant protein *antithrombin III* are also synthesized in the liver. Some of the *coagulation factors* (II, VII, IX, and X) require vitamin K for post-translational carboxylation within the hepatocyte. *Protein C* is also synthesized in the liver by a vitamin K–dependent plasma zymogen. Parenchymal liver disease of sufficient severity to impair protein synthesis and obstructive disease sufficient to impair intestinal absorption of the fat-soluble vitamin K are therefore potential causes of bleeding disorders.[6] Because of the great functional reserve of the liver, failure of hemostasis is rarely a complication except in liver disease of long standing or of great severity. Thus, testing for a coagulation defect is not a screening procedure but rather a means of following progress of the disease or of assessing risk of bleeding.

The *prothrombin time* (PT) is prolonged in coagulation defects due to liver disease, because it is affected by deficiencies of more than one factor. Partial thromboplastin time (PTT) or assays of individual factors do not ordinarily give any additional diagnostic information. An important use of coagulation tests in liver disease patients is in the evaluation of the clotting process before liver biopsy. A typical battery of coagulation tests for this purpose is a PT and platelet count.

Lipid and Lipoprotein Synthesis

The liver is the major source of plasma lipoprotein production and metabolism. In acute hepatocellular injury, levels of hepatic enzymes such as lecithin-cholesterol acyltransferase (LCAT) and hepatic triglyceride lipase (H-TGL; triacylglycerol lipase) are decreased.[9] Patients with acute liver disease, therefore, have increased levels of plasma triglycerides, a decreased percentage of cholesteryl esters, and abnormal lipoprotein electrophoretic patterns. In acute viral or alcoholic hepatitis, there is an absence of α- and pre-β-bands and presence of a wide, densely staining β-band. The hypertriglyceridemia characteristic of acute hepatocellular injury is mild; there is, however, accumulation of a triglyceride-rich low-density lipoprotein (LDL), which migrates in the β-region and gives rise to the broad β-bands that appear on electrophoresis.

Lipoprotein abnormalities are common in chronic intra- or extrahepatic cholestasis. There are marked elevations in the plasma levels of cholesterol and phospholipids. In addition, in patients with obstructive jaundice, an abnormal lipoprotein (Lp-X) is present.

Metabolic Function

The liver has central importance in metabolic and regulatory pathways, including the metabolism of drugs (activation and detoxification) as well as the disposal of exogenous and endogenous metabolites such as galactose and ammonia. As a corollary, multiple metabolic abnormalities due to specific, inherited enzyme deficiencies can primarily or secondarily affect the liver.

Urea Synthesis

Patients with end-stage cirrhosis may have decreased urea synthesis and thus low levels of urea in plasma and urine. In addition, plasma levels of the urea precursors ammonia and amino acids are elevated, and the specific activities of the enzymes involved in urea synthesis are low. The rate of hepatic urea synthesis is dependent upon exogenous intake of nitrogen as well as endogenous protein catabolism. In cirrhotic patients, the pathway is unable to incorporate excess ammonia nitrogen into urea.

AMMONIA

Excess ammonia exerts toxic effects on the central nervous system. The major source of circulating ammonia is the gastrointestinal tract, where ammonia is derived from the action of bacterial proteases, ureases, and amine oxidases on the contents of the colon as well as from the hydrolysis of glutamine in both the small and large intestines. Under normal circumstances most of the portal vein ammonia load is metabolized to urea in hepatocytes in the Krebs-Henseleit urea cycle during the first pass through the liver.

Inherited deficiencies of urea cycle enzymes are the major causes of hyperammonemia in infants. Two other types of inherited metabolic disorders produce hyperammonemia; namely, those involving the metabolism of the dibasic amino acids, lysine and ornithine, and those involving the metabolism of organic acids such as propionic acid, methylmalonic acid, and isovaleric acid.

Of the possible acquired causes of hyperammonemia, the most common is advanced liver disease. Severe liver disease, whether acute (as in toxic or fulminant viral hepatitis and Reye's syndrome) or chronic (as in cirrhosis), can lead to a significant impairment of normal ammonia metabolism. Hepatic encephalopathy in the cirrhotic patient can be precipitated by gastrointestinal bleeding, which enhances ammonia production by bacterial metabolism of the blood proteins in the colon and subsequently increases blood ammonia levels. Other precipitating causes of encephalopathy include excess dietary protein, constipation, infections, drugs, or electrolyte and acid-base imbalance. When cirrhosis is accompanied by impaired venous drainage from the intestine into the liver via the portal vein, collateral vessels shunt ammonia of intestinal origin away from the liver and into the general circulation, causing increased blood ammonia levels. Impaired renal function may also accompany severe liver disease. As urine output decreases, blood urea concentration increases, leading to increased excretion of urea into the intestine, where it is converted to ammonia.

Methods for the Determination of Ammonia

Methods for measuring ammonia can be divided into two major classes: *two-stage procedures* and *one-stage (direct) procedures.* Flow-injection analysis is an example of a two-stage procedure. The specimen is injected into a flowing stream of alkaline solution from which the ammonia diffuses through a membrane into a pH-sensitive indicator stream. The method is fast, requires minimal specimen, and carries minimal risk of ambient ammonia contamination but does require dedicated equipment. A cation-exchange resin is used in a two-stage assay that removes NH_4^+ from its matrix.[8] The ammonium ion is then eluted from the resin and the eluate is reacted with Berthelot (phenolhypochlorite) reagent prior to spectrophotometric quantitation.

One-stage methods do not separate NH_4^+ from the specimen before the analytic step. The enzymatic assay used in these procedures is most frequently based on the reaction:

$$\text{2-Oxoglutarate} + NH_4^+ + \text{NADPH} \xrightleftharpoons[\text{pH 7.4}]{\textit{Glutamate dehydrogenase}} \text{Glutamate} + \text{NADP}^+ + H_2O$$

Change in absorbance at 340 nm is measured, as NADPH is transformed to $NADP^+$.

Reliable ammonia levels can be achieved only by meticulous precautions to avoid false elevations.

1. *Smoking* is a source of ammonia contamination and should be avoided by patient and phlebotomist.

2. *Laboratory atmosphere* is a source of ammonia contamination. Therefore, blood collection as well as the performance of the ammonia analysis should ideally be done in a laboratory with restricted traffic. Glassware should be soaked in a solution of hypochlorite, 52.5 g/L, and rinsed thoroughly with deionized water.

3. *Poor venipuncture* technique may result in increased ammonia levels. Use of a heparin lock, probing for a vein, drawing blood into a syringe and transferring it to an anticoagulated tube, or partial fill of the evacuated tube allowing subsequent entry of air may cause elevation of ammonia levels by 100–200 μg/L. EDTA and nonammonium heparin salts are acceptable anticoagulants. Ammonia values in serum are significantly but variably higher than the corresponding plasma values.

4. *Metabolism of nitrogenous constituents* in the specimen is a source of ammonia contamination that can be minimized by placing the specimen on ice immediately, by centrifuging without delay, and by performing the analysis immediately.

Reference values. For the enzymatic method, the range is 140–490 μg/L (11–35 μmol/L) as ammonia nitrogen. The reference range for the resin method is 220–470 μg/L (16–33 μmol/L) as ammonia nitrogen. Reference values should be determined in the laboratory performing the assay. If values in healthy subjects are significantly higher than those noted, consideration should be given to the existence and correction of sources of error identified above.

Carbohydrate Metabolism

Because the liver is a major processor of dietary and endogenous carbohydrates, liver disease can be expected to affect carbohydrate metabolism (see also Chapter 13). For example, many aspects of the metabolism of galactose are unique to the liver, and thus the determination of the rate of galactose elimination is a helpful, but not widely used, measure of hepatic metabolic capacity.[13] When galactose concentrations are relatively low (<30 mg/dL), hepatic extraction averages ~90%.

Mechanisms of Enzyme Release from Diseased Liver Tissue

In various forms of liver disease, serum levels of numerous cytosolic, mitochondrial, and membrane-associated enzymes are increased; the degree of elevation varies with the type of disease. Alanine and aspartate aminotransferases, alkaline phosphatase, and γ-glutamyltransferases are the enzymes most often measured for evaluation of liver disease.

Cytosolic and mitochondrial enzymes. The exact mechanism by which enzymes are released from the cytosol and mitochondria of hepatocytes into the blood stream is not known. However, it has been shown that impairment of cellular energy production, as in viral hepatitis, increases permeability of the hepatocyte membrane and allows cytosolic isoenzymes of the aminotransferases to spill into the sinusoids and from there into the peripheral blood.[10] Permeability of mitochondrial membranes may also increase, and mitochondrial isoenzymes will then be released as well.

Membrane-bound enzymes. The increased activities of γ-glutamyltransferase and alkaline phosphatase in diseased human liver are believed to be due to enhanced synthesis. The elevation of serum activity may be due to the release of these enzymes from the membrane matrix to which they are normally anchored.

Dynamics of Liver Enzyme Changes in Liver Disease

The degree of increase of serum enzyme activities in liver disease and the pattern of changes can vary markedly throughout the clinical course of the disease (see Chapter 12, *Enzymes*). Factors that may affect the degree of elevation and the pattern of changes are (1) specific activity (i.e., the units of catalytic activity per gram of tissue) of the enzymes in the liver cell at the time they are released into the blood stream; (2) the pattern of relative specific activities of the different enzymes in the liver at the time serum activities are determined; and (3) the relative rates of catabolic clearance of the enzymes from the blood stream.

Clearance of different liver enzymes from plasma occurs at greatly different rates. The average $t_{1/2}$ of alanine aminotransferase is 47 h, of cytosolic aspartate aminotransferase 17 h, of alkaline phosphatase 9.7 d, and of γ-glutamyltransferase 4.1 d.

DISEASES OF THE LIVER

Mechanisms and Patterns of Liver Cell Injury

Hepatic injury can occur by various mechanisms and may be either diffuse or localized. Sherlock has suggested that there are three specific patterns of liver cell injury: direct, immunologic, and cholestatic.[24]

In the *direct pattern of liver injury,* individual cell death occurs in specific areas such as the centrilobular zone. Localization of injury may be due to the lobular vascular gradient; for example, with hypoxia the central location of necrosis can be attributed to a lower oxygen content of sinusoidal blood in the central areas. Certain toxins will cause injury to specific hepatocyte organelles. For example, following chronic alcohol ingestion there may be predominantly mitochondrial damage eventuating in cell death; this sequence has been attributed to oxidation products of ethanol, such as acetaldehyde. Metabolic conversion in the hepatocyte may activate drugs or produce potentially toxic agents.[15] Certain toxic metabolites, such as those formed from acetaminophen, may bind covalently to liver cell macromolecules that are essential to hepatocyte viability; therefore, liver cell death may occur.

Immunologically mediated hepatocyte injury may be more complex and enigmatic. Cytotoxic lymphocytes are capable of attacking liver cell membrane antigens. The classic example of immunological hepatic injury is chronic active hepatitis, in which it is postulated that a hepatocyte membrane protein serves as an antigen to which immunologically mediated activity is directed. Drugs may also serve as haptens and may combine with normal liver cell membrane antigens, thus providing a focus for immunologically directed hepatocyte injury. Primary biliary cirrhosis has also been cited as an example of immunologically mediated liver disease. In this situation, there is a marked infiltration of lymphocytes and plasma cells and injury to the bile ducts.

Cholestatic liver injury can be due to either extrahepatic or intrahepatic obstruction. Extrahepatic obstruction is most frequently due to a mechanical cause such as atresia, stenosis, or obstruction of the bile ducts secondary to carcinoma or gallstones. Intrahepatic cholestatic liver injury may be due to functional disturbances in the bile secretory mechanism.

Table 22-3. CLINICAL FEATURES OF ACUTE VIRAL HEPATITIS

	Hepatitis A	Hepatitis B
Incubation period	2–6 weeks	6 weeks–6 months
Onset	Usually acute	Usually insidious
Mode of transmission		
Oral (fecal)	Usual	Infrequently oral; not fecal
Parenteral	Infrequent	Usual; intimate contact
Epidemiology		
Sporadic cases	Mainly children	Males predominate
Outbreaks	Food or water borne	Contaminated blood products
Polyarthralgia	Rare	Common
Rashes	Occasional	Common
Duration	Brief (14–21 d)	Prolonged
Sequelae		
Carrier state	No	10% of cases
Chronic hepatitis	Rare	Frequent (~10% of cases)

Acute Viral Hepatitis

Viral hepatitis is a primary infection of the liver that is most commonly caused by *hepatitis A* virus (HAV) or *hepatitis B* virus (HBV). However, other viruses, tentatively grouped under the name *non-A, non-B,* can produce clinical symptoms similar to those produced by the classic HBV. Of less importance as agents of viral hepatitis are cytomegalovirus, herpes simplex, varicella-zoster, Epstein-Barr, rubella, and Coxsackie viruses. Clinical features of acute viral hepatitis caused by HAV or HBV are summarized in Table 22-3.

Antigens related to HAV infection are not detectable in serum. The humoral immune response of HAV produces, in the initial phase of the infection, IgM type anti-HAV and, in the later phase, IgG type anti-HAV (Figure 22-10). The latter antibodies persist for a prolonged period, possibly indefinitely.

HBV is a double-shelled structure; the outer coat, called the HBV surface antigen (HB_sAg), surrounds the HBV core antigen (HB_cAg), double-stranded DNA, and the enzyme DNA polymerase. A soluble antigen (HB_eAg) is closely associated with HB_sAg. HB_sAg appears in serum before clinical features of hepatitis B infection appear and persists throughout the clinical course of the disease (Figure 22-11). The humoral immune response to HBV infection produces antibodies to the surface antigen (anti-HB_s) that first appear during the resolution and convalescent phases of the disease. Anti-HB_s antibodies are considered essential for the resolution of HBV infection and are believed to provide prolonged immunity to patients in whom they persist. The newly discovered

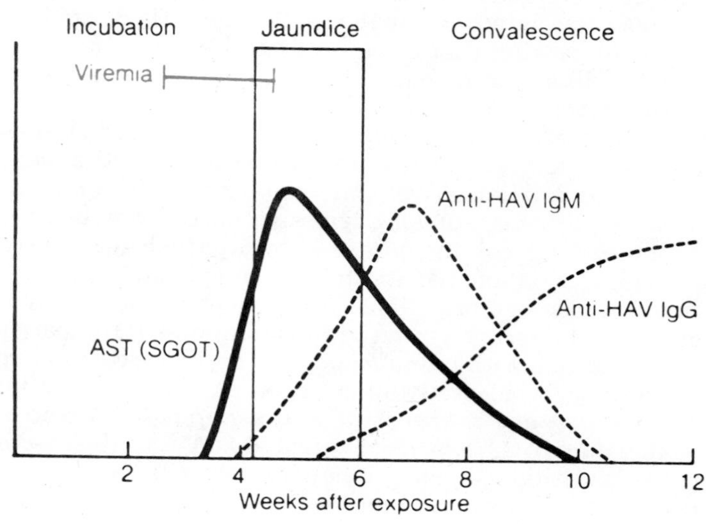

Figure 22-10. In *hepatitis A* infection, jaundice occurs 14–35 d (median, 32 d) after exposure. AST elevations precede the development of jaundice and usually remain abnormal after the serum bilirubin returns to normal values. Viremia and fecal excretion of virus usually occur during the preicteric stage of the disease. The limited period of viremia makes hepatitis A virus a very uncommon cause of post-transfusion hepatitis. The appearance and disappearance of serologic markers to hepatitis infection (anti-HAV) are depicted by dotted lines. (From Balistreri, W. F.: Consultant, *24*:131–153, 1984.)

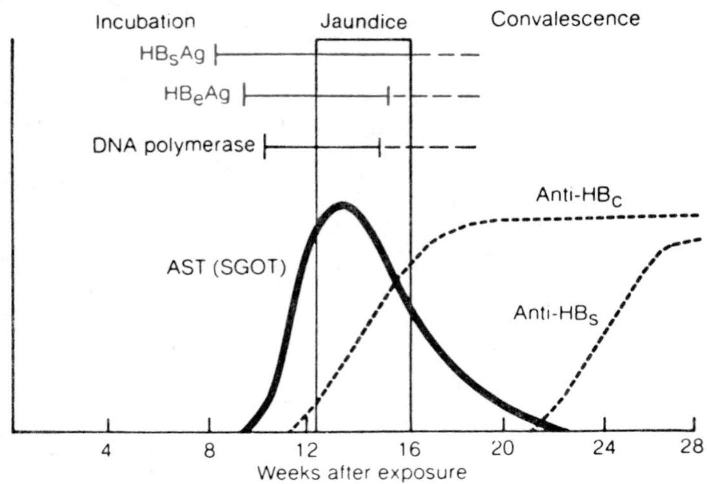

Figure 22-11. Course of *acute Type B hepatitis* with recovery. Illustration indicates: (1) Onset of hepatitis with jaundice three months after exposure. (2) Detection of hepatitis B surface antigen (HB_sAg) ~2–8 weeks after exposure, followed by appearance of its antibody (anti-HB_s) ~2–4 weeks after HB_sAg is no longer detectable. (3) Detection of hepatitis B_e antigen (HB_eAg) shortly after appearance of HB_sAg and its disappearance shortly after HB_sAg disappears; this is usually followed by the appearance of antibody to HB_eAg (anti-HB_e) that persists. (4) Detection of hepatitis B core antibody (anti-HB_c) at the time of onset of disease two to three months after exposure; anti-HB_C IgM will be detectable in high levels for ~5 months. (From Balistreri, W. F.: Consultant, *24*:131–153, 1984.)

delta virus may coinfect patients who are HB_sAg positive. This may be responsible for worsening of a previous stable clinical condition (e.g., exacerbation of chronic hepatitis).

Those patients who do not develop anti-HB_s become chronic carriers of an HBV, which is typified by the prolonged presence in serum of HB_sAg and anti-HB_c only. The carrier state occurs in ~10% of patients infected with HBV. Chronic carriers whose serum is also persistently positive for HB_eAg are at an especially high risk for the development of chronic active liver disease. Assessment of HB_eAg and its antibodies may be an important predictor of which chronic carriers of HBV will later develop chronic active liver disease. The screening of donated blood for HB_sAg is now an essential laboratory activity to minimize the risk of transmitting HBV infection to patients undergoing blood transfusions.

The *epidemiological aspects* of HAV and HBV are quite different. Hepatitis A is transmitted by the fecal-oral route or by contaminated food, water, or shellfish; parenteral transmission is uncommon. In contrast, hepatitis B is transmitted most commonly by parenteral injection or by the exchange of body secretions as in sexual intercourse. There is no known carrier state for HAV; however, chronic carriage of HBV is associated with significant implications for both the individual and his contacts. High risk for hepatitis B correlates with exposure to blood and body fluids, as is frequent with health care personnel. Certain lifestyles, such as male homosexuality and drug addiction, are also associated with high risk for hepatitis B. Epidemiology of non-A, non-B hepatitis seems to parallel that of hepatitis B. With the institution of testing for HB_sAg among blood donors, non-A, non-B hepatitis has become the most frequent transfusion-transmitted form of viral hepatitis.

The spectrum of *pathological changes* caused by these viruses may vary from a mild bout of acute hepatitis that resolves without further consequence to a more severe, even fatal disease. Alternatively, there may be resolution with scarring or chronic liver disease of either a benign or an aggressive nature. Cirrhosis and subsequent development of hepatocellular carcinoma are known pathological outcomes of chronic hepatitis B infection. Fulminant hepatic failure complicates acute viral hepatitis of any cause in a small percentage of patients. Treatment of the infected patient is supportive. A major goal is to interrupt spread of infection among the patient's intimates either by passive immunization with immune globulin or, in the case of hepatitis B, with high-titer specific immune globulin. Active immunization against HBV infection should be considered.

Laboratory features of acute hepatitis A and hepatitis B. The ranges of the peak values of ALT, AST, conjugated and unconjugated bilirubin, ALP, and GGT among patients with a diagnosis of hepatitis A or B are similar. Peak values for ALT and AST range from ten to one hundred times normal, with ALT higher than AST activity at the peak and throughout the course of abnormal values. Peak total bilirubin concentrations vary widely. In the anicteric form of the disease, values range from normal to slightly above normal, whereas in icteric viral hepatitis, values range from about 50 to 200 mg/L. In ~10% of the patients with icteric viral hepatitis, intrahepatic cholestasis is present. In the absence of histologically significant intrahepatic cholestasis, acute viral hepatitis is associated with peak alkaline phosphatase values that are up to approximately two times and GGT values that are up to approximately five times the upper limit of the reference range. In contrast, in the presence of intrahepatic cholestasis, both enzyme activities are significantly higher. Peak values for alkaline phosphatase of about five times normal and for GGT of about ten times normal are a characteristic finding.

The sequence of events associated with *hepatitis A* infection is shown in Figure 22-10. Within a week after onset of clinical illness, ALT and AST levels rise to a peak of more than ten times normal; they may fall to normal as early as 1–2 weeks later. Serum bilirubin levels rise later, if at all, but return to normal before transaminases; levels of both unconjugated and conjugated bilirubin may be elevated to varying degrees. Viremia and fecal excretion of the virus usually occur in the preicteric stage of the disease. Because the period of viremia is short, HAV is rarely a cause of post-transfusion hepatitis.

The course of events following infection with *hepatitis B* is depicted in Figure 22-11. Approximately 8–10 weeks after infection there is a gradual rise in aminotransferases, followed by the onset of clinical icterus. The peak increase in serum aminotransferase levels occurs at ~14–15 weeks after infection; the jaundice abates before the aminotransferases become normal. The typical sequence of viral markers is depicted in the figure—antibody to the hepatitis B core (anti-HB_c) appears early, followed by appearance of antibody to the surface antigen (anti-HB_s); the latter is protective. In the chronic carrier state, HB_sAg remains in the patient's serum and there is no *detectable* anti-HB_s rise.

Chronic Hepatitis

The clinical presentation of patients with chronic active liver disease is highly variable and may not be heralded by a clinically apparent acute episode. The operational definition of chronic hepatitis is hepatic inflammation with histological or biochemical documentation that persists for more than six months (Figure 22-12). Regardless of the initiating event, the classification, prognosis, and treatment are based on the histologic features. There are three classifications: (1) chronic persistent hepatitis; (2) chronic lobular hepatitis; and (3) chronic active (aggressive) hepatitis. The two major *etiologic*

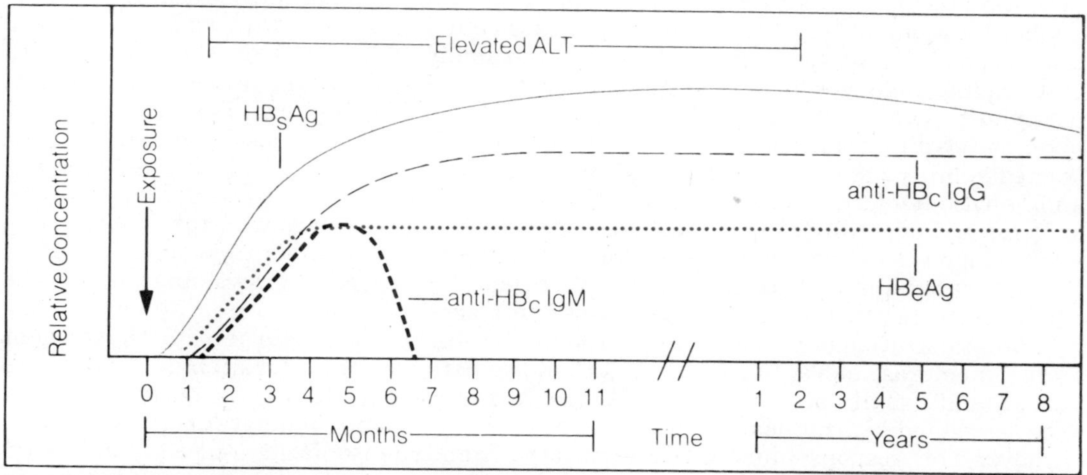

Figure 22-12. *Chronic active type B hepatitis.* Illustration shows (1) prolonged period of abnormal serum alanine aminotransferase (ALT) levels, (2) persistence of HB_sAg and HB_eAg, (3) high levels of anti-HB_c IgM for about five months, and (4) no detectable anti-HB_e or anti-HB_s. (Courtesy of Abbott Laboratories, North Chicago, IL 60064.)

subclassifications are (a) hepatitis B–related, in which HB_sAg is positive, and (b) autoimmune. The latter group of patients may have a variety of associated findings, such as positive antinuclear antibody (ANA) and smooth muscle antibody (SMA), as well as increased γ-globulin.

Alcoholic Liver Disease

Depending upon the degree and duration of alcohol consumption, significant abnormalities in liver structure and function can occur.[3] These may range from mild changes, such as fatty liver, to moderate disease (alcoholic hepatitis), or to full-blown hepatic fibrosis or alcoholic cirrhosis (Laennec's cirrhosis).

The chronic consumption of alcohol will induce a rise in serum γ-glutamyltransferase. However, levels of the enzyme will return to normal if the patient abstains from alcohol. Persistently abnormal values in the absence of continuing exposure suggest the presence of underlying structural liver disease. In the absence of liver disease, the serum γ-glutamyltransferase is useful as an index to monitor alcohol consumption. The increase in GGT activity is a more sensitive and specific index for detecting liver injury in alcoholics than are increases in aminotransferases or alkaline phosphatase.[2,3,14] Hyperbilirubinemia and other manifestations of cholestasis may complicate severe alcoholic liver disease, and there may be bile duct obstruction secondary to pancreatitis. Depending upon the severity of the liver disease and the nutritional status, hypoalbuminemia or hypergammaglobulinemia or both may be present. In some cases of alcoholic liver disease, as in cirrhosis and alcohol-induced fatty liver, acute hemolysis accompanied by marked elevation of serum lipids may be found (Zieve's syndrome).[14] In chronic alcoholics serum AST may be elevated owing to skeletal muscle injury (alcoholic myopathy) or alcoholic cardiomyopathy.

Monitoring serum GGT activity is helpful in distinguishing chronic alcoholics with liver disease from those without liver disease.[16,22] Liver disease is assumed if GGT activity initially is 8–10 times normal and if the elevation persists after 6–8 weeks of abstention from alcohol. On the other hand, if initial GGT levels are only 2–3 times normal and return to normal after abstention, the patient is assumed to be free of liver disease.[2,3,14]

Drug-Induced Liver Diseases

Certain drugs or their metabolites are able to produce a direct, predictable injury to liver cells and to elicit a dose-related hepatotoxic reaction. Other drugs are less predictable in their hepatotoxic potential; they may produce a pattern of injury similar to that seen in either viral hepatitis or cholestatic liver injury. The mechanism in these cases may be an immune-hypersensitivity reaction.

Two classic examples of hepatotoxins are acetaminophen and isoniazid. Acetaminophen is metabolized by the hepatic microsomal P_{450} system to produce small concentrations of a reactive metabolite that is rapidly inactivated by conjugation with glutathione. Ingestion of acetaminophen in toxic doses, however, overwhelms the capacity of the normal elimination mechanisms. In the face of glutathione depletion, the toxic metabolite accumulates and causes necrosis. Clinical manifestations include protracted vomiting and nausea, followed by an apparent recovery phase; marked jaundice and hepatic tenderness ensue and deterioration of liver function is rapid. There may be extreme increases in transaminase activity; prothrombin times are grossly abnormal. Early diagnosis of acetaminophen-induced hepatotoxicity is imperative since initiation of therapy within 16 h of ingestion with drugs such as *N*-acetylcysteine lessens the potential for hepatic injury and decreases the mortality rate.

Administration of isoniazid for treatment of tuberculosis is associated with transient aminotransferase elevations in up to 20% of the patients. This elevation is ~5–10 times normal and usually occurs within the first 2–3 months of therapy; it does not appear to be associated with clinical hepatitis and apparently does not mandate drug withdrawal. However, this response must be differentiated from true isoniazid-induced liver injury, which occurs in ~1% of all isoniazid recipients in whom there is also an increase in serum alkaline phosphatase, bilirubin, and prothrombin time. The injury may develop during the second and third months or may have its onset much later in the course of drug treatment.

Cirrhosis

Cirrhosis refers to diffuse fibrosis with nodular regeneration of hepatocytes. Cirrhosis is accompanied by decreased albumin synthesis, increased prothrombin time, decreased cholesterol synthesis, and insulin resistance. Cholestasis may also be present as evidenced by a mild increase in ALP and GGT, as well as total and conjugated bilirubin. The mild degree of liver cell necrosis that is usually present in cirrhosis may be responsible for a slight increase in AST and ALT as well as LDH-5. Any type of cirrhosis may cause an increase in blood pressure within the portal vein and its tributaries; i.e., portal hypertension, which leads to development of portal-systemic collateral venous drainage, splenomegaly, and ascites. Collateral vessels, which develop in the mucosa of the gastric fundus and esophagus (esophageal varices), may bleed profusely.

Primary Biliary Cirrhosis

Primary biliary cirrhosis (PBC) is characterized by progressive destruction of the intrahepatic bile ducts. The disease primarily affects middle-age women; the incidence in females is nine times higher than in males. The clinical picture is characterized by intractable itching that precedes onset of jaundice and leads to excoriated, hyperpigmented skin. The progressive degree of cholestasis is associated with a decrease in bile excretion; therefore, bile acids do not reach the duodenum, and fat malabsorption and fat-soluble vitamin deficiency can occur.

The classic biochemical alteration early in PBC is a marked elevation in the alkaline phosphatase level; transaminases are only modestly elevated, and bilirubin elevations are variable. Antimitochondrial antibody (AMA) and antinuclear (ANA) and smooth muscle (SMA) autoantibodies may also be present.

Hepatic Tumors

Benign and malignant lesions in the liver include primary and secondary neoplasms, cysts, and abscesses. Metastatic lesions are much more common (20:1) than primary liver tumors. Factors such as pre-existing cirrhosis or chronic infection (carrier state) of hepatitis B virus are frequently associated with primary liver cell carcinoma. World-wide, the incidence of hepatocellular carcinoma correlates with the prevalence of HB_sAg carrier states. The suspicion of a tumor should be high when abnormalities in serum biochemical tests occur without an obvious reason. The transaminases may rise slowly initially, with more marked increases as the mass spreads throughout the parenchyma.

Assay of serum alkaline phosphatase activity is widely used as a biochemical screening test and is frequently elevated when hepatic parenchyma is displaced by tumor (either primary or metastatic), cysts, or granulomas. This elevation may occur even when the serum bilirubin level remains normal. 5′-Nucleotidase, leucine aminopeptidase, and γ-glutamyltransferase activities may also be elevated. A marked rise in GGT activity may be noted in patients with metastatic disease with masses that are small and not easily detected. A more than twofold increase in serum total LDH activity has been noted in the presence of both primary and secondary hepatic tumors. Serum α_1-fetoprotein (AFP) concentration may be used as an indicator of hepatic cell carcinoma. A resurgence of serum AFP levels or of another oncofetal protein such as carcinoembryonic antigen (CEA) to >400 ng/mL in later life is highly suggestive of the presence of primary liver cancer. These markers are useful in serially monitoring the course of the disease and its treatment.

Metabolic Liver Diseases

Iron storage diseases. The liver plays an important role in iron metabolism; it synthesizes transferrin, to which much of circulating iron is bound. Like many other tissues, the liver also synthesizes and contains ferritin, which binds and stores iron (see Chapter 24). Impaired hepatic synthesis of transferrin leads to decreased serum concentrations and low values for total iron-binding capacity (TIBC). In acute hepatic disease with liver cell necrosis, both serum iron and serum ferritin concentrations increase since the damaged hepatocyte cannot take up iron or retain ferritin.

Pathologic iron overload (hemochromatosis) may be either primary (i.e., idiopathic or familial) or secondary (acquired) as a consequence of alcoholism, thalassemia, or repeated transfusion. Idiopathic familial primary hemochromatosis is a genetic disorder in which iron absorption is inadequately regulated; therefore, progressive iron overload occurs with deposition of iron in the hepatic, cardiac, and pancreatic parenchymal cells, and in the skin.

Secondary hemochromatosis cannot be differentiated from the primary form by laboratory findings alone. The hepatic histopathologic findings in both are that of increased iron stores within the hepatocytes and reticuloendothelial cells, in association with variable fibrosis and macronodular cirrhosis. The pattern of liver cell damage is the same regardless of the cause of the iron overload, but the degree of fibrosis is proportional to the load. Hepatocyte damage may be due to either lipid peroxidation with attendant membrane damage or lysosomal rupture with cell damage. Clinical symptoms are due to massive iron deposition in the heart, pancreas, and skin, as well as in the liver. Abuse of ethanol may exacerbate the liver damage or accelerate iron accumulation. The eventual hepatic lesion may be fibrosis and cirrhosis leading to portal hypertension. There may be an attendant cardiomyopathy, onset of diabetes, and various joint and skin manifestations.

The diagnosis can be made by documenting elevation of serum ferritin levels in conjunction with increased serum iron and per cent of transferrin saturation, together with decreased transferrin and TIBC. Definitive diagnosis may require measurement of iron in a liver biopsy. It is important to identify relatives at risk for hemochromatosis so that early treatment can be instituted. Human leukocyte antigen (HLA) typing may be useful for this purpose.

Wilson's disease. Wilson's disease, or hepatolenticular degeneration, is a disorder of copper metabolism. It is inherited in the autosomal dominant mode and is characterized by copper deposition in brain, liver, kidney, and cornea. Clinical manifestations include neurologic degeneration, cirrhosis, and typical corneal deposits called Kayser-Fleischer rings. In younger patients liver disease is predominant, but in patients over 20 years of age neurological symptoms predominate. The hepatic manifestations may vary from hepatomegaly to subacute or chronic hepatitis, cirrhosis, and fulminant failure. The effect on the liver and other tissues is essentially that of copper toxicity.

The defect in Wilson's disease is poorly characterized; however, it appears that there is an alteration in the excretory pathways of ceruloplasmin-bound copper. Impaired biliary excretion of copper causes the metal to accumulate in the liver.

The disease is fatal if untreated, but treatment by copper chelation (e.g., D-penicillamine) is effective. Because of the genetic implications, the presentation of an index case should lead to screening of the whole family. Institution of appropriate treatment in asymptomatic cases can prevent development of hepatic and other manifestations.

The best screening test for Wilson's disease is serum ceruloplasmin, since 96% of patients with Wilson's disease have levels <30 mg/dL. Total serum copper is high in early Wilson's disease and is present in serum unbound to ceruloplasmin. Urinary copper excretion exceeds 100 μg/d. An extremely reliable diagnostic finding is increased copper content of liver tissue.

Cholestasis. Cholestasis implies functional and morphologic disturbances in hepatic excretory processes.[18] Cholestasis can result from interference with bile excretion at any level from the hepatocyte to the ampulla of Vater. *Intrahepatic cholestasis* implies a defect in hepatic excretory function as a result of defective liver cell function due to damage to the bile secretory apparatus, transport mechanisms, or to altered metabolism. Diseases such as viral hepatitis and alcoholic hepatitis can be associated with cholestasis. The pattern of hepatotoxicity induced by certain drugs (e.g., chlorpromazine) may also be cholestatic. In *extrahepatic cholestasis,* there is an apparent mechanical blockade in the biliary tree, as may develop from biliary atresia, common duct stone, strictures, or carcinoma of the biliary tract or head of the pancreas.

Clinical consequences of prolonged cholestasis of any nature are due to (1) failure of bile to reach the duodenum and consequent malabsorption of fat and the fat-soluble vitamins A, D, E, and K; and (2) accumulation in the liver of biliary constituents—bile acids, bilirubin, cholesterol, and trace elements normally excreted in bile—and their regurgitation into serum.

Neonatal cholestasis. Cholestasis in the neonatal period can be associated with a wide variety of infectious, metabolic, or endocrine disorders. Jaundice is also a frequent concomitant of various syndromes or inborn diseases such as cystic fibrosis, α_1-antitrypsin deficiency, and Down's syndrome.

In evaluating the infant with jaundice, the first goal is to differentiate quickly neonatal cholestasis from its more benign counterpart, physiologic hyperbilirubinemia. The latter is characterized by an increase in unconjugated bilirubin because of immaturity of metabolic processes. Serum bilirubin can rise to high levels with the attendant risk of the development of kernicterus (bilirubin encephalopathy); however, in most cases the jaundice will resolve as the liver matures. On the other hand, jaundice in infants with an increased *conjugated* bilirubin fraction is never benign (i.e., some form of hepatobiliary disease is implied).

The second goal is to differentiate *neonatal hepatitis* from *biliary atresia.* Children affected by biliary atresia may be aided by early surgical intervention and creation of bile drainage, but neonatal hepatitis requires medical management for the consequences of cholestasis. A simple test is the examination of stool for the presence of bile pigments. Consistent absence of bilirubin or bile pigments suggests bile duct atresia; conversely, pigmented stools are strong evidence against this disease. Patients with neonatal hepatitis may have acholic or normal stools intermittently. The ultimate discrimination is by liver biopsy.

Reye's syndrome. Acute encephalopathy in combination with fatty degeneration of the viscera was initially described by Reye and colleagues in Australia in 1963.[20] The syndrome is characterized by an abrupt onset of protracted vomiting, which usually follows a prodromal, febrile viral illness caused by either varicella or influenza B. The vomiting is associated with neurological changes, such as lethargy and confusion, which may deteriorate rapidly into stupor and coma. At the same time, the liver enlarges and marked abnormalities in liver function become evident. This disease most frequently strikes children in the age range of 6–11 years.

The etiology of Reye's syndrome is unknown; however, it is apparent that there is a generalized mitochondrial dysfunction due either to a virus or a toxin or to viral potentiation of chemical toxins. The role of host susceptibility and the effect of exogenous agents such as aspirin need to be more fully evaluated.

Laboratory findings include a marked elevation in ALT and in AST from liver or muscle; an increase in plasma ammonia concentration, with the degree of increase having possible prognostic significance; normal total bilirubin concentration; hyperaminoacidemia; increase in plasma fatty acids; hyperuricemia; hypoprothrombinemia unresponsive to vitamin K; and, in infants, a tendency toward hypoglycemia.

Management consists of judicious treatment of the increased intracranial pressure secondary to cerebral edema with osmotic diuretics such as mannitol. Cerebral edema is the major factor contributing to the high mortality rate.

Acute cholecystitis. Acute cholecystitis usually develops in association with gallstones in the cystic duct, resulting in obstruction and a subsequent rise in biliary pressure, which in turn causes direct damage to the gallbladder. Less commonly, cholecystitis can develop without coincident gallstones as a complication of bacteremia, vascular disease such as polyarteritis, treatment with drugs such as steroids, or trauma.

The clinical features are variable depending upon the degree of inflammation and ischemia. The most common symptoms are (1) right upper quadrant abdominal pain, which may be initiated by the ingestion of heavy or fatty-food-containing meals; (2) increased temperature associated with bacterial infection of the gallbladder wall; (3) abdominal rigidity; (4) tender mass palpable in the gallbladder area; (5) elevated white count; and (6) jaundice.

The most common type of gallbladder disease is *chronic cholecystitis*, secondary to gallstone formation. Obesity, female gender, a family history of gallstones, multiparity, or a previous attack of cholecystitis are predisposing factors. Abdominal distention and epigastric pain, especially postprandially, along with belching and flatus may be present.

Techniques such as *ultrasound* and *endoscopic visualization* (retrograde cholangiography) are very helpful in documenting the presence and pathogenesis of biliary obstruction.

Gallstones. The most common type of gallstone is composed predominantly of cholesterol. Lithogenic (supersaturated) bile contains excess cholesterol relative to the other components of the mixed micelles (i.e., bile salts and phospholipids) due to either increased secretion of cholesterol or a decrease in bile acid pool size. In susceptible patients, lithogenic bile is produced by the liver, and the gallstones are formed in the gallbladder by intravesicular factors such as impaired gallbladder contractility and stasis, with or without infection, and mucin alterations that favor cholesterol crystallization.

Pigment gallstones, although much less frequent than cholesterol gallstones, are characterized by a cholesterol content of $<25\%$ and a black or dark brown color; they are composed of calcium bilirubinate, phosphate, and carbonate. They are most commonly found in association with hemolysis or cirrhosis and are a frequent complication in sickle cell anemia.

Systemic Conditions that Produce Liver Disease

Heart disease. Liver damage is a common complication of acute or chronic heart failure. Hepatocytes are especially vulnerable to lowered oxygen tensions; therefore, in the face of heart failure, diminished blood pressure, and reduced hepatic blood flow, there is congestion of centrilobular areas with hemorrhage, focal necrosis, sinusoidal dilatation, and impairment in bile secretion.

Mild changes in serum ALT, AST, total bilirubin, and ALP occur in chronic right ventricular failure. GGT values are usually elevated to about twice the upper limit of the reference range. On the other hand, when acute right ventricular failure occurs, there is a rapid release of enzymes (ALT, AST, and LDH-5) into the blood stream. Mildly abnormal unconjugated and conjugated bilirubin as well as increased ALP and GGT values are also typical in acute right heart failure.

Gastrointestinal disease. Hepatobiliary disease is a frequent complication of chronic inflammatory intestinal diseases such as Crohn's disease (regional ileitis) and ulcerative colitis. In addition, the incidence of gallstones is increased in patients with ileal dysfunction.

Effect of pregnancy on liver function. The altered hormonal milieu present during pregnancy may be associated with a rise in serum ALP late in gestation. It is important to note that the source of marked elevations in ALP may be the placenta, not the liver. Therefore, evaluation of the pregnant patient for liver disease may be more reliably carried out by assessing serum total and conjugated bilirubin, GGT, and aminotransferase levels.

The liver in diabetes mellitus. Diminished tissue utilization of glucose and overproduction of glucose by the liver occur in diabetes mellitus. The liver plays a major role in insulin metabolism; therefore, alterations in blood glucose levels may be seen in liver disease because of failure of hepatic insulin degradation. In patients with well-controlled diabetes mellitus, there are no detectable abnormalities in liver function.

Use of the Laboratory in the Diagnosis and Management of Liver Disease

Rational and judicious use of a carefully selected group of tests should allow the clinician to answer the specific questions that arise when confronted with a patient with suspected liver disease: (1) Is liver disease present? (2) What *type* of disturbance is present? (3) How *severe* is the liver disease? (4) How can the disease be *monitored* during treatment? and (5) What is the *prognosis* for this individual patient? None of the currently available tests or procedures can, by themselves, answer these questions.

Is liver disease present? The presence of liver disease may be suspected on the basis of a careful history and physical examination. In other cases, routine chemical screening may identify a biochemical abnormality.[5,25] The combination of laboratory tests most frequently used in confirming this suspicion includes serum bilirubin, AST, and ALT, as well as ALP or GGT, or both (see also Chapter 12, *Enzymes*). These tests complement each other and may suggest specific abnormalities such as cell damage and cholestasis. Normal values on all of these screening tests can, with substantial certainty, rule out significant liver or biliary tract disease. There are important exceptions, however, such as asymptomatic cholelithiasis, compensated inactive cirrhosis, or small space-occupying lesions. Further workup will be based on the results of these tests if any are abnormal.

What type of disturbance is present? When the presence of liver disease has been suggested by physical and laboratory findings, then the common questions that arise in clinical practice are: (1) Does the patient have hepatic *parenchymal* disease or *cholestasis?* (2) If cholestasis is present, is the *site* of the obstruction intrahepatic or extrahepatic? (3) In an alcoholic, is this nonalcoholic liver disease or alcohol-related hepatitis?

Acute liver cell injury (*parenchymal disease*) can be seen in viral hepatitis, alcoholic liver disease, drug- or toxin-induced liver disease, hypoxemia, shock, or metabolic liver diseases. The cell injury in all of these conditions, except alcoholic hepatitis, is reflected by a marked increase in aminotransferase levels in direct relationship to the degree of hepatocyte injury.

Cholestasis is reflected by an accumulation in serum of the components of bile. The total *and* conjugated serum bilirubin levels are elevated to some degree in all forms of cholestasis except unilateral or *focal duct obstruction*, as occurs with tumors, cystic masses, or granulomas. Alkaline phosphatase and, to a greater degree, γ-glutamyltransferase activities are increased in most forms of cholestasis. Therefore, the disparity of a normal serum bilirubin level with increased alkaline phosphatase and γ-glutamyltransferase activities may be seen in patients with this type of lesion. Precise localization of an intrahepatic obstruction is made through the use of ultrasonography, percutaneous transhepatic cholangiography (PTC), endoscopic retrograde cholangiopancreatography (ERCP), or computed axial tomography (CAT), or by operative intervention.

The biochemical feature of acute *alcoholic hepatitis* is a markedly elevated GGT in serum; this is not true of acute hepatitis. Alcoholic hepatitis has also been characterized by an AST/ALT ratio of >1.0 when the AST is <300 U/L.

A positive test for antimitochondrial antibody (AMA) in the presence of elevated alkaline phosphatase, cholesterol, and serum IgM strongly supports the diagnosis of primary biliary cirrhosis instead of extrahepatic obstruction. These tests can then be supplemented by ancillary studies such as liver biopsy, PTC, ERCP, and other radiographic or scintigraphic techniques.

How severe is the liver disease and what is the prognosis? Specific measurements of hepatic function can be of benefit in determining the severity of the liver disease in an individual patient. Fulminant hepatic failure, which carries a very high mortality rate, is most commonly seen with hepatitis B infection but can occur as a complication of any type of acute or chronic hepatic injury. There may be ascites, hypoglycemia, hypokalemia, hyponatremia, and hypoalbuminemia as well as anemia secondary to gastrointestinal bleeding. The most informative biochemical indicators are serum bilirubin, prothrombin time, and serum albumin. A uniformly poor prognosis has been associated with continued elevation of bilirubin, marked decrease in serum albumin, and prolonged prothrombin time (>10 s beyond control) unresponsive to parenteral vitamin K administration. The previously elevated aminotransferases may decrease as hepatic parenchymal mass shrinks. Blood ammonia levels may also become elevated. A rise in α_1-fetoprotein (AFP) levels is a *favorable* index of recovery.

How can liver disease be monitored? Serial determination of serum ALT and AST levels should be sufficient to chart the course of a patient with acute viral hepatitis. ALT and AST activities usually return to normal 2–3 months after clinical recovery, indicating that the liver has healed. However, abnormal transaminase values that persist for six months or more are indicative of chronic hepatitis. Measurement of ALT and AST is usually not helpful in monitoring the course of cirrhosis, since their values are typically normal or only slightly abnormal; however, onset of hepatic decompensation is usually indicated by a falling albumin level, hyperbilirubinemia, and increasing PT values.

Laboratory tests used to diagnose hepatic dysfunction may be affected by other factors. Drugs may interfere by causing either false-positive or false-negative results; nonhepatic diseases may affect levels of bilirubin and liver enzymes. In acute pancreatitis, aminotransferase elevations up to 3–5 times normal are not uncommon. Systemic illnesses, such as sepsis due to either specific bacteria or overwhelming fungal infections, can be associated with an elevation of serum bilirubin and alkaline phosphatase activity. There may also be a mild rise in the aminotransferase levels. Septic shock due to gram-negative bacteremia or other causes can result in an extreme rise of the aminotransferase levels to several thousand units per liter, whereas serum bilirubin may rise only moderately. Congestive heart failure of varying degrees will cause hepatic enlargement and abnormalities of hepatic function, but the increased aminotransferase levels will rapidly return to normal as the cardiac status improves.

A number of systemic diseases that have no apparent hepatic manifestations may cause an abnormality in the standard indices of liver function. A situation that commonly arises is an elevated alkaline phosphatase value in diseases such as Hodgkin's disease, myeloid metaplasia, or other malignancies.

A final caution to consider is the fact that significant liver disease can exist in the presence of *normality* of a multitude of tests of hepatic structural and functional integrity. For example, in small, space-occupying lesions such as tumors or in congenital hepatic fibrosis, there may be no detectable abnormality in the results of standard laboratory tests.

References

1. Balistreri, W. F., Suchy, F. J., Farrell, M. K., et al.: Pathologic versus physiologic cholestasis: Elevated serum concentration of a secondary bile acid in the presence of hepatobiliary disease. J. Pediatr., *98*:399–402, 1981.
2. Baraona, E., Leo, M. A., Borowsky, S. A., et al.: Pathogenesis of alcohol-induced accumulation of protein in the liver. J. Clin. Invest., *60*:545–554, 1977.
3. Belfrage, P., Berg, B., Cronholm, T., et al.: Prolonged administration of ethanol to young, healthy volunteers: Effects on biochemical, morphological and neurophysiological parameters. Acta Med. Scand. (Suppl.), *552*:5–55, 1973.
4. Branch, R. A.: Drugs as indicators of hepatic function. Hepatology, *2*:97–105, 1982.
5. Corless, J. K., and Middleton, H. M., III: Normal liver function: A basis for understanding hepatic disease. Arch. Intern. Med., *143*:2291–2294, 1983.
6. Corrigan, J. J., Jeter, M., and Earnest, D. L.: Prothrombin antigen and coagulant activity in patients with liver disease. JAMA, *248*:1736–1739, 1982.
7. Doumas, B. T., Perry, B. W., Sasse, E. A., et al.: Standardization in bilirubin assays: Evaluation of selected methods and stability of bilirubin solutions. Clin. Chem., *19*:984–993, 1973.
8. Forman, D. T.: Rapid determination of plasma ammonia by an ion-exchange technique. Clin. Chem., *11*:1–9, 1964.
9. Freeman, M., Kuiken, L., and Ragland, J. B.: Hepatic triglyceride lipase deficiency in liver disease. Lipids, *12*:443–445, 1977.
10. Friedel, R., Diederichs, F., and Lindena, J.: Release and extracellular turnover of cellular enzymes. *In:* Advances in Clinical Enzymology. E. Schmidt, F. W. Schmidt, I. Trauschold, et al., Eds. Munich, S. Karger, 1979.
11. Kahn, S.: Personal Communication.
12. Lauff, J. J., Kasper, M. E., and Ambrose, R. T.: Separation of bilirubin species in serum and bile by high performance reversed-phase liquid chromatography. J. Chromatog., *226*:391–402, 1981.
13. Lindskov, J.: The quantitative liver function as measured by the galactose elimination capacity. Acta Med. Scand., *212*:295–302, 1982.
14. Mendenhall, C. L.: Alcoholic hepatitis. Clin. Gastroenterol., *10*:417–441, 1981.
15. Mitchell, J. R., Nelson, S. D., Thorgeirsson, S. S., et al.: Metabolic activation: Biochemical basis for many drug-induced liver injuries. Prog. Liver Dis., *5*:259–279, 1976.
16. Moussavian, S. N., Becker, R. C., Piepmeyer, J. L., et al.: Serum gamma-glutamyl transpeptidase and chronic alcoholism: Influence of alcohol ingestion and liver disease. Dig. Dis. Sci., *30*:211–214, 1985.
17. Perry, B. W., Doumas, B. T., Bayse, D. D., et al.: A candidate reference method for determination of bilirubin in serum. Test for transferability. Clin. Chem., *29*:297–301, 1983.
18. Popper, H.: Cholestasis: The future of a past and present riddle. Hepatology, *1*:187–191, 1981.
19. Rashid, S. A., Axon, A. T. R., Bullen, A. W., et al.: Serum β_2-microglobulin in hepato-biliary diseases. Clin. Chim. Acta, *114*:83–91, 1981.
20. Reye, R. D. K., Morgan, G., and Baral, J.: Encephalopathy and fatty degeneration of the viscera: A disease entity in childhood. Lancet, *2*:249–252, 1963.
21. Rudman, B., DiFolco, T. J., Galambos, J. T., et al.: Maximal rate of urea excretion and synthesis in normal and cirrhotic subjects. J. Clin. Invest., *52*:2241, 1973.

22. Ryback, R. S., Eckardt, M. J., Felsher, B., et al.: Biochemical and hematologic correlates of alcoholism and liver disease. JAMA, *248*:2261–2265, 1982.
23. Schmid, R.: Bilirubin metabolism: State of the art. Gastroenterology, *74*:1307–1312, 1978.
24. Sherlock, S.: Patterns of hepatocyte injury in man. Lancet, *1*:782–786, 1982.
25. Skrede, S., Solberg, H. E., Ritland, S., et al.: Diagnostic and prognostic value of laboratory tests assessed in a follow-up study of 200 patients with liver disease. Clin. Chem., *28*:1177–1181, 1982.
26. Sundberg, M. W., Lauff, J. J., Weiss, J. S., et al.: Estimation of unconjugated, conjugated, and "delta" bilirubin fractions in serum by use of two coated thin films. Clin. Chem., *30*:1314–1315, 1984.
27. Weiss, J. S., Gautam, A., Lauff, J. J., et al.: The clinical importance of a protein-bound fraction of serum bilirubin in patients with hyperbilirubinemia. N. Engl. J. Med., *309*:147–150, 1983.
28. Wu, T.-W., Dappen G. M., Powers, D. M., et al.: The Kodak Ektachem clinical chemistry slide for measurement of bilirubin in newborns: Principles and performance. Clin. Chem., *28*:2366–2372, 1982.

Additional Reading

Balistreri, W. F.: Viral hepatitis: Implications to pediatric practice. Adv. Pediatr., *32*:287–320, 1985.

Balistreri, W. F., and Schubert, W. K.: Liver disease in infancy and childhood. *In*: Diseases of the Liver. L. Schiff and E. Schiff, Eds. Philadelphia, J.B. Lippincott Co., 1982.

Balistreri, W. F., and Soloway, R. D.: Clinical Guide to Bile Acid Physiology and Alterations in Disease States. North Chicago, IL, Abbott Laboratories, July, 1978.

Demers, L. M., and Shaw L. M.: Evaluation of liver function: A multifaceted approach to clinical diagnosis. Baltimore, Urban and Schwarzenberg, 1978.

Koff, R. S.: Viral Hepatitis. New York, John Wiley & Sons, 1978.

Wallnofer, H., Schmidt, E., and Schmidt, F. W.: Diagnosis of Liver Diseases. Stuttgart, Georg Thieme Publishers, 1977.

23 GASTRIC, PANCREATIC, AND INTESTINAL FUNCTION

NORBERT W. TIETZ, Ph.D.
ALAN D. RINKER, M.S.
A. RALPH HENDERSON, M.B., Ch.B., Ph.D.

Efficient digestion and absorption of nutrients are the results of coordinated functions of the stomach, intestinal tract, and pancreas. Coordination and regulation are in large measure dependent on hormones that stimulate or inhibit secretion of fluids containing hydrochloric acid (HCl), bile acids, bicarbonate, and digestive enzymes.

The ***human stomach*** consists of three major zones: the cardiac zone, the body, and the pyloric zone (Figure 23-1). The upper *cardiac zone* contains mucus-secreting *surface epithelial cells.* The *body of the stomach* contains cells of four different types: (1) the surface epithelial cells, which secrete mucus; (2) the *parietal cells,* which are the main and possibly only source of HCl and intrinsic factor; (3) the *chief* or *peptic cells*, which secrete a considerable amount of pepsinogen; and, finally, (4) the *neck chief cells* or mucus cells, which secrete mucus and pepsinogen. The third portion of the stomach, the *pyloric zone*, is subdivided into the *antrum,* the *pyloric canal*, and the *sphincter.* Its cells secrete mucus, some pepsinogen, and gastrin but no HCl.

The three phases of the digestive process are the cephalic, gastric, and intestinal phases. The *cephalic phase* is initiated by the sight, smell, and taste of food that stimulate the cerebral cortex and subsequently the vagal nuclei, and result in secretion of pepsinogen, HCl, and gastrin. Hydrogen ion secretion takes place against a millionfold concentration gradient, an energy-dependent process catalyzed by K^+, H^+-ATPase.

The distention caused by food entry into the stomach initiates the *gastric phase* of digestion. HCl release is caused by (1) direct stimulation of the parietal cells by the vagus nerve; (2) local distention of the antrum and stimulation of antral cells by the

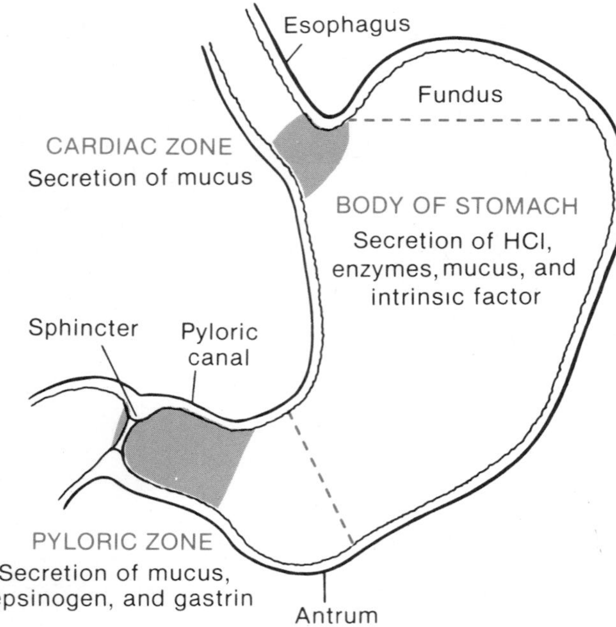

Figure 23-1. Schematic drawing of the stomach, with major zones.

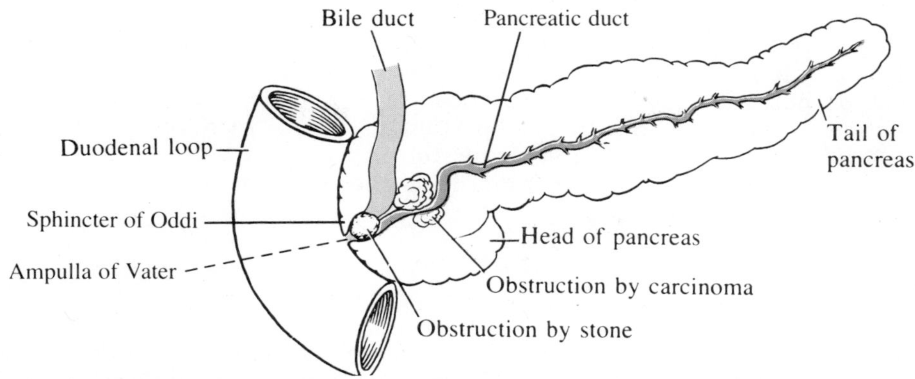

Figure 23-2. Cross section through pancreas with obstruction of the pancreatic duct by stone or carcinoma.

vagus nerve to secrete gastrin, which in turn causes HCl release from parietal cells; and (3) release of gastrin, stimulated by the near neutralization (pH 5–7) of gastric HCl by ingested food entering the pyloric zone. Gastrin also stimulates antral motility, the secretion of pepsinogen and of pancreatic fluid rich in enzymes, and the release of a number of gastrointestinal hormones (secretin, insulin, acetylcholine, somatostatin, and pancreatic polypeptide). As a result of the initially acid environment, pepsinogen is rapidly converted to the active proteolytic enzyme pepsin.

The *intestinal phase* of digestion begins when the weakly acidic digestive products of proteins and lipids enter the duodenum. Four of the peptides secreted by the intestine are recognized as legitimate hormones—namely, gastrin, cholecystokinin (CCK), secretin, and gastric inhibitory peptide. Others are considered candidate hormones, and their exact role is not completely understood. Gastrin, secretin, and cholecystokinin are released as the upper duodenal mucosa comes in contact with gastric HCl and chyme (partially degraded food, mixed with mucus). The principal actions of these hormones are described later.

The ***pancreas*** lies across the posterior wall of the abdomen. Its head is located in the duodenal curve (loop) and its body and tail extend to the spleen (Figure 23-2). Pancreatic digestive enzymes, in a bicarbonate-rich juice, are collected in the pancreatic duct and pass through the ampulla of Vater, which is controlled by the sphincter of Oddi, and then mix with the food bolus in the duodenum. During the passage through the small intestine (duodenum, jejunum, ileum), carbohydrates are broken down by amylase and oligo- and disaccharidases into monosaccharides, which are then actively absorbed into the blood stream. Protein is degraded in the duodenum by trypsin, chymotrypsin, and carboxypeptidase from the pancreas, and aminopeptidases from the small intestine. The resulting dipeptides and amino acids are absorbed in the jejunum and ileum by specialized absorptive mechanisms in the mucosal surface. Dietary fats are emulsified in the duodenum by the action of bile; are hydrolyzed by lipase (aided by colipase) to individual fatty acids, monoacylglycerols (monoglycerides), and glycerol; and are then absorbed in the remainder of the small intestine. Most nutrients, including vitamins and minerals, have been absorbed by the time the food passes into the large intestine (cecum, colon, rectum), where water is actively absorbed, electrolyte balance is regulated, and bacterial actions take place. These processes result in the formation of feces.

Gastrointestinal Hormones

The gastrointestinal (GI) tract is both a major endocrine organ and a target for many hormones released locally and at a distance. GI hormones are released by endocrine cells found throughout the gut mucosa and are so numerous that the gut is recognized as the largest endocrine organ in the body. Collectively, the GI hormones influence motility, secretion, digestion, and absorption in the gut. They regulate bile flow as well as secretion of pancreatic hormones and affect tonicity of vascular walls, blood pressure, and cardiac output. Many of these hormones are present both in the gut and in the central nervous system, making them important in the neuroendocrine control of the gut. In addition, they may have a role as neurotransmitters when released from nerve synapses. A listing of some gastrointestinal hormones is given in Table 23-1.

GASTRIN

Chemistry and Physiology

Three molecular forms of gastrin that stimulate gastric acid secretion are known to exist in blood and tissues: *big gastrin,* a linear polypeptide of 34 amino acids designated as G-34; *little gastrin,* or G-17; and *minigastrin,* or G-14. These polypeptides circulate in both nonsulfated and sulfated forms and have similar biologic activities. Derivatives of gastrin, containing the same terminal tetrapeptide residue, are physiologically active, but the potency of each derivative differs. A synthetic derivative, *pentagastrin,* is used for maximal stimulation of HCl secretion in gastric function testing.

Gastrin is produced and stored mainly by special endocrine cells (G cells) of the antral mucosa of the stomach, and to a lesser extent by G cells of the proximal duodenum and D cells of the pancreatic islets. After secretion, the principal circulating form of gastrin in healthy individuals and in patients with hypergastrinemia is G-34. It stimulates the secretion of gastric acid by fundal parietal cells, secretion of gastric pepsinogen and intrinsic factor by the gastric mucosa, release of secretin by the small intestinal mucosa, and secretion of pancreatic HCO_3^- and enzymes as well as hepatic bile. G-34 also increases gastric and intestinal motility, mucosal growth, and blood flow to the stomach. Gastrin is secreted in response to antral distention and the presence of partially digested protein products (peptides, polypeptides, and amino acids) in the stomach. Other stimuli of gastrin include alcohol, caffeine, insulin-induced hypoglycemia (see page 771), and vagal stimulation initiated by smelling, tasting, chewing, and swallowing of food.

Maximum secretion of gastrin from the fundus occurs at an antral pH of 5–7 but is inhibited by lower pH. This negative feedback regulation appears to be a safeguard against overacidification.

Clinical Significance

Zollinger-Ellison (Z-E) syndrome. Knowledge of blood gastrin levels is helpful in the diagnosis of Zollinger-Ellison syndrome, caused by gastrinomas (duodenal or pancreatic endocrine tumors) that produce and secrete large quantities of gastrin. The syndrome consists of fulminant peptic ulcers, massive gastric hypersecretion, and non-B islet cell tumors of the pancreas. Hypergastrinemia, diarrhea, steatorrhea, and other endocrinopathies are also frequent characteristics of this syndrome. In Z-E patients, fasting gastrin levels are usually markedly elevated, ranging from 2 to 2000 times the normal level. Concentrations >1000 pg/mL with gastric acid hypersecretion are virtually diagnostic of gastrinoma. Unfortunately, there is substantial overlap in gastrin levels between patients with Z-E syndrome and patients with common peptic ulceration and a variety of other diseases. The differential diagnosis is aided by the gastrin response to secretin infusion. Management of the Z-E patient requires surgical removal of the tumor, gastrectomy, or therapy with histamine H_2-receptor antagonists.

The *secretin challenge* is the most useful provocative test for the diagnosis of Z-E syndrome. Serum gastrin is measured before and at intervals following the intravenous injection of secretin. A positive result, consistent with the diagnosis of gastrinoma, is indicated by an increase in gastrin concentration of ≥100 pg/mL over the basal level. In normal individuals and in patients with other disorders, secretin causes a slight decrease in serum gastrin.

Miscellaneous conditions. Increased fasting serum gastrin concentrations are often seen in patients >60 years of age and in patients with atrophic gastritis who may

Table 23-1. CHARACTERISTICS OF PRINCIPAL GASTROINTESTINAL HORMONES

Hormone	Molecular Weight	Number of Amino Acids	Tissues (Principal tissue[s] are listed first)	Half-Life in Blood	Serum Reference Range	Principal Actions
Little gastrin	2098 (I) 2178 (II)	17	Stomach antrum; also vagal fibers, duodenum	~5 min	<100 pg/mL*	Stimulates secretion of gastric acid, pepsinogen, intrinsic factor, secretin, pancreatic enzymes and HCO_3^-, and hepatic bile; increases gastric and intestinal motility and mucosal growth
Big gastrin	3839 (I) 3919 (II)	34	Duodenum, also stomach antrum	~42 min	<100 pg/mL*	
Secretin	3055	27	Duodenum	~4 min	0–500 pg/mL	Increases pancreatic secretion of HCO_3^-, enzymes, and insulin; promotes pancreatic growth; increases HCO_3^- and water secretion from liver and Brunner's glands; stimulates gallbladder contractions; reduces gastric and duodenal motility; inhibits gastrin release and gastric acid secretion
Cholecystokinin-pancreozymin (CCK-PZ)	3884	33	Duodenum and jejunum; also CNS	~3 min	5–800 pg/mL	Regulates gallbladder contraction; increases motility of small intestine; stimulates secretion of pancreatic enzymes, insulin, glucagon, pancreatic polypeptide, and HCO_3^-; stimulates pancreatic growth and secretion from Brunner's glands; slightly stimulates gastric HCl and pepsinogen
Vasoactive intestinal polypeptide (VIP)	3326	28	Nervous system; also all areas of GI tract	~1 min	0–100 pg/mL; about 10-fold higher in CSF	Causes relaxation of smooth muscles of circulatory system, gut, and genitourinary system; increases water and electrolyte secretion from pancreas and gut; releases hormones from pancreas, gut, and hypothalamus; stimulates lipolysis, glycolysis, and bile flow; inhibits gastrin and gastric acid secretion
Gastric inhibitory polypeptide (GIP)	5104	43	Duodenum and jejunum	~15 min	15–100 pmol/L	Stimulates insulin release in hyperglycemia; inhibits gastric acid, pepsin, and gastric secretion; reduces gastric and intestinal motility; increases fluid and electrolyte secretion from small intestine

*This value represents gastrin immunoreactivity as measured in "G-17 equivalents" because gastrin assays are standardized by using synthetic human G-17. Other forms of gastrin (e.g., G-34, G-14) have variable cross-reactivities with G-17 antibodies used in these assays.

have a reduced rate of gastric acid secretion. Patients with *gastric ulcer disease* usually demonstrate elevated basal and postprandial serum gastrin levels, which are most likely due to decreased inhibition of gastrin secretion. Fasting serum gastrin levels of patients with *duodenal ulcer disease* do not differ from those of normal subjects, but postprandial serum gastrin responses are usually greater than normal.

Increased basal gastrin concentrations may be classified as "appropriate" or "inappropriate" according to their association with decreased or increased gastric acid secretion. In patients with very low or absent acid secretion and a functionally intact gastric antrum, such as in pernicious anemia, an increase in serum gastrin is physiologically appropriate and is expected. The increase is due to hyperplasia of antral G cells as the result of continuous stimulation in the absence of negative feedback. Inappropriate hypergastrinemia does not respond to the normal negative feedback by gastric HCl.

Determination of Gastrin in Serum

Specimen. Draw a venous blood sample, without additives, after the patient has fasted 12 h or more. Plasma is unsuited for analysis. Gastrin is unstable in serum, even at refrigerator temperatures, due to the action of proteolytic enzymes in serum. Therefore, centrifuge specimens promptly and keep frozen prior to assay. Assay specimens immediately after thawing; avoid repeated freezing and thawing.

Principle. Most of the commercial kit procedures available for the measurement of gastrin utilize a double-antibody radioimmunoassay technique. A sample size of either 100 or 200 μL of serum is required.

Comments. Standards in most gastrin assays consist of synthetic human G-17. Since G-34 is the predominant form of gastrin in gastrinoma patients, measurement of serum gastrin in these patients should be interpreted in terms of G-17 equivalents.

Reference ranges.[54] Reference values for serum gastrin in fasting patients range up to 100 pg/mL. Values may be higher in elderly individuals. Concentrations fluctuate throughout the day, according to a circadian rhythm (lowest at 0300–0700 h, highest during daytime) and in response to meals.

CHOLECYSTOKININ-PANCREOZYMIN (CCK-PZ; CCK)

CCK is an intestinal hormone that activates gallbladder contraction and stimulates secretion of pancreatic enzymes. Because of its dual action it is sometimes called CCK-PZ. It is a polypeptide containing 33 amino acids, of which the five *C*-terminal amino acids are identical to those of gastrin. CCK has multiple molecular forms; the biologically active molecule contains a sulfated tyrosyl residue.

CCK is found in the I cells of the upper small intestinal mucosa, mainly in the duodenum. Mixtures of polypeptides and amino acids from partially digested protein stimulate CCK secretion, but pure undigested protein does not. Secretion is also stimulated by gastric hydrochloric acid entering the duodenum and by fatty acids with nine-carbon chains or longer, especially fatty acids in the form of micelles.

CCK regulates the contraction of the gallbladder and increases the motility of the duodenum and the small intestine. Because it possesses the same terminal amino acid tetrapeptide as gastrin, it has a slight degree of gastrin activity and also competes with gastrin for the receptor sites on the HCl-secreting cells. This action may contribute to the termination of gastric secretion after a meal. Gastrin and CCK activity are additive in their stimulation of the pancreas. In addition, CCK stimulates pancreatic growth, relaxes the sphincter of Oddi, and stimulates secretions from Brunner's (duodenal) glands.

Methods. Little information is available on plasma CCK concentrations in various disorders, and no reliable and convenient methods are currently available for the measurement of CCK. Various radioimmunoassays have been developed, but methodological problems have resulted in considerable disagreement over values for circulating CCK. Reference values for fasting serum from healthy individuals are 5–800 pg/mL but are highly dependent on the method in use.[38]

SECRETIN

Secretin is a polypeptide with structural similarities to glucagon, vasoactive intestinal polypeptide (VIP), and gastric inhibitory polypeptide (GIP). It stimulates the flow of pancreatic juice into the duodenum. The intact secretin molecule is required for its biological activity.[38,58]

Secretin is secreted by the mucosal granular S cells located in greatest concentration in the duodenum but present throughout the length of the small intestine. It is released primarily on contact of the S cells with gastric HCl. Alcohol appears to increase secretin release by stimulation of gastric acid secretion with subsequent lowering of duodenal pH. The only known inhibitor of secretin release is somatostatin.

The primary physiological role of secretin appears to be the stimulation of the pancreas to secrete an increased amount of juice with a high bicarbonate content. This is the reason for its use in pancreatic function tests. Secretin inhibits gastrin release (except in Z-E syndrome) and therefore gastric acid secretion. There is currently little knowledge of abnormal secretion of secretin in pathological conditions.

VASOACTIVE INTESTINAL POLYPEPTIDE (VIP)

VIP is a linear polypeptide consisting of 28 amino acids; it has structural similarities to secretin, GIP, and glucagon. VIP is present throughout the body and is found in highest concentrations in the nervous system and gut. Unlike other GI hormones, VIP is distributed along the entire length of the gut from esophagus to colon but is more concentrated in the jejunum, ileum, and colon.

VIP has a large number of ill-defined physiological actions including neurotransmission; increase of water and electrolyte secretion from the pancreas and gut; release of hormones from the pancreas, gut, and hypothalamus; stimulation of lipolysis, glycogenolysis, and bile flow; and inhibition of gastrin and gastric acid secretion. There is no evidence that VIP is released during digestion; however, vagal stimulation can cause its release.

Measurement of VIP concentrations is a very useful screening test for the diagnosis of VIP-secreting tumors that cause intractable diarrhea, for detecting occult metastases, and for evaluating the effect of therapy in patients with VIP-secreting tumors.

Methods. Radioimmunoassays have been developed to measure circulating plasma concentrations in pathological conditions in which VIP is elevated; however, many RIA methods are not sensitive enough to measure VIP levels accurately in normal individuals.[38] Although variations in values between measurement methods are large, reported values for normal individuals are generally <100 pg/mL.[2,38,54] Patients with VIP-secreting tumors have plasma VIP concentrations of 600–9000 pg/mL.[38]

GASTRIC INHIBITORY POLYPEPTIDE (GIP)

Gastric inhibitory polypeptide is a linear peptide consisting of 43 amino acids. Its *N*-terminal end has a close resemblance to glucagon and secretin. GIP is synthesized and released by K cells located in the duodenal and jejunal mucosa. The plasma level is increased by oral administration of glucose or triacylglycerols and by intraduodenal infusions of solutions containing a mixture of amino acids. In order for food components to stimulate GIP release, they must be absorbed by the intestinal mucosa.

The biological actions of GIP include (1) stimulation of insulin secretion in the presence of hyperglycemia, (2) reduction of intestinal motility with stimulation of small intestinal fluid and electrolyte secretions, and (3) in supraphysiological concentrations, inhibition of gastric acid, pepsin, and gastrin secretion. The insulinotropic action of GIP appears to be the most important of its biological actions, and, as a result, this hormone has recently been called *glucose-dependent insulinotropic peptide* (GIP). Currently, the diagnostic or therapeutic value of GIP has yet to be established.

GIP is measured by radioimmunoassay. Reported levels of GIP in fasting human plasma are 15–100 pmol/L[3] but depend on the method used.

OTHER HORMONES OF THE GUT

Somatostatin is one of the most potent known inhibitors of endocrine secretions. Somatostatin inhibits the release of growth hormone, thyroid-stimulating hormone, insulin, glucagon, gastrin, CCK, secretin, VIP, GIP, motilin, pancreatic polypeptide, enteroglucagon, neurotensin, substance P, and others; it also inhibits the effect of these hormones on their target tissues.

Bombesin is present in nerves and in special endocrine cells in the gut; the highest concentrations are found in the stomach, antrum, and duodenum. Bombesin appears to function as a local releasing factor for some gastrointestinal hormones.

Motilin is a strong stimulant for contraction of smooth muscles of the upper gastrointestinal tract. It increases the motility of the fundus, antrum, and duodenum as well as contractions of the lower esophageal sphincter. Motilin is unique in that its actions are generally restricted to the fasting state.

Pancreatic polypeptide (PP) is derived mainly from the pancreas. The exact physiological role of PP is not understood. PP has a biphasic effect; it initially increases and then inhibits the secretion of pancreatic enzymes, water, and electrolytes. Increased PP concentrations can be used as biochemical markers for pancreatic endocrine tumors.[58]

Glucagon-like immunoreactivity (GLI) or ***enteroglucagon*** consists of a group of substances that have biological actions similar to those of pancreatic glucagon but are chemically different. GLI is found in all parts of the small intestine and colon.

Enzymes of the Gastrointestinal Tract

PEPSIN AND PEPSINOGEN

The chief and peptic cells, the neck chief cells or mucus cells, and, to a lesser degree, the cells of the pyloric zone of the stomach secrete *pepsinogen,* which is activated to *pepsin* at the acid pH provided by gastric hydrochloric acid. Pepsins catalyze the hydrolysis of proteins to polypeptides.

Pepsinogen A is a proteolytic enzyme precursor (proenzyme) that, after secretion, is converted by acid to the active enzyme, *pepsin A* (EC 3.4.23.1). Pepsin A is capable of converting additional pepsinogen to pepsin (autocatalysis). It preferentially splits peptide bonds adjacent to phenylalanine, tyrosine, tryptophan, or leucine.

Of the pepsinogen released from the gastric mucosa, about 99% is secreted into the stomach to become part of the gastric fluid. The remaining 1% diffuses into the interstitial fluid surrounding the chief cells and eventually enters the blood. The portion that reaches the kidneys passes through the glomeruli and is excreted into the urine. At the mildly acid pH of urine, part or all of the proenzyme, now called *uropepsinogen,* may be converted (activated) to ***uropepsin.*** Daily excretion is fairly constant for any one individual, although there is a diurnal variation. If the pH of the urine specimen is kept between 5.0 and 6.5, the urine enzyme is stable at room temperature for 2–3 d and in the refrigerator for two weeks. No inhibitor of the enzyme is present in the urine; uropepsin is destroyed at alkaline pH.

Pepsinogen secretion is stimulated by the vagus nerve and by some GI hormones (gastrin, secretin, CCK). Its secretion is inhibited by gastric inhibitory polypeptide (GIP), anticholinergics, histamine H_2-receptor antagonists, and vagotomy.

Serum concentrations of pepsinogen reflect the parietal cell mass and correlate well with the maximum acid-secreting capacity. In general, uropepsinogen concentrations in the urine parallel those of pepsinogen in the blood.

Measurement of Pepsinogen and Pepsin

Pepsin, as well as pepsinogen after activation to pepsin, has been measured in mucosal cells, in gastric contents, in serum, and in urine. These measurements are rarely used in clinical laboratories.

Methods include (1) measurement of enzymatic activity by hydrolysis of specific protein or peptide substrates followed by measurement of the amount of substrate remaining or product formed,[21] (2) radioimmunoassay,[14,42] and (3) diffusion of the enzyme through a turbid mixture of agarose and dye-linked protein with subsequent measurement of cleared zones as the dye is split off from the protein by proteolytic action.[46] The methods vary in their ease, specificity, and stoichiometry; reference ranges depend on the method used.

Enzymes Derived From the Pancreas

The normal pancreas secretes several enzymes that pass almost entirely into the duodenum; only a small fraction of these enzymes reaches the blood. Pancreatic enzymes synthesized by the acinar cells are stored there in the granules, some in the form of zymogens (e.g., amylase). Enzymes of predominant clinical interest are *amylase, lipase, phospholipase A, ribonucleases*, and a group of *proteolytic enzymes* that includes trypsin, chymotrypsin, carboxypeptidase A and B, elastase, and collagenase.

Upon stimulation by CCK and, to a lesser extent, by gastrin and secretin, pancreatic enzymes are released by the acinar cells of the pancreas into a fluid containing HCO_3^-. These enzymes pass through the ductules into the main pancreatic duct, which empties into the duodenum through the ampulla of Vater. The bile duct joins the pancreatic duct just proximal to or at the ampulla of Vater (see Figure 23-2). The pancreas may have a second or accessory duct (duct of Santorini), which empties separately or jointly with the main duct into the duodenum. In the duodenum, the pancreatic juice mixes with the food material coming from the stomach. The combination of enzymes secreted by the pancreas, at the proper pH provided by the bicarbonate, can digest virtually any food material.

In some disorders, such as pancreatitis or obstruction of the pancreatic duct due to stones or carcinoma, the flow of enzymes and bicarbonate into the duodenum is impeded and an increased amount of pancreatic enzymes reaches the blood stream (back flooding). The mechanism by which the enzymes enter the circulation may involve changes in pressure in the pancreatic duct and ductules, changes in permeability of acinar cells, or disruption of the acinar-limiting membrane.

Pancreatic Enzymes in Serum

The pancreatic enzymes most commonly measured in serum are *amylase, lipase,* and *trypsin.* The determination of these enzymes and their role in the diagnosis of pancreatic diseases are discussed in more detail in Chapter 12. Here we comment briefly on the role of these enzymes and then discuss elastase.

Amylase

Amylase in serum. Measurements are requested chiefly for the diagnosis of acute pancreatitis, in which serum amylase activity is increased 2–12 h after an acute attack and reaches a peak 12–72 h after the attack. The test is simple and widely available; however, elevations of serum amylase are relatively nonspecific since they are observed in a variety of nonpancreatic diseases as well (Table 12-4). Determination of S- and P-type isoenzymes increases specificity and aids in the differentiation of hyperamylasemia of pancreatic origin from that due to other causes (see also Chapter 12).

Amylase in urine. Amylase is filtered by the glomeruli and excreted in the urine. In pancreatitis, an increased urinary excretion rate accounts for the earlier, more frequent, and more prolonged elevation of the enzyme in urine, as compared with blood. Increased clearance rate is not a specific response to pancreatitis but instead is due to competition for absorption of low-molecular-weight proteins by the renal tubules. Increases in amylase clearance and in the ratio of amylase to creatinine clearance are observed in a number of nonpancreatic disorders (see Chapter 12).

Lipase

Assays for lipase in *serum* are used chiefly for the diagnosis of acute pancreatitis, in which levels change in a manner similar to those of amylase. The elevations are frequently more pronounced, more prolonged, and more specific for the disease than are amylase elevations. For diagnostic purposes, lipase and amylase evaluations complement one another; one study reports clinical sensitivity of 98% for the diagnosis of acute pancreatitis when the two enzymes are used together[29] (see Chapter 12). Lipase in *urine* is undetectable with most assay methods.

Trypsin

Recently developed immunochemical techniques have shown that elevation of *serum* trypsin is a highly sensitive and relatively specific indicator of pancreatic disease.[19] Very high levels in acute pancreatitis contrast sharply with low or normal levels in chronic pancreatitis and with essentially normal levels in hepatic jaundice. Levels in carcinoma of the pancreas depend on the degree of obstruction involved and the amount of remaining acinar tissue. Since trypsin is excreted by the kidneys, serum levels are also elevated in renal disease. Recently, immunoreactive trypsin determinations have been suggested as screening tests for cystic fibrosis in newborns.[16]

Chymotrypsin

Although chymotrypsin has been determined in the serum of patients with various forms of pancreatic disease, its application for diagnostic use is limited (see Chapter 12). Direct and indirect chymotrypsin determinations on duodenal contents and feces, however, have proved to be of diagnostic value (see also *Trypsin and Chymotrypsin in Stool,* later in this chapter).

Elastase

Elastase (EC 3.4.21.11), unlike trypsin, chymotrypsin, and pepsin, rapidly hydrolyzes elastin, the yellow scleroprotein that is the basic ingredient of yellow elastic connective tissue. Recent reports point out that elastase is increased in acute and relapsing chronic pancreatitis to a greater degree than is serum amylase activity.[49] Elevations also persist for a longer time and are said to reflect the clinical course better than amylase activities. Furthermore, no or only minor elevations in elastase are observed in cases of hyperamylasemia of nonpancreatic origin. Elevations of elastase have also been observed in carcinoma of the pancreas, especially carcinoma of the head of the pancreas.

The recent development of a radioimmunoassay for elastase 1 has generated new clinical interest in this enzyme and avoids the problems associated with measurement of enzymatic activity.[49] The *reference range* is 80–320 ng/dL.

Enzymes in the Intestinal Mucosa

Although the intestinal mucosa contains a wide variety of enzymes, most of these are not measured for routine diagnostic purposes. *Disaccharidases* will be discussed in a later section of this chapter.

Gastric Function Tests

The various motor and secretory activities of the stomach can be evaluated by a number of clinical laboratory tests, among which are the analysis of gastric residue, the determinations of basal and stimulated secretion rates, and the determinations of intrinsic factor and pepsinogen. These tests, however, are more helpful in evaluating specific functions of the stomach than in providing specific diagnostic data.[12,13] Detection of compounds not normally seen in gastric contents, such as blood or lactic acid, is also of clinical significance.

Gastric Residue and Its Major Constituents

Gastric residue is defined as the entire contents of the stomach after a fast of ~12 h. The specimen is obtained by aspiration through a gastric tube. The *normal total volume* is between 20 and 100 mL, but usually below 50 mL. Among the causes for *increased volume* are delayed emptying, increase in gastric secretion, and admixture of material regurgitated from the duodenum. In this last case, the gastric residue generally contains bile, which can be confirmed by chemical tests for bilirubin (e.g., Ictotest, Ames Laboratories, Elkhart, IN 46525). The *consistency* of gastric residue is rather fluid but may be viscous in the presence of excessive amounts of mucus. The residue is usually colorless, but in the presence of regurgitated bile its color may be slightly yellow or green. This is the case in ~25% of normal individuals and in the majority of patients after partial gastrectomy or gastroenterostomy. A red or brown color in the gastric residue is usually due to blood, which can readily be confirmed by chemical tests, such as the Gastroccult test (SmithKline Diagnostics, Inc., Sunnyvale, CA 94086).

Free hydrochloric acid. Hydrochloric acid is secreted by the parietal cells at a concentration of ~155 mmol/L. The hydrochloric acid secreted into the lumen of the stomach is derived from the blood. Rapid secretion of gastric hydrochloric acid after a meal is accompanied by a decrease in blood chloride and an increase in blood pH and HCO_3^-, the so-called *alkaline tide.* The pH of parietal secretion is ~0.9, but mixture with other gastric constituents increases the pH to 1.5–3.5. The secretion of hydrochloric acid by the mucosa is continuous, but the volume fluctuates considerably depending upon the degree and type of stimulation.

The *concentration* of free acid in gastric residue of normal individuals varies from 0–40 mmol/L (without stimulation). Approximately 4% of young healthy individuals may have no free hydrochloric acid in the fasting stomach. This percentage increases with increase in age and is about 25% in individuals at age 60. Absence of free hydrochloric acid in gastric residue is considered abnormal only if the condition persists after maximal stimulation with pentagastrin. *False achlorhydria* is a term used for those cases in which free hydrochloric acid is secreted, but in which the hydrogen ions are subsequently partially or fully neutralized. A determination of chloride might help avoid misinterpretation in these cases (normal: 45–155 mmol/L).

The determination of the concentration of free hydrochloric acid is generally performed by titration, but measurement of the pH of the gastric residue is the preferred method according to some investigators.[36]

Total acidity in gastric residue includes hydrogen ions occurring as (1) free HCl, (2) mucoprotein, (3) acid salts, and (4) organic acids such as lactic and butyric acid, which are formed in gastric content by bacterial action when food is retained in the stomach for long periods (>6 h) at neutral or slightly alkaline pH. Such a condition is usually associated with carcinoma of the stomach or pyloric stenosis and may be due to either decreased secretion of HCl or neutralization of HCl by gastric content. These organic acids are not found when free HCl is present. The total acid concentration is usually between 10 and 50 mmol/L.

Mucus is present in only small amounts in normal gastric contents; however, increased amounts are sometimes found in gastric carcinoma, in gastritis, and in cases of mechanical irritation due to passage of a stomach tube.

Blood. In some pathological conditions, such as carcinoma of the stomach, peptic ulcer, gastritis, esophageal varices, or bleeding gums, blood may be present in gastric residue. At the strongly acid pH in the stomach, acid hematin is formed; it has a brownish appearance, resembling coffee grounds. Fresh red blood may be due to accidental trauma from the gastric tube or to underlying lesions, such as ulcer or carcinoma.

Food. Normal gastric residue does not contain any appreciable amount of food. Excessive accumulation of food particles indicates decreased motor activity of the stomach or pyloric obstruction.

Intrinsic factor. Gastric residue contains intrinsic factor (IF), which is secreted by the parietal cells of the fundus and combines with vitamin B_{12} to make absorption of this vitamin possible. Lack of this intrinsic factor causes vitamin B_{12} deficiency with resultant arrest in the development of erythrocytes. Intrinsic factor can be measured indirectly by the Schilling test (see Chapter 24).

Stimuli of Gastric Secretion

Test meals, caffeine, and alcohol. In the past, stimulation of gastric secretion was effected by oral administration of meals, caffeine, sodium benzoate, or ethanol. These stimuli give essentially no undesirable side reactions, but the degree of stimulation is relatively weak.

Gastrin and pentagastrin. Gastrin is the natural and most powerful stimulus for gastric HCl secretion. Pure gastrin is administered subcutaneously, intramuscularly, or by intravenous infusion. The synthetic product *pentagastrin* is also a potent stimulus for gastric HCl secretion, although a higher dose is required for maximal stimulation. The response to the stimulus is highly reproducible, and side reactions are extremely rare. Pentagastrin is commercially available (e.g., Peptavlon, Ayerst Laboratories, Division of American Home Products Corp., New York, NY 10017) and is currently the *stimulus of choice.*

Insulin. Hypoglycemia in response to insulin administration stimulates the vagal medullary nuclei and causes increased gastrin release and HCl secretion. The dose of insulin administered is chosen to decrease blood glucose to <45 mg/dL (see *Hollander Test,* below).

ANALYSIS OF GASTRIC RESIDUE

Preparation of the patient. The patient should be in the fasting state (no food or liquids for 12 h). The patient is not permitted to smoke the morning of or during the test and should avoid any form of exercise. Gastric residue is collected by oral or nasal intubation under fluoroscopic guidance.

Screening test for free acid. Measure the pH of aspirated gastric juice or place a drop of Toepfer's reagent into the gastric residue. Do not use pH indicator paper. A pH below 3.0 or a red color after the addition of Toepfer's reagent indicates the presence of free acid. A pH above 3.0 or a yellow color of Toepfer's reagent indicates absence of free acid. For some purposes (e.g., exclusion of pernicious anemia), the test may be discontinued at this point, as presence of HCl excludes this diagnosis. If no HCl is found, a gastric stimulus is administered (see *Stimuli of Gastric Secretion*, above).

Determination of Free Hydrochloric Acid in Gastric Residue

Principle. A known volume of gastric residue is titrated with NaOH, 0.10 mol/L, to a pH of 3.5 by using a pH meter or until Toepfer's reagent (0.5 g diethylaminoazobenzene/100 mL ethanol) added to gastric juice takes on a salmon color. The concentration of free HCl is then calculated:

$$\text{Free HCl, mmol/L} = \frac{\text{mL NaOH} \times \text{concentration of NaOH, mol/L} \times 1000}{\text{Gastric specimen titrated, mL}}$$

Comments. Since HCl is the only strongly ionized acid in gastric content, titration to a pH of 3–3.5 detects essentially all free hydrochloric acid. Titration beyond a pH of 3.5 will overestimate the HCl concentration to varying degrees, depending on the composition of the gastric residue. On the other hand, titration to pH 3.5 may underestimate the amount of free H^+ secreted if some of these ions are bound to or have reacted with other constituents of gastric content. Thus, no fully satisfactory procedure is available to measure accurately the true total amount of free acid secreted by the gastric mucosa.

The appearance of gastric content and its volume should be noted on the report since it may suggest performance of additional tests, such as a test for blood (e.g., Gastroccult test) in case of brownish gastric content, a test for bile (e.g., Ictotest) if the gastric content appears yellowish or green, or a test for lactic acid if a foul smell is noted.

Determination of Total Titratable Acidity in Gastric Residue

For the determination of total acidity in gastric residue (or other gastric specimens such as basal secretions or secretions after stimulation) the specimen is titrated to a pH of 7.0 by using a pH meter or phenol red as indicator. Calculation of total acidity is done in the same way as the calculation of free hydrochloric acid, shown previously. The ***reference range*** for total acidity of gastric residue (without stimulation) is 10–60 mmol/L.

Determination of Basal Acid Output (BAO) and Peak Acid Output (PAO) of Gastric Acid[9]

Determination of the Basal Secretion Rate of Gastric HCl

A basal condition, in the context of gastric analysis, is one in which the patient is at complete rest and not exposed to any visual, auditory, or olfactory stimuli. Since such a condition is maintained during sleep, many clinicians prefer a 12-h nocturnal collection for the determination of the basal acid output. However, this causes the patient significant discomfort. A reasonably satisfactory alternative is the collection of a specimen for a shorter period of time (e.g., 30 min) after the patient has had a satisfactory night's sleep in a quiet private room.

Twelve-hour nocturnal secretion rate. During the day of the test, withhold after 0800 h medications that influence gastric secretion (antacids, anticholinergic or adrenergic drugs, reserpine, corticosteroids). Give the patient a clear liquid evening meal the day before the test; thereafter allow no food or drink until completion of the collection. Intubate, aspirate, and discard the gastric residue at ~2000 h. Collect all gastric secretions during the subsequent 12 h by using a continuous-operating pump. At the end of 12 h, immediately send the specimen to the laboratory for analysis.

Procedure

Measure and record the volume (mL) and titrate the amount of free acid in the specimen (see page 772). Then calculate the mmol of free acid secreted per hour.

$$\text{Free acid, mmol/h} = \frac{\text{mmol free acid/L} \times \text{total volume of specimen, mL}}{1000} \times \frac{60}{\text{Collection period of specimen, min}}$$

Thirty-minute basal secretion rate test. On the day preceding the test, prepare the patient as described for the 12-h test but do not intubate. Awaken the patient in the early morning of the test day and then intubate; aspirate and discard gastric residue. Then collect gastric fluid for the 30-min period. Request the patient to expectorate saliva during the collection period. The specimen is analyzed for free HCl as described above and the result is reported in mmol of free acid/h.

Peak Acid Output (PAO) and Maximum Acid Output (MAO)

After completing the basal specimen collection (12 h or 30 min), leave the tube in place and give the patient a gastric stimulant (e.g., pentagastrin). Collect specimens during the next four 15-min intervals into separate, time-labeled containers for the calculation of the MAO. Collection of six 15-min specimens is required for determining the PAO, since maximal stimulation in some individuals may start after 30 min and continue beyond 1 h. Determine the amount of free HCl in each specimen by titration as described above.

Calculation

Maximum acid output. Calculate the free acid output (mmol/h) in each of the first four 15-min post-stimulation specimens, by using the formula above. Average the acid output values for all four post-stimulation specimens. This gives the MAO/h.

Peak acid output. Calculate the acid output per hour for all six 15-min post-stimulation specimens. Average the results of the two specimens with the highest acid output. This value is used to calculate the peak acid output (PAO/h) by the formula for free acid given above.

BAO/PAO ratio. Additional diagnostic information may be obtained by calculating the ratio of basal acid output to peak acid output or the per cent of BAO relative to PAO as follows:

$$\frac{\text{BAO}}{\text{PAO}} \times 100 = \%\text{BAO/PAO}$$

Interpretation and Reference Ranges[54]

Gastric residue **(after 12-h fast)**

Volume:	20–100 mL (generally below 50 mL)
pH:	1.5–3.5
Free acid (without stimulation):	0–40 mmol/L
Free acid (after stimulation with pentagastrin):	10–130 mmol/L

12-h Gastric secretion

Volume:	150–1000 mL
pH:	1.5–3.5

Gastric acid secretion rate

1. *Basal Acid Output (BAO)*	
Normal; gastric ulcer:	0–5 mmol/h
Possible duodenal ulcer:	5–15 mmol/h
Zollinger-Ellison (Z-E) syndrome:	>20 mmol/h (at times >60 mmol/h)
2. *Peak Acid Output (PAO)*	
Normal; gastric ulcer:	5–20 mmol/h
High normal; duodenal ulcer; possible Z-E syndrome:	20–60 mmol/h
Z-E syndrome:	>60 mmol/h (generally not more than twice basal acid output)

3. *BAO/PAO*	%	Fraction
Normal; gastric ulcer or gastric carcinoma:	<20	<0.20
Gastric or duodenal ulcer:	20–40	0.20–0.40
Possible duodenal ulcer or Z-E syndrome:	40–60	0.40–0.60
Z-E syndrome:	>60	>0.60

Note: Data related to the acid output are not diagnostic by themselves; due to wide overlap, borderline values are of little diagnostic significance.

Confirmation of Complete Vagotomy (Hollander Test)[11,20,34]

Surgery to deal with peptic ulcer usually involves section or division of the vagus nerve (vagotomy), which innervates gastrin- and HCl-releasing cells in the stomach. Completeness of vagotomy is not always anatomically apparent. Hypoglycemia of <45 mg/dL, produced by insulin administration, strongly stimulates gastric HCl and pepsinogen secretion. Such a response to hypoglycemia is not observed in *completely* vagotomized subjects. The insulin (Hollander) test is therefore performed to confirm the completeness of the nerve division (vagotomy).

The optimal time for performance of the Hollander test is between three and six months after the surgical procedure. The test should not be performed in the first two weeks immediately after surgery.

Procedure

Patient preparation and placement of the gastric tube are the same as outlined under *Analysis of Gastric Residue* (page 772).

1. Aspirate the gastric residue completely and discard the specimen.
2. Collect a 30-min basal gastric secretion for laboratory analysis.
3. Draw blood specimens for glucose determination before and 30, 60, and 90 min after insulin injection.
4. Administer insulin intravenously immediately after the collection of the basal gastric sample. (The insulin must be administered by a physician and the patient must be continually observed for adverse reactions.)
5. Collect eight 15-min post-insulin gastric specimens (total 120 min) for laboratory analysis.
6. Measure free acid output per hour for each sample (basal and post-insulin).

Interpretation. Vagotomy is considered incomplete if (1) acid concentration increases >20 mmol/L (or >10 mmol/L if basal secretion was not acid), or (2) total acid output is >2 mmol/h. Test results are invalid if serum glucose has not fallen below 45 mg/dL.

Detection of Lactic Acid in Gastric Contents

Lactic, butyric, and acetic acids in gastric contents are most likely the products of bacterial fermentation of food stagnated in the stomach in the absence of free hydrochloric acid. Testing for lactic acid has relatively little clinical value, but it is occasionally requested because the presence of lactic acid, together with gastric retention and hypochlorhydria, is a common finding in carcinoma of the stomach.

Principle. A portion of gastric contents is extracted with ether. An aliquot of this ether extract is treated with ferric chloride, 10 g/dL, which gives a slight yellow-greenish color with low concentrations (50–100 mg/dL) of lactic acid and an intense yellow-green color with high concentrations (>100 mg/dL). Lack of color indicates absence of lactic acid.

Tests Measuring the Exocrine Function of the Pancreas and Intestinal Absorption

The exocrine function of the pancreas is the production and secretion of pancreatic juice that is rich in enzymes and bicarbonate. Normal pancreatic juice is colorless and odorless; it has a pH of 8.0–8.3 and a specific gravity of 1.007–1.042. The total 24-h secretion volume is 800–3000 mL.

A number of laboratory tests are available to measure exocrine functions either directly or indirectly. These tests should be viewed as measurements of pancreatic function or intestinal absorption and not as tests designed to diagnose a specific disorder. The interpretation of results must be tempered by the realization that great overlap occurs between results observed in normal individuals and those found in patients with pancreatic disorders. Furthermore, clinical signs of pancreatic insufficiency often do not appear until loss of 90% of acinar tissue has occurred.[4]

The concentration and total output of pancreatic juice, bicarbonate, and enzymes can be measured directly on duodenal contents or pancreatic juice obtained by cannulation. Measurement of enzymes, especially chymotrypsin, in feces is simpler and noninvasive and is preferred by some clinicians.

Measurements of the total volume of pancreatic juice, amount or concentration of bicarbonate, or rates of enzyme secretion are generally preceded by stimulation of the pancreas by one of several test meals or stimulants. Examples are the *Lundh meal* and injection of either or both secretin or cholecystokinin. Stimulation is followed by collection of duodenal contents and by the determination of HCO_3^-, trypsin, amylase, lipase, or a combination of these.

Trypsin and Chymotrypsin in Stool

Trypsin and chymotrypsin secreted by the pancreas and present in duodenal contents mix with food material and, except for the fraction digested in the intestinal tract, are excreted in the feces. Measurement of the activity of these enzymes in stool has been used for the diagnosis of pancreatic insufficiency, as, for example, in cystic fibrosis.

For many years, a *screening test for trypsin* has been used in which a drop of a dilution of emulsified stool is placed on an X-ray film. This procedure is described in Chapter 12. Digestion of the gelatin layer of the film by an appropriate dilution of stool suggests presence of normal amounts of tryptic activity. Although very fast and simple, the test is at best only a screening procedure. It is subject to interferences from proteolytic enzymes present in intestinal bacteria or in contaminating urine. The test is applicable only to children, since lower intestinal motility in adults results in partial or total autodigestion of the enzyme and therefore falsely low or negative values. Several synthetic low-molecular-weight substrates are now available for estimating trypsin or chymotrypsin activity.

Of these two enzymes, chymotrypsin is significantly more stable,[18,48] and its output in stool correlates well with chymotrypsin secretion. In addition, the chymotrypsin test has a much higher sensitivity. The incidence of falsely low results (falsely positive for diminished pancreatic function) is ~10%. Falsely normal results in patients with mild pancreatic insufficiency may be as high as 25%.[41]

A recently introduced procedure for chymotrypsin is simple and fast and eliminates inconsistency of results caused by particle-bound enzyme.[32] In this procedure, a fecal extract is mixed with a synthetic pentapeptide conjugated with 4-nitroaniline and is hydrolyzed by chymotrypsin to produce free yellow 4-nitroaniline, which can be measured spectrophotometrically at 405 nm.[32] The reagent kit is presently available only in Europe.

Bz-Ty-PABA Test (PABA Test, BTP Test) for Chymotrypsin

This test for pancreatic chymotrypsin uses the synthetic peptide *N*-benzoyl-L-tyrosyl-*p*-aminobenzoic acid (BTP), which is administered orally together with a test meal to stimulate pancreatic secretion. BTP is specifically hydrolyzed by chymotrypsin in the duodenum to release *p*-aminobenzoic acid (PABA). PABA is subsequently absorbed in the intestinal tract and metabolized in the liver to hippurate and to PABA glucuronide and PABA acetylate. These arylamines are then excreted by the kidneys into the urine, where they can be measured spectrophotometrically. Low chymotrypsin levels, as found in pancreatic insufficiency, result in less chromogen excretion in the urine.

The peptide substrate BTP is available commercially (Adria Laboratories, Inc., Columbus, OH 43215) in kit form, and the method is described in detail in the kit insert. The test is affected by various medications, especially antibiotics, sulfonamides, and diuretics. Foods such as prunes and cranberries contain hippurate precursors that interfere with the chemical determination of aromatic amines. Test results are also affected by the rate of intestinal absorption and the degree of renal excretion.

The sensitivity of the test in the detection of chronic pancreatitis is ~86%. The specificity of the test (i.e., the number of true normal results in patients with nonpancreatic disease) is 93%. There is a high correlation between the results of this test and those of the direct test for chymotrypsin in stool.

Indirect Tests of Pancreatic Function

This group of tests is based on the fact that proper pancreatic function is essential for normal intestinal absorption. Since the pancreas is the major source of amylase, lipase, and proteolytic enzymes, significant decreases in pancreatic functions cause decreased digestion of starch, fats, and protein and a parallel increase in excretion of these food materials in stool. In pancreatic deficiency, microscopic examination of stool demonstrates large amounts of undigested cell nuclei and meat fibers (creatorrhea), increased fat (steatorrhea), and increased starch (amylorrhea). Demonstration of these conditions is suggestive of impaired absorption.

FECAL LIPIDS

Fecal lipids are composed of mono-, di-, and triacylglycerols; fatty acids and fatty acid salts (soaps); phospholipids and glycolipids; and sterols and cholesteryl esters. Most fecal lipids are nonabsorbed dietary fats, but a significant amount (~2 g) is derived from intestinal bacteria and from epithelial cells that have sloughed normally from the intestinal mucosa. The quantitation of fecal fat is the most definitive test for determining the presence of steatorrhea (increased presence of lipids) but not for determining its cause.

Unfortunately, fecal fat and other indirect tests for pancreatic function have no diagnostic value in mild cases of pancreatic disease or in those cases in which the acute phase has subsided. On the other hand, stools of patients with celiac disease or sprue may be similar to those from patients with pancreatic insufficiency because increased intestinal motility decreases the time that food remains in the intestinal tract where digestive enzymes can act.

Determination of Total Fecal Lipids[7,31]

Principle. In the *gravimetric* method for the determination of total fecal lipids, a preweighed, emulsified stool specimen is acidified to decrease the ionization of fatty acids. The lipids are then extracted from the stool specimen into an organic solvent, the solvent is evaporated, and the residue is quantitated by gravimetry.

In the *titrimetric* method, a weighed stool specimen is subjected to alkaline hydrolysis to convert esterified lipids to fatty acids and then acidified to facilitate the transfer of fatty acids into an organic solvent. The solvent is evaporated, and the residue is dissolved in ethanol and titrated with a standard solution of sodium hydroxide. The result represents ~80% of total fecal lipids.

Specimen. The patient should be on a diet containing 50–150 g of fat per day for 3 d prior to this study and during the 72-h collection of the stool specimen. The patient should not have had castor oil or other laxatives prior to the specimen collection.

Stool specimens are conveniently collected in a one-gallon paint can, and the container should be refrigerated during the collection period. Collection of 24- or 48-h specimens is not recommended, since results on such specimens are subject to more variability. Any obvious foreign matter should be removed from the specimen prior to processing. Contamination of feces with urine should be avoided.

Reference range for children up to six years is <2 g/d; adults, 2–6 g/d. Values >7 g/d indicate steatorrhea.

Serum Carotene, Vitamin A, and Vitamin A Tolerance Tests

These tests are based on the fact that carotene and vitamin A are fat soluble and thus can be absorbed to a significant extent only if fats are first hydrolyzed by pancreatic lipase. See also Chapter 16, *Vitamins.*

$^{14}CO_2$ Breath Test[6,47]

In order to avoid the difficulties in obtaining 72-h stool collections for fecal fat, some investigators have administered orally triglycerides containing ^{14}C-fatty acids to determine the degree to which fat was absorbed. The test is based on the assumption that the oxidation of ^{14}C-fatty acids to $^{14}CO_2$ is proportional to the rate of absorption of fatty acids. An extension of this test is the comparison of results obtained after administration of labeled free fatty acids and labeled triglycerides. It was hoped that this approach would distinguish between disorders of fat digestion and fat absorption.

This test appears to be an easy approach to establishing fat malabsorption; however, the necessity of performing this test over a 6-h period and the need to administer isotopes are clear disadvantages of this procedure and may be responsible for its limited use. In addition, there is uncertainty about the effect of a variety of disorders of fat metabolism, such as hyperlipemia or obesity, on test results.[47]

D-Xylose Absorption Test[1]

Xylose is a pentose not normally present in significant amounts in blood. Pancreatic digestive enzymes are not needed for its absorption. When given orally, ~60% is passively absorbed in the proximal small (duodenojejunal) intestine, and most is subsequently excreted by the kidneys. The amount of xylose recovered in the urine or blood, in a specified time interval after administration of a measured dose, is used to evaluate mucosal absorption ability. Low absorption of xylose is observed in intestinal malabsorption. However, low values are also observed in other intestinal and gastric disorders as well as in circulatory, renal, and other diseases. In addition, excretion decreases with age as a reflection of decreased kidney function. In malabsorption due to pancreatic insufficiency, the absorption of xylose will be essentially normal provided that there is no significant increase in intestinal motility; on the other hand, >80% of patients with jejunal malabsorption show low values. Therefore, the test is of some help in distinguishing between these two types of malabsorption and in evaluating the response to therapy.

Principle. A 25-g dose of xylose in water (or 5 g if so specified by the physician) is given to a fasting patient, and all of the urine voided over the next 5 h is collected. An anticoagulated blood specimen is taken ~2 h after the xylose is given (in children, after 1 h). The xylose concentrations in the 5-h urine specimen and in the blood specimen are determined by treating the diluted urine and a protein-free filtrate of the blood with *p*-bromoaniline in an acid medium. When heated with acid, xylose is dehydrated to furfural, which in turn reacts with *p*-bromoaniline. A pink color (A_{max} = 520 nm) is produced in this reaction. Thiourea minimizes formation of interfering colored compounds. The concentrations of xylose in blood and urine are calculated on the basis of absorbance readings obtained with a standard xylose solution.

Reference Ranges

	Plasma, mg/dL	Urine
Child, 0.5-g/lb dose (max. 25 g):	>30	16–33% of ingested dose/5h
Adult, 5-g dose:	>20	>1.2 g/5 h
Adults, 25-g dose:	>25	>4.0 g/5 h

Some investigators feel that a 2-h urine collection is more sensitive. The reference range for a 2-h collection is 14.2–29% of dose.[1,33]

Comments on Method and Sources of Error

The accuracy of the procedure depends not only on the rate of absorption but also on the rate of excretion of xylose by the kidneys. Thus, patients with renal insufficiency will excrete a decreased amount of xylose. In order to eliminate misinterpretations as a result of renal retention, a blood determination of xylose is carried out along with the determination of xylose in urine. A normal post-challenge blood xylose level in the presence of decreased urine xylose excretion suggests renal retention. False-positive (low) results on plasma may also occur in the presence of vomiting, gastric stasis, dehydration, myxedema, massive ascites, edema, or massive bacterial overgrowth of the intestines.

Haeney and colleagues[28] performed a study on 216 subjects with ages ranging from 14 to 92 and found that administration of 5 g of D-xylose, followed by measurement of blood xylose concentrations 1 h after oral administration and after correction of results for body surface, shows maximum discrimination between normal and abnormal subjects. The blood test circumvents many of the conditions that lead to low urinary excretion, and values are not affected by age.

Several drugs[60] cause impaired absorption of xylose and can lead to false-positive (low) results. Failure to fast can have the same effect. The presence of glucose, galactose, or phenazopyridine in the urine can cause false-negative results due to chemical interference.

The Sweat Test

The determination of the chloride concentration of sweat is considered the most reliable single test in the diagnosis of cyctic fibrosis. Sodium and chloride elevations in sweat are seen even in the absence of gastrointestinal or respiratory symptoms and when pancreatic insufficiency cannot be demonstrated. In most affected infants, the test becomes positive between 3 and 5 weeks of age. Only 1–2% of cystic fibrosis patients have sweat chloride values <60 mmol/L and only 1 in 1000 has a value <50 mmol/L.[17] Some laboratories also measure the sodium concentration of the sweat sample.

Collection of Sweat by Iontophoresis

1. Use forceps to place two prewashed 2-inch-square gauzes (or two Whatman No. 42 filter papers, diameter 5.5 cm) into a weighing bottle, close with a stopper, and determine the combined weight accurately, using an analytical balance. Handle the weighing bottle with tissue or gauze to avoid direct contact with fingers.

2. Wash and dry the anterior (inside) surface of the patient's forearm and place two-inch-square gauzes (not the preweighed gauze) over the washed area. Moisten the gauzes well with a freshly prepared solution of pilocarpine nitrate, 4 mg/dL. Place two gauzes saturated with saline on the posterior (outside) surface of the arm. The positive electrode of a direct-current power supply designed for iontophoresis is placed over the gauze with pilocarpine, and the negative electrode is placed on the saline-soaked gauze. Ensure good contact and secure the electrodes to the arm. If the arm is too small, as in infants, use the thigh or the interscapular area of the back of the patient.

3. Apply a current of 0.16 mA/cm^2 for 5 min.[23] The current will tend to increase during this time interval but should be adjusted to stay at the stated setting. After 5 min take off the electrodes, clean the skin area well with distilled water, and dry the area. If the patient complains about discomfort, discontinue the test. A tickling sensation and a reddish skin color at the site of the electrode are common findings and should be disregarded. Complete kits for iontophoretic stimulation (either battery-operated or AC-operated) are available commercially from Farrall Instrument Company, Grand Island, NE 68801.

4. Use forceps to place the preweighed gauze or filter paper over the skin area that was exposed to pilocarpine. Place a plastic sheet (~4 inch square) over this area and seal airtight with surgical tape. Allow sweat to accumulate on the gauze or filter paper. This takes ~20–30 min, but the time of sweat collection may be extended as long as necessary. In general, the appearance of droplets on the plastic sheet indicates that enough sweat has accumulated. These droplets must be included in the collection.

5. Remove the gauze or filter paper with forceps, place it immediately into the weighing bottle, and close with a stopper.

Recovery of Sweat

Accurately (within 1 mg) determine the weight of the bottle or Petri dish containing the sweat-soaked gauze or filter paper and calculate the amount of sweat by difference. One gram of sweat is assumed to be 1 mL of sweat. The minimum amount of sweat required for reliable quantitation is 100 mg. Very low sweat rates may lead to falsely low chloride results in cystic fibrosis patients.

Determination of Sweat Chloride

Sweat is eluted from the gauze into a measured volume of deionized water. The amount of chloride in the sweat sample is then measured by standard laboratory techniques (e.g., chloridometry). The concentration of chloride (mmol/L) is calculated from the chloride content and volume of the sweat sample. The sodium concentration may also be measured.

Reference Ranges

	Healthy Children	Cystic Fibrosis Homozygotes
Na^+, mmol/L:	10–40	70–190
Cl^-, mmol/L:	5–35	60–160

Mean values of Na^+ and Cl^- for normal children are 27 mmol/L; in normal adults means are slightly higher: Na^+, 33; Cl^-, 39. Chloride values >70 mmol/L are suggestive of cystic fibrosis. In adults with cystic fibrosis, mean values are Na^+, 110 and Cl^-, 118.

Sources of Error

False-high results can be obtained if the skin is not cleaned properly or if contaminated sponges are used. A blank determination using clean gauze and deionized water should be performed to assure that the gauze is low in sodium and chloride content.

False-normal results can be obtained in cystic fibrosis patients with salt depletion, which is common in these patients during hot weather. Additional electrolyte studies on serum help to avoid misinterpretation of results in such instances. Inconsistently elevated levels of Na^+, Cl^+, or K^+ in sweat have also been reported in several other disorders, but most of these conditions are rare and clinically distinct from cystic fibrosis.[50]

The diagnosis of cystic fibrosis is very critical and difficult. It is therefore of utmost importance that the laboratory either follow a procedure that conforms to the guidelines given by the Cystic Fibrosis Foundation[23] or refer the patient to a designated center supported by the Cystic Fibrosis Foundation.

Gastric, Pancreatic, and Intestinal Diseases

Peptic ulcer disease. This is the collective name given to chronic duodenal ulcer and chronic gastric (benign) ulcer. Although both of these diseases have many features in common, they differ with respect to incidence, pathogenesis, medical history, outcome, and management.

Chronic duodenal ulcer. A duodenal ulcer is a round, sharply punched-out defect in the mucosa of the duodenum or in the distal antrum and pyloric channel of the stomach. In time, chronic duodenal ulcers heal by the formation of fibrous scar tissue and the inward growth of the duodenal mucosal epithelium. Apart from the morbidity due to the ulcer itself, duodenal ulcer has four important, although uncommon, complications: *hemorrhage* due to the ulcer's eroding a blood vessel, *perforation* through the anterior wall of the duodenal cap, causing peritonitis; *penetration* of the ulcer into adjacent structures such as pancreas, biliary tract, liver, colon, abdominal wall, and even lung; and *luminal obstruction* (pyloric stenosis) caused by the gradual contraction of the ulcer scar.

Diagnosis is made primarily by radiologic techniques or endoscopy. Serum levels of pepsinogen group I may be of value since they reflect the gastric chief cell mass and correlate with the peak acid output (PAO). One-third to one-half of patients with duodenal ulcer have PAO of about 30 mmol/h[5] (see also page 774). The most important diseases to exclude during the diagnostic process are gastric ulcer and gastric carcinoma.

Chronic benign gastric ulcer. Gastric ulcers occur anywhere in the stomach, although most are found along the lesser curvature. Some patients with gastric ulcer have a coexistent duodenal ulcer. Gastric ulcers heal with the formation of fibrous tissue and the inward growth of the gastric epithelium. As with duodenal ulcers, gastric ulcers may cause hemorrhage or may perforate. Chronic benign ulcers must be differentiated from acute gastric erosion and ulceration (see *Gastritis*, below).

Diagnosis. It is important to establish that the ulcer is indeed benign; radiological characteristics are important. The PAO of gastric ulcer disease is not significantly different from that of healthy controls[5] (see also page 774).

Zollinger-Ellison syndrome. The pathophysiology of the Zollinger-Ellison syndrome has already been described. *Gastrinomas* (gastrin-secreting tumors) are most frequently found in the head or tail of the pancreas. About half of all gastrinomas are multiple, about two thirds are malignant, and one-fourth are part of the multiple endocrine neoplasm syndrome Type I (MEN I), with associated tumors in pancreatic islets and in parathyroid and pituitary glands. Diarrhea often precedes ulcer symptoms and occurs in about one-third of all patients with the syndrome. Steatorrhea also occurs due to the lowering of the small intestinal pH, but it is less common than diarrhea. Lipase is inactivated at low pH, bile salts are rendered insoluble, and vitamin B_{12} is not absorbed by the distal ileum; gastric secretion of intrinsic factor is unaffected.

Diagnosis. Gastric acid hypersecretion is a common finding; however, there is a substantial overlap in gastric acid production with duodenal ulcer disease. Serum gastrin levels are usually markedly increased in the Z-E syndrome but sometimes only marginal elevations are found; additional testing such as the secretin challenge is then necessary. The 12-h overnight secretion of HCl and the basal HCl secretion are >100 mmol/L (see also pages 773–774).

Gastritis[57] is a term used to denote mucosal inflammation. It can be classified as erosive, nonerosive, or specific; specific gastritis is very rare.

Erosive gastritis (synonyms: stress lesions, stress erosions, or acute gastritis) occurs in four well-defined clinical situations:[45] following severe trauma; following severe burns (Curling's ulcer); in patients with intracranial disease (Cushing's ulcer) and following craniotomy or traumatic head injuries; and in patients who chronically ingest drugs such as corticosteroids, aspirin, or ethanol. Endoscopy is usually the definitive technique to establish the diagnosis.

Nonerosive gastritis (synonym: chronic gastritis) is associated with gastric ulcer disease or gastric carcinoma, following partial gastrectomy, in pernicious anemia, and in the healthy elderly person. Serum gastrin levels are elevated in achlorhydric patients because there is no negative feedback by HCl.

Gastric cancer (adenocarcinoma) arises from the mucus cells of the gastric mucosa. It is the third most common gastrointestinal cancer in the United States. *Diagnosis*[5] is made by biopsy. The majority of patients have fasting achlorhydria, but less than a quarter of all cases have achlorhydria following maximal stimulation. The volume of the gastric residue may be increased, and in some cases blood may be found in the stomach. Lactate is often present.

Postgastrectomy syndrome.[35] Annually, more than 50 000 Americans are treated for peptic ulcer disease by gastric resection or denervation (vagotomy) of the stomach. Of these, about 5000 will suffer some functional incapacity to which the name *postgastrectomy dysfunction* or *syndrome* has been given. This term includes the dumping syndrome, diarrhea, weight loss, anemia, and bone disease. It is caused by decreased absorption of vital nutrients.

The dumping syndrome occurs because the stomach, after gastrectomy, empties more rapidly than normal. The patient experiences abdominal pain and diarrhea together with vagomotor phenomena such as flushing, lightheadedness, and cardiac palpitations 30–60 min after eating. Because the proximal jejunum contains a hypertonic solution of partially digested food, rapid osmotic equilibration with blood results in a slight but rapid decrease in the blood volume, with concomitant intestinal distention. Between 90 and 180 min after a meal, cardiac palpitations, lassitude, mental confusion, and even loss of consciousness are experienced and may be due to reactive hypoglycemia. Diarrhea may occur 1–2 h after eating as a result of rapid filling of the jejunum after meals.

Weight loss and specific nutritional deficiencies are not uncommon after gastric surgery. Anemia develops as a result of decreased absorption of iron, vitamin B_{12}, and folate. Maldigestion may occur due to rapid gastric emptying (resulting in less gastric dispersion of solid food and reduced digestion time), small intestinal "hurry," and reduced pancreatic secretion. There is also asynchrony between the arrival of pancreatic and biliary secretions and the entry of food into the small intestine.

An increased prevalence of metabolic bone disease occurs following ulcer surgery. A common defect is osteoporosis, which is characterized by bone atrophy. Serum calcium and serum alkaline phosphatase activities are generally normal. Osteomalacia, failure to mineralize normal bone matrix, is also present in about half of the patients with osteoporosis. In osteomalacia, the skeleton is susceptible to stress; serum calcium values are low but alkaline phosphatase activity is increased. It is estimated that 5–15% of patients who have had ulcer surgery develop metabolic bone disease.

Pancreatic and Intestinal Diseases

The pancreas is inaccessible to direct examination, and thus disease processes may be far advanced before pancreatic disease becomes manifest.

Acute pancreatitis. Acute pancreatitis is the name given to a discrete, sudden episode of more or less diffuse enzymatic destruction of pancreatic tissue as a result of the escape of active pancreatic enzymes into the glandular parenchyma. It is distinguished from chronic pancreatitis in that, upon elimination of the causative factor, normal endocrine and exocrine functions are restored and there is no permanent functional damage.[52] *Relapsing pancreatitis* is a term used to characterize recurrent occasions of either acute or chronic pancreatitis. The milder form of acute pancreatitis is edematous; the more severe form is hemorrhagic. Up to 20% of alcoholic patients with pancreatitis have hypertriglyceridemia (Type V pattern). A similar association occurs in some patients with idiopathic pancreatitis. Patients with Type I, IV, or V hyperlipoproteinemias frequently develop acute pancreatitis that often progresses to chronic pancreatitis. The morphology of the pancreas in acute pancreatitis indicates that proteolysis, lipolysis, and hemorrhage are caused by the pancreatic enzymes.

Alcohol-induced changes have an important role in the genesis of acute pancreatitis. Alcohol increases pancreatic secretion; it may also cause partial ductal obstruction or may cause hyperlipemia in alcoholics after a fatty meal.

Diagnosis.[37,52] The diagnosis of acute pancreatitis is principally a diagnosis of exclusion involving both radiologic and laboratory procedures. The laboratory is of value in providing a number of chemical and hematologic assays, as later shown in Table 23-3. The enzyme findings in acute pancreatitis are discussed in Chapter 12. Simultaneous determinations of serum amylase and lipase are recommended, although the severity of the attack does not correlate with the degree of elevation of serum enzyme activity. The four most valuable indicators of the severity of pancreatitis are serum albumin <3.0 g/dL, hematocrit <30%, serum creatinine >2 mg/dL, and serum calcium ≤8 mg/dL.[37]

Chronic pancreatitis. Chronic pancreatitis is defined as progressive functional damage to the pancreas that persists even when the causative factor is eliminated.[27] Patients with chronic pancreatitis may suffer episodes of exacerbation indistinguishable from acute pancreatitis. About half of all patients with chronic pancreatitis suffer attacks of acute pancreatitis. *Etiology and pathogenesis*[27] are largely unknown, but associated factors that may be causative are well recognized.

Ethanol abuse is clearly associated with chronic pancreatitis. It is essential, in view of the close association of chronic pancreatitis with alcoholism, to check liver and biliary function by determination of serum bilirubin and alkaline phosphatase levels. Pancreatic exocrine function should be examined by stimulation tests and fecal fat estimations. Radiological investigations and other techniques are of considerable value in confirming suspected chronic pancreatitis. Serum and urine pancreatic enzyme levels are of limited assistance because the acinar cell mass may be markedly decreased or mild inflammation may not increase serum enzyme activities.

Tumors of the pancreas. *Carcinoma of the pancreas* is the major tumor in this organ. *Adenocarcinoma of pancreatic ductal epithelium*[15] is a common malignancy. It develops insidiously, progresses relentlessly, and is invariably fatal.[15] There are no characteristic clinical features that lead to the diagnosis. About half of all patients will have jaundice due to a tumor in the head of the pancreas that obstructs or encircles the distal common bile duct. Radiological tests are used to investigate the possible presence of pancreatic neoplasms.

A wide range of *laboratory tests* for carcinoma of the pancreas have been proposed, with widely varying sensitivities and specificities. These include the serum markers galactosyltransferase isoenzyme II,[43] carcinoembryonic antigen, α_1-fetoprotein, ferritin, and ribonuclease. In some patients results of conventional tests for serum amylase, lipase, trypsin, or trypsin/creatinine clearance may be elevated. At present, reliance must be placed on clinical suspicion and on invasive imaging techniques.

Tumors of the pancreatic islet tissues[59] include insulinoma, glucagonoma, gastrinoma, VIP-secreting tumors, somatostatinoma, and multiple endocrine neoplasia (MEN) syndromes.

Insulinoma.[59] These B cell islet tumors of the pancreas cause variable degrees and symptoms of hypoglycemia, especially after an overnight fast. The *diagnosis* is normally established by finding a low serum glucose (male <55, female <35 mg/dL). The diagnosis can be confirmed by finding an inappropriately elevated plasma insulin level of >15 mU/L and a plasma proinsulin level >40 fmol/L. However, these findings may be mimicked by the surreptitious use of oral sulfonylureas.

Glucagonoma. The A cell islet tumors of the pancreas are mostly malignant and demonstrate progressive metastatic growth. The *diagnosis* is established by the appearance of clinical symptoms and by the finding of a plasma glucagon level exceeding 500 pg/mL.

Somatostatinoma. High circulating levels of somatostatin are found in patients with this tumor, but somatostatin may also be secreted ectopically by a variety of neoplasms. The pancreatic tumor may also secrete adrenocorticotropic hormone and calcitonin. A mild diabetic syndrome is usually present. Gallbladder contraction is inhibited, which, together with inhibition of pancreatic juice secretion, can lead to diarrhea and steatorrhea. Delayed gastric emptying and inhibition of acid secretion produce dyspepsia and achlorhydria.

Multiple endocrine neoplasia (MEN) syndrome. A number of syndromes are associated with hyperplasia or tumors in two or more endocrine organs (see also Chapter 21). These have been described as the multiple endocrine neoplasia syndromes. One of these syndromes, MEN I (Wermer's syndrome), consists of a highly variable combination of hyperplasia or tumors of the non-B islet cells of the pancreas, the parathyroids, and the anterior pituitary, with variable degrees of hyperfunction of these systems. The tumors may also be "nonfunctioning" in the sense that insufficient hormone is secreted to produce symptoms. The most important relationship between pancreatic tumor and MEN I is in gastrinoma, where up to a quarter of all gastrinoma patients have MEN I.

Cystic fibrosis.[25] This disease is the most common lethal genetic defect of Caucasian populations. It is inherited in an autosomal recessive fashion, and its incidence in Caucasian populations ranges from 1:1900 to 1:15 000. These data suggest a heterozygote carrier incidence of about 1:20. Unfortunately, at present heterozygote carriers cannot be identified. Cystic fibrosis is a major cause of malabsorption in infants and children and of chronic pulmonary disease in childhood. All cases of cystic fibrosis will eventually develop chronic pulmonary disease, and, indeed, the respiratory disorder is responsible for much of the morbidity and nearly all of the mortality associated with this disease. Cystic fibrosis is a systemic disease affecting many types of exocrine glands. It is associated with a unique sweat gland defect, with abnormalities of mucus secretion, and with abnormalities of the male genital tract. The nature of the primary defect is not known.

The disease should be suspected in a child with chronic or recurrent respiratory tract infections, but there are many other presentations, among them pancreatic exocrine and endocrine dysfunctions and functional and anatomic intestinal abnormalities.

Nearly 90% of patients with cystic fibrosis present with exocrine pancreatic dysfunction caused by obstruction of the pancreatic small ducts. Abnormal electrolyte secretion is a characteristic and constant finding in cystic fibrosis. The sodium and chloride and less often the potassium content of sweat are elevated, and stimulated pancreatic juice shows an abnormal electrolyte pattern. The defect appears to be in the reabsorption of sodium and chloride in the sweat ducts.

The *diagnosis*[10,30,56] rests primarily upon the findings of the sweat test, although these results must be integrated with the clinical findings and family history.

Diarrhea[8,55] is defined as an increase in fecal water excretion. Normally, the colon absorbs more than 90% of the fluid entering it from the small gut. A fluid load exceeding the reabsorptive capacity of the colon results in diarrhea.

There are four main types of diarrhea.

1. *Secretory diarrhea* results from stimulated active ion (predominantly sodium) secretion that is caused by bacterial toxins.

2. *Diarrhea due to decreased solute absorption* is caused by loss of mucosal transport functions as the result of altered small intestinal mucosal morphology (celiac disease) or isolated transport defects.

3. *Osmotic diarrhea* is caused by the accumulation of osmotically active molecules that draw water into the gut. This type of diarrhea is caused by laxatives, maldigestion due to enzyme deficiencies, malabsorption of sugars in mucosal disease, and transport defects of sugars.

4. *Diarrhea due to malabsorption* causes increased hydrostatic pressure and results in luminal distention and stimulation of fluid and electrolyte secretion.

The loss of diarrhea fluid results in water depletion (hypovolemia), Na^+ and K^+ depletion, and HCO_3^- loss that may cause a hyperchloremic metabolic acidosis with a normal anion gap.

Maldigestion, Malabsorption, and Related Disorders

The total quantity of fluid absorbed each day by the gut is estimated to be ~9 L, of which 1.5 L is ingested liquids and 7.5 L is gastrointestinal secretions. The majority of secretions enter the gut in the upper portion of the tract—from the mouth, stomach, pancreas, and bile duct. More than 90% of the fluid is absorbed from the small intestine. The efficiency of absorption is due to unique features of the epithelial cell lining and its relation to the underlying rich vascular plexus. Folds, villi, and microvilli together present an absorptive surface of about 250 m^2, the surface area of a tennis court!

Several hundred grams of carbohydrates, over 100 g of fat, and 50–100 g of amino acids are absorbed daily in the small gut, but maximal absorptive capacity is believed to be at least 10 times greater. The maximal absorptive capacity for fluid is probably

Table 23-2. BRUSH BORDER OLIGOSACCHARIDASES

Enzyme	Substrate	Product
Lactase	Lactose	Glucose and galactose
Sucrase*	Sucrose	Fructose and glucose
	Maltose	Glucose
Trehalase	Trehalose	Glucose
α-Glucosidase	Short linear oligosaccharides	Glucose
α-Dextrinase* (isomaltase)	α-Dextrins containing some α-1,6 linkages	Glucose

*Sucrase and isomaltase occur as a single complex in the brush border.

at least 20 L. This considerable reserve of capacity may compensate for mild to moderate degrees of dysfunction induced by disease processes, at least in the early phases.

It is necessary to distinguish between maldigestion and malabsorption. *Maldigestion* is a dysfunction of the *digestive* process that can occur at a number of sites in the gastrointestinal tract, for example, hypoacidity in the stomach, hyperacidity of the duodenum, loss of brush border enzymes in the small intestine, or pancreatic insufficiency. On the other hand, *malabsorption* is a dysfunction of the *absorptive* process by the small gut; it is due to loss of absorptive epithelial cells as may be caused by gluten, inflammation, infection, surgical resection, infiltrations, and so on.

Carbohydrates

Normal digestion in the small intestine. After the action of salivary and pancreatic α-amylase on dietary starch and glycogen, carbohydrate content in the gut consists of newly formed maltose, ingested monosaccharides, dietary disaccharides, oligosaccharides, and indigestible polysaccharides such as cellulose.[40] This mixture is osmotically active and draws water into the gut. The osmotic load is increased as hydrolases of the brush border (Table 23-2) split di- and oligosaccharides into their constituent monosaccharides, but it is decreased as the transport mechanisms for the monosaccharide products move them into the epithelial cells (absorption by active transport).[40,44] For most oligosaccharides (except lactose), hydrolytic rate is greater than transport rate. Mucosal lactase has the lowest activity of all the disaccharidases and is not increased (induced) by feeding large amounts of lactose (as are maltase and sucrase, if the diet contains large amounts of maltose and sucrose, respectively).

Maldigestion and malabsorption of carbohydrates. Generalized impairment of carbohydrate absorption occurs following a number of diseases that cause mucosal damage or dysfunction, such as celiac disease, tropical sprue, and acute gastroenteritis.[51] Absorption defects can be caused by a deficiency in a single or all brush border oligosaccharidases.[40,44] The most common defect is lactose intolerance (lactase deficiency). Physiological change is responsible for *acquired lactase deficiency.* Lactase activity declines as a child ages, but the age at which decline begins differs among ethnic groups.

Tests for the diagnosis of disaccharidase deficiency.[39] The most direct diagnostic test for an intestinal enzyme deficiency is histochemical examination of the brush border of the intestinal epithelium. A more practical approach is an oral challenge of the patient (50 g in adults and 2 g/kg in children) with one disaccharide at a time followed by measurement of serum glucose levels. If the appropriate disaccharidase is present, the disaccharide will be hydrolyzed at the brush border and its component monosaccharides absorbed. Increase of >30 mg/dL over the fasting glucose level indicates adequate disaccharidase activity; an increase of <20 mg/dL indicates deficiency. Verification of absorptive ability is made by performing a glucose tolerance test prior to the disaccharide challenge test. A normal rise above the fasting glucose level (see Chapter 13, *Carbohydrates*) verifies absence of a transport defect.

Lipids[7]

Normal digestion in the small intestine. The average daily dietary intake in North America contains over 100 g of fat, mostly triglycerides.[40] Less than 5% of this amount is recovered in the feces. Although pancreatic lipase is quantitatively the most important hydrolytic enzyme, lipase secreted by the tongue and soft palate and by the gastric mucosa can hydrolyze up to 30% of fats in the stomach.[24,40] This nonpancreatic lipase may have a significant role in lipid digestion when pancreatic function is impaired. The lingual and gastric lipases do not require bile salts to function, and they are active at gastric pH. Liberation of fatty acids in the stomach serves to stabilize the surface of the triglyceride emulsion and promotes the binding of pancreatic colipase; in addition, the liberated fatty acids stimulate release of cholecystokinin from the duodenal mucosa.

In the duodenum and jejunum, pancreatic lipase in the presence of bile salts, colipase, and the triglyceride emulsion produces fatty acids and monoglycerides. These products form micelles with bile salts and lysophosphoglycerols; the micelles convey the nonpolar lipid molecules from the lumen to the epithelial cell surface and dissociate there to produce high concentrations of monoglycerides, lysophosphoglycerols, and fatty acids that are absorbed into the mucosal cell. This absorption is facilitated by the presence of a fatty acid–binding protein in the cytosol of the cell. Within the cell, fatty acids are immediately incorporated into triglycerides. The triglycerides, together with phospholipids, cholesterol and its esters, and a specific apolipoprotein, are formed into spherical chylomicrons that are then released by exocytosis into the lymphatic system of the small gut. From the lymphatics, chylomicrons enter the blood stream and are distributed to the liver, adipose tissue, and other organs.[40]

Conditions associated with maldigestion and malabsorption of lipids include diseases that affect the pancreatic stage, biliary stage, or cellular and delivery stage, or combinations of these.

Amino Acids and Proteins

The average daily dietary intake of protein in North America is ~100 g. Another 70 g of so-called endogenous protein enters the intestinal lumen daily as protein from digestive juices and shed mucosal cells. Of the total protein load, only ~10 g is lost in the feces.[40]

Normal digestion and absorption. Protein digestion is initiated in the stomach by the action of pepsin in a highly acid medium. The acidity also denatures the protein, unfolding the polypeptide chains for better access by the gastric, pancreatic, and intestinal proteolytic enzymes. Additionally, the polypeptides and amino acids produced in the stomach by action of pepsin are potent secretagogues for the hormones that stimulate the pancreas and intestine. Stimulated pancreatic secretion contains the proenzyme forms of trypsin, chymotrypsin, elastase, exopeptidases, carboxypeptidases, and other enzymes. Stimulation of the intestine by gastrointestinal hormones also liberates several proteolytic enzymes from the brush border. One of them, an enteropeptidase called enterokinase, selectively cleaves trypsinogen to form trypsin. Trypsin then activates more trypsin (autocatalysis) and other pancreatic proenzymes as well. Action of the pancreatic enzymes on the partly digested proteins within the lumen produces peptides, 2–6 amino acid residues in length, and also single amino acids. The peptides are hydrolyzed to single amino acids by the aminopeptidases and other hydrolytic enzymes of the brush border. At least seven carrier systems with overlapping specificities for different amino acids are involved in transport of amino acids into the cells. Small quantities both of whole proteins and of peptides are probably absorbed into the cell, but the major portion of digested protein is taken up by the cell in the form of amino acids, which are passed into the portal circulation, on to the liver, and then to the systemic circulation.

Maldigestion and malabsorption of proteins. Pancreatic and small intestinal diseases are major causes of protein maldigestion and malabsorption. Note, however, that the reserve capacity of the pancreas is very great, and fecal loss of protein may not become significant in pancreatic insufficiency states until trypsin has fallen to ~10% of normal.[22] *Mucosal diseases* may affect protein assimilation by reduction in the number of mucosal cells, increased turnover of intestinal cells, or increased losses of plasma proteins from damaged intestinal surface. *Surgical resection* of the intestine not only reduces the total intestinal absorptive surface but may remove a segment of the gut that is specialized for absorption of certain nutrients. Another cause for reduced protein

Table 23-3. SELECTED TESTS USED IN DIAGNOSIS OF PANCREATIC AND INTESTINAL DISEASES

Pancreatitis, Pseudocyst
- Amylase, serum
- Amylase, urine
- Isoamylase, serum
- Amylase/creatinine ratio, urine
- Lipase, serum
- Immunoreactive trypsin, serum
- Miscellaneous tests: calcium, serum; glucose, serum; creatinine, serum; albumin, serum; hematocrit
- Visualization techniques*

Pancreatic Exocrine Insufficiency
- Pancreatic juice or duodenal content after stimulation with Lundh meal, secretin, or secretin-CCK for volume, concentration of HCO_3^-, trypsin, chymotrypsin, lipase, and amylase
- Bz-Ty-PABA test after stimulation with Lundh meal
- Stool for trypsin or chymotrypsin, with or without Lundh meal
- Stool for fat

Malignant Disease
- Carcinoembryonic antigen, serum
- Pancreatic oncofetal antigen, serum
- α_1-Fetoprotein, serum
- Special enzyme tests, pancreatic juice or duodenal contents
- Pancreatic exocrine function tests, with cytologic examination
- Visualization techniques*

Malabsorption
- Hydrogen test, breath
- Disaccharide tolerance tests
- Xylose absorption test
- Fat, stool
- Tests for pancreatic exocrine insufficiency (above)
- Intestinal biopsy, histochemical study for intestinal enzymes
- Iron, folate, vitamin B_{12} in serum
- Visualization techniques*

*Visualization techniques include angiography, computed axial tomography (CAT scan), endoscopic retrograde cholangiopancreatography (ERCP), ultrasound scan, and radiologic examination of intestine with contrast media.

assimilation is *protein-losing enteropathy,* in which inflamed or ulcerated mucosa exudes considerable quantities of plasma proteins, which, if not subsequently digested and reabsorbed, represent large fecal losses of nitrogen; these losses will cause hypoproteinemia.

Diagnosis of Malabsorption and Maldigestion

Pancreatic and Intestinal Disease

The laboratory plays an important part in detecting or ruling out malabsorption and in elucidating its cause. Table 23-3 illustrates the many diagnostic modalities that are available.

The diagnostic process should begin with the determination of total fecal lipids.[26,53] If steatorrhea is not present, hydrogen breath tests and selected disaccharide tolerance tests are utilized to evaluate brush border oligosaccharidase deficiencies. If steatorrhea is present, the total fecal lipid value is a measure of the degree of malabsorption of lipids. The xylose absorption test is then carried out, along with intestinal X-rays; if xylose absorption is normal, as is usual in pancreatic exocrine insufficiency, tests of pancreatic function are required. Jejunal biopsy for histochemical and morphological examination is helpful to confirm congenital oligosaccharidase deficiencies and, in other conditions, is necessary to evaluate the nature and degree of damage to the intestinal mucosa. X-rays and other visualization techniques are required to demonstrate abnormal motility or anatomic features of the gut. A wide variety of serum and urine tests (Table 23-3) are employed in specific contexts to refine or to confirm possible diagnoses or to assign the cause for the presenting malabsorption syndrome.

References

1. Adria Laboratories: Xylo-pfan (D-xylose) kit insert. Adria Laboratories, Inc., Columbus, OH 43215.
2. American Bio-Science Laboratories, Van Nuys, CA 91405, 1984.
3. Anderson, D., Elahi, D., Brown, J. C., et al.: Oral glucose augmentation of insulin secretion. J. Clin. Invest., *62*:152–161, 1978.
4. Arvanitakis, C., and Cooke, A. R.: Diagnostic tests of exocrine pancreatic function and disease. Gastroenterology, *74*:932–948, 1978.
5. Baron, J. H.: Clinical Tests of Gastric Secretion: History, Methodology, and Interpretation. London, Macmillan Press, 1978.
6. Bartlett, K., Dobson, J. V., and Eastham, E.: A new method for the detection of hydrogen in breath and its application to acquired and inborn sugar malabsorption. Clin. Chim. Acta, *108*:189–194, 1980.
7. Bin, T. L., Stopard, M., Anderson, S., et al.: Assessment of fat malabsorption. J. Clin. Pathol., *36*:1362–1366, 1983.
8. Binder, H. J.: Absorption and secretion of water and electrolytes by small and large intestine. *In:* Gastrointestinal Disease: Pathophysiology, Diagnosis, Management. 3rd ed. M. H. Sleisenger and J. S. Fordtran, Eds. Philadelphia, W. B. Saunders Co., 1983.
9. Blackman, A. H., Lambert, D. L., Thayer, W. R., et al.: Computed normal values for peak acid output based on age, sex, and body weight. Am. J. Digest. Dis., *15*:783–789, 1970.
10. Blyth, S. A., and Farrell, P. M.: Advances in the diagnosis and management of cystic fibrosis. Clin. Biochem., *17*:277–283, 1984.
11. Bouchier, I. A. D.: Gastric and pancreatic function tests. Ann. Clin. Biochem., *7*:122–125, 1970.
12. Brooks, F. P.: Clinical usefulness of gastric acid secretory tests. Postgrad Med., *51*:189–193, 1972.
13. Brooks, F. P., and O'Neill, F.: Gastric analysis for hydrochloric acid. *In*: Practice of Medicine, Vol. 2. Hagerstown, Md., Harper & Row, 1973.
14. Broughton, A: Pepsinogen-I and gastrin radioimmunoassay in the diagnosis of gastric pathology. Lab. Man., April, 1980, pp. 46–48.
15. Cello, J. P.: Carcinoma of the pancreas. *In*: Gastrointestinal Disease: Pathophysiology, Diagnosis, Management. 3rd ed. M. H. Sleisenger and J. S., Fordtran, Eds. Philadelphia, W. B. Saunders Co., 1983.
16. Crossley, J. R., Smith, P. A., Edgar, B. W., et al.: Neonatal screening for cystic fibrosis, using immunoreactive trypsin assay in dried blood spots. Clin. Chim. Acta, *113*:111–121, 1981.
17. Davis, P. B., Hubbard, V. S., and di Sant'Agnese, P. A.: Low sweat electrolytes in a patient with cystic fibrosis. Am. J. Med., *69*:643–646, 1980.
18. Dürr, H. K., Schneider, R., Bode, C., et al.: Fecal chymotrypsin: Study on some characteristics of the enzyme. Digestion, *17*:396–403, 1978.
19. Elias, E., Wood, T., and Redshaw, M.: Diagnostic importance of changes in circulating concentrations of immunoreactive trypsin. Lancet, *2*:66–68, 1977.
20. Farrar, G. E., and Bower, R. J.: Gastric juice and secretion: Physiology and variations in disease. Ann. Rev. Physiol., *29*:141–168, 1967.
21. Ford, T. F., Hermon-Taylor, J., and Grant, D. A. W.: A sensitive fluorometric assay for the simultaneous estimation of pepsin and pepsinogen in gastric mucosa. Clin. Chim. Acta, *126*:17–23, 1982.
22. Freeman, H. J., Sleisenger, M. H., and Kim, Y. S.: Human protein digestion and absorption: Normal mechanisms and protein-energy malnutrition. Clin. Gastroenterol., *12*:357–378, 1983.
23. Gibson, L. E., di Sant'Agnese, P. A., and Shwachman, H.: Procedure for the quantitative iontophoretic sweat test for cystic fibrosis. Cystic Fibrosis Foundation, 6000 Executive Blvd., Suite 510, Rockville, MD 20852, 1985.
24. Glickman, R. M.: Fat absorption and malabsorption. Clin. Gastroenterol., *12*:323–334, 1983.
25. Grand, R. J., Park, R. W., and Stafford, R. J.: Pancreatic disorders in childhood. *In*: Gastrointestinal Disease: Pathophysiology, Disease, Management. 3rd ed. M. H. Sleisenger and J. S. Fordtran, Eds. Philadelphia, W. B. Saunders Co., 1983.
26. Gray, G. M.: Maldigestion and malabsorption: Clinical manifestations and specific diagnosis. *In*: Gastrointestinal Disease: Pathophysiology, Diagnosis, Management. 3rd ed. M. H. Sleisenger and J. S. Fordtran, Eds. Philadelphia, W. B. Saunders Co., 1983.
27. Grendell, J. H., and Cello, J. P.: Chronic pancreatitis. *In*: Gastrointestinal Disease: Pathophysiology, Diagnosis, Management, 3rd ed. M. H. Sleisenger and J. S. Fordtran, Eds. Philadelphia, W. B. Saunders Co., 1983.
28. Haeney, M. R., Culank, L S., Montgomery, R. D., et al.: Evaluation of xylose absorption as measured in blood and urine: A one-hour blood xylose screening test in malabsorption. Gastroenterology, *75*:393–400, 1978.
29. Hansen, W.: Akute Pankreatitis. Der Informierte Artz, *8*:62–69, 1980.
30. Heeley, A. F., and Watson, D.: Cystic fibrosis—its biochemical detection. Clin. Chem., *29*:2011–2018, 1983.
31. Jeejeebhoy, K. N., Ahmad, S., and Kozak, C.: Determination of fecal fats containing both medium and long chain triglycerides and fatty acids. Clin. Biochem., *3*:157–163, 1970.
32. Kasper, P., Möller, G., and Wahlefeld, A.: A new photometric assay for chymotrypsin in stool. Clin. Chem., *30*:1753–1757, 1984.
33. Kendall, M. J., Nutter, S., and Hawkins, C. F.: Bacteria and the xylose test. Lancet, *1*:1017–1018, 1972.
34. Korman, M. G., Soveny, C., and Hansky, J.: Radioimmunoassay of gastrin. The response of serum gastrin to insulin hypoglycaemia. Scand. J. Gastroenterol., *6*:71–75, 1971.
35. Meyer, J. H.: Chronic morbidity after ulcer surgery. *In*: Gastrointestinal Disease: Pathophysiology, Diagnosis, Management. 3rd ed. M. H. Sleisenger and J. S. Fordtran, Eds. Philadelphia, W. B. Saunders Co., 1983.

36. Moore, E. W.: The terminology and measurement of gastric acidity. Ann. N.Y. Acad. Sci., *140*:866–874, 1967.
37. Moosa, A. R.: Current concepts: Diagnostic tests and procedures in acute pancreatitis. N. Engl. J. Med., *311*:639–643, 1984.
38. Morgan, L. M., and Marks, V.: The gastrointestinal hormones. *In*: Hormone Analysis: Methodology and Clinical Interpretation. G. W. Pennington and S. Naik, Eds. Boca Raton, FL, CRC Press, 1981, Vol. 2.
39. Newcomer, A. D.: Disaccharidase deficiencies. Mayo Clin. Proc. *48*:648–652, 1973.
40. Newsholme, E. A., and Leech, A. R.: Biochemistry for the Medical Sciences. New York, John Wiley & Sons, 1983.
41. Otte, M.: Pankreasfunktionsdiagnostik. Der Internist, *20*:331–340, 1979.
42. Plebani, M., DiMario, F., Vianello, F., et al.: Pepsinogen group I radioimmunoassay and total serum pepsinogen colorimetric determination: A comparative study in normal subjects and in peptic ulcer patients. Clin. Biochem., *16*:20–22, 1983.
43. Podolosky, D. K., McPhee, M. S., Alpert, E., et al.: Galactosyltransferase isoenzyme II in the detection of pancreatic cancer: Comparison with radiologic, endoscopic, and serologic tests. N. Engl. J. Med., *304*:1313–1318, 1981.
44. Ravich, W. J., and Bayless, T. M.: Carbohydrate absorption and malabsorption. Clin. Gastroenterol., *12*:335–356, 1983.
45. Robert, A., and Kauffman, G. L.: Stress ulcers. *In*: Gastrointestinal Disease: Pathophysiology, Diagnosis, Management. 3rd ed. M. H. Sleisenger and J. S. Fordtran, Eds. Philadelphia, W. B. Saunders Co., 1983.
46. Rudolph, C., and Tauschel, H. D.: Fast determination of pepsin activity in a colored radial diffusion test at pH optimum. Anal. Biochem., *119*:49–54, 1982.
47. Ryan, M. E., and Olsen, W. A.: A diagnostic approach to malabsorption syndromes, a pathophysiological approach. Clin. Gastroenterol., *12*:533–550, 1983.
48. Sale, J. K., Goldberg, D. M., Thjodleifsson, B., et al.: Trypsin and chymotrypsin in duodenal aspirate and faeces in response to secretin and cholecystokinin-pancreozymin. Gut, *15*:132–138, 1974.
49. Satake, K.: Pancreatic disease and the case for serum elastase 1 RIA. Diagn. Med., September/October, 1983, pp. 47–55.
50. Shwachman, H., and Mahmoodian, A.: The sweat test and cystic fibrosis. Diagn. Med., June, 1982, pp. 61–77.
51. Sleisenger, M. H., Ed.: Malabsorption and nutritional support. Clin. Gastroenterol., *12*:1–610, 1983.
52. Soergel, K. H.: Acute pancreatitis. *In*: Gastrointestinal Disease: Pathophysiology, Diagnosis, Management. 3rd ed. M. H. Sleisenger and J. S. Fordtran, Eds. Philadelphia, W. B. Saunders Co., 1983.
53. Theodossi, A., and Gazzard, B. G.: Have chemical tests a role in diagnosing malabsorption? Ann. Clin. Biochem., *21*:153–165, 1984.
54. Tietz, N. W., Ed.: Clinical Guide to Laboratory Tests. Philadelphia., W. B. Saunders Co., 1983.
55. Walmsely, R. N., and Guerin, M. D.: Digestive tract disorders. *In*: Disorders of Fluid and Electrolyte Balance. Bristol, J. Wright, 1984.
56. Webster, H. L.: Laboratory diagnosis of cystic fibrosis. CRC Crit. Rev. Clin. Lab. Sci., *18*:313–338, 1982.
57. Weinsten, W. M.: Gastritis. *In*: Gastrointestinal Disease: Pathophysiology, Diagnosis, Management. 3rd ed. M. H. Sleisenger and J. S. Fordtran, Eds. Philadelphia, W. B. Saunders Co., 1983.
58. Williams, R. H.: Gastrointestinal hormones. *In*: Textbook of Endocrinology. 6th ed. R. H. Williams, Ed. Philadelphia, W. B. Saunders Co., 1981.
59. Yamada, T.: Secretory tumors of the pancreas. *In*: Gastrointestinal Disease: Pathophysiology, Diagnosis, Management. 3rd ed. M. H. Sleisenger and J. S. Fordtran, Eds. Philadelphia, W. B. Saunders Co., 1983.
60. Young, D. S., Pestaner, L. C., and Gibberman, V.: Effects of drugs on clinical laboratory tests. Clin. Chem., *21*:1D–432D, 1975.

BIOCHEMICAL ASPECTS OF HEMATOLOGY

24

VIRGIL F. FAIRBANKS, M.D.
GEORGE G. KLEE, M.D., Ph.D.

Mature erythrocytes lack nuclei, mitochondria, and ribosomes and therefore are incapable of biosynthesis of proteins. Thus, they depend principally on anaerobic glycolysis for their energy needs. Reticulocytes contain ribosomes that make them capable of some protein synthesis, but these very early forms of circulating erythrocytes lose their reticulum (i.e., their ribosomes) in the course of ~48 hours and thereafter have no capacity to synthesize proteins.

Genetically determined abnormalities of erythrocyte proteins cause several important diseases. Among such abnormalities are those affecting structural proteins of the erythrocyte membrane; those affecting hemoglobin structure, function, or stability; and those affecting important erythrocyte enzymes. Deficiency of vitamin B_{12}, folic acid, or iron causes anemia by impairing the ability of the bone marrow to form erythrocytes.

The clinical consequences of abnormalities of erythrocyte structure, function, and metabolism include *anemia*, decreased number of erythrocytes or decreased hemoglobin content of blood; *polycythemia*, greater than normal number of erythrocytes in the blood; *cyanosis*, a bluish discoloration of skin and mucous membranes that results from deoxygenation of hemoglobin; and *hemolysis*, more rapid than normal destruction of erythrocytes. Some of the causes of these conditions and some of the laboratory methods required for their investigation are the subjects of this chapter.

Disorders of Erythrocyte Metabolism[4]

Principal Pathways of Erythrocyte Metabolism

The major pathways related to erythrocyte glycolysis are shown in Figures 24-1 and 24-2. The *Embden-Meyerhof pathway* produces ATP and NADH. The hexose monophosphate pathway (or pentose shunt) provides both reduced NADPH and pentose phosphates and is in turn linked with the glutathione cycle. The *Rapoport-Luebering cycle* (RLC) regulates O_2 affinity of hemoglobin by increasing or decreasing the amount of 2,3-diphosphoglycerate (2,3-DPG). Abnormalities of these metabolic pathways may result in hemolytic anemia.

Embden-Meyerhof pathway. Inherited deficiency of hexokinase (EC 2.7.1.1), glucosephosphate isomerase (EC 5.3.1.9), phosphofructokinase (EC 2.7.1.11), triosephosphate isomerase (EC 5.3.1.1), phosphoglycerate kinase (EC 2.7.2.3), or pyruvate kinase (EC 2.7.1.40) may cause hemolytic anemia. Of these, pyruvate kinase (PK) deficiency is the most common, even though quite rare.

Hexose monophosphate pathway (HMP). Normally, ~10% of glucose is catabolized through the hexose monophosphate pathway, but this fraction may be markedly increased when there is oxidative stress. The HMP reduces $NADP^+$ to NADPH, which in turn maintains glutathione in the reduced state and generates seven-, five-, four-, and three-carbon sugar phosphate intermediates that may be recycled into the Embden-Meyerhof pathway of glycolysis (see Figure 24-1).

Deficiency of glucose-6-phosphate dehydrogenase (G-6-PD) in erythrocytes is widely prevalent throughout the world, particularly among people of Southeast Asia, in blacks, in people of the Mediterranean littoral, and in people of India. The disorder is X-chromosome-linked.

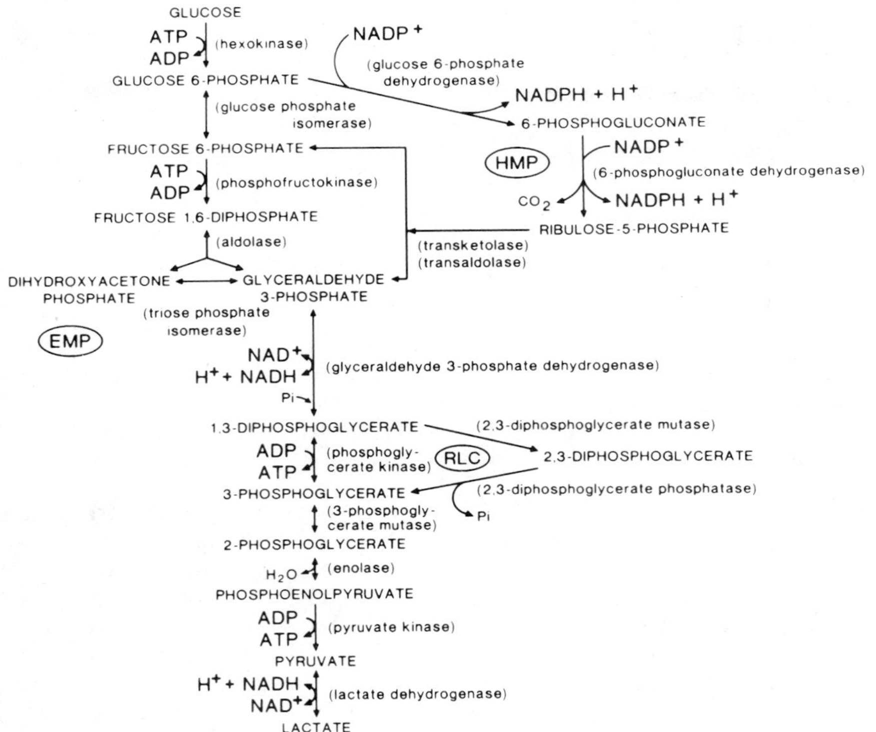

Figure 24-1. Major glycolytic pathways of the erythrocyte. EMP = the Embden-Meyerhof pathway; HMP = hexose monophosphate pathway or pentose shunt; RLC = the Rapoport-Luebering cycle.

Substrates are in upper case type, enzymes in parentheses. ADP = adenosine diphosphate; ATP = adenosine triphosphate; NAD^+ = nicotinamide–adenine dinucleotide; NADH = reduced nicotinamide–adenine dinucleotide; $NADP^+$ = nicotinamide–adenine dinucleotide phosphate; NADPH = reduced nicotinamide–adenine dinucleotide phosphate. The step from ribulose-5-phosphate that is shown as catalyzed by transketolase and transaldolase is an abbreviation of this portion of the HMP.

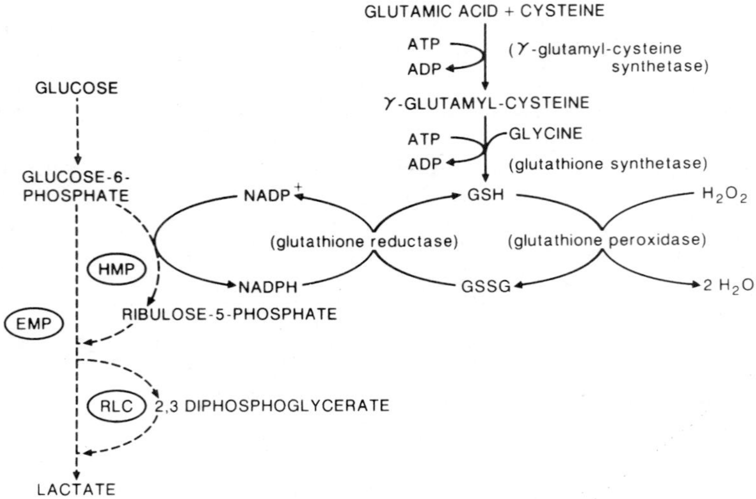

Figure 24-2. Interrelationship of hexose monophosphate and glutathione pathways. GSH = reduced glutathione; GSSG = oxidized glutathione; other abbreviations as in Figure 24-1.

G-6-PD variants have been classified on the basis of severity of enzyme deficiency:

Class I	Deficiency associated with chronic hemolytic anemia (e.g., G-6-PD Chicago)
Class II	Severe deficiency, usually without spontaneous hemolytic anemia (e.g., G-6-PD Mediterranean, Mahidol, Union)
Class III	Moderate to mild deficiency (e.g., G-6-PD A−)
Class IV	Very mild or no deficiency (e.g., G-6-PD A)
Class V	Increased activity (only one such variant has been described, G-6-PD Hektoen)

Subclassification of the variants may be made according to the electrophoretic mobility and kinetic properties of G-6-PD variants.

Glutathione pathway. Glutathione (GSH) protects critical erythrocyte proteins from peroxidative injury. Glutathione reductase links this pathway to the HMP through the reversible oxidation and reduction of $NADP^+$ (see Figure 24-2). Glutathione peroxidase (GSH-Px, EC 1.11.1.9) converts H_2O_2 to water, thus reducing the likelihood of peroxidative damage. Glutathione must be continually synthesized in the erythrocyte. This requires two enzymes: γ-glutamyl cysteine synthetase (γ-GC-S) and glutathione synthetase (GSH-S). Deficiency of each of these erythrocyte enzymes has been found in a few cases associated with congenital hemolytic anemia.

The Rapoport-Luebering cycle (RLC). This cycle, which is unique to mammalian erythrocytes, is the principal physiological means for regulating the O_2 affinity of hemoglobin. It does this by its product, 2,3-diphosphoglycerate (2,3-DPG), which is quantitatively the major glycolytic intermediate of the erythrocyte. 2,3-DPG content of the erythrocyte is increased in all anemias, usually in inverse proportion to the hemoglobin concentration. This compensatory increase in 2,3-DPG concentration results in a decrease in O_2 affinity of hemoglobin so that O_2 is more readily transferred to tissues. The RLC is also sensitive to pH, and this pH-sensitivity also facilitates change in 2,3-DPG concentration to "fine-tune" the O_2 affinity of hemoglobin.

Two enzymes involved in the RLC are *2,3-diphosphoglycerate mutase (2,3-DPGM)* and *2,3-diphosphoglycerate phosphatase.** The only well-documented case of deficiency of either enzyme resulted in erythrocytosis.

Methods for the Determination of Erythrocyte Enzymes

Very few clinical laboratories need to measure the activity of erythrocyte enzymes other than G-6-PD and pyruvate kinase. However, methods for the determinations of glutathione and 2,3-diphosphoglycerate will be mentioned as well.

Specimens. With few exceptions, venous blood specimens for these assays should be anticoagulated either with Na_2EDTA, acid-citrate-dextrose (ACD) solution, or citrate-phosphate-dextrose (CPD) solution. ACD or CPD anticoagulant is essential if glutathione is to be assayed. With use of any of these anticoagulants, erythrocyte enzyme activity is stable at 4 °C for three weeks. Plasma should not be removed until shortly before the assay is performed. Special conditions of specimen collection and handling are required for the assay of 2,3-DPG. Because leukocytes have significantly greater enzyme activity than do erythrocytes, assays are performed using saline-washed, packed erythrocytes that are essentially leukocyte-free. The same hemolysate may be used for the assay of any of the erythrocyte enzymes.

---

*These enzymes, unique to the mammalian erythrocyte, have not been assigned EC designations.

Preparation of Hemolysate[5]

Reagents and Supplies

1. NaCl, 0.155 mol/L; keep at 4 °C.
2. Hemolyzing solution. Place 10 mL of Na_2EDTA, 0.27 mol/L, pH 8.0, and 0.05 mL of 2-mercaptoethanol in a 1-L volumetric flask and bring to volume with reagent grade water. Store at 4 °C.

Procedure

1. Centrifuge 2 mL of whole blood at 1500 × *g* for 10 min. Remove and discard the plasma and "buffy coat" (the cream-colored layer of leukocytes between plasma and packed erythrocytes).
2. Resuspend the erythrocytes in 10 mL of cold NaCl, 0.155 mol/L. Mix thoroughly.
3. Centrifuge at 1500 × *g* for 10 min at 4 °C.
4. Remove and discard the supernatant and topmost layer of cells, including any residual buffy coat.
5. Repeat the suspension of cells in cold NaCl, 0.155 mol/L, centrifugation, and discarding of supernatant and uppermost layer of cells (to a depth of about 0.5 mm) a total of three times.
6. Resuspend the washed cells in an equal volume of cold NaCl, 0.155 mol/L.
7. Pour about 150 mL of acetone into a 250-mL beaker.
8. Using tongs or insulated gloves, place a small piece of dry ice into the beaker. While waiting for frothing of the dry ice–acetone mixture to subside, add 0.2 mL of the erythrocyte-NaCl suspension to 1.8 mL of hemolyzing solution in a glass tube (10 × 150 mm). Cap the tube.
9. Place the tube into the dry ice–acetone mixture until frozen, usually a few minutes.
10. Thaw at room temperature. (Do not use hot water to thaw; some enzymes are very thermolabile.)
11. Examine for turbidity. The hemolysate should be clear upon visual inspection. If it is turbid, centrifuge at 2250 × *g* for 30 min at 4 °C.

Determination of Glucose-6-Phosphate Dehydrogenase Activity[5]

Principle. The reaction catalyzed by G-6-PD is:

$$\text{Glucose-6-phosphate} + \text{NADP}^+ \xrightarrow{G\text{-}6\text{-}PD} \text{6-Phosphogluconate} + \text{NADPH} + \text{H}^+$$

The enzyme activity is determined by measurement of the rate of increase in NADPH concentration by measuring the rate of increase in absorbance at 340 nm.

Reagents

1. Tris-HCl buffer. Place 121.14 g of tris(hydroxymethyl)methylamine (Tris, THAM, Trizma) into a 2-L beaker and add ~800 mL reagent grade water. Adjust pH to 8.0 by adding HCl, 1 mol/L, while constantly stirring. After pH adjustment, transfer to a 1-L volumetric flask and add reagent grade water to bring to volume.
2. $MgCl_2$, 0.1 mol/L.
3. $NADP^+$, 2 mmol/L (Sigma Chemical Co., St. Louis, MO 63178).
4. Stock reagent mixture. The above reagents may be premixed in the following proportions:

Reagent	Proportion
Tris-HCl buffer	0.50
$MgCl_2$ solution	0.50
$NADP^+$ solution	0.50
H_2O	2.90

The mixture may be prepared in advance and stored frozen for as long as one month.

Procedure

1. For each specimen, prepare blank and assay cuvets as follows. Include at least one normal control with each set of specimens.

	Blank Cuvet, mL	Assay Cuvet, mL
Stock reagent mixture	0.88	0.88
H_2O	0.10	—
Hemolysate (see page 792)	0.02	0.02

2. Invert twice to mix.
3. Incubate at 37 °C for 10 min.
4 Add 0.1 mL of glucose-6-phosphate solution to assay cuvets only.
5. Rapidly invert twice to mix, then place in spectrophotometer and measure the increase in absorbance at 340 nm for 10 min, at 37 °C.
6. While absorbance is being recorded, measure the hemoglobin concentration of the hemolysate by using a standard cyanmethemoglobin method (see page 803).

Calculation

$$\text{Net } \Delta A_{340}/\text{min} = \frac{A_{10\text{ min}} - A_{0\text{ min}}}{10}$$

$$\text{G-6-PD activity, U/g Hb} = \frac{(\text{Net } \Delta A_{340}/\text{min}) \times 804}{C_{Hb}}$$

where C_{Hb} = hemolysate hemoglobin concentration (g/dL)
Net ΔA_{340}/min = the rate of change in absorbance of the assay cuvet minus the rate of change in absorbance of the blank cuvet

Example:

A_{340}, initial = 0.230
A_{340} at 10 min = 0.500
ΔA_{340}/10 min = 0.270
ΔA_{340}/min = 0.027

ΔA_{340}/min for blank = 0.001
Net ΔA_{340}/min for specimen = 0.026
C_{Hb} = 1.47

$$\text{G-6-PD, U/g Hb} = \frac{0.026 \times 804}{1.47} = 14.2$$

Reference values are 8–18 U/g Hb (mean ± 2 SD); values >18 U/g Hb are encountered in any condition associated with younger than normal erythrocytes (as in hemolytic anemias not due to G-6-PD deficiency) but are of no clinical significance.

Comments and precautions. The absorbance change in the G-6-PD assay outlined above reflects NADPH generated in both the G-6-PD and the 6-phosphogluconate dehydrogenase (6-PGD) reactions. However, in practice it is not necessary to correct for 6-PGD activity, since deficiency of erythrocyte 6-PGD has not been reported as a cause of hemolytic disease, nor will normal 6-PGD activity obscure G-6-PD deficiency.

In black persons with the common A− G-6-PD variant, a few days following onset of a hemolytic episode, so many young erythrocytes with normal G-6-PD activity enter the blood that G-6-PD activity may rise into the normal range and G-6-PD deficiency cannot be demonstrated until 6–8 weeks have elapsed. This interval is required to allow circulating erythrocytes to become older and G-6-PD activity to decline to levels that permit diagnosis. However, in Caucasians who have the Mediterranean G-6-PD variant, even the youngest erythrocytes have little or no G-6-PD activity and diagnosis can be made by G-6-PD assay at any time. The Mahidol and Union variants, which probably account for severe G-6-PD deficiency in ~25% of the nearly 1 000 000 Southeast Asians now residing in North America, appear to be similar in this respect to G-6-PD Mediterranean. G-6-PD variants of Class I that cause chronic hemolytic anemia can also be detected at any time by this assay.

Determination of Pyruvate Kinase (PK) Activity[5]

Principle. The PK reaction is linked with the lactate dehydrogenase (LDH) reaction in which NADH is oxidized to NAD^+.

$$\text{Phosphoenolpyruvate} + \text{ADP} \xrightarrow{PK} \text{Pyruvate} + \text{ATP}$$

$$\text{Pyruvate} + \text{NADH} + \text{H}^+ \xrightarrow{LDH} \text{Lactate} + \text{NAD}^+$$

Since lactate dehydrogenase is present in excess, the rate of NADH oxidation is limited by the activity of PK. The reaction rate is measured by the rate of decrease in absorbance at 340 nm. Assays are performed at high and low substrate concentrations with and without addition of fructose-1,6-diphosphate (FDP), because some PK variants associated with hemolysis have atypical reaction kinetics (and thus may exhibit "normal" activity at usual substrate concentrations) or may show absence of enhancement by FDP. A normal control must be run concurrently with every specimen assayed.

The ***reference ranges*** (mean ± 2 SD) are 11–19 U/g Hb for the high-substrate assay, 1.4–3.0 U/g Hb for the low-substrate assay without FDP, and 4.7–7.3 U/g Hb for the low-substrate assay with FDP stimulation.

Determination of Erythrocyte Glutathione (GSH) Concentration[5]

Clinical significance. Erythrocyte glutathione concentration is diminished in many persons who have defects in the hexose monophosphate or glutathione synthesis pathways.

Principle. Virtually all of the nonprotein sulfhydryl groups of erythrocytes are in the form of reduced glutathione. 5,5′-Dithiobis(2-nitrobenzoic acid), or DTNB, is a disulfide chromogen that is readily reduced by sulfhydryl compounds to an intensely yellow compound. The absorbance of the reduced chromogen is measured at 412 nm and is directly proportional to the GSH concentration.

Specimen. Whole blood, 3 mL, anticoagulated with ACD.

Reagents

1. Precipitating solution. Place in a 100-mL volumetric flask 1.67 g of glacial metaphosphoric acid, 0.20 g of disodium or dipotassium ethylenediaminetetraacetic acid (EDTA), and 30.00 g of NaCl, and bring to volume with distilled water. This solution is stable for three weeks at 4 °C. A fine precipitate may form owing to EDTA, but this does not interfere with the test.
2. Phosphate solution, 0.30 mol/L. Place in a 1-L volumetric flask 42.59 g of Na_2HPO_4, and bring to volume with distilled water. The solution is stable indefinitely at 4 °C. If crystals form, dissolve by heating.
3. DTNB reagent. Place in a 100-mL volumetric flask 40.0 mg of 5,5′-dithiobis(2-nitrobenzoic acid; Sigma Chemical Co., St. Louis, MO 63178). Bring to volume with a solution of sodium citrate, 1 g/dL. This solution is stable for at least 13 weeks at 4 °C.
4. GSH standards. Place 100 mg GSH (Sigma Chemical Co., St. Louis, MO 63178) in a 100-mL volumetric flask and bring to volume with reagent grade water. Invert repeatedly until GSH is completely dissolved. Prepare 50 mg/dL and 10 mg/dL standards by diluting 5 mL of the 100 mg/dL standard with 5 mL and 45 mL, respectively, of reagent grade water. The GSH standards are not stable and must be freshly prepared on the day of the assay.

Procedure

1. Place 0.2 mL of whole blood into a 10-mL test tube and add 1.8 mL distilled water. Mix to hemolyze.
2. Promptly add 3.0 mL of precipitating solution. Mix.
3. Allow to stand 5 min at room temperature and then filter through coarse-grade filter paper.
4. Prepare these mixtures for each sample:

Reagent	Blank, mL	Assay, mL
Filtrate	—	2.0
Precipitating reagent	1.2	—
H_2O	0.8	—
Phosphate solution	8.0	8.0
DTNB solution	1.0	1.0

5. Invert three times to mix.
6. Read absorbance at 412 nm within 4 min of preparing the mixtures.
7. Measure the hematocrit of the original whole blood specimen.
8. Assay the GSH standards, omitting the filtration step.
9. Graph the standard curve. Determine GSH concentration of the blood specimen from the graph.
10. Calculate GSH concentration:

$$\text{GSH, mg/dL of erythrocytes} = \frac{\text{GSH conc. (from standard curve)}}{\text{Hematocrit}}$$

The ***reference range*** is 47–100 mg/dL erythrocytes.

Comments and precautions. The ability of the erythrocyte to maintain a normal GSH concentration when stressed by an oxidant substance may be tested by incubation of blood with 1-acetyl-2-phenylhydrazine (APH) and glucose for 2 h at 37 °C. In this GSH stability test, GSH concentration is measured before and after incubation of blood with APH. Erythrocytes deficient in G-6-PD, GSH-S, or γ-GC-S exhibit marked decrease in GSH concentration after such stress. GSH concentration declines in blood anticoagulated with heparin or EDTA; for such specimens the assay should be carried out on the same day. Include at least one normal control specimen with each assay.

Determination of 2,3-DPG in Erythrocytes[5]

Clinical significance. 2,3-Diphosphoglycerate (2,3-DPG) is the predominant glycolytic intermediate of the erythrocyte and plays a vital role in modulating the affinity of hemoglobin for oxygen (see pages 797–798).

Principle. 2,3-DPG is formed by the so-called Rapoport-Luebering pathway; the relevant reactions are:

$$\text{Phosphoenolpyruvate} \xrightleftharpoons{\textit{Enolase}} \text{2-Phosphoglycerate} \xrightleftharpoons[\textit{MPGM}]{\textit{2,3-DPG}} \text{3-Phosphoglycerate}$$

In the reaction mixture used in this assay, enolase and monophosphoglycerate mutase (MPGM) establish an equilibrium between phosphoenolpyruvate (PEP) and 2-phosphoglycerate. The addition of 2,3-DPG shifts the equilibrium so that more 2-phosphoglycerate is converted to 3-phosphoglycerate. This in turn results in the conversion of PEP to 2-phosphoglycerate. The rate of decrease in PEP concentration is measured by the decrease in absorbance at 240 nm, which is proportional to the amount of 2,3-DPG in the specimen.

The ***reference range*** is 13.4–17.4 μmol/g Hb (mean ± 2 SD).

Specimen. Immediately after drawing blood (without anticoagulant), expel two drops of whole blood from the syringe tip into 2 mL of ice cold distilled water. Mix to hemolyze. Deliver the specimen to the laboratory packed in ice (0–4 °C) within 30 min. Since 2,3-DPG concentration declines rapidly once blood is drawn, it is critical that blood for this assay be lysed and chilled immediately.

Hemoglobin and its Degradation[7]

Hemoglobin (Hb) is a spheroidal protein that consists of four subunits: two pairs of identical polypeptide chains, each with a cleft or pocket on its exterior. The cleft contains a heme (iron-protoporphyrin) group that is the site of O_2 uptake and release. Hemoglobin generally contains as subunits a pair of α- or α-like globin chains and a pair of identical non-α chains that may be β, γ, δ, or ϵ. The two α and two non-α chains are arranged around the central cavity, as shown in Figure 24-3. The hemoglobin tetramer opens and closes slightly as O_2 is taken up or released.

The Structure of Normal Hemoglobins

In normal human adults, at least 96% of the hemoglobin is Hb A, which is composed of two normal α- and two normal β-chains and may thus be represented as $\alpha_2^A\beta_2^A$, or more simply as $\alpha_2\beta_2$. *Hb* A_2 is usually about 2.5–3.0% of total hemoglobin; it contains

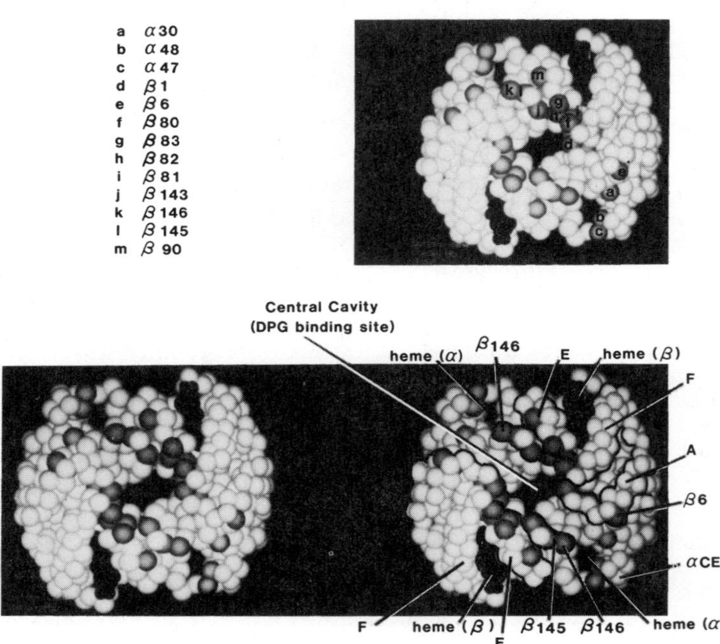

Figure 24-3. Model of the hemoglobin tetramer with the β-chain subunits facing the reader. Amino acids are represented as spherules. Those that are shaded affect hemoglobin mobility in citrate agar gel. Spherules labeled d, j, k, and l in the upper panel constitute the 2,3-DPG binding site and are important in regulation of the oxygen affinity of hemoglobin. The lower panel is a stereo pair for three-dimensional viewing. To obtain the three-dimensional effect, hold the illustration at arm's length. The β-globin chains appear to be between the reader and the more distant α-chains. On the right, some of the segments of the β-globin chain have been outlined, and segments and critical components have been marked. β146 is the COO^- end of the β-chain; β6 is the site of amino acid substitutions responsible for hemoglobins S and C. (Reproduced by permission from Winter, W. P., and Yodh, J.: Interaction of human hemoglobin and its variants with agar. Science, *221*:175-178, 1983. Copyright 1983 by the Am. Assoc. for the Adv. of Science.)

two δ-chains, and may thus be designated $\alpha_2\delta_2$. *Fetal hemoglobin (Hb F)* predominates during fetal life but rapidly diminishes during the first year of postnatal life. In normal adults <1% of hemoglobin is Hb F. It consists of two α- and two γ-chains, and thus is $\alpha_2\gamma_2$. Three embryonic hemoglobins are *Hb Portland* ($\zeta_2\gamma_2$), *Hb Gower-1* ($\zeta_2\epsilon_2$), and *Hb Gower-2* ($\alpha_2\epsilon_2$). These three normal embryonic hemoglobins have little import for clinical laboratories since they are not encountered during postnatal life.

Each of the globin chains is a long strand of amino acids, 141 in each α-chain and 146 in each β-, δ-, and γ-chain. Each chain contains helical segments separated by short noncoiled segments (Figure 24-4).

Normally, the heme pocket is formed entirely by nonpolar (hydrophobic) amino acids. The heme moiety is suspended within this pocket by an attachment of its iron atom to the imidazole group of the proximal histidine—position 92 of the β-chain (β92) or position 87 of the α-chain (α87). The imidazole group of the distal histidine (β63 or α58) also contacts the iron of heme, but it appears to swing in and out of this position to permit the ingress and egress of O_2. The iron atom in each heme is in the divalent state, whether hemoglobin is oxygenated or deoxygenated. *Methemoglobin* is a nonfunctional derivative with iron in the trivalent state.

Hemoglobin Degradation

Hemoglobin degradation normally occurs within phagocytic cells of the reticuloendothelial system. Aged and metabolically run-down erythrocytes can no longer provide the energy to sustain the Na^+,K^+-pump and therefore leak cations at an increasing rate. The aged cells are ingested by phagocytes, in which the degradation of hemoglobin to bilirubin begins (see Chapter 22). When there is accelerated extravascular erythrocyte destruction, hemoglobin degradation is accelerated. Under certain circumstances (for example, mechanical hemolysis due to faulty intracardiac valvular prosthesis or the episodic hemolysis of *paroxysmal nocturnal hemoglobinuria*), erythrocyte destruction may be predominantly intravascular. In this case, the free plasma hemoglobin is bound by the plasma protein *haptoglobin*. The hemoglobin-haptoglobin complex formed is rapidly removed by parenchymal cells of the liver. In hemolytic disorders the unbound

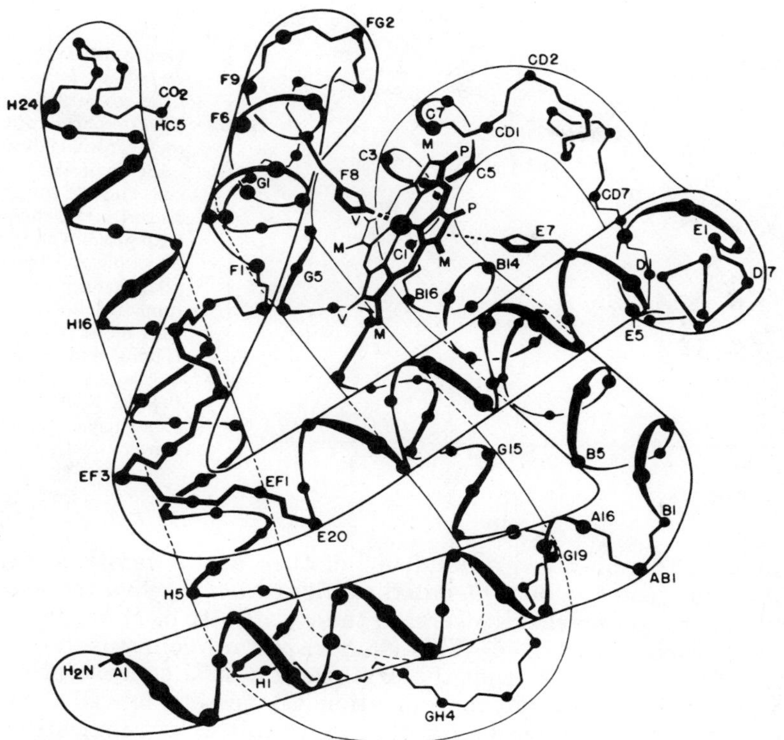

Figure 24-4. Structure of the hemoglobin subunit. Chains of amino acids in spiral or helical segments are linked by short nonhelical segments. The helical segments are designated A through H. In this illustration, amino acids are designated in accordance with the helical or nonhelical segment in which they occur. (From Dickerson, R. E.: X-ray analysis and protein structure. *In*: The Proteins: Composition, Structure and Function, Vol. 2. 2nd ed. H. Neurath, Ed. New York, Academic Press, 1964, pp. 603–778. Reproduced with permission.)

haptoglobin therefore is usually decreased or absent. Free hemoglobin, in excess of the binding capacity of haptoglobin, is excreted in the urine, to which it imparts a chocolate-brown color. *Hemoglobinuria* is identified by the occult blood test, which is described later in this chapter. The plasma haptoglobin concentration returns to normal after 2–3 d unless the hemolysis persists. Some of the heme that dissociates from hemoglobin in plasma may be taken up by *hemopexin*, a heme-binding protein present in normal plasma, and some is taken up by serum albumin to form an albumin-heme complex called *methemalbumin*.

Mechanisms of Hemoglobin Function

Changes in Conformation States of Hemoglobin with Uptake and Release of O_2

Hemoglobin is uniquely adapted to take up and release O_2. This is the result of subunit-to-subunit interaction whereby an alteration in spatial relationships between portions of the molecule facilitates the uptake or release of O_2. The O_2 dissociation curve of normal blood hemoglobin is sigmoidal. The uptake of the first molecule of O_2 by the hemoglobin molecule facilitates the uptake of the second, third, and fourth molecules of O_2 with only a modest increment in oxygen pressure. Conversely, a molecule of hemoglobin with all four heme sites oxygenated readily gives up O_2 at three of these sites. This highly coordinated response to slight changes in oxygen tension, which is absolutely vital for effective O_2 delivery, appears to depend in part on (1) reversible binding of certain amino acids across the $\alpha_1\beta_2$ contact point, permitting the $\alpha\beta$-dimers to snap back and forth as the molecule is oxygenated and deoxygenated; and (2) the

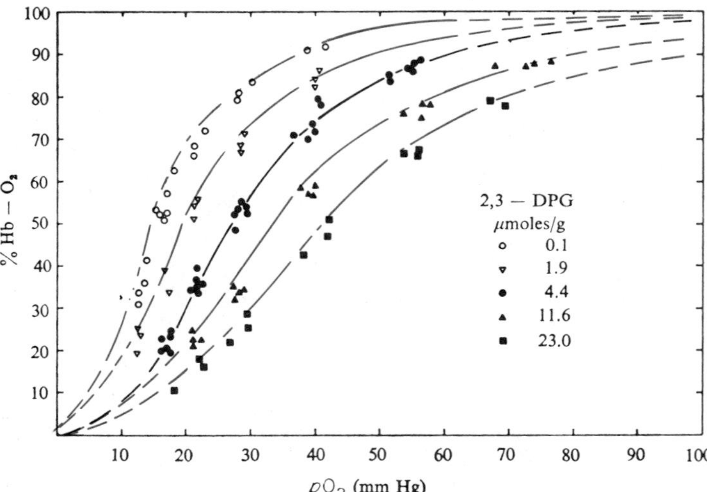

Figure 24-5. Normal oxygen dissociation curve of hemoglobin. Changes in 2,3-diphosphoglycerate (DPG) concentration in the erythrocyte markedly influence the position of the curve. As the concentration of 2,3-DPG increases, the curve shifts to the right (Reproduced with permission from Duhm, J.: The effect of 2,3-DPG and other organic phosphates on the Donnan equilibrium and the oxygen affinity of human blood. *In*: Oxygen Affinity of Hemoglobin and Red Cell Acid Base Status [Alfred Benzon Symposium, IV]. M. Rorth and P. Astrup, Eds. Copenhagen, Denmark, Alfred Benzon Foundation, 1972.)

movement of a DPG molecule into and out of the central cavity of the hemoglobin molecule; DPG stabilizes the deoxygenated configuration. In vivo increases in erythrocytic DPG concentration markedly decrease the O_2 affinity of hemoglobin and increase O_2 delivery to tissues (Figure 24-5). This is an important compensatory mechanism in anemia, and it has been shown that the erythrocytic DPG concentration varies reciprocally with the blood hemoglobin concentration. Conversely, low DPG concentration is associated with a shift of the O_2 dissociation curve to the left, indicating increased O_2 affinity of hemoglobin and decreased O_2 delivery to tissues. This is a problem in blood banking, because transfusion of large amounts of DPG-deficient stored blood may compromise a patient's ability to oxygenate his tissues for several hours until more DPG is synthesized.

Changes in pH of blood also influence the affinity of hemoglobin for oxygen. A decrease in blood pH or an increase in pCO_2 causes the dissociation curve to shift to the right and hemoglobin to become less avid for O_2. This favors increased delivery of O_2 to tissues. Conversely, an increase in blood pH or a decrease in pCO_2 causes the curve to shift to the left and hemoglobin to become more avid for O_2. This favors increased uptake of O_2 by hemoglobin and less delivery of O_2 to tissues. These changes in O_2 affinity of hemoglobin that are caused by pH changes in the range of 7.2–7.6 are known as the *alkaline Bohr effect.*

Thalassemias and Related Conditions

Thalassemias are conditions in which there is a reduction in the rate of synthesis of one type of globin chain. If there is a reduction in α-chain synthesis, the condition resulting is an α-thalassemia; if reduction is in β-chain synthesis, the resulting condition is a β-thalassemia. Thalassemias occur worldwide but with highest frequency in persons of Asian, African, or Mediterranean origin. Alpha-thalassemia appears to be the most common genetic disorder of man. Approximately 30% of black Americans have a gene for α-thalassemia. A similarly high prevalence of α-thalassemia is observed in persons of Southeast Asian origin. Alpha-thalassemia also occurs in persons of Mediterranean origin, although at a lower prevalence. Beta-thalassemia occurs frequently in Mediterranean and Asian peoples. Clinical expressions of thalassemia genes range from none (in the mildest type of α-thalassemia trait) to very severe. The most severe forms of thalassemia are incompatible with life. The terms *thalassemia minor* and *major* indicate the severity of the clinical disorder; the term *thalassemia trait* signifies the *heterozygous* state for a thalassemia gene. Thalassemias are summarized in Table 24-1.

The α-thalassemias. Most of the α-thalassemias are the result of deletions of one or more α-gene loci. The mildest form of α-thalassemia is the result of a single α-gene

Table 24-1. THE THALASSEMIAS AND ASSOCIATED HEMOGLOBIN ABNORMALITIES

Abnormality	Microcytosis	Anemia	Hemoglobin F	Hemoglobin A_2	Hemoglobin H	Other Features
α-Thalassemia						
1 α-chain deleted	0	0	Normal	Normal	0	Small quantities (~2%) hemoglobin Barts (γ_4) detectable at birth
2 α-chains deleted	0	0	Normal	Normal	0	Hemoglobin Barts (~5%) detectable at birth
3 α-chains deleted	+	Mild to severe	May be increased	Normal	10–15%	Unstable hemoglobin hemolytic anemia (hemoglobin H)
4 α-chains deleted	—	Severe	0*	0*	0*	Fetal death with severe pallor and edema (hydrops fetalis)
β-Thalassemia and δβ-Thalassemia						
Heterozygous						
β	+	+	Normal	4–10%	0	—
δβ	+	+	5–15%	Decreased	0	—
Homozygous						
β	Marked	Marked	70–90%	Up to 20%	0	Skeletal abnormalities, cutaneous ulcers, growth retardation
δβ	Moderate	±	100%	0	0	—
Doubly heterozygous						
Hb S-β-thal.	+	Moderate to marked	Increased	Increased	0	May exhibit all typical features of sickle cell disease
Hb C-β-thal.	+	Moderate	Increased	Increased	0	Often marked splenic enlargement†
Hb E-β-thal.	+	Moderate to severe	Increased	Increased	0	
Hereditary Persistence of Fetal Hemoglobin						
Heterozygous	0	0	10–20%	Decreased	0	Asymptomatic
Homozygous	0	0	100%	0	0	Asymptomatic
Hb S-HPFH	0	0	10–20%	Decreased	0	Usually asymptomatic

* Hemoglobin Barts, 100%.
† May be present (in β^+ or β^{++}-thalassemia, not in β^0-thalassemia).

deletion, leaving three functional α-genes. Persons who have this condition are said to have α-thalassemia-2 trait, a harmless condition. Two α-genes may be deleted on the same chromosome (α-thalassemia-1 trait) or one may be deleted from each of the two chromosomes (homozygous α-thalassemia-2). Both conditions cause mild microcytosis but are otherwise innocuous. Persons with three deleted α-genes have a moderately severe thalassemia, Hb H (β_4) disease. Homozygosity for the doubly deleted α-thalassemia-1 results in inability of the fetus to make any but embryonic hemoglobins or Hb Barts (γ_4); extreme anemia results, followed by fetal death in utero.

The β-thalassemias. The β-thalassemias are classified as β^0, β^+, and β^{++}, depending on the degree of deficiency of β-chain synthesis. In β^0-thalassemia, β-chain synthesis is absent and no Hb A is demonstrable; in β^+-thalassemia there is a slight synthesis; and in β^{++}-thalassemia there is moderately reduced β-chain synthesis. The amount of Hb A demonstrable in β^+- and β^{++}-thalassemia varies accordingly. Heterozygotes for each of these types of β-thalassemia have thalassemia minor, which is rarely of any clinical consequence. Homozygotes or compound heterozygotes have severe thalassemia major.

Delta-beta thalassemia. Considerably less common than β-thalassemia is *δβ-thalassemia trait,* a condition that exhibits the same mild clinical and hematological features as β-thalassemia trait but is not associated with an increase in Hb A_2. Hemoglobin F concentration is characteristically increased to 5–20% of total hemoglobin.

Lepore. *Hemoglobin Lepore* is the prototype of a large number of hemoglobinopathies and thalassemias that are a consequence of nonhomologous meiotic crossing over.

This results in deletions of portions of the gene loci or in gene fusions. Hemoglobin Lepore trait is not uncommon; it causes microcytosis without anemia.

Hereditary Persistence of High Fetal Hemoglobin

Several types of *hereditary persistence of high fetal hemoglobin (HPFH)* have been defined. This condition is the result of deletion of an entire $\delta\beta$-gene region. HPFH trait is harmless and generally considered unassociated with clinical or hematologic features. In HPFH (African type), Hb F concentration is usually 15–30% of total. In some cases there may be very slight microcytosis, but anemia is not observed.

Combinations of Thalassemias

Compound heterozygotes for α- and β-thalassemias (those who have inherited both an α- and a β-thalassemia gene) have a very mild anemia and microcytosis. The same appears to be true of compound heterozygotes for β- or $\delta\beta$-thalassemia and Hb Lepore.

Since the hemoglobin β^E-gene is also expressed as a mild thalassemia, compound heterozygotes for Hb E and β-thalassemia have severe thalassemia that may completely mimic classic β-thalassemia major. This is the most frequently encountered severe thalassemia disorder in Southeast Asia. In contrast, a compound heterozygote for α-thalassemia-1 and Hb E has moderate microcytosis but little or no anemia. The proportion of Hb E in such cases is usually reduced to 10–20% of total hemoglobin (in contrast to 30–35% in the usual simple Hb E trait). Persons who have inherited the α-thalassemia-1 gene (2 α-loci deleted), the α-thalassemia-2 gene (1 α-locus deleted), and one β^E-gene have 10–15% Hb E and clinical features of Hb H disease but, paradoxically, they may show no Hb H upon electrophoresis. This is because the deficit in β-chain synthesis that results from the presence of a β^E-gene balances the deficit in α-chain synthesis; there are not enough surplus β-chains to form Hb H (β_4-tetramers) in large quantities.

Hemoglobinopathies

Disorders of hemoglobin that result from structural abnormality of a globin chain are called hemoglobinopathies. Approximately 400 structural variants have been described. By far the majority of these are entirely unattended by clinical or hematologic manifestations and are discovered only incidentally. Some of the abnormal hemoglobins are listed in Table 24-2, together with their clinical and laboratory manifestations. By far the largest number of hemoglobin variants are due to *point mutations* in the code for globin chains. *Nonhomologous crossing over* may result in large deletions or fusion genes (Hb Lepore) or in the plucking out or insertion of single nucleotides. In such events, the primary structure of the globin chain peptide is altered.

Clinical Effects of Hemoglobinopathies

Sickling disorders. When deoxygenated, *hemoglobin S* forms long rope-like polymers (tactoids) that distort the shape of the erythrocyte, thus forming sickle cells. The formation of irreversibly sickled erythrocytes is the cause of severe manifestations that accompany homozygous Hb S disease. In contrast, *Hb S trait*, the heterozygous state, is usually innocuous.

Thalassemias also have important interactions with Hb S trait. When Hb S trait is inherited together with β^0-thalassemia trait, a severe sickling disorder results, since most of the hemoglobin is S. Somewhat milder sickling disorders occur in those who have inherited both Hb S trait and β^+- or β^{++}-thalassemia. In such cases Hb A is 10–25% of the total; most of the rest is Hb S. Those who inherit Hb S trait together with HPFH trait have no sickling disorder and are not anemic, although Hb S is about 70% of the total Hb (the remainder is Hb F). Alpha-thalassemias are protective against severe sickling, so that persons who have homozygous Hb S disease and also α-thalassemia trait have less severe sickling disorders than those who are simple Hb S homozygotes. This phenomenon may account both for the high frequency of α-thalassemia trait in blacks and for many persons with "mild homozygous sickle cell disease."

Table 24-2. CLINICALLY IMPORTANT HEMOGLOBINOPATHIES*

Clinical Manifestation	Hemoglobin Designation	Substitution	Comments
Sickling disorders†	S	β(A3)6 Glu→Val	
	C	β(A3)6 Glu→Lys	Target cells
	D-Punjab	β(GH4)121 Glu→Gln	
	O-Arab	β(GH4)121 Glu→Lys	
Unstable hemoglobin	Köln	β(FG5)98 Val→Met	Compensated hemolytic anemia
Hemolytic anemia	Zürich	β(E7)63 His→Arg	Sulfonamide-induced hemolysis
	Hasharon	α(CE5)47 Asp→His	No hemolysis or mild hemolysis
Cyanosis	M-Iwate	α(F8)87 His→Tyr	This group of rare abnormal hemoglobins is characterized by abnormal absorption spectra and by normal methemoglobin values when tested by the usual spectrophotometric measurement of methemoglobin (see methemoglobin assay)
	M-Boston	α(E7)58 His→Tyr	
	M-Milwaukee	β(E11)67 Val→Glu	
	M-Hyde Park	β(F8)92 His→Tyr	
	M-Saskatoon	β(E7)63 His→Tyr	
	Kansas	β(G4)102 Asn→Thr	Rightward displacement of O_2 dissociation curve
Erythrocytosis	Chesapeake	α(FG4)92 Arg→Leu	This group of abnormal hemoglobins is characterized by marked leftward displacement of the O_2 dissociation curve
	J-Camptown	α(FG4)92 Arg→Gln	
	Malmö	β(FG4)97 His→Gln	
	Kempsey	β(G1)99 Asp→Asn	
	Ypsilanti	β(G1)99 Asp→Tyr	
	Yakima	β(G1)99 Asp→His	
	Rainier	β(HC2)145 Tyr→Cys	
	Bethesda	β(HC2)145 Tyr→His	
Thalassemic blood picture	E	β(B8)26 Glu→Lys	In Southeast Asians
	Lepore	δβ Fusion	In Mediterraneans
	H	β_4	In Asians
"Physiologic anemia"	Yoshizuka	β(G10)108 Asn→Asp	Decreased O_2 affinity facilitates release of O_2 to tissues

*This is a selected list of abnormal hemoglobins with significant clinical effects.

†Hb S and variants which, when inherited together with Hb S (in double heterozygotes), cause sickling disorders.

Abbreviations for amino acids are Glu, glutamic acid; Val, valine; Lys, lysine; Met, methionine; His, histidine; Arg, arginine; Tyr, tyrosine; Gln, glutamine; Asn, asparagine; Leu, leucine; Thr, threonine; Asp, aspartic acid; Cys, cysteine.

Unstable hemoglobin hemolytic anemia. These variants are usually the result of substitution of a polar (hydrophilic) amino acid for a nonpolar (hydrophobic) amino acid that lines the "pocket" in which the heme group is located. The instability conferred by these substitutions results in accelerated denaturation of hemoglobin and in a hemolytic disorder that may be mild and fully compensated (as in *Hb Köln* disease), or in severe hemolysis (as in *Hb Santa Ana disease*). In some of these disorders, hemolysis may be induced by ingestion of sulfonamides. Most of these conditions are quite rare.

Erythrocytosis. The hemoglobin variants with high oxygen affinity often cause moderate to marked increase in hemoglobin concentration and hematocrit as well as in erythrocyte count. This is because such hemoglobin variants do not readily release O_2 to tissues and are thus functionally inert. Each of the more than three dozen known high O_2-affinity hemoglobins is quite rare.

Cyanosis. Lifelong cyanosis may be the result of a hemoglobin variant in which the iron atom of the heme cannot be reduced by physiological means from the trivalent state. Several of these "*M-hemoglobin*" *methemoglobinemias* have been defined. All but one (Hb M-Milwaukee) are the result of tyrosine's substituting for one of the heme-binding histidines. The M-hemoglobin methemoglobinemias are extremely rare. Furthermore, most are not identified by the usual assay for methemoglobin.

Classification of Hemoglobin Variants*

Hemoglobin variants are classified according to the globin chains affected and according to the sites of substitution in the globin chains. They are further classified as to whether they represent point mutations (single amino acid substitutions), deletion, fusion, or chain extension. A few hemoglobin variants have two amino acid substitutions. The best known of these is *Hb C-Georgetown (C-Harlem).* Although Hb C-Georgetown behaves electrophoretically like Hb C at alkaline pH, in every other respect it behaves like Hb S, and the person who has Hb C-Georgetown trait really has Hb S trait.

Genetics

Inheritance of structurally abnormal hemoglobinopathies follows simple Mendelian laws. Most abnormal hemoglobins cause little or no clinical manifestations in the heterozygote and thus are autosomal recessive traits. On the other hand, a person who is heterozygous for an unstable hemoglobin, for a Hb M, or for a high O_2-affinity hemoglobin has the clinical effects indicated in Table 24-2. These conditions are transmitted as autosomal dominant traits.

Since the structure of α-chains is controlled by four gene loci, abnormal hemoglobins that have substitutions in the α-chain generally make up ~25% of the total hemoglobin. If one of the two β-chain genes contains an altered sequence of DNA (a mutation), ~50% of the β-chains formed will be abnormal.

Abnormal β-chains are not produced as rapidly as normal β-chains. The proportion of Hb S in a person with sickle cell trait is usually ~40%. Similarly, in Hb C trait, there is about 40% Hb C. Hemoglobin E is ~30% of total hemoglobin in Hb E trait, and Hb O-Arab is also ~30–35% of total hemoglobin. If a mutation in only one of the pair of genes that controls β-chain structure results in an unstable hemoglobin (for example, Hb Köln), the unstable hemoglobin is often ~20% of total hemoglobin. This may be due in part to accelerated denaturation and preferential removal of circulating erythrocytes that contain a significant amount of unstable hemoglobin.

The inheritance of an identical mutation in both of the genes that determine β-chain structure results in a homozygous hemoglobinopathy. Thus, if both genes direct the synthesis of β^S-chains, only Hb S ($\alpha_2\beta_2^S$) can be formed, and no Hb A will be found. This gives rise to homozygous Hb S disease or sickle cell disease. The simultaneous inheritance of different mutations in both of the β-genes may also lead to serious disease, as exemplified by Hb S-C disease. In this disorder, which exhibits all of the features of sickle cell disease, both β^C- and β^S-chains are formed, leading to production of Hb S ($\alpha_2\beta_2^S$) and Hb C ($\alpha_2\beta_2^C$). Again, no Hb A is formed, inasmuch as there are no normal β^A-genes.

Hemoglobin Derivatives

The term *hemoglobin* implies that the iron atom is in the reduced (divalent or ferrous) state, whether hemoglobin is in the *oxygenated form* (*oxyhemoglobin*) or *deoxygenated form* (*deoxyhemoglobin*). When the iron is trivalent, the brown pigment is called *methemoglobin. Carboxyhemoglobin* is a hemoglobin–carbon monoxide complex. (Measurement of carboxyhemoglobin is reviewed in Chapter 27.) *Sulfhemoglobin, verdohemoglobin,* and *choleglobin* are degradation products of hemoglobin. In strongly basic or strongly acidic solution, hemoglobin is denatured to alkaline hematin or acid hematin, respectively. Spectral absorbance maxima for the more important hemoglobin derivatives are given in Table 24-3.

---

*Tables of hemoglobin variants, classified in the manner described, are available from the International Hemoglobin Information Center, Comprehensive Sickle Cell Center, Augusta, GA 30912.

Table 24-3. ABSORBANCE MAXIMA OF HEMOGLOBIN AND HEMOGLOBIN DERIVATIVES

Compound	Absorbance Maxima (nm)				
Deoxyhemoglobin	428–430		555		
Oxyhemoglobin	412–415		541	576–578	
Carboxyhemoglobin	417–418		537	568–572	
Methemoglobin (neutral pH)	404–407	500	540	578	630
Cyanmethemoglobin	413–418	480	541	580–590	

Measurement of Hemoglobin Concentration in Whole Blood

The measurement of hemoglobin concentration in blood is one of the most frequently performed clinical laboratory tests. The principle of the method given here has been adopted almost universally for both manual and automated procedures.

Clinical significance. Blood hemoglobin concentration may be diminished as a consequence of hemorrhage or hemolysis or as a result of impaired blood formation in the bone marrow. Conversely, blood hemoglobin concentration may be increased when there is impaired gas exchange through the lungs; it also is increased in a variety of other disorders.

Principle. The Fe(II) of hemoglobin is oxidized to Fe(III) by ferricyanide and then converted into stable cyanmethemoglobin by addition of KCN:

$$\mathrm{HbFe(II)} + \mathrm{Fe(III)(CN)_6^{3-}} \longrightarrow \mathrm{HbFe(III)} + \mathrm{Fe(II)(CN)_6^{4-}}$$

$$\mathrm{HbFe(III)} + \mathrm{CN^-} \longrightarrow \mathrm{HbFe(III)CN}$$

where HbFe(II) represents hemoglobin; HbFe(III), methemoglobin; and HbFe(III)CN, cyanmethemoglobin. Nonreactive ions are omitted. The absorbance of cyanmethemoglobin is measured at 540 nm, where it exhibits a broad absorbance peak (Figure 24-6).

Specimen. Blood may be anticoagulated with Na_2(EDTA) or may be taken directly from a finger or heel puncture without use of an anticoagulant.

Reagents

1. Drabkin's solution. Dissolve, in succession, 0.20 g of $K_3Fe(CN)_6$, 0.05 g of KCN, and 1.0 g of $NaHCO_3$ in distilled H_2O and dilute to 1000 mL. Store the reagent in a dark bottle at 4 °C. Stable for at least four months.

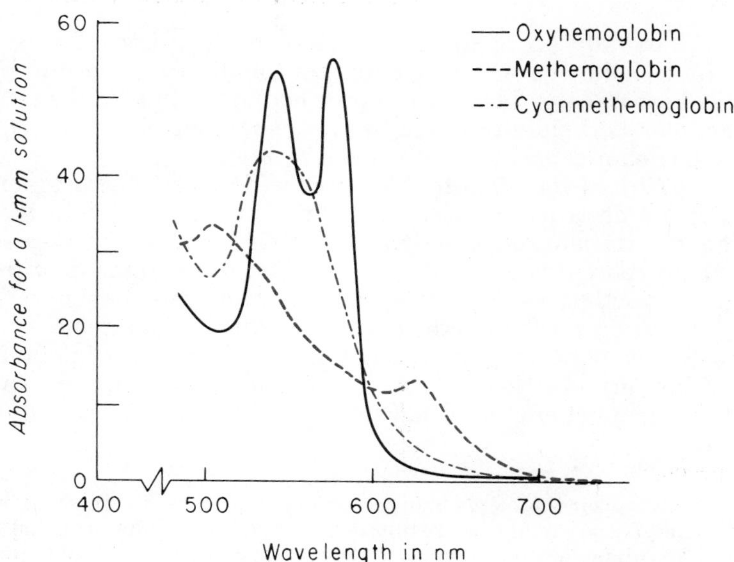

Figure 24-6. Spectrophotometric absorption curves for oxyhemoglobin, methemoglobin, and cyanmethemoglobin (authors' data). Oxyhemoglobin and cyanmethemoglobin are used in measuring the hemoglobin concentration. The peak at 630–635 nm, which is distinctive for methemoglobin, is abolished by addition of cyanide, and the resultant decrease in absorbance is directly proportional to the methemoglobin concentration. All heme proteins exhibit their maximal absorbance in the Sorét band region of 400–440 nm. Because the absorbance of hemoglobin in the Sorét region is approximately 10 times the absorbance at 540 nm, the Sorét peaks have been omitted from this diagram. The absorbance curve for methemoglobin is markedly influenced by small changes in pH. The curve given here is at pH 6.6.

2. Cyanmethemoglobin standard, usually 80 mg/dL. The concentration of the standard preparation is certified by an appropriate certifying agency.

Procedure

1. Set up a series of cuvets labeled blank, standard, control, sample 1, sample 2, sample 3, and so forth.
2. Pipet 6.0 mL of standard into the standard cuvet and 6.0 mL of Drabkin's solution into all other cuvets.
3. Thoroughly mix all blood specimens to be tested by repeated inversion immediately before testing.
4. Transfer 0.020 mL of whole blood from each sample to its respective cuvet.
5. Mix each sample cuvet and then let stand for ~5 min to ensure complete cell lysis and conversion of hemoglobin to cyanmethemoglobin.
6. Set spectrophotometer wavelength to 540 nm.
7. Place the blank in the spectrophotometer and adjust the meter to read 100% transmittance (0 absorbance). Then replace the blank cuvet with the standard cuvet and read its absorbance (A_S).
8. Measure the absorbance (A_U) of each of the sample cuvets.

Calculations

The dilution of blood in Drabkin's solution is 0.02 mL to 6.02 mL (1/301). The hemoglobin concentration of each sample (C_U) is calculated as:

$$C_U, \text{g/dL} = 301\left(\frac{A_U \times C_S}{A_S}\right) \times \frac{1}{1000} = 0.301\left(\frac{A_U \times C_S}{A_S}\right)$$

Reference ranges. For persons residing at or near sea level, the normal values of hemoglobin concentration are 14–18 g/dL in adult males and 12–16 g/dL in adult females. In healthy pregnant women, hemoglobin concentrations may be as low as 11.0 g/dL. In children, hemoglobin concentrations may be normally as low as 10.5 g/dL, and in adult males of age ≥60 years, as low as 12.5 g/dL.

Comments and precautions. The cyanmethemoglobin method has been adopted internationally as the reference method for hemoglobin measurement. Although cyanide is a lethal chemical, its concentration in Drabkin's solution is so low that it does not constitute any significant hazard to personnel. Nonetheless, reasonable care must be exercised in preparing and handling the solution and its subsequent disposal.

Measurement of Plasma Hemoglobin[15]

Virtually all of the hemoglobin in blood is contained within the erythrocytes. A minute quantity of hemoglobin is normally released into plasma by the destruction of the erythrocytes but is promptly bound by haptoglobin. The haptoglobin-hemoglobin complex is rapidly removed by parenchymal cells of the liver. Thus, the normal plasma hemoglobin concentration is close to zero.

Clinical significance. An increase in plasma hemoglobin concentration is indicative of acute destruction of erythrocytes (hemolysis) within the vascular system, as in hemolytic transfusion reactions, or in elevation of the degree of hemolysis occurring in extracorporeal treatment of blood. The measurement of free hemoglobin in plasma is of no practical value in the diagnosis of chronic hemolytic disorders.

Principle. The method is based on the measurement of oxyhemoglobin at 415 nm, the Sorét band of maximal absorbance of oxyhemoglobin. Measurements at 380 and 450 nm are also required for correction of turbidity and other nonspecific absorbance. A spectrophotometer with 1-nm spectral resolution is required.

Specimen

Measurement of plasma hemoglobin requires a minimum volume of 1 mL of heparinized plasma. Blood should be freshly drawn by a technique designed to minimize in vitro hemolysis from mechanical injury to erythrocytes. Since some degree of hemolysis always occurs when blood clots, the measurement of serum hemoglobin has no diagnostic value whatever.

1. Attach a 30-cm long polyvinyl tubing to the hub of a sterile 18-gauge needle or use an 18-gauge needle with attached tubing from an infusion set. Clamp the other end with a hemostat.
2. Prepare two plastic (or silicone-coated glass) centrifuge tubes, labeled 1 and 2, for collection of blood from the free end of the tubing.
3. Lightly coat the interior of tube 2 by spraying it with a fine mist of sodium heparin solution, 1000 units/mL. Only a small amount (0.05 mL) of heparin solution should be applied. Any large droplets should be wiped away.
4. Loosely tie a tourniquet around the upper arm. With minimal delay, perform a clean puncture of the antecubital vein. Release the tourniquet, and permit ~5 mL of blood to flow freely into tube 1; then collect 4 mL of blood in tube 2. Be careful that droplets of blood do not adhere to the test tube above the meniscus. Discard the blood in tube 1.
5. Promptly centrifuge tube 2 at 1000 × *g* for 10 min; carefully draw off the supernatant plasma with a transfer pipet and deliver it into a second centrifuge tube.
6. Recentrifuge plasma at 1600 × *g* for 20 min; carefully draw off the supernatant plasma and transfer it to a third tube.
7. Proceed promptly to the hemoglobin measurement described below or freeze at −20 °C until the test is performed.

Reagents

1. Na_2CO_3 stock solution, 1 g/dL.
2. Na_2CO_3 working solution, 0.01 g/dL. Dilute 1 mL of stock solution to 100 mL with distilled water.

Procedure

1. Prepare a blank by adding 4 mL of Na_2CO_3, 0.01 g/L, to a cuvet.
2. To an additional cuvet add 3 mL of Na_2CO_3, 0.01 g/L, and 0.3 mL of specimen. Cover with Parafilm and invert three times to mix.
3. Measure absorbance at 380, 415, and 450 nm.

Calculations

$$\text{Plasma Hb, mg/dL} = \frac{2A_{415} - (A_{380} + A_{450})}{1.655} \times \frac{1000\ D}{E}$$

where $\frac{2}{1.655}$ and $\frac{A_{380} + A_{450}}{1.655}$ are corrections for "nonspecific absorption,"
1000 converts g/dL to mg/dL,
D is the dilution of specimen in working solution (0.3 mL in 3.0 mL = 1:11), and
E is the absorptivity at 415 nm of a 1 g/dL solution of oxyhemoglobin (79.46).

Reference values. Reference values are difficult to define since the concentration of hemoglobin depends on the care with which samples are obtained and processed. With casually collected blood samples, plasma hemoglobin concentration is usually <5 mg/dL. With meticulous care in obtaining plasma, normal plasma hemoglobin should be <1.0 mg/dL. Even this amount is principally the result of in vitro hemolysis. Very lipemic plasma may give spuriously high results despite measurements at 380 and 450 nm. The method is not reliable when serum bilirubin concentration is increased.

Comments and precautions. Plasma Hb assays with benzidine derivatives, such as 3,3′,5,5′-tetramethylbenzidine, are much more sensitive than that described above. However, they are inaccurate because normal plasma contains an inhibitor of the peroxidase (or pseudoperoxidase) reaction that is the basis for benzidine-type assays of hemoglobin concentration. Thus, such assays underestimate plasma hemoglobin concentration by about 50%. The nature of the inhibitor(s) is unknown. Unsubstituted benzidine cannot be used because it is a carcinogen.

Determination of Methemoglobin and Sulfhemoglobin

When the iron in hemoglobin is oxidized to the trivalent state, the resulting brownish pigment is *methemoglobin. Sulfhemoglobin* is believed to contain sulfur bonded to pyrroles of the porphyrin ring.

Clinical significance. The normal concentration of *methemoglobin* is <1.5% of total hemoglobin. In congenital methemoglobinemia that is due to methemoglobin reductase deficiency, the methemoglobin concentration in blood is increased to 10–20% of the total heme pigment, although a few cases have been reported in which the methemoglobin concentration has been as high as 40–55%. Methemoglobin concentrations >60% are usually lethal regardless of cause. Methemoglobinemia also occurs in persons who have ingested nitrites, nitrates, or certain drugs or chemicals such as sulfones or aniline dyes. *Sulfhemoglobin* is not a normal constituent of blood. It is sometimes found in concentrations of 1–10% of total hemoglobin after excessive use of certain drugs such as phenacetin or dapsone.

Principle. The absorbance spectrum of methemoglobin exhibits a small, characteristic peak at 630–635 nm (Figure 24-6). Addition of cyanide eliminates this peak by converting methemoglobin to cyanmethemoglobin. The decrease in absorbance is proportional to the methemoglobin concentration.

The normal absorbance spectrum of oxyhemoglobin shows very little absorbance above 600 nm. However, if sulfhemoglobin is present in a hemolysate, there is a broad increase in the absorption curve in the range of 600–620 nm. This sulfhemoglobin plateau is not affected by treatment with cyanide.

Specimen. Blood should be fresh and may be anticoagulated with heparin, EDTA, or ACD (acid-citrate-dextrose) solution. No fluid or food restriction is needed.

Reagents

1. Potassium ferricyanide. Dissolve 2.0 g of $K_3Fe(CN)_6$ in distilled H_2O and dilute to 10.0 mL. If stored in a brown bottle at 4 °C, this solution is stable for at least one year.
2. Potassium cyanide solution (CAUTION: lethal poison). Dissolve 500 mg of KCN in distilled H_2O and dilute to 10 mL. Label "POISON." Stable at 20 °C for at least four months.
3. Potassium phosphate buffer, 0.15 mol/L, pH 6.6 (20 °C). Dissolve 17.1 g of $K_2HPO_4 \cdot 3\ H_2O$ (or 13.2 g anhydrous) in 500 mL of distilled H_2O and 10.2 g of KH_2PO_4 in 500 mL of distilled water. Transfer 200 mL of the KH_2PO_4 solution to a 1-L beaker and add an equal volume of K_2HPO_4 solution. Place a pH electrode in the beaker. Then slowly add more K_2HPO_4 solution, with constant stirring, until the mixture has a pH of 6.6. (The ratio of volumes required is about 9:16, respectively.) Store at 4 °C. Discard whenever the solution appears turbid. New buffer should be prepared at least once every three months.

Procedure (Methemoglobin Only)[3]

1. Prepare a blank cuvet containing 1.5 mL of phosphate buffer and 1.5 mL of H_2O. Designate this cuvet C_1.
2. Pipet 0.1 mL of whole blood into a test tube containing 3.9 mL of distilled H_2O; swirl to mix.
3. Add 4.0 mL of potassium phosphate buffer and mix thoroughly.
4. Transfer 3 mL of hemolysate to each of two cuvets; designate these C_2 and C_3.
5. To cuvet C_3, add 0.1 mL of $K_3Fe(CN)_6$ solution. Cover with Parafilm, mix by inverting three times, and measure its absorbance at 630 nm after 2 min using C_1 as a blank. Record as A_{3a}.
6. Measure the absorbance at 630 nm for cuvet C_2 by using C_1 as blank. Record as A_{2a}.
7. Add 0.1 mL of KCN to all cuvets. (Use a safety pipet or add 2 drops from a transfer pipet fitted with a rubber bulb.) Mix by inverting three times, and allow to stand for 5 min.
8. Measure absorbance at 630 nm for cuvets C_2 and C_3, with C_1 as blank. Record as A_{2b} and A_{3b}.

Calculations

$$\text{Methemoglobin, \% of total Hb} = 100\left(\frac{A_{2a} - A_{2b}}{A_{3a} - A_{3b}}\right)$$

The *reference range* for methemoglobin is 0–1%. A methemoglobin concentration >1.5% is abnormal.

Solubility Test for Hemoglobin S[10,12,19]

This simple and rapid method detects the presence of any sickling hemoglobin, such as Hb S or Hb C-Georgetown, but does not quantitate the amount of abnormal hemoglobin and therefore does not differentiate Hb S trait from homozygous Hb S disease. Furthermore, other hemoglobin variants may be present in the blood whether results are positive or negative.

Principle. Hemoglobin S, when deoxygenated, is insoluble in concentrated phosphate buffer and produces a visible turbidity. Almost all other hemoglobins, including hemoglobins A, F, C, E, and D, are soluble in such solutions. A reducing substance, sodium hydrosulfite ($Na_2S_2O_4$, sodium dithionite), is used to deoxygenate the hemoglobin, and saponin is used to lyse the erythrocytes.

Interpretation. The test is positive if there is sufficient turbidity to prevent reading of print through the suspension. Normal specimens show only a faint haziness; most are completely transparent. A positive test is presumptive evidence for Hb S but should be confirmed by hemoglobin electrophoresis, since rare sickling hemoglobins will also give positive solubility tests, as will Hb Barts (when present as 100% of hemoglobin) and specimens that contain numerous normoblasts or Heinz bodies, as is seen after splenectomy in patients with severe anemias.

Comments and precautions. Because it does not identify other deleterious hemoglobin variants or thalassemias, the solubility test is best used not in screening programs but instead in determining whether a hemoglobin variant that appears to be hemoglobin S, D, or G by electrophoresis is, in fact, hemoglobin S.

False-negative results are very rare but may occur when Hb S concentration is <15% of total. Increasing the volume of erythrocytes added to the test will then result in a positive test if Hb S is present. Reagent kits for Hb S screening are commercially available from several sources. If the test is done using whole blood rather than packed erythrocytes, false-positive results may result from lipemia or dysproteinemia.

Electrophoretic Separation of Hemoglobins on Cellulose Acetate[6,10]

Principle. At pH 8.6 hemoglobins are negatively charged ions that move toward the anode in an electrical field. However, hemoglobins differ in the amount of charge per molecule if their component amino acids differ in the number of amino groups. Hemoglobin S has two more and Hb C has four more positive charges per molecule than does Hb A. Therefore, Hb S moves toward the anode more slowly than does Hb A, and Hb C moves even more slowly. Conversely, hemoglobins H, I, and J, which have additional negative charges, all move toward the anode more rapidly than Hb A.

Cellulose acetate is convenient, provides sharp resolution of hemoglobin bands in a relatively short time, and may be stored permanently.

Specimen. This procedure requires a minimum of 0.5 mL of whole blood. Any anticoagulant may be used. A specimen may be stored at 4 °C for one week if the anticoagulant is heparin and for three weeks if collected with EDTA or ACD (acid-citrate-dextrose) solution. However, if an unstable hemoglobin (for example, H or Köln) is suspected, electrophoresis should be carried out within 24 h of collection.

Reagents

1. Tris-EDTA-borate (TEB) buffer, pH 8.6 (25 °C). Dissolve 12 g of tris(hydroxymethyl)-aminomethane (Tris), 1.56 g of disodium ethylenediaminetetraacetate (EDTA), and 0.92 g of boric acid in distilled H_2O and dilute to 1000 mL. Store at 4 °C. This solution is stable indefinitely.
2. Stain. Dissolve 20 g of ponceau S, 30 g of trichloroacetic acid, and 30 g of sulfosalicylic acid in distilled H_2O and dilute to 1000 mL. Make a new solution at least once every two months.
3. NaCl, 0.155 mol/L. Store at 4 °C.

Procedure

1. Place 0.5–1 mL of whole blood or, if plasma has been removed, 0.2–0.5 mL of sedimented cells into a centrifuge tube. Suspend the cells in 10–15 mL of cold NaCl solution.

2. Centrifuge at 5 °C for 10 min at 900 × *g*.
3. Aspirate and discard the supernatant saline.
4. Add 9 vol of cold distilled H_2O per 1 vol of packed cells, followed by 0.2 mL of xylene or toluene. Resuspend the packed cells and then thoroughly mix. Hemoglobin H is denatured and removed by this step. If an unstable hemoglobin is suspected, omit addition of xylene or toluene and do not use vortex mixer.
5. Centrifuge at 1600 × *g* for 15 min.
6. Aspirate and discard the supernatant xylene or toluene.
7. The clear hemolysate lying above the stromal precipitate is removed for electrophoresis. It may be stored at 4 °C for use the same day or at −20 °C for use in subsequent studies.
8. Prepare a cellulose acetate membrane as described by the supplier. Apply samples at a template position near the cathode in alternating positions with control specimens containing hemoglobins of known type, such as J, A, F, S, or C. Place TEB buffer in both buffer chambers, making sure that the ends of the membrane are in contact with the buffer and that the buffer level is the same in each chamber.
9. Perform electrophoresis at room temperature for 30 min at 450 *V*.
10. Stain with ponceau S. Then destain, clear the background, and dry membrane for storage.

Interpretation

The order of electrophoretic mobility of the more commonly encountered hemoglobin variants is shown in Figures 24-7 and 24-8. Usually, Hb A is preceded by a fainter band that merges with it. This band is a mixture of glycosylated hemoglobins.

Hemoglobins A_2, C, O-Arab, and E cannot be differentiated on the basis of electrophoretic mobility in this medium, nor can hemoglobins D, G, and Lepore be distinguished from Hb S. Hemoglobin F may be poorly resolved from Hb A. Hemoglobins H and I are indistinguishable. However, the relative amounts found are useful in distinguishing between some of the hemoglobins that have the same mobility. For example, Hb A_2 is never as high as 10% of the total hemoglobin. Thus, if a very slowly moving band constitutes 10% or more of the total hemoglobin, it may be presumed to be hemoglobin C, O-Arab, or E. Hemoglobins C and O-Arab are virtually limited to persons of Central African ancestry, whereas Hb E is virtually limited to those of Southeast Asian ancestry. Hemoglobin Lepore constitutes only 10–15% of the total hemoglobin, whereas hemoglobin S, D, or G constitutes 25–45%. Similarly, Hb H constitutes ~5–15% of the total hemoglobin, whereas Hb I (an α-chain variant) is usually ~25%. Very rarely, Hb H may comprise 30–40% of total hemoglobin in acquired Hb H disease associated with erythroleukemia.

Thus, by measuring the proportions of these hemoglobin bands densitometrically and by knowing the ethnic origin of a patient, one may deduce the probable identity of an abnormal hemoglobin. Additional confirmatory tests are often necessary, such as the hemoglobin solubility

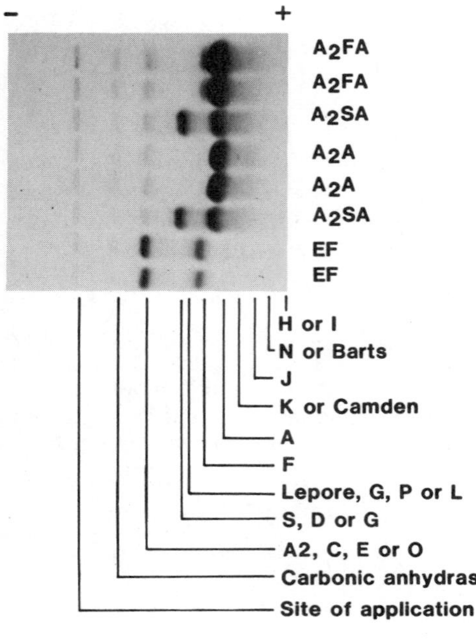

Figure 24-7. Hemoglobin electrophoresis on cellulose acetate at pH 8.6 following staining with ponceau S. In the first two tracks, beginning at the top, is a specimen from a patient with β-thalassemia minor. Note prominent bands for Hb A_2 and F. The third and sixth tracks contain a specimen from a patient with Hb S trait. Hemoglobins A and S are present in nearly equal proportions. The fourth and fifth tracks contain specimens from a normal adult. The two lowermost tracks contain specimens from a patient with Hb E/β^0-thalassemia. Note the absence of Hb A, and that Hb E and Hb F are present in proportions of about 2 and 3 to 1, respectively. Hemoglobin C/β^0-thalassemia or O/β^0-thalassemia would result in the same pattern as in the two lowermost tracks.

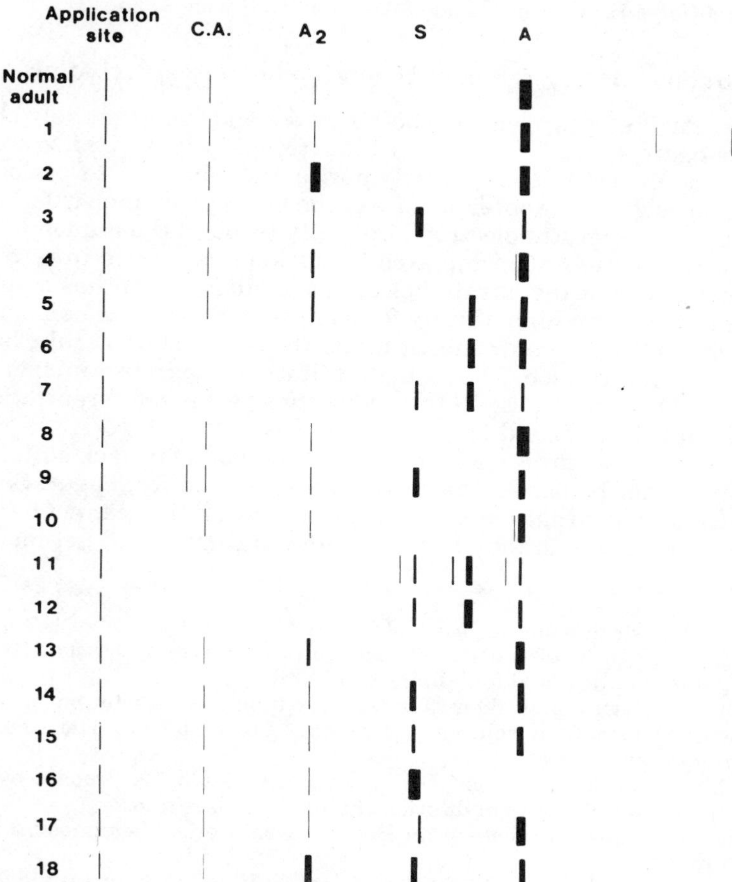

Figure 24-8. Hemoglobin electrophoresis on cellulose acetate, pH 8.6. Some abnormal patterns that are commonly encountered in clinical laboratories are shown in tracks 1-18. C.A. = carbonic anhydrase band. Relative proportions of each band are indicated by width.

Track 1. Hemoglobin H disease. Two minor bands are seen anodal to Hb A. These are Hb H (most anodal) and Hb Barts. Hb H is 10%. Hb Barts is seen typically in umbilical cord blood of neonates with α-thalassemia; Hb H is not seen in neonates.

Track 2. Hb C trait. Hemoglobins O and E cannot be distinguished from Hb C in this medium. However, in Hb C trait, Hb C is about 40% of total hemoglobin, whereas in Hb E trait or O trait, the abnormal band is usually 25-30% of total hemoglobin.

Track 3. Hb S/β^+-thalassemia. Whenever the proportion of Hb S exceeds that of Hb A, concurrent β-thalassemia is likely. The same pattern would characterize a specimen from a patient with homozygous Hb S disease who had been transfused with blood that contains Hb A. Information about recent transfusion is critical to interpretation of hemoglobin electrophoresis results.

Track 4. β-thalassemia minor with increase in the proportion of Hb A_2 to 3.5–9% of total hemoglobin. Hemoglobin F is 2–5% of total hemoglobin in about half the cases of β-thalassemia minor. When Hb F is present in small proportion, it migrates just cathodal to Hb A, but when present in high proportion it migrates nearly midway between Hb A and Hb S, as in tracks 5–7.

Track 5. Homozygous β^+-thalassemia or homozygous β^0-thalassemia following transfusion of normal blood. Hb F is markedly increased in all homozygous β-thalassemias, and increase in Hb A_2 may also be observed.

Track 6. Umbilical cord blood from a normal newborn. Characteristic are high proportion of Hb F (50–70% of total) and absence of Hb A_2 and carbonic anhydrase.

Track 7. Umbilical cord blood from a neonate who has Hb S trait. Note approximately equal proportions of hemoglobins A and S, preponderance of Hb F, and absence of Hb A_2 and carbonic anhydrase.

Track 8. Normal adult or child >1 year of age. However, several abnormal hemoglobins co-migrate with Hb A; a normal pattern therefore does not exclude the presence of an abnormal hemoglobin. An example is Hb Malmö that causes erythrocytosis. This pattern would be observed in Hb Malmö trait. This pattern is also typical for β-thalassemia minor or Hb H disease when either is associated with iron deficiency, since iron deficiency reduces the proportion of Hb A_2 and may result in marked diminution in Hb H. The correct diagnosis is made by repeating electrophoresis following treatment of iron deficiency.

(*continued*)

test for Hb S, or a test for an unstable hemoglobin (e.g., Köln, E, or H). Agar gel electrophoresis at pH 6.2 differentiates Hb C from E and O-Arab, or Hb S from D and G.

Electrophoretic Separation of Hemoglobins in Agar Gel at pH 6.2[10,21]

Principle. Purified agar contains both *agarose* and the negatively charged sulfated polysaccharide *agaropectin*. In an agar gel, agarose is polymerized to constitute an immobile matrix; agaropectin does not polymerize and can bind reversibly with a small number of amino acids on the outer surface of the hemoglobin molecule. The hemoglobin-agaropectin complex migrates electrophoretically through the immobile agarose matrix toward the anode, whereas noncomplexed hemoglobin is carried toward the cathode by electroendosmotic flow of the citrate buffer. Thus Hb S, which has a substitution near the external surface, has a high affinity for agaropectin, whereas hemoglobins D-Punjab and G-Philadelphia, with similar substitutions deeper in the molecule, have low affinity for agaropectin. Similarly, Hb C has a high affinity for agaropectin, but Hb E, with like charge, does not. This simple, useful technique thus permits differentiation of Hb S from D and G, and Hb C from E and O.

Gels prepared with agarose, a refined derivative of agar, lack sulfated agaropectin; thus, such gels cannot be used in place of those prepared from agar. Some recent commercial batches of purified agar also lack agaropectin and thus cannot be used. Therefore, clinical laboratory personnel may have to resort to commercial gel membranes.

Reagents

1. Citrate buffer, stock solution, 0.5 mol/L, pH 5.8 (25 °C). Dissolve 147 g of trisodium citrate ($Na_3C_6H_5O_7 \cdot 2\ H_2O$) in distilled H_2O and dilute to 800 mL. Dissolve 10.5 g of citric acid ($H_3C_6H_5O_7 \cdot H_2O$) in distilled H_2O and dilute to 100 mL.

Add the citric acid solution (~68 mL) to the trisodium citrate solution to adjust the mixture to pH 5.8. Add H_2O to a final volume of 1000 mL. Store at 4 °C. The solution is stable for approximately one year.

2. Citrate buffer, working solution, 0.05 mol/L, pH 6.2 (25 °C). Prepare by diluting 1 vol of stock solution with 9 vol of H_2O. On dilution the pH increases to 6.2.

3. Agar gel membrane. Purchase from Helena Laboratories, Beaumont, TX 77704, or other commercial source.

4. Naphthol Blue-Black stain. Mix 0.6 g of Naphthol Blue-Black dye (Amido Black 10B; Sigma Chemical Co., St. Louis, MO 14508), 6.0 g of trichloroacetic acid, 50.0 mL of 95% ethanol, and

---

Figure 24-8 (*Continued*)

Track 9. Hemoglobin S trait together with Hb A_2' trait. Hb A_2' is seen cathodal to carbonic anhydrase. It is found in 1% of blacks and is harmless. The proportion of Hb A_2' is equal to that of Hb A_2.

Track 10. Slight increase in Hb F, as typically seen in many chronic anemias such as hereditary spherocytosis or myelofibrosis.

Track 11. Improperly handled umbilical cord blood from a neonate with Hb S trait. Failure to maintain the specimen at 4 °C during several days that preceded electrophoresis resulted in formation of methemoglobins A, F, and S, shown as minor bands.

Track 12. Same specimen as in Track 11 following addition of KCN to convert methemoglobins to cyanmethemoglobins.

Track 13. Hemoglobin E trait in a patient who also has either iron deficiency or α-thalassemia. In either circumstance, the proportion of Hb E may be diminished to 10–20%, and Hb E may be mistaken for Hb A_2.

Track 14. Hemoglobin S trait, Hb D trait, or Hb G trait. These conditions cannot be distinguished by electrophoresis in this medium.

Track 15. Hb G-Philadelphia trait. Hb G-Philadelphia is seen in the Hb S position in this medium and is usually about 25–35% of total hemoglobin.

Track 16. Homozygous Hb S disease or Hb S/Hb D disease or Hb S/β^0-thalassemia. Hemoglobin F proportion may also be increased. Note the absence of Hb A.

Track 17. Hb Lepore trait. Hemoglobin Lepore is about 10–15% of total hemoglobin.

Track 18. Hemoglobins A, S, and C in a blood specimen? Not genetically possible because β^A, β^C, and β^S are alleles for the same chromosomal locus. Usually this pattern is due to the fact that the patient has Hb S trait together with Hb G-Philadelphia trait, resulting in the combinations $\alpha_2^A\beta_2^A$ or Hb A, $\alpha_2^G\beta_2^A$ or Hb G-Philadelphia (in the Hb S position), $\alpha_2^A\beta_2^S$ or Hb S, and Hb $\alpha_2^G\beta_2^S$ or the hybrid Hb S-G-Philadelphia that migrates in the Hb C position. Other possible explanations are (1) the patient has Hb S trait and was transfused with blood from a donor with Hb C trait (or has Hb S/C disease and was transfused with blood from a normal donor), or (2) specimens from patients with Hb S trait and Hb C trait were inadvertently mixed in the laboratory.

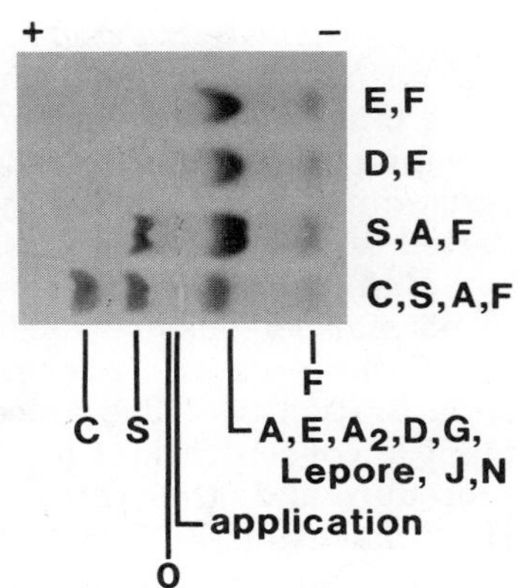

Figure 24-9. Electrophoresis of hemoglobin on citrate agar gel, pH 6.2. Cathode is to right, anode to left. Characteristic hemoglobin positions are indicated below the gel. Letters on the right margin indicate the hemoglobins present in four specimens. The specimen in the uppermost track is from a patient with homozygous Hb E and contains 98% Hb E and 2% Hb F. The second track is from a patient with Hb D-Punjab trait. It contains 51% Hb A and 49% Hb D (co-migrating) and 1% Hb F. The third track is a specimen from a patient who has Hb S trait and α-thalassemia trait. It contains hemoglobins S, A, and F. Hb S is 20% of total hemoglobin, a low proportion that indicates concurrent α-thalassemia, consistent with a low mean corpuscular volume of 68 fL. The lowermost track is the control, prepared by mixing blood of hemoglobins C trait and S trait. It contains hemoglobins C, S, A, and F.

0.75 mL of glacial acetic acid; dilute with distilled H_2O to 2000 mL. Prepare a fresh solution once a month.

5. Acetic acid rinse, 2%. Dilute 10 mL of glacial acetic acid with distilled H_2O to 500 mL.
6. Sodium chloride, 0.9 g/dL.

Preparation of Hemolysates

1. Centrifuge 1 mL of anticoagulated whole blood at 700 × *g* for 10 min.
2. Remove and discard plasma, and then resuspend erythrocytes in 10 mL of NaCl, 0.9 g/dL.
3. Centrifuge at 700 × *g* for 10 min. Remove and discard supernatant.
4. Lyse erythrocytes by alternately freezing and thawing twice.
5. Measure the hemoglobin concentration of the hemolysate and adjust it to 10 g/dL by adding working buffer.

Procedure. Follow manufacturer's instructions.

Interpretation. The relative order of hemoglobin migration in this medium is shown in Figure 24-9. Hemoglobin F shows the most marked cathodal movement of the common hemoglobins, whereas Hb C shows anodal migration. Hemoglobin S remains close to the site of application or may be slightly anodal. Hemoglobins D, G, and A migrate slightly cathodal and thus separate well from Hb S. Hemoglobin E also travels with A and thus is easily differentiated from C. Hemoglobin O-Arab has a slightly slower cathodal migration than A, and this serves to distinguish it from C as well as from A, D, G, and E. Hemoglobin A_2 has mobility similar to that of Hb A. Hemoglobin Lepore also exhibits mobility identical to that of Hb A, thus differentiating it from Hb S.

Comments and precautions. The concentration of hemoglobin applied to the gel affects mobility. Thus, if there is a larger quantity of Hb A or of a variant hemoglobin, the apparent electrophoretic mobility toward the cathode will be increased.

Test for Unstable Hemoglobins[8,10]

Clinical significance. More than 100 unstable hemoglobin variants have been described, and most of these are associated with hemolytic anemia. Although heat tests have been widely used in the past, chemical stress tests are more practical for the demonstration of unstable hemoglobin.

Principle. If an unstable hemoglobin is present in a hemolysate, incubation in dilute isopropanol solution at 37 °C results in marked turbidity within 5–10 min, whereas hemolysates containing normal hemoglobins remain transparent much longer.

Determination of Fetal Hemoglobin (Alkali-Resistant Hemoglobin)[2,20]

Clinical significance. In δβ-thalassemia minor, the fetal hemoglobin concentration is usually in the range of 5–15% of the total hemoglobin, whereas in β-thalassemia major the fetal hemoglobin may be 70–90% or more of the total hemoglobin. Slight increases in Hb F concentration are found in a variety of unrelated hematological disorders such as aplastic anemia and acute leukemia. In homozygous sickle cell disease, the Hb F concentration is often slightly increased. Higher concentrations of Hb F occur in Hb S-thalassemia and in patients who are doubly heterozygous for the Hb S gene and for a gene for hereditary persistence of fetal hemoglobin (HPFH).

Principle. Most hemoglobins denature readily at alkaline pH and can then be precipitated by addition of ammonium sulfate to 40% saturation. Fetal hemoglobin is resistant to denaturation by alkali and remains soluble. All hemoglobins are first converted to cyanmethemoglobin before denaturation.

Reference ranges. Hb F is usually <1% and never >2% of the total hemoglobin in normal persons over one year of age.

Comments and precautions. There are several published variations of the alkali denaturation test. The procedure referenced is the most sensitive method for measurement of concentrations of fetal hemoglobin when it is <10% of total hemoglobin. This procedure and other rapid tests for fetal hemoglobin give erroneously low results when Hb F makes up 30% or more of the total hemoglobin in a specimen. In order to obtain reproducible results, the final cyanmethemoglobin concentration measured in this method should not be <480 mg/mL.

Determination of Hemoglobin A_2[1,10,14]

Principle. Hemoglobins are ionically bound to DEAE cellulose and then eluted by changes in pH or salt concentration. Hb A_2 is eluted by a solution of glycine-KCN. Then, hemoglobins A and F are eluted with a solution of glycine, KCN, and NaCl. Absorbances of the eluted fractions are measured at 415 nm.

Specimen. The specimen is prepared as for electrophoresis on cellulose acetate (see page 807), or 50 μL of whole blood are hemolyzed by mixing with 200 μL of distilled H_2O and letting stand for 10 min.

Reagents. Columns and reagents can be purchased from commercial sources (Isolab Inc., Akron, OH 44321; Helena Laboratories, Beaumont, TX 77704; and others). These kits include:

1. Chromatographic "minicolumns" of DEAE cellulose, prepoured and pre-equilibrated with pH-adjusted glycine-KCN solution. The resin is held in place by a plastic disk.
2. Hemoglobin elution solutions:
   Solution I (for elution of Hb A_2) contains glycine, 0.2 mol/L, and KCN, 1.54 mmol/L, adjusted to pH ~7.5 depending on the batch of DEAE cellulose.
   Solution II (elution of Hb A and F) contains glycine, 0.2 mol/L; KCN, 1.54 mmol/L; and NaCl, 0.2 mol/L.

Procedure. Follow manufacturer's instructions and include a quality control specimen with each group of assays.

Reference range. Results in normal adults are in the range of 1.5–3%, median 2.2%. In β-thalassemia minor, Hb A_2 values range from 3.5 to 6%. A very small proportion of specimens yields values for Hb A_2 in the indeterminate range between 3.0 and 3.5%.

Comments and precautions. In contrast to other hemoglobins, Hb A_2 must be measured with precision because the diagnosis of β-thalassemia minor may hinge on a Hb A_2 value only 1–2% above the reference range. Alternative procedures suffer from unacceptable imprecision.

Hemoglobins C, E, and O co-elute with A_2 in this procedure. Thus, the presence of these hemoglobins precludes Hb A_2 quantitation by this technique. On the other hand, when Hb C, E, or O is present, an assay of Hb A_2 is not needed for diagnosis. Any measured values for A_2 of ≥10% should be assumed to represent hemoglobin C, E, or O.

Detection of Hemoglobin in Sputum, Urine, and Feces (Occult Blood)

Clinical significance. Hemoglobin is not normally present in sputum, urine, or feces except in minute amounts; the demonstration of the presence of hemoglobin in these materials is an important diagnostic clue that usually implies bleeding within the respiratory, urinary, or gastrointestinal tracts, respectively. ("Occult" refers to the fact that the amount of blood present may be too small to be noticeable upon visual inspection.)

Principle. These tests for the detection of hemoglobin depend on the fact that heme proteins act as peroxidases, catalyzing the reduction of hydrogen peroxide to water, a reaction that requires a hydrogen donor. If guaiac or *o*-tolidine (a benzidine derivative) is used as a hydrogen donor, the oxidation of the donor chromogen results in a blue color, and the intensity of the final color is primarily a function of the hemoglobin concentration. Commercially available kits are suitable for this test.

Comments and precautions. *o*-Tolidine is used in a commercially available dipstick method (Hemastix, Ames Company, Elkhart, IN 46514) to test for hemoglobin or myoglobin in urine. Hematest (Ames Co.) detects blood in feces. The *o*-tolidine tests (Hemastix and Hematest) are sensitive to ~2 mg of hemoglobin/dL of urine and to ~100 mg of hemoglobin/100 g of feces, the equivalent of ~10 erythrocytes/μL of urine or 1 mL of blood/100 g of feces, respectively.

A false-positive urine test may occur if the urine container has been contaminated with hypochlorite solution, if peroxidase-containing bacteria are present in the urine, or if myoglobin is present. The *o*-tolidine-based test (Hematest) is so sensitive that false-positive results are very common (e.g., in patients who are taking oral iron medications). A guaiac-based test for fecal occult blood is the Hemoccult test (SmithKline Diagnostics, Inc., Sunnyvale, CA 94086), in which false-positive results are very infrequent. However, false-negative reactions are quite common when the rate of blood loss in feces is ~10 mL/d.

A recently developed test (Hemoquant; SmithKline Bioscience Laboratories, Inc., King of Prussia, PA 19406) for occult gastrointestinal bleeding is based on measurement of porphyrins in feces. This test may have greater sensitivity and specificity than the Hemoccult test.

MYOGLOBIN[18]

Myoglobin is an oxygen-binding protein of striated (cardiac and skeletal) muscle. It closely resembles the α-, β-, γ-, and δ-chains of hemoglobin, but in contrast to hemoglobin, myoglobin exists only as a monomer and thus is unable to release oxygen except at extremely low oxygen tensions. Its physiological role is uncertain but may be an oxygen reservoir of last resort under circumstances of extreme hypoxia. Increases in serum myoglobin concentrations occur following trauma to either skeletal or cardiac muscle, as in crush injury or myocardial infarction. A radioimmunoassay method for serum myoglobin is commercially available.

Measurement of serum myoglobin concentration has been advocated for the diagnosis of myocardial infarction. However, it is diagnostically less sensitive than the creatine kinase assay and is subject to false-positive results arising, for example, from seemingly minor injury to skeletal muscle.

Following injury to skeletal or cardiac muscle, myoglobin may be released and excreted in the urine (*myoglobinuria*). Pronounced myoglobinuria, sufficient to cause darkening of the urine, may occur after strenuous physical exercise or as the result of an accident in which there is crushing of muscle. Copious amounts of myoglobin in the urine may contribute to impairment of renal function.

Methods. Several physicochemical properties permit differentiation of hemoglobin and myoglobin. The nephelometric *immunoprecipitation* method is the procedure of choice. Alternatively, *ultrafiltration* of clear, centrifuged urine through a Millipore filter 8–12 nm in pore diameter holds back hemoglobin but allows myoglobin to pass. The presence of heme protein in the ultrafiltrate is indicative of myoglobin.

A *solubility test* for myoglobin in urine is based on the fact that hemoglobin is completely precipitated in 80% saturated ammonium sulfate solution, whereas myoglobin remains in solution. The supernatant is tested for heme (Hemastix is convenient for this purpose). If the test for heme protein is positive, this is presumptive evidence for the presence of myoglobin. Unfortunately, this simple and widely used test frequently gives erroneous results.

Reference ranges. Serum myoglobin concentrations in adult males are <80 μg/L and in adult females <60 μg/L and increase with age; children have very low values. The mean serum myoglobin concentration of black men is about 10% higher than that of white men. Results of qualitative urine tests are negative in normal subjects.

VITAMIN B_{12}[9]

Structure and Terminology

The vitamin B_{12} group of substances are physiologically active cobalamins. They are composed of tetrapyrrole rings surrounding central cobalt atoms, with nucleotide side chains attached to the cobalt. The cobalamin tetrapyrrole ring, exclusive of cobalt and other side chains, is called a *corrin*. The cobalt-corrin complex is termed *cobamide*. *Cyanocobalamin* is a stable compound that forms dark red, needle-like crystals; it is the reference compound used for measuring serum cobalamin concentration. Less stable serum cobalamins may be converted to this compound for quantitation. The predominant physiological form of cobalamin in serum is *methylcobalamin*, whereas that in cytosols is *5′-deoxyadenosylcobalamin* (Figure 24-10). Cyanocobalamin is gradually destroyed on exposure to light. Aqueous solutions of cyanocobalamin exhibit a distinctive absorption spectrum with maxima at 278, 361, and 550 nm. The spectrum is independent of pH but changes when cyanocobalamin binds to intrinsic factor.

Biochemical function. Vitamin B_{12} is a coenzyme for two physiologically important functions in humans: (1) the synthesis of methionine, which is a required step in the production of the active folate coenzyme, and (2) the conversion of methylmalonic acid to succinic acid. Deficiency of vitamin B_{12} in humans is associated with megaloblastic anemia and neuropathy. The metabolic lesion in these disorders is not fully understood, but studies indicate that impairment in methionine synthesis is most directly related to clinical manifestations.

Metabolism. In humans, the daily vitamin B_{12} requirement is $\sim 0.5\ \mu g$. The only significant dietary sources of vitamin B_{12} are meats, milk or milk products, and eggs. In the stomach, vitamin B_{12} forms a complex with intrinsic factor (IF), which is absorbed by cells in the distal ileum. Within the mucosal epithelial cells, the vitamin B_{12}–IF complex is dissociated, and the vitamin then passes into the plasma. Almost all of the vitamin B_{12} in plasma is taken up by hepatocytes and is stored within the liver until it is again released into plasma to meet physiological demands. Normally the body stores of ~1 mg far exceed the daily metabolic requirement. It is for this reason that symptoms of physiologic vitamin B_{12} deficiency do not become evident for five years or more after dietary deficiency begins.

Three classes of vitamin B_{12} transport proteins are found in humans: (1) intrinsic factor, which occurs only in gastric juice, (2) transcobalamin II, which occurs in plasma and other body fluids, and (3) cobalophilins or "R proteins," which are ubiquitous. *Intrinsic factor*, secreted by the parietal cells of the stomach, is required for the intestinal absorption of vitamin B_{12} in the distal ileum. No other known substance can transport vitamin B_{12} across the intestinal wall. Pernicious anemia, the most important disorder

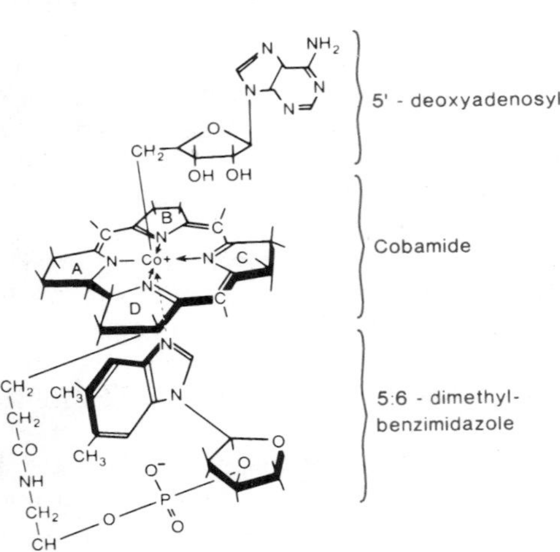

Figure 24-10. The structure of 5′-deoxyadenosylcobalamin, a physiologically important form of vitamin B_{12}. All of the active cobalamins contain a corrin ring (here labeled cobamide), a purine nucleotide (5:6-dimethylbenzimidazole), and an organic ligand, in this case 5′-deoxyadenosine. The various cobalamins differ principally in the moiety linked to the cobalt above the plane of the corrin nucleus in this illustration. Methylcobalamin is the simplest of these with a methyl group bound to the cobalt atom. Cyanocobalamin, commercial vitamin B_{12}, has CN^- at this same position. (Reprinted from Chanarin, I.: The Megaloblastic Anaemias. 2nd ed. Oxford, Blackwell Scientific, 1979; with permission.)

of vitamin B_{12} metabolism, is due to a defect in the secretion of IF. The most important vitamin B_{12} transport protein in plasma is *transcobalamin II* (*TC II*), a β-globulin synthesized mainly in the liver. TC II transports vitamin B_{12} to receptors on cell membranes throughout the body.

Two types of vitamin B_{12} binders occur in human gastric juice, one with slow and one with rapid mobility in zone electrophoresis. The slow component is *intrinsic factor* and the rapid component is designated "R proteins" (R for rapid). R proteins, known as transcobalamin I and transcobalamin III, occur in plasma, amniotic fluid, milk, saliva, ascitic fluid, and granulocytes. Collectively, these binders are called *cobalophilins*. They also bind vitamin B_{12} analogs to varying extent. No definite physiological function for these proteins is known.

Clinical significance. Deficiency of vitamin B_{12} causes serious and often irreversible neurologic disorders, such as burning pain or loss of sensation in the extremities, weakness, spasticity and paralysis, confusion, disorientation, and dementia. Neurologic symptoms may occur without any discernible hematological changes. However, when hematological changes are observed, they are usually accompanied by some neurologic changes. The classic morphological changes in the blood, in approximate order of appearance, are hypersegmentation of neutrophils, macrocytosis, anemia, leukopenia, and thrombocytopenia, accompanied by *megaloblastic* changes in bone marrow. Decreased concentrations of serum vitamin B_{12} are frequently, but not always, associated with increased mean corpuscular volume (MCV) of erythrocytes.

Vitamin B_{12} deficiency may result from a variety of causes. Most common are lack of intrinsic factor secretion by the stomach and intestinal malabsorption. In the absence of intrinsic factor (*pernicious anemia*), vitamin B_{12} cannot be absorbed from food. Absence of IF also results from total gastrectomy. Malabsorption of vitamin B_{12} may be caused by gastrectomy, ileal resection, tropical sprue, inflammatory disease of the small intestine, intestinal stasis with overgrowth of colonic bacteria, or infestation of the small intestine by the fish tapeworm *Diphyllobothrium latum*.

FOLIC ACID[9]

Structure and terminology. *Folate* and *folic acid* are general terms for a family of compounds related to pteroic acid. Pteroic acid is composed of a pteridine ring joined to a *p*-aminobenzoic acid residue (Figure 24-11). Pteroic acid, when conjugated with one molecule of L-glutamic acid (pterolyglutamic acid), can be reduced to dihydrofolic acid or tetrahydrofolate. Only the reduced forms are biologically active. Although various forms of folic acid are normally present in human serum and other body fluids, the principal form is N^5-methyltetrahydrofolate. Folate polyglutamates are quantitatively and functionally the major forms of folate within the cytosol.

Biochemical functions. Folate coenzymes are essential for the transfer of single carbon units. Five of the major reactions are conversion of serine to glycine, catabolism of histidine, and synthesis of thymidylate, methionine, and purine. Different folates are involved in these reactions, depending on the chemical state of the single fragments transferred.

Metabolism. Folate is absorbed from dietary sources such as green, leafy vegetables. Polyglutamate forms present in food are converted to monoglutamates in the intestinal mucosa. After cellular uptake, most of the folate is reduced and methylated and enters the circulation as *N^5-methyltetrahydrofolate*. Usual dietary sources provide ~500 μg of folate/day. The human daily requirement for folate is ~50 μg. Body stores of folate amount to ~5 mg. Interruption of absorption of folate may, therefore, result in deficiency within a few months.

Folic acid and B_{12} metabolism are thought to be linked by the reaction that transfers a methyl group from N^5-methyltetrahydrofolate to cobalamin. In cases of cobalamin deficiency, folate is "trapped" as N^5-methyltetrahydrofolate, cannot be recycled back into the functional folate polyglutamate pool, and diffuses out of the cell. The folate deficiency thus arising eventually results in megaloblastic anemia and neuropathies.

Both high- and low-affinity folate binders are found in plasma. The functional role

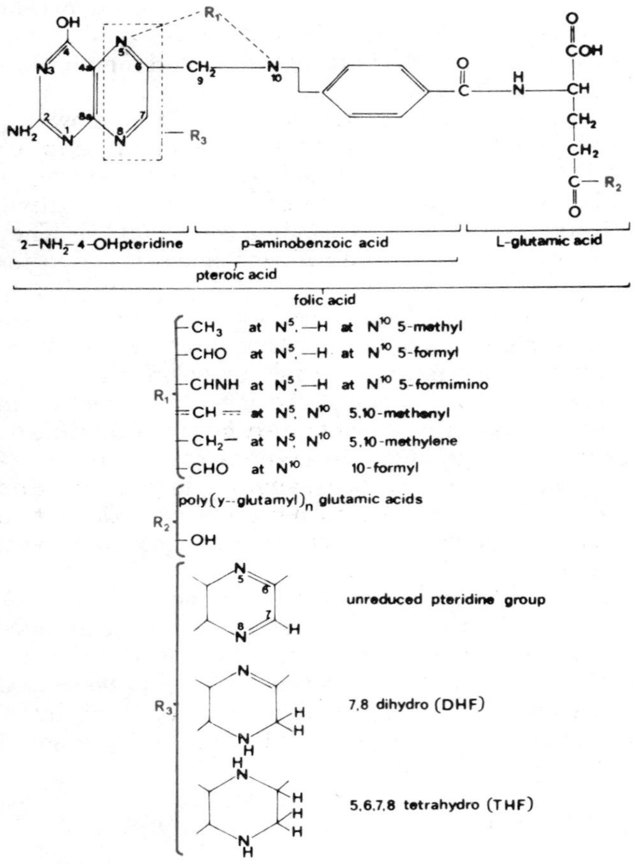

Figure 24-11. Structure and relationships of folic acid and its derivatives. The three basic subunits of folic acid are indicated. Biologically active derivatives have substituent groups at positions R_1, R_2, or R_3. For example, the physiologically active compound N^5N^{10}-methylene-tetrahydrofolate heptaglutamate has a methylene bridge between N^5 and N^{10}; four additional protons at positions 5, 6, 7, and 8 in the pteridine ring (R_3); and six additional glutamate residues attached at position R_2 in the glutamic acid side chain. (Reprinted from Row, P. B.: Inherited disorders of folate metabolism. *In*: The Metabolic Basis of Inherited Disease. 5th ed. J. B. Stanbury, J. B. Wyngaarden, and D. S. Fredrickson, Eds. New York, McGraw-Hill, 1983; with permission.)

of these binders is unknown, but they may serve to reduce loss of folate by renal filtration, and they possibly could have a role in bacterial defense. Unlike the situation with vitamin B_{12}, no clinical disorders have been reported that are related to the absence of folate-binding proteins.

Clinical disorders. Deficiency of folate may result from the absence of intestinal microorganisms (gut sterilization), poor intestinal absorption (e.g., after surgical resection or in celiac disease or sprue), insufficient dietary intake, excessive utilization (as in pregnancy, liver disease, and malignancies), and administration of antifolate drugs (e.g., methotrexate).

Megaloblastic anemia is the major clinical manifestation of folate deficiency, although sensory loss and neuropsychiatric changes also occur. Combined deficiency of both folate and iron commonly occurs in malnourished people. In such cases, macrocytosis of erythrocytes, otherwise typical of folic acid deficiency, is not observed.

Determination of Vitamin B_{12} and Folic Acid

Methods. Two types of methods for the quantitation of vitamin B_{12} and folic acid are microbiological assays and competitive protein binding (CPB) radioassays. The latter have largely replaced the less convenient microbiological assays.

Several commercial kits are available for the *competitive protein binding assays* of vitamin B_{12} and folic acid. Most provide for the simultaneous measurement of both substances in the same reaction tube by using ^{125}I-labeled folate and ^{57}Co-labeled cobalamin. The binder used in the folate assay is a protein that occurs naturally in milk, β-lactoglobulin. The vitamin B_{12} binder is nonhuman intrinsic factor (IF), usually obtained from hog stomach. If the IF is not highly purified, it may contain R proteins, which will bind vitamin B_{12} as well as related metabolically inactive compounds (vitamin B_{12} analogs that may be present in the sample), thereby causing artificially elevated vitamin B_{12} results. To measure only metabolically active vitamin B_{12} ("true B_{12}") in CPB assays, either the IF must be highly purified or *cobinamide* (B_{12} analog) must be added to less-pure IF to saturate all binding sites on the contaminating R proteins. Cobinamide is not bound by IF since IF is highly specific for true B_{12}. This elimination of R protein binding sites prevents interferences in the assay caused by B_{12} analogs that may be present in the sample or by other binders that may be present in the IF preparation.

In the most widely used CPB assays for simultaneous determinations of folate and vitamin B_{12} levels, folate (principally N^5-methyltetrahydrofolate) competes with ^{125}I-labeled folate for a limited number of binding sites on the β-lactoglobulin binder, and B_{12} (cobalamin) competes with ^{57}Co-labeled cobalamin for a limited number of binding sites on IF. As a preliminary step, the specimen containing folate and vitamin B_{12} is boiled in a buffered solution containing dithiothreitol, potassium cyanide, and the ^{125}I- and ^{57}Co-labeled tracers. Boiling releases folate and vitamin B_{12} from endogenous binding-proteins and inactivates these proteins to prevent their interference in the reaction. Also, folate is reduced and stabilized by dithiothreitol, and serum vitamin B_{12} is converted by reaction with potassium cyanide to the more stable cyanocobalamin. Alternatively, some procedures irreversibly denature endogenous binding-proteins by increasing the pH to 12–13 and then readjusting the pH to 9.3 before the binding reagent is added, thereby avoiding a boiling step to release folate and vitamin B_{12}. Next, buffered binder reagent is added and the reaction mixture is incubated at pH 9.3 to allow the competitive reactions to take place. Subsequent separation of bound and free folate and vitamin B_{12} is achieved by contact with dextran-coated charcoal, which adsorbs the free (unbound) molecules, leaving protein-bound folate and vitamin B_{12} in the solution. Some procedures eliminate charcoal separation by attaching the folate binder and IF to solid-phase particles such as glass or polyacrylamide beads. The free folate and vitamin B_{12} then remain in the supernatant, and the bound analytes become part of the solid-phase suspension. Both charcoal and solid-phase particles can be separated from the reaction mixture by centrifugation. For methods employing charcoal, radioactivity of the supernatants is measured, whereas in solid-phase radiosorbent assays, the radioactivity of the pellet is measured. For the performance of the simultaneous folate/vitamin B_{12} measurement, a gamma-scintillation counter must be used that discriminates between the disintegrations produced by the ^{57}Co and ^{125}I tracers present in the samples to be counted.

Specimen collection and storage. Serum specimens are recommended for this procedure; however, plasma specimens collected in EDTA may also be used. Heparin should not be used as an anticoagulant since it has a modest vitamin B_{12}–binding ability. Specimens should be collected from fasting individuals because recent food intake may appreciably increase the serum folic acid concentration, although not the vitamin B_{12} concentration. Unhemolyzed cell-free serum must be protected from light to prevent folate deterioration. Store at 2–8 °C for up to 4 h or store frozen (−20 °C) for longer periods.

For *whole blood hemolysates* used in the erythrocyte folate assay, collect the blood in a tube containing EDTA. Record the hematocrit value and prepare a hemolysate in ascorbic acid solution (100 μL blood in 2 mL ascorbic acid solution, 0.2 g/dL; incubate 90 min to hydrolyze folate polyglutamates). Protect from light. Hemolyzed specimens may be stored at 2–8 °C for up to 3 h or stored frozen (−20 °C) for longer periods.

Reference ranges for adults in the fasting state are: serum folate, 3–16 μg/L; RBC folate, 150–800 μg/L; and serum vitamin B_{12}, 140–700 ng/L. Serum vitamin B_{12} concentration declines significantly as a function of age in healthy adult males.

Comments and precautions. Persons with untreated pernicious anemia usually have vitamin B_{12} concentrations of <100 ng/L. Those with folate deficiency usually have fasting serum folate levels of <1.0 μg/L. Indeterminate results (B_{12} of 100–400 ng/L or serum folate of 2.0–3.0 μg/L) are common. Because these deficiency disorders develop gradually, patients with results in the indeterminate range should have serum reassayed in a few months, except in cases in which a diagnosis can be clearly established by other means.

Vitamin B_{12} Absorption (Schilling) Test[9,22]

Principle. The vitamin B_{12} absorption test permits differentiation between pernicious anemia and intestinal malabsorption as causes of vitamin B_{12} deficiency. The proportion absorbed from orally administered ^{57}Co- or ^{58}Co-labeled vitamin B_{12} can be measured by determining the radioactivity in feces, urine, or serum, or by externally scanning the liver. The usual procedure is to measure radioactivity in a 24-h urine specimen that is collected after oral administration of 0.5 μg of radioactive cobalt–labeled vitamin B_{12} following an overnight fast. The patient is also injected intramuscularly with a "flushing dose" of 1000 μg unlabeled vitamin B_{12} to saturate the binding capacity of plasma and liver for vitamin B_{12} and to cause urinary excretion of the absorbed vitamin B_{12}.

Interpretation. When the oral test dose of ^{57}Co-vitamin B_{12} is 0.5 μg, normal individuals excrete in the urine 8% or more of the dose administered, whereas in persons with pernicious anemia, <7% and most often ~3% is excreted. A confirmatory test for lack of IF requires ingestion of vitamin B_{12} with IF and repeat of the flushing dose, one or two days after the initial vitamin B_{12} absorption test. Oral administration of IF simultaneously with vitamin B_{12} increases the absorption of vitamin B_{12} by the intestinal tract and thus increases the urinary excretion to ≥8% of the dose administered. In patients with malabsorption, oral administration of IF does not increase the per cent of the dose excreted. However, ~25% of patients with pernicious anemia also have intestinal malabsorption that is secondary to vitamin B_{12} deficiency and is reversible following weeks or months of vitamin B_{12} therapy. Thus, a low value for vitamin B_{12} absorption when administered with IF does not exclude pernicious anemia.

Iron Metabolism and Iron Compartments[11]

The total iron content of the body is ~3–3.5 g. Of this amount, ~2.5 g is in hemoglobin, virtually all of which is contained within erythrocytes or their precursors in the bone marrow. Most of the remainder is storage iron. However, numerous cellular enzymes and coenzymes require iron, notably *peroxidases* and *cytochromes* that are also heme proteins, many of the enzymes of the Krebs cycle, and monoamine oxidase, which is involved in neurotransmission. Plasma contains only ~2.5 mg of iron.

Iron is transported from one organ to another as Fe(III) bound to a plasma protein called *apotransferrin*. The apotransferrin-Fe(III) complex is called ***transferrin***. Transferrin occurs within the cytosol of many cells and may serve as an intracellular iron transport protein. Iron in storage form in various cells is ~800 mg in men and 0–200 mg in women. ***Ferritin*** is the major iron-storage compound; it occurs in nearly all body cells but particularly in hepatocytes. The *apoferritin* shell surrounds an interior ferric oxyhydroxide crystalline core that contains 2000–4000 Fe(III) atoms. Nonenzymatic reduction of Fe(III) to Fe(II) occurs as iron is released from ferritin. Thus, ferritin is both a very efficient iron trap and a readily available source of iron for metabolic requirements and for formation of hemoglobin and other heme proteins. The minute concentration of ferritin in serum is an indicator of total body storage iron. Liver injury results in release of relatively enormous amounts of ferritin into plasma, raising the serum ferritin concentration several thousand times and the serum iron concentration by thousands of μg/dL.

Hemosiderin, the other form of storage iron, consists of aggregated, partially deproteinized ferritin. In contrast to ferritin, hemosiderin is insoluble in aqueous solutions, and iron is only slowly released from hemosiderin. Like ferritin, hemosiderin normally occurs predominantly in cells of the liver, spleen, and bone marrow.

Absorption, Transport, and Excretion

The average American diet provides 10–15 mg of iron daily, mostly in the form of heme proteins in meat. Normally ~1 mg of iron is absorbed each day, principally in the duodenum. Heme is absorbed as such, whereas inorganic iron must be in the ferrous state. Both ferritin and transferrin are present in the absorptive cells of the intestinal mucosa and are believed to act together to regulate iron absorption. When body iron stores are high, the ferritin content of mucosal epithelium is also high and the transferrin content is low. Iron that enters mucosal cells is trapped in ferritin and lost when the mucosal cells are sloughed into the intestinal lumen. Thus, iron absorption is minimized when body stores of iron are increased. Conversely, when there is iron deficiency, mucosal cell content of ferritin is diminished, apotransferrin content is increased, and iron absorption is accelerated.

The major pathway of iron metabolism (Figure 24-12) is a virtually closed cycle in which iron passes from plasma transferrin to the erythrocyte precursors in the bone marrow, where it is incorporated into hemoglobin. These cells then enter the circulation as mature erythrocytes, where they remain for nearly four months before they become metabolically "worn out" and are engulfed by phagocytes. The iron is released from hemoglobin and returns to plasma transferrin, thus completing one cycle and beginning another. From this iron cycle, small quantities of iron are diverted for use in other iron compartments; there is also a slow, continual exchange between the storage and transport compartments.

Each day ~1–2 mg of iron absorbed from the intestinal tract enters this cycle to compensate for the 1–2 mg of iron lost each day from the body. Most of the iron loss is due to minute quantities of iron present in epithelial cells and erythrocytes lost in urine and feces. With each menstrual cycle, women lose ~40–80 mL of blood, which is equivalent to 20–40 mg of iron. This loss must be made up by increased absorption of iron by the intestinal mucosa. Similarly, ~600–900 mg of iron is diverted from the maternal body pool to the fetus as a consequence of each pregnancy. Compensation for iron loss from pregnancy and menstruation is often difficult, particularly since many women consume iron-poor diets. Thus, iron deficiency is very common in women, even affluent women whose diets seem adequate.

Disorders of Iron Metabolism

Iron deficiency. Iron deficiency is one of the most prevalent disorders of humans. It is particularly a disease of children, of young women, and of older persons. In children, iron deficiency is frequently due to dietary deficiency because milk has a low iron content. In adults it almost always indicates chronic blood loss.

Measurements of *serum iron concentration* and *total iron-binding capacity* (TIBC) are widely used in the diagnosis of iron deficiency. However, assay of *serum ferritin concentration* is a much more sensitive and reliable means of demonstrating this disorder. *Free erythrocyte protoporphyrin* (FEP) concentration is increased in patients with iron deficiency or lead poisoning, whereas it is normal in thalassemia minor, a condition

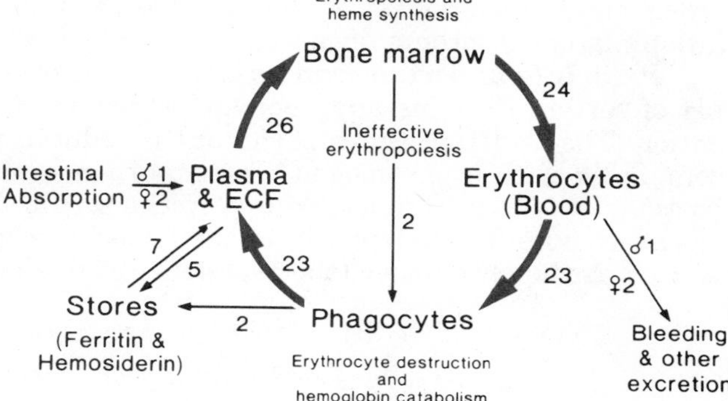

Figure 24-12. Principal pathways of iron metabolism in humans. Numbers indicate approximate milligrams that enter or leave the various compartments daily.

easily confused with iron deficiency. The most reliable method for diagnosis of iron deficiency is microscopic evaluation of bone marrow aspirate stained with Prussian blue, which demonstrates presence or absence of hemosiderin.

Iron overload. *Hemosiderosis* is a term that implies iron overload without associated tissue injury. *Hemochromatosis* implies iron overload with injury to involved organs, as manifested by cellular degeneration and fibrosis. Iron overload most commonly results from chronic excessive absorption of iron from a normal diet. It is rarely the result of protracted ingestion of iron medication over many years or inappropriate injections of iron, as in iron dextran or blood transfusion.

Hereditary hemochromatosis, the classic disorder of iron overload, is due to an inborn error of iron absorption. Approximately 10% of Caucasians of Northern European origin are carriers of the gene, although only a portion of homozygotes develop the clinical disease. This condition, of lifelong duration, is usually manifested by hyperpigmentation of skin, diabetes mellitus, cirrhosis of the liver, pituitary insufficiency, hypogonadism, arrhythmias of the heart or congestive heart failure, and arthritis. The name *sideroblastic anemia* is given to a group of iron-overload disorders of unknown cause.

Diagnosis of iron-overload disorders is best made by measurement of serum iron concentration and total iron-binding capacity. Usually the serum iron concentration is >200 μg/dL (35.8 μmol/L) and transferrin saturation is $>55\%$. Serum ferritin assay is a somewhat less sensitive test for ascertainment of early iron overload, whereas the free erythrocyte protoporphyrin test has no role at all in the diagnosis of iron overload.

The Determination of Serum Iron and Iron-Binding Capacity[16,17]

Clinical Significance

Serum iron concentration connotes principally the Fe(III) bound to serum transferrin. Serum iron concentration is decreased in many, but not all, patients with *iron deficiency anemia* and in *acute or chronic inflammatory disorders* such as *acute infection, immunization,* and *myocardial infarction.* Serum iron concentration diminishes markedly in patients who are beginning to respond to specific therapy for anemias of other causes, for example, *treatment of pernicious anemia* with vitamin B_{12}. Acute or recent *hemorrhage,* including that due to blood donation, results in low serum iron concentration. Serum iron concentration also drops at the time of *menstruation.* Use of hormonal contraceptives raises serum iron concentration, but on cessation of contraceptive hormone intake, serum iron concentration decreases as much as 30% concurrent with uterine bleeding. Greater than normal concentrations of serum iron occur in iron-overload disorders such as *hemochromatosis,* in *acute hepatitis,* in *acute iron poisoning* in children, and following *oral ingestion of iron medication* or *parenteral iron administration.*

Since normally only about one-third of the iron-binding sites of transferrin are occupied by Fe(III), serum transferrin has considerable reserve iron-binding capacity. This is called the serum unsaturated (or latent) iron-binding capacity (UIBC). The ***total iron-binding capacity*** (TIBC) is a measurement of the maximum concentration of iron that serum proteins, particularly transferrin, can bind. The serum TIBC varies in disorders of iron metabolism. It is often increased in iron deficiency and decreased in chronic inflammatory disorders or malignancies. See also *Transferrin* in Chapter 11 (page 333).

Principle of serum iron assay. Iron is released from transferrin by a decrease in pH of serum. Proteins and apotransferrin are removed by precipitation and centrifugation. The Fe(III) of the supernatant is reduced to Fe(II) with thioglycolic acid and is complexed with a chromogen. The absorbance of the iron-chromogen complex is proportional to the iron concentration in the serum. Buffer is added to adjust pH to 4.5 to ensure maximum absorptivity of the iron-chromogen complex. If bathophenanthroline is used as the chromogen, the reactions involved are:

$$\text{Transferrin-Fe(III)} + \text{HCl-trichloroacetic acid} \longrightarrow \text{Fe(III)} + \text{transferrin}\downarrow$$

$$\text{Fe(III)} + \text{thioglycolic acid} \longrightarrow \text{Fe(II)}$$

$$\text{Fe(II)} + \text{bathophenanthroline}_{\text{(colorless)}} \longrightarrow \text{Bathophenanthroline-Fe(II)}_{\text{(colored)}}$$

Principle of total iron-binding capacity assay. Excess ammonium ferric citrate is incubated with serum to saturate all iron-binding sites of transferrin. The unbound Fe(III) is removed by addition of $MgCO_3$ powder [which adsorbs excess Fe(III)] and a buffer. The mixture is centrifuged and the iron content of the supernatant (modified serum) is assayed.

Specimen. Sufficient blood to provide 3.5 mL serum or heparinized plasma is required. Plasma specimens collected with EDTA, oxalate, or citrate are unsatisfactory, since they bind iron, preventing its reaction with the chromogen. Specimens should be collected in the morning in order to avoid low results due to diurnal variation. Markedly hemolyzed specimens should be rejected.

Reagents

Reagents must be of highest purity and virtually iron-free. All glassware must be rendered iron-free by immersion overnight in HCl, 6 mol/L, followed by repeated rinsing in iron-free (Class I) water.

1. Trichloroacetic acid (Fisher Scientific Co., Pittsburgh, PA 15219) or iron-free trichloroacetic acid solution, 6.1 mol/L (Sigma Chemical Co., St. Louis, MO 63178). ACS-certified trichloroacetic acid may contain 200 μg Fe/g, an amount sufficient to raise the Fe concentration of standards and specimens by 10 μg/dL.
2. Protein precipitant solution. Place in a 1-L volumetric flask 33.3 mL of thioglycolic acid (mercaptoacetic acid), 98 g of trichloroacetic acid (or 98.4 mL of trichloroacetic acid solution, 6.1 mol/L), and 400 mL of water. Swirl to dissolve. Slowly add 2 mol of HCl (e.g., 192 mL of concentrated HCl, if assayed at 38% HCl). Add H_2O to volume. Store in a dark brown bottle. Stable for two months.
3. Chromogen solution. Place in a 1-L volumetric flask 250 mg of disodium bathophenanthroline disulfonic acid (ICN K and K Laboratories, Inc., Plainview, NY 11803) and 84.04 g of sodium acetate. Add H_2O to volume.
4. Iron standards, 100 and 200 μg/dL. Commercial standards may be purchased or standards may be prepared by dissolving pure iron wire in concentrated HCl.
5. Ammonium ferric citrate, ~1.25 mmol Fe(III)/L. Place 400 mg ammonium ferric citrate in a 1-L volumetric flask and add ~950 mL H_2O. Add a few crystals of citric acid and shake to dissolve. Adjust pH to 7.0 with dilute NH_4OH and dilute to volume with water. Store at 4 °C and avoid prolonged exposure to light. Solution should be a light yellow-green and free of rust-colored precipitate. Prepare fresh every three months.
6. Barbital-NaCl buffer, 40 mmol barbital/L, pH 7.5. Place 12 g of diethylbarbituric acid, 4.5 g of sodium diethylbarbiturate, and 12.8 g of NaCl in a 2-L beaker. Add 1950 mL of H_2O and stir to dissolve. Add NaOH, 1 mol/L, in 1-mL increments until all reagents have dissolved. Place a pH electrode in the solution, adjust pH to 7.5 by addition of HCl, 0.1 mol/L, transfer to a 2-L volumetric flask, and bring to volume with H_2O.
7. Magnesium carbonate, light powder (MCB Manufacturing Chemists, Cincinnati, OH 45212).

Procedure

Serum Iron

1. Mix 2 mL serum with 2 mL of protein precipitant solution in a 15-mL centrifuge tube, and allow to stand 5 min.
2. Centrifuge at 1500 × *g* for 10 min, and transfer 2 mL of the clear supernatant to a labeled cuvet.
3. To a cuvet labeled blank, add 1 mL of water and 1 mL of protein precipitant solution. Mix and allow to stand 5 min.
4. To another cuvet marked standard, add 1 mL of standard solution and 1 mL of protein precipitant solution. Mix and allow to stand 5 min.
5. Add 2 mL of chromogen solution to all cuvets and mix. Allow to stand 5 min.
6. Measure absorbances of specimen, blank, and standard cuvets at 534 nm, using water to set zero absorbance.

Calculation

$$\text{Serum iron concentration, } \mu g/dL = \frac{(A_u - A_b)}{(A_s - A_b)} \times C_s$$

where A_u is the absorbance of the specimen, A_b is the absorbance of the blank, A_s is the absorbance of the standard, and C_s is the concentration of the standard.

Total Iron-Binding Capacity

1. To 1 mL of serum add 0.1 mL of ammonium ferric citrate solution. Mix and allow to stand 10 min. Then add 2 mL of barbital-NaCl buffer and 0.4–0.5 g of light magnesium carbonate. Cap with Parafilm.
2. Mix on rotary mixer for at least 15 min.
3. Centrifuge for 10 min at 1500 × *g*.
4. Transfer 2 mL of supernatant to another tube and proceed to protein precipitation as for step 1 of serum iron determination.

Calculation

$$\text{TIBC, } \mu\text{g/dL} = 3.1 \times \frac{(A_u - A_b)}{(A_s - A_b)} \times C_s$$

The factor 3.1 corrects for dilution of the specimen with buffer and ammonium ferric citrate solutions.

$$\text{Transferrin saturation, \%} = \frac{\text{Serum iron concentration} \times 100}{\text{TIBC}}$$

Reference ranges. The *serum iron* concentrations in fasting morning specimens from healthy adults range from 70 to 180 μg/dL (12.5–32.2 μmol/L) in men and from 60 to 180 μg/dL (10.7–32.2 μmol/L) in women. Serum iron concentrations in normal neonates range from 95 to 225 μg/dL (17–40.3 μmol/L). After one month of age the range closely approximates that of adults unless there is iron deficiency.

The *TIBC* for adults is 250–450 μg/dL (44.8–80.6 μmol/L). Transferrin saturation in men is 20–50% and in women 15–50%.

Many factors influence serum iron concentration and TIBC. Day-to-day variation is quite marked in healthy persons. Diurnal variation causes serum iron concentrations to be distinctly lower in the afternoon than in the morning and quite low in the evening (as low as 10–20 μg/dL in healthy persons). Many persons with iron deficiency have normal values for serum iron concentration and TIBC.

Comments and precautions. Because free hemoglobin is precipitated with other serum proteins, slight hemolysis has very little effect on the serum iron assay results. However, when serum specimens show marked hemolysis, a small amount of iron may be liberated from hemoglobin. Such sera should be rejected.

Because of the great quantities of iron in the environment, scrupulous care is needed to ensure that glassware, water, and reagents are not contaminated with iron.

Serum transferrin concentration may be estimated from the TIBC by the following relationship:

$$\text{Serum transferrin, mg/dL} = 0.70 \times \text{TIBC}$$

Methods are also available for immunoassay of serum transferrin concentration.

SERUM FERRITIN[13]

Clinical significance. The quantity of iron in the storage compartment is reflected in plasma ferritin concentration. The plasma ferritin concentration declines very early in the development of iron deficiency. Thus, measurement of serum ferritin concentration is a very sensitive indicator of an iron deficiency that is uncomplicated by other acute concurrent diseases. On the other hand, a large number of chronic diseases result in increased serum ferritin concentration. These diseases include chronic infections, chronic inflammatory disorders such as rheumatoid arthritis or renal disease, Gaucher's disease, and numerous types of malignancies, especially lymphomas, leukemias, breast cancer, and neuroblastoma. In patients who have any of these chronic disorders together with iron deficiency, serum ferritin concentrations are very often normal. Increase in plasma ferritin concentration also occurs in viral hepatitis or following toxic liver injury, as a result of release of ferritin from damaged liver cells. Plasma ferritin concentration is also increased with increases of iron stores, as seen in patients with hemosiderosis or hemochromatosis.

Principle. Serum ferritin assay may be performed by radioimmunoassay (RIA), immunoradiometric assay (IRMA), or enzyme-linked immunosorbent assay (ELISA). In most laboratories, serum ferritin assays are performed with commercially available kits.

Reference Ranges

	μg/L
Newborn infants:	25–200
1 month:	200–600
2 to 5 months:	50–200
6 months to 15 years:	7–142
Adults,	
Males:	20–300
Females:	10–120

Comments and precautions. Considerable variation in reference values has been observed with different methods. To avoid serious misinterpretation, reference values must be determined for each laboratory. Some IRMA and ELISA assays give incorrect information, which may even be in the normal range, when the true concentration of ferritin in a specimen is extremely high (for example, >10 000 μg/L). This phenomenon is referred to as the "high-dose hook effect." The problem usually can be prevented if specimens with assayed ferritin concentrations >800 μg/L are diluted and reassayed.

FREE ERYTHROCYTE PORPHYRINS (FEP)[23]

Clinical significance. The free erythrocyte porphyrins (FEP) are porphyrins that are readily extracted from erythrocytes with organic solvents. The principal free porphyrin is *protoporphyrin IX*. FEP concentration is moderately increased in *iron deficiency* and *lead poisoning*, since both conditions prevent the final step of heme synthesis and, consequently, porphyrin precursors accumulate in the erythrocyte. Very marked increases in FEP concentration also occur in *lead poisoning* and in *erythropoietic protoporphyria*. On the other hand, thalassemic disorders are not associated with increased FEP concentrations. The FEP assay is used as a screening test, especially in pediatric practice, for detection of iron deficiency and lead poisoning. Whenever increased concentrations of FEP are observed, more specific tests are required, such as measurement of serum iron or serum ferritin concentration and measurement of erythrocyte lead or whole blood lead concentration. A description of the quantitative determination of erythrocyte protoporphyrin is given in Chapter 25.

References

1. Abraham, E. C., Reese, A., Stallings, M., et al.: Separation of human hemoglobins by DEAE-cellulose chromatography using glycine-KCN-NaCl developers. Hemoglobin, *1*:27–44, 1976–1977.
2. Betke, K., Marti, H. R., and Schlicht, I.: Estimation of small percentages of foetal haemoglobin. Nature, *184*:1877–1878, 1959.
3. Betke, K., Steim, H., and Tönz, O.: A family with congenital methaemoglobinaemia due to reductase deficiency. German Medical Monthly, *7*:217–219, 1962.
4. Beutler, E.: Energy metabolism and maintenance of erythrocytes. *In*: Hematology, 3rd ed. W. J. Williams, E. Beutler, A. J. Erslev, et al., Eds. New York, McGraw-Hill, 1983, pp. 331–345.
5. Beutler, E.: Glucose-6-phosphate dehydrogenase (G-6-PD) and 6-phosphogluconate dehydrogenase (6-PGD). *In*: Red Cell Metabolism. A Manual of Biochemical Procedures, 2nd ed. E. Beutler, Ed. New York, Grune & Stratton, 1975, pp. 66–69.
6. Briere, R. O., Golias, T., and Batsakis, J. G.: Rapid qualitative and quantitative hemoglobin fractionation: cellulose acetate electrophoresis. Am. J. Clin. Pathol., *44*:695–701, 1965.
7. Bunn, H. F., Forget, B. G., and Ranney, H. M.: Human Hemoglobin. Philadelphia, W. B. Saunders Co., 1977.
8. Carrell, R. W., and Kay, R.: A simple method for the detection of unstable haemoglobins. Br. J. Haematol., *23*:615–619, 1972.

9. Chanarin, I.: The Megaloblastic Anaemias, 2nd ed. Oxford, England, Blackwell Scientific Publications, 1979.
10. Fairbanks, V. F., Ed.: Hemoglobinopathies and Thalassemias: Laboratory Methods and Case Studies. New York, Brian C. Decker, Division of Thieme-Stratton Inc., 1980.
11. Fairbanks, V. F., and Beutler, E.: Iron metabolism. *In*: Hematology, 3rd ed. W. J. Williams, E. Beutler, A. J. Erslev, et al., Eds. New York, McGraw-Hill, 1983, pp. 300–310.
12. Fairbanks, V. F., and Klee, G. G. Biochemical aspects of hematology. *In*: Textbook of Clinical Chemistry. N. W. Tietz, Ed. Philadelphia, W. B. Saunders Co., 1986, pp. 1540–1541.
13. Fairbanks, V. F., and Klee G. G.: Ferritin. *In*: Progress in Clinical Pathology, Volume 8. M. Stefanini, and E. S. Benson, Eds. New York, Grune & Stratton, 1981, pp. 175–203.
14. Hamilton, S. R. , Miller, M. E., Jessop, M., et al.: Comparison of microchromatography and electrophoresis with elution for hemoglobin A_2 (Hb A_2) quantitation. Am. J. Clin. Pathol., *71*:388–396, 1979.
15. Harboe, M.: A method for determination of hemoglobin in plasma by near-ultraviolet spectrophotometry. Scand. J. Clin. Lab. Invest., *11*:66–70, 1959.
16. International Committee for Standardization in Haematology: Recommendations for measurement of serum iron in human blood. Br. J. Haematol., *38*:291–294, 1978.
17. International Committee for Standardization in Haematology: The measurement of total and unsaturated iron-binding capacity in serum. Br. J. Haematol., *38*:281–290, 1978.
18. Kagen, L. J.: Myoglobin: methods and diagnostic uses. *In*: CRC Critical Reviews in Clinical Laboratory Sciences. J. Batsakis, and J. Savory, Eds. Cleveland, OH, CRC Press, 1978, pp. 273–302.
19. Nalbandian, R. M., Nichols, B. M., Camp, F. R., et al.: Dithionite tube test—a rapid, inexpensive technique for detection of hemoglobin S and non-S sickling hemoglobin. Clin. Chem., *17*:1028–1032, 1971.
20. Pembrey, M. E., McWade, P., and Weatherall, D. J.: Reliable routine estimation of small amounts of foetal haemoglobin by alkali denaturation. J. Clin. Pathol., *25*:738–740, 1972.
21. Robinson, A. R., Robson, M., Harrison, P., et al.: A new technique for differentiation of hemoglobin. J. Lab. Clin. Med., *50*: 745–752, 1957.
22. Schilling, R. F.: Intrinsic factor studies. II. The effect of gastric juice on the urinary excretion of radioactivity after the oral administration of radioactive vitamin B_{12}. J. Lab. Clin. Med., *42*:860–866, 1953.
23. Stockman, J. A., Weiner, L. S., Simon, G. E., et al.: The measurement of free erythrocyte porphyrin (FEP) as a simple means of distinguishing iron deficiency from β-thalassemia trait in subjects with microcytosis. J. Lab. Clin. Med., *85*:113–119, 1975.

Additional Reading

Stanbury, J. B., Wyngaarden, J. B., and Fredrickson, D. S., Eds.: The Metabolic Basis of Inherited Disease, 5th ed. New York, McGraw-Hill, 1983.
Williams, W. J., Beutler, E., Erslev, A. J., and Lichtman, M. A., Eds. Hematology, 3rd ed. New York, McGraw-Hill, 1981.
Wintrobe, M. M., Lee, G. R., Boggs, D. R., et al.: Clinical Hematology, 8th ed. Philadelphia, Lea & Febiger, 1981.

PORPHYRINS AND DISORDERS OF PORPHYRIN METABOLISM

25

ROBERT F. LABBE, Ph.D.
JOEL M. LAMON, M.D.

Although the primary (inherited) disorders of porphyrin metabolism are relatively uncommon, several secondary (induced) disorders are very common in some groups of the population. All of these disorders are rather complex metabolic diseases with overlapping clinical and biochemical features. Systematic approaches to diagnoses are now possible by utilizing combinations of recently developed methods for measuring metabolite concentrations and enzyme activities. As management and therapy become increasingly effective, early and accurate diagnosis assumes added importance.

Chemistry of Porphyrins

Porphyrins are derivatives of porphin, a macrocyclic, highly unsaturated structure composed of four pyrrole rings bonded by four methene bridges (−CH=). Porphin itself has no side-chain substituents on the pyrrole rings. Porphyrins are differentiated by the kind and order of substituents occupying the eight peripheral positions on the four pyrrole rings of porphin (Figure 25-1). Four basic isomer series are possible, but only the Type III isomers occur in normal metabolism and heme synthesis. Protoporphyrin IX is derived from Type III precursor porphyrins.* Certain Type I porphyrins form insignificant amounts as by-products of a rare biochemical defect, but they have no metabolic function.

Of the many kinds of porphyrins found in nature, only *uroporphyrin, coproporphyrin*, and *protoporphyrin* have clinical significance. These metal-free porphyrins have no biological functions in man; porphyrins are metabolically active only in the form of chelates. The iron chelates of porphyrins are termed *hemes; protoheme* is by far the most common and, in a quantitative sense, the most significant (Figure 25-2). Heme always functions as a prosthetic group of a protein. In mammals, hemoproteins participate in a variety of biochemical processes, all of which are associated with some aspect of oxidative metabolism such as oxygen transport (hemoglobin), cellular respiration (cytochromes), or hydrogen peroxide degradation (catalase). A cobalt chelate, cobalamin or vitamin B_{12}, and a magnesium chelate, chlorophyll, are other forms of naturally occurring tetrapyrroles, although greatly modified in structure from the porphyrin from which they are derived. Zinc protoporphyrin also occurs in trace amounts in normal metabolism; it increases markedly with impaired iron utilization.

Crystalline porphyrins and their concentrated solutions are very dark red or purple in color. In acid solution porphyrins are known for their intense orange-red fluorescence (620–630 nm) on exposure to long-wavelength ultraviolet light, a property that is used in most porphyrin analyses. In general, free porphyrins tend to be more stable in acid solution and in the dark. Dilute hydrochloric acid solutions of coproporphyrin are very stable and are most often used as a standard in porphyrin assays by fluorometric techniques.

*There are 15 possible isomers of protoporphyrin. The naturally occurring isomer has been designated IX, although it is now known to be derived biochemically from precursors of the Type III isomer.

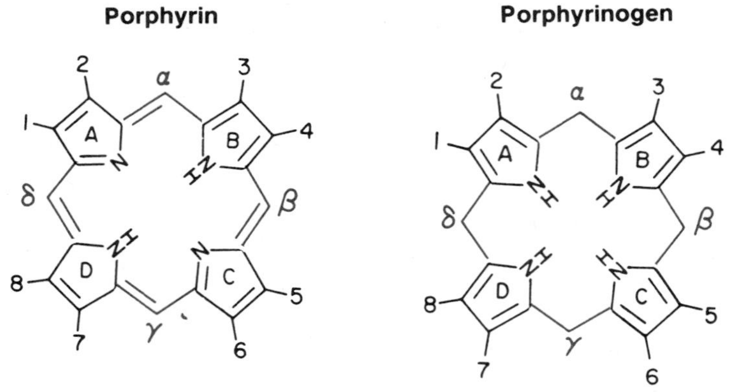

Porphyrin

Porphyrinogen

SUBSTITUENTS IDENTIFYING DIFFERENT PORPHYRINS

Ring	Position	URO	COPRO	PROTO
		Porphyrin III or Porphyrinogen III		
A	1	Acetate	Methyl	Methyl
	2	Propionate	Propionate	Vinyl
B	3	Acetate	Methyl	Methyl
	4	Propionate	Propionate	Vinyl
C	5	Acetate	Methyl	Methyl
	6	Propionate	Propionate	Propionate
D	7	Propionate	Propionate	Propionate
	8	Acetate	Methyl	Methyl
		Porphyrin I or Porphyrinogen I		
D	7	Acetate	Methyl	Does not occur in nature
	8	Propionate	Propionate	

Figure 25-1. Structure of porphyrin and porphyrinogen and a listing of the substituents that identify the different porphyrins.

The degree of polarity of the porphyrins, and thus their water solubility, is influenced by the number of carboxyl groups in the pyrrole substituents. **Uroporphyrin**, having eight carboxyl groups, is the most soluble at physiological pH. **Protoporphyrin**, having only two carboxyl groups, is quite insoluble at this pH, but is very soluble in lipid solvents. **Coproporphyrin**, with its four carboxyl groups, has intermediate solubility. These differing solubility properties and associated degrees of ionization form the basis for the separation and assay of the individual porphyrins. Uroporphyrin is excreted, for all practical purposes, exclusively in the urine, protoporphyrin exclusively in the feces, and coproporphyrin by either route depending upon its rate of formation and the pH of the urine. Alkalinity in urine favors coproporphyrin excretion.

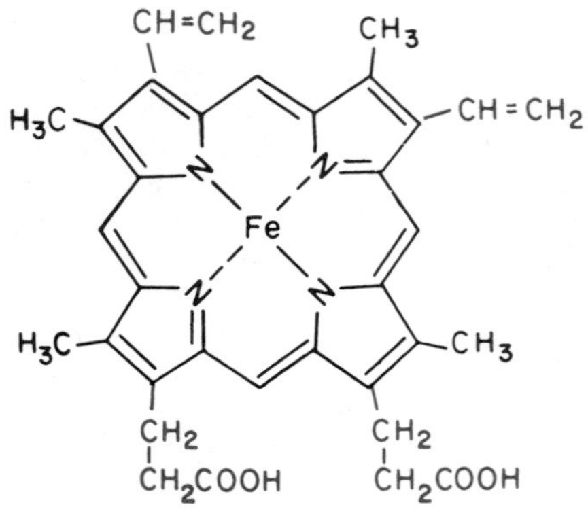

Figure 25-2. The structural formula of protoheme, a chelate of ferrous iron and protoporphrin IX. The vinyl groups occur in positions 2 and 4 of rings A and B, and the propionic acid (carboxyethyl) groups occur in positions 6 and 7 of rings C and D. (Compare with Figure 25-1.)

Porphyrinogens are reduced forms of porphyrins containing six additional hydrogen atoms, one at each of the four methene bridge carbons, and one at each of the two nonhydrogenated pyrrole nitrogens (Figure 25-1). Porphyrinogens are functional precursors of heme. Oxidation of a porphyrinogen to the corresponding porphyrin irreversibly removes that molecule from the heme biosynthetic pathway. Protoporphyrin IX, in which side-chain alteration is complete, is the only nonporphyrinogen intermediate in heme biosynthesis.

The porphyrinogens are colorless and nonfluorescent. Porphyrinogens are also highly unstable, especially in an acid medium, and rapidly oxidize to porphyrins, making their analysis somewhat impractical for laboratory diagnosis. Therefore, porphyrins are clinically the most significant compounds in the diagnosis of disorders of porphyrin metabolism.

Physicochemical and Pharmacological Effects of Porphyrins and Porphyrin Precursors

Porphyrins deposited in skin and then exposed to long-wavelength ultraviolet radiation can cause considerable skin damage. The different lesions that are caused by uro-, copro-, or protoporphyrin probably relate to their respective solubilities. Uroporphyrin and coproporphyrin typically cause delayed bullous lesions, whereas protoporphyrin causes an almost immediate burning sensation and inflammatory reaction in sun-exposed areas of the skin.

The porphyrin precursors, δ-aminolevulinic acid and porphobilinogen, have very low renal thresholds that contribute to low blood concentrations. The expression of neurologic symptoms is always accompanied by the excretion of excessive amounts of porphyrin precursors. This precursor excess may be a significant factor in the pathogenesis of neurologic abnormalities, but the correlation between the level of precursor excretion and the severity of neurologic symptoms is poor.

Biosynthesis of Porphyrins and Heme

Porphyrin and heme biosynthetic activity is quantitatively most prominent in bone marrow and liver; however, porphyrins and heme are synthesized in all mammalian cells. The series of reactions leading to heme (Figure 25-3) begins with the condensation of succinyl coenzyme A and glycine. Pyridoxal phosphate (B_6PO_4) is a cofactor in this condensation. In the same reaction sequence, the glycine moiety is decarboxylated by δ-aminolevulinate (ALA) synthase, forming δ-aminolevulinate. Two molecules of δ-aminolevulinate are then condensed and cyclized through the action of the zinc-containing enzyme porphobilinogen synthase to form the monopyrrole porphobilinogen. Porphobilinogen is condensed and cyclized through the concerted action of two enzymes, uroporphyrinogen I synthase and uroporphyrinogen III cosynthase. Subsequently, the four acetate side-chains of uroporphyrinogen III are decarboxylated through the action of uroporphyrinogen decarboxylase to form coproporphyrinogen III. Next, coproporphyrinogen oxidase decarboxylates and dehydrogenates the propionic acid side-chains in positions two and four, converting these to vinyl groups and yielding protoporphyrinogen IX. Protoporphyrinogen oxidase then oxidizes protoporphyrinogen to protoporphyrin. Finally, ferrochelatase catalyzes the chelation of a ferrous ion by protoporphyrin to form heme.

Control of this biosynthetic pathway occurs principally by modulation of the activity of the first and rate-limiting enzyme, δ-aminolevulinate synthase. Heme has two effects on this enzyme. First, heme can act as a repressor to control the rate of formation of δ-aminolevulinate synthase. Second, excess cellular heme can act by feedback inhibition to control the activity of formed δ-aminolevulinate synthase. Thus, any metabolic change

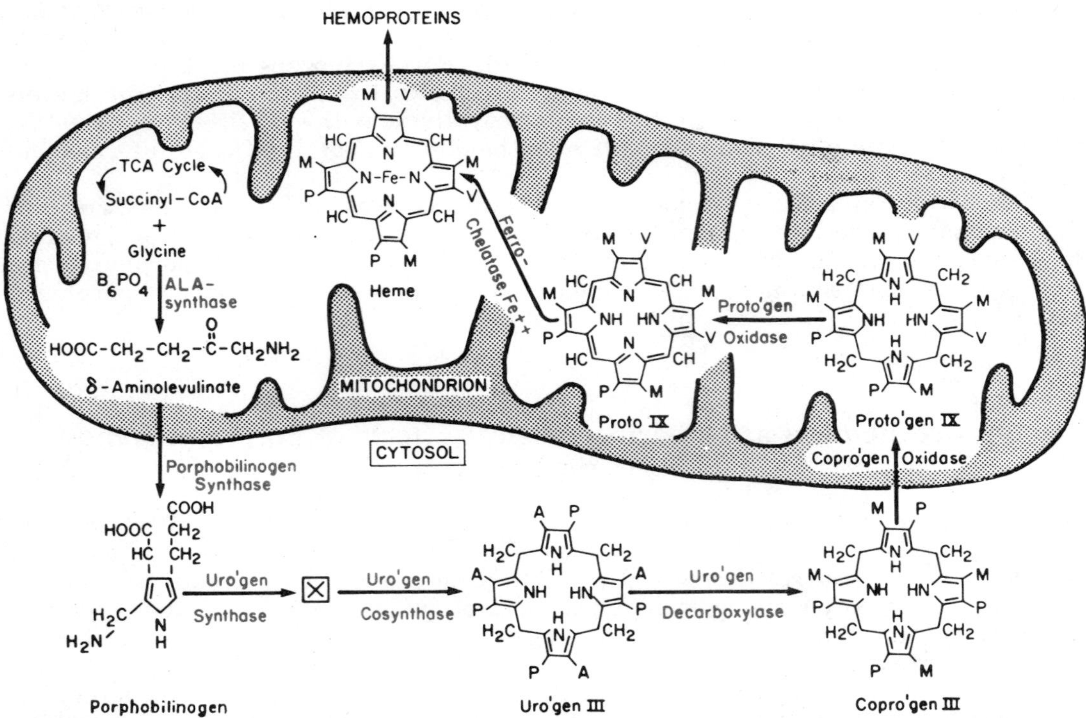

Figure 25-3. The heme biosynthethic pathway showing the distribution of enzymes between the mitochondria and cytoplasm. Intermediates between porphobilinogen and urporphyrinogen, designated by (X), remain unidentified. B_6PO_4 = pyridoxal phosphate.

that affects the rate of either formation or degradation of heme might also affect its biosynthesis.

Free porphyrins found in body fluids and tissues arise as nonenzymatic oxidation by-products of porphyrinogens from the heme biosynthetic pathway. Normally, only trace amounts of porphyrins escape the biosynthetic processes, because control mechanisms maintain the pathway in a delicate state of balance to meet the requirements of each cell. In erythrocytes, for example, about one molecule of excess protoporphyrin accumulates for each 30 000 molecules of heme formed in the process of hemoglobin synthesis. However, a number of pathological conditions are associated with stimulated or inhibited heme biosynthesis leading to abnormal tissue levels or excretion rates of the porphyrins and their precursors. From a diagnostic point of view, abnormalities in porphyrin metabolism are considered only in terms of *increased* levels of porphyrins and their precursors. No known clinical significance can be ascribed to the occurrence of low concentrations.

Primary Disorders of Heme Biosynthesis

The metabolic abnormalities observed in the porphyrias result from inherited deficiencies of specific enzymes in the heme biosynthetic pathway (Table 25-1). Most classifications of the porphyrias categorize the diseases into hepatic and erythropoietic groups, based on the presumed major sites of excess porphyrin and porphyrin precursor production. In clinical practice, however, a classification relating to symptomatology may be more useful (Table 25-2). Classifying the porphyrias primarily into broad clinical categories (i.e., neurological and cutaneous) simplifies the clinical and chemical correlations in these disorders and improves the efficacy of the diagnostic evaluations of porphyria.

Table 25-1. ENZYME DEFECTS IN PORPHYRINOPATHIES AND THEIR RELATIONSHIPS TO LABORATORY FINDINGS

Specific Enzyme Defect	Porphyrinopathy	Diagnostic Laboratory Findings
δ-Aminolevulinate synthase	None known; probably lethal lesions if primary	May increase secondary to impaired heme synthesis in some disorders
Porphobilinogen synthase	Symptoms not established	Limited to decreased enzyme activity
Uroporphyrinogen I synthase	Acute intermittent porphyria	↓ Uroporphyrinogen I synthase ↑ Urinary δ-aminolevulinic acid* ↑ Urinary porphobilinogen* ↑ Urinary uroporphyrin*
Uroporphyrinogen III cosynthase	Congenital erythropoietic porphyria	↑ Urinary uroporphyrin ↑ Urinary coproporphyrin ↑ Blood porphyrins ↑ Fecal porphyrins
Uroporphyrinogen decarboxylase	Porphyria cutanea tarda	↑ Urinary uroporphyrin ↑ Urinary 7-COOH porphyrin ↑ Fecal isocoproporphyrin
Coproporphyrinogen oxidase	Coproporphyria†	↑ Urinary δ-aminolevulinic acid* ↑ Urinary porphobilinogen* ↑ Urinary coproporphyrin ↑ Fecal coproporphyrin
Protoporphyrinogen oxidase	Porphyria variegata	↑ Urinary δ-aminolevulinic acid* ↑ Urinary porphobilinogen* ↑ Urinary coproporphyrin ↑ Fecal protoporphyrin, coproporphyrin
Ferrochelatase	Protoporphyria	↑ Blood protoporphyrin ↑ Fecal protoporphyrin

*Typical finding during acute neuropathy; may be normal in remission.
† The name *hereditary coproporphyria* is commonly used, but it is redundant because all porphyrias by definition are hereditary.

Table 25-2. CLASSIFICATION OF PORPHYRINOPATHIES

Primary (inherited):
- *Neurologic (acute attack)*
  - Acute intermittent porphyria (uroporphyrinogen I synthase deficiency)
- *Cutaneous (photosensitive)*
  - Congenital erythropoietic porphyria (uroporphyrinogen cosynthase deficiency)
  - Porphyria cutanea tarda (uroporphyrinogen decarboxylase deficiency)
  - Protoporphyria (ferrochelatase deficiency)
- *Mixed (neurologic and cutaneous)*
  - Coproporphyria (coproporphyrinogen oxidase deficiency)
  - Porphyria variegata (protoporphyrinogen oxidase deficiency)
- *Asymptomatic* (?)
  - Porphobilinogen synthase deficiency

Secondary (induced):
- *Coproporphyrinuria*
  - (Examples: tyrosinemia, lead poisoning, alcoholism)
- *Protoporphyrinemia*
  - (Examples: iron deficiency, lead poisoning, inflammation)

Neurological (Acute Attack) Forms of Porphyria

Excess excretion of the porphyrin precursors δ-aminolevulinic acid and porphobilinogen are the characteristic chemical abnormalities in acute intermittent porphyria during symptomatic episodes. The acute attack or neurological forms of porphyria may be considered disorders of porphyrin precursors. Each of the diseases in this group reflects a specific enzyme defect that is inherited as an autosomal dominant trait; they include ***acute intermittent porphyria, variegate porphyria***, and ***coproporphyria***. The acute attack manifestations of each are identical, but symptomatic illness can be latent for indeterminate periods. During the acute phase, δ-aminolevulinic acid and porphobilinogen are excreted in the urine in excess, but during asymptomatic intervals these chemical abnormalities may resolve. Thus, although levels of δ-aminolevulinic acid and porphobilinogen may be diagnostic at any particular time, normal results during quiescent periods do not exclude a diagnosis. Although each of these disorders has a unique enzyme abnormality, urine and fecal porphyrin excretion patterns are generally used to establish the specific form of porphyria (Table 25-1).

Symptoms of an acute attack may include abdominal pain, back pain, nausea, paresthesia, weakness, inability to think clearly, and self-destructive thoughts. Signs of acute attacks often include hypertension and tachycardia. Acute attack episodes generally have a relatively brisk onset of hours to days and may last several days to several weeks. The spectrum of symptoms and signs as well as the duration and intensity of symptoms is quite variable among patients. An acute attack may be precipitated by drugs and other factors; therefore, prevention is a primary mode of patient management. After a protracted illness, neurological disease may develop.

Porphobilinogen synthase deficiency is a recently described disorder with an autosomal dominant inheritance. Individuals can be totally asymptomatic, but identification and study of further cases are necessary in order to elucidate the true clinical significance of this enzyme deficiency. Symptoms that have been noted are neurological in nature.

Cutaneous (Photosensitive) Forms of Porphyria

The common feature linking this group of diseases is the characteristic excess porphyrin production and excretion (Table 25-1). A bullous dermatosis occurs in ***porphyria cutanea tarda***, which is a cutaneous disorder only. Lesions are confined to the dorsum of the hands, face, and ears. These lesions are related to photosensitivity and occur primarily during the sunnier months of the year. Acute attacks do not occur, and neither neurological nor psychotic symptoms are present.

Porphyria cutanea tarda has been considered an acquired disease more often than other forms of porphyria. However, autosomal dominant transmission of a deficiency of uroporphyrinogen decarboxylase has been confirmed in both affected and nonaffected family members. The toxic or acquired cases may result from inhibition of the same enzyme or from irreversible oxidation of porphyrinogens to porphyrins. Excessive alcohol consumption is commonly associated with this disorder. However, iron accumulation has been shown to be the major etiological factor due to an inhibitory effect of iron on uroporphyrinogen decarboxylase. Estrogen exposure is another etiological setting. Porphyria cutanea tarda is probably the most common inherited porphyrin disorder in North America.

Congenital erythropoietic porphyria is the rarest form of inherited porphyrin disorders. It is unique among the porphyrias for having an autosomal recessive inheritance. The basic defect is a deficiency of uroporphyrinogen cosynthase in the bone marrow and accumulation of series I porphyrin isomers. Dark red urine and severe photosensitivity are typically observed in the neonatal period, but adult onset has been reported. Porphyrins stain the bones and teeth (erythrodontia) of these individuals, which results in a bright fluorescence of the stained areas on exposure to long-wavelength ultraviolet light. Ineffective erythropoiesis, hemolytic anemia, and splenomegaly complicate the clinical course.

Protoporphyria is the result of a ferrochelatase deficiency that leads to protoporphyrin accumulation (Table 25-1). This defect is associated with severe photosensitivity.

The cutaneous reaction occurs within 15–20 min of sun exposure and begins with burning and itching, followed by eventual painful swelling and erythema that may last for 48 h or longer. Beta-carotene provides protection against photosensitivity.

A discussion of clinical features of protoporphyria would be incomplete without noting the occurrence of hepatobiliary disease in a subgroup of these patients. Liver decompensation may occur abruptly with jaundice and evidence of portal hypertension (ascites, esophageal varices). The major route of protoporphyrin excretion is via the liver. Impaired liver function and bile excretion in this group of patients connote hepatic injury secondary to protoporphyrin accumulation in the liver parenchyma. The large enterohepatic circulation of protoporphyrin can be a very ominous development, since further hepatic protoporphyrin accumulation may then occur. Often, cholelithiasis occurs in protoporphyria at an unusually early age. In some instances, protoporphyrin deposition has been documented in stones that were removed surgically.

Secondary Disorders of Heme Biosynthesis

A variety of clinical conditions besides porphyrias are accompanied by excess accumulation and excretion of porphyrins or porphyrin precursors. Symptoms in these conditions may be indistinguishable from those of the porphyrias. In such cases, the disturbance of porphyrin metabolism is a result of a superimposed disorder or toxin and is usually not a result of an inherited defect in the heme biosynthetic pathway. It is interesting to note that those disorders that produce the clinical picture of an acute attack of porphyria exhibit increased amounts of only δ-aminolevulinic acid in the urine, whereas porphobilinogen excretion is not increased. This demonstrates the specificity of porphobilinogen for acute intermittent porphyria and may offer a clue with regard to an important etiological factor in the clinical illness. The two diseases in this category are *lead poisoning* and *hereditary tyrosinemia*. Lead reversibly inhibits both the activity of porphobilinogen synthase and the incorporation of iron into heme. In addition to increased urine excretion of δ-aminolevulinic acid, erythrocyte protoporphyrin concentration (as a zinc chelate) is increased. Urine coproporphyrin is elevated as a delayed response; in fact, chronic lead poisoning is one of the main causes of significant coproporphyrinuria. Despite the excess coproporphyrin in the urine and protoporphyrin in erythrocytes, individuals with lead poisoning do not experience photosensitivity. *Heavy metal intoxication* may produce a skin change unrelated to secondary coproporphyrinuria. However, metals such as mercury, bismuth, copper, gold, silver, and arsenic do not cause elevation of urinary δ-aminolevulinic acid or erythrocyte protoporphyrin. Hereditary tyrosinemia also produces an acute porphyria-like illness. One of the metabolites accumulating in excess (succinyl acetone) is a potent inhibitor of porphobilinogen synthase. Interestingly, coproporphyrinuria occurs in this disorder as well.

All conditions that produce an imbalance between protoporphyrin formation and iron availability can lead to accumulation of erythrocyte protoporphyrin or its zinc chelate. This abnormal metabolite is markedly increased in chronic lead exposure. Zinc protoporphyrin also increases in iron deficiency. This abnormal porphyrin chelate circulates in the erythrocyte attached to a heme site on globin, where it serves no biological function. The mechanism of formation is unknown. Sideroblastic anemia, in which iron is not properly utilized; hemolytic anemia, in which erythropoiesis is greatly accentuated; secondary polycythemia, in which there is a stimulus to erythropoiesis extrinsic to the marrow; and the commonly seen inflammatory block are all associated with elevated erythrocyte protoporphyrin or its zinc chelate. Anemia accompanying chronic disease has also been stated to cause elevated erythrocyte protoporphyrin. The conditions of heme biosynthesis associated with secondary protoporphyrinemia typically have protoporphyrin concentrations below 400 μg/dL of erythrocytes, whereas erythropoietic protoporphyria typically produces erythrocyte protoporphyrin concentrations in considerable excess of those occurring in these secondary disorders.

Coproporphyrin is by far the most common porphyrin excreted in secondary porphyrinuria. Hexachlorobenzene, alcohol, sedatives, and hypnotics can also cause coproporphyrinuria. Neoplasia, liver disease, myocardial infarction, and thalassemia have on occasion been associated with coproporphyrinuria.

Diagnosis of Porphyrin Disorders

Porphyrins and Porphyrin Precursors

Essentially, the quantitation of three porphyrins (uroporphyrin, coproporphyrin, and protoporphyrin) and two precursors (δ-aminolevulinic acid and porphobilinogen) comprises the basis for the laboratory diagnosis of most disorders affecting porphyrin metabolism (Table 25-1 and Figure 25-1).

Selection and Interpretation of Laboratory Tests

If an acute neurological porphyria is suspected, the primary screening procedure is a porphobilinogen determination. Increased urinary porphobilinogen is the characteristic chemical abnormality of acute intermittent porphyria, variegate porphyria, and coproporphyria during an acute attack. Since porphobilinogen concentration in the urine may or may not be increased at nonsymptomatic times, screening procedures should be reserved for determining whether a patient's acute symptoms are the result of a neuropathic porphyria. False-negatives are very rare and false-positives are uncovered by assay of porphobilinogen in a 24-h urine specimen. A negative screening test is usually reliable in a symptomatic patient. When porphyria is strongly suspected on clinical grounds and quantitative urine studies are performed, it is worthwhile to determine both δ-aminolevulinic acid and porphobilinogen concentrations. These two measurements allow a differentiation of lead poisoning and tyrosinemia, two disorders that resemble acute porphyria clinically but typically exhibit normal urinary porphobilinogen and elevated urinary δ-aminolevulinic acid levels. After a diagnosis of porphyria has been confirmed, screening tests are not appropriate for following a patient's course.

Since δ-aminolevulinic acid is most stable at acid pH and porphobilinogen at alkaline pH, organizing a proper urine collection procedure is difficult. This problem is compounded when uroporphyrin and coproporphyrin measurements are also requested. A practical approach is the collection of a 24-h urine specimen that is kept in the cold and protected from light, with sodium bicarbonate added to maintain pH near neutrality. This provides maximum stability of both precursors as well as the porphyrins in the same specimen.

Urine screening for total porphyrins has only limited specificity, and urine porphyrin composition can vary greatly. Nevertheless, a total porphyrin screen is often requested to identify a suspected neuropathic porphyria.

Uroporphyrinogen I synthase deficiency, unlike the level of urine porphobilinogen excretion, is a constant finding in acute intermittent porphyria, and latent cases (i.e., without clinical or urine abnormalities) can be identified. The two other acute attack forms of porphyria may have an identical clinical and chemical picture during the acute phase, but the uroporphyrinogen I synthase activity will be normal in these porphyrin disorders.

The photosensitive porphyrias are best characterized chemically as disorders of porphyrin excess, as opposed to porphyrin precursor excess. In addition to urine porphyrins, erythrocyte protoporphyrin assay is essential for diagnosis of photosensitive porphyria; protoporphyrin is not excreted in the urine. Fecal porphyrin assays can be helpful in diagnosing coproporphyria, and these assays are often considered essential to confirm a diagnosis of porphyria variegata.

Analysis of Porphyrin Precursors

Two porphyrin precursors, δ-aminolevulinic acid and porphobilinogen, accumulate or are overproduced in the neuropathic porphyrin disorders. When used in clinical diagnosis, porphyrin precursors are measured almost exclusively in urine. Both screening tests and quantitative assays are used for porphobilinogen, but no screening test exists for δ-aminolevulinic acid. Serum concentrations can provide clinically useful information, but the much lower concentrations of the porphyrin precursors restrict serum assays to research laboratories.

Determination of δ-Aminolevulinic Acid[9]

Principle. δ-Aminolevulinic acid is condensed with ethyl acetoacetate to form a pyrrole. This derivative is purified by extraction into ethyl acetate. The extracted derivative is then reacted with Ehrlich's reagent to give a cherry-red colored compound that is measured spectrophotometrically.

Specimen. A 24-h urine collection is obtained and the total volume recorded. During the urine collection period, the specimen should be refrigerated and should contain 2 g barbituric acid as preservative. If porphyrins or porphobilinogen is also to be assayed, substitute 4–5 g of sodium bicarbonate for barbituric acid to ensure a near neutral pH.

Reagents

1. Acetate buffer, pH 4.6. To 700 mL of water, add 57 mL of glacial acetic acid and 136 g of sodium acetate trihydrate. Dilute to 1 L with water.
2. Ethyl acetoacetate, reagent grade.
3. Ethyl acetate, reagent grade.
4. Ehrlich's reagent.* To ~30 mL of glacial acetic acid in a 50-mL cylinder, add 1.0 g of *p*-dimethylaminobenzaldehyde, 5 mL of 70% perchloric acid, and 5 mL water. Dilute to 50 mL with glacial acetic acid.
5. δ-Aminolevulinic acid standard. Stock δ-aminolevulinic acid standard is prepared by dissolving 0.128 g of the commercially available hydrochloride in 100 mL of water to obtain 1000 μg δ-aminolevulinic acid/mL. Dilute 0.10 mL of this stock solution to 10 mL with water to obtain 10 μg/mL. A standard curve is prepared by further diluting this solution as follows: 0.10 mL stock solution to 1.0 mL with water, 0.20 mL stock solution to 1.0 mL with water,. . . continuing in 0.10-mL increments to a maximum of 1.0 mL undiluted stock solution. This gives a standard curve ranging from 1.0 to 10 μg/mL. The stock solution is stable for two months at 4 °C.

Procedure

1. Place 1.0 mL urine into each of three glass-stoppered centrifuge tubes for each urine specimen (duplicate specimens plus one blank).
2. Add 1.0 mL acetate buffer to each tube.
3. Add 0.20 mL ethyl acetoacetate to each specimen tube and to a water blank but NOT to the patient specimen blank. Mix by vortexing, stopper, and place in a boiling water bath for 10 min.
4. Allow the tubes to cool, or hasten cooling by placing tubes into ice or cold water.
5. Add 3.0 mL ethyl acetate to each tube, stopper, and shake vigorously 50 times.
6. Remove the glass stoppers and centrifuge lightly (about 1000 × *g*) to separate the phases. A short (1 min) centrifugation will not cause significant evaporation.
7. Using a volumetric pipet, withdraw 2.0 mL of the upper ethyl acetate phases and transfer to separate tubes.
8. Add 2.0 mL Ehrlich's reagent to each tube. Vortex-mix the reactants and allow to stand 10 min.
9. Measure the absorbance immediately at 553 nm, using water to zero the spectrophotometer.

Calculation. Subtract the blank reading from each specimen reading, then determine the μg δ-aminolevulinic acid/mL urine from the standard curve. Calculate the 24-h excretion as follows:

$$\text{mg } \delta\text{-Aminolevulinic acid/d} = \frac{\mu\text{g } \delta\text{-Aminolevulinic acid/mL}}{1000} \times \text{TV}$$

where 1000 = conversion of μg to mg and TV = total volume (mL) of urine collected in 24 h.

The ***reference range*** for δ-aminolevulinic acid is 1.5–7.5 mg/d.

---

*Several modifications of Ehrlich's reagent have been used, each varying in the kind and concentration of acid solvent and the concentration of *p*-dimethylaminobenzaldehyde. This is the result of attempts to obtain a stabilized chromophore and to enhance its absorbance. The particular Ehrlich's reagent indicated here is that described in the original method.

METHODS FOR THE DETECTION OF PORPHOBILINOGEN

Principle. Porphobilinogen condenses with *p*-dimethylaminobenzaldehyde in acid solution (Ehrlich's aldehyde reagent) to form a magenta-colored product.

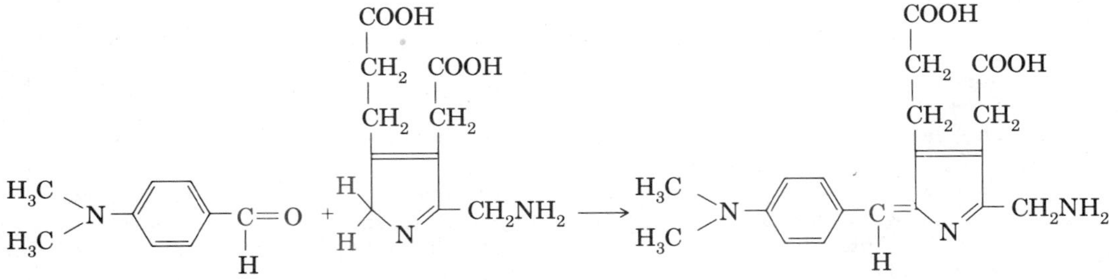

p-Dimethylamino-benzaldehyde + **Porphobilinogen (PBG)** → **PBG-Aldehyde Chromophore**

Since this reaction can occur with other urinary constituents, the two screening tests use pH adjustment and solvent extractions to remove interfering substances, thus making the tests reasonably specific. For quantitative analysis, porphobilinogen is purified by adsorption to an ion-exchange resin. Color-producing interfering substances, such as urobilinogen, methyldopa, or chlorpromazine, as well as indole and related compounds that interfere by reacting with the chromophore to produce colorless derivatives, may be removed by repeated washings of the resin with water prior to the elution of porphobilinogen with acetic acid.

Specimen. Screening tests for porphobilinogen are carried out preferably on a fresh morning specimen. Quantitative analyses should be performed on 24-h urine collections. If the pH of the urine is adjusted to near neutrality (pH 6–8) with sodium bicarbonate, the specimen can be stored frozen for periods up to two weeks, although assays should be performed as soon as possible.

Reagents. Prepare either reagent 1, 2, or 3 for the method desired. Reagents 7, 8, and 9 are only required for the quantitative assay.

1. Ehrlich's reagent, qualitative (Watson-Schwartz test).* Dissolve 0.7 g *p*-dimethylaminobenzaldehyde in 150 mL of concentrated HCl and add 100 mL water. Store in a brown bottle. Although the reagent becomes increasingly yellow with storage, it can be used for a period of several months.
2. Ehrlich's reagent, qualitative (Hoesch test).* Dissolve 2.0 g of *p*-dimethylaminobenzaldehyde in 100 mL HCl, 6 mol/L. The reagent can be stored without loss of activity for at least nine months.
3. Ehrlich's reagent, quantitative.* Dissolve 2.0 g of *p*-dimethylaminobenzaldehyde in 25 mL of concentrated HCl plus 75 mL of glacial acetic acid. The reagent becomes increasingly yellow on prolonged storage. These changes can be retarded by refrigeration in the dark.
4. Sodium acetate, saturated. An excess of sodium acetate $\cdot$ 3 H_2O (~1 g/mL) is added to water and the reagent is stored at room temperature.
5. Chloroform, reagent grade.
6. *n*-Butanol, reagent grade.
7. Acetic acid, 1.0 mol/L.
8. Ammonia, reagent grade, 28 g NH_3/100 g.
9. Dowex 2-X8 resin (200-400 mesh), acetate form. Suspend 100–200 g of Dowex 2-X8 resin in about 4 volumes of sodium acetate, 10 g/dL. Mix the suspension well and allow it to settle by gravity until the supernatant solution can be easily decanted. Wash the resin 8–10 times, each with about 4 volumes of deionized water, decanting each time to remove both the fines and the excess sodium acetate. A stock suspension of resin is prepared by adding an equal volume of water to the washed, packed resin. Immediately before use, the suspension should be shaken thoroughly, since the resin settles rapidly. This resin preparation is stable for several months.

---

*Several modifications of Ehrlich's reagent have been used, each varying in the kind and concentration of acid solvent and the concentration of *p*-dimethylaminobenzaldehyde. This is the result of attempts to obtain a stabilized chromophore and to enhance its absorbance. The particular Ehrlich's reagent indicated here is that described in the original method.

Screening Tests for Porphobilinogen in Urine

Procedure (Watson-Schwartz Test)[7]

1. To 2.0 mL of urine in a test tube, add 2.0 mL of Ehrlich's reagent, qualitative (Watson-Schwartz), and mix.
2. Add 4.0 mL of saturated sodium acetate and mix again. Check with pH indicator paper to confirm that the solution is in the range of pH 4–5.
3. Add 5 mL of chloroform, stopper, and shake vigorously for 1 min. Permit the phases to separate; centrifuge if necessary.
4. Transfer the upper (aqueous) phase to another tube, to which is added 2 mL of *n*-butanol. Stopper and shake vigorously for 1 min. Allow the phases to separate; centrifuge if necessary.
5. Examine the lower (aqueous) layer for a magenta color. If this color is present, the test is positive for porphobilinogen, indicating a concentration several times normal. A magenta color in the upper (butanol) layer indicates increased urobilinogen. If the interpretation is ambiguous, proceed to the Hoesch test.

Procedure (Hoesch Test)[3]

1. Place 2 mL of Ehrlich's reagent, qualitative (Hoesch), into a test tube, then overlay with 3 drops of urine.
2. If increased porphobilinogen is present, an immediate cherry-red color will develop at the point of mixing.

Comments (Watson-Schwartz and Hoesch tests). Despite modifications to improve the specificity of these two tests, an inexperienced person can have difficulties with interpretation of some results with either the Watson-Schwartz or Hoesch test. An effective solution to the uncertainties of interpretation is the combined use of both tests. Since the Hoesch test does not react with urobilinogen, it serves primarily to confirm the results of the Watson-Schwartz test.

In the Watson-Schwartz test, porphobilinogen and the chromogen that it forms always remain in the aqueous phase. The extractions with chloroform and butanol are essential for removing frequently occurring substances that interfere with the test. If no magenta color is observed in the upper (aqueous) phase following the chloroform extraction, the *n*-butanol extraction may be omitted and the test considered negative. The most common interfering substance is chloroform-soluble urobilinogen. Other substances sometimes present in urine can give a variety of colors, including yellow, orange, and red. All of these tend to make a positive identification of porphobilinogen difficult. Under these circumstances the quantitative procedure for porphobilinogen should be performed, because this method removes the most common interfering substances, as well as the decolorizing indoles, and permits positive identification of porphobilinogen.

Quantitative Determination of Porphobilinogen in Urine[5]

Procedure

1. Pipet 4 mL of well-mixed Dowex 2-X8 resin suspension into a 12-mL centrifuge tube and centrifuge ~1 min; discard the supernatant solution. Pipet 1 mL of urine onto the packed resin, followed by 0.1 mL of concentrated ammonium hydroxide (28 g NH_3/100 g). Stir the mixture thoroughly and centrifuge ~1 min; discard the supernatant solution.
2. Wash the resin, to which the porphobilinogen has been adsorbed, four times with 5-mL aliquots of deionized water to remove interfering substances. Centrifuge ~1 min after each wash.
3. Elute the porphobilinogen from the washed resin with four 2-mL aliquots of acetic acid, 1 mol/L. After each acetic acid addition, mix and then centrifuge the resin for 1 min and decant the supernatant solution into a 12-mL graduated centrifuge tube. Adjust the final volume to exactly 10.0 mL with acetic acid, 1 mol/L; mix the solution and centrifuge briefly to remove any traces of resin that may have been carried over during decanting.
4. Treat 2 mL of the acetic acid eluate with 2 mL of Ehrlich's reagent (quantitative). Prepare a reagent blank, substituting 2 mL of acetic acid, 1 mol/L, for the eluate. Thoroughly mix the reactants. After allowing the tubes to stand 6–8 min, read them against the reagent blank at 555 nm and 525 nm in cuvets with a 10-mm light path, using a spectrophotometer with a resolution of ≤ 2 nm.

Calculation. Porphobilinogen excretion is calculated as follows:

$$\text{Porphobilinogen, mg/d} = \frac{A_{555} \times 10 \times \text{TV}}{0.114 \times 1000}$$

where 0.114 = absorbance of the reaction product in a 10-mm light path when the porphobilinogen concentration is 1 μg/mL
10 = dilution of the original urine aliquot with acetic acid
TV = total volume (mL) of urine collected in 24 h
1000 = μg to mg conversion factor.

Comments. The factor 0.114 was determined empirically according to the conditions outlined and using a pyrrole derivative of δ-aminolevulinic acid as a standard. Alternatively, porphobilinogen can now be obtained commercially, but it is somewhat more expensive and unstable.

If porphobilinogen is present in significant amounts, the color developed by Ehrlich's reagent should be rose to crimson, and the ratio of A_{525}/A_{555} should be near 0.83. A ratio >1.00 is rare and indicates that interfering substances are still present and that the result should not be interpreted as an abnormal concentration of porphobilinogen. Alternatively, an Allen correction can be applied by measuring the absorbance at 535, 555, and 575 nm, and then utilizing the formula $A_{corrected} = 2A_{555} - (A_{535} + A_{575})$. Calculating with this corrected absorbance helps to eliminate the effects of interfering chromogens, but a chromatographic purification remains essential. With a porphobilinogen concentration two to three times normal, which is clinically significant, no difficulty is likely to be encountered in its quantitation.

The reaction of porphobilinogen with Ehrlich's reagent yields a product whose absorbance follows Beer's law up to an absorbance of at least 0.750. If the absorbance is too high, less of the original sample should be used and a correction made in the dilution factor of the calculation.

Upon addition of Ehrlich's reagent, maximum color develops within 6 min. The color remains stable for 2–3 min and then begins to fade slowly. Within 20 min absorbance decreases by about 10%.

The optimum amount of resin is 2 volumes of packed resin to 1 volume of urine. The concentration of acetic acid required to elute porphobilinogen from the resin is not critical, but four elutions are essential to obtain quantitative recovery of the porphobilinogen. The assay has a CV of 7.0%. Purification of porphobilinogen can be done by utilizing commercially available chromatographic columns* as a substitute for the batch process. The remainder of the procedure can be carried out as described.

The ***reference range*** for the normal urinary excretion rate of porphobilinogen is < 1 mg/d; however, caution should be exercised in interpreting slightly elevated values as diagnostic of porphyria.

METHODS FOR PORPHYRIN ANALYSES

Principle. Virtually all porphyrin analyses are based on the isolation of one or more porphyrins from the specimen, separation of the individual porphyrins, and observation or measurement of the porphyrins by fluorometry or spectrophotometry. Porphyrins are usually isolated by extraction into an acidified organic solvent. For screening purposes, minimal purification is required. For quantitation, the individual porphyrins are separated by selective solvent extraction or by some type of chromatography. Their characteristic orange-red fluorescence (620–630 nm) when irradiated with long-wavelength ultraviolet light (398–408 nm) allows porphyrins in acid solutions to be detected fluorometrically at concentrations below 10^{-8} mol/L. When the concentrations of porphyrins are sufficiently high, they can be measured spectrophotometrically.

Specimens. *Whole blood* specimens are collected using any common anticoagulant. *Urine* for screening purposes preferably should be a first morning specimen, but random specimens can be used. Quantitative analyses should be performed on 24-h urine specimens collected in a container to which 4–5 g of sodium bicarbonate has been added to maintain the pH at 6–8. If *feces* are used only for qualitative tests, a small specimen (1 g) is adequate.

When porphyrin analyses cannot be performed soon after collection, specimens should be stored in the dark at 4 °C, or kept frozen if more than 1–2 d will elapse before analysis.

---

*Prepared chromatographic columns for porphobilinogen assay are supplied by Bio-Rad Laboratories, Richmond, CA 94804.

Screening Tests for Porphyrins

All screening tests are performed in glass apparatus. Plastic tubes must be avoided because fluorescence quenching by many synthetic materials will interfere with observation of the porphyrin fluorescence.

Reagents

1. Hydrochloric acid, 3 mol/L. Dilute 25 mL of concentrated HCl, reagent grade, to 100 mL with water.

2. Ethyl acetate: glacial acetic acid (4:1). Mix reagent grade solvents in the indicated volume ratio.

Equipment. For all porphyrin screening tests (blood, urine, feces), use a long-wavelength ultraviolet or Wood's lamp of adequate intensity; a 100-*W* mercury spot bulb (7000 $\mu W/cm^2$ at 15 inches) fitted with a red-purple filter is recommended.

Qualitative Determination of Urinary Porphyrins[2]

Procedure. Place 5 mL of urine and 3 mL of ethyl acetate:glacial acetic acid into a centrifuge tube and shake thoroughly. Allow the phases to separate by standing, or hasten the process by a brief centrifugation. Irradiate the upper (organic) layer with ultraviolet light and observe for porphyrin fluorescence. This layer fluoresces lavender with moderately elevated concentrations of porphyrins, then pink, and finally red with markedly increased porphyrin concentration. Since drugs, abnormal metabolites, or unknown materials will occasionally impart a fluorescence or color to the organic layer, this upper layer should be removed carefully with a pipet and transferred to a second tube to which is added 0.5 mL of HCl, 3 mol/L. With mixing, the porphyrins are extracted into the lower acid phase, whereas most interfering substances remain in the organic phase. In an acid medium, porphyrin fluorescence is greatly intensified; moreover, the bluish tint of common urine components is removed by the extraction to give the orange-red fluorescence typical of porphyrins. Urine specimens from healthy subjects can be used as a guide for normal levels of porphyrins.

Qualitative Determinations of Fecal Porphyrins[2]

Procedure. Transfer a small portion of feces (~50 mg or a volume about the size of a large drop of water) from the end of a glass rod into a 12-mL centrifuge tube. Add 3 mL of ethyl acetate:acetic acid and stir vigorously. Centrifuge briefly and decant the supernatant solution into a second centrifuge tube containing 0.5 mL of HCl, 3 mol/L. Shake or vortex the solutions well for 1 min and allow to separate, or centrifuge briefly. Observe the lower (aqueous) phase in ultraviolet light for porphyrin fluorescence. Normally, this acid solution shows little or no fluorescence; porphyrins, when present in increased concentration, will impart a clearly distinguishable orange-red fluorescence.

The occasional presence of dietary chlorophyll in fecal matter may cause confusion because it also has a red fluorescence not easily distinguishable from that of porphyrins. Even though chlorophyll is insoluble in aqueous solution and remains in the upper organic layer, it can exhibit a deep red fluorescence that might cause uncertainty in interpretation of the test. When high levels of chlorophyll appear to be present, the upper organic layer can be drawn off and discarded and a fresh 3-mL aliquot of ethyl acetate:acetic acid used to extract the remaining HCl layer. This procedure will remove the chlorophyll interference, and any remaining orange-red fluorescence in the aqueous phase can be considered the result of porphyrins.

Qualitative Determination of Blood Porphyrins[2]

Procedure. Place 1 mL of whole blood into 3 mL of ethyl acetate:acetic acid in a centrifuge tube. Stir the mixture thoroughly, centrifuge, and decant the supernatant solution into a second centrifuge tube containing 0.5 mL of HCl, 3 mol/L. Shake or vortex the two phases thoroughly and allow to separate. Illuminate the lower HCl layer with an ultraviolet light. An orange-red fluorescence indicates porphyrins are present. With these volumes of solutions, normal blood will usually show just a trace of porphyrin fluorescence. This will be equivalent to about 60 μg of porphyrin/dL erythrocytes, or near the upper end of the normal range. A distinct porphyrin fluorescence is seen at concentrations of two to three times normal, becoming deep orange-red as the concentration increases further.

Virtually all of the porphyrins found in blood are within the erythrocytes; the plasma contains only traces of porphyrins. Therefore, an analysis of whole blood actually reflects the porphyrin content of the erythrocytes. Increased cellular porphyrin concentrations can also be easily observed by fluorescence microscopy of an unfixed, unstained blood smear. However, the intense ultraviolet irradiation focused on the slide quickly destroys the protoporphyrin, causing its fluorescence to fade rapidly in the field of view. By contrast, uroporphyrin and coproporphyrin are quite stable under these conditions.

Identification of Porphyrins

When abnormal levels of porphyrins are seen or suspected in a screening test, identification of the specific porphyrin or porphyrins that are elevated can aid in diagnosis. Identification can be accomplished very simply with thin-layer chromatography. Cut Gelman Instant TLC Sheets Type SA (Gelman Sciences, Inc., Ann Arbor, MI 48106) with a sharp knife on a clean flat surface to approximately 1.1 × 6.6 cm; handle the strips with forceps to avoid contamination. Store the strips in a clean test tube or Petri dish. To separate the porphyrins, transfer ~30 μL of the ethyl acetate layer from the screening test to the chromatographic strip, placing the spot ~1.5 cm from one end. (*Note*: Depending upon the estimated fluorescence, as little as 10 μL or as much as 60 μL can be spotted.) Apply the solutions slowly and dry with a hair dryer during the course of addition to minimize dispersion of the spot. Place the TLC strip into a 10-mL screw-topped vial (21 × 70 mm) containing ~1.0 mL of developing solvent prepared by mixing 1.2 mL chloroform, 1.4 mL methanol, and 0.3 mL ammonium hydroxide. Allow the solvent to ascend to near the top of the strip (5–10 min). Remove the strip from the vial and allow to dry briefly. While supporting the strip on a clean firm surface (white filter paper serves well), examine the fluorescence under ultraviolet light and note the pattern. Uroporphyrin is normally visible as a faint round spot at the origin. Coproporphyrin will appear as a sharp band at 1.5–2.0 cm from the origin. Other fluorescent compounds usually move near the solvent front. Significantly elevated uroporphyrin and coproporphyrin are easy to identify. A unique feature of porphyria cutanea tarda is a sharp band of 7-carboxyl porphyrin appearing slightly above the uroporphyrin spot. Other intermediate porphyrins of five and six carboxyl groups are rarely seen and need not be utilized for diagnosing porphyrinopathies (see Table 25-1).

Urinary porphyrins may be quantified, as well as identified, by high performance liquid chromatography.[8] *Reference ranges* are 4.0–20 μg/d for uroporphyrin and 13–179 μg/d for coproporphyrin.

Quantitative Determination of Erythrocyte Protoporphyrin[6]

Principle. Several simplified, rapid micro-methods for erythrocyte protoporphyrin measurement have been developed. Typically, all erythrocyte porphyrins are first effectively removed from the blood by adsorption or extraction. In this instance Celite is the adsorbant. The described method does not discriminate among uro-, copro-, or protoporphyrin, but this has little clinical consequence, since the predominant extractable porphyrin is protoporphyrin. The adsorbed porphyrin is further purified by eluting it into acidified ethyl acetate. It is then measured fluorometrically, using coproporphyrin as a standard.

Specimen. Whole blood is anticoagulated with EDTA or heparin. The specimen is stable for several weeks when stored at 4 °C. It should *not* be frozen, since this can affect porphyrin extractability.

Reagents

1. Celite (No. C211, Fisher Scientific Co., Pittsburgh, PA 15219). Suspend 5 g in 100 mL sodium chloride, 0.156 mol/L. Shake well immediately before removing an aliquot. The suspension is stable for at least two months.
2. Ethyl acetate:acetic acid (4:1). Mix reagent-grade solvents in the indicated volume ratio. The solution is stable indefinitely.
3. Hydrochloric acid, 1.5 mol/L. Dilute 12.5 mL of concentrated HCl to 100 mL.

4. Coproporphyrin I standard, 5 μg (Sigma Chemical Co., St. Louis, MO 63178). Dissolve the coproporphyrin standard in 5 mL of HCl, 1.5 mol/L, by heating in a hot water bath according to the package insert. This standard should be stable for many months if stored in a well-sealed container in the dark at 4 °C. To determine accurately the concentration of this coproporphyrin reference standard, measure absorbance of the HCl solution in a cuvet with a 10-mm light path. At the Sorét peak (399 nm), the molar absorptivity is 4.70×10^5. Note that near 399 nm all porphyrins have a very sharp spectral peak and high molar absorptivity, which means that slight variations in instrument adjustments can have a relatively large influence on the reading. Hence, it is essential that the concentration be calculated from the amount of absorbance at the Sorét peak, whether this occurs slightly above or below 399 nm. For a standard curve, prepare concentrations of 10–50 ng/mL. This concentration range should give a linear response through zero. Coproporphyrins I and III have the same spectral properties; thus, either can be used as a standard.

Procedure

1. Place 750 μL of 5% Celite suspension into a 12-mL centrifuge tube. Add 50 μL of whole blood and mix by vortexing for 10 s.
2. Add 5 mL of ethyl acetate:acetic acid and vortex for 10 s to solubilize the porphyrins.
3. Centrifuge for 5 min at about $1000 \times g$ to pack the Celite-protein precipitate.
4. Decant the supernatant solution into another tube and add 5 mL of HCl, 1.5 mol/L. Vortex mix for 30 s to extract the porphyrins. Centrifuge briefly to separate the layers; the lower layer is the porphyrin-containing HCl fraction.
5. A simple fluorometer fitted with an excitation filter having a peak transmission at 405 nm and emission filter with a cutoff below 600 nm is blanked with HCl, 1.5 mol/L. The fluorescence of the HCl layer is then measured. Calculate the concentration using the standard curve. When the fluorescence reading is too high, the porphyrin solution may be diluted in HCl, 1.5 mol/L, a neutral density filter may be placed behind the secondary filter, or the assay may be repeated using less specimen.

Calculation. Calculation of erythrocyte protoporphyrin must take into account the fact that a coproporphyrin standard is used and that the result is typically normalized to 100 mL of erythrocytes. (Calculation per gram of hemoglobin can also be used.)

$$\text{Protoporphyrin, } \mu\text{g/dL erythrocytes} = \frac{\text{Fluorescence (unkn)} \times \text{conc (std)} \times 2.6 \times 100}{\text{Fluorescence (std)} \times 0.05 \times 1.11 \times \text{hematocrit}}$$

where 2.6 = final volume (mL) of porphyrins in HCl
0.05 = volume (mL) of blood used for assay
1.11 = correction for use of a coproporphyrin fluorescence standard to measure protoporphyrin fluorescence
100 = conversion of hematocrit from per cent to a decimal fraction.

Comments. In this assay, acid treatment of the sample breaks the zinc protoporphyrin chelate and the determination reflects total protoporphyrin. The assay therefore detects protoporphyrinemia without reference to cause, which may be either primary as in photosensitive protoporphyria or secondary to chronic lead poisoning, hereditary tyrosinemia, or iron deficiency.

The ***reference range*** for erythrocyte protoporphyrin is 17–77 μg protoporphyrin/dL erythrocytes. These values are highly method dependent. The units and values also differ depending on whether porphyrin content is expressed per deciliter whole blood, deciliter erythrocytes, gram hemoglobin, or mole of heme.

Determination of Zinc Protoporphyrin by Hematofluorometry[1]

Zinc protoporphyrin forms during states of iron-deficient erythropoiesis from a variety of causes, including poor diet, excessive blood loss, chronic inflammation, or chronic lead exposure. Prior to the discovery of zinc protoporphyrin and the development of hematofluorometry, erythrocyte porphyrins were assayed by acid extraction procedures. This resulted in the loss of zinc and the formation of metal-free protoporphyrin; hence, the term *free erythrocyte protoporphyrin* or *FEP* that has come into common usage. Now, these increased levels of zinc protoporphyrin can be determined directly with a hematofluorometer. This is an application of front-surface fluorometry. Incident (exciting) light is directed onto a drop of blood on a glass slide at an angle of 37°. The emitted light is then detected at an angle to the incident beam. The equipment as originally described uses a tungsten-halogen lamp and excitation or primary filters that transmit wavelengths of 420–430 nm. The emission or secondary filters transmit at wavelengths of 580–680 nm. Commercial equipment is available.

The test requires one drop (~50 μL) of anticoagulated whole blood, venous or capillary. Zinc protoporphyrin is stable in a refrigerated specimen for at least one week. Since hemolysis adversely affects test results, specimens should not be frozen prior to analysis. Blood must be thoroughly mixed and completely oxygenated when on the slide for placement into the instrument, since a partially deoxygenated sample gives erroneously low results. Bilirubin can be a positive interferent, but its effect has been significantly diminished in contemporary instruments by appropriate choice of filters.

Zinc protoporphyrin results by hematofluorometry are linear from normal to grossly elevated levels (11 000 μg/L) and correlate well with results obtained by extraction methods. The coefficient of variation is ~2%. *Reference ranges* depend on the instrument and method, but a concentration of 17–77 μg/dL erythrocytes is a reasonable expectation in most circumstances.

Measurement of Enzymes of Heme Biosynthesis

Assays for all of the enzymes of the heme biosynthetic pathway have been described. However, only uroporphyrinogen I synthase activity is measured regularly in the clinical laboratory.

Determination of Uroporphyrinogen I Synthase Activity[4]

Principle. The enzyme catalyzes the formation of uroporphyrinogen from the monopyrrole precursor, porphobilinogen. The uroporphyrinogen is rapidly oxidized to uroporphyrin during acid deproteinization of the incubation mixture. Uroporphyrin fluorescence is then measured directly.

Specimen. A whole blood specimen is collected with either heparin or EDTA. The specimen can be stored at 4 °C for one week without significant loss of activity.

Reagents

1. Tris, 0.05 mol/L, pH 8.2. Dissolve 600 mg tris(hydroxymethyl)aminomethane in 90 mL water. Adjust to pH 8.2 with HCl, 6 mol/L, and bring volume to 100 mL.
2. Porphobilinogen, 2 mmol/L. Dissolve 4.5 mg in 10.0 mL deionized water. Stable for one month if stored at −20 °C.
3. Trichloroacetic acid (TCA), 25% (25 g/dL). Dissolve 25 g TCA in 50 mL water and bring volume to 100 mL.
4. Coproporphyrin standards (5–50 ng/mL). Prepare as described for quantitative determination of erythrocyte protoporphyrin, except that dilutions should be in 12.5% TCA.

Equipment. A simple fluorometer is fitted with an excitation filter with a peak transmission at 405 nm and an emission filter with a cutoff below 606 nm.

Procedure

1. Prepare an erythrocyte hemolysate by adding 50 μL of whole blood to 950 μL of Tris buffer. Leave this solution on ice for 10 min before proceeding.
2. Measure the hemoglobin concentration of the hemolysate (step 1) for calculation on the basis of activity per gram of hemoglobin. The cyanmethemoglobin (Drabkin) or any comparable method is suitable.
3. Combine in two separate tubes (test and blank) 0.9 mL of Tris buffer and 0.5 mL of porphobilinogen substrate. Add 5 μL of the hemolysate. Stop the blank reaction immediately by adding 1 mL of trichloroacetic acid.
4. Incubate the reaction tubes in the dark with gentle agitation at 37 °C. After 60 min, stop the reaction in the test by adding 1.0 mL of 25% trichloroacetic acid.
5. Centrifuge the reaction tubes at 1000 × *g* for 10 min or as necessary for clarification. Measure the fluorescence of the supernatant after blanking the instrument with 12.5% TCA.

Calculations. Results are expressed in enzyme units of activity per unit of hemoglobin. One unit of uroporphyrinogen synthase activity is equivalent to the formation of 1 μmol of uroporphyrin/min.

$$\text{mU/g Hb} = \frac{\text{Fluorescence (unkn)}}{\text{Fluorescence (std)}} \times \text{std conc} \times \frac{1.09}{0.05} \times \frac{1}{\text{Hb}} \times \frac{1}{60}$$

where std conc = concentration of coproporphyrin standard in nmol/mL
0.05 = mL of blood added (1:20 hemolysate)
Hb = hemoglobin concentration in the 1:20 hemolysate (g/mL)
1.09 = correction factor to express porphyrin formation as uroporphyrin after measurement against a coproporphyrin standard
60 = minutes of incubation

Comments. The erythrocyte uroporphyrinogen I synthase activity varies with cell age. Therefore, a shift in the erythrocyte population toward younger cells might lead to an increased enzyme activity that does not accurately represent the true condition. An elevated reticulocyte percentage would indicate that such a situation could be present.

The ***reference range*** for uroporphyrinogen I synthase activity is 1.27–2.01 mU/g hemoglobin.

References

1. Blumberg, W. E., Eisinger, J., Lamola, A. A., et al.: Zinc protoporphyrin level in blood determined by a portable hematofluorometer: A screening device for lead poisoning. J. Lab. Clin. Med., *89*: 712–723, 1977.
2. Haining, R. G., Hulse, T., and Labbe, R. F.: Rapid porphyrin screening of urine, stool and blood. Clin. Chem., *15*:460–466, 1969.
3. Lamon, J., With, T. K., and Redeker, A. G.: The Hoesch test: Bedside screening for urinary porphobilinogen in patients with suspected porphyria. Clin. Chem., *20*:1438–1440, 1974.
4. Magnussen, C. R., Levine, J. B., Doherty, J. M., et al.: A red cell enzyme method for the diagnosis of acute intermittent porphyria. Blood, *44*:857–868, 1974.
5. Moore, D. J., and Labbe, R. F.: A quantitative assay for urinary porphobilinogen. Clin. Chem., *10*:1105–1111, 1964.
6. Piomelli, S.: Free erythrocyte porphyrins in the detection of undue absorption of Pb and of Fe deficiency. Clin. Chem., *23*:264–269, 1977.
7. Schwartz, S., Berg, M. H., Bossenmaier, I., et al.: *In*: Methods of Biochemical Analysis, Vol. VIII. D. Glick, Ed. New York, Interscience Publishers, Inc., 1960, p. 249.
8. Schreiber, W. E., Raisys, V. A., and Labbe, R. F.: Liquid-chromatographic profiles of urinary porphyrins. Clin. Chem., *29*:527–530, 1983.
9. Tomokumi, K., and Ogata, M.: Simple method for determination of urinary δ-aminolevulinic acid as an index of lead exposure. Clin. Chem., *18*:1534–1536, 1972.

Additional Reading

Bishop, D. F., and Desnick, R. J., Eds.: Assays of the heme biosynthetic enzymes. *In*: Enzyme, Vol. 28, 2–3. New York, F. Karger Publishers, Inc., 1982.
Elder, G. H.: Haem synthesis and breakdown. *In*: Iron in Biochemistry and Medicine II. A. Jacobs and M. Worwood, Eds. New York, Academic Press, 1980, pp. 245–292.
Falk, J. E.: Porphyrins and Metalloporphyrins. New York, Elsevier, 1964.
Goldberg, A., and Moore, M. R., Eds.: The Porphyrias. Clin. Haematol., *9*:225–444, 1980.
Kappas, A., Sassa, S., and Anderson, K. E.: The Porphyrias. *In*: The Metabolic Basis of Inherited Disease. 5th ed. J. B. Stanbury, J. B. Wyngaarden, and D. S. Fredrickson, Eds. New York, McGraw-Hill, 1983, pp. 1301–1384.
Meyer, U. A.: Porphyrias. *In*: Harrison's Principles of Internal Medicine. 9th ed. K. J. Isselbacher, R. D. Adams, E. Braunwald, et al., Eds. New York, McGraw-Hill, 1980, pp. 494–500.
Romeo, G.: Enzymatic defects of hereditary porphyrias: An explanation of dominance at the molecular level. Human Genet., *39*:261–276, 1977.
With, T. K.: A short history of porphyrins and the porphyrias. Int. J. Biochem., *11*:189–200, 1980.

26 THERAPEUTIC DRUG MONITORING

THOMAS P. MOYER, Ph.D.
CHARLES E. PIPPENGER, Jr., Ph.D.
ROBERT V. BLANKE, Ph.D.
ROBERT A. BLOUIN, Pharm.D.

Therapeutic drug monitoring (TDM) is a rapidly developing area in the clinical laboratory. To be effective, TDM requires the acquisition of valid specimens followed by routine, timely, and reliable determinations of drug concentrations in these specimens. Results may be reported individually or collated so that, in conjunction with the dosing schedule, they may be interpreted in a pharmacokinetic context. TDM calls for a multidisciplinary approach since it relies on cooperative efforts of the physician, nurse, pharmacologist, pharmacist, and clinical chemist.

The recent rapid evolution of TDM stems from a fortuitous conjunction of several factors: (1) development of more numerous and more effective drugs with therapeutic potential and application; (2) development of analytical techniques that are accurate, precise, sensitive, and specific for measurement of drugs in biological fluids; and (3) utilization of computers in analytical technology, in pharmacokinetic data analysis, in dose regimen design, and in the collection and storage of patient data.

The mechanism of action of a drug refers to the actual biochemical or physical process that occurs to elicit the biological response. Drug action is generally mediated through a receptor on the cell surface or within the cell. Cellular proteins, such as enzymes, as well as structural or transport proteins, are important examples of drug receptors. Nonprotein macromolecules may also bind drugs, resulting in altered cellular function such as membrane permeability. Some drugs are chemically similar to natural endogenous substances and may compete with these for binding sites. Some drugs may block formation, release, uptake, or transport of essential substances. Although the exact molecular interactions remain obscure for most drugs, theoretical models have been developed to explain them. One concept postulates that if the drug-receptor complex is sufficiently stable and able to modify the target system, an observable pharmacological response will occur. As Figure 26-1 illustrates, the response is dose dependent until a maximal effect is reached. A plateau at this point may be due to saturation at the receptor or to a variety of other effects.

For many drugs, a strong correlation exists between the serum drug concentration and the observed pharmacological effect. Nevertheless, one must always keep in mind

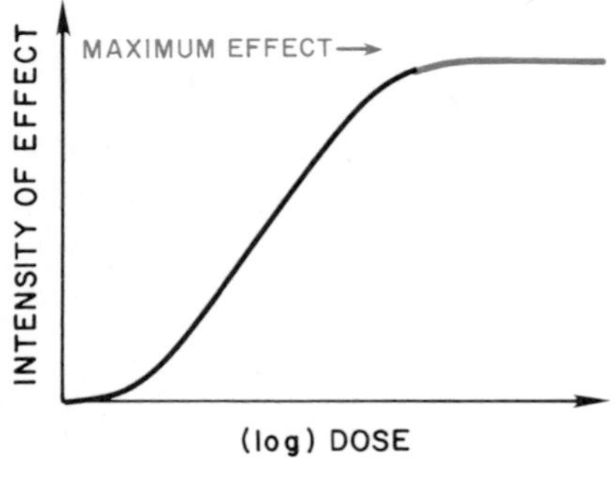

Figure 26-1. The log dose-effect relationship. The plateau (maximum effect) may be due to saturation at the receptor or to a variety of other effects.

that a serum drug concentration does not necessarily equal the concentration at the receptor; it may merely reflect it.

Changes in serum drug concentrations versus time may mirror changes in local concentrations at the receptor site or in body tissues. This effect is sometimes called the *property of kinetic homogeneity* and is applicable to all pharmacokinetic models in post-absorptive and postdistributive phases of the time course.

The property of kinetic homogeneity is an important assumption in TDM because it is the basis upon which all therapeutic and toxic concentration ranges are established. The ranges collectively define the *therapeutic window*, which is illustrated in Figure 26-2 as a relationship between minimum effective concentration (MEC) and minimum toxic concentration (MTC). Ideally, the *trough blood concentration* (the lowest concentration achieved just before the next dose) within each dosing cycle should not fall below the MEC, and the *peak blood concentration* (the highest concentration achieved within the dosing cycle) should not rise higher than the MTC (i.e., the concentrations should stay within the therapeutic range). Blood concentrations above the MTC put the patient at risk for toxicity; concentrations below the MEC have him at risk for the disorder the drug is supposed to treat. Dosage must be planned to achieve therapeutic concentrations and these must be monitored to allow adjustment of dosage if necessary. The smaller the difference between MEC and MTC, the more likely TDM will be necessary.

Pharmacology is involved with therapeutic agents and their effects on living processes.

Pharmacodynamics is the study of the biochemical and physiological effects of drugs and their mechanisms of action.[22] For many drugs, the ultimate effect or mechanism of action at the molecular level is understood poorly if at all. However, effects at the cellular or organ system level or in the whole body can be related to dosage of the drug.

Pharmacokinetics is the study of the rate processes of absorption, distribution, biotransformation, and excretion of drugs. It is this aspect of pharmacology that most strongly influences the interpretation of TDM results and that is dealt with in more detail in this chapter.

Toxicology can be considered an aspect of pharmacology concerned with adverse effects of chemicals on living systems (see also Chapter 27). Toxic effects and mechanisms of action may be different from therapeutic effects and mechanisms for the same drug. Similarly, at the high dosage of drugs at which toxic effects may be produced, rate processes are frequently altered in comparison with those at therapeutic dosages. For more complete discussions of these basic concepts the reader is encouraged to review standard textbooks.[4,22,35,41]

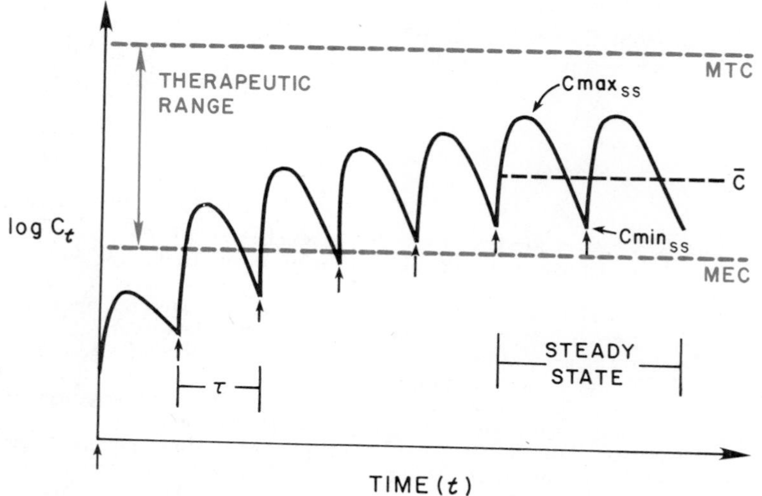

Figure 26-2. Sequence of drug concentration changes with multiple identical doses. Note that at steady state, peak and trough concentrations lie within the therapeutic range (or therapeutic window) and that 5–7 half-lives are required to reach steady state. $C_{max_{ss}}$ and $C_{min_{ss}}$ = maximum and minimum steady-state concentrations; $\overline{C}$ = average steady-state concentration; τ = dosing interval; ↑ = dose; MTC = minimum toxic concentration; MEC = minimum effective concentration. (Modified from Gilman, A. G., Goodman, L., and Gilman, A., Eds.: The Pharmacological Basis of Therapeutics. 6th ed. New York, Macmillan, 1980.)

Pharmacokinetics

Pharmacokinetics is the study of the rate processes of absorption, distribution, biotransformation, and excretion of drugs. *Clinical pharmacokinetics* applies the principles of pharmacokinetics to therapeutic management of an individual patient. Figure 26-3 depicts in detail the many factors affecting drug concentration and pharmacological response.

Pharmacokinetics utilizes mathematical approaches to predict or describe certain

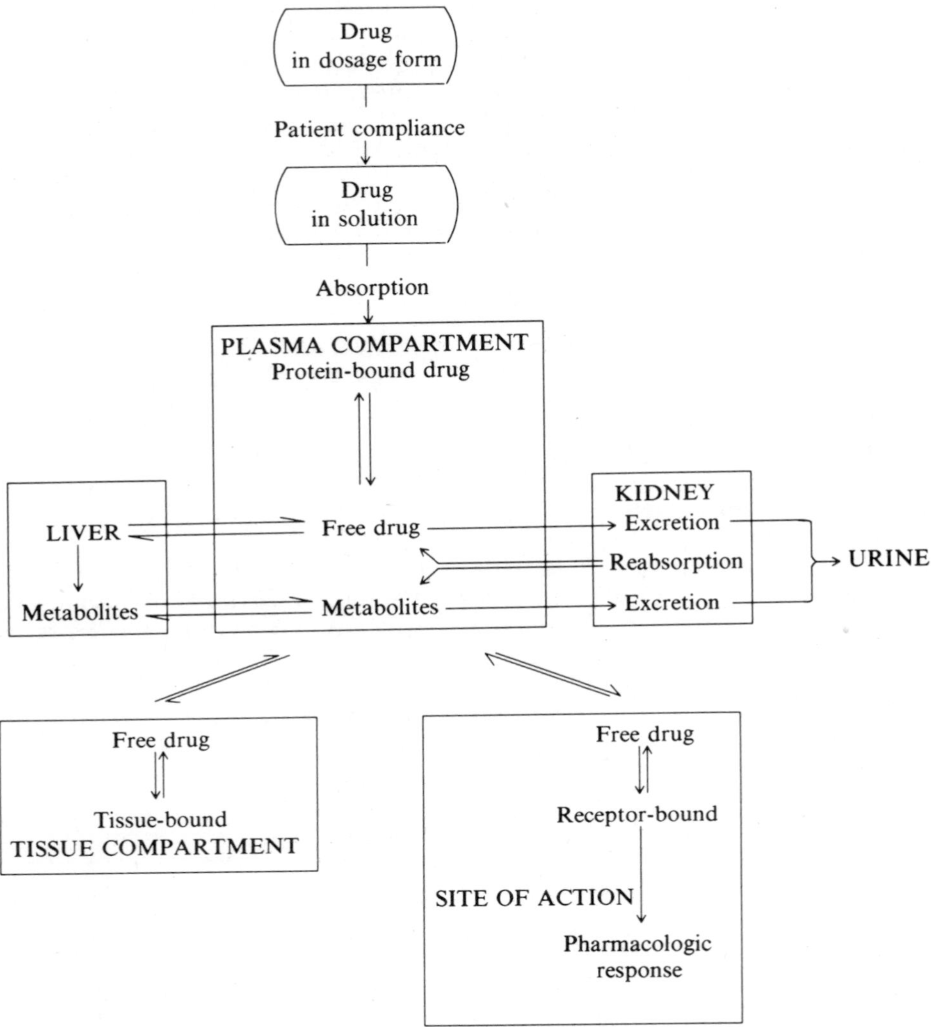

Figure 26-3. Factors affecting plasma drug concentration.

Absorption. Drug must be formulated to assure bioavailability for absorption from the GI tract or other administration site.

Metabolism. Drug is converted to a more soluble compound (metabolite) that may be pharmacologically active or inactive. Metabolism may occur in tissues other than the liver.

Excretion. The more water-soluble drugs and their metabolites are usually excreted in urine. Excretion may also occur via bile, feces, saliva, and expired air.

Tissue storage. The drug may be stored in tissues that exhibit no pharmacologic response to it; side effects may occur from drug reaction with a specific physiological system.

Site of action. Free drug binds to the receptor and produces pharmacologic response. The number and type of receptors to which the drug is bound determine the intensity and duration of the observed response.

(Adapted from Pippenger, C. E.: TDM: Principles of drug utilization. Syva Monitor, Nov., 1978, pp. 1, 3–5. Syva Co., Palo Alto, CA.)

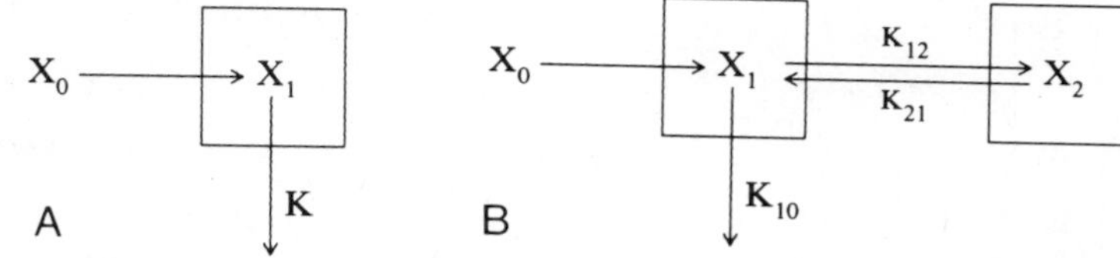

Figure 26-4. *A,* A one-compartment pharmacokinetic model. *B,* A two-compartment pharmacokinetic model. See text for details.

events, usually for calculating a dosing regimen or predicting the serum drug concentration following a given drug dose. The mathematical tools most often employed in clinical pharmacokinetics are *compartmental models* and *model-independent relationships*. Both of these approaches will be discussed.

In ***compartmental models***, blood drug concentrations and time data determine or define the model. The number and values of compartments assigned to the model have no true physiological meaning or anatomical reality. Usually the intravascular fluid (blood, plasma) is the anatomical central reference compartment. The advantage of intravascular fluid as the central compartment is the ease with which it may be sampled in order to provide a definitive profile of blood concentration of drug versus time. The number of compartments can be quite extensive; however, one- and two-compartment models are most often employed.

In a *one-compartment model*, the body is considered as a single compartment. It is assumed that after introduction of a drug, the substance is rapidly and uniformly distributed throughout the body. In contrast, the *two-compartment model* accounts for an initial decline of drug concentration in the central compartment due to distribution from the plasma into tissues (the peripheral compartment). Figure 26-4,*A* depicts a one-compartment model and Figure 26-4,*B* a two-compartment model of the result of an intravenous, bolus administration of a drug. In these figures, X_0 represents the drug dose given and therefore the amount of drug in the system at zero time; X_1, the amount of drug in the central compartment; and X_2, the amount of drug in the peripheral compartment in the case of the two-compartment model. The Ks are first-order rate constants. K and K_{10} are *elimination* rate constants; i.e., they describe the rates at which the drug leaves the central compartment and is lost from the system. K_{12} and K_{21} are *transfer* rate constants describing, for the two-compartment model only, rates at which the drug is exchanged between compartments within the system. Figure 26-5 graphically illustrates the relationship between log of concentration and time for the one-compartment model. Figure 26-6 illustrates the considerably more complex kinetics demonstrated by the two-compartment model.

The elimination rate constant can be related to another variable, *half-life*($t_{1/2}$), by the following equation:

$$t_{1/2} = \frac{0.693}{K} \tag{1}$$

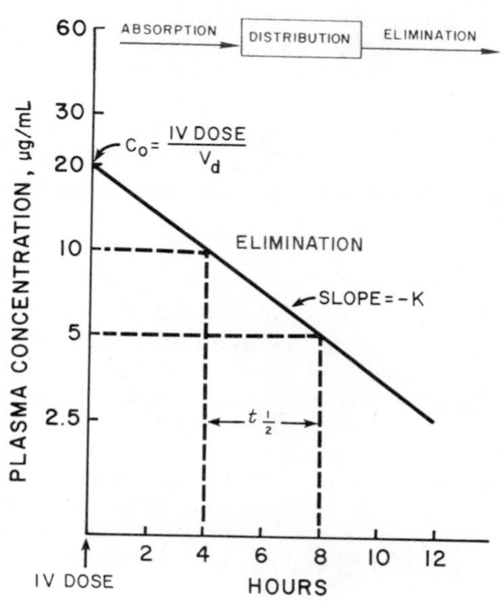

Figure 26-5. Drug concentration in plasma after intravenous administration of a single dose. The data show a monoexponential decline that can be described by a one-compartment model. V_d = volume of distribution; K = elimination constant; C_0 = blood concentration at time 0.

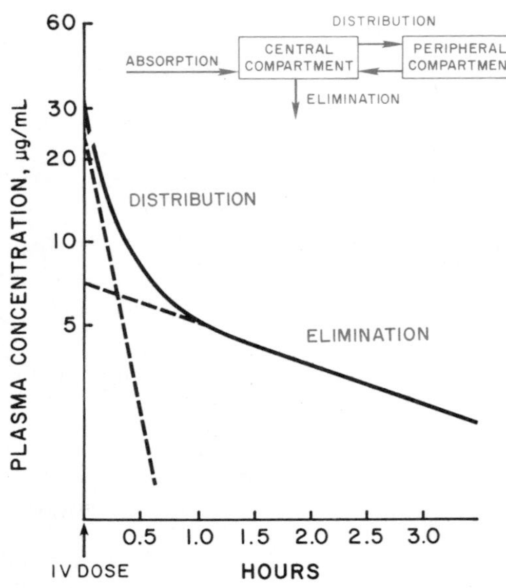

Figure 26-6. Decline of plasma concentration versus time after intravenous administration for a drug best characterized by a two-compartment model.

Half-life can be defined as the time required for the amount of drug in the system or for the blood drug concentration to decline to one-half of the previously measured value. Figure 26-5 demonstrates how the half-life can be rapidly determined from a semilog plot of drug concentration versus time. If half-life differs markedly from reported values for the drug in question, a component of elimination (K) may be the cause (e.g., renal impairment or liver dysfunction).

The apparent *volume of distribution* (V_d) of a drug can be described as a proportionality constant to translate the absolute amount of drug present in the body into its concentration relative to a volume. The units of V_d are usually liters (L) or liters per kilogram of body weight (L/kg). Although V_d is a mathematical term and not a real physiological parameter, it is useful for contrasting degrees to which different types of drugs distribute. For instance, the polar, hydrophilic drug acetylsalicylic acid has a V_d of 0.15 L/kg, whereas the nonpolar, lipophilic drug desipramine has a V_d of 34 L/kg.

Total plasma clearance (Cl_T) is defined as the theoretical total volume of plasma completely cleared of drug per unit of time and is usually expressed in units of mL/min, L/h, mL/min/kg, or L/h/kg. Cl_T is the sum total of all the clearances contributed by each elimination route: $Cl_T = Cl_R + Cl_B + Cl_M + ...$, where Cl_R, Cl_B, and Cl_M represent renal, biliary, and metabolic contributions to clearance, respectively. Clearance is an important measurement since it provides an assessment of the body's ability to eliminate a drug. In addition, Cl_T is readily used to relate the dosing rate to steady-state concentration.

Clearance is mathematically related to V_d and K as

$$Cl_T = V_d K \tag{2}$$

A different and preferred way to express Cl_T is

$$Cl_T = \frac{X_0}{AUC} \tag{3}$$

where AUC is the area under the curve described by the plasma concentration versus time and integrated over time from zero to infinity and X_0 is the dose administered. The reader will see later the particular advantage of equation (3).

Therapeutic drug monitoring is rarely concerned with a drug administered as a single, one-time intravenous bolus. Drugs administered chronically pose the usual therapeutic situations. Figure 26-2 shows that a drug, repetitively administered at a fixed dosing interval, will accumulate in the body until a steady-state condition exists. The symbol τ stands for dosing interval. *Steady state* can be defined as that point in a dosing scheme when the amount entering the circulation (governed by dosing rate) equals the amount eliminated (governed by elimination rate).

Assuming first-order elimination, 5–7 half-lives of a drug are required to reach 97–99% of a steady-state value. At steady state, and assuming τ is constant, the profiles of concentration versus time should be superimposable from dose to dose.

Model-independent relationships are becoming increasingly popular in clinical pharmacokinetics. As implied earlier, anything beyond a one-compartment model is impractical for clinical application. The fact that most drugs actually possess multicompartmental characteristics is reason enough to consider noncompartmental alternatives. The main advantages of model-independent relationships are that there are fewer relationships to remember, fewer restrictive assumptions, a more general insight into elimination mechanisms, and easier computations. However, model-independent relationships are not without their disadvantages; conceptualization of compartments or physiological spaces may be lost, specific information that may be clinically relevant or pertinent to mechanisms of distribution or elimination can be lost, and difficulty in constructing profiles of concentration versus time can be increased.

The most frequently used model-independent approach utilizes the concept of the average steady-state plasma concentration ($\overline{C}$; see Figure 26-2). The $\overline{C}$ following chronic dosing can be calculated by using equation (4):

$$\overline{C} = \frac{X_0 F}{Cl_T \times \tau} \tag{4}$$

where F is the bioavailability (fraction of dose available to reach the systemic circulation) of the drug. Equation (4), rearranged, can be used to calculate an appropriate dosing regimen:

$$X_0 = \frac{\overline{C} \times Cl_T \times \tau}{F} \tag{5}$$

The equations (4) and (5) are most useful when the half-life of a drug is much greater than τ, the dosing interval.

Principles of Drug Disposition

Absorption. Most drugs administered chronically to patients are given orally. The absorption process depends on the drug's dissociating from its dosage form, dissolving in gastrointestinal fluids, and then diffusing across biological membranes into the blood stream. The rate and extent of drug absorption may vary considerably.

The fraction of a drug that is absorbed into the systemic circulation is referred to as its *bioavailability*. The bioavailability of a particular drug, if the drug is to be useful, must be great enough so that the active component can pass in sufficient amount and in a desirable time from the gut into the systemic circulation. An exception would be the case in which the lumen of the gastrointestinal tract is the site of drug action (e.g., as with antibiotics used to sterilize the gut). Then, low bioavailability would be considered advantageous.

Some drugs that are rapidly and completely absorbed nevertheless have low bioavailability to the systemic circulation. This is true of drugs with a high hepatic extraction rate. After oral administration, drugs that are absorbed from the lumen of the small intestine are carried via the portal vein directly to the liver. Thus, a drug with a high hepatic extraction rate may be extensively metabolized by the liver before it reaches the systemic circulation. This phenomenon is called the *first pass effect*.

In addition to the *extent* of absorption, the *rate* of absorption is important. The absorption rate constant of a drug is usually much greater than its elimination rate constant. Efforts are now being made in the pharmaceutical industry to decrease the apparent rate of absorption of many drugs by manipulating their formulations. Dosage formulations that provide *sustained release* permit drugs to be taken at less frequent intervals. Conditions that may influence the extent or rate of drug absorption include abnormal gastrointestinal motility, diseases of the stomach as well as of the small and large intestine, radiation, food, and interaction with other substances in the gastrointestinal tract. One should be particularly aware of drugs that directly affect gut absorption, such as antacids, cholestyramine, and anti-ulcer medications.

Distribution. Following the entrance of a drug into the vascular compartment, the drug will be carried to all organs and fluid compartments. The overall process is referred to as *distribution*. Factors determining the distribution pattern of a drug are the binding of the drug to circulating blood components or to fixed receptors and the ability of the drug to pass through membrane barriers or to dissolve in structural or storage lipids.

Once a drug enters the systemic circulation it will rapidly distribute and come to equilibrium with many of the blood components. The most significant are plasma proteins. Figure 26-3 shows the equilibrium that exists between *free* and *protein-bound* drugs. It is believed that only the drug unbound to plasma protein (free) is available for distribution beyond the intravascular space and for elimination. Additionally, only the free drug is available to elicit a biological response. Therefore, changes in the protein-binding characteristics of a drug can have a profound influence on the distribution and elimination of a drug as well as on the manner in which steady-state concentrations are interpreted. Each drug has its own characteristic protein-binding pattern.

A drug may be either tightly or loosely bound to plasma proteins. A weakly bound drug can be displaced by a drug or endogenous compounds with greater affinity for the plasma protein-binding sites. Even though the total drug concentration may remain unchanged, displacement of a drug from its plasma protein-binding sites elevates free drug concentrations and can result in clinical toxicity.

Anything that alters the concentration of *free* drug in the plasma will ultimately alter the amount of drug available to enter the tissues and interact with specific receptor systems. Different disease states can alter free drug concentration. For example, in uremia, the composition of plasma is characterized by an increase in nonprotein nitrogen compounds, by acid-base and electrolyte imbalances, and often by a decrease in albumin. There may even be structural changes of some proteins that alter their binding capacities. In uremic patients, then, free drug concentrations are frequently elevated. If one monitors *total* plasma drug concentrations in these patients, it is possible that marked alerations of free drug concentration will not be detected since the total drug concentration may not be dramatically different from that observed in nonuremic patients. For example, phenytoin is 90% bound and 10% free in healthy subjects. In uremic patients, 20–30% of the total plasma concentration of phenytoin may be free. If one considers the clinical situation in which a nonuremic patient has a total plasma phenytoin concentration of 15 μg/mL, the free phenytoin concentration will be 1.5 μg/mL. If a uremic patient has a total concentration of 15 μg/mL, the free drug concentration may be 4.5 μg/mL, which may be high enough to precipitate severe phenytoin side effects.

Some drugs exhibit saturation of the available plasma protein–binding sites at optimal total drug concentrations. For example, valproic acid is a drug that shows saturation at concentrations >100 μg/mL. Thus, an increase of total plasma valproate concentration from 100 to 125 μg/mL represents a significant increase in the free valproate concentration.

Any change in normal physiological status can alter free drug concentrations. Geriatric patients often exhibit hypoalbuminemia; thus, there is a marked decrease in protein-binding sites for drugs. In the elderly, the clinical symptoms of drug intoxication may be manifested as impaired cognitive function, particularly confusion. An elderly patient may often be considered senile, when in reality an increased free drug concentration is affecting his or her cognitive ability. Reduction of drug dosage to decrease the free drug concentrations can result in dramatic improvements in the personalities of these patients. At present, total blood drug concentrations (the sum of protein-bound and free drug) are routinely analyzed, but ultrafiltration and dialysis methods for routine quantitation of free drugs are currently being evaluated.

In ***equilibrium dialysis,*** serum or plasma is placed on one side of a dialysis membrane and an isotonic buffer solution on the other side. The pore size of the membrane is selected to retain binding proteins on the specimen side of the membrane (retentate) while unbound drug passes through the membrane into the buffer solution (dialysate). At equilibrium, the free drug concentrations in the retentate and dialysate are equal. Depending on the drug, equilibrium may take from 1 to 24 h. Free drug concentration in the dialysate may be measured by any routine method, provided that sensitivity is adequate. Equilibrium dialysis for estimation of free drug concentration is not yet sufficiently fast or convenient for clinical use.

Estimation of free drug concentration by ***ultrafiltration*** is more practical. In this procedure, a cup in which the ultrafiltrate is collected is connected to a reservoir tube containing the serum or plasma. A semipermeable membrane is placed between the two compartments and the entire device is centrifuged at about $3000 \times g$. The free drug concentration is measured in the water-clear ultrafiltrate by a sensitive procedure. As serum or plasma water is driven through the membrane together with any free drug, protein-bound drug and free protein are concentrated at the same rate above the membrane. Under ideal conditions the free drug concentrations in the filtrate will remain constant throughout the filtration. However, each drug and filtration system used should be evaluated for nonspecific binding of the drug to the membrane or apparatus, protein leakage of the membrane, and other sources of error. Commercially available devices appear to be satisfactory for some drugs but not for all.

Estimation of the free drug concentration will continue to be of interest to TDM. Ultrafiltration techniques promise to be useful in satisfying this need. However, it should be remembered that laboratory measurements now only *estimate* the free drug concentration in circulating blood. Artifacts introduced in drawing, processing, and storing blood can modify dissociation equilibria for some drugs.

Metabolism. The rate of enzymatic processes that metabolize most drugs is usually characterized by the Michaelis-Menten equation and follows first-order reaction kinetics. In other words, the rate of metabolism is proportional to the concentration of the drug. First-order kinetics are characteristic of the metabolism of numerous drugs.

There are several drugs, notably phenytoin, salicylates, and ethanol, that cannot be characterized by simple first-order kinetics. Instead, the rate of metabolism of these compounds is said to be capacity-limited. The term *capacity-limited* or *nonlinear pharmacokinetics* means that clearance or the apparent half-life changes with changes in concentration. Figure 26-7, curve B, shows how a dose-response curve may be linear (first order) until a capacity limit of some pathway is reached and then may become nonlinear. Important clinical considerations arise when a patient is treated with a drug that displays nonlinear kinetics. First, changes in dosing result in disproportionate changes in steady-state drug concentrations so that titration to appropriate serum levels must be approached conservatively. Second, since both clearance and apparent half-life of the drug change with increasing drug concentration, the length of time required to reach a new steady-state concentration will be prolonged.

Biotransformation. The liver is the organ principally responsible for drug metabolism. Its role is to convert lipophilic nonpolar molecules to more polar, water-soluble forms. The drug molecule can be modified by reactions that alter its chemical structure (oxidation, reduction, or hydrolysis), or by reactions that conjugate the drug (glucuronidation, sulfation), or by both. Many environmental chemicals and endogenous biochemicals are processed by the same mechanisms.

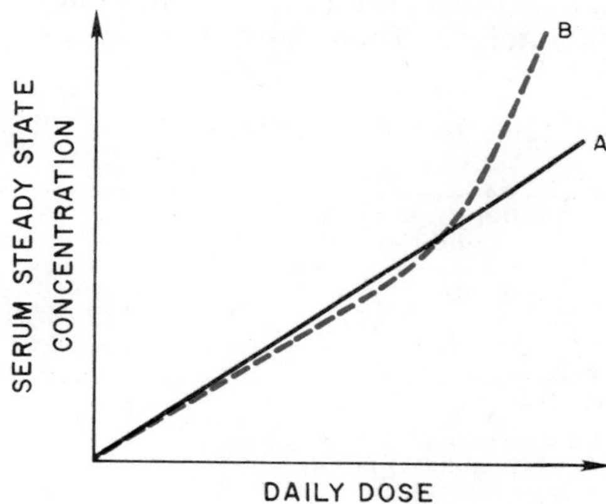

Figure 26-7. Linear and nonlinear dose-response curves. Line A illustrates the linear relationship between serum drug concentration and total daily dose of a drug that displays first-order kinetics. Line B illustrates the dose-response relationship for a drug that displays capacity-limited kinetics because of a saturable enzyme or transport mechanism; in this situation, serum concentration becomes independent of total daily dose, and the relationship of drug concentration to dose becomes nonlinear. (Adapted from Pippenger, C. E.: Practical pharmacokinetic applications. Syva Monitor, Jan., 1979, pp. 1–4. Syva Co., Palo Alto, CA.)

The enzymes of the hepatic microsomal system can be induced or inhibited. Microsomal *enzyme induction* leads to an increase in the amount of enzyme present, most commonly the amounts of the oxidizing enzymes. A classic example is phenobarbital. Obviously, when a patient is on a drug with a narrow therapeutic index, the dosage regimen would need to be adjusted should a known enzyme-inducing drug be added to or deleted from therapy. Since the drug-metabolizing enzymes of the liver interact with a wide variety of endogenous and exogenous substances, it is not very suprising to witness the *metabolic inhibition* of one drug by another. Most drug-drug interactions probably fall into the categories of substrate competition or competitive or noncompetitive inhibition. Examples of drugs that have been shown to inhibit drug metabolism significantly include chloramphenicol, cimetidine, valproic acid, allopurinol, and erythromycin. As with enzyme inducers, the addition or deletion of an inhibitory drug to a patient's drug therapy requires appropriate therapeutic drug monitoring and probable dosage adjustment.

The value of therapeutic monitoring is nowhere more evident than for drugs that are principally metabolized by the liver. There are two cogent reasons for this. First, wide variability in the rate of metabolism of any given drug exists not only in different patients but also in the same patient at different times. The variability is due to factors such as age, weight, gender, genetics, exposure to environmental substances, diet, coadministered drugs, and disease. Second, there is no good endogenous biochemical marker by which hepatic function, and consequently hepatic capability for drug clearance, can be routinely assessed before drug therapy is initiated.

The biotransformation of drugs may produce *metabolites* that are pharmacologically active. In such instances, the metabolite should also be measured since it is contributing to the effect of the drug on the patient. If the metabolite is inactive, it need not be measured, but steps should be taken to ensure that it does not interfere in the analytical process.

Renal excretion is a major pathway for the elimination of drugs and their metabolites. Therefore, decreased renal function will cause elevated serum drug concentrations and increase the pharmacologic response. Three processes are involved in the elimination of the drug via the kidney: *glomerular filtration (GF)*, *tubular secretion*, and *tubular reabsorption*.

Kidney function is readily and reliably evaluated by estimation of *creatinine clearance*. A strong correlation has been shown to exist between creatinine clearance and the total body clearance and elimination rate constant of those drugs primarily dependent upon the kidney for their elimination. Examples of drugs whose therapeutic use is adjusted to account for changes in creatinine clearance include gentamicin, tobramycin, amikacin, digoxin, vancomycin, and cyclosporine.

Summary of Factors Influencing Drug Disposition

A large number of factors are now recognized to have a profound influence on a patient's pharmacologic response to a drug. Consideration of the patient's history and adjunct drug therapy is essential at the initiation of drug therapy and therapeutic drug monitoring. Table 26-1 summarizes the influence various altered pathophysiological

Table 26-1. PATHOPHYSIOLOGICAL STATES THAT USUALLY DECREASE TOTAL BODY CLEARANCE OF PARTICULAR DRUGS

Pathophysiological Condition	Drugs Affected
Renal disease	Amikacin, gentamicin, tobramycin, kanamycin, netilmicin, digoxin, procainamide, *N*-acetylprocainamide, methotrexate, lithium, vancomycin, cimetidine, chloramphenicol*
Liver disease	Phenytoin,* phenobarbital,* valproic acid,* lidocaine, quinidine, verapamil, nifedipine, propranolol, theophylline, chloramphenicol*
Cardiovascular disease	Verapamil, nifedipine, disopyramide, propranolol
Congestive heart failure	Lidocaine, quinidine, theophylline
Hypothyroidism	Digoxin

*Decrease or no change.

states have on clearance mechanisms of several drugs frequently monitored by therapeutic drug monitoring programs.

Clinical Utility of Therapeutic Drug Monitoring

Therapeutic drug monitoring (TDM) is most valuable when the gravity of an acute clinical situation demands aggressive therapy with a drug that has a narrow therapeutic index. Laboratory data interpreted in the context of the dose regimen and clinical situation allow the physician to recognize and document the patient with an unusual pattern of drug disposition and thus to tailor therapy and to reduce risks of over- or under-medication. As the regimen is continued, laboratory results provide the basis for timely and rational adjustment of drug dosage and scheduling as the pathological effects of the primary or intercurrent disease progress or regress. TDM allows the physician to establish a baseline drug concentration at which the patient responds well. This knowledge is helpful if a certain therapeutic regimen must be reinstituted at a later time, since such knowledge increases the chance that the treatment will be efficacious at the outset of renewed treatment. Continuation of therapeutic drug monitoring will also identify the need for changes in the regimen in response to physiological changes such as puberty, pregnancy, or aging. Finally, TDM is useful in the detection of noncompliance, which is important to avoid failure of treatment, especially in those conditions that are frequently free of pain or unusual discomfort, such as epilepsy, asthma, hypertension, and mild heart disease.

Factors Unique to the TDM Laboratory

The value and quality of a TDM service are only as good as the data produced. Issues of continuing concern include the following:

1. Each laboratory should *inform the medical staff* as to therapeutic and toxic concentration ranges, analytical method (when appropriate), action values, required sample volume, and specifications for collection tubes.

2. Guidelines should be available for *ideal sample schedules* for each individual drug monitored. Usually, steady-state trough concentrations are most desirable; however, other sample schedules are often invoked, depending on the properties of the drug or individual needs of the patient.

3. The *time and date of collection* of the drug sample and of the *last dose* should be noted. To assess steady-state conditions, the length of time the patient has been on a particular regimen should be known. Numerous factors have been shown to influence the quantitation of a patient's drug level in vitro following specimen collection. Penicillin and heparin, for example, may inactivate aminoglycoside antibiotics during specimen storage. Additionally, hemolysis of a patient's drug blood specimen may alter the specificity of a particular analytical method.

4. Since *laboratory reports* become part of a patient's chart, it is useful to devise a reporting form that incorporates all of the data necessary for interpretation (dosage form, frequency and amount, plasma concentration, trough or peak value, time of dosage, time of draw, and other drugs coadministered).

SPECIFIC DRUG GROUPS

Drugs that are routinely monitored are conveniently classified by the kind of therapy they support (e.g., control of epilepsy, management of respiratory or cardiac function). An analytical method for one drug in a grouping is often applicable to other drugs in the same grouping. The following discussion is organized in accordance with classifications commonly recognized. Note that some drugs such as salicylate are discussed in Chapter 27, *Analysis of Toxic Substances*.

Table 26-2. PHARMACOKINETIC VALUES FOR ANTICONVULSANT DRUGS

Antiepileptic Drugs	Therapeutic Range, per mL of Plasma	Half-Life, h	Volume of Distribution, L/kg	Bioavailability, Oral, %	Protein Bound, %
Bromide	1000–1500 μg	290	*		0
Carbamazepine	8–12 μg	27 ± 4	1.4 ± 0.2	>70	82 ± 5
Clonazepam	10–50 ng	39 ± 12	3.2 ± 1.1	70–100	47
Ethosuximide	40–100 μg	33 ± 6	0.7 ± 0.1		0
Mephobarbital	1–5 μg	30 ± 10			50
Methsuximide	20–40 ng	3			
(Normethsuximide)	(20–40 μg)	(30)			
Phenobarbital	15–40 μg	86 ± 7	0.9 ± 0.3	90	50
Phenytoin	10–20 μg	†	0.6 ± 0.1	90–100	89 ± 23
Primidone	5–10 μg	8 ± 5	0.6 ± 0.5	92 ± 18	20
Valproic acid	50–100 μg	16 ± 3	0.1 ± 0.05	100 ± 10	93 ± 4‡

*Where data are missing, they are either not applicable or not available.
†Phenytoin does not obey first-order kinetics at therapeutic concentrations; therefore, a true half-life cannot be calculated. In adults, the apparent half-life is 18–30 h; in children it is 12–22 h.
‡Concentration dependent.

Anticonvulsant Drugs

Anticonvulsant drugs are those used to treat epilepsy as well as seizure disorders secondary to other diseases. The most common anticonvulsants are phenobarbital and phenytoin (for major motor seizures) and valproic acid (for absence seizures). Other drugs are used to control more complicated or mixed types of seizures. Pharmacokinetic values of common anticonvulsants are listed in Table 26-2.

PHENOBARBITAL

Phenobarbital is used in the treatment of all types of seizures except absence seizures (petit mal). Absorption of oral phenobarbital is slow but complete. The time at which *peak plasma concentrations* are reached is widely variable and ranges from 4–10 h after the dose. The *elimination half-life* is age dependent (children average 70 h, geriatric patients 100 h). Since hepatic metabolism is one of the prime routes of elimination, reduced liver function results in prolonged half-life. The predominant side effect observed in adults at blood concentrations >40 μg/mL is sedation, although tolerance to this effect develops with chronic therapy.

Phenobarbital is metabolized in the liver to *p*-hydroxyphenobarbital, which is largely excreted as the glucuronide or sulfate ester. When renal and hepatic function are decreased, clearance of the drug decreases. Elimination of phenobarbital may be decreased during chronic administration of either valproate or salicylate; the concentration of phenobarbital may increase 10–20%, and a dosage adjustment may be necessary to avoid intoxication. Phenobarbital induces mixed-function oxidative enzymes, resulting in increased metabolism of other xenobiotics after ~1–2 weeks of therapy.

Because of the long elimination half-life of phenobarbital, the blood concentration does not change rapidly. Therefore, a *serum specimen* collected late in the dose interval (trough) is representative of the overall effect. Results from specimens collected 2–4 h after the dose can be misleading, because they may be construed to be the peak concentration when in actuality they are not.

PHENYTOIN

Phenytoin (diphenylhydantoin) is used in the treatment of primary or secondary generalized tonic-clonic seizures, elementary-partial or complex-partial seizures, and status epilepticus. The drug is not effective for absence seizures. Phenytoin is not readily soluble in aqueous solutions, and absorption of oral phenytoin is slow and sometimes incomplete. Once absorbed, the drug is highly protein bound (90–95%). As with all drugs, the pharmacological effect of phenytoin is directly related to the amount present in the free (unbound) state. The degree of protein binding can be reduced by the presence of other drugs, anemia, and the hypoalbuminemia that can occur in the elderly. In these conditions, an increased effect is observed at the same total drug concentration as in plasma from normal patients.

The optimal *therapeutic concentration* for seizure control also serves as a reasonable guideline when the drug is used as a cardiac *antiarrhythmic* agent. Total phenytoin concentrations in excess of 20 μg/mL do not usually enhance seizure control and are often associated with nystagmus and ataxia. Total phenytoin plasma concentrations in excess of 35 μg/mL have been shown actually to precipitate seizure activity.

Phenytoin is metabolized by hepatic microsomal hydroxylating enzymes. Hepatic metabolism of phenytoin may become nonlinear within the therapeutic range. Once metabolism is nonlinear, small dosage increments result in large changes in blood concentration; this phenomenon partially explains the wide variation in dosage among patients that is required to accomplish a therapeutic effect.

The time to collect the *specimen* is dictated by the reason for monitoring. If the patient displays any symptoms of intoxication, then the peak blood concentration is of interest. This specimen is collected 4–5 h after the dose, although the peak level may be delayed up to 8 h if the drug is given in conjunction with substances that increase stomach acidity. If the principal question at hand is adequate therapy, then the trough concentration is more useful, and the specimen is collected just before the next dose is given.

A number of *drug interactions* alter the disposition of phenytoin. Alcohol, barbiturates, and carbamazepine induce oxidative enzymes, resulting in increased metabolism of phenytoin. Drugs such as chloramphenicol, disulfiram, isoniazid, and dicumarol compete with phenytoin metabolism, resulting in an increase in both total and free phenytoin concentrations. Salicylate, valproic acid, phenylbutazone, and sulfonylureas compete with phenytoin for serum protein-binding sites. The end result is diminished total serum concentration of phenytoin while the free phenytoin concentration and the pharmacological effect remain approximately the same. The interest in monitoring the free phenytoin concentration is in response to these altered disposition states.

VALPROIC ACID

Valproic acid is used for treatment of absence seizures. It has been shown to be useful against tonic-clonic and partial seizures when used in conjunction with other antiepileptic agents. Valproic acid is rapidly and almost completely absorbed after oral administration. *Peak concentrations* occur 1–3 h after an oral dose. The principal metabolite has anticonvulsant activity comparable with that of valproic acid, but it does not accumulate in plasma. The single dose $t_{1/2}$ is 16 h in healthy adults but reduces to 12 h on chronic therapy, and the $t_{1/2}$ may be as short as 8 h in children. In neonates and in hepatic disease, when metabolism is reduced, the $t_{1/2}$ becomes prolonged. Valproic acid is highly protein bound (93%). In circumstances when competition for protein binding increases, such as in uremia, cirrhosis, or concurrent drug therapy, the percentage of free valproic acid increases.

Clearance of valproic acid is rapid, which presents a dosing dilemma. Dosage must be adequate to provide a plasma concentration >40 μg/mL but avoid concentrations >100 μg/mL. The ideal *specimen* for monitoring the blood concentration is that drawn just prior to the next dose, usually early in the morning, to confirm that an adequate dose has been prescribed prior to bedtime. Dosage is particularly problematic in young children who might sleep for more than one complete half-life of the drug. Valproic acid modulates the action of a variety of other common antiepileptic drugs. Other common antiepileptic drugs that induce hepatic oxidative enzymes result in increased valproic acid clearance, thus requiring a higher dose to maintain effective therapeutic levels.

Analysis of valproic acid can be accomplished by either GLC or immunoassay, such as EIA and fluorescence polarization. In GLC analysis, valproic acid (VPA) is extracted from acidified serum into an organic solvent. A chemical analog, cyclohexane carboxylic acid, is used as an internal standard. The two acids are then extracted into an aqueous alkaline medium that is again acidified and the two acids are re-extracted into an organic phase (chloroform). A small aliquot of the chloroform extract is subjected to GLC analysis and quantitated by using a flame ionization detector.[20]

PRIMIDONE

Primidone is effective in the treatment of tonic-clonic and complex-partial seizures. The action of this drug is due partially to the accumulation of its major metabolite, *phenobarbital*. A second metabolite of primidone, *phenylethylmalonamide* (PEMA), also has some antiepileptic activity. Primidone is rapidly and completely absorbed after oral administration. Disposition of the drug is not known to be significantly altered by other disease states or other drugs.

Since phenobarbital is an active metabolite of primidone, concurrent analysis of phenobarbital is required for complete interpretation of results. The phenobarbital concentrations rise gradually over a period of 1–2 weeks after therapy is initiated. *Toxicity* due to accumulation of primidone is usually associated with symptoms of sedation, nausea, vomiting, diplopia, dizziness, ataxia, and a phenobarbital level >40 μg/mL. Specimen collection is dictated by the same rules that apply for phenobarbital.

Coadministration of acetazolamide with primidone will result in decreased gastrointestinal absorption of primidone and subsequent diminished plasma concentrations. Primidone administered in association with phenytoin will produce a modest elevation of the phenobarbital/primidone ratio since phenytoin competes with the hepatic hydroxylating enzymes associated with phenobarbital's metabolism.

CARBAMAZEPINE

Carbamazepine is used in the treatment of generalized tonic-clonic, partial, and partial-complex seizures and for the treatment of pain associated with trigeminal neuralgia. After oral administration, carbamazepine is rapidly absorbed but shows wide individual variability. The *elimination half-life* early in therapy is ~1 d. With chronic therapy, hepatic enzymes responsible for metabolism are induced, and the elimination half-life is reduced to 15–20 h. Any reduction in liver function will result in drug accumulation.

Toxicity associated with excessive carbamazepine ingestion is characterized by blurred vision, nystagmus, ataxia, drowsiness, and diplopia. Side effects unrelated to plasma concentration include development of an urticarial rash (which usually disappears upon discontinuation of the drug) and hematological depression (leukopenia, thrombocytopenia, or aplastic anemia).

The active *metabolite* of carbamazepine, carbamazepine-10,11-epoxide, may contribute to symptoms of intoxication in children who have a therapeutic plasma concentration of the parent drug. Drugs that induce the hepatic oxidative enzyme system (phenytoin, phenobarbital) increase the rate of clearance of carbamazepine.

Because of carbamazepine's relatively long half-life, the *trough specimen* yields the most useful information, although in the case of suspected mild intoxication, the peak value of the plasma concentration correlates better with toxicity. The peak specimen should be collected 2–4 h after the oral dose.

ETHOSUXIMIDE

Ethosuximide has been used for the treatment of absence seizures (petit mal) characterized by brief loss of consciousness. Ethosuximide is readily absorbed from the gastrointestinal tract. The drug is cleared mainly by metabolism as either the hydroxyethyl compound or the glucuronide ester of the hydroxyethyl metabolite. The *trough specimen* yields the most useful information regarding therapeutic efficacy. Toxicity related to an excessive blood concentration of ethosuximide is rare.

Methods for the Determination of Common Anticonvulsant Drugs

Phenobarbital, phenytoin, primidone, carbamazepine, and ethosuximide can be coanalyzed by either gas-liquid chromatography (GLC) or high performance liquid chromatography (HPLC) or individually analyzed by immunoassay. The advantage of HPLC or GLC is that simultaneous analysis can be accomplished. Immunoassay procedures, such as EIA and fluorescence polarization, are less labor intensive and are usually quicker than HPLC or GLC for a single analyte. The principle of an HPLC method is presented here as an example of a reliable method that simultaneously measures several anticonvulsants. The principles of EIA and fluorescence polarization are discussed in Chapter 2, Section Two.

Principle.[21] Phenobarbital, phenytoin, carbamazepine, and ethosuximide are extracted from serum into acetonitrile. The extract is injected into the mobile phase; drugs separated on the column are detected and quantitated by using the absorbance at 200 nm. An internal standard is included to correct for losses in extraction or for chromatographic variability.

Cardioactive Drugs

Subgroups of the cardioactive drugs are the so-called cardiac glycosides (digoxin and digitoxin) and the antiarrhythmic drugs. Pharmacokinetic values of cardioactive drugs are listed in Table 26-3.

DIGOXIN

Digoxin, one of a group of cardiac glycosides obtained from digitalis plants, restores the force of cardiac contraction in congestive heart failure. It is also used in the management of supraventricular tachycardias. The drug affects myocardial transport of Na^+, K^+, and Ca^{2+} and improves cardiac contractility.

Disagreement over the clinical value of digoxin measurements and the failure of digoxin concentration to correlate with clinical toxicity are usually related to aberrations in serum and tissue concentrations of sodium, potassium, magnesium, and calcium. An increased sensitivity to digoxin can be noted in states of hypokalemia, hypomagnesemia, and hypercalcemia, which make establishment of the true therapeutic concentration of digoxin difficult.

Table 26-3. PHARMACOKINETIC VALUES FOR CARDIOACTIVE DRUGS

Cardioactive Drugs	Therapeutic Range, per mL of Plasma	Half-Life, h	Volume of Distribution, L/kg	Bioavailability, Oral, %	Protein Bound, %
Amiodarone	1.0–2.0 μg	24 ± 12	12 ± 3	60–90	90
Digoxin	0.5–2.0 ng	42 ± 19	5–12	75	25
Disopyramide	2.0–4.5 μg	8 ± 2	0.8 ± 0.3	83 ± 11	*
Flecainide	0.2–1.0 μg	12 ± 4	†		
Lidocaine	2–5 μg	1.8 ± 0.4	1.1 ± 0.4	35 ± 11	51 ± 8
Mexiletine	0.75–2.0 μg	11–26	†		
Procainamide (plus NAPA)	4–8 μg (30 μg)	3 ± 0.6 (NAPA: 8)	1.9 ± 0.3	83 ± 16	16 ± 5
Propranolol	50–100 ng‡	3.9 ± 0.4	3.9 ± 0.6	36 ± 10	94 ± 2
Quinidine	2–5 μg	6.2 ± 1.8	2.7 ± 1.2	80 ± 15	71 ± 11
Tocainide	8–12 μg	12–16	1.4 ± 1.6	100	50
Verapamil	50–200 ng	3.5–7	5	20	90

*Blood-concentration dependent; see text.

†Where data are missing, they are not applicable or not available.

‡This range provides for maximal reduction of exercise-induced tachycardia. No therapeutic range exists for antihypertensive use. Concentrations as high as 1000 ng/mL have been observed without toxic effects.

Absorption of digoxin is variable and dependent upon the dosage form. At steady state the concentration of digoxin in cardiac tissue is 15–30 times that of plasma. The *peak plasma concentration* is reached 2–3 h after the oral dose, whereas the *peak tissue concentration* occurs 6–10 h after an oral dose. Pharmacological effects and toxicity correlate with tissue concentration, rather than plasma concentration, and the effective and safe *therapeutic plasma concentration* range of digoxin is not determined at the peak plasma concentration but rather at the time of peak tissue concentration.[12] Thus, to ensure a correlation between plasma concentration and tissue concentration, the appropriate time to collect the *specimen* is 8 h or more after the dose. Results from specimens collected earlier than 8 h after the dose are inappropriate and often misleading.

Digoxin *toxicity* is characterized by nonspecific symptoms of nausea, vomiting, anorexia, and predominance of green-yellow visual distortion. Cardiac symptoms include a rapid and irregular heart rate.

Digoxin is excreted unchanged or in the form of digoxigenin mono- or disaccharides in the urine. As a result, digoxin toxicity develops more frequently and lasts longer in patients with renal impairment. Dosage requirements are decreased in patients with renal disease.[12]

Decreased gastrointestinal absorption occurs with sprue and small intestinal resections, high-fiber diets, hyperthyroidism, and situations of increased gastrointestinal motility. A more dangerous situation due to the interaction of quinidine and digoxin develops,[26] which results in an increase in the digoxin concentration.

Method for the Determination of Digoxin

Immunoassay has been the only practical method for clinical use, since any other method would require prohibitively large specimen volumes. Fluorescence polarization[31] and fluorescence-quenching immunoassays now compete in popularity with radioimmunoassay for routine use, although the first method is more affected by variation in protein content of the specimen. Most of these methods require <100 μL serum and can be completed in 1–1½ h.

Digitoxin, another cardiac glycoside, is also given therapeutically, although less commonly than digoxin. RIA kits for the specific assay of digitoxin are available. The drug cross-reacts to various degrees with most of the antibodies provided in commercial immunoassays for digoxin. When ordering drug assays on digitalized patients, personnel involved in direct patient care do not always distinguish between digoxin and digitoxin, but these are, nevertheless, different drugs whose quantitation requires different assays.

Note: Digitoxin is infrequently prescribed today; however, serum levels should be evaluated in patients suspected of digitalis intoxication but who have nondetectable digoxin levels. RIA procedures for digoxin cross-react only minimally with digitoxin.

LIDOCAINE

Lidocaine is the drug of choice for the initial therapy of premature ventricular contractions and the prevention of ventricular arrhythmias. Because lidocaine undergoes nearly complete first-pass hepatic metabolism when administered orally, it is administered only by intravenous or intramuscular injection. Clearance of lidocaine is very rapid. Reduced hepatic function will impair clearance, prolong elimination, and cause accumulation of the drug.

Blood concentrations ranging from 1.5 to 4.0 μg/mL are usually effective and are rarely associated with any form of central nervous system or cardiovascular toxicity. Concentrations ranging from 4 to 6 μg/mL may be needed for suppressing arrhythmias but may be associated with mild central nervous system depression and slight QRS widening on the electrocardiogram. Concentrations ranging from 6 to 8 μg/mL are acceptable only if alternative therapy is not possible, since these levels have been associated with significant central nervous system depression and atrioventricular node blockage.

Lidocaine has two *metabolites* commonly found in plasma, monoethylglycinexylidide

(MEGX) and glycinexylidide (GX). MEGX and lidocaine have nearly identical toxic equivalency.

Since lidocaine is most commonly administered as a constant infusion following a loading dose, the time to collect the specimen is determined by the reason for monitoring. If the blood concentration is intended to document an adequate concentration early in therapy, the specimen should be collected 30 min after the loading dose, or 5–7 h after therapy is initiated if no loading dose is given (five half-lives after start of therapy). If the patient shows toxic symptoms, the specimen should be collected as close to the episode as possible and analyzed immediately, since these symptoms present a potentially life-threatening situation.

The principal binding protein of lidocaine, α_1-acid glycoprotein, has been demonstrated to accumulate after myocardial infarction. The result of accumulation of this protein is reduction of free lidocaine, which reduces the pharmacological effect of the drug.[38] Another interaction of interest is the decreased clearance of lidocaine associated with concomitant dosage of propranolol, which could result in toxicity.

QUINIDINE

Quinidine, available as either quinidine sulfate or quinidine gluconate, is used in the treatment of certain types of irregular and rapid heart rates. *Peak serum concentrations* are reached in 1.5–2 h of oral intake, unless the slow release preparation (quinidine gluconate) is used. *Peak plasma concentrations* are reached 4–5 h after quinidine gluconate administration, and the trough concentration occurs 1–2 h after the next administration. Clearance of quinidine is dependent on adequate hepatic and renal function. Reduction of either of these two functions will result in accumulation of the drug.

The optimal *therapeutic concentration* is method dependent because of the interference in some tests by quinidine metabolites and the presence of dihydroquinidine, a naturally occurring analog of quinidine found in all pharmaceutical preparations. Single-extraction fluorescence methods display interference due to coadministration of triamterene (Diazide) and the accumulation of inactive quinidine metabolites in patients with impaired renal function. Double-extraction fluorescence techniques are more specific for quinidine and a few of its active metabolites [(3s)-3-hydroxyquinidine and quinidine-*N*-oxide]. The antibody used in the commercially available EMIT and fluorescence polarization procedures for quinidine also appears to cross-react with these compounds; double extraction fluorescence procedures and the EMIT assay yield similar results corresponding to an optimal *therapeutic concentration* of 2–5 μg/mL.[33] Liquid chromatography is a more specific procedure for analysis of quinidine and can separate dihydroquinidine from quinidine; the therapeutic range by liquid chromatography is 1.5–4.5 μg/mL. Quinidine *toxicity* is usually observed at concentrations >8 μg/mL and is commonly associated with cinchonism, tinnitus, lightheadedness, giddiness, and cardiovascular toxicity. The most prominent toxic effect is on the gastrointestinal tract. Hypersensitivity reactions associated with quinidine are not related to blood concentration.

Routine monitoring of quinidine is most commonly an effort to document adequate dosage. In this case the *trough specimen* is appropriate. When symptoms of intoxication are observed, then a *peak specimen* may be more useful. Clearance of quinidine is dependent upon an active mixed-function oxidase system in the liver. Induction of this system by barbiturates will lead to enhanced clearance of quinidine.

PROCAINAMIDE

Procainamide is used for therapy of premature ventricular contractions, ventricular tachycardia, atrial fibrillation, and paroxysmal atrial tachycardia. *Peak plasma concentrations* after oral administration are reached within 0.75–1.5 h if the drug is given in capsule form, or within 1–3 h if given in tablet form. Excretion of procainamide is dependent upon hepatic metabolism and renal clearance; therefore, alteration of function in either organ will lead to accumulation of procainamide and its metabolites. The concentration at which procainamide will block premature ventricular contractions and inhibit ventricular tachycardia ranges from 4 to 8 μg/mL,[23] although most patients can tolerate concentrations higher than this, and some may require higher doses.[28]

The issue of the ideal therapeutic concentration for procainamide is complicated by the fact that one of its metabolites, *N*-acetylprocainamide (NAPA), has antiarrhythmic activity similar to that of procainamide. The compound has been shown to accumulate in patients with impaired renal function and in fast acetylators.[42] Coanalysis of NAPA is necessary to provide a complete assessment of therapy. Fast acetylators have NAPA concentrations equal to or exceeding those of procainamide in a specimen collected 3 h after administration, whereas slow acetylators have procainamide present at greater than twice the NAPA concentration in specimens collected during the same time interval. Because the effects of procainamide and NAPA are cumulative, peak concentrations of procainamide should be limited to 8–12 μg/mL, and peak concentrations of procainamide plus NAPA should not exceed 30 μg/mL.

Symptoms of *intoxication* include bradycardia, prolongation of the QRS interval, AV block, and induced arrhythmias. The development of systemic lupus erythematosus (SLE) associated with procainamide therapy is not related to plasma concentration but is associated with the acetylator status of the patient; slow acetylators predominate in the group that develops SLE.

DISOPYRAMIDE

Disopyramide is used for suppression of unifocal, multifocal, and premature ventricular contractions, ventricular tachycardia, and ventricular and supraventricular arrhythmias; it is also used in the treatment of Wolff-Parkinson-White syndrome. The drug is commonly used as replacement therapy for quinidine when quinidine's side effects are intolerable. Elimination is by renal clearance and hepatic metabolism. With renal insufficiency, clearance is prolonged, causing accumulation of disopyramide.

The usual *therapeutic range* is 2–4.5 μg/mL, but plasma levels up to 7.5 μg/mL are sometimes required to control ventricular arrhythmias. The relationship between clinical response and optimal therapeutic concentration is confused by a decrease in protein binding that occurs as plasma concentration of disopyramide rises.[17] Thus, free disopyramide blood concentrations would be a better index of therapeutic response. The principal metabolite is nordisopyramide, but little additional therapeutic information is gained by monitoring it.

The predominant side effects of disopyramide are anticholinergic (i.e., dry mouth, urinary hesitancy, and constipation). These symptoms occur at plasma concentrations >4.5 μg/mL. Cardiac *toxicity* is usually associated with blood concentrations >10 μg/mL and is characterized by atrioventricular node blockage, bradycardia, and asystole. Because of the wide degree of variability of protein binding, interindividual variability of the blood concentration at which these symptoms develop is large.

Methods. *Lidocaine, procainamide, NAPA, quinidine,* and *disopyramide* can be analyzed by HPLC, GLC, or immunoassay. Quinidine can also be accurately analyzed by double-extraction direct-fluorescence analysis.[33]

For HPLC the antiarrhythmic drugs are extracted from alkalinized serum or plasma into a polar organic solvent. After solvent evaporation, the residue is dissolved in a mobile phase, chromatographed, and quantitated with an ultraviolet detector.[25]

PROPRANOLOL

Propranolol is used in the treatment of arrhythmias of atrial and ventricular origin, angina pectoris, myocardial infarction, and hypertension. Its principal effect is to reduce the heart rate, thus relieving angina, and to reduce the ventricular rate in patients with atrial fibrillation.

Propranolol undergoes such a high degree of first-pass hepatic metabolism that its bioavailability is widely variable between individuals and is dose dependent. Reduced hepatic function or reduced blood flow to the liver will cause accumulation of propranolol.

A close relationship between slowing of heart rate and blood concentration has been reported.[34] This study also suggests that the heart rate serves as an adequate means to monitor therapeutic efficacy, precluding blood measurements in most patients.

Methods. Propranolol as well as other β-blockers such as *metoprolol*, *aprindolol*, and *nadolol* can be analyzed by EIA, fluorescence polarization, and HPLC by using the native fluorescence of the molecule for detection.

For HPLC, propranolol and an internal standard (pronethalol) are extracted from alkalinized serum into an organic solvent (butanol/hexane). The drugs can be back-extracted into an acidified medium and subjected to reverse-phase chromatography. Fluorescence detection allows for quantitation by peak height determination.[30]

BRONCHODILATORS

Theophylline relaxes bronchial smooth muscle to relieve or prevent asthma. The drug is principally eliminated by hepatic metabolism. Theophylline is readily absorbed after oral, rectal, or parenteral administration. If the drug is taken orally without food, the blood concentration peaks within 2 h. If it is administered with food or as the slow release formula, peak concentrations occur 3–5 h after the dose. In children and in adults who smoke, the $t_{1/2}$ ranges from 3 to 4 h. Nonsmoking adults in good health has an *elimination half-life* of about 9 h. The $t_{1/2}$ in neonates and in adults with congestive heart failure can be prolonged to 20–30 h, depending on the degree of liver immaturity or loss of liver function. Coadministration of erythromycin or troleandomycin also has been shown to reduce the clearance of theophylline significantly. Cimetidine reduces hepatic metabolism of theophylline, resulting in significant and potentially life-threatening increases in serum concentrations.

Optimal *therapeutic effect* of theophylline, as evaluated by forced expiratory volume measurements, appears to occur at plasma concentrations of 5–20 μg/mL. Suppression of exercise-induced bronchospasm in the asthmatic patient is optimal at 15 μg/mL. Neonatal apnea treated with theophylline responds to slightly lower concentrations.

Theophylline *clearance* is dose dependent. At serum concentrations >20 μg/mL, small dosage increases lead to disproportionately large increases in serum concentrations and risk of intoxication. Symptoms of theophylline *toxicity* include nausea, vomiting, headache, diarrhea, irritability, and insomnia. Serious toxicity, characterized by cardiac arrhythmias and seizures, is usually associated with serum concentrations in excess of 30 μg/mL. Once seizure activity begins, the final prognosis is very poor. Mortality can be as high as 50%.[47]

A minor metabolite of theophylline in adults, *caffeine*, has been shown to accumulate to significant concentrations in neonates.[10] Caffeine itself is an effective inhibitor of apnea,[2] which may explain the lower therapeutic concentration required in the neonate. The case has been made for concurrent determination of both theophylline and caffeine concentrations in neonates treated with theophylline.[43] Therapy with caffeine in the treatment of neonatal apnea is popular because of caffeine's long half-life in the neonate (>30 h).

Pharmacokinetic details of bronchodilators are summarized in Table 26-4.

Methods. A variety of immunoassays, especially fluorescence polarization and EIA, are available for the determination of theophylline.[13] Theophylline, caffeine, and

Table 26-4. PHARMACOKINETIC VALUES FOR BRONCHODILATOR DRUGS

Bronchodilators	Therapeutic Range, per mL Plasma	Half-Life, h	Volume of Distribution, L/kg	Bioavailability, Oral, %	Protein Bound, %
Caffeine	8–14 μg*	30*	†		
Theophylline	10–20 μg 5–10 μg*	6–8	0.5 ± 0.16	96 ± 8	56 ± 4

*For neonatal apnea.
†Where data are missing, they are either not applicable or not available.

dyphylline can be measured by HPLC. Theophylline, caffeine, or dyphylline, along with an internal standard, β-hydroxyethyltheophylline, are extracted into ethyl acetate. After solvent evaporation, the redissolved residue of the extract can be separated by HPLC and the concentration determined from the resultant chromatogram.[32]

ANTIBIOTICS

The **aminoglycosides** are a class of antibiotics that act by inhibiting bacterial protein synthesis. The aminoglycoside drugs include *streptomycin*, *neomycin*, *netilmicin*, *sisomicin*, *gentamicin*, *tobramycin*, *kanamycin*, and *amikacin*. They are used for treating serious infections due to aerobic and facultatively anaerobic gram-negative bacilli.

Therapy with antimicrobial agents differs from the approach used for other drugs. The goal is to accomplish a minimal inhibitory concentration (MIC) in plasma such that the bacterial organism is killed while the host remains undamaged. Because the organisms treated are variable and can become resistant to certain drugs, treatment with specific aminoglycoside agents should always be directed by susceptibility testing.

The aminoglycosides are a very polar group of drugs and are thus poorly absorbed from the intestinal tract. They are routinely administered intravenously or intramuscularly to bypass this problem. When administered directly into the blood, aminoglycosides are rapidly distributed in the extracellular fluid but do not cross cell membranes, nor do they bind to plasma proteins. *Elimination half-lives* are short, ranging from 2 to 3 h. Since clearance is almost completely dependent on renal function, impairment of glomerular filtration causes accumulation of these drugs.

There is considerable variability reported regarding the relationship of blood concentration to later onset of toxicity, mainly renal tubular necrosis and degeneration of the auditory nerve. The *trough specimen* has been suggested as the best specimen to use for monitoring toxicity. Table 26-5 identifies maximum trough serum concentrations (together with data on peak and toxic values) for various antibiotics. In this mode of monitoring, the intent of therapy should be dosing the patient in such a manner that the trough concentration does not exceed these limits. An alternative and equally convincing argument can be made for utilizing the *peak serum concentrations* (Table 26-5) as a therapeutic index.[45] Dosage corrections must be made in patients with compromised renal function because in these patients the aminoglycosides have prolonged half-lives and slower elimination.

Toxicity associated with aminoglycosides shows up as delayed-onset vestibular and cochlear sensory-cell destruction and acute renal tubular necrosis. The degree and severity of cell damage are variable among the various drugs, but all will cause cell damage if the concentrations exceed the limits identified in Table 26-5. Unfortunately, some patients will develop toxicity regardless of the concentration, although these cases are rare. Irreparable loss of vestibular, cochlear, or renal function usually correlates with grossly elevated blood concentrations of any one of the aminoglycosides.

Heparin has been implicated as a deactivator of gentamicin by formation of an inactive complex.[40] This complex, although biologically inactive, will cross-react with

Table 26-5. EFFECTIVE CONCENTRATIONS OF ANTIBIOTICS

	μg/mL		
	Peak	*Trough*	*Toxic*
Amikacin	20–25	5–10	>35
Chloramphenicol	20–30	10	>50
Gentamicin	5–8	1–2	>10
Kanamycin	20–25	5–10	>35
Sulfonamides	125	>75	>300
Tobramycin	5–8	1–2	>10
Trimethoprim	5–8	—	>50
Vancomycin	20–30	10	>50

Table 26-6. PHARMACOKINETIC VALUES FOR ANTIBIOTIC DRUGS

Antibiotics	Therapeutic Range, per mL of Plasma	Half-Life, h	Volume of Distribution, L/kg	Bioavailability, Oral, %	Protein Bound, %
Amikacin	10–25 μg	2 ± 0.6	0.2 ± 0.08	*	
Chloramphenicol	10–20 μg	2.7 ± 0.8	0.9	75–90	53 ± 5
Gentamicin	2–8 μg	2–3	0.25		<10
Kanamycin	10–25 μg	2.1 ± 0.2	0.26 ± 0.05		0
Sulfonamides, all	75–125 μg	6–9	0.1–0.2	~100	60–90
Tobramycin	2–8 μg	2.2 ± 0.1	0.26 ± 0.09		<10
Trimethoprim	5–8 μg	11 ± 1.4	1.8 ± 0.2	~100	70 ± 5
Vancomycin	10–30 μg	5–6	0.4		<10

*Where data are missing, they are either not applicable or not available.

antibodies to the specific aminoglycoside. Specimen collection tubes containing heparin (1000 units/mL) may lead to such complex formations.

Pharmacokinetic details of antibiotics are summarized in Table 26-6.

Methods. The aminoglycoside antibiotics have been analyzed for years by bioassay techniques. These methods, however, are variable and subject to interference by numerous drugs. Recently both liquid chromatographic and immunochemical methods have become available.[21] The enzyme immunoassay (EMIT) is discussed here as an example of drug assay by this technique, but fluorescence polarization (see Chapter 2, Section Two) is also widely used.

Principle. The EMIT assays used for the measurement of drugs are homogeneous enzyme-labeled immunoassays. The products of the reaction are measured directly in the reaction mixture, eliminating the requirement for separation of bound and free label. The reagent system consists of (1) the drug bonded covalently to an appropriate enzyme, (2) an antibody specific for the drug assayed, and (3) reaction substrates and cofactors for the enzyme used as label. The specimen is mixed with these three reagents and allowed to react during an incubation period. In the absence of drug in the specimen, the enzyme-labeled drug binds to the antibody and forms a complex that alters the conformation or blocks the active site of the enzyme, thus preventing the enzyme reaction from occurring. If the drug is present in the specimen, it will displace enzyme-labeled drug from the antibody; the free enzyme-labeled drug can then react with enzyme substrate to produce a measurable product. The amount of product resulting from the enzymatic activity increases as the concentration of unlabeled drug in the specimen increases. The concentration of the drug can be calculated by interpolation from a standard curve.

EMIT methods for gentamicin, amikacin, and tobramycin use glucose-6-phosphate dehydrogenase (G-6-PD) as the enzyme label along with glucose-6-phosphate (G-6-P) and $NADP^+$ as substrate. The type of enzyme used for labeling and its specific substrate may differ in EMIT assay systems for other drugs; however, the principles of the reaction remain the same. The free G-6-PD–labeled drug catalyzes the following reaction:

$$\text{G-6-P} + \text{NADP}^+ \rightarrow \text{6-Phosphogluconolactone} + \text{NADPH} + \text{H}^+$$

Increase in absorbance at 340 nm is measured with a spectrophotometer capable of accurate measurements in the range of 0.02–0.5 A, of accurate temperature control at 30 ± 0.1 °C, and of the recording of absorbance versus time.[45] The standard curve is nonlinear; therefore, specialized curve-fitting data reduction methods must be used.

CHLORAMPHENICOL

Chloramphenicol is used as an antimicrobial agent. It acts by binding to the 50S ribosomal subunit of bacteria and inhibiting protein synthesis in prokaryotic organisms.

Use of this drug is dependent upon its relative toxicity against the microorganism versus the host. The drug is used against gram-negative bacteria.

Chloramphenicol is rapidly absorbed in the gastrointestinal tract. Peak serum concentration occurs 1–2 h after the oral dose with the free base form of the drug and 2–3 h after the dose with the palmitate salt. Times of peaks after intravenous administration of the succinate salt are highly variable. Chloramphenicol is distributed to all tissues and is concentrated in the cerebrospinal fluid. Since the drug is actively metabolized by the liver, chloramphenicol will accumulate in cases of hepatic disease. Renal disease, however, does not dramatically reduce clearance. Host *toxicity* displayed after chloramphenicol therapy includes blood dyscrasias and cardiovascular collapse; both show a modest relationship to blood concentration. Other blood concentration–related toxicities include anemia, characterized by maturation arrest in the marrow; reticulocytopenia; and increases in both serum iron and serum iron-binding capacity. These symptoms are all associated with serum concentrations in excess of 25 μg/mL.[36] Development of aplastic anemia, the most serious complication, is not related to dose or blood concentration but appears to be an idiosyncratic reaction. Cardiovascular collapse, which occurs primarily in newborns, has been related to a total serum chloramphenicol concentration in excess of 50 μg/mL.[27]

Methods for the determination of chloramphenicol concentrations in blood serum include bioassay, EIA, gas chromatography, and HPLC with UV detection.[21] Methods for chloramphenicol determination must allow differentiation between the pro-drug forms, chloramphenicol palmitate or succinate, and chloramphenicol.

VANCOMYCIN

Vancomycin is a glycopeptide that is bactericidal against gram-positive bacteria and some gram-negative cocci. Although the drug is poorly absorbed when given orally, a 1-g dose given intravenously every 12 h accomplishes a *peak blood concentration* of 20–30 μg/mL and a *trough concentration* of 5–10 μg/mL. Blood concentration–dependent *toxicity* involves the auditory nerve but occurs rarely with concentrations <30 μg/mL. In patients with impaired renal function, the serum concentration may increase to toxic levels. Vancomycin can be analyzed by HPLC,[21] by fluorescence polarization, and by commercially available RIA kits.

Psychotrophic Drugs

Drugs commonly used and monitored in psychiatric care include *lithium*, *antidepressants*, and *neuroleptics*. Pharmacological values for lithium and some antidepressants are shown in Table 26-7.

Table 26-7. PHARMACOKINETIC VALUES FOR ANTIPSYCHOTIC DRUGS

Antipsychotic Drugs	Therapeutic Range, per mL of Plasma	Half-Life, h	Volume of Distribution, L/kg	Bioavailability, Oral, %	Protein Bound, %
Amitriptyline	70–225 ng (Ami + Nor)†	15 ± 5	8.3 ± 2	*	97 ± 0.8
Amoxapine	200–500 ng (Amox + 8-OH Amox)†	8 (8-OH Amox = 30)			
Desipramine	150–300 ng	18 ± 5	35 ± 8	68	92 ± 1
Doxepin	150–250 ng (Dox + Nordox)†	17 ± 6	20 ± 8	27 ± 10	
Imipramine	150–250 ng (Imi + Desip)	13 ± 3	15 ± 6	47 ± 21	89–94
Lithium	0.8–1.2 mmol/L	22 ± 8	0.79 ± 0.34	100	0
Maprotiline	200–600 ng	27–58			
Nortriptyline	50–150 ng	31 ± 13	18 ± 4	51 ± 5	94 ± 0.6
Trazodone	0.8–1.6 μg	13			

*Where data are missing, they are either not applicable or not available.
†Indicates that therapeutic range is the sum of the concentration of the parent drug and its metabolite.

LITHIUM

Lithium, administered as lithium carbonate, is used for treatment of manic-depressive illness. Absorption of lithium from the gastrointestinal tract is complete, and the *peak plasma concentration* is reached in 2–4 h after an oral dose. The lithium cation does not bind to protein. Clearance is predominantly by renal excretion.

The *optimal therapeutic response* to lithium has not been related to a specific serum concentration; however, serum lithium concentrations are monitored to ensure patient compliance and to avoid intoxication. A standardized 12-h post-dose serum lithium concentration is recommended to assess adequate therapy in patients taking sustained-release preparations. A concentration in excess of 1.5 mmol/L in a specimen drawn 12 h after the dose indicates an increased risk of intoxication. Early symptoms of intoxication include apathy, sluggishness, drowsiness, lethargy, speech difficulties, irregular tremors, myoclonic twitchings, muscle weakness, and ataxia. These symptoms, although not life-threatening, are uncomfortable and indicate that the onset of life-threatening seizures is imminent.

Lithium excretion parallels that of sodium. It readily passes the glomerular membrane and is reabsorbed in the proximal convoluted tubules. In situations in which the patient is vulnerable to dehydration (fever, watery stools, vomiting, loss of appetite, hot weather), the potential for lithium intoxication is increased. In dehydration, the proximal tubular response to reabsorption of sodium (and lithium) is reduction of clearance. Severe intoxication, characterized by muscle rigidity, hyperactive deep tendon reflexes, and epileptic seizures, is usually associated with lithium concentrations >2.5 mmol/L.

Method for the Determination of Lithium

The concentration of lithium in serum, plasma, urine, or other body fluids can be determined by either flame emission photometry or atomic absorption spectrophotometry.[1]

Principle. Following dilution of the specimen with a solution containing sodium and potassium (internal standard), the sample is aspirated into the flame of an emission photometer. (In some procedures, cesium is used as an internal standard.) Thermal energy of the flame raises the energy level of both lithium and potassium to an unstable excited state. As the molecules revert to ground state, they emit energy in the visible region of the energy spectrum. The energy lines most commonly monitored are 671 nm for lithium and 768 nm for potassium. The emission intensities of lithium and potassium from the sample are compared as a ratio, and quantitation is performed by comparing this ratio with the ratio observed with standard reference solutions. The standard and sample are diluted with a solution containing a constant amount of sodium because sodium suppresses lithium emission, and consistency of analysis medium in standard and sample reduces this factor as a variable.

Antidepressants

The **tricyclic antidepressants** (e.g., amitriptyline, desipramine, doxepin, imipramine, nortriptyline) or **tetracyclic antidepressants** (e.g., maprotiline) are used to treat endogenous depression characterized by depressed mood, feelings of guilt, appetite suppression, insomnia, weight change, and decreased sexual drive. In more severe cases, depersonalized behavior, paranoid behavior, and suicidal tendencies become apparent. Endogenous depression implies that there is no apparent organic or societal cause for these behavioral changes.

The antidepressants are nearly completely absorbed from the gastrointestinal tract but undergo first-pass hepatic metabolism, so their ultimate *bioavailability* is moderate and variable. Because these drugs slow gastrointestinal activity and gastric emptying, their absorption may be delayed. *Peak plasma concentrations* are reached from 2 to 12 h after the oral dose. Metabolism is via *N*-demethylation and aromatic-ring hydroxylation, followed by conjugation with glucuronic acid. When the drug administered is the tertiary amine (amitriptyline, doxepin, imipramine), metabolism causes accumulation of the respective secondary amine (nortriptyline, nordoxepin, desipramine). These substances have generally equal pharmacological activity and accumulate to concentrations approximately (but variably) equal to the parent drug. Taking these factors into consideration (i.e., variable bioavailability, high volume of distribution, variable metabolic activity, and generation of pharmacologically active metabolites), it is not surprising that patient response to these drugs is widely variable. Determination of serum concentration is used primarily in assessing initial pharmacokinetic values to ensure adequate initial dosage.

There is a linear relationship between clinical improvement and serum concentration for most of these drugs; the exception is nortriptyline, which has a specific therapeutic window. A serum concentration of nortriptyline below or above the concentration range of 50–150 ng/mL correlates with worsening of mood.[3] The other antidepressants do not display this effect; the upper limit of the optimal blood concentration for these other antidepressents is limited by the onset of mild toxicity. *Toxicity* is expressed as dry mouth and perspiration, signs that may occur also with depression. Thus, it is difficult to differentiate between signs of mild toxicity due to the drug and those due to the disease that is being treated. More serious toxicity is expressed as atrioventricular node block, which occurs at *serum concentrations* ranging from 800 to 1200 ng/mL. The severity of intoxication is related to the serum concentration.[7]

Methods. Numerous methods have been published for the analysis of tricyclic (TCAs) and tetracyclic (TTAs) antidepressants. The choice of method made by a clinical laboratory must take into consideration several points: (1) the method must be sensitive to concentrations as low as 25 μg/mL; (2) it should distinguish and measure both parent drug and metabolites; and (3) it should discriminate the TCAs and TTAs from structurally similar, potential interferents such as mild sleep-inducers, antihistamines, and appetite suppressants available as over-the-counter compounds.

Analysis by gas-liquid chromatography and mass spectrophotometry using selected ion monitoring (GLC/MS-SIM) is the reference method; the electron impact mode[6] or the chemical ionization mode[14] can be used. Chemical ionization, because of its high sensitivity and selectivity, is the method of choice. EIA is available, and gas chromatography with a nitrogen/phosphorus detector (GC-N/P) is a commonly used technique in clinical laboratories because it is well characterized and has adequate sensitivity; its principle is given below. HPLC has also proved to be useful.[44]

Principle.[18] Antidepressants, along with appropriate internal standards, are extracted from alkalinized plasma into a polar organic solvent (isoamyl alcohol/heptane). The TCAs and TTAs are then back-extracted into an acidic aqueous phase (dilute HCl) to purify the sample further. Finally, after alkalinization, the TCAs and TTAs are extracted again into an organic phase (ethyl ether), which is evaporated. An aliquot of the reconstituted residue is then chromatographed. The effluent, which passes into a nitrogen/phosphorus detector, causes a response with characteristic retention time for each antidepressant. The intensity of that response, when compared with the response of the internal standard, can be used to quantitate the TCA or TTA in the sample.

Neuroleptic Drugs

The psychotic patient is most often treated with one of the **phenothiazines**, **haloperidol**, or **lithium**, or a combination of these drugs. Because response to these drugs is unpredictable and the patients are difficult to control, monitoring serum concentration should assist in more appropriate treatment.

The procedures for phenothiazine determination suffer from one drawback—they are too specific. The major problem encountered with monitoring the phenothiazines is the number of *metabolites* of each, most of which have some pharmacological activity.[15] Thus, therapeutic monitoring of these drugs requires coanalysis of numerous metabolites, and standards for most of the metabolites are not available.

Neuroleptic activity can also be monitored by using a competitive protein-binding assay. Haloperidol and phenothiazines compete with dopamine binding to the dopamine receptor in central nervous system tissue. Using a radiolabeled ligand, Creese and Snyder[16] demonstrated that these drugs could displace this ligand from the receptor, and that the degree of displacement was proportional to the concentration of the active neuroleptic drug. Furthermore, the many active metabolites of the phenothiazines also displace the radiolabeled ligand. Thus, it may be possible, using this one procedure, to assess the degree of dopamine displacement accomplished by a particular drug regimen. The procedure is still under investigation; it may prove to be clinically useful in the future.

Antineoplastic Drugs

METHOTREXATE

Methotrexate (MTX) has proved useful in the management of acute lymphoblastic leukemia in children, choriocarcinoma and related trophoblastic tumors in women, and carcinomas of the breast, tongue, pharynx, and testes. High-dose methotrexate administration followed by leucovorin rescue may result in regression of both carcinoma of the lung and osteogenic sarcoma.

Methotrexate inhibits DNA synthesis by competitively inhibiting the enzyme dihydrofolate reductase, thus decreasing the rate of pyrimidine nucleotide synthesis. Leucovorin, a folate analog, is used to rescue host cells from methotrexate inhibition. Methotrexate is a nonspecific cytotoxin, and prolongation of blood levels appropriate to killing tumor cells may lead to severe, unwanted cytotoxic effects such as myelosuppression, gastrointestinal mucositis, and hepatic cirrhosis.

Serum concentrations of methotrexate may be monitored during high-dose therapy (>50 mg/m^2) in order to identify the time at which active intervention of leucovorin rescue should be initiated. Criteria for methotrexate blood concentration indicative of a potential for *toxicity* after single-bolus, high-dose therapy are: (1) 10 μmol/L, 24 h after dose; (2) >1 μmol/L, 48 h after dose; and (3) >0.1 μmol/L, 72 h after dose.

Characteristically, blood concentrations are monitored at 24, 48, and 72 h after the single dose and leucovorin is administered when methotrexate levels are inappropriately high for a post-dose phase. The route of elimination for methotrexate is primarily renal excretion. During the period of high blood levels, particular attention must be paid to maintaining a large output volume of alkaline urine. The pK_a of methotrexate is 5.5; thus, small decreases in urine pH result in reduction in its solubility with the risk of intratubular precipitation of the drug and obstructive nephropathy. Monitoring blood levels therefore provides the basis for decisions for timing of initiation and continuance of leucovorin treatment and for managing urinary pH. Table 26-8 lists pharmacokinetic values for methotrexate.

Methods. Methotrexate (MTX) has been measured in biological specimens by a wide variety of techniques. Fluorescence polarization, enzyme immunoassay, radioimmuno-

Table 26-8. PHARMACOKINETIC VALUES FOR METHOTREXATE

Volume of distribution	0.4 L/kg
Bioavailability, oral	65%
Protein bound	45 ± 14%
Half-life, at concentrations:	
>100 μmol/L	1.8 h
<100 μmol/L	8.4 h
<0.1 μmol/L	>10 h
Potentially toxic concentrations:	
24 h after dose (single-bolus)	>10 μmol/L
48 h after dose	>1 μmol/L
72 h after dose	>0.1 μmol/L

assay, and folate reductase inhibition techniques are most commonly used, although nonisotopic homogeneous immunoassays are becoming popular. Liquid chromatographic procedures have also been developed that allow for coanalysis of the drug and its metabolites. However, metabolite concentrations have been shown to be of minor importance in toxicity. Because reagents for enzyme immunoassay are generally available and the method does not require specialized equipment, this technique is commonly utilized by clinical laboratories. The principle of the MTX assay by EMIT technique is identical to that given for aminoglycoside antibiotics. The difference in the assays consists only in the antibody and labeled drug present in the reagent system.[24]

OTHER ANTINEOPLASTIC DRUGS

For various reasons, other antineoplastic drugs are not routinely monitored outside of clinical trials. *Cytosine arabinoside* and *5-fluorouracil* are antimetabolites similar to methotrexate, but in neither case has the relationship between circulating blood level and therapeutic efficacy been sufficiently firm to justify routine monitoring. Alkylating agents such as *cyclophosphamide* are metabolically converted to active compounds with life spans of only seconds before they interact with tissue and are destroyed. *Actinomycin* and *doxorubicin* have toxic effects that appear to relate not to a circulating drug concentration but to dosage mass and length of exposure. *Cisplatin*, easily measurable by platinum analysis, has a renal toxicity that may be related to both blood levels and length of exposure; although monitoring is not now common, it may be in the future.[37]

Immunosuppressants

Many drugs have immunosuppressant effects but two, azathioprine and cyclosporine, are particularly used for suppressing the immune response responsible for allograft rejection. Of the two drugs, cyclosporine is currently considered the more valuable drug for this purpose.

CYCLOSPORINE

Cyclosporine, formerly cyclosporin A, is a cyclic peptide that is composed of 11 amino acids and isolated from the fungus *Trichoderma polysporum*. The compound has been shown effective in suppressing host-versus-graft rejection in organ transplants. Cyclosporine acts by inhibiting proliferation of lymphocytes, resulting in suppression of both humoral and cellular immunity.[8,9]

There is a poor relationship between dose and blood concentration, but the relationship between whole blood concentration, immunosuppression, and toxicity is predictable.[5] Immunosuppression requires trough whole blood concentrations of 100–300 ng/mL as measured by HPLC. Trough whole blood concentrations >400 ng/mL are associated with hepatic, renal, neurologic, and infective complications.[11,19]

Because cyclosporine is soluble only in an oil-based medium, it is given orally as a suspension. Bioavailability is variable but averages 27%, and peak concentrations are reached within 1–2 h. Cyclosporine is 90% protein bound and concentrated in erythrocytes. The degree of concentration in erythrocytes is temperature dependent in vitro; thus, measurement of plasma concentration requires strict attention to specimen temperature prior to separation of plasma by centrifugation if reproducible results are to be obtained.[46] Because of this effect, the best *specimen* for analysis is whole blood. The average $t_{1/2}$ of cyclosporine is 12 h. Most metabolites of cyclosporine are inactive.

Several drugs are reported to alter the disposition of cyclosporine. Ketoconazole, melphalan, amphotericin B, and aminoglycoside antibiotics all prolong metabolism of cyclosporine sufficiently to increase the risk of nephrotoxicity.[29]

Methods. The first procedure available for analysis of cyclosporine was a radioimmunoassay developed by the drug's producer. The immunoassay may exhibit nearly 100% cross-reactivity with inactive metabolites. Recently, several HPLC methods have been developed. The methods differ from most HPLC procedures because they require heating of the column. Cyclosporine is a peptide and like other peptides interacts with the packing material used in the chromatographic column. The end effect, at room temperature, is a very broadened peak, found with all brands of chromatography columns. Peak-broadening can be eliminated by maintaining the column temperature at 75 °C.

Principle.[39] Whole blood or plasma, spiked with cyclosporin D as an internal standard, is acidified and subjected to ether extraction. The cyclosporines, soluble in the organic phase, are transferred to a clean tube and the solvent is evaporated. The residue is redissolved in dilute acid and methanol and washed with hexane to remove interfering substances; hexane is then discarded. The aqueous medium is made alkaline and the cyclosporins are again extracted into ether. The organic phase is transferred to a clean tube and the ether is evaporated. The remaining residue is reconstituted and chromatographed, and quantitation is performed by peak area analysis.

References

1. Amdisen, A.: Serum lithium determinations for clinical use. Scand. J. Clin. Lab. Invest., *20*:104–108, 1967.
2. Aranda, J. V., Gorman, W., Bergsteinsson, H., et al.: Efficacy of caffeine in treatment of apnea in the low-birth-weight infant. J. Pediatr., *90*:467–472, 1977.
3. Asberg, M., Cronholm, B., Sjoquist, F., et al.: Relationship between plasma level and therapeutic effect of nortriptyline. Br. Med. J., *3*:331–334, 1971.
4. Baer, D. M. and Dito, W. R., Eds.: Interpretations in Therapeutic Drug Monitoring. Chicago, American Society of Clinical Pathologists, 1981.
5. Beveridge, T., Gratwohl, A., Michot, F., et al.: Cyclosporin A: Pharmacokinetics after a single dose in man and serum levels after multiple dosing in recipients of allogenic bone-marrow grafts. Current multiple dosing in recipients of allogenic bone-marrow grafts. Curr. Ther. Res., *30*:5–18, 1981.
6. Biggs, J. T., Holland, W. H., Chang, S., et al.: Electron beam ionization mass fragmentographic analysis of tricyclic antidepressants in human plasma. J. Pharm. Sci., *65*:261–268, 1976.
7. Biggs, J. T., Spiker, D. G., Petit, J. M., et al.: Tricyclic antidepressant overdose: Incidence of symptoms. JAMA, *238*:135–138, 1977.
8. Borel, J. F., Feurer, C., Gubler, H. U., et al.: Biological effects of cyclosporin A: A new antilymphocytic agent. Agents Action, *6*:468–475, 1976.
9. Borel, J. F., Feurer, C., Magnee, C., et al.: Effects of the new antilymphocytic peptide cyclosporin A in animals. Immunology, *32*:1017–1025, 1977.
10. Bory, C., Baltassat, P., Porthault, M., et al.: Metabolism of theophylline to caffeine in premature newborn infants. J. Pediatr., *94*:988–993, 1979.
11. Bowers, L.: University of Minnesota Hospital, Minneapolis, personal communication.
12. Bresnahan, J. F., and Vlietstra, R. E.: Digitalis glycosides. Mayo Clin. Proc., *54*:675–684, 1979.
13. Chang, J., Gotcher, S., and Gushaw, J. B.: Homogeneous enzyme immunoassay for theophylline in serum and plasma. Clin. Chem., *28*:361–367, 1982.
14. Chinn, D. M., Jennison, T. A., Crouch, D. J., et al.: Quantitative analysis for tricyclic antidepressant drugs in plasma or serum by gas-chromatography-chemical-ionization mass spectrometry. Clin. Chem., *26*:1201–1204, 1980.
15. Cohen, B. M., Herschel, M., and Aoba, A.: Neuroleptic, antimuscarinic, and anticholinergic activity of chlorpromazine, thioridazine and their metabolites. Psychiatry Res., *1*:199–208, 1979.
16. Creese, I., and Snyder, S. H.: A simple and sensitive radioreceptor assay for antischizophrenic drugs in blood. Nature, *270*:180–182, 1977.
17. David, B. M., Madsen, B. W., and Ilett, K. F.: Plasma binding of disopyramide. Br. J. Clin. Pharmacol., *9*:614–618, 1980.
18. Dawling, S., and Braithwaite, R. A.: Simplified method for monitoring tricyclic antidepressant therapy using gas-liquid chromatography with nitrogen detection. J. Chromatogr., *146*:449–456, 1978.
19. Dietrich, B. A., Krauss, A. M., Reidenberg, M., et al.: Alterations in state in apneic pre-term infants receiving theophylline. Clin. Pharmacol. Ther., *24*:474–478, 1978.
20. Dusci, L. J., and Hackett, L. P.: Gas chromatographic determination of valproic acid in human plasma. J. Chromatogr., *132*:145–147, 1977.
21. Gerson, B., and Anhalt, J. P.: High-Pressure Liquid Chromatography and Therapeutic Drug Monitoring. Chicago, Am. Soc. Clin. Path., 1980.
22. Gilman, A. G., Goodman, L. S., and Gilman, A.: The Pharmacological Basis of Therapeutics. 6th ed. New York, Macmillan, 1980.
23. Koch-Weser, J.: Serum procainamide levels as therapeutic guides. Clin Pharmacokinet., *26*:389–402, 1977.
24. Lafara, R. L.: Computer Methods for Science and Engineering. Rochelle Park, N.J., Hayden Book Co., 1973, pp. 153–156.
25. Lagerström, P. O., and Persson, B. A.: Liquid chromatography in the monitoring of plasma levels of antiarrhythmic drugs. J. Chromatogr., *149*:331–340, 1978.

26. Leahey, E. B., Reiffel, J. A., Drusin, R. E., et al.: Interaction between quinidine and digoxin. JAMA, *240*:533–534, 1978.
27. Lietman, P. S.: Chloramphenicol and the neonate—1979 view. Clin. Perinatol., *6*:151–162, 1979.
28. Meyerburg, R. J., Kessler, K. M., Kiem, I., et al.: Relationship between plasma levels of procainamide suppression of premature ventricular complexes and prevention of recurrent ventricular tachycardia. Circulation, *64*:280–290, 1981.
29. Morgenstern, G. R., Powles, R., Robinson, B., et al.: Cyclosporin interaction with ketoconazole and melphalan. Lancet, *2*:1342, 1982.
30. Nation, R. L., Peng, G. W., and Chiou, W. L.: High-pressure liquid chromatographic method for the simultaneous quantitative analysis of propranolol and 4-hydroxypropranolol in plasma. J. Chromatogr., *145*:429–436, 1978.
31. Nelson, K., and Bowers, L.: Evaluation and service experience with a fluorescence polarization immunoassay for digoxin. Clin. Chem., *29*:1175, 1983.
32. Ou, C. N., and Frawley, V. L.: Concurrent measurement of theophylline and caffeine in neonates by an interference-free liquid-chromatographic method. Clin. Chem., *29*:1934–1936, 1983.
33. Pape, B. E.: Enzyme immunoassay and two fluorometric methods compared for the determination of quinidine in serum. Ther. Drug Monit., *3*:357–363, 1981.
34. Pine, M., Favrot, L., Smith, S., et al.: Correlation of plasma propranolol concentration with therapeutic response in patients with angina pectoris. Circulation, *52*:886–889, 1975.
35. Pippenger, C. E.: Pediatric clinical pharmacology of antiepileptic drugs: A special consideration. *In*: Antiepileptic Drugs, Quantitative Analysis and Interpretation. C. E. Pippenger, J. K. Penry, and H. Kutt, Eds. New York, Raven Press, 1977, pp. 315–319.
36. Polin, H. B., and Plant, M. E.: Chloramphenicol. NY State J Med., *77*:378–381, 1977.
37. Powis, G., Ames, M., and Kovach, J.: Dose-dependent pharmacokinetics and cancer chemotherapy. Cancer Chemother. Pharmacol., *6*:1–9, 1981.
38. Routledge, P. A., Stangel, W. W., Wagner, G. S., et al.: Alpha-1-acid glycoprotein (AAG) as a determinant of lidocaine (L) disposition in myocardial infarction (MI). Clin. Res., *27*:776A, 1979.
39. Sawchuck, R. J., and Cartier, L. L.: Liquid-chromatographic determination of cyclosporin A in blood and plasma. Clin. Chem., *27*:1368–1371, 1981.
40. Schmelter, R. F., Dirksen, J. W., Stalp, J. T., et al.: Effect of heparin on radioimmunoassay of gentamicin. Am. J. Hosp. Pharm., *38*:534–537, 1981.
41. Sjoqvest, F., Borga, O., and Orme, M. L. E.: Fundamentals of Clinical Pharmacology. *In*: Drug Treatment. 2nd ed. G. S. Avery, Ed. Sydney, ADIS, 1980, pp. 1–61.
42. Soonhag, C., and Karlsson, E.: Comparative antiarrhythmic efficacy of intravenous N-acetylprocainamide and procainamide. Eur. J. Clin. Pharmacol., *15*:311–317, 1979.
43. Standefer, J. C., and Callaway, S.: Biotransformation of theophylline in neonates and infants. Clin. Chem., *27*:1086, 1981.
44. Thoma, J. J., Bondo, P. B., and Kozak, C. M.: Tricyclic antidepressants in serum by a Clin-elut column extraction and high pressure liquid chromatography analysis. Ther. Drug. Monit., *1*:335–358, 1979.
45. Tracy, R. P., Ebnet, L. E., and Moyer, T. P.: Use of decreased reagent volumes in enzyme immunoassay. Clin. Chem., *25*:1868–1869, 1979.
46. Wenk, M., Follath, F., and Abisch, E.: Temperature dependency of apparent cyclosporin A concentrations in plasma. Clin. Chem., *29*:1865, 1983.
47. Zwillich, C. W., Sutton, F. D., Neff, T. A., et al.: Theophylline-induced seizures in adults. Correlation with serum concentrations. Ann. Intern. Med., *82*:784–787, 1975.

ANALYSIS OF TOXIC SUBSTANCES

27

ROBERT V. BLANKE, Ph.D.
WALTER J. DECKER, Ph.D.

Analytical toxicology, effectively performed, is an important and frequently essential part of the sequence of events leading from diagnosis to treatment of chronic and acute poisonings. Therefore, an increasing number of clinical laboratories are establishing sections devoted primarily to the measurement of drugs and other toxic substances. The overlap of pharmacological principles and methodologies between toxicology and therapeutic drug monitoring (Chapter 26) often places the analytical responsibility for both in the same section of a large laboratory. This arrangement allows for efficient utilization of the same facilities, personnel, and analytical equipment.

In order to fulfill its role in dealing with emergency situations, the toxicology laboratory must provide 24-h service and utilize reliable methods with short turnaround times. Whatever the circumstances, the extent of service offered must necessarily be a pragmatic compromise between the service demanded for adequate clinical response and the resources that can be marshalled to meet the demand. The compromise is often achieved by a well-chosen mix of qualitative and quantitative methods representing a wide variety of analytical principles and modes.

Emergency Toxicology

There are two major reasons for the difficulty and complexity of diagnosis and treatment of toxic episodes. First, there are many thousands of chemical substances used as drugs, as household products, and for industrial and farming applications. Although substances that are encountered relatively frequently in acute poisoning incidents are, in most instances, successfully identified, information as to their toxic effects and the most effective approach to treatment remains largely incomplete. Second, there is rarely a single individual with the necessary knowledge, training, and skills adequate to respond effectively to all types of poisoning incidents. As a result, the response to the toxicological emergency must be a team effort.

The response involves at least three components: the clinical findings, the laboratory results, and the coordination of data. First, the patient must undergo a thorough clinical assessment—including both physical examination and medical, occupational, and social histories—by a physician. Life-threatening effects must be aggressively and promptly treated, and further exposure of the patient to the toxic substances must be prevented or minimized. For diagnostic purposes, the history is one of the most important parts of the examination. Information sought includes drugs or chemicals with which the patient may have been in direct contact; those available in the home, workplace, or other environment; and possible drug interactions. The history elicited from the patient may be a confused recollection or garbled information and may not match the clinical state of the patient or the laboratory results. Therefore, it has become axiomatic to encourage the clinician to "treat the patient and not the poison."[10] This is to say, if the clinical state is either more severe or less severe than would be indicated by the history or the laboratory results, the treatment is geared to the clinical state rather than to the other, possibly misleading, factors.

The second component of the response—the laboratory—should then be brought into play. Here, not only specific assays but screening tests are conducted. Many toxic substances can be ruled out of, or into, a differential diagnosis by demonstration of the presence or absence of a particular chemical in specimens removed from the patient or in substances available to the patient. Although the history, together with the clinical state of the patient, may indicate a specific substance as a causative agent, screening tests are best employed at this point. The results of the laboratory analysis may confirm a previous impression of the clinician or may raise the possibility of a toxic substance not previously suspected. A decision may be made to quantitate the substance in specimens obtained from the patient to gauge the severity of the poisoning. In some cases, quantitative results may determine the type of treatment procedure to be instituted. The laboratory also has a role in follow-up determinations for evaluation of effectiveness of treatment or for prognosis.

A third component, a toxicology data coordinator, has in recent years been recognized as an important facilitator of communication between the clinician and the laboratory. This person serves by correlating a specific patient's history and clinical state with information on toxic substances and their known effects. The coordinator then can set priorities for the laboratory for testing procedures offering the greatest potential for useful analytical results, interpret the results in a meaningful way, and communicate both the results and the informational context to the physician. Without this informed communication, the laboratory may waste valuable time in meaningless manipulations or the clinician may be tempted to treat the laboratory result rather than the patient. In some successful emergency toxicology applications, the coordinator is a trained and experienced toxicologist. In others, this role may be filled in part by a clinician or a laboratory worker who has acquired capability through training and experience.

Information resources are many and varied. Poison Information Centers assemble information on common toxic chemicals. The Toxicology Data Bank at the National Library of Medicine (available through MEDLARS at most libraries) permits rapid retrieval of specific information by means of a telephone link to the computer at the National Library of Medicine. Commercial information resources are also available. All three components of response must be reviewed periodically to assure optimal interaction.

Chronic Poisoning: Industrial and Environmental Monitoring

The problem of chronic poisoning has always been present, but only in recent years has it received attention. Most of the data upon which decisions regarding action levels by regulatory agencies are based draw on experimental animal studies or retrospective epidemiological studies in which subjects exposed to various concentrations of chemicals developed signs and symptoms of poisoning. Little evidence exists correlating concentrations of these chemicals and their metabolites in plasma or target organs with the observed effect.

The nature of toxic substances that give rise to chronic poisoning varies widely, ranging from elements (particularly metals) through complex organic and inorganic compounds. These substances may be encountered as drugs, pesticides, industrial chemicals, food additives, and environmental chemicals and pollutants. Thus, they constitute a spectrum of substances present in a variety of states in a multiplicity of matrices at extremely low concentrations. They represent an enormous challenge both to the analytical and the clinical toxicologist.

Action levels are those concentrations of toxic chemicals in food, air, or water that may be expected to exert adverse effects after prolonged exposure in humans. Data correlating action levels and length of exposures to fluid and tissue concentrations of toxic substances are scanty. Currently, the clinical laboratory is seldom required to perform toxicological analyses on nontraditional samples, although breath and fat analyses have been found to be useful. Current concern for the prevention and detection of chronic poisoning in the workplace, the development of technology and the means for monitoring by toxicological analyses, and a growing base of clinical information all forecast for the clinical laboratory an increasing role in detection and follow-up in chronic poisonings.

Responsibilities of the Clinical Toxicology Laboratory

Selection of methods that are reliable and sensitive enough for assays of biological specimens of limited size and that can be completed in time to be of practical use to the clinician is important. Depending on the nature of the analyte and the available laboratory facilities, turnaround time can range from minutes to hours, but a turnaround time >24 h is often of little use to the clinician. Defining those analytes for which tests should be available is also important, since tens of thousands of chemicals can conceivably be encountered by potential patients. Some drugs and chemicals are encountered rather commonly by almost all individuals. Thus, ethanol, carbon monoxide, sedatives, analgesics, and tranquilizers represent analytes of general concern. Furthermore, geographical location and the industries in the area affect the variety of toxic chemicals that a laboratory might be called upon to analyze.

Qualifications of personnel and choice of equipment for toxicological analyses are dependent on the level of technical sophistication and expertise demanded by the repertoire of tests made available.

Reporting results. Many of the test procedures are *screening tests* employed for preliminary assessments of poisonings. By their very nature, screening tests are designed to rule in or rule out substances in a qualitative manner. Frequently, screening tests are not highly specific; a "positive" result may be the result of interferences. On the other hand, a negative result may effectively rule out a toxic agent or group of substances. If results of a drug screen of urine are reported as negative, the clinician must know what drugs were tested for in order to appreciate which drugs were absent in that particular specimen. It is misleading to imply in the report that the specimen analyzed was negative for all possible drugs when only a limited number were actually tested for. If a standard array of drugs is routinely included in the urine drug screen, those drugs must be indicated on the report. "Presumptive" or "preliminary" test results must be clearly labeled as such, since definitive or confirmatory procedures subsequently applied to specimens may negate, validate, or supplement the initial information.

Reporting should also indicate the *sensitivity* of the analysis used. Some procedures that detect drugs in overdose concentrations may not be sensitive enough to detect them in therapeutic concentrations. Processing a less than adequate but irreplaceable sample by a standard procedure may make the analyte concentration fall below the detection limit of the method. The clinician should be informed of the sensitivity limit of the analytical method so that the meaning of a negative result is made clear.

Complications raised by *drug nomenclature* must also be resolved. For clarity, results should be routinely reported in terms of generic names or chemical names rather than trade names.

Units for reporting toxic substances have not been standardized in a way comparable with those for biological substances. Currently, the most popular units for drug concentrations for therapeutic monitoring and pharmacokinetic calculations are μg/mL or mg/L. For other substances, notably metals, parts per million (ppm) are sometimes used, but, strictly speaking, this unit usually requires conversion to the more familiar mass/volume form for clinical applications.

Screening Procedures

The most readily available and least expensive tests are those using the analyst's eyes and nose as detectors of skin color or staining and of characteristic odors.[29] Even with limited resources, however, some additional laboratory support can be furnished for the clinician.[33]

Prime requirements for laboratory screening tests are selectivity, sensitivity, and rapid turnaround time. Tests described in this section require a minimum of sophisticated equipment and can usually give at least presumptive evidence of exposure to specific toxic substances. Two points must always be borne in mind. First, false-negative results are not tolerable; they could lead to detrimental or suboptimal treatment of the poisoned patient. Second, and for the same reason, positive results of screening tests are generally presumptive only and must be supported by the use of specific (and ideally independent) confirmatory tests. Although there are few safe, specific antidotes available,[13] the benefit to the patient is great if a laboratory result, quickly obtained, can support administration of a specific antidote.

Spot Tests

Spot tests are analytical procedures that are simple and rapid and make use of the analyst's eye as the instrument of detection. A number of chemical spot tests are applicable to bedside or near-bedside rapid screening for poisoning. Some of these are quantitative, some are semiquantitative, and others are strictly qualitative. When used with discretion, they are especially helpful when adequate diagnostic laboratory facilities are not immediately available.

Decker and Treuting[14] reviewed spot tests available in the literature and selected a number of these that were found to be reliable when used in their laboratories. Some have been slightly modified by these authors; they are listed here in alphabetical order.

Other lists of spot tests have been compiled by Sunshine,[42] Curry,[12] and Kaye.[29]

Acetaminophen. See *p-Aminophenol*.

Acetone (for detecting uncontrolled diabetes or the metabolite of isopropyl alcohol poisoning). See Chapter 13, page 439.

Alcohols. See *Volatile reducing substances*.

Aldehydes. See *Volatile reducing substances*.

p-Aminophenol (metabolite of phenacetin or acetaminophen). Add 3 drops HCl, 1.0 mol/L, to 1 mL urine; place in ice bath. When cold, add 3 drops freshly prepared sodium nitrite solution, 0.14 mol/L, and 3 drops α-naphthol reagent (100 mg α-naphthol in 5 mL NaOH, 2.0 mol/L). A red color (semiquantitative) is indicative of *p*-aminophenol.

Amphetamine. See *Sympathomimetic amines*.

Antimony. See *Heavy metals*.

Arsenic. See *Heavy metals*.

Barbiturates, glutethimide, ethchlorvynol. Acidify 1 mL serum, urine, or filtered stomach contents or washings with HCl, 1.0 mol/L. Extract with 5 mL chloroform. Evaporate the chloroform layer in a 100 °C water bath. Take up the residue in a few drops of chloroform and spot on a piece of filter paper. Allow the chloroform to evaporate and add a few drops of aqueous mercurous nitrate, 0.02 mol/L. A dark blue to purple color is indicative of a number of acid hypnotic drugs, including barbiturates, glutethimide, and ethchlorvynol.

Bromate. See *Oxidizing agents*.

Carbamates. See *Meprobamate*.

Carbon monoxide. Add 20 mL distilled water to 0.5 mL whole blood and add 1 mL NaOH, 1.0 mol/L, and mix. Blood containing over 20% carbon monoxide hemoglobin (HbCO) will have a light cherry-red color; normal blood will turn brown.

Carbon tetrachloride. See *Halogenated hydrocarbons*.

Chloral hydrate. See *Halogenated hydrocarbons* and *Volatile reducing substances*.

Chlorate. See *Oxidizing agents*.

Chloroform. See *Halogenated hydrocarbons*.

Cholinesterase (activity is depressed by organophosphorus insecticides). Make up a fresh aqueous solution of bromthymol blue, 0.008 mol/L, containing 100 mg/mL acetylcholine bromide. Dip a piece of filter paper in this solution and air dry. Apply two drops of any pH 5 buffer (control) to one piece of impregnated paper; to another apply two drops of cell-free plasma (unknown). The color of the control paper will change from blue to yellow in ~5 min at room temperature. If the unknown paper takes significantly longer to change color (10 min or more), the cholinesterase activity is depressed; the degree of depression is roughly proportional to time.

Codeine. See *Morphine*.

Desipramine. See *Imipramine*.

Ethchlorvynol. See *Barbiturates*.

Ethylene glycol. See *Oxalic acid*.

Glucose (to rule out drug-induced symptoms or to implicate insulin misadventure). See Chapter 13, page 431.

Glutethimide. See *Barbiturates*.

Halogenated hydrocarbons (chloroform, chloral hydrate, carbon tetrachloride). Extract 1 mL blood with 2 mL ether. Pipet off 1 mL of the ether layer into a test tube; add to it 1 mL pyridine and 1 mL NaOH, 5.0 mol/L. Place the tube in a boiling water bath for 1 min. A red color in the pyridine layer is positive for halogenated hydrocarbons. Urine (1 mL) may be used without extraction. A water blank should also be run.

Note: Not all halogenated hydrocarbons will react positively with this test.

Heavy metals (Reinsch test for antimony, arsenic, bismuth, and mercury). Clean a 10-cm length of heavy-gauge copper wire with sandpaper or emery cloth. Without touching the wire with bare fingers, coil two-thirds of it in a loose spiral. To 5–10 mL stomach contents or urine, add an equal volume of HCl, 2.0 mol/L, and place the copper spiral into the solution. Place in a boiling water bath for 10 min; examine for a stain on the copper. If no stain appears in 1 h, the test may be considered negative. Sensitivity is ~5 μg.

Antimony: blue or purple-black	Bismuth: shiny black
Arsenic: dull black	Mercury: silver gray

Heroin. See *Morphine*.

Imipramine and desipramine. Add 1 mL urine to 1 mL of reagent consisting of equal volumes of potassium dichromate, 0.007 mol/L; H_2SO_4, 5.4 mol/L; perchloric acid 2.0 mol/L; and nitric acid, 5.0 mol/L. A green color indicates imipramine; blue indicates desipramine.

Insulin. See *Glucose*.

Iron. To 1 mL filtered stomach contents or washings, add 1 drop concentrated HCl and 0.5 mL potassium ferricyanide, 0.03 mol/L. A blue color is indicative of ferrous ion. To another 1-mL aliquot, add 1 drop concentrated HCl and 0.5 mL potassium ferrocyanide, 0.027 mol/L. A blue color is indicative of ferric ion.

Isopropyl alcohol. See *Acetone*.

Meprobamate. Extract 2 mL serum with 5 mL ether. Pipet off 3 mL of the ether layer and evaporate in a 50-mL beaker. Add 3 mL of a reagent containing 2 g hydroquinone in 100 mL of H_2SO_4:H_2O at a ratio of 15:85. Heat for 25 min at 100 °C. Under ultraviolet illumination (preferably under a Wood's light, 365 nm), a rose fluorescence is positive for meprobamate.

Extract another sample and evaporate the extract as above. Take up residue in 3 drops ether; place on a filter paper. Add 1 drop 10% furfural in ethanol. After the ethanol evaporates, expose the paper to fumes from concentrated HCl. A deep purple color is positive for meprobamate.

Other carbamates, including insecticides such as *N*-methylcarbamate (carbaryl, Sevin), will also react if present in sufficiently high concentrations. Filtered stomach contents or washings are preferred over serum when these compounds are suspected.

Mercury. See *Heavy metals*.

Methamphetamine. See *Sympathomimetic amines*.

Methanol. 1 mL protein-free filtrate of serum is added to 0.1 mL $KMnO_4$ solution (5 g/dL in H_2O). Swirl the test tube gently, and after 5 min add sufficient powdered sodium bisulfite to decolorize the permanganate. Add 0.2 mL freshly prepared chromotropic acid solution (0.5 g/dL in H_2O) and 6 mL concentrated H_2SO_4. Mix and heat in a boiling water bath for 5 min. A red-violet color is positive and specific for methanol. Quantitation can be accomplished by colorimetry at 570 nm.

Morphine. Adjust 20 mL urine to pH 9.0–9.5 with NH_4OH and extract with 10 mL chloroform:ethyl acetate (1:4). Evaporate half of the extract to dryness, add 0.5 mL concentrated HCl and 1 drop concentrated H_2SO_4, and place test tube in a boiling water bath. A reddish-purple color is indicative of morphine. After all HCl is evaporated, add 0.5 mL fresh HCl and neutralize with solid Na_2CO_3. Morphine is indicated by a violet color that changes to green upon dropwise addition of a 57% HI solution to which 0.5 g I_2/dL in H_2O has been added.

Evaporate the other half of the chloroform:ethyl acetate extract to dryness; acidify with 0.5 mL HCl, 2.0 mol/L. Add 0.5 mL Lugol's solution diluted 1:100 with water. The slow formation of reddish-brown crystals is indicative of morphine.

Note: Several other opium alkaloids and synthetic opioids will also react positively. Also, the presence of morphine in urine results from the biotransformation of codeine or heroin.

Nitrate. See *Oxidizing agents*.

Organophosphorus insecticides. See *Cholinesterase*.

Oxalic acid, oxalates (metabolite of ethylene glycol). Acidify urine or filtered stomach contents or washings to pH 1.5 with concentrated HCl, filter, adjust the pH of a 5-mL aliquot to approximately 8 with NH_4OH, and acidify with acetic acid to pH <7.0. Add saturated $CaCl_2$ solution dropwise. A precipitate is indicative of oxalate. To confirm, add concentrated HCl dropwise.

If the precipitate dissolves, oxalate is very probably present. Warm the mixture and add $KMnO_4$ solution (0.1 g in 100 mL H_2O) dropwise; the violet color will disappear if oxalate is present.

Oxidizing agents (bromate, chlorate, nitrate, nitrite). To 1 mL filtered stomach contents or washings, add 2 mL of a solution of diphenylamine in concentrated H_2SO_4 (0.5 g in 100 mL). A blue color is indicative of bromate, chlorate, or nitrate. The latter ion may be a result of ingestion of nitrite.

To 2 mL whole blood, add 1 mL HCl, 2.0 mol/L, centrifuge, and decant. Add 0.1 mL Evans blue solution (0.01 g in 100 mL H_2O) to the supernatant; decolorization of the dye indicates the presence of one or more of these ions.

Paraquat. To 5 mL filtered stomach contents or urine, add 0.1 g sodium bicarbonate and 0.1 g sodium dithionite. The resulting blue color can be quantitated by colorimetry at 625 nm.

Phenacetin. See *p-Aminophenol*.

Phenols (cresols, naphthols, thymols). Extract 2 mL stomach contents with 10 mL ether. Evaporate the ether layer at room temperature in a small white crucible. Add 1 drop freshly prepared sodium nitrite, 0.14 mol/L, in concentrated H_2SO_4, 1 drop water, and a NaOH or KOH pellet in that order. Positive reactions are as follows:

Cresols:	dark brown	Phenols:	reddish green to blue
Naphthols:	green	Thymols:	green to purple

Phenothiazine antipsychotic drugs. See also *Salicylates*. Add 1 mL urine to 1 mL FPN reagent (ferric chloride solution, 5 g/dL; perchloric acid, 2.0 mol/L; nitric acid, 5.0 mol/L; 1:9:10 mixture). Pink-to-violet color indicates a phenothiazine; the color depends upon dose and metabolites (the latter are predominantly pink). Shake 5 mL urine with 10 mL chloroform; a red metabolite (if present) is extracted. To the chloroform layer, add 5 mL HCl, 2.0 mol/L, and a small piece of tin; a colorless solution will then result. Filter the chloroform layer and add 1 mL 30% H_2O_2; a red color indicates a positive reaction.

Salicylates (and phenothiazines). Place 1 drop urine or serum on a Phenistix reagent stick (Ames Division, Miles Laboratories, Elkhart, IN 46515). A violet color indicates a salicylate. The test is semiquantitative using the color chart provided.

Note: Phenothiazine antipsychotic drugs also give a violet color that will persist after H_2SO_4:H_2O (1:1) is added; the color due to salicylates will disappear.

Add 1 mL urine or serum to 1 mL ferric chloride solution (5 g in 100 mL H_2O). A violet color indicates a salicylate.

Strychnine. To 5 mL filtered stomach contents or washings add 1 mL concentrated HCl and a small piece of metallic zinc. Boil the mixture for 1 min, cool, and filter. Hold a test tube containing 2 mL concentrated H_2SO_4 at approximately a 45° angle, and allow 2 mL filtrate to run down the side very slowly. A rose-red ring at the interface of the two liquids indicates strychnine.

Sulfonamides (and other compounds containing a primary aromatic amino group such as a *p*-aminophenol and *p*-aminosalicylate). Place 0.5 mL urine or filtered stomach contents on a piece of newspaper. Add 1 drop concentrated HCl. An intense yellow-to-orange color is a positive result. Positive and negative control urine specimens must also be carried through the procedure.

Sympathomimetic amines (amphetamine, ephedrine, hydroxyamphetamine, methamphetamine, phenylephrine, synephrine). To 10 mL urine or filtered stomach contents or washings, add sufficient NaOH, 2.0 mol/L, to raise the pH to ~8. Extract with 10 mL chloroform. To the chloroform layer add 2 mL HCl, 5.0 mol/L, and extract. Remove 1 mL of the aqueous layer and add 0.5 mL Sanchez reagent (100 mg dimethylaminobenzaldehyde in 20 mL ethanol acidified with 4 drops concentrated H_2SO_4). Evaporate to dryness. Note the color of the residue. Dissolve the residue in water and again note the color.

Compound	Residue	Solution
Amphetamine	Yellow-green	Yellow-green
Ephedrine	Dark green	Pale green
Hydroxyamphetamine	Red-brown	Yellow-brown
Methamphetamine	Dark green	Pale green
Phenylephrine	Orange	Pale orange
Synephrine	Yellow-brown	Dark yellow

Note: Certain opium alkaloids will yield a red or violet residue; aqueous solution of the residue will be colorless or pale yellow.

Thallium. To 1 mL urine or filtered stomach contents or washings, add 2 drops concentrated HCl and 5 drops bromine water. Five minutes later, add 5 drops sulfosalicylic acid solution (20 g/dL), 0.5 mL toluene, and 5 drops aqueous methyl violet solution (0.1 g/dL) and shake. A blue to blue-green color in the upper toluene layer is indicative of thallium.

Volatile reducing substances (alcohol, aldehydes, chloral hydrate). Place 1 mL blood, urine, or stomach contents or washings into a small screw-capped test tube. Wipe the inside of the tube dry. Place a loose plug of glass wool about halfway down the tube. Moisten the wool with a drop of reagent (0.5 g potassium dichromate in 100 mL H_2SO_4, 21.6 mol/L). Screw on the cap tightly and immerse the bottom of the tube in hot (almost boiling) water. The reagent will change color from orange to green in less than 10 min if a significant amount of volatile reducing substance is present.

Comments. The results of spot tests must be evaluated with a considerable amount of discretion. As a rule, false-positives occur much more frequently than do false-negatives, since most spot tests lack absolute specificity. On the other hand, a false-negative may result from a number of conditions, including low sensitivity of the method used, binding of the agent or its metabolites to protein or other high-molecular-weight substances, complexation of the agent or its metabolites with highly adsorptive or reactive compounds, formation of metabolites that do not react the same as the parent compound, volatilization of or chemical change in the agent during processing, masking of color by other substances present, outdated or unstable reagents, and inefficient solvent extraction of the compound. Some of these pitfalls can be overcome by carrying positive and negative controls through the test.

Thin-Layer Chromatography (TLC)

A rapid TLC screening method for the detection of a significant number of commonly abused drugs in urine is presented.[47] Blood may also be assayed by this method, but urine usually contains a higher concentration of most drugs, and it is therefore the specimen of choice. The method has been modified slightly in the authors' laboratories.

Materials and Reagents

1. TLC plates. Soft-surfaced plates generally give the best resolution in the system; a plate such as the Analtech Silica Gel GF is recommended. Many analysts prefer to use channeled or preadsorbent plates or both. Score the plates horizontally 10 cm above the spotting line to provide a uniform front and to simplify calculation of R_f values.
2. Standards. A composite aqueous standard consists of amphetamine, codeine, meperidine, methadone, methaqualone, morphine, phenobarbital, quinine, and propoxyphene, each in a concentration of 3 mg/L. Prepare fresh monthly; store refrigerated in a brown bottle.
3. Extraction reagents:
   a. *Buffer*. Adjust saturated ammonium chloride solution to pH 9.2 with concentrated ammonium hydroxide solution.
   b. *Solvent*. Mix chloroform:isopropanol in a ratio of 96:4. *Caution*: Mix and use in a fume hood. Chloroform is a demonstrated carcinogen in animals.
4. Developing solvent mixture:
   a. Mix ethyl acetate:methanol:concentrated NH_4OH:water in a ratio of 150:18:0.5:10. Keep tightly stoppered.
   b. Replace solvent after 2 plates have been developed in the tank, or when R_f values are not reproducible if solvent has lost ammonia.
5. Spray reagents:
   a. Fluorescamine (Fluram, Roche), 20 mg in 100 mL dry acetone. Keep refrigerated. Prepare fresh monthly.
   b. Ninhydrin, 100 mg in 100 mL acetone. Keep refrigerated. Prepare fresh monthly.
   c. Iodoplatinate, 5 mL platinic chloride, 5.0 g/dL, and 45 mL potassium iodide (10 g/dL) added to 100 mL deionized water. Stable for 6 months.
   d. Mercuric sulfate. Dissolve 2 g mercuric oxide in 20 mL concentrated sulfuric acid; dilute to 100 mL with deionized water. Stable indefinitely.
   e. Diphenylcarbazone, 100 mg in 100 mL of a 50:50 acetone-chloroform solution. *Caution*: Use fume hood; keep refrigerated; prepare fresh monthly.
   f. Dragendorff reagent:
      (1) Dissolve 2.2 g bismuth subnitrate in a solution of glacial acetic acid:water, 25:100.

(2) Dissolve 50 g potassium iodide in 125 mL water.

(3) Combine 10 mL each of solutions (1) and (2) with 20 mL glacial acetic acid and 100 mL water. Stable for one year if kept tightly stoppered in the dark.

Procedure

1. To ~20 mL urine, add 1.5 mL pH 9.2 buffer, mix, and centrifuge. Extract the clear supernatant liquid with 30 mL extraction solvent by shaking for 10 min. Centrifuge and place the lower (solvent) layer into a 100-mL beaker. Evaporate until the solvent is about half evaporated, add 1 drop concentrated HCl (to reduce volatilization of amines such as amphetamine), and continue evaporation to dryness. Carry 1 mL of the standard plus 19 mL water through the extraction and evaporation procedure.
2. Take up the residue in about 20 μL methanol; spot the entire solution 2 cm above the bottom edge of the TLC plate (or in a preabsorbent layer). Do the same with the extract residue of the standard.
3. Immerse the plate in the solvent tank; develop to the 10-cm scored line. Allow the solvent to equilibrate at this line for 10–15 min. Remove the plate from tank and air-dry in the fume hood.
4. Examine the plate under short- and long-wavelength ultraviolet light (e.g., in a Chromatovue cabinet); several drugs such as quinine or quinidine will exhibit characteristic fluorescence. Others will quench fluorescence if a plate incorporating an inorganic fluorophore is used. With a sharp pencil, outline the spot(s) observed and calculate the R_f value(s) in this and the following steps.
5. Spray the plate with fluorescamine and re-examine under long-wavelength ultraviolet light. This will reveal drugs containing primary amine groups, such as amphetamine. Perform this and all subsequent sequential sprays in a fume hood.
6. Spray with ninhydrin; heat the plate in an oven for 10 min at 85 °C, then activate the plate for 3 min under simultaneous long- and short-wave ultraviolet light in an ultraviolet view box (e.g., a Chromatovue). Primary and secondary amines are seen as pink to pinkish-tan spots (the latter more characteristic of methamphetamine).
7. Spray with mercuric sulfate. Barbiturates, glutethimide, and phenytoin appear as dry or white areas; in this system, all three drugs will have approximately the same R_f values.
8. Spray with diphenylcarbazone for further confirmation of these drugs. Glutethimide will be blue, for example, and amobarbital will be red on a violet background.
9. Spray the same plate with iodoplatinate; alkaloids particularly, but also many other nitrogen-containing compounds, become evident. Many drugs will have characteristic colors with this reagent.[9]
10. Spray the same plate with Dragendorff reagent; methaqualone is detected as a bright orange spot; benzodiazepines, such as chlordiazepoxide, diazepam, and flurazepam, will appear as less bright spots. Previously seen spots will usually turn tan or brown.

Interpretation and Comments

Approximate R_f values of a number of commonly encountered drugs appear in Table 27-1. The table is intended to be only a guide; it is highly recommended that each laboratory calculate its own R_f values and include results from additional drug standards.

Table 27-1. R_f VALUES FOR SOME COMMONLY ENCOUNTERED DRUGS ON ANALTECH GF TLC PLATES[47]

Drug	R_f Value
Morphine	0.14
Codeine	0.22
Amphetamine	0.99
Quinine	0.38
Meperidine	0.53
Amitriptyline	0.63
Methadone	0.67
Pentazocine	0.73
Phencyclidine	0.79
Propoxyphene	0.81
Xylocaine	0.82
Phenobarbital	0.77
Methaqualone	0.87

It is important that all positive results be confirmed by at least one independent method of analysis.

Additional thin-layer chromatographic procedures have been described.[3,21,40,43]

Gas-Liquid Chromatographic (GLC) Screening Procedures

Screening by GLC is practical in those laboratories with appropriate equipment and expertise. A method using two columns and detecting acidic, neutral, and basic drugs has been published by the authors.[5] A single-column, temperature-programmed screening approach[36] can be extremely useful when the laboratory has limited equipment. Finkle and colleagues[18] developed a highly comprehensive GLC-based method that uses four columns and three liquid phases; the publication also lists relative retention data for almost 600 toxic substances and metabolites in humans.

Criteria for a practical GLC screening method include:

1. An extraction system that recovers from a blood, urine, or gastric specimen a majority of drugs most frequently abused or overdosed. Generally, this consists of a two-stage extraction, one stage for basic drugs, the second for acidic and neutral drugs.
2. Chromatographic conditions that will detect a significant amount of a majority of the drugs extracted without prior derivatization.

Screening Test for Metals

The Reinsch test for detection of antimony, arsenic, bismuth, and mercury has been described in the section on *Spot Tests*, as has a test for thallium.

A differential procedure for six metals (bismuth, copper, gold, lead, mercury, and zinc) is based upon the fact that ions of each of these metals will result in a specific color when complexed with dithizone (diphenylthiocarbazone) and adjusted to a specific pH.[29] This procedure is quite tedious; steps must be taken to remove contaminant metals from most of the reagents employed. Analyses based on thin-layer chromatography[48] are more convenient. Detection limits are for the most part high (5–100 μg); thus, a 24-h urine collection is generally the specimen of choice. A typical procedure has been described.[5]

Screening Tests for Volatile Substances

A number of toxic substances, such as the lower alcohols, acetaldehyde, formaldehyde, acetone, lower-molecular-weight hydrocarbons, inhalation anesthetics, freons, and carbon monoxide, are volatile at room or body temperature. Others, such as cyanide or fluoride salts, can be converted to the gaseous state by pH adjustment. Screening methods commonly used to detect volatiles include gas-liquid chromatography, osmometry, spectrophotometry, and microdiffusion analyses. In the *headspace method* for volatiles, a sample mixed with an internal standard is incubated in a sealed vial. The gas above the liquid phase (in the headspace) is then sampled and subjected to gas-liquid chromatography (see page 880). *Direct on-column injection* of a blood or urine specimen can also be used,[27] but many analysts find that plugging of the syringe needle, due to heat precipitation of proteins or nonvolatile compounds, may occur.

Osmometry is also useful in detecting overdose of ethanol and other polar, water-soluble substances.[35] In the absence of conditions that cause high serum osmolality (diabetic acidosis and coma, renal failure, hypernatremia) a workable approximation of blood ethanol concentration in mg/dL can be achieved by subtracting an average "normal" value of 290 mOsm/kg H_2O from the observed osmolality of the serum specimen and then multiplying the difference by the empirically derived factor 4.24.

Screening Test for Organochlorine Pesticides[45]

Screening for organochlorine pesticides or environmental pollutants can be done on serum or whole blood. Since many of these substances are persistent and ubiquitous in

Table 27-2. "NORMAL" CONCENTRATIONS OF INSECTICIDES IN HUMAN BLOOD[45]

	Mean, ppb*	Range, ppb
p,p'-DDE	26	2–500
p,p'-DDT	12	2–100
p,p'-TDE	11	2–70
Dieldrin	4.5	1–25
Methoxychlor	19.0	10–40
Lindane	3.5	1–20
α-BHC	3.5	1–10
Heptachlor epoxide	4	2–20

*ppb = parts per billion

the environment, it is not unusual to detect low concentrations in serum of normal patients. It is important, therefore, to discriminate between "normal" and elevated concentrations (Table 27-2). A GLC method described by the authors minimizes interferences.[5]

Other Screening Procedures

A number of tests, often indirect in nature, can be useful in screening for toxic substances. Many of these are particularly useful when the physician has already made a provisional diagnosis based on clinical signs, symptoms, known exposure to a substance, information from friends or relatives, and medical history. Examples include the determination of cholinesterase activity when poisoning by organophosphate or carbamate pesticide is suspected, and the screening test for ketones when it is likely that the patient has consumed isopropanol. These tests have been described in the section on *Spot Tests*.

Abnormal results of several tests commonly performed in the clinical chemistry laboratory may indicate the need to pursue more specific procedures to confirm or rule out poisoning. Table 27-3 gives some examples. Extensive descriptions appear in the literature.[7,25,30]

Liquid chromatography (high performance chromatography, HPLC) is emerging as a general screening technique for the detection of drugs in blood.[28] Twenty common drugs can be determined simultaneously in a single run in ~35 min. The primary difficulty in adapting liquid chromatography to a screening method is that no single column packing or solvent system (unlike gas-liquid chromatography) will effectively resolve widely different chemical classes of drugs. Perhaps in the near future ternary (or even quaternary or higher) solvent delivery systems will overcome this problem.

Determination of Toxic Substances in Biological Matrices

In most chemical analytical procedures, the analyte must be separated from the biological matrix prior to the application of an instrumental analysis. Historically, analytical toxicologists have tended to classify toxic substances according to the manner by which they are separated from the biological specimen. For this reason, some of the toxicological reference textbooks have categorized poisons as (1) gases, (2) volatile substances, (3) corrosives, (4) metals, (5) nonmetals, and (6) nonvolatile organic substances. In contemporary analytical toxicology there has been a tendency to combine categories that naturally overlap. The following discussion will include those separation procedures that are commonly utilized, but the discussion will not attempt to deal with all separation procedures.

Methods for the Determination of Volatile Substances

Volatile toxic compounds consist of gases or liquids that have boiling points of <100 °C. *Microdiffusion* can be used to separate many members of this group; it has

Table 27-3. EXAMPLES OF BIOCHEMICAL ABNORMALITIES THAT MAY RESULT FROM TOXIC AGENTS[30]

Acidosis: Methanol, ethylene glycol, iron salts, phenytoin, salicylates, tricyclic antidepressant drugs
Albuminuria: Many toxic substances
Aminoaciduria: Cadmium, phosphorus
Bilirubinemia: Ethylene glycol or its monomethyl ether
Copper deficiency: Molybdenum
Delta-aminolevulinic dehydratase activity (decreased): Lead
Hematemesis: Acetonitrile, arsenic, boron, tetrachloroethane
Hematuria: Anticoagulant drugs, benzidine, carbon tetrachloride, naphthalene, picric acid, sulfonamides, turpentine
Hemoglobinuria: Acetic acid, arsine, naphthalene, stibine
Hemolysis: Acetic acid, arsine, lead, naphthalene
Hemoptysis: Chlorine
Hypercalciuria: Beryllium, cadmium, phosphorus
Hyperglycemia: Acetone, insulin, isoniazid, phenytoin, salicylates (early stages)
Hyperkalemia: Digitalis glycosides, ethylene glycol, isoniazid
Hypoglycemia: Ethanol, methanol, salicylates (late stages)
Hypokalemia: Tricyclic antidepressant drugs, diuretics
Hypoxia: Aniline, dinitrotoluene
Jaundice: Arsine, chlorinated hydrocarbons, dinitrophenol, gold, naphthalene, nitrobenzene, stibine, sulfonamides
Lactic acid dehydrogenase (increased): Zinc oxide
Liver damage: Many toxic substances
Methemoglobinemia: Aniline, chlorates, mercaptans, nitrated and chlorinated benzenes and toluenes, oxides of nitrogen, phenacetin, pyridium, sulfonamides
Oxygen saturation (decreased): Beryllium
Porphyrinuria: Benzene, chlorinated hydrocarbons, lead, mercury
Proteinuria: Phosphorus, thallium
Renal damage: Many toxic substances
Stools, bloody: Anticoagulant drugs, antimony, arsenic, castor beans, copper, iron, oxalic acid, methaqualone, zinc chloride
Sulfhemoglobinemia: Nitrobenzene, nitrochlorobenzene
Uremia: Diethylene glycol, isopropanol, lead, paraquat

the advantages of small sample size, minimal and inexpensive equipment, and simplicity of operation. In this technique, the specimen containing the substance to be separated and a "trapping" solution are placed in separate compartments inside a sealed container, where they share the same vapor space. The substance to be separated, because of its vapor pressure, leaves the specimen and enters the common vapor space, from which it is absorbed by the trapping solution. Thus, by gaseous diffusion, the substance to be separated is continuously transferred from the specimen to the trapping solution until an equilibrium is reached. If the trapping solution contains a reagent that converts the separated substance into a different compound, a quantitative transfer results. The entire operation can be carried out with a small specimen by using a Conway unit. This unit consists of two round, concentric chambers molded into a dish that can be sealed. The trapping solution is placed in the center well and the specimen (and in some cases a "releasing" agent) in the outer compartment (Figure 27-1).

The time required for completion of the diffusion process depends on the vapor pressure of the substance to be separated, the volume of the specimen solution, the

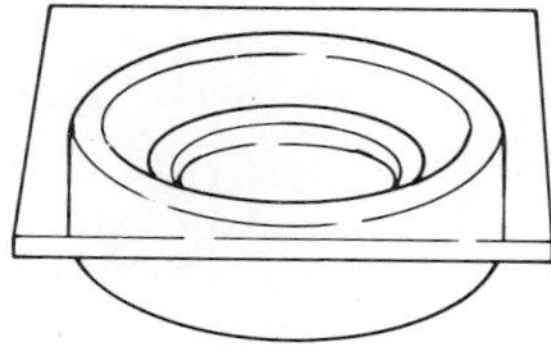

Porcelain diffusion unit

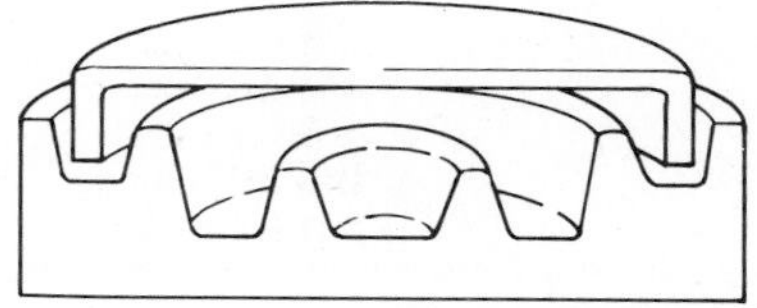

Sealable porcelain diffusion unit, showing cover sealed in liquid.

Figure 27-1. Microdiffusion chambers of the Conway type.

nature of the trapping solution, and the temperature at which the process is conducted. In general, the diffusion time is shortened by higher temperatures, by small volumes of specimen solutions, and by substances of high vapor pressure. If diffusion is carried out at an elevated temperature, precautions must be taken to ensure against loss of expanding gases, which tend to lift the cover. It is advisable to place a 300-g weight on the lid at room temperature, and at elevated temperatures the lid should be clamped in place.[11]

A number of toxicologically important volatile substances have been measured by this technique. Since most of these procedures are colorimetric in nature, all that is required besides the necessary reagents and the Conway unit is a spectrophotometer. Feldstein and Klendshoj[17] have applied this method to a number of solvents and gases as shown in Table 27-4.

A second general approach to the separation of volatile substances is by the use of *headspace gas chromatography*. This general approach is applicable to more substances than the microdiffusion procedure and is more convenient for those laboratories equipped with a gas chromatograph. The principle of this method is to decrease the solubility of the volatile substance in the biological sample by saturating the aqueous phase with salt.

Generally, the specimen is measured into a vial fitted with a septum stopper and a quantity of salt is added to the specimen to achieve a saturated solution. After ap-

Table 27-4. MICRODIFFUSION TECHNIQUES

	Outer Compartment			
Analysis	***Sample***	***Liberating Agent***	**Inner Compartment**	**Diffusion Time**
Acetaldehyde	3 mL blood 5 mL tissue*	3–4 drops H_2SO_4, 10%	3.3 mL $NaHSO_3$, 0.15 mol/L	3 h
Acetone	3 mL blood 5 mL tissue*	3–4 drops H_2SO_4, 10%	3.3 mL $NaHSO_3$, 0.15 mol/L	3 h
Carbon monoxide	1.0 mL blood	1 mL H_2SO_4, 10%	2.0 mL palladium chloride, 0.005 mol/L, in HCl, 0.01 mol/L	1 h
Cyanide	2–4 mL blood 5 mL tissue*	3–4 drops H_2SO_4, 10%	3.3 mL NaOH, 0.1 mol/L	3 h
Ethanol	0.8 mL blood 4 mL tissue*	1 mL K_2CO_3, sat. soln.	2 mL acid-dichromate, $K_2Cr_2O_7$, 0.02 mol/L in H_2SO_4, 7.76 mol/L	3 h
Fluoride	1.0 mL blood 1.0 mL urine	$HClO_4$, conc.	Zr SPADNS†	24 h at 60 °C
Formaldehyde	3 mL blood 5 mL tissue*	3–4 drops H_2SO_4, 10%	3.3 mL $NaHSO_3$, 0.15 mol/L	3 h
Halogenated hydrocarbons	1–4 mL blood 1–4 mL tissue*	—	1.0 mL toluene	3 h
Isopropanol	2.0 mL blood 5 mL tissue*	1 mL K_2CO_3, sat. soln.	3.3 mL H_2SO_4, 10%	3 h
Methanol	2.0 mL blood 5 mL tissue*	1 mL K_2CO_3, sat. soln.	3.3 mL H_2SO_4, 10%	3 h
Phenols	2–4 mL blood 5 mL tissue*	3–4 drops H_2SO_4, 10%	3.3 mL NaOH, 0.1 mol/L	3 h
Sulfide	2–4 mL blood 5 mL tissue*	3–4 drops H_2SO_4, 10%	3.3 mL NaOH, 0.1 mol/L	3 h

* Tissue homogenate.

† Zirconyl chloride · 8 H_2O, 0.4 mmol/L, and 4,5-dihydroxy-3-(*p*-sulfophenylazo)-2,7-naphthalene disulfonic acid trisodium salt, 4.6 mmol/L, in HCl, 2.0 mol/L.

After Feldstein, M., and Klendshoj, N.: The determination of volatile substances by microdiffusion analysis. J. Forensic Sci., *2*:39–58, 1957.

Table 27–5. RELATIVE RETENTION TIMES OF VOLATILE SUBSTANCES

Compound	Relative Retention Time*	Compound	Relative Retention Time*
Petroleum ether	0.19	Benzene	1.22
Acetaldehyde	0.20	Isopropanol	1.24
n-Pentane	0.21	Methyl cyclohexane	1.32
Diethyl ether	0.25	Allyl ether	1.39
Propionaldehyde	0.36	Butanol, tertiary	1.40
Methylacetate	0.38	*n*-Propyl acetate	1.63
Acetone	0.40	Chloroform	1.64
Carbon disulfide	0.40	Allyl acetate	1.69
Acrolein	0.41	1-Chloro-3-methyl butane	1.72
n-Hexane	0.44	Trichlorethanol	1.80
Methyl iodide	0.44	1,4-Dioxane	1.81
Isopropyl ether	0.47	Ethylene dichloride	1.84
Diethylamine	0.57	Trichlorethylene	2.10
Methanol	0.60	*n*-Octane	2.40
Acetonitrile	0.65	*n*-Propanol	2.50
Methylene chloride	0.69	2,3-Dichloropropane	2.63
Ethyl acetate	0.71	Methyl isobutyl ketone	2.75
Cyclohexane	0.72	Butanol, secondary	3.04
Tetrahydrofuran	0.78	Toluene	3.15
Acrylonitrile	0.80	Paraldehyde	3.45
Butanone	0.84	*n*-Butylacetate	4.00
n-Heptane	0.98	Isobutanol	4.68
1,1,1-Trichloroethane	0.98	*n*-Nonane	5.95
Ethanol	1.00	Pyridine	6.08
Carbon tetrachloride	1.01	*n*-Butanol	6.60
Propionitrile	1.14	Xylene	7.57

*Relative to ethanol, whose absolute retention time is 1.9 ± 0.1 min on a 6′ × 1/4″ O.D. column packed with 5% Carbowax K-600 and 3% Halcomid 18 on 60- to 80-mesh Teflon 6HC.

Reprinted with permission, from Dubowski, K.: Organic volatile substances. *In*: Methodology for Analytical Toxicology, I. Sunshine, Ed. Boca Raton, FL, CRC Press, Inc., 1975.

propriate mixing, the vial is placed in a water bath maintained at a specific temperature and, after equilibration, a measured sample of the vapor space in the vial above the aqueous medium is sampled and injected into an appropriate gas chromatograph. Dubowski has reported a number of column packings and retention times that are appropriate for this type of analysis[15] together with relative retention times of many volatile substances that can be measured in this fashion (Table 27-5).

Methods for the Determination of Nonvolatile Organic Substances

This is the largest group of substances encountered in the clinical toxicology laboratory and includes most drugs and alkaloids, pesticides, and occupational and environmental hazards and pollutants. Extraction methods utilizing water-immiscible solvents must usually be employed to separate the substance from the specimen. However, extractions are frequently not quantitative or may result in troublesome emulsions. Some toxic substances are rapidly metabolized, excreted, or bound to protein, making their detection difficult. Many substances are chemically similar to naturally occurring compounds and must be differentiated from them by purification steps. Because of the frequent importance of measuring metabolites as well as the parent compound, analysis may require additional separation procedures. Finally, contaminants that may interfere with the subsequent analysis are frequently encountered and must be separated from the analyte of interest.

Those substances that are acids or bases are usually water soluble when they are in the form of salts. Conversion of the compound to the free acid or base makes it less water soluble but more soluble in solvents used in extraction procedures. In Figure 27-2, note that if the pH of the specimen is made acidic, organic acids are converted to the

$$\underset{\substack{\text{Relatively nonpolar}\\\text{(Solvent soluble)}}}{RCOOH} \underset{H^+}{\overset{OH^-}{\rightleftharpoons}} \underset{\substack{\text{Polar}\\(H_2O\text{ soluble})}}{RCOO^-}$$

$$\underset{\substack{\text{Relatively nonpolar}\\\text{(Solvent soluble)}}}{R\text{-}NH_2} \underset{OH^-}{\overset{H^+}{\rightleftharpoons}} \underset{\substack{\text{Polar}\\(H_2O\text{ soluble})}}{R\text{-}\overset{+}{N}H_3}$$

$$\underset{\substack{\text{Nonpolar}\\\text{(Solvent soluble)}}}{RH} \underset{OH^-}{\overset{H^+}{\longleftrightarrow}} \underset{\substack{\text{Nonpolar}\\\text{(Solvent soluble)}}}{RH}$$

$$\underset{\substack{\text{Polar}\\(H_2O\text{ soluble})}}{H_3\overset{+}{N}\text{-}R\text{-}COOH} \underset{H^+}{\overset{OH^-}{\rightleftharpoons}} H_2N\text{-}R\text{-}COOH \underset{H^+}{\overset{OH^-}{\rightleftharpoons}} \underset{\substack{\text{Polar}\\(H_2O\text{ soluble})}}{H_2N\text{-}R\text{-}COO^-}$$

Figure 27-2. Effects of manipulating pH of solvent for separation of nonvolatile organic substances by solvent extractions. R-COOH, R-NH_2, RH, and HN_2-R-COOH refer to acidic, basic, neutral, and amphoteric drugs.

relatively nonpolar, free acid form that is soluble in an appropriate organic solvent. Conversely, if the specimen is rendered alkaline, organic bases are converted to the relatively nonpolar free base form that is soluble in organic solvents. Neutral compounds, obviously, exist in the same relatively nonpolar form regardless of the pH of the medium. Generally, these are more soluble in the organic solvent used for extraction. The most difficult organic substances to extract effectively are amphoteric in nature. As can be seen in Figure 27-2, either acidification or alkalinization may result in polar, water-soluble forms that are not appreciably soluble in organic solvents. For these substances a specific pH must be found at which the compound predominantly exists in the relatively nonpolar, solvent-soluble form.

Utilizing the pH-dependent solubility properties of nonvolatile organic substances, it is possible to separate chemical species in a specimen by differential extraction (Figure 27-3).

Clean-up Procedures

Differential extractions, as described above, constitute a partial removal of interfering substances. Most analytical procedures currently in use that require a preliminary extraction step rely on a chromatographic step in order to remove interferences and quantitate the analyte simultaneously. Thus, in analytical procedures based on TLC, HPLC, or GLC, additional clean-up procedures may not be necessary.

Toxic substances that must be measured at extremely low concentrations (in the parts per billion or parts per trillion range) need additional clean-up steps. Some nonvolatile, organic substances have become ubiquitous in the environment; they are found as contaminants in reagents and glassware as well as in the patient's specimen. Phthalate esters used as plasticizers, polychlorinated biphenyls, pesticides, and other pollutants in the environment fall into this category. Most of the common clean-up procedures are chromatographic in nature and, indeed, some utilize serial chromatographic separations.

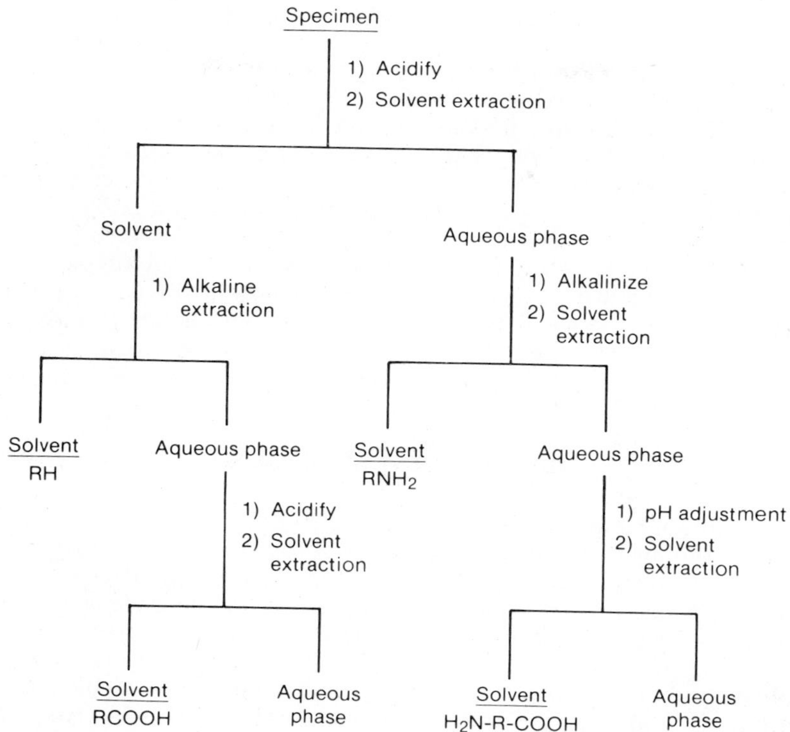

Figure 27-3. A scheme of separation for nonvolatile organic substances by differential solvent extraction. For abbreviations see Figure 27-2.

Separation of Inorganic Toxic Agents

For those inorganic toxic substances that are stable at elevated temperatures, separation can be accomplished relatively simply by burning away the organic material in the specimen and measuring the component of interest in the inorganic residue. The combustion process may be either a wet digestion with strong, oxidizing acids or a dry ashing procedure in a flame or furnace. Some metals and their salts are volatile at high temperatures, and special precautions must be taken to avoid loss during the ashing step. Others are present only in trace quantities and are subject to loss or contamination. Many instrumental procedures, such as atomic absorption spectrometry or anodic stripping voltammetry, do not require extensive specimen manipulation, since the method itself accomplishes either destruction of organic material at elevated temperatures or a separation of the analyte from the matrix. Application of atomic absorption is discussed in Chapter 17. Some specimens of importance in toxicological analysis require specific preparation procedures. This is particularly true of solid specimens such as hair, nails, bone, or other tissues. *Dry ashing procedures* are restricted to those elements that exist in chemical forms and that are not volatile at temperatures of ~500 °C. Generally, the specimen is weighed into a silica crucible, treated with appropriate reagents in order to ensure that the element is in the proper oxidation state, and then placed into a cold muffle furnace. The temperature is slowly raised and held at a high level until all organic material is oxidized. The residue is then analyzed. The *wet digestion procedure* is applicable to those elements that may be volatilized at high temperature. Generally, the specimen is weighed into a quartz or borosilicate glass flask and treated with a mixture of nitric acid and a small amount of perchloric and sulfuric acids (4:1:1). The mixture is heated slowly to the boiling point of sulfuric acid with occasional addition of oxidizing acids until no further charring or darkening of the liquid occurs. By connecting the outlet of the flask to a water aspirator pump, acid fumes can safely be diverted to the drain. The resulting solution is neutralized prior to further analysis.

Dialysis and Membrane Filtration

Some toxic substances are difficult to separate by standard procedures, either because of their physical properties or because the substance itself is so labile that it becomes chemically modified during the separation process. Some toxic anions fall into this category, and dialysis has been used successfully for these separations. Conventional dialysis across a semipermeable membrane is effective but time consuming. More recently, a technique has been applied that separates compounds by centrifugation or vacuum filtration through a membrane or special filter that permits passage of substances of limited molecular size. This technique has also been suggested for the separation of protein-bound substances in plasma from those in the free unbound state. The rapidity with which this type of procedure can be carried out makes this means of separation particularly attractive for the clinical laboratory. Few studies have been conducted, however, on the recovery of analytes by this separation procedure.

Immunoassay Procedures

The application of immunoassay procedures to the detection of drugs in biological specimens has developed into an important area of toxicological analysis. Small volumes of specimens and a minimum of specimen preparation, frequently with no separation step, are required. Since few manipulations are involved, the methods are easily adaptable to automation for screening large numbers of patients. The principles of this technique are described more completely in Chapter 3. Although immunoassay procedures are extremely valuable in laboratories engaged in therapeutic drug monitoring, they have had only limited application in toxicological analysis. This is primarily due to the following factors:

1. The nature of the toxic substances must be known and an immunoassay kit must be available for such an assay.
2. Although immunoassays are highly specific, frequently there are cross reactions with structurally similar compounds or metabolites.
3. When immunoassays are performed on a single specimen, the cost of this assay may be high because standards and controls are included with each assay. If the test is performed infrequently, the limited shelf life of reagents also increases the cost.

For some toxic substances, immunoassays are the only practical approach for the clinical laboratory. This applies to such substances as cardiac glycosides, lysergic acid diethylamide (LSD), and Δ^9-tetrahydrocannabinol (THC). LSD and THC are only rarely encountered as causes of acute toxic emergencies.

Detection of Toxic Substances by Specific Procedures

Recommending specific procedures for every toxic substance is impractical, since hundreds of methods are available and for each substance several analytical methods may be equally meritorious. In addition, methods may be quickly outdated as new knowledge is gained and technical advancements are made. Therefore, references should be consulted when specific procedures are needed.[1,2,42,44]

The methods outlined or cited in this chapter have the merit that they perform as described and have been used in the authors' laboratories. In addition, they have been selected in order to illustrate a variety of analytical or instrumental applications.

CARBON MONOXIDE

Carbon monoxide, the product of incomplete combustion of organic substances, is the most common of the gaseous poisons. It is present in the free state in manufactured gas (coal gas) but not in natural gas. Furnaces and other appliances, as well as internal combustion engines, release carbon monoxide as one of the combustion products and are frequently causes of carbon monoxide poisoning when used under conditions of poor ventilation or improper operation.

Carbon monoxide combines reversibly with hemoglobin in a manner almost identical to oxygen, but with a bond that is about 210 times stronger. As a result, carbon monoxide is not readily displaced from hemoglobin except at high oxygen tension; accidental poisonings can occur even with prolonged exposure to low levels of carbon monoxide in the atmosphere. In these instances the carbon monoxide level in blood builds up slowly until toxic levels are reached.

The detection and estimation of carbon monoxide in biological specimens can be approached in two general ways: (1) release of the gas from the hemoglobin complex with subsequent direct or indirect measurement of the gas, or (2) estimation of carboxyhemoglobin by its typical color or absorption bands. The first approach can be carried out by gasometric techniques, gas chromatography,[22] microdiffusion (pages 878–881), or infrared spectrophotometry.[41] The second approach utilizes spectrophotometric[46] or spectrographic analysis or simple color comparison.

Specimen. Regardless of the analytical method used, the specimen to be analyzed must contain hemoglobin. Relatively little carbon monoxide dissolves in the aqueous or lipid fractions of tissue, compared with that bound to hemoglobin. This rather obvious statement is made since some clinicians request serum, spinal fluid, or even urine carbon monoxide levels. The most satisfactory specimen is whole blood. Clotted blood is less desirable and must be homogenized before analysis with a minimum exposure to air.

Determination of Carbon Monoxide by Spectrophotometry

Hemoglobin and its derivatives have characteristic absorption bands in the visible region that can be utilized to detect carboxyhemoglobin and to measure the quantity present. In alkaline solution the absorption maxima for oxygenated hemoglobin are 576–578 and 540–542 nm; for carboxyhemoglobin they are 568–572 and 538–540 nm (Figure 27-4,*A*). Deoxygenated hemoglobin has a single broad band at 555 nm (Figure 27-4,*B* and *C*). If a weakly alkaline dilution of blood is treated with sodium hydrosulfite (dithionite), oxygenated hemoglobin (and any methemoglobin present) is converted to deoxygenated hemoglobin (Figure 27-4,*C*). Carboxyhemoglobin is unaffected by such treatment (Figure 27-4,*D*).

This is the basis of several methods for the determination of percent saturation of hemoglobin by carbon monoxide. The method to be described[46] works satisfactorily with fresh, oxalated blood but is not satisfactory with postmortem blood or specimens containing denatured hemoglobin.

Principle.[46] A dilute hemolysate of blood is treated with sodium dithionite, which reduces methemoglobin and oxyhemoglobin but does not affect carboxyhemoglobin. The absorbance of this solution is measured at 541 and 555 nm, the absorbance ratio A_{541}/A_{555} is calculated, and the per cent carboxyhemoglobin is determined from the standard curve.

Reagents

1. NH_4OH, 0.12 mol/L. Dilute 15.9 mL of concentrated NH_4OH to 1.0 L with deionized water. This solution is stable.
2. Sodium hydrosulfite (sodium dithionite), reagent grade. Preweigh 10-mg portions of sodium dithionite into individual small tubes. Stopper the test tubes or cover with Parafilm.
3. Carbon monoxide. Lecture bottle (Matheson Gas Products, Division of Will Ross, Inc., East Rutherford, NJ 07073).
4. Oxygen, CP.
5. Nitrogen.

Special apparatus. A narrow-bandpass (<2 nm) spectrophotometer with 10-mm cuvets is required, although the use of a recording spectrophotometer with the same specifications is desirable.

It is imperative that the spectrophotometer be checked regularly for wavelength and spectrophotometric accuracy with appropriate calibrating filters (e.g., NBS Reference Material 930) and with liquid photometric standards (e.g., NBS Reference Material 931).

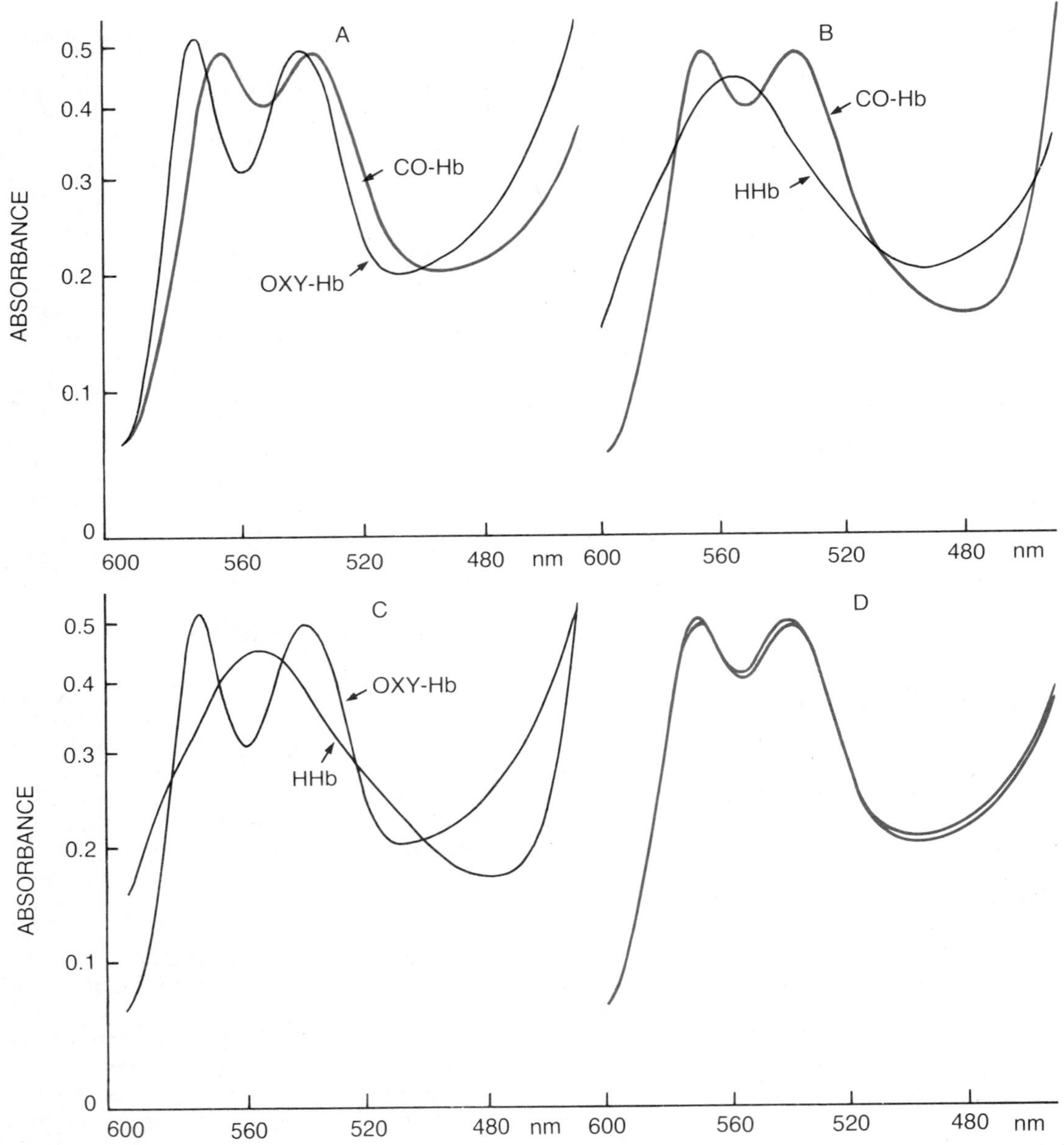

Figure 27-4. Spectral curves (overtraced) of (*A*) 100% oxyhemoglobin (OXY-Hb) and 100% carboxyhemoglobin (CO-Hb) before treatment with sodium dithionite; (*B*) 100% oxyhemoglobin (now HHb) and 100% carboxyhemoglobin after treatment of both with sodium dithionite; (*C*) 100% oxyhemoglobin before and after treatment with sodium dithionite; and (*D*) 100% carboxyhemoglobin before and after treatment with sodium dithionite. (Reproduced by permission of the Institute for Clinical Science, from Tietz, N. W., and Fiereck, E. A.: The spectrophotometric measurement of carboxyhemoglobin. Ann. Clin. Lab. Sci., *3*:36–42, 1973.)

Procedure

1. Add 100 μL of whole heparinized blood to 25 mL of NH_4OH, 0.12 mol/L. Mix the solution and allow it to stand for 2 min. This is the hemolysate.
2. Transfer 3.0 mL of NH_4OH and 3.0 mL of the hemolysate into separate 10-mm cuvets. (Analyze the sample in triplicate.)
3. At timed intervals add 10 mg of sodium dithionite to each cuvet. Cover the cuvets with Parafilm and invert gently 10 times. If a number of hemolysates are analyzed, space the addition of the reducing agent so that each can be read after exactly 5 min.

4. Exactly 5 min after the addition of dithionite to the hemolysates, read the absorbance at 541 and 555 nm against the NH_4OH blank.

5. Calculate the ratio of the absorbance at 541 nm to that at 555 nm, A_{541}/A_{555}, and determine the per cent carboxyhemoglobin from the calibration curve. *Note*: For confirmation and for the purpose of record, the sample without and with dithionite (steps 1 and 3, respectively) may be scanned between 450 and 600 nm (see Figure 27-4).

Preparation of the Standard Curve

Caution: Use a fume hood when working with carbon monoxide gas.

1. Collect 20 mL of heparinized blood from a healthy person who does not smoke.

2. Transfer a 4.0-mL portion of the fresh, heparinized blood sample into each of two 125-mL separatory funnels. Allow pure oxygen to flow through one funnel and pure carbon monoxide through the other for 15 min while the funnels are gently rotated. After the addition of the gases, close the separatory funnels and rotate them gently for an additional 15 min. Analyze the fully saturated samples immediately, in triplicate, according to the procedure given above. Use these results to establish the 0 and 100% carboxyhemoglobin calibration points. These samples may not be used to establish the intermediate calibration points.

Plot the ratio of the absorbance at 541 nm to that at 555 nm for the 0% and for the 100% carboxyhemoglobin samples and draw a line between the two points (see Figure 27-5).

3. Fill the funnel containing the 100% carboxyhemoglobin sample with nitrogen gas and rotate it for 5 min. Treatment with nitrogen removes the physically dissolved CO from the sample, but a small amount of CO will also dissociate from hemoglobin. Determine the exact carboxyhemoglobin content of this sample by the method described, using the standard curve just prepared. Prepare intermediate standards by mixing appropriate proportions of the nitrogen-treated sample with the oxygen-treated sample.

4. Analyze each of the diluted blood samples from step 3 in triplicate, according to the procedure given above.

5. Plot the calculated concentrations against the absorbance ratios obtained. These points should fall on the line drawn for the fully saturated samples, since the curve is linear over the entire range (see Figure 27-5).

Interpretation. Normal carbon monoxide levels depend on the degree of exposure to this gas without signs and symptoms of poisoning being produced. For example, in smokers levels may reach up to 8–9% saturation of hemoglobin with carbon monoxide, whereas non-smokers may show only 0.5–1.5% saturation.

In our experience, patients can survive brief periods of 70–75% saturation. Prolonged periods at these high levels can, of course, be fatal. Interpretation of lethal levels must be related to other factors in each case; i.e., time of exposure, normal hemoglobin level of the patient, age and general health of the patient, degree of activity, and so on.

In patients treated with oxygen, carbon monoxide is fairly rapidly released from hemoglobin. Frequently, a patient is treated with oxygen while being transported to the hospital. By the time a blood sample is drawn and analyzed, the carbon monoxide level may be close to normal. These patients should be kept quiet with good oxygenation to ensure that all tissue-bound carbon monoxide (e.g., with myoglobin and heme-containing enzymes) is dissipated before the patient is discharged. Effects of carbon monoxide toxicity at various carboxyhemoglobin concentrations are shown in Table 27-6.

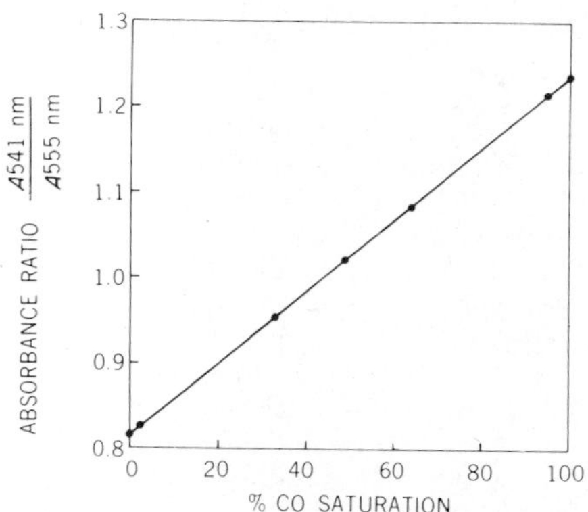

Figure 27-5. Example of a standard curve for conversion of absorbance ratio, A_{541}/A_{555}, to per cent carboxyhemoglobin saturation. (Reproduced by permission of the Institute for Clinical Science, from Tietz, N. W., and Fiereck, E. A.: The spectrophotometric measurement of carboxyhemoglobin. Ann. Clin. Lab. Sci., *3*:36–42, 1973.)

Table 27-6. CARBOXYHEMOGLOBIN EFFECTS

Carboxyhemoglobin, %	Response
10	Shortness of breath on vigorous muscular exertion
20	Shortness of breath on moderate exertion, slight headache
30	Decided headache, irritation, ready fatigue, disturbance of judgment
40–50	Headache, confusion, collapse, and fainting on exertion
60–70	Unconsciousness, respiratory failure, and death if exposure is continued
80	Rapidly fatal
Over 80	Immediately fatal

From Deichmann, W. B., and Gerarde, H. W.: Symptomatology and Therapy of Toxicological Emergencies. New York, Academic Press, 1964.

Determination of Volatile Substances

Among the volatile substances are many compounds that are present in industrial and household products. The most common of these are often amenable to specific, quantitative assay by microdiffusion technique (pages 878–880, Table 27-4) or are ideally suited to gas chromatographic assay (Table 27-5) after separation from their matrix by headspace techniques (pages 880–881). In some GC approaches, blood or urine is injected directly after dilution with internal standard.

In the following procedure, the three most commonly ingested alcohols (ethanol, methanol, and 2-propanol) and acetone, a metabolite of 2-propanol, can be determined simultaneously. Use of this procedure is particularly appropriate, since "alcohol" ingestion may frequently be missed if only ethanol is measured.

Determination of Alcohols by Gas Chromatography[42]

Principle. Poisoning by alcohols almost always results in blood or serum levels that are quite high compared with those of most other toxic agents. This fact, coupled with the low boiling points of alcohols, permits gas chromatographic analysis at low temperatures, using specimens that can be injected directly or after simple dilution. Normal constituents of biological specimens do not interfere. For example, if a flame ionization detector is used, the large excess of water in the specimen does not interfere significantly, since it will elicit only a minimal response by the detector. For greatest accuracy, the use of a recorder with an integrator is recommended. If ±10% accuracy is tolerable, measurement of peak height can be used for quantitation. 1-Propanol is used in this procedure as an internal standard.[42]

Specimen. For specimen collection see page 890.

Apparatus. Gas chromatograph, equipped with a flame-ionization detector and a glass column, 2 m by 2 mm, packed with 5% Carbowax 20 M on 60/80 Carbopack B. Column temperature, 85 °C; injection port, 100 °C; detector, 125 °C. Place a small plug of glass wool in the end of the column nearest the injection port to trap solids. The plug is easily removed and should be replaced periodically with a clean plug.

Reagents

All chemicals used should be anhydrous, analytical reagent grade.

1. Diluent stock solution, 23.4 g/L. Dilute 30.0 mL of 1-propanol to 1 L with water at 20 °C.

2. Diluent working solution, 234 mg/L. Dilute 10.0 mL of the diluent stock solution to 1.00 L with water.

3. Alcohol stock solutions. Dilute 3.00 mL each of anhydrous methanol, ethanol, and 2-propanol, respectively, to 100.0 mL with water. At 20 °C, this will result in concentrations as follows:
   a. Methanol, 23.7 g/L
   b. Ethanol, 23.7 g/L
   c. 2-Propanol, 23.6 g/L

4. Alcohol standard solutions. Dilute 1.00, 2.00, 4.00, 8.00, and 16.00 mL of each stock solution to 100.0 mL with diluent working solution. This will result in standard solutions as follows:
   a. Methanol, 237, 474, 948, 1900, and 3790 mg/L.

b. Ethanol, 237, 474, 948, 1900, and 3790 mg/L.
c. 2-Propanol, 236, 472, 944, 1890, and 3780 mg/L.

Procedure

1. Dilute the specimen 10-fold with diluent working solution.
2. Inject duplicate 1-μL aliquots of the prepared samples into the gas chromatograph, making sure that the stated operating conditions are met.
3. Inject 1-μL aliquots of each alcohol standard solution, suitably diluted with the diluent working solution.

Calculation. Calculate the ratio of the peak area or peak height of the unknown alcohol to that of the 1-propanol internal standard. Compare this ratio with similar ones obtained from the corresponding alcohol standard solutions. The ratios obtained from the reference solutions are linear with concentrations from 0–4000 mg/L.

Accuracy and precision. This method can easily detect concentrations of 100 mg alcohol/L blood. Lower concentrations can be detected by injecting larger volumes of diluted specimen into the gas chromatograph. The standard deviation of the method is ±30 mg/L when using peak-area ratios.

Interpretation. Under the stated conditions, 1-propanol has a retention time of about 5 min. The retention times for other volatile substances relative to 1-propanol are as follows: acetaldehyde, 0.16; methanol, 0.25; acetone, 0.34; ethanol, 0.5; 2-propanol, 0.66 (Figure 27-6).

ETHANOL

Ethanol (ethyl alcohol) is the most common toxic substance encountered. In the case of a patient brought to the hospital in coma, the effect of alcohol, if any is present, must be ruled out in a differential diagnosis of the cause of coma.

There are probably more published methods for the determination of ethanol in blood than for any other toxic substance. The ideal methods, as far as specificity and rapidity are concerned, are based on the use of gas chromatography. Even this procedure is not absolutely specific by itself, but it is extremely unlikely that an interfering substance would be encountered in biological specimens. A commonly used method is an enzymatic one utilizing alcohol dehydrogenase.[31] An ultraviolet spectrophotometer is necessary for this test; however, this method lends itself well to automation.

Determination of Ethanol with Alcohol Dehydrogenase (ADH)[31]

Principle. Ethanol is oxidized in the presence of ADH to acetaldehyde. In the course of this reaction, the coenzyme NAD^+ is reduced:

$$C_2H_5OH + NAD^+ \xrightleftharpoons{ADH} CH_3CHO + NADH + H^+$$

The increase in NADH is measured by the increase in absorbance at 340 nm. The equilibrium for this reaction lies strongly to the left. However, the reaction can be driven almost completely to the right by maintaining a high pH and removing the acetaldehyde as it is formed by reacting it with semicarbazide.

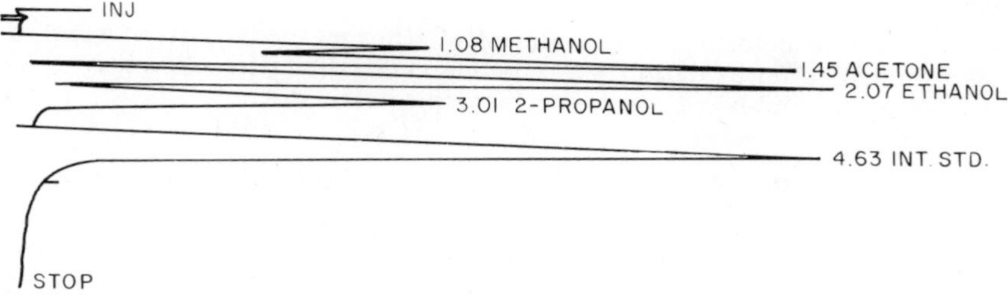

Figure 27-6. Gas chromatographic separation of some commonly encountered volatile compounds. Numbers refer to retention times.

The ADH method is not specific; other alcohols can interfere at high concentrations. Despite this drawback, it is still used in many clinical laboratories because of its convenience and the relatively rare occurrence of other alcohol intoxications.

Many modifications of this method have been reported, and several commercial kits are available. Some vendors claim to have minimized or eliminated interferences due to alcohols other than ethanol (DuPont *aca*).

Specimen. Blood, serum, urine, and saliva are all appropriate specimens for this test. The site of venipuncture, finger puncture, or ear puncture should be cleansed and disinfected with aqueous Zephiran (benzalkonium chloride), aqueous Merthiolate (thimerosal), or other suitable disinfectant. Never use alcohol or other volatile disinfectants. If reusable containers, syringes, and needles are employed, they must not be cleaned or stored with alcohol or other volatile solvents. Fluoride, 5 mg/mL blood, is the best preservative, but citrate, oxalate, and heparin can be used as anticoagulants. All specimens must be stored under refrigeration until they are analyzed.

Procedure. For details, see the insert with the kit to be used.

Comments. This is an extremely sensitive test. Alcohol from outside sources must be carefully avoided as a contaminant.

Temperature and reaction time are not critical since the reaction proceeds to completion. However, if the temperature is lower than 30 °C, the reaction may not be complete after 10 min. This is one reason for checking a standard with each group of tests.

Whole blood is the specimen of choice (see below). Serum or plasma will have much higher ethanol levels than blood. The saliva ethanol concentration is 1.1–1.2 times greater than that of blood, whereas the urine concentration is quite variable.

Although ethanol is the preferred substrate for this enzyme, the rate of dehydrogenation of 2-propanol is 6%, that for 1-propanol is 36%, and that for 1-butanol is 17.5% relative to that of ethanol. The rate of dehydrogenation of methanol is negligible.

Interpretation. With blood ethanol levels of 500–1000 mg/L, various signs of intoxication may be observed: flushing, loquaciousness, slowing of reflexes, impairment of visual acuity, and so on; however, there is much individual variation. Individuals with levels >1000 mg/L are considered under the influence of alcohol, and depression of the central nervous sytem is more apparent.

With higher blood alcohol levels, central nervous system impairment is more pronounced and coma may appear at levels of 3000 mg/L. Death may occur with levels above 4000 mg/L.

In some areas, state laws require that blood alcohol levels, when measured for legal purposes, be stated in terms of per cent by weight. In these cases, weighing the sample is mandatory. For clinical work, pipetting the sample is more convenient. Interpretation of blood levels is the same in each case. When serum or plasma is the specimen analyzed, concentrations of ethanol average 20% higher than in whole blood, since alcohol distributes into the aqueous phase.

METHANOL

Methanol (methyl or wood alcohol) is a widely used solvent in paints, varnishes, and paint removers. It is used alone as an antifreeze fluid and with ethanol and soap as a solid canned fuel. Poisonings are usually due to accidental ingestion by children or by alcoholics. In some areas, methanol may be a contaminant in "moonshine."

Methods for the Determination of Methanol

The most convenient and reliable method for methanol determination is gas chromatography.

Methanol can also be determined by a variety of other methods, most of which involve measuring the color intensity after oxidation of methanol to formaldehyde, followed by the development of a color by reacting formaldehyde with chromotropic acid (CTA):

$$CH_3OH + MnO_4^- + 2H^+ \longrightarrow CH_2O + MnO_2 + 2\,H_2O$$

$$CH_2O + \text{Chromotropic acid (CTA)} \longrightarrow CH_2O\text{—CTA (Unknown structure)}$$

Violet color

These methods work well since chromotropic acid is a specific reagent for formaldehyde. The microdiffusion method referred to earlier is useful for the determination of methanol; it also utilizes CTA for color development.

The CTA colorimetric procedure for methanol has two major drawbacks. First, methanol is not quantitatively oxidized to formaldehyde. After formation of formaldehyde by the oxidation reaction just noted, the formaldehyde itself can be oxidized to formic acid and further to carbon dioxide as follows:

$$CH_2O + MnO_4^- \longrightarrow CO_2 + MnO_2 + H_2O$$

Conditions must be chosen such that constant proportions of methanol are oxidized. The set conditions must be rigidly observed before quantitative results can be achieved.

Second, the presence of reducing substances other than methanol will affect the system so that the procedure is not quantitative. The presence of ethanol invalidates a methanol procedure based on oxidation followed by CTA color development, if the calibration curve has been set by using pure methanol standards. In the procedure of Hindberg and Wieth,[26] an excess of ethanol is added to both standards and unknowns. This results in a constant "interference" of a magnitude much greater than would ever be encountered in practice.

Interpretation. Methanol poisoning is considerably more dangerous than poisoning due to ethanol since methanol is metabolized in humans to formaldehyde and formic acid. The accumulation of formic and other acids severely reduces the alkali reserve, resulting in a metabolic acidosis. Furthermore, necrosis of the pancreas and serum amylase elevations have been demonstrated. Therefore, in addition to blood methanol levels, plasma carbon dioxide content, serum amylase determination, and electrolyte studies are useful laboratory tests for determining the severity of the poisoning and for following the progress of treatment.

As little as 2 teaspoonsful (10 mL) of methanol is considered toxic; fatal results have been reported with dosages between 50 and 200 mL (2 and 8 ounces). A blood level greater than 800 mg/L is dangerous to life. Metabolites of methyl alcohol can damage the optic nerve, resulting in either temporary or permanent blindness.

Treatment is twofold. First, the acidosis is treated, generally with sodium bicarbonate given both intravenously and orally. Second, ethanol is administered to saturate the alcohol dehydrogenase enzyme system. Since ethanol is the preferred substrate for this enzyme, this prevents the conversion of methanol to its toxic metabolites.

Corrosives

This group includes those strong mineral acids or fixed alkalis that produce chemical burns on contact. There are no good tests that can be carried out on blood, serum, or urine by which the type of acid or alkali can be detected and the ingested quantity estimated.

The only specimen that can be examined profitably is gastric contents. Frequently, this specimen is not available unless the patient has vomited, since gastric lavage is contraindicated in this type of poisoning. If gastric contents are available, the pH should be measured. Ions such as Na^+, K^+, Cl^-, SO_4^{2-}, and PO_4^{3-} can be demonstrated by routine methods used in the clinical laboratory. Obviously, most of the common ions would be present normally in gastric contents. To be of significance in this type of case, a large excess must be present. Since many compounds in the group of corrosives can cause major disturbances in acid-base balance, it is advisable to perform electrolyte studies on blood. Usually the clinician has evidence from lesions in the mouth and esophagus that a corrosive substance has been ingested.

Metals

All metals are toxic if a sufficient quantity is absorbed. Generally they are not encountered in their toxic form in the elemental or free state, but rather in the form of salts. The degree of toxicity of a given metal is dependent on the solubility of the salt; the greater the solubility, the more likely it is that it will be absorbed and the greater will be its toxicity. For example, barium chloride is soluble and extremely toxic, but barium sulfate is insoluble enough to be given orally as a radiopaque medium prior to X-ray examinations of the gastrointestinal tract.

ARSENIC

Arsenic, despite its reputation, is not a common poison. However, since arsenic is an ingredient in some herbicides and insecticides, accidental poisonings, both acute and chronic, may be encountered on occasion.

The symptoms of both acute and chronic arsenic poisoning can easily be confused with those of a variety of other conditions. Arsenic combines readily with proteins because of its great affinity for sulfhydryl groups. The important toxic effects of arsenic are the result of protein interactions that produce gastrointestinal irritation and irreversible inhibition of important enzyme systems. The strong affinity of arsenic for tissue proteins is responsible for its rapid removal from the blood. Blood, therefore, may not be an appropriate specimen for detection of arsenic poisoning except in those cases in which very large doses have been ingested. The specimen of choice is urine, even if 2–3 weeks have elapsed since an ingestion may have occurred.

Arsenic is a trace element normally present in all human tissues. Its concentration may be increased in specimens from persons who are occupationally or environmentally exposed to it. Accumulations associated with chronic poisoning due to exposure are best evaluated by analysis of hair. For early confirmation of acute poisoning, urine is the preferred specimen.

Determination of Arsenic

Colorimetric determination has been described by the authors elsewhere.[5] The method utilizes a previously ashed specimen that is subjected to Zn-HCl arsine generation following conversion of arsenic to the trivalent form. The arsine formed is chelated with diethyldithiocarbamate (DDC) and the absorbance of the chelate is measured spectrophotometrically.

Atomic absorption spectrophotometry[5] is more specific and more sensitive and is the method of choice in laboratories equipped with the required instrumentation.

Interpretation. Normal arsenic levels in urine are <0.05 mg/L. In cases of chronic poisoning, arsenic levels in urine will rise to 0.1 mg/L; in acute poisoning, 10.0 mg/L or more may be present.

Since arsenic is readily bound by sulfhydryl groups of protein, considerable arsenic is bound by keratin and subsequently deposited in hair and nails. This phenomenon has led to the analysis of hair and nails in an effort to detect previous exposure to arsenic. Interpretation of these analyses is difficult because of the problem of differentiating between surface contamination of the hair and endogenous arsenic. If such an examination is required, a minimum of 1.0 g of clean hair (a large handful), clipped close to the scalp, should be collected. For specimen collection, see also Chapter 17.

LEAD

Lead is one of the most serious of the metallic poisons. The usual cause of poisoning in adults is industrial exposure to inorganic or organic lead compounds; strict controls in industry and the education of workers have been helpful in minimizing these poisonings. Children are particularly susceptible to lead poisoning; unfortunately, their exposure to lead-containing paint and plaster, especially in dilapidated housing, has continued despite regulations, labeling laws, and attempts to educate the public. Lead encephalopathy in the severely poisoned child has a high mortality rate; children who survive frequently show evidence of permanent damage to the nervous system.

Lead poisoning is demonstrated by an increased concentration of lead in blood. Lead is a ubiquitous element and is normally present in trace amounts in biological fluids. Analytical procedures must be extremely sensitive and precise and be conducted with great care in order to discriminate between normal and increased concentrations or to detect change during treatment. These requirements may make lead analysis a function only of a laboratory with special experience and equipment for exacting analysis of trace metals.

Table 27-7. RELATIONSHIP OF BLOOD LEAD AND FREE ERYTHROCYTE PROTOPORPHYRIN (FEP) CONCENTRATIONS

	Interpretation	Blood Concentrations	
		Lead, mg/L	***FEP, mg/L***
I.	Normal	<0.29	<0.59
Ia.	Probable Fe deficiency	<0.29	0.60–1.89
Ib.	Probable transient or declining Pb level	0.30–0.49	<0.59
II.	Minimal elevation of Pb	0.30–0.49	0.60–1.09
III.	Moderate elevation of Pb	0.50–0.79	1.10–1.89
IV.	Extreme elevation of Pb	>0.80	>1.90
	Erythropoietic protoporphyria	<0.29	>1.90
	Probable Pb contamination—repeat Pb assay	>0.50	<1.10

Since lead inhibits the heme synthetic pathway, determination of δ-aminolevulinic acid, coproporphyrin III, or protoporphyrin IX (Chapter 25) can be used to screen adult and pediatric populations at risk. Both bench and field fluorometric methods are available for detection of elevated levels of free erythrocyte protoporphyrin. Table 27-7 illustrates relationships between blood lead and free erythrocyte protoporphyrin levels in various conditions.

Methods for Determination of Lead

Lead is best determined by atomic absorption spectrophotometry (AAS) or by anodic stripping voltammetry (ASV).

Specimen. An overriding consideration in collection of specimens for lead analysis is prevention of contamination. For blood specimens, special lead-free evacuated tubes are commercially available and recommended. Heparinized whole blood is the appropriate specimen since most of the lead in blood is located in the erythrocytes. However, lead content of the anticoagulant used should be known so that proper correction can be made.

For collection of a 24-h urine specimen, the patient should void directly into a lead-free container (a borosilicate glass or polyethylene container from which surface lead has been removed by washing, then by rinsing with hot nitric acid, 1 mol/L, and by rinsing twice with metal-free water). A preservative should not be added because it might contaminate the specimen. The entire specimen (or a minimum of 100 mL after recording the total volume) is submitted to the toxicological laboratory for analysis.

Urine stored without a preservative in clear polystyrene containers, exposed to light at room temperature, showed no significant lead change for a period of 10 days.

Atomic Absorption Spectrophotometry[1,5]

Conventional AAS necessitates processing of blood specimens prior to the actual assay. The processing may include ashing the specimen, or precipitation of protein with trichloroacetic acid followed by direct measurement of lead in the supernatant, or concentration of lead by solvent extraction followed by measurement of lead in the extract. All these approaches suffer from poor sensitivity and imprecision, require rather large samples, and are susceptible to contamination from reagents. Use of flameless (graphite furnace) AAS circumvents some of the problems since excellent results can be obtained on as little as 0.5 μL of packed erythrocytes or 1 μL of diluted whole blood with a minimum of preparation.

Anodic Stripping Voltammetry[1,5]

In ASV, a mercury electrode in contact with the solution to be analyzed is held at a negative potential greater than necessary to reduce the metal ion to the metal. Plating of the analyte metal then occurs on the mercury of the electrode. The potential is then rapidly and linearly changed in the positive (anodic) direction. When the oxidation-reduction potential of a metal/metal ion couple is reached, the metal is reoxidized to the ion and passes into solution, causing a measurable flow of current. The potential at which the peak current occurs is characteristic of the metal, and the magnitude of the current flowing at that potential is proportional to the amount of the metal ion in the solution. ASV for lead analysis has excellent sensitivity and is rapid, simple, and precise. Sample size can be small and sample manipulation can be minimized.

Stability of Standards

Lead standards deteriorate rapidly in polystyrene or borosilicate glass containers exposed to light, probably because of adsorption of lead ions on the glass or plastic surface. More concentrated solutions saturate the container surface and, after losing about 0.5 μg Pb/L, remain stable for several days. Borosilicate adsorbs more lead than polystyrene. To prevent errors, solutions containing <0.2 μg Pb/L should be prepared in a darkened room and analyzed immediately.

Reference ranges. Normal lead levels are <0.08 mg/L of urine or 0.3 mg/L of blood. Levels higher than normal indicate increased absorption of lead compounds; levels >0.1 mg/L of urine or >1 mg/L of blood are usually associated with signs and symptoms of lead poisoning. Normal blood lead levels in children are 0.15–0.20 mg/L. In this age group, levels of 0.4 mg/L represent an abnormal exposure to lead compounds. Urine levels should be reported in mg/L together with the total volume of the 24-h specimen to allow correlation of 24-h excretion of lead with clinical factors that may be related to an excessively high or low urinary output.

BROMIDE

Bromides are used in both organic and inorganic forms in medicine, chiefly as sedatives. These drugs are sometimes abused or may be taken in overdosage accidentally. The nonprescription status of drugs containing bromide makes them easily available to the patient predisposed to drug abuse. Bromide intoxication occurs easily because of slow excretion. In toxicity, delirium, delusions, hallucinations, mania, lethargy, or coma may occur; EEG changes accompany the intoxication.

Methods for estimating the blood concentration of bromide[24] have been based on the spectrophotometric properties of a gold bromide complex. Bromide, after deproteinization of plasma by trichloroacetic acid, yields a colored complex with gold chloride that has an absorption maximum at 440 nm.

$$AuCl_3 + 3\,Br^- \longrightarrow AuBr_3 + 3\,Cl^-$$

The formation of gold tribromide may also be accompanied by the formation of $AuBrCl_2$ and $AuBr_2Cl$. The resulting brown color is very stable in acid solution and can be read quantitatively in a spectrophotometer at 440 nm.

Reference ranges. Normal bromide levels in serum are 8–15 mg/L. Therapeutic levels for serum bromide are ~1000 mg/L, and toxic levels are usually >1500 mg/L.

BARBITURATES

In cases associated with drug overdose, barbiturates are leading offenders. These drugs are commonly prescribed as treatment for a variety of conditions. Because of their availability, they are frequently the cause of accidental poisoning. Since barbiturates are hypnotics (sleep-producing drugs), they are used in suicide attempts. There are many individual drugs in this group, but all are chemically similar, substituted, pyrimidine-ring compounds. All bring about sleep, and in overdose they cause coma and death.

Methods for Determination of Barbiturates

These drugs are readily extractable from blood, serum, or urine. Gas-liquid chromatography, preferably using a nitrogen-sensitive detector[6] to minimize nonbarbiturate peaks on the chromatogram, is a widely used method for detection and quantitation. Gas chromatography is particularly advantageous because a specific barbiturate can be identified in a single analytical run. Ultraviolet spectrophotometry, the classical and still very valuable alternative, can be used effectively to detect and quantitate the barbiturate group of compounds but has only limited application to identification of specific drugs in the group.

The ***ultraviolet spectrophotometric methods*** for determination of 5,5-substituted barbiturates in biological material depend on the fact that the compounds exist in three forms in solution: a nonionized form in acid solution, with almost no absorption in the range 230–270 nm; the first ionized form at pH 9.8–10.5 with an absorption maximum at 240 nm; and the second ionized form at pH 13–14 with an absorption maximum at 252–255 nm and a minimum at 234–237 nm. The three forms of the 5,5-disubstituted drugs can be represented as follows:

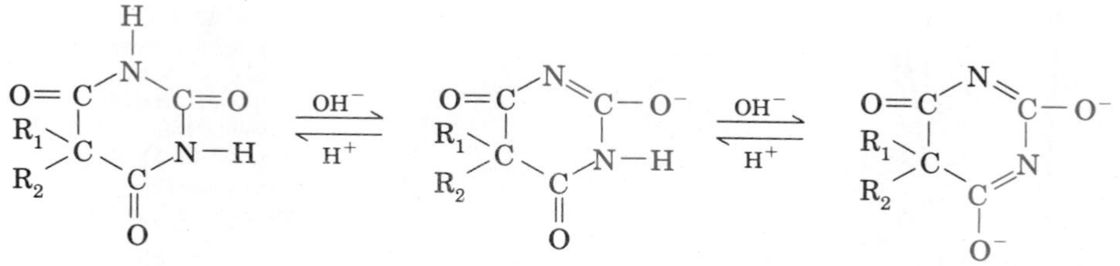

Acid form *First ionized form* *Second ionized form*

The 1,5,5-trisubstituted barbiturates exist only in two forms in solution, since they lack one enolizable hydrogen. The nonionized form in acid solution has almost no absorption in the range of 230–270 nm; the first and only ionized form at pH 9.8–14 has an absorption maximum at 245 nm.

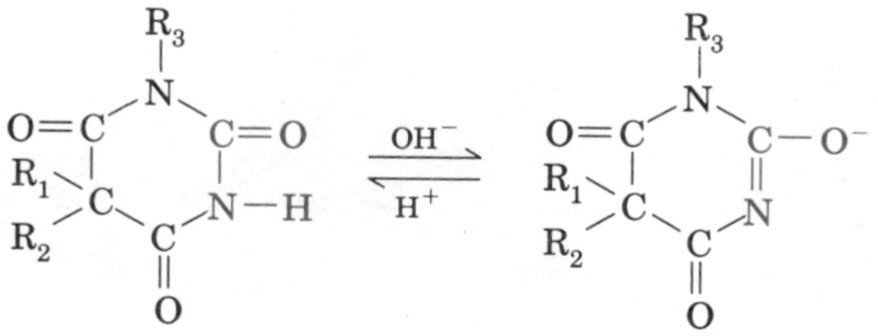

Acid form *Ionized form*

In the acid form, barbiturates are relatively water insoluble, but they are soluble in organic solvents. In both the first and second ionized forms these drugs are very water soluble but are insoluble in organic solvents. Thus, they can be extracted from blood or serum at physiological pH values or from acidified urine by organic solvents. By washing the organic solvent with an aqueous phosphate buffer of pH 7.4, some interfering impurities (e.g., salicylates) can be removed, although some loss of barbiturates occurs. Shaking the organic solvent with dilute alkaline solution converts the free acid form of barbiturate into its salt, resulting in the transfer of the barbiturate into the aqueous phase. This aqueous extract is used for scanning in the ultraviolet spectrophotometer (pH 14); equal volumes of this solution and H_3BO_3-KCl, 0.6 mol/L, result in a pH 9.9 solution for the second UV scan (see Figure 27-7).

The criteria for the presence of barbiturates are as follows:

1. In pH 9.9 solution: a maximum absorbance at 238–240 nm.
2. In pH 14 solution: a maximum absorbance at 252–255 nm.
3. In pH 14 solution: a minimum absorbance at 234–247 nm.
4. Isosbestic points: 227–230 nm and 247–250 nm.

Note: Isosbestic points are points on a spectral curve at which two substances have equal absorbance at the same wavelength.

Dilute solutions of barbiturates may not give all of the characteristic points, and

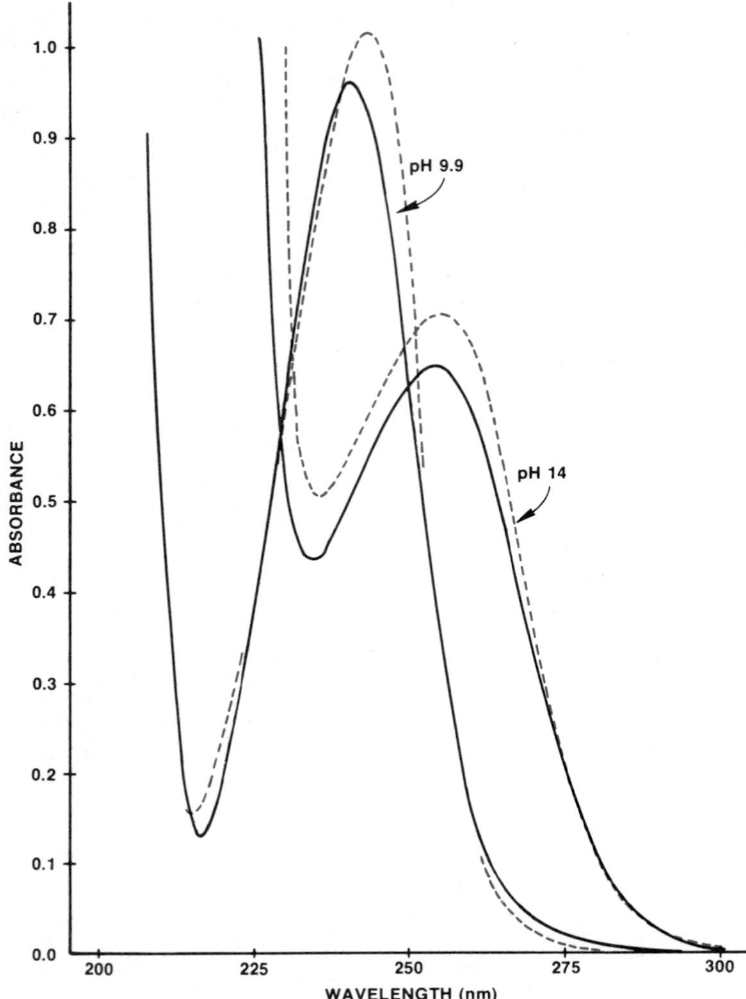

Figure 27-7. Typical ultraviolet absorption curves of secobarbital in alkaline extract before (dashed lines) and after (solid lines) hydrolysis. Note that a pH change from 9.9 to 14 shifts absorbance maxima upscale for both the hydrolyzed and unhydrolyzed species.

the absorption peaks of salicylates and sulfonamides may obscure some of them. The phosphate wash of the extract (described in the complete method)[5] should remove most of the interferences.

Proper interpretation of a blood or serum barbiturate level cannot be done unless the type of barbiturate present is known. For example, a 10 mg/L concentration of barbital is not too serious, but the same level of secobarbital is close to a lethal level. Also, 20 mg/L of phenobarbital in blood is a therapeutic level, but the same level of secobarbital can be lethal.

Since the prognosis in a given case of overdose is influenced by the type of barbiturate involved, it is important to identify the drug or to determine the type of barbiturate present. This is done by treatment of the barbiturate with hot alkaline solution, by thin-layer chromatography, or by GLC.

About 29 different barbituric acid derivatives are, or have been, used clinically. Of these, only about six or eight are commonly prescribed and available to the general public.

1. *Long-acting*: barbital, phenobarbital, mephobarbital, diallylbarbituric acid.
2. *Intermediate-acting*: amobarbital, aprobarbital, butabarbital, hexethal.
3. *Short-acting*: cyclobarbital, pentobarbital, secobarbital.
4. *Ultrashort-acting*: hexobarbital, thiamylal, thiopental.

The ultrashort-acting barbiturates are used exclusively as anesthetic agents and consequently are rarely encountered in cases of overdosage. Thus, the balance of this

Table 27-8. TOXIC LEVELS IN BLOOD FOR COMMON BARBITURATES

Generic Name (Trade Name)	Blood Level When Consciousness Regained, mg/L
Barbital (Veronal)	80
Phenobarbital (Luminal)	50
Butabarbital (Butisol)	30
Amobarbital (Amytal)	30
Pentobarbital (Nembutal)	10
Secobarbital (Seconal)	10

discussion is restricted to the classification of the first three groups: long-acting, intermediate-acting, and short-acting barbiturates.

Short-acting barbiturates may be distinguished from long- and intermediate-acting compounds by heat treatment of the alkaline extract. The short-acting compounds are more stable than the others to heat-alkali treatment because they do not hydrolyze as readily to malonic acid and urea. Absorbance scans measured before and after heating are compared. The extent of the decrease of UV absorbance after heat-alkali treatment is evidence of the presence of long-, intermediate-, or short-acting barbiturates.

Interpretation of results. Blood barbiturate levels must be interpreted cautiously for the following reasons:

1. There is variation in the response of different individuals to a given dose of any drug.

2. A given concentration of barbiturate has a more profound effect when the blood level is rising than when the blood level is falling.

3. The effects observed in the patient will be more severe if other depressant drugs, particularly alcohol, are present in addition to the barbiturates.

4. Some individuals who are tolerant of or addicted to barbiturates may have high blood levels without obvious effects.

Table 27-8 shows approximate blood levels at which consciousness is regained when the blood levels of various barbiturates are declining. Prognosis is poor for patients with concentrations above the indicated levels because such concentrations are associated with fatalities. Except in epileptics treated with phenobarbital, therapeutic barbiturate levels in blood rarely exceed 1–2 mg/L.

SALICYLATES

Aspirin is responsible for more cases of accidental poisoning in children than any other substance. Toxic doses of salicylates initially produce a stimulation of the central nervous system. This may be reflected by hyperventilation, flushing, and fever. Unfortunately, an unrecognized case of salicylate poisoning may be thought to be a case of infection and further aspirin given in a vain attempt to control the fever. Central nervous system stimulation is followed by depression and a complex disturbance of acid-base balance. Initially a respiratory alkalosis caused by severe hyperventilation occurs, but this may be followed, especially in infants, by a metabolic acidosis due in part to the absorbed drug and in part to accumulation of other metabolic acids.

Methods for the Determination of Salicylates

A common and convenient ***spectrophotometric determination*** of salicylate in urine, serum, or other specimens is based on the formation of a violet-colored complex between Fe(III) ions and phenols.

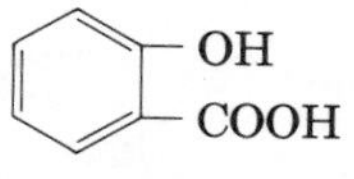

$$C_6H_4(OH)(COOH) + Fe^{3+} = \text{Violet color complex}$$

Salicylic acid

The test is not specific for salicylates, but false-negatives do not occur at toxic levels.

HPLC methods are more specific, and most can simultaneously detect and quantitate acetaminophen and phenacetin. An example is the method[5] in which salicylic acid and acetaminophen are extracted into an organic solvent containing 8-chlorotheophylline as internal standard. The concentrated extract is analyzed by liquid chromatography with ultraviolet absorbance detection at 248 nm. Calculation of concentration is based on a response factor determined from a standard curve. Normal plasma constituents and eight other commonly used drugs do not interfere with the assay, but theophylline co-elutes with salicylate and thus interferes.

Interpretation of results. Therapeutic levels of salicylic acid rarely rise above 200 mg/L in blood or serum. Above 300 mg/L, toxic symptoms such as headache, tinnitus, flushing, and hyperventilation may be seen. Lethal salicylate levels are usually >600 mg/L.

PROPOXYPHENE

Propoxyphene exists in two optically active forms. The dextrorotatory form is a narcotic analgesic marketed as Darvon, whereas the levorotatory form is available as Novrad, an antitussive. The isomers cannot be differentiated by usual chemical means; therefore, levels are reported as propoxyphene. Quantitative methods for assaying blood and tissue levels of this drug have been plagued by many difficulties. Unlike most organic bases, the hydrochloride salt of propoxyphene is soluble in certain organic solvents. Normal metabolism of this drug converts it to norpropoxyphene. Although this metabolite would not be expected to differ greatly from the parent drug, strong alkalinization of norpropoxyphene causes it to rearrange, forming norpropoxyphene amide. Once the amide is extracted from an alkaline medium, it cannot be reextracted from the organic solvent by acid solutions.[32] These factors, plus protein binding of the drug, have contributed to the difficulties associated with assays of propoxyphene in biological specimens.

Determination by gas-liquid chromatography[5] also measures norpropoxyphene, an active metabolite. Therapeutic concentrations of propoxyphene are 0.1–0.4 mg/L. The metabolite, norpropoxyphene (measured as norpropoxyphene amide), has about one-half the analgesic potency of propoxyphene, and its therapeutic concentration is 0.1–1.5 mg/L.

AMPHETAMINES

Amphetamine and its homolog methamphetamine are members of a group of phenylethylamine derivatives that share two features: they are potent central nervous system stimulants and they are widely abused. Although amphetamine itself is no longer extensively used therapeutically, other phenylethylamines are encountered as decongestants, appetite suppressants, and antidepressants. Many are abused for their real or fancied hallucinogenic properties.

Most of these substances can be detected by similar methods. The following procedure, although specifically applied to amphetamine and methamphetamine, will detect most phenylethylamines.

Determination of Amphetamines[5]

Amphetamine, methamphetamine, and other phenylethylamines are extracted from a basic solution into toluene. The compounds are derivatized in the toluene using trichloroacetylchloride. The derivatizing reagent is removed with sodium hydroxide. A part of the toluene layer is injected into a gas chromatograph equipped with an electron-capture detector.

The following levels are seen after a single dose:

	Plasma, mg/L	Urine, mg/L
Amphetamine	0.02–0.12	1–5
Methamphetamine	0.01–0.05	0.5–4

ACETAMINOPHEN

This widely used analgesic is available without a prescription and is frequently recommended when aspirin may present problems to a patient, particularly in pediatrics or after surgery. When taken in overdosage, it may produce liver toxicity due to its conversion to a toxic metabolite. Normally, the metabolite is conjugated with glutathione and excreted, but if glutathione is depleted for any reason, excess metabolite will bind covalently to hepatic cellular components. Analysis not only confirms a suspected overdosage but, by additional analysis of timed specimens, can offer a means of estimating the half-life of the drug. A lengthened half-life indicates that liver toxicity may occur.

HPLC is the favored method for acetaminophen assay. A typical method[5,23] involves addition of an internal standard to a serum sample, which is then deproteinized with trichloroacetic acid. The supernatant is injected into a mobile phase of methanol:acetic acid:water and separated on a reverse-phase column.

Toxic manifestations of acetaminophen are observed if the serum concentration 4 h after an overdose ingestion is > 300 mg/L, or if the $t_{½}$ exceeds 2 h. Antidotal treatment is then recommended. Serum concentrations > 100 mg/L may be toxic when the time of ingestion is not exactly known.

BENZODIAZEPINES

This class of compounds comprises a therapeutically useful group of structurally related substances that are used as sedative, hypnotic, or anticonvulsant drugs. Therapeutic concentrations are low, and many of these drugs are transformed to active metabolites. These in turn are also used as drugs. The HPLC method described is applicable to many of the currently used drugs of this group.

Determination of Benzodiazepines by HPLC[5]

Serum or urine is extracted under basic conditions into an organic solvent. The solvent is evaporated and the residue dissolved in methanol and injected into an HPLC equipped with a reverse phase column. The mobile phase is 70% methanol in water. Constituents are detected at 240 nm. Quantitation is accomplished by calculating the ratio of peak height to an internal standard and by comparing it with a standard curve.[34]

Since chlordiazepoxide is rather unstable, the standard curve should be checked with each sample run.

Drug	Therapeutic Range, mg/L	Half-Life, h
Diazepam	0.5–2.0	21–37
Nordiazepoxide	0.1–0.5	48
Chlordiazepoxide	0.5–3.0	7–14
Norchlordiazepoxide	0.2–0.7	Not known

MORPHINE, CODEINE, AND HYDROMORPHONE

Opioid analgesics comprise a large group of substances that control pain by central nervous system depression. Since the isolation of morphine in 1803, a large number of naturally derived or synthetic compounds for therapeutic analgesia have been described. Abuse of these drugs has adverse effects, even including fatalities. It is estimated that in the early 1970s, opiates were responsible for 40–50% of all drug-related fatalities annually.

The isolation, characterization, and quantitation of these drugs have been a continuing challenge. Virtually every analytical technique has been applied to this problem, but all suffered from insufficient sensitivity or poor specificity, or both. Immunoassay techniques provide the necessary sensitivity but are applicable only when the identity of the drug is known, and positive results must be confirmed by an alternative assay because of cross-reaction by metabolites.

Recently, derivatization followed by gas-liquid chromatography using nitrogen-phosphorus detectors (NPD) or electron capture detectors (ECD) has been reported. These techniques, together with gas chromatography/mass spectroscopy (GC/MS), have been reviewed elsewhere.[19] The principle of a rapid, simple, and specific GC/MS procedure for the simultaneous quantitation of morphine, codeine, and hydromorphone is described below. The procedure uses selected ion monitoring (SIM). Hydromorphone (Dilaudid) was included, since it has proved elusive in many other published assays.

Determination of Opiates

Principle. Serum or urine, previously hydrolyzed to release conjugated opioids, is adjusted to pH of 9.9. Extraction of the drugs together with an added internal standard is followed by derivatization of hydroxyl groups to fluorinated esters. GC/MS is carried out while selected ions characteristic of each compound are monitored (Figures 27-8 and 27-9).[38]

Calculation. The ions (m/z) are monitored at specific retention times. Standard curves for each drug are established with nalorphine as the internal standard by plotting peak area ratios against concentration. Analytes in unknown specimens are identified on the basis of their retention times and ion characteristics, then quantitated from peak area ratios..

Comments. The extraction efficiency for all the opiates has been found to be between 55 and 60%. The quantitative use of the internal standard nalorphine, coupled with the sensitivity of the GC/MS, obviates further extraction.

Monitoring characteristic ions at specific retention times makes selected-ion monitoring (SIM) a powerful analytical tool. The use of nalorphine and the sensitivity of response of the system make this procedure quantitative from <0.02 mg/L to at least 8 mg/L for morphine and codeine.

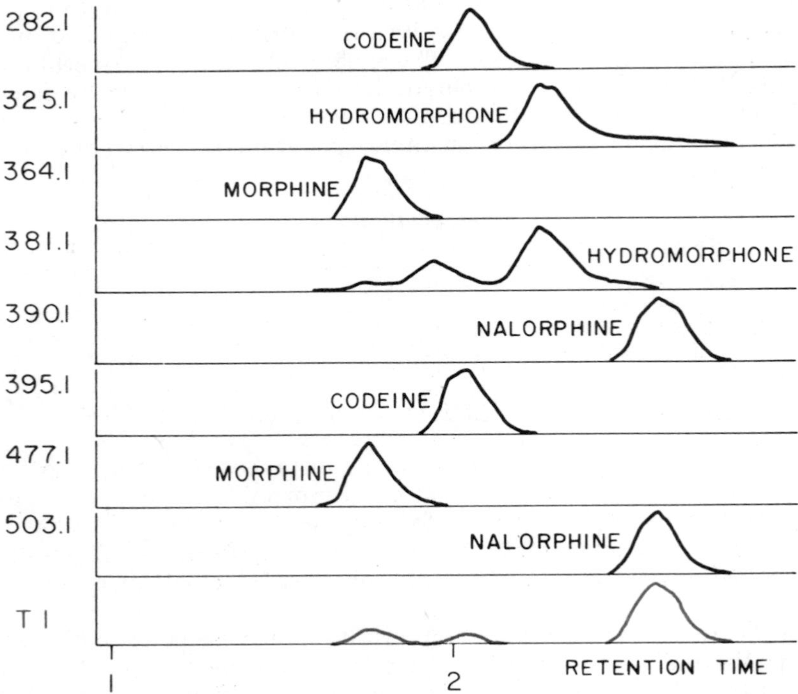

Figure 27-8. Typical selected ion and total ion chromatograms of a plasma extract containing morphine, codeine, hydromorphone, and nalorphine by using SIM/GC/MS. (Reproduced by permission of Preston Publications, Inc., from Saady, J. J., Narasimhachari, N., and Blanke, R.: Rapid simultaneous quantification of morphine, codeine, and hydromorphone by GC/MS. J. Anal. Toxicol., *6*:235–237, 1982.)

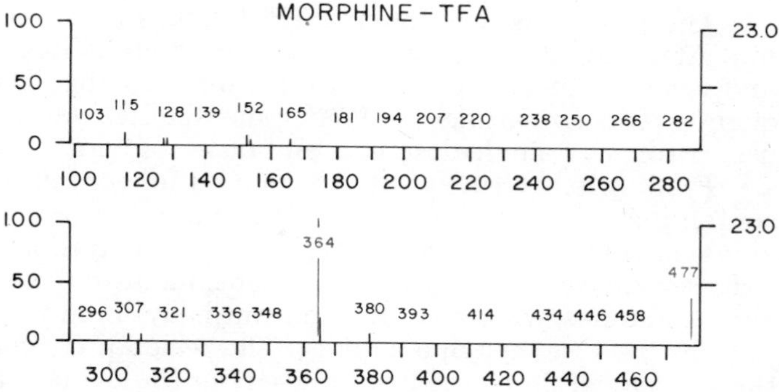

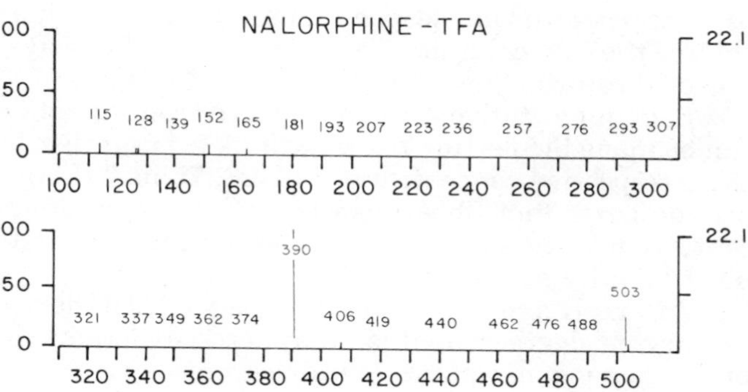

Figure 27-9. Individual mass spectra of trifluoroacetyl derivatives of morphine and nalorphine (internal standard) by electron ionization GC/MS. "Relative abundance" is given on the ordinate, and *m/z* values are seen on the abscissa. (Reproduced by permission of Preston Publications, Inc., from Saady, J. J., Narasimhachari, N., and Blanke, R.: Rapid, simultaneous quantification of morphine, codeine, and hydromorphone by GC/MS. J. Anal. Toxicol., *6*:235–237, 1982.)

Interpretation. *Therapeutic concentrations* of morphine range from 0.01 to 0.07 mg/L in serum and from 0.5 to 10 mg/L in urine. Codeine has similar concentrations in serum, but urine concentrations may range up to 30 mg/L. Hydromorphone is more potent; therapeutic concentrations in serum are <0.03 mg/L but may reach 1 mg/L in urine. Depending on the degree of tolerance, toxic concentrations generally exceed 0.1 mg/L in serum for each of the three opiates.

CANNABINOIDS

By definition, marijuana is a mixture of the dried leaves and the flowering tops of the plant *Cannabis sativa*. The primary active ingredient in marijuana is Δ^9-tetrahydrocannabinol (THC); street marijuana usually contains <5% THC. Hashish, the resinous coating from the flowers and leaves of *Cannabis sativa*, contains up to 15% THC and may be smoked as a mixture of the resin and tobacco.

Overdosage of THC is rarely severe enough to be life threatening. Furthermore, THC is almost completely metabolized. The resulting metabolites are held in fat stores and excreted in urine and feces over a long time period. Thus, to distinguish between active and inactive forms in a quantitative manner requires difficult and time-consuming procedures relying on GC/MS.[20] There is a lack of practical methods for the identification of the active ingredient in human serum or urine.

For medical-legal reasons or to assess the clinical state of a patient, it may be useful to detect recent use of THC. A homogeneous enzyme immunoassay procedure can be applied in these cases, provided that results are interpreted cautiously. The method assays the major urinary metabolites of THC and is most sensitive to 11-nor-Δ^9-THC-9-carboxylic acid. One kit is designed to detect at a sensitivity of ~20 ng/mL and to discriminate reliably between a negative specimen and one containing 50 ng/mL cannabinoids.

The peak effect of smoking THC occurs in 20–30 min and the duration is 90–120 min after one cigarette. Elevated urinary metabolites are found within hours of exposure and remain detectable for 3–10 d after smoking. The rate of metabolism and, consequently, concentrations of urinary metabolites vary widely depending on frequency of use, chronic or single-dose use, and other factors.

Principle.[37] Urine diluted with a buffer containing malate is mixed with a reagent containing the coenzyme NAD^+ and antibodies to a THC derivative. Binding occurs to any substance in the urine recognized by the antibody. A second reagent containing a THC derivative labeled with the enzyme malate dehydrogenase is then added. The labeled drug combines with any unfilled antibody–binding sites and thereby proportionally reduces the total enzyme activity. The residual enzymatic activity is directly related to the concentration of the drug present in the urine. The active enzyme converts NAD^+ and NADH, resulting in an absorbance change that is measured by a spectrophotometer at 340 nm.

Procedure. Refer to the insert in the EMIT-d.a.u. kit for details (Syva, Palo Alto, CA 94303).

Interpretation. Response of an unknown urine specimen equal to or greater than that of the low calibrator is interpreted as "positive"; a response less than that of the low calibrator is "negative."

Semiquantitative estimates of THC concentrations, in terms of analyte equivalents, can be made by plotting the A_{340} units for negative, low, and medium calibrators against their respective concentrations (0, 20, 75 ng/mL) on semilog paper. The resulting calibration curve (not linear) can be used to approximate the cannabinoid component concentration in the specimen. For appropriate interpretation, the limitations of the method must be fully appreciated.

Interferences. Although a variety of THC derivatives and metabolites will respond to varying degrees in this assay, cross-reactivity is low or absent for amphetamine, propoxyphene, morphine, benzoylecgonine, methaqualone, aspirin, amitriptyline, diazepam, meperidine, phencyclidine, and secobarbital. Recently ibuprofen has been reported as an interferent. Urine specimens high in salt content may yield false-negative results. Positive results must be confirmed by an alternative method specific for Δ^9-THC or its metabolites.

Quality Assurance

Although the topic of quality assurance is dealt with elsewhere (Chapter 8), some aspects unique to the toxicology laboratory should be stressed.

Standards and Controls

The problem of acquiring reliable and verifiable standards for drugs, metabolites, and other toxic chemicals is a serious one that does not seem to have a simple solution. Some substances are available that can be considered primary standards, but generally one must rely on commercial sources for the preparation of standards. The testing of scheduled drugs further requires that the analytical laboratory be licensed by federal and state agencies for the use of these compounds. Some drug standards, available commercially in small quantities, do not require licensure, but these place the analyst in a position of complete reliance on the distributor for authenticity and purity of these substances. Even when drug and chemical standards are acquired in the purest form available, the analyst must still confirm or determine their state of purity prior to use.[39] Many drugs have a limited shelf life and may become modified during storage. One common source of error in this regard is the uncertain amount of water of hydration of very pure crystalline material.

The most common approach to resolving the purity problem is to measure a physical property of the standard (such as melting point) and compare it with literature values. The presence of impurities can frequently be detected by a simple TLC or other chromatographic assay. Many compounds have significant ultraviolet absorbance properties. Comparing the extinction coefficient with values found in the literature is a useful check of purity or proper preparation. This offers a further advantage in that the standard stock solution can be checked periodically in order to detect any changes that might have occurred owing to evaporation or degradation of the compound.[16] In any event, it is of extreme importance to be aware that purity of standard is one of the most vulnerable factors in any assay.[8]

Suitable controls for use in clinical toxicology laboratories also present a problem. Laboratories that have a relatively high workload for a particular type of assay can prepare controls by pooling sera or urines. The mean and standard deviation can then be calculated from results obtained from this pool.

Frequently, fortifying outdated blood bank plasma with the drug or chemical to be measured and then using this as a control is the only alternative to a true control. This is obviously not identical to a "physiological" specimen, since metabolites or other interferences may be absent from this type of control. When other sources are lacking, however, this artificial control is better than none. Many commercially available controls are prepared this way. In addition, many controls contain a variety of analytes that, depending on the assay in use, may pose problems of interferences and thus must be used cautiously.

The procedures for which it is most difficult to provide adequate control material are screening tests. In these situations, the nature of the analyte may not be known until the screening procedure has been carried out on a specimen from the patient. It is impractical then to obtain or prepare a control specimen. A judicious selection of the most frequently encountered, representative analytes can be added to pooled, analyte-free specimens at concentrations near the sensitivity limit of the test in order to provide control material for these procedures. It is most important to incorporate a control material that confirms the test is operating effectively close to its limit of sensitivity.

References

1. Baselt, R. C.: Analytical Procedures for Therapeutic Drug Monitoring and Emergency Toxicology. Davis, CA, Biomedical Publ., 1980.
2. Baselt, R. C.: Biological Monitoring Methods for Industrial Chemicals. Davis, CA, Biomedical Publ., 1980.
3. Bastos, M. L., Kananen, G. E., Monforte, J. R., et al.: TLC of basic organic drugs. *In*: Methodology for Analytical Toxicology. I. Sunshine, Ed. Boca Raton, FL, CRC Press, 1975, pp. 434–442.
4. Blanke, R. V.: Correlation of urinary coproporphyrin III with lead intoxication. J. Forensic Sci., *1*(3):79–87, 1956.
5. Blanke, R. V., and Decker, W. J.: Analysis of toxic substances. *In*: Textbook of Clinical Chemistry, N. W. Tietz, Ed. Philadelphia, W. B. Saunders Co., 1986.
6. Blanke, R. V., and Saady, J. J.: Rapid simultaneous determination of six anti-convulsant drugs by GLC. Presented at the 26th SE Regional Meeting, Am. Chem. Soc., 1974.
7. Block, J. B.: The Signs and Symptoms of Chemical Exposure. Springfield, IL, Charles C Thomas, 1980.
8. Booth, E., Drofton, P., and Roberts, L. B.: The influence of standards on interlaboratory quality control programs. Clin. Chim. Acta, *55*:367–375, 1974.
9. Clarke, E. G. C.: Isolation and Identification of Drugs. London, W. Clous and Sons, 1969.
10. Comstock, E. G.: Treat the patient not the poison. Tex. Med., *65*:25–28, 1969.
11. Conway, E. J.: Microdiffusion Analysis and Volumetric Error. 5th ed. London, Crosby Lockwood & Sons, 1962.
12. Curry, A. S.: Poison Detection in Human Organs, 3rd ed. Springfield, IL, Charles C Thomas, 1976.
13. Czajka, P., and Duffy, J. P.: Poisoning Emergencies—A Guide for Emergency Medical Personnel. St. Louis, C. V. Mosby, 1980.
14. Decker, W. J., and Treuting, J. J.: Spot tests for rapid diagnosis of poisoning. Clin. Toxicol., *4*:89–97, 1971.
15. Dubowski, K. M.: Organic volatile substances. *In*: Methodology for Analytical Toxicology. I. Sunshine, Ed. Boca Raton, FL, CRC Press, 1975, pp. 407–411.
16. Elion-Gerritzen, W. E.: The absorbance of the standard: A valuable tool in quality control of spectrophotometric methods. Am. J. Clin. Pathol., *60*:493–498, 1973.
17. Feldstein, M., and Klendshoj, N. C.: The determination of volatile substances by microdiffusion analysis. J. Forensic Sci., *2*:39–58, 1957.
18. Finkle, B. S., Cherry, E. J., and Taylor, D. M.: A GLC based system for the detection of poisons, drugs and human metabolites encountered in forensic toxicology. J. Chromatogr. Sci., *9*:393–419, 1971.
19. Foltz, R. L., Fentiman, A. F., and Foltz, R.: GC/MS assays for abused drugs in body fluids. NIDA Research Monograph 32, DHHS publication number (ADM) 80–1014, 1980.

20. Foltz, R. L., McGinnis, K. M., and Chinn, D. M.: Quantitative measurement of Δ^9-tetrahydrocannabinol and two major metabolites in physiological specimens using capillary column GC-negative ion chemical ionization mass spectrometry. Biomed. Mass Spectrom., *10*:316–323, 1983.
21. Forney, R., Jr.: Modified Davidow TLC drug screening procedure. *In*: Methodology for Analytical Toxicology. I. Sunshine, Ed. Boca Raton, FL, CRC Press, 1974, pp. 443–446.
22. Goldbaum, L. R., Schoegel, E. L., and Dominguez, A. M.: Application of gas chromatography to toxicology. *In*: Progress in Chemical Toxicology, Vol. 1. A. Stohman, Ed. New York, Academic Press, 1963, pp. 11–52.
23. Gotelli, G. W., Kabra, P. M., and Marton, L. J.: Determination of acetaminophen and phenacetin in plasma by HPLC. Clin. Chem., *23*:957–959, 1977.
24. Gray, M. G., and Moore, M.: Blood bromide determinations: Their use and interpretation. J. Lab. Clin. Med., *27*:680–686, 1941.
25. Handbook of Common Poisonings in Children. HEW Publication No. (FDA) 76–7004, 1976.
26. Hindberg, J., and Wieth, J. O.: Quantitative determination of methanol in biologic fluids. J. Lab. Clin. Med., *61*:355–362, 1963.
27. Jain, N. C.: Direct blood injection method for gas chromatographic determination of alcohols and other volatile compounds. Clin. Chem., *17*:82–85, 1971.
28. Kabra, P. M., Stafford, B. E., and Marton, L. J.: Rapid method for screening toxic drugs in serum with liquid chromatography. J. Anal. Toxicol., *5*:177–182, 1981.
29. Kaye, S.: Emergency Toxicology. 5th ed. Springfield, IL, Charles C Thomas, 1970.
30. Key, M. M., et al., Eds.: Occupational Diseases: A Guide to Their Recognition. NIOSH Publication No. 77–181, 1977.
31. Lundquist, F.: Ethyl alcohol in blood and tissue. *In*: Methods of Biochemical Analysis, Vol. 7. D. Glick, Ed. New York, Interscience Publ., 1959, p. 217.
32. McMahon, R. E., Ridolfo, A. J., Culp, H. W., et al.: The fate of radiocarbon-labeled propoxyphene in rat, dog and human. Toxicol. Appl. Pharmacol., *19*:427–444, 1971.
33. Meade, B. W., Widdop, B., Blackmore, D. J., et al.: Simple tests to detect poisons. Ann. Clin. Biochem., *9*:35, 1972.
34. Peat, M. A., and Kopjak, L.: The screening and quantitation of diazepam, fluorazepam, chlordiazepoxide and their metabolites in blood and plasma by electron capture gas chromatography and high pressure liquid chromatography, J. Forensic Sci., *24*:46–54, 1979.
35. Pedetski, H. M., Koerner, T. A., Hughes, J. R., et al.: Osmometry in the evaluation of alcohol intoxication. Clin. Toxicol., *5*:343–363, 1972.
36. Peel, J. W., and Perrigo, B. A.: Practical gas chromatographic screening procedure for toxicological analysis. Can. Soc. Forensic Sci. J., *9*:69–74, 1976.
37. Rodgers, R., Crowl, C. P., Einstad, W. M., et al.: Homogeneous enzyme immunoassay for cannabinoids in urine. Clin. Chem., *24*:95–100, 1978.
38. Saady, J. J., Narasimhachari, N., and Blanke, R. V.: Rapid simultaneous quantification of morphine, codeine and hydromorphone by GC/MS. J. Anal. Toxicol., *6*:235–237, 1982.
39. Saady, J. J., Narasimhachari, N., and Friedel, R. O.: Unsuspected impurities in imipramine and desipramine standards and pharmaceutical formulations. Clin. Chem., *27*:343–344, 1981.
40. Smith, R. V., and Stewart, J. T.: Chromatographic methods. *In*: Textbook of Biopharmaceutic Analysis. Philadelphia, Lea & Febiger, 1981, pp. 51–56.
41. Stewart, R. D., and Erley, D. S.: Detection of toxic compounds in humans and animals by rapid infrared techniques. J. Forensic Sci., *8*:31–45, 1963.
42. Sunshine, I., Ed.: Methodology for Analytical Toxicology. Boca Raton, FL, CRC Press, 1975.
43. Sunshine, I.: TLC for weak acids, neutrals and weak bases. *In*: Methodology for Analytical Toxicology. I. Sunshine, Ed. Boca Raton, FL, CRC Press, 1975, pp. 412–413.
44. Sunshine, I., and Jatlow, P., Eds: Methodology for Analytical Toxicology, Vol. II, Boca Raton, FL, CRC Press, 1982.
45. Taylor, J. R., Calabrese, R. P., and Blanke, R. V.: Organochlorine and other insecticides. *In*: Intoxications of the Nervous System. Part I. Vol. 36 of Handbook of Clinical Neurology, P. J. Vinken and G. W. Bruyn, Eds. New York, North-Holland Publ. Co., 1979, pp. 391–455.
46. Tietz, N. W., and Fiereck, E. A.: The spectrophotometric measurement of carboxyhemoglobin. Ann. Clin. Lab. Sci., *3*:36–42, 1973.
47. Wahl, K., and Rejent, T.: Identification of drugs of abuse in urine using single development thin-layer chromatography. J. Anal. Toxicol., *3*:216–217, 1979.
48. Zweig, G., and Sherma, J., Eds.: Handbook of Chromatography, Vols. I and II. Boca Raton, FL, CRC Press, 1972.

Additional Reading

Quality Control and Proficiency Testing:

Boone, D. J., Guerrant, G. O., and Knouse, R. W.: Proficiency testing in clinical toxicology: Program sponsored by the Center for Disease Control. J. Anal. Toxicol., *1*:147–150, 1977.
Field, P. H.: Quality assurance and proficiency testing. *In*: Introduction to Forensic Toxicology. R. H. Cravey and R. C. Baselt, Eds. Davis, CA, Biomedical Publications, 1981.

Jain, N. C.: Blind proficiency testing in urine drug screening: The need for an effective quality control program. J. Anal. Toxicol., *1*:142–146, 1977.
Jatlow, P.: Proficiency testing in clinical toxicology: An overview. J. Anal. Toxicol., *1*:109–110, 1977.
Walberg, C. B.: Proficiency assessment programs in toxicology. J. Anal. Toxicol., *1*:105–108, 1977.

Medical-Legal Considerations:
Baselt, R. C., and Cravey, R. H.: Chapter 26, Forensic toxicology. *In*: Casarett and Doull's Toxicology: The Basic Science of Poisons, 2nd ed. New York, Macmillan, 1980.
Cravey, R. H., and Baselt, R. C., Eds.: Introduction to Forensic Toxicology. Davis, CA, Biomedical Publications, 1981.

28 BIOCHEMICAL ASPECTS OF PREGNANCY

MICHAEL F. GREENE, M.D.
MONTSERRAT DEM. FENCL, Ph.D.
DAN TULCHINSKY, M.D.

Biochemical Changes in Normal Pregnancy

Hormones produced throughout pregnancy are derived from maternal as well as fetal endocrine glands.[29] In addition, the placenta produces a variety of steroid and protein hormones that are secreted into both the fetal and maternal circulations. Thus, hormones and other substances found in the maternal circulation may originate in the fetal compartment.

The Placenta

The placenta is an incomplete endocrine organ and is capable of producing hormones only by utilizing either fetal or maternal precursors. As Figure 28-1 shows, maternal cholesterol is the main precursor for placental progesterone production. Dehydroepiandrosterone sulfate (DHEAS), derived from both the fetal and maternal adrenals, is the major precursor for placental estrone and estradiol production, whereas fetal 16-α-DHEAS is the main precursor for estriol production. Precursors for placental protein hormone synthesis are derived mainly from the maternal blood supply that nourishes the placenta.

Included among the major protein hormones produced by the placenta are human chorionic gonadotropin (hCG) and human placental lactogen (hPL; see Figure 28-1). Generally, hormone production by the placenta is proportional to its mass; hCG, which is produced maximally at the end of the first trimester (10-12 weeks of pregnancy), is an exception.

Although the placental protein hormones are secreted mainly into the maternal circulation, smaller amounts also reach the fetal circulation, where they can exert an effect on both the maternal and fetal endocrine systems. However, with few exceptions, protein and thyroid hormones of either fetal or maternal origin do not cross the placenta. In contrast, certain steroid sulfates and unconjugated steroids have the ability to cross from one circulation to the other. One of the exceptions is the steroid conjugate DHEAS.

Maternal Endocrine Changes

The maternal pituitary. Increasing estrogen levels throughout pregnancy significantly increase the number of pituitary prolactin-producing cells (lactotropes); prolactin (PRL) secretion increases up to 10-fold. Conversely, the high estrogen levels during pregnancy suppress secretion of luteinizing hormone (LH) and follicle-stimulating hormone (FSH) to undetectable levels. Baseline levels of other pituitary hormones such as thyroid-stimulating hormone (TSH) and adrenocorticotropic-stimulating hormone (ACTH) remain unchanged (see Table 28-1), but growth hormone (GH) response to provocative stimuli is blunted.

The maternal thyroid and parathyroid. No clear effect of pregnancy on thyroid function has been demonstrated. Although high levels of thyroxine-binding globulin (TBG) during pregnancy raise the concentrations of total thyroxine (T_4) and triiodothyronine (T_3) in blood by ~50%, no change is noted in the free T_4 and T_3 concentrations, which represent the active forms of these hormones. In contrast, throughout pregnancy there is an ~30–50% increase in plasma level of parathyroid hormone (PTH) with almost

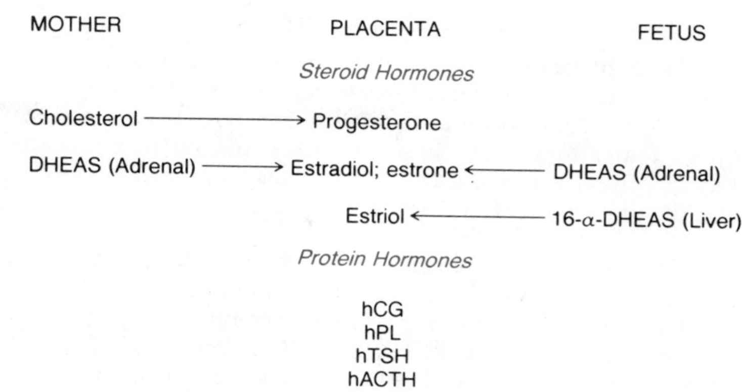

Figure 28-1. Schematic representation of steroid and protein hormone production by the placenta. The placenta is known to produce many other protein hormones.

no change in plasma ionized calcium, thus suggesting a new "set point" for the secretion of PTH. Calcitonin does not increase predictably during pregnancy, whereas vitamin 1,25-$(OH)_2D_3$ is increased during pregnancy to promote increased intestinal calcium absorption.

The maternal adrenal. Although some of the increase in plasma cortisol is due to increased CBG and a decreased metabolic clearance rate of cortisol, there also appears to be an increased production of cortisol and a twofold increase in production of dehydroepiandrosterone sulfate (DHEAS). The diurnal rhythm of cortisol with higher AM than PM levels is maintained, but the absolute plasma level of cortisol doubles in pregnancy, mainly because of an increased plasma concentration of sex hormone–binding globulin (SHBG). Increased plasma aldosterone as well as increased deoxycorticosterone concentrations are also observed.

The maternal ovaries. The maintenance of early pregnancy is dependent on the secretion of adequate amounts of progesterone by the corpus luteum. The continuous function of the corpus luteum of early pregnancy (less than ~50 days) is ensured by hCG production and stimulation until enough progesterone is produced by the placenta to assume this function.

Table 28-1. PROTEIN, STEROID, AND THYROID HORMONE CHANGES AND ASSOCIATED LABORATORY FINDINGS IN NORMAL PREGNANCY

	Nonpregnant Women	Pregnant Women at Term
LH, mIU/mL*	5–25	<2*
FSH, mIU/mL	2–15	<2
PRL, ng/mL	5–25	100–300
GH, ng/mL	<5	<7
TSH, μU/mL	<5	<5
ACTH, pg/mL	7.8–120	12–60
T_4 (total), μg/dL	5–12	7–18
T_3 (total), ng/dL	50–250	120–280
Cortisol, μg/dL	5–25	10–40
Cortisone, μg/dL	1–5	1–10
Deoxycorticosterone, ng/dL	4–16	50–210
Aldosterone, ng/dL	2–10	40–150
Estradiol, ng/mL	0.07–0.3	5–25
Progesterone, ng/mL	1–25	70–250
Parathyroid hormone, μLEq/mL†	20–60	20–120
Calcitonin, pg/mL	50–350	50–450
1,25-$(OH)_2$ D_3, pg/mL	10–90	20–200

*Since some commercially available LH assays do not distinguish LH from hCG, the "LH" values obtained in the presence of hCG may be extremely high.

†Ranges vary with the laboratory.

Changes in the Fetal Endocrine Systems

Both placental and fetal pituitary hormones are required for the normal development of the fetal endocrine system. During the *first trimester*, the fetal pituitary gland is not fully functional and pituitary-hypothalamic communications are not well developed. Thus, during the first trimester, placental human chorionic gonadotropin plays a key role in stimulating the fetal testes to produce testosterone, which is responsible for the development of the male external genitalia. No similar effect is noted on female gonadal development.

The *second trimester* is characterized by fetal pituitary hormone hypersecretion, presumably as a result of the insensitivity of the hypothalamic-pituitary axis to negative feedback inhibition by circulating hormones. The result is increased pituitary secretion of gonadotropins (LH, FSH), GH, TSH, and ACTH. At this stage of development, the secretion of FSH and LH results in oocyte development, and the secretion of ACTH and TSH ensures stimulation of fetal adrenal and thyroid gland functions.

The *third trimester* of pregnancy is marked by maturation of the fetal hypothalamic-pituitary feedback system. The presence of higher than normal levels of circulating hormones causes a decrease in the secretion of hypothalamic releasing hormones. As a result, the concentrations of most circulating pituitary hormones return to normal.

Screening of the Pregnant Patient

Effects of Pregnancy on Routine Laboratory Tests

The large amounts of estrogens, progesterone, human placental lactogen, and corticosteroids produced during pregnancy affect various metabolic, physiologic, and endocrine systems. The secretion of estrogens and progesterone throughout pregnancy ensures appropriate development of the endometrium and uterine growth, adequate uterine blood supply, and preparation of the uterus for the onset of labor. An increase in resistance to angiotensin, predominance of lipid metabolism over glucose utilization, and increased synthesis by the liver of steroid-binding proteins, fibrinogen, and other proteins are characteristic of pregnancy. As a result of such changes, many of the laboratory reference ranges for the nonpregnant state are not applicable during the pregnant state (Table 28-1).

Hematological changes. Maternal *blood volume* increases during pregnancy by an average of 45%. Plasma volume increases more rapidly than erythrocyte mass. Therefore, in spite of augmented erythropoiesis, the concentration of *hemoglobin*, the *erythrocyte count*, and the *hematocrit* commonly decrease during normal pregnancy. Hemoglobin concentrations at term average 12.1 g/dL compared with 13.3 g/dL for the nonpregnant state.

The concentrations of several blood *coagulation factors* are increased during pregnancy. Plasma *fibrinogen* increases ~50% and ranges from 300 to 600 mg/dL; the increase contributes to the increase in sedimentation rate. Other clotting factors also increase, resulting in slightly shortened one-stage prothrombin time and partial thromboplastin time. There does not appear to be a significant decrease in the number of platelets, and the clotting time of whole blood remains unchanged.

Chemical changes. During pregnancy, serum triglycerides, cholesterol, phospholipids, and free fatty acids increase by 40%. Plasma albumin is decreased to an average of 3.0 g/dL in late pregnancy as compared with the mean of 4.3 g/dL in the nonpregnant state; plasma globulins increase slightly. Alkaline phosphatase activity in serum approximately doubles, mainly due to an increase in heat-stable placental alkaline phosphatase.

Renal function. An increase in glomerular filtration rate (GFR) results in an increase in creatinine clearance rate and decreased concentrations of creatinine and urea in serum. The failure of pregnant women to excrete concentrated urine after fluids are withheld does not necessarily indicate renal damage; it may be due to altered fluid distribution. Glucosuria may be present owing to increased GFR.

Effect of Pregnancy on Endocrine Test Results

HUMAN CHORIONIC GONADOTROPIN (hCG)

Chemistry. Human chorionic gonadotropin is a glycoprotein composed of two nonidentical, noncovalently bound glycoprotein subunits, the alpha (α-) and beta (β-) subunits. When the dimer is dissociated, most of the hormone activity is lost. The amino acid sequence of α-hCG exhibits considerable homology with the α-subunits of human LH, TSH, and FSH. Some homology also exists between the peptide portions of β-hCG and β-LH subunits; 80% of the first 115 amino acids in both β-subunits are similar in sequence, but the additional 30 amino acid residues of β-hCG determine the specificity of the hormone.

Biochemistry and physiology. Human chorionic gonadotropin is synthesized in the syncytiotrophoblast cells of the placenta. The synthesis of α- and β-subunits of hCG may be independently controlled; the β-subunit is the limiting factor in the appearance of complete hCG chains. Between the first and third trimesters, the ratio of α-subunits to β-subunits increases twofold, which coincides with increasing amounts of free α-chains and decreasing amounts of free β-chains in maternal plasma as pregnancy progresses.

From the physiological point of view, hCG plays an important role in maintaining the function of the corpus luteum during the first weeks of pregnancy (i.e., until the luteoplacental shift of progesterone production has occurred). hCG appears also to promote steroidogenesis in the fetoplacental unit and plays an important role in stimulating fetal testicular secretion of testosterone.

Methods for the Determination of hCG

Several assays involving either biological or immunological techniques are available for the measurement of hCG in biological fluids (Table 28-2). Bioassays, however, have now been replaced by simpler, more rapid, accurate, and specific techniques for the direct measurement of hCG in biological fluids.

Immunochemical Techniques

Rapid Agglutination-Inhibition Tests

Specimen. For qualitative pregnancy tests, urine specimens are collected in a clean container at any time of the day. However, the first voided morning specimen is generally the most concentrated, and therefore preferred. To ensure a concentrated specimen, the patient should be instructed to restrict fluid intake from 2000 h until the morning collection. The specimen should be clear; if there is turbidity or a sediment, the specimen should be filtered or centrifuged.

Methods. In recent years, commercial kits for pregnancy tests based on hemagglutination (or latex agglutination) inhibition have become available. These may be divided into two categories: slide tests and tube tests.

The *slide test* is performed on a slide and the results are obtained within 2 min. Antiserum to hCG is added to the urine, followed by the addition of hCG-coated latex particles. If there is no hCG in the urine (nonpregnant patient), the antiserum will react with the hCG-coated latex particles and agglutination will occur (agglutination = negative test). However, in case of pregnancy, the presence of sufficient quantities of hCG in the urine will neutralize the hCG antibodies and no agglutination will occur within 2 min (no agglutination = positive test).

In the *tube test*, urine is mixed with an optimal amount of hCG antiserum in a test tube. If hCG is present in the urine specimen (pregnant patient), hCG will combine with and neutralize the hCG antibodies. When hCG-coated erythrocytes are added to the system and left undisturbed for 2 h, no hemagglutination will occur, since all hCG antibodies have been neutralized during step one. As a result, the erythrocytes will settle to the bottom of the tube in a ring pattern (positive test). The size and shape of the ring

Table 28-2. METHODS FOR hCG MEASUREMENTS

Test Type	Sensitivity, mIU/mL	Time Required for Assay	Specimen	Comments
1. In vivo bioassay	500–1000	1–5 d	Urine	Measures biological activity of hCG. Does not distinguish LH from hCG. Expensive, time-consuming.
2. Immunologic methods				
Agglutination-inhibition	1000–2000	2 min–2 h	Urine	Nonquantitative, simple, rapid; up to 5% false-positives.
β-hCG radioimmunoassay (RIA)	1–10	3–24 h	Serum	Quantitative, specific. Requires RIA equipment and provision for disposal of radioactive waste.
Rapid β-hCG RIA	25–150	2–3 h	Serum	Quantitative, specific. Requires RIA equipment and provision for disposal of radioactive waste.
Enzyme immunoassay (EIA)	5–25	3–5 h	Serum	Quantitative. Nonisotope labeling. Sensitivity depends on antibody and type of enzyme detection.
Immunoradiometric, monoclonal antibody (IRMA)	1.5–5	4 h	Serum	Quantitative, high specificity. Requires equipment for counting of radioactivity and provision for disposal of radioactive waste.
3. In vitro bioassay system				
Radioreceptor	100–200	1–2 h	Serum	Quantitative; measures biological activity of hCG. Does not distinguish LH from hCG. Requires equipment for counting and provision for disposal of radioactive waste.
Steroid response in target cells	0.1–0.5	3–5 h	Serum	Quantitative; measures biological activity of hCG. Does not distinguish LH from hCG. Requires RIA equipment and provision for disposal of radioactive waste.
Cytochemical assay	0.5	5 d	Serum	Quantitative; does not distinguish LH from hCG. Too cumbersome for routine clinical use.

formation is generally compared with a control tube prepared simultaneously with the urine sample. In the absence of hCG in the urine sample (nonpregnant patient), the hCG antibodies added are not neutralized; they will react with the hCG-coated erythrocytes, and agglutination produces a mat-like precipitate (negative test). For a semiquantitative assay by hemagglutination-inhibition, the urine is serially diluted until further dilution will not cause inhibition of hemagglutination (end point). A direct comparison with an hCG standard gives the approximate amount of hormone present in the unknown specimen.

Comments. False-positive results occur in 2–5% of cases due to the presence in urine of interfering substances such as proteins, drugs, bacteria, erythrocytes, and leukocytes. Cross-reactivity with pituitary gonadotropins may give false-positive test results (e.g., LH surge during the menstrual cycle). Conditions that denature the antibody to hCG (heat, pH, or long storage) will yield false-positive results in an agglutination inhibition test. False-negative results in the agglutination inhibition tests are common since the test usually does not detect hCG at concentrations below 1000–2000 mIU/mL. These qualitative tests, therefore, will not become positive until 8–14 d after the first missed menstrual period. They may not be sensitive enough to diagnose a normal pregnancy after the second trimester or detect abnormal or ectopic pregnancies when hCG concentrations are low. The use of known positive and negative controls and standardization of the reagents is therefore extremely important in order to obtain reliable results.

Accuracy of the various commercially available pregnancy tests ranges from 93.8 to 98.8%.

Determination of the hCG β-Subunit by Radioimmunoassay (RIA)

RIA provides a very specific assay for the measurements of hCG in blood or urine *if* the antibody used in the RIA has been raised against the β-subunit of hCG or against its carboxyl terminal peptide. This antiserum discriminates between hCG and LH, whereas most antisera produced against intact hCG do not. Moreover, the β-subunit RIA is also 100 to 200 times more sensitive than the urine slide test, and levels of hCG as low as 1 mIU/mL can be measured in serum (plasma). The sensitivity and specificity of the β-subunit RIA permit the diagnosis of very early pregnancy (6–8 d after conception) and allow the detection of low levels of hCG in patients with abnormal intrauterine or ectopic pregnancy, as well as the monitoring of patients with gestational trophoblastic diseases or hCG-secreting tumors.

Specimen. Blood specimens for the hCG immunoassays are collected into suitable tubes without anticoagulants, allowed to clot at room temperature, and centrifuged to obtain clear serum. All specimens not tested within 48 h of collection should be stored at −20 °C. As with most biological materials, repeated freezing and thawing should be avoided. Serum specimens showing gross hemolysis, gross lipemia, or turbidity may give false results.

Methods. Several commercial kits for the determination of hCG are available. These kits contain β-hCG-antibody, ^{3}H- or ^{125}I-hCG tracer, hCG standards, hCG controls, and precipitating antibodies (or other separating agents), together with detailed instructions. Most of the available kit procedures are similar. Some require a preincubation of unlabeled hormone with antibody to enhance assay sensitivity, followed by an overnight incubation (equilibration) after labeled hCG is added. Alternatively, a short incubation of 30 min, in which all the components of the reaction are incubated simultaneously in a nonequilibrium competitive-binding assay, can be used. The choice of a long or short incubation will depend upon the user's requirement for sensitivity, reproducibility, and turnaround time for reporting results.

Comments. High levels of LH should not significantly cross-react with the hCG antibody used. Assay sensitivity should be such that small amounts of hCG will be detected while false-positive results due to the presence of LH will be minimized.

Commercially available kits for hCG immunoassay are calibrated against either the Second International Standard hCG for bioassay (2nd IS-hCG), established in 1964 by the Committee on Biological Standardization of the World Health Organization (WHO), or the most recent WHO International Reference Preparation of hCG (1st IRP-hCG), established in 1974. Because the two standards are not equally pure, laboratories and clinicians must be aware of the implications of using different standards. Results obtained from methods referenced to each of these international standards are not the same and cannot be directly compared. On a weight basis, the 2nd IS-hCG standard has about one-half the biological potency of the more recently prepared 1st IRP-hCG; that is, 1 ng of 2nd IS-hCG ≃ 5 mIU, and 1 ng of the 1st IRP-hCG ≃ 10 mIU. Therefore, tests standardized with the 2nd IS-hCG standard have about one-half the value (in mIU/mL) of tests standardized with the 1st IRP-hCG. Ultimately, the reported values of hCG will have meaning only in relation to the specificity of the assay system and the reference material used to calibrate it.

Immunoradiometric assay (IRMA). Commercial kits, incorporating a mixture of monoclonal antibodies into a specific and sensitive IRMA system, are now available for the hCG assay. Some of the kits are specific for either the intact hCG molecule (Tandem hCG, Hybritech Inc., San Diego, CA 92121), the free α- or β-subunit, or a combination of both intact hCG and β-subunit (MAIA clone, Serono Diagnostics, Inc., Braintree, MA 02184). For tumor detection, this difference can be important because some tumors secrete only free β-subunits or free α-subunits. In case of pregnancy detection, a good correlation has been reported between results by IRMA kits detecting the intact hCG molecule and conventional β-subunit RIAs. The Tandem hCG incorporates two monoclonal antibodies recognizing different regions of the hCG molecule. The first antibody is coated on a plastic bead and the second (soluble) antibody is labeled with ^{125}I. Following formation of the solid-phase antibody/hCG/labeled antibody sandwich, the bead is washed to remove unbound labeled antibody. The radioactivity bound to the bead is measured; it is directly proportional to the concentration of hCG present in the specimen. The assay can detect as little as 1.5 mIU/mL of hCG (1st IRP, WHO) and it shows a linear response up to a concentration of 400 mIU/mL. Cross-reactivity with LH is stated to be <1%.

Enzyme Immunoassay (EIA) Techniques

EIA methods, which utilize enzymes as labels instead of radioactive isotopes, have proved to be a suitable alternative to RIAs. Disposal of radioactive waste is avoided and expensive equipment for measuring isotopes is not required since enzyme activity is usually determined by photometry. EIAs using monoclonal antibody technology have been introduced recently for the *qualitative* measurement of hCG in urine or serum. The Tandem ICON pregnancy test (Hybritech, Inc., San Diego, CA 92121) is a sensitive solid-phase EIA in which two monoclonal antibodies recognizing different regions of the hCG molecule are used. The procedure consists of passing the urine or serum specimen containing hCG through a membrane saturated with the first monoclonal antibody. The time required to gain maximal binding of hCG is merely the time required to pass the specimen through the membrane. The second antibody, conjugated to the enzyme alkaline phosphatase, is then added and binds to the hCG captured on the membrane. Excess unbound antibody is washed out, and substrate for the enzyme is then added. A distinct blue spot in the center of the membrane appears when the specimen contains hCG (blue spot = positive test). The lower limit of detectability of hCG by this filter membrane method is 50 mIU/mL; the total assay time is 5 minutes.

In Vitro Bioassay Systems

In traditional bioassays, the activity of the hormone is analyzed after its administration to an animal. Recently, several simplified assays have been introduced for measuring the biological activity of hCG in cell target tissues using in vitro systems. Among these techniques are radioreceptor and cytochemical assays.

The principle of the ***radioreceptor assay (RRA)*** is based on the ability of radiolabeled hCG to bind to specific receptors prepared from cell membranes of gonadal tissues and on the inhibition of this binding by unlabeled hCG present in the biological samples (e.g., Biocept G, Wampole Laboratories, Cranbury, NJ 08512). Advantages of RRA are that results can be obtained in less than 2 h and that they reflect the bioactivity of the hormone rather than its immunological properties. However, the RRA cannot distinguish hCG from hLH. High concentrations of hLH in blood at midcycle and in postmenopausal women (up to 200 mIU/mL) may give false-positive results in this assay. To circumvent this problem, the lower limit of this assay is therefore set at 200 mIU/mL. This limits the usefulness of this test for the diagnosis of very early or abnormal pregnancies.

The principle of the ***cytochemical assay***[6] is based on a biochemical response induced by hCG on target cells maintained in vitro (e.g., depletion of ascorbic acid from luteinized ovaries). By measuring such cytochemical reactions it is possible to assess very minute concentrations of hCG (e.g., 0.5 mIU/mL).

Clinical Applications

Diagnosis of pregnancy. Qualitative tests for hCG in blood or urine are primarily used for the confirmation of pregnancy. Urinary hCG determinations are usually sufficient for diagnosis of normal pregnancy when it has progressed beyond the second week following the first missed period. However, a pregnancy test using serum possesses greater sensitivity and may be advantageous for the early diagnosis of pregnancy. Moreover, quantitative and serial determinations of hCG in serum can be helpful whenever abnormalities of pregnancy are suspected and for diagnosis or monitoring treatment in patients with gestational trophoblastic neoplasm (GTN) or other hCG-producing tumors.

In normal pregnancies, the first significant rise in hCG concentration in maternal peripheral blood can be detected between 8 and 10 d after ovulation and soon after ovum

Table 28-3. REFERENCE RANGES FOR SERUM hCG IN PREGNANCY

Length of Gestation		
After Fertilization	***After LMP***	**mIU/mL**
2nd week	4th week	30–100
3rd week	5th week	100–1000
4th week	6th week	1000–10 000
2nd–3rd month	7–14th week	30 000–100 000
2nd trimester	15–26th week	10 000–30 000
3rd trimester	27–40th week	5000–15 000

The samples were assayed using the specific β-hCG assay. The Second International Standard for hCG was used as the reference preparation. LMP = last menstrual period.

implantation. Thereafter, serum and urine concentrations of hCG rise, reaching levels of 30 000–100 000 mIU/mL in serum and 100 000–600 000 mIU/mL in urine after 60–80 days of pregnancy. Subsequently, concentrations of hCG start to decline slowly in serum and urine, and by the end of the second trimester a 90% fall from its peak concentration will usually have occurred (Table 28-3).

Ectopic pregnancy. Measurements of β-hCG in serum can be useful in identifying ectopic pregnancies or abnormal intrauterine pregnancies. Patients with ectopic pregnancies often have persistently low serum levels of hCG (50–500 mIU/mL); however, concentrations of hCG depend on the size and viability of the trophoblastic tissue. Not infrequently, high levels of hCG (and a positive pregnancy test) can be associated with ectopic pregnancy. On the other hand, an undetectable hCG serum level (<1 mIU/mL) in a woman who has missed her period is quite convincing evidence that pregnancy has not occurred. However, relatively low levels of hCG can be associated with a normal early stage or abnormal intrauterine pregnancy. To further distinguish between the various possibilities, repeat measurements of hCG may be needed. In normal intrauterine pregnancy, a rapid increase of hCG will follow, whereas in cases of ectopic pregnancy or abortion, hCG concentrations will fall. However, since hCG is cleared from serum very slowly, it may be detectable in serum or urine up to four weeks after abortion. For these reasons there are occasional difficulties in interpreting results, particularly when levels of β-hCG range from 5 to 20 mIU/mL.

Trophoblastic disease. Serum hCG determinations are very useful for monitoring patients with gestational trophoblastic neoplasm (GTN) or other hCG-producing tumors, such as some lung carcinomas in male and female patients. These GTN patients characteristically have high hCG levels and low human placental lactogen concentrations. After the tumor is removed or treated with chemotherapy (or both) or after the uterus has been emptied, hCG should fall to undetectable levels within eight weeks.

ESTRIOL

Biogenesis during pregnancy. The biosynthesis of estrogens during pregnancy differs qualitatively and quantitatively from that of women who are not pregnant. In the nonpregnant state, the ovaries are the main site of estrogen synthesis; estradiol (E_2) is secreted in microgram quantities and estriol (E_3) is only a by-product of E_2 metabolism. During pregnancy, the placenta is the major source of estrogens; E_3 is produced in milligram quantities and estrone (E_1) and E_2 are produced in microgram amounts. However, the placenta, as opposed to the ovary, cannot accomplish de novo synthesis of estrogens from precursors such as acetate, cholesterol, or progesterone and has to be provided with adrenal C-19 steroid precursors of either maternal or fetal origin.

The fetus and placenta cooperate in E_3 biosynthesis during pregnancy. The fetal adrenals provide DHEAS, which is 16α-hydroxylated in the fetal liver. The fetal 16α-OH-DHEAS is then hydrolyzed by placental sulfatase and aromatized to E_3. The E_3 is then secreted into both the fetal and maternal circulations, where its measurement, mostly by RIA, was formerly used as an indicator of fetoplacental status. However, in recent years the trend in antepartum assessment has been away from biochemical evaluation and toward biophysical evaluation, such as ultrasound examinations.

Complications of Pregnancy

Erythroblastosis Fetalis

Hemolytic disease of the newborn, or erythroblastosis fetalis, is a disease of the fetus and newborn caused by an immunological incompatibility between fetal and maternal blood. Any of a large number of erythrocyte antigens may be responsible for isoimmune disease. The most common cause of severe disease is sensitization of an Rh_o-negative woman to the Rh_o(D) antigen.

Pathophysiology

If an Rh_o-negative woman, for example, is inadvertently transfused with Rh_o(D)-positive blood, she may respond by producing anti-D antibodies. With modern transfusion techniques, however, the source for sensitization in women is more likely due to exposure to Rh_o(D)-positive fetal blood from a pregnancy. Although the fetal and maternal blood compartments are generally considered to be separate during normal gestation, it is now known that fetal erythrocytes continually have access to the maternal circulation. This antigenic challenge is sufficient in some women to provoke an antibody response. Substantially larger antigenic exposures may result from disruptions in the integrity of the fetal compartment that accompany spontaneous or induced abortion, ectopic pregnancy, or normal delivery of an infant. The larger the fetomaternal hemorrhage, the more likely it is that the mother will respond to the challenge with an antibody. Other antigens of the Rh system (C, c, E, e) may also stimulate antibody formation but appear to be only about 1/30th as potent at stimulating a response as the D antigen.

The immunoglobulins produced are usually of the IgG class and cross the placenta readily. If the sensitized woman has another Rh-positive fetus, antibodies will cross the placenta and cause destruction of fetal erythrocytes. Furthermore, during pregnancy or at the time of delivery, repeat exposure to Rh-positive erythrocytes causes an augmented response resulting in higher antibody titers. This augmented response gives rise to the frequent observation that subsequent pregnancies are more severely affected than earlier ones.

The basic pathophysiological effects of all types of isoimmune hemolysis in the fetus are the same for all antigens. Differences in severity of disease are due to differences in the expression of the antigen on the surface of the cells, the intrinsic immunogenicity of the antigen, and peculiarities of the immune response of the mother.

The anemia, when severe, leads to congestive heart failure and generalized fetal edema with ascites, as well as pleural and pericardial effusions. When the fetal condition has progressed to this degree, it is referred to as hydrops fetalis and carries a very grave prognosis. The edema and effusions are readily observable by ultrasound examination. When these changes are observed and there is no therapeutic intervention, intrauterine demise will follow in a relatively short time.

With the fetus in utero, the functions of respiration and removal of bilirubinoid pigments resulting from hemolysis are performed by the placenta and mother. When a fetus is delivered and must assume these functions for itself as a neonate, hydrops is a severe handicap. The lungs are edematous, and pleural effusions and ascites physically restrict their ability to expand. The damaged liver is unable to conjugate and excrete bilirubin adequately. When unconjugated bilirubin accumulates in the blood to excessive concentrations, it passes through the blood-brain barrier to stain and destroy basal ganglia. This form of brain damage, termed *kernicterus*, may lead to cerebral palsy and deafness.

Evaluation of Erythroblastosis Fetalis

To care for sensitized women, it is first necessary to distinguish them from the general obstetric population. For this purpose, it is routine at the first prenatal visit to determine each patient's blood group and Rh status as well as to screen the patient's serum for irregular (i.e., other than anti-A or anti-B) antibodies against a standard panel of erythrocytes. If an antibody that agglutinates erythrocytes is found, it must be identified and titered. If the father possesses the antigen in question, and the maternal antibody is of the IgG class and of sufficiently high titer, then there may be a problem for the fetus. Under these circumstances, further evaluation is required.

Liley[17] noted that the degree of hemolysis in sensitized pregnancies could be assessed by spectrophotometrically measuring the absorbance of bilirubinoid pigments in amniotic fluid, which he named ΔOD_{450}, a term now replaced by ΔA_{450}. Therefore, when the titer indicates a problem, amniocentesis for quantitation of bilirubin in amniotic fluid is indicated. Since the earliest intervention by intrauterine transfusion can be done at 22 weeks of gestation, the first amniocentesis should be performed at about that time. Unless the first amniocentesis shows a very severely affected fetus requiring immediate intervention, amniocentesis will be repeated at intervals to monitor the pregnancy. The appropriate interval between amniocenteses is determined by the degree of severity of the disease and the trend with time.

Interpretation of the ΔA_{450} depends upon knowing the gestational age of the pregnancy (Figure 28-2). Values that fall into Liley's bottom zone I represent an unaffected or very mildly affected fetus. Values in the middle zone are still compatible with a minimally affected fetus, but as values rise within zone II, it is increasingly likely that the fetus is suffering moderate to marked hemolysis. Values in the top zone (zone III) indicate severe disease. Without intervention, a fetus whose values fall in the top zone will most likely die.

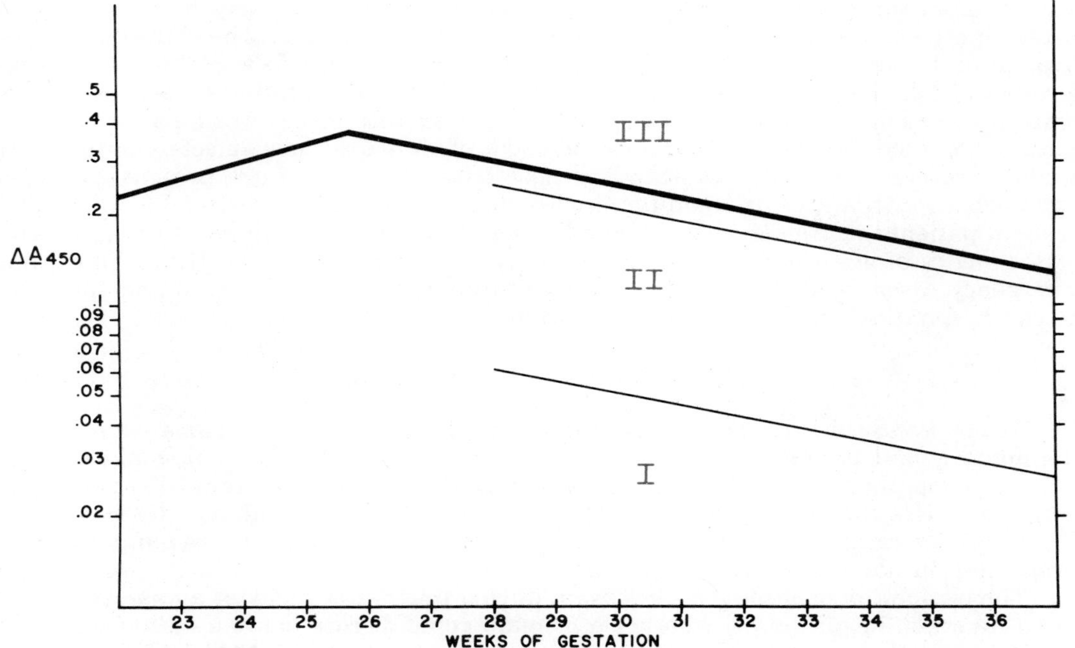

Figure 28-2. Liley's three-zone chart (with modification) for interpretation of amniotic fluid ΔA_{450}. Note that the original three zones (divided by the two lighter lines) extended only to 28 weeks of gestation. Uppermost (heavy) line is the upward revision of the "danger line" by Dr. Irving Umansky. (Redrawn from Reid, D. E., Ryan, K. J., and Benirschke, K.: Principles and Management of Human Reproduction. Philadelphia, W. B. Saunders Co., 1972.)

Therapeutic options for fetuses severely affected by erythroblastosis fetalis are limited. Intrauterine transfusion and early delivery are the two most important modes of management. Both are associated with considerable potential for fetal and neonatal morbidity and mortality. An intrauterine transfusion is performed by passing a needle, under ultrasound guidance, through the maternal abdominal wall and uterus into the fetal peritoneal cavity. Compatible erythrocytes are infused. They are absorbed into the fetal circulation across peritoneal surfaces. Once transfusions have begun, they must be continued until the fetus has reached a gestational age that ensures a high likelihood for survival in the neonatal intensive care unit. Early delivery and exchange transfusion may be a preferable alternative, depending upon the particular circumstances.

Diabetes Mellitus During Pregnancy

Several of the hormonal changes of pregnancy, especially the increased production of glucocorticoids and the production of human placental lactogen, tend to exacerbate some of the pathological changes in diabetes mellitus (DM), including hyperglycemia and the tendency toward ketosis. Modern prenatal care with emphasis on scrupulous maintenance of euglycemia has dramatically improved the outlook for pregnant diabetics. Perinatal survival of their infants now closely approximates that in the nondiabetic population. Control is established through careful diet and insulin therapy and monitored by capillary blood glucose determinations performed several times daily at home by the patient.

A particularly difficult problem in the care of pregnant diabetics is the high incidence of *congenital anomalies*. Recent work by Miller and coworkers[21] has shown that an elevated level of glycosylated hemoglobin early in pregnancy, which indicates poor control of DM, correlates significantly with a substantially increased risk for major congenital anomaly.

The *routine monitoring* and management of pregnant diabetics is somewhat more complex than for nondiabetics. In addition to day-to-day monitoring of capillary blood sugars, a serum and 24-h urine specimen should be collected early in pregnancy to determine baseline creatinine clearance and total protein loss. The diabetic patient population in our institution has an incidence of neural tube defects of 20/1000. Therefore, all of our diabetic patients are screened at 16 weeks' gestation by a quantitative maternal serum α-fetoprotein determination and an ultrasound examination. As discussed below, this will detect approximately 90% of all neural tube defects. Chromosome studies are done only if other standard obstetrical indications are present. As term approaches, pregnancies are monitored with electronic fetal monitoring. Elective deliveries in patients with otherwise uncomplicated diabetes are not done prior to 38 completed weeks of gestation. Fetal lung maturity is checked prior to delivery, in all but emergency cases, by determining the lecithin-sphingomyelin (L/S) ratio and the saturated phosphatidylcholine (SPC) concentration.

Renal Disease and Hypertension

Renal disease during pregnancy may originate from all of the same causes as in the nonpregnant patient, e.g., chronic pyelonephritis, chronic glomerulonephritis, diabetic nephropathy, and lupus nephritis. Similarly, the evaluation of renal disease during pregnancy uses the same basic laboratory determinations as in the nongravid patient. The reference ranges for renal function tests in pregnancy, however, are quite different from those in normal nonpregnant women.

The evaluation of *chronic hypertension* during pregnancy includes an assessment of renal function. Regardless of whether a primary renal disease has caused hypertension or vice versa, the implications for the pregnancy are the same. Patients with decreased creatinine clearance, azotemia, and hypertension are at significantly increased risk for premature separation of the placenta (*abruptio placentae*), intrauterine growth retardation, fetal distress, and fetal demise.

Pre-eclampsia is a syndrome peculiar to pregnancy and consists of hypertension, edema, and proteinuria. It is more common among primigravidas, but women with chronic hypertension also are at risk regardless of their parity. Pre-eclampsia is a systemic disorder whose cause is unknown. It is frequently associated not only with decreased renal function but also with hepatic dysfunction, coagulation abnormalities, and microangiopathic hemolysis. The central nervous system is often irritable and if left untreated, pre-eclampsia may progress to eclampsia with the development of seizures. There is no laboratory test or set of tests that alone establishes or rules out the diagnosis of pre-eclampsia. In the appropriate clinical setting, however, the following laboratory data help to support the diagnosis: (1) falling creatinine clearance; (2) rising serum creatinine and urea nitrogen; (3) rising serum uric acid; (4) hemoconcentration with rising hematocrit; (5) proteinuria in excess of 400 mg/d; (6) evidence of intravascular hemolysis, such as schistocytes on peripheral smear or hemoglobinuria; (7) evidence of intravascular coagulation, such as falling platelet count or hypofibrinogenemia; and (8) abnormal liver function studies.

The time-honored prophylaxis against convulsions in the peripartum period is IM or IV administration of magnesium sulfate. Therapeutic serum levels are 3–4 mmol/L; respiratory depression can be seen with levels in excess of 5 mmol/L and complete paralysis at 7–7.5 mmol/L. Intravenous calcium should be available as therapy for magnesium toxicity. Magnesium ions cross the placenta freely, and newborns delivered to mothers who are maintained on magnesium therapy may be lethargic and develop hypocalcemia.

Antenatal Diagnosis

The field of antenatal diagnosis has undergone an explosion of knowledge, especially within the last 12–15 years. This has been due to the application of established techniques (e.g., cell culture and enzyme determinations) and to the development of new methods, such as high speed ultrasonic imaging and recombinant DNA techniques. A recent review contains over 180 conditions that have been diagnosed in utero.[27] These conditions range from twins to fatal chromosomal and metabolic disorders.

Ultrasonographic Examination

Basic to all modern antenatal diagnoses is visualization of the uterine contents by the use of ultrasound. Several types of imaging devices are in common use. The transducer emits very high frequency sound waves, which are transmitted through the tissues and reflected back to the transducer from interfaces of differing tissue densities. The reflected sound waves picked up by the transducer are then processed and assembled into an image on a cathode ray tube. Amniocentesis should always be done with the assistance of ultrasound, which aids in the visualization of the uterine contents and helps localize the fetus, placenta, and amniotic fluid. This reduces the incidence of blood-contaminated specimens, multiple needle passes to obtain fluid, and ultimately unsuccessful anmiocenteses. It also helps to avoid trauma to the fetus and the mistaken sampling of urine from the maternal bladder.

Amniotic Fluid

Throughout intrauterine life, the fetus lives within a fluid compartment. The amniotic fluid provides a medium in which the fetus can readily move, it cushions the fetus against injury, and it helps maintain a constant temperature for the fetus. This fluid is a dynamic medium whose volume and chemical composition are controlled within relatively narrow limits.

Volume. The volume of amniotic fluid increases progressively to a mean of 984 mL at the 34th week of gestation and drops off slightly until the 40th week and more sharply toward the 42nd week. Total fluid fluxes between maternal serum and amniotic fluid are large (~450 mL/h) and result in complete exchange of the amniotic fluid volume every 2–3 h.

Composition. Early in gestation the composition of the amniotic fluid very closely resembles a complex dialysate of the maternal serum. As the fetus grows, the amniotic fluid changes in several ways. Most notably, the Na^+ concentration and osmolality drop, whereas the concentrations of urea, creatinine, and uric acid rise.[2] The activities of many enzymes in amniotic fluid have been studied with respect to both gestational age and fetal status but have not been found to be clinically useful. The major lipids of interest are the phospholipids whose type and concentrations reflect fetal lung maturity; these will be discussed further below. Numerous steroid and protein hormones are present in amniotic fluid.[31] The rare syndrome of congenital adrenal hyperplasia due to 21-hydroxylase deficiency has been diagnosed antenatally by measuring 17-hydroxyprogesterone and pregnanetriol in the amniotic fluid at midterm.[12]

Early in pregnancy, there is little or no particulate matter in the amniotic fluid. By 16 weeks of gestation, large numbers of cells are present that have been shed from the surfaces of the amnion, skin, and tracheobronchial tree. They have proved to be of great utility in antenatal diagnosis and will be discussed below. As pregnancy continues to progress, scalp hair and lanugo (very fine body hair) are also shed into the fluid and contribute to its turbidity.

Determination of ΔA_{450} in Amniotic Fluid

The severity of intrauterine hemolysis (e.g., in Rh isoimmune disease) can be estimated by determining the amount of bilirubin in the amniotic fluid. The concentration of bilirubin is normally so low (~10–30 μg/dL) that the elevated level of bilirubin can be rapidly and accurately detected by absorption spectrophotometry. In the absence of significant amounts of bilirubin, the absorbance curve for the amniotic fluid between 365 and 550 nm is nearly a straight line (Figure 28-3). The maximal absorbance of bilirubin is at 450 nm. The degree to which the curve deviates from a straight line at 450 nm is linearly proportional to the concentration of bilirubin in the amniotic fluid.

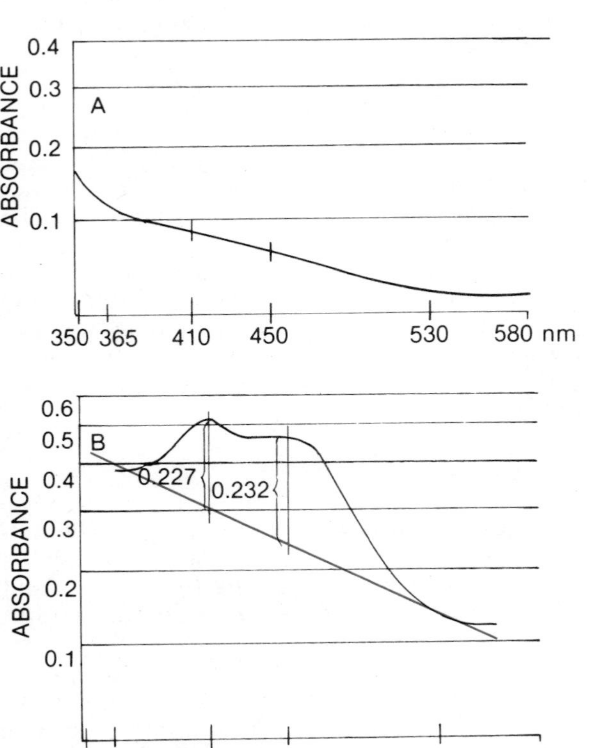

Figure 28-3. *A*, Normal amniotic fluid. Note near linearity of the curve. *B*, Amniotic fluid showing bilirubin peak at 450 nm and oxyhemoglobin peak at approximately 410 nm. Note baseline drawn between linear parts of the curve, from 550 to 365 nm.

This is the change in absorbance at 450 nm (ΔA_{450}). There is normally a small amount of bilirubin in amniotic fluid, and this amount changes with gestational age (see Figure 28-2). To properly interpret the ΔA_{450}, therefore, it is necessary to know the gestational age with reasonable precision.

Specimen. The amniotic fluid specimen is obtained by amniocentesis with a 22-gauge needle under ultrasound guidance; care should be taken not to contaminate the specimen with blood. A total of ~10 mL should be withdrawn. Since bilirubin is unstable in light, the specimen should be protected from light during transport to the laboratory and during storage. This may be done conveniently by wrapping the specimen in aluminum foil. Liley[18] found that the absorbance peak had a $t_{½}$ of 10 h in laboratory daylight and 12–18 min in winter sunlight. When stored in the dark, however, the peak is stable for 30 d at room temperature and for at least nine months under refrigeration (see also Chapter 10).

Instrumentation. A recording spectrophotometer is most conveniently used and the fluid is scanned between 350 and 560 nm. As described by Liley,[17] the original procedure used a semilogarithmic plot of absorbance against wavelength. Subsequently, others have used linear plots with insignificantly different results.[23] Since the major contaminant of concern is oxyhemoglobin with an absorbance peak of 410 nm, it is important that the instrument have a narrow bandpass to minimize interference by this compound when making the A_{450} reading.

Procedure

1. Clear the specimen of particulate matter by centrifugation at ~2000 × *g* for 5 min.
2. Fill cuvet with undiluted amniotic fluid.
3. Zero the spectrophotometer according to the directions for that particular instrument by using a NaCl solution, 0.15 mol/L.
4. Scan the fluid from 350–560 nm. If a manual spectrophotometer is used, make individual absorbance readings at the following wavelengths: 350, 365, 380, 390, 400, 410, 415, 420, 430, 440, 450, 460, 470, 485, 500, 515, 530, 540, 555. Plot the absorbance readings against wavelengths on semilog paper with the absorbance values on the semilog scale (y-axis).
5. On the absorbance plot, draw a straight line (the baseline) from the linear area of the curve at 350–375 nm to the linear area of the curve at 525–550 nm.
6. Construct a line perpendicular to the x-axis from the absorbance peak at 450 nm to the baseline. This line represents the ΔA_{450} (see Figure 28-3).
7. If there is an oxyhemoglobin peak at 410 nm, determine the ΔA_{410} in a similar fashion.

Calculations. In the absence of other pigment peaks, the ΔA_{450} derived from the graphic method above should be reported directly. If there is oxyhemoglobin present as a peak at 410 nm, then 5% of the ΔA_{410} should be subtracted from the ΔA_{450} and that number reported as the ΔA_{450}.

Sources of error. Contamination of amniotic fluid with blood should be obvious from the red color and presence of erythrocytes in the button after centrifugation. When a blood-contaminated specimen is obtained, it is important to minimize hemolysis of the erythrocytes since this is the main source of oxyhemoglobin. A bloody specimen, therefore, should be refrigerated immediately and processed as soon as possible. Contamination of an amniotic fluid specimen with fetal blood from an affected fetus with a high serum bilirubin could introduce a substantial error, the magnitude of which cannot accurately be determined. Whenever a "bloody tap" is obtained, a Kleihauer-Betke test should be done on the erythrocytes to establish their origin. If they are found to be in large part fetal, this should be kept in mind when the fluid results are interpreted. If the amniotic fluid remaining within the amniotic cavity has been contaminated with blood by a traumatic amniocentesis, these erythrocytes will lyse and contaminate the fluid with hemoglobin and bilirubinoid pigments. It will take 2–3 weeks for these pigments to be cleared and for the fluid to return to the original state.

After steroid treatment, the ΔA_{450} will be lowered relative to the level found immediately before the steroid administration.[5] Further observation and amniocenteses will show that the ΔA_{450} returns to the previous level or higher within a week.

Meconium is a substance found in the fetal intestine and contains a number of pigments. Meconium staining of amniotic fluid will cause a substantial rise in the ΔA_{450}, with a broad and variable peak at 400–415 nm. There is no way to compensate quantitatively for meconium contamination. A single episode of meconium passage into the amniotic fluid will require about three weeks to clear.

Tests for the Determination of Fetal Lung Maturity

Respiratory distress syndrome (RDS), or hyaline membrane disease, is the most common immediate complication that a premature infant is likely to encounter. Normally, adequate concentrations of pulmonary surfactant reduce the surface tension in the alveoli so that at end-expiration, the walls of the alveoli do not collapse and do not stick together. When surfactant is not present, the alveoli collapse and considerable negative pressure must be generated with each inspiration to reopen them. Depending upon the degree of severity, an infant may not be able to reopen the alveoli, or may do so only with great physical effort. Affected infants require supplemental oxygen and mechanical ventilation to remain properly oxygenated.

Pulmonary surfactant is a complex mixture of lipids and proteins with less than 5% carbohydrates. Most of the lipid is phospholipid and the majority of that is lecithin.[1] Up to 85% of the lecithin has saturated fatty acids at both the α- and β-positions and about 60% of the fatty acids is palmitic acid. Other lipids present are phosphatidylglycerol (PG), phosphatidylinositol (PI), phosphatidylethanolamine, and sphingomyelin.

The most common situation in which a test of fetal lung maturity is sought is when early delivery is anticipated due to some medical or obstetrical indication. Results indicating immaturity of the fetal lungs might cause postponement of elective delivery or prompt, active intervention to suppress premature labor.

Numerous tests of amniotic fluid for fetal maturity have been proposed. Some of these, such as the absorbance at 650 nm, creatinine concentration, lipid staining of cells, and "amniocrit," may correlate with gestational age, but either they do not measure lung maturity specifically or they cannot be used in the presence of blood or meconium. Two methods that directly measure surface active materials are presented below.

Foam Stability Index (FSI)[26]

Principle. When pulmonary surfactant is present in amniotic fluid in sufficient concentration, the fluid is able to form a highly stable surface film that can support the structure of a foam. There are other substances in the fluid, such as proteins, bile salts, and salts of free fatty acids, that are also capable of forming a stable foam, but these can be excluded from the film by ethanol, which competes with these other substances for a position in the surface film. This is the principle of the test originally proposed by Clements and colleagues.[8] A quantitative test, the *foam stability index*, has been devised and is described below. This test makes use of the principle that more surfactant activity is necessary to support a stable foam as the fraction of ethanol in the mixture is increased. A fixed volume of undiluted amniotic fluid is mixed with increasing volumes of ethanol, and the largest fraction of ethanol in which the amniotic fluid is still capable of supporting a foam is noted.

Specimen. Amniotic fluid is obtained by amniocentesis. Fluid may be obtained by transvaginal puncture of the bulging membranes, but it should not be grossly contaminated with vaginal secretions. Whenever possible, the fluid should be tested immediately. If there is to be a delay of a few hours, the fluid should be centrifuged and refrigerated at 4 °C. If the fluid is not to be tested the same day, it should be centrifuged and stored frozen at −20 °C. Immediately before testing the fluid, it should be gently inverted several times to obtain a uniform suspension without shaking and creating a foam.

Procedure

1. Centrifuge the amniotic fluid at 1000 × *g* for 3 min. Decant the supernatant.
2. Gently mix the supernatant without creating a foam by inverting several times.
3. Set up six 12 × 75-mm screw-top plastic tubes and label them 0.50, 0.48, 0.47, 0.46, 0.45, and 0.44 to indicate final ethanol concentrations.
4. Carefully pipet the following volumes of 95% ethanol into the respective tubes: 0.55, 0.51, 0.49, 0.47, 0.45, and 0.43 mL.
5. Add exactly 0.50 mL of the amniotic fluid supernatant to each tube.
6. Cap tubes and shake vigorously for 30 s. Allow to settle for 15 s and then examine.

7. Using a bright overhead light and placing each tube against a flat, black background, examine for the presence of bubbles at the edge of the air-fluid meniscus. A ring of bubbles 360° around the tube indicates a "positive" result. Record the highest concentration of ethanol at which a positive reading is obtained; this is the foam stability index.

Interpretation

In two series, a combined total of 229 fetuses with an FSI of 0.48 or greater were reported. None of these developed RDS and were termed *mature*. Of 41 fetuses with an FSI of 0.47, two (5%) developed RDS. For 110 fetuses with an FSI of 0.46 or less, 44 (40%) developed RDS. These were termed *immature*.

Comments

1. Tubes and stoppers must be clean and free of detergents, to avoid false-positive readings.
2. The final concentration of the ethanol is critical. A 95% solution of ethanol is hygroscopic; therefore, careful attention must be paid to the preparation of the ethanol reagent from absolute ethanol and water and to its storage in a tightly stoppered container. The 95% ethanol must be accurately pipetted into the amniotic fluid.
3. The test should be conducted at 20–25 °C. Temperatures substantially below this may retain a relatively unstable foam, whereas higher temperatures may cause the deterioration of an otherwise stable foam.[13]
4. Specimens contaminated with either blood or meconium will tend to give a falsely "mature" test result. Since a significant percentage of fluid samples submitted to a clinical laboratory may be contaminated, this is an important limitation of this test. If the test result is "immature" or "intermediate" in the face of obvious contamination, it is probably reasonable to conclude that the fetal lungs are immature. If possible, the procedure should be repeated on an uncontaminated specimen.
5. Recently, a kit for determining the foam stability index (Lumadex-FSI Test, Beckman Instruments, Brea, CA 92621) has been introduced.[25]

Methods for the Simultaneous Determination of the Lecithin/Sphingomyelin (L/S) Ratio and Saturated Phosphatidylcholine (SPC)[28]

The major surface-active component of the lung surfactant is saturated phosphatidylcholine (SPC), which is a subfraction of the total lecithins. Its concentration in amniotic fluid tends to rise with increasing gestational age. The concentration of SPC in amniotic fluid can be measured directly and correlated with fetal lung maturity, or its concentration can be related to another lipid, sphingomyelin, and that ratio correlated with lung maturity. Sphingomyelin has surface-active properties and is found in the lung, but it is ubiquitously distributed in cell membranes and in plasma. Sphingomyelin is believed to play no role in the surfactant system in the lung; however, it is used as a convenient marker against which lecithin is measured. The concentration of lecithin relative to sphingomyelin rises suddenly at 34–36 weeks of gestation, which correlates with the development of fetal lung maturity.[10]

The quantitative assay of SPC requires the removal of unsaturated lecithin from the sample. Osmium is a very powerful oxidizing agent that reacts with double bonds in fatty acids and renders lecithins containing unsaturated fatty acids insoluble.[19] This reaction proceeds rapidly to completion and is the principle upon which the present procedure is based.

Specimen. A minimum of 3 mL of amniotic fluid should be obtained, preferably by transabdominal amniocentesis. Specimens obtained from a pool of fluid in the vagina immediately after rupture of the membranes are acceptable (but not for the shake test). Specimens contaminated with blood or meconium are also acceptable, although the results must be interpreted with care. Specimens that cannot be analyzed immediately are quite stable, but refrigeration is recommended.[30] Uncentrifuged fluid should not be frozen.

Materials and Reagents

1. Chromatogram sheets, Eastman 6061 Silica Gel without fluorescent indicator, catalog # 13179. Sheets must be activated before use by heating for ≥2 h in a 90 °C oven.

2. Developing solvent. 65 parts chloroform: 25 parts methanol: 4 parts distilled water.

3. Bromothymol blue dye solution. Add 64 mL of sodium hydroxide solution (1 mol/L) to 896 mL distilled water. Dissolve in this solution 10 g boric acid powder. Then add 500 mg bromothymol blue dye and swirl until dissolved (a magnetic stir bar is helpful).

4. Osmium tetroxide
Stock solution: 1 g OsO_4 + 10 mL CCl_4
Working solution: 0.7 mL stock + 9.3 mL CCl_4

5. Saturated phosphatidylcholine (SPC; Sigma Chemical Co., St. Louis, MO 63178) standard, 100 ng/μL. Mix 9 mL chloroform plus 1 mL of methanol as the solvent, and add 1 mg of SPC. Store in refrigerator.

6. L/S standard. Dissolve 30 mg each of lecithin and sphingomyelin in 10 mL chloroform. Store in refrigerator.

Procedure

Preparation of the Amniotic Fluid Specimen

1. Centrifuge the fluid in a 15-mL tube at 500 × *g* for 10 min at 4 °C.
2. Pipet 3 mL of the supernatant into a 50-mL tube.
3. Add 3 mL methanol and vortex well.
4. Add 6 mL chloroform and vortex for 30 s.
5. Centrifuge at 500 × *g* for 5 min.
6. Aspirate off the lower (chloroform) phase and transfer into a 15-mL tube and evaporate to dryness under flowing N_2 in a 60 °C water bath.
7. Rinse down the sides of the tube with a small amount of chloroform and evaporate again as above.
8. Redissolve residue in 150 μL chloroform.
9. Transfer 50 μL to a 12 × 75-mm tube and evaporate tubes containing 100 and 50 μL chloroform extract to dryness. Chill on ice.

Determination of the L/S Ratio

10. Add 2-3 drops of cold acetone to the residue in the 15-mL tube containing 100 μL amniotic fluid residue and agitate to dissolve.
11. Add 0.75 mL cold acetone and let stand on ice for 5 min.
12. Centrifuge at 800 × *g* for 1 min at 4 °C. Decant off and discard acetone.
13. Dry the tube under flowing N_2 and chill.
14. Redissolve residue in 100 μL chloroform and spot these aliquots onto activated chromatogram sheets: two 2-μL, two 5-μL, two 10-μL, and two 20-μL bands. Spot a 10-μL and 20-μL band of the standard for reference. Leave blank lanes between the samples so that the sheet can be read in the reflectance spectrodensitometer.

Determination of SPC

15. Place the 12 × 75-mm tube containing the 50-μL amniotic-fluid residue prepared above in Step 9 in a rack under a hood.
16. Add 0.5 mL of the OsO_4 working solution and allow to stand at room temperature for 15 min.
17. Evaporate under N_2 in the hood and redissolve the residue in 50 μL chloroform.
18. Place three 10-μL spots onto the TLC sheet, leaving blank lanes between samples.
19. Warm the SPC standard solution to room temperature and place two 10-μL spots (1 μg) onto the TLC sheet.
20. Dry sheet in oven. If spots are particularly dark (usually cases of heavy blood contamination), leave in oven for 10–15 min.
21. Develop, stain, and read chromatograms as below in steps 22–26.

Chromatography

22. Dry sheet in a 90 °C oven for 2 min.
23. Develop in a tank with 1 cm of solvent and allow the front to migrate to within 0.5–1 cm of top of sheet.
24. Dry in a 90 °C oven for 2 min and stain by immersion in a tank with bromothymol blue dye. Do not spray.
25. Blot off excess dye and dry in a 90 °C oven for 5 min.
26. Cool to room temperature and read in a reflectance spectrodensitometer.

Calculation

$$\text{SPC in sample, } \mu g/dL = \frac{A_{sample} \times 1.0 \times 100}{A_{std} \times 0.2} = \frac{A_{sample}}{A_{std}} \times 500$$

where 1.0 = amount of standard (std) used, in μg
100 = converts to dL
0.2 = amount of original amniotic fluid in mL

Example. If sample spots give absorbance readings of 33, 35, and 37 ($\bar{x} = 35$) and the standard spots read 20 and 22 ($\bar{x} = 21$), then

$$\text{SPC, } \mu g/dL = \frac{35}{21} \times 500 = 833$$

Calculation of L/S Ratio

The lecithin and sphingomyelin spots are identified by their R_f values as determined from the L/S standard. Their areas are derived from scanning with the densitometer. L/S ratio is calculated from the values obtained for the areas.

Interpretation

The L/S ratio indicates mature lungs if it exceeds 2.0/1 and the SPC indicates maturity if it exceeds 500 μg/dL. Both of these tests predict fetal lung maturity more reliably than they predict immaturity. Less than 2–3% of babies delivered within 24 h of obtaining an L/S ratio of 2.0 or greater can be expected to develop RDS; i.e., >97–98% of babies predicted to be mature will, in fact, be mature. Almost one-half of the infants with L/S ratios between 1.5 and 2.0, however, will not develop RDS. Thus, in contrast to the high degree of reliability of a mature prediction, only half of the babies predicted to be immature will, in fact, be immature. Corresponding numbers for the SPC are ~1% falsely indicated to be mature and <20% falsely indicated to be immature.

A number of investigators have reported that infants of diabetic mothers (IDMs) more frequently develop RDS in spite of "mature" L/S ratios of greater than 2.0/1.[9] For this reason, the value for lung "maturity" for an IDM has been raised progressively over the past several years to an L/S ratio of 3.5/1 and an SPC of 1000 μg/dL. Fetuses with isoimmune hemolytic disease also seem to develop RDS frequently despite "mature" indices.

Comments

1. During the initial centrifugation of the amniotic fluid to remove particulate matter, it is important to avoid excessive *g* force or prolonged centrifugation. At *g* forces in excess of ~500 there is a progressive fall in both the L/S ratio[7] and total phospholipid content.[30]

2. In the initial extraction procedure it is important to create first a monophasic system with the addition of methanol and then to add the chloroform to form the biphasic system. Insoluble proteins will be left at the interface. Failure to mix thoroughly with methanol first will lead to an inadequate extraction and loss of lipid in the protein layer.[3]

3. While ammonium-sulfate charring of lecithin and sphingomyelin spots has been popular, charring is less intense as the degree of saturation of the lipid being examined increases. The phospholipids most important for surfactant activity are saturated lipids. Thus, the paradox develops that as the lecithin composition "matures," the lecithin chars less intensely. Staining with bromothymol blue is a saturation-insensitive technique. Lecithin, which is highly saturated, stains with the same intensity as relatively unsaturated lecithins. The major disadvantage of bromothymol blue staining is that it fades rather rapidly, and thus chromatograms should be scanned promptly after staining. Dipping the TLC sheet is preferable to spraying because color development is more uniform.

4. The major disadvantage of this one-dimensional technique is that it fails to separate phosphatidylinositol (PI) from lecithin. Thus, it may slightly overestimate the L/S ratio.

5. The L/S ratio of whole blood is probably between 1.4/1 and 1.6/1. The values of blood-contaminated specimens will therefore tend to be drawn to that range from either higher or lower values. "Borderline" values of bloody specimens must be interpreted with care, as they may represent either very "immature" values drawn up or mature values of just over 2.0 drawn down.

6. There is essentially no SPC in either blood or meconium. Thus, contamination of amniotic fluid samples with either of these will not change the SPC value.

7. Vaginal contamination of specimens obtained from a pool of fluid that has been present in the vagina for a relatively brief period does not affect either test.[22]

The Determination of the Lung Profile[14,15]

The lung profile, also referred to as the "two-dimensional" lung profile, provides the simultaneous determination of the L/S ratio, PG, and PI as percentages of total phospholipid and the saturated lecithin as a percentage of total lecithin. By using the two-dimensional system, this technique separates the acidic phospholipids PG and PI from the lecithin fraction so that the lecithin is not overestimated. This procedure depends upon a cold acetone precipitation to separate the highly saturated lecithins important for surfactant activity from the soluble, less saturated lecithins (which are less important for surfactant activity).

Congenital Malformations of the Central Nervous System

During embryonic development, the area that is to form the central nervous system (brain and spinal cord) differentiates into a plate of cells. The flat plate then rolls up and its edges fuse into a hollow neural tube, which drops into the embryo to develop just under what will become the skin of the back. Neural tube formation is normally complete very early in gestation (by the end of the fourth week following fertilization). Failure of the embryo to complete neural tube fusion successfully will lead to permanent developmental defects of brain or spinal cord such as anencephaly, spina bifida, or myelomeningocele.[11] Approximately 90–95% of these defects are "open"; i.e., there is a direct communication between the neural tube and the outside world (which is amniotic fluid for the fetus) without intact overlying skin. Under this circumstance, proteins that are normally relatively restricted to the fetal blood and cerebrospinal fluid gain access to the amniotic fluid in large quantities. Furthermore, these proteins in the amniotic fluid gain access, in smaller but significant and measurable amounts, to the maternal circulation. It is on this principle that present biochemical techniques for the antenatal diagnosis of neural tube defects (NTDs) are based.

Alpha$_1$-Fetoprotein and Acetylcholinesterase

Alpha$_1$-fetoprotein (AFP) is a glycoprotein of MW ~70 000. It is produced initially by the fetal yolk sac and then by the fetal liver as the yolk sac degenerates. Considerably smaller amounts are probably also produced in the fetal gut and kidneys. Early in embryonic life this protein functions as the major serum protein. During development, the relative concentration of AFP in fetal serum decreases as the liver switches over to production of serum albumin. Maximal concentration in the fetal serum of 2–3 mg/mL is reached at ~16 weeks of gestation and declines steadily to ~0.07 mg/mL by term. The rise and fall in concentration of AFP in the amniotic fluid roughly parallels that in the fetal serum, but is two orders of magnitude lower in concentration (~0.015 mg/mL at 16 weeks). The relationship with respect to maternal serum concentration is slightly more complicated owing to the additional consideration of fetal membrane and placental permeability. While the fetal serum concentration of AFP falls progressively from 16 weeks' gestation, the membranes and placenta become increasingly "leaky," leading to a progressive rise in maternal serum AFP from 14 weeks to term.[16] Methods for AFP determination are listed in Chapter 11.

It has also been demonstrated that amniotic fluid in pregnancies with NTDs (and omphaloceles) contains a second isoenzyme of acetylcholinesterase (AChE) of central nervous system origin that is not found in amniotic fluid of normal pregnancies.[5]

Table 28-4. CONDITIONS ASSOCIATED WITH ELEVATED AFP

Anencephaly	Nuchal bleb
Congenital nephrosis	Omphalocele
Duodenal atresia	Rh isoimmune disease
Encephalocele	Sacrococcygeal teratoma
Esophageal atresia	Spontaneous abortion
Fetal blood contamination	Trisomy 13
Fetal death	Turner's syndrome with cystic hygroma
Gastroschisis	Twins (AFP increased in maternal serum; amniotic fluid AFP usually normal)
Hydrocephalus	Urinary obstruction
Meckel's syndrome	
Myelomeningocele	
Necrosis of fetal liver secondary to herpes virus infection	

Testing for Neural Tube Defects

Any patient who has had a child with an NTD of a type that has significant risk for recurrence should be offered, in subsequent pregnancies, antenatal diagnosis consisting of a combination of ultrasound examination and amniotic fluid analysis. Testing should be done at ~16 weeks' gestation for several reasons: (1) This is a good time for an ultrasound examination because a detailed fetal examination can be done. (2) The AFP concentration in the amniotic fluid is maximal. (3) The amniocentesis is relatively easy technically. (4) If the AFP is elevated there is still time before 24 weeks, the limit for elective termination, to complete the fetal evaluation. When an amniocentesis is done at 16 weeks of gestation for cytogenetic studies, most centers are also performing an AFP assay on the fluid. If the amniocentesis reveals an elevated AFP (>3 SD above $\bar{x}$) and no NTD or other explanation for the elevation is found by ultrasound examination (Table 28-4), then further steps are indicated. The possibility of contamination of the specimen with fetal blood should be excluded, and an AChE assay should be performed.

The fact remains, however, that 90% of all infants with NTDs are delivered to totally unsuspecting parents who have no recognized risk factor for NTD. Thus, it is argued that if a significant impact is to be made on NTDs as a public health issue, a simple, reliable, noninvasive, inexpensive screening test should be made available to all women. Maternal serum AFP screening has been suggested for this purpose and has been shown to be effective in numerous demonstration projects around the world. Maternal serum AFP values greater than two multiples of the median (MoM) deserve investigation for the possibility of an NTD. Low maternal AFP values have a significant association with Down syndrome.[20]

Diagnostic Uses of the Cellular Elements in Amniotic Fluid

Amniotic fluid contains numerous cells that are shed from various surfaces. The main sources are thought to be skin and amnion, with minor contributions from the genitourinary, respiratory, and alimentary tracts. The number of cells per unit volume of fluid tends to rise with increasing gestational age. However, the percentage of these cells that are viable in culture decreases with advancing gestational age. Generally, between 15 and 20 weeks of gestation the number of cells and their viability are such that cultures can successfully be established with a high degree of reliability.

The incidence of major and minor chromosomal abnormalities in all live births is approximately 1 in 200. The majority of second trimester diagnostic amniocenteses are performed for the purpose of detecting these major abnormalities. The anomalies are not uniformly distributed among all pregnant women. It is well known that there is a markedly increased tendency toward these problems with increasing maternal age. For Down syndrome, for example, the risk at maternal age 18 is approximately 1:2000 whereas at age 40 it is 1:80. Thus, the most common indication for a "genetic amniocentesis" is advanced maternal age.

Cell cultures. Amniotic fluid cells to be cultured must be collected and handled by using *sterile* technique. Approximately 20 mL of fluid should be obtained by amniocentesis, and the fluid should be kept at body or room temperature and transported to the laboratory as rapidly as possible. It should *never* be frozen or exposed to toxins or fixatives. The specimen may conveniently be stored and transported in the disposable syringe used for the amniocentesis. The cells tend to adhere to glass, and therefore glass syringes and tubes should not be used. Upon arrival in the cytogenetics laboratory, samples are centrifuged and the cell pellet is used for preparing cell cultures. For details of preparing the cultures, the reader is referred elsewhere.[24]

Chromosome analysis. Actively growing and dividing amniotic fluid cells in culture are prepared for chromosome analysis by arresting cell division at metaphase. This is done by exposing the cells to an antimitotic agent such as colchicine for several hours. The arrested cells are fixed to a microscope slide and ruptured so that the chromosomes are spread out with little or no overlap of one another. They can then be stained and photographed for analysis. Simple chromosome counting and identification can lead to diagnosis of conditions involving lost chromosomes (45,X Turner's syndrome) or extra chromosomes (47 trisomy 21). More sophisticated staining techniques that demonstrate banding permit the identification of subtle deletions, inversions, and translocations. By making use of minor variations in staining patterns (polymorphisms), one can trace an extra chromosome or set of chromosomes back to one parent or the other.

Detection of Inborn Errors of Metabolism

Cultured amniotic fluid cells are generally the preferred specimen for studying inborn metabolic errors. For most metabolic studies (e.g., assays of enzyme activity), larger numbers of cells are needed than for cytogenetic studies. Usually, therefore, cultures must be maintained for somewhat longer periods of time. In some instances, enzyme activity assays have been performed directly on amniotic fluid, but this is generally less reliable, and cultured cells are the preferred specimens.

Most of the diseases that can be diagnosed are inherited as autosomal recessive disorders. Thus, in most situations, the first affected individual is born to a couple unexpectedly. It is therefore the subsequent pregnancy that is monitored by amniocentesis for the possible recurrence of the disease. In this case, the parents each carried a relatively rare gene that caused them no harm as heterozygotes but, when passed to their offspring in the homozygous condition, could cause serious disease or death. This unexpected carrier state has changed somewhat in recent years with the advent of the ability to test for "carriers" (heterozygotes) among adults. Thus, it is now possible for persons of ethnic groups at high risk for carriage of a deleterious gene to be tested for the presence of that gene. The gene for Tay-Sachs disease, for example, is carried by 1 in 30 Ashkenazic Jews. A Jewish couple anticipating having children might be tested in advance, using peripheral blood. Specific inborn errors of metabolism are discussed in more detail in Chapters 11, 12, and 14 of this book.

Until recently, studies of amniotic fluid cells were limited to metabolic activities that are expressed by the cells. Recombinant DNA techniques, however, have permitted the direct examination of the genome without requiring its expression.

References

1. Abramovich, D. R., Keeping, J. D., and Thom, H.: The origin of amniotic fluid lecithin. Br. J. Obstet. Gynecol., *82*:204–207, 1975.
2. Benzie, R. J., Doran, T. A., Harkins, J. L., et al.: Composition of the amniotic fluid and maternal serum in pregnancy. Am. J. Obstet. Gynecol., *119*:798–810, 1974.
3. Brown, L. M., and Duck-Chong, C. G.: Methods of evaluating fetal lung maturity. CRC Crit. Rev. Clin. Lab. Sci., *16*:85–159, 1982.
4. Buamah, P. K., Evans, C., and Ward, A. M.: Amniotic fluid acetylcholinesterase isoenzyme patterns in the diagnosis of neural tube defects. Clin. Chim. Acta, *102*:147–151, 1980.

5. Caritis, S. N., Mueller-Henbach, E., and Edelstone, D. I.: Effect of betamethasone on analysis of amniotic fluid in the rhesus-sensitized pregnancy. Am. J. Obstet. Gynecol., *127*:529–532, 1977.
6. Chayen, J., Daly, J. R., Loveridge, N., et al.: The cytochemical bioassay of hormones. Recent Prog. Horm. Res., *32*:33–72, 1976.
7. Cherayil, G. D., Wilkinson, E. J., and Borkowf, H. I.: Amniotic fluid lecithin/sphingomyelin ratio changes related to centrifugal force. Obstet. Gynecol., *50*:682–688, 1977.
8. Clements, J. A., Platzker, A. C. G., Tierney, D. F., et al.: Assessment of the risk of the respiratory distress syndrome by a rapid test for surfactant in amniotic fluid. N. Engl. J. Med., *286*:1077–1081, 1972.
9. Cruz, A. C., Buhi, W. C., Birk, S. A., et al.: Respiratory distress syndrome with mature lecithin/sphingomyelin ratios: Diabetes mellitus and low Apgar scores. Am. J. Obstet. Gynecol., *126*:78–82, 1976.
10. Gluck, L., and Kulovich, M.: Lecithin/sphingomyelin ratios in amniotic fluid in normal and abnormal pregnancy. Am. J. Obstet. Gynecol., *115*:539–546, 1973.
11. Holmes, L. B., Driscoll, S. G., and Atkins, L.: Etiologic heterogeneity of neural tube defects. N. Engl. J. Med., *294*:365–369, 1976.
12. Jeffcoate, T. N. A., Fliegner, J. R. H., Russell, S. H., et al.: Diagnosis of adrenogenital syndrome before birth. Lancet, *2*:553–555, 1965.
13. Keniston, R. C., Noland, G. L., and Pernoll, M. L.: The effect of blood, meconium and temperature on the rapid surfactant test. Obstet. Gynecol., *48*:442–446, 1976.
14. Kulovich, M. V., and Gluck, L.: The lung profile. II. Complicated pregnancy. Am. J. Obstet. Gynecol., *135*:64–70, 1979.
15. Kulovich, M. V., Hallman, M. B., and Gluck, L.: The lung profile. I. Normal Pregnancy. Am. J. Obstet. Gynecol., *135*:57–63, 1979.
16. Lau, H. L., and Linkins, S. E.: Alpha-fetoprotein. Am. J. Obstet. Gynecol., *124*:533–554, 1976.
17. Liley, A. W.: Liquor amnii analysis in the management of the pregnancy complicated by rhesus sensitization. Am. J. Obstet. Gynecol., *82*:1359–1370, 1961.
18. Liley, A. W.: Errors in the assessment of hemolytic disease from amniotic fluid. Am. J. Obstet. Gynecol., *86*:485–494, 1963.
19. Mason, R. J., Nellenbogen, J., and Clements, J. A.: Isolation of disaturated phosphatidylcholine with osmium tetroxide. J. Lipid Res., *17*:281–284, 1976.
20. Merkatz, I. R., Nitowsky, H. M., Macri, J. N., et al.: An association between low maternal serum α-fetoprotein and fetal chromosomal abnormalities. Am. J. Obstet. Gynecol., *148*:886–894, 1984.
21. Miller, E., Hare, J. W., Cloherty, J. P., et al.: Elevated maternal HbA_{1C} in early pregnancy and major congenital anomalies in infants of diabetic mothers. N. Engl. J. Med., *304*:1331–1334, 1981.
22. Phillippe, M., Acker, D., Torday, J., et al.: The effects of vaginal contamination on two pulmonary phospholipid assays. J. Reprod. Med., *27*:283–286, 1982.
23. Queenan, J. T.: Modern Management of the Rh Problem. Hagerstown, Md., Harper & Row, 1977.
24. Sandstrom, M. McH., Beauchesne, M. T., Gutashaw, K. M., et al.: Prenatal cytogenetic diagnosis. Meth. Cell. Biol., *26*:35–66, 1982.
25. Sher, G., and Statland, B. E.: Assessment of fetal pulmonary maturity by the Lumadex foam stability index test. Obstet. Gynecol., *61*:444–449, 1983.
26. Sher, G., Statland, B. E., and Freer, D. E.: Clinical evaluation of the quantitative foam stability index tests. Obstet. Gynecol., *55*:617–620, 1980.
27. Stephenson, S. R., and Weaver, D. D.: Prenatal diagnosis—a compilation of diagnosed conditions. Am. J. Obstet. Gynecol., *141*:319–343, 1981.
28. Torday, J., Carson, L., and Lawson, E. E.: Saturated phosphatidylcholine in amniotic fluid and prediction of the respiratory-distress syndrome. N. Engl. J. Med., *301*:1013–1018, 1979.
29. Tulchinsky, D., and Ryan, K. J., Eds.: Maternal-Fetal Endocrinology. Philadelphia, W. B. Saunders Co., 1980.
30. van Voorst tot Voorst, E. J. G. M.: Effects of centrifugation, storage, and contamination of amniotic fluid on its total phospholipid content. Clin. Chem., *26*:232–234, 1980.
31. Wu, C. H., Mennuti, M. T., and Mikhail, G.: Free and protein-bound steroids in amniotic fluid of midpregnancy. Am. J. Obstet. Gynecol., *133*:666–672, 1979.

29 APPENDIX

NORBERT W. TIETZ, Ph.D.
NANCY M. LOGAN, B.A.

CONTENTS

Miscellaneous Constants and Information:

Buffers Commonly Used in the Laboratory:

Statistical Tables:

Nomograms:

Reference Ranges:

MISCELLANEOUS CONSTANTS AND INFORMATION

Table 29-1. METRIC UNITS

Prefix Name	Prefix Symbol	Equivalent	Units of Length	Units of Mass*	Units of Capacity
kilo-	k	10^3	kilometer (km)	kilogram (kg)	kiloliter (kL)
—	—	1	meter (m)	gram (g)	liter (L)
deci-	d	10^{-1}	decimeter (dm)	decigram (dg)	deciliter (dL)
centi-	c	10^{-2}	centimeter (cm)	centigram (cg)	centiliter (cL)
milli-	m	10^{-3}	millimeter (mm)	milligram (mg)	milliliter (mL)
micro-	μ	10^{-6}	micrometer (μm)	microgram (μg)	microliter (μL)
nano-	n	10^{-9}	nanometer (nm)	nanogram (ng)	nanoliter (nL)
pico-	p	10^{-12}	picometer (pm)	picogram (pg)	picoliter (pL)

*According to the SI system, *mass* is the preferred term; *weight* is more commonly used in the United States.

Table 29-2. TEMPERATURE CONVERSIONS (CELSIUS → FAHRENHEIT)

Temp., °C	0	1	2	3	4	5	6	7	8	9
−10	14.0	12.2	10.4	8.6	6.8	5.0	3.2	1.4	−0.4	−2.2
−0	32.0	30.2	28.4	26.6	24.8	23.0	21.2	19.4	17.6	15.8
0	32.0	33.8	35.6	37.4	39.2	41.0	42.8	44.6	46.4	48.2
10	50.0	51.8	53.6	55.4	57.2	59.0	60.8	62.6	64.4	66.2
20	68.0	69.8	71.6	73.4	75.2	77.0	78.8	80.6	82.4	84.2
30	86.0	87.8	89.6	91.4	93.2	95.0	96.8	98.6	100.4	102.2
40	104.0	105.8	107.6	109.4	111.2	113.0	114.8	116.6	118.4	120.2
50	122.0	123.8	125.6	127.4	129.2	131.0	132.8	134.6	136.4	138.2
60	140.0	141.8	143.6	145.4	147.2	149.0	150.8	152.6	154.4	156.2
70	158.0	159.8	161.6	163.4	165.2	167.0	168.8	170.6	172.4	174.2
80	176.0	177.8	179.6	181.4	183.2	185.0	186.8	188.6	190.4	192.2
90	194.0	195.8	197.6	199.4	201.2	203.0	204.8	206.6	208.4	210.2
100	212.0	213.8	215.6	217.4	219.2	221.0	222.8	224.6	226.4	228.2

°C = 5/9 × (°F − 32); °F = (9/5 × °C) + 32.

Table 29-3. GREEK ALPHABET (UPRIGHT AND SLOPING TYPES)

Name	Upright		Sloping		Name	Upright		Sloping	
alpha	Α	α	*Α*	*α*	nu	Ν	ν	*Ν*	*ν*
beta	Β	β	*Β*	*β*	xi	Ξ	ξ	*Ξ*	*ξ*
gamma	Γ	γ	*Γ*	*γ*	omicron	Ο	ο	*Ο*	*ο*
delta	Δ	δ	*Δ*	*δ*	pi	Π	π	*Π*	*π*
epsilon	Ε	ε, ϵ	*Ε*	*ε, ϵ*	rho	Ρ	ρ	*Ρ*	*ρ*
zeta	Ζ	ζ	*Ζ*	*ζ*	sigma	Σ	σ	*Σ*	*σ*
eta	Η	η	*Η*	*η*	tau	Τ	τ	*Τ*	*τ*
theta	Θ	ϑ, θ	*Θ*	*ϑ, θ*	upsilon	Υ	υ	*Υ*	*υ*
iota	Ι	ι	*Ι*	*ι*	phi	Φ	φ, ϕ	*Φ*	*φ, ϕ*
kappa	Κ	ϰ, κ	*Κ*	*ϰ, κ*	chi	Χ	χ	*Χ*	*χ*
lambda	Λ	λ	*Λ*	*λ*	psi	Ψ	ψ	*Ψ*	*ψ*
mu	Μ	μ	*Μ*	*μ*	omega	Ω	ω	*Ω*	*ω*

Table 29-4. ORGANIC SOLVENTS COMMONLY USED FOR EXTRACTION OF AQUEOUS SOLUTIONS

Solvent	BP, °C*	Flammability†	Lighter/Heavier Than Water‡	Characteristics
Ethyl ether	34.6	++++	L	Absorbs 1.5% water; strong tendency to form peroxides
Methylene chloride	40	−	H	May form emulsions; easily dried
Petroleum ethers	35–80	+++	L	Easily dried; poor solvents for polar compounds
Ethyl acetate	77.1	+++	L	Good for polar compounds; absorbs up to 3.3% water
2-Butanol	99.5	+	L	Good for extraction of highly polar water-soluble compounds from buffer solutions; easily dried
Carbon tetrachloride	76.8	−	H	Good for nonpolar compounds; is easily dried
Chloroform	61.7	−	H	Easily dried; tends to form emulsions
Diisopropyl ether	69	+++	L	Tends to form peroxides
Benzene	80.1	+++	L	Prone to form emulsions; good for extraction of alkaloids and phenols from buffered solutions

*BP = Boiling point.
† − to ++++ indicates degree of flammability.
‡ L = lighter than; H = heavier than.

Modified from Shugar, G. J., Shugar, R. A., Bauman, L., et al.: Chemical Technicians' Ready Reference Handbook, 2nd ed. New York, McGraw-Hill, 1981; and Gordon A. J., and Ford, R. A.: The Chemist's Companion. New York, Wiley-Interscience, 1972.

Table 29-5. PROPERTIES OF COMMONLY USED SOLVENTS

Name	Alternative Name	Molecular Weight*	Boiling Point,† °C	Specific Gravity, $d_4^{20‡}$	Soluble in Water
Acetic acid	Ethanoic acid	60.05	118	1.0492	yes
Acetone	2-Propanone	58.08	56.2	0.7899	yes
Aniline	Aminobenzene	93.13	184	1.02173	1 g in 28.6 mL
Benzene		78.12	80.1	0.8765	1 part in 1430
n-Butanol	1-Butanol	74.12	117.5	0.8098	yes
Carbon disulfide		76.14	46	1.2632	no
Carbon tetrachloride	Tetrachloromethane	153.81	76.8	1.5867	1 part in 2000
Chloroform	Trichloromethane	119.37	61.7	1.4832	1 part in 200
Ethanol	Ethyl alcohol	46.07	78.5	0.7893	yes
Ethyl acetate	Acetic acid ethyl ester	88.10	77	0.9003	yes
Ethyl ether	Diethyl ether	74.12	34.6	0.7134	slightly
Ethylene dichloride	1,2-Dichloroethane	98.96	83.5	1.2569	1 part in 120
Heptane		100.20	98.4	0.6837	no
Isoamyl acetate	Acetic acid 3-methyl-butyl ester	130.18	142	0.8670	1 part in 400
Isoamyl alcohol	3-Methyl-1-butanol	88.15	131	0.8092	2 g/dL at 14 °C
Isobutyl alcohol	2-Methyl-1-propanol	74.12	108.4	0.8018	1 part in 20
Isopropyl alcohol	2-Propanol	60.09	82.4	0.785	yes
Methanol	Carbinol; Methyl alcohol	32.04	65	0.7914	yes
Methyl isobutyl ketone	4-Methyl-2-pentanone	100.16	117	0.801	moderately
Methylene chloride	Dichloromethane	84.93	40	1.326	1 part in 50
Nitrobenzene		123.11	211	1.2037	1 part in 500
Petroleum ether		varies with fraction	35–80; varies with fraction	0.625–0.660	no
Pyridine		79.10	115.5	0.9780§	yes
Toluene	Methylbenzene	92.13	110.6	0.8669	very slightly
o-Xylene	1,2-Dimethylbenzene	106.16	144	0.8802	no
m-Xylene	1,3-Dimethylbenzene	106.16	139	0.8642	no
p-Xylene	1,4-Dimethylbenzene	106.16	138	0.8611	no

* See footnote for Table 29-1, page 930.
† Boiling points were measured at or near 760 mm Hg.
‡ The density of the solvent was measured at 20 °C; the density of water was measured at 4 °C.
§ d_4^{25}
Information compiled from Weast, R. C., Ed-in-Chief. CRC Handbook of Chemistry and Physics, 65th ed. Boca Raton, FL, CRC Press, Inc., 1984–85; Windholz, M., Ed. The Merck Index, 10th ed. Rahway, NJ, Merck and Co., Inc., 1983.

Table 29-6. PRIMARY AND SECONDARY STANDARDS

SRM*	Standard	Formula	Purity,† Assay,† or pH(S) at 25 °C†	Molecular Weight (Also, Gram Quantity Needed for 1 Liter of Solution Containing 1 Mole)
186IIc	Disodium hydrogen phosphate	Na_2HPO_4	pH 7.415	141.98
	Hydrochloric acid	HCl	36.5–38%	36.46
	Nitric acid	HNO_3	69–71%	63.01
	Oxalic acid	$H_2C_2O_4$		90.04
	Oxalic acid, dihydrate	$H_2C_2O_4 \cdot 2H_2O$	pH 6.863	126.07
186Ic	Potassium dihydrogen phosphate	KH_2PO_4	≥ 99.5%	136.09
185e	Potassium hydrogen phthalate	$KHC_8H_4O_4$	pH 4.004	204.22
188	Potassium hydrogen tartrate	$KHC_4H_4O_6$	pH 3.557	188.18
189	Potassium tetroxalate	$KHC_2O_4 \cdot H_2C_2O_4$	pH 1.679	218.16
	Potassium iodate	KIO_3	~99%	214.02
191a	Sodium bicarbonate	$NaHCO_3$	pH 10.011	84.00
192a	Sodium carbonate	N_2CO_3	≥ 99.95%	106.00
40h	Sodium oxalate	$Na_2C_2O_4$	99.972%	134.01
187b	Sodium tetraborate decahydrate (Borax)	$Na_2B_4O_7 \cdot 10H_2O$	pH 9.183	381.27
	Succinic acid	$HOOC\text{–}(CH_2)_2\text{–}COOH$	≥ 99.0%	118.09
	Sulfuric acid	H_2SO_4	95–98%	98.08
922	Tris (hydroxymethyl)-aminomethane (tromethamine, THAM or Tris)	$NH_2C(CH_2OH)_3$	99.9%	121.14
923	Tris HCl	$NH_2C(CH_2OH)_3\ HCl$	99.7%	157.59

*Standard Reference Material available from the National Bureau of Standards, Washington, D.C.
†May vary with lot or manufacturer.

Table 29-7. pH INDICATORS

Common Name	Chemical Name	pH Range	Color Change	Commonly Used Concentration
Alizarin	1,2-Dihydroxyanthraquinone	5.5–6.8	Yellow to red	0.50 g/dL in ethanol
Bromcresol green (bromocresol green)	3,3′,5,5′-Tetrabromo-*m*-cresolsulfonphthalein	3.8–5.4	Yellow to blue-green	0.10 g in 7.15 mL NaOH, 0.02 mol/L; dil. to 250 mL with water
Bromcresol purple (bromocresol purple)	5,5′-Dibromo-*o*-cresolsulfonphthalein	5.2–6.8	Yellow to purple	0.10 g in 9.25 mL NaOH, 0.02 mol/L; dil. to 250 mL with water
Bromphenol blue (bromophenol blue)	3,3′,5,5′-Tetrabromophenolsulfonphthalein	3.0–4.6	Yellow to purple	0.10 g in 7.45 mL NaOH, 0.02 mol/L; dil. to 250 mL with water
Bromphenol red (bromophenol red)	Dibromophenolsulfonphthalein	5.2–6.8	Yellow to red	0.04 g/dL in water (salt form)
Bromthymol blue (bromothymol blue)	3,3′-Dibromothymolsulfonphthalein	6.0–7.6	Yellow to blue	0.10 g in 8.0 mL NaOH, 0.02 mol/L; dil. to 250 mL with water
Chlorphenol red (chlorophenol red)	3,3′-Dichlorophenolsulfonphthalein	5.2–6.8	Yellow to red	0.10 g in 23.6 mL NaOH, 0.01 mol/L; dil. to 250 mL with water
Congo red	Sodium diphenyldiazo-bis-α-naphthylaminesulfonate	3.0–5.0	Blue-violet to red	0.10 g/dL in water
Cresol red	*o*-Cresolsulfonphthalein; α-hydroxy-α,α-bis(4-hydroxy-*m*-tolyl)-*o*-toluenesulfonic acid γ-sultone	7.2–8.8	Yellow to red (orange to amber at pH 2–3)	0.04 g/100 mL of NaOH, 1.1 mmol/L
Litmus	Lacmus; tournesol; turnsole; lacca musica; lacca coerulea (blue coloring matter of various lichens)	4.5–8.3	Red to blue	
Metachrome yellow (alizarine yellow GG)	Sodium *m*-nitrobenzeneazosalicylate	10.2–12.0	Colorless to yellow	0.10 g/dL in 50% ethanol
Metacresol purple	*m*-Cresolsulfonphthalein	1.2–2.8 7.4–9.0	Red to yellow Yellow to purple	0.10 g in 26.2 mL NaOH, 0.01 mol/L + 223.8 mL water
Methyl orange	Sodium *p*-dimethylaminoazobenzenesulfonate	3.1–4.4	Red to yellow	0.10 g/dL in water
Methyl red	4′-Dimethylaminobenzene 2-carboxylic acid	4.4–6.2	Red to yellow	0.10 g/dL in ethanol
Methyl violet	Pentamethylbenzyl-*p*-rosaniline HCl	0.0–1.6	Yellow to blue	0.01–0.05 g/dL in water
Neutral red	Aminodimethylaminotoluaminozine hydrochloride	6.8–8.0	Red to yellow	0.10 g/dL in 60% alcohol
Phenol red	Phenolsulfonphthalein; α-hydroxy-α,α-bis(*p*-hydroxyphenyl)-*o*-toluenesulfonic acid γ-sultone	6.8–8.4	Yellow to red	0.02–0.05 g/dL in ethanol
Phenolphthalein	3,3-Bis(*p*-hydroxyphenyl)-phthalide	8.3–10.0	Colorless to red	0.10 g/dL in 95% ethanol
Thymol blue	Thymolsulfonphthalein; α-hydroxy-α,α-bis(5-hydroxycarbacryl)-*o*-toluenesulfonic acid γ-sultone	1.2–2.8 8.0–9.6	Red to yellow Yellow to blue	If sodium salt is used: 0.04 g/dL in 95% ethanol If acid form is used: 0.04 g/dL NaOH, 1.0 mol/L
Thymolphthalein	5,5′-Diisopropyl-2,2′-dimethylphenolphthalein	9.3–10.5	Colorless to blue	0.10 g/dL in 95% ethanol
Toepfer's reagent	Dimethylaminoazobenzene	2.9–4.0	Red to yellow	0.50 g/dL in 95% ethanol

Table 29-8. IONIZATION CONSTANTS K AND pK_a FOR COMMON ACIDS AND BASES IN WATER*

	K	pK_a
Acetic acid	1.75×10^{-5}	4.76
Acetoacetic acid	2.62×10^{-4}	3.58 (18 °C)
Ammonia	5.6×10^{-10}	9.25
Boric acid†	6.4×10^{-10}	9.19
Carbonic acid	4.47×10^{-7}	6.35
	4.68×10^{-11}	10.34
Citric acid	7.4×10^{-4}	3.13
	1.7×10^{-5}	4.77
	4.0×10^{-7}	6.40
Diethylbarbituric acid (Veronal)	3.7×10^{-8}	7.43
Ethylenediamine	1.4×10^{-7}	6.85
	1.12×10^{-10}	9.93
Ethylenediamine-tetraacetate	1.00×10^{-2}	2.00
	2.16×10^{-3}	2.67
	6.92×10^{-7}	6.16
	5.50×10^{-11}	10.26
Formic acid	1.76×10^{-4}	3.75
Glycine	4.5×10^{-3}	2.35
	1.7×10^{-10}	9.77
Glycylglycine	7.24×10^{-4}	3.14
	5.62×10^{-9}	8.25
Hydroxylamine	9.1×10^{-9}	8.04
Imidazole	1.01×10^{-7}	6.95
Isocitric acid	5.13×10^{-4}	3.29
	1.99×10^{-5}	4.70
	3.98×10^{-7}	6.40
p-Nitrophenol	7.0×10^{-8}	7.15
Oxalacetic acid	2.75×10^{-3}	2.56
	4.27×10^{-5}	4.37
Oxalic acid	6.5×10^{-2}	1.19
	6.1×10^{-5}	4.21
Phosphoric acid	7.5×10^{-3}	2.12
	6.2×10^{-8}	7.21
	4.8×10^{-13}	12.32
Phosphorous acid	1.0×10^{-2}	2.00
	2.6×10^{-7}	6.59
Pyruvic acid	3.23×10^{-3}	2.49
Succinic acid	6.2×10^{-5}	4.21
	2.3×10^{-6}	
Sulfuric acid	>>1	—
	1.2×10^{-2}	1.92
Tartaric acid	1.1×10^{-3}	2.96
	6.9×10^{-5}	4.16
Triethanolamine	1.26×10^{-8}	7.90
Tris(hydroxymethyl)-aminomethane	8.32×10^{-9}	8.08

*Temperature at or near room temperature (25 °C) unless otherwise indicated.
† Boric acid acts as a monotropic acid in aqueous solution.

Table 29-9. COMMON CONCENTRATED ACIDS AND BASES

Acid or Base	Spec. Grav. at 25 °C*	% by Wt., w/w*	g/L*	M.W.	Approx. Molarity	Approx. Normality	Approx. mL Required to Make 1 L of 1 N Soln.
Acetic acid (CH_3COOH)	1.06	99.5	1060	60.05	17.6	17.6	57
Ammonium hydroxide (NH_4OH)	0.880	29	252 (NH_3)	17.03 (NH_3)	14.8	14.8	67
Hydrochloric acid (HCl)	1.19	37	440	36.46	12.1	12.1	83
Nitric acid (HNO_3)	1.42	70	990	63.02	15.7	15.7	64
Perchloric acid ($HClO_4$), 60%	1.53	60–62	934	100.46	9.3	9.3	108
72%	1.68	70–72	1195	100.46	11.9	11.9	84
Phosphoric acid (H_3PO_4)	1.7	85	1441	98.00	14.7	44.1	23
Potassium hydroxide, saturated† (KOH)	1.55	~50	785	56.11	14	14	71
Sodium hydroxide, saturated† ($NaOH$)	1.50	~50	750	40.00	19	19	53
Sulfuric acid (H_2SO_4)	1.84	96	1765	98.08	18	36	28

*May vary with lot or manufacturer.
† Saturated solutions made from the usual C.P. sodium or potassium hydroxide will vary in strength, chiefly because of the variable amount of carbonate that such solutions contain.

Table 29-10. ATOMIC WEIGHTS*

Name	Atomic Number	Symbol	International Atomic Mass	Oxidative States
Aluminum	13	Al	26.98	+3
Antimony (Stibium)	51	Sb	121.8	+3, +5, −3
Argon	18	Ar	39.95	0
Arsenic	33	As	74.92	+3, +5, −3
Barium	56	Ba	137.3	+2
Beryllium	4	Be	9.012	+2
Bismuth	83	Bi	209.0	+3, +5
Boron	5	B	10.81	+3
Bromine	35	Br	79.90	+1, +5, −1
Cadmium	48	Cd	112.4	+2
Calcium	20	Ca	40.08	+2
Carbon	6	C	12.01	+2, +4, −4
Cerium	58	Ce	140.1	+3, +4
Cesium	55	Cs	132.9	+1
Chlorine	17	Cl	35.45	+1, +5, +7, −1
Chromium	24	Cr	52.00	+2, +3, +6
Cobalt	27	Co	58.93	+2, +3
Copper	29	Cu	63.55	+1, +2
Fluorine	9	F	19.00	−1
Gold (Aurum)	79	Au	197.0	+1, +3
Helium	2	He	4.003	0
Hydrogen	1	H	1.008	+1, −1
Iodine	53	I	126.9	+1, +5, +7, −1
Iron (Ferrum)	26	Fe	55.85	+2, +3
Lanthanum	57	La	138.9	+3
Lead (Plumbum)	82	Pb	207.2	+2, +4
Lithium	3	Li	6.941	+1
Magnesium	12	Mg	24.31	+2

Name	Atomic Number	Symbol	International Atomic Mass	Oxidative States
Manganese	25	Mn	54.94	+2, +3, +4, +7
Mercury	80	Hg	200.6	+1, +2
Molybdenum	42	Mo	95.94	+6
Neon	10	Ne	20.18	0
Nickel	28	Ni	58.69	+2, +3
Nitrogen	7	N	14.01	+1, +2, +3, +4, +5, −1, −2, −3
Oxygen	8	O	16.00	−2
Palladium	46	Pd	106.4	+2, +4
Phosphorus	15	P	30.97	+3, +5, −3
Platinum	78	Pt	195.1	+2, +4
Potassium (Kalium)	19	K	39.10	+1
Selenium	34	Se	78.96	+4, +6, −2
Silicon	14	Si	28.09	+2, +4, −4
Silver (Argentum)	47	Ag	107.9	+1
Sodium (Natrium)	11	Na	22.99	+1
Strontium	38	Sr	87.62	+2
Sulfur	16	S	32.06	+4, +6, −2
Tellurium	52	Te	127.6	+4, +6, −2
Thallium	81	Tl	204.4	+1, +3
Thorium	90	Th	232.0	+4
Tin (Stannum)	50	Sn	181.7	+2, +4
Titanium	22	Ti	47.88	+2, +3, +4
Tungsten (Wolfram)	74	W	183.8	+6
Uranium	92	U	238.0	+3, +4, +5, +6
Vanadium	23	V	50.94	+2, +3, +4, +5
Xenon	54	Xe	131.3	0
Zinc	30	Zn	65.38	+2

*According to the SI system, *mass* is the preferred term; *weight* is more commonly used in the United States. Values have been updated according to Holden, N.E., and Martin, R. L.: Atomic weights of the elements, 1981. Pure and Appl. Chem., *55*(7):1011–1118, 1983. Values are based on carbon-12 and rounded to four significant figures.

BUFFERS COMMONLY USED IN THE LABORATORY

Table 29-11. SØRENSON'S PHOSPHATE BUFFER, 0.1 MOL/L

Na_2HPO_4 (0.1 mol/L; 14.2 g/L), mL	KH_2PO_4 (0.1 mol/L; 13.6 g/L), mL	pH at 20 °C*
0.25	9.75	5.29
0.5	9.5	5.59
1.0	9.0	5.91
2.0	8.0	6.24
3.0	7.0	6.47
4.0	6.0	6.64
5.0	5.0	6.81
6.0	4.0	6.98
7.0	3.0	7.17
8.0	2.0	7.38
9.0	1.0	7.73
9.5	0.5	8.04

Mix solutions in the volumes indicated to achieve 10 mL of buffer at the desired pH.

*~ 0.3 pH unit lower at 37 °C.

Table 29-12. TRIS(HYDROXYMETHYL)-AMINOMETHANE BUFFER

Tris(hydroxymethyl)-aminomethane (~0.08 mol/L; 10.114 g/L), mL	HCl (0.1 mol/L), mL	pH at 23 °C*
50	5.0	9.10
50	7.5	8.92
50	10.0	8.74
50	12.5	8.62
50	15.0	8.50
50	17.5	8.40
50	20.0	8.32
50	22.5	8.23
50	25.0	8.14
50	27.5	8.05
50	30.0	7.96
50	32.5	7.87
50	35.0	7.77
50	37.5	7.66
50	40.0	7.54
50	42.5	7.36
50	45.0	7.20

Mix solutions in the volumes indicated and dilute to 100 mL with deionized water to achieve 100 mL of buffer at the desired pH.

*~ 0.15 pH unit lower at 37 °C.

Table 29-13. DELORY AND KING'S CARBONATE-BICARBONATE BUFFER

Na_2CO_3 (0.1 mol/L; 10.6 g/L), mL	$NaHCO_3$ (0.1 mol/L; 8.4 g/L), mL	pH at 25 °C*
1.1	8.9	9.1
1.4	8.6	9.2
2.2	7.8	9.4
2.7	7.3	9.5
3.9	6.2	9.7
5.1	4.9	9.9
6.4	3.6	10.1
7.4	2.5	10.3
7.9	2.1	10.4
8.3	1.6	10.5
8.8	1.2	10.6

Mix solutions in the volumes indicated to achieve 10 mL of buffer at the desired pH.

*~ 0.1 pH unit lower at 37 °C.

Table 29-14. ACETIC ACID–SODIUM ACETATE BUFFER

CH_3COOH (0.2 mol/L; Dilute 11.5 mL Reagent Grade Glacial Acetic Acid to 1 L), mL	CH_3COONa (0.2 mol/L; 16.4 g/L), mL	pH at 25 °C*
92.5	7.5	3.6
88.0	12.0	3.8
82.0	18.0	4.0
73.5	26.5	4.2
63.0	37.0	4.4
52.0	48.0	4.6
41.0	59.0	4.8
30.0	70.0	5.0
21.0	79.0	5.2
14.0	86.0	5.4
9.0	91.0	5.6
6.0	94.0	5.8

Mix solutions in the volumes indicated to achieve 100 mL of buffer at the desired pH.

*~0.05 pH unit lower at 37 °C.

STATISTICAL TABLES

Table 29-15. UPPER TAIL AREAS UNDER THE STANDARD GAUSSIAN CURVE FOR POSITIVE VALUES OF *z*

z	0.00	0.01	0.02	0.03	0.04	0.05	0.06	0.07	0.08	0.09
0.0	0.5000	0.4960	0.4920	0.4880	0.4840	0.4801	0.4761	0.4721	0.4681	0.4641
0.1	0.4602	0.4562	0.4522	0.4483	0.4443	0.4404	0.4364	0.4325	0.4286	0.4247
0.2	0.4207	0.4168	0.4129	0.4090	0.4052	0.4013	0.3974	0.3936	0.3897	0.3859
0.3	0.3821	0.3783	0.3745	0.3707	0.3669	0.3632	0.3594	0.3557	0.3520	0.3483
0.4	0.3446	0.3409	0.3372	0.3336	0.3300	0.3264	0.3228	0.3192	0.3156	0.3121
0.5	0.3085	0.3050	0.3015	0.2981	0.2946	0.2912	0.2877	0.2843	0.2810	0.2776
0.6	0.2743	0.2709	0.2676	0.2643	0.2611	0.2578	0.2546	0.2514	0.2483	0.2451
0.7	0.2420	0.2389	0.2358	0.2327	0.2296	0.2266	0.2236	0.2206	0.2177	0.2148
0.8	0.2119	0.2090	0.2061	0.2033	0.2005	0.1977	0.1949	0.1922	0.1894	0.1867
0.9	0.1841	0.1814	0.1788	0.1762	0.1736	0.1711	0.1685	0.1660	0.1635	0.1611
1.0	0.1587	0.1562	0.1539	0.1515	0.1492	0.1469	0.1446	0.1423	0.1401	0.1379
1.1	0.1357	0.1335	0.1314	0.1292	0.1271	0.1251	0.1230	0.1210	0.1190	0.1170
1.2	0.1151	0.1131	0.1112	0.1093	0.1075	0.1056	0.1038	0.1020	0.1003	0.0985
1.3	0.0968	0.0951	0.0934	0.0918	0.0901	0.0885	0.0869	0.0853	0.0838	0.0823
1.4	0.0808	0.0793	0.0778	0.0764	0.0749	0.0735	0.0721	0.0708	0.0694	0.0681
1.5	0.0668	0.0655	0.0643	0.0630	0.0618	0.0606	0.0594	0.0582	0.0571	0.0559
1.6	0.0548	0.0537	0.0526	0.0516	0.0505	0.0495	0.0485	0.0475	0.0465	0.0455
1.7	0.0446	0.0436	0.0427	0.0418	0.0409	0.0401	0.0392	0.0384	0.0375	0.0367
1.8	0.0359	0.0351	0.0344	0.0336	0.0329	0.0322	0.0314	0.0307	0.0301	0.0294
1.9	0.0287	0.0281	0.0274	0.0268	0.0262	0.0256	0.0250	0.0244	0.0239	0.0233
2.0	0.0228	0.0222	0.0217	0.0212	0.0207	0.0202	0.0197	0.0192	0.0188	0.0183
2.1	0.0179	0.0174	0.0170	0.0166	0.0162	0.0158	0.0154	0.0150	0.0146	0.0143
2.2	0.0139	0.0136	0.0132	0.0129	0.0125	0.0122	0.0119	0.0116	0.0113	0.0110
2.3	0.0107	0.0104	0.0102	0.0099	0.0096	0.0094	0.0091	0.0089	0.0087	0.0084
2.4	0.0082	0.0080	0.0078	0.0075	0.0073	0.0071	0.0069	0.0068	0.0066	0.0064
2.5	0.0062	0.0060	0.0059	0.0057	0.0055	0.0054	0.0052	0.0051	0.0049	0.0048
2.6	0.0047	0.0045	0.0044	0.0043	0.0041	0.0040	0.0039	0.0038	0.0037	0.0036
2.7	0.0035	0.0034	0.0033	0.0032	0.0031	0.0030	0.0029	0.0028	0.0027	0.0026
2.8	0.0026	0.0025	0.0024	0.0023	0.0023	0.0022	0.0021	0.0021	0.0020	0.0019
2.9	0.0019	0.0018	0.0018	0.0017	0.0016	0.0016	0.0015	0.0015	0.0014	0.0014
3.0	0.0013	0.0013	0.0013	0.0012	0.0012	0.0011	0.0011	0.0011	0.0010	0.0010
3.1	0.0010	0.0009	0.0009	0.0009	0.0008	0.0008	0.0008	0.0008	0.0007	0.0007
3.2	0.0007	0.0007	0.0006	0.0006	0.0006	0.0006	0.0006	0.0005	0.0005	0.0005
3.3	0.0005	0.0005	0.0005	0.0004	0.0004	0.0004	0.0004	0.0004	0.0004	0.0003
3.4	0.0003	0.0003	0.0003	0.0003	0.0003	0.0003	0.0003	0.0003	0.0003	0.0002
3.5	0.0002	0.0002	0.0002	0.0002	0.0002	0.0002	0.0002	0.0002	0.0002	0.0002
3.6	0.0002	0.0002	0.0001	0.0001	0.0001	0.0001	0.0001	0.0001	0.0001	0.0001
3.7	0.0001	0.0001	0.0001	0.0001	0.0001	0.0001	0.0001	0.0001	0.0001	0.0001
3.8	0.0001	0.0001	0.0001	0.0001	0.0001	0.0001	0.0001	0.0001	0.0001	0.0001
3.9	0.0000	0.0000	0.0000	0.0000	0.0000	0.0000	0.0000	0.0000	0.0000	0.0000

Because of the symmetry of the standard Gaussian curve, the entries in this table are also the lower tail areas for the negative value of *z*.

Table 29-16. PROBABILITY POINTS OF THE STUDENT'S t DISTRIBUTION WITH ν DEGREES OF FREEDOM; α IS THE UPPER TAIL PROBABILITY

ν \ α	0.40	0.25	0.10	0.05	0.025	0.01	0.005	0.0005
1	0.325	1.000	3.078	6.314	12.706	31.821	63.657	636.619
2	0.289	0.816	1.886	2.920	4.303	6.965	9.925	31.598
3	0.277	0.765	1.638	2.353	3.182	4.541	5.841	12.941
4	0.271	0.741	1.533	2.132	2.776	3.747	4.604	8.610
5	0.267	0.727	1.476	2.015	2.571	3.365	4.032	6.859
6	0.265	0.718	1.440	1.943	2.447	3.143	3.707	5.959
7	0.263	0.711	1.415	1.895	2.365	2.998	3.499	5.405
8	0.262	0.706	1.397	1.860	2.306	2.896	3.355	5.041
9	0.261	0.703	1.383	1.833	2.262	2.821	3.250	4.781
10	0.260	0.700	1.372	1.812	2.228	2.764	3.169	4.587
11	0.260	0.697	1.363	1.796	2.201	2.718	3.106	4.437
12	0.259	0.695	1.356	1.782	2.179	2.681	3.055	4.318
13	0.259	0.694	1.350	1.771	2.160	2.650	3.012	4.221
14	0.258	0.692	1.345	1.761	2.145	2.624	2.977	4.140
15	0.258	0.691	1.341	1.753	2.131	2.602	2.947	4.073
16	0.258	0.690	1.337	1.746	2.120	2.583	2.921	4.015
17	0.257	0.689	1.333	1.740	2.110	2.567	2.898	3.965
18	0.257	0.688	1.330	1.734	2.101	2.552	2.878	3.922
19	0.257	0.688	1.328	1.729	2.093	2.539	2.861	3.883
20	0.257	0.687	1.325	1.725	2.086	2.528	2.845	3.850
21	0.257	0.686	1.323	1.721	2.080	2.518	2.831	3.819
22	0.256	0.686	1.321	1.717	2.074	2.508	2.819	3.792
23	0.256	0.685	1.319	1.714	2.069	2.500	2.807	3.767
24	0.256	0.685	1.318	1.711	2.064	2.492	2.797	3.745
25	0.256	0.684	1.316	1.708	2.060	2.485	2.787	3.725
26	0.256	0.684	1.315	1.706	2.056	2.479	2.779	3.707
27	0.256	0.684	1.314	1.703	2.052	2.473	2.771	3.690
28	0.256	0.683	1.313	1.701	2.048	2.467	2.763	3.674
29	0.256	0.683	1.311	1.699	2.045	2.462	2.756	3.659
30	0.256	0.683	1.310	1.697	2.042	2.457	2.750	3.646
40	0.255	0.681	1.303	1.684	2.021	2.423	2.704	3.551
60	0.254	0.679	1.296	1.671	2.000	2.390	2.660	3.460
120	0.254	0.677	1.289	1.658	1.980	2.358	2.617	3.373
∞	0.253	0.674	1.282	1.645	1.960	2.326	2.576	3.291

Table 29-17. PROBABILITY POINTS OF THE F DISTRIBUTION WITH ν_1 AND ν_2 DEGREES OF FREEDOM*

ν_2 \ ν_1	1	2	3	4	5	6	7	8	9	10	12	15	20	24	30	40	60	120	∞
1	161.4	199.5	215.7	224.6	230.2	234.0	236.8	238.9	240.5	241.9	243.9	245.9	248.0	249.1	250.1	251.1	252.2	253.3	254.3
2	18.51	19.00	19.16	19.25	19.30	19.33	19.35	19.37	19.38	19.40	19.41	19.43	19.45	19.45	19.46	19.47	19.48	19.49	19.50
3	10.13	9.55	9.28	9.12	9.01	8.94	8.89	8.85	8.81	8.79	8.74	8.70	8.68	8.64	8.62	8.59	8.57	8.55	8.53
4	7.71	6.94	6.59	6.39	6.26	6.16	6.08	6.04	6.00	5.96	5.91	5.86	5.80	5.77	5.75	5.72	5.69	5.66	5.63
5	6.61	5.79	5.41	5.19	5.05	4.95	4.88	4.82	4.77	4.74	4.68	4.62	4.56	4.53	4.50	4.46	4.43	4.40	4.36
6	5.99	5.14	4.76	4.53	4.39	4.28	4.21	4.15	4.10	4.06	4.00	3.94	3.87	3.84	3.81	3.77	3.74	3.70	3.67
7	5.59	4.74	4.35	4.12	3.97	3.87	3.79	3.73	3.68	3.64	3.57	3.51	3.44	3.41	3.38	3.34	3.30	3.27	3.23
8	5.32	4.46	4.07	3.84	3.69	3.58	3.50	3.44	3.39	3.35	3.28	3.22	3.15	3.12	3.08	3.04	3.01	2.97	2.93
9	5.12	4.26	3.86	3.63	3.48	3.37	3.29	3.23	3.18	3.14	3.07	3.01	2.94	2.90	2.86	2.83	2.79	2.75	2.71
10	4.96	4.10	3.71	3.48	3.33	3.22	3.14	3.07	3.02	2.98	2.91	2.85	2.77	2.74	2.70	2.66	2.62	2.58	2.54
11	4.84	3.98	3.59	3.36	3.20	3.09	3.01	2.95	2.90	2.85	2.79	2.72	2.65	2.61	2.57	2.53	2.49	2.45	2.40
12	4.75	3.89	3.49	3.26	3.11	3.00	2.91	2.85	2.80	2.75	2.69	2.62	2.54	2.51	2.47	2.43	2.38	2.34	2.30
13	4.67	3.81	3.41	3.18	3.03	2.92	2.83	2.77	2.71	2.67	2.60	2.53	2.46	2.42	2.38	2.34	2.30	2.25	2.21
14	4.60	3.74	3.34	3.11	2.96	2.85	2.76	2.70	2.65	2.60	2.53	2.46	2.39	2.35	2.31	2.27	2.22	2.18	2.13
15	4.54	3.68	3.29	3.06	2.90	2.79	2.71	2.64	2.59	2.54	2.48	2.40	2.33	2.29	2.25	2.20	2.16	2.11	2.07
16	4.49	3.63	3.24	3.01	2.85	2.74	2.66	2.59	2.54	2.49	2.42	2.35	2.28	2.24	2.19	2.15	2.11	2.06	2.01
17	4.45	3.59	3.20	2.96	2.81	2.70	2.61	2.55	2.49	2.45	2.38	2.31	2.23	2.19	2.15	2.10	2.06	2.01	1.96
18	4.41	3.55	3.16	2.93	2.77	2.66	2.58	2.51	2.46	2.41	2.34	2.27	2.19	2.15	2.11	2.06	2.02	1.97	1.92
19	4.38	3.52	3.13	2.90	2.74	2.63	2.54	2.48	2.42	2.38	2.31	2.23	2.16	2.11	2.07	2.03	1.98	1.93	1.88
20	4.35	3.49	3.10	2.87	2.71	2.60	2.51	2.45	2.39	2.35	2.28	2.20	2.12	2.08	2.04	1.99	1.95	1.90	1.84
21	4.32	3.47	3.07	2.84	2.68	2.57	2.49	2.42	2.37	2.32	2.25	2.18	2.10	2.05	2.01	1.95	1.92	1.87	1.81
22	4.30	3.44	3.05	2.82	2.66	2.55	2.46	2.40	2.34	2.30	2.23	2.15	2.07	2.03	1.98	1.94	1.89	1.84	1.78
23	4.28	3.42	3.03	2.80	2.64	2.53	2.44	2.37	2.32	2.27	2.20	2.13	2.05	2.01	1.96	1.91	1.86	1.81	1.76
24	4.26	3.40	3.01	2.78	2.62	2.51	2.42	2.36	2.30	2.25	2.18	2.11	2.03	1.98	1.94	1.89	1.84	1.79	1.73
25	4.24	3.39	2.99	2.76	2.60	2.49	2.40	2.34	2.28	2.24	2.16	2.09	2.01	1.96	1.92	1.87	1.82	1.77	1.71
26	4.23	3.37	2.98	2.74	2.59	2.47	2.39	2.32	2.27	2.22	2.15	2.07	1.99	1.95	1.90	1.85	1.80	1.75	1.69
27	4.21	3.35	2.96	2.73	2.57	2.46	2.37	2.31	2.25	2.20	2.13	2.06	1.97	1.93	1.88	1.84	1.79	1.73	1.67
28	4.20	3.34	2.95	2.71	2.56	2.45	2.36	2.29	2.24	2.19	2.12	2.04	1.96	1.91	1.87	1.82	1.77	1.71	1.65
29	4.18	3.33	2.93	2.70	2.55	2.43	2.35	2.28	2.22	2.18	2.10	2.03	1.94	1.90	1.85	1.81	1.75	1.70	1.64
30	4.17	3.32	2.92	2.69	2.53	2.42	2.33	2.27	2.21	2.16	2.09	2.01	1.93	1.89	1.84	1.79	1.74	1.68	1.62
40	4.08	3.23	2.84	2.61	2.45	2.34	2.25	2.18	2.12	2.08	2.00	1.92	1.84	1.79	1.74	1.69	1.64	1.58	1.51
60	4.00	3.15	2.76	2.53	2.37	2.25	2.17	2.10	2.04	1.99	1.92	1.84	1.75	1.70	1.65	1.59	1.53	1.47	1.39
120	3.92	3.07	2.68	2.45	2.29	2.17	2.09	2.02	1.96	1.91	1.83	1.75	1.66	1.61	1.55	1.50	1.43	1.35	1.25
∞	3.84	3.00	2.60	2.37	2.21	2.10	2.01	1.94	1.88	1.83	1.75	1.67	1.57	1.52	1.46	1.39	1.32	1.22	1.00

*α = 0.05 (upper tail)

NOMOGRAMS

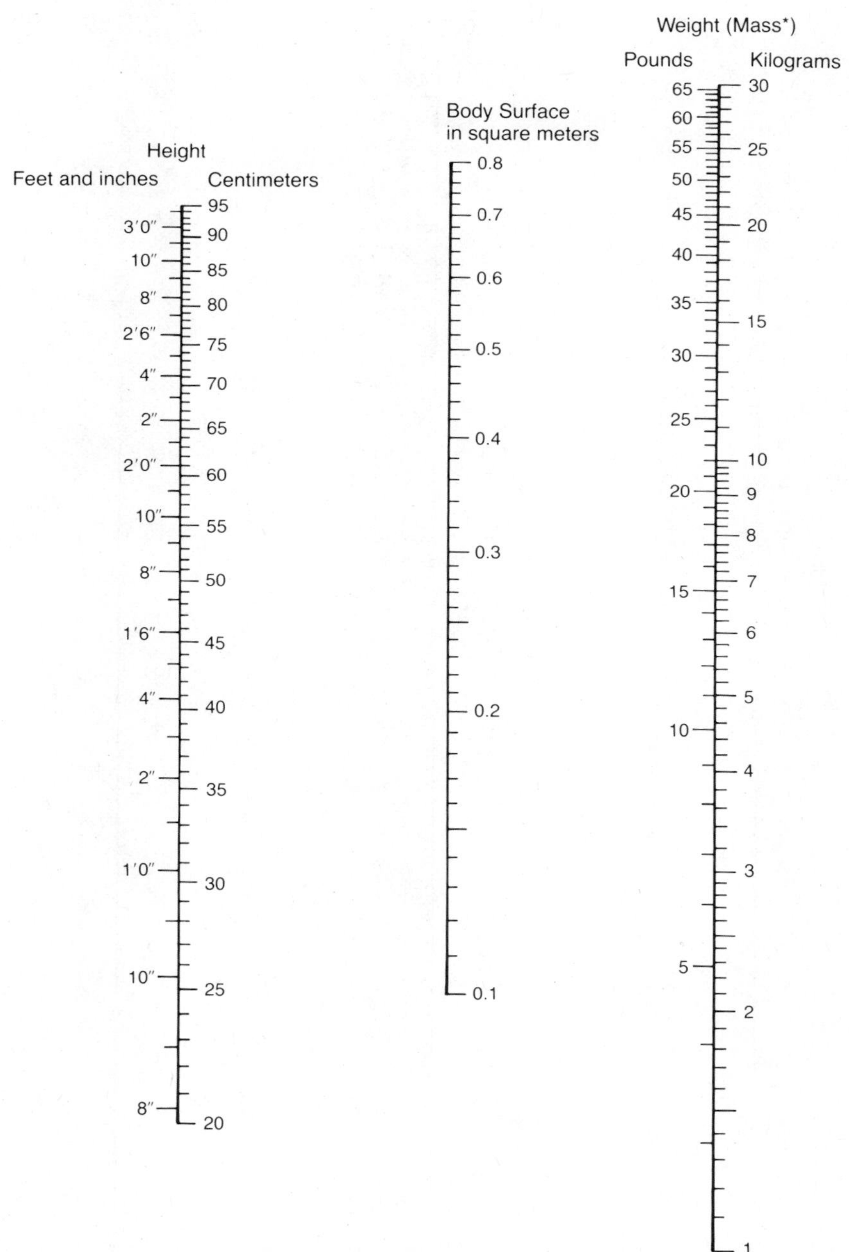

Figure 29-1. Nomogram for the determination of body surface area of children. (From DuBois, E. F.: Basal Metabolism in Health and Disease. Philadelphia, Lea & Febiger, 1936.)
*According to the SI system, *mass* is the preferred term; *weight* is more commonly used in the United States.

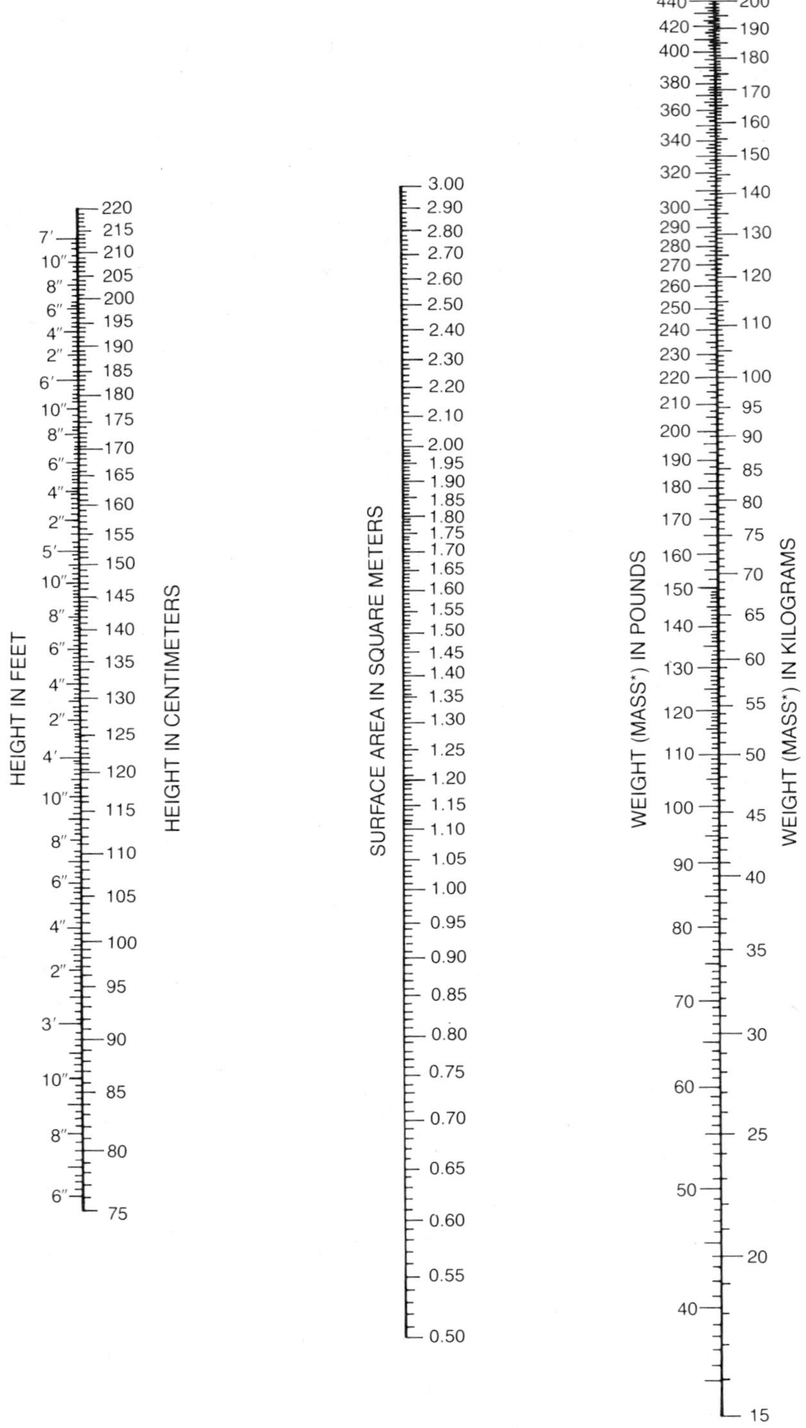

Figure 29-2. Nomogram for the determination of body surface area of children and adults. (From Boothby, W. M., and Sandiford, R. B.: Boston M. & S. J. *185*:337, 1921.)

*According to the SI system, *mass* is the preferred term; *weight* is more commonly used in the United States.

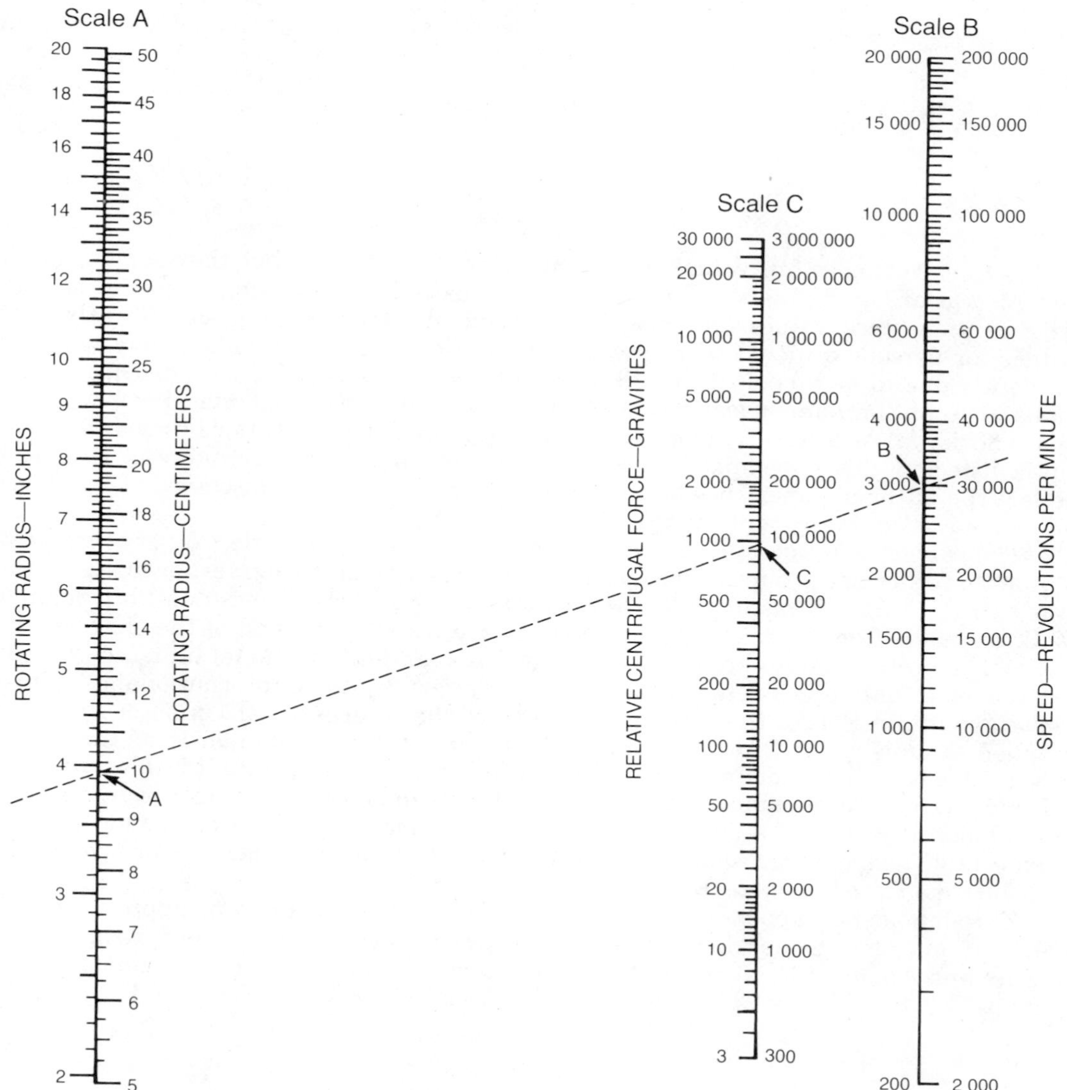

Figure 29-3. Nomogram for calculating relative centrifugal force. (Modified from the IEC [a division of Damon Corporation, Needham Heights, MA 02194] Relative Centrifugal Force Nomograph.)

Instructions

To find, for example, the relative centrifugal force (RCF) at a radial distance of 10 cm from the center of rotation when the centrifuge is operated at a speed of 3000 rpm, place a straightedge on the chart, connecting the 10-cm point on Scale A (rotating radius) with the 3000-rpm point on Scale B (speed). Read the point at which the straightedge intersects Scale C (relative centrifugal force); for this example, that point is $1000 \times g$. Use the right side of Scales B and C for high-speed and ultracentrifuges.

If the desired RCF is known, the necessary speed for a particular rotating radius may be determined by connecting appropriate points on Scales A and C, and reading the speed from Scale B.

Equation for calculating RCF:

$$RCF = 1.118 \times 10^{-5} \times r \times n^2$$

where RCF is obtained in gravities ($\times g$),
r is rotating radius in centimeters, and
n is rotating speed in revolutions per min.

REFERENCE RANGES

INTRODUCTION

Reference ranges are valuable guidelines for the clinician, but they should not be regarded as absolute indicators of health and disease. Reference ranges should be used with caution since values for "healthy" individuals often overlap significantly with values for persons afflicted with disease. In addition, laboratory values may vary significantly due to methodological differences and mode of standardization. This is especially true for immunological tests, which utilize antibodies that may have different characteristics. As a result, laboratory values in individual institutions may differ from those listed in this appendix. Reference ranges given in other chapters of this book reflect the authors' experience and are specific for the methods discussed. Thus, these values may also differ from those listed here.

The values in this chapter are for adults in the fasting state unless otherwise stated. Values for other age groups, when included, are clearly identified. Most of this material was extracted from *Clinical Guide to Laboratory Tests,* N. W. Tietz, Ed.; Philadelphia, W.B. Saunders Company, 1983. Other sources are listed at the end of the chapter.

All laboratory values are given in conventional and international units. In general, the international units given conform to the SI system (Système International d'Unités). However, in some cases the recommendations of the International Union of Pure and Applied Chemistry (IUPAC) and the Commission on World Standards of the World Association of Societies of Pathology (COWS of WASP) are used, since it is felt that these have found wider acceptance in clinical laboratories and offer advantages over the units recommended in the SI system. If the exact molecular weight of a compound has not been established, e.g., for some proteins, values in the conventional unit are converted to values expressed in mass/L.

Throughout this appendix, we have used the prefixes for units as approved by the CGPM (Conférence Générale des Poids et Mésures), 1964, and the International Union of Pure and Applied Chemistry (IUPAC) and International Federation of Clinical Chemistry (IFCC). See Clinica Chimica Acta, *96*:157F–183F, 1979, and Chapter 1.

Abbreviations

For convenience and to preserve space, we have used standard abbreviations commonly used in laboratory medicine. Less common abbreviations and some nonstandard abbreviations are given below.

ACD	acid-citrate-dextrose
Amf	amniotic fluid
Ascf	ascitic fluid
Br	breath
Cit	citrate
CSF	cerebrospinal fluid
EDTA	ethylenediaminetetraacetate, edetic acid
F	feces
F^-	fluoride
F^-/Ox	fluoride and oxalate
Gast Cont	gastric contents
Gastf	gastric fluid
Gast Res	gastric residue
H	hair
Hep	heparin
Occup. exp.	occupational exposure
Ox	oxalate
P	plasma
Pericf	pericardial fluid
Placent	placental
Plf	pleural fluid
Plt	platelets
post-stim	post-stimulation
pp	postprandial
RBC	red blood cells
S	serum
Sal	saliva
Semf	seminal fluid
Serf	serous fluid
Swt	sweat
Synf	synovial fluid
T	tissue
U	urine
WB	whole blood

Table 29-18. CLINICAL CHEMISTRY AND TOXICOLOGY

Test	Specimen	Reference Range		Conversion Factor	Reference Range (International Units)
Acetoacetate					
Semiquant.	S or P (F^-/Ox)	Negative (< 3 mg/dL)		× 0.098	Negative (< 0.3 mmol/L)
	U	Negative (< 50 mg/dL)			Negative (< 4.9 mmol/L)
Acetone					
Semiquant.	S or P (F^- or Ox)	Negative (< 3 mg/dL)		× 0.172	Negative (< 0.52 mmol/L)
			mg/dL		*mmol/L*
Quant.			0.3–2.0	× 0.172	0.05–0.34
Semiquant.	U	Negative			Negative
Quant.			0.3	× 0.172	0.05
Acetylcholinesterase in Erythrocytes	WB (ACD, EDTA, or Hep)	36.9 ± 3.83 U/g Hb (SD)		× 0.0645	2.38 ± 0.23 MU/mol Hb
		1070 ± 111 U/10^{12} RBC		× 10^{-3}	1.07 ± 0.11 nU/RBC
		12.5 ± 1.30 U/mL RBC		× 1	12.5 ± 1.30 kU/L RBC
		Lower in newborns			
Adrenocorticotropic Hormone (ACTH)	P (EDTA)		*pg/mL*		*ng/L*
		0800 h:	25–100	× 1	25–100
		1800 h:	< 50		< 50
Adrenocorticotropic Hormone Stimulation Test (Prolonged Infusion)	U, 24 h	17-KGS:	2–4-fold rise		Same
		17-KS:	2-fold rise		
		17-OHCS:	2–5-fold rise		
Dose: 500 μg Cortrosyn/d × 3	S	Cortisol:	25–50 μg/dL	× 0.028	Cortisol: 0.7–1.4 μmol/L
Adrenocorticotropic Hormone Stimulation Test (Rapid Test) *Dose: 250 μg Cortrosyn, IM*	S: baseline, 30 and 60 min post-stim	Cortisol,			
		Baseline:	> 5.0 μg/dL	× 0.028	Baseline: > 0.14 μmol/L
		After Cortrosyn:	2 × baseline		Same
Alanine Aminotransferase (ALT, GPT)			*U/L*		
Henry, optimized, 30 °C	S	Newborn/Infant:	5–28		Same
		Adult:	8–20		
		> 60 y, M:	7–24		
		F:	7–16		
IFCC, with P-5′-P, 30 °C			< 30		
Albumin					
Nephelometric, colorimetric	S	Adult:	3.5–5.0 g/dL	× 10	35–50 g/L
		> 60 y:	3.4–4.8 g/dL		34–48 g/L
		Avg. ~0.3 g/dL higher in upright individuals			Avg. ~3 g/L higher in upright individuals
Nephelometric, rate	CSF	10–30 mg/dL		× 10	100–300 mg/L
	U	At rest:	< 80 mg/d		Same
		Ambulatory:	50–150 mg/d		
Aldosterone	S or P (Hep, EDTA)	Adult, *average sodium diet*			
			ng/dL		*nmol/L*
		Supine:	3–10	× 0.0277	0.083–0.28
		Upright, F:	5–30		0.138–0.83
		M:	6–22		0.166–0.609
		2–3 × higher during pregnancy			
		Low sodium diet:			
		Increases 2–5-fold			
		Florinef suppression:	< 4 ng/dL		< 0.11
		ACTH or angiotensin stimulation, 1 h: 2–5-fold increase			
	U, 24 h	*Total Urinary Na, mEq/d* — *Plasma Renin Activity, (ng AI*/h)/mL*	*Urinary Aldosterone, μg/d*		*Urinary Aldosterone, nmol/d*
		< 20 — 5–24	35–80	× 2.77	97–222
		50 — 2–7	13–33		36–91.4
		100 — 1–5	5–24		13.8–66.5
		150 — 0.5–4	3–19		8.3–52.6
		200	1–16		2.7–44.3
		250	1–13		2.7–36
		(Assuming normal serum Na, K, and extracellular vol)			

*Angiotensin I

(continued)

Table 29-18. CLINICAL CHEMISTRY AND TOXICOLOGY *(Continued)*

Test	Specimen	Reference Range		Conversion Factor	Reference Range (International Units)
Aluminum (Al)			*μg/L*		*μmol/L*
	S		3.5–8.5	× 0.0371	0.13–0.32
		Patients on Al medication:	< 30		< 1.11
δ-Aminolevulinate Dehydratase *Colorimetric, 38 °C*	RBC (Hep)	139–211 U/mL erythrocytes		× 1	139–211 kU/L
δ-Aminolevulinic Acid (δ-ALA)	S	15–23 μg/dL; lower in children		× 0.076	1.14–1.75 μmol/L
	U	1.3–7.0 mg/d		× 7.626	9.9–53.4 μmol/d
Ammonia Nitrogen *Resin or enzymatic*	S or P (NaHep)		*μg N/dL*		*μmol N/L*
		Newborn:	90–150	× 0.714	64–107
		0–2 wk:	79–129		56–92
		> 1 mo:	29–70		21–50
		Adult:	15–45		11–32
Amniotic Fluid Analysis $\Delta A_{450\ nm}$ *L/S Ratio, see Lecithin/sphingomyelin Ratio*	Amf	28 wk:	0–0.048 ΔA		Same
		40 wk:	0–0.02 ΔA		
Amylase *UV-method, Beckman; BMD*	S	Adult:	25–125 U/L		Same
		> 70 y:	20–160 U/L		
	U, timed	1–17 U/h			1–17 U/h
Amylase/Creatinine Clearance Ratio	S and U, random	1–4%		× 0.01	Clearance fraction: 0.01–0.04
Angiotensin I	P (KEDTA), periph-vein	11–88 pg/mL		× 1	11–88 ng/L
Angiotensin II	P (KEDTA)	Arterial:	2.4 ± 1.2 ng/dL (SD)	× 10	24 ± 12 ng/L (SD)
		Venous:	50–75% of arterial conc.	× 0.01	Fraction of arterial conc.: 0.50–0.75
Anion Gap [**Na** − (**Cl**$^-$ + **HCO**$_3^-$)]	P (Hep)	7–16 mEq/L		× 1	7–16 mmol/L
α_1-**Antitrypsin** *Nephelometric*	S	Newborn:	145–270 mg/dL	× 0.1	1.45–2.70 g/L
		Adult:	78–200 mg/dL		0.78–2.00 g/L
Arsenic	WB (Hep)		*μg/dL*		*μmol/L*
			0.2–6.2	× 0.133	0.03–0.83
		Chronic poisoning:	10–50		1.33–6.65
		Acute poisoning:	60–930		7.98–124
	U, 24 h	5–50 μg/d		× 0.0133	0.07–0.67 μmol/d
Ascorbic Acid, *see Vitamin C*					
Aspartate Aminotransferase (AST, SGOT), *30 °C*	S		*U/L*		
		Newborn/Infant:	25–75		Same
		Adult:	8–20		
		> 60 y, M:	11–26		
		F:	10–20		
		With P-5′-P:	12–29		
Base Excess	WB (Hep)		*mmol/L*		Same
		Newborn:	(−10) to (−2)		
		Infant:	(−7) to (−1)		
		Child:	(−4) to (+2)		
		Adult:	(−2) to (+3)		
Bicarbonate (**HCO**$_3^-$)	S	Art.:	21–28 mmol/L		Same
		Ven.:	22–29 mmol/L		
			mEq/L		*mmol/L*
	WB, arterial (Hep; anaerobic)	Newborn:	17.2–23.6	× 1	17.2–23.6
		Infant:	19.0–23.9		19.0–23.9
		2 mo–2 y:	16.3–23.9		16.3–23.9
		Adult:	18–23		18–23

Table 29-18. CLINICAL CHEMISTRY AND TOXICOLOGY (Continued)

Test	Specimen	Reference Range		Conversion Factor	Reference Range (International Units)
Bile Acids, Total	S	0.3–2.3 μg/mL		× 2.45	0.74–5.64 μmol/L
GLC	S, 1 h pp	1.8–3.2 μg/mL		(Conv. factor based on	4.41–7.84 μmol/L
	F	120–225 mg/d		cholic acid, M.W. 408.6)	294–551 μmol/d
Bilirubin					
Total	S	*Premature mg/dL*	*Full Term mg/dL*		*Premature* / *Full Term* *μmol/L*
		Cord: <2.0	<2.0	× 17.1	<34.2 / <34.2
		0–1 d: <8.0	<6.0		<137 / <103
		1–2 d: <12.0	<8.0		<205 / <137
		3–5 d: <16.0	<12.0		<274 / <205
		>5 d: <2.0	0.2–1.0		<34.2 / 3.4–17.1
	U	Negative			Negative
	Amf	28 wk:	0.075 mg/dL (or ΔA_{450} <0.048)	× 17.1	1.28 μmol/L
		40 wk:	0.025 mg/dL (or ΔA_{450} <0.02)		0.43 μmol/L
Conjugated (direct)	S	0–0.2 mg/dL		× 17.1	0–3.4 μmol/L
Blood Volume	WB (Hep)	M:	52–83 mL/kg	× 0.001	0.052–0.083 L/kg
		F:	50–75 mL/kg		0.050–0.075 L/kg
C-Peptide	S		*ng/mL*		*μg/L*
		Adult:	≤4.0	× 1	≤4.0
		>60 y, M:	1.5–5.0		1.5–5.0
		F:	1.4–5.5		1.4–5.5
C-Reactive Protein	S	Cord blood:	1–35 μg/dL	× 10	10–350 μg/L
		Adult:	6.8–820 μg/dL		68–8200 μg/L
Calcium, Ionized (Ca_I)	S, P, or WB (Hep)		*mg/dL*		*mmol/L*
		Cord, venous:	5.5 ± 0.3 (SD)	× 0.25	1.38 ± 0.08 (SD)
		Newborn,			
		3–24 h:	4.3–5.1		1.08–1.28
		24–48 h:	4.0–4.7		1.00–1.18
		Adult:	4.48–4.92		1.12–1.23
		or	2.24–2.46 mEq/L	× 0.5	1.12–1.23
Calcium, Total	S		*mg/dL*		*mmol/L*
		Cord:	9.0–11.5	× 0.25	2.25–2.88
		Newborn,			
		3–24 h:	9.0–10.6		2.25–2.65
		24–48 h:	7.0–12.0		1.75–3.0
		Child:	8.8–10.8		2.20–2.70
		Adult:	8.4–10.2		2.10–2.55
	CSF	2.1–2.7 mEq/L or		× 0.50	1.05–1.35 mmol/L
		4.2–5.4 mg/dL		× 0.25	1.05–1.35 mmol/L
	U, 24 h	*Ca in diet*	*mg/d*		*mmol/d*
		Calcium-free:	5–40	× 0.025	0.13–1.00
		Low to average:	50–150		1.25–3.75
		Average (800 mg/d):	100–300		(20 mmol/d): 2.50–7.50
Cannabinoids	P (Hep or EDTA)	Negative	0.019–0.026 mg/L	× 3.18	0.060–0.083 μmol/L
		THC after smoking 10 mg:			
	U	Negative			Negative
Carbon Dioxide, Partial Pressure (pCO_2)	WB, arterial (Hep)		*mm Hg*		*kPa*
		Newborn, infant:	27–40	× 0.133	3.59–5.32
		Adult, M:	35–48		4.66–6.38
		F:	32–45		4.26–5.99
Carbon Dioxide, Total (tCO_2)	S or P (Hep)		*mmol/L*		Same
		Newborn:	13–22		
		Premature, 1 wk:	14–27		
		Child:	20–28		
		Adult:	22–29		
Carbon Monoxide	WB (EDTA)		*% HbCO*		*HbCO Fraction*
		Nonsmokers:	0.5–1.5	× 0.01	0.005–0.015
		Smokers,			
		1–2 pack/d:	4–5		0.04–0.05
		>2 packs/d:	8–9		0.08–0.09
		Toxic:	>20		>0.20
		Lethal:	>50		>0.5

(continued)

Table 29-18. CLINICAL CHEMISTRY AND TOXICOLOGY *(Continued)*

Test	Specimen	Reference Range		Conversion Factor	Reference Range (International Units)
Carboxyhemoglobin, *see Carbon Monoxide*					
Carcinoembryonic Antigen (CEA)	S	Nonsmokers:	0–3.0 ng/mL	× 1	0–3.0 μg/L
		Smokers:	0–5.0 ng/mL		0–5.0 μg/L
β-Carotene	S		*μg/dL*		*μmol/L*
Petroleum ether extraction		Infant:	20–70	× 0.0186	0.37–1.30
		Child:	40–130		0.74–2.42
		Adult:	60–200		1.12–3.72
Catecholamines	U, 24 h				
HPLC		<110 μg/d		× 5.91	<650 nmol/d
Fluorometric		<280 μg/d			<1655 nmol/d
				(Conv. factor based on norepinephrine, M.W. 169.18)	
Catecholamines, Fractionated	U, 24 h	Norepinephrine,	*μg/d*		*nmol/d*
		1–4 y:	0–29	× 5.91	0–171
		4–10 y:	8–65		47–384
		10–15 y:	15–80		89–473
		Adult:	0–100		0–591
		Epinephrine,	*μg/d*		*nmol/d*
		1–4 y:	0–6.0	× 5.46	0–32.8
		4–10 y:	0–10.0		0–54.6
		10–15 y:	0.5–20		2.7–109
		Adult:	0–15		0–81.9
		Dopamine,	*μg/d*		*nmol/d*
		1–4 y:	40–260	× 6.53	261–1698
		>4 y:	65–400		425–2610
Catecholamines, Free	P (EDTA and Na metabisulfite)		*pg/mL*		*pmol/L*
		Epinephrine, random:	<88	× 5.46	<480
		Norepinephrine, random:	104–548	× 5.91	615–3239
		Dopamine, random:	<136	× 6.53	<888
		Lower values with patient supine			
Cerebrospinal Fluid Volume	CSF	Child:	60–100 mL	× 0.001	0.06–0.10 L
		Adult:	100–160 mL		0.10–0.16 L
Ceruloplasmin	S		*mg/dL*		*μmol/L*
RID		Newborn:	1–30	× 10	10–300
		Adult:	15–60		150–600
Chloride	S or P (Hep)		*mmol/L*		Same
		Newborn:	96–110		
		Thereafter:	98–106		
	CSF	Infant:	111–130		
		Adult:	118–132		
	U, 24 h		*mmol/d*		Same
		Infant:	2–10		
		Child:	15–40		
		Thereafter:	110–250		
		(vary greatly with Cl intake)			
	Swt		*mmol/L*		Same
		Normal:	0–30		
		Cystic fibrosis:	60–200		
Cholesterol, Total	S or P (EDTA)		*mg/dL*		*mmol/L*
		Infant:	70–175	× 0.0259	1.81–4.53
		Adolescent:	120–210		3.11–5.44
		Adult:	140–310		3.63–8.03
		Recommended (desirable) range for adults:	140–220		3.63–5.70
Cholinesterase II (S-Pseudocholinesterase)					
RID	S	0.5–1.5 mg/dL		× 10	5–15 mg/L
DuPont aca		8–18 U/mL			8–18 kU/L
Dibucaine:		83.6 ± 1.3% (SD) inhibition		× 0.01	Fraction of activity inhibited: 0.84 ± 0.013 (SD)
Fluoride:		79.7 ± 1.2% (SD) inhibition			0.80 ± 0.012 (SD)

Table 29-18. CLINICAL CHEMISTRY AND TOXICOLOGY *(Continued)*

Test	Specimen	Reference Range		Conversion Factor	Reference Range (International Units)
Chorionic Gonadotropin, β-Subunit (β-hCG)	S or P (EDTA)		*mIU/mL*		*IU/L*
		M and nonpregnant F:	<3.0	× 1	<3.0
		F, post-conception,			
		7–10 d:	>3.0		>3.0
		30 d:	100–5000		100–5000
		40 d:	>2000		>2000
		10 wk:	50 000–140 000		50 000–140 000
		14 wk:	10 000–50 000		10 000–50 000
		Trophoblastic disease:	>100 000		>100 000
	U, 24 h	Pregnancy (wk),	*IU/d (mean)*		Same
		6th:	13 000		
		8th:	30 000		
		12–14th:	105 000		
		16th:	46 000		
		Thereafter:	5000–20 000		
Chymotrypsin	F	120–1265 μg/g stool (mean: 290)		× 1	120–1265 mg/kg stool (mean: 290)
Complement Components					
Total hemolytic complement activity	P (EDTA)	75–160 U/mL or		× 1	75–160 kU/mL
		>33% of plasma CH50		× 0.01	Fraction of CH50: >0.33
Total complement decay rate (functional)	P (EDTA)	~10–20%			Fraction decay rate: ~0.10–0.20
		Deficiency: >50%			>0.50
		mg/dL			*mg/L*
C3 (β_1 C-globulin)	S	80–155		× 10	800–1550
C4 (β_1 E-globulin)	S	13–37			130–370
Factor B (C3 proactivator)	S	20–45			200–450
Properdin	S	2.8 ± 0.4 (SD)			28 ± 4 (SD)
C1 inhibitor (Esterase inhibitor)	P (EDTA)	17.4–24.0			174–240
Copper	S		*μg/dL*		*μmol/L*
		Birth–6 mo:	20–70	× 0.157	3.14–10.99
		Child:	80–190		12.56–29.83
		Adult, M:	70–140		10.99–21.98
		F:	80–155		12.56–24.34
		Pregnancy at term:	118–302		18.53–47.41
		Values for Blacks ~8–12% higher			Blacks ~8–12% higher
	U, 24 h	15–30 μg/d		× 0.0157	0.24–0.47 μmol/d
Coproporphyrin	U, 24 h	34–234 μg/d		× 1.5	51–351 nmol/d
	F, 24 h	<30 μg/g dry wt		× 1.5	<45 nmol/g dry wt
		400–1200 μg/d			600–1800 nmol/d
Cortisol	S or P (Hep)	0800 h:	5–23 μg/dL	× 27.6	138–635 nmol/L
		1600 h:	3–15 μg/dL		83–414 nmol/L
		2000 h:	≤50% of 0800 h	× 0.01	Fraction of 0800 h: ≤0.50
Cortisol, Free	U, 24 h		*μg/d*		*nmol/d*
		Child:	2–27	× 2.76	5.5–74.5
		Adolescent:	5–55		13.8–152
		Adult:	10–100		27.6–276
Creatine	S or P	M: 0.17–0.70 mg/dL		× 76.3	M: 13–53 μmol/L
		F: 0.35–0.93 mg/dL			F: 27–71 μmol/L
	U, 24 h	M: 0–40 mg/d		× 7.63	M: 0–305 μmol/d
		F: 0–80 mg/d			F: 0–609 μmol/d
Creatine Kinase (CK)	S				
			U/L		
Total, CK-NAC, 30 °C		Newborn:	68–580		Same
		Adult, M:	25–130		
		F:	10–115		
		Higher after exercise			
Isoenzymes		Fraction 2 (MB):	<4–6% of total (method dependent)	× 0.01	Fraction of total: <0.04–0.06
Creatinine	S or P		*mg/dL*		*μmol/L*
Jaffe, kinetic or enzymatic		Cord:	0.6–1.2	× 88.4	53–106
		Newborn, 1–4 d:	0.3–1.0		27–88
		Infant:	0.2–0.4		18–35
		Child:	0.3–0.7		27–62
		Adult, M:	0.6–1.2		53–106
		F:	0.5–1.1		44–97

Table 29-18. CLINICAL CHEMISTRY AND TOXICOLOGY *(Continued)*

Test	Specimen	Reference Range	Conversion Factor	Reference Range (International Units)
Creatinine *(cont'd.)*				
Jaffe, manual	S or P	0.8–1.5 mg/dL	× 88.4	71–133 μmol/L
	Amf	After 37 wk gestation: >2.0 mg/dL		After 37 wk gestation: >177 μmol/L
	U, 24 h	*(mg/d)/kg*		*μmol · d^{-1} · kg^{-1}*
		Child: 8–22		71–195
		Adult, M: 14–26		124–230
		F: 11–20		97–177
		Declines with age to (10 mg/kg)/d at age 90		
		or M: 800–1800 mg/d	× 0.00884	7.1–15.9 mmol/d
		F: 600–1600 mg/d		5.2–14.1 mmol/d
Creatinine Clearance (Endogenous)	S or P, and U	<40 y, M: (97–137 mL/min)/1.73 m^2	× 0.00963	(0.93–1.32 mL/s)/m^2
		F: (88–128 mL/min)/1.73 m^2		(0.85–1.23 mL/s)/m^2
		Decreases (∼6.5 mL/min)/1.73 m^2 per decade		Decreases (∼0.06 mL/s)/m^2 per decade
Cyclic AMP	P (EDTA)	M: 5.6–10.9 ng/mL	× 3.04	M: 17–33 nmol/L
		F: 3.6–8.9 ng/mL		F: 11–27 nmol/L
	U, 24 h	<3.3 mg/d	× 344	<10 μmol/d
		or <1.64 mg/g creatinine		or <565 μmol/mol creatinine
Cystine and Cysteine				
Qual.	U, random	Negative		Negative
Quant.	U, 24 h	10–100 mg/d	× 8.254 (Conversion based on cysteine, M.W. 121.16)	83–830 μmol/d
Dehydroepiandrosterone (DHEA)				
Total	S	*ng/mL*		*nmol/L*
		Adult, M: 1.7–4.2	× 3.47	6–15
		F: 2.0–5.2		7–18
Dehydroepiandrosterone Sulfate (DHEA-SO$_4$)	S or P (Hep or EDTA)	*μg/mL*		*μmol/L*
		Child: 0.60–2.54	× 2.6	1.6–6.6
		Adult, M: 1.99–3.34		5.2–8.7
		F, Premenopausal: 0.82–3.38		2.1–8.8
		Postmenopausal: 0.11–0.61		0.3–1.6
		Pregnancy, term: 0.23–1.17		0.6–3.0
Deoxycholic Acid, Total	S	0.22 ± 0.13 μg/mL (SD)	× 2.55	0.56 ± 0.33 μmol/L (SD)
11-Deoxycorticosterone (DOC)	S or P (Hep, EDTA, or Ox)	Ad lib diet, 0800 h: 4.3–12.3 ng/dL	× 0.0303	0.13–0.37 nmol/L
11-Deoxycortisol (Compound S)	P (Hep, EDTA, or Ox)	*μg/dL*		*μmol/L*
		<1 without metyrapone	× 0.029	<0.029
		>7 after metyrapone		>0.203
Dexamethasone Single Dose Overnight Suppression Test *Dose: 1 mg orally at 2300 or 2400 h*	S, 0800 h	Cortisol: Suppression to 5–10 μg/dL or to <50% of baseline	× 27.59	Cortisol: Suppression to 138–276 nmol/L or fraction of baseline: <0.50
Dexamethasone Suppression Test (Standard) *Low dose, adult: 0.5 mg q 6 h × 8*	S, 0800 h Control, day 2 and 3	Cortisol: Suppression on day 3 to <50% of baseline or to <5 μg/dL	× 0.01 × 27.59	Cortisol: Suppression on day 3; fraction of baseline: <0.50 or <138 nmol/L
	U, 24 h Control, day 1, 2, and 3	17-KGS: Suppression on day 2 to <7.5 mg/d	× 3.467	17-KGS: Suppression on day 2 to <26 μmol/d
		17-OHCS: Suppression on day 2 to <4.5 mg/d	× 2.76	17-OHCS: Suppression on day 2 to <12.4 μmol/d
		Free cortisol: <50% of baseline	× 0.01	Free cortisol: Fraction of baseline: <0.50

Table 29-18. CLINICAL CHEMISTRY AND TOXICOLOGY (Continued)

Test	Specimen	Reference Range			Conversion Factor	Reference Range (International Units)	
Dexamethasone Suppression Test (Standard) *(cont'd.)*							
High dose, adult *2.0 mg q 6 h × 8*		Cortisol, 17-KGS, 17-OHCS: Suppression on day 3 to <50% of baseline			× 0.01	Cortisol, 17-KGS, 17-OHCS: Suppression on day 3; fraction of baseline: <0.50	
Dihydrotestosterone (DHT)	S		*ng/dL*			*nmol/L*	
		Pubertal	*M*	*F*		*M*	*F*
		stage I:	<10	<10	× 0.0344	<0.34	<0.34
		II:	<20	<15		<0.69	<0.52
		III:	<35	<25		<1.20	<0.86
		IV–V:	<75	<25		<2.58	<0.86
		Adult:	60–300	10–40		2–10.3	0.34–1.4
Disaccharide Absorption Test	S		*mg/dL*			*mmol/L*	
		Change in glucose from fasting value:	>30		× 0.055	Change in glucose from fasting value: >1.67	
Estradiol (E_2), Total	S or P (Hep or EDTA)		*pg/mL*			*pmol/L*	
		M, Pubertal stage I:	2–8		× 3.67	7–29	
		II:	11			40	
		III:	>20			>73	
		Adult, M:	8–36			29–132	
		F, Pubertal stage I:	0–23			0–84	
		II:	0–66			0–242	
		III:	0–105			0–385	
		IV:	20–300			73–1100	
		Adult, F:					
		Follicular:	10–90			37–330	
		Midcycle:	100–500			367–1835	
		Luteal:	50–240			184–881	
		(Lower on oral contraceptives)					
		Postmenopausal:	10–30			37–110	
Estriol (E_3), Free	S	Pregnancy (wk),	*μg/L*			*nmol/L*	
		30–32:	4.5–16.0		× 3.47	15.6–55.5	
		34:	5.5–18.5			19.1–64.2	
		36:	7.0–25.0			24.3–86.8	
		37:	8.0–28.0			27.8–97.2	
		38:	9.0–32.0			31.2–111.0	
		39:	10.0–34.0			34.7–118.0	
		40–41:	10.5–25.0			36.4–86.8	
Estrogen Receptor Assay (ERA)	T		*fmol/mg Protein*			*nmol/kg Protein*	
		Negative:	<3.0		× 1	<3.0	
		Borderline positive:	3–20			3–20	
		Positive:	>20			>20	
Estrogens, Total	S		*pg/mL*			*ng/L*	
		Child:	<30		× 1	<30	
		Adult, M:	40–115			40–115	
		F, cycle-days,					
		1–10 d:	61–394			61–394	
		11–20 d:	122–437			122–437	
		21–30 d:	156–350			156–350	
		Prepubertal and postmenopausal:	≤40			≤40	
	U, 24 h		*μg/d*			Same	
		Child:	1.0 (mean)				
		M, pubertal stage I:	2.5 (mean)				
		II:	5.9 (mean)				
		III:	6.2 (mean)				
		Adult, M:	5–25				
		F, Preovulation:	5–25				
		Ovulation:	28–100				
		Luteal peak:	22–80				
		Pregnancy:	<45 000				
		Postmenopause:	<10				
Ethanol	WB (Ox) or S		*mg/dL*			*mmol/L*	
		Toxic:	50–100		× 0.217	11–22	
		Depression of CNS:	>100			>21.7	
		Fatalities reported:	>400			>86.8	

Table 29-18. CLINICAL CHEMISTRY AND TOXICOLOGY *(Continued)*

Test	Specimen	Reference Range	Conversion Factor	Reference Range (International Units)
Fat, Fecal	F, 72 h	*g/d*		Same
		Infant, breast-fed: <1		
		0–6 y: <2		
		Adult: <7		
		Adult (fat-free diet): <4		
Fatty Acids, Nonesterified (Free)	S or P (Hep)	*mg/dL*		*mmol/L*
		Adult: 8–25	× 0.0354	0.30–0.90
		Child and obese adults: <31	(Conv. factor based on oleic acid, M.W. 282.47)	<1.10
Fatty Acids, Total	S	190–420 mg/dL	× 0.0352	7–15 mmol/L
Ferric Chloride Test	U, random	Negative		Negative
Ferritin	S	*ng/mL*		*µg/L*
		Newborn: 25–200	× 1	25–200
		1 mo: 200–600		200–600
		2–5 mo: 50–200		50–200
		6 mo–15 y: 7–140		7–140
		Adult, M: 15–200		15–200
		F: 12–150		12–150
α_1**-Fetoprotein** (AFP)	S	Adult: <30 ng/mL	× 1	<30 µg/L
		Fetal: Peak of 200–400 mg/dL in first trimester; falls to 1% of peak (<5 mg/dL) in cord blood	× 0.01	Peak of 2–4 g/L in first trimester; falls to 1% of peak (<0.05 g/L) in cord blood
		1 y: <30 ng/mL	× 1	<30 µg/L
	S, maternal	*µg/dL*		*µg/L*
		Pregnancy (wk) / *10th–90th Percentile* / *Median*		*10th–90th Percentile* / *Median*
		6–9 / <1.0–1.5 / <1.0	× 10	<10–15 / <10
		10–12 / <1.0–2.5 / <1.0		<10–25 / <10
		13–15 / <1.0–4.4 / 1.9		<10–44 / 19
		16–18 / <1.0–7.1 / 3.6		<10–71 / 36
		19–21 / 2.1–9.6 / 5.9		21–96 / 59
		22–24 / 5.1–15.0 / 7.6		51–150 / 76
		25–27 / 5.6–17.0 / 9.9		56–170 / 99
		28–30 / 7.9–19.8 / 14.0		79–198 / 140
		31–33 / 8.4–34.4 / 14.9		84–344 / 149
		34–36 / 6.0–33.5 / 12.3		60–335 / 123
		37–40 / 6.3–16.5 / 13.5		63–165 / 135
	Amf	*mg/dL*		*mg/L*
		Pregnancy (wk) / *Median* / *±2 log SD*		*Median* / *±2 log SD*
		11–12 / 2.4 / 1.0–5.0	× 10	24 / 10–50
		13–14 / 2.3 / 1.3–4.1		23 / 13–41
		15–16 / 1.8 / 0.9–3.5		18 / 9–35
		17–18 / 1.5 / 0.6–3.3		15 / 6–33
		19–20 / 1.0 / 0.5–2.5		10 / 5–25
		21–25 / 0.7 / 0.4–1.4		7 / 4–14
		26–30 / 0.6 / 0.3–1.0		6 / 3–10
		31–35 / 0.2 / 0.05–0.7		2 / 0.5–7
		36–40 / 0.1 / 0.02–0.3		1 / 0.2–3
Fibrinogen	P (NaCit)	200–400 mg/dL	× 0.01	2.00–4.00 g/L
FIGLU *Dose: 5 g histidine q 4 h × 3*	U, 24 h, after initial dose	<35 mg/d	× 5.74	<201 µmol/d
Fluoride		*µg/mL*		*µmol/L*
	P (Hep)	0.01–0.2	× 52.6	0.5–10.5
	U	0.2–1.1		10.5–57.9
		Occup. exp.: 4–5		210–263
Folate	S	3–16 ng/mL	× 2.265	6.8–36.2 nmol/L
	RBC (EDTA)	130–628 ng/mL packed cells		294–1422 nmol/L packed cells
Folate Absorption Test	U, 24 h	45 ± 7% of dose	× 0.01	Fraction of dose: 0.45 ± 0.07

Table 29-18. CLINICAL CHEMISTRY AND TOXICOLOGY (*Continued*)

Test	Specimen	Reference Range		Conversion Factor	Reference Range (International Units)
Follicle-Stimulating Hormone (hFSH)	S or P (Hep)		*mIU/mL (IRP-2-hMG)**		*IU/L*
		Adult, M:	4–25	× 1	4–25
		F,			
		Premenopause:	4–30		4–30
		Midcycle peak:	10–90		10–90
		Pregnancy:	Low to undetectable		Low to undetectable
		Postmenopause:	40–250		40–250
		*2nd International Reference Preparation of Human Menopausal Gonadotropin			
	U, 24 h		*IU/d (IRP-2-hMG)**		
		Adult, M:	4–18		Same
		F:	3–12		
		Higher in males >60 y			
		*2nd International Reference Preparation of Human Menopausal Gonadotropin			
Free Thyroxine Index (FT_4I)	S		*FT_4 Index* 1.2–5.0		Same
With normalized T_3RU		1–3 d:	9.3–26.6		
		1–4 wk:	7.6–20.8		
		>6 y:	5.5–10.0		
		Adult:	5.0–12.0		
Free Triiodothyronine, *see Triiodothyronine, Free*					
Free Triiodothyronine Index (FT_3I)	S	1–5 y:	165		Same
		5–10 y:	150		
		10–15 y:	130		
Gastric Acid	Gast Res		*Without Stimulation mmol/L*		Same
		Free:	0–40		
		Combined:	10–20		
		Total:	10–50		
Gastric Content	Gast Res	Volume:	20–100 mL	× 0.001	0.02–0.1 L
		pH:	1.5–3.5		H^+ concentration: 32–316 μmol/L
Gastric Secretion Rate	Gast Cont, total; six 15-min spec.	BAO:	0–5 mmol/h		Same
		PAO:	5–20 mmol/h (post Pentagastrin)		
		BAO/PAO:	0.20		
Gastrin	S		*pg/mL*		*ng/L*
		Cord, venous:	64 ± 12.5 (SEM)	× 1	64 ± 12.5 (SEM)
		4 d:	151 ± 15.8		151 ± 15.8
		<3 y:	32.7 ± 22.3		32.7 ± 22.3
		>3 y:	14.8 ± 14.3		14.8 ± 14.3
		<60 y:	<100		<100
		>60 y:	<100 (15% have values of 100–800 pg/mL)		<100 (15% have values of 100–800 pg/mL)
Gastrin-Secretin Stimulation Test *IV dose: 5 U secretin/kg*	S: baseline and 15-min intervals for 1 h	Gastrin: No response or slight suppression Z-E syndrome: Increase >110 pg/mL if baseline is 80–500 pg/mL; increase >1400 pg/mL if baseline is high			Same Z-E syndrome: Increase >110 ng/L if baseline is 80–500 ng/L; increase >1400 ng/mL if baseline is high
Glomerular Selectivity Test (IgG/Albumin Ratio)	U	<0.16 indicates high selectivity			Same
Glucose	S, fasting		*mg/dL*		*mmol/L*
		Cord:	45–96	× 0.0555	2.50–5.33
		Premature:	20–60		1.11–3.33
		Neonate:	30–60		1.67–3.33
		Newborn,			
		1 d:	40–60		2.22–3.33
		>1 d:	50–80		2.78–4.44

(continued)

Table 29-18. CLINICAL CHEMISTRY AND TOXICOLOGY *(Continued)*

Test	Specimen	Reference Range		Conversion Factor	Reference Range (International Units)	
Glucose *(cont'd.)*						
	S, fasting		*mg/dL*	$\times$ 0.0555	*mmol/L*	
		Child:	60–100		3.33–5.55	
		Adult:	70–105		3.89–5.83	
		> 70 y:	83–110		4.61–6.10	
	CSF	Infant, child:	60–80		3.33–4.44	
		Adult:	40–70		2.22–3.89	
Quant., enzymatic	U	< 0.5 g/d		$\times$ 5.55	< 2.8 mmol/d	
Qual.	U	Negative			Negative	
Glucose-6-Phosphate Dehydrogenase (G-6-PD) in Erythrocytes *WHO and ICSH*	WB (ACD, EDTA, or Hep)	12.1 $\pm$ 2.09 U/g Hb (SD) 351 $\pm$ 60.6 U/10^{12} RBC 4.11 $\pm$ 0.71 U/mL RBC		$\times$ 0.0645 $\times$ 10^{-3} $\times$ 1	0.78 $\pm$ 0.13 MU/mol Hb 0.35 $\pm$ 0.06 nU/RBC 4.11 $\pm$ 0.71 kU/L RBC	
Glucose, 2 h Postprandial	S	< 120 mg/dL Diabetes: *see Glucose Tolerance Test, Oral*		$\times$ 0.0555	< 6.7 mmol/L	
Glucose Tolerance Test (GTT)	S		*mg/dL*		*mmol/L*	
		Adults,	*Normal* / *Diabetic*		*Normal*	*Diabetic*
Oral		Fasting:	70–105 / > 140	$\times$ 0.0555	3.9–5.8	> 7.8
		60 min:	120–170 / $\geq$ 200		6.7–9.4	$\geq$ 11.1
		90 min:	100–140 / $\geq$ 200		5.6–7.8	$\geq$ 11.1
		120 min:	70–120 / $\geq$ 140		3.9–6.7	$\geq$ 7.8
IV		5 min:	< 250		< 13.88	
		90 min:	at or below fasting concentration			
		or K > 1.5%				
Glucose Tolerance Test (GTT) with Cortisone *Dose: 50 mg, 8.5 and 2 h before test*	P (F^-/Ox): baseline, 1, 1½, 2 h after glucose	Glucose,	*mg/dL*		*mmol/L*	
		Fasting:	70–105	$\times$ 0.0555	3.89–5.83	
		1 h:	< 200		< 11.10	
		1½ h:	< 200		< 11.10	
		2 h:	< 140		< 7.77	
γ-Glutamyltransferase (GGT) *37 °C, aca*	S	M: F:	9–50 U/L 8–40 U/L		Same	
Glutathione, Reduced (GSH)	WB (ACD, EDTA, or Hep)	6.57 $\pm$ 1.04 μmol/g Hb (SD) 190 $\pm$ 30.16 μmol/10^{12} RBC 2.23 $\pm$ 0.35 μmol/mL RBC		$\times$ 0.0645 $\times$ 1 $\times$ 1 (Conv. factor based on hemoglobin, M.W. 64 500)	0.42 $\pm$ 0.07 mol/mol Hb 190 $\pm$ 30 amol/RBC 2.23 $\pm$ 0.35 mmol/L RBC	
Glutathione Reductase in Erythrocytes *ICSH, 37 °C*	WB (ACD, EDTA, or Hep)	Adult (reaction without added flavin adenine dinucleotide, FAD): 7.18 $\pm$ 1.09 U/g Hb (SD) 208 $\pm$ 31.6 U/10^{12} RBC 2.44 $\pm$ 0.37 U/mL RBC		 $\times$ 0.0645 $\times$ 10^{-3} $\times$ 1	 0.46 $\pm$ 0.07 MU/mol Hb 0.21 $\pm$ 0.03 nU/RBC 2.44 $\pm$ 0.37 kU/L RBC	
		Adult (reaction using FAD): 10.4 $\pm$ 1.5 U/g Hb (SD) 302 $\pm$ 43.5 U/10^{12} RBC 3.54 $\pm$ 0.51 U/mL RBC		 $\times$ 0.0645 $\times$ 10^{-3} $\times$ 1	 0.67 $\pm$ 0.10 MU/mol Hb 0.30 $\pm$ 0.04 nU/RBC 3.54 $\pm$ 0.51 kU/L RBC	
Glycerol, Free	P (EDTA)	3–10 y: 11–80 y:	0.56–2.14 mg/dL 0.29–1.72 mg/dL	$\times$ 0.1086	0.061–0.232 mmol/L 0.032–0.187 mmol/L	
Gonadotropins, *see Pregnancy Tests and Chorionic Gonadotropin, β-Subunit*						
Growth Hormone (hGH, Somatotropin)	S or P (EDTA or Hep)		*ng/mL*		*μg/L*	
		Newborn:	10–40	$\times$ 1	10–40	
		Child:	< 1–10		< 1–10	
		(Occasional values up to 20)				
		Adult, M:	< 2		< 2	
		F:	< 10		< 10	

Table 29-18. CLINICAL CHEMISTRY AND TOXICOLOGY *(Continued)*

Test	Specimen	Reference Range	Conversion Factor	Reference Range (International Units)
Haptoglobin (Hp)	S			
RID		83–267 mg/dL	× 10	830–2670 mg/L
Hemoglobin binding capacity		40–180 mg Hb/dL	× 0.155	6.20–27.9 μmol Hb/L
Nephelometry		26–185 mg/dL	× 10	260–1850 mg/L
HDL-Cholesterol (HDL-C)	S or P (EDTA)	*mg/dL*		*mmol/L*
		M *F*		*M* *F*
		Mean: 45 55	× 0.0259	1.17 1.42
		Range,		
		0–19 y: 30–65 30–65		0.78–1.68 0.78–1.68
		> 20 y: 30–70 30–85		0.78–1.81 0.78–2.20
		Values for Blacks ~10 mg/dL higher		
		CHD Risk — *HDL-C, % of Total Chol*		*Fraction HDL-C of Chol*
		M F		M F
		Dangerous: < 7 < 12	× 0.01	< 0.07 < 0.12
		High: 7–15 12–18		0.07–0.15 0.12–0.18
		Average: 15–25 18–27		0.15–0.25 0.18–0.27
		Below average: 25–37 27–40		0.25–0.37 0.27–0.40
		Protection probable: > 37 > 40		> 0.37 > 0.40
Hemoglobin (Hb)	WB (EDTA)	*g/dL*		*mmol/L*
		Cord: 13.5–20.0	× 0.155	2.09–3.10
		1–3 d (cap): 14.5–22.5		2.25–3.49
		1 wk: 13.5–21.5		2.09–3.33
		2 wk: 12.5–20.5		1.94–3.18
		1 mo: 10.0–18.0		1.55–2.79
		2 mo: 9.0–14.0		1.40–2.17
		3–6 mo: 9.5–13.5		1.47–2.09
		0.5–2 y: 10.5–13.5		1.63–2.09
		2–6 y: 11.5–13.5		1.78–2.09
		6–12 y: 11.5–15.5		1.78–2.40
		12–18 y, M: 13.0–16.0		2.02–2.48
		F: 12.0–16.0		1.86–2.48
		18–49 y, M: 13.5–17.5		2.09–2.71
		F: 12.0–16.0		1.86–2.48
	P (EDTA, ACD, or Hep)	1–4 mg/dL	× 0.155	0.16–0.62 μmol/L
	S	< 3 mg/dL with butterfly setup and 18-g needle		< 0.47 μmol/L with butterfly setup and 18-g needle
	U, fresh, random	Negative		Negative
Hemoglobin A_{1c}	WB (Hep, EDTA, or Ox)			
Electrophoresis		5.6–7.5% of total Hb	× 0.01	Fraction of Hb: 0.056–0.075
Column		6–9% of total Hb		0.06–0.09
Hemoglobin A_2 (HbA_2)	WB (EDTA or Ox)	Adult: 1.5–3.5% (2 SD)	× 0.01	0.015–0.035 (2 SD)
		Lower in infants < 1 y		
		β-Thalassemia trait: 3.75–6.5% (2 SD)		0.038–0.065 (2 SD)
Hemoglobin (Hb) Electrophoresis	WB (EDTA, Cit, or Hep)			Hb Fraction
		HbA: > 95%	× 0.01	> 0.95
		HbA_2: 1.5–3.5%		0.015–0.035
		HbF: < 2%		< 0.02
Hemoglobin F	WB (EDTA)	*% HbF*		*Mass Fraction HbF*
Alkali denaturation (White)		1 d: 77.0 ± 7.3	× 0.01	0.77 ± 0.073
		5 d: 76.8 ± 5.8		0.768 ± 0.058
		3 wk: 70.0 ± 7.3		0.70 ± 0.073
		6–9 wk: 52.9 ± 11.0		0.529 ± 0.11
		3–4 mo: 23.2 ± 16.0		0.232 ± 0.16
		6 mo: 4.7 ± 2.2		0.047 ± 0.022
		8–11 mo: 1.6 ± 1.0		0.016 ± 0.010
		Adult: < 2.0		< 0.020
Hemoglobin H (HbH) *Isopropanol precipitation*	WB (ACD, EDTA, or Hep)	No precipitation at 40 min		No precipitation at 40 min

(continued)

Table 29-18. CLINICAL CHEMISTRY AND TOXICOLOGY (Continued)

Test	Specimen	Reference Range		Conversion Factor	Reference Range (International Units)
Hemopexin	S	Fetus, term:	> 30% of mean adult conc. or 18% of maternal conc.	× 0.01	Fraction of adult conc.: > 0.30; Fraction of maternal conc.: 0.18
		Maternal:	> 1.5 times that of nonpregnant adults		Same
		Adult:	50–115 mg/dL	× 0.01	0.50–1.15 g/L
	U	Mean:	0.2 mg/d		0.2 mg/d
Homocystine	U, random	Negative			Negative
Homogentisic Acid	U, random	Negative			Negative
Homovanillic Acid (HVA)	U, 24 h	Child:	3–16 μg/mg creatinine	× 0.621	1.86–9.94 mmol/mol creatinine
		Adult:	< 15 mg/d	× 5.49	< 82 μmol/d
β-Hydroxybutyric Acid *Qual.*	S or P	Undetectable			Undetectable
17-Hydroxycorticosteroids (17-OHCS)	U, 24 h		*mg/d*		*μmol/d*
		0–1 y:	0.5–1.0	× 2.76	1.4–2.8
		Child:	1.0–5.6		2.8–15.5
		Adult, M:	3.0–10.0	(Conversion based on	8.2–27.6
		F:	2.0–8.0	hydrocortisone, M.W. 362)	5.5–22
		or 3–7 mg/g creatinine		× 0.312	or 0.9–2.5 mmol/mol creatinine
	Amf	Increases 4–6-fold throughout pregnancy			Same
5-Hydroxyindoleacetic Acid (5-HIAA) *Qual.*	U, random	Negative (< 25 mg/d)		× 5.2	Negative (< 130 μmol/d)
Quant.	U, 24 h	2–8 mg/d		× 5.2	10.4–41.6 μmol/d
17-Hydroxyprogesterone (17-OHP)	S		*ng/mL*		*nmol/L*
		M, Pub. stage I:	0.1–0.3	× 3.03	0.30–0.91
		Adult:	0.2–1.8		0.61–5.45
		F, Pub. stage I:	0.2–0.5		0.61–1.52
		Follicular:	0.2–0.8		0.61–2.42
		Luteal:	0.8–3.0		2.42–9.10
		Postmenopausal:	0.04–0.5		0.12–1.52
		Values are highest at birth but decrease rapidly during 1st week of life			
Hydroxyproline	S		*mg/dL*		*μmol/L*
		6–18 y, M:	0–0.66	× 76.3	0–50
		F:	0–0.58		0–44
		Adult, M:	0–0.55		0–42
		F:	0–0.46		0–34
Immunoglobulin A (IgA) *Nephelometric*	S		*mg/dL*		*g/L*
		Cord:	0–5	× 0.01	0–0.05
		Newborn:	0–2.2		0–0.02
		4–6 mo:	3–82		0.03–0.82
		6 mo–2 y:	14–108		0.14–1.08
		2–6 y:	23–190		0.23–1.90
		6–12 y:	29–270		0.29–2.70
		12–16 y:	81–232		0.81–2.32
		Adult:	76–390		0.76–3.90
Immunoglobulin D (IgD)	S	Newborn:	< 1.0 mg/dL	× 10	< 10 mg/L
		Adult:	0–8 mg/dL		0–80 mg/L
Immunoglobulin E (IgE)	S		*IU/mL* 0–380	× 1	*kIU/L* 0–380

Table 29-18. CLINICAL CHEMISTRY AND TOXICOLOGY (Continued)

Test	Specimen	Reference Range		Conversion Factor	Reference Range (International Units)
Immunoglobulin G (IgG) *Nephelometric*	S		*mg/dL*		*g/L*
		Cord:	760–1700	× 0.01	7.6–17.0
		Newborn:	700–1480		7.0–14.8
		½–6 mo:	300–1000		3.0–10.0
		6 mo–6 y:	500–1300		5.0–13.0
		6–16 y:	700–1650		7.0–16.5
		Adult:	650–1500		6.5–15.0
		(higher in Blacks)			
RIA	CSF		*mg/dL*		*mg/L*
		15–20 y:	3.5 ± 2.0 (SD)	× 10	35 ± 20 (SD)
		21–40 y:	4.2 ± 1.4		42 ± 14
		41–60 y:	4.7 ± 1.0		47 ± 10
		61–87 y:	5.8 ± 1.6		58 ± 16
Immunoglobulin G/ Albumin Ratio	CSF and S	0.3–0.6			0.3–0.6
Immunoglobulin G Synthesis Rate	CSF and S	−9.9 to +3.3 mg/d			−9.9 to +3.3 mg/d
Immunoglobulin M (IgM) *Nephelometric; Std: Beckman; Atlantic Antibodies; WHO*	S		*mg/dL*		*mg/L*
		Cord:	4–24	× 10	40–240
		Newborn:	5–30		50–300
		½–6 mo:	15–109		150–1090
		6 mo–6 y:	43–239		430–2390
		6–16 y:	45–260		450–2600
		Adult:	40–345		400–3450
		Results vary with std. preparation			
	CSF		*mg/dL*		*mg/L*
		15–20 y:	0.02 ± 0.009 (SD)	× 10	0.2 ± 0.09 (SD)
		21–40 y:	0.016 ± 0.003		0.16 ± 0.03
		41–87 y:	0.017 ± 0.005		0.17 ± 0.05
Insulin (12 h Fasting)	S		*μIU/mL*		*mIU/L*
		Newborn:	3–20	× 1.0	3–20
		Adult:	6–24		6–24
		> 60 y:	6–35		6–35
	Amf	< 16 wk:	Undetectable		Undetectable
		Term:	11.3 (mean)	× 1	11.3 (mean)
Insulin Antibodies	S	Undetectable			Undetectable
Insulin and Glucose Suppression Test	S, every 6–12 h	> 50 mg glucose/dL during a 72-h period of fasting, with values slightly lower in females		× 0.0555	> 2.8 mmol glucose/L during a 72 h period of fasting, with values slightly lower in females
		Insulin: < 4 μIU/mL or undetectable		× 1	Insulin: < 4 mU/L or undetectable
		Normal fasting insulin-glucose ratio: < 0.3			Normal fasting insulin-glucose ratio: < 5.4
Insulin with Oral Glucose Tolerance Test	S	*Min*	*Insulin, μIU/mL*		*mIU/L*
		0:	6–24	× 1	6–24
		30:	25–231		25–231
		60:	18–276		18–276
		120:	16–166		16–166
		180:	4–38		4–38
Insulin Tolerance Test *Dose: 0.1–0.15 U/kg, IV*	S	*Glucose*: Decrease ~50% of the fasting level by 30 min and return to normal fasting limits by 90–120 min		× 0.01	Fractional decrease in glucose ~0.50 of the fasting level by 30 min and return to normal fasting limits by 90–120 min
		hGH: Increase of > 5 ng/mL within 60 min of hypoglycemia		× 1	*hGH*: Increase of > 5 μg/L within 60 min of hypoglycemia
		Cortisol: Increase of > 6 μg/dL with peak of > 20 μg/dL		× 27.59	*Cortisol*: Increase of > 165 nmol/L with peak of > 552 nmol/L

(continued)

Table 29-18. CLINICAL CHEMISTRY AND TOXICOLOGY (Continued)

Test	Specimen	Reference Range		Conversion Factor	Reference Range (International Units)
Intrinsic Factor, *see Vitamin B_{12} Intrinsic Factor*					
Iron	S		*μg/dL*		*μmol/L*
ICSH		Newborn:	100–250	× 0.179	17.90–44.75
		Infant:	40–100		7.16–17.90
		Child:	50–120		8.95–21.48
		Adult, M:	65–170		11.64–30.43
		F:	50–170		8.95–30.43
		Strongly method dependent			
Iron-Binding Capacity, Total (TIBC)	S	Infant:	100–400 μg/dL	× 0.179	17.90–71.60 μmol/L
		Thereafter:	250–450 μg/dL		44.75–80.55 μmol/L
Iron Saturation	S	20–55%		× 0.01	Fraction saturation 0.20–0.55
Isocitrate Dehydrogenase (ICD), 30 °C	S	1.2–7.0 U/L			1.2–7.0 U/L
17-Ketogenic Steroids (17-KGS)	U, 24 h		*mg/d*		*μmol/d*
		0–1 y:	<1.0	× 3.467	<3.5
		1–10 y:	<5	(Conversion based on	<17
		11–14 y:	<12	dehydroepiandros-	<42
		Adult, M:	5–23	terone, M.W. 288)	17–80
		F:	3–15		10–52
		>70 y, M:	3–15		10–52
		F:	3–13		10–45
Ketone Bodies					
Qual.	S	Negative			Negative
	U, random	Negative			Negative
Quant.	S	0.5–3.0 mg/dL		× 10	5–30 mg/L
17-Ketosteroids (17 KS), Total	U, 24 h		*mg/d*		*μmol/d*
Zimmerman reaction		14 d–2 y:	≤1	× 3.467	≤3.5
		2–6 y:	≤2	(Conversion based on	≤7
		6–10 y:	1–4	dehydroepiandros-	3.5–14
		10–12 y:	1–6	terone, M.W. 288)	3.5–21
		12–14 y:	3–10		10–35
		14–16 y:	5–12		17–42
		Adult,			
		M:	8–22		28–76
		F:	6–15		21–52
		Decreases with age			Decreases with age
GLC		Adult, M: 5.0–12.0			17–42
		F: 3.0–10.0			10–35
L-Lactate			*mg/dL*		*mmol/L*
	P (NaF)	Venous:	4.5–19.8	× 0.111	0.5–2.2
		Arterial:	4.5–14.4		0.5–1.6
	WB (Hep)	At bed rest,			
		Venous:	8.1–15.3		0.9–1.7
		Arterial:	<11.3		<1.25
	U, 24 h	496–1982 mg/d		× 0.0111	5.5–22 mmol/d
	CSF	<25.2 mg/dL		× 0.111	<2.8 mmol/L
Qual.	Gastf	Negative			Negative
Lactate Dehydrogenase (LDH)					
Total (L → P), 30 °C	S		*U/L*		Same
		Newborn:	160–450		
		Neonate:	300–1500		
		Infant:	100–250		
		Child:	60–170		
		Adult:	45–90		
		>60 y:	55–100		
Total (P → L), 30 °C			150–320		
	CSF	~10% of serum value		× 0.01	~0.10 fraction of serum value

Table 29-18. CLINICAL CHEMISTRY AND TOXICOLOGY (Continued)

Test	Specimen	Reference Range	Conversion Factor	Reference Range (International Units)
Lactate Dehydrogenase (LDH) *(cont'd.)*				
Isoenzymes, Agarose	S	*% of Total* Fraction 1: 14–26 Fraction 2: 29–39 Fraction 3: 20–26 Fraction 4: 8–16 Fraction 5: 6–16	× 0.01	*Fraction of Total* 0.14–0.26 0.29–0.39 0.20–0.26 0.08–0.16 0.06–0.16
Lactate/Pyruvate Ratio	WB (Hep)	10/1		10/1
LDL-Cholesterol (LDL-C) *Calculated*	S or P (EDTA)	*mg/dL* M / F Cord blood: 10–50 / 10–50 0–19 y: 60–140 / 60–150 20–29 y: 60–175 / 60–160 30–39 y: 80–190 / 70–170 40–59 y: 90–205 / 80–190 >60 y: 90–215 / 100–235 *Recommended* (desirable) range for adults: 65–175 mg/dL	× 0.0259	*mmol/L* M / F 0.26–1.30 / 0.26–1.30 1.55–3.63 / 1.55–3.89 1.55–4.53 / 1.55–4.14 2.07–4.92 / 1.81–4.40 2.33–5.31 / 2.07–4.92 2.33–5.57 / 2.59–6.09 1.68–4.53
Lead	WB (Hep)	*µg/dL* Child: <25 Adult: <40 Toxic: ≥100	× 0.0483	*µmol/L* <1.21 <1.93 ≥4.83
	U, 24 h	<80 µg/L	× 0.00483	<0.39
Lecithin/Sphingomyelin (L/S) Ratio	Amf	2.0–5.0 indicates probable fetal lung maturity; >3.0 in diabetics		Same
Lecithin Phosphorus	Amf	>1.0 mg/dL indicates probable adequate fetal lung maturity	× 0.3229	>0.33 mmol/L indicates probable adequate fetal lung maturity
Lipase	S			
Tietz method		<1.0 unit/mL	× 278	<278 U/L
BMD turbidimetric		Adult: 10–150 U/L >60 y: 18–180 U/L		Same
β-Lipoprotein (LDL, Low-Density Lipoprotein)	S, 12–14 h fasting	28–53% of total lipoproteins	× 0.01	Mass fraction of total: 0.28–0.53
Lipoprotein Electrophoresis	S	Distinct β-band; negligible chylomicron and pre-β bands		
Long-Acting Thyroid-Stimulating Hormone (LATS)	S	Undetectable		Undetectable
Luteinizing Hormone (hLH)	S or P (Hep); avoid EDTA	*mIU/mL* Child: <5 Adult, M: 2–12 F: Follicular: 2–15 Midcycle: 30–100 Luteal: 2–15 Postmenopausal: 20–100	× 1	*IU/L* <5 2–12 2–15 30–100 2–15 20–100
Lysozyme (Muramidase)	S or P (EDTA)	5–15 µg/mL	× 1	5–15 mg/L
Macroamylase	S	Present in ~1% of healthy subjects with normal serum amylase activity and in 2.5% of patients with abnormal activity		Same
Magnesium	S	*mEq/L* Newborn, 2–4 d: 1.0–1.8 Adult: 1.3–2.1 (Higher in females during menses)	× 0.5	*mmol/L* 0.5–0.9 0.65–1.05
	U, 24 h	6.0–10.0 mEq/d		3.00–5.00 mmol/d

(continued)

Table 29-18. CLINICAL CHEMISTRY AND TOXICOLOGY (*Continued*)

Test	Specimen	Reference Range		Conversion Factor	Reference Range (International Units)
Melanin	U, random	Negative			Negative
Mercury	WB (EDTA)	<5.0 μg/dL		× 0.0499	<0.25 μmol/L
	U, 24 h	<20 μg/L		× 0.00499	<0.1 μmol/L
		Toxic: >150 μg/L			>0.75 μmol/L
Metanephrine, Total	U, 24 h		*μg/mg creatinine*		*mmol/mol creatinine*
		<1 y:	0.001–4.60	× 0.574	0.0006–2.64
		1–2 y:	0.27–5.38		0.15–3.09
		2–5 y:	0.35–2.99		0.20–1.72
		5–10 y:	0.43–2.70		0.25–1.55
		10–15 y:	0.001–1.87		0.0006–1.07
		15–18 y:	0.001–0.67		0.0006–0.38
		Adult:	0.05–1.20		0.03–0.69
Methanol	WB (F^-/Ox)		*mg/L*		*mmol/L*
			<1.5	× 0.0312	<0.05
		Toxic:	>200		>6.24
	U	Occup. exp.:	<50		<1.56
	Br		<0.8 ppm	× 0.0312	<0.02
		Occup. exp.:	<2.5 ppm		<0.08
Methemoglobin (MetHb, Hemiglobin)	WB (EDTA, Hep, or ACD)	0.06–0.24 g/dL or 0.78 ± 0.37% of total Hb (SD)		× 155* × 0.01 (*Conv. factor based on hemoglobin, M. W. 64 500)	9.3–37.2 μmol/L Mass fraction of total Hb: 0.008 ± 0.0037 (SD)
Metyrapone (Metopyrone) Stimulation Test *Dose, adult: 750 mg q 4 h × 6; child: 300 mg/m²*	S	11-Deoxycortisol:	>7.0 μg/dL	× 28.86	11-Deoxycortisol: >200 nmol/L
		Cortisol:	<8 μg/dL	× 27.59	Cortisol: <220 nmol/L
	U, 24 h	17-KGS:	2.5–3-fold rise, but at least 10 mg/d	× 3.467* (*Conv. factor based on DHEA, M.W. 288)	17-KGS: 2.5–3-fold rise but at least 35 μmol/d
		17-KS:	>2 × base level		Same
		17-OHCS:	3–5 × base level		Same
Single Dose Metyrapone Test Dose: 30 mg/kg orally with milk or snack at midnight	S, 0800 h	11-Deoxycortisol:	>7 μg/dL	× 28.86	>200 nmol/L
Microsomal Antibodies, Thyroid, *see Thyroid Microsomal Antibodies*					
Mucopolysaccharide Screen (MPS)	U, random	Negative			Negative
Myelin Basic Protein	CSF	<4 ng/mL		× 1	<4 μg/L
Myoglobin	S	M: 49 ± 17 μg/L (SD) F: 35 ± 14 μg/L Increases slightly with age			Same
	U, random	Negative			Negative
Niacin	U, 24 h	0.3–1.5 mg/d (mean: 0.6)		× 8.113	2.43–12.17 μmol/d (mean: 4.87)
Nitrites	U	Negative Toxic: Nitrites detected			Same
Nitrogen, Total	F	Infants: 0.11–0.52 g N/d Adults: <2 g N/d		× 71.4	7.9–37 mmol N/d <143 mmol N/d
Nonprotein Nitrogen (NPN)	WB (Ox)	<50 mg/dL		× 0.714	<35.7 mmol/L
	S	<35 mg/dL			<25.0 mmol/L
Normetanephrine, Total	P (EDTA and Na metabisulfite)	Normotensive: 1.2 ± 0.1 ng/mL (SEM)		× 5.46	6.55 ± 0.55 nmol/L
		Primary hypertensive: 2.5 ± 0.2 ng/mL (SEM)			13.65 ± 1.09 nmol/L

Table 29-18. CLINICAL CHEMISTRY AND TOXICOLOGY *(Continued)*

Test	Specimen	Reference Range	Conversion Factor	Reference Range (International Units)
Occult Blood	F, random	Negative (<2 mL blood/150 g stool per d)	× 6.67	Negative (<13.3 mL blood/kg stool per d)
Qual.	U, random	Negative		Negative
Orosomucoid, *see α₁-Acid Glycoprotein*				
Osmolality	S	Neonate: May be as low as 266 mOsmol/kg Child, adult: 275–295 mOsmol/kg >60 y: 280–301 mOsmol/kg		Same
	U, random	50–1400 mOsmol/kg, depending on fluid intake		
	U, random	After 12 h fluid restriction: >850 mOsmol/kg		Same
	U, 24 h	~300–900 mOsmol/kg		
Osmolality Ratio, Urine/Serum	U and S	1.0–3.0 >3.0 after 12 h fluid restriction		Same
Oxalate	S	1.0–2.4 μg/mL	× 11.4	11–27 μmol/L
		Ethylene glycol poisoning: >20 μg/mL		>228 μmol/L
	U, 24 h	8–40 mg/d	× 11.4	91–456 μmol/d
		Ethylene glycol poisoning: >150 mg/d		>1710 μmol/d
Oxygen, P_{50}	WB (Hep)	Newborn: 18–24 mm Hg	× 0.133	2.39–3.19 kPa
		Adult, adjusted to pH (P) 7.4: 25–29 mm Hg		3.33–3.86 kPa
Oxygen, Partial Pressure (pO_2)	WB, arterial (Hep)	*mm Hg*		*kPa*
		Birth: 8–24	× 0.133	1.1–3.2
		5–10 min: 33–75		4.4–10.0
		30 min: 31–85		4.1–11.3
		>1 h: 55–80		7.3–10.6
		1 d: 54–95		7.2–12.6
		Thereafter: 83–108		11.04–14.36
		(Decreases with age and high altitude)		
Oxygen Saturation	WB, arterial (Hep)			Fraction saturated:
		Newborn: 40–90%	× 0.01	0.40–0.90
		Thereafter: 95–98%		0.95–0.98
Parathyroid Hormone (hPTH)	S	Varies with laboratory,		
		SmithKline: *pg/mL*		*ng/L*
		C-terminal 500–1500	× 1	500–1500
		Midmolecule: 70–330		70–330
		Nichols Institute:		
		N-terminal 11–24		11–24
		Midmolecule: 50–330		50–330
		Immuno Nuclear:		
		Midmolecule 29–85 pmol/L		Same
		Higher in pregnancy		
pH (37 °C)	WB, arterial (Hep)	Premature, 48 h: 7.35–7.50		Same
		Birth, full term: 7.11–7.36		
		5–10 min: 7.09–7.30		
		30 min: 7.21–7.38		
		>1 h: 7.26–7.49		
		1 d: 7.29–7.45		
		Thereafter: 7.35–7.45		
		or H^+ conc.: 36–44 nEq/L	× 1	36–44 nmol/L
	U, random	Newborn/neonate: 5–7		Same
		Thereafter: 4.5–8		
		Average ~6		
	Serf (Plf, Pericf, Ascf)	6.8–7.6		
	Synf	Parallels serum		
	CSF	7.35–7.40		
	Semf	7.2–8.0 (average: 7.8)		
	F	7.0–7.5 (May be acid with high lactose intake)		

(continued)

Table 29-18. CLINICAL CHEMISTRY AND TOXICOLOGY *(Continued)*

Test	Specimen	Reference Range		Conversion Factor	Reference Range (International Units)
Phenylalanine	S		*mg/dL*		*mmol/L*
		Premature:	2.0–7.5	× 0.06054	0.12–0.45
		Newborn:	1.2–3.4		0.07–0.21
		Adult:	0.8–1.8		0.05–0.11
Phenylpyruvic Acid *Qual.*	U, random	Negative $FeCl_3$ test			Negative by $FeCl_3$ test
Phosphatase, Acid					
Prostatic (RIA)	S	< 3.0 ng/mL		× 1	< 3.0 μg/L
Roy, Brower, and Hayden, 37 °C		0.11–0.60 U/L			0.11–0.60 U/L
Phosphatase, Alkaline	S		*U/L*		Same
p-Nitrophenyl phosphate, carbonate buffer, 30 °C		Infant:	50–165		
		Child:	20–150		
		Adult:	20–70		
		> 60 y:	30–75		
Bowers and McComb, 30 °C		Adult:	25–100		
IFCC, 30 °C		M:	30–90		
		F:	20–80		
Phosphatidylglycerol (PG)	Amf	Absent:	Fetal immaturity		Same
		Present:	Fetal maturity		
Phospholipids, Total	S or P (EDTA)		*mg/dL*		*g/L*
		Newborn:	75–170	× 0.01	0.75–1.70
		Infant:	100–275		1.00–2.75
		Child:	180–295		1.80–2.95
		Adult:	125–275		1.25–2.75
		> 65 y:	196–366		1.96–3.66
Phosphorus, Inorganic	S		*mg/dL*		*mmol/L*
		Cord:	3.7–8.1	× 0.323	1.20–2.62
		24 h:	3.5–8.6		1.13–2.78
		24–48 h:	5.5–9.5		1.78–3.07
		Infant:	4.5–6.5		1.45–2.10
		Child:	4.5–5.5		1.45–1.78
		Thereafter:	2.7–4.5		0.87–1.45
		> 60 y, M:	2.3–3.7		0.74–1.20
		F:	2.8–4.1		0.90–1.32
	U, 24 h	Adults on diet containing 0.9–1.5 g P and 10 mg Ca/kg: < 1.0 g/d		× 32.3	Adults on diet containing 29–48 mmol P and 0.25 mmol Ca/kg: < 32.3 mmol/d
		On nonrestricted diet:	0.4–1.3 g/d		On nonrestricted diet: 12.9–42.0 mmol/d
Plasma Volume	P (Hep)	M: 25–43 mL/kg		× 0.001	0.025–0.043 L/kg
		F: 28–45 mL/kg			0.028–0.045 L/kg
Porphobilinogen (PBG)					
Quant.	U, 24 h	0–2.0 mg/d		× 4.42	0–8.8 μmol/d
Qual.	U, random	Negative			Negative
Postheparin Lipolytic Activity	P (Hep)		*(μmol FFA/min)/mL*		*U/L*
		TG < 200 mg/dL:	0.24–0.57	× 1000	240–570
		TG > 200 mg/dL:	0.06–0.71		60–710
Potassium	S		*mEq/L*		*mmol/L*
		Premature,			
		Cord:	5.0–10.2	× 1	5.0–10.2
		48 h:	3.0–6.0		3.0–6.0
		Newborn,			
		Cord:	5.6–12.0	× 1	5.6–12.0
		Newborn:	3.7–5.9		3.7–5.9
		Infant:	4.1–5.3		4.1–5.3
		Child:	3.4–4.7		3.4–4.7
		Thereafter:	3.5–5.1		3.5–5.1

Table 29-18. CLINICAL CHEMISTRY AND TOXICOLOGY (Continued)

Test	Specimen	Reference Range		Conversion Factor	Reference Range (International Units)
Potassium (cont'd.)					
	P (Hep)	Adult: 3.4–4.5 mEq/L		× 1	3.4–4.5 mmol/L
	U, 24 h	25–125 mEq/d; varies with diet			25–125 mmol/d; varies with diet
	CSF	70% of plasma level; 2.5–3.2 mEq/L, rises with plasma hyperosmolality		× 0.01 × 1	0.70 plasma level fraction; 2.5–3.2 mmol/L, rises with plasma hyperosmolality
	RBC	~105 mEq/L		× 1	~105 mmol/L
	Swt	5–17 mEq/L		× 1	5–17 mmol/L
Pregnancy Tests					
Chorionic Gonadotropin (hCG) Tube Test					
Qual.	S or U	Negative Positive by 4th–8th d after expected menstrual period			Same
Semiquant.	S or U, 24 h	Peak values up to 120 000 mIU/mL		× 1	Peak values up to 120 000 IU/L
Chorionic Gonadotropin (β-hCG), see Chorionic Gonadotropin, β-Subunit					
Radio Receptor Assay (RRA) *Qual.*	S	Negative; pregnancy can be detected < 10 d after conception			Same
Pregnanediol	U, 24 h		*mg/d*		*μmol/d*
		< 2 y:	< 0.1	× 3.12	< 0.3
		3–5 y:	< 0.3		< 0.9
		6–9 y:	< 0.5		< 1.6
		M, Adult:	0.1–0.7		0.3–2.2
		F, 10–15 y:	0.1–1.2		0.3–3.7
		Adult,			
		Follicular:	< 1.0		< 3.1
		Luteal:	2–7		6.2–21.8
		Postmenopausal:	0.2–1.0		6.2–3.1
Pregnanetriol	U, 24 h		*mg/d*		*μmol/d*
		2 wk–2 y:	0.02–0.2	× 2.97	0.06–0.59
		2–5 y:	< 0.5		< 1.49
		5–15 y:	< 1.5		< 4.46
		> 15 y:	< 2.0		< 5.94
Progesterone	S		*ng/mL*		*nmol/L*
		M, Pub. stage I:	0.11–0.26	× 3.18	0.35–0.83
		Adult:	0.12–0.3		0.38–0.95
		F, Pub. stage I:	0–0.3		0–0.95
		II:	0–0.46		0–1.46
		III:	0–0.6		0–1.91
		IV:	0.05–13.0		0.16–41.3
		Follicular:	0.02–0.9		0.06–2.86
		Luteal:	6.0–30.0		19.1–95.4
Progesterone Receptor Assay (PRA)	T, tumor		*fmol/mg protein*		*nmol/kg protein*
		Normal, or benign and nonresponsive tumor:	≤ 5	× 1	≤ 5
		Positive:	> 10		> 10
Proinsulin	S	< 30% of total immunoreactive insulin-like material, or < 0.2 ng/mL		× 0.01	Fraction of immunoreactive insulin-like material: 0.30, or < 0.2 μg/L
Prolactin (hPRL)	S		*ng/mL*		*μg/L*
		Adults, M:	< 20	× 1	< 20
		F, Follicular:	< 23		< 23
		Luteal:	5–40		5–40
		Pregnancy,			
		1st trimester:	< 80		< 80
		2nd trimester:	< 160		< 160
		3rd trimester:	< 400		< 400
		Newborn:	> 10-fold adult levels		

(continued)

Table 29-18. CLINICAL CHEMISTRY AND TOXICOLOGY *(Continued)*

Test	Specimen	Reference Range		Conversion Factor	Reference Range (International Units)
Properdin	S		*mg/dL ± SD*		*mg/L ± SD*
		Cord:	1.5 ± 0.1	× 10	15 ± 1
		1 mo:	1.4 ± 0.4		14 ± 4
		6 mo:	1.9 ± 0.3		19 ± 3
		Adult:	2.8 ± 0.4		28 ± 4
Protein					
Total	S		*g/dL*		*g/L*
		Premature:	3.6–6.0	× 10	36–60
		Newborn:	4.6–7.0		46–70
		Cord:	4.8–8.0		48–80
		1 wk:	4.4–7.6		44–76
		7 mo–1 y:	5.1–7.3		51–73
		1–2 y:	5.6–7.5		56–75
		≥3 y:	6.0–8.0		60–80
		Adult,			
		Ambulatory:	6.4–8.3		64–83
		Recumbent:	6.0–7.8		60–78
		>60 y: lower by	~0.2		~2
Electrophoresis		Albumin,	*g/dL*		*g/L*
		Adult:	3.5–5.0		35–50
		>60 y:	3.7–4.7		37–47
		α_1-Globulin,			
		Adult:	0.1–0.3		1–3
		α_2-Globulin,			
		Adult:	0.6–1.0		6–10
		β-Globulin,			
		Adult:	0.7–1.1		7–11
		γ-Globulin,			
		Adult:	0.8–1.6		8–16
Total	U, 24 h	1–14 mg/dL		× 10	10–140 mg/L
		50–80 mg/d (at rest)			Same
		<250 mg/d after intense exercise			
Electrophoresis			*Average % of Total Protein*		*Fraction of Total Protein*
		Albumin	37.9	× 0.01	0.379
		α_1-Globulin	27.3		0.273
		α_2-Globulin	19.5		0.195
		β-Globulin	8.8		0.088
		γ-Globulin	3.3		0.033
Total	CSF, lumbar		*mg/dL*		*mg/L*
Column Turbidimetry			8–32	× 10	80–320
		Premature:	15–130		150–1300
		Full-term newborn:	40–120		400–1200
		<1 mo:	20–80		200–800
		Thereafter:	15–40		150–400
Electrophoresis			*% of Total Protein*		*Fraction of Total Protein*
		Prealbumin:	2–7	× 0.01	0.02–0.07
		Albumin:	56–76		0.56–0.76
		α_1-Globulin:	2–7		0.02–0.07
		α_2-Globulin:	4–12		0.04–0.12
		β-Globulin:	8–18		0.08–0.18
		γ-Globulin:	3–12		0.03–0.12
Protoporphyrin	WB (Hep or EDTA)	<50 μg/dL RBC		× 0.0178	<0.89 μmol/L RBC
	F, 24 h	≤60 μg/g dry wt or <1500 μg/d		× 0.00178	≤0.11 mmol/kg dry wt or <2.67 μmol/d
Pseudocholinesterase (PChE), *see Cholinesterase II*					
Pyruvic Acid	WB (Hep)	0.3–0.9 mg/dL		× 0.114	0.03–0.10 mmol/L
	U	88.1 mg/d		× 0.0114	1 mmol/d
	CSF	0.5–1.7 mg/dL		× 0.114	0.06–0.19 mmol/L
Renal Plasma Flow (RPF)	P and U	M: 560–830 mL/min		× 0.01667	9.34–13.84 mL/s
		F: 490–700 mL/min			8.17–11.67 mL/s
		or (390 mL/min)/m² body surface			or (6.50 mL/s)/m² body surface
		>40 y: decreases ~75 mL/decade			

Table 29-18. CLINICAL CHEMISTRY AND TOXICOLOGY *(Continued)*

Test	Specimen	Reference Range		Conversion Factor	Reference Range (International Units)	
Renin (PRA)	P (EDTA)	*Normal sodium diet,*	*(ng/h)/mL (mean ± SE)*		$\mu g \cdot h^{-1} \cdot L^{-1}$ (mean ± *SE*)	
		Supine:	1.6 ± 1.5	× 1	1.6 ± 1.5	
		Standing (4 h):	4.5 ± 2.9		4.5 ± 2.9	
		Low-sodium diet,				
		Supine:	3.2 ± 1.1		3.2 ± 1.1	
		Standing (4 h):	9.9 ± 4.3		9.9 ± 4.3	
Reverse Triiodothyronine (rT_3)	S		*ng/dL*		*nmol/L*	
		1–5 y:	15–71	× 0.0154	0.23–1.09	
		5–10 y:	17–79		0.26–1.22	
		10–15 y:	19–88		0.29–1.36	
		Adults:	30–80		0.46–1.23	
Schilling Test (Intrinsic Factor Test) *Dose: 0.5–1.0 μCi ^{58}Co-Vitamin B_{12}*	U, 24 h	> 7.5% of dose		× 0.01	Fraction of dose: 0.075	
Serotonin	WB (EDTA)	39–361 ng/mL		× 0.00568	0.22–2.05 μmol/L	
	Plt	~314 ng/10^9 platelets		× 0.00568	~1.78 amol/platelet	
Sodium	S or P (Hep)		*mEq/L*		*mmol/L*	
		Premature, Cord:	116–140	× 1	116–140	
		48 h:	128–148		128–148	
		Newborn, Cord:	126–166		126–166	
		Newborn:	134–144		134–144	
		Infant:	139–146		139–146	
		Child:	138–145		138–145	
		Thereafter:	136–146		136–146	
	U, 24 h	40–220 mEq/d (diet dependent)		× 1	40–220 mmol/d	
	Swt	10–40 mEq/L		× 1	10–40 mmol/L	
		Cystic fibrosis:	> 70 mEq/L		> 70 mmol/L	
Somatomedin C	P (EDTA)	0.4–2.0 IU/mL		× 1000	400–2000 IU/L	
Specific Gravity	U, random	Newborn:	1.012		Same	
		Infant:	1.002–1.006			
		Adult:	1.002–1.030			
		After 12 h fluid restriction:	> 1.025			
	U, 24 h	1.015–1.025				
Sulfhemoglobin	WB (EDTA, Hep or ACD)	≤ 1.0% of total Hb		× 0.01	≤ 0.010 of total Hb (mass fraction)	
Testosterone, Free	S	*ng/dL (mean ± SE)*	*% of Total*		*pmol/L (mean ± SE)*	*Fraction of Total*
		Prepubertal,				
		M: 0.04 ± 0.01	0.7 ± 0.2	× 34.7	1.4 ± 0.35	0.007 ± 0.002
		F: 0.04 ± 0.01	0.7 ± 0.1		1.4 ± 0.35	0.007 ± 0.001
		Adult,		(× 0.01 for fraction)		
		M: 7.9 ± 2.3	1.4 ± 0.3		274 ± 80	0.014 ± 0.003
		F: 0.31 ± 0.07	0.9 ± 0.2		10.8 ± 2.4	0.009 ± 0.002
Testosterone, Total	S		*ng/dL ± SD*		*nmol/L ± SD*	
		Prepubertal, M:	6.6 ± 2.5	× 0.0347	0.23 ± 0.09	
		F:	6.6 ± 2.5		0.23 ± 0.09	
		Adult, M:	572 ± 135		19.85 ± 4.68	
		F:	37 ± 10		1.28 ± 0.35	
		Pregnancy:	114 ± 38		3.96 ± 1.32	
		(Level unrelated to sex of fetus)				
Thyroglobulin (Tg)	S	< 50 ng/mL		× 1	< 50 μg/L	
Thyroid Antibodies	S	Adult:	≤ 1:10 dilution		Same	
		Child:	≤ 1:4			
Thyroid Microsomal Antibodies	S	Nondetectable (hemagglutination) or < 1:10 (IFA)			Same	

(continued)

Table 29-18. CLINICAL CHEMISTRY AND TOXICOLOGY (*Continued*)

Test	Specimen	Reference Range		Conversion Factor	Reference Range (International Units)
Thyroid-Stimulating Hormone (hTSH)	S or P		*μIU/mL*		*mIU/L*
		Cord:	3–12	× 1	3–12
		Newborn:	3–20		3–20
		Child:	4.5 ± 3.6 (SD)		4.5 ± 3.6 (SD)
		Adult:	<10		<10
		>60 y, M:	2–7.3		2–7.3
		F:	2–16.8		2–16.8
Thyroid-Stimulating Hormone — Response to TRH *Dose, adult: 400–500 μg TRH (Thypinone), IV, rapid; child: 7 μg/kg*	S	30 min after stimulation,	*μIU/mL*		*mIU/L*
		Child:	11–35	× 1	11–35
		Adult, M:	15–30		15–30
		F:	20–40		20–40
Thyrotropin-Releasing Hormone (hTRH)	P	Newborn, 30 min after delivery:	78 pg/mL (mean), falling to normal at 24 h	× 1	78 ng/L (mean)
		Adult:	5–60 pg/mL		5–60 ng/L
	U, 24 h	Adult, M:	195 ng/d (mean)		Same
		F:	119 ng/d (mean)		
Thyrotropin-Releasing Hormone Stimulation Test *Dose, adult: 500 μg TRH, IV*	S	TSH within 30 min,			TSH within 30 min,
		<40 y:	>6 μIU/mL rise		<40 y: >6 mIU/L rise
		>40 y, M:	>2 μIU/mL rise		>40 y, M: >2 mIU/L rise
		hPRL:	3–5-fold rise above baseline (diminishes with age)		hPRL: 3–5-fold rise above baseline (diminishes with age)
Thyroxine-Binding Globulin (TBG)	S		*mg/dL*		*mg/L*
		Cord:	1.4–9.4	× 10	14–94
		1–4 wk:	1.0–9.0		10–90
		1–12 mo:	2.0–7.6		20–76
		1–5 y:	2.9–5.4		29–54
		5–10 y:	2.5–5.0		25–50
		10–15 y:	2.1–4.6		21–46
		Adult:	1.5–3.4		15–34
		Pregnancy, last 5 mo:	5.6–10.2		56–102
		As T_4-binding capacity, Adult: 10–25 μg/dL			As T_4-binding capacity, Adult: 100–250 μg/L
Thyroxine, Free (FT_4)	S	0.8–2.4 ng/dL		× 12.9	10.3–31.0 pmol/L
Thyroxine Index, Free, *see Free Thyroxine Index*					
Thyroxine Ratio, Effective (ETR)		0.86–1.13		× 1	0.86–1.13
Thyroxine/TBG Ratio	S	0.2–0.5 T_4 (μg/dL)/TBG (μg/mL)		× 12.9	2.6–6.4 T_4 (nmol/L)/TBG (mg/L)
Thyroxine (T_4), Total	S		*μg/dL*		*nmol/L*
		Cord:	8–13	× 12.9	103–168
		Newborn:	11.5–24		148–310
		Neonate:	9–18		116–232
		Infant:	7–15		90–194
		1–5 y:	7.3–15		94–194
		5–10 y:	6.4–13.3		83–172
		Adult:	5–12		65–155
		>60 y, M:	5–10		65–129
		F:	5.5–10.5		71–135
		Pregnancy, last 5 mo:	6.1–17.6		79–227
Transferrin *Nephelometric*	S		*mg/dL*		*g/L*
		Newborn:	130–275	× 0.01	1.30–2.75
		Adult:	220–400		2.20–4.00
		>60 y:	180–380		1.80–3.80
Transketolase *Ribose/sedoheptulose, 37 °C*	WB (Hep)	(9–12 μmol/h)/mL whole blood		× 16.67	150–200 U/L whole blood
		If Hct is abnormal: (2.1–2.4 μmol/h)/10^9 red cells		× 0.0167	0.035–0.040 nU/red cell

Table 29-18. CLINICAL CHEMISTRY AND TOXICOLOGY (Continued)

Test	Specimen	Reference Range			Conversion Factor	Reference Range (International Units)	
Triglycerides (TG)	S, ≥ 12 h		*mg/dL*			*mmol/L*	
Fluorometric;	fast		*M*	*F*		*M*	*F*
colorimetric		Cord blood:	10–98	10–98	× 0.0113	0.11–1.11	0.11–1.11
		0–5 y:	30–86	32–99	(Conv. factor based on	0.34–0.97	0.36–1.12
		6–11 y:	31–108	35–114	triolein, M.W. 885)	0.35–1.22	0.40–1.29
		12–15 y:	36–138	41–138		0.41–1.56	0.46–1.56
		16–19 y:	40–163	40–128		0.45–1.84	0.45–1.45
		20–29 y:	44–185	40–128		0.50–2.09	0.45–1.45
		30–39 y:	49–284	38–160		0.55–3.21	0.43–1.81
		40–49 y:	56–298	44–186		0.63–3.37	0.50–2.10
		50–59 y:	62–288	55–247		0.70–3.25	0.62–2.79
		Values decrease slightly after age 60				Levels for Blacks: 0.11–0.23 mmol/L lower	
		Levels for Blacks: 10–20 mg/dL lower					
		Recommended (desirable) levels for adults:					
		Male:	40–160 mg/dL			0.45–1.81 mmol/L	
		Female:	35–135 mg/dL			0.40–1.53 mmol/L	
Triiodothyronine, Free	S			*mean pg/dL*		*mean pmol/L*	
		Cord:		130 ± 10 (SE)	× 0.0154	2.0 ± 0.15 (SE)	
		1–3 d:		410 ± 20		6.31 ± 0.31	
		6 wk:		400 ± 20		6.16 ± 0.31	
		Adult (20–50 y):		230–660 pg/dL		3.54–10.16 pmol/L	
				% of Total		*Fraction of Total*	
Triiodothyronine Resin Uptake Test (T_3RU; T_3U)	S	Newborn:		25–37	× 0.01	0.25–0.37	
		Adult:		24–34		0.24–0.34	
		> 60 y, M:		24–32		0.24–0.32	
		F:		22–32		0.22–0.32	
				(Method dependent)			
Triiodothyronine, Total T_3 (RIA)	S			*ng/dL*		*nmol/L*	
		Cord:		30–70	× 0.0154	0.46–1.08	
		Newborn:		65–275		0.98–4.26	
		1–5 y:		100–260		1.54–4.00	
		5–10 y:		90–240		1.39–3.70	
		10–15 y:		80–210		1.23–3.23	
		Adult:		120–195		1.85–3.00	
		> 60 y, M:		105–175		1.62–2.69	
		F:		108–205		1.66–3.16	
Tubular Reabsorption of Phosphate (TRP)	U, 4 h (0800–1200 h) and S	82–95%			× 0.01	Fraction: 0.82–0.95	
Urea, *see Urea Nitrogen*							
Urea Nitrogen	S			*mg/dL*		*mmol urea/L*	
		Cord:		21–40	× 0.166	3.5–6.6	
		Premature (1 wk):		3–25		0.5–4.2	
		Newborn:		4–12		0.7–2.0	
		Infant/child:		5–18		0.8–3.0	
		Adult:		7–18		1.2–3.0	
		> 60 y:		8–21		1.3–3.5	
				Higher after protein intake			
	U	12–20 g/d			× 16.66	200–333 mmol urea/d	
Urea Nitrogen/Creatinine Ratio	S	12/1 to 20/1				12/1 to 20/1	
Uric Acid							
Phosphotungstate	S			*mg/dL*		*mmol/L*	
		Adult, M:		4.5–8.2	× 0.059	0.27–0.48	
		F:		3.0–6.5		0.18–0.38	
		> 60 y, M:		4.2–8.0		0.25–0.47	
		F:		3.2–7.3		0.19–0.43	
Uricase		Child:		2.0–5.5		0.12–0.32	
		Adult, M:		3.5–7.2		0.21–0.42	
		F:		2.6–6.0		0.15–0.35	
	U, 24 h			*mg/d*		*mmol/d*	
		Low purine diet, M:		< 480	× 0.0059	< 2.83	
		F:		< 400		< 2.36	
		Average diet:		250–750		1.48–4.43	

(continued)

Table 29-18. CLINICAL CHEMISTRY AND TOXICOLOGY (Continued)

Test	Specimen	Reference Range		Conversion Factor	Reference Range (International Units)
Urine Volume	U, 24 h		*mL/d*		Same
		Newborn (1–2 d):	30–60		
		Adult, M:	800–1800		
		F:	600–1600		
		(Varies with water intake and other factors)			
Urobilinogen	U, 2 h	0.1–0.8 EU*/2 h			Same
	U, 24 h	0.5–4.0 EU/d			
	F	75–275 EU/100 g		× 10	750–2750 EU*/kg
		75–400 EU/d			Same
		40–280 mg/d		× 1.69	67–473 μmol/d
		*Ehrlich Unit			
Uroporphyrin	U, 24 h	< 50 μg/d		× 1.2	< 60 nmol/d
	F, 24 h	10–40 μg/d			12–48 nmol/d
	RBC (Hep or EDTA)	Negative			Negative
Vanillylmandelic Acid (Vanilmandelic Acid)	U, 24 h		*mg/d*		*μmol/d*
		Newborn:	< 1.0	× 5.05	< 5.1
		Neonate:	< 1.0		< 5.1
		Infant:	< 2.0		< 10.1
		Child:	1–5		5.1–25.3
		Adolescent:	1–5		5.1–25.3
		Thereafter:	2–7		10.1–35.4
		or 1.5–7 μg/mg creatinine		× 0.571	or 0.86–4.00 mmol/mol creatinine
Viscosity	S	1.10–1.22 Centipoise		× 1	1.10–1.22 Centipoise
Vitamin A	S		*μg/dL*		*μmol/L*
		Newborn:	35–75	× 0.0349	1.22–2.62
		Child:	30–80		1.05–2.79
		Adult:	30–65		1.05–2.27
Vitamin B_{12}, True	S	100–700 pg/mL		× 0.738	74–516 pmol/L
		> 60 y: 110–800 pg/mL			81–590 pmol/L
Vitamin B_{12} Intrinsic Factor	Gastf	50–400% enhancement of $^{57}Co\text{-}B_{12}$ uptake by GPIMH*		× 0.01	Fractional increase in $^{57}Co\text{-}B_{12}$ uptake by GPIMH*: 0.50–4.00
		*Guinea pig intestinal mucosal homogenate			
Vitamin C *Dinitrophenylhydrazine, colorimetric*	P (Ox, Hep, or EDTA)	0.6–2.0 mg/dL		× 56.78	34–114 μmol/L
	Buffy coat (Hep)	20–53 μg/10^8 WBC		× 0.568	11.4–30.1 amol/cell
Vitamin C Saturation Test *Dose: 0.5–2.0 g ascorbate orally over a period of 4 d*	U, 24 h	60–80 % of test dose excreted		× 0.01	Fraction test dose excreted: 0.60–0.80
Vitamin D_3, 1,25-dihydroxy	S	25–45 pg/mL		× 2.4	60–108 pmol/L
Vitamin D_3, 25-hydroxy	P (Hep)	Summer: 15–80 ng/mL		× 2.496	37.4–200 nmol/L
		Winter: 14–42 ng/mL			34.9–105 nmol/L
Xylose Absorption Test	WB (NaF)		*mg/dL*		*mmol/L*
		Child, 1 h (5 g dose):	> 20	× 0.0666	> 1.33
		Adult, 2 h (25 g dose):	> 25		> 1.67
	U, 5 h	Child: 16–33% of ingested dose		× 0.01	Fraction ingested dose: 0.16–0.33
		Adult,	*g/5 h*		*mmol/5 h*
		5 g dose:	> 1.2	× 6.66	> 8.00
		25 g dose:	> 4.0		> 26.64
		> 65 y:	> 3.5		> 23.31
Zinc	S	70–150 μg/dL		× 0.153	10.7–22.9 μmol/L

Table 29-19. DRUGS—THERAPEUTIC AND TOXIC

Doses for drugs listed as part of the protocol for laboratory tests, as well as their therapeutic and toxic ranges, are those commonly accepted on the basis of present knowledge. The constant acquisition of new knowledge makes it necessary to confirm and compare information supplied in this appendix with data furnished by manufacturers; additional experience with these drugs may lead to revised values. It should also be considered that there are significant differences in drug response among individuals. Where no therapeutic or toxic level is given, it was felt by the authors that there are presently no reliable data available.

Data collected for therapeutic monitoring require careful coordination between the time of sampling and drug administration. Therefore, recommendations have been made regarding the optimal time for sampling of many drugs. Hours in parenthesis indicate the minimum sampling times.

For convenience and to preserve space, we have used standard abbreviations commonly used in laboratory medicine. Less common abbreviations and some nonstandard abbreviations are given below.

Abbreviations

EDTA	ethylenediaminetetraacetate, edetic acid	prem.	premature
		S	serum
Hep	heparin	Therap.	therapeutic
Occup. exp.	occupational exposure	U	urine
P	plasma	WB	whole blood

Drug	Specimen		Reference Range	Conversion Factor	Reference Range (International Units)
Acetaminophen	S or P (Hep or EDTA)	Therap.:	10–30 μg/mL	× 6.62	66–199 μmol/L
		Toxic:	> 200 μg/mL		> 1324 μmol/L
***N*-acetylprocainamide**	S or P (Hep or EDTA); trough	Therap.:	5–30 μg/mL	× 3.61	18–108 μmol/L
		Toxic:	> 40 μg/mL		> 144 μmol/L
Amikacin	S or P (EDTA)	Therap.,	*μg/mL*		*μmol/L*
		Peak:	25–35	× 1.71	43–60
		Trough (less severe infection):	1–4		1.7–6.8
		(life-threatening infection):	4–8		6.8–13.7
		Toxic,			
		Peak:	> 35–40		> 60–68
		Trough:	> 10–15		> 17–26
ε-Aminocaproic Acid	S or P (Hep or EDTA); trough	Therap.:	> 130 μg/mL	× 7.62	> 991 μmol/L
Amitriptyline	S or P (Hep or EDTA); trough (> 12 h after dose)	Therap.:	120–250 ng/mL	× 3.61	433–903 nmol/L
		Toxic:	> 500 ng/mL		> 1805 nmol/L
Amobarbital	S	Therap.:	1–5 μg/mL	× 4.42	4–22 μmol/L
		Toxic:	> 10 μg/mL		> 44 μmol/L
Amphetamine	S or P (Hep or EDTA)	Therap.:	20–30 ng/mL	× 7.40	148–222 nmol/L
		Toxic:	> 200 ng/mL		> 1480 nmol/L
Bromide	S	Therap.:	750–1500 μg/mL	× 0.0125	9.4–18.7 mmol/L
		Toxic:	> 1250 μg/mL		> 15.6 mmol/L
Caffeine	S or P (Hep or EDTA)	Therap.:	3–15 μg/mL	× 5.15	15–77 μmol/L
		Toxic:	> 50 μg/mL		> 258 μmol/L
Carbamazepine	S or P (Hep or EDTA); trough	Therap.:	8–12 μg/mL	× 4.23	34–51 μmol/L
		Toxic:	> 15 μg/mL		> 63 μmol/L
Carbenicillin	S or P (Hep or EDTA)	Therap:	Dependent on minimum inhibition conc. of specific organism		Same
		Toxic:	> 250 μg/mL (neurotoxicity)	× 2.64	> 660 μmol/L

(continued)

Table 29-19. DRUGS—THERAPEUTIC AND TOXIC (*Continued*)

Drug	Specimen	Reference Range	Conversion Factor	Reference Range (International Units)
Chloral Hydrate	S	As trichloroethanol: Therap.: 2–12 μg/mL Toxic: > 20 μg/mL	× 6.69	13–80 μmol/L > 134 μmol/L
Chloramphenicol	S or P (Hep or EDTA); trough	Therap.: 10–25 μg/mL Toxic: > 25 μg/mL	× 3.09	31–77 μmol/L > 77 μmol/L
Chlordiazepoxide	S or P (Hep or EDTA); trough	Therap.: 700–1000 ng/mL Toxic: > 5000 ng/mL	× 0.0033	2.34–3.34 μmol/L > 16.7 μmol/L
Chlorpromazine	S or P (Hep or EDTA); trough	Therap.: 50–300 ng/mL Toxic: > 750 ng/mL	× 3.14	157–942 nmol/L > 2355 nmol/L
Cimetidine	S or P (Hep or EDTA); trough	Therap.: > 1.0 μg/mL	× 3.96	> 4.0 μmol/L
Clonazepam	S or P (Hep or EDTA); trough	Therap.: 15–60 ng/mL Toxic: > 80 ng/mL	× 3.17	48–190 nmol/L > 254 nmol/L
Clonidine	S or P (Hep or EDTA)	Therap.: 1.0–2.0 ng/mL	× 4.35	4.4–8.7 nmol/L
Clorazepate	S or P (Hep or EDTA)	As desmethyldiazepam: Therap.: 0.12–1.0 μg/mL	× 3.01	0.36–3.01 μmol/L
Cocaine	S or P (Hep or EDTA); on ice	Therap.: 100–500 ng/mL Toxic: > 1000 ng/mL	× 3.3	330–1650 nmol/L > 3300 nmol/L
Desipramine	S or P (Hep or EDTA): trough (≥ 12 h after dose)	Therap.: 75–160 ng/mL Toxic: > 1000 ng/mL	× 3.75	281–600 nmol/L > 3750 nmol/L
Desmethylmeth-suximide	S	Therap.: 10–40 μg/mL Toxic: > 40 μg/mL	× 5.29	53–212 μmol/L > 212 μmol/L
Diazepam	S or P (Hep or EDTA); trough	Therap.: 100–1000 ng/mL Toxic: > 5000 ng/mL	× 0.0035	0.35–3.51 μmol/L > 17.55 μmol/L
Digitoxin	S or P (Hep or EDTA); ≥ 6 h after dose	Therap.: 20–35 ng/mL Toxic: > 45 ng/mL	× 1.31	26–46 nmol/L > 59 nmol/L
Digoxin	S or P (Hep or EDTA); ≥ 12 h after dose	*ng/mL* Therap., CHF: 0.8–1.5 Arrhythmias: 1.5–2.0 Toxic, Adult: > 2.5 Child: > 3.0	× 1.28	*nmol/L* 1.0–1.9 1.9–2.6 > 3.2 > 3.8
Diphenylhydantoin, *see Phenytoin*				
Disopyramide	S or P (Hep or EDTA); trough	Therap., *μg/mL* Arrhythmias, Atrial: 2.8–3.2 Ventricular: 3.3–7.5 Toxic: > 7	× 2.95	*μmol/L* 8.3–9.4 9.7–22 > 20.7
Doxepin	S or P (Hep or EDTA); trough (≥ 12 h after dose)	Therap.: 30–150 ng/mL Toxic: > 500 ng/mL	× 3.58	107–537 nmol/L > 1790 nmol/L

Table 29-19. DRUGS—THERAPEUTIC AND TOXIC (*Continued*)

Drug	Specimen	Reference Range	Conversion Factor	Reference Range (International Units)
Ethchlorvynol	S or P (Hep or EDTA)	Therap.: 2–8 μg/mL Toxic: > 20 μg/mL	× 6.92	14–55 μmol/L > 138 μmol/L
Ethosuximide	S or P (Hep or EDTA); trough	Therap.: 40–100 μg/mL Toxic: > 150 μg/mL	× 7.08	283–708 μmol/L > 1062 μmol/L
Fenoprofen	P (EDTA)	Therap.: 20–65 μg/mL	× 4.12	82–268 μmol/L
Furosemide	S	Therap.: 1–2 μg/mL 30 min after last dose	× 3.023	3–6 μmol/L
Gentamicin	S or P (EDTA)	Therap., *μg/mL* Peak (less severe infection): 5–8 (severe infection): 8–10 Trough (less severe infection): < 1 (moderate infection): < 2 (severe infection): < 2–4 Toxic, Peak: > 10–12 Trough: > 2–4	× 2.09	*μmol/L* 10.4–16.7 16.7–20.9 < 2.1 < 4.2 < 4.2–8.4 > 21–25 > 4.2–8.4
Glutethimide	S	Therap.: 2–6 μg/mL Toxic: > 5 μg/mL	× 4.60	9–28 μmol/L > 23 μmol/L
Imipramine	S or P (Hep or EDTA); trough (≥ 12 h after dose)	Therap.: 125–250 ng/mL Toxic: > 500 ng/mL	× 3.57	446–893 nmol/L > 1785 nmol/L
Isoniazid	S or P (Hep or EDTA)	Therap.: 1–7 μg/mL Toxic: 20–710 μg/mL	× 7.29	7–51 μmol/L 146–5176 μmol/L
Kanamycin	S or P (EDTA)	Therap., *μg/mL* Peak: 25–35 Trough (less severe infection): 1–4 (life-threatening infection): 4–8 Toxic, Peak: > 35–40 Trough: > 10–15	× 2.06	*μmol/L* 52–72 2–8 8–16 > 72–82 > 21–31
Lidocaine	S or P (Hep or EDTA); ≥ 45 min following bolus dose	*μg/mL* Therap.: 1.5–6.0 Toxic, CNS, cardiovascular depression: 6–8 Seizures, obtundation, decreased cardiac output: > 8	× 4.27	*μmol/L* 6.4–26 26–34.2 > 34.2
Lithium	S or P (Hep or EDTA); ≥ 12 h after last dose	Therap.: 0.6–1.2 mEq/L Toxic: > 2 mEq/L	× 1	0.6–1.2 mmol/L > 2 mmol/L
Lorazepam	S or P (Hep or EDTA)	Therap.: 50–240 ng/mL	× 3.11	156–746 nmol/L
Lysergic Acid Diethylamide	P (EDTA) U	After hallucinogenic dose: 0.005–0.009 μg/mL 0.001–0.050 μg/mL	× 3090	15.5–27.8 nmol/L 3.1–154 nmol/L
Meperidine	S or P (Hep or EDTA)	Therap.: 70–500 ng/mL Toxic: > 1 μg/mL	× 4.04 × 4043	283–2020 nmol/L > 4043 nmol/L
Meprobamate	S	Therap.: 6–12 μg/mL Toxic: > 60 μg/mL	× 4.58	28–55 μmol/L > 275 μmol/L

(continued)

Table 29-19. DRUGS—THERAPEUTIC AND TOXIC (*Continued*)

Drug	Specimen	Reference Range	Conversion Factor	Reference Range (International Units)
Methaqualone	S or P (Hep or EDTA)	Therap.: 2–3 μg/mL Toxic: >10 μg/mL	× 4.00	8–12 μmol/L >40 μmol/L
Methotrexate	S or P (Hep or EDTA)	Therap.: variable Toxic, 1–2 wk after low-dose therapy: >9.1 ng/mL 48 h after high-dose therapy: 454 ng/mL	× 2.20	variable >20 nmol/L >1000 nmol/L
Methsuximide (*N*-desmethyl methsuximide)	S	Therap.: 10–40 μg/mL Toxic: >40 μg/mL	× 5.29	53–212 μmol/L >212 μmol/L
Methyldopa	P (EDTA)	Therap.: 1–5 μg/mL Toxic: >7 μg/mL (Not well defined)	× 4.73	4.7–23.7 μmol/L >33 μmol/L
Methyprylon	S	Therap.: 8–10 μg/mL Toxic: >50 μg/mL	× 5.46	43–55 μmol/L >273 μmol/L
Morphine	S or P (Hep or EDTA)	Surgical anesthesia: 65–80 ng/mL Toxic: 200–5000 ng/mL	×3.50	227–280 nmol/L 700–17 500 nmol/L
Nitroprusside	S or P (EDTA)	As thiocyanate: Therap.: 6–29 μg/mL	× 17.2	103–499 μmol/L
Normethsuximide	S	Therap.: 10–40 μg/mL Toxic: >40 μg/mL	× 5.29	53–212 μmol/L >212 μmol/L
Nortriptyline	S or P (Hep or EDTA); trough (≥12 h after dose)	Therap.: 50–150 ng/mL Toxic: >500 ng/mL	× 3.80	190–570 nmol/L >1900 nmol/L
Oxazepam	S or P (Hep or EDTA)	Therap.: 0.2–1.4 μg/mL	× 3.49	0.70–4.9 μmol/L
Paraldehyde	S or P (Hep or EDTA)	Therap., *μg/mL* Sedation: 10–100 Anesthesia: >200 Toxic: 200–400 Lethal: >500	× 7.57	*μmol/L* 76–757 >1514 1514–3028 >3785
Paraquat	 WB (EDTA) U	*μg/mL* Toxic: 0.1–1.6 Occup. exp.: 0.3 Toxic: 0.9–64	× 3.89	*μmol/L* 0.39–6.2 1.17 3.50–249
Pentobarbital	S or P (Hep or EDTA); trough	Therap., *μg/mL* Hypnotic: 1–5 Therap. coma: 20–50 Toxic: >10	× 4.42	*μmol/L* 4–22 88–221 >44
Phenacetin	P (EDTA)	Therap.: 1–20 μg/mL Toxic: 50–250 μg/mL	× 5.58	6–112 μmol/L 279–1395 μmol/L
Phencyclidine	S or P (Hep or EDTA)	Toxic: 90–800 ng/mL	× 4.11	370–3288 nmol/L
Phenobarbital	S or P (Hep or EDTA); trough	*μg/mL* Therap.: 15–40 Toxic, Slowness, ataxia, nystagmus: 35–80 Coma with reflexes: 65–117 Coma without reflexes: >100	× 4.31	*μmol/L* 65–170 151–345 280–504 >430

Table 29-19. DRUGS—THERAPEUTIC AND TOXIC (*Continued*)

Drug	Specimen	Reference Range		Conversion Factor	Reference Range (International Units)
Phensuximide (both parent and *N*-desmethyl metabolites)	S or P (Hep or EDTA)	Therap.:	40–60 μg/mL	× 5.71	228–324 μmol/L
Phenylbutazone	P (EDTA)	Therap.:	50–100 μg/mL (Not well defined)	× 3.08	162–324 μmol/L
		Toxic:	> 100 μg/mL		> 324 μmol/L
Phenytoin	S or P (Hep or EDTA); trough	Therap.:	10–20 μg/mL	× 3.96	40–79 μmol/L
		Toxic:	> 20 μg/mL		> 79 μmol/L
Primidone	S or P (Hep or EDTA); trough	Therap.:	5–12 μg/mL	× 4.58	23–55 μmol/L
		Toxic:	> 15 μg/mL		> 69 μmol/L
Procainamide	S or P (Hep or EDTA); trough	Therap.:	4–10 μg/mL	× 4.23	17–42 μmol/L
		Toxic:	> 10–12 μg/mL		> 42–51 μmol/L
		Also consider effect of metabolite (NAPA)			
Propoxyphene	P (EDTA)	Therap.:	0.1–0.4 μg/mL	× 2.95	0.3–1.2 μmol/L
		Toxic:	> 0.5 μg/mL		> 1.5 μmol/L
Propranolol	S or P (Hep or EDTA); trough	Therap.:	50–100 ng/mL	× 3.86	193–386 nmol/L
Protriptyline	S or P (Hep or EDTA); trough (≥ 12 h after dose)	Therap.:	70–250 ng/mL	× 3.80	266–950 nmol/L
		Toxic:	> 500 ng/mL		> 1900 nmol/L
Quinidine	S or P (Hep or EDTA); trough	Therap.:	2–5 μg/mL	× 3.08	6–15 μmol/L
		Toxic:	> 6 μg/mL		> 18 μmol/L
Salicylates	S or P (Hep or EDTA); trough	Therap.:	150–300 μg/mL	× 7.24	1086–2172 μmol/L
		Toxic:	> 300 μg/mL		> 2172 μmol/L
Secobarbital	S	Therap.:	1–2 μg/mL	× 4.20	4.2–8.4 μmol/L
		Toxic:	> 5 μg/mL		> 21.0 μmol/L
Theophylline	S or P (Hep or EDTA)	Therap.,	*μg/mL*		*μmol/L*
		Bronchodilator:	8–20	× 5.55	44–111
		Premature apnea:	6–13		33–72
		Toxic:	> 20		> 110
Thiocyanate	S or P (EDTA)		*μg/mL*		*μmol/L*
		Nonsmoker:	1–4	× 17.2	17–69
		Smoker:	3–12		52–206
		Therap. after nitroprusside infusion:	6–29		103–499
	U	Nonsmoker:	1–4 mg/d	× 17.2	17–69 μmol/d
		Smoker:	7–17 mg/d		120–292 μmol/d
Thiopental	S or P (Hep or EDTA); trough		*μg/mL*		*μmol/L*
		Hypnotic:	1.0–5.0	× 4.13	4.1–20.7
		Coma:	30–100		124–413
		Anesthesia:	7–130		29–536
		Toxic conc.:	> 10		> 41
Thioridazine	S or P (Hep or EDTA)	Therap.:	1.0–1.5 μg/mL	× 2.70	2.7–4.1 μmol/L
		Toxic:	> 10 μg/mL		> 27 μmol/L
Tobramycin	S or P (Hep or EDTA)	Therap.,	*μg/mL*		*μmol/L*
		Peak (less severe infection):	5–8	× 2.14	11–17
		(severe infection):	8–10		17–21

(continued)

Table 29-19. DRUGS—THERAPEUTIC AND TOXIC (*Continued*)

Drug	Specimen	Reference Range	Conversion Factor	Reference Range (International Units)
Tobramycin (*cont'd*)	S or P (Hep or EDTA)	*μg/mL*		*μmol/L*
		Trough (less severe infection): <1		<2
		(moderate infection): <2		<4
		(severe infection): <2–4		<4–9
		Toxic,		
		Peak: >10–12		>21–26
		Trough: >2–4		>4–9
Valproic Acid	S or P (Hep or EDTA); trough	Therap.: 50–100 μg/mL	× 6.93	347–693 μmol/L
		Toxic: >100 μg/mL		>693 μmol/L
Vancomycin	S or P (Hep or EDTA); trough	Therap: Not well established		
		Toxic: >80–100 μg/mL	× 1	>80–100 mg/L
Warfarin	S or P (Hep or EDTA)	Therap.: 1–10 μg/mL	× 3.24	3–32 μmol/L

References

Beutler, E.: Hemolytic Anemia in Disorders of Red Cell Metabolism. New York, Plenum Publishing Company, 1978.

Brown, S. S., Mitchell, F. L., and Young, D. S., Eds.: Chemical Diagnosis of Disease. Amsterdam, Elsevier/North-Holland Biomedical Press, 1979.

Gilman, A. G., Goodman, L., and Gilman, A., Eds.: The Pharmacological Basis of Therapeutics, 6th ed. New York, Macmillan, 1980.

Mabry, C. and Tietz, N. W.: Tables of normal laboratory values. *In:* Nelson Textbook of Pediatrics, 12th ed. R. E. Behrman and V. C. Vaughan, III, Eds. Philadelphia, W. B. Saunders Co., 1983.

Tietz, N. W., Ed.: Clinical Guide to Laboratory Tests. Philadelphia, W. B. Saunders Co., 1983.

Tietz, N. W.: Reference ranges and laboratory values of clinical importance. *In:* Cecil Textbook of Medicine, 17th ed. J. B. Wyngaarden and L. H. Smith, Jr., Eds. Philadelphia, W. B. Saunders Co., 1985.

Tietz, N. W., and Blackburn, R. H., Eds.: Reference Ranges and General Information. Clinical Laboratories, A. B. Chandler Medical Center, University of Kentucky, Lexington, Kentucky, 1984.

INDEX

Page numbers in *italic* type indicate figures; page numbers followed by (t) indicate tables.

% TRANSMISSION-ABSORBANCE CONVERSION CHART

%T	A	%T	A	%T	A	%T	A
1	2.000	1.5	1.824	51	0.2924	51.5	0.2882
2	1.699	2.5	1.602	52	0.2840	52.5	0.2798
3	1.523	3.5	1.456	53	0.2756	53.5	0.2716
4	1.398	4.5	1.347	54	0.2676	54.5	0.2636
5	1.301	5.5	1.260	55	0.2596	55.5	0.2557
6	1.222	6.5	1.187	56	0.2518	56.5	0.2480
7	1.155	7.5	1.126	57	0.2441	57.5	0.2403
8	1.097	8.5	1.071	58	0.2366	58.5	0.2328
9	1.046	9.5	1.022	59	0.2291	59.5	0.2255
10	1.000	10.5	0.979	60	0.2218	60.5	0.2182
11	0.959	11.5	0.939	61	0.2147	61.5	0.2111
12	0.921	12.5	0.903	62	0.2076	62.5	0.2041
13	0.886	13.5	0.870	63	0.2007	63.5	0.1973
14	0.854	14.5	0.838	64	0.1939	64.5	0.1905
15	0.824	15.5	0.810	65	0.1871	65.5	0.1838
16	0.796	16.5	0.782	66	0.1805	66.5	0.1772
17	0.770	17.5	0.757	67	0.1739	67.5	0.1707
18	0.745	18.5	0.733	68	0.1675	68.5	0.1643
19	0.721	19.5	0.710	69	0.1612	69.5	0.1580
20	0.699	20.5	0.688	70	0.1549	70.5	0.1518
21	0.678	21.5	0.668	71	0.1487	71.5	0.1457
22	0.658	22.5	0.648	72	0.1427	72.5	0.1397
23	0.638	23.5	0.629	73	0.1367	73.5	0.1337
24	0.620	24.5	0.611	74	0.1308	74.5	0.1278
25	0.602	25.5	0.594	75	0.1249	75.5	0.1221
26	0.585	26.5	0.577	76	0.1192	76.5	0.1163
27	0.569	27.5	0.561	77	0.1135	77.5	0.1107
28	0.553	28.5	0.545	78	0.1079	78.5	0.1051
29	0.538	29.5	0.530	79	0.1024	79.5	0.0996
30	0.523	30.5	0.516	80	0.0969	80.5	0.0942
31	0.509	31.5	0.502	81	0.0915	81.5	0.0888
32	0.495	32.5	0.488	82	0.0862	82.5	0.0835
33	0.482	33.5	0.475	83	0.0809	83.5	0.0783
34	0.469	34.5	0.462	84	0.0757	84.5	0.0731
35	0.456	35.5	0.450	85	0.0706	85.5	0.0680
36	0.444	36.5	0.438	86	0.0655	86.5	0.0630
37	0.432	37.5	0.426	87	0.0605	87.5	0.0580
38	0.420	38.5	0.414	88	0.0555	88.5	0.0531
39	0.409	39.5	0.403	89	0.0505	89.5	0.0482
40	0.398	40.5	0.392	90	0.0458	90.5	0.0434
41	0.387	41.5	0.382	91	0.0410	91.5	0.0386
42	0.377	42.5	0.372	92	0.0362	92.5	0.0339
43	0.367	43.5	0.362	93	0.0315	93.5	0.0292
44	0.357	44.5	0.352	94	0.0269	94.5	0.0246
45	0.347	45.5	0.342	95	0.0223	95.5	0.0200
46	0.337	46.5	0.332	96	0.0177	96.5	0.0155
47	0.328	47.5	0.323	97	0.0132	97.5	0.0110
48	0.319	48.5	0.314	98	0.0088	98.5	0.0066
49	0.310	49.5	0.305	99	0.0044	99.5	0.0022
50	0.301	50.5	0.297	100	0.0000		